American College of Veterinary Internal Medicine (ACVIM) Forum 2017

National Harbor, Maryland, USA
8 - 10 June 2017

Volume 1 of 2

ISBN: 978-1-5108-4279-3

Printed from e-media with permission by:
Curran Associates, Inc.
57 Morehouse Lane
Red Hook, NY 12571

Some format issues inherent in the e-media version may also appear in this print version.

Printed by Curran Associates, Inc. (2017)

Address correspondence to:
American College of Veterinary Internal Medicine (ACVIM)
8301 E. Prentice Avenue, Suite 300
Greenwood Village, CO 80111
www.acvim.org

Additional copies of this publication are available from:

Curran Associates, Inc.
57 Morehouse Lane
Red Hook, NY 12571 USA
Phone: 845-758-0400
Fax: 845-758-2633
Email: curran@proceedings.com
Web: www.proceedings.com

TABLE OF CONTENTS

VOLUME 1

SMALL ANIMAL INTERNAL MEDICINE

SMALL ANIMAL INTERNAL MEDICINE & LARGE ANIMAL INTERNAL MEDICINE

TRANSFUSION

RESEARCH ABSTRACTS – ORAL PRESENTATIONS

RESEARCH ABSTRACTS – POSTER PRESENTATIONS

RESEARCH ABSTRACTS – ORAL PRESENTATIONS

RESEARCH ABSTRACTS – POSTER PRESENTATIONS

RESEARCH ABSTRACTS – ORAL PRESENTATIONS

RESEARCH ABSTRACTS – POSTER PRESENTATIONS

RESEARCH ABSTRACTS – ORAL PRESENTATIONS

RESEARCH ABSTRACTS – POSTER PRESENTATIONS

RESEARCH ABSTRACTS – ORAL PRESENTATIONS

RESEARCH ABSTRACTS – POSTER PRESENTATIONS

RESEARCH ABSTRACTS – ORAL PRESENTATIONS

RESEARCH ABSTRACTS – POSTER PRESENTATIONS

RESEARCH ABSTRACTS – ORAL PRESENTATIONS

RESEARCH ABSTRACTS – POSTER PRESENTATIONS

RESEARCH ABSTRACTS – ORAL PRESENTATIONS

RESEARCH ABSTRACTS – POSTER PRESENTATIONS

RESEARCH ABSTRACTS – ORAL PRESENTATIONS

RESEARCH ABSTRACTS – POSTER PRESENTATIONS

RESEARCH ABSTRACTS – ORAL PRESENTATIONS

RESEARCH ABSTRACTS – POSTER PRESENTATIONS

RESEARCH ABSTRACTS – ORAL PRESENTATIONS

RESEARCH ABSTRACTS – POSTER PRESENTATIONS

RESEARCH ABSTRACTS – ORAL PRESENTATIONS

RESEARCH ABSTRACTS – POSTER PRESENTATIONS

RESEARCH ABSTRACTS – ORAL PRESENTATIONS

RESEARCH ABSTRACTS – POSTER PRESENTATIONS

RESEARCH ABSTRACTS – ORAL PRESENTATIONS

RESEARCH ABSTRACTS – POSTER PRESENTATIONS

RESEARCH ABSTRACTS – ORAL PRESENTATIONS

RESEARCH ABSTRACTS – POSTER PRESENTATIONS

RESEARCH ABSTRACTS – ORAL PRESENTATIONS

RESEARCH ABSTRACTS – POSTER PRESENTATIONS

AMERICAN ASSOCIATION OF VETERINARY NUTRITION/AMERICAN COLLEGE OF VETERINARY NUTRITION

AMERICAN COLLEGE OF VETERINARY CLINICAL PHARMACOLOGY/AMERICAN ACADEMY OF VETERINARY PHARMACOLOGY AND THERAPEUTICS

AMERICAN SOCIETY FOR CLINICAL PATHOLOGY

EVIDENCE BASED VETERINARY MEDICINE ASSOCIATION

INTERNATIONAL SOCIETY FOR COMPANION ANIMAL INFECTIOUS DISEASES

LIVER STUDY GROUP

VOLUME 2

ONE MEDICINE

SOCIETY FOR COMPARATIVE ENDOCRINOLOGY

VETERINARY COMPARATIVE RESPIRATORY SOCIETY (LARGE ANIMAL)

VETERINARY COMPARATIVE RESPIRATORY SOCIETY (SMALL ANIMAL)

VETERINARY EMERGENCY AND CRITICAL CARE SOCIETY

CARDIOLOGY

NEUROLOGY

ONCOLOGY

SMALL ANIMAL INTERNAL MEDICINE

EQUINE

acvim001

Reflections of the Heart: Echocardiography for Veterinary Technicians

H. Edward Durham Jr., CVT, LATG, VTS (Cardiology)
St. Kitts, West Indies

INTRODUCTION

Echocardiography is a method of diagnosing cardiac disease utilizing ultrasound to create an image of the heart. With today's devices, these images are in motion and nearly real time. Echocardiography specifically refers to an ultrasound of the heart and great vessels. Prior to the advent of echocardiography, most cardiac conditions were diagnosed with cardiac catheterization, requiring an invasive procedure. Ultrasound is a noninvasive diagnostic tool, which is simpler to perform and less painful. The first clinical ultrasound images of the heart were done with a single "ice pick" beam of ultrasound and displayed on a moving paper roll, much like electrocardiography. As advances were made in ultrasound transducer design, a "cross-sectional" image could be generated by utilizing a fan-shaped beam of multiple single beams. This is the most common modality of echocardiogram used today and is referred to as 2D or two-dimensional echocardiography. The newest technology generates a three-dimensional image, displayed in real time and is referred to as "4D." As part of the standard information of the echocardiogram, the single ultrasound beam view of the heart, known as M-Mode, is still used. From the M-Mode images, precise measurements of cardiac wall thickness and chamber sizes can be obtained. When combined with a Doppler evaluation, a complete picture of cardiac morphology and function can be obtained.

BASIC TECHNOLOGY

In brief, the ultrasound wave is generated by electrically exciting a piezoelectric crystal within the transducer causing it to expand. When the electrical stimulation is applied to the piezoelectric crystal in an alternating current, the crystal expands and contracts to generate an ultrasonic wave. A piezoelectric crystal will also convert ultrasound wave it senses to an electric signal; thereby serves as either transmitter or receiver. The number of waves generated per second is known as the frequency or hertz (Hz). Diagnostic medical transducers generate millions of hertz or megahertz (MHz); in veterinary medicine, transducers range from 2.0 to 12 MHz. The ultrasound waves are transmitted through the patient and reflect off body tissue, returning to the transducer to be received by more crystals. The electrical signals generated by the crystals are transformed by a computer into an image displayed on a monitor. Ultrasound images are displayed in black, white, and shades of grey. In abdominal ultrasound, the more shades of grey presented the better the interpretation, but in echocardiography, less shades of grey are used and increased edge detection is preferred.

The different frequencies of transducers allow for penetration to different depths. There is no transducer frequency that is suitable for all situations. Lower-frequency transducers have greater penetration, but have less ability to show detail. Higher-frequency transducers have better image resolution, but less penetration.

There are other factors affecting how the ultrasound waves behave in the body. Attenuation is the tendency for the waves to be blocked or lose energy from transmission and refraction is the tendency of the waves to be scattered off their original course. These concepts are important because attenuation is related to tissue density, thus bone attenuates the signal so severely that ultrasound waves do not pass through them. Air, on the other hand, has such a profound refraction effect on ultrasound waves that nothing can be seen beyond it. For this reason, a coupling gel must be used between the patient and the transducer to eliminate air and facilitate imaging. Some of these problems can be overcome by changing the machine setting and it is incumbent on the sonographer to learn the complete operation of their machine.

INDICATIONS

Echocardiography provides two broad categories of information about the heart: morphology and (systolic and diastolic) function. Morphology of the heart allows the ultrasonographer to see what the heart looks like. Information, such as chamber enlargement, evidence of effusions, presence of cardiac tumors, and appearance of the cardiac valves, can be assessed. Systolic function is what most people call contractility; however, contractions of the heart are influenced by many variables, such as Frank-Starling forces, afterload and sympathetic nervous tone. Consequently, systolic function is the preferred term for cardiac motion. Indications for performing an echocardiogram include heart wall thickness evaluation, assessment of chamber size and (systolic and diastolic) function, definitive diagnosis of heart murmurs, diagnosis of congenital heart disease, diagnosis of cardiac tamponade with pericardial effusion, and identification of cardiac tumors as a cause of pericardial effusion.

PATIENT POSITIONING AND PREPARATION

There are many ways of performing an echocardiogram. In very large patients, such as horses or cattle, standing is the preferred position. Many echosonographers in Europe prefer this position for dogs as well. The most common position, however, is a lateral recumbency with the hair clipped over the precordium. The echocardiographic table has openings in it to allow the operator to reach under the table to access the patient through the hole. This position allows the heart to fall close to the thoracic wall to increase image quality. For imaging the opposite hemithorax, the operator may reach over the patient and image from the top, or more likely, the patient is turned to the opposite lateral recumbency. An electrocardiogram is run during the echocardiogram to indicate specific times in the cardiac cycle.

STANDARD IMAGE PLANES

Traditionally, even though the transducer is placed on the underside of the patient, the transducer anchor point is oriented at the top of the viewing screen (Figure 1). The image is displayed upside down from how it is generated. In this orientation, objects closest to the transducer are at the top of the screen or in the near field. Objects at the bottom of the screen are in the far field and furthest away from the transducer. There are several standard image planes recognized in veterinary medicine. These planes of view allow a set of standard measurements to be performed to assess cardiac chamber dimensions and systolic function. Echocardiograms usually begin with views from the right hemithorax. The heart is viewed in either its long axis or in its short axis. One way to understand these planes is to imagine a pineapple. If the pineapple is sliced from top to bottom making a pineapple boat, then the long axis is represented. If the pineapple is sliced transversely making pineapple rings, that represents the short axis. When viewing the heart from the right hemithorax, the first view commonly acquired is the right parasternal long-axis four-chamber view. In this view, the heart is typically situated horizontally across the screen with the left atrium (LA) and ventricle (LV) in the far field and the right ventricle (RV) and atrium (RA) in the near field. From this position all four chambers of the heart may be seen together. By making a slight adjustment in the angle and rotation of the transducer, the aorta may be brought into view, the so-called "five" chamber view. Here, the LA is much smaller in appearance, but the aorta is in full view and represents the fifth chamber. From this position the transducer is rotated ninety degrees and the heart is seen in the short axis.

There are three standard short-axis views: 1) LV at the level of the papillary muscles, 2) LV at the level of the mitral valve (MV), 3) the heart base showing the aortic valve, LA, RA, RV and RV outflow tract (RVOT). In the first, the LV is circular, with the LV lumen shaped somewhat like a mushroom, and the RV appears as a crescent-shaped chamber in the near field. The papillary muscles are seen projecting into the LV lumen creating the "stalk" of the mushroom. In the second short-axis view, the transducer angle is tilted toward the base of the heart until the MV is seen opening and closing. This view is commonly referred to as the "fish mouth" view since the MV resembles a fish at the surface of the water opening and closing its mouth. By angling the transducer more basilar, the beam passes through the heart base. The LV is no longer visible, but the aortic valve is seen crosswise, with the LA in the far field, the RA to one side, the RV in the near field, and the RVOT opposite the RA on the screen. Sometimes the pulmonic valve can also be seen in this view, and if not, a slight cranial adjustment in transducer angle often will bring it into view.

To complement these 2D views, a series of M-Mode views are also acquired. Because M-Mode uses only one ultrasound beam, requiring less raw data to be processed as in 2D, M-Mode allows for greater temporal resolution and finer detail along the single beam. M-Mode therefore allows for greater detail in assessing cardiac wall motion and delineating the edges of the chambers compared to the chamber lumens. M-Mode views are still the standard views for making many of the measurements of chamber dimensions and systolic function. The M-Mode views can be derived from either the long or the short axis 2D images, with the first usually taken from the short axis at the papillary muscle level. A cursor is typically placed across the 2D image, such that the cursor crosses the interventricular septum (IVS) passing through the LV lumen between the papillary muscles and out the LV free or posterior wall (LVPW). The data received by the computer **only** along the cursor line are then displayed over time. The resulting image is a graphic representation of the heart with time along the X axis and depth along the Y axis. The RV and IVS will be in the near field and the LV lumen will appear in the middle, with the LVPW in the far field (Figure 2). From this image, measurement of the LV can be made in diastole and systole and calculations of systolic function can be determined. From the "fish mouth" short-axis view, the motion of the MV can be recorded as it moves during the cardiac cycle. This image is similar to the previous LV view with triangular lines of the MV in the midst of the LV lumen. As the operator moves toward the base of the heart to the short-axis basilar view, a subsequent M-Mode can be recorded with the aorta in the center of the screen and the LA in the far field. In this image, a comparison of the LA to the aorta diameters is made. A ratio of LA to aorta (LA/Ao ratio) is calculated.

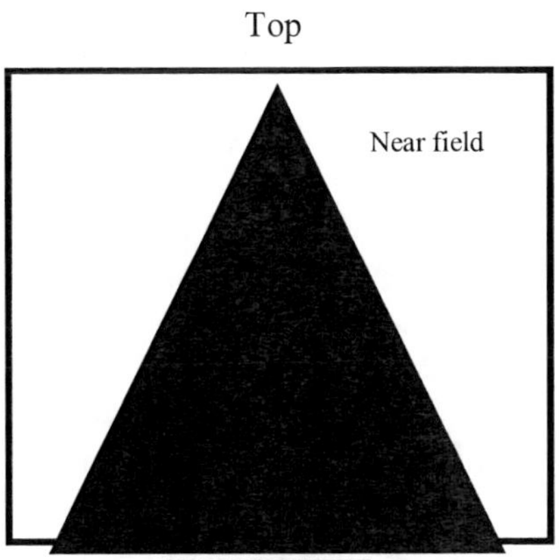

Figure 1

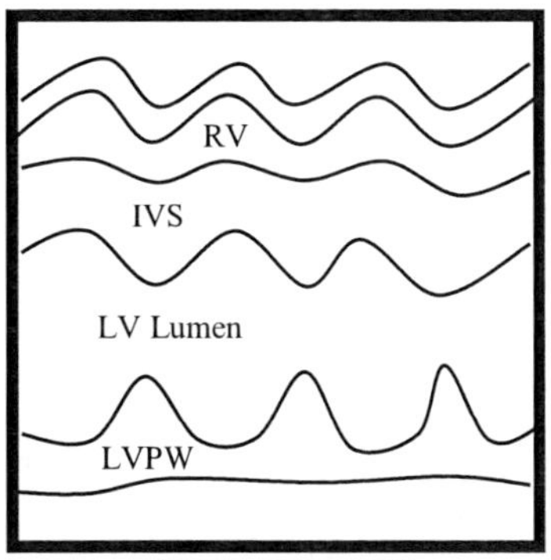

Figure 2

In addition to these standard views, other image plane may be acquired. Views of the main pulmonary artery and its branches may be seen by making angle changes from the short-axis basilar view. This view is useful for the visualization of heartworms, if present. From the long-axis four-chamber view, RA tumor may be seen well by sliding the transducer toward the sternum to tip the heart on the screen, bringing the RA more into view. From the left hemithorax, the heart is seen in a four-chamber view with the atria in the far field and the apex of the ventricles in the near field, the so-called left apical view. From this orientation, accurate Doppler velocity measurement may be acquired as well as potentially increased views of RA tumors. Other short-axis views of the RA and a long-axis view of the pulmonary artery may also be acquired from the left hemithorax. Representative images of all the planes discussed will be shown during the presentation.

Measurements and Interpretation

A series of standard measurements is performed in virtually all echocardiograms. From the 2D long-axis four-chamber view, a measurement of the LA diameter is done at the end of systole, when the LA is at its largest. Assessment for pericardial effusion and subjective impressions of the LV and MV function may also be performed. From the 2D long-axis five-chamber view, the aortic diameter is measured, often at several places and timings. Most other measurements come from the M-Mode views. One of the most important is from the LV papillary muscle level M-Mode. In this view, measurements are made of the IVS thickness, LV lumen diameter, and LVPW thickness, in systole and diastole. Diastolic wall thicknesses in cats are used to diagnose HCM. The difference between the LV lumen in systole and diastole is used to determine percent fractional shortening. Only one measurement is taken from the MV M-Mode, the "E" point to septal separation (EPSS). The EPSS measures the distance from the tip of the MV at its first opening after the T wave on the electrocardiogram, to the IVS. This distance is known to increase in the face of systolic dysfunction. This measurement is specific but may not be sensitive. As mentioned earlier, the LA/Ao ratio is calculated from the short-axis basilar M-Mode image. This measurement takes the diameter of the aorta and the diameter of the left atrium and compares the two. In the normal patient, the LA and the aorta are approximately the same size. As congestive heart failure occurs, the LA will increase in size, yet the aortic size remains static, giving each patient its own normal to compare any left atrial enlargement to.

Assessment of systolic function is perhaps one of the most important measurements performed by the echosonographer. No single measurement provides a comprehensive or foolproof method of assessing systolic function. The most common measures used are percent fractional shortening (%FS), shortening area (%SA), and ejection fraction (%EF). Each of these measurements determines the percent change in the LV from diastole to systole. Fractional shortening is a percent change in the linear measurement of the LV diameter from an M-Mode image using the following formula:

LVIDd – LVIDs/ LVIDd x 100 = %FS

where LVIDd = LV internal dimension in diastole and LVIDs = LV internal dimension in systole.

Fractional shortening is the easiest and most common determination of systolic function performed, but it has important limitations. The fractional shortening only measures the change in the LV lumen at two points, one on the IVS and another on the LVPW. The heart is a three-dimensional structure and moves in all three directions during systole. Thus, %FS may underestimate true systolic function. The %SA is a measurement taken from the short-axis 2D image of the LV papillary muscle level. The area of the LV lumen is traced in systole and diastole. The area measurements used the same formula used for %FS and percent change in the LV area is calculated. For a three-dimensional evaluation of the LV systolic function, the %EF is used. Ejection fraction is similar to %SA, but a volume of the left ventricle is measured in systole and diastole and a percent change calculated. To measure the LV volume requires some mathematical assumptions since the LV is not a perfect cylinder, but more of a truncated ellipse. Using a formula called the Modified Simpson's rule, the area of the LV is traced from the long-axis four-chamber view. The computer measures the maximal length and then fills this area with a series of disks approximating the area at various levels of the LV. The area of these disks is summed to generate an approximate volume. The advantage is that by calculating the volume, the operator takes as many data points into account as possible, but because the LV is irregularly shaped, a perfect volume cannot be measured. Another advantage of the %EF is that the difference between the LV volume in diastole and systole is equal to the stroke volume, which allows for calculations of cardiac output by multiplying stroke volume by heart rate. In most echocardiographic laboratories, if the %FS is normal, then the patient is considered normal. If the %FS is not normal, then the other assessments of systolic function are utilized. However, it is important to keep in mind the most important limitation of all of these measures: they are simply derived from the ratio of the starting point in diastole to the ending point in systole; abnormalities of either of these dimensions will alter the final product. You cannot assume because a patient has an ejection fraction below 50% that systolic dysfunction is present; the diastolic dimensions could be too low due to volume depletion. Similarly, a normal ratio does not guarantee that systolic function is adequate: a large increase in diastole due to increased preload can result in this ratio being "normal" in a patient with decreased contractility. The bottom line: treat your patient, not the numbers!

Interpretation of an echocardiogram requires a detailed knowledge of cardiac disease conditions. Simply acquiring images and performing measurements is usually insufficient for making accurate diagnoses of heart disease. The echosonographer needs to have an excellent understanding of the cardiac cycle as well. It is imperative to know what to expect from a normal heart to recognize abnormal function or anatomy. One method of understanding the echocardiogram is to understand what question about the patient is being answered by the echocardiogram. For example, if the patient is a Doberman pinscher, a breed in which dilated cardiomyopathy (DCM) is common, then the operator must understand how DCM affects the heart and what changes in the echocardiogram are to be expected. Also, issues of systolic function become critical in the diagnosis of DCM. Conversely, if the patient is a cat, which rarely acquires DCM but commonly develops hypertrophic cardiomyopathy, the sonographer needs to understand that pathophysiology. In this instance, systolic function is less of an issue and the thickness of the ventricular walls becomes critical. By understanding the pathophysiology of cardiac disease, the echosonographer is never surprised by their findings, and the echocardiogram does not simply become the "heart tester," cast like a net seeking an unknown heart condition.

acvim002

How Fast Is Fast: Doppler Echocardiography

H. Edward Durham Jr., CVT, LATG, VTS (Cardiology)
St. Kitts, West Indies

INTRODUCTION

Doppler echocardiography is a non-imaging modality of cardiac ultrasound that is used to measure the velocity of blood flow through the heart. Doppler studies may be performed in any vessel of the body, but for the purposes of this discussion the topic will be limited to the heart and great vessels. Doppler echocardiography is another application of ultrasound technology and is typically performed in conjunction with an echocardiogram. Where the echocardiogram creates an image of the heart and provides information about the morphology and systolic function of the heart, the Doppler portion provides hemodynamic information and allows assessment of diastolic function and estimates of chamber pressures. The Doppler study gives the cardiologist information about the direction and velocity of blood flow. It can also be used to indicate turbulent flow in the heart.

BASIC TECHNOLOGY

Doppler studies are simply an advanced form of ultrasound that is specifically for velocity measurements of blood. This allows the cardiologist to make some conclusions about physiology and prognosis. Doppler echocardiography works on the principle of the Doppler shift, first described by Johann Christian Doppler in 1864. According to the Doppler principle, a sound wave emanating from a moving object as it moves toward a stationary observer will become compressed and increase in frequency; conversely, if the object is moving away from an observer, the sound waves get **rarefied** or stretched out and decrease in frequency. This is a phenomenon that we all have witnessed when a siren on a firetruck passes by our standing position. The degree of frequency shift is directly related to how fast the object is moving. For a Doppler echocardiogram, the sonographer use this principle in the same way a bat uses sonar to hunt for food. The bat transmits an ultrasonic sound at a known frequency. When the signal reflects off a moth, it returns to the bat with the frequency changed depending on the moth's direction in relation to the bat; increased if the moth is moving toward the bat, decreased if moving away. The degree of frequency shift tells the bat how fast the moth is moving. In the heart, a known frequency signal is transmitted by the transducer into the blood pool. From the reflected signal a calculation is made using the frequency shift to calculate the blood velocity and direction. The following formula is used by the computer to make the calculation:

$$\mathbf{V = C \times f_d / 2(f_o) \times \cos\theta}$$

where V=velocity in m/sec, C=the average speed of ultrasound waves in soft tissue (1540 m/sec), f_d=frequency shift, $2(f_o)$=twice the transmitted velocity, and cos θ=the cosine of the angle of interrogation. The frequency shift is the difference between the transmitted frequency and the reflected frequency.

Because all the variables in the equation are known, except for the cos θ, the angle of interrogation, also called intercept angle, is of critical importance. It is the angle that the ultrasound beam strikes its target, in this case the blood pool (Figure 1). The most accurate measurement is made when the beam strikes the target at perfect parallel (zero or 180 degrees) and the cosine is equal to 1 or -1. As the intercept angle increases, the cosine moves away from 1, and the accuracy of the calculation decreases. The angle of interrogation is the only variable in the equation that the operator has any control over and it is critical the angle closest to parallel be achieved for the most accurate velocity. As a result, the operator cannot overestimate a true velocity, but can underestimate one with misalignment outside of pure parallel. Generally, angles of intercept greater than 15 degrees are considered unacceptable in cardiology. To facilitate alignment, the Doppler study is performed on the left apical 5-chamber view from the left hemithorax. In this view, the atria are at the bottom of the screen, the apex of the heart is at the top and the aorta heading out of the left ventricle (LV) toward the bottom. This orientation allows for alignment of the Doppler cursor with blood flow heading straight up toward the transducer from the atria and straight down out of the aorta. The left ventricular outflow tract velocity may also be obtained with the subcostal view in the dog. The pulmonary artery (PA) can be interrogated from a right parasternal short axis basilar view.

MODALITIES

Three modalities are commonly used for cardiac Doppler measurements. Spectral Doppler includes pulsed-wave Doppler and continuous-wave Doppler. The information generated is displayed on a graph with time along the X axis and velocity on the Y axis. In the center of the graph is a baseline indicating zero flow. Returning ultrasound signals of flow toward the transducer are displayed above the baseline, and signal moving away from the transducer are displayed below the baseline. The baseline can be shifted up or down to create more graph above or below depending on the predominate direction of blood flow being recorded at the time (Figure 2). Color-flow Doppler uses the same principles but the information gained is displayed as a color signal superimposed atop a 2D echocardiographic map. To use spectral Doppler a cursor is turned on over the 2D echocardiographic image and aligned with the blood flow to be interrogated and the Doppler of choice activated.

Pulsed-wave Doppler (PW) is a system that allows for velocity measurements at a specific location in the heart. A sample volume or gate is positioned along the cursor line over the area for velocity measurement. When the PW Doppler is turned on, a piezoelectric crystal transmits a short burst of ultrasound waves that travel to the designated depth while the transducer waits to receive, and the reflected data are received by another crystal. This pattern is repeated many thousands of times a second. The number of times the crystal can transmit, shut off, wait for the returning signal and then transmit again is the **pulse repetition frequency** (PRF) and represents how many times it can sample per second. The PRF is important in that the faster the PRF or sampling rate, the faster a velocity that can be recorded. If the returning signal has a frequency shift greater than twice the PRF, ambiguous information is returned to the computer. The farther the sample gate is from the transducer, the lower the PRF due to distance and the slower a velocity can unambiguously be recorded. In essence, the returning signal is faster than the transducer's ability to sample, and **aliasing** occurs. A similar phenomenon occurs when watching a carriage with spoke wheels move. When the carriage is moving slowly, the spokes turn at a rate slower than the human eyes' ability to see them move, but as the carriage speeds up the spokes begin to appear to move backward. Our minds know the carriage is moving forward, yet because our eyes can only sample images at set speed, the spokes appear to move backward. With PW Doppler an aliased signal is displayed "wrapped around" into the opposite side of the spectral graph (Figure 3). The velocity at which aliasing occurs is variable due to depth, but the average is 1.5 m/sec. So the advantage of PW is the ability to record velocities at the specific depth, but because the transducer must stop transmitting for a split second between pulses to receive returning information, PW has a limited velocity range it can detect.

Continuous-wave Doppler (CW) allows for measurement of very fast velocities by continuously transmitting and receiving, and is displayed the same as PW Doppler (Figure 2). Unlike PW Doppler that records velocities in one site, the CW Doppler records velocities all along the length of the ultrasound beam. It is impossible for the frequency shift to be greater than the sample rate which is continuous, so extremely fast velocities can be measured. The point along the beam that the highest velocity occurs however is unknown. Continuous-wave Doppler therefore provides a method of measuring the high velocity blood flow of congenital heart disease, but is range ambiguous. In a practical sense, high-velocity jets can be located with the PW Doppler and then accurately quantified by the CW Doppler.

Color-flow Doppler (CF), also known as color-flow mapping, is an advanced form of PW, but rather than the returning signal being displayed on a graph, the information is color coded and displayed atop the 2D image. In CF Doppler there are a series of PW sample gates alongside each arranged on a plane. When the CF feature is initiated, a box appears over the 2D echocardiographic image. This box can be moved over the region of interest and size adjusted. The colors of red and blue are traditionally used to display directions of blood flow; blue for away from the transducer and red for toward the transducer. The acronym BART (Blue Away; Red Toward) is often used. Color-flow Doppler allows the speedy identification of both normal and abnormal flow. It aids in the alignment of the PW or CW Doppler cursor for velocity measurements. It also helps locate and estimate the severity of valvular regurgitation or cardiac shunts. An added feature of CF Doppler is **variance**. Variance is the term used for turbulent blood flow, which is normally laminar in the heart. Turbulence is simply blood moving in multiple directions at multiple velocities. To demonstrate turbulence, most echocardiographs add a yellow-green mosaic color map with the standard red or blue signal. The limitation of CF Doppler is that it is also subject to aliasing. As with PW when the returning frequency shift is greater than twice the PRF, then the signal gets "wrapped." In CF Doppler that means a change in color. The colors selected by the computer occur on a color wheel, with shades of blue on one half and shades of red on the other. As the recorded velocity increases in CF, the intensity of the color increases in that direction or color. If the recorded velocity exceeds the PRF, then the computer will take the next color on the color wheel. For example, if the CF signal is looking at flow away from the transducer, it will code the returning signal in blue of intensifying color until it reaches the maximum velocity it can record in the direction (the Nyquist limit), then it will use the next color in the wheel which will be a shade of red. So even though the operator may know that the flow is moving away from the transducer, the color displayed on the screen will be red. Adjustments to the depth of the color box or some changes in PRF may correct this problem, but only to a point.

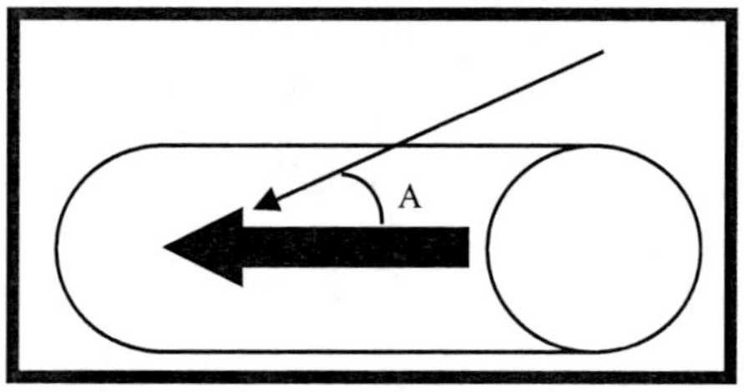

Figure 1
A= Angle of Intercept or θ
Heavy arrow = blood flow
Light arrow = ultrasound beam

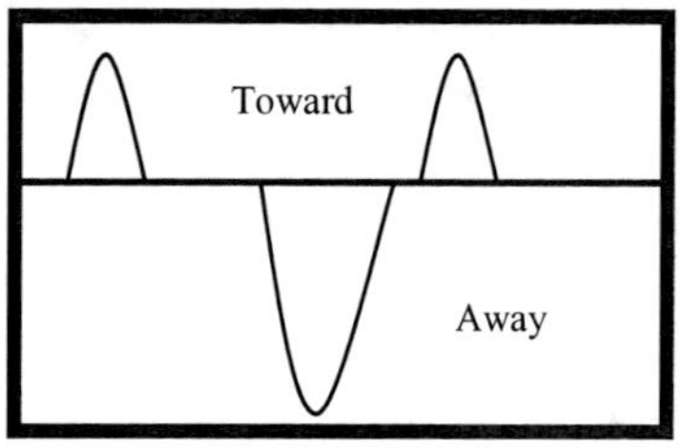

Figure 2

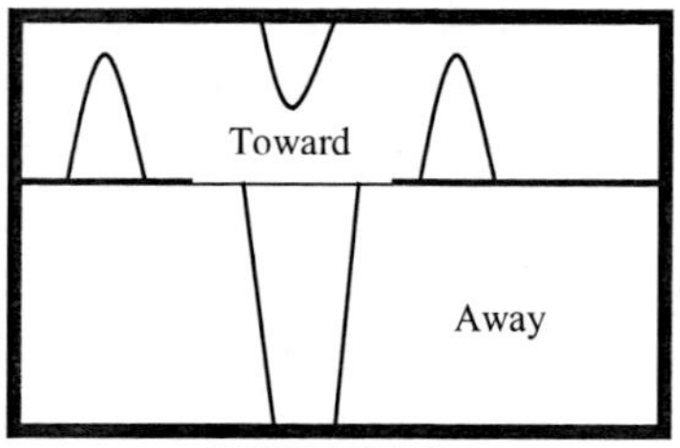

Figure 3

INTERPRETATION

The practical use of Doppler in echocardiography is for the determination of velocity from one chamber of the heart to the next to calculate a pressure gradient (PG). All movement of blood within the body, from a physics point of view, is based on pressure gradients; the increase in pressure in one chamber that drives blood to the next. The velocity of blood moving from one chamber to the next is directly related to the pressure gradient by the modified Bernoulli equation: **PG = 4 x V²** where V is velocity in m/sec. The pressure gradient is expressed in mm Hg. The operator must understand normal cardiac physiology to interpret the findings. For example, it is known that a normal velocity out the aorta is 1.0–1.7 m/sec. Using the Bernoulli equation indicates PG from the LV to the aorta is approximately 4–12 mm Hg. Left ventricular outflow tract obstruction (i.e., systolic anterior motion of the mitral valve, subaortic stenosis) may be present if this velocity is significantly increased. The velocity and calculated PG are then used to assign severity. If the velocity recorded by spectral Doppler is 5 m/sec, then a PG of 100 mm Hg exists. From a physiology standpoint, a 100 mm Hg PG means the LV must then generate an extra 100 mm Hg of pressure to move blood across the aortic valve. This represents a marked increase in the work load of the ventricle indicating severe outflow tract obstruction (the normal 125 mm Hg + the 100 mm Hg of the stenosis). The same function is used for the pulmonic valve to diagnose and assign severity to pulmonic stenosis.

Another application of Doppler-derived PGs is the estimation of normal or abnormal chamber pressures. In the case of a ventricular septal defect (VSD), Doppler can be used to detect equilibration in the ventricles. Again, an understanding of normal physiology is critical. In the normal heart, the LV systolic pressure is approximately 125 mm Hg, and the right ventricular systolic pressure is approximately 25 mm Hg. The pressure difference between the two ventricles is approximately 100 mm

Hg. When the VSD is interrogated by CW Doppler, the operator will record a velocity at or close to 5 m/sec if the ventricles still have their normal pressure, implying normal chamber pressures and therefore a compensated heart. If the velocity recorded is significantly below 5 m/sec, then decompensation may be occurring. Likewise with mitral regurgitation (MR), the velocity of the MR jet should be 5 m/sec or above since the left atrial (LA) pressure should be 5–10 mm Hg during ventricular systole, and the LV pressure should be 125 mm Hg. If the LV to LA pressure gradient is significantly below 5 m/sec, then the left ventricle is not able to generate its normal pressure or there is development of pulmonary hypertension.

In cases of patent ductus arteriosus (PDA), both CW and CF Doppler are of great benefit. Patient with PDA, flow is continuously flowing from the aorta into the main PA. By placing a CF Doppler box over the main PA, a continuous turbulent color signal will be seen and is diagnostic for PDA. The CW cursor can then be placed in the PDA aperture and a velocity through the PDA measured. If the patient is well compensated, then this PG should be close to 100 mmHg, since the normal aortic pressure is 120 mm Hg and the normal PA pressure is 20–25 mm Hg. If the patient is decompensating, then the trans-PDA velocity will be below 5 m/sec, typically indicating lower than normal aortic pressure and a failing left ventricle or development of pulmonary hypertension.

A third application for Doppler is the estimation of chamber pressures "downstream" of two adjacent chambers. Pulmonary arterial pressures can otherwise only be determined by invasive means of direct catheterization and pressure transducers. By using Doppler velocities and pressure gradients of tricuspid regurgitation (TR), the echosonographer can estimate pulmonary arterial pressure noninvasively. First, the operator must determine the PG between the right ventricle (RV) and the pulmonary artery. The PG from the RV to the PA is normally approximately 4 mm Hg, provided there is no pulmonic stenosis which would increase the RV pressure, and thus the RV to PA pressure gradient. With a RV to PA pressure gradient of 4 mm Hg, the operator can stipulate that the RV pressure and the PA pressure are essentially the same (RV mm Hg = PA mm Hg) because 4 is so close to zero. At this point the operator looks for TR and, if present, measures its velocity. If no TR is present, then this application of Doppler is unavailable. The PG of the TR tells the operator the pressure difference between the RV and the right atrium (RA), so if the TR velocity is 4 m/sec then the PG is 64 mm Hg between the RV and the RA. The normal pressure of the RA is also about 4 mm Hg, so the operator can likewise stipulate it to be zero also (technically the 4 mmHg from the PV to PA pressure gradient and the 4 mm Hg RV to RA gradient cancel each other out). The conclusion is that the RV pressure is approximately 64 mm Hg systolic. As stipulated earlier, the RV mm Hg is equal to the PA mm Hg, so the operator knows the PA pressure is also approximately 64 mm Hg. Normal pulmonary arterial pressure in systole is reported to be 20–25 mm Hg, therefore a pressure of 64 mm Hg is indicative of pulmonary hypertension.

In addition to these uses of Doppler echocardiography, assessments of diastolic function can be made. By looking at filling patterns of the left ventricle, and flow through the pulmonary veins as they enter the left atrium, the cardiologist can make some observations about relaxation of the ventricle. Doppler can also be used to look at the isovolumic relaxation time (IVRT), the time between the closing of the aortic valve and the opening of the mitral valve as a measure of cardiac relaxation. As the ventricle fails it does not fill as well and IVRT will increase in length. There are many other methods of assessing diastolic function that are beyond the scope of this proceeding that the interested echosonographer may study.

ABBREVIATIONS

Left ventricle	LV	Pulse-wave Doppler	PW
Right ventricle	RV	Continuous-wave Doppler	CW
Left atrium	LA	Color-flow Doppler	CF
Right atrium	RA	Pulse repetition frequency	PRF
Pulmonary artery	PA	Pressure gradient	PG
Ventricular septal defect	VSD	Meters per second	m/sec

Patent ductus arteriosus	PDA	Millimeters of mercury	mm Hg
Mitral regurgitation	MR	2D	2 dimensional echocardiogram
Tricuspid regurgitation	TR	Isovolumic relaxation time	IVRT

CONCLUSION

Doppler echocardiography is an extremely valuable tool for the diagnosing and prognosticating of cardiac disease. When combined with 2D and M-Mode echocardiography, Doppler can provide a complete picture of acquired and complex congenital heart disease. Color-flow Doppler adds speed and accuracy to the Doppler study by allowing the operator to locate and align with the blood flow much faster. It is incumbent on the echosonographer to become well versed in the operation and limitations of each type of Doppler modality and have a detailed knowledge of cardiac physiology.

acvim003

Understanding the Development of Arrhythmias

Carol Fitzpatrick, CVT, VTS (Cardiology)
Levittown, PA, USA

Understanding the specialized cells, the elements of the cardiac conduction system, and the cardiac action potential is key to understanding the development of arrhythmias.

Two types of cardiomyocytes generate and conduct electrical impulses:

- Cells that generate the stimulus to trigger a heartbeat. Examples are the SA and AV node.
- Cells that respond to the stimulus by converting it into a physical contraction helping to eject blood out of the heart. Examples include atrial or ventricular myocytes.

The Cardiac Conduction System

Five major elements make up the cardiac conduction system (see diagram below). This system incorporates the specialized muscle cells and conducting fibers responsible for initiation and transmission of the impulse throughout the heart.

The five elements are:

- SA node
- AV node
- Bundle of His
- Left and right bundle branches
- Purkinje fibers

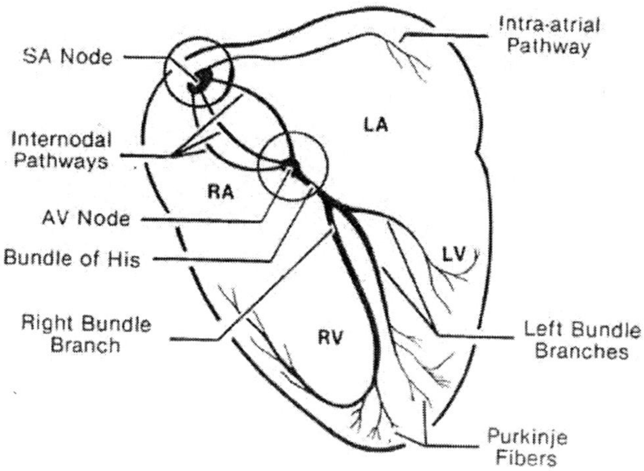

The SA node is located in the wall of the right atrium and can spontaneously generate electrical impulses (action potentials) which spread throughout atria and the heart to generate the heartbeat. As the electrical impulse spreads through the atria, they become excited and will contract simultaneously which ejects blood out of the atria and into the ventricles.

The AV node is located in the center of the heart, at the base of right atrium between the atria and ventricles. The electrical impulse generated at the SA node will travel across the atria and arrive at the AV node. The AV node slows conduction and delays transmission of the electrical impulse. This delay is critical because it allows for maximum filling of the ventricles allowing for optimal cardiac output.

From the AV node the electrical signal travels rapidly to the **bundle of His, the left and right bundle branches and the Purkinje fibers**.

The Purkinje fibers penetrate the left myocardium to aid in coordinated depolarization of the ventricles. Depolarization of the ventricles results in contraction and ejection of blood out of the heart.

THE CARDIAC ACTION POTENTIAL

An action potential is defined as the change in electrical potential due to the passage of an electrical impulse along the membrane of a muscle cell. Below is a diagram showing the action potential curve and a corresponding sample ECG. When there is a net movement of positive charges into a cell, the transmembrane potential becomes more positive. When there is a net movement of positive charges out of a cell, the transmembrane potential becomes more negative.

Surface ECG Recording

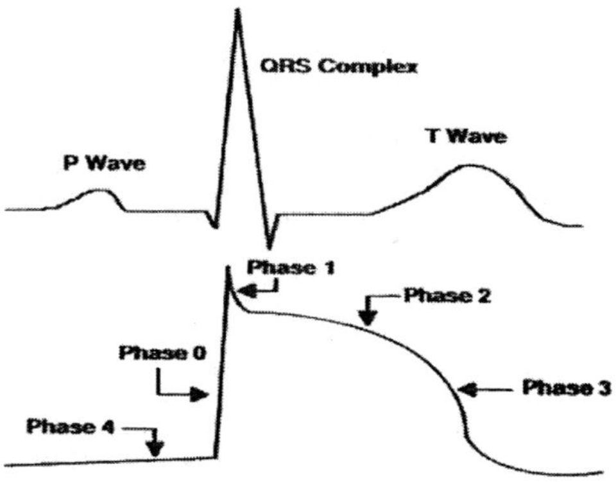

Ventricular Muscle Cell Action Potential

The five phases of the ventricular muscle cell (non-nodal cell) action potential are as follows:

- **Phase 0**: rapid depolarization - slow and fast sodium channels open and ions rush into the cell; slower calcium channels open and calcium begins to move into the cell.
- **Phase 1**: early repolarization - the fast sodium channels close and potassium channels open allowing a rapid efflux of potassium out of the cell.
- **Phase 2**: plateau phase - some sodium and calcium are moving into the cell while potassium is moving out of the cell. Positive charges are moving in and out of the cell at the same time, so the charge inside the cell remains relatively positive. Calcium ions move more slowly than the sodium ions and this prevents the cell from repolarizing too rapidly. This helps extend the refractory period and the rate at which the cardiac tissue can depolarize.
- **Phase 3**: repolarization - the calcium and sodium channels are closed and the potassium channels are still open, making the cell more negative.
- **Phase 4**: resting potential - the potassium channels are closing slowly and the cell is repolarizing so that the cycle can begin again at phase 0.

Notice how the phases of the action potential correspond to the ECG. The movements of sodium, potassium, and calcium ions in and out of the cell in the sequence described are responsible for the action potential, which in turn causes cardiomyocytes to contract and a heartbeat to occur. Ventricular arrhythmias occur when there are changes in phase 2 and phase 3 (delayed after depolarizations). Phase 2 and 3 are phases of the refractory period. The refractory period is a protective mechanism in the heart to prevent multiple action potentials from occurring, thus limiting excessive heart rates.

The SA and AV nodes have their own action potential. These cells are also known as pacemaker cells and are capable of generating regular spontaneous action potentials, which is termed automaticity. In other words, these cells are capable of spontaneous depolarization. This is different from the non-nodal cells which require an external stimulus to depolarize and generate an action potential. Nodal (pacemaker) cells will spontaneously depolarize and their electrical impulse spreads to non-nodal cells to trigger depolarization of the rest of the myocardium to generate the heartbeat.

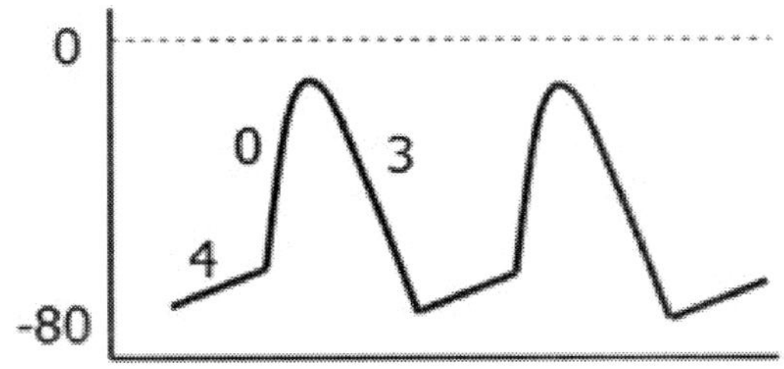

Pacemaker cell action potential

The three phases of the pacemaker cell (nodal) action potential are as follows:

- **Phase 4:** spontaneous depolarization - the slow influx of sodium ions causes the membrane potential to begin to spontaneously depolarize; calcium channels then begin to open and the depolarization progresses.
- **Phase 0:** depolarization - because the movement of calcium ions is slower than the fast sodium channels in the ventricular muscle cell, the rate of depolarization is slower.
- **Phase 3:** repolarization - the calcium channels become inactivated and close and potassium exits the cell. The cycle begins again at phase 4.

Spontaneous depolarization of the pacemaker cells can be affected by autonomic nerves, hormones, drugs, and ischemia/hypoxemia.

Three broad mechanisms of arrhythmogenesis are defined:

- Abnormal automaticity
- Triggered activity
- Reentry

ABNORMAL AUTOMATICITY

In the normal heart, only certain cells have the property of spontaneous diastolic depolarization. With disease, non-nodal cells can become partially depolarized and spontaneous diastolic depolarization can occur. For example, in ischemic tissue the cells can become partially depolarized affecting the movement of sodium and calcium ions, which causes a repetitive impulse initiation, thus generating a tachyarrhythmia.

TRIGGERED ACTIVITY

Triggered activity refers to the development of an impulse due to **afterdepolarizations** (increases in membrane potential that occur during or immediately after an action potential). Afterdepolarizations are classified as early or delayed depending on where in the action potential they occur. They can only exist in the presence of a previous action potential, and if they generate sufficient increase in membrane potential they can trigger a new action potential, which may result in tachyarrhythmias.

REENTRY

A reentry arrhythmia occurs when the electrical impulse loops back on itself. This looping takes additional time and the impulse will find cells that have repolarized and are able to be re-excited. This can set up a continuous cycle resulting in a tachyarrhythmia.

Alterations in timing and the refractory state of the tissue are critical for reentry to occur. Many antiarrhythmic drugs alter the refractory period or conduction velocity in an attempt to abolish the reentrant mechanisms.

Part two of this lecture will focus on the treatment of arrhythmias. We will review the classification of antiarrhythmic medications, their uses with particular arrhythmias, and their side effects.

References

1. Fox PR, Sisson D, Moise NS. *Textbook of Canine and Feline Cardiology: Principles and Clinical Practice*. Philadelphia, PA: WB Saunders; 1999.
2. Kittleson MD, Kienle RD. *Small Animal Cardiovascular Medicine*. St. Louis, MO: Elsevier/Mosby; 1998.
3. Silverstein DC, Hopper K. *Small Animal Critical Care Medicine*. 2nd edition. St. Louis, MO: Elsevier Saunders; 2015.

acvim004

Understanding the Treatment of Arrhythmias

Carol Fitzpatrick, CVT, VTS (Cardiology)
Levittown, PA, USA

In veterinary medicine the most common treatment for arrhythmias is medication. Different medications alter the action potential in different ways. A classification scheme organizes the different medications based on how they alter the action potential.

Vaughan-Williams classification

Class			Mechanism	Example
I	A	Sodium channel blockers	Moderate Na channel blockade, ↑ ERP, variable anticholinergic effects	Procainamide, quinidine
	B		Weak Na channel blockade, ↑ ERP	Lidocaine, mexiletine
	C		Strong Na channel blockade, → ERP	Flecainide, propafenone
II		Beta blockers	Depress phase 4	Atenolol, esmolol, propanolol
III		Potassium channel blockers	Prolong phase 3	Sotalol, amiodarone
IV		Calcium channel blockers	Prolong phase 2	Diltiazem, verapamil
V		Other		Digoxin, atropine

Class I Drugs - Sodium Channel Blockers

Sodium channel blockers block fast sodium channels responsible for rapid depolarization (Phase 0) of non-nodal cardiomyocytes. Phase 0 relies on activation of fast sodium channels and rapid entry of sodium into the cell, so sodium channel blockers decrease the slope of phase 0 and decrease the amplitude of the action potential, which in turn decreases conduction velocity. By blocking the sodium channels, the slope of Phase 0 and the action potential is decreased. This decreases conduction velocity. By slowing the rate of depolarization, the rate of conduction can be decreased and tachyarrhythmias suppressed. In addition, sodium channel blockers may also affect the effective refractory period related to drug actions on the potassium channels.

Class I antiarrhythmics are subdivided into three classes (IA, IB, IC). They each affect the sodium channel slightly differently and have different kinetics.

- Class IB drugs (such as lidocaine) have fast kinetics and will only block the sodium channels at **high heart rates**. At normal heart rates, these drugs do not affect the sodium channels. These drugs decrease the effective refractory period.
- Class IC drugs are not common in veterinary medicine and have slow kinetics and affect the sodium channel at all heart rates. These drugs do not significantly alter the action potential duration.
- Class IA drugs (such as procainamide) have intermediate kinetics. These drugs increase the effective refractory period.

Sodium channel blockers may also change the effective refractory period which is due to the drug actions on the potassium channels involved in phase 3 - repolarization. Increasing the effective refractory period increases or decreases arrhythmogenesis depending on the underlying cause of the arrhythmia. For example, in a reentrant arrhythmia the drug may help terminate a tachyarrhythmia by prolonging the time that the tissue is unexcitable. However, increasing the effective refractory period increases the action potential and can precipitate afterdepolarizations which can trigger tachyarrhythmias such as Torsades de Pointes.

By unknown mechanisms, sodium channel blockers can suppress abnormal automaticity by decreasing the slope of Phase 4.

Side effects of class I drugs: all have the potential to be proarrhythmic, that is, they have the potential to generate a drug-related arrhythmia. In addition, they can cause nausea, vomiting, lethargy, tremors, seizures. These typically resolve with discontinuing the drug.

CLASS II - BETA-BLOCKERS

Beta blockers are drugs that bind to beta adrenergic receptors and block the binding of norepinephrine and epinephrine. The beta receptors are located in the cardiac nodal tissue (the SA node and AV node), the conducting system and contracting myocytes. The heart has both β1 and β2 receptors but the predominant receptor is β1. These receptors primarily bind norepinephrine which is released from sympathetic nerves.

Beta-blockers can blunt the sympathetic effects and decrease heart rate, conduction velocity, contractility, relaxation rate and can inhibit aberrant pacemaker activity.

Side effects of class II drugs: bradycardia, hypotension, AV block, bronchoconstriction.

CLASS III - POTASSIUM CHANNEL BLOCKERS

Potassium channel blockers bind to and block the potassium channels that are responsible for Phase 3 - repolarization. Blocking these channels slows repolarization, which leads to an increase in action potential and an increase in the effective refractory period. Because the action potential is prolonged, the heart rate is reduced.

By increasing the effective refractory period, these drugs are useful in suppressing both supraventricular and ventricular tachyarrhythmias caused by reentry mechanisms.

It is important to note that all class III antiarrhythmics have effects in other classes as well. For example, sotalol is a potassium channel blocker and has beta blocker activity.

Side effects: proarrhythmic, hypotension (especially amiodarone).

CLASS IV - CALCIUM CHANNEL BLOCKERS

Calcium channel blockers are divided into two groups: dihydropyridines (amlodipine) and non-dihydropyridines (diltiazem).

While amlodipine is included in the antiarrhythmic classification scheme, its primary use is for treatment of systemic hypertension.

The non-dihydropyridines are calcium channel blockers that are relatively selective for the myocardium and are useful for treating arrhythmias. By reducing the slope of phase 4 which decreases the rate of spontaneous depolarization, thus reducing the rate of pacemaker firing. These drugs are mainly used to treat supraventricular tachyarrhythmias.

Side effects of Class IV drugs: excessive bradycardia, AV node block, and negative inotropy, hypotension, bradycardia.

CLASS V - MISCELLANEOUS

Digoxin decreases conduction of electrical impulses through the AV node. It also increases parasympathetic activity to the heart which decreases SA node firing. It is used to treat supraventricular tachyarrhythmias, most commonly atrial fibrillation.

Side effects of digoxin: digoxin has a relatively narrow therapeutic safety window and must be monitored. Too much can cause new arrhythmias. When the therapeutic window is exceeded, the animal may develop decreased appetite, vomiting, and first-degree AV block. These side effects generally resolve with a decrease in dose. Toxicity more likely with renal dysfunction, hypokalemia, advancing age, chronic lung disease, hypothyroidism.

Atropine inhibits the effects of excessive parasympathetic/vagal tone on the heart which is manifested as sinus bradycardia or AV nodal block. Atropine prevents acetylcholine from binding to its

receptors which are found principally on cells in the SA and AV node, resulting in an increase in HR and conduction velocity.
Side effects of atropine: tachycardia, mydriasis, dry mouth, urinary retention, inhibition of sweating, and constipation.

REFERENCES

1. Fox PR, Sisson D, Moise NS. *Textbook of Canine and Feline Cardiology: Principles and Clinical Practice*. Philadelphia, PA: WB Saunders; 1999.
2. Kittleson MD, Kienle RD. *Small Animal Cardiovascular Medicine*. St. Louis, MO: Elsevier/Mosby; 1998.

acvim005

The Use of Anti-Arrhythmic Drugs vs. Electrical Cardioversion to Treat Tachyarrhythmias in the Canine and Feline Patient

Anne Myers, RVT, VTS (Cardiology)
Apex, NC, USA

Introduction

When a canine or feline patient with congenital or acquired heart disease presents to a veterinary hospital with a serious disturbance of their cardiac conduction system resulting in an abnormally high heart rate and rhythm, correct diagnosis and treatment relies on the ability to perform a diagnostic quality electrocardiogram (ECG) for interpretation. Many cases are in need of immediate triage, such as the hemodynamically unstable patient with sustained runs of ventricular tachycardia and "R on T" complexes which can progress to ventricular fibrillation and cardiac arrest. Sustained runs of supraventricular tachycardia (SVT) or high heart rate atrial fibrillation (AF) should also be treated as soon as possible, but are not as life-threatening. A canine patient with lone A-fib presents with the arrhythmia, but not the rapid rate, so there is more time to make a treatment plan. The goals for treating a tachyarrhythmia are to lower heart rate and try to correct the rhythm in order to decrease workload on the heart muscle, improve cardiac output and make the patient feel better. If the patient is unresponsive to administration of antiarrhythmic drugs, in some cases, another option is to deliver a transthoracic electrical shock to the heart in order to restore a normal rate and rhythm. This can be performed as an emergency procedure, or as a planned outpatient procedure, depending on the type of tachyarrhythmia.

Pharmacologic Treatment of Ventricular Tachycardia

The first line of treatment for any tachyarrhythmia is usually the administration of anti-arrhythmic drugs. For sustained runs of ventricular tachycardia (VT) in the canine and feline, immediate treatment with an intravenous bolus of **lidocaine** is warranted. Lidocaine is a class Ib antiarrhythmic drug that acts to block sodium channels in the myocardium, slowing ventricular depolarization. A dose of 2 mg/kg can be repeated up to 4 times for a total of 8 mg/kg in the canine; however, after that; lidocaine can cause gastrointestinal and neurologic toxicity in the form of vomiting and seizures. Felines are extremely sensitive to the side effects of lidocaine and a lower dose of 0.2–0.75 mg/kg IV can be used up to 3 times. Once a patient has reverted back to a sinus rhythm, many veterinarians will also administer an oral dose of **sotalol** (canine 1–2.5 mg/kg q12h, feline 10 mg q12h). This class II-III non-selective beta blocker can control ventricular arrhythmias by blocking potassium channels in the myocardium, slowing ventricular repolarization and prolonging the action potential and refractory period at the AV node. If stable, the patient can be sent home on sotalol and should return for a 24-hour Holter monitor in 1 week to evaluate efficacy of drug therapy. The addition of the oral sodium channel blocker mexiletine may be indicated.

Unsynchronized Electrical Cardioversion of Ventricular Fibrillation

If the VT is unresponsive to lidocaine or other anti-arrhythmic medications (i.e., procainamide, esmolol) and progresses to ventricular flutter or fibrillation, the next step, if the equipment is available, would be unsynchronized electrical cardioversion, also called **defibrillation** in which a strong electrical impulse of 3 to 5 joules/kg is delivered to the heart with hand held paddles or electrode pads placed on the chest wall at the level of the ventricles. This "shock" causes myocardial depolarization and asystole, allowing the sinoatrial node to begin the normal depolarization/repolarization sequence. If successful, the patient is usually monitored in hospital for 24 hours. Discharge medications and follow up are similar to primary pharmacologic treatment.

Supraventricular Tachycardia

Treatment to break sustained runs of SVT should be started as soon as possible to avoid myocardial failure due to rapid pacing. This can start with a vagal maneuver such as carotid artery massage or ocular compression. If there is transient or no response, a slow intravenous bolus of the calcium channel blocker **diltiazem**, can be administered at a dose of 0.1–0.25 mg/kg. This drug works by slowing conduction through the SA and AV node. Other intravenous drugs used to treat SVT include the calcium channel blocker **verapamil**, or the beta blocker **esmolol**. If there is no response to IV meds, synchronized electrical cardioversion is a last option and will be discussed in detail in the lone A-fib section of this paper.

Atrial Fibrillation

Atrial fibrillation (AF) is a type of SVT in which the atria send rapid, irregular impulses to the AV node, triggering the ventricles to contract. AF comes in two forms; the first due to structural heart disease and dilation of the atria from cardiomyopathy, congenital heart defects such as AV valve dysplasia, or from chronic AV valve degeneration. Many of these patients are in heart failure and sympathetic tone is high, resulting in an elevated ventricular response and heart rates ranging from 160 to over 250 beats per minute. The other type is "**lone A-fib**" occurring in large and giant breed dogs, humans, and horses that by the nature of their size have large atrial chambers. Lone A-fib can be idiopathic, or the result of an anesthetic episode and may be transient. Heart rate is normal or only mildly elevated, because these patients are not in heart failure and the AV node can act to filter some of the fibrillation waves. However, these patients are lacking the "atrial kick" part of normal sinus node conduction which will affect cardiac output and chronically, may cause structural and functional remodeling of the heart. Most lone A-fib canine patients are asymptomatic, with discovery occurring on a routine physical or pre anesthetic exam, or they may have a history of mild exercise intolerance. Many of these patients are good candidates for a procedure known as **synchronized electrical cardioversion** in which an electrical impulse is delivered to the atria, resetting the rhythm to sinus and restoring a normal conduction pathway.

Pharmacologic Treatment of Atrial Fibrillation in the Heart Failure Patient

Because the duration of sinus rhythm after synchronized electrical cardioversion is significantly less for dogs with structural heart disease,[6] and anesthesia is more of a risk, antiarrhythmic medications are used to control heart rate in these patients. The goal is to slow conduction through the AV node to lower the ventricular response rate. Oral **diltiazem** is frequently used at a dose of 0.5 to 1.5 mg/kg q8h for the canine and 0.5 to 2 mg/kg q8h for the feline. It also comes in an extended release form (XR) at a dose of 3–4 mg/kg q12h in the canine and 30 mg q12–24h in the feline. Oral **digoxin** is a cardiac glycoside that prolongs cardiac depolarization, by slowing AV node conduction time, thus slowing heart rate. It also enhances cardiac muscle contractility by increasing the availability of intracellular calcium and with careful monitoring of blood levels, makes it a good choice for heart failure patients in AF. Dosing in the canine is 0.003–0.005 mg/kg q12h and ¼ of a 0.125 mg tablet every 24 to 48 hours in the feline. Treatment with an ACE inhibitor, diuretic and an inodilator is added.

Synchronized Electrical Cardioversion for Lone A-Fib

Pharmacologic treatment with oral diltiazem can also be tried first in the lone A-fib patient; however, long-term complications such as cardiomyopathy may still occur. Using the **LIFEPAK 20/20e** defibrillator/monitor device, cardioversion to a sinus rhythm can be performed by a veterinary cardiologist and staff as a hospital day case. The canine patient is placed under general anesthesia as the shock does cause discomfort. Before induction, the right and left lateral/ventral chest is clipped and cleaned. Two adult-sized QUIK-COMBO therapy electrode pads are placed on the chest over the atria, or hand-held paddles with a metallic plate surface covered with a saline based conductive gel can be used to deliver the shock. The patient is positioned in right lateral recumbency and attached to the ECG of the LIFEPAK which analyzes patient rhythm and identifies the R wave of the QRS complex. The system allows the operator to deliver a brief, high-energy **biphasic** pulse of direct current electricity measured in joules. A biphasic shock allows the current to be applied from two different directions by changing the polarity of the electrodes in the latter portion of the shock. This uses less energy and makes it more efficient than a monophasic shock in which the electrical current enters from just one electrode.

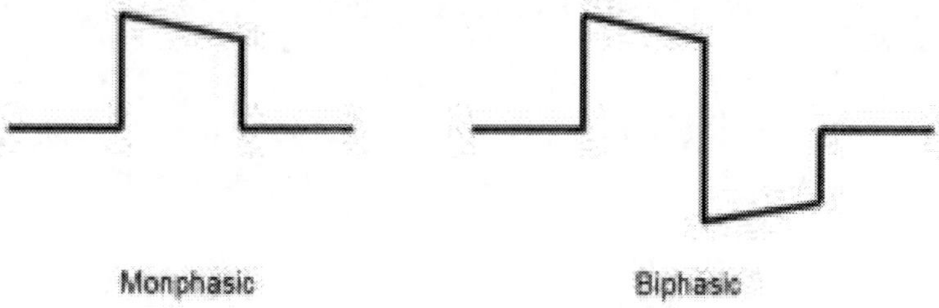

!!!!!!!!!!!!!!!!!!!!!VIN editor: Insert title/caption!!!!!!!!!!!!!!!!!!!!!Successful synchronized electrical cardioversion in a canine patient with lone A-fib. Notice that synchronization to the R wave is disabled after the shock is delivered

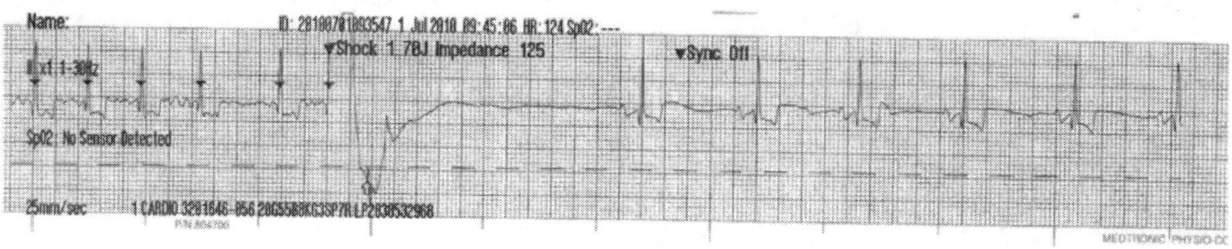

The most important aspect of the procedure is the ability to deliver the electrical impulse at the optimal moment of the cardiac cycle, which corresponds to the R wave of the QRS complex. **Synchronizing the shock to the R wave (depolarization) prevents delivery during the T wave (repolarization) which could induce a chaotic reentry circuit and ventricular fibrillation.**

If more than one shock needs to be administered, along with an increase in joules, the operator must resynchronize the ECG monitor to make sure the system is set to deliver another shock on the R wave, as most will default to an unsynchronized defibrillation mode after each shock in case emergency defibrillation is needed. A successful shock of 1 to 2 joules/kg usually depolarizes sufficient atrium to disrupt fibrillation waves and restore sinus rhythm. The patient is recovered from anesthesia, monitored to ensure stability of the rhythm, and can go home the same day. Heart rate and rhythm should be rechecked periodically by ECG or Holter monitor, which can be performed by a veterinary technician for review by the cardiologist.

References

1. Atkins CE *et al. J Vet Intern Med.* 1995;100:807.
2. Bright JM *et al. J Vet Intern Med.* 2008;22:114.
3. Estrada EH *et al. J Vet Card.* 2009;11:137.
4. LIFEPAK 20/20e device description manual on CD 2010
5. McGurrin K. Therapeutic options in atrial fibrillation. 2010 ACVIM Forum. June 9–12. Anaheim, CA.
6. Romain P. Atrial Fibrillation: An Overview of the Most Common Arrhythmia. 2010 ACVIM Forum. June 9-12. Anaheim, CA.
7. Tilley LP, Smith Jr. WKF, Oyama M, Sleeper MM. Chapter 16. In: *Manual of Canine and Feline Cardiology.* Saunders Elsevier. 2008;315–332.
8. Ware WA. *Cardiovascular Disease in Small Animal Medicine.* Manson Publishing Ltd. 2007;205–208.

acvim007

Cardiac Effects of Systemic Disease: Structural, Functional, and Electrical Heart Changes of Noncardiac Origin

Jessica Ward, DVM, DACVIM (Cardiology)
Ames, IA, USA

Introduction

When evaluating veterinary patients who present with clinical signs suggestive of a cardiac problem, it is important to remember that not all cardiac changes are the result of **primary** structural heart disease. Many systemic diseases can have **secondary** effects on the heart. In such cases, the heart is not the underlying problem, but rather a "target" or "bystander" damaged by the primary disease. Prompt recognition of the true underlying disease is critical for proper management. This presentation discusses 5 common cardiac manifestations of systemic disease in dogs and cats.

1. Left Ventricular Hypertrophy in Cats Secondary to Systemic Hypertension or Hyperthyroidism

Differential Diagnoses for Left Ventricular Hypertrophy in Cats

Left ventricular (LV) hypertrophy is the most common cardiac disease "phenotype" in cats (the most common echocardiographic abnormality), and can result from several different disease processes. A mnemonic to remember the 4 main causes of LV hypertrophy in cats is "The 3 HYPERs": HYPERtrophic cardiomyopathy (HCM), HYPERtensive heart disease, and HYPERthyroid heart disease. **Hypertrophic cardiomyopathy** (HCM) is the most common **primary** cardiomyopathy in cats and is caused by abnormalities in the contractile elements of the heart muscle (the sarcomere). Abnormal structure and function leads to impaired relaxation (diastolic dysfunction), and the heart wall responds by getting thicker (concentric hypertrophy). **Systemic hypertension** is a common vascular disease of older cats, most commonly caused by chronic kidney disease or hyperthyroidism. The resulting pressure overload causes the left ventricle to get thicker (concentric hypertrophy). **Hyperthyroidism** is a common endocrine disease in older cats caused by a functional thyroid adenoma. The overproduction of thyroid hormone increases the heart's workload by upregulating the body's metabolism and oxygen demand; additionally, thyroid hormone directly stimulates heart muscle growth (hypertrophy). Among these 3 diseases, HCM is most common in middle-aged (and particularly male) cats; hyperthyroidism and systemic hypertension are much more common in cats over the age of 10 years.

Diagnosis and Treatment of Secondary Left Ventricular Hypertrophy in Cats

Echocardiography establishes the presence of LV hypertrophy. Thoracic radiographs and point-of-care thoracic ultrasound can identify pulmonary edema and/or pleural or pericardial effusion suggestive of congestive heart failure (CHF). However, additional testing is required to diagnose (or rule out) a secondary cause of LV hypertrophy. Specifically, systolic blood pressure measurement and thyroid hormone level are recommended in any cat over the age of ~5–6 years diagnosed with LV hypertrophy. If systemic hypertension or hyperthyroidism are identified, these diseases **must** be treated in addition to the secondary cardiac changes (and CHF, if present). Depending on the severity at time of diagnosis, some of the cardiac changes may "reverse remodel" (return to more normal appearance) with adequate treatment of the underlying disease.

For a patient diagnosed with CHF secondary to...	What diagnostic testing is recommended?	What treatments are recommended?	Will cardiac changes reverse?
Hypertrophic cardiomyopathy	Echocardiogram (Assess treatment side effects: renal panel, blood pressure)	Furosemide Pimobendan Enalapril/benazepril Clopidogrel	No

Hyperthyroidism	Echocardiogram Thyroid level (T4) Blood pressure CBC/Chem/UA	(Same as HCM) **plus** Methimazole or I-131	Maybe
Systemic hypertension	Echocardiogram Blood pressure (Investigate cause: CBC/Chem/UA/T4)	(Same as HCM) **plus** Amlodipine	Maybe

2. High-Output Heart Failure Secondary to Anemia

Anemia and the Heart

Severe anemia can affect the heart in 2 main ways. First, anemic patients can develop a **physiologic heart murmur** from turbulent blood flow due to decreased blood viscosity. This heart murmur itself is benign and does not require specific treatment. Of greater clinical significance, chronic severe anemia can lead to **high-output heart failure** (HOHF), or congestive heart failure in the setting of **increased** cardiac output. Echocardiographically, HOHF patients have 4-chamber enlargement with no other abnormalities. Besides severe anemia, other causes of HOHF include pregnancy, arteriovenous fistulas, and thiamine deficiency. The mechanism by which anemia leads to HOHF relates to systemic vascular resistance. Recall that blood pressure (BP) = cardiac output (CO) x systemic vascular resistance (SVR). In severe anemia, systemic vascular resistance is **decreased** due to a combination of (1) decreased blood viscosity and (2) peripheral vasodilation induced by tissue hypoxia. In order to maintain blood pressure in the face of decreased systemic vascular resistance, the body expands intravascular volume to increase cardiac output. In the short term, this compensatory mechanism maintains blood pressure and improves oxygen delivery; however, the long-term consequence is volume overload and HOHF.

Diagnosis and Treatment of Cardiac Changes Secondary to Anemia

HOHF is a relevant differential diagnosis for tachypnea/dyspnea in severely anemic patients. Echocardiography can detect 4-chamber enlargement and to rule out other causes of heart murmurs. Thoracic radiographs and point-of-care thoracic ultrasound can identify CHF. If HOHF is diagnosed, short-term treatment for CHF is indicated in addition to treatment of the underlying anemia. Blood transfusions are critical in restoring appropriate blood viscosity and systemic vascular resistance. Once anemia resolves, cardiac changes typically reverse.

For a patient diagnosed with heart disease secondary to...	What diagnostic testing is recommended?	What treatments are recommended?	Will cardiac changes ever reverse?
Anemia	Echocardiogram Thoracic radiographs (Workup underlying cause of anemia)	Blood transfusion Treat underlying cause of anemia If patient is in CHF, short-term: Furosemide + pimobendan +/- ACEi, Clopidogrel	Yes

3. Sepsis-Induced Myocardial Dysfunction

Sepsis and the Heart

Sepsis-induced myocardial dysfunction (SIMD) is a syndrome of cardiac dysfunction caused by systemic inflammation and/or infection. The phenotypic appearance of SIMD is LV systolic dysfunction (impaired contractility) and LV dilation, or a "dilated cardiomyopathy phenotype." The pathophysiology of SIMD is likely multifactorial, involving a combination of myocardial ischemia, proinflammatory cytokines, structural damage to cardiomyocyte mitochondria and/or cytoskeletons, and dysregulation of signaling pathways. SIMD is common in critically ill humans and has been reported in dogs. Clinical manifestations of SIMD can include presence of a new heart murmur, hypotension due to poor cardiac output, and even CHF.

Diagnosis and Treatment of Sepsis-Induced Myocardial Dysfunction

SIMD is an important differential diagnosis for weakness/syncope, hypotension, new heart murmurs, or tachypnea/dyspnea in patients with sepsis or SIRS. Echocardiography will identify LV systolic dysfunction and LV dilation (as well as rule out vegetative endocarditis), and thoracic radiographs can diagnose CHF, if present. Cardiac troponin I, a cardiac biomarker that increases with cardiomyocyte damage, can also be useful. Positive inotropes are indicated for treatment of hemodynamically significant SIMD. Typically, oral pimobendan is sufficient, but IV dobutamine can be used in very ill animals incapable of taking oral medications. Cardiac changes associated with SIMD typically resolve within 1 week of resolution of sepsis/SIRS.

For a patient diagnosed with cardiac changes secondary to...	What diagnostic testing is recommended?	What treatments are recommended?	Will cardiac changes ever reverse?
Sepsis-induced myocardial dysfunction	Echocardiogram Thoracic radiographs Troponin I (Workup underlying cause of sepsis)	Treat septic shock and underlying cause of sepsis Pimobendan (if able to receive PO medications) Dobutamine (if severe)	Yes

4. Noncardiac Causes of Ventricular Arrhythmias

Differential Diagnoses for Ventricular Arrhythmias

Ventricular premature complexes (VPCs) and ventricular tachycardia (VT) are often associated with structural heart diseases, particularly dilated cardiomyopathy (DCM) or arrhythmogenic right ventricular cardiomyopathy (ARVC). However, ventricular arrhythmias can also be caused by noncardiac disease. The acronym HEART categorizes differential diagnoses for ventricular arrhythmias: **H**eart disease; **E**lectrolyte abnormalities (particularly hypokalemia and hypomagnesemia); **A**utonomic imbalance (increased sympathetic tone); **R**andom intra-abdominal diseases (particularly splenic disease and gastric dilatation-volvulus); and **T**oxins/drugs or **T**rauma. The pathophysiologic mechanism for ventricular arrhythmias associated with intra-abdominal disease is poorly understood, but may involve cardiomyocyte electrical instability (abnormal repolarization), myocardial depressant factors (cytokines), and/or myocardial ischemia.

Diagnosis and Treatment of Ventricular Arrhythmias

An electrocardiogram establishes presence of ventricular ectopy. If echocardiography fails to identify structural heart disease, noncardiac causes of ventricular arrhythmias should be investigated with an electrolyte panel, abdominal imaging, and historical questioning about possible toxicity or trauma. Regardless of etiology, ventricular arrhythmias which are either hemodynamically significant (ventricular tachycardia) or which predispose patients to electrical instability and sudden death (R-on-T phenomenon) require anti-arrhythmic treatment. Lidocaine IV is the appropriate first-line therapy. If continued anti-arrhythmic treatment is needed in hospital, a lidocaine CRI or oral sotalol can be considered. Generally, ventricular arrhythmias caused by noncardiac disease resolve rapidly with treatment of the underlying abnormality.

For a patient with ventricular arrhythmias secondary to...	What diagnostic testing is recommended?	What treatments are recommended?	Will ventricular arrhythmias resolve?
Structural heart disease	Echocardiogram Thoracic radiographs	Lidocaine → sotalol ACEi, pimobendan, +/- furosemide	No; likely will require lifelong anti-arrhythmic treatment
Electrolyte abnormalities (↓K^+, ↓Mg^{++})	Electrolyte panel	Electrolyte supplementation	Yes
Intra-abdominal disease	Abdominal imaging (XR,	Lidocaine +/- short-	Yes

(splenic disease, GDV, sepsis)	US) CBC/Chemistry/UA	term sotalol Treat intra-abdominal disease	
Toxins/drugs	CBC/Chemistry/UA +/- blood drug/toxin levels	Lidocaine Decontamination	Yes

5. VAGALLY MEDIATED BRADYARRHYTHMIAS

Vagal Tone and the Heart

The vagus nerve provides parasympathetic innervation to the heart, predominately to the sinoatrial and atrioventricular (AV) nodes. Increased parasympathetic outflow (increased vagal tone) decreases heart rate and slows AV conduction. Exaggerated vagal tone can lead to bradyarrhythmias including sinus bradycardia, profound sinus arrhythmia, low-grade AV block, and periods of sinus arrest. Certain systemic diseases can cause abnormally high vagal tone, including disorders of the gastrointestinal, respiratory, ocular, neurologic, or endocrine systems.

Diagnosis and Treatment of Vagally Mediated Bradyarrhythmias

An electrocardiogram will characterize the bradyarrhythmia. An atropine response test is the diagnostic test of choice to confirm that a bradyarrhythmia is mediated by high vagal tone. Atropine 0.04 mg/kg is given SQ and an ECG is repeated 20–30 minutes later. If injection of this parasympatholytic converts the rhythm to sinus (generally >150 bpm), then the cause of the bradyarrhythmia is high parasympathetic tone. (If atropine has no effect, then the bradyarrhythmia is likely due to structural disease [most commonly fibrosis] of the cardiac conduction system.) No long-term treatment is required for vagally mediated bradyarrhythmias.

For a patient with a bradyarrhythmia due to...	What diagnostic testing is recommended?	What treatments are recommended?	Will the bradyarrhythmia ever resolve?
High vagal tone	Electrocardiogram Atropine response test +/- Echocardiogram (if murmur) Consider working up underlying cause of high vagal tone	None required Treat concurrent heart disease (if present) Treat underlying cause of high vagal tone	Maybe

CONCLUSION

Many systemic diseases can have negative effects on the heart. Practitioners and technicians should be aware of the mechanisms by which noncardiac diseases can affect the heart, because diagnosis and treatment of the underlying disease is imperative for effective treatment of secondary cardiac changes.

REFERENCES

1. Brown *et al. J Vet Intern Med.* 2007;21:542–558.
2. Reusch *et al. Vet Clin North Am Small Anim Pract.* 2010;40(2):335–352.
3. Chetboul *et al. J Vet Intern Med.* 2003;17(1):89–85.
4. Snyder *et al. J Vet Intern Med.* 2001;15:52–56.
5. Liu *et al. Am J Pathol.* 1981;102(3):388–395.
6. Bond *et al. J Am Vet Med Assoc.* 1988;192(11):1546–1549.
7. Syme. *Vet Clin North Am Small Anim Pract.* 2007;37(4):723–743.
8. Sangster *et al. J Vet Intern Med.* 2014;28:465–472.
9. Yaphe *et al. J Am Vet Med Assoc.* 1993;202:961–964.
10. Spotswood *et al. Vet Radiol Ultrasound.* 2006;47:358–365.
11. Wilson *et al. J Feline Med Surg.* 2010;12(1):904–909.
12. Morelli *et al. Intensive Care Med.* 2005;31(5):638–644.
13. Nelson & Thompson. *J Am Anim Hosp Assoc.* 2006;42(5):344–349.

14. Dickinson *et al. J Vet Intern Med.* 2007;21:1117–1120.
15. Langhorn *et al. J Vet Intern Med.* 2013;27(4):895–903.
16. Celes *et al. Pathobiology.* 2013;80(2):70–86.
17. Schober *et al. J Am Vet Med Assoc.* 2002;221(3)L381–388.
18. Macphail *et al. J Vet Emerg Crit Care.* 2006;16(3):192–198.
19. Hamlin. *Pharmacol Ther.* 2007;113(2):276–295.
20. Bulmer. *Vet Clin North Am Small Anim Pract.* 2011;41(4):717–726.
21. Bruchim *et al. J Vet Emerg Crit Care.* 2012;22(4):419–427.
22. Ward *et al. J Vet Cardiol.* 2016;18(3):199–212.

acvim008

Recent Trends in Diagnosis and Treatment of Congestive Heart Failure in Dogs and Cats

Jessica Ward, DVM, DACVIM (Cardiology)
Ames, IA, USA

INTRODUCTION

Management of congestive heart failure (CHF) in companion animals has come a long way since the days of furosemide monotherapy. Although diuretics remain important, veterinarians today have many more "tools in our toolbox" to both diagnose and treat CHF in dogs and cats. This presentation will cover 5 recent trends in management of CHF in small animals.

1. NT-PROBNP FOR DIAGNOSIS OF CHF IN CATS

Theory and Evidence

B-type natriuretic peptide (BNP) is a hormone secreted by the heart. In normal animals, BNP is secreted in small amounts in the atria; in patients with heart disease (specifically in response to ventricular wall stress), BNP is produced in larger amounts in the ventricles. BNP secretion is an adaptive response; BNP counteracts the renin-angiotensin-aldosterone system (RAAS) by causing diuresis, natriuresis (excretion of sodium), and vasodilation. Because BNP is hypersecreted in patients with heart disease, it acts as a circulating biomarker of cardiac disease. BNP has many possible clinical applications in the diagnosis and monitoring of heart disease in veterinary patients. Of practical importance, the prohormone BNP is rapidly cleaved to an active portion (C-BNP) and an inactive portion (NT-proBNP). IDEXX Laboratories owns proprietary rights to NT-proBNP testing in dogs and cats, while Antech Diagnostics owns rights to C-BNP testing in dogs.

One of the clinical settings in which BNP testing has proved most useful is differentiating CHF from noncardiac disease in cats with respiratory distress. Elevated NT-proBNP (above a cutoff threshold) has shown high (~90%) sensitivity and specificity for diagnosis if CHF in dyspneic cats, whether the biomarker is evaluated in blood[1-3] or pleural effusion[4]. The original studies documenting the diagnostic utility of NT-proBNP to detect CHF were published nearly 10 years ago; however, the original NT-proBNP tests were send-out tests requiring specific shipping and handling instructions with turnaround times of 1–2 working days, making the test impractical for emergency management of dyspneic patients. A major logistical advancement in cardiac biomarker testing has been the development of a point-of-care NT-proBNP test for cats (SNAP® Feline proBNP, IDEXX Laboratories). With a colorimetric cutoff ("positive" sample spot) of ~200 pmol/L, this test has shown excellent sensitivity (95.2%) and specificity (87.5%) for diagnosis of CHF in cats with pleural effusion.[5]

Clinical Experience and Unanswered Questions

Point-of-care NT-proBNP measurement is becoming part of the standard emergency workup for cats with respiratory distress. This is a patient population notorious for (1) rapid clinical decompensation and (2) equivocal thoracic radiographic findings. A small blood sample (3–4 drops) for a point-of-care NT-proBNP test can greatly assist the triage process for dyspneic cats in an emergency or general practice setting. (Ironically, NT-proBNP testing is less clinically useful for cardiologists, since echocardiography is the gold standard to which biomarker testing is compared.)

There are several challenges involved with BNP testing in general. As with any diagnostic test, sensitivity and specificity depend on the index of suspicion for the disease in the tested population, so widespread screening of asymptomatic animals using BNP is unlikely to be rewarding. BNP and NT-proBNP assays are species-specific, and no SNAP Canine proBNP has yet been developed. Furthermore, each laboratory uses its own reference values, and assays (and their reference values) frequently change. Another limitation to BNP testing, particularly longitudinal testing or evaluation of animals with more mild disease, is relatively high inter- and intra-individual variation.[6,7] Finally, BNP can be affected by several noncardiac diseases, including chronic kidney disease, pulmonary and systemic hypertension, and critical illness.[8,9]

There remain many unanswered questions about BNP, and cardiac biomarkers are an active field of research. Some studies have suggested that NT-proBNP may be helpful in predicting prognosis for dogs with asymptomatic heart disease[10] or CHF;[11] however, echocardiographic measures of heart size have shown similar predictive value, and an added benefit of serial NT-proBNP testing has yet to be established. Other researchers are investigating whether BNP-targeted therapy of CHF may improve survival.[12]

2. Point-of-Care Lung Ultrasound for Diagnosis of CHF

Theory and Evidence

Point-of-care lung ultrasound (LUS) is an emerging imaging modality which can suggest the presence of alveolar or interstitial pulmonary pathology by identifying ultrasound artifacts known as B-lines.[13,14] B-lines (also called "ring-down artifacts," "comet tails," or "lung rockets") are created by the high-impedance gradient between small fluid-filled alveoli and surrounding air. Stacks of horizontal ring-down artifacts coalesce to create the ultrasonographic appearance of narrow hyperechoic vertical lines which extend from the pleural-pulmonary interface to the far aspect of the ultrasound screen without fading, and which move synchronously with respiration.[14,15]

LUS has shown high diagnostic accuracy in differentiating CHF (characterized by multiple bilateral B-lines) from noncardiac causes of respiratory distress in both humans[16] and veterinary patients[17-21]. In a study of 100 dyspneic dogs and cats, LUS had a sensitivity of 84% and specificity of 74% for identifying CHF (similar to the accuracy of thoracic radiographs).[20] In another study of 368 patients, absence of any B-lines on LUS ruled out CHF with specificity of 88% in dogs and 96% in cats.[21]

Clinical Experience and Unanswered Questions

Point-of-care LUS has become part of my standard emergency workup for dogs and cats with respiratory distress. This rapid, noninvasive test is particularly useful for initial triage of unstable patients (to determine if furosemide is indicated), and can be performed cageside (or even through a portal into an oxygen cage). LUS has become an extension of a "FAST" ultrasound examination already performed in many emergency centers during initial triage examination (prior to thoracic radiographs). Diagnostic accuracy for detection of CHF in cats is excellent because the most relevant differential diagnosis for a dyspneic cat (feline asthma) is not associated with any alveolar or interstitial fluid. In contrast, LUS results in some false positive results in dogs with severe diffuse alveolar or interstitial diseases, such as ARDS, noncardiogenic pulmonary edema, or diffuse pneumonia.

Many avenues of LUS research are open to investigation. Our research group is currently investigating the utility of LUS for diagnosis of CHF in dog with cough, as well as monitoring resolution of CHF over time. Additionally, the possibility that LUS may be **more** sensitive than thoracic radiographs at detecting very mild pulmonary edema warrants examination. Finally, B-lines are not the only artifact seen on LUS; additional findings such as "shred," "tissue," and "nodule" signs may have diagnostic significance.

3. Pimobendan in Cats with CHF

Theory and Evidence

Pimobendan (Vetmedin®) is a combined positive inotrope (calcium sensitizer) and vasodilator (phosphodiesterase-III inhibitor). The drug is currently FDA approved for use in dogs with CHF and has shown survival benefit in dog with CHF secondary to mitral valve disease[22] or dilated cardiomyopathy[23]. Pimobendan has also shown benefit in prolonging time to CHF in dogs with asymptomatic mitral valve disease[24] or dilated cardiomyopathy[25]. Although cardiologists generally agree on the utility of pimobendan in dogs with heart disease, there is less consensus about use of pimobendan in cats. The most common cause of CHF in cats is hypertrophic cardiomyopathy (HCM). In this disease, myocardial relaxation (diastolic function) is impaired, but systolic function (contractility) is generally normal or even hyperdynamic. Thus, positive inotropes are theoretically counterintuitive in HCM and may seem contraindicated in cats with outflow tract obstruction (hypertrophic **obstructive** cardiomyopathy, HOCM). However, there are also theoretical benefits to pimobendan in HCM: calcium-sensitizing properties may improve myocardial relaxation as well as contraction; balanced vasodilation may be beneficial in cats with CHF and elevated left ventricular filling pressures; and pimobendan's adjunctive anti-cytokine effects may improve appetite and attitude.[26]

Several retrospective studies of cats with CHF **and LV systolic dysfunction** (i.e., dilated cardiomyopathy or end-stage HCM with severe remodeling) have shown that pimobendan is well tolerated in cats,[26–28] though these studies lacked control groups needed to assess efficacy. A total of 229 cats were evaluated in these 3 studies, including 4 cats with HOCM, none of whom experienced adverse effects from pimobendan. A single cat with multiple congenital cardiac defects (including an outflow tract obstruction) did experience hypotension after receiving pimobendan, which resolved with discontinuation of the drug.[26] More recently, a retrospective case control study of 54 cats with CHF and **normal systolic function** (i.e., "typical" HCM) demonstrated a significant survival benefit of pimobendan.[29] Cats who began receiving pimobendan within 24 hours of diagnosis of CHF survived a median of 626 days, compared to 103 days for cats who did not receive pimobendan during treatment for CHF. Five cats in each group (pimobendan vs. control) had HOCM, and there were no adverse effects of pimobendan treatment.[29]

Clinical Experience and Unanswered Questions

On the basis of published evidence and clinical experience, I give pimobendan as a first-line treatment to every cat diagnosed with CHF, regardless of underlying disease, systolic function, or presence of outflow tract obstruction. Over the past 5 years, I have treated over 100 cats with pimobendan throughout the course of their CHF, with clear and consistent (though admittedly anecdotal) positive outcomes. I have also found that oral pimobendan can obviate the need for dobutamine in cats presenting with CHF and cardiogenic shock. The typical dose for pimobendan in cats is the same as dogs (label dose of 0.2–0.3 mg/kg PO q 12 h), which in most average-sized cats equates to 1.25 mg per cat PO q 12 h. As with dogs, I "dose-escalate" pimobendan in cats along with furosemide dose to maintain optimal CHF control as the disease progresses; the highest dose I have prescribed in a cat is 1 mg/kg PO q 8 h (6x starting dose). Side effects are exceedingly rare (occasional vomiting or diarrhea), and I have not personally experienced any adverse effects (hypotension, etc.) in cats with outflow tract obstruction. The major difficulty with pimobendan administration in cats relates to drug formulation: tablet sizes are relatively large for pilling cats, and cats with CHF rarely eat the flavored tablets willingly. I often recommend that owners crush the tablets, mix with water, and administer via syringe.

There are still many unanswered questions regarding the use of pimobendan in cats. Some cardiologists remain concerned about use of pimobendan in cats with outflow tract obstructions, particularly given the small total number of cats with HOCM receiving pimobendan in published studies; future studies should address efficacy of pimobendan specifically in this subgroup of cats. The ideal dose of pimobendan for cats has yet to be established; pharmacokinetic studies suggest that pimobendan reaches higher maximal concentrations and has a longer half-life in cats versus dogs,[30] but that conversion of pimobendan to the active metabolite may be reduced in cats compared to dogs[31]. Additional future directions include developing efficacious alternative drug formulations (such as transdermal or oral compounded formulations) to ease administration of pimobendan to cats.

4. Spironolactone in Dogs with CHF

Theory and Evidence

Neurohormonal dysregulation, particularly activation of the RAAS, is an important component of the pathophysiology and progression of CHF. The end products of the RAAS system, angiotensin II and aldosterone, have negative long-term cardiovascular consequences, including vasoconstriction, sodium and water retention, and myocardial remodeling and fibrosis. The RAAS can be pharmacologically downregulated at several levels. The most common first-line drugs used for this purpose in veterinary medicine are angiotensin-converting enzyme inhibitors (ACEi), which inhibit formation of angiotensin II (and thus also production of aldosterone). However, experimental and clinical models have shown that patients with RAAS activation who receive ACEi can have paradoxically increased levels of aldosterone (termed "aldosterone escape") due to upregulation of systems other than systemic RAAS. Spironolactone is a direct aldosterone receptor blocker (and, by virtue of this aldosterone blockade, a potassium-sparing diuretic) which can provide adjunctive suppression of the RAAS system.

The current accepted standard treatment for CHF in dogs is "triple therapy" (furosemide, an ACEi, and pimobendan).[32] However, in both people and dogs, there is evidence of improved outcomes when spironolactone is added to standard therapy of CHF, even in patients whose CHF is well controlled (as opposed to refractory). The human RALES trial demonstrated a 30% reduced risk of cardiac death and 25% reduced risk of re-hospitalization in patients receiving spironolactone.[33] A multicenter European canine trial[34] similarly demonstrated a 55% reduced risk of cardiac death using aldosterone in dogs with CHF and mitral valve disease. (However, at the time of that study, pimobendan was not yet in widespread use; the comparison "standard therapy" was furosemide, ACEi, and digoxin.) In Europe, where ACEi are generally expensive (sometimes cost-prohibitive), spironolactone is approved for the treatment of CHF in dogs with mitral valve disease.

Clinical Experience and Unanswered Questions

"Quadruple therapy" (furosemide, ACEi, pimobendan, and spironolactone) is my standard of care for chronic treatment of CHF in dogs. I generally begin spironolactone at the first recheck visit after initial onset of CHF (usually ~1 week following CHF episode), so that I am able to recheck renal values and electrolytes after initiating furosemide (prior to starting spironolactone). Assuming that patients are nonazotemic and either normokalemic or hypokalemic at that recheck, I begin spironolactone 2 mg/kg PO q 24 h (the dose used in the multicenter European trial).[34] I generally withhold spironolactone, at least initially, in azotemic or hyperkalemic patients. Spironolactone is never indicated as a monotherapy (except for primary hyperaldosteronism [Conn's syndrome]). I am more hesitant about the use of spironolactone in cats due to reports of ulcerative facial dermatitis in a relevant proportion (~30%) of Maine Coons receiving spironolactone in one study,[35] and thus I tend to reserve spironolactone for cats with refractory CHF.

Many unknowns remain regarding the optimal use of spironolactone in CHF. Several study design issues were apparent in the multicenter European trial, including a high dropout rate, low percentage (~20%) of study participants reaching the primary endpoint (cardiac death), and questionable severity of mitral valve disease in the study population (stage B2 vs. stage C). It is also not known **when** in the progression of heart disease is the most appropriate time to begin spironolactone: at first onset of CHF, at first recheck after diagnosis of CHF, or perhaps even during the asymptomatic phase (stage B2) when ACEi are first initiated? Finally, is it necessary to recheck renal values and electrolytes after starting spironolactone? Although mechanistically a potassium-sparing diuretic, spironolactone actually causes no diuresis in healthy dogs[36] and no significant azotemia or electrolyte derangements in dogs with mitral valve disease and CHF[37].

5. Sildenafil in Dogs with CHF and Pulmonary Hypertension

Theory and Evidence

Pulmonary hypertension (PHT), or elevated pulmonary arterial blood pressure, is common in patients with CHF. Pathophysiologically, PHT can arise from two sources. **Precapillary** PHT results from disease of the pulmonary arteries or arterioles causing direct vasoconstriction and/or obstruction (etiologies include pulmonary fibrosis, heartworm disease, and pulmonary thromboembolic disease). **Postcapillary** PHT results when elevated left atrial and pulmonary venous pressures "back up" across the capillary bed, indirectly increasing pulmonary arterial pressures. In general, all dogs with left-sided CHF have some mild degree of postcapillary PHT (since, by definition, they have severely elevated left atrial pressures). However, some dogs with CHF develop more severe PHT that can lead to clinical signs (including syncope and exercise intolerance) despite good control of pulmonary edema. Pulmonary artery pressures can be estimated noninvasively via echocardiography using calculations based on velocity of tricuspid or pulmonic regurgitation.

The targeted treatment for PHT is sildenafil, a phosphodiesterase-V inhibitor and selective pulmonary arterial vasodilator. Sildenafil has been shown to improve clinical signs in dogs with PHT secondary to various precapillary and postcapillary causes,[38,39] including CHF[40]. Additionally, treatment with pimobendan (a balanced systemic vasodilator) can decrease postcapillary PHT by lowering left atrial pressure and has been shown to improve quality of life and survival in dogs with mitral valve disease and PHT.[41]

Clinical Experience and Unanswered Questions

I consider PHT as a differential diagnosis for dogs with historical CHF and worsening clinical signs (syncope, weakness, exercise intolerance) despite good control of CHF (i.e., no evidence of radiographic pulmonary edema). In this patient subgroup, I offer recheck echocardiography to screen for moderate to severe PHT. If identified, I either (1) increase the current dose of pimobendan (generally my first line for moderate PHT), or (2) prescribe sildenafil as an adjunct treatment (if PHT is severe or if there is clinical suspicion for a precapillary cause of PHT). Dose of sildenafil is 1–3 mg/kg PO q 8–12 h; I generally begin at 2 mg/kg PO q 12 h and titrate dose based on clinical response. Side effects of sildenafil are rare, though hypotension is possible when used with other vasodilators. The major challenge in prescribing sildenafil is prohibitive cost of the brand-name product (Viagra®). I can generally find pharmacies selling sildenafil 20 mg, either generic or the less well-known brand name Revatio®, for more reasonable prices.

As with the other trends in CHF management, more research is needed regarding use of sildenafil in veterinary patients. The majority of dogs with stage C (and likely B2) mitral valve disease have mild, subclinical PHT; would these dogs benefit from adjunct sildenafil treatment even if they are not overtly symptomatic for PHT? What is the optimal dose and frequency of sildenafil treatment? Finally, how can we best monitor response to sildenafil therapy, since some studies document clinical benefit of sildenafil without a significant decrease in echocardiographically-estimated pulmonary artery pressures?[38,39]

Conclusion

There have been many advances in diagnosis and treatment of CHF in dogs and cats within the last 5 years. Additional emerging trends in CHF treatment (not covered in today's talk) include (1) use of the ultrapotent loop diuretic torsemide (rather than furosemide) in refractory CHF, and (2) use of furosemide constant rate infusions (rather than intermittent boluses) for in-hospital management of acute CHF. Cardiologists frequently disagree about certain aspects of CHF management; individual clinical preference is guided by a combination of published literature and personal experience, both of which have informed this presentation.

References

1. Oyama, *et al. J Am Vet Med Assoc.* 2009;235:1319–25.
2. Hassdenteufel, *et al. J Vet Emerg Crit Care.* 2013;23(4):416–22.
3. Connolly, *et al. J Vet Intern Med.* 2008;22:96–105.
4. Humm, *et al. J Small Anim Pract.* 2013;54:656–61.
5. Hezzell, *et al. J Vet Intern Med.* 2016;30(2):536–542.
6. Harris, *et al. J Feline Med Surg.* 2017;19(2):216–223.
7. Kellihan, *et al. J Vet Cardiol.* 2009;11:S93–7.
8. Kellihan, *et al. J Vet Cardiol.* 2011;13:171–82.
9. Raffan, *et al. J Vet Intern Med.* 2009;23:1184–9.
10. Reynolds, *et al. J Vet Cardiol.* 2012;14:193–202.
11. Moonarmart, *et al. J Small Anim Pract.* 2010;51:84–96.
12. Wolf, *et al. J Vet Cardiol.* 2012;14:399–408.
13. Lichtenstein, *et al. Am J Respir Crit Care Med.* 1997;156:1640–6.
14. Volpicelli, *et al. Intensive Care Med.* 2012;38:577–91.
15. Lichtenstein. *Minerva Anestesiol.* 2009;75:313–7.
16. Al Deeb, *et al. Acad Emerg Med.* 2014;21:843–52.
17. Lisciandro, *et al. Vet Radiol Ultrasound.* 2014;55(3):315–322.
18. Lisciandro, *et al. J Vet Emerg Crit Care.* 2016 (abstract).
19. Rademacher, *et al. Vet Radiol Ultrasound.* 2014;55(4):447–452.

20. Ward, *et al. J Am Vet Med Assoc.* In press.
21. Lisciandro, *et al. J Vet Emerg Crit Care.* In press.
22. Häggström, *et al. J Vet Intern Med.* 2008;22:1124–35.
23. O'Grady, *et al. J Vet Intern Med.* 2008;22:897–904.
24. Boswood, *et al. J Vet Intern Med.* 2016;30(6):1765–1779.
25. Summerfield, *et al. J Vet Intern Med.* 2012;26(6):1337–49.
26. Gordon, *et al. J Am Vet Med Assoc.* 2012;241:89–94.
27. Macgregor, *et al. J Vet Cardiol.* 2011;13:251–60.
28. Hambrook, *et al. J Feline Med Surg.* 2012;14:233–9.
29. Reina-Doreste, *et al. J Am Vet Med Assoc.* 2014;245:534-539.
30. Hanzlicek, *et al. J Vet Cardiol.* 2012;14:489–96.
31. Yata, *et al. J Vet Cardiol.* 2016;18:310–25.
32. Atkins, *et al. J Vet Intern Med.* 2009;23:1142–50.
33. Pitt, *et al. N Engl J Med.* 1999;341:709–17.
34. Bernay, *et al. J Vet Intern Med.* 2010;24:331–41.
35. MacDonald, *et al. J Vet Intern Med.* 2008;22:335–41.
36. Jeunesse, *et al. J Vet Cardiol.* 2007;9:63–8.
37. Thomason, *et al. J Vet Cardiol.* 2007;9:103–8.
38. Bach, *et al. J Vet Intern Med.* 2006;20:1132–5.
39. Kellum, *et al. J Vet Intern Med.* 2007;21:1258–64.
40. Brown, *et al. J Vet Intern Med.* 2010;24:850–4.
41. Atkinson, *et al. J Vet Intern Med.* 2009;23:1190–6.

1.
acvim009

Targeted Nutritional Management of Horses at Risk of Laminitis

Harry B. Carslake, MA, VetMB, DACVIM (LAIM), MRCVS
Liverpool, UK

Introduction

Targeted nutritional management involves identifying horses that are in a group of interest, in this case at risk of laminitis, and then recommending appropriate nutritional plans which are tailored to the horse and to the facilities available. Veterinary technicians are often heavily involved in the nutritional management of horses under veterinary care, as veterinarians can feel uncomfortable with nutrition and are often less familiar with feeding practices.

Identifying Horses at Increased Risk of Laminitis

There are three major causes of laminitis, each with associated risk factors: 1. Supporting limb laminitis (SLL). Horses at risk have severe or chronic unilateral lameness and increased loading of the contralateral sound limb; 2. Sepsis related laminitis (SRL): This occurs in horses suffering from diseases that cause a systemic inflammatory response and multiple organ failure. In a hospital setting a common example of this is colitis; 3. Endocrinopathic laminitis. Equine metabolic syndrome (EMS) and/or pituitary par intermedia dysfunction (PPID) are associated with an increased risk of laminitis. Horses that have received corticosteroids are also included in this group.

Severe lameness and sepsis are relatively simple to identify, and in these cases if laminitis occurs it is usually severe and shortly after the risk factor becomes apparent. Clinical signs of endocrinopathic risk factors are more subtle and variable, and onset of laminitis is more variable both in timing and severity. Usually the three categories of laminitis are considered in isolation, as there are differences in their pathophysiology, but what is not known is whether risk factors are additive, for example, whether supporting limb laminitis is more likely to occur in a horse with pre-existing EMS. As endocrinopathic disease is the most common cause of laminitis, accounting for 90% of cases in one study,[1] and is most influenced by nutritional management, it will be the focus of this session.

Equine Metabolic Syndrome (EMS)

This term was first coined in 2002,[2] because of the similarities noted between EMS and metabolic syndrome in people. EMS was further defined by an ACVIM consensus statement in 2010,[3] when it was described as a collection of endocrine and metabolic risk factors that caused an increased susceptibility to laminitis. The definition of EMS continues to evolve as its pathophysiology is better understood. Horse factors associated with EMS include age, breed, season, obesity, regional adiposity (e.g., cresty neck), reproductive abnormalities, inactivity, diet, and evidence of current or previous laminitis, however, these do not define the disease. For example, EMS can occur in non-obese horses. A definitive diagnosis is established by metabolic and endocrine testing, normally involving detection of insulin dysregulation (ID). Insulin dysregulation refers to increased serum concentration of insulin, either before or after feeding, and/or resistance of cells to the normal action of insulin. Additional tests for adiponectin and leptin, which are hormones produced by adipose tissue can also add supportive information. Management of EMS is largely based on dietary modification and increased exercise.

Pituitary Pars Intermedia Dysfunction (PPID)

Previously termed equine Cushing's disease, PPID is a neurodegenerative disorder that results in an overproduction of several pituitary hormones and is common in older (>15 years) horses. Clinical signs are variable but can include generalized or localized hypertrichosis (long hair coat), muscle wasting, excessive sweating, abnormal fat distribution, lethargy, infertility, and recurrent laminitis. Several diagnostic tests are available to confirm a diagnosis. Not all horses with PPID develop laminitis, and ID secondary to PPID is considered to be an important risk factor.[4]

Use of Exogenous Corticosteroids

The risk of laminitis following administration of corticosteroids is prominent in the minds of most veterinarians and horse owners, despite several retrospective studies indicating that the risk is low.[5,6] The causative mechanism is unknown, but is likely to be at least partly linked to the inhibitory effects corticosteroids have on the actions of insulin. Assessing ID in horses with risk factors for EMS or PPID before administering corticosteroids is advocated by some veterinarians. When considering administration of corticosteroid to a horse with EMS or PPID, using an alternative medication, or nutritional management to minimize insulin response to feeds and resolve ID is a reasonable precaution.

Role of Insulin Dysregulation in Endocrinopathic Laminitis

Multiple experimental and clinical studies have demonstrated an association between ID and susceptibility to laminitis. Laminitis has been induced in normal horses by infusing high concentrations of insulin;[7] however, the causative mechanism is still not fully understood. ID is a central feature of EMS. Not all cases of PPID have ID, but those that do are more likely to develop laminitis and have a poorer prognosis for survival.[4] Measures to reduce circulating insulin concentration and resolve ID are essential in the nutritional management of these cases.

NUTRITIONAL MANAGEMENT OF HORSES AT RISK OF ENDOCRINOPATHIC LAMINITIS

The short, medium, and long term goals for nutritional management of EMS and PPID are:

- Short: Minimize serum glucose and insulin concentrations after feeding.
- Intermediate: Normalize body condition and insulin dysregulation.
- Long term: Maintenance on a low glycemic/insulinemic diet, while returning the horse to its intended use.

Baseline Data

After establishing a diagnosis of EMS or PPID, baseline data is required to tailor a nutritional plan.

- History and clinical examination. Hoof examination and baseline radiographic assessment of the feet. Concurrent medical conditions (e.g., respiratory/dental disease) also need to be considered.
- Assessment of insulin dysregulation, ideally with an oral sugar/glucose test.
- Details of current feeding. Measurements should be by weight, not scoops/nets/armfuls. Nutritional analysis can be obtained from packaging for commercial feeds, or for forages from labs such as Equi-Analytical, Ithaca. Estimating rate of pasture intake is difficult, will vary considerably according to factors such as level of work, grass type, other feeds offered, and time at pasture. Ponies grazing fall pastures consumed an average 0.8% body weight as dry matter (DM) in 3 hours in one study.[8]
- Body weight, by weighbridge if available, or estimation by girth circumference or a formula combining girth and length.[9]
- Body condition score (BCS). Scores of 1–9 or 1–5 are commonly employed. There are useful online videos for clients showing how to perform a BCS.[10]
- Girth and belly circumference. These are more sensitive indicators of early weight loss than BCS[11] and are an easy way for owners to monitor weight loss/gain. Snipping small areas of hair helps ensure a consistent location.
- Facilities and preferences of the owner. In most cases the "ideal" nutritional plan needs to be tailored to considerations such as horse yard protocols, grazing and forage availability, and preferences of owners and horses.
- The use or intended use of the horse.

Calculation of Nutrient Requirements and Balancing Dietary Supply

NRC guidelines (https://nrc88.nas.edu/nrh/) give estimates of requirements based on weight and activity of a horse; however, there is considerable variation between individuals. It also calculates total dietary supply according to nutritional analysis and weights fed. It is useful to compare previous and new diets and check them against recommendations.

General Advice for Nutritional Management of PPID and EMS

High non-structural carbohydrate (NSC) (sugar/starch) feeds should be avoided. Both PPID and EMS horses are prone to exaggerated serum insulin responses to feeds, which contributes to laminitis risk. Forages vary considerably in NSC content, however, <10–12% DM is recommended, which can be achieved through forage analysis or soaking of hay. Haylage (ensiled grass) normally has a lower NSC compared to hay, but can be associated with marked post-feeding insulin responses[12] and is not recommended.

Calorie restricted diets, especially if feeding only preserved forage, can be deficient in certain nutrients, and a proprietary feed balancer is recommended to ensure adequate provision of protein, minerals, and vitamins.

Any changes to diets should be introduced over 1–2 weeks to allow the gastrointestinal tract to adapt. Caloric restriction usually results in reduced time feeding, and can result in boredom and stereotypies. Dividing daily rations into several meals, using small holed hay nets, hanging nets in the center of the barn, and using grazing muzzles will help increase forage consumption time. Feeding regimens that allow plenty of exercise are preferable.

Feeding Underweight PPID Horses

Muscle loss is common in PPID horses and requires provision of additional calories and amino acids. If there is concurrent insulin dysregulation or a history of laminitis, then extra caution with pasture and higher NSC feeds is required. In addition to provision of a high-quality forage, suitable supplementary feeds to promote weight gain include: vegetable oil gradually introduced to a maximum of 1 ml/kg/day; soaked un-molassed sugar beet pulp, 0.05–0.2% BW (presoaked)/day; and alfalfa chaff and proprietary feeds specifically formulated for PPID. Supplementary vitamin E is recommended, especially if there is no access to pasture.

Feeding the Overweight PPID Horse or EMS Horse

Obesity is very common in horses with EMS, and is a risk factor for the development of laminitis.[13,14] Based on studies in other species, the current recommendations for safe weight loss in horses is 0.5–1% BW/week. Rates >1% BW/week are considered too high. The goal is to reduce BCS to <7/9 or <4/5, whilst minimizing serum insulin responses to feeds and ensuring that the protein, vitamin, and mineral content of the diet is balanced.

- Feeding only a low NSC forage (soaked or analyzed) is recommended during periods of weight loss and laminitis, supplemented with a balancer.
- Feed according to current, not "ideal" bodyweight (BW).
- Start by feeding 1.5–2% BW/day fresh hay, depending on BCS, previous level of feeding and level of work.
- If NSC content of hay is >10–12% then initially feed 1.75–2.25% BW/day dry hay, which is then soaked.
- Soaking hay for 8–16 hours in cold water significantly reduces NSC but also mineral content.
- A previous recommendation that hay should be fed at an initial rate of 1.5% of "ideal" BW, and soaked if NSC is >10%[3] can result in excessive rates of weight loss[15] and is not advised as an initial diet.
- There is considerable variation in weight loss between individuals.

Pasture and Grazing Muzzles

Turning horses out to pasture is a source of cheap forage and provides exercise and socialization, so is popular with owners and horses alike. Unfortunately pasture forage intake is very hard to measure and the NSC content of grass can be very high.

- Zero grazing should be recommended in cases with, or at high risk of, laminitis as pasture can have a very NSC content and intake is very difficult to regulate.
- NSC content is highest in pasture from mid-morning to late afternoon, and during the spring and autumn.

- Strategies to reduce pasture intake include strip grazing, grazing after other horses/livestock, mowing pastures (with removal of clippings), and turning out onto dirt yards/schools, taking precautions against sand and toxic weed ingestion.
- A pony can ingest up to 2/3 of its daily DM requirements in 3 hours of unmuzzled grazing on good pasture.
- Muzzles significantly reduce pasture intake and are useful for allowing turnout. Owners should ensure that a horse can still drink through the muzzle and potential complications such as changes in herd hierarchy, skin rubs, and abnormal incisor wear are monitored.

Monitoring and Maintaining Contact with Owner

Initially, the owners should perform weekly assessment of weight, BCS, and girth and belly measurements. A repeat visit by the veterinarian/technician 4–6 weeks after starting a dietary plan to discuss progress, improvements in clinical signs/BCS, and adherence to the dietary plan is recommended. Repeated oral glucose/sugar tests after 10–12 weeks will allow monitoring of insulin dysregulation. Nutritional recommendations need to be regularly revised according to clinical response, alterations in weight, BCS, measures of ID, and level of work.

Non-Endocrinopathic Risk Factors for Laminitis

Horses at risk of sepsis-related or supporting limb laminitis usually have nutritional requirements specifically related to the primary disease. To the author's knowledge there is no evidence demonstrating an increased risk of supporting limb or sepsis related laminitis in horses with pre-existing EMS or PPID. Despite this, in horses with evidence of pre-existing EMS or PPID, it would be advisable to offer low NSC where possible, and to use corticosteroids judiciously.

References

1. Karikoski NP, *et al. Domest Anim Endocrinol.* 2011;41(3):111–117.
2. Johnson PJ. *Vet Clin North Am Equine Pract.* 2002;18:271–293.
3. Frank N, *et al. J Vet Intern Med.* 2010;24(3):467–475.
4. McGowan CM, *et al. Equine Vet J.* 2004;36(3):295–298.
5. Jordan VJ, *et al. Equine Vet J.* 2017;49(1):19–25.
6. McCluskey MJ, *et al. Equine Vet Educ.* 2004;16(2):86-89.
7. Asplin KE, *et al. Vet J.* 2007;174(3):530-535.
8. Longland AC, *et al. J Equine Vet Sci.* 2016;40:26–33.
9. Carroll CL, *et al. Equine Vet J.* 1988;20(1):41–45.
10. Hollands T. www.youtube.com/watch?v=Zih1jT_pUgQ. 2013.
11. Dugdale AH, *et al. Equine Vet J.* 2010;42(7):600-610.
12. Carslake HB, *et al.* ECEIM Congress; Utrecht, Netherlands. 2015.
13. Carter RA, *et al. Equine Vet J.* 2009;41(2):171–178.
14. Treiber KH, *et al. J Am Vet Med Assoc.* 2006;228(10):1538–1545.
15. Argo CM, *et al. Vet J.* 2015;206(2):170–177.

acvim010

Biosecurity Measures to Prevent Spread of Equine Infectious Transmissible Diseases

Angelika Schoster, DVM, DVSc, PhD, DACVIM (Large Animal Internal Medicine), DECEIM
Zurich, Switzerland

INTRODUCTION

'Bio' means life, and 'security' means protection. Biosecurity is an important concept in veterinary hospitals to protect patients, visitors, and staff. Understanding the concepts behind why certain biosecurity measures are taken, leads to better compliance. This talk aims to give an overview on important equine infectious transmissible diseases. It will cover pathophysiology of the disease and how this leads to development of specific biosecurity guidelines. A focus will be put on zoonotic diseases and how to prevent staff infection.

EQUINE INFECTIOUS DISEASES

Infectious diseases are associated with microorganisms, including bacteria, viruses, and parasites. Not all infectious diseases are transmissible. Examples of infectious but non-transmissible diseases include West Nile virus and equine alpha virus encephalitides (EEE, VEE, WEE). Some infectious diseases are transmissible between animals of the same species while others are transmissible from one species to another (e.g., *Salmonella*). Some transmissible infections can be transferred to humans, so called zoonoses. Infectious diseases often vary by geographic region, therefore it is important to know which diseases are prevalent in the practice region and biosecurity concepts have to be tailored accordingly.

Some infections are spread by direct contact between horses, while others are spread via vectors such as insects and ticks and by fomites. Important fomites in the environment of horses include water and feed buckets, stall cleaning equipment, riding equipment, as well as brushes and other horse cleaning supplies. The hands and clothes of the caregivers, riders, visitors, farriers, and veterinarians are equally important to spread infections. The importance of hand hygiene cannot be emphasized enough.[1]

Table 1. Overview of equine transmissible infectious diseases

Disease	Other species in danger of transmission	Zoonotic potential
Bacterial		
Salmonellosis	Warm- and cold-blooded animals	x
Enteric clostridial disease		
Streptococcus equi equi (strangles)		
Leptospirosis	Mammals	x
Burkholderia mallei (glanders)	Felines, dogs, camels, wolves	x
Lawsonia intracellularis		
Viral		
Equine herpesvirus 1	New World camelids	
Equine infectious anemia		
Equine viral arteritis		
Equine influenza		
Hendra virus	Dogs, bats	x
Rotavirus		
Coronavirus		
Vesicular stomatitis	Ungulates	x
African horse sickness		

Rabies	Warm-blooded animals	x
Parasitic		
Cryptosporidiosis	Vertebrates	x

General Biosecurity Measures in Hospitals

With regards to biosecurity, any equine hospital differs from a horse barn. Similar to human hospitals, sick animals with different epidemiologic backgrounds (age, vaccination status, immune competence, deworming status, etc.), concurrent diseases, and immunosuppression are stabled in a small area. Human traffic differs between a horse barn and equine hospitals, facilitating the spread of diseases. A biosecurity concept should be available in writing, easily accessible and updated regularly. The biosecurity guidelines should be tailored to the specific needs of the hospital, taking into account common regional infections, patient population, and building facilities. Ideally, every hospital should have isolation facilities where animals with known or suspected infections can be individually isolated. General recommendations for cleaning, disinfection and sterilization, hand hygiene, isolation guidelines, and antimicrobial guidelines should be followed. One person should be the infection control officer, where all information is gathered and distributed when necessary. Daily rectal temperatures should be taken and recorded from all horses in hospital. If one or more horses develop fever without an obvious cause, further infection control measures should be taken, depending on the specific hospital guidelines.[2]

Isolation Guidelines

It is advisable to isolate horses with clinical signs or history of a potential transmissible infection until an infectious disease is ruled out. Definitive diagnosis often takes several days, over which spread of the disease may ensue. Rules when a horse has to be placed in isolation should therefore be based on common clinical signs of infectious disease instead of positive diagnostic tests. Once a diagnosis has been established, further recommendations should be available regarding handling of the horse and all related samples and equipment, treatment, etc. and when quarantine can be lifted. In hospital surroundings, it is advisable to discharge a horse from isolation without lifting the quarantine to reduce risk for other patients and facilitate cleaning and disinfection.[3]

Pathophysiology and Biosecurity Measures for Selected Diseases

Streptococcus equi equi, Strangles

Streptococcus equi equi, the causative agent of strangles, is a highly contagious bacterium with a worldwide distribution. It is circulating in the equine population, but does not survive for prolonged periods outside the host. Infection most commonly is spread from an acutely affected animal with clinical signs. Some animals show very mild signs of disease and will shed the bacterium. Another important epidemiologic group are horses that have had a strangles infection in the past, which had affected the retropharyngeal lymph nodes and then broke into the guttural pouch resulting in empyema. The infection usually resolves, but occasionally purulent material remains in the guttural pouch, may dry out but could remain infectious. These horses no longer show clinical signs but may remain infectious for years. Transmission occurs by direct contact and through fomites. The bacterium enters the host through the mucous membranes of the upper respiratory tract, especially the pharynx area, and spreads locally. The most common clinical form of the disease affects young horses and causes fever, nasal discharge, lymph node abscessation, and general malaise. Once infection is established, the host starts shedding the bacterium via respiratory secretions and is now contagious. Fever is the first clinical sign and occurs before shedding starts. This important feature can be used to stop the spread of the disease in an outbreak. If the affected horse is isolated right after the fever onset, the spread can be stopped. Biosecurity measures in a strangle outbreak, therefore, include the formation of 3 groups: diseased horses, healthy horses who were in contact with diseased horses, and healthy non-contact horses. Barrier precautions have to be built between the groups, as the bacterium is highly contagious. Twice-daily temperature measurements should be taken in all groups and horses moved to the diseased group once they are febrile. In a hospital setting, fever of unknown origin without an obvious non-infectious cause in a horse should raise the suspicion of strangles and biosecurity measures should be considered.[4,5]

Salmonellosis

Salmonella are a group of bacteria that inhabit the gastrointestinal tract of humans and animals worldwide. Cold- and warm-blooded animals can be infected with *Salmonella.* It is an important cause of disease in humans and animals, including horses. Equine hospital outbreaks occur frequently. Acutely infected animals with clinical signs also shed *Salmonella* in large numbers. Some healthy animals and people carry and shed *Salmonella* and pose a risk for others. Particularly, horses with gastrointestinal disease, such as colic, are prone to shed *Salmonella* without showing signs of salmonellosis. Transmission occurs via the feco-oral route through contact with the feces of infected animals. The main clinical signs in horses are fever, general malaise, and, often but not always, diarrhea. Clinical signs in humans are similar and include diarrhea and fever. Biosecurity measures should include isolation of all horses with diarrhea and particularly suspected *Salmonella* cases. Some horses do not develop diarrhea but show other signs of colitis (fever, hypoproteinemia, leucopenia). Therefore, a horse showing two or more of the aforementioned signs should also be considered a *Salmonella* suspect and be isolated. An infected horse can shed *Salmonella* for several weeks after the clinical signs have resolved; testing is therefore mandatory before declaring an animal free. *Salmonella* is shed intermittently; one negative fecal sample is not enough to declare an animal *Salmonella* free. A vaccine is not available. Biosecurity measures are very important for this disease and should focus on prevention. Horses with gastrointestinal disease (colic) should be housed in a separate ward, particularly separate from patients with a weakened immune system or patients undergoing elective surgery. Some hospitals rely on fecal testing of all patients entering the hospital, or at least testing high-risk patients.[6]

Leptospirosis

Leptospirosis is one of the most widespread zoonotic infections, but an uncommon disease in horses. The bacterium can infect many different mammals and infected animals shed the bacterium in their urine. Transmission to humans occurs through contact with contaminated water or urine of infected animals. Particularly, rodents, skunks and other wildlife are often infected and shed the bacterium. Clinical signs in horses include fever, liver failure, pulmonary hemorrhage, and renal failure. Equine recurrent uveitis is sometimes associated with *Leptospira;* however, this is a sequel to infection and does not indicate current systemic infection. In humans, flu-like symptoms are commonly seen, but the disease can also affect the lungs and the heart. Pregnant women are at risk for miscarriage. Treatment requires antimicrobial administration; in humans, the disease is fatal in 1–5%. A vaccine for animals is available, however does not contain all serovars of the bacterium and therefore is not 100% protective. There are no leptospiral vaccines for people in North America. Biosecurity measurements to prevent the spread of disease to veterinary hospital staff should particularly focus on preventing contact with urine of horses. Rodent control and prevention of stagnant waters are important to reduce the risk of exposure for hospital patients. Pets should not be allowed to roam free and urinate in horse barns to prevent spread.[7]

Equine Herpesvirus 1 Infection

There are several different equine herpesviruses (EHV). EHV-1 and EHV-4 are associated with equine respiratory disease. They have a worldwide distribution and once an infection has occurred, latency is established and in most cases the virus cannot be cleared. The virus goes into a latent phase where no clinical signs occur, but the virus can be reactivated during periods of stress. Disease can result from new infection or reactivation of latent ones. Acutely infected animal showing clinical signs can spread the virus through respiratory secretions. Additionally, animals with a latent infection are intermittently shedding the viruses with no signs or very mild clinical signs. EHV-1 is able to cause a viremia and spread to other tissues after infecting the respiratory epithelium and can then cause abortion and equine myeloencephalopathy. The aborted fetus and all placental fluids also contain active virus in large numbers and pose a risk for other animals. Clinical signs in horses depend on the form. The respiratory form is characterized by high fever, nasal discharge, and coughing. Myeloencephalopathy is characterized by signs consistent with neurologic disease. Typical signs include ataxia and difficulties urinating and defecating. Treatment is symptomatic. In an outbreak situation, horses should be divided into 2 groups, diseased and non-diseased. The virus is highly contagious and barrier precautions should be established between the groups. Daily temperatures should be taken from the healthy horses and horses moved to the disease group once they develop fever. Shedding of the virus usually starts after the first fever spike; therefore spread can be prevented if the horse is isolated immediately after the fever is detected. The virus can be shed intermittently, therefore a single negative sample does not rule out disease and is not enough to ensure that an animal is no longer shedding the virus. A vaccine for horses exists; the vaccine, however, cannot fully protect the horse from infection, but usually clinical signs are milder. Vaccination is recommended nevertheless, as increased antibodies in the horse population aid in preventing spread of the disease. Equine herpes virus outbreaks have occurred in equine hospitals and are much-feared occurrences. Horses with suspected EHV-1 infection, based on clinical signs such as fever of unknown origin, abortion, and fever or neurologic signs and fever should be isolated until a definitive diagnosis is established.[8]

Rabies

Rabies is a viral disease affecting the central nervous system of warm-blooded animals including humans. The inciting agent, a *Lyssavirus*, causes encephalitis, which is usually fatal. The virus has a worldwide distribution; however, some countries have managed to control the virus to the extent to be considered rabies-free. Cases are still encountered in North America, particularly in wild animals but occasionally also in domestic dogs and occasionally in horses. An eradication program is underway. Transmission occurs through animal bites, as the saliva of infected animals is infectious. Transmission from horses to humans can occur when inspecting the mouth or other contact with the saliva of an infected horse in addition to being bitten. Both in humans and animals the clinical signs are consistent with neurologic disease. In horses, clinical signs are similar to other causes of encephalitis, which makes clinical diagnosis difficult. A definitive diagnosis can only be reached postmortem through tissue samples from the brain of the diseased animal. Pre-exposure vaccination is recommended in some countries for employees in the veterinary sector. A vaccinated horse is less likely to be affected by rabies but infection is still possible, but in North America horses should be vaccinated to reduce the risk of disease transmission. Biosecurity measures should focus particularly on preventing contact with the saliva, CSF of affected animals.

Summary

Equine infectious diseases are common, and a variety of microorganisms have zoonotic potential. It is important to establish a biosecurity protocol tailored to the specific veterinary clinic or hospital, depending on common diseases, patient population, and available facilities. Emphasis should be put on prevention of spread rather than dealing with outbreaks. People in the veterinary field are at high risk of exposure to zoonotic microorganisms and disease and should take specific preventative precautions to avoid such infections.

Resources

- OIE World Organization for Animal Health www.oie.int
- CDC Center for Disease Control and Prevention www.cdc.gov
- ECDC European Centre for Disease Prevention and Control http://ecdc.europa.eu/en/Pages/home.aspx

References

1. Anderson ME. Contact precautions and hand hygiene in veterinary clinics. *The Veterinary Clinics of North America: Small Animal Practice*. 2015;45:343–360, vi.
2. Weese JS. Infection control and biosecurity in equine disease control. *Equine Veterinary Journal*. 2014;46:654–660.
3. Traub-Dargatz JL, Dargatz DA, Morley PS, Dunowska M. An overview of infection control strategies for equine facilities, with an emphasis on veterinary hospitals. *The Veterinary clinics of North America: Equine Practice*. 2004;20:507–520, v.
4. Waller AS. New perspectives for the diagnosis, control, treatment, and prevention of strangles in horses. *The Veterinary Clinics of North America: Equine Practice*. 2014;30:591–607.

5. Sweeney CR, Timoney JF, Newton JR, Hines MT. *Streptococcus equi* infections in horses: guidelines for treatment, control, and prevention of strangles. *Journal of Veterinary Internal Medicine*. 2005;19:123–134.
6. Burgess BA, Morley PS. Managing *Salmonella* in equine populations. *The Veterinary Clinics of North America: Equine Practice*. 2014;30:623–640.
7. Verma A, Stevenson B, Adler B. Leptospirosis in horses. *Veterinary Microbiology*. 2013;167:61–66.
8. Lunn DP, Davis-Poynter N, Flaminio MJ, Horohov DW, Osterrieder K, Pusterla N, Townsend HG. Equine herpesvirus-1 consensus statement. *Journal of Veterinary Internal Medicine*. 2009;23:450–461.

acvim011

A Technician's Guide to Radiography in Equine Medicine

Richard Hepburn, BVSc, MS, CertEM (IntMed), DACVIM (LAIM), MRCVS
Gloucestershire, UK

INTRODUCTION

Radiography is commonly used in equine medicine; however, most technicians only feel comfortable imaging the appendicular skeleton for orthopaedic cases. Obtaining diagnostic quality radiographs of the head, cervical and thoracolumbar spine, and thorax is an important part of the diagnostic workup of many equine medicine cases. Enabling technicians to image these regions effectively requires an appreciation of how to position the horse and orient the X-ray beam, as well as an understanding of how to adjust exposure settings to create a diagnostic image. In order to image these areas higher exposures are required and so an appreciation of how to minimise the risk of radiation exposure is also essential. A consistent approach to radiography of the equine medical patient will improve standards of clinical care, enhance workflow within a clinic, increasing productivity and job satisfaction.

RADIOGRAPHIC SAFETY[1]

Individual technicians are advised to refer to legislative acts and regulations, which refer to the use of ionising radiation in the location where they work. Local radiation rules should be in place reflecting risk assessment of the use of radiographic imaging both in a clinic setting and when using mobile equipment. Safety is required as X-rays are high-power electromagnetic waves that have the ability to pass through tissue causing biological changes to that tissue: somatic damage (becomes evident within a person's lifespan) such as cataract, anemia, cancer, sterility; and genetic damage (gene mutations detected in future generations). The primary X-ray beam is strongest being directed through the tissue being imaged onto the imaging plate; however, it also then passes on to whatever is in line behind the plate. Secondary scattered radiation is weaker being produced as the primary beam interacts with matter; however, it forms the majority of occupational exposure. The direction of this scatter during imaging of equine medicine cases is predominantly back towards the X-ray generator and hence the imager themselves.

Control measures should be in place to restrict X-ray exposure by the use of:

1. **Controlled areas** - Radiography at a clinic should be performed in a controlled area, with sufficient shielding to stop the primary beam leaving in any direction (remember doors), with personnel warning signs/systems to limit entry during imaging. When using mobile equipment, a suitable location should be chosen that can be controlled to prevent entrance of unprotected personnel and maximise containment of the primary beam beyond the imaging plate (aiming toward a solid brick wall with no one behind is preferable to aiming towards a door onto a passageway). Suitable signs should be placed to let others know that X-rays are being used.
2. **Equipment maintenance** - X-ray-generating equipment should be regularly checked for leakage. If any damage occurs (e.g., from a horse kicking the generator), the equipment should not be used again until its safety has been assured.
3. **Personal protection** - People present within the controlled area should wear personal protective equipment - suitably sized double-sided aprons (0.25–0.5 mm thickness of lead), gloves or open lined mittens (0.35–0.5 mm lead), thyroid collars and sleeves (the specifications will be determined by your local regulations). Personal radiation dose monitors should be used (central and peripheral monitors may be required) following a prescribed monitoring plan. Personal protective clothing (gowns, etc.) should be routinely checked for performance, again especially after any suspected damage.
4. **Good technique** - Minimising exposure of personnel to X-rays requires effective patient restraint (a compliant patient will have fewer undiagnostic images taken), effective positioning of the patient and imaging plate with minimal handling and good technique to minimize scatter.

Employees should receive training in the radiation safety protocols and rules of an individual practice, and acknowledge their duties of never knowingly exposing themselves or anyone else, informing their employer if pregnant, using and caring for personal protective equipment and dosimeters, and reporting any over-exposure or incidents including equipment damage.

Principles of Radiography and Exposure[2]

The X-ray tube generates the X-rays in the form of the unidirectional primary beam. Collimator shields control the size of the primary beam, a light source allows visualisation of this area. The tube stand (or gantry) supports the tube head. The control panel has an on/off switch, exposure factor controls (mA, kV, time), the exposure button and an exposure indicator, and the light source on/off button. kV (kilovolts) defines the penetrating power of the X-ray beam. mA (milliamperes) defines the quantity (intensity) of X-rays produced. mAs is mA times by time. X-ray intensity follows the inverse square law - if you double the distance from the tube head, you reduce the X-ray intensity four-fold. The distance between the X-ray tube and the imaging plate is the source-image distance (SID, or what used to be called film-focal distance, FFD). If the SID is increased the exposure must be increased - **(e.g., if 10 mAs is used with a SID of 75 cm, when the SID is increased to 150 cm 40 mAs is required**. Shortening the SID without changing the exposure settings will act to increase the exposure. Differential absorption of the primary beam is caused by type of tissue: with increased absorption occurring from air-fat-water-bone-metal; and varying thickness of tissue. Less absorption equals more X-rays on the imaging plate, which appears more black (termed radiolucent). More absorption equals fewer X-rays on the imaging plate, which appears more white (termed radiopaque). The dynamic range of the image (also called latitude) is the range of exposures over which an image is formed.

Exposure factors affect the opacity and contrast of the X-ray image: the quantity of X-rays reaching the imaging plate is a result of the mAs and SID, which affects the radiopacity of the image; the quality and intensity of the X-ray reaching the imaging plate is a result of the kV, which affects both the radiopacity and the contrast (difference between radiopaque and radiolucent areas). High mA, low kV creates high contrast but low dynamic range (seen as a very black and white image); whereas low mA and high kV produces low contrast but high dynamic range (seen as a very grey image). In general, a higher kV setting allows a lower mAs setting and will produce an image of comparable density. If the optimum kV is exceeded, the image will suffer from an overall increase in density with loss of contrast as the effect of differential tissue absorption is reduced. Increased scatter will also occur, further reducing contrast and image quality. Collimation is essential to reducing scatter as reducing the primary beam size to the area of interest will limit scatter from neighbouring tissues.

Factors that affect exposure selection include the type of X-ray machine and its output, the type of imaging screen used, the patient details (body composition) and the area imaged. For consistent good-quality images an exposure chart should be used that relates to the anatomical area to be imaged and the position/projection to be taken; relies upon consistent imaging plate, SID and if necessary grid usage; and identifies the kV and mA (mAs) required for a specific patient/view combination.

Projections and Positioning[2]

When describing a projection, the terminology refers to the direction of the X-ray beam from "point of entry" to "point of exit" (e.g., left to right laterolateral). When imaging equine medicine cases, a large imaging plate (35 x 43 cm) is preferable as this allows fewer exposures and better anatomical description. Overlapping fields should be taken, with 4 collimated margins within the plate boundaries. When using smaller imaging plates (especially with portable equipment), more exposures will be required to cover the same anatomic region, and radiopaque markers may be required to allow anatomic localisation. Poor positioning describes poor localisation to the area of interest, poor tube-imaging plate alignment, excessive collimation and movement blur (which can relate to the X-ray tube, the patient and/or the imaging plate). Plate holders are vital to safe, effective imaging of equine medicine cases as they reduce or eliminate exposure of an assistant compared to hand holding.

Prior to positioning, the patient should be checked for any mud/dirt on the coat or areas of wet hair that will cause artefact, any found should be removed/dried off. For imaging the head, axial skeleton and thorax, true laterolateral projections are essential, which requires that the horse be standing square and perpendicular to the primary beam. This author recommends having a permanent long axis marker on the floor of the X-ray room, which allows easy positioning. When X-raying in a temporary location, this marker can readily be created using chalk.

Head

For laterolateral projections the head should be supported in a neutral position with the muzzle at the level of the shoulder joint; this can easily be achieved by resting the muzzle on a small trolley. The head should be vertical and not tilted to either side (ear tips level). The head collar should be removed, and the ears gently secured forward using adhesive tape (this allows imaging of the caudal surface poll). When imaging the cheek teeth, it is helpful to open the mouth slightly. For ventrodorsal (or dorsoventral) views, the head is extended forward from the poll by a handler holding the muzzle, this is difficult to achieve in a conscious horse as any degree of rotation can make interpretation difficult.

Cervical Spine

The horse should be stood square, with the head in a neutral position (muzzle level with the shoulder joint). This is best achieved by a handler standing in front of the horse with the muzzle held in their cupped hands, which allows the handler to hold the neck straight by ensuring the ear tips are level and that there is an equal amount of shoulder and rump visible between each ear and the midline of the neck (this limits rotation and yaw).

Thoracolumbar Spine

The horse should be stood square, ensuring that they are not resting either hindlimb (as this causes rotation of the spine).

Thorax

The horse should be stood square, with the front limbs slightly forward to reduce muscle mass over the cranial thorax. Left and right laterolaterals may be required to locate a lesion.

The Head[2,3]

- **Cranium** - Laterolateral view, using a large imaging plate (portrait orientation), centre the X-ray beam 5 cm caudal to the orbit at right angles to the midline. Ventrodorsal view centres on the larynx and generally requires general anaesthesia. Oblique views of the caudal skull are challenging to obtain.
- **Frontal and maxillary sinuses, maxilla** - Laterolateral view centres midway between the orbit and the lateral opening of the infra-orbital canal, about 1" dorsal to the facial crest at right angles to the midline. Ventrodorsal (or D-V) view is centred on the midline between the horizontal rami of the mandible in the middle third of the face, with the primary beam aligned perpendicular to the imaging plate.
- **Cheek teeth and mandible** - Laterolateral views centred on the point where the upper and lower 4th cheek teeth meet (rostral end of the facial crest), at right angles to the midline. It is helpful to slightly open the mouth with a bite block to separate the occlusal surfaces of the cheek teeth. Dorsoventral view can be obtained in a heavily sedated horse with their head resting on a low support, with the beam centred on the nasal septum at a level of the rostral end of the facial crest. Offset mandible views can be obtained this way, with the mandible positioned laterally with a bit block. Oblique views image the roots of the cheek teeth, the tooth of interest should be against the cassette, the X-ray beam is then aligned dorsally 30° to image the maxillary roots, or ventrally 30° for the mandibular roots, with the beam centred on the tooth of interest. A separate maxillary lateral 40–60° ventrodorsal oblique view can be obtained with the mouth wide open to obtain further apical detail.
- **Canines and incisors** - Extraoral horizontal slightly caudorostral oblique views image the canines. Intraoral views can image both the canines (perpendicular to the long axis of the head, 45° to the imaging plate), and the incisors (90° to the imaging plate).

CERVICAL SPINE[2,4-6]

Sequential laterolateral views from the occiput to T1 are required to image the entire cervical spine. Radiolucent skin markers should be positioned starting at the cranial margin of the wing of the atlas, then each sequential marker should be positioned at a level approximately 3 cm dorsal to the palpable cranial tubercle of the transverse process of the cervical vertebrae (which corresponds with the location of the vertebral canal), at a distance that allows the imaging plate to overlap by ¼ for each view. C7 has no transverse processes and so the final marker should be positioned approximately 10 cm caudal and horizontal to the cranial tubercle of the transverse process of C6. The imaging plate (landscape orientation) should be positioned at an angle parallel to the long axis of the vertebral column, the tube should be perpendicular to the long axis of the neck. Interpretation of cervical radiographs requires true laterolateral projection, the adequacy of obtained images should be assessed by looking for complete overlap of the cranial tubercles of the left and right transverse processes of each vertebra. If dorsal-ventral distraction is present, the horse's neck is rotated (seen by the handler as unlevel ear tip position). If craniocaudal distraction is present, the neck is displaced left to right (seen by the handler as unequal shoulder/rump between the ear tips). Distracted images should be repeated. When using portable radiography equipment to obtain laterolateral images, it is helpful to reduce the SID, but maintain an air gap between the horse and the imaging plate as this can control scatter.

Paired left and right dorsal to ventral (or ventral to dorsal) oblique views should be obtained if an abnormal appearance to articular process joints is found on laterolateral projections. Again the horse is stood square, slightly retracting the forelimb on the side of the imaging plate can help with positioning, with the head held as for the laterolateral view. A radiolucent skin marker is positioned horizontally at the level of the articulation of interest, midway dorsoventrally up the neck. The primary beam is perpendicular to the long axis of the neck, and positioned DV (or VD) at an angle of 50–55° for C4-C5 and 45–55° for C5-T1, the imaging plate is in landscape orientation.

THORACOLUMBAR SPINE[2,7]

To obtain laterolateral images of the dorsal spinous process, a series of different radiopaque markers should be positioned along the dorsal midline, at a distance apart of ½ the width of the imaging plate (landscape orientation). The X-ray beam is centred horizontally 10–20 cm ventral to the top of the back (lowest at the withers where the top of the plate should be just above the highest point), and perpendicular to the long axis of the trunk. A series of overlapping images is required to identify any affected vertebrae, which are then located relative to the radiopaque markers. It is often helpful to start in the mid-thoracic region (T9–T15) and then image cranially and caudally.

Paired left and right oblique projections are required to image the vertebral bodies and their articular process joints. Radiolucent skin markers should be positioned approximately 15–25 cm ventral to the dorsal midline (roughly the level of the vertebral canal), again allowing for overlap of the imaging plate (landscape orientation). The X-ray beam is positioned perpendicular to the long axis of the trunk and angled 20° ventrodorsally. Effective collimation is essential to image quality.

THORAX[2]

Laterolateral radiographs are obtained during peak inspiration. Four overlapping views are taken: dorsocaudal, dorsocranial, and ventrodorsal (all in landscape orientation) and ventrocranial (in portrait orientation). Radiolucent skin markers should be placed that correspond to the middle of the imaging plate. For the two dorsal views, the top of the imaging plate is at the level of the thoracic vertebral canal. This is most easily estimated in the caudodorsal thorax where it is approximately 15 cm ventral to dorsal midline. The caudal margin of the imaging plate of the caudodorsal view should be at approximately the 15/16th rib. The craniodorsal view should overlap the cranial margin of the caudodorsal view by 1/3 of the imaging plate. For the caudoventral view, the bottom margin of the imaging plate should be the dorsal margin of the olecranon and the caudal margin at approximately the 8/9th rib. The cranioventral image is positioned 5 cm more ventrally and should overlap the cranial margin of the caudoventral view by ¼ of the imaging plate. Structures closest to the imaging plate will have sharp margins, whereas those on the contralateral side will be magnified and less distinct. Paired left and right laterals should be taken if pathology is suspected in one hemithorax. Mediastinal pathology will appear largely identical on left and right views.

Copies of the lecture images showing the position of all the radiographic views described above will be available at: www.dropbox.com/sh/lz7xqjyjnr0vdf8/AAD17WNmxFRzwxOH3y8MVN5pa?dl=0

References

1. Pilsworth R. *Radiation Safety in Equine Practice.* DVD British Equine Veterinary Association; 2006.
2. Butler J, *et al. Clinical Radiology of the Horse.* 4th ed; 2016.
3. Baratt R. *Vet Clin North Am Equine Pract.* 2013;29(2):367–395.
4. Down S. *Equine Vet J.* 2009;41(6):518–524.
5. Withers J, *et al. Equine Vet J.* 2009;41(9):895–902.
6. Tambaschi. *Equine Vet J.* 2014;S47:10–11.
7. Girodroux M, *et al. Equine Vet J.* 2009;41(2):130–138.

acvim012

Common Clinical Presentations and Diseases of Neonatal Ruminants and Camelids

Melissa G. Ackerman, DVM, MS
Pullman, WA, USA

Introduction

Neonatal ruminants are vulnerable to illness and death. It is estimated that 5–10% of the calf crop and 15–20% of the lamb crop each year dies before weaning in the USA; the majority of these deaths happen during the first 3 days of life.[3] A number of factors impact neonatal health, including the health and nutrition status of the dam in the periparturient period, events during labor and delivery, transfer of colostral immunity, management on the farm, and weather and other environmental factors. Sick ruminant and camelid neonates may have a variety of clinical signs. Some musculoskeletal congenital abnormalities may be easily identified at birth. Other problems can occur such as hypoxia, sepsis, pneumonia, gastrointestinal (GI) ulcers, abomasal bloat, and failure of passive immune transfer, and may not be readily recognized at first. Identifying risk factors and understanding how they lead to disease will help the veterinary technician or assistant to identify neonates at high risk of developing disease or of dying. Early diagnosis and treatment is important in giving the sick neonate the best possible chance at recovery.

Risk Factors for Neonatal Disease Development

Events that occur before birth impact the neonate, therefore it is important to gather a thorough history about the dam's specific diet, environment, health, and labor and delivery. Proper nutrition to appropriately meet increased metabolic requirements associated with pregnancy and lactation are essential for normal fetal and neonatal development. Mineral excess or deficiencies in the dam directly impact the mineral status in the fetus. Energy and protein deficiency in the dam, pregnancy toxemia, vitamin A deficiency, copper deficiency or excess, iron excess, iodine excess or deficiency, and selenium deficiency are all common causes of stillbirth and perinatal death in ruminants.[3] When inquiring about feeding, it is important to have the client provide supplement labels, hay analyses, and amount fed per head per day to properly evaluate the diet.

Extreme weather during pregnancy can negatively affect the fetus. Heat stress in the third trimester interferes with normal fetal growth and with the dam's production of high quality colostrum, also increasing the risk for failure of passive transfer, or FPT (discussed later). This is primarily because dry matter intake decreases and this promotes negative energy balance, mobilization of fat, ketogenesis, and pregnancy toxemia.[3] In fact, anything that negatively affects the dam's health or ability to deliver will negatively impact the fetus, including obesity, protein energy malnutrition, hypocalcemia and other micronutrient deficiencies.[3]

Maternal rejection, prematurity, FPT, prolonged or induced labor, cesarean section, weakness/poor appetite, disease or debilitation of the dam, and adverse weather conditions are all risk factors that contribute to high risk of neonatal disease and death. Primiparous dams, multiple offspring, and cesarean section are all associated with failure of maternal-offspring bond development. Starvation and hypothermia is the second leading cause of death in neonatal lambs. As the amniotic fluid dries, the neonate experiences evaporative heat loss. Neonates have little energy reserves and have to shiver and metabolize brown adipose tissue for thermoregulation. If they are born during rain, wind, or cold temperatures there are no energy reserves to meet the increased (up to 500%) metabolic requirements that weather stress demands. If the animal does not eat or continues to be exposed it will die from hypothermia or cerebral hypoglycemia. Additionally, with all of these conditions the absorption of immunoglobulins from ingested colostrum is decreased in the affected neonate.[3]

Forty to sixty percent (40–60%) of stillbirths are associated with dystocia. Dystocia commonly causes edema in the head and tongue, making it difficult for the neonate to nurse and ingest colostrum until the edema resolves, making FPT likely. Another common sequela is severe respiratory and metabolic acidosis. Calves that survive dystocia are at higher risk of becoming sick in the first 45 days of life than calves delivered normally. Fetopelvic incompatibility is a common cause of dystocia in heifers, which is why appropriate breeding time and feeding to achieve appropriate growth (and therefore size) by calving time in primiparous cattle is essential for dystocia prevention.[3]

If dystocia impairs blood flow to the placenta or if the umbilical cord ruptures during dystocia, the fetus can become deprived of oxygen. The resultant ischemia and anaerobic metabolism can result in acid-base derangements and sets the tissue up for reperfusion injury. In these cases, the neonate may require treatment for metabolic acidosis, and may suffer ischemic organ injury like renal tubular necrosis, gastrointestinal injury, and encephalopathy. Any process that interferes with gas exchange to the placenta can also cause fetal hypoxia and ischemia; for example, infectious placentitis.[3]

Colostrum and Passive Immunity

Neonatal ruminants are born agammaglobulinemic because immunoglobulins do not transfer across the placenta. Therefore, it is imperative for the neonate to ingest high-quality colostrum for antibody protection. Colostrum contains many components important for nutrition and immunity including maternal immunoglobulins, leukocytes, cytokines, and various antimicrobial proteins and peptides.[4] For this discussion we will focus on the importance of maternal immunoglobulins.

At birth, neonates are exposed to numerous environmental microorganisms. Common routes of infection include the umbilicus and the oro/nasopharynx into the respiratory and GI tract. The neonate's naïve immune system needs colostral immunity to help protect it from severe infection. Successful passive transfer of immunity occurs when the neonate has ingested and absorbed adequate colostral immunoglobulins. This absorption is a passive, non-selective process which ceases by 24 hours old, and is due to the short-lived neonatal enterocytes of the newborn. In calves, adequate passive transfer is defined as >10 mg/ml serum IgG1 by 48 hours old (IgG1 comprises the majority of immunoglobulins in ruminant colostrum). Neonates should consume 6–10% of their body weight as colostrum in the first 24 hours of life, and the first feeding should occur in the first 6 hours. Neonates with FPT are at higher risk of developing disease, especially bacterial septicemia, but other important interacting factors include management, hygiene, environment, infection pressure, organism virulence, and antibody specificity. Therefore, neonates with low levels of serum IgG1 do not always contract disease, and neonates with adequate serum IgG1 levels can still develop disease, especially when exposed to an environment with a high pathogen load, as in a crowded maternity pen with multiple cows.[3]

Diagnosis of Failure of Passive Transfer

Passive transfer of immunity may be measured directly or indirectly. Most tests are optimally performed at 24–48 hours old, and the greatest accuracy is obtained within the first week of life, before the neonate makes significant amounts of immunoglobulins. Single radial immunodiffusion and enzyme-linked immunosorbent assay tests can directly measure serum IgG concentration. Other available tests rely on total proteins for which passive immunity of IgG is significantly associated. Zinc sulfate turbidity, sodium sulfite precipitation, glutaraldehyde coagulation, gamma-glutamyl transferase activity, and refractometer of total serum solids all estimate serum IgG concentration based on total proteins in the blood or serum.[3] Perhaps the most practical test in the clinical setting is using total serum solids to estimate IgG concentration. This test does not have to be sent to a diagnostic lab, most clinics have a refractometer, and results are nearly immediate. Serum protein >5.0 g/dL is associated with successful passive transfer, 4.5–5.0 g/dL is uncertain, and <4.5 g/dL is consistent with failure of passive transfer.[3]

Milestones Immediately After Birth

Neonatal resuscitation is an important topic in neonatal care and the at-risk neonate, but is beyond the scope of this discussion. This topic is discussed in some of the references provided.

Calves and lambs are born with a head-righting reflex that should be present just after birth. Within the next few minutes, the neonate should be able to achieve sternal recumbency.[3] Time to standing and nursing is variable between species and breed. Calves should be attempting to stand within 15 to 30 minutes after birth.[1] Beef calves will stand by about 35 minutes and nurse by 81 minutes and dairy calves can take up to 70 minutes to stand and over an hour and a half to nurse.[3] Most lambs will stand by 30 minutes and nurse by 90 minutes. Llama and alpaca crias often stand and nurse within the first hour of life, but should by standing by 60 minutes and should be nursing by 2–4 hours old.[5]

Physical Exam and Identification of Common Congenital Abnormalities and Common Neonatal Diseases

Observation

Completion of the physical exam will allow for identification of some common congenital abnormalities. An important part of the physical exam is observation of body condition, behavior, posture, movement, and attitude. Sick neonates may hold their head and ears dropped, spend more time recumbent or be reluctant to move or nurse, and may be dull and uninterested in interacting with their surroundings. Meconium-staining indicates a stressed newborn, and hypoxic neonates may have difficulty maintaining sternal recumbency, be slow to rise, weak, lack a suckle reflex, and develop depressed mentation within a few hours.[3]

Musculoskeletal System

Anything that prevents a neonate from standing and nursing soon after birth could be fatal. Arthrogryposis, tendon laxity, and obvious congenital deformations such as scoliosis, kyphosis, lordosis, and atlantoaxial malformations may be obvious during observation. Examination for signs of trauma, like rib or limb fractures from birth, strenuous manipulation during dystocia, or being stepped on by an adult is important, as is identifying flexural and angular limb deformities. Some limb deformities may be correctable with splints. Others may be caused by nutrient deficiencies such as vitamin D, calcium, phosphorous, or "rickets" that result in insufficient mineralization of bone that is easily seen on radiographs and also results in spontaneous fractures. These patients can respond well to supplementation and be normal, but also may have residual angular limb deformity and tendon laxity. Animals that are reluctant to move, have a stiff, painful gait, but are otherwise bright and alert may have vitamin E and/or selenium deficiency-associated myopathy. These neonates are usually born to vitamin E and/or selenium deficient dams, and often do not respond to supplementation. The prognosis is poor because the myopathy is irreversible. The joints should be palpated for signs of heat, swelling, and pain, which could be due to septic arthritis. In this case, several joints are often affected and systemic antibiotics and joint lavage are necessary for recovery.[3]

Body Temperature

Sick neonates may be febrile, especially with infectious septicemia. Hypothermia and hypoglycemia often are present if the animal is anorexic, as neonates have little glycogen stores. Drying, warming, and tube-feeding colostrum can revive a weak neonatal lamb or kid if recognized early enough.[3] Table 1 indicates normal body temperatures and other physical exam parameters for neonatal ruminants.

Head, Eyes, and Ears

The eyes should be clear; cloudy eyes may mean the animal has uveitis due to a uterine or systemic infection acquired soon after birth; in this case fibrin can often be seen in the anterior chamber. Corneal abrasions or ulcers and congenital cataracts can also cause the eyes to look cloudy. Entropion and ectropion are often easily identifiable as well. Entropion is often easily corrected with a subconjunctival injection of a small volume of penicillin procaine G or with eyelid sutures. The ears should be held erect, if one is drooping it may be due to otitis and facial nerve paralysis, and sometimes purulent exudate can be seen in the ear. The exudate may be cultured with antimicrobial sensitivity testing. Concurrent respiratory disease is often seen with otitis, and *Mycoplasma*, *Pasteurella*, and *Haemophilus* species are often cultured from otitis cases. These animals often benefit from systemic antibiotics. Calves with hydranencephaly due to bovine viral diarrhea virus (BVDV) may have a dome-shaped head. Brachygnathism is often easily identified here and can interfere with normal nursing if severe.[3]

Oral Cavity
In the oral cavity the incisors should be erupted and the hard palate should be examined, as cleft palate is the most common congenital defect of ruminants.[3] With cleft palate the animal often is seen with milk dribbling out of the nose, and aspiration of milk results in pneumonia. The mucous membranes should be pink and moist. Tacky or dry mucous membranes are consistent with dehydration. Sick neonates may be pale when anemic, hypothermic, hypovolemic, and in shock. Septic neonates may have dark pink or red mucous membranes and a visible "toxic line" along the incisors. Petechiae along the mucous membranes is consistent with sepsis. A complete blood count (CBC), chemistry panel, and blood gas is often helpful in characterizing these abnormalities and guiding fluid therapy and treatments.

Cardiovascular System
Congenital heart defects like tetralogy of Fallot and Eisenmenger's complex that result in right to left shunting of blood result in cyanotic mucous membranes. Cardiovascular compromise during shock can also cause a blue or grey color to the mucous membranes. Normal capillary refill time is <2 seconds.[3]

Cardiac rhythm should be regular; arrhythmias may be consistent with illness due to diarrhea, cardiomyopathies, ionophore toxicosis, and hypomagnesemia. The most common cardiac defect in ruminants is ventricular septal defect. This may or may not be accompanied by a pansystolic murmur heard on the left and right sides during cardiac auscultation.[3] Patent foramen ovale and patent ductus arteriosus also occur in ruminants and often a heart murmur may be auscultated. Detailed descriptions of the heart murmurs associated with VSD and other congenital cardiac defects can be found in (3). Peripheral pulses may be palpated on the coccygeal and brachial arteries (ventral aspect of the tail and the medial aspect of the elbow, respectively) and should be strong and regular. Extremities in the normal neonate are warm.[3]

Respiratory Tract
Malodorous breath, increased respiratory rate and effort, wheezes, stridor, coughing, and nasal discharge are all consistent with respiratory disease. Over 20% of preweaned calves and 50% of weaned heifers die due to respiratory disease.[1] Chest auscultation can help identify neonates with pneumonia. Pneumonia may be viral and/or bacterial in etiology. Neonatal septicemia can result in hematogenous spread of bacteria to the lungs. Bronchoalveolar lavage with cytology and/or bacterial culture and sensitivity testing may be helpful in guiding treatment, and blood culture may be helpful in cases of sepsis. However, broad-spectrum antimicrobial treatment should be started when pneumonia is first recognized, rather than waiting for diagnostic test results. Treatment may be adjusted if needed.

Aspiration pneumonia occurs when liquids like milk or solids like meconium are inhaled. This can be caused by improper esophageal tube feedings, or by hypoxia during birth causing reflex inhalation of meconium. Aspiration pneumonia can result in immediate death or chronic bronchopneumonia that may ultimately be fatal. Treatment is usually with long-term antimicrobial and anti-inflammatory therapy.[1]

Choanal atresia is the most common cause of respiratory problems in the cria.[2] It is a congenital abnormality that is often recognized at birth in llama and alpaca crias by dyspnea, open-mouth breathing, and difficulty nursing. The oronasal cavity and the choanae, which are normally continuous, are separated by a membranous or osseous structure due to abnormal embryogenesis. Because camelids are obligate nasal breathers, this causes dyspnea. This condition can be confirmed with contrast radiography or CT. The prognosis is poor and euthanasia is often recommended.[5]

Gastrointestinal Tract
Abnormal abdominal contour, appetite, and fecal consistency often signifies GI disease. Congenital defects like atresia ani, atresia coli, and atresia recti occur and these neonates will have a history of declining appetite, colic, and distended abdomen. Intestinal atresia is the primary cause of abdominal distension in calves up to 7 days old; the spiral colon is the most common site.[3] Radiographs and exploratory abdominal surgery may help with diagnosis and prognosis as far as feasibility of surgical correction. Diarrhea staining on the tail is important to note, as infectious diarrhea is the primary reason for mortality in calves 3–21 days of age. Though small amounts of blood may be present in normal calf feces, blood and fibrin passed in feces is associated with inflammation of the bowel and pathogens like *Salmonella* and coronavirus should be considered as likely etiologic agents. Enterotoxigenic *E. coli* causes

secretory diarrhea, and therefore alkaline fecal pH, which is quickly distinguished with pH paper.[3] If diarrhea is causing electrolyte imbalances, acid/base disturbances, and/or sepsis evident on blood gas, CBC and chemistry panel, then broad-spectrum antimicrobial treatment is warranted with fluid therapy and treatments to correct the abnormalities.

Putrefactive indigestion occurs when large amounts of milk are delivered to the abomasum and it refluxes back to the rumen, or when milk escapes the esophageal groove and is delivered into the rumen. Clinical signs include diarrhea, poor growth, rough haircoat, and recurrent bloat. Passage of an orogastric tube allows for rumen fluid collection and examination; pH is usually ≥7 with rumen putrefaction. Reducing milk volume per feeding and introducing a creep feed or calf starter will help promote ruminal development.[3]

Abomasal ulcers may not result in clinical signs in young calves, but peritonitis, shock, and death can rapidly occur with perforating ulcers. Abdominal distension and pain with palpation may occur with abomasal ulcers in calves. Abomasal ulcers reportedly appear after adverse weather. Perforating ulcers are associated in calves with copper deficiency, dietary changes, mycotic infections, abomasal bezoars, and *Clostridium perfringens* abomasitis. With abomasal obstruction and inflammation, the abomasal contents will reflux into the rumen; pH is low to normal. Radiographs and ultrasound can be helpful in diagnosing and distinguishing problems of abdominal distension.[3]

Umbilicus

Infectious pathogens can ascend the umbilicus and cause infection and septicemia, usually during the first 2 weeks of life. Palpation of the umbilicus should be part of the examination, noting size, discharge, discomfort on palpation either caudally as the umbilical arteries course toward the bladder, or cranially as the umbilical vein courses toward the liver. Though patent urachus is uncommon in calves, umbilical infection sequelae include sepsis, peritonitis, abdominal adhesions, umbilical hernias, abscesses, septic arthritis, and osteomyelitis. Ultrasound of the suspected umbilical cord infection can be helpful to identify the structures and the extent of involvement.[3] Because of this potential route of infection, the umbilicus should be dipped with 7% iodine tincture or 0.5% chlorhexidine solution after birth.

Summary

Neonatal ruminants and camelids are vulnerable to infectious disease and death. This is due to a number of factors including the health and nutrition status of the dam in the periparturient period, events during labor and delivery, transfer of colostral immunity, management on the farm, and weather and other environmental factors. It is important for the veterinary technician and assistant to gather relevant information in the history to help identify potential risk factors or differentials when a sick neonate presents to the clinic. Additionally, common congenital defects and neonatal problems and diseases may be identified at birth or soon thereafter during physical examination.

Table 1. Normal physical exam parameters in ruminants and camelids

Species/age	Body temperature °F	Heart rate beats/min	Respiratory rate breaths/min
Cattle adult	100.5–102.5	40–80	12–36
Calf	101.5–103	100–140	30–60
Sheep adult	102–103.5	60–120	12–72
Lamb	102.5–104	120–160	30–70
Goat adult	101.5–103.5	70–110	15–40
Kid	102–104	120–160	40–65
Llama/alpaca adult	99.5–102	60–90	10–30
Cria	100–102	60–90	20–40

Table modified from Smith, *Large Animal Internal Medicine*, 4th ed. and Fowler, *Medicine and Surgery of South American Camelids*, 2nd ed.

References

1. Poulsen KP, McGuirck SM. Respiratory disease of the bovine neonate. *The Veterinary Clinics of North America Food Animal Practice*. 2009;25:121–137.
2. Smith B, Timm K, Long P. *Llama and Alpaca Neonatal Care*. Corvallis, OR: Bixby Press; 1996.
3. Smith BP. *Large Animal Internal Medicine*. 4th ed. St Louis, MO: Mosby Elsevier; 2009.
4. Barrington GM, Parish SM. Bovine neonatal immunology. *The Veterinary Clinics of North America Food Animal Practice*. 2001;17(3):463–476.
5. Whitehead CE. Management of neonatal llamas and alpacas. *The Veterinary Clinics of North America Food Animal Practice*. 2009;25:353–366.
6. Fowler ME. Clinical diagnosis: examination and procedures. In: *Medicine and Surgery of South American Camelids*. 2nd ed. Ames, IA: Iowa State Press Blackwell; 1989:69–88.

acvim016

Advanced MRI Safety for Veterinary Professionals

James J. Stuppino, BS, RT (R) (MR)
Bethlehem, PA, USA

MRI Use in Veterinary Medicine

MRI technology has only been around since 1980 and high costs originally prohibited the use of MRI in animals until the mid-1990s. This phenomenal diagnostic tool has not been readily available to the veterinary world until recently. Even in today's market, dedicated veterinary MRI facilities are not very common. Most states have maybe 2–3 Veterinary MRI centers available, and owners are sometimes traveling up to 5 hours or more when their pet needs an MRI. The demand for dedicated independent veterinary MRI centers with certified staff continues to grow across the United States.

MRI technology allows veterinarians the ability to see more detailed images of injuries and diseases within pets' bodies than traditional diagnostic tests, such as X-rays or even CT. MRI is now taking the guesswork out of veterinary medicine by enabling veterinarians to see exactly why their patient may be experiencing symptoms. It is an extremely useful tool for veterinarians when it comes to an animal with a mysterious or undiagnosed problem as well but it's a technology that does not come without risks. The invisible risk of the magnetic field and liquid helium should not be taken lightly.

MRI Safety

MRI has gained much press over the years from dangerous accidents but probably the most infamous was the Michael Colombini case in August of 2001. Michael was a 6-year-old boy who died after undergoing an MRI exam at a New York-area hospital when the machine's powerful magnetic field pulled a metal oxygen tank out of the hands of a nurse and accelerated it across the room, crushing the child's head. Accidents like these are not very common but the machines strength cannot be taken lightly. It's the invisible danger of the magnetic field and the common misconception of "I didn't know the machine was on." Many individuals who do not routinely work around a MRI machine think if it's not making noise, there is no magnetic field. It's this misconception that causes injuries. There has been a dramatic increase in adverse MRI events in recent years, from 40 reported adverse events in 2004 to 164 reported events in 2012.

The difference between discovering a potentially dangerous item and it going flying is usually a matter of where the person is standing in the room. Someone saying, "Oh, I just realized that I had my pocket knife in my pocket," can easily turn into a much more serious situation if that person is a foot closer to the magnet and someone else is between them and the magnet. Providers need to recognize that any time a ferromagnetic object makes it through the screening process undetected and into the controlled access of the MRI suite that all the conditions required for a projectile accident are in place. The following diagram (Figure 1) represents magnetic fringe fields of a 1.5T MRI Scanner. The diagram is in (mT) and the 0.5 gauss line labeled with the yellow arrows is very significant. This is the pacemaker line and it should be kept inside the MRI scanner room. Magnetic shielding should be considered if this line goes past the walls of the scan room. As you get closer to the scanner, the stronger these gauss lines get and the more significant the pull on a metal object. Every scanner installation should have a field plot like this completed for it and it should be taken very seriously to determine where the safety zones around the MRI should be considered.

Figure 1

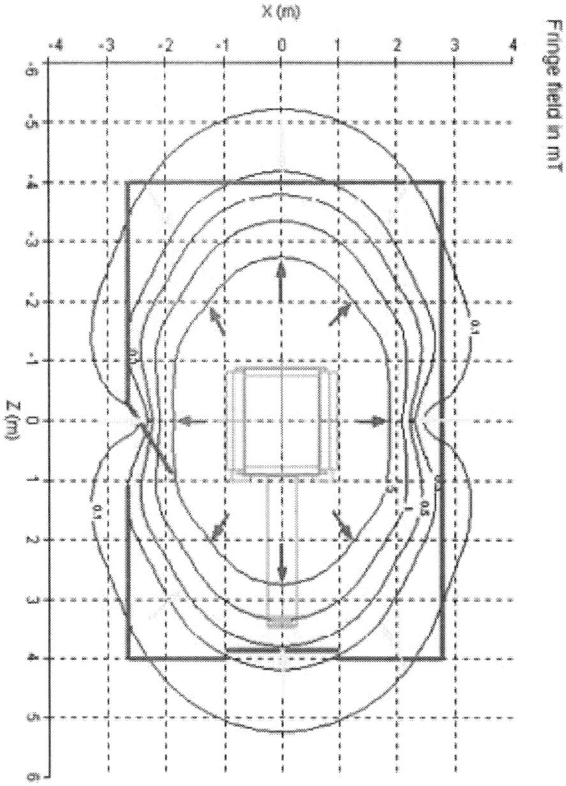

Projectile incidents accounted for only 12% of reported injuries. Excessive heating and burns account for another large percentage of mostly avoidable injuries. Facility based standards should be established in veterinary medicine and restricting access to the MRI site and staff education should be the main priorities of any facility using a MRI Scanner. Continuous staff education should be provided on a quarterly basis with updates to discuss any potential incidents and solutions how to correct them. Veterinary imaging facilities should be designed using the American College of Radiology in human medicine's four-zone design concept, where Zone 4 is the actual scan room and no admittance without documented training and screening.

Some recent events in veterinary MRI have brought out questions concerning another dangerous part of MRI Technology; liquid helium and quenching of the scanner. A recent accident concerning a very old veterinary magnet and a decision to "quench" the magnet instead of ramping it down to eliminate the magnet field. So it brings up a question "When do you quench the magnet?" This process can occur spontaneously if there is a manufacturing fault with the scanner. It can also occur in a fire situation. If the scanner heats up internally by just 1°C this is sufficient for a quench to occur. A quench is more likely to be initiated manually, however, in an emergency situation. The purpose of initiating a quench is to make the magnet lose its magnetic field quickly. This may be done for the following reasons:

- Someone is trapped against or in the magnet by a ferromagnetic object.
- There is a fire in the scanning room or one likely to get out of control in the controlled access area.

Emergency Caused by a Ferromagnetic Object Endangering Life

Ferromagnetic objects can be attracted towards the magnet with a force that may be many times the original weight of the object. This makes something as small as scissors difficult to remove and something larger, like an oxygen trolley impossible to get off the scanner. Therefore, if someone was trapped by a large object, the only way to free them quickly would be to remove the magnetic field; i.e., by quenching the magnet.

Non-Emergency

If a large ferromagnetic object - e.g., a trolley - gets stuck on the magnet and no one is trapped or could be potentially hurt by the accident then the magnet does not need to be quenched as MRI engineers can 'ramp down' the magnetic field with specialized equipment in a more controlled manner than quenching. A quench initiated by anyone other than an MRI engineer is normally a chargeable call and can cost up to 60–80k.

MR Authorized'/'Responsible Persons are the only staff who should quench the magnet as they should know and have been trained in which precise situations require such a drastic measure to be taken. To quench, press the large red Quench button under the glass cover in the control room or on the wall in the scanner room. After pressing the quench button be aware that the magnetic field will dissipate within 20–60 seconds and so be ready because the object pinned to the magnet will fall to the ground. A trapped individual should be helped at this point and immediately removed from the scan room.

MRI Contrast Media

Another area of safety concern is MRI Contrast media. "Adverse side effects from the administration of contrast media vary from minor physiological disturbances to rare severe life-threatening situations. Preparation for prompt treatment of contrast media reactions must include preparation for the entire spectrum of potential adverse events and include prearranged response planning with availability of appropriately trained personnel, equipment, and medications." MRI contrast has always been considered to be one of the safest injectable agents but over the past few years this ideology has completely changed. Gadolinium contrast agents are paramagnetic agents that act indirectly to provide contrast enhancement by facilitating proton T1 and T2 relaxation processes through alteration of the local magnetic environment. The effect of contrast agents depends on their biodistribution as well as on pulse sequence parameters. Defined risks of gadolinium include allergic reactions, adverse reactions, and in those with renal failure, the development of nephrogenic systemic a fibrosis (NSF). Clear evidence has now been discovered that gadolinium based contrast agents can be found in varied levels of accumulation of residual gadolinium in the brain, bones, and muscles of patients even with normal renal function. The use of a gadolinium based contrast agent as part of every standard MRI scan in veterinary medicine must now be considered. One must now consider the unknown risks of previously unanticipated residual gadolinium as to whether administer a gadolinium contrast agent at all.

Burns

The most common patient injuries in the MRI suite are burns and the most common objects to undergo significant heating are wires and leads. Other objects associated with burns are pulse oximeter sensors and cables, cardiorespiratory monitor cables, safety pins, metal clamps, drug delivery patches (which may contain metallic foil), and tattoos (which may contain iron oxide pigment). It is "possible" for patients to get 1st, 2nd, or even 3rd degree burns in an MRI if items such as ECG cables are looped and are touching the patients skin (even if these devices are MRI compatible). All "cables" should not touch the animal's skin directly, and should **not** be in a looped configuration.

- Coil cables, such as ECG/VCG cables should be separated and parallel.
- Cables should be parallel to the main magnetic field.
- Layer of insulation/padding between the patient and the coils.
- Never use damaged coils or equipment

References

1. ACR Manual on Contrast Media - Version 10.1, 2015: V8. 2012;83.
2. Tobias Gilk, Emanuel Kanal, MD2. *Journal of Magnetic Resonance Imaging*. Interrelating Sentinel Event Alert #38 with the ACR Guidance Document on MR Safe Practices: 2013. An MRI Accreditation Safety Review Tool. 2013;37:531–543.
3. Preventing accidents and injuries in the MRI suite. The Joint Commission. 2008;38.
4. Patient Safety & Quality Healthcare. http://psqh.com/component/content/article/137-november-december-2011/992-mri-safety-10-years-later.html (VIN editor: Link was not accessible on 4/19/2017).

acvim017

Holistic Epilepsy Management

Holger A. Volk, PhD, DECVN, MRCVS
London, UK

The most common chronic neurological brain disease in people and dogs is epilepsy with an estimated prevalence of 0.4–1% (Cowan 2002; Sander, Shorvon 1996) and 0.6–2%, respectively (Schwartz-Porsche 1986; Kearsley-Fleet *et al.* 2013; Heske *et al.* 2014). A bit more than twenty per cent of human patients with epilepsy will continue to have seizures despite chronic medical treatment with two adequate antiepileptic drugs (AEDs) (Picot *et al.* 2008). Around 20–80% of dogs with suspected idiopathic epilepsy (depending on the examined dog population [hospital vs. primary care]) will not get into remission and continue to have recurrent seizures (Heynold *et al.* 1997; Berendt *et al.* 2002; Arrol *et al.* 2012; Boothe *et al.* 2012) and around 20–30% will remain inadequately controlled (less than 50% reduction of seizure frequency) despite adequate treatment with the standard AED phenobarbital (PB) and/or potassium bromide (KBr) (Trepanier *et al.* 1998; Schwartz-Porsche *et al.* 1985; Podell, Fenner, 1993). Epilepsy is more than a just pure seizure disorder (De Risio *et al.* 2015). Dogs suffering from epilepsy, especially drug-refractory epilepsy frequently characterized by a high seizure frequency and cluster seizures, have an increased risk of developing behavior changes, premature death, and a reduced owner-perceived quality of life (Berendt *et al.* 2007; Shihab *et al.* 2011; Chang *et al.* 2006; Wessmann *et al.* 2012). Behavioural changes, the effects of recurrent seizure activity and anti-epileptic drugs can affect significantly the quality of life (QoL) of affected dogs (e.g., injury, disorientation, fear, anxiety, restlessness, polyphagia and polydipsia/polyuria). There is also a potentially high burden upon the owners of dogs with epilepsy that may negatively affect their QoL, including unpredictable and uncontrollable seizure activity, medicating their pets several times a day and emergency vet visits (Wessmann *et al.* 2012; Chang *et al.* 2006). Owner's perceived dog's QoL is especially affected by high seizure frequency and if the dog is on more than 2 drugs (Wessmann *et al.* 2016). The QoL of the owner is on the other hand mainly influenced by the sedation and ataxia level of their AED-treated dog with epilepsy. It is not surprising, but relevant, that reductions in the perceived dog's QoL scores are associated with reductions in owner's QoL, and vice versa. Understanding what influences QoL is of uttermost importance. Podell *et al.* (2016) state in the ACVIM Small Animal Consensus statement on seizure management that the most important outcome measure is QoL, as this will ultimately affect the owner's decision making in looking after a pet with epilepsy or ultimately can force a decision if the pet has a life worth living for. Owners will elect euthanasia if caring for a pet with epilepsy carries too much emotional stress, becomes financially challenging and it has an impact on the owner's own psychosocial environment. It is therefore important to understand how different management options are tolerated and influence the QoL of the affected pet and owner.

Key for reducing stress and improving QoL for the owner and the dog is to consider the influence of the environment, diet, comorbidities and antiepileptic treatment. Some reports have indicated the importance of diets, such as the ketogenic diet, hypoallergenic diet and fatty acid supplementation as new or alternative treatment strategies for canine epilepsy. Food supplementation with omega-3 showed inconclusive results (Matthews *et al.* 2012). There is some anecdotal evidence that a hypoallergenic diet might improve seizure control in dogs with gastrointestinal hypersensitivity (Lujan *et al.* 2004). A traditional high-fat low-carb/protein ketogenic diet failed to improve seizure control (Patterson *et al.* 2005). A more promising ketogenic diet is based on medium-chain triglycerides which improved seizure control in the majority of cases (Law *et al.* 2015).

Pet owner education is another key to successful management of the patient with epilepsy. As better the pet owner is educated about epilepsy, its comorbidities and antiepileptic treatment (side effects, pharmacodynamics and pharmacokinetic aspects), the more the owner will learn to live with the condition successfully and help you in the care of the patient (e.g., active monitoring of seizure frequency [e.g., paper or electronic seizure diary using an APP]) (Bhatti *et al.* 2015). The owner can also help to reduce environmental stress factors and help with the individualization of the drug treatment depending on the seizure status of the dog (Packer *et al.* 2015). The principle of comprehensive epilepsy care is that 'every little helps' to increase the seizure threshold and therefore improve the management of epilepsy.

Conceptually, epilepsy is not one disease with one unique and simple pathophysiology. The term embraces a heterogeneous group of chronic conditions with seizures being its clinical manifestation. Complex simplicity - the brain is composed by a complex structure; however, it can only function and dysfunction in certain limited ways. This explains why a wide variety of disturbances of brain structure and function can result in seizures and epilepsy. It is thought that similar pathophysiological pathways cannot only lead to epilepsy, but also to other neurodevelopmental and neurobehavioral disorders (Johnson, Shorvon 2011; Shihab *et al.* 2011). These disorders share pathophysiological commonalities at various levels. Understanding the common and discrete pathophysiological pathways can help improve our understanding of these diseases and how to diagnose, monitor and treat them appropriately. Furthermore, pharmacologically characterizing these disorders and understanding why treatment does fail in some cases cannot only improve our knowledge of the mechanisms behind drug resistance and how to overcome them, but potentially even more importantly deepen our understanding of their pathophysiology and natural clinical course. Furthermore, understanding comorbidities of epilepsy such as behavior abnormalities might also give us novel treatment angles, which could improve QoL of the affected dog and potentially seizure control.

Shihab *et al.* (2011) first described that dogs with idiopathic epilepsy have an increase in fear/anxiety and defensive aggression behavior. In the Lagotto Romagnolo dog a benign "childhood" or juvenile epilepsy form has been described with the first seizures occurring when they are around 6 weeks of age (Jokinen *et al.* 2007). Nearly all of the affected dogs come into remission at the age of 10 weeks generally without receiving medical treatment. Jokinen and colleagues (2015) looked at long-term outcome of these dogs and found despite these dogs remaining seizure free they had an increase in inattention and excitability/impulsivity behavior compared to controls. ADHD-type behavior has been also described in a recent study from the UK (Packer *et al.* 2016), which could be in part improved when a medium-chain triglyceride diet was fed.

Drug-resistant epilepsy continues to be a major challenge for veterinary medicine and further studies are needed to unravel its pathophysiology, determine clinical and behavior factors, which might provide new treatment angles and so that new treatment strategies can be developed which are individually tailored for the patient to maximize response to AEDs. AEDs will remain the mainstay for epilepsy management, but environmental factors, diet management and treatment of comorbidities might also need to be considered to improve ultimately the QoL of the patient and the owner.

REFERENCES

1. Arrol L. *et al. Vet Rec.* 2012;170:335.
2. Berendt M, *et al. J Vet Intern Med.* 2002;16:262.
3. Berendt M, *et al. J Vet Intern Med.* 2007;21:754.
4. Boothe DM, *et al. J Am Vet Med Assoc.* 2012;240:1073.
5. Chang Y, *et al. J Small Anim Pract.* 2006;47:574.
6. Cowan LD. *Mental Retard Dev Dis Res Rev.* 2002;8:171.
7. De Risio L, *et al. BMC Vet Res.* 2015.
8. Heske L, *et al. Vet J.* 2014;202:3.
9. Heynold V, *et al. J Small Anim Pract.* 1997;38:7.
10. Johnson MR, Shorvon SD. *Epilepsy Behav.* 2011;22:421.
11. Jokinen TS, *et al. J Vet Intern Med.* 2007;21:464.
12. Jokinen TS, *et al. J Vet Intern Med.* 2015;10.1111.
13. Kersley-Fleet L, *et al. Vet Rec.* 2013;171(13):338.
14. Packer RMA, *et al. PloS One.* 2015.
15. Packer RMA, *et al. Epilepsy Behav.* 2016;55.
16. Podell M, Fenner WR. *J Vet Intern Med.* 1993;7:318.
17. Podell, *et al. J Vet Intern Med.* 2016.
18. Sander JW, Shorvon SD. *J Neurol Neurosurg Psychiatry.* 1996;61:433.

19. Schwartz-Porsche D, *et al. J Vet Pharmacol Ther*. 1985;8:113.
20. Schwartz-Porsche D. In: Proceedings ACVIM 1986;1161.
21. Shihab N, *et al. Epilepsy Behav*. 2011;21:160.
22. Trepanier LA, *et al. J Am Vet Med Assoc*. 1998;213:1449.
23. Wessmann A, *et al. J Vet Intern Med*. 2012;26:1.
24. Wessmann A, *et al. Vet Rec*. 2016;10:1136.

acvim018

Caring for the Down Patient

Laurie Cook, DVM, DACVIM (Neurology)
Columbus, OH, USA

Introduction

Veterinary patients may become recumbent for a variety of reasons. Neurologic diseases, particularly those affecting the spinal cord and peripheral nervous systems, commonly cause patients to have difficulty or inability to rise. Orthopedic diseases and other systemic illnesses may also cause patients to have trouble rising. These diseases also may be painful and affected patients may be less willing to move. There are many consequences to being recumbent. Larger patients in particular may be predisposed to decubital ulcers particularly if they lie on soiled or poorly padded bedding. Urine retention and urinary incontinence in recumbent patients may lead to urine scald, pyoderma, and cutaneous ulcers. Recumbent patients are also more predisposed to urinary tract infections and aspiration pneumonia. They also may develop muscle atrophy either as a direct consequence of neurologic disease or due to disuse of affected limbs. Proper care of these patients including pain control, fastidious patient care and monitoring, and physical therapy can all help improve quality of life and hasten recovery.

Pain Control

Many neuromuscular and orthopedic diseases are painful. Many of these diseases can be treated long term with pain medications. Treatment with anti-inflammatory medications such as carprofen, meloxicam, or firocoxib can provide substantial relief long term. If treatment with an anti-inflammatory medication alone is insufficient, other pain medications such as tramadol, gabapentin, or amantadine can be added. In cases of severe pain, oral morphine can be very helpful. The author uses oral morphine often for hospice type care situations. Improvement in pain control can help patients maintain mobility, quality of life, and musculoskeletal strength.

Patients may experience pain following spinal surgeries. Irritation of the nerve roots and meninges from a disc may be present for a few hours to days following surgery. Muscle dissection to approach the spinal column during surgery can also be painful. Anxiety is also often a factor as patients may be nervous or fearful in the hospital setting, and pain can also produce anxiety. There are many protocols for pain control, and the author prefers using a multimodal approach following spinal surgeries.

Immediately following surgery, patients may be treated with opioids. Fentanyl at a constant rate infusion is useful because it is effective, the dose can be easily adjusted based on the patients needs due to its short half-life, and it is relatively safe. Some patients can become dysphoric and vocal on fentanyl, and it can suppress the patient's appetite. Methadone is also effective for pain control and has sedative properties as well making it a good choice for anxious patients. Hydromorphone appears to be effective for pain control. In the author's experience, patients may become dysphoric and vocal with this drug.

Anxiety may be addressed by using sedative/anxiolytic medications. Acepromazine has sedative but no analgesic properties. When used in conjunction with other pain medications, it can be effective for reducing anxiety. Trazadone may be more effective as an anxiolytic medication and is useful in dogs that can take medications orally. If trazadone and tramadol are used together, caution should be taken as there is some risk of serotonin syndrome. Gabapentin also has anxiolytic properties and can be used safely with other pain medications.

Anti-inflammatory medications are very beneficial. Carprofen, meloxicam, firocoxib, or prednisone used at an anti-inflammatory dose, are some of the more common choices. Gabapentin may be useful for alleviation of neurogenic pain. Tramadol is often used in conjunction with other medications. Its efficacy for pain control has not yet been proven, but anecdotally some patients seem to respond.

Ice applied to the incision every 6–8 hours for the first 1–2 days postoperatively can improve pain control in conjunction with use of pain medications and can also help to reduce swelling and bleeding postoperatively.[1]

In-Hospital Patient Care

Turning the Patient

Recumbent patients should be turned either to a sternal position or to the opposite lateral position every 4–6 hours when possible.[2] Maintaining recumbence in the same position for prolonged periods of time can be deleterious to the patient's health. Pulmonary atelectasis may occur on the down side and can predispose patients to hypoxemia or pneumonia. Venous stasis may cause dependent edema and can increase risk for thrombosis. Pressure sores may develop in patients lying in the same position for long periods of time.

Clean and Dry Bedding

Recumbent patients should be kept on clean and dry bedding that is well padded. They should be checked frequently for urination/defecation on bedding and soiled bedding should be changed as soon as possible. It is particularly important to keep these patients dry to prevent urine scald and decubital ulcers.[2]

Nutrition and Hydration

Care should be taken to ensure the dog has access to water and food. Paraparetic/plegic dogs may not be able to get to food/water bowls on their own. Consequently if the dog is eating and drinking, food and water may need to be held up for the dog. If the dog is not eating or drinking, IV fluid therapy can help to maintain hydration until they are able to take water orally.[2]

Bladder Care

Dogs with thoracolumbar or lumbosacral diseases are prone to inability to urinate and urinary incontinence. Urine retention due to loss of upper motor neuron (UMN) control in dogs with thoracolumbar spinal cord injuries. UMN bladder dysfunction generally occurs when the affected animal loses motor function, so these patients are usually paraplegic. They tend to regain bladder function when motor function returns. As long as the spinal cord injury can recover, the prognosis for recovery of urination is good. However, proper bladder management is crucial to facilitate recovery.

Urine retention and incontinence due to lower motor neuron (LMN) injury from loss of innervation by the pelvic and pudendal nerves occurs with lumbosacral injuries. LMN bladder dysfunction can occur in dogs that are still ambulatory since the sacral spinal segments are caudal to the segments responsible for walking ability. Since there is direct damage to the pelvic nerve, prognosis for recovery is more guarded than with UMN bladder dysfunction.

If urine retention is persistent, the bladder wall can become stretched and atonic and will not recover function even if the neurologic injury recovers. In addition, patients are prone to recurrent urinary tract infections, pyelonephritis, urine scald and bed sores. Therefore, these bladders should be emptied a few times per day. Manual expression can be used particularly with UMN bladders. Urethral sphincter tone can make manual expression difficult, so prazosin or phenoxybenzamine can be prescribed to reduce sphincter tone. Manual expression is less effective at keeping LMN bladders empty due to poor bladder muscle tone. If the bladder cannot be emptied by manual expression, indwelling or intermittent urethral catheterization may be necessary.[2]

At-Home Care Following Disc Surgery

Following surgery for acute disc extrusions, restricted activity for a period of about 4 weeks is important. The reason for this activity restriction is to prevent further disc extrusion at the site of injury where the annulus fibrosis may still be torn. For most dogs, restricted activity means confinement to a crate or a small room (such as a laundry room) where the dog cannot jump on and off furniture. Owners are instructed to take the dog outside on a leash a few times per day for urination and defecation, but no walks for exercise. It is permissible for the dog to be out with the owner under supervision on their lap as long as they are not following the owner around the house or jumping on and off furniture. For some patients, the author instructs owners to keep the dog on a leash with them in the house when the owner is home. For patients that are very anxious being confined, medical therapy with an anxiolytic medication such as trazadone may be beneficial during the time when the dog needs to be confined.

Following the 4 weeks of restricted activity, the owner is instructed to allow slow return to normal activity level with leash walks of increasing length over 2–3 weeks.

Sling support for the hind end may be helpful to the owner when walking a paraparetic dog. There are many types of commercial slings available for purchase. Alternatively, owners can use a towel or other soft cloth type material to serve as a sling.

Some dogs may scrape their toenails as they try to walk particularly on asphalt or other hard surfaces. Use of protective booties can help to prevent injury to the toes while the patient recovers.

Physical Therapy

Physical therapy offers potential benefits to the recovery of dogs following surgery for disc extrusions as well as dogs with other causes of neuromuscular weakness. However overly aggressive physical therapy soon after spinal surgery could be detrimental. Therefore a 5-step approach has been suggested:[1]

Figure 1. 5-step approach to physical therapy following hemilaminectomy for disc extrusions

Physical therapy step	Protocol
Step 1: Immediately post op	Ice packing surgical incision Passive range of motion exercises Massage of affected limb muscles
Step 2: Patient can bear weight	Standing exercises Neuromuscular stimulation
Step 3: Patients limb motion returns	Weight-shifting exercises Walking - assisted or water treadmill Swimming
Step 4 and 5: Patient begins to walk	Sit-to-stand exercises Balance and coordination exercises Increased intensity walking and swimming

Access to veterinary physical therapists is becoming more and more common, so referral to a physical therapy center if available should be discussed with owners if appropriate.

References

1. Kerwin S, Levine J, Hicks D. Thoracolumbar spine. *Veterinary Surgery: Small Animal*. S.J. Karen Tobias, ed. vol. 1. 2011: Saunders.
2. Tefend MB, Dewey CW. Nursing care and physical therapy for patients with neurologic disease. *A Practical Guide to Canine and Feline Neurology*. Dewey CW, da Costa RC, eds. 2016, Ames, Iowa: Wiley-Blackwell.

acvim019

Feline Neurology in Practice

Laurie Cook, DVM, DACVIM (Neurology)
Columbus, OH, USA

INTRODUCTION

Cats are seen much less frequently than dogs with neurologic diseases in clinical practice. One reason for this discrepancy is the nature of the diseases affecting cats compared with dogs. Two of the most common neurologic diseases treated in dogs are intervertebral disc extrusions and epilepsy. These diseases are much less common in cats comparatively. Another reason may be cost associated with diagnosis and treatment of neurologic diseases in cats. Consequently, neurologic diseases in cats often are not discussed as often as neurologic diseases in dogs. We will discuss pitfalls and unique aspects of neurologic examinations in cats compared with dogs and some of the more common neurologic diseases encountered in cats.

NEUROLOGIC EXAMINATION

Neurologic lesion localization in cats is very similar to that in dogs. Performing the neurologic exam in cats, however, can pose some unique challenges depending on the temperament of the cat. Some cats are very tolerant, but others have less patience for an extensive neurologic exam than dogs do. With these cats, the neurologic exam may need to be adjusted accordingly to minimize the stress on the cat and to try to get the most important information out of the exam.

Gait exams in cats can be challenging if the cat is nervous and apprehensive in the clinic environment. Often these cats will walk crouched to the floor making assessment of weakness, ataxia, and posture difficult. Cats that are scared may run to the nearest hiding place also making gait assessment difficult. Taking these cats to a quiet room with few people around can be helpful in some cases. Also, allowing the cat to walk in the room with the owner present may alleviate some stress for the cat allowing for a more accurate gait exam.

Proprioceptive testing also can be more challenging in cats compared with dogs. Though there are many postural reaction tests, many neurologists rely primarily on proprioceptive placing (knuckling) in dogs. Proprioceptive placing can be reliable in cats, but in the author's experience, many cats are not amenable to this test. The author will usually attempt proprioceptive placing first. If it is unsuccessful, many cats are more amenable to hopping or tactile placing. Wheelbarrowing and extensor postural thrust testing are also usually easier to perform in cats compared with placing.

Menace testing to assess vision can be slightly less reliable in cats compared with dogs. The menace test is a learned behavioral response. Cats that are scared and apprehensive may be less likely to respond to a menace test even when their vision is normal. If menace response is questionable, a cotton ball test or navigation in a room can be used to try to assess vision in these cats.

NEUROLOGIC DISEASES

This review is not intended to encompass all neurologic diseases in cats. Rather it will focus on some of the more commonly encountered neurologic conditions, their etiologies, diagnoses, treatments and prognosis.

Cerebellar Hypoplasia

Cerebellar hypoplasia, underdevelopment of the cerebellum, occurs in kittens exposed to feline panleukopenia virus (FPV). Kittens infected with FPV in utero or as neonates may develop cerebellar hypoplasia due to destruction of the external germinal cell layer of the cerebellum. This cell layer is actively growing up through the first few weeks of life. Affected kittens usually start showing signs when they begin to walk. Clinical signs include intention tremors, hypermetria and cerebellar ataxia. Usually these kittens feel well otherwise and are energetic and playful. The condition is non-progressive, and most kittens can adapt to maintain a good quality of life and are excellent pets. Clinical signs often lessen with age as they learn to adapt, but clinical signs usually don't resolve completely. Some kittens are affected so severely that ambulation and prehension of food are very difficult limiting quality of life, and euthanasia may be appropriate for these kittens.[1]

Diabetic Neuropathy

Diabetic cats may present with signs of a peripheral neuropathy causing weakness. The typical clinical signs are symmetric pelvic limb weakness, plantigrade stance, poor reflexes and proprioception and atrophy. The neuropathy can also lead to weakness in the thoracic limbs with a palmigrade stance and Horner's syndrome. Abnormal Schwann cell and myelin function, and possibly abnormal axon function are thought to be responsible for the clinical signs. There are multiple theories as to the cause of this dysfunction:

1. Microvascular theory: Microvascular dysfunction occurs in diabetics that can lead to many metabolic consequences. Ischemic injury to the peripheral nerves may lead to axonal and myelin degeneration.
2. Metabolic theory: Diabetics have increase in activity of a polyal pathway due to excess glucose leading to accumulation of sorbitol. This leads to depletion of myoinositol which is necessary for Na/K ATPase pump function. These pumps are necessary for energy production. A depletion of energy production may interfere with axonal and Schwann cell function. Free-radical production can lead to membrane damage as well.
3. Auto-immune theory: There is evidence of auto-immune dysfunction in diabetic patients. This dysfunction may attack myelin and lead to vascular thrombosis, leading to further microvascular nerve injury.

Diagnosis is usually based on identification of a peripheral neuropathy concurrently with diabetes mellitus and ruling out other causes of polyneuropathy. Prognosis is guarded but some cats may improve with improved control of diabetes mellitus. There is no specific treatment, but maximizing glycemic control may lead to partial or complete resolution of signs over weeks to months.[1]

Hepatic Encephalopathy (HE)/Portosystemic Shunts

Signs of diffuse forebrain dysfunction often occur in animals with severe liver disease, most commonly in animals with congenital or acquired portosystemic shunts (PSS). PSS is an abnormal blood vessel that bypasses the hepatic portal system. Therefore, blood from the GI tract that contains gut-derived toxins bypasses the liver and these toxins enter systemic circulation. In severe hepatic failure, the liver is unable to clear gut derived toxins. In both cases, the complete pathogenesis of HE is not completely understood. Endogenous gut-derived toxins including ammonia, skatoles, indoles, and short-chain fatty acids are thought to be primarily involved. In addition, alterations of brain neurotransmitters and circulating benzodiazepine-like substances that act on brain gamma-aminobutyric acid (GABA) receptors are probably involved.

Clinical signs usually include profoundly altered mentation, abnormal behavior, compulsive pacing, head pressing, blindness and seizures. Affected cats often have other signs of hepatic dysfunction including small stature (congenital PSS), failure to gain weight, anorexia, vomiting, diarrhea, ptyalism and PU/PD.[1,2] Cats with congenital PSS may have amber or copper colored irises noted on general physical exam.[2]

Diagnosis is made by documenting signs of diffuse forebrain dysfunction in a patient with hepatic dysfunction on bloodwork. Serum bile acids and ammonia levels are often markedly elevated. Abdominal ultrasound, nuclear scintigraphy and mesenteric portography may demonstrate the presence of a PSS.

Treatment of hepatic encephalopathy is directed at reducing the circulating toxins. Oral or rectal lactulose helps to decrease colonic bacterial toxin production and increases transit time through the colon. Oral antibiotics such as amoxicillin, metronidazole or neomycin may help reduce urease producing bacteria. Low protein diets can help reduce ammonia production by gut bacteria.[1,2] Long-term treatment of PSS or hepatic failure is directed at correcting the underlying cause if possible and the reader is directed to other resources for an in depth discussion of these treatments.

CNS Neoplasia (Meningioma, Lymphoma)
Cats are commonly affected with neoplasia affecting the central nervous system. Meningiomas grow from the arachnoid layer of the meningeal lining and are the most common primary brain tumor in cats. These generally benign, well-encapsulated, slow growing masses do not invade the surrounding neuropil. This is in contrast to meningiomas in dogs that do invade surrounding brain tissue and can be more aggressive. Diagnosis is made with cross-sectional imaging. MRI is preferred, but many meningiomas can be seen with CT as well. Because these masses are benign and encapsulated, they generally are very amenable to surgical removal depending on the location with a good prognosis. Lymphoma is the most common spinal tumor in cats and usually affects the epidural space but can be intradural or intramedullary. Signs are usually attributed to a myelopathy in the region that is affected, and it is common for it to be multifocal. Diagnosis is made with MRI and CSF analysis, which may or may not show neoplastic cells. Treatment usually includes corticosteroids, radiation therapy and/or chemotherapy.[1]

Feline Infectious Peritonitis (FIP)
FIP meningoencephalitis is the most common infectious cause of CNS disease in cats. It is usually associated with the dry or non-effusive form of the disease. Affected cats are usually young to middle age. Signs of multifocal brain disease including mentation changes, seizures and/or vestibular signs are most common. Often signs of systemic illness including fever, weight loss, inappetence, etc., are present as well. Ocular changes such as chorioretinitis and corneal precipitates may be present. Confirming a diagnosis can be challenging. Bloodwork changes are non-specific but may include anemia, inflammatory leukogram, hyperglobulinemia, and elevated liver enzymes. Brain MRI may show enlarged ventricles with multifocal periventricular contrast enhancement. CSF analysis is typically abnormal with a neutrophilic pleocytosis and elevated protein. Serum coronavirus titers are not specific for CNS FIP. Immunohistochemical confirmation of virus can be confirmed on biopsy samples or fluids. Unfortunately often fluid samples are not cellular enough to provide enough antigen for immunohistochemical confirmation. Therefore, often a presumptive diagnosis is made antemortem. Diagnosis can usually be confirmed with necropsy.[1]

Middle Ear Inflammatory Polyps/Otitis Media-Interna
Inflammatory nasopharyngeal polyps can affect the middle ears of cats unilaterally or bilaterally. Otitis media-interna can occur secondary to inflammatory polyps, due to hematogenous spread, or secondary to otitis externa. Signs of peripheral vestibular syndrome including head tilt, nystagmus, strabismus, ataxia, nausea and inappetence are common in affected cats. They may also have concurrent facial paralysis and Horner syndrome as these nerves run near the middle and inner ear and are commonly affected with peripheral vestibular syndromes. Occasionally the disease affects both middle ears resulting in bilateral peripheral vestibular syndrome. Since the vestibular systems work relative to each other, disease on both sides "cancels out" the typical signs of vestibular syndrome. These cats do not have a head tilt or pathologic nystagmus. However, they walk with a crouched tentative gait and make very wide head excursions - "Stevie Wonder head movements". They also do not have a normal physiologic nystagmus (oculocephalic reflex or tracking). Occasionally cats with otitis media-interna will have extension of infection into the brain resulting in a secondary meningoencephalitis with cerebellar and/or central vestibular signs.

Diagnosis is made by cross sectional imaging with CT or MRI. Myringotomy can help to sample fluid for culture. Antibiotic treatment can be tried for otitis media-interna but often fails to completely resolve the infection. Ventral bulla osteotomy surgery may be necessary to treat refractory otitis media-interna and polyps.[1]

Idiopathic Epilepsy
Idiopathic or **primary** epilepsy is much more common in dogs but can occur in cats. Any breed of cat may be affected and they are usually young adult to middle age when they first start having seizures. Typically these cats are normal between seizures, have normal neurologic exam, and are healthy otherwise. Seizures in cats are more commonly associated with underlying disease processes than

epilepsy, so idiopathic epilepsy is diagnosed by ruling out other causes of seizures. Treatment is directed at controlling seizures with anti-convulsant therapy if necessary.[1]

Pyrethrin/Permethrin Toxicity

Permethrin and other pyrethroid exposure in cats usually comes from well-meaning owners applying canine topical flea medications to their cats. These insecticides interfere with sodium channels and GABA receptors in the CNS. Affected cats present usually within hours of exposure. Clinical signs include severe tremors, hypersalivation, ataxia, vomiting, CNS depression or excitability, seizures and in severe cases, death. Prognosis is generally good with symptomatic treatment. The skin at the site of application should be washed. IV fluids are administered. Methocarbamol IV can be very beneficial to minimizing the tremors. If seizures are present, anticonvulsant therapy should be administered. Atropine can minimize hypersalivation and diarrhea. Most cats with appropriate treatment will recover within 72 h.[1]

Hypertensive Encephalopathy

Hypertensive encephalopathy occurs in cats often secondary to hypertension with chronic renal disease. Other diseases that lead to hypertension include hyperthyroidism, hyperadrenocorticism, and pheochromocytoma. Neurologic signs have been reported in 29–46% of hypertensive cats. It is more common when there is a sudden increase in blood pressure or a systolic blood pressure >180 mm Hg. In people it is characterized by white matter edema, vascular changes, hemorrhage or infarction. Clinical signs are usually attributed to diffuse intracranial disease including dull mentation, behavior changes, seizures, vestibular signs, or focal signs due to a focal ischemic episode. Diagnosis is made by documentation of persistent hypertension (2 or more accurate blood pressure readings) and ruling out other causes of central nervous system disease. Prognosis is fair to good with control of blood pressure. Medications to control blood pressure include amlodipine, enalapril, benazepril, atenolol, and others.[3]

References

1. Dewey CW, da Costa RC. *A Practical Guide to Canine and Feline Neurology*. 2016, Ames, Iowa: Wiley-Blackwell.
2. Johnson SE. What's new with congenital PSS? CVC in San Diego Proceedings. November 1, 2010.
3. Brown S, Atkins C, Bagley R *et al*. ACVIM Consensus Statement: Guidelines for identification, Evaluation, and Management of Systemic Hypertension in Dogs and Cats. *J Vet Intern Med*. 2007;21:542–558.

1.
acvim020

Neurology Technician's Role in Management of Patients with Intracranial Trauma

Karen Dyer Inzana, DVM, PhD, DACVIM (Neurology)
Blacksburg, VA, USA

There are 3 major tissues within the calvaria: neural parenchyma, blood, and cerebrospinal fluid. Following injury, hemorrhage, contusions, and subsequent edema compromise the area allocated for neural tissue and increase intracranial pressure. CNS trauma consists of a primary phase where the neural tissue is directly injured by blunt force or penetrating injuries. Unfortunately, this is only the beginning of the problem. A secondary phase of neural destruction follows this, which is mediated by a series of biochemical events. The most important treatment for both phases of neural trauma is to maintain cerebral blood flow while reducing intracranial pressure. Unfortunately, this is easier said than done.

Intracranial pressure is affected by:

1. **Systemic hypotension**: If the mean arterial blood pressure falls significantly (as in shock), cerebral blood vessels dilate to try to maintain cerebral blood flow. This increases intracranial pressure.
2. **Elevated $PaCO_2$ levels** that happen when a patient is hypoventilating due to lung contusions cause cerebral vasodilatation, which increases intracranial pressure. Decreasing $PaCO_2$ levels rapidly reduces intracranial pressure. Unfortunately, it does so by causing cerebral vasoconstriction, which decreases blood flow.
3. **Venous occlusion** (holding off jugular veins, putting pressure on abdomen) causes dilation of cerebral veins and increased intracranial pressure.

Specific Therapy for Intracranial Trauma

1. Address extracranial complications first. Treat shock, hemorrhage, etc.
    - **High-volume resuscitation**
        - Isotonic crystalloid 60–90 mg/kg bolus; 40–60 mg/kg bolus cats
    - **Low-volume resuscitation**
        - Colloids
            - Hetastarch - 10–20 mg/kg up to 40 mL/kg/h administered in 5 mL/kg boluses over 5–10 minutes.
            - Dextran-70 - 10–20 mL/kg for dogs and 5 mL/kg for cats (up to 40–50 mL/kg/h dogs and 20 mL/kg cats). May increase glucose levels, which may actually increase secondary trauma.
        - Hypertonic saline 7% (4–5 mL/kg) or 3% (5.3 mL/kg administered over 3–5 minutes. Human studies have also tried CRI of 2% administered at a rate between 0.1–1 mL/kg/h to maintain serum osmolarity around 320 mOsm/L. Do not administer to hyponatremic patients. Need to monitor sodium and prevent hypernatremia.
2. Maintain $PaCO_2$ between 25–35 mm Hg
3. Elevate the head (15–30 degrees), which decreases intracranial pressure by improving venous drainage.
4. Furosemide 2–5 mg/kg IV. This will act as a diuretic, which may directly decrease cerebral edema. Most believe it is beneficial to administer furosemide prior to mannitol, because it a) increases the effect of mannitol, and b) may blunt the effects of volume overload associated with mannitol administration.
5. Mannitol 0.5–1 g/kg over 20 minutes. Mannitol should not be given more than 3 times over a 24-hour period. It is best to keep serum osmolality at or below 320 mOsm/L to prevent renal failure.

Human protocols recommend also monitoring Na+ and discontinuing of the drug if serum sodium exceeds 160 mmol/L. Beneficial effects include not only reduction in intracranial pressure, but mannitol also reduces blood viscosity, improving cerebral perfusion.

6. Hypothermia: Core body temperature between 33–35 degrees C (91.4–95 degrees F) for up to 48 hours with gradual rewarming. Must abolish shivering with benzodiazepines, opioids, or barbiturates. Experimental results still mixed regarding benefit/cost ratio. Hypothermia may mitigate many of the mediators of secondary neural injury, but may contribute to hypovolemia, arrhythmias, coagulopathy, electrolyte disorders, metabolic acidosis, and increased insulin resistance.
7. Barbiturates: Are directed at counteracting excitotoxicity and other biochemical mediators of secondary neural injury. They also decrease energy demands of cerebral tissue. Most institute this therapy only when other therapies have failed. Phenobarbital 2–15 mg/kg initially, followed by a CRI of 0.2–1 mg/kg/h. Cardiovascular and respiratory depression are serious consequences as well as inability to monitor changes in neurological status.
8. Steroids: Current recommendations are to avoid steroids in intracranial trauma patients unless they are deteriorating rapidly and you have little to lose.

Neuroimaging Intracranial Trauma

Skull radiographs may show depressed fractures. However, unless they are significantly depressed, these are rarely repaired. Since the dura is adhered to the periosteum in dogs and cats, epidural hematomas are rare. Subarachnoid hematomas will not typically cause an expanding mass effect, because the blood should mix with CSF and diffuse away. The incidence of hematomas between the dura and arachnoid (subdural hematomas) has traditionally been considered quite low. However, the increased accessibility of CT/MRI has raised the possibility that hematomas may be more frequent than realized. CT/MRI evaluation with potential surgical decompression should be considered in a patient that is deteriorating despite appropriate medical treatment.

Neurological Evaluation in Patients with Altered Mentation

1. Level of consciousness is evaluated by trying to arouse the patient. Clearly coma is worse than semi-coma, which is worse than stupor.
2. Body posture
    - Injury to different portions of the brain may cause abnormal postures.
    - Marked head turn (nose turned to side of injury) is typical of cerebral lesions.
    - Marked head tilt with rolling is typical of caudal brainstem (vestibular) disorders. These animals usually have marked nystagmus and strabismus.
    - Opisthotonus (head and neck extended) may be seen with injury to either the diencephalon (decerebrate posturing) or cerebellum (decerebellate posturing).
    - Rigid extension of all 4 limbs may be seen with either decerebrate or decerebellate posturing. Decerebellate posturing may also cause the rear limbs to be flexed rather than extended. Both of these postures are usually associated with other signs of either cerebellar or brainstem dysfunction.
3. Pupillary light reflexes
    - Normal, reactive pupils are obviously the best.
    - Miotic pupils have an intermediate prognosis. Localization of the injury is difficult with miotic pupils. This is probably the equivalent of a central nervous system-induced Horner's syndrome and indicates injury to the "upper motor control" of the sympathetic innervation of the eye.
    - Mydriatic, unresponsive pupils are the worst scenario, indicating injury to the parasympathetic portion of cranial nerve III, usually as a consequence of brain herniation.
4. Other cranial nerve reflex function

- Oculocephalic reflex: Turn the head from side to side and induce normal vestibular nystagmus. Instilling cold or warm water into the ears may also induce an oculocephalic reflex. Absence of this reflex indicates either bilateral injury to the vestibular system or injury to the medial longitudinal fasciculus - a large, peripheral tract in the brainstem that links the vestibular system to cranial nerves III, IV, and VI.
- Corneal or palpebral reflex: Touch the cornea or palpebral fissure to induce an eye blink. Absence indicates injury to either cranial nerves V or VII.

It is important to note that no one portion of the exam indicates whether the outcome will be positive or not. The most important prognostic sign is how the patient is changing over time and in response to your treatments. The following Coma Scale has been modified from human medicine and is a useful tool to monitor changes over time.

Modified Glasgow Coma Scale

Motor activity	
Normal gait, normal spinal reflexes	6
Hemiparesis, tetraparesis, or decerebrate activity	5
Recumbent, intermittent extensor rigidity	4
Recumbent, intermittent extensor rigidity with opisthotonus	3
Recumbent, constant extensor rigidity with opisthotonus	2
Recumbent, hypotonia of muscles, depressed or absent spinal reflexes	1
Brainstem reflexes	
Normal pupillary light reflexes and oculocephalic reflexes	6
Slow pupillary light reflexes and normal to reduced oculocephalic reflexes	5
Bilateral unresponsive miosis with normal to reduced oculocephalic reflexes	4
Pinpoint pupils with reduced or absent oculocephalic reflexes	3
Unilateral, unresponsive mydriasis with reduced or absent oculocephalic reflexes	2
Bilateral, unresponsive mydriasis with reduced or absent oculocephalic reflexes	1
Level of consciousness	
Occasional periods of alertness and responsive to environment	6
Depression or delirium, capable of responding to environment but response may be inappropriate	5
Stupor, responsive to visual stimuli	4
Stupor, responsive to auditory stimuli	3
Stupor, responsive only to repeated noxious stimuli	2
Coma, unresponsive to repeated noxious stimuli	1

Assessment

Patients with a score of 15–18 have a good prognosis; patients with a score of 9–14 would be guarded; while patients with a score of 8 or less have a poor prognosis. While these numbers are helpful, whether the patient is improving or deteriorating is much more indicative of final outcome.

Future Therapies

Most future therapies will be directed toward specific mediators of secondary neural trauma. This is a complex, interrelated event, but from the perspective of future therapeutics, it can be broken down into 4 major components: excitotoxicity, oxidative stress, mitochondrial breakdown, and edema formation. Multimodal drugs, or drugs that mitigate more than one of these factors, include progesterone, dexanabinol, magnesium, and cyclosporin A.

REFERENCES

1. Algattas H, Huang JH. *Int J Mol Sci*. 2014;15:309.
2. Sheriff FG, Hinson HD. *Semin Neurol*. 2015;35:42.
3. Sande A, West C. *J Vet Emer Crit Care*. 22010,20:177.
4. Finnie JW, Blumbergs PC. *Vet Path*. 2002;39:679.

acvim021

How to Perform a Neurologic Exam

Maureen Sroufe, LVT, VTS (Neurology)
Blacksburg, VA, USA

Tools Needed

You only need a few simple tools to perform a neurologic (neuro) exam on a dog or cat. The basic things you need are a reflex hammer, bright light source, cotton balls, a non-slip surface, and an exam form. It is preferred that you use a reflex hammer over other things such as a hemostat handle or end of your stethoscope. These may work, but can look unprofessional in front of the client. A bright light, such as a transilluminator, can be helpful over a penlight. Most penlights are not bright enough to get a good pupillary light response. A non-slip surface is very helpful for gait analysis. It can also help animals feel more secure when walking or standing than the slick tile floor, especially if they are nervous or ataxic. A carpeted area, yoga, or other type of mat would be good inside. You could also go outside to a cement or grassy area. If you have space, a long carpeted area or mat inside is preferred for gait analysis. An exam form is extremely helpful to document your findings and keep them organized. There are numerous types of forms. You can put together one that suits your practice preferences. There is a copy of a form in the handouts.

Order of the Neurologic Exam

The best way to organize your neuro exam is to come up with a system that works for you and always follow it. This will allow you to form a routine and not miss anything. It is always best to look at the whole animal first through observation at a distance. You can often gain a lot of information watching your patient interact with their environment. If your patient will not be amenable to having the neuro exam done, this may be all you are able to do, but it will tell you a lot about your patient. You can then move on to cranial nerves, postural reactions, and spinal reflexes. These can be done in any order, but it is usually helpful to do postural reactions before reflexes because reflexes are not reliable in the absence of postural reaction deficits. Always exercise good judgement when doing exams on patients that are fractious, critical, or extremely painful. In these patients it is not about completeness, but getting the most information without stress or further trauma or pain.

Gait and Strength

When we observe the patient, walking around the exam room or down the hall, we want to assess several things about their gait. First, are they ambulatory or nonambulatory? To be ambulatory they need to be able to walk under their own power, with little or no assistance. If they cannot lift their own weight, but do have motor to the affected limbs, we consider this nonambulatory. Weakness or paresis is when the patient has motor, plegia is when the patient has no motor to the affected limb. If only the rear limbs are affected, this is para- paretic/plegia. In all four limbs, is tetra- paretic/plegia. Only one limb is mono-paretic/plegia and only the left or right side, front and rear, hemi- paretic/plegia. We also look at stride length. A short stride is indicative of a lower motor neuron problem to the affected limbs, while a long stride or ataxic gait can point to an upper motor neuron problem to the affected limbs. We need to observe their body stance, such as if they lean or curl their body to one side or stand with a wide-based stance.

Mentation

We often observe mentation while taking a history and watch as they interact with their environment. Keeping in mind normal behavior for a dog or cat in the hospital setting. They may be scared, withdrawn, nervous, or fractious. Ask the owner what is normal for their pet coming into the hospital. Mentation can be described as alert, obtunded, stuporous, or comatose. Obtunded or dull animals are somewhat unresponsive. They will interact when talked to or touched and respond to stimuli. Stuporous animals may seem unresponsive, but will respond to noxious stimuli, but still may be an inappropriate response. Comatose animals will not respond even to noxious stimuli.

Head Position

The most common abnormal head position is a head tilt. We may also see a head turn or ventral or dorsal flexion of the head and neck.

Cranial Nerves

There are numerous tests for the cranial nerves (CN). Several of these test more than one nerve as some nerves provide the sensory pathway and others are the motor pathway.

Olfactory nerve (CN I) is often not tested in dogs and cats. Use a cotton ball with alcohol or a small amount of cat food to elicit a behavioral response of aversion or pleasure. Cover the animal's eyes and hold in front of their nose or you can hide food under a cup and have a second cup with nothing and see if they can pick which cup has the food.

Menace response tests the optic nerve (CN II) and the facial nerve (CN VII). Make a gesture with your hand toward one eye of the animal. The animal should blink or move their head in aversion. This will assess the sensory path of vision in the optic nerve and the motor of the facial nerve. **Use caution not to create air flow or touch hair or whiskers of the animal.** The optic nerve can also be tested by dropping a cotton ball in front of the animal to see if they follow it.

Pupil light response (PLR) will test CN II and oculomotor nerve (CN III). The optic nerve will be the sensory to vision and oculomotor will be motor for the reflex. Shine a bright light source into the animal's eye and the pupil should constrict. We then look in the opposite eye to see if there is a consensual response. The test is then repeated on the other eye.

Strabismus is an abnormal eye position. This can be normal in some breeds, especially brachycephalic breeds. Lift the animal's head up and look for the eye position. Eye position should be the same in both eyes. This will test oculomotor (CN III), trochlear (CN IV), abducent (CN VI), and vestibulocochlear (CN VIII) nerves. Trochlear and abducent nerves innervate the muscles of the eye, along with the oculomotor nerve are responsible for normal eye movement and position. Vestibulocochlear nerve helps with input to where the eye should be.

Oculocephalic test is the normal physiologic nystagmus as you move the head from side to side. It tests cranial nerves III, IV, VI, and VIII. The eye movements should stop when the head stops; if not, then CN VIII is affected and the animal has positional nystagmus.

The trigeminal nerve (CN V) is motor to the muscles of mastication and sensory to the face. Muscle mass and jaw tone are tested by palpating the muscles of mastication for atrophy or asymmetry. The facial nerve (CN VII) provides motor to the muscles for facial expression and sensory for taste and inner part of the pinna. The branches of CN V and VII are usually tested together. Looking for any asymmetry to the face, lips, eye fissures, or ears. The three branches, ophthalmic, maxillary, and mandibular, are tested by touch. This includes palpebral or blink reflex, touching the inside of the ear and the lips. Start with a light touch and the animal should flinch or blink. If needed, a pinch with your finger or hemostats may be used, especially if the nerve is damaged.

The vestibulocochlear nerve (CN VIII) has two branches, cochlear and vestibular. The cochlear branch is tested by making a squeak, whistle, or loud noise. The gold standard for testing hearing is a brain auditory evoked response (BAER), but that isn't feasible for every patient. The vestibular branch shows distinct characteristics if there is a lesion. The lesions are usually unilateral. Deficits seen are ataxia, head tilt, and nystagmus. The head tilt is usually in the direction of the lesion and the fast phase of the nystagmus is away from the lesion. There are three different directions the nystagmus can be: horizontal, vertical, or rotatory. The direction of the nystagmus is based on the direction the eye is moving. Horizontal the eye is moving left to right, vertical the eye is moving up and down, rotary the eye is moving either clockwise or counterclockwise. If there is a change in the direction of the fast phase, the lesion is most likely a central vestibular lesion.

The glossopharyngeal (CN IX), vagus (CN X), accessory (CN XI), and hypoglossal nerve (CN XII) are tested together as they innervate the pharynx and tongue. Test swallow and gag by palpating pharynx and observing a swallow. You may also palpate the laryngeal area to elicit a swallow. You will also see movement of the tongue. Look for any asymmetry in the tongue. History from the owner may also help with assessment as they will know of historical changes in swallow or voice.

Postural Reactions

Postural reactions are helpful to do before reflexes. These tests help determine which limbs are affected. In the absence of postural reaction deficits, our reflexes are hard to interpret because by definition, those limbs should be normal. Proprioceptive reactions should be done in a standing position. Supporting the weight of the animal, turn over the paw so the dorsal surface is touching the ground. The animal should quickly place the foot correctly. A delay or not turning over can indicate a problem with sensory or motor function. Care should be taken to ensure that you are not getting a withdrawal response. If the patient will not let you place their foot when you are touching the distal paw, try holding slightly higher on the limb and moving the limb to turn the paw over. When testing the thoracic limbs, support the animal's weight from their chest. When testing the pelvic limbs, support their pelvis. If you support in front of the limb, you may inhibit the patient from being able to move the limb and giving an appropriate response.

Hopping response is testing one limb at a time. Lift the dog so that only one limb is touching the ground, then shift their weight laterally. When they sense a shift in their center of gravity, they should initiate a step. Abnormal responses can include no step at all, leaning too far before stepping, very short steps, or an over-exaggerated step. It is important to compare left to right and front to back responses.

Wheelbarrow reaction is testing the front limbs. The animal is lifted so to bear all its weight on the front limbs and walked forward. If the response is normal, repeat the procedure with the animal's head lifted or eyes covered. This causes the animal to depend solely on proprioceptive information and can bring out subtle abnormalities. Care should be exercised in patient showing extreme neck pain or if you suspect a luxation in the cervical spine.

Extensor postural thrust response tests the rear limbs. As the animal is lowered, the animal should extend their pelvic limbs to reach for the floor, then symmetric steps backward should be seen. The reaching for the floor tests the vestibular reaction, whereas the stepping backward will test their proprioception.

Hemiwalk and hemistand response test the left or right side of the dog. Lift the thoracic and pelvic limbs on one side and walk laterally. Very subtle signs can be seen with the hemiwalk/stand.

Spinal Reflexes

Spinal reflexes help us with lesion localization. We use the response to determine if there is an upper motor neuron (UMN) or lower motor neuron (LMN) problem. UMN reflexes can appear normal to exaggerated, while LMN reflexes can look decreased to absent. We also use muscle tone to help with localization. The most reliable reflex in the thoracic limb is the flexor or withdrawal. Even in normal dogs the other reflexes can be difficult to elicit. In the pelvic limb the most reliable reflexes are the patellar and withdrawal. Practice and seeing lots of variants of normal and abnormal are helpful in becoming proficient.

The flexor, or withdrawal reflex, for the thoracic and pelvic limbs are similar. A noxious stimulus is applied to the lateral toe and the animal should flex all the joints of the limb. The medial toe should also be tested with the same results. In the pelvic limb the animal should flex the hock, stifle, and hip. In the thoracic limb they should flex the carpus, elbow, and shoulder. Start with the least amount of pressure on the toes and increase if more is needed to elicit the reflex. The important thing to remember is that the flexor reflex is a spinal reflex and does not require the brain to flex the limb. Which means that the flexor reflex does not equal deep pain sensation. If the animal shows some form of conscious recognition when eliciting the withdrawal, then the animal does have pain sensation to the limb.

Pelvic Limb Reflexes

Patellar reflex tests the femoral nerve. Support the limb and percuss the patellar ligament with a reflex hammer. The response should be one quick extension of the stifle. Multiple "kicks" are considered exaggerated, where a slower response may indicate decreased response.

Sciatic reflex tests the sciatic nerve. Place your finger on the sciatic notch and percuss your finger or you can percuss directly on the sciatic notch. The response should be one quick flexion of the stifle. Often you may just see the muscle flex.

Cranial tibial reflex tests the peroneal branch of the sciatic nerve. Support the limb and percuss the belly of the cranial tibial muscle. The response should be flexion of the hock.

Gastrocnemius reflex tests the tibial branch of the sciatic nerve. This can be tested several ways. One is to hold the gastrocnemius muscle where it inserts and percuss your thumb. You may see movement in the distal limb or you may feel the muscle flinch in your hand. The other way to test gastroc is to slightly flex the hock to stretch the muscle and percuss the tendon. You may feel extension of the hock. The gastroc reflex can be difficult to elicit.

Perineal reflex tests the pudendal nerve. Stimulation to the perineum should produce contraction of the anal sphincter or flexion of the tail.

Thoracic Limb Reflexes

Biceps reflex tests the musculocutaneous nerve. Place a finger on the brachialis tendon above the elbow and slightly extend the elbow and percuss your finger. You may feel flexion of the elbow or the skin over the biceps brachii muscle may be seen.

Triceps reflex tests the radial nerve. This can be tested by holding the tendon above the olecranon and seeing movement in the limb or feeling the muscle twitch. You can also percuss the triceps brachii muscle and may see contraction of the muscle or movement of the limb. This reflex is often difficult to elicit.

Extensor carpi radialis reflex test the radial nerve. Support the limb with the elbow and carpus in flexion and percuss the extensor carpi radialis muscle just below the elbow. You should see extension of the carpus.

Odds and Ends

Superficial and deep pain are assessed in the same manner as the withdrawal reflex looking for conscious recognition such as turns to look, cries, or stops panting. Again, only apply as much pressure as needed to elicit the response without causing undue pain. Testing a lateral and medial digit is often done.

Cutaneous trunci is typically tested in a standing position. Usually tested caudal to cranial, pinch the skin on either side of the vertebral column with fingers or hemostats. The skin should flinch on both sides. The reflex may be absent caudal to a lesion in the spinal cord.

Spinal hyperpathia is typically reserved for last as to preserve cooperation of your patient that may be painful. Again working caudal to cranial, you can press on the spinous process or the transverse process of the vertebral bodies. Start with gentle pressure and increase to deep tissue palpation. This should not be painful in the normal animal. Often animals that are extremely painful may react everywhere making localization of pain more difficult.

References

1. Lorenz MD, *et al.* In: *Handbook of Veterinary Neurology.* Chapter 1. 5th ed. St. Louis, MO: Elsevier Saunders; 2011:15.
2. Dewy CW, *et al.* In: *Practical Guide to Canine and Feline Neurology.* 3rd ed. Wiley Blackwell; 2015:9

acvim022

Spinal Cord Localization and Diseases

Maureen Sroufe, LVT, VTS (Neurology)
Blacksburg, VA, USA

LOCALIZATION

We look at our neurologic exam findings to help with lesion localization. We divide the spinal cord into functional segments, C1-C5, C6-T2, T3-L3, and L4-S3. Each segment will show different signs if there is a problem in that segment. We look at gait, postural reactions, spinal reflexes, muscle tone, and muscle atrophy.

We can use the following table to help determine if each segment falls into upper motor neuron (UMN) signs or lower motor neuron (LMN) signs.

	Upper motor neuron lesion	**Lower motor neuron lesion**
Gait	Long-strided, stiff, spastic	Short-strided, looks lame, weakness
Postural reactions	Reduced to absent	Reduced to absent
Spinal reflexes	Normal to increased	Decreased to absent
Muscle tone	Normal to increased	Decreased to absent
Muscle atrophy	Minimal to none	Moderate to severe

When we look to localize, the following table can be useful:

	Thoracic limbs	**Pelvic limbs**
C1-C5	UMN	UMN
C6-T2	LMN	UMN
T3-L3	Normal	UMN
L4-S3	Normal	LMN

Once we have our localization, we look at the different diseases affecting the spinal cord. Using localization, signalment, and history, we can compile a list of differential diagnoses.

DIFFERENTIAL DIAGNOSIS

We use the DAMNNIITT scheme to help us rule in or out different diseases affecting the spinal cord. These categories break down the numerous choices into manageable categories.

- Degenerative
- Anomalous
- Metabolic and nutritional
- Neoplastic
- Infectious and Inflammatory
- Ischemic or vascular
- Toxic
- Traumatic

1.
acvim023

Pain Management Options for Dogs and Cats

Alix Partnow, DVM, MS, DACVIM (Neurology)
Lynnwood, WA, USA

INTRODUCTION

Recognition and treatment of pain are central to diagnosis and management of numerous conditions in cats and dogs. Not only does appropriate management of pain improve patient outcomes, but it also enhances quality of life for the pet and his/her owner. The field of pain management is always evolving, and we in the veterinary profession must continue to adjust our protocols to maintain a high standard of care for our patients.

In order to effectively treat pain, we must first be able to identify pain in a clinical setting and counsel pet owners on how to recognize signs of pain in the animal's natural environment. Once pain is detected, standard diagnostic techniques should be implemented to determine the source, as direct therapy of the underlying condition will typically yield the most complete resolution of signs. However, as we strive to achieve the aforementioned goal, it is critical that we recognize both pharmacologic and non-pharmacologic/"complementary" modalities for management of pain.

PHYSIOLOGY OF PAIN

The physiology of pain involves initial stimulation of peripheral nociceptors (mechanical, chemical, thermal) followed by transduction of the stimulus into an electrical impulse sent through the afferent peripheral sensory nerve to the level of the dorsal horn of the spinal cord. Within the dorsal horn there is transmission of the electrical stimulus through the spinal cord into the brainstem, the thalamus, and ultimately into the cerebral cortex where pain is consciously perceived. There can be physiologic modulation of pain, either through release of excitatory or inhibitory neurotransmitters by descending pathways within the spinal cord.

While historically pain was classified as acute or chronic based upon the duration of signs, more recently the paradigm has shifted to become more clinically relevant. To this end, it is important to distinguish adaptive from maladaptive/"wind-up" pain. Adaptive pain includes the inflammatory response and is considered the normal tissue reaction to a noxious stimulus. Once the stimulus is removed (e.g., inflammation resolved), the pain will subside. Maladaptive pain develops as the result of either intense acute or uncontrolled chronic pain resulting in changes to central nervous system (spinal cord and/or brain). Examples of maladaptive pain include allodynia, where a non-noxious stimulation of the peripheral nervous system triggers central transmission of pain or where the thalamus, which typically relays signs from the spinal cord to the cerebral cortex, becomes a spontaneous pain generator leading to perception of pain in absence of any stimulation (i.e., thalamic pain syndrome).

Adaptive and maladaptive pain may exist concurrently in the same patient. It is crucial to recognize these two distinct physiologic forms of pain, as the treatment protocols vary.

CLINICAL RECOGNITION OF PAIN

While trauma, osteoarthritis, and surgical pain are commonly recognized, it is important to consider pain can develop as a result of disease in nearly any organ system. Some common but often overlooked sources of pain include constipation/obstipation, cystitis, and metabolic conditions such as diabetic neuropathy. For the comfort of the patient as well as for the safety of all members of the veterinary team, it is important to obtain a complete history from the owner regarding any potential signs of pain seen prior to presentation. It is also necessary to anticipate how planned procedures such as restraint, venipuncture, etc. may either cause or exacerbate pain in a particular dog or cat.

Common signs of pain among dogs include change in gait (e.g., lameness), decrease in activity/mobility, increased respiratory rate and/or panting, vocalization, aggression or avoidance of physical interaction with other pets or humans, restlessness, and decreased appetite. Additionally, dogs may lick excessively or chew at painful spots on the body, their third eyelids can sometimes be elevated, and certain individuals will demonstrate needy/attention-seeking behavior. In cats, pain is unfortunately often overlooked. However, cats may display virtually any of the signs listed above. They may also develop changes in litter box habits.

When tracking a patient's pain over time, serial examination findings and thorough history-taking are both equally important. In the clinical setting, physiologic changes such as tachycardia, tachypnea, hyperthermia, hypertension, and mydriasis can all be associated with pain. There are also multimodal, semiquantitative tools such as the Colorado State University Canine and Feline Acute Pain Scales that provide a guideline for tracking physiologic and behavioral signs as well as response to palpation and body tension. Additionally, there are validated owner questionnaires, such as the Canine Brief Pain Inventory developed by Dr. Dorothy Cimino Brown at University of Pennsylvania, to quantify pain and response to therapy.

Pharmacologic Intervention

Before selecting the appropriate analgesic medications for a given patient, one must consider the source and severity of pain as well as any concurrent medical conditions such as renal or hepatic failure that may influence metabolism and safety of various classes of drugs. Route of administration (injectable vs. oral), available size/concentration, and necessary frequency of administration are also crucial when designing a pain protocol that can be easily maintained. For example, it may be necessary to have certain medications compounded to accommodate dosing for small dogs and feline patients. Additionally, careful attention must be paid to palatability; if an owner cannot easily administer the prescribed medication, then compliance will drop dramatically.

Analgesia may be prescribed either preemptively, such as with premedication prior to induction of anesthesia for elective surgery such as ovariohysterectomy, or in reaction to pre-existing pain. Classes of drugs to consider include α-2 agonists (e.g., dexmedetomidine), corticosteroids (e.g., prednisone and dexamethasone), local anesthetics (e.g., bupivacaine and lidocaine), NMDA receptor antagonists (e.g., ketamine and amantadine), nonsteroidal anti-inflammatories (e.g., carprofen, meloxicam), opioids (e.g., tramadol, codeine, hydromorphone, methadone), and nonclassified drugs such as gabapentin and pregabalin. Anxiolytics and tricyclic antidepressants may also be used adjunctively to enhance analgesic activity.

Complementary Therapies

In conjunction with pharmacologic therapy, non-pharmacologic techniques such as application of heat and ice, use of acupuncture, cold laser therapy, massage, and other physical therapy can enhance comfort and improve patient outcome.

Conclusion

Improved understanding of the physiology of pain in conjunction with recognition of clinical signs in our canine and feline patients enables us as veterinary professionals to better advocate for our patients and provide them with multimodal pain control consisting of both pharmacologic and complementary therapies.

References

1. Brown DC, Boston RC, Coyne JC, Farrar JT. Development and psychometric testing of an instrument designed to measure chronic pain in dogs with osteoarthritis. *Am J Vet Res*. 2007;68(6):631–637.
2. Epstein M, Rodan I, Griffenhagen G, Kadrlik J, Petty M, Robertson S, Simpson W. 2015 AAHA/AAFP pain management guidelines for dogs and cats. *J Am Anim Hosp Assoc*. 2015;51(2):67–84. doi: 10.5326/JAAHA-MS-7331.
3. Harman R, Carlson K, Gaynor J, *et al*. A prospective, randomized, masked, and placebo-controlled efficacy study of intraarticular allogeneic adipose stem cells for the treatment of osteoarthritis in dogs. *Front Vet Sci*. 2016;3:81. Published online 2016 Sep 16. doi: 10.3389/fvets.2016.00081. PMCID: PMC5025432.

acvim036

Chronic Diarrhea

Nicola Ackerman, BSc (Hons), RVN, CertSAN, CertVNECC, VTS (Nutrition)
Plymouth, UK

The causes of many gastrointestinal disorders in dogs can be difficult to initially identify. Treatments initially start based on the clinical symptoms, which do tend to resolve over a couple of days. In situations where clinical symptoms persist (vomiting, diarrhea and weight loss) a diagnosis may be required in order to accurately treat the patient (Table 1). Electrolyte imbalances can quickly develop in animals with severe diarrhea, with oral hydration or intravenous supplementation being provided when required, intravenous routes being required if the animal is vomiting or clinically dehydrated.[1]

Primary causes of chronic diarrhea	Secondary causes of chronic diarrhea
Infectious Endoparasites, enteropathogens, small intestinal dysbiosis	**Exocrine pancreatic disease** Exocrine pancreatic insufficiency Chronic pancreatitis
Inflammatory Idiopathic inflammatory bowel disease	**Hepatic disease** Liver failure
Neoplastic Intestinal lymphoma Other	**Kidney disease** Chronic kidney disease
Mechanical Sliding intussusception	**Adrenal disease** Hypoadrenocorticism
Toxic	**Thyroid disease** Hyperthyroidism in cats Hypothyroidism in dogs
Other rare conditions	**Cardiovascular disease** **Central nervous system disease**

Table 1. Causes for chronic diarrhea[3]

The gastrointestinal system requires nutrition, and therefore requires to be fed, even when the disease/disorder still persists. With there being numerous gastrointestinal disorders and with each individual reacting differently to different diets, it is important to obtain a full nutritional / diet history from owners about their pets. Talking to the owner and finding out what the pet has traditionally ate, how well does the pet eat, does it only eat a home cooked diet or a specific flavour of food is not only important to find out whether the gastrointestinal symptoms are related to what the pet is being fed (or dietary indiscretion), but also where to go to next in the recommendation of a new diet.

Dietary Management

Dietary manipulation can aid in a number of nutrient responsive conditions, but it can be a process of elimination to find the one that suits the individual the best. What may have suited one animal may not suit another. Some animals may require a combination of diets; some may have concurrent conditions which mean that the normal recommended diet may not be suitable, or even cause more problems. Many diets aimed at conditions affecting the gastrointestinal system work on the bases of manipulation of fat content, fibre content, digestibility or novel protein source (Table 2). Monitoring of the animal whilst dietary change is occurring is important, weight, body condition, faecal quality and output, and the monitoring of blood parameters (folate, B12) can all indicate the success of nutrition (Picture 1). Gaining a diagnosis from the veterinary surgeon is an important starting point; though there are many occasions where nutrition can also be used as a diagnostic tool (for example dietary sensitivities).

Diet type	Dietary characteristics
Novel protein source	Contains one protein and one carbohydrate source that need to be novel

	to the animal.
Hydrolyzed protein diet	Contains protein that has been broken down into smaller peptides or amino acids. The diets differ in degree of hydrolyzation, and protein source.
Easily digestible diet	Low-residue diet will contain a good source of prebiotics.
High-fiber diets	Useful with fiber-responsive colitis.
High-fat/low-carbohydrate or moderate-fat/moderate-carbohydrate diet	Mechanism of action unknown, chosen diets need to digestible.

Table 2. Dietary types that may be effective for a dietary trial with chronic diarrhea in cats and dogs[3]

Picture 1: Fecal scoring chart for cats and dogs

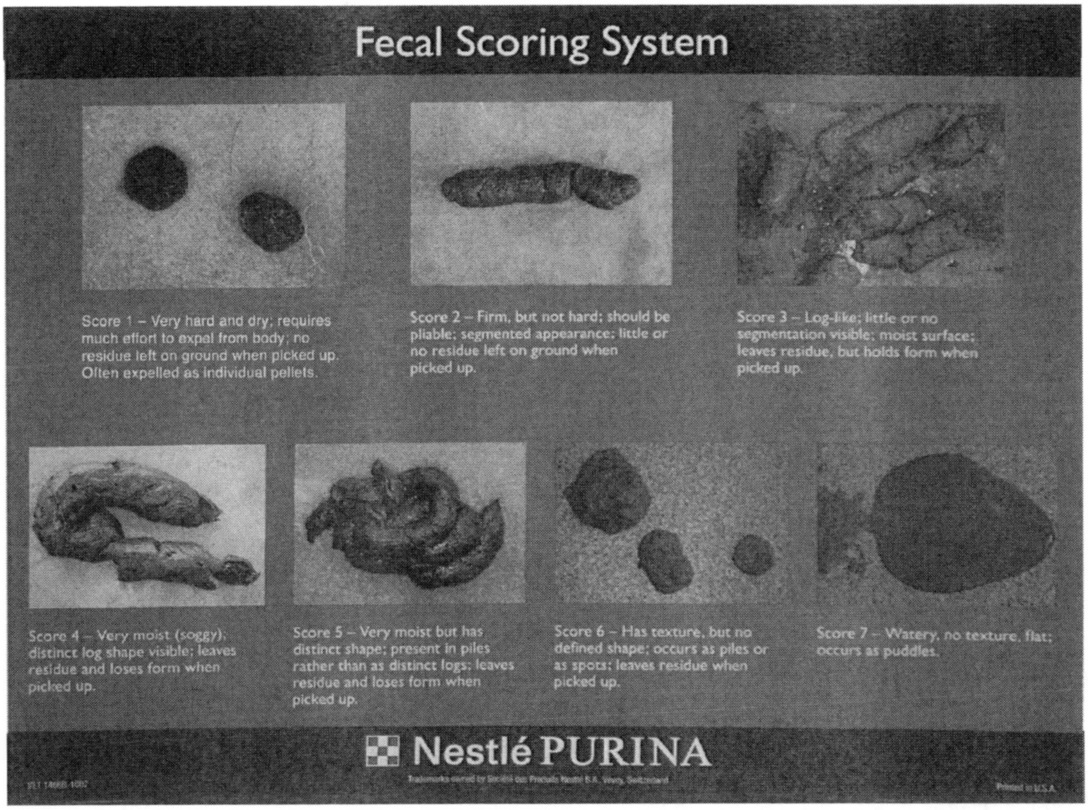

For those cases that are suffering from severe unretractable vomiting for over 48 hours total parenteral nutrition must be provided, alongside very small amounts of nutrition being given by mouth (microenteral nutrition), and the necessary medications prescribed by the veterinary surgeon.[2] The use of micro-enteral nutrition has shown to improve prognosis of many cases that were traditionally starved. Dietary therapy for dog's suffering from acute pancreatitis was traditionally nil-by-mouth, until the clinical symptoms ceased, and in some practices this could be up to five to seven days. Evidence has shown that nil-by-mouth in these cases has little benefit to the animal. The oral nutritional support is required to prevent gastroduodenal ulceration, bacterial translocation from the gut and septicaemia. Small amounts of watered down baby rice or cottage cheese can be used before moving onto a commercial low-fat, highly digestible diet, oral hydration fluids can also be utilised.[2]

In human medicine there is a continuing bank of evidence indicating that "feeding through" diarrhea with an appropriate diet can be beneficial. Comparisons can be difficult, as many cases of diarrhea in man tend to be secretory, whereas osmotic diarrhea is more common in dogs and cats. The advantages of "feeding through" include maintaining mucosal health, reducing the risk of bacterial translocation. Feeding through should not be used in animals that are vomiting or severely dehydrated, unless the hydration and electrolyte status is initially correctly prior to feeding. The obvious disadvantage of "feeding through" is the risk of "accidents" in the house, consideration should be taken for owners that may not be able to care for the pet themselves when the diarrhea is present. Whether food is withheld or not, unlimited access to water must be maintained at all times. The use of oral rehydration solutions containing simple sugar, soluble fibres, peptides and electrolytes should be advocated in these cases. When using this method feeding a diet, which won't cause an inflammatory reaction in the future is important. Dietary antigens can cross the compromised gastrointestinal mucosa and set up a hypersensitivity reaction. Use of a novel protein and/or carbohydrate source should be recommended in selected cases. Once the animal is then well, and transferred to its original diet, it is unlikely to have any reactions to the original protein source.

Dietary transitions are as important as the diet itself. When the gastrointestinal system is dysfunctioning transitions may need to be over a longer period of time. Palatability of the diet can be affected by the animal feeling ill, nausea being the biggest cause. Treatment to correct any factors of illness that effect palatability is important, alongside correction of any hydration issues.

Pro- and Prebiotics

Pro- and prebiotics are widely used in the treatment of gastrointestinal disorders. Prebiotics are substances that are able to alter the gastrointestinal flora in a manner to benefit the microorganisms, these are normally types of fibres (e.g., manno-oligosaccharides, fructooligosaccharides) or amino acids like glutamine. Probiotics however, are a live microbial feed supplement, which benefits the host animal by improving the gastrointestinal microbial population. Probiotics generally used are comprised of lactic acid bacteria such as *Lactobacillus, Streptococcus* and *Bifidobacterium.* Use of probiotics in cases of chronic diarrhea has shown to be beneficial. Hart *et al.* (2012) showed that seventy-two percent of owners perceived an improvement in their cat's diarrhea following a 21-day course of synbiotic supplementation. Honneffer *et al.* (2014) review piece demonstrated the complex role of the microbiome and its influence in gastrointestinal disease processes, with the positive effects of pro- and prebiotics[6].

Pharmaceuticals

Many different pharmaceuticals can be utilised in the treatment of different gastrointestinal problems that can be seen in veterinary practice. It is important when conducting nursing clinics to read the clinical history and to make note of the medication regimes that the veterinary surgeon has prescribed for the animal. In some chronic cases drug dosages are tapered to effect and therefore support from the veterinary nurse may be required in order to help clients achieve this.

Vitamin B12 (cobalamin) injections are commonly used in animals with low blood cobalamin, which is commonly seen in animals with gastrointestinal problems. Weekly injections in some cases may be required for a number of weeks, and this is a role for the veterinary nurse to undertake. Oral supplementation is possible in some cases, and there are now nutritional supplements which contain Vitamin B12.

References

1. Battersby I, Harvey A. Differential diagnosis and treatment of acute diarrhea in the dog and cat. *In Practice*. 2006; 28:480–488.
2. Watson PJ. Managing canine pancreatitis. In: BSAVA Congress 2005 Scientific Proceedings. 2005; 97–99.
3. Steiner J. A step-wise approach to dogs and cats with chronic diarrhea. *Veterinary Focus*. 2013;23(2):54–56.
4. Westermarck E, Skrzypczak T, Harmoinen J, Steiner JM, Ruaux C, Williams DA, Eerola E, Sundbäck P, Rinkinen M. Tylosin-responsive chronic diarrhea in dogs. *J Vet Intern Med*. 2005;19:177–186.
5. Hart ML, Suchodolski JS, Steiner JM, Webb CB. Open-label trial of a multi-strain synbiotic in cats with chronic diarrhea .*J Feline Med Surg*. 2012;14(4):240–5.
6. Honneffer JB, Minamoto Y, Suchodolski JS. Microbiota alterations in acute and chronic gastrointestinal inflammation of cats and dogs. *World J Gastroenterol*. 2014;20(44):16489–16497.

acvim037

Food Intolerances

Nicola Ackerman, BSc (Hons), RVN, CertSAN, CertVNECC, VTS (Nutrition)
Plymouth, UK

Introduction

Dietary-induced skin problems include food intolerances, primary and secondary nutrient deficiencies, and nutrient toxicities. A full dietary history needs to be taken and owners should include specific commercial foods, all snacks and treats, supplements, chewable/palatable medications and vitamins, chew toys, human foods and any food that the animal may have access to. Actively encourage owners to keep a food diary, as with obese animals and diabetics. The classification and diagnosis of nutrition-related skin disorders are initially based on a detailed dietary history and food evaluation. Veterinary nurses and technicians are in an ideal position to help owners make an accurate representation of what the pet consumes.

Food Intolerances and Allergies

The terms food allergy and food hypersensitivity should be reserved for those adverse reactions to food that have immunologic basis. Food intolerance refers to adverse food reactions due to non-immunologic mechanisms. Dermatologists suggest that only 1–6% of all dermatoses seen in practice relate to adverse food reactions and that food allergies constitute 10–20% of allergic responses in dogs and cats, (Figure 1). A full dermatological workup is required before a nutritional factor can be entirely confirmed. It can be commonplace, however, for owners to self-diagnose their pet's food allergy or intolerance, even before seeing the veterinary surgeon. There are many diets available that are gluten free, grain free, hypoallergenic and/or natural; and when the pet improves on these diets the owners assume that their pet had an allergy to one of the ingredients. In many cases it is purely a coincidence.

Figure 1. The difference between adverse food reactions

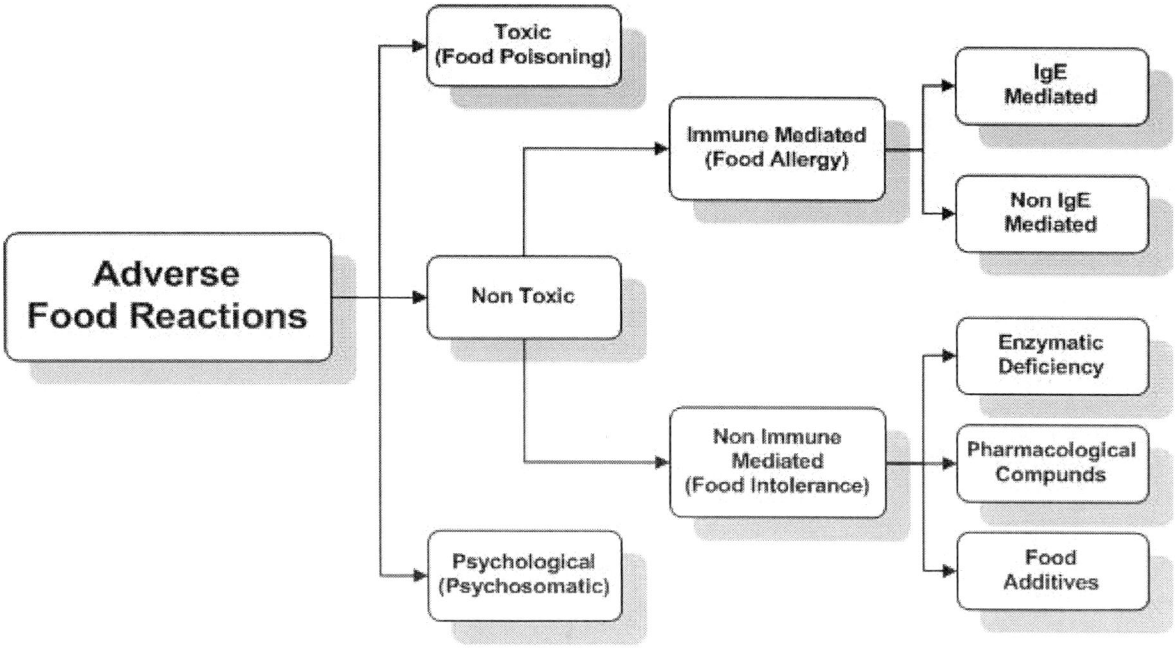

The skin is the largest organ of the body and has heavy demand on bodily nutrient supply. When factors become suboptimal, it is usually the skin which becomes the first organ to demonstrate deterioration. The owner can easily detect signs of deterioration, loss of coat sheen, coat shedding, and the coat can become greasy and scurf at the skin surface can be evident, but in many cases the signs of deterioration occur at such a slow rate that owners don't notice. It is not until the issue is resolved, or the animal is placed on a different diet that the animal's coat returns to 'healthy' and the owner notices a difference.

Food intolerance is not so clear cut and is a more controversial area. Although not life-threatening, it can and often does make the sufferer feel extremely unwell and can have a major impact on life. Some people actually lack the enzymes needed to break foods down; for example lactose intolerance where the enzyme lactase is not produced in large enough amounts to break down the lactose (milk sugars) in milk. Food intolerance reactions do not involve IgE and the mechanisms are unclear; however, it is known that reactions are more likely to be delayed with the reaction occurring several hours and sometimes up to several days after eating the offending food. The symptoms caused by these reactions are numerous, but have been associated with gut symptoms such as bloating, diarrhoea, constipation, and IBS, and skin problems such as eczema. Specific food additives, which are known to cause problems, include onions and propylene glycol, which can cause haematological abnormalities in cats. Lactose intolerance is a relatively common metabolic adverse reaction in dogs and cats. Diarrhoea can develop when given cow's or even goat's milk, due to the lactose content being higher than that of bitch's or queen's milk. Gluten-sensitive enteropathy has been well documented in Irish Setters.

Inflammatory Bowel Disease

The role of food allergies in canine and feline inflammatory bowel disease/disorder (IBD) is unknown. It is thought that hypersensitivity to food is involved in the pathogenesis of this syndrome. The role of elimination food trials often alleviate the signs of IBD, which does seem to imply that food allergies or food intolerances play a role in this syndrome.

Elimination Diets

When designing or recommending an elimination diet, certain characteristics do need to be taken into consideration. Review the diet history sheets and diaries to give you an indication of what the animal has been exposed to in the past. A limited number of protein sources are required, along with the protein being of a novel source. Excessive levels of protein should be avoided; the protein present needs to be of a digestibility greater than 87%. Hydrolysed protein can be used, as digestibility is high. Digestibility is an important factor as free amino acids and small peptides make poor antigens. The diet needs to be free of additives and excessive levels of vasoactive amines, whilst still being nutritionally adequate for the animal's life stage and body score.

Hydrolysed diets are advantageous as they remove the absolute necessity of finding the primary causal agent of the food allergen. Hydrolysed diets work by taking a protein and reducing it in size through hydrolysis. When the hydrolysed protein is then ingested, the body no longer recognises the protein as an antigen, and the allergic immune response is not initiated.

The use of carbohydrates in the diet also needs to be limited. In many elimination diets the use of a novel carbohydrate source is also utilised. Many carbohydrates sources used include rice, potato, and tapioca.

Feeding an Elimination Food (Dermatological and Gastrointestinal Disease)

The main diagnosis method for nutritional adverse food reactions is dietary elimination trials. Either a commercial product or a homemade diet can achieve these trials. It must be noted that homemade diets are likely to be nutritionally inadequate for maintenance requirements. Most homemade diets lack a source of calcium, essential fatty acids, vitamins, and other micronutrients. These diets can also contain excessive levels of protein. If the owner is insistent on the use of a homemade diet, guidance on constructing a balanced diet is required from the veterinary practice.

In order to make a diagnosis of nutritional adverse food reactions, only the elimination food can be used. The trial needs to be performed for 2–12 weeks, and full dedication from the owner is required. In some cases, resolution of clinical signs can take a couple of weeks, though in some cases this can take longer.

Elimination trials can prove to be extremely useful when diagnosing the causal nutritional agent in both dermatological and gastrointestinal disease. A strict elimination trial protocol does need to be adhered to, and the owner and anyone in contact with the animal need to be aware of this. No other foodstuffs, other than the elimination food, should be ingested over this period. This includes treats, flavored vitamins, chewable/palatable medications, fatty acids supplements, and chew toys. Advice must be clearly conveyed to the owner when discussing the use of an elimination food trial. The patient is fed the controlled diet for four to 12 weeks for dermatological disease. A shorter period of two to four weeks is usually satisfactory in gastrointestinal disease. The degree of clinical improvement during the elimination trial will be 100%, only if food sensitivity is the sole causal agent. These trials are often difficult to interpret in dermatological disease. This is due to concurrent allergic skin disease. These patients may only partially respond to an elimination trial. Confirmation of a diagnosis can be made to an adverse food reaction when the clinical signs reoccur within 10–14 days after the animal is challenged with its original diet. Always give the client written instructions describing the requirements of the elimination trial and feeding quantities. Details on how to contact an advising veterinary nurse can prove useful if the owners have any questions during the trial.

Prolonged Elimination Can Build Tolerance

Weeks or months of elimination of the reactive food may well lead to reintroduction of the food without reaction. This is known as tolerance, and its maintenance depends on establishing the threshold of both frequency and quantity for that person - in other words, eating the food occasionally may be tolerated, but reintroducing it in large quantities or on a very regular basis (e.g., every day) might lead to symptoms recurring. This is purely individual so working this out and not restricting the diet more than is necessary is a major consideration.

Figure 2. Flow diagram of protocol of elimination diets

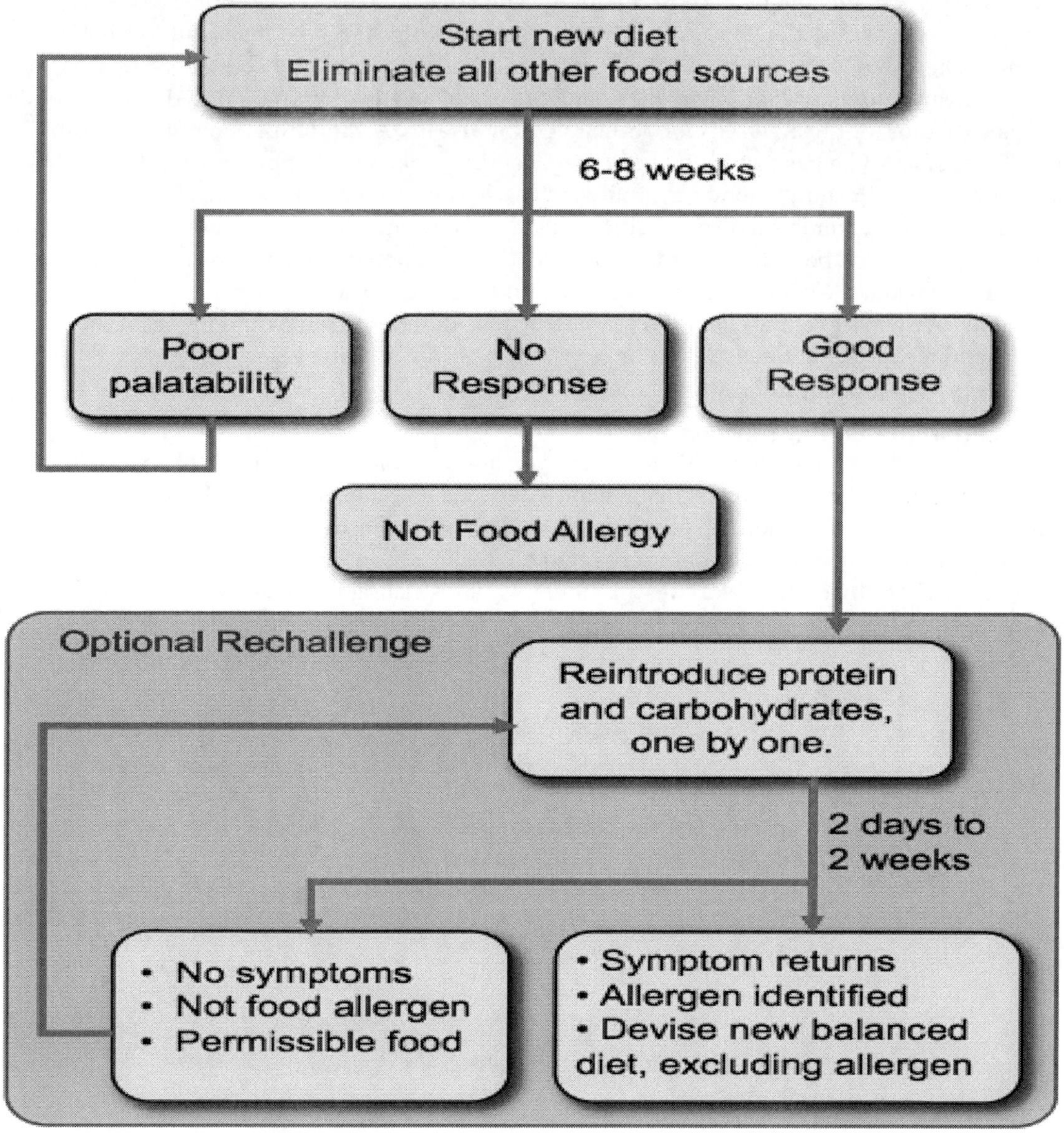

Conclusion

Food intolerances can cause a variety of clinical signs, from skin issues to gastrointestinal signs. A good diagnostic workup by the veterinary surgeon is required, part of this can include a food elimination trial. These trials can prove to be useful - but interpretation of them can be difficult. Exceptionally good client guidance is needed in diet trials and veterinary nurses should take a pivotal role in aiding clients.

acvim038

What Do Prebiotics and Probiotics Really Do?

Ann Wortinger, BIS, LVT, VTS (ECC, SAIM, Nutrition)
Belleville, MI, USA

Introduction

It seems that everywhere we look these days we are seeing products that include either probiotics or prebiotics. This can be in the form of supplements, known as nutraceuticals, or in over-the-counter foods and therapeutic diets. According to the hype, these products can improve your immune system, improve your gastrointestinal function, and any number of other things just short of emptying the litter box for you!

According to the currently adopted definitions by the Food and Agriculture Organization (FAO) and the World Health Organization (WHO), probiotics are "living microorganisms which, when administered in adequate amounts, confer a health benefit on the host" (Weese 2003), whereas prebiotics are "non-digestible food ingredients that selectively stimulate the growth and activities of specific bacteria in the gastrointestinal tract and exert beneficial effects on the host" (Pan 2009). The Food and Drug Administration (FDA) defines nutraceuticals as "nondrug substances produced in a purified or extracted form and administered orally to provide agents required for normal body structure and function with the intent of improving health and wellbeing" (Lerman 2007).

As we can see by all these definitions we are either talking about bacteria or substances that benefit bacteria in the intestinal tract. We all know that there are millions of bacteria that normally reside in the small and large intestines of normal, healthy animals. These bacteria help with digestion of food, maintain intestinal mucosal integrity, participate in metabolism, as well as local and systemic immune stimulation (Kelly 2006).

The Intestines

The intestines are seen as the largest immune system in the body, comprising ~70% of the total system. The mucosal barrier found in the intestines helps block the entrance of most pathogenic bacteria into the body while allowing the entrance of permeable nutrients (Kelly 2006). Since most pathogens enter the body through the mouth and then the intestinal tract, these intestinal defenses must be working optimally to cope with the onslaught of foreign substances and pathogens to which it is constantly exposed (Kelly 2006).

This defense involves the coordination of three different systems within the intestines: the first is the resident intestinal microflora which provides an environment which favors the growth and function of beneficial bacteria; the second is the intestinal mucosa which provides a protective barrier against pathogenic bacteria; and lastly, the gut-associated lymphoid tissue (GALT) (Kelly 2006).

It would be difficult for us to affect the intestinal mucosa or the GALT, but we can help modulate the intestinal microflora to improve the environment the bacteria are living in allowing a positive impact on the dog or cat. There is substantial research being done looking at how enteral nutrition can be used to help improve both the GALT but also the intestinal mucosa barrier. Hopefully, we will see the benefits of that research in the next few years. Currently, we are using prebiotics and probiotics to help with this modulation.

Prebiotics

Typically, prebiotics are a type of carbohydrates called oligosaccharides, though by definition non-carbohydrates can be classified as prebiotics. The ones used most frequently are classified as soluble fibers. These were first identified as a functional food in 1995 by Marcel Roberfroid. In 2007, Dr. Roberfroid clarified that only two classes of fructooligosaccharides fully meet the definition of a prebiotics. These would include oligofructose and inulin (Roberfroid 2007).

Oligofructose is a 2–8 chain fructose-based saccharide molecule that undergoes fermentation fairly quickly in the colon providing nourishment to the bacteria in that area. Examples of sources for fructooligosaccharides (FOS) would include soybeans, oats, beets, and tomatoes.

Inulin is a longer 9–64 chain fructose-based saccharide that tends to be fermented more slowly benefiting bacteria farther down the colon. It can also be broken down into the FOSs by intestinal bacteria to provide both FOS and inulin. Inulin can be found in Jerusalem artichoke, jicama, and chicory root.

These two fibers are seen as minimally digestible because of their beta-bond base connecting the fructose molecules. Dogs and cats lack the enzyme in their intestines to break down the beta bond, instead they usually breakdown alpha saccharide bonds. The resident bacteria in these animals do have the ability to break the beta bond producing short chain fatty acids (SCFA). The SCFAs seen most commonly include acetate, propionate and butyrate.

Plants do not have only one type of carbohydrate in them, so both FOS and inulin can be found in varying amounts in the same plant, while none of either may be found in others. Just adding fiber to the diet may or may not provide prebiotic effects, dependent on the types of fibers found in the product.

The SCFAs produced by the breaking of the beta bond through bacterial fermentation serve as an energy source for the colonocytes, act to lower the colonic pH, and stimulate sodium and water absorption (Pan 2009). One SCFA in particular, butyrate serves as a primary source of energy for colonocytes, but may also directly enhance cell proliferation of normal cells, while suppressing proliferation of transformed cells. Because both oligofructose and inulin are fibers, there can be negative side effects to the addition to the diet. Both FOS and inulin have been shown to cause decrease in fecal protein digestibility when used at higher levels (Pan 2009).

Studies have shown there are generally positive effects on gut microflora, host health as evaluated through gut integrity and bacterial colonization and animal performance as evaluated through digestion, body weight gain and feed efficiency when FOS and fructose based inulins are added to the diet (Verdonk 2009).

PROBIOTICS

Probiotics, rather than encouraging the growth of beneficial bacteria and suppressing the growth of pathogenic bacteria, actually introduce these beneficial bacteria into the environment. To be able to function as a prebiotic certain criteria must be met by the bacteria being used. They must be able to:

- Survive the acid and bile found in the gastrointestinal tract
- Adhere to the intestinal cells or transiently colonize various areas within the GI tract
- Exclude or reduce pathogenic bacterial adherence
- Produce acids, hydrogen peroxide and/or bacteriocins that antagonize the growth of pathogens
- Co-aggregate to help achieve normal balanced microflora population
- Be safe, noninvasive, noncarcinogenic and nonpathogenic to the animal (Kelly 2006).

Initially, the gastrointestinal tract of a newborn is sterile but is colonized with bacteria within hours of birth. These bacteria find their individual niches within the intestinal tract and reach a state of equilibrium. Once this neonate "grace period" is over though, introduction of bacteria is significantly more difficult due to the gastric acidity and the introduction of bile acids to the chime leaving the stomach. Though yogurt products contain beneficial bacteria, as do a number of other bacterial fermented products, these bacteria seldom survive the stomach and therefore do not provide any benefit to the intestinal bacteria (Weese 2003).

The primary bacterial populations included in prebiotics that are seen to benefit the cat and dog are lactic acid bacteria, especially *Lactobacillus, Bifidobacterium* and *Enterococcus*. These bacteria use fermentation to transform some sugars into organic acids, particularly lactic and acetic acids. These acids serve to lower the pH in the intestinal tract, and inhibit the growth of pathogenic bacteria (Kelly 2006).

Probiotics can benefit the intestinal microflora in a number of ways. They can increase fecal bacteria count of good bacteria while decreasing the numbers of pathogenic bacteria. Some prebiotics have been shown to minimize adherence to the intestinal epithelial cells and establishment of pathogenic bacterial populations. Compared to the large intestines, the small intestines have poor microflora colonization and limited barrier protection against pathogens. Therefore, probiotics can exert an important beneficial effect in the small intestines (Kelly 2006).

Table 1.1. Intestinal bacteria

Beneficial	Pathogenic
Lactobacilli	*Pseudomonas aeruginosa*
Eubacteria	*Proteus* species
Bifidobacteria	Staphylococci
Enterococci	Clostridia
	Veillonella

(Kelly 2006)

Probiotics can also help control diarrhea caused by bacterial overgrowth or parasitic infection. This uses the principles of competitive exclusion, competition for nutrients and binding sites and an increase in specific and nonspecific immune response (Kelly 2006).

One of the biggest challenges for manufacturers of probiotic products is survivability of the bacteria. Most commercially available products suffer tremendous loss of activity during storage. After 5-6 months of storage, virtually no live organisms are present. Those bacteria that are present still have to survive the gastric pH and duodenal bile acids before they can colonize the small intestines. To address these concerns, some manufacturers have developed microencapsulation to protect the bacteria while others provide a variety of bacterial species to cover more bases.

CONCLUSION

Since prebiotics are used as both a source of nutrition for the colonocytes and as a fiber source for the animal, inclusion in the diet is fairly easy. The fiber source used should be listed on the ingredient panel, but may not state whether it is a source of FOS or inulin. Increasing your familiarity with the different fiber sources use, and their relative FOS and inulin contents can help when evaluating a diet. When in doubt, contact the manufacturer for further information.

Unlike prebiotics, the use of probiotics is more of a transient process based on need rather than a long-term process. Ideally, the body will supply the bacterial population to support the best intestinal health. But when this doesn't occur, or when challenges are expected that could affect intestinal health, probiotics can be easily added into the diet to help support the bacterial populations there. When evaluating products to use for either prebiotics or probiotics, ensure that there is research to support the claims given, that the product is proven to contain the stated levels of additives and that they have been proven to promote normal intestinal microflora. A review of probiotics in commercial dog and cat foods by Dr. Scott Weese, demonstrated that few products meet these guidelines (Weese 2003). Alternately, evaluation in a controlled clinical setting can provide firsthand knowledge of the effectiveness of the products used. By looking at these facts, you can ensure that you are providing the best product to support a happy, healthy intestinal tract with a hardworking population of beneficial bacteria.

REFERENCES

1. Kelly M. The role of probiotics in GI tract health. Nestle Purina PetCare Company. 2006.
2. Lerman A, Lockwood B. Nutraceuticals in veterinary medicine. *Pharmacy Journal*. 2007;278:51.
3. Pan X, Chen F, Wu T, Tang H, Zhao Z. Prebiotic oligosaccharides change the concentrations of short-chain fatty acids and the microbial population of mouse bowel. *Journal of Zhejiang University Science B*. 2009;10(4):258–263.
4. Roberfroid M. Prebiotics: the concept revisited. *Journal of Nutrition*. 2007;137:830S.
5. Verdonk JMAJ, Shim SB, van Leeuwen P, Verstegan MWA. Application of inulin-type fructans in animal feed and pet food. *British Journal of Nutrition*. 2005;93(supplement 1):s125–s138.
6. Weese, JS, Arroyo L. Bacteriological evaluation of dog and cat diets that claim to contain probiotics. *Canadian Veterinary Journal*. 2003;44:212–216.

acvim015

Flexible GI Endoscopy

Liza Wysong Rudolph, BAS, CVT, VTS (CP-CF, SAIM)
Aberdeen, MD

Introduction

Endoscopy of the gastrointestinal tract (GIT) continues to be popular in small animal veterinary medicine and its incorporation into private practices is becoming well established. This necessitates that veterinary technicians have familiarity with endoscopic equipment and procedures in order to anticipate clinician needs, maximize patient care, and assist in providing diagnosis and proper treatment options.

Flexible Endoscopes

Due to their versatility, flexible endoscopes are the mainstay of small animal GI endoscopy. Flexible endoscopes have a variety of beneficial capabilities including mechanical deflection, insufflation, irrigation, and suction. Endoscopes come in a variety of lengths and diameters and offer high quality optics and the ability to perform interventions (e.g., cytology, biopsy, cautery).

There are several human and veterinary endoscopy manufacturers. A human pediatric endoscope with four-way directional tip deflection, insufflation, irrigation and interventional capabilities is a good choice for a general all-purpose small animal endoscopy of the GIT. Most human pediatric endoscopes are approximately 100 cm or slightly more, whereas endoscopes specific to the veterinary industry can be up to 240 cm. The advantage of a longer length is the ability to reach the duodenum in large and giant breed dogs. Unfortunately, they can be unwieldy and awkward for use in small dogs and cats. The diameter of the insertion tube that is usually used in small animals ranges from about 5.9 to 11 mm. Larger diameters will allow for larger biopsy specimens to be obtained, but increase the difficulty of pyloric and duodenal intubation in small dogs and cats, which may lead to an incomplete examination. Conversely, smaller-diameter, more flexible scopes may curl up in the stomach, making it more difficult to pass the pyloric sphincter into the duodenum.

Two types of flexible endoscopes are available, fiber and video. Fiberscopes transmit images from the endoscope tip and light via optical glass fibers. Images may be viewed through a magnifying eyepiece or, more commonly, processed onto a video monitor. These scopes are more affordable, have decent image quality, and are available in a wide range of diameters and lengths. A video endoscope has a video chip located behind the objective lens that transmits the image via a digital signal. The main benefits of a videoscope is superior image quality, resolution, and color. Historically the diameters available for video endoscopes were somewhat more limited than those of the fiberscopes, but this is no longer the case. Video endoscopy is more expensive because the endoscopes are more expensive and a processor is required.

The flexible endoscope is comprised of three parts: the umbilical cord, the handpiece, and the insertion tube. The umbilical cord connects the light source to the endoscope. It also has the connectors for air, water, and suction. This section should be handled carefully in order to avoid damage to the fiber optics and/or electronics.

The handpiece controls the flexible endoscope. Mechanical deflection of the distal insertion tube can either be two-directional or four-directional and the deflection cables are controlled manually by knobs on the handpiece. Endoscopes with two-directional deflection have a single knob to control up/down motion. A four-way directional tip is preferred for GIT endoscopy to maximize visualization, facilitate entrance to the small intestine, and is most appropriate for proper biopsy procurement. Four-directional endoscopes have two knobs. The larger knob is used for up/down deflection while the smaller knob is used for right/left motion. A lock is attached to the small knob to freeze right/left motion if desired. To avoid potential damage, always be sure this is unlocked before attempting to move the right/left knob.

There are two valves on the handpiece of a GI endoscope. The first valve, usually marked red, is the suction valve. Fluid and air can be suctioned through the endoscope by depressing the valve, but large particles can cause the channel to become obstructed. The second valve, marked in blue, allows control of air and water. By covering the hole in the center of the valve with a finger, air insufflation can occur. This will allow for distension of the GIT for proper visualization. If the valve is depressed fully, water flows down to the tip of the endoscope to clear away mucus and debris that may obscure the image. It is recommended that prior to and after all procedures these valves are checked for proper function. The entrance to the biopsy port or channel is also located on the handpiece. Instruments such as biopsy forceps, cytology brushes, foreign-body graspers, cautery, laser fibers, or injection/aspiration needles can be passed through the channel.

The insertion tube contains the method of transmitting the image (fiber optics or electric transmission), deflection cables, and multiple channels. The distal end of the insertion tube allows for mechanical deflection, and at its most terminal point, houses the objective lens. The insertion tube is the most fragile section of the endoscope and damage should be avoided. Common forms of damage to insertion tubes include lens breakage from striking hard surfaces and channel perforation from forceful equipment insertion. A variety of problems can also result from crushing (due to patient bite), tight coiling, bending, or twisting of the tube.

The biopsy channel diameter ranges approximately 2.0–2.8 mm in GI endoscopy. The diameter of flexible instruments passing through the channel is slightly smaller (1.8–2.6). For ease of use, it is recommended that the total working length of endoscopic instruments be about 50 cm longer than the channel itself.

Endoscopic Instrumentation

There is a wide variety of endoscopic instrumentation available. Biopsy retrieval instruments (oval, cup, smooth, serrated, spiked, etc.) are likely those most commonly encountered, but foreign-body retrievers (graspers, snares, loops, baskets, etc.), cytology brushes, and injection or aspiration needles may all be used in the small animal patient.

When using endoscopic instruments, always be sure that the instrument is closed before inserting it into the channel and that it remains closed when continuing down the channel. If any resistance is met, do not force the instrument or damage could occur. Once the end of the instrument is seen protruding from the endoscope it is safe to open. If FB retrieval is being performed, do not remove foreign objects through the channel. Firmly grasp the foreign body and then remove the instrument and endoscope from the patient as a single unit.

When obtaining biopsy samples, the operator should push the open forceps against the area to be biopsied before closing. Be sure to close the instrument firmly, but not too forcefully. Using too much force will prematurely dull the instrument and put excess strain on the mechanism. Once the sample is obtained, retract the closed forceps through the biopsy channel. If the endoscope is needed to be retroflexed 180 degrees to obtain a sample, it is recommended to insert the forceps when the scope is in a neutral position to minimize undue wear and tear on the biopsy channel.

Biopsy samples should be handled minimally. Often the use of small-gauge needles and/or small amount of saline flush is recommended to coax the sample out of the forceps and onto your biopsy cassette. The samples should be kept moist to avoid desiccation and placed in formalin as soon as possible.

GI Endoscopy

Gastrointestinal endoscopy enables the clinician to visualize the mucosal surfaces of the GIT and obtain biopsies without surgical intervention. A flexible endoscope suitable for examining the GIT is also useful for foreign-body (FB) retrievals, stricture dilation, and feeding-tube placement.

The technician should set up and test the endoscope prior to every procedure and before patient anesthesia. All ancillary equipment should be gathered before the start of the procedure. This may include a mouth gag, sterile lube, gloves, biopsy forceps, cassettes, formalin jars, gauze sponges, needles, saline, etc. The suction, air/water, light source, image quality, and white balance should all be evaluated and deemed to be in proper working order. The light source should be turned off until the start of the procedure to maximize lamp life and prevent the distal end of the insertion tube from getting too warm.

Endoscopy of the GIT typically involves examining the following five main areas: the esophagus, stomach, duodenum, proximal jejunum, ileum, and colon. For simplicity, these areas are generally divided into two procedures: esophagogastroduodenoscopy (esophagus, stomach, and duodenum) and colonoscopy (colon and ileum). In some cases, rigid endoscopy may be employed for proctoscopies (rectum and descending colon) and esophagoscopies for FB retrieval, but this is beyond the scope of this topic.

GASTRODUODENOSCOPY

Patient preparation for esophagogastroduodenoscopy, often called an "upper GI," is fairly straightforward. A 12- to 24-hour food fast is usually sufficient for most patients, but longer periods may be required in patients with gastric retention and foreign bodies. As general anesthesia is necessary, routine premedication and induction agents that effect GIT motility or intestinal secretions may be altered according to clinician preference. At induction, the anesthetist should be sure to note any abnormalities during intubation. The patient is typically placed in left lateral recumbency to aid entry into the duodenum. A mouth gag is placed and an oral exam takes place prior to the procedure. The endoscope passes through the oropharynx and upper esophageal sphincter. Insufflation is required for proper visualization of the esophagus and may require manual compression of the esophagus so air does not readily escape the proximal esophagus into the oral cavity. The mucosa should appear smooth, pale, and glistening. After the esophagus is examined, the scope passes through the lower esophageal sphincter (usually closed and darker red in color) into the stomach.

The stomach is comprised anatomically of the cardia, fundus, body, antrum, and pylorus. The greater and lesser curvature are used to describe the external and internal areas of the body and antrum that create the larger (greater) and more acute (lesser) curves of the stomach. Normal stomach mucosa can be described as smooth, glistening, and pink with longitudinal rugal folds. As described previously for esophagoscopy, pressure may need to be placed around the cervical esophagus by an assistant to prevent the escape of air. Regarding insufflation, it is important to monitor for overdistension of the stomach, as this will make it more difficult to obtain biopsies, cause gastric mucosa ischemia, and compromise the overall cardiopulmonary status of the patient.

In order to proceed to the duodenum, the endoscope must pass through the pylorus and pyloric sphincter. Intubation of the pylorus is often the most difficult part of the procedure. The endoscope is usually advanced slowly while keeping the center of the pylorus in view. If that is unsuccessful, some endoscopists will attempt to use closed biopsy forceps as a stylet before passing through the pylorus. Repositioning of the patient and external massage may help facilitate intubation. The duodenum generally appears paler, but still pink, smooth, and uniform. The mucosa appears more granular and not as smooth as the gastric mucosa. Closely packed lymph follicles involved with immune response, called Peyer's patches, may also be observed. In some instances, the duodenal papillae may be visualized. It is very difficult to differentiate the duodenum from jejunum endoscopically.

After complete examination of the esophagus, stomach, and proximal small intestine, the endoscope will be drawn back and biopsy samples will be obtained. Biopsies should always be taken when the scope is being withdrawn. Excessive bleeding during biopsy procurement may obscure visualization, impeding a proper examination. Additionally, there is an increased risk of perforation if the endoscope is advanced after obtaining biopsies. At the completion of the procedure, all air should be suctioned from the stomach (inadequate removal can lead to gastric dilation and volvulus formation), as well as any fluid observed in the esophagus.

COLONOSCOPY

Unlike upper-GI tract endoscopy, the patient preparation for colonoscopy is more complex and varied based on clinician preference. Generally, most clinicians will recommend 24 to 48 hours of withholding food, gastrointestinal lavage, and enemas. Electrolyte solutions such as Colyte® or GoLYTELY® (polyethylene glycol) can be offered to the patient or administered via nasoesophageal (CRI) or orogastric intubation or OsmoPrep® (sodium phosphate) can be administered orally. These medications cause large amounts of water to be pulled from the body into the colon (osmotic diarrhea), which promotes evacuation. Water should be available to these patients and intravenous or subcutaneous fluid administration should be considered in patients receiving any of the above medications. Administration of gastrointestinal lavage inducing agents is contraindicated in patients with confirmed or suspected intestinal strictures or obstruction.

Before beginning colonoscopy, the patient is placed under general anesthesia and usually positioned in left lateral recumbency. A digital palpation should be performed before the procedure to aid in identification of any gross abnormalities. The endoscope is inserted into the rectum and the endoscope is advanced. Insufflation is necessary for visualization and may require manual (digital) occlusion of the rectum around the endoscope. Although there are some angles and turns that will obscure the view, a central view can usually be obtained for proper mucosal examination. Submucosal vessels are visible in the colon and the surface is usually smooth, pink, and glistening. The ileum is entered with the same techniques described for the duodenum. The ileocecal valve often has a button-like appearance and is located in proximity to the entrance into the cecum. The ileum is similar in appearance to the duodenum: pink, smooth, and uniform. Again, biopsies are taken as the scope is withdrawn.

Foreign Body Retrieval

Another common use for flexible endoscopy is FB retrieval. Esophageal foreign bodies may be removed through the oral cavity or advanced into the stomach for removal by gastrotomy. If neither can be accomplished, surgical intervention will be necessary to remove the object from the esophagus as forceful endoscopic removal is contraindicated.

Endoscopy can be used to retrieve gastric FBs as well. Any object that could lead to an obstruction, perforation, or toxicity should be removed. Endoscopic removal of gastric foreign material is contraindicated if foreign objects will cause severe trauma to the lower esophageal sphincter, esophagus or upper esophageal sphincter upon removal. Smaller objects may not require removal as they will likely pass. After FB removal, the mucosa should be examined for any secondary trauma.

Reprocessing

One of the most important duties of the technician is proper care and cleaning (reprocessing) of the endoscope. Proper reprocessing is critical for appropriate maintenance of the endoscope as well as prevention of cross contamination and nosocomial infections. Manufacturer reprocessing guidelines should be followed after every endoscopy procedure. It is important that the reprocessing begin immediately after the procedure is completed. The first steps that are performed at patient side are termed precleaning. The endoscope is first wiped clean with enzymatic detergent followed by suctioning of the detergent through the endoscope until it runs clear. It is also important to use an all-channel irrigator valve as part of the precleaning process.

A leak test must be performed prior to manual cleaning and immersion of the endoscope to prevent water damage. Placement of any recommended protective cap for the endoscope electrodes should occur prior to the leak test and endoscope immersion. In the event that the endoscope fails a leak test, follow the manufacturer's recommendations on how to proceed. After testing, the endoscope is completely immersed in enzymatic detergent for manual cleaning and the exterior is cleaned with a soft cloth or sponge. A cleaning brush is then fed through the appropriate channels per manufacturer instructions. The channels are then flushed with enzymatic detergent and rinsed with clean or distilled water using the manufacturer's channel cleaning adapters. The water is then purged from the channels. The purpose of manual cleaning is to remove any protein-rich material and reduce the quantity of bacteria.

High-level disinfection is the next step in the endoscopic reprocessing procedure. Be sure to reference the manufacturer guidelines regarding compatible high-level disinfectant (HLD) use. It is also important to follow the HLD directions regarding temperature, dilution, contact time, and the use of appropriate personal protective equipment (PPE). The endoscope will be immersed in the disinfectant and the solution flushed through all channels. All endoscopic accessories including valves, water bottles, cleaning brushes, and mouth gags will also be submerged. After the recommended contact time, the endoscope should be thoroughly flushed of the HLD, rinsed thoroughly, and dried. A final flush of the channels with 70% isopropyl followed by forced air will expedite the drying process and reduce the likelihood of bacterial contamination during the drying process.

Automated endoscopy reprocessors (AERs) are available, but are not commonly utilized in private small animal veterinary practices. These devices increase efficiency as they automatically wash, disinfect, and rinse the channels of the endoscope and should be considered in any practice with a high endoscopy case load. Automatic reprocessors provide a more reliable method of endoscope reprocessing due to their consistent channel irrigation and precise disinfectant soaking times. AERs also reduce the risk of personnel exposure to hazardous HLD.

Apart from fragile ultra-thin endoscopes that are typically stored in cases, it is recommended that endoscopes be hung vertically to promote a more thorough drying and minimize the likelihood of bacterial growth. Proper care of endoscopes will maximize equipment life, minimize repair costs, and contribute to performing consistent high-quality endoscopy.

As veterinary endoscopy use continues to increase, veterinary technicians will be required to have advanced knowledge of endoscopy equipment and procedures. Skilled technicians that are comfortable with endoscopy are assets to their patients, doctors, and practice.

References

1. Chamness CJ. Endoscopic instrumentation and documentation for flexible endoscopy. In: Tams TR, Rawlings CA, eds. *Small Animal Endoscopy.* 3rd ed. St. Louis, MO: Elsevier; 2011:3–26.
2. Cox S. Lower gastrointestinal tract endoscopy. In: *Endoscopy for the Veterinary Technician.* Ames, IA: Wiley; 2016:83–94.
3. Cox S. Upper gastrointestinal tract endoscopy. In: *Endoscopy for the Veterinary Technician.* Ames, IA: Wiley; 2016:63–82.
4. Fornan M. Endoscopy. In: *Small Animal Internal Medicine for Veterinary Technicians and Nurses.* Ames, IA: Wiley-Blackwell; 2012:487–495.
5. Leib MS. Colonoscopy. In: Tams TR, Rawlings CA, eds. *Small Animal Endoscopy.* 3rd ed. St. Louis, MO: Elsevier; 2011:217–244.
6. Tolbert MK. Gastrointestinal endoscopy. In: *Textbook of Veterinary Internal Medicine.* 8th ed. St. Louis, MO: Elsevier; 2017:437–440.
7. Sherding RG, Johnson SE. Esophagoscopy. In: Tams TR, Rawlings CA, eds. *Small Animal Endoscopy.* 3rd ed. St. Louis, MO: Elsevier; 2011:41–96.
8. Sum S, *et al.* Flexible endoscopy in small animals. *Vet Clin North Am Small Anim Pract.* 2009;39:881–902.
9. Tams TR, Spector, DJ. Endoscopic removal of gastrointestinal foreign bodies. In: Tams TR, Rawlings CA, eds. *Small Animal Endoscopy.* 3rd ed. St. Louis, MO: Elsevier; 2011:293–310.
10. Tams TR. Gastrointestinal endoscopy: instrumentation, handling technique, training, and implementation in practice. In: Tams TR, Rawlings CA, eds. *Small Animal Endoscopy.* 3rd ed. St. Louis, MO: Elsevier; 2011:27–40.
11. Tams TR. Gastroscopy. In: Tams TR, Rawlings CA, eds. *Small Animal Endoscopy.* 3rd ed. St. Louis, MO: Elsevier; 2011:97-172.
12. Walker V, *et al.* Care of endoscopic equipment. In: *Textbook of Veterinary Internal Medicine.* 8th ed. St. Louis, MO: Elsevier; 2017:332–335.
13. Walker V. Endoscope care and cleaning. In: *Endoscopy for the Veterinary Technician.* Ames, IA: Wiley; 2016:15–32.
14. Walker V. Endoscopy equipment. In: *Endoscopy for the Veterinary Technician.* Ames, IA: Wiley; 2016:1–14.

1.
acvim027

The Incidentalomas: Splenic and Adrenal Masses

Anne Corrigan[1], DVM, MS, DACVIM (SAIM); Sandra (Axiak) Bechtel[2], DVM, DACVIM (Oncology)
[1]Grenada, West Indies; [2]Columbia, MO, USA

INTRODUCTION

Advances in diagnostic imaging - ultrasound, CT scan, and MRI - have provided us with powerful diagnostic tools and therefore the ability to better evaluate our veterinary patients for various disease processes. However, we are also able to detect additional abnormalities, some of which are unanticipated and can dramatically change the therapeutic options and prognosis for our patients. When unexpected masses are found, particularly in the abdomen, the result is a diagnostic and therapeutic dilemma. What do we do with incidental masses - coined incidentalomas - in the clinical veterinary patient? How do we counsel clients and best guide them through the decision-making process with a minimum of stress? Today, we will cover the most common abdominal incidentalomas - splenic and adrenal masses, differential diagnoses, common diagnostic recommendations, and potential outcomes.

SPLENIC MASSES

Splenic masses may originate from blood vessels, lymphoid tissues, smooth muscle, or the connective tissue that makes up the fibrous stroma. The double two-thirds rule defines risk of malignancy in dogs - two-thirds of splenic tumors are malignant, and two-thirds of malignant splenic tumors are hemangiosarcoma. Other malignant splenic tumors are lymphoma, fibrosarcoma, histiocytic sarcoma, and other round cell or mesenchymal tumors. Benign tumors may also occur and hematomas and hemangiomas are a common type of benign splenic mass in dogs; lipoma, myelolipoma, and leiomyoma may also occur. Additional non-neoplastic splenic masses are abscesses and granulomas. Important to note is that extramedullary hematopoiesis may also present as splenic nodules on abdominal imaging. In cats, malignant tumors are more common than benign, and the two most common malignancies are mast cell tumor and lymphoma.[1]

Diagnosis of Splenic Masses

If appropriate, a fine needle aspirate and cytology may help define and diagnose a splenic mass. Many times, if a mass is cavitated and fluid filled, cytology will not be attempted due to risk of mass rupture and bleeding. Solid masses, nodules, and infiltrative disease of the spleen can be safely aspirated in many cases.[1]

Treatment

Therapy of a splenic mass is directed at the underlying cause. Splenectomy is the treatment of choice for a variety of solid tumors, including sarcomas and, in cats, mast cell tumor. If cytology was not considered safe, then surgery can be utilized as both a diagnostic and therapeutic tool with histopathology for definitive diagnosis. Lymphoma and canine mast cell tumor is best treated with chemotherapy in the dog and cat.[1]

ADRENAL MASSES

When an incidental adrenal mass is found, determining whether or not the mass is functional is an important diagnostic step. Blood pressure and fundic exam to investigate for a pheochromocytoma and endocrine testing to investigate a functional adrenal adenoma/carcinoma resulting in adrenal dependent hyperadrenocorticism should be considered. Staging for other tumor types is important as metastatic cancers to adrenal glands are common, especially in cats. In general, benign adenomas are more common than carcinomas and are non-invasive, and pheochromocytomas and carcinomas are often fixed and invasive. Functioning adrenal carcinomas secrete cortisol independent of pituitary control; secretion may intermittent and random, occasionally other hormones maybe produced. Similarly, pheochromocytoma secretion of epinephrine or norepinephrine is intermittent and random.[2]

Differentiation of Benign and Malignant

Adrenal adenomas tend to be small (less than 2 cm), well encapsulated or at least partially encapsulated, and well differentiated on histopathology. In contrast, carcinomas tend to be greater than 2 cm in diameter with peripheral fibrosis and invasion into the capsule. Adrenal carcinomas may also metastasize with the liver and lungs; other sites reported include kidney, ovary, mesenteric lymph node, peritoneal cavity, and thyroid gland. The metastatic rate reported ranges from 14–50%.[3,4] Pheochromocytomas are functional tumors of the adrenal medulla derived from chromaffin cells. Normal chromaffin cells are stimulated by sympathetic nerve fibers to secrete hormones into the blood stream - epinephrine, norepinephrine, and dopamine. Pheochromocytomas are not innervated, so catecholamine release is not initiated by neural impulses and their release is variable and unpredictable. Pheochromocytomas are uncommon in dogs (0.01–0.1% of all canine tumors) and rare in cats (only 3 case reports). Differentiation of origin of the mass may require histopathology; however, numerous pre-operative tests can be performed.[2-4]

Diagnosis of Adrenal Dependent Hyperadrenocorticism (ADH) - Functional Tumors of the Adrenal Cortex

Dogs tend to be older with a median age of 11 years with clinical signs consistent with hyperadrenocorticism. These signs include polyuria, polydipsia, polyphagia, weakness, lethargy, dermatologic changes, abdominal enlargement, and panting.[3,4]

Common diagnostics for ADH include screening tests, specific diagnostic tests for hyperadrenocorticism, and differentiating tests to distinguish ADH from pituitary dependent (PDH). Imaging is required for diagnosis as well as for preparing for treatment and to evaluate for caval invasion or metastasis. One screening test is the urine cortisol:creatinine ratio. The diagnosis of hyperadrenocorticism can be achieved with ACTH stimulation tests or low-dose dexamethasone suppression test (LDDST), however, both of these tests can have both false positive and false negative results. Differentiation tests include endogenous ACTH levels (which can be difficult to interpret) and advanced imaging; abdominal ultrasound, or abdominal CT or MRI. The LDDST can often act as a differentiating test but only for identifying PDH, you must use additional imaging to confirm ADH.[5]

Diagnosis of Pheochromocytoma - Functional Tumors of the Adrenal Medulla

Physical examination may be normal, or findings non-specific: tachypnea, panting, tachycardia, weakness, pallor, arrhythmias, and hypertension may be noted. No consistent findings are reported for complete blood count, chemistry panel, or urinalysis. Hypertension (greater than 160 mm Hg) is documented in 40–50% of dogs with pheochromocytoma. Note that a normal blood pressure does not rule out a pheochromocytoma, as catecholamine release is episodic in nature. Ocular examination may reveal mydriasis, retinal hemorrhage, or, less commonly, retinal detachment. Neurologic examination may reveal spinal pain, rear limb ataxia, or head tilt.[6,7] Measurement of plasma or urine catecholamine metabolites are superior to measurement of the catecholamine in human medicine due to the short half-life of catecholamines in circulation and the sporadic and unpredictable release of catecholamines from the pheochromocytoma. Membrane bound catechol-O-methyltransferase (COMT) on the tumor cells is responsible for the metabolism of tumor released catecholamines - since catecholamines are only intermittently released by the tumor, the continuous release of the metabolites metanephrine and/or normetanephrine leads to more consistent test results. In fact, one study found plasma metanephrine to have moderate sensitivity with excellent sensitivity, and normetanephrine to have excellent sensitivity and specificity, in differentiating dogs with pheochromocytoma from healthy dogs, dogs with non-adrenal disease, and dogs with adrenal cortical tumors. Urine normetanephrine:creatinine ratio may also be helpful in the diagnosis of pheochromocytoma. False positive results for both serum and urine metanephrine and normetanephrine measurements can be obtained if the dog is on phenoxybenzamine.[8,9] Definitive diagnosis requires histopathology and immunohistochemistry.

Treatment for Adrenal Carcinomas

Adrenalectomy should be considered for functional tumors, locally invasive tumors, or those larger than 2.5 cm. If the mass is determined to be nonfunctional and is less than 2 cm with no invasion, repeating an abdominal ultrasound regularly is recommended. With adrenal carcinomas a median survival time of

230–778 days is reported with surgery. Potential complications of surgery are cardiovascular events and hemorrhage; dogs are also at risk for adrenocortical insufficiency, pulmonary thromboembolism, pancreatitis, renal failure, and wound dehiscence. Perioperative mortality rates range from 9–60%. Post-operatively, ACTH stimulation testing within 24 to 48 hours should be considered, in addition to monitoring of glucose and electrolytes. Heparin and glucocorticoids may be indicated in addition to mineralocorticoid supplementation in some dogs.[2] Medical therapy for functional adrenal tumors include mitotane, although much higher doses are required than when used for PDH and relapses are common (up to 60% of dogs) and untoward side effects are also common in >50% of dogs, not including dogs who develop mineralocorticoid deficiencies. The median survival time reported is 16.4 months. The median survival time in dogs with adrenocortical carcinomas treated with surgery alone is 778 days; in dogs that survive the peri-and post-operative period and are discharged from the hospital, the median survival time is reported as 992 days in one study and 17.5 months in another.

Treatment for Pheochromocytomas

Adrenalectomy is the treatment of choice for pheochromocytoma in both the dog and cat. Prior to consideration of treatment for a suspected pheochromocytoma, phenoxybenzamine, a non-selective alpha-1 and -2 receptor blocker, is recommended to decrease blood pressure, decrease ventricular arrhythmias, and support blood volume expansion. A study in dogs found that pre-treatment with phenoxybenzamine reduced the overall mortality rate; treated dogs were six times more likely to survive than dogs that did not receive phenoxybenzamine as a pre-treatment (mortality rate 13% versus 48% in untreated dogs). Potential complications of adrenalectomy include hypertension, hypotension, cardiac arrhythmias, and hemorrhage.[4,5,10]

Overall prognosis following surgery is a median survival time of 1 year, with some dogs surviving 2 and 3 years postoperatively. Actual metastatic rates are difficult to determine as, histologically, the appearance of pheochromocytomas are variable and not predictive of malignancy. Dogs with tumors larger than 5 cm, evidence of metastasis, or thrombosis have a worse prognosis than dogs without these factors. Adjuvant chemotherapy has not been studied in dogs and is not thought to be helpful in people; radiation therapy has also not been studied.[4-6] Medical management is indicated for dogs in which surgical resection is not a viable treatment option. Phenoxybenzamine is recommended long term for medical management of catecholamine release. Beta blockers, such as propranolol or atenolol, may be considered for severe tachycardia to prevent unopposed alpha stimulation and uncontrolled hypertension.[2]

REFERENCES

1. de Morais A, *et al.* In: Ettinger SJ, Feldman EC, eds. *Textbook of Veterinary Internal Medicine.* 7th ed. 2010;810.
2. Reusch CE, Feldman EC, *et al.* eds. *Canine & Feline Endocrinology.* 4th ed. 2015;5213.
3. Labelle P, *et al. Vet Pathol.* 2004;41:490.
4. Anderson CR, *et al. J Am Anim Hosp Assoc.* 2001;37:93–97.
5. Melian, *et al.* In: Ettinger SJ, Feldman EC, eds. *Textbook of Veterinary Internal Medicine.* 7th ed. 2010;1816–1840.
6. Barthez PY, *et al. J Vet Intern Med.* 1997;11:272–279.
7. Gilson SD, *et al. J Vet Intern Med.* 1994;8:228–232.
8. Gostelow R, *et al. J Vet Intern Med.* 2013;27:83–90.
9. Quante S, *et al. J Vet Intern Med.* 2010;24:1093–1097.
10. Herrera MA, *et al. J Vet Intern Med.* 2008;22:1333–1339.

acvim028

Prevention, Management and Treatment of Chemotherapy-Induced Gastrointestinal Toxicity in Veterinary Medicine

Emma Warry, BVSc, MS, DACVIM (Oncology)
Columbus, OH, USA

Introduction

Gastrointestinal toxicity is the most commonly reported side effect following chemotherapy administration in companion animals. It may manifest as vomiting, nausea, reduced appetite, anorexia, diarrhea, and rarely constipation. The severity of these side effects varies greatly among patients, and can vary in an individual from one treatment to the next. For most patients, gastrointestinal toxicity is mild to moderate and can be treated and managed on an outpatient basis. However, for some individuals toxicity can be severe and require hospitalization and aggressive supportive care. Rarely death can occur due to complications from gastrointestinal toxicity.

The development of chemotherapy-induced gastrointestinal toxicity impacts quality of life for the patient, and for many clients it becomes emotionally distressing. In addition, treatment and management of side effects, particularly if hospitalization is required, can become an unforeseen financial burden for the client and may impact the ability to continue with additional chemotherapy treatment. For many patients who develop gastrointestinal toxicity, dose delays and/or dose reductions for subsequent treatments may be indicated. Changes to the treatment protocol that impact dose intensity may negatively influence disease-free interval and median survival time.

In human cancer patients being treated with chemotherapy, multiple risk factors have been identified for the development of gastrointestinal toxicity. These include: the emetic potential of the chemotherapy drug being administered, female gender, less than 50 years of age, anxiety, nausea prior to treatment, fatigue, chemotherapy administration on an outpatient basis, low alcohol intake, and a history of motion sickness or vomiting during pregnancy; whereas inpatients and patients with severe alcohol consumption have a decreased risk for the development of gastrointestinal toxicity. Unfortunately, risk factors in veterinary patients are yet to be clearly established. One exception to this is the identification of the MDR-1 mutation in heading breeds such as the Rough Collie, Australian shepherd, and the Shetland sheepdog. Dogs with the mutation have an increased sensitivity to chemotherapy drugs that are considered MDR-1 substrates, and thus more likely to develop severe toxicity.

In order to reduce the negative impact associated with the development of gastrointestinal toxicity following chemotherapy treatment, it is imperative that a focus be placed on prevention and prophylaxis rather than simply waiting for patients to develop toxicity before intervening. By prophylactically treating for gastrointestinal toxicity the success and tolerability of treatment are improved.

Classification and Pathophysiology of Chemotherapy-Induced Nausea and Vomiting

Chemotherapy-induced nausea and vomiting (CINV) encompasses three distinct symptoms, vomiting, nausea and retching, and while they frequently occur simultaneously they are independent consequences of chemotherapy administration. CINV can be further defined as either acute, delayed or anticipatory. Acute CINV occurs within 24 hours of chemotherapy administration, with peak intensity most commonly reported 5 to 6 hours after treatment. Delayed CINV occurs 24 hours post-chemotherapy administration, with maximal intensity occurring 48 to 72 hours after treatment and can persist for up to 5–7 days. This is the most commonly encountered and most debilitating form of CINV encountered in veterinary medicine. Anticipatory CINV occurs prior to chemotherapy administration and is an expected or conditioned response. This form of CINV is rarely appreciated in veterinary medicine. It is now understood that acute and delayed CINV have different pathophysiology.

The single most important risk factor for the development of CINV is the emetic potential of the drug or protocol that is prescribed. The classification of emetic potential in humans is based on the percentage of patients that can be expected to develop CINV with treatment, and this information is then used to determine the most appropriate antiemetic medications to be used.

Table 1. Classification of emetic potential and risk for CINV

Level	Emetic potential	Acute CINV	Delayed CINV	Example
1	Minimal (0–10%)	-	-	Vincristine, vinblastine, bleomycin
2	Low (10–30%)	+	-	Mitoxantrone, gemcitabine, paclitaxel
3	Moderate (>30–90%)	++	+	Carboplatin, doxorubicin, ifosfamide
4	High (>90%)	++	++	Cisplatin, dacarbazine, streptozocin

Acute Chemotherapy-Induced Nausea and Vomiting

Following the administration of chemotherapy, free radicals are generated within the gastrointestinal tract, causing excessive release of serotonin from the enterochromaffin cells (EC). The released serotonin binds to the 5-HT3 receptors, located at the afferent vagal terminals, conducting information to the brain. This results in stimulation of the chemotherapy trigger zone (CTZ) and vomiting center (VC), thus triggering the vomiting reflex.

Delayed Chemotherapy-Induced Nausea and Vomiting

Substance P (SP) is a neurotransmitter distributed throughout the peripheral and central nervous systems and binds to neurokinin-1 (NK-1). SP release is also mediated by chemotherapy administration; however, as opposed to serotonin, it binds mostly to centrally located NK-1 receptors, passing the signal to the CTZ and ultimately the VC inducing delayed CINV.

Treatment Options and Strategies for Chemotherapy-Induced Nausea and Vomiting

Standard guidelines and recommendations exist in human oncology for the prophylactic management and treatment of CINV. These guidelines are primarily based on the emetic potential level that is assigned to a particular drug or protocol. When these guidelines are adhered to by doctors and patients, the incidence and severity of CINV are significantly reduced, with patients reporting an improved quality of life, and are more likely to be able to continue with routine daily activities. Similar standardized guidelines do not currently exist in veterinary medicine; however, we can use the human experience to help us better manage CINV in our patients.

Table 2. Guidelines for the prophylactic management of acute CINV in humans

Emetic potential (level)	Recommended antiemetic drugs
1	No routine prophylaxis
2	5-HT3 antagonist or DEX or DOP antagonist
3 (except carboplatin)	5-HT3 antagonist + DEX
3 (carboplatin)	5-HT3 antagonist + NK-1 antagonist + DEX
4	5-HT3 antagonist + NK-1 antagonist + DEX

Table 3. Guidelines for the prophylactic management of delayed CINV in humans

Emetic potential (level)	Recommended antiemetic drugs
1	No routine prophylaxis
2	No routine prophylaxis
3	NK-1 antagonist or 5-HT3 antagonist + DEX
4	NK-1 antagonist + DEX +/- 5-HT3 antagonist

5-HT3 Receptor Antagonists

This class of antiemetic drugs function to suppress vomiting and nausea by preventing binding of serotonin to the 5-HT3 receptors located within the gastrointestinal tract (afferent vagal nerve), CTZ and the VC. 5-HT3 receptor antagonists are used frequently in human oncology for the management and treatment of CINV, particularly in combination with other antiemetic medications for prevention when a moderately to highly emetic chemotherapy protocol is being used. They are particularly important for the management of the acute phase due to their activity at the level of the gastrointestinal tract. In veterinary medicine, the most commonly prescribed 5-HT3 receptor antagonist is ondansetron, which can be administered subcutaneously, intravenously, or orally every 8 to 12 hours. I routinely administer injectable ondansetron at the time of chemotherapy administration if a moderately to highly emetic drug is being administered, such as carboplatin or doxorubicin.

NK-1 Receptor Antagonists

This class of drugs prevent binding of SP to the NK-1 receptor, blocking the preventing activation of the VC and thus the vomiting reflex. These drugs have been shown to reduce the severity of both acute and delayed CINV. In addition there is also evidence to suggest that they increase the activity of other antiemetic medications such as 5-HT3 and dopamine antagonists. In veterinary medicine, maropitant is the NK-1 receptor antagonist that is used. This drug can be administered intravenously, subcutaneously or orally. A clinical trial performed by Rau *et al.* (2010) demonstrated that when maropitant was administered intravenously to dogs at the time of doxorubicin administration, and then orally for four consecutive days, fewer episodes of vomiting and diarrhea and reduced severity were reported. This supports the use of prophylactic administration of maropitant to patients being treated with doxorubicin. I will also administer the same maropitant protocol to patients receiving carboplatin, those that have experienced gastrointestinal toxicity following previous chemotherapy treatments, or if the development of any gastrointestinal side effect will negatively affect the clients' willingness to continue with treatment.

Dopamine Receptor Antagonists

This class of drugs has many different indications, one of which is use as an antiemetic. They function by blocking the dopamine receptors at the CTZ. In addition, they are also able to improve gastric emptying, increase lower esophageal pressure, and improve intestinal motility. Metoclopramide is the most commonly used drug from this class. It is generally used only for patients who are receiving chemotherapy with low emetic potential, or in patients with breakthrough nausea and vomiting. In veterinary medicine, metoclopramide is used mainly in the breakthrough setting and in patients with severe or refractory ileus.

Alternative Treatment Strategies

Several alternative strategies are described for control and treatment of CINV. These include behavioral therapies, such as hypnosis and systematic desensitization, and acupuncture. These strategies are anecdotally reported to be highly successful by some clinicians; however, phase 3 clinical trials are yet to be performed or do not demonstrate benefit over current standard of care. The efficacy of these strategies is yet to be investigated in veterinary patients.

Pathophysiology and Management of Chemotherapy-Induced Diarrhea

Chemotherapy-induced diarrhea (CID) is a frequently encountered complication of chemotherapy treatment, and in human oncology is reported in 50 to 80% of patients. Unlike CINV, risk factors have not been clearly identified. In addition to chemotherapy, diarrhea associated with the administration of tyrosine kinase inhibitors is also frequently reported.

The underlying mechanism for CID is unclear; however, it is likely a combination of inflammation within the GI tract, secretory dysfunction, GI dysmotility, and alterations in GI innervation. Mucositis manifests as a combination of inflammation and ulceration resulting in altered gastrointestinal microflora and secretions. The sequential phases of mucositis development include: (i) Initiation which occurs via the direct or indirect effects of chemotherapy on rapidly dividing epithelial cells within the gastrointestinal tract; (ii) upregulation of proinflammatory cytokines; (iii) signaling and amplification; (iv) inflammation and ulceration within the mucosal epithelium; (v) the healing phase occurs when proliferation and differentiation of the gastrointestinal epithelium returns; however, functional changes may remain.

Prophylactic and treatment strategies for CID are less well described than CINV, with many conflicting reports and studies. Therapies are considered to be either nonpharmacologic or pharmacologic. Nonpharmacologic treatments described in human oncology include glutamine, activated charcoal, and probiotics. Pharmacologic treatments include prophylactic use of antibiotics such as neomycin, agents to reduce inflammation such as budesonide and NSAIDs, opioids such as loperamide or octreotide.

In veterinary oncology, the most commonly used treatments for the management of CID include administration of metronidazole, loperamide, probiotics, dietary modification, and RX Clay. For patients with severe diarrhea that results in perianal scolding, topical therapies and analgesia, as well as appropriate husbandry are required.

CONCLUSION

Appropriate management of CINV and CID in veterinary oncology is a very important aspect of cancer care. Appropriate prophylactic management can help reduce the occurrence and severity of side effects, maintaining patient quality of life and client willingness to proceed with treatment. In addition, successful management of CINV and CID may allow for dose intensification and potentially improved disease control and outcome.

REFERENCES

1. Hawkins R, *et al. Clin J Oncol Nurse*. 2009;13(1):54.
2. Janelsins MC, *et al. Expert Opin Pharmacother*. 2013;14(6):757.
3. McQuade RM, *et al. Front Pharmacol*. 2016;7:414.
4. Navari RM, *et al. N Engl J Med*. 2016;374(14):1356.
5. Rau SE, *et al. J Vet Intern Med*. 2010;24(6):1452.
6. Stein A, *et al. Ther Adv Med Oncol*. 2010;2(1):51.
7. Tageja N, *et al. Postgrad Med J*. 2016;92:34.

acvim029

Bronchoscopy for the Endoscopy Technician

Susan Cox, RVT, VTS (SAIM)
Davis, CA, USA

As the veterinary profession grows and becomes more specialized, many veterinary specialists are being trained in advanced procedures such as bronchoscopy and integrating this specialized procedure into their treatment plans. We will be examining airway evaluation techniques from the technician's point of view, which includes comprehensive patient care, endoscope and instrument selection, and diagnostic sampling techniques.

Developing a team approach to the bronchoscopic examination is highly recommended. These patients are often dyspneic and stressed, so it's imperative for the procedure to be efficient to minimize anesthetic time. If possible, a technician should be dedicated to anesthesia and another assist with the bronchoscopy. Depending on the patient and differentials, individuals from the surgery team may also need to be available which should be scheduled ahead of the procedure. Emergency personnel should also be notified. Distractions such as phone calls, pagers or irrelevant conversation should be discouraged so the team may focus on the procedure. All instrumentation should be tableside, and the bronchoscope "plugged in" and ready to go.

Respiratory Examination

A complete airway examination at the UC Davis Veterinary Medical Teaching Hospital includes a thorough examination of the larynx as well as the lungs (bronchoscopy). Nasopharyngoscopy may also be performed if indicated.

Laryngoscopy is the examination of the laryngeal area and is most often used to assess upper airway structures as well as laryngeal function. Laryngoscopy is most commonly performed at anesthetic induction prior to endotracheal intubation. Indications include respiratory distress, exercise intolerance, voice change and increased respiratory effort. The laryngeal exam is often accomplished at anesthetic induction - drugs are titrated until the mouth can be opened safely. The tongue is then pulled forward so the laryngeal area can be observed. If evaluating arytenoid function, an additional assistant may be needed to inform the endoscopist when the patient is inspiring so that respiratory phase is correlated with arytenoid movement. Doxapram administration is recommended to stimulate respiration and maximize the diagnostic utility of this exam. Doxapram is administered at 0.5–1mg/kg IV. Instrumentation needed include a tongue depressor, cotton-tipped swabs, and a laryngoscope. Rigid telescopes can also be used with a camera attached for documentation or teaching purposes. If a foreign body or mass is suspected, biopsy or alligator forceps should be available. A thorough oral exam should also be performed after laryngoscopy - the tonsils should be observed if in or out of tonsillar crypts, and the crypts probed for foreign bodies. The underside of the tongue should also be probed for foreign objects. The soft and hard palate should be palpated for masses, and a cursory look at the teeth should be performed.

Bronchoscopy is a systematic endoscopic examination of the trachea and bronchial tree. Bronchoscopy is commonly performed to evaluate the airways, retrieve diagnostic samples and perform therapeutic interventions (tracheal stent placement). Bronchoscopy helps with lesion location and identification. As discussed for patients undergoing laryngoscopy, these patients are often unstable, so bronchoscopy should be performed as efficiently as possible to limit anesthesia time. These patients are often preoxygenated prior to anesthesia. Indications include workup of acute and chronic cough, abnormal breathing pattern, tracheal collapse, airway neoplasia, foreign body, and airway obstruction. A bronchoscopy may also be performed to aid in selective bronchial intubation and to help surgeons with lesion localization in the operating room.

INSTRUMENTATION

Unlike gastroscopes that possess 4-way flexion, bronchoscopes are limited to two-way deflection for easier maneuverability within the lungs. For dogs over 25 kg, an endoscope that has a working length of 90 cm with a 2-mm biopsy channel is suitable. For smaller canines, an endoscope of 5–5.5 mm outer diameter and 55 cm working length should be adequate. For felines and dog breeds such as the Papillon and Chihuahua, a flexible cystoscope with an outer diameter of 2.9 mm and working length of 55 cm will be needed (instrumentation may be unavailable for the small 1.2-mm instrument channel). These endoscopes have a biopsy channel for bronchoalveolar lavage (BAL), foreign body retrieval, cytology brushings, and biopsy forceps. When purchasing a bronchoscope, make sure that the instrument channel port will accommodate a syringe hub snugly - this creates a better "seal" for BAL sampling. Foreign bodies should be removed when first seen with biopsy forceps, and BALs performed before biopsies and brushings to avoid contamination. Pulmonary biopsy forceps have a spike or needle within the biopsy cups for better purchase on tough bronchial bifurcations, and should be used carefully to avoid complications. Bronchial brushes are also used to obtain cytology samples, and their use will be further detailed.

Having an adequate oxygen delivery system during bronchoscopy can be challenging. Customized short endotracheal tubes can be used for canines that can accommodate a 7.5 Fr ET tube or larger (ET tubes smaller than 7.5 Fr may damage the endoscope). The shorter length helps the endoscopist visualize more of the trachea than with conventional endotracheal tubes. Adapters attach to the ET tube and allow the endoscope to pass safely through the tube. Ports for attaching inhalation anesthesia/O_2 delivery and scavenge attach to ports on the sides of the adapter. For smaller patients, jet ventilators can be utilized to pulse oxygen through a large-bore (14–16 g) catheter placed in the trachea, with the endoscope carefully placed alongside. The downside with jet ventilators is that they can be expensive. In patients that are extubated for the procedure, adapting the biopsy channel to administer oxygen is recommended. A constant rate of infusion of propofol will ensure an adequate anesthetic plane in extubated patients.

A typical tableside bronchoscopy setup includes warmed saline, saline-soaked gauze squares, non-bacteriostatic lube packets, mouth speculae, 20-ml syringes for BAL samples, and the endoscopy report for recording observations during the procedure.

PROCEDURE

Patients should be placed in sternal recumbency with a rolled towel or bolster under the chin and 2 mouth speculae in place. Mechanical suction should be avoided as it may damage delicate airways and instrument channels. The insertion tip is carefully placed through the arytenoids and into the trachea, or through the membrane of the adapter and out the endotracheal tube. Submucosal vessels and a rounded appearance are normal for the trachea. Next, the carina divides the right and left bronchial tree. Each bronchus should be examined and observed for color change, surface texture, contents, size, and patency. If BAL sampling is indicated, sites should be identified at this time. To prevent contamination of BAL samples, remove the bronchoscope, flush the instrument channel with saline, and wipe the insertion tube with saline-soaked gauze sponges. Also clean the endotracheal tube adapter with saline-soaked cotton-tipped applicators.

Bronchoalveolar lavage is the most common airway sampling technique. This can be done using a syringe or BAL catheter. BAL volumes vary for cats (3–5 ml per aliquot), small canines <6 kg (5–10 ml per aliquot) and large canines (15–20 ml per aliquot). For the procedure, saline is instilled into the lower airways and retrieved for culture and cytology. The endoscope is wedged into the bronchus, and saline is slowly flushed down the biopsy channel in a 20-ml syringe or BAL catheter. Follow with 3–4 ml of air to clear the channel. Use short pulls with the syringe or BAL catheter to retrieve the sample as the endoscopist moves the tip slightly up and down and back and forth within the airway. Negative pressure may occur within the syringe if the bronchoscope's tip is against mucosa. Inform the endoscopist when this occurs to adjust the tip of the bronchoscope. A diagnostic (deep) sample has a foamy top, termed surfactant, which means that the sample has had contact with the alveoli. A sample that appears bloody or has excessive mucopurulent material should be noted. A retrieval of about 60% of the instilled amount is normal. Each sample should be submitted for cytological examination, and multiple samples may be combined for culture. Saline may also be used to flush mucopurulent discharge away from a bronchus for better visualization.

Guarded bronchial brushes are used after BALs. Once the brush end is out of the endoscope and deployed at the desired site, the brush and endoscope move as a unit to obtain the sample. The brush end is then brought back into the guard and removed from the instrument channel. The brush may be rubbed onto a slide and submitted for cytology, or cut into a RTT or repackaged and submitted for culture.

When the procedure is complete, have suction and an anesthesia mask on hand for compromised patients. The endoscopist may choose to intubate smaller patients, although intubation may cause further minor tracheal trauma. The endoscope should be cleaned and processed as soon as the patient is stable.

References

1. Cox S. *Endoscopy for the Veterinary Technician.* 2016:107–119.
2. Johnson LR. *Clinical Canine and Feline Respiratory Medicine.* 2010:27–37.

1.
acvim030

License to Pill - What You Need to Know to Keep Your Job and Your Cancer Patients Safe

Carolyn J. Henry, DVM, MS, DACVIM (Oncology)
Columbia, MO, USA

Introduction

The field of veterinary oncology is reliant upon administration of established anticancer drugs that are not approved for use in animals, the use of conditionally approved drugs that are still undergoing evaluation, and the participation in clinical trials that include new agents not yet fully evaluated in pets with cancer. For the safety of our patients and clients and the legal and ethical practice of veterinary medicine, it is vitally important to have a working knowledge of the legal and safety issues associated with dispensing and administering anticancer drugs. This session will provide specific examples from veterinary oncology practice and test participants' knowledge of these issues and the ramifications of not appropriately addressing them.

Dispensing Medications

When dispensing drugs from a veterinary practice, appropriate containers and labeling are necessary. Childproof containers are recommended and in some states, are required. Labels must contain complete information as outlined by the AVMA guidelines:

- **The actual container must bear the veterinarian's name, address, name of the drug (active ingredient), identification of the animal(s) to be treated, adequate directions for proper use, and cautions/precautions including milk and meat withdrawal times. This information may be on the label applied by the manufacturer, or on a label attached to the product by the veterinarian.[1]**

Whereas not all of the above guidelines are relevant to small animal oncology, others are absolutely essential. Proper identification of the drug's active ingredients should be listed on the label and must be distinct from the trade name of the product to avoid confusion. A further consideration with anticancer drugs is the potential for harm to the client who will be administering the medication. Proper label warnings and additional informational handouts should be standardized and dispensed at each visit. This is especially important for drugs that may result in side effects after chronic use. For example, sterile hemorrhagic cystitis may result after administration of cyclophosphamide, but this is often after months of use (median time to onset = 92 days in a case-control study).[2] If warnings are provided only at the initiation of the treatment protocol, owners may forget this potential side effect and be less vigilant about monitoring for signs of cystitis. Clients must also be counseled about the proper storage of any dispensed medications. Chemotherapy drugs dispensed for use in pets should be stored separate of human drug prescriptions to avoid inadvertent use of the wrong product. When drugs such as chlorambucil, which requires refrigeration and protection from light, are dispensed, bottles should be amber, labeled for refrigeration, and ideally placed in a bag to further protect from light.

The prior two examples include chemotherapy drugs (cyclophosphamide and chlorambucil) approved for use in people for a specific indication. But what about use in an extra-label manner in veterinary patients? What constitutes extra-label use and is it legal? The guiding principles for these questions come from the 1994 Congressional Animal Medicinal Drug Use Clarification Act (AMDUCA).[3]

The act states:

- **Extra-label use is legal only when ordered by a veterinarian and within the context of FDA's VCPR (Veterinary Client-Patient Relationship) found within 21 CFR 530.3 and state VCPR requirements.[4]**
- **Veterinarians may utilize drugs in an extra-label manner in their regular course of practice when the health of an animal is threatened or death may result from failure to treat. Under AMDUCA regulations, extra-label use means the actual or intended use of a drug, by or on the order of a**

veterinarian, in a manner that is not in accordance with approved labeling. Any deviation from the label, by veterinarians or lay persons is an illegal use, unless the use meets all the requirements of FDA's extra-label drug use rules. Deviations from the label include, but are not limited to:

- **Use in a species or production class not on the label.**
- **Use of a different route of administration, indication, frequency, dose, or duration.**

Because we do not currently have a veterinary-labeled form of most anticancer drugs, extra-label use is the norm in veterinary oncology. However, there is renewed interest in seeking FDA approval for chemotherapy agents in veterinary medicine. As approvals occur, the conditions under which substitution of an approved human drug can be used in animals must be understood. The situations where a human drug may be used even when a veterinary-approved drug exists include:

- The animal drug does not contain the active ingredient needed.
- The approved animal drug is not in the required dosage form (liquid vs. tablet, for example).
- The approved drug is not in the required concentration.
- In the context of a valid veterinarian-client-patient relationship, the veterinarian has found that the approved drug is not clinically effective when used as labeled.

FDA vs. USDA Approval

When using the term, "approved" animal drug, one may be referring to either USDA or FDA approval. According the DHHS FDA website, the USDA approves "veterinary biologicals which can include: viruses, serums, toxins, or analogous products of natural or synthetic origin which are intended for use in the treatment of animals and which act primarily through the direct stimulation, supplementation, enhancement, or modulation of the immune system or immune response. Such products include, but are not limited to vaccines, bacterins, allergens, antibodies, diagnostics, antitoxins, toxoids, immunostimulants, antigenic, or immunizing components of microorganisms intended for use in the prevention, diagnosis, management, or cure of disease in animals." www.fda.gov/AboutFDA/Transparency/Basics/ucm193868.htm. The distinction has been most notable in veterinary oncology, where tumor vaccines and immune modulatory compounds can blur the lines between the two agencies. In the case of tumor vaccines, depending on the source and content of the vaccine, approval could either be through the USDA (as with Oncept Canine Melanoma vaccine) or the FDA (as with newer human tumor vaccines), or may not require approval from either agency (autologous vaccines in veterinary medicine).

The first FDA-approved anticancer drug in veterinary medicine was Palladia, first approved in 2009. This has opened the door for subsequent FDA approval of new anticancer drugs, with the norm being conditional approval, followed by full licensure after additional clinical evaluation is completed. In some cases (i.e., Kinavet-CA1 and Paccal-Vet-CA1) conditional approval may be withdrawn if the requirements for full licensure are not met. Drugs that are under conditional approval will have their name followed by "CA1" to indicate "conditional approval". For example, Tanovea-CA1 is currently conditionally approved for treatment of canine lymphoma.

Drugs that are under conditional approval can only be used for the specific label indication for which the conditional licensure was approved. Alternative use is not permitted until the drug is fully licensed unless it is in the context of an investigational study. For this reason, it is imperative to read and understand the parameters for any conditionally-approved drug and ensure that your practice is not suggesting or promoting the illegal off-label use of that drug. Once a drug is given full licensure, off-label use for other indications is permitted.

Compounded Drugs

The use of compounded drugs in veterinary medicine is especially popular for oncology drugs, given the paucity of antineoplastic agents approved for animal use and in sizes and formulations suitable for administration to pets. Compounding may entail combining two or more approved drugs, enhancing flavor for the species of interest, or producing a formulation or size that is more suitable for a given patient than what is commercially available. Over 6 million prescriptions are filled by compounding pharmacies for veterinarians in the USA each year.[5] While the necessity for these drugs has driven the demand for more compounded products, the quality and safety of compounded products is often questioned. A recent study in the state of Missouri revealed that roughly 20% of compounded prescription products do not contain the amount of active drug indicated on the label.[6] A study published this year evaluated the potency and stability of compounded chlorambucil, cyclophosphamide, and melphalan and demonstrated considerable variability in both drug potency and stability for these oral medications.[7] How is this possible, given the strict FDA guidelines that govern activities in veterinary medicine? Actually, pharmacies in the USA are regulated by state boards of pharmacy and the FDA has limited jurisdiction over compounding pharmacies unless their activity falls under the category of manufacturing. This means that regulation of veterinary compounding varies widely from state to state. Furthermore, pharmacies can register with more than one state board of pharmacy and participate in interstate commerce involving veterinary compounded products.

Because compounding pharmacies are not regulated by the FDA, practitioners and technicians need to provide clients with the appropriate information to make an informed decision about the risks of using compounded drugs. In general, the use of compounded drugs is recommended only when a label-approved product is not available and the active ingredient is necessary for humane treatment of the pet. The use of a compounded drug simply as a cheaper alternative to an approved product is inappropriate, with ramifications varying by state. With the recent dramatic price increase for chlorambucil, some veterinarians are turning to compounding pharmacies to purchase a more economically feasible formulation. Compounding pharmacies may be unable to sell capsules of the same strength as the commercially-available FDA-approved human tablet (2 mg), but can market capsules of alternative strength (i.e., 1.9 mg). Again, clients should be informed of the differences between FDA-approved and compounded products and practitioners must make product safety and efficacy a priority.

Conclusions

The practice of veterinary oncology presents unique challenges in terms of providing the highest quality of care within the limitations of client financial resources, drug tolerability, and convenience, and a paucity of FDA - and USDA - approved drugs for the veterinary oncology market. An understanding of the regulatory terminology, legal ramifications of clinical decisions, and the anticipated health risks vs. potential benefits of various human and veterinary anticancer drugs is essential for the reasoned, safe, and ethical practice of veterinary oncology. As the liaison between the veterinarian and client, veterinary oncology nurses and technicians should be knowledgeable of best practices and remain current in their training as new drugs become available.

References

1. www.avma.org/KB/Policies/Pages/Guidelines-for-Veterinary-Prescription-Drugs.aspx.
2. Gaeta R, *et al. Vet Comp Oncol.* 2014;12(4):277.
3. www.fda.gov/AnimalVeterinary/GuidanceComplianceEnforcement/ActsRulesRegulations/ucm085377.htm.
4. www.fda.gov/AnimalVeterinary/ResourcesforYou/ucm380135.htm#Valid_Veterinarian-Client-Patient_Relationship.
5. Davidson G. *Pharmaceutics.* 2017:5.
6. Missouri Board of Pharmacy, Annual Reports from 2006–2014. http://pr.mo.gov/pharmacists-annual-reports.asp. (Accessed on 5 January 2017).
7. Burton JH, *et al. Vet Comp Oncol.* 2017:1.

acvim032

The Long Goodbye: Hospice and Palliative Care

Liz Hughston, Med, RVT, CVT, VTS (SAIM, ECC)
San Jose, CA, USA

Introduction

Terminal diagnoses may be made rapidly and suddenly in the veterinary practice, stunning owners who have to make difficult decisions in a short amount of time. Or patients can see the veterinary team for their entire life, and begin a slow decline into the end of life. Offering palliative and hospice care in your practice can help ease the journey to the end of a pet's life for both your patients and their owners. These proceedings will discuss setting up a hospice program, pain management options, and at-home care that can be provided as owners help their pets at the end of their lives.

There are three common life stages: puppy/kittenhood, adulthood, and senior. As pets near the end of their life, we can consider this time a fourth stage: the end-of-life stage. For some pets, this stage comes at the end of a long life. For others, it can be due to a sudden illness or terminal diagnosis. In either case, both clients and patients need support through what can be an emotionally difficult time and the veterinary healthcare team plays a primary role in this support. Recognizing the importance and strength of the human-animal bond is incumbent on the veterinary practitioner and providing hospice and palliative care can demonstrate that recognition.

Benefits

A hospice program can benefit the practice in many ways. First, there are direct income benefits to the practice in the form of extended consultation visits, charges for telephone consultations, medication dispensed, and house-call charges. There are also indirect income benefits from a hospice program including referrals, and creating a client that is very bonded to the practice. The benefits of a hospice program also extend to the staff as it gives technicians and veterinarians the opportunity to interact with clients and their pets on a deep, emotional level that many in our profession find very rewarding. A complete hospice program would also include grief services for clients that the practice can open to include staff participation. Supporting staff in their grieving - whether for their own pets, or for client pets they have bonded with - can lead to less burn-out and compassion fatigue, and higher levels of employee satisfaction.

Getting Started

As with any new offering, it is ideal to start hospice with a small group of clients to enable your practice to easily measure the success of the program, and to assess the areas that need improvement. The biggest part of any hospice program is customer service and this will be the largest area of time investment. Clients need lots of coaching and counseling, in addition to learning how to administer medications and monitor their pet's condition and quality of life. Regular - weekly or even daily - consulting on the pet's quality of life and adjusting the hospice plan can also be very time consuming. It is important for the practice owner or manager to realize that starting this program will mean a dedicated staff: another good reason to start small at first. Staff involved with the hospice program can come from any area of your practice but should include at least one registered/licensed/credentialed veterinary technician and a veterinarian. It makes sense to have a technician manage the program and consult with the veterinarian regarding treatment plans and prescriptions. This leaves the more time-consuming aspects in the hands of the veterinary technician, freeing the veterinarian to see other patients.

It is important to consult your state's Veterinary Practice Act to clarify what veterinarians and technicians may do in a client's home. Some states have very strict regulations about the prescription or transportation of controlled drugs as well. Keeping very detailed records is very important, with special attention paid to the Drug Enforcement Agency (DEA) log. The legal ramifications are, unfortunately, beyond the scope of my experience, and of these proceedings. If you are considering establishing a hospice program, a consultation with your practice's legal counsel is certainly warranted. Depending on your state's regulations, it is likely that veterinary technicians, in addition to the clients themselves, will perform the bulk of the hospice care. This requires dedicated training prior to beginning a hospice program. There is veterinary training available, but human hospice training can also be helpful. Additional communication training may also be warranted, depending on staff experience.

An important aspect of any hospice program is the need for emotional support for all parties involved. Plan to have regular debriefing meetings with the staff of the hospice program to give them an opportunity to decompress and process their feelings after losing a pet, or having difficult client interactions. Regular meetings also give the staff the opportunity to share experiences and best practices, in order to improve the hospice program. Clients must be consulted with regularly, and scheduled home visits are warranted to ensure that the pet is being appropriately cared for and to help address any unstated client needs that may be evident in person. While the patient is the focus of the hospice program, every patient comes with people that must also be supported.

ENROLLING PATIENTS

Patients enrolled in your practice's hospice program will be geriatric and moving into the final stage of their lives, in the absence of an active disease process, or - more likely - those pets that have received a terminal diagnosis. In either case, it is important for the hospice program coordinator to develop a comprehensive care plan for each patient. The first step is to discuss the owner's goals for their pet, as well as their beliefs, and needs. Cultural sensitivity is a crucial skill in this step, as owners may have philosophical or religious beliefs that will affect the end-of-life care plan for the pet. Many pet owners may lack sufficient time or space to care for their pet at home. In some cases, detailed discussions about entering hospice may lead to euthanasia, as owners realize they are unable to perform the necessary hospice care. Once the owner has decided to enter the hospice program, they should also be fully educated on the disease - if any - their pet has been diagnosed with. This should include handouts for the client to read and refer to. The veterinary healthcare team can then work together with the client to develop a personalized care plan for the patient, taking into account the disease process (if any) and the needs, beliefs, and goals of the client for their pet. Care plans will be discussed later in these proceedings.

Once the plan has been developed, hospice staff must take the time to train the clients on aspects of care they will be providing in the home. This may include administration of medication, fluids subcutaneously, and body care (e.g., massage, passive range of motion, monitoring skin integrity). Training may also need to take place in the home to ensure that the environment is aligned with the goals for the pet, and the needs of the client. One of the most important part of the hospice plan is the ongoing support of the veterinary healthcare team to the client. It can be both physically and emotionally taxing to provide hospice care for a pet and owners need to know that the veterinary hospice program can support them either with educational materials, staff help, or other resources to help them succeed in reaching their goals. The final step in hospice care is providing grief counseling and support after the pet has passed. This can be done in-house with your own trained staff, or the practice can hired a trained grief counselor to manage the process. Alternatively, referrals are also a possibility.

HOSPICE CARE PLANS

Any hospice care plan developed by your staff should include attention to the following points:

Pain Management

This is perhaps the most important component of a successful hospice program. It is not acceptable to allow patients, especially those with terminal illnesses, to remain in pain. It is imperative that the hospice care team train owners and caregivers to watch for signs of pain in their pet and give them the tools necessary to treat that pain. Depending on your state's laws and regulations, it may be possible for owners to administer opioid pain medication at home, which is indicated for the painful hospice patient. Opioids are generally safe for the patient's cardiovascular system and can be titrated to effect. The primary side effect is respiratory depression. Fentanyl patches can be applied to the patient and can be replaced every 3–5 days as needed to maintain pain control. Depending on underlying disease process, the addition of non-steroidal anti-inflammatory agents can be highly beneficial in managing the patient's pain as well. If the patient is geriatric, or has long-standing orthopedic or other neuropathic or chronic pain conditions, consider adding agents like gabapentin or amantadine. While many terminal conditions are not painful, if you are in doubt as to whether your patient may be painful, treat for pain. Pet owners generally are not very skilled at recognizing pain in their companions and frequently under-report signs of pain. Training and resources for the client are very important to help them understand dog and cat pain and monitor response to medication. A list of pain indicators for owners is attached as Appendix B to these proceedings. Owners can also be trained to perform massage, which can help alleviate pain.

Nutrition

Proper and adequate nutrition is key to maintaining critical physical functions and structures. Dogs and cats that are unable to meet their nutritional needs on their own may require hand feeding. If hand feeding is still not adequate, or the owner is unable to feed by hand, or the pet is uninterested in eating when hand fed, placement of a feeding tube is warranted. Syringe feeding is no longer recommended as it is quite stressful and can lead to food aversion, worsening a nutritional crisis. The easiest type of tube to place that the owners can manage at home is an esophagostomy tube (e-tube). Blenderized diets and crushed or liquid medications can be given through the tube and, with proper maintenance, the tubes can be left in place for weeks to months.

Grooming and Sanitation

Pets that are unable to move themselves away from their feces or urine are not only at risk of urine scald or skin maceration, but also may suffer from depression or anxiety at being soiled. Using beds or mats that wick away moisture can be helpful if a pet is incontinent or has mobility issues. Owners should be trained to clean and completely dry their pet after any elimination accident and to examine the skin and hair coat regularly for signs of pressure sores, or other wounds or irritated areas. Additionally, if the pet is unable to groom him or herself, owners can be trained to perform grooming duties and keep the coat free of mats. Matted fur can cause skin irritation and pain; regular grooming will remove mats and keep the fur and skin healthy.

Physical Therapy

Physical therapy in the form of passive range of motion can help keep patients comfortable by improving their flexibility and maintaining muscle tone; this is especially helpful for patients that are not mobile. Massage can also be performed to stimulate lymphatic drainage. If the pet is able to rise, stand, and walk, gentle walks on high-traction surfaces are important to help maintain quality of life and mobility.

Quality of Life

Clients must evaluate their pet's quality of life at regular intervals to determine if hospice is still meeting the goals they have for the patient, as well as their needs and beliefs. One example is Dr. Alice Villalobos' Quality of Life Scale (attached as Appendix I) - additional scales and examples will be covered in the lecture. Many clients may find it easier to keep track of "good" and "bad" days; however, the client chooses to define those terms. Whichever tool is used, the hospice and palliative care team should provide instruction on tracking their pet's quality of life and the scale or journal or other tracking method results should be reviewed at every hospice visit or consultation. Maintaining a pet's quality of life, as identified during the hospice care planning process, is one of the key purposes of hospice. Therefore, it is a topic of conversation during each consultation and the client must complete the scale (or other assessment) on a regular (i.e. monthly or weekly) basis. A useful tool to help owners complete the scale or to track good vs. bad days is a Care Journal. In the journal, owners can keep track of what interventions the patient likes or doesn't like, responses to medications, pain levels, and overall well-being.

End-of-Life Care

Upon entering hospice, the owner should be encouraged to think about how they would like to handle the remains of their pet after he or she is deceased. Most practices offer cremation services, but information should be provided on other options including burial at home (if allowed in your municipality) or in a pet cemetery. The availability of special urns, or other customized remembrance items such as clay or ink paw prints should also be considered as part of the plan.

Once the plan is developed, regular consultations should be scheduled with the hospice program team. Depending on disease process and the client's needs and goals, meetings may be as frequent as two to three times a week, or may be monthly or every other month. Consultations can occur either by phone or in the home or both. It can be difficult for clients to bring their patient to the hospital, and every effort should be made to keep the consultations as convenient as possible. When visiting the home, take the time to check the surroundings and the pet's environment and bring any concerns to the hospice caregiver. Seeing pets in their own homes can give the hospice program a more complete view of the patient and more information about what is important to the owners as well. Adjustments to medications or to other aspects of the plan should be approved by the supervising veterinarian and agreed to by the client. Follow-up is key to ensure therapies are working as intended.

When Hospice Ends

Hospice care ends in one of two ways: either the pet passes naturally or the pet is euthanized. In many cases, the client and their wishes for their pet's passing will determine the outcome. This can be a very difficult time for some clients, similar to the euthanasia experience in the hospital or clinic setting. However, the added investment of time, energy, and effort can make the decision to euthanize even more difficult and painful for an owner.

It is important to teach owners the signs that a pet is moving towards an unassisted passing so they can be watched for and anticipated. Some signs include moving away from warmth and seeking a cold area to rest, restlessness, rising of the third eyelid with a vacant stare, and short shallow breathing. Some pets will want to be by themselves, others may seek attention from family members. With good education and planning, owners can know what to look for and be ready.

More often, because of difficulties in care or declining quality of life, the decision is made to euthanize. If your practice act allows it, in-home euthanasia is a fitting end to hospice and can allow owners to say goodbye when they feel the time is right and in the environment their pet feels most comfortable. If you do perform at home euthanasia, it is key to be sure you bring extras of everything you will need, including euthanasia solution! Plan to sedate the patient to the point of unconsciousness prior to administering the euthanasia solution. Some practitioners will use propofol for this purpose: if given fast enough IV it will induce apnea which can also reduce the gasping that sometimes occurs after euthanasia. If venous access is not an option, many sedatives and tranquilizers can be given via the subcutaneous route and have good effect. In my experience, the intraperitoneal route of administration for euthanasia solution works well in those cases where venous access is not an option. It is virtually painless and cessation of cardiac activity is achieved well within ten minutes and very peacefully. Each practitioner will have his or her own protocols and methods for performing this vital hospice service.

After euthanasia or natural death, it is imperative that the hospice care team reaches out to the owners and support them in their grief. As mentioned previously, in-hospital grief counseling or referrals should be offered to every hospice client. Recognition of the passing should be commemorated with a card or letter; also consider donating to a cause important to the client, or related to the patient's disease process. The end of life care plan should be followed meticulously, after confirming the owner's wishes. Regular check-ins with the client after the pet has passed can help with grief and further bonds the owner to the practice. Any additional grief resources can also be provided when checking in. Consider sending a remembrance to the family on the pet's birthday the year following their passing, or any other special date.

Conclusion

A hospice program can benefit the veterinary practice in several ways and requires a commitment from practice ownership and management to ensure success. Proper training of staff members is key, as is support of the client and their pet throughout the difficult fourth stage of life.

APPENDIX I

Dr. Alice Villalobos' Quality of Life (QOL) Scale
(a.k.a., the HHHHH-MM Scale)
Rate each aspect of your pet's life on a scale of 1 (poor) to 10 (best):

- **Hurt**
    - Is your pet painful?
    - Can your pet breathe easily, with minimal effort?
        - Difficulty breathing is reported by human patients as the worst kind of pain and discomfort.
    - Does your pet need supplemental oxygen?
- Score: ______
- **Hunger**
    - Is your pet eating enough?
    - Can your pet eat on his or her own?
    - Does hand feeding help?
    - Does your pet need a feeding tube?
- Score: ______
- **Hydration**
    - Is your pet drinking water?
    - Does your pet's skin feel "sticky" if you pinch it between your fingers?
    - Do your pet's gums feel dry?
    - Are you able to give fluids under the skin?
- Score: ______
- **Hygiene**
    - Will your pet tolerate brushing, grooming, or cleaning?
    - Is your pet able to move him or herself away from waste (urine, feces)?
    - Is your pet able to groom him or herself?
    - Does your pet have any pressure sores?
- Score: ______
- **Happiness**
    - Is your pet able to do the things that he or she enjoys?
    - Is your pet responsive to you and other family members or pets?
    - Does your pet seem lonely, depressed, or anxious?
    - Can the pet's bed be near household activities?
- Score: ______
- **Mobility**
    - Can your pet get up by him or herself?
    - Does your pet like to go for walks?
    - Does your pet fall down often?
    - Does your pet have seizures?
    - Can your cat get to, in, and out of the litterbox?
- Score: ______
- **More Good Days Than Bad**
    - If your pet has more bad days than good days, quality of life may be compromised.
    - Can you interact with your pet as you'd like to?
    - Do you have difficulty providing the care your pet needs?
- Score: ______

A total score of 35 or higher indicates that hospice care should continue. A score less than 35 must prompt a thorough quality of life discussion with the hospice care team.

Appendix II

Signs or indicators of pain in dogs and cats, in order of severity:

- Droopy head or ears
- Tucked tail
- Lack of social interaction
- No interest in play or games
- Subtle lack of alertness gradually increasing to a deep apathy (early sign will be subtle)
- Diminished appetite
- Tense body
- Tense face
- Accepting treats or food gingerly (particularly if your pet used to accept them enthusiastically)
- Lack of interest in walks
- Won't come when called (particularly if this is unusual for your pet)
- Worried or sad facial expression
- Ears pulled back or flattened
- Unable to get comfortable or lie down
- Frequent shifting of position when resting
- Difficulty getting up
- Excessive panting (particularly when it is not hot)
- Shivering/trembling/shaking
- Pacing
- Difficulty moving after a long rest
- Slow or unusual gait
- Limping
- Hunched back
- Compulsive licking or rubbing of a certain body part
- Looking at sides or other body part suddenly and/or worriedly
- Suddenly running away from "nothing in particular"
- Inability jump on couch or bed
- Positioning to avoid contacting certain body parts
- Any change in normal sleep patterns
- "Guards" a particular body part
- Reluctance to be touched in a certain area
- Reluctance to be picked up
- Isolating self from other family members or activities
- Unusual attention seeking
- Aggressive behavior to protect a particular area
- Aggressive behavior on a usually docile pet
- Crying when a particular area is touched
- Refusing to go on walks
- Moaning
- Whimpering
- Refusing to eat
- Pressing head against wall (if head pain is present)
- Unable to get up
- Crying in pain
- Teeth clenched, biting down on an object (particularly cats)
- Howling/screaming uncontrollably

Other types of pain or discomfort include:

- Difficulty breathing

- Fear or anxiety
- Fever or cold
- Inability to empty bladder
- Hunger pains
- Intestinal cramping (i.e., from diarrhea)
- Constipation
- Constant itchiness
- Nausea

RESOURCES

1. AVMA Hospice Guidelines: www.avma.org/KB/Policies/Pages/Guidelines-for-Veterinary-Hospice-Care.aspx
2. Lap of Love Veterinary Hospice (http://www.lapoflove)
3. Pawspice (www.pawspice.com)
4. The International Association of Animal Hospice and Palliative Care (www.iaahpc.org)
5. The International Veterinary Academy of Pain Management (www.ivapm.org)
6. The Veterinary Society for Hospice and Palliative Care (www.vethospicesociety.org)
7. Veterinary Wisdom (www.veterinarywisdom.com)
8. The Ralph Site Pet Loss Support (www.theralphsite.com)

REFERENCES

References are available upon request.

1.
acvim043

Cystitis or Transitional Cell Carcinoma? Differentiation of Bacterial Infection and Cancer of the Bladder

Sandra (Axiak) Bechtel[1], DVM, DACVIM (Oncology); Anne Corrigan[2], DVM, MS, DACVIM (SAIM)
[1]Columbia, MO, USA; [2]Grenada, West Indies

INTRODUCTION

The diagnosis of cystitis versus bladder cancer can be a frustrating one in many cases. Severe cystitis can lead to a thickened bladder wall, and, if severe, sometimes cause a mass effect within the bladder. Bladder cancer, specifically transitional cell carcinoma (TCC), can cause clinical signs very similar to cystitis, and, in fact, many cases of TCC are complicated by underlying infection. Given the difference in prognosis and therapy, it is important to understand how to differentiate between these two diseases and appropriately advise clientele.

CYSTITIS

Presenting Clinical Signs

The most common clinical signs of cystitis in both dogs and cats are pollakiuria, stranguria, periuria, dysuria, and hematuria. These signs are consistent regardless of the inciting cause; inflammation, infection, urinary calculi, neurologic disease and neoplasia, or any combination thereof; additionally, animals with urinary tract infection (UTI) can be non-clinical.[1] However, the specific diseases that cause these signs are often very different when comparing dogs and cats. Developing an appropriate diagnostic testing protocol to help determine the inciting cause is necessary to appropriately treat and assess prognosis.

Urinary Tract Infections

Urinary tract infections (UTI) occur in dogs and, less frequently, cats. Female dogs have the highest incidence and male cats have the lowest, with the incidence in cats increasing after the age of 10 years.[1] The most common bacterial isolates in small animals are *E. coli*, *Staphylococcus* spp., *Streptococcus* spp., and *Enterococcus* spp., followed by *Proteus*, *Klebsiella*, *Pasteurella*, *Pseudomonas*, and *Corynebacterium*. Identifying bacteriuria from urine collected via cystocentesis or catheterization is highly suggestive of UTI, with pyuria commonly seen. Culture and susceptibility testing should be performed to appropriately treat UTIs; this becomes much more important for complicated infections including male patients with UTI, relapsing, refractory/persistent, reinfection, or superinfections. Urolithiasis is both a cause and an effect of bacterial UTI and must be evaluated and monitored.[1,5,6]

Diagnostics

Normal host defenses prevent bacterial colonization, so evaluating for any abnormalities is an important initial diagnostic step. Evaluation points include normal anatomy, micturition, urine, mucosal barriers and host immunocompetence, and these will assist in appropriately treating as well as possibly creating opportunities for direct correction of any abnormalities and the prevention of complicated UTIs and/or progression of infection e.g., pyelonephritis.[2] Diagnostic testing for UTI depends on the patient. A first occurrence simple UTI can be empirically treated with no diagnostic work up, however, once a UTI becomes complicated a full multisystemic diagnostic work up including complete blood count, chemistry panel, urinalysis including sediment evaluation, urine culture and susceptibility, and imaging should be performed to identify likely causes and prevent even more complicated infections as described above.[1,2]

Intact male dogs with prostatitis will commonly present with signs of lower urinary tract disease, usually due to ascending bacterial infection; they will often have concurrent prostatic abnormalities that should be investigated and identified including benign prostatic hyperplasia, prostatic cysts, or neoplasia. Diagnostics should include thoracic radiograph and abdominal imaging to evaluate for metastatic disease along with a full systemic work up.[2]

Fungal infections are much more uncommon; *Candida albicans, C. glabrata* and *C. tropicalis, Aspergillus* spp., *Blastomycosis* spp., and *Cryptococcus* organisms have all been identified. Urine cytology and specific fungal testing should be performed.[1,2,6]

Additional canine cystitis conditions that are much rarer but should be identified and evaluated include polypoid cystitis and emphysematous cystitis. Polypoid cystitis is a rare disease that can present identically to TCC. Diagnostics should include ultrasound imaging of the bladder to identify pedunculated to polypoid mass/es. Histopathology and identifying inflammation, most commonly in the cranioventral bladder mucosa, will always be necessary to identify and differentiate this from TCC. Hematuria is a common sign in dogs with polypoid cystitis, and in one report half of dogs had concurrent UTIs at the time of diagnosis; culture and susceptibility should be part of the diagnostic work up.[3] Emphysematous cystitis is most commonly caused by *E. coli, Proteus* spp., *Clostridium* spp., and *Aerobacter aerogenes.* Emphysematous cystitis is usually a result of concurrent diabetes mellitus; high concentrations of glucose in the urine allows bacterial invasion of the bladder mucosa and gas production in the wall and lumen of the bladder from microbial glucose fermentation.[1,2] Often radiographs will identify air in the bladder wall, and diagnosis is confirmed with a systemic work up and appropriate testing.

Feline Idiopathic Cystitis

Feline idiopathic cystitis (FIC) is the most common cause of lower urinary tract signs in cats, has an unknown etiology, and remains a diagnosis of exclusion. Aggressive research has identified many factors, both host and environmental, that interact to cause a variety of clinical signs as well as a variety of responses to therapeutics. It is likely multisystemic with cardiovascular, gastrointestinal, dermatologic, respiratory, central nervous system, and immune system comorbidities reported, as well as both acute and chronic episodes. Consequences can include secondary UTI, urinary tract obstruction, and recurrent episodes of cystitis. Diagnostics must include ruling out anatomic abnormalities of the urinary tract as well as environmental evaluation, trauma, UTI, and neoplasia.[4] A full multisystemic diagnostic work up including complete blood count, chemistry panel, urinalysis including sediment evaluation, urine culture and susceptibility, and imaging as well as a full evaluation of the environment, and ultimately bladder wall histopathology, and special testing may be required. Treatment is variable and patient specific; options include increasing water intake, altering the cats' environment (household and animal interaction), increasing litter box availability, weight loss, pain control, glycosaminoglycans, behavioral medications, pheromones, and sedatives. Some cats respond some do not, but most cats will get better with time regardless of treatment modality.[1,2,4]

BLADDER CANCER

Presenting Clinical Signs

The most commonly diagnosed bladder cancer in dogs and cats is TCC. Common clinical signs are stranguria, hematuria, pollakiuria, and dysuria. Oftentimes recurrent urinary tract infections are the first sign of disease; initial improvement in clinical signs are reported with appropriate antibiotic therapy, however, signs return upon cessation of therapy or do not completely resolve. Breeds at greater risk for development of TCC are Scottish Terrier, Shetland Sheepdog, Beagle, Wire-Haired Fox Terrier, and West Highland White Terrier. Secondary urinary tract infections are common.[7,8]

Behavior

Transitional cell carcinoma is a locally invasive tumor that infiltrates the bladder wall. While the trigone of the bladder is most often affected, any region of the bladder and urethra may be involved. Multi-focal disease is common. Over half of bladder TCCs involve the urethra, and 29% of male dogs have involvement of the prostate. In addition to local invasiveness, TCCs are highly metastatic. Carcinomas metastasize through both the lymphatic and hematogenous route, meaning metastasis can be widespread. Lymph node and lungs are commonly involved, and other metastatic sites include bone, liver, spleen, and skin. Approximately 16% of dogs have local lymph node metastasis and 14% of dogs have distant metastasis at the time of diagnosis of TCC.[7,8]

Diagnosis

In many cases of recurrent UTIs, and ultrasound of the bladder will be performed to examine for bladder stones or masses. Ultrasound of the bladder will reveal in many cases a distinct mass, usually in the trigone of the bladder (although any area of the bladder may be affected). Masses may be multifocal or singular in nature. Diffuse thickening of the bladder wall is common. If obstruction is present, hydroureter and hydronephrosis may also be evident.

Cytology is an important tool for diagnosis of TCC. It is important to note that background inflammation can interfere with cytologic interpretation; inflammation will result in reactive epithelial cells that are indistinguishable from cancerous cells. Therefore, an underlying UTI can complicate diagnosis of TCC. Neoplastic cells can be shed in the urine, and a diagnosis of carcinoma may be made based on the urine sediment if no underlying inflammation is present. Diagnostic catheterization can be utilized, with or without ultrasound guidance, to collect tumor cells for evaluation. It is important to note that, in this scenario, inflammation can also interfere with diagnostic interpretation. Cystoscopy can also be utilized to collect biopsy samples for histopathology, in addition to visualizing the urethra and bladder. Cystocentesis and percutaneous biopsy sampling of urine or tumor tissue should be avoided, as seeding of tumor cells in the needle tract has been reported.[7,8]

Staging

Given the metastatic behavior of TCC, full staging should be undertaken prior to treatment recommendations. Thoracic radiographs and abdominal ultrasound are standard for imaging of the thorax and abdomen. A minimum database (complete blood count, chemistry panel, and urinalysis) should be performed to assess the overall health of the dog or cat, and to investigate for underlying urinary tract infection. If evidence of a urinary tract infection is found (pyuria, bacteriuria) a culture should be performed. By definition, a UTI with underlying tumor is a complicated UTI, and antibiotic therapy is warranted for 4–6 weeks, with a recheck culture after 1 week of antibiotics and 1 week after completion of therapy to confirm resolution of the UTI.[1,7,8]

Therapy

Therapy is two-fold, aimed at both local control of disease for improvement of quality of life, and systemic disease with the goal of extending lifetime. Local control may be obtained with surgical excision of the mass if: the lesion is singular and discrete, if the mass is in the apex of the bladder and therefore excisable, and/or if surgical debulking is necessary to restore urine flow by relieving ureteral obstruction. As most TCCs tend to be large, infiltrative, and multifocal, surgical excision is not often a viable option. Even with surgical excision, a cure is not affected due to the infiltrative nature of the disease. Radiation therapy, although not well studied, can also be utilized with the goal of local disease control. Urethral and ureteral stents may be placed in cases of obstruction - while these do not treat the tumor, they do provide relief of obstruction.

Chemotherapy is utilized with the goal of decreasing bladder mass size and delaying or treating metastatic disease. Nonsteroidal anti-inflammatory drugs can be used alone or in combination with chemotherapy. Many combinations of chemotherapy have been studied, and the most common utilized are carboplatin or mitoxantrone with a nonsteroidal anti-inflammatory drug. Vinblastine and chlorambucil have also been studied. Response rates (complete or partial remission) of up to 36% with these drugs has been reported.[7,8]

Prognosis

The prognosis of TCC is variable and related both to stage of disease and response to therapy. It is not possible to directly compare protocols from different studies, and the optimal treatment is therefore not yet determined. Median survival times for dogs treated with chemotherapy range from 130–300 days. Many dogs will progress sequentially through chemotherapy protocols based on their response to treatment.[7,8]

References

1. Pressler B, Bartges JW. *Ettinger and Nelson Textbook of Veterinary Internal Medicine.* 7th ed. 2010:2036.
2. Bartges JW. *Ettinger and Nelson Textbook of Veterinary Internal Medicine.* 7th ed. 2010:164.

3. Takiguchi M, Inaba M. *J Vet Med Sci.* 2005;67(1):67, 57.
4. Forrester SD, Towell T. *Vet Clin North Am Small Anim Pract.* 2015;45,:783.
5. Ling GV, *et al. J Vet Intern Med.* 2001;15:341.
6. Norris CR, *et al. J Am Anim Hosp Assoc.* 2000;36:484.
7. Fulkerson CM, Knapp DW. *Vet J.* 2015;205:217.
8. Henry CJ. *Vet Clin North Am Small Anim Pract.* 2003;33:597.

acvim024

Protein-Losing Nephropathy: The Technician's Role

Dawn Terrill, CVT, VTS (SAIM)
Feeding Hills, MA, USA

Pathophysiology

The glomerulus is a complex filtration bed that allows water and solutes to pass. Size is the main determiner of what gets through. Macromolecules, such as albumin and ATIII, are too large to pass. When damage to the glomerulus occurs, larger molecules are allowed to pass. This damage may be transient, or persistent and progressive.

Pathophysiology of Protein-Losing Nephropathy (PLN)

Glomerulonephritis is the most common cause and is due to immune complex deposition in the glomeruli. These immune complexes trigger inflammation along filtration membranes. Amyloidosis is due to a pathologic deposition of polymerized proteins, called amyloids, in the kidneys.

Both glomerulonephritis and amyloidosis are triggered by inflammation elsewhere in the body. The kidney is an innocent bystander. Proteinuria can be the first sign of kidney damage. It occurs earlier than changes in urine specific gravity or blood work. **Persistent proteinuria with an inactive urine sediment = renal disease, even if all other values are normal.**[1]

Causes

Infection, such as pyelonephritis, tickborne diseases, heartworm disease, and leptospirosis. Inflammation from chronic periodontal disease, chronic skin disease, systemic lupus erythematous or other immune-mediated diseases. Corticosteroid excesses from Cushing's disease or long-term steroid use. Neoplasia, although not a common cause. The most common cause is idiopathic. There are also some familial causes such as Shar Pei amyloidosis and the hereditary PLN and PLE in soft-coated Wheaten terriers.

Clinical Signs

Generally, no clinical signs if diagnosed early. Polyuria and polydipsia are very common signs, and often the first signs owners note. Decreased appetite and vomiting may occur as the disease progresses. If nephrotic syndrome develops, peripheral edema will be noted. Nephrotic syndrome is severe, advanced PLN and is characterized by proteinuria, hypoalbuminemia, hypocholesteremia and peripheral edema. The prognosis for nephrotic syndrome is very guarded.

Diagnostics

A minimal database should consist of a CBC, chemistry panel, urinalysis, +/- urine culture and urine protein:creatinine ratio (UPC). A three-day urine collection by the client prior to their appointment is recommended. According to one study, 50% of samples collected at the hospital had a higher UPC than those collected at home by the owner.[2] It is recommended to mix 2 ml of urine from each sample and submit the 6 ml. A study performed in 2010 notes that this helps alleviate the day-to-day variance that can occur from just one sample.[3]

A UPC cannot be interpreted without a urinalysis, as the presence of bacteria, RBC, and WBC may falsely elevate the UPC. If infection is suspected, a urine culture should be done first. A urinalysis with reflex UPC is not recommended as the veterinarian knows the patient better and they, not the lab, should be the one to determine if a UPC is run or not.

It is very important that a urinalysis be done with every annual screening. This may give insight into early renal disease due to the proteinuria or a decreased urine specific gravity. Albumin and creatinine values may be normal until very late stages of renal disease. In one study, 19% of apparently healthy elderly dogs had persistent proteinuria, which is associated with risk of death and progression of chronic kidney disease.[4]

In addition to the minimal database, testing for tickborne diseases and leptospirosis should be performed, as both cause inflammation and damage to the kidney. As leptospirosis is a zoonotic disease, it is also a danger to staff and the pet owner. Thoracic radiographs and abdominal ultrasound to look for sources of inflammation are recommended. Ultrasound also allows the veterinarian to assess kidney structure.

Importance of Doing Blood Pressures

Systemic hypertension (> 160 mm Hg) has been reported in 80% of dogs with glomerular disease.[5] Increased proteinuria is associated with an increase in systemic hypertension.[6] Blood pressures should be taken yearly on young healthy dogs and at every visit for at-risk patients, such as those with heart disease, renal disease, hyperthyroidism, and for those on hypertensive medications.

When taking a blood pressure, the technician should be consistent and use the same size cuff and same leg on the patient each time. The leg should be level with the heart and the patient should be in lateral recumbency whenever possible. Three to five measurements should be taken and averaged. The patient's demeanor should be noted and all vitals should be recorded in the patient's record.[7] A Doppler unit provides a good-quality, low-cost method of obtaining blood pressure. It is accurate in small animals, and the pulse rate and quality are easily assessed.

General Treatment

- Transition to a prescription renal diet to decrease stress on the kidneys.
- Ace inhibitors to help control hypertension within the kidney. Enalapril is started at a low dose (0.25 mg/kg SID or BID) and slowly increased to a maintenance dose of (0.5 mg/kg BID). A creatinine level is checked after each increase to be certain the blood flow within the kidney is not being reduced too much, which would adversely affect the GFR.
- Anticoagulants to help prevent clotting – aspirin (0.5 mg/kg SID) or clopidogrel (2 mg/kg)
- Omega 3 FA to reduce inflammation - (EPA 40 mg/kg) (DHA 25 mg/kg)
- Antihypertensive medications if needed - amlodipine (0.1 mg/kg SID or BID)
- Antinausea medications if needed - Cerenia (2 mg/kg SID)
- Appetite stimulants if needed - mirtazapine (0.25–0.5 mg/kg SID)
- Gastroprotectants if needed - famotidine (0.25–0.5 mg/kg BID), omeprazole (1 mg/kg BID)

Goals of Treatment

- Reduce proteinuria, thereby slowing destruction of renal cells
- Reduce inflammation that can lead to further renal damage
- Control systemic hypertension which can lead to further proteinuria, as well as damage to other target organs.
- Prevent nausea
- Reduce risk of clotting that can lead to PTE
- Improve overall QOL

Consultations

Technicians are often the first person the client meets at an appointment. The history that is obtained from the client will help the veterinarian provide better care for the patient. Travel history may impact testing, as different areas of the country have different diseases that may cause inflammation that can lead to PLN. Information on previous tickborne disease illnesses, leptospirosis, as well as chronic inflammatory conditions should be obtained. Polyuria and polydipsia are often the first clinical signs a client will see in a dog suspected of PLN. Changes in appetite or weight, or any GI upset, such as vomiting or diarrhea, should be noted.

Owners should bring all medications that the pet is on to the appointment. Many clients feel that supplements, herbal and holistic medications are not the same as prescribed medications. Many of these types of medications have no quality control and may actually adversely interact with the prescribed medications. In a drug-induced liver injury (DILI) study done on humans, 15.5% of the injuries were due to supplements.[8]

A diet history should also be obtained, including brand(s) of food being eaten, dry and/or canned, amounts fed and how often. In addition, it should also be noted what human foods and treats the pet is being fed.

CLIENT COMMUNICATIONS

Client communications are an essential part of the technician's job. PLN is a complex disease, often requiring multiple medications, diet transition, and frequent visits to the veterinarian for exams and blood work. Clients may become overwhelmed or uncertain about what to do. In order to help clients and patients, technicians should:

- Explain the disease in terms the client can understand, by using drawings, pictures or analogies.
- Teach medical terms and provide handouts for the client to read.
- Explain the purpose of each medication prescribed. This helps improve compliance as owners understand the importance of each medication to the treatment of PLN.
- Explain the purpose of the lab work being performed. This helps educate the owner and improves compliance.
- Discuss the importance of the prescription renal diet. Diet alone can significantly improve QOL and slow down progression of renal disease.[9] Educate the owner on the difference between a commercial low-protein diet and a prescription renal diet.
- Formulate a diet plan for each patient. Provide handouts with information on prescription renal diets, as well as kcal/cup and kcal/can of each diet.
- Teach owners how to calculate kcal for the pet with the information provided.
- Discuss how to transition to the renal diet and provide support by way of weekly phone updates.

CONCLUSION

Protein losing nephropathy is a manageable, but progressive disease. As technicians, we can educate our clients in the benefits of:

- Correct diet
- Medical therapy
- Regular exams
- Follow-up blood work

If you have questions or would like any of the handouts shown in the presentation, please email me at internalmedicine@veshdeerfield.com.

REFERENCES

1. *J Vet Intern Med*. 2005;19:377–385.
2. *J Vet Intern Med*. 2015;29:1029–035.
3. *Vet Clin Path*. 2010;39(1):53–56.
4. *J Vet Intern Med*. 2017;31:93–101.
5. *Textbook of Veterinary Internal Medicine*. Vol 2, Chap 325.
6. *Textbook of Veterinary Internal Medicine*. Vol 2, Chap 157.
7. ACVIM Consensus Statement. *J Vet Intern Med*. 2007;21:542–558.
8. *Hepatology*. 2014;60(4):1399–408.
9. *J Am Vet Med Assoc*. 2002;220:1163–1170.

acvim033

Bartonellosis: A One Health Perspective on an Emerging Zoonotic Infectious Disease

Edward B. Breitschwerdt, DVM, DACVIM
Raleigh, NC, USA

Bartonella species are fastidious Gram-negative bacteria that are highly adapted to a mammalian reservoir host and within which the bacteria usually cause a long-lasting intra-erythrocytic bacteremia. These facts are of particular importance to veterinarians and physicians, as an increasing number of animal reservoir hosts have been identified for various *Bartonella* species. Among numerous other examples, *Bartonella henselae* has co-evolved with cats, *Bartonella vinsonii* subsp. *berkhoffii* has co-evolved with dogs and wild canines, and *Bartonella bovis* has co-evolved with cattle. Importantly, the list of reservoir-adapted *Bartonella* species, including a large number of rodent species that might serve as "pocket pets," continues to grow exponentially, as new *Bartonella* spp. are discovered. Prior to 1990, there were only two named *Bartonella* species, whereas there are now at least 36 named and numerous unnamed or candidatus species, based upon deposited Gen Bank sequences or preliminary reports, respectively.

In the natural reservoir host, chronic bacteremia with a *Bartonella* species can frequently be detected by blood culture or PCR in outwardly healthy individuals. In contrast, the diagnostic detection of a *Bartonella* spp. in a non-reservoir adapted host can be extremely difficult. Most, although not all diseases caused by *Bartonella* spp. occur in accidental hosts and these organisms are being increasingly implicated as a cause of zoonotic infections. Until recently, mechanisms that facilitate persistent *Bartonella* bacteremia in mammals were not well understood. Recent reports have identified an intra-endothelial and intra-erythrocytic localization for these bacteria, which represents a unique strategy for bacterial persistence. Non-hemolytic intracellular colonization of erythrocytes and endothelial cells would preserve the organisms for efficient vector transmission, protect *Bartonella* from the host immune response, and potentially contribute to decreased antimicrobial efficacy. Other *in vitro* studies indicate that *Bartonella* spp. can infect dendritic cells, microglial cells, pericytes, monocytes, and $CD34^+$ bone marrow progenitor cells.

EPIDEMIOLOGY

Bartonella henselae was initially isolated from an HIV-infected human and subsequently from cats, dogs, horses, marine mammals and other small terrestrial mammals. *Bartonella vinsonii* subsp. *berkhoffii* was initially isolated from a dog with endocarditis in our North Carolina laboratory in 1993 and subsequently from cats, coyotes and human patients. Retrospectively, long-term administration of immunosuppressive doses of corticosteroids for a presumptive diagnosis of systemic lupus erythematosus with cutaneous vasculitis may have facilitated the isolation of the original type strain of *B. vinsonii (berkhoffii)* from this dog that subsequently developed endocarditis. Due to the relatively recent recognition that dogs can be infected with *B. vinsonii* (*berkhoffii*), *B. henselae* and potentially other *Bartonella* spp., seroprevalence data are somewhat limited. Seroprevalence was determined in 1,920 sick dogs from North Carolina or surrounding states that were evaluated at a veterinary teaching hospital. Using a reciprocal titer of >32, only 3.6% of sick dogs had antibodies to *B. vinsonii (berkhoffii)*. Risk factors that could be associated with seroreactivity included heavy tick exposure (Odds ratio 14.2), cattle exposure (OR 9.3), rural vs. urban environment (OR 7.1), and heavy flea exposure (OR 5.6). These data were interpreted to support the possibility that exposure to *B. vinsonii (berkhoffii)* was more likely in dogs in rural environments that were allowed to roam. In addition, these dogs were likely to have a history of heavy tick and flea infestations. Experimental flea transmission of *B. henselae* to dogs has now been confirmed in the laboratory (Lappin and Breitschwerdt, unpublished data). Also, crossreactivity to *Bartonella* antigens was not detected when testing sera from dogs experimentally infected with *R. rickettsii* or *Ehrlichia canis*. However, 36% of serum samples derived from dogs naturally infected with *E. canis* were reactive to *B. vinsonii* antigens. As *E. canis* is transmitted by *Rhipicephalus sanguineous*, this tick may be involved in the transmission of *B.*

vinsonii. The possibility of tick transmission was further supported by two additional studies involving dogs infected with one or more *Ehrlichia* spp. from the same geographic region, in which seroreactivity to *B. vinsonii (berkhoffii)* antigens was 30% and 89%, respectively. Seroprevalence, using *B. vinsonii (berkhoffii)* antigens, was 10% (4/40 dogs) in dogs with suspected tick-borne illness from Israel and 36% in dogs with fever and thrombocytopenia from Thailand. Using an ELISA assay, 35% of 869 samples, derived from coyotes in California, contained antibodies to *B. vinsonii (berkhoffii)* antigens. Current data indicate that canine and human exposure to *B. henselae* and *B. vinsonii (berkhoffii)* can be found throughout much of the United States and most tropical and subtropical regions of the world.

Studies from Hawaii, the United Kingdom and Japan identified *B. henselae* seroprevalences of 6.5% (2/31dogs), 3.0% (3/100 dogs) and 7.7% (4/52). In our laboratory, *B. henselae* is the most common *Bartonella* species found in sick dogs using the BAPGM enrichment blood culture platform. The pathogenicity of all *Bartonella* spp. in dogs is poorly understood; however, like humans, dogs can be infected with numerous *Bartonella* spp. *B. henselae* DNA has been amplified and sequenced from the livers of dogs with peliosis hepatis, a unique pathological lesion also reported in *B. henselae*-infected people. *B. henselae* DNA was also amplified from a dog and a horse with granulomatous hepatitis, a histopathological lesion reported with some frequency in *B. henselae*-infected children and adults. *Bartonella clarridgeiae* DNA has been amplified and sequenced from the liver of a Doberman pincher with copper storage disease and from the aortic valve of a dog with vegetative valvular endocarditis. *Bartonella elizabethae*, a species that infects rodents, was found in a dog that had experienced chronic weight loss culminating in sudden unexplained death. Based upon a large seroepidemiological, controlled study from the University of California (Davis), dogs that were seroreactive to either *B. henselae, Bartonella clarridgeiae* or *B. vinsonii (berkhoffii)* were referred for evaluation of lameness, neutrophilic polyarthritis, nasal discharge, epistaxis, and splenomegaly.

In cats, *B. henselae* and *B. clarridgeiae* have been amplified or grown most frequently from cats and the fleas collected from cats. Flea-associated transmission has been well documented amongst cats. In flea endemic areas, *Bartonella* spp. seroprevalence rates in cats can be greater than 90% and bacteremia rates can be greater than 50%. Granulomatous myocarditis has been reported in cats, naturally or experimentally infected with *B. henselae*. *B. vinsonii* (*berkhoffii)* caused recurrent osteomyelitis in a cat.

PATHOGENESIS

Although as yet unproven, *B. vinsonii (berkhoffii)* may be transmitted to dogs by the bite of an infected flea or tick. Based upon antidotal evidence, dogs may also become infected with *B. henselae* by a cat bite or scratch, analogous to cat scratch disease in people. *B. vinsonii* appears to cause chronic intra-erythrocytic and endothelial cell infections in dogs for extended periods of time, potentially resulting in vasoproliferative pathologies. Similar to other highly adapted intracellular vector-transmitted pathogens, the factors that ultimately result in these bacteria causing disease manifestations are likely multifactorial and as yet to be defined. If similar to babesiosis, another intra-erythrocytic pathogen, stress, hard work, parturition, concurrent infection with other organisms or therapeutic immunosuppression may contribute to the development of pathology. Following experimental inoculation of SPF dogs with culture grown *B. vinsonii (berkhoffii)*, there was sustained suppression of peripheral blood CD8+ lymphocytes, accompanied by an altered cell surface phenotype and an increase in CD4+ lymphocytes in the peripheral lymph nodes. Therefore, infection with *B. vinsonii (berkhoffii)* appears to induce a degree of chronic immunosuppression that might predispose dogs to other infectious agents, resulting in a wide array of clinical manifestations in naturally infected dogs. The pathogenesis of disease (myocarditis or endocarditis) in cats may ultimately be associated with the virulence of the specific strain. Cats infected with *Bartonella* spp. are commonly co-infected with hemoplasmas and at times, more than one *Bartonella* sp. However, whether co-infections magnify disease manifestations of either genera is unclear and in most studies co-infections did not appear to potentiate illness. From an evolutionary perspective, it is obvious that vectors, vector-borne organisms, and animal and human hosts have developed a highly adapted form of interaction. In general, vectors need blood for nutrition; bacterial, rickettsial and protozoal organisms need an intracellular environment to survive, and immunologically, most hosts appear to be able to support chronic infection with vector-borne organisms for months to years without obvious deleterious effects. These factors serve to illustrate the potential difficulty in establishing causation in cats, dogs or people infected with a single or co-infected with multiple tick-transmitted pathogens. Recently, we proposed an addition to Koch's postulates entitled the postulate of comparative infectious disease causation. By satisfying this postulate, *Bartonella* species appear to be able to cause endocarditis, granulomatous inflammatory diseases, particularly involving heart, liver, lymph nodes, and spleen, persistent intravascular infections and the induction of vasoproliferative tumors in animals and human patients.

Clinical Findings

The spectrum of disease associated with *Bartonella* infection in dogs and most other animal species is currently unknown. Endocarditis has been reported in cats, cows, dogs, humans, and wildlife infected with a spectrum of *Bartonella* spp. In some dogs, intermittent lameness, bone pain, epistasis or fever of unknown origin can precede the diagnosis of endocarditis for several months, whereas other dogs will present with an acute history of cardiopulmonary decompensation. Cardiac arrhythmias secondary to myocarditis can be detected in cats and dogs without echocardiographic evidence of endocarditis. Granulomatous lymphadenitis has been associated with *B. vinsonii (berkhoffii)* and *B. henselae* in dogs. *B. vinsonii (berkhoffii)* and other *Bartonella* species appear to contribute to the development of dermatologic lesions indicative of a cutaneous vasculitis, panniculitis, as well as anterior uveitis, polyarthritis, meningoencephalitis, and immune-mediated hemolytic anemia. Additional research efforts, using carefully designed case-controlled studies are necessary to establish the frequency and extent to which *Bartonella* spp. contribute to dermatological, ocular, orthopedic, neurological or hematological abnormalities in dogs (and humans). Clinically, many disease manifestations have also been attributed to *Bartonella* spp. infections in cats. However, it is very difficult to prove disease associations in cats in the field because of the high prevalence rates in non-clinical carriers. In research cats that are infected by exposure to *C. felis*, fever, endocarditis, and myocarditis are the most common disease manifestations. As discussed for dogs, additional case-controlled studies are needed in cats.

Diagnosis

Thrombocytopenia, anemia, which frequently can be immune-mediated, and neutropenia or neutrophilic leukocytosis are the hematological abnormalities in dogs that are seroreactive or BAPGM enrichment blood culture/PCR positive. Thrombocytopenia is found in approximately half, eosinophilia approximately one third of infected dogs, and monocytosis frequently occurs in *Bartonella* endocarditis. Hematological abnormalities have been rarely reported in cats, but similar to dogs, a subset of *Bartonella*-infected cats are neutropenic. Serum biochemical abnormalities are usually very mild or nonexistent in both cats and dogs. In cats, *Bartonella* spp. antibodies have correlated with polyclonal hyperglobulinemia and hypoglycemia. As *B. henselae, B. koehlerae* and *B. vinsonii (berkhoffii)* antibodies are infrequently detected (<4%) in sick dogs in North America, detection of *Bartonella* spp. antibodies in a sick dog provides diagnostic support for prior exposure and potentially active infection. For this reason, treatment of seroreactive dogs or dogs from which any *Bartonella* spp. DNA is detected in blood or tissue samples would be recommended.

Isolation and Molecular Detection of *Bartonella* Species

Because conventional microbiological isolation techniques lack sensitivity, bartonellosis is usually diagnosed by PCR amplification of organism specific DNA sequences and/or through serological testing. Recently, the development of a more sensitive enrichment culture approach, using BAPGM (*Bartonella* alpha *Proteobacteria* growth medium) followed by real time PCR has greatly facilitated the molecular detection or isolation of *Bartonella* species from the blood of sick or healthy animals, including dogs, horses, and human beings. Obviously, the relative sensitivity of the diagnostic methods used to detect *Bartonella* species infection greatly influences an investigator's ability to establish disease causation or a clinician's ability to initiate appropriate treatment. Specifically, the use of this optimized microbiological approach has facilitated the recognition of blood-borne *Bartonella* spp. infections in dogs, horses, human beings, and porpoises. Diagnostic testing (animals and humans) for *Bartonella* species (serology, PCR and

BAPGM Enrichment Blood Culture/PCR) is available through Galaxy Diagnostics, Inc. (contact@galaxydx.com). In cats, serology, PCR or culture combined with serology is recommended and can be procured at Galaxy Diagnostics Inc. and Colorado State University (www.dlab.colostate.edu)

Pathologic Findings

In dogs (and humans), pathologic findings associated with *Bartonella* spp. infection include endocarditis, myocarditis, granulomatous lymphadenitis, granulomatous hepatitis, osteomyelitis, bacillary angiomatosis, and peliosis hepatitis. Multifocal areas of severe myocardial inflammation can be found in dogs with *B. vinsonii* (*berkhoffii*) endocarditis. Although not specific for *Bartonella* infections, organisms can be detected in diseased tissues using silver stains, particularly in acute *Bartonella* infections, such as acute regional lymphadenitis (cat scratch disease). During chronic infections, organisms are often too few in number to be detected in tissues by silver staining, unless a fulminate infection is localized to heart valves. The cardiac abnormalities noted in cats to date are similar to those described for dogs. It seems likely that the spleen plays an important immunomodulatory role in controlling persistent *Bartonella* spp. bacteremia in animals and people.

Therapy

To date, an optimal protocol has not been established for the treatment of *Bartonella* infections in cats, dogs, or people. Regardless of the antibiotic(s) that is used for treatment, a long duration of antibiotic administration (at least 4–6 weeks) may be necessary to eliminate the infection. Due to the rapid development of resistance to macrolides (azithromycin), I no longer recommend these antibiotics as a sole or first-line antibiotic for treating *Bartonella* infections. Fluoroquinolones in combination with doxycycline are currently being used by the author to treat clinical cases of bartonellosis, while exploring antibiotic efficacy following natural or experimental infections. Doxycycline alone does not appear to eliminate *B. vinsonii (berkhoffii), B. henselae* or *B. clarridgeiae* in cats, dogs or other animal species. Serum antibody titers often decrease rapidly (3–6 months) and are generally no longer detectable in dogs that recover following antimicrobial therapy. Therefore, post-treatment serology may be a useful adjunct to BAPGM/PCR to determine if therapeutic elimination of *Bartonella* infections has been achieved. Whether there is clinical benefit to follow serologic or molecular assay results in cats has not been widely studied, but most treated cats do not become seronegative in the short term. However, bacteremia can resolve after treatment or resolve spontaneously in some cats, whereas other cats remain bacteremic despite four to six weeks of antibiotic (documented for several antibiotic regimens) administration, despite resolution of clinical abnormalities (such as lethargy, inappetance, and fever).

Prevention

Although somewhat circumstantial, there is increasing evidence that *Bartonella* species can be transmitted by fleas and ticks to cats, dogs, and human beings. Based upon scientific evidence generated during the past several decades, vector-transmitted pathogens can induce clinical manifestations ranging from acute fatal illness (i.e., Rocky Mountain spotted fever, ehrlichiosis, babesiosis, and bartonellosis) to chronic debilitating disease states (ehrlichiosis, babesiosis, borreliosis, and bartonellosis). Therefore, minimizing or eliminating flea and tick exposure is perhaps of greater veterinary and public health importance today, than during any previous time in history. When rigorous flea and tick control measures are instituted, it is highly probable that transmission of *Bartonella* species to pets and their owners will be greatly reduced or eliminated.

Public and Occupational Health Considerations

There is increasing evidence to support an important role for *Bartonella* species as a cause of a spectrum of disease manifestations in human patients. Due to extensive contact with a variety of animal species, veterinary professionals appear to be at occupational risk for infection because of frequent exposure to *Bartonella* spp.; therefore these individuals should exercise increased precautions to avoid arthropod bites, arthropod feces (i.e., fleas and lice), animal bites or scratches and direct contact with bodily fluids from sick animals. As *Bartonella* spp. have been isolated from cat, dog or human blood, cerebrospinal fluid, joint fluid, aqueous fluid, seroma fluid and from pleural, pericardial and abdominal effusions, a substantial number of diagnostic biological samples collected on a daily basis in veterinary practices could contain viable bacteria. The increasing number of defined *Bartonella* spp., in conjunction with the high level of bacteremia found in reservoir-adapted hosts, which represent the veterinary patient population, ensures that all veterinary professionals will experience frequent and repeated exposure to animals harboring these bacteria. Therefore, personal protective equipment, frequent hand washing and avoiding cuts and needle sticks have become more important as our knowledge of this genus has improved and various modes of transmission have been defined. Physicians should be educated as to the large number of *Bartonella* spp. in nature, the extensive spectrum of animal reservoir hosts, and the diversity of confirmed and potential arthropod vectors, current limitations associated with diagnosis and treatment efficacy, and the ecological and the medical complexities induced by these highly evolved intravascular, endotheliotropic bacteria.

REFERENCES

An extensive reference list is available to readers in the following article:

Breitschwerdt EB. Bartonellosis, One Health and all creatures great and small. *Vet Dermatol*. 2017;28:96-e21. http://onlinelibrary.wiley.com/doi/10.1111/vde.12413/abstract

acvim034

The Axis of Evil: Triaditis in the ICU

Ashley DiPrete, RVT, VTS (SAIM)
Dublin, CA, USA

INTRODUCTION

Triaditis (triad syndrome) is the concurrent state of pancreatitis, cholangiohepatitis/cholangitis, and inflammatory bowel disease.

Liver

The liver is the second largest abdominal organ whose function includes aiding in lipid metabolism, production of bile acids and albumin, to produce glucose (gluconeogenesis), and to store glucose as glycogen. In cats, unlike other species, the bile duct fuses with the pancreatic duct prior to terminating in the small intestines, which has been proven to be a source of problems.

Pancreas

The pancreas has both endocrine and exocrine functions. For this talk, we will focus on the exocrine function. The exocrine functions are performed in the pancreatic acinar cells and duct system that opens directly into the proximal duodenum. These functions include synthesis and secretion of digestive enzymes, zymogens of digestive enzymes, trypsinogen, phospholipase. These enzymes are essential for digestion of proteins, triglycerides, and complex carbs. The pancreas also synthesizes bicarbonate, intrinsic factor, and colipase.

Small Intestines/Gastrointestinal (GI) Tract

The function of the small intestines includes the absorption of electrolytes, water, vitamins, and, after being chemically altered by digestive enzymes secreted by the pancreas, the absorption of carbohydrates, proteins, and fats.

PANCREATITIS

While pancreatitis has many triggers, there is no definitive cause. Bacterial infections have been documented in the pancreas, as well as rarely pancreatic and hepatic flukes. Trauma (ex: high rise syndrome), complications of surgery or minimally invasive treatment have also been implicated as risk factors. Anticonvulsive medications have not been proven to be a true risk factor and ingestion of toxins, such as Easter Lilies, can have a pancreotoxic effect, but this is limited to degeneration of acinar cells without inflammation. Pancreatitis is a cause of diabetes mellitus, rather than a consequence of it.

The key factor of pancreatitis is the blockage of the acinar cell apex in the pancreatic duct leading to colocalization and fusion of zymogen and lysosomal granules. In much more basic terms, the pancreatic ducts become blocked (via inflammation) and the digestive enzymes leak out, thereby effectively self-digesting. Oxidative stress and hypotension are also factors that can lead to pancreatitis. The self-defense mechanism of the pancreas against self-digestion is overwhelmed when >10% of intracellular trypsin is activated.

Signs

The signs of feline pancreatitis are vague and can include hyporexia, lethargy, vomiting, dehydration. It is very common for canines to present with cranial abdominal pain; however, this is difficult to assess in cats. Due to the vagueness of signs and symptoms, pancreatitis should be on every list of differential diagnoses with cats that present with any combination of these signs and symptoms.

Diagnosis and Treatment

For a definitive diagnosis, there must be documentation of inflammation of the pancreatic parenchyma via histopathology. Despite complications, which include possible increased inflammation to the pancreas itself, it is a relatively safe procedure. The ideal candidate may be the stable chronic pancreatitis patient. In canines, pancreatic lesions tend to have a patchy distribution, and since there are no studies that prove otherwise, we must assume this is the case with felines as well.

Other diagnostic testing includes complete blood count (CBC)/serum biochemistries, pancreatic specific tests, radiographs, abdominal ultrasound, endoscopic ultrasounds, CT, biopsy and histopathology. There are no pathognomonic changes on bloodwork that are specific to pancreatitis. Hypocalcemia has been associated with severe disease and with necrotizing pancreatitis. Most of these abnormalities are indicative of comorbidities and will require additional and specific treatment. It is important to note that amylase and lipase have little to no diagnostic ability in the feline.

The pancreas is the only organ to make the following proteins: feline trypsinogen (fTLI), trypsin-activation peptide (TAP), pancreas-specific lipase (fPLI and spec fPLI). Since there are normally low levels of these proteins in the serum, any elevation points to acinar cellular dysfunction and therefore confirms the diagnosis of pancreatitis. There is a point-of-care test available, called the SNAP® PLi, made by IDEXX. This test can be run on whole blood and is 100% specific and 87% sensitive. This test should be followed up by reference laboratory findings. The best test currently available is the quantitative fPLI which is 67–79% sensitive and 67–82% specific. In cases of moderate to severe pancreatitis it has been shown to be up to 100% specific, and values of >20 mcg/L have been associated with poor prognosis.

Plain radiographs may show subtle and nonspecific findings. Abdominal ultrasound (AUS) can be quite helpful in finding supporting evidence for pancreatitis, such as pancreatic enlargement, hypoechoic pancreatic region (acute pancreatitis) or hyperechoic pancreatic region (chronic pancreatitis), abdominal effusions, and dilation of the pancreatic duct. Where AUS excels, is scanning for comorbidities, such as cholangitis/cholangiohepatitis.

Treatment

The severity of signs of the patient will dictate how aggressive treatment needs to be. Patients commonly present dehydrated and will require IV fluid resuscitation. Pancreatic ischemia is a significant contributor to necrotizing pancreatitis and holds a poorer prognosis. Treatment specific to pancreatitis includes colloids and fresh frozen plasma, which likely provides oncotic support while replacing coagulation cascade proteins. Once vomiting and nausea has been addressed, it is reasonable to start to feed these patients. New research shows that an early return to enteral nutrition is associated with improved outcomes, shorter hospital stays, and a decrease in frequency of side effects. Pancreatitis is a painful condition, and therefore adequate pain control must be at the forefront of treatment.

Cholangiohepatitis/Cholangitis

Cholangiohepatitis is inflammation of the liver, the gall bladder, and the biliary tree, whereas cholangitis is inflammation solely of the gall bladder and biliary tree. Histologically, neutrophils will be seen within the lumen of the bile duct. Generally, this occurs when an infection travels from the small intestines via the common bile duct into the gall bladder. Patients presenting for cholangiohepatitis or cholangitis may present with any combination of the following: fever, weight loss, lethargy, vomiting, hyporexia, icterus, and hepatomegaly. Cats are also prone to nonsuppurative cholangiohepatitis, which does not respond to antibiotics, and often must be treated with immunosuppressants.

Lab Findings

Some of the diagnostic testing for cholangiohepatitis/cholangitis include CBC and serum biochemistry panels. Abnormalities on CBC may include leukocytosis with band neutrophils and anemia. Patients may have normal liver enzymes on chemistry panels; however, 98% of patients will have an elevated AST. With inflammation of the bile duct or gall bladder, serum total bilirubin may be elevated. It is important to monitor clotting times, as liver dysfunction may negatively affect them. The gold standard for testing liver function is via serum bile acids. Elevations in a fasting and post-prandial sample (2 hours after feeding) of >25 µmol/L are indicative of decreased liver function and warrant further investigation via liver biopsy or fine needle aspirate. Another diagnostic test is abdominal ultrasound (AUS). During AUS, the liver parenchyma may be hyperechoic, supporting the diagnosis of inflammation within the liver/gall bladder. Other concurrent findings are hepatomegaly, gall bladder distension, and increased gall bladder sediment. With the assistance of AUS, a fine needle aspirate of the liver may be obtained. Fine needle aspirates are quick, easy, and generally do not require the testing of coagulation times prior to procedure. However, aspirates yield a small sample size and do not reflect morphology of parenchymal architecture and will therefore limit the ability to identify the primary hepatic disorder. Cytology of fine needle aspirates may not detect infiltrative lesions, particularly if they are nodular, multifocal or periportal. An ultrasound-guided biopsy can be obtained, which allows for a larger sample. The most common adverse side effect to an ultrasound-guided biopsy is hemorrhage, which may be seen up to 10 hours post-procedure. A surgical or laparoscopic liver biopsy is superior for diagnostic purposes. When the biopsy is taken surgically, there is an opportunity to sample a visibly abnormal area, and will allow for other procedures to be done at the same time, such as aspiration and culture of the gall bladder, and the opportunity to obtain other non-liver biopsies.

Treatment

The mainstay of treatment for neutrophilic cholangitis/cholangiohepatitis is antibiotics, which should be chosen based on culture and sensitivity results obtained during the diagnostic process. The antibiotic that is chosen must be effective against anaerobic and aerobic enteric bacteria, and must be excreted into the bile in its active form. To treat the nonsuppurative form of cholangitis/cholangiohepatitis, the use of immunosuppressant drugs may be needed. Other treatments include diet changes to include a highly digestible moderate fat diet. Unless signs of overt hepatic encephalopathy are present, protein restriction is not recommended. Unless there is a biliary obstruction, the use of choleretics is recommended. Although there has not been much in the way of research regarding the use of S-adenosylmethionine (SAMe) and its uses specifically for cholangiohepatitis/cholangitis, it has been shown to be beneficial for other hepatic disease processes by reducing inflammation and negative side effects of free radicals within the liver itself. Depending on the severity of signs, hospitalization may be warranted.

Inflammatory Bowel Disease

Inflammatory bowel disease (IBD) is a collective term that includes disorders of the GI tract characterized by persistent or recurrent GI signs and histologic evidence of intestinal inflammation. There is no age or gender predisposition and the cause of IBD is unknown. In humans, the breakdown to immunologic tolerance to luminal antigens (bacterial/dietary) is thought to be critical.

Nomenclature reflects the anatomic distribution and the predominant cell type, with the most common being lymphocytic-plasmacytic enteritis (LPE). IBD may be overdiagnosed due to ease of obtaining endoscopic biopsies, difficulties in interpretation of small samples, and failure to eliminate other causes of mucosal inflammation prior to obtaining samples.

Presentation

Patients suffering from IBD may present with waxing and waning of signs and symptoms, which may resolve spontaneously. In some instances, there may be a precipitating event, such as acute gastroenteritis. Often, the signs that are present can help point to the location of the GI tract that is being affected. In cats, the most predominate sign is vomiting, which points to gastric or upper small intestinal involvement. Patients may have large or small bowel diarrhea. Patients may also present with increased borborygmus and flatulence, weight loss and altered appetite. Abdominal palpation may reveal pain and thickened bowel loops. Pain after eating can be a significant symptom of IBD. Some of the symptoms of more severe disease include bloody vomit or diarrhea, anorexia, and polyphagia in the face of significant weight loss.

Diagnosis

To diagnose IBD, an intestinal biopsy is necessary. Diagnosis should be limited to cases where histologic inflammation is found without an underlying cause. It is imperative to exclude all infectious responses, diet responses, and antibacterial responses prior to final diagnosis. Intestinal biopsies can be obtained either endoscopically or via laparotomy. In cats, where there is a tendency for multiorgan involvement, laparotomy may be the better option. Endoscopic biopsies are generally easy and quick to obtain; however, their small sample size, tendency to be superficial samples, and possibility of crush artifact are the major drawbacks. Endoscopy is limited to the proximal SI tract, which makes obtaining biopsy samples of the ileum, where samples tend to be more diagnostic, difficult to obtain. A laparotomy provides an opportunity to obtain full-thickness samples of the entire GI tract, as well as any other organ that may be abnormal (pancreas, liver).

There are no pathognomonic findings on CBC or chemistry profiles. Abnormalities on complete blood count (CBC) can include neutrophilia +/- left shift. Anemia may be due to chronic inflammation or blood loss via the gut. Thrombocytosis is seen with chronic GI bleeding. Some abnormalities that may be seen on chemistry profiles are hypocholesteremia, which is suggestive of malabsorption. Hypoalbuminemia is associated with a poorer prognosis. In felines, any increase in serum activities of liver enzymes are more likely to be results of primary liver disease as they have a shorter half-life in this species.

A fecal should be checked to eliminate other reasons for mucosal inflammation, such as parasites and bacterial pathogens. Due to the potential for occult *Giardia*, empirical treatment is recommended.

Serum vitamin concentrations can also provide information about the disease status for patients. Low cobalamin and folate are potentially due to intestinal malabsorption. The degree of hypocobalaminemia correlates with histologic damage and a poorer prognosis. Survey radiographs are generally unremarkable for cases of IBD. Abdominal ultrasound can document focal or diffuse disease, while allowing for the evaluation of intestinal wall thickness and appearance, and evaluation for other abdominal structure abnormalities.

Treatment

Despite which form of IBD the patient is suffering from, all treatment generally involves some combination of dietary modification, antibacterial therapy, +/- immunosuppressive therapy. Mild cases of IBD generally respond to diet change +/- antibacterial therapy, whereas severe cases may necessitate the use of immunosuppressive medications.

Dietary modifications include feeding an easily digestible diet that lowers the intestinal antigenic load and reduces mucosal inflammation (novel protein diet). Supplementation of folate and cobalamin is advised if serum levels are below normal.

Antimicrobial therapy is used to treat secondary small intestinal bacterial overgrowth (SIBO) and to reduce the bacterial antigens that cause IBD symptoms. Metronidazole is the go-to treatment for IBD in felines due to the potential immunomodulatory effects it may have on cell-mediated immunity, in addition to its antibacterial properties.

The last resort of treatment for IBD is immunosuppressive medications, which should be reserved for patients who have not responded to diet modifications and antimicrobial therapy, or for patients who present with severe signs. Cytotoxic drugs may be added if clinical signs recur as these medications are tapered, or to spare their side effects.

Prebiotics and probiotics may prove useful in the treatment of IBD, as they help replace the beneficial gut flora and thereby reduce the intestinal inflammation. Species-specific prebiotics and probiotics are ideal.

Generally, hospitalization for IBD patients is brief and revolves around supportive care.

TECHNICIAN'S ROLE

As technicians, it is our responsibility to be the clinician's eyes and ears, as well as being advocates for our patients. We can use Kirby's Rule of Twenty as a guide for keeping our patients as content as possible while hospitalized. It is important to remember that feline patients also present their own set of challenges while being hospitalized.

Patients who are hospitalized for triaditis have a myriad of issues that need to be addressed simultaneously. Luckily, treatment for the 3 concurrent disease processes are relatively similar. Many times, these patients will require IV fluids. IV fluids should be aimed to correct hydration status over 12–24 hours, and to keep up with ongoing fluid losses (e.g., diarrhea and vomiting). These patients may be hypoproteinemic, therefore fluid therapy should be administered with caution. It is important to remember that with these disease processes, patient's electrolytes may be subject to significant imbalances. It is the role of the nurse to anticipate these needs and help correct them, as instructed by the managing clinician. For instance, in cases of severe pancreatitis, patients may become hypokalemic.

Pain control is highly important in these patients, and many patients will show clinical improvement once their pain is managed. Opioids are recommended for these cases. Buprenorphine or a fentanyl CRI work very well in these patients. Buprenorphine falls victim to the ceiling effect, and therefore it does not serve well as on-going, severe pain management medication. However, as the patient improves, the need for pain control will also diminish. Buprenorphine may be good choice of medication to send home for continued low-moderate pain management.

Antiemetics should be administered prior to reintroducing food. Maropitant may be beneficial for these conditions as it has antinociceptive properties in the viscera. In cases of protracted vomiting or nausea, the addition of a 5-HT3 receptor antagonist may be warranted to gain adequate control.

It is always recommended to entice the patient to eat on their own; however, it is **not** recommended to force feed/syringe feed these patients. Cats are notorious for developing food aversions, and the act of force feeding them may reinforce a negative correlation between eating and negative side effects. If the feline patient refuses to eat on their own, a nasoesophageal (NE) or nasogastric (NG) feeding tube may easily be placed and used for enteral nutrition. Reintroduction of calories should be slow and tailored to the patient needs. Nutrition can be administered as a trickle feeding via syringe pump, or as bolus feedings every few hours. Due to the narrow diameter of the NG/NE tubes, liquid enteral nutrition is required (e.g., CliniCare® and Vivonex®). Nutrition chosen will be based on clinician's choice, preferences, and patient's needs. Another option of feeding tubes is the esophagostomy tube. This is a larger-bore feeding tube (14– 20 Fr) that, when properly placed, will allow for calorie-dense nutrition to be administered. During all feedings, patients need to be monitored for adverse side effects, such as nausea and vomiting. If the patients start showing any of these signs, feedings should be halted and the managing clinician should be informed.

Creature comforts are as important as medications and nursing care. As best as possible, all felines should be kept as content as possible. Stress is a contributing factor to the morbidity during hospitalization stays. If possible, all felines should be housed away from canines, in a quiet room. If this is not feasible, the next best option is to place cats on the opposite side of the ICU. Patients should be given clean, dry, and warm bedding, and should be checked frequently and regularly to ensure they remain clean and dry. Some cats who are extra-nervous may appreciate an area where they can hide. A blanket over half of the kennel, or a small box inside of their kennel can accomplish this. Placing feeding bowls in the opposite corner from the litter box is also important, as cats tend not to eat where they eliminate. The use of pheromones to reduce anxiety has also proven to help with hospital visits. A plug-in diffuser, or a blanket with calming pheromones sprayed onto a towel and placed in the patient's kennel should help to reduce anxiety.

REFERENCES

1. Ettinger SJ, Feldman EC, Côté E. (2016). *Textbook of Veterinary Internal Medicine, Expert Consult.* 8th ed. Saunders.
2. Nelson RW, Couto CG. (2009). *Small Animal Internal Medicine.* St. Louis, MO: Mosby/Elsevier.
3. Odunayo A. (2014). Management of potassium disorders. *Clinician's Brief.* 69–72
4. *The Merck Veterinary Manual electronic version.* (2016). Retrieved January 29, 2017, from www.merckvetmanual.com/
5. Linklater A. (2014). Feline pancreatitis. *Clinician's Brief.* 17–21.

acvim035

Atypical Endocrine Diseases and the Role of the Veterinary Technician

Colleen Ruderman, RVT, VTS (SAIM)
Atlanta, GA, USA

Summary/Introduction

Endocrine diseases are common in dogs and cats. The technician plays an important role in client education and disease management in these cases. Technicians are integral as the primary contact for pet owners, to provide follow-up and to ensure continuity in treatment. While some endocrine diseases are more common, such as diabetes mellitus and feline hyperthyroidism, others are much less common or subtle in their presentation. They can be a point of frustration for owners and veterinary staff alike, whether there are difficulties in diagnosis or treatment. Technicians in specialty and general practice should be aware of the presentation, expected clinical findings, diagnostics required, and different treatment options for less common endocrine disorders in order to optimize client satisfaction and patient care.

Canine Atypical Hypoadrenocorticism

Primary or 'typical' hypoadrenocorticism (Addison's disease) is most commonly due to the immune-mediated destruction of the adrenal cortex, leading to glucocorticoid and mineralocorticoid deficiencies. The result is a clinical syndrome related to inadequate cortisol production that includes gastrointestinal signs (anorexia, vomiting, and diarrhea), low blood pressure, and weakness. Mineralocorticoid deficiency leads to electrolyte derangements and hypotension. Treatment is supportive for vomiting and hypotension and includes supplementation of glucocorticoids and mineralocorticoids. Atypical hypoadrenocorticism is due to a glucocorticoid deficiency only, and patients are not affected by the often life-threatening electrolyte imbalance that comes with primary hypoadrenocorticism. Approximately 10–30% of dogs with hypoadrenocorticism have glucocorticoid deficiency only.[6] It is suspected to be partial immune destruction of the adrenal cortex causing some zones to be affected, so it is primarily adrenal in origin; however, acute presentation of the disease can be located to the pituitary, typically after a trauma. However, the most common cause is abrupt cessation of higher dose long-term glucocorticoids. Another form of poor cortisol secretion is critical illness-related corticosteroid insufficiency (CIRCI), which is a syndrome of consumption, rather than immune destruction of the adrenals or pituitary trauma.[7]

Clinical symptoms are similar to primary hypoadrenocorticism with weight loss, gastrointestinal signs, lethargy, and weakness that may be waxing and waning in nature. Complete blood counts will usually show lack of a stress leukogram due to the cortisol deficiency. Sodium, potassium, and chloride will be normal on the biochemistry panel. Hypoglycemia, hypercalcemia, hypoalbuminemia, and hypocholesterolemia may also be found.

A definitive diagnosis for complete hypoadrenocorticism is made with the ACTH stimulation test. Results of the pre- and post-cortisol levels should be below reference range. Diagnostic testing is the ACTH stimulation test for atypical hypoadrenocorticism; however, results are low rather than unreadable, ranging from 1–5 µg/dl on the post-sample.[6,7] It is important that no form of steroid be administered 2 weeks prior to testing.[6] Plasma aldosterone levels may also be tested and identify early mineralocorticoid deficiency. Lower than normal levels may be demonstrated; however, with no clinical symptoms or electrolyte imbalance, these dogs typically respond to treatment with glucocorticoid supplementation alone.[1] Plasma endogenous ACTH concentrations can also help locate the origin of the disease. If low to low normal ACTH, the origin is likely pituitary but may be hypothalamic. With adrenal origin disease, the ACTH should be high. Treatment is administration of physiologic doses of prednisone (0.1–0.25 mg/kg/day) and response should be quick resolution of clinical symptoms. Initial doses might be higher and given parenterally in more severe cases. Monitoring of aldosterone levels may be performed to monitor for development into primary hypoadrenocorticism. Alternatively, monitoring of electrolyte levels and sodium potassium ratio may be performed.

Canine Atypical Hyperadrenocorticism

Hyperadrenocorticism, or Cushing's disease, is the overproduction of cortisol as a result of a pituitary or adrenal tumor. Atypical Cushing's disease is a poorly defined condition in which clinical manifestations of hyperadrenocorticism are present with normal cortisol levels when an ACTH stimulation test and low-dose dexamethasone suppression test are performed. The theory is that elevations in steroid hormones other than cortisol cause similar clinical findings. While the exact hormones responsible are unknown, the hormones currently measured are 17-hydroxyprogesterone (17OHP), progesterone, deoxycorticosterone, estradiol, and androstenedione, and can be associated with adrenal tumors.[10] Atypical Cushing's disease is controversial because of the variability in clinical signs, clinical findings, steroid hormones involved, and response to treatment.

Clinical presentation is often similar to the more common form of hyperadrenocorticism and includes increased thirst, increased urination, increased appetite, potbellied appearance, systemic hypertension, and symmetrical alopecia. General bloodwork for these patients often has elevated liver enzymes, elevated cholesterol, and a low urine specific gravity. It is important to remember that these dogs may have some, but not all of these physical and laboratory findings. In addition, on standard diagnostic testing with LDDS testing **and** ACTH stimulation testing are normal or give questionable. Ultrasonography can aid diagnosis by identifying an adrenal mass.

Once a patient is suspected of having atypical hyperadrenocorticism, additional testing by submission of a sex hormone panel (done with the same protocol as the ACTH stim for cortisol) can be performed. If a patient has normal cortisol levels and elevations in sex hormones with clinical symptoms the diagnosis can be presumed. Alternatively, the ACTH stim or LDDS test can be repeated at a later date. It is important to note that many patients with non-adrenal illness and typical hyperadrenocorticism can have elevations in sex hormone levels. It is also suspected that hormones we are unable to measure may be responsible for the clinical symptoms in question. Treatment is still aimed at relief of clinical symptoms and improvement of quality of life. The medication chosen depends on the steroid hormone in excess. Certain medications fail to lower certain steroid hormones and may actually increase them. Trilostane and mitotane can be used; however, it is cautioned until true cortisol excess has been demonstrated as there is a risk for iatrogenic hypoadrenocorticism with those medications. The initial treatment typically involves melatonin and flaxseed lignin. Melatonin is involved in neuroendocrine control which can improve clinical symptoms such as increased thirst and urination, and may also improve hair growth. Side effects of melatonin given orally are uncommon and approximately 4 months of treatment may be needed before improvement in clinical symptoms is seen.[11] Flaxseed lignin inhibits 3 beta HSD enzymes which can reduce cortisol levels, and is suspected to have effects on some sex hormones that may contribute to clinical symptoms. Side effects of flaxseed lignin are typically gastrointestinal and many preparations of the compound have low palatability. Melatonin and flaxseed doses and monitoring is based on changes in clinical symptoms. Mitotane and trilostane may also be used in cases where melatonin and lignans have shown little response, but is not the first line in treatment for atypical hyperadrenocorticism. Trilostane can be used as it can drop cortisol secretion and may increase sex hormones like progesterone and 17OPH; however, many dogs treated will still show clinical improvement. Monitoring with trilostane and mitotane should follow standard protocols. Overall, the outcomes of treatment with mitotane and trilostane are varied and inconsistent with atypical hyperadrenocorticism.

Feline Hyperadrenocorticism

Feline Cushing's is rare but similar in pathophysiology to canine hyperadrenocorticism. It is also most commonly caused by a pituitary adenoma. Although the etiology is similar, the clinical symptoms in feline hyperadrenocorticism are typically more profound. The majority of cases have poorly controlled or insulin-resistant diabetes mellitus. Clinical symptoms also include polyuria/polydipsia, urinary tract infections, upper respiratory infections, and hair loss, are also similar to the canine version. However, the primary difference is in less common cutaneous manifestations such as extreme friability of the skin, poor wound healing, and curled ear tips. General bloodwork can show the expected abnormalities (elevated liver enzymes, low urine concentration, leukocytosis); however, glucosuria is present in almost all cases due to concurrent poorly controlled diabetes mellitus. Anemia may also be present on complete blood counts.

Diagnostic testing is also different than in dogs because cats respond poorly to ACTH in the form of synthetic corsynotropin or ACTH gel. The low-dose dexamethasone suppression test is preferred, but because cats are more resistant to glucocorticoids than dogs, the dose is higher at 0.1 mg/kg IV of dexamethasone and it does not differentiate between pituitary and adrenal illness. Endogenous ACTH levels, a HDDS test or diagnostic imaging can help distinguish adrenal from pituitary origin. Cats are more resistant to the therapeutic effects of mitotane, which is why, historically, bilateral adrenalectomy was performed for pituitary origin disease. There is limited evidence cats respond to trilostane, so it is also recommended for management of pituitary disease. Iatrogenic hypocortisolism can occur at any time during treatment and will lead to signs of hypocortisolism. In addition, due to the abrupt drop in cortisol, which is the primary cause for insulin resistance in these cases, all cats with diabetes mellitus should be monitored carefully for insulin overdosage. If a primary adrenal tumor is found to be the cause (rare), then treatment with trilostane for 2 weeks prior to surgical removal is the treatment of choice. However, glucocorticoid and mineralocorticoid supplementation may be needed following surgery.

Feline Acromegaly/Hypersomatotropism

Acromegaly is characterized by a pituitary adenoma, which causes excessive levels of growth hormone to be produced. It is being identified with increased prevalence in cats. Excessive levels of growth hormone cause overgrowth of bone, muscles, and organs. In some cats the changes in the skull can be characterized by widening of the head and jaw, increased interdental spaces, and protrusion of the lower jaw. In addition, weight gain, increased muscle mass, enlarged kidneys, and heart disease may be seen. The tumor will also cause an increase in insulin-like growth factor (IGF-1), which has anabolic effects and causes insulin resistance and most cats have diabetes mellitus at the time of diagnosis.

Clinical symptoms, aside from changes in appearance, are associated with insulin resistance and consist of polyuria, polydipsia, and polyphagia. Weight loss can be seen, but many of these cats are actually able to maintain or gain weight. Due to the effect of growth hormone on organs, cats may present in cardiac failure as a result of hypertrophic cardiomyopathy, or may have secondary renal disease or other organomegaly. The diagnosis is made by measuring IGF-1 levels, as a test is not readily available for measuring growth hormone directly. Computed tomography or magnetic resonance imaging can confirm the presence of a pituitary tumor and help determine treatment options.

Medical management is a conservative option and involves treatment of diabetes mellitus and other manifestations such as heart failure. Typically, high-dose insulin and possible combination of long-term and short-term insulin will be needed, but requires frequent blood glucose curves and monitoring of clinical symptoms. Glucose is difficult to impossible to control with insulin in these cats but they often do better with higher glucose than non-acromegalic diabetic cats. A more aggressive option for treatment is radiation therapy which can help shrink tumors and improve the response to insulin. Pituitary irradiation may slow progression of cardiac disease. Unfortunately, complete remission is uncommon with radiation therapy. The emerging gold standard in treatment is the transsphenoidal hypophysectomy, which is recommended for tumors 1 cm in diameter or less. Monitoring postoperatively can be complex and there are temporary and permanent side effects to this procedure; however, remission for diabetes is possible with this procedure. At this time, only a small number of institutions are performing this surgery. Another emerging treatment is the use of somatostatin analogs. Existing medications have been unsuccessful; however, recent studies have shown promise with the drug pasireotide in long-acting forms. As with any treatment involving changes in insulin resistance, decreased insulin doses are recommended prior to the start of treatment. If no treatment is pursued, then progressive diabetes mellitus with diabetic ketoacidosis, cardiac failure, and renal disease often occur leading to euthanasia or death within 1–3 years from the time of diagnosis.

Feline Hyperaldosteronism - Conn's Syndrome

Hyperaldosteronism is a disease most commonly seen in cats caused by excess mineralocorticoid (aldosterone) production by one or both adrenals. The overproduction causes increase in blood pressure and decreases potassium, leading to clinical symptoms. Tumors are typically unilateral, and due to their vascular nature often involve the vena cava. Bilateral cortical hyperplasia has also been described. Clinical symptoms are often blindness secondary to retinal detachment as a result of severe hypertension. If severe hypokalemia is present, muscle weakness, lethargy, cervical ventroflexion, and dropped hocks can also be seen. Treatment often requires hospitalization to correct the hypokalemia with IV supplementation, and then oral potassium gluconate can be administered. ACE inhibitors such as benazepril, in addition to amlodipine and spironolactone, are used to decrease hypertension. For adrenal tumors, surgical adrenalectomy is recommended. Because this is a disease of older cats with severe hypertension, renal disease/failure may be present or develop in these cats.

Pheochromocytoma

Pheochromocytoma is a catecholamine-secreting adrenal tumor, most commonly found in canines and is extremely rare in cats. This tumor can cause hypertension and arrhythmias due to the secretion of hormones such as epinephrine, norepinephrine, and dopamine. Bloodwork is typically unremarkable; however, imaging will usually identify a tumor. They are typically malignant in their behavior with frequent vascular involvement and less frequent metastasis. Clinical symptoms are associated with severe hypertension or arrhythmias (weakness, collapse, pu/pd, exercise intolerance) and, if tumors rupture, acute hemorrhage can occur. The diagnosis is suspected in a patient with clinical symptoms and identification of an adrenal tumor on imaging. Definitive diagnosis can only be made on histopathology. Treatment recommendation is typically surgical if there is no extensive invasion or metastasis. Phenoxybenzamine is used prior to surgery to decrease hypertension and risk of cardiac arrhythmias and variable blood pressure during surgery.

References

1. Feldman E, *et al. Canine and Feline Endocrinology*. Saunders;2004:69–80, 263, 352, 358–389, 394–439, 440–448.
2. Mellett K, *et al*. Trilostane therapy for spontaneous hyperadrenocorticism in cats. *J Vet Intern Med*. 2013;3;27:1471–1477.
3. Scott-Moncrieff JCR. Diagnosis and management of feline Cushing's. In: Proceedings of the British Small Animal Veterinary Congress; 2011.
4. Ward C, *et al*. Emerging endocrine diseases in the cat. In: Proceedings of the NAVC Conference; 2015.
5. Baumstark M, *et al*. Evaluation of aldosterone concentrations in dogs with hypoadrenocorticism. *J Vet Intern Med*. 2014;28:154–159.
6. Reusch C. Atypical Addison's disease. In: Proceedings of the 20th ECVIM-CA Congress; 2010.
7. Nichols R. Primary, secondary, atypical, and pseudo hypoadrenocorticism. In: Proceedings of the ACVIM Forum; 2015.
8. Greco D. Hyperadrenocorticism associated with sex steroid excess. *Clin Tech Small Anim Pract*. 2007;10:1053.
9. Gostelow R, *et al*. Pasireotide long-acting release treatment for diabetic cats with underlying hypersomatotropism. *J Vet Intern Med*. 2017;31:355–364.
10. Behrend E. Clinical approach to adrenal incidentalomas. In: Proceedings of the ACVIM Forum; 2016.
11. Clinical Endocrinology Service *et al*. Treatment option considerations for hyperadrenocorticism in dogs. College of Veterinary Medicine, University of Tennessee.

acvim045

Semi-Elemental Nutrition in Internal Medicine Patients

Kara M. Burns, MS, MEd, LVT, VTS (Nutrition)
Wamego, KS, USA

Only recently has nutritional support of critically ill patients been considered a supportive measure of high priority. Advances in both human and veterinary medicine have demonstrated that nutritional support is an important therapeutic modality and does benefit the management of diseases. In diseased states, the inflammatory response triggers alterations in cytokines and hormone concentrations and shifts metabolism toward a catabolic state. When food intake is insufficient, the predominant energy source is derived from accelerated proteolysis, which alone is an energy-consuming process. Consequently, critically ill animals may actually preserve fat deposits in the face of lean muscle tissue loss. The goal of nutritional support in these catabolic patients is to feed the catabolism with exogenous sources of protein and fat, therefore sparing endogenous protein which is critical to recovery.

Malnutrition in veterinary patients is thought to increase morbidity and mortality. In the GI tract, transit times increase, absorptive capabilities decrease, there is villous atrophy and an increased risk of bacterial translocation. In the kidneys, excretion of urinary calcium and phosphorus increases, ability to excrete acid decreases, gluconeogenesis increases, and glomerular filtration rate decreases. Malnutrition has been documented to decrease humoral immunity and barrier function (skin and mucosal surfaces), inflammatory response, leukocyte motility, and bactericidal activity. Patients are at risk for pulmonary complications due to decreased response to hypoxia, decreased lung elasticity, and secretion production, altered permeability, and decreased tidal volume. Cardiovascular complications include increased incidence of arrhythmias and decreased weight of the heart muscle. Protein malnutrition may also alter the normal or expected metabolism of certain drugs, which may increase or decrease their therapeutic effect, even when given at recommended dosages.

Historically, nutritional support was not considered necessary until animals had inadequate intake for 10 days. This concept is outdated and unjustified. Recent evidence suggests a more appropriate goal for most cases is to initiate nutritional support within 3 days of hospitalization (at the latest). Evaluating nutritional status may be challenging for healthcare team members. Historical weight loss may provide some evidence of inadequate intake, and body condition scoring may be helpful to assess fat loss but is not as sensitive for muscle wasting. Muscle condition scoring should also be implemented in compromised patients. Laboratory abnormalities may not be present or may be nonspecific. Because of the limitations in assessing nutritional status, early risk factors must be identified that predispose patients to malnutrition. The risk factors to consider are a history of inadequate nutritional intake lasting more than 3–5 days; serious underlying disease (e.g., severe trauma, sepsis, peritonitis, acute pancreatitis, major gastrointestinal surgery); and large protein losses (e.g., protracted vomiting, diarrhea, protein-losing nephropathies, draining wounds, burns). Patients with these risk factors are candidates for nutritional support.

The gastrointestinal tract needs to be fed. The gut receives an overwhelming percentage of its nutrition from the chyme passing through it. In the small intestine, enterocytes utilize lumenal glutamine preferentially as their source of metabolic fuel. The colonocytes prefer butyrate, a short-chain fatty acid formed by fermentation of lumenal carbohydrates. In the absence of these fuel sources, the gut epithelium slows growth and replication resulting in atrophy, necrosis, and increased risk of bacterial translocation across the now abnormal gut barrier. A key to prevention of this potentially serious problem is providing nutrition support to the gut. The rule of nutritional support is "if the gut works use it."

Placing an indwelling feeding tube is the preferred method if enteral-assisted feeding is necessary for more than two days. After an indwelling tube has been placed, feeding is easier and less stressful. Nasoesophageal, esophagostomy, gastrostomy, and jejunostomy feeding tubes are the most commonly used. In animals undergoing laparotomy, placing gastrotomy or jejunostomy feeding tubes should be considered. The decision to use one tube over another is based on the anticipated duration of nutritional support (e.g., days vs. months), the need to circumvent certain segments of the gastrointestinal tract (e.g., oropharynx, esophagitis, pancreatitis), clinician experience, and the patient's ability to withstand anesthesia (very critical animals may only tolerate placement of nasoesophageal feeding tubes).

Nutritionists debate the exact formula for determining daily energy requirements in ill patients. A good starting point is to calculate the resting energy requirements for the patient's current weight. If a patient's ideal weight is used, underweight animals may be at greater risk for complications from overfeeding. Resting energy requirements can be calculated using the following equation: $RER = 70 \times (\text{current body weight in kg})^{0.75}$. Another equation that may be used is RER = (kg x kg x kg, √, √) x 70.

The nutritional status of the patient must be monitored vigilantly as caloric intake may need to be adjusted to prevent weight loss or unintended weight gain. Food selection depends on tube size and location within the GI tract, the availability and cost of products, and the experience of the clinician. Commercial foods available for enteral use in veterinary patients can be divided into two major types: 1) liquid or modular products and 2) blended pet foods. Nasal and jejunostomy tubes usually have a small diameter (<8 Fr.), which requires use of liquid foods. Orogastric, pharyngostomy esophagostomy and gastrostomy tubes have large diameters (>8 Fr.) and are suitable for blended pet foods. The majority of this discussion moving forward will center on semi-elemental nutrition.

Generally, human liquid foods cost more than veterinary liquid products. Most human liquid foods may be adequate for adult dogs but are too low in protein for cats, puppies, and adult dogs with increased protein losses (e.g., protein-losing enteropathies, drains). Human liquid enteral products may not contain adequate concentrations of protein, taurine, arginine, and arachidonic acid for long-term feeding of cats, but are satisfactory for fewer than seven days. It is recommended that veterinary liquid products be used when managing veterinary patients. It is ideal to utilize nutritional formulations specifically designed for the species you are managing.

Liquid foods are of two basic types: 1) elemental or monomeric and 2) polymeric. Foods said to be "elemental" are not truly elemental - rather they are semi-elemental and contain nutrients in small hydrolyzed absorbable forms and are best described as monomeric. The proteins are usually present as free amino acids, small dipeptides or tripeptides or larger hydrolyzed protein fractions. The fat source is often an oil of mixed (medium- and long-chain) fatty acids and the carbohydrate sources are mono-, di- and trisaccharides. Semi-elemental nutrition provides easier digestion and rapid absorption of nutrients, designed to support the critical systems of the body.

The gastrointestinal tract is lined with cells called enterocytes, which aid in digestion, absorption, and transportation of nutrients into the body. Vital nutrients absorbed by enterocytes include amino acids and peptides, carbohydrates (complex and simple saccharides), lipids, water, vitamins, and ions (or minerals). Enteral nutrition is preferred whenever possible, because enterocytes undergo atrophy without luminal nutrient stimulation. By focusing on the nutritional needs of the enterocyte, semi-elemental diets provide the maximum amount of nutrition that can be absorbed with the least energy expenditure.

In ill patients, the body's main goal is to support key organs like the heart, brain, liver, and lungs. Blood flow to the gastrointestinal tract is reduced, thus slowing gastrointestinal motility. As enterocytes form a tight seal to protect the body from harmful bacteria or toxins, digestion and absorption of nutrients are similarly altered. Unfortunately, these physiologic changes make it difficult for nutrients to be absorbed when the body needs nutrition the most. Thus, weakened and debilitated patients benefit from a predigested or a semi-elemental diet.

Semi-elemental diets contain ingredients that are purified and hydrolyzed, and are usually combined with a small amount of complex highly digestible ingredients. Semi-elemental diets that are nourishing the enterocytes of the GI tract help maintain an osmotic balance, which minimizes the risk of diarrhea by preventing the loss of water and other vital nutrients from the gastrointestinal tract.

There are several liquid foods on the human medical market that are positioned as monomeric or hydrolyzed diets; however, they may not be suitable for the canine or feline species. There are feline- and canine-specific semi-elemental diets that are recommended for the critically ill cat, dog, avian, and exotic species. These products can be fed through any feeding tube including a J-tube. Monomeric foods are indicated in disease conditions such as inflammatory bowel disease, lymphangiectasia, refeeding parvoviral enteritis and pancreatitis cases and any other condition in which a patient's digestive capabilities are questionable.

Polymeric products contain mixtures of more complex nutrients. Protein is supplied in the form of large peptides (e.g., casein or whey). Carbohydrates are usually supplied as corn starch or syrup, and fats are provided by medium-chain triglycerides (MCT) or vegetable oil. These foods require normal digestive processes and are appropriate for most veterinary clinical situations, especially when a small tube (<8 Fr.) has been placed and particular nutrient profiles are needed (e.g., low sodium, high protein, soluble fiber).

Module products are concentrated powdered or liquid forms of nutrients and are primarily supplemental. These products may be added to a liquid product to increase the concentration of a specific nutrient. There are protein, fat and carbohydrate modules (e.g., casein powder, vegetable oil or corn syrup). For example, a protein modular product may be added to a human liquid product for an animal with high protein requirements. Soluble fiber can be added to these foods using psyllium husk fiber or pectin; however, these fibers may block the small side ports in 8-Fr. and smaller tubes.

The feeding schedule is often determined by the patient's ability to tolerate food and the logistics of feeding. Feeding an amount equal to the patient's RER during the first 24 hours of food re-introduction, if physically tolerated, is recommended. Feeding one-third of the RER and then increasing the amount by one-third every 24 hours is a more cautious approach to initial feeding. Foods should be warmed to room temperature, but not higher than body temperature before feeding.

Food boluses must be infused slowly (over approximately one minute per bolus) to allow gastric expansion. Daily food dosage should be divided into several meals according to the expected stomach capacity. Capacities for cats and dogs are 5 to 10 ml/kg body weight during initial food reintroduction. Maximum capacities as high as 45 to 90 ml/kg body weight have been measured in cats and dogs when fully re-alimented. Most often, meeting the patient's RER can be done in volumes far less than these maximums. Salivating, gulping, retching, and even vomiting may occur when too much food has been infused or when the infusion rate is too fast.

Research in people has demonstrated that the stomach does not "shrink" during a prolonged fast as was once thought to be the case; but rather the stretch receptors are more sensitive and stimulated by a smaller volume when refeeding occurs. Feeding should be stopped at the first sign of retching or salivating, the meal size reduced by 50% for 24 hours and then increased by 25% gradually. Foods provided via J-tubes must be infused slowly and often in either very small quantities or by a slow gravity drip or enteral pump with an hourly rate equal to RER/24 hours because the jejunum is volume sensitive. An additional strategy to reduce complications when feeding a liquid diet would be to utilize a continuous trickle feeding.

Remember to follow each meal with water flush to clear the feeding tube of food residue. When the patient is volume sensitive, it is important to know the minimum volume required to flush the tube. The patient's daily fluid requirement must also be met and additional tap water may be administered through the feeding tube to meet that requirement. Liquid oral medications may also be administered easily through feeding tubes. Plugged feeding tubes can be cleared by filling the tube with water or a nonalcoholic carbonated beverage and allowing time for the food plug to dissolve. In general, endport tubes are easier to maintain than sideport tubes because food tends to become trapped in the blind end of sideport tubes. All tubes except orogastric and nasoesophageal tubes require standard every-other-day bandage care.

Summary

The consequences of malnutrition in all patients, especially sick or injured patients, are decreased immunocompetence, decreased tissue synthesis and repair, and altered drug metabolism. Nutritional management in ill internal medicine patients require constant monitoring and careful calculations by the veterinary technician to address malnutrition. Semi-elemental diets provide the maximum amount of nutrition that can be absorbed with the least energy expenditure, an important combination when feeding critical patients. Constant vigilance and exceptional nursing care will help the patient manage their malnutrition and facilitate healing and recovery.

References

1. Kerl ME, Johnson PA. Nutritional plan: matching diet to disease. *Clin Tech Small Anim Pract.* 2004:19(1):9–21.
2. Thomovsky E, *et al.* Parenteral nutrition: uses, indications, and compounding. *Compend Contin Educ Vet.* 2007;29:76–8, 80–5.
3. Thomovsky E, *et al.* Parenteral nutrition: formulation, monitoring, and complications. *Compend Contin Educ Vet.* 2007;29:88–102.
4. Chan DL, Freeman LM. Nutrition in critical illness. *Vet Clin North Am Small Anim Pract.* 2006;36(6):1225–41.
5. Proulx J. Nutrition in critically ill animals. In: *The Veterinary ICU Book.* Jackson Hole, WY: Teton NewMedia; 2000:202–217.
6. Saker K, Remillard RL. Critical care nutrition and enteral-assisted feeding. In: *Small Animal Clinical Nutrition.* 5th ed. Topeka, KS: Mark Morris Institute; 2010:439–476.
7. Chandler ML, *et al.* Use of peripheral parenteral nutritional support in dogs and cats. *J Am Vet Med Assoc.* 2000;216:669–73.
8. Remillard RL, Saker K. Parenteral-assisted feeding. In: *Small Animal Clinical Nutrition.* 5th ed. Topeka, KS: Mark Morris Institute; 2010:477–498.
9. Eirmann L, Michel K. Enteral nutrition. In: *Small Animal Critical Care Medicine.* St. Louis, MO: Saunders; 2009: 53–58.
10. Ford RB, Mazzaferro E. Kirk and Bistner's Handbook of Veterinary Procedures and Emergency Treatment. 9th ed. St. Louis, MO: Elsevier; 2012.
11. Martindale RG, Shikora SA, Nishikawa R, Siepler JK. In: American Society for Parenteral and Enteral Nutrition, ed. *Nutritional Considerations in the Intensive Care Unit: Science, Rationale, and Practice.* Dubuque, IA: Kendall/Hunt Publishing Company; 2002.
12. Remillard RL. Nutrition: feeding hospitalized patients for best clinical outcome: a guide to oral, tube feeding and estimating caloric requirements. 2009. *DVM360.*

acvim046

Parathyroid Disorders in Dogs and Cats

Tracey Nowers, CVT, VTS
Sandwich, MA, USA

There are 4 parathyroid glands located within the neck region of dogs and cats. They are located near the thyroid gland, but have no functional relation to it. Parathyroid glands consist of mainly chief cells. The chief cells secrete parathyroid hormone (PTH). PTH regulates serum calcium, which in return regulates PTH, within a negative feedback loop.

When blood calcium decreases, PTH is secreted by the parathyroid glands. PTH stimulates osteoclasts. These are the cells that break down bone and send calcium into blood stream. This causes blood calcium to increase, which stops PTH secretion. The increase in blood calcium also causes calcitonin to stimulate osteoblasts to take calcium from blood and build bone. This action decreases blood calcium. The feedback loop begins again when blood calcium decreases.

When there is an increase in PTH, calcium excretion by the kidney decreases, which causes hypercalcemia. In turn, the body excretes more phosphorus which leads to hypophosphatemia.

In addition, when blood calcium is low, PTH instructs kidneys to start making more vitamin D. There are vitamin D stores in the body that are ready for the kidney to produce when needed. When the kidney starts producing vitamin D, it helps the gastrointestinal tract absorb more calcium. When vitamin D increases and calcium homeostasis is achieved, PTH secretion stops.

This interaction between calcium, phosphorus, parathyroid hormone and vitamin D is happening constantly.

HYPERPARATHYROIDISM

Hyperparathyroidism is the unregulated overproduction of parathyroid hormone (PTH) resulting in abnormal calcium homeostasis.

There are 2 types in cats and dogs: primary and secondary hyperparathyroidism.

Primary Hyperparathyroidism

Primary hyperparathyroidism (PHPT) is not related to another disease. Excessive PTH production is usually caused by a parathyroid adenoma, parathyroid carcinoma, or hyperplasia. Multiple parathyroid adenomas are possible, but rare.

It is seen more commonly in dogs, 6 years and older with median age of 11 years at diagnosis. There is no gender predisposition. The highest breed incidence is in Keeshond. In cats, the median age is 13 years, with no apparent gender predisposition.

Many patients are asymptomatic. However, most clinical signs are attributed to the hypercalcemia. One of the most popular clinical signs of PHPT is polyuria and polydipsia from decreased renal tubular response to antidiuretic hormone.

Hematuria, stranguria, pollakiuria, and urinary tract obstruction can occur. Dilute urine-specific gravity predisposes patients to urinary tract infections. Calcium phosphate or oxalate urolithiasis may result from calciuria and phosphaturia. Lethargy, muscle wasting, and stiffness can occur from decreased neuromuscular tissue excitability. Inappetence, vomiting, and constipation are caused by decreased excitability and motility of GI smooth muscle.

Diagnosis

When a blood chemistry shows an increased total calcium, hypercalcemia can be confirmed with an ionized calcium. There are several different forms of calcium in blood. These include ionized calcium, calcium bound to other minerals called anions, and calcium bound to proteins like albumin. Ionized calcium, also known as free calcium, is the most active form.

Additionally, a malignancy panel can be sent to Michigan State to evaluate ionized calcium level, vitamin D level, parathyroid hormone level, and a protein called parathyroid-related protein that can be secreted by some cancers. PTH values that are increased or in the upper half of reference range are consistent with PHPT in hypercalcemic patients.

Imaging may also be helpful to help diagnosis. A cervical ultrasound can be performed to try to locate the nodule. But they are generally small. This would also aid the surgeon in locating a parathyroid tumor, if one was visualized. Nuclear imaging can also aid in diagnosing PHPT, while abdominal imaging or thoracic radiographs can point to other causes of hyperparathyroidism.

In cats, a mass can usually be palpated.

Treatment

If hypercalcemia is severe, diuresis using a non-calcium-containing electrolyte solution is suggested prior to surgery. A diuretic, such as furosemide, may be needed to enhance the excretion of calcium in the urine. Glucocorticoids may also be given to enhance urinary excretion of calcium. Additionally, bicarbonate may be given, in conjunction with intravenous saline and diuretics, to shift the amount of ionized calcium (the dangerous form) to unionized calcium (a safer form).

Surgery is required to remove the abnormal parathyroid gland(s). They can be difficult to locate during operation because they often look like just another lobule of thyroid gland.

Removing 1 or 2 glands in dogs usually requires no or short-term management for hypocalcemia, but other patients require permanent therapy. For clinically stable patients, therapy may include oral vitamin D therapy and/or calcium supplementation until the function of the remaining parathyroid glands returns to normal.

If hypocalcemia is severe, intravenous calcium supplementation may be necessary. Removing all 4 glands requires lifelong supplemental therapy.

Postoperative monitoring of ionized calcium is recommended every 12–24 hours for 7 days. The decline in PTH and ionized calcium occurs over 1 to 7 days. Monitoring should be continued until the calcium level has returned to normal. If 1 or 2 glands were removed, weaning of supplementation should begin in a couple weeks. Levels should be monitored during weaning process, then checked every 6 months.

Secondary (Tertiary) Hyperparathyroidism

Parathyroid glands are overproducing parathyroid hormone in response to hypocalcemia in an attempt to increase blood calcium. It occurs most commonly secondary to chronic kidney disease and severe gastrointestinal disease The commonality for patients with secondary hyperparathyroidism is that the intestines do not absorb enough calcium, whether the problem is with the intestinal wall or not having enough vitamin D from the kidneys, or nutritionally.

Diagnosis

Diagnosis includes high parathyroid hormone levels, hyperphosphatemia and low blood calcium.

Treatment

Treatment includes supportive care for the underlying disease process, and vitamin D and calcium supplementation. Hyperphosphatemia is treated with fluid therapy to correct electrolyte imbalances and oral phosphorus binders (e.g., aluminum hydroxide) are also administered.

HYPOPARATHYROIDISM

Hypoparathyroidism is a condition of parathyroid hormone (PTH) deficiency. It is usually diagnosed as idiopathic but can occur as a complication secondary to thyroidectomy (accidental removal or injury) or parathyroidectomy. Transient hypoparathyroidism may occur with rapid correction of long-standing hypercalcemia (hypercalcemia causing temporary atrophy of the gland), as can occur with surgical removal or ablation of a parathyroid adenoma causing primary hyperparathyroidism.

Clinical Signs

Clinical signs include seizures, muscle tremors, muscle cramping, muscle weakness, stiff gait, and behavioral changes (e.g., restlessness, aggression, hypersensitivity to stimuli, disorientation). Less commonly, they include facial rubbing, biting and licking paws, excessive panting, vomiting, diarrhea, weight loss. Loss of the normal membrane-stabilizing effect of ionized calcium on nerve cells causes increased excitability of central and peripheral nervous tissue.

Diagnosis

Diagnosis is made by evaluating a serum PTH concentration and a concurrent ionized calcium concentration. Patients will have decreased PTH levels, hypocalcemia, and hyperphosphatemia.

Treatment

Hospitalization is necessary for monitoring of seizures, hyperthermia, and other clinical signs of hypocalcemia. Hyperthermia in hypocalcemic patients usually derives from muscle fasciculations, not true fever. Therefore, cooling measures may be appropriate (fan/cooling paw pads), but anti-pyrexic drugs are not.

The goal of therapy is to increase serum calcium concentrations above the threshold for clinical signs and maintain it just below or at the low end of the reference range. Therapy should include use of calcium supplements and vitamin D. Parenteral supplemental administration is usually needed initially while waiting for the oral form to take effect.

If hypocalcemic tetany or seizures are present, calcium should be administered slowly intravenously immediately. Calcium gluconate is available in both oral and injection formulations. Calcium chloride is more irritating and is more likely to cause tissue necrosis with extravasation, making calcium gluconate the preferred option When the hypocalcemia is slow evolving, oral calcium gluconate or carbonate is suitable for use. However, if the hypocalcemia is severe, the injectable formulation is the preferred means of administration. Monitoring an electrocardiogram (ECG) for potential cardiac arrhythmias is important during use of IV calcium supplementation.

For oral vitamin D supplementation, calcitriol comes in a form that does not need an activation process. It helps to regulate the amount of 2 minerals in your body: calcium and phosphate.

SUMMARY

The parathyroid hormone controls extracellular calcium and phosphate concentrations by regulating intestinal reabsorption, renal excretion and exchange between the extracellular fluid and bone of these ions. Not only is calcium homeostasis important to bone density, but it also affects muscle contraction, heart contractibility, neurons in nervous system to send signals and coagulation.

REFERENCES

1. Stephen J. Ettinger, Edward C. Feldman. *Textbook of Veterinary Internal Medicine*. Sixth edition. 2004.
2. John E. Hall. *Guyton and Hall Textbook of Medical Physiology*. Thirteenth edition. 2016.
3. Vincent A. Ziglioli, David Grant. Interpretation of parathyroid abnormalities. *Clinician's Brief*. December 2015.

acvim047

Understanding Diabetic Remission in Cats

Laura Van Vertloo, DVM, MS, DACVIM (SAIM)
Ames, IA, USA

Introduction

Many cats treated for diabetes mellitus later go on to develop resolution of clinical signs and maintenance of normoglycemia without administration of exogenous insulin, a state referred to as diabetic remission. This has been an area of active research over the past decade and the goals of therapy in managing feline diabetics have subsequently shifted. While treatment objectives used to consist of maintaining reasonable glycemic control in order to eliminate clinical signs and avoid the complications of untreated diabetes mellitus, it is now increasingly common to treat feline diabetics with the additional goal of attaining (and maintaining) a state of diabetic remission.

Pathophysiology

The majority of cases of feline diabetes mellitus are thought to resemble type II diabetes mellitus in humans. Type II diabetics (and, therefore, most diabetic cats) have a relative insulin deficiency due to the combined effects of one or more sources of insulin resistance and varying degrees of dysfunctional β-cells within the pancreatic islets. In contrast, type I diabetics have an absolute deficiency of endogenous insulin production and dependence on lifelong exogenous insulin administration. This is the most common form seen in dogs and is thought to be rare in the cat.

Insulin Resistance

Insulin resistance describes a state in which the action of insulin is impaired at the level of the insulin-dependent tissues (mainly liver, muscle, and adipose), leading to a decrease in glucose uptake by these tissues, and a lack of suppression of hepatic gluconeogenesis and glycogenolysis. Causes of insulin resistance are diverse and include sources of inflammation, administration of diabetogenic drugs, and obesity.

β-Cell Dysfunction

It is important to note that some cats with insulin resistance respond appropriately with increased insulin secretion in order to maintain normoglycemia. In order for overt diabetes mellitus to develop, cats with insulin resistance must also have some degree of β-cell dysfunction. Those cats whose β-cells cannot keep up ultimately become hyperglycemic and develop clinical signs of diabetes mellitus. The cause of β-cell failure in cats is not clear but is likely multifactorial with amyloid deposition and inflammation potentially playing a role.

Glucotoxicity, which is impaired insulin secretion in response to hyperglycemia, is an important factor contributing to β-cell dysfunction in diabetic cats. Sustained, experimental hyperglycemia from IV glucose administration in healthy cats resulted in a reduction of endogenous insulin production to levels consistent with type I diabetes. Following discontinuation of IV glucose, these cats had a transient dependence on exogenous insulin administration to maintain normoglycemia as their β-cells recovered. Glucotoxicity has also been associated with the development of variable pancreatic islet lesions in cats with experimental hyperglycemia and naturally occurring diabetes mellitus. Although glucotoxicity is not the triggering cause of feline diabetes mellitus, since overt diabetes with hyperglycemia has to be present for it to occur, the ability to reverse or partially reverse β-cell dysfunction with treatment is thought to be a major explanation for how and why some cats go into diabetic remission.

Managing Feline Diabetics with the Goal of Remission

Which Insulin Is Best?

Administration of insulin is the most important aspect of treatment for the diabetic cat. Because of the influence of glucotoxicity on progressive β-cell failure, early, effective glycemic control is critical for reaching diabetic remission and insulin is the most effective way to achieve this. There have been several studies in recent years looking at the effect of different insulin types on rates of diabetic remission; however, due to inconsistencies between study methods, it is difficult to conclude that one type of insulin is superior to the others for feline diabetic management.

The three types of insulins currently used most commonly in diabetic cats are porcine lente insulin (Vetsulin), protamine zinc insulin (ProZinc) and insulin glargine (Lantus). NPH is frequently problematic in cats because of its potency and short duration of action and is not recommended as a first-line insulin, although it has been associated with remission in one study.

Porcine lente insulin (Vetsulin) has been FDA approved for use in cats and is classified as having an intermediate duration of action. Published remission rates for this insulin type have varied from 11–43%, depending on the study design. The highest remission rates were reported in studies done on newly diagnosed diabetics that were concurrently fed a standardized low-carbohydrate diet.

Protamine zinc insulin (PZI) is a long-acting recombinant human insulin that is also FDA approved for use in cats. Although PZI has not yet been extensively studied specifically pertaining to remission in cats, reported remission rates in two studies were 38% and 68% in cats fed a low-carbohydrate/high-protein diet.

Glargine, a long-acting insulin analogue, has received the most literature attention pertaining to diabetic remission, with 5 total studies reporting remission rates from 17–100%. In a study comparing newly diagnosed diabetic cats receiving a low-carbohydrate diet and either lente, PZI or glargine insulin, the glargine group had the highest rate of remission. Another long-acting insulin analogue, insulin **detemir**, has recently been evaluated and shown to be associated with remission rates similar to glargine.

Diabetic Monitoring

It has been established in a number of studies that cats are more likely to achieve remission if they have had a shorter duration of diabetes mellitus and if glycemic control is achieved rapidly, supporting the role of glucotoxicity reversal in the development of remission. Thus, in order to increase the chance of remission it is essential to not only start treatment with insulin early but to provide effective monitoring to optimize glycemic control.

Owners of newly diagnosed diabetic cats should be prepared for 3–4 months of frequent veterinary visits to allow time to find an optimal insulin dose and also to allow more careful assessment for hypoglycemia during the period of time when cats most frequently experience remission.

Home monitoring is gaining more popularity in cats due to the potential for stress-induced hyperglycemia to interfere with accurate interpretation of blood glucose curves generated in the hospital. Additionally, home monitoring may improve chances of diabetic remission. Some of the studies reporting the highest rates of remission with glargine and detemir involved a German Internet-based diabetes forum for cat owners. Cats enrolled in these studies had intensive home monitoring and dose adjustments with the goal of targeting normoglycemia. Owners were instructed to frequently measure blood glucose and adjusted insulin dose at home while following a strict dosing protocol. While remission rates were high in these groups of cats, the requirements for cat owners may not be realistically applied to the general pet-owning population.

Nutrition

A low-carbohydrate/high-protein diet has been shown in multiple studies to be associated with improvement in glycemic control and a higher rate of remission when compared to traditional high-fiber diets. When such a diet is not contraindicated due to concurrent disease processes, cats with diabetes mellitus should be transitioned to one of these diets. Owners of overweight or obese cats should also be counseled on weight loss and specific caloric recommendations should be made for these patients with a goal rate of weight loss of 0.5–1% of body weight per week.

Other Factors Predicting Remission

Other than achieving rapid glycemic control in a cat with a short duration of diabetes mellitus, there are a few other factors that have been associated with an increased chance of diabetic remission.

- **Recent corticosteroid treatment**. This is presumably due to the removal of a source of insulin resistance (exogenous corticosteroid) that likely contributed to the development of overt diabetes mellitus.
- **Lack of hypercholesterolemia** has been shown to be associated with remission. Hypercholesterolemia may reflect poor glycemic control. Additionally, excess lipid may also decrease β-cell function (lipotoxicity), although the evidence for this is not as strong as for glucotoxicity.
- **Lack of peripheral neuropathy**. Presumably, cats with peripheral neuropathy have experienced more prolonged hyperglycemia than cats without, making them less likely to achieve remission.

Relapse

While it can be tempting to say that a cat in diabetic remission has been cured of its diabetes, this is not the case. Approximately 25–50% of cats in remission will ultimately relapse and go back to needing exogenous insulin for control of clinical signs. The majority of cats that have gone into remission have impaired glucose tolerance and can be considered "prediabetic." Cats that have achieved remission should continue on dietary management (low-carbohydrate/high-protein). Veterinarians should be vigilant about reducing sources of insulin resistance, including maintaining or targeting an ideal body condition and avoiding exogenous glucocorticoids.

Summary Points

- The majority of cats with diabetes mellitus have a form resembling human type II diabetes and have developed overt signs of diabetes mellitus due to a combination of insulin resistance and β-cell dysfunction.
- Chronic, sustained hyperglycemia in an untreated diabetic contributes to impaired insulin release (glucotoxicity), which is at least partially reversible with treatment and likely explains why remission can occur after initiating treatment.
- Early treatment with insulin and effective glycemic control are key to reversing glucotoxicity and improving chances of remission.
- All insulins used in cats have the potential to result in diabetic remission if effective glycemic control is obtained.
- While comparison of insulin remission rates between studies with a wide variety of confounding factors is difficult, the evidence available suggests that glargine may be more likely to result in remission, but additional studies are needed to confirm this.
- Careful monitoring with frequent visits to assess clinical signs, glucose curves, and fructosamine is recommended following diagnosis to optimize glycemic control early on, and home monitoring may facilitate this and improve chances of remission.
- Cats with diabetic remission can relapse and should be treated as prediabetic.

References

1. Gottlieb S, *et al. J Vet Intern Med.* 2015;29(1):184–192.
2. Gostelow R, *et al. Vet J.* 2014; 202(2):208–221.
3. Marshall RD, *et al. J Feline Med Surg.* 2009;11:683–691.
4. Kley S *et al. J Am Vet Med Assoc.* 2004;225;261–266.
5. Roomp K, Rand J. *J Feline Med Surg.* 2009;11:668–682.
6. Roomp K, Rand J. *J Feline Med Surg.* 2012;14:566–572.
7. Bennett N, *et al. J Feline Med Surg.* 2006;8:73–84.
8. Link KR, Rand JS. *J Vet Intern Med.* 1996; 10: 185.
9. Nelson RW, *et al. J Vet Intern Med.* 1999;13(1):28–35.
10. Michiels L, *et al. J Feline Med Surg.* 1998;10:439–451.

acvim048

Stenting for the Veterinary Technician; Procedure, Equipment and Techniques

Meri Hall, LVT, RVT, LATG, VTS (SAIM)
Palm Beach Gardens, FL, USA

Stenting, by definition, is the placement of a tube inserted into the lumen of a vessel. In veterinary medicine, we use stents in many conditions. Collapsing trachea and urethral obstruction are among the most common. The general principles for stenting are similar for all uses.

There are several different types of stents with the self-expanding metal stents being the most common. They are used for respiratory, urinary, gastrointestinal, hepatobiliary, and cardiovascular stenting procedures. Self-expanding metal stents are a fine wire mesh that is preloaded on an introducer. These stents may be repositioned prior to complete deployment, but, once placed, are difficult to remove. The self-expanding metal stents do have foreshortening, which is where they overexpand during placement and then will return to a predetermined length and diameter. A woven metal self-expanding stent is similar to the self-expanding metal with the exception that the metal strands are woven and are more cloth like. These stents are more flexible and may not be as rigid in placement.

Laser cut stents are compressed and when placed the body temperature causes the stent to expand. They are typically used in interventional radiology and in the urethra and cardiovascular system. Balloon-expandable metallic stents are preloaded over a balloon deployment system. They are typically utilized in nasopharyngeal stenosis. They are stainless steel stents that will maintain their expanded shape unless manually compressed.

In all cases, premeasurements of the affected lumen must be made prior to performing the procedure. Many times, this can be achieved with digital radiology. A marker catheter is placed in a natural orifice adjacent to the lumen to be stented to allow for calibration due to magnification that naturally occurs. A contrast agent is injected into the lumen to be stented allowing for accurate measurements.

Tracheal Stenting

For tracheal collapse, the patient is placed under general anesthesia and an image is obtained in lateral recumbency while 20 cm H_2O positive pressure is performed. The trachea is then measured at the cervical, thoracic, and intrathoracic levels with a marker catheter placed in the esophagus. The diameter of the stent selected is 10–20% greater than the maximal tracheal measurement. The length is measured 1 cm caudal to the cricoid cartilage and 1 cm cranial from the carina. The goal is to span the entire portion of the trachea that is collapsed. For facilities that do not maintain a stock of stents, this can be done the week prior to placement.

The day of placement, the patient is placed under general anesthesia and placed in lateral recumbency on the fluoroscopy table. The endotracheal tube is pulled to just caudal to the larynx. The tracheal stent is placed through the endotracheal tube and, utilizing fluoroscopy, is placed at the premeasured location. The stent is then slowly deployed while confirming placement with the fluoroscope. Postplacement radiographs are taken to confirm proper placement. Postprocedural complications include stent shortening, collapse of the trachea cranial or caudal to the stent, stent fracture, granulation tissue formation, and bacterial tracheitis.

Intravascular Stenting

Central venous obstruction stenting is performed in some facilities. Fluoroscopy is utilized and the stent is deployed over a hydrophilic guide wire. Access is made via the jugular vein or femoral vein.

Nasopharyngeal Stenting

In cases of nasopharyngeal stenosis, a balloon-expandable metallic stent is most commonly used. Prior balloon-expanding procedures may be utilized prior to placement of the stent. The patient is again placed under general anesthesia to obtain the measurements. The size of stent chosen is 10–20% larger than the vessel. The stent should not extend within 1 cm of the caudal aspect of the soft palate. Complications are uncommon.

URETHRAL STENTING

Urethral stenting is used many times in the case of transitional cell carcinoma. The placement can be done with surgical assistance or fluoroscopy or a combination. Contrast media is utilized to highlight the area of obstruction. A weasel wire is used to assist in the placement of the stent by first bypassing the obstruction before the stent is placed over the guidewire. The stent diameter is based on the maximal diameter of the surrounding normal urethra and the length measured to be 1 cm cranial and caudal into normal tissue surrounding the obstruction. The stent is passed over the guidewire and slowly deployed under fluoroscopy. Postprocedural radiographs are obtained to confirm placement.

ESOPHAGEAL STENTING

For esophageal stenting, lateral radiographs are obtained to measure the stricture. Balloon expansion is done prior to placement of the stent. The stent should be 1–2 cm cranial and caudal to the stricture. The stent is placed and is sutured in place. The complication for esophageal stenting is esophageal tearing.

In all stenting procedures, the luminal obstruction is measured prior to stent placement. The type and size of the stents used are veterinarian driven. The technician should be familiar with the procedure, the specific equipment required based on which stent is being used and be comfortable with the postprocedural care required.

REFERENCES

1. Weisse C. In: Ettinger SJ, Feldman EC, Côté E, eds. *Textbook of Veterinary Internal Medicine*. 8th ed. Elsevier; 2017: 459–466.
2. Beal MW. In: Ettinger SJ, Feldman EC, Côté E, eds. *Textbook of Veterinary Internal Medicine*. 8th ed. Elsevier; 2017: 466–472.
3. Weisse C, Berent A, eds. *Veterinary Image-Guided Interventions*. Wiley-Blackwell; 2015:12–14.

acvim013

Acquisition and Analysis of Joint Fluid

Carmela Pratt, MS, DVM, DACVIM
Saint Louis, MO, USA

Joint fluid acquisition and analysis is an important part of identifying various joint abnormalities and diseases. These vary from infectious, inflammatory, and neoplastic conditions. Arthropathies can be classified as erosive vs. non-erosive, and infectious vs. non-infectious. Local joint swelling can occur with an infection of a solitary joint that occurred secondary to trauma. Polyarthropathy can occur from an infection or non-infectious disease. The indication for acquiring joint fluid is based off clinical signs, which vary from ADR, fever of unknown origin, lameness, shifting leg lameness, back or neck pain, or inability to walk. These clinical signs can also indicate other disease states and would have to be ruled out along with a polyarthropathy. Many times a stepwise approach is used ranging from the least invasive to the most invasive diagnostics. Various diagnostics may be needed to look for an underlying cause for the joint disease, which need to be ruled out so the proper treatments can be provided. Joint taps are considered less invasive due to their simple acquisition and the minimal anesthesia needed to acquire them. There is low risk for contaminating infection providing sterile technique is followed.

Sites for Joint Fluid Acquisition

There are several sites that can be used to acquire joint fluid samples. In general, unless only one limb is affected, at least 3 joint taps are performed on different joints. This ensures that all the joints are affected, such as with immune-mediated polyarthropathy, one of the most common reasons for obtaining joint fluid. Sites that are commonly used are the stifle, carpus, hock, elbow, and hip. The joint fluid can be obtained from the same joint but on a different limb to count as a second joint sample.

Preparation of Sites

The selected sites should be shaved approximately 1 inch on each side from the sampling site. The site is then scrubbed with chlorhexidine and ethanol interchangeably at least 3 times. After the site is cleaned then it is considered sterile and nothing should come in contact with the area. Sterile gloves are placed out for the doctor performing the procedure along with needles and syringes placed on the sterile field when the glove pack is opened. The technician may be asked to hold the limb in position to allow the doctor to remain sterile with both gloves. Alternatively, the doctor may hold the limb with one hand, considered the dirty hand, and palpate and aspirate with the clean hand. This is the doctor's preference. If the technician is required to hold the limb it is important that the area remains sterile. You will likely be asked to flex and extend the joint for the doctor for them to locate the proper site to stick the needle to obtain a sample. During the procedure the doctor should avoid bumping the bone too much as this can cause blood contamination. Several attempts may be needed to obtain a clean sample.

Fluid Sample

Analysis of the fluid sample is performed by placing a sample in a red top tube (without serum separator), in a purple top EDTA tube, and/or a culture swab. The fluid in the EDTA tube or directly from the syringe is used to make a few slides. Ideally the slide would be made directly from the syringe. The remainder can be submitted to the lab if they wish to make more slides. The red top tube and culture swab are used to culture the synovial fluid.

In a normal joint there is no more than ¼ to ½ mL of clear fluid depending on the joint being sampled. The fluid will be viscous and clear. The cytology of normal joint fluid may reveal a few mononuclear cells, but is otherwise fairly clean and boring.

In a non-erosive inflamed joint (non-infectious), such as with idiopathic immune-mediated polyarthropathy, there would contain excess fluid in the joint compared to a normal joint. The viscosity of the fluid would be decreased and will be discolored (yellow, pale brown). The cellular content of the fluid would contain numerous non-degenerative neutrophils. A few mononuclear cells may be present. This is diagnostic for immune-mediated polyarthropathy when diagnosed in 3 joints.

Infectious arthropathy is similar to inflammatory immune-mediated arthropathy except there are generally degenerative neutrophils with bacteria. You cannot always visualize the bacteria. Infectious arthropathies tend to affect one joint if related to trauma or multiple joints is related to a systemic infection. You will be able to collect up to a few milliliters of fluid depending on how affected the joint is. A culture is necessary to diagnose these cases.

Erosive immune-mediated polyarthritis is caused by immune destruction of the bone and joint region. Diseases such as rheumatoid arthritis in dogs or periosteal proliferative polyarthritis in cats are responsible for this. Radiographs would show periosteal proliferation and lysis of bone of the joint. The joint fluid will reveal neutrophilic inflammation is seen but other inflammatory cells will also be visualized (lymphocytes, macrophages).

Neoplasia can be diagnosed by visualizing neoplastic cells. The types of neoplasia that can occur are synovial cell sarcoma, fibrosarcoma, rhabdomyosarcoma, osteosarcoma, malignant fibrous histiocytoma (MFH), liposarcoma, hemangiosarcoma, myxoma, malignant giant cell tumor of soft tissue, and undifferentiated sarcomas. It is not often that the cells of these tumors exfoliate very well. Most often these are diagnosed with more aggressive means such as with bone marrow biopsies or surgical biopsies and can often be seen with radiographs.

Treatment

The treatment is variable among the different diseases. For immune-mediated polyarthropathy, immunosuppressive treatment is used. This includes a combination of prednisone, azathioprine, mycophenolate, or cyclosporine. The recommendation is to repeat joint taps approximately 3–4 weeks after control of disease is performed to determine response to treatment. Alternatively, some base control of disease solely on clinical signs. Once the joints are normal or the animal is not clinical anymore then you begin to taper the medication. Rechecks should be performed every 3–4 weeks until the dog is completely off its medications.

For infectious arthropathy the treatment is based off a culture and sensitivity testing. Some antibiotics that might be started empirically would be cephalexin, doxycycline, or Clavamox. These encompass many of the typical pathogens that infect joints. These pathogens include *Staphylococcus*, *Streptococcus*, and *Mycoplasma*.

Neoplastic arthropathies need further treatment with an oncologist. Some options that would exist would be radiation therapy, some chemotherapies, and amputation. It would be imperative to stage the patient to make sure there is no spread to other organs before pursuing treatment.

Prognosis

The prognosis would depend on the type of arthritis present. Immune-mediated polyarthropathy and infectious arthropathy have a good prognosis for recovery. Erosive polyarthropathies lead to permanent arthritis and joint pain after the underlying immune reaction is corrected. They may need intermittent NSAID therapy to keep them comfortable. Depending on the tumor identified, neoplasia can have a fair to poor prognosis for recovery.

acvim014

Antibiotics: Pearls for the Veterinary Technician

Dennis Spann, DVM, DACVIM (Small Animal Internal Medicine)
Roseville, CA, USA

INTRODUCTION

Sir Alexander Fleming identified penicillin the first Western natural antibiotic in 1928. That discovery ushered in an era of unprecedented improvement in medical care. Penicillin was followed by more discoveries including cephalosporins - in 1948, carbapenems - in 1976 and fluoroquinolones - in 1980. These tools allowed us to fight natural infections and complications of medical procedures. Unfortunately antibiotic resistance became apparent almost immediately and has become increasingly prevalent. Healthcare acquired infections (HAIs) have emerged as a major health care risk. Antibiotic resistance is an increasingly prevalent aspect of HAIs. Inappropriate use of antibiotics encourages the development of bacterial resistance. The rate of new antibiotic development has slowed dramatically, so replacement is only part of the solution to combat antibiotic resistance. Proper and optimal use of our current antibiotics is essential to preserving these valuable tools. Effective use of antibiotics requires knowledge of their chemistry, spectrum of activity, and mechanism of action. Specific use of appropriate antibiotics decreases antibiotic resistance development.

CLASSIFICATION OF ANTIBIOTICS

Antibiotics can be classified by a number of different methods. Important categories include: chemical class or family of antibiotic, spectrum of activity, bactericidal or bacteriostatic. Antibiotics can also be categorized as dose dependent and time dependent killing. The choice of which antibiotic to use when depends on what is know or suspected about the infection that is being treated. It is important to consider which antibiotics should be readily available for use and which antibiotics might need reservation for special infections.

Families of Antibiotics

Beta-lactams

Beta-lactams antibiotics include penicillins, cephalosporins, carbapenems, monobactam (Aztreonam). The beta-lactams are cell wall inhibitors that bind penicillin-binding proteins and prevent cell wall cross-linking. These antibiotics are bactericidal and work in a time dependent fashion. Killing bacteria is dependent on how long the serum concentration is above the MIC. Beta-lactams are ineffective against atypical intracellular organisms like mycoplasma, etc. The SPICE-A organisms *(Serratia, Pseudomonas,* indole-positive *Proteus, Citrobacter, Enterobacter* and *Acinetobacter*) often have inducible chromosomal beta-lactamases leading to resistance to all beta lactams except some penems.

Aminopenicillins - amoxicillin and ampicillin are used for UTIs, some skin some and susceptible enterococcal infections. Aminoglycosides are sometimes used synergistically for enterococcal infections. Amoxicillin has much better oral absorption, than ampicillin. Piperacillin and ticarcillin are antipseudomonal beta-lactams. Combination with beta-lactamase inhibitors improves activity. Amoxicillin/clavulanate has a broad spectrum against gram positives, some gram negatives, and anaerobes. (SPICE-A and pseudomonas organisms are usually resistant). Amoxicillin-clavulanate is used for abscesses, urinary and respiratory tract and dental infections. Ampicillin/sulbactam is an IV formulation with some activity against *Acinetobacter. E. coli* has a high rate of resistance to ampicillin sulbactam, so suspected polymicrobial infections are better treated with other antibiotics. Ticarcillin/clavulanate - IV. Broad spectrum is currently discontinued.

Cephalosporins are similar to penicillins. The spectrum and pharmacology is determined by the side chains. Cephalosporins are divided into generations (first through fourth). First generation cephalosporins have good gram-positive activity. Moving from first to third generation generally increases the gram-negative spectrum.

1st generation - cefazolin - IV, cephalexin - PO. Spectrum: Excellent gram-positive (methicillin sensitive and streptococci), some gram-negative coverage for *Proteus*, *E. coli*, *Klebsiella*.

2nd generation - cefuroxime (PO and IV) spectrum: Gram positive and increased gram negatives - notably *Enterobacter*. Uses include: respiratory infections (upper and lower tract), UTIs, Lyme disease (alternative to doxycycline).

3rd generation cephalosporins: Ceftriaxone - IV, cefotaxime - IV, cefpodoxime - PO. Third generation spectrum: Good gram-positive and excellent gram-negative coverage (*E. coli*, *Proteus*, *Klebsiella* and most SPICE organisms, but not *Pseudomonas*), little anaerobic coverage.

3rd/4th generation ceftazidime (IV) (3rd/4th generation cephalosporin) spectrum: only has gram-negative coverage (including *Pseudomonas*).

4th Generation - cefepime (IV) spectrum: broad gram positive (methicillin-sensitive staphylococci, strep) and gram negative including *Pseudomonas*, but poor coverage of anaerobes.

Carbapenems - imipenem/cilastin, Meropenem, ertapenem, doripenem (all IV) spectrum: Broadest spectrum antibiotics, cover gram positive, gram negative including *Pseudomonas* (except ertapenem) and ESBL (extended spectrum beta-lactamase producers), also anaerobes. This class is reserved for serious infections due to resistant gram negatives including hospital-associated pneumonia, intra-abdominal infections, complicated soft tissue infections, meningitis and others. Carbapenems are most reliable class of antibiotics against ESBL organisms and the SPICE-A organisms. Penems have penetration virtually everywhere, including CSF.

Protein Synthesis Inhibitors

All the protein synthesis inhibitors work by binding to either 30S or 50S ribosomal subunits. Most are bacteriostatic except for aminoglycosides.

Macrolides - erythromycin, clarithromycin, azithromycin - these are all 50S ribosomal inhibitors. Spectrum includes atypical organisms (*Chlamydia*, *Mycoplasma*), also some activity vs. gram-positive cocci and some gram negatives.

Azithromycin - mycoplasma infections primarily.

Tetracyclines - doxycycline, tetracycline, minocycline - 30S inhibitors (PO and IV). These antibiotics have fairly broad spectrum with some staph and MRSA and some gram-negative coverage as well as atypicals infections including: *Rickettsia*, Lyme disease, tularemia, *Brucella*, Q fever, and anthrax. Doxycycline has excellent bioavailability.

Clindamycin - 50S inhibitor (PO and IV) It has excellent activity vs. anaerobes and gram-positive cocci - strep and staph, including ~50% of community-acquired MRSP/MRSA but not *Enterococci*. Clindamycin is often used in skin/soft tissue and dental infections. It is excellent for toxoplasmosis (with pyrimethamine) and has an antitoxin effect in necrotizing fasciitis due to Group A Strep. Associated with higher rates of clostridial infections.

Aminoglycosides - gentamicin, amikacin - 30S inhibitor (all IV), extremely efficacious against aerobic gram negatives including *Pseudomonas*. NO activity against gram positives (except when used for synergy) or anaerobes. Aminoglycosides are used for serious gram negative and pseudomonas infections. Used with beta-lactams against gram-positive organism for broad spectrum or synergistic effects. Aminoglycosides exhibit concentration-dependent killing

Chloramphenicol - 50S ribosomal inhibitor. PO or IV. Chloramphenicol has broad spectrum against gram positives and gram negatives, anaerobes, and is useful against unusual pathogens including spirochetes, *Rickettsia*, *Erhlichia*, *Coxiella*, and *Salmonella*. Limited use due to potential toxicity. Toxicity includes 1. Bone marrow suppression - direct, dose-related effect that is reversible. 2. Aplastic anemia - rare, idiosyncratic reaction (occurs at rate of 1/40,000), but generally fatal event. Usually occurs several weeks after drug is stopped. 3. Marrow effects in cats

Topoisomerases

The quinolones are drugs that inhibit DNA gyrase (Gram -) and topoisomerase IV (Gram +) to prevent proper replication of DNA. They are increasingly broad spectrum with gram-positive and -negative coverage. Spectrum varies by drug. They are effective against intracellular pathogens. Newer fluroquinolones have some coverage of anaerobes. Distribution is high in lung, kidney, prostate, urine and stool. Toxicities include: CNS affects, arthropathies, retinopathies and GI side effects.

Sulfonamides

Sulfonamides are derivatives of para-aminobenzenesulfonamide. They prevent the formation of folic acid. These antibiotics have broad gram-positive and -negative spectrum. They bacteriostatic in most cases but bactericidal in the urinary tract. Sulfonamides have high bioavailability. The urinary system is the main route of excretion. Sulfonamides are useful for urinary pathogens, nocardia, and toxoplasmosis. Effective against some MRSP/MRSA. Sulfasalazine is used to treat colitis, silver sulfadiazine is used to prevent wound and burn infections. Drug reactions include: KCS, hemolytic anemia, agranulocytosis, aplastic anemia, TEN, skin rashes, liver necrosis, and gastrointestinal side effects.

Optimal Usage of Antibiotics

Antibiotics should be used against infection. Signs of infection include: fever, redness, swelling, pain, discharge. Specific signs may depend on the organ system affected. Cough and tachypnea with pulmonary infections, arrhythmias and weakness with endocarditis, diarrhea, pain, hematochezia with gastrointestinal infections, discoloration, swelling, discharge with skin and soft tissue infections, frequent urination, fever, renal failure with urinary tract infections. Signs of serious infection and sepsis include depression changes in mentation, tachypnea, increased or decreased body temperature, tachycardia, hypotension, hypoglycemia, leukocytosis or leukopenia and more. Suspected infections should be cultured to identify the responsible organism and their sensitivity. Culture optimizes the use of antibiotics. By identifying the organism's sensitivity the antibiotic use can be refined and antibiotics can be used at the appropriate dosage for an appropriate length of time. Prior to the receipt of culture results, empirical choices should be broader spectrum and reflect typical organisms at that site.

Pharmacology

Antibiotic choice can be further refined though use of a local antibiogram if one is available. Antibiotic sensitivities should be used as guidance for antibiotic selection. These are based on the *in vitro* minimum inhibitory concentration, knowledge of the plasma levels obtained at standard dosing regimens and the knowledge of time and concentration dependence of the antibiotic.

Time-dependent antibiotics have optimal bactericidal effect when drug concentrations are maintained above the minimum inhibitory concentration (MIC). The goal is to maintain concentrations are maintained at 2 to 4 times the MIC throughout the dosing interval. Higher concentrations do not result in greater kill of organisms. These antibiotics have less post-antibiotic effect. Concentration-dependent antibiotics have increasing bacterial kill with increasing levels of drug and a significant post antibiotic effect. Peak concentration and area under the concentration curve (AUC) determine efficacy of these antibiotics. For this group of drugs, concentrations of at least 10 times the MIC are needed for optimal bactericidal effect. Some bacteriostatic antibiotics show characteristics of time and concentration dependence.

Examples of time-dependent antibiotics: penicillins, cephalosporins

Examples of concentration dependent antibiotics: fluoroquinolones, aminoglycosides

Precautions with particular antibiotic/bacterial combinations:

- *Salmonella* - Host flora inhibition - antibiotics inhibit the inoculum of salmonella required to create infection. Use caution in animals an around humans on antibiotics.
- *Pseudomonas* - Inappropriate use of antibiotics induces resistance factors in Pseudomonas spp. Carbapenems, fluoroquinolones, aminoglycosides induce biofilm formation. Ciprofloxacin enhance the frequency of mutation to carbapenem resistance Tetracycline induce enhanced cytotoxicity.
- *E. coli* and *Klebsiella* ssp. species are the most common enterics to have extended spectrum betalactamase expression. Resistance confers high frequency resistance to: enrofloxacin (92%), doxycycline 82%, gentamicin 44%, Chloramphenicol 44%, TMS 49%. Sheehan *et al*

References

1. Jed F. Fisher and Shahriar Mobashery. Endless resistance, endless antibiotics? *MedChemComm*, 2016Jan1; 7(1) 37–49.
2. Kong *et al*. Beta-lactam antibiotics: from antibiosis to resistance and bacteriology. *APMIS*, January 2010:118(1) 1–36.
3. *Infectious Diseases of the Dog and Cat*. Craig Greene, ed. Fourth ed. Elsvier, 2012.
4. *Canine and Feline Infectious Diseases*. Jane E. Sykes, ed. Elsvier, 2014.

acvim025

The Nursing Process

Liza Wysong Rudolph, BAS, CVT, VTS (CP-CF, SAIM)
Aberdeen, MD, USA

Introduction

The nursing process is an organizational tool that was first described in human medicine in 1955 to guide nursing students and promote patient care. It is currently included in the American Nursing Association's (ANA) standards of practice. Utilization of the nursing process in veterinary medicine allows veterinary technicians to function in a similar capacity to that of human nurses. The process employs a systemic approach to the delivery of care to every patient. In order to correctly use the nursing process, the veterinary technician must be able to apply critical thinking and veterinary nursing knowledge to institute a high standard of care and enhance patient care. Veterinary knowledge refers to a thorough understanding of common disease processes, diagnostics, treatments, and associated procedures - as well as potential complications or risk factors associated with these diseases and procedures.

The core of the veterinary nursing process includes patient assessment, planning, implementation, and evaluation of results. Each step of the nursing process is addressed separately, but depending on the patient case dynamic, these steps may occur concurrently, overlap, or occur out of sequence.

Assessment

The veterinary technician should evaluate the individual needs of the patient to provide proper patient care. A large part of this part of the process is data collection and analysis. Data collection is finding the clues, and analysis is making sense of the clues. Information should be collected from the direct observations of the veterinary technician as well as information from the client and other team members. These data may come from patient history (owner, patient medical records), staff members (rounds, patient medical records), or the patient status (attitude, pain score, laboratory values, physiologic parameters, physical examination). It is important for the veterinary technician to become comfortable and skilled at evaluating their patients and performing physical examinations. Ideally, every patient should be examined toward the beginning of every nursing shift. This will help establish a baseline for comparison in evaluating ongoing nursing or medical interventions.

Diagnosis (Problem List)

Unlike registered nurses who have the ability to assign a diagnosis from a database to their patient, veterinary technicians are limited by Veterinary Practice Acts and it is considered unlawful for a technician to assign a diagnosis to a patient. However, veterinary technicians can compile a patient problem list. Defining the actual (current) and potential (future) problems of a patient, will encourage the technician to think critically and will directly impact the quality of patient care.

Planning

Once information has been gathered during the assessment phase, the next part of the nursing process is the planning stage. Planning should be broken into identifying actual problems versus potential problems. Actual problems are those that interfere with the patient's well-being and for which further diagnostics and treatments are required. Potential problems are those that have not yet occurred, but the patient is at risk to develop them. It is important to try to solve the solvable and lessen the impact of the unsolvable patient problems. It is also essential to monitor patients for the re-emergence of previously managed issues that may present themselves in a decompensating patient.

Once a problem list is composed, it must be prioritized and goals should be set. Goals should be stated in terms of outcomes that are able to be observed, measured, or tested so that effective evaluation can then be carried out; distinctions should be made between short-term and long-term goals. In circumstances in which the patient's problem(s) cannot be solved or alleviated, maintaining quality of life should be a defined as a goal. Steps should be taken to help make the patient as comfortable as possible - even when death is inevitable.

Implementation

Implementation, or nursing intervention, is addressed after the planning stage. It is important to make clear what decision-making has taken place to justify the nursing intervention. Veterinary technicians should understand the reason for the action (intervention) and properly document the actions in the medical record.

Evaluation

A vital and sometimes very difficult part of the nursing process is the evaluation of the nursing plan. Ideally, a positive response should be observed in the patient, confirming that the nursing goals have been achieved. The nursing plan should be constantly reevaluated in order to provide high-quality patient care. It is difficult to justify planning and implementing nursing interventions if you are unsure that there has been a benefit to the patient or even the client in some way. The plan should be flexible and dynamic so that it can be adapted to the changes in the patient's needs. If the plan is not effective, the following questions should be asked:

- How has the problem responded?
  - Has there been partial improvement?
  - Is the problem unchanged or worsened? If so, should the nursing intervention be altered or stopped? Should the plan include something completely different?
- Was the goal inappropriate when first set?
- Is there additional intervention required?
  - Is more information from the clinician or the client needed to decide the next step in the nursing care?
  - Does the goal require interventions from other members of the veterinary team?

Conclusion

The nursing process is generally a cooperative venture between the doctor and the technician. Having a systematic approach encourages the practice team as a whole to be fully involved with patient care and promotes good case communication between different technicians, veterinarians, and even different clinics. Nursing care plans can save time and allow better communication in larger practices. In smaller practices, where inpatient care may be less common, the veterinary practice team may benefit from the structure and prompts in a plan. When used appropriately, the veterinary nursing process allows the veterinary technician to be utilized to the extent of their ability, and can free the veterinarian to diagnose and prescribe. This can improve hospital efficiency and economy, as well as job satisfaction for the entire veterinary team.

References

1. Brown C. Experience of designing and implementing a care plan in the veterinary environment. *The Veterinary Nurse*. 2012;9:534-542.
2. Davis H. A new approach to veterinary nursing: the nursing process. In: Proceedings of the 4th Annual Back To School Veterinary Technician and Assistant CE Seminar. Davis: University of California Davis, School of Veterinary Medicine; 2011:40-44.
3. Hughston L. Critical thinking in critical care. In: Proceedings of the 19th Annual International Veterinary Emergency and Critical Care Symposium. San Diego, CA; 2013.
4. Jeffery A. Advanced nursing practice. In: *BSAVA Manual of Canine and Feline Advanced Veterinary Nursing*. 2nd ed. Gloucester, UK: BSAVA; 2008:1-8.
5. Lock K. *The Veterinary Nurse*. 2011;2:272.
6. Main C. *Veterinary Nursing Journal*. 2011;26:149.
7. Main C. *Veterinary Nursing Journal*. 2011;26:207.
8. Orpet H. *Veterinary Nursing Journal*. 2011;26:242.

9. Rockett J, *et al. Patient Assessment, Intervention, and Documentation for the Veterinary Technician.* Clifton Park, NJ: Delmar; 2009.
10. Wagner C. *The Veterinary Nurse.* 2011;9:328.

acvim040

Alternative Use of Blood Products

Kenichiro Yagi, BS, RVT, VTS (ECC, SAIM)
Los Altos, CA, USA

INTRODUCTION

RBCs for anemia, plasma for coagulation factor deficiency, platelet products for platelet disorders, albumin concentrate for hypoalbuminemia, and immunoglobulin concentrate for immune-mediated diseases are among the list of conditions for which blood products are used for therapy. In addition, blood products are traditionally administered intravenously for these conditions. Autologous serum eye drop application, autologous blood patch pleurodesis (ABPP), and administration of platelet-rich concentrate (PRC) for joints and tendons constitute non-traditional forms of blood product use.

AUTOLOGOUS SERUM EYE DROPS

Autologous serum can be used as eye drops, most commonly used in dogs with ulcerative keratitis, especially when melting corneal ulcers are present. Melting corneal ulcers result when there is an imbalance between proteinases and anti-proteinases, leading to higher levels of proteinases.[2] Proteinase inhibitors, such as plasmin, α1-proteinase inhibitor, α2-macroglobulin, and metalloproteinase inhibitors, are present in normal tears and counteract the effects of proteolytic enzymes. Proteolytic enzymes are necessary for normal turnover and remodeling of the corneal stroma. When proteinase inhibitor levels are decreased, there is a pathologic degradation of the corneal stroma, resulting in keratomalacia (corneal melting).[2] Autologous serum contains the same proteinase inhibitors that are found in tears, and serve as a replacement to restore the balance between anti-proteinase and proteinase activity. In addition to the proteinase inhibitors, autologous serum contains epithelial growth factor (EGF), transforming growth factor beta (TGF-β), vitamin A, albumin, α2-macroglobulin, fibronectin, substance P, insulin-like growth factor 1, and immunoglobulins.[1] Both EGF and TGF-β are found in higher concentration in serum than in tears. EGF accelerates the migration of epithelial cells and prevents apoptosis.[2] TGF-β mediates epithelial and stromal repair.[2]

Autologous serum eye drops are initially instilled in the eye every 1–2 hours. When improvement in the ulcer is seen, or ocular discomfort is reduced, the frequency can be reduced to every 4 to 6 hours.[2] The total duration of therapy typically does not exceed 72 hours.[3]

Autologous serum is prepared through centrifugation of blood and extraction of the serum (Box 1). The amount of blood to be collected is dictated by the amount of serum desired to be yielded, however, not to exceed 10% of total blood volume. Before preparation and use of the eye drops, informed consent should be obtained from the owner, as the use of autologous serum is considered extra-label. The patient should be screened for infectious diseases blood donors are screened for.[2] Filtration of the serum to remove fibrin strands is thought to reduce the efficacy of the drops and not recommended. It is unclear whether dilution of the serum to reduce the potential for irritation from the viscosity of the solution is beneficial, and simultaneous use of antibiotic eye drop solutions also may reduce the efficacy.[2]

Box 1. Sample protocol for autologous serum production
1. Perform venipuncture to collect blood (typically no more than 5 mL/kg from small dogs and 2.5 mL/kg from cats).
2. Transfer the blood into a sterile container and allow it to clot for 30 minutes at room temperature.
3. Centrifuge at 3000 x g for 15 minutes (or according to the centrifuge recommendations for separation of serum).
4. Remove the serum and aliquot into sterile dropper bottles or syringes.
5. Label with the name and date of production.
6. Store at -20°C for a maximum of 6 months.
7. Ideally use a new aliquot daily, stored in the refrigerator (4°C) between uses.
Adapted from Holowaychuk MK and Yagi K 2016[2]

Contamination during collection and processing of the serum is of concern. Sterile technique should be employed during collection and production of the serum. An aliquot can be submitted for culture at time of production to assure sterility.

The serum aliquots are frozen at -20°C until needed. Thawed aliquots are to be stored at 4°C for up to 24 hours. Recommendations for use after thawing vary between 16 hours to 1 week.[2] Care should be taken to avoid exposure to light during storage to avoid loss of vitamin A and proper aseptic technique when handling to prevent contamination by and proliferation of bacteria. A study observed that efficacy was maintained after 6 months of storage, suggesting sufficient anti-proteinase activity for up to 6 months post-collection.[2]

The use of autologous serum eye drops is contraindicated in animals where blood collection puts the patient at risk (e.g., severe anemia, congestive heart failure) or if harmful substances may be in the plasma. Animals with active infections, or positive screening tests for infectious diseases, as well as patients with hyperbilirubinemia, or hyperproteinemia are not acceptable candidates. Animals receiving medications irritating or toxic to the cornea are also not ideal candidates for autologous serum.[4] Complications have been reported in 13% of human patients which are most commonly an inflammatory response or infection. Scleral vasculitis, immune complex deposition, peripheral corneal infiltrate, corneal ulcer, and discomfort and irritation were other reported complications.[2]

There was the same degree in efficacy seen between autologous serum and a commercial ointment containing vitamin A, E, and hydrolyzed casein for the treatment of corneal ulceration in dogs, which indicates some benefit to the use of autologous serum eye drops.[5]

AUTOLOGOUS BLOOD PATCH PLEURODESIS

An autologous blood patch pleurodesis (ABPP) procedure involves the instillation of autologous blood into the pleural space for the treatment of a persistent pulmonary air leak (PAL) or pneumothorax. Options for correcting persistent PAL include surgical repair or pleurodesis. Pleurodesis uses irritating substances such as talc, silver nitrate, antibiotics, or antineoplastic agents to incite an inflammatory response which seals the airway leak, while having potential harmful effects for these substances to be persistently left in the pleural space. With ABPP, harmful residual effects are eliminated because the irritant is a natural substance originating from the same individual and has been observed to bring the shortest air leak cessation time with the least amount of complications when compared with chemical pleurodesis.[6]

Autologous blood patch pleurodesis has been described for use in people for relapse of spontaneous pneumothorax, PAL after lung lobectomy, spontaneous pneumothorax secondary to chronic obstructive pulmonary disease or interstitial lung disease, postoperatively after lung lobectomy, and in acute respiratory distress syndrome.[2,12] ABPP has been used to treat persistent pneumothorax in dogs due to various causes such as bite wounds, blunt trauma, bullous emphysematous disease, chronic bronchitis, grass awn migration, and postoperative diaphragmatic hernia repair.[2,6,13]

To prepare for the procedure, any existing air should be evacuated through the thoracostomy tube. Collection of blood for ABPP is performed through a sampling catheter or aseptically from the jugular or peripheral veins with a syringe and needle. Infusion of 3–5 mL/kg of blood, and not exceeding 10 mL/kg, is recommended to be infused in dogs.[2] Infusion of 1–3 mL/kg in people and 2–3 mL/kg in rats have been observed to be effective.[2] Blood is collected into a syringe without any anticoagulant and is immediately infused into the thoracostomy tube and repeated as necessary to achieve the calculated volume to be infused. Blood is allowed to stay in the pleural space for 3–4 hours with intermittent repositioning of the patient to distribute the blood in the pleural space. The patient is monitored for respiratory compromise from the infusion of blood through observation of respiratory rate and effort and other pulmonary monitoring capabilities (Box 2).

Box 2. Autologous blood patch pleurodesis procedure
Preparation
1. The procedure is recommended to be performed with three people: one person to monitor the patient, one person

to collect the blood, and one person to instill the blood.

2. Consider administration of sedation if the patient is likely to stress from handling. The patient can be positioned in sternal or lateral recumbency.

3. Placement of a blood sampling catheter is recommended. An aseptically prepared jugular or a large peripheral vein can be used.

4. Disconnect the thoracostomy tube from continuous suction, if applicable. Ensure the thoracic cavity is completely evacuated of air by aspirating the thoracostomy tube until negative suction is obtained.

5. Calculate the desired autologous blood volume (3–5 mL/kg). Divide this volume in half (for each hemithorax) if bilateral thoracostomy tubes are present.

Collection and infusion

6. Collect the blood in 20–60-mL syringes without the use of anticoagulant solution. The collection should be rapid and atraumatic to prevent clotting.

7. Instill the collected blood into the pleural space via the thoracostomy tube rapidly to ensure instillation prior to clot formation. The use of an assistant for the injection is recommended so there is no delay in collecting further volume(s) needed to reach the total volume desired.

8. Repeat steps 6 and 7 until the entire calculated blood volume is instilled via one or two thoracostomy tubes.

9. Instill 10–20 mL of 0.9% NaCl or air into the thoracostomy tube(s) to remove residual blood from inside the tubes.

10. Leave the thoracostomy tube(s) clamped for 3–4 hours, then aspirate the tubes to remove the instilled blood. Resume aspiration of the tubes sooner if the patient demonstrates worsening tachypnea, respiratory effort, or SpO_2 measurements.

Adapted from Holowaychuk MK and Yagi K 2016[2]

The success rate of ABPP in resolving pneumothorax in dogs has been reported as a 67% resolution rate after one treatment and 89% resolution rate after additional ABPP procedures.[2,6] Similar results have been reported with the use of ABPP in people, with 76.6% resolution rate on the first procedure and 92.7% resolution overall with additional ABPP procedures. Complications, such as transient fever, pyothorax, infections, and hypercapnia (presumably from space-occupying nature of blood), can occur. ABPP is contraindicated with active infection. Overall, ABPP is considered a simple, inexpensive, and safe procedure to aid in the resolution of persistent pneumothorax.

Platelet-Rich Concentrates

Platelet-rich concentrates (PRC) have gained attention due to their use in sports-related injuries in humans and animals. Platelets contain platelet-derived growth factor (PDGF), TGF-β, and vascular endothelial growth factor (VEGF). These growth factors support anabolic activity by stimulating cell proliferation, extracellular matrix synthesis, and vascular invasion. Growth factors also suppress cytokines and, as a result, suppress catabolic activity.[2] By providing a higher platelet concentration solution, the concentrations of these factors are increased. PRCs come in two different forms as (PRP) and autologous conditioned plasma (ACP), the latter having reduced leukocyte and erythrocyte numbers and, theoretically, less inflammation and better for healing.[7]

Autologous PRP

The use of autologous PRP to promote the healing of flap closures of acute and chronic wound has been evaluated in a variety of species with benefits including improved microcirculation, better tissue perfusion, and greater skin flap viability post-surgery. The effect of autologous PRP injections in the dog on flap incisions was compared to flap incisions without PRP injections used as a control.[14] The PRP-treated flaps showed a reduced rate of necrosis, increased mean percentage of survival area, improved vascularity and perfusion, and reduced edema when compared to the control.

Autologous PRP has also been used to promote healing of wounds that are unable to be closed and must heal by second intention.[15] Evaluation of the effect of autologous PRP for wounds shows varied results. There are beneficial effects noted with the use of autologous PRP in the healing of decubital ulcers in the dog. A 93.5% wound reduction was seen in PRP dogs when compared to 13.2% in the control group.[8] A case study reported a necrotic wound unresponsive to conventional therapy healing after PRP injection.[9] Another case study reported accelerated epithelial differentiation and organized collagen formation in the horse.[10]

The effect of autologous PRP has also been studied in tendinopathies in the dog.[16] Evaluation of subjective measures such as lameness scores and canine brief pain inventory and objective measures such as gait analysis, and ultrasound lesion size and appearance has been conducted, with subjective measures indicating improvement. More evidence related to improvements in objective measures would be ideal to determine the benefit of autologous PRP in this manner.

Platelet-rich plasma is obtained through centrifugation of whole blood and extraction of the platelet-rich portion of the plasma. While an ordinary centrifuge can be used for this purpose, commercial PRP production systems are available with variable yields in platelet concentration. The effects of variability in platelet number on diseases are unknown.

Autologous Conditioned Plasma

Autologous conditioned plasma has been used in the treatment of tendon and ligament lesions in horses. A case series evaluating the effect of ACP treatment of acute severe tendinitis in the superficial or deep digital flexor tendons documented improvements in cross-sectional area of injury, percentage of the maximal injury zone, echogenicity score, and fiber alignment scores.[17] Each of the horses were performing at previous workload or in full training with no re-injury reported.[2]

A comparison of the effect of hyaluronan and ACP for the treatment of elbow osteoarthritis in the dog through intra-articular injections resulted in equal levels of reduction in lameness grade between the groups treated with hyaluronan and ACP. Owner survey also reflected improvement in both groups, leading the authors to conclude that ACP was as effective as hyaluronan and corticosteroids in the alleviation of osteoarthritis. Neither group experienced any adverse effects.[2,11]

Autologous conditioned plasma is prepared through the use of commercial systems which eliminate red and white blood cells from the mixture which reduces detrimental inflammatory effects and elevates concentration of PDGF, EGF, and VEGF, when compared to plasma.

Conclusion

Blood products are used commonly in veterinary medicine. In addition to the traditional uses for blood products as cell and component therapy, there are other components of blood that appear to be beneficial in treating certain conditions in animals. These emerging treatment methods have limited but supportive evidence for their use and additional studies are recommended to evaluate their efficacy.

References

1. Quinto GG, *et al. Arq Bras Oftalmol.* 2008;71:47–54.
2. Holowaychuk MK, *et al. The Manual of Veterinary Transfusion Medicine and Blood Banking.* 2016;103–116.
3. Ollivier FJ, *et al. Vet Ophthalmol.* 2007;10:199–206.
4. Yamada C, *et al. Transfusion.* 2008;48:1245–1255.
5. Ortiz JF, *et al. Rev Colomb Cienc Pec.* 2012;25:90–96.
6. Oppenheimer N, *et al. J Vet Emerg Crit Care.* 2014;24(2):215–220.
7. Sief M, *et al. Vet Comp Orthop Traumatol.* 2011;24:122–125.
8. Tambella AM, *et al. Vet Surg.* 2014;43:726–733.
9. Kim JH, *et al. Vet Dermatol.* 2009;20:123–126.
10. Carter CA. *Exp Mol Pathol.* 2003;74:244–255.
11. Franklin SP, *et al. Can Vet J.* 2013;54:881–884.
12. Martinez-Escobar S, *et al. J Crit Care.* 2006;21:209–216.
13. Merbl Y, *et al. J Am Vet Med Assoc.* 2010;237:299–303.
14. Karayannopoulou M, *et al. Vet Comp Orthop Traumatol.* 2014;27:379–386.
15. Karayannopoulou M, *et al. Vet Comp Orthop Traumatol.* 2015;28:172–178.
16. Ho LK, *et al. Can Vet J.* 2015;56:845–849.
17. Georg R, *et al. J Vet Sci.* 2010;11:173–175.

acvim041

Evans Syndrome: A Technician's Role

Yvonne R. Brandenburg, RVT, VTS (Small Animal Internal Medicine)
Livermore, CA, USA

Introduction

As technicians, either in specialty practice or on the front lines in general practice, knowing how to recognize patients with hematologic disorders is important in preventing complications and improving long-term survivability of patients. Evans syndrome is a rare autoimmune hematologic disorder that affects both red blood cells and platelets.

Learner Outcomes

The goal of this lecture is to 1) define Evans syndrome; 2) review basic hematologic cytology so able to recognize specific markers for Evans; 3) review patient care and client education; 4) discuss treatment options available; and 5) discuss the long-term management and risk reduction for patients.

Review of Evans Syndrome

Evans syndrome is a rare autoimmune disease where both primary immune-mediated hemolytic anemia (IMHA) and immune-mediated thrombocytopenia (IMTP) are present. The disease was originally described by Robert Evans in 1951 when he monitored 24 human patients with IMHA and IMTP and how they reacted similarly to treatments including splenectomy.[1]

Primary IMHA is defined by the antibody-mediated destruction of red blood cells and is characterized by a moderate to marked regenerative anemia, spherocytes, and autoagglutination.[2] Typically, patients present with severe anemia (PCV of 13–21%) that is most often regenerative, although this may not always be the case. Antibodies stick to the surface of the red blood cell and label the cells as 'foreign intruders' which activates the immune system to target and destroy them. This can be done either as extravascular hemolysis (more common) with macrophages breaking down the cells primarily in the spleen, or as intravascular hemolysis (less common) which happens when the red blood cells are broken down in the blood stream. Patients typically present pale, lethargic, depressed, weak, and anorexic.

Primary IMTP is described as an autoimmune disorder with production of antibodies directed against normal platelet antigens.[3] This can happen either at the level of the bone marrow with inadequate production or as platelet destruction once they are in circulation. Thrombocytopenia is described as platelet counts less than 200,000/mcL. Once a patient has <50,000/mcL, abnormal hemostasis may occur, and once <20,000/mcL spontaneous bleeding is common, which can be seen in the oral cavity, nose, or urinary tract. It has been reported that 25–75% of IMHA patients have concurrent thrombocytopenia at the time of diagnosis and are thought to have Evans.[2] Patients typically present with multiple petechiae and spontaneous bleeding, such as epistaxis.

It is important when diagnosing Evans to rule out underlying diseases. The typical diagnostic workup consists of a full chemistry, CBC, vector-borne disease testing, thoracic radiographs, and an abdominal ultrasound. Performing a PT, PTT and D-Dimer (or TEG) is warranted in case DIC is also present or a cause of the anemia and thrombocytopenia. DIC should not have a positive Coombs anemia, spherocytes or platelet antibodies. A bone marrow biopsy and cytology can help with ruling out leukemia, multiple myeloma, lymphoma or other myeloproliferative disorders that can cause anemia and thrombocytopenia. Severe thrombocytopenia is not a contraindication for performing a bone marrow biopsy.[4] Ensuring that the patient is stable for the procedure is indicated and a blood transfusion may be warranted if clinical signs associated with hypoxia are present.

Published reports of prognosis of IMHA alone are that 50–88% will survive to release from the clinical setting.[2] Of those patients that are released from the hospital, death or euthanasia from ongoing complications typically occurs by 2 months.[2] Variables that have been seen to increase mortality include hyperbilirubinemia, auto-agglutination, increased band neutrophil counts, thrombocytopenia, and hypoalbuminemia.[2] At this time, there are no specific studies about the prognosis with Evans, but it is inferred that Evans has a 50–70% survival to release, with similar variables for mortality.

The most common complication and cause of death for IMHA and Evans is confirmed or suspected thromboembolic disease, including pulmonary thromboembolism (PTE) and disseminated intravascular coagulation (DIC).[2] The systemic inflammatory response in the body from this disease promotes a hypercoagulable state and is the cause of significant morbidity and mortality.[2]

Review of Hematologic Cytology

The CBC should be reviewed microscopically by a clinical pathologist. However, this is not readily available for most and a turnaround time of 12 to 24 hours is expected. Because of this, technicians should have the training and be able to review a blood smear slide to look for classic cytological changes.

For IMHA, spherocytes are suggestive of red blood cell damage and destruction. Spherocytes are round in shape, smaller in diameter, and have decreased central pallor. They are formed when partial phagocytosis occurs of the cellular membrane and the red blood cell reseals itself and reenters circulation. Another important cell characteristic to look for is whether the anemia is regenerative or non-regenerative. Regeneration is suggested on the hemogram (cell volume, hemoglobin concentration) but cytologically anisocytosis (different RBC cell sizes), polychromasia (multiple colors), and increased nucleated red blood cells may be seen. Reticulocytes are released early from the bone marrow when stimulated by decreased oxygenation and can be identified by staining with new methylene blue.[2] Reticulocytes are more sensitive and specific indicators of regeneration. A non-regenerative anemia occurs when the bone marrow does not respond to stimulation and does not produce or release young RBCs to replenish the supply in the blood stream. It is also possible that the red cell destruction occurs within the bone marrow, so no young red cells (reticulocytes) are produced and a non-regenerative anemia is seen.

A saline agglutination test should also be performed to evaluate the presence of RBC autoagglutination. One drop of whole EDTA blood and 1 drop of 0.9% saline is placed on a microscope slide. Once the sample is prepared, it is important to review the slide both macroscopically and microscopically. If there is persistent agglutination, it is suggestive of antibody-mediated clumping of the cells. It is important to review microscopically to ensure that it is autoagglutination and not rouleaux formation.[2]

Cytological characteristics of IMTP are severe thrombocytopenia (<50,000/mcL), normal to increased megakaryopoiesis, and microthrombocytosis. Performing manual platelet counts is important to monitor for response to immunotherapy. Typically a blood smear is done and viewed under high-power magnification in oil immersion. The platelet count can be estimated by counting platelets in 10 high-power fields, adding together and dividing by 10, and then multiplying by 15,000–20,000. It is important to scan the feathered edge and the slide to look for platelet clumping as this will alter the actual platelet count. It is common for megakaryocytes to be present early once the body starts producing and releasing platelets, as these are young immature cells.[3]

Daily monitoring of PCV and cytology is common during treatment to evaluate response to treatment. This includes fresh blood smears looking at morphology of RBCs and platelets.

In the photo below, an Evans patient's blood smear was reviewed on day 5 after initial presentation.[5] Notice that spherocytes are still present. There is also moderate-severe anisocytosis, polychromasia, and megakaryocytes present, all indicating that the bone marrow is responding appropriately and immunosuppressive therapy is starting to work, although not completely resolved.

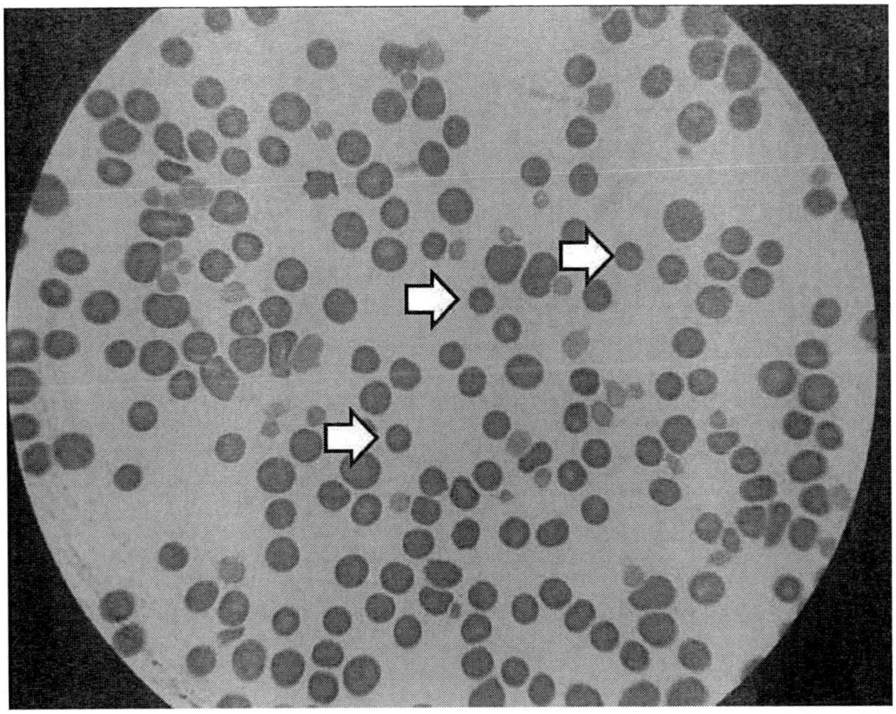

Fig. 1 Arrows pointing to spherocytes present on day 5 of hospitalization

REVIEW TECHNICIAN'S ROLE IN CLINICAL SETTING

Initial triage can prevent serious complications with these patients. Recognition of petechia or other suggestions of hemorrhage is critical to prevent obtaining blood samples from inappropriate vessels such as the jugular vein. If a technician suspects a bleeding disorder, it is vital to minimize the amount of venipuncture performed on a patient and take samples form sites that are easier to apply pressure to minimize hemorrhage. Initial blood samples can be acquired through placing intravenous catheters to minimize the number of times a blood vessel is breached.

Typically, the technicians are involved in performing the initial workup for Evans patients. The technicians ensure that the patient is admitted into the patient care ward and completes the treatments recommended by the veterinarian of record. Once the veterinarian has given the preliminary diagnosis, it is vital that a senior technician performs the patient care for these patients as there are significant risks associated with the disease.

These patients are typically severely anemic when they arrive and the first 3–5 days in the hospital is when blood transfusions typically occur. There may be multiple transfusions needed as it can take up to 5 days for the immune system to be suppressed enough to allow for red blood cells (RBCs) and platelets to be formed and not attacked. Increased platelets typically occur 2–5 days after starting immunosuppressive therapy.[3] For RBCs, once the bone marrow has been stimulated it takes 2–5 days for reticulocytes to be identified in circulation. Peak numbers occur 4–7 days after stimulation.[2]

Blood transfusions are indicated for patients with PCV of <15% as oxygen transport is severely impaired at that point and can lead to hypoxia to the body's tissues. Blood typing and crossmatching the patient to the donor prior to starting a transfusion is often performed by technicians. Blood typing is often done with cards or kits but crossmatching is often performed manually with repeated cell washes. Alternatively, kits are available. Once a compatible blood product is found, technicians administer the blood product, typically packed red blood cells, and monitor closely for any potential transfusion reactions. These reactions can be broken down into four categories: immunologic or non-immunologic; and acute or delayed.[7]

While the patient is hospitalized, it is important to ensure that the patient is comfortable and stable. Supportive care with fluid therapy is recommended to maximize tissue perfusion and minimize kidney damage associated with hemoglobin excretion and systemic inflammation.[7] Jugular venipuncture, central catheters, cystocentesis are all contraindicated in these patients due to the inability to clot. Gentle handling and wet food are also indicated. Careful venipuncture of a peripheral vein with small-gauge needles is recommended with low platelet counts. Applying a pressure wrap is recommended for 30–60 minutes with careful monitoring of the site once the wrap is removed to ensure the patient does not continue to bleed from the venipuncture site.

These patients are immune compromised and should be handled with gloves to prevent the acquisition of a nosocomial infection. A dedicated patient care technician should be assigned to these patients, and these technicians should not handle any potentially infectious patients to help minimize the risk.

Close monitoring for pulmonary thromboembolism (PTE) is important both in the hospital and once the patient is released. Due to the severe inflammatory response associated with the disease, as well as the hypercoagulable state associated with glucocorticoids, microembolism is common complication and a cause of death in these patients.[2] Close monitoring of respiratory effort and function is crucial to early recognition of PTE. Collecting blood for a D-Dimer or TEG tracing can help support the diagnosis of a thrombus as it represents the breakdown of a clot.[8]

Treatment Options Available to Patients

During the acute phase, transfusion therapy is warranted depending on the severity of the anemia. As red blood cells carry oxygen, low red blood cell count will result in decreased oxygen delivery to the tissues or hypoxemia. Oxygen therapy does not address the primary problem as the red blood cells are already saturated with oxygen. To treat hypoxemia associated with anemia, packed red blood cells are given to increase the number of red blood cells available to carry oxygen. Arterial blood gas analysis (PaO_2) is preferred to determine hypoxemia rather than SpO_2 monitors. PaO_2 measures oxygen delivery to the tissues whereas SpO_2 measures red blood cell oxygen saturation. Clinical signs of hypoxemia include increased heart rate, increased respiratory rate, lethargy, and pale or muddy mucous membranes. Prior to any transfusion, blood typing and a crossmatch is recommended to minimize transfusion reactions. Auto-agglutination can make interpretation of crossmatches difficult.

Immunosuppressive therapy is the cornerstone of treatment for Evans.[2] The first line of drug therapy is the use of glucocorticoids such as or prednisone orally or dexamethasone SP injections if the patient is not tolerating oral medications. Glucocorticoids at immunosuppressive doses are the foundation of treatment therapy for Evans patients.[2] The standard starting dose of prednisone is 1 to 2 mg/kg q12 h in dogs, with some internists limiting the dose for larger dogs to decrease the amount of side effects seen.[9] Monitoring PCV and platelet counts determines when it is appropriate to taper the prednisone. This is typically done slowly to prevent a relapse as long as the patient is tolerating the glucocorticoids.

Human intravenous immunoglobulin (IVIg) and plasmapheresis can also be used in these patients. Human IVIg has been shown to decrease the platelet recovery time, therefore decreasing the amount of time spent in hospital.[10] IVIg has not been shown to be beneficial in cases of IMHA alone.[10] IVIg can be cost-prohibitive for some clients. Plasmapheresis has been shown to help with IMHA but not necessarily in IMTP.[10]

Vincristine has been shown to stimulate thrombopoiesis in healthy dogs.[10] There have been some concerns raised about the function of these early released platelets. There are contradictory studies regarding platelet function. However, it was speculated that platelet function in patients with IMTP that had >40,000 platelets per mcL did not show signs of active hemorrhage, indicating that vincristine may help.[10]

Secondary immunosuppressive drugs are recommended for patients with Evans to help facilitate the tapering of the glucocorticoids. Azathioprine was the standard for treating IMHA, IMTP and Evans for years. The initial dose recommended is 2 mg/kg PO q24 h. Azathioprine has fallen out of favor due to its bone marrow suppression effects as well as the potential for acute liver toxicity.[2] Careful monitoring of liver chemistries is recommended in addition to CBCs. Per Plumb[11] the starting dose is 2 mg/kg PO q24h, with long-term therapy being 0.5–1 mg/kg PO every other day.

A common drug to use is cyclosporine modified. Cyclosporine is a 'potent immunosuppressive drug indicated for the treatment of inflammatory and immune-mediated diseases, and for organ transplantation.'[12] The dose is 5–10 mg/kg PO q12–24h. Variability in absorption has been reported between formulations and using the brand Neoral™ (human formula) or Atopica™ (veterinary formula) versus generic forms has been shown to have better bioavailability and control of disease.[12]

Monitoring drug through plasma concentrations at 2–4 weeks to maintain the concentrations between 100–300 mg/mL is recommended.[2] More recently, assays of lymphocyte activity are available for cyclosporine and thought to be a better biologic indicator of disease control than blood levels. The test is now available at Mississippi State University.[13]

Mycophenolate mofetil has been used in humans for patients receiving organ transplants. This is an option for treating veterinary patients, but there are very limited data regarding efficacy. Reports suggest that the dose required may be toxic to the gastrointestinal system and caution should be used with this drug.[13] The current recommended dose of mycophenolate is 12–17 mg/kg PO per day, once or divided in two doses.[11]

Prophylactic anticoagulant therapy to prevent PTE is typically done by platelet inhibitory drugs such as low-dose aspirin at 5 mg/kg PO q24h, or clopidogrel at 10 mg/kg loading dose, with 2 mg/kg PO q24h afterward.[9]

A limited case series was done with melatonin used in association with IMTP.[10] It is thought to 'stimulate platelet generation, probably by promoting megakaryocyte fragmentation.'[10] The dose per Plumb[11] for dogs under 20 kg is 3 mg PO BID, and >20 kg dogs at 6 mg PO BID. Some reports have a dose up to 20 mg PO daily.[10]

Splenectomy can also be a treatment option for these patients. This is indicated when the first-line therapy has failed and is thought to be due to removing the primary organ response for destruction of antibody-sensitized platelets and red blood cells.[10]

Review Technician's Role in Client Education

One of the biggest responsibilities of the technician is to work with the client to ensure that they fully understand the medication dosing instructions, as well as the side effects they should be aware of. Compliance with following medication recommendations is crucial to minimize relapse and potentially life-threatening conditions.

Explaining what they should expect from their pets as they transition from the hospital to home can help alleviate unnecessary frustrations. Patients typically are released with immunosuppressive doses of glucocorticoids and other immunosuppressants that are associated with side effects, so setting up realistic expectations regarding adverse reactions is important. Explaining to clients the significant clinical changes such as polyuric (PU), polydipsic (PD), and polyphagic (PD) that are common with glucocorticoids and that they can be quite extreme. Warning clients that the pet will typically need to urinate every 1–2 hours, even overnight is indicated. Clients should also be aware of potential behavioral changes associated with steroids including aggression, food scavenging, and begging. Ensuring that food items are out of reach of pets is important, including garbage cans to prevent food-related gastrointestinal upset. Muscle wasting is also something that clients should be made aware of. This can be very dramatic, especially on the temporalis muscle and the spine. Explaining that this is temporary until glucocorticoid doses are reduced can also alleviate client concerns. Larger dogs can have more pronounced muscle wasting and can have issues with lameness due to muscle loss. These muscle-wasting side effects, increased susceptibility to infection, susceptibility to clot formation, and insulin resistance are reasons why glucocorticoid therapy is minimized and alternative immunosuppressants used.

Clients should be informed that wet food is recommended during the initial phase until platelet numbers are above 80–100 k/mcL. It is also recommended that the pet not be allowed to chew on toys, bones or other objects that can exacerbate gingival bleeding. Patients should also have restrictions of no running, or strenuous activity to help prevent undue petechia and/or hemorrhage.

These patients need close CBC monitoring every 1–2 weeks to ensure that they are not relapsing during the tapering of drug therapy to the lowest effective dose.[2] After release from the clinical setting, chemistries are monitored a minimum of every 6 months, or more frequently if the patient had abnormalities while in hospital. The goal is to wean patients off of the steroids as safely and quickly as possible to minimize the adverse side effects. Evans syndrome patients have a higher risk of relapse and are more likely to require lifelong immunosuppressive drugs, so multimodal drug therapy is typically done.

Educating clients on how to recognize a thromboembolic event, specifically PTE, is key to catching an event early and potentially treating if possible. Anticoagulant therapy can be initiated once platelet numbers are significant enough to not cause spontaneous bleeding. See treatment options for more information.

The doctor and technicians should also explain that a pet's immune system is suppressed and they are more susceptible to infectious diseases. Clients and dogs should avoid dog parks, grooming centers, and other places with high animal traffic. Limiting other dogs from contacting their pet is appropriate too, especially young animals whose vaccine status is unknown.

The final aspect to discuss is the financial implications of this disease. With weekly rechecks, lab work, and the drug therapies associated with maintaining these patients, the financial commitment can be a limiting factor for clients. Pet insurance is helping to bridge the financial gap for owners fortunate to have insurance prior to diagnosis, but once a pet is diagnosed insurance will not cover this disease.

Help clients know that they are not alone. The emotional roller-coaster that clients feel is understandable and normal. Being an advocate for both the pet and the client can build a lasting partnership for the patient's health. There are few resources for clients specifically about Evans but owners can try Facebook[8] or other support pages.

Technician's Role in Long-Term Case Management and Risk Reduction

Client communication and follow-up is crucial to long-term survivability of patient with Evans syndrome. Facilitating follow-up appointments and checking in with clients to ensure that they do not have questions can reduce the chance of non-compliance. Answering client's questions as they come up and letting them know that they should feel comfortable asking these questions, can prevent a major relapse, or worse, other potential life-threatening complications. Reminder calls for rechecks and recommended lab work, can be beneficial for clients struggling with compliance.

Maintenance of these patients involves slow, gradual tapering of immunosuppressive therapy, with minimal exposure to unnecessary drugs, vaccination, or stress conditions to avoid relapse.[3] Clients should always check with their veterinarian before changing medications or adding over-the-counter medications as some of these can interact with prescribed medications. Flea and tick prevention is warranted to prevent infectious agents as these patients are immune suppressed. Vaccinations are not recommended because of the inability of the animal to mount an effective response to the vaccine.

Conclusion

Evans syndrome is a combination of both immune-mediated hemolytic anemia and immune-mediated thrombocytopenia. Veterinary technicians play a crucial role in performing diagnostics including basic hematologic cytology, patient treatment, and performing client education which facilitates successful outcomes of this life-threatening disease.

References

1. Norton A, *et al. British Journal of Haematology*. 2006;132(2):125.
2. Mitchell, K, *et al. Vet Int Med Dz Dog & Cat.* 2010;1:761.
3. Brooks MB, *et al. Vet Int Med Dz Dog & Cat.* 2010;1:772.
4. Neel JA, *et al. Kirk's Current Veterinary Therapy XV*. 2014;15:280.
5. Brandenburg YR. 2016.

6. Bistner SI, *et al. Kirk & Bist Han Vet Proc & ER Tx.* 2010;7:665.
7. Garcia J, *et al. SAIM for Vet Tech & Nurs.* 2012;161.
8. Facebook: https://www.facebook.com/groups/6228146980/ *IMHA*, 2017
9. Johnson V, *et al. Kirk's Cur Vet Ther.* 2014;15:275.
10. Nakamura RK, *et al. JVECC.* 2012:22(1):59.
11. Plumb D. *Plumb's Veterinary Drug Handbook.* 2011.
12. Mackin A. *2016 ACVIM Forum.* 2016;1113.
13. MSU College of Veterinary Medicine. 2014 www.cvm.msstate.edu/animal-health-center/pharmacodynamic-laboratory
14. West LD, *et al. JVECC.* 2014;24(2):226.

acvim042

Introduction to Transfusion Medicine for the ER and ICU Technician

Jordin Karalunas, LVT, VTS (ECC)
Shelby Township, MI, USA

Transfusion medicine has certainly come a long way and is continually changing with our growing advancements in veterinary medicine. A crucial understanding of transfusion indications, component therapy, contraindications, blood typing, monitoring, and transfusion reactions, can determine the outcome of a successful transfusion. Product handling and administration protocols also play an important role in patient outcome. A thorough comprehension of why, how, and what we should transfuse is pivotal to each ICU case.

COMPONENT THERAPY

Knowing what disorder we are treating, and the blood product required to treat is the basis of component therapy. Administering only what the patient needs prevents unnecessary transfusion reactions, circulatory overload, and additional blood product waste. Products, dosages, indications, and administration protocols are listed below. It is recommended all blood products be administered through a dedicated, separate IV catheter without additional fluids or medications to prevent complications.

Fresh Whole Blood

Blood collected within 6–8 hours. Contains red cells, white blood cells, plasma proteins, a small amount of platelets, and coagulation factors. Dosage is 10–20 ml/kg, but is patient-dependent. The "Rule of 1s," though not proven, may apply. It states that 1 ml of transfused blood/1 lb of body weight will raise the patient's PCV by 1%. Indications include trauma with massive hemorrhage, hypovolemia, hemoabdomen, rodenticide poisoning, sepsis, hypoxemia due to anemia, and coagulopathy or thrombopathia. Administration requires a filter drip set (free drip), or a microaggregate (ex. Hemo-Nate®) filter and syringe pump.

Stored Whole Blood

Blood that has been collected and stored for longer than 8 hours. Contains red blood cells, plasma proteins, and some coagulation factors. Dosage is 10–20 ml/kg, but is patient-dependent and the "Rule of 1s" may apply. Indications include trauma with massive hemorrhage, hypovolemia, hemoabdomen, and hypoxemia due to anemia. Administration is the same as fresh whole blood.

Packed Red Blood Cells

Blood collected and separated (removing plasma from whole blood). Contains red blood cells, a small amount of platelets, white blood cells, and a minimal amount of plasma. Dosage is patient-dependent, approximately 6–10 ml/kg over 2–4 hours. Indications include trauma with hemorrhage (including surgical), hemoabdomen, sepsis, hypoxemia due to anemia, immune-mediated hemolytic anemia (IMHA), and immune-mediated thrombocytopenia (ITP). Keep in mind the "10/30" rule is not proven in indicating transfusion. This rule states that a hemoglobin of 10 g/dl and a PCV of 30% are markers for transfusion requirement. However, many patients do not show clinical signs of transfusion requirement until their PCV is below 20%. Transfusion indications should not be based on patient numbers and are always patient-dependent! Administration requires a filter drip set (free drip) or a microaggregate (ex. Hemo-Nate®) filter and syringe pump.

Fresh Frozen Plasma

Blood that has been separated within 6 hours of collection (removing blood cells from plasma). Contains plasma, albumin, globulins, coagulation factors (V, VIII, IX, vWF), and electrolytes. Dosage is patient-dependent, approximately 6-10 ml/kg over 2–4 hours. Indications include coagulation disorders (congenital vs. acquired), hemophilia A & B, von Willebrand disease, disseminated intravascular coagulation (DIC), sepsis, envenomation, liver disease, and pancreatitis. Administration requires a filter drip set (free drip), or a microaggregate (ex. Hemo-Nate®) filter and syringe pump. Filter drip sets are available that are compatible with IV pumps, allowing more precise administration.

Frozen Plasma

Blood that has been separated after 6 hours of collection (removing blood cells from plasma), FFP that was thawed and re-frozen, or fresh frozen plasma that is greater than 1 year old. Contains a decreased amount of coagulation factors with varying stability, plasma, and albumin. Dosage is patient-dependent, approximately 6–10 ml/kg. Indications include some clotting factor deficiencies and colloid necessity. Administration requires a filter drip set (free drip), or a microaggregate (ex. Hemo-Nate®) filter and syringe pump. Filter drip sets are available that are compatible with IV pumps, allowing more precise administration.

Cryoprecipitate

Portion of plasma that precipitates after fresh frozen plasma has been slowly thawed. Contains heavily concentrated von Willebrand's factor and factor VIII, fibrinogen/fibronectin. Dosage is patient-dependent, approximately 1 unit/10 kg over 2 hours. Indications include complications due to von Willebrand's disease, or hemophilia A. Administration requirements include a micro-aggregate filter (ex. Hemo-Nate®) and syringe pump.

Frozen Platelet Concentrate

A concentrated platelet product, made from the cryopreserving and freezing of fresh platelet concentrate. Contains platelets, has been leukoreduced, and preserved in DMSO (dimethyl sulfoxide). A dose of approximately 1 unit/10 kg over 1–2 hours may increase the platelet count 20,000/μL 1–2 hours post-transfusion. Though efficacy data on this product have not been published, potential indications include thrombocytopenia/pathia, DIC, sepsis, and ITP. Administration requirements include a filter drip set, allowing the product to free drip. Gentle handling of the product is a must! Because this product contains DMSO, it must be transfused slowly to avoid bradycardia.

Human Albumin

Derived from human blood, it contains human serum albumin (HSA). Dosage is dependent on patient, not to exceed 1 ml/min over 4 hours and not to exceed 2 g/kg/day. Indications include hypoalbuminemia, sepsis, systemic inflammatory response syndrome (SIRS), and burn patients. Administration requirements include dilution of HSA into normal saline to maintain a 10% solution. Some protocols call for no dilution. Use of a regular fluid administration set is required. It is recommended to wear gloves when handling, as this is derived from human blood. Severe transfusion reactions with this product have been published and caution should be taken when transfusing this product.

Lyophilized Canine Albumin

Derived from canine blood using a heat shock process, it contains pure canine albumin. Dosage is patient-dependent; however, a formula courtesy of Animal Blood Resources Intl. has been provided (Body weight (kg) x 90 ml/kg x (2 g/dl - Patient albumin) x 0.2 dL/g). No maximum dose has been identified. Indications include hypoalbuminemia, sepsis, SIRS, and burn patients. Administration requires a filter drip set (free drip), or a microaggregate (ex. Hemo-Nate®) filter and syringe pump. This product is currently not available due to the high incidence of transfusion reaction.

Typing and Crossmatching

Typing should be done on every patient prior to transfusion, especially in felines. This greatly reduces the risk of transfusion reaction. Most canines can typically tolerate their first transfusion of universal type blood, as they have not yet built up antibodies. Felines, however, have naturally occurring antibodies which require typing prior to transfusion. In-house typing of the recipient allows the option of type-specific transfusion [i.e., D.E.A. (dog erythrocyte antigen) 1.1 positive or negative and type A, B, or AB in cats]. In-house recipient blood typing takes only a few minutes. Alvedia typing cartridge, RapidVet typing cards, and RapidVet-H IC are available for canine and feline typing while in hospital. Though blood typing can be difficult with agglutinating patients, causing false type results, the Alvedia typing cartridge allows typing in spite of agglutination for accurate type-specific transfusion. It is possible to transfuse a canine with a D.E.A. of 1.1 positive with a D.E.A. 1.1 negative unit. However, it is contraindicated to transfuse a D.E.A. 1.1 negative dog with a D.E.A. 1.1 positive unit. This could lead to severe transfusion reaction. With cats, blood transfusions must be type specific (i.e., a type A cat must receive type A blood and a type B cat must receive type B blood). It is acceptable practice, though not ideal, to transfuse a type AB cat with type A blood when AB blood is not available. It is important to also note that in regard to ferrets, they appear to have no identifiable blood type, and thus do not require typing or crossmatching.

Crossmatching should be performed in every patient who has had previous transfusions, or if it has been longer than 4 days since the animal's first transfusion, as the animal has had time to build up antibodies which could lead to transfusion reaction. It has been recommended to crossmatch cats with every transfusion, as cats with potential lymphoproliferative disorders (ex. FeLV/FIV) are difficult to type and may yield a false blood type when tested in house. Major crossmatching is the most common method performed and uses recipient serum or plasma against donor red blood cells. Minor crossmatching is less commonly performed and uses recipient red blood cells and donor plasma. Crossmatching takes approximately 40 minutes and can be performed in house using a testing kit (ex. RapidVet) for both canine and feline. Results will yield a positive or negative crossmatch via evidence of agglutination/hemolysis. A negative crossmatch likely means a safe transfusion between donor and recipient.

TRANSFUSION MONITORING

Any patient receiving a transfusion requires close monitoring. Monitoring parameters include temperature, heart rate and pulse quality, respiration rate and quality, mucous membrane (MM) color and capillary refill time (CRT), blood pressure, drip rate, IV catheter patency, and patient attitude. Typically, a dedicated technician and detailed monitoring sheet are recommended. Monitoring forms can also be used at a later date to determine patient compatibility with donor, transfusion rate, and blood product used. It is not recommended to feed a patient during a transfusion, as vomiting can occur with transfusion reactions. Though not proven, it may be beneficial to administer diphenhydramine to a patient prior to transfusion if they have had a history of immunologic allergic-type reactions with previous transfusions. Diphenhydramine is not indicated in febrile non-hemolytic transfusion reactions.

A typical transfusion monitoring protocol includes a baseline set of vitals including blood work, while taking into account any abnormalities such as hypothermia/hyperthermia, tachycardia, tachypnea, and mucous membrane color. These abnormalities should be expected in critical patients requiring transfusion, and may resume to normal parameters during and after transfusion. Once a baseline set of vitals is achieved and transfusion is initiated, a full set of vitals (as indicated above) should be taken routinely, and protocols differ from hospital to hospital. Initial protocol should be baseline vitals, a set of vitals every 15 minutes for a half hour, then every 30 minutes for 1 hour, then hourly during the remainder of the transfusion. It is important to keep in mind that patients can have delayed transfusion reactions. Some reactions can be delayed days.

TRANSFUSION REACTIONS

There are different types of transfusion reactions. A complete understanding of the types of transfusion reactions and their cause can aid in the reduction of reaction progression and patient risk. There are four types of transfusion reactions: acute immunologic (non-hemolytic and hemolytic), acute non-immunologic, delayed immunologic, and delayed non-immunologic.

Acute Immunologic Reactions

This type of transfusion reaction is due to the antibodies in the recipient's blood reacting to foreign antigens in the transfused blood. These reactions can be acute or delayed by days.

Febrile non-hemolytic transfusion reactions are the most common. This reaction typically occurs because of antibodies reacting to donor IgA or donor white blood cells, platelets, and plasma proteins. Symptoms of this reaction often entail vomiting, facial swelling, pruritis, and fever. Treatment for this reaction may include slowing down the transfusion rate, and supportive care. If the temperature continues to climb, transfusion should be stopped.

Acute immunologic type 1 hypersensitivities/anaphylaxis reactions are rare. This type of transfusion reaction typically occurs due to mast cell release. Pruritis, urticarial (hives), edema, elevated temperature, and anaphylaxis with shock-type symptoms can occur. Treatment plan includes immediate cessation of transfusion, anti-histamines, IV fluids, and shock treatment plan if indicated (i.e., oxygen, vasopressors, epinephrine, glucocorticoids, etc.).

Acute hemolytic transfusion reactions are extremely rare and very serious. Antibody sensitivity to antigens being transfused can cause this reaction. Hemolysis occurs because the immune system recognizes the antigens being transfused as foreign. Symptoms noted during this reaction include hemoglobinuria, hemoglobinemia, tremors/seizures, vomiting, weakness, pyrexia, tachypnea, and tachycardia/bradycardia, and death. This reaction is most severe in cases of a type B cat receiving type A blood, because of their naturally occurring antibodies. This reaction can also be seen in cases where a D.E.A. 1.1 negative dog is transfused with D.E.A. 1.1 positive blood. Treatment involves discontinuing the transfusion, supportive care with IV fluids, shock treatment plan if indicated, urine output monitoring, hemoglobin and bilirubin monitoring. (Delayed hemolysis can also occur. Patients should be closely monitored after transfusion.)

Acute Non-immunologic Reactions

This reaction type can occur for a variety of reasons aside from blood type mismatch. Tainted blood products, patient illness, and technician error can all induce a non-immunologic transfusion reaction. Some examples include thromboembolic complications, bacterial contamination, circulatory overload, citrate toxicity, and elevations in the patient's ammonia levels.

Transfusion-Associated Circulatory Overload (TACO)

This non-immunologic transfusion reaction can occur in patients with compromised cardiovascular systems. This can also occur in patients who are transfused too rapidly, due to technician error, causing circulatory overload as well. The colloid components of some blood products and the risk for rapid infusion of large volumes of blood products can lead to vascular overload with cardiogenic pulmonary edema. Signs of reaction include dyspnea, coughing, hypertension, tachycardia, vomiting, elevations in central venous pressures and systemic blood pressures, and cyanosis. Treatment for transfusion-related circulatory overload is supportive care, discontinuation of transfusion, oxygen therapy, and diuretics if indicated. The use of component therapy vs. whole blood transfusion can reduce the risk of TACO.

Transmission of Infectious Disease or Bacterial Contamination

Inappropriate donor screening for infectious disease can lead to transmission of disease to the recipient, and delayed symptoms. Poor handling of blood products during donation and transfusion can also lead to bacterial contamination, thus compromising the recipient. It is important to have strict aseptic technique during donation, blood product separation and transfusion. Signs of reaction can include GI signs, fever, dyspnea, hypoglycemia, hypotension, and tachycardia. Typically, signs occur relatively early during the transfusion. Treatment involves immediate cessation of the transfusion, with blood cultures of the recipient and the blood product. Antibiotics and supportive care should be initiated.

Hyperammonemia

This reaction can occur in patients with liver dysfunction, when older blood products are transfused which contain large amounts of ammonia. Ammonia levels rise in the unit during blood product storage. It may be recommended to use a newer unit of blood to transfuse patients with compromised liver function. Neurologic signs are to be expected with elevated levels of ammonia. Such signs include depressed mentation, ataxia, head pressing, seizures, and circling. Immediately stop the transfusion if signs occur and treat supportively to decrease ammonia levels. Oral lactulose, lactulose enemas, monitoring of mentation, serum ammonia levels, and decreased dietary protein intake may be considered during treatment.

Citrate Toxicity

This occurs when large volumes of citrated blood products are administered to patients with liver dysfunction, causing hypocalcemia. Citrate is used in stored blood products as an anticoagulant which chelates calcium. Tremors, seizures, facial twitching, decreased cardiac output, and arrhythmias will be seen with this form of reaction. Treatment includes discontinuing the transfusion, calcium supplementation, supportive care, and calcium monitoring.

Transfusion-Related Acute Lung Injury (TRALI)

Though more prevalent in human medicine, this complication can be seen in veterinary transfusion medicine as well. Signs are respiratory in nature, including dyspnea, pulmonary infiltrates, hypoxemia, and fever. Though the exact pathophysiology of TRALI is unknown, it is thought to involve recipient leukocyte antigens which react to donor leukocyte antibodies, as well as cytokine involvement allowing fluid to enter the alveoli creating lung injury. Signs can be delayed up to 6 hours after transfusion and are respiratory in nature. Treatment includes oxygen supplementation and ventilation. Leukoreduction may be beneficial in prevention.

Transfusion-Related Immunomodulation (TRIM)

Thought to occur as a result of immunosuppression post-transfusion. Can allow increased chance of organ survival post-transplant, but also increase risk of infection. It is thought that leukoreduction may be beneficial in reduction of immunosuppression and thus infection.

Delayed Immunologic Transfusion Reaction

These reactions can occur 4–21 days post-transfusion and can result from development of antibodies that shorten the transfused RBC lifespan.

Delayed Non-immunologic Transfusion Reaction

This results from the contamination of transfused blood. For example, retroviruses such as FeLV/FIV. These reactions can be prevented through complete donor screening.

New findings to consider in transfusion medicine include the emergence of leukoreduced blood products. It is thought that the use of leukoreduction filters can decrease risk, and possibly eliminate transfusion reaction by removing leukocytes which have inflammatory effects and can sensitize the immune system. Leukoreduction has proved to be beneficial in humans by reducing frequency and severity of febrile non-hemolytic transfusion reaction. The use of leukoreduction filters may also increase red cell survival. There are few studies in veterinary medicine to warrant mandatory animal blood banking protocol changes; however, leukoreduction in veterinary medicine is emerging in a positive light.

The DAL antigen is newer and requires attention when transfusing Dalmatians in particular. This antigen exists commonly in most dogs, but is not found in some Dalmatians. Thus transfusing a dog that does not have the antigen, with the DAL antigen can lead to severe transfusion reaction. It is strongly recommended to cross-match any Dalmatian before transfusion. More recently, this antigen is found to be lacking in Dobermans as well. Cross-matching Dobermans prior to transfusion is also suggested.

MiK is an antigen that has been recently identified in cats. MiK is found to be present in most cats; however, some have naturally occurring antibodies against this antigen. If a cat has naturally occurring antibodies to MiK and is transfused with this antigen, hemolytic transfusion reaction will occur. The importance of cross-matching each cat prior to transfusion must be stressed.

Xenotransfusion is the transfusion of blood from one species to another. This has been studied in the past with K9 to feline transfusions. However these studies are dated and patients had severe and often fatal transfusion reactions. This practice is not currently recommended and requires further study.

Cryopreservation (freezing) of red blood cells is currently performed in human medicine, in cases of mass casualty, by some of our armed forces. Through a process of glycerolization and deglycerolization of red blood cells, the product can be considered to have a stable shelf life of 10 years. This process does require special equipment and tools. However, by removing the plasma and washing the white blood cells, there is a significant decrease in transfusion reactions. Some articles show evidence of virtually no transfusion reactions with this product. There are no current studies using this cryopreservation process in veterinary medicine.

The importance of knowing patient requirements, component therapy, potential complications and monitoring protocols is absolutely necessary when transfusion medicine is indicated. Safe transfusions allow better patient outcome and blood product preservation. While transfusion medicine is growing, and the emergence of safer products is made available, transfusion outcome still rests on the knowledge and nursing skills of the veterinary team.

C01

Right Ventricular Systolic Function in Dogs with Preclinical and Clinical Myxomatous Mitral Valve Disease (Abstract C01)

Emily Hopkin Chapel[1]; Brian A. Scansen[2]; Karsten E. Schober[1]; John D. Bonagura[1]

[1]College of Veterinary Medicine, The Ohio State University, Columbus, OH, USA; [2]Colorado State University, Fort Collins, CO, USA

Right ventricular (RV) dysfunction strongly and independently predicts outcomes in people with myxomatous mitral valve disease (MMVD). The purpose of this prospective pilot study was to assess RV systolic function (RVSF) in dogs with different stages of MMVD. We hypothesized that RVSF would differ among stages of disease, and decrease in decompensated MMVD.

Client-owned dogs with MMVD not previously receiving cardiovascular medications and without moderate-or-severe pulmonary hypertension were eligible for enrollment. Dogs presenting in acute congestive heart failure received a single dose of furosemide prior to echocardiography. Five validated indices of RVSF were measured: tricuspid annular plane systolic excursion (TAPSE), fractional area change, lateral tricuspid annular longitudinal peak systolic velocity (S'), and longitudinal global strain (Strain) and strain rate of the RV free wall. Groups were compared using one-way ANOVA and Sidak's test. Frequencies of cases with cardiac remodeling falling outside previously-established reference intervals were compared using Fisher's exact test.

Thirty-five dogs at 3 stages of MMVD were recruited based on statistical power analysis: B1 (n=12), B2 (n=12), and C (n=11). Dogs in stage B2 had higher values for TAPSE, S', and Strain ($p \leq 0.05$) compared to stage B1; and TAPSE compared to stage C ($p=0.019$). TAPSE exceeded reference limits in 67% of B2 dogs compared with 9% in stage C ($p=0.009$). This study indicates that compensated MMVD with remodeling is associated with hyperdynamic RVSF, but with cardiac decompensation, TAPSE, S' and Strain often normalize. These results demonstrate that RVSF differs across stages of MMVD.

C02

High-Pressure Balloon Valvuloplasty for Severe Pulmonic Stenosis: A Prospective Observational Study in 25 Dogs (Abstract C02)

Catherine Belanger[1]; Catherine T. Gunther-Harrington[1]; Satoko Nishimura[1]; Maureen S. Oldach[2]; Lance C. Visser[3]; Joshua A. Stern[3]

[1]Veterinary Medical Teaching Hospital, University of California-Davis, Davis, CA, USA; [2]University of California-Davis, Davis, CA, USA; [3]Department of Medicine and Epidemiology, University of California-Davis, Davis, CA, USA

We aimed to evaluate the safety and efficacy of high-pressure balloon valvuloplasty (HPBVP) for treatment of canine pulmonic stenosis.

Twenty-five dogs with severe valvular PS (echocardiographically-derived transpulmonic peak/maximum pressure gradient (EDPG) ≥80 mm Hg) were recruited. All patients underwent complete echocardiography before and 16–24 hrs after balloon valvuloplasty (BVP) was performed using a high-pressure balloon (Atlas Gold PTA Dilatation Catheter) with rated burst pressures ranging from 14–18 mm Hg. Balloon sizes were 12–24 mm and 4 cm in length. The pulmonic valve annulus diameter was multiplied by a mean sizing factor of 1.4±0.15 to determine balloon size. Procedural success was defined as a post-HPBVP EDPG reduction of at least 50% or reduction into at least the moderate category of PS (50–79 mm Hg). Optimal result was defined as a post-procedural EDPG ≤30 mm Hg.

Initial median (IQR) EDPG for all dogs was 96 (88, 127) mm Hg with a post-operative median of 48 (36, 65) mm Hg. HPBVP produced a median pressure gradient reduction of 63 (39, 68)% and procedural success rate of 92% (23 dogs). Optimal results were achieved in 56% (14 dogs). There were no significant correlations between pressure gradient reduction and valve morphology (Type I and Type II) or severity of right ventricular hypertrophy. Pulmonic valve annulus diameter was the only tested variable that was significantly correlated to pressure gradient reduction ($p=0.02$; $r2=0.22$). No patients experienced suicide right ventricle and all patients survived the procedure.

These findings suggest that HPBVP is an appealing treatment option for routine cases of severe PS. Considering the procedural success rate and high number of optimal results, further studies comparing HPBVP to traditional BVP are warranted.

C03

Exit Block as a Mechanism of Sinus Node Dysfunction Evidenced by Geometric Heart Rate Variability (Abstract C03)

Flavia Giacomazzi[1]; Romain Pariaut[2]; Roberto Santilli[3]; Niels Otani[4]; N. Sydney Moise[5]

[1]Cornell University, Ithaca, NY, USA; [2]College of Veterinary Medicine, Cornell University, Ithaca, NY, USA; [3]Clinica Veterinaria Malpensa, SAMARATE, Lombardia, Italy; [4]College of Science, Rochester Institute of Technology, Rochester, NY, USA; [5]Section of Cardiology, Department of Clinical Sciences, College of Veterinary Medicine, Cornell University, Ithaca, NY, USA

The purpose of this study was to determine if beat-to-beat patterning as judged by geometric heart rate variability (HRV) was compatible with the hypothesis of impulse exit block as a mechanism for sinus node dysfunction (SND) in dogs.

Twenty-four dogs with SND and 24 age/size matched control dogs were studied. Geometric HRV was assessed using Poincaré plots and tachograms from Holter recordings. Autonomic tone was assessed using time and frequency domain HRV. A computer model was designed to assess beat-to-beat behavior. Inputs included (1) sinusoidal oscillations of vagal tone with a 4-second period (respiration), slowly varying component of sleep-wake cycle, and randomly varying noise; (2) oscillator of sinus node with variation dependent on vagal tone and a random component based on previous exit block; (3) probability of exit block with threshold dependent on vagal tone.

SND dogs had higher indices of HRV than control dogs. In contrast to control dogs, recordings from SND dogs showed >2 beat-to-beat clusters that approximated multiples. Depending on input, control and SND dog beat-to-beat patterns could be mimicked by the model. The computer model verified a pattern representing exit block (Frames A,B,C). SND dogs had Poincaré plots that failed to reveal an inherent sinus node rate (90–100 bpm) as did the control dogs. Instead, the shortest intervals were faster (120–170 bpm) with multiple bifurcations as the rate slowed.

Beat-to-beat patterning of SND dogs is consistent with block in the sinus node supporting the hypothesis of impulse block in the exit pathways from the sinus node.

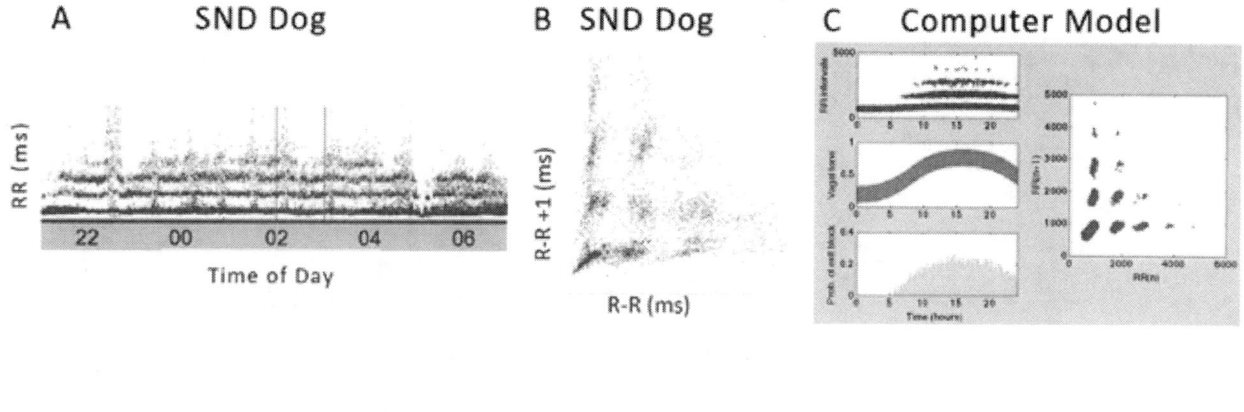

C04

Is the Left Atrium Prothrombotic in Cats with Aortic Thromboembolism? (Abstract C04)

Tsumugi Kurosawa[1]; Wan Ching Cheng[2]; Lois Wilkie[2]; Julia Sargent[3]; David J. Connolly[2]; Virginia Luis Fuentes[1]

[1]Royal Veterinary College, Hertfordshire, England, UK; [2]Royal Veterinary College, Hatfield, England, UK; [3]Royal Veterinary College/Southern Counties Veterinary Specialists, Bath, UK

Left atrial (LA) endothelial damage or stretch results in a procoagulable environment in association with down-regulation of thrombomodulin expression. The role of the endothelium in aortic thromboembolism (ATE) in cats is unknown. The aim of our study was to compare LA thrombomodulin gene expression in normal cats, cats with congestive heart failure (CHF) and cats with ATE.

Myocardium was collected from 5 normal cats, 8 cats with CHF and 12 cats with ATE within 15 minutes of euthanasia. LA myocardium was placed in RNA later and frozen in -80°C until batch analysis. The rest of the heart was formalin-fixed for histopathology. RNA was extracted and GeXP eXpress Profiler software was used to design primers. Multiplex qRT-PCR was performed to quantify thrombomodulin gene expression. All multiplex results were normalized to ribosomal protein S7. Results were expressed as median (range) and compared between groups using the Kruskal-Wallis and Mann Whitney tests.

There was no difference in sex (p=0.22) or age between groups (p=0.38). Median age was 79 months (4–184 months) in normal cats, 77 months (7–130 months) in CHF cats and 103 months (44–222 months) in ATE cats. Normalized thrombomodulin concentration in the ATE group was 0.10 (0.01–0.19) compared with 0.14 (0.08–0.32) in the CHF group and 0.2 (0.08–0.22) in the normal group (p=0.09).

Although no significant difference in thrombomodulin expression was identified, further investigations with a larger sample size would be justified.

C05

Characterization of Gene Expression Profiles Linked to Degenerative Mitral Valve Disease in Small Breed Dogs (Abstract C05)

Eva M. Oxford[1]; N. Sydney Moise[1]; Ed J. Rice[2]; Adam R. Boyko[3]; Charles G. Danko[2]

[1]Section of Cardiology, Department of Clinical Sciences, College of Veterinary Medicine, Cornell University, Ithaca, NY, USA; [2]Department of Biomedical Sciences, Baker Institute for Animal Health, Ithaca, NY, USA; [3]Department of Biomedical Sciences, College of Veterinary Medicine, Cornell University, Ithaca, NY, USA

The purpose of this study was to use a genome-wide association study (GWAS) and precision nuclear run-on and sequencing (PRO-seq), to identify novel gene candidates and expression profiles linked to DMVD in small breed dogs.

Blood samples from 249 (95 control, 154 DMVD) client-owned dogs (<20 kg) were used for the GWAS. Tissue samples from the anterior mitral valve leaflets were available from 2 DMVD and 5 control dogs for Western blot and PRO-seq analysis. DMVD dogs had a murmur and mitral regurgitation associated with leaflet thickening and/or prolapse on echocardiography. Control dogs were >9 years of age with no murmur, and no abnormalities on echocardiography.

A GWAS was performed with genomic DNA using a custom single nucleotide polymorphism (SNP) genotyping array containing 185,850 markers. Results were analyzed by calculating the odds ratio and P value for each SNP. A gene of interest was identified and overlapping primer pairs were designed to incorporate exon segments between 300–400 base pairs in length. Polymerase chain reactions (PCR) were performed using a touchdown program. Sanger sequencing was performed on a total of 20 samples (10 DMVD, 10 controls). Sequence alignments were performed using Clustal Omega software.

A total of 5 control and 2 DMVD valves were available for Western blot analysis. Samples were probed for antibodies recognizing alpha-catenin and beta-catenin (P-Y142 isoform), and normalized to GAPDH signal.

PRO-seq was performed on one control and one DMVD valve, in order to directly detect changes in gene transcription (expression) through the genome-wide mapping of RNA polymerase. Data were analyzed for genome wide up-or-down regulation using the PANTHER classification system (pantherdb.org). Genes were grouped based on their biological function.

The GWAS revealed the strongest association signal on Chromosome 17 ($P=7.7 \times 10^{-6}$), near the CTNNA2 gene, which codes for alpha-catenin. Fine mapping of CTNNA2 did not detect a mutation within the coding regions of the gene. However, Western blot analysis revealed a 53% decrease in normalized alpha-catenin levels and an 86% decrease in beta-catenin (P-Y142 isoform) compared to control samples. PRO-seq experiments did not reveal a clear signal for regulatory enhancers upstream of alpha-catenin near the highest association signal. The PRO-seq data additionally revealed genome-wide expression profiles for genes significantly up-or-down regulated ($p<0.05$ after Bonferroni correction) in the DMVD sample. A >2-fold change was noted in 187 genes; with functions such as wound healing, regulation of muscle contraction, and muscle structure development.

This study identified a candidate gene (CTNNA2), which codes for alpha-catenin, an essential protein of the adherens junctions and modulator of the canonical Wnt pathway. Although a mutation was not detected within the coding regions of the gene or within upstream enhancers of CTNNA2, Western blot revealed decreased levels of alpha-catenin (and the phosphorylated form of beta-catenin) in DMVD valves. These preliminary data may support the presence of a mutation located within the non-coding regions controlling CTNNA2 transcription and further investigation is warranted.

C06

Prognostic Value of MAPSE and TAPSE in Feline Hypertrophic Cardiomyopathy (Abstract C06)

Ilaria Spalla[1]; Jessie Rosie Payne[2]; Kieran Borgeat[2]; Virginia Luis Fuentes[3]; David J. Connolly[1]

[1]Royal Veterinary College, Hatfield, England, UK; [2]Royal Veterinary College/Highcroft Veterinary Referrals, Bristol, England, UK; [3]Royal Veterinary College, Hertfordshire, England, UK

Hypertrophic cardiomyopathy (HCM) has a variable prognosis depending on many factors, including left atrial size, the presence of clinical signs and left ventricular (LV) systolic function. Mitral annular plane systolic excursion (MAPSE) and tricuspid annular plane systolic excursion (TAPSE) assess longitudinal ventricular systolic function and may have prognostic value in cats with HCM.

We sought to investigate whether MAPSE and TAPSE were associated with survival time in cats with HCM.

Medical records were searched for cats diagnosed with HCM by echocardiography (LV wall thickness ≥6 mm), images of adequate quality to measure MAPSE/TAPSE and survival information. Of 185 included cats, 77 were showing clinical signs (53 CHF, 12 aortic thromboembolism [3 with concurrent CHF], 9 syncope, 3 exertional dyspnoea). MAPSE and TAPSE were measured by anatomical M-Mode at the atrioventricular annular level of the LV free wall (MAPSE-FW), interventricular septum (MAPSE-IVS) and right ventricular wall (TAPSE).

Cats with MAPSE-FW <3.6 mm had a median survival time (MST) of 352 days vs 2171 days in MAPSE-FW ≥3.6 mm ($p<0.01$). Cats with MAPSE-IVS <3.2 mm had a MST of 352 days vs 2171 days for MAPSE-IVS ≥3.2 mm ($p<0.001$) and cats with TAPSE <6.4 mm had a MST of 380 days vs 2171 days for TAPSE ≥6.4 mm.

Lower MAPSE and TAPSE were associated with worse survival, suggesting systolic longitudinal dysfunction has prognostic value in cats with HCM.

C07

Echocardiographic Assessment of Aortic Root Rotation in Dogs with Congenital Heart Disease (Abstract C07)

Nicole R. Wyatt[1]; Brian A. Scansen[2]

[1]Colorado State University Veterinary Teaching Hospital, Fort Collins, CO, USA; [2]Colorado State University, Fort Collins, CO, USA

Rotation of the aortic root, defined as abnormal position of the aortic valvar sinuses relative to surrounding cardiac structures, has been described in children with congenital heart disease but not dogs. Children with tetralogy of Fallot (TOF) show clockwise aortic root rotation and have an increased prevalence of coronary artery anomalies. Coronary artery anomalies are also common in brachycephalic dogs with pulmonary valve stenosis (PS). This study evaluated aortic root rotation in dogs with normal hearts, PS, subaortic stenosis (SAS), and TOF. Aortic root rotation in normocephalic compared to brachycephalic PS dogs was also investigated.

Echocardiograms were retrospectively evaluated from normal (n=31), PS (31), SAS (16), and TOF (5) dogs. Theta was defined as the angle between interatrial septum and commissure of left coronary and noncoronary sinuses, while rho was the angle between interatrial septum and commissure of right coronary and left coronary sinuses; each angle was measured in triplicate for all dogs and averaged. Groups were compared by ANOVA or Welch's test with alpha set at 0.05.

Differences in angles theta ($p<0.001$) and rho ($p=0.021$) were apparent between groups. Dogs with PS showed mild clockwise aortic root rotation, while dogs with TOF showed severe rotation, compared to normal or SAS dogs. In PS dogs, clockwise rotation was greater in brachycephalic compared to normocephalic dogs ($p=0.0004$).

We report clockwise aortic root rotation in dogs with congenital heart disease, particularly PS and TOF. Ongoing studies are investigating the relationship between aortic root rotation and coronary artery anomalies in brachycephalic PS dogs.

C08

Echocardiographic Right Atrial Volumes in Cats with Cardiac Disease and Pleural Effusion Versus Pulmonary Edema (Abstract C08)

Melissa Wilson; **Katherine F. Scollan**

Oregon State University, Corvallis, OR, USA

Cardiomyopathies, including hypertrophic and restrictive, are the most common acquired cardiac diseases in cats. Many affected cats develop pulmonary edema secondary to left-sided congestive heart failure (L-CHF). In addition, cats with primarily left-sided cardiomyopathy can develop pleural effusion without the presence of pulmonary edema. It is well known, but not well understood, why cats with left-sided CHF can develop both pulmonary edema and pleural effusion. A previous study identified that cardiomyopathic cats with pleural effusion had larger right atrial (RA) diameters than cats with pulmonary edema. The aim of this study was to compare right atrial volumes in cats in L-CHF with pleural effusion, L-CHF with pulmonary edema, and a group of healthy cats.

Medical records and echocardiographic studies from 2008–2015 were reviewed and measured. Cats were categorized into three groups: 1) L-CHF cats with pleural effusion 2) L-CHF cats with pulmonary edema, and 3) healthy cats. Right atrial volumes were measured from both the right long-axis four-chamber (RLA) and left apical four-chamber (LAP) views. Volumes were calculated using both the modified Simpson's method of discs (MOD) and the area-length formula (A-L) and indexed to body to yield mL/kg values. The LA:Ao ratio was measured from the right parasternal short-axis view. The RA volume methods (MOD versus A-L) and views (RLA versus LAP) were compared using Wilcoxon matched-pairs signed rank tests. Group medians of the MOD RLA volumes and LA:Ao ratios were compared using Kruskal Wallis tests. Receiver operator curves were generated to assess sensitivity and specificity of various cut-off values of indexed RA volumes for cats to develop pleural effusion.

Sixty-nine cats were included in the study, 27 cats with L-CHF and pleural effusion, 21 cats with L-CHF and pulmonary edema, and 21 healthy cats. The median indexed RA volume of the pleural effusion group (0.41 mL/kg, range 0.12–2.4) was significantly larger than the pulmonary edema (0.16 mL/kg, range 0.09–0.6; $p<0.0001$) and normal (0.23 mL/kg, range 0.12–0.39; $p=0.007$) groups. RA volumes were not significantly different between the pulmonary edema and normal groups. Using the MOD method, RA volumes were not different from the RLA and LAP views, however values obtained by MOD and A-L methods were significantly different ($p=0.0013$). Median LA:Ao values were not statistically different between the pleural effusion (2.17, range 1.63–3.14) and pulmonary edema groups (2.23, range 1.59–3.2), though both were different than the normal group. Using RA volumes to differentiate pleural effusion from pulmonary edema CHF, the area under the curve was 0.83. Sensitivity (Se %) and specificity (Sp %) for selected cut-off values of indexed RA volumes were: 0.3 mL/kg, Se 74, Sp 82; 0.35 mL/kg, Se 63, Sp 87; and 0.4 mL/kg, Se 56, Sp 95.

The results of this study indicate that cats that developed pleural effusion had significantly larger indexed RA volumes than those that developed pulmonary edema and healthy cats. Cats with CHF in both groups had LA enlargement, though this study would indicate that concurrent RA enlargement is more likely to cause pleural effusion as a manifestation of CHF than LA enlargement alone. RA volumes greater than 0.3 mL/kg offer adequate sensitivity and specificity for the development of pleural effusion in cardiomyopathic cats. Measurement of RA volumes are useful in estimating the risk of developing pleural effusion in cats with cardiomyopathy and normal values of RA volumes from a larger population should be investigated.

C09

Reader Performance of Radiographic Left Atrial Enlargement in Dogs and Comparison to Echocardiographic Left Atrial Assessment (Abstract C09)

Laetitia Duler; Nicole LeBlanc; Katherine F. Scollan; Stacy Cooley; Sarah Nemanic

Oregon State University, Corvallis, OR, USA

Accurate assessment of left atrial (LA) size is integral to the evaluation of left-sided cardiac disease in dogs and provides risk stratification for the development of left-sided congestive heart failure (CHF). Thoracic radiographs are routinely used to subjectively assess the presence and severity of left atrial enlargement (LAE). The primary aim of this study was to assess interreader agreement of subjective radiographic LAE between readers with different levels of experience and expertise. A second aim was to compare subjective degrees of radiographic LAE with echocardiographic measurements of LA size.

This retrospective study identified dogs that underwent thoracic radiographs and an echocardiogram on the same day at the Oregon State University Veterinary Teaching Hospital between 2012 and 2014. All thoracic radiographic views were anonymized and reviewed by 2 board-certified cardiologists, 2 board-certified radiologists, and 2 small animal rotating interns. Radiographs were evaluated for the presence or absence of LAE, severity of LAE if present (mild, moderate, or severe), and presence or absence of 7 pre-defined subjective LAE criteria. On the lateral projection these included elevation of the left caudal mainstem bronchus, loss of the normal gentle curvature of the caudal margin of the cardiac silhouette, straightening of the caudodorsal margins of the cardiac silhouette, and increased height of the caudodorsal border of the heart. On the ventrodorsal and dorsoventral projections, these criteria included divergence (or splaying) of the principal bronchi, a double opacity sign on 5.00–7.00 o'clock position on the cardiac silhouette, and evidence of left auricular enlargement at the 2:30–3 o'clock position.

Echocardiographic LA size was evaluated objectively by the left atrial-to-aortic ratio (LA:Ao) and LA volume via the monoplane modified Simpson's method of discs (MOD). LA volumes (LAV) were indexed to body weight as mL/kg. Objective measures of LA size were stratified into categories of normal (LA:Ao <1.5; LAV <1.3), mild (1.5≤LA:Ao<1.7; 1.3≤LAV<2.35), moderate (1.7≤LA:Ao<2; 2.35≤LAV<3.4), or severely enlarged (LA:Ao ^{3} 2; LAV ^{3} 3.4). Interreader agreement of radiographic LAE evaluation was assessed by linearly weighted kappa (k) and intraclass correlation (ICC). Sensitivity (Se) and specificity (Sp) of identifying varying degrees of echocardiographic LAE on radiographs was assessed using both LA:Ao and LAV.

101 dogs were included with a median weight of 8.6 kg (range 1–65 kg) and a median age of 8 years old (range 0.2–15 years). Echocardiographic diagnoses included normal (n=13), chronic degenerative valve disease (n=43), dilated cardiomyopathy (n=12), pulmonic stenosis (n=13), left-to-right patent ductus arteriosus (n=11), subaortic stenosis (n=4), and single cases of other cardiac diseases. Agreement for the presence or absence of radiographic LAE was excellent between all readers (ICC=0.7828) The interreader agreement for the degree of radiographic LAE ranged from moderate to substantial (k=0.493–0.701) dependent on the combination of readers. For the 7 distinct subjective LAE criteria, the interreader agreement ranged from fair to good (ICC=0.264–0.586) with the strongest interreader agreement for increased height of the caudodorsal heart margin, splaying of the mainstem bronchi, and enlargement of the left auricle. Agreement between specialists was higher than agreement between all readers for each individual criterion, with the exception of elevation of the left caudal mainstem bronchus.

Overall there was strong agreement between methods of echocardiographic LAE stratification: The agreement between LAV and subjective LAE was almost perfect (k=0.939), while LA:Ao had moderate agreement with the LAV (k=0.587). The agreement between subjective radiographic LAE and objective echocardiographic LAE was moderate and varied slightly with echocardiographic method of classification (LAV, k=0.573; LA:Ao, k=0.428). The sensitivity and specificity of thoracic radiographs to identify LAE was slightly better using LAV than LA:Ao (LAV Se=84% and Sp=70%; LA:Ao: Se=78% and Sp=71%).

The results of this study confirm the hypothesis that interreader agreement for radiographic classification of LAE is strong, even across specialty disciplines and levels of experience. There was variable agreement between readers for application of distinct, subjective criteria of LAE and this study indicated that increased height of the caudodorsal heart margin, splaying of the mainstem bronchi, and enlargement of the left auricle were the most consistent criteria used by cardiologists, radiologists, and interns to identify LAE. Subjective radiographic LAE severity had only moderate agreement with echocardiographic LAE severity regardless of objective LA size measure, however LAV and subjective LAE showed stronger agreement than LA:Ao.

C10

Echocardiographic Effects of Butorphanol and Atenolol in Dogs with Pulmonary Valve Stenosis (Abstract C10)

Satoko Nishimura[1]; Lance C. Visser[2]; Catherine Belanger[1]; Catherine T. Gunther-Harrington[1]; Maureen S. Oldach[3]; Joshua A. Stern[2]

[1]Veterinary Medical Teaching Hospital, University of California-Davis, Davis, CA, USA; [2]Department of Medicine and Epidemiology, University of California-Davis, Davis, CA, USA; [3]University of California-Davis, Davis, CA, USA

We sought to determine the effects of butorphanol and atenolol on maximum transpulmonary pressure gradient (maxPG), pulmonary valve effective orifice area indexed to body size (EOAi), and right ventricular (RV) systolic function in dogs with pulmonic stenosis (PS).

Dogs (n=18) with PS (maxPG >50 mm Hg) were prospectively recruited. Dogs had 3 echocardiograms: baseline, 5-minutes post-butorphanol (0.2–0.25 mg/kg IV), and 2–3 weeks post-atenolol (1–1.5 mg/kg q 12). In a blinded fashion, maxPG, EOAi, pulmonic-to-aortic velocity time integral ratio (VTI ratio) and RV function (assessed via tricuspid annular plane systolic excursion, RV fractional area change, and RV systolic velocity [S′]) were measured. MaxPG, VTI ratio, EOAi and RV function indices were compared pre- and post-drug using a paired t-test. Percent changes post-drug were compared with an unpaired t-test.

Post-butorphanol, there were no significant differences in indices of RV function or severity of PS but heart rate was significantly reduced (mean % change ±SD = -21%±16; $p<0.001$). Post-atenolol, heart rate and maxPG was significantly ($p\leq0.01$) reduced with mean %changes ±SD of -17%±13 and -15%±9, respectively. There were no significant changes in EOAi, VTI ratio and RV function post-atenolol. Percent change in VTI ratio post-atenolol (6%±12) was significantly ($p=0.01$) different than maxPG (-15%±9). Three dogs previously diagnosed with severe PS (maxPG >80 mm Hg) were diagnosed with moderate PS (maxPG 50–80 mm Hg) post-atenolol.

Atenolol reduces PS severity assessment via maxPG. EOAi and VTI ratio may be useful flow-independent methods for PS severity assessment, particularly post-drug or intervention.

C11

Prospective Clinical Trial Evaluating Spironolactone in Dobermans with Congestive Heart Failure Due to Dilated Cardiomyopathy (Abstract C11)

Lynne O'Sullivan[1]; Sonja Fonfara[1]; Andrew Laskary[2]; Heidi Chambers[1]

[1]University of Guelph, Guelph, ON, Canada; [2]Ontario Veterinary College, University of Guelph, Guelph, ON, Canada

The aldosterone antagonist spironolactone was recently shown to improve survival in dogs with congestive heart failure (CHF) due to chronic mitral valvular disease. Whether spironolactone has beneficial effects in dogs with dilated cardiomyopathy (DCM) is not known. The primary objective of the study was therefore to evaluate the effect of spironolactone on survival time in Dobermans with CHF due to DCM.

The trial design was prospective, randomized, single-blinded, and placebo-controlled. 67 dogs were randomized to receive 50 mg of spironolactone twice daily (34 dogs) or a placebo (33 dogs), in addition to standard therapies. Clinical follow up was every 1–4 weeks until endpoint. Quality of life questionnaire and physical examination were performed at every recheck, while renal biochemistry, ECG, echocardiography, and thoracic radiography were reassessed as needed. The primary endpoint was time to cardiac death, defined as death from CHF, euthanasia (euth) for CHF, or sudden death (SD). Kaplan-Meier survival estimates were computed and compared.

The 67 dogs comprised 20 females and 47 males with average (range) age 7.9 years (2.0–14.3). One dog in the spironolactone group was withdrawn by the owner. 59 dogs (89%) reached the primary endpoint (5 CHF death, 26 CHF euth, 28 SD). 7 dogs (11%) were censored (6 euth non-heart, 1 lost to follow-up). The median survival time in the spironolactone group (183 days) was not statistically significantly different than that for the placebo group (124 days) (Log-Rank p=0.35, Wilcoxon p=0.19). Results need to be verified in larger trials given the relatively small sample size.

C12

Heart-Fatty Acid Binding Protein - A Potential Marker for Canine Cardiac Diseases (Abstract C12)

Christopher Lam[1]; Domingo Casamian Sorrosal[2]; Gabrielle Monteith[1]; Sonja Fonfara[3]

[1]Ontario Veterinary College, University of Guelph, Guelph, ON, Canada; [2]Dick White Veterinary Referrals, Cambridge, UK; [3]University of Guelph, Guelph, ON, Canada

Heart-Fatty Acid Binding Protein (hFABP) is produced primarily in cardiomyocytes, mainly responsible for fatty acid homeostasis, and has been proposed as an early and sensitive biomarker in human cardiac diseases. Its potential use as a cardiac biomarker in veterinary medicine has yet to be investigated. This study aims to assess hFABP as a prospective biomarker for canine cardiac diseases compared to cardiac Troponin-I (cTnI).

Twenty-one healthy dogs (control), 23 dogs with degenerative mitral valve disease (MVD), and 13 dogs with dilated cardiomyopathy (DCM) of various cardiac disease stages were studied. Serum measurements of HFABP and cTnI, routine blood work and echocardiography were performed at initial presentation for all dogs and at 1 and 3 months for the diseased dogs, and followed up until 12 months to verify survival.

Levels recorded during the initial presentation showed significantly elevated hFABP with highest levels found in ACVIM stage C disease regardless of cause, followed by stage B and controls. cTnI differed stage C from both B and controls, but was unable to differentiate between stage B and controls. DCM patients had higher hFABP and cTnI levels over MVD patients, but cTnI was indistinguishable between MVD dogs and controls. Reduction of hFABP and cTnI during the follow-up period was also observed from initial presentation amongst both DCM and stage C dogs. hFABP was the most significant predictor for death when multivariate study was performed.

This study showed the potential value of hFABP as a cardiac marker to differentiate stages and disease type in dogs.

C13

Non-Gated Computed Tomography Angiography for Evaluation of Coronary Artery Anomalies in Bulldogs with Pulmonic Stenosis (Abstract C13)

Catherine T. Gunther-Harrington[1]; Kathryn Phillips[2]; Lance C. Visser[3]; Joshua A. Stern[3]

[1]Veterinary Medical Teaching Hospital, University of California-Davis, Davis, CA, USA; [2]University of California-Davis, Davis, CA, USA; [3]Department of Medicine and Epidemiology, University of California-Davis, Davis, CA, USA

Prepulmonic coronary anomalies (CAs) have been reported in bulldogs and represent an increased risk when performing balloon valvuloplasty (BVP). Identification of CAs have been reported using high-slice cardiac-gated computed tomography angiography (CTA). However, the utility of the more accessible, non-gated, 16-slice CTA is unknown. The purpose of this study was to evaluate the feasibility of non-gated CTA to identify CAs in anesthetized bulldogs with severe pulmonic stenosis (PS) prior to BVP.

In this prospective observational study, bulldogs and bulldog mixes with severe PS were recruited. Full echocardiograms were performed to diagnose PS. Dogs were placed under general anesthesia and a non-gated 16 slice CTA was performed. CTAs were reviewed by a board-certified radiologist and cardiologist and coronary anatomy identified. When normal coronary artery anatomy was noted on CTA, a right ventricular outflow tract angiogram was performed and evaluated during levophase to confirm normal coronary anatomy prior to BVP. Dogs with prepulmonic CAs were recovered from anesthesia and no BVP was performed.

Five bulldogs and 2 bulldog crosses were enrolled. All dogs had clear, interpretable images from the non-gated CTA. Coronary anomalies were identified in 4 dogs based on non-gated CTA, all of which were consistent with R2A anomaly. Three dogs had normal coronary anatomy based on non-gated CTA which was confirmed during levophase angiography. Successful BVP was performed without incident in these 3 dogs.

We conclude that non-gated CTA represents an accessible, noninvasive method for diagnosing CAs in bulldogs with PS.

C14

Ventricular Remodelling Marker Levels Differ in Cats with Hypertrophic Cardiomyopathy with and without Atrial Thrombus (Abstract C14)

Sonja Fonfara[1]; Sarah Kitz[2]; Anja Kipar[2]

[1]University of Guelph, Guelph, ON, Canada; [2]University of Zurich, Zurich, Switzerland

Atrial thrombosis (AT) is a known complication in cats with hypertrophic cardiomyopathy (HCM). Slow blood flow, endothelial damage and hypercoagulability, caused by an enlarged left atrium (LA) and cardiac dysfunction are suspected to be involved in thrombus formation. However, not all cats with HCM and LA enlargement develop AT, and how cats with AT differ from cats without AT is not known.

We therefore compared the myocardial transcription of cytokines (interleukin [IL]-1, -2, -4, -6, -8, -10, -12, -18, tumor necrosis factor-α, interferon-γ, transforming growth factor [TGF]-β), matrix metalloproteinases (MMP)-2, -3, -9, -13, and tissue inhibitors of metalloproteinase (TIMP)-1, -2, and -3 from atria and ventricular samples from 10 cats with HCM and 5 cats with HCM and AT using quantitative real time PCR. The cats had been clinical patients, and diagnosis of the disease was obtained by echocardiography.

Cats with HCM showed generally higher atrial than ventricular marker transcription, which applied for cats with HCM and AT only for IL-1, TGF-β, MMP-2, MMP-9 and TIMP-1. Cats with HCM and AT exhibited also significantly lower atrial IL-12 and lower ventricular MMP-2 and TIMP-2 transcription than atria and ventricular samples, respectively, from cats with HCM without AT.

These results suggest different remodelling processes primarily in the ventricles of cats with HCM and AT, which might contribute to cardiac dysfunction and AT formation in cats with HCM. Lower IL-12 and TIMP-2 have been reported in association with increased left ventricular hypertrophy, fibrosis and dysfunction.

C15

Normal Echocardiographic Parameters in Clinically Healthy Adult Captive Tigers with and without Medetomidine Sedation (Abstract C15)

Kiira T. Rodriguez; Andrew Cushing; Ed Ramsay; Rebecca Gompf

Department of Small Animal Clinical Sciences, The University of Tennessee, Knoxville, TN, USA

This study aimed to describe echocardiographic findings in healthy adult tigers and determine the difference between echocardiographic variables under sedation with medetomidine and after reversal with atipamezole. Eight adult tigers (*Panthera tigris*) from an exotic cat sanctuary were included.

Each tiger underwent a complete echocardiographic examination by the same operator following sedation with medetomidine, midazolam and ketamine and induction into general anesthesia. Following completion of the first exam, the medetomidine was reversed with atipamezole. A second echocardiographic examination was then performed 30 minutes later by the same operator. Measurements were tested for statistical differences between the two examinations and between gender and weight.

One tiger was excluded from analysis due to the finding of a ventricular septal defect, leaving a total of 7 tigers. Physiologic regurgitation of the tricuspid and pulmonic valves was seen in 86% and 100% of the tigers. There was a statistically significant difference in left ventricular internal dimension at end-diastole based on weight and interventricular septum thickness at end-systole between pre and post-echoes. Fractional shortening and left ventricular internal dimension at end-systole approached, but did not reach, statistical significance between pre and post-echoes.

In conclusion, physiologic tricuspid and pulmonic regurgitations are common findings in healthy adult tigers. The echocardiographic measurements obtained during this study can be used as references for future examinations in adult tigers. The effects of medetomidine on these measurements, specifically the left ventricular internal dimension at end-diastole and interventricular septum thickness at end-systole, should be taken into account.

C16

Evaluating Clopidogrel Effect in Cats Using Plateletworks (Arachidonic Acid) and Platelet Function Analyzer 200 (P2Y) - (Abstract C16)

Anthony Abrams-Ogg[1]; Lynne O'Sullivan[1]; Sonja Fonfara[1]; Shari Raheb[1]; Sophie Saati[1]; Kimberly Ho[2]; Shauna Blois[1]; R. Darren Wood[1]

[1]University of Guelph, Guelph, ON, Canada; [2]Central Toronto Veterinary Referral Clinic, Toronto, ON, Canada

Platelet function testing may be useful to identify if clopidogrel resistance occurs in cats. Plateletworks® (PW) consistently detects clopidogrel effect in normal cats using ADP as agonist. In canine PW studies, arachidonic acid (AA) is as useful an agonist as ADP to detect clopidogrel effect, but there are no data for cats. Platelet Function Analyzer® (PFA) inconsistently detects clopidogrel effect in normal cats using the collagen-ADP cartridge. This also occurs in humans, prompting development of the INNOVANCE® P2Y cartridge, which consistently detects clopidogrel effect in humans and dogs. The objective of this study was to investigate PW-AA and PFA-P2Y in cats. Nine normal cats and 13 cats with cardiovascular diseases were studied. All had normal hematocrits and platelet counts. Blood samples were collected by jugular venipuncture. The 9 normal cats showed PW-AA %Aggregation (%Agg)=83–97% and PFA-P2Y Closure Time (CT)=38–111 sec. For cats with cardiovascular diseases (one or both tests performed): a) 7 cats already receiving clopidogrel: PW-AA %Agg=4–19% (n=3) and PFA-P2Y CT >300 sec (n=6). b) 6 cats prior to starting clopidogrel: PW-AA %Agg=21–96% (n=5) and PFA-P2Y CT=50–118 sec (n=3). For 3 of these 6 cats one or both tests were repeated while receiving clopidogrel: PW-AA %Agg=0–12% (n=3) and PFA-P2Y CT >300 sec (n=1). For both tests, differences between normal cats and cats being treated with clopidogrel were significant ($p<0.05$). PW with AA agonist and PFA with P2Y cartridge may be highly sensitive to detect clopidogrel effect in cats.

C17

Postoperative Changes in Aortic Characteristics Affect the Low Fractional Shortening in Canine Patent Ductus Arteriosus (Abstract C17)

Takuma Aoki; Yoko Fujii; Hiroshi Sunahara; Keisuke Sugimoto; Takuya Uehara; Yoshito Wakao

Azabu University, Sagamihara, Kanagawa, Japan

Our objective is to elucidate the influence of changes in diastolic aortic pressure and aortic compliance on the fractional shortening (FS%) in dogs that undergo patent ductus arteriosus (PDA) occlusion. 62-client owned dogs that underwent percutaneous PDA occlusion were retrospectively used for our study. The dogs were evaluated using echocardiography and a catheter for the measurement of their direct blood pressure during PDA occlusion. The diameter of descending aorta was measured using digital fluoroscopic cine loops. After PDA occlusion, the systolic blood pressure did not change; however, pulse pressure significantly decreased ($p<0.01$) and diastolic blood pressure (DAP) significantly increased ($p<0.01$). In a univariate analysis ($n=49$), the low postoperative FS% was associated with an older age ($p<0.05$), the increased postoperative diastolic blood pressure, and decreased pulse pressure (both with $p<0.01$). In a multivariate analysis ($n=49$), low postoperative FS% was associated with an older age ($p<0.05$) and a decreased postoperative pulse pressure ($p<0.05$). The diameter of the descending aorta significantly increased both in systole and diastole (both with $p<0.01$) and the aortic FS% was significantly decreased ($p<0.05$) just after PDA occlusion. In conclusion, left ventricle might not be able to contract after PDA occlusion because of increased DAP and decreased aortic compliance, i.e., the aortic valve more strongly closed and the aorta was rapidly dilated like a rupturing balloon due to trapping a large amount of blood into aorta.

C18

Congestive Heart Failure Associated with DOCP Therapy in 12 Dogs with Hypoadrenocorticism (Abstract C18)

Christophe Bourguignon[1]; Teresa C. DeFrancesco[1]; Sandra Tou[1]; Rita Hanel[2]; Steven Marks[1]; Bruce Keene[1]

[1]North Carolina State University, Raleigh, NC, USA; [2]BluePearl Veterinary Partners, Raleigh, NC, USA

Steroid-associated congestive heart failure (CHF) has been documented in both humans and cats. Episodes of possible DOCP-associated CHF in dogs with hypoadrenocorticism (HA) prompted a search of medical records from 2003–2016 to evaluate the potential relationship between DOCP administration and CHF.

Of 146 canine HA cases identified during this time period, 12 (8%) also experienced CHF. Eight of these dogs had chronic mitral valve disease; the other 4 had dilated cardiomyopathy. Five of the 12 dogs experienced CHF after the first injection of DOCP; 3 of these had known pre-existing heart disease. The median time between the DOCP and CHF was 11 days (range 1–29 days).

The 12 HA/DOCP/CHF cases were matched to historical controls (CHF without HA, matched by age, breed, sex, cardiac diagnosis and year of diagnosis). The number and time interval of documented CHF episodes and survival after CHF was evaluated in both groups.

Median survival after the onset of CHF was 448 days in controls vs. 953 days in the HA/DOCP group, but this difference was not significant (p=0.43). The number and time interval of CHF episodes was likewise not different between groups.

The small number of cases and the large variability in the parameters measured limited our ability to identify differences between groups. Despite the negative findings in this small, retrospective study, further study of the relationship between DOCP administration and CHF seems warranted.

C19

An Impaired Oculocardiac Reflex Identifies the Autonomic Imbalance in Dogs with Symptomatic Mitral Valve Disease (Abstract C19)

Bruna C. Brüler[1]; Thayane C. S. Vieira[1]; Marcela Wolf[2]; Stephany B. Lucina[2]; Marlos Gonçalves de Sousa[3]

[1]Federal University of Paraná - UFPR/Department of Veterinary Science, Curitiba, Parana, Brazil; [2]Federal University of Paraná - UFPR, Curitiba, Parana, Brazil; [3]Univ. Estadual Paulista UNESP/Jaboticabal, Jaboticabal, São Paulo, Brazil

The oculocardiac reflex is the heart's physiological response to digital compression of the eyeballs, producing negative chronotropic and inotropic responses in healthy subjects. In this study, we investigated the autonomic imbalance in dogs with CHF due to myxomatous mitral valve disease (MMVD) by means of their individual response to that vagal maneuver. Fifty-two client-owned MMVD dogs (stages B1/B2/C/D) and twelve healthy dogs admitted for regular cardiac evaluation were recruited for a prospective observational study. Every dog had an ECG recorded for 5 minutes, and during the last minute both globes were compressed into the orbits using digital pressure until the 5 minute-recording was over. Twenty RR intervals were used to calculate the mean RR interval, the standard deviation of RR intervals, the root mean square of the successive differences in RR intervals and the vasovagal tonus index before eyeball compression and at approximately twenty seconds after compression started. The percent change of every surrogate was calculated. Although a significant difference was only documented for VVTI ($p=0.0419$), the other surrogates did behave differently in stages C and D dogs as compared to B1 and B2 animals (Figure 1). An explanation for this might be the parasympathetic withdraw in CHF, in addition to a subtle sympathetic activation due to discomfort, which was likely suppressed in dogs with preserved vagal response. Also, the sustained increased sympathetic activation rather than transitional sympathetic activation might be involved with the distinct response to OCR in symptomatic dogs as compared to the asymptomatic ones.

Figure1: Median percent change of the surrogates of autonomic balance after digital ocular compression in healthy dogs and dogs with MVD (B1/B2/C/D).

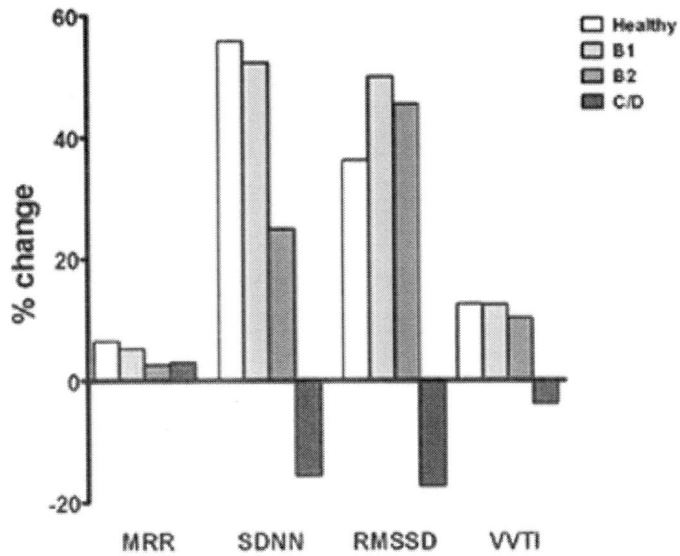

C20

Instability of the QT Interval in Dogs with Myxomatous Mitral Valve Disease (Abstract C20)

Bruna C. Brüler[1]; Flavio S. Jojima[1]; Gustavo Dittrich[1]; Amalia T. Giannico[1]; Marlos Gonçalves de Sousa[2]

[1]Federal University of Paraná - UFPR/Department of Veterinary Science, Curitiba, Parana, Brazil; [2]Univ. Estadual Paulista UNESP/Jaboticabal, Jaboticabal, São Paulo, Brazil

In people, QT prolongation and instability are known to relate to ventricular arrhythmias. Since dogs with myxomatous mitral valve disease are prone to developing ventricular arrhythmias, this study investigated whether indices derived from QT interval could be used to predict an increased likelihood of arrhythmias regardless of clinical signs. 167 client-owned MMVD dogs (stages B1/B2/C) admitted for cardiac evaluation were recruited for a prospective cross-sectional study. Every dog underwent an echocardiogram and ECG recording. Fifty consecutive QT intervals were measured for each dog. QTs were corrected for heart rate and used to calculate the average QT (QTa), QT variance (QTv), total instability (TI), short term (STI) and long term (LTI) instabilities. A significant difference between asymptomatic and symptomatic animals was documented for every parameter ($p<0.0001$), with higher values found for stage C dogs. Significant difference between B1 and B2 also existed for QTa. An AUC >0.5 was found when QTa and STI were used to discriminate dogs with or without ventricular arrhythmias (Figure 1). Dogs with QTa >310 (sensitivity: 56%/specificity: 71%), and STI >11 (sensitivity: 56%/specificity: 85%) were shown to be slightly (15%) and moderately (30–45%) more prone to developing ventricular arrhythmias, respectively. Also, an AUC >0.5 was documented when every QT index was used to distinguish animals with remodeled (B2/C) and non-remodeled hearts (B1). QTa and QTv best discriminated B1 from B2/C dogs and B1/B2 from C dogs, respectively. Because an increased QT instability was shown in symptomatic patients, QT indices might translate into prognostic surrogates for arrhythmias and CHF in MMVD dogs.

Figure 1: Receiver operating characteristic (ROC) curves obtained when differentiating MMVD dogs with ventricular arrhythmias from dogs without ventricular arrhythmias on the standard ECG recording.

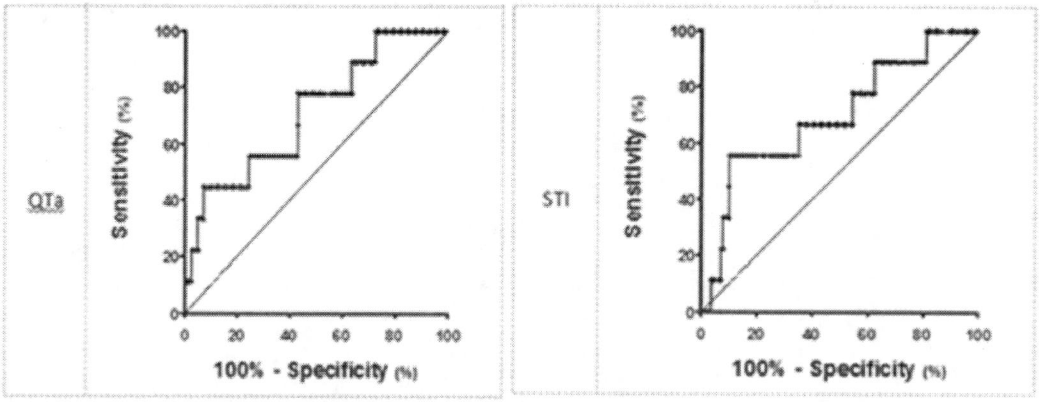

C21

Cardiovascular Changes in Brachycephalic Dogs (Abstract C21)

Raphaela Marques Canola; Jaislane Braz; Evandro Zacché; Alejandro Zamora Restan; Jorge Cardoso Silva Filho; Bruno Cristian Rodrigues; Michelli Fenerich; Elizabeth Regina Carvalho; **Marlos Gonçalves de Sousa**; Aparecido Antonio Camacho

UNESP - Jaboticabal, São Paulo, Brazil

Brachycephalic syndrome is characterized by primary and secondary upper respiratory tract abnormalities, which may result in significant upper airway obstruction. Cardiovascular and hemodynamic changes may also be present. However, unlike upper respiratory abnormalities which have been intensely studied, there is a paucity of information regarding cardiovascular and hemodynamic changes related to the syndrome.

To that purpose, one evaluated 19 animals selected in two distinct groups: Control Group (CG) comprised of 6 beagles with mean age of 2 years, and Brachycephalic group (BG) consisted of 13 french bulldog with brachycephalic syndrome, aging 2.5 years. Thoracic radiographs, arterial blood gas analysis and echocardiographic examination were performed. Data were submitted to normality test and statistical differences in between groups were assessed by using t-test ($p \leq 0.05$).

BG presented P_aO_2 (89.4±9.9 mm Hg) and P_aCO_2 (34.7±2.1 mm Hg) below and above CG (104.9±5.2 mm Hg; 32.4±1.7 mm Hg), respectively ($p \leq 0.05$). VHS index in BG (11.9±0.9) was higher when compared to CG (10.9±0.2) ($p \leq 0.05$). BG showed better performance on M-mode regarding LVIDs (15.4±3.5 mm) and FS (44.8±9.5%) when compared to CG (19.1±3.0 mm; 33.2±4.7%, respectively) ($p \leq 0.05$). The LA/Ao ratio was also higher in BG (1.43±0.1) than in CG (1.26±0.1) ($p \leq 0.05$).

Based on our results, brachycephalic dogs presented signs of cardiorespiratory disorders, demonstrated by the lower P_aO_2 values. Furthermore, their heart silhouette was considered bigger than CG (higher VHS index), with signs of right-side cardiac remodeling. However, no significant variations were observed on echocardiographic examination which could indicate signs of cardiac remodeling in both groups. The best performance obtained by the BG based on LVIDs and FS% variables could be explained by catecholamine release since there was no chemical restraint during examination procedure and these animals were more stressed at the examination table. It is believed that, since the animals on BG group were all young adults, with a mean age of 2.5 years, they did not have enough time to manifest all the complications, including cardiovascular and hematological, related to the syndrome. It is our belief that with aging thus, with progression of the condition, right-side cardiac remodeling, and alterations in blood gas and arterial blood pressure would be more evident. Even so, depending on severity, development of pre-capillary pulmonary arterial hypertension may also occur. Our results suggest brachycephalic syndrome may have effects over the cardiovascular system. Therefore, we encourage the need of a thorough investigation to better understand the cardiovascular changes related to brachycephalic syndrome in order to distinguish them from specific heart conditions, especially when attaining predisposing dog breeds.

C22

Dobutamine Stress Test in Dogs with Degenerative Mitral Valve Disease (Abstract C22)

Bruno Cristian Rodrigues; Evandro Zacché; Jorge Cardoso Silva Filho; Raphaela Marques Canola; Jaislaine Braz; Michelli Fenerich; Elizabeth Regina Carvalho; Alejandro Zamora Restan; **Marlos Gonçalves de Sousa**; Aparecido Antonio Camacho

UNESP - Jaboticabal, São Paulo, Brazil

Mitral valve myxomatous degenerative disease (MVMD) is the main heart disease in dogs, whose evolution leads to significant secondary hemodynamic changes with activation of cardiac and neurohormonal compensatory mechanisms. Pharmacological stress with dobutamine was performed aiming detection of possible changes in the myocardial reserve and systolic dysfunction, which are usually not detected during the conventional echocardiographic exam. Therefore might help veterinarians with early diagnosis of these disorders and thus, establishing a prognosis and appropriate treatment for the patient.

For this purpose, 18 beagles, male and female, older than 6 years old were evaluated. These animals were selected in 3 groups: control group (G1) comprised of 6 animals and 2 other groups composed by dogs with MVMD. The second group (G2) with 7 animals, had the left ventricular internal diameter ratio in diastole indexed to aorta (DIVEd/Ao) less than 2.8 and the third group (G3) with 5 animals and DIVEd/Ao above 2.8.

Echocardiography was performed before and after pharmacological stress with dobutamine. The protocol for pharmacological stress was a continuous infusion of 5 μg/kg/minute during 5 minutes, followed by 10 μg/kg/minute for 10 minutes. The results were based on the difference between pre and post pharmacological stress values of shortening fraction (FS%) and ejection fraction (EF%) variables, obtaining means of each groups and comparing them to each other.

Statistical analysis was performed by variance analysis followed by the Tukey's test. Regarding FS%, there was no statistical difference ($p<0.05$) among G1 and G2 (0.44 ± 0.14 and 0.49 ± 0.13, respectively). However, G1 and G3 (0.44 ± 0.14 and 0.19 ± 0.14) and G2 and G3 (0.49 ± 0.13 and 0.19 ± 0.14) showed statistical difference. Along with EF% variable, there was no statistical difference among G1 and G2 (0.25 ± 0.08 and 0.30 ± 0.08) and G1 and G3 (0.25 ± 0.08 and 0.13 ± 0.09), but there was difference between G2 and G3 (0.30 ± 0.08 and 0.13 ± 0.09).

Based on our results, it was possible to infer that pharmacological stress test with dobutamine did not evidence changes in systolic function in G2 animals with mild cardiac remodeling when compared to G1. However, animals with severe remodeling (G3) the systolic function differed when compared to other groups, suggestive of systolic dysfunction.

C23

Speckle Tracking in Dogs with Myxomatous Mitral Valve Disease Submitted to Pharmacological Stress Test (Abstract C23)

Jorge Cardoso Silva Filho; **Marlos Gonçalves de Sousa**; Evandro Zacché; Bruno Cristian Rodrigues; Raphaela Marques Canola; Jaislane Braz; Michelli Fenerich; Elizabeth Regina Carvalho; Alejandro Zamora Restan; Aparecido Antonio Camacho

UNESP - Jaboticabal, São Paulo, Brazil

Myxomatous mitral valve disease (MMVD) is a more common heart disease in dogs, leading to valve heart failure and congestive heart failure.

The evaluation of myocardial function through two-dimensional speckle tracking has been shown to be an effective method for the early identification of ventricular dysfunction, being safe and feasible for evaluation of myocardial function in dogs with suspected of some form of latent myocardial dysfunction not evidenced by conventional echocardiographic examination. Stress echocardiography with dobutamine is commonly indicated to detect cardiac dysfunction in asymptomatic patients, evaluating myocardial contractile reserve and viability. Pharmacological stress test was performed with dobutamine to evaluate the myocardial function via two-dimensional speckle tracking in healthy dogs and in different stages of the MMVM. The study included 15 Beagle dogs, distributed in 3 groups and selected according to the internal diameters of the left ventricle in diastole and the aortic ratio (DIVEd/Ao). The groups were constituted: Group I (Control) - 5 healthy dogs without MMVD; Group II - 5 dogs with MMVD (DIVEd)/Ao ratio lower than 2.8; Group III - 5 dogs with MMVD and DIVEd/Ao ratio higher than 2.8. The animals were submitted to the echocardiographic examination performed by a single operator and evaluation was performed before and after dobutamine infusion. Dobutamine was diluted in 5% glucose solution, following the progressive scheme: 5 µg/kg/minute for 3 consecutive minutes; 7.5 µg/kg/minute for 3 consecutive minutes; 10 µg/kg/minute for 3 consecutive minutes. The images were analyzed using the Syngo Velocity Vector Imaging (VVI) algorithm optical flow algorithm (SIEMENS®). Data were submitted to the normality test and to repeated measures analysis of variance and Tukey's test.

Table 1. Comparison of the mean values of the Strain and Strain rate radial values before and after the Dobutamine pharmacological stress test in healthy beagle dogs and dogs with DMVM.

	Group 1	Group 2	Group 3	p Value
Peak systolic radial strain (%)	44,75±3,87	17,45±10,87	19,81±8,21	<0,001
Peak systolic radial strain rate (s^{-1})	1,90±0,18	0,88±0,62	0,81±0,51	0,018

C24

Cardiovascular Echocardiography Aspects During Pregnancy in Female Dogs (Abstract C24)

Vivian Tavares de Almeida; Ricardo Andres Ramirez Uscategui; Victor José Correia dos Santos; Aparecido Antonio Camacho; **Marlos Gonçalves de Sousa**; Raphaela Marques Canola; Ana Paula Rodrigues Simões; Michele Lopes Avante; Wilter Ricardo Russiano Vicente

UNESP - Jaboticabal, São Paulo, Brazil

The aim this study was to evaluate the adaptations involving the left ventricle, across the gestational process and its diastolic function, using tissue Doppler echocardiography. Animals: 15 pregnant bitches, healthy aged 2 to 6 years and weighing between 5 and 15 kg. Heart rate and blood pressure clinical evaluation was performed, as well as echocardiography M-mode, spectral and tissue Doppler, in the estrous (EZ), every 15 days during pregnancy (PD 15, 30, 45 and 60) and 45 days postpartum (PP). The variables from the evaluation were submitted to statistical analysis. Heart rate (HR) ($p=0.149$) and systolic blood pressure (SAP) ($p=0.564$) did not show significant changes during pregnancy. The parameters that have undergone significant changes were EF ($p=0.0016$) and FS ($p=0.0019$). AVVmax, PVmax, wave E'(s) S' (s) and S' (lat) being influenced by the gestational period ($p<0.05$). These changes could be explained by cubic regression model. The E:A ($p=0.041$) IVRT (0.044) relation and A '(lat) (0.023) also suffered influence of pregnancy, but had a nonparametric residual distribution. Pregnancy produces changes in maternal hemodynamics also influencing diastolic function. Myocardial contractility increases as a result of an improvement in systolic performance, while diastolic function changes significantly revealing a more compliant ventricle, which is the physiological response to increased volume and metabolic requirements due to gestation.

C25

TAPSEidx Echocardiographic Evaluation in Dogs with Pulmonary Hypertension Pre-Capillary and Post-Capillary (Abstract C25)

Jaislane Braz; Raphaela Marques Canola; Alejandro Zamora Restan; Jorge Cardoso Silva Filho; Evandro Zacché; **Marlos Gonçalves de Sousa**; Aparecido Antônio Camacho

UNESP - Jaboticabal, São Paulo, Brazil

Pulmonary hypertension (PH) is a hemodynamic and pathophysiologic condition defined as increase in systolic pulmonary artery pressure greater than 30 mm Hg. Classically, this condition can lead to right ventricular dysfunction, which is associated to poor outcomes, especially in patients with chronic congestive heart failure. Unfortunately, right ventricular function in dogs with pulmonary arterial hypertension has been poorly investigated. Thus, the aim of the present study was to assess the right ventricular systolic function in dogs with tricuspid regurgitation and pulmonary hypertension by echocardiographic evaluation.

To that purpose, one evaluated 35 dogs with echocardiographic diagnosis of tricuspid regurgitation and PH, calculating the transtricuspid gradient regurgitation through Bernoulli modified equation and then adding an assumed right atrial pressure. Animals were selected into 5 groups: Pos-capillary PH group, composed by dogs with myxomatous mitral valve disease and clinical or echocardiographic congestion signs; Pre-capillary PH group including dogs with pulmonary disease. Each group was subdivided in mild PH group, including dogs with pulmonary arterial pressure gradient estimated below 50 mm Hg; and moderate/severe PH group, comprised by animals with pulmonary arterial pressure gradient over 50 mm Hg; Pre-capillary mild group was composed by 14 dogs; pre-capillary PH moderate/severe group 6 dogs; pos-capillary PH mild group 6 dogs and pos-capillary PH moderate/severe group 9 dogs. All PH groups were comprised by animals between 5 and 16 years, 2.8 and 27.6 kg and pulmonary arterial pressure estimated between 36 and 96 mm Hg. These animals did not receive antihypertensive treatment until the time of examination. Control group with 6 clinically healthy dogs with no echocardiographic abnormality, 2 years old and weight between 9.1 and 12.7 kg.

Echocardiographic evaluation of RV was performed in all dogs and followed the recommendation of the Guidelines for the echocardiographic assessment of the right heart endorsed by the European Association of Echocardiography. The parameter assessed was tricuspid annular plane systolic excursion (TAPSE), acquired by placing an M-mode cursor through the tricuspid annulus and measuring the amount of longitudinal motion of the annulus at peak systole an apical 4-chamber window. Once TAPSE varies according patient's weight it was indexed by aortic diameter (TAPSEidx). The data was submitted to normality test, the analysis between groups was performed by one way ANOVA test ($p<0.05$) and means were compared by Tukey test.

There was statistical difference between control and pre-capillary moderate/severe groups ($p<0.05$), and from pos-capillary moderate/severe and control groups ($p<0.05$) when evaluating TAPSEidx parameter. Results are summarized in Table 1.

Statistical difference detected by TAPSEidx among control group and pre and pos-capillary moderate/severe groups may suggest that patients with moderate/severe PH have right ventricular systolic dysfunction.

Table 1. TAPSE echocardiographic parameters of RV systolic function (mean ± SD) of dogs with PH (n = 35) and clinically healthy dogs (n = 6).

	Control	PHpre mild	PHpre m s	PHpos mild	PHpos m s
TASPEidx	1.34 ± 0.27	0.89 ± 0.22	0.64 ± 0.47	0.97 ± 0.32	0.77 ± 0,46

PHpre mild = Pulmonary hypertension pre-capillary mild group; PHpre m s = Pulmonary hypertension pre-capillary moderate/severe group; PHpos mild = Pulmonary hypertension pos-capillary mild group; PHpos m s = Pulmonary hypertension pre-capillary moderate/severe group;

C26

Pre Clinical Results of LASSBIO294: An Inodilator Compound (Abstract C26)

Rosangela de Oliveira Alves Carvalho[1]; Ana Paula Araujo Costa[1]; Priscilla Regina Nasciutti[1]; Rogério Vieira Silva[1]; Bruna Dantas Matos[1]; Nayara Gonçalves Melo[1]; Gabriel Moreno Matos de Andrade[1]; Eliezer de Jesus Barreiro[2]; Valéria de Oliveira[1]

[1]Federal University of Goias, Goiania, Goias, Brazil; [2]Federal University of Rio de Janeiro, Rio de Janeiro, Brazil

LASSBio 294 is a compound, developed by the Laboratory of Evaluation and Synthesis of Bioactive Substances (LASSBio®) of the Federal University of Rio de Janeiro (UFRJ) made from the substrate safrole, a compound extracted from sassafras oil, found in Brazilians plants such as the cinnamon-white (*Ocotea pretiosa*). The inodilator effect of the compound has been demonstrated in bioassays, using isolated and intact aortic rings, and in whole hearts and fractions of myocardial of rats, suggesting that the mechanism of action for its vasodilatory effects is related to inhibition of phosphodiesterases, and the inotropic effect on the increased Ca 2+ uptake by the sarcoplasmic reticulum. In rats with experimentally induced myocardial infarction, LASSBio 294 decreased collagen deposition, presumably regulating inflammation mediators. In addition, it increased cardiac lusitropism, decreasing left ventricular filling pressure, modulating the action of the SERCA protein, increasing Ca 2+ uptake by the sarcoplasmic reticulum, inducing myocardial relaxation and improving diastolic function. In previous studies performed by these authors on healthy beagle dogs, the oral administration of the compound lowered blood pressure, without causing hypotension or other toxic effects and increased the shortening fraction by 19.1%. Faced with the need to test the compound in a non-rodent mammal with myocardial disease, it was proposed in this study to verify the action of LASSBio 294 (2 mg/kg, n=7) on the cardiovascular parameters of New Zealand rabbits with dilated cardiomyopathy induced by doxorubicin, using pimobendan (3 mg/kg, n=6) as a positive control, both treatments were administered every 12 hours for 30 days. DCM was induced by intravenous administration of 1 mg/kg of doxorubicin, twice a week for three weeks and weekly thereafter until FS=25% or less was achieved. The animals were evaluated weekly by: electrocardiography, echocardiography, blood pressure measurement, chest X-ray and biomarkers of renal and hepatic function and myocardial injury. This study was approved by the Ethics Committee of Animal Use of the Federal University of Goias (UFG) under the protocol number 006/14. The data were submitted to analysis of variance and compared by means of Tukey test, adopting 5% of significance. Under these conditions, LASSBio 294 increased systolic function, increasing FS by 35% in the second week, and improved diastolic function in a superior manner to pimobendan, maintaining the E/A ratio <1 and E'/ A' <1 and reducing IVRT, without altering the arterial pressure or producing proarrhythmogenic or toxic effects and also, reduced the serum concentration of creatinine. However, it did not prevent the evolution of the congestive condition, since two animals presented pulmonary edema and effusions. Thus, the pre-clinical *in vitro* and *in vivo* testing stages of LASSBio 294 was concluded, and it was tested in rodent and non-rodent mammals, healthy and with induced DCM, with a positive inodilator and lusitropic effect. Therefore, it can be concluded that LASSBio 294 is a promising compound, which needs to be used in clinical trials with individuals with naturally occurring DCM, so that it can complete the necessary trials for its rise as drug, allowing an increase in the existing therapeutic protocols.

C27

Clinical and Echocardiographic Features, and Survival of French Bulldogs with Pulmonic Stenosis (66 Cases: 2005–2016) - (Abstract C27)

Cécile Damoiseaux[1]; Loic Desquilbet[2]; Vassiliki Gouni[1]; Marie Lavennes[1]; Camille Poissonnier[1]; Sandrine Bertrand[3]; Maud Chenebaux[3]; Jean-Louis Pouchelon[1]; Valérie Chetboul[1]

[1]Alfort Cardiology Unit, National Veterinary School of Alfort, Maisons-Alfort, Ile-de-France, France; [2]Department of Biostatistics and Clinical Epidemiology, National Veterinary School of Alfort, Maisons-Alfort, Ile-de-France, France; [3]National Veterinary School of Alfort, Maisons-Alfort, Ile-de-France, France

Pulmonic stenosis (PS) is one of the most common congenital heart defects in dogs, affecting various canine breeds including English and French Bulldogs (FB). PS has been extensively described in English bulldogs. However, to the best of our knowledge, no specific study has been dedicated to PS in FB. The objective of this prospective observational study was therefore to document the epidemiological, clinical, and echocardiographic features, and risk factors for death in FB with PS. The study population consisted of 66 FB. In most cases (53/66, 80%), at least 2 obstructive lesions were observed (most commonly [42/66, 64%] valvular and supravalvular), using color-flow Doppler mode and 3-dimensional echocardiography. The median Doppler-derived peak transvalvular pulmonary gradient (DG) was high: 170 mm Hg (lower-upper quartiles: 139–223 mm Hg), with a DG ≥200 mm Hg in 33% of FB. Among the 51 FB for which a follow-up was available, 21/51 (41%) died, 67% (14/21) of all deaths being cardiac-related (CD) at a median age of 3.5 years. In univariate analyses, age (HR=2.2 per 1 year; p=0.02), clinical signs at presentation (HR=3.7; p=0.03), DG (HR=1.2 per 10 mm Hg; p=0.01), right ventricle dilation (HR=5.0; p=0.04), moderate to severe tricuspid regurgitation (HR=7.6; p<0.01), and right-sided congestive heart failure (HR=4.8; p=0.05) were significantly associated with time to CD. After adjustment for age and DG, tricuspid regurgitation remained significantly associated with time to CD (HR=5.1; p=0.02). In conclusion, PS in FB are commonly severe and complex, with tricuspid regurgitation being an independent predictor of CD.

C28

Comparative Multidimensional Imaging of Patent Ductus Arteriosus in Dogs (Abstract C28)

Kelley R. Doocy; Ashley Saunders; Sonya Gordon

College of Veterinary Medicine & Biomedical Sciences, Texas A&M University, College Station, TX, USA

Accurate measurements are integral for device selection and sizing to minimize complications associated with minimally invasive transcatheter closure of patent ductus arteriosus (PDA), in particular regarding minimal ductal diameter (MDD) to device ratio depending on method of measurement. Standard measurements are typically obtained with two dimensional (2D) angiography and echocardiography. The aims of this study were to characterize PDA dimensions including MDD, ampulla diameter 3 mm above the MDD and morphology based on right lateral angiography, transthoracic echocardiography (TTE) from the right and left imaging windows and 2D, biplane, and 3D transesophageal echocardiography (TEE).

The study population consisted of 25 dogs with left-to-right shunting PDA; 20 had minimally invasive closure and 5 had surgical ligation (see table).

The 2D TTE and angiographic MDD measurement is acquired in the same orientation as 3D TEE measurement A. A was the smallest of the 2 perpendicular 3D measurements in 23/24 (96%) dogs. The ratio of A/B indicated a circular (0.05 mm difference between A and B) shaped ampulla in 5/24 (21%) and MDD in 7/24 (29%). Real-time 3D TEE documented a dynamic change in the pulmonary ostium dimensions throughout the cardiac cycle. In 21 dogs with angiography performed prior to catheter-based closure, morphology was classified as type IIA (16), type IIB (3), type III (1, subsequently ligated), and other (1, similar to Krichenko type D). All 7 dogs with circular MDD on 3D TEE were classified as type IIA on angiography. Median Amplatz® canine duct occluder (ACDO) size used in 20 dogs was 7 (range, 3–10). Median (range) CDO to MDD ratio was 2.04 (1.48–3.07) with angiography and 2.09 (1.25–4.00) with 2D TEE. Median (range) CDO proximal disc to ampulla ratio was 1.68 (1.26–2.78) with angiography and 1.76 (1.07–2.63) with 2D TEE. The measurement deemed most accurate including 3D TEE was used to make device selection in individual cases. No complications occurred in any dog.

In conclusion, 3D TEE provides a clear cross sectional image of the PDA pulmonary ostium and documented an elliptical shape MDD in the majority (71%) with the minor axis (A) similar in location to the angiographic and 2D TEE measurement which could contribute to device under sizing. Based on standard 2D TEE and angiographic imaging, the ACDO was oversized by approximately 2 times the MDD as previously reported, and the proximal disc to ampulla ratio was always at least 1.

Measurements are reported in mm as median (range).

	Angiography	TTE – Right	TTE – Left	TEE	3D TEE*
MDD	3.0 (1.7-4.9)	4.0 (1.0-8.0)	3.0 (2.-0-4.0)	3.0 (1.0-6.0)	A: 2.0 (0.4-3.6) B: 3.0 (0.9-5.8)
Ampulla diameter	8.0 (3.2-11.3)	9.0 (4.0-15.0)	9.0 (3.0-12.0)	7.0 (4.0-15.0)	A: 5.7 (1.4-8.3) B: 7.6 (2.6-11.8)

*Two perpendicular measurements obtained in a cross sectional view

C29

Changes in Neurohumoral Factors in 9 Dogs that Received Carperitide and Presented with Pulmonary Edema (Abstract C29)

Masayuki Enokizono[1]; Nobuyuki Kanno[2]; Haruhiko Suzuki[3]; Shusaku Yamada[4]; Katsuhiro Matsuura[5]

[1]Japan Small Animal Medical Center, Tokorozawa, Saitama, Japan; [2]Nihon University, Fujisawa, Kanagawa, Japan; [3]Animal Cardiovascular and Thoracic Surgery Center, Sayama, Saitama, Japan; [4]Animal Cardiovascular and Thoracic Surgery, Fuchuu, Tokyo, Japan; [5]Animal Cardiovascular and Thoracic Surgery Center, Sayama-shi, Saitama, Japan

The purpose of this study was to clarify the treatment effects of carperitide in veterinary cases. Nine dogs presenting with pulmonary edema due to mitral valve insufficiency were treated with a continuous infusion of carperitide (0.05 μg/kg/min). Dogs with hypotension were excluded from the study. Blood sampling was performed twice: once before the administration of carperitide (Pre) and then after the administration of carperitide (Post). Neurohumoral factors, such as atrial natriuretic peptide (ANP), N-terminal pro b-type natriuretic peptide (NT-proBNP), and aldosterone levels, were measured at both time points. In addition, chest radiography, echocardiography, and blood pressure measurements were performed depending on the specific cases. Seven of the 9 dogs showed improvements post-treatment, although 2 cases did not show any improvement. ANP levels increased post-treatment in all cases (mean ± standard deviation: pre = 292±163 vs. post = 908±866). NT-proBNP levels increased in 4 cases and decreased in 5 cases (pre = 6563±2061 vs. post = 5627±3026). Aldosterone level increased only in one case (pre = 232±172 vs. post = 124±130) In addition, the left atrial aortic root ratio (LA/Ao), E-wave, vertebral heart score (VHS), and blood pressure decreased. Carperitide is a human ANP agent, which is why we expected higher ANP levels post-treatment. We observed a decrease in NT-proBNP levels in all cases that were treated for more than 7 hours. This is thought to be due to a time lag until the half-life period of NT-proBNP and the stretch stimulation of the ventricular muscle, which are reflected in the blood concentration of NT-proBNP. In addition, the ventricular muscle stretch was inhibited by the decrease in left atrial pressure following the administration of carperitide, which lasted for more than 7 hours. Lastly, the decrease in aldosterone levels was thought to be due to the inhibitory effects of carperitide on the renin-angiotensin-aldosterone system. This study suggested that carperitide administration is effective since pulmonary edema is improved and aldosterone is suppressed.

C30

Myocardial Remodeling in Canine Dilated Cardiomyopathy (DCM) - (Abstract C30)

Stefania Gasparini[1]; **Sonja Fonfara**[2]; Udo Hetzel[3]; Anja Kipar[1]

[1]University of Zurich, Zurich, Switzerland; [2]University of Guelph, Guelph, ON, Canada; [3]Institute of Veterinary Pathology, Vetsuisse Faculty, University of Zurich, Zurich, Switzerland

Dilated cardiomyopathy (DCM) is one of the most common cardiac diseases in dogs. However, its pathogenesis is still not well understood. Based on the detailed histological and transcriptional assessment of the myocardium in DCM, the present study aimed to investigate the involvement of remodeling processes in the pathogenesis of DCM.

The myocardium of 16 dogs with DCM was examined by histology and immunohistology. Alongside, quantitative reverse transcriptase PCR (qPCR) for cytokines and remodeling markers was performed in four of these cases, in relation to their transcription in six dogs with non-cardiac diseases. Laser microdissection with subsequent qPCR served to confirm the cardiomyocytes as the source of these markers.

The histological examination confirmed the occurrence of degenerative myocardial changes ("fatty infiltration-degenerative type" and "attenuated wavy fiber type"), but also revealed a general increase in interstitial cells. In addition, focal lesions were observed. These comprised accumulations of fibroblast-like cells, mild fibrosis and edema, with neovascularization and degenerating cardiomyocytes. The majority of interstitial cells appeared to be of macrophage origin and were often found to express intracellular adhesion molecule (ICAM-1), transforming growth factor (TGF)-b and vascular endothelial growth factor, which also cardiomyocytes expressed frequently. The transcription of ICAM-1, matrix metalloproteinase (MMP)-2, tissue inhibitor of MMP (TIMP)-1, TIMP-2, lysyl oxidase and tumor necrosis factor (TNF)-a was significantly increased, and cardiomyocytes were shown transcribe most relevant markers.

The results confirm remodeling processes as a major component in the pathogenesis of canine DCM and suggest that cardiomyocytes at least contribute to the pathogenic processes.

C31

Retrospective Evaluation of Effect of Heart Rate on Survival in Dogs with AF (Abstract C31)

Brigite Pedro[1]; Joanna Dukes-McEwan[2]; Mark Oyama[3]; Marc S Kraus[3]; **Anna R Gelzer**[3]

[1]Willows Veterinary Centre and Referral Service, Solihull, UK/Small Animal Teaching Hospital, University of Liverpool, Solihull, England, UK; [2]Small Animal Teaching Hospital, University of Liverpool, Neston, England, UK; [3]Department of Clinical Studies, School of Veterinary Medicine, University of Pennsylvania, Philadelphia, PA, USA

Atrial fibrillation (AF) is a common arrhythmia in dogs, usually associated with a rapid ventricular response rate, and typically managed with pharmacological rate control. This retrospective study aimed at assessing the effect of heart rate (HR) on survival time in dogs with AF. Forty-six dogs from 2 sites with AF and at least one 24 h Holter recording were included from records reviewed between 2008–2016. The mean heart rate (meanHR, 24 h average), minimum HR (minHR, 1 minute average), maximum HR (maxHR, 1 minute average) were recorded at the time when adequate rate control was presumed and no further adjustments of rate control therapy were made. Long-term outcome was ascertained and the primary study endpoint was all cause mortality.

All dogs had structural heart disease, 31/46 were in congestive heart failure, 44/46 were receiving anti-arrhythmic drugs. MeanHR was 125 bpm (range 62–203 bpm), minHR was 82 bpm (range 37–163 bpm) and maxHR was 217 bpm (range 126–307 bpm). Sex, age, weight, cardiac diagnosis, and study site had no significant impact on survival. For every 10 bpm increase in meanHR, the risk of all-cause mortality increased by 35% (hazard ratio 1.35; 95% CI 1.17–1.55; $p<0.001$). Median survival time of dogs with meanHR <125 bpm (n=23) was significantly longer (1037 days; range 524-open) than dogs with meanHR ≥125 bpm (n=23) (105 days; range 67–267days) ($p=0.0012$). Only meanHR was independently associated with survival time ($p<0.003$) by multivariate regression. Our results suggest that aggressive rate control in dogs with AF leads to reduced mortality and prospective studies targeting meanHR of 125 bpm or lower are indicated.

C32

Left Atrial Volume Can Detect Congestive Heart Failure in Dogs with Chronic Mitral Valve Disease (Abstract C32)

Matheus M. Mantovani[1]; Jacqueline Ribeiro de Castro[1]; **André Martins Gimenes**[1]; Lilian C. Petrus[1]; Caio N. Duarte[1]; Mariana Yukari Ueda[1]; Guilherme T. Goldfeder[2]; Maria Helena M.A. Larsson[2]; Denise S. Schwartz[1]

[1]School of Veterinary Medicine and Animal Science, University of São Paulo (USP), São Paulo, Brazil; [2]University of São Paulo, São Paulo, Brazil

This study aimed to evaluate left atrial volume (LAV) in dogs in different stages of chronic mitral valve disease (CMVD), and to evaluate the association of these values with congestive heart failure (CHF). Our hypothesis was that LAV increases with severity of CMVD and can be used to detect CHF in dogs with CMVD. Eigthy dogs were included in a prospective cross-sectional clinical study, grouped according to the staging of CMVD proposed by the American College of Veterinary Internal Medicine (A, B1, B2 and C), based on clinical signs and echocardiographic evaluation. Left atrial volume was calculated using the biplane area-length method from the left apical two and four chamber views at the end of ventricular systole. The LAV was indexed to body weight. LAV/kg showed high values ($P_{25\%-75\%}$=3.91–6.81) compared with groups A (0.82 mL/kg, $P_{25\%-75\%}$=0.64–0.95), B1 (0.91 mL/kg, $P_{25\%-75\%}$=0.47–1.29) and B2 (2.23 mL/kg, $P_{25\%-75\%}$=1.93–2.85). Analysis of ROC curve for CHF diagnosis showed area under curve of 0.92 for LAV/kg. A LAV/kg >3.26 mL/kg was the optimal cut-off identified for detecting CHF in dogs with CMVD with a sensitivity and specificity of 90%. In addition, when LAV is greater than this cut-off, it increases to 90% the probability to detect CHF (an increase of 40% in diagnostic capacity than pre-test), which allows for the decision to treat for CHF. In conclusion, left atrial volume increases with severity of CMVD and helps the detection of CHF in these dogs.

C33

Serum Levels of ST2 Soluble Fraction in Clinically Healthy Dogs (Abstract C33)

André Martins Gimenes[1]; Leticia Beccari[1]; Jacqueline Ribeiro de Castro[1]; Matheus M. Mantovani[1]; Caio N. Duarte[1]; Lilian C. Petrus[1]; Maria Helena M.A. Larsson[2]; Denise S. Schwartz[1]

[1]School of Veterinary Medicine and Animal Science, University of São Paulo (USP), São Paulo, Brazil; [2]University of São Paulo, São Paulo, Brazil

The soluble fraction of ST2 (sST2) is the circulating isoform of ST2, a member of the interleukin-1 (IL-1) receptor family. The sST2 is induced when cardiomyocytes and cardiac fibroblasts undergo mechanical strain, making it a promising biomarker of fibrosis and cardiac remodeling. sST2 has been widely studied in human medicine, and characterized as an important prognostic biomarker and an independent mortality predictor. However, there are no studies determining reference values or evaluating the influence of gender and age on sST2 in healthy dogs. Therefore, this study aimed to measure serum levels of sST2 in healthy dogs, in order to determine normal values for the study population and to evaluate the influence of gender and age on this biomarker concentration for future assessment of dogs with heart disease. With this purpose, a prospective cross-sectional clinical study was conducted, and included a sample of 60 healthy small-to-medium size dogs, males and females, adults of different ages. There was no significant difference in sST2 serum levels between males (22) and females (38) ($p=0.980$) or between age groups in the study population ($p=0.358$). The reference values found for the serum levels of sST2 in healthy dogs ranged from 36.94 to 44.67 pg/mL (CI 95%), with a mean value of 40.80 pg/mL. There was no influence of gender or age group on serum levels of sST2 for the assessed dog population. It was possible to suggest a reference range for the biomarker in healthy dogs.

C34

Soluble ST2 Biomarker in Heart Failure Secondary to Chronic Mitral Valve Disease in Dogs (Abstract C34)

André Martins Gimenes[1]; Jacqueline Ribeiro de Castro[1]; Matheus M. Mantovani[1]; Caio N. Duarte[1]; Lilian C. Petrus[1]; Mariana Yukari Ueda[1]; Guilherme T. Goldfeder[2]; Maria Helena M.A. Larsson[2]; Denise S. Schwartz[1]

[1]School of Veterinary Medicine and Animal Science, University of São Paulo (USP), São Paulo, Brazil; [2]University of São Paulo, São Paulo, Brazil

The soluble fraction of ST2 (sST2) is considered a cardiac biomarker, which is induced mechanically by the stretch of cardiomyocytes and fibrobasts. This biomarker has prognostic value in human cardiology, and it is considered an important independent mortality predictor. However, studies assessing sST2 in dogs are still lacking. Therefore, this study aimed to determine the utility of sST2, compared to NT-proBNP and cardiac troponin I (cTnI), for assessment of dogs with chronic mitral valve disease (CMVD). With this purpose, 151 dogs in different stages of CMVD were submitted to clinical exam, laboratory tests, cardiologic exams and measurement of sST2, NT-proBNP and cTnI. The results demonstrate that sST2 was higher in dogs in CHF compared to non-CHF dogs. However, when dogs were stratified according to CMVD stages, it was not possible to differentiate the stages based on sST2 concentrations. NT-proBNP and cTnI were higher in CHF dogs, as well as in more advanced stages of CMVD. There was correlation between sST2 and NT-proBNP and cTnI. Analysis of ROC curve for CHF diagnosis showed areas under the curve (AUC) of 0.66 for sST2, 0.86 for NT-proBNP and 0.76 for cTnI. As for prognosis assessment, sST2 concentrations higher than 40 pg/mL indicated higher mortality compared to groups with lower sST2 values. Results suggest that sST2's performance as a standalone diagnostic test is limited and inferior to the other studied biomarkers. However, sST2 seems to be a useful independent prognostic marker, and mortality predictor in dogs with CMVD.

C35

Prevalence of Heart Disease in Dogs at a Teaching Hospital: A Retrospective Populational Study (Abstract C35)

Vinicius Dayoub; Paula Itikawa; **Guilherme T. Goldfeder**; Ana Elisa M. Gonçalves; Bruno Silva; Rebecca B. Pessoa; Nathalia Cardoso; Suzana Enumo; Maria Helena M.A. Larsson

University of São Paulo, São Paulo, Brazil

Chronic valvular heart disease (CVHD) is the most common morbidity, of great relevance in the internal medicine of small animals. The scope of the present study was to evaluate the main diagnoses performed in an 11-year (2005–2015) period at the cardiology unit of a teaching hospital. In the period of the study, 11,921 dogs were attended, with gender balance: 51% males and 49% females. Mongrel dogs and Poodles represented both 25% of the animals. The mean age of the patients was 129 months, of which 45% (5,387) sought the unit for pre-anesthetic evaluation. Excluding pre-anesthetic evaluation, CVHD accounted for 64% of the diagnoses, healthy dogs 19%, congenital anomalies 3%, arrhythmias 3%, pericardial effusion 3% (178), being 148 (83%) with cardiac base neoplasia, dilated cardiomyopathy 2%, and 5% left including tracheal collapse, bronchitis, and dirofilariosis. Between 2005 and 2010, 54 animals were diagnosed with cardiac base neoplasia, while between 2010–2015, 94 dogs were diagnosed, representing an increment of 74% of cardiac neoplasias. From 141 cases with dilated cardiomyopathy, 97 were attended between 2005–2010, and 44 between 2010–2015, showing a decrease of 121% in the last period. The data obtained are in agreement with the literature that reveals CVHD and congenital anomalies as the most frequent cardiopathies. The significant decrease in the prevalence of CMD may be caused by despopularization of the breeds predisposed to DCM, while the increase of cardiac base neoplasias may be theorized by the improvement in canine nutritional management increasing survival of companion animals.

C36

Assessment of Left Atrial Function by Speckle-Tracking Echocardiography in Dogs with Myxomatous Mitral Valvular Disease (Abstract C36)

Min Hee Kang; **Sang-Woo Han**; Hee-Myung Park

College of Veterinary Medicine, Konkuk University, Seoul, Seoul-t'ukpyolsi, Republic of Korea

Left atrial function is well known prognostic factor in dogs with myxomatous mitral valvular disease (MMVD). Left atrial functional analysis with advanced echocardiographic technique such as speckle-tracking echocardiography (STE) have not been performed in affected dogs. The purpose of this retrospective study was to evaluate left atrial STE variables associated with disease severity and to compare it with conventional left atrial and ventricular echocardiographic parameters.

Retrospective review of medical records for dogs with MMVD was performed. Client-owned forty-eight dogs were included in this study. Every dog was presented for evaluation of heart disease: 34 dogs with MMVD (according to mitral regurgitation (MR) severity: mild, n=12; moderate, n=11; and severe groups, n=11) and 14 healthy age and weight-matched controls. The conventional echocardiographic parameters including fractional shortening (FS), ejection fraction (EF), E velocity, E/A ratio, E/E′ ratio, left atrium to aorta ratio (LA/Ao), end-systolic volumes indexed to body surface area in B-mode ($EDVI_B$), end-diastolic volumes indexed to body surface area in B-mode ($ESVI_B$) and PW TDI parameters were recorded. The left atrial speckle tracking imaging including global peak atrial longitudinal strain (PALS) and global peak atrial contractile strain (PACS) was acquired from two-dimensional cine loops of apical four-chamber view.

Global PALS was increased in the mild MR group, reduced in the moderate MR group and further reduced in the severe MR group. Global PACS was reduced in moderate and severe MR groups. Global PALS and PACS correlated with LA/Ao, $EDVI_B$, $ESVI_B$, FS, EF, E velocity, and E/E′. Dogs with diastolic dysfunction had reduced global PALS when plotted for left atrial size.

The STE is repeatable and reproducible tool and it can have more important role in managing advanced heart disease in dogs than the left atrial size parameters, which may underestimate the severity of disease in end-stage heart disease.

C37

Acute Echocardiographic Effects of Sotalol on Ventricular Function and Ectopy in Dogs with Ventricular Tachyarrhythmias (Abstract C37)

Joanna L. Kaplan[1]; Lance C. Visser[2]; Satoko Nishimura[3]; Catherine T. Gunther-Harrington[3]; Catherine Belanger[3]; Mikaela S. Mueller[4]; Joshua A. Stern[2]

[1]University of California-Davis, Davis, CA, USA; [2]Department of Medicine and Epidemiology, University of California-Davis, Davis, CA, USA; [3]Veterinary Medical Teaching Hospital, University of California-Davis, Davis, CA, USA; [4]University of California-Davis, Davis, CA/Blue Pearl Specialty and Emergency Pet Hospital, Renton, WA, USA

We sought to determine the acute effects of sotalol (administered orally) on left ventricular (LV) systolic function and ventricular ectopy in dogs with ventricular tachyarrhythmias.

Dogs (n=27) diagnosed with stable ventricular tachyarrhythmias were prospectively enrolled. Dogs had an echocardiogram and 5-minute ECG before and 2–3 hours post-sotalol. Systolic function was assessed during sinus rhythm via fractional shortening (FS), LV shortening area (LVSA), and modified Simpson's ejection fraction (EF). Number of ventricular premature complexes (VPCs)/5-min were tabulated and ventricular arrhythmias were graded: 0=no VPCs, 1=single VPCs, 2=accelerated idioventricular rhythm, 3=bigeminy/trigeminy, 4=couplets/triplets, and 5=ventricular tachycardia or R-on-T phenomenon. Echocardiographic and ECG measurements were performed in a blinded fashion. Number of VPCs/5-min, complexity grade, indices of LV systolic function and sinus heart rate were compared using a paired t-test (or Wilcoxon signed-rank test).

Baseline heart rate, FS, LVSA, and EF were all significantly ($p<0.001$) decreased post-sotalol (2.5±0.2 mg/kg PO) with mean percent changes ± SD of -18±16%, -14±11%, -10±12%, -13±16%, respectively. Median, IQR of VPCs/5-min (18, 9–32 vs 4, 0–17) and arrhythmia grade (1, 1–4 vs 1, 0–1) were also significantly ($p<0.01$) decreased post-sotalol.

The reduction in LV systolic function post-sotalol is relatively mild in most dogs and percent change post-sotalol may not exceed previously reported within-day coefficients of variation for echocardiographic LV systolic function indices. Based on a 5-minute ECG, sotalol acutely improved ventricular ectopy.

C38
Speckling-Tracking Echocardiographic Alterations in Dogs with Chronic Mitral Valvular Disease and Concurrent Renal Disorders (Abstract C38)

Yunhye Kim[1]; Chul Park[2]; Junil Kim[3]; Mi-Yeon Kim[4]

[1]College of Veterinary Medicine, Chonbuk National University, Curi, Kyonggi-do, Republic of Korea; [2]College of Veterinary Medicine, Chonbuk National University, Iksan, Cholla-bukto, Republic of Korea; [3]Taegu Plus Animal Medical Center, Buk-gu, Taegu-jikhalsi, Republic of Korea; [4]Chonbuk National University, Chonbuk, Taejon-jikhalsi, Republic of Korea

Chronic mitral valvular disease (CMVD) is the most common acquired canine heart disease, characterized by resultant volume overload firstly leading to left atrial enlargement. Occasionally, dogs with CMVD concomitant with renal insufficiency are found. Pathological interactions exist between cardiovascular diseases and renal disorders, which was recently defined as cardiovascular-renal disorders (CvRD) in veterinary medicine. Although the mechanisms of CvRD are not yet fully understood, cardiac diseases and renal dysfunction have recognized as both important determinant of prognosis to each other. Of tools for evaluation of cardiac function, speckle-tracking echocardiography (STE) is newer technique that allows more accurate assessment of left atrial (LA) as well as left ventricular (LV) functions. In the present study, the aims were to evaluate LA and LV function using STE, to assess the correlation between the class of cardiac and renal disease, and to investigate the correlation among the class of renal disease, echocardiographic variables, and laboratory variables in different stage of CMVD dogs and CMVD dogs concomitant with renal disorders.

Forty-one client-owned dogs with CMVD were included. All the dogs underwent physical examination, blood works, radiographs, abdominal ultrasound, urinalysis, electrocardiogram, and echocardiography. Dogs were categorized according to American College of Veterinary Internal Medicine (ACVIM) consensus statement and International Renal Interest Society (IRIS) staging system. 2D, M-mode, and Doppler examinations were performed for conventional echocardiography. Using STE, strains of circumferential, radial, and longitudinal motions, LV twist, LA volume, and LA strain were assessed. Systolic twist increased to moderate stage (ACVIM C class) and decreased in severe stage (ACVIM D class). Dogs in advanced stage had increased circumferential and radial strain, whereas decreased longitudinal strain compared with earlier stages. LA maximal and minimal volume increased, and left atrial active emptying index, LA-FACact and peak LA strain rate during atrial contraction were decreased with severity. Higher concentrations of serum creatinine and serum urea were found in ACVIM class D. IRIS stage had statistically significant correlation with ACVIM class and several echocardiographic variables including early diastolic untwisting rate and E/A.

In this study, the significant correlation among IRIS stage, serum creatinine concentration, echocardiographic variables, and ACVIM class was shown. This finding suggests connection between chronic mitral valve disease and renal disorder.

Table1. Values of Characteristics and clinical and echocardiographic variables in CMVD dogs (n=41).

	ACVIM B1(n=14)	ACVIM B2 (n=5)	ACVIM C (n=15)	ACVIM D(n=7)
Percentage of over IRIS stage 2 (%)	7.14	0	33.3	57.14
LA/Ao ratio	1.3 (0.9-1.45)	1.64 (1.45-1.72)	1.45 (1.19-2.39)	1.99 (1.72-2.53)
Fractional Shortening (%)	45.6 (31.4-59.8)	56 (44.2-67.1)	53.9 (32.1-71.9)	56.6 (36.8-66.7)
E/A ratio	0.93 (0.73-1.16)	0.97 (0.75-1.05)	0.99 (0.74-2.26)	1.54 (1.02-1.9)
Systolic twist (°)	4.21 (2.25-12.41)	6.86 (4.61-7.26)	6.85 (3.08-9.17)	4.63 (2.62-7.93)
Circumferential Strain (%)	-18.14 (-13.18--24.32)	-14.17 (-10.14--21.51)	-15.07 (-13.49--20.17)	-30.95 (-17.24--38.65)
Radial Strain (%)	22.74 (11.52-36.29)	13.81 (6.4-42.76)	20.06 (10.48-26.69)	40.12 (16.63-58.11)
Longitudinal Strain (%)	-23.39 (-16.10--28.76)	-14.78 (-12.31--25.44)	-17.97 (-11.76--29.47)	-17.50 (-12.18--26.50)
Left Atrial Active Emptying Index (%)	43.22 (25.19-67.59)	41.90 (22.26-52.49)	59.73 (21.39-55.28)	32.89 (25-45.20)

C39

Comparative Evaluation of Dogs with Chronic Mitral Valve Regurgitation Treated with Amlodipine or Pimobendan (Abstract C39)

Orson Kamakura; Paula Itikawa; Guilherme T. Goldfeder; Rebecca B. Pessoa; **Maria Helena M.A. Larsson**

University of São Paulo, São Paulo, Brazil

This prospective, randomized, double blinded study has the purpose to evaluate the echocardiographic parameters in dogs with chronic mitral valve degeneration (CMVD) stage C treated with pimobendan or amlodipine in addition to conventional therapy. The use of pimobendan in stage C dogs is well stablished, but has been debated due to the apparent preservation of the myocardial inotropism in these patients. Amlodipine is a vasodilator drug with no inotropic effects, usually used in stage D dogs. Ten dogs were randomized to group A and eleven to group B. Both groups were treated with furosemide and enalapril maleate, but group A dogs also received pimobendan (0.25 mg/kg BID) and the group B ones, amlodipine (0.1 mg/kg BID). All dogs were evaluated before the beginning of the medications (T0), thirty days later (T30) and sixty days after T0 (T60), observing changes in echodopplercardiographic and electrocardiographic examinations, as well as systolic blood pressure. There were no significant differences in electrocardiographic or systolic blood pressure measurements between the groups. The echodopplercardiographic examinations showed a statistically significant difference in speeds in Em waves, between groups at T0 and in the amlodipine group between T0 and T60 and speed of Am wave, between T30 and T60 times. These parameters may indicate a change in early myocardial relaxation. This is the first prospective randomized double blinded study comparing pimobendan and amlodipine in stage C CMVD dogs. Despite the relative small sample size, with the findings of this study, we cannot conclude superiority of one drug over another.

C40

Systolic Blood Pressure Obtained by Doppler Method in Dogs: Comparison Between Anatomical Positions and Evaluators (Abstract C40)

Ana Elisa M. Gonçalves[1]; Paula Itikawa[1]; Guilherme T. Goldfeder[1]; André Martins Gimenes[2]; **Maria Helena M.A. Larsson**[1]

[1]University of São Paulo, São Paulo, Brazil; [2]School of Veterinary Medicine and Animal Science, University of São Paulo (USP), São Paulo, Brazil

Because obtaining accurate systolic blood pressure (SBP) measurement is important in clinical practice, a comparison of SBP measurements obtained by Doppler method in dogs when cuff was placed on the left forelimb and tail, by three examiners with different degrees of experience, was made in order to evaluate the correlation of values between the anatomical positions and reliability of the measurements of the evaluators. Agitation of the animals in each situation was also considered, in order to correlate it to different evaluators and anatomical cuff positions. For this purpose, 36 dogs, males and females, up to 20 kg were used. Each patient had their SBP measured by Doppler method by three evaluators, both on the left forelimb as the tail, in randomly selected order. Each evaluator classified the animal as calm or agitated. The results showed no difference in SBP measured between forelimb (136.8±26.37, p=0.3549) and tail (133.2±25.64, p=0.3062), with concordance between the anatomical positions. Measures obtained between evaluators with different degrees of experience did not differ (Forelimb: evaluator A: 136.9±31.52, p=0.3928; evaluator B: 134.8±24.44, p=0.1257; evaluator C: 138.8±31.14, p=0.1322. Tail: evaluator A: 132.9±28.33, p=0.5157; evaluator B: 127.2±24.50, p=0.4182; evaluator C: 132.4±21.09, p=0.6715). Agitation of patients did not change between the anatomical positions and evaluators. SBP values and agitation decreased as the dogs get used to the measurements in forelimb. It was concluded that the measurement of SBP by Doppler method can be performed reliably in both forelimb and tail, and adequate instruction is enough to examiners measured precisely SBP.

C41

Effects of Anti-Inflammatory Glucocorticoids on Hemodynamics and Cardiac Function in Clinically Healthy Dogs (Abstract C41)

Allison Masters[1]; Darren J. Berger[1]; Wendy A. Ware[1]; Johann F. Coetzee[2]; Natalie Langenfeld[3]; Jessica L. Ward[1]

[1]Iowa State University, Ames, IA, USA; [2]College of Veterinary Medicine, Kansas State University, Manhattan, KS, USA; [3]University of Iowa, Centralia, IA, USA

The use of glucocorticoids in veterinary patients with heart disease is presently limited due to concern for possible precipitation of congestive heart failure. The objective of this study was to investigate previously proposed mechanisms by which glucocorticoids could cause progression of heart disease in dogs, including fluid retention from mineralocorticoid effect, transient hyperglycemia causing intravascular fluid shift, increased afterload due to systemic hypertension, and direct cardiac remodeling.

In this prospective clinical trial, clinically healthy dogs with allergic dermatitis (n=11) were given anti-inflammatory doses of oral prednisone (1 mg/kg/day) for 14 days, followed by a taper and washout period. Hemodynamic, biochemical, and echocardiographic variables were measured prior to prednisone administration (d 0) and at days 7, 14, and post-washout (d 35). A matched control population was evaluated concurrently. We hypothesized that prednisone administration would cause no changes in measured variables at any time point. Linear mixed models were used to compare changes in variables from baseline to d 7, d 14, and d 35 between treatment groups.

Results demonstrated no significant changes in sodium/potassium ratio, blood glucose, or echocardiographic variables. A significant increase in blood pressure from baseline occurred in the treatment group when compared to the control group at d 7 ($p=0.014$). Expected changes in hematologic and biochemical variables due to prednisone administration (neutrophilia, eosinopenia, isosthenuria, and elevated ALP and ALT) occurred in the treatment group. Overall, this study demonstrated that anti-inflammatory doses of glucocorticoids in clinically healthy dogs have the potential to impact cardiac function by causing elevated blood pressure and thus increased afterload.

C42

Effect of Ultrafiltration During Cardiopulmonary Bypass in Dogs Undergoing Mitral Valve Plasty (Abstract C42)

Katsuhiro Matsuura[1]; Shusaku Yamada[2]; Masayuki Enokizono[3]; Haruhiko Suzuki[1]; Dai Nagakubo[4]; Nobuyuki Kanno[5]

[1]Animal Cardiovascular and Thoracic Surgery Center, Sayama-shi, Saitama, Japan; [2]Animal Cardiovascular and Thoracic Surgery, Fuchuu, Tokyo, Japan; [3]Japan Small Animal Medical Center, Tokorozawa, Saitama, Japan; [4]The University of Tokyo, Sayama, Saitama, Japan; [5]Nihon University, Fujisawa, Kanagawa, Japan

Cardiopulmonary bypass (CPB) is necessary to perform mitral valve plasty (MVP) in dogs. Ultrafiltration (UF) has major advantages such as reducing the amount of blood transfusion and removal of various cytokines in human medicine. We performed MVP for dogs with spontaneous mitral valve regurgitation (MR) under CPB and investigated the effect of ultrafiltration. Twenty two dogs diagnosed with severe MR that underwent MVP under CPB were included. Depending on whether or not ultrafiltration was used, it was divided into a non-UF group and a UF group. We performed CUF (Conventional Ultrafiltration) and MUF (Modified Ultrafiltration). The lowest hematocrit (Ht) value and Ht after the end of CPB in both groups were compared. The average of the lowest Ht during the operation was 23.0% in the non-UF (n=11) group and 24.0% in the UF group (n=11), and no significant difference was observed between the two groups. The average Ht after the end of the CPB was 29.0% in the non-UF group and 35.5% in the UF group. Thus the average Ht was significantly higher in the UF group (p=0.01). We also analyzed cases with low body weight (5 kg or less) separately. There were 6 such cases in the non-UF group and 9 low body weight cases in the UF group. The lowest Ht and the Ht after the end of the CPB was significantly higher in the UF group. Notably, the average Ht after the CPB was 10% higher in UF group (36.0%) than in non-UF group (26.0%). The amount of cardioplegia sufficient to produce cardiac arrest varies depending on the specific case. This often results in excessive hemodilution. Since it helps to drain excess fluids under such circumstances, CUF is thought to contribute to the difference in Ht in this study. In addition, during MUF, residual blood in the extracorporeal circuit can be returned to the patient's body. As a result, the Ht after the end of the CPB were higher in the UF group than in the non-UF group. We conclude that UF during CPB increases the lowest Ht value during CPB and increases Ht after end of the CPB in dogs undergoing MVP. These differences are particularly noticeable in small dogs.

C43

Right Ventricular Dysfunction and Dyssynchrony Assessed by Echocardiography in Dogs with Pre-Capillary Pulmonary Hypertension (Abstract C43)

Tomoya Morita[1]; Kensuke Nakamura[2]; Tatsuyuki Osuga[1]; Keitaro Morishita[2]; Noboru Sasaki[2]; Hiroshi Ohta[2]; Mitsuyoshi Takiguchi[2]

[1]Hokkaido University, Sapporo, Hokkaido, Japan; [2]Department of Veterinary Clinical Sciences, Hokkaido University, Sapporo, Hokkaido, Japan

Right ventricular (RV) dysfunction and dyssynchrony have been associated with decreased cardiac function and clinical worsening in human patients with pre-capillary pulmonary hypertension (PH). However, few reports are available on RV function and dyssynchrony in dogs with pre-capillary PH. The objective of this retrospective study was to investigate RV morphological change, function and dyssynchrony assessed by echocardiography in dogs with pre-capillary PH.

Medical records were retrospectively reviewed in 75 client-owned dogs; 21 dogs with pre-capillary PH, and 54 healthy dogs. RV morphological variables, including RV internal diameter in end-diastole (RVIDd), RV end-diastolic area (RVEDA) and RV wall thickness in end-diastole (RVWTd), and echocardiographic indices of RV function, including RV strain by speckle tracking echocardiography, and dyssynchrony index, the standard deviation of the time to peak longitudinal strain of RV (RV-SD) by speckle tracking echocardiography, were evaluated. In addition, the relationship between echocardiographic variables, QRS duration and RV-SD was validated in dogs with pre-capillary PH.

The dogs with pre-capillary PH had larger RVIDd, RVEDA, RVWTd, lower RV strain, and higher RV-SD compared with healthy dogs. There was a positive correlation between RV-SD and RVIDd, RVEDA and TR velocity (RVIDd, $r=0.55$, $p=0.012$; RVEDA, $r=0.81$, $p<0.001$; TR velocity, $r=0.73$, $p<0.001$), but RVWTd and QRS duration were not related with RV-SD.

In conclusion, the dogs with pre-capillary PH had RV dysfunction and dyssynchrony, and RV dyssynchrony may be caused by RV dilation and elevated pulmonary arterial pressure in dogs with pre-capillary PH.

C44

Short-term Results of Quality-of-Life and Right-Ventricle Function After Palatoplasty and Rhinoplasty in Brachycephalic Dogs (Abstract C44)

Jeff Munoz-Pérez[1]; Jorge E. Rojas-Aguirre[1]; Sergio E. Linares-Villalba[1]; Sebastian Cardona-Ramirez[2]

[1]Departamento de Salud Animal, Universidad de Caldas, Manizales, Caldas, Colombia; [2]Department of Pathobiology, University of Missouri, Columbia, MO, USA

Brachiocephalic syndrome (BS) is the result of a combination of anatomical abnormalities in predisposed breeds including stenotic nares and elongated soft palate among others. The signs of BS include laryngeal stridor, rales, exercise intolerance, and higher risk of heat stroke, cyanosis, vomiting, pulmonary hypertension, and right ventricle congestive heart failure. The English bulldog has been reported as a natural model for the study of sleep obstructive apnea syndrome in people, because this breed show alterations like snoring, fragile sleep and hypoxemia. In addition, upper respiratory tract obstruction and inflammation are known to have direct effects on the cardiovascular system and health-related quality of life (HRQoL). The aim of this study was to compare quality of life and right ventricular global function before and after corrective surgery for upper respiratory tract obstruction.

For this longitudinal prospective study, all patients with clinical evidence of Upper Airway Obstruction candidates for corrective Palatoplasty (PP) and Rhinoplasty (RP) were recruited during December 2015 and December 2016. To evaluate the HRQoL, the Spanish version of the "FETCH™ Questionnaire" was used (FETCHSV2-Q™). The global right ventricular function was evaluated with parameters such as the Tricuspid Annular Plane Systolic Excursion "TAPSE". Because there is a curvilinear relationship between TAPSE and weight, the following correlation equation was used. TAPSE = 0.17+0.52 Weight$^{1/3}$. The maximal velocity of the S′ wave and E′ early diastolic wave of the tricuspid annulus were evaluated by tissue Doppler. In addition, systolic time intervals and the Index of myocardial performance (RVIMP) were evaluated from the spectral signal of the transtricuspid and pulmonary blood flow. All patients were evaluated between 10 and 2 days before surgery and 1 to 6 months afterward. Student t-test and the Wilcoxon rank test were used to compare parametric and non-parametric variables, respectively. The results were considered statistically significant with a p-value <0.05.

Twenty dogs of different brachiocephalic breeds with signs of upper airway obstruction were recruited. No difference was found before and after surgery in the total score obtained with the FETCHSV2-Q™ instrument: 30.1±11.2 vs 29.3±6.7 (p>0.05). In this sample, TAPSE values were reduced compared to the established reference ranges in dogs (population of this research: 0.9±0.3 cm vs reference range: 1.5–2.0 cm). In addition, there was difference between TAPSE obtained 10 and 2 days before and 2–6 months after corrective surgery: 0.9±0.3 vs 1.4±0.9 (p<0.05). The maximum velocity of the S′ wave of the tricuspid annulus before surgery was 8.4±2.3 cm/s and after surgery was 9.25±0.9 cm/s, being a statistically significant difference (p-value <0.05). The values of the E′ early wave velocity was similar and before the procedure was 7.75±1.09 cm/s vs. 8.50±0.89 cm/s afterward (p<0.05). RVIMP was 0.19±0.7 before surgery and similar after intervention 0.25±0.2 (p-value >0.05). The Pre-ejection time (PEP) was 81±7.9 ms before surgery and markedly diminished post-recovery 64±11 ms (p-value <0.05).

In conclusion, corrective surgery for upper airway obstruction may release the right ventricle from an increased afterload and stabilize systolic function values. However, apparently does not improve the quality of life of patients undergoing to these interventions.

C45

Prognostic Factors in Cats with Restrictive Cardiomyopathy: A Retrospective Study of 92 Cases (2001–2015) - (Abstract C45)

Peggy Passavin[1]; Valérie Chetboul[1]; Vassiliki Gouni[1]; Cécile Damoiseaux[1]; Jean-Louis Pouchelon[1]; Emilie Trehiou-Sechi[1]; Amandine Petit[1]; Camille Poissonnier[1]; Marie Lavennes[1]; Loic Desquilbet[2]

[1]Alfort Cardiology Unit, National Veterinary School of Alfort, Maisons-Alfort, Ile-de-France, France; [2]Department of Biostatistics and Clinical Epidemiology, National Veterinary School of Alfort, Maisons-Alfort, Ile-de-France, France

Restrictive cardiomyopathy (RCM) is the second most common feline primary cardiomyopathy. However, few data are available regarding prognostic variables in large feline RCM populations. The aim of this retrospective study was therefore to analyze the clinical course of cats diagnosed with RCM between 2001 and 2015 at the Alfort Cardiology Unit, and to identify factors associated with cardiac death (CD). RCM diagnosis was based on the following echocardiographic criteria: left atrial (LA) or biatrial enlargement associated with normal or mildly altered left ventricle (LV). Cats with mildly altered LV should also have a restrictive LV filling pattern. Median survival time to CD and adjusted hazard ratios (HR) were estimated by Kaplan-Meier method and multivariable Cox models, respectively. The study population consisted of 92 cats with RCM. Most RCM cats (64/92, 70%) were symptomatic at the time of diagnosis, with dyspnea related to congestive heart failure in 57/64 cats (89%). Fifty-eight of the 72 RCM cats for which follow-up was available died (81%). CD represented 64% of all death causes, with a median survival time from diagnosis to CD of 667 days. Independently of age, biatrial enlargement and arrhythmia, the risk of CD was increased by both the LA on aorta (Ao) ratio and the presence of severe LA enlargement (LA:Ao ratio ≥2): HR=4.2 for a 1-unit increase, 95% confidence interval (CI)=[1.7–10.7] (p=0.002) and HR=2.5, 95% CI=[1.1–5.6] (p=0.003), respectively. In conclusion, CD is common in RCM cats, and LA enlargement is significantly associated with decreased survival time.

C46

Interindividual Variability in Clopidogrel Active Metabolite Concentrations in Feline Plasma (Abstract C46)

Michele C.L. Reinelt[1]; Pamela M. Lee[1]; Michael H. Court[2]

[1]Washington State University, Pullman, WA, USA; [2]Program in Individualized Medicine (PrIMe), Washington State University, Pullman, WA, USA

Arterial thromboemboli (ATE) are a devastating consequence of heart disease in cats. Although the antiplatelet medication clopidogrel has shown superior efficacy to aspirin for prevention of cardiogenic ATE in cats, treatment failures with clopidogrel still occur. In humans, treatment failures are often attributed to variation in metabolism of clopidogrel to the clopidogrel active metabolite (CAM). The primary objectives of this project were to quantify variation in plasma CAM concentrations between cats and determine if these differences are correlated with variability in clopidogrel metabolism and/or body weight.

Thirty-five cats were screened, and 19 healthy cats were enrolled.

A single oral dose of 18.75 mg clopidogrel was administered to each cat. Blood was collected 2 hours post-administration. Plasma concentrations of clopidogrel, CAM, and clopidogrel acid (major inactive metabolite) were measured using high performance liquid chromatography with tandem mass spectrometry.

A significant correlation was appreciated between the percentage of drug circulating as CAM (CAM%; an indicator of the extent of clopidogrel metabolism to CAM) and CAM concentrations (simple linear regression: $r^2=0.47$, $p=0.001$; multiple linear regression including weight as covariate: $r^2=0.55$, $p=0.001$). No significant correlation was seen between weight and CAM concentrations (simple linear regression: $r^2=0.12$, $p=0.147$; multiple linear regression including CAM% as covariate: $r^2=0.55$, $p=0.121$). One cat with high CAM% and high CAM concentration and four cats with low CAM% and low CAM concentrations were identified.

Substantial variation in circulating CAM concentrations exists between cats, and 47% of this variation could be attributed to individual differences in clopidogrel metabolism to CAM.

C47

Biological Variation of Galectin-3 in Clinically Healthy Dogs (Abstract C47)

Jacqueline Ribeiro de Castro[1]; Bruna B. Real[1]; André Martins Gimenes[1]; Matheus M. Mantovani[1]; Caio N. Duarte[1]; Lilian C. Petrus[1]; Guilherme T. Goldfeder[2]; Maria Helena M.A. Larsson[2]; **Denise S. Schwartz**[1]

[1]School of Veterinary Medicine and Animal Science, University of São Paulo (USP), São Paulo, Brazil; [2]University of São Paulo, São Paulo, Brazil

Galectin-3 (Gal-3) is novel serum biomarker associated with fibrosis and inflammation of heart failure in human patients. However, there are no studies determining reference values or evaluating the influence of gender and age on Gal-3 in healthy dogs. This study aimed to establish reference intervals for Gal-3 in a healthy dog population and to evaluate the influence of gender and age on this biomarker concentration. A prospective cross sectional clinical study was conducted, and included a sample of 60 healthy small-to-medium size dogs, males and females, adults of different ages. Considering the gene homology of approximately 85% (www.ncbi.nlm.nih.gov/homologene) between canine and human Gal-3, human ELISA kit (KitRD#DGAL30, Quantikine ELISA, Fenix, USA) and canine (dog Galectina-3, EIAab®, Wuhan EIAAB Science, China) were tested. Gal-3 concentration recovery for healthy dogs obtained with canine kit was significantly lower, with low repeatability and reproducibility, compared to the human kit, suggesting lower sensitivity of the canine Gal-3 kit used. Reference values obtained for human Gal-3 were 7.548 ng/mL ($P_{25\%-75\%}$=8.933–10.960). There was no significant difference in Gal-3 serum levels between males and females ($p=0.2297$) or between age groups ≤6 years and >6 years in the study population ($p=0.6936$). The evaluation of the biological variation of this biomarker showed that concentrations of Gal-3 were unaffected by age and gender. These data suggest a reference range for the biomarker human Gal-3 in healthy dogs.

C48

Galectin-3 as Biomarker in Heart Failure Secondary to Chronic Mitral Valve Degeneration in Dogs (Abstract C48)

Jacqueline Ribeiro de Castro[1]; André Martins Gimenes[1]; Matheus M. Mantovani[1]; Caio N. Duarte[1]; Mariana Yukari Ueda[1]; Lilian C. Petrus[1]; Guilherme T. Goldfeder[2]; Maria Helena M.A. Larsson[2]; **Denise S. Schwartz**[1]

[1]School of Veterinary Medicine and Animal Science, University of São Paulo (USP), São Paulo, Brazil; [2]University of São Paulo, São Paulo, Brazil

Chronic mitral valve degeneration (CMVD) is a highly prevalent heart disease seen mainly in older small breed dogs. In order to follow the progression of heart failure (HF), galectin-3 (Gal-3) has been applied as a biomarker to identify pre-clinical cardiac diseases, progression and decompensation in human patients. This study aimed to establish the utility of this new biomarker, isolated or in association with Type B natriuretic pro-peptide (NT-proBNP) and cardiac troponin I to estimate short term prognosis in dogs with HF caused by CMVD. It was designed as a prospective cross-sectional clinical study with a longitudinal arm. One hundred thirty nine dogs were distributed among five groups with rigorous selection criteria, according to ACVIM CMVD staging (Control group: stage A- 60 healthy small breed dogs, predisposed to CMVD; 28 dogs in stage B1, 20 dogs in stage B2, 20 dogs in stage C and 11 dogs in stage D), recruited from a Veterinary Teaching Hospital. Groups B1, B2, C and D had a second blood sampling at day 60. Measurements were obtained for human Gal-3, NT-proBNP and cTnI. Reference values obtained for group A for human Gal-3 were 7.548 ng/mL ($P_{25\%-75\%}$=8.933–10.960). We concluded that the magnitude and variation observed in Gal-3 did not allow for detection of differences between stages of CMVD nor were capable of identifying patients in HF, compared to the other measured biomarkers, NT-proBNP and cTnI, already established for canine HF evaluation.

C49

Assessment of Longitudinal Systolic Function Using Tissue Motion Annular Displacement in Healthy Dogs (Abstract C49)

Marcela Wolf[1]; Bruna C. Brüler[2]; Giovana Lais Ruviaro Tuleski[1]; Stephany B. Lucina[1]; Ana Paula Sarraff Lopes[3]; Flavio S. Jojima[2]; **Marlos Gonçalves de Sousa**[4]

[1]Federal University of Paraná - UFPR, Curitiba, Parana, Brazil; [2]Department of Veterinary Science, Federal University of Paraná - UFPR, Curitiba, Parana, Brazil; [3]Pontifical Catholic University of Parana, Curitiba, Parana, Brazil; [4]Universidade Estadual Paulista UNESP/Jaboticabal, São Paulo, Brazil

Although not assessed in standard echocardiography, longitudinal myocardial fibers play an active role in ventricular contraction. Longitudinal Strain (LSt) and Tissue Motion Annular Displacement (TMAD) are techniques that evaluate these fibers. TMAD, which has not been investigated in dogs, is calculated as the dislodgement of a virtual point automatically determined between mitral's septal and lateral annulus towards the left ventricular apex. One of the reported advantages of TMAD over LSt is the possibility of having it calculated using images not suitable for strain analysis, as well as the reduced processing time. In this prospective cross-sectional observational study twenty-seven healthy dogs (1–14 y/2–32 kg) underwent an echocardiogram. Apical 4-chamber (AP4) and 2-chamber (AP2) images were obtained, which allowed the calculation of TMAD and LSt obtained from both images. Data underwent the Shapiro-Wilk test to check for a normal distribution. Results are shown in Table 1. Similar TMAD and LSt was calculated from AP4 and AP2 images. No differences existed between males and females for TMAD-AP4, TMAD-AP2 and LSt. A positive correlation existed between TMAD-AP2 and LSt-AP2, Global LSt (combined AP4+AP2 LSt) and body weight. Interestingly, TMAD correlated significantly with echocardiographic surrogates of volume and LV filling pressure, but LSt did not. In addition to being technically easier to perform, TMAD required less image enhancement and operator experience as compared to LSt. Further studies are warranted to clarify whether TMAD might be a reliable substitute for global LSt in the assessment of longitudinal systolic function in dogs.

Table 1 – Tissue motion annular displacement and longitudinal strain of the left ventricle measured in healthy dogs. Either means (SD) or medians (IQR) are shown depending on the parameter attaining a normal distribution or not on the Shapiro-Wilk normality test.

		Females (n=16)	Males (n=11)	All (n=27)	*p*
TMAD (mm)	AP4	6.7 (1.6)	7.1 (1.1)	6.8 (1.4)	0.7794
	AP2	6.9 (1.4)	7.7 (1.5)	7.2 (1.5)	0.3552
TMAD (%)	AP4	13.6 (2.1)	13.1 (1.9)	13.4 (2.0)	0.8479
	AP2	13.5 (2.3)	13.9 (1.2)	13.7 (1.9)	0.9093
LSt*	AP4	24.5 (20.8-26.0)	24.5 (23.3-26.6)	24.5 (22.5-26.0)	0.8831
	AP2	23.6 (22.0-26.5)	23.9 (22.4-26.2)	23.8 (22.4-26.2)	0.9694
	Global	24.3 (21.5-25.0)	24.9 (24.0-25.3)	24.4 (22.9-25.0)	0.6686

* Nonparametric data; TMAD: tissue motion annular displacement calculated as the dislodgement of a virtual point automatically determined between mitral's septal and lateral annulus towards the left ventricular apex; TMAD%: percent displacement of the mitral annulus towards the heart apex as compared with the left ventricular length; LSt: left ventricular longitudinal strain.

C50

Coupling Interval and Prematurity Index in Dogs with Mitral Valve Disease and Ventricular Arrhythmias (Abstract C50)

Elizabeth Regina Carvalho; Roberto Andrés Navarrete Ampuero; Aparecido Antonio Camacho; **Marlos Gonçalves de Sousa**

Universidade Estadual Paulista UNESP/Jaboticabal, São Paulo, Brazil

Ventricular arrhythmias (VA) were already demonstrated to be a concern in dogs with myxomatous mitral valve disease (MMVD). The coupling interval (CI) and the prematurity index (PI) (Figure 1) were shown to accurately differentiate benign and malignant VA in human beings, in which VA are known to be associated with an increased risk to either evolve into signs of heart failure or die suddenly. In this study, we investigated how the CI and PI perform in dogs with MMVD.

In a retrospective, cross-sectional investigation that included dogs with either symptomatic (stages C/D; n=41) or asymptomatic (stages B1/B2; n=29) MMVD, we reviewed the electrocardiographic tracings to calculate both the CI and PI. In eight dogs we also compared these indices obtained from both a Holter recording and a standard ECG tracing and no statistical differences existed (CI, p=0.97; PI, p=0.17). Even though CI and PI were determined in all animals enrolled in the study, VPCs' characteristics were only compared between symptomatic and asymptomatic dogs when a Holter recording was available (n=49).

The PI was different (p=0.01) between symptomatic (0.65±0.17) and asymptomatic (0.56±0.18) dogs, but CI was considered similar (p=0.91). Also, the symptomatic dogs had more polymorphic VPCs (p=0.002) and supraventricular arrhythmias (p=0.0002) than the asymptomatic animals.

In conclusion, VPCs in dogs with symptomatic MMVD are more premature and more commonly associated with supraventricular arrhythmias and polymorphic ventricular premature complexes than asymptomatic animals.

Figure 1. Schematic representation of the calculation of the CI and PI.

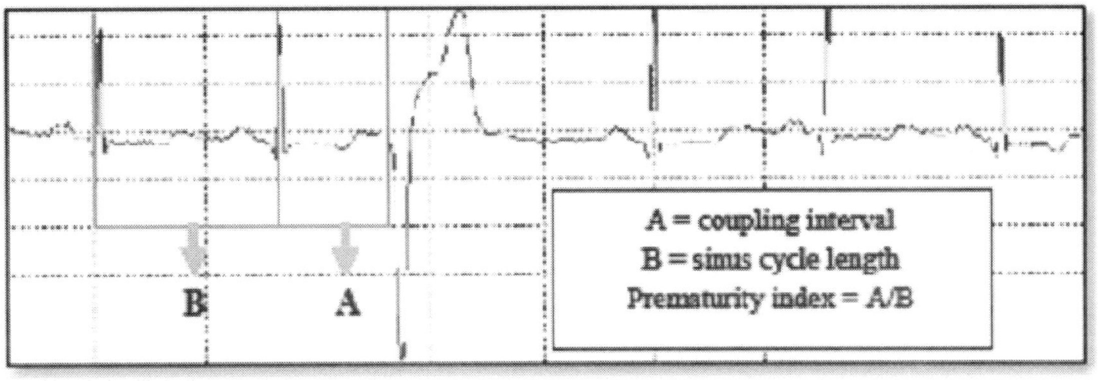

C51

Computed Tomographic Angiography and 3D Printing of Aberrant Coronary Arteries in Dogs with Pulmonic Stenosis (Abstract C51)

Susanne M. Stieger-Vanegas[1]; Rachel Cunningham[1]; Katherine F. Scollan[2]

[1]College of Veterinary Medicine, Oregon State University, Corvallis, OR, USA; [2]Oregon State University, Corvallis, OR, USA

Pulmonic stenosis is one of the most common congenital cardiac defects in dogs and can occur concurrently with type R2A aberrant coronary artery anatomy, most commonly reported in English Bulldogs and Boxers. Computed tomography angiography (CTA) allows to non-invasively assess coronary artery anatomy. Three dimensional (3D) models improve the understanding and visualization of complex cardiac anatomy.

Our hypothesis is that CTA a can be used to evaluate the presence and anatomy of anomalous coronary arteries in dogs with pulmonic stenosis and that 3D models of anomalous and normal coronary arteries can be created using specialized software and printed.

Retrospective study. CTA images of the coronary arteries from 6 dogs with confirmed R2A coronary artery anomalies and 5 dogs with normal coronary artery anatomy were segmented using two different 3D modeling software programs and printed using a 3D printer.

Normal and aberrant coronary arteries could be visualized using CTA and segmented using both 3D software programs. All R2A cases had a single coronary ostium in the right aortic sinus that gave rise to a normal right main and circumpulmonary left main coronary artery with all other visualized branches following normal anatomic courses. All normal cases had normal coronary artery anatomy. 3D printed models were of excellent quality and allowed direct visualization of normal and abnormal coronary anatomy.

CTA is effective for identification of R2A coronary artery anomaly. Non-gated CTA is not suitable for creating printable 3D models. 3D models increase understanding of the complex anatomy of normal and anomalous coronary arteries.

C52

Radial Strain and Strain Rate Assessed by Speckle-Tracking Echocardiography in Turkish Kangal Dogs (Abstract C52)

Osman Safa Terzi; Hasan Albasan

Department of Internal Medicine, Faculty of Veterinary Medicine, University of Ankara, Diskapi, Ankara, Turkey

Cardiac diseases especially dilated cardiomyopathies are common in dogs. Quantitative analysis of left ventricular deformation by two-dimensional speckle tracking echocardiography provides better diagnosis. The purpose of this study was to specify radial strain and strain rate values of left ventricle by two-dimensional speckle-tracking echocardiography in Turkish Kangal dogs.

Study groups of A and B consisted of 3 dogs with dilated cardiomyopathy and 4 dogs with valve disease, respectively. Control group (C) consisted of 7 healthy dogs. The mean ages of 3.8±2.9, 5.5±2.4, and 2.1±1.7 years in groups of A, B and C, respectively were not significantly ($p>0.05$) different.

Right parasternal short-axis view of the left ventricle at the level of the papillary muscles was used to determine radial strain and strain rates by two-dimensional speckle-tracking echocardiography. Results of this study revealed that group A was significantly ($p<0.05$) different from group C and group B. However, there was no significant difference between group C and group B.

Two-dimensional speckle tracking echocardiography revealed the decreased left ventricular myocardial systolic performance in dogs with dilated cardiomyopathy while similar strain and strain rates between healthy dogs and dogs with valve disease. Therefore, this procedure can be applied to detect dilated cardiomyopathy in dogs with cardiac disease.

C53

CT-Based Anatomical Features of Large Airway and Heart Volume in Dogs with Different Body Size (Abstract C53)

Takuya Uehara; Yoko Fujii; Keisuke Sugimoto; Hiroshi Sunahara; Takuma Aoki

Azabu University, Sagamihara, Kanagawa, Japan

Chronic valvular heart disease is commonly seen in small-breed dogs and a cough is a common clinical sign. We hypothesized that anatomical features of the heart and large airway were different, influenced by body size, and would relate to coughing when the heart enlarged. The purpose of this study was to compare relative heart volume and anatomical features of large airway in different body sizes using CT scan. Retrospective analyses performed on medical records of dogs without heart and lung disease that underwent CT scanning. Dogs (n=209) were divided into 3 groups on the basis of body weight; small (<7 kg), medium (7 kg–20 kg), and large (≥20 kg). The following parameters were calculated using OsiriX: thoracic volume, heart volume, relative heart volume to thoracic volume, heart length, distance from main stem bronchi to vertebra, tracheal diameter, and angle of bronchus. Relative heart volume was significantly large in small dogs (0.29±0.070) and medium dogs (0.28±0.061), compared with large dogs (0.23±0.075). The relative distance from main stem bronchi to vertebra was significantly short in small dogs (0.10±0.037), compared with medium (0.15±0.043) and large dogs (0.19±0.050). There was no significant difference in the angle of bronchus among the groups. Findings revealed that small dogs had a large heart relative to the thorax, and the distance from the heart to vertebra was relatively short, compared with larger dogs. This anatomical feature may be potentially predisposed to occurrence of cough.

C54

Long-term Outcome of Irish Wolfhound Dogs with Preclinical Cardiomyopathy Treated with Pimobendan, Methyldigoxin, or Benazepril (Abstract C54)

Andrea Vollmar[1]; Philip Fox[2]

[1]Private practice, Bonn and Wissen, Germany; [2]Animal Medical Center, New York, NY, USA

Dilated cardiomyopathy (DCM) is a common cause of morbidity and mortality in Irish Wolfhounds (IW), but clinical data reporting treatment efficacy are limited. We aimed to evaluate long-term pimobendan, benazepril, and methyldigoxin therapy in 121 IW with preclinical DCM.

This retrospective, open label, cohort study evaluated IW receiving pimobendan (n=25; 0.24–0.26 mg/kg PO q 12 h); benazepril (n=24; 0.25–0.5 mg/kg PO q 12 h); methyldigoxin (n=21; 0.005 mg/kg PO q 12 h); methyldigoxin+benazepril (n=40); or pimobendan+benazepril (n=11), for up to 7.8 years. Time to events were estimated by Kaplan Meier method for first onset of CHF, cardiac death, and all-cause mortality using pair-wise comparisons between groups. Pooled data from 86 IW with echocardiographic follow-up were dichotomized based upon improvement in %FS, LV end-systolic, and LV end-diastolic dimensions into references ranges ('responders'), versus no echo improvement ('nonresponders'), to compare cardiac survival.

Time to CHF was not significantly different between groups (all p>0.2). Time to cardiac death was longer with benazepril-vs-methyldigonin (p=0.034), pimobendan-vs-methyldigoxin (p=0.007), benazepril-vs-pimobendan+benazepril (p=0.005), pimobendan-vs-pimobendan+benazepril (p=0.004), and methyldigoxin+benazepril-vs-pimobendane+benazepril (p=0.027), respectively. Survival from all-cause mortality was significantly longer for pimobendan-vs-methyldigoxin (0.029) and benazepril-vs-pimobendan+benazepril, respectively (p=0.045). Time to cardiac death was longest for 'responders' (median-4.8 years) vs (median-2.9 years) (p=002); time to all-cause death was not different (p=0.067).

Larger, prospective, blinded studies are needed to clarify clinical benefits of these drugs.

C55

Evaluation of Lung Ultrasound for the Diagnosis of Congestive Heart Failure in Dogs with Cough (Abstract C55)

Jessica L. Ward[1]; Gregory R. Lisciandro[2]; Wendy A. Ware[1]; Teresa C. DeFrancesco[3]

[1]Iowa State University, Ames, IA, USA; [2]Hill Country Veterinary Specialists, Spicewood, TX, USA; [3]North Carolina State University, Raleigh, NC, USA

Lung ultrasound (LUS) is an emerging imaging technique that can suggest presence of cardiogenic pulmonary edema (CHF) by identifying certain artifacts (B-lines) caused by interstitial or alveolar fluid. This study aimed to determine the accuracy of LUS for diagnosing CHF in dogs with cough.

Sixty coughing dogs were enrolled. Patients underwent protocolized LUS quantifying the presence and number of B-lines at 4 sites on each hemithorax. Medical records, including thoracic imaging, were evaluated for final diagnosis.

Seven dogs were diagnosed with CHF and 53 with noncardiac (NC) causes of cough (most commonly dynamic airway disease). Dogs with CHF had a greater number of B-lines compared to NC dogs ($p<0.0001$) and more sites containing B-lines ($p<0.0001$). Using the cutoff criterion that at least 4 out of 8 sites contain B-lines, sensitivity and specificity of LUS for diagnosing CHF were 100% and 92%, respectively. Diagnostic accuracy was higher among the subset of dogs with heart murmurs (n=33; sensitivity 100%, specificity 96%) and dogs with moderate to severe left atrial enlargement (n=19; sensitivity and specificity 100%). Notably, 23% (12/53) of dogs with NC cough had been treated previously with diuretics due to prior CHF misdiagnosis.

LUS was useful in predicting CHF as the cause of cough in this study population, particularly among dogs with concurrent heart disease. LUS is a potentially valuable diagnostic tool to identify (and avoid misdiagnosis of) CHF in coughing dogs.

O01

Systematic Optimization of Therapy for Canine Intracranial Neoplasia Using PAC-1, a Novel and Potent Procaspase-3 Activator (VCS Award Winner) – (Abstract O01)

Lisa J. Schlein[1]; H.C. Podenis[1]; E.J. Roy[1]; Stephanie Lezmi[2]; J.S. Looper[3]; Michael Podell[4]; Peter J. Dickinson[5]; Timothy M. Fan[6]; Paul J. Hergenrother[7]

[1]University of Illinois, Urbana, IL, USA; [2]Ipsen Pharmaceuticals, Urbana, IL, USA; [3]Chicago Veterinary Specialty Group, Chicago, IL, USA; [4]MedVet Chicago, Chicago, IL, USA; [5]University of California-Davis, Davis, CA, USA; [6]Department of Veterinary Clinical Medicine, University of Illinois at Urbana-Champaign, Urbana, IL, USA; [7]University of Illinois at Urbana-Champaign, Urbana, IL, USA

Canine central nervous system tumors can develop deeply within the brain parenchyma, precluding surgical resection and limiting therapeutic options. PAC-1 is a novel, blood-brain barrier penetrant, pro-apoptotic small molecule activator of procaspase-3 (PC-3), with orphan drug status for the treatment of human glioblastoma multiforme. PC-3 is frequently overexpressed in malignantly transformed tissues, providing an opportunity to selectively induce apoptosis in cancer cells with dysregulated upstream apoptotic circuitry. This study evaluates the *in vitro* activity of PAC-1 against a panel of brain tumor cells, and the feasibility of combining PAC-1 with conventional therapies in dogs with spontaneous brain cancer.

Immunohistochemical characterization of PC-3 was performed in 21 normal canine brains and over 450 canine and human intracranial neoplasms. Murine, canine, and human glioma cell lines were evaluated for PC-3 expression and *in vitro* sensitivity to PAC-1 and radiation monotherapies. Dogs with spontaneously-arising glioma and meningioma were treated with oral PAC-1 in combination with conventional therapies to determine tolerability, with serial clinical and MRI outcome assessments.

PC-3 is overexpressed in canine intracranial neoplasms and in high grade human astrocytomas relative to normal brain tissues. Immortalized canine, human, and murine glioma cell lines show *in vitro* sensitivity to PAC-1 and radiation monotherapies at biologically relevant exposures. Six dogs (3 gliomas 3 meningiomas) received PAC-1 combinatorial therapies and achieved objective responses (1 complete response, 3 partial responses, 2 with stable disease).

PAC-1 shows therapeutic promise in the treatment of canine intracranial neoplasia, as well as high grade human astrocytoma. Investigation of therapeutic approaches that combine PAC-1 with radiation therapy and/or temozolomide will further elucidate its therapeutic potential in murine models and canine patients.

O02

Development of a Histopathologic and Flow Cytometric Subclassification System in Canine B-Cell Lymphoma (VCS Award Winner) - (Abstract O02)

Amber Wolf-Ringwall[1]; Lynelle Lopez[1]; Brad Charles[1]; Robyn Elmslie[2]; Brooke Fowler[3]; Janet Lori[4]; Gabriella Sfiligoi[5]; Anne Skope[6]; Erin Arnold[7]; Douglas H. Thamm[8]; E.J. Ehrhart[1]; Anne Avery[1]; Susan Lana[1]

[1]Colorado State University, Fort Collins, CO, USA; [2]Veterinary Referral Center of Colorado, Englewood, CO, USA; [3]Aspen Meadow Veterinary Specialists, Longmont, CO, USA; [4]Animal Emergency and Specialty Clinic, Parker, CO, USA; [5]Wheat Ridge Animal Hospital and Veterinary Specialists, Wheat Ridge, CO, USA; [6]VCA Alameda East Veterinary Hospital, Denver, CO, USA; [7]Animal Hospital Specialty Center, Highlands Ranch, CO, USA; [8]Flint Animal Cancer Center, Colorado State University, Fort Collins, CO, USA

Canine B-cell lymphoma is a common hematologic malignancy and a clinically heterogeneous disease. Little progress has been made to improve remission duration and survival. The primary goal of this study was to correlate modified WHO histopathologic subclassification and flow cytometry (FC) with clinical outcome. We hypothesized that histologically distinct subclassifications would have unique cell surface markers as determined by FC, and these subcategories of disease would have variable outcomes when treated with a standardized treatment protocol.

In this prospective study, dogs with naïve multicentric B-cell lymphoma were treated with a 19-week CHOP chemotherapy protocol at the Flint Animal Cancer Center and six specialty clinics. All patients had lymph node biopsies and FC collected prior to therapy. Outcome was correlated with patient characteristics, treatment, histopathologic subclassification, and FC features.

A total of 74 dogs were enrolled. Flow cytometry diagnosis of B-cell lymphoma was consistent with histopathologic diagnosis in all cases. Seventy-seven percent of dogs were diagnosed with diffuse large B-cell lymphoma (DLBCL). There were no phenotyping differences between the subcategories of lymphoma detectable by our FC panel. All DLBCLs fell above a minimum forward light scatter value, indicating that cell size, as established by FC, may be useful to rule out this disease. The median overall survival time (MST) for DLBCL (316 days) was statistically significant compared to that of all other histopathologic subgroups combined (114 days, $p=0.0231$).

Prognostic factors identified on univariate analysis as significant for median progression free survival time and/or MST included cell size on FC, presence of lymphocytosis, presence of hypercalcemia, need for dose reduction, experiencing complete remission as best response to therapy, and completing CHOP. Further studies are needed to determine factors that may allow for further patient stratification to eventually modify therapy to improve outcome.

O03

Incidence and Prognostic Significance of HER2/neu Expression in Canine Osteosarcoma (Abstract O03)

Martha MaloneyHuss[1]; Carolyn Gross[2]; Julie Engiles[1]; Nicola Mason[1]

[1]University of Pennsylvania, Philadelphia, PA, USA; [2]Pieper Memorial Veterinary Center, Middletown, CT, USA

Canine osteosarcoma (OSA) is a highly aggressive mesenchymal cancer accounting for 85% of canine bone tumors. OSA commonly arises in the long bones of large breed dogs, and 95% of patients are thought to have micrometastases at diagnosis. Standard of care treatment for canine OSA consists of amputation followed by chemotherapy, with median survival times ranging from 207–366 days. Development of metastatic disease limits long-term survival, with 9.7%–28% of patients alive two years from diagnosis. Treatment of macroscopic metastasis is unrewarding and given this, therapy is focused on prevention of metastatic disease.

HER2/neu is an epidermal growth factor receptor tyrosine kinase that is expressed in human cancers of the prostate, breast, ovary, stomach, and colon. HER2/neu expression is reported in up to 60% of human pediatric osteosarcomas, where increased expression correlates with aggressive biologic behavior and an increased risk of metastasis. Contrary to the membranous expression seen in carcinomas, HER2/neu expression in OSA is principally cytoplasmic. In canine patients, one previous retrospective study reported HER2 expression in 4/10 primary tumor samples, with a suggestion that HER2 expression correlated with decreased survival. We therefore hypothesized that HER2 expression in canine OSA denotes a more aggressive disease phenotype.

To evaluate this hypothesis, we reviewed the medical records of 115 OSA patients with sufficient formalin-fixed tumor tissue available for immunohistochemical (IHC) analysis. Samples with at least 10% of malignant cells staining positive for HER2/neu were classified as HER2/neu+. Based on these criteria, 95/115 (82.6%) primary tumor samples were HER2/neu+. Additionally, 54 metastatic lesions from a total of 13 dogs were evaluated and 39/54 (72.2%) of these lesions were HER2/neu+. There was no correlation between HER2/neu expression by the primary tumor and HER2/neu expression by metastatic foci.

Samples were then scored based on the percentage of neoplastic cells staining for HER2/neu (distribution score; 50% = 3) and the intensity of HER2/neu staining (weak = 1, moderate = 2, strong = 3). A combined score was obtained by multiplying the distribution score by the intensity score. Based on the combined score, patients were grouped into negative/low (combined score of 1–4), intermediate (>4, <7.5), or high (7.5 or greater) categories of HER2/neu expression for Time To Progression (TTP) and Overall Survival (OS) analysis.

To determine whether expression of HER2/neu has prognostic significance, TTP and OS of dogs with negative/low, medium or high combined HER/neu scores were determined. Only dogs treated with standard of care amputation and chemotherapy (n=55) were included in this analysis. The median TTP and OS of dogs with negative/low HER2/neu expression (n=22) was 190 and 299 days respectively compared with a TTP and OS of 154 and 253 days respectively for dogs with intermediate HER2/neu expression (n=23). TTP and OS for dogs with high HER2/neu expression (n=10) was 127 days and 187 days respectively.

When previously reported (ACVIM 2014), the original data set showed no correlation between HER2/neu expression and TTP or OS. However, in this report, only patients that underwent standard of care amputation and chemotherapy were included in the TTP and OS analysis, revealing a trend towards shorter TTP and OS for dogs with tumors that highly express HER2/neu.

Prospective studies will be required to determine whether a statistically significant correlation exists between HER2/neu status and patient outcome. Nevertheless, taken together, these data show that the majority of canine OSA express HER2/neu, and that dogs with tumors that have high HER2/neu combined scores have a tendency towards earlier metastasis and decreased OS. These data support the evaluation of HER2/neu targeted immunotherapies in both the prevention and possible treatment of metastatic OSA.

O04

MicroRNA-34a Regulates the Invasive Capacity of Canine Osteosarcoma Cell Lines (Abstract O04)

Cecilia M. Lopez[1]; Peter Yu[2]; Xiaoli Zhang[3]; Cheryl A. London[4]; Joelle M. Fenger[5]

[1]Department of Veterinary Clinical Sciences, The Ohio State University, Columbus, OH, USA; [2]Wexner Medical Center, College of Medicine, The Ohio State University Columbus, OH, USA; [3]Center for Biostatistics, The Ohio State University, Columbus, OH, USA; [4]The Ohio State University, Tufts University, Columbus, OH, USA; [5]The Ohio State University, Columbus, OH, USA

Osteosarcoma (OSA) is the most common bone tumor in children and dogs; however, despite aggressive treatment, little improvement in survival times have been achieved in either dogs or people over the past 15 years. MicroRNAs (miRNAs) are small non-coding RNAs that regulate gene expression and play a fundamental role in cancer. Our laboratory previously identified a unique miRNA expression signature associated with canine OSA and found miR-34a expression to be significantly decreased in canine OSA tumors compared to normal canine osteoblasts. These data are concordant with recent studies demonstrating that miR-34a functions as a tumor-suppressor gene in human OSA and that loss of miR-34a is associated with a poor prognosis in human OSA patients. We hypothesize that loss of miR-34a generates a pattern of gene dysregulation that contributes to the pathogenesis of canine OSA.

MiR-34a expression was evaluated in primary canine OSA tumors, OSA cell lines and normal canine osteoblasts (Cell Applications, Inc.) using Taqman miRNA assays. To investigate the effects of miR-34a on OSA cell behavior, canine OSA8 and Abrams cell lines were stably transduced with empty control or pre-miR-34a lentiviral constructs (Systems Biosciences) and cell proliferation, cell cycling, cell invasion and migration was evaluated. RNA sequencing of canine OSA8 cells overexpressing miR-34a was performed and changes in gene expression were validated using real-time PCR.

Real time PCR demonstrated that miR-34a expression levels were significantly lower in primary canine OSA tumors and OSA cell lines compared to normal canine osteoblasts. Furthermore, miR-34a expression was found to be significantly decreased in distant metastases compared to primary OSA tumors, suggesting that loss of miR-34a is directly involved in the metastatic process. To address the effects of miR-34a on OSA cell behavior, the canine OSA8 and Abrams cell lines which express low basal levels of miR-34a were transduced with miR-34a lentiviral constructs resulting in high levels of miR-34a. In transformed canine OSA8 and Abrams cell lines, miR-34a overexpression reduced cellular invasion and migration but had no effect on cell proliferation or cell cycling. Transcriptional profiling of OSA8 cells overexpressing miR-34a demonstrated dysregulation of several genes, including down-regulation of putative miR-34a target genes associated with cellular invasion. Consistent with these findings, enhanced expression of these genes correlates with decreased miR-34a expression in primary canine OSA tumors.

Taken together, these findings suggest that miR-34a may play an important role in regulating OSA invasion and loss of miR-34a *in vivo* may promote the aggressive behavior of canine OSA.

O05

Outcome of Cats with Intermediate or Large Cell Gastrointestinal Lymphoma Treated with Surgical Mass Resection (Abstract O05)

Kathleen Tidd[1]; Dorothy Brown[1]; Amy Durham[2]; Sridhar Veluvolu[3]; Erika L. Krick[3]

[1]Department of Clinical Studies, School of Veterinary Medicine, University of Pennsylvania, Philadelphia, PA, USA; [2]Department of Pathobiology, School of Veterinary Medicine, University of Pennsylvania, Philadelphia, PA, USA; [3]School of Veterinary Medicine, University of Pennsylvania, Philadelphia, PA, USA

Standard of care treatment for feline gastrointestinal lymphoma consists of administration of systemic chemotherapy. The use of surgery except in cases of intestinal obstruction or perforation/peritonitis is largely unknown. The study purpose was to determine the outcome of cats with intermediate and large-cell gastrointestinal lymphoma following surgical resection.

Records of 34 cats diagnosed with intermediate or large cell gastrointestinal lymphoma based on biopsy of a mass from 2005–2015 were retrospectively reviewed. Cases were included if the cat survived surgery and had a known two-week post-operative status. Data collected included anatomic site, surgical margins, surgical complications, chemotherapy, and post-operative and long-term outcome.

Within the included cases, affected sites were small intestines (n=18), large intestines (n=7), and stomach (n=5). Complete margins (n=13) were unrelated to anatomic site of lymphoma. Thirty cats survived to hospital discharge, and 26 cats survived to suture removal approximately two weeks after surgery. Reported causes of post-operative death included peritonitis, sepsis, and persistent or worsening clinical signs. Overall MST for cats surviving to suture removal was 213 d (95% CI:48, 465). MST for cats with large intestinal masses was significantly longer than for cats with small intestine or stomach masses for all cats assessed (675 d, 51 d, 96 d; p=0.03) and for cats that survived to suture removal (675 d, 83 d, 96 d; p=0.03). Adjunctive systemic chemotherapy (n=16) and completeness of the surgical margins did not significantly affect survival.

Serious complications were overall uncommon following resection of gastrointestinal lymphoma with the majority of cases found to survive the peri-operative period. Given the significantly longer survival times of cats with large intestinal lymphoma, it may be reasonable to recommend surgical removal for cats with solitary lymphoma tumors in this location.

O06

Retrospective Evaluation of Toceranib Phosphate (Palladia®) Use in the Treatment of Canine Gastrointestinal Stromal Tumors (Abstract O06)

Erika P. Berger; Chad M. Johannes; Albert E. Jergens; Karin Allenspach; Leslie E. Fox

College of Veterinary Medicine, Iowa State University, Ames, IA, USA

The purpose of this study was to solicit data regarding current clinical use of toceranib in dogs with gastrointestinal stromal tumors (GIST) to provide initial assessment of biologic activity. The American College of Veterinary Internal Medicine Oncology and Small Animal Internal Medicine listservs were used to solicit data pertaining to cases in which toceranib was used to treat canine GIST. Case data from 28 dogs with histopathogically confirmed GIST were received from 17 participating institutions. Dogs undergoing treatment other than surgical excision prior to toceranib were excluded. Clinical benefit (CB) was observed in 93% (26/28; 24 complete response [CR]/no evidence of disease [NED], 1 partial response [PR], 1 stable disease [SD]). All dogs with microscopic disease experienced CB (21/21; 21 NED), while 71% (5/7; 3 CR, 1 PR, 1 SD) with gross disease experienced CB. Median duration of treatment in dogs experiencing CB was 36 weeks (1–159). At the time of data submission, 8/26 responding dogs were still receiving treatment and an additional 4/26 were experiencing continued NED following completion of prescribed therapy. In dogs with CB, the median toceranib dose was 2.6 mg/kg (0.5–3.5) and in 81% (21/26) was given three days per week. Treatment was well-tolerated with 57% (16/28) experiencing adverse events, a majority of which were Grade 1 or 2 gastrointestinal toxicities. Biologic activity of toceranib is evident in studied dogs; however, prospective studies are needed to refine the role of toceranib in treatment of GIST in microscopic and gross disease settings.

O07

Association of Tumor-Infiltrating Regulatory T Cells With Adverse Outcome in Dogs With Solid Tumors (Abstract O07)

Kosei Sakai; Shingo Maeda; Tomohiro Yonezawa; Naoaki Matsuki

Graduate School of Agricultural and Life Sciences, The University of Tokyo, Bunkyo-ku, Tokyo, Japan

Regulatory T cells (Tregs) play an integral role in suppression of inflammation and the maintenance of immunological tolerance. In addition, Tregs infiltrate into a variety of tumor tissues and thereby suppress antitumor immunity. However, data on the association between tumor-infiltrating Tregs and prognosis in dogs with various tumors are limited. The purpose of this study was to examine the number of tumor-infiltrating Tregs and their association with prognosis in dogs with solid tumors.

This study included 113 canine tumors that were surgically excised in the Veterinary Medical Center of the University of Tokyo between May 2009 and December 2015. On the basis of the histopathology, the following diagnoses were determined: squamous cell carcinoma in 14 dogs; hepatocellular carcinoma in 28 dogs; oral malignant melanoma in 29 dogs; lung adenocarcinoma in 16 dogs; and transitional cell carcinoma in 26 dogs. Normal tissue samples were obtained from 6–13 healthy dogs. Foxp3-positive Tregs in the normal and tumor tissues were examined by immunohistochemistry. Cells with clear lymphocyte morphology, distinct nuclear staining for Foxp3, and an unstained cytoplasm were classified as Tregs. Foxp3-positive Tregs were quantified in different compartments (intratumoral and peritumoral areas). For survival analysis, cases were classified as having a high or low density of Foxp3-positive Tregs according to the median Treg number in each tumor. Overall survival was compared between cases with a high Treg number and those with a low Treg number using log-rank test.

Compared to the normal tissues, the number of intratumoral and peritumoral Tregs in the tumor tissues were significantly increased in dogs with squamous cell carcinoma ($p<0.0001$), oral malignant melanoma ($p<0.0001$), lung adenocarcinoma ($p<0.0001$), and transitional cell carcinoma ($p<0.0001$). No significant difference was observed in Foxp3-positive Tregs between normal liver and hepatocellular carcinoma. The overall survival for cases with a high number of intratumoral Tregs was significantly shorter than for cases with a low number of intratumoral Tregs in dogs with lung adenocarcinoma ($p=0.0074$), squamous cell carcinoma ($p=0.0145$), and transitional cell carcinoma ($p=0.0011$). In dogs with hepatocellular carcinoma and oral malignant melanoma, the number of Foxp3-positive Tregs was not significantly related to the prognosis.

This study demonstrated that Foxp3-positive Tregs infiltrated into various canine tumor tissues. The higher number of intratumoral Tregs was associated with negative outcomes in dogs with lung adenocarcinoma, squamous cell carcinoma, and transitional cell carcinoma, suggesting that tumor-infiltrating Tregs contribute to the suppression of antitumor immunity in dogs as well as in humans.

O08

Adoptive T Cell Transfer in Healthy Dogs, and Dogs with Osteosarcoma (Abstract O08)

Brian K. Flesner[1]; Gary W. Wood[2]; Pamela A. Gayheart-Walsten[3]; F. Lynn. Sonderegger[2]; Carolyn Henry[4]; Sandra M. Bechtel[1]; Kim A. Selting[1]; Debbie J. Tate[1]; **Jeffrey N. Bryan**[1]; Noe Reyes[5]

[1]University of Missouri, Columbia, MO, USA; [2]TVAX Biomedical Inc., Olathe, KS, USA; [3]Xenometrics LLC, Stillwell, KS, USA; [4]Professor of Oncology, University of Missouri, Columbia, MO, USA; [5]ELIAS Animal Health, Olathe, KS, USA

Osteosarcoma is the most common primary bone malignancy in the dog. Based on favorable findings in infected canine limb-spare patients, stagnant survival times despite a plethora of adjuvant chemotherapy options, and promising recent preclinical and human immunotherapy outcomes, novel immunotherapy approaches have become attractive treatment options.

We hypothesized that a T cell infusate made up of peripheral blood mononuclear cells (PBMCs) activated and stimulated to proliferate *ex vivo* by exposure to superantigens could be safely administered to healthy dogs and dogs with osteosarcoma. Additionally, we evaluated the safety of a course of low-dose human interleukin-2 (IL-2) following T cell infusion.

Four dogs were used in the healthy dog study (2 female hounds and 2 male mongrels). Three dogs with spontaneously occurring osteosarcoma have been enrolled in an initial phase I/II clinical trial at the University of Missouri. Two underwent the course of autologous vaccinations and leukapheresis, with one also receiving *ex vivo*-activated T cells.

Healthy dogs underwent leukapheresis to isolate PBMCs. T cells were selectively grown *ex vivo* using IMDM media and superantigens. Following culture, superantigens were removed and the product was then infused into the dogs after premedication; they also received five subcutaneous doses of IL-2. Dogs' vitals were analyzed, aerobic/anaerobic cultures were performed, and cytokines were analyzed.

Minimal side effects were noted in the healthy dogs. Vitals showed normal variation during the study; gastrointestinal signs (most often noted as random emesis) were documented in three of four dogs. Pro-inflammatory cytokines (IL-6, IFNγ, TNFa, and CCL2) and anti-inflammatory cytokine IL-10 were produced, although significant inflammatory-type symptoms were not noted in the healthy dogs. Both clinical patients were discharged the same day of the leukapheresis procedure. The only clinical patient who received *ex vivo*-activated T cells had not been pre-medicated and experienced a severe inflammatory response and became septic. The dog recovered and is still alive 11 months after diagnosis with no signs of metastasis and with no further cancer therapy.

Based on the encouraging safety data in our healthy canine patients, this adoptive T cell transfer approach has the potential to change treatment recommendations and outcome in dogs diagnosed with osteosarcoma. A phase I/II clinical trial is ongoing to determine if ECI-OSA-1 after amputation is indeed safe and results in similar or enhanced efficacy compared to surgery and chemotherapy.

O09

Dysregulation of Healthy Osteoblasts by Canine Osteosarcoma Exosomes (Abstract O09)

Andrew Schlueter[1]; Cheri Goodall[1]; Ryan Troyer[1]; Kallan Bathke[1]; Liping Yang[2]; Shay Bracha[1]

[1]College of Veterinary Medicine, Oregon State University, Corvallis, OR, USA; [2]Oregon State University, Corvallis, OR, USA

Exosomes are small lipid vesicles that mediate cell-to-cell communication. Cancer cells exploit exosomes to promote metastasis, evade host immune responses, and facilitate angiogenesis. Presently, the effects of osteosarcoma (OSA) exosomes on healthy osteoblasts have not been studied. This is the first study to investigate the proteomic composition of canine OSA exosomes and their impact on osteoblasts. We hypothesized that OSA exosomes carry a unique cargo which manipulates healthy osteoblasts, resulting in alteration of bone turnover.

Exosomes were extracted from OSA (POS, HMPOS, COS) and healthy osteoblast (CnOb) supernatants. The size and concentration of the exosomes were validated via nanoparticle tracking analysis. Exosomal proteomic content was determined by mass spectrometry and validated by Western blot. *In vitro* assays were performed by co-culturing OSA exosomes with CnOb. Treated CnOb, untreated CnOb, and exosomes from untreated CnObs were also subjected to proteomic analysis. Fluorescent calcification assays for assessment of hydroxyapatite deposition were performed with OSA cells, osteoblasts alone, and osteoblasts treated with OSA exosomes.

OSA exosomes contained over 200 proteins. In comparison to CnOb derived exosomes, OSA exosomes revealed additional proteins associated with ossification (gelsolin, osteocalcin, vitamin D binding protein). CnOb treated with OSA exosomes showed an upregulation of 129 proteins and a downregulation of 42 proteins compared to untreated CnOb.

OSA cells exhibited osteoblastic characteristics on fluorescent calcification assays. When CnOb were challenged with OSA exosomes, their secretion of hydroxyapatite was increased significantly ($p=0.018$).

These findings demonstrate that OSA exosomes contain pro-osteoblastic cargo which promotes healthy osteoblasts to increase bone deposition. Our study provides a preliminary understanding of the role of OSA-secreted exosomes and their impact on the bone microenvironment.

O10

Quantification and Prognostic Significance of PD-L1 Expression in Canine Diffuse Large B-cell Lymphoma (VCS Award Winner) - (Abstract O10)

Lisbeth Ambrosius; Deepika Dhawan; Audrey Ruple; Jose Ramos-Vara; Michael Childress

Purdue University, West Lafayette, IN, USA

Increased expression of programmed cell death ligand 1 (PD-L1) has been noted in a subset of human lymphomas and recently has been targeted with antibody therapy. The purpose of this study was to measure the expression of PD-L1 in canine diffuse large B-cell lymphoma (DLBCL) utilizing quantitative real time polymerase chain reaction (qPCR), and to determine if PD-L1 expression is correlated with progression free survival (PFS).

A biorepository maintained by the Purdue Comparative Oncology Program was searched for RNA samples of histopathologically-confirmed cases of canine DLBCL diagnosed between 2010 and 2015. Archived lymph node biopsy samples, processed in TRIzol reagent, from treatment-naïve dogs were retrieved. Paired RNA samples from lymph node from clinically healthy dogs were retrieved to serve as controls. Multiplex qPCR was performed using TaqMan probes and primers against canine PD-L1 and GAPDH. The data were analyzed using the $2^{-\Delta\Delta CT}$ method to calculate the fold change in PD-L1 expression in DLBCL relative to normal lymph node.

Forty-two dogs met inclusion criteria. Median progression free survival time was 252 days (range 23–466) days. PD-L1 expression in the DLBCL samples ranged from 0.21 to 7.45 fold different from that measured in normal canine lymph node. 21/42 (50%) dogs had PD-L1 fold change less than 1.0, and 21/42 (50%) had PD-L1 fold change more than 1.0 (compared to normal lymph node average). When PD-L1 expression was dichotomized as fold change greater than or less than 1.0, it was not found to be significantly associated with PFS using the log-rank test. However, when modeled continuously, PD-L1 expression was found to be negatively correlated with PFS in both univariate and multivariate Cox proportional hazards models.

These findings demonstrate that PD-L1 is variably expressed in canine DLBCL, and that higher expression of PD-L1 correlates with reduced PFS in affected dogs.

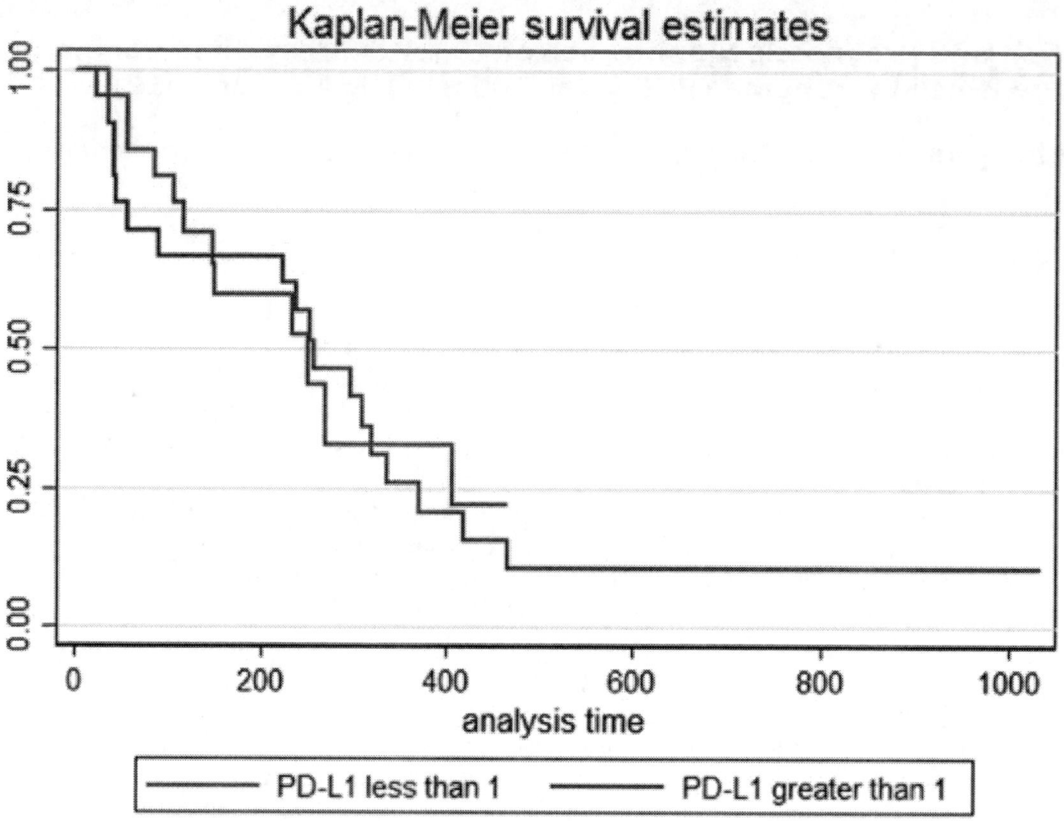
Kaplan-Meier survival estimates
1.00
0.75
0.50
0.25
0.00
0
200
400
600
800
1000
analysis time
PD-L1 less than 1
PD-L1 greater than 1

O11

In vitro Inhibition of Monocarboxylate Transporters Display Synergistic Activity with Doxorubicin or Metformin in Canine Osteosarcoma (VCS Award Winner) - (Abstract O11)

Heather L. Gardner[1]; Joelle M. Fenger[1]; Vincent Sandanayaka[2]; Cheryl A. London[3]

[1]The Ohio State University, Columbus, OH, USA; [2]Nirogyone Therapeutics, Boston, MA, USA; [3]The Ohio State University, Columbus, OH, USA and Tufts University, Medford, MA, USA

Introduction: Monocarboxylate transporters (MCT) transport lactic acid across plasma membranes, providing a mechanism for tumor cells to meet their bioenergetic needs during glycolysis in a variety of microenvironments. The objective of this work is to explore the therapeutic potential of modulating cellular metabolism through inhibition of MCT1 and MCT4 in canine osteosarcoma, with the ultimate goal of providing proof of principle data to support future clinical studies.

Materials & Methods: MCT1 and MCT4 expression in canine osteosarcoma were assessed with qRT-PCR, western blotting and IHC. Proliferation, apoptosis and oxygen consumption were evaluated using shRNA and small molecule inhibitors directed against MCT1 and MCT4. Metformin and doxorubicin were evaluated for their ability to synergize with MCT1/4 small molecule inhibitors.

Results: MCT1 is expressed at higher levels relative to MCT4 in canine osteosarcoma cell lines. Small molecule inhibitors directed against MCT1/4 do not significantly decrease cell proliferation. Synergistic activity was observed when cells were treated with doxorubicin or metformin in the presence of MCT1 and MCT4 small molecule inhibitors.

Conclusions: MCT1 and MCT4 are expressed in canine osteosarcoma. Drug combinations with synergistic activity are necessary to realize the full potential of MCT1/4 inhibition.

O12

Expression of PDGFR-β, VEGFR2, and KIT in Canine Transitional Cell Carcinoma (Abstract O12)

Lark S. Walters; Olya Martin; Mee-Ja Sula

College of Veterinary Medicine, University of Tennessee, Knoxville, TN, USA

Transitional cell carcinoma (TCC) is the most common neoplasia of the urinary tract in dogs. It tends to be locally invasive and has a moderate metastatic rate. Systemic medical therapy with a combination of a non-steroidal anti-inflammatory drug (NSAID) and a chemotherapeutic agent is the current mainstay of treatment. Receptor tyrosine kinases (RTKs) play an important role in promoting cell growth, differentiation and regulation of cell function. RTK inhibitor Palladia has been used anecdotally to treat TCC.

The goal of this study was to evaluate archived normal urinary bladder, TCC, and cystitis bladder samples for expression of Palladia targets: VEGFR2, PDGFR-β, and stem cell factor receptor (KIT).

Formalin-fixed paraffin wax-embedded samples of 30 canine urinary bladders with transitional cell carcinoma, 10 canine urinary bladders with cystitis and 10 normal canine urinary bladders were obtained from the University of Tennessee, Department of Pathobiology histologic archives for use in this study. Staining for PDGFR-β, VEGFR2, and KIT was carried out using standard immunohistochemical procedures developed for use in formalin fixed tissues. Staining intensity was evaluated at 20x in representative sections. Overall staining intensity was scored from 0–3 for the entire section. Immunohistochemical staining (% cells stained) was evaluated over ten 40x fields and given a score from 1–4. A final scoring product was calculated by multiplying the intensity score by the immunohistochemical staining score.

A significant number of TCC samples (30/30) expressed PDGFR-β compared to cystitis (5/10) and normal bladder samples (3/10). A large number of the TCC samples (11/30) also had mild to moderate staining for KIT compared to cystitis (0/10) and normal bladder (0/10) samples, but the difference was not statistically significant.

Based on this proof of target study, further investigation is warranted to determine clinical response of TCC to Palladia.

O13

Clinical Survey of Veterinary Specialists Evaluating Management of Chemotherapy Induced Vomiting and Inappetence in Dogs (Abstract O13)

Natalie Patricia Punt[1]; Chad M. Johannes[2]; Lauren R. Hackbarth[3]; Leslie E. Fox[2]

[1]Kansas State University, Manhattan, KS, USA; [2]College of Veterinary Medicine, Iowa State University, Ames, IA, USA; [3]Iowa State University, Ames, IA, USA

Chemotherapy-induced vomiting and inappetence (CIVI) are common adverse events experienced by cancer patients. However, the incidence and recommended treatment following administration of specific chemotherapeutics has not been established in dogs. The aim of this study was to assess perceived emetogenic and inappetence potential of chemotherapeutics and evaluate the clinical use of antiemetics and appetite stimulants in dogs receiving chemotherapy.

A survey was issued on the American College of Veterinary Internal Medicine Oncology and Small Animal Internal Medicine listservs to assess the perceived incidence and clinical management of acute and delayed CIVI associated with 20 chemotherapeutics as defined to VCOG-CTCAE v1.1.

Eighty-five completed surveys were evaluated with responses from oncologists (83.5%) in private practice (73%) with >80% of their practice being medical oncology (70.6%). The oral chemotherapeutics evaluated were associated with a minimal/low (<50%) risk of acute and delayed vomiting independent of VCOG grade. Injectable chemotherapeutics most associated with moderate/high (>50%) risk of causing clinically significant (>Grade 2) acute or delayed vomiting included cisplatin, dacarbazine, epirubicin, paclitaxel, and streptozocin. Of the nine antiemetics evaluated, maropitant and ondansetron were most frequently administered. Inappetence following chemotherapy administration was most commonly treated with maropitant and mirtazapine.

The perceived incidence and management of CIVI in dogs widely differs. Dogs receiving chemotherapeutics with currently perceived lower potential for CIVI may also benefit clinically from treatment with antiemetics and/or appetite stimulants. Further research to understand the true incidence and clinical impact of CIVI will contribute to development of guidelines to appropriately manage these effects in dogs.

O14

Immunohistochemical Expression of P-Glycoprotein in Dogs with Epitheliotropic and Nonepitheliotropic Cutaneous Lymphoma (Abstract O14)

Sílvia Lucas[1]; Ana Luiza Alves[1]; Valter M. Winkel[2]; Lucas Rodrigues[1]; Bruno Cogliati[1]

[1]University of São Paulo, São Paulo, São Paulo, Brazil; [2]School of Veterinary Medicine and Animal Science, University of São Paulo, São Paulo, São Paulo, Brazil

Cutaneous lymphoma represents 1% of skin tumors in dogs and is classified in epitheliotropic and nonepitheliotropic. Unlike the other anatomical forms of lymphoma, cutaneous lymphomas have a low response to the treatment, and dogs with nonepithelitropic lymphoma have a lower survival time when compared to dogs with epitheliotropic form. The resistance of neoplastic cells to cytotoxic drugs is the major cause of failure of chemotherapy in cancer and one of multiple drug resistance mechanisms (MDR) is the active drug efflux mediated by ATP-Binding Cassette superfamily (ABC). Among these transporters is P-glycoprotein (ABCB1). The aim of this study was to investigate ABCB1 immunoexpression in cutaneous epitheliotropic and non-epitheliotropic lymphoma. Immunohistochemistry was performed in 18 paraffin-embedded tumor samples of cutaneous lymphoma in dogs (11 nonepitheliotropic and 7 epitheliotropic) using CD20 and CD3 markers for immunophenotyping and ABCB1 antibody (C494 - Enzo Life Sciences®). It was found 5 epitheliotropic T-cells and 2 B-cells and 10 nonepitheliotropic T-cells and 1 B-cell lymphoma. Two samples of epitheliotropic lymphoma were negative for ABCB1 and the median percentage of labeled cells was 20% (0–40%) whereas in nonepitheliotropic lymphoma, all cases were positive for ABCB1 and the median percentage of labeled cells was 50% (20–90%) and statistical analysis showed a significant difference expression of this marker ($p=0.0162$). In conclusion, the majority of cases with cutaneous lymphoma expressed P-glycoprotein and the subtype nonepitheliotropic showed higher labeled cells than epitheliotropic.

O15

Cell Proliferation Markers Ki-67 and Mitotic Index in Cats with Intestinal Lymphoma (Abstract O15)

Valter M. Winkel[1]; Ana Luiza Alves[2]; Bruno Cogliati[2]; Archivaldo Reche[2]; Sílvia Lucas[2]

[1]School of Veterinary Medicine and Animal Science, University of São Paulo, São Paulo, São Paulo, Brazil; [2]University of São Paulo, São Paulo, São Paulo, Brazil

Intestinal lymphoma is the main anatomical form of lymphoma diagnosed in cats. Cell proliferation indicators as Ki-67 expression and mitotic index may be important in assessing the clinical behavior of the tumor and to determine the degree of malignancy. The objectives of this study were to determine the expression of Ki-67 and mitotic index in intestinal lymphoma of cats and its relation with overall survival time. Forty-seven samples of cats with intestinal lymphoma were studied and immunohistochemical was performed for: CD3, PAX5, and Ki-67. The analysis of Ki-67 was made by counting 1,000 infiltrating lymphocytes in the tumor microenvironment and mitotic index was ranked as mild, moderate, or high, by the evaluation of 10 microscopic fields at 40X. The results showed a significant difference in the immunohistochemical expression of Ki-67 in B and T-cell Lymphoma ($p=0.001$) with a median of 47.2% and 8.5%, respectively. The same was observed when comparing the histological type (lymphoblastic and lymphocytic), with a median of 40.5% and 8.2%, respectively ($p=0.002$). Regarding mitotic index, two cats with lymphoblastic B-cell and one with lymphoblastic T-cell lymphoma showed high expression, while three animals showed moderate expression and the majority of cases, 41 animals (87.2%) had mild expression. In conclusion, Ki-67 expression was a better proliferative marker when compared to mitotic index in cats with intestinal lymphoma. Higher expression of Ki-67 was observed in those animals affected by lymphoblastic or B-cell and is related to low therapeutic response and shorter survival time.

N01

Common Cytogenetic Alterations Define Choroid Plexus Tumors in Dogs (Abstract N01)

Devin Ancona[1]; Miriam Aguilar[2]; Danika L. Bannasch[3]; Peter J. Dickinson[3]

[1]UC Davis Veterinary Medical Teaching Hospital, Davis, CA, USA; [2]UC Davis Population Health and Reproduction, Veterinary Medicine, Davis, CA, USA; [3]University of California-Davis, Davis, CA, USA

Defining commonly occurring cytogenetic abnormalities in canine choroid plexus tumors (CPTs) is essential to an understanding of oncogenesis, and the definition of appropriate therapeutic strategies. We used Illumina 173K HD SNP arrays to determine copy number alterations (CNAs) and allelic imbalances in 12 histologically confirmed spontaneous canine choroid plexus tumor samples (3 papillomas/CPP, 8 carcinomas/CPC, 1 recurrent CPP) and tumor matched normal tissue. Copy number calls and allelic events were determined by matched paired analysis using BioDiscovery Nexus Copy Number™ software.

Hierarchical clustering by aberration profile did not distinguish between CPPs and CPCs. Choroid plexus tumors were characterized by highly recurrent single copy, whole chromosomal losses with specific losses involving chromosomes 2, 5, 8 and 20, seen in 90–100% of tumors. The pattern of predominantly whole chromosomal losses was most similar to findings in human CPCs compared to CPPs, and several recurrent chromosomal losses in dog tumors were syntenic to commonly reported regions within whole chromosomal losses in human CPTs. Highly recurrent losses in dog CPTs encompassed key tumor suppressor genes and genes associated with chromosomal instability including TP53 and VHL.

The strikingly high prevalence of specific CNAs in canine tumors is supportive of their biological relevance in CPT oncogenesis and progression. If the characteristic CNAs of whole chromosomal losses reflects a common oncogenic mechanism in human and canine CPTs, comparative cytogenetics and more detailed analysis of candidate genes within orthologous regions may provide insight into both novel and common pathways in choroid plexus tumor biology.

N02

Pullout Properties of Monocortical and Bicortical Pins and Screws in Canine Lumbar Vertebral Bodies (Abstract N02)

Devin Ancona[1]; Beverly Sturges[2]; Amy Kapatkin[2]; Tanya Garcia-Nolen[3]; Susan Stover[3]

[1]UC Davis Veterinary Medical Teaching Hospital, Davis, CA, USA; [2]Surgical and Radiological Sciences, Veterinary Medicine, Davis, CA, USA; [3]Anatomy, Physiology and Cell Biology, Veterinary Medicine, Davis, CA, USA

Implants used in surgical vertebral stabilization are typically placed bicortically due to a theoretical increase in pullout strength and construct stiffness. Bicortical implants, however, have an increased risk of impingement of regional neurovascular structures. Recent studies involving cervical vertebrae suggested that monocortical pin/screw plate constructs tested in bending are biomechanically equivalent to bicortical constructs. There is no data directly comparing the pullout strength of each implant type/method in canine lumbar vertebrae.

A full factorial design was used to assess pullout strength of 3.2 mm x 4.0 mm positive profile pins, 3.5 mmg cortical screws, and 3.5 mm locking screws, inserted monocortically or bicortically in predrilled holes in 6 lumbar vertebrae (L1-L6) from 6 canine cadavers of similar body weight and size. Implants were distracted in a mechanical testing system in a direction along their long axis at a constant rate of 0.5 mm/s until screw/pin pullout. Displacement and load data were acquired at 128 Hz. The effects of implant and insertion depth on implant pullout were assessed with a repeated measures ANOVA that accounted for cadaver and vertebra.

Pullout strength was greater with bicortical (4531 N±459 N) than monocortical implants (1783 N±459 N) regardless of implant type. There were no significant differences in pullout strength between implant type or vertebra; however, this may be due to low sample number. These data provide evidence that bicortical implant placement is a greater factor than the nature of the implant with respect to pullout strength in lumbar vertebral bodies, for those implants tested.

N03

Pharmacokinetic Analysis of Single Dose Extended Release Levetiracetam per OS in Healthy Cats (Abstract N03)

Laura R. Barnard[1]; Heidi L. Barnes Heller[1]; Dawn M. Boothe[2]

[1]University of Wisconsin-Madison, Madison, WI, USA; [2]Auburn University, Auburn, AL, USA

The purpose of this study was to determine the pharmacokinetic profile and side effects following a single dose (500 mg) of extended release levetiracetam (XRL) in healthy cats. Seven healthy cats, weighing ≥ 5 kg were included. Health was determined by normal physical and neurologic examination, PCV/ TP and serum biochemistry evaluation. XRL was administered per os, with food, at time 0. Timed blood collection occurred over 36 hours via jugular catheter. Serum levetiracetam was quantitated by high performance liquid chromatography using a technique previously validated in cats. Data is reported as mean±SD. C_{max} = 89.7±25.7 μg/ml, adjusted C_{max} = 1.03±0.27 μg/ ml, T_{max} = 291.4±94.4 minutes, MRT = 534.0±61.7 minutes, CLss (compensated for F) = 0.001±0.0005, C 12 hour = 43.7±18.4 μg/ ml, C 24 hour = 4.9±3.4 μg/ml, Vd (adjusted for F) = 0.51±0.11 L/ kg, dose adjusted AUC = 676.5±205.0 μg h/ ml. All cats had levetiracetam concentrations above the minimum human interval at 12 and 18 hours; 1 cat maintained levetiracetam concentrations above the interval at 24 hours. No adverse effects were noted throughout the study period and all cats maintained normal neurologic examinations. In conclusion, side effects appear to be minimal and the mean concentration at 24 hours was below the minimum human therapeutic interval by 0.1 μg/ml suggesting that once daily dosing may be possible however individual peak and trough serum concentrations should be considered when administered once daily.

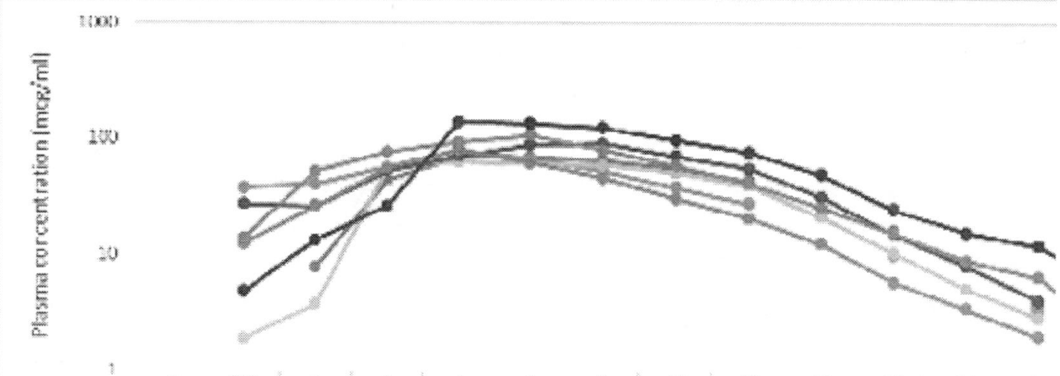

	0	0.5	1	2	4	6	8	10	12	15	18	21	24
Cat 1	0	0	7.8	44.1	69.3	68.4	65.3	56.3	43.5	0	16.3	0	3.7
Cat 2	0	37.8	40.1	56.7	78.8	67.3	52.7	38.1	27.5	0	10	0	3.4
Cat 3	0	1.9	2.7	56.6	60.3	60.8	53.0	50.7	30.9	21.5	10.5	5.2	2.9
Cat 4	0	22.2	26.1	[illegible]	[illegible]	87.7	90.9	70.8	55.4	[illegible]	15.3	8.3	4.1
Cat 5	0	12.4	26	55.7	80.9	61.8	45.6	30.6	20.3	12.6	5.9	3.5	2
Cat 6	0	13.5	51.9	76.8	92.9	106.5	79.5	58.5	41.8	25.7	15.6	5	6.5
Cat 7	0	4.8	13.2	26.2	138.6	130.9	123.5	96.5	76.3	49.5	24.9	15.7	12

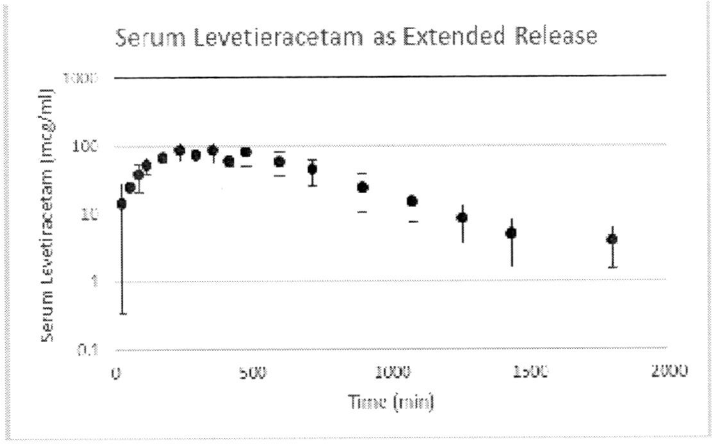
Serum Levetieracetam as Extended Release
Serum Levetiracetam (mcg/ml)
Time (min)
0.1
1
10
100
1000
0
500
1000
1500
2000

N04

Surgical Decompression, With or Without Adjunctive Therapy, for Treatment of Primary Vertebral Osteosarcoma in Dogs (Abstract N04)

Sasha Dixon[1]; Annie Chen-Allen[1]; Beverly Sturges[2]; Karen Vernau[3]; Jonathan Levine[4]; Arturo J. Otamendi[4]; Peter J. Early[5]; Lara Curtis[6]; Alix Partnow[7]; Stephanie A. Thomovsky[8]; Hillary Greatting[1]; Tom Jukier[1]

[1]Washington State University, Pullman, WA, USA; [2]Surgical and Radiological Sciences, Veterinary Medicine, Davis, CA, USA; [3]College of Veterinary Medicine, University of California-Davis, Davis, CA, USA; [4]College of Veterinary Medicine, Texas A&M University College Station, TX, USA; [5]North Carolina State University, Raleigh, NC, USA; [6]VCA Veterinary Specialty Center of Seattle, Lynnwood, WA, USA; [7]VCA Veterinary Specialty Center of Seattle, Kenmore, WA, USA; [8]College of Veterinary Medicine, Purdue University, West Lafayette, IN, USA

Osteosarcoma is the most common primary vertebral tumor in dogs, however previous studies examining outcome after surgical decompression of these tumors are limited. The goal of this study was to determine survival time after decompressive surgery alone and combined with adjunctive therapies such as definitive radiation or chemotherapy.

Records were collected retrospectively from six academic and private practice institutions, and were reviewed for data regarding surgical procedure, neurologic function, adjunctive therapy and survival. Sixteen cases were included, all of which received decompressive surgery. Of those cases, 1 dog was treated with chemotherapy and 3 dogs were treated with both chemotherapy and definitive radiation therapy. All dogs either improved neurologically following surgery or remained stable neurologically with improved pain control. All were large breed dogs with an average age of 9 years. Median survival time for dogs treated with surgery alone (n=12) was 43.5 days (range 3–1333). One outlier was still alive 4 years after surgery alone. The one dog treated with chemotherapy survived 56 days. Dogs treated with radiation therapy and chemotherapy (n=3) had a median survival time of 257 days (rage 223–270). In most cases where follow up was available, dogs were euthanized due to recurrence of the primary tumor or return of original clinical signs.

In conclusion the results of the study indicate that overall survival time with surgical decompression alone is poor. Adjunctive therapy appears to prolong survival in cases of primary vertebral osteosarcoma that have had decompressive surgery.

N05

The Effects of Stabilizing Agents, Fetal Calf Serum and VetStarch, on Canine Cerebrospinal Fluid Analysis (Abstract N05)

Lindsey N. Peterson[1]; Stephanie A. Thomovsky[1]; John A. Christian[2]; George E. Moore[1]; R. Timothy Bentley[1]

[1]College of Veterinary Medicine, Purdue University, West Lafayette, IN, USA; [2]Department of Comparative Pathobiology, Purdue University, West Lafayette, IN, USA

A common challenge of CSF analysis is time-dependent rapid degradation of nucleated cells. This degradation leads to an inability to accurately interpret CSF, diagnose disease and institute treatment. CSF additives, fetal calf serum and the hydroxyethyl starch, Hetastarch, have been used to delay cell degradation but some alter microprotein levels. The purpose of this research is to determine if a more readily available and less expensive hydroxyethyl starch, VetStarch™, is effective at preserving CSF nucleated cell morphology without altering microprotein levels.

CSF samples were collected from 9 dogs undergoing neurologic workup to date with a target sample size of 28. Samples were divided into 4 aliquots. One aliquot was analyzed immediately (control). The remaining three aliquots received one stabilizing agent each [saline, fetal calf serum or VetStarch™] before storage at 4°C. Samples were analyzed at 48 hours; total nucleated cell count, microprotein, and cell morphology scores were recorded. A morphology score of 1 indicated no observed cellular degeneration; a score of 4 indicated severe degeneration.

Samples stored in VetStarch™ and fetal calf serum exhibited improved morphology scores, median 2.46 (1.42–3.20) and 2.25 (1.13–3.33), respectively, as compared to those in stored in saline, 3.0 (1.14–4.00). Microprotein levels for samples stored in VetStarch™ were similar to those recorded in control samples, 18.6±4.50 and 18.8±2.81 (mean±standard deviation), respectively. Fetal calf serum significantly altered microprotein levels, 464.3±109.75.

VetStarch™ is a commercially available, inexpensive, effective stabilizing agent for canine CSF without erroneous protein elevation.

N06

GM2 Gangliosidosis in Shiba Inu Dogs with an In-Frame HEXB Deletion and Autofluorescent Storage Granules (Abstract N06)

Natalie A. Villani[1]; Ana Kolicheski[1]; Gary S. Johnson[1]; Dennis P. O'Brien[2]; Tendai Mhlanga-Mutangadura[1]; David A. Wenger[3]; Jeremy F. Taylor[4]; Robert D. Schnabel[5]; Martin L. Katz[6]

[1]Department of Veterinary Pathobiology, University of Missouri, Columbia, MO, USA; [2]Department of Veterinary Medicine and Surgery, University of Missouri, Columbia, MO, USA; [3]Department of Neurology, Jefferson Medical College, Philadelphia, PA, USA; [4]Division of Animal Sciences, University of Missouri, Columbia, MO, USA; [5]Division of Animal Sciences and Informatics Institute, University of Missouri, Columbia, MO, USA; [6]Mason Eye Institute, University of Missouri, Columbia, MO, USA

This study was performed to identify the mutation causing a progressive neurodegenerative disease in Shiba Inu dogs with storage body autofluorescence. Two related Shiba Inu dogs were suspected to have neuronal ceroid lipofuscinosis (NCL) based on clinical signs, antemortem diagnostics, disease progression and autofluorescence of storage material on histopathology. A whole genome sequence (WGS) was generated with DNA from one affected dog using previously described techniques. The data were screened for potentially causal mutations in known NCL genes but no candidate variants were found. However, the WGS contained a novel homozygous deletion of a conserved codon in the *HEXB* gene (which encodes the beta subunit of beta-hexosaminidase). The deletion was not found in the WGS from 105 unaffected dogs of various breeds. The other affected Shiba Inu was also homozygous for this variant. Mutations in *HEXB* have been implicated in GM2 gangliosidosis (Sandhoff disease) in humans and in other dog breeds. Beta-hexosaminidase activity was measured in brain tissue from one of the affected dogs using previously described synthetic substrates. Activity against two different substrates was markedly reduced compared to a normal dog, consistent with Sandhoff-like disease (HEXB deficiency). Thus, we conclude that the *HEXB* mutation in these dogs caused GM2 gangliosidosis and led to their clinical signs. To our knowledge, this is the first report that GM2 gangliosidosis-associated storage material is autofluorescent. This study also demonstrates the utility of WGS for the efficient diagnosis of potentially heritable diseases in veterinary medicine by identifying a mutation that abolishes HEXB activity in Shiba Inu dogs.

N07

Exosome-Associated Integrins as Liquid Biopsy Biomarkers for Canine Glioma (Abstract N07)

Chai-Fei Li[1]; Matt Mellema[2]; Randy Carney[2]; Beverly Sturges[3]; Daniel York[2]; Ruiwu Liu[4]; Kit Lam[5]; Peter J. Dickinson[2]

[1]VMTH, UC-Davis, Davis, CA, USA; [2]University of California-Davis, Davis, CA, USA; [3]Surgical and Radiological Sciences, Veterinary Medicine, Davis, CA, USA; [4]University of Davis, Sacramento, CA, USA; [5]Harold Albin Johnson Endowed Chair in Biomedical Research, Department of Biochemistry & Molecular Medicine, Division of Hematology & Oncology, UC Davis Cancer Center, Sacramento, CA, USA

Easily accessible, validated biomarkers for gliomas are needed to aid in diagnosis and therapeutic monitoring. Exosomes are extracellular membranous vesicles (30–100 nm) that are released by cells into the microenvironment and may carry signatures of the cell of origin. We hypothesized that integrin targeting cyclic nonapeptides (LXW7/αvβ3, LXY30/α3β1) derived from screening of human and canine glioma cells would identify canine glioma cell lines and cell line derived exosomes *in vitro*.
Three glioma cell lines (J3TBg, G06A, SDT3G) were used to validate targeting of peptides to live cells using flow cytometry and immunofluorescent labelled peptide. Canine glioma cell derived exosomes were isolated from serum free media using ExoQuick precipitation and centrifugation, and characterized using nanoparticle tracking analysis (NTA) technology (NanoSight® LM10HS) and peptide conjugated quantum dots. Non-glioma targeting peptide (LLP2A/α4β1) was used as a negative control.
All cell lines were labelled by the LXW7 and LXY30 peptides with no LLP2A labelling. Exosomes were isolated from media collected from all cell lines and confirmed by NTA analysis and CD9 immunocytochemistry. Quantum dot conjugated labelling of exosomes was demonstrated for LXW7 and LXY30 peptides, but not for LLP2A.

Expression of cell surface αvβ3, and α3β1integrin in canine glioma cells was demonstrated using integrin specific peptide targeting, consistent with human cell lines and expression in dog tumor samples. Importantly, these integrin surface markers are also expressed in cell derived exosomes. Characterization of LXW7 and LXY30 binding exosomes in CSF and serum is therefore a rational strategy to define accessible biomarkers for gliomas *in vivo*.

N08

Investigation of Immune or Infectious Etiologies in Canine Meningitis/Meningoencephalitis/Meningoencephalomyelitis (Abstract N08)

George E. Moore[1]; Michael Podell[2]; Rebecca Windsor[2]; Zachary Niman[2]; Jennifer Ochsenhirt[2]; Edward Dubovi[3]

[1]College of Veterinary Medicine, Purdue University, West Lafayette, IN, USA; [2]MedVet Chicago, Chicago, IL, USA; [3]Cornell University, Ithaca, NY, USA

Meningitis, meningoencephalitis, and meningoencephalomyelitis (M/ME/MEM) are inflammatory conditions which may be of immune or infectious origins. A perceived increase in these conditions at a specialty referral practice led to two investigative studies. The first study was a retrospective survey of MRI diagnoses made during a 3-year period (2013–2015) to assess increased prevalence and/or temporal trends. The second study was a 7-month prospective study of possible immune or infectious causes of M/ME/MEM diagnosed at the practice.

In the first study, monthly incidence of M/ME/MEM was calculated for the study period using computer searches for "immune", "immuno", "meningitis", "meningoencephalitis", or "meningoencephalomyelitis" within recorded MRI diagnoses. Of the total MRIs, 229 (9.3%) were annotated as possible cases. A possible seasonal trend was seen as 143 of 229 (62.4%) occurred April–September.

In the second study, 64 dogs were considered M/ME/MEM cases based on MRI and/or CSF results. Median age was 7.8 yr (0.2–16.0 yr) and median weight 7.7 kg (1.8–56.3 kg); more than 30 breeds were represented. Infectious diseases (one each *Bartonella, Anaplasma, Ehrlichia*) were identified via routine tests in 3 dogs. Vaccination histories were available on 61/64 dogs; 21/61 (34.4%) of cases had received one or more vaccinations within 35 days of the onset of clinical signs. Multivalent *Bordetella* vaccines (IN) were most commonly administered in the 35-day window, followed by *Leptospira* vaccines. The number of vaccines administered most recently did not differ between dogs vaccinated.

A cause of seasonal increases in M/ME/MEM was not identified. An immune origin may play a role, mediated in some dogs by undefined mechanisms related to immunological booster from multivalent or monovalent vaccinations.

N09

Imaging Findings at 3T and Clinical Outcome in Deep-Pain Negative Dogs with Intervertebral Disc Herniation (Abstract N09)

Arturo J. Otamendi; Beth Boudreau; Lindsey Gilmour; Jay Griffin; Jonathan Levine; Nicholas Jeffery

College of Veterinary Medicine, Texas A&M University, College Station, TX, USA

Non-invasive prognostic indicators are minimal for dogs with absent pelvic limb deep-pain perception resulting from acute thoracolumbar intervertebral disc herniation. We examined dogs presented to one veterinary hospital between 2011 and 2015 with paraplegia and absent pelvic limb deep-pain perception undergoing 3T MRI. Our purpose is to correlate MRI findings in these dogs with clinical outcome. Medical records were reviewed to identify dogs meeting the above criteria. Dogs were included if their MRI study comprised T2-weighted sagittal and transverse images covering the T3-L3 region, and acute disc herniation within this region was confirmed at surgery or necropsy. Surviving dogs were prospectively followed through follow-up neurological examinations 4–6 weeks post-operatively. MRI studies were reviewed by two board-certified radiologists blinded to outcome.

Fifty-two dogs met our inclusion criteria, and outcome data were available for 47. Twelve developed myelomalacia, 10 failed to recover deep-pain perception, and 25 recovered deep-pain perception and voluntary pelvic limb movement by 4–6 weeks post-operatively. Estimates of relative intramedullary T2-weighted hyperintensity length in the sagittal plane were well-correlated between observers ($r=0.76$, $p< 0.001$). Median length of average T2-weighted hyperintensity did not differ among outcome groups. 45% of dogs with T2-weighted hyperintensity > 6 x L2 body length had a good recovery. 42% of dogs that developed myelomalacia had T2-weighted hyperintensity < 6 x L2 body length.

We conclude that sagittal length of intramedullary T2-weighted hyperintensity associated with acute thoracolumbar disc herniation in dogs with absent deep-pain perception, as apparent on 3T MRI scans, does not correlate with the likelihood of functional recovery or the development of ascending-descending myelomalacia.

N10

Efficacy of a Therapeutic Diet on Dogs with Signs of Cognitive Dysfunction Syndrome (Abstract N10)

Yuanlong Pan[1]; Gary Landsberg[2]; Isabelle Mougeot[2]; Stephanie Kelly[3]; Hui Xu[1]; Sandeep Bhatnagar[1]; Norton W. Milgram[2]

[1]Nestlé Purina Petcare Global Resources, St. Louis, MO, USA; [2]CanCog Technologies, Toronto, ON, Canada; [3]Vivocore Inc, Fergus, ON, Canada

The objective of this clinical study was to evaluate the effects of dietary supplementation with medium chain triglyceride oil (MCT) and a proprietary brain nutrient blend (BPB containing DHA, EPA, and enhanced levels of B vitamins, antioxidants and arginine) on pet dogs with cognitive dysfunction syndrome (CDS). Previous studies have shown that diets supplemented with MCT or BPB enhance learning, memory and problem-solving ability in senior dogs.

Participating veterinary clinics screened senior dogs for signs of CDS as determined by a Senior Canine Behavior and Health Questionnaire, which included six categories (disorientation, altered social interaction, sleep-wake cycle disturbance, loss of house training, anxiety and altered activity). Out of 102 dogs screened, after ruling out potential medical causes, 87 were randomly enrolled into one of three diet groups with 29 dogs per group: Control (0% MCT), 6.5% MCT, 9% MCT. Each dog's signs were re-evaluated at day 30 and day 90.

All 6 categories of the CDS signs were significantly ($p<0.05$, Fisher least significant difference test) improved in the dogs given the 6.5% MCT oil and BPB diet at the end of the 90-day study. By contrast, the control diet did not significantly improve disorientation and social interactions scores. The 9% MCT diet was observed to produce significant improvements in only those dogs whose owners had positive feedback on the diet, suggesting possible palatability issues.

These findings support the use of the 6.5% MCT+BPB supplemented diet for treatment of clinical signs in senior dogs with signs consistent with CDS.

N11

Cognitive Impairment, Health Status and Overall Adaptation in Aged Dogs: An Epidemiological Study (Abstract N11)

Anna Salas Mani[1]; Jonathan Bowen[2]; Jaume Fatjó[2]; Celina Torre[1]

[1]Affinity Petcare, Barcelona, Catalonia, Spain; [2]Ethometrix, East Sussex, England, UK

Previous epidemiological studies using validated scales have provided very useful information about changes in behaviour and cognitive competence in senior dogs. However, they usually present 3 limitations: 1) They don't include information on the overall health status of the dog, 2) they target a pre-selected population of senior dogs, thus missing information from other age groups with whom comparisons could be made, and 3) behavioural measurements are focused on very specific behaviour changes linked to the aging process, as opposed to the impact that cognitive impairment could have on overall adaptation and functioning.

A study was designed to explore the impact of the aging process in a population of owned dogs, including information on the overall health status and the overall adaptation to the physical and social environment.

An online questionnaire was made available to dog owners through social media and a selected group of veterinary clinics. To avoid an age-related selection-bias, the study was presented as an online tool for all dog owners to estimate the overall level of adaptation of a dog to its environment, with each participant being given a score for their dog (which we called the "global functioning score"). Questions were divided into 4 sections: demographic data (including age and weight, a validated scale to assess cognitive dysfunction ("Canine Cognitive Dysfunction Rating scale", Salvin *et al.*), physical health conditions and how worried owners were about them, and how well the dog performed in a wide range of common situations (such interacting with people, family members and other dogs).

To estimate the prevalence of mild, moderate and severe cases of cognitive dysfunction, two methods were used. Firstly, scores from the cognitive dysfunction rating scale were calculated according to the authors' scheme. Secondly a two-step cluster analysis was performed on the cognitive function scores. Kruskal-Wallis was used to explore differences in cognitive dysfunction scores between weight groups and to compare Global Functioning Score (GFS) between age groups.

Two thousand one hundred and fourteen valid questionnaires were obtained. More than 28% of the dogs were 8-years-old or older (n=605). The overall prevalence of cognitive dysfunction in that population was 19.7%. No severe cases of cognitive dysfunction were detected below 7 years of age. However, global functioning score increased until the age of 7 and then declined, with the lowest scores being seen in dogs aged 14 and above. The reported prevalence of cognitive dysfunction by the owners was just 2.3%, being ranked last in a list of 10 health conditions reported by the owner, including arthritis, tumours, hair loss, deafness and breathing problems. When owners were asked to rate the seriousness of each of a list of 5 health conditions, cognitive dysfunction ranked third, behind cancer and heart disease but above arthritis. Older owners ranked cognitive dysfunction as more serious than younger owners.

Our findings indicate that although cognitive dysfunction is common, it is often not detected by the owner. This may be due to a combination of a lack of knowledge of the signs of the condition and a false perception that it is less serious than other some other health problems. Both will need to be tackled if we want to improve take up of treatments for this condition. Global functioning score, as an indicator of the dog's everyday performance as a pet, offer added information that could be refined to help in early detection of cognitive dysfunction and other age-related conditions.

References

1. Salvin HE, McGreevy PD, Sachdev PS, Valenzuela MJ. 2011. The canine cognitive dysfunction rating scale (CCDR): a data-driven and ecologically relevant assessment tool. *Vet J*. 188:331–336.

N12

Non-Surgical Treatment of Intracranial Empyema and Abscesses in 5 Dogs (Abstract N12)

Michelle Tensley[1]; Benjamin Young[2]; Scott Schatzberg[3]

[1]The Animal Neurology and Imaging Center, Rio Rancho, NM, USA; [2]VCA Alameda East Veterinary Hospital, Denver, CO, USA; [3]The Animal Neurology and Imaging Center, Algodones, NM, USA

The gold standard treatment recommendations for both veterinary and human patients with intracranial infection includes surgical drainage and antibiotic therapy. The purpose of this retrospective case series was to describe the clinical and MRI findings, as well as the medical treatment protocols, for dogs with confirmed or presumed intracranial bacterial infections that were treated without surgical intervention. Five dogs that were treated successfully between July 2011 and May 2015 are described. Medical treatment decisions were based upon the diagnoses of intracranial bacterial infection and/or owner reluctance for surgical intervention. MRI images for each patient were reviewed by a board-certified veterinary neurologist and radiologist. Success was defined by post-treatment MRI confirming resolution of the primary lesion(s) or a survival time of greater than 6 months past the discontinuation of treatment without relapse of clinical signs.

All dogs received enrofloxacin in combination with a second antimicrobial, most frequently amoxicillin-clavulanic acid. Prednisone also was prescribed at a variable dosage and duration. Treatment in 3 animals included an initial period of intravenous medications, while the other 2 dogs were treated with oral medication only. Treatment duration ranged from 6 weeks to 4 months. Initial clinical signs varied in severity from mild neurological deficits noted on examination to severe clinical signs such as altered mental status and evidence of increased intracranial pressure. Four dogs were alive at the time of manuscript composition, 10 to 28 months after diagnosis. The remaining dog died from unrelated cause 36 months after diagnosis.

We concluded that medical treatment without surgery is a feasible option for patients with presumed intracranial infection. Early and aggressive therapy with broad-spectrum antibiotics is recommended. Clinical severity did not affect outcome in this group of dogs.

N13

Cytarabine Constant Rate Infusion as Treatment for Meningoencephalomyelitis of Unknown Etiology in 22 Dogs (Abstract N13)

Rachel L. Arnold[1]; Randolph L. Winter[2]; Jill Narak[3]; Amanda R. Taylor[4]

[1]Tuskegee University/Auburn University/Southeast Veterinary Neurology, Miami, FL, USA; [2]Depatment of Clinical Sciences, College of Veterinary Medicine, Auburn University, Auburn, AL, USA; [3]Veterinary Referral Surgical Practice, Roswell, GA, USA; [4]Auburn University, Auburn, AL, USA

Meningoencephalomyelitis of unknown etiology (MUE) is a noninfectious inflammatory disease of the central nervous system. Cytarabine, an antimetabolite, has been used in combination with glucocorticoids to treat MUE in the dog.

Medical records of 22 dogs (age 6.27±2.83 years) with MUE that received cytarabine were reviewed. MRI and CSF findings were used to confirm the diagnosis of MUE. Data was reported as mean±SD or median (range) as appropriate. Student's t-test or Wilcoxon rank sums were used for comparisons, and correlation was assessed by Spearman rank correlation and simple linear regression.

All dogs received at least 1 CRI treatment of cytarabine. Fourteen dogs also received at least 1 SQ treatment. The number of CRI treatments per dog was median of 5.5 (1–38). Survival of all dogs was median of 258.5 days (61–1312). Survival was not affected by age at diagnosis ($p=0.289$), duration of clinical signs ($p=0.825$), or sex ($p=0.413$). There was a positive and significant correlation with number of CRI treatments and survival ($r=0.87$, $p\leq0.001$). The number of SQ treatments did not affect survival time ($r=-0.05$, $p=0.819$). In dogs with a relapse, there was a positive and significant correlation between number of CRI treatments and the longest period between clinical signs relapse ($r=0.82$, $p=0.002$).

Dogs with MUE may live longer and have longer periods of time between clinical relapse with increasing number of cytarabine CRI treatments.

N14

Hereditary Myotonia in Pigs Associated with Chloride Voltage-Gated Channel 1 Gene Partial Deletion

Alexandre S. Borges[1]; César Araújo[2]; José Barbosa[3]; Molly E. McCue[4]; Carlos Oliveira[3]; Luiz A. Resende[5]; José P. Oliveira-Filho[2]

[1]Sao Paulo State University - FMVZ - UNESP, Botucatu, Sao Paulo, Brazil; [2]FMVZ-UNESP-Botucatu, Botucatu, Sao Paulo, Brazil; [3]UFPA, Castanhal, Para, Brazil; [4]University of Minnesota, St Paul, MN, USA; [5]FMB-UNESP, Botucatu, Sao Paulo, Brazil

Mutations in the *CLCN1* gene were previously associated with hereditary myotonia in several species but not in pigs. The aim of this study is to characterize hereditary myotonia in pigs caused by a *CLCN1* gene mutation.

Five pigs from the same litter, presenting clinical myotonia and two non-myotonic pigs were submitted to clinical examination, electromyography (EMG), muscle biopsy (hematoxylin and eosin), and molecular investigation (muscle and blood samples were respectively used for RNA and DNA extractions). RT-PCR, PCR-sequencing reactions and alignments of obtained sequences from all clinically affected, clinically normal, and the NCBI reference pig sequence (GenBank: XM_003134570.3) were performed.

All affected animals presented muscle hypertrophy and stiffness. Action myotonia episodes, especially after startle and warm-up phenomenon were observed. The EMG of the affected animals evidenced spontaneous myotonic discharges with a characteristic sound. Dystrophic muscle abnormalities were not observed. All affected animals had a 348 base pairs (bp) deletion (c.[1582_1930del]) corresponding to exons 15, 16, and 17 in the mRNA. Using the DNA, an absence of a 4165 bp fragment (g. [7153448_7157612del]) including the aforementioned exons, introns 15 and 16 entirely, and introns 14 and 17 partially was verified. *In silico* analysis suggested that this mRNA deletion results in a chloride voltage-gated channel protein without 116 amino acids.

Pedigree analysis indicated an autosomal recessive mode of inheritance of this form of hereditary myotonia associated with a CLCN1 gene partial deletion in an animal species not yet reported with this disease.

N15

Population Pharmacokinetics of Extended-Release Levetiracetam in Epileptic Dogs When Administered Alone, with Phenobarbital or Zonisamide (Abstract N15)

Karen Munana[1]; Arturo J. Otamendi[2]; Julie A. Nettifee[3]; Mark G. Papich[3]

[1]College of Veterinary Medicine, North Carolina State University, Raleigh, NC, USA; [2]College of Veterinary Medicine, Texas A&M University, College Station, TX, USA; [3]North Carolina State University, Raleigh, NC, USA

The extended release formulation of levetiracetam (LEV-XR) has gained acceptance as an antiepileptic medication in dogs, but there are no studies evaluating its disposition in dogs with epilepsy. The objective of this study was to evaluate the pharmacokinetics of LEV-XR in epileptic dogs when administered alone, or concurrently with other AEDs. A population pharmacokinetic approach and non-linear mixed effects modeling (NLME) was used to examine the variation and source of variation in our population. Eighteen client-owned dogs on maintenance therapy with LEV-XR only (Group L, n=6), LEV-XR and phenobarbital (Group LP, n=6) or LEV-XR and zonisamide (Group LZ, n=6) were enrolled. All drugs were examined at steady state concentrations. Blood samples were collected at 0, 2, 4, 8 and 12 hours after the morning dose of LEV-XR was administered with food. The LP group had significantly lower C_{MAX} compared to the L and LZ groups (13.48±9.48 versus 33.93±29.84 and 34.13±9.01 µg/mL, respectively), significantly lower AUC (134.15±153.67 versus 353.10±284.11 and 452.80±116.89 h·µg/mL, respectively), and significantly higher CL/F (0.17±0.20 versus 0.08±0.07 and 0.07±0.02 L/kg/hr, respectively), demonstrating that concurrent administration of phenobarbital significantly lowers the LEV concentration in epileptic dogs. Elimination half-life was similar among groups (L, 1.46±2.08 h; LP, 1.54±1.97 h; LZ, 1.57±1.13 h), but shorter than previously reported in healthy dogs. These findings warrant consideration when utilizing LEV-XR as a treatment for epilepsy in dogs.

N16

Steroid-Responsive Myelitis in Five Dogs: Different Characteristics in Comparison with Steroid-Responsive Meningitis-Arteritis (Abstract N16)

Seo-Jin Park[1]; Dong-Woo Chang[1]; Ji-Houn Kang[2]; Mhan-Pyo Yang[3]; Byeongteck Kang[3]

[1]College of Veterinary Medicine, Chungbuk National University, Chenogju, Ch'ungch'ong-bukto, Republic of Korea; [2]College of Veterinary Medicine/Veterinary Medical Center, Chungbuk National University, Cheongju, Ch'ungch'ong-bukto, Republic of Korea; [3]College of Veterinary Medicine, Chungbuk National University, Cheongju, Ch'ungch'ong-bukto, Republic of Korea

The most frequently occurring non-infectious inflammatory lesion of the canine spine is steroid-responsive meningitis-arteritis (SRMA), while pure myelitis is a rare disease in dogs. Although meningomyelitis of unknown cause and granulomatous meningomyelitis are the most common diagnoses in canine cases, non-infectious myelitis responsive to glucocorticoid treatment has not been well described. This report describes the clinical and diagnostic features of steroid-responsive myelitis (SRM), which were different to those of SRMA, in five dogs.

Five dogs with SRM were identified retrospectively. Signalment, history, body temperature, neurological signs, diagnostic workup, treatment, and prognosis were analyzed. Affected breeds were small- to medium-sized: Miniature Pinscher (n=2), Maltese (n=1), Toy Poodle (n=1), and Cocker Spaniel (n=1). The mean age of the dogs was 4.2 years (range 2 to 8 years). Paraspinal hyperesthesia, paresis, and paralysis were acutely observed according to the affected region of the spinal cord. However, all dogs were non-febrile, and only one dog had neutrophilic leukocytosis on hematological analysis. Magnetic resonance imaging (MRI) of each dog revealed diffuse spinal lesions with T2 and FLAIR-hyperintense and T1-hypo- or isointense signal intensities at different regions: cervical (n=1), thoracolumbar (n=1), lumbosacral (n=1), and multifocal (n=2) spinal parenchyma. Contrast enhancement was not detected in the meninges. Four dogs had cerebrospinal fluid (CSF) pleocytosis: mononuclear (n=3) and mixed (n=1). In both serum and CSF, the levels of C-reactive protein (CRP) and immunoglobulin A (IgA) were within respective normal limits, except in two dogs with elevated serum CRP. No dog tested positive on CSF for infectious disease. A presumptive diagnosis of non-infectious inflammatory myelitis was established based on the clinical signs, MR scanning, CSF analysis, and laboratory tests.

Treatment was initiated with prednisolone (4 mg/kg/day orally) for immunosuppression, and then doses were tapered according to follow-up results. The mean duration of immunosuppressive therapy was 202 days (range 131 to 264 days), and medication was ceased in four dogs. Improvement in spinal lesions was confirmed in all dogs by neurological evaluation, CSF analysis, and MR follow-up scans. Because clinical signs did not recur after discontinuation of glucocorticoid, the type of myelitis present in all cases was considered to be steroid-responsive.

In general, SRMA is characterized by fever, neck pain, neutrophilia, increased levels of IgA and CRP, neutrophilic pleocytosis, and contrast enhancement of the meninges on MR scanning. The present cases did not have these features of SRMA, while mononuclear or mixed pleocytosis, diffuse spinal lesions on MR images, and excellent responsiveness to glucocorticoid were commonly noted in all dogs.

N17

The Pharmacokinetics of Cytarabine Administered Subcutaneously to Dogs with Meningoencephalomyelitis of Unknown Etiology (Abstract N17)

Bethany Pastina[1]; Peter J. Early[1]; Robert L. Bergman[2]; Karen Munana[3]; Julie A. Nettifee[1]; Katherine Bray[3]; Aude Castel[3]; Rennie Waldron[4]; Akshaya Maller[3]; Mark G. Papich[1]; Kristen Messenger[3]

[1]North Carolina State University, Raleigh, NC, USA; [2]Carolina Veterinary Specialist, Charlotte, NC, USA; [3]College of Veterinary Medicine, North Carolina State University, Raleigh, NC, USA, [4]Upstate Veterinary Specialist, Greenville, SC, USA

The objective of this study was to evaluate the pharmacokinetics (PK) of cytarabine (CA) after subcutaneous (SC) administration to dogs with meningoencephalomyelitis of unknown etiology (MUE). Twelve client-owned dogs, from 22 months to 10 years old, and weighing 2.5 to 39.2 kg, received a single SC dose of CA at 50 mg/m^2 for treatment of MUE. A sparse sampling technique was used to collect pharmacokinetic data. Four samples were collected from each dog to cover the time period from 0 to 360 minutes after administration. All dogs were concurrently on an oral prednisone dose of 0.5 to 2 mg/kg/day. Plasma CA concentrations were measured by HPLC and pharmacokinetic parameters were estimated using nonlinear mixed effects modeling (NLME). Plasma drug concentrations ranged from 0.05–2.8 µg/ml. The population estimate (CV%) for elimination half-life and T_{max} of cytarabine in dogs was 0.7 (46.2) h and 0.8 (11.3) h, respectively. The volume of distribution per fraction absorbed was 0.7 (38.4) L/kg. Mean plasma concentration of CA for all dogs was at or above 1.0 µg/mL at the 30, 60, 90, and 120-minute time points. In this study, the pharmacokinetics of CA in dogs with MUE after a single 50 mg/m^2 SC injection in dogs was similar to what has been previously reported; there was moderate variability in the population estimates in this small clinical population of dogs.

N18

Cavernous Sinus Syndrome in Dogs and Cats: 15 Cases (2002–2015) - (Abstract N18)

Helena Rylander[1]; Ellison Bentley[2]; Aslynn Jones[3]

[1]University of Wisconsin Madison, Madison, WI, USA; [2]School of Veterinary Medicine, UW Madison, Madison, WI, USA; [3]College of Veterinary Medicine and Biomedical Sciences, Veterinary Teaching Hospital, Fort Collins, CO, USA

The purpose of this study was to identify the presenting complaints, neurologic findings, diagnosis, and outcomes in dogs and cats with cavernous sinus syndrome (CSS).
Medical records between 2002 and 2015 were reviewed. Inclusion criteria were neurologic signs consistent with CSS and advanced imaging and/or post mortem examination.
Thirteen dogs and 2 cats were included. Twelve dogs received advanced imaging. Post mortem examination was performed on 2 cats and 3 dogs.

Dogs were 6–13 years (mean=10.8 years) of different breeds. Both cats were male neutered domestic shorthair, 3 and 14 years respectively.

Presenting complaints included mydriasis (N=4), behavior changes (N=3), hyporexia (N=3), ptosis (N=2), ataxia (N=2), pain (N=2), weakness (N=2), lethargy (N=2), and one each of epiphora, ocular swelling, polydipsia, seizures, facial muscle atrophy, dysphagia, and head tilt.

Neurologic signs included ophthalmoparesis/plegia (N=13), reduced/absent pupillary light response (N= 11), mydriasis (N= 10), reduced/absent corneal sensation (N= 7), ptosis (N= 6), reduced facial sensation (N= 2), and enophthalmos (N=1).

Thirteen patients had a mass lesion within the cavernous sinus, 6 of which were neoplastic on histopathology.

Median survival time for 4 patients treated with radiation therapy was 1035 days (range 150–2280), for 4 patients receiving medical treatment 360 days (range 7–1260 days), and for 5 non-treated patients 14 days (range 0–90 days).

In conclusion mydriasis and ophthalmoplegia are common signs of CSS. A mass lesion is the most common cause. Survival time may improve with radiation therapy.

N19

Effects of Chloride in the Diet on Serum Bromide Concentrations in Dogs

Go Togawa[1]; Miyoko Saito[1]; Shinsuke Iwata[2]; Naoyuki Watanabe[3]; Kensuke Orito[1]

[1]School of Veterinary Medicine, Azabu University, Sagamihara, Kanagawa, Japan; [2]School of Veterinary Medicine, Azabu University, Sagamihara, Kagawa, Japan; [3]Watanabe Animal Hospital, Shimada, Shizuoka, Japan

Potassium bromide (KBr) is an antiepileptic drug that is widely used for seizure control in dogs. Diets with high contents of Cl^- have been shown to decrease serum Br^- concentrations in dogs. However, the numerical relationship between dietary Cl^- intake and serum Br^- concentrations currently remains unknown. Therefore, the aim of the present study was to quantify serum Br^- concentrations in dogs fed diets with different contents of Cl^- and reveal the underlying relationship in order to contribute to the therapeutic drug monitoring of KBr.

Steady-state serum Br^- concentrations were measured using a gold chloride method in 23 dogs treated with KBr. The content of Cl^- (per 1g) was measured using Mohr's method or calculated from the ingredients label provided by the manufacturer. A regression analysis was used to evaluate the relationship between daily Cl^- intake (mg/kg/day) and the serum Br^- concentration per dose (µg/ml per mg/kg).

Dogs with a higher Cl^- intake had lower serum Br- concentrations. A strong negative correlation was observed between daily Cl^- intake (mg/kg/day) and the serum Br^- concentration per dose (µg/ml per mg/kg) ($p < 0.01$).

These results suggest the importance of considering the dietary content of Cl^- for KBr-treated dogs and may also assist with the selection of KBr doses and appropriate diets by clinicians for dogs treated with KBr.

EN01

Molecular Analyses of Pituitary Somatostatin and Dopamine Receptors in Feline Hypersomatotropism (ESVE Award Winner) - (Abstract EN01)

Christopher Scudder[1]; S. Mirczuk[2]; K. Richardson[2]; Ruth Gostelow[1]; Yaiza Forcada[1]; I. McGonnell[2]; David Church[1]; P. Kenny[2]; M. Korbonits[2]; R. Fowkes[2]; Stijn Niessen[3]

[1]Royal Veterinary College, Hatfield, England, UK; [2]Royal Veterinary College, North Myms, Hatfield, England, UK; [3]Newcastle Medical School, Royal Veterinary College, Hatfield, England, UK

Feline hypersomatotropism (HS) (which results in acromegaly) is typically caused by a growth hormone (GH) secreting pituitary adenoma and is thought to be a common cause of feline diabetes mellitus. Medical pituitary inhibition has been challenging in some humans and cats with this condition. This could be related to variable expression of somatostatin (SSTR) and dopamine (DR) receptor expression. There are five different SSTR subtypes (SSTR1-5) and five different subtypes of DRs, with D2R being the predominant DR subtype in the human pituitary. Pasireotide, a somatostatin analogue with high binding affinity to SSTR1, SSTR2 and SSTR5, is effective at controlling GH secretion in feline acromegalics but somatostatin analogues with high binding affinity to SSTR2 (octreotide and lanreotide) are not.

The aim of this study was to document the receptor expression in feline somatotrophinomas and to compare this to the healthy feline pituitary. We utilised a novel customised multiplex RT-qPCR assay to investigate the expression profile of the somatostatin and dopamine receptors in GH-secreting pituitary adenomas. Pituitary adenoma RNA was extracted from 20 GH-secreting adenomas and eight control cats followed by multiplex RT-qPCR analyses for multiple genes, including somatostatin receptors (SSTR1-5), dopamine receptor (D2R), growth hormone receptor, growth hormone secretagogue receptor and the housekeeper RPL18.

The normalised gene expression of SSTR1-5 in acromegalic cats was SSTR1=SSTR5>SSTR2 vs SSTR5=SSTR1=SSTR2 in control cats. SSTR3 and SSTR4 were undetectable. Expression of SSTR1, SSTR2 and SSTR5 were upregulated in acromegalics vs controls (SSTR1:p=0.006, SSTR2:p=0.005, SSTR5:pSSTR1 increased x9 [95% CI 4–12], SSTR2 increased x6.5 [95% CI 2.5–8.5] and SSTR5 increased x4 [95% CI 2–4]. D2R gene expression was not significantly different between tumorous or normal pituitary tissue; within the hypersomatotropism group D2Rexhibited a moderate negative correlation with pituitary volume (r=-0.62, p<0.01, Rsquare 0.39).

Feline GH-secreting adenomas expressed SSTRs are similar to human GH-secreting adenomas, having higher SSTR1, SSTR2 and SSTR5 expression than SSTR3 and SSTR4. However, the higher expression of SSTR1 and SSTR5 vs SSTR2 than in most humans may explain the greater biochemical response to pasireotide than octreotide. Decreased D2R expression in larger tumours may be a mechanism allowing unchecked adenoma growth. The receptor expression profiles of feline GH-secreting adenomas will help the search for improved medical management options of feline hypersomatotropism.

EN02

Comparison of Two Doses for ACTH Stimulation Testing in Dogs Suspected of Hypoadrenocorticism (Abstract EN02)

Annabel S. Botsford; Hollie P. Lee; Robert Kemppainen; Ellen N. Behrend

College of Veterinary Medicine, Auburn University, Auburn, AL, USA

The ACTH stimulation test, using cosyntropin at 5 mcg/kg IV, is the preferred method for diagnosis of hypoadrenocorticism. Previous studies have shown maximal stimulation of the adrenal glands using 1 mcg/kg cosyntropin in normal dogs. No studies have evaluated the efficacy of the lower dose in dogs suspected of hypoadrenocorticism. Our objective was to compare 1 mcg/kg (low-dose) to 5 mcg/kg (high-dose) cosyntropin IV to determine if both doses result in a similar adrenocortical response.

Testing was prospectively performed in dogs suspected of hypoadrenocorticism. Dogs suspected of having hypoadrenocorticism received 1 mcg/kg cosyntropin IV followed four hours later by 5 mcg/kg cosyntropin IV. Preliminary studies confirmed the validity of performing two ACTH stimulation tests on the same day. Blood samples were obtained pre- and one hour post-ACTH for each dose (4 measurements total). Cortisol was measured using a previously validated radioimmunoassay. To detect differences in cortisol concentration between cosyntropin doses (1 and 5 mcg/kg) and between time points (baseline and 60-minutes), data were analyzed using a paired t-test and equivalence testing by commercial statistical computer programs. Significance was set at the $p \leq 0.05$ level.

Twelve dogs were included. Administration of both ACTH doses caused a significant increase in cortisol concentration ($p<0.001$ for both). A significant difference was detected between the post-ACTH cortisol concentrations. The mean post-ACTH cortisol concentration after the high-dose was significantly greater than the mean post-ACTH cortisol concentration obtained after the low-dose (one-tailed p value, $p=0.061$). Using equivalence testing, the results of the two tests were not equivalent ($p>0.05$).

Therefore, the 1 mcg/kg IV dose of cosyntropin does not cause maximal adrenal response as does the standard 5 mcg/kg IV dose in dogs suspected of hypoadrenocorticism. The lower dose is not sufficient for ACTH stimulation testing in those patients suspected of hypoadrenocorticism.

EN03

Performance Evaluation of a Veterinary-Specific Enzyme-Linked Immunosorbent Assay for Measurement of Canine Serum Cortisol Concentrations (Abstract EN03)

Michael Lane[1]; Bente Flatland[1]; Shelly Olin[1]; Kellie Fecteau[1]; Markus Rick[2]; Luca Giori[1]

[1]College of Veterinary Medicine, University of Tennessee, Knoxville, TN, USA; [2]College of Veterinary Medicine, Michigan State University, East Lansing, MI, USA

A veterinary-specific enzyme-linked immunoassay (ELISA) (SNAP Cortisol Test Kit, IDEXX, USA) is available to veterinarians for in-house diagnosis and management of canine hypo- and hyperadrenocorticism. However, analytical performance of the assay has not been reported in peer-reviewed literature. The purpose of this study was to investigate the precision of the ELISA, agreement of the ELISA with two chemiluminescence assays (CAs) at two reference laboratories, and the clinical implications of any identified bias.

Stored, frozen canine serum samples were used. ELISA precision was prospectively evaluated by measuring three canine serum pools having low, medium, and high cortisol concentrations, respectively, in triplicate on five consecutive days. A prospective method comparison study was performed by measuring cortisol concentrations in sixty-three independent canine serum samples using the ELISA and two CA. Methods were compared using Pearson 's Correlations, Passing-Bablok regressions and Bland-Altman analyses. Clinical implications of identified bias were investigated using an error grid.

Within-laboratory coefficients of variation for the low, medium, and high concentration pools were 21%, 28.9%, and 13.0%, respectively. Constant and proportional biases between the ELISA and CA methods were present at all concentrations. There was good overall correlation with both CA (r=0.92 and r=0.94). Despite identified biases, there was clinical agreement for results in 74.6% of samples.

There is clinical agreement between the ELISA and CA in most cases. Clinical disagreement would occur in 25.4% of cases. Equivocal results reported by the ELISA should be confirmed by a reference laboratory.

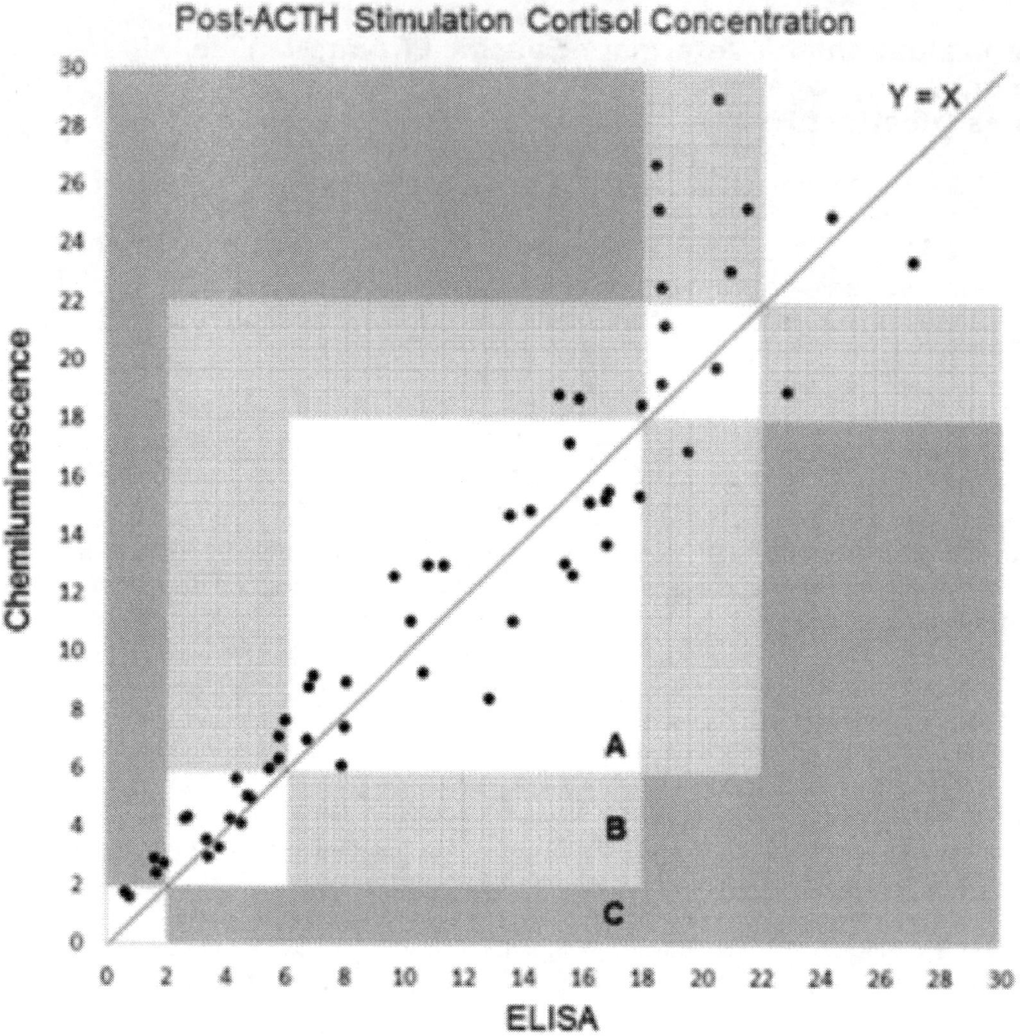
Post-ACTH Stimulation Cortisol Concentration
Chemiluminescence
ELISA
Y = X
A
B
C
0
2
4
6
8
10
12
14
16
18
20
22
24
26
28
30

EN04

Investigation of a Novel Modified Fixed Dose Determination Protocol for Radioiodine Treatment of Feline Hyperthyroidism (Abstract EN04)

Wendy A. Morre; David Pancier; Gregory Daniel; William Monroe

Virginia-Maryland College of Veterinary Medicine, Blacksburg, VA, USA

While radioiodine is the treatment of choice for feline hyperthyroidism because of its safety and efficacy, dose determination of ^{131}I in cats remains controversial. The goal is to achieve a euthyroid state with one treatment, avoiding persistent hyperthyroidism or hypothyroidism. No method of dose determination has been able to consistently achieve this goal. We tested the hypothesis that a radioiodine dose determination based on thyroid scintigraphy would be associated with reduction in hypothyroidism without an increase in persistent hyperthyroidism at 6 months post radioiodine treatment compared with a fixed dose of ^{131}I.

The purpose of this study was to compare a novel method of radioiodine dose determination that utilized thyroid scintigraphy (novel dose group) with a standard fixed dose method (fixed dose group) for the treatment of hyperthyroidism in cats. Specifically, this study focused on determining radioiodine dose by measuring thyroid gland size and percent of $^{99m}TcO_4^-$ uptake using thyroid scintigraphy. Hyperthyroidism was diagnosed in all cats by finding a serum T4 concentration above the reference interval and characteristics diagnostic of hyperthyroidism on nuclear scintigraphy. Scintigraphy was performed by obtaining images 30 minutes after IV administration of 3.0 mCi $^{99m}TcO_4$. In cats in the novel dose group, percent dose uptake was calculated by drawing a region of interest around the thyroid gland lobe(s) and dividing the corrected counts (CPM) in the thyroid gland by the $^{99m}TcO_4$ dose (CPM). Cats with a percent dose uptake <5%, 5–10%, and >10% were calculated to receive 3 mCi, 3.5 mCi, or 4.5 mCi of ^{131}I, respectively administered subcutaneously. Thyroid size was assessed by a ratio of the thyroid ROI size (# of pixels) to the salivary ROI size (# of pixels) (TS). Cats with a TS ratio of <5, 5–10, and >10 were calculated to receive 3 mCi, 3.5 mCi, and 4.5 mCi of ^{131}I. The dose of ^{131}I to be given was first determined by measuring percent dose uptake. If the thyroid size fell into a higher dosing category than percent uptake, the dose was increased accordingly. Cats in the fixed dose group received a mean of 4.3 mCi of ^{131}I subcutaneously. Cats were evaluated 6 months after treatment to determine the functional status of the thyroid gland. Hypothyroidism was defined as having a serum T4 concentration below the reference interval and persistent hyperthyroidism was defined as having a serum T4 concentration above the reference interval 6 months post treatment. Fifty-eight cats were included in the novel dose group and the fixed dose group consisted of 25 cats treated prior to those in the novel dose group. Univariate analysis using Chi-square and Fisher 's exact analysis was used to determine associations between dose determination method and outcome with the level of significance set at $p<0$ 0.05.

Hypothyroidism developed in 32% of cats in the fixed dose group compared to 21% in the novel dose group ($p=0.27$). Persistent hyperthyroidism occurred in 8% of cats in the fixed dose group compared to 16% of cats in the novel dose group ($p=0.35$).

Results of this study showed no statistical difference in the number of cats remaining persistently hyperthyroid or becoming hypothyroid at 6 months post radioiodine treatment using the novel modified fixed dose protocol compared to the standard fixed dose protocol. More studies investigating this novel radioiodine dose determination are needed that incorporate additional doses of ^{131}I in addition to the 3mCi, 3.5mCi, and 4.5mCi described in this study.

EN05

Comparison Between Glucose Measurements in Canine Whole Blood, Serum and Plasma (Abstract EN05)

Ellen N. Behrend; Natalie D. Suchowersky; Elizabeth A. Carlson; Hollie P. Lee

College of Veterinary Medicine, Auburn University, Auburn, AL, USA

Blood glucose concentration is frequently measured in dogs as many diseases are associated with hypoglycemia and hyperglycemia such as diabetes mellitus. Use of whole blood, serum or plasma can affect results when using a human glucometer, with serum and plasma yielding more accurate results (Tauk *et al.*, J Am Vet Med Assoc. 2015). The purpose of this study was to investigate the correlation between glucose concentrations in canine whole blood, serum and plasma measured by the AlphaTrak2 (Zoetis, Parsippany, NJ), a veterinary glucometer, and glucose concentration measured by a biochemical analyzer.

In this prospective study, blood samples were collected from 31 dogs. Reason for presentation was not an absolute entrance criteria, but samples from diabetic dogs were sought. Immediately following each blood draw, blood was placed in a heparinized tube for submission to the laboratory and a drop of blood was analyzed by an AlphaTrak2 specifically designated for the study; blood was also used to fill a heparinized capillary tube and a nonheparinized capillary tube. The capillary tubes were centrifuged within 2 minutes after collection to acquire serum (nonheparinized tube) and plasma (heparinized tube) to obtain a glucose concentration measured by the AlphaTrak2. Within 10 minutes following each blood draw, blood was submitted to the Auburn University Clinical Pathology Lab for determination of the glucose concentration by an automated biochemical analyzer. The blood glucose concentration in whole blood, serum, and plasma measured by the AlphaTrak2 were compared to the blood glucose concentration measured by the automated analyzer. The Lin concordance correlation coefficient (pc) and Bland-Altman plots were used to determine correlation and agreement between the results obtained in the different sample types.

Glucose concentration in whole blood measured by the AlphaTrak2 was more strongly correlated with the glucose concentration measured by the biochemical analyzer (pc=0.92) compared to serum and plasma glucose concentrations (pc=0.59 and 0.61 for canine serum and plasma, respectively). The mean difference between the glucose concentrations in whole blood, serum, and plasma measured by the AlphaTrak2 and the glucose concentration determined by the biochemical analyzer were 15, 119, and 113 mg/dL, respectively.

Our findings suggest that when using an AlphaTrak2, the accuracy and reliability of a glucose measurement obtained may be increased when using whole blood compared to serum or plasma. Thus, use of whole blood would allow for a more correct assessment and diagnosis which are necessary for appropriate therapeutic intervention.

EN06

Stereotactic Irradiation of Canine ACTH-Secreting Pituitary Tumors: Clinical and MRI Time-Related Response Evaluation (Abstract EN06)

Mario Dolera[1]; Luca Malfassi[2]

[1]La Cittadina Fondazione Studi e Ricerche Veterinarie, Romanengo, Lombardia, Italy; [2]La Cittadina Fondazione Studi e Ricerche Veterinarie, Crema, Lombardia, Italy

Optimal irradiation schedule for canine adrenocorticotropic hormone (ACTH) secreting pituitary tumors has not been defined and response evaluation criteria has not been established.

The aim of this study was to evaluate time-related clinical and magnetic resonance imaging (MRI) response in dogs suffering from pituitary hyperadrenocorticism with a measurable mass at MRI treated with stereotactic radiotherapy (RT).

A single institution prospective single arm observational study was conducted. Dogs naive of any endocrine treatment suffering from pituitary hyperadrenocorticism diagnosed by standard laboratory and imaging tests were irradiated with 6 MV LINAC using Volumetric Modulated Arc Therapy with a dose of 38 Gy delivered in 10 fractions on alternated days basis. Concomitant administration of trilostane were started. Tumor and endocrine response were evaluated by repeated MRI of pituitary and adrenal glands at 0–2–4–6–8–12–18–24 months. Basal and ACTH-stimulated cortisol were serially measured and the dose of trilostane to achieve eucortisolism was recorded.

Fourteen dogs were enrolled. Median pituitary tumor volume was 1 cmc (range 0.5–8.5 cmc). Mean basal and stimulated cortisol at the diagnosis were 7.8 mcg/dL and 36.1 mcg/dL. Mean initial dose of trilostane was 2.0 mg/kg/12h. Complete regression of the tumor was observed in 10/14 dogs at 6–18 months (median 8 months), partial regression in 4/4 at the end of the study. No progressions were observed. Normalization of the thickness of adrenal glands was observed in all dogs at 8–18 months (median 12 months). Eucortisolism was obtained in all except 1 dogs. Trilostane was progressively tapered and withdrawn at 14–18 months (median 16.5 months). No radio toxicity were encountered.

The RT schedule used in this study allowed optimal tumor and endocrine control. A gradual progression in the response was observed, starting with a volumetric response of the pituitary tumor, delayed of several months after the treatment, then a response of the adrenal glands and ultimately a reduction of the serum cortisol levels was achieved. This pattern of response must be taken in consideration during follow up evaluation to adjust the dose of trilostane and to establish evaluation criteria.

EN07

Pharmacodynamics of Insulin Detemir and Insulin Glargine (300 U/ml) in Healthy Dogs (Abstract EN07)

Heidi Fink[1]; Chen Gilor[2]; Claire Herbert[3]

[1]Veterinary Medical Center, The Ohio State University, Columbus, OH, USA; [2]University of California-Davis, Davis, CA, USA; [3]The Ohio State University, Columbus, OH, USA

Insulin glargine 300 U/ml (Toujeo®) and insulin detemir (Levemir®) are synthetic long-acting insulin analogs associated with minimal intra- and inter-patient variability or episodes of hypoglycemia in people. Toujeo® is more predictable and longer acting compared to glargine 100 U/ml (Lantus®) in people. The purpose of this study was to assess the pharmacokinetics or pharmacodynamics of Toujeo® compared to Levemir® in healthy dogs.

Eight healthy purpose-bred dogs each received 2.4 nmol/kg of Levemir® and Toujeo® (0.1 U/kg and 0.4 U/kg, respectively, SC injection) on two different days, at least one week apart, in random order. Endogenous and exogenous insulin were measured hourly with the Mercodia Isoinsulin and the Mercodia canine insulin ELISA's. Blood glucose (BG) was measured every 5 minutes and glucose was administered intravenously at a variable rate with the goal of maintaining BG within 10% of baseline BG ("Isoglycemic clamp"). Glucose infusion rate and a decline in endogenous insulin were used to define exogenous insulin action.

Median onset of action was significantly delayed with Toujeo® compared to Levemir® (4.0 h [3.3–5.8 h] versus 0.6 h [0.6–1.2 h], p=0.002). There was no significant difference in peak (median [range]=6.3 h [5.0–21.3 h] versus 4.3 h [2.9–7.4 h], p=0.15) or duration (16.3 h [6.1–20.1 h] versus 10.8 h [8.8–14.8 h], p=0.21) of action between Toujeo® or Levemir®, respectively. Toujeo® demonstrated a peakless time-action profile in 4/8 dogs.

Toujeo® might be better suited than Levemir® as a once-daily insulin in some dogs based on its long duration of action and peakless time-action profile. Day-to-day variability in insulin action should still be assessed for both formulations.

EN08

Impact of an Auditory Stimulus on Baseline Cortisol Concentrations in Clinically Normal Dogs (Abstract EN08)

Taylor Gin[1]; Melanie Puchot[2]; Audrey K. Cook[2]

[1]College of Veterinary Medicine, Texas A&M University, Bryan, TX, USA; [2]College of Veterinary Medicine, Texas A&M University, College Station, TX, USA

Baseline cortisol concentrations are routinely used to screen dogs for hypoadrenocorticism (HOC). A result 20%. We hypothesized that exposure to an unpleasant auditory stimulus prior to sample collection would trigger cortisol production and increase concentrations above the 55 nmol/L cut-off in healthy dogs.

Twenty-eight healthy client-owned dogs were evaluated; median age was 4 years (range 2–9 years) and median weight was 20 kg (range 10–27 kg). Dogs were ineligible for inclusion if they had received short- or long-acting glucocorticoids within the previous 30 and 90 days, respectively.

Dogs were randomly assigned to control (n=7) and test (brief noise: n=10; and long noise: n=11) groups. Each dog and owner waited in a secluded area for 10–15 minutes. "Brief noise" dogs were exposed to the sound of a wet-dry vacuum for 3 minutes at the start of this period. "Long noise" dogs were exposed to random bursts of wet-dry vacuum noise for 15 minutes. Blood was then collected in an adjacent room. Samples were processed immediately; serum was frozen at -80°C prior to submission.

Median cortisol concentrations for the three groups were similar; control: 79 nmol/L (range 33–201 nmol/L); brief noise: 64 nmol/L (range 36–115 nmol/L), and long noise: 65 nmol/L (range 40–177 nmol/L) (Figure 1).

Exposing dogs to an unpleasant noise stimulus for 3–15 minutes prior to sample collection did not increase serum cortisol concentrations and is unlikely to improve the specificity of baseline cortisol determination as a screening test for HOC.

Figure 1: Serum cortisol concentration for the 3 groups

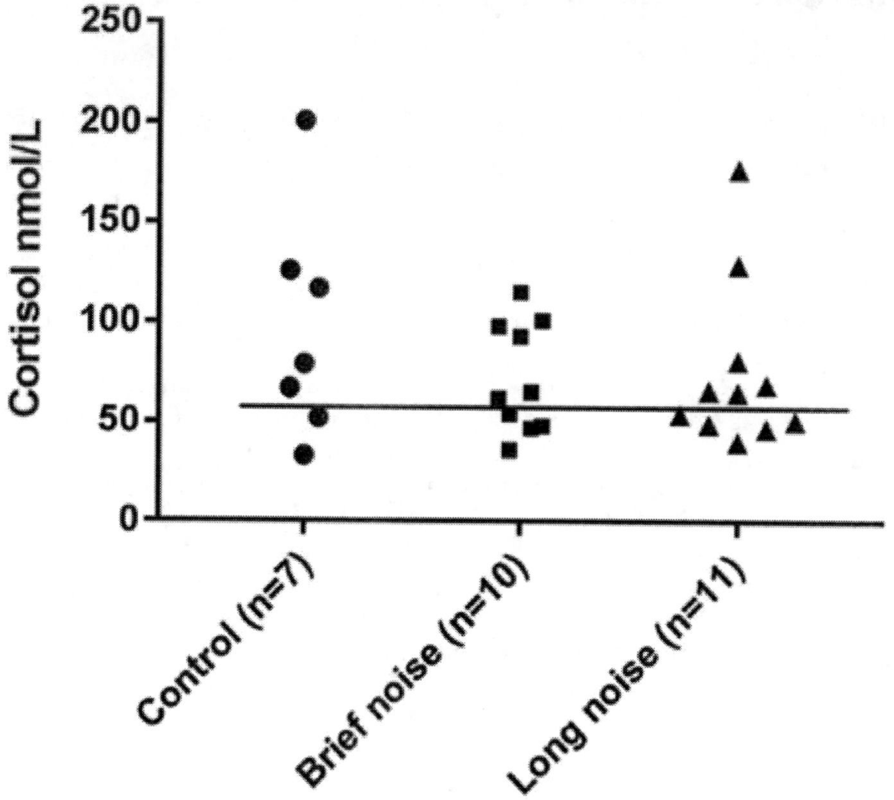

EN09

Diabetic Remission in Cats Examined in UK Primary-Care Veterinary Practices: Occurrence and Risk Factors (Abstract EN09)

Ruth Gostelow[1]; Dan O'Neill[1]; David Brodbelt[1]; David Church[1]; Oliver Dunham[1]; Stijn Niessen[2]

[1]Royal Veterinary College, Hatfield, England, UK; [2]Newcastle Medical School, Royal Veterinary College, Hatfield, England, UK

Identifying predictors of feline diabetic remission aids veterinarians when advising on diabetic cat's prognosis, and might provide insight into the etiology of feline diabetic remission. Previous investigations have often studied referral patients or utilized relatively small sample sizes, whereas examining larger, diverse populations might identify new predictors. This study aimed to document the remission rate among a large cohort of diabetic cats examined in UK primary-care veterinary practices, and identify predictors of remission.

An observational, retrospective cohort study was conducted using electronic patient record data collected through the VetCompass Companion Animal Surveillance Programme from 1st September 2009 to 31st August 2014. Cats with newly-diagnosed diabetes mellitus (DM) were identified and occurrence of remission was recorded. Cumulative probability of remission was assessed using the Kaplan-Meier estimator, and predictors of remission were identified using multivariable, time-dependent Cox proportional hazards modelling. Mortality within 30 days of DM diagnosis, and predictors of death, were also assessed using the Kaplan-Meier estimator and multivariable binary logistic regression, respectively. Significance was set at <0.05. Average values are given as median (interquartile range).

Five hundred and eighty-three cats (347 male, 235 female, 1 unknown sex) were recruited from 103 practices. Five hundred and seventeen (88.2%) were crossbred, with Burmese (n=24; 4.1%) being the most common pure breed. Median age was 13.0 (10.4–15.0) years, median bodyweight 4.5 (3.8–5.3) kg, and median follow-up time 232 (42–596) days. One-year cumulative probability of remission was 13.1% (95% confidence interval [CI] 10–17%). Multivariable analysis revealed glucocorticoid therapy in the 90 days before diagnosis (hazard ratio [HR] 5.9; 95% CI 3.3–10.4; p<0.0001), and Burmese breed compared to crossbred status (HR 3.7; 95% CI 1.6–8.1; p=0.001) to be independent predictors of remission. Cumulative probability of survival 30 days after diagnosis was 83% (95% CI 79–85%). Lack of medical anti-hyperglycemic therapy (odds ratio (OR) 14.7; 95% CI 7.5–29.0; p<0.0001) and lower bodyweight (kg) at diagnosis (OR 1.3; 95% CI 1.0–1.7; p=0.03) were independently associated with increased odds of death within 30 days of DM diagnosis.

Probability of remission was lower than often documented in referral populations. High mortality following DM diagnosis and owner reluctance to pursue medical therapy present substantial barriers to achieving greater diabetic remission rates in primary-care practice. Increased likelihood of diabetic remission has not been previously identified in Burmese cats. This could result from differences in the pathogenesis of DM in Burmese compared to other breed types and warrants further investigation.

EN10

One-Year Prospective Randomized Trial Comparing Efficacy of Glargine and Protamine Zinc Insulin in Diabetic Cats (Abstract EN10)

Ruth Gostelow[1]; Christopher Scudder[1]; Katarina Hazuchova[1]; Yaiza Forcada[1]; David Church[1]; Stijn Niessen[2]

[1]Royal Veterinary College, Hatfield, England, UK; [2]Newcastle Medical School, Royal Veterinary College, Hatfield, England, UK

Long-acting insulin treatment is recommended in feline diabetes mellitus (DM) and insulin glargine has frequently been proposed to offer the greatest efficacy and likelihood of diabetic remission. However, a recent systematic review identified that unbiased, adequately powered, long-term comparisons of insulin types are lacking. A one-year, prospective, randomized clinical trial was conducted to compare the glycemic control and remission rates associated with glargine (Lantus®, Sanofi) or human-recombinant PZI (PZIR; ProZinc®, Boehringer Ingelheim) therapy in diabetic cats.

Recently-diagnosed diabetic cats (duration 6 weeks–5 months) were recruited through a dedicated feline DM clinic. Cats were transitioned to a low-carbohydrate diet 10 days before first examination, with overweight cats (body condition score >5/9) receiving caloric intake aimed at weight loss. Cats were excluded if pre-recruitment screening identified ketoacidosis, clinical pancreatitis, glucocorticoid/progestogen administration, hyperthyroidism, hypersomatotropism, or other conditions causing likely insulin resistance. Diabetic control was assessed using fructosamine, a validated diabetic clinical score (DCS), mean blood glucose (MBG) during 24-hour blood glucose curves (BGCs) and twice-daily (BID) insulin dose (U/kg). Pet and owner quality of life was assessed using the validated DIAQoL-pet score. Cats were randomized to receive BID glargine or PZIR and underwent a one-year, standardized re-examination program. Insulin dose was adjusted according to a single, nadir-led protocol based on BGC results.

Two-sample t-tests or Mann-Whitney U-tests assessed for differences in MBG, fructosamine, BID insulin dose, DIAQoL-pet score, and DCS between groups at baseline and at 1, 3, 6, and 12 month(s). Proportion of cats entering remission was compared using Kaplan-Meier product limit estimates and log rank test, and predictors of remission were identified using multivariable Cox proportional hazards modelling. Significance was set at $p<0.05$.

Forty-six cats were recruited (glargine=24, PZIR=22). No significant difference in MBG, fructosamine, BID insulin dose, DIAQoL-pet score, or DCS was identified between groups at any time point. Eight glargine-treated cats (33.3%) and five PZIR-treated cats (23%) achieved remission ($p=0.42$). Baseline bodyweight >5.2 kg (hazard ratio (HR) 2.7; 95% confidence interval (CI) 0.9–8.3; $p=0.07$), baseline feline body mass index ≥25% (HR 3.3; 95% CI 1.1–10.0; $p=0.04$) and ≥2% weight loss at the 1-month time point (HR 12.9; 95% CI 3.5–47.4; $p=0.0002$) were identified as possibly associated with remission during univariate analysis (criterion: $p<0.1$). Multivariable analysis revealed a minimum of 2% weight loss at the 1-month time point to be the only significant predictor of remission (HR 13.5; 95% CI 2.9–62.5; $p=0.0009$).

The current study is unprecedented in terms of size and control of bias. It revealed that glargine and PZIR produced comparable glycemic control and no detectable difference in remission rate among recruited cats. Clinicians should consider implementing a controlled weight loss program as a priority in overweight diabetic cats in order to promote remission.

EN11

Biochemical Markers of Bone Turnover in Dogs with Naturally Occurring Hyperadrenocorticism (Abstract EN11)

Eilidh Gunn[1]; Robert Shiel[2]; Karolina Jankowska[2]; Elizabeth Matthews[2]; Mary Sekiya[2]; Carmel T. Mooney[2]

[1]University of Glasgow, Glasgow, Scotland, UK; [2]University College Dublin, Dublin, Ireland

The effect of hyperadrenocorticism (HAC) on bone in dogs is controversial. The aim of this study was to investigate its effect on bone turnover by assessing intact and whole parathyroid hormone (PTH) concentrations and markers of both osteoblastic (osteocalcin (OC)) and osteoclastic (carboxyterminal cross-linked telopeptide of type 1 collagen (ICTP)) activity.

Blood samples were collected from dogs with HAC and a hospital control population. Intact and whole PTH by IRMA (Scantibodies), OC by ELISA (Immunodiagnostic Systems) and ICTP by RIA (Orion Diagnostica) were measured using commercial human kits. Data were reported as median (range) with differences between groups assessed by Mann Whitney testing.

Intact and whole PTH concentrations were significantly higher ($p<0.001$ in each case) in 41 dogs with HAC (82.27 [12.18–867.73] pg/mL and 41.9 [5.09–1138.18] pg/mL, respectively) compared with 24 control dogs (35.00 [18.27–183.27] pg/mL and 18.27 [8.00–113.55] pg/mL, respectively). The median OC concentration in 35 dogs with HAC was not significantly different from 20 control dogs (3.3 [0.7–118.8] ng/mL and 4.0 [2.0–15.1] ng/mL, respectively). The median ICTP concentration in 34 dogs with HAC was significantly lower ($p<0.001$) than in 19 control dogs (3.2 [1.2–12.8] ng/mL and 7.3 [3.7–21.3] ng/mL, respectively).

Whole and intact PTH concentrations are higher in dogs with HAC. This does not appear to be associated with a decrease in bone formation or an increase in bone resorption as assessed by OC and ICTP measurements, respectively.

EN12

A Comparison of Intact Parathyroid Hormone Assays for Use in Dogs (Abstract EN12)

Eilidh Gunn[1]; Robert Shiel[2]; Karolina Jankowska[2]; Elizabeth Matthews[2]; Mary Sekiya[2]; Carmel T. Mooney[2]

[1]University of Glasgow, Glasgow, Scotland, UK; [2]University College Dublin, Dublin, Ireland

Several assays are available for measurement of intact parathyroid hormone (iPTH) concentrations but there are few reports comparing assays in dogs.

Intact PTH concentrations were measured using an immunoradiometric assay (IRMA) (Scantibodies) and a chemiluminescent assay (CLIA) (Immulite 2000) designed for human use, and a canine specific enzyme-linked immunosorbent assay (ELISA) (Immutopics). Samples, in duplicate, were assessed using each technique within 4 hours of each other.

45 and 48 samples, respectively, were available for comparison between IRMA and ELISA and IRMA and CLIA. Correlation between ELISA and IRMA was good ($r=0.90$ [95% CI 0.83–0.95], $p<0.001$). Bland-Altman difference plot revealed a bias with ELISA results greater than IRMA values by a mean + sd of 80.74 + 89.51% (95% CI -94.7-256.2). Using CLIA, 39 (81.3%) samples had iPTH concentrations at or below the reported limit of detection of the assay (3.0 pg/mL). The remaining nine dogs had detectable but low concentrations ranging from 4.82–452.0 pg/mL. By comparison these values were higher by IRMA (range, 125.55–867.73 pg/mL).

The CLIA cannot be recommended for measuring canine iPTH as most values were undetectable. This may be related to the antibodies used within the assay. The non-isotopic ELISA conferred advantages over IRMA with regard to ease of use but yielded higher values that would have impacted clinical decision-making. Additionally, a high prevalence of duplicate errors and gel formation impacted reliability. Whilst more technically demanding, the IRMA proved the most reliable and robust method evaluated.

EN13

Predicted Ionized Calcium Can Accurately Confirm Hypo/Hypercalcemia in Dogs, Even with Missing Biochemistry Values (Abstract EN13)

Kevin Le Boedec[1]; Julie Danner[2]

[1]Centre Hospitalier Veterinaire FREGIS, Arcueil, Ile-de-France, France; [2]College of Veterinary Medicine, University of Illinois, Urbana, IL, USA

Abnormal total calcium (tCa) values do not necessarily imply hypo/hypercalcemia based on measured ionized calcium (miCa). Unfortunately, miCa is not readily available in practice.

In dogs with increased or decreased tCa, a multivariate adaptive regression splines model was used to predict miCa from creatinine, albumin, tCa, phosphorus, sodium, potassium, chloride, alkaline phosphatase (ALP), triglycerides, and age. Sensitivity and specificity of predicted ionized calcium (piCa) were compared to those of corrected tCa for true hyper/hypocalcemia (McNemar 's test). The impact of missing biochemistry values on predictive model accuracy was assessed by replacing missing biochemistry variables by their mid-reference interval value. Accuracy was considered altered if piCa became significantly different from miCa (Wilcoxon test).

Out of 53 dogs with elevated tCa, 26 were truly hypercalcemic based on miCa. Sensitivity of piCa (92%) and of corrected tCa (100%) were equivalent to detect hypercalcemia ($p=0.5$). Specificity of piCa (81%) was significantly higher than specificity of corrected tCa (29%) to diagnose hypercalcemia ($p<0.001$). Out of 19 dogs with decreased tCa, 10 were truly hypocalcemic based on miCa. Sensitivity of piCa (90%) was significantly higher from sensitivity of corrected tCa (20%) to detect hypocalcemia ($p=0.01$). Specificity of piCa (78%) and of corrected tCa (100%) were equivalent to diagnose hypocalcemia ($p=0.5$). Without missing values, piCa was not significantly different from miCa ($p=0.12$). Triglycerides, phosphorus, sodium, and ALP could be missing without altering piCa accuracy.

In dogs with abnormal tCa values, true hyper/hypocalcemia may be assessed by piCa even if triglycerides, phosphorus, sodium, and ALP measurements are not available.

EN14

Serum Symmetric Dimethylarginine Concentrations in Hyperthyroid Cats With and Without Azotemic Chronic Kidney Disease (Abstract EN14)

Tim L. Williams

University of Cambridge, Cambridge, England, UK

Hyperthyroidism decreases body muscle mass and serum creatinine concentrations which can confound diagnosis of concurrent chronic kidney disease (CKD). Serum symmetric dimethylarginine (SDMA) concentrations are influenced to a lesser extent by muscle mass, and might be a better biomarker of CKD in hyperthyroidism. This study aimed to evaluate serum SDMA concentrations in hyperthyroid cats with and without azotemic CKD.

Hyperthyroid and euthyroid (without azotemic CKD [control]) cats >9 years old were included. Hyperthyroid cats received anti-thyroid medication and were monitored for azotemia (serum creatinine concentration >1.7 mg/dL) during a three month follow up period. Hyperthyroid cats that were azotemic before treatment or during follow up were included in the hyperthyroid-CKD group. Kruskal-Wallis with Dunn 's multiple comparisons test was used to compare baseline serum creatinine and SDMA concentrations (SDMA measured by IDEXX Laboratories) between the control, hyperthyroid-CKD and hyperthyroid non-azotemic groups. Median (range) values are presented.

Serum creatinine concentrations in control (1.4 [1.1–1.8] mg/dL, n=8) and hyperthyroid groups were not significantly different, however, serum creatinine was significantly higher in the hyperthyroid-CKD group (1.9 [1.1–2.4] mg/dL, n=5) than the hyperthyroid non-azotemic group (1.1 [0.7–1.3] mg/dL, n=4; $p<0.05$). Serum SDMA concentrations were not significantly ($p>0.05$) different between the control (11 [9–13] µg/dL), hyperthyroid-CKD (14 [7–17] µg/dL) and hyperthyroid non-azotemic groups (13 [11–19] µg/dL).

Serum SDMA concentrations did not distinguish hyperthyroid cats with and without azotemic CKD. The influence of hyperthyroidism on serum SDMA concentrations warrants further investigation.

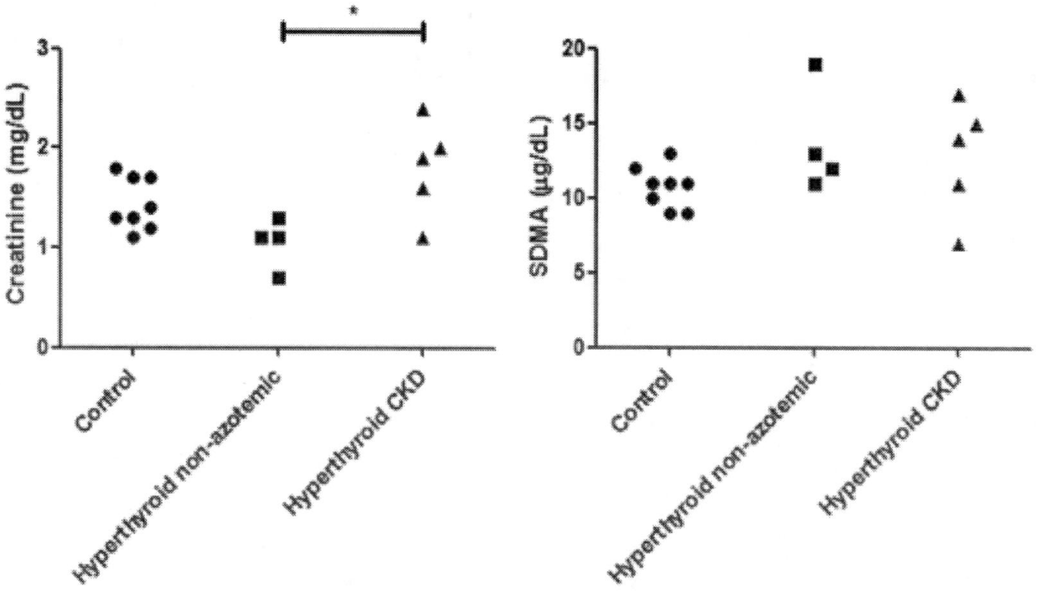

EN15

Increased Adipocyte Size and Leukocyte Infiltration in Adipose Tissue of Obese Dogs (Abstract EN15)

Carolina Zaghi Cavalcante[1]; Rayane Sgarbossa[1]; Vanessa Gmyterco[1]; Luiz Capriglione[1]; Thais Casagrande[2]; Taise Fuchs[2]; Jorge Castro[1]; Jair Engracia[1]; Marconi Farias[1]; Anita Nishiyama[3]

[1]Pontifical Catholic University of Parana, Curitiba, Parana, Brazil; [2]Universidade Positivo, Curitiba, Parana, Brazil; [3]Federal University of Parana, Curitiba, Parana, Brazil

Obesity is a frequent nutritional disorder in dogs and it is associated with several metabolic alterations and represents a risk factor for several systemic comorbidities. However, there is controversy about which of the adipose tissues contributes more to these complications in the canine species. The present study aimed to evaluate the subcutaneous and visceral adipose tissue of ten obese female dogs and ten female dogs with eutrophic body condition, regarding adipocyte size and presence of leukocyte infiltration. The study was conducted with routine animals from the Veterinary Hospital, and all of the animals were submitted to a complete clinical examination, classification based on the body condition score, and blood exams to exclude concomitant diseases. Fragments of subcutaneous and visceral adipose tissue were obtained during electives ovariosalpingohisterectomy procedures, the samples were stained with Hematoxylin-Eosin and analyzed with Zeiss Microscope software. It was observed that obese dogs presented remarkably larger adipocyte size when compared to animals with ideal body condition, in a significance level of 5% through Student 's T-test. Likewise, it was noted the presence of significantly higher leukocyte infiltrate in obese animals, in a significance level of 5% in the Kruskal-Wallis test, compared to animals with normal body condition that did not have leukocyte infiltration detected. However, the size of adipocytes and leukocyte infiltrate in obese animals did not differ significantly between the visceral and subcutaneous adipose tissue samples. It was concluded, therefore, that obesity in dogs is related with the increase of adipocytes and with presence of leukocyte infiltration, but it was not possible to state in this study that different fat deposits have differentiated metabolic and endocrine levels, which could interfere in a specific way in the inherent processes in obese and diabetic patients.

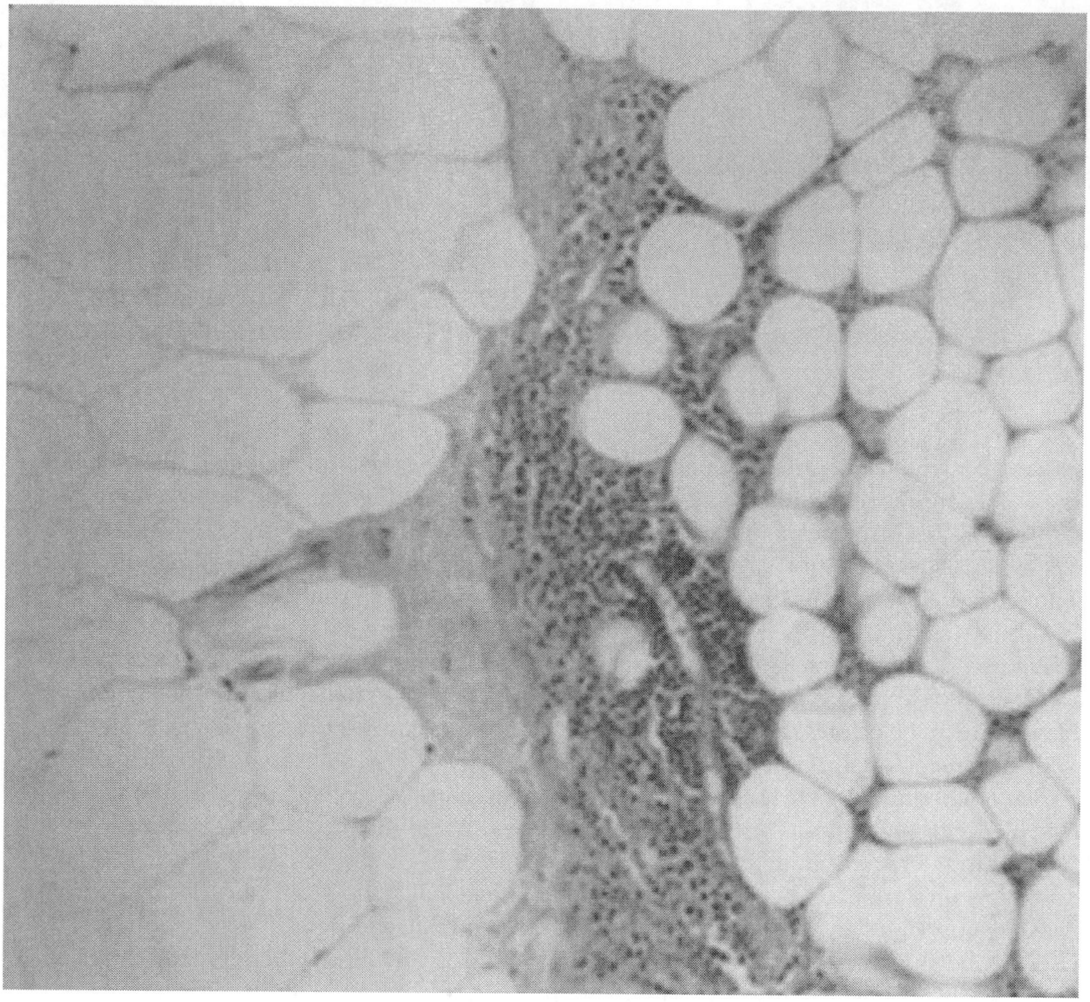

EN16

Lipid Peroxidation in Obese Dogs Before and After the Weight Loss Program (Abstract EN16)

Carolina Zaghi Cavalcante[1]; Andressa Roncoski[1]; Rayane Sgarbossa[1]; Vanessa Gmyterco[1]; Ana Paula Sarraff Lopes[1]; Gisele Sechi[1]; Anita Nishiyama[2]

[1]Pontifical Catholic University of Parana, Curitiba, Parana, Brazil; [2]Federal University of Parana, Curitiba, Parana, Brazil

Obesity is a frequent nutritional disorder in dogs and cats and it results in oxidative stress. The evaluation of oxidative stress can be performed through the process of lipid peroxidation, and malonate is one of the most abundant aldehydes. The evaluation method is called "test for substances that react with thiobarbituric acid" (TBARS). Objective of the present study is to evaluate oxidative stress by measuring the TBARS present in samples of obese dogs before and after the weight loss program. The work was carried out with routine animals of the Veterinary Hospital of the Pontifical Catholic University of Parana (located in Parana, Brazil). The animals were submitted to a complete clinical examination, blood count, biochemistry, urinalysis, and systolic blood pressure measurement. Subsequently, they were classified as obese on the body scale; body mass index, and percentage of body fat. The animals were divided into group 1 (pre-diet) and group 2 (post-diet). Ten obese animals were selected and submitted to a hypocaloric diet loss program. TBARS was performed using the Elisa technique. All animals in groups 1 and 2 presented TBARS values above the reference value, and there was a statistically significant reduction of the values from group 1 to group 2 ($p=0.034$). Based on the results, it was concluded that obese dogs present higher production of lipid peroxidation than animals with ideal body scale, leading to the development of oxidative stress and elevation of TBARS concentrations.

EN17

Evaluation of Corticotropin (Acthelea) Effect in the ACTH Stimulation Test in Dogs with Hyperadrenocorticism (Abstract EN17)

Priscila Viau Furtado[1]; Debora Cattaruzzi Rodini[2]; Rogério Soila[2]; Patricia Lopes[1]; Raquel Fukumori[1]; Claudio Alvarenga Oliveira[1]

[1]FMVZ/USP, São Paulo, São Paulo, Brazil; [2]PROVET, São Paulo, São Paulo, Brazil

One of the most important endocrine tests for diagnosis and control of hyperadrenocorticism (HAC) in dogs is the ACTH stimulation. The objective of this study was to evaluate the efficacy of corticotropin inducing cortisol production in dogs. This possibility can be used as an alternative to tetracosactrin (Synacthen) which is, currently, the most used drug in this adrenal challenge test, mainly in Latin America.

Thirty ACTH stimulation tests were performed on 15 dogs divided into 2 groups: 5 control animals (2 females and 3 males, healthy, mean age 2.5±1.5 years) and 10 treatment animals (6 females and 4 males, previously hyperadrenocorticism diagnosed, mean age 11.7±2.3 years). All animals were controlling them, in other words, first were challenged with Synacthen and then with Acthelea. For dogs on trilostano (1 to 2 mg/kg/day), ACTH stimulation tests were initiated 4–6 hours post-pill administration as per manufacturer instructions. The first stimulation tests were performed using Synacthen as the drug of election to induce cortisol response and the second, which were separated by 15 days period, using Acthelea. Blood samples were obtained just before injection and 60 minute post-ACTH. The first to assess the cortisol basal concentration of control group (CSynbas and CEleabas) and Treated group (TtoSynbas and TtoEleapos) and the second collection for assessment of cortisol concentration post-ACTH in the control group (CSynpos and CEleapos) and treated group (TtoSynpos and TtoEleapos). The doses were: 0.04 mL/Kg for Synacthen and 0.08 mL/Kg for Acthelea, IV. Serum cortisol concentrations were measured at the PROVET/BR Hormone Laboratory using previously validated radioimmunoassay (MP Biomedicals).

Results were expressed for their medium-sized ± SEM. For the control group, the results obtained for the baseline values were 0.89±0.08 µg/dL (CSynbas) and 0.96±0.12 µg/dL (CEleabas), with p=0.68; and post-ACTH values were 9.14±0.46 µg/dL (CSynpos) and 9.00±0.71 µg/dL (CEleapos), with p=0.58 There was no significant difference between the mean baseline and post-stimulation in the control group, when comparing the results among drugs. For the treated group, the baseline values were 1.71±0.34 µg/dL (TtoSynbas) and 1.82±0.48 µg/dL (TtoEleabas), with p=0.82; and after ACTH values were 7.85±2.2 µg/dL (TtoSynpos) and 7.90±2.1 µg/dL (TtoEleapos), with p=0.98. Therefore, no significant difference was observed between the values obtained when comparing the use of these two drugs applied in the tests.

According to the results obtained in this study, we can conclude that the Acthelea was effective in stimulating the production of cortisol in dogs and can be used as an option in the ACTH stimulation test, particularly in places where Synacthen is not available.

EN18

Evaluation of Triglycerides, Cholesterol and Glucose in Dogs with Mastocytoma Submitted to Surgery and Chemotherapy (Abstract EN18)

Eric Vieira Januário[1]; Samanta Rios Melo[2]; Bruna Castro Miranda[2]; Erika Brito Marco Zanutto[2]; Julia Maria Matera[2]

[1]Faculty of Veterinary Medicine and Animal Science, University of São Paulo, São Paulo, São Paulo, Brazil, [2]University of São Paulo, São Paulo, São Paulo, Brazil

Humans show metabolic changes after chemotherapy, due to the action mechanism of chemotherapy drugs and/or concomitant corticosteroid therapy, with health consequences. The aim of this study is to evaluate similar changes in dogs.

Laboratory tests of 19 dogs diagnosed with cutaneous mastocytoma treated at the Small Animal Surgery Department - Faculdade de Medicina Veterinária e Zootecnia, Universidade de São Paulo, submitted to different chemotherapy protocols after tumor excision were analyzed.

We compared the levels of triglycerides, cholesterol, and fasting glucose of these patients submitted to chemotherapy according to the protocols: Intravenous I-vinblastine (2 mg/m 2-8 weekly sessions, 4 fortnightly sessions); metronomic dose of oral II-lomustine (2.84 mg/m 2 SID) for 16 weeks associated with oral prednisone (2 mg/kg SID); Intravenous synthetic III-phosphoethanolamine (70 mg/m 2, 4 weekly sessions) without glucocorticoids. Measurements were performed in a period before (P0); in three periods during (P1, P2, P3: monthly in the Protocols I and II and weekly in Protocol III) and 30 days after completion of chemotherapy (C).

Nineteen dogs were assessed: 11 (57.89%) females and eight (42.11%) males with a mean age of 8.7 years. At P0 the dogs showed mean values of triglycerides (MVTRI) of 100.28 mg/dL (standard deviation = SD=55.58); cholesterol (MVCHOL) 222.84 mg/dL (SD=59.97) and glucose (MVGLUC) 74.5 mg/dL (SD=24.14). P1, MVTRI 199.728 mg/dL (SD=187.51), MVCHOL=230.2 mg/dL (SD=63.19), MVGLUC=91.5 (SD=1206.14). In period P2, MVTRI=148.47 mg/dL (SD=138.84), MVCHOL=217.57 mg/dL (SD=108.93), MVGLUC=123.24 (SD=8.42). In P3, MVTRI=1293.55 mg/dL (SD=1471.78), MVCHOL=279.45 mg/dL (SD=104.32). In period C, MVTRI=139.4 (SD=63.26) and MVCHOL=242.8 (SD=88.63).

Hypertriglyceridemia and elevated serum glucose were observed in dogs treated with chemotherapy associated with glucocorticoids. Further studies on these changes are necessary.

EN19

Retrospective Incidence of Hyperthyroidism in Domestic Cats at a Veterinary Hospital of São Paulo, Brazil (Abstract EN19)

Livia Scalize; **Marcia Jericó**; Tatiana Pacini; Renata Castilho

Anhembi Morumbi University, São Paulo, São Paulo, Brazil

In recent decades, the prevalence of feline hyperthyroidism (FHT) has increased significantly worldwide and is currently a major endocrine disease that causes high morbidity in middle-aged cats. Due to the lack of studies on this endocrine disease in Brazil, a retrospective study of stored serum samples from all cats older than eight years treated at the Veterinary Hospital of Anhembi Morumbi University (São Paulo, Brazil) was carried out from January 2013 to January 2015, regardless of the main complaint that led them to be treated at the service. Thyroid function was evaluated through total thyroxine (T4T) measurements. A total of 212 cat serum samples were sent to a private laboratory, where T4T was measured by radioimmunoassay (RIA). Of the 212 assessed serum samples, 3.30% confirmed feline hyperthyroidism, 59.91% were within the normal range and 36.79% were below the normal value, suggesting sick euthyroid syndrome. In 2.35% of the samples, T4T values were close to the upper limit value, which raises the possibility that these cats had FHT, but whose comorbidities at the time of blood collection reduced T4T values in blood serum. We conclude that even though the prevalence of FHT in the assessed group was shown to be discreet, this clinical suspicion should be part of the differential diagnosis in cats aged ≥eight years.

EN20

Alterations in the Hormones Regulating Blood Glucose Concentrations in Healthy Dogs Undergoing Hydrocortisone Administration (Abstract EN20)

Hakhyun Kim[1]; Ji-Houn Kang[2]; Chan-Joo Sung[1]; Jiyeon Yun[1]; Purum Son[1]; Seungji Woo[1]; Byeong-Teck Kang[1]; Mhan-Pyo Yang[1]

[1]College of Veterinary Medicine, Chungbuk National University, Cheongju, Ch'ungch'ong-bukto, Republic of Korea; [2]College of Veterinary Medicine, Veterinary Medical Center, Chungbuk National University, Cheongju, Ch'ungch'ong-bukto, Republic of Korea

Endo- and exogenous hypercortisolemia is well known to associate with insulin resistance in dogs. However, the types of changes in the levels of hormones involved in maintaining the blood glucose levels of dogs with hypercortisolemia are unknown. The objective of this study was to assess whether insulin sensitivity and the factors regulating blood glucose levels in dogs are altered by chronic administration of hydrocortisone.

Six laboratory dogs received hydrocortisone (8.5 mg/kg) orally every 12 hours for 90 days. All variables were serially examined during the 90 days of hydrocortisone administration. Insulin sensitivity was evaluated by homeostatic model assessment $(HOMA)_{\text{insulin sensitivity}}$. Visceral fat was quantified by computed tomography.

With hydrocortisone administration, $HOMA_{\text{insulin sensitivity}}$ was significantly decreased, although the fasting glucose concentrations were maintained within the reference range; and the visceral fat mass was significantly increased. In addition, $HOMA_{\text{insulin sensitivity}}$ was significantly correlated with visceral fat mass. The changes in serum concentrations of leptin, adiponectin, insulin, glucose-dependent insulinotrophic polypeptide (GIP), pancreatic polypeptide, glucagon, and ghrelin were significant. Except for KC-like chemokine and interleukin-7, differences between the concentrations of inflammatory cytokines were significant. The correlations between visceral fat mass, leptin, adiponectin, ghrelin, GIP, insulin concentrations, and $HOMA_{\text{insulin sensitivity}}$ were significant.

These results suggest that insulin sensitivity decreased by hydrocortisone administration is likely associated with increased visceral fat mass and changes in serum concentrations of the hormones regulating glucose levels in dogs. Further study clarifying our results in naturally occurring hyperadrenocorticism in dogs is needed.

EN21

Persistent Hypertonicity in Diabetic Dogs (Abstract EN21)

Cristian Perez; Tera Brandt; **Thomas Schermerhorn**

Kansas State University, Manhattan, KS, USA

Tonicity is a function of the serum concentrations of effective osmoles. Sodium (NA), chloride (CL), and glucose (GLU) are the major endogenous osmoles that affect tonicity, but GLU contributes significantly only in hyperglycemic states. The pathophysiological consequences of hypertonicity (HT) are not completely understood but persistent HT may affect disease progression in diabetic humans. Many diabetic dogs experience persistent hyperglycemia but the role of HT in these dogs has not been investigated. The study objective was to evaluate diabetic dogs for evidence of persistent HT.

A retrospective evaluation was performed to compare pertinent solute concentrations and other tonicity indicators between insulin-treated diabetic dogs (DD; n=37) dogs and non-diabetic dogs (NDD; n=164). For data analysis, comparisons between groups were made using Student 's T test, correlations were performed using Spearman 's rho test and ROC analysis was used to study the relationship between the mean corpuscular volume difference (dMCV) and fructosamine.

In DD, tonicity (315 vs. 311 mOsm/L; p=0.02) and GLU (295 vs. 110 mg/dl; p=0.0001) were elevated while NA and CL were decreased compared with NDD. HT (tonicity >315 mOsm/L) occurred more frequently in DD (45.9% vs. 31.1% for NDD). The dMCV, a marker for HT, was significantly increased in DD (3.9 vs. 1.7 fL for NDD). Increased dMCV predicted an elevated fructosamine (>450 umol/L) with 82.6% sensitivity and 42.8% specificity. The results show that stable, insulin-treated diabetic dogs frequently have HT and that GLU is the major osmole that contributes to HT in DD. An increase in dMCV is a modest predictor for serum fructosamine, suggesting the latter is an indirect marker for hypertonicity in DD. In conclusion, dogs with moderate/poor glycemic control show evidence of persistent HT, although additional studies are needed to investigate the role of persistent HT and its possible consequences for diabetic dogs.

EN22

Serum 25-Hydroxyvitamin D Concentrations in Dogs with Acute Pancreatitis (Abstract EN22)

Dong-In Kim[1]; Ji-Houn Kang[2]; **Purum Son**[1]; Hakhyun Kim[1]; Jiyeon Yun[1]; Seungji Woo[1]; Chan-Joo Sung[1]; Byeong-Teck Kang[1]; Mhan-Pyo Yang[1]

[1]College of Veterinary Medicine, Chungbuk National University, Cheongju, Ch'ungch'ong-bukto, Republic of Korea; [2]College of Veterinary Medicine, Veterinary Medical Center, Chungbuk National University, Cheongju, Ch'ungch'ong-bukto, Republic of Korea

An imbalance in vitamin D status has recently been suggested to be associated with disease progression in humans with pancreatitis. To date, however, no studies have investigated the association between vitamin D status and acute pancreatitis (AP) in dogs. Therefore, the purpose of this study was to examine differences between the concentrations of serum 25-hydroxyvitamin D (25-[OH]D) in healthy dogs versus AP dogs, and, if so, to determine the correlations between 25-(OH)D, ionized calcium (iCa), and C-reactive protein (CRP) levels.

This prospective observational cohort study included 20 client-owned dogs with newly diagnosed AP and 22 otherwise healthy dogs with similar body-condition scores. The circulating concentrations of serum 25-(OH)D, blood iCa, and serum CRP were measured.

The serum 25-(OH)D and blood iCa concentrations were significantly lower, and the serum CRP concentration was significantly higher in dogs with AP than in healthy dogs. There were significant differences between the concentrations of serum 25-(OH)D in AP survivors and non-survivors. The differences between the concentrations of blood iCa and serum CRP in AP survivors and non-survivors were not significant. Concentrations of 25-(OH)D were positively correlated with the concentrations of iCa and albumin in dogs with AP. Conversely, a negative correlation was observed between serum concentrations of CRP and iCa in the AP group.

This study found that dogs with AP exhibited decreased 25-(OH)D levels, which might be the result of inflammatory responses rather than a cause of AP. Moreover, alerations in serum 25-(OH)D concentrations might be associated with disease severity or mortality in dogs with AP.

EN23

Clinical Relationship Between Cholestatic Disease and Pituitary-Dependent Hyperadrenocorticism in Dogs: A Retrospective Analysis (Abstract EN23)

Kun-Ho Kim[1]; **Woo-Jin Song**[1]; Qiang Li[2]; Min-Ok Ryu[1]; Sang-Chul Park[1]; Hyeon-Jin Kim[1]; Su-Yeon Kim[1]; Hye-Mi Yang[1]; Hwa-Young Youn[1]

[1]Seoul National University, Seoul, Republic of Korea; [2]Seoul National University, Seoul, China (People's Republic)

A high prevalence of cholestatic disease, including gall bladder mucocele (GBM) has been reported in dogs with naturally-occurring pituitary-dependent hyperadrenocorticism. Differences exist in the clinical features of dogs with PDH and concurrent cholestatic disease, and also is the management of these dogs with trilostane. In this study, 65 client-owned dogs with naturally-occurring PDH were analyzed. Each dog was treated with trilostane for at least 3 months before the study, and had a good clinical response, as determined by owners. Statistical comparisons of clinical signs, results of routine blood tests, basal and post-ACTH cortisol concentration, and optimal trilostane dosage were made after dogs were separated into the following 3 groups by ultrasonographic imaging: normal on ultrasound (NOU) group, cholestasis group, and GBM group. The GBM group had more severe clinical signs, and significantly different total serum cholesterol concentration and post-ACTH stimulation cortisol concentration at the time of diagnosis. Dogs that weighed <6 kg had a significantly higher prevalence of cholestatic disease than did the other dogs (p=0.003). The optimal trilostane dosages for the GBM and cholestasis groups were 2.5 and 1.5 times the dosage of the NOU group, respectively (p<0.001). Gall bladder disease associated with cholestatic disease is correlated with PDH in dogs, both in its clinical features and drug management. These findings may be associated with hypercholesterolemia, unidentified genetic factors, and the hydrophobic nature of trilostane.

EN24

Usefulness of Thyrotropin Stimulation Test in Dogs with Hypothyroidism, Non-Thyroidal Illness, or Receiving Hydrocortisone (Abstract EN24)

Chan-Joo Sung[1]; Ji-Houn Kang[2]; Hakhyun Kim[1]; Seungji Woo[1]; Purum Son[1]; Jiyeon Yun[1]; Ki-Jeong Na[1]; Mhan-Pyo Yang[1]

[1]College of Veterinary Medicine, Chungbuk National University, Cheongju, Ch'ungch'ong-bukto, Republic of Korea; [2]College of Veterinary Medicine, Veterinary Medical Center, Chungbuk National University, Cheongju, Ch'ungch'ong-bukto, Republic of Korea

Hypothyroidism is a common endocrinopathy in dogs; however, the diagnosis of hypothyroidism may be difficult because nonthyroidal illness (NTI) and medication might affect the results of thyroid function tests. The objective of this study was to evaluate the usefulness of a thyrotropin (TSH) stimulation test for diagnosing hypothyroidism.

This case-controlled prospective study included 41 dogs, 35 client-owned dogs who were suspected to have altered thyroid function and 6 dogs who had received oral hydrocortisone for 90 days. The TSH stimulation test was performed in each dog. Based on the results of the test, the 35 client-owned dogs were divided into 3 groups as follows: dogs with hypothyroidism, dogs with NTI, and dogs without hypothyroidism.

The mean basal serum TT4 concentration of dogs with hydrocortisone administration was similar to that of dogs with hypothyroidism (0.32 µg/dL and 0.27 µg/dL, respectively). The mean basal serum TT4 concentrations of dogs with NTI and dogs without hypothyroidism were 0.78 µg/dL and 1.56 µg/dL, respectively. The mean post-rhTSH TT4 concentrations of dogs administered hydrocortisone, dogs with NTI, and dogs without hypothyroidism were increased to 2.17 µg/dL, 2.19 µg/dL, and 3.49 µg/dL, respectively; while the post-rhTSH TT4 concentration of dogs with hypothyroidism remained <1.0 µg/dL. The peak post-rhTSH TT4 concentration in dogs administrated hydrocortisone occurred at 4 hours after stimulation. Those of dogs with NTI and dogs without hypothyroidism occurred at 6 hours after injection of rhTSH.

These results indicate that a TSH stimulation test using rhTSH could be a useful method for diagnosing hypothyroidism, even in dogs receiving long-term, high-dose hydrocortisone. In clinical practice, the TSH stimulation test is valuable for the diagnosis of hypothyroidism in dogs with a variety of NTIs, hyperadrenocorticism, or dogs administered glucocorticoids.

EN25

Effect of Freezing on Biological Activity of Depot Tetracosactide in Healthy Dogs (Abstract EN25)

Thatiana Leite Pinto[1]; **Aline Bomfim Vieira**[2]; Renata Simoes Barros[1]; Vitor Nowosh[1]; Mário Felipe Alvarez Balaro[1]; Marcia Carolina Salomão[1]; Nayro Xavier de Alencar[1]

[1]College of Veterinary Medicine, Universidade Federal Fluminense, Niteroi, Rio de Janeiro, Brazil; [2]School of Veterinary Medicine, Ross University, Basseterre, Saint George Basseterre, Saint Kitts and Nevis

The adrenocorticotropic hormone (ACTH) stimulation test is an excellent tool for assessing adrenal gland function in dogs. Among the various presentations of ACTH available for testing, the Depot Tetracosactide (Synacthen® Depot) is economically advantageous in some countries, because of the volume and concentration of its ampoule (1.0 mg/1.0 mL). This advantage depends on the reuse of the surplus volume of the ampoule, as the indicated dose for dogs is 0.25 mL per dog. Our aim was to evaluate the effect of freezing on the biological activity of Depot Tetracosactide for the stimulation of adrenal gland in healthy dogs. Thirteen healthy adult dogs were submitted to ACTH stimulation test (0.25 mL/dog/IM) four times at intervals of two months between each test (0, 2, 4, and 6 months). The first stimulation test was done in every dog with the first aliquot of a fresh, refrigerated ampoule, opened at the time of application. The remaining three aliquots were stored at -20°C in plastic syringes of 50 units protected from light. Samples for determination of serum cortisol by radioimmunoassay were obtained before, one and two hours after administration of the ACTH. All samples were thawed once and analyzed in duplicate by ANOVA. Serum cortisol levels did not differ significantly ($p<0.01$) before or after freezing for all sampling times. We conclude that Depot Tetracosactide can be stored in plastic syringes at -20°C for up to six months without interfering with the biological stability of the hormone.

EN26

Evaluation of the Clinical Accuracy of 6 Portable Blood Glucose Meters in Dogs (Abstract EN26)

Min-Keun Shin[1]; Ji-Houn Kang[2]; **Jiyeon Yun**[1]; Bohye Shin[1]; Hakhyun Kim[1]; Seungji Woo[1]; Purum Son[1]; Chan-Joo Sung[1]; Woon-Bum Baek[1]; Byeong-Teck Kang[1]; Ki-Jeong Na[1]; Mhan-Pyo Yang[1]

[1]College of Veterinary Medicine, Chungbuk National University, Cheongju, Ch'ungch'ong-bukto, Republic of Korea; [2]College of Veterinary Medicine, Veterinary Medical Center, Chungbuk National University, Cheongju, Ch'ungch'ong-bukto, Republic of Korea

Portable blood glucose meters (PBGMs) are widely used because of their usefulness in the control of blood glucose levels. However, the accuracy of PBGMs has been questioned repeatedly, and factors that can interfere with blood glucose measurements have been identified. The objective of this study was to evaluate the factors associated with deviations in the measurements obtained by PBGMs from the blood glucose reference range in dogs.

This prospective study included 67 client-owned dogs presenting to the veterinary medical center for various diseases, including diabetes mellitus. All of the blood samples (N=219) were collected from jugular veins. Blood glucose concentrations measured by 6 PBGMs (AlphaTrak2, iPet, OneTouch Ultra, Cerapet, VetMate, Optium Xceed) were compared to those measured by the reference method, and accuracies were evaluated according to ISO 15197:2013. Statistical agreements of PBGM measurements were determined by Spearman correlation analysis, Passing-Bablok regression analysis, the Wilcoxon signed-rank test, Bland-Altman plotting, and error grid analysis. Mann-Whitney and Kruskal-Wallis tests were performed to assess the variables expected to affect glucose measurements.

Spearman correlation analysis revealed high correlations between the values obtained by the PBGMs and the reference method. However, Bland-Altman plotting showed that deviations of the PBGM values from the reference range increased when actual glucose concentrations increased. Decreased packed cell volume (PCV) resulted in overestimation or underestimation of blood glucose values from iPet and OneTouch Ultra, respectively. Hypertriglyceridemia resulted in overestimation of glucose values from AlphaTrak2, iPet, Cerapet, and VetMate. Hyperproteinemia resulted in overestimation of glucose values from OneTouch Ultra and Optium Xceed. Error grid analysis revealed that AlphaTrak2, OneTouch Ultra, Cerapet, and Optium Xceed were adequate, based on minimum accuracy criteria of ISO 15197:2013.

The results suggest that hypertriglyceridemia, hyperproteinemia, and decreased PCV can significantly affect the results obtained by various PBGMs. Clarification of the causes of errors in glucose measurements should lead to more accurate assessment of disease and appropriate glucose control and treatment for diabetic dogs.

GI01

Relationship of the Mucosal Microbiota to Inflammation and Disease Activity in Cats with Intestinal Lymphoma (Abstract GI01)

Kayode Garraway[1]; Chad M. Johannes[2]; Angela Bryan[2]; John Peauroi[3]; Giacomo Rossi[4]; Todd Atherly[5]; Curtis Mosher[6]; Karin Allenspach[2]; Albert E. Jergens[2]

[1]Small Animal Internal Medicine, Veterinary Clinical Sciences, College of Veterinary Medicine, Iowa State University, Ames, IA, USA; [2]College of Veterinary Medicine, Iowa State University, Ames, IA, USA; [3]VDx Veterinary Diagnostics and Preclinical Research Services, Davis, CA, USA; [4]School of Biosciences and Veterinary Medicine, University of Camerino, Camerino, Italy; [5]Department of Oral Biology, University of Florida, Gainesville, FL, USA; [6]Iowa State University, Ames, IA, USA

Previous studies have shown a relationship between mucosal bacteria and host responses in cats with inflammatory bowel disease (IBD) and lymphoma (LSA). However, these earlier studies have not investigated the specific microbial etiologic factors nor the selective recruitment of mucosal tumor-infiltrating immune (CD11b+) cells which may promote tumor progression. The objective of the present study was to investigate the relationship between pro-inflammatory microbiota (including *Fusobacterium* spp., *Helicobacter* spp., Enterobacteriaceae) and mucosal CD11b+ cells in cats with small cell LSA versus IBD. A retrospective analysis of cats diagnosed with intestinal LSA (n=14) or IBD (n=14) from 2010–2015 was performed. Following detailed diagnostic evaluation and confirmation of a histopathologic diagnosis, the mucosal microbiota of endoscopic intestinal biopsies of IBD and LSA cats was evaluated by fluorescence *in situ* hybridization (FISH) targeting the 16S rRNA genes of select pro-inflammatory microbiota. Host responses to microbial dysbiosis were studied using immunohistochemistry to identify cells of the gastrointestinal mucosa expressing intestinal NF-kB (p65) and CD11b+. Results indicated that most IBD/LSA cats were middle aged or older having chronic gastrointestinal signs of several months duration. Analysis by FISH showed that mucosal bacteria (EUB) were uncommonly observed in the duodenum of diseased cats. The numbers of mucosal *Bacteroides* spp. and *Clostridium* spp. were increased ($p<0.05$) in IBD cats as compared to cats with LSA. Conversely, total numbers of *Fusobacterium* spp. were increased ($p<0.05$) in colonic tissues of cats with LSA versus IBD (LSA: median 30 [range 0–378]; IBD:median 9 [range 0–346]). Moreover, the numbers of *Fusobacterium* spp. found within adherent mucus of colons in cats with LSA were increased ($p\leq0.05$) relative to other mucosal compartments. Mucosal cells expressing NF-kB and CD11b+ were increased ($p<0.05$) in intestines of cats with LSA versus IBD suggesting the presence of a pro-inflammatory microenvironment. In conclusion, alterations exist in the composition of the mucosal microbiota in cats with LSA and IBD. The intestine of cats with LSA is selectively enriched with bacterial species, including *Fusobacterium* spp. These microbiota may trigger recruitment of tumor-infiltrating immune (CD11b+) cells contributing to a pro-inflammatory environment that is conducive for intestinal LSA progression.

GI02

Effects of *Akkermansia muciniphila* Supplementation on Markers of Intestinal Permeability in Dogs Following Antibiotic Treatment (Abstract GI02)

Maria C. Jugan[2]; Adam Rudinsky[2]; Oleg Paliy[3]; Alex Gordon[3]; Joshua Daniels[2]; Prosper Boyaka[1]; Chen Gilor[4]

[1]The Ohio State University, Columbus, OH, USA; [2]Veterinary Clinical Sciences, The Ohio State University, Columbus, OH, USA; [3]Wright State University, Dayton, OH, USA; [4]University of California-Davis, Davis, CA, USA

Diarrhea and other sequelae of gastrointestinal hyperpermeability are common complications of antibiotic therapy. *Akkermansia muciniphila* is a mucin-degrading bacterium, positively associated with gastrointestinal epithelial health and decreased permeability. The objectives of this study were to measure effects of *Akkermansia* administration on markers of gastrointestinal permeability following antibiotic administration.

Eight healthy, purpose-bred dogs were randomized to receive either *Akkermansia* (109 CFU/kg; N=4) or vehicle (N=4) for 6 days following a 7-day course of metronidazole. After a 20-day washout, dogs were crossed-over to the alternate treatment. After an additional 20-day washout, the experiment was repeated with amoxicillin-clavulanate. Fecal *Akkermansia* qPCR and plasma concentrations (measured by ELISA) of cytokeratin-18, lipopolysaccharide, and glucagon-like peptides (GLP-1, GLP-2) were measured at baseline (T0), post-antibiotic (T1), and post-treatment (vehicle or *Akkermansia*) (T2). For each antibiotic, absolute or delta concentrations were compared between time-points using repeated measures ANOVA.

Akkermansia was detected in feces in 7/8 dogs following supplementation (T2) but not at T0 or T1. Delta (T2-T1) cytokeratin-18 after metronidazole was significantly lower on vehicle (-0.27 ng/ml) versus *Akkermansia* (2.4 ng/ml; $p=0.03$). Cytokeratin-18 concentrations tended to decrease from T0 to T1 on amoxicillin-clavulanate ($p=0.05$). Post-prandial GLP-1 concentrations (38.2 pM/ml) were higher than pre-prandial (15.5 pM/ml) concentrations. No adverse side-effects or other significant biomarker alterations were noted.

Akkermansia muciniphila PCR detection suggested successful gastrointestinal transit following oral supplementation in dogs, with an effect on gastrointestinal epithelium based on plasma cytokeratin-18 alterations. Further study is needed to determine impact in dogs with naturally occurring disease.

GI03

The Impact of Oral Tylosin on the Fecal Microbiota of Healthy Dogs (Abstract GI03)

Alison C. Manchester[1]; Craig Webb[1]; Jonathan A. Lidbury[2]; Jörg M. Steiner[2]; Jan S. Suchodolski[2]

[1]Department of Clinical Sciences, Colorado State University, Fort Collins, CO, USA; [2]Gastrointestinal Laboratory, College of Veterinary Medicine, Texas A&M University, College Station, TX, USA

Dysbiosis is thought to play a major role in the pathogenesis of intestinal disease. Treatment of intestinal disease may involve the use of antimicrobials, but their effects on the microbiota are not fully understood and several antibiotics have been shown to promote dysbiosis. The aim of this study was to prospectively evaluate the impact of tylosin administration on the fecal microbiota of healthy dogs.

Ten healthy pet dogs were randomly assigned to one of 2 groups in a double-blinded fashion: 6 dogs treated with oral tylosin at 20 mg/kg and 4 dogs treated with a placebo capsule, each given q12h for 7 days. All dogs were maintained on their usual diet. Fecal samples were collected on day 0. A fecal score (range: 1 to 7) was noted daily during drug administration, with additional fecal samples collected on days 7, 10, 21, and 63. Fecal samples were assessed using qPCR for 7 bacterial groups: *Faecalibacterium, Turicibacter, E. coli, Streptococcus, Blautia, C. hiranonis,* and *Fusobacterium*. Results were analyzed for each bacterial group and also compiled mathematically into a single numeric value, the Dysbiosis Index (DI). Parameters were compared using a Friedman 's test, followed by Dunn 's post-test. A p-value <0.05 was considered statistically significant.

None of the dogs in either group developed diarrhea. The DI increased significantly on day 7 (median [range]: 2.9 [+1.7 to +3.8]) compared to day 0 (-6.5 [-8.3 to +1.7]; p=0.03) with tylosin treatment. The DI remained significantly increased on day 10 (p=0.03) but not on day 21 (p=0.06). The abundance of *Faecalibacterium* was significantly decreased after tylosin treatment. No significant changes in the abundance of any bacterial group or in the DI were noted in dogs receiving placebo.

Tylosin had a profound effect on the fecal microbiota, which in some dogs persisted for at least 2 months. Further studies are warranted to determine if dysbiosis secondary to antimicrobial administration increases patient risk of developing disease, intestinal or otherwise, in the future.

GI04

Prognostic Indicators in Dogs with Extrahepatic Bile Duct Obstruction Secondary to Pancreatitis (Abstract GI04)

Shannon Palermo; Dorothy Cimino Brown; Mark Rondeau

School of Veterinary Medicine, University of Pennsylvania, Philadelphia, PA, USA

Pancreatitis may cause extrahepatic bile duct obstruction secondary to local inflammation in the vicinity of the common bile duct. To date, there has been limited evidence to support the optimal treatment for this condition. The purpose of this study was to describe the clinical presentation and outcome of dogs that develop extrahepatic bile duct obstruction secondary to pancreatitis, and assess factors associated with survival.

A retrospective review of medical records from the Matthew J. Ryan Veterinary Hospital at the University of Pennsylvania from 1989–2016 was performed. Cases were included if clinical signs and ultrasonographic findings were consistent with pancreatitis and extrahepatic bile duct obstruction was suspected based on ultrasonographic common bile duct dilation of greater than 3 mm and presence of hyperbilirubinemia. Cases were excluded if another cause for bile duct obstruction was identified, or if the medical record was incomplete. Medical records were searched for physical examination findings, clinicopathologic data, diagnostic imaging findings, treatment and outcome. Referring veterinarians were contacted when necessary for information regarding survival. Median survival times were determined by the use of the Kaplan-Meier product limit method and Cox multivariable survival methods were employed to determine which factors were associated with survival time following diagnosis.

Complete medical records were available for 45 dogs. 30 dogs were managed medically and 15 dogs were managed surgically. The overall median survival time was 241 days (95% CI 25, 631). Dogs 9 years of age or older with azotemia had a 9.9 greater hazard for death (95% CI 2.5, 38.1; $p=0.001$) compared to dogs younger than 9 years old without azotemia. Dogs with a temperature greater than 102.5°F at admission had a 3.1 greater hazard for death (95% CI 1.3, 7.7; $p=0.013$) than dogs with a body temperature of less than 102.5°F at admission. Dogs without gallbladder distension on ultrasound had a 4.4 greater hazard for death (95% CI 1.3, 15.4; $p=0.018$) compared to dogs with gallbladder distension.

In dogs with extrahepatic bile duct obstruction secondary to pancreatitis, age of 9 years or older, azotemia at admission, body temperature greater than 102.5°F at admission and lack of gallbladder distension on ultrasound are associated with an increased risk of death. Medical vs. surgical management, total bilirubin at admission, change in total bilirubin during hospitalization, and magnitude of increased liver enzyme activity were not significantly associated with outcome in this group of dogs.

GI05

IL-1beta Protein Expression is Increased in Biopsies of Dogs with Severe Inflammatory Bowel Disease (Abstract GI05)

Martin Hawes[1]; Anna Riddle[1]; Joseph Kirk[1]; **Karin Allenspach**[2]

[1]Royal Veterinary College, London, North Mymms, England, UK; [2]College of Veterinary Medicine, Iowa State University, Ames, IA, USA

Canine Inflammatory Bowel Disease (IBD) is believed to be a multifactorial disease due to an abnormal immune response to intestinal microbes in genetically susceptible individuals. IL-1β, a pro-inflammatory cytokine, has previously been shown to be elevated in biopsies of dogs with food responsive disease in a small number of cases. The purpose of this study was to investigate duodenal levels of IL-1β protein in a larger number of dogs with different clinical severity of IBD. Duodenal biopsies from ten dogs each with mild IBD (Canine Chronic Enteropathy Clinical Activity Index, CCECAI 0–5), moderate IBD (CCECAI 6–8), severe IBD (CCECAI ≥9) and dogs with protein-losing enteropathy (PLE), as well as 10 healthy Beagle dogs were included in the study. Biopsy samples were weighed prior to manipulation. Samples were homogenized at 20 Hz for 2 minutes in 500 μL homogenization buffer (PBS containing 0.05% Tween-20) with protease inhibitor cocktail. Tissue homogenates were centrifuged at 12500 RCF for 2 minutes and the supernatants frozen at -80°C until the day of ELISA. Quantification of IL-1β protein expression was determined using a commercially available canine-specific IL-1β ELISA (Kingfisher Biotech Inc) following the manufacturer's assay instructions. All samples were analysed in duplicate. Median IL-1β expression (pg/ml/mg tissue) was 121.6 (range 38.3–219.4) in the severe IBD group; 19.7 (range 10.9–24.8) in the moderate severity IBD group; 13.3 (range 8.0–29.8) in the mild severity IBD group, 13.0 (range 8.0–27.9) in the PLE group, and 0.9 (range 0.3–1.4) in the healthy control dogs. There was a statistically significant difference in IL-1β expression between all IBD dogs and the healthy controls (p2=0.47, p<0.001). The results of this study support previous findings that IL-1β may be involved in the pathogenesis of canine IBD.

GI06

Histologic Findings in Endoscopically Derived Intestinal Biopsies From Clinically Healthy Cats (Abstract GI06)

Sina Marsilio[1]; Jörg M. Steiner[1]; Jonathan A. Lidbury[1]; Jan S. Suchodolski[1]; Mark Ackerman[2]

[1]Gastrointestinal Laboratory, College of Veterinary Medicine, Texas A&M University, College Station, TX, USA; [2]Department of Veterinary Pathology, College of Veterinary Medicine, Iowa State University, Ames, IA, USA

Criteria developed by the World Small Animal Veterinary Association (WSAVA) are commonly used for the assessment of feline intestinal biopsy specimens. However, the histological characteristics of intestinal biopsy specimens collected from healthy cats have not been well characterized using this scoring system. The aim of this study was to describe the histologic findings in endoscopically derived intestinal biopsy specimens from clinically healthy cats.

Twelve clinically healthy, client-owned cats underwent gastroduodenoscopy at the Veterinary Medical Teaching Hospital at Texas A&M University (Animal Use Protocol 2015-0276 CA). Tissue specimens were collected from the stomach (10 cats) and the duodenum (12 cats) and were evaluated according to WSAVA criteria by a board-certified pathologist (MRA) who was blinded to the health status of the cats.

The median age of the cats was 9.5 years (min–max: 3 to 18 years). Sample quality was reported as very good in all cases. Histologic evaluation of the stomach revealed lymphoplasmacytic infiltration in all specimens evaluated, which was graded as minimal (1 cat), minimal to mild (2 cats), mild (3 cats), mild to moderate (2 cats), or moderate (2 cats). Histologic evaluation of duodenal biopsies revealed evidence of lymphoplasmasmacytic infiltration in 11 of 12 cats, which was described as minimal to mild (3 cats), mild (1 cat), mild to moderate (4 cats), or moderate (3 cats). On assessment of duodenal specimens 1 cat was diagnosed as a having diffuse low-grade lymphoma. Architectural changes were present in 8 of 10 gastric and 12 of 12 duodenal sections. These consisted of fibrosis (8 cats) in gastric specimens and lacteal dilation (1 cat), crypt abscesses (1 cat), fibrosis (3 cats), and crypt hyperplasia (11 cats) in duodenal specimens.

In conclusion, intestinal biopsies from clinically healthy cats commonly show mucosal changes that are considered abnormal when evaluated according to WSAVA guidelines.

GI07

Effect of the Bile Acid Sequestrant Cholestyramine on Fecal Bile Acid Concentrations in Healthy Dogs (Abstract GI07)

Blake C. Guard[1]; Celeste Alexander[2]; Julia B. Honneffer[1]; Jonathan A. Lidbury[1]; Jörg M. Steiner[1]; Kelly S. Swanson[3]; Jan S. Suchodolski[1]

[1]Gastrointestinal Laboratory, College of Veterinary Medicine, Texas A&M University, College Station, TX, USA; [2]Division of Nutritional Sciences, University of Illinois, Urbana, IL, USA; [3]Department of Animal Sciences and Division of Nutritional Sciences, Department of Veterinary Clinical Medicine, Urbana, IL, USA

Cholestyramine is a bile acid sequestrant that acts in the gastrointestinal tract by binding bile acids, thus preventing their reabsorption. In human patients with inflammatory bowel disease (IBD), cholestyramine is often used to treat individuals with symptoms of diarrhea. Preliminary studies in canine patients with IBD suggest that their fecal bile acid concentrations are altered, typified by an increase in primary bile acids and a decrease in secondary bile acids compared to healthy controls. The purpose of this study was to determine the effect of cholestyramine on the fecal bile acids profile in healthy dogs.

Baseline fecal samples were collected from healthy Beagle dogs (n=12) two weeks before administration of cholestyramine. Additional fecal samples were collected after two weeks of daily cholestyramine administration and again two weeks later after a washout period. All dogs were maintained on the same maintenance diet during the study and were fed 11.4 g/day of cholestyramine powder (8 g active ingredient) suspended in 75 mL of water during the cholestyramine administration period. Fecal concentrations of primary bile acids (i.e., cholic acid and chenodeoxycholic acid) and secondary bile acids (i.e., lithocholic acid, deoxycholic acid, and ursodeoxycholic acid) were evaluated using gas chromatography coupled with mass spectrometry. Data were assessed for normality using the Shapiro-Wilk test and differences in bile acid concentrations were compared using the Friedman's test. The Dunn's post-test was used where appropriate. Statistical significance was set at $p<0.05$.

Total bile acids concentrations were significantly increased after two weeks of cholestyramine administration (median [min–max]: 14.2 μg/mg [5.7–25.5 μg/mg]) when compared to baseline values (median [min–max]: 6.6 μg/mg [5.6-17.7 μg/mg]; p=0.0062). Total secondary bile acids concentrations were significantly increased after two weeks of cholestyramine administration (median [min–max]: 13.1 μg/mg [5.5–23.3 μg/mg]) when compared to baseline values (median [min–max]: 6.2 μg/mg [5.4–17.1 μg/mg]; p=0.0183). After two weeks of washout there were no parameters that remained significantly different compared to the original baseline values.

In conclusion, the fecal bile acids profile is altered in healthy dogs after cholestyramine administration. Further studies are needed to understand the potential clinical utility of cholestyramine as a therapeutic option in canine patients with gastrointestinal disease.

GI08

Fecal Fatty Acid Concentrations in Dogs with Exocrine Pancreatic Insufficiency Receiving Enzyme Supplementation (Abstract GI08)

Julia B. Honneffer[1]; Amanda B. Blake[1]; Joseph C. Parambeth[1]; Olesia C. Kennedy[2]; Blake C. Guard[1]; Jonathan A. Lidbury[1]; Jörg M. Steiner[1]; Jan S. Suchodolski[1]

[1]Gastrointestinal Laboratory, College of Veterinary Medicine, Texas A&M University, College Station, TX, USA; [2]Epi4Dogs Foundation, Inc., Farmville, VA, USA

Exocrine pancreatic insufficiency (EPI) is a disease characterized by insufficient synthesis and secretion of pancreatic enzymes by the exocrine pancreas, resulting in malassimilation of macro-nutrients. For example, insufficient pancreatic lipase prevents normal digestion of dietary fat. Consequently, EPI would be expected to be associated with excessive fat (e.g., fatty acids) remaining in the feces. Treatment of EPI includes oral supplementation of pancreatic digestive enzymes and is often effective at decreasing severity of clinical signs, but it is unclear if assimilation normalizes concomitantly. This study evaluated fecal fatty acid (FA) concentrations in dogs with EPI undergoing enzyme supplementation. The hypothesis of this study was that fecal fatty acid concentrations would be increased in dogs with EPI compared to those of healthy dogs, even when being treated with enzyme supplementation.

Fatty acid concentrations were quantified in fecal samples from 34 dogs diagnosed with EPI that were being treated with pancreatic enzyme supplements and from 82 healthy control dogs using an in-house gas chromatography/mass spectrometry (GC/MS) assay. Target analytes included palmitic acid (16:0), stearic acid (18:0), oleic acid (18:1ω9), linoleic acid (18:2ω6), α-linolenic acid (18:3ω3), gondoic acid (20:1ω9), and erucic acid (22:1ω9). A Mann-Whitney U test was used for comparison between groups. P-values were adjusted for multiple comparisons and statistical significance was set at $p<0.05$.

Fecal FAs were significantly increased in the feces of dogs with EPI (all $p<0.001$). Concentrations of FAs for dogs with EPI vs. healthy control dogs were (median [min-max] μg/mg of lyophilized feces): palmitic acid (12.0 [1.6–48.4] vs. 4.2 [1.3–13.4]), stearic acid (6.6 [1.1–43.2] vs. 2.3 [0.9–14.4]), oleic acid (13.8 [1.8–70.2] vs. 4.0 [0.3–16.9]), linoleic acid (10.3 [1.7–34.5] vs. 4.0 [0.4–29.7]), α-linolenic acid (1.0 [0.2–5.5] vs. 0.4 [0.1–3.5]), gondoic acid (0.69 [0.10–2.36] vs. 0.19 [0.03–0.61]), and erucic acid (0.07 [0.02–0.96] vs. 0.03 [0.01–0.27]). The sum of measured fecal FAs was 46.8 [9.0–174.6] vs. 15.3 [4.0–61.3].

Fecal fatty acid concentrations were increased in dogs with EPI, even while being treated with pancreatic enzyme supplementation. These data are consistent with malassimilation of fat in these patients.

GI09

Ki-67/CD3 Index in Canine Inflammatory Bowel Disease (Abstract GI09)

Sonja Karlovits[1]; Anita Manz[1]; Karin Allenspach[2]; Ingrid Walter[3]; Stefan Kummer[3]; Ursula Reichart[3]; Alexander Tichy[4]; Barbara Richter[5]; Andrea Fuchs-Baumgartinger[5]; Iwan A. Burgener[6]; Nicole Luckschander-Zeller[1]

[1]Department for Companion Animals and Horses, Clinic for Internal Medicine, University of Veterinary Medicine, Vienna, Austria; [2]College of Veterinary Medicine, Iowa State University, Ames, IA, USA; [3]VetCore Facility for Research, University of Veterinary Medicine, Vienna, Austria; [4]Department of Biomedical Sciences, University of Veterinary Medicine, Vienna, Austria; [5]Department for Pathobiology, Institute of Pathology and Forensic Veterinary Medicine, University of Veterinary Medicine, Vienna, Austria; [6]Clinic for Small Animals and Infectious Diseases, VetMedUni Vienna, Austria, Vienna

Hypersensitivity and proliferation of T-cells play a key role in the pathogenesis of IBD. In human medicine, Ki-67 is an indicator for cell growth, but there are only few studies in canine medicine.

The aim of this study was to investigate Ki-67 in relation to T-cells as a marker for canine IBD.

According to clinical signs and histopathologic results eleven privately owned dogs were grouped as IBD (IBD). Six healthy beagles served as controls (CO). All dogs were clinically assessed using the Canine Chronic Enteropathy Activity Index (CCECAI) scoring system. Duodenal mucosal biopsy samples were endoscopically retrieved for histopathological examination using the World Small Animal Veterinary Association-scheme (WSAVA). Double-marked immunofluorescence microscopy was used to investigate Ki-67 for identifying proliferating cells in combination with CD3 to detect T-cells in 4 different areas (epithelium (E) and lamina propria (LP) of villi and crypts, respectively).

In IBD-patients there was a significant increase in the clinical scoring ($M = 5.2 \pm 2.7$) versus CO (constant 0) ($p<0.001$). Furthermore, double-positive cells were significantly increased in the LP of the crypt region of IBD-dogs ($M = 1.23 \pm 1.33$ cells/mm^2) versus CO ($M = 0.120 \pm 0.143$ cells/mm^2) ($p=0.022$). In this area, a significant correlation was found between CCECAI and the Ki-67/CD3 index ($r=0.670$).

In summary, the Ki-67/CD3 index is significant upregulated in the LP, specifically in the crypt area of IBD dogs indicating a metabolically active region. In conclusion, Ki-67/CD3 could be a useful marker for canine IBD activity.

GI10

Videofluoroscopic Swallow Study Features of Obstructive Lower Esophageal Sphincter Disorders in Dogs (Abstract GI09)

Megan Grobman[1]; James Schachtel[1]; Teresa Lever[2]; Carol Reinero[1]

[1]College of Veterinary Medicine, University of Missouri, Columbia, MO, USA; [2]Department of Otolaryngology - Head and Neck Surgery, School of Medicine, University of Missouri, Columbia, MO, USA

In people, lower esophageal sphincter (LES) disorders causing functional obstruction have been identified as rare causes of megaesophagus (ME). Obstructive LES disorders have been identified in dogs but lack clear diagnostic criteria. Because these conditions may respond to targeted therapy, correct diagnosis is critical. Videofluoroscopic swallow study (VFSS) is the gold-standard for the diagnosis of dysphagia in dogs. Our objective was to characterize VFSS features of LES achalasia-like syndrome and dyssynchrony in dogs. We hypothesized that LES achalasia-like syndrome (LES-AS) and LES dyssynchrony (LES-D) could be distinguished from each other and normal dogs using standardized VFSS criteria.

Retrospective study. Sixty dogs presented to the University of Missouri Veterinary Health Center between April 2015–November 2016 for VFSS using a free-feeding standing protocol using at least 2 consistencies of iohexol or barium impregnated food (thin liquid, puree, and kibble). Dogs with a functional LES obstruction (n=11) were further evaluated using subjective and objective metrics: presence of ME; distal esophageal bolus height relative to T12; maximal LES opening; esophageal contractility (in the face of physiologic distention), propulsions (aboral bolus movement), and primary or secondary peristalsis; "bird beak"; and timing of LES opening/closure.

Using VFSS, dogs were determined to be healthy (n=6), or had LES-AS (n=6), LES-D (n=5), and "other" (primary esophageal dysmotility/GERD; n=1). All dogs with LES-AS and 2/5 dogs with LES-D had ME, with the ratio of the esophageal bolus column height:height of T12 vertebral body being significantly greater in LES-AS compared to healthy dogs (p=0.021). Dogs with LES-AS had a significantly lower ratio of distal esophageal column height:LES opening height compared to normal (p=0.010) and LES-D (p=0.043) dogs. Subjective contractility of the esophagus was normal (n=3 LES-D), hypocontractile (n=2 LES-D, n=1 LES-AS) or acontractile (n=5 LES-AS). Propulsions were completely effective (n=3 LES-D) or ineffective (n=2 LES-D). Presence of any primary and/or secondary peristaltic waves were observed in all dogs with LES-D (n=5) compared to LES-AS (n=1). A "bird-beak" sign was seen in all dogs with LES-AS. All dogs with LES-D had abnormal timing of LES opening and closing.

Functional obstruction of the LES due to LES-AS and LES-D represent discrete disease states distinguishable by VFSS.

GI11

Effect of an Extruded Vegetable Diet on Fecal Microbiota of Dogs with Food-Responsive Enteropathy (Abstract GI11)

Francesca Bresciani[1]; **Yasushi Minamoto**[2]; Jan S. Suchodolski[2]; Giorgia Galiazzo[1]; Carla Vecchiato[1]; Carlo Pinna[1]; Giacomo Biagi[1]; Marco Pietra[1]

[1]Department of Veterinary Medical Sciences, University of Bologna, Ozzano dell'Emilia, Emilia-Romagna, Italy; [2]Gastrointestinal Laboratory, College of Veterinary Medicine, Texas A&M University, College Station, TX, USA

Intestinal dysbiosis and adverse food reactions are involved in the pathogenesis of food responsive enteropathy (FRE) in dogs. Various options for an elimination diet are available, and a vegetable dry food is one alternative. Dietary interventions are thought to alter gut microbial communities in healthy individuals and the resolution of dysbiosis is expected in diseased animals concurrent with remission of clinical signs. Therefore, the aim of this study was to evaluate changes in fecal microbiota in dogs with FRE before and after an elimination dietary trial with a vegetable diet. The same vegetable diet trial was performed in healthy control dogs (HC) to evaluate changes in fecal microbiota before and after the trial, and to compare them to FRE dogs.

Dogs with FRE (n=10) and HC (n=14) were fed the vegetable diet for 60 days. Fecal samples were collected before and after the dietary trial. Fecal genomic DNA was extracted and used for Illumina sequencing of 16S rRNA genes. Sequence data were analysed using the QIIME pipeline. The dysbiosis index of the sequence data was calculated using a published mathematical model, and a score >0 was considered as dysbiotic. Statistical significance was set at $p<0.05$.

Significantly lower alpha diversity was observed in dogs with FRE-baseline compared to HC-baseline and FRE-after trial. Distinct microbial communities were observed in dogs with FRE-baseline compared to HC-baseline (ANOSIM p=0.001) and dogs with FRE-after trial (ANOSIM p=0.032). Microbial communities were still different in FRE-after trial compared to HC-baseline (ANOSIM p=0.001). The calculated dysbiosis index was higher in dogs with FRE-baseline compared to HC-baseline (p=0.022), but no significance difference was observed between FRE-after trial and HC-baseline. The fecal microbiota in HC did not show any significant differences before vs. after the vegetable dietary trial.

Results of this study suggest that in FRE dogs, treatment with the vegetable elimination diet led to partial recovery of the fecal microbiota by significantly increasing microbiota richness, which was significantly closer to healthy microbiota after treatment. In contrast, no changes were detected in fecal microbiota of healthy control dogs fed with the same vegetable diet.

GI12

Evaluation of Hydromorphone and Dexmedetomidine for Emesis Induction in Cats (Abstract GI12)

Mike Nystrom; Adesola Odunayo; Chika Okafor

University of Tennessee, Knoxville, TN, USA

Induction of emesis after ingestion of a toxin can be challenging in cats due to the variable efficacy and adverse effects associated with the use of alpha-2 agonists as emetics. Hydromorphone may be a more effective alternative with fewer adverse effects. The objective of this study was to compare the efficacy of hydromorphone and dexmedetomidine at inducing emesis in cats.

This was a prospective, blinded, randomized crossover study utilizing 12 healthy purpose-bred cats. Cats were randomly assigned to receive hydromorphone (0.1 mg/kg [0.045 mg/lb], subcutaneously) or dexmedetomidine (7 mcg/kg [3.2 mg/lb], intramuscularly). Following administration, the incidence of emesis, number of emetic events, signs of nausea, temperature, heart rate, respiratory rate and sedation score were recorded.

Emesis was successful in 9 of 12 (75%) cats when treated with hydromorphone and in 7 of 12 (58%) cats when treated with dexmedetomidine ($p=0.67$). Dexmedetomidine was more likely to cause sedation than hydromorphone ($p<0.001$). Heart rate in cats were significantly decreased at 1 and 2 hours post-hydromorphone ($p=0.003$, 0.014 respectively) and at 1, 2, 3, 5, 6 hours post-dexmedetomidine ($p=0.001$, 0.003, 0.038, 0.013, 0.001 respectively). Cats were more likely to develop an increase in body temperature with hydromorphone administration.

Results of this study indicate that hydromorphone is an effective alternative to dexmedetomidine for the induction of emesis in cats. Sedation and decrease in heart rate were observed less with hydromorphone administration. Further investigation into the most adequate dose of hydromorphone for optimizing emesis is warranted.

GI13

Identification of Jejunal Lesions in Dogs Using Capsule Endoscopy (Abstract GI13)

Jill S. Pomrantz; Jeffrey A. Solomon

Infiniti Medical, Menlo Park, CA, USA

Imaging the jejunum in dogs is problematic. Ultrasound frequently misses mucosal lesions and traditional endoscopy does not typically reach the jejunum. ALICAM is a capsule endoscopy system that images the entire bowel. The purpose of this study was to retrospectively assess the frequency of jejunal lesions in dogs given ALICAM for gastrointestinal signs and/or laboratory abnormalities. A board-certified internist interpreted each study. Eight of 131 (6.1%) dogs had complete studies with lesions identified in the jejunum that were considered significant. Of the eight dogs with positive studies, 6 were given ALICAM for a regenerative anemia, while 2 had signs of overt gastrointestinal bleeding (melena). Five of 8 dogs had a normal gastrointestinal tract on ultrasound prior to ALICAM administration. One dog had a normal gastroduodenoscopy and a second dog had a normal gastroduodenoscopy and laparoscopy prior to ALICAM. ALICAM findings included ulcerated masses in 2/8 dogs, large erosions or ulcers in 4/8 dogs, and focal fluid, blood or mucosal abnormalities in 2/8 dogs. Following ALICAM, two dogs underwent manual push enteroscopy. In one case, the ulcerative lesions seen on ALICAM were not found. In the second case, the suspected lesion was identified and resected. A third dog had an exploratory laparotomy with subsequent identification and resection of the mass seen with ALICAM. ALICAM is useful in imaging the jejunum and should be considered an important diagnostic step in the work up of gastrointestinal disease, especially in cases of unexplained anemia where traditional endoscopy or ultrasound failed to identify a cause.

GI14

Platelet Function and Endoscopic Changes after Clopidogrel, Aspirin, Prednisone, or Combination Therapy in Dogs (Abstract GI14)

Jacqueline Whittemore[1]; Allison Mooney[1]; Dianne Mawby[1]; John Thomason[2]

[1]University of Tennessee, Knoxville, TN, USA; [2]College of Veterinary Medicine, Mississippi State University, MS, USA

The purpose of this study was to determine the effects of sustained therapy with antiplatelet drugs and glucocorticoids on platelet function tests and the upper gastrointestinal tract.

Fifteen healthy dogs were randomized to 1 of 5 groups: aspirin 2 mg/kg/d (A), clopidogrel 2–3 mg/kg/d (C), prednisone 2 mg/kg/d (P) or combined prednisone with aspirin (PA) or prednisone with clopidogrel (PC) therapy PO for 28 days. Platelet count, PFA-100® analysis (collagen/epinephrine or collagen/ADP as appropriate), and endoscopy were performed at baseline, 14 days (platelet tests only), and 28 days. The presence of ≥15 hemorrhages, ≥5 punctate erosions, ≥1 invasive erosion, or ≥1 ulcer was considered clinically-significant bleeding. Dogs were categorized as antiplatelet responders if closure times were >300 sec (aspirin) or increased >30% compared to baseline (clopidogrel). Gastrointestinal bleeding and antiplatelet responder status were compared among groups by Fisher's exact test; $p<0.05$ was considered significant.

One PA dog was removed after 14 days for severe weight loss; invasive gastric erosions were identified via capsule endoscopy. Endoscopic lesions were present at day 28 in 0/3 C dogs, 1/3 A dogs (>20 hemorrhages, 4 hemorrhagic tracts), 1/3 P dogs (≥12 invasive erosions), 2/2 remaining PA dogs (≥10 invasive erosions, each), and 2/3 PC dogs (1: 8 invasive erosions, diffuse hemorrhage; 1: 1 deep ulcer). All platelet counts were normal. There were 3/3 and 3/3 C dogs, 2/3 and 1/3 A dogs, 3/3 and 1/3 PC dogs, and 0/3 and 0/2 PA dogs classified as responders based on PFA closure times on days 14 and 28, respectively. Frequency of bleeding did not differ significantly among groups. Responder status differed significantly among groups at day 14 ($p=0.03$) but not 28 ($p=0.6$)

Clinically significant gastrointestinal bleeding did not occur with sustained clopidogrel therapy but occurred in all other treatment groups. Efficacy of antiplatelet therapy decreased over time, particularly in dogs receiving concurrent steroid therapy. Further evaluation in a larger population of dogs is warranted.

GI15

Prevalence of Hypoadrenocorticism in Dogs with Chronic Gastrointestinal Signs (Abstract GI15)

Stefan Unterer[1]; Christina Schiessl[1]; Iwan A. Burgener[2]; Barbara Kohn[3]; Jörg Lechner[4]; Reto. Neiger[5]; Thomas Rieker[6]; Astrid Wehner[1]; Silke Salavati Schmitz[5]

[1]Clinic of Small Animal Medicine, LMU Munich, Munich, Bayern, Germany; [2]Clinic for Small Animals and Infectious Diseases, VetMedUni Vienna, Vienna, Wien, Austria; [3]Clinic of Small Animals, Faculty of Veterinary Medicine, Freie Universität Berlin, Berlin, Germany; [4]Head of Tierärztliche Klinik Nürnberg Hafen, Nürnberg, Bayern, Germany; [5]Small Animal Hospital (Internal Medicine), Justus Liebig University Gießen, Giessen, Hessen, Germany; [6]AniCura Kleintierspezialisten Ravensburg, Ravensburg, Baden-Wurttemberg, Germany

Dogs with hypoadrenocorticism (HA) frequently show gastrointestinal signs (GIS). However, the proportion of dogs presented for chronic GIS as their main complaint ultimately suffering from HA is unknown. The aims of this study were to determine the prevalence of HA in dogs with chronic GIS and to identify clinical and laboratory parameters that help to diagnose HA in these dogs.

A standardized work-up was performed in all dogs with GIS >3 weeks duration presented to one of the specialists of the German gastroenterology working group between November 2014 and December 2015. This included a basal serum cortisol measurement, and an ACTH stimulation test (ACTHST) if basal serum cortisol was <3 μg/dL.

151 dogs were prospectively included. Basal serum cortisol concentration was <3 μg/dL in 80/151 (53%) and <2 μg/dL in 42/151 (28%) dogs. In 6/151 dogs HA was diagnosed based on ACTHST (stimulated serum cortisol concentration <2 μg/dL), representing a prevalence of 4%. No hyperkalemia/hyponatremia was observed in any dog with HA. In 5/6 HA dogs, melena or hematochezia was noted. No historical information or laboratory parameter was able to separate HA from other disorders causing chronic GIS.

In conclusion, the prevalence of HA amongst dogs with chronic GIS is considerable. No differentiation between HA and other causes leading to chronic GIS is possible based on history, physical examination or baseline blood work. Determination of basal serum cortisol concentration should be performed as a screening test for HA, followed by an ACTHST in suspicious cases.

GI16

Oral Cobalamin Supplementation in Dogs with Exocrine Pancreatic Insufficiency (Abstract GI16)

Linda Toresson[1]; Jörg M. Steiner[2]; Jan S. Suchodolski[2]; Thomas Spillmann[3]

[1]Evidensia Specialist Animal Hospital, Helsingborg, Sweden and University of Helsinki, Finland, Helsingborg, Skane Lan, Sweden; [2]Gastrointestinal Laboratory, College of Veterinary Medicine, Texas A&M University, College Station, Texas, USA; [3]Faculty of Veterinary Medicine, Department of Equine and Small Animal Medicine, Helsinki, Finland

Exocrine pancreatic insufficiency (EPI) is associated with decreased production of intrinsic factor (IF) in dogs and cats. Since IF is required for cobalamin (CBL) absorption in the ileum, CBL deficiency is a prevalent sequel to EPI in dogs and is associated with a negative prognosis.

Current supplementation protocols for CBL in dogs call for parenteral (PE) supplementation. In humans, several studies have reported equal efficacy of PE and oral (PO) supplementation of CBL. Recently, this group has reported successful PO CBL supplementation in dogs with CE and hypocobalaminemia in a retrospective observational study and in a prospective study comparing oral and parenteral CBL supplementation. The purpose of this retrospective study was to evaluate the effect on PO CBL supplementation in dogs with EPI and subnormal or low-normal serum CBL concentrations. A computerized database search for dogs treated at Evidensia Specialist Animal Hospital, Helsingborg, Sweden during 2010–2016 was performed. Inclusion criteria were dogs diagnosed with EPI based on subnormal serum canine trypsin-like immunoreactivity concentration (cTLI; reference range 5–35 µg/L), an initial serum cobalamin concentration ≤350 ng/L (reference interval: 234–811 ng/L), and daily treatment with CBL tablets (cyanocobalamin 1 mg/tablet; dogs < 20 kg ¼ tablet/10 kg, dogs ≥ 20 kg 1 tablet). Exclusion criteria were PE CBL supplementation in parallel with PO or failure to comply with the treatment protocol.

Ten dogs aged 1.7–9.2 (median 4.0) years matched the inclusion criteria. Serum CBL for follow-up was analyzed 28–199 (median 44) days after start of supplementation using an automated chemiluminescence immunoassay (Immulite 2000, Siemens for 8/10 dogs and ADVIA Centaur, Siemens for 2/10 dogs). The most common breeds included were German Shepherds (n=3) and mixed breed dogs (n=2). Median serum cTLI concentration was 1.6 µg/L (range < 1 to 3 µg/L).

At inclusion, the median (range) serum CBL concentration was 204 ng/L (150–350 ng/L). It increased significantly to 1,113 ng/L (794–2,385 ng/L) after supplementation. This difference was statistically significant (p=0.002; Wilcoxon matched-pairs signed rank test). Due to the retrospective nature of the study, metabolic markers of CBL deficiency could not be analyzed. Despite this limitation, our results suggest that oral CBL supplementation appears effective in treating dogs with EPI and subnormal or low-normal serum cobalamin concentrations. Since, according to the manufacturer, only traces of IF should be present in the PEz, this finding suggests that dogs, as has been demonstrated in humans, may have an IF-independent pathway for CBL absorption. Whether such an alternative pathway does exist in dogs requires further studies. Additionally, further studies comparing cellular cobalamin status after PO or PE supplementation in dogs with EPI are warranted.

GI17

Development of a Canine Intestinal Organoid Model (Abstract GI17)

Andre Meneses[1]; Bart Spee[1]; Loes A. Oosterhoff[1]; Manabu Sakai[1]; Kerstin Schneeberger[1]; Louis C. Penning[1]; Iwan A. Burgener[2]

[1]Faculty of Veterinary Medicine, Utrecht University, Utrecht, Netherlands; [2]Clinic for Small Animals and Infectious Diseases, VetMedUni Vienna, Vienna, Wien, Austria

Three-dimensional organ-buds, or organoids, are a novel and organ mimicking *in vitro* model based on LGR5-positive stem cells that show realistic micro-anatomy. Recently, intestinal organoid cultures from various species including mice and humans have been described. However, canine intestinal crypt-derived organoids are not yet described. Since organoids can be infinitely expanded in culture, remain genetically stable, and can be differentiated to all cell types of the intestine, they represent an ideal *in vitro* model for intestinal diseases such as inflammatory bowel disease, virus infections, and drug screening. The aim of this study was to establish small intestinal and colon-based organoid culture systems from dogs, followed by molecular and cellular characterization.

Fresh cadaveric samples were obtained from duodenum, jejunum and colon of seven individual dogs, followed by crypt isolation. Organoids were propagated in 3D-culture using RPSO1-based expansion media. Differentiation was achieved by removing proliferation inducing components. We established a crypt isolation protocol and kept organoid lines from three areas of the intestine (duodenum, jejunum and colon) from seven dogs in culture during at least 12 passages. The gene-expression levels of stem cell markers (*LGR5, CD133, ASCL2* and *OLFM4*) were stable in expansion media. After differentiation, expression of the goblet cell marker *MUC2* and Paneth cell marker *NEUROG3* were increased, whereas expression levels of stem cell marker LGR5 was decreased.

This novel robust *in vitro* model of small intestinal and colonic organoids can be applied as a model to study canine intestinal diseases in the future.

GI18

Administration of *Enterococcus faecium* NCIMB 10415 to Normal Dogs Does Not Change Their Fecal Microbiota (Abstract GI18)

Linda Alison Matthewman[1]; Jan S. Suchodolski[2]; Dirk Werling[3]

[1]Royal Veterinary College, Hatfield, England, UK; [2]Gastrointestinal Laboratory, College of Veterinary Medicine, Texas A&M University, College Station, TX, USA; [3]Department of Pathology and Pathogen Biology, Royal Veterinary College, London, Hertfordshire, England, UK

Claims for probiotics include that they colonise the gastrointestinal tract and support, stabilise or maintain the balance of a healthy microbiota. There is evidence of the effect of administration of *Enterococcus faecium* SF68 to unhealthy dogs but no information yet on any effect, particularly on the microbiota, of healthy dogs.

Twelve healthy dogs with no history of gastrointestinal disease and normal faecal examinations, received a 2 week course of a preparation containing 2 x 10^9 CFU of *Enterococcus faecium* NCIMB 10415 (SF68) per dose, taken once daily. Normally voided faeces were collected before, during, at the end of treatment and 4 days after the end of treatment. The DNA extracted from the faecal samples, was used to detect and quantify the probiotic bacteria by qPCR, and to analyse the faecal microbiota by Illumina sequencing.

The probiotic bacteria were detected in the faeces whilst the dog received the product but were not detectable 4 days after treatment stopped. There was no effect of the probiotic on the microbiota of the faeces of the 12 normal dogs. Principle co-ordinate analysis showed the microbiota of the faeces to be more similar to the individual dog, than to the treatment and there was no change in species richness or diversity ($p>0.05$). There was a small insignificant increase in *Lactobacillus* spp. ($p>0.05$) and there was no effect on the amounts of *E. coli* or other *Enterobacteriaceae*. A cladogram showed that the only difference between all the faecal samples of these dogs taken over the trial period was a 100-fold increase in the amount of *E. faecium* on day 14 ($p<0.01$). Our data clearly show that this probiotic is a transient coloniser in the gastrointestinal tract of healthy dogs and does not appear to affect the faecal microbiota after 14 days of administration.

Cladogram of the microbiota of the faeces of 12 dogs which took Synbiotic D-C

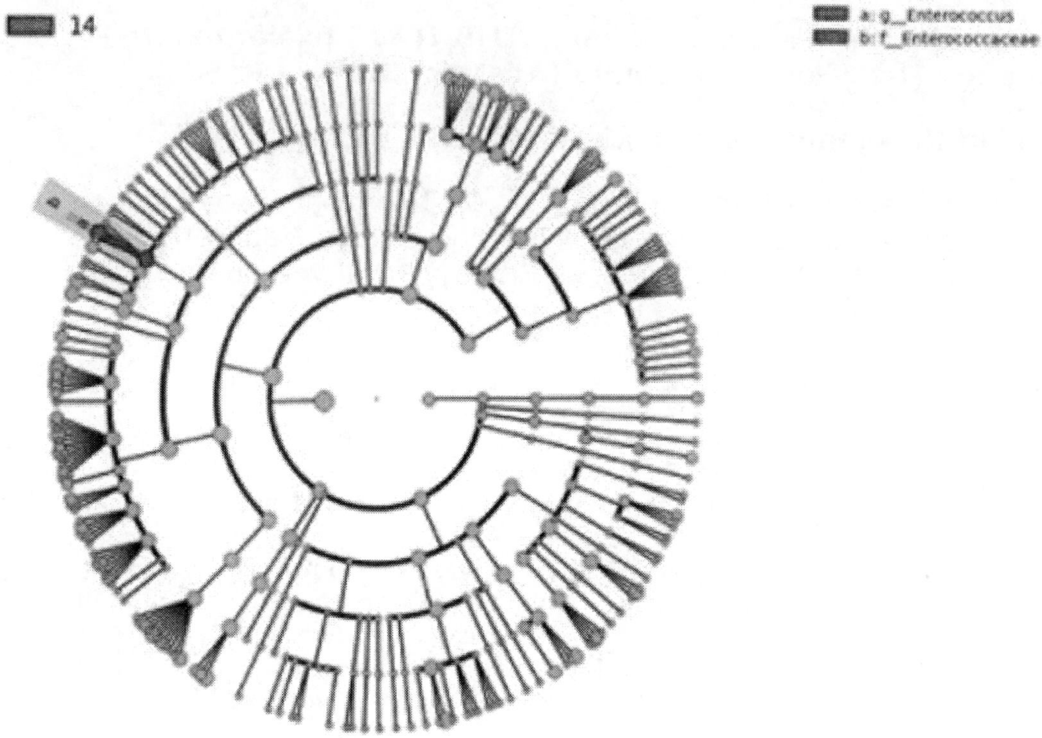

Cladogram of the microbiota of the faeces of 12 dogs which took Synbiotic D-C. Samples include those taken before treatment, on day 14 at the end of treatment and 4 days after the end of treatment. The only difference in the microbiota of these dogs over this time period is the presence of the probiotic bacteria *Enterococcus faecium* NCIMB 10415 on day 14.

GI19

Comparison of Maropitant to Metoclopramide Prior to Orogastric Administration of Polyethylene Glycol Solution in Dogs (Abstract GI19)

David C. Grant; Michael S. Leib

Virginia Maryland College of Veterinary Medicine, Blacksburg, VA, USA

We hypothesized maropitant is superior to metoclopramide in preventing vomiting and regurgitation when administered prior to orogastric tube administration of polyethylene glycol solution (PEG). Dogs (n=46) with naturally occurring diseases for which PEG was administered for bowel cleansing prior to colonoscopy were prospectively enrolled. Dogs were randomly assigned to be administered maropitant or metoclopramide subcutaneously 30 minutes prior to orogastric tube administration of 66 ml/kg of PEG by or under supervision of the blinded authors. The PEG dose was repeated 2 hours later, but the anti-emetic was not. The incidences of vomiting or regurgitation were 25.8% and 44.9%, respectively. Regurgitation occurred more often in dogs with lower body weight (15.5 kg vs. 24.3 kg, p=0.021), in dogs with more rapid administration of PEG (Dose 1: 73.7 seconds vs. 112 seconds, p=0.005; dose 2: 71.6 seconds vs. 111.8 seconds, p=0.015.), and in female dogs (13/21 vs. 7/25, p=0.021). When author 2 inserted the orogastric tube, vomiting occurred less often than when Author 1 did so (5/34 vs. 5/9, p=0.008). There was no difference in incidence of vomiting (Dose 1: 16.7% vs. 23.8%, p=0.713; Dose 2: 12.5% vs. 21.1%, p=0.681) or regurgitation (Dose 1: 25.0% vs. 52.4%, p=0.059; Dose 2: 12.5% vs. 31.6%, p=0.153) in dogs that received maropitant or metoclopramide, respectively.

Vomiting and regurgitation occur commonly following PEG administration in dogs with naturally occurring large intestinal diseases. Maropitant was not superior in preventing these. PEG administration technique may be important.

GI20

The Role of Pore-Forming Toxins in Dogs with Acute Hemorrhagic Diarrhea Syndrome (Abstract GI20)

Kathrin Busch[1]; Natalie Sindern[1]; Katrin Hartmann[1]; Christian Leutenegger[2]; John Prescott[3]; Lena Proksch[1]; Ralf Müller[1]; Stefan Unterer[1]

[1]Clinic of Small Animal Medicine, LMU Munich, Munich, Bayern, Germany; [2]IDEXX Laboratories, Inc., West Sacramento, CA, USA; [3]Department of Pathobiology, University of Guelph, Guelph, ON, Canada

Recently, novel pore-forming toxins designated as NetE and NetF were identified in a *Clostridium (C.) perfringens* type A strain isolated from a dog with acute hemorrhagic diarrhea. Pore-forming toxins, which have cytotoxic activity, could be responsible for the necrotizing mucosal lesions typically present in dogs with acute hemorrhagic diarrhea syndrome (AHDS). Thus, this study aimed to determine the prevalence of *C. perfringens* encoding *netE* and *netF* in feces of dogs with AHDS and to evaluate any association between severity of clinical signs and the presence of these toxin genes.

Fecal samples of 54 dogs with AHDS, 54 dogs with canine parvovirus (CPV) infection and 66 healthy dogs were tested by real-time PCR for *netE* and *netF* genes. Severity of clinical signs, time to recovery, mortality rate, and selected laboratory parameters were compared between toxin gene-positive and -negative dogs with AHDS.

There was a significant difference between the three groups in the prevalence of the pore-forming toxins *netE* and *netF* (dogs with AHDS: 26/54 [48.1%]; dogs with CPV-infection: 0/54; [0%]; healthy dogs: 8/66 [12.1%]) ($p<0.01$). In dogs with AHDS, no significant difference was detected in any parameter evaluated between netE- and *netF*-positive and *netE*- and *netF*-negative dogs.

The results suggest that overgrowth of *C. perfringens* type A strains encoding for *netE* and *netF* is not a general sequel of acute hemorrhagic diarrhea, since these genes were not detected in dogs with CPV infection. A primary pathogenic role of these toxins in AHDS is likely.

GI21

Serum 3-Bromotyrosine Concentrations in Dogs with Chronic Enteropathy (Abstract GI21)

Panpicha Sattasathuchana[1]; Karin Allenspach[2]; Rosana Lopes[3]; Jonathan A. Lidbury[3]; Jan S. Suchodolski[3]; **Jörg M. Steiner**[3]

[1]Gastrointestinal Laboratory, College of Veterinary Medicine, Texas A&M University, College Station, TX, USA and Department of Companion Animal Clinical Sciences, Faculty of Veterinary Medicine, Kasetsart University, Bangkok, Donmaung, Krung Thep, Thailand; [2]College of Veterinary Medicine, Iowa State University, Ames, IA, USA; [3]Gastrointestinal Laboratory, College of Veterinary Medicine, Texas A&M University, College Station, TX, USA

Peripheral eosinophilia has limitations as a marker of eosinophil-mediated disease. The measurement of 3-bromotyrosine (3-BrY) in serum has been reported to serve as a monitoring tool for eosinophil-mediated diseases in humans. In a previous study, we showed serum 3-BrY concentrations to be increased in dogs with eosinophilic gastroenteritis as well as lymphocytic-plasmacytic enteritis. However, serum 3-BrY concentrations have not previously been evaluated more specifically in dogs with food-responsive diarrhea (FRD) and steroid-responsive diarrhea (SRD). Therefore, the objectives of this study were to compare peripheral eosinophil counts in dogs with FRD, SRD, and healthy control dogs and to evaluate and compare serum 3-BrY concentrations in dogs with FRD, SRD, and healthy control dogs.

The association of peripheral eosinophilia with group (FRD [n=38], SRD [n=14], and healthy control dogs [n=46]) was determined using Fisher's exact test. Serum 3-BrY concentrations were measured using electron impact ionization gas chromatography/mass spectrometry. Serum 3-BrY concentrations in dogs with FRD, dogs with SRD, and healthy control dogs were compared using the Kruskal-Wallis test. Post-test analysis was used to determine the differences of serum 3-BrY concentrations among groups. Significance was set at $p<0.05$.

There was no significant association between peripheral eosinophilia and the group of dogs ($p=0.0692$). The median (min–max) serum 3-BrY concentrations in dogs with FRD, SRD, or healthy control dogs were 0.99 (0–8.82) μmol/L, 3.27 (0.9–26.23) μmol/L, and 0.62 (0.62–1.79) μmol/L, respectively. Serum 3-BrY concentrations in dogs with SRD were significantly higher than those in dogs with FRD ($p=0.0070$) or in healthy control dogs ($p<0.0001$). Serum 3-BrY concentrations in dogs with FRD were also significantly higher than those in healthy control dogs ($p=0.0247$).

In conclusion, serum 3-BrY concentrations were higher in dogs with SRD or FRD than in healthy control dogs. Also, dogs with SRD had higher serum 3-BrY concentrations than dogs with FRD. This suggests that eosinophil-meditated inflammation plays a role in the pathogenesis of these disorders, especially SRD. Further research is needed to determine whether the measurement of 3-BrY is useful to sub-classify dogs with chronic enteropathy as having FRD or SRD.

GI22

Eosinophil Peroxidase Monoclonal Antibody Labels Intact and Degranulated Eosinophils in the Gastrointestinal Tract (Abstract GI22)

Robert J. Washabau[1]; Idil Bastan[2]; Aaron Rendahl[3]; Davis Seelig[2]; Michael Day[4]; Edward Hall[5]; Savita P. Rao[3]; Sriram Rao[3]

[1]University of Minnesota, West Chester, PA, USA; [2]College of Veterinary Medicine, University of Minnesota, St. Paul, MN, USA, [3]University of Minnesota, St. Paul, MN, USA; [4]University of Bristol, Langford, England, UK; [5]University of Bristol, Bristol, England, UK

We have previously shown that a monoclonal antibody against eosinophil peroxidase (EPX) detects intact and degranulated eosinophils in canine jejunal mucosa (AJVR, January 2017). The aim of the present study was to compare EPX labelling and haematoxylin and eosin (H&E) staining of intact and degranulated eosinophils in endoscopically-derived samples of canine duodenal mucosa. Our objective was to determine whether EPX could improve the detection of intact and degranulated eosinophils in the duodenal mucosa of dogs with inflammatory bowel disease (IBD).

Tissue samples were obtained from the histopathology archives of the University of Bristol, School of Veterinary Sciences, and dogs were included in the study if they met WSAVA standards for IBD (i.e., clinical signs, upper G.I. endoscopic imaging and biopsy, and histopathology). Samples were divided into three groups as dogs affected by: i) lymphocytic-plasmacytic enteritis (LPE), ii) eosinophilic enteritis (EE), and iii) a control group based on WSAVA criteria. Serial sections of paraffin wax-embedded tissues were immunolabelled with EPX mAb or stained by H&E. Intact and degranulated eosinophils in consecutive microscopic fields (x400 magnification) of the upper (villus tips) and lower (between the muscularis mucosae and crypts) regions of duodenal mucosa were counted manually. Inflammation was classified as mild, moderate, or severe by one blinded investigator (MD), and eosinophil counts were carried out by two investigators blinded to the identity of the cases (IB, DS).

EPX mAb readily detected degranulated and intact eosinophils in the duodenal mucosa and there was good agreement in eosinophil quantification between the two observers in both EPXmAb and H&E stained sections. While H&E staining readily detected intact eosinophils in the duodenum, it was ineffective in detecting degranulated eosinophils. Further, across all study groups, EPX-mAbIHC (immunohistochemistry) detected three times more eosinophils compared to H&E-stained slides. In terms of disease severity, degranulated and intact eosinophil numbers were up to two-fold greater in moderate vs. mild disease.

We conclude that H&E staining not only fails to detect degranulated eosinophils effectively but grossly underestimates eosinophil counts in the duodenal mucosa, while EPX IHC detects both intact and degranulated eosinophils. Results of the present study suggest that EPX IHC may serve as a potential diagnostic method for identifying IBD, monitoring disease activity, and evaluating the success of treatment in dogs. It may further constitute a first step toward understanding the role of eosinophil degranulation in the pathogenesis of IBD in dogs.

GI23

Pilot Evaluation of the Effect of Omeprazole on the Feline Fecal Microbiome and Metabolome (Abstract GI23)

Sarah M. Schmid[1]; Jan S. Suchodolski[2]; Emily N. Gould[1]; Josh Price[1]; M. Katherine Tolbert[1]

[1]University of Tennessee, Knoxville, TN, USA; [2]Gastrointestinal Laboratory, College of Veterinary Medicine, Texas A&M University, College Station, TX, USA

Proton pump inhibitor administration in humans is associated with an increased risk of enteric, particularly *Clostridium difficile*, infections. Omeprazole also alters the canine microbiome; however, no studies have explored its effect on the feline fecal microbiome and metabolome. We hypothesized that prolonged oral omeprazole administration would alter the fecal microbiome and metabolome in cats.

A within-subjects, before and after, study was performed whereby all cats received 60 days of consecutive treatment with placebo (250 mg lactose PO q 12 h), followed by 5 mg (0.83–1.6 mg/kg PO q 12 h) omeprazole. On days 0, 30, and 60 of placebo and omeprazole therapy, free catch fecal samples were collected from each cat. The fecal microbiome was characterized using 16S rRNA gene sequencing. qPCR assays were also used to specifically target *Clostridium perfringens*, *Clostridium hiranonis*, and *Bifidobacterium* spp. The metabolome was characterized by untargeted mass spectrometry-based methods. Two-way repeated-measures ANOVA tests were utilized to determine the effects of phase of treatment, time, and the interaction of phase and time.

No global changes in the microbiome and metabolome were observed based on multivariant analysis (ANOSIM, $p>0.05$). However, some bacterial genera were altered following omeprazole administration including an increased relative abundance of *Streptococcus*, *Lactobacillus*, *Clostridium*, and *Faecalibacterium* spp. and decreased *Bifidobacterium* spp. in all cats. The results of this pilot study suggest that omeprazole might have little impact on the feline fecal microbiome and metabolome when administered for 60 days or less. Larger studies are needed to confirm these results.

GI24

Serum Biomarkers of Oxidative Stress as a Marker of Inflammatory Bowel Disease in Dogs (Abstract GI24)

Camila Rubio[1]; Silvia Martinez-Subiela[1]; Jh Hernandez-Ruiz[1]; Asta Tvarijonaviciute[2]; Jose Ceron[1]; **Karin Allenspach**[3]

[1]University of Murcia, Espinardo, Murcia, Spain; [2]University Autonoma of Barcelona, Barcelona, Catalonia, Spain; [3]College of Veterinary Medicine, Iowa State University, Ames, IA, USA

There is evidence that oxidative stress plays an important role in the pathogenesis of IBD in human patients particularly in the initiation and perpetuation of inflammation and in subsequent tissue damage. Oxidative stress occurs when there is a marked imbalance between the production of Reactive Oxygen Species (ROS) and their removal by antioxidants. Recent publications suggest that oxidative stress could also represent a significant factor in the pathogenesis of IBD in dogs. The objective of this study was to compare a panel of various serum biomarkers evaluating both the antioxidant response and the oxidative damage in dogs with inflammatory bowel disease (IBD) and healthy control dogs. Eighteen dogs with IBD and 20 healthy dogs were enrolled in the study. Trolox equivalent antioxidant capacity (TEAC), cupric reducing antioxidant capacity (CUPRAC), ferric reducing ability of the plasma (FRAP), total thiol concentrations, and paraoxonase 1 (PON1) activity were evaluated in serum to determine antioxidant response. To evaluate oxidative status, ferrous oxidation-xylenol orange (FOX), thiobarbituric acid reactive substances (TBARS) and reactive oxygen species production (ROS) concentrations in serum were determined. Mean concentrations of all antioxidant biomarkers analyzed, with exception of FRAP, were significantly lower ($p<0.0001$) in serum of dogs with IBD than healthy dogs. The oxidant markers studied were significantly higher ($p<0.0001$) in serum of dogs with IBD than in healthy dogs. These findings support the hypothesis that oxidative stress plays an important role in the pathogenesis of IBD in dogs.

GI25

Analytical Validation of an Enzymatic Assay for Measurement of Fecal D- and L-Lactate in Dogs (Abstract GI25)

Amanda B. Blake; Yasushi Minamoto; Jonathan A. Lidbury; Jörg M. Steiner; Jan S. Suchodolski

Gastrointestinal Laboratory, College of Veterinary Medicine, Texas A&M University, College Station, TX, USA

L-lactate is generated from pyruvate via lactate dehydrogenase during normal cell metabolism. Lactobacillales (lactic acid bacteria) in the intestinal tract can synthesize both D- and L-lactate. Currently, serum lactate concentration is used as a marker of tissue hypoperfusion and hypoxia. Cases of D-lactic acidemia have been reported in dogs and humans with intestinal disease. Fecal lactate has also been shown to be increased in humans with short bowel syndrome and calves with diarrhea. Fecal lactate concentrations might serve as a marker of the microbial activity in the intestine. Therefore, the aim of this study was to analytically validate an enzymatic method for the measurement of both D- and L-lactate in canine feces.

A commercial kit was adapted for use in feces (D-/L-lactic acid kit, R-Biopharm). Surplus homogenized canine fecal samples from seven dogs were used for method validation. Validation variables included lower limit of detection, dilutional parallelism, spiking recovery, and inter- and intra-assay variability.

The lower limit of detection for D- and L-lactate concentrations were 0.007 and 0.003 mM, respectively. Observed-to-expected ratios for dilutional parallelism ranged from 89% to 117% for D-lactate and 87% to 112% for L-lactate. Recovery for spiking sample extracts ranged from 95.9% to 103.4% for D-lactate, 96.4% to 118.5% for L-lactate, and 97.8% to 112.9% for total lactate. Average intra-assay coefficients of variation (%CVs) for D-, L-, and total lactate were 5.0%, 5.1%, and 4.1%, respectively. Average inter-assay %CVs for D-, L-, and total lactate were 23.5%, 20.1%, and 19.4%, respectively.

This assay was shown to be linear, precise, accurate, and reproducible. Future studies are underway to evaluate the utility of fecal lactate measurement in dogs with gastrointestinal diseases.

GI26

Dogs with Exocrine Pancreatic Insufficiency Have Dysbiosis and Abnormal Fecal Lactate and Bile Acid Concentrations (Abstract GI26)

Amanda B. Blake[1]; Blake C. Guard[1]; Julia B. Honneffer[1]; Fayth G. Kumro[1]; Olesia C. Kennedy[2]; Jonathan A. Lidbury[1]; Jörg M. Steiner[1]; Jan S. Suchodolski[1]

[1]Gastrointestinal Laboratory, College of Veterinary Medicine, Texas A&M University, College Station, TX, USA; [2]Epi4Dogs Foundation, Inc., Farmville, VA, USA

It has been reported that dogs with exocrine pancreatic insufficiency (EPI) commonly have intestinal dysbiosis. However, the effects of EPI on microbial metabolism are poorly understood. The aim of this study was to compare fecal dysbiosis as well as fecal lactate and bile acid concentrations between dogs with EPI and healthy control dogs.

Fecal samples were collected from eleven dogs with EPI that had not received antibiotics for at least 3 weeks and had been on enzyme supplementation for 0.5–10 years (median 5 years). Fecal samples from healthy dogs (n=18), collected for three consecutive days and pooled, served as control samples. DNA was extracted and analysed by qPCR for selected bacterial groups and data expressed as Dysbiosis Index (as previously reported). Fecal lactate was measured by enzymatic methods (D-/L-lactic acid kit, R-Biopharm) and bile acids were quantified with gas chromatography/mass spectrometry from lyophilized feces. The Mann-Whitney U test was used to compare the Dysbiosis Index and fecal lactate and bile acid concentrations between dogs with EPI and healthy control dogs. Correlations were assessed using Spearman's correlation coefficient and significance was set at $p<0.05$.

Dogs with EPI had a higher Dysbiosis Index (median [min–max]: +3.08 [-7.29 to +7.62]) than healthy control dogs (-3.81 [-7.57 to +3.32]; p=0.0232). Total fecal lactate concentrations were increased in dogs with EPI (3.44 mM [0.71–158.30 mM]) compared to healthy control dogs (1.14 mM [0.54–6.64 mM]; p=0.0037). The proportion of secondary bile acid was lower in dogs with EPI (70% [6–96%]) compared to healthy control dogs (93% [12–97%]; p=0.0431). There was no correlation between any measurements and duration of enzyme therapy.

In conclusion, this study identified differences in the fecal microbiota as well as fecal lactate and bile acid concentrations between dogs with EPI and healthy control dogs.

GI27

Analytical Validation of a Radioimmunoassay for the Measurement of Soluble CD25 Concentrations in Canine Serum (Abstract GI27)

Agostino Buono; Jonathan A. Lidbury; Jan S. Suchodolski; Jörg M. Steiner

Gastrointestinal Laboratory, College of Veterinary Medicine, Texas A&M University, College Station, TX, USA

CD25, also known as interleukin-2 receptor α (IL-2Rα), is a 55 kDa transmembrane cytokine receptor, primarily expressed by activated T cells, regulatory T cells (Tregs), and some types of cancer cells. In humans, increased expression of membrane-bound CD25 has been reported in patients with autoimmune or inflammatory diseases, during transplant rejection, and some types of neoplasia. A soluble form of CD25 (sCD25) is generated by proteolytic cleavage and has a molecular mass of 45 kDa. Elevated serum concentrations of sCD25 have been reported in human patients with the aforementioned conditions. A positive correlation has been reported between the membrane expression of CD25 and the serum concentration of sCD25. *In vitro*, activated T cells and Tregs have been shown to shed sCD25, therefore high concentrations of serum sCD25 are thought to be reflective of a sustained immune activation. Moreover, sCD25 has been used as a biomarker to help characterize disease progression, prognosis, and treatment in humans. Thus, the aim of this study was to establish and analytically validate a radioimmunoassay for the measurement of sCD25 in canine serum.

Canine specific CD25 protein and antibodies were commercially available and used to establish this assay. A competitive, liquid-phase ^{125}I radioimmunoassay (RIA) was set-up and analytically validated by determination of lower limit of detection (LLOD), dilutional parallelism, spiking recovery, and intra- and inter-assay variability. Surplus canine serum samples were pooled and used for validation. In addition, due to the inability to find canine serum samples with naturally-occurring high concentrations of sCD25, three samples were spiked with different concentrations of pure canine CD25 and treated as unknown samples to test the assay across the entire working range. Recovery was evaluated by spiking three samples with five different concentrations of CD25 (5, 10, 25, 35, and 40 ng/mL).

LLOD was calculated to be 0.49 ng/mL. Mean (± SD) observed-to-expected ratio (O/E) for serial dilutions was 95.8 ± 14.4% for 6 samples, and mean (± SD) O/E for spiking recovery was 93.2 ± 4.2% (mean ± SD) for 3 samples. Coefficients of variation for intra-assay variability for 8 serum samples were ≤12.5% (mean ± SD: 7.5 ± 4.2%), whereas inter-assay coefficients of variation for 6 samples were ≤15.7% (mean ± SD: 11 ± 4.4%), respectively.

In conclusion, the RIA described here is linear, accurate, precise, and reproducible for the measurement of soluble CD25 in canine serum. Further studies are warranted to evaluate the clinical utility of measuring serum sCD25 concentrations for evaluation of dogs with immune-mediated and inflammatory conditions.

GI28

Fecal Microbial Transplantation Decreases the Dysbiosis Index in Dogs Presenting with Chronic Diarrhea (Abstract GI28)

Jennifer Chaitman[1]; Blake C. Guard[2]; Fatima Sarwar[2]; Jonathan A. Lidbury[2]; Jörg M. Steiner[2]; Jan S. Suchodolski[2]

[1]Veterinary Internal Medicine and Allergy Specialists, New York, NY, USA; [2]Gastrointestinal Laboratory, College of Veterinary Medicine, Texas A&M University, College Station, TX, USA

The intestinal microbiome is important for maintaining gastrointestinal health and alterations may contribute to gastrointestinal disease. It is currently unclear whether fecal microbial transplantation (FMT) may present an alternative approach to standard therapy for gastrointestinal disease. The purpose of this study was to evaluate the fecal microbiota before and after FMT in patients that presented for chronic diarrhea.

Fresh or frozen stool from an infection free donor dog on a hypoallergenic diet was administered as an FMT using 5 g/kg of stool diluted with 60 mL saline administered per rectum via a red rubber catheter on an out-patient basis to unsedated patients. Stool samples were collected from 7 patients with gastrointestinal disease immediately before and one week after administration of FMT. Microbial dysbiosis was determined using a previously described Dysbiosis Index which targets several key bacteria typically associated with maintaining health including: Universal 16S rRNA, *Faecalibacterium, Turicibacter, Streptococcus, E. coli, Blautia, Fusobacterium,* and *Clostridium hiranonis*. Data was analyzed by a Wilcoxon signed rank test for matched pairs. Significance was set at $p<0.05$.

The Dysbiosis Index significantly decreased in dogs post FMT (median [min–max]: -5.1 [-7.1 to -0.8]) compared to dogs before FMT (median [min–max]: -1.3 [-4.8 to +8.1]; $p=0.0156$).

This study suggests that FMT alters the abundance of some commonly targeted bacterial groups typically associated with gastrointestinal disease. Further studies are warranted to investigate how FMT alters fecal microbiota in specific enteropathies and its long-term effect on the microbiome.

GI29

Fecal Short Chain Fatty Acid Collection, Storage and Variability in Healthy Dogs (Abstract GI29)

Anne-Claire Duchaussoy[1]; Kevin Le Boedec[2]; Caroline Mansfield[3]

[1]UVET Hospital, Faculty of Veterinary and Agricultural Sciences, University of Melbourne, Moorabin, VIC, Australia; [2]Centre Hospitalier Veterinaire FREGIS, Arcueil, Ile-de-France, France; [3]UVET Hospital, Faculty of Veterinary and Agricultural Sciences, University of Melbourne, Werribee, VIC, Australia

Genetic, environmental, defective immune system and dysbiosis have been identified as main factors involved in the pathophysiology of chronic enteropathies. Short Chain Fatty Acids (SCFA) are volatile compounds that are directly correlated to the quantitative and qualitative bacterial population of the gastrointestinal tract. An increasing number of publications assessing gastrointestinal health and inflammation include SCFA measurements in fecal matter. However, there is scant information about the repeatability and reliability of SCFA analysis in dogs and cats. As SCFA are highly volatile and unstable, and our preliminary results in feces from both dogs and cats showed high variability for the same animal and between healthy animals, we investigated repeatability and variability of SCFA measurement.

We started with a prospective study of 4 healthy dogs (normal body condition). Fecal samples were collected upon voiding and divided into aliquots immediately after collection. The aliquots (in duplicate) were stored in ambient temperature or +4C before freezing at -20°C versus -80°C. We also varied the duration in time between collection and freezing. In all samples, SCFA were measured using a dilution gas chromatography-mass spectrometry (GC-MS) assay as previously described. All samples were thawed by batch, and kept on ice until the internal standard solution was added. We focused on the three most prevalent short chain fatty acids identified in dogs, namely propionic acid, butyric acid and valeric acid. Results showed a linear increase in the amount of SCFA over time in between collection and the freezing process when left at room temperature. This was not observed when sampled were stored at +4°C. There was no significant difference in between samples stored at -20°C and -80°C results.

The next component of the study obtained fecal samples from 17 healthy dogs monthly. Samples were frozen immediately after collection at -80°C, and SCFA then measured using GC-MS assay. Results showed wide variability of results for normal dogs, and also a noticeable degree of individual variability overtime. Potential causes of this variability include variation within a stool sample, as well as the possible presence of undigested debris in the fecal materials.

In conclusion, we showed a high degree of variability within individual dogs as well as with storage methods. This data improves our understanding of SCFA analysis, and the potential interpretation of SCFA in studies of chronic enteropathies in dogs.

GI30

Immunohistochemical Distribution of TGR5 Bile Acid Receptors in the Gastrointestinal Tract of Dogs (Abstract GI30)

Paula R. Giaretta; Raquel R. Rech; Anna K. Blick; Jörg M. Steiner; Jonathan A. Lidbury; Blake C. Guard; Jan S. Suchodolski

Gastrointestinal Laboratory, College of Veterinary Medicine, Texas A&M University, College Station, TX, USA

Beyond their role in digestion, bile acids also act as systemic signaling modulators of several pathways involving glucose and lipid metabolism and play an anti-inflammatory role through the activation of nuclear and G-protein coupled receptors (i.e., GPBAR-1 or TGR5). TGR5 expression has been demonstrated in many gastrointestinal tissues of humans and laboratory animals. The aim of this study was to evaluate the immunohistochemical distribution of TGR5 receptors in the gastrointestinal tract of dogs.

Samples were collected from six adult dogs (3 females and 3 males; median: 4 years, range: 1 to 13 years) that were humanely euthanized due to causes not related to this study. These dogs did not have any clinical signs or histopathological lesions suggesting gastrointestinal disease. Sections of duodenum, jejunum, ileum, colon, cecum, rectum, liver, gallbladder, and pancreas were fixed in 10% formalin, followed by routine histologic processing. For immunohistochemistry, heat-induced antigen retrieval was performed. Tissues were incubated with a polyclonal TGR5 antibody (directed against amino acids 264 to 330 of human TGR5) that was diluted 1:100 or rabbit IgG as a negative control for 1 hour at room temperature.

In the intestinal mucosa, TGR5 labeling was identified as a granular staining pattern in the membrane of goblet cells, enterocytes, and colonocytes. Enteroendocrine cells in the crypts of the small and large intestine showed granular labeling in the cytoplasm and membrane, as did resident macrophages in the lamina propria and Peyer's patches. Granular and homogeneous labeling was present in the cytoplasm and membrane of ganglion cells in the submucosal and muscularis plexus of the small and large intestine. In the muscularis, both layers of smooth muscle expressed TGR5 in the membrane. In the liver, the membrane of the biliary epithelium and the canalicular surface of the hepatocellular membrane were positive, as were the cytoplasm and membrane of Kupffer cells. The epithelium of the gallbladder showed granular labeling in the membrane and cytoplasm. In the pancreas, membranous labeling was evident in the epithelial cells of the ducts, and a membranous/cytoplasmic pattern was identified in approximately 40% of islet cells. The cytoplasm and membrane of endothelial cells were positive in all the examined tissues.

Our findings show that, based on antigenic expression, TGR5 is ubiquitously distributed in the gastrointestinal tract of dogs, expressed predominantly in the membrane of epithelial cells and both in the membrane and cytoplasm of ganglia, histiocytes, and enteroendocrine cells.

GI31

Immunolocalization of TGR5 Bile Acid Receptors in the Colon of Dogs with Chronic Enteropathy (Abstract GI31)

Paula R. Giaretta; Raquel R. Rech; Anna K. Blick; Jörg M. Steiner; Jonathan A. Lidbury; Blake C. Guard; Yasushi Minamoto; Jan S. Suchodolski

Gastrointestinal Laboratory, College of Veterinary Medicine, Texas A&M University, College Station, TX, USA

Canine chronic enteropathy (CE) is suspected to have a multifactorial etiology, with an interplay of genetics and environmental factors and an altered gut microbiota and host immune system. Dogs with CE have alterations in the serum and fecal metabolome, including changes in fecal bile acids. Bile acids are important signaling molecules that act through the interaction with their receptors: G protein-coupled bile acid receptor 1 (TGR5) and farnesoid X receptor (FXR). TGR5 agonists have been shown to suppress TNF-α production in macrophages from the lamina propria of human patients with Crohn's disease. The aim of this study was to compare the antigenic expression of TGR5 receptors in colonic samples as well as the fecal concentration of bile acids between dogs with CE and control dogs.

The CE group was comprised of seven dogs (3 females and 4 males, median: 7 years, range: 6 months to 10 years) with a history of gastrointestinal signs for more than three weeks, mild to moderate inflammatory infiltration of the colonic mucosa, and no other identified cause for gastrointestinal (GI) inflammation. The control group consisted of six adult dogs (3 females and 3 males; median: 4 years, range: 1 to 13 years) that were humanely euthanized due to causes not related to this study. These dogs did not have any GI signs or histopathologic lesions. None of the dogs in this study received antibiotic or corticosteroid therapy within 3 weeks of colonic and fecal sampling. Formalin-fixed and paraffin-embedded sections of colon were used for immunohistochemistry. Heat-induced antigen retrieval was performed and after blocking, tissues were incubated with a polyclonal TGR5 antibody (directed against amino acids 264–330 of human TGR5) that was diluted 1:100 or rabbit IgG as a negative control for 1 hour at room temperature. Concentrations of fecal bile acids were measured by gas chromatography/mass spectrometry. The Mann-Whitney U-test was used to test for statistical significance ($p < 0.05$).

In both groups, immunolabeling was identified in the membrane of colonocytes, goblet cells, and smooth muscle cells of the muscularis; and in the membrane and cytoplasm of enteroendocrine cells in the crypts, resident macrophages, ganglion cells, and endothelial cells. The infiltrating inflammatory cells in CE dogs were mainly composed of lymphocytes and plasma cells, neither of which expressed TGR5. Immunohistochemical expression of TGR5 in the colon did not differ between control and CE dogs. However, dogs with CE had an increased proportion of primary bile acids in their feces ($p=0.012$).

Although dogs with CE showed altered proportions of fecal bile acids, differences in the expression of TGR5 in colonic samples could not be detected by immunohistochemistry. Further investigations using other techniques, including gene arrays, are needed and might contribute to the understanding of bile acid dysmetabolism in dogs with CE.

GI32

Enteropathogenic *Escherichia coli* is Associated with Diarrheal Mortality and Intestinal Inflammation in Foster-Age Kittens (Abstract GI32)

Victoria Watson[1]; James Flowers[1]; Sandra Strong[2]; Chitrita DebRoy[3]; Jody Gookin[1]

[1]College of Veterinary Medicine, North Carolina State University, Raleigh, NC, USA; [2]Wake County Animal Center, Raleigh, NC, USA; [3]The *E. coli* Reference Center, Pennsylvania State University, University Park, PA, USA

It is estimated that ~15% of foster kittens die or are euthanized due to illness before 8 weeks of age. Most reportedly have clinical signs or post-mortem evidence of gastrointestinal disease. In children, enteropathogenic *E. coli* (EPEC) is a common cause of diarrhea, is associated with higher hazard of death, and has proven difficult to detect or experimentally model in animals. We previously reported EPEC in a series of kittens that died with diarrhea. The purpose of this study was to determine the prevalence of EPEC infection in kittens and its association with diarrhea, diarrhea-associated mortality, specific intestinal tract pathology, and possible risk factors.

Live kittens (n=61) with and without diarrhea and deceased kittens (n=54) that were euthanized in apparent good health or were euthanized/died due to severe diarrhea were obtained from 2 different shelter facilities. All kittens were ≤12 weeks of age and ≤1 kg body weight. Fecal culture for EPEC and extraction of fecal DNA for qPCR amplification of EPEC enterocyte-attaching-effacing gene (*eae*) was performed. Deceased kittens underwent autopsy and sections of the stomach, small intestine, and colon were obtained for histopathology, scoring of lesions, and fluorescence *in situ* hybridization (FISH) to detect enteroadherent *E. coli*. Medical records were examined for variables of potential association with EPEC infection.

Based on fecal culture the prevalence of EPEC infection was 18% in both live and deceased kittens. Based on qPCR amplification of *eae* from feces prevalence was 37% in live kittens and 48% in deceased kittens. EPEC was observed in kittens with and without diarrhea. However, kittens that died or were euthanized due to severe diarrhea harbored a significantly greater quantity of live EPEC and *eae* compared to kittens without diarrhea. Kittens diagnosed with EPEC had significantly greater severity of inflammatory infiltrates in the small intestine and colon and increased epithelial injury in the small intestine. Enteroadherent *E. coli* was observed by means of FISH in 2 kittens. When live and deceased kitten populations were examined on the basis of EPEC diagnosis, significant associations were observed between EPEC infection and medical history of subcutaneous fluid administration.

These findings identify EPEC as a prevalent infection in kittens and implicate EPEC as a significant primary or contributing cause of intestinal inflammation, dehydration, and diarrhea-associated mortality in kittens. A high prevalence of EPEC in healthy kittens and children complicates interpretation of EPEC infection in patients with diarrhea. Future efforts to identify mechanisms responsible for susceptibility to EPEC-induced diarrhea will lead to advanced understanding of EPEC pathogenesis in kittens and children.

GI33

Development and Analytical Validation of an Assay for the Quantification of Canine Fecal Bile Acids (Abstract GI33)

Blake C. Guard; Michelle M. Jonika; Julia B. Honneffer; Jonathan A. Lidbury; Jörg M. Steiner; Jan S. Suchodolski

Gastrointestinal Laboratory, College of Veterinary Medicine, Texas A&M University, College Station, TX, USA

The gut microbiota is important in maintaining intestinal health. Bile acids are increasingly appreciated to play a role in regulation of gut microbial composition and intestinal health. Bile acids are synthesized from cholesterol, conjugated in the liver, and once secreted into the gastrointestinal tract (GIT), undergo modification by certain members of the intestinal microbiota. Numerous bile acid receptors (e.g., farnesoid X receptor and G protein-coupled membrane receptor) have been identified along the GIT and are responsible for regulating metabolism and maintaining an anti-inflammatory environment in the gut. The aim of this study was to develop and analytically validate a gas chromatography/mass spectrometry (GC/MS) assay for the identification and quantification of bile acids in canine feces.

Fecal bile acids (cholic acid [CA], chenodeoxycholic acid [CDCA], lithocholic acid [LCA], deoxycholic acid [DCA], and ursodeoxycholic acid [UDCA]) were measured in their unconjugated form after undergoing butyl esterification for chromatographic separation. A capillary DB-1ms Ultra Inert column was used with a 20:1 split sample injection ratio. Validation parameters included the lower and upper limits of quantification (LLOQ and ULOQ, respectively). Additionally, precision of the assay was calculated by assaying 6 aliquots taken from a single fecal sample from 4 dogs on the same run/day followed by calculating intra-assay coefficients of variation (CV%). Reproducibility of the assay was determined by analyzing 6 aliquots taken from a single fecal sample from 4 dogs on 6 consecutive days followed by calculating inter-assay variation (CV%).

The LLOQ and ULOQ in μg/mL were as follows for each compound: cholic acid (3.9 and 1000), chenodeoxycholic acid (6.25 and 200), lithocholic acid (1.9 and 500), deoxycholic acid (31.3 and 1000), and ursodeoxycholic acid (0.78 and 50). For intra-assay variability, the average CV%s were: 6.0, 5.6, 7.1, 7.3, and 8.8% for CA, CDCA, LCA, DCA, and UDCA, respectively. For inter-assay variability, the average CV%s were: 8.3, 8.0, 4.8, 8.6, and 13.2% for CA, CDCA, LCA, DCA, and UDCA, respectively.

In conclusion, the present assay was found to be both reproducible and precise for the quantification of select bile acids in canine feces.

GI34

Longitudinal Characterization of the Fecal Metabolome in Dogs with Inflammatory Bowel Disease (Abstract GI34)

Blake C. Guard[1]; Michelle M. Jonika[1]; Julia B. Honneffer[1]; Jonathan A. Lidbury[1]; Jörg M. Steiner[1]; Albert E. Jergens[2]; Jan S. Suchodolski[1]

[1]Gastrointestinal Laboratory, College of Veterinary Medicine, Texas A&M University, College Station, TX, USA; [2]College of Veterinary Medicine, Iowa State University, Ames, IA, USA

Canine inflammatory bowel disease (IBD) is a multifactorial disease, the pathogenesis of which includes alterations in gut microbiota and improper activation of the immune system. Recent studies have used untargeted metabolomics of serum and feces to investigate patients with chronic gastrointestinal (GI) disease. However, evidence is lacking about how GI inflammation and ongoing microbial dysbiosis affect metabolites long-term in patients that undergo immunosuppressive therapy. Therefore, the purpose of this study was to characterize the fecal metabolome in dogs with IBD upon initial diagnosis and after therapy, using an untargeted approach.

Nine dogs that were non-responsive to dietary or antibiotic therapy and had histologically confirmed intestinal inflammation were enrolled. Fecal samples were collected at baseline, 3 weeks, and 8 weeks. Patients received immunosuppressive therapy after initial diagnosis and clinical signs were scored according to the canine IBD activity index (CIBDAI). Fecal samples were also collected from healthy dogs (n=13) to serve as controls. Fecal samples were analyzed by an untargeted metabolomics platform using gas chromatography coupled with mass spectrometry. Data were found to be non-parametric. Therefore, comparisons were made across time points using a Friedman's test for repeated measures. A Dunn's post-test was used where appropriate. P-values were adjusted for multiple comparisons using the Benjamini and Hochberg False Discovery Rate and significance was set at $p<0.05$.

Multivariate testing consisted of principal component analysis that revealed significant differences between healthy controls and dogs with IBD at baseline, 3 weeks, and 8 weeks after therapy. The median CIBDAI score for patients 8 weeks following initial diagnosis was 0 compared to a median CIBDAI score of 7 at enrollment. Univariate analysis revealed that of the 233 metabolites measured, trans-4-hydroxyproline, sinapinic acid, glycine, 2-methylglyceric acid, stearic acid, heptadecanoic acid, myo-inositol, and dehydroabietic acid were found to be significantly different among the repeated time points sampled from dogs with IBD ($p<0.05$).

In conclusion, distinct changes in metabolic profiles were observed between healthy dogs and dogs with IBD. Despite improvement in clinical activity scores, several metabolites remained altered at 8 weeks follow up.

GI35

Longitudinal Assessment of Fecal Sterol and Fatty Acid Concentrations in Dogs with Diarrheal Diseases (Abstract GI35)

Julia B. Honneffer[1]; Blake C. Guard[1]; Stefan Unterer[2]; Francesca Bresciani[3]; Sara A. Wennogle[4]; Jonathan A. Lidbury[1]; Jörg M. Steiner[1]; Albert E. Jergens[5]; Jan S. Suchodolski[1]

[1]Gastrointestinal Laboratory, College of Veterinary Medicine, Texas A&M University, College Station, TX, USA; [2]Clinic of Small Animal Medicine, LMU Munich, Munich, Bayern, Germany; [3]Department of Veterinary Medical Sciences, University of Bologna, Ozzano dell'Emilia, Emilia-Romagna, Italy; [4]College of Veterinary Medicine, Colorado State University, Fort Collins, CO, USA; [5]Iowa State University, College of Veterinary Medicine, Ames, IA, USA

Sterols and fatty acids play an essential role as building blocks for structural components, as signaling molecules, and in energy metabolism, yet can also be toxic to cells at increased concentrations. Increased cholesterol and decreased phytosterol concentrations in the feces of dogs with chronic enteropathy have previously been reported. This study aimed to explore concentrations of fecal sterols and fatty acids in dogs exhibiting a wide variety of gastrointestinal disease phenotypes.

Baseline fecal samples were collected from dogs with acute hemorrhagic diarrhea syndrome (AHDS, n=22), food-responsive diarrhea (FRD, n=10), steroid-responsive diarrhea (SRD, n=24), and healthy control dogs (n=82). In a subset of diseased dogs, follow-up samples were collected 2–3 months after the baseline sample (AHDS, n=7; FRD, n=6; SRD, n=9). Diagnoses of FRD and SRD were based on response to therapy. All dogs with FRD were successfully managed with the same vegetarian diet. Feces were analyzed by gas chromatography/mass spectrometry (GC/MS) using an in-house assay. At each of the two timepoints, a Kruskal-Wallis test and the Benjamini-Hochberg step-up method to adjust for multiple comparisons were used to identify significantly altered compounds. Dunn's test was used to compare groups, and statistical significance was set at $p<0.05$.

At baseline, dogs with AHDS exhibited decreased fecal phytosterol (i.e., β-sitosterol, campesterol, sitostanol, fucosterol, and stigmasterol, all $p<0.001$), increased cholesterol ($p<0.001$), and increased fatty acid (i.e., stearic, erucic, gondoic, palmitic, and α-linolenic, all $p<0.001$) concentrations when compared to healthy control dogs. There were no statistically significant differences in any of the compounds between FRD and SRD dogs at baseline. There were several significant differences observed between dogs with SRD and healthy control dogs at baseline: decreased β-sitosterol, sitostanol, campesterol, and fucosterol ($p<0.001$, $p<0.001$, $p=0.002$, and $p=0.013$, respectively), and increased erucic, gondoic, and stearic acids ($p=0.008$, $p=0.012$, and $p=0.038$, respectively). However, at the 2–3 months follow-up, several sterols were altered between dogs with SRD and FRD: sitostanol, β-sitosterol, and fucosterol were decreased in SRD ($p<0.001$, $p=0.004$, and $p=0.033$, respectively), and α-linolenic, erucic, and gondoic acids were also decreased in dogs with SRD ($p<0.001$, $p=0.004$, and $p=0.004$). At the 2–3 months follow-up campesterol, stigmasterol, cholesterol, stearic acid, and palmitic acid were no longer significantly different between dogs with AHDS and healthy control dogs.

The most profound changes in fecal sterols observed were at baseline between dogs with AHDS and healthy control dogs, but many of these changes normalized rapidly. The observed differences between dogs with SRD and FRD at follow-up after 2–3 months may be largely due to the effect of diet. Future studies are required to evaluate the utility of these compounds as markers of disease phenotype or response to therapy.

GI36

Validation of an Assay for Quantification of Sterol and Fatty Acid Concentrations in Dog Feces (Abstract GI36)

Julia B. Honneffer; Blake C. Guard; Jonathan A. Lidbury; Jörg M. Steiner; Jan S. Suchodolski

Gastrointestinal Laboratory, College of Veterinary Medicine, Texas A&M University, College Station, TX, USA

Cholesterol is the most abundant sterol in mammalian tissues, playing a critical role in cell membrane function and serving as a precursor molecule for hormones, vitamins, and bile acids. Structurally similar to cholesterol, phytosterols have been associated with anti-inflammatory effects in rodent models of colitis. Previous studies have shown altered fecal sterol concentrations in dogs with chronic enteropathy. Fatty acids are an important source of energy, and along with sterols, they are also critical for intestinal epithelial cell membrane structure and function. The aim of this study was to analytically validate a gas chromatography/mass spectrometry (GC/MS) assay for the measurement of several sterols (i.e., cholesterol, coprostanol, cholestanol, β-sitosterol, stigmasterol, campesterol, sitostanol, and fucosterol) and fatty acids (i.e., palmitic, oleic, linoleic, α-linolenic, stearic, gondoic, and erucic acids) in canine fecal samples.

A fecal sample was collected from eight dogs, lyophilized, and used for assessment of intra- and inter-assay variability (%CV). Sample and standard preparation were adapted from published methods, using deuterated analogs of cholesterol, sitostanol, and stearic acid for internal standards. Validation parameters also included lower and upper limits of quantitation (LLOQ and ULOQ, respectively). Quality control procedure for each sample run consisted of blanks, continuing calibration verification standards, and a laboratory control sample of pooled feces.

Intra- and inter-assay variability (median %CV) were 4% and 6% for cholesterol, 5% and 11% for coprostanol, 4% and 9% for cholestanol, 4% and 6% for β-sitosterol, 4% and 7% for stigmasterol, 4% and 6% for campesterol, 4% and 11% for sitostanol, 6% and 10% for fucosterol, 4% and 7% for palmitic acid, 4% and 9% for oleic acid, 8% and 10% for linoleic acid, 14% and 12% for α-linolenic acid, 4% and 6% for stearic acid, 5% and 11% for gondoic acid, and 10% and 18% for erucic acid. Limits of quantitation for each compound in μg/mL (LLOQ-ULOQ) were as follows: palmitic acid (6.25–800), linoleic acid (6.25–800), α-linolenic acid (12.5–400), oleic acid (12.5–800), stearic acid (6.25–800), gondoic acid (1.56–200), erucic acid (1.56–200), cholesterol (1.56–1600), campesterol (6.25–400), stigmasterol (3.13–100), fucosterol (1.56–200), β-sitosterol (3.13–800), coprostanol (1.56–200), cholestanol (0.78–200), and sitostanol (3.13–400).

In conclusion, the GC/MS method validated here is precise, reproducible, and sensitive for the quantification of several sterols and fatty acids in canine fecal samples.

GI37

Longitudinal Assessment of Fecal Dysbiosis in Dogs with Acute and Chronic Diarrhea (Abstract GI37)

Hannah L. Klein[1]; Stefan Unterer[2]; Albert E. Jergens[3]; Jonathan A. Lidbury[1]; Jörg M. Steiner[1]; Anna Lena Ziese[4]; Jan S. Suchodolski[1]

[1]Gastrointestinal Laboratory, College of Veterinary Medicine, Texas A&M University, College Station, TX, USA; [2]Clinic of Small Animal Medicine, LMU Munich, Munich, Bayern, Germany; [3]College of Veterinary Medicine, Iowa State University, Ames, IA, USA; [4]Ludwig Maximilian University, Munich, Germany, Munich, Bayern, Germany

Previous studies have identified dysbiosis in dogs with idiopathic IBD (IBD) and dogs with acute hemorrhagic diarrhea syndrome (AHDS). The aim of this study was to evaluate whether dysbiosis patterns differ between these disease groups, and over time.

Fecal samples were analyzed from healthy dogs (n=27), dogs with IBD (n=19), and dogs with AHDS (n=23). Follow-up samples were collected over multiple time-points up to 8 weeks from 8 dogs in each group. None of the dogs was treated with antibiotics. Microbiota was analyzed by PCR for *E. coli, Streptococcus, Turicibacter, Faecalibacterium, Fusobacterium, Blautia,* and *C. perfringens.* Data was expressed individually and numerically combined as Dysbiosis Index (DI). Kruskal-Wallis and Friedman's tests were used to compare data between groups and over time with significance set at $p<0.05$.

At baseline, dogs with IBD ($p<0.001$) and AHDS ($p<0.01$) had higher DI compared to healthy dogs. While the CIBDAI score decreased in dogs with SRE significantly ($p<0.05$), the DI remained significantly increased compared to healthy dogs even at 8 weeks. In dogs with AHDS, the DI was significantly increased at day 7 ($p=0.002$), but not at day 14 ($p=0.59$). Differences in dysbiosis patterns were also observed between IBD and AHDS, primary an increase in *C. perfringens* in dogs with AHDS but not in IBD.

In conclusion, although dogs with SRE improved clinically, the DI remained increased. In contrast, dogs with AHDS showed a rapid normalization in microbiota within 14 days, suggesting differences in dysbiosis patterns due to the underlying disease.

GI38

Fecal Microbiota and Body Condition in Growing Beagle Dogs Fed Unrestrictedly for 3.5 H Daily (Abstract GI38)

Lucie Leclerc[1]; Chantal Thorin[2]; Anitha Isaiah[3]; John Flanagan[4]; Vincent Biourge[4]; Jan S. Suchodolski[3]; Patrick Nguyen[1]

[1]Nutrition & Endocrinology Unit, Oniris - Vet Nantes, Nantes, Pays de la Loire, France; [2]Animal Physiopathology and Functional Pharmacology Unit, Oniris - Vet Nantes, Nantes, Pays de la Loire, France; [3]Gastrointestinal Laboratory, College of Veterinary Medicine, Texas A&M University, College Station, TX, USA; [4]Royal Canin - R&D Discover, Aimargues, Languedoc-Roussillon, France

Several human and animal studies have reported that the gut microbiota (GM) is altered in adult obese compared to non-obese subjects, but few studies have been conducted in growing ones. The aim of this study was to identify differences in GM in a homogenous population of growing dogs and to identify when these alterations occur.

We evaluated the composition of the fecal microbiota in 24 growing female Beagle dogs 7 times from 5 to 21 months of age by Illumina sequencing and quantitative PCR of the 16S rRNA gene. The population of dogs was homogenous in terms of age, age at neutering (8 months), environmental and feeding conditions. They were fed the same diet *ad libitum* for 3.5 hours daily. Up to 10.5 months of age, they were fed a puppy growth diet (fat=20%, total dietary fiber=8%, crude protein=29%, on a dry matter basis [DM]) and subsequently an adult maintenance diet (fat=11%, total dietary fiber=17%, crude protein=28% on as DM basis). The dogs were allotted a posteriori into 3 groups according to fat-free mass, fat mass percentage and pelvic circumference at 24 months: optimal body weight (IW, n=9), moderately overweight (OW1, n=6) and overweight (OW2, n=9). The data were analyzed by a linear mixed-effect model with the animal as a random term and corrected for multiple comparisons.

The most abundant phyla were *Firmicutes, Bacteroidetes, Fusobacteria* and *Proteobacteria*, accounting for 97 to 99% of sequences. During the pre-neutering period, *Proteobacteria* were significantly less abundant ($p<0.05$) in the OW2 group than in the IW group. However, throughout the study, the relative abundance of *Proteobacteria* increased significantly more in the OW1 and the OW2 groups than in the IW group ($p<0.05$), without any significant difference in abundance at 21 months between groups.

These results suggest that the early low and then increase abundance of *Proteobacteria* could be related to overweight development. However, *Proteobacteria* have highly diverse metabolisms and further analyses at genus level and/or metagenomics analyses may explain the underlying link between the microbiota modulation and excessive weight gain.

GI39

Evaluation of the Fecal Microbiota in Cats with Chronic Enteropathy Before and After Treatment (Abstract GI39)

Sina Marsilio[1]; Betty Chow[2]; Steve L. Hill[2]; Jonathan A. Lidbury[1]; Jörg M. Steiner[1]; Jan S. Suchodolski[1]

[1]Gastrointestinal Laboratory, College of Veterinary Medicine, Texas A&M University, College Station, TX, USA; [2]Veterinary Specialty Hospital, San Diego, CA, USA

The gastrointestinal (GI) microbiota is considered to play a crucial role in the onset, progression, response to treatment, and outcome in many disease processes including chronic enteropathy (CE) in cats. However, only limited information on the fecal microbial composition in cats with CE is available. The aim of this study was to characterize the fecal microbiota in cats with histologically confirmed CE before and after treatment.

Fecal samples were collected from clinically healthy cats (n=52) and from cats with CE (n=14) before and 4 weeks after medical treatment. Chronic enteropathy was defined as the display of signs of gastrointestinal disease for ≥3 weeks. Cats that received antibiotics or steroids within 2 weeks prior endoscopy and fecal sampling were excluded from the study. All cats with CE underwent standard medical work-up before collecting intestinal biopsies endoscopically. Fecal DNA was evaluated using quantitative PCR (qPCR) analysis for: total bacteria, *Fusobacteria*, *Streptococcus* spp., *Blautia* spp., *E. coli*, *Faecalibacterium* spp., *Turicibacter* spp., and *Clostridium hiranonis*. Differences in bacterial abundance among healthy cats and cats with CE (before treatment) were evaluated using a Mann Whitney test. Samples collected before and after treatment from each affected cat were compared using a Wilcoxon signed rank test. The Feline Chronic Enteropathy Activity Index (FCEAI) was compared in cats with CE before and after treatment using a Wilcoxon signed rank test. Statistical significance was set at $p<0.05$.

Histologic evaluation of intestinal biopsies was consistent with IBD (n=10) or small cell lymphoma (n=4). Treatment consisted of prednisolone, chlorambucil, various antibiotics, cyanocobalamin, or a combination of these. Before treatment, the abundance of *Faecalibacterium* spp. and *Turicibacter* spp. was significantly decreased in cats with CE (p=0.004 and 0.007, respectively) compared to healthy control cats, while *Streptococcus* spp. and *Fusobacteria* were significantly increased in cats with CE (p=0.037, 0.028, respectively). There were no significant differences in the abundance of the other bacterial groups or the total bacterial counts between groups. FCEAI was significantly decreased after treatment in cats with CE (median [range]: 3 [0–4]) compared to before treatment (6 [4–8]), p=0.0312). However, no significant differences were found in the abundance of any bacterial group before and after treatment in cats with CE.

In conclusion, fecal dysbiosis is present in cats with CE. Although the FCEAI decreased significantly after 4 weeks of medical treatment, the abundance of selected bacterial groups did not change significantly. Results might suggest that a longer period of treatment or a different approach to medical therapy (i.e., pre- or probiotics) may be needed for cats to recover from fecal dysbiosis associated with CE.

GI40

Platelet Indices as Markers of Disease Presence and Severity in Dogs with Inflammatory Bowel Disease (Abstract GI40)

Susan O. Mehain; Jillian M. Haines; Pamela M. Lee

College of Veterinary Medicine, Washington State University, Pullman, WA, USA

In human medicine, recent studies have shown that individuals with inflammatory bowel disease (IBD) have statistically significant decreases in mean platelet volume (MPV) and platelet distribution width (PDW) and increases in platelet crit (PCT) compared to healthy controls. In remission, PDW and PCT have been found to be significantly greater than in controls. No studies to date have investigated the association between platelet indices in dogs with IBD compared to healthy controls and to determine if platelet indices could be useful predictors of disease presence or severity. Therefore, the purpose of this retrospective study was to determine if any platelet indices are significantly different between dogs diagnosed with IBD and healthy matched control dogs, or if any indices are predictive of severity based on canine chronic enteropathy clinical activity index (CCECAI) scores assigned retrospectively to dogs diagnosed with IBD.

Dogs with a histopathologic diagnosis of inflammatory bowel disease (IBD) that had no other concurrent illness and had not received immunosuppressant therapy were eligible for inclusion in this study. Twenty dogs diagnosed with IBD and 20 healthy matched control dogs were included in the analysis. Variables compared between the two groups included MPV, PDW, and PCT, as well as mean platelet number (PLT), mean platelet component concentration (MPC), mean platelet component concentration distribution width (PCDW), mean platelet mass (MPM), and platelet dry mass distribution width (PMDW). Significant differences between dogs diagnosed with IBD and healthy controls were evaluated using a t-test (reported as the mean ± standard deviation) and a Receiver Operating Characteristic (ROC) curve. Additionally, the correlation between platelet indices and the CCECAI scores was evaluated using bivariate scatterplots and linear regression models to determine if any variables were predictive of disease severity.

The results indicate that none of the variables compared between the two groups were statistically significant (p-value<0.05) in predicting the presence of disease. Although not significant, those variables close to a p-value of 0.10 all trend towards significance, and could represent possible predictive indices for disease: mean PLT and PCT were noticeably higher in IBD dogs and mean PCDW was noticeably lower in IBD dogs than in healthy controls. The ROC analysis was performed to determine predictive value of each variable using area under the curve (AUC). Based on AUC values, no platelet indices are strongly correlated with disease. However, PLT, PCT, PCDW, MPM, and PMDW all have AUC values between 0.6–0.7, indicating they may be somewhat predictive of disease presence. Bivariate scatterplots with moving average demonstrate that only PLT (p-value 0.02) and PCT (p-value 0.03) were statistically significant in predicting disease severity. Linear regression showed mild correlation with CCECAI score for these two indices with PLT r^2 value of 0.28 and PCT of 0.25.

In conclusion, a strong correlation between platelet indices and presence or severity of disease in dogs with inflammatory bowel disease is not statistically supported. In spite of this, many indices show potential value as predictors of disease in patients with IBD. A larger study population is likely needed to determine true significance.

	IBD dogs	Healthy controls		
Indices	**Mean ± SD**	**Mean ± SD**	**AUC**	**p-value**
Mean platelet number ($x10^3$cells/μL)	292 ± 136	237 ± 76	0.61	0.13
Mean platelet volume (fL)	10.77 ± 1.87	11.0 ± 1.74	0.54	0.69
Platelet distribution width (%)	57.56 ± 17.64	61.48 ± 5.38	0.45	0.35
Platelet crit (%)	0.31 ± 0.03	0.25 ± 0.01	0.6	0.12
Mean platelet component concentration (g/dL)	23.05 ± 1.39	22.63 ± 1.69	0.55	0.42
Mean platelet component concentration distribution width (g/dL)	5.06 ± 1.28	5.71 ± 1.09	0.69	0.11
Mean platelet mass (pg)	2.29 ± 0.27	2.2 ± 0.25	0.61	0.29
Platelet dry mass distribution width (pg)	0.92 ± 0.14	0.88 ± 0.12	0.62	0.27

GI41

Comparison of Serum Lipoprotein Profiles Between Cats with Hepatic Lipidosis and Healthy Control Cats (Abstract GI41)

Tomomi Minamoto[1]; Alexandra Hamilton[2]; Steve L. Hill[2]; Jan S. Suchodolski[1]; Jörg M. Steiner[1]; Jonathan A. Lidbury[1]

[1]Gastrointestinal Laboratory, College of Veterinary Medicine, Texas A&M University, College Station, TX, USA; [2]Veterinary Specialty Hospital, San Diego, CA, USA

Lipoproteins are conglomerates of triglycerides, cholesterol, phospholipids, and apolipoproteins that transport lipids in the blood stream and are classified into 5 classes based on their density: chylomicrons, very low density lipoproteins (VLDL), intermediate density lipoproteins (IDL), low density lipoproteins (LDL), and high density lipoproteins (HDL). Based on the human literature, altered proportions of lipoprotein classes have been described in certain disease states, such as cardiovascular disease or metabolic disease. Few published reports are available in veterinary medicine regarding the characteristics of lipoproteins in diseased animals. Feline hepatic lipidosis (FHL) is characterized by excessive accumulation of lipid within hepatocytes, which can lead to intrahepatic cholestasis and severe hepatic dysfunction. We hypothesized that cats with FHL show an altered serum lipoprotein profile due to intensive lipolysis in adipose tissue and alterations in hepatocellular lipid metabolism. The aim of this study was to compare serum lipoprotein profiles as well as serum cholesterol and triglyceride concentrations between cats with FHL and healthy control cats.

Analysis of serum lipoprotein profiles using density gradient ultracentrifugation was performed in 20 cats with FHL (confirmed by cytological and/or histological findings) and 20 healthy control cats. The area under the curve for triglycerides-rich-lipoproteins (TRL), LDL, and HDL was calculated. In addition, serum cholesterol and triglyceride concentrations were measured using a clinical chemistry analyzer. Data were analyzed using a Wilcoxon rank sum test. Significance was set at $p<0.05$.

LDL content was significantly higher in cats with FHL than in healthy control cats ($p=0.0001$), while HDL content was significantly lower in cats with FHL than in healthy control cats ($p=0.0032$). TRL content was not significantly different between the two groups ($p=0.0699$). Also, serum cholesterol and triglyceride concentrations were not significantly different between the two groups ($p=0.5075$ and $p=0.2541$, respectively).

In our study, serum lipoprotein profiles were altered in cats with FHL even though serum cholesterol and triglyceride concentrations were not significantly different compared to healthy control cats. The clinical importance of lipoprotein profiling in cats with hepatic lipidosis warrants further study.

GI42

Ileal and Colonic Mucosal Microbiota in Dogs with Steroid Responsive Chronic Enteropathy (Abstract GI42)

Francesca Bresciani[1]; **Yasushi Minamoto**[2]; Jan S. Suchodolski[2]; Giorgia Galiazzo[1]; Carla Vecchiato[1]; Carlo Pinna[1]; Giacomo Biagi[1]; Marco Pietra[1]

[1]Department of Veterinary Medical Sciences, University of Bologna, Ozzano dell'Emilia, Emilia-Romagna, Italy; [2]Gastrointestinal Laboratory, College of Veterinary Medicine, Texas A&M University, College Station, TX, USA

Exact aetiology for inflammatory chronic enteropathies in dogs remains unknown. Accumulating evidence suggests a pivotal role for intestinal dysbiosis in disease pathogenesis. Many studies have evaluated the alteration of fecal microbiota in canine chronic gastrointestinal (GI) disease, and less research is focused on mucosal microbiota, especially in the ileum and colon. The objectives of the current study were to evaluate ileal and colonic mucosal microbiota in dogs with steroid responsive enteropathy (SRE) before and after 4 months of treatment, and to compare them to control dogs (CD).

A total of 10 dogs diagnosed with SRE were enrolled. Complete GI endoscopy was performed and samples were collected by a cytology brush at diagnosis (SRE-Baseline, n=10) and after 4 month of treatment (SRE-After, n=8). Oral laxative and 2–4 water enemas were performed before endoscopy. A total of 6 CD that were euthanized for reasons unrelated to this study, with no GI disease, were included. Samples from CD were obtained during necropsy within 3 hours of death. Mucosal genomic DNA was extracted and used for Illumina sequencing of 16S rRNA genes. Sequence data were analysed using the QIIME pipeline. Statistical significance was set at $p<0.05$.

Clinical signs improved significantly after 4 month of treatment in SRE, but no improvement was seen on endoscopic or histological evaluation. Significant differences in microbial communities between SRE-baseline and CD were observed in the colon (ANOSIM $p=0.002$), but not in the ileum (ANOSIM $p=0.180$). In dogs with SRE, both ileal and colonic microbial communities remained similar after 4 month of treatment (ANOSIM $p=0.189$ and $p=0.637$, respectively), and were different from CD (ANOSIM $p=0.001$ and $p=0.004$, respectively).

Results of this study suggest that the mucosal microbiota in the colon of dogs with SRE is different from that of CD. Although clinical signs improved, colonic mucosal dysbiosis was still present after 4 months of treatment.

GI43

Effects of Prednisolone Administration on Pancreatic Tissue and Pancreatic Lipase Immunoreactivity Concentration in Healthy Dogs (Abstract GI43)

Hiroshi Ohta; Noboru Sasaki; Keitaro Morishita; Kensuke Nakamura; Mitsuyoshi Takiguchi

Department of Veterinary Clinical Sciences, Hokkaido University, Sapporo, Hokkaido, Japan

The canine pancreatic lipase immunoreactivity (cPLI) concentration is considered the most specific biomarker for the diagnosis of canine pancreatitis. In one report, 4 weeks of oral administration of prednisone (2.2 mg/kg/day) to six healthy dogs had no significant effect on the serum cPLI concentration (Steiner *et al.*, 2009). The aim of this study was to determine the effect of high-dose prednisolone (4 mg/kg/day) on the pancreatic tissue and serum cPLI concentration in healthy dogs. Six healthy beagles received prednisolone (4 mg/kg/day subcutaneously) for 2 or 3 weeks. Measurement of the cPLI concentration and ultrasonographic examination of the pancreas were performed before and after treatment. Laparoscopic biopsy of the liver and right pancreatic lobe was also performed before and after treatment. The serum cPLI concentration was compared before and after treatment using the Wilcoxon signed-rank test. The serum cPL concentration was significantly higher after than before treatment. The post-treatment cPLI concentration was classified as normal (≤200 μg/L) in five dogs and abnormal (≥400 μg/L) in one dog.

Ultrasonographic examination of the pancreas showed no significant change after treatment. There was no histological evidence of pancreatitis in the right pancreatic lobe after treatment. Administration of prednisolone at 4 mg/kg/day for 2 or 3 weeks increased the serum cPLI concentration in six healthy dogs, although a clinically significant increase (≥400 μg/L) was observed in only one dog. Histological evidence of pancreatitis was not observed in any of the six dogs examined.

GI44

Corticosteroid Therapy as Initial Treatment for Dogs with Acute Pancreatitis: 65 Cases (2011–2016) - (Abstract GI44)

Hiroki Okanishi[1]; Takayuki Nagata[2]; Shinsuke Nakane[2]; Toshihiro Watari[1]

[1]Laboratory of Veterinary Internal Medicine, Department of Veterinary Medicine, College of Bioresource Sciences, Nihon University, Fujisawa, Kanagawa, Japan; [2]Nakane Animal Hospital, Kitaku, Tokyo, Japan

In humans with acute pancreatitis (AP), treatment within 48 hours of onset is very important and can profoundly influence the prognosis. In dogs with AP, the efficacy of corticosteroid treatment is unknown. The purpose of this study was to investigate the responsiveness to initial corticosteroid therapy and prognosis in dogs with AP.

Sixty-five dogs were included in this study. Dogs were divided into the prednisolone (PDL) or non-prednisolone (NPDL) groups. Prednisolone was administered at a dose of 1 mg/kg/day in the PDL group. Enrofloxacin, maropitant, famotidine, fentanyl, and IV fluids were used as treatments in both groups. The number of days until the CRP level <2 mg/dl, the clinical sign score (APCS) <2, duration of hospitalization, mortality, and relapse rate were analyzed. Treatment effect was assessed using relative and absolute risk reduction (RRR, ARR) and number needed to treat (NNT).

CRP value was significantly lower in the PDL group than in the NPDL group after day 3 of hospitalization (4.2 vs. 12 mg/dl, p<0.05). Days taken for CRP level <2 mg/dl and APCS <2 and the duration of hospitalization were significantly shorter in the PDL group than the NPDL group (4 vs. 8 days, 5 vs. 7 days, and 5 vs. 8 days, p<0.01, respectively). Mortality in the PDL group was significantly lower than in the NPDL group (11.3 vs. 46.1%), and had a therapeutic effect (RRR, 0.76; ARR, 34%; NNT, 2.9).

In conclusion, initial corticosteroid therapy may be effective for dogs with AP.

GI45

Fecal Composition in Dogs with Naturally Occurring Exocrine Pancreatic Insufficiency (Abstract GI45)

Joseph C. Parambeth; Anitha Isaiah; Jonathan A. Lidbury; Jan S. Suchodolski; Jörg M. Steiner

Gastrointestinal Laboratory, College of Veterinary Medicine, Texas A&M University, College Station, TX, USA

Exocrine pancreatic insufficiency (EPI) is a syndrome characterized by the inadequate synthesis and secretion of pancreatic digestive enzymes, resulting in maldigestion. Dietary trials in dogs with EPI have not demonstrated a benefit for any specific diet. Determination of fecal composition can be used to determine nutrient digestibility. The aim of this study was to evaluate the fecal composition [dry matter (DM), organic matter (OM), ash, crude protein (CP), and total lipid content (acid hydrolyzed fat, AHF)] in dogs with naturally occurring EPI.

Surplus fecal samples from 17 dogs (9 dogs were receiving enzyme supplementation) enrolled in an unrelated clinical trial involving dogs with EPI were utilized. To be included into the study, the dogs had to be at least 1 year of age, have clinical signs of EPI (i.e., polyphagia, weight loss, steatorrhea, and/or loose, voluminous, and/or malodorous stools), have a serum cTLI concentration ≤2.5 µg/L, and be free from any clinically apparent disease other than EPI. Twelve healthy dogs were enrolled as healthy controls. Three naturally voided fecal samples were collected by the owners at home and stored frozen until shipment. Upon receipt, samples were pooled, aliquoted, and stored at -80°C until shipment to the Animal Sciences Laboratory, Department of Animal Sciences and Division of Nutritional Sciences, University of Illinois. Fecal samples were analyzed for DM, OM, ash, CP, and AHF. Fecal composition parameters were compared between the three groups; healthy control dogs, dogs with EPI on enzyme supplementation (EPI-S), and dogs with EPI without enzyme supplementation (EPI-NS) using the Kruskal-Wallis test. Where significant differences were identified, post-hoc testing was performed using a Dunn's multiple comparison test. Significance was set at $p<0.05$.

Median DM content of feces (min–max) were 36% (30–46.1%), 29.8% (26.5–32.4%), and 28.6% (21.2–34.7%) in the healthy, EPI-S, and EPI-NS dogs, respectively. Median OM content of feces on DM basis (min–max) were 68.3% (60.9–71.5%), 83.3% (72.9–89.9%), and 89.2% (86–95.9%) in the healthy, EPI-S, and EPI-NS dogs, respectively. Median ash content of feces on DM basis (min–max) were 31.7% (28.6–39.1), 16.7% (10.1–27.1), and 10.8% (4.2–14) in healthy, EPI-S, and EPI-NS dogs, respectively. Median CP content of feces on DM basis (min–max.) were 25.1% (21.7–28.1), 34.9% (24.5–40.2), and 32.3% (16.3–51.6) in the healthy, EPI-S, and EPI-NS dogs, respectively. Median AHF content of feces on DM basis (min–max) were 7.2% (5.4–9.5), 11.4% (8–20.1), and 20.2% (9–34.1) in the healthy, EPI-S, and EPI-NS dogs, respectively. DM, OM, ash, and AHF were significantly different between healthy dogs and dogs with EPI. However, no significant differences were observed in the fecal composition of EPI dogs, regardless of whether they were treated with enzyme supplementation.

This study showed that fecal composition is altered in dogs with naturally occurring EPI even when managed with enzyme supplementation. Larger prospective clinical studies are necessary to elucidate the mechanisms leading to this altered fecal composition despite enzyme replacement therapy.

GI46

Genome-Wide Association Study in German Shepherd Dogs with IBD Reveals Possible Role of Th2 Cytokines (Abstract GI46)

Atiyeh Peiravan[1]; Francesca Bertolini[2]; Max Rothschild[2]; Albert E. Jergens[3]; Kenneth Simpson[4]; Dirk Werling[5]; Karin Allenspach[3]

[1]Department of Veterinary Clinical Sciences and Services, Royal Veterinary College, London, Hertfordshire, England, UK; [2]Department of Animal Science, College of Agriculture and Life Sciences, Iowa State University, Ames, IA, USA; [3]Iowa State University, College of Veterinary Medicine, Ames, IA, USA; [4]College of Veterinary Medicine, Cornell University, Ithaca, NY, USA; [5]Department of Pathology and Pathogen Biology, Royal Veterinary College, London, Hertfordshire, England, UK

German shepherd dogs (GSD) in the UK are at increased risk of developing Inflammatory Bowel Disease (IBD). Similar to IBD in humans, understanding the genes involved in canine IBD is believed to reveal insights into disease pathogenesis in the dog.

The aim of the present study was to identify single nucleotide polymorphisms using a genome wide association study (GWAS) which may confer genetic susceptibility or resistance to IBD. Genomic DNA was extracted from EDTA blood or saliva samples of 96 cases and 98 controls. Genotyping of cases and controls was performed on the Canine Illumina HD SNP array and data generated were analyzed using PLINK. Using Fst and logistic association analyses, we identified 16 candidate genes that have been previously reported to be associated with human IBD or are involved in the pathways that are linked to the disease. Eight of these genes, one (PTPRC; Protein Tyrosine Phosphatase, Receptor Type C) on chromosome 7 and seven (IL4, IL5, IL13, CSF2; Colony Simulating Factor 2, SLC22A4 and A5; Solute Carrier Family22 member 4 and 5, IRF1; Interferon Regulatory Factor 1) on chromosome 11 have been identified in both approaches. In conclusion we identified 16 genes associated with IBD in GSDs. Our results suggest that Th2 cytokines could be implicated in the pathogenesis of IBD in GSDs. Targeted re-sequencing of the genes identified in our study will help to identify causative SNPS and consequently functional analysis of the causal SNPs reveal insights into mechanisms involved in pathogenesis on canine IBD.

GI47

Capsule Endoscopy Findings in Dogs with Hypoalbuminemia (Abstract GI47)

Jill S. Pomrantz; Jeffrey A. Solomon

Infiniti Medical, Menlo Park, CA, USA

The purpose of this study was to analyze capsule endoscopy (CE) findings in hypoalbuminemic dogs. The records of 178 dogs given ALICAM for gastrointestinal signs and/or laboratory abnormalities were retrospectively assessed. Seventeen dogs were hypoalbuminemic and had complete studies. The mean ± SD albumin was 1.9 ± 0.4 g/dL, mean ± SD age was 7.5 ± 2.8 years, and mean ± SD weight was 20.5 ± 11.4 kg. Eight of 17 dogs had concurrent anemia. Prior to CE, 15 of 17 dogs had ultrasound examinations. The gastrointestinal tract (stomach, small intestine [SI], and colon) was normal (n=8) or characterized by nonspecific wall thickening (n=7). Seventeen of 17 CE studies were abnormal. Findings consisted of irregular/thickened mucosa with or without erosions (n=6), numerous dilated lacteals (n=4), severe erosions or ulcers with active bleeding (n=2), mass (n=2), hookworms (n=1), SI nodules (n=1), and lesions consistent with colonic vascular ectasia (n=1). Lesions seen on CE (dilated lacteals, colonic vascular ectasia) were confirmed in two dogs that subsequently underwent traditional endoscopy. A third dog had gastroduodenoscopy prior to CE, which showed SI hemorrhage, but missed the SI nodules later identified with CE. A diagnosis of Strongyloides was made based on endoscopic biopsies and fecal Baermann test. A fourth dog with severe melena had a negative gastroduodenoscopy and laparoscopy, yet CE subsequently identified severe gastrointestinal erosions and ulcers thought to be secondary to NSAID administration. CE can be informative in dogs with hypoalbuminemia even when lesions are not identified with traditional endoscopy, surgery, or ultrasound.

GI48

Canine Adipose Tissue-Derived Mesenchymal Stem Cells Reduce Severe Acute Pancreatitis by Regulating T Cells (Abstract GI48)

Woo-Jin Song[1]; Qiang Li[2]; Min-Ok Ryu[1]; Sang-Chul Park[1]; Hyeon-Jin Kim[1]; Su-Yeon Kim[1]; Hye-Mi Yang[1]; Hwa-Young Youn[1]

[1]Seoul National University, Seoul, Republic of Korea; [2]Seoul National University, Seoul, China (People's Republic)

Severe acute pancreatitis (SAP) is associated with systemic complications and high mortality rate in dogs. Mesenchymal stem cells (MSCs) have been investigated for their therapeutic potential in several inflammation models. This study investigated the effects of canine adipose tissue-derived (cAT) MSCs in a rat model of SAP induced by retrograde injection of 3% sodium taurocholate solution into the pancreatic duct. cATMSCs labeled with dioctadecyl-3,3,3′-tetramethylindo-carbocyanine perchlorate (1×10^7 cells/kg) were systemically administered to rats and pancreatic tissue was collected 3 days later for histopathological, quantitative real-time PCR, and immunocytochemical analyses. Greater numbers of infused cATMSCs were detected in the pancreas of SAP as compared to sham-operated rats. cATMSC infusion reduced pancreatic edema, inflammatory cell infiltration, and acinar cell necrosis, and decreased pancreatic expression of the pro-inflammatory cytokines tumor necrosis factor-α, interleukin (IL)-1β, -6, -12, -17, and -23 and interferon-γ, while stimulating expression of the anti-inflammatory cytokines IL-4 and IL-10 in SAP rats. Moreover, cATMSCs decreased the number of cluster of differentiation 3-positive T cells and increased that of Forkhead box P3-positive T cells in the injured pancreas. These results indicate that cATMSCs can be effective as a cell-based therapeutic strategy for treatment of SAP in dogs.

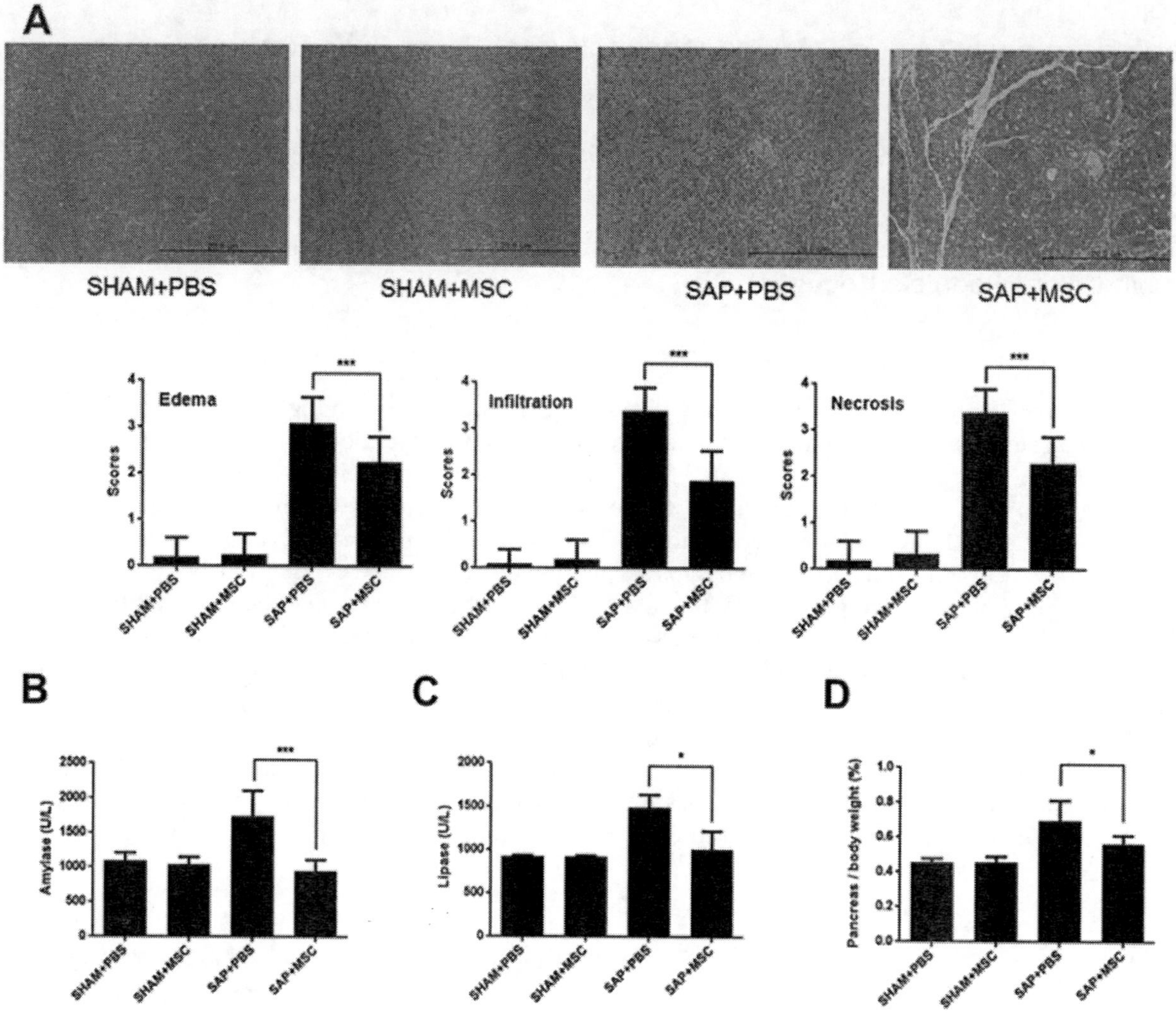

A
SHAM+PBS
SHAM+MSC
SAP+PBS
SAP+MSC
Edema
Infiltration
Necrosis
Scores
***
B
C
D
Amylase (U/L)
Lipase (U/L)
Pancreas / body weight (%)
*

GI49

Likelihood and Outcome of Esophageal Perforation Associated with an Esophageal Foreign Body in Dogs (Abstract GI49)

Allyson Sterman; Audrey K. Cook; Kelley Thieman-Mankin

College of Veterinary Medicine, Texas A&M, College Station, TX, USA

There is limited information regarding the likelihood and outcome of esophageal perforation secondary to an esophageal foreign body (EFB) in dogs.

The medical records databases at two veterinary hospitals identified 125 dogs with EFBs from Jan 2005–Jan 2014. Bones (44%) and fishhooks (37%) predominated. Fifteen dogs (12%) were diagnosed with a perforation based on radiography (n=4), esophagoscopy (n=10) or surgical exploration (n=1). Ten of these had a fishhook EFB; five had a bone. Esophageal perforation was not influenced by patient size. Esophageal perforation was more likely in dogs with a fishhook (10/15 vs. 27/110; odds ratio 6.1; p=0.002) than those with other types of EFB. For fishhook and bone EFB, the median time from ingestion to presentation was substantially longer for dogs with a perforation (12 hours and 96 hours, respectively) than those without (1 hour [p=0.007] and 24 hours [p=0.003], respectively). Endoscopic EFB removal was attempted in 13 dogs; eight recovered without surgical intervention and one was euthanized because of esophageal necrosis. Six dogs were treated surgically; five survived and one was euthanized intraoperatively. The two non-survivors both had long-standing bone EFBs. Survival rate for dogs with esophageal perforation was 87%.

Esophageal perforation secondary to an EFB is relatively uncommon (12%). However, ingestion of a fishhook markedly increases the likelihood of perforation, as does a delay between EFB ingestion and presentation to the hospital. All ten dogs with perforation secondary to a fishhook were discharged alive, and the majority of the survivors (8/13; 62%) did not require surgical intervention.

GI50

The Effect of Prolonged Ambient Temperature Exposure on the Feline Fecal Microbiota (Abstract GI50)

Moran Tal Gavriel[1]; Diego Gomez-Nieto[2]; Charlotte Chau[1]; J Scott Weese[2]; Adronie Verbrugghe[1]

[1]Department of Clinical Studies, Ontario Veterinary College, University of Guelph, Guelph, ON, Canada; [2]Department of Pathobiology, Ontario Veterinary College, University of Guelph, Guelph, ON, Canada

Research involving fecal collection from client-owned cats can be very challenging, due to the time difference between defecation and collection, as well as the owners' compliance to refrigerate/freeze samples immediately following collection. This study examined the effect of prolonged ambient temperature exposure on the feline fecal microbiota composition.

Fecal samples were collected within 15 minutes after defecation from 12 healthy cats, at a boarding facility in Guelph, ON, Canada. Samples were aliquoted and the first aliquot was frozen at -80°C within one hour of defecation (time point [T] 0). Remaining aliquots were maintained at ambient temperature (20 to 23°C) and frozen at -80°C at 6, 12, 24, 36, 48, 72 and 96 h after collection. DNA was extracted from all aliquots, and the V4 region of the 16S rRNA gene was amplified using polymerase chain reaction (PCR). The PCR products were sequenced with next-generation sequencing (Illumina MiSeq). Mothur software was used for bioinformatics analysis.

A total of 9,118,609 sequences passed all filters, with a median of 93,909 sequences per sample (range: 50,315 to 207,127 sequences/sample). Relative abundance of different taxa was calculated, and compared between time points, with no significant differences ($p>0.05$ for all taxa between time points, following the Benjamini-Hochberg adjustment). Alpha-diversity indexes (Shannon diversity, Simpson diversity and Chao1) were used to evaluate evenness, diversity and richness respectively, and no significant differences were identified ($p>0.05$ for all comparisons). Beta-diversity measurements were applied to assess similarity in community membership and structure. Phylogenetic trees were created, to visually present community membership and structure (Jaccard and Yue & Clayton indexes, respectively) over time between cats, and no significant differences were identified (parsimony and unifrac-unweighted tests, $p>0.05$). Principal coordinate analyses were performed and showed that samples cluster mainly by cat, with no significant differences between time points (AMOVA, $p>0.05$; HOMOVA, $p>0.05$). Linear discriminant analysis (LDA) effect size (LEfSe) method was performed to identify genomic features, such as enriched taxa, between time points. LEfSe analysis failed to detect any differences between groups. Random forest algorithm analysis identified a 92% error rate, indicating homogeneity across time points.

Contradictory evidence exists in human studies in regards to the effect of storage at different temperatures on the fecal microbiota. A recent study by Weese and Jalali showed limited impact of refrigeration for 14 days on the feline and canine microbiota. This concurs with the current study, which demonstrated that maintaining feline fecal samples at ambient temperature up to 96 hours (4 days) has no demonstrable effect on the bacterial membership and structure. Results from LEfSe and random forest analysis also provides supportive evidence for the lack of difference between time points.

GI51

Effects of Donor Sex and Pre-Conditioning on Cytokine Production of Canine Adipose-Derived Mesenchymal Stem Cells (Abstract GI51)

Tracy Webb; Stephanie Smith; Rebecca Timmons; Craig Webb

Colorado State University, Fort Collins, CO, USA

Current research suggests that the use of mesenchymal stem cell (MSC) therapy to treat clinical diseases, such as chronic enteropathy, may be optimized through donor selection and pre-conditioning strategies. We therefore investigated the effects of several pre-conditioning methods and donor sex on *in vitro* cytokine production of adipose-derived canine MSC (cMSC). Passage 6 cMSC were generated from cryopreserved subcutaneous adipose tissue from 6 healthy, adult, intact dogs, 3 female and 3 male. cMSC were subjected to 6 different conditions: control, poly I:C, IFNγ, TGFβ, serum-free media, and a mixed hypoxic environment. Supernatants from the cMSC were harvested after 1, 6, 12, 18, and 24 hours. Levels of IL-6, IL-8, IL-10, TGFβ, VEGF, and MCP-1 were determined by ELISA. IL-10 was below the level of detection in all samples. TGFβ was measured in the wells that were treated with TGFβ but was generally otherwise undetectable. IL-6 was only detected in relatively low levels in some samples from one male and one female donor. cMSC produced small to moderate amounts of IL-8 constitutively, which reached peak levels by 18 hours and was altered by pre-conditioning. cMSC produced large amounts of MCP-1 and VEGF, which were altered by pre-conditioning. Generally, the serum-free and the mixed hypoxic environment decreased production of all measured cytokines. Samples from the male donors showed significant donor variability compared to females suggesting that treatment success might be influenced by careful donor selection as well as pre-treatment conditioning, depending on the desired outcome.

GI52

Effects of Proviable-DC® on Clinical Signs, Fecal Microbiomes, and Metabolomic Profiles in Cats Receiving Clindamycin (Abstract GI52)

Jacqueline Whittemore[1]; Jennifer E. Stokes[2]; Nikki Laia[3]; Josh Price[1]; Jan S. Suchodolski[4]

[1]University of Tennessee, Knoxville, TN, USA; [2]College of Veterinary Medicine, University of Tennessee, Knoxville, TN, USA; [3]Nashville Vet Specialists, Nashville, TN, USA; [4]Gastrointestinal Laboratory, College of Veterinary Medicine, Texas A&M University, College Station, TX, USA

The purpose of this study was to compare the incidence of antibiotic-associated gastrointestinal signs (AAGS) and alterations in the fecal microbiome and metabolome of 16 healthy cats randomized to receive 150 mg clindamycin with 1 capsule placebo or synbiotic (Proviable-DC®) once daily for 21 days with reevaluation 603 days after antibiotic discontinuation.

Daily food consumption, occurrence of vomiting, and fecal score were recorded. Fecal samples were collected on days 5–7 (baseline), 26–28, and 631–633. Sequencing of 16S rRNA genes was performed, and mass spectrometry was used to determine metabolomic profiles. $P<0.05$ was considered significant, with the Benjamini & Hochberg's False Discovery Rate (fdr) used to adjust for multiple comparisons.

All cats developed AAGS. Vomiting was less in cats receiving a synbiotic, but the difference was not significant. Alpha and beta diversity changed significantly during treatment. Decreases in *Actinobacteria* (*Bifidobacterium, Collinsella, Slackia*), *Bacteriodetes* (*Bacteroides*), *Lachnospiraceae* (*Blautia, Coprococcus, Roseburia*), *Ruminococcaceae* (*Faecilobacterium, Ruminococcus*), and *Erysipelotrichaceae* (*Bulleidia,* [*Eubacterium*]) and increases in *Clostridiaceae* (*Clostridium*) and *Proteobacteria* (*Aeromonadales, Enterobacteriaceae*) occurred, with incomplete normalization by days 631–633. Derangements in short-chain fatty acid, bile acid, indole, sphingolipid, polyamine and cinnaminic acid profiles occurred, some of which persisted through days 631–633 and differed between groups.

In cats, AAGS are accompanied by short- and long-term dysbiosis and alterations in metabolomic profiles. Metabolomic profiles, but not AAGS or bacterial abundances, differed between groups over time. Further investigation is warranted to determine whether antibiotic-induced dysbiosis is associated with an increased risk of AAGS in cats and whether prior synbiotic administration ameliorates this risk.

GI53

Effects of Staggered Proviable-DC® Administration on Fecal Microbiome and Metabolomic Profiles of Cats Receiving Clindamycin (Abstract GI53)

Jacqueline Whittemore[1]; Jennifer E. Stokes[2]; Josh Price[1]; Jan S. Suchodolski[3]

[1]University of Tennessee, Knoxville, TN, USA; [2]College of Veterinary Medicine, University of Tennessee, Knoxville, TN, USA; [3]Gastrointestinal Laboratory, College of Veterinary Medicine, Texas A&M University, College Station, TX, USA

The purpose of this study was to compare the fecal microbiome and metabolome of 16 healthy cats randomized to receive 75 mg clindamycin with food, followed 1 hour later by 2 capsules of a placebo or synbiotic (Proviable-DC®) once daily for 21 days with reevaluation 6 weeks after treatment.
Fecal samples were collected on days 5–7 (baseline), 26–28, and 68–70. Sequencing of 16S rRNA genes was performed, and mass spectrometry was used to determine metabolomic profiles. $P<0.05$ was considered significant, with Benjamini & Hochberg's False Discovery Rate used to adjust for multiple comparisons.

Alpha and beta diversity were altered significantly during treatment, and alterations in the Shannon and dysbiosis indices persisted at days 68–70. Age was significantly associated with the Shannon and dysbiosis indices and relative bacterial abundances. Decreases in *Actinobacteria* (*Adlercreutzia, Bifidobacterium, Collinsella, Slackia*), *Bacteriodetes* (*Bacteroides, Prevotella*), *Ruminococcaceae* (*Faecilobacterium, Ruminococcus*), *Veillonellaceae* (*Megamonas, Megasphaera, Phascolarctobacterium*) and *Erysipelotrichaceae* ([*Eubacterium*]) and increases in *Clostridiaceae* (*Clostridium*), *Proteobacteria* (*Enterobacteriaceae*) occurred, with incomplete normalization by days 68–70. Metabolite profiles differed significantly by time, group and time x group for 106, 5 and 10 metabolites, respectively. Microbiome changes were accompanied by derangements in short-chain fatty acid, bile acid, indole, sphingolipid, and cinnaminic acid profiles, some of which persisted through days 68–70 and differed between groups.

In cats, antibiotic administration causes sustained dysbiosis. Age-associated changes in the microbiome could affect sensitivity to antibiotic-induced gastrointestinal effects. Metabolomic profiles differed between treatment groups over time, suggesting a potential mechanism for decreased antibiotic-induced gastrointestinal effects in cats administered synbiotics.

HM01

Serum IL-17 in Dogs with Immune-Mediated Hemolytic Anemia (Abstract HM01)

Benoit Cuq[1]; Shauna Blois[2]; R. Darren Wood[2]; Anthony Abrams-Ogg[2]; Christian Bédard[3]; Geoffrey A. Wood[1]

[1]Ontario Veterinary College, University of Guelph, Guelph, ON, Canada; [2]University of Guelph, Guelph, ON, Canada; [3]Faculté de Médecine Vétérinaire, Université de Montréal, St Hyacinthe, QC, Canada

Canine immune mediated hemolytic anemia (IMHA) has a high mortality rate, up to 70%, despite aggressive immunosuppressive therapy and supportive care. Development of immune-mediated disease is likely a multifactorial process leading to a loss of normal immune self-tolerance. Humans affected by autoimmune hemolytic anemia show several cytokine derangements, including a newly defined subset of T cells, Th17 cells secreting IL-17. IL-17 appears to have a critical role in the pathogenesis of AIHA in people, and work in mouse models of the disease indicates IL-17 as a potential treatment target. It was hypothesized that dogs with IMHA would display significant elevations in IL-17 concentrations compared to healthy controls.

Serum IL-17 was measured in thirty dogs with primary IMHA at diagnosis (D0); and 2 (D2) and 4 (D4) days after initiation of immunosuppressive therapy as well as in 25 clinically healthy dogs. IL-17 was significantly increased at the time of presentation (D0) in dogs with IMHA ($p=0.0158$) compared to healthy dogs as well at D2 ($p=0.0404$). Over the course of hospitalization, IL-17 significantly decreased between D0 and D2 ($p=0.0146$) and D0 and D4 ($p=0.0010$) for the discharged patients ($n=19$). No changes in IL-17 levels were observed between D0, D2 and D4 among the dogs that were euthanized in hospital. There was no correlation between IL-17 concentration and bilirubin levels, lactate or any of the CBC parameters at the time of presentation or throughout the hospitalization. IL-17 is elevated in dogs with IMHA and seems to decrease when response to therapy is observed.

HM02

Glutathione Peroxidase, Total Antioxidant Capacity, and Urinary F2-Isoprostane Concentrations in Anemic Dogs (Abstract HM02)

Allison Kendall; Andrew Woolcock; Aimee Brooks; George E. Moore

College of Veterinary Medicine, Purdue University, West Lafayette, IN, USA

Oxidative stress plays a considerable role in the pathophysiology of multiple diseases and has been well documented as a contributor to disease in both the human and veterinary literature. Oxidative stress results from the production of reactive oxygen species that cause cellular injury through direct and indirect mechanisms. One at-risk cell type is the erythrocyte, however, the role of oxidative stress in canine anemia has not been investigated. Oxidative stress is primarily assessed through the activity of antioxidant enzymes, the concentration of endogenous antioxidants, and byproducts of oxidative damage. The purpose of this study was to determine if oxidative stress is present in anemic dogs. We hypothesized that anemic dogs would have an alteration in the activity of glutathione peroxidase (GPx), a reduction of total antioxidant capacity (TAC), and an increased concentration of urinary F2-isoprostanes (F2IsoP, a byproduct of lipid peroxidation) when compared to healthy dogs.

In this prospective, cross-sectional study, we recruited 40 client-owned dogs with anemia (PCV<30%) due to any cause and 40 client-owned age-matched healthy control dogs. All dogs were evaluated for whole blood GPx activity, plasma TAC and urinary F2-isoprostane concentrations. Anemic dogs were further stratified into those with a hemolytic cause and those with a non-hemolytic cause.

Anemic dogs had a significantly lower GPx activity (43.1×10^3 U/L $\pm$ 1.6×10^3) than dogs in the control group (76.0×10^3 U/L $\pm$ 1.9×10^3) ($p<0.01$) indicating inactivation of GPx activity and depletion of antioxidant defense mechanisms. Further, dogs with a hemolytic anemia had lower mean GPx activity when compared to dogs with non-hemolytic anemia, but this difference did not reach significance ($p=0.51$). However, this may suggest that the degree of oxidative stress may be more severe with hemolysis. TAC concentrations ($p=0.10$) and urinary F2-isoprostanes ($p=0.75$) did not significantly differ between groups.

This is the first study in veterinary medicine demonstrating a glutathione peroxidase deficiency and subsequent oxidative stress in anemic dogs. Further studies are warranted to determine if antioxidant supplementation would improve survival and overall outcome as part of a therapeutic regimen for anemic states, especially those with hemolysis.

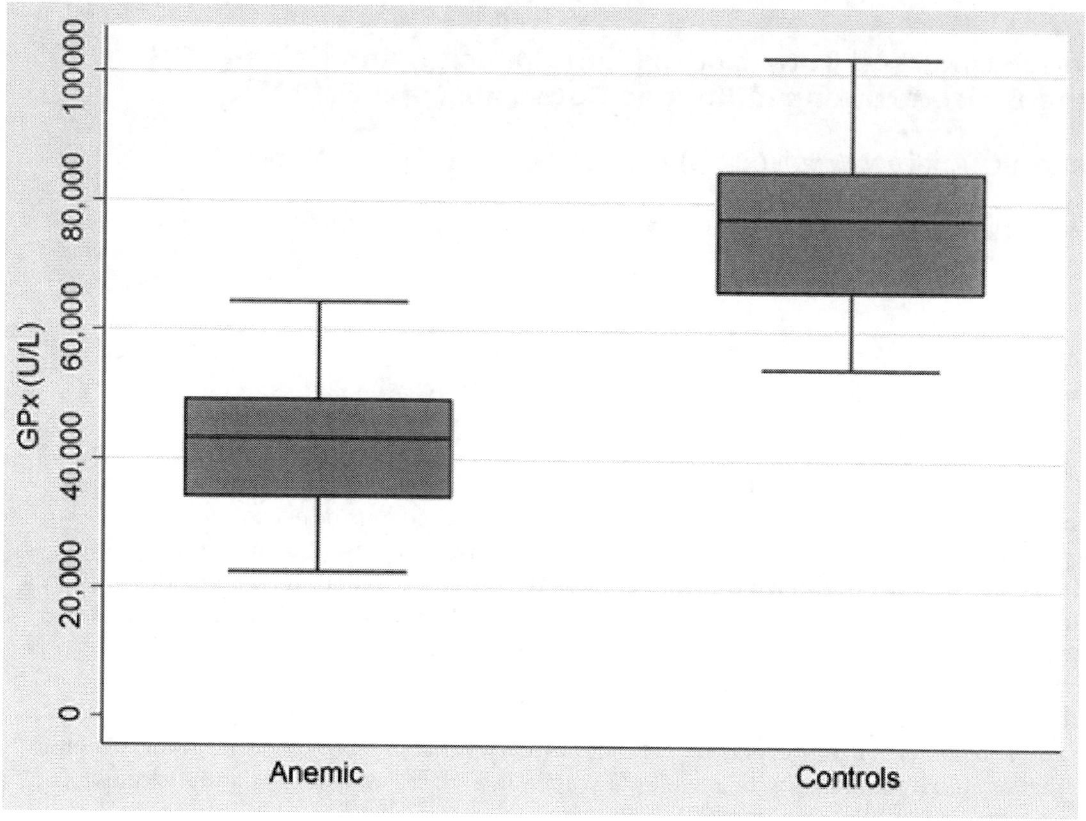
GPx (U/L)
100000
80,000
60,000
40,000
20,000
0
Anemic
Controls

HM03

The Effect of Yunnan Baiyao on the Kinetics of Hemostasis in Healthy Dogs (Abstract HM03)

Robin MacRae; Anthony Carr

Western College of Veterinary Medicine, Saskatoon, SA, Canada

Yunnan Baiyao is a popular Chinese herbal medicinal agent that has been used for its anti-inflammatory properties, hemostatic properties and wound healing properties in people for approximately a century. It has been recommended for use in dogs to control hemorrhage such as epistaxis or with splenic hemangiosarcoma. However there have been no studies evaluating the safety of Yunnan Baiyao for use in dogs and there have been no *in vivo* studies on the effect of Yunnan Baiyao on coagulation parameters in healthy dogs. The objectives of this study were to evaluate the effects of Yunnan Baiyao on the kinetics of hemostasis and to determine if there are any adverse effects associated with the administration of Yunnan Baiyao to healthy dogs. This study used six healthy purpose bred dogs. Baseline evaluation included a complete blood count, serum biochemistry profile, urinalysis, prothrombin time, partial thromboplastin time, and hemostatic kinetics using ROTEM thromboelastrometry (EXTEM, INTEM, FIBTEM). ROTEM parameters assessed included clotting time, clot formation, clot stability and clot lysis. The dogs were administered oral Yunnan Baiyao at a dose of 25 mg/kg orally once daily for 6 days. On day 2, day 4 and day 6 ROTEM thromboelastometry was again performed on all dogs. On day 6 a complete blood count, serum biochemistry profile, urinalysis, prothrombin time and partial thromboplastin time was also performed on each dog. No significant changes were noted with treatment in any of the parameters monitored. No significant change in clotting times ($p>0.2$), clot formation time ($p>0.2$), and clot formation rate ($p>0.2$) were noted when pre-treatment ROTEM values and post treatment ROTEM values were compared. From the results of this study we can conclude that Yunnan Baiyao is safe for administration to dogs and does not result in significant changes to baseline health parameters or coagulation parameters in healthy dogs.

HM04

Prestorage Leukoreduction Attenuates Neutrophil Extracellular Trap Formation in Canine Stored Packed Red Blood Cell Units (Abstract HM04)

Erin R. McQuinn[1]; Stephanie Smith[2]; Austin K. Viall[1]; Chong Wang[1]; Dana N. LeVine[1]

[1]Iowa State University, Ames, IA, USA; [2]Colorado State University, Fort Collins, CO, USA

Neutrophil extracellular traps (NETs), webs of DNA and citrullinated histones extruded from activated neutrophils to trap invading organisms, are a newly recognized cause of transfusion-related acute lung injury. NETs form during storage of human red blood cells (RBCs); formation is attenuated by prestorage leukoreduction (LR). Inflammatory cytokines that form in stored RBCs may promote NETosis in the neutrophils of the transfusion recipient.

We hypothesized that NETs form during storage of canine RBCs, that prestorage LR reduces NETosis, and that application of supernatant from stored, non-leukoreduced (NLR) RBC units would induce NETosis in healthy canine neutrophils.

A blood unit was collected from 6 healthy dogs, processed, then divided into two equal subunits of RBCs, one of which was leukoreduced with a commercial filter. Units were stored at 4°C for 42 days and sampled biweekly. NET markers citrullinated histone H3 and cell-free DNA were quantified in unit supernatants via western blot and fluorescence with DNA dye, respectively. Supernatants were applied *ex vivo* to neutrophils from healthy dogs and release of extracellular DNA representing NETosis was assessed. Data was analyzed by repeated measures ANOVA with treatment and time as fixed variables.

Both markers of NETs increased during RBC storage and were higher in NLR than LR units (Figure 1). Isolated neutrophils did not form NETS in response to supernatant from stored canine RBC.

NETosis occurs in stored canine NLR RBC units, and is attenuated by prestorage LR. NETs may be mediators of transfusion reactions; the clinical benefits of prestorage LR should be further explored.

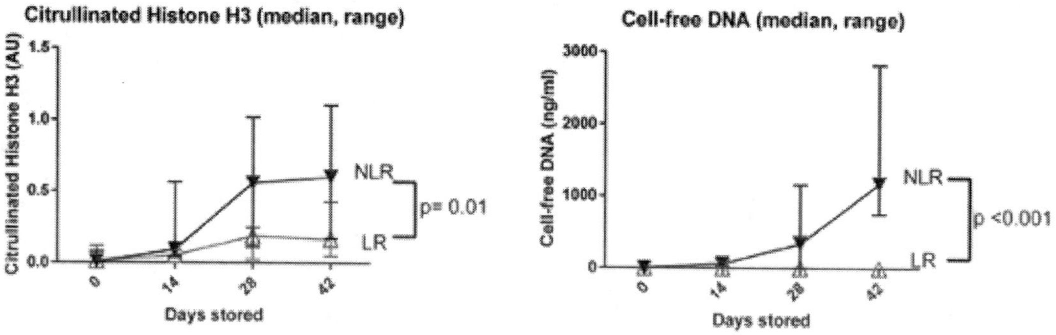

Figure 1. NET markers in stored non-leukoreduced (NLR) and leukoreduced (LR) RBC units.

HM05

Retrospective Analysis of Length of Glucocorticoid Treatment in 77 Cases of Canine Autoimmune Blood Disorders (Abstract HM05)

Nicole Puza[1]; Sean Hillock[2]

[1]Veterinary Medical Center of Long Island, West Sayville, NY, USA; [2]Veterinary Medical Center of Long Island, West Islip, NY, USA

The purpose of this study was to describe the length of time glucocorticoids are utilized in the treatment of autoimmune hemolytic anemia (AIHA), immune-mediated thrombocytopenia (ITP), and Evans syndrome in canines treated at the authors' institution.

A retrospective analysis was performed of patients that achieved remission and were weaned from glucocorticoid treatment for which full treatment data was available. Seventy-seven cases were included and analyzed for median dosage, tapering timeline, and length of glucocorticoid therapy. The collected data is presented alongside anecdotal data obtained by survey of Diplomates of the American College of Veterinary Internal Medicine in the specialty of Internal Medicine.

The median starting glucocorticoid dose was 2.3 mg/kg/day (prednisone equivalent). All cases were started on a secondary immunosuppressant. The median time between taper steps was 7 days. The median taper of the original dose at each step was 25%. Median length of glucocorticoid treatment was 49 days (range 0–922). Thirteen cases (16.8%) experienced relapse within the study period. Two occurred within 6 months of discontinuing prednisone while on a secondary immunosuppressant, and two within 6 months of discontinuing secondary immunosuppressive therapy. The remainder occurred 8–26 months after discontinuing secondary immunosuppressive therapy.

Survey data indicated 85.1% of respondents initiate glucocorticoid treatment at 2 mg/kg/d. 80.5% begin to taper glucocorticoids at >2 weeks, with each additional step at 2–4 week intervals (97%). 85.4% of respondents decrease by 25% of the original dose at each step, with 79% of responses indicating glucocorticoid treatment for greater than three months. Estimated relapse rates within six months of discontinuing immunosuppressive therapy was <20% (85.7% of respondents).

The results of this study indicate success in maintaining remission using a rapid glucocorticoid tapering schedule with use of a secondary immunosuppressant. These data are similar to that anecdotally achieved by prolonged treatment with glucocorticoids with a presumed decrease in severity of side effects of prolonged glucocorticoid therapy.

HM06

Erythrocyte Damage and In-Line Pressure Changes Associated with Transfusion of Canine Blood Through Microaggregate Filters (Abstract HM06)

Katie Cooley-Lock[1]; Peyton Williams[1]; Matthew Williams[1]; Steven Elder[2]; Todd Archer[1]; Andrew Mackin[1]; John Thomason[1]

[1]College of Veterinary Medicine, Mississippi State University, MS, USA; [2]Agricultural and Biological Engineering, Mississippi State University, MS, USA

Blood transfusions are common, potentially life-saving therapies in small animal medicine. Unfortunately, in dogs, previous studies have shown that red blood cells (RBCs) transfused via syringe pump have a significantly shorter post-transfusion survival time compared to RBCs transfused via gravity flow. One explanation for this reported decrease in RBC survival is that, during a transfusion, there is an increase in shearing stresses that damage RBCs as blood is forced through an in-line microaggregate filter.

The goal of this study was to determine if blood passing through a microaggregate filter, via a syringe pump, would damage canine RBC. Our hypothesis was that, over time, the filter would become clogged with fibrin and cellular debris, requiring an increase in force to maintain the preferred flow rate and thereby causing an increase in damaged RBCs.

Whole blood (50 mL) was collected from 8 healthy dogs and, using a syringe pump, blood was passed through a standard microaggregate filter at three standard rates used in the clinical setting (12.5, 25, and 50 mL/h). Pre- and post-filter blood samples were collected at the beginning and end of a simulated transfusion. Samples were analyzed to determine the erythrocyte osmotic fragility (mean corpuscular fragility [MCF]), RBC count, hemoglobin concentration, RBC distribution width (RDW), and RBC morphology. The force required to pass blood through the microaggregate filter was also measured continuously during the simulated transfusion.

Regardless of the transfusion rate, there was no change in the MCF, RBC count, hemoglobin concentration, or RDW between pre- and post-filter samples. Additionally, in 5 dogs, no significant change in pressure within the filter was seen over time. Echinocytes were the most common abnormal erythrocyte morphology noted at the 25 and 12.5 mL/h rates at the end of the simulated transfusion in both the pre- and post-filter samples. Schistocytes were identified in one sample (post-filter at 12.5 mL/h).

Our study suggests that, regardless of the transfusion rate, the microaggregate filter does not alter the fragility of the canine RBC, but may alter the morphology. RBC morphology alterations could contribute to a decrease in RBC survival when administrating blood via a syringe pump.

HM07

Analytical Validation of a Flow Cytometric Protocol for Circulating Platelet Microparticle Quantification in Dogs (Abstract HM07)

Signe E. Cremer[1]; Anne K.H. Krogh[1]; Matilda E.K. Hedström[1]; Liselotte B. Christiansen[2]; Inge Tarnow[3]; Annemarie T. Kristensen[1]

[1]Department of Veterinary Clinical and Animal Sciences, University of Copenhagen, Frederiksberg C, Denmark; [2]Department of Veterinary Disease Biology, University of Copenhagen, Frederiksberg C, Denmark; [3]Savara Pharmaceuticals A/S, Hørsholm, Hoersholm, Denmark

Circulating platelet microparticles (PMP) are subcellular vesicles released from platelet and possibly megakaryocyte membranes. In people, altered PMP concentrations are associated with pro-thrombotic, inflammatory, neoplastic and cardiovascular conditions, but only few canine studies exist. Blood-processing pre-analytically influences the PMP concentration. The study purpose was to analytically validate PMP quantification in dogs with a perspective of clinical application.

Using flow cytometry, dually CD61 and Annexin positive PMP of 0.3–1.0 µm were quantified in citrated whole blood (WB) and platelet-poor plasma (centrifuged at 4500 g for 3 minutes). The coefficient of variance (CV) for duplicate - (intra-assay) and parallel (inter-assay) PMP analyses and the detection limit (DL) were calculated, accepting maximally 20% variation.

Data obtained from 13 healthy dogs demonstrated acceptable intra-assay CV for WB (2.5%, 0.3–10.8% [median, range]) and plasma (5.7%, 0.0–11.1%) and acceptable inter-assay CV for WB (11.1%, 3.9–21.2%) but not for plasma (24.2%, 4.1–105.0%). Plasma PMP concentrations equaled the noise level of negative controls, and antibody centrifugation (plasma protocol) markedly reduced this noise with a DL of 1560 PMP/µL. Antibody centrifugation was implemented in another seven dogs. The results demonstrated acceptable WB intra-assay (2.4%, 1.2–6.7%) - and inter-assay (6.0%, 0.0–12.3%) CV, except in one dog (58.2% inter-assay CV). Unfortunately, hemolysis and/or lipidemia interfered with plasma PMP quantification in five of the seven dogs.

In conclusion, PMP concentrations were quantified reliably in WB with acceptable analytical precision, suggesting an application in clinical settings. Plasma PMP concentrations were invalid or unobtainable due to DL overlap or hemolysis and/or lipidemia, respectively.

HM08

Flow Cytometric Platelet Microparticle Quantification in Cats with Mild Cardiomyopathy (Abstract HM08)

Signe E. Cremer; Jørgen Koch; Nanna Graversen; Anne S. Gravgaard; Louise Bochsen; Rebecca Langhorn; Jakob Willesen; Annemarie T. Kristensen; Lise N. Nielsen

Department of Veterinary Clinical and Animal Sciences, University of Copenhagen, Frederiksberg C, Denmark

Cardiogenic embolism (CE) in cats is a devastating condition primarily associated with cardiomyopathy. Hypercoagulability may pose a risk for thrombus formation; however, no single test can predict development of CE. Platelet microparticles (PMP) are subcellular vesicles released from platelet membranes and are associated with thrombosis in humans. The purpose of this study was to: i) analytically validate flow cytometric PMP quantification in cats and ii) in a pilot study, evaluate the concentration of procoagulant PMP in healthy cats and cats with cardiomyopathy.

The cats underwent physical examination, hematology, serum biochemistry, an extended coagulation screen, and a full cardiac examination including blood pressure measurement, electrocardiography, and echocardiography. Using flow cytometry, dually CD61 and Annexin positive PMP of 0.3–1.0 μm were quantified in citrated whole blood (WB) and platelet-poor plasma. The coefficient of variance (CV) for duplicate - (intra-assay) and parallel (inter-assay) PMP analyses were calculated.

Six healthy cats and five cats with mild hypertrophic cardiomyopathy were examined. PMP concentrations could be quantified with acceptable intra-assay CV for WB (3.8%, 0.1–12.5% [median, range]) and plasma (7.4%, 0.5–15.3%) and acceptable inter-assay CV for WB (7.1%, 0.7–22.2%) but not for plasma (27.8%, 8.4–77.1%). The group-wise comparisons of WB PMP concentrations demonstrated no differences between the healthy cats (362.000 PMP/μL, 224.900–677.100 PMP/μL) and the cats with mild cardiomyopathy (383.600 PMP/μL, 4.700–477.400 PMP/μL, $p>0.99$).

In conclusion, WB PMP concentrations could be reliably quantified in cats in a clinical setting. The PMP concentration however, did not differ between healthy cats and cats with mild cardiomyopathy.

HM09

Diagnosis and Clinical Outcome Associated with a Leukemoid Response in Dogs: 81 cases (2006–2016) - (Abstract HM09)

Blakeley W. Janacek[1]; Leah Cohn[1]; Megan Grobman[2]

[1]University of Missouri, Columbia, MO, USA; [2]College of Veterinary Medicine, University of Missouri, Columbia, MO, USA

The purpose of the study was to review cause and outcome of leukemoid response in dogs. A computerized search for dogs seen over the last 10 years with neutrophil counts ≥60,000 cells/µL at the University of Missouri was completed. Demographic data, hemogram, diagnoses and clinical outcome were extracted from records. Based on recorded diagnosis, leukemoid response was deemed to be due to focal or systemic infection or inflammation, immune-mediated disease, neoplastic disease, tissue damage/necrosis, or cause undetermined.

A total of 81 dogs were identified, including 43 male (9 intact, 33 castrated, 1 of unknown status) and 38 female (8 intact, 30 spayed) dogs with a mean age of 6.8 years (minimum 0.17, maximum 15). Nine were mixed breed, and German shepherd dogs (7/81; chi square p=0.035) were over-represented. The median neutrophil count was 70,940 cells/µL (IQR ± 24,080; maximum 197,680 cells/µL). A definitive diagnosis was available for 56 dogs, and a presumptive diagnosis for 17. Forty-two (51.9%) had infection/inflammation, 22 (27.2%) neoplastic disease (7 lymphoma or leukemia), 5 (6.2%) immune-mediated disease, 4 (4.9%) tissue damage, and 8 dogs (9.9%) went undiagnosed. Forty-four dogs survived to discharge, 27 were euthanized, 10 died in hospital. The median segmented neutrophil counts did not differ between disease group based on Kruskal-Wallis One Way Analysis of Variance on Ranks. Similarly, survival was not related to neutrophil count.

Magnitude of leukemoid response alone may not predict disease cause or prognosis.

HM10

Immediate Post-Transfusion Survival of Biotinylated Canine Red Blood Cells with Various Transfusion Techniques (Abstract HM10)

Pamela Y. Martin[1]; Brittany Thames[2]; Charlee Mulligan[2]; John Thomason[1]; Andrew Mackin[1]

[1]College of Veterinary Medicine, Tuskegee University, Tuskegee, AL, USA; [2]College of Veterinary Medicine, Mississippi State University, Mississippi State, MS, USA

The use of infusion or syringe pumps during canine blood transfusion is controversial, because pumps can potentially damage red blood cells (RBC), rendering the transfusion less effective. A previous study in healthy dogs reported a complete absence of detectable biotinylated RBCs 24 hours post-transfusion in many animals when blood was given through syringe or infusion pumps. Complete loss of transfused RBCs within one day of transfusion is, however, inconsistent with anecdotal clinical experiences.

The objectives of our current study were to confirm whether a number of different transfusion methods (one syringe and one infusion pump brand, compared to gravity drip) resulted in premature loss of transfused cells, and to explore the timing of early RBC loss if it occurred. Eight healthy hound dogs were phlebotomized (54 ml of whole blood in CPDA-1), and blood was centrifuged, with supernatant plasma collected and stored. Remaining RBCs were incubated with NHS-biotin dissolved in 100% DMSO, and then washed in saline. The RBCs were then resuspended in the stored autologous plasma, and 5 mL was retained as a positive biotinylated control for each sample, with the remainder transferred to either a 100 mL dry blood bag (infusion pump and gravity drip) or 60 mL syringe (syringe pump). Assigned transfusion treatments were as follows: gravity drip (2 dogs), infusion pump (Heska Vet/IV 2.2; 3 dogs), and syringe pump with an in-line filter (Medfusion 2001; Hemo-Nate Blood Filter; 3 dogs). Autologous whole blood transfusion was performed at a rate of 30 mL per hour. Blood samples were collected in sodium citrate anticoagulant at time 0 (pre-transfusion, used as negative non-biotinylated control), and at hours 1, 6, 12, 18, and 24, and days 2, 3, and 4 post-transfusion. Samples were labeled with phycoerythrin-streptavidin, and percentage of biotinylated RBCs analyzed by flow cytometry with RBCs gated according to their forward and side light scatter at 10,000 gated events per sample.

Immediate post-transfusion (1 hour) percent biotinylated RBCs (percent of total RBCs; mean ± SEM) was 1.43 ± 0.23% (gravity), 1.45 ± 0.05% (infusion) and 1.55 ± 0.10% (syringe), consistent with minimal loss of RBCs during transfusion, since estimated post-transfusion mean target percent biotinylated RBCs was 1.12% (range 1.02–1.24%). Over the first 24 hours, and the subsequent 3 days, there was no significant difference between transfusion techniques, with approximately 1.5% of biotinylated cells present up to 4 days following transfusion (Figure 1).

Our study found acceptable post-transfusion survival of labeled RBCs regardless of method, a finding that is discrepant with the prior study documenting unmeasurable levels of biotinylated RBCs one day after transfusion using various pumps.

Subsequent exploration of reasons for discrepant results suggests several possibilities. DMSO is the standard solvent used for biotin, since biotin is insoluble in aqueous solutions. Biotin precipitates in DMSO solutions of 50% or less, and high DMSO concentrations are therefore needed for biotin labeling studies. Inadvertent dilution of DMSO in residual plasma during RBC labeling therefore has the potential to render biotinylation ineffective. Additionally, we found that different concentrations of DMSO cause variable lysis of RBCs: 60% DMSO (diluted in PBS) caused marked hemolysis, 50 and 100% caused moderate hemolysis, 30, 40 and 90% caused mild hemolysis, and 10, 20, 70 and 80% caused no hemolysis. Our findings suggest that transfusion studies using biotinylated RBCs may be susceptible to technique artifacts, with subtle variations in DMSO concentrations leading to either failure of biotinylation or inadvertent RBC damage.

Figure 1. Mean percent biotinylated RBCs post-transfusion

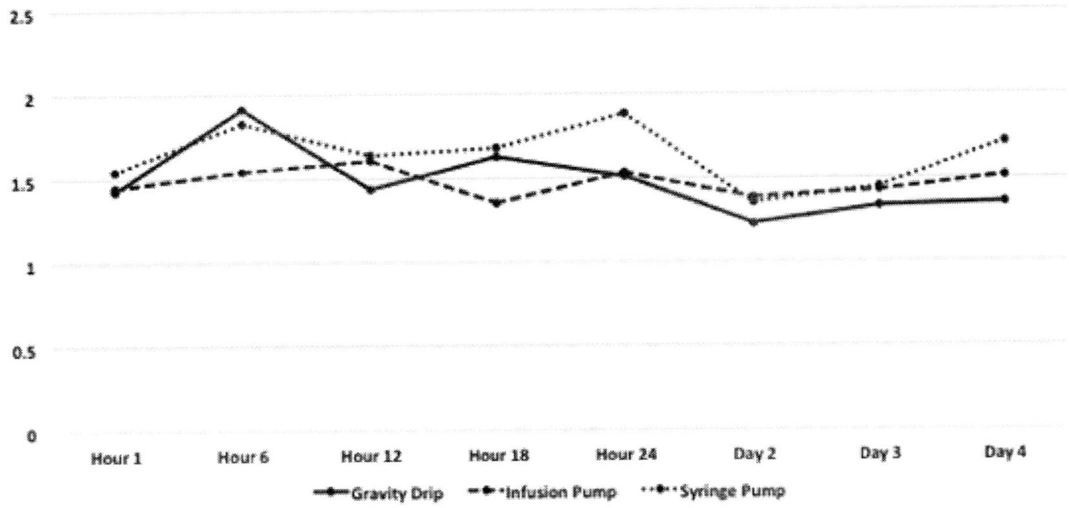

HM11

Markers of Oxidative Damage During Storage of Feline Packed Red Blood Cells (Abstract HM11)

Jennifer M. Reinhart; Julie M. Walker

Department of Medical Sciences, School of Veterinary Medicine, University of Wisconsin-Madison, Madison, WI, USA

Biochemical and immunologic changes consistent with erythrocyte storage lesion have been described in feline packed red blood cell (pRBC) units. Oxidative damage occurs in stored human erythrocytes as evidenced by decreased glutathione concentrations, but this has yet to be investigated in cats. The purpose of this study was to determine the effect of storage on glutathione, osmotic fragility, and methemoglobin reductase activity in feline pRBCs. Whole blood (60 mL) was collected from five healthy cats into blood bags containing CPDA-1. Each bag was split in half and pRBC units were prepared. For each cat, additive solution 3 (AS-3) was added to one of the units and pRBCs were stored at 4°C. Aliquots of each unit were sampled on days 1, 14, 28, 35, and 42 of storage and assayed for glutathione, osmotic fragility, and methemoglobin reductase activity. Friedman's test was used to assess changes over time. Glutathione decreased significantly with time in both AS-3-treated and untreated pRBCs. Osmotic fragility increased significantly with time in untreated pRBCs, but did not change from baseline in pRBCs treated with AS-3. Methemoglobin reductase activity was unchanged in both treated and untreated samples. Storage of feline pRBCs over 42 days leads to glutathione depletion and increased osmotic fragility. These findings are further evidence for storage lesion in feline pRBCs. However, the addition of AS-3 mitigates the some of these changes and should be considered in the development of future blood product protocols.

HM12

Adherence to Human Guidelines for Plasma Transfusions in Dogs and Cats: 154 Cases (Abstract HM12)

Tom Taney; Elizabeth A. Rozanski

Cummings School of Veterinary Medicine, Tufts University, North Grafton, MA, USA

Fresh frozen plasma is used to treat a variety of conditions in critical illness. Plasma transfusion has the potential for causing adverse reactions and is a limited resource. There are currently no evidence-based plasma transfusion recommendations in dogs and cats. These include documented coagulopathy with bleeding or a planned invasive procedure (DC), warfarin reversal (WR), massive transfusion (MT), and factor deficiency (FD). The aim of this study was to determine if plasma transfusions in animals met guidelines used in human medicine.

The medical record system was retrospectively searched for dogs and cats that had received plasma transfusion. The indication(s) for the transfusion recorded. The percentage in each category was recorded, as well as those cases with no clear indication. Survival was recorded.

Dogs (n=136) and cats (n=18) were identified. Seventy-six percent (n=117/154) of patients had plasma transfusions that could be classified under an indication in the proposed guidelines for human patients. 15 patients had indications that could be classified into multiple categories. The most common indications were DC at 61% (n=94/154), MT at 17 % (n=27/154), WR at 5.8 % (n=9/154), and FD at 1.2% (n=2/154). Twenty-four percent (n=37/154) of animals were administered plasma for other reasons, most frequently hypoproteinemia. Survival to discharge was 48% for dogs (n=65/136) and 44% for cats (n= 8/18).

The majority of plasma transfusions were considered indicated based guidelines for human patients, however, 24% of transfusions were not considered indicated. Similar guidelines should be considered for companion animals.

HM13

Antiplatelet Effects of Generic Clopidogrel Compared to Trade Name Clopidogrel (Plavix®) in Cats (Abstract GI48)

Mays Malkawi; Andrew Woolcock; Daniel Hogan

College of Veterinary Medicine, Purdue University, West Lafayette, IN, USA

Cats with hypertrophic cardiomyopathy (HCM) are at an increased risk for formation of an aortic thromboembolism (ATE) due to blood stasis and platelet hyper-reactivity in enlarged cardiac chambers, among other prothrombotic factors. Anti-platelet medications are commonly used in cats considered at risk of developing thromboembolism. Clopidogrel bisulfate, a thienopyridine drug, causes irreversible inhibition of platelet aggregation in response to ADP and collagen, common initiators of clot formation. Trade name clopidogrel (Plavix®) was previously investigated in healthy cats and was associated with a significant inhibition of platelet activity. In 2012, the FDA approved the generic form of clopidogrel, but its inhibitory effects on platelet aggregation have not been investigated. We hypothesized that generic clopidogrel is associated with equal anti-platelet effects to those associated with trade name clopidogrel in healthy cats, and that both formulations are associated with minimal oral/gastrointestinal adverse events.

In this randomized, double-blind, crossover study, five skeletally mature cats were determined to be healthy based on physical examination and diagnostics including CBC, serum chemistry profile, urinalysis, and coagulation times. The cats participated in three treatment phases in which they were randomly assigned to receive either: trade name clopidogrel 18.75 mg PO SID for 7 days, generic clopidogrel 18.75 mg PO SID for 7 days, or a placebo (cellulose) PO SID for 7 days. On day 0, prior to starting the treatment, baseline platelet aggregation was assessed on a multiplate platelet aggregometer. Briefly, blood was collected into a lithium heparin tube, and platelets were activated for the assay using ADP. Platelet aggregation was repeated on day 8 to determine the percentage of platelet inhibition achieved. Information regarding potential adverse effects were recorded daily for each cat during each phase. A minimum four-week washout period occurred between each phase.

All cats maintained a normal platelet count at each time point of the study. No adverse effects were noted with the administration of trade name or generic clopidogrel in any cat. Two cats reliably showed significant inhibition of platelet aggregation in response to both trade name and generic clopidogrel (full responders). One cat showed slight inhibition of platelet aggregation in response to generic clopidogrel, but not trade name clopidogrel (partial responder). Two cats did not show any inhibition of platelet aggregation in response to both trade name and generic clopidogrel (non-responders).

While we determined that both the trade name and generic formulations of clopidogrel are safe to administer in cats, there was not reliable inhibition of platelet aggregation in all cats. Cats tended to have a similar response to both formulations of clopidogrel implying that intrinsic factors affect the efficacy of the drug, as opposed to human or assay error. This variable response to treatment may be due to alterations in the cytochrome P450 gene coding for enzymes responsive for clopidogrel metabolism, as several polymorphisms have been identified in humans. This is the first study to investigate the effect of the generic formulation of clopidogrel on platelet aggregation, but due to the low sample size, meaningful conclusions cannot be made about the varying response to the drug. Further studies should aim to increase sample size, and investigate metabolic, genetic, and other intrinsic factors that could affect drug efficacy.

HM14

In Vivo Effects of 7.5% Hypertonic Saline Solution on Canine Whole Blood Coagulation Using Thromboelastography (Abstract HM14)

Hyeyoung Kim[1]; **Suyeon Baek**[1]; Aryung Nam[2]; Kyoung won Seo[3]

[1]College of Veterinary Medicine, Chungnam National University, Daejeon, Taejon-jikhalsi, Republic of Korea; [2]Department of Veterinary Internal Medicine, College of Veterinary Medicine, Seoul National University, Seoul, Seoul-t'ukpyolsi, Republic of Korea; [3]Chungnam National University, Daejeon, Taejon-jikhalsi, Republic of Korea

The objective of this study is to evaluate the effect of 7.5% hypertonic saline solution (HSS) on whole blood coagulation in healthy dogs and to compare the effects of *in vivo* and *in vitro* dilution of HSS. This prospective, controlled, experimental study included twelve healthy beagles. All twelve dogs were included in the treatment group with five of them randomly selected to serve as the control group. The dogs in the treatment group (n=12) were given a 5 mL/kg 7.5% HSS IV over 1 mL/kg/min. Following a 2-week washout period, the dogs in the control group (n=5) received 5 mL/kg 0.9% normal saline solution. Subsequently, 4-mL blood samples were collected at 0, immediately after infusion, 30, 60 and 90 min after infusion for thromboelastography (TEG) measurement.

The *in vitro* study was performed with diluted samples, and 2-mL blood samples were collected at baseline and diffused with 7.5% HSS at a ratio of 1:18, which was equivalent to 5 mL/kg infusion *in vivo* situation.

Measurements using ANOVA were carried out repeatedly for analysis of the *in vivo* study, and there were no statistically significant changes ($p<0.01$) in any TEG variables between the treatment group and control group. For comparison between *in vivo* and *in vitro* studies, paired t-tests were used.

The comparison between *in vitro* and *in vivo* dilution was performed only at a point of time immediately after infusion, which is most similar situation with *in vitro* dilution because HSS is immediately diffused *in vivo*.

There were significant changes in maximum amplitude ($p=0.04$) and G values ($p=0.04$), suggesting that HSS affects whole blood coagulation differently, according to the study design. Unlike in previous *in vitro* studies, which revealed that 7.5% HSS has anticoagulant effects, a clinically relevant dose (5 mL/kg) of 7.5% HSS had no effect on whole blood coagulation when compared with 0.9% normal saline *in vivo*.

Table 1. Reference intervals for thromboelastography variables for citrated kaolin-activated whole blood and 90% CIs of upper and lower reference limits obtained. (n = 40 healthy Beagle dogs)

Variable	Unit	Sample size	Lower 2.5% reference limit	Upper 97.5% reference limit	90% CI interval lower	90% CI interval upper
R	min	40	2.3	5.1	1.7-2.9	4.5-5.7
K	min	40	1.0	2.3	0.7-1.3	2.0-2.6
α	degree	39†	59.6	77.5	55.9-63.4	73.8-81.3
MA	mm	39†	54.9	71.2	51.7-58.1	67.1-73.5
G	Kdyn/cm^2	39†	5.5	11.5	4.3-6.7	10.3-12.7

† Values excluded were from 2 dogs (1 for α and 1 other for MA and G values) for which values were > 3 SD from the mean.

R = reaction time; K = clotting time; α = angle; MA = maximum amplitude; G = global clot strength (G)

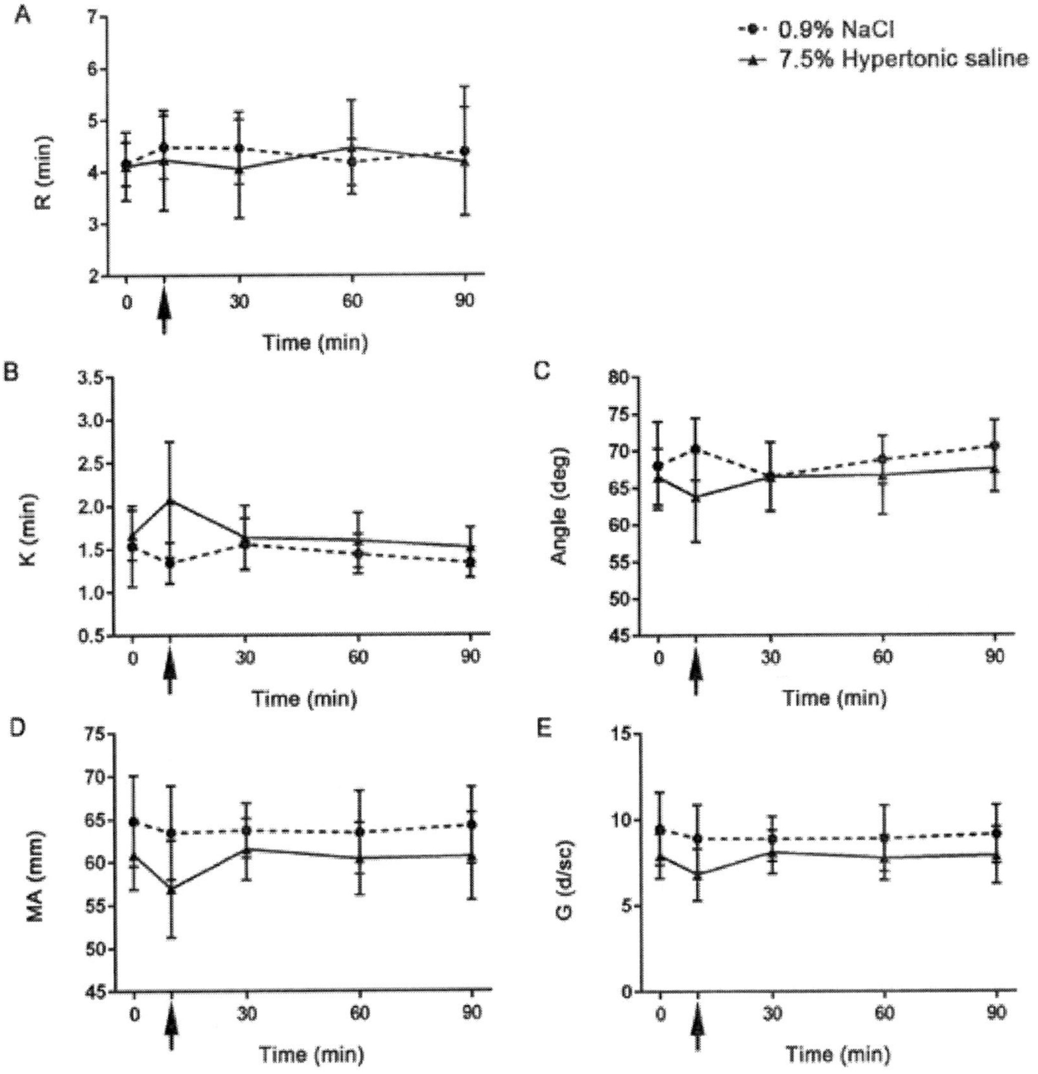
A
B
C
D
E
0.9% NaCl
7.5% Hypertonic saline
R (min)
K (min)
Angle (deg)
MA (mm)
G (d/sc)
Time (min)

HM15

In Vitro Inhibition and Detection of Canine Complement Activation in Canine Immune-Mediated Hemolytic Anemia (Abstract HM15)

Daniela Hernandez Muguiro; Erica Behling-Kelly; Robert Goggs

[1]College of Veterinary Medicine, Cornell University, Ithaca, NY, USA

The purpose of this study is to evaluate the ability of two human complement inhibitors to inhibit canine complement-mediated hemolyisis *in vitro* and to validate the use of an ELISA for detection of canine complement activation in dogs.

Human complement inhibitors, C3-inhibitor (Compstatin) and human recombinant C1-esterase Inhibitor (C1-INH) were evaluated using an *in vitro* hemolytic assay. Antibody-coated sheep erythrocytes were exposed to canine complement with and without inhibitors. Dose curves were generated by plotting the percentage of hemolysis against inhibitor concentration. Complement activation in samples from 11 healthy and 13 dogs with IMHA was determined using an ELISA for detection of Terminal Complement Complex C5b-9 (TCC C5b9). The concentration of TCC C5b9 in patient samples was extrapolated from a standard curve generated using the supplied protein standard. Samples treated with snake venom were used as a positive control for complement activation.

C1-INH showed dose-dependent inhibition of canine complement-mediated hemolysis. Compstatin was minimally effective. TCC C5b9 concentrations tended to be higher in control samples compared to dogs with IMHA.

C1-INH inhibited canine complement activation *in vitro*. Compstatin showed minimal inhibition of canine complement-mediated hemolysis at the doses evaluated. The ELISA tested did not consistently detect complement activation.

Figure 1. Inhibition of canine complement-mediated hemolysis of antibody-coated sheep erythrocytes (Ab-SRBCs) by two human complement inhibitors. The calculated percentages (%) of hemolysis of canine complement serum (1:16, y-axis) are plotted against the inhibitor concentration (x-axis). Hemolysis was measured by spectrophotometry at 540 nm (OD_{540}). (A) ab-SRBCs were incubated with canine complement serum and decreasing concentrations of human recombinant C1-esterase Inhibitor (C1-INH, 500 μg/mL to 31.25 μg/mL). (B) Ab-SRBCs were incubated with canine complement serum and decreasing concentrations of C3-inhibitor (Compstatin, 100 μM to 0.001 μM).

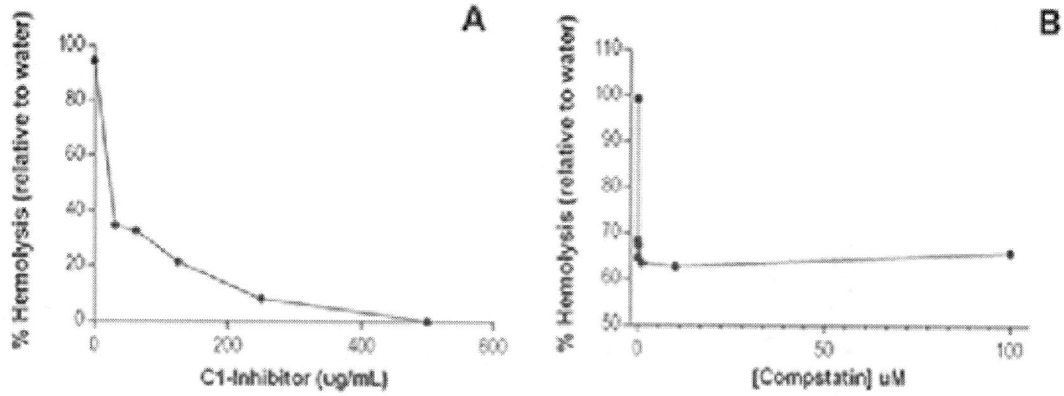

HM16

Novel Thrombogenicity Examination of Whole Blood: Total Thrombus-Formation Analysis System in Dogs (Abstract HM16)

Tomoko Iwanaga[1]; Shusaku Yamada[2]; Ryuji Fukushima[3]; Tomoka Nagasato[1]; Ikuro Maruyama[1]; Naoki Miura[1]

[1]Kagoshima University, Kagoshima, Kagoshima, Japan; [2]Animal Cardiovascular and Thoracic Surgery, Fuchuu, Tokyo, Japan; [3]Tokyo University of Agriculture and Technology, Fuchuu, Tokyo, Japan

Blood clotting is controlled by platelet activation and the coagulation system. Virchow's three conditions must be met to form a clot *in vivo*. However, current blood coagulation examinations are fragmentary. The Total Thrombus-formation Analysis System (T-TAS) is a new laboratory procedure to evaluate the blood clotting ability in which whole blood flows into a microflow chamber imitating blood vessels.

T-TAS utilizes two microchips: the PL-chip and AR-chip. The PL-chip predominantly evaluates platelet function, and the AR-chip evaluates both platelet activity and coagulation simultaneously. The blood clotting ability is evaluated by measuring the internal pressure in the chip, which rises with clot formation. The measurement values are the occlusion start time (OST: time to reach 10 kPa), occlusion time (OT: time to reach 80 kPa), and area under the pressure curve (AUC). In this study, we assessed the reproducibility of each chip in 8 healthy dogs, established the reference range in 54 dogs (PL, n=19; AR, n=35), and confirmed the effect of aspirin in 3 dogs.

The reproducibility of each chip was confirmed based on the coefficient of variation (CV) (PL: CV_{OST}=8.69%, CV_{OT}=13.58%, CV_{AUC}=6.45%; AR: CV_{OST}=10.01%, CV_{OT}=7.65%, CV_{AUC}=1.57%). The reference ranges of the OST and OT for the PL-chip (average ± standard deviation) were 2.26±0.81 and 4.73±1.21 min, respectively (AUC, 406.77±54.36), and those for the AR-chip were 5.02±1.83 and 6.04±2.14, respectively (AUC, 1955.35±156.67). In all dogs, the graph shifted to the right for the PL-chip after aspirin administration, but not for the AR-chip; thus, apparent extension of PL was observed, indicating successful evaluation of platelet suppression by aspirin.

The results of this study indicate that evaluation of blood coagulation in dogs using the T-TAS is useful and that the blood clotting ability in dogs with various diseases might be evaluated more subtly by T-TAS in the future.

HM17

Vaccination Behavior and Adverse Events in Dogs Treated for Primary Immune-Mediated Hemolytic Anemia (Abstract HM17)

Alaina K.B. Moon[1]; Julia Veir[2]

[1]Oregon State University, Corvallis, OR, USA; [2]Colorado State University, Fort Collins, CO, USA

The purpose of this study was to describe the frequency with which dogs treated for primary immune-mediated hemolytic anemia (IMHA) at veterinary teaching hospitals are vaccinated after diagnosis, and investigate whether an association exists between vaccination and vaccine reactions (including but not limited to relapse of IMHA) in this population. A retrospective descriptive observational study was performed involving 66 dogs at 2 veterinary teaching hospitals: Colorado State University (57) and Oregon State University (9). Medical records databases were searched for cases meeting inclusion and exclusion criteria for primary IMHA. Owners and referring veterinarians of identified dogs were contacted with a survey via postal mail to gather information regarding vaccination of the dog following diagnosis. Referring veterinarians were contacted by phone regarding vaccination history for each dog prior to diagnosis of IMHA. At least one survey was returned in 44 cases. Results indicated 22 dogs did not receive vaccinations after diagnosis, and 21 dogs did receive vaccinations after diagnosis. For the remainder of dogs, no survey was returned or the survey was excluded for inconsistency of information. Of the vaccinated dogs, only 2 possible vaccination reactions were reported (vomiting and urticaria in 1 dog, and worsening anemia in 1 dog). A date of vaccination prior to diagnosis was available for 46 dogs. The mean number of days between vaccination and initial diagnosis of IMHA was 351 days. These results indicate that vaccination occurred among nearly half of dogs diagnosed with IMHA, with rabies being the most common vaccine administered (19 of 21 dogs). This may reflect the legal requirement of rabies vaccination in most locations. Adverse events were uncommon overall, but comparison to the rate of adverse events in the general population was not performed. Finally, this study did not document a temporal relationship between vaccination and initial diagnosis of IMHA, although comparison to a control population is necessary to confirm this finding.

HM18

Evaluation of Thromboelastography in Dogs with Heartworm Infection (Abstract HM18)

Seri Park[1]; Yunhye Kim[2]; Chul Park[1]; Junghyun Kim[3]

[1]College of Veterinary Medicine, Chonbuk National University, Iksan, Cholla-bukto, Republic of Korea; [2]College of Veterinary Medicine, Chonbuk National University, Curi, Kyonggi-do, Republic of Korea; [3]College of Veterinary Medicine, Konkuk National University, gwangjin-gu, Seoul-t'ukpyolsi, Republic of Korea

Canine heartworm disease due to *Dirofilaria immitis* is an important disease causing hemostatic disorders. Although there are several studies showing the relationship between coagulation pathway and heartworm infection, most of studies have used conventional methods of laboratory assessment of hemostasis. There is a limitation in using conventional coagulation test to evaluate hypercoagulable state in heartworm infected dogs. Thromboelastography (TEG) provides a more truthful reflection of *in vivo* hemostasis compared to conventional coagulation tests in the diagnosis and monitoring of patients. In this study, we describe TEG findings in dogs with heartworm infection and compare their coagulation status as determined by TEG to clinical presentation, heartworm grade, and conventional coagulation tests.

This study consisted of fifteen healthy dogs and twenty-five dogs with heartworm infection. Heartworm infection was diagnosed using commercial ELISA test kit. Dog with heartworm infection were graded to four groups according to blood work, radiography, echocardiography. Platelet count, prothrombin time, activated partial thromboplastin time, plasma D-dimer concentration, plasma fibrinogen concentration, and thromboelastography were performed in the all dogs. In the results, the value of alpha showed a tendency to increase as the grade increased (table 1, p=0.03), which indicates higher level of fibrinogen activation in more advanced grades. In the rank coefficient correlation between conventional coagulation tests and TEG, platelet had a significantly positive coefficient correlation with MA and G value in heartworm infected grade 1. An interesting result is that a dog diagnosed femoral arterial thromboembolism showed significantly hypercoagulable state in TEG, compared with conventional tests (figure 1). Also, the serial TEG analysis of a dog with disseminated intravascular coagulation represented more hypocoagulable state over time (figure 2). This study suggests that application of TEG is helpful for diagnosis and treatment of heartworm infection in dogs.

Table 1. Mean values and SD for TEG variables in heartworm infected dogs and control dogs.

Variables	Control	HW grade 1	HW grade 2	HW grade 3	HW grade 4	*P* value
R (min)	3.7±1.55	3.5±1.43	4.7±0.87	4.1±0.57	3.7±1.55	0.599
K (min)	1.7±0.6	1.2±0.27	1.6±0.65	1.3±0.23	1.7±0.6	0.168
α (°)	66.3±6.07	72.8±3.47	68.7±8.95	71.9±3.37	66.3±6.07	0.030*
MA (mm)	61.4±6.65	67.0±6.39	70.3±9.90	68.8±7.79	61.4±6.65	0.109
G (dynes/cm²)	8260.0±1994.56	10700.0±3306.90	13350.0±5500.82	12000.0±4599.13	8466.7±6934.94	0.106
LY30 (%)	0.4±0.65	0.0±0.0	0.0±0.04	0.1±0.29	0.4±0.65	0.330

* *P* value for comparison of TEG variables in dogs with heartworm infection. Jockheere-terpstra adjusted P < .05.

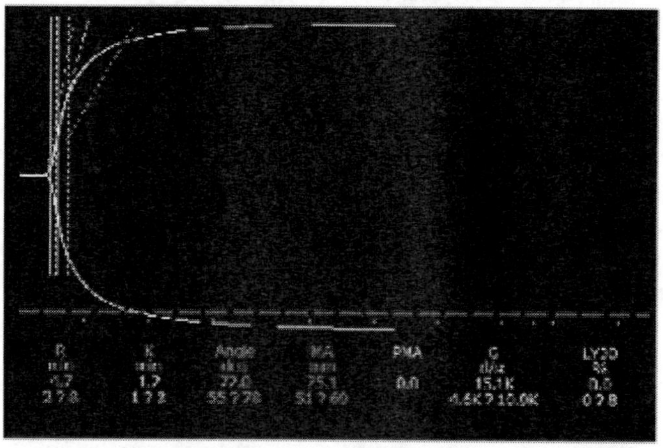

Figure 1. Thromboelastogram of a heartworm infected dog with atrial thromboembolism. It suggested that the coagulation status of dog was hypercoagulable

HM19

Characteristics of Hemostasis During Experimental *Ehrlichia canis* Infection in Laboratory Beagles (Abstract HM19)

Sarah Shropshire; Christine Olver; Michael R. Lappin

Colorado State University, Fort Collins, CO, USA

Ehrlichia canis infection in dogs can manifest with clinical signs related to bleeding and also commonly causes thrombocytopenia. The exact mechanisms of bleeding and thrombocytopenia are unknown but are thought to be related to processes such as vasculitis and immune and non-immune processes on platelets. It is also unknown why some dogs show signs of bleeding while other dogs do not despite significant thrombocytopenia. It is possible that platelets become activated during infection and/or blood clots become resistant to fibrinolysis which could prevent a bleeding phenotype. The purpose of this study was to assess the platelet indices of activation, platelet function via whole blood impedance platelet aggregometry, percentage of immunoglobulin associated platelets (percent IgG), and thromboelastography (TEG) measurements including velocity curve variables in dogs experimentally infected with *E. canis*.

Six healthy laboratory beagles were experimentally infected with *E. canis* via needle inoculation. Four dogs were inoculated from one naturally infected dog and two dogs were inoculated from a second naturally infected dog. Four dogs became positive for *E. canis* via serology and PCR but the remaining two dogs never became positive despite repeated testing. As a result, the four dogs were designated as positive and the two dogs were designated as negative. Platelet indices; platelet count, mean platelet volume (MPV), plateletcrit (PCT), platelet volume distribution width (PDW), mean platelet mass (MPM), mean platelet component concentration (MPC) and platelet component distribution width (PCDW) were measured. For platelet aggregometry, the area under the curve (AUC) was recorded using a saline control (AUC_{saline}) and for each agonist (adenosine diphosphate (ADP) [AUC_{ADP}] and arachidonic acid (AA) [AUC_{AA}]). Percent IgG was assessed by direct flow cytometry. Tissue factor (TF)-activated TEG and TF-activated TEG + tissue plasminogen (tPA) were performed. TEG variables (reaction time [R], clotting time [K], rate of clot formation [α], and maximum amplitude [MA], LY30 [percent clot lysis 30 minutes after MA], CL30 [clot lysis 30 minutes after MA], LY60 [percent clot lysis 60 minutes after the MA], and CL60 [clot lysis 60 minutes after MA]) were recorded in addition to the velocity curve variables: maximal rate of thrombus generation (MRTG), time to maximum rate of thrombus generation (TMRTG), total thrombus generated (TG), maximum rate of lysis (MRL), time to maximal rate of lysis (TMRL), and total lysis (L). Testing was performed in both the positive and negative dogs prior to inoculation and then on weeks 1, 2, 3, 4, and 5. Due to clinical signs (weight loss, hyporexia, lethargy, fever), the positive dogs were administered doxycycline at 10 mg/kg, PO, daily between week 3 and week 4. For each variable, a repeated measures analysis (2-way ANOVA) with *post hoc* analysis was performed comparing the positive and negative dogs. For all analysis, $p<0.05$ was considered significant.

There were no significant differences between groups (positive versus negative) for the variables PDW, MPC, PCDW, AUC_{saline}, R, K, α, CL60, TMRTG and MRL. During infection, the platelet count, PCT, AUC_{ADP}, AUC_{AA}, LY30, and LY60 were significantly lower and the MPV, MPM, percent IgG, MA, CL30, MRTG, TG, TMRL, and L were significantly higher in the positive group. No overt signs of bleeding were noted during the study.

The positive dogs had significantly lower platelet counts but had evidence of activated platelets based on an increased MPV and MPM. There was evidence of decreased platelet function in the positive dogs but this could have been related to the degree of thrombocytopenia. The positive dogs also showed evidence of anti-platelet antibodies during infection. Overall, the positive dogs appeared hypercoagulable based on an increased MA, MRTG, and TG and hypofibrinolytic based on a lower LY30, LY60 and higher

CL30 and TMRL. The characteristics of hemostasis during experimental *E. canis* infection are complicated but platelet indices of activation and evidence of a hypercoagulable, hypofibrinolytic state may explain the lack of a bleeding phenotype in some dogs despite significant thrombocytopenia.

HM20

Whole Blood Impedance Platelet Aggregometry in Healthy Greyhounds Compared to Healthy Beagles (Abstract HM20)

Sarah Shropshire; Christine Olver

Colorado State University, Fort Collins, CO, USA

Whole blood impedance platelet aggregometry can be affected by hematocrit and platelet count in humans. We sought to evaluate whether this phenomenon exists in dogs by comparing blood from Greyhounds, a breed known to have high hematocrits and low platelet counts, to Beagles, a breed with "normal" hematocrits and platelet counts. We hypothesized that aggregometry in Greyhounds may result in erroneous interpretations of platelet function. The purpose of this study was to compare whole blood impedance platelet aggregometry and other hemostatic assays in healthy Greyhounds to healthy Beagles.

Whole blood impedance platelet aggregometry and tissue factor-activated thromboelastography (TF-TEG) were performed in six healthy client-owned Greyhounds and six healthy laboratory Beagles. For platelet aggregometry, the area under the curve was recorded for each agonist (adenosine diphosphate [ADP] and arachidonic acid [AA]) and a saline control. For TF-TEG, reaction time (R), clotting time (K), rate of clot formation (α), maximum amplitude (MA), percent clot lysis 30 minutes after MA is reached (LY30), amount of clot lysis 30 minutes after MA is reached (CL30), percent clot lysis 60 minutes after the MA is reached (LY60), and amount of clot lysis 60 minutes after MA is reached (CL60) were recorded. The hematocrit, platelet count, and fibrinogen were also recorded. An unpaired t-test was performed on ADP, AA, hematocrit, platelet count, fibrinogen, R, α, MA, and a Mann-Whitney U test was performed on the saline control, K, LY30, CL30, LY60, and CL60. Pearson correlation was also assessed between hematocrit and ADP, AA, and MA, platelet count and ADP, AA, and MA, and fibrinogen and MA. For all tests, $p<0.05$ was considered significant.

The ADP, AA, platelet count, fibrinogen, MA, CL30, and CL60 were significantly lower and the hematocrit, LY30, and LY60 were significantly higher in the Greyhounds. There were no statistical differences for the saline control, R, K, and α between groups. In all twelve dogs, there was correlation between hematocrit and MA (r=-0.70), ADP (r=-0.91), and AA (r=-0.81), and platelet count and ADP (r=0.9), AA (r=0.72), and MA (r=0.68) and fibrinogen and MA (r=0.85).

In agreement with previous studies, healthy Greyhounds had significantly higher hematocrits and lower platelet counts. Based off of the TEG tracings, Greyhounds appeared significantly hypocoagulable (decreased MA) and hyperfibrinolytic (increased LY30, LY60, and decreased CL30, CL60). Future studies should investigate whether hyperfibrinolysis contributes to the propensity to bleed in Greyhounds. The platelet aggregometry results were significantly lower for both agonists in Greyhounds as compared to Beagles which could be erroneously interpreted as decreased platelet function in this breed. Therefore, just as hematocrit and platelet count can affect TEG tracings, the same considerations should be applied to whole blood impedance platelet aggregometry. Both the high hematocrit and low platelet counts in healthy Greyhounds can potentially contribute to lower aggregometry readings due to suspected *in vitro* effects rather than reflecting true *in vivo* effects since other studies have demonstrated normal platelet function in Greyhounds. Therefore, it is not recommended to compare whole blood impedance platelet aggregometry measurements of Greyhounds to non-Greyhounds. Subject-based or population-based reference intervals should be established in this breed for accurate interpretation of results.

HM21

Thawing Times and Hemostatic Proteins in Canine Fresh Frozen Plasma Using a Commercial Thawing Device (Abstract HM21)

Laurie Torkildsen[1]; Medora Pashmakova[1]; Micah Bishop[2]; James Barr[1]

[1]Texas A&M University, College Station, TX, USA; [2]WAVE Veterinary Internal Medicine, Naples, FL, USA

The objectives of this study were to compare thawing times between a 37°C warm water bath (WWB), running water bath (RWB), and Plasmatherm and compare hemostatic protein function in canine fresh frozen plasma (FFP) thawed with WWB or Plasmatherm.

Thawing time was measured for a 240 mL bag of FFP thawed in WWB, RWB, or Plasmatherm 10 times. Fresh canine donor plasma samples were split into 120 mL bags and 3 mL control aliquots prior to freezing. Bags were thawed by WWB or Plasmatherm and aliquots equilibrated to room temperature. Concentrations of hemostatic proteins, albumin, D-dimers, prothrombin time (PT) and partial thromboplastin time (PTT) were obtained. Repeated measures ANOVA, or Friedman, as appropriate with corresponding post tests were used to compare groups. Alpha was set to 0.05.

There was an overall difference in thawing times between the three groups ($p=0.0274$). Post testing revealed significantly shorter thawing time with RWB as compared to Plasmatherm ($p=0.0417$). A significant difference in PTT was noted overall ($p<0.001$) and post testing revealed longer PTT in Plasmatherm and WWB groups compared to control ($p=0.0010$, $p=0.0052$, respectively). There was an overall significant difference in factor VII concentration ($p=0.0007$) and post testing revealed decreased factor VII in Plasmatherm and WWB groups compared to control($p=0.0010$, $p=0.0041$, respectively). A significant difference in factor X concentration was seen overall ($p=0.0317$) and post testing revealed decreased factor X in Plasmatherm compared to control ($p=0.0417$).

The RWB had the fastest thawing time. Significant differences were detected in hemostatic protein function but these differences are not considered clinically significant.

HM22

Clinical Evaluation of the QuickVet/RapidVet Feline Blood Typing Test (Abstract HM22)

Christiane Weingart; Janina Assmann; Barbara Kohn

Clinic of Small Animals, Faculty of Veterinary Medicine, Freie Universität Berlin, Berlin, Germany

Blood transfusions have become an integral part of feline intensive care treatment. For patient safety blood typing is an important pretransfusion test.

Objective of this prospective study was evaluation of an automated feline cartridge blood typing method (QuickVet®/RapidVet®(QVRV) SMB, Farum, Denmark) and comparison of results with slide and tube (reference) methods (not commercially available). Samples with discordant results were additionally tested with Alvedia® (Quick Test A+B, Alvedia, Limonest, France). To determine precision, blood typing of 3 A and B samples were repeated 5 times within 24 hours and over 5 consecutive days.

83 EDTA-anticoagulated blood samples routinely collected for diagnostic purposes and 16 provided by outside laboratories were available. (One sample with severe hemolysis could not be typed with QVRV.) Slide/tube tests yielded the same results on all 98 samples. 78 samples were blood type A, 14 B and 6 AB. QVRV identified 75/78 of the A and 14/14 of the B samples. Three A samples with low hematocrit (≤ 10%) could only be typed after diluting them. Six AB samples gave discordant results compared to the reference method: 1 AB sample was typed as blood type A with QVRV (and Alvedia®) and 5 as B with QVRV (and AB with Alvedia®). All repeats yielded the same results. Based on all samples run, QVRV agreed in 90.8% with the reference method (exclusively for A and B cats, the agreement was 96.5%). Blood samples with type B should be confirmed by back typing and samples with type AB by tube method.

HM23

Characterization of Thrombocytosis in 715 Dogs (2011–2015) - (Abstract HM23)

Andrew Woolcock; Alessandra Keenan; Cheryl Cheung; George E. Moore

College of Veterinary Medicine, Purdue University, West Lafayette, IN, USA

Thrombocytosis is an uncommon hematologic abnormality in canine patients that has been associated with various neoplastic, metabolic, and inflammatory conditions. The aim of this study was to characterize thrombocytosis in dogs by its severity and the diseases to which it is associated.

Medical records of all dogs with thrombocytosis between 2011–2015 were reviewed. The thrombocytosis for each patient was characterized as primary or secondary, and then further classified by severity and diagnosis. Diagnostic categories included neoplasia, endocrine disease, inflammatory disease, or miscellaneous. Inflammatory disease was further classified by the body system(s) involved. Serum and plasma potassium concentration, if available, were evaluated for the possibility of pseudohyperkalemia.

A total of 1254 complete blood counts with thrombocytosis from 715 dogs were included in the study. Median platelet count in this population was 582 x 10^3/μL (500–1810 x 10^3/μL). No correlation between severity of thrombocytosis and diagnosis was identified. Secondary causes included neoplasia (55.7%), endocrine disease (12.0%), and inflammatory disease (46.6%). Of the dogs with neoplasia, carcinoma (52.0%) and lymphoma (27.2%) were the most common diagnoses. Immune-mediated disease was common (22.2%), associated with frequent glucocorticoid administration, and had a significantly higher median platelet count when compared to the other body systems ($p<0.001$).

Thrombocytosis is commonly associated with neoplasia and immune-mediated disease diagnoses in dogs. Further research is warranted to investigate the role of thrombocytosis in hepatobiliary and renal disease.

HP01

Biomarkers of Oxidative Stress and Redox Status of the Liver in Dogs with Liver Disease (Abstract HP01)

Caitlin Barry-Heffernan[1]; Joanne L. Ekena[2]; Sarah Dowling[3]; Katrina Viviano[1]

[1]University of Wisconsin, Madison, WI, USA; [2]Department of Medical Sciences, School of Veterinary Medicine, University of Wisconsin-Madison, Madison, WI, USA; [3]Pet Emergency Treatment and Specialties, Inc., Lancaster, PA, USA

Oxidative stress plays a significant role in canine liver disease. The main intracellular antioxidant, reduced glutathione (rGSH), is decreased in dogs with naturally occurring liver disease. It is unknown whether circulating rGSH levels correlate with liver concentrations. The aims of this study were to determine whether circulating rGSH correlates with liver rGSH in dogs with naturally occurring liver disease and to evaluate whether other markers of systemic oxidative stress, plasma vitamin E and urine 8-isoprostanes/creatinine (8-IsoP), correlate with liver rGSH.

Dogs with liver disease presenting for liver biopsies as part of their clinical evaluation were included. Urine and blood were collected within 24 hours of liver biopsy. Plasma, erythrocyte, and liver rGSH were measured using high performance liquid chromatography (HPLC); vitamin E was measured by HPLC, and 8-IsoP was measured by gas chromatography/mass spectrometry.

Thirty-one dogs, of the targeted enrollment of 34, have completed the study. The median rGSH in the liver, plasma, and erythrocytes were 5.78 mM, 3.90 μM, and 1.88 mM, respectively. No correlations were found between liver rGSH and erythrocyte rGSH, vitamin E, or 8-IsoP. There was a modest negative correlation between liver and plasma rGSH (ρ - 0.40; P 0.043). Dogs with liver disease had significantly higher urine 8-IsoP compared with healthy controls (5885 vs. 2980 pg/mg Cr; $P < 0.0001$). The initial results of this study demonstrate that in dogs with naturally occurring liver disease only plasma rGSH correlates with liver rGSH, albeit negatively. In addition, dogs with naturally occurring liver disease have evidence of increased systemic oxidative stress compared to healthy controls.

HP02

Canine Cholangitis/Cholangiohepatitis: A Descriptive Study in 54 Dogs (2004–2014) (Abstract HP02)

Jennifer L. Harrison[1]; Bradley Turek[2]; Dorothy Cimino Brown[3]; Charles W. Bradley[1]; Julie Callahan Clark[4]

[1]School of Veterinary Medicine, University of Pennsylvania, Philadelphia, PA, USA; [2]Antech Diagnostics, Irvine, CA, USA; [3]University of Pennsylvania, Philadelphia, PA, USA; [4]New England Animal Medical Center, West Bridgewater, MA, USA

Canine cholangitis appears to be more common than previously thought; yet our understanding of the disease remains incomplete. The study aim was therefore to describe a population of dogs with biopsy confirmed cholangitis.

Medical records and biopsies of fifty-four dogs with cholangitis or cholangiohepatitis confirmed via histopathology between January 2004 and December 2014 were retrospectively reviewed.

Presenting clinical signs included vomiting (72.2%), lethargy (70.4%) and inappetence (64.8%). The majority of dogs (49/50) had liver enzyme elevations, hyperbilirubinemia (32/50) and hypercholesterolemia (24/43). Ultrasonographic abnormalities of the liver and/or biliary tree were seen in 84% of cases. Histopathology revealed that 53/54 cases had neutrophilic cholangitis/cholangiohepatitis while 1 case had lymphocytic cholangitis. The majority (42/54) were chronic. Histopathology revealed concurrent gallbladder disease in 46.2% of cases and biliary tract obstruction in 42.6% of cases. Seventeen of 36 biliary and 11/25 liver cultures were positive for bacterial growth; *Escherichia coli* and *Enterococcus* spp. were most common. The median patient survival was 671 days (CI 95% 114, 1426). Controlling for age, dogs that did not have a cholecystectomy had a 2.1 greater hazard ratio for death ($p=0.001$, CI 95% 1.0, 4.3) compared to dogs that had a cholecystectomy. Controlling for cholecystectomy, dogs greater than 13 years of age had an increased risk of death ($p=0.037$, CI 95% 1.9, 13.2) compared to those less than 13 years.

In this population of dogs, chronic neutrophilic cholangitis/cholangiohepatitis was most common. Gallbladder disease, specifically cholecystitis and biliary tract obstruction, occurred often in conjunction with cholangitis. In this study dogs with cholangitis and biliary tract disease that underwent cholecystectomy had a reduced risk of death, thus cholecystectomy may improve patient outcome.

HP03

Hepatic Fibrosis is Associated with Shortened Survival Times in Dogs with Chronic Hepatitis (Abstract HP03)

Christopher R. Dolan[1]; Aline Rodrigues Hoffmann[2]; Randi M. Gold[2]; Jan S. Suchodolski[1]; Jörg M. Steiner[1]; Jonathan A. Lidbury[1]

[1]Gastrointestinal Laboratory, Texas A&M University, College Station, TX, USA; [2]Department of Veterinary Pathobiology, Texas A&M University, College Station, TX, USA

Chronic hepatitis is a common condition in dogs and fibrosis is an important component of this disease process. In human patients with chronic hepatitis prognosis is related to the stage of hepatic fibrosis. Therefore, the aim of this study was to determine if there is a relationship between the stage of hepatic fibrosis and survival times in dogs with chronic hepatitis.

Dogs histologically diagnosed with chronic hepatitis between January 2003 and January 2013 at the Veterinary Medical Teaching Hospital, Texas A&M University were retrospectively enrolled. Only dogs for which sufficiently detailed follow-up medical records were available were included in the study. For each dog the time of liver biopsy, survival time, and cause of death or euthanasia were recorded. Hepatic fibrosis was staged by a board-certified veterinary pathologist (ARH) using a 5-point scoring system adapted from the Ishak scoring system (absent, mild, moderate, marked, very marked). The survival of dogs with absent to moderate and marked to very marked hepatic fibrosis was compared using Kaplan-Meyer survival curve analysis and the Mantel-Cox test. Statistical significance was set as $p<0.05$. Medical records from 112 dogs with chronic hepatitis were reviewed. Sufficiently detailed follow-up records were available for 47 dogs. Median survival times were 2,438 days for dogs with absent to mild fibrosis ($n=19$) and 675 days for dogs with marked to very marked hepatic fibrosis ($n=28$; $p=0.0126$). The median survival time of dogs with absent to mild fibrosis were significantly longer than that of patients with marked to very marked fibrosis.

HP04

Proteomic Analysis of Liver Tissue from Dogs with Chronic Hepatitis (Abstract HP04)

Yuri A. Lawrence[1]; Lawrence J. Dangott[2]; Jörg M. Steiner[1]; Jan S. Suchodolski[1]; Jonathan A. Lidbury[1]

[1]Gastrointestinal Laboratory, Texas A&M University, College Station, TX, USA; [2]Department of Biochemistry and Biophysics, Texas A&M University, College Station, TX, USA

Chronic hepatitis is a common cause of morbidity and mortality in dogs. The identification of proteins that are differentially expressed in dogs with chronic hepatitis could contribute to the development of diagnostic markers of this disease and provide insights into its pathogenesis. The objective of this study was to identify proteins that are differentially expressed in the liver of dogs with chronic hepatitis.

Hepatic tissue was collected from 8 healthy dogs during ovariohysterectomy and also from 8 dogs with histologically confirmed chronic hepatitis. The proteome of the liver tissue was extracted by mechanical disruption and detergent-based cell lysis and was analyzed by 2-dimensional fluorescence difference gel electrophoresis. Proteins were separated in the first dimension by isoelectric focusing on a 3–10 nonlinear immobilized pH gradient and by 12% sodium dodecyl sulfate-polyacrylamide gel electrophoresis in the second dimension. DeCyder 2-D Differential Analysis Software® was utilized for data analysis. Proteins with an absolute fold change value >2.0 were selected for further analysis. Protein identification was achieved by nanoflow liquid chromatography tandem mass spectrometry (NANO LC-MS/MS). Post-hoc testing was performed using Bonferroni's multiple comparisons test and significance was set at $p<0.05$.

Twenty-eight protein spots showed an increased abundance and 12 showed a decreased abundance in liver tissue from dogs with chronic hepatitis as compared with liver tissue from healthy control dogs. Sixteen proteins were selected for further analysis. Proteins significantly increased in the liver of dogs with chronic hepatitis included protein disulfide-isomerase A4, carbamoyl-phosphate synthase, senescence marker protein 30, formimidoyltransferase-cyclodeaminase, and complement C3. These proteins are involved in a variety of processes including inflammation, redox homeostasis, as well as detoxification.

Further studies are needed to determine whether some of the proteins identified may prove useful as a biomarker of chronic hepatitis in dogs.

HP05

Characterization of the Urine Metabolome of Dogs with Chronic Hepatitis (Abstract HP05)

Yuri A. Lawrence[1]; Blake C. Guard[1]; Jörg M. Steiner[1]; Jan S. Suchodolski[1]; Jonathan A. Lidbury[1]

[1]Gastrointestinal Laboratory, College of Veterinary Medicine, Texas A&M University, College Station, TX, USA

Canine chronic hepatitis can present a diagnostic challenge with various etiologies leading to similar clinical signs and laboratory findings. The histopathological assessment of a liver biopsy specimen is often required for a definitive diagnosis to be made and few prognostic markers exist. The emerging science of metabolomics is used to detect changes in endogenous low molecular weight metabolites in biological samples and offers the possibility of identifying non- or minimally-invasive surrogate markers of disease. The objective of this study was to characterize metabolite changes in the urine of dogs with chronic hepatitis using an untargeted methodology.

Urine samples were collected by cystocentesis from 10 adult dogs with chronic hepatitis and 10 healthy adult control dogs. Urine samples were analyzed by gas chromatography time-of-flight mass spectrometry (GC-TOF). Data were found to be non-parametric. Principal component analysis was used to make multivariate comparisons between dogs with chronic hepatitis and healthy control dogs. A Wilcoxon rank-sum test was used to make univariate comparisons between groups. p values were adjusted for multiple comparisons using the Benjamini and Hochberg False Discovery Rate and significance was set at $p<0.05$.

Principal component analysis revealed significant differences between dogs with chronic hepatitis and healthy control dogs. The univariate analysis demonstrated statistically significant differences between dogs with chronic hepatitis and healthy control dogs for 25 of 220 named metabolites. The most clinically relevant of these included glutamine, glycylproline, putrescine, and ornithine, which are important stimulators of M2 macrophages that play a role in tissue repair and proliferation.

Significant differences between the urine metabolites of dogs with chronic hepatitis and healthy control dogs were identified. Further targeted assessment of the altered metabolites is needed to validate these observations.

HP06

Hepatic Leptospiral Infections Without Prominent Renal Involvement (Abstract HP06)

Katie E. McCallum[1]; Fernando Constantino-Casas[1]; Andre J. Kortum[1]; Niamh Lingley[2]; Alex J. Sterritt[3]; Harry Swales[4]; James Warland[5]; Tristan Cogan[6]; Penny J. Watson[1]

[1]University of Cambridge, Cambridge, England, UK; [2]Suffolk, England, UK; [3]Highcroft Veterinary Group, Bristol, England, UK; [4]University of Liverpool, Neston, England, UK; [5]Animal Health Trust, Suffolk, England, UK; [6]University of Bristol, Bristol, England, UK

Leptospiral infection is an infectious disease seen in dogs that has been shown to cause dysfunction in multiple organ systems but typically demonstrates a predilection to the liver and kidneys. Reports of chronic hepatitis caused by *Leptospira* spp. with absence of renal involvement are confined to small case series in individual kennels. To date, the role of this organism in idiopathic chronic hepatitis remains unknown. However, with the recent availability of Fluorescence In-Situ Hybridisation (FISH), accurate identification of spirochaetes in liver samples is possible. Consequently, this may help elucidate the role of the organism in cases of chronic canine hepatitis. The aim of this study was to describe the clinico-histopathological presentation of cases of hepatic leptospirosis diagnosed by FISH, with an absence of renal involvement.

Cases included in this study consisted of cases presented to the Queen's Veterinary School Hospital, University of Cambridge from 2013–2016 or cases consulted for telephone advice. Dogs were included in the study if they had a histological diagnosis of hepatitis and the presence of leptospires identified by FISH, with clusters suggesting living, reproducing organisms. Liver histology in all cases was reviewed by a single pathologist. Cases without histopathological diagnosis were excluded. The presence of clinical evidence of renal involvement: - azotaemia (serum creatinine >140 mmol/L - based on IRIS guidelines) and glycosuria were further criteria of exclusion.

Ten dogs were included in the study. Age at presentation ranged from 2.8 years to 10 years (median 8.5 years). Six out of ten were spaniels: breeds represented included Cavalier King Charles Spaniel (n=1), Labrador Retriever (n=2), Cocker Spaniel (n=2), English Springer Spaniel (n=3), West Highland White Terrier (n=1) and Jack Russell Terrier (n=1). Eight of the cases were female and two were male. Clinical signs varied but most commonly included anorexia (n=6), abdominal pain (n=2), vomiting (n=2), lethargy (n=4), gingivitis (n=2), pyrexia (n=3), and polydipsia/polyuria (n=2). Two dogs were siblings that lived together and developed clinical signs at the same time. Three dogs were asymptomatic but had persistent elevations in liver enzymes. Mild thrombocytopenia was reported in four cases and all cases had an elevation in liver enzymes. Histopathology of the liver documented chronic hepatitis in all cases with granulomatous hepatitis in seven cases, one of which showed sheets of macrophages, and three cases demonstrated lymphoplasmacytic/neutrophilic hepatitis. Three dogs underwent repeat liver biopsy 3, 15 and 16 months following initial diagnosis, respectively. Fibrosis and vacuolar hepatopathy was diagnosed in the first case and cirrhosis in the second case (both of these cases remained positive for *Leptospira* by FISH). The case re-biopsied at 16 months was diagnosed with granulomatous hepatitis and had become negative by FISH. One patient with a histopathological diagnosis of neutrophilic, lymphoplasmacytic gingivitis and vasculitis demonstrated the presence of *Leptospira* by FISH. Leptospirosis speciation in one case of granulomatous hepatitis documented the presence of *Leptospira interrogans*.

All dogs were treated with a minimum of 10 mg/kg doxycycline PO SID for 14 days. Clinical outcome was known for all cases: six dogs were alive at the time of writing and four dogs were euthanised as a result of progressive liver disease. This study suggests that the presence of leptospiral organisms within the liver may be associated with chronic hepatitis (often granulomatous) without clinical evidence of renal involvement. We intend to undertake further studies that perform sequencing of *Leptospira* spp. organisms as well as performing FISH on control liver biopsies with an absence of hepatitis.

HP07

Demographics and Histological Changes in a Cohort Group of Dogs with Abnormal Hepatic Copper Concentrations (Abstract HP07)

Tarini Ullal; David Twedt

Colorado State University, Fort Collins, CO, USA

Liver disease associated with abnormal hepatocyte copper (Cu) concentrations occurs commonly in dogs. Excessive hepatic Cu results in hepatocyte damage from oxidative stress, resulting in cell death and subsequent necroinflammatory changes. Genetic mutations in Cu metabolism, impaired biliary excretion of Cu, and increased dietary Cu are thought to contribute to excessive hepatic Cu.

The aim of this study was to document trends in age, sex, and breed for dogs having abnormally elevated hepatic Cu and to then correlate the abnormal copper content to the degree of histopathological inflammatory change. We hypothesized that abnormal hepatic Cu concentrations would be more common in middle aged to older female dogs with certain breeds being overrepresented. We also hypothesized that the higher the Cu content, the more severe the inflammatory liver changes would be.

We examined our diagnostic laboratory records for cases having both liver histopathology and hepatic Cu quantitation in the year 2014. Fresh liver tissue was analyzed for Cu via flame atomic absorption spectroscopy and expressed as μg/g dry weight liver ([dwl], normal 120–400). We elected to evaluate a subset of dogs having hepatic Cu >1000 μg/g because those cases would more likely reflect a primary copper storage disorder. Histopathologic diagnoses were grouped as inflammatory or non-inflammatory. Inflammatory samples were sub-classified as mild, moderate or severe. Non-inflammatory samples were characterized as hepatocellular swelling, hyperplasia, lipidosis, nonspecific reactive hepatopathies, fibrosis, or neoplasia.

628 cases met the criteria of having both liver histology and Cu quantitation. 125 cases (20%) had hepatic Cu >1000 (median 1670 μg/g dwl). Ages ranged from 1–14 years (median 7.0 years). 56% were female. 47 different breeds were represented. The Labrador retriever (21.6%), Doberman pinscher (8%), West Highland white terrier (5.6%), and standard poodle (4%) were the four most common pure breeds. Median age for Labrador retrievers was 8 compared to 5 years for Doberman pinschers. 52% of Labrador retrievers, 70% of Doberman pinschers, 80% of standard poodles, and none of the West Highland white terriers were female. Of the 326 cases having normal copper levels, the Labrador retriever (10%), Chihuahua (5.2%), Maltese (4.9%), standard poodle (4.6%), shih tzu (4.3%), and Yorkshire terriers (4%) comprised the most cases.

108 cases (86%) having hepatic Cu >1000 also had inflammatory changes on histopathology. The inflammation was classified as mild (16.7%), moderate (58.3%), or severe (24%). Dogs with inflammatory changes were far more likely to have elevated hepatic Cu ($p<0.001$). Specifically, Cu dogs with moderate to severe inflammation had significantly higher hepatic Cu concentrations ($p<0.0001$). The Labrador retriever was the breed most represented in the moderate and severe groups.

Our findings indicate that the Labrador retriever is a common breed with copper associated liver disease. However, this may be an overrepresentation due to the popularity of the breed. More female cases had elevated hepatic Cu in all breeds with the exception of the West Highland white terrier, in which only male cases were identified. Affected dogs in this study were on average 7–8 years old at the time of diagnosis except for Doberman pinschers, which were on average younger. Higher levels of hepatic Cu were more likely associated with more severe inflammatory changes on histopathology.

HP08

Coagulation Abnormalities and their Correlation to Use of Gelfoam® to Aid in Hemostasis During Laparoscopic Liver Biopsy in Dogs (Abstract HP08)

Sara A. Wennogle[1]; Christine Olver[2]; David Twedt[2]

[1]College of Veterinary Medicine, Colorado State University, Fort Collins, CO, USA; [2]Colorado State University, Fort Collins, CO, USA

Coagulation abnormalities are common in human and veterinary patients with hepatobiliary disease. Due to the liver's complex and varied influence on hemostasis these derangements can result in a multitude of outcomes in the patient, including a propensity towards bleeding and/or thrombosis. Hepatic biopsy is commonly indicated in the diagnostic work-up of hepatobiliary disease in dogs, and increasingly performed via laparoscopy. Knowledge of a patient's hemostatic status prior to biopsy is considered essential to assess bleeding risk. However, conventional tests are not always accurate predictors of clinical bleeding. Additionally, newer techniques for evaluating the coagulation system have recently become available.

The objective of our study was to evaluate both conventional and modern tests of the coagulation system in a group of dogs undergoing laparoscopic liver biopsy for diagnostic work-up of hepatobiliary disease. Results of these tests were then correlated with the use of intervention to control clinical bleeding during laparoscopic liver biopsy.

27 dogs undergoing laparoscopic liver biopsy at Colorado State University were prospectively enrolled. Blood was collected prior to the procedure for chemistry profile, platelet count, platelet aggregometry (Multiplate® analyzer), thromboelastography, prothrombin time, activated partial thromboplastin time, antithrombin activity, fibrinogen, d-dimers, and fibrin degradation products. Bleeding was scored based on whether Gelfoam® (absorbable gelatin sponge, USP) was utilized by the laparoscopist to aid in control of bleeding. Gelfoam® was applied when active bleeding persisted for >30 seconds at one or more biopsy sites.

Histology confirmed hepatic disease in all dogs (Idiopathic chronic hepatitis; n=12; Copper-associated hepatopathy: n=5; Hydropic/Vacuolar: n=5; Non-specific reactive: n=3; Vascular: n=1; Lymphoma: n=1). 18/27 (67%) dogs had at least one coagulation abnormality. 5/27 (19%) dogs had Gelfoam® placed to aid hemostasis in one or more biopsy sites. Hypoalbuminemia (p=0.009), decreased antithrombin activity (p=0.003), and older age (p=0.038) were associated with the use of Gelfoam®. 14/27 (52%) dogs had abnormal platelet aggregometry, including all 5 dogs who received Gelfoam®. Abnormal platelet aggregometry was associated with the use of Gelfoam® to aid in hemostasis (p=0.041). Prolonged PT or APTT were not associated with use of Gelfoam®. Hypofibrinogenemia was also not associated with Gelfoam® use, but was correlated with hypoalbuminemia (p=0.042). Bleeding was controlled with Gelfoam® intervention alone in all dogs. One dog sustained a splenic laceration during entry into the abdomen, but hemostasis was achieved quickly. Otherwise, no significant complications occurred.

Thrombocytopathia may have contributed to clinical bleeding during laparoscopic liver biopsy in this group of dogs with hepatobiliary disease. Additionally, all dogs with low antithrombin had Gelfoam® placed at biopsy sites to aid hemostasis. Low antithrombin may reflect decreased liver function and/or be due to consumption as a consequence of low-grade disseminated intravascular coagulation. As hypofibrinogenemia was correlated with hypoalbuminemia, it is possible fibrinogen could be assessed as an additional marker of hepatic function. In these 27 dogs, laparoscopic liver biopsy was safe and adequate samples to achieve a histologic diagnosis were obtained in all cases.

HP09

Bleeding Risk Assessment for Percutaneous Ultrasound Guided Hepatic Biopsy in Dogs and Cats (Abstract HP09)

Michelle K. Pavlick[1]; Cynthia R. Leveille-Webster[2]; Dominique G. Penninck[2]

[1]Cummings School of Veterinary Medicine, Tufts University, Worcester, MA, USA; [2]Cummings School of Veterinary Medicine at Tufts University, North Grafton, MA, USA

Hepatic biopsy is often necessary for a definitive diagnosis of liver disease. Because the liver plays a central role in hemostasis, bleeding after biopsy procedure is a concern. The aim of this study was to determine risk factors for bleeding following percutaneous ultrasound-guided liver biopsy. The records of 104 dogs and 30 cats that had a percutaneous ultrasound-guided liver biopsy were retrospectively reviewed. Based on human guidelines for bleeding, blood loss was classified as minor, moderate and severe by an absolute decrease in the hematocrit of ≤6%, 7–14% and ≥15%, respectively. Complications were defined as physiologic compromises that necessitated an intervention (transfusion or resuscitative fluids) or death. The relationship between change in hematocrit/complications and initial hematocrit, PT, aPTT, platelet count, serum parameters (liver enzymes, bilirubin) as well as the number of biopsies, biopsy needle gauge, final histological diagnosis and ultrasonographic findings (focal vs. diffuse, nodularity, ascites) were assessed using Chi square tests, One Way Analysis of Variance or Pearson's correlation coefficient with a p value <0.05 considered significant.

Mild, moderate and severe bleeding occurred in 61 (58.7%), 33 (31.7%) and 10 (9.6%) dogs, respectively and complications in 2 (1.9%) dogs. Dogs with a higher initial hematocrit had a greater decrease in hematocrit. Dogs with neoplasia were more likely to have complications. Mild, moderate and severe bleeding occurred in 13 (43.3%), 16 (53.3%) and 1 (3.3%) cat(s), respectively and complications in 5 (16.7%) cats. Moderate/severe bleeding was more likely in cats with a nodular liver. Cats with a higher initial platelet count had a greater decrease in hematocrit. Complications were associated with a lower pre-biopsy hematocrit in cats. In both dogs and cats, there was no correlation between the change in hematocrit/complications and PT, aPTT, bilirubin, serum liver enzyme activity, the number of biopsies, biopsy needle gauge, histological diagnosis, vitamin K administration or the presence of ascites, focal or diffuse disease.

Ultrasound-guided liver biopsy is a relatively safe procedure. The risk of complications is higher in cats and is associated with a lower pre-biopsy hematocrit. The reason for the paradoxical findings that higher platelet number and hematocrit in the cat and dog respectively are associated with greater decrease in hematocrit will require further investigation. Typical indicators of hypocoagulability (prolonged PT, aPTT and low platelet count) did not predict the occurrence of complications or the magnitude of change in hematocrit post-biopsy. This likely reflects the complex state of coagulation that exists in liver disease.

HP10

High-Throughput RNA Sequencing and Differential Gene Expression Analysis in Dogs with Chronic Hepatitis (Abstract HP10)

Vera M. Eulenberg[1]; Yuri A. Lawrence[1]; Jan S. Suchodolski[1]; Jörg M. Steiner[1]; Jonathan A. Lidbury[1]

[1]Gastrointestinal Laboratory, College of Veterinary Medicine, Texas A&M University, College Station, TX, USA

Hepatic fibrosis is the result of chronic hepatic injury and can lead to loss of functional hepatic mass and portal hypertension and thus is an important event in the progression of chronic hepatitis. The pathogenesis of hepatic fibrosis in dogs has not yet been fully characterized. The aim of this study was to identify differentially expressed genes in dogs with chronic hepatitis to further elucidate the molecular mechanisms involved in the development and regulation of hepatic fibrosis.

RNA extraction and high-throughput RNA sequencing were performed on liver biopsy specimens collected from 16 dogs with chronic hepatitis and 12 healthy control dogs undergoing ovariohysterectomy. The sequencing data were analyzed using CLC Genomics software (CLC bio, QIAGEN) and the included empirical analysis of differential gene expression. Genes with an absolute fold change-value between groups >2.5 and a multiple comparisons adjusted p value <0.05 were considered to be differentially expressed. Differentially expressed genes and transcripts potentially related to the pathogenesis of hepatic fibrosis were selected and subcategorized using the genecards.org databases (GeneCards, PathCards, and GeneAnalytics).

Six-hundred-and-fifty-one genes (1,136 transcripts) were differentially expressed in dogs with chronic hepatitis. A connection with hepatic fibrosis or related pathways was identified for 190 of these genes (269 transcripts). One-hundred-and-sixty-five of these genes (222 transcripts) were upregulated and 25 (47 transcripts) were downregulated. They were grouped into the following subcategories: extracellular matrix degradation, extracellular matrix organization, hepatic stellate cell activation, apoptosis, detoxification of reactive oxygen species, and components of fibrotic mechanisms mentioned in the human literature.

A large number of genes were found to be differentially expressed in dogs with chronic hepatitis compared to healthy control dogs and some of these genes may potentially be related to the pathogenesis of hepatic fibrosis. Further study of this dataset and confirmatory testing of a subset of the identified genes with quantitative RT-PCR might provide further insights into the pathogenesis of canine hepatic fibrosis.

IM01

In Vitro Effects of Vitamin D on Phagocytosis, TLR4, and Cytokine Production in Healthy Dogs (Abstract IM01)

Jared Jaffey; Juliana Amorim; Amy DeClue

University of Missouri Veterinary Health Center, Columbia, MO, USA

Hypovitaminosis D has been identified in dogs with sepsis. In people, calcitriol reverses the immunopathology during sepsis, yet little is known about the effects of calcitriol on innate immune function in the dog. We hypothesized that calcitriol will reduce canine leukocyte TNF production, increase IL-10 production, augment PMN phagocytosis and reduce TLR4 expression.

Blood was collected from eight healthy, adult dogs and incubated with calcitriol (10^{-7} M) or control for 24h. Following incubation, LPS-stimulated leukocyte production of TNF and IL-10 were measured using a canine-specific multiplex assay and PMN phagocytosis of opsonized *E. coli* and constitutive TLR4 expression were evaluated using flow cytometry. In three dogs, a concentration response curve was performed to evaluate if the cytokine modulation was concentration dependent. Comparisons were made using paired T-test or Wilcoxon Signed Rank test, with a $p<0.05$ considered significant.

Calcitriol significantly reduced LPS-stimulated leukocyte TNF production (mean±SD, calcitriol, 823±862; control, 1694±1023 pg/mL; p=0.022) and TNF:IL-10 ratio (median, Q1, Q3, calcitriol, 0.9, 0.2, 1.1; control, 3.5, 1.4, 6.9; p=0.039) but not IL-10 production (calcitriol, 791±386, control, 551 ± 323 pg/mL; p=0.114). Blunting of TNF production was concentration dependent (concentration range: 10^{-9} to 10^{-7} M). Phagocytosis and TLR4 expression were not significantly altered by calcitriol.

These data indicate that calcitriol induces an anti-inflammatory shift in leukocytes exposed to LPS without altering phagocytosis or TLR4 expression, *in vitro*, which may be due to alterations in post-receptor signaling or protein translation. Thus, calcitriol could be a novel candidate immunomodulatory therapy for sepsis in dogs.

IM02

Urban Environment and Odds of Canine Immune-Mediated Disease (Abstract IM02)

Unity Jeffery

Texas A&M University, College Station, TX, USA

Urban environment and exposure to air pollution are associated with human auto-immune diseases including systemic lupus erythematosus and multiple sclerosis. Specific environmental triggers for canine immune-mediated disease remain largely undiscovered, but their identification would improve the management of high-risk individuals.

The aim of this study was to determine if dogs living in urban areas have increased odds of immune-mediated disease compared with those in rural areas.

Case records from a referral hospital in the American Mid-West were retrospectively reviewed to identify dogs meeting pre-specified criteria for immune-mediated hemolytic anemia (IMHA) and/or thrombocytopenia (ITP); immune-mediated polyarthritis (IMPA) or meningoencephalitis of unknown origin (MUO). Breed-matched controls without immune-mediated disease were randomly selected. The census tract in which the dog lived and its population density were retrieved from a government database. Urban environment was classified as >1000 people per square mile, as defined by the US census bureau.

137 cases and 137 controls meet the inclusion criteria. The odds ratio for any immune mediated disease was 0.94 (95% confidence interval: 0.58–1.52, p=0.81) for urban versus rural dogs. For specific immune-mediated diseases, odds ratios were 0.85 (0.45–1.61, p=0.63, n=79 cases) for IMHA/ITP; 0.83 (0.25–2.77, p=0.76, n=21 cases) for IMPA and 1.26 (0.49–3.23, p=0.63, n=37 cases) for MUO.

This study did not reveal significant associations between urban environment and canine immune-mediated disease. Larger studies, or studies in other geographical areas might still reveal a relationship, but the current study suggests exposures common to both rural and urban dogs merit further investigation as triggers of immune-dysregulation.

IM03

Immune Cells in the Nose and Oropharynx of Cats and Response to Mucosal Immune Stimulation (Abstract IM03)

Steven Dow[1]; William Wheat[1]; Elena T. Contreras[1]; Michael R. Lappin[1]

[1]Colorado State University, Fort Collins, CO, USA

Little is known regarding the normal immune cell populations resident in the nose and oropharynx of healthy cats. Nor have previous studies investigated the cellular responses of these mucosal tissues to non-specific immune stimulation. Such information could provide important insights into the nature and timing of mucosal immune responses in the upper respiratory and GI tracts of cats.

Nasal lavage and oropharyngeal swabs were used to obtain resident cells from 9 healthy adult cats. The cell populations were phenotyped and enumerated using flow cytometry. After baseline samples were collected, cats were treated by intranasal and oral administration of a liposome-TLR mucosal immune stimulant (MucosImmune), and changes in immune cell populations were assessed by nasal lavage and oral swabs obtained at 6h, 24h and 72h after treatment.

We found that the normal cat nasal passages contained a mixed population of CD5+ T cells and CD14+ monocytes, with relatively abundant neutrophils. In contrast, the oropharynx contained numerous CD5+ T cells and few monocytes or neutrophils. Following administration of MucosImmune, there was a sustained influx of neutrophils into the nose over 72h, while neutrophils only increased in the oropharynx at 6h after treatment. There was a significant increase in CD14+ monocytes in both sites by 6h, with gradual declines over 72h. In both sites, there was a transient disappearance of CD5+ lymphocytes at 6h, which rebounded to normal numbers by 24 to 72h.

These data indicate that nasal passages and the oropharynx in cats both contain a relatively abundant population of immune cells capable of responding rapidly to TLR activation, with the oropharynx containing a dense T cell population and the nose containing more neutrophils and likely B cells. Both sites responded rapidly to mucosal delivery of TLR agonists, with early appearance of innate immune cells and disappearance and then reappearance of T lymphocytes. These results indicate that mucosal administration of a liposome-TLR agonist rapidly activates local immune responses in the feline upper respiratory tract, which may be associated with non-specific protection from viral or bacterial pathogens.

IM04

The Use of a Combination-Drug Protocol versus Glucocorticoids Alone for Treatment of Idiopathic Immune-Mediated Polyarthritis (Abstract IM04)

Sara Ravicini[1]; Andrew C.C. Kent[1]; Mark D. Dunning[2]; Fergus Allerton[1]

[1]Willows Referral Centre, Solihull, England, UK; [2]University of Nottingham, Sutton Bonington, England, UK

Idiopathic immune-mediated polyarthritis (IMPA) is an inflammatory non-infectious joint disease caused by immune-complex deposition in the synovium due to a type-III hypersensitivity reaction. Oral administration of immunosuppressive drugs is the standard treatment for this condition. The most commonly used drug is prednisolone, often in combination with other immunosuppressive agents. The benefit of combination immunosuppressive therapy introduced at the time of diagnosis has not been previously evaluated. The aim of this retrospective study was to compare outcomes for dogs with IMPA treated with prednisolone alone or in combination with other immunosuppressive drugs introduced at the time of diagnosis. We hypothesised that the use of combination immunosuppressants would not improve outcome in canine IMPA.

The medical records from a single tertiary referral centre were reviewed between January 2014 and May 2016 for dogs diagnosed with idiopathic IMPA on the basis of consistent joint fluid analysis. Cases were excluded if an underlying disease was identified. Thirty-six dogs met the inclusion criteria: 17 females and 19 males of various breeds, with a median age of 8 years (range 2–12) and median bodyweight of 19kg (6.3–46kg). 16/36 cases were presented for lameness or reduced mobility with 20 dogs displaying systemic signs (e.g., lethargy, decreased appetite, pyrexia) of which 7 manifested no mobility issues prior to diagnosis. Median time to diagnosis was 100 days (range 5–560) with no seasonal bias. Group 1 (23 dogs) received prednisolone, whilst group 2 (13 dogs) received prednisolone with azathioprine (7) or ciclosporin (6). Outcome data was compared using non-parametric tests.

13/23 dogs in group 1 relapsed during tapering of medication compared to 4/13 in group 2 ($p=0.07$). Three dogs in group 1 were euthanased due to suspected complications arising from IMPA or therapy, (one dog died after developing pyothorax and was excluded from calculations). All 13 dogs in group 2 are still alive ($p=0.17$). In group 1 the median treatment duration was 397 days (104–1250) compared to a median of 158 days (81–260) in group 2 ($p=012$).

There was no significant difference between groups 1 and 2 for duration of therapy, relapse rate or mortality. Thus the hypothesis was accepted. The observations that dogs started on a multi-drug protocol from the outset required a shorter treatment duration, had lower mortality and a lower relapse rate may reflect a type II error due to low group numbers. A prospective study with greater numbers is ongoing to investigate this further.

IM05

Activation of Innate Immune Responses in Cat Leukocytes by a Liposomal TLR Ligand Immune Stimulant (Abstract IM05)

Steven Dow[1]; William Wheat[1]; Jonathan Coy[1]; Lyndah Chow[2]; Michael R. Lappin[1]

[1]Colorado State University, Fort Collins, CO, USA; [2]Fort Collins, CO, USA

Previous studies have demonstrated the ability of liposome-TLR ligand complexes (LTLC) to potently activate innate immune responses in several species, but they have not been fully evaluated for activity in cats. There is currently an unmet need for a broad spectrum immune stimulant for prevention or early treatment of viral and bacterial respiratory tract pathogens in cats. The purpose of the present study was to investigate *in vitro* responses of cat lymphocytes and monocytes to stimulation by LTLC.

Cat peripheral blood mononuclear cells (PBMC) obtained from healthy cats were used to assess immune activation by LTLC. PBMC were incubated with LTLC for 24 to 72 hours, and immune activation was assessed by cytokine release assays, RT-PCR, and flow cytometry.

We found that LTLC triggered IFN-g production by PBMC within 24h of incubation. In addition, upregulated expression of IL-12 and TNF-a mRNA was demonstrated using RT-PCR. With flow cytometry, LTLC induced activation of B cells and T and monocytes, as reflected by upregulated expression of MHCII expression on B cells and monocytes and upregulated expression of OX40 on T cells and B cells. In addition, LTLC induced T cell proliferation.

In a companion abstract, we demonstrate the mucosal immune stimulatory properties of an LTLC-based immune stimulant administered to healthy cats. Therefore, taken together with the *in vitro* findings presented here, these results suggest that LTLC may be used as an effective non-specific immune stimulant for induction of anti-viral and anti-bacterial immunity in cats.

IM06

Validation of an Electrochemiluminescent Immunoassay for Measurement of Tumor Necrosis Factor-Alpha in Cat Serum (Abstract IM06)

Caleb C. Thomas[1]; Sina Marsilio[1]; Jan S. Suchodolski[1]; Jörg M. Steiner[1]; **Jonathan A. Lidbury**[1]

[1]Gastrointestinal Laboratory, College of Veterinary Medicine, Texas A&M University, College Station, TX, USA

Tumor necrosis factor-alpha (TNF-α) is a cytokine released mainly by macrophages in response to contact with infectious agents, tissue injury, tumor growth, and autoimmune conditions. Thus, measurement of TNF-α warrants evaluation as a diagnostic and prognostic marker in cats with inflammatory diseases. However, currently no analytically validated assay is commercially available for the measurement of this cytokine in cats. The aim of this study was to develop and analytically validate an electrochemiluminescent immunoassay for the measurement of TNF-α in cat serum.

Using a commercially available platform (UPLEX, Meso Scale Diagnostics) a feline TNF-α electrochemiluminescent assay was developed from recombinant feline TNF-α and polyclonal rabbit anti-feline TNF-α antibodies. Validation variables included the lower limit of detection, dilutional parallelism, spiking recovery, intraassay variability, and interassay variability.

The lower limit of detection was 1.3 ng/L. Observed-to-expected ratios for dilutional parallelism (n=6) ranged from 43.3%–133.3% with a mean (±SD) of 102.9%±25.4%. Recoveries for spiked TNF-α in serum samples (n=4) ranged from 72.2%–112.1% with a mean±s.d. of 93.1%±12.5%. Intraassay coefficients of variation (%CVs; n=4) were ≤4.3% with a mean±s.d. of 3.4%±0.7%. Interassay coefficients of variation (n=4) were ≤17.6% with a mean±s.d. of 12.5%±4.9%.

This assay was shown to be sensitive, precise, and accurate for the measurement of TNF-α in cat serum. However, assay linearity was suboptimal. Further studies are needed to determine whether linearity can be improved with the current assay system. Also, the clinical implications of serum TNF-α concentrations need to be evaluated in this species.

IM07

A Flow Cytometric Study of CD4+CD25+FoxP3+ Regulatory T cells Changes in Canine Atopic Dermatitis (Abstract IM07)

Dong-jae Lee[1]; **Chul Park**[1]; Ha-jung Kim[2]; Yunhye Kim[3]; Sung tae Yoon[1]

[1]College of Veterinary Medicine, Chonbuk National University, Iksan, Cholla-bukto, Republic of Korea; [2]College of Veterinary Medicine, Chonnam National University, Kwang-ju, Cholla-namdo, Republic of Korea; [3]College of Veterinary Medicine, Chonbuk National University, Curi, Kyonggi-do, Republic of Korea

Canine atopic dermatitis (CAD) is a common, chronic, genetically-predisposed, and relapsing inflammatory skin disease. CAD is a multifaceted disease resulting from a complex interaction between environmental and genetic factors. Both of these factors can shape skin barrier function and the immunological response of predisposed patients. CD4+CD25+FoxP3+ Regulatory T cells (Tregs) become known to prevent excessive immune responses through their immunoregulatory properties and therefore play a leading role in the maintenance of immunological homeostasis. There was only few researches that was performed to analyze the relationship between CAD and proportion of Tregs in periphery blood. In the present study, the aims were to assess the proportion of circulating Tregs (CD4+CD25+FoxP3+, CD4+FoxP3+) in the healthy and the atopic dogs to analyze implication of Tregs to CAD and to investigate the correlation of Tregs with CAD severity and age.

The animals for analysis consisted of eight healthy dogs and nine atopy dogs that had never been treated. The diagnosis of CAD was made with fulfillment of clinical criteria (Favrot's criteria) and exclusion of differential diagnosis including ectoparasitic and infectious dermatoses. Expression of Tregs in peripheral blood was evaluated by flow cytometry. Tregs were identified as CD4+CD25+PoxP3+ T cells and CD4+PoxP3+ T cells. There were statistically significant increase in proportion of both CD4+CD25+FoxP3+ T cell and CD4+FoxP3+ T cell in periphery blood of atopic dogs compared to healthy dogs (mean 0.91% versus 4.2%, p=0.012; mean 1.76% versus 10.06%, p=0.004). The circulating regulatory T cells (Tregs) of atopic dogs had negative correlation with their disease severity (r=-0.758, p=0.05) and also negative with age (r=-0.666, p=0.05).

Both CD4+CD25+FoxP3+ Tregs and CD4+FoxP3+ Tregs may have a relevance to the pathogenesis of CAD in course of atopic dermatitis. This study demonstrated negative association between Treg frequency and disease severity. The numbers of circulating Tregs of atopic dogs decreased as they got older.

IM08

12/15 Lipoxygenase Deficiency Enhances the Development of Pneumonic Tularemia (Abstract IM08)

Carmela L. Pratt[1]; Jerod Skyberg[2]; Charles R. Brown[2]

[1]University of Missouri, Saint Louis, MO, USA; [2]University of Missouri, Columbia, MO, USA

Francisella tularensis, a pathogenic Gram negative intracellular bacteria that causes pneumonia, skin lesions, and systemic illness known as tularemia that can affect cats in certain regions of the United States. 12/15 lipoxygenase (LO) is an enzyme that utilizes arachidonic acid to produce anti-inflammatory mediators. Deletion would be expected to produce an increase in the severity of disease. Since 12/15-LO has a role in regulating inflammation, our hypothesis is pneumonic tularemia in mice would produce increases in morbidity and mortality. *Francisella tularensis* Live Vaccine Strain (Ft LVS) is pathogenic to mice but not to people. C3H wild type (WT) and 12/15 LO knock out (KO) mice were infected intratracheally with a sublethal dose of Ft LVS and their weights were monitored for 10 days. On day 10 post-infection the mice were euthanized and the lungs, spleens, and liver were collected for analysis. The cellular phenotype of infiltrating inflammatory cells, leukotriene B4 (LTB4) and KC levels were determined from lung tissue. Histopathology was assessed from the lung and liver, and splenic weights were determined. 12/15 LO KO mice had increased morbidity compared to WT mice indicated by their significant weight loss. KC, a chemoattractant for neutrophils, was significantly ($p<0.05$) higher in the lung tissue of KO mice. Inflammatory cells (neutrophils, T cells, and macrophages) and pulmonary histopathology scores were significantly higher ($p<0.05$) in KO mice. In contrast LTB4, which assists in host defense was significantly decreased ($p<0.05$) in 12/15 LO KO mice. Splenic weights were significantly ($p<0.05$) higher than WT spleens correlating with the severity of disease. Given this information, 12/15 LO plays a role in regulating inflammation in pneumonic tularemia. Eicosanoids are highly conserved among species and It is possible that 12/15 LO agonists may be used as natural therapeutics to aid in the resolution of pneumonia in domestic animals, but further studies are needed.

IM09

Feline Adipose Tissue-Derived Mesenchymal Stem Cells Reduce Inflammation by Secreting Soluble Factors (Abstract IM09)

Hyung-Kyu Chae[1]; Bo-Yeon Lee[1]; **Woo-Jin Song**[1]; Qiang Li[2]; Min-Ok Ryu[1]; Sang-Chul Park[1]; Hyeon-Jin Kim[1]; Su-Yeon Kim[1]; Hye-Mi Yang[1]; Hwa-Young Youn[1]

[1]Seoul National University, Seoul, Republic of Korea; [2]Seoul National University, Seoul, China (People's Republic)

Mesenchymal stem cells (MSCs) have immunomodulatory functions and differentiation capacity, conferring them with various clinical uses. Although MSCs have been applied in the treatment in various inflammatory diseases, mechanistic research on feline MSCs is lacking. Accordingly, in this study, we aimed to analyze the immunomodulatory mechanisms of MSCs isolated from feline adipose tissue (fAT-MSCs). Cytokine expression in RAW264.7 murine macrophages and allogeneic feline peripheral blood mononuclear cells (fPBMCs) was measured using quantitative real-time polymerase chain reaction (qRT-PCR) and compared according to the presence of fAT-MSCs. We also extracted RNA and collected supernatants from fAT-MSCs for measurement of soluble factor expression levels by qRT-PCR and enzyme-linked immunosorbent assay. The levels of pro-inflammatory cytokines, e.g., tumour necrosis factor-α (TNF-α), inducible nitric oxide synthase and interleukin (IL)-1β, were significantly decreased in coculture of mitogen-stimulated RAW264.7 cells with fAT-MSCs compared with that in RAW264.7 cells. Additionally, the expression of immunomodulatory factors from fAT-MSCs, including cyclooxygenase-2 (COX-2), transforming growth factor (TGF)-β, indoleamine-2,3-dioxygenase (IDO) and hepatocyte growth factor, increased in the presence of RAW264.7 cells. Expression of pro-inflammatory TNF-α, interferon-γ and IL-6 decreased, and expression of anti-inflammatory IL-10 increased during coculture of mitogen-stimulated allogeneic fPBMCs with fAT-MSCs. TGF-β, COX-2 and IDO expression and prostaglandin E-2 production from fAT-MSCs increased in the presence of allogeneic fPBMCs. Our data suggested that soluble factors secreted from fAT-MSCs play an important role in the immunomodulatory effects of these cells. These findings may be helpful in the application of fAT-MSCs to feline patients with immune-related diseases.

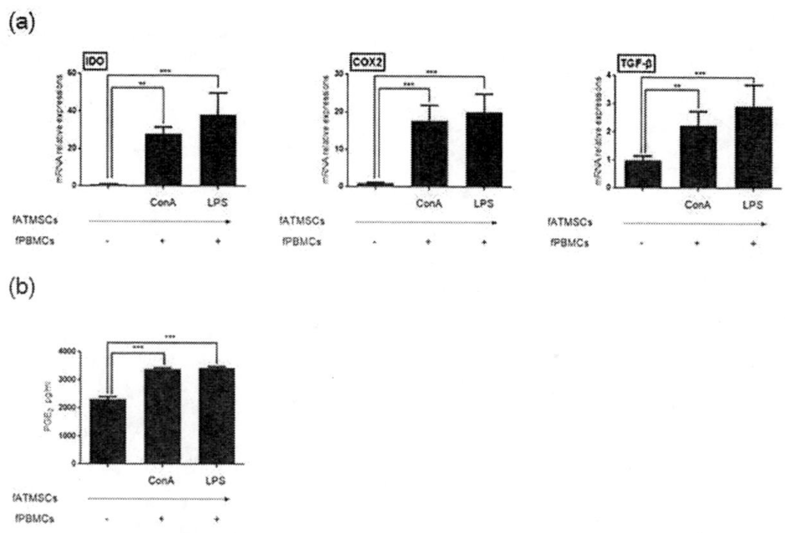

IM10

Presence of Naturally Occurring Circulating Anti-AAV8 and AAV9 Antibodies in Pet Dogs and Cats (Abstract IM10)

Brittany Lucchetti[1]; **Andrew Specht**[2]; Kristen Coleman[3]; Thomas Conlon[3]

[1]BluePearl Veterinary Partners Arizona, Scottsdale, AZ, USA; [2]College of Veterinary Medicine, University of Florida, Gainesville, FL, USA; [3]Department of Pediatrics/Powell Gene Therapy Center, University of Florida, Gainesville, FL, USA

Gene therapy is an emerging treatment modality in human medicine and animal models including dogs and cats are often used in preclinical studies to demonstrate efficacy and safety. Research into gene therapies for some conditions in animals has started as well. Adeno-associated virus (AAV) based vectors are among the most common delivery mechanisms. It is well established that antibodies to AAV vectors can impact the success of gene transfer. The primary objective of this study was to determine if pet populations have circulating antibodies to two AAV vectors used for gene therapy.

An ELISA technique was utilized to assess the titer level of circulating antibodies to serotypes AAV8 and AAV9 in 115 canine and 85 feline patients presenting to a university hospital. Results were variable, but mean values for canines were $2.3x10^3$ (anti-AAV8) and $1.4x10^4$ (anti-AAV9). Feline mean values were $1.2x10^5$ (anti-AAV8) and $4.1x10^5$ (anti-AAV9). There was very weak, positive correlation between serotypes within species based on a two-sided Pearson's product-moment correlation test (Canine r^2=0.18, Feline r^2=0.37).

Based on these results, it appears dogs and cats in the pet population have some circulating antibodies capable of binding to AAV8 and AAV9 particles. However, without concurrent neutralizing antibody levels, the clinical significance of this finding is unknown. If a correlation between circulating antibody titers and neutralizing antibody titers exists, it would suggest that potential candidates for gene therapy drawn from a pet population might require screening for antibodies or immunosuppressive therapy prior to receiving AAV vector based treatment.

IM11

Serum from Some Pet Dogs and Cats Neutralizes AAV8 and AAV9 Mediated Gene Transduction (Abstract IM11)

Thomas Conlon[1]; Sanford Boye[2]; Wei Li[2]; Kristen Coleman[1]; Brittany Lucchetti[3]; **Andrew Specht**[4]

[1]Department of Pediatrics/Powell Gene Therapy Center, University of Florida, Gainesville, FL, USA; [2]Department of Ophthalmology, College of Medicine, University of Florida, Gainesville, FL, USA; [3]BluePearl Veterinary Partners Arizona, Scottsdale, AZ, USA; [4]College of Veterinary Medicine, University of Florida, Gainesville, FL, USA

With the increasing interest in gene therapy as a potential treatment modality for a variety of human diseases, large animal models including dogs and cats are often used in preclinical studies. There has also been interest in using these techniques to treat diseases that affect pet populations. Adeno-associated virus (AAV) based vectors are among the most common utilized delivery mechanisms and circulating antibodies capable of binding to AAV8 and AAV9 particles have been identified in pet dogs and cats. The primary objective of this study was to identify whether some of these same pets might have neutralizing antibody titers against AAV8 and AAV9 transduction.

Serum samples from seven pet dogs and seven pet cats with varying levels of circulating anti-AAV8 and anti-AAV9 antibodies were utilized. Neutralizing antibody assays were performed using a Self-complementary c-smCBA-mCherry reporter packaged in either AAV8 or AAV9 and permissive cells (661W and Lec2, respectively) at a multiplicity of infection of 10,000 vg/cell. Positive controls were serum samples from dogs in a research colony that had been previously treated with AAV8 and AAV9 vectors and negative controls had no serum. Vectors were pre-incubated with series of serum dilutions (1:10, 1:40, 1:160, 1:640 and 1:2560) before infecting cells. Three days post infection, transduction was assessed by measuring mCherry-mediated fluorescence. Neutralization of transduction greater than 50% relative to no serum control in the lowest dilution (1:10) was identified in 2/7 cats and 1/7 dogs for AAV8 and 1/7 cats and 5/7 dogs for AAV9. There was no obvious positive correlation between circulating antibody titers and neutralization titers. There was a trend for the presence of feline serum to have a positive impact on transduction efficiency of both vectors.

The clinical significance of these results remains unknown, but they suggest the need for further evaluation to determine if candidates for gene therapy from a pet population might require screening tests or immunosuppressive therapy prior to receiving AAV vector based treatment.

ID01

Bartonella Seroepidemiology in Dogs from North America, 2008–2014 (Abstract ID01)

Erin Lashnits[1]; Maria Correa[1]; Barbara Hegarty[1]; Adam Birkenheuer[1]; Edward Breitschwerdt[2]

[1]North Carolina State University, Raleigh, NC, USA; [2]NCSU-CVM, Raleigh, NC, USA

Improved understanding of *Bartonella* spp. serological patterns in dogs may aid clinical decision making and enhance current understanding of naturally-occurring arthropod vector transmission. We analyzed a large diagnostic laboratory database to better understand *Bartonella* seroepidemiology of dogs in North America, identifying demographic groups in which *Bartonella* exposure may be more likely, describing variations in chronologic and geographic patterns of *Bartonella* seroreactivity, and examining co-exposure to other vector-borne pathogens. We hypothesize that there will be no significant demographic risk factors for *Bartonella* exposure, and that *Bartonella* exposure will be more common in regions and seasons with high vector prevalence and among dogs with other canine vector-borne disease (CVBD) exposure.

CVBD serology results (15,451 diagnostic submissions) generated between January 1, 2008 and December 31, 2014 at the North Carolina State University, College of Veterinary Medicine Vector Borne Disease Diagnostic Laboratory were retrospectively reviewed. *Bartonella henselae, B. vinsonii* subsp. *berkhoffii*, and/or *B. koehlerae* seroreactivity was determined by indirect immunofluorescent antibody (IFA) testing. For a subset of accessions a more comprehensive diagnostic CVBD serology panel (*Ehrlichia* spp., *Anaplasma* spp., *Rickettsia* spp., *Borrelia burgdorferi*, *Babesia* spp. and *Dirofilaria immitis*) was performed using SNAP® 4Dx® (IDEXX Laboratories) or IFA.

Overall, 3.26% of accessions were *Bartonella* spp. seroreactive, a percentage that is comparable to seroreactivity for other CVBDs among US canine population-wide serosurveys. Seroreactivity to *B. henselae* (2.13%) and *B. koehlerae* (2.39%) antigens was detected more frequently than seroreactivity to *B. vinsonii* subsp. *berkhoffii* (1.42%, p<0.0001) antigen. Intact male dogs were more likely to be seroreactive (5.04%) than neutered males (2.87%, p<0.0001; OR 1.80, 95% CI 1.37–2.35, p<0.0001) or intact or spayed females (3.22%, p=0.0003; OR 1.59, 95% CI 1.23–2.05, p=0.0004). Mixed or non-AKC breed dogs were more likely to be seroreactive (4.45%) than purebred dogs (3.02%, p=0.0002; OR 1.49, 95% CI 1.21–1.85, p=0.0002). There was no seasonal seroreactivity trend for any of the three *Bartonella* spp. Geographic patterns of seroreactivity did not correspond with other regional CVBD patterns. Rather, seroreactivity was distributed broadly across North America with some areas of apparent higher exposure risk (*B. henselae* in WA and CT, *B. koehlerae* in MO). Co-exposure to other vector-borne pathogens was common, with associations between *B. henselae* and *B. vinsonii* subsp. *berkhoffii* and multiple other CVBD pathogens. *B. koehlerae* seroreactivity was associated with *Ehrlichia* spp. (OR 3.31; 95% CI 2.43–4.51, p<0.0001) and *Rickettsia* spp. (OR 2.72; 95% CI 1.57–4.71, p=0.0002) exposures, but was not associated with the pathogens transmitted predominantly by *Ixodes scapularis* (*B. burgdorferi* and *Anaplasma* spp.).

We conclude that *Bartonella* spp. exposures may occur throughout North America and may be seen at any time of year. Dogs exposed to other CVBD, male intact dogs, and mixed breed dogs are at higher risk for *Bartonella* exposure. Fleas and multiple tick species are proposed vectors for canine bartonellosis; our seroepidemiological analyses suggest that there may be multiple vectors for *Bartonella* infection in dogs, or that the major vector may depend on local geographic, environmental, or host factors.

ID02

Impact of Immune Suppression on Dogs with Treated Select Vector Borne Disease Agents (Abstract ID02)

Masahiko Sato; Julia Veir; Sarah Shropshire; Michael R. Lappin

Colorado State University, Fort Collins, CO, USA

A 4-week course of doxycycline is the recommended treatment for *Ehrlichia canis* infection in dogs. Although some studies report clearance of the infection after antibiotic therapy, others have documented persistent infection even after treatment. Therefore, concerns for recrudescence of disease arise when an immunosuppressive drug for treatment of an immune mediated disease in a dog previously infected with *E. canis*. The purpose of this study was to determine whether administration of prednisolone and cyclosporine at a standard dose would induce clinical or laboratory evidence of reactivated *E. canis* in dogs previously treated for the infection.

Seven beagles that were experimentally infected with *E. canis* and administered doxycycline at 10 mg/kg, PO daily for 4 weeks were included in this study. Three of the dogs were also infected with *Anaplasma platys* and *Babesia vogeli* and were administered two doses of imidocarb at 6.6 mg/kg, IM 2 weeks apart prior to enrollment. Each dog was administered prednisolone (Dose range 0.94–1.25 mg/kg, PO, q12 hours) in conjunction with cyclosporine whose dosages were adjusted based on the trough levels (Final dose range 5.3–8.3 mg/kg, PO, q12 hours) for 6 weeks. Clinical signs (fever, lethargy, and anorexia), complete blood count (CBC), polymerase chain reaction (PCR) for DNA of *E. canis, A. platys,* and *B. vogeli* in blood, anti-*E. canis* indirect fluorescent antibodies (IFA) titers, and flow cytometry for antiplatelet antibodies were monitored.

All dogs were negative on PCR but positive on IFA prior to immunosuppression and completed the immunosuppressive protocol. No recrudescence of *E. canis* infection was detected based on clinical signs or results of CBC, PCR, IFA and flow cytometry for anti-platelet antibodies. Anti-*E. canis* IFA titers were negative in 5/7 dogs at the end of immunosuppressive protocol and were negative 6 months after the protocol in 5/5 dogs. All dogs were negative via PCR for *E. canis, A. platys* and *B. vogeli* throughout immunosuppression.

A 4-week course of doxycycline with or without supplemental imidocarb cleared *E. canis* infection for at least 6 months after immunosuppression for 6 weeks. A risk of reactivation of *E. canis* infection with immune suppression in previously treated dogs was considered minimum.

ID03

Immune Complex Dissociation Methods for Detecting Heartworm Antigen and Their Utility in Assessing Patient Samples (Abstract ID03)

Rick Alleman[1]; Andrea Arguello-Marin[2]; Jan Drexel[2]; Joe Liu[2]; Melissa Beall[2]; Ramaswamy Chandrashekar[2]

[1]Lighthouse Veterinary Consultants, Alachua, FL, USA; [2]IDEXX Laboratories, Inc., Westbrook, ME, USA

Heartworm antigen testing is routinely performed to diagnose heartworm infections in dogs. The original diagnostic tests employed immune complex dissociation (ICD) methods to release antigen that was bound by endogenous canine antibodies. However, ICD methods were discontinued in the majority of commercially available assays due to improvements in assay technology (O'Connor T, *et al.* "Immunoassay applications in veterinary diagnostics." In: Wild D, ed. *The Immunoassay Handbook.* 4th ed. Great Britain: Elsevier, 2013: 623–645). Interference from immune complexes is not unique to heartworm diagnostics and various methods are available for ICD. The purpose of this study was to evaluate different methods of ICD for the detection of free heartworm antigen by microtiter plate ELISA and evaluate the performance of ICD in pet dogs. The original, USDA licensed PetChek® Heartworm Test (IDEXX Laboratories, Inc.) utilized pepsin at an acidic pH for ICD prior to antigen testing. Using a set of experimental samples, the pepsin method was compared to heat treatment, and acid treatment. All methods were found to release complexed antigen and produce positive results when the post-treatment samples were tested for heartworm antigen by microtiter plate ELISA. While heat treatment required a minimum of 600 µl of either serum or plasma, the pepsin and acid methods needed only 50 µl of the patient sample. To extend this to a field population, samples from 1115 patients were submitted to IDEXX Laboratories between 2014–2016 for investigation of discrepant heartworm results. Samples were evaluated with and without pepsin ICD using PetChek® Heartworm. A total of 112 patients became antigen positive with the ICD protocol only, 122 were positive with and without ICD but demonstrated at least a 50% increase in plate absorbance (OD), and 109 were positive both with and without ICD but had minimal change in plate OD following ICD. The remaining 772 patients did not have detectable antigen with and without ICD. In a prospective study, dogs receiving pre-adulticide heartworm treatment consisting of a macrocyclic lactone in combination with doxycycline were evaluated in the same way. Serial samples from 12 dogs revealed that 3 had a reduction in free circulating antigen within 4 weeks of initiating treatment. In all cases, free antigen levels could be recovered with ICD. Heartworm antigen testing with ICD can be a valuable adjunct diagnostic for those patients who have discrepant results, have intermittently used a macrocyclic lactone, or are receiving treatment.

ID04

Fecal Shedding of Parvovirus Following Modified Live Feline Panleukopenia Virus Vaccination in Adult Healthy Cats (Abstract ID04)

Michèle Bergmann[1]; Stephanie Schwertler[1]; Uwe Truyen[2]; Stephanie Speck[2]; Sven Reese[3], Katrin Hartmann[1]

[1]Clinic of Small Animal Medicine, LMU Munich, Munich, Bayern, Germany; [2]Veterinary Faculty (Virology) Leipzig, Leipzig, Bayern, Germany; [3]Institute for Anatomy, Histology and Embryology, LMU Munich, Munich, Bayern, Germany

Fecal samples positive for canine parvovirus (CPV) have been reported in apparently healthy dogs after modified-live vaccination. There is no information about fecal shedding of feline panleukopenia virus (FPV) in cats after modified-live vaccination. Aim of this study was to provide information about parvovirus DNA-shedding in recently vaccinated, adult, healthy cats and to assess related factors using Chi-squared and Fisher´s exact test. Forty cats were included in the study and were vaccinated with a commercial FPV modified-live vaccine. Feces of cats were tested for the presence of parvovirus DNA on day 7, 14, 21, and 28 by quantitative real-time PCR. Pre- and post-vaccination serum antibody titers were measured by hemagglutination inhibition on day 0, 7, and 28. Almost one third of the cats (12/40; 29.3%; 95% CI: 18.0–45.6) shed parvovirus DNA during the post-vaccination period. Parvovirus DNA was detectable up to 28 days after immunization. Parvovirus DNA-shedding was significantly associated with a lack of protective pre-vaccination antibody titers (p=0.016; odds-ratio (OR): 6.44; 95% CI: 1.44–28.89) and with post-vaccination titer increases (p=0.029; OR: 5.00; 95% CI: 1.17–21.39). Vaccinated cats can shed parvovirus DNA at least up to 4 weeks after vaccination and shedding occurs more commonly in cats without protective titers. Post-vaccinal shedding thus can interfere with diagnosis in recently vaccinated cats that are presented with clinical signs suspicious for of panleukopenia.

ID05

Effects of a Liposome-TLR Mucosal Immune Stimulant on Kittens Infected with Feline Herpesvirus 1 (Abstract ID05)

Elena T. Contreras[1]; Steven Dow[1]; Michael R. Lappin[1]

[1]Colorado State University, Fort Collins, CO, USA

Feline herpesvirus 1 (FHV-1) is the most common feline infectious disease in cats, with over 90% of cats housed in group settings being infected. Liposome-TLR ligand complexes (LTLC) have been shown in mice to enhance innate immunity and induce non-specific protection against both viral and bacterial diseases. A new formulation of LTLC to be used as a mucosal immune stimulant (MucosImmune) has been shown to be safe and capable of stimulating local immunity in cats. The objectives of this study were to determine whether MucosImmune administered prior to FHV-1 experimental inoculation or at the first signs of clinical illness post-inoculation in kittens could lessen clinical signs and shedding of FHV-1 when compared to untreated controls.

A total of 12 female and 13 male, 12-week-old purpose-bred kittens that were negative for FHV-1 antibodies in serum and for FHV-1 DNA in cells collected from the oropharynx by swab were divided into three groups housed in three separate rooms. Thirteen kittens were randomized into Group A (n=7; prevention) or Group B (n=6; treatment). The 12 kittens assigned to Group C served as untreated controls. Group A was administered MucosImmune (0.2 ml each nostril, 0.6 ml into the oropharynx) 24 hours prior to FHV-1 inoculation. On Day 0, all kittens were sedated and administered $10^{5.3}$ $TCID_{50}$ of FHV-1 divided between the nares and oropharynx. Group B kittens were administered one dose of MucosImmune the first day that clinical signs were noted with a second dose administered 24 hours later. Two trained, masked observers applied a standardized clinical scoring system detailing ocular and respiratory clinical signs to each room for 30 minutes each day for a total of 28 days. Mucosal cells were collected from the caudal pharynx of each kitten from Groups A and C on Days 7, 14, 21, and 28, and from each kitten from Group B after clinical signs were recognized but prior to the first MucosImmune dose, 7, 14, and 21 days after the first treatment, and then on Day 28 after FHV-1 inoculation. Total DNA was extracted from the oropharyngeal swabs, and quantitative PCR assays were performed for FHV-1 DNA and GAPDH with results presented and compared amongst groups as a ratio. The Wilcoxon rank-sum test was used for comparisons amongst groups within the time periods Days 1 to 14, Days 15 to 28, and Days 1 to 28. Time to resolution of illness (total score ≤1) was also compared amongst groups.

No significant differences were found in total combined scores in Group A as compared to Group C kittens. When compared to Group C, total ocular scores in Group A kittens were significantly lower in Days 15 to 28 ($p=0.04$) and lower conjunctivitis scores were detected in all periods ($p=0.01$; $p<0.001$; $p<0.001$). When compared to Group A, total respiratory scores in Group C kittens were significantly lower in Days 15 to 28 ($p<0.001$) and Days 1 to 28 ($p<0.001$). Clinical illness was of shorter duration in Group A kittens (median 17.5 days, range 10 to over 28 days) as compared to Group C kittens (median 22, range 20 to 27 days) but the result was not significantly different ($p=0.13$). On Days 21 and 28, FHV-1/GAPDH ratios for Group A were significantly lower ($p=0.01$) than those for Group C. When Group B results were compared to those of Group C, no significant differences were found in total combined scores, total ocular scores, total respiratory scores, or time to resolution. However, Group B kittens had significantly less pyrexia on Days 1 to 28 ($p=0.03$) as compared to Group C.

Administration of MucosImmune lessened some clinical signs of FHV-1 in both experiments in this model and decreased FHV-1 shedding when administered 24 hours prior to inoculation with FHV-1.

ID06

Cell Block Cytology Applied to Sporotrichosis Diagnosis (Abstract ID06)

Fernanda Fidelis. Gonsales[1]; Natalia Coelho Couto de Azevedo Fernandes[2]; Juliana Mariotti. Guerra[2]; Amanda Moreto[1]; Juliana Takahashi[2]; Silvana Mello Pereira. da Silva[2]; Nilson Benites[1]

[1]School of Veterinary Medicine, University of São Paulo, São Paulo, Brazil; [2]Adolfo Lutz Institute, São Paulo, Brazil

Sporotrichosis is the most frequent subcutaneous mycosis detected in urban areas, it is caused by zoonotic pathogenic species of *Sporothrix schenckii* complex that affects especially cats. In Brazil, due to a large outbreak of sporotrichosis in the city of Guarulhos (State of São Paulo), the disease has become notifiable. Clinically the disease is characterized as a subcutaneous mycosis, with papular lesions with nodular feature or ulcero-gummy. The definitive diagnosis is obtained by the isolation of the fungus in culture. However, the result can take weeks, resulting in a delay to start the appropriate treatment and disease could spread through other community members. The aim of this study was standardization of the diagnosis of feline sporotrichosis by Cell Block Cytology (CB). CB consists of immediate wet cell fixation in preservative liquid based cytology solution (Surepath, BD), and histological processing. We compared two collecting methods: fine needle aspiration (FNA) and exfoliative brushing (EB) with an endocervical brush. Ten samples of skin lesions of nine suspects cats were collected, (five samples with each method). EB allowed a cell thick sediment, and production of several histological sections, with a slide stained with periodic acid Schiff method. All samples collected by this method showed positive results for fungal structures. FNA cases revealed low cellularity and impossible histological processing. This standardization shows that CB processing is possible with EB of lesions and fixation in cytology preservative solution, allowing storage for up to a month after harvest and further testing, as immunohistochemistry. CB is an alternative for sporotrichosis diagnosis.

ID07

A Rapid In-Clinic Test Detects IgM Antibodies to Four Different *Leptospira* Serovars (Abstract ID07)

Josh Lizer[1]; Michael Loenser[2]; Angela Weber[1]; Patrick Meeus[1]; Matthew Krecic[3]

[1]Zoetis, Kalamazoo, MI, USA; [2]Zoetis, Parsippany, NJ, USA; [3]Zoetis, Morristown, NJ, USA

Canine leptospirosis, a disease of worldwide distribution and significant zoonotic potential, is an acute bacterial infection of dogs caused by pathogenic spirochetes belonging to the genus *Leptospira*. Timely diagnosis of canine leptospirosis is important because early initiation of antibiotic therapy provides the greatest opportunity for a successful patient outcome. However, diagnosis is hindered by vague clinical signs and by diagnostic methods that can be difficult to interpret and require a lag time for results to be reported. A lateral flow test (WITNESS® Lepto, Zoetis) has been developed for the detection of IgM antibodies to *Leptospira*, offering the advantage of a result in 10 minutes at the point-of-care without the need for specialized equipment or expertise. Our hypothesis is WITNESS Lepto can detect IgM to four significant *Leptospira* serovars and will yield positive results earlier or concurrent to that of the microscopic agglutination test (MAT), the serological reference method for diagnosis of leptospirosis.

Thirty-two beagle dogs approximately six months of age were experimentally infected on days zero, one, and two with serovars Canicola, Grippotyphosa, Icterohaemorrhagiae, or Pomona (n=8 dogs/serovar). Sera were collected and evaluated for leptospiral antibodies with WITNESS Lepto and MAT at three to four day intervals. A minimum four-fold increase of antibody titer (≥1:400) for any serovar of a 7-serovar MAT panel indicated seroconversion. This study was approved by the Zoetis Institutional Animal Care and Use Committee.

Seroconversion was detected in all dogs by days 10 and 14 for WITNESS Lepto and MAT, respectively, and as early as day 4 for both methods. At day 7, WITNESS Lepto yielded positive results for 28/32 dogs (87.5%) whereas MAT yielded positive results for only 22/32 dogs (69%). In conclusion, WITNESS Lepto detects IgM to the four serovars included in this study and detects seroconversion earlier or concurrent to that of MAT. At the point-of-care, WITNESS Lepto may be useful as a screening test for acutely ill dogs suspected to have leptospirosis.

Test	Serovar	Day 0	Day 4	Day 7	Day 10*	Day 14*
WITNESS Lepto	C	0/8	0/8	8/8	8/8	4/4
	G	0/8	8/8	8/8	6/6	3/3
	I	0/8	0/8	4/8	8/8	8/8
	P	0/8	0/8	8/8	7/7	6/6
MAT	C	0/8	0/8	2/8	7/8	4/4
	G	0/8	6/8	7/8	5/6	3/3
	I	0/8	0/8	5/8	8/8	8/8
	P	0/8	0/8	8/8	7/7	6/6

*Some dogs removed from study early

ID08

Relationship Between *Histoplasma capsulatum* Urine EIA Results and Outcome in Cats Treated for Histoplasmosis (Abstract ID08)

Jennifer McGill[1]; Gary Norsworthy[2]; Andrew Hanzlicek[3]; Katherine KuKanich[4]; Audrey K. Cook[5]

[1]College of Veterinary Medicine, Louisiana State University, Baton Rouge, LA, USA; [2]Alamo Feline Health Center, San Antonio, TX, USA; [3]College of Veterinary Medicine, Oklahoma State University, Stillwater, OK, USA; [4]College of Veterinary Medicine, Kansas State University, Manhattan, KS, USA; [5]College of Veterinary Medicine, Texas A&M University, College Station, TX, USA

Infection with *Histoplasma capsulatum* is routinely diagnosed in cats using a quantitative enzyme immunoassay (EIA) performed on urine +/or serum. Any level of antigenuria/antigenemia is considered indicative of active infection, and urine EIA has a >90% sensitivity. The objective of this study was to determine if urine EIA results at the time of diagnosis (prior to the administration of antifungal agents) were predictive of outcome (i.e., survival to 6 months) for cats with histoplasmosis. Pre-treatment urine EIA results were available for 50 cats examined at four institutions between April 2012 and December 2015; 35 cats were alive at Day 180, 12 had died (median survival time: 24 days; range 2–124 days) and three were lost to follow up. For statistical purposes, positive results below the level of quantification (19 ng/ml) were listed as 19.1 ng/ml. The median EIA at the time of diagnosis for cats alive at 6 months was 5 ng/ml (range: 0–19.1); this was similar to findings for the non-survivors (median: 7.29 ng/mL; range of 0.78–19.1; p=0.54) [Figure 1]. Surviving cats were significantly younger (mean age: 6.9 yrs) than non-survivors (mean age 9.9 yrs; p=0.03) but median body weights (3.8 kg v 3.6 kg) and rates of pulmonary involvement (22/35 v 9/12) were similar. These findings indicate that urine EIA results at the time of diagnosis are not predictive of outcome in cats with histoplasmosis and should not be used as a prognostic indicator.

Figure 1: Urine EIA result for 47 cats with Histoplasmosis
(solid bars indicate the median value)

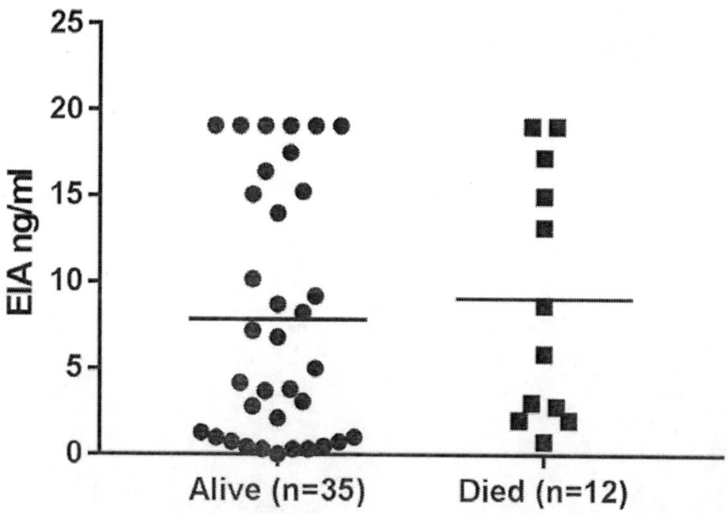

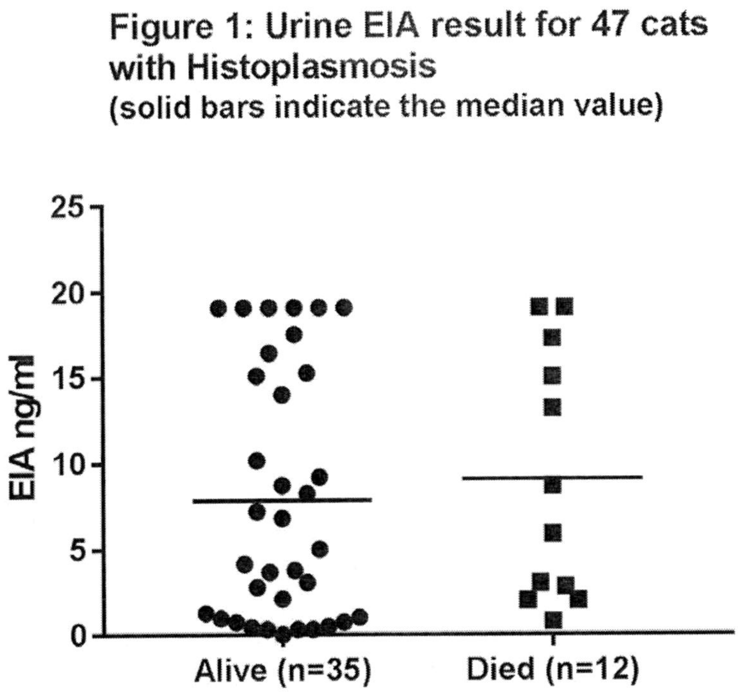

ID09

Prevalence and Phylogeography of Methicillin Resistant *Staphylococcus* spp. Carriage in Companion Animals in Brisbane, Australia (Abstract ID09)

Erika Meler; Ricardo Soares Magalhaes; Rochelle Rochelle; Trisha Farry; Emma Bennett; Tina Maguire; Justine Gibson

The University of Queensland, Gatton, QLD, Australia

Methicillin resistant *Staphylococcus* spp. (MRS) carriage rates in companion animals have been well described in North America and Europe but not investigated in Australia. The aims of this study were to quantify the overall carriage rate of MRS in companion animals presenting to two major veterinary hospitals in the Brisbane area during 2016 and to identify geographical differences in carriage rates and risk factors for MRS carriage between a rural (R) and urban (U) veterinary hospital.

Nasal and rectal swabs were taken from animals within 24 h of hospital admission. Following growth on selective media (Brilliance Staph Agar™), PCR for *mec-A* and *nuc* genes were performed to confirm methicillin resistance and differentiate *S. aureus* (MRSA) from *S. pseudintermedius* (MRSP). Medical data collected included residential address, signalment, previous drug use, surgical procedures, and veterinary visits/hospitalisations over the prior year. Risk factors were evaluated using Stata Corp™. Geographical variation in carriage analysis used semivariograms and OpenBUGS™.

Overall prevalence was 7.2% (19/265) in dogs, 1.6% (1/61) in cats. No (0/45) horses carried MRS. MRSP carriage was found in 19 dogs and none of the cats. MRSA carriage was identified in one dog (also MRSP carrier) and in one cat. We found a statistically significant difference in MRSP carriage rate between site R (3.2%) and site U (24.4%). Significant spatial variation was found in MRSP carrier dogs explained by clinical history and contextual factors of dog's residence.

In conclusion, major difference in canine MRSP carriage rates may exist between urban and rural areas.

ID10

Prevalence of Vector-borne Diseases in Healthy Dogs in Oklahoma (Abstract ID10)

Emma Rupert[1]; Laura A. Nafe[1]; Paul DeMars[1]; LaRinda Holland[1]; Lara Sypniewski[1]; Susan Little[1]

[1]Center for Veterinary Healthy Sciences, Oklahoma State University, Stillwater, OK, USA

Vector-borne diseases are common in dogs and result in a multitude of clinical signs with variable severity between individual patients. The purpose of this study was to determine the PCR prevalence of *Ehrlichia* spp., *Anaplasma* spp., *Rickettsia* spp., *Babesia* spp., *Hepatozoon* spp., *Mycoplasma* spp., *Babesia* spp., *Leishmania* spp., *Neorickettsia* spp., and *Bartonella* spp. in presumptively healthy dogs presenting to the Oklahoma State University VTH for annual wellness evaluation. A secondary objective was to correlate PCR results with historical information, physical examination, and clinicopathologic results. All dogs enrolled in the study had a client survey completed, SNAP® 4Dx Plus® (IDEXX), CBC, chemistry panel, and vector-borne disease PCR. One hundred and twelve dogs (61 females, 51 males) with a median age of 5 years (range: 0.75–14 years) were enrolled in the study between July 2016 and November 2016. SNAP® 4Dx Plus® results showed that 13 dogs (11.6%) were positive for *Ehrlichia* spp., while only 1 of 13 was *Ehrlichia* PCR positive. Eight dogs (7.1%) were PCR positive for a vector-borne pathogen, with 6/8 infected with Candidatus *Mycoplasma haematoparvum,* 1 with *Mycoplasma haemocanis,* and 1 with *Ehrlichia ewingii.* Four of 8 infected dogs had normal CBC and chemistry results, while 2/8 dogs had clinicopathologic abnormalities potentially attributed to vector-borne disease (thrombocytopenia 62,000/uL and hyperglobulinemia 4.0 g/dL). Anemia was not documented in any dog infected with *Mycoplasma* spp. The prevalence of vector-borne disease in healthy dogs presenting to the Oklahoma State University VTH is fairly low, with *Mycoplasma* spp. infection being most common.

ID11

Prevalence of *Anaplasma phagocytophilum* Infection in Feral Cats in Massachusetts (Abstract ID11)

Erin Galemore; Mary Labato; Elizabeth O'Neil

Tufts University, North Grafton, MA, USA

Anaplasma phagocytophilum infection has been commonly associated with clinical disease in humans and dogs. However, *A. phagocytophilum* infection has not routinely been associated with clinical disease in cats. Recent reports have shown that *Anaplasma* infection can cause significant illness in cats. The objective of this study was to determine the prevalence of *A. phagocytophilum* infection in adult feral cats in Massachusetts, an endemic area for *A. phagocytophilum* and its tick vector *Ixodes scapularis*.

Blood samples were collected between June and December 2015 from 175 apparently healthy adult feral cats that were presented to trap and release spay/neuter centers in Massachusetts. Complete blood count, blood smear evaluation, SNAP® 4Dx® (IDEXX), and *A. phagocytophilum* PCR were performed on all samples to document acute infection (PCR positive and/or inclusions observed on blood smear) and exposure to *A. phagocytophilum* (SNAP® 4Dx® positive for *A. phagocytophilum* antibodies).

The prevalence of exposure to *A. phagocytophilum* in feral cats in Massachusetts was 9.71%, whereas the prevalence of acute infection was 6.94%. All blood smears were negative for *Anaplasma* inclusions; therefore, acute infection was defined as testing positive on PCR analysis.

The prevalence of *A. phagocytophilum* exposure in feral cats approaches 10% and is significantly higher than the previously reported prevalence of 4.3% in the United States ($p<0.001\%$). *A. phagocytophilum* infection may be an emerging infectious disease in felines. Further research is needed to determine the prevalence of clinical illness associated with *A. phagocytophilum* infection in cats living in endemic areas.

ID12

Opportunistic Fungal Infections Associated with Cyclosporine Use in Dogs with Immune-Mediated Diseases (Abstract ID12)

Brigitte B. McAtee[1]; Kevin Cummings[2]; Audrey K. Cook[3]; Jonathan A. Lidbury[4]; Johanna Heseltine[1]; Mike Willard[1]

[1]Texas A&M Veterinary Teaching Hospital, College Station, TX, USA; [2]Dept. of Veterinary Integrative Biosciences, Texas A&M University, College Station, TX, USA; [3]College of Veterinary Medicine, Texas A&M University, College Station, TX, USA; [4]Gastrointestinal Laboratory, Texas A&M University, College Station, TX, USA

Opportunistic fungal infections (OFIs) have previously been reported in dogs undergoing immunosuppressive therapy. However, the incidence of OFIs and risk factors for their development in dogs undergoing immunosuppressive treatment are unknown. The aim of this study was to determine the proportion of dogs diagnosed with certain immune-mediated diseases treated with immunosuppressive drugs that develop OFIs during follow-up and to determine if particular drug(s) were a risk factor for OFIs.

Medical records of dogs presenting to the Texas A&M University, Veterinary Teaching Hospital between January 2008 and December 2015 diagnosed with IMHA, IMPA, ITP, and/or Evans' syndrome were retrospectively reviewed. Age, sex, and breed of patients were evaluated. Dogs were excluded if they were euthanized, died, or lost to follow up within 30 days of initiation of immunosuppressive treatment. For each dog the immunosuppressive agent(s) that they were treated with and whether or not they developed an OFI during follow up was recorded. Multivariable logistic regression was used to determine the association between particular immunosuppressive agents and the development of OFI. Statistical significance was set at $p<0.05$.

Two hundred and forty six dogs diagnosed with selected immune-mediated diseases were retrospectively included in the study: 105 with immune mediated hemolytic anemia, 62 with immune mediated thrombocytopenia, 61 with immune mediated polyarthritis, and 18 with Evans' syndrome. Sixteen patients (6.5%) were diagnosed with an OFI. Multivariable logistic regression analysis showed that the odds of developing an OFI were significantly higher among dogs that received cyclosporine (OR, 4.4; 95% CI, 1.2–16.2; $p=0.02$), relative to dogs that did not receive cyclosporine and that OFI were significantly more likely in male dogs than in female dogs (OR, 4.1; 95% CI, 1.4–12.4; $p=0.01$).

OFIs were an important complication in these dogs with select immune-mediated diseases. Patients were significantly more likely to develop an OFI if they had been treated with cyclosporine and if they were male.

ID13

Staging Feline Leukemia Virus Infections Using Quantitative p27 ELISA and Real-Time PCR Results (Abstract ID13)

Melissa Beall[1]; Jesse Buch[1]; Roberta Cahill[1]; Genevieve Clark[1]; Jancy Hanscom[1]; Christian Leutenegger[2]; Ramaswamy Chandrashekar[1]

[1]IDEXX Laboratories, Inc., Westbrook, ME, USA; [2]IDEXX Laboratories, Inc., West Sacramento, CA, USA

Feline leukemia virus (FeLV) is an oncogenic retrovirus of cats. Studies of FeLV infected cats have demonstrated a correlation between outcome of infection and viral RNA and proviral DNA loads. It has been proposed that a similar correlation exists for p27 antigen concentration. The purpose of this study was to utilize a next-generation microtiter plate ELISA to quantify p27 antigen concentrations and compare that with real-time PCR results for a population of samples obtained from the field. A total of 353 feline samples submitted between 2010 and 2015 to IDEXX Laboratories for RealPCR™ testing that included FeLV were utilized for the study. Matched plasma that had been stored frozen was tested using the PetChek® FeLV 15 ELISA (IDEXX Laboratories, Inc.) for p27 antigen. Recombinant p27 protein was used to generate a standard curve for quantifying the results. Of the 89 FeLV PCR positive samples, 27 had crossing points (Cp) less than or equal to 21 and all had p27 concentrations greater than 50 ng/ml (average 286 ng/ml). Twenty-eight PCR positive samples had Cp greater than 30 and all had p27 concentrations less than 10 ng/ml, with 19 having no detectable antigen. Of the 264 PCR negative samples, 24 samples had p27 antigen concentrations from 1–30 ng/ml. This study demonstrates a potential correlation between proviral DNA loads and the concentration of circulating p27 antigen. Results from the combined diagnostic methodologies may provide more objective information for staging (progressive or regressive-latent or regressive-focal) and monitoring FeLV infections.

ID14

Worldwide Clinic-Based Serologic Survey of FIV Antibody and FeLV Antigen in Cats (Abstract ID14)

Jesse Buch; Melissa Beall; Tom O'Connor; Ramaswamy Chandrashekar

IDEXX Laboratories, Inc., Westbrook, ME, USA

Feline Immunodeficiency Virus (FIV) and Feline Leukemia Virus (FeLV) are globally distributed retroviruses capable of inducing fatal disease in cats. FeLV is spread primarily via saliva among cats in close casual contact whereas FIV is most commonly transmitted between combative cats through deep bite wounds. Rapid test methods for detection of FIV antibody and FeLV p27 antigen in blood are well established as standard practice among veterinarians to screen cats for these diseases. The purpose of this observational study was to evaluate macrogeographic trends in FIV and FeLV test results in pet cats over a 9-year period between 2008 and 2016. The data were mined from a comprehensive international database of SNAP® Feline Triple®, SNAP FIV/FeLV Combo, and SNAP FIV/FeLV Combo Plus field results and used to assess the frequency of singular and co-infections of these retroviruses in cats. Results were obtained from more than 2.9 million samples distributed across more than 26,000 unique postal codes from 68 countries. Countries were grouped geographically into 7 global regions: North America (3 countries, n=2,538,792); Caribbean (8 countries, n=6,882); Latin America (9 countries; n=9,984); Northern Europe (20 countries, n=95,800); Southern Europe (12 countries, n=206,157); Middle East-Africa (6 countries, n=4,787 samples); and Asia-Pacific (10 countries, n=81,201).

The weighted average and range for FIV positive rates by region were: North America (Avg=5.0%, Range=3.5–5.3%); Caribbean (12.6%, 7.7–27.7%); Latin America (7.4%, 2.5–19.7%); Northern Europe (7.5%, 1.6–19.4%); Southern Europe (12.4%, 5.6–20.2%); Middle East-Africa (14.2%, 9.2–24.1%); and Asia-Pacific (13.0%, 3.4–18.8%). The weighted average and range for frequency of FeLV positives by region were: North America (Avg=3.9%, Range=1.6–4.1%); Caribbean (9.2%, 0.5–18.7%); Latin America (12.6%, 2.5–19.7%); Northern Europe (7.5%, 1.6–19.4%); Southern Europe (12.4%, 5.6–20.2%); Middle East-Africa (14.2%, 7.1–42.9%); and Asia-Pacific (6.1%, 2.7–20.7%). The weighted average and range for incidence of FIV-FeLV co-infections by region were: North America (Avg=0.5%, Range=0.2–0.5%); Caribbean (1.9%, 0.0–8.5%); Latin America (1.6%, 0.5–9.2%); Northern Europe (0.6%, 0.0–1.9%); Southern Europe (1.7%, 0.0–5.4%); Middle East-Africa (2.0%, 0.2–6.3%); and Asia-Pacific (1.8%, 0.6–4.6%).

This study represents the broadest worldwide feline retroviral serologic survey reported to date utilizing data from a single test platform. These results along with future monitoring of this data may be useful in developing a global view of the efficacy of various control and eradication strategies being employed for FeLV and FIV across different regions of the world.

ID15

Evaluation of a Rapid Immunoassay for Point-of-Care Detection of Bacteria in Cat Urine (Abstract ID15)

Joshua Daniels[1]; Nancy Martin[2]; **Julie Byron**[2]

[1]The Ohio State University, Columbus, OH, USA; [2]College of Veterinary Medicine, The Ohio State University, Columbus, OH, USA

Lower urinary tract signs in cats (LUTS) are more frequently due to sterile inflammation rather than infection. Practitioners may forgo microbiological culture due to cost, and prescribe antimicrobial agents empirically, incurring unnecessary selection pressure and side effects on most patients. Therefore, a rapid screening test for detection of bacteria in cat urine could direct appropriate therapy.

A "dip-stick" style immunoassay to detect bacteria was performed on feline urine (n=100) submitted for routine bacterial culture in a Veterinary Teaching Hospital. The immunoassay was performed according to manufacturer instructions and culture was performed quantitatively on Columbia agar with 5% sheep's blood, with a maximum 72 hour incubation time at 35°C in ambient air. Bacteria were identified using matrix-assisted laser-desorption time-of-flight mass spectrometry.

Eleven samples yielded ≥100,000 cfu/ml in culture, ten of which (90.9%) were positive on the immunoassay, and consisted of *E. coli* (n=5), *Enterococcus* spp. (n=5), and *Staphylococcus felis* (n=1). Three samples yielded ≤200 cfu/ml, one of which (33.3%) was positive on the immunoassay, and consisted of *E. coli* (n=2), and *Staphylococcus felis* + *E. coli* (n=1). The immunoassay was positive for two (2.3%) culture negative samples. The immunoassay correctly categorized the bacteria as gram-negative or gram-positive for eleven (100%) of the concordant samples.

When samples with ≤1000 cfu/ml were considered negative, sensitivity and specificity of the immunoassay were 90.9% and 96.6%, respectively, with positive and negative predictive values of 76.9% and 98.9%. Given the low prevalence of bacterial cystitis in cats, these data suggest that the immunoassay is a useful screening tool for cats presenting with LUTS.

ID16

Enterococcus faecium Inhibits Feline *T. foetus* Growth and Adhesion to Intestinal Epithelial Cells (Abstract ID16)

Rachel E. Dickson[1]; David A. Bemis[1]; M. Katherine Tolbert[1]

[1]University of Tennessee, College of Veterinary Medicine, Knoxville, TN, USA

Tritrichomonas foetus (Tf) causes chronic diarrhea in cats. Limited treatment options and increasing resistance support investigation of alternative treatments. Probiotics reduce proliferation and pathogenicity of other intestinal pathogens. However, little work has been done to investigate their efficacy against Tf infection. Demonstrating that probiotics reduce Tf infection could have important clinical implications in treating feline trichomonosis.

Enterococcus faecium (Efm) was cultured from a commercially available probiotic. Its effect on Tf proliferation was evaluated throughout log phase growth. The potential of Efm to reduce Tf adhesion to the intestinal epithelium was analyzed with a co-culture model using porcine intestinal epithelial cells (IPEC-J2). Tf and IPEC-J2 were pre-treated with Efm prior to co-culture to evaluate both prophylactic and post-infection treatment with Efm. Tf adhesion and cytotoxicity were evaluated using fluorescent microscopy and crystal violet spectrophotometric analysis. Data were analyzed using SigmaStat.

Efm-induced inhibition of Tf growth was observed at concentrations as low as 10:1 Tf:Ef ($p<0.001$) and was determined to be largely pH dependent ($p<0.01$). Tf-induced cytopathogenicity, when co-cultured with Efm prior to infection of IPEC, was unaltered. However, pretreating IPEC monolayers with Efm reduced Tf adhesion ($p=0.002$), suggesting a benefit of pre-treatment with Efm for cats at risk of Tf infection.

These results demonstrate that Efm inhibits Tf growth. Pre-treatment of intestinal epithelial cells with Efm, but not treatment at the time of Tf infection, inhibits adhesion of Tf to intestinal epithelial cells *in vitro*. Studies investigating other *Enterococcus* spp., as well as multi-strain probiotics, in combination with ronidazole are warranted.

ID17

A Molecular Epidemiological Survey of Gammaherpesviruses in Domestic Cats in Japan (Abstract ID17)

Morihiro Tateno; Masashi Takahashi; **Yasuyuki Endo**

Lab Small Anim Intern Med, Joint Faculty of Vet Med, Kagoshima Univ, Kagoshima, Japan

Gammaherpesviruses (GHVs) are members of an emerging subfamily of the Herpesviridae. The two major human GHVs, Epstein-Barr virus (EBV) and Kaposi's sarcoma-associated herpesvirus (KSHV), are clinically important pathogens because those viruses possess tumorigenesis potential especially in immune-compromised patients. In feline practice, we often experience tumor cases, especially lymphomas. Feline leukemia virus (FeLV) was a major cause for the development of lymphoma in the past days, however, the majority of lymphoma cases are shifting to FeLV-free cats. But the mechanism of tumorigenesis in these spontaneously developed lymphomas has not been well understood. Under this situation, GHV became a candidate which explain the lymphomagenesis in domestic cats. Recent study identified a novel GHV in domestic cats (*Felis catus* GHV, FcaGHV), and epidemiological surveys found that FcaGHVs are distributing in several countries including the United States, Australia, Singapore, Germany and Austria. In the present study, a nation-wide molecular epidemiological study was conducted to investigate the prevalence of GHVs in Japan.

GHV-derived DNA in blood samples was detected by degenerate nested pan-GHV PCR targeting gB gene as previously reported. Nucleotide sequences of obtained DNA fragments were determined and BLAST analysis was carried out. Phylogenic analysis was also conducted. Risk factors relating to GHV infection on clinical parameters, including age, sex, status of retrovirus infections, bite wound history and clinical signs, were statistically analyzed using univariable (Fisher's exact test) and multivariable (Logistic regression model) analyses.

GHV-derived DNA was detected in 1.4% (23/1,738) of Japanese domestic cats. BLAST analysis revealed that all of them showed high similarity with FcaGHVs which were isolated from domestic cats in the United States. Phylogenic analysis also revealed that all Japanese isolates formed one cluster with FcaGHV. In the univariable analysis, older age (OR, 4.93; 95% CI, 1.43–16.99; $p<0.01$), male (OR, 2.51; 95% CI, 0.98–6.40; $p=0.05$), FIV infection (OR, 5.35; 95% CI, 2.30–12.46; $p<0.01$), FeLV infection (OR, 2.02; 95% CI, 0.74–5.50; $p=0.18$) and sick cat (OR, 2.37; 95% CI, 0.80–7.02; $p=0.12$) were statistically selected as risk factors for GHV infection. In the following multivariable analysis, older age (OR, 3.84; 95% CI, 1.06–13.86; $p=0.04$) and FIV infection (OR, 4.11; 95% CI, 1.52–11.11; $p<0.01$) were found to be significant and independent risk factors for GHV infection.

In this study, the prevalence of GHVs in Japanese domestic cats was surveyed and risk factor for GHV infection was determined. FcaGHV prevalence in Japan (1.4%) was lower than previously reported prevalence in other countries and areas (US, 16%; Europe, 16.2%). It was unable to determine the cause of this difference. However, Japanese isolates were identified to have close genetic relationship to domestic cat derived FcaGHV. This finding suggested that the virus is highly conserved in each animal species and GHV is harboring in host specific manner as well as other herpesviruses. The risk factor for GHV infection in Japanese cats was FIV infection and it was same as previously reported. But we have to clarify how FIV enhances the GHV infection and/or viremia. The clinical importance of GHVs infection in feline practice is still unknown. Further investigation will be necessary to identify the pathogenicity of GHVs in domestic cats.

ID18

Minimal Inhibitory Concentrations of Antimicrobials Against *Escherichia coli* Isolated from Companion Animal Urinary Tract Infections (Abstract ID18)

Lacie Johansen[1]; Cindy Lindeman[2]; Don Bade[3]; George Rodgers[4]

[1]Zoetis, Kalamazoo, MI, USA; [2]Formerly with Zoetis, Kalamazoo, MI, Apple Valley, MN, USA; [3]Microbial Research Incorporated, Fort Collins, CO, USA; [4]Zoetis, Parsippany, NJ, USA

Zoetis has been conducting an ongoing surveillance program to evaluate the susceptibility trends of antimicrobial agents against canine and feline *Escherichia coli* (*E. coli*) pathogens isolated from Urinary Tract Infections (UTIs) since 2011. The data were analyzed annually and the first five years of susceptibility data for amoxicillin/clavulanic acid, ampicillin, cefovecin, cefpodoxime, ceftiofur, enrofloxacin, marbofloxacin, orbifloxacin, trimethoprim/sulfamethoxazole and cephalexin are presented here. A total of 3,300 *E. coli* strains, isolated as the etiological agent in cats and dogs presenting with naturally occurring UTIs and being seen at primary/general care practices were submitted to this surveillance program by 16 veterinary diagnostic laboratories throughout the United States (US) and Canada. The Minimal Inhibitory Concentrations (MICs) for all isolates were tested using a broth microdilution system (Thermo Fisher Scientific, Oakwood Village, OH) that conforms to the Clinical and Laboratory Standards Institute (CLSI) standards.

Data collected during the first 5 years of this program showed that the modes and MIC_{50} values remained stable (within one doubling dilution) for each antimicrobial tested against *E. coli* from both dogs and cats for each year of testing; MIC_{90} values were stable over the years tested with some exceptions. For amoxicillin/clavulanic acid and ampicillin, mode and MIC_{50} values remained the same (4 µg/mL) and MIC_{90} values ranged from 8–>16 µg/mL. During this time, the *in vitro* testing for cefovecin, cefpodoxime, and ceftiofur were steady, the mode, MIC_{50}, and MIC_{90} values ranged from 0.5–2 µg/mL, with a few single drug, single year MIC_{90} values testing outside that range. The mode and MIC_{50} values for the fluoroquinolones (enrofloxacin, marbofloxacin and orbifloxacin) all tested at ≤0.12 µg/mL, while MIC_{90} values ranged from 0.06–4 µg/mL. For trimethoprim/sulfamethoxazole, mode and MIC_{50} values were stable (≤0.12 µg/mL) and MIC_{90} values ranged from ≤0.25–1 µg/mL. *In vitro* testing of cephalexin was included in 2015. The mode and MIC_{50} values for this single year of testing *E. coli* from both dogs and cats were both 8 µg/mL, while the MIC_{90} value was 16 µg/mL in cats and >64 µg/mL in dogs. Screening and phenotypic confirmation only of Extended-Spectrum β-Lactamase (ESBL) production in *E. coli* were done following CLSI standards. Over the 5 years of testing, 1.2% (23 of 1933) of the *E. coli* isolated from dogs and 0.8% (11 of 1367) of the *E. coli* isolated from cats were confirmed as ESBL-producing strains.

Overall, antimicrobial susceptibility data were stable from 2011–2015 and did not show progression of resistance during that time frame.

ID19

Effect of Doxycycline Administration on Antibodies Against Different *Borrelia burgdorferi* Peptides (Abstract ID19)

Michael R. Lappin[1]; Scott Moroff[2]

[1]Colorado State University, Fort Collins, CO, USA; [2]Antech Laboratories, Lake Success, NY, USA

Borrelia burgdorferi is transmitted by *Ixodes* spp. and is associated with inflammatory arthritis or nephritis in some dogs (Lyme disease). A number of diagnostic modalities have been evaluated to aid in the diagnosis of Lyme disease in dogs, with serological testing proving to be the most beneficial. Antibodies against outer surface protein A (OspA), OspC, OspF and antibodies against the C6 peptide are currently used most frequently in commercially available serological assays. Antibodies against OspC are often the first detected in serum after a new exposure to *B. burgdorferi*. In humans with *Ixodes* spp. infestation, doxycycline treatment has been shown to block progression of *B. burgdorferi* infection but whether this occurs in dogs is unknown. The purpose of this study is to report the effect of doxycycline administration on serological responses to OspA, OspC, OspF, and C6 peptide in dogs experimentally infested with wild-caught *I. scapularis*.

A total of 23 young adult dogs were infested with 25 female and 15 male *I. scapularis* wild caught in Rhode Island (48% *B. burgdorferi* carriage rate). All dogs had multiple fed ticks present when removed from the chambers at Day 7. Serum collected prior to infestation and weekly after infestation were assayed in a commercially available in clinic assay (SNAP4DXPlus; IDEXX Laboratories; C6 peptide) and at a commercial laboratory (Accuplex; Antech Diagnostics; OspA, OspC, and OspF). The first 3 dogs to be positive for OspC antibodies but negative for antibodies against the other 3 peptides were administered doxycycline at approximately 10 mg/kg, daily for 28 days.

By Week 10 after *I. scapularis* infestation, 21 of 23 dogs had developed serological evidence of *B. burgdorferi* infection. Of the 3 dogs with OspC antibodies alone that were administered doxycycline for 28 days, 2 dogs never developed antibodies against the other peptides and were negative for OspC antibodies by Week 2 or Week 6 after starting doxycycline administration. The other treated dog became positive to the C6 peptide alone during week 1 of doxycycline administration, was positive for 2 of the next 3 weeks, and then was negative for all *B. burgdorferi* antibodies for 4 weeks after cessation of treatment. In contrast, only 1 of the 18 infected, untreated dogs transiently (1 week) developed OspC antibodies but then stayed negative for all antibodies for 6 weeks, suggesting a self-limited infection or a false positive OspC antibody titer.

This model induced serological evidence of *B. burgdorferi* infection of 21 dogs with 17 of 18 untreated dogs developing serological evidence of chronic infection characterized by antibodies against OspF and C6 peptide. While only 1 dog may have spontaneously limited antibody production against all 4 peptides, all 3 dogs administered doxycycline during the acute phase limited antibody production. These results should be confirmed in larger numbers of dogs, but may suggest that treatment when OspC antibodies are first detected may block development antibody responses associated with chronic *B. burgdorferi* infection.

ID20

Performance of a Leishmanin Skin Test for the Diagnosis of Feline Leishmaniosis (Abstract ID20)

Ludmila S.V. Sobrinho[1]; Juliana P. Vides[1]; Karina Y. Hirata[1]; Mariana Aschar[1]; Vânia Matta[1]; Marcia D. Laurenti[1]; **Mary Marcondes**[1]

[1]São Paulo State University, Aracatuba, São Paulo, Brazil

A natural resistance of cats to leishmaniosis, primarily due to a cellular immune response, is widely suggested. Leishmanin skin test (LST) has been successfully used in humans and dogs and a positive LST result indicates exposure to *Leishmania* and a state of cell mediated immunity. A total of 96 cats, seronegative for FeLV and FIV, presenting to a Veterinary Teaching Hospital in an endemic area in Brazil, were evaluated by LST. *Leishmania* infection was confirmed in 56 cats by bone marrow qPCR, and 40 non-infected cats composed the control group. Among infected cats, 23/56 presented clinical signs and 33/56 were asymptomatic. Each cat was intradermally injected with a suspension of $4x10^7$ *L. chagasi* promastigotes (inactivated in thimerosal solution), in the inner surface of the right thigh, and thimerosal solution was injected in the opposite limb. All control and infected cats tested negative to LST at any time reading (24, 48 and 72 h after inoculation). We expected that at least part of the 26 asymptomatic (and seronegative) cats presented a positive LST. In canine leishmaniosis (CanL) LST can detect a high proportion of asymptomatic dogs, although some of them have negative LST even in the presence of a cellular immune response. The method employed was previously used for CanL, and its potential immunogenicity was confirmed by promoting positive LST in five naturally infected and asymptomatic dogs from the same endemic area. However, the LST tested do not seem to be an adequate tool for the diagnosis of feline leishmaniosis.

ID21

Evaluation of Dried Blood on Filter Paper for Amplification of *Ehrlichia canis* DNA from Dogs (Abstract ID21)

Yunus H. Ozekin; Cody Minor; Jennifer Hawley; Melissa Brewer; Michael R. Lappin

Colorado State University, Fort Collins, CO, USA

Transportation of blood from remote areas can be problematic. While DNA is generally stable and a number of studies have evaluated DNA extracted from dried blood of mammals, there is minimal information determining the sensitivity of different techniques using dog blood. The objective of this study was to determine the optimal technique for extracting DNA from filter paper dried blood spots using *Ehrlichia canis* DNA as the target.

Blood samples from dogs were collected with owner permission for use in a vector borne agent prevalence study in Mexico. After collection, the aliquot of blood was placed in EDTA, and then 300 μl was blotted onto filter paper (VWR® Grade 413), allowed to dry, and stored in individual sterile bags. The filter paper samples were stored at room temperature, and the EDTA blood was stored at -80°C until assayed at the central laboratory. A glyceraldehyde 3-phosphate dehydrogenase (GAPDH) quantitative PCR was used to estimate DNA yields in the different experiments. Different buffers and incubation periods were assessed in experiments using the entire filter paper in different washing procedures, and it was determined that an overnight PBS wash gave the best results (Method 1). In Method 2, a commercially available kit (QIAamp® DNA Blood Mini Kit Dried Blood Spot (DBS) protocol) was performed as per the manufacturer's instructions. Once Method 1 was optimized, 12 samples previously shown to be positive for *E. canis* DNA in blood by conventional PCR assay and 3 known negative samples were thawed and used to make 2 filter paper blots which were allowed to dry overnight before being subjected to DNA extraction Methods 1 and 2. The whole blood and filter paper derived DNA extracts were then assayed in the GAPDH qPCR and the *Ehrlichia* spp. PCR assay. A total of 11 whole blood samples were PCR positive with bands characterized as strong (n=4), medium (n=3), or weak (n=4). One sample using Method 1 was excluded due to the washing of an incorrect blot.

When the GAPDH qPCR results from whole blood were compared to those from the 2 filter paper extract Methods, it was shown that both Method 1 (74 fold) and Method 2 (15 fold) had lower DNA yields. *Ehrlichia canis* DNA was amplified from 5 of 11 whole blood positive samples (45.5%) using Method 2 and 2 of 10 samples (20.0%) using Method 1. The filter paper samples that had been dried for several months gave the same results as those filter paper samples that had been dried for 24 hours.

While specific DNA can be extracted from blood dried on filter paper, the sensitivity will be decreased compared to whole blood. If this technique is used, the commercial kit is likely to be more sensitive than an overnight PBS wash.

ID23

Clinical and Laboratory Findings in 30 Dogs Naturally Infected with *Ehrlichia ewingii* (Abstract ID23)

Barbara Qurollo[1]; Ramaswamy Chandrashekar[2]; Jesse Buch[2]; Melissa Beall[2]; Caroline A. Yancey[1]; Alexander Caudill[1]; Alaire Comyn[1]; Edward Breitschwerdt[1]

[1]NCSU-CVM, Raleigh, NC, USA; [2]IDEXX Laboratories, Inc., Westbrook, ME, USA

E. ewingii is the most prevalent tick-transmitted *Ehrlichia* spp. identified in dogs from the Southern US. Fever, lameness, neutrophilic polyarthritis and neurologic abnormalities are considered common clinical findings in dogs naturally-infected with *E. ewingii*; however, nonclinical dogs can be PCR+. To further assess the clinicopathologic features in *E. ewingii* PCR+ cases, we analyzed the medical records of 30 naturally-infected dogs. To be included in this study, dogs had clinical data, archived whole blood available from North Carolina State University-College of Veterinary Medicine, and were *E. ewingii* PCR (+) and PCR (-) for other vector-borne pathogens including *Ehrlichia* spp. (*E. canis*, *E. chaffeensis*, and Panola Mountain *Ehrlichia*), *Anaplasma* spp., *Babesia* spp., *Bartonella* spp., Hemotropic *Mycoplasma* spp., and *Rickettsia* spp. Based upon diagnostic and/or retrospective research testing, all 30 dogs satisfied the inclusion criteria. Frequently reported physical exam findings included joint pain and effusions 10/30 (33.3%); heart murmur 8/30 (26.75); organomegaly 7/30 (23.3%); vomiting 6/30 (20%) and fever 5/30 (16.7%). Of the 30 *E. ewingii* PCR+ dogs, 28, 26 and 20 had corresponding CBCs, biochemistry panels and urinalyses, respectively. Hematological and biochemical abnormalities included increased abnormal lymphocytes 18/28 (64.3%); neutrophilia 15/28 (53.6%), of which 9/15 (60%) were characterized by a left shift; anemia 14/28 (50%), of which 8/14 (57%) were characterized as regenerative; monocytosis 13/28 (46.4%); lymphopenia 12/28 (42.9%); thrombocytopenia 11/28 (39.3%); elevated ALP 13/26 (50%) and ALT 10/26 (38.5%), of which only two dogs were receiving corticosteroids; hypoalbuminemia 9/26 (34.6%); and elevated creatinine 6/26 (23.1%). Urinalyses abnormalities included albuminuria 11/20 (55%); proteinuria with inactive sediment 6/20 (30%); and bilirubinuria 6/20 (30%). Of the 7 dogs with urine/protein/creatinine ratios, 6 (85.7%) were greater than 1.5 (range: 1.74–11.7). Canine vector-borne disease IFA serology results were available for 28/30 (93.3%) dogs (no sera submitted for 2 dogs), of which 13/28 (46.4%) were *E. canis* IFA seroreactive (titers ≤1:256) and 5/28 (17.8%) were *R. rickettsii* (spotted fever group rickettsia) IFA seroreactive (titers ≤1:256). All dogs were *Babesia* spp., and *Bartonella* spp. seronegative. Using the SNAP®4Dx® Plus test kit, 23/29 (79.3%) serum or blood samples were *Ehrlichia* (+), 1/29 (3.4%) was Lyme (+) and all dogs were *Anaplasma* antibody (-) and *Dirofilaria immitis* antigen (-). The most frequent medical record diagnoses included renal disease 6/30 (20%); IMHA 5/30 (16.7%); neurologic disease 4/30 (13.3%); polyarthritis 3/30 (10%); hepatic disease 3/30 (10%); cancer 3/30 (10%); and other or undiagnosed 3/30 (30%) illnesses. Ehrlichiosis was the sole diagnosis for only 1/30 (0.033%) dogs and was listed concurrently with other diagnoses for 8 (26.6%) dogs. We hypothesize that infection with *E. ewingii* may contribute to co-morbidities or function as a precipitating factor in a spectrum of clinical syndromes in dogs. This hypothesis will be tested using inferential statistics and a large diagnostic data set from IDEXX Laboratories, Inc.

NU01

Fluorescent *In Situ* Hybridization in Tissues from Dogs with Proliferative Urethritis (Abstract NU01)

Moria Borys[1]; Sean Hulsebosch[1]; Jane Sykes[1]; F. Charles Mohr[1]; Katherine Watson[1]; Kenneth Simpson[2]; Jodi L. Westropp[1]

[1]School of Veterinary Medicine, UC Davis, Davis, CA, USA; [2]College of Veterinary Medicine, Cornell University, USA, Ithaca, NY, USA

The cause of proliferative urethritis (PU) has not been elucidated. Chronic bacterial infections and immune-mediated processes have been proposed. Aerobic bacterial urine cultures (ABUC) are often positive at diagnosis, but whether this is a cause or consequence of PU is unknown.

The objectives of this retrospective study were to 1) describe the clinicopathologic data of cases with PU, 2) utilize fluorescent *in situ* hybridization (FISH) to determine whether mucosally associated bacteria are present, and 3) describe whether an association exists between the type of inflammation and the presence or absence of bacteria on FISH evaluation.

Twenty-two cases of biopsy-confirmed PU were identified (mean age 8 years; range 1–15 years). There were 14 females, 8 males, and 15 breeds represented. Four dogs had positive ABUC perioperatively, 4 had positive mucosal bacterial cultures, and 11 had negative ABUC. Urine cultures were unavailable or not performed in 7 dogs. FISH analysis was performed in 13 cases; adherent/invasive bacteria were noted in 5 dogs, and tissue-associated bacteria in 2 dogs. Of these 7 dogs, 3 had positive ABUC at the time of biopsy. The 6 FISH analyses that were negative for bacteria also had negative ABUC. One of the dogs with adherent/invasive bacteria visualized had lymphoplasmacytic and histiocytic urethritis; the remaining 12 had neutrophilic, lymphoplasmacytic, or mixed inflammation reported on histopathology. These results suggest that primary or secondary bacterial invasion may be associated with PU, despite negative ABUC; this knowledge may alter treatment of this disease.

NU02

Vitamin D Metabolism in Dogs with Calcium Oxalate Urolithiasis (Abstract NU02)

Elizabeth Groth[1]; Jody Lulich[1]; Dennis Chew[2]; Valerie Parker[3]; Eva Furrow[1]

[1]University of Minnesota, St. Paul, MN, USA; [2]The Ohio State University, Columbus, OH, USA; [3]The Ohio State University Veterinary Clinical Sciences, Columbus, OH, USA

The aim of this study was to measure vitamin D metabolites in dogs with and without a history of CaOx urolithiasis.

Twenty Miniature Schnauzers, 10 Bichons Frise, and 10 Shih Tzus were included in the study. Cases consisted of 20 dogs with a history of CaOx urolithiasis and fasting urinary calcium to creatinine ratios (UCa/Cr) >0.05. Controls consisted of 20 age-, sex-, and breed-matched dogs without urolithiasis (as determined by screening radiography) and with UCa/Cr 3, 24,25(OH)$_2$ vitamin D$_3$, and total 1,25 (OH)$_2$ vitamin D measured through a commercial laboratory (Heartland Assays, Ames, Iowa).

There were no statistically significant differences between cases and controls when comparing 25(OH) D$_3$, 24,25(OH)$_2$ D$_3$, total 1,25(OH)$_2$, or the ratio of total 1,25(OH)$_2$ to 25(OH) D$_3$. However, cases had a significantly higher ratio of 25(OH) D$_3$ to 24,25(OH)$_2$ D$_3$ compared to controls (median=1.43 and 1.17, respectively; p=0.012). Nine of the 10 highest ratios were found in cases, but there was substantial overlap with more than half the cases falling within the range observed in controls. There was a moderate positive correlation between the ratio of 25(OH) D$_3$ to 24,25(OH)$_2$ D$_3$ and UCa/Cr (r=0.37, p=0.017).

This data suggests that decreased conversion of 25(OH) vitamin D$_3$ to the inactive 24,25(OH)$_2$ vitamin D$_3$ occurs in a subset of dogs with CaOx urolithiasis. Abnormalities in vitamin D metabolism may contribute to stone risk in some dogs.

NU03

Risk Factors for Urinary Tract Infections in Cats After Ureteral Stent and Ureteral Bypass Placement (Abstract NU03)

Lucy Kopecny[1]; Carrie A. Palm[1]; Kenneth J. Drobatz[2]; Ingrid M. Balsa[1]; William T.N. Culp[1]

[1]University of California-Davis, Davis, CA, USA; [2]University of Pennsylvania, Philadelphia, PA, USA

Ureteral stenting and bypass are commonly used for managing benign ureteral obstructions. Objectives of this case-control study were: 1) to determine post-operative rate of urinary tract infection (UTI) in cats undergoing these procedures and 2) to identify risk factors associated with UTI development. Forty-two client-owned cats undergoing 46 renal decompressive surgeries were included. Cats were included if they had urine cultures performed prior to and after surgery; variables were compared statistically to identify risk factors.

Median duration of follow-up was 203 days (range 9–2184 days). Of 137 post-operative urine cultures, 20 (15%) were positive in 14/42 (33%); 3/42 cats had more than one UTI. Nine different bacteria were identified; *Enterococcus faecium* (n=5), and *E. coli* (n=5) predominated. Cats undergoing bilateral surgery were more likely (OR: 4, 95% CI: 1,15; p=0.03) to develop a UTI postoperatively compared to cats with unilateral surgery. Anuric cats were also more likely to develop a UTI compared to non-anuric cats (OR: 5, 95% CI 1,27; p=0.03). Pre-existing chronic kidney disease and post-operative urinary catheter or abdominal drain placement were not significantly associated with infection, either prior to discharge or at long-term follow-up.

Post-operative UTIs occurred relatively commonly in this cohort of cats. Post-operative urethral catheterization has been speculated to increase UTI risk, but this was not evident in this cohort of cats. Greater case numbers will be required before definitive conclusions can be made. Future work should focus on gathering greater case numbers and on the clinical significance of these infections.

NU04

Concentrated Alkaline Urine is Hostile to Uropathogenic *E. coli* Growth (Abstract NU04)

Laura Thornton[1]; Richard Burchell[1]; Stacey Burton[1]; Nicolas Lopez-Villalobos[1]; Danielle Pereira[1]; Izzy MacEwan[1]; Ares Fong[1]; Antonius Hatmodjo[1]; Mark Nelson[1]; Alex Grinberg; Niluka Velathanthiri; Arnon Gal[1]

[1]Institute of Veterinary, Animal, and Biomedical Sciences, Massey University, Palmerston North, Manawatu-Wanganui, New Zealand

Uropathogenic *E. coli* (UPEC) are common bacteria involved in lower urinary tract infections in dogs. This study tested the hypothesis that acidic diluted urine promotes UPEC growth.

The authors obtained early morning urine samples from nine healthy bitches. Following urine filtration through a 0.2 μm biofilter, we obtained a matrix of three levels of pH (5.5, 7.0, 8.5) by three levels of urine specific gravity (USG; 1.010, 1.020, 1.030) by titrating the urine with HCl and NaOH. We inoculated the urine with known quantities of three UPEC strains, incubated at 37°C for four hours, made serial dilutions from 10^{-4} to 10^{-7}, plated on nutrient agar for 24 hours and subsequently counted the numbers of colonies forming units (CFU). We only counted plates from dilutions that grew 30–300 CFU. All experiments had negative and positive controls.

The number of UPEC CFU was analyzed with a mixed linear model that includes the fixed effect of strain, pH, concentration and the interaction between the three, the random effect of dog, and inoculated dose of UPEC as a covariate. Least Square Means were obtained and used for comparisons. For all comparisons, the alpha probability error was $p=0.05$.

There were significant differences in the mean (±SE) log(CFU) between the different pH 7.0>5.5>8.5 and USG 1.010>1.020>1.030. The least bacteria grew in pH 8.5 USG 1.030 and the most in acidic and neutral diluted urine. The strain of UPEC had a significant effect on growth. Differences between strains were substantial in concentrated alkaline urine.

The impact that acidifying pet food diets may have on UPEC infections should be determined.

NU05

Correlation Between Canine Urine Cortisol-Creatinine and Protein-Creatinine Ratios Collected at Home and in Hospital (Abstract NU05)

Lindsey E. Citron[1]; Nicole M. Weinstein[1]; Meryl P. Littman[1]; J.D. Foster[2]

[1]School of Veterinary Medicine, University of Pennsylvania, Philadelphia, PA, USA; [2]Friendship Hospital for Animals, Washington, DC, USA

Recently, it has been demonstrated that urine-protein-creatinine (UPC) ratios were higher in canine urine samples collected in hospitals (IH) settings than those collected at home (AH). Stress associated in-hospital procedures and travel to the hospital could play a role in the higher UPC observed. The purpose of this study was to evaluate urine cortisol-creatinine ratios (UCCR), as a metric of stress, and to correlate to UPC in urine samples obtained at home and in hospital.

This was a prospective, non-masked study. Clients obtained a free-catch urine sample at home or off-campus (<12 hours prior to their appointment). Time of urine collection and total hospital visit time were recorded. A second urine sample was obtained at the conclusion of the dog's appointment in hospital. Urine creatinine, protein, and cortisol levels, and full urinalysis were measured on both urine samples. Urine protein creatinine ratios and UCCR were compared between AH and IH settings via Wilcoxon signed rank test. Correlation between UCCR and UPC, in any setting, and effect of travel/hospitalization time on change in UCCR and UPC were evaluated with Spearman's analysis.

Thirty-six dogs were enrolled. The majority (78%) of dogs were nonproteinuric (UPC<0.2) in both urine samples. The UCCR was significantly lower (p<0.0001) in urine samples obtained AH (median 17.5, range 1–41) compared to IH (median 20.5, range 10–189). The UPC was not statistically different (p=0.144) in urine samples obtained AH (median 0.02, range 0.02–2.87) compared to IH (median 0.03, range 0.01–2.64). UCCR was not significantly associated with UPC in all samples and in any collection setting. Travel time, time in hospital, and total time prior to IH urine collection was not associated with change in UCCR or UPC. In the 8 patients that were borderline proteinuric (UPC 0.2–0.5) or proteinuric (UPC>0.5), there was no statistical difference in UPC between settings (p=0.25), however the UCCR was significantly higher (p=0.0312) in samples collected IH (median 17.5, range 11–159) compared to AH (median 13, range 1–41).

Contrary to an earlier study, this data demonstrated that UPC was not higher when measured in urine samples collected IH compared to AH. Dogs had higher UCCR IH compared to AH, however the UCCR was not associated with UPC. Stress, as estimated through UCCR, did not affect proteinuria in these patients.

NU06

Application of Peritoneal Dialysis in Companion Animals in Turkey (Abstract NU06)

Hasan Albasan[1]; Osman Safa Terzi[1]; Sule Duzlu Arun[1]; Erdal Kara[1]; **Nevra Keskin**[1]; Irem Gul Sancak[2]

[1]Department of Internal Medicine, Faculty of Veterinary Medicine, University of Ankara, Diskapi, Ankara, Turkey; [2]Department of Surgery, Faculty of Veterinary Medicine, University of Ankara, Diskapi, Ankara, Turkey

Peritoneal dialysis (PD) is a renal replacement therapy that has been commonly used in human and veterinary medicine for the treatment of kidney insufficiency. The objective of this study was to assess the renal function in animals with renal failure after PD.

Renal cases presented to veterinary teaching hospital at Ankara University were treated with PD between June 30, 2014 to June 29, 2015. A total of 5 cases consisted of 3 dogs with acute renal failure of ethylene glycol toxicity (n=1) and ingestion of unknown plants (n=2) and of the remaining one dog and one cat were with chronic renal failure. PD was applied once daily ranged between 4 to 10 days.

Concentration of serum urea in all patients significantly decreased after the second attempt of PD but, not serum creatinine. When PD procedures completed, the concentrations of serum urea and creatinine were significantly lowered in all cases. However, those values were still above the reference range except in one dog with acute renal failure.

Complications identified during PD were peritonitis (n=4), hypoalbuminemia (n=4), hypokalemia (n=3), dialysate retention or leakage from the catheter site (n=3), edema (n=2) and pleural effusion/chylothorax (n=2).

This is the first report of PD procedure successfully applied as life-saving, easily applicable, and cost effective clinical procedure to companion animals with renal failure that could be adapted by veterinary practitioners in Turkey.

NU07

Predicted Ionized Calcium Better Reflects Calcium Homeostasis Than Total Calcium in Dogs with Renal Azotemia (Abstract NU07)

Kevin Le Boedec[1]; Julie Danner[2]

[1]Centre Hospitalier Vétérinaire Frégis, Arcueil, Ile-de-France, France; [2]College of Veterinary Medicine, University of Illinois, Urbana, IL, USA

Total calcium (tCa) and corrected tCa are considered unreliable predictors of calcium homeostasis, especially in dogs with renal azotemia. Unfortunately, measured ionized calcium (miCa), which truly reflects calcium homeostasis, is not readily available in practice.

Dogs with renal azotemia (i.e., creatininemia >2.0 mg/dL and urine specific gravity <1.030) were retrospectively identified from 751 records for inclusion in the study. In included dogs, tCa was corrected for total protein or albumin concentrations by use of published equations. Predicted ionized calcium (piCa) was calculated from routine biochemical values and age via the Canine Ionized Calcium Calculator website (available at: http://vetmed.illinois.edu/study/mars-model/VetMed.php). Sensitivity and specificity of piCa were compared to those of tCa and corrected tCa for true hypercalcemia and hypocalcemia, as determined by miCa (McNemar's test).

Out of 40 azotemic dogs included in the study, 7 were hypercalcemic and 14 were hypocalcemic based on miCa. Sensitivities of piCa (86%), tCa (86%) and corrected tCa (100%) were equivalent to detect hypercalcemia ($p=1.0$). Specificities of piCa (94%) and tCa (88%) were not statistically different ($p=0.5$), but specificity of piCa was significantly higher than specificity of corrected tCa (70–76%) to diagnose hypercalcemia ($p=0.008$–0.03). Furthermore, sensitivity of piCa (86%) was significantly higher from sensitivities of tCa (29%) and corrected tCa (7%) to detect hypocalcemia ($p<0.01$). Specificities of piCa (96%), tCa (100%) and corrected tCa (100%) were equivalent to diagnose hypocalcemia ($p=1.0$).

In dogs with renal azotemia, piCa seems more accurate to diagnose true hypercalcemia and more sensitive to detect true hypocalcemia than corrected tCa.

NU08

Urine Protein to Creatinine Ratios of Proteinuric Canine Urine Samples Stored in Multiple Container Types (Abstract NU08)

Patrick Moyle[1]; Andrew Specht[1]; Richard Hill[1]

[1]College of Veterinary Medicine, University of Florida, Gainesville, FL, USA

Adsorption of protein to the surface of polymer plastic or glass could affect urine protein:creatinine ratios. Homopolymer polypropylene specimen containers mitigate protein adsorption but urine samples collected by owners are sometimes stored in propylene copolymer plastic food storage or glass containers either at room temperature, refrigerated, or frozen. The primary aim of this study was to determine whether protein:creatinine ratios (UPC) are affected by these storage conditions.

Urine was collected over a 24 hour period from one mildly proteinuric and one severely proteinuric dog each with negative urine cultures and a unremarkable urine sediment and stored at 4°C in a homopolymer propylene collection jug designed for 24-hour urine protein analysis in humans. Five aliquots of each urine sample were subjected to each of nine storage conditions: in three types of container (glass, propylene copolymer, and homopolymer propylene) at either 24°C for 4 hours, 4°C for 12 hours, or -20°C for 72 hours. An additional five aliquots were stored at -80°C in homopolymer as a control. Following experimental storage, samples were then stored at -80°C in homopolymer until analysis. To determine the minimum size of each of the study groups, a power analysis was performed a priori. With a power of 80%, an alpha of 5%, and a standard deviation of 0.3 for severely proteinuric animals and 0.1 for mildly proteinuric animals (based on pilot study), 4 samples would be needed to detect an approximate 10% decrease in UPC in a severely proteinuric sample (from 10 to 9) and an approximate 33% decrease in the UPC in a mildly proteinuric animal (from 0.6 to 0.4). The latter was chosen as it would alter recommendations based on the 2005 ACVIM guidelines. Shapiro-Wilk test indicated that values were normally distributed. For each urine sample, ratios were compared using a one-way ANOVA with storage condition as the only factor in the model, and ratios were compared among treatments post-hoc with a Bonferroni correction (experimentwise type 1 error <0.05).

Ratios differed slightly but significantly ($p<0.03$) among aliquots stored under different conditions for both urine samples. For mildly proteinuric urine, UPC values were slightly higher (mean 0.68 vs. 0.66) in aliquots stored in copolymer containers at 24°C than at -20°C, and were slightly lower (0.66 vs. 0.68) in aliquots stored in glass containers at 24°C than at -4°C, but were not different from controls (0.67 +/- 0.01). For severely proteinuric urine, UPCs were higher in aliquots stored in copolymer at 24°C (9.6 +/- 0.9) and glass at -20°C (9.5 +/- 0.5) than in controls (8.4 +/- 0.2). Nevertheless, most storage conditions did not affect ratios and the detected differences would not have led to different clinical recommendations based on the ACVIM consensus statement or IRIS guidelines.

The results suggest that collecting and storing urine in clean commonly used household containers such as propylene copolymer food storage containers and glass jars is unlikely to lead to changes of sufficient magnitude to be of clinical importance. However, further study would be needed to determine if there might be specific circumstances or variables (i.e., patients with dysproteinemia, different amounts of surface area contact) in which container type might still cause clinically significant changes in measured UPC.

NU09

Evaluation of Gastric pH and Serum Gastrin Concentrations in Cats with Chronic Kidney Disease (Abstract NU09)

M. Katherine Tolbert[1]; Shelly Olin[1]; Sally MacLane[2]; Jörg M. Steiner[3]; Shelly Vaden[4]; Emily N. Gould[6]; Josh Price[6]

[1]College of Veterinary Medicine, University of Tennessee, Knoxville, TN, USA; [2]Appalachian Animal Hospital, Piney Flats, TN, USA; [3]Gastrointestinal Laboratory, College of Veterinary Medicine, Texas A&M University, College Station, TX, USA; [4]North Carolina State University, Raleigh, NC, USA; [5]University of Tennessee, Knoxville, TN, USA

Chronic kidney disease (CKD) is a prevalent condition in cats. Advanced CKD is associated with hyporexia and vomiting. These signs are typically attributed to uremic toxins and gastric hyperacidity and underlie IRIS recommendations for acid suppressant use with advanced CKD in cats. However, gastric pH has not been measured in cats with CKD. Accordingly, the central objective of this study was to determine if cats with moderate to severe CKD have decreased gastric pH compared to age-matched, healthy cats. Based on previously published work demonstrating an association of hypergastrinemia and CKD, we hypothesized that cats with CKD would have a reduced gastric pH compared to healthy, aged-matched cats.

All cats with concurrent disease were excluded on the basis of history, physical examination, CBC, serum chemistry profile, urinalysis and urine culture, serum total thyroxine, and SDMA (controls only) obtained within 24 hours of pH monitoring and assessment of serum gastrin concentrations. Ten cats with IRIS stage II–IV CKD (mean±SD age: 14.1±5.7 yrs) and nine healthy mature adult cats (mean age: 12.3±1.7 yrs) were prospectively enrolled. Continuous, 12hr gastric pH monitoring was performed by oral administration of a pH capsule. Serum gastrin concentrations, mean gastric pH, and mean percentage of time gastric pH was strongly acidic (pH<2) were compared between groups using an independent 2-sample t-test.

pH parameters including mean gastric pH (CKD: 1.8±0.5; healthy: 1.6±0.3; p-value=0.23) were not significantly different between cats with CKD and healthy cats. Mean serum gastrin concentrations (CKD: 110±210; healthy: 57±33 ng/dL) were higher and highly variable in cats with CKD, as previously described, but were not significantly different between groups (p-value=0.72). pH data suggest that cats with CKD do not have gastric hyperacidity compared to aged-matched healthy cats. Thus, further study investigating the benefit of acid suppression in cats with CKD is warranted.

NU10

Effect of High Sodium Diet on Urine Characteristics in Healthy Adult Dogs (Abstract NU10)

Hui Xu[1]; Xuemei Si[1]; Sandeep Bhatnagar[1]; Dorothy Laflamme[2]

[1]Nestlé Purina Petcare Global Resources, St. Louis, MO, USA; [2]Purina Institute, St. Louis, MO, USA

Increasing urine volume and decreasing urine specific gravity (USG) are considered important in managing dogs with crystal-associated lower urinary tract diseases. The objective of this study was to evaluate the impact of increased dietary salt (sodium chloride) on water intake, urine volume, USG, and mineral excretion in dogs.

Two diets, identical except for salt content, were fed to 12 healthy adult dogs using a 2 x 2 cross-over design. Control Diet (CON) contained 0.11 g sodium/100 kcal ME, and High Sodium Diet (HNA) contained 0.37 g sodium/100 kcal ME. All dogs were fed CON for 2 weeks as the baseline, then randomly allocated to receive CON or HNA for 2 weeks. Following a 2-week washout when all dogs were fed CON, a final 2 week feeding period was used with dogs fed the alternate diet from phase 1. Water consumption was measured daily, total voided urine was collected on days 10, 12, and 14 of each phase for determination of volume, USG and mineral excretion. Generalized linear mixed models detected differences among diets with $p<0.05$ considered statistically significant.

The increase in water consumption did not reach statistical significance. However, urine volume (ml/kg/day) increased and USG decreased (both $p<0.05$) in dogs fed HNA (28.60 ± 13.28 and 1.027 ± 0.008, respectively) vs. CON-fed dogs (15.66 ± 14.81 and 1.035 ± 0.005, respectively). Total daily excretion of calculogenic components (calcium, magnesium, oxalate, and phosphate) were increased in dogs fed HNA ($p<0.05$) but urine concentration of these minerals were not different from baseline or CON (all $p>0.10$). Total excretion of citrate was increased in HNA dogs ($p<0.05$ vs. baseline) but urine concentration differences did not reach significance ($p=0.079$) vs. baseline, with no difference vs. CON. Among urine concentrations, only Na and Cl were increased and K decreased in HNA vs. baseline and CON, all $p<0.05$.

In conclusion, increased sodium intake resulted in greater urine volume and lower USG without changing urine concentrations of calculogenic components. Moderately increased dietary sodium could be a viable way to increase urination in dogs with lower urinary tract conditions.

NU11

SDMA Is a More Reliable Marker for Kidney Function in Cats and Dogs with Cancer (Abstract NU11)

Murthy V. Yerramilli[1]; Mahalakshimi Yerramilli[1]; Giosi Farace[1]; Jane Robertson[1]; Roberta Relford[1]; Dennis Jewell[2]; Jean Hall[3]

[1]IDEXX, Westbrook, ME, USA; [2]Hill's Pet Nutrition, Topeka, KS, USA; [3]College of Veterinary Medicine, Oregon State University, Corvallis, OR, USA

While the prevalence of functional impairment and injury in the kidney is high in cancer patients, it is significantly underdiagnosed using creatinine. In a study of lymphoma patients, 4 out of 10 had comprised kidney function but, only 1 out of 4 of this cohort was identified using creatinine. The remainder needed kidney biopsies for diagnosis. The most common causes of compromised kidneys include infiltration, nephrotoxic chemotherapeutics, volume depletion, tumor lysis syndrome, obstruction and sepsis. The ineffectiveness of creatinine is likely due to reduced production as a result of cachexia, reduced protein intake and therapeutics to mention few.

In a French epidemiological study with 5000 human cancer patients with different types of solid tumors it was reported that 52% had a reduced GFR with 12% being stages 3-5. While such detailed information is nonexistent for veterinary patients at this time, there is no reason to think that the situation is any different. Aside from the increasing rate of cancer due to longevity, many of the chemotherapeutics are nephrotoxic so there is a significant need for effective and sensitive biomarkers to diagnose kidney disease and effectively manage interventions.

To understand the effectiveness of SDMA over creatinine to diagnose kidney impairment in cancer patients, we performed a retrospective study in cats and dogs using banked samples and correlated results to histopathology. The study included 19 patients (10 dogs and 9 cats) with a variety of malignancies. SDMA ranged from 8-26 µg/dL with 7 patients being above the upper reference limit. Necropsies and histopathology were performed on all patients.

All the patients with increased SDMA had histopathological findings indicating kidney disease or neoplastic infiltration. No findings were observed in patients with normal SDMA. In all patients, creatinine was below the upper reference limit and ranged from 0.4 to 1.3 mg/dL

In an additional retrospective evaluation using 50 cats and dogs with neoplasia, 15% had increased SDMA while only 2% had elevated creatinine.

In conclusion, the increase in SDMA correlated with histopathologic findings in patients with neoplasia. Thus, SDMA is a more reliable estimate of kidney function in cats and dogs with cancer.

NU12

Increased Water Intake and Hydration in the Domestic Cat Ingesting a Nutrient-Enriched Water (Abstract NU12)

Brian Zanghi[1]; Cari Gardner[2]; Arleigh Reynolds[1]

[1]Nestle Purina PetCare, St. Louis, MO, USA; [2]Nestle Purina PetCare Research Statistics, St. Louis, MO, USA

This study investigated water intake and hydration in healthy cats and evaluated a nutrient-enriched water on hydration. Domestic shorthair cats (N=18; mean age 9.8yrs±SD 2.5; BCS 5–7 on 9-point scale) received either tap-water (W; N=9) or test-water containing whey protein and glycerol (TW; N=9) for 56 days. Prior to treatment, a 7-d baseline established daily W and food consumption using a 2-bowl monitoring system with *ad libitum* water and dry food. Blood and urine samples were collected on days -1, 8, 15, 30, and 56 for urine specific gravity (USG), creatinine, and osmolality measurements. Endogenous creatinine-based glomerular filtration rate (GFR) was calculated from sample collection on days 28–30 or 31–33. Baseline W and calorie intake were similar between groups. Although no difference was observed between groups during the treatment phase, TW group had a daily liquid increase of 52%, with a median intake across all days that ranged from 5% to 254% increase. Variance was controlled post hoc with cats characterized as responders (R:consumption >25%) and non-responders (NR) (W: 2R/7NR versus TW: 7R/2NR; p=0.02). Daily urine volume (mL/kg BW) was higher (p=0.02) in TW (15.2±SE 1.8) versus the W treated cats (10.3±SE 0.7), but no difference (p=0.60) between groups for GFR (1.8±SE 0.1 versus 1.9±SE 0.2, respectively). Baseline USG was similar between groups (1.052 g/mL), but declined (ANOVA p<0.01) with TW (1.037g/mL±SE 0.005) versus W (1.053g/mL±SE 0.002). Serum osmolality was similar for both groups over the entire trial. This study suggests that specific nutrients added to a cat's water can improve liquid intake to significantly improve overall hydration, as determined by greater urine output and dilution, which may offer a health benefit to some cats in need of greater water consumption.

NU13

Feline Serum Cytokines and Renal Alpha-Enolase Immunohistochemical Staining After Repeated Administration of a FVRCP Vaccine (Abstract NU13)

Stacie Summers[1]; Shannon M. McLeland[2]; Jessica J. Quimby[3]; Steven Dow[3]; Jennifer Hawley[3]; Jonathan Coy[3]; Michael R. Lappin[3]

[1]College of Veterinary Medicine, Colorado State University, Fort Collins, CO, USA; [2]College of Veterinary Medicine, Iowa State University, Ames, IA, USA; [3]Colorado State University, Fort Collins, CO, USA

Progressive interstitial nephritis (IN) is the primary cause of feline chronic kidney disease (CKD) which affects as many as 50% of elderly cats and prevalence increases with age. Vaccination has been shown to be a risk factor for development of feline CKD. The Crandell Rees feline kidney (CRFK) cell line is commonly used to grow viruses for use in feline vaccine production. Repeated administration of a vaccine containing CRFK remnants as a model for interstitial nephritis was shown to induce strong antibody responses against CRFK lysates and against alpha-enolase, an immunodominant glycolytic pathway enzyme found in all mammalian cells that has been studied as a marker of inflammatory diseases in people. The primary objectives of this study were to further evaluate this potential model for IN by assessing samples from the cats for cell-mediated immune responses to alpha-enolase, by evaluating serum cytokine concentrations using a commercially available kit, and by describing changes in renal alpha-enolase staining patterns.

In the IN model, serum, urine, peripheral blood mononuclear cells, and a renal biopsy were collected from six, one year old, healthy purpose bred, unvaccinated kittens (three male; three female). After sample collection on Week 0 and again on Weeks 2, 4, 6, 8, 10, 12, and 14, all cats were administered a commercially available parenteral FVRCP vaccine shown previously to induce anti-CRFK and anti-enolase antibodies. A second renal biopsy was collected on Week 16, avoiding the previous biopsy sites. Blood and urine samples were collected on all sample dates. A single lymphocyte proliferation assay was performed on 4 of 6 hyperinoculated cats and 2 unvaccinated healthy control cats using 4 antigens (kidney, spleen, alpha-enolase, and CRFK). Sera were assayed using a commercially available kit that detects sFas, Flt-3 ligand, GM-CSF, IFN-gamma, IL-1beta, IL-2, IL-4, SDF-1, KC, IL-6, MCP-1, IL-13, IL-8, IL-12 (p40), PDGF-BB, RANTES, SCF, IL-18, and TNF-alpha (Milliplex MAP Cytokine/Chemokine Feline Panel, Millipore, Darmstadt, Germany). Alpha-enolase immunohistochemical staining was performed on kidney biopsy samples (week 0 and week 16) and then evaluated and scored (0–4) by a board-certified pathologist that was masked to the timing of the biopsies.

Lymphocytes from hyperinoculated cats proliferated significantly more in the presence of alpha-enolase compared to the other three antigens confirming the induction of a cell-mediated immune response toward alpha-enolase contained in the vaccine. Glomerular and tubular alpha-enolase expression was significantly ($p<0.05$) greater at week 16 (mean score 3.37 and 1.95, respectively) compared to week 0 (0.37 and 1.01, respectively). Increased cytokine concentrations over the hyperinoculation period were noted for GM-CSF (3 cats), IFN-gamma (5 cats), IL-1beta (3 cats), IL-4 (4 cats), SDF-1 (5 cats), IL-6 (4 cats), MCP-1 (3 cats), IL-13 (4 cats), IL-8 (5 cats), PDGF-BB (3 cats), RANTES (6 cats), IL-18 (3 cats), and TNF-alpha (3 cats).

Hyperinoculation with a market leading FVRCP parenteral vaccine induced both antibody and cell-mediated immune response towards alpha-enolase, increased alpha-enolase expression in both renal tubules and glomeruli, and led to elevations in select inflammatory cytokines and chemokines within this 16-week model. Although there has been no direct link made between parenteral FVRCP vaccines and feline interstitial nephritis to date, results of urinary cytokines and further assessment of alpha-enolase expression within feline kidneys will be used to continue evaluation of this potential IN model.

NU14

Tamm-Horsfall Protein in Dogs with Natural Occurring Chronic Kidney Disease: A Marker of Progression? (Abstract NU14)

Fernanda Chacar[1]; Márcia Kogika1; Talita Sanches[2]; Douglas Caragelasco[1]; Cinthia Martorelli[1]; Camila Rodrigues[2]; José Capcha[2]; Lucia Andrade[2]

[1]School of Veterinary Medicine and Animal Science/Internal Medicine, University of Sao Paulo, Sao Paulo, Brazil; [2]School of Medicine, University of Sao Paulo, Sao Paulo, Brazil

Tamm-Horsfall Protein (THP) is exclusively synthesized by the distal nephron under physiological conditions. Therefore, it is expected larger amounts of this protein in the urine of healthy individuals. In chronic kidney disease (CKD) dogs, a few investigations have reported about the quantities of THP in urine. Lower urinary excretion of THP in CKD dogs might be associated with distal tubular injuries and/or to the loss of renal mass, and then, THP may act as a marker of disease progression. The aim of this study was to assess the role of THP as a marker of CKD progression in dogs. Nine clinically healthy dogs were recruited as control group. Forty CKD dogs were enrolled (IRIS staging, Stage 1, n= 10; Stage 2, n =10; Stage 3, n= 10; Stage 4, n= 10). Western blotting of THP was performed, and the results of densitometry were expressed in mean±SEM. Urinary immunodetection of THP was 100.0±7.7% in control dogs, and 117.2 ±10.8%, 86.0±9.0%, 65.4±8.7% and 46.7±9.0% in CKD dogs in stages 1, 2, 3 and 4, respectively. Differences among all groups were observed, particularly between Stages 1 and 4 ($p \leq 0.0001$). It is likely that progressive reduction in urinary excretion of THP may contribute to detect gradual loss of nephrons, and it may act as a marker of disease progression in CKD dogs.

NU15

Light and Electron Microscopic Analysis of Kidneys from Dogs Infected with *Angiostrongylus vasorum* (Abstract NU15)

Bernhard Gerber[1]; Luca Aresu[2]; Manuela Schnyder[3]

[1]Clinic for Small Animal Internal Medicine, Vetsuisse Faculty University of Zurich, Zurich, Switzerland; [2]Department of Comparative Biomedicine and Food Science, Università di Padova, Legnaro (PD), Veneto, Italy; [3]Institute of Parasitology, Vetsuisse Faculty University of Zurich, Zurich, Switzerland

Infection with *Angiostrongylus vasorum* (Av) is an emerging disease in dogs with a wide range of clinical signs. To date, it is not clear whether Av is able to cause glomerular immune complex deposition or other renal damage.

The aim of the study was to evaluate the kidneys of dogs experimentally infected with Av.

Sixteen beagle dogs 1 to 1.5 years of age were orally infected with about 266 third stage larvae each. Eight dogs were treated with an anthelmintic against Av after 30 days and the other eight were left untreated. All dogs were euthanized 56 to 59 days after infection. Renal samples were obtained and fixed in neutral-buffered formalin and 2.5% glutaraldehyde for light- (LM) and electron microscopy (EM) analysis.

Of the untreated dogs, 3 showed no significant changes at LM, 2 had occasional fetal glomeruli (<5%). In 3 dogs multifocal granulomas were observed in the cortex.

Of the treated dogs 4 showed no significant changes at LM, 3 had occasional fetal glomeruli (<5%), and 1 had focal and segmental mesangial sclerosis and mild increase of mesangial cells associated with occasional mesangial interposition.

Overall, EM confirmed the histological diagnosis and no glomerular immune deposits were found.

We demonstrated that kidneys of dogs infected with Av for 56 to 59 days have only mild LM changes and no immune mediated disease.

NU16

Effect of Weight Gain in Dogs on Renal Function and Perfusion Assessed by Contrasted-Enhanced Ultrasound (Abstract NU16)

Daisy Liu[1]; Myriam Hesta[2]; Emmelie Stock[1]; Siska Croubels[3]; Mathias Devreese[3]; Bart Broeckx[2]; Evelien Bogaerts[1]; Sylvie Daminet[4]; Katrien Vanderperren[1]

[1]Department of Veterinary Medical Imaging and Small Animal Orthopedics, Faculty of Veterinary Medicine, Ghent University, Merelbeke, Oost-Vlaanderen, Belgium; [2]Department of Nutrition, Genetics and Ethology, Faculty of Veterinary Medicine, Ghent University, Merelbeke, Oost-Vlaanderen, Belgium; [3]Department of Pharmacology, Toxicology and Biochemistry, Faculty of Veterinary Medicine, Ghent University, Merelbeke, Oost-Vlaanderen, Belgium; [4]Department of Medicine and Clinical Biology of Small Animals, Faculty of Veterinary Medicine, Ghent University, Merelbeke, Oost-Vlaanderen, Belgium

Studies in obese humans and animal models of obesity revealed that complex interactions between metabolic, inflammatory and hemodynamic factors result in altered renal perfusion, vascular dysfunction and histological changes in the kidneys. Early detection of renal dysfunction is crucial in reducing morbidity and mortality. Contrast-enhanced ultrasound (CEUS) has been suggested to be a sensitive method to study renal microvascular perfusion. Therefore, the objective of this study was to assess the effect of weight gain in dogs on kidney function and perfusion using CEUS.

Sixteen healthy lean adult beagles (2–8 years) with ideal body weight (BW) (average 11.39±1.76 kg) and ideal body condition scores (BCS) (range 4–5/9, median: 4/9) were equally divided into a control group (group C) and a treatment group (group T), based on age and sex. After four weeks adaptation to a commercial adult maintenance diet, group T received 1.3 x maintenance energy requirements to induce weight gain for 47 weeks, whereas group C was fed to maintain an ideal BW and BCS for 47 weeks. All dogs were weighed and BCS was assessed on a weekly basis. At the end of the adaptation (week 0) and at week 47, systolic blood pressure; serum urea, creatinine, and albumin; urinary specific gravity, urinary protein:creatinine ratio (UPC) were measured. Glomerular filtration rate (GFR) was measured by plasma clearance of exo-iohexol after an intravenous bolus injection of iohexol (Omnipaque 300®, 64.7 mg/kg BW). While sedated (butorphanol, 0.04 ml/kg), CEUS of the left kidney was performed using a bolus injection with Sonovue®. Time intensity curves reflecting tissue perfusion were generated based on manually drawn region of interests on the renal cortex and medulla. Perfusion parameters (rise/fall time, area under curve (AUC), time to peak, wash-in/out rate (WiR/WoR), wash-in/out AUC, wash-in perfusion index (WiPI), mean transit time, and peak enhancement (PE)) were calculated and evaluated. Parametric and non-parametric tests were used to compare groups and paired data with significance set at $p<0.05$.

At week 0, no significant differences between group C and group T were found. At week 47, group T gained on average 37% more weight; and consequently had a higher BCS ($p<0.001$) than group C. Despite the weight gain, no significant effect on blood pressure and GFR was observed in group T at week 47. However, a significantly higher median UPC ($p=0.031$) was observed at week 47 compared to week 0 in group T. CEUS perfusion parameters of the cortex were not significantly different between the two groups and between week 0 and 47. In the medulla, however, PE, a measure related to blood volume, was significantly higher in group T than in group C at week 47 ($p=0.021$). Furthermore, WiR, WiPI and WoR of the medulla, parameters related to blood velocity, were also significantly higher or close to significant when compared to group C at week 47 ($p=0.038$, 0.021 and 0.065, respectively) or to week 0 (WiR: $p=0.039$; WoR: $p=0.039$).

A weight gain of 37% in dogs does not appear to affect most routine markers of kidney function or GFR. Increased UPC does suggest that weight gain results in higher urinary excretion of protein as seen in humans. Additionally, CEUS was able to detect differences in blood volume and velocity of the renal medulla, which may indicate that structural and functional changes may have occurred as a result of weight gain. Further studies utilizing non-invasive markers or techniques to localize abnormalities in kidney function or injury in obese dog are warranted. Additional studies are also required to evaluate whether these changes are reversible after weight loss.

NU17

Evaluation of FGF-23 and Urinary Fractional Excretion of Phosphorus Ratio in Chronic Kidney Disease Dogs (Abstract NU17)

Cinthia Martorelli[1]; Márcia Kogika[1]; Fernanda Chacar[1]; Douglas Caragelasco[1]; Dennis Chew[2]; Ramiro Toribio[3]; Katarzyna Dembek[4]; Camila Rodrigues[5]; Talita Sanches[5]; Lucia Andrade[5]

[1]School of Veterinary Medicine and Animal Science/Internal Medicine, University of Sao Paulo, Sao Paulo, Brazil; [2]The Ohio State University, Columbus, OH, USA; [3]The Ohio State University Veterinary Clinical Sciences, Columbus, OH, USA, [4]Iowa State University, Ames, IA, USA; [5]School of Medicine, University of Sao Paulo, Sao Paulo, Brazil

Fibroblast growth factor 23 (FGF-23) is a bone-derived hormone that regulates phosphorus (P) and vitamin D metabolism. It was recently shown that dogs with chronic kidney disease (CKD) have increased FGF-23 concentrations. FGF-23 increases urinary fractional excretion of phosphorus (uFEP) in order to limit increases in serum phosphorus. The phosphaturic activity of FGF-23 is based on suppression of renal P reabsorption in the proximal tubules. Recently human studies showed that uFEP/FGF-23 ratio may reflect functional nephrons and the phosphaturic actions of FGF-23, and can be used as prognostic marker of CKD. A low uFEP/FGF-23 ratio may indicate a reduction in the number of functional nephrons and/or impaired FGF-23 signaling from Klotho (co-receptor for FGF-23) expression. The goal of this study was to evaluate whether uFEP/FGF-23 ratio reflect severity of CKD and could predict prognosis of CKD dogs. Sixteen CKD dogs were followed-up for 13 months or until death (IRIS stage 2, n=5; stage 3, n=11). Urinary and serum phosphorus as well as serum FGF-23 were measured at the first visit. The uFEP/FGF-23 ratio was determined in fifteen healthy dogs (mean±SEM; 0.083±0.024). Normophosphatemia (3.54±0.07 mg/dL; range 2.3 to 5.4 mg/dL) was detected in all CKD dogs in stage 2 most of the times during the follow-up, and hyperphosphatemia detected in 8 out of 11 dogs in stage 3 (5.68±0.23 mg/dL; range 2.5 to 16.0 mg/dL). The uFEP/FGF-23 ratio was 0.027±0.014 in stage 2 CKD, and 0.006±0.003 in stage 3 CKD (p=0.0122). Serum FGF-23 concentrations were higher in stage 3 and even though FGF-23 was increased, there was no increment in phosphaturia. CKD dogs in stage 3 with lower uFEP/FGF-23 ratios had a poorer prognosis. Decreased survival was predicted when uFEP/FGF-23 ratios were less than 0.006 (survival up to 300 days). The worst prognosis occurred when uFEP/FGF-23 ratio was <0.003 (less than 130 days survival). In conclusion, a low uFEP/FGF-23 ratio associated with normal to high serum FGF-23 concentrations supports a reduction in functional nephrons and/or impaired FGF-23 signaling from renal injury since increased FGF-23 should promote sufficient phosphaturia. Evaluation of Klotho (serum, urine, renal tissue) will be valuable to this question. Studies with larger population of dogs in various IRIS CKD stages will confirm the prognostic value of uFEP/FGF-23 in dogs with spontaneous chronic kidney disease.

NU18

Alpha-Enolase Staining Patterns in the Renal Tissues of Cats with and without Chronic Kidney Disease (Abstract NU18)

Shannon M. McLeland[1]; Jessica J. Quimby[2]; Michael R. Lappin[2]

[1]College of Veterinary Medicine, Iowa State University, Ames, IA, USA; [2]Colorado State University, Fort Collins, CO, USA

Alpha-enolase, a ubiquitous, glycolytic enzyme known for catalyzing the conversion of 3-phosphoglycerate to phosphoenolpyruvate, has been implicated as a self-antigen in various autoimmune diseases. Autoantibodies in serum are capable of eliciting renal injury and ultimately can contribute to patient morbidity and mortality. Furthermore, endogenous renal alpha-enolase has been shown to be overexpressed in human patients with systemic lupus nephritis and mixed cryoglobulinemia. Chronic kidney disease (CKD) is common in companion animals and affects a significant proportion of geriatric cats. While antibodies against alpha-enolase have been shown to be present in sera of cats, patterns of alpha-enolase protein expression in the kidney have not been determined in healthy or diseased cats. The purpose of this study was to identify and characterize alpha-enolase expression in renal tissues of cats with and without CKD.

None of the cats from which tissues were collected were euthanized because of the study. Renal tissues were collected from 29 cats with IRIS Stage II–IV CKD (mean age 14.7 years; range 8–20.5 years of age) and 12 cats with a creatinine concentration <1.6 mg/dl, and a urine specific gravity of >1.035. The cats with normal creatinine concentrations were divided into those <1.5 years of age (n=8 cats) and >10 years of age (n=4 cats). Interstitial nephritis was the predominant histopathological finding in the CKD group and was minimal to absent in the 12 cats with normal creatinine concentrations. An immunohistochemistry technique for semi-quantitative measurement of alpha-enolase protein expression in feline tissue was optimized utilizing a commercially available monoclonal alpha-enolase antibody. Each section of kidney was evaluated for positive immunoreactivity and assigned a score based on signal intensity - immunoreactivity - ranging from 0 to 3. Scoring was defined as follows: 0=no staining, 1=light brown/tan, 2=moderate brown staining, 3=dark brown which may obscure the nucleus. Mean scores between young, geriatric and CKD were compared using Kruskal-Wallis with Dunn's post hoc analysis.

Non-azotemic cats <1.5 years of age had moderate alpha-enolase expression in tubular epithelium (median score=2; range 2–3) but staining was absent to mild in glomeruli (median score=1; range 0–2). In contrast, alpha-enolase expression was moderate to high in tubular epithelium (median score=2; range 1.95 to 2.05) and glomeruli (median score=2; range 1.8 to 2.6) in non-azotemic cats >10 years of age. Lastly, evaluation of kidneys of cats with CKD showed that alpha-enolase expression was decreased in tubules that were degenerative or atrophic (median score=1; range 0–3), similar to the other groups of cats in normal tubules (median score=2; range 1–3), and moderate to highly expressed in glomeruli (median score=2; range 0–3). While no statistical difference was found between tubular immunoreactivity scores, there was a significant difference between glomerular immunoreactivity scores between non-azotemic cats >10 years of age and non-azotemic cats <1.5 years of age (p=0.04) and CKD cats and non-azotemic cats <1.5 years of age (p=0.009).

Alpha-enolase expression decreases in damaged tubules and increases in the glomeruli of older cats prior to the development of detectable CKD when compared to young, non-azotemic cats. Further studies will be required to determine whether these findings relate to the pathogenesis of CKD or could be used as an early diagnostic test for CKD.

NU19

Metabolomic Evaluation of Feline Urinary Diseases (Abstract NU19)

Nathalie Priymenko[1]; Bastien Autuoro[2]; Celina Torre[3]; Isabelle Jouanin[1]; Marie Tremblay-Franco[1]; **Isabelle Jeusette**[4]

[1]Toxalim, Université de Toulouse, INRA, ENVT, Toulouse, Midi-Pyrénées, France; [2]École Nationale Vétérinaire de Toulouse, Toulouse, Midi-Pyrénées, France; [3]Affinity Petcare, Barcelona, Catalonia, Spain; [4]Affinity Pet Care SA, L'Hospitalet de Llobregat, Catalonia, Spain

Metabolomics may be used for providing description of endogenous metabolites present in biological samples in response to pathophysiological or external stimuli. The purpose of the present study was to use metabolomics to compare cats with feline idiopathic cystitis (FIC) or struvite plug (SP) with healthy cats.

Male cats presenting either SP (n=6) or FIC (n=19) and healthy (n=10) were recruited through veterinary practices. Urine samples were collected by cystocentesis. NMR spectra were acquired on Bruker DRX-600 Avance spectrometer operating at 600.13 MHz ^1H resonance frequency and equipped with a cryoprobe. After data reduction and normalisation, 751 buckets RMN were analysed with multivariate statistical analyses.

Inspection of the spectra revealed variation in urine composition between the three groups and enabled identification of metabolites responsible for some peaks. After OSC filtration, a valid and robust PLS-DA model explaining the variability between the three groups was built. Urine from FIC cats contained higher concentrations of indoxyl sulfate, hippurate, creatinine, felinine, betaine and TMAO compared to SP or healthy cats. Urine from SP cats contained more guanidinoacetate and dimethylamine compared to FIC or healthy cats and more creatine and TMAO than healthy cats. Urine of healthy cats contained higher quantities of carnitine, but less of TMAO and creatinine.

In this study, the use of metabolomics was able to describe biochemical variations between urines of healthy cats and cats presenting FIC or SP. This novel application of metabolomics opens new hypothesis on pathophysiology and treatment in such cats.

NU20

Comparison of Single, Averaged, and Pooled Urine Protein:Creatinine Ratios in Proteinuric Dogs Undergoing Medical Therapy (Abstract NU20)

Sarah Shropshire[1]; Robert Cerda[2]; Jessica J. Quimby[1]

[1]Colorado State University, Fort Collins, CO, USA; [2]Arapahoe Animal Hospital, Boulder, CO, USA

Monitoring the urine protein:creatinine (UPC) ratio in dogs with protein-losing nephropathy (PLN) to determine response to therapy can be challenging due to day-to-day variation in their UPC values. The purpose of this study was to determine if a single sample, averaged sample, or pooled sample are comparable for determination of the UPC ratio, regardless of the degree of proteinuria, in PLN dogs receiving medical therapy.

Twenty-five client-owned dogs diagnosed with PLN currently receiving ACE inhibitors and/or ARBs for treatment of proteinuria were enrolled into the study; 16 dogs had UPC ratios less than 4 and 9 dogs had UPC ratios ≥4. UPC ratios were measured in each dog utilizing three separate methods; a single sample (day 3), an average sample (day 1–3), and a pooled sample (equal pooling of urine from day 1–3). Bland-Altman analysis was performed to evaluate a single UPC versus a pooled UPC, a single UPC versus an average UPC, and a pooled UPC versus an average UPC. The correlation coefficient for each of the three analyses was also calculated followed by simple linear regression. Bland-Altman analysis was also performed on subgroups (UPC ≤4 or UPC >4) to compare the methods in differing degrees of proteinuria.

When evaluating all of the dogs, there was a strong correlation between a single UPC compared to a pooled UPC ($r=0.96$, $p<0.0001$), between a single UPC and an average UPC ($r=0.98$, $p<0.0001$), and between the pooled UPC and average UPC ($r=0.99$, $p<0.0001$). For each Bland Altman comparison 95% limits of agreement (95% of the time the difference between two methodologies will fall within this range) are described below:

There was good agreement between the three methods particularly for dogs with UPC ≤4. Poorer agreement was seen between single UPC and pooled or average UPC, with the single Day 3 UPC typically being higher than the pooled or average UPC.

The results of this study show that the UPC ratio from a single sample, averaged sample, and pooled sample are comparable in PLN dogs receiving medical therapy when UPC ≤4. In PLN dogs with UPC >4, more variability between methods exists but a previous study in X-linked nephropathy dogs not receiving medical therapy showed this degree of variability may not be clinically relevant. However, future studies are needed to verify if this is true in all dogs with PLN while receiving medical therapy.

	Pooled vs Single	Average vs Single	Average vs Pooled
UPC ≤ 4 95% Limits of Agreement	-0.4 to 0.8	-0.4 to 0.8	-0.3 to 0.3
UPC > 4 95% Limits of Agreement	-0.2 to 2.4	-0.4 to 2.2	-0.9 to 0.4

NU21

Retrospective Study of the Efficacy of Oral Potassium Supplementation in Cats with Chronic Kidney Disease (Abstract NU21)

Liberty Sieberg[1]; Jessica Quimby[2]

[1]Colorado State University, Fort Collins, CO, USA; [2]The Ohio State University, Columbus, OH, USA

The purpose of this retrospective study was to gain insight into the effect of three oral potassium supplements: potassium gluconate tablets (PGT), potassium gluconate powder (PGP) and potassium citrate granules (PCG) on hypokalemia and serum bicarbonate.

Colorado State University Veterinary Teaching Hospital medical records (2006–2016) were retrospectively searched for cats that had been prescribed an oral potassium supplement for management of their CKD-associated hypokalemia. Data collected included signalment, bodyweight, age, supplement type, dose, directions, concurrent medications, diet, indication owner was unable to give the supplement, and serum chemistry and urinalysis values. For inclusion, laboratory work had to be available at the time of hypokalemia diagnosis, and at recheck within one to six weeks. Treatment response was defined in three ways: any increase in potassium, an increase in potassium to within the normal reference range, and an increase to greater than 4 mEq/L. Not all data was normally distributed thus Wilcoxon Sign Rank test was used to compare pre and post-supplementation values. Spearman Rank was used to assess correlation between dose and change in serum potassium and bicarbonate.

A total of 205 cats were identified in the search; 167 were ineligible and 38 met inclusion criteria. Of the 38 included cats, 16 were given PGT, 11 were given PGP, and 11 were given PCG. 5/38 had to change the dose of supplement, and 5/38 changed the supplement type. Dosing ranged from 0.21–1.6 mEq/kg (49–374.4 mg/kg) for PGT, 0.25 mEq/kg (59–347 mg/kg) for PGP, and 0.04–1.41 mEq/kg (4–144 mg/kg) for PCG. 36/38 cats had an increase in potassium after supplementation (14/16 for PGT, 11/11 for PGP, and 11/11 for PCG). 34/38 cats had an increase to within the potassium reference range (13/16 for PGT, 10/11 for PGP, and 11/11 for PCG). 24/38 cats had an increase in potassium to >4 mEq/L (8/16 for PGT, 8/11 for PGP, and 8/11 for PCG). Pre-supplementation potassium mean +/ SD for PGT was 3.3 +/- 0.5 and 4.1 +/- 0.6 post-supplementation. Pre-supplementation potassium mean +/ SD for PGP was 3.3 +/- 0.4 and 4.2 +/- 0.7 post-supplementation. Pre-supplementation potassium mean +/ SD for PCG was 3.3 +/- 0.2 and 4.1 +/- 0.4 post-supplementation. There was a statistically significant difference in serum potassium post supplementation for all three medications: PGT (p=0.0001), PGP (p=0.001), PCG (p=0.001). There was a positive correlation between PGT dose and change in potassium levels (p=0.04), but no significant correlation for PGP or PCG. There was no significant difference between pre and post supplementation serum bicarbonate nor a correlation between potassium supplement dose and serum bicarbonate level. There was not enough data to assess effect on urine pH.

All three potassium supplements were effective in treating hypokalemia secondary to CKD in the majority of cats despite quite variable dosing, but treatment response depending on the definition of response used. No indication of alkalinizing effect was seen, and prospective studies are needed for more thorough assessment.

NU22

Serum Concentrations of Symmetric Dimethylarginine in Dogs with Dehydration (Abstract NU22)

Woojoo Choi[1]; Ji-Houn Kang[2]; **Seungji Woo**[1]; Purum Son[1]; Jiyeon Yun[1]; Chan-Joo Sung[1]; Hakhyun Kim[1]; Byeong-Teck Kang[1]; Mhan-Pyo Yang[1]

[1]College of Veterinary Medicine, Chungbuk National University, Cheongju, Ch'ungch'ong-bukto, Republic of Korea; [2]College of Veterinary Medicine/Veterinary Medical Center, Chungbuk National University, Cheongju, Ch'ungch'ong-bukto, Republic of Korea

Symmetric dimethylarginine (SDMA) and neutrophil gelatinase-associated lipocalin (NGAL) have been suggested as useful biomarkers for evaluating renal disease. The objective of this study was to assess the circulating SDMA and NGAL concentrations in dogs before and after correction of dehydration.

Sixty-nine client-owned dogs with dehydration and 21 healthy dogs were included in this case-controlled observational study. The dogs with dehydration were divided into 2 groups, based on the presence of azotemia. In addition, the dogs with dehydration were divided into 3 groups, according to the severity of dehydration. All dehydrated dogs were rehydrated within 24 hours by standardized fluid therapy, and serum concentrations of SDMA and NGAL were measured before and after correction of dehydration.

There were significant differences between the concentrations of SDMA and NGAL measured before and measured after the correction of dehydration in dogs with azotemia, but not in dogs without azotemia. In addition, SDMA and NGAL concentrations showed significant decreases after the correction of dehydration in dogs with moderate or severe dehydration, but not in dogs with mild dehydration. Notably, in 16 of 69 dogs, serum SDMA concentrations before correction of dehydration were higher than the reference interval and were normalized to the reference interval after rehydration. The correlation between serum SDMA and creatinine concentrations was significant, regardless of the presence of azotemia; whereas the serum NGAL concentration was significantly correlated with creatinine concentration only in dogs with azotemia.

This study found that the circulating SDMA concentration can be affected by hydration status, indicating that measurement of SDMA concentration in dehydrated dogs might need to be repeated after correction of dehydration. However, further studies are needed to clarify our results.

NM01

Effect of Antioxidant Supplementation on Markers of Oxidative Stress and Clinical Outcome in Hospitalized Dogs (Abstract NM01)

Danielle Hagen[1]; Joanne L. Ekena[2]; Brian Geesaman[3]; Katrina Viviano[1]

[1]University of Wisconsin, Madison, WI, USA; [2]Department of Medical Sciences, School of Veterinary Medicine, University of Wisconsin-Madison, Madison, WI, USA; [3]Carolina Veterinary Specialists, Winston Salem, NC, USA

Oxidative stress is central to the pathogenesis of many systemic diseases. In humans, antioxidant depletion and lipid peroxidation correlate with disease severity and outcome. We have previously shown ill dogs experience oxidative stress, including decreased glutathione and vitamin E, and increased urinary 8-isoprostane/creatinine (8-IsoP/Cr). Short-term N-acetylcysteine supplementation did not impact clinical score or outcome but prevented further glutathione depletion. The purpose of this study was to evaluate longer-term, combination antioxidant supplementation on markers of oxidative stress, clinical score, and survival in ill dogs.

Hospitalized client-owned dogs were eligible for inclusion. Dogs were randomized to no supplementation (NS) or oral Denamarin and vitamin E supplementation (AS) for 30 days. Antioxidants (glutathione, cysteine, vitamin E, and selenium), 8-IsoP/Cr, and clinical SPI2 scores were determined at enrollment and 30 days. Glutathione, cysteine, and vitamin E were quantified by high-performance liquid chromatography. Selenium and 8-IsoP/Cr were measured using atomic absorption spectroscopy and gas chromatography/mass spectrometry, respectively.

Thirty-two ill dogs (target enrollment 40) have completed the study to date (AS, n=17; NS, n=15). Ill dogs had significantly elevated 8-IsoP/Cr versus healthy controls (3660 vs. 2980 pg/mg Cr; $p=0.021$). Antioxidant supplementation did not result in significant differences in median endogenous glutathione, cysteine, vitamin E, or selenium concentrations, 8-IsoP/Cr, SPI2 score, or 30 day survival. Interestingly, the NS group had significantly decreased plasma cysteine concentrations at 30 days when compared to baseline (9.62 to 6.95 µM; $p=0.039$). These preliminary findings suggest, in this heterogeneous population of ill dogs, combination antioxidant supplementation did not alter redox state or clinical outcome.

NM02

Voluntary Food Ingestion in Dogs Following Intestinal Surgery (Abstract NM02)

Christopher Kennedy[1]; Juliet Gladden[1]; Jonathan Babyak[1]

[1]Tufts University, North Grafton, MA, USA

Eating after intestinal surgery in dogs is variable. In humans, early enteral feeding has been shown to decrease morbidity and reduce the duration of post-operative ileus (POI); POI increases the length of hospitalization. This study evaluated the duration of anorexia and time to discharge in dogs following simple enterotomy (SA), multiple enterotomy (linear foreign body) (LFB) and resection and anastomosis (R&A). Additionally, prokinetic use was evaluated.

The medical records were searched: A convenience sample of fifty cases was identified. Data recorded included: signalment, the first time voluntary consumption of food occurred, use of a prokinetic and the time of discharge. The time until eating (TUE) and time until discharge (TUD) following surgery were calculated from the time of extubation. Incomplete records, dogs with a history of previous intestinal surgery, or cases in which dehiscence occurred, were excluded. A multivariable regression was performed to model the effect of variables after adjusting for population characteristics.

Nineteen dogs had SA, seventeen had LFB and fourteen had R&A. The mean age was 5.26 (+/- 4.37), 5.09 (+/- 4.53) and 6.79 (+/- 2.92) years, respectively. All dogs received opioids: 96% received full mu-agonists (n=48). Forty-two percent (n=21) received metoclopramide (six SA [32%], nine LFB [53%], six R&A [43%]): these dogs remained in hospital thirty hours longer on average compared to dogs that did not receive metoclopramide (p=0.007). Reasons for metoclopramide use were not clearly documented. All dogs survived to discharge. The mean TUE was 11.26 (+/- 6.31), 13.88 (+/- 7.65), 11.64 (+/- 7.1) hours, respectively. TUE significantly extended TUD independent of surgery type: each additional hour of not-eating delayed discharge by 0.5 hours (p=0.01). The mean TUD was 47.16 (+/- 22.06), 50.0 (+/- 16.14), 85.14 (+/- 53.02) hours respectively. R&A dogs were discharged on average thirty hours later than SA (p=0.005). R&A dogs that received metoclopramide were discharged sixty hours later compared to SA dogs that did not receive metoclopramide.

Eating decreased the time of hospitalization following intestinal surgery in dogs. Early enteral feeding should be encouraged. Metoclopramide was associated with prolonged hospitalization. Its use suggested real or anticipated POI in this study. The use of prokinetics in POI is contested in human medicine: they do not consistently decrease hospitalization, though they may improve feeding tolerance. Resection and anastomosis is expected to lengthen hospitalization.

NM03

Measurement of Body Composition and Body Water Using Quantitative Magnetic Resonance in Preweaning Puppies (Abstract NM03)

Julie K. Spears[1]; Brian Zanghi[2]

[1]Nestle Purina, Saint Louis, MO, USA; [2]Nestle Purina PetCare, St. Louis, MO, USA

The period from birth through weaning represents a critical period in puppy development. Body composition during this preweaning period has not been well-studied because sedation was previously required for analysis. Quantitative Magnetic Resonance (QMR) is a noninvasive technology that does not require anesthesia and has been validated as a method to assess body fat, lean, and total body water (TBW) in companion animals. The objective of this study was to determine if QMR can be used to measure body composition in puppies during the preweaning period. Healthy newborn puppies (Labrador retrievers and beagles; n=16/breed) were selected for this study. Body weights were collected and QMR scans were performed on awake and unrestrained puppies within 24 hours of birth, and then at 2, 4, 6, and 8 weeks of age. Differences in weight, percent lean, body fat, TBW, and lean mass hydration were assessed. Body weight was higher for Labrador retrievers at all timepoints ($p<0.05$). Body fat percentage was significantly higher in beagle puppies at birth, 2, 4, and 6 weeks of age ($p<0.05$). Although lean mass percentage was similar between breeds at birth, Labrador retrievers had significantly higher proportion at Weeks 2 and 4 ($p<0.05$). No breed differences were observed in TBW percentage at any time period, but beagles had significantly higher lean mass hydration at birth ($p<0.05$). These results indicate QMR can be used to detect differences in body composition during the critical preweaning period. Further research evaluating differences in body composition among dog breeds during growth is required.

NM04

Hypocholesterolemia in Cats: A Retrospective Study of 48 Cases (Abstract NM04)

Claudia A. Bowman; Austin K. Viall; Jean-Sébastien Palerme

Iowa State University, Ames, IA, USA

A retrospective study was performed to describe the underlying diseases, clinicopathologic abnormalities and prognosis associated with hypocholesterolemia in a population of cats presenting to a tertiary referral hospital between 2004 and 2016. Case inclusion criteria were: hypocholesterolemia, physical exam, CBC, serum biochemistry and urinalysis. When available, serum total T4, retroviral tests, imaging reports and necropsy reports were reviewed. For analysis, cats were categorized according to their disease by body system or etiology: neoplastic, infectious, sterile inflammatory, hepatobiliary, cardiac, gastrointestinal, malnutrition, toxin, neurologic and unknown.

Forty-eight cats met inclusion criteria. Neutered males comprised 54% of the cats, intact males 4%, spayed females 38% and intact females 4%. Median age was 5 years (range 0.5–14 years). Domestic mixed breed cats comprised 75% of the study population, 6% were of unknown breed. Diagnosis categories were distributed as follows: 23% infectious, 23% gastrointestinal, 17% neoplastic, 11% hepatobiliary, 6% malnutrition, 6% inflammatory, 6% unknown, 4% cardiac, 2% neurologic and 2% toxin. Anemia was the most common CBC abnormality, present in 37.5% of cats. The most common biochemical abnormalities were hyperglycemia (45.8%), increased BUN (22.9%), hyperproteinemia (22.9%) and increased ALT (22.9%). The median cholesterol concentration was 67.5 mg/dl (range 50–74 mg/dl). Comorbidities were present in 37% of cats with renal and cardiac disease being the most prevalent at 27% and 16%, respectively.

No significant difference was found in the serum cholesterol concentrations between diagnosis categories. No significant difference was found between the serum cholesterol concentrations of survivors and nonsurvivors. Hypocholesterolemia did not appear to be associated with survival in this study.

NM05

Effect of a Commercially Available Probiotic on Immune Responses in Healthy Dogs (Abstract NM05)

Michael R. Lappin; Jonathan Coy; Jennifer Hawley; Steven Dow

Colorado State University, Fort Collins, CO, USA

In a previous study, *Enterococcus faecium* strain SF-68 (FortiFlora®, Purina) was fed to puppies from weaning to 1 year of age and differences in some immune responses were detected when compared to a control group. In that study, the first data collection point was after 10 weeks of supplementation and so it is unknown if any potentially beneficial effects may occur earlier. In addition, whether the probiotic induces changes in immune responses of adult dogs is unknown and since the first study was performed, there have been advances in techniques for evaluating the immune system of dogs. The objective of the current study was evaluate select immune responses of adult dogs before and after supplementation with *E. faecium* strain SF-68.

A total of 7 age-matched clinically healthy laboratory reared young adult beagles housed in a research facility were chosen for study. Blood in EDTA and serum were collected from each dog prior to supplementation and then monthly for 12 weeks. Flow cytometry was used to measure percentage of B cells expressing surface bound IgG or MHCII, the percentage of CD4+ T cells expressing MHCII, and the percentage of CD8 T cells expressing MHCII or CD11a. For each of these molecules, the amount of surface expression was also determined by calculating the geometric mean fluorescence index (GMFI). T lymphocyte proliferative responses of all dogs to a non-specific mitogen (Concanavalin A) were assessed by flow cytometry and the ratio between proliferation of unstimulated cells and stimulated cells determined. Sera were assessed for a panel of cytokines using a commercially available kit (Milliplex MAP Canine Cytokine/Chemokine Magnetic Bead Panel; EMD Millipore). Differences between groups was assessed with significance defined as $p<0.05$.

T lymphocyte proliferative responses to mitogen activation in the dogs on Week 12 (Range=2.12–3.9; median=2.25) was statistically greater ($p=0.03$) than Week 0 (1.47–2.47; median=1.92). Compared to Week 0, statistically significant increases in B cells expressing surface bound IgG or MHCII and increased B cell MHCII GMFI were first noted on Week 4 and were maintained through Week 12. Compared to Week 0, numerically increased percentages of CD4+ T cells expressing MHCII, percentages of CD8 T cells expressing MHCII, increased T cell CD4+ MHC Class2 GMFI, and increased T cell CD8+ MHCII GMFI were detected Weeks 4, 8 and 12, but the differences were not statistically significant. Compared to Week 0, T cell CD8+ CD11a+ and T cell CD8+ CD11a+ GMFI were significantly greater on Weeks 4 and 8 but the effect was not apparent on Week 12. Of the cytokines/chemokines, at least 50% of the sera assessed had concentrations within the range of the assay for KC-like, IP-10, IL-7, IL-8, IL-18, and MCP-1; statistically significant differences after supplementation with the probiotic were not noted.

Supplementation of adult beagles with *Enterococcus faecium* strain SF-68 can induce immunomodulation as early as 4 weeks in dogs. Additional studies are ongoing to detail clinical benefits.

NM06

Behavior and Health in Aged Dogs Fed a Food with Antioxidants, Phytonutrients and Fatty Acids (Abstract NM06)

Heidi Shiefelbein[1]; Jennifer M. MacLeay[1]; Inke Paetau-Robinson[1]; Dennis Jewell[1]

[1]Hill's Pet Nutrition, Inc., Topeka, KS, USA

Two out of 3 pet parents do not feed their older pet an age-appropriate food. In 2005 AAHA published guidelines for care of older dogs which recommend focus areas for client education that includes wellness care and nutrition. Oxidative stress from free radicals and inflammatory processes are considered leading causes of degenerative diseases and cellular aging. Therefore, the purpose of this study was to evaluate the efficacy of a food designed to counter these changes in senior dogs compared to a non-nutritionally controlled cohort.

This 6-month study was reviewed and approved by an animal care and use committee and veterinarians and owners were masked as to the study sponsor. To be enrolled small and medium breed dogs must have been ≥9 years of age and large breed dogs must have been ≥7 years of age with no overt evidence of major organ disease. Dogs from 17 North American veterinary hospitals were screened (physical examination, CBC, serum chemistry, urinalysis) by attending veterinarians. Dogs were randomized to the test or control groups. The control group could consume any dry non-therapeutic food, excluding raw foods, of the owner's choosing and were able to switch foods at their discretion. The test food contained controlled minerals, added antioxidants including vitamin E, ascorbic acid, beta carotene and lipoic acid, sources of n6 and n3 fatty acids including chicken fat, fish oil and flax seed, carnitine and phytonutrients supplied by the inclusion of carrots, citrus pulp, spinach, tomato pomace and plant extracts. History, physical examination forms, medication records, dietary information and owner questionnaires were filled out electronically at 0, 1, 2, 3 and 6 months. Blood and urine samples were obtained at screening (T0), month 3 (T3) and month 6 (T6). Blood work results but not questionnaire results have been previously reported (Hall *et al. PLoS One*. 2016). The test group had significantly decreased blood urea nitrogen (BUN) from T0 vs. T3 and T6 and these were significantly lower than the control group at T3 and T6. BUN was significantly lower in the test compared to the control group at T3 and T6.

Three hundred dogs were screened and 255 were enrolled in the study: 129 Test (59m, 70f, mean age 9.54y, range 7–15y), 126 Control (58m, 68f, mean age 9.86y, range 7–15y); 103 Test and 111 Control dogs completed the study. Statistical analysis was a repeated measures regression model with value of interest as the dependent variable and diet, time point and diet interacted with time point as independent variables. Differences in least square means were used to test for significant differences. p value of 0.05 or less was used to determine significance.

Previous foods consumed by enrolled dogs included 44 dry and 8 canned brands of food. The majority of foods were not labeled as being indicated specifically for senior pets. Pet owners reported that the dogs in the test group exhibited significantly more positive responses and greater improvements in vitality, seeking the owner's attention, coat shine, coat softness, decreased stiffness and improved ability to rise from rest compared to the control group and compared to baseline at Month 6.

This study design did not use a standardized control food and therefore may have captured bias toward and/or away from the test food due to loyalty to owner's chosen food(s) or the novelty of the test food. Improvements in BUN are consistent with controlled studies where pets were transitioned onto a masked senior or control food. This study does provide a real world scenario reflecting the pet parent's overall positive experience in transitioning from their current food to the test food. Dogs benefitted from the test food as shown by the decrease in BUN, positive changes in appearance and aging-associated behaviors such as rising from rest, increased activity and interacting more with family members.

NM07

Behavior and Health in Aged Cats Fed a Food with Antioxidants, Phytonutrients and Fatty Acids (Abstract NM07)

Heidi Shiefelbein[1]; Jennifer M. MacLeay[1]; Inke Paetau-Robinson[1]; Dennis Jewell[1]

[1]Hill's Pet Nutrition, Inc., Topeka, KS, USA

Two out of 3 pet parents do not feed their older pet an age-appropriate food. In 2005 AAHA published guidelines for care of older cats which recommends focus areas for client education that includes wellness care and nutrition. Oxidative stress from free radicals and inflammatory processes are considered leading causes of degenerative diseases and cellular aging. Therefore, the purpose of this study was to evaluate the efficacy of a food designed to counter these changes in senior cats compared to a non-nutritionally controlled cohort.

This 6-month study was reviewed and approved by an animal care and use committee and veterinarians and owners were masked as to the study sponsor. To be enrolled cats must have been ≥9 years of age with no overt evidence of major organ disease. Cats from 11 North American veterinary hospitals were screened (physical examination, CBC, serum chemistry, urinalysis) by attending veterinarians. Cats were randomized by age to test or control groups. The control food could be any dry non-therapeutic food, excluding raw foods, of the owner's choosing and they were able to switch foods at their discretion. The test food contained controlled minerals, added antioxidants including vitamin E, ascorbic acid, beta carotene and sources of n6 and n3 fatty acids including chicken fat and fish oil, carnitine and phytonutrients supplied by the inclusion of broccoli, tomato pomace and plant extracts. History, physical examination forms, medication records, dietary information and owner questionnaires were filled out electronically at 0, 1, 2, 3 and 6 months. Blood and urine samples were obtained at screening (T0), month 3 (T3) and month 6 (T6). Blood work results but not questionnaire results have been previously reported (Hall *et al. PLoS One*. 2016). The test group had significantly decreased creatinine and blood urea nitrogen (BUN) from T0 vs. T3 and T6 and these were significantly lower than the control group at T3 and T6.

One hundred twenty-nine cats were screened, 105 were enrolled in the study: 51 Test (19m, 32f, mean age 11.61y, range 8–18y), 54 Control (20m, 34f, mean age 11.56y, range 9–17y). 33 Test and 49 Control cats completed the study. Statistical analysis was a repeated measures regression model with value of interest as the dependent variable and diet, time point and diet interacted with time point as independent variables. Differences in least square means were used to test for significant differences. p value of 0.05 or less was used to determine significance.

Previous foods consumed included 20 dry and 12 canned brands of food. The majority were not labeled as being indicated specifically for senior pets. At 6 months the Test group pet owners reported significantly improved scores in vitality, coat texture, less matting, increased interactions with family members and improved ability to run and play than the control group and compared to baseline.

This study design did not use a standardized control food and therefore may have captured bias toward and/or away from the test food due to loyalty of the owner to their chosen food(s) or the novelty of the test food. Improvements in BUN and creatinine are consistent with controlled studies where pets were transitioned onto a masked senior or control food. This study does provide a real world scenario reflecting the pet parent's experience in transitioning from their current food to the test food. Cats benefitted from the senior food as shown by the improved BUN and creatinine concentrations, positive changes in appearance and aging associated behaviors such as increased ability to run and play, increased activity and greater interaction with the family.

OT01

Factors in Successful IV Catheterization of Patients by Veterinary Students in the Emergency Room (Abstract OT01)

Shannon Babyak[1]; Jonathan Babyak[2]; Elizabeth A. Rozanski[3]

[1]Tufts University/Becker College, North Grafton, MA, USA; [2]Tufts University, North Grafton, MA, USA; [3]Cummings School of Veterinary Medicine, Tufts University, Grafton, MA, USA

Successful placement of IV catheters is essential for treatment of ill and injured pets. Veterinary students often struggle to gain skills in IV insertion. The purpose of this prospective study was to evaluate factors in student success for surrounding placement of IV catheters.
Students were asked to complete a survey following each catheter placement attempt in the ER. The responses were anonymous. The IRB reviewed the study for student participation. Students were asked to self-assess their level of experience (high, medium and low), and then characteristics of the patient, restrainer, catheter location and type were recorded as well as success or failure in catheter placement. Degree of "coaching" was also recorded. The chi-square test was used to evaluate differences between insertion in dogs and cats. A univariate logistic regression analysis was used to assess the relationship between success and all other variables.

One hundred and thirty-five catheters were attempted, 117 in dogs and 18 in cats. 93% of students reported low or moderate prior experience. 65% percent were successful. Veterinary technicians restrained for the majority of the catheter placements (54%), and provided some coaching in 60% of the attempts. There was no relationship between successful catheter placement and prior experience, species, indication for catheterization, brand of catheter, vessel, and restrainer. No student was successful when a faculty member was restraining the patient, although this was not significant due to the small sample size.

Catheterization is vital clinical skill; on-going efforts in how to improve teaching during clinical instruction are still necessary.

OT02

Vasopressor Use in Critically Ill Cats 41 Cases (2007–2015) (Abstract OT02)

Nikki Licht[1]; Elizabeth A. Rozanski[2]; John E. Rush[1]

[1]Tufts University, North Grafton, MA, USA; [2]Cummings School of Veterinary Medicine, Tufts University, Grafton, MA, USA

Prolonged hypotension may have negative impact on survival in critically ill cats. Hypotension refractory to fluid therapy is an indication for continuous rate infusion (CRI) of vasoactive medications (VM). The purpose of this study was to review the indications and outcome of cats receiving VM.
The electronic medical record database was searched using the term "cat" and and any of "dopamine", "epinephrine", "norepinephrine" and "vasopressin". Records were reviewed; cats were excluded if they received VM during CPR or immediately following ROSC or if they received VM only during anesthesia or immediately post-operatively. The underlying disease, initial BP, the VM(s) used, the rate, duration of VM use and outcome was recorded. Descriptive statistics were used and Fisher exact test to compare groups.

Forty-one cats met the inclusion criteria for study enrollment. Initial BP was not detectable or <50 mm Hg (Doppler) in all cats. Sepsis (49%), neoplasia (29%) or DKA (10%) were common underlying conditions. The VM used most commonly was dopamine, with cats typically receiving between 6 and 20 mcg/kg/min. Norepinephrine was the second most common VM and cats typically received between 1.1 and 2 mcg/kg/min. Sixteen cats survived. 27 cats received one VM and 10 cats survived, while 14 of cats received 2 vasopressors, and 6 of these cats survived ($p=0.22$). Thirty cats received VM for <24 hours with 13 surviving, while 11 received VM>24 hours, with 3 surviving ($p=0.48$). No adverse affects were attributed to VM. Vasopressors may be used in cats with critical illness.

OT03

Evaluation of Middle Ear in Cats with Eustachian Tube Obstruction (Abstract OT03)

Nevra Keskin[1,2]; Patricia A Schachern[1]; Irem Gul Sancak[1,3]; Hasan Albasan[1,2]; Michael M. Paparella[1,4]; Sebahattin Cureoglu[1]

[1]Department of Otolaryngology-Head and Neck Surgery, University of Minnesota, Minneapolis, MN, USA; [2]Department of Internal Medicine, Faculty of Veterinary Medicine, University of Ankara, Diskapi, Ankara, Turkey; [3]Department of Surgery, Faculty of Veterinary Medicine, University of Ankara, Diskapi, Ankara, Turkey; [4]Paparella Ear Head and Neck Institute, Minneapolis, MN, USA

Eustachian tube obstruction (ETO) is one of the leading causes of inflammation in the middle ear (ME). Undiagnosed and untreated otitis media can progress and result in serious pathological changes. The purpose of this study was to assess the middle ear in cats with ETO. Study materials consisted of a total of 15 temporal bones of cats with ETO at progressive time intervals of 2 days, 1-week, 2-week and 4-week.

The results revealed moderate hyperplasia equally developed throughout the epithelial layer surrounding the ME and neutrophil-rich mild inflammatory cell infiltration; more pronounced at one week obstruction compared to 2 days. Moreover, the majority of the ME cleft was filled with serous effusion. At 2-weeks post-obstruction, hyperplasia, edema and hyper vascularization of the ME epithelial layer were severe and there was an increased number of inflammatory cells within the serous effusion. After 4 weeks, there was a transition of ME effusion from serous to mucoid with an increased infiltration of plasma and other mononuclear cells, as well as very severe epithelial hyperplasia. In addition, granulation tissue appeared to originate from the extracellular matrix in the ME cavity and round window niche.

ME epithelial hyperplasia and characteristics of ME effusion were directly proportional to the duration of ETO. Therefore, patients presented with long-lasting complaints of ocular and nasal discharge, epistaxis, sneezing, otorrhea, ear itching, head tilt, facial nerve paralysis, and noisy and difficult breathing should be examined for dysfunction or blockage of the Eustachian tube by diagnostic techniques such as tympanometry.

OT04

Radial and Coccygeal Artery Indirect Doppler Systolic Blood Pressure Measurements in Privately-owned, Conscious Dogs (Abstract OT04)

Allison Mooney[1]; Dianne Mawby[1]; Josh Price[2]; Jacqueline Whittemore[1]

[1]University of Tennessee, Knoxville, TN, USA

The purpose of this study was to assess the effects of age, body condition score (BCS) and muscle condition score (MCS) on indirect radial and coccygeal Doppler systolic arterial blood pressure (SAP) measurements in dogs.

Sixty-two privately-owned dogs were enrolled between June and July 2016. BCS and MCS were determined by 2 investigators. Blood pressure was measured per published guidelines using headphones, and the order of measurement site was randomized. Dogs were positioned in right lateral recumbency for radial measurements and sternal recumbency or standing for coccygeal measurements. Associations between SAP and other variables were assessed by correlation coefficients and analysis of covariance.

Radial and coccygeal SAP measurements were moderately correlated ($r=0.45$, $p<0.01$). Radial SAP was significantly higher than coccygeal SAP (mean difference 9 mm Hg, $p<0.01$), but discordance occurred in both directions. No difference was observed between the first measurement taken, the average of measurements 2–6, or the average of all 6 measurements for either the radial (128, 129, and 129 mm Hg; $p=0.36$) or coccygeal (121, 122, and 122 mm Hg; $p=0.82$) site. Associations were not found between SAP measurements for either site and age, weight, BCS, MCS, anxiety score, or cuff size. Heart rate decreased significantly from the start of acclimation to the end of the first data collection series regardless of site ($p<0.01$).

Measurement site can be based on patient and operator preference given lack of associations with patient variables, but the same site should be used for serial SAP measurements given discordant results between sites.

OT05

Hemodynamic and Cardio-Respiratory Effects of an Infusion of Fentanyl, Ketamine and Lidocaine in Canines (Abstract OT05)

Luis A. Vergara Saldarriaga[1]; Sonia Orozco Padilla[2]; Sandra Acevedo Toro[3]; Luis F. Restrepo Betancur[1]

[1]Universidad de Antioquia, Medellín, Antioquia, Colombia; [2]Director Veterinary Hospital of the University of Antioquia, Medellín, Antioquia, Colombia; [3]Ocularvet, Medellín, Antioquia, Colombia

To evaluate the cardio-respiratory and hemodynamic effects produced by the continuous infusion of Fentanyl, Lidocaine and Ketamine (FLK), and the use of Isoflurane as a general anesthetizer in ASA I patients undergoing surgery. Study design. Prospective experimental study. Animals. Nine female dogs. The variables evaluated in the study were heart rate (HR), respiratory rate (RR), non-invasive blood pressure (NIBP), cardiac output (CO), cardiac index (CI), arterial oxygen saturation (SaO_2), end tidal CO_2 ($EtCO_2$), temperature (T). Arterial pressure of carbon dioxide ($PaCO_2$), arterial pressure of oxygen (PO2), bicarbonate, pH, total carbon dioxide (TCO_2) and presence of lactate in the blood were also evaluated. In order to assess the influence of the medications in the cardiorespiratory function, hemodynamic function and acid-base status, measurements were taken at two different time points: time point one (T1) corresponded to anesthesia with Isoflurane and time point two (T2), corresponds to anesthesia with Isoflurane and continuous infusion of FLK (0,08, 30 y 15 $\mu g\ kg^{-1}\ min^{-1}$).

There is a statistically significant difference between the effects of medications on temperature and acid-base status at T1 and T2 ($p<0.05$). In contrast, there is no statistically significant difference between the effects of medications on the cardiorespiratory and hemodynamic variables at T1 and T2 ($p>0.05$).

The use of FLK as a continuous infusion at dosage of 0,08, 30 and 15 $\mu g\ kg^{-1}\ min^{-1}$ in patients anesthetized with Isoflurane generated statistically significant changes in the blood pH and body temperature ($p<0.05$). Regarding the cardiovascular and hemodynamic variables, there were no statistically significant difference changes ($p>0.05$) at the different time points although it was possible to observe a tendency to improve the tissue perfusion and hemodynamic stability.

Keywords: Anesthesia, blood gases, cardiac output, I STAT, multimodal analgesia.

Figures

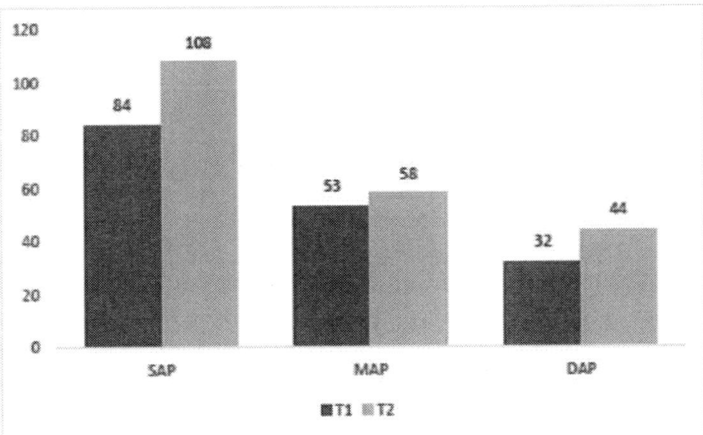

Figure 1. Systolic arterial pressure (SAP), mean arterial pressure (MAP) and diastolic arterial pressure (DAP) (mmHg) at times T1 and T2

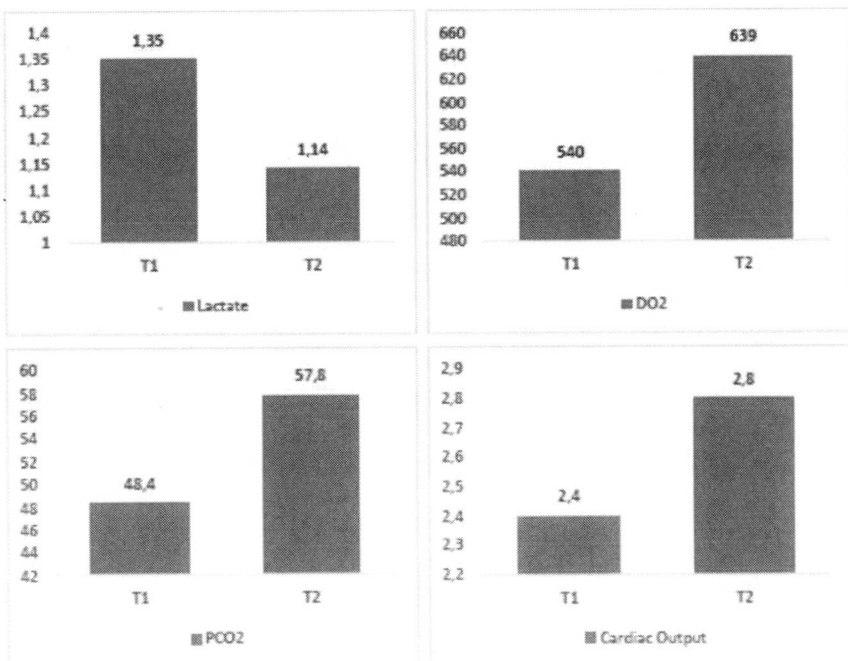

P01

Effect of Corticosteroids on Plasma Protein Binding of Mycophenolic Acid in Dogs and Cats (Abstract P01)

Alice Morassi[1]; S.M. Rivera[1]; Jennifer E. Slovak[2]; Michael H. Court[3]; Nicolas Villarino[1]

[1]Washington State University, Pullman, WA, USA; [2]College of Veterinary Medicine, Washington State University, Pullman, WA, USA; [3]Program in Individualized Medicine (PrIMe), Washington State University, Pullman, WA, USA

Mycophenolic acid, the active form of mycophenolate mofetil (MMF), is effective in controlling dysregulated lymphocyte responses associated with immune-mediated diseases. This drug has been shown to be promising for the treatment of multiple autoimmune diseases in dogs and cats. In humans, dose optimization and plasma drug concentration monitoring (TDM) are recommended in order to maximize the MPA effect, particularly in patients that require sustained immunosuppression. In humans, MPA is a highly protein bound drug (~97%). It is recommended to monitor free drug plasma concentration, as the free (unbound) MPA correlates with its immunosuppressive effect. It is likely that dogs and cats would also need TDM and optimization of dosage regimens. However, the extent of MPA plasma protein binding in dogs and cats are unknown. Importantly, the extent of MPA protein binding cannot be extrapolated from other species due to inherent differences in albumin structure and drug binding capacity.

We hypothesized that the extent of plasma protein binding in dogs and cats was >90%. Furthermore, considering that MMF is likely to be co-administered with corticosteroids, we also evaluated the effect of prednisolone and dexamethasone on the plasma protein binding of MPA. The extent of plasma protein binding of MPA (15 µg/mL) and the potential interaction with dexamethasone and prednisolone in plasma were evaluated using a high throughput equilibrium dialysis device and a validated chromatographic method. The concentration of drugs used in this study corresponds to plasma concentrations reported for clinical patients treated with therapeutic doses. For determination of plasma protein binding, we used individual plasma samples from clinically healthy dogs (n=14) and cats (n=13). For evaluating the impact of corticosteroids on MPA plasma protein binding, we pooled plasma from 6 healthy dogs and 6 cats. All the experiments were run in triplicate, and were repeated on 3 different days. The data from the groups were analyzed by Student t-test, and a $p \leq 0.05$ was considered significant.

The extent of plasma protein binding (% mean and range) was 96 (95–97) and 92 (90–93) for dogs and cats, respectively. The unbound plasma concentration (µg/mL) ranged between 0.51 and 0.94, and 1.15 and 1.61 in dogs and cats, respectively. The mean unbound plasma concentration of MPA was statistically different ($p<0.001$) between the two species. Results of the effect of corticosteroids on plasma protein binding of MPA are presented in tables 1 and 2. The mean unbound plasma concentration of MPA was higher when combined with prednisolone ($p<0.01$) but not with dexamethasone ($p>0.05$).

In conclusion, results of this study suggest that MPA was highly bound to plasma proteins. Furthermore, prednisolone had an apparent *in vitro* effect on the extent of binding of MPA in plasma from dogs and cats. These results will not be necessarily reflected on higher unbound plasma concentration of MPA *in vivo* due to the effect of drug distribution and elimination on plasma concentration.

Table: 1 Effect of prednisolone (PRED) on the extent of plasma protein of MPA

MPA	DOG range	CAT range
MPA		
% bound	95-96	90-90
% unbound	4-5	10-10
Bound (µg/mL)	14.1-14.7	16.-16.4
Unbound (µg/mL)	0.63-0.65	1.6-1.63
MPA + LOW DOSE PRED (0.6µg/mL)		
% bound	94-95	88-89
% unbound	5-6	11-12
Bound (µg/mL)	14.9-18.6	15.-16.0
Unbound (µg/mL)	0.79-1.14	1.75-1.81
MPA + HIGH DOSE PRED (6µg/mL)		
% bound	94-95	(88-89)
% unbound	5-6	11-12
Bound (µg/mL)	14.2-17.6	(4.4-16.1
Unbound (µg/mL)	0.77-1.10	1.62-1.84

Table: 2 Effect of dexamethasone (DEX) on the extent of plasma protein of MPA

MPA	DOG range	CAT range
MPA		
% bound	95-95	89-89
% unbound	5-5	11-11
Bound (µg/mL)	18.34-18.98	15.36-16.50
Unbound (µg/mL)	0.90-0.94	1.75-1.81
MPA + LOW DOSE DEX (0.6µg/mL)		
% bound	94-94	88-89
% unbound	6-6	11-12
Bound (µg/mL)	16.6-17.8	14.4-16.
Unbound (µg/mL)	1-1.1	1.62-1.84
MPA + HIGH DOSE DEX (6µg/mL)		
% bound	94-94	88-89
%unbound	6-6	11-12
Bound (µg/mL)	17.4-18.3	14.4-15.9
Unbound (µg/mL)	1.00-1.08	1.62-1.81

P02

Assessment of Safety, Toxicity, and Pharmacokinetics of Cannabidiol in Healthy Dogs (Abstract P02)

Lisa Bartner[1]; Linda Hyatt[2]; Luke Wittenburg[3]; Sangeeta Rao[4]; Stephanie McGrath[5]

[1]Colorado State University, Fort Collins, CO, USA; [2]College of Veterinary Medicine, University of Tennessee, Knoxville, TN, USA; [3]University of California-Davis, Davis, CA, USA; [4]Adjunct Faculty of Epidemiology, Colorado School of Public Health, University of Colorado, Denver, Fort Collins, CO, USA; [5]Department of Clinical Sciences, College of Veterinary Medicine and Biomedical Sciences, Fort Collins, CO, USA

Our purpose was to determine the pharmacokinetics (PK) and tolerability of cannabidiol (CBD) in healthy dogs. We hypothesized oral and transdermal delivery would be well tolerated and provide measurable systemic levels of CBD.

Thirty, healthy Beagle dogs were randomly assigned to receive one of three CBD formulations (capsule, oil, or transdermal cream), twice daily for a total dose of 150 or 300 mg/day (approximately 10 and 20 mg/kg/day), over 6 weeks. Blood was collected for 12 hours following the first dose for CBD PK. At 2-, 4-, and 6-weeks, blood was collected for CBC, chemistry panel, bile acids, and CBD plasma levels. Daily observational notes and weekly physical examinations were also performed.

Pharmacokinetic analysis demonstrated that the oil formulation resulted in higher plasma concentrations and systemic exposure, with less variability. Cannabidiol exposure was dose proportional. Elimination half-life of the oil formulation following a dose of 75 and 150 mg was 199.7±55.9 and 127.5±32.2 minutes, respectively. Twice-daily administration maintained plasma CBD levels until study completion.

There were no significant abnormalities on CBC, urinalysis, or pre- and postprandial bile acids. Mild elevations in serum alkaline phosphatase (ALP) occurred at 4 and 6 weeks, more frequently with capsule and oil formulations at 300 mg/day. All dogs experienced diarrhea; presence, onset, and severity did not correlate with formulation or dose.

Cannabidiol appears to be well tolerated in dogs. Exposure is proportional to dose and oil formulation provides more favorable PK. These results provide a framework for future efficacy studies of CBD in dogs.

P03

Concentration-Dependent Effects of Mycophenolic Acid on Canine Lymphocytes (Abstract P03)

Charlee Mulligan[1]; Keun-Seok Seo[2]; Kari Lunsford[2]; John Thomason[2]; Todd Archer[2]; Barbara Kaplan[2]; Nogi Park[2]; Lakshmi Narayanan[3]; Benjamin Lee[4]; Andrew Mackin[2]

[1]College of Veterinary Medicine, Mississippi State University, Mississippi State, MS, USA; [2]College of Veterinary Medicine, Mississippi State University, Starkville, MS, USA; [3]Mississippi State University, Starkville, MS, USA; [4]Colorado State University, Fort Collins, CO, USA

Mycophenolate mofetil (MMF) is the prodrug of mycophenolic acid (MPA), a potent immunosuppressive agent. MPA inhibits inosine monophosphate dehydrogenase (IMPDH), an enzyme in the *de novo* purine nucleotide synthesis pathway. Lymphocytes lack a purine salvage pathway as an alternative source of nucleotides, and are therefore especially susceptible to the effects of MPA. MMF is reported to inhibit both B and T cell proliferation. MMF was originally developed for use in human transplant patients but, since inexpensive generic versions have become available, the drug has become popular in veterinary medicine for the treatment of immune-mediated disorders in dogs. While it has been assumed that MMF will have effects in dogs that are similar to established effects in humans, to date very limited data are available regarding the effects of MPA on canine lymphocytes.

The objective of our current project was to evaluate the concentration-dependent effects of MPA on canine lymphocyte function utilizing a CD4 cell activation assay, and B and T cell lymphocyte proliferation assays. Peripheral blood mononuclear cells (PBMCs) were isolated from six healthy hound dogs using Histopaque® density gradient centrifugation. Once isolated, the PBMCs were plated, incubated in escalating concentrations of MPA (1.25, 2.5, 5, 10, 20, 40, 80, 160, and 320 ng/mL), vehicle (DMSO) control, and media only control, and then stimulated with either pokeweed mitogen (PWM) or concanavalin A (ConA). The CD4 cell activation assay was modeled on a commercially available human assay of activated CD4 positive T cell ATP synthesis that is marketed as a non-specific test of the efficacy of immunosuppressive agents. Briefly, CD4 positive cells were magnetically sorted from PBMCs following activation with ConA, cells were lysed to release intracellular ATP, a luminescence reagent (luciferin/luciferase) activated by ATP was added, and resultant bioluminescence was measured using a luminometer. B and T cell proliferation were evaluated by measuring incorporation of tritiated thymidine via a scintillation beta-counter following activation with PWM and ConA, respectively.

Our study revealed that MPA had no concentration-dependent effect on the CD4 cell activation assay, with no significant difference between DMSO vehicle control (376±25, mean relative fluorescence units ± SEM) and even the highest MPA concentration evaluated of 320 ng/mL (352±24). Lymphocyte proliferation assays, in contrast, revealed that MPA reliably inhibited both B and T cell proliferation in a concentration-dependent manner, with an IC50 (half maximal inhibitory concentration) of between 5 and 10 ng/mL for both assays (Figure 1).

This study confirmed that MPA reliably inhibits canine B and T cell proliferation in a concentration-dependent fashion, and that IC50 is achieved at MPA concentrations that are readily achievable at standard MMF dose rates. Based on the results of this *in vitro* study, oral MMF would be anticipated to have the potential to inhibit both B and T cell proliferation in dogs. Based on the results of our study, the CD4 cell activation assay was not significantly affected by exposure to MPA, suggesting that the assay, although marketed in human medicine as a non-specific test of drug-induced suppression of CD4 positive T cells, may not be of clinical value in dogs receiving MMF.

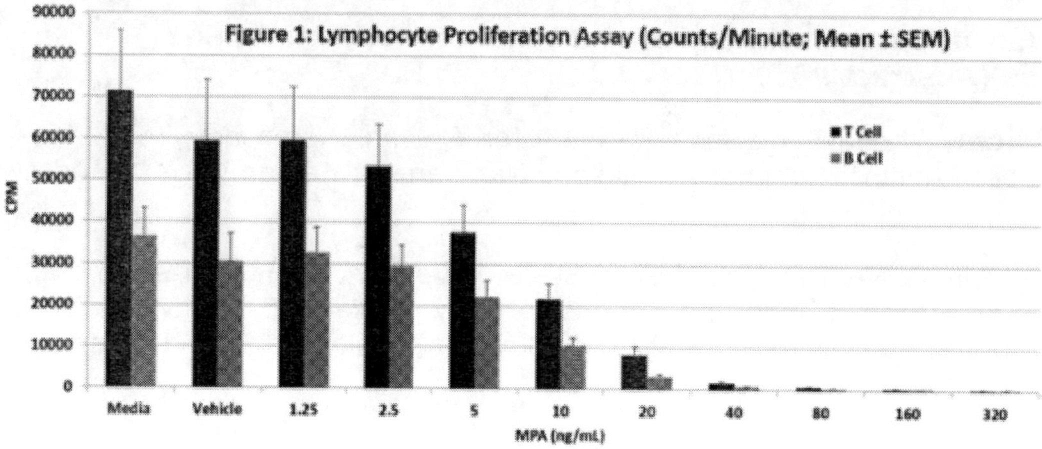
Figure 1: Lymphocyte Proliferation Assay (Counts/Minute; Mean ± SEM)
90000
80000
70000
60000
50000
40000
30000
20000
10000
0
CPM
T Cell
B Cell
Media
Vehicle
1.25
2.5
5
10
20
40
80
160
320
MPA (ng/mL)

P04

Pharmacokinetics of Mycophenolic Acid after Multi-Oral Dose Administration of Mycophenolate Mofetil in Healthy Cats (Abstract P04)

Jennifer E. Slovak[1]; S.M. Rivera[2]; J.K. Hwang[2]; Michael H. Court[3]; Nicolas Villarino[2]

[1]College of Veterinary Medicine, Washington State University, Pullman, WA, USA; [2]Washington State University, Pullman, WA, USA; [3]Program in Individualized Medicine (PrIMe), Washington State University, Pullman, WA, USA

Mycophenolate mofetil (MMF) is an attractive, alternative, commercially available immunosuppressant with a rapid onset of action and minimal side-effects. Our goal in this study was to evaluate the pharmacokinetics (PK), and various clinical parameters after 1 week of oral dosing of MMF in healthy cats. The PK of mycophenolic acid (MPA), the active form of MMF, was evaluated following oral administration of 10 mg/kg PO BID (n=3), 15 mg/kg PO BID (n=3), and 15 mg/kg PO TID (n=4) of MMF (CellCept® oral suspension, Roche Lab Inc.) for 1 week to a healthy population of cats. Plasma samples were taken prior to the first administration of MMF and at 0.25, 0.5, 1, 1.5, 3, 6, 8, 10, 12 h after the first MMF administration. In addition, plasma samples were taken 24, 72, 121, 145, 157 h after the first administration (before the next dose).

Concentration of MPA and its derivatives in plasma were determined using a validated chromatographic method. Non-compartmental analysis was done to estimate PK parameters of MPA. All cats bio-transformed MMF into MPA. Results suggest that cats bio-transformed MMF to MPA relatively fast, as observed in an intravenous PK experiment done in our laboratory. There was significant variation in the absorption phase of the concentration time profile of the drug as reflected in the dispersion of the Tmax values. In addition, there was significant inter-individual variability in MPA exposure. The shape of concentration vs. time profiles of MPA after the first dose of MMF suggests that MPA undergoes enterohepatic recycling. Following multiple administrations, the trough MPA levels do not reflect a significant accumulation of drug in plasma for the BID dosage regimens evaluated. There was a small accumulation over time of MPA after the TID administration of MMF. Three metabolites of MPA were detected in cats; phenol glucuronide, phenol glucoside and acyl glucuronide after the administration of 15 mg/kg. No acyl glucuronide metabolites were identified in the 10 mg/kg PO BID group.

No cats vomited during the study, 2 of 10 cats in the 15 mg/kg TID group were hyporexic, but they recovered at discontinuation of MMF. A mild reduction in serum protein levels, PCV, and platelet counts occurred between day 1 and day 7 of the study. No other values were altered by MMF use in the cats. This study describes the clinical parameters and PK of oral MMF administration in healthy cats. Future studies will evaluate the pharmacodynamics of MPA in patients receiving these doses of MMF.

P05

Limited Sampling Strategy to Determine Mirtazapine Pharmacokinetics in Cats with Liver Disease and Age-Matched Controls (Abstract P05)

Rikki Fitzpatrick[1]; **Kellyi Benson**[2]; Luke Wittenburg[3]; Ryan Hansen[2]; Paul Lunghofer[2]; Jessica Quimby[4]

[1]VCA Advanced Veterinary Care Center, Fishers, IN, USA; [2]Colorado State University, Fort Collins, CO, USA; [3]University of California-Davis, Davis, CA, USA; [4]The Ohio State University, Columbus, OH, USA

The purpose of this study is to use a limited sampling strategy to determine if pharmacokinetics of oral mirtazapine is altered in cats with liver disease compared to healthy age matched control cats (AMC).

Twenty cats were enrolled; 10 with liver disease and 10 apparently healthy age-matched controls. Cats with liver disease did not have evidence of chronic kidney disease or hyperthyroidism based on complete blood count, serum biochemistry, urinalysis and total T4 as well as no recent exposure to mirtazapine. Apparently healthy AMC cats were screened with a physical examination, complete blood count, serum biochemistry, urinalysis and total T4. Blood was drawn from all cats 1 hour and 4 hours following one 1.87 mg oral dose of mirtazapine. In a subset of 6 cats from each group, blood was also obtained 24 hours post-administration. Mirtazapine concentrations were measured by liquid chromatography coupled to tandem mass spectrometry. Drug exposure (AUC) and half-life was predicted using a limited sampling approach based on multiple linear regression analysis from previous full sampling studies, and estimated using non-compartmental methods. The data was not normally distributed and a Mann Whitney test was used to compare parameters between the two groups and Spearman Rank was used to analyze correlation between liver values and pharmacokinetic parameters.

There was a statistically significant difference in Tmax between liver disease cats and AMC cats (p=0.03). Tmax was 4 hours in 7 liver cats versus 2 AMC cats and 1 hour in 3 liver cats versus 8 AMC cats. AUC and was not significantly different between the two groups; however, the calculated half-life of liver disease cats was significantly increased (*p<0.004) compared to healthy AMC cats. There was a correlation between serum ALP (p=0.0003; r=0.89), ALT (p=0.003; r=0.80) and total bilirubin (p=0.001; r=0.84) when compared to the serum half-life of mirtazapine.

Liver disease in cats prolongs the half-life of mirtazapine. Dosing with a decreased frequency should be considered in cats with liver disease. As this effect was quite variable, additional work will need to be done to determine which cats are most effected.

	Liver Group		Age-Matched Control	
	Mean +/- SD	Median (range)	Mean +/- SD	Median (range)
Dose mg/kg	0.46 +/- 0.08	0.45 (0.30-0.58)	0.45 +/- 0.10	0.45 (0.28 – 0.59)
ALT (IU/l)	480 +/- 311.9	391 (138 – 1031)	50.7 +/- 13.4	47 (30-72)
ALP (IU/l)	217.7 +/- 163.5	226 (24-640)	25.6 +/- 9.7	28 (8-43)
T Bili (mg/dl)	5.8 +/- 6.8	3.9 (0.1 – 21.1)	0.03 +/- 0.04	0 (0-0.1)
C_{max} (ng/ml)	44.3 +/- 19.7	39.1 (26 – 87.3)	54.0 +/- 16.6	49.7 (32.2 – 80.2)
AUC (ng/ml•hr)	442 +/- 242	384 (215 – 1075)	423 +/- 98	422 (277 – 571)
Half- life*	15.1 +/- 7.5	12.6 (7.9 – 27.9)	7.4 +/- 0.67	7.3 (6.7 – 8.6)

P06

Antiacidic Effect of Intravenous Esomeprazole Twice a Day with Two Different Doses in Beagle Dogs (Abstract P06)

Do-hyun Seo[1]; **Yoora Cho**[1]; Kyoung won Seo[2]

[1]College of Veterinary Medicine, Chungnam National University, Daejeon, Taejon-jikhalsi, Republic of Korea; [2]Chungnam National University, Daejeon, Taejon-jikhalsi, Republic of Korea

Esomeprazole is an S-enantiomer of omeprazole that has been reported to have more favourable pharmacokinetics and efficacious antacid properties in humans. However, there is little information about the dosage of esomeprazole and pharmacokinetic property in conscious dogs. The aim was to evaluate the efficacy of intravenous esomeprazole q12h and to compare the efficacy between 1 mg/kg and 0.5 mg/kg in healthy dogs.

Six healthy adult colony beagle dogs were enrolled. Randomized, 2-way crossover design. In each group, the dogs received 1 mg/kg or 0.5 mg/kg, respectively intravenously q12h for 48 hours. The continuous intragastric pH was measured by Bravo® pH monitoring system and recorded as mean percentage time (MPT) that pH was ≥3 and ≥4 for 24 hours in each group. The comparison between groups was analysed by repeated measured analysis of variance with a *post hoc* Bonferroni 's method.

The MPT values that pH was ≥3 and ≥4 for first 24 hours were 89.78±9.08, 84.93±11.30 for 1 mg/kg group and 88.18±7.47 and 81.16±9.43 for 0.5 mg/kg group, respectively. There are no statistical differences in MPT between groups. Both groups showed no critical adverse effects.

It is the first study to evaluate intravenous esomeprazole twice daily in non-anesthetized healthy dogs. The results suggest that 0.5 mg/kg IV q12h administration of esomeprazole had the same efficacy of 1 mg/kg IV q12h in healthy dogs.

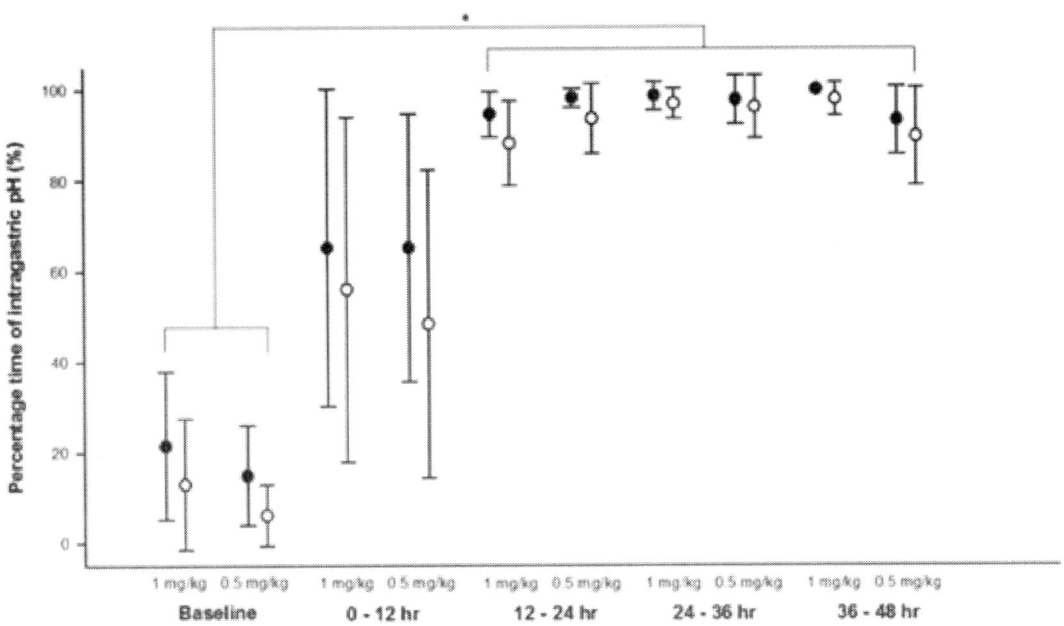

Table 1. Pharmacokinetic parameters of 2 dosing regimens, 1 mg/kg and 0.5 mg/kg q12h in 6 dogs. The mean ± SD are presented. *Comparison when dose normalized

Parameters	Dosing regimens		
	1 mg/kg (n = 6)	0.5 mg/kg (n = 6)	P value
T_{max} (h)	0.05 ± 0.00	0.06 ± 0.02	0.394
$T_{1/2}$ (h)	0.90 ± 0.32	0.72 ± 0.09	0.485
C_{max} (μg/mL)	5.39 ±1.92	2.03 ± 0.47	*0.156
AUC_{last} (μg·h/mL)	5.05 ± 2.24	1.44 ± 0.51	*0.094
V_{ss} (L/kg)	0.25 ± 0.10	0.30 ± 0.05	0.240
CL (L/h/kg)	0.24 ± 0.13	0.39 ± 0.14	0.065

P07

Pharmacokinetics and Antacid Efficacy of Esomeprazole After Intravenous, Oral and Subcutaneous Administration to Beagle Dogs (Abstract P07)

Ji-hye Hwang[1]; **Jae geum Cho**[1]; Seula Lim[1]; TaeSung Koo[2]; Kyoung won Seo[3]

[1]Veterinary Internal Medicine, Chungnam National University, Daejeon, Taejon-jikhalsi, Republic of Korea; [2]Graduate School of New Drug Discovery and Development, Chungnam National University, Daejeon, Taejon-jikhalsi, Republic of Korea; [3]Chungnam National University, Daejeon, Taejon-jikhalsi, Republic of Korea

Esomeprazole is an S-enantiomer of omeprazole that has been reported to have more favourable pharmacokinetics and efficacious antacid properties in humans. However, the pharmacokinetics and effects on intragastric pH of esomeprazole in dogs have not been reported. To determine the pharmacokinetics of esomeprazole administered via various routes (PK study) and to investigate the effect of esomeprazole on intragastric pH using a Bravo pH monitoring system (PD study).

Twelve healthy adult male Beagle dogs were included in this study. Both studies used an open, randomized, crossover design. In the PK study, 7 dogs received intravenous (IV), subcutaneous (SC), and oral doses (PO) of esomeprazole (1 mg/kg). Each treatment period was separated by a washout period of at least 10 days. Esomeprazole plasma concentrations were measured by HPLC/MS/MS. In the efficacy study, intragastric pH was recorded without medication (baseline pH) and following IV, SC, and PO esomeprazole dosing regimens (1 mg/kg) in 5 dogs.

The bioavailability of esomeprazole administered as PO enteric-coated granules and as SC injections was 71.43% and 106.61%, respectively. The half-life was approximately 1 hour. The mean percent time with intragastric pH was ≥3 or ≥4 was significantly increased; these results indicate that esomeprazole administration increased intragastric pH regardless of the dosing route ($p<0.05$). The PK parameters of for PO and SC esomeprazole administration were favourable and esomeprazole significantly increased intragastric pH following IV, PO, and SC administration.

IV and SC administration of esomeprazole may be useful when PO administration is not possible. No significant adverse effects were observed.

Table 1. The pharmacokinetic parameters of esomeprazole (n = 7) following IV, SC and PO administration of 1 mg/kg doses. All parameters were calculated as the mean (geometric mean) ± SD. T_{max}, time until maximum concentration; C_{max}, peak plasma concentration; $T_{1/2}$, terminal elimination half-life; *AUC*, area under the plasma concentration-time curve; *CL*, systemic plasma clearance; V_{ss}, steady state volume of distribution; *BA*, bioavailability

	Esomeprazole IV	Esomeprazole SC	Esomeprazole PO Day 1	Esomeprazole PO Day 5
T_{max} (h)	0.05 ± 0.00	0.38 (0.37) ± 0.13	1.81 (1.51) ± 1.23	0.81 (0.69) ± 0.50
C_{max} (μg/ml)	4.68 (4.51) ± 1.39	2.62 (2.53) ± 0.79	1.34 (0.96) ± 0.91	1.75 (1.69) ± 0.49
$T_{1/2}$ (h)	0.73 (0.71) ± 0.17	0.90 (0.87) ± 0.24	0.98 (0.97) ± 0.17	1.03 (1.01) ± 0.22
AUC_{last} (μg·h/ml)	3.82 (3.55) ± 1.65	4.07 (3.99) ± 0.89	2.72 (2.00) ± 2.05	2.89 (2.70) ± 1.34
$AUC_{0-\infty}$ (μg·h/ml)	3.82 (3.55) ± 1.65	4.07 (3.99) ± 0.89	2.73 (2.01) ± 2.05	2.89 (2.70) ± 1.35
CL (L/h/kg)	0.30 (0.28) ± 0.11			
Vss (L/kg)	0.27 (0.27) ± 0.07			
BA (%)		106.61	71.43	75.75

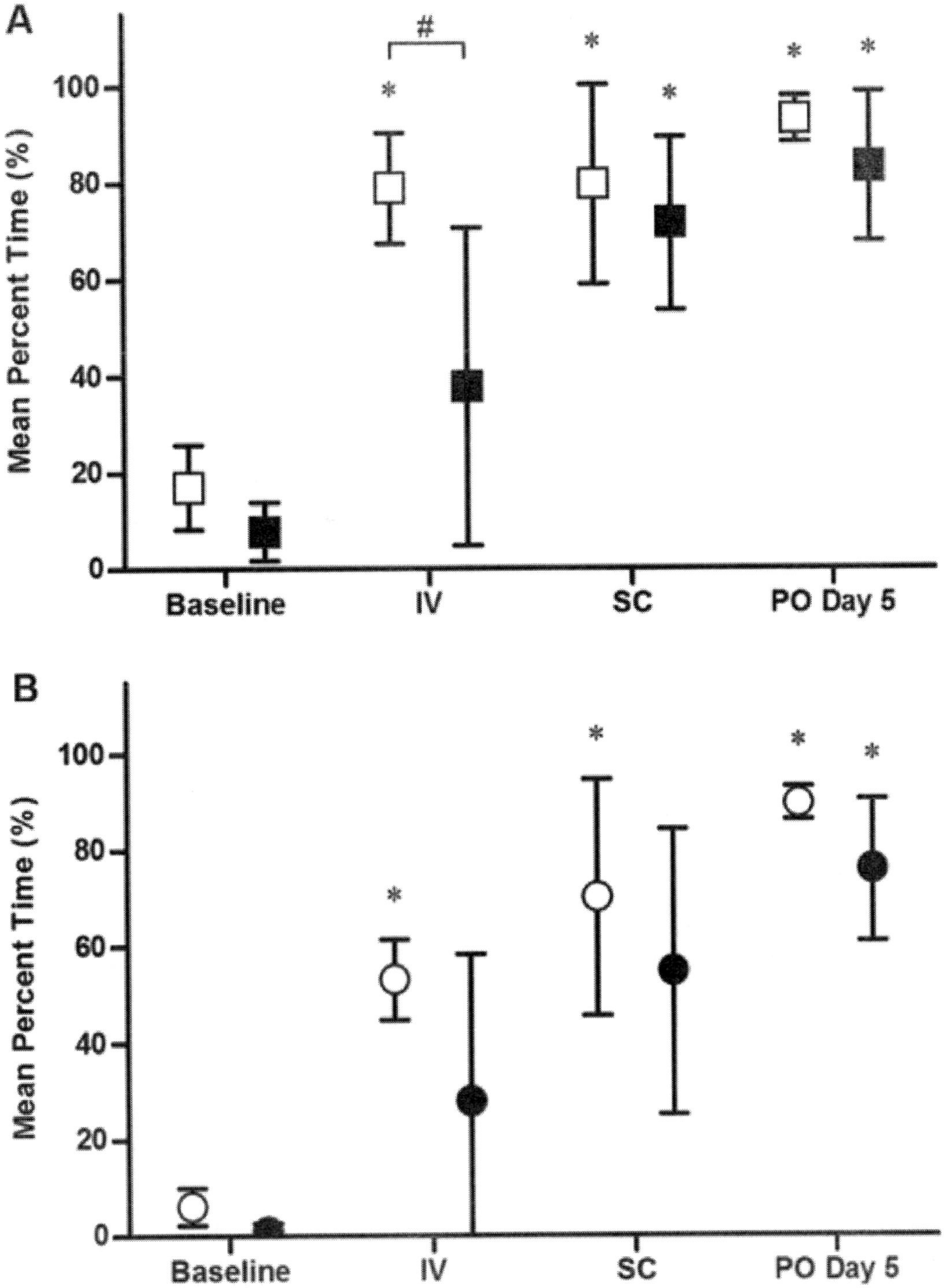
A
#
*
Mean Percent Time (%)
100
80
60
40
20
0
Baseline
IV
SC
PO Day 5
B
Mean Percent Time (%)
100
80
60
40
20
0
Baseline
IV
SC
PO Day 5

P08

Prediction of Platelet Responsiveness to Aspirin and Fluoride Effects on Esterase Activity in Healthy Dogs (Abstract P08)

Jillian M. Haines[1]; Rachel Hegedus[1]; Pamela M. Lee[1]; Michael H. Court[2]

[1]Washington State University, Pullman, WA, USA; [2]Program in Individualized Medicine (PrIMe), Washington State University, Pullman, WA, USA

Acetylsalicylic acid (ASA), commonly known as aspirin, is a cyclooxygenase inhibitor frequently used to inhibit platelet function and formation of thromboemboli in at risk patients. A potential phenomenon of resistance to these antiplatelet effects has been documented in people and potentially in dogs. Aspirin resistance may be a contributing factor to the development of life threatening thrombi thus increasing the rate of morbidity and mortality in several canine diseases. Understanding the phenomenon of aspirin resistance and developing methods for predicting it could lead to improvements in survival in these patients. The cause of aspirin resistance remains unclear. Individual variability in esterase activity and therefore hydrolysis of aspirin has been suggested as a potential cause of poor bioavailability of aspirin in blood in people.

The objectives of our study were to determine whether the ASA concentration needed to cause 50% inhibition of platelet aggregation (IC50) *in vitro* in healthy dogs was predictive of the IC50 value measured in the same dogs after administration of oral aspirin (*in vivo* IC50). Furthermore, we wanted to evaluate whether the addition of fluoride, known to cause non-specific esterase inhibition, altered the concentrations of ASA or its metabolite, salicylic acid (SA), in blood samples collected after oral aspirin administration.

Twenty-one healthy client-owned dogs were enrolled. First, *in vitro* IC50 testing was performed twice (trial 1 and 2) at least 1 week apart to determine individual repeatability. Whole blood samples were drawn and incubated with 6 different aspirin concentrations. Aggregation was measured in AUC (area under the curve) units using a multiple electrode impedance aggregometer (Multiplate). For *in vivo* IC50 testing, dogs were given a single oral dose of 4 mg/kg aspirin. Aggregation was assessed prior to aspirin administration then 20, 40, and 180 minutes after administration. At each time point, samples were also collected into EDTA and EDTA-fluoride blood tubes for measurement of ASA and SA plasma concentrations using liquid chromatography with mass spectrometry.

Based on calculation of *in vitro* IC50, dogs were separated into 3 different response groups. Eighteen of the 21 dogs (86%) were consistent as aspirin responders with *in vitro* IC50 values below 250 umol/L (median 66.7 umol/L, range 34.8–151.4). One dog showed a consistently poor response with *in vitro* IC50 values between 250–500 umol/L for both trials. Two dogs had marked variation in their *in vitro* IC50 values between the trials and were classified as variable responders. No dogs were found to be aspirin non-responders consistently for both trials. Between trial variation in *in vitro* IC50 for all dogs and the 18 dogs with IC50 values below 250 umol/L was mean (range) of 51.9% (3.4–194.3) and 38.6% (3.4–78.8), respectively. After administration of 4 mg/kg aspirin orally, 8 of 21 dogs did not show >50% platelet inhibition by 3 hours. *In vivo* IC50 values for the remaining 13 dogs for samples collected in EDTA and EDTA fluoride were median (range) of 31.7 umol/L (0.1–107.6) and 30.9 umol/L (0.1–92.4), respectively. All three dogs identified as poor or variable responders *in vitro* had >50% inhibition of platelet aggregation *in vivo*. The addition of fluoride to blood samples resulted in a mean percent reduction in SA concentrations of 17.3% and mean percent increase in ASA concentrations of 41.4%.

This study suggests individual variation in aspirin response does exist in dogs as assessed by multiple electrode aggregometry. A single low dose (4 mg/kg) of aspirin was not sufficient to cause measurable inhibition of platelet aggregation in all dogs and resulted in ASA concentrations that were substantially less than those used *in vitro*. Aspirin responsiveness *in vitro* did not correlate with responsiveness *in vivo* in this scenario. The use of EDTA fluoride anticoagulant resulted in inhibition of conversion of ASA to SA suggesting successful inhibition of esterase activity. When storing dog plasma samples containing aspirin, the use of EDTA containing fluoride may prevent esterase hydrolysis of the ASA. Further investigation into the effects of esterase activity as a cause of aspirin resistance in dogs is warranted.

P09

Pharmacokinetics of an Orally-Disintegrating Sildenafil Film After Oral and Rectal Routes Administration to Beagle Dogs (Abstract P09)

Hyuckjoo Yang[1]; **Hwaran Hong**[1]; Kyoung won Seo[2]

[1]College of Veterinary Medicine, Chungnam National University, Daejeon, Taejon-jikhalsi, Republic of Korea; [2]Chungnam National University, Daejeon, Taejon-jikhalsi, Republic of Korea

Sildenafil citrate, a highly selective phosphodiesterase type V inhibitor, is used to treat pulmonary hypertension (PH) in veterinary medicine. Sometimes, patients with PH experience respiratory distress, and the oral administration of any type of drug to these patients can be fatal. For such patients, the rectal route should be carefully considered as an alternative means of administering drugs.

The objective of this study was to compare the pharmacokinetic characteristics of an orally disintegrating film (ODF) formulation of sildenafil with those of a conventional film-coated tablet (FCT) formulation in healthy dogs. Additionally, the pharmacokinetic profile of the ODF formulation of sildenafil was analyzed after its rectal administration.

Twelve healthy Beagle dogs were divided into four groups and given sildenafil per the following specifications: FCT formulation 2 mg/kg PO (per oral), ODF formulation 2 mg/kg PO, ODF formulation 2 mg/kg diluted in distilled water and administered rectally, and ODF formulation 10 mg/kg dissolved in water and administered rectally. Blood samples were collected over the course of 24 h and the concentrations of sildenafil were assayed.

There were no significant differences in T1/2 and Tmax across the four arms. The Cmax of the rectally administered 10 mg/kg dose did not differ significantly from those of the orally administered 2 mg/kg doses. In contrast, the Cmax of the rectally administered 2 mg/kg dose was significantly lower than those of the orally administered 2 mg/kg doses and rectally administered 10 mg/kg dose. The AUC after rectally administering the 2 mg/kg dose was significantly smaller than those obtained after sildenafil was administered orally. The AUC after rectally administering the 10 mg/kg dose was significantly higher than those for the oral administration arms. There were no notable systemic adverse effects in any of the groups.

During emergencies, it may be helpful for sildenafil to be administered rectally to respiratory distress patients. With the exception of an increased AUC, the pharmacokinetic characteristics of the rectally administered 10 mg/kg dose have been shown to be comparable to those of an orally administered 2 mg/kg dose.

Table 1

Pharmacokinetic parameters of sildenafil in beagle dogs after administration of each arm. C_{max} = Maximum plasma concentration. T_{max} = Time at the maximum concentration. $T_{1/2}$ = Elemination half-life. AUC = Area under the curve.

Pharmacokinetic parameters	FCT PO 2 mg/kg	ODF PO 2 mg/kg	Rectal 2 mg/kg	Rectal 10 mg/kg
C_{max} (μg/mL)	0.21 ± 0.07	0.25 ± 0.10	0.09 ± 0.04	0.26 ± 0.08
T_{max} (hours)	2.08 ± 1.30	1.88 ± 1.65	3.28 ± 2.49	4.30 ± 2.91
$T_{1/2}$ (hours)	5.46 ± 2.39	4.62 ± 1.49	4.82 ± 1.91	4.20 ± 1.40
AUC (μg·h/mL)	1.52 ± 0.47	1.51 ± 0.51	0.77 ± 0.31	3.12 ± 1.39

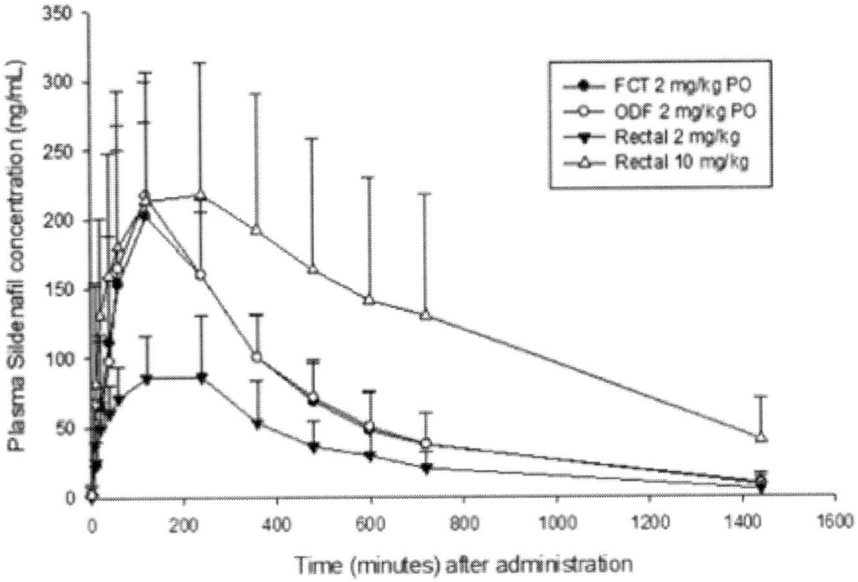

Fig. 1. Mean plasma sildenafil concentrations for each arm. Plasma concentration-time profiles of sildenafil in Beagle dogs after single-dose oral and rectal administration. FCT, Film-coated tablet PO; ODF, Orally disintegrating film PO.

P10

Safety of Mirtazapine 2% Transdermal Ointment Administered Topically to Cats at 5 mg/kg for 28 Days (Abstract P10)

Allyson Avenatti[1]; Williams Buhles[2]; Jessica J. Quimby[3]; Daizie Labelle[4]; Melinda Poole O'Banion[1]

[1]Kindred Biosciences Inc., Burlingame, CA, USA; [2]Kindred Biosciences Inc., Fairfax, CA, USA; [3]Colorado State University, Fort Collins, CO, USA; [4]Kindred Biosciences Inc., Atherton, CA, USA

Mirtazapine 2% transdermal ointment (2% MZP) achieves clinically relevant serum concentrations in cats and results in increased body weights at a dose of 0.5 mg/kg. The purpose of this study was to evaluate the safety profile at an exaggerated dose.

In this unmasked study, 2% MZP was applied topically to the inner pinna of five healthy adult intact female cats and two intact female cats with chronic kidney disease (CKD) (serum creatinine 1.7 mg/dL and 1.5 mg/dL; urine specific gravity 1.023 and 1.037 respectively) daily for 28 consecutive days at 5 mg/kg. Cats ranged in age from 1.4 to 3.2 years. 2% MZP volume was split equally between ears at each administration. Behavioral observations were conducted twice daily throughout the study and scored on interaction, activity, vocalization, and tremors. Physical examination, ECG, and doppler blood pressure were conducted on study days -5, 13, and 27. Scanning echocardiogram and abdominal ultrasound were conducted on study days -6 and 26. Hematology, and serum chemistry were conducted on study days -6, 13, and 27. Body weights were measured on study days -5, -1, 6, 13, 20, and 27. Body weight, blood pressure, and serum ALT were compared between time points using a Wilcoxon Signed Rank Test for paired samples. Sum of pre-dose mean behavior scores were compared to mean behavior scores at Day 27 using a Wilcoxon Signed Rank Test for paired samples. Statistical significance was set at $p<0.05$ for all tests.

Erythema of the inner pinna was a common finding in multiple cats at various time points with one cat showing an excoriation of one ear on day 5 which subsequently healed uneventfully. One cat with CKD exhibited vomiting between study days 16 and 21 and recovered without treatment. The same cat showed intermittent bouts (2–3 sec.) of spasms and ataxia with tail curled over the back on day 20 only. All cats gained a significant amount of weight between Day 0 and Day 28 ($p=0.016$). Blood pressure was analyzed in the healthy cats at a single time point and there was no significant difference compared to baseline at any time point. Hematology and serum chemistry showed no clinically significant changes in either apparently healthy cats or the two cats with CKD. There were no clinically significant changes in cardiac findings. There was no significant difference in behavior scores between the beginning and end of the study despite a 5x dose of mirtazapine.

Mirtazapine 2% transdermal ointment was well tolerated in cats when administered at 5 mg/kg for 28 consecutive days.

P11

Double-Blind, Placebo-Controlled, Randomized Study of Transdermal Mirtazapine Ointment for the Management of Feline Weight Loss (Abstract P11)

Kristy Longpre[1]; Williams Buhles[2]; Ming Tin[1]; Tianhua Hu[1]; Jessica J. Quimby[3]; Daizie Labelle[4]; Melinda Poole O'Banion[1]; Jessica Lee[1]; Stephanie Markey[1]

[1]Kindred Biosciences Inc., Burlingame, CA, USA; [2]Kindred Biosciences Inc., Fairfax, CA, USA; [3]Colorado State University, Fort Collins, CO, USA; [4]Kindred Biosciences Inc., Atherton, CA, USA

Mirtazapine 2% Transdermal ointment (2% MZP) has been demonstrated to achieve clinically relevant serum concentrations in cats. The purpose of this study was to demonstrate the effectiveness and safety of 2% MZP for the management of weight loss in cats.

This study was a multi-center, double-blind, placebo-controlled, randomized study. Cats at least 1 year of age, weighing ≥2 kg, with a documented medical history of ≥5% body weight loss that was deemed clinically significant were enrolled. Cats could not have a serum creatinine >5.0 mg/dL or diagnosed neoplasia. Randomization was 1:1 (mirtazapine:placebo). 2% MZP was dosed at 2 mg per cat (equivalent to 0.1 mL volume of ointment) regardless of body weight. The placebo ointment was identical to treatment ointment but without active ingredient mirtazapine, and was dosed at the same volume. The ointments were administered to the inner pinna daily for 14±3 days. Effectiveness was demonstrated by comparison of mean percent change in body weight between the Day 1 and Week 2 visits using a two-sample t-test with significance set at $p<0.05$ where the percent change in body weight in the 2% MZP group was greater than or equal to 0. Clinical pathology data from all cases receiving the 2% MZP or placebo were included in the safety analysis and each variable was analyzed using Analysis of Covariance, with treatment as a fixed effect and the baseline value as a covariate. Adverse events for safety analysis included any observations, whether or not considered to be product related, that were unfavorable and unintended, and that occurred after the use of study drug on Day 1 through study termination.

Two-hundred thirty-one (231) client-owned cats were enrolled in the study with 230 being administered at least one dose of 2% MZP or placebo and included in the safety analysis. One-hundred eighty-seven (187) cats were determined to be evaluable for the effectiveness analysis. Of those, 90 cats were in the 2% MZP group and 97 cats were in the placebo group.

The results are summarized in Table 1. The difference in mean percent change in body weight (95% C.I.) between the 2% MZP group and the placebo group was 3.78% (2.49%, 5.07%), which was statistically significant ($p<0.0001$). No clinically relevant statistically significant changes in clinical pathology parameters were seen. Adverse events reported in the 230 cat safety population were mild with the most common being application site reactions observed in 31.3% of placebo treated cats and 14.8% of 2% MZP treated cats. Nine cats had serious adverse events while on study, related to their underlying disease, of which, 3 cats received 2% MZP and 6 cats received placebo.

Mirtazapine 2% transdermal ointment is effective in managing weight loss in cats. Use of 2% MZP resulted in a mean weight gain of 4.07% compared to 0.29% for placebo after two weeks of transdermal treatment. The product was generally well tolerated.

Table 1. Summary of Results

	2 mg of 2% MZP (n = 90)		Placebo (n = 97)	
	Mean +/- SD	Median (range)	Mean +/- SD	Median (range)
Age (years)	14.3 +/- 3.64	14.9 (2.8-24.6)	13.5 +/- 2.87	14.0 (6.0-19.0)
Dose (mg/kg)	0.53 +/- 0.145	0.51 (0.22-0.95)	0	0
Baseline Weight (kg)	4.1 +/- 1.20	3.9 (2.1-9.2)	4.3 +/- 1.00	4.3 (2.3-7.5)
Weight change (%)	4.07 +/- 5.313	4.5(-10.7-16.7)	0.29 +/- 3.472	0 (-14.6-7.7)
Weight change (kg)	0.16 +/- 0.215	0.2 (-0.6-0.5)	0.01 +/- 0.147	0 (-0.7-0.3)

P12

Single Dose Pharmacokinetics Following Transdermal and Oral Administration of Mirtazapine 2% Transdermal Ointment in Cats (Abstract P12)

Beasley Mason[1]; Williams Buhles[2]; Jessica J. Quimby[3]; Daizie Labelle[4]; Donald Jung[5]; Melinda Poole O'Banion[1]

[1]Kindred Biosciences Inc., Burlingame, CA, USA; [2]Kindred Biosciences Inc., Fairfax, CA, USA; [3]Colorado State University, Fort Collins, CO, USA; [4]Kindred Biosciences Inc., Atherton, CA, USA; [5]Pharmaceutical Research Services, Cupertino, CA, USA

Mirtazapine has been shown to increase body weight in cats. The purpose of this study was to evaluate single dose pharmacokinetics (PK) of mirtazapine in a novel formulation of mirtazapine 2% transdermal ointment (2% MZP) in cats.

This study was a randomized, masked, cross-over PK study. Eight apparently healthy adult cats 6–12 years of age were studied. All cats were healthy as determined by daily clinical observations through a 7 day acclimation period, physical examination, and CBC/serum chemistry. All cats wore Elizabethan collars throughout the course of the study. On Day 0, four cats were treated once with 0.5 mg/kg of 2% MZP applied to the inner pinna of the right ear, and four cats were treated once with 0.5 mg/kg of 2% MZP given orally by syringe to the back of the mouth. On Day 5 the cats crossed over to the opposite treatment. Plasma was collected pre-dose and at 1, 2, 4, 6, 8, 12, 24, 48, 72, and 96 hours after administration and frozen at -20 °C prior to shipping and analysis. Plasma mirtazapine concentrations were measured using a validated LC-MS/MS method. PK parameters were calculated using standard methods WinNolin Professional 5.3. PK parameters were statistically compared between treatments using Wilcoxon Signed Rank test with significance set at $p<0.05$.

The PK parameters are summarized in Table 1. PK parameters were significantly different between the orally and transdermally administered 2% MZP (T_{max} p=0.02; C_{max} p=0.04; Half-life p=0.03) with the exception of $AUC_{0-\infty}$ (p=0.09).

Transdermal application of 2% MZP to the inner pinna achieves measurable plasma concentrations in cats. Absorption of 2% MZP was faster and more consistent following oral administration as compared to topical application. Additionally, the mean extent of absorption following oral administration was about 2-fold higher than following topical application of 2% MZP. The longer apparent terminal half-life observed following topical as compared to oral administration (26.8 vs. 10.1 h) suggests the presence of flip-flop pharmacokinetics, where the rate of absorption is significantly slower than the rate of elimination.

Table 1. Summary of PK parameters

Parameter	0.5 mg/kg Transdermal 2% MZP (n=8)		0.5 mg/kg Oral 2% MZP (n=8)	
	Mean ± SD	Median (range)	Mean ± SD	Median (range)
T_{max} (hr)	15.9 ± 15.6	9 (1-48)	1.13 ± 0.35	1 (1-2)
C_{max} (ng/mL)	21.5 ± 43.5	5.6 (2.1-129)	83.1 ± 31.2	83.2 (43.4-128)
$AUC_{0-\infty}$ (ng-h/mL)	270 ± 61.5*	250 (204-397)*	434 ± 149	494 (208-590)
Half-life (hr)	26.8 ± 6.0*	27.4 (19-34.5)*	10.1 ± 4.2	9.1 (4.7-17.0)

*n=6

P13

Multiple Dose Pharmacokinetics of Mirtazapine 2% Transdermal Ointment in Cats (Abstract P13)

Melinda Poole O'Banion[1]; Williams Buhles[2]; Jessica J. Quimby[3]; Daizie Labelle[4]; Donald Jung[5]

[1]Kindred Biosciences Inc., Burlingame, CA, USA; [2]Kindred Biosciences Inc., Fairfax, CA, USA; [3]Colorado State University, Fort Collins, CO, USA; [4]Kindred Biosciences Inc., Atherton, CA, USA; [5]Pharmaceutical Research Services, Cupertino, CA, USA

Mirtazapine has been shown to increase body weight in cats and is transdermally absorbed. The purpose of this study was to evaluate multiple dose pharmacokinetics (PK) of mirtazapine 2% transdermal ointment (2% MZP) in cats.

This study was a masked, randomized, three-arm parallel study to determine the plasma PK of two dosages (0.5 and 2.0 mg/kg) of 2% MZP applied topically once daily for 14 days to healthy adult cats (6–13 years old) as determined by CBC/serum chemistry and physical exam. Twenty cats were randomized: eight to 0.5 mg/kg, eight to 2.0 mg/kg, and four cats to no treatment. Cats underwent acclimation for 7 days, a dosing phase of 14 days, and a post-dosing phase of 3 days for PK sampling. Cats wore Elizabethan collars for the duration of the study. Cats were assessed daily for general health and application site abnormalities. Physical examinations and body weights were recorded at the beginning and end of the study. Plasma samples for mirtazapine concentrations were collected within one hour prior to treatment on the last dosing day, and at 1, 2, 4, 6, 8, 12, 24, 48, 72, and 96 h post final dosing and analyzed using a validated LC-MS/MS method. Noncompartmental pharmacokinetic analysis was performed with WinNonlin Professional ver 5.3.

The PK profile is summarized in Table 1. Mean ± SD body weight for cats that received 0.5 mg/kg 2% MZP was 5.4 kg ± 1.1 kg prior to treatment and 5.7 ± 1.2 kg after 14 days of treatment. Mean ± SD body weight for cats that received 2.0 mg/kg 2% MZP was 5.3 kg ± 1.1 kg prior to treatment and 5.7 ± 1.2 kg after 14 days of treatment. Mean ± SD body weight for control cats was 5.8 kg ± 1.2 kg at baseline and 6.1 ± 1.2 kg after 14 days. Mild redness of the pinna (application site) was noted following administration but there was no pinnal excoriation or ulceration observed in any cat.

After once daily dosing for 14 days, mirtazapine 2% transdermal ointment achieves clinically relevant serum concentrations in cats.

Table 1. PK parameters following 14 days of transdermal dosing of 2% MZP

Parameter	0.5 mg/kg 2% MZP (n=8)		2.0 mg/kg 2% MZP (n=8)	
	Mean +/- SD	Median (range)	Mean +/- SD	Median (range)
T_{max} (hr)	2.1 ± 1.3	2 (1.0-4.0)	3.0 ± 1.1	3.0 (2.0-4.0)
C_{max} (ng/mL)	39.6 ± 9.7	40.9 (27.2-56.5)	98.2 ± 55.0	73.9 (56.6-212)
$AUC_{0-\infty}$ (ng-hr/mL)	647 ± 225	590 (453-1167)	2045 ± 525	1778 (1576-2909)
Half-life (hr)	20.7 ± 4.0	21.0 (15.8-25.5)	28.4 ± 8.6	27.2 (20.3-47.5)

P14

Oral Mycophenolate Mofetil Dose Escalation Trial Assessing Adverse Effects and Pharmacodynamic Responses in Normal Dogs (Abstract P14)

Charlee Mulligan[1]; Keun Seok Seo[2]; Barbara Kaplan[2]; Kari Lunsford[2]; John Thomason[1]; Todd Archer[1]; Nogi Park[2]; Lakshmi Narayanan[2]; Kayla Alexander[2]; Andrew Mackin[1]

[1]College of Veterinary Medicine, Mississippi State University, Mississippi State, MS, USA; [2]College of Veterinary Medicine, Mississippi State University, Starkville, MS, USA

Mycophenolate mofetil (MMF) is a purine synthesis inhibitor and immunosuppressive agent that was originally developed for use in human organ transplantation, but has recently been increasingly used in veterinary medicine for treatment of canine immune-mediated disorders. In dogs, however, there are very limited experimental data available to enable establishment of an effective dose rate, and dose rates vary widely in published case series and anecdotal clinical reports. Adverse effects, particularly diarrhea, are frequently reported in both human and canine patients, and often lead to either dose reduction or drug discontinuation.

The objectives of our study were to, utilizing a dose escalation study in healthy dogs, determine the maximum tolerated dose (MTD) of MMF in each animal, and determine the effects of this dose on pharmacodynamic assays (B and T cell proliferation) focused on the established drug mechanism of action. Eight adult hound dogs were given oral MMF at three different escalating dose rates (10 mg/kg bid, 15 mg/kg bid, and 20 mg/kg bid) for one week at each dose, with a 2-week washout period between doses, and monitored for significant diarrhea, vomiting, or inappetence. Once MTD was determined, dogs were administered this dose rate for 2 weeks, and blood was collected on Days 0, 7, and 14 for B and T cell proliferation assays measuring incorporation of tritiated thymidine following activation with pokeweed mitogen and concanavalin A, respectively. A complete blood count (CBC), serum biochemistry, and urinalysis were also performed at the beginning and end of the study.

All 8 dogs tolerated oral MMF dose rates of 10 and 15 mg/kg bid for one week without significant side effects. Three dogs developed significant diarrhea (one of which also developed vomiting) at 20 mg/kg bid for one week, leading to a final starting MTD of 15 mg/kg bid (3 dogs) and 20 mg/kg bid (5 dogs) for the subsequent 2-week trial. However, 2 of the dogs on 15 mg/kg bid then developed significant diarrhea at the start of the second week of drug administration, necessitating a further dose reduction to 10 mg/kg bid. Therefore, by the end of the second week, 5 dogs were on 20 mg/kg, 1 dog was on 15 mg/kg, and 2 dogs were on 10 mg/kg bid. There were no significant changes in CBC, serum biochemistry or urinalysis results over the study period.

B and T cell proliferation results after one week of oral MMF at MTD were not significantly different than Day 0 results (pretreatment baseline) (Figure 1). However, after 2 weeks of oral MMF, both B and T cell proliferation were significantly reduced: B cell proliferation (counts/min) was on average reduced to 49% of baseline, and T cell proliferation to 42% of baseline. Interestingly, in all 3 dogs on reduced MMF doses, both B and T cell proliferation were still decreased by more than 50%.

Our study confirmed that oral MMF at maximally tolerated dose rates significantly inhibits both B and T cell proliferation in dogs. This effect, however, appears to be delayed for several weeks, suggesting that MMF may not be the ideal treatment choice for dogs with severe or emergent immune-mediated diseases that require rapid immunosuppression. The mechanism for this delay in the immunosuppressive effects of MMF has not been elucidated, although another purine synthesis inhibitor, azathioprine, is also reported to have a delayed effect, the cause of which has also not yet been determined. Diarrhea was, as has been previously reported, the most common dose-limiting side effect in our study. The exact mechanism of MMF-associated diarrhea has not been determined in either humans or dogs. Interestingly, in some dogs, drug-induced diarrhea, as with drug effects on lymphocyte proliferation, appears to be delayed until after the first week of therapy. Since a lower oral dose rate (10 mg/kg bid) of MMF still causes significant suppression of lymphocyte proliferation in some dogs, without associated diarrhea, this reduced dose rate warrants exploration in future pharmacodynamic studies.

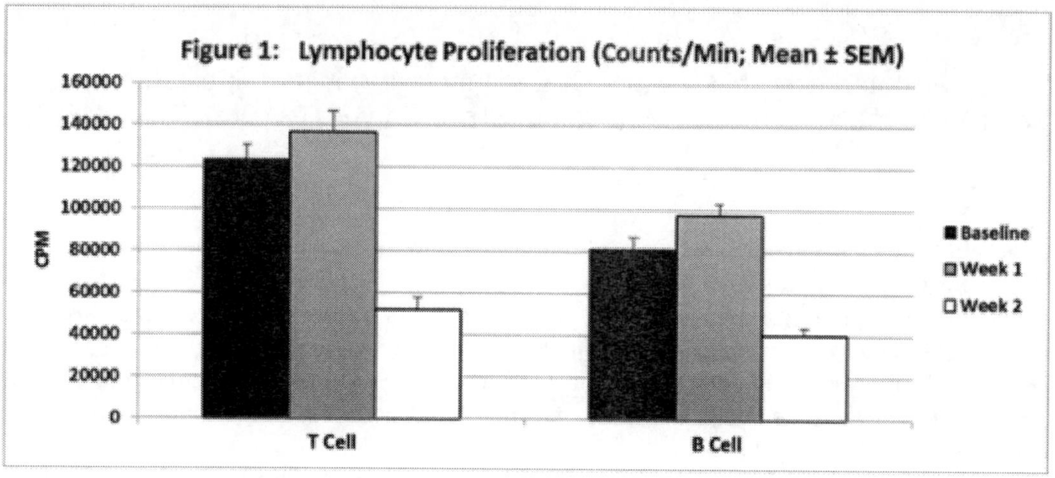

P15

Functional Evaluation of the Canine *CYB5R3* A729G Single Nucleotide Polymorphism Associated with Sulfonamide Hypersensitivity (Abstract P15)

Jennifer M. Reinhart; Joanne L. Ekena; Lauren A. Trepanier

Department of Medical Sciences, School of Veterinary Medicine, University of Wisconsin-Madison, Madison, WI, USA

The enzyme cytochrome b_5 reductase (encoded by *CYB5R3*) detoxifies reactive metabolites of sulfonamide antibiotics. The single nucleotide polymorphism (SNP) *CYB5R3* 729A>G was recently associated with sulfonamide hypersensitivity in dogs. This synonymous coding SNP may partially explain risk for this idiosyncratic drug reaction, however, the functional significance of this variant is unknown. Based on in silico predictions, we hypothesized that the 729A>G SNP would decrease protein synthesis and total cellular cytochrome b_5 reductase activity without decreasing mRNA expression in an *in vitro* model. HepG2 cells were transfected with an expression vector containing either wildtype (A) or variant (G) *CYB5R3* cDNA and incubated for 48 hours. Cells were harvested and *CYB5R3* mRNA, protein, and function were measured with qRT-PCR, immunoblots, and cytochrome c reduction assay, respectively. There was no significant difference in protein expression between the A and G alleles. mRNA quantification and cytochrome b_5 reductase activity assays are underway. Preliminary results do not support *CYB5R3* 729A>G as a functional variant leading to decreased cytochrome b_5 reductase expression. However, immunoreactivity is only a semi-quantitative measure of protein expression; results of mRNA and enzyme activity assays may better inform the mechanistic significance of 729A>G SNP in sulfonamide hypersensitivity in dogs.

P16

Recovery of T-Cell Function in Healthy Dogs Following Cessation of Cyclosporine Therapy (Abstract P16)

John Thomason[1]; Charlee Mulligan[1]; Lakshmi Narayanan[2]; Lisa Durso[1]; Brittany Thames[1]; Andrew Mackin[1]; Todd Archer[1]

[1]College of Veterinary Medicine, Mississippi State University, Mississippi State, MS, USA; [2]Mississippi State University, Starkville, MS, USA

Cyclosporine is a potent immunosuppressive agent commonly used to treat immune-mediated disorders in dogs. Cyclosporine targets T lymphocytes by decreasing cytokine expression (IL-2, IL-4, and IFN-γ) of activated T cells. Unfortunately, significant complications, such as life-threatening secondary infections, can sometimes arise as a result of cyclosporine administration, necessitating cessation of therapy. Once cyclosporine is stopped, the expectation is that the patient's immune system will completely recover, and commence functioning appropriately. Currently, it is unknown how quickly the immune system and T-cell function recover following the cessation of cyclosporine therapy. The objective of this study was to utilize a pharmacodynamic bioassay to quantify the average time it takes for T-cell function to return to pre-treatment levels in healthy dogs after cessation of an immunosuppressive dose of cyclosporine.

Six healthy adult dogs were given cyclosporine (10 mg/kg BID, PO) for 7 days. Whole blood gene expression of IL-2 and IFN-γ was evaluated before treatment and again on Day 7 of drug administration, two hours after giving the medication, using quantitative reverse transcription polymerase chain reaction (qRT-PCR). Samples were activated for 5 hours with phorbol 12-myristate 13-acetate and ionomycin, RNA was extracted, and relative gene quantification was performed comparing gene expression after drug administration to pre-treatment levels. Additionally, for one week after cessation of cyclosporine, daily blood samples were collected for evaluation of IL-2 and IFN- γ gene expression.

After 7 days of oral cyclosporine, both IL-2 and IFN- γ were markedly suppressed, reaching 1% and 2% of pre-treatment values, respectively. Mean IL-2 and IFN-γ expression remained decreased below pre-treatment levels on Days 1 and 2 of washout (WO), but by WO Day 3 were 115% (SEM, 25%) and 103% (23%) of pre-treatment values, respectively. By WO Day 4, mean IL-2 and IFN-γ expression were 156% (25%) and 125% (25%) of pre-treatment values, respectively.

Our results suggest that, following cessation of immunosuppressive doses of cyclosporine, T-cell expression of IL-2 and IFN-γ will begin to return to pre-treatment values within the initial 24 hours. In some dogs, IL-2 and IFN-γ expression return to pre-treatment values within 3 days of stopping cyclosporine, but other dogs may require an additional recovery day. Interestingly, upon recovery, both IL-2 and IFN-γ expression exceed pre-treatment values, suggesting a potential rebound effect following the cessation of cyclosporine therapy. Additional studies are required to determine the timing of post-cyclosporine recovery of T-cell function in dogs with naturally occurring immune-mediated disorders.

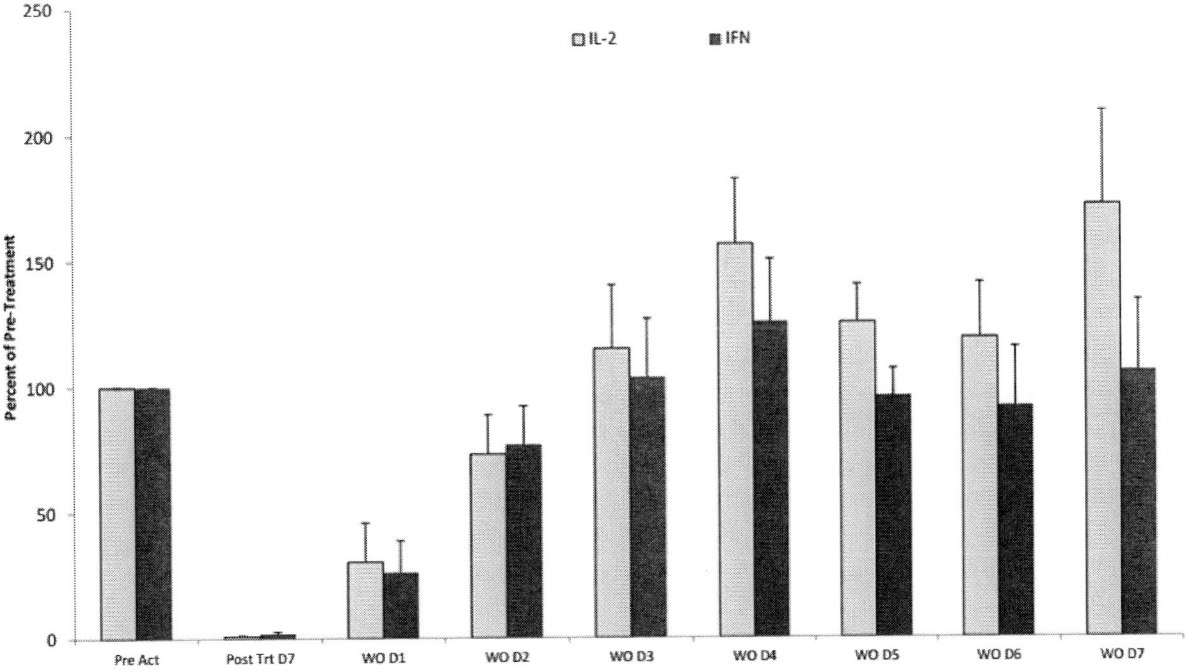

250
200
150
100
50
0
Percent of Pre-Treatment
IL-2
IFN
Pre Act
Post Trt D7
WO D1
WO D2
WO D3
WO D4
WO D5
WO D6
WO D7

R01

Effect of Oral Yunnan Baiyao on Periprocedural Hemorrhage and Coagulation in Dogs Undergoing Nasal Biopsy (Abstract R01)

Lauren Adelman; Shelly Olin; Christine M. Egger; Jennifer E. Stokes

College of Veterinary Medicine, University of Tennessee, Knoxville, TN, USA

The hemostatic efficacy and safety of *Yunnan baiyao* (YB) has been demonstrated across multiple species. Despite the widespread anecdotal use of YB in veterinary medicine for dogs with bleeding tendencies, few studies have evaluated its efficacy in dogs. The primary objective of this study was to evaluate the effect of YB on clinical hemorrhage following nasal biopsy in dogs as measured by blood loss and time to stop bleeding. A secondary objective was to evaluate the effect of YB on measures of coagulation including buccal mucosal bleeding time (BMBT) and thromboelastography (TEG).

Nineteen client-owned dogs who were presented to the UT-VMC internal medicine service for nasal biopsy between March 2015 and August 2016 were enrolled prospectively. Dogs were randomized to treatment (YB) or placebo control group. Owners, statisticians and primary investigators were blinded to which treatment the dog received. Each dog received 2 doses of drug, one dose at 6 pm the evening before and one dose at 6 am the morning of the procedure according to body weight. All dogs had baseline blood pressure, platelet count, prothrombin time, and activated partial thromboplastin time performed at presentation and a BMBT performed before and after capsule administration. Dogs were excluded from the study if systolic blood pressure was >160 mm Hg, platelet count 13 sec, PTT >90.6 sec or BMBT >4 minutes. In addition to BMBT, 9/19 had both TEG and blood loss measurements performed, 6/19 had only TEG performed, and 4/19 had only blood loss measurements performed. Measurements of blood loss included total blood loss, blood loss as a percentage of body weight and time to stop bleeding. A standardized anesthetic protocol was followed. A final diagnosis was determined by histopathologic evaluation of biopsy samples, CT findings and/or cytological exam on an individual case basis. Statistical analysis of blood loss measures was performed using a two-way ANCOVA. A repeated measures mixed effects ANCOVA was used to analyze BMBT and TEG variables (alpha angle, K, R, MA). Backwards variable elimination was performed to evaluate the effect of covariates. A P value <0.05 was considered statistically significant.

Time to stop bleeding was significantly shorter in the YB group (300±12 sec) compared to the control group (367±9 sec) and was not influenced by underlying disease. There was no significant effect of body weight, weight of biopsies, or weight of biopsies as a percentage of body weight on time to stop bleeding. Age and presence of epistaxis at presentation were significantly associated with a longer time to stop bleeding while blood pressure and biopsy number were significantly associated with a shorter time to stop bleeding. Treatment effect remained significant even when the effect of these covariates were considered. Total blood loss and blood loss as a percentage of body weight were not significantly different between groups. There was no effect of YB on measures of coagulation including BMBT or TEG. No adverse effects were noted during or after treatment in either group.

In this study, pre-procedural oral administration of YB to dogs undergoing nasal biopsy resulted in significantly shorter time to stop bleeding following biopsy. The results of this study support the use of YB prior to nasal biopsy in dogs. Additional studies are needed to determine the clinical use of YB for other procedures or medical conditions associated with hemorrhage in dogs. Furthermore, pharmacokinetic studies are needed to determine optimal oral dosing of YB in dogs.

FIGURE 1

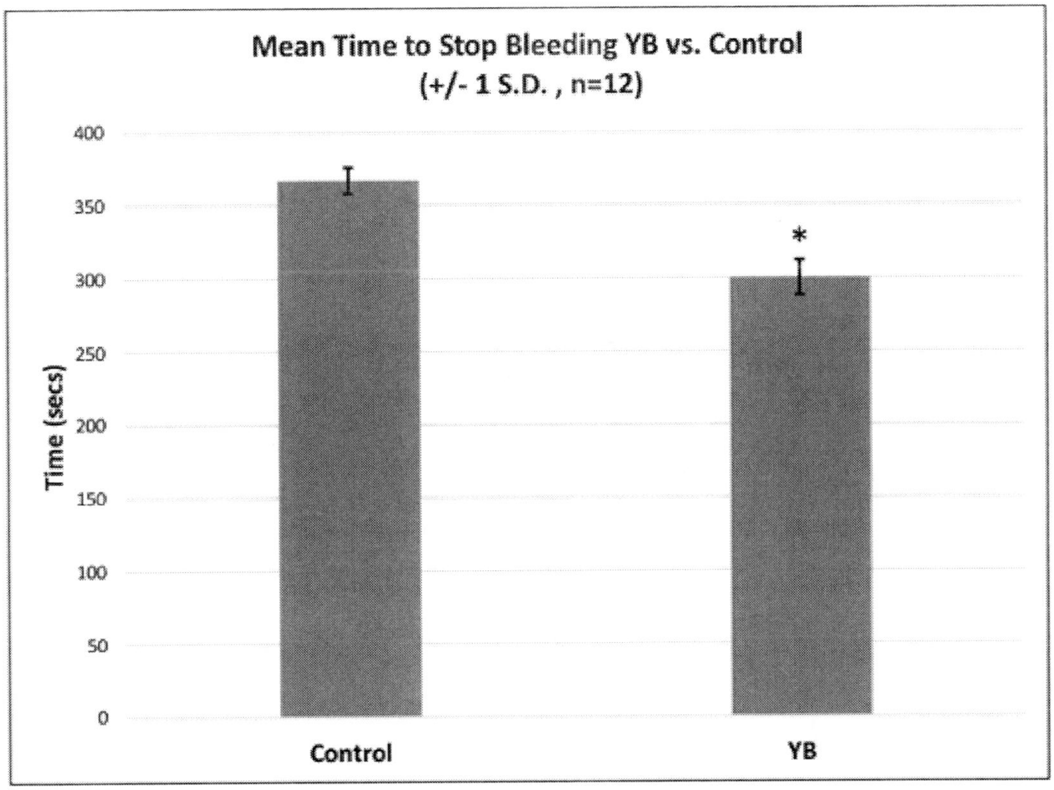

*A P value < 0.05 was considered statistically significant

Table 2: Total blood loss, blood loss as a percentage of body weight and time to stop bleeding in dogs receiving YB versus placebo control

Parameter	YB N=6			Control N=6			P Value	
	Neoplasia N = 1	Inflammatory N = 5	All	Neoplasia N = 2	Inflammatory N = 4	All	Disease	Treatment
Total Loss (grams)	26	40 ± 30 (9-73)	38 ± 27	40 ± 11 (33-48)	20 ± 12 (4-32)	27 ± 15	0.83	0.86
Loss % BW (%)	6	15 ± 9 (4-27)	14 ± 9	28 ± 25 (10-45)	23 ± 26 (7-62)	25 ± 23	1.00	0.69
Bleeding Time (sec)	411	284 ± 151 (114-493)	300 ± 12	479 ± 257 (297-660)	326 ± 161 (142-500)	367 ± 9	0.16	0.01*

All values are expressed as mean +/- standard deviation followed by the range in parentheses
Total Loss = Total blood loss
Loss % BW = Blood loss as a percentage of body weight
Bleeding Time = Time to stop bleeding
P Value category "disease" represents the difference between parameters with regards to underlying disease category (neoplastic vs inflammatory)
P Value category "treatment" represents the difference between parameters between YB and control groups
*A P value < 0.05 was considered statistically significant

R02

Bacterial Infection in Canine Tracheal Collapse Syndrome Patients Prior to, and Following, Tracheal Stent Placement (Abstract R02)

Sylvia Lesnikowski[1]; Chick Weisse[2]; Allyson Berent[2]; Alexandre Le Roux[2]

[1]ASPCA, New York, NY, USA; [2]Animal Medical Center, New York, NY, USA

The purpose of this study was to determine the incidence of positive pathogenic pre-stent and post-stent cultures in dogs with tracheal collapse (TC).

This was a retrospective study with review of records of 26 traditional-type TC and 27 malformation-type TC dogs receiving tracheal stents at the Animal Medical Center from 2009–2014. All dogs with tracheal stent placement and concurrent endotracheal culture were included as pre-stent cultures. Any subsequent cultures obtained were considered post-stent cultures. Positive pathogenic cultures were defined as positive bacteriologic cultures with compatible cytologic and/or radiographic evidence of infection.

Positive pathogenic cultures occurred in 25/38 (66%) pre-stent and 20/31 (65%) post-stent dogs. Traditional-type TC dogs had positive pathogenic pre-stent and post-stent cultures in 16/26 (62%) and 7/26 (27%), respectively. Malformation-type TC dogs had positive pathogenic pre-stent and post-stent cultures in 9/27 (33%) and 13/27 (48%), respectively. The most common isolates included *Escherichia* spp. (16%), *Pseudomonas* spp. (16%), and *Staphylococcus* spp. (16%) in pre-stent patients and *Escherichia* spp. (29%) and *Pseudomonas* spp. (12%) in post-stent patients.

Positive pathogenic cultures were present in a similar majority of dogs with tracheal collapse prior to, and following, tracheal stent placement. Malformation-type TC dogs had a higher incidence of pathogenic post-stent cultures compared to traditional-type TC dogs. The risk of a positive post-stent culture increased in malformation-type TC but decreased in traditional-type TC patients. Tracheal stents may not increase the risk of airway infections in all tracheal collapse patients. Bacterial cultures in TC dogs may reflect poor clearance of pharyngeal and enteric organisms.

R03

Oral Probiotic Bacterial Species Can Colonize the Healthy Feline Airway Microbiota (Abstract R03)

Aida I. Vientos Plotts[1]; Aaron Ericsson[2]; Hansjorg Rindt[3]; Megan Grobman[1]; Carol Reinero[1]

[1]College of Veterinary Medicine, University of Missouri, Columbia, MO, USA; [2]Metagenomics Center, College of Veterinary Medicine, University of Missouri, Columbia, MO, USA; [3]Comparative Internal Medicine Laboratory, College of Veterinary Medicine, University of Missouri, Columbia, MO, USA

While probiotics may beneficially alter the gastrointestinal microbiota, no studies in cats have determined if oral probiotics affect the composition of the airway microbiota. Using 16S rRNA amplicon sequencing of bacterial DNA extracted from oropharyngeal (OP) swabs, bronchoalveolar lavage (BAL), feces, and blood, our objective was to characterize how oral probiotics changed the composition of the upper and lower airway, fecal, and blood microbiota. We hypothesized that probiotics would modulate the respiratory microbiota in healthy cats, demonstrated by the detection of the probiotic bacterial species present in fecal samples and altered "core" airway microbiome communities in the respiratory tract.

Six healthy research cats <1-year-old had OP, BAL, fecal, and blood samples collected at baseline and 4 weeks after receiving oral probiotics. Extracted DNA underwent PCR of the 16S rRNA gene. Once sequenced, coverage (number of reads/sample), richness, and relative abundance of representative operational taxonomic units (OTUs) were determined. Principal component analysis (PCA) demonstrated relatedness of samples.

After receiving probiotics, coverage in BAL (mean ± SD, 269 ± 23 to 11,122 ± 23,546 reads/sample; $p = 0.002$), and blood (269 ± 43 to 1,513 ± 631 reads/sample; $p = 0.002$) significantly increased. There was no significant difference in coverage in OP and fecal swabs ($p = 0.56$, and $p = 0.59$, respectively). Richness in the upper (from 105 ± 8 to 124 ± 13; $p = 0.030$) and lower airways (from 27 ± 4 to 68 ± 31; $p = 0.002$) and blood (from 15 ± 6 to 28 ± 11; $p = 0.041$) significantly increased. In feces richness did not significantly differ after probiotics ($p = 0.485$). The probiotic capsule resolved to the OTU level consisted of the following taxa: genus *Bifidobacterium* (36%), *Bifidobacterium animalis* (23%), genus *Streptococcus* (26%), *Lactobacillus zeae* (12%), and genus *Lactobacillus* (3%). *B. animalis* and *L. zeae* were not detected in any sample site at baseline. *L. zeae* was detected in all sites post-treatment, and *B. animalis* was detected in OP, BAL, and feces, but not blood. The overall composition of the upper and lower airway microbiota underwent the most changes in regards to relative abundance of predominant OTUs with 4.8% and 6.9% of OTUs changing significantly in OP and BAL samples, respectively. In contrast only 0.5% of OTUs in feces changed significantly from baseline. PCA revealed distinct microbiomes in each site; samples clustered more tightly at baseline and had more variation after probiotic administration.

This is the first study describing the effect of oral probiotics on the respiratory microbiota of healthy cats. Finding bacterial species present in the oral probiotics in the upper and lower airways provides pilot data suggesting that oral probiotics could be investigated as a tool to target dysbiosis occurring in inflammatory airway diseases such as feline asthma.

R04

Thoracic Radiograph Findings in Feline Histoplasmosis (Abstract R04)

Lauren Kuzimski[1]; Laura A. Nafe[1]; Kate Sippel[2]; Andrew Hanzlicek[1]

[1]College of Veterinary Medicine, Oklahoma State University, Stillwater, OK, USA; [2]Gulf Coast Veterinary Specialists, Stillwater, OK, USA

Histoplasmosis is a common fungal infection that has the potential to result in severe clinical disease, with the majority of cats having pulmonary involvement. The purpose of this study was to evaluate thoracic radiograph findings in cats diagnosed with histoplasmosis. Medical records of cats presenting to the Oklahoma State University VTH between 2005–2015 were reviewed. Cats included in the study had orthogonal thoracic radiographs performed, a complete medical record, and a definitive diagnosis of histoplasmosis based on identification of *H. capsulatum*, positive urine antigen, and/or positive fungal culture. A radiologist blinded to the clinical severity of each patient evaluated all radiographs to determine the predominant pulmonary pattern and severity score (0-normal, 1-mild, 2-moderate, 3-severe) for each lung lobe. Forty-seven cats (27 female, 20 male) were included in the study with a median age of 7 years (range: 0.7–19 years) of which 32 (68%) were presumed to be alive at the time of this abstract. The predominant pulmonary pattern was bronchial in 38 cats (81%) and 6 cats (13%) had a mixed pulmonary pattern (combination of bronchial, interstitial, and/or alveolar). Three cats (6%) had bilateral pleural effusion and 3 cats (6%) had normal radiographs. All lung lobes were abnormal in the majority of cats and the median severity score of all lung lobes was 2 out of 3 (moderate). Feline histoplasmosis results in pulmonary abnormalities in the majority of cats infected, and the predominant pulmonary pattern is a moderate bronchial pattern with or without an interstitial component.

R05

Frequency of Bronchoalveolar Lavage Hemosiderosis in 171 Dogs and 34 Cats with Respiratory Disease (2007-2016) - (Abstract R05)

Stipe Jelovcic[1]; Alice Defarges[1]; **Kimberly Hooi**[1]; Stephanie Nykamp[1]; J. Scott Weese[2]; Dorothee Bienzle[2]

[1]University of Guelph, Guelph, ON, Canada; [2]Department of Pathobiology, Ontario Veterinary College, University of Guelph, Guelph, ON, Canada

To retrospectively evaluate the presence of hemosiderin in canine and feline bronchoalveolar lavage (BAL) samples and possible association with signalment, pre-BAL treatment, diagnosis, blood parameters, transthoracic fine-needle aspiration, BAL procedure time, and prior immune-mediated disease or neoplasia.

Medical records of dogs and cats with respiratory disease that underwent BAL between 2007 and 2016 at the Ontario Veterinary College were reviewed.

Records from 171 dogs and 34 cats met inclusion criteria. Hemosiderin was reported in BAL cytology from 13/171 (7.6%) dogs and 18/34 (52.9%) cats ($p<0.001$). Cats had a higher risk of BAL hemosiderosis than dogs (OR=7.5). Dogs that had a history of cancer were at higher risk for BAL hemosiderosis (OR=3.64). None of the other factors were associated with BAL hemosiderosis, except for prolonged BAL procedure in dogs ($p=0.044$). An increased proportion of BAL eosinophils in cats was associated with absence of hemosiderin ($p=0.0463$). Hemosiderin was identified in the BAL of 4/72 (5.5%) dogs with pneumonia, 6/56 (10.7%) dogs with bronchitis and 2/9 (22.2%) dogs with neoplasia. One of 16 non-diagnostic BAL samples in dogs contained hemosiderin. Hemosiderin was identified in the BAL of 4/6 (66.7%) cats with asthma, 7/12 (58.3%) cats with pneumonia, 5/12 (41.6%) cats with bronchitis and 2/4 cats (50%) with neoplasia. Hemosiderin was present in 2/6 (33.3%) dogs and 1/2 cats (50%) that underwent transthoracic fine needle aspiration prior to BAL.

Hemosiderosis is more common in BAL samples from cats than from dogs, and does not appear to be associated with specific respiratory conditions.

R06

Lung Lobe Torsion in 39 Dogs and Cats (2004-2016) - (Abstract R06)

Kathryn Benavides[1]; Elizabeth A. Rozanski[2]; Trisha Oura[1]

[1]Tufts University, North Grafton, MA, USA; [2]Cummings School of Veterinary Medicine, Tufts University, Grafton, MA, USA

Lung lobe torsion (LLT) is an uncommon pulmonary disease where the lung twists around its pedicle, resulting in necrosis of the lobe. The primary objective of this study was to evaluate the characteristics of animals identified with LLT, with a secondary objective to evaluate any prognostic indicators.

The medical record system was searched using for "Lung lobe torsion", and individual cases were reviewed for confirmation of LLT either at the time of surgical intervention or on necropsy. Data recorded included signalment, large or small breed, imaging modality(ies), presence of pre-operative pleural effusion, lobe affected, and outcome. Fisher 's exact test was used to compare groups.

Thirty-nine cases were identified including 35 dogs and 4 cats. Pugs (n=10) were most common, with the remaining dogs being small breed (n=7), or large breed (n=18). Three cats were DSH and one was a minskin. One small breed dog and one cat were euthanized at the time of diagnosis. Thoracic radiographs were performed in all cases, CT in 19, and bronchoscopy in 5. Lobes affected included right middle (n=18), left cranial (n=18), right cranial (n=2), and accessory (n=1) Of those treated, 86.5% survived to discharge. All 16 treated pugs, other small breed dogs and cats survived, while only 12/18 large breeds survived. (p=0.02). Thirty-one animals had pre-operative pleural effusion, 26 survived; 6 animals had no pre-operative effusion, all survived (p=0.57).

Small breeds and cats with LLT have an excellent prognosis for survival to discharge; the prognosis in large dogs is more guarded.

R07

Fluoroscopic Estimation of Thoracic Dimensional Changes in Healthy Dogs (Abstract R07)

Jennifer Chan; Lynelle Johnson; Craig Brown; Rachel Pollard

School of Veterinary Medicine, University of California-Davis, Davis, CA, USA

There are limited non-invasive methods available to evaluate lung mechanics. We hypothesized that inspiratory to expiratory change in lung area as assessed by fluoroscopy would result in reproducible measures that could be used for future evaluation of lung function and mechanics in dogs with respiratory disease. Forty-four conscious, unsedated dogs with no evidence of respiratory disease were minimally restrained in sternal recumbency. Tidal respiration was recorded fluoroscopically for 10 respiratory cycles, and maximal inspiratory and expiratory images from 3 respiratory cycles were evaluated. The number of pixels in the entire thoracic cavity was measured for both inspiration and expiration, and the average percent change in thoracic dimension was determined for each dog. This process was repeated using a hemithorax measurement that excluded the mediastinum and cardiac silhouette. Reference ranges with 95% confidence were computed using a non-parametric percentile distribution. Median percent change in thoracic dimension for the total thorax measurement was 12.5% (confidence interval of 8.91–23.97%). Median percent change for the hemithorax measurement (20.8%) was significantly different ($p<0.001$) with a larger confidence interval of 14.32–37.61%, rendering it more responsive to lung area changes. Neither measurement technique was correlated with age, sex, thoracic conformation, BCS, or breed but varied with body weight above and below 10 kg. In conclusion, fluoroscopy provides a noninvasive and reliable measure of lung area changes during respiration in small and large dogs. Further investigations of lung area changes are required in dogs with respiratory disease to assess its utility in detecting dysfunction.

R08

Exposure to Second-Hand Smoke in Pets from Different Socioeconomic Backgrounds (Abstract R08)

Nichole Smith[1]; Elizabeth A. Rozanski[2]; Diane Welsh[1]; Gregory Wolfus[3]

[1]Tufts University, North Grafton, MA, USA; [2]Cummings School of Veterinary Medicine, Tufts University, Grafton, MA, USA; [3]Tufts University, Worcester, MA, USA

Environmental tobacco smoke (ETS) exposure is more common in children from lower socioeconomic classes. ETS exposure has been linked with a variety of health risks. Urine cotinine may be used to estimate ETS exposure. The purpose of this study is to evaluate the rate of ETS exposure in dogs and cats from different socioeconomic classes in Massachusetts.

Urine samples submitted left-over from routine testing were collected from two university laboratories. Samples from animals living in Worcester (two zip codes), Webster/West Brookfield, Westborough/Walpole, or Wellesley/Weston were eligible for inclusion, representing low income, lower middle income, upper middle income, and high income areas. No further owner information was recorded. Informed consent was waived by the Clinical Sciences Review Committee. Urine was tested for cotinine and classified as positive or negative. Results were compared using Chi-square analysis.

One hundred and seventy-eight dogs and cats were enrolled. Seventeen were positive for urine cotinine (9.5%). Worcester (low-income) had a significantly higher rate of positive animals. ($p=0.00093$). Worcester animals had a 21% rate of ETS exposure, Webster/West Brookfield 11%, Westborough/Walpole 1% and Wellesley/Weston 0%.

Dogs and cats, similar to children, have a higher rate of exposure to ETS in lower socioeconomic settings, and may be at risk of more smoke-associated diseases.

E01

Quantification of Left Atrial Wall Motion in Healthy Horses Using Two Dimensional Speckle Tracking (Abstract E01)

Christina Eberhardt[1]; Katharyn J. Mitchell[2]; Colin C. Schwarzwald[3]

[1]University of Zurich, Vetsuisse Faculty, Zurich, Switzerland; [2]University of Zurich, Zurich, Switzerland; [3]Clinic for Equine Internal Medicine, Vetsuisse Faculty, University of Zurich, Zurich, Switzerland

The objective of this study was to validate the use of two-dimensional speckle tracking (2DST) echocardiography for assessment of left atrial (LA) mechanical function in healthy horses. We aimed to establish test-retest reliability and calculate normal reference intervals.

26 healthy Warmblood horses were included. 2DST analyses of LA wall motion were performed on digitally stored cineloop recordings of a standardized right-parasternal 4-chamber view focusing on the left atrium. A p-wave trigger was used to define the start of a cardiac cycle. Global longitudinal strain (S_L), global longitudinal strain rate (SR_L) and time to peak left atrial contraction (TPLC) were measured to characterize LA contractile, reservoir and conduit function. Measurements of 10 randomly selected horses were repeated in a blinded fashion by the same observer and by a second observer on different days.

The mean, SD and lower and upper limit of the reference interval were calculated using Reference Value Advisor software (www.biostat.envt.fr/spip). Intra-observer and inter-observer measurement variability was quantified using the coefficient of variation (CV).

Reference intervals (mean ± SD, lower and upper limit) were S_L (%): -11.82±3.52, -19.20 to -4.44 and SR_L (s^{-1}): -1.31±0.43, -2.23 to -0.40 for LA contractile function; S_L (%): 15.33±2.94, 9.17 to 21.50 and SR_L (s^{-1}): 0.78±0.12, 0.52 to 1.03 for LA reservoir function; and S_L (%): -5.29±1.74, -8.95 to -1.62 and SR_L (s^{-1}): -1.53±0.23, -2.01 to -1.06 for LA conduit function. Normal TPLC (ms) was 264±23; 215 to 313. Day-to-day intra-observer and inter-observer variability (CV%), respectively, for LA contractile function were S_L: 9.2 and 10.7 and SR_L: 7.6 and 14.2; for LA reservoir function were S_L: 7.5 and 13.0 and SR_L: 10.4 and 6.4; for conduit function were S_L: 11.1 and 15.4; SR_L: 6.1 and 12.3; and for TPLC were 2.0 and 5.1.

This study shows that 2DST is a feasible and reliable method to quantify left atrial wall motion, with very low to low day-to-day intra-observer and inter-observer variability. Reference intervals for healthy Warmblood horses are reported. Future studies will focus on left atrial wall motion in horses with cardiac disease to proof the concept of using 2DST for LA wall motion analysis.

E02

Effect of Flecainide on Atrial Fibrillatory Rate in Horses with Induced Atrial Fibrillation (Abstract E02)

Eva Z. Hesselkilde[1]; Helena Carstensen[1]; Maria M. Haugaard[1]; Jonas Carlson[2]; Thomas Jespersen[3]; Pyotr G. Platonov[2]; Rikke Buhl[1]

[1]Department of Large Animal Sciences, Faculty of Health and Medical Sciences, University of Copenhagen, Taastrup, Hovedstaden, Denmark; [2]Department of Cardiology, Clinical Sciences, Lund University, Lund, Skane Lan, Sweden; [3]Department of Biomedical Sciences, Faculty of Health and Medical Sciences, University of Copenhagen, Hovedstaden, Denmark

Atrial fibrillation (AF) is the most important arrhythmia in equine medicine and traditionally, characteristics of AF have been assessed by intra-cardiac measurements of atrial fibrillation cycle length. Atrial fibrillatory rate (AFR) retrieved from surface ECGs can however, be used as a biomarker for atrial electrical remodeling. The aim of this study was to study characteristics of induced AF and its modification by flecainide in healthy horses.

Fourteen horses were electrically stimulated into AF by a multipolar catheter placed in the right atrium and were treated with either isotonic saline (n=7) or flecainide 2 mg/kg (n=7) IV. A single horse, with spontaneous persistent AF lasting more than three weeks, was also treated with flecainide. ECGs from all 15 horses were analyzed using spatiotemporal cancellation of QRST complexes and calculation of AFR from the residual atrial signal.

At drug administration AFR was 322±30 fibrillations per minute (fpm) in the flecainide-treated horses and 299±12 fpm in the control group (ns). Flecainide infusion was followed by significant ($p<0.05$) AFR decrease that preceded restoration of sinus rhythm in all induced horses (Fig. 1A). Time to cardioversion after flecainide infusion start was 4.3±2.5 min. In the control animals, mean AF duration was 205 min (range 152–273 min) and no alteration in AFR was observed following saline administration. The last 30 min before spontaneous cardioversion, a gradual decrease in AFR was observed ($p<0.05$) (Fig. 1B). In the horse with spontaneous persistent AF, AFR decreased from 450 fpm to 300 fpm following flecainide administration without restoring sinus rhythm.

Flecainide caused a rapid decrease in AFR that resulted in restoration of sinus rhythm while AFR gradually decreased in the horses that spontaneously cardioverted. AFR in the horse with naturally occurring AF was markedly higher which suggest AFR to be a non-invasive method to study atrial remodeling and the effect of antiarrhythmic compounds.

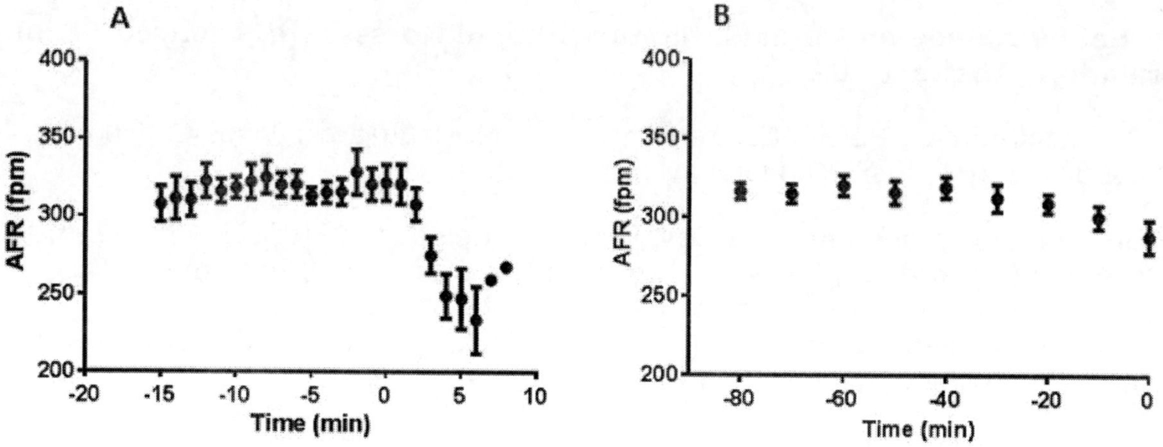

Fig. 1A+B. AFR trend preceding cardioversion in 14 horses with induced AF. Mean ± SEM. A Represent horses treated with intravenous flecainide at time = 0. Note how AFR rapidly decreases until restoration of sinus rhythm (P<0.05). B Represent the control group where AFR slowly decreases before cardioversion at time = 0 (P<0.05).

E03

Reduced Exercise Capacity in Horses Due to Atrial Fibrillation (Abstract E03)

Helena Carstensen[1]; Eva Z. Hesselkilde[1]; Merle Frederike Fenner[1]; Bjørg Z. Klein[1]; Steen Pehrson[2]; Thomas Jespersen[3]; **Rikke Buhl**[1]

[1]Department of Large Animal Sciences, Faculty of Health and Medical Sciences, University of Copenhagen, Taastrup, Hovedstaden, Denmark; [2]Department of Cardiology, The Heart Centre, Copenhagen University Hospital, Copenhagen, Hovedstaden, Denmark; [3]Department of Biomedical Sciences, Faculty of Health and Medical Sciences, University of Copenhagen, Hovedstaden, Denmark

Atrial fibrillation (AF) is the most common pathological arrhythmia resulting in poor performance in horses and the purpose of this study was to quantify the impact of AF on exercise capacity.

Nine unfit Standardbred mares were included in the study; six AF horses and three time-matched controls. All horses performed a maximal exercise treadmill test (Test 1) before (day 0) and after (day 57/58) (Test 2) induction of AF by pacemaker stimulation in the right atrium. During the tests all horses were equipped with surface ECG and a venous catheter for blood sampling.

Maximum speed in the AF group was 1.2±0.4 m/s slower in Test 2 compared to Test 1 whereas the control horses ran 0.7±0.6 m/s faster in their second test ($p<0.05$). In the AF group speed at HR 200 bpm was 8.6±0.7 m/s in Test 1 opposed to 6.5±0.8 m/s in Test 2 ($p<0.05$) and time for the HR to reach 100 bmp after cessation of the test was 20±22 and 65±44 minutes in Test 1 and 2 respectively ($p<0.05$). For horses in AF maximum HR reached 311±27 bpm (range 276–346 bpm) during exercise compared to 226±11 bpm (range 207–242 bpm) for the same horses in sinus rhythm ($p<0.05$). No significant differences in blood parameters were found between Test 1 and 2.

Exercise capacity was significantly reduced in horses after developing AF characterized by decreased velocity, increased HR at lower speed, increased maximum HR and prolonged cardiac recovery after exercise.

E04

Antiarrhythmic Effects of Flecainide on Experimentally Induced Atrial Fibrillation of Different Duration in Horses (Abstract E04)

Helena Carstensen[1]; **Eva Z. Hesselkilde**[1]; Merle Frederike Fenner[1]; Ameli V. Loft-Andersen[1]; Jørgen K. Kanters[2]; Steen Pehrson[3]; Thomas Jespersen[2]; Rikke Buhl[1]

[1]Department of Large Animal Sciences, Faculty of Health and Medical Sciences, University of Copenhagen, Taastrup, Hovedstaden, Denmark; [2]Department of Biomedical Sciences, Faculty of Health and Medical Sciences, University of Copenhagen, Copenhagen N, Hovedstaden, Denmark; [3]Department of Cardiology, The Heart Centre, Copenhagen University Hospital, Copenhagen, Hovedstaden, Denmark

Atrial fibrillation (AF) is the most common clinical arrhythmia in horses. Only a limited number of drugs are available for pharmacological treatment and their use can be a challenge due to adverse reactions. Flecainide has been tested in equine AF with variable efficacy. The purpose of this study was to test whether the efficacy of flecainide is dependent on AF duration.

Six AF horses and three controls were included in the study. The study was performed during a period of 55 days. On day 0 all horses had flecainide (2 mg/kg IV) administered in sinus rhythm for baseline measurements. AF was then induced in the AF group by means of pacemaker stimulation in the right atrium. On days 3, 9, 27 and 55 flecainide was administered again to all horses.

All horses in AF cardioverted to sinus rhythm on day 3 and 9. Five out of six horses cardioverted on day 27 whereas only 2/6 horses cardioverted on day 55. Flecainide infusion prolonged QRS duration from 0.10±0.01 s to 0.13±0.02 s ($p<0.001$) and the corrected QT interval from 0.44±0.03 s to 0.49±0.04 s ($p<0.001$). Four out of six horses in the AF group experienced ventricular arrhythmias following flecainide infusion on days 9, 27 and/or 55 whereas none of the controls experienced side effects.

Flecainide efficacy decreases with the length of AF duration but could be used in horses with AF of recent onset. The drug can cause ventricular arrhythmias and should be used with caution.

E05

Beat-to-Beat Heart Rate Variability During Ridden Exercise in Healthy Horses (Abstract E05)

Patricia Egli[1]; Olivia Lorello[2]; Cristobal Navas de Solis[2]; Colin C. Schwarzwald[3]; Katharyn J. Mitchell[4]

[1]Vetsuisse Faculty, University of Zurich, Zurich, Switzerland; [2]Department of Veterinary Clinical Sciences, University of Bern, Bern, Switzerland; [3]Clinic for Equine Internal Medicine, Vetsuisse Faculty, University of Zurich, Zurich, Switzerland; [4]University of Zurich, Zurich, Switzerland

The aim was to quantify beat-to-beat heart rate variability (HRV) in healthy horses during a standardised ridden exercise test (SRET) and determine if level of fitness influences the HRV.

SRETs were performed in 13 horses used for pleasure riding (Group P) and 23 used for medium level eventing competitions (Group E). Velocity, HR, blood lactate concentration and ECGs were recorded. The ECGs were corrected for artefacts and arrhythmias noted. HRV analysis was performed on the gallop segments and percent (%) beat-to-beat variability was calculated. Performance indices (velocity at blood lactate 4 mmol/L [V_4] and velocity at HR 200 min^{-1} [V_{200}]) were calculated.

Group P had lower V_{200} and V_4 compared with Group E (mean V_{200}±SD: Group P: 10±1 m/s, E: 11±0.78 m/s, p=0.0372; mean V_4±SD: Group P: 8.3±0.77 m/s, E: 9.7±0.73 m/s, p<0.0001). Four Group P and 4 Group E horses had arrhythmias during the gallop phase of the exercise test. The % beat to beat variation during the gallop phase was very low (median [1% and 99% percentiles] Group P: 0% [-3.5 to 3.3%], E: 0% [-3.6 to 3.2%]) and there was no difference between groups (p>0.9999).

Horses used for pleasure riding had lower performance indicators (V_{200}, V_4) than horses used for higher level eventing competition. Fitness level did not influence the very low beat-to-beat variability seen in this population of horses. Arrhythmias detected in healthy horses during ridden exercise were infrequent and did not influence beat-to-beat HRV. This group of ridden horses had similar-beat to-beat HRV to that seen in an additional study (using a different data set) during exercise on a high-speed treadmill.

E06

Corticotropin Releasing Hormone in Foals with Adrenocortical Insufficiency (Abstract E06)

Katarzyna Dembek[1]; Stephanie Vijan[2]; Barry David[3]; Bonnie Barr[4]; Teresa Burns[2]; Ramiro Toribio[5]

[1]College of Veterinary Medicine, Iowa State University, Ames, IA, USA; [2]The Ohio State University, Columbus, OH, USA; [3]Hagyard Equine Medical Institute, Lexington, KY, USA; [4]Rood and Riddle Equine Hospital, Lexington, KY, USA; [5]Department of Veterinary Clinical Sciences, The Ohio State University, Columbus, OH, USA

Dysfunction of the hypothalamic-pituitary-adrenal axis (HPAA), manifested as relative adrenal insufficiency (RAI) or critical illness related corticosteroids insufficiency (CIRCI) has been associated with sepsis and mortality in newborn foals. HPAA dysfunction at the adrenal gland (cortisol) level is well documented in critically ill foals, but information at the hypothalamic (corticotropin releasing hormone [CRH]) and pituitary (adrenocorticotropin [ACTH]) levels is lacking. We hypothesized that in most septic foals the HPAA will be activated at all three levels and the degree of activation will be linked to disease severity and mortality. We proposed that CRH and ACTH deficiency will occur in few critically ill foals and will be associated with disease severity and mortality.

Blood samples were collected on admission from 37 septic (sepsis score >12), 38 sick non-septic (SNS), and 18 healthy foals of <4 days of age. Blood concentrations of CRH, ACTH, and cortisol were determined by radioimmunoassays. CRH and ACTH deficiency was defined as values below the upper level of the 95% confidence interval from values of healthy foals. Hormone values in the normal range in critically ill foals were considered inappropriate.

CRH, ACTH and cortisol concentrations were higher in septic and SNS than healthy foals ($p<0.01$). Nonsurviving foals had higher CRH and ACTH concentrations than survivors ($p<0.05$). Twenty one percent, 11% and 13% of septic foals had CRH, ACTH and cortisol deficiency, respectively. Septic foals with hypocortisolemia also had decreased CRH and ACTH concentrations but increased CRH/ACTH ratios compared to septic foals with adequate cortisol release ($p<0.01$).

All components of the HPAA were activated in most critically ill foals. Some septic foals demonstrated HPAA dysfunction with hypocortisolemia as well as ACTH, and CRH insufficiency. High CRH/ACTH ratio in septic foals with hypocortisolemia suggested the pituitary dysfunction in addition to adrenocortical insufficiency.

E07

Immunohistochemical Expression of Insulin, Glucagon and Somatostatin in Pancreatic Islets from Insulin Resistant Horses (Abstract E07)

Kim Newkirk[1]; Gordon Ehrensing[1]; Agricola Odoi[1]; **Nicholas Frank**[2]

[1]College of Veterinary Medicine, The University of Tennessee, Knoxville, TN, USA; [2]Cummings School of Veterinary Medicine, Tufts University, North Grafton, MA, USA

Hyperinsulinemia and insulin resistance are associated conditions in horses and insulin secretion increases as tissue insulin sensitivity decreases. We hypothesized that insulin-resistant (IR) horses would have increased insulin staining of pancreatic islets when compared to insulin-sensitive (IS) horses.

Twenty-three horses were assigned to IS (n=13) and IR (n=10) groups on the basis of frequently-sampled intravenous glucose tolerance tests and minimal model analysis, or fasted insulin concentrations. Humane euthanasia was performed for reasons other than this study, and pancreas samples were obtained for immunohistochemical analysis. Pancreas tissues were stained for insulin, glucagon and somatostatin and digital images were analyzed to determine expression of each hormone relative to total islet area.

No significant differences in insulin staining were detected between groups and our hypothesis was not supported, but IR horses had a significantly less glucagon within pancreatic islets than IS horses. Median (interquartile range) percentage of total islet area staining positive for glucagon was 12% (5%, 13%) of total islet area in IR horses, compared to 18% (16%, 22%) in IS horses (p=0.001). Immunoreactive glucagon, insulin, and somatostatin were present in approximately 15%, 63%, and 8% of total islet area, respectively.

It was concluded that insulin expression did not differ between IR and IS groups and there was no evidence that insulin resistance increases insulin secretion within pancreatic islets. Lower glucagon expression in IR horses may be a result of compensatory down-regulation of hormone secretion in response to hyperinsulinemia or hyperglycemia.

E08

Equine Breed Specific Insulin/Glucose Dynamics During Frequently Sampled Intravenous Glucose Tolerance and Oral Sugar Tests (Abstract E08)

Jane M. Manfredi[1]; Raymond J. Geor[2]; Patty Sue D. Weber[1]; L. Jill McCutcheon[2]; Molly E. McCue[3]

[1]Michigan State University, East Lansing, MI, USA; [2]Massey University, Palmerston North, Auckland, New Zealand; [3]University of Minnesota, St. Paul, MN, USA

Insulin dysregulation (ID), specifically hyperinsulinemia, has been identified as a cause of laminitis. Certain breeds are seemingly more susceptible. Understanding breed differences in insulin responses and measures of lipid metabolism/adipokines may be critical for identifying truly "at risk" individuals. Insulin-modified frequently sampled intravenous glucose tolerance tests (FSIGTT) (n=90) and oral sugar tests (OST) (n=82) were performed along with assessment of markers for lipid metabolism/adipokines (n=90) in 5 breeds (Quarter Horses [QH], Arabians, Morgans, Welsh Ponies [WP], and Thoroughbreds). Minimal model analyses of the FSIGTT for insulin sensitivity (SI), the acute insulin response to glucose (AIRg), disposition index (DI), glucose mediated glucose disposal (Sg), the lowest glucose value (Gmin), and the deflection of glucose below baseline (dGB) were assessed. OST insulin thresholds, trajectories, and area under the curve (AUC) were assessed. Statistics included: multilevel regression analysis, regression modeling of time trajectories, one-way ANOVA, Kruskal-Wallis test, ROC curve analysis (significant at $p<0.05$). QH had higher SI than all other breeds, lower AIRg than WP and Arabians, and higher DI than Morgans. Arabians had higher AIRg than Morgans, and lower Sg than WP. Morgans had lower AIRg than WPs. Arabians, Morgans, and WP had lower Gmins than Thoroughbreds. Morgans had a greater dGB than QHs. Different OST insulin thresholds for ID existed between breeds. OST glucose and insulin trajectories and AUC were lowest in QH. Significant breed differences existed in nonesterified fatty acids, high molecular weight adiponectin, leptin, and triglycerides. Breed specific differences are important to consider when evaluating metabolic health.

E09

Equine Breed Histologic Differences in Adipocytes and Muscle Related to Insulin Dynamics and Body Fat (Abstract E09)

Jane M. Manfredi[1]; Raymond J. Geor[2]; Patty Sue D. Weber[1]; Molly E. McCue[3]; L. Jill McCutcheon[2]

[1]Michigan State University, East Lansing, MI, USA; [2]Massey University, Palmerston North, Auckland, New Zealand; [3]University of Minnesota, St. Paul, MN, USA

Insulin dysregulation, specifically hyperinsulinemia, has been identified as a cause of laminitis. Muscle and adipose tissue have large roles in insulin regulation and in the pathology of human metabolic syndrome, but these roles have not been well interrogated in horses. Tailhead adipose tissue (n=76) and middle gluteal muscle (n=28) biopsies were performed (4 equine breeds: Quarter Horses [QH]; Arabians, Morgans, Welsh Ponies [WP]) to relate adipocyte area (AA) and muscle fiber type percent area and proportions to body condition score (BCS), total body fat mass (TBFM) (using the deuterium dilution technique) and minimal model parameters from a frequently sampled intravenous glucose tolerance test (FSIGTT). Statistics included: Kolmogorov-Smirnov analysis, MANOVA, and Spearman correlation. Significance was set at $p<0.05$. Overall BCS was weakly to moderately correlated to insulin sensitivity (SI), the acute insulin response to glucose (AIRg), and AA. TBFM was not correlated to SI and weakly correlated to AIRg and AA overall, but moderately correlated to AA in QH. QH had a significantly smaller AA than Arabians and WP but not Morgans. AA was weakly related to SI and moderately to AIRg. Baseline insulin concentrations were moderately correlated to Type 1 muscle fiber percent area and proportion. No breed differences existed between muscle fiber type area or proportion. There were breed differences in adipocyte, but not muscle histology. The weak correlations between BCS, TBFM and SI and AIRg suggest that adiposity may not be a key factor in determining metabolic fitness in horses.

E10

Shared Genetic Loci Underlying Equine Metabolic Syndrome in Welsh Ponies and Morgan Horses (Abstract E10)

Elaine Norton[1]; Nichol Schulz[1]; Raymond J. Geor[2]; James Mickelson[3]; Molly E. McCue[1]

[1]University of Minnesota, St. Paul, MN, USA; [2]Massey University, Palmerston North, Auckland, New Zealand; [3]College of Veterinary Medicine, University of Minnesota, St. Paul, MN, USA

Equine metabolic syndrome (EMS), a clustering of metabolic disturbances resulting in insulin dysregulatoin and derangements in fat metabolism, is the most common cause of laminitis. Previous work from our lab has confirmed that the pathophysiology of EMS is complex, with both genetic and environmental factors contributing to the phenotypic variability in metabolic traits. The objective of this project was to identify genomic regions contributing to EMS by performing a genome-wide association study (GWAS) in a cohort of 232 Welsh ponies and 286 Morgan horses phenotyped for 11 metabolic traits. Genotyping was performed on one of three SNP arrays (54,000, 670,000 to 1,800,000 SNPs); genotype imputation allowed for generation of a uniform set of makers across platforms (~1.8 million SNPs). GWAS was performed using a mixed linear regression model with sex and age included as covariates. Within breeds, a total of 67 (Welsh ponies) and 116 (Morgans) significant loci were identified for the metabolic traits, of which 4 were shared between breeds. Loci were defined as shared if SNPs with a p-value <1.00e-05 were within a 500 kb window of each other. Specifically, shared loci were identified on ECA10, ECA18, ECA20, and ECA3 for insulin levels post oral sugar test, and baseline non-esterified fatty acids, triglycerides, and ACTH, respectively, with a total of 75 candidate genes identified. These data confirm that EMS is a polygenic trait with risk loci unique to individual breeds as well as those shared between breeds. Future directions include interrogation of these regions through whole genome sequencing.

E11

Effect of Age and Diet on Insulin and Glucose Dynamics in Horses (Abstract E11)

Sarah Jacob[1]; Raymond J. Geor[2]; Patricia Harris[3]; Molly E. McCue[4]

[1]Michigan State University, East Lansing, MI, USA; [2]Massey University, Palmerston North, Auckland, New Zealand; [3]WALTHAM Centre for Pet Nutrition, Leicestershire, England, UK; [4]University of Minnesota, St. Paul, MN, USA

The relationship between insulin dysregulation, dietary adaptation and aging is poorly understood yet critical to making dietary recommendations for horses at risk of insulin dysregulation and associated laminitis. In this study, the effect of age and diet on insulin and glucose dynamics was examined in 16 healthy Thoroughbred and Standardbred horses, divided into two groups: Adult (8.8±2.9 years; mean±SD) and Aged (20.6±2.1 years). Four concentrate isocaloric diets: Base (low starch, low sugar), Starch (base plus kibbled corn), Fiber (base plus beet pulp and soybean hulls), and Sugar (base plus dextrose powder) were each fed for seven weeks (balanced four-way crossover). Horses were group fed low non-structural carbohydrate hay once daily, and individually fed concentrate meals twice daily. In the seventh week, insulin-modified frequently sampled intravenous glucose tolerance tests and oral sugar tests (OST) (0.25 mL/kg corn syrup) were performed. Data were analyzed with multivariable linear mixed regression (significance set at $p \leq 0.05$). Acute insulin response to glucose (AIRg), OST_peak insulin, insulin area under the curve, and $OST\_insulin_{75\ minutes}$ were higher in Aged than Adult independent of diet. Aged had higher AIRg and disposition index on Fiber and Sugar, and glucose effectiveness on Sugar than Adults adapted to the same diet. Adults had lower $OST\_insulin_{basal}$ on Fiber compared to Sugar. Aged horses had higher OST_peak insulin than Adults on Fiber. Age and diet had an influence on insulin and glucose dynamics, and should be considered when evaluating a horse's metabolic status.

E12

Fecal Bile Acids, Sterols and Fatty Acid Concentrations in Healthy and Obese Horses (Abstract E12)

Rodolfo G. Madrigal[1]; Michelle Coleman[1]; Noah Cohen[1]; Jan S. Suchodolski[2], Blake C. Guard[2]; Julia B. Honneffer[2]

[1]College of Veterinary Medicine, Texas A&M University, College Station, TX, USA; [2]Gastrointestinal Laboratory, Department of Small Animal Clinical Sciences, College of Veterinary Medicine, Texas A&M University, College Station, TX, USA

Diagnosing and managing horses with metabolic disorders including obesity is challenging due to a poorly understood pathogenesis and lack of well-characterized risk factors and therapeutic options. In other species, differences in fecal bile acids (BA), fatty acids (FA), and sterols have been recognized in metabolic disease; however, these have not been reported for horses. Thus, the objective of this study was to compare fecal BA, sterols, and FA between obese and non-obese horses.

Fecal samples were collected from obese horses with a body condition score (BCS) of ≥7 and healthy horses with a BCS of ≥3 and ≤5. Fecal BA, FA, and sterol concentrations were obtained using gas chromatography coupled with mass spectrometry. Variables were assessed for normality and analyzed using t-test or Wilcoxon rank sum tests to assess differences between groups.

A total of 40 samples were analyzed, including 21 obese horses and 19 healthy horses. Signalment was similar between groups. Seven FA, 9 sterols and 5 BA were analyzed. No statistical difference in bile acids was noted between groups. Results showed statistically significantly lower concentrations of linoleic acid (p=0.024), oleic acid (p=0.006), stearic acid (p=0.009) and gondoic acid (p=0.021) and increased cholesterol concentration (p=0.020) in obese horses.

Understanding the role of fecal FA and sterols in horses may improve our understanding of the pathogenesis of metabolic disease and lead to the development diagnostic tests or novel therapeutic agents.

E13

Eosinophilic Peritonitis in 8 Horses (1995–2015) - (Abstract E13)

Fauna Leah Smith[1]; K Gary Magdesian[2]; Mary Beth Whitcomb[3]

[1]William R. Pritchard Veterinary Medical Teaching Hospital, University of California-Davis, Davis, CA, USA; [2]Department of Medicine and Epidemiology, School of Veterinary Medicine, University of California-Davis, Davis, CA, USA; [3]Department of Surgical and Radiological Sciences, School of Veterinary Medicine, University of California-Davis, Davis, CA, USA

Eosinophilic peritonitis has previously not been described in horses. The objective of this study was to identify clinicopathologic features, treatment and outcome of horses with eosinophilic peritonitis.

A retrospective study of medical records (1995–2015) at the Veterinary Medical Teaching Hospital, University of California-Davis identified 8 horses with ≥10000/µl total nucleated cell count and ≥1000 eosinophils/µl in abdominal fluid samples.

Arabians were overrepresented (p=0.01, OR 8.6 [CI: 2.1-34.4]). The most common presenting complaint was colic (6/8). The median peritoneal total nucleated cell and eosinophil counts were 21,040/µl (range 10,000–75,000/µl) and 3786 cells/µl (range 1329–35,343/µl), respectively. Peritoneal fluid total protein concentration was increased (median 4.5, range 2.3–6.7 g/dl). All horses had peripheral eosinophilia (964, range 540–3144 cells/µl). Abdominal ultrasonography revealed gastrointestinal thickening (5/5) and lymphadenopathy (4/5). Microscopic evidence of eosinophilic infiltration of abdominal organs was present (5/6). Quantitative fecal egg counts were negative in 5/6 horses.

Treatment included anthelmintics (8/8), antimicrobials (7/8), corticosteroids (4/8), flunixin meglumine (3/8) and cetirizine (1/8).

All horses were discharged. Two horses were euthanized for colic 1 and 9 months after discharge. Five horses were alive at follow up (2–24 months). Horses treated with corticosteroids with serial abdominocenteses and CBC had improvement of the peritonitis and eosinophilia (3/3). Only 1/3 horses not treated with corticosteroids with serial abdominocenteses had improvement in abdominal counts, and 2/2 did not have improvement of eosinophilia.

Eosinophilic peritonitis is a newly described hypereosinophilic syndrome of horses, which may involve abdominal organ infiltration. Treatment with corticosteroids may result in remission.

E14

The Effects of Metronidazole on the Equine Cecal and Fecal Microbiome and Metabolome (Abstract E14)

Carolyn E. Arnold[1]; Anitha Isaiah[2]; Josie Coverdale[3]; Jörg M. Steiner[2]; Jan S. Suchodolski[2]

[1]Department of Large Animal Clinical Sciences/Gastrointestinal Laboratory, Texas A&M University, College Station, TX, USA; [2]Gastrointestinal Laboratory, College of Veterinary Medicine, Texas A&M University, College Station, TX, USA; [3]Department of Animal Science, Texas A&M University, College Station, TX, USA

Antimicrobial administration is considered a risk factor for colitis in horses. The subsequent alteration of the resident population of gastrointestinal bacteria may result in dysmetabolism. The purpose of this study was to investigate the effects of metronidazole, a commonly prescribed antibiotic for anaerobic infections, on the cecal and fecal microbiome and metabolome of horses.

Metronidazole (15 mg/kg PO q 12 h) was administered to adult horses (n=5) with indwelling cecal cannulas (AUP 2013-0123). Fecal and cecal samples were taken on days -52, -28, -14, 0, 3 (fecal only), 7, and 14. Illumina sequencing of 16S rRNA genes was performed on extracted DNA. Data were analyzed using Quantitative Insights Into Molecular Ecology (QIIME) and analysis of similarities (ANOSIM) was performed. Univariate analysis of bacterial groups was performed using a Friedman's test and Dunn's post-test. Linear discriminant analysis effects size (LEfSe) was used to find bacterial taxa associated with study day. Short chain fatty acids (SCFA) and untargeted metabolomics were analyzed using mass spectrometry platforms.

Alpha diversity was significantly decreased between days -28 and 14 for cecal samples (Chao1, p=0.043; Observed OTUs, p=0.001; Shannon, p=0.002) and between days 0 and 3 for fecal samples (Shannon, p=0.021). Beta diversity, as measured by weighted Unifrac distances, showed significant differences between days -52, -28, -14, 3 and 7 in cecal and fecal samples (ANOSIM, p=0.001). The LEfSe analysis revealed that several bacterial taxa (*Christensenellaceae*, *Campylobacteriaceae*, and *Spirochaetaceae*) were significantly altered due to antibiotic administration. SCFA were not significantly affected by metronidazole. Of the 553 metabolites analyzed, 223 were named and 330 were unnamed. Analysis of the named fecal metabolites found 100 that were significantly different amongst study days 0, 3, 7, and 14. Those metabolites that were most statistically significant ($p<0.05$, $q=0.041$) were active in the biosynthesis of fatty acids, steroids, proteins, carbohydrates, sucrose, and starches. Of the 223 named cecal metabolites, 93 were statistically different on study days 0, 7, and 14 ($p<0.05$) but failed to remain significant after adjustment for multiple comparisons ($q=0.091$). In conclusion, metronidazole administration changed the cecal and fecal microbiome and metabolome in horses.

E15

Effect of Age and Cortisol on Equine Dendritic Cells After Exposure to Bacteria *ex Vivo* (Abstract E15)

Brina Gorham; David Hurley; Steeve Giguere; Natalie Norton; Shyla Giancola; Kelsey Hart

College of Veterinary Medicine, University of Georgia, Athens, GA, USA

Dendritic cells (DCs) are important in the immune response to bacteria, but foal DC function is not well characterized. Monocyte-derived DCs (MoDCs) were generated from 8 foals at 1, 7 and 30 days-of-age and 9 adult horses. MoDCs were incubated for 48 hours with or without killed *Escherichia coli* or *Staphlococcous aureus* antigen in the presence or absence of cortisol (15 μg/dl). MoDC expression of MHCII, CD86, and CD14 was measured with flow cytometry, and supernatant cytokine concentrations (IL-4, IL-17, IFN-gamma, IFN-alpha, and IL-10) were quantified with a fluorescent bead-based immunoassay. Effects of age and cortisol on MoDC variables were determined with mixed-effects linear models (significance $p<0.05$). The percentage of cells expressing MHCII and CD86 was lower and CD14 was higher in 1-day-old-foal MoDCs compared to horse MoDCs ($p\leq0.006$). Bacterial exposure alone did not alter surface marker expression in foal or horse cells, but exposure to bacteria with cortisol further reduced MHCII expression in foal cells ($p<0.001$). Following bacterial exposure, 1-day-old-foal MoDCs produced significantly less IL-4, IL-17, IFN-gamma, and IFN-alpha and more IL-10 than horse MoDCs ($p<0.001$). Following bacterial exposure, 1-day-old-foal MoDCs produced significantly less IL-4, IL-17, IFN-gamma, and IFN-alpha and more IL-10 than horse MoDCs ($p<0.001$). Foal MoDCs exhibit phenotypic and functional immaturity that persists during the first month of life and is compounded by exposure to cortisol. Thus, foal DCs may be less effective at activating lymphocyte responses, which could contribute to foals' increased susceptibility to bacterial sepsis.

E16

Blood Neutrophil Activation Markers in Equine Asthma Syndrome and Impact of Dexamethasone (Abstract E16)

Nicolas Herteman[1]; Carolyn Grimes[2]; Jean-Pierre Lavoie[3]

[1]University of Zürich, Zürich, Switzerland; [2]Département de Pathologie et Microbiologie, Faculté de Médecine Vétérinaire, Université de Montréal, Saint-Hyacinthe, QC, Canada; [3]Equine Internal Medicine Professor, Université de Montréal, Saint-Hyacinthe, QC, Canada

Pulmonary neutrophilia is a characteristic finding in severe equine asthma (SEA or heaves) and is also common in moderate equine asthma (MEA, or inflammatory airway disease). MPXI, NeutX and NeutY are blood neutrophil activation parameters (activity of myeloperoxidase and neutrophils' size) automatically provided by the ADVIA 120 Hematology Analyzer. The aim of this study was to evaluate these indices in healthy and asthmatic horses, and how corticosteroids can affect them.

We retrospectively studied MPXI, NeutX and NeutY in the blood of 26 with MEA, 14 horses with SEA, and 12 controls. These parameters were also studied in horses with SEA and 5 controls after the administration of dexamethasone (0.06 mg/kg, PO, SID). Blood was collected by jugular venipuncture in EDTA vacutainers and analyzed with ADVIA 120 Hematology Analyzer. Data were compared using Student t-tests and one way ANOVA.

All three parameters were significantly increased in the blood of horses with SEA when compared to controls and to MEA. Dexamethasone significantly decreased NeutX values in SEA and in controls, but MPXI and NeutY were only significantly reduced in SEA.

This study describes for the first time the use of MPXI, NeutX and NeutY in equine asthma. Our results indicate that neutrophils are more activated in SEA than in MEA and controls. However, the diagnostic value of these parameters for equine asthma is limited due to data's overlapping between the groups. Furthermore, the effects of dexamethasone on MPXI, NeutX and NeutY should be taken into account when monitoring the parameters in horses.

E17

Use of Serum Amyloid A to Differentiate Between Infectious and Non-Infectious Diseases in Horses (Abstract E17)

Fiona M. Wensley[1]; Kaitlyn James[2]; Julie Burges[3]; Danielle Carrade-Holt[3]; Nicola Pusterla[1]

[1]Department of Medicine and Epidemiology, School of Veterinary Medicine, University of California-Davis, Davis, CA, USA; [2]UC Davis Graduate Group in Epidemiology, Boston, MA, USA; [3]Clinical Diagnostic Laboratories, School of Veterinary Medicine, University of California-Davis, Davis, CA, USA

Stall-side serum amyloid A (SAA) assays have recently increased in availability resulting in a more widespread use of SAA to identify disease processes requiring further investigation and diagnostic modalities. Treatment decisions are often made on the basis of a high or change in SAA value. The objectives of this study were to improve correlations of serum amyloid A with plasma fibrinogen and neutrophil count, and to assess the use of SAA as a screening tool to identify infectious disease.

A randomized study on 314 equine patients presented to the UC Davis Veterinary Medical Teaching Hospital with the inclusion criteria of a complete blood count (CBC) submission. SAA samples were run on whole blood using the StableLab Lateral Flow Immunoassay. The horses were classified into 5 groups using clinical and diagnostic modalities from the VMTH Medical records, with no reference to the CBC or SAA results to aid in classification. The 5 groups included healthy (n=36); healthy recently vaccinated (n=10); inflammatory, non-infectious (n=106); inflammatory, infectious (n=121); and non-healthy, non-inflammatory (n=41). The median SAA value for inflammatory non-infectious and inflammatory infectious were significantly different at 9.5 mg/mL and 181 mg/mL, respectively. To refine the model, the SAA values were categorized into normal, low, medium and high ranges. SAA was normal for 86% of the healthy group and for 80% of the non-healthy, non-inflammatory group. High elevations between 1,000 and 3,000 mg/mL were mainly distributed in the inflammatory infectious group (78%), with none in the healthy group. SAA values in the medium or high range correlated well with abnormalities to the white blood cell count and fibrinogen.

The data set was further statistically analyzed amongst only the three unhealthy groups. The inflammatory infectious (n=121) compared with the two groups of non-infectious (n=147). When comparing the infectious to the non-infectious groups, the SAA value was 2.6 times more likely to be within the medium range and 8 times more likely to be within the high range of 1,000–3,000 mg/mL. The sensitivity of SAA alone was 60% with an overall accuracy of 46%. The combination of an elevated SAA, abnormal neutrophil count and hyperfibrinogenemia, further improves the sensitivity (70%) and overall accuracy (59%) of correctly identifying an underlying infectious inflammatory process.

In conclusion, Serum Amyloid A performance held a greater overall accuracy when compared to neutrophil count and fibrinogen, both individually and combined, for defining infectious versus non-infectious diseases. When used in conjunction with neutrophil count and fibrinogen, SAA further enhances the accuracy of differentiating disease status. The stall side assay was found to be an excellent screening tool however, a normal SAA does not rule out inflammation, nor is a high SAA definitive of infection, as with all hematological parameters, there is high variability among individuals.

E18

Fecal Shedding Prevalence of Equine Coronavirus in Hospitalized Horses (Abstract E18)

So young Kwon[1]; Jenifer R. Gold[2]; James Evermann[3]; Fairfield Bain[4]; Macarena Sanz[2]

[1]Washington State University, Pullman, WA, USA; [2]Department of Veterinary Clinical Sciences, Washington State University, Pullman, WA, USA; [3]Washington Animal Disease and Diagnostics Laboratory, Washington State University, Pullman, WA, USA; [4]Merck Animal Health, Lexington, KY, USA

Coronavirus infection has been recognized in foals and adult horses with pyrogenic and enteric disease but its significance as an etiologic agent remains unclear. Recently Equine Coronavirus (ECoV) has been shown in feces of horses without clinical illness suggesting that inapparent carriers may shed the virus.

The objective of this current, ongoing study is to investigate the presence of ECoV in feces of horses hospitalized for enteric disease (n=35) or orthopedic conditions (n=35). Fecal samples are collected at admission and 48–72 h thereafter and the presence of ECoV is evaluated using electron microscopy (EM) and reverse transcription polymerase (RT-PCR). Categorical analyses will be performed using Kruskal-Wallis test to determine the association between observations (age, breed, sex, clinical signs, clinicopathological data, EM and RT-PCR results).

Our preliminary data shows 17% (3/18) of horses with enteric disease were positive for ECoV during hospitalization. These horses had the following clinicopathologic abnormalities; fever (2/3), elevated liver values (1/3), hepatic lipidosis (1/3), and sand enteropathy (1/3). So far, ECoV has not been detected in the orthopedic group (n=16). Interestingly, a discrepancy between EM and RT-PCR results has been noted in our ongoing study and is currently being investigated further.

The results of the current study show that ECoV is detected in a small percentage of horses with enteric disease and is not a common finding in horses with orthopedic conditions. Also, inconsistent agreement between EM and RT-PCR results emphasizes diagnosis of ECoV is confounded by limited sensitivity and specificity data of currently available diagnostic tests.

E19

Investigation of an Experimental Infection Model of Equine Coronavirus in Adult Horses (Abstract E19)

Emily A. Schaefer[1]; Samantha Barnum[2]; Corey Harms[3]; Molly Viner[3]; Nicola Pusterla[2]

[1]University of California-Davis, Davis, CA, USA; [2]Department of Medicine and Epidemiology, School of Veterinary Medicine, University of California-Davis, Davis, CA, USA; [3]William R. Pritchard Veterinary Medical Teaching Hospital, School of Veterinary Medicine, University of California-Davis, Davis, CA, USA

To investigate the clinical, hematological, molecular and serological parameters of adult horses experimentally infected with ECoV.

Four horses were experimentally infected with frozen feces containing approximately 10^9 genome equivalents of ECoV/g of feces collected from the intestinal lumen of a horse diagnosed with ECoV at necropsy. Four additional horses were exposed daily to the feces from the experimentally challenged horses. Monitoring included twice daily physical examinations, daily nasal swab, blood, and fecal collection for quantitative molecular detection of ECoV. Blood was collected every other day for hematologic analysis and prior to experimental infection and weekly thereafter for serological analysis.

All eight horses showed evidence of fecal shedding of ECoV by qPCR. Six of the eight horses (75%) exhibited mild, clinical disease of soft-formed manure; only one horse exhibited transient pyrexia. All horses maintained a total white cell count within normal limits, though 3 horses developed a mild, transient lymphopenia. There were no statistically significant differences ($p=0.2$) in quantity of fecal shedding between experimentally infected horses and horses exposed to ECoV-containing feces.

The experimental infection of adult horses with ECoV was associated with mild and self-limiting clinical signs, transient lymphopenia, and fecal shedding of ECoV. The experimental infection mimics natural infection, where morbidity rates vary between 20–80%. Although the number of horses in each group was relatively small, there were no differences between experimentally infected horses and horses being exposed to ECoV-containing feces. The study supports a fecal-oral route of transmission.

E20

Investigation of Shedding Frequency of Respiratory Pathogens in Horses Recently Imported to the United States (Abstract E20)

Fauna Leah Smith[1]; Nicola Pusterla[2]; Joie Watson[2]; Isabelle Kilcoyne[3]; Claudia Sonder[4]

[1]William R. Pritchard Veterinary Medical Teaching Hospital, University of California-Davis, Davis, CA, USA; [2]Department of Medicine and Epidemiology, School of Veterinary Medicine, University of California-Davis, Davis, CA, USA; [3]Department of Surgical and Radiological Sciences, School of Veterinary Medicine, University of California-Davis, Davis, CA, USA; [4]Center for Equine Health, School of Veterinary Medicine, University of California-Davis, Davis, CA, USA

Imported horses may represent a significant risk for spreading infectious respiratory pathogens into equine populations. They are significantly stressed, and have been comingled with other horses increasing the likelihood of viral recrudescence and novel exposure. The study objective was to investigate the frequency of shedding of respiratory pathogens in recently imported horses.

168 horses undergoing CEM quarantine between October 2014 and June 2016 had a veterinary exam and nasal secretions collected at the time of entry and if they developed acute respiratory signs during quarantine. Samples were assayed for equine influenza virus (EIV), equine herpes virus 1, 2, 4, and 5 (EHV-1, -2, -4, -5), equine rhinitis virus A and B (ERVs) and *Streptococcus equi* subspecies *equi* using qPCR.

EHV-2 and EHV-5 were found in 28.6% and 42.2% of nasal swabs, respectively. EHV-4 was identified in 4 horses and EHV-1 (N_{752}) was detected in 2 horses. 34 horses were qPCR positive for multiple EHV types. All EHV-1 qPCR positive horses were symptomatic at sampling. Horses that were qPCR-positive for EHV-4, -2 and -5 were asymptomatic 57.1% 52.8% and 59.0% of the time, respectively. Clinical respiratory abnormalities were seen 49.1% of qPCR negative horses. EIV and ERVs were not detected in any samples; however, 25% of samples failed quality control. *S. equi* was not recovered from any horse.

Clinical signs were poorly correlated with being qPCR negative or qPCR positive for EHV-2, -4, -5. All EHV-1 positive horses were clinically affected. While EHV-2 and EHV-5 were commonly shed, their significance in acute respiratory disease is questionable. EHV-1 or EHV-4 were shed by 3.6% of horses, which could have serious consequences if such animals were introduced into a large mixed population of susceptible horses upon entry into the USA.

E21

Does Vaccinating Mares Prior to Parturition Increase Colostral Antibody Transfer to Foals? (Abstract E21)

John Haffner[1]; Steven Grubbs[2]; Jeremy Carlton[3]; Amanda Adams[4]; Stephanie Reedy[4]

[1]Middle Tennessee State University, Murfreesboro, TN, USA; [2]Boehringer Ingelheim, Stewartsville, MO, USA; [3]MTSU Horse Science Center, Murfreesboro, TN, USA; [4]Gluck Equine Research Center, Lexington, KY, USA

Foals are immunologically naïve at birth, meaning they have had no exposure to foreign antigens and have therefore not yet mounted any type of protective immune response or accumulated significant levels of immunoglobulins (Ig). Ingestion and absorption of immunoglobulin-rich colostrum are the sole means of passive transfer in foals providing temporary protection from infection for the first few months of life. Current recommendations for optimizing colostral antibody transfer from mares to foals includes vaccination of the mare 4 to 6 weeks prior to parturition. This practice has been recommended to increase the serum antibody level in the dam; therefore the mare would concentrate a higher level of Ig in colostrum during the final two weeks of gestation. Although this management practice is widely accepted, it has not been adequately substantiated using evidenced-based medicine from adequately controlled studies. The purpose of this study was to determine if vaccinating pregnant mares approximately 4 to 6 weeks prior to foaling increases the IgG antibody transferred to foals specific for West Nile Virus (WNV) and Equine Influenza Virus (EIV). Twenty-seven pregnant adult Tennessee Walking Horse (TWH) mares ranging in age from 5 to 23 years were enrolled in the study. All mares were confirmed in foal via rectal palpation and confirmed via ultrasound. Mares were randomized into three treatment groups: G0 naïve control group (no previous vaccinations; no vaccinations administered during the study); G1 vaccinated group (initial 2 vaccination series and revaccinated 4 to 6 weeks prior to foaling); and G2 vaccinated group (initial 2 vaccination series; no pre-foaling revaccination). Mares in G1 and G2 were administered Vetera® 4XP + WNV (eastern and western encephalitis, equine influenza clades 1 and 2 + WNV) labeled safe for use in pregnant mares. Following parturition, blood was collected and processed from each foal to determine IgG levels (passive immunity) and serum aliquots frozen at -80°C. Serum was shipped to Gluck Equine Research Center, Lexington, KY for measurement of serum hemagglutination-inhibiting (HI) equine influenza virus antibody levels against Kentucky/14 (Florida clade 1) and Ayrshire/13 (Florida clade 2) reference strains and samples were shipped to Cornell Animal Health Diagnostic Center, Ithaca, NY for analysis of WNV IgG antibody levels measured by ELISA. The mean antibody level was reported accompanied by standard deviation and 95% confidence intervals. Serological titer values below the quantifiable limit were reported as '< 10' and the value of 5 was substituted for these analyses, which provided a conservative estimate. The null hypothesis there was no vaccination effect was tested with a mixed model repeated measures methodology with subject as a random effect. All pairwise comparisons were reported between vaccination groups. Models were run separately for each reference strain. P values < 0.05 were considered statistically significant.

Foals from G1 mares (initial 2 vaccination series and revaccinated 4 to 6 weeks prior to foaling) had significantly higher WNV antibody levels (p=0.0012) compared to foals from non-vaccinated G0 mares. G1 foals also had significantly (p=0.0321) higher ELISA WNV antibody levels compared to foals from G2 mares only administered the initial two vaccination series. There was no significant difference (p=0.4429) in ELISA WNV antibody levels in foals from G2 mares and foals from unvaccinated G0 mares. Foals from G1 mares had significantly (p=0.0012) higher KY/14 (EIV Florida clade 1) antibody levels compared to foals from non-vaccinated G0 mares. Although the arithmetic means of EIV (Florida clade 1) antibody levels were 24% greater in foals from G1 mares compared to foals from G2 mares, there was no significant difference (p=0.3891). There was a significant (p=0.0389) difference in KY/14 (EIV Florida clade 1) antibody levels in foals from G2 mares compared to foals from unvaccinated G0 mares. Foals from G1 mares had significantly (p=0.0012) higher Ayrshire/13 (EIV Florida clade 2) antibody levels compared to foals from non-vaccinated G0 mares. Although the arithmetic means of Ayrshire/13 (EIV Florida clade 2) antibody levels in G1 mares were 13% greater than foals from G2 mares, no significant difference (p=0.6896) was identified. There was a significant (p=0.0430) difference in Ayrshire/13 (EIV Florida clade 2) antibody levels in foals from G2 mares compared to foals from unvaccinated G0 mares.

The foals of mares vaccinated with the initial series and revaccinated 4 to 6 weeks prior to parturition had significantly higher levels of ELISA WNV and EIV (clades 1 and 2) passively transferred antibodies 24 to 72 hours following parturition. The results of this study supports the importance and recommendation to vaccinate broodmares 4 to 6 weeks before foaling not only for their own protection, but most importantly to maximize concentrations of immunoglobulins in their colostrum to be passively transferred to their foals.

E22

Utility of Acute Phase Proteins in Equine Protozoal Myeloencephalitis and Other Equine Nervous System Diseases (Abstract E22)

Neil S. Mittelman[1]; Amy L. Johnson[2]

[1]New Bolton Center, University of Pennsylvania, Kennett Square, PA, USA; [2]New Bolton Center, School of Veterinary Medicine, University of Pennsylvania, Kennett Square, PA, USA

The purpose of this pilot study was to determine if C-reactive protein (CRP) or serum amyloid A (SAA) increase with equine protozoal myeloencephalitis (EPM) compared to other neurologic diseases. Accurate antemortem EPM diagnosis requires evidence of intrathecal antibody production (e.g. serum:CSF SnSAG2, 4/3 titer ratio < 100). Some advocate the use of acute phase proteins in addition to serology, which alone results in substantial false positives. Serum and CSF CRP and SAA were measured in 25 cases of equine neurologic disease: EPM (10), cervical vertebral stenotic myelopathy (CVSM) (10), neuroborreliosis (2), equine motor neuron disease (1), degenerative myelopathy (1), and leukoencephalomalacia (1). Nine of 10 EPM cases had serum:CSF SnSAG2, 4/3 titer ratios < 25. The untested EPM case was confirmed postmortem, as were 4 other EPM cases. Nine of the 10 CVSM cases were confirmed postmortem, and the other case had a positive myelogram and SNSAG2, 4/3 titer ratio of 200. The remaining cases were diagnosed on postmortem exam. Serum CRP was not elevated in any case. Serum SAA was only elevated in the two cases of neuroborreliosis (132 mg/L and 460 mg/L; reference 0–20 mg/L) suggesting SAA may aid its diagnosis, but further testing is warranted. Neither serum CRP nor SAA increased in cases of EPM or CVSM. Cerebrospinal fluid CRP and SAA also failed to differentiate cases of EPM (CRP median 3.35 mg/L, range 0.19–13.43 mg/L; SAA median 0.1 mg/L, range < 0.1–2.4 mg/L) from CVSM (CRP median 4.015 mg/L, range 0.16–9.62 mg/L; SAA median 0.62 mg/L, range < 0.1–2.91 mg/L).

E23

Expression and Sequence of Calcium Regulatory Micropeptides in Equine Skeletal Muscle: Implications for Exertional Rhabdomyolysis (Abstract E23)

Kaitlin M. Soave[1]; Melissa Schott[1]; Keri Gardner[1]; Carrie Finno[2]; Sudeep Perumbakkam[1]; **Stephanie Valberg**[1]

[1]Michigan State University, East Lansing, MI, USA; [2]University of California-Davis, Davis, CA, USA

Recurrent exertional rhabdomyolysis (RER) affects 5–7% of racing Thoroughbreds, Standardbreds and potentially Quarter Horses, A genetic basis has been proposed but not substantiated. Dysregulation of sarcoplasmic reticulum calcium cycling has been suggested to cause RER. We hypothesized that calcium dysregulation in RER horses could be impacted by genes encoding newly discovered micropeptides such as sarcolipin (*SLN*), myoregulin (*MRLN*), *DWORF* and phospholamban (*PLN*) that modify the activity of sarcoplasmic reticulum calcium ATPase (SERCA). The purpose of this study was to determine: 1) gene expression of *SLN, MRLN, DWORF* and *PLN* in equine muscle, 2) the coding sequence of these micropeptides and 3) potential differences in their coding sequence between controls and RER Thoroughbreds, RER Standardbreds and RER Quarter Horses. Gene expression was determined in gluteal muscle from 11 horses using Next generation sequencing. Sanger sequencing of *SLN, MRLN* and *DWORF* was performed on DNA isolated from blood or muscle tissue of a minimum of 9 RER horses per gene and 9 control horses selected from the Neuromuscular Diagnostic Laboratory.

SLN was the predominant regulatory micropeptide in gluteal muscle with several hundred fold higher gene expression than *PLB* or *MRLN*. Unlike all other studied mammalian species, equine SLN encoded a truncated protein (29 vs. 31 amino acids) missing key regulatory residues (Ser4, Thr5, Cys9, Tyr31) that are proposed to mediate controlled, reversible SERCA inhibition. In contrast, equine *MLN* had one conservative amino acid substitution, *DWORF* 2 non-conservative amino acid substitutions and *PLN* was 100% orthologous to other species. Non-synonymous mutations in *SLN, MLN* or *DWORF* were not identified in RER horses compared to control horses.

We conclude that SLN is a key regulator of SERCA in equine muscle. Conservation of *SLN*'s unique regulatory peptides in the horse suggests an evolutionary advantage - slightly elevated myoplasmic calcium - which can enhance muscle power output. RER, however, does not appear to arise from distinct mutations in *SLN, MRLN* or *DWORF*.

E24

Identification of Candidate Genes for Recurrent Exertional Rhabdomyolysis in Thoroughbreds and Standardbreds (Abstract E24)

Samantha K. Beeson[1]; James Mickelson[1]; Matthew Binns[2]; Sarah Blott[3]; Paul Caputo[4]; Cajsa Isgren[5]; Alison Moore[6]; Richard Piercy[7]; June Swinburne[8]; Mark Vaudin[9]; Molly E. McCue[10]

[1]College of Veterinary Medicine, University of Minnesota, St. Paul, MN, USA; [2]Equine Analysis Systems, LLC, Midway, KY, USA; [3]School of Veterinary Medicine and Science, University of Nottingham, Leicestershire, UK; [4]Pompano Beach, FL, USA; [5]University of Liverpool, Wirral, UK; [6]Moore Equine Services, Cambridge, AB, Canada; [7]The Royal Veterinary College, Hertfordshire, UK; [8]Animal DNA Diagnostics LTD, Cambridge, UK; [9]Animal Health Trust, Newmarket, UK; [10]University of Minnesota, St. Paul, MN, USA

5–10% of Thoroughbred (TB) and 6% of Standardbred (STDB) racehorses suffer from painful cramping and muscle damage following exercise (recurrent exertional rhabdomyolysis [RER]). Gender, temperament, diet, and activity level influence expression of the clinical phenotype, and heritability ranges from 0.34–0.42% in TBs and 0.39–0.45% in STDBs. The objective of this project is to identify variants associated with RER. In both breeds, total phenotypic variance is explained by < 1% of the SNPs tested. These SNPs were further partitioned based on estimated effect sizes. In the TB, SNPs of small, moderate, and large effect account for 45%, 35%, and 20% of the phenotypic variability, respectively. Similarly, in STDBs, small, moderate, and large effect SNPs explain 47%, 30%, and 23% variability, respectively. Genome-wide association studies (GWAS) identified seven genomic regions of interest (ROIs) significantly associated with RER in TB (n=491) and STDB (n=476) horses: two in TBs, four in STDBs, and one shared locus. Of the 274 genes within these ROIs, 34 were computationally predicted to be candidate genes for rhabdomyolysis in humans. Variants in each ROI were identified using whole genome sequences from 10 RER and 10 controls in each breed. 189,610 total variants were discovered in the ROIs, with an average of 7.5 variants per kb. Of these variants, 47.5% were within genes, and 1.1% had putative functional effects. Variants will be prioritized based on segregation between RER and control horses and predicted functional effect. High-priority variants will be used to create an assay for genotyping in the entire GWAS population.

E25

Foal Survival Score Can Predict Disease Severity in Neonatal Foals (Abstract E25)

Katarzyna Dembek[1]; Jillian Minuto[2]; Nathan Slovis[3]; Stephen Reed[4]; Teresa Burns[2]; Ramiro Toribio[5]

[1]The Ohio State University, Ames, IA, USA; [2]The Ohio State University, Columbus, OH, USA; [3]Hagyard Equine Medical Institute, Lexington, KY, USA; [4]Rood and Riddle Equine Hospital, Lexington, KY, USA; [5]Veterinary Clinical Sciences, The Ohio State University Columbus, OH, USA

The sepsis score (SS) has been used to diagnose or assume the presence of sepsis in newborn foals over the last thirty years. However, clinical experience and recent validation studies suggest that the SS is not as accurate to predict sepsis as in the original study. Therefore, a novel scoring method - the foal survival score (FSS) was recently developed to predict likelihood of survival in sick newborn foals. Based on sensitivity and specificity, the FSS has good ability to predict survival; however, the association of the FSS with severity of illness remains to be determined. The lower the FSS the more likely is the foal to die. High SS indicates severe disease. We hypothesized that the FSS will be a good predictor of disease severity (compared to SS) and outcome in hospitalized foals.

Foals (< 4 days of age) were classified into 4 type of disease groups based on clinical and laboratory findings: systemic inflammatory response syndrome [SIRS] (n = 45), sepsis (n = 14), multiple organ dysfunction syndrome [MODS] (n = 37), and healthy foals (n = 65). Multinomial regression models and odds ratios (OR) were calculated to predict each type of disease.

The median FSS was inversely associated with the SS; and foals with MODS, sepsis and SIRS had low FSS compared to healthy foals ($P < 0.01$). The probability of SIRS, sepsis and MODS decreased by 77%, 84% and 86% with increases in FSS compared to healthy foals, respectively ($P < 0.01$). Increases in SS were associated with higher likelihood of SIRS (OR = 3.5), sepsis (OR = 4.7) and MODS (OR = 5.7) compared to healthy foals ($P < 0.01$). Foals that survived had higher FSS and lower SS than non-survivors ($P < 0.01$). Likelihood of survival increased with higher FSS (OR = 1.9) and decreased with higher SS (OR = 0.74) in hospitalized foals ($P < 0.01$).

Our study demonstrated that FSS can be used as a scoring method to determined disease severity in addition to predicting survival in critically ill foals. Future research is needed to establish specific cutoff values for FSS to diagnose each type of neonatal disease.

E26

Pharmacokinetic and Biochemical Profiles After Single Administration of Oral Torsemide in Healthy Horses (Abstract E26)

Gustavo Ferlini Agne[1]; Seung Woo Jung[1]; Anne A. Wooldridge[1]; Aime K. Johnson[1]; William Ravis[2]; Sue H. Duran[1]

[1]Department of Clinical Sciences, College of Veterinary Medicine, Auburn University, Auburn, AL, USA; [2]Department of Pharmacal Sciences, Auburn University, Auburn, AL, USA

Torsemide is a loop diuretic agent more potent than furosemide. Its oral administration has demonstrated high bioavailability across species. The aim of the study was to determine the pharmacokinetic profiles of oral torsemide in the horse.

Torsemide was administered intragastrically at a single dose of 6 mg/kg to six healthy adult mares. Blood samples were collected at predetermined time points over a period of 48 hours. Plasma torsemide concentrations were measured by the use of high performance liquid chromatography, and serum biochemical profiles were examined. Student's t-test with Bonferroni correction was used to identify differences in pharmacokinetic and biochemical parameters between baseline (pre) and 24 hour post drug administration (post). Statistical significance was set at $p < 0.05$.

Pharmacokinetic analysis revealed peak concentration (C_{max}) of 11.05±5.26 µg/mL, time of maximum concentration (T_{max}) of 2.83±1.33 h, area under the curve (AUC) of 92.2±48.9 µg*h/mL, and an elimination half-life ($T_{1/2}$) of 9.11±1.5 h. Significantly increased concentrations of creatinine (pre: median of 1.65 mg/dL and range of 1.5–1.8 mg/dL, post: median of 2 mg/dL and range of 1.9–2.3 mg/dL; p=0.02) and bicarbonate (pre: median of 22.8 mmol/L and range of 17.6–25.3 mmol/L, post: median of 30.7 mmol/L and range of 27.6–34 mmol/L; p=0.0007) were noted. Chloride concentrations were markedly decreased (pre: median of 99 mmol/L and range of 96–103 mmol/L, post: median of 87.5 mmol/L and range of 85–88 mmol/L; p=0.0007).

Oral torsemide successfully reached therapeutic concentrations in blood, and resulted in mild pre-renal azotemia and electrolyte changes. These results suggest that oral torsemide may be employed as a clinically useful diuretic agent for management of congestive heart failure and fluid retention in the horse.

E27

Pharmacodynamic Properties of Oral Torsemide in Healthy Horses (Abstract E27)

Gustavo Ferlini Agne[1]; SeungWoo Jung[1]; Anne A. Wooldridge[1]; William Ravis[2]; Sue H. Duran[1]; Aime K. Johnson[1]

[1]Department of Clinical Sciences, College of Veterinary Medicine, Auburn University, Auburn, AL, USA; [2]Department of Pharmacal Sciences, Auburn University, Auburn, AL, USA

Diuretic therapy has been limited to injectable formulations in the horse due to poor absorption of oral furosemide. The purpose of this study was to determine whether oral torsemide produced clinically significant diuresis in healthy horses.

Torsemide was administered orally at 4 mg/kg/day to 6 healthy adult mares for 6 days. A 28 Fr Foley urinary catheter with a urine collection bag was used to measure urine output and urine specific gravity (USG) over 12 hours on days 0 and 6. Blood samples were collected daily for serum biochemical analysis. Non-invasive mean arterial pressure (MAP) was measured daily with a tail cuff. Student's t-test with the Bonferroni correction was used to analyze for differences in laboratory parameters between Day 0 (pre) and Day 6 (post).

Urine output (median and range) significantly increased with torsemide [pre: 2.2 L, (1.7–4.4 L); post: 11.5 L (8.7–15.4 L); p=0.0054]. USG markedly decreased [pre: 1.046 (1.032–1.049); post: 1.010 (1.008–1.011); p=0.002]. Significant alterations in biochemical parameters included hyponatremia [pre: 137 mmol/L (136–141 mmol/L); post: 129 mmol/L (128–134 mmol/L); p=0.001], hypokalemia [pre: 3.75 mmol/L (3.5–4.1 mmol/L); post: 1.95 mmol/L (1.6–2.4 mmol/L); p=0.001], and elevated creatinine [pre:1.4 mg/dL (1.3–1.6 mg/dL); post: 1.95 mg/dL (1.8–2.2 mg/dL); p=0.006]. MAP significantly decreased after torsemide [pre: 76.5 mm Hg (70–87 mm Hg); post: 56 mm Hg (48–74 mm Hg); p=0.02].

Torsemide administered orally induced clinically significant diuresis with mild pre-renal azotemia and electrolytes changes. These results suggest that oral torsemide therapy could help overcome current therapeutic limitations for horses with excessive fluid retention.

E28

Diarrhea in Horses Treated for Lower Respiratory Tract Disease with Macrolides Versus Other Antimicrobials (2000–2015) - (Abstract E28)

Emily John[1]; Gaelle Hirsch[2]; Henry Staempfli[2]

[1]Atlantic Veterinary College, Charlottetown, PI, Canada; [2]Ontario Veterinary College, Guelph, ON, Canada

Macrolide antimicrobials have historically been associated with severe adverse gastrointestinal effects when administered to adult horses. However, they are desirable antimicrobials for use in lower respiratory tract disease, and recent anecdotal results suggest efficacy in adult horses without a high incidence of adverse effects.

The objective of this study was to investigate the association between the development of diarrhea in both adult horses and foals administered macrolide antimicrobials versus other classes of antimicrobials for treatment of lower respiratory tract disease.

One hundred and fifty-eight horses (96 adults, 62 foals) admitted to a referral centre between 2000 and 2015, diagnosed with pneumonia, pleuropneumonia, or pleuritis, and administered systemic antimicrobials were included in this retrospective study.

Information obtained from the medical records included age at admission, breed, sex, antimicrobials administered while in hospital, clinical diagnosis, clinical outcome, and presence or absence of diarrhea. Macrolide treatment (erythromycin, azithromycin, clarithromycin) both with and without rifampin was compared to treatment with any other combination of antimicrobial(s). The relationship between age, antimicrobial(s) used, and development of diarrhea was investigated. Pearson's Chi-square and Fisher's exact test were applied ($P < 0.05$), including odds ratios with 95% confidence intervals.

Macrolide administration to adult horses ($P = 0.104$; OR 3.4 [0.78, 14.38]) and foals ($P = 0.729$; OR 1.48 [0.38, 5.74]) showed no statistically significant difference in adverse gastrointestinal effects compared to administration of other classes of antimicrobials. Macrolides are potentially safe to administer to adult horses for the treatment of lower respiratory tract disease.

E29

Pharmacokinetics of Intravenous Lithium Chloride and Agreement Between Two Methods of Lithium Measurement in Horses (Abstract E29)

Lynn M. Martin[1]; Alex Bukoski[1]; Dorothy Whelchel[2]; Tim Evans[1]; Charles Wiedmeyer[1]; Philip Johnson[2]

[1]University of Missouri, Columbia, MO, USA; [2]College of Veterinary Medicine, University of Missouri, Columbia, MO, USA

Pharmacokinetic evaluation of lithium chloride (LiCl), administered to healthy adult horses as an intravenous (IV) bolus, has not been reported. Nor does there exist a validated rapid procedure for determining plasma lithium (Li^+) concentrations in horses in order to monitor therapeutic levels and avoid toxicity. The goals of this study were to determine the IV bolus pharmacokinetic parameters of Li^+ for healthy adult horses and compare agreement between two methods of measurement of plasma Li^+ concentration: spectrophotometric enzymatic assay (SEA) and inductively coupled plasma mass spectrometry (ICP-MS; reference method).

Lithium chloride was administered as a 0.15 mmol/kg bwt IV bolus. Lithium concentrations were measured at times -30, -15, 0 (pre LiCl administration), 2.5, 5, 10, 20, 40 min and 1, 2, 4, 8, 12, 24, and 48 h (post LiCl administration) in healthy adult horses (n= 8). Performance outcomes of the SEA were assessed using the following criteria: linearity-of-dilution, inter-day coefficient of variation (CV), and spike-and-recovery. The two methods (SEA and ICP-MS) of plasma Li^+ concentration measurement were compared for agreement. Pharmacokinetic parameters were determined based on the reference ICP-MS data.

The IV LiCl bolus was well tolerated and side effects were not observed. The SEA showed acceptable linearity, $R^2 = 0.9752$; inter-day CV, 2.5%; and recovery, 96.3%. Both non-compartmental and compartmental analyses (traditional two-stage and nonlinear mixed-effects [NLME] modeling) were performed. Geometric mean values of non-compartmental parameters were: plasma Li^+ concentration at time zero, 2.19 mM; terminal elimination half-life, 1541 min; area under the plasma concentration-time curve from time zero to infinity, 550 mM min; clearance, 0.273 mL/min/kg; mean residence time, 1875 min; and volume of distribution at steady state, 511 mL/kg. Results of the traditional two-stage analysis showed good agreement with the NLME modelling approach. Bland-Altman analyses demonstrated poor agreement between the SEA and ICP-MS methods over the therapeutic Li^+ concentration range (95% limits of agreement = 0.14 ± 0.13 mM).

Lithium chloride dosed at 0.15 mmol/kg bwt IV bolus was well tolerated by healthy adult horses and displayed pharmacokinetic parameters similar to that reported in other species. The SEA displayed acceptable internal validity but did not agree well with ICP-MS. The results of this study support further evaluation of LiCl for clinical use.

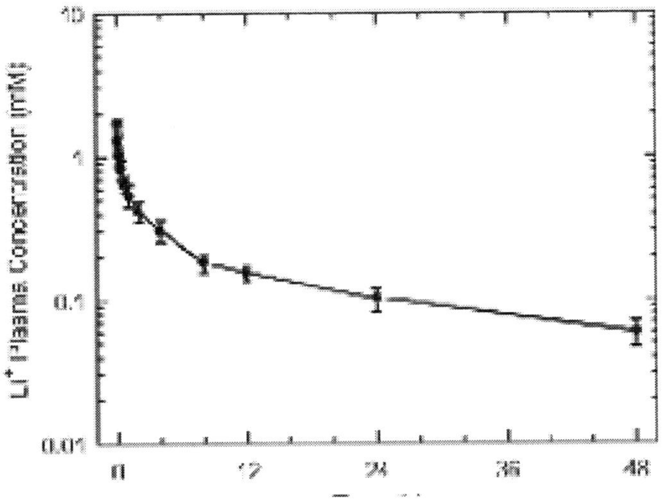

Li⁺ Plasma Concentration (mM)
10
1
0.1
0.01
0
12
24
36
48

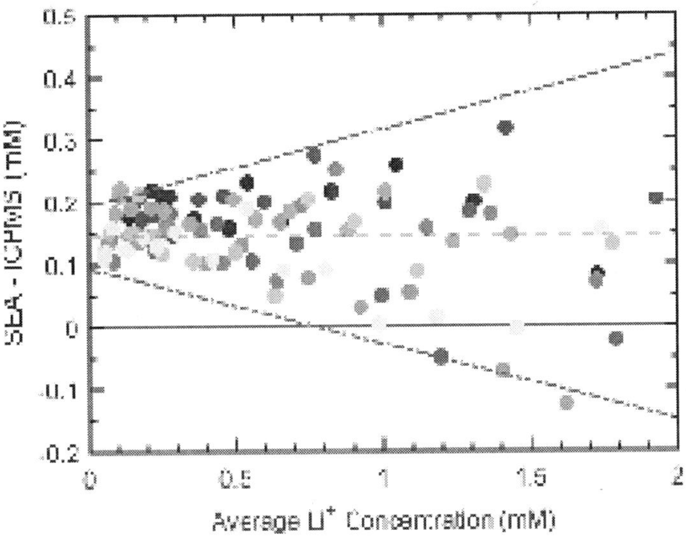

SEA - ICPMS (mM)
0.5
0.4
0.3
0.2
0.1
0
-0.1
-0.2
0
0.5
1
1.5
2
Average Li⁺ Concentration (mM)

E30

Ocular Tissue Concentrations and Safety After Subconjunctival Injection of a Voriconazole Thermogel in Horses (Abstract E30)

Mariano Mora Pereira[1]; Eva M. Abarca[2]; Anne A. Wooldridge[1]; Sue H. Duran[1]; William Ravis[3]

[1]Department of Clinical Sciences, College of Veterinary Medicine, Auburn University, Auburn, AL, USA; [2]Department of Clinical Sciences, Auburn University College of Veterinary Medicine, Bern, Bern, Switzerland; [3]Department of Pharmacal Sciences, Auburn University, Auburn, AL, USA

Previous *in vitro* studies using a voriconazole-poly(lactic-co-glycolic acid)-b-poly(ethylene glycol)-b-poly(lactic-co-glycolic acid) thermogel (VORI-GEL) demonstrated sustained-release of voriconazole above therapeutic concentrations for most equine ocular fungal pathogens. The objective was to evaluate safety and voriconazole concentrations in ocular tissues and aqueous humor (AH) following subconjunctival injection of the VORI-GEL in horses.

Six normal horses received topical 1% voriconazole (VORI-TOP) every 4 hours for 2 days, and then subconjunctival injection of 1.5% VORI-GEL after washout in alternate eyes. Tears and AH were collected day 2 in VORI-TOP and days 2, 7, 14, and 23 in VORI-GEL. To evaluate ocular tissue concentrations, 2 horses received a subconjunctival injection of 1.5% VORI-GEL 2 days and 2 hrs before euthanasia in alternate eyes. Voriconazole concentrations were measured by high performance liquid chromatography, and evaluated using Student's t-tests and descriptive statistics. Ocular inflammation was assessed using a modified Hackett-McDonald scoring system.

Neither VORI-GEL nor VORI-TOP caused changes in ocular inflammatory scores. Voriconazole concentrations in tears on day 2 were not different between treatments. Voriconazole concentrations in AH for VORI-GEL were detectable, but not quantifiable on days 2 and 7. Voriconazole concentrations above 0.5 μg/g were detected in the anterior segment tissues 2 hrs and 2 days after VORI-GEL injection.

Subconjunctival injection of VORI-GEL is safe. In normal horses, low AH and tear concentrations are likely due to intact ocular barriers and high drug lipophilicity. There is sustained release of therapeutic concentrations of voriconazole from the VORI-GEL to the anterior segment tissues for at least 2 days following injection.

E31

Comparison of Respiratory Inductive Plethysmography and Forced Oscillatory Mechanics to Measure Airway Hyperreactivity in Horses (Abstract E31)

Claire E. Dixon; Daniela Bedenice; Melissa Mazan

Tufts Cummings School of Veterinary Medicine, North Grafton, MA, USA

A diagnosis of inflammatory airway disease (IAD) is commonly based on abnormal bronchoalveolar lavage fluid (BALF) cytology or airway hyperreactivity (AHR) in horses with poor performance or chronic cough. Our study purpose was to determine whether two modalities of non-invasive lung function testing (FOM-forced oscillatory mechanics vs. RIP-Respiratory Inductive Plethysmography) establish the same clinical diagnosis of AHR, using histamine bronchoprovocation. AHR was defined by the histamine dose needed to double FOM baseline resistance, or to achieve a 35% increase in RIP delta flow.

Nineteen horses (3–25 years, 335–650kg) with clinical signs suggestive of IAD were randomly assigned to undergo FOM and RIP testing on consecutive days. BALF was subsequently collected and stained with modified Wright's and toluidine blue stains. Categorical data were compared between groups using Chi-square analysis.

Abnormal BALF cytology confirmed IAD in 14/19 (73.7%) horses. Both FOM and RIP revealed AHR in 7/14 (50%) of these IAD horses. An additional 4/19 (21.1%) horses showed AHR based solely on RIP testing, including 2 horses with normal BALF cytology. A diagnosis of AHR was more often associated with RIP than FOM testing ($P = 0.013$), although the prevalence of AHR was significantly higher in IAD vs non-IAD horses, regardless of testing methodology. The phase angle between thoracic and abdominal components of breathing did not differ between patient groups.

In conclusion, RIP diagnosed AHR more frequently than did FOM, including horses with no other diagnostic evidence of IAD. Without further evaluation, these two testing modalities of AHR cannot be used interchangeably.

E32

Effect of Valacyclovir on EHV-5 Viral Kinetics in Horses Diagnosed with Equine Multinodular Pulmonary Fibrosis (Abstract E32)

Charlotte A. Easton-Jones[1]; John E. Madigan[2]; Samantha Barnum[2]; Nicola Pusterla[2]

[1]School of Veterinary Medicine, University of California-Davis, Davis, CA, USA; [2]Department of Medicine and Epidemiology, School of Veterinary Medicine, University of California-Davis, Davis, CA, USA

Previous studies have revealed that equine herpesvirus-5 (EHV-5) is commonly isolated from the lungs of horses diagnosed with equine multinodular pulmonary fibrosis (EMPF), suggesting an etiological link. There are currently no studies assessing the impact of valacyclovir treatment on viral kinetics of EHV-5 in EMPF cases.

The objective of the study was to determine the effect of valacyclovir treatment on viral load of EHV-5 measured by quantitiative PCR in blood, nasal swabs and broncho-alveolar lavage fluid. Four horses with EMPF were treated for 10 days with oral valacyclovir. The diagnosis of EMPF was based on histopathological findings and EHV-5 detection by qPCR on lung tissue. Sequential blood, nasal swabs and broncho-alveolar lavage fluid samples were collected to analyze the viral kinetics of EHV-5 during the 10 days of treatment.

There was no statistical difference in median EHV-5 viral load between the two time points for all of the three sample types tested. EHV-5 viral load for blood on day 0 ranged from 5,830 to 18,635 (median 7,738) gB gene copies/million cells, and on day 10 ranged from 2,625 to 18,635 (median 12,138) gB gene copies/million cells. EHV-5 viral load for nasal secretions on day 0 ranged from 23,639 to 3.394×10^9 (median 1.522×10^6), and on day 10 ranged from 251,186 to 1.697×10^7 (median 452,957). EHV-5 viral load for broncho-alveolar lavage fluid on day 0 ranged from 614 to 315,655 (median 33,080) gB gene copies/million cells, and on day 10 ranged from 3,562 to 542,417 (median 115060) gB gene copies/million cells.

Valacyclovir is a relatively expensive drug that is routinely used to treat horses with EMPF. This study revealed that 10 days of valacyclovir treatment did not significantly alter viral kinetics of EHV-5 as measured by qPCR in nasal swabs, blood and broncho-alveolar lavage fluid samples.

E33

Relationship Between Tracheobronchoscopic EIPH Score and Bronchoalveolar Lavage Erythrocyte Count in Horses (Abstract E33)

Carolina Lopez; Warwick Bayly; Jenifer R. Gold; Macarena Sanz; Debra Sellon

Department of Veterinary Clinical Sciences, Washington State University, Pullman, WA, USA

Exercise-induced pulmonary hemorrhage (EIPH) is diagnosed and severity assessed by post-exercise tracheobronchoscopic examination (TBE), and/or enumeration of erythrocytes (RBCs) in bronchoalveolar lavage fluid (BALF). TBE EIPH scores range from 0-4, 0 indicating no EIPH. BALF RBC < 1000/µl indicates no EIPH. There is scant information regarding the relationship of TBE EIPH score to BALF RBC number. We hypothesized that BALF RBC was weakly correlated ($r^2 < 0.20$) with TBE score.
TBE followed by BAL was performed 223 times on 151 different horses after treadmill (8 horses, 45 runs), racetrack (6 horses, 24 runs), or barrel racing (143 horses, 155 runs) exercise. TBE EIPH score ranged from 0 (n = 83) to 4 (n = 5). RBC number ranged from 0 to 994,000/µl. Eleven TBE scores were 0 despite RBC count > 1000/µl. There were 61 exams with TBE scores ≥ 1 but BALF RBC counts < 1000/µl. The relationship between TBE score and BALF RBC number was significant but weak ($p < 0.001$; $r^2 = 0.19$) after Spearman's evaluation. TBE EIPH score was weakly correlated with BALF RBC number and there were large ranges of BALF RBCs associated with each TBE score. An EIPH score of 0 does not rule out the occurrence of EIPH, but neither does a BALF RBC count < 1000/µl. The latter finding was unexpected and may be explained by fast clearance of the respiratory tract or lavage of an airway segment other than the one bleeding. The possibility of technical errors counting RBCs in some samples cannot be discounted.

E34

Prevalence of Exercise-Induced Pulmonary Hemorrhage in Barrel Racing Horses in the Pacific Northwest (Abstract E34)

Jenifer R. Gold[1]; Donald P. Knowles[2]; Todd Coffey[3]

[1]Department of Veterinary Clinical Sciences, Washington State University, Palouse, WA, USA; [2]Animal Disease Research Unit, USDA-Pacific West Area, Washington State University, Pullman, WA, USA; [3]Center for Interdisciplinary Statistical Education and Research, Washington State University, Pullman, WA, USA

Exercise-induced pulmonary hemorrhage (EIPH) refers to presence of blood in the airways caused by strenuous exercise. It has been documented in racing, polo and eventing horses, but little is known about EIPH in barrel racing horses. The purpose of this study was to test for the presence of EIPH in barrel racing horses, estimate its prevalence in the Pacific Northwest and determine whether a relationship between EIPH and performance exists.

This study enrolled 158 barrel racing horses competing at events in WA, ID, and MT, after obtained owner consent. Data collected included signalment, illness/respiratory history, race division, and race day medications. Tracheobronchoscopy was scored based on quantity of blood in the trachea (0=no blood to 4=large amounts throughout the trachea). Erythrocyte counts were obtained from bronchoalveolar lavage fluid. Statistical analysis included logistical regression, Chi squared, Fisher's Exact T-test, linear regression and calculation of correlation coefficient. Significance was set at $P < 0.05$.

The prevalence of EIPH in our study was 54%. Horses that ran faster had a significantly greater likelihood of bleeding $P < 0.014$. Bleeding did not significantly affect performance, although horses that bled finished lower than horses that didn't bleed. Significant ($P < 0.001$) positive linear relationship between the tracheal score and BAL erythrocyte count was shown. Correlation coefficient between these 2 tests was poor ($r^2 = 0.14$).

This study shows EIPH is present in barrel racing horses in the Pacific Northwest and may impact performance. The long term effects of EIPH and other variables requires further study.

E35

Prevalence of Cardiac Arrhythmias in Competition Draft Horses (Abstract E35)

Lisanne Gallant[1]; Sarah Jacob[1]; Elizabeth M. Tadros[2]; Jane Woodrow[3]; Melissa Hines[3]; Susan Ewart[1]; Harold C. Schott, II[1]

[1]Michigan State University, East Lansing, MI, USA; [2]Diagnostic Center for Population and Animal Health, Michigan State University, Lansing, MI, USA; [3]University of Tennessee, Knoxville, TN, USA

Cardiac arrhythmias, most notably atrial fibrillation, have been linked to poor performance. In a recent study Draft and Warmblood breeds were more likely to be presented with atrial fibrillation in comparison to most light breeds, with the exception of Standardbreds. To our knowledge there is no data describing a population of competition ready Draft horses. Our objective was to determine the prevalence of silent cardiac arrhythmias in a population of Draft horses. A convenience sample of 245 Draft horses was studied at the 2015 Michigan Great Lakes International Draft Horse Show. Information gathered on each horse included signalment, history of poor performance, upper airway noise, myopathy, and cardiac arrhythmias or murmurs. All horses were auscultated on each side of the thorax for 30 seconds. A 30-second electrocardiographic (ECG) recording was then taken using a handheld recording device that transmitted to a smart phone (AliveCor®). ECG recordings were evaluated for rhythm and heart rate, P-R interval, and R-R interval were measured. The study population included 82 Percherons, 69 Belgians, 69 Clydesdales, 11 Belgian mules, 8 Shires, and 5 Percheron mules. Arrhythmias were found in four horses: atrial fibrillation (1), ventricular premature depolarizations (1), second degree atrioventricular block (2). None of these horses had a history of poor performance or known cardiac arrhythmias. We concluded that the prevalence of silent arrhythmias in well-conditioned, competitive Draft horses is low (2/245; 0.8%).

E36

Beat-to-Beat Heart Rate Variability During Treadmill Exercise in Healthy Horses and Horses with Cardiac Disease (Abstract E36)

Ladina Frick[1]; Colin C. Schwarzwald[2]; Katharyn J. Mitchell[3]

[1]Vetsuisse Faculty, University of Zurich, Zurich, Switzerland; [2]Vetsuisse Faculty, Clinic for Equine Internal Medicine, University of Zurich, Zurich, Switzerland; [3]University of Zurich, Zurich, Switzerland

The aim was to quantify beat-to-beat heart rate variability (HRV) in healthy horses during a high-speed standardised treadmill exercise test (HSET) and compare with HRV in horses diagnosed with cardiac disease or other causes of poor performance.
HSET were performed in 30 healthy horses (Group H), 9 horses with cardiac disease (Group C) and 11 horses with non-cardiac causes of poor performance (Group O). Velocity, heart rate, blood lactate and ECGs recorded. The ECGs were corrected for artefacts and arrhythmias noted. HRV analysis was performed on trot and canter segments and % beat-to-beat variability was calculated. Performance indices (velocity at blood lactate 4 mmol/L [V_4] and velocity at HR 200 min^{-1} [V_{200}]) were calculated. Group C had lower V_{200} and V_4 compared with Group H and Group O (meanV_{200} ± SD: Group C: 7.6 ± 0.7 m/s, H: 8.8 ± 1.2 m/s, O: 8.7 ± 0.9 m/s, p =0.0053; meanV_4 ± SD: Group C: 6.5 ± 0.8 m/s, H: 7.3 ± 1.0 m/s, O: 7.5 ± 1.0 m/s, p = 0.0383). Seven Group C horses had arrhythmias during exercise, while 4 Group H horses had arrhythmias only during the canter phase. No Group O horses had arrhythmias during exercise. The % beat to beat variation during both the trot and canter phases was low (median [1% and 99% percentiles] Trot Group H: 0% [-4.0 to 3.7%], C: 0% [-5.1 to 3.7%], O: 0% [-3.4 to 3.3%]; Canter Group H: 0% [-5.5 to 5.3%], C: 0% [-4.9 to 4.6%], O: 0% [-5.1 to 4.8%]). However, the minimum and maximum % inter-beat variation was greater in Group C during the trot phase (minimum % variation Group H: -14.9%, C: -28.1%, O: -9.2%, p=0.0085; maximum % variation Group H: 19.8%, C: 78.1%, O: 14.9%, p=0.0014).

Horses with cardiac disease have lower performance indicators (V_{200}, V_4) than healthy horses or horses with other causes of poor performance. During HSET, horses have very little beat to beat variation. Infrequent arrhythmias are detected in horses during HSET and do not modify the median beat-to-beat HRV but can influence the minimum and maximum beat-to-beat HRV indices.

E37

Simpson's Method of Discs in Argentina-Chair Horse: Echocardiographic Reference-Intervals for End-Diastolic and End-Systolic Left-Ventricle Volumes (Abstract E37)

Jeff Munoz-Pérez[1]; Gerhard Wess[2]; Sergio E. Linares-Villalba[1]; Carlos E. Giraldo[3]

[1]Departamento de Salud Animal, Universidad de Caldas, Manizales, Colombia; [2]Ludwig-Maximilians-Universität München, Munich, Berlin, Germany; [3]Universidad de Caldas, Manizales, Caldas, Colombia

Frequently the measurement of ventricular volumes by echocardiography in horses has been performed using parameters obtained through the M-mode. The Teichholz formula has being the most applied for the calculation of the ventricular volumes and the ejection fraction. The American Society of Echocardiography recently does not recommend its use in clinical settings. Instead, the Simpson's method of discs (SMOD) was recommended in the updated of cardiac chamber quantification guidelines with high levels of evidence. In veterinary medicine, the diagnostic superiority of SMOD versus M-mode measurements has been similar in different breeds of dogs, some as Doberman pinscher, Boxer, Giant Breeds, and Sighthounds. The aim of this study was established the breed-specific reference intervals for the End-diastolic volume (EDV) and End-systolic volume (ESV) by Biplanar Simpson's method of discs (SMOD) in healthy Argentina Chair horses, hypothesizing that in horses the apparent diagnostic success of SMOD is similar, especially in high performance breeds with sports aptitudes.

In this study, 17 Argentina Chair horses with an appropriate health state confirmed by clinical examination, ECG, echocardiography and laboratory test were recruited. Horses with evidence of systemic disease or cardiovascular alterations were excluded. Horses were evaluated without sedation and 2D and M-mode measurements were performed in the right para-sternal and left para-sternal views. The ventricular volumes obtained by SMOD and M-mode were indexed to body surface area (BSA). Reference intervals for SMOD with and without weight normalization were established using the right para-sternal and left para-sternal views. The volumes obtained in each view were compared, as well as between genders and ages using a paired t-test when normal distribution was apparent. The volumes obtained by SMOD and M-mode were compared using a T-test. Reference intervals were reported with 95% confidentiality. A p-value $< .05$ was considered statistically significant.

All indexed volumes were of normal distribution according to Kolmogorov-smirnov test ($p=0.50$). The volumes obtained with the SMOD in both views demonstrated modest agreement in the student T-test ($p<0.05$). The breed-specific references intervals were 98,93 ml/m^2 and 47,67 ml/m2 for indexed end-diastolic volume (EDVI) and indexed end-systolic volume (ESVI), respectively. There were no difference between the indexed volumes obtained in mares and males ($p > 0.05$). As well as between the age groups ($p=0.07$). However, there was difference between the volumes obtained with SMOD method and One-dimensional M-mode ($p=0.03$). The volumes derived from M-mode had significant difference between left and right para-sternal views ($p=0.001$).

In conclusion, this study provided echocardiographic reference intervals for SMOD-derived left ventricular volumes for these breed with aptitude for equestrian sports. Because some of these sports are under stressful activity, the cardiovascular effort of the athlete horse is greater. The indexed-volumes provided with this study may possibly have the ability to recognize early, horses with a diminished athletic performance associated to cardiovascular comorbidities.

E38

Equine Pituitary Pars Intermedia Dysfunction (PPID) in the Southern Hemisphere, What Is Different? (Abstract E38)

Francois-Rene Bertin

The University of Queensland, Gatton, QLD, Australia

PPID is a well characterized endocrine disorder in the temperate climates of the Northern Hemisphere; however, its description in tropical and subtropical climates of the Southern Hemisphere is scarce. This analysis reviews the cases of PPID reported in the Southern Hemisphere and describes cases from The University of Queensland that reflect a different presentation. Clinical features of the disease and variables associated with laminitis, insulin dysregulation and survival are reported.

Ninety-nine cases including 34 ponies and 65 horses were diagnosed with PPID with either a basal ACTH (91%) or a dexamethasone suppression test (9%).

The most common clinical signs were tachypnoea (67%), anhidrosis (55%), hypertrichosis/hirsutism (40%), weight loss (38%), polyuria/polydipsia (35%) and lameness (34%). The most common clinicopathologic abnormality was hypophosphataemia (36%). When investigated, laminitis was diagnosed in 82% of cases, insulin dysregulation in 35% and gastrointestinal parasitism in 100%.

Pergolide was administered in 83% of cases at a dose of 0.001 [0.001–0.005] mg/kg increased to 0.002 [0.001–0.015] mg/kg over a period of 1 [0–48] months. Eighty percent showed a clinical improvement and 65% survived.

Factors associated with laminitis were younger age ($p=0.007$) and absence of weight loss ($p=0.030$). Factors associated with insulin dysregulation were younger age ($p=0.026$), duration of clinical signs ($p=0.044$) and need for higher doses of pergolide ($p=0.017$). The only factor associated with survival was clinical improvement ($p=0.007$).

This study shows that commonly described features of PPID are found in the Southern Hemisphere but that more unusual findings also warrant investigation of endocrine disorders.

E39

Hair Cortisol Concentration as a Stress Biomarker in Horses (Abstract E39)

Maria Carolina Duran Graeff[1]; David Janz[2]; Cheryl Waldner[3]; John Campbell[4]; Fernando Marques[3]

[1]WCVM, University of Saskatchewan, Saskatoon, SK, Canada; [2]WCVM, Toxicology Centre, University of Saskatchewan, Saskatoon, SA, Canada; [3]WCVM, University of Saskatchewan, Saskatoon, SA, Canada; [4]WCVM, University of Saskatchewan, Saskatoon, SA, Canada

Cortisol is commonly used as an indicator of physiological stress. Traditional samples (e.g., blood, urine, saliva) to measure cortisol secretion have proven to adequately assess short term/acute response to stress, while hair cortisol concentration (HCC) is increasingly being used to assess long-term stress in mammals. The purpose of the study was to analyze HCC in whole hair and hair segments to investigate potential effects of body location and surgical procedure (castration) on HCC in horses.
Mane and tail hair from healthy broodmares (n=10) was collected, segmented and analyzed. Additionally, yearling male (n=14) and female horses (n=14) were used. Separate mane and tail hair samples were collected at the beginning of the study (before castration) and 3 months after from male and female yearlings. Cortisol was extracted into methanol from ground hair and quantified using ELISA. HCC in mane samples from broodmares was greater than in tail samples. In all studied groups, segments further away from the hair root contained lower HCC. The decrease in hair cortisol concentrations between 1 and 2-months post-castration was significantly lower in castrated males than intact females. Cortisol accumulation in the month after castration could explain the significantly smaller decrease in HCC concentrations in males than females during the same time period. Results suggest that HCC analysis in horses can be used to address basal and elevated HPA activity in horses. Nevertheless, other factors such as segment location and collection site may also have an effect on measured HCC.

E40

The Association Between Endocrine Disrupting Chemicals and Equine Metabolic Syndrome (Abstract E40)

Sian Durward-Akhurst[1]; Nichol Schulz[2]; Elaine Norton[2]; Raymond J. Geor[3]; James Mickelson[4]; Molly E. McCue[2]

[1]University of Minnesota, Center City, MN, USA; [2]University of Minnesota, St. Paul, MN, USA; [3]Massey University, Palmerston North, Auckland, New Zealand; [4]College of Veterinary Medicine, University of Minnesota, St. Paul, MN, USA

Our group has demonstrated that up to 49% of Equine Metabolic Syndrome (EMS) phenotypic variability is explained by shared environment; however only 4–18% of this environmental variability is explained by diet, exercise and season, suggesting other environmental factors play a role in EMS. Endocrine Disrupting Chemicals (EDCs) are associated with metabolic syndrome and other endocrine abnormalities in humans, and our preliminary data demonstrated that horses from farms < 31 miles of EDC disposal sites (SF) were more likely to have had laminitis and had higher post oral-sugar-challenge insulin concentrations (INS-OST). These data suggested that EDC exposure is an EMS risk factor.
We have previously reported the measurement of the concentration of EDCs capable of activating the aryl hydrocarbon receptor via a bioassay (AHR-EDC); and the correlations between these EDCs and metabolic measurements in 301 horses from 32 farms. Here we extend these findings by presenting correlations between EDCs capable of activating the estrogen receptor (ER-EDC). Significance was $p < 0.05$. Mean (range) ER-EDC concentrations were 279.66 (4.35–15,000) pg/ml plasma. EDCs were below the detection limit in approximately half of the horses. AICc statistics were used to determine the best linear multivariable regression model for EDC association with 9 EMS phenotypic variables. ER-EDC was associated with glucose after an oral sugar test (OST), glucose, insulin and insulin-OST. Overall, these results suggest that some of the unexplained environmental variance in individuals with EMS is due to EDC exposure, which is mediated in part through the estrogen and arylhydrocarbon receptors.

E41

Clinical Signs Associated with PPID in the Equine Athlete (Abstract E41)

Marc Baus[1]; **Johanna Kremburg**[1]; Kara Kirchherr[2]; Steven Grubbs[3]; Dwana Neal[4]; Tom Keefe[5]

[1]Grand Prix Equine, Hawleyville, CT, USA; [2]Boehringer Ingelheim, Belchertown, MA, USA; [3]Boehringer Ingelheim, Stewartsville, MO, USA; [4]Boehringer Ingelheim, Platte City, MO, USA; [5]Enviro stats, Windsor, CO, USA

Pituitary Pars Intermedia Dysfunction (PPID) has been described as one of the most common diseases of horses and ponies 15 years of age and older. Recently, the clinical signs of PPID have been divided into early and advanced clinical signs. Establishing a diagnosis of PPID in horses with early clinical signs is currently a difficult challenge facing equine veterinarians. Particularly difficult may be the diagnosis of horses with PPID in the equine athlete. Many of the same clinical signs identified in early or advanced PPID may be recognized in the sport horse along with tendon or suspensory ligament degeneration. Suspensory ligament injuries have been considered a common cause of lameness, in the equine athlete involved in competitive events. A recent histopathological study concluded that an association exists between PPID and suspensory ligament (SL) degeneration. The objective of this study was to identify the most common clinical signs associated with PPID in the sport horse.

Sport horses >10 years of age, any breed, and sex were eligible for study enrollment as long as they were documented to be exhibiting one or more of the early or advanced clinical signs of PPID including suspensory ligament desmitis. Forty-nine horses were evaluated and included in the final data analysis with at least one clinical sign of early or advanced PPID. Demographic data, signalment, and a physical examination was conducted and each horse was tested for PPID using the TRH stimulation test measuring ACTH at 0 (T0ACTH) and 10 (T10ACTH) min. Insulin and glucose levels were also determined. Normal horses were excluded from the study. Blood samples were shipped overnight to the Animal Health Diagnostic Center, Cornell University, Ithaca, NY for analysis. The association between PPID status, based on ACTH and insulin results, and each of the demographic variables (age, sex, and breed), clinical signs, the two test result variables insulin and glucose were statistically evaluated individually using the Pearson chi-square test. Odds ratios for significant predictors of PPID status were computed using corresponding 95% confidence intervals when applying multiple logistic regression analysis.

Of the 49 horses, 19 (39%) were $PPID^+$ and 8 (16%) were IR^+. Of the 19 $PPID^+$ horses, only 2 (11%) were $PPID^+$ at the T0ACTH time point, whereas 19/19 (100%) were $PPID^+$ at T10ACTH. Six (32%) of 19 $PPID^+$ horses were hyperinsulinemic. Horses' ages ranged from 11 to 25 years of age (arithmetic mean 17 years) in the $PPID^+$ group. The most common clinical signs observed in the $PPID^+$ horses were delayed regional shedding, loss of epaxial muscle mass, regional adiposity, skeletal muscle atrophy and suspensory desmitis. $PPID^+$ was significantly ($p=0.023$) associated with lameness (suspensory desmitis, tendon laxity, superficial digital flexor tendonitis). Of the horses that were lame, 70% were $PPID^+$.

Based on the results of this study, the TRH stimulation procedure was required in 89% of enrolled horses for laboratory confirmation of PPID. In the sport horse, suspensory desmitis was significantly associated with $PPID^+$ status. Veterinarians should include PPID in the list of differential diagnoses when examining sport horses with suspensory desmitis along with early and advanced clinical signs of PPID.

E42

Assessment of Prolactin Concentration as a Screening Test for PPID (Abstract E42)

Harold C. Schott, II[1]; Bo Norby[1]; Donald Thompson[2]

[1]Michigan State University, East Lansing, MI, USA; [2]Louisiana State University, Baton Rouge, LA, USA

To assess measurement of prolactin (PRL) concentration as a screening test for PPID, serum PRL concentration was compared between 31 horses that were enrolled in and 26 horses that failed inclusion criteria for the clinical efficacy study for pergolide. Enrollment required horses to have a hypertrichosis score ≥1 and either an elevated plasma ACTH concentration (≥50 pg/mL) or abnormal overnight dexamethasone suppression test results. PRL was measured again after 3 and 6 months of treatment with pergolide in the horses enrolled in the study. PRL concentrations were compared by rank sum analysis and an ANOVA on ranks for repeated measures. A ROC curve was constructed to assess sensitivity and specificity of PRL at various cut-off values as a screening test for PPID. PRL was greater ($p=0.02$) in enrolled (7.0±8.2 ng/mL, mean±SD) as compared to excluded (4.3±2.6 ng/mL) horses. Area under the ROC curve was 0.68 (95% CI: 0.55–0.80). Using cut-off values of ≥ 3.5 and ≥ 4.0 ng/mL, sensitivity and specificity were 77 and 54% and 65 and 54%, respectively. Curiously, PRL tended to decrease after 3 months of pergolide treatment (3.8±1.8 ng/mL) but the only significant difference was a higher ($p<0.01$) PRL after 180 days of pergolide treatment (9.7±6.4 ng/mL) as compared to the 90 day treatment value. In this cohort of horses, measurement of PRL did not appear to be a useful screening tool for PPID. Further, treatment of PPID-affected horses with pergolide did not decrease PRL.

E43

Adrenocorticotropic Hormone Levels in Equine Plasma After Centrifugation at Multiple Time-Points From Collection (Abstract E43)

Kayla Shepard[1]; John Haffner[2]; Steven Grubbs[3]; Dwana Neal[4]; Greg Pierce[5]

[1]Boehringer Ingelheim, St. Joseph, MO, USA; [2]Middle Tennessee State University, Murfreesboro, TN, USA; [3]Boehringer Ingelheim, Stewartsville, MO, USA; [4]Boehringer Ingelheim, Platte City, MO, USA; [5]Innovative Data Resources, Asheville, NC, USA

Pituitary pars intermedia dysfunction (PPID) has been considered the most common endocrinologic disorder of aged horses. The diagnosis of PPID is typical based on history, clinical signs and demonstration of endocrinologic dysfunction via diagnostic assay results. The most common diagnostic test utilized by veterinary practitioners for the diagnosis of PPID is currently measurement of basal plasma adrenocorticotropic hormone (ACTH) concentration. Determination of resting plasma ACTH concentration has been the most practical diagnostic test for PPID likely due to the ease of collection and the availability of seasonal reference ranges that allow for testing any time of the year. Multiple recommendations exist concerning the *in vitro* stability of ACTH. Most all recommendations state samples should be chilled promptly after collection. Multiple recommendations exist for the timing of centrifugation of chilled samples from sample collection. In the US, a discordance exists between laboratory recommendations concerning centrifugation time from sample collection; anywhere from 2 hours up to 4 hours from sample collection. Realistically, many equine ambulatory practitioners do not have the ability to centrifuge the sample within 2 to 4 hours. Based on this dilemma, the purpose of this study was to determine the stability of ACTH levels in blood samples with increasing duration from collection prior to centrifugation.

Ten horses, 5 $PPID^+$ and 5 $PPID^-$ (non-PPID) horses were enrolled in the study during the autumn time-period. On Day 0, 5mL of whole blood was collected from each horse into each of six EDTA tubes and immediately chilled. One tube was centrifuged within 15 minutes of collection (time 0) followed by centrifugation of one tube from each horse at 4, 8, 12, 24 and 36 hours following collection. At each time-point, plasma was centrifuged and separated into 1.5 mL polypropylene tubes and stored at -80C. Plasma was shipped frozen with cold packs overnight to the Animal Health Diagnostic Center, Cornell University, Ithaca, NY for analysis. The percent change from baseline (PCFB) was reported to standardize the data due to differing baseline values. Mean PCFB, standard deviation and 95% confidence intervals (CI) were reported. The null hypothesis that there was no time effect and was tested with a mixed model repeated measures methodology with subject as a random effect. P-values < 0.05 were considered statistically significant.

The absolute changes over time revealed no pattern of change over time. The mean PFCB was 2.8% ± 7.96% (CI -2.88%, 6.99%). There was not significant evidence (p-value of 0.5056) of a time effect using the repeated measures model. There was no evidence of any significant time effect on the level of ACTH in $PPID^+$ or $PPID^-$ (non-PPID) horses. There did not appear to be any systematic degradation of ACTH level due to delay in measurement through 36 hours post collection. The results of this study supports the recommendation that immediately chilled EDTA blood samples (properly stored) may be centrifuged and plasma separated within 36 hours of collection for shipment to the respective laboratory for analysis without degradation of the ACTH level. Further research is needed to determine ACTH stability in a larger number of horses ($PPID^+$ and $PPID^-$) from separate sites and different geographic regions.

E44

Effects of Dexamethasone on Circulating ACTH and Cortisol Concentrations in Healthy and PPID-Affected Horses (Abstract E44)

Dorothy Whelchel[1]; Philip Johnson[1]; Amanda Trimble[2]; Loren Schultz[1]

[1]College of Veterinary Medicine, University of Missouri, Columbia, MO, USA; [2]College of Veterinary Medicine, Kansas State University, Manhattan, KS, USA

The purpose of this study was to investigate effects of systemic corticosteroid administration on circulating ACTH and cortisol concentrations in healthy horses and horses with Pituitary Pars Intermedia Dysfunction (PPID). The hypothesis of this study was that administration of synthetic corticosteroids would decrease circulating ACTH and cortisol concentrations in healthy horses while significant changes would not be detected in horses with PPID.

Seventeen horses meeting the study's inclusion criteria were tested for PPID, using endogenous ACTH plasma concentrations and the thyrotropin releasing hormone (TRH) stimulation test. The study included six healthy horses (group 1: plasma ACTH concentrations <110 pg/ml post TRH administration) and 6 horses with PPID (group 2: plasma ACTH concentrations above seasonal reference ranges or >110 pg/ml post TRH administration). All 12 subjects were stabled in box stalls with free choice access to grass hay and water for the duration of the study. The study was conducted in accordance of the University of Missouri's Animal Care and Use Committee and between February to July. After a 48-hour acclimation period, blood samples were collected to measure baseline ACTH and cortisol concentrations in all horses. Dexamethasone (0.05 mg/kg) was subsequently administered IV once. For the next 48 hours, ACTH and cortisol were measured every 6 hours. Repeated measures ANOVA analysis followed by *post hoc* testing determined significant effects of dexamethasone treatment on ACTH and cortisol concentrations over time in both groups of horses and compared differences between groups.

Significant changes in circulating ACTH ($p<0.0017$) and cortisol concentrations ($p<0.0001$) were detected at numerous time points after dexamethasone treatment in both groups of horses; however, significant differences between groups were not evident. In conclusion, dexamethasone administration appeared to suppress plasma ACTH concentrations for up to 36 hours and serum cortisol concentrations for at least 48 hours in healthy horses and horses affected by PPID. Studies with larger sample sizes may better distinguish differences in the response of the endocrine axis to dexamethasone treatment in healthy horses compared to horses with PPID.

E45

Evaluation of Dexamethasone to Prevent Post-Operative Ileus in 61 Horses With and Without Intestinal Resection (Abstract E45)

Kate F. McGovern[1]; Bruce M. Bladon[2]

[1]Donnington Grove Veterinary Group, Hermitage, England, UK; [2]Donnington Grove Veterinary Group, Newbury, England, UK

Inflammation is key in the development of post-operative ileus (POI) in rodents, with a similar pathogenesis likely occurring in other species. It is well established that dexamethasone reduces inflammation and therefore may reduce POI in horses. A retrospective study was performed to determine if dexamethasone reduces POI in horses with small intestinal (SI) disease, and to assess effects on incisional health and short-term survival. Data from all horses that underwent SI colic surgery and received 0.1 mg/kg dexamethasone intravenously during surgery (DEX) was extracted from clinical records (61 horses). Thirty-two horses underwent (SI) resection and 29 did not. Data from 61 horses that did not receive dexamethasone (NoDEX) was collected sequentially from clinic records for comparison. Horses were matched for the type of resection performed (if any). Fisher's Exact and Student's t tests were used for data analysis. Fewer horses produced gastric reflux when dexamethasone was administered (DEX 13/61; NoDEX 21/61, $p=0.1570$). Nasogastric reflux produced (litres) was lower when horses received dexamethasone, (DEX 12.06±35.73, NoDEX 16.85±38.90, $p=0.4841$). Differences were not statistically significant. There was no difference in survival to discharge (53/61 DEX, 52/61 NoDEX, $p=1.0000$). Fewer surgical incisions became infected in the DEX group with no statistical significant difference (6/61 DEX, 8/61 NoDEX, $p=0.7774$). Dexamethasone may have a beneficial effect on the incidence of POI but no statistically significant differences were detected. A single dose of dexamethasone does not have a detrimental effect on short-term survival or increase the risk of incisional infections. A randomized controlled trial is required.

E46

Acid-base Disorders in Diarrheic Calves and Horses: Comparison Between the Traditional and Physicochemical Approaches (Abstract E46)

Cheryl Faso[1]; J. Scott Weese[2]; William Sears[3]; Henry Staempfli[4]; Diego Gomez-Nieto[2]

[1]Department of Clinical Studies, University of Guelph, ON, Canada; [2]Department of Pathobiology, Ontario Veterinary College, University of Guelph, Guelph, ON, Canada; [3]Department of Population Medicine, University of Guelph, Guelph, ON, Canada; [4]Ontario Veterinary College, Guelph, ON, Canada

The objectives of this study were to evaluate the capacity of 3 different approaches (Henderson-Hasselbalch (HH), base excess (BE) and strong ion difference (sSID)) to identify acid-base (AB) disorders in diarrheic calves and horses with normal venous pH and pCO_2, and to describe the sSID variables in a subgroup of diarrheic calves and horses without evidence of any AB disorder according to the traditional approach (normal venous pH, pCO_2 and BE).

Medical records were reviewed for all calves and horses presented to a teaching hospital between 2010 and 2016 (calves) and 2010 to 2012 (horses). Data of pCO_2 (mm Hg), pH, and plasma [Na^+] (mM/L), [K^+] (mM/L), [Cl^-] (mM/L), [L-lactate$^-$] (mM/L), and total protein [TP] (g/L) were recorded. All venous blood gases, electrolytes, and pH measurements were performed using a Radiometer 800 Flex machine. TP was measured as total solids by refractometry. By the HH, BE and sSID approaches, AB disorders in calves were defined when any of the respective variables were outside of the reference ranges: [HCO_3] (20 to 30 mM/L), anion gap (AG) (< 20 mM/L), BE (-2 to 6 mM/L), Strong ion difference (SID_4) (38 to 48 mM/L), Total plasma concentration of non-volatile weak acids (A_{tot}) (15 to 22 mM/L) and unmeasured strong ions (USI) (-2 to 3 mM/L). sSID variables for calves were calculated as $SID_4 = (Na^+ + K^+) - (Cl^- + L\text{-}Lactate^-)$; $A_{tot} = 0.343 \times TP(g/L)$ and $USI = A_{tot}/(1+10^{(7.08-pH)}) - SID_4 - HCO_3$. AB disorders in horses were defined when any of the following variables were outside of the following reference ranges: BE (1 to 7 mM/L), [HCO_3] (24 to 34 mM/L), AG (< 16 mM/L); SID_4 (38 to 47), A_{tot} (11 to 16 mM/L) and USI (-2 to 3 mM/L). sSID variables for horses were calculated as $SID_4 = (Na^+ + K^+) - (Cl^- + L\text{-}Lactate^-)$; $A_{tot} = 0.22 \times TP(g/L)$ and $USI = A_{tot}/1+10^{(6.65-pH)} - SID_4 - HCO_3^-$. McNemar's test was used to evaluate whether pairs of approaches are equally likely in finding an AB disorder. The level of agreement between pairs of approaches in detecting AB disorders was assessed using Kappa coefficient test.

For the first part of the study, 37 calves and 50 horses hospitalized for diarrhea with normal venous pH (mean±SD; 7.39±0.03 and 7.41±0.03, respectively) and $pvCO_2$ (41±4.5 mm Hg and 42±4.5 mm Hg, respectively) were included. In these groups, HH approach detected AB disorders in 11 (30%) calves and 15 (30%) horses, BE in 9 (25%) calves and 20 (40%) horses, and sSID in 25 (68%) calves and 33 (66%) horses. McNemar's test demonstrated that HH/BE are equally likely to detect AB disorders in horses and calves, but BE/sSID were not. Kappa agreement test revealed a moderate agreement between HH/BE, and a fair agreement between HH/sSID and BE/sSID, for detection of AB disorders in horses (Table 1)

For the second part of the study, a subgroup of 27 calves and 30 horses with normal venous pH, $pvCO_2$, and BE was identified. 47% (17/27) of those calves had sSID AB disorders: USI acidosis (6/27), A_{tot} acidosis (4/27), A_{tot} alkalosis (4/27), SID_4 acidosis (1/27), SID_4 and USI acidosis (1/27), USI acidosis and A_{tot} acidosis (1/27). 47% (14/30) of those horses had sSID AB disorders: SID acidosis (6/14), SID acidosis and A_{tot} alkalosis (5/14), USI acidosis (1/14), USI acidosis and A_{tot} alkalosis (1/14), and USI acidosis and A_{tot} alkalosis (1/14).

Diarrheic calves with apparently normal AB state had underlying mixed metabolic disorders characterized by the acidifying effects of low SID_4, USI and total plasma proteins, whereas underlying mixed metabolic AB disorders in horses were characterized by the acidifying effect of low SID_4, with concurrent alkalinizing effect of hypoproteinemia. This results suggest that, in addition to venous pH, $PvCO_2$, HCO_3 and BE, clinicians should investigate the quantitative effect of the SID_4, A_{tot}, and USI on the AB status of diarrheic calves and horses.

Table 1. McNemar's (McNem) test and Kappa coefficient results

	HH / BE			HH / sSID			BE / sSID		
	Calves	Horses	Comb.	Calves	Horses	Comb.	Calves	Horses	Comb.
McNem*	0.50	0.18	0.81	0.02	<0.001	<0.001	<0.001	<0.001	<0.001
Kappa	0.4	0.6	0.52	0.18	0.36	0.24	0.17	0.43	0.31
95% CI**	0.08 to 0.72	0.38 to 0.83	0.33 to 0.71	-0.05 to 0.41	0.18 to 0.54	0.14 to 0.43	-0.01 to 0.37	0.22 to 0.64	0.16 to 0.46
p-value#	0.01	<0.001	<0.001	0.15	<0.001	<0.001	0.12	<0.001	<0.001

*McNemar's test p-value **CI, confidence interval. Comb, combined calves and horses. #p-value Kappa test.

E47

Radiographic Evaluation of Compounded and Illegal Over-the-Counter Omeprazole Products (Abstract E47)

Mark Wallace

Carolina Equine Hospital, Browns Summit, NC, USA

Compounded and illegal omeprazole paste products often have production quality control problems as documented by incomplete filling of syringes, air pockets and variations in homogeneity. There is one FDA-approved omeprazole product (GastroGard®) for the treatment of gastric ulcers in horses. Lack of FDA-oversight can lead to undetermined and unregulated production quality. The objective of this study was to compare the production quality of compounded and illegally manufactured omeprazole products compared to GastroGard® paste.

Syringes of compounded omeprazole paste from 5 pharmacies, 8 illegally manufactured over-the-counter omeprazole products and 2 syringes of GastroGard® were acquired. Syringes were radiographed using digital radiographic equipment, labelled to ensure specimen identity in the film. Syringe fill, air pocket presence and product consistency were evaluated. Results were reported using qualitative descriptors for levels of syringe fill and product consistency.

All compounded and over the counter products had incomplete fill and air pockets clearly visible. Eight showed variability in homogeneity. GastroGard® paste showed no visible air pockets and had consistent homogeneity.

The use of compounded omeprazole and illegally produced over the counter omeprazole products in the equine industry is widespread. The lack of an FDA finding on production quality leaves the potential for significant production quality problems. Compounded and illegal omeprazole paste products often have production quality control problems as documented by incomplete filling of syringes, air pockets and variations in homogeneity.

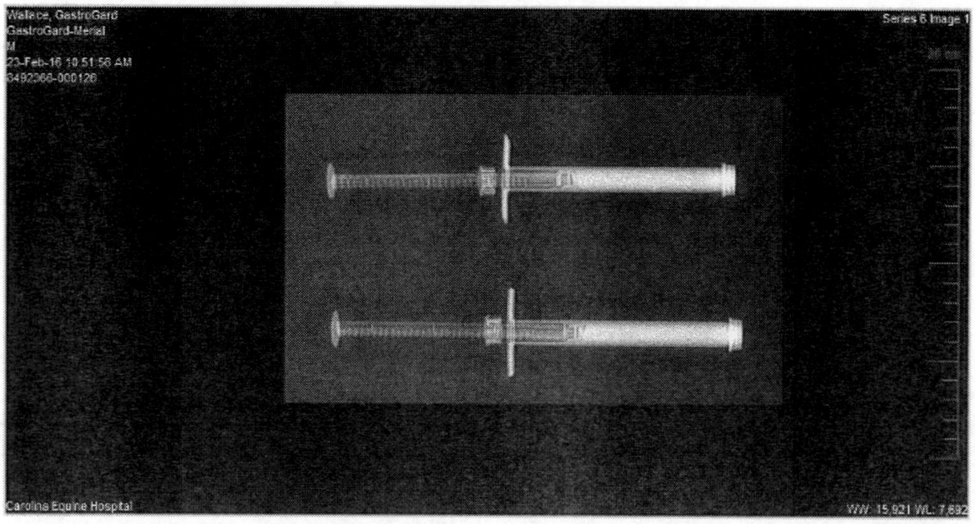

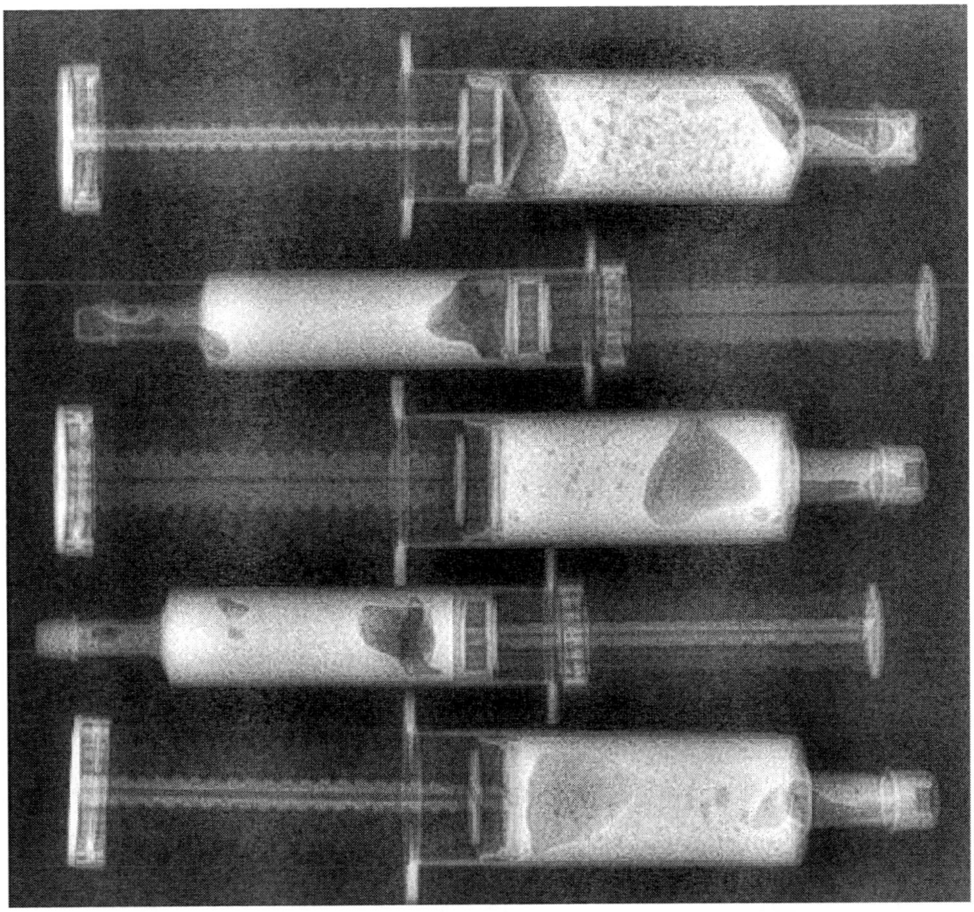

E48

Effects of TLR Agonist and Clarithromycin on Growth of *Rhodococcus equi* in Adult Equine Macrophages (Abstract E48)

Tasha Likavec[1]; Angela Bordin[2]; Steeve Giguere[3]; Noah Cohen[2]

[1]Texas A&M University, Bryan, TX, USA; [2]College of Veterinary Medicine, Texas A&M University, College Station, TX, USA; [3]College of Veterinary Medicine, University of Georgia, Athens, GA, USA

The increasing prevalence of macrolide-resistant *Rhodococcus equi* and the lack of alternative antimicrobials necessitates identifying novel therapies. Stimulating host innate immune responses can enhance clearance of infectious agents. The objectives of this study were: 1) to determine whether an agonist for Toll-like receptors (TLR) 2/6 and 9 (*viz.*, PUL-042) could increase the killing capacity of adult equine monocyte-derived macrophages (MDMs) against macrolide-resistant and macrolide-susceptible *R. equi*; and, 2) to determine whether PUL-042 had a synergistic effect with clarithromycin on the killing capacity of MDMs against *R. equi*. The null hypotheses were that PUL-042 would not increase the killing capacity of MDMs against either strain of *R. equi* and that PUL-042 would not have a synergistic effect on the killing capacity of MDMs when combined with clarithromycin.

To achieve these objectives, MDMs were harvested from 8 apparently healthy, University-owned adult horses. The MDMs of each horse were then separately infected with a macrolide-resistant and a macrolide-susceptible strain of *R. equi*. Phagocytosis of the bacteria was determined by counting colony forming units (CFUs) from lysed MDMs at baseline (T0 = 45 minutes post-infection). The CFUs were also determined at 48 hours post-infection (T48) with 1 of 4 treatments: no treatment (negative control), PUL-042 alone (PUL), clarithromycin alone (CLR), or a combination of PUL-042 and clarithromycin (PULC). The percentage of CFU reduction was determined by calculating the ratio of CFUs between T48 and T0 for each treatment and multiplying by 100%. Results showed that phagocytosis varied significantly ($P < 0.05$) among horses. Additionally, there was no significant difference in phagocytosis (i.e., T0 CFU) between strains ($P = 0.27$). For both macrolide-susceptible and macrolide-resistant *R. equi*, treatment with CLR or PULC had marked and significant ($P < 0.05$) reduction of CFU relative to either negative control or treatment with PUL; for both strains, there were no differences between control and PUL or between CLR and PULC. Although the macrolide-resistant strain tended to be less effectively reduced, this difference was not significant ($P = 0.14$).

In conclusion, PUL-042 alone did not effectively kill either a macrolide-resistant or a macrolide-susceptible strain of *R. equi* in equine MDMs. Additionally, there was not a detectable synergistic effect of PUL-042 in combination with clarithromycin. A similar experiment using adult alveolar macrophages is in progress to determine if more specialized macrophages are affected by PUL-042 and results will be reported. It is possible that PUL-042 might exert indirect effects on macrophages through stimulation of respiratory epithelial cells.

E49

Outcome of a Novel Vertebral Interbody Fusion Technique in Horses with Cervical Vertebral Compressive Myelopathy (Abstract E49)

Yvette Nout-Lomas; Jeremiah Easley; Ellison Aldrich; Rosemary Bayless; Howard Seim

Colorado State University, Fort Collins, CO, USA

Current surgical techniques for cervical stabilization are technically challenging and can result in significant complications. Also, they rely solely on compression and do not provide stabilization in tension, which can result in micromotion, potential catastrophic failure, and implant migration. Here we evaluated a novel technique for cervical vertebral interbody fusion in 6 horses with cervical vertebral compressive myelopathy (CVCM). Six horses diagnosed with CVCM with pre-operative ataxia grades ranging from 2–4/5 on the modified Mayhew scale underwent this novel approach. Two horses were compressed at 1 site and 4 were compressed at 2 sites (Table). The surgical procedure consisted of creation of an intervertebral disc-space defect and placement of an intervertebral fusion device (IFD), 4 titanium polyaxial pedicle screws and 2 connecting rods secured with set-screws (Figure). All horses were re-evaluated at 7 days, 6 months, and 1-year post-operative. The procedures were approved by the Colorado State University Institutional Animal Care and Use Committee. Surgery was performed safely and post-operative radiographs showed good placement of all IFDs, screws and rods in all horses. No implant failure or IFD displacement occurred up to 1-year post-operative. All horses improved within 3 months and by 1-year post-operative ataxia had reduced by 1–2 grades. The use of this novel fusion system requires less disk space removal, provides stabilization in compression and tension, and was technically straight-forward to perform. Surgery was performed successfully with improvement of clinical signs in all horses. This technique is a promising alternative to current surgical procedures for treatment of CVCM.

HORSE	Pre-op Grade	Site(s) Affected	Post-op Grade	Months post-op
1 Joker	3/5	C3-C4 C6-C7	2/5	18
2 Mac	3/5	C4-C5 C5-C6	2/5	18
3 Weeman	4/5	C5-C6 C6-C7	2/5	12
4 Francesca	2/5	C4-C5	1-2/5	6
5 Lucky	2/5	C4-C5	1-2/5	6
6 Reina	2/5	C3-C4 C4-C5	1-2/5	6

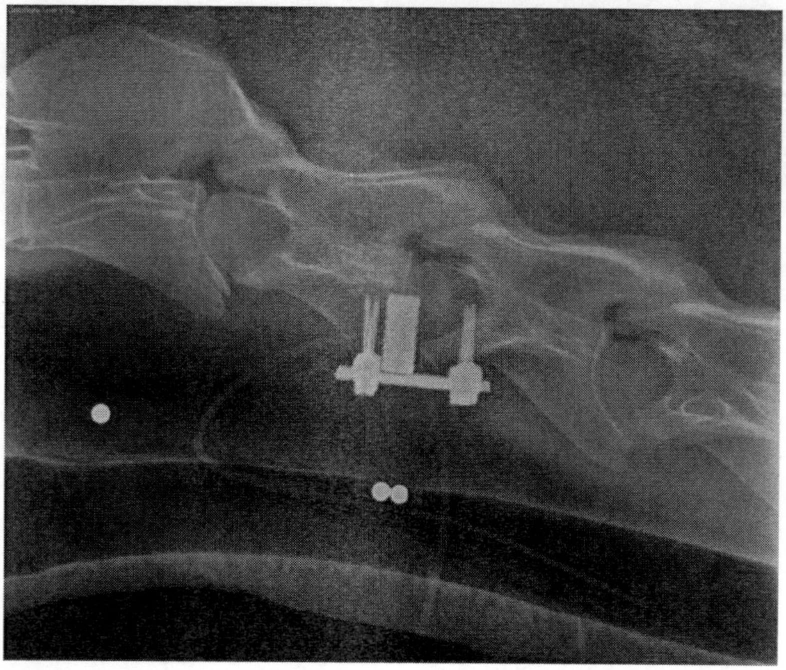

E50

Accuracy of Two Patient Registration Techniques for Neuronavigation-Guided Brain Biopsy in the Horse (Abstract E50)

Lawrence Santistevan[1]; Rebecca A. Packer[1]; Jeremiah Easley[2]; Elissa Randall[1]; Audrey Ruple[3]; Fred Wininger[4]

[1]College of Veterinary Medicine and Biomedical Sciences, Colorado State University, Fort Collins, CO, USA; [2]Colorado State University, Fort Collins, CO, USA; [3]Purdue University, West Lafayette, IN, USA; [4]Veterinary Specialty Services, Manchester, MO, USA

Neuronavigation-guided intracranial surgery is increasingly common in veterinary medicine, although its use has not yet been described in equine species. The objective of this study was to determine the accuracy of neuronavigation-guided intracranial procedures in the horse using a standard fiducial array compared to anatomic landmarks for patient registration. Six equine cadaver heads were used for the study. CT images of each specimen were acquired, with the fiducial array rigidly secured to the frontal bone. Six targets were selected in each specimen. Patient registration was performed separately for 3 targets using the fiducial array, and 3 targets using anatomic landmarks. In lieu of biopsy, 22-gauge wire seeds were placed at each targeted site. Postoperative images were coregistered with the planning scan to calculate offset from the tip of the seed to the target. No statistical difference in accuracy between registration techniques was identified. The impact of surgeon experience was examined separately for each technique using a two-sample t-test and the experienced surgeon was significantly more accurate (1.35 mm, SD 0.69) than novice surgeons (4.04 mm, SD 1.78) using the fiducial array ($P < 0.01$). Although not statistically significant ($P = 0.41$), for the experienced surgeon the mean distance to target was smaller when registering with the fiducial array (1.35 mm) than natural landmarks (2.24 mm). In conclusion, surgeon experience impacts degree of accuracy. Data also suggest that registration using anatomic landmarks is feasible, but perhaps should be reserved for cases where a separate planning scan with fiducials cannot be performed.

E51

Causes of Fatalities During FEI Three-Day Eventing (Abstract 51)

Imogen Comyn[1]; Andrew Bathe[1]; Alastair Foote[2]; Celia Marr[3]

[1]Rossdales LLP, Exning, England, UK; [2]Rossdales LLP, Newmarket, England, UK; [3]Rossdales LLP/Equine Veterinary Journal/University of Glasgow/Horserace Betting Levy Board, Exning, England, UK

Data on equine deaths in Three-Day Eventing (3DE) is lacking. We aimed to report the prevalence of sudden death and fatal musculoskeletal injury in 3DE and describe causes of sudden death. Numbers of starts, fatalities (death/euthanasia) and fatal musculoskeletal injuries between 2008–2014 were obtained from the International Federation for Equestrian Sports (FEI). Post-mortem reports were reviewed and sudden deaths were classified using categories defined for Thoroughbred racing.[1]

There were 112,958 starts and 66 fatalities thus a prevalence of 5.8 deaths/10,000 starts. There were 41 (62%, 3.6/10,000 starts) musculoskeletal injuries; 16 (24%, 1.4/10,000 starts) sudden deaths, 9 cases (14%) where further details were not provided. There was considerable variation in available information on post-mortem examination. A definitive cause of death was identified in 6 of the 16 sudden death cases: 3 central nervous system trauma and 3 blood vessel rupture. Presumptive diagnoses were proposed in 5/16 cases, all cardiopulmonary failure. Cause of sudden death was unidentified in 5/16. This distribution of musculoskeletal injury versus sudden death is similar to fatalities in racing. The distribution amongst categories was also similar to Thoroughbred racing fatalities but definitive diagnoses were reached less frequently. A more standardised system of reporting fatalities would be valuable.

References

1. Lyle CH *et al.* (2011). Sudden death in racing Thoroughbred horses: an international multicentre study of post mortem findings. *Equine Veterinary Journal*. 43:324-331.

E52

Effects of Endothelial Colony Forming Cell/PEG-Fibrinogen Scaffolds on Wound Healing in Horses (Abstract E52)

Randolph L. Winter[1]; Yuan Tian[2]; Wen J. Seeto[2]; Fred J. Caldwell[1]; David D. Pascoe[3]; Jey W. Koehler[4]; Elizabeth A. Lipke[2]; Anne A. Wooldridge[1]

[1]Depatment of Clinical Sciences, College of Veterinary Medicine, Auburn University, Auburn, AL, USA, [2]Department of Chemical Engineering, Auburn University, Auburn, AL, USA; [3]School of Kinesiology, Auburn University, Auburn, AL, USA; [4]Department of Pathobiology, Auburn University, Auburn, AL, USA

Endothelial colony forming cells (ECFCs) are progenitor cells which function in neovascularization and may be useful therapeutically in conditions with poor blood supply, such as distal limb wounds in horses. An injectable ECFC/hydrogel scaffold may ensure cell survival and localization to improve neovascularization and healing.

Autologous ECFCs were isolated from 6 adult horses, labeled with quantum nanodots (QD), and a subset of cells encapsulated in poly(ethylene) glycol fibrinogen microspheres (MS). Full-thickness dermal wounds were created in duplicate on each distal limb and randomly assigned to injections with empty MS, serum, ECFCs, and ECFCs encapsulated into MS (ECFC-MS). Wounds were either biopsied weekly or at week 4 only. Analysis included wound surface area (WSA), granulation tissue scoring (GS), thermography, and immunohistochemical staining for von Willebrand factor (VWF).

Treatments were well tolerated in all horses. QD labeled cells were identified in wound biopsies up to 3 weeks post-injection. There were no significant effects of treatment on GS, but the WSA was influenced by treatment ($p=0.0002$), with ECFCs alone having the smallest WSA at 4 weeks. GS were greater for wounds biopsied ($p=0.0009$) and for hindlimbs ($p\leq0.0001$). Thermographic analysis revealed that the ECFC-MS group had the greatest decrease in limb temperature over time ($p=0.011$). The vWF immunostaining density was lowest in the ECFC-MS group.

Treatment effects on wound temperature, vessel density, and wound size may indicate that ECFCs affect wound healing in equine distal limbs. Despite variability within the model, ECFC treated wounds had a significantly positive effect on wound healing.

E53

Measured and Calculated Parameters of Global Oxygenation in Healthy Foals (Abstract E53)

David M. Wong[1]; Kate L. Hepworth-Warren[2]; Joan Howard[3]; Beatrice Sponseller[3]

[1]Iowa State University, Ames, IA, USA; [2]Farmingdale, NJ, USA; [3]Ames, IA, USA

Septicemia remains a common clinical entity in neonatal foals with varying degrees of illness associated with the process, including cardiovascular shock. In some instances, clinical parameters such as heart rate, blood pressure and urine output may be unreliable or late indicators of inadequate tissue perfusion and poor global oxygenation status, thus providing limited guidance for timely therapy. Conversely, more invasive monitoring such as measurement of cardiac output (CO) and mixed venous oxygen saturation of hemoglobin have been advocated in some studies to better direct therapy (i.e. early goal-directed therapy) and improve outcome in people that are critically ill or in septic shock.

Very few investigations have evaluated global oxygenation in foals, despite the frequency of equine neonatal sepsis. Consequently, equine clinicians primarily rely on physical exam parameters and blood lactate to guide therapy. Because placement of a pulmonary artery catheter in neonatal foals to measure mixed venous blood (SvO_2) is clinically impractical, the objective of this study was to evaluate parameters associated with global oxygen from central venous blood in healthy neonatal foals. Specifically, in this study, the following parameters were measured from 12 healthy neonatal foals from central venous (cv) and arterial (a) blood: saturation of hemoglobin with oxygen ($S_{cv}O_2$, SaO_2), partial pressure of oxygen ($P_{cv}O_2$, PaO_2), partial pressure of CO_2 ($P_{cv}CO_2$, P_aCO_2), pH (pH_{cv}, pH_a) and L-lactate (Lac_{cv}, Lac_a). Additionally, the following parameters reflecting global oxygenation were calculated: venous-to-arterial CO_2 gap (dCO_2), oxygen extraction ratio (O_2ER), ratio between P_aO_2 and the fraction of inspired oxygen (P_aO_2:FIO_2 ratio) and central venous and arterial bicarbonate (HCO_{3cv}, HCO_{3a}), base excess (BE_{cv}, BE_a), hemoglobin (Hb_{cv}, H_{ba}) and oxygen content (O_2Ct_{cv}, O_2Ct_a).

No complications were observed with catheter placement or data collection in this study and in summary, this study reports central venous and arterial blood parameters associated with global oxygenation in healthy foals to establish baseline values at different time points within the neonatal period. These parameters, many utilized in human critical care, might allow better hemodynamic monitoring in critically ill foals. The forthcoming objective is to compare similar global oxygen parameters from critically ill foals, with the hope of improving patient monitoring, therapy and prognostication in equine neonatal sepsis and septic shock.

E54

A Randomized, Blinded, Controlled Crossover Study of Dipyrone to Control Fever in Horses (Abstract E54)

Ryan Avenatti[1]; Ming Yin[2]; Craig Reinemeyer[3]; Melinda Poole O'Banion[1]; Emily Sundman[1]; Tianhua Hu[1]

[1]Kindred Biosciences, Inc, Burlingame, CA, USA; [2]Kindred BioSciences, Inc, Burlingame, CA, USA; [3]East Tennessee Clinical Research, Inc, Rockwood, TN, USA

Dipyrone has well-described anti-pyretic effects. This study evaluated the effectiveness of dipyrone administered intravenously to control fever in horses compared to placebo.

Thirty-one horses with fever (rectal temperature ≥ 102.0°F) from naturally occurring disease were randomized to two groups. Fifteen horses were treated with a single dose of 30 mg/kg dipyrone intravenously (IV), and sixteen horses were treated with a single dose of saline (placebo) IV. Medications or procedures that may have affected the rectal temperature were prohibited. Rectal temperature was assessed six hours following treatment. Twenty-four hours following dose administration, febrile horses were crossed over to the other treatment group under the same procedures. Treatment response was defined as a temperature decrease of either ≥2.0°F from baseline or a return to normothermia (≤101.0°F) six hours after administration of dipyrone or saline. Statistical comparison between treatments was performed using the Fisher's Exact test, two-sided.

Thirty-one (31) horses completed the first phase and 27 horses completed the second phase. Results are presented in Table 1.

Table 1. Treatment response rate

	Treatment response rate (%)		P-value
	Dipyrone responder/N (%)	Placebo responder/N (%)	
First phase (N=31)	10/15 (66.7%)	2/16 (12.5%)	0.0032
Second phase (N=27)	8/14 (57.1%)	0/13 (0%)	0.0019

The majority of adverse events were due to infection from underlying disease. Three horses in the dipyrone group had episodes of soft manure, of which two were pre-existing prior to treatment. One horse in the dipyrone group had several serious adverse events related to severe pleuropneumonia, and was removed from the study following the first phase to receive treatment.

A single administration of dipyrone IV effectively controlled fever 6 hours post administration and was superior to placebo.

E55

A Controlled Study of the Clinical Safety of an Oral Formulation of Dipyrone in Horses (Abstract E55)

Emily Sundman[1]; Craig Reinemeyer[2]; Melinda Poole O'Banion[1]

[1]Kindred Biosciences, Inc, Burlingame, CA, USA; [2]East Tennessee Clinical Research, Inc, Rockwood, TN, USA

Dipyrone is available as an injectable formulation in Europe; however, no oral formulation has been developed for use in horses. The commonly accepted dose of dipyrone is 30 mg/kg administered intravenously. This study evaluated the clinical safety of a high dose oral formulation of dipyrone in horses.

Seven horses were randomized in two treatment groups; 5 horses to dipyrone treatment and 2 horses to saline treatment (placebo). The dipyrone group was treated with 80 mg/kg dipyrone per os (PO) twice daily for 15 days. Placebo horses received saline on the same schedule. Gastroscopy was performed and blood samples for hematology and serum chemistry were collected at baseline, Day 8 and Day 15.

All horses tolerated high dose oral dipyrone administration well. Two dipyrone-treated horses and one placebo horse showed oral mucosal trauma that was related to syringe dosing. One placebo horse had a mild episode of colic on study day 9. Gastroscopy results are summarized in Table 1, below. One dipyrone-treated horse had a clinically significant increase in creatine kinase (CK) and aspartate aminotransferase (AST) beginning on day 8 and remained unchanged throughout the study. One dipyrone-treated horse had a clinically significant increase in CK at study termination.

Initial assessment of high dose dipyrone was well tolerated in horses when administered twice daily for 15 days.

Table 1. Gastric ulcer observations

Horse	Treatment	Gastric Ulcer Number / Description		
		Baseline	Day 8	Day 15
458	Dipyrone	0	0	1-2 lesions / mucosa only
577	Dipyrone	0	0	0
802	Dipyrone	0	0	0
818	Dipyrone	0	0	0
821	Dipyrone	1-2 lesions / mucosa only	0	1-2 lesions / mucosa only
634	Saline	0	0	0
814	Saline	0	0	0

E56

In Vitro Effects of Tamoxifen on Equine Neutrophil Respiratory Burst and Phosphatidylserine Expression (Abstract E56)

Carla Olave; Natalia Morales; Claudio Henriquez; Benjamin Uberti; Jose Sarmiento; Hugo Folch; Gabriel Moran

Universidad Austral de Chile, Valdivia, Los Rios, Chile

Neutrophils participate in innate immunity as the first line of host defense against microorganisms. However, exacerbated neutrophil activity can be harmful to surrounding tissues; this is important in a range of diseases, including allergic asthma and chronic obstructive pulmonary disease in humans, and equine asthma. Tamoxifen (TX) is a non-steroidal estrogen receptor modulator with effects on cell growth and survival. TX has been shown to induce early apoptosis of peripheral blood and bronchoalveolar lavage (BAL) neutrophils from horses with acute lung inflammation (Perez *et al.*, 2016). Our study group previously showed that neutrophil respiratory burst and IL-8-induced chemotaxis also decrease after treatment with TX, confirming a direct action of the drug on the cell type (Borlone *et al.*, 2017).

We studied the pharmacological effect of TX on equine neutrophils obtained by jugular venipuncture from 6 healthy horses, and purified by Ficoll gradient centrifugation. Respiratory burst was evaluated by chemoluminescence induced by opsonized zymosan in neutrophils. Luminol-emitted luminescence was measured at 37°C for 3 hours using a luminometer. The Different concentrations of TX and its combination with selective estrogen receptor modulators (SERM) were evaluated (TX 0,5; 1; 2,5; 5; 7,5; 10; 15 and 20 µM; ICI 182,780 [high affinity estrogen receptor antagonist] 1, 10 and 100 nM; G15 [selective G-protein estrogen receptor antagonist] 1, 10 and 100 nM; β-estradiol 1, 10 and 100 nM; G1 [selective G-protein estrogen receptor agonist] 1, 10 and 100 nM; ICI 100 nM + TX 11,5 µM; G15 100 nM + TX 11,5 µM; β-estradiol 100 nM + TX 11,5 µM; and G1 100 nM + TX 11,5 µM). Phosphatidylserine expression was quantified by flow cytometry using a commercial Annexin V-FITC kit (BD Biosciences). Cells were incubated at 37°C in a humidity chamber (CO_2 5%) for 20 minutes with: DMSO 0,1% (vehicle), TX 11,5 µM, ; β-estradiol 100nM; ICI 100nM; G15 100nM and G1 100nM.

Results showed that TX inhibits neutrophil respiratory burst induced by opsonized zymosan in a dose dependent manner. Nuclear (17-β-Estradiol) and GPR30 cell membrane (G1) estrogen receptor agonists and their antagonists (ICI 182,780 and G15 respectively) do not block nor reproduce TX´s effect. Thus, TX does not inhibit respiratory burst through estrogen receptors. TX 11,5 µM also produced an increase in phosphatidylserine translocation, which was not observed with any of the SERM used. This shows a direct pro-apoptotic effect on neutrophils. In conclusion, TX´s inhibitory and pro-apoptotic effect on neutrophils seems to be independent of estrogen receptors.

E57

Effects of Magnesium Sulfate Infusion on Lung Function and Clinical Scores in Severe Asthmatic Horses (Abstract E57)

Ludovic Tanquerel[1]; Gabrielle Fillon-Berttrand[2]; Jean-Pierre Lavoie[3]; Mathilde Leclère[3]

[1]Université de Montréal/Resident in Equine Internal Medicine, Saint-Hyacinthe, PQ, Canada; [2]Université de Montréal, Saint-Hyacinthe, PQ, Canada; [3]Université de Montréal, Equine internal Medicine Professor, Saint-Hyacinthe, PQ, Canada

We hypothesized that magnesium sulfate ($MgS0_4$) has bronchodilator effects in severe asthmatic horses and potentiates the effects of inhaled salbutamol.

Six severe asthmatic horses were studied using a crossover experiment with a wash-out period between treatments. Clinical scores and lung function measured by impulse oscillometry (IOS) were recorded before and after administration of salbutamol (800 µg) and mock procedure (experiment Ia, to document reversibility). The same parameters were also recorded before and after $MgSO_4$ infusion (2.2 mg/kg/min IV over 20 min) and an equivalent volume of saline, adjusted to the same osmolarity (experiment Ib), and before and after $MgSO_4$ infusion and salbutamol, compared to salbutamol alone (experiment II).

Salbutamol significantly improved the lung function in studies Ia and II. $MgSO_4$ significantly improved clinical scores when administered alone or with salbutamol ($P < 0.05$), with a trend for lung function improvement only observed when administered in combination with salbutamol ($P = 0.06$). $MgSO_4$ also increased the tidal volume (experiment II). Saline administration and the mock procedure had no effect on lung function or clinical scores.

$MgSO_4$ alone or in combination with salbutamol had a mild positive effect on clinical scores of severe asthmatic horses. It had more effect on clinical scores than IOS parameters, which could be due to an undetected bronchodilatory effect or a change in the breathing strategy.

F01

Comparison of Oral, Intravenous and Subcutaneous Fluid Therapy for Resuscitation of Calves with Diarrhea (Abstract F01)

Vincent Dore[1]; Derek M. Foster[1]; Geof W. Smith[2]

[1]College of Veterinary Medicine, North Carolina State University, Raleigh, NC, USA

Neonatal diarrhea is a major source of economic loss to the cattle industry and the leading cause of calf mortality in most countries. The major goals of treatment are to correct dehydration, increase blood pH, restore electrolyte concentrations and provide nutritional support. It is quite possible that the use of oral electrolytes in calves with moderate dehydration and acidosis is of equal or perhaps even greater benefit than giving large volumes of fluids either intravenously or subcutaneously.

The objective of this study was to compare the ability of a commercially available oral electrolyte solution (OES) alone or in combination with hypertonic saline in comparison to the administration of intravenous or subcutaneous fluid therapy to resuscitate calves with osmotic diarrhea.

Thirty-two Holstein bull calves 5–27 days of age were used in this study. Calves were acclimated to their housing and normal diet for two days. Osmotic diarrhea and dehydration were induced by adding sucrose to the milk replacer. In addition, spironolactone and hydrochlorothiazide were given orally and furosemide intramuscularly three times per day. Clinical hydration scores, depression status, fecal consistency and body weight were recorded at each feeding period. Treatment was started when the calf had a hydration score of 2, depression score of 1 and fecal score of 3. Calves were randomly divided in four treatment groups of 8 calves each: 1) OES alone (mixed according to label directions); OES with hypertonic saline (4 ml/kg, IV); 3) intravenous fluids (LRS, 2 liters) or 4) subcutaneous fluid (LRS, 2 liters). Treatment was administered at 0 and 12 hours. Changes in plasma volume, acid-base status, electrolytes concentrations and physical examination parameters were recorded before fluid therapy and again at 1, 2, 4, 8, 12 hours after each treatment.

Decreases in hematocrit, total protein, albumin concentration as well as increase in blood pH between time 0 and 2 were observed in all group. The oral electrolytes group showed greater increase of blood pH at time 1 and 2. All calves showed improvement of their hydration scores, plasma volume, attitude score and fecal score over the period of study. Fecal score did not appear to be affected by treatment.

In conclusion, oral electrolytes with or without hypertonic saline appeared to be an excellent alternative for resuscitation of calves with moderate diarrhea and acidosis and often performed better than small volumes of IV or SC fluids.

F02

Variability of Bovine Leukemia Virus Bulk Tank Milk Antibody Levels Over Different Sample Collection Intervals (Abstract F02)

Emily John[1]; Omid Nekouei[2]; J.T. McClure[1]; Marguerite Cameron[1]; Greg Keefe[1]; Henrik Stryhn[1]

[1]Atlantic Veterinary College, Charlottetown, PI, Canada; [2]Faculty of Veterinary Medicine, University of Calgary, Calgary, AB, Canada

Bulk tank milk (BTM) samples are used to determine the infection status and estimate dairy herd prevalence for Bovine Leukemia Virus (BLV) using an antibody ELISA assay. BLV ELISA variability between samples from the same herd or from different herds has not been investigated. The objective was to determine the within-herd and between-herd variability of a BTM BLV ELISA assay over 1-month, 3-month, and 3-year sampling intervals.

Canadian Maritime region dairy herds (n=523) contributing to bulk tank milk samples in 2013 and 2016 participated. BLV antibody levels were measured in three BTM samples collected at 1-month intervals in early 2013 as well as two BTM samples collected over a 3-month interval in early 2016. Random-effects models were used to estimate the variability between BTM samples from the same herd and between herds for 3 sampling intervals.

The majority of variability of BTM samples was seen between herds (81.14–92.66%). Unexplained variance between samples from the same herd, on square-root scale, was greatest for the 3-year (0.983±0.061), followed by the 1-month (0.615±0.027) then the 3-month (0.560±0.035) intervals.

Variability of BTM sample antibody levels within the same herd is present but is much smaller than the variability between herds, and is greatest for the 3-year sampling interval. The 3-month sampling interval resulted in the least variability and is appropriate to use to determine the baseline level of within-herd prevalence for BLV control programs and to determine the effectiveness of control programs when repeated at regular intervals.

F03

Comparison of Six On-Farm Tests to Estimate Somatic Cell Count at Dry-Off in Dairy Cattle (Abstract F03)

Sahar A. Kandeel; Ameer A. Megahed; Peter D. Constable

University of Illinois at Urbana-Champaign, Urbana, IL, USA

Somatic cell count (SCC) is the most commonly used method to monitor udder health in dairy cattle. Our objective was to evaluate the clinical performance of different cow-side tests for SCC (California Mastitis Test {CMT}, Somaticell® test and three PortaSCC® tests) for diagnosing subclinical intramammary infection (IMI) at dry-off.

Quarter foremilk samples were collected from 117 dairy cows at dry-off. Quarter SCC was measured using Delaval® cell counter (reference method; IMI=SCC>200,000 cells/ml), CMT, Somaticell®, and three PortaSCC® tests. Spearman correlation coefficient, area under the receiver operating curve (AUC), and kappa coefficient (κ) were calculated. P<0.05 was considered significant.

Compared to the reference method, CMT had the highest correlation (rs=0.89), and at optimal cut-point (score ≥ trace), AUC=0.88 and κ=0.77. PortaSCC® color test was the second best performing test (rs=0.82), but required 45 minutes to produce a result, and at the optimal cut-point (50,098 SCC/ml), AUC=0.91 and κ=0.74. PortaSCC® quick test was the third best performing test (rs=0.80), and at the optimal cut-point (54,482 SCC/ml), AUC=0.88 and κ=0.62. PortaSCC® reader (rs=0.68) was inferior to PortaSCC® color test, and at the optimal cut-point (73,419 SCC/ml), AUC=0.82 and κ=0.50. The Somaticell® test didn't perform well (rs=0.44; AUC=0.68, κ=0.24) with an optimal cut-point of 123,864 SCC/ml.

We conclude that the CMT provides the most accurate, practical, and least cost on-farm screening test to predict IMI at dry-off. However, because the CMT is a semi-quantitative test, we recommend using the PortaSCC color test® if a more quantitative on-farm screening test is required.

F04

Ability of Milk pH to Predict Increased Somatic Cell Count at Dry-Off in Dairy Cattle (Abstract F04)

Sahar A. Kandeel; Ameer A. Megahed; Peter D. Constable

University of Illinois at Urbana-Champaign, Urbana, IL, USA

Milk pH is increased in lactating dairy cattle with subclinical mastitis (SCM). Milk pH testing may therefore provide an economic, rapid, and practical method for diagnosing SCM in the field. Our objective was to evaluate the clinical utility of measuring milk pH using a PICCOLO plus® pH meter, Multistix® 10 SG Reagent Strips for Urinalysis (Multistix® strips), and pH Hydrion® paper as on-farm screening methods for diagnosing SCM in dairy cattle at dry-off.

Quarter foremilk samples were collected from 117 dairy cows at dry-off. Quarter somatic cell count (SCC) was measured using a Delaval® cell counter with SCC≥200,000 cells/ml as the reference method for SCM. Milk pH was measured using the pH meter, Multistix® strips, and pH Hydrion® paper. Spearman correlation coefficient, area under the receiver operating curve (AUC), and kappa coefficient (κ) were calculated. $P<0.05$ was considered significant.

Compared to the reference method, Multistix® strips had the highest association (rs=0.61) with Delavel® SCC, and at the optimal cut-point (pH≥7.0), AUC=0.77 and κ=0.42. The pH meter (rs=0.52) was inferior to Multistix® strips, and at the optimal cut-point (pH≥6.69), AUC=0.74 and κ=0.34. pH Hydrion® paper performed poorly at an optimal cut-point of ≥6.2 (rs=0.36; AUC=0.59, κ=0.12).

We conclude that milk pH does not provide a clinically useful primary method for diagnosing SCM at dry off in dairy cattle. Combining milk pH with other diagnostic tests may improve the clinical utility of milk pH to detect quarters with increased SCM.

F05

Clinical Utility of Four Point-of-Care Enzymatic Tests in Predicting Somatic Cell Count in Dairy Cattle (Abstract F05)

Sahar A. Kandeel; Ameer A. Megahed; Peter D. Constable

University of Illinois at Urbana-Champaign, Urbana, IL, USA

Inflammation of the mammary gland increases the activity of more than 20 enzymes in the glandular secretion. Our primary objective was to characterize the ability of three enzymes in quarter milk samples, catalase, lactate dehydrogenase (LDH), and esterase, to predict the presence of a high somatic cell count (SCC) and intramammary infection (IMI) in lactating dairy cattle.

Quarter foremilk samples were collected from 117 dairy cows at dry-off and 109 fresh cows within 4–7 days post calving. A Delaval® cell counter was used as the reference method to measure quarter SCC, with IMI being defined as SCC≥200,000 cells/ml. Milk catalase activity was measured using the Accutest Uriscreen® test. Milk LDH activity was measured using Udder-Check® strips. Milk esterase activity was measured using Multistix® and Serim PeriScreen® strips. Spearman correlation coefficient, area under the receiver operating curve (AUC), and kappa coefficient (κ) were calculated. P<0.05 was considered significant.

Compared to the reference method, Udder-Check® strips had the strongest association (rs=0.79) with IMI, and at the optimal cut-point score≥1, AUC=0.88 and κ=0.73. The Accutest Uriscreen® test had the second highest association (rs=0.72) with IMI, and at the optimal cut-point foam height≥4 mm, AUC=0.85 and κ=0.54. Serim PeriScreen® strips had the third highest association (rs=0.69), and at optimal cut-point score≥1, AUC=0.72 and κ=0.42. Multistix® strips had the lowest association (rs=0.45; AUC=0.70, κ=0.19) at an optimal cut-point score≥1.

We conclude that Udder-Check® strips and the Accutest Uriscreen® test may provide clinically useful on-farm enzymatic screening tests for predicting the presence of IMI.

F06

Persistence of Coagulase Negative Staphylococcal Intramammary Infection in Dairy Goats (Abstract F06)

Véronique Bernier Gosselin; John R. Middleton; Pamela R. F. Adkins

University of Missouri, Columbia, MO, USA

The objective of this study was to describe the persistence of species-specific coagulase negative staphylococcal (CNS) intramammary infection (IMI) in goats from a large dairy goat herd in Missouri.

Udder-half level milk samples were aseptically collected for bacterial culture, somatic cell count (SCC), and milk components (% fat and % protein) determination from all lactating goats (n=909) on a commercial goat dairy. Aerobic bacterial culture was performed on Columbia blood agar for 24 h at 37 °C. Goats with at least one udder-half yielding a single colony morphology of CNS were enrolled for two additional once monthly udder-half level milk samplings. Staphylococcal isolates were identified to the species-level using matrix assisted laser desorption/ionization time of flight (MALDI-ToF) mass spectrometry or PCR amplification and sequencing of the *tuf* gene when MALDI-ToF yielded an indeterminate result. Intramammary infection status was defined based on the presence of the same species of CNS in 1, 2, or 3 of the samples from each udder half. Differences in prevalence of chronic IMI (CNS-CNS-0, CNS-0-CNS or CNS-CNS-CNS) versus only an IMI on the first sample (CNS-0-0) were assessed within species using the Chi-square test or Fisher's Exact test as applicable (Table 1).

Table 1. Association between prevalence of a given coagulase negative *Staphyloccus* spp. (CNS) and persistence of IMI for 217 udder-halves (n=63 with single IMI and n=154 with chronic imi). Parenthetic data represents the within column percentage of IMI

Species	Single IMI (CNS-0-0)	Chronic IMI (CNS-CNS-0; CNS-0-CNS; CNS-CNS-CNS)	P-value
S. simulans	9 (14.3%)	73 (47.4%)	< 0.01
S. xylosus	15 (23.8%)	32 (20.8%)	0.62
S. caprae	4 (6.3%)	19 (12.3%)	0.23
S. chromogenes	2 (3.2%)	13 (8.4%)	0.24
S. epidermidis	3 (4.8%)	11 (7.1%)	0.56
S. arlettae	12 (19.0%)	1 (0.6%)	< 0.01
S. lentus	11 (17.5%)	1 (0.6%)	< 0.01
Other CNS*	7 (11.1%)	4 (2.6%)	0.02
Total	63 (100%)	154 (100%)	

*Other CNS include *S. warneri* (n=3 chronic), *S. equorum* (n=3 single IMI), *S cohnii* (n=1 chronic), and *S. capitis, S. haemolyticus, S. hominia,* and *S. saprophyticus* (each 1 single IMI).

Significance was declared at P<0.05. For species with ≤ 3 IMI within species, data were combined into a single category (Other CNS) for analysis.

Bacterial culture of milk samples from the initial sampling yielded 250 udder-halves with a CNS IMI on 219 of the 909 goats. Of these 250 udder-half IMI, 18 were lost to follow-up after the 1st sample. An additional 15 halves were excluded due to unsuccessful identification of at least one of the CNS isolates from the 3 samplings. The remaining 217 IMI were classified based on the number of samples positive for the same CNS species (Table 1).

The most commonly identified species were *S. simulans, S. xylosus, S. caprae, S. chromogenes,* and *S. epidermidis,* and the majority of these IMI persisted over the 3-month sampling period. On the other hand, most of *S. arlettae* and *S. lentus* IMI did not persist after the first sample. Unlike previous studies, *Staphylococcus simulans* was found to be associated with chronic IMI in the goats in the present study herd. These results suggest that similar to cows, some CNS species might be better host-adapted whereas other CNS species might have an environmental niche.

F07

Use of a Minimally Invasive Tube Cystostomy for the Management of Obstructive Urolithiasis in Goats (Abstract F07)

Leslie F. Wagner[1]; Robert Streeter[1]; Melanie Boileau[1]; Rachel Oman[2]; Jennifer L. Halleran[1]

[1]Oklahoma State University, Stillwater, OK, USA; [2]University of Missouri, Columbia, MO, USA

Obstructive urolithiasis is the most economically significant urologic disorder in male goats. Depending on the urolith type present, medical or surgical treatment options may be utilized. Medical therapy alone has a high rate of recurrence, while surgical options can be costly and typically require general anesthesia in an already compromised patient. A few low cost urine diversion options have been described, however many of them are associated with significant complications. An initial description of a minimally invasive tube cystostomy technique showed promise in the management of urolithiasis in small ruminants. The following report details our implementation and adaptations of this technique to patients in the United States.

A mixed population of Boer, Boer-cross and Pygmy bucks and wethers (n=10) presented to Oklahoma State University for obstructive urolithiasis between July 2015 and October 2016 and underwent a minimally invasive tube cystostomy (MITC) procedure. They ranged in age from 3 months to 8 years (mean=56 months) and in weight from 7 kg to 90 kg (mean=53 kg).

Briefly, the MITC procedure was performed with local anesthesia and sedation. A small keyhole incision was performed in the upper left paralumbar fossa, the dorsal aspect of the bladder was punctured with a trocar and a 10F silicone Jackson-Pratt drain (n=8) or a 16F pig-tail latex gastrostomy tube (n=2) was advanced into the bladder. The body wall was closed in a standard fashion and the tube was secured with a Chinese finger trap ligature where it exited the body wall. A one-way valve was placed on the end of the tubing. Medical therapy, including urine acidification, antibiotics and non-steroidal anti-inflammatory drugs, was instituted after cystostomy tube placement. Depending on the overall health status of the patient, budget, and owner availability for continued care and monitoring, patients were either hospitalized or sent home following tube placement.

Following MITC, resolution of urolithiasis was achieved in 70% of the cases reviewed. The cystostomy tube was in place for an average of 8 days (range of 6–13 days). Complications were associated with the presence of calcium carbonate uroliths (n=2), displacement of the tube from the bladder and accompanying uroabdomen (n=3), or hemoabdomen (n=1). Identifying the urolith type and utilizing a suitable tube and trocar may mitigate potential complications. Based on these cases, MITC shows promise to be a low cost yet effective method of managing obstructive urolithiasis due to phosphatic uroliths in male goats.

F08

Clinical Utility of Measuring Plasma Fructosamine Concentration During Early Lactation in Dairy Cattle (Abstract F08)

Ameer A. Megahed[1]; Mark Hiew[2]; Peter D. Constable[1]

[1]University of Illinois at Urbana-Champaign, Urbana, IL, USA; [2]Universiti Putra, Faculty of Veterinary Medicine, Putrajaya, Putrajaya, Malaysia

Fructosamine is widely used as a long term hyperglycemic biomarker in humans and dogs, but the clinical usefulness of fructosamine as a hypoglycemic biomarker is uncertain. Our primary objective was to evaluate the clinical utility of fructosamine in quantifying the magnitude of hypoglycemia and negative energy balance (NEB) during early lactation in dairy cattle.

Plasma samples were collected weekly from 106 Holstein-Friesian cattle (34 primiparous, 72 multiparous) on days 4–34 postpartum. Plasma concentrations of glucose ([gluc]), β-hydroxybutyrate ([BHB]), total protein ([PP]), and other clinicopathologic indices of energy status were determined. Backfat thickness and longissimus dorsi muscle thickness ([LDT]) were measured ultrasonographically. Plasma fructosamine concentration ([FRA]) was measured at approximately 28 days postpartum. Associations between [FRA] and study variables were evaluated using Spearman's rho and stepwise linear regression. $P<0.05$ was considered significant.

A positive association was detected between plasma [FRA] and mean plasma [gluc] for days 4–28 postpartum ($r_s=+0.34$, $p=0.0016$), and between plasma [FRA] and LDT ($r_s=+0.31$, $p=0.0039$). Plasma [FRA] was negatively associated with mean plasma [BHB] for days 4–28 postpartum ($r_s=-0.31$, $p=0.019$). Stepwise linear regression identified a positive association between plasma [FRA] and mean [PP] for days 4–28 postpartum ($r_s=+0.24$, $p=0.015$). After correcting plasma [FRA] for [PP] at approximately day 28 of lactation, there was only a marginal improvement in the association between plasma [FRA] and mean postpartum plasma [gluc] ($r_s=+0.37$, $p=0.0009$).

We conclude that plasma [FRA] is not sufficiently sensitive or specific to provide a clinically useful method for quantifying the magnitude of hypoglycemia or NEB in early lactation.

F09

Evaluation of a Portable Ion-Selective Electrode Meter for Measuring Sodium Concentration in Cattle Body Fluids (Abstract F09)

Ameer A. Megahed[1]; Mark Hiew[2]; Peter D. Constable[1]

[1]University of Illinois at Urbana-Champaign, Urbana, IL, USA; 2 Faculty of Veterinary Medicine, Universiti Putra, Putrajaya, Putrajaya, Malaysia

A low cost ion-selective electrode (ISE) hand-held meter (LAQUAtwin B-722; Horiba, US) has recently become available for measuring the sodium concentration ([Na^+]) in biological fluids. The objective of this study was therefore to characterize the analytical performance of the ISE meter in measuring [Na^+] in bovine plasma, urine, milk, and abomasal fluid.

Four method comparison studies were completed using 152 plasma, 80 milk, and 206 urine samples from 16 lactating Holstein-Friesian cows with experimentally-induced free water, electrolyte, and acid-base imbalances, and 94 abomasal fluid samples from 6 healthy male Holstein-Friesian calves fed fresh milk with or without an oral electrolyte solution. Deming regression and Bland-Altman plots were used to characterize the meter performance against reference methods (indirect ISE, Hitachi 911 and 917).

The LAQUAtwin ISE meter applied directly in plasma measured [Na^+] 11.0 mEq/L (7.9%) lower than the indirect ISE reference method, consistent with the recommended adjustment of +7.5% when indirect ISE methods are used to analyze plasma. The LAQUAtwin ISE meter run in indirect mode accurately measured urine [Na^+]. The LAQUAtwin ISE meter run in direct mode measured milk [Na^+] 3.1 mEq/L (15.0%) lower than the indirect ISE reference method. The LAQUAtwin ISE meter run in direct mode measured abomasal fluid [Na^+] 4.5 mEq/L (8.6%) lower than the indirect ISE reference method.

We conclude that the LAQUAtwin ISE meter provides a practical, low cost, rapid, and accurate point-of-care instrument for the on farm measurement of [Na^+] in plasma, milk, urine, and abomasal fluid samples from cattle.

F10

Ultrasonographic Determination of Longissimus Dorsi Muscle Thickness and Intramuscular Fat Percentage in Periparturient Dairy Cattle (Abstract F10)

Ameer A. Megahed[1]; Mark Hiew[2]; Peter D. Constable[1]

[1]University of Illinois at Urbana-Champaign, Urbana, IL, USA; [2]Universiti Putra, Faculty of Veterinary Medicine, Putrajaya, Putrajaya, Malaysia

Several studies have examined the rate of backfat mobilization in periparturient dairy cattle, but the rate of intramuscular fat (IMF) mobilization relative to backfat mobilization and muscle tissue mobilization does not appear to have been studied. Our objective was therefore to characterize the changes in IMF relative to changes in muscle thickness and backfat thickness (BFT) in the physiologic response to negative energy balance in periparturient dairy cattle.

One hundred and six periparturient Holstein-Friesian cattle (34 primiparous, 72 multiparous) at -3 and +28 days relative to parturition were examined. A 5 MHz linear ultrasound probe was used to measure BFT and the maximum thickness and IMF percentage (determined using a thresholding-segmentation method) of the longissimus dorsi muscle in the thoracic region (LD-thoracic). Spearman's correlation coefficient (r_s) and mixed models analysis were used for statistical analysis and $P<0.05$ was declared significant.

The mean decrease in BFT over the 31 day period was similar for primiparous (-46%) and multiparous cows (-45%). The mean decrease in IMF of the LD-thoracic muscle was similar for primiparous (-22%) and multiparous cows (-30%). Likewise, the mean decrease in LD-thoracic muscle thickness was similar for primiparous (18%) and multiparous cows (21%). The decrease in IMF was not associated with BFT decrease (r_s=0.03), but was associated with the decrease in LD-thoracic muscle thickness (r_s=0.59, P=0.0011).

We conclude that ultrasonographic measurement of the LD-thoracic muscle thickness and IMF complements ultrasonographic measurement of BFT when quantifying negative energy balance in periparturient dairy cattle.

F11

Plasma Potassium-Lowering Effect of Different Hypertonic Infusion Solutions in Hyperkalemic Neonatal Diarrheic Calves (Abstract F11)

Florian M. Trefz[1]; Peter D. Constable[2]; Ingrid Lorenz[3]

[1]Clinic for Ruminants with Ambulatory and Herd Health Services, LMU Munich, Oberschleißheim, Bayern, Germany; [2]University of Illinois at Urbana-Champaign, Urbana, IL, USA; [3]Bavarian Animal Health Service (Tiergesundheitsdienst Bayern e.V.), Poing, Bayern, Germany

Hyperkalemia is a frequently observed electrolyte imbalance in dehydrated and acidemic calves with neonatal diarrhea that can result in skeletal muscle weakness and life-threatening cardiac conduction abnormalities. This study aimed to compare the plasma potassium-lowering effect of hypertonic sodium bicarbonate, sodium chloride, and glucose infusion solutions in the initial treatment of hyperkalemic calves. A total of 22 diarrheic calves (age < 21 days) with a plasma potassium concentration (cK) > 5.8 mmol/L were included in the study. Calves randomly received one of the following intravenous infusions over a period of 5 minutes: 8.4% $NaHCO_3$ (6.4 ml/kg BW; n=7), 7.5% NaCl (5 ml/kg BW; n=8), or 46.2% glucose (5 ml/kg BW; n=7), and were subsequently allowed to suckle 2 liters of an electrolyte solution. Infusions with $NaHCO_3$ and NaCl provided an identical sodium load of 6.4 mmol/kg BW.

Infusions with hypertonic glucose resulted in marked hyperglycemia and hyperinsulinemia but cK remained unchanged during the first 20 minutes after initiation of treatment. Between 30 and 120 min after start of treatment the most marked decrement of cK was observed in hypertonic $NaHCO_3$, with values that differed significantly ($p<0.05$) from hypertonic NaCl, but not from hypertonic Glucose. After 120 min the observed decrements were equivalent to -26±10%, -9±8%, and -22±6% of baseline cK in groups $NaHCO_3$, NaCl, and Glucose, respectively. Results indicate a treatment advantage of hypertonic $NaHCO_3$-infusions over the use of hypertonic NaCl or glucose in hyperkalemic diarrheic calves, with alkalinization providing the most effective potassium-lowering mechanism.

F12

Effect of Age on Distribution of Danofloxacin and Tulathromycin in Dairy Calves (Abstract F12)

Danielle A. Mzyk[1]; Jennifer L. Davis[2]; Ronald E. Baynes[3]; Geof W. Smith[3]

[1]Food Animal Residue Avoidance Databank, College of Veterinary Medicine, North Carolina State University - Cary, NC, USA; [2]VA-MD College of Veterinary Medicine, Blacksburg, VA, USA; [3]College of Veterinary Medicine, North Carolina State University Raleigh, NC, USA

Calfhood diseases have a major economic impact on beef and dairy operations, due to the costs associated with mortality, treatment and the long term effects on growth and performance. With very few approved drugs for use in veal calves and young heifers, questions remain about the distribution and effectiveness of these drugs in young animals. The objective of this study was to compare the distribution of active drug concentrations in the plasma vs. different effector compartments including interstitial fluid (ISF) and pulmonary epithelial lining fluid (PELF) of healthy 3-week-old and 6-month-old calves. 8 calves in each age group were given a single subcutaneous (SC) dose of either danofloxacin (8 mg/kg) or tulathromycin (2.5 mg/kg). Plasma, ISF, and bronchoalveolar lavage (BAL) fluid were collected and analyzed for drug concentration. Plasma protein binding was measured using a microcentrifugation system in individual animals at 1, 7, and 21 days, 8 weeks and 6 months of age.

6-month old calves maintained higher ISF/plasma concentration ratios throughout the study period compared to that observed in 3-week old calves for both danofloxacin and tulathromycin. Non-linear mixed effect modeling revealed age as a significant covariate in plasma concentration differences seen for calves administered tulathromycin but not for danofloxacin treated calves. Plasma protein binding was assessed *in vitro* in individual dairy calves (n=5) for 5 drugs, including danofloxacin and tulathromycin. No significant differences were seen in calves from the ages of 1, 7 and 21 days of age but trends towards changes in 8-week and 6-month old calves indicate the need to assess protein binding changes for future drug approvals in young calves.

In conclusion, administration of danofloxacin and tulathromycin SC at label doses in 3-week old and 6-month old calves demonstrated differences in distribution that are affected by age. More research is needed to determine the impacts of physiological differences and disease on the distribution and efficacy of commonly used drugs in neonatal calves.

F13

Cross-Sectional Study of Teat and Inguinal Skin Colonization of Dairy Heifers with Various Staphylococcal Species (Abstract F13)

Pamela R.F. Adkins[1]; John R. Middleton[1]; Simon Dufour[2]

[1]University of Missouri, Columbia, MO, USA; [2]University of Montreal, Saint-Hyacinthe, PQ, Canada

The objective of this project was to describe the ecology of host-adapted *Staphylococcus* spp. on the teat and inguinal skin of dairy heifers. A cross-sectional study including Holstein heifers (n=106) at the University of Missouri Foremost Dairy was conducted. Heifers were randomly selected based on housing type and age group and enrolled in the following groups; preweaned heifers (n=22), weaned heifers age 2–6 months (n=15), young pasture raised heifers age 6–13 months (n=26), pasture raised heifers age 13–18 months (n=22), and breeding age heifers >18 months (n=21). Body site swabbing samples, which included a composite sample of all four teats and a second composite sample from the inguinal regions of each heifer, were collected. Swabbing samples were mixed with 10 ml of sterile saline, agitated, and plated on mannitol salt agar and incubated at 37°C for 24 h. Up to 10 staphylococcal colonies, including at least one of each morphologically distinct colony type, were saved for analysis. Staphylococcal isolates were speciated using matrix-assisted laser desorption ionisation-time of flight (MALDI-ToF) mass spectrometry or PCR amplification and sequencing of the *rpoB* or *tuf* gene. The Chi square or Fisher's exact test (as applicable) was used to investigate whether the prevalence of each staphylococcal species differed between the inguinal and teat region. Logistic regression models were used to investigate the relationship between heifer's age (continuous variable) and the probability of having a given body site colonized with a certain staphylococcal species.

A total of 2,042 staphylococcal isolates from 212 body sites of 106 heifers were evaluated. Of the 2,042 isolates, 1,359 were eliminated from the data analysis due to replication of a single species on the same body site, leaving 683 isolates. A total of 17 different staphylococcal species were identified. *Staphylococcus aureus* was more prevalent on teat skin than in the inguinal region ($p<0.01$). All other staphylococcal species were equally distributed between the teat and inguinal region. For most species, colonization of the inguinal and teat region was associated with age of the heifer. *Staphylococcus equorum, Staphylococcus lentus, Staphylococcus sciuri,* and *Staphylococcus vitulinus* were more likely to be found on younger heifers, while *Staphylococcus chromogenes, Staphylococcus devriesei,* and *Staphylococcus haemolyticus* were more likely to be found on the older heifers. A diverse population of staphylococcal species was found on the teat and inguinal region of heifers, and staphylococcal populations tended to change with age. Some important host-adapted species frequently associated with intramammary infection, e.g. *S. chromogenes,* become more prevalent with age.

F14

Quantifying Stress in Primiparous and Multiparous Dairy Cattle Around Parturition (Abstract F14)

Ameer A. Megahed[1]; Mark Hiew[2]; Peter D. Constable[1]

[1]University of Illinois at Urbana-Champaign, Urbana, IL, USA; [2]Faculty of Veterinary Medicine, Universiti Putra, Putrajaya, Putrajaya, Malaysia

Stress can negatively impact the health and productivity of lactating dairy cattle. The primary objective of this study was to determine the degree of stress experienced by dairy cattle around calving, and whether primiparous cattle were more stressed than multiparous cattle in the periparturient period.
One hundred and six Holstein-Friesian cattle (34 primiparous, 72 multiparous) were studied. Animals were moved indoors into separate calving pens approximately 3 days before the anticipated calving date. Heart rate, respiratory rate, and mean arterial pressure (MAP) were determined daily from 3 days prepartum to 3 days postpartum. The MAP was measured using indirect oscillometry by application of a cuff to the coccygeal artery, and pressures were corrected for the height difference above the scapulohumeral joint. Plasma samples were collected and concentrations of glucose ([gluc]) and cortisol ([cortisol]) determined. Mixed models ANOVA was used for statistical analysis and $p<0.05$ was considered significant.

Least squares mean daily heart rates were consistently 10 bpm higher in primiparous (97 bpm) than multiparous cattle (87 bpm, $p<0.0001$). Least squares MAP was consistently higher in primiparous (97 mm Hg) than multiparous cattle (91 mm Hg, $p=0.0004$). Least squares mean daily respiratory rates were consistently 8 breaths/min higher in primiparous (52 breaths/min) than multiparous cattle (44 breaths/min, $p<0.0001$) Primiparous cattle had higher plasma [gluc] than multiparous cattle on days -2, -1, and 0.5 to 3.0 relative to parturition. Plasma [cortisol] was higher ($p=0.028$) in primiparous (10.3 ng/mL) than multiparous (7.7 ng/mL) cattle the day before parturition.

We conclude that primiparous cattle are more stressed around parturition than multiparous cows.

F15

Is Plasma L-Lactate a Reliable Predictor of Non-Survival in Critically Ill Neonatal Calves? (Abstract F15)

Florian M. Trefz[1]; Ingrid Lorenz[2]; Annette Lorch[1]

[1]Clinic for Ruminants with Ambulatory and Herd Health Services, LMU Munich, Oberschleißheim, Bayern, Germany; [2]Bavarian Animal Health Service (Tiergesundheitsdienst Bayern e.V.), Poing, Bayern, Germany

Both in human and veterinary medicine, L-lactate is a well-established biomarker of tissue hypoxia, sepsis, disease severity and mortality and has also attracted increasing attention in bovine medicine due to the availability and validation of cheap and portable L-lactate analyzers. The aim of the present retrospective analysis was therefore to assess the prognostic relevance of increased plasma L-Lactate concentrations (cL-lactate) in a large study population of hospitalized neonatal calves. For this purpose, the data of 10,060 neonatal calves admitted to the Clinic for Ruminants, LMU Munich, were retrospectively analyzed. L-Lactate was determined as part of a routinely performed clinical biochemistry panel on admission.

The overall survival rate in this study population was 65.5% and increased cL-lactate (> 2.2 mmol/L) were evident in 42.2% of calves. The survival-rate of calves in the 9th (4.14-6.70 mmol/L) and 10th decile (> 6.70 mmol/L) of cL-lactate was 52.3% respective 41.5%, opposed to 74.5% in calves of deciles 1–5 which had cL-lactate in the reference range. A binary logistic regression analysis indicated that the odds for non-survival increased by a factor of 1.16 (95% CI: 1.15–1.18) for every mmol/L increase of cL-lactate ($P < 0.001$). The area under the ROC curve for cL-lactate was 0.64 (95% CI: 0.63–0.66).

In conclusion, increased cL-Lactate are associated with an increased risk of non-survival in hospitalized neonatal calves, but they are not sufficiently reliable for the prediction of a negative outcome. Further work is however necessary to assess whether the prognostic relevance is dependent from the underlying disorder.

acvim190

Management Strategies for the EMS Horse

Megan Shepherd, DVM, DACVN (Comparative)
Blacksburg, VA, USA

Outcomes

Upon completion, participants will be able to

1. Describe nutritional management for the obese horse.
2. Describe nutritional management for the hyperinsulinemic/insulin-resistant horse.
3. Describe the validity of nutraceuticals for managing the metabolic horse.

Introduction

Equine metabolic syndrome (EMS) represents obesity/regional adiposity, insulin resistance/hyperinsulinemia and past/present laminitis.

Managing Obesity

Overweight condition and obesity are common in horses. Body condition score (BCS) is positively associated with hyperinsulinemia and negatively associated with insulin sensitivity.[1,2] Therefore, weight loss should be a priority for the overweight horse. A weight loss ration should meet the horse's requirements for protein, minerals and vitamins (http://nrc88.nas.edu/nrh/), while concurrently restrict calories. Strict dry matter intake restriction (1.0–1.3% BW per day) has been prescribed to promote weight loss in horses and ponies.[3] The author's observation is that this degree of restriction isn't necessary for all cases and depends on the energy density of the forage fed. A DM goal of 1.5% ideal body weight (BW) is recommended to help maintain satiety.[4] Maintenance of satiety, by maximizing calories from forage, is important for gut and mental health.[5] Ideally, forage would be fed to meet energy requirements (30.3 kcal/kg ideal body weight),[6] but this requires a forage nutrient analysis.

Forage quality greatly determines the digestible energy; mature forage will be lower in digestible energy. A visual assessment may provide a rough idea of the forage quality. Forages with a low leaf:stem ratio (i.e., relatively stemmy hays) will be more appropriate for the obese horse than a soft leafy hay. Ideally, a forage analysis would be conducted and should include analyses of digestible energy (DE), neutral detergent fiber (NDF) and crude protein (CP). For example, if a horse is fed 20 lb grass hay per day, calorie/energy provision would be 14 Mcal from a poor-quality grass hay (0.7 Mcal per pound, as fed) versus 18 Mcal from a high-quality grass hay (0.9 Mcal per pound); a difference of 4 Mcal per day or 120 Mcal per month!

Time-restricted grazing is not a solution to reduce intake because it increases grazing rate.[7] Grazing muzzles reduce pasture intake by 30–80%.[8,9] Grazing muzzles should be checked for wear and changed at least once within a grazing season. Grazing muzzles enable horses to remain on pasture, where they can cover larger areas and expend more energy than in a dry lot with a hay feeder.[10]

Hay nets slow rate of intake and hole size is negatively correlated with hay intake rate.[11,12] Hampson and collegues[13] increased activity by providing hay free choice in a feeder with two doors that opened in alternating 5-minute intervals. The horses had to walk around a 10-meter-long fence to access the other door of the feeder. Alternately, food toys may be used to promote activity and slow consumption. However, food toys work best with pelleted feeds. The obese horse does not need extra calories; however, grass hay cubes and/or pellets should be considered if a food toy is to be implemented.

Horses should lose 0.5–1% body weight per week or 0.5–1 BCS point on a 9-point scale per month.[4,14] After weight loss, previously obese horses will likely have lower energy requirements than their lean counterparts. Therefore, free-choice forage feeding, without modification by grazing muzzle or hay net, is contraindicated in the long term.

Managing Hyperinsulinemia/Insulin Resistance

Diets high in starch and sugar diets lead to insulin resistance/altered insulin sensitivity/hyperinsulinemia.[15,16] Forage is lower in non-structural carbohydrates (NSC=sugar+starch+fructan; 6–20% DM) than grains (e.g., barley, oats; 37–67% DM).[17] Furthermore, the higher the forage quality, the more likely the forage will meet the horse's energy requirement, negating the need for grains and grain-based concentrates. Forage quality matters and conducting a forage nutrient analysis is helpful in fine-tuning a horse's ration. For example, an 1,100-lb adult average horse needs 16.7 Mcal digestible energy per day. If feeding to meet energy requirements, a 14% NSC hay with 1.1 Mcal DE per pound will provide 967 g NSC per day, which is less than the 985 g NSC provided by a 11% NSC hay with 0.85 Mcal DE per pound. Furthermore, feeding an 18% NSC Orchard grass hay to overweight Arabian geldings did positively influence fasted serum insulin, but only during the 1st week and serum insulin remained within the reference range.[18] Hay may be soaked to reduce the NSC content; however, other nutrients are lost during soaking.[19]

Laminitis has been associated with pasture consumption and inulin (fructan in chicory root) has been used to induce laminitis. Orchard grass NSC, specifically the water-soluble carbohydrate (WSC) fraction, fluctuates throughout the day and season, with season having the greatest effect.[20] Furthermore, horses prefer short grass,[21] which may concentrate NSC in cool-season grasses. However, to date, there is no evidence that forage fructans (i.e., phlein) are linked to changes in the hindgut microbiota and laminitis in the horse.

If the available forage is not meeting the energy needs for the horse, the author suggests energy dense, low NSC complete feeds. Complete feeds are relatively high in fiber (i.e., >15% crude fiber DM), but NSC content varies and often isn't published by the manufacturer. Beet pulp is more energy dense than grass hay and has a low NSC content (4–20% NSC DM).[17] In contrast, rice bran is an energy dense feed/ingredient, but has a higher NSC content (16–34% DM).[17]

Supplements

Horse-specific vitamin-mineral products or balancer pellets should be used to complement essential nutrients missing in the forage. Balancer pellets are commercial products highly fortified in protein, vitamins, and minerals and thus to be fed in small quantities (i.e., 1 lb per 1,100-lb horse; see manufacturer's instructions for details).

A variety of nutraceuticals are marketed for managing the horse with hyperinsulinemia, insulin resistance, metabolic syndrome, and PPID. Nutraceuticals are numerous, under-regulated and under-studied. Nutraceuticals often do not contain what the label states and may even contain molecules/compounds not stated on the label.

Dietary oils (e.g., corn, soybean, canola oil) may be used to add calories for thin horse without increasing dietary NSC. Before adding dietary oils, ensure that the ration meets the essential nutrient needs of the horse. Add oils to the diet slowly to allow the small intestine to adapt to increased dietary fat and prevent diarrhea. Dietary oil should be complemented with additional dietary vitamin E, an antioxidant that specifically protects polyunsaturated fatty acids.

Omega-3 fatty acids (39 mg DHA/kg BW/day) appear to augment the negative effect of dexamethasone on insulin sensitivity.[22,23] However, the benefit of omega-3 supplementation in the face of naturally occurring hyperinsulinemia/insulin resistance has not been determined.

The author suggests starting with a complete and balanced ration before reaching for the supplements/nutraceuticals. Evidence of third-party auditing of products and/or manufacturing facilities, such as with US Pharmacopoeia Convention (USP) or National Animal Supplement Council (NASC), can provide some peace of mind with regard to the product quality. Furthermore, peer-reviewed published, ideally double-blinded placebo control studies provide some evidence of efficacy and safety.

Conclusion

Institute a forage-based weight loss plan for the obese horse. Maximize forage in the hyperinsulinemic/insulin-resistant horse's diet. If forage alternatives are needed, choose low-NSC options, like beet pulp or low-NSC complete feeds. Balance the ration with equine-specific vitamin-mineral products (e.g., loose/granular horse vitamin-mineral, horse vitamin-mineral blocks or balancer pellets). Ensure that the ration is complete and balanced before reaching for the supplements/nutraceuticals.

Questions?

Contact our **Nutrition Service**; vetnutrition@vt.edu; 540-231-1775; 540-231-6448 (fax).

REFERENCES

1. Hoffman RM, *et al. J Anim Sci.* 2003;81:2333.
2. Geor RJ, *et al. J Vet Intern Med.* 2007;21:601.
3. Argo CM, *et al. Vet J.* 2012;194:179.
4. Kienzle E, Fritz J. *Tierarztl Prax Ausg G Grosstiere Nutztiere.* 2013;41:257.
5. Stanley SO, Cant JP, Osborne VR. *J Equine Vet Sci.* 2015;35:973.
6. NRC; The National Academies Press, 2007.
7. Glunk EC, Pratt-Phillips SE, Siciliano PD. *J Equine Vet Sci.* 2013;33:421.
8. Glunk EC, *et al. J Equine Vet Sci.* 2014;34:930.
9. Longland AC, Barfoot C, Harris PA. *J Equine Vet Sci.* 2016;40:26.
10. Shingu Y, *et al. J Equine Sci.* 2000;11:69.
11. Glunk EC, *et al. J Equine Vet Sci.* 2014;34:986.
12. Aristizabal F, *et al. Equine Vet J.* 2014;46:484.
13. Hampson BA, *et al. Equine Vet J.* 2013;45:481.
14. McGowan CM, *et al. Vet J.* 2013;196:153.
15. Treiber K, *et al. J Anim Physiol Anim Nutr.* 2005;89:427.
16. Treiber K, *et al. J Anim Sci.* 2005;83:2357.
17. Equi-Analytical Common Feed Profiles; 2016: http://equi-analytical.com/common-feed-profiles/ (VIN editor: the original link http://www.equi-analytical.com/CommonFeedProfiles/GetResults.asp gives a 404 "Page not found" error).
18. Shepherd ML, *et al. J Anim Physiol Anim Nutr (Berl).* 2012;96:428.
19. Martinson KL, *et al. J Equine Vet Sci.* 2012;32:776.
20. Kagan IA, *et al. J Equine Vet Sci.* 2011;31:488.
21. Fleurance G, Duncan P, Mallevaud B. *Anim Res.* 2001;50:149.
22. Brennan KM, *et al. Am J Vet Res.* 2015;76:889.
23. Hess TM, *et al. J Equine Vet Sci.* 2013;33:446.

acvim191

The Equine Microbiome in Adults and Foals

J. Scott Weese, DVM, DVSc, DACVIM
Guelph, ON, Canada

INTRODUCTION

Historically, thoughts of the host-microbial interaction have typically focused on a host-pathogen approach, whereby microbes cause disease and must often be eliminated. While that is certainly true in many situations, it is an overly simplistic approach that ignores the complex and critical host-microbial interaction that shapes, fosters and maintains life. Microbes can certainly cause disease. At the same time, we cannot live in a microbe-free environment and there is clear evidence that a lack of proper host-microbial interactions is detrimental to health, and even survival.

THE COMMENSAL MICROBIOTA

The horse (like any other species) harbors a vast microbial population consisting of an array of bacteria, Archaea, viruses, fungi, and parasites. This microbial population (the microbiota) outnumbers the hosts own cells by an estimated factor of ~10, meaning that what we identify as a single individual (a horse) is comprised of only 10% horse cells. The sum of the genetic components of the microbiota (the microbiome) exceeds the genetic material of the host by a factor of ~100 (or more), and this has been referred to as our 'second genome.'[1] While numbers of cells or DNA copies do not necessarily define an individual or its function, it is clear that the horse is subject to profound potential influences from this endogenous microbial population, let alone the broader range of microorganisms that it encounters on a daily basis. However, the host typically escapes unscathed from this constant microbial and antigenic exposure. While medicine focuses on the harmful aspects of microbial encounters, the microbiota is critical for normal development and maintenance of health of the host, and the absence of the commensal microbiome is largely incompatible with life. At the same time, there are various potential negative influences of the microbiota and its by-products that can result in illness and death. An inability of the host to develop adequate immunological tolerance of the accompanying massive immunological onslaught, while at the same time to preventing the opportunistic pathogens amongst the microbiota (a rather large component) and externally microorganisms from causing disease are at the root of myriad infectious and inflammatory conditions. The influences of the microbiota in equine disease have only been superficially studied, but in other species, there is abundant evidence of important roles in those types of diseases, as well as other conditions such as allergy, metabolic disease, neoplasia, and obesity.[2-4]

There is probably not one specific 'equine microbiota.' Every body site that communicates with the external world will have a microbiota, the composition of which can vary both between and within sites (e.g., the skin microbiota is different from the respiratory tract microbiota, and the composition of the duodenal microbiota is different than that of the cecum[5]). Understanding the composition of microbiotas, their function, how they vary between and within individuals normally and how they are affected by disease and various endogenous and exogenous factors is an important area of study.

THE GASTROINTESTINAL MICROBIOTA

In terms of biomass and probably influence on the host, the gastrointestinal microbiota is dominant and has received most of the attention to date. The intestinal tract harbors one of the most dense microbial populations on the planet,[6] and alterations in this microbiota have been associated with disease for some time. In humans, it has been estimated that there are 1012–1014 bacteria per gram of feces,[6,7] with estimates of the number of different bacterial species ranging 1000 to many thousands. These are somewhat moving targets, largely as a result of changes in sequencing technology and bioinformatics. Equine estimates are less well developed but there is no reason to believe this would be different in other animal species. Absolute numbers are not highly relevant; the key is there is a massive microbial load in the gut microbiota and it contains a very large number of genetically and functionally diverse microorganisms.

Assessment of the Microbiota

It is only with very recent advances in technology that microbiotas are starting to be understood. Traditionally, assessment of the bacterial microbiota relied on bacterial culture. While providing some useful information, culture has severe limitations because of the inability of a large percentage of microorganisms to grow under conventional culture conditions and corresponding underestimation of bacterial richness (number of different species present). Ready growth by a small number of bacteria leads to overestimation of the relative abundance of those organisms, with corresponding underestimation of diversity. Even if most or all organisms were culturable, culture is completely impractical for description of a population with billions of organisms per gram and 1000 or more different species. Molecular methods can detect organisms that are not culturable using conventional methods, as well detect unknown microorganisms, and offer a distinct advance to culture. Early methods such as denaturing gradient gel electrophoresis (DGGE) provided insights into microbiota variation,[8] but had limited ability to determine components that account for the variation, or to provide specific information about the overall microbial composition. Quantitative PCR can be used to quantify components of the microflora,[9] but this method is not useful for large-scale and broad studies because each PCR reaction can only be used to assess one specific organism or group, and target organisms must be known to create appropriate primer sets. Group-specific oligonucleotide probes have been used for quantitative analysis of the microbiota,[10] but only provide information about limited, targeted organisms. Fluorescent *in situ* hybridization (FISH) can provide detailed information about specific organisms or groups and can be very important in some situations, but is not able to provide broad microbiota assessment and has had little application in horses. Ready availability of sequencing then lead to major advances. In particular, recent advances in next generation sequencing, first with the 454 sequence platform then through Illumina sequencing, have allowed for sequencing of complex microbial populations at unprecedented sequence depth. Paired with advances in bioinformatics, assessment of complex microbial populations can now be cost-effective and routine, generating massive amounts of data.

The 'Normal' Bacterial Microbiota

While powerful laboratory and analytical tools are available, care must be taken to ensure that there is truly a proper understanding of the 'normal' microbiota, through investigation of adequate numbers of animals, study of animals of different breeds and under different management, and evaluation of the spectrum of microbiota compositions (metacommunities) that may occur in healthy individuals. Care must also be taken to ensure that proper testing and analytical tools are used, as massive datasets like this can be markedly mismanaged and manipulated, resulting in erroneous interpretations.

While what truly constitutes the 'normal' microbiota is probably not completely understood (and likely varies with many factors such as diet, management, age, and myriad other factors), general conclusions can be made. In most, if not all, mammalian species, a few bacteria phyla predominate, and while thousands of bacterial species can be present, a small number account for the bulk of the microbiota. In horses, the phyla Firmicutes, Bacteroidetes, Proteobacteria, Actinobacteria and Verrucomicrobia dominate.[11-17] Differences in proportions of these phyla vary between studies, likely as a result of different study populations and different methodologies; however, Firmicutes tend to dominate in healthy individuals, as is the case with most other mammalian species. This phylum of gram-positive bacteria (including Clostridia, one of the dominant classes, and Clostridiales, a prominent order) is receiving increasing attention as its role in health becomes apparent. This can run somewhat counter to typical (superficial) perceptions of the microbiota, particularly the 'bad' nature of clostridia and the desire to limit their populations.

The composition of the microbiota is not static over the horse's life. The foal is presumably born with a sterile gastrointestinal tract, yet it rapidly becomes colonized after birth and pronounced richness and diversity are evident within 24 hours of life.[18]

Archaeal Microbial

Little is known about Archaea, members of a Kingdom of single celled prokaryotes previously classified as bacteria, with limited study of both the composition of the archaeal microbiota and its role in health and disease. Specific study of Archaea in horses has been lacking, with bacterial microbiota studies sometimes briefly reporting data (as some Archaea are detected using universal bacterial primers) and others removing Archaea. Archaea can be found in most (and likely all) equids, but at a low relative abundance compared to bacteria and with limited study of individual members and communities.[19,20] Archaeal richness is dwarfed by that of bacteria, as shown by a study that reported over 86,000 unique bacterial operational taxonomic units (OTUs) but only 63 for Archaea, even when including Archaea-specific primers.[20] Despite the low relative abundance and species diversity compared to bacteria, this group should not be ignored, Of particular interest in methanogenic Archaea, particularly *Methanobrevibacter* and *Methanocorpusculum*, which are dominant members of the archaeal microbiota.[19,20] Limited study also suggests that horses harbor various novel methanogens.[21] The clinical relevance of Archaea is unknown but given the potential effects of methane, including physical effects of gas accumulation and direct impacts on intestinal motility, study of this group in gastrointestinal disease is warranted.

Viral Microbiota

The viral microbiota (virome) is likely vast, with a combination of bacteriophages and mammalian viruses. Assessment of the virome is more complicated and costly compared to relatively straightforward bacterial assessment. As technologies become easier and more affordable, study of the virome has expanded in different species. Certain viruses are known pathogens, but how the broader viral microbiota influences horse health, either directly or through influences on the bacterial microbiota, is not known. It is likely that studies on the equine virome will be published in upcoming years as this area gains increased interest and methods become easier and lower cost.

Fungal Microbiota

Fungi are part of the commensal microbiota, yet their role in health and disease is poorly studied in most species. Equine study from a broader microbiota context is lacking.

Parasitic Microbiota

Parasites are often ignored when the microbiota is considered; however, this broad group contains many members that are commonly present in horses and can influence health and disease. These can range from single-celled protozoal to complex helminthes. Investigation of intestinal parasites has typically focused on population prevalence and the role of parasites in disease, rather than in a host-pathogen mindset. A study targeting ciliated protozoa identified a surprisingly abundant protozoal population, with protozoal sequences accounting for 15% of sequences (compared to 72% for bacteria and 13% for Archaea).[20] Fifteen genera, all belonging to the Litostomatea, were identified, comprised of a variety of species whose activities are largely unknown.

The influence on parasitic populations on aspects such as inflammation and development of immunotolerance is of increasing interest in various species. The interrelationship between the bacteria and parasitic microbiotas is another area of interest that needs further study.

Structure vs. Function

While our ability to describe the microbiota has profoundly changed in recent years, our ability to understand the results continues to lag behind. Even if we can fully define what is present (an elusive goal, given current methods and databases), presence and function are not necessarily the same. There is evidence of profound functional redundancy (the ability of genetically diverse populations to achieve the same end goal) in the human microbiota.[22] Studies of normal horses that have shown marked differences between healthy horses perhaps relate to this. Thus, while understanding of composition and changes in composition are useful, microbial community structure does not always clearly indicate its functional potential or its function. Functional aspects can be investigated through 'shotgun' sequencing approaches, whereby sequencing is performed on whole genome DNA fragments, not 16S rRNA gene PCR products. This eliminates potential PCR bias and provides more information about functional genes;

however, it requires much larger sequence depth (and corresponding cost) and is hampered by limitations in databases. Further, presence of genes or gene fragments does not indicate if or how they are being expressed. Transcriptomics, sequencing of cDNA created from RNA from a sample, can provide a better understanding of the function of the sample (at that time), but is more technically challenging and much more costly because of the required sequence depth. Limitations in databases (i.e., understanding the function of the transcribed gene) also exist, but it can be assumed that continued advances in technology will make these techniques more accessible. Additional methods such as metabolomics can help provide an assessment of certain functions, albeit with some limitations. Ultimately, combinations of many different 'omics' approaches (e.g., metagenomics, transcriptomics, proteomics...) will be used to more accurately assess the microbiota and its function, and to properly understand the role of the microbiota in health and disease.

REFERENCES

1. Zhao L. *Nature*. 2010;465:879–880.
2. Schwiertz A, *et al*. *Obesity*. 2010;18:190–195.
3. Sanz Y, *et al*. *Proc Nutr Soc*. 2010;69:434–441.
4. Bisgaard H, *et al*. *J Allergy Clin Immunol*. 2011;128:646–652.e641–645.
5. Costa MC, *et al*. *Vet J*. 2015;205:74–80.
6. Whitman WB, *et al*. *Proc Natl Acad Sci USA*. 1998;95:6578–6583.
7. Furrie E. *Gut*. 2006;55:141–143.
8. Grønvold A-MR, *et al*. *Vet Microbiol*. 2010;145:366–372.
9. Grønvold A-MR, *et al*. *Vet Microbiol*. 2010.
10. Daly K, *et al*. *FEMS Microbiol Ecol*. 2003;44:243–252.
11. Costa MC, *et al*. *PLoS One*. 2012;7:e41484.
12. Costa MC, *et al*. Characterization of the bacterial microbiota of different gastrointestinal tract compartments in healthy horses. In: Proceedings from the American College of Veterinary Internal Medicine Forum; 2014; Nashville, TN.
13. Moreau MM, *et al*. *Vet Microbiol*. 2013.
14. O' Donnell MM, *et al*. *Lett Appl Microbiol*. 2013;57:492–501.
15. Proudman CJ, *et al*. *Equine Vet J*. 2014.
16. Steelman SM, *et al*. *BMC Vet Res*. 2012;8:231.
17. Costa MC, *et al*. *BMC Vet Res*. 2015;11:19.
18. Costa MC, *et al*. Intestinal colonization of newborn foals characterized by next generation sequencing. In: Proceedings from the American College of Veterinary Internal Medicine Forum; 2013; Seattle, WA.
19. Liu X, *et al*. *Curr Microbiol*. 2014;68:657–662.
20. Fernandes KA, *et al*. *PLoS One*. 2014;9:e112846.
21. Lwin K-O, *et al*. *Archaea*. 2014;2014:483574.
22. Consortium HMP. *Nature*. 2012;486:207–214.

acvim192

Role of the Equine Microbiota in Health and Disease

J. Scott Weese, DVM, DVSc, DACVIM
Guelph, ON, Canada

INTRODUCTION

As investigation of the body's microbiotas increases, it is becoming clear that the complex microbial populations that inhabit different ecological niches on and in the host play important roles in health and have the potential to cause or influence various disease processes. Despite profound advances over the past decade, understanding of these microbial communities and their impacts on health and disease is still in its infancy. Regardless, there are various diseases in horses that are, or are plausibly, influenced by the microbiota. It is almost certain that more equine diseases will be identified as microbiota-associated in coming years and that mechanisms will be discerned. Currently, understanding of the influence of the microbiota on health and disease in horses is superficial, but interesting information is emerging about various conditions.

Colitis

It should not be surprising that an altered fecal microbiota is present in horses with colitis,[1] although broad study is lacking. A study comparing horses with idiopathic colitis vs. controls identified significant decreases in Actinobacteria and Spirochetes in horses with colitis, along with an increase in Fusobacteria.[1] Additionally, while not statistically significant, there was a numerical increase in Proteobacteria, a phylum widely associated with 'dysbiosis.' Changes in community membership (species present) and structure (species present and their relative abundances) were identified. Further, diarrheic horses had significantly lower relative abundances of some key families that are increasingly associated with 'gut health' in other species, particularly Lachnospiraceae and Ruminococcaceae. Those two families are particularly noteworthy given their increasing recognition as being important for health in a variety of animal species.[2,3] Since 'colitis' is not a specific disease, further study of different causes (e.g., *Salmonella, Clostridium difficile*) and differentiation of cause vs effect are required.

Colic

Colic is a multifactorial disease, with many known mechanisms potentially influenced by the microbiota. Among the most plausible effects include influences on gas production and intestinal motility, and many risk factors for colic (e.g., diet change) could directly relate to microbiota alterations. Study has been limited, but one prospective longitudinal study provided important insights. In that study, alterations in the microbiota that occurred before the onset of colic in post-partum mares were observed.[4] This included a marked shift in the relative abundances of Firmicutes and Proteobacteria (with significant increases in Proteobacteria in mares that colicked), as well as depletion of various members of the Ruminococcaceae and Lachnospiraceae families in mares that developed large colon volvulus. In addition to bacteria, Archaea, a strange kingdom of prokaryotic organisms, are of interest because this group includes various methanogens. Methane-producing bacteria could be of relevance, both in terms of the physical effects of gas production and the impact of methane on intestinal motility.

Laminitis

The gut microbiota appears to play a role in at least some cases of another multifactorial disease, laminitis. Influences of diet have long been recognized in the pathophysiology of pasture-associated and grain-overload laminitis, particularly through increases in streptococci and other lactic acid-producing bacteria.[5-7] Changes in the cecal bacterial microbiota have been identified following carbohydrate infusion and experimentally produced early laminitis,[8] including significant increases in *Lactobacillus, Streptococcus, Veillonella,* and *Serratia*. Another study identified increased bacterial richness in horses with laminitis, along with significant differences in relative abundances of two unclassified Clostridiales.[9] As is common, cause and effect are hard to discern, and the changes in these taxa may represent both. For example, *Lactobacillus* and *Streptococcus* are lactic acid bacteria that may be associated with disease through production of lactic acid in response to carbohydrate load, with corresponding local and systemic effects. However, *Veillonella* have been shown to increase as lactate levels decrease in an *in vitro* model of the equine gut microbiota, suggesting this could be a rebound growth in response to changes in the local environment after an inciting microbiota shift.[10] Defining the broad changes that occur during development of laminitis and the range of bacteria that are involved is a challenge.

Gastric Ulcer Disease

The pathophysiology of gastric ulcer disease is complex, multifactorial, and not well understood. There is a distinct gastric microbiota,[11,12] and it can be influenced by management factors such as feeding[11]. Whether there is a direct or indirect role of the microbiota in gastric ulcer disease is unknown.

Metabolic Syndrome

There is great interest in the role of the intestinal microbiota in obesity and metabolic disease in humans, and obesity is increasingly recognized as (in part) an infectious disease. A study of horses with equine metabolic syndrome (EMS) identified a decrease in microbial diversity in horses with EMS, enrichment of Verrucomicrobia and depletion of *Fibrobacter*.[13] How these relate to metabolic disease induction or maintenance or, more broadly, their effects on obesity, remain to be discerned.

Periodontal Disease

The oral microbiota is known to play an important role in dental disease in other species, and differences in the oral microbiota of horses with periodontitis and healthy horses were identified in one study.[14] The microbiota in horses with oral disease was more diverse and dominated by *Prevotella* and *Veillonella*, compared to controls, where *Gemella* and *Actinobacillus* were most abundant. However, the comparison was weakened by use of subgingival swabs for the diseased horses compared to gingival swabs from controls. Whether the changes reflect disease or sampling site is unknown, as is whether any real changes, if present, represent cause or effect.

Other

The number of conditions that the microbiota has been shown to (or to potentially) influence in horses is limited compared to other species, but this is likely a function of limited study. The range of diseases that may have a microbiota basis in humans and other species in impressive, and many of these have equine analogues. These include diverse conditions such as inflammatory bowel disease, atopy, *Clostridium difficile* infection, neurocognitive disorders, obesity, metabolic diseases, colorectal cancer, periodontal disease, and pneumonia.

LIMITATIONS

Associations between the microbiota and health or disease must be interpreted for what they are, associations, not necessarily causation. Differentiating cause from effect is difficult and requires more vigorous study. This can be difficult, particularly when there may be different influences on both the disease process and microbiota, including differences in diet or antimicrobial exposure.[15-17] Other factors such as exercise,[18] as well as fasting, transportation and anesthesia[19] can also impact the equine intestinal microbiota. As a result of the various potential influencing factors and the degree of normal inter-horse variation, large scale or longitudinal studies (or both) may be required to help identify causation. For example, while decreases in Lachnospiraceae in horses with gastrointestinal disease suggest that this family might play an important role in gut health in the horse, particularly given their importance in other animal species, cross-sectional studies cannot differentiate cause or effect. More informative was subsequent study of post-partum mares that identified significant increases in Proteobacteria and depletion of Lachnospiraceae prior to the onset of large colon volvulus compared to controls that did not go on to develop colic.[4] While not necessarily proof of causation, prospective study can provide stronger support for the clinical relevance of observed changes, albeit with much greater effort in study execution. For many sporadic conditions, this may be impractical (e.g., massive numbers of animals would need to be enrolled to result in an adequate population of diseased animals for most conditions).

CONCLUSIONS

It is clear that the bacterial microbiota of the intestinal tract plays an important role in development, immune regulation, and maintenance of health. It is also evident that discernable changes in the microbiota occur with certain diseases, and that some of these changes may be identifiable before the onset of clinical disease. Therefore, it is likely that targeted approaches could be used to identify horses at increased risk of developing diseases such as large colon volvulus or perhaps a broader range of colic types. Technological advances will probably make rapid microbiota analysis feasible in the near future, increasing the chance that this could be a clinically relevant surveillance tool for high-risk groups. Yet, there is currently inadequate information about what changes are relevant and how those vary across different horse populations. Broader study of the microbiota as a predictor of different diseases and in different horse populations is required to be able to develop the evidence needed to interpret microbiota assessment results, since it is clear that **technology** currently outpaces **knowledge**.

REFERENCES

1. Costa MC, *et al. PLoS One*. 2012;7:e41484.
2. Antharam VC, *et al. J Clin Microbiol*. 2013;51:2884–2892.
3. Suchodolski JS, *et al. PLoS One*. 2012;7:e51907.
4. Weese JS, *et al. Equine Vet J*. 2014;doi: 10.1111/evj.12361.
5. Milinovich GJ, *et al. ISME J*. 2008;2:1089-1100.
6. Milinovich GJ, *et al. Environ Microbiol*. 2007;9:2090–2100.
7. Milinovich GJ, *et al. Environ Microbiol*. 2006;8:885–898.
8. Moreau MM, *et al. Vet Microbiol*. 2014;168:436–441.
9. Steelman SM, *et al. BMC Vet Res*. 2012;8:231.
10. Biddle AS, *et al. PLoS One*. 2013;8:e77599.
11. Dong H-J, *et al. J Microbiol Biotechnol*. 2016;26:763–764.
12. Perkins GA, *et al. Appl Environ Microbiol*. 2012.
13. Elzinga SE, *et al. J Equine Vet Sci*. 2016;44:9–16.
14. Kennedy R, *et al. Vet Res*. 2016;47:49.
15. Dougal K, *et al. PLoS One*. 2014;9:e87424.
16. Harlow BE, *et al. J App Microbiol*. 2015;119:1234–1244.
17. Fernandes KA, *et al. PLoS One*. 2014;9:e112846.
18. Almeida MLMd, *et al. PLoS One*. 2016;11:e0167108.
19. Schoster A, *et al. Equine Vet J*. 2015;48:595–602.

1.
acvim310

Animal Supplement Regulations in the USA: Separating Facts from Fiction

Bill Bookout, BS, MBA
Valley Center, CA, USA

Regulatory Landscape for Animal Supplements

Under current law in the United States, only two legal categories exist for animal products that are similar to human dietary supplements: animal food/feed or animal drugs. Categorically - meaning legally - there is no other choice. Most people are unaware that a legal category does not exist for animal supplements, in addition to not knowing that many commonly utilized ingredients are not approved for nutritional purposes in animals. Examples of ingredients unapproved in animal food/feed include those often found in joint products, such as glucosamine, chondroitin sulfate, and methylsulfonylmethane (MSM), as well as virtually all herbal ingredients if they are in products for purposes other than flavorings.

When the Dietary Supplement Health Education Act of 1994 (DSHEA) became law, the legislation amended the Federal Food, Drug and Cosmetic Act (FD&C) to create a specific legal category under Food for Human Dietary Supplements and allowed the labeling and marketing of dietary supplements for human use. Unfortunately, when the issue was debated in Congress, companion animals (dogs, cats, and horses) were not considered and language was not included in the legislation to allow similar products for animals to be regulated as "dietary supplements," as was provided for people.

Examination of the Congressional Record when the bill was finalized showed Congress neither specifically included nor excluded the application to animals. Although it may have been reasonable to at least include animals not intended for human consumption, such as dogs, cats, and horses, the topic of animals was overlooked, probably because the animal supplement industry did not exist to any significant degree in 1994 and the purpose of DSHEA was to address the increasing consumer demand for dietary supplements for people.

The primary agency responsible for the regulation of animal food and drugs is the Food and Drug Administration, Center for Veterinary Medicine (FDA-CVM), which works closely with the states through regulatory associations like the Association of American Feed Control Officials (AAFCO) to help ensure both federal and state laws are followed and companies remain in compliance. In 1996, FDA-CVM published a notice in the Federal Register explaining why the agency believes DSHEA does not apply to animals, stating that upon thorough review of the legislation it was evident that Congress was concerned only with human products and did not consider animal products, and for this reason the agency concluded that Congress did not intend the law to apply to animal products. Accordingly, the agency stated they did not intend to apply DSHEA to animal products.

This opinion places products similar to human dietary supplements but marketed for animals with only two possible legal categories under United States law: animal food/feed or animal drugs. The FD&C defines food (animal feed) and a drug as:

- **Food** - "articles used for food or drink for man or other animals...and articles used for components of any such article." There is no requirement that animal foods have pre-market approval by FDA-CVM. The act does require that animal foods, like human foods, be pure and wholesome, contain no harmful or deleterious substances, and be truthfully labeled.
- **Drug** - "an article intended for use in the diagnosis, cure, mitigation, treatment or prevention of disease, or an article intended to affect the structure or function of the body other than food." In the drug definition, the courts have interpreted "food" as something that provides nutrition, taste, or aroma. If a food affects the structure or function of the body, it does so by these properties (e.g., a food may provide nutrients such as calcium for proper bone structure). However, if a substance affects the structure or function of the body apart from its nutritive value, such as improvement in joint function, it may be considered a drug. Structure/function effects extending beyond the "food"

umbrella also include claims for improved or increased production and performance, and alteration or improvement in function.

When a substance, including one considered food, is intended to be used for the treatment or prevention of disease or for a "non-food" structure/function effect, FDA considers it a drug. Under the FD&C, a new animal drug must be shown to be safe and effective for its intended use by adequate data from controlled scientific studies as part of a New Animal Drug Application (NADA). If a product on the market is not approved, it may be deemed an adulterated drug and subject to regulatory action.

Nevertheless, with the dramatic expansion of the human dietary supplement industry, the natural extension was to make similar products for animals in response to consumer demand recognizing the benefits for companion animals. This was a great idea except for one thing: the products are technically not approved and, in 2002, were very close to being removed from the marketplace.

AAFCO is a not-for-profit, private volunteer association whose members must be regulatory officials from a state, federal, or foreign government. As an association AAFCO has no regulatory authority or enforcement ability, yet a basic goal of AAFCO is to provide a mechanism for developing and implementing uniform and equitable laws, regulations, standards, and enforcement policies for regulating the manufacture, distribution, and sale of animal feeds; resulting in safe, effective, and useful feeds. AAFCO thereby promotes new ideas and innovative procedures, and urges adoption by member agencies, for uniformity.

Recognizing the growing problem, evidenced by the increasing number of companion animal products entering the market, AAFCO attempted to be proactive and formed the following committees to try and address the issue:

- Feed Labeling Subcommittee on Nutritional Health and Nutrient Function Claims (1998)
- Nutraceutical Regulatory Advisory Panel (NRAP, 1998–1999)
- Novel Ingredients Regulatory Framework Task Force (NIRFTF, 1999–2000)
- Botanicals and Herbs Committee (2000–2002)

Each committee failed to identify a viable pathway for dealing with "unapproved" ingredients, non-nutritional uses, and the rapidly increasing number of products, and were all disbanded. Ultimately this led to the formation of the Enforcement Strategy for Marketed Ingredients (ESMI, 2001–2004). The ESMI charter was simple: to identify and recommend enforcement policy and actively remove products from the marketplace which contained ingredients not approved for use in animal feed.

NASC Response and Accomplishments

The National Animal Supplement Council (NASC) was formed in direct response to ESMI announcements. NASC is a non-profit industry trade association. While the organization was founded in 2001, the objectives and approach were defined in April 2002 when principals and senior executives from approximately 20 companies met to formalize tactical actions and strategic objectives addressing the issue for animal health products.

The strategic objectives of NASC are:

1. To identify and define a viable pathway which allows responsible companies to purvey products to the marketplace without the threat of broad sweeping regulatory policy or actions by cooperatively engaging the regulatory agencies, if possible, at both state and federal levels.
2. To evaluate, define, and implement guidelines which are fair, reasonable, responsible, and nationally consistent. This system of industry driven self-regulation may provide a platform for a long-term solution.
3. To differentiate NASC members by their conduct and influence buying decisions though education.
4. To take any actions the Board of Directors deems appropriate to protect and defend our industry.

Between 2002 and 2005, NASC tried diligently to navigate the food/feed pathway knowing that submitting a NADA to FDA-CVM was not possible for these types of products. At some point, it was realized that many drugs currently marketed are not approved for use in specific animal species, yet are allowed to be sold by FDA-CVM. One should understand what FDA can and cannot do: the agency can allow products to be marketed as unapproved drugs under enforcement discretion, provided they have a high degree of confidence the company is acting responsibly; however, the agency cannot create or change the law. This realization was a turning point for all involved.

Therefore, NASC developed two approaches for feed supplements and dosage form animal health products that are consistent with both federal and state laws.

1. Feed supplements - NASC follows the recommendations of AAFCO or the laws of the state in which the product is sold. NASC does not represent products that are intended as a sole source of nutrition.
2. Dosage form animal health products - Utilizing the approach from over-the-counter drug products, NASC developed a separate set of labeling templates and quality parameters that apply these types of products.

NASC redefined its plan, communicated its intentions and reasoning to FDA-CVM and AAFCO, and offered its commitment to act completely in a transparent manner while requesting input as we progressed. Looking at commitments and accomplishments of the industry gives a good overview of where NASC is today:

- Have developed and implemented Good Manufacturing Practice Standards (cGMPs) which are similar, but not identical to FDA's 21 CFR Part 111 for Human Dietary Supplements.
- Labeling products properly, consistent with claims are within those supporting the normal health, structure, and/or function of the animal.
- Review of ingredients by Scientific Advisory Committee.
- Randomly testing finished products to help ensure these products meet label claims.
- Continued vigilance and effective risk management through NAERS™ (NASC Adverse Event Reporting System), which is available to regulatory agencies.
- Routine meetings with both FDA-CVM and AAFCO representatives to discuss progress, issues, outcomes, etc.
- Conducting routine joint training programs with invitations for participation from federal and state regulators.
- Establishing and requiring independent quality audit programs to provide verification that quality systems requirements have been implemented.
- Registration of products and facilities with FDA-CVM.
- Supplier qualification programs for raw materials suppliers and contract manufacturers.

So far, the approach has worked and NASC has no reason to think this will not continue. Since that time when strong enforcement policies were recommended, the situation has progressed from one of genuine survival of the industry to one of cooperation with regulatory agencies at both the state and federal levels. Representing the majority of responsible participants in the industry, NASC has achieved solutions that are in the best interests of everyone influenced by animal supplements: consumers, manufacturers, dealers, distributors, veterinarians, regulators, raw material suppliers, stores, breeders, trainers, and, not least importantly, the animals themselves.

ASC Quality Seal

As part of its ongoing effort to improve and standardize the animal health supplement industry, NASC initiated its Quality Seal program. When a consumer sees a product with the Quality Seal it means two things: the company is reputable and voluntarily agreed to adhere to NASC's quality standards, which include submitting to an independent audit to ensure conformance with quality system requirements;

and the company is contributing to a goal greater than the company itself, which is helping to ensure these products will continue to be available to animal owners.

Use of the Quality Seal is a privilege that cannot be purchased. Before they are granted permission to display the Quality Seal on their products, websites, and marketing materials, NASC member companies are required to demonstrate compliance with the following criteria:

1. The company must have a **quality control manual** in place that provides written standard operating procedures for production process control. This helps ensure the company is providing a consistent and quality product.
2. The company must have an **adverse event reporting/complaint system** in place to continually monitor and evaluate products and ingredients.
3. The company must follow proper **label guidelines** for all products and all forms of labeling.
4. The company's product labels must include any specific **warning and caution statements** recommended by the Food & Drug Administration's Center for Veterinary Medicine and the NASC Scientific Advisory Committee for ingredients.

The importance of NASC and its comprehensive audit program is to help minimize and control risk, statistically track all product information, keep the consumer educated with correct labeling information, and make the consumer aware of any possible ingredient complexities with caution and warning statements. Ultimately, the entire program is designed to help elevate the quality of the animal health supplement industry for the benefit of all stakeholders including consumers and, more importantly, the animals themselves.

CALL TO ACTION FOR VETERINARIANS

How can veterinarians contribute to the progress of NASC's mission? When clients turn to you for product advice, seize the opportunity to tell them about the Quality Seal, then recommend products that display the Quality Seal. In addition, educate them about the importance of purchasing products from NASC members, as it is the best way they can support efforts to improve the quality of animal health supplements and ensure the continued availability of these products in the U.S. market.

And when researching products to carry in your practice, recommend to clients, or to use with your own animals, evaluate every one by asking the following questions:

1. **Who formulates the products?**
   It is important that animal supplements are formulated by professionals who have the expertise and knowledge necessary to successfully formulate a product. Call the maker of the product and ask who formulated it and what expertise they have.
2. **Who can I call if I have questions or need expert advice?**
   The maker of a product you're considering should have employees, technicians, or veterinarians on staff who are knowledgeable and have the expertise to answer any question you may have. It is important to establish faith in the companies you deal with, and if they can't answer your questions to your satisfaction, consider choosing a different product.
3. **What sort of testing and quality standards do the companies have?**
   All companies should have testing and quality standards in place for their products, as well as someone on staff who is well qualified to answer any questions you may have about their testing and quality standards.
   Is the company an NASC member and have they passed an NASC audit?
   Products which display the Quality Seal come from companies that have been audited by NASC and have had the following verified:
   a. The company has a written quality control manual that meets NASC cGMPs, which are modeled after FDA's 21 CFR Part 111 for Human Dietary Supplements.
   b. The company complies with labeling guidelines from FDA and AAFCO.

c. The company enters all products into a National Adverse Event Reporting System, which FDA has access to.
d. The company complies with any recommendations from the NASC Scientific Advisory Committee.
e. The company has passed independent testing for meeting label claims.
f. The company has signed a written code of conduct contract.
g. The company attends routine meetings for quality and regulatory issues.
h. The company is in good standing with NASC.

4. **Has the product been independently tested by NASC?**
It is important that the company's products are independently purchased (to avoid sample bias) and tested to verify label claim or potential contamination issues such as the presence of lead or *Salmonella*. This is exactly what NASC does with member products.
5. **Are there any red flags that make me think twice about the products?**
Always keep this in mind: Companies that make claims that sound too good to be true probably are, and cheap products are usually cheap for a reason. Never hesitate to investigate a company you don't feel quite right about, and never hesitate to walk away from a company that won't directly answer your questions.

When the majority of responsible industry participants are working cooperatively with regulatory agencies and are supported by downstream business partners, including veterinarians and animal owners purchasing their products, true and positive change is possible. This is precisely the result NASC is helping to achieve. The next time you evaluate or purchase an animal supplement remember that NASC member companies are driving and funding the solutions that apply to all industry participants to make the industry better, and to contribute to a cause greater than their own individual company interests.

References

1. Nolen RS. Facing crackdown, dietary supplement companies promise changes. *J Am Vet Med Assoc.* 2002;221(4):479–481, 483.
2. Burns K. Supplement companies made changes after facing crackdown. *J Am Vet Med Assoc.* 15 January 2017.
3. Burns K. Assessing pet supplements. *J Am Vet Med Assoc.* 15 January 2017.
4. Natural Products Insider. Regulatory considerations for animal supplements in North America. www.naturalproductsinsider.com/digital-issues/2015/03/the-science-of-supplements.aspx.
5. Natural Products Insider. Regulatory update, claims for animal supplements, and important changes. www.naturalproductsinsider.com/digital-issues/2016/09/animal-joint-health.aspx.
6. Association of American Feed Control Officials (AAFCO). www.aafco.org/.
7. FDA's Regulation of Pet Food. www.fda.gov/animalveterinary/resourcesforyou/ucm047111.htm.
8. Food Safety Modernization Act. www.fda.gov/food/guidanceregulation/fsma/default.htm.
9. US Food & Drug Administration. Guidance for Industry: Structure/Function Claims, Small Entity Compliance Guide.
10. North American Compendiums, LRVNHP Program. naccvp.com.
11. National Animal Supplements Council. (VIN editor: Original link was modified on 3-29-2017).

acvim312

Immunologic, Structural and Functional Response to Intestinal Abstinence

Korinn E. Saker, MS, DVM, PhD, DACVN
Raleigh, NC, USA

What Is Meant by Intestinal Abstinence?

Typically, abstinence is referred to in association with human proclivities such as alcohol consumption or other personal questionable activities, and is defined as the practice of not doing or having something that is wanted or enjoyable. Following that thread, intestinal abstinence would then be defined as the practice of purposefully not feeding or withholding oral provision of nutrients to a patient via the enteral route. Not to be confused with anorexia, a self-imposed nutrient deprivation, intestinal abstinence is nutrient deprivation imposed by the attending clinician for any number of reasons. To better understand why this concept is of interest to veterinarians, an overview of how the gastrointestinal tract (GIT) responds to nutrients, as well as to the absence of nutrients, is a great starting point.

How Do Nutrients, or Lack Thereof, Impact the GIT?

Adequate nutrition, as well as specific individual nutrients are integral in enterocyte functionality and gastrointestinal immunity. Enterocytes are constantly being exposed to foreign material (antigens) from diet, environment, drugs, etc. To process these antigens, the enterocytes require adequate and appropriate nourishment for optimal development to provide the functions associated with nutrient digestion and absorption. Additionally, enteral provision of nutrients to the GIT improves blood flow to the enterocytes maintaining splanchnic perfusion and prevents structural and functional alterations of the gut barrier. Preservation of epithelial tight cell junctions is necessary for barrier structure in prevention of bacterial translocation, while optimizing the environment of villous and crypt cells helps ensure secretion of necessary brush-border enzymes and preserves mucosal mass and microvilli height. As well, a number of factors related to intramural nutrients and the intestinal microbiota affect gut integrity and the risk for increased permeability. Under normal circumstances with adequate enteral nutrient intake, one can expect an adequate number of protective probiotic commensal flora and dietary short-chain fatty acids, all of which stimulate heat shock proteins (HSPs) in the gut epithelial cells. These processes stabilize barrier function and maintain gut integrity. When there are adequate nutrients present to feed the commensal bacteria, they can keep virulent bacteria at bay.[1]

Nutrients have a well-documented role in the gut-associated lymphoid tissue (GALT) aspect of the immune system. Specifically, protein, lipid, and selected minerals and vitamins influence T- and B-cell function, delayed hypersensitivity reactions, macrophage and neutrophil function, and complement. Limitations on nutrient delivery to the GIT negatively impact the GALT's ability to protect against antigen-driven insults that can ultimately result in single or multi-organ failure, severely compromising patient morbidity and mortality. **Table 1** provides an overview of the effect of malnutrition on immune function.

Table 1. Effect of inadequate nutrient provision on immune function

Type of nutrient deficiency	T-cell function	DST	B-cell function	Macrophage function	Neutrophil function	Complement
Macro-nutrients						
Chronic P–E	↓↓	↓↓	↓ to variable	↓	↓	↓ esp C2, C3
Acute protein	N or ↓	↓	↓	↓↓	↓	↓ esp C2, C3
Lipid	↓	-		-		-
Micro-minerals						
Zinc, iron, copper, selenium	↓↓	↓	N	Variable	↓	-
Vitamins						

Bs A C	↓ Variable	↓ ↓ ↓	↓↓ ↓ N	↓ - -	↓ - Variable	↓ C3

Modified from Chandra; DST=delayed-type hypersensitivity skin test; N=normal; esp =especially; (-) indicated conflicting data.

In the absence of enteral nutrient delivery, the gut is not the sole concern, as the body elicits both systemic physiologic and metabolic responses to nutrient deprivation. The responses differ depending on the general health status of the patient faced with intestinal abstinence. In simple starvation, nutrient deprivation generally occurs in the healthy individual. Shifts in nutrient utilization occur based on endogenous availability and the innate instinct to survive in the face of nutrient deprivation. Fat or adipose stores become the primary source of energy sustaining the individual until they are depleted, at which time lean muscle is catabolized for energy via gluconeogenesis. The initial sparing effect of protein stores from lean muscle mass prolongs survival when food deprivation persists. Eventually, protein stores from lean muscle mass will diminish to the point of respiratory or cardiac failure.

The scenario is quite different when critical illness, trauma, surgery is a part of the scenario, where a characteristic three-phase (acute, stable, and recovery) **stress response** occurs. This stressor, compounded by nutrient deprivation from anorexia or prescribed NPO, results in depletion of both endogenous adipose (fat) stores for energy and lean muscle (protein) stores to support this illness-induced stress response. Increased circulating levels of stress hormones (i.e., cortisol, glucagon, catecholamines, TSH, growth hormone, etc.), inflammatory cytokines (i.e., TNF-α, IL-1, IL-2, additional inflammatory interleukins), and acute phase proteins drain endogenous protein stores, while fat remains the primary energy source. The resultant negative nitrogen balance from catabolism of lean muscle can severely compromise immunocompetence, delay wound healing, and increase risk for postoperative complications. Metabolic ramifications of this stress response along with nutrient deprivation to the GIT promote severe gut mucosal cell compromise, and thus barrier breakdown; diminish functions of GALT; and promote a protein-energy malnourished state.[2]

Is Intestinal Abstinence a Beneficial Practice for the Critical Patient?

This is the million-dollar question! A clinician's decision to feed or to withhold food from the critical patient is, in many cases, multifactorial. Reasons include the desire to follow traditional dogma of "resting the gut"; fear of metabolic intolerance (i.e., sepsis, neurological impairment, vasopressor support, etc.) or intestinal intolerance to enteral nutrition support (i.e., dysmotility, vomiting, regurgitation, bowel anastomosis, etc.); uncertainty as to what diet and/or nutrient levels to provide for any particular patient (i.e., complicated clinical cases); inexperience in translating diagnostic results into necessary nutrient and diet characteristic needs; the belief that "they will eat when they feel better"; and so on. More recently, veterinary nutritionists appear to be encouraging early nutritional support for critical patients. This belief is supported by clinical guidelines established by an exhaustive review of available human studies provided through national and international human nutrition organizations including the American Society of Enteral and Parenteral Nutrition (ASPEN), the European Society of Enteral and Parenteral Nutrition (ESPEN), and associated working groups.

That trauma and surgery patients are commonly in a catabolic state is not surprising;[3] but patients with chronic and acute critical illness should also be looked at in this vein, as they commonly experience the stress response described above. This stress-associated skeletal muscle breakdown can only be partially mitigated by aggressive enteral macronutrient delivery and can last for weeks post-discharge (Dante *et al.* 2016). Appropriate pre-op and early post-op nutritional support have overwhelmingly been linked to improved patient outcomes after clinical illness, including reduction in postoperative infectious complications; increased ventilator-free days; shorter hospital stays; and lower mortality.[4-8] The traditional NPO approach for patients diagnosed with a primary or secondary pancreatitis, has recently been challenged showing improved patient outcomes with early enteral feeding.[9] The prevailing argument for early enteral feeding focuses on maintaining a healthy GI mucosal environment, and how this can minimize adverse single- or multi-organ complications to the patient. To this end, early vs.

delayed enteral nutrition intervention has recently become the recommended standard of care in both human and veterinary medicine for most GI and pancreatitis patients.

The flip side of this argument, for earlier and aggressive (targeted) enteral nutrition support, suggests that anorexia is a preserved evolutionary response that may be beneficial during acute illness. Over tens of thousands of years, in the absence of antibiotics and critical care medicine, intestinal abstinence (starvation) appears to have provided a selective advantage to host defense mechanisms.[10] This early observation has been investigated over the years and researchers have identified that starvation promotes autophagy which plays a key role in promoting host defenses and the immune response to intracellular pathogens. This is supported by studies following both *Listeria monocytogenes*-infected mice and a large cohort of primary famine victims that reported a decreased survival and increased risk of infection in those force-fed or allowed unmanaged re-feeding, respectively. The authors concluded that the autophagy associated with intestinal abstinence (starvation) destroys intracellular pathogens, thus the starvation approach improved survival rates. The most important initiator of cellular autophagy is nutrient starvation which inhibits activity of the kinase, rapamycin.[11] Unarguably, long-term nutrient starvation is not compatible with full recovery of the critical patient, nor the healthy patient challenged with an acute condition. As well, these studies do not reflect the protected environment of an ICU, so this muddies the waters as to the benefit of starvation to patients managed in "modern medicine" environments. Recently, several clinical trials compared trophic, permissive underfeeding, and normocaloric feeding to critically ill patients, and results of all trials dispelled the notion that short-term starvation is harmful and failed to demonstrate any benefit from the aggressive feeding approach.[10] Clearly, evidence-based arguments can be made for and against the practice of gastrointestinal abstinence (starvation) as a means of managing the critical patient.

Is There a Standard Patient-Safe Approach to Nutritional Support for the Critical Patient?

Consider this…the neuroendocrine, immunologic and metabolic responses to severe illness, trauma, or surgery involve three defined phases: acute, stable, and recovery. Each phase is actually identified with specific metabolic, immunologic, and physiologic changes which necessitate different macronutrient intakes. Timing (hours, days, weeks) of transition from phase to phase differs for each patient. This implies that a "standard" protocol based on day post-op, or hours hospitalized in ICU, may be challenging to develop for our veterinary patients. A more realistic and patient-beneficial approach would be matching the nutrient formulation and intake to the patient's imminent needs. The practice of utilizing a "PRN" nutritional assessment as part of each patient examination can provide direction as to when, what, how to "feed the gut," as enteral nutrient support has been well documented to ensure better clinical outcomes for our patients.

Balancing the divide between "feeding vs. starving the gut" can lead to the development of patient-appropriate nutrition support protocols for your practice. Whether your patient presents for an acute or chronic disease state, with a history of anorexia or hyporexia; an elective surgery or in need of an emergency procedure; as septic or inflammatory; or any combination of these conditions, addressing their specific nutrient needs will increase the chance of successful recovery. Remember, the gastrointestinal tract is the **most** physiological route for nutrient delivery. The longer it is "starved," the higher the risk for metabolic and physiologic complications. Although human nutrition has published specific nutrition support guidelines (i.e., A.S.P.E.N. Guidelines), we need to extrapolate those guidelines with some caution. In particular, the "time to re-feeding" guidelines published by human nutrition organizations, should be considered from the perspective that unlike most human patients, our veterinary patients often present with a prolonged history of adverse clinical signs, particularly anorexia/hyporexia. Withholding enteral nutrition support (i.e., intestinal abstinence) based on human guidelines may be putting your veterinary patients at a much higher risk of serious complications due to malnutrition.

Accurate determination of daily protein needs is an aspect of nutrition support in veterinary patients that is poorly defined. Human nutrition utilizes various calculations along with assessment of protein status biomarkers and more recently quantitative lean muscle mass monitoring to best meet the patient's daily protein needs. Recently, a push for increased protein intake for critically and recovering patients has shown measurable benefit. Likewise, the delivery of select amino acids (i.e., glutamine), prebiotics, and probiotics to provide the mucosal cells with the nutrients and protective mechanisms necessary to hasten recovery and minimize medical or surgical treatment complications is being closely evaluated in human patients. This level of nutrient specifics, utilizing evidence-based research, is not yet fully available for our veterinary patients. Veterinary nutritionists are more commonly extrapolating from the human literature for these specific levels of guidelines; but we are definitely making progress for the vast array of species that our veterinary colleagues care for each day. As more research is completed to support forward progress for the finer details of nutrient support, do not void the obvious. Become comfortable regarding how and when to utilize the gastrointestinal tract to help ensure successful patient outcomes.

References

1. Chilloux J, *et al. Curr Opin Clin Nutr Metab Care.* 2016;19(4):250.
2. Joosten KF, *et al. Curr Opin Clin Nutr Metab Care.* 2016;19(3):226.
3. Finnerty CC, *et al. JPEN J Parenter Enteral Nutr.* 2013;37(5 Suppl):21S.
4. McClave SA. *JPEN J Parenter Enteral Nutr.* 2012;36(1):24.
5. Yeh DD, *et al. JPEN J Parenter Enteral Nutr.* 2016;40(1):37.
6. Bruns BR, *et al. Nutr Clin Pract.* 2016;31(1):14.
7. Miller KR, *et al. JPEN J Parenter Enteral Nutr.* 2013;37(5 Suppl):39S.
8. Fremont RD, *et al. Nutr Clin Pract.* 2015;30(3):344.
9. Harris, *et al. J Vet Emerg Crit Care.* 2017 (in press).
10. Marik PE. *Curr Opin Clin Nutr Metab Care.* 2016;19(2):155.
11. Choi AM, *et al. N Engl J Med.* 2013;368:651.

1.
acvim313

The Intestinal Microbiome: The Potential and the Promise

Matthew D. Koci, PhD; Anne L. Ballou, PhD
Raleigh, NC, USA

INTRODUCTION

Gut bacteria have been recognized as an essential part of animal growth, development, and health for centuries;[1] however, only recently have we developed the tools to begin to understand the complex ecosystem living inside animals and how changes in their composition can affect host physiology.[2] These new high throughput sequencing, proteomic, and metabolomic tools have allowed for a better understanding of the bacterial populations present, the factors they produce, and how these change under different stages of animal life or disease state. While these studies have garnered much attention in the scientific and lay press; with all the various aspects of host biology it seems to affect, we are just now beginning to develop the insights that will enable us to begin to exploit these interactions clinically. A key component to this success will be merging animal phenotype data from centuries of research with our ability to now know what bacteria are involved.

GUT DEVELOPMENT

The best example of the role gut bacteria play in animal health can be seen in the comparisons of gnotobiotic and conventionally raised animals.[3-7] Germ-free animals have altered size and shape of different intestinal segments, reduced villus surface area, fewer slower dividing crypt cells, fewer goblet cells, reduced production of mucus, and impaired barrier function.[8,9] Various studies have demonstrated that various *Lactobacillus* species, *Bacteroides thetaiotaomicron*, *E. coli*, as well as other bacteria, were able to restore or a more normal gut structure and function to gnotobiotic animals.[10-13]

NUTRITION AND METABOLISM

Germ-free mice appear to be resistant to diet-induced obesity.[14] This difference is not only related to the loss of microbial enzymes needed to breakdown complex carbohydrates, as germ-free animals on a high-fat, high-sugar diets still have difficulties utilizing the available nutrients.[15] Studies in humans and mice demonstrate the hosts microbiome increases levels of leptin, glucose, insulin, and hepatic lipogenesis as compared to germ-free animals.[14,16] This suggests the gut microbes signal to and have significant influence over the metabolic and nutritional status of the hosts.

This influence appears to work in both directions as studies in various animals demonstrates that changes in the host's diet can have profound impact on the composition of the microbiome.[17-19] These studies typically compare diets with considerable differences in energy level or composition. For example, changes in the microbiomes of people after consuming isocaloric vegan or meat-and-dairy diets demonstrate substantial changes between diets as well as relative to the subjects' baseline biome.[19] In addition to changes in the phylogenetic profile, there were also changes in levels of secondary bile salts and microbial markers of gut inflammation.[17]

IMMUNE FUNCTION

The association between the intestinal microbiota and host immune function has been one of the most studied, but as yet least understood aspect of microbiome-host interaction. There are numerous studies in several species that demonstrate altering the microbiome can impact immune function.[16,20] Typically, these studies demonstrate increased antibody responses, changes in cytokine expression, and/or changes in inflammation.[21-24] The mechanisms behind these changes are poorly understood; however, it seems the changes are linked to how the mucosal immune system differentiates microbial pathogen from non-pathogen.

Pattern recognition receptors (PRRs), found on a variety of cell types, particularly antigen-presenting immune cells like dendritic cells and macrophages, and some enterocytes and include the toll-like receptors (TLR), nod-like receptors (NLR), and C-type lectin receptors.[25-27] These PRRs detect microbial cells through microbe-associated molecular patterns (MAMPs) such as flagellin, LPS, microbial DNA, glycoproteins, and peptidoglycans.[25] Binding of MAMPs to PRRs induce various activating or inhibitory responses, leading to different immune responses.[26] There is growing evidence that the MAMPs produced by commensal and symbiotic bacteria promote more of a tolerant host response, while those from pathogens induce more of a classical anti-bacterial response.[28] *Bacteroides fragilis* polysaccharide A, for example, interacts with TLR2 on regulatory T-cells (Tregs) to promote tolerance and inhibit the proinflammatory TH17 type of response.[28,29]

The regulation these bacteria play on host immune function can be even more fundamental. Germ-free mouse studies demonstrate that certain gut bacteria may be necessary for normal gut-associated lymphoid tissue (GALT) development. For example, the aggregated lymphoid tissue in the intestine known as Peyer's patches, are smaller in gnotobiotic models.[30,31] The introduction of *Bifidobacterium infantis* to these mice was enough to restore normal IgE-mediated oral tolerance relative to gnotobiotic controls.[32] However, some aspects of GALT development may be time sensitive. Introducing a normal gut microbiota to a young gnotobiotic mice can restore normal oral tolerance, but does not in adults.[32] Postnatal colonization of the gut also appears to be important for generating a balance between pro- and anti-inflammatory immune responses in the gut.[33,34]

Cancer

One of the most surprising studies into the impact that the host microbiome can have on overall health has to do with cancer. While there have been numerous studies associating changes in intestinal bacterial populations and increased rates of cancer, most notably colon cancer, these studies were mostly correlative.[35] However, several groups have since demonstrated that the constituent populations of the host intestinal microbiome can predict a patient's response rate to anti-cancer therapies.[36] This suggests the host microbiome metabolizes or otherwise augments the therapy increasing or even allowing for its therapeutic effect.[36,37]

Behavior

The interaction between the host and microbiome is providing more scientific support for the saying "gut feeling." The past several years we have seen an explosion of research studying how the enteric nervous system communicates with and in many cases may even regulate the central nervous system.[38] Add to this the ever-expanding list of neurotransmitter, hormones, and related mimics that gut bacteria seem able to produce have led many to begin to develop strategies for controlling stress, depression, and other disorders through manipulation of the intestinal bacterial populations.[39,40]

Next Steps

The intestine is one of, if not the most complex organ of any animal. It has more immune cells than any other organ. It has more neuro-endocrine cells than any other organ. It represents the largest surface area for pathogens to invade the body that it must exclude, while at the same time allow for the digestion and uptake of nutrients. Add to this the fact that it harbors more microbial cells than there are cells in the animal itself, resulting in a collective microbial genetic potential ~150 times that of the host genome.[41] This rich diversity provides countless possibilities for regulating all aspects of animal health and physiology, and given the fact that this interaction is happening in the gut and the microbiome is heavily influenced by diet, animal nutrition will potentially have an even larger role to play in the future of animal health.

References

1. Metchnikoff E, *et al. The prolongation of life*. New York, NY: G. P. Putnam's Sons; 1910:xxviii, 1 l.
2. Gilbert JA, *et al. Stand Genomic Sci*. 2010;3:249–253.
3. Bruckner G, *et al. Adv Nutr Res. 1984;*6:271–332.
4. Coates ME. *Proc Nutr Soc*. 1973;32:53–58.
5. Dufty JH. *Aust Vet J*. 1976;52:355–361.
6. Kubelkova K, *et al. Cell Mol Life Sci*. 2016;73:3961–3969.

7. Thompson GR, *et al. Gut*. 1971;12:230–235.
8. Gordon HA, *et al. Am J Physiol*. 1961;201:175–178.
9. Sommer F, *et al. Nat Rev Microbiol*. 2013;11:227–238.
10. Alam M, *et al. Scand J Gastroenterol*. 1994;29:445–451.
11. Banasaz M, *et al. Appl Environ Microbiol*. 2002;68:3031–3034.
12. Hooper LV, *et al. Science*. 2001;291:881–884.
13. Ulluwishewa D, *et al. J Nutr*. 2011;141:769–776.
14. Backhed F, *et al. Proc Natl Acad Sci U S A*. 2007;104:979–984.
15. Backhed F, *et al. Proc Natl Acad Sci U S A*. 2004;101:15718–15723.
16. Kau AL, *et al. Nature*. 2011;474:327–336.
17. David LA, *et al. Nature*. 2011;505:559–563.
18. Muegge BD, *et al. Science*. 2011;332:970–974.
19. Zimmer J, *et al. Eur J Clin Nutr*. 2012;66:53–60.
20. Buntyn JO, *et al. Annu Rev Anim Biosci*. 2016;4:335–355.
21. Kumar S, *et al. Probiotics Antimicrob Proteins*. 2017.
22. Qiu R, *et al. J Anim Sci*. 2012;90:2639–2651.
23. Guttsches AK, *et al. Innate Immun*. 2007;18:204–216.
24. Balcazar JL, *et al. FEMS Immunol Med Microbiol*. 2007;51:185–193.
25. Lebeer S, *et al. Nat Rev Microbiol*. 2010;8:171–184.
26. Medzhitov R, *et al. Nature*. 2007;449:819–826.
27. Hausmann M, *et al. Gastroenterology*. 2002;122:1987–2000.
28. Round JL, *et al. Science*. 2011;332:974–977.
29. Round JL, *et al. Proc Natl Acad Sci U S A*. 2010;107:12204–12209.
30. Maeda Y, *et al. Immunobiology*. 2001;204:442–457.
31. Pollard M, *et al. Infect Immun*. 1970;2:96–100.
32. Sudo N, *et al. J Immunol*. 1997;159:1739–1745.
33. Bouskra D, *et al. Nature*. 2008;456:507–510.
34. Eberl G. *Nat Rev Immunol*. 2005;5:413–420.
35. Schwabe RF, *et al. Nat Rev Cancer*. 2013;13:800–812.
36. Sivan A, *et al. Science*. 2015;350:1084–1089.
37. Perez-Chanona E, *et al. Curr Opin Immunol*. 2016;39:75–81.
38. O'Mahony SM, *et al. Behav Brain Res*. 2015;277:32–48.
39. Zheng P, *et al. Mol Psychiatry*. 2016;21:786–796.
40. Mulle JG, *et al. Curr Psychiatry Rep*. 2013;15:337.
41. Qin J, *et al. Nature*. 2010;464:59–65.

1.
acvim217

Integrating Diagnostic Test Results Using Bayesian Thinking

Sébastien Buczinski[1], Dr Vét, MSc, DACVIM; Renaud Léguillette[1], DMV, PhD, DACVIM; Edouard Timsit[2], DVM, PhD, DECBHM
[1]St-Hyacinthe, QC, Canada; [2]Calgary, AB, Canada

Introduction

Large animal internists make decisions every day; should I treat this animal? Should I perform additional tests to obtain a diagnosis? Etc. Most of these decisions are based on signalment, history, clinical examination, and ancillary/diagnostic tests. Access to diagnostic tests has increased exponentially in the last 10 years (imaging, blood markers, other fluid analysis), and it is crucial to use and interpret them adequately. Typically, the internist has an intuitive approach which consists in updating his/her diagnosis based on what he/she thought before and after performing a diagnostic test and integrating its result. This decision-making process can be described as Bayesian thinking, (i.e., updating the diagnosis based on new information obtained from test results).

The objective of this presentation is to formalize this approach based on the use of likelihood ratios and Bayesian theorem.

Diagnostic Test Accuracy

All diagnostic tests used in practice should have a known accuracy, defined in terms of sensitivity (Se; ability to detect diseased animals) and specificity (Sp: ability to detect non-diseased animals). Ideally, a test should have a perfect Se and Sp. Unfortunately, very few tests have such accuracy. Attempts can be made to improve Se or Sp of a test by either increasing or decreasing its threshold for positivity. However, one should remember that Se and Sp vary in opposite direction; improving Se is always done at the "cost" of decreasing Sp (and reciprocally). Typically, a sensitive test enables ruling-out a diagnosis (i.e., few false negatives) whereas a specific test enables us to rule-in a diagnosis (i.e., few false positives).

Likelihood Ratio of Diagnostic Tests

Se and Sp cannot be directly translated into practice, as they are determined based on disease status (diseased [Se] and healthy [Sp]) and not based on test results. What an internist wants to know is: what is the proportion of test positive (or test negative) animals that are truly diseased (or truly healthy)? And not, what is the proportion of healthy or sick animals that are correctly classified by a test? In other words, the internist wants to rule-in or rule-out a specific disease based on a diagnostic test result. The proportions/probabilities of test positive (or test negative) animals that are truly diseased (or truly healthy) are typically defined at the herd level in terms of positive (PPV) or negative (NPV) predictive values. Unfortunately, predictive values cannot be easily translated at the individual level (as they are often based on prevalence of disease within a population). To overcome this limitation, the internist can use the likelihood ratios (LR) of a test instead of its Se, Sp, NPV, or PPV. The positive and negative likelihood ratios of a test are quantifying how frequent is a positive (T+) or negative (T-) test in diseased versus non-diseased cases.[1]

$$LR+ = P_{T\ +/\text{Diseased}} / P_{T\ +/\text{Non-diseased}} = Se/(1\text{-}Sp)$$

$$LR- = P_{T\ -/\text{Diseased}} / P_{T\ -/\text{Non-diseased}} = (1\text{-}Se)/Sp$$

The LR is independent of disease prevalence and is therefore portable from one population to another (or one animal to another).

Bayesian Theorem for Quantifying the Added Value of a Diagnostic Test

The LR+ and LR- translated into practice through the Bayes theorem (also called Bayes formula).[1,2] This theorem links the pre-test odd of a disease with the post-test odd of the disease using the LRs (positive or negative based on the test result).

Post-test odd = pre-test odd * LR which when replacing the odds by "Probability of disease/(1-Probability of disease)" can also be written as:

$$P_{\text{Post-test}} = P_{\text{Pre-test}}*LR/(1-P_{\text{Pre-test}} + P_{\text{Pre-test}}*LR)$$

The clinician can use this theorem to update his/her pre-test probability based on the test results and LRs. The probability of disease updated by the test result is called post-test probability. It is noteworthy that many phone apps or online calculators can ease the work of internist in calculating post-test probabilities.[3] The difference between pre- and post-test probabilities can be interpreted as the added value of a test (increased probability when the test is positive or decreased probability when the test is negative).

Determination of Pre-Test Probability

Bayes' theorem introduces the need of accurately assessing the pre-test probability. Each internist has his/her own way of defining pre-test probability based on his/her prior experience, the clinical examination performed, etc. The need to explicitly state this pre-test probability is important and has been extensively investigated in medical studies.[4,5]

To assess how much a post-test probability is affected by a pre-test probability, the clinician can make several "what-if" scenario by testing different pre-test probabilities and observing their effects on the post-test probability.

Integration of Uncertainty Using Stochastic Modeling

Test accuracy (Se, Sp, and derived LRs) and post-test probability are often not known with certainty. For example, different Se and Sp may have been reported for a particular test. Se and Sp estimates are also typically defined within a margin of error (e.g., credible or confidence intervals). To take into account these uncertainties, a stochastic approach can be used. With this approach, uncertainties are translated in term of prior distributions and post-test probabilities are calculated using a stochastic model with Monte-Carlo sampling algorithm. This approach is more time consuming and cannot be performed without specific software. However, it significantly improves the accuracy of diagnosis as it provides the internist with a distribution of post-test probabilities, which includes all uncertainties (versus a single value for post-test probability).

Conclusion

The use of Bayes' theorem and likelihood ratios can enable the internist to easily measure the added value of a diagnostic test and better interpret a test results. For this approach, it is important to know the test accuracy and accurately define a pre-test probability of disease.

References

1. Gardner IA, Greiner M. Receiver-operating characteristic curves and likelihood ratios: improvements over traditional methods for the evaluation and application of veterinary clinical pathology tests. *Vet Clin Pathol.* 2006;35(1):8–17.
2. Akobeng AK. Understanding diagnostic tests 2: likelihood ratios, pre- and post-test probabilities and their use in clinical practice. *Acta Paediatr.* 2007;96(4):487–491.
3. Buczinski S, O'Connor AM. Specific challenges in conducting and reporting studies on the diagnostic accuracy of ultrasonography in bovine medicine. *Vet Clin North Am Food Anim Pract.* 2016;32(1):1–18.
4. Phelps MA, Levitt MA. Pre-test probability estimates: a pitfall to the clinical utility of evidence-based medicine? *Acad Emerg Med.* 2004;11:692–694.

5. Bianchi MT, Alexander BM, Cash SS. Incorporating uncertainty into medical decision making: an approach to unexpected test results. *Med Decis Making*. 2009;29:116–124.

acvim218

Intramammary Therapy: Impact of Cow and Farm Management Factors

Ted Whittem, BVSc, PhD, DACVCP, FANZCVS; Andrew Woodward, BVSc (Hons), BAnSci (Hons)
Melbourne, VIC, Australia

INTRODUCTION

Drug Residues in Milk

Intramammary therapies (IMT) for mastitis were first used in the 1930s, and from then have required focus to ensure that milk is produced free from drugs.[1] Because milk is regarded as wholesome food for infants and as a pure raw material for food processing, milk must remain free of antibiotics.[2-4] Regulation of labelled milk discard time (MDT) is the primary method used to ensure drug-free milk.[6]

However, the methods prescribed for determining MDT involve estimation of drug elimination in approximately 20 healthy cows. Data which are not on a log-linear are censored, and concentration estimates are then made and inappropriately regarded as 'observations'; dependent estimates of the population means and variance are derived with confidence intervals to determine MDT. These methods underestimate the variance of population parameters,[7] have been statistically criticized,[8] ignore cow disease and farm management factors,[9,10] yet are still required to be used by the regulators.

Treatments tested in mastitic cows in 1970 showed that 16 of 21 (76%) had persistent residues beyond the MDT.[11] Subsequently, mastitis was demonstrated to delay elimination from milk of penicillin G and dihydrostreptomycin.[12]

Because the discard of milk during treatment is the largest cost associated with mastitis, even if not including milk discarded after bulk tank rejection[13,14] and because there is a probability of unnecessary milk discard because of conservative MDTs, and because MDT calculations are irrelevant to the population and management factors wherein the drugs are used, a new approach to individualised MDT determination is needed.

Efficacy of Antimicrobials After IMT

Research on efficacy of intramammary therapy has allowed a general conclusion that an unacceptable proportion of cows do not respond to intramammary antimicrobial therapy. The overall bacteriological cure-rate is approximately 75% based on trial results showing from 25% to 100% cures. Cure rates determined vary because of the animal numbers treated, the severity and etiology of the mastitis, the experimental model, the drug chosen and its dose regimen and the formulation.[15-21] A major factor responsible for suboptimal efficacy is short duration of therapy and non-optimal drug-concentration profiles. Dose regimens need to be reviewed.

Population Approaches for Normal Cows

In 1999, the relatively new pharmacokinetic tool of nonlinear mixed effects modelling (NLME) allowed re-evaluation of the milk discard recommendations for pirlimycin (Pirsue®, Zoetis). The result for MDT in healthy animals was similar to the labelled MDT. However, NLME identified that individual pirlimycin clearance in milk is strongly correlated to the cow's productivity mastitis had a formidable impact on the variability of drug clearance. This population modelling also showed that several other factors influence the pirlimycin milk concentration profile: milk protein, body condition score, frequency of rumen contractions, number of quarters treated, milk lactose, body temperature, milk urea nitrogen, milk fat content and somatic cell count.[22]

However, this modelling violated a number of fundamental assumptions for traditional pharmacokinetic models; sampling of milk directly increased the rate of drug elimination, the sample space is not well mixed, the volume of the sample compartment is not constant, so an improved new approach for data analysis was needed.[5]

A New Physiologically Based Pharmacokinetic Model

Addressing each assumption violation, we introduced a new physiologically based pharmacokinetic model; the physiological compartments do change in volume during the milking and inter-milking interval, the sample space can be unmixed, non-first order transfer of drug between compartments is modelled.[5]

The model was validated using cefuroxime data from healthy cattle. Predictions of cefuroxime milk concentration in the secretory compartment where infections occur were modelled for allowing recommendations for optimization of milk concentration for achievement of efficacy with non-labelled dose and milking regimes. However, this initial PBPK model poorly predicted pirlimycin milk concentrations but with further development we have now validated a PB-PK model for pirlimycin, penicillin, cephapirin and cefuroxime.

This new model has allowed us to identify or confirm several cow and farm management factors that alter persistence of drug residues in milk. Based on this identification, new recommendations are now possible to minimise the risk of bulk tank contamination and to improve efficacy.

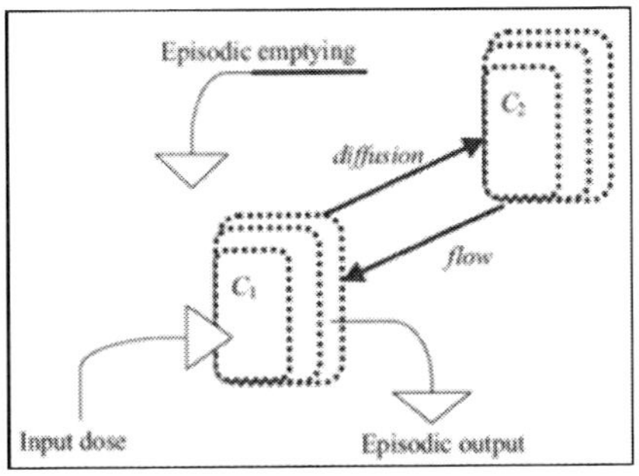

Figure 1. A simplified PBPK model of the udder, where dotted boxes indicate changing volume[5]

Management Factors

The efficiency of milking determines the post-milking residual volume of the udder. Smaller residual volume results in shorter drug retention in the udder and vice versa. This is the most important management factor for its impact on MDT.

Increasing the single-quarter dose has little effect on MDT whereas increasing the number of treated quarters increases the required MDT by 1–2 milkings. The duration of therapy can be varied by farmers, going off-label to increase the duration of therapy with an aim to increase efficacy. Increasing the duration of therapy does not change the MDT after last dose for most drugs, but for cephapirin there is some doubt due to the non-linearity of PK for its degradation products.

The inter-milking interval (4–24 h) influences both the duration that antibiotic concentration in milk remains above the minimum inhibitory concentration (MIC) and the required MDT. Variation in the duration of the first post-dose inter-milking interval can alter the duration of time above MIC and possibly alter the needed MDT.

Cow Factors

Low milk productivity causes prolongation of the drug residues in milk and increases the necessary MDT. This is the most important cow factor for its impact on MDT.

For pirlimycin, Gram-positive infections increase the systemic absorption of active drug and is likely to cause prolongation of meat drug residues.

Modelling confirms that MDT should be calculated based on number of milkings, not hours post-last dose, although very short inter-milking intervals will result in greater drug retention after each milking because small dilution factors result in greater mass retention in the residual volume. This finding directly contradicts current recommendations in regulatory guidelines.

Conclusions

The development of fast digital computers and sophisticated software for data analysis have led to the development of PB-PK modelling as a tool, to improve our understanding of factors which alter MDT and milk drug-concentration profiles. This more detailed understanding is timely. The dairy industry is undergoing rapid change in management practices and this change impacts both MDT and antibiotic efficacy. It is important that mandated regulatory methods and guidelines be revised accordingly, and that such new approaches are considered as tools for improving the sophistication of farming practice.

References

VIN editor: References 5 and 9 are the same.

1. Johnson. *Cornell Vet.* 1941;31:127.
2. Jones, *et al. J Dairy Sci.* 1988;71:1691–1699.
3. Reigart JR. *Environ Health Perspect.* 1998;106(6):A263–A264.
4. Jurdi DA, *et al. J Food Protection.* 1981;44:674–676.
5. Whittem T, *et al. J Vet Pharmacol Ther.* 2012;35(5):460–471.
6. Hentschel, *et al. J Am Vet Med Assoc.* 1991;199:1–23.
7. Katz. *Progress in Food and Nutrition Science.* 1988;12:325–338.
8. Concordet D, *et al. J Vet Pharmacol Ther.* 1997;20:380–386.
9. Whittem T, *et al. J Vet Pharmacol Ther.* 2012;35(5):460–471.
10. Whittem T. *J Vet Pharmacol Ther.* 1999;22(1):41–51.
11. Mercer HD, *et al. Am J Vet Res.* 1970;31(9):1549–1560.
12. Mercer HD, *et al. Am J Vet Res.* 1974;35(9):1191–1196.
13. Bartlett PC, *et al. J Dairy Sci.* 1991;74:1561–1572.
14. Shim EH, *et al. J Dairy Sci.* 2004;87(8):2702–8.
15. McDougall S. *N Z Vet J.* 1998;46(6):226–32.
16. McDougall S. *N Z Vet J.* 2003;51(3):111–6.
17. McDougall S, *et al. N Z Vet J.* 2007;55(4):161–70.
18. Owens WE, *et al. J Dairy Sci.* 1997;80(2):313–7.
19. Owens WE, *et al. J Dairy Sci.* 1988;71(11):3143–7.
20. Serieys F, *et al. J Dairy Sci.* 2005;88(1):93–9.
21. Spencer, *et al. Am J Vet Res.* 1947;8(29):325–9.
22. Whittem. "Reducing the persistence of antimicrobial residues in milk." In: Proceedings of the Seventeenth Annual Veterinary Medical Forum of the American College of Veterinary Internal Medicine. American College of Veterinary Internal Medicine, Chicago, IL, 1999.

1.
acvim314

Population Pharmacokinetics and Antiepileptics in Dogs and Cats

Dawn Merton Boothe, DVM, PhD, DACVIM (Small Animal Internal Medicine), ACVCP;
Roy Harmon, BS
Auburn, AL, USA

Introduction

The safe and effective use of a drug in the individual patient depends, in part, on using the most appropriate drug at the most appropriate dosing regimen. Determining the most appropriate dosing regimen for a patient relies on being able to predict individual patient variability in response to the drug. The dose-response-concentration relationship is composed of a pharmacodynamic factor (that is, what the drug does to the body) and a pharmacokinetic component (or what the body does to the drug - drug disposition). Pharmacodynamic responses generally follow the drug receptor theory, which indicates that the greater the number of drug molecules that interact with the receptors, the greater the response to that drug. As such, response is anticipated to be positively related to tissue drug concentrations, which generally parallel the more easily measured plasma drug concentrations (PDC). The time course of PDC after the administration of a given dose is determined by four major drug movements in the body: absorption, distribution, metabolism, and excretion. Individual variability of drug response among patients presumably might be predicted based on those factors that influence movement of each of these drugs. These might be categorized, for example, as physiologic (e.g., species, breed [ethnicity in humans], gender, or age); pathologic (especially the impact of renal, hepatic, or gastrointestinal disease); or pharmacologic (i.e., drug interactions, particularly those impacting drug transport systems or pumps and drug-metabolizing enzymes).

Pharmacokinetics

Traditional

The term "pharmacokinetics" refers to the mathematical modeling of the time course of plasma drug concentrations following any given dosing regimen. Traditional pharmacokinetic studies generally involve multiple samples collected at fixed intervals from healthy volunteers and a two-stage process of pharmacokinetic analysis. In the first stage of modeling, concentration is modeled as a function of time (nonlinear regression), and key parameters that describe the model are identified. In the second stage, descriptive statistics are estimated for each of the parameters and then generally reported as mean ± standard deviation. Among the limitations of such fixed models are the small size of the sample population and, because such studies are generally performed in healthy subjects, relevance to target populations (including diseased subjects). In veterinary medicine, relevance also is impacted by differences among species. Variability in such modeling can be great not only because of variability in the subjects due to factors impacting disposition, but also variability of the estimates. Further, while standard pharmacokinetic modeling is relatively easy, such models are not able to handle sparse data and cannot discriminate among the potential sources of pharmacokinetic variability nor quantify their effects (Charles 2014).

Population

Population pharmacokinetics (PPK) has emerged as a potentially effective alternative to individual drug monitoring and has become a key tool for predicting the impact of factors on drug response in humans (Teuscher; Duffull 2011). With population pharmacokinetic data, the information is collected from patients receiving the drug therapeutically. Data from patients may be collected as part of a clinical trial or as routine therapeutic drug monitoring (TDM). Generally, only a limited number of samples (2 to 4) are collected. Data from each subject is characterized by a set of PK parameters. However, the population model expands on individual analysis by relating the individual parameters to a set of theoretical, but typical, parameters and quantifies the impact of information known about the patient on the variability in the individual parameters. The data is subjected to a mixed effect model: fixed effects are the PK parameters (e.g., clearance) and the factors that significantly impact those parameters (e.g., weight, age, which may be added as covariates), whereas random effects account for unexplained between-individual variability. For PPK, the typical estimated value for each parameter is the mode; as patient numbers increase, the mode approaches the true population mean. However, individuality of patient information is then used to estimate the most likely value of the parameter for each patient (Charles 2014). Thus, each patient "lends" data to the model, but in turn receives information back.

As with all models, those generated from population data are wrong, but they may be among the most useful when compared to traditional pharmacokinetic modeling. Population pharmacokinetic analysis can be used to expand an understanding of the relationship between the individual and drug movement in a variety of clinical and nonclinical (including experimental) settings (Teuscher). Advantages of PPK include the collection of data in patients taking the drug, flexibility in modeling that allows incorporation of data from subjects participating in different studies (Hussein 1997), inclusion of patients for which data is sparse, the potential for collection of covariate data that might explain model variability, and the ability to distinguish between inter- and intraindividual variability. Population pharmacokinetic data does require at least 2 samples per patient if the interindividual variability is to be identified, and a larger number of patients (≥ 40). In contrast to traditional PK studies, a large amount of data is collected and must be analyzed. Modeling is complex, tedious, and labor and time intensive. It is negatively impacted if data is missing (especially covariates in all patients) (Charles 2014).

THERAPEUTIC DRUG MONITORING

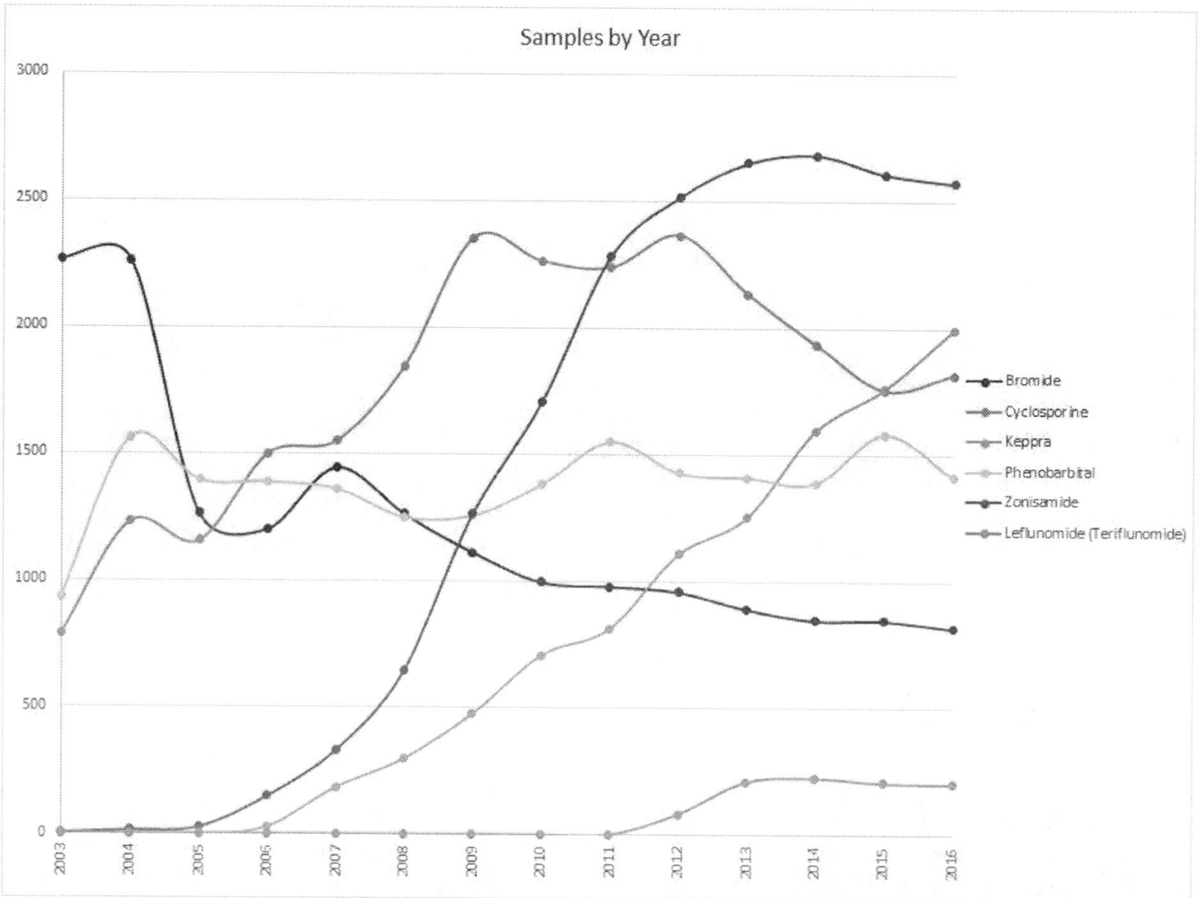

Therapeutic drug monitoring (TDM) has become routine practice for some drugs as a means of improving the design of dosing regimens for the individual patient. The goal of TDM should not be to make sure that a patient is within a therapeutic range, but rather to determine the individual patient's therapeutic range. Whereas a 2010 Cochrane review of TDM in human epileptics, which was based only on one study (n = 180; 5 drugs), found no evidence that TDM was helpful in reaching predefined target ranges in new epileptics on monotherapy, it did acknowledge that TDM might be useful in selected patients, polytherapy, or special situations. Indeed, a position paper of the ILAE Commission on Therapeutic Strategies (Patsalos 2008) delineates 6 circumstances under which TDM is likely to be useful. These include the establishment of an individual therapeutic range, to diagnose toxicity, to assess compliance in refractory patients, to guide dosage adjustment in the face of factors that contribute to individual variability in pharmacokinetics, in the face of anticipated pharmacokinetic changes such as drug interaction, and when using a dose-dependent anticonvulsant. Although no paper has summarized the general role of TDM data in generating PPK for anticonvulsants, a number of studies have addressed the PPK of selected drugs (Graves 1998; Hussein 1997 [carbamazepine], Serrano 1999 [valproate]; Hashimoto 1994, 6 [zonisamide, phenytoin]).

The Clinical Pharmacology Laboratory at Auburn University provides TDM services to over 4000 veterinary practices in the USA and throughout the world. Among the drugs tested are anticonvulsants, including bromide, phenobarbital, zonisamide, and levetiracetam. The applicability of population pharmacokinetics for these drugs will vary in part because of their disposition, and most notably their elimination half-life, which determines, in turn, the number of samples collected during TDM. The information that generally accompanies our samples includes signalment (including weight), whether or not the patient is controlled, additional drugs (including the anticonvulsant the patient is receiving), the source of the drug (compounded, and specific manufacturers), any side effects that might be attributed to the drug, and any disease-related factors. Regarding sample collection, for drugs with a short half-life, for which drug concentrations are anticipated to fluctuate more than 30% during a dosing interval, we ask for two samples to be collected (levetiracetam, gabapentin; generally a 2- to 3-h "peak" and a just-before-the-next-dose "trough"). This allows determination of a disappearance rate constant and thus elimination half-life ($t_{1/2}$ = 0.693/kd, where kd = rise/run or ln (C1/C2)/t1–t2, where C1, t1 and C2, t2 are the peak and trough concentrations, respectively) (Morris 2002). For drugs that do not fluctuate, a trough sample is requested after the drug is anticipated to reach steady-state.

Table 1. Percent responders to antiepileptic drugs used as monotherapy or combination therapy.(-1 indicates the drug was used, 0 that it was not. LEViteracetam, ZoNiSamide, Phenobarbital, Bromide, Immediate Release, eXtended Release)											
	DRUG						Controlled?				
	LEV	ZNS	PH	BR	LEV IR	LEV XR	Yes	No	Unk	% CON	Rank
1	-1	-1	-1	-1			108	131	0	0.45	17
2	-1	-1	-1	0			172	271	0	0.39	18
3	0	-1	-1	-1			235	211	0	0.53	11
4	-1	-1	0	-1			146	138	0	0.51	13
5	-1	0	-1	-1			209	176	0	0.54	9
6	-1	0	0	-1			156	149	0	0.51	14
7	0	0	-1	-1			493	346	0	0.59	5
8	0	-1	0	-1			392	204	0	0.66	4
9	0	-1	-1	0			529	476	0	0.53	12
10	-1	-1	0	0			262	257	0	0.50	15
11	-1	0	-1	0			379	439	0	0.46	16
12	0	0	0	-1			851	402	0	0.68	3
13	-1	0	0	0			967	723	0	0.57	7
XR	-1	0	0	0	0	-1	51	41	0	0.55	8
IR	-1	0	0	0	-1	0	76	64	0	0.54	9
Unknown	-1	0	0	0	0	0	840	618	0	0.58	6
14	0	0	-1	0			2155	919	0	0.70	2
15	0	-1	0	0			3030	1221	0	0.71	1

Concentration	Controlled?		
(mcg/ml)	Yes	No	Unspecified
Keppra	22.7	25.6	23.5
Zonisamide	27.9	29.8	28.2
Phenobarbital	23.3	25.7	24.1
Bromide (mg/ml)	1.6	1.4	1.6

POPULATION-DESCRIPTIVE STATISTICS

For the purposes of this manuscript, population data provided through both traditional and population pharmacokinetic analysis will be presented, with the latter based on a comparison of immediate versus extended-release levetiracetam. Without statistical analysis, the percent responders appear to be greatest to either phenobarbital or zonisamide (approximately 70%), followed by bromide (68%) and then levetiracetam (57% as either extended or immediate release) when used as sole agents. Patients on combination therapy generally respond at a rate of 40 to 50%. Data that summarizes drug disappearance half-life (Table 2) indicates that LEV-XR does have a longer half-life compared to IR in dogs, which is supportive of q 12 h dosing. Interestingly, in contrast to previous reports, the half-life of neither LEV nor ZNS was shorter in patients also on PH compared to monotherapy (Munana 2015).

Table 2. Half-life (hrs) of patients receiving anticonvulsants as overall, as mono or sole therapy and in combination with other antiepileptics.

Levetiracetam (LEV)					Zonisamide (ZNS)				Phenobarbital (PH)			
Overall	n	Mean	SD	CV	OVERALL	Mean	SD	CV	OVERALL	Mean	SD	CV
XR Half-life:	46	9.91	9.43	0.95	(n) 815	32.99	15.08	0.46	(n) 285	36.01	15.68	0.44
IR Half-life:	52	4.70	2.68	0.57	MONO				MONO			
MONO					500	34.37	15.08	0.44	183	34.67	15.68	0.45
XR Half-life:	22	9.91	9.43	0.95	LEV				LEV			
IR Half-life:	33	4.70	2.68	0.57	90	30.46	13.34	0.44	45	34.84	14.49	0.42
ZNS					PH				ZNS			
XR Half-life:	8	18.94	16.12	0.85	223	29.61	14.93	0.50	41.00	43.55	17.64	0.41
IR Half-life:	7	4.21	1.33	0.32	BR				BR			
PH					90	32.16	16.04	0.50	41	39.89	20.46	0.51
XR Half-life:	17	8.19	6.26	0.76	LEV & PH				LEV & ZNS			
IR Half-life:	15	4.29	2.88	0.67	35	28.59	10.58	0.37	11	38.85	11.78	0.30
					LEV & BR				LEV & BR			
					16	28.68	13.61	0.47	10	41.22	13.24	0.32
					PH and BR				ZNS & BR			
					42	28.55	15.52	0.54	7	49.79	32.09	0.64

An interesting finding is the proportion (15 to 20%) of patients on phenobarbital that have serum that is lipemic (lipemic index of 4 or higher) (see figure below). We do not see this level of lipemia in dogs not receiving phenobarbital. Lipemia does not interfere with the assay. It is not clear if this data suggests that patients on phenobarbital are more likely to have lipemic serum compared to patients that are not.

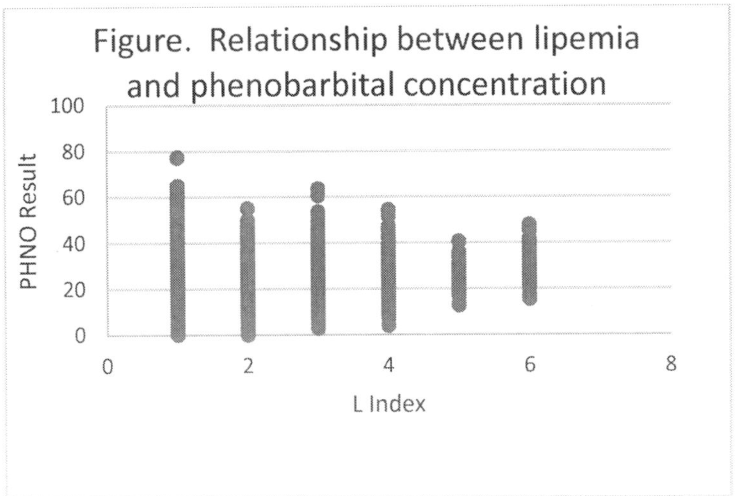

Figure. Relationship between lipemia and phenobarbital concentration

Levetiracetam is among the anticonvulsants with a half-life that is short enough that drug concentrations fluctuate significantly during a dosing interval. As such, many patients have 2 samples submitted, allowing for population kinetic analysis. Table 3 reflects population pharmacokinetic analysis using a nonlinear effects model.

Table 3. Pharmacokinetic parameters of serum levetiracetam versus time based on population non-compartmental analysis of therapeutic drug monitoring data collected from epileptic dogs receiving either regular or extended release levetiracetam.

Immediate Release (n=121)	Units	Mean	Std Dev	95% LCI	95% UCI	Minimum	Maximum	Median	P value
C_{max}	mcg/ml	30.0	20.3	26.4	33.7	2.5	142.7	25.5	
T_{max}	hr	2.59	0.77	2.45	2.73	1.25	4.75	2.00	
C_{max}/Dose	mcg/ml / (mg/kg)	1.27	0.6	1.17	1.38	0.18	3.4	1.2	
C_{last}	mcg/ml	10.7	11.9	8.6	12.9	1.0	80.1	8.4	
T_{last}	hr	8.2	1.7	7.9	8.5	5.0	14.0	8.0	
AUC_{Inf}	mcg*hr/ml	268	307	213	323	22	1664	175	
AUC_{Inf}/Dose	mcg*hr/ml	12.2	16.1	9.3	15.1	1.06	118.9	7.7	
K_d	hr(-1)	0.25	0.13	0.22	0.27	0.004	0.59	0.23	
$t_{1/2}$	hr(-1)	6.1	16.3	3.2	9.0	1.2	170	3.0	
Vd/F	ml/kg	701	532	605	796	217	3981	569	
CL/F	ml/hr/kg	161	127	138	183	8	947	130	
MRT_{Inf}	hr	9.8	23.6	5.6	14.1	2.2	246.0	5.3	
Extended Release (n=51)		**Mean**	**Std Dev**	**95% LCI**	**95% UCI**	**Minimum**	**Maximum**	**Median**	
C_{max}	mcg/ml	25.7	15.4	21.4	30.1	5.3	93.3	24.2	
T_{max}	hr	2.9	1.2	2.5	3.2	1.0	8.3	2.5	0.066
C_{max}/Dose	mcg/ml / (mg/kg)	0.85	0.38	0.75	0.96	0.20	1.85	0.85	<0.001
C_{last}	mcg/ml	11.3	7.9	9.1	13.6	1.0	36.0	9.7	
T_{last}	hr	11.2	1.7	10.7	11.6	5.5	14.5	11.8	<0.001
AUC_{Inf}	mcg*hr/ml	428	363	325	530	39	1622	303	<0.001
AUC_{Inf}/Dose	mcg*hr/ml	14.8	14.0	10.9	18.8	1.3	80.7	9.8	0.0037
K_d	hr(-1)	0.13	0.10	0.10	0.15	0.01	0.48	0.08	0.037
$t_{1/2}$	hr	11.5	12.4	8.0	14.9	1.5	78.9	8.4	<0.001
Vd/F	ml/kg	1202	746	993	1412	325	3893	1013	<0.001
CL/F	ml/hr/kg	125	118	92	158	12	746	102	0.09
MRT_{Inf}	hr	18	18	13	23	2.6	115	13	0.03

References

1. Charles B. Population pharmacokinetics: an overview. *Australian Prescriber*. 2014;37(6):210–213.
2. Duffull SB, *et al.* Interpreting population pharmacokinetic-pharmacodynamic analyses - a clinical viewpoint. *Br J Clin Pharmacol*. 2011;71(6):807–814.
3. Graves NM, *et al.* Population pharmacokinetics of carbamazepine in adults with epilepsy. *Pharmacotherapy*. 1998;18(2):273–281. PubMed PMID: 9545146.
4. Hashimoto Y, *et al.* Population analysis of the dose-dependent pharmacokinetics of zonisamide in epileptic patients. *Biol Pharm Bull*. 1994;17(2):323–326.
5. Hussein Z, Posner J. Population pharmacokinetics of lamotrigine monotherapy in patients with epilepsy: retrospective analysis of routine monitoring data. *Br J Clin Pharmacol*. 1997;43(5):457–465.
6. Morris RG, *et al.* Comparison of trough, 2-hour, and limited AUC blood sampling for monitoring cyclosporine (Neoral) at day 7 post-renal transplantation and incidence of rejection in the first month. *Ther Drug Monit*. 2002;24(4):479–486.
7. Munana KR, *et al.* Effect of chronic administration of phenobarbital, or bromide, on pharmacokinetics of levetiracetam in dogs with epilepsy. *J Vet Intern Med*. 2015 Feb 25.
8. Odani A, *et al.* Population pharmacokinetics of phenytoin in Japanese patients with epilepsy: analysis with a dose-dependent clearance model. *Biol Pharm Bull*. 1996;19(3):444–448.
9. Patsalos PN, *et al.* Antiepileptic drugs - best practice guidelines for therapeutic drug monitoring: a position paper by the subcommission on therapeutic drug monitoring, ILAE Commission on Therapeutic Strategies. *Epilepsia*. 2008;49(7):1239–1276.
10. Serrano BB, *et al.* Valproate population pharmacokinetics in children. *J Clin Pharm Ther*. 1999;24(1):73–80.

11. Teuscher NS. Introducing population pharmacokinetic analysis into your early drug development efforts. White Paper, Phoenix NLME.
http://www.pharsight.com/library/sign_in/Introducing_Population_Pharmacokinetic_Analysis_White_Paper.pdf. (VIN editor: Link could not be accessed as of 4-17-17.)

acvim315

Medical Cannabinoids: Prophecy or Prayer?

Dawn Merton Boothe, DVM, PhD, DACVIM (Small Animal Internal Medicine), ACVCP
Auburn, AL, USA

INTRODUCTION

The legalization of medical marijuana (*Cannabis* spp.) for treatment of human diseases has been accompanied by an increased use in animals. Much of this use is implemented without veterinary supervision by pet owners able to purchase animal dietary supplements derived from marijuana or its constituents. However, this use is largely accomplished by clients who access them through internet sites marketing products specifically for use in dogs and cats. Although such use is not currently evidence-based, support for medical use of marijuana-based products is increasing emerging in human medicine and most of the indications should extrapolate to animal diseases. The purpose of this manuscript is to describe the types of products being marketed, the regulations surrounding their use, and provide a scientific bases for their use based on presumed mechanisms of action. Finally, evidence supporting use for various indications will be summarized.

MARIJUANA INGREDIENTS

Marijuana refers to the dried leaves and tops of the hemp plant (*Cannabis sativa*) (Svienska 2008). It has been a part of recreational, religious and medical activities of a variety of cultures for over 5,000 years and was among the most commonly prescribed medications in the United States Pharmacopeia until declared illegal in the 1930s (Krietzer 2009; Burns 2006). *Cannabis* sp is a pharmacologically (and toxicologically) diverse herb, containing at least 480 distinct compounds with their proportions varying between each subspecies, the part of the plant, and how that product is cured or prepared. Plant products include, in addition to marijuana, hashish and hashish oil, formed from the resin secreted by the plant. Hemp is commonly used to refer to the stem of the marijuana plant. However, marijuana is one of several varieties of hemp plants grown and harvested specifically for the stem which is used for a variety of products such as ropes, animal bedding. Unique to *Cannabis* are close to 70 different terpene phenolic compounds referred to as cannabinoids. Phytocannabinoids are unique to *Cannabis*. These lipophilic, low-molecular-weight compounds (300 Da) (Hosking 2008) are structurally similar to the **eicosanoid arachidonic acid**, the precursor of prostaglandins and leukotrienes. The most important of the phytocannabinoids are: Δ^9-tetrahydrocannabinol (Δ^9-THC), cannabidiol (CBD), cannabichromene (CBC), and cannabigerol (CBG) (Grotenhermen 2003). In addition to phytocannabinoids, marijuana contains approximately 140 different terpenoids. These compounds are responsible for a variety of actions as well as its scent. The specific terpenoids yielded from a particular marijuana plant depend on the type of *Cannibis* (determining the fiber content), the part of the plant, its sex and age, whether or not it is cultivated in or outdoors, when it is harvested and the conditions at harvest, and how it is dried and stored. The serotonergic effects of marijuana (5-HT1A and 2A) may reflect the impact of these essential oils, contributing to analgesia and mood modification. Other components in the plant include nitrogen containing compounds (n=70: alkaloids, amines); carbohydrates, including common monosaccharides (n=13: fructose, glucose, mannose), selected disaccharides (sucrose, maltose), and several polysaccharides (e.g., cellulose, pectin) as well as several sugar alcohols (n=12; mannitol, sorbitol, glycerol). A number of flavonoids also are present (n=23); among them, apigenin has a wide variety of effects, including interaction with benzodiazepine receptors, resulting in an anxiolytic effect. Other ingredients include fatty acids (n=33) and others.

PROPOSED MECHANISMS OF ACTION

The Endocannabinoid System

Among the phytocannabinoids, Δ^9-THC is the most understood as it is the main property of psychogenic producing behavior and pharmacological activity against pain (Grotenhermen 2003; Di Marzo 2007). Its discovery and elucidation of its role in the human body paralleled that of the discovery of the opioid receptors, leading to a description of the endocannabinoid system, including endogenous cannabinoid ligands (endocannabinoids) and their respective endocannabinoid (eCB) receptors (Hosking 2006; Di Marzo 2006; Di Marzo 2007). The major **cannabinoid receptors**, CB_1R and CB_2R, are G-protein coupled receptors found within the cytoplasm of the cell. The CB_1R receptor is ubiquitous and commonly found within the central and peripheral nervous tissue as well as in peripheral tissues associated with the immune system (e.g., tonsils and spleen) (Burns 2006; Grotenhermen 2003; Hosking 2008). Similar to opioids, endogenous endocannabinoid ligands are capable of acting as agonists or antagonists on their corresponding receptors (Di Marzo 2006). CB2r are located principally on immune cells, but this includes microglia. The cannabinoid receptors are influenced by both endocannabinoids and phytocannabinoids. At least 5 **endogenous cannabinoids** have been described, with anandamide (CB1 and 2 agonist, but higher affinity for CB1) being the most thoroughly studied. It is synthesized by post-synaptic neurons, acting as a retrograde messenger to influence neurotransmitter, and particularly GABA, release. It is extremely unstable, being rapidly hydrolyzed to ethanolamine (an antihistamine) and arachidonic acid. Cannabinoids are able to disrupt short-term memory, impair cognition and time perception, alter mood while enhancing body awareness, discoordination, sleepiness, and reduce attention focus and the ability to "filter" irrelevant information.

Although interaction with cannabinoid receptors is unique among plants to hemp, cannabinoids do not necessarily cause their effects by direct interaction with CBR. Other receptors are also targeted (e.g., benzodiazepines, serotonin, others). Cannabinoids can influence the release of other neurotransmitters.

Pharmacodynamic Effects

The endocannabinoid system is a known contributor to physiology, but has been recognized for only about 25 years. In general, it contributes to homeostasis (*Relax, Eat, Sleep, Forget and Protect*; McParland 2014).

Endocannabinoids appear to be important as neuroprotectants (e.g., antioxidants, inhibition of calcium influx and excessive glutamate production), for example, that associated with CNS ischemia or hypoxia, or the presence of neurotoxicants. These effects appear to be mediated predominantly by CB1 (located particularly in the dorsal horn of the spinal cord) although CB_2 also plays a role, depending on the tissue (Svizenska 2008). Cannabinoids also inhibit neuroinflammation (see therapeutic indications). Although not all effects of cannabinoids are mediated by CBR, their extensive distribution contributes to a variety of physiologic responses. The dopaminergic reward pathway is stimulated by CB1 receptors, motivating eating, smoking and substance abuse. A variety of clinical effects occur, including but not limited to inhibition of nociception (sensation and pain), decreased anxiety and emesis, manipulation of gastrointestinal and cardiovascular function, and stimulation of appetite.[4] The CB_2R receptors are principally found in cells and organs of the immune system, including leukocytes, monocytes, B and T cells, spleen, and tonsils. Activation of CB_2R receptor receptors do not cause the effects on mentation that activation of CB_1R produces, and has become a target for therapeutic use in human medicine by reducing inflammation, immune suppression, and as a chemotherapeutic (Burns 2006; Grotenhermen 2003; Hohmann 2006).CB_1R receptors often modulate other signals. For example, they inhibit voltage-activated calcium channels, decreasing excitatory (acetylcholine) and increasing inhibitory (GABA) neurotransmitters. CB_2 also is located on neurons where it may be associated with cell differentiation (Svizenska 2008).

Specific Cannabinoids

Cannabidiol (CBD) was the first isolated phytocannabinoid to be isolated from the *Cannabis* plant in 1930–1940s. In 1960s, CBD was employed as an anticonvulsant due to having similar pharmacologic effects as phenobarbital and diphenylhydantoin (DPH) (Mechoulam 2002). CBD, has very low affinity for the cannabinoid receptors; however, it serves as an antagonist for CB_1R and CB_2R agonists (Mechoulam 2007). The natural CBD (-) does not bind to CB_1R; however, the synthetic (+) has been shown to bind to both CB_1R and CB_2R (Mechoulam 2002). Besides anti-convulsant effects, CBD has been used for its anxiolytic, antipsychotic, and antinausea effects (Mechoulam 2002). Interestingly, after oral administration of CBD in a murine model for rheumatoid arthritis, researchers saw a diminished interferon gamma (IFN-γ), decreased release of tumor necrosis factor alpha (TNF-a) and nitrous oxide (Mechoulam 2002; Malfait 2000). CBD also works as a potent antioxidative agent, showing greater protective nature against glutamate neurotoxicity than either ascorbate (vitamin C) or α tocopherol (vitamin E) (Mechoulam 2007). These antioxidative effects may have explained why CBD was successful in correcting hypermotility in mice, with no effects on the control population (Capasso 2008). On a smaller scale, CBD has been shown to stimulate mesenchymal stem cells responsible for bone formation and fracture healing, while also controlling bone resorption (Izzo 2009). Although CBD has very low toxicity on rhesus monkeys after IV administration, it has been reported to have very low oral bioavailability (9 h), which may be due to first pass metabolism (Mechoulam 2002). Cannabichromene (CBC), one of the non-psychotropic cannabinoids, has been shown to have strong anti-inflammatory properties through indirect activation of CB_1R, through inhibition of the endocannabinoid inactivation (Shinjyo 2013). Most recently, CBC was determined to normalize the intestinal motility of an experimental model of intestinal inflammation in mice, but not alter the rate of transit in control animals (Izzo 2001; Izzo 2012). **Cannabigerol** (CBG), whose mechanism of action has not been completely elucidated, has a wide variety of therapeutic targets from antitumor activity as well as potent antibacterial effects towards selected microbes, including methicillin resistance staphylococci (MIC of 0.5 to 2 mcg/ml) (Appendino 2008; Rock 2011; Izzo 2009). Traditionally, cannabinol (CBN), the primary product of Δ^9-THC breakdown, was been used to predict the age of the marijuana plant. CBN has recently been discovered to have an immunosuppressive effect by decreasing the production of interleukin-2 (IL-2) by decreasing T cell activation (Faubert 2000).

Untoward Pharmacologic Effects

As with many CNS active drugs, marijuana is associated with both tolerance (higher concentration needed to impart a similar pharmacologic effect) and withdrawal (a clinical syndrome of nervousness, tension, restlessness, sleep disturbance and anxiety). However, the long elimination half-life of the most active ingredient, THC (and others) appears to preclude a clear cut abstinence syndrome (Svizenska 2008). As with other addictive agents, laboratory rodents have been demonstrated to self-medicate, suggesting an addictive component. Tolerance also should be expected: dogs exhibit a unique ataxic response to IV CBD. However, tolerance to this effect rapidly emerges within one week of repetitive treatment.

Cannabinoids in Dogs or Cats

Cannabinoid receptors have been studied in a limited fashion in dogs. Initial studies focused on relevance to humans and provide evidence that dogs may react with unique behaviors. **Receptors**: In 1975, tritium-labeled Δ-9 THC (0.5 mg/kg IV) radioactivity was distributed throughout canine cerebellum and cerebral cortex, with increased concentrations in grey matter versus white matter noted; up to 50% of the signal reflected metabolites (28). Peripherally, radioactivity occurred in all organs save the vitreous humor. Peripheral tissues with the highest concentrations (relative) were bile (8), adrenal gland (3, 5), liver, auricle and ventricle of the heart, renal cortex, and pancreas (1), with the least concentration in the fat, trachea, and testis. The canine CB_2R has been relatively cloned and characterized and shows 76% homology with other species (31). CB_1R is located in the apical region of the striated cells of parotid and mandibular salivary glands (12). Both CB_1R and CB_2R were demonstrated in various cells of canine epidermis and dermis of dogs; both receptors increased in atopic dogs (8).

Marijuana and Pets

Legalization in states has yet to include veterinary medicine. Legalization of medical or recreation marijuana among the states is likely to be associated with an increased incidence of toxicity, with a 4-fold increase cited in one study although toxicity may reflect additional ingested foods (e.g., chocolate) (Meola 2012). THC is among the compounds cited as a toxicologic hazard in detection (police) dogs (Llera 2008). It is the most common drug to which detection dogs are exposed. Both dogs and cats may become intoxicated with smoke inhalation as well as ingestion of food containing marijuana (or hashish). It is absorbed rapidly following either oral or inhalant administration with clinical signs evident within 30 to 60 minutes of ingestion, although one reference (Osweiler 2008) indicates onset as long as 12 hours after exposure. Cannabinoids of medical significance appear to undergo first pass metabolism and as such, the risk of toxicity with inhalant products is much greater compared to oral. The implication for medical use is that oral administration may not be cost effective. The drug is eliminated by hepatic metabolism and biliary excretion with elimination being complete in 5 days in dogs; duration of toxicity ranges from 30 minutes to 3 days, but 18–24 is the average. Enterohepatic circulation may contribute to the prolonged half-life. The most common signs of toxicity following ingestion in dogs include tachycardia, hypotension, depression, ataxia, vomiting (inducing emesis is not recommended in clinically depressed dogs because of the risk of aspiration), altered behavior, bradycardia, hypersalivation, weakness, hypothermia and seizures. Treatment is largely supportive, with sedation with benzodiazepines or phenothiazines as needed. Antiemetic therapy may be indicated.

Pharmacologic Manipulation

The system can be manipulated by interfering with endogenous receptor ligands with cannabinoid or cannabinoid-like drugs, and enzymes responsible for endocannabinoid synthesis and degradation.

Regulatory Considerations

Currently, at least 24 (PRO-CON). States have approved marijuana in some form. According to NORML (www.norml.org/states), a site dedicated to law reformation, 34 states have some type of conditional use, 15 states of decriminalized use, 14 states of medical marijuana laws. Several states have passed Industrial Hemp bills, with such plants being legal under some conditions as long as they contain less than 0.3% THC (DMW). However, cannabinoids themselves, included the oil CBD concentrate from these plants remains as a Class 1 Schedule substance, meaning it has a high risk of abuse potential and no recognized medical benefit. 10 different pharmaceutical cannabis products (including synthetic) have or are undergoing some level of approval (http://medicalmarijuana.procon.org/view.resource.php?resourceID=000883).

Medical Uses

The proposed indications for medical marijuana have included, but are not limited to behavioral, sleep and gastrointestinal disorders, neuroprotection, antispasmodic but prokinetic, anorexia, nausea, glaucoma, diabetes, immunosuppression, malaria, anti-inflammatory and, of course, pain (Table 1, Izzo 2009). A proposed advantage of medical marijuana compared to a single drug (e.g., dronabinol, a synthetic THC [Marinol®]), is the multiple compounds contained in the plant. Two advantages are offered: 1. The compounds might act synergistically (a "synergistic" shotgun or entourage effect) to provide an enhanced desired pharmacologic effect, while 2. at the same time, mitigating (one compound acting on another) undesirable effects. However, evidence for a synergistic benefit is lacking based on the lack of differences when THC is consumed as marijuana, versus Marinol® (humans) (Brenneisen 200X). Presumably, because marijuana contains so much THC, it may not be the most effective portion of the plant and it may contribute to more side effects. Cannabinoid deficiency has been linked as an etiology of a variety of illnesses: ("eCB deficiency syndrome") as an etiology in migraine, fibromyalgia, irritable bowel syndrome, psychological disorders, and others (McParland 2014). However, finding evidence to support either the negative or positive effects of *Cannabis* can be difficult because such information is often tainted with emotionally mediated opinion. PRO-CON (http://medicalmarijuana.procon.org/view.resource.php?resourceID=000881) is a useful site that provides links to evidence using a categorical approach, as well as information on approval status among the states.

Pain Management

Cannabinoid use has been best prescribed for its use in controlling neuropathic pain. Peripherally, CB2R receptors, and to a much lesser extent CB1R, have been effective in modulating the inflammatory response as well as tissue and nerve injury (19, 33). CB2R, present on both mast cells and leukocytes, play multiple key roles in the modulation of the local inflammatory response including preventing mast cell degranulation, diminishing neutrophil migration, and decreasing the release of nitric oxide from macrophages (33). Cannabinoids have an analgesic effect on neuropathic pain in rodent models (thought to be mediated via the CB1 and CB2 receptors) in addition to other receptors, such as transient receptor potential vanilloid type 1 (TRPV1). Secondary to nerve injury, cannabinoid induced antinociception is more effective in alleviating pain than opioid drugs by suppressing wind up and noxious stimulus induced central sensitization (Hohmann 2006). The mechanism of action is hypothesized to be due to loss of opioid receptors along the traumatized nerve tissue being either removed or compromised secondary to the injury, where CB_1R remain intact (Bridges 2001; Rice 2002). Recently, studies of the interactions between the cannabinoid and the opioid systems indicates that co-administration of two agents may produce favorable synergistic effects, and may offer a new treatment strategy for multi-modal analgesia (MacPherson 2000; Ripamonti 2001). Recent findings have also suggested that NSAIDs may also owe some of their therapeutic success to their interaction with the endocannabinoid system either by inactivation of proteins or by encouraging biosynthesis. Rofecoxib, a cyclooxygenase-2 (COX-2) selective nonsteroidal anti-inflammatory, synergizes with anandamide (the endogenous agonist of CB_1R and CB_2R), in a positive-feedback loop to further elevate levels of anandamide as well as other analgesic fatty acid ethanolamide levels (Di Marzo 2007).

Neurologic

Suppression of convulsions/seizures: While the exact mechanisms resulting in suppression of epileptic seizures by cannabinoids are unknown, there are many receptors for cannabinoids (particularly CB1) in areas of the brain known to be sites where partial seizures originate CBD reduces calcium oscillations in hippocampal neurons *in vitro*, and may exert its antiepileptic action by reducing calcium available for cell excitation. CB1 receptors have been detected with strong immunoreactivity in the hippocampi, skin and salivary glands of normal dogs. Experimentally, CBD attenuates experimentally-induced seizures in animals; this may reflect reduced calcium fluxes (Izzo 2009). THCV also has been associated with some anticonvulsant effects by virtue of its inhibitory effects on CB1. **Anxiolytic**: These effects have been demonstrated in healthy human volunteers (Izzo 2009). CBD exerts benzodiazepine independent effects, possibly by activating post synaptic 5-HT1A receptors. **Neuroprotection**: CBD is an antioxidant and as such has been proposed for treatment of Alzheimer's disease, Parkinson's disease and Huntington's disease. Restoration of calcium homeostasis may prevent apoptosis (Izzo 2009). In rodents, CBD reverses brain damage associated with ischemia. **Emesis and appetite**: Control of emesis and approved appetite are among the approved indications for FDA-approved cannabinoids. Emesis involves, among other signals, release of serotonin and subsequent stimulation of 5HT that activate neurons in the area postrema. CB1 receptors in the cerebrum, vestibular nuclei, and other brainstem nuclei involved in emesis suppress vestibular nuclei signals associated with nausea. Among the mechanisms of improved appetite is facilitated olfaction. Appetite is an approved indication for FDA-approved cannabinoid products.

Other

Cancer: In addition to control of adverse clinical signs associated with cancer and its treatment, a number of the cannabinoids have antiproliferative-antiapoptotic effects in a number of tumor cell lines. The National Cancer Institute has a link describing ongoing studies: www.cancer.gov/about-cancer/treatment/cam/patient/cannabis-pdq. **Diabetes mellitus**: CBD inhibits development of diabetes in non-obese diabetic mice, including ameliorating clinical signs of disease. This appears to reflect, in part, control of pancreatic inflammation, but also reduction of oxidative stress in target tissues (e.g., retina). **Bone formation**: A number of cannabinoids (essentially all in Table 1) stimulate mesenchymal stem cells responsible for bone formation and fracture healing. CBD Aldo controls bone resorption, reducing bone loss (Izzo 2009). Antimicrobial: CBC and CBG have demonstrated potent antibacterial effects towards selected microbes, including methicillin resistance staphylococci (MIC of 0.5 to 2 mcg/ml).

References

1. Brenneisen R. Chemistry and analysis of phytocannabinoids and other cannabis constituents. In: Elshohly MA, ed. *Forensic Science and Medicine: Marijuana and the Cannabinoids*. NY: Humana Press Inc. 17–49.
2. Izzo AA. *Non-Psychotropic Plant Cannabinoids: New Therapeutic Opportunities from an Ancient Herb*. Cell Press, Elsevier. 2009;515–5273.

3. Kreitzer FR, Stella N. The therapeutic potential of novel cannabinoid receptors. *Pharm Ther.* 2009;122:83–96.
4. Llera RM. Toxicologic hazards for police dogs involved in drug detection. *J Am Vet Med Assoc.* 2006;228:1028–1032.
5. Meola SD *et al.* Evaluation of trends in marijuana toxicosis in dogs living in a state with legalized medical marijuana: 125 dogs (2005–2010). *J Vet Emerg Crit Care.* 2012;22:690–696.
6. Osweiler GD *et al. Small Animal Toxicology.* Wiley-Blackwell, 224–229.
7. Svizenska I *et al.* Cannabinoid receptors 1 and 2 (CB1 and CB2), their distribution, ligands and functional involvement in nervous system structures: A short review. *Pham Biochem Beha.* 2008;90:501–511.
8. McPartland JM *et al.* Care and feeding of the endocannabinoid system: a systematic review of potential clinical interventions that upregulate the endocannabinoid system. *PLoS One.* 2014;9(3):e89566.

acvim316

Planning a Pharmacokinetic Study Without Elementary Mistakes

Ted Whittem, BVSc, PhD, DACVCP, FANZCVS
Melbourne, VIC, Australia

INTRODUCTION

A well-planned and executed pharmacokinetic (PK) study can provide valuable information, both for drug use and hypothesis testing. Comprehensive understanding of the pharmacokinetics of a drug enables rational dose design and appropriate clinical use. Often, PK studies involve short intensive bursts of researcher activity which can be scheduled into available timeslots. Therefore these studies are appealing for House Officers and clinical research investigators. There are several elementary steps in the design and conduct of pharmacokinetic experiments which, if overlooked or ignored, result in poor outcomes and rejection for publication.

STEPS FOR PLANNING AND CONDUCT OF PHARMACOKINETIC EXPERIMENTS

Ask a Question

Many PK studies are purely descriptive, rather than inquisitive. Without a formulated hypothesis and a suitably inquisitive study design, such descriptive PK studies are often not worthy of full manuscripts, and many high-impact journals do not accept them. PK should be thought of as a research tool for answering hypotheses, not as a quick route to rapid publication.

Build a Team

A publishable outcome is unlikely if a study is designed without *a priori* expectation of results. You must estimate the expected PK behavior and parameters from the literature if possible, potentially using interspecies scaling.[1] In particular, you need to develop an estimate of the target plasma concentration. Ideally, conduct a pilot study in 2 or 3 animals with sparse collection points and, perhaps, a range of doses. The samples will then need to be analyzed for drug concentration and the data evaluated. To satisfactorily navigate this study design stage, assemble a research team with expertise in interspecies allometry, data modelling and analytical methodology. Failure to get expert input at the design stage often results in unsatisfactory and unpublishable outcomes.

Measure Plasma Concentration

The target plasma concentration dictates the necessary sensitivity of the analytical assay. The assay must accurately measure concentrations to at least a tenth of the minimum effective concentration. Before any animal work is done, you must develop and validate the analytical assay. For successful publication, assay validation parameters which are required include: the inter- and intra-day coefficients of variation, statistical evaluation of the standard curve, the stability of stock and standard solutions and of plasma samples at -20°C and possibly -80°C, the analyte recovery through the sample preparation process, and the lower limit of quantification (LLOQ).[2] The minimum sample volume must be calculated from these assay performance characteristics as this impacts the study's IUCAC application!

Choose the Dosing Strategy: Doses, Frequency, Formulation, Routes of Administration

These study design questions are driven by the hypothesis to be tested. However, additional considerations are important including availability of formulations and active moiety, cost versus budget and the potential for acute toxicity. To adequately describe PK of any route or formulation, the PK for intravenous administration is needed. If intravenous PK are not available, then plan to study both the intravenous and the intended routes of administration. If an intravenous formulation is not readily available, plan to make one and consult a pharmacist or formulation chemist.

Choose the Sample Frequency, Sample Processing and Storage

Sample collection time points must be both frequent and prolonged and the assay must be sufficiently sensitive to measure over the entire duration. Mathematical methods are available to help estimate the most efficient time points for sample collection. Alternatively, if clinical responses for drug effects can be and have been measured, then the focus of the sample schedule should be primarily during the period of clinically detectable activity. It is not possible to accurately estimate the half-life of a drug unless 4 or more samples are collected over at least 2 to 3 terminal half-lives, so to plan the experimental design an estimate of the elimination rate constant is needed which determines the minimum duration of sample collection.

Choose the Study Design and Mathematical Approaches

The background understanding gained from the preparation for the study will determine the optimum study design and data processing method. Study types which are commonly used include cross-over or parallel comparisons, with stratified random block allocation of animals. PK data processing approaches that are commonly used include the naive pooled data set, the standard two-stage approach, a pharmacokinetic-pharmacodynamic (PK-PD) study, or a population approach (POP-PK).[3] The study type will influence the chosen experimental design.

The naïve pooled study combines all the data from all the animals together for data analysis. This loses the individual animal differences and can result in incorrect population estimates. This naïve pooled approach is used when each individual animal is sampled on one or two occasions, for example for wildlife species or where the sampling itself might change the PK by markedly altering circulating blood volume.

The standard two-stage approach is most usual. Typically, healthy and similar animals are enrolled, paying attention to blocking for gender, body weight and age, as these factors can be correlated to PK parameter estimates. The data from each individual are independently analyzed. The sample population of estimates for individual parameters in each treatment group allows calculation of population estimates. Statistical comparisons can be made between the results to evaluate the effect of the manipulation. Typically, the statistical method used is a general linear model with factors for dose, sequence, animal, period, sequence plus animal interactions, and other manipulations.

For PK-PD studies data are collected to measure the drug concentration and at the same time the effect of the drug is measured either directly or by surrogate, such as biological markers such as an intermediate enzyme activity. The STS approach is usually used, but with more complex modelling compared to PK-only studies. PK-PD studies are useful to correlate dose and effect and to better predict ideal doses in the presence of pathophysiological processes. Because of the complexity of the modelling and the larger number of parameters to be estimated, a large number of samples per animal is needed.

The POP-PK approach avoids the need for experimental animals, allowing the use of clinical cases. Only a few samples, usually from 2 to 5, are taken from each case and the sampling can be at times convenient for that individual case. Indeed, for POP-PK it is better if animals are sampled at different times from one another. Factors that may impact the pharmacokinetics of the drug are also measured and recorded. With only 2–3 time points per animal, from 30 to 50 cases are needed to adequately describe the PK. Although POP-PK depends upon an *a priori* knowledge of the suitable PK structural model for the drug, the approach generates evidence for adaption of clinical dosing regimens based on individual animal characteristics.

Conduct the Study

Preparation before the study is the key to success. All equipment, sample vials and personnel need to be assembled, labelled and trained. The in-life phase of these studies has intensely busy times interspersed by inactivity. Planning is key to ensure adequate study personnel are available for the hectic periods, and that everyone knows their role. Protocol training sessions are essential. Having one person in charge of all sample labelling and processing is a good idea because a simple switching of two samples can be sufficient error to prevent any conclusions, because the sequence of the samples is important.

Animal welfare is an important consideration: the environment and the comfort of the animals must be maintained since excitement, stress, eating and hydration status can all be factors that markedly alter PK.

Sample management is critical. Many drugs are not stable, or are metabolized in blood or plasma, so cooling and separation of cells from plasma needs to be rapid, yet consistent between samples. Hemolysis needs to be avoided as it often interferes with analytical processing. The type of sample collection tube or anticoagulant can also influence assay results, so must be consistent with the assay validation requirements.

Analyze the Results

Off-the-shelf pharmacokinetic analysis software allows the pharmacokinetically naïve to produce apparent PK results without any real intellectual input. Unfortunately, this often produces results which are erroneous. Mathematical discipline is necessary, first to review the data for consistency and then to iterate between possible solutions, seeking the best structural modelling approach. The ideal statistical model to adequately describe the data is selected after the structural model is defined. Although both these steps require in-depth PK knowledge, there are usually multiple pathways which lead to alternative correct procedures and the 'correct' approach can be, therefore, somewhat subjective.

Accordingly, after data processing one must review the results for plausibility and reiterate through data analysis decisions, changing methodology as needed to establish robust PK parameter estimates. Finally, a review of all modelling assumptions is needed and sometimes the results show that the experimental design was inappropriate and the experiment must be repeated after redesign.

Classical Compartmental Models

Classically, pharmacokinetic models include the one, two, and three compartment open mammillary models with first-order elimination from the central compartment after administration to the central compartment. The purpose of any modeling is to derive estimates of pharmacokinetic variables which can be used to make predictions of suitable dosing regimens.

Compartmental models may be suitable if the concentration versus time data appear to draw one or more straight lines when plotted on semi-logarithmic axes. The first step, therefore, is to plot the time and concentration data on semi-logarithmic axes for each individual, and subjectively evaluate it for linearity. First, check to see if the data draw a single straight line: a one-compartment model is likely appropriate. If a single straight line cannot fit to data, the residual differences between the measured and predicted concentrations for each time point are not randomly distributed around zero. The researcher then evaluates the data within a two-compartment model and, if needed, additional compartments are considered subsequently. This non-mathematical subjective approach helps the analyst to make decisions about the modelling and statistical methodology to be used and to correctly direct the PK software. The PK software varies the model parameters to best fit the data and compares the goodness of fit for each structural model, often using Akaike's Information Criterion (AIC) as the test statistic.[4] These comparisons must be conducted within the same statistical model. If the researcher varies the statistical model, such as by changing weighting of error, then structural models cannot be compared directly across different statistical models.

The statistical model describes the relationship between plasma concentration and the distribution of its residual variance. If the distribution of the residuals is homoscedastic, the residuals are not weighted. If a plot of the residuals versus time or concentrations shows a pattern, or non-random distribution, and if there is confidence that the structural model is appropriate, then the statistical model should be changed. This change can markedly alter the resulting parameter estimates.

Different initial estimates for each parameter can also induce different final model parameter estimates. Therefore, the initial estimates used are very important. Some PK software requires manual input of initial estimates and some derive their own initial estimates. You should always test your modelling conclusions for robustness to variation of the initial estimates. If the results are dependent on the initial estimates used, then either the pharmacokinetic model or the statistical model may be unsuitable. Software for PK modelling often offers options to be chosen by the user. Decisions regarding the mathematical approach, the steps size and the number of iterations can result in changes to the resulting parameter estimates.

Non-Compartmental Models

An alternative is to use non-compartmental modeling. It should be noted that this is not "model-independent." Non compartmental analysis provides estimates of "mean time" parameters, which can be thought of as the average total time spent by molecules in a kinetic space after introduction to the space. A homogeneous kinetic space is analogous to a single compartment in a compartmental model. Non-compartmental analysis, however, is equally applicable to situations where there is a heterogeneous kinetic space, which can be thought of as a collection of two or more compartments of a compartmental model. The mathematics used for the non-compartmental modeling are simple but their derivation and ease of understanding is less so. The formulae used rely on the assumption of linear pharmacokinetics.[5,6]

The two important calculated variables in non-compartmental analyses which are used for parameter estimation are the Area Under the concentration versus time Curve ($AUC_{(0\text{-inf})}$), and the Area Under the first Moment of the concentration versus time Curve ($AUMC_{(0\text{-inf})}$) which are calculated using the linear or log-linear trapezoidal methods.[7] These variables must be accurately calculated to avoid misleading pharmacokinetic parameter estimates. If the pharmacokinetic curve is not measured for an adequate time period, the extrapolated portion of $AUC_{(0\text{-inf})}$ is large: if the percent of $AUC_{(0\text{-inf})}$ that was extrapolated is greater than 20%, then the experiment is flawed and needs to be redesigned and repeated. The duration and frequency of sampling and the lower limit of quantification of the assay should be reassessed.

Common Problems in PK Due to the Drug

For drugs with complex formulations and nonlinear absorption, data analysis is more complex and a deconvolution approach may be needed. Frequently it is helpful to define the intravenous pharmacokinetics in each individual, then progress to modeling of the absorption phases. Drugs which alter their own pharmacokinetics, for example through either upregulation or inhibition of their own metabolism or through changing their own renal clearance, may have non-linear and/or non-dose-proportionality of their PK, and the simple approaches to data analysis described above become inappropriate. Delayed absorption can result in a lag between dosing and changes in concentration, and there are many drugs where absorption is altered by food, ambient temperature, surface area of application, site of application, and so on. Lag times can be modelled in compartmental analysis, but this is more difficult for non-compartmental analysis. Reabsorption after clearance from the kinetic space, such as occurs with enterohepatic recycling, causes a violation of primary assumptions for both compartmental and non-compartmental methods described above, and requires much more complex data analysis to result in meaningful findings.

Conclusions

Pharmacokinetic studies can provide meaningful and useful clinically relevant information, helping to understand dosing schedules and their adjustment for individuals and disease. However, PK studies have many pitfalls for the unwary.

References

1. Mordenti. *Journal of Pharmaceutical Sciences*. 1986;75(11):1028–1040.
2. Ng. *Reviewer Guidance: Validation of Chromatographic Methods*. Food and Drug Administration: Center for drug evaluation and research, Washington DC; 1994.
3. Sheiner. *Drug Metabolism Reviews*. 1984;15(1–2):153–171.
4. Yamaoka, *et al. Journal of Pharmacokinetics and Biopharmaceutics*. 1978;6(2):165–175.
5. Veng-Pedersen. *Clinical Pharmacokinetics*. 1989;17(5):345–366.
6. Gibaldi, *et al. Pharmacokinetics*. New York: Marcel Dekker Inc.; 1982.
7. Purves. *Journal of Pharmacokinetics and Biopharmaceutics*. 1992;20:211–226.

acvim317

Pharmacogenetics of Idiosyncratic Drug Reactions: Sulfonamide Hypersensitivity and Beyond

Jennifer M. Reinhart, DVM, MS, DACVIM (SAIM)
Madison, WI, USA

Idiosyncratic Drug Reactions

In human and veterinary medicine, adverse drug reactions (ADRs) are classified as either dose-dependent or idiosyncratic. In dose-dependent reactions, most members of a population will experience toxicity at high enough dosages. Conversely, idiosyncratic reactions only occur in a small subset of individuals and escalating drug doses do not increase the likelihood of an ADR in the remainder of patients. Thus, idiosyncratic reactions are considered independent of dose in the general population. More severe toxicity may occur in susceptible individuals at higher doses. However, from a practical perspective, **exposure to the drug** is much more important than the magnitude of that exposure.

A variety of drugs have been reported to cause idiosyncratic reactions in veterinary patients. Some of the better-known culprits include sulfonamide antimicrobials[1-3] and carprofen[4] in dogs, and methimazole[5] and oral diazepam[6] in cats. Clinical manifestations vary with the drug and individual patient, but common target organs include liver, bone marrow, and skin. Typically, a "sensitization period" of approximately 5 days to several weeks occurs between initiation of therapy and development of clinical signs. Because the presence of the drug and not the amount of drug determines whether the reaction occurs, dose reduction is often insufficient to resolve these signs and complete cessation of drug therapy is required.

As with most clinical traits, risk for idiosyncratic drug reactions can be categorized as environmental or heritable. By definition, all ADRs have at least one environmental risk factor because, without drug exposure, the reaction cannot occur. However, idiosyncratic reactions only occur in a small number of exposed individuals, so other factors must contribute proportionately more risk. The influence of genetic variation is of particular interest because some idiosyncratic reactions are overrepresented in certain population subgroups (e.g., breed, sex). Because of low disease incidence, genetic risk for idiosyncratic reactions can be difficult to study. However, genes involved in the pathogenesis of these reactions provide a logical starting place.

Most idiosyncratic drug reactions involve the generation of reactive metabolites, so metabolic enzymes are prime pharmacogenetic targets. The cytochrome P450 (CYP) superfamily is often implicated as these enzymes produce electrophilic, intermediate compounds from a wide range of substrates through Phase I oxidation.[7] Furthermore, many CYP enzymes exhibit significant genetic variation, which influences metabolite production and thus clinical phenotype. For example, in humans, aspirin-intolerant urticaria has been associated with polymorphisms in *CYP2C9*.[8] Phase II enzymes may also contribute to genetic risk by modulating enzymatic detoxification of drug intermediates. Variants of the *N*-acetyltransferase 2 and glutathione-*S*-transferases genes predispose to the development of asthma in patients treated with paracetamol[9] and diclofenac-associated hepatopathy has been associated with polymorphisms in UDP-glucuronosyltransferase isoforms[10]. Once generated, these reactive metabolites can cause direct oxidative damage and necrosis of susceptible tissues. Therefore, enzymes involved in the production and maintenance of the cellular antioxidant system (e.g., glutathione synthetase and reductase) may also influence the development of idiosyncratic drug reactions.[11]

In addition to causing oxidative damage, some drug metabolites can covalently bind endogenous proteins. These haptens elicit a self-directed immune response leading to inflammation and tissue damage. In this situation, idiosyncratic drug toxicities are considered hypersensitivity reactions. Similar to other immunologic disorders, genetic variation in antigen processing and presentation pathways can confer risk. In particular, the major histocompatibility complex (MHC) genes have been associated with a variety of drug hypersensitivity reactions in humans.[12] Genetic polymorphisms in cytokines, cell ligands, and other proteins mediating propagation and maintenance of the immune response may also play a role.

SULFONAMIDE HYPERSENSITIVITY

Sulfonamide hypersensitivity is uncommon in the general dog population, with a reported incidence of only 0.25% in one study.[13] However, in comparison with other antimicrobials, it is one of the most reported and best characterized systemic idiosyncratic drug reactions in veterinary medicine. Signs usually begin 5–14 days after initiation of drug therapy. A variety of clinical manifestations can occur, but the most common include fever, polyarthropathy, thrombocytopenia, and hepatopathy, with the latter two being possible negative prognostic indicators.[2] Both sex (female) and breed (Doberman Pinscher, Samoyed, Schnauzer) predilections have been suggested for sulfonamide hypersensitivity[2,3] and the condition appears heritable in humans[14]. These form the basis for a suspected genetic predisposition in both species.

The proposed pathogenesis of sulfonamide hypersensitivity is similar to that of other drug hypersensitivity reactions in that both a metabolic and immunologic phase are necessary to elicit a response (**Figure 1**). Sulfonamide antimicrobials are unique among sulfonamide drugs in that they contain an arylamine group, which can undergo N-hydroxylation to a hydroxylamine, reactive intermediate.[15] In humans, this reaction is primarily catalyzed by CYP2C9, which is polymorphic and thus could affect the development of the hypersensitivity reaction.[16,17] In oxidizing conditions, the hydroxylamine spontaneously degrades to an electrophile, which adducts to self-proteins and elicits an adaptive immune response. This response is primarily cell-mediated and is dependent on presentation of drug adducts by cellular MHC molecules. The MHC genes represent a major source of genetic diversity in all mammalian species. In humans, a specific MHC haplotype (HLA-A30 B13 Cw6) has been associated with sulfonamide hypersensitivity.[18] MHC pharmacogenetics has yet to be specifically evaluated in canine sulfonamide hypersensitivity, but multiple associations exist for other immune-mediated diseases in dogs.[19-22]

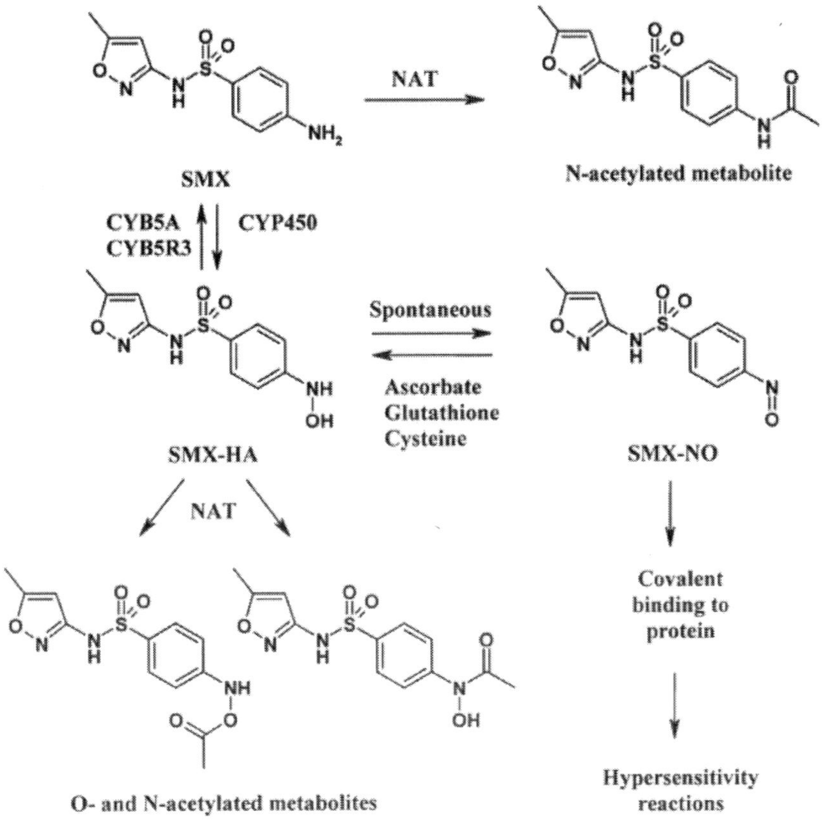

Figure 1. Biotransformation of sulfamethoxazole (SMX) as a representative sulfonamide antibiotic. NAT = N-acetyltransferase, CYB5A=cytochrome b5, CYB5R3 = cytochrome b5

reductase, CYP450 = cytochrome P450, SMX-HA = sulfamethoxazole-hydroxylamine, SMX-NO = sulfamethoxazole-nitroso. Note: NAT is not expressed in dogs.

In addition to those that generate reactive metabolites, enzyme systems also exist that detoxify these intermediates. In humans, the N-acetyltransferase enzymes (NAT1/2) convert both the parent sulfonamide drug and the hydroxylamine metabolite to an acetylated form, which facilitates renal excretion. *NAT2* is genetically polymorphic, but studies conflict as to whether these variants affect clinical phenotype.[23-25] Interestingly, dogs do not express a functional NAT enzyme,[26] which may partially explain why sulfonamide hypersensitivity occurs more commonly in dogs relative to veterinary species. This lack of activity also may force canine metabolism to rely more heavily on other detoxification systems such as the cytochrome b5 (CYB5A) and cytochrome b5 reductase (CYB5R3) system.[27] CYB5A/CYB5R3 convert the hydroxylamine metabolite back to the parent drug preventing further oxidation to the nitroso form and adduction.[27,28] In one large human study, no *CYB5A* or *CYB5R3* polymorphisms were significantly associated with sulfonamide hypersensitivity.[25] However, a parallel study in dogs did identify an association with a single nucleotide polymorphism (SNP) in *CYB5R3*, A729G.[29]

Genome-Wide Approaches and Precision Medicine for Drug Hypersensitivities

In the past ten years, the study of pharmacogenetics, including idiosyncratic drug reactions, has shifted from candidate gene studies to whole genome analysis. This is accomplished largely through the use of genome-wide association studies (GWAS). GWAS is particularly useful for the study of complex traits, such as drug hypersensitivity, because multiple associations can be identified simultaneously. At its core, GWAS is a case-control study with SNP markers examined as potential risk factors for a clinical trait of interest. In a GWAS evaluating drug hypersensitivity, allele frequencies are compared between patients with the hypersensitivity and patients tolerant to that drug. Multivariate regression is used to estimate the risk of developing the reaction associated with each SNP and significance is adjusted for multiple comparisons. This last point should be emphasized because hundreds of thousands or even millions of SNPs are included on genotyping arrays. Thus, to reach sufficient power to detect an association, large sample sizes are generally required, which can be problematic for rare traits such as drug hypersensitivity.

Using candidate gene approaches, a variety of gene types have been associated with hypersensitivity reactions in humans. However, at a genome level, an interesting trend emerges: almost all identified associations are with MHC loci. These associations include both MHC class I (HLA-A, B, C)[30,31] and class II (HLA-DP, DQ, DR)[31,32] genes. This trend highlights the importance of genetic variability in the immune response to xenobiotics - the genetic effects of these loci are strong enough to persist in pharmacogenetic genome-wide studies, which usually have modest sample sizes. Moreover, because these genes have large effects, MHC genotyping can be useful in assessing individual risk for idiosyncratic drug reactions.

One of the overarching goals of pharmacogenetics is to develop methods to predict and prevent adverse drug reactions in clinical patients. To date, the most widespread and successful screening program for drug hypersensitivity is HLA genotyping for the antiviral drug abacavir. A strong link has been demonstrated between the HLA-B*57:01 allele and abacavir-induced liver injury in people.[33] Several studies have confirmed the clinical utility of this screening program and the FDA label for abacavir carries a black box warning strongly recommending genotyping prior to initiation of therapy.[34,35] Similar programs exist for other drugs associated with hypersensitivities such as carbamazepine (HLA-A*31:01, HLA-B*15:02) and allopurinol (HLA-B*58:01).[35]

Array-based GWASs have greatly expanded our understanding of adverse drug reaction pharmacogenetics. With the advent of rapid and cost-effective whole-genome sequencing techniques, the number and complexity of detectable variants are increasing exponentially, as is our ability to identify links between these variants and clinical phenotype. Other medical fields have already incorporated these techniques into the study of complex disease traits and pharmacology will soon follow.

REFERENCES

1. Twedt DC, *et al. J Vet Intern Med*. 1997;11(1):20.
2. Trepanier LA, *et al. J Vet Intern Med*. 2003;17(5):647.
3. Giger U, *et al. J Am Vet Med Assoc*. 1985;186(5):479.
4. MacPhail CM. *J Am Vet Med Assoc*. 1998;212(12):1895.
5. Trepanier LA. *Vet Clin North Am Small Anim Pract*. 2007;37(4):775.
6. Center SA. *J Am Anim Hosp Assoc*. 1996;209(3):618.
7. Court MH. *Vet Clin North Am Small Anim Pract*. 2013;43(5):1027.
8. Palikhe NS, *et al. Allergy Asthma Immunol Res*. 2011;3(4):273.
9. Kang SH, *et al. Allergy Asthma Immunol Res*. 2013;110(5):364.
10. Daly AK, *et al. Gastroenterology*. 2007;132(1):272.
11. Trepanier LA. *Vet Clin North Am Small Anim Pract*. 2013;43(5):1055.
12. Pavlos R, *et al. Pharmacogenomics*. 2012;13(11):1285–306.
13. Noli C, *et al. Vet Q*. 1995;17(4):123.
14. Mohanasundaram J, *et al. J Indian Med Assoc*. 1998;96(1):21.
15. Cribb AE, *et al. Drug Metab Dispos*. 1991;19(5):900.
16. Cribb AE, *et al. Drug Metab Dispos*. 1995;23(3):406.
17. Gill HJ, *et al. Pharmacogenetics*. 1999;9(1):43.
18. Ozkaya-Bayazit E, *et al. Eur J Dermatol*. 2000;10(4):288.
19. Kennedy LJ, *et al. Tissue Antigens*. 2006;68(6):467.
20. Peiravan A, *et al. Vet Immunol Immunopathol*. 2016;182:101.
21. Friedenberg SG, *et al. Immunogenetics*. 2015;68(3):205.
22. Threlfall AJ, *et al. Canine Genet Epidemiol*. 2015;2:8.
23. Zielińska E, *et al. Eur J Clin Pharmacol*. 1998;54(9-10):779.
24. Wolkenstein P, *et al. Pharmacogenetics*. 1995;5(4):255.
25. Sacco JC, *et al. Pharmacogenet Genomics*. 2012;22(10):733.
26. Trepanier LA, *et al. Biochem Pharmacol*. 1997;54(1):73.
27. Trepanier LA, *et al. Xenobiotica*. 2000;30(12):1111.
28. Kurian JR, *et al. J Pharmacol Exp Ther*. 2004;311(3):1171.
29. Funk-Keenan J, *et al. J Vet Intern Med*. 2012;26(5):1126.
30. Daly AK, *et al. Nat Genet*. 2009;41(7):816.
31. Lucena MI, *et al. Gastroenterology*. 2011;141(1):338.
32. Fernandez CA, *et al. Blood*. 2014;124(8):1266.
33. Tangamornsuksan W, *et al. J Pharm Pharm Sci*. 2015;18(1):68.
34. Ruiz-Iruela C, *et al. Pharmacogenet Genomics*. 2016;26(8):390.
35. Plumpton CO, *et al. Pharmacoeconomics*. 2016;34(8):771.

acvim319

Individualizing the Use of NSAIDs in Cats: A Pharmacometabolomics Approach

Nicolas F. Villarino, VM, DVSc, PhD, DACVCP
Pullman, WA, USA

Introduction

Cats are the most common companion animals in the United States, with approximately 70 million cats living in American households. Unfortunately, a significant number of cats suffer from various painful and inflammatory conditions that require pharmacological intervention. The primary goal of the pharmacological control of pain and inflammation is to prevent suffering, avoid further tissue damage, and promote the recovery of normal tissue structure and organ function(s). Nonsteroidal anti-inflammatory drugs (NSAIDs), such as meloxicam, are essential to treat pain and inflammation in veterinary species, including cats. Nevertheless, one of the greatest challenges in feline medicine is the effective and safe use of NSAIDs.

One of the key factors that defy the optimal treatment of pain and inflammation is that clinicians in the US only have access to two NSAIDs licensed by the U.S. Food and Drug Administration (FDA) for use for cats, meloxicam and robenacoxib. This situation is in clear contrast with the situation for dogs, for which 10 FDA-licensed NSAIDs are available, providing veterinarians with the tools to offer more optimal pain control and anti-inflammatory therapy. One of the greatest challenges in feline medicine is the effective and safe administration of nonsteroidal anti-inflammatory drugs (NSAIDs), particularly for treating conditions such as trauma and peritonitis that require their long-term administration at high doses. Each year, thousands of cats unnecessarily suffer from pain and inflammation when they could have benefitted from the long-term administration of NSAIDs such as meloxicam.

NSAIDs and Kidney Damage

Cats are exquisitely sensitive to NSAIDs. The long-term administration of relatively high doses of NSAIDs results in severe kidney changes associated with acute injury, including glomerular and tubular damage and interstitial lymphocytic infiltration. The sustained damage to the glomerulus or tubule interstitium can lead to progressive damage and an expected loss of functional kidney mass. Regrettably, these events remain clinically occult and undetected until there is a significant decrease in the functional renal mass, ultimately detectable by traditional clinical markers.

What Do We Not Know About NSAIDS in Cats?

In spite of the fact that NSAIDs are critical for feline medicine, the exact mechanism(s) of the adverse effects of these drugs in cats remain poorly understood. Unfortunately, most of our current knowledge about the mechanism(s) of action and toxicity of NSAIDs in cats is extrapolated from other species. However, data generated in other species, like dogs, humans, and mice, are not always transferable to cats. In order to improve feline medical care, substantial progress must be made in understanding the effects of NSAIDs in this species. There is an urgent need for safer NSAIDs and strategies to control painful and inflammatory conditions in cats. One of the strategies that could be implemented to control pain and inflammation is the individualization of the administration of NSAIDs.

Why Is It Critical to Individualize NSAIDS in Cats?

It is crucial to individualize the administration of NSAIDs for cats because it provides the best chances of maximizing the anti-inflammatory effects, while minimizing the risk of causing kidney damage or favoring the progression of an asymptomatic underlying kidney disease. Preventing NSAID-induced kidney damage is necessary because acute kidney injury is now recognized as a potential mediator of progressive chronic kidney disease and end-stage kidney disease.[1-3]

Individualized Medicine in Veterinary Patients

Individualization of the use of NSAIDs in cats is doable. The best example of individualized medicine in animals is the testing for the mutation of the multidrug resistance gene in dogs. Approximately 60,000 dogs were tested for MRD-1 worldwide. Testing for the mutation of the multidrug resistance gene helps to prevent severe adverse drug reactions to several drugs in dogs, including ivermectin and related drugs, the antidiarrheal agent loperamide, and several anticancer drugs (vincristine, doxorubicin). Considering the important role of individualized veterinary medicine, WSU has recently established the Program in Individualized Medicine (PrIME) to improve diagnostics and optimize pharmacologic interventions for dogs and cats (https://prime.vetmed.wsu.edu/). One of the objectives of scientists of the PrIME is to individualize the use of NSAIDs in cats because it will help improve the feline medical care.

Biomarkers for Individualized Medicine

Individualization of treatments can be accomplished by using biomarkers for drug safety and/or efficacy. A biomarker is a characteristic that is objectively measured and evaluated as an indicator of normal biological processes, pathogenic processes, or pharmacologic responses to a therapeutic intervention. A contributing factor to NSAID-induced kidney injury is the delayed detection of this condition due to the lack of noninvasive sensitive diagnostic tests. For example, serum creatinine concentrations do not become elevated above the normal range until approximately >75% of renal function is lost. More recently, symmetric dimethylarginine (SDMA) has become available.[4] However, although this metabolite seems to be very helpful for diagnosing and staging chronic kidney disease, recent preliminary studies done in our laboratory suggest that SDMA is not a sensitive biomarker for predicting or detecting NSAID-induced kidney damage in its early stages. Early detection of kidney damage is critical for individualizing the administration of NSAIDs for cats. Kidney-specific clusterin, cystatin B, and inosine are molecules that can become biomarkers for the diagnosis of active kidney injury in cats[5] in the future.

What Do We Need to Individualize the Administration of NSAIDs in Cats?

Biomarkers for detecting early non-symptomatic acute kidney injury and predicting and monitoring drug toxicity are necessary to avoid NSAID-induced kidney damage, maximize the benefits of NSAIDs, and improve the overall medical care and welfare of cats in need of long-term analgesic or anti-inflammatory therapy. Unfortunately, biomarkers for detecting NSAID-induced kidney damage in cats remain to be discovered. Important characteristics of kidney biomarkers for cats are described comprehensively in several recent reports.[2,5] Biomarker for individualizing the administration of NSAIDS in cats should at least:

- Be noninvasive, easily measured, inexpensive.
- Be detectable in urine and/or plasma, such that it can be assessed routinely and serve as an indicator of kidney function or dysfunction or response to injury.[2]
- Predict which cat is predisposed to NSAID-induced kidney damage.[2]
- Be highly sensitive and specific to ensure timely diagnosis.
- Reflect kidney-specific events, be unique and specific to the kidney, and reflect very early and potentially sustained phases of the pathogenesis and repair processes.[2]
- Reflect the extent of ongoing processes, changes in these processes in response to therapeutic intervention over time, and potentially the etiologic insult.[2]

Current evidence suggests that it is highly unlikely to find a single ideal biomarker for individualizing the administration of NSAIDs in cats. However, it is probable that a panel of biomarkers will be necessary to assist the use of NSAIDs in cats.

Pharmacometabolomics

Metabolomics is the nonbiased identification and characterization of "all" the small molecular compounds (metabolites) in a biological system, using highly sensitive analytical techniques, in combination with bioinformatics.[6] The metabolome is a collective term for all the metabolites in a specific biological system/sample. In metabolomics studies, samples are collected and then metabolites are extracted from the sample using different procedures. Finally, metabolites are separated and detected in the samples based on solubility and mass by using sophisticated analytical instruments (HPLC/MS/MS). Mass and the retention times (times of detection after introduction in the system) of the molecules are used for metabolite identification using open access metabolite libraries databases.

Since its first applications in the 1960s and 1970s,[7] the use of metabolomics as a research tool has expanded into fields such as physiology,[8] nutrition,[9] infectious diseases[10]. During the last decade, metabolomics has been successfully applied to unveil new therapeutic targets and to discover new biomarkers and mechanism(s) of drug toxicity. This emerging field is known as pharmacometabolomics. Pharmacometabolomics complements epigenomic, genomic, transcriptomic, and proteomic, and contributes to a wide-ranging understanding of drug effects by considering both intrinsic and extrinsic contributions to interindividual variation in drug response.[12] Changes in metabolite concentrations have served as the basis for the development of more than a hundred different chemical or metabolite biomarker tests that are commonly used today. It, therefore, presents a very promising approach for gaining a new understanding of the effects of NSAIDs on numerous metabolic processes in cats. Pharmacometabolomics could also help identify a panel of biomarkers for detecting early non-symptomatic NSAID-induced kidney damage and predicting and monitoring drug toxicity.

In our laboratory, we have applied a pharmacometabolomics approach for evaluating metabolites changes in urine and plasma from cats treated with meloxicam. The resulting metabolite profile is analogous to a chemical fingerprint left behind by NSAID-induced alterations in cellular processes. By associating upregulated and/or downregulated small molecules with a reference chemical pathway database, we were able to identify, for the first time, key feline-specific metabolic pathways that are stimulated or repressed by the NSAID meloxicam.

Feline's Urine Metabolome

Urine is commonly used in metabolomics studies because it contains hundreds of low-molecular-weight substances. Low-molecular-weight substances in urine include metabolic breakdown products from foods, drugs, environmental contaminants, endogenous waste metabolites, and bacterial end products.[13-15] Using an untargeted metabolomics approach, we performed a comprehensive analysis of low-molecular-weight substances in feline urine from healthy domestic cats. We detected 319 metabolites in urine. Strikingly, these metabolites are key components of at least 100 chemical pathways.[16] We identified the presence of multiple metabolites that were previously not known to be present in feline urine. Of the 319 metabolites, 126 are considered known (when the spectra and retention time are matched to library entries). Most of these metabolites were carbohydrates and carbohydrate conjugates. However, the feline urine also contained organic acids and derivatives, heterocyclic compounds, amino acids, peptides and analogues, phenols and derivatives, nucleosides, nucleotides and analogues, alcohols and polyols, lipids and lipid-like molecules, and amines.

Feline's Plasma Metabolome

We performed a comprehensive analysis of 412 low-molecular-weight substances in plasma from domestic cats. One hundred and thirty-two of those metabolites are considered known. The feline plasma metabolites detected in plasma belong to multiple chemical groups including carbohydrates, organic acids and derivatives, heterocyclic compounds, amino acids, nucleosides, nucleotides and analogues, alcohols and polyols, lipids and lipid-like molecules, and amines.

Feline Urine and Plasma Metabolome in Cats Treated with Meloxicam

Combining principal component analysis with univariate statistical analysis, we evaluated the plasma and urine metabolomes from cats treated with meloxicam.

In urine, there were no noticeable changes in the metabolome following a single administration of meloxicam in all of the cats. However, the repeated administration of meloxicam resulted in an alteration of several low-molecular-weight substances. Some of these substances were upregulated while others were downregulated. The same trend was observed in the plasma metabolome. However, there was a markedly larger number of metabolites altered in plasma than in urine.

Metabolomics analysis reveals that repeated administration of meloxicam seems to perturb critical cellular processes, including the beta-oxidation, collagen synthesis, and the cyclic acid cycle. Importantly, there is scientific information that links the alteration of some of these altered metabolic pathways with cellular dysfunction and tissue kidney damage.[17-23]

These original results open a window for discovering unknown or poorly understood, metabolic pathways affected by the administration of meloxicam in cats, which could lead to the discovery of noninvasive biomarkers.

Future Directions

Discovery of noninvasive robust biomarkers for predicting the development and progression of kidney damage is necessary for individualizing the use of NSAIDs in cats. Noninvasive biomarkers could assist clinicians in making therapeutic decisions according to individual needs; for example, by helping the categorization of patients based on the risk for suffering NSAID-induced kidney damage. Furthermore, biomarkers could aid the selection of optimal dose intervals and extent of washout periods between drugs and/or the selection of intermittent therapy or daily administration.

References

1. Lakhmir S. *N Engl J Med*. 2014;371:58–66.
2. Cowgill LD. *Vet Clin North Am Small Anim Pract*. 2016;46(6):995–1013.
3. Jepson RE. *Vet Clin North Am Small Anim Pract*. 2016;46(6):1015–48.
4. Relford R, *et al*. *Vet Clin North Am Small Anim Pract*. 2016;46(6):941–60.
5. Yerramilli M, *et al*. *Vet Clin North Am Small Anim Pract*. 2016;46(6):961–93.
6. Nadia K, *et al*. *Adv Protein Chem Struct Biol*. 2016;102:53–78.
7. Hoult DI, *et al*. *Nature*. 1974;252:285–287.
8. Dallmann R, *et al*. *Proc Natl Acad Sci USA*. 2012;109:2625–2629.
9. Piccolo BD, *et al*. *J Nutr*. 2015;145:681–690.
10. Villarino N, *et al*. *Proc Natl Acad Sci U S A*. 2016;113:2235–2240.
11. Jun Zou. *Biomed Res Int*. 2013;2013:742835.
12. Burt T, Nandal S. *Clin Transl Sci*. 2016;9:128–138.
13. Kind T, *et al*. *Anal Biochem*. 2007;363:185–195.
14. Issaq HJ, *et al*. *J Urol*. 2008;179:2422–2426.
15. Want EJ. *Nature Protocols*. 2010;5:1005–1018.
16. Kanehisa M, Goto S. *Nucleic Acids Res*. 2000;28:27–30.
17. Dzúrik R, *et al*. *Nephron*. 1992;61(1):64–67.
18. Suhre K. *et al*. *Nature*. 2011;477:54–60.
19. van Guldener C, *et al*. *Semin Vasc Med*. 2005;5(2):201–8.
20. Bánhegyi G, *et al*. *Free Radic Biol Med*. 1997;23(5):793–803.
21. Sharma K, *et al*. *J Am Soc Nephrol*. 2013;24(11):1901–12.
22. de Gramont A, *et al*. *Nat Rev Clin Oncol*. 2015;12(4):197–212.
23. Gronwald W, *et al*. *Kidney Int*. 2011;79(11):1244–53.

acvim320

Canine Body Cavity Effusions: Classification and Differentials

Andrea A. Bohn, DVM, PhD, DACVP (Clinical Pathology)
Fort Collins, CO, USA

Introduction

Classification of an effusion can be very helpful in prioritizing differentials and directing additional diagnostics for determining the underlying cause for fluid accumulation. The underlying pathogeneses for fluid accumulation will be reviewed and the different classifications of effusions and how these relate to differential diagnoses will be discussed. A slightly simplified variation from the traditional classification system will be utilized and its usefulness demonstrated by retrospective analysis of clinical cases.

Pathogenesis

Body cavity effusion is a very general term meaning that there is an accumulation of fluid within the cavity. There are many things that can lead to an effusion, although the number of underlying mechanisms is fewer. Evaluation of the characteristics of effusion fluid can occasionally lead to a diagnosis, but more often primarily aids in developing a diagnostic plan. Therefore, consideration of the underlying pathogenesis for effusion formation is a very important aspect of fluid evaluation.

In normal states, a small amount of fluid escapes from submesothelial capillaries (comprised of water, electrolytes, and small molecules), based on a balance of oncotic and hydrostatic pressures between the capillaries and interstitial tissue. This fluid is low in proteins, as the permeability for proteins is low in endothelium. Any of this fluid that is not reabsorbed by submesothelial lymphatics easily passes through the mesothelial lining into the body cavity. Stomas are direct openings of the lymphatic system within the mesothelial lining that are important for reabsorption of cavity fluid, including cells, particles, and proteins.

Anything that disrupts the oncotic or hydrostatic pressure balance between vasculature and surrounding tissues can result in effusion, as well as anything that prevents the reuptake of fluid by lymphatics. In addition, vessels can become more leaky due to inflammatory mediators or compromise of vessel wall integrity, which allows the larger components of blood or lymphatics to escape as well, including protein and cells. Keeping these underlying mechanisms in mind, rather than categorizing fluids based on numerical cutoffs alone, is necessary to effectively prioritize differentials and determine the most appropriate diagnostic plan.

Classification

Body cavity effusions have been traditionally classified as transudates, modified transudates, and exudates depending on various cell counts and protein concentrations. It is not clear how the original recommendations were established and there is some disagreement on whether this terminology is appropriate. A retrospective look at over 100 canine abdominal fluids suggests that a simpler method of categorizing abdominal effusions may be more effective.[1] This method first looks at the cell concentration. Greater than 3,000 cells/µl is considered a fluid of increased cellularity and less than 3,000 cells/µl is a transudate. Transudates are then categorized as having low (<2.5 g/dl) or high (≥2.5 g/dl) protein concentration. The next critical step is microscopic examination of the sample to differentiate the cells present and look for abnormal cells or cell morphology, microorganisms, or foreign substances.

Transudates are effusions of low cellularity (<3,000 cells/µl), which are typically associated with decreased oncotic pressure, increased hydrostatic pressure, or decreased lymphatic removal of fluid. An exception of an effusion that numerically may fall into this category is uroabdomen, where there is leakage of urine into the abdomen without an increase in inflammatory cell concentrations. Transudates can be of high or low protein content, the difference being which tissues are involved and, possibly, chronicity and severity of lesions, the precise underlying disease processes, along with potential comorbidities.

Effusions of high cellularity (traditionally called exudates) are most commonly associated with inflammatory processes, where cells migrate from the vasculature into body cavities due to inflammatory mediators. The predominant cell is typically a neutrophil and macrophages are also often present in low, moderate, or high numbers, depending on cause. Inflammation can be a fairly nonspecific finding, unless infectious agents, inciting substances, or neoplastic cells are seen. Leakage of lymphatic fluid into body cavities often results in increased cell numbers, although lymphocytes are not always the predominant cell seen. Since leukocyte concentrations are normally higher in the peripheral blood than in normal cavity fluid, hemorrhagic effusions typically have a cell count >3,000 cells/μl. In addition, neoplastic cell populations are occasionally shed into body cavity effusions, resulting in increased cell counts.

Differentials

Characterizing canine abdominal fluids, using a 3,000/μl cell count to differentiate transudates from effusions of high cellularity and a protein concentration of 2.5 g/dl to differentiate between low and high protein transudates, was retrospectively evaluated in a recent study.[1] This simplified method of fluid evaluation performed well in helping prioritize differentials in that study as well as within an additional population of 108 dogs. There are a few disease processes that have consistent effusion characteristics and others that can result in a spectrum of changes for which more complicated numerical cut-offs did not provide any improved diagnostic value.

Based on over 200 dogs that were patients at Colorado State University Veterinary Teaching Hospital that had peritoneal effusion between July 1, 2013, and July 1, 2015, neoplasia was the most common cause for the effusion. Approximately half of the cases with effusions having >3000 cells/μl and slightly less than a quarter of transudates with high or low protein were due to neoplasia. The most common cause of low protein transudates was severe liver disease (approximately 30% of low protein transudates). Other fairly common causes were protein-losing enteropathy (20%), enteritis (14%) and vasculitis/thrombosis (11%). The most common cause of high protein transudates, comprising >40% of cases, was cardiac disease. Other causes of high protein transudates that involved more than one dog were vasculitis/thrombosis (14% of high protein transudates), organ displacement (7%), and a few cases of enteritis and GI foreign body. Exudates that weren't due to neoplasia were septic (28%) or associated with pancreatitis (8%), bile leakage (4%) or uroabdomen (<3%). There were also several fluid samples collected post-operatively that all contained elevated inflammatory cell concentrations except for a few cases that had Jackson-Pratt drains in place.

Prioritizing Differentials and Work-Up for Canine Abdominal Effusions

Neoplastic effusion (can fall within any of the subsequently listed effusion categories):

- All body cavity fluid samples must be microscopically evaluated for atypical cell populations.
- If obvious or suspected neoplastic cells are present, further evaluation may include:
    - Diagnostic radiology imaging or endoscopic examination for localization of the neoplasm
    - Aspiration or biopsy; possible immunocytochemistry or immunohistochemistry
    - Polymerase chain reaction for antigen receptor rearrangements (PARR) or flow cytometry for suspected lymphoproliferative disorders (large/immature lymphocytes are generally neoplastic; Increased suspicion for small cell lymphoproliferative disorder with lymphocyte counts >10,000/μl).

Low protein transudates (cells <3,000/μl, refractometer protein <2.5 g/dl):

- Top differentials should include conditions causing hypoalbuminemia and/or portal hypertension. Chronic liver disease and protein-losing enteropathies are common causes.
    - Further diagnostics:
        - Biochemical panel; assess albumin, liver enzymes
        - Urinalysis
        - Possible gastrointestinal evaluation/testing; possible liver function tests

- Diagnostic radiology imaging as indicated

High protein transudates (cells <3,000/µl, refractometer protein >2.5 g/dl):

- A top differential should include heart failure or cardiac tamponade. Organ displacement (torsion, etc.) is another consideration.
    - Further diagnostics:
        - Cardiac evaluation
        - Other diagnostic radiology imaging as indicated

Transudates of either low or high protein:

- Neoplasia is a common cause of all types of effusions (see above).
- Enteritis (serum albumin may be normal) and vasculitis or thrombosis are other causes of transudates that can be of low or high protein concentration.
    - Further diagnostics:
        - CBC
        - Coagulation testing; D-dimers
        - Diagnostic radiology imaging as indicated

Effusions of high cellularity:

- Inflammatory exudates may be associated with septic abdomen, pancreatitis, bile peritonitis, neoplasia, necrosis, and uroabdomen.
    - Further diagnostics:
        - Culture
        - Further testing for bile peritonitis, uroabdomen, pancreatitis, or sepsis, as indicated
        - Diagnostic radiology imaging as indicated
- Hemorrhagic effusions (defined here as >500,000 erythrocytes/µl) are often associated with neoplasia; a few cases were also associated with pancreatitis and anaphylaxis.
    - Further diagnostics:
        - Diagnostic radiology imaging as indicated
        - CBC, coagulation testing for possible coagulopathy
        - Possible further testing for pancreatitis

EFFUSIONS THAT CAN BE CLASSIFIED AS EITHER A TRANSUDATE *OR* AN EFFUSION WITH INCREASED CELLULARITY:

- Neoplasia; with or without neoplastic cells present (see above)
- Uroabdomen; if suspected, measure creatinine in fluid and serum (fluid creatinine >2x serum creatinine)
- Lymphangiectasia; small lymphocytes typically comprise >30% of cells
- Chylous effusion; triglycerides >100 mg/dl; lymphocytes may not be the predominant cell; underlying conditions are generally similar to those for transudates, possibly idiopathic.

OTHER BODY CAVITIES

Mechanisms for effusions within peritoneal, pericardial, and thoracic body cavities are similar, although the most common underlying disease processes can differ, depending on location and species. Data obtained from analysis of canine thoracic and pericardial fluids using the cut-offs of 3,000 cells/µl to differentiate between transudates and effusions of high cellularity and of 2.5 g/dl protein to differentiate between low and high protein transudates will also be discussed.

REFERENCES

1. Bohn AA. Analysis of canine peritoneal fluid analysis. *Vet Clin North Am Small Anim Pract*. 2017;Jan;47(1):123.

acvim321

Feline Body Cavity Effusions: Classification and Differentials

Andrea A. Bohn, DVM, PhD, DACVP (Clinical Pathology)
Fort Collins, CO, USA

INTRODUCTION

Classification of an effusion can be very helpful in prioritizing differentials and directing additional diagnostics for determining the underlying cause for fluid accumulation. The underlying pathogeneses for fluid accumulation will be reviewed and the different classifications of effusions and how these relate to differential diagnoses will be discussed. A slightly simplified variation from the traditional classification system will be utilized and its usefulness demonstrated by retrospective analysis of clinical cases.

PATHOGENESIS

Body cavity effusion is a very general term meaning that there is an accumulation of fluid within the cavity. There are many things that can lead to an effusion, although the number of underlying mechanisms is fewer. Evaluation of the characteristics of effusion fluid can occasionally lead to a diagnosis, but more often primarily aids in developing a diagnostic plan. Therefore, consideration of the underlying pathogenesis for effusion formation is a very important aspect of fluid evaluation.

In normal states, a small amount of fluid escapes from submesothelial capillaries (comprised of water, electrolytes, and small molecules), based on a balance of oncotic and hydrostatic pressures between the capillaries and interstitial tissue. This fluid is low in proteins, as the permeability for proteins is low in endothelium. Any of this fluid that is not reabsorbed by submesothelial lymphatics easily passes through the mesothelial lining into the body cavity. Stomas are direct openings of the lymphatic system within the mesothelial lining that are important for reabsorption of cavity fluid, including cells, particles, and proteins.

In general, effusions are almost always associated with vascular effects, including blood vasculature and/or lymphatics. Anything that disrupts the oncotic or hydrostatic pressure balance between vasculature and surrounding tissues can result in effusion, as well as anything that prevents the reuptake of fluid by lymphatics. In addition, vessels can become leakier due to inflammatory mediators or compromise of vessel wall integrity, which allows the larger components of blood or lymphatics to escape as well, including protein and cells. Keeping these underlying mechanisms in mind, rather than categorizing fluids based on numerical cutoffs alone, is necessary to effectively prioritize differentials and determine the most appropriate diagnostic plan. The different primary pathogeneses leading to body cavity effusion are listed in the following table:

Body cavity effusions

Pathogenesis	Causes	Typical effusion characteristics
Changes in osmotic or hydrostatic vascular pressures or if normally formed fluid is not reabsorbed as usual	Hypoalbuminemia: - Liver insufficiency, protein-losing enteropathy, protein-losing nephropathy.	Low cells. Low protein.
	Increased blood or lymphatic vascular pressure: - Systemic changes due to cardiovascular disease, portal hypertension, or anything that can compress or occlude vasculature: neoplasia or other mass lesions, organ displacement or swelling, thrombosis. (Given the close association of blood and lymphatic vessels, differentials for lymphatic effusions do not differ from other transudates, other than lymphangiectasia and idiopathic chylous effusions should be added.)	Low cells. Low or high protein, depending on underlying cause, location, duration.

Increased permeability of vessels	Vasculitis/anaphylaxis Inflammation - Septic - Non-septic: FIP, pancreatitis, bile peritonitis, neoplasia	Low cells. Possible increased RBCs. Low or high protein. High cellularity. FIP cellularity may be low.
Leakage from vessels or viscus	Hemorrhage Lymph - Triglycerides >100 mg/dl is diagnostic for chylous effusion; lower triglycerides do not rule out lymph effusion Uroabdomen - Urine is typically of low cell count and low protein, but cellularity can rise due to an inflammatory response.	Usually high cells and protein. High erythrocytes. Low or high cellularity. Few to many lymphocytes. Low or high protein. Low protein. Low or high cells.
Exfoliation of cells	Neoplasia (often causes effusion secondarily via other pathogeneses listed rather than this mechanism)	Low or high cells. Low or high protein.

CLASSIFICATION

The traditional classification of body cavity effusions has categorized effusions as transudates, modified transudates, and exudates depending on various cell counts and protein concentrations. It is not clear how the original recommendations were established and there is some disagreement on whether this terminology is appropriate. According to most sources, transudates are supposed to have fewer cells than modified transudates and there are variable recommendations as to the cut-off between, or overlap of, modified transudate and exudate cellularity. There are also variable and overlapping protein concentration recommendations for differentiating transudates, modified transudates, and exudates. Because there is overlap and omission in the traditional classification categories for body cavity effusions, and because there are causes for effusion that result in fluid characteristics that span across categories, it is important not to focus too much on numbers but to focus on determining the underlying pathogenesis and potential causes for the effusion.

A recent article retrospectively assessed canine abdominal fluids that utilized simplified cut-off values for categorizing effusions compared to those traditionally recommended that were as, if not more, effective.[1] This method first looks at the cell concentration. Greater than 3,000 cells/μl is considered a fluid of increased cellularity and less than 3,000 cells/μl is a transudate. Transudates are then categorized as having low (<2.5 g/dl) or high (≥2.5 g/dl) protein concentration. Of course microscopic examination of the sample to differentiate resident cells and look for abnormal cells or cell morphology, microorganisms, or foreign substances is critical in the analysis, as it is for all classification schemes.

Transudates are effusions of low cellularity, which are typically associated with decreased oncotic pressure, increased hydrostatic pressure, or decreased lymphatic removal of fluid. An exception of an effusion that numerically may fall into this category is uroabdomen, where there is leakage of urine into the abdomen without an increase in inflammatory cell concentrations. Transudates can be of high or low protein content, the difference being which tissues are involved and, possibly, chronicity and severity of lesions, the precise underlying disease processes, along with potential comorbidities.

Effusions of high cellularity (traditionally called exudates) are most commonly associated with inflammatory processes, where cells migrate from the vasculature into body cavities due to inflammatory mediators. The predominant cell is typically a neutrophil and macrophages are also often present in low, moderate, or high numbers, depending on cause. Inflammation can be a fairly nonspecific finding, unless infectious agents, inciting substances, or neoplastic cells are seen. Leakage of lymphatic fluid into body cavities often results in increased cell numbers, although lymphocytes are not always the predominant cell seen. Since leukocyte concentrations are normally higher in the peripheral blood than in normal cavity fluid, hemorrhagic effusions typically have an increased cell count. In addition, neoplastic cell populations are occasionally shed into body cavity effusions, resulting in increased cell counts.

Although there is some disagreement in what numbers and terminology should be used on how best to classify effusions, there seems to be a general agreement that consideration of the underlying pathogenesis is the most important factor. From there, a differential list and diagnostic plan can be best formulated. In this session, retrospective analysis of feline body cavity effusions will be used to evaluate whether modifications of the traditional effusion classification system can make body cavity effusion analysis less complicated. Disease processes that cause body cavity effusions will also be reviewed, the most common of which are included in the following table:

DIFFERENTIALS

Common causes of body cavity effusions in cats

Cause	Comments	Typical effusion characteristics
Cardiovascular disease	One of the most common causes of thoracic effusion. One of the most common causes of abdominal effusion.	Low cells. Low protein. Low cells. High protein.
Neoplasia	Common cause of effusions (especially in old cats). Diagnostic if neoplastic cells present. Effusions are often due to vascular congestion, inflammation, hemorrhage, or blocking lymphatic stoma.	Highly variable. Low or high cells. Low or high protein.
Bacterial infection	Multiple possible causes: penetrating wounds, foreign body migration, pneumonia or lung abscess, hematogenous spread, previous surgery, decreased integrity of digestive tract.	High cells. Usually high protein.
FIP	Vasculitis is the predominant cause of effusion. Typically high protein (often >4.5 g/dl) with albumin:globulin <0.9. Cells typically <20,000/μl, often <5,000).	High protein. Low to high cells.
Idiopathic chylous effusions	Other causes of chylous effusion must be ruled out. Differentials are similar to transudative effusions (heart cardiovascular disease and neoplasia are common causes; also lymphangiectasia, organ displacement, trauma). Triglycerides >100 mg/dl is diagnostic for chylous.	Low or high cells. Low or high protein. Small lymphocytes often, but not always, predominant cell.
Trauma	Could result in hemorrhagic or chylous effusions from ruptured vessels/organs. Possible transudates from displaced organs, herniation. Possible inflammation from damaged or leaking viscera.	Variable, depending on process
Less common causes	Uremia, vasculopathy, hypoalbuminemia, coagulopathy; diaphragmatic hernia, organ displacement/torsion, pancreatitis	Variable, depending on process

DIAGNOSTIC PLANNING

The following box offers recommended steps to take based on results from body cavity fluid analysis using the simplified method of classification. Of course signalment, history, and clinical exam findings should also be kept in mind when prioritizing differentials and further diagnostic procedures. The effectiveness of this simplified classification scheme will be discussed.

Effusions Cell Count ≥3,000/µl

- If neoplastic cells are present:

1. Diagnostic radiology imaging for localizing the neoplasm; aspiration or biopsy of mass lesion, as appropriate.

- If mostly neutrophils +/- macrophages:

1. Look for microorganisms or foreign material; culture
2. Consider FIP, especially if protein >3 g/dl (often ≥5.0), albumin:globulin ratio is <0.9, and cellularity is <5,000 cells/µl (cellularity almost always <20,000 with rare cases up to 30,000). If cellularity is determined by a Sysmex XT-2000iV hematology analyzer, a >1.7 difference in cell counts between the DIFF and BASO channels has 90% sensitivity and 93.5% specificity for FIP (Giordano *et al.* 2015). Additional supportive testing on the fluid may include immunofluorescence staining of coronavirus antigen in macrophages, anti-coronavirus antibody detection, Rivalta's test.
3. Consider whether surgery or diagnostic radiology imaging for assessment of abdominal or thoracic structures is indicated. -Neoplasia is a common cause of effusion - aspiration or biopsy of mass lesions, as appropriate.
4. Consider whether additional testing for bile peritonitis, uroabdomen, pancreatitis, or sepsis is indicated.

- If high numbers of erythrocytes are present (determine erythrocyte concentration or PCV of fluid):

1. Rule out blood contamination (platelets present with no erythrophagocytosis or hemosiderin)
2. Evaluate for neoplasia or trauma
3. Less likely considerations for hemorrhagic effusion include organ torsion, septic effusion, coagulopathies. Diagnostic radiology imaging or CBC with platelet count and coagulation testing, as indicated.

- If a high proportion of lymphocytes is present or fluid appears milky:

1. Small lymphocytes: probable lymphatic effusion - chylothorax is common in cats: first, rule out cardiac disease, then neoplasia, trauma, lung-lobe torsion, lymphangiectasia. Idiopathic chylous effusion if no underlying disease process evident.
2. Immature lymphocytes are indicative of a lymphoproliferative disorder.
3. If a small cell lymphoproliferative disorder is of concern, consider PARR (polymerase chain reaction for antigen receptor rearrangements) or flow cytometry, especially if lymphocytes are greater than 10,000/µl.
4. In general, differentials for lymphatic effusion are similar to those of transudates, with the addition of idiopathic chylous effusion and lymphangiectasia.

Effusions with <3,000 cells/µl and protein <2.5 g/dl

- If neoplastic cells are present (Diagnostic radiology imaging, etc., as indicated)
- Cardiac evaluation is indicated for thoracic effusion

- No cardiac disease or obvious neoplasia: serum biochemistry panel is indicated

1. Hypoalbuminemia: follow-up with urine analysis, liver function tests, GI testing as indicated
2. Liver enzyme elevation: portal hypertension
3. Compare fluid creatinine to serum creatinine if uroabdomen possible
4. Consider other differentials as discussed with high protein transudates, below

Effusions with <3,000 cells/µl and protein ≥2.5 g/dl

- If neoplastic cells are present (Diagnostic radiology imaging, etc., as indicated)
- Cardiac evaluation is indicated for abdominal effusion
- Consider FIP, especially if protein >3 g/dl (often ≥5.0), albumin:globulin ratio is <0.9 (see above).

- Diagnostic radiology imaging, if not previously performed
1. Neoplasia is a common cause of a transudate: if mass lesion detected, aspiration or biopsy of mass lesion, as appropriate
2. Organ displacement or distension, GI foreign body, assessment of spleen, pancreas, hepatobiliary, genitourinary, and/or respiratory systems, mediastinal lesions

- Evaluate GI system for enteritis
- Consider vasculitis or thrombosis

References

1. Beatty J, Barrs V. *J Feline Med Surg*. 2010;12:693.
2. Davies C, Forrester SD. *J Small Anim Pract*. 1996;37:217.
3. Dempsey SM, Swing PJ. *J Am Anim Hosp Assoc*. 2011;47(1):1.
4. Epstein SE. *Vet Clin North Am Small Anim Pract*. 2014;44(1):16.
5. Giordano A, *et al*. *Vet Clin Pathol*. 2015;44(2):295.
6. Hartmann K, *et al*. *J Vet Intern Med*. 2003;17(6):781.
7. Wright KN, *et al*. *J Am Vet Med Assoc*. 1999;214(3):375.

acvim322

Synovial Fluid Review

Paula M. Krimer, DVM, DVSc, DACVP (Clinical Pathology)
Athens, GA, USA

Introduction

Synovial fluid (SF) can be a valuable tool in the diagnosis of some musculoskeletal diseases. This presentation will discuss normal physiology, important aspects of collecting and processing an adequate diagnostic sample, and interpretation of results. Promising new directions in synovial fluid diagnostics will be presented.

Normal Physiology

Synovial fluid is present in joints, bursas, and tendon sheaths. It is necessary to increase the viscosity and elasticity of articular cartilage, to provide lubrication between synovial surface and between synovium and cartilage, and to protect joint tissues from friction and other damage. Unlike synovium, articular cartilage does not have a blood, nerve, or lymphatic supply and also relies on synovial fluid for to provide nutrients, remove waste products, and for the exchange of regulatory cytokines.

Synovial fluid is composed of an ultrafiltrate of passively dialyzed serum that passes through the synovial membrane with the addition of substances actively secreted from the synovial subintimal layer.[1] The ultrafiltrate is dependent on difference between intracapillary and intra-articular hydrostatic pressures and between colloid osmotic pressures of capillary plasma and synovial tissue fluid. Changes in plasma composition such as hyperuricaemia are reflected in the joint fluid. In normal joints, intra-articular pressures are slightly subatmospheric at rest.

The joint intimal layer is composed of macrophage type A synoviocytes and fibroblast type B synoviocytes without a basement membrane. Macrophage type A are derived from blood monocytes and are resident phagocytes (like Kupfer cells in the liver). Most proteins in synovial fluid are actively secreted by type B synoviocytes including collagens, hyaluronic acid, aggrecan, fibronectin, lubricin, proteoglycans (e.g., proteoglycan 4; PRG4), surface-activated phospholipids (SAPL), lubricin, proteinases, collagenases, and prostaglandins. Hyaluronic acid is the best studied. It is required to separate soft tissue surfaces (synovium on synovium, or synovium on cartilage) and allows them to slide over each other for boundary lubrication.[1]

The subintimal (subsynovial) layer does not have a basement membrane and is composed of loose connective tissue (type II collagen), adipose, abundant lymphatics, vasculature, and nerves. The intimal layer contains extracellular matrix gaps composed of collagen I, III, V, HA, chondroitin sulphate, biglycan and decorin proteoglycans. These gaps allow passage of ultrafiltrate and small molecules (<100,000 Daltons) from subintimal layer into circulation within joint space. Synovial fluid is cleared through lymphatics in the synovium assisted by joint movement.

Protein concentration in the joint is inversely proportional to molecular size and overall is ~25%–33% of plasma. Glucose and electrolytes are similar in synovial fluid as in serum or plasma, but SF has approximately half the ions. The calcium serum:synovial ratio is ~1.18 due to the base combining power of mucin for calcium. The normal pH in dogs is considered 7.0–7.8 and decreases with infection. Enzymes enter synovial fluid directly from the plasma, may be produced locally, or are released by synovial fluid macrophages. Alkaline phosphatase, acid phosphatase, lactic dehydrogenase, and other enzymes are present in detectable quantities.

The viscosity of synovial fluid does not directly relate to the quality of lubrication for soft tissues. Cartilage-on-cartilage lubrication does not rely on hyaluronates but is dependent on glycoproteins and a fluid film. At low loads the lubricating action of glycoproteins (especially aggrecan and lubricin) is due to sliding. At higher loads the cartilage surfaces are kept apart by a fluid film composed of fluid and interstitial fluid wept from the articular cartilage itself. The elasticity of articular cartilage may potentiate the fluid-film lubricating mechanisms at high loads.

Sample Collection & Submission

Collecting a synovial fluid sample should be considered in animals with joint pain or effusion, unexplained lameness, gait abnormalities, ambulatory difficulties such as stiffness or immobilization, and fever of unknown origin. Sufficient quantities of synovial fluid can be difficult to obtain in normal joints and varies by location. Less than 0.25 mL is expected in most joints of cats and small dogs, less than 0.5 mL from the stifle of large dogs, while dogs with hip dysplasia can have 2–6 mL of synovial fluid in the hip joint.[2] Joint selection and approach will vary by presentation and suspected disease process. Ultrasonography can be useful to determine ideal needle placement. For suspected systemic conditions, especially non-infectious polyarthritis, multiple joints should be aspirated; the stifles, carpi, and tarsi are generally recommended.[3]

Joints should be shaved and aseptically prepared similar to other surgical procedures and patients restrained and either heavily sedated or anaesthetized. All instruments and supplies should be sterile. As iatrogenic hemorrhagic contamination is common, a needle-only technique can be considered prior to active aspiration. A 22- or 25-gauge needle is generally used, with a 1-, 3-, or 6-mL syringe. The depth of the needle will also vary by location. In general, smaller instrumentation is used for smaller joints and helps reduce blood contamination.[4]

During aspiration great care must be taken to prevent hemorrhage, with which most of the synovial fluid analysis becomes futile. The plunger should be completely released and no negative pressure should be present when withdrawing the needle from the joint.

The amount of fluid collected and suspected disease process determine sample processing. A direct smear should always be made immediately at the time of collection for best cellular preservation. Cells can only be evaluated in areas sufficiently thin for cytoplasmic dispersal, and is generally best at the edges of the smear. For this reason, thin smears are critical for accurate evaluation and less material should be used in slide preparation than with most other fluids. However, diseased joints often have decreased viscosity and slides are easier to prepare. If necessary synovial fluid can be mixed with 150 IU/mL of hyaluronidase in a 1:1 ratio for several minutes prior to slide preparation.[2]

If only a minimal amount of synovial fluid is collected, preparation of direct smears and culture or PCR should be prioritized over other laboratory analyses. If there is sufficient fluid, an EDTA tube should be used for most synovial analysis to prevent clotting; the smallest size for the volume aspirated should be used to avoid dilution. EDTA is bacteriostatic so an aliquot can be reserved in a tube without additive or several drops can be placed on an aerobic culturette swab directly from the needle. Blood culture media should be used if available to allow both aerobic and anaerobic culture, and samples allowed to incubate 24 hours to reduce false negative culture results.[5] PCR does not appear to be more accurate than culture in synovial fluid.[6] EDTA samples cannot be used to evaluate pH, viscosity or mucin clot formation; a red top or heparin tube must be used.

Fluid samples can be shipped on ice without significant loss of integrity. Cytology slides should be completely dry and unstained, avoiding exposure to formalin fumes and refrigeration to avoid cell lysis after condensation. If fluid samples are being sent with slides, do send with ice packs but try to package so slides are as far away from ice as possible.

Routine Analysis

Routine synovial fluid analysis includes evaluation of physical properties, protein, nucleated cell count, and differential. Normal synovial fluid is pale yellow and clear. Increased turbidity (cloudier) samples indicate increased cellularity or rarely crystalluria; if 12 font writing cannot be read behind the fluid then the cell count is usually >30 x 10^9/L. White, red, or brown coloration indicates acute to subacute hemorrhage, while dark yellow (xanthochromic) samples indicate chronic hemorrhage and the presence of hemoglobin breakdown products. Thought it will not clot, synovial fluid can resemble a gel that liquefies upon agitation, a property termed thixotropism. Actual coagulation occurs in the presence of fibrinogen such as with hemarthrosis or a damaged synovial membrane.

SF viscosity can be estimated and reflects the amount of hyaluronic acid present as well as its degree of polymerization, but is not a reliable indicator of inflammation in dogs with polyarthritis.[7] The subjective "string test" is performed by placing a small drop between the fingers or on an applicator stick, and extending them until the string of fluid breaks; >2 cm is considered acceptable. On smears, windrowing is evidence of viscosity, with randomly distributed cells suggesting reduced sample viscosity.

The mucin clot test is rarely performed in small animals, requiring a heparin sample and sufficient quantity of fluid. It estimates the integrity of joint lubrication by adding a small amount of synovial fluid in a 1:4 ratio with 2.5% acetic acid and evaluation of mucin precipitation into a clot.[8]

The pH can be measured but should be performed immediately with narrow-range pH paper. Handheld devices are not reliable for measuring pH.

Protein can be estimated with a refractometer, but a quantitative biochemical assay on an automatized analyzer also used for serum/plasma biochemistry is recommended for accurate results. Normal joints are expected to have 1.5–3 g/dL of protein.[9] Protein can be increased due to inflammation and leakage from damaged vessels, or decreased with marked effusions or after joint lavage/injection. In horses, total protein concentration greater than 40 g/L indicates infection.[10]

Nucleated cell counts are generally performed on a hemacytometer or an automated analyzer, though the thickness of the fluid can interfere with even cell distribution. For automated analyzer evaluation, samples are diluted with specific quantities of saline and sometimes hyaluronidase to prevent clotting. In our laboratory, synovial fluid is incubated for 5 minutes in a 37°C water bath after dilution 1:2 with hyaluronidase prior to being counted in an automated impedance analyzer. In dogs, a cell count of <3000 cells/μL and in cats <1000 cells/μL is expected.[4] In horses, <1000 cells/μL is generally considered normal.[10] Tendon sheath fluid in horses has a reported normal range of 200 to 3500 cells/μL.[11]

Cellularity can be estimated from a slide, though there is a tendency to over-estimate.[12] A more recent study indicated visual review of a properly prepared direct smear correctly categorized samples within, mildly to moderately increased, or markedly increased in cellularity in 78% of samples.[13] Samples with increased viscosity and cell clumping are more likely to be incorrectly classified.

Evaluating a direct cytology smear is likely the most critical test in laboratory evaluation. Determining the presence of iatrogenic hemorrhage is critical before proceeding. The absence of hemosiderin-laden macrophages or erythrophagia with significant numbers of erythrocytes suggests blood contamination. If visible, the presence of platelets is a strong indicator of acute/iatrogenic contamination. The presence of windrowing (cells aligned in rows), pink granularity, crescenting, and color of the background should be noted for subjective analysis of sample viscosity. The absence of a pink granular background suggests reduced glycoproteins or hyaluronic acid concentration. Cells should only be evaluated in areas sufficiently thin for cytoplasmic dispersal, and is generally best at the edges of the smear. For this reason, thin smears are critical for accurate evaluation and less material should be used in slide preparation than with most other fluids.

The most common cell type is large mononuclear cells; these are macrophages, monocytes, synovial lining cells and small lymphocytes. Neutrophils are infrequent and should be well preserved; normal percentages vary but less than 10% is accepted as normal.[4] Macrophages can phagocytose cells, cellular debris, and organisms. Neutrophil morphology should be carefully examined for cytoplasmic inclusions, vacuolation, indistinct cells margins, nuclear degenerative changes, and organisms. Ragocytes are neutrophils with small round cytoplasmic vacuoles of phagocytosed nuclear proteins easily confused for bacteria; they confirm an immune-mediated joint disease and are relatively common. LE cells, or lupus erythematosus cells, are very rare neutrophils with a marginated nucleus and large cytoplasmic homogenous inclusion of nucleoproteins that also indicate an immune-mediated condition. Eosinophils are very infrequent. Rarely multinucleated giant cells or osteoclasts are also found in SF samples and confirm cartilaginous damage.

BIOCHEMICAL TESTING

Biochemical testing is useful when joint sepsis is suspected and can usually be analyzed before culture or PCR results are available. Lactic acid and glucose can be measured patient-side with a point of care device with similar findings to values determined in reference laboratories.[14] Glucose is normally no lower than 10 mg/dl less than serum glucose levels but can be decreased in joint disorders. A >20 mg/dl decrease difference from serum is suggestive of infection. Lactate is increased in septic arthritis in horses while pH is decreased. In dogs, a low lactate level ruled out infection in a recent study.[15]

Lactate dehydrogenase can be elevated in synovial fluid in a variety of conditions including rheumatoid arthritis, infectious arthritis, and gout. LDH appears to originate through cartilage abrasion.[16] The increase may be partially due to the increase of neutrophils and should be evaluated in comparison with serum levels.

Myeloperoxidase is an enzyme released from stored granules by neutrophils in the presence of bacteria. It is a good biomarker of joint sepsis in horses.[17-18] In dogs it has only been evaluated for osteoarthritis where it appears to be useful in confirming disease.[19]

Serum amyloid A (SAA) is an acute phase inflammatory protein that can be measured in serum and in synovial fluid. It is higher in SF & serum of horses with septic joints compared to healthy joints, osteoarthritis, and non-septic synovitis (WBC<$0.5x10^9$/L).[20] Unlike protein, it does not appear to change in response to repeated arthrocentesis or joint lavage.[21-22]

Novel Biomarkers

SF contains collagen and other matrix components are degraded and synthesized in synovial disease, and measuring degradation products, synthesized elements, or enzymes involved in these processed in synovial fluid may become clinically useful. Hundreds have been evaluated in humans, however, assays are not commercially available and currently these are investigational only.

Chondroitin sulfate is a glycosaminoglycan component of proteoglycans abundant in the joint. Levels in SF have been shown to be elevated in experimental and natural degenerative joint conditions in multiple species including cruciate rupture in the dog. It has also been shown to be a predictor of future joint disease in polo ponies and shows promise to detect asymptomatic osteoarthritis in horses.[23-24]

Matrix metalloproteinases are proteolytic enzymes that are up-regulated with inflammation and active degradation of cartilage. The most commonly investigated in veterinary species are MMP-9 and MMP-2. MMP-9 is predominantly produced by neutrophils and is proportionally increased with the total nucleated cell count. The assay is accurate but technically difficult to perform and prolonged, limiting its practical use.[25] MMP-2 is produced by chondrocytes and synovial fibroblasts. The ratio of MMP-9 to MMP-2 in SF of horses is a useful predictor of survival after a septic joint, with a ratio of 3.01 in survivors and 14.5 in non-survivors.[25] In dogs, pro-MMP-2 activity in SF is increased with cranial cruciate ligament rupture.[26-27] MMP-3, also known as stromelysin, is also significantly increased in osteoarthritis in dogs and horses.[28]

References

1. Tamer TM. *Interdiscip Toxicol*. 2013;6(3):111–125.
2. Martinez CR, Santangelo KS. *Vet Clin North Am Small Anim Pract*. 2017;47(1):111–122.
3. Ellison RS. *Semin Vet Med Surg (Small Anim)*. 1988;3(2):133–139.
4. Pacchiana PD, *et al*. *J Am Vet Med Assoc*. 2004;225(12):1866–1870.
5. Montgomery RD, *et al*. *Vet Surg*. 1989;18(4):300–303.
6. Scharf VF, *et al*. *Aust Vet J*. 2015;93(6):204–207.
7. Jacques D, *et al*. *Vet Surg*. 2002;31(5):428–434.
8. Boon GD. *Vet Med*. 1997;92(5):443, 451–443, 451.
9. MacWilliams PS, Friedrichs KR. *Vet Clin North Am Small Anim Pract*. 2003;33(1):153–178.
10. Steel CM. *Vet Clin North Am Equine Pract*. 2008;24(2):437–454, viii.
11. Malark JA, *et al*. *Am J Vet Res*. 1991;52(8):1292–1294.
12. Gibson NR, *et al*. *Vet Rec*. 1999;144(17):463–465.
13. Dusick A, *et al*. *Vet J*. 2014;202(3):550–554.
14. Dechant JE, *et al*. *Vet Surg*. 2011;40(7):811–816.
15. Proot JL, *et al*. *Vet Comp Orthop Traumatol*. 2015;28(5):301–305.
16. Walter EL, *et al*. *Histochem Cell Biol*. 2007;128(1):7–18.
17. Fietz S, *et al*. *Res Vet Sci*. 2008;84(3):347–353.
18. Wauters J, *et al*. *Equine Vet J*. 2013;45(3):278–283.
19. Hurlbeck C, *et al*. *Res Vet Sci*. 2014;96(3):429–435.
20. Jacobsen S, *et al*. *Am J Vet Res*. 2006;67(10):1738–1742.

21. Sanchez-Teran AF, *et al. Vet Surg*. 2016;45(2):223–230.
22. Sanchez-Teran AF, *et al. Vet J*. 2016;210:30–33.
23. Baccarin RY, *et al. Can J Vet Res*. 2014;78(1):50–60.
24. Machado TS, *et al. Equine Vet J*. 2012;44(4):404–411.
25. Kidd JA, *et al. Vet Rec*. 2007;161(10):329–334.
26. Murakami K, *et al. J Vet Med Sci*. 2016;78(6):1051–1054.
27. Boland L, *et al. Vet Comp Orthop Traumatol*. 2014;27(3):210–215.
28. Fox DB, Cook JL. *J Am Vet Med Assoc*. 2001;219(6):756–761.

acvim323

Cerebrospinal Fluid Review

Paula M. Krimer, DVM, DVSc, DACVP (Clinical Pathology)
Athens, GA, USA

Introduction

Cerebrospinal fluid (CSF) is an important ancillary diagnostic tool in the diagnosis of most neurological diseases. A normal CSF can rule-out many disease processes though it can still be normal in up to 10% of dogs with confirmed inflammatory disease[1] and can detect inflammation despite normal MRI findings[2]. This presentation will discuss current understanding of normal physiology, important aspects of collecting and processing an adequate diagnostic sample, and issues in interpretation of results.

Normal Physiology

The CSF fluid serves multiple purposes. It provides nourishment, acts as a chemical buffer, and removes waste products, but also protects the brain by regulating intracranial pressure and providing buoyancy. CSF fluid is 10% of the cranium's volume, letting the brain float to reduce its weight and pressure on the brainstem as well as acting as buffering physical impacts. Intracranial pressure can change with posture, exercise, respiration, or trauma such as intracranial hemorrhage, and can be accommodated by dislocation of CSF into the subarachnoid space (acute response) as well as reduced production or increased absorption (chronic response).[3]

There are four fluid compartments in the brain: cerebrospinal fluid, interstitial fluid, intracellular fluid, and blood vessels. Our understanding of the relationship between these four compartments, CSF production, flow, and absorption, have changed over time.

The classical theory of CSF circulation is based on the active creation of CSF fluid by the choroid plexus with unidirectional bulk flow to arachnoid villi in the dural venous sinuses for passive absorption.[4] However, recent studies contradict this theory; for example removal of the choroid plexus does not seem to impact the volume of CSF produced or its content.[5-6] Current theories suggest CSF and interstitial (extracellular) fluid are a single unit and the entire central nervous system vasculature continuously contributes to the formation of CSF, including the choroid plexus.[7] Corollary to this newer theory, flow is not unidirectional but allows for local pulsatile mixing and diffusion.[7] Most absorption of CSF fluid is thought to occur through what is coined the glymphatic system (a portmanteau of glial and lymphatic) under low pressure throughout the central nervous system. The arachnoid space previously thought to be the major egress of CSF fluid is more important in draining the brain under higher pressure conditions.[7]

Arteries traversing the pia and bathed in CSF fluid produced by the choroid plexus descend into the brain parenchyma surrounded by a space called the Virchow–Robin space. This space is composed of layered endothelium, pial matter, and glial (astrocyte) cell layers each with its own basement membrane.[8-9] The space narrows as the arteries branch into smaller vessels, carrying CSF fluid with it, and eventually disappears. Astrocytes completely surround brain capillaries, forming a "neurovascular unit" and boundary to the perivascular space. The astrocytes contain significant numbers of dynamically regulated aquaporins (water-transporting pores, especially aquaporin 4) at their end feet to control bilateral movement of water, solutes, and ions in and out of the capillaries into the interstitial space.[9,10] In this perivascular space, CSF and filtered interstitial continuously mix.

There fluid moves from ingressing capillaries and their perivascular space through the brain parenchyma and into egressing perivascular spaces, then capillaries and venules. This movement of fluid from vasculature, through parenchyma, and back to vasculature, allows for the removal of waste products analogous to the lymphatic system, hence the moniker "glymphatics."[9,11] This system is partially dependent on sleeping position, with lateral positioning being superior to supine or prone for clearance along the glymphatic system, and may explain the need for sleep.[12] Its implications for conditions such as Alzheimer's disease, characterized by the accumulation of waste products, are beginning to be studied.[8]

The choroid plexus is still an important source of CSF production. This is a villous structure of highly folded ependymal epithelium lining the lateral, third, and fourth ventricles. The underlying endothelium is fenestrated to allow passage of a plasma ultrafiltrate with assistance from hydrostatic pressure. As CSF pressure is increased, ultrafiltrate production is decreased. This ultrafiltrate is then actively transported through the choroidal epithelium into the ventricles via ion transport and is the main determinant of choroidal CSF production. The net efflux of bicarbonate, chloride, and potassium enter the CSF via channels in the epithelial cell apical membranes, creating an osmotic gradient for the movement of water from blood vessels to the ventricles via aquaporin.[1]

Eighty to 90% of CSF protein is albumin, and the ratio of CSF and serum albumin, called the albumin quotient (AQ), can be used to evaluate blood-brain barrier disruption. Minor concentrations of transthyretin (TTR or prealbumin), retinol-binding protein (RBP) and transferrin are also synthesized by choroid plexi. Traces of beta and gamma globulins, tau protein (a fraction of modified transferrin), glial fibrillary acidic protein, and myelin basic protein are produced intrathecally. The movement of proteins in and out of the CSF is determined primarily by molecular radius but also depends on electrical charge, plasma concentration, and the status of the blood-brain barrier.

Sample Collection

CSF samples should not be collected in animals with known or suspected intracranial pressure, cerebral edema, hemorrhagic diathesis or spinal luxation. A mass effect on imaging should also be considered a contraindication. If the cerebellomedullary cistern is absent, a tap should not be attempted. CSF analysis should occur prior to contrast studies as dye can cause inflammation and may also be less accurate immediately post-ictal.[13]

A complete analysis requires >0.5 mL (preferably 1–2 mL) which in addition to visual determination of transparency and color includes a total nucleated cell count (TNCC), red blood cell count, microprotein evaluation, and cytospin preparation for cytology. Minimizing blood contamination is important to prevent an inaccurate TNCC. Approximately 1 mL of fluid per 5 kg of body weight can be collected safely.[14]

Samples should be submitted and analyzed immediately after collection. Due to its low protein content, cell morphology begins to deteriorate within 30 minutes. Though monocytes deteriorate more quickly than neutrophils, analysis up to 8 hours after collection should not impact diagnostic interpretation and higher protein samples better preserve cell morphology.[15] The use of hetastarch, fetal calf serum (3.7 g/dL added at a concentration of 20% volume), and autologous serum have all been shown to improve cellular morphology when sample analysis is delayed.[15-16] Fetal calf serum 3.7 g/dL (Biuret method) can be added to CSF at concentration of 20% volume, or hetastarch 1:1 (vol:vol). Consider collecting samples into two aliquots; one for the cell count and cytology to which a preservative can be added, and the other for measuring protein. Samples should always be labeled as to location.

Routine Analysis

Normal CSF is clear and colorless. Turbidity is noticeable at >500 WBC/μL.[3] Pink or red samples can be attributed to iatrogenic, acute, or chronic hemorrhage. Samples with iatrogenic or very acute hemorrhage (<3 hours old) have a clear supernatant after centrifugation and lack erythrophagia, while chronic *in vivo* hemorrhage will have a yellow-pink supernatant and erythrophagia or hemosiderin-laden macrophages are expected.

The impact of iatrogenic hemorrhage on the TNCC is controversial with conflicting results in the literature. Some papers indicate no correlation between the RBC count and TNCC count,[17] while others indicate no impact when RBC counts are below 500 RBC/μL but increased neutrophils and protein when it is >500 RBC/μL[18]. One study showed that CSF samples with RBC counts <5000/μL all had TNCC counts below 6 WBC/μL.[19] Two studies have created formulas for correcting the TNCC based on the number or erythrocytes detected; one contaminating TNCC could be accounted for by 500 RBC/μL in dogs[20] and 100 RBC/μL in cats[21].

The total nucleated cell count is typically manually performed on a hemacytometer and requires experience to accurately differentiate cell types. Automated hematology analyzers (ADVIA 120 and 2120) have been validated for measuring CSF total nucleated and red blood cells counts in dogs. Limitations for this method include inability to classify a variety of cells including macrophages, plasma cells, neoplastic cells, and nucleated erythrocytes. Technical issues such as the need for multiple wash steps to eliminate background cell counts, minimum 300 μL volume, and costly quality control reagents with short shelf lives are also a considerable impediment to their routine use in a veterinary setting.[15,22] Reference intervals for the TNCC from cerebellomedullary samples are generally considered <5 cells/μL in dogs and slightly lower in cats at <3 cells cells/μL; lumbar samples are expected to have lower cellularity. The TNCC in healthy rabbits has been reported to be ≤4 cells/μL,[23] and in ferrets <8 cells/μL[24].

Cytospin evaluation for cell differential and morphology is arguably the most important aspect of the CSF analysis. The precision of the cell yield on cytospins of low cellularity samples has been shown to be over 30% with both a calibrated and manual pipetting technique.[25] The majority of cells, >70% should be monocytoid with up to 30% small lymphocytes and 3–10% neutrophils. Other occasionally seen cell types in normal samples include eosinophils, mitotic figures, phagocytic cells, and ependymal cells, as well as cells of the choroid plexus and myelin. The coefficient of variation for neutrophil percentage has been reported as 146% in low cellularity samples, indicating tremendous variation in the neutrophil percentage on a single cytospin; however the same study demonstrated CSF samples with >3% neutrophils had neutrophils present on a single cytospin.[25] Calibrated pipetting had significantly fewer foamy macrophages than the manual technique,[25] which questions their value in interpretation of disease processes.

Protein measurements are best performed on commercial biochemical analyzers. Turbidimetric and dye-binding methods (Pyrogallol Red-Molybdate and Coomassie Brilliant Blue) each yield different reference intervals; laboratories should establish their own reference intervals.[26] Urine test strips are specific for albumin and have also been used but are semiquantitative; 1+ is generally considered normal and >2+ is specific for increase protein. Increased protein is a non-specific indicator of pathology that can be attributed to disruptions in the blood brain barrier, intrathecal immunoglobulin production, or necrosis. When TNCC is not concurrently increased, consider extradural compressive lesions, intramedullary masses, CNS necrosis, trauma, or vasculitis.

Biochemical Testing

Enzymes in the CSF can originate in the blood, neural tissue, or cells within the CSF.

Creatine kinase (CK) has a unique brain isoenzyme and is independent of serum values. Increases in creatine kinase may reflect a poorer prognosis in intervertebral disc disease and have been reported in canine cervical spondylomyelopathy.[27-28] AST and CK can be increased with myelin degeneration but do not appear to increase in chronic non-inflammatory conditions.[29]

Though lactate dehydrogenase can be increased with CNS lymphoma and other inflammatory diseases, and in people is useful to differentiate bacterial and sterile meningitis, it is a labile enzyme deteriorating >10% within 24 hours regardless of storage methods.[30-31] Further study into isoenzymes of this enzyme is warranted.[31]

Glucose enters the CSF by selective transport (e.g., GLUT1). CSF glucose varies with blood glucose, rate of transport into CSF, and metabolic rate of the central nervous system. Values are normally 60–80% of plasma. Low CSF glucose is considered useful to diagnose infectious meningitis.

Glutamine is produced from ammonia and alpha-ketoglutarate, allowing removal of the toxic metabolic waste product ammonia. Normal concentration of ammonia is 8–18 mg/dl, which is elevated in liver disorders.

Recent reference intervals for lactic acid in canine CSF were investigated (1.1 to 2.0 mmol/L) and findings were similar to humans values of 1.8±0.06 mmol/L.[32] Destruction of tissue within the CNS owing to oxygen deprivation (hypoxia) causes increased CSF lactic acid levels proportional to the severity of brain injury in cats and were reported higher in dogs with worse clinical neurological scores.[32-33]

Other CSF Testing

Additional diagnostic procedures on CSF fluid are limited by volume and availability. Table 1 summarizes the more common additional tests requested and their suggested use. Investigative biomarkers are summarized in Table 2.

Table 1. Common additional laboratory diagnostic tests for CSF fluid

Test	Diagnostic capture	Condition
IgA	Elevated levels	Steroid-responsive meningitis-arteritis
Protein electrophoresis	BBB disturbance, immunoglobulin production	Intrathecal IgG production
Flow cytometry	Lymphocytic neoplasms	Immunophenotyping
CSF culture	Bacterial MEM (low sensitivity)	Bacteria, fungal
PCR	Assess infectious disease (poor sensitivity, specificity affected by blood contamination)	Rickettsia, fungal, toxoplasmosis, neosporosis, FIP, CDV
CSF-specific antibody	Increase in Ab, accuracy improves when combined with PCR	FIP, CDV, EPM, toxoplasmosis, neosporosis, rickettsia, fungus

Table 2. Investigative biomarkers evaluated in CSF fluid

Test	Description	Applied diseases	Utility
Neuron-specific enolase (NSE)	Cytoplasm of neurons, oligodendrocytes, blood, and neuroendocrine cells	Detect neuronal damage in CNS; disease study (GM1 gangliosidosis, MEM)	Lacks specificity; diagnostic utility undetermined
Myelin basic protein[28,34]	Produced by oligodendrocytes; marker of WM injury	Dogs with MBP >3 ng/mL had poorer changes of ambulation after acute disk herniation	78% sensitive; 76% specific. ROC 0.688
Glial fibrillary acidic protein (GFAP) - serum[35-37]	Intermediate filament in astrocytes; evaluate BBB	Investigated in myelomalacia, inflammatory disease but lacks specificity in DDx inflammation	NME (Pugs): 67% sensitive; 100% specific; myelomalacia: 75% sensitive; 97% specific
Cleaved-tau (c-tau)[38]	Microtubule structure localized in axons	Investigated prognosis in plegic dogs with IVDH	Sensitivity 86%; specificity 83%. ROC 0.887
Phosphorylated neurofilament-heavy (pNF-H) - serum[39]	Major cytoskeletal component of axons	Paraplegic dogs absent DPP prognosis for ambulation	34.8% sensitive; 100% specific. ROC 0.613
Matrix-metalloproteases (MMP-9; MMP-2)[40-42]	Endopeptidases that degrade ECM as role in tissue remodeling. 4 groups based on substrate specificity: collagenases, stromelysin, gelatinases, and membrane-type MMPs	TL IVDH - paraplegia; detected in dogs with brain neoplasia	Prognosis undetermined in IVDH paraplegic dogs; DDx in brain tumors?

References

1. Tipold A. *J Vet Intern Med.* 1995;9(5):304–314.
2. Lamb CR, *et al. Vet Radiol Ultrasound.* 2005;46(1):17–22.
3. Di Terlizzi R, Platt SR. *Vet J.* 2009, 180(1):15–32.

4. Oreskovic D, Klarica M. *Prog Neurobiol.* 2011;94(3):238–258.
5. Milhorat TH, *et al. J Neurosurg.* 1976;44(6):735–739.
6. Milhorat TH. *Int Rev Cytol.* 1976;47:225–288.
7. Miyajima M, Arai H. *Neurol Med Chir (Tokyo).* 2015;55(8):647–656.
8. Jessen NA, *et al. Neurochem Res.* 2015;40(12):2583–2599.
9. Nedergaard M. *Science.* 2013;340(6140):1529–1530.
10. Oreskovic D, Klarica M. *Brain Res Rev.* 2010;64(2):241–262.
11. Plog BA, *et al. J Neurosci.* 2015;35(2):518–526.
12. Lee H, *et al. J Neurosci.* 2015;35(31):11034–11044.
13. Creevy KE, *et al. Am J Vet Res.* 2013;74(8):1118–1125.
14. Elias A, Brown C. *Lab Anim (NY).* 2008;37(10):457–458.
15. Fry MM, *et al. Vet Clin Pathol.* 2006;35(1):72–77.
16. Bienzle D, *et al. J Am Vet Med Assoc.* 2000;216(11):1761–1764.
17. Hurtt AE, Smith MO. *J Am Vet Med Assoc.* 1997;211(7):866–867.
18. Doyle C, Solano-Gallego L. *Vet Clin Pathol.* 2009;38(3):392–396.
19. Wilson JW, Stevens JB. *J Am Vet Med Assoc.* 1977;171(3):256–258.
20. Bailey CS, Higgins RJ. *Am J Vet Res.* 1985;46(5):1162–1165.
21. Rand JS, *et al. Am J Vet Res.* 1990;51(7):1044–1048.
22. Ruotsalo K, *et al. Vet Clin Pathol.* 2008;37(2):242–248.
23. Jass A, *et al. Vet Rec.* 2008;162(19):618–622.
24. Platt SR, *et al. Am J Vet Res.* 2004;65(6):758–760.
25. Krimer PM, *et al. J Vet Diagn Invest.* 2016;28(2):158–164.
26. Riond B, *et al. Vet Clin Pathol.* 2014;43(1):78–88.
27. Martin-Vaquero P, *et al. Spine (Phila Pa 1976).* 2015;40(9):601–612.
28. Witsberger TH, *et al. J Am Vet Med Assoc.* 2012;240(5):555–562.
29. Rand JS, *et al. Can Vet J.* 1994;35(3):174–181.
30. Abate O, *et al. Zentralbl Veterinarmed B.* 1998;45(2):73–85.
31. Houle JC, *et al. Vet Clin Pathol.* 2015:44(2):253–261.
32. Caines D, *et al. Can J Vet Res.* 2013;77(4):297–302.
33. Choi SC, *et al. J Neurotrauma.* 1998;15(10);771–776.
34. Levine GJ, *et al. J Vet Intern Med.* 2010;24(4):890–896.
35. Sato Y, *et al. J Vet Med Sci.* 2013;75(7):949–953.
36. Toda Y, *et al. Vet Rec.* 2007;161(8):261–264.
37. Miyake H, *et al. J Vet Med Sci.* 2013;75(11):1543–1545.
38. Roerig A, *et al. Vet J.* 2013;197(2):253–258.
39. Nishida H, *et al. Vet Surg.* 2014;43(3):89–93.
40. Levine JM, *et al. Am J Vet Res.* 2006;67(2):283–287.
41. Mariani CL, *et al. Am J Vet Res.* 2013;74(1):122–129.
42. Mandara MT, *et al. Vet Pathol.* 2009;46(5):836–845.

acvim324

Hematology, Biochemistry and Iron of Aging

Lauren B. Radakovich, DVM, DACVP (Clinical)
Fort Collins, CO, USA

Introduction

Just as improved health care has led to an increasing lifespan for human beings, our dogs and cats are also experiencing longer life spans. In fact, it is estimated that anywhere from 30–49% of canine companions are 7 years old and older.[1] This is great news for pet owners but, along with an increasing life span, there are a host of aging-associated health problems that veterinarians must face. Aging is linked with an increased incidence of many diseases including cancer, musculoskeletal disease, neurodegenerative diseases, renal disease, and cardiovascular dysfunction. Aging is also associated with an increased propensity for obesity, which can cause or exacerbate many health conditions on its own. Obesity is especially common amongst our spayed and neutered pets.

Routine complete blood counts (CBC) and serum biochemical profiles are a mainstay of clinical practice and are often one of the first tools used by veterinarians to assess patient health. They are also more objective and sensitive at detecting disease than physical exams, alone.[2] These clinical pathology tests become even more important in aging pets, as they may detect emerging or occult disease. This talk discusses hematologic and biochemical changes that may be noted in **clinically healthy** senior and geriatric dogs. As obesity is intricately linked with aging, its effects on CBC and serum biochemical tests are also discussed. As iron is emerging as a key player in aging and its associated diseases, there will be an overview of iron homeostasis as well as discussion of how iron becomes dysregulated with obesity and aging.

Canine Aging - CBC and Serum Biochemistry

Because CBCs and serum biochemical profiles are commonly used to detect disease, and our pets are living longer, there is great interest in how aging might affect these routine tests. There are several manuscripts discussing hematologic and biochemical changes associated with aging; however, these have been predominantly done in a highly controlled laboratory setting. As such, many of the studies have examined a single or only a few specific breeds, mainly beagles and Labrador retrievers. We wanted to determine how aging might affect baseline physiology and, thus, affect routine clinical pathology tests in clinically healthy, aging dogs. All dogs evaluated were client-owned, so a variety of breeds, diet, and lifestyles contributed to the data.

We retrospectively collected CBC and serum biochemical data from dogs ≥1 year of age. Dogs were classified as "clinically healthy" based on physical exam findings documented in the medical record. Dogs were grouped as adult (1–7.9 years old), senior (8–11.9 years old), and geriatric (12+ years old) for comparisons. There was no statistical difference in the frequency of large- versus small- to medium-breed dogs between the groups. We developed exclusion criteria based on what we believed a practicing veterinarian would typically use to form clinical treatment decisions. After exclusions, 302 dogs were identified to be included in the study.

Senior and geriatric dogs exhibited several hematologic and biochemical differences when compared to the younger adults. One of the most intriguing findings of our study was that aging dogs have evidence of potential gastrointestinal bleeding, which is commonly reported in elderly humans. This was supported by an increased urea concentration in the geriatric group while serum creatinine values did not differ among the three groups. Additionally, the urea-to-creatinine ratio (U:C) was significantly higher in geriatric dogs and exceeded the reported normal value of 20:1.[3] The most common causes of an increased U:C in small animals are reported to be gastrointestinal bleeding followed by dehydration. While neither dehydration nor chronic kidney disease can be ruled out as causes for this increase in U:C, they are considered less likely due to clinical and biochemical findings. It is also possible that a lower muscle mass in aged dogs masked an increase in creatinine. Gastrointestinal bleeding with concurrent iron deficiency could also explain our findings of decreased hematocrit, decreased erythrocyte mean cell volume, lower serum iron, and increased platelet counts in older dogs. A recently published review states that the most common cause of anemia in elderly people in Western societies is iron deficiency due to gastrointestinal bleeding.[4] Future studies examining sources of occult blood loss are indicated to further explore the possibility of underlying gastrointestinal bleeding in aged dogs.

In addition to possible gastrointestinal bleeding, older dogs also had evidence of "inflammaging." This term refers to an accumulation of inflammatory mediators over time that predisposes individuals to chronic diseases and cancer.[5] Although the retrospective nature of our study did not allow for the measurement of specific inflammatory cytokines, there was biochemical evidence to suggest low-grade systemic inflammation was present in older dogs. Geriatric dogs had significantly higher refractometric plasma protein concentrations, which was mirrored by a similar trend in serum total protein. This increase in protein was attributed to a higher serum globulin level, as albumin concentration was lower in older dogs. Increased globulins coupled with decreased albumin represents a classic inflammatory pattern, as albumin is a negative acute phase protein. The presence of increased platelets in our cohort, while possibly secondary to iron deficiency, could also represent a reactive thrombocytosis consequent to inflammation. Likewise, the decreased hematocrit, MCV, and serum iron could also be attributed to inflammation, often referred to as anemia of chronic disease (discussed in Basic Iron Metabolism, below). It is not uncommon for elderly humans to experience anemia due to concurrent absolute iron deficiency (due to blood loss or iron-poor diet) and functional iron deficiency. Functional iron deficiency refers to limited iron delivery to the bone marrow that occurs with inflammation due to the iron sequestration effects of the hormone hepcidin.[6] Unfortunately, diagnostic tests available in veterinary medicine are unable to differentiate absolute iron deficiency, functional iron deficiency, and the combination of these two entities.

Other biochemical changes noted in older dogs included increased liver enzyme activity, increased cholesterol concentrations, and evidence for decreased serum water fraction. Specifically, alkaline phosphatase (ALP) and alanine aminotransferase (ALT) activities were increased with age. There are numerous possible explanations for these increased liver enzymes. The increase in ALP raises concern for smoldering hyperadrenocorticism while increasing cholesterol raises the possibility of underlying hypothyroidism. Future studies would include endocrine testing for these disorders. Findings that may be attributed to decreased serum water fraction or dehydration include a higher anion gap and serum potassium in aged dogs. Both of these analytes are often increased with metabolic acidosis. Unexpectedly, serum chloride decreased with age, while serum sodium remained similar among the three groups. The significance of this finding is not well understood and deserves further investigation.

Canine Obesity - CBC and Serum Biochemistry

Obesity is commonly associated with aging in people, and this trend is mirrored in our pet canine population as well. Obesity is a risk factor for many diseases in dogs, including endocrinopathies, musculoskeletal disorders, insulin resistance, exercise intolerance, and some cancers, among others. Obesity is often accompanied by a low-grade, chronic inflammatory state that may eventually lead to more serious health problems down the road. In a retrospective study similar to the one described above in aging dogs, we investigated how obesity affects CBC and serum biochemical values in clinically healthy dogs.

Dogs were sorted into 2 basic weight categories based on a 9-point body condition score (BCS): control (BCS 4–6) and overweight/obese (BCS 7–9). After exclusions, there were 240 dogs in the control group and 116 in the overweight/obese group. Overweight and obese dogs had a higher total white blood cell count characterized by higher neutrophil counts, which may indicate a higher degree of inflammation in these dogs. However, increased levels of corticosteroids in overweight/obese dogs cannot be ruled out, and future studies should include measurement of inflammatory cytokines. Overweight/obese dogs also had higher plasma and serum protein levels with increased globulins and albumin. These changes in proteins may be due to either inflammation or decreased serum water fraction in overweight dogs. Similar to the canine aging study, overweight dogs had an increased anion gap coupled with higher sodium, higher potassium, and lower chloride values. The relative hypochloremia and disconnect with serum sodium in overweight dogs is intriguing and not well understood. This finding has been reported in obese humans, however, and deserves further exploration, as its cause remains elusive.

Basic Iron Metabolism and Regulation

Iron is an element essential to many biologic processes. Importantly, it is required for the synthesis of hemoglobin, the protein that carries oxygen in red blood cells. Iron is also needed for the production of myoglobin and plays a critical role in oxidative phosphorylation within mitochondria.[7] Like many key micronutrients, iron is critical for survival and optimal organism function, but a delicate balance must be maintained. A deficiency of iron can eventually lead to anemia, but iron is also deleterious in excess as it can promote free radical damage and inflammation via Fenton chemistry.[8]

Iron Absorption, Transport, Uptake, and Recycling

Because iron is required for life, but is also toxic in excess, mammals have evolved a complex system for its regulation. Interestingly, there is no dedicated mechanism for iron excretion. As most iron used by the body on a daily basis is recycled within macrophages, only a small portion must be obtained from the diet. This is to replace iron that is lost through routine epithelial cell shedding. Iron can be absorbed in heme and non-heme forms. Each form enters the enterocyte via its own transporter. Once inside the enterocyte, iron is either stored as ferritin (if the body is iron replete) or absorbed into the bloodstream (if there is a need for iron). Iron exits the basolateral side of the enterocyte via ferroportin, the only known cellular iron exporter.[9]

After crossing into the bloodstream, iron is quickly picked up by transferrin to prevent any potential oxidant injury while in the bloodstream. Transferrin shuttles iron to its destination, which is mainly immature erythroid cells in the bone marrow. Transferrin-bound iron binds to the transferrin receptor, TfR1, on erythroid cells. The complex is then internalized, iron is released from transferrin to be used by the cell to synthesize hemoglobin, and transferrin is recycled to the cell surface to return to the bloodstream.[10]

As mentioned above, the vast majority of iron utilized for hemoglobin synthesis is derived from macrophage iron recycling. As red blood cells become senescent, splenic macrophages remove them from the circulation via erythrophagocytosis. Macrophages also have receptors to scavenge free hemoglobin and heme. Within macrophages, iron is released from either heme or transferrin, where it has two potential fates. If there is a need for iron, it will be exported through ferroportin so that it may go the bone marrow or other target tissues. Another possible outcome is for it to be incorporated into ferritin, the main storage form of iron within macrophage cytoplasm. Ferritin acts as a protein cage that encases iron in its ferric state, which is less prone to engage in the Fenton reaction.[11] If ferritin becomes saturated within a cell, it will denature into hemosiderin.

Iron stores are present primarily in the liver. Stores are also present in the spleen and bone marrow, serving as a reserve source of iron, should there be an unexpected shortage (e.g., blood loss). Of note, cats lack marrow iron stores normally. The body taps into these iron stores to support hemoglobin synthesis when iron becomes limited, such as iron-poor diets or, more commonly, chronic external blood loss.

Iron Regulation via Hepcidin

When iron stores reach a critical level, and further absorption bears the risk of unmitigated free radical formation, the liver is stimulated to produce hepcidin. Hepcidin, the "master iron regulator," is a peptide hormone made primarily in response to high total body iron load. Once produced, hepcidin binds to ferroportin, promoting its internalization and degradation within the cell. This results in a block of iron bioavailability: it can neither be absorbed through enterocytes nor exported from hepatocytes or macrophages. Systemic inflammation is also a stimulus for hepcidin production. Interleukin-6 (IL-6), a proinflammatory cytokine, induces the liver to produce hepcidin. This is an attempt by the innate immune system to sequester iron from potential invading pathogens. If inflammation is chronic, a condition called "functional iron deficiency" can develop. This term refers to ongoing iron sequestration within macrophages and enterocytes, thereby limiting its availability to the bone marrow. If this iron inaccessibility goes on long enough, the animal may develop anemia. We often call this "anemia of chronic disease" or "anemia of inflammation." Inhibitors of hepcidin include iron deficiency, tissue hypoxia, and anemia. There is also the newly discovered erythroferrone, a hormone produced by erythroblasts that inhibits hepcidin, allowing for recovery from blood loss and inflammation.[12,13]

Iron Dysregulation in Aging

As mentioned above, elderly patients have a higher incidence of anemia that is commonly due to iron deficiency. Iron deficiency may result from poor diet, gastrointestinal bleeding, and/or inflammatory disease. Often inflammatory disease and bleeding may occur concurrently, such as with a bleeding gastrointestinal tumor, or as seen in a patient with bleeding secondary to nonsteroidal anti-inflammatory medication used to manage chronic pain. On the other side of the coin, iron accumulation throughout life is emerging as a cause for a multitude of health problems. Iron has powerful pro-oxidant potential, and without a designated mechanism for excretion of excess iron, it tends to accumulate in tissues over time. This is an area of research that is being actively investigated in rodent laboratory models, and iron chelation therapy has been successfully applied to a multitude of conditions where iron excess is implicated.

Iron Dysregulation in Obesity

Aberrations in iron metabolism have been described in humans, rodent models, and companion dogs. Obese patients often experience iron deficiency, which can be attributed to a variety of causes. Iron-poor diet is often cited in people but is rare in dogs fed commercial diets. The most commonly cited cause for iron deficiency in obesity is due to decreased absorption secondary to elevated levels of hepcidin.[14] As mentioned above, hepcidin is a peptide hormone produced mainly by the liver. The liver produces hepcidin in response to inflammation, specifically IL-6. As IL-6 is commonly increased in obese individuals, they also tend to have higher levels of hepcidin. Hepcidin binds to and degrades its target - ferroportin, the only known iron export protein. Ferroportin is predominantly found on duodenal enterocytes, macrophages, and hepatocytes. Its degradation prevents iron absorption in the gastrointestinal tract. It also halts the iron cycle in macrophages, thereby limiting iron availability to the bone marrow. This iron restriction can eventually lead to anemia in some patients, as a form of anemia of chronic disease. While iron is restricted from the bone marrow, its sequestration in other tissues, primarily adipose, can lead to local oxidant damage and subsequent inflammation, thus causing a vicious cycle.[15]

Conclusion

In summary, there are several CBC and serum biochemistry differences in clinically healthy aging dogs. Further studies are warranted to further explore gastrointestinal bleeding and inflammaging in senior and geriatric patients and to determine if dogs may be a suitable model for anemia in elderly people.

References

1. AVMA. *US Pet Ownership and Demographic Sourcebook*. 2007:6–23.
2. Metzler F, *et al. Vet Clin North Am Small Anim Pract*. 2012;42(4):615–29.
3. Meuten D. *Veterinary Hematology and Clinical Chemistry*. 2nd ed. 2012:326–28.
4. Goodnough L, *et al. Am J Hematol*. 2014;89(1):88–96.
5. Ostan R, *et al. Neuroimmunomodulation*. 2008;15(4–6):224–40.
6. Nemeth E, *et al. Science*. 2004;306(5704):2090–93.
7. Pantopoulos K, *et al. Biochemistry*. 2012;51(29):5705–24.
8. Walling C. *Acc Chem Res*. 1975;8(4):125–31.
9. Drakesmith H, *et al. Cell Metab*. 2015;22(5):777–87.
10. Ponka O, *et al. Int J Biochem Cell Biol*. 1999;31(10):1111–37.
11. Finazzi D, *et al. Arch Toxicol*. 2014;88(10):1787–1802.
12. Kautz L, *et al. Blood*. 2014;124(16):2569–74.
13. Camaschella C, *et al. Int J Hematol*. 2016;38(Suppl 1):20–6.
14. Citelli M, *et al. Nutrients*. 2015;7(1):335–48.
15. Orr J, *et al. Diabetes*. 2014;63(2):421–32.

acvim325

Effects of Paraneoplastic Syndromes on Complete Blood Counts and Serum Chemistry Profiles

Amy L. MacNeill, DVM, PhD, DACVP
Fort Collins, CO, USA

INTRODUCTION

Paraneoplastic syndromes are caused by tumor production of functional hormones or cytokines, or by tumor-induced alterations of the immune response. Paraneoplastic syndromes most commonly manifest as endocrinopathies, neuropathies, musculoskeletal diseases, dermatopathies, or hematologic abnormalities. **Table 1** includes examples of paraneoplastic syndromes involving all of these systems. The proceedings and talk will focus on alterations in results of hematologic data and biochemical profiles caused by paraneoplastic syndromes.

Table 1. Common paraneoplastic syndromes

System affected	Paraneoplastic syndrome	Pathogenesis	Associated cancer
Endocrine	Acromegaly	Growth hormone releasing hormone	Pituitary adenoma
	Hyperadrenocorticism	Adrenocorticotropic hormone	Adrenal adenocarcinoma Primary lung tumor
	Hyperaldosteronism	Aldosterone	Adrenal adenoma Lymphoma
	Hypercalcemia	1,25-dihydroxyvitamin D	Lymphoma
		Bone lysis	Multiple myeloma Osteosarcoma
		Parathyroid hormone (PTH)	Parathyroid gland tumor
		PTH-related protein	Adenocarcinoma Apocrine gland adenocarcinoma Bronchogenic adenocarcinoma Lymphoid leukemia Lymphoma Mammary gland tumor Melanoma Osteosarcoma Primary lung tumor Renal angiomyxoma Squamous cell carcinoma Thyroid tumor
		Protein/cytokine (e.g., prostaglandin E1/2, IL-1β, TGFβ, receptor activator of NF-κB ligand, G-CSF & other osteoclast activating factors, other PTH-like hormone)	Varied cancer types
	Hyperestrogenism	Estrogens	Sertoli cell tumor
	Hypergastrinemia	Gastrin	Gastrinoma Pancreatic non-B islet cell neoplasia
	Hyperhistaminemia	Histamine	Mast cell tumor
	Hyperthyroidism	Thyroxin	Thyroid gland tumors

	Hypoglycemia	Insulin	Pancreatic islet cell tumor
		Insulin-like growth factor	Adenocarcinoma Hepatocellular carcinoma Mammary carcinoma Primary lung tumor Other solid tumors
		Undetermined	Lymphocytic leukemia Renal adenocarcinoma Renal lymphoma
	Syndrome of inappropriate antidiuretic hormone	Antidiuretic hormone Atrial natriuretic peptide	Carcinoma Primary lung tumor Sarcoma
Musculoskeletal	Hypertropic osteopathy	Possible hypoxia Possible growth factor (VEGF or PDGF) release from platelets in cases with thrombocytosis	Primary lung tumor Renal transitional cell carcinoma Sarcoma
	Osteomalacia	Fibroblast growth factor	Sarcoma
	Polymyositis	Immune-mediated damage	Varied cancers
Neurologic	Myasthenia gravis	Antibody vs. acetylcholine receptor or skeletal muscle proteins	Thymoma
		Possible antibody formation but may have been coincidental	Apocrine gland adenocarcinoma Cholangiocarcinoma Cutaneous lymphoma Osteogenic sarcoma
	Polyneuropathy	Antibody vs. myelin-associated glycoprotein, gangliosides, glycosphingolipids, other	Mammary tumor Multiple myeloma
Dermatologic	Epidermal necrosis	Probable immune-mediated damage	Lymphoma Pancreatic carcinoma
	Erythema multiforme	Probable immune-mediated damage	Thymoma
	Erythroderma	Probable immune-mediated damage	Thymoma
	Exfoliative dermatitis	Probable immune-mediated damage	Thymoma
	Feline paraneoplastic alopecia	Undetermined	Cholangiocarcinoma Hepatocellular carcinoma Intestinal adenocarcinoma Pancreatic tumor Plasma cell tumor
	Hepatocutaneous syndrome	Glucagon	Gastric tumor Pancreatic carcinoma
	Nodular dermatofibrosis	Probable immune-mediated damage	Renal cystadenoma/adenocarcinoma
	Pemphigus	Anti-plakin IgG	Sarcoma
		Undetermined	Lymphoma
Hematologic	Anemia	Alteration of the bone marrow microenvironment	Varied cancer types
		Chronic disease	Most cancers
		Immune-mediated hemolysis	Varied cancer types

		Microangiopathic hemolysis	Varied cancer types
		Phagocytosis	Hemophagocytic histiocytic sarcoma
	Eosinophilia	GM-CSF IL-5	Carcinoma Lymphoma Mast cell tumor
	Myelodysplasia	Immune-mediated damage	Broncho-alveolar carcinoma
		Cytokine dysregulation (e.g., TNFα, TGFβ, IRF-1)	Varied cancer types
	Neutropenia	Alteration of the bone marrow microenvironment	Broncho-alveolar carcinoma
	Neutrophilia	G-CSF GM-CSF IL-6	Carcinoma Renal carcinoma Renal transitional cell carcinoma Squamous cell carcinoma
	Polycythemia	Erythropoietin	Leiomyosarcoma Renal lymphoma T cell lymphoma
	T cell lymphocytosis	Aberrant T cell function	Thymoma
	Thrombocytopenia	Alteration of the bone marrow microenvironment	Leukemia
		Disseminated intravascular coagulopathy	Lymphocytic leukemia Mucin-producing adenocarcinomas Prostatic adenocarcinoma Renal carcinoma Thyroid carcinoma Visceral cancers
	Thrombocytosis	IL-6	Broncho-alveolar carcinoma Hepatocellular carcinoma Lymphoma
Other	Fever	Pyrogenic cytokines, prostaglandins	Varied cancer types
	Hyperalbuminemia	Possible increased production or defect in receptivity to increased osmolality	Hepatocellular carcinoma
	Hypoalbuminemia	Acute phase protein response; may be due to increased IL-6	Hemophagocytic histiocytic sarcoma
		Increased osmolality	Multiple myeloma
		Loss	Melanoma Renal carcinoma
	Monoclonal gammopathy	Immunoglobulin	Lymphoma Lymphocytic leukemia Multiple myeloma

Paraneoplastic Syndromes That Affect Complete Blood Counts

Many neoplastic diseases can alter blood cell counts by damaging the circulatory system directly (e.g., leukemia, stage V lymphoma, solid tumor metastasis to bone marrow). Paraneoplastic syndromes are observed when tumors affect blood cell counts indirectly. Understanding the underlying pathogenic mechanisms of indirect damage to hematopoietic cells can help direct treatment strategies for individual patients. Patients with paraneoplastic hematologic syndromes typically have a poor prognosis.[1]

Anemia

Anemias in cancer patients are often normocytic, normochromic (non-regenerative) anemias. This type of anemia is classified as anemia of chronic disease or anemia of chronic inflammation. Prolonged increased concentrations of circulating pro-inflammatory cytokines, such as interleukin (IL)-1 and IL-6, are thought to inhibit hematopoiesis in affected patients. In cancer patients, these cytokines may be produced by the tumor cells or by surrounding tissue that is damaged from tumor growth.

Hemolytic anemias also have been induced by cancers. Often hemolysis is immune-mediated due to tumor-directed production of anti-erythrocyte antibodies or abnormal T cell or B cell function. In some patients, hemolysis occurs secondary to microvascular disease caused by products produced by the tumor or damaged surrounding tissues. Direct phagocytosis of erythrocytes occurs with some erythrophagocytic tumors, particularly histiocytic sarcomas. Increased tumor production of IL-6 has be implicated as a mechanism for enabling phagocytosis of the red blood cells to occur.

Differentiation of non-regenerative and hemolytic anemia can be determined from results of a complete blood count (CBC) and microscopic evaluation of a blood smear.

Polycythemia

Cancer growth (particularly in the lungs or kidney) may lead to hypoxia which causes physiologic release of erythropoietin and secondary polycythemia. Additionally, some tumors can produce erythropoietin and cause polycythemia without evidence of hypoxia. Tumor types reported to produce erythropoietin include renal tumors, lymphoma, and leiomyosarcoma. Unfortunately, measurement of serum erythropoietin is not readily available in veterinary patients. If the polycythemia is due to a paraneoplastic syndrome, the patient's hematocrit should return to reference intervals following tumor removal.

Eosinophilia

There are a few types of tumors that induce peripheral eosinophilia and often eosinophilic inflammation within the tumor. The most common tumor types associated with paraneoplastic eosinophilia in veterinary patients are mast cell tumors, lymphomas, and carcinomas. Cytokine production by tumor cells (IL-5 and/or GM-CSF) is thought to cause the eosinophilia.[1] Eosinophil counts can be monitored with CBCs and are expected to decrease if the tumor is successfully treated.

Neutropenia

Neutropenia and thrombocytosis secondary to tumor-induced myelodysplasia was reported in a cat with bronchoalveolar carcinoma.[2] Hematopoietic complications that occur secondary to paraneoplastic myelodysplasia are relatively common in human patients and are thought to be immune-mediated conditions caused by tumor-driven production of auto-antibodies or altered lymphocyte function (see Myelodysplasia).

Neutrophilia

Neutrophilia may be observed in cancer patients for many reasons, including secondary bacterial infection, corticosteroid administration, and systemic inflammatory response syndrome. If these disease processes are not occurring, paraneoplastic neutrophilia should be considered. In patients with paraneoplastic neutrophilia, neutrophil counts >100,000 cells/μL with a left shift can be seen. Carcinomas are the most common tumor type observed to induce this condition. Tumor production of G-CSF, GM-CSF, or IL-6 stimulates hematopoiesis and leads to marked myeloid hyperplasia and peripheral neutrophilia.[1] If production of G-CSF is the underlying mechanism of the neutrophilia, concurrent hypercalcemia of malignancy may be present (see Hypercalcemia).

T Cell Lymphocytosis

Rarely, mixed populations of circulating $CD4^+$ and $CD8^+$ T cells may be increased in patients with thymomas.[3] This is thought to be due to abnormal T cell regulation by neoplastic thymic epithelial cells and is not a leukemia or lymphoma. Flow cytometry and PCR for antigen receptor rearrangement are particularly helpful diagnostic tools for determining if a lymphocytosis is a neoplastic or reactive process.

Thrombocytopenia

Proteins secreted by tumors that alter the bone marrow microenvironment or activate the coagulation cascade may lead to thrombocytopenia. Tumors known to cause paraneoplastic thrombocytopenia include lymphocytic leukemia and (adeno)carcinomas. Mucins from adenocarcinoma have been shown to activate Factor X which induces release of tissue factor and cysteine proteases from the tumor.[1] This leads to excessive clot formation, rapid utilization of platelets, and thrombocytopenia. Hypoxia also may contribute by inducing release of tissue factor and plasminogen activator inhibitor-1 by damaged cells. The uncontrolled activation of the coagulation cascade often leads to disseminated intravascular coagulopathy (DIC). In other tumor types, thromboplastin release secondary to vascular stasis has be implicated in leading to thrombocytopenia and DIC.[4]

In some humans with visceral neoplasms, Trousseau's syndrome is recognized which includes recurrent thrombophlebitis, arterial emboli, and hemorrhage. Hematologic abnormalities in these patients include thrombocytopenia, hypofibrinogenemia, increased prothrombin time, increased fibrin degradation products, and microangiopathic hemolytic anemia.[1]

Thrombocytosis

Thrombocytosis may be present in patients with chronic inflammation or infection. Less commonly, paraneoplastic thrombocytosis is diagnosed. As with other paraneoplastic hematologic syndromes, carcinomas and lymphomas are the most common tumor types associated with thrombocytosis. Paraneoplastic thrombocytosis is believed to be induced by tumors that produce IL-6.

Myelodysplasia

Autoimmune diseases associated with myelodysplastic syndromes (MDS) include vasculitis, polyarthritis, polyneuropathy, glomerulonephritis, and immune-mediated hemolytic anemia. Abnormal immunologic function causing abnormal antigen presentation or T cell responses or B and T cell interactions have been reported to be due to defective B or T cell function, decreased T cell or natural killer cell numbers, hypergammaglobulinemia, or hypogammaglobulinemia.[5] In some patients, immune dysfunction may lead to myelodysplasia, while in others, myelodysplasia is thought to initiate paraneoplastic autoimmune disease.

Paraneoplastic Syndromes That Affect Serum Biochemical Profiles

Hyperadrenocorticism

Production of excess adrenocorticotropic hormone (ACTH) by a pituitary tumor is the cause of hyperadrenocorticism in 80–85% of dogs and cats with the disease. Adrenocortical tumor production of cortisol is found in approximately 15–20% of dogs and cats. Pituitary pars intermedia dysfunction is the most common cause of paraneoplastic hyperadrenocorticism in horses.

In all species, patients with hyperadrenocorticism often present with polydipsia (PD), polyuria (PU), polyphagia, lethargy, and weakness. In dogs, other complaints include a pendulous abdomen, lameness, panting, and hepatomegaly. Dermatopathies associated with canine hyperadrenocorticism include endocrine alopecia, pyoderma, demodicosis, and calcinosis cutis. Cats often present with concurrent diabetes mellitus. Alopecia, and fragile skin also have been associated with hyperadrenocorticism in cats. In horses, hirsutism and muscle wasting are common signs of the disease.

A stress leukogram is often seen on a CBC of patients with hyperadrenocorticism. In dogs, increased alkaline phosphatase due to an increase in the corticosteroid-induced isoenzyme is common. Diagnostic tests for hyperadrenocorticism include a urine cortisol to creatinine ratio and ACTH stimulation test in dogs, and dexamethasone suppression tests and endogenous ACTH measurement in dogs, cats, and horses. Abdominal ultrasound can aid in diagnosis of adrenocortical tumors in dogs and cats.

Hyperaldosteronism

Aldosterone-producing adrenal gland adenomas are the most common cause of hyperaldosteronism in humans. These patients may present with hypertension and hypokalemia. Rare cases of ectopic aldosterone-producing (adeno)carcinomas and lymphomas have also be reported.[6] Secondary hyperaldosteronism due to a non-neoplastic disease process is more commonly observed in veterinary patients.

Hypercalcemia

Hypercalcemia is one of the more common paraneoplastic syndromes diagnosed in veterinary patients. Malignancy is the most common cause of hypercalcemia in dogs and is second only to idiopathic hypercalcemia in cats. In horses, hypercalcemia of malignancy is rare. Symptoms associated with hypercalcemia include PU/PD, renal failure, anorexia, nausea, vomiting, abdominal pain, lethargy, dehydration, constipation, muscle weakness, arrhythmia, and coma. Additionally, confusion, fatigue, irritability, depression, anxiety, and cognitive dysfunction have been reported in hypercalcemic people.

Pathologic causes of hypercalcemia include renal failure (occasional in dogs and cats, common in horses), hypoadrenocorticism, osteolysis, and increased vitamin D, parathyroid hormone (PTH), or PTH-related protein (PTHrp). Increased PTHrp has been detected in patients with many types of tumors and in patients with granulomas. In humans, increased production of other proteins that affect calcium mobilization from bone also have been documented to cause hypercalcemia of malignancy. Examples of these proteins include prostaglandin E_1, prostaglandin E_2, interleukin (IL)-1β, transforming growth factor (TGF) β, receptor activator of nuclear factor (NF)-κB ligand, granulocyte colony stimulating factor (G-CSF), and other osteoclast activating factors including PTH-like hormones.

Measurement of serum calcium is complicated by the fact that serum biochemistry profiles report a total calcium concentration, which includes free (ionized) and anion-bound calcium. Approximately 45% of serum calcium is bound to proteins (mostly albumin), therefore total calcium concentration may be artificially decreased by concurrent hypoalbuminemia. Also, calcium concentration may be falsely elevated in lipemic samples and in patients with hyperproteinemia. Determination of ionized calcium is highly recommended, particularly in patients with suspected alterations in calcium concentration. In patients with increased ionized calcium, measurement of PTH, PTHrp, and 25-hydroxyvitamin D can help determine the underlying cause of hypercalcemia. Patients with hypercalcemia of malignancy are expected to have a low or low normal PTH, elevated PTHrp, and normal 25-hydroxyvitamin D concentrations in serum.

Paraneoplastic hypercalcemia is often associated with apocrine gland adenocarcinoma, lymphoma, and multiple myeloma in dogs. However, several types of epithelial tumors and lymphoproliferative diseases as well as osteosarcoma and melanoma have been reported to cause hypercalcemia of malignancy. Likewise, carcinomas and lymphomas are the most common types of tumors that induce hypercalcemia of malignancy in cats, but case reports of other tumor types, including osteosarcoma, have been published. In horses, rare cases of lymphoma, multiple myeloma, squamous cell carcinoma, parathyroid adenoma, and ameloblastoma have been documented to induce hypercalcemia.

Hyperestrogenism

Sertoli cell tumors can secrete large amounts of estrogens. Clinical signs associated with hyperestrogenism include dermatologic changes (alopecia and pigmentation), feminization, squamous metaplasia of the prostate, and bone marrow suppression.

Hypergastrinemia

Gastrinomas of the pancreas are rare in veterinary species. The tumor induces excess secretion of hydrochloric acid from the stomach leading to ulceration of the esophagus, stomach, and duodenum. Clinical signs may include severe weight loss, anorexia, abdominal pain, vomiting, and diarrhea.

Hyperhistaminemia

Increased concentration of or acute release of histamine can be a complication of mast cell tumors and acute myelocytic leukemias with basophil differentiation. Clinical signs of hyperhistaminemia can range from swelling and erythema around the tumor to gastrointestinal ulceration and shock. Dogs with mast cell tumors often have a peripheral eosinophilia.

Hyperthyroidism

Hyperthyroidism in veterinary species is often caused by excess thyroxin (T_4) by a functional thyroid tumor. This disease is the most common endocrinopathy in cats. The classic clinical signs of feline hyperthyroidism are polyphagia with weight loss, PU/PD, increased activity, restlessness, diarrhea, and vomiting.

Increased liver enzymes (ALT, ALP) are the most consistent and significant abnormality. Other laboratory abnormalities may include erythrocytosis, an excitement leukogram, increased BUN with normal creatinine (due to increased protein catabolism), and hypocholesterolemia.

Total T_4 is considered diagnostic for hyperthyroidism, if elevated. If total T_4 is normal to high normal, measurement of free T_4 by equilibrium dialysis may be necessary. Free T_4 concentration is slightly more sensitive than total T_4, but it is less specific. Free T_4 within the reference interval rules out hyperthyroidism. If free T_4 is elevated, it likely means hyperthyroidism, however 6–12% of euthyroid cats can have an increased free T_4 concentration. A triiodothyronine (T_3) suppression test also can be used to diagnose cats with hyperthyroidism. In hyperthyroid cats, there is minimal to no reduction in total T_4 after repeated administration of synthetic T_3.

Hypoglycemia

Many cancer patients have abnormal carbohydrate, protein, and lipid metabolism due to impaired insulin sensitivity, abnormal glucose oxidation, decreased gluconeogenesis, or increased glucose recycling. Tumors use glucose anaerobically which creates lactose. Lactose is converted back to glucose by the lactic acid (Cori) cycle, which requires energy, leading to a net energy loss. This process can lead to paraneoplastic hypoglycemia and cachexia.[7]

Syndrome of Inappropriate Antidiuretic Hormone (SIADH)

Tumor production of antidiuretic hormone or atrial natriuretic peptide can lead to SIADH. Carcinomas in the lungs, gastrointestinal tract, and urogenital tract are most often associated with this syndrome, although lymphomas and some sarcomas (Ewing's tumor and leiomyosarcoma) have been implicated as well. Non-neoplastic diseases that lead to SIADH should be ruled out before a paraneoplastic syndrome is diagnosed. Non-neoplastic causes of SAIDH induce release of antidiuretic hormone from the posterior pituitary and include central nervous system damage, hypothyroidism, adrenal insufficiency, volume depletion, pulmonary infection, and several pharmaceutical agents. Abnormalities observed in patients with SIADH include hyponatremia, hypo-osmolality, increased urine osmolality, and increased fractional excretion of sodium.

References

1. Kanaji N, *et al. World J Clin Oncol.* 2014;5(3):197–223.
2. Hogan DF, *et al. J Am Anim Hosp Assoc.* 1999;35(6):483–486.
3. Batlivala TP, *et al. J Small Anim Pract.* 2010;51(9):491–494.
4. Weller RE. *Vet Clin North Am Small Anim Pract.* 1985;15(4):805–816.
5. Enright H, Miller W. *Leuk Lymphoma.* 1997;24(5–6):483–489.
6. Aronova A, *et al. World J Cardiol.* 2014;6(5):227–233.
7. Ogilvie GK, *et al. Am J Vet Res.* 1997;58(3):277–281.

acvim326

Biochemical Abnormalities with Cytologic/Hematologic Correlations in Dogs

Rose E. Raskin, DVM, PhD, DACVP (Clinical Pathology)
West Lafayette, IN, USA

Introduction

Cases presented will demonstrate biochemical abnormalities affecting hepatic and renal blood parameters, electrolytes and acid-base status particularly. Where pertinent, specific cytologic and hematologic changes will be demonstrated from these cases to assist in the diagnosis. While some abnormalities may be specific for certain diseases, the emphasis will be placed on using all available data to best evaluate these cases. In addition, specific instrument artifacts relative to some of the cases will be highlighted. As a review, these proceedings will help establish an appropriate baseline for understanding the abnormalities presented in these canine cases.

Urinary System

Tests for Glomerular Function

Glomerular function evaluation involves serum measurements of urea and creatinine along with a complete urinalysis to determine the presence and character of the azotemia as indicated by an elevated urea and/or creatinine. Other test results for impaired glomerular function include increased urine protein and hyperphosphatemia. Increased dietary protein or gastrointestinal bleeding will increase urea levels.

Creatinine is produced from muscle creatine and as such decreased muscle mass as in emaciation or severe starvation may cause reduced creatinine levels. Compared with urea, creatinine is less affected by dietary protein.

Pancreatic enzymes are released by leakage from pancreatic acini following injury by trauma, neoplasia or inflammation and then are absorbed by the blood. Both amylase and lipase are degraded and excreted by the kidney. Therefore, chronic renal insufficiency will produce increased levels of one or both of the enzymes. These enzymes will increase to a mild or moderate degree with decreased glomerular filtration, such as dehydration or shock. Lipemia minimally affects the enzyme measurements since they are kinetic assays.

Hepatobiliary System

Enzymes of Hepatocellular Injury

Alanine aminotransferase (ALT) is a cytoplasmic enzyme of the hepatocyte that is considered highly liver specific for dogs and cats, while very specific increases have been detected with very severe muscle damage. The half-life in the dog is 2–3 days but shorter (hours) in the cat. The upper end of the reference range is used to determine fold increases in activity. Persistent mild to moderate increases (over months) suggest smoldering inflammation (e.g., chronic hepatitis). Phenobarbital may induce increased ALT concentration.

Aminotransferase (AST) is present in both cytoplasm and mitochondria of the hepatocyte. Due to mitochondrial location, severe damage must occur to the cell to release enzyme, therefore levels are often lower than ALT. Half-life in dogs is 1–2 days but shorter (hours) in the cat. It is not as specific as ALT as AST is present in both liver **and** muscle (cardiac, skeletal) as well as erythrocytes so it may increase with hemolysis.

Indicators of Hepatic Function

Liver function is reflected by several analytes including albumin, alpha and beta globulins, urea, glucose, bilirubin, and cholesterol. Erythrocyte size and shape, serum bile acids, blood ammonia, and most coagulation factors are also affected. Microcytosis may be seen in portosystemic shunt animals or those with decreased hepatic function. This likely relates to a relative iron deficiency from decreased transferrin production which transports iron to erythroid precursors.

Urea is produced by the liver as part of the urea cycle, therefore decreased levels often indicate reduced liver function. Decreased levels are often seen in portosystemic shunt cases related to reduced blood flow through the hepatic parenchyma.

Bilirubin is the result of hemoglobin breakdown. The heme component loses its iron portion and is converted to bilirubin in mammals, which is bound to albumin for transport to the liver. Within hepatocytes, the bilirubin is conjugated with glucuronic acid in most species and can be measured in the blood in a direct reaction. The water-soluble conjugated form is secreted into the bile or excreted into the urine. Total bilirubin includes the unconjugated form known as indirect and the direct or conjugated form of bilirubin. With rapid or severe hemolysis, the amount of unconjugated bilirubin will increase, but if liver function is normal, an increase in conjugated bilirubin will soon occur as well. Intrahepatic disease and hemolysis can cause elevations in both unconjugated and conjugated forms, while extrahepatic cholestasis mostly results in conjugated bilirubin. Bilirubinuria may precede hyperbilirubinemia.

Hypoproteinemia can result from hypoalbuminemia, hypoglobulinemia, or both concurrently. Concurrent decrease may indicate mild or early blood loss or loss through gastrointestinal tract (bleeding, inflammatory, infectious or neoplastic GI disease). Greyhounds reportedly have significantly lower globulin levels. Hypoalbuminemia may reflect decreased synthesis due to liver damage or secondary to increased oncotic pressure in addition to inflammation. As albumin is a negative acute phase protein, its production is reduced in acute inflammation. Pure albumin loss may be renal (glomerular) or due to vasculitis.

Hypoglycemia can reflect reduced production from liver damage or lack of glycogen stores (young, hunting dogs, extreme exertion). Other causes of hypoglycemia involve seizures, which can cause reduction in brain glycogen as well as hypoadrenocorticism, which can be related to primary adrenal disease or to hypopituitarism. Hypoglycemia can result from *in vitro* consumption of glucose from blood cells if left in contact too long before serum/plasma separation. Bacteria can also utilize glucose so septic conditions must be considered.

Cholesterol, when decreased in concentration, can be associated with a lack of hepatic production.

Musculoskeletal System

The muscle enzyme creatine kinase (CK) is very specific in animals and is used concurrently with AST to determine whether liver or muscle damage is most likely. With severe muscle damage, ALT has been shown to increase. Countering the increases from muscle injury are the effects of muscle atrophy which produces reductions in these enzyme concentrations. The largest increases in CK result from seizures but expect lesser increases with injections or recumbency.

Acid-Base Disturbances

Anion gap is a calculated value $[Na^+ + K^+] - [Cl^- + HCO3^-]$ that is highly dependent on bicarbonate concentration. Finding an elevated anion gap supports a titrational metabolic acidosis likely due to an increase in renal acids, ketoacids, lactic acids, or exogenous unmeasured anions such as ethylene glycol, etc. These will typically reduce the bicarbonate concentration. However, when the total CO_2 measurement is at the upper reference limit or higher, a mixed disturbance with bicarbonate retention is occurring. This increase in bicarbonate higher than expected may reflect a loss of negative ions such as chloride, which often occurs through gastric vomiting creating a metabolic alkalosis. One can calculate the difference between sodium and chloride and the mean of their respective reference interval to determine if they are following each other, which should be within 5 units. It is best to confirm the presence of acidemia or alkalemia through blood gas measurements of arterial blood.

Tumor Lysis Syndrome

Marked cytoreduction and lysis of neoplastic cells following chemotherapeutic administration or by spontaneous cell rupture may result in the syndrome known as acute tumor lysis. Laboratory findings in this case often include hyperphosphatemia, hyperkalemia, and increased purine metabolites in blood and urine. Hypocalcemia has also been observed as a result of metastatic calcification when the calcium x phosphorus product exceeds 70. Typically two or more laboratory abnormalities occurs within 3 days or up to 7 days after therapy has begun. Neoplastic cells contain high concentrations of nucleic acids, phosphorus, and potassium and upon rupture release these materials into the bloodstream.

Hyperphosphatemia and hypocalcemia: Phosphorus is predominantly an intracellular ion that can be 4–6 times higher in human neoplastic lymphocytes than in normal lymphocytes. Hyperphosphatemia may precipitate the calcium and result in crystalline deposition in the kidney. Reduction in vitamin D production from renal disease can also lead to hypocalcemia.

Hyperkalemia: Potassium is a major intracellular cation in leukocytes and platelets. Dogs do not have high potassium in red blood cells, except for certain breeds of dogs (e.g., Akita, Shiba Inu); however, reticulocytes in all canine breeds are rich in potassium. Concurrent acute renal failure secondary to obstructive hyperuricemia or nephrocalcinosis can enhance elevated potassium levels through decreased GFR and excretion. Concurrent metabolic acidosis may have also contributed to hyperkalemia.

Hyperuricemia: Purine metabolism occurs in dogs within the small intestine to release adenine and guanine, which are further catabolized to xanthine. The enzyme, xanthine oxidase, found mostly in blood, liver, lungs, and intestines converts xanthine to uric acid. In dogs (breeds other than the Dalmatian or English bulldog), uric acid is metabolized to allantoin by the enzyme uricase found in peroxisomes within hepatocytes and then excreted in the urine. Xanthine and allantoin crystalluria have been observed in canine patients following allopurinol administration and spontaneously with acute myeloid leukemia, respectively.

Normoblastosis/Neutrophil Hypersegmentation

Normoblastosis without evidence of regeneration suggests inappropriate rubricytosis. This often results from splenic changes such as splenectomy, splenic contraction, or neoplasia (e.g., hemangiosarcoma). Another major explanation is bone marrow damage with injury to small vessels from heat, hypoxic damage, necrosis, inflammation or neoplasia. Lead toxicosis produces abnormal red cell related to incomplete hemoglobin production.

Hypersegmented neutrophils most often result from accelerated aging of cells from endogenous or exogenous corticosteroids. Other explanations include artifact from old or heated samples as well as primary or secondary dysplasia resulting from neoplasia or nutritional deficiencies.

Sample-Related Causes of Instrument Error

Certain instruments will give a false concentration of sodium and chloride in blood related to marked increases in lipids. Most analyzers determine electrolyte ion activity by ion-selective electrodes and potentiometry methods. However, it is significant when the ion activity is measured in a nondiluted sample (direct potentiometry) versus measured in a diluted sample (indirect potentiometry). The reported electrolyte concentration reflects its activity in the aqueous phase, so indirect potentiometry will reduce the electrolyte concentration due to displacement of the water content of the sample volume by increased triglycerides or cholesterol and the subsequent altered dilution factor. Note that hypercholesterolemia does not result in visible lipemia as seen with hypertriglyceridemia.

Changes in erythrocyte size, as measured by some automated analyzers, may be affected by the osmotic effects of sodium. In the absence of reticulocytosis, macrocytosis and hypochromasia may be caused by artifactual cell swelling (from EDTA or hypernatremia). Cells from the hypertonic blood swell *in vitro* when exposed to the hypotonic diluent fluid from the analyzer. The reverse appears to occur with hyponatremia responsible for microcytic erythrocytes as measured by the analyzer. Blood cells do not appear smaller on blood films.

Marked hyperbilirubinemia is expected to alter results measured by spectrophotometer, such as creatinine by the Jaffe method, but this will differ between test methodologies. Using a higher wavelength at 570 nm on icteric samples will avoid the interference. In another study, marked hyperbilirubinemia is associated with a false decrease in serum total protein and therefore globulin concentrations when using a bichromatic spectrophotometric biuret assay.

References

1. Brunson D, *et al. Vet Clin North Am Small Anim Pract.* 2004;34:1419.
2. Gupta *et al. Vet Clin Pathol.* 2014;43:422
3. Mastrorilli C, *et al. Vet Clin Pathol.* 2013;42(2):145.
4. Meichner K, *et al. Vet Clin Pathol.* 2015;44:613.
5. Mylonakis M, *et al. Aust Vet J.* 2007;85:206.
6. Roberts M, *et al. Anesthesiology.* 2001;95:716.
7. Stockham S, *et al. Fundamentals of Veterinary Clinical Pathology.* 2nd ed. Ames, IA: Blackwell Pub.; 2008.
8. Torrente C, *et al. J Vet Emerg Crit Care.* 2014;24:509.
9. Vickery K, *et al. J Vet Intern Med.* 2007;21:1401.

acvim327

Biochemical Abnormalities with Cytologic/Hematologic Correlations in Cats

Rose E. Raskin, DVM, PhD, DACVP (Clinical Pathology)
West Lafayette, IN, USA

Introduction

Cases presented will demonstrate biochemical abnormalities affecting hepatic and renal blood parameters, electrolytes and proteins particularly. Where relevant, specific cytologic and hematologic changes will be demonstrated from these cases to assist in the diagnosis. While some abnormalities may be specific for certain diseases, the emphasis will be placed on using all available data to best evaluate these cases. In addition, specific instrument artifacts relative to some of the cases will be highlighted. As a review, these proceedings will help establish an appropriate baseline for understanding the abnormalities presented in these feline cases.

Urinary System

Tests for Renal Tubular Function or Damage

Common serum tests for renal tubular function include electrolyte abnormalities and calculated anion gap. Impaired renal function affects such electrolytes as sodium, potassium, chloride, bicarbonate, calcium, phosphorus, and magnesium. The acid-base conditions are frequently determined through bicarbonate, sodium, chloride, potassium, anion gap measurement, and the urine pH.

Phosphorus in the serum is measured as inorganic phosphate. Extracellular shifts occur with acidosis similar to potassium and insulin causes intracellular shifts of phosphorus similar to potassium. Decreased renal excretion occurs with renal disease (decreased GFR) and with young animals due to the effects of growth hormone to increase renal reabsorption of phosphate by kidney tubules.

Calcium may be reduced from a decrease in the protein-bound fraction secondary to hypoalbuminemia and a decrease in the free fraction due to increased calcium binding such as with oxalate crystal formation or soft tissue mineralization related to a high calcium-phosphorus product, generally over 70. In chronic kidney disease, the decrease in 1,25 dihydroxyvitamin D inhibits synthesis of a calcium-binding protein leading to reduced distal renal tubular reabsorption of calcium that results in hypocalcemia. Increased excretion of calcium from the kidney during acute pancreatitis is thought to be related to a calcitonin effect from the release of glucagon which enhances calcitonin activity.

Increased renal tubular reabsorption or decreased excretion of magnesium (e.g., renal disease in cats or obstructive uropathy in cats) results in hypermagnesemia.

Changes in sodium with hyponatremia result from increased renal loss of sodium due to reduced tubular resorption as a result of tubular disease, decreased aldosterone (hypoadrenocorticism), or persistent polyuria due to osmotic diuresis. Redistribution of sodium from third-space fluid (e.g., urinary bladder rupture) may result in hyponatremia. Hypernatremia is seen with increased renal retention at the proximal tubules due to influence of hyperaldosteronism or mineralocorticoid therapy or with disturbances to thirst regulation.

Chloride is the most abundant extracellular anion whose changes often follow those of sodium. Hyperchloremia results from increased retention by the kidney (e.g., decreased bicarbonate resorption as occurs in renal tubular acidosis). Dilution of electrolytes due to redistribution of fluids from third space may result in hypochloremia.

Increases in potassium are commonly seen with reduced urine excretion due to obstruction, anuric or oliguric renal failure or less commonly with uroabdomen. With polyuric renal failure, there is increased renal excretion of potassium. Hypokalemia is also present in cats with early stages of chronic renal failure.

Total carbon dioxide (TCO_2) is typically found on the routine chemistry profile. It can be used as an approximation for bicarbonate as TCO_2 is generally 1 mmol/L higher than bicarbonate. Secretional loss of bicarbonate occurs through saliva, duodenal vomiting, or diarrhea and reduced bicarbonate resorption (e.g., renal tubular acidosis [hyperchloremic]).

Increases in anion gap due to an increase in unmeasured anions will alter the concentrations of the prevalent anions, usually HCO_3^-, causing a loss of body buffer base. This titrational effect on bicarbonate results in a metabolic acidosis. Endogenous sources includes uremic, lactic, and keto-acids. Exogenous sources include drugs (e.g., salicylates) and toxins (e.g., ethylene glycol), which can increase the anion gap and significantly lower bicarbonate and/or chloride concentrations. Calculation of anion gap is a helpful diagnostic tool for ethylene glycol toxicosis in small animals.

HEPATOBILIARY SYSTEM

Indicators for Cholestasis

Alkaline phosphatase (ALP) is located in cell membranes of the bile canaliculi and bile ducts. Elevations in serum levels result from induction of synthesis. The ALP enzyme consists of many isoenzymes arising from several sites such as bone, intestine, placenta, kidney, and liver. Increases in the bone isoenzyme are due to young growing animals with increased bone turnover or in some cancer patients with osteosarcoma. Levels of total ALP and bone isoenzyme have been determined to be important prognostic factors for appendicular osteosarcoma in dogs. Only in dogs are levels induced by glucocorticoids. Both glucocorticoid and drug (anticonvulsants) induction affect the steroidal isoenzyme, a specific hepatic isoenzyme, formed possibly by a change in glycosylation of an intestinal isoenzyme, giving it new physical properties including reduced clearance rate. Due to the very short half-life for ALP isoenzymes from the placenta, intestine, and kidney, these sources are not clinically relevant causes of increased ALP. Peak rise takes about one week, increases are not immediate, especially with the steroidal isoenzyme. Increases in ALP often precede hyperbilirubinemia. Levels in cats are often low due to the very short half-life (6 h) and low production level of enzyme. Any rise in ALP in cats indicates significant cholestasis or possibly decreased enzyme clearance as may occur in feline hepatic lipidosis. Levels in dogs are higher than cats due to the longer half-life (3 days) and higher levels of enzyme production. Hyperthyroidism in cats is commonly associated with increased ALP levels due to unknown mechanisms.

Gamma-glutamyltransferase (GGT) is an inducible enzyme located in membrane of biliary epithelial cells. Similar to ALP, there is some induction in the dog from corticosteroids and anticonvulsants but to a lesser extent. High ALP with low GGT can be a useful indicator for hepatic lipidosis in cats. Higher GGT increases compared with ALP in the cat may suggest extrahepatic bile duct obstruction or cholangiohepatitis. In addition to the liver, GGT is found in kidney tubules. Therefore tubular damage produces an increase in urine GGT, such as occurs from gentamicin toxicity. Colostrum absorption by puppies causes marked increases in serum GGT within 10 days of age.

Organic analytes increased with cholestasis include bilirubin, predominately the conjugated form, serum bile acids, and cholesterol related to reduced bile flow.

GASTROINTESTINAL SYSTEM/EXOCRINE PANCREAS/PROTEIN

Losses of fluids and nutrients will affect electrolytes and proteins with shifts in sodium, chloride, bicarbonate, potassium, and anion gap. It is expected that both albumin and globulin will be lost with a protein-losing enteropathy or gastrointestinal bleeding. Albumin may also be downregulated in synthesis secondary to hyperglobulinemia. It is thought that increased immunoglobulins cause increased oncotic pressure and secondarily decrease albumin synthesis at the level of gene expression. Increased oncotic pressure may also affect concentrations of hepatocyte nuclear factor-1, which, when decreased, will lower expression of the albumin synthesis gene.

Serum amylase initially rises quickly, then remains elevated for one week (experimental pancreatitis - dogs). There is poor sensitivity and specificity in cats; poor specificity in dogs for pancreatic disease except with moderate to marked elevations (greater than three-fold increase above high end of reference range. The enzyme may be increased in GI disease, renal disease, and hepatobiliary disease.

Serum lipase may be more sensitive than amylase. Corticosteroids have been reported to increase lipase values 2–5-fold above the high end of the reference range. Lipase is normally present in gastric mucosa and pancreas and increased lipase levels can occur in liver disease and abdominal disease.

Endocrine Conditions

Hypoadrenocorticism results in changes particularly with sodium and potassium. Evaluation should include the remaining electrolytes, anion gap, and the urinary system as biochemical data often mimic renal disease with azotemia.

Hyperadrenocorticism and hyperaldosteronism will exhibit severe hypernatremia and hyperchloremia. Additionally, this will produce hypertension and increased circulating volume.

Hypoparathyroid or hyperparathyroid disease will affect calcium and phosphorus levels whose influence is most noted in the kidney. Tremors or generalized weakness may occur clinically from calcium imbalances.

Pituitary disease may arise from genetic, neoplastic, or trauma-induced conditions and result in deficits in ACTH, TSH, LH, GH, FSH from the anterior lobe (pars distalis) and deficits in ADH and oxytocin from the posterior lobe (pars nervosa). Panhypopituitarism may lead to lack of ADH, TSH and ACTH, which results in polydipsia and polyuria, secondary hypothyroidism and atypical hypoadrenocorticism, respectively.

Sample-Related Causes of Instrument Error

Relative to the use of the hand-held refractometer for the measurement of total solids or plasma proteins, it is known that certain non-protein solids, when increased, can cause factitiously elevated results by 0.5 to 1.0 g/dL (5 to 10 g/L). The errors can be produced by such materials as cholesterol, urea, lipoproteins, and glucose. Excess EDTA in plasma samples due to underfilled blood collection tubes may also falsely increase the plasma protein concentration by refractometry. Hemolysis, while it interferes with accurate reading, does not elevate protein concentration by refractometer.

Certain instruments will give a false concentration of sodium and chloride in blood related to hyperproteinemia. Most analyzers determine electrolyte ion activity by ion-selective electrodes and potentiometry methods. However, it matters whether the ion activity is measured in a nondiluted sample (direct potentiometry) versus measured in a diluted sample (indirect potentiometry). The reported electrolyte concentration reflects its activity in the aqueous phase so indirect potentiometry will reduce the electrolyte concentration due to displacement of the water content of the sample volume by increased proteins or lipids and the subsequent altered dilution factor.

References

1. Botha CJ, *et al. J S Afr Vet Assoc.* 2009;80:63.
2. George J. *Vet Clin Pathol.* 2001;30:201.
3. Heiene R, *et al. J Feline Med Surg.* 2009;11:474.
4. Mellor P, *et al. Vet Pathol.* 2008;45:159.
5. Smith J, *et al. J Small Anim Pract.* 2004;45:405.
6. Stockham S, *et al. Fundamentals of Veterinary Clinical Pathology.* 2nd ed. Ames, IA: Blackwell Pub.; 2008.

acvim329

Feline Calcium Oxalate Urolithiasis: What We Do - Can We Do Better?

Joe Bartges, DVM, PhD, DACVIM (SAIM), DACVN
Athens, GA, USA

Formation of uroliths is not a disease, but rather a complication of several disorders. A common denominator of these disorders is that they can from time to time create oversaturation of urine with one or more crystal precursors resulting in formation of crystals.[1] Urolith formation, dissolution, and prevention involves complex physical processes. Major factors include: 1) supersaturation resulting in crystal formation, 2) effects of inhibitors of crystallization and inhibitors of crystal aggregation and growth, 3) crystalloid complexors, 4) effects of promoters of crystal aggregation and growth, and 5) effects of noncrystalline matrix.[1]

Calcium Oxalate Crystal and Urolith Formation

CaOx accounts for 40–50% of feline uroliths. Certain breeds of cats are predisposed to CaOx urolith formation including long-haired cats.[2-4] CaOx urolith formation occurs when urine is oversaturated with calcium (Ca) and oxalate (Ox).[1] In addition to these alterations in activities of ions, large molecular weight proteins occurring in urine influence CaOx formation.[5] Hypercalciuria may occur with hypercalcemia or with normocalcemia. Hypercalciuria is a significant risk factor, but not necessarily the cause of CaOx urolith formation.[6] Hypercalciuria has not been defined in normocalcemic cats with CaOx uroliths but is thought to occur. Although excessive dietary intake of Ca may result in hypercalciuria, studies in human beings refute this. Hypercalciuria may also occur with administration of loop diuretics, glucocorticoids, urinary acidifiers, and vitamin D and/or C. Hypercalcemia is associated with increased risk of CaOx urolith formation. In cats with CaOx uroliths, hypercalcemia was observed in 35% of the cases.[7] Conversely, uroliths developed in 35% of cats with idiopathic hypercalcemia.[8] Hypercalcemia results in increased Ca fractional excretion and hypercalciuria. Oxalic acid forms soluble salts with sodium and potassium ions, but a relatively insoluble salt with Ca ions; therefore, any increased urinary concentration of oxalic acid may promote CaOx formation. Hyperoxaluria has been observed experimentally in kittens consuming Vitamin B_6 deficient diets, but has not been associated with naturally occurring CaOx urolith formation in adult cats[9] and in cats with an inborn error of metabolism.[10] *Oxalobacter formigenes*, an enteric bacterial organism that metabolizes oxalic acid in the gastrointestinal tract, has been shown to increase risk of CaOx formation in dogs that have low enteric colonization.[11] Compared with human beings, urinary Ox appears to have a lesser role in CaOx formation in cats and urinary Ca appears to have a greater role.[12,13] Metabolic acidosis promotes hypercalciuria by promoting bone turnover (release of Ca with buffers from bone), increasing serum ionized Ca concentration resulting in increased urinary Ca excretion, and decreased renal tubular reabsorption of Ca. Consumption of diets by cats supplemented with ammonium chloride has been associated with increased urinary Ca excretion.[14] Aciduria decreases urinary citrate concentration and to impair function of macromolecular protein inhibitors. In cats, magnesium supplementation as magnesium chloride was associated with increased urinary Ca excretion and aciduria, while magnesium supplementation as magnesium oxide was associated with alkaluria and a lesser degree of urinary Ca excretion.[15] Feeding a lower protein diet (35% versus 57% crude protein dry matter basis) was associated with decreased urinary Ca excretion and urinary saturation for CaOx.[16] In a study of healthy cats fed similar diets that differed in only their acidifying or alkalinizing properties, urinary saturation with CaOx was lower when the urine pH was >7.2 and greater when urine pH was <6.5.[17] Decreased urine volume results in increased Ca and oxalic acid saturation and an increased risk for urolith formation. Many cats affected with uroliths have a USG >1.040.[6]

Management of Calcium Oxalate Urocystoliths

Lower Urinary Tract Uroliths

Uroliths may result in clinical signs of lower urinary tract disease including urethral obstruction. Treatment for urethral obstruction involves relieving the obstruction and correcting the metabolic imbalances as quickly as possible.[18] Medical protocols that dissolve CaOx uroliths are not currently available; therefore, uroliths must be removed physically.[19,20] Uroliths that cause clinical signs should be removed with minimally invasive procedures when possible.[19] Detection of urocystoliths does not necessarily warrant surgical intervention; however, obstruction of urine outflow, an increase in size and/or number of calculi, persistent clinical signs, and a lack of response to therapy are indications for calculi removal.[21] Traditional open surgical options are available for treatment of urolithiasis including cystostomy and urethrostomy. There are several minimally-invasive treatment options for retrieval of bladder and urethral stones. These include voiding urohydropropulsion, transurethral cystoscopic stone removal with or without use of laser lithotripsy, and percutaneous cystolithotomy (PCCL).[19,20,22-25]

Upper Urinary Tract Uroliths

Only those nephroliths contributing to outflow obstruction, recurrent infection, pain, and those enlarging to the point of causing renal parenchymal compression should be considered for removal in cats. CaOx nephroliths can be removed by endoscopic nephrolithotomy, pyelolithotomy, or bisection nephrotomy.[26] If hydroureter and hydronephrosis is present, then removal of CaOx ureteronephroliths or placement of a ureteral stent or subcutaneous ureteral bypass device should be performed.[19,26-30] If hydroureter and hydronephrosis are not present with an ureterolith, then medical expulsion therapy may be attempted and if unsuccessful then additional procedures as described should be considered.[19,26-30]

Prevention of Calcium Oxalate Uroliths

A significant number of cats with CaOx uroliths will have recurrence within 2 years of removal if preventative measures are not undertaken.[31] Because the cause(s) of CaOx urolith formation is/are not completely known, no treatment has been shown to be completely effective. If possible, metabolic factors known to increase CaOx risk should be corrected or minimized. Goals of dietary prevention include: 1) reducing urine Ca and Ox concentration, 2) promoting high concentrations and activity of urolith inhibitors, 3) reduce urine acidity, and 4) promote dilute urine. Epidemiologic studies consistently identify acidifying diets as a risk factor for CaOx urolithiasis.[2,3,32] There is a linear relationship between increasing CaOx solubility and urine pH in healthy cats.[17] In 5 cats with hypercalcemia and CaOx uroliths, discontinuation of acidifying diets or urinary acidifiers was associated with normalization of serum Ca concentration.[3] A target urine pH of 6.6 to 7.5 is suggested.[33] Potassium citrate is often included in diets designed for CaOx prevention. In urine, citric acid combines with Ca to form soluble complexes, directly inhibits nucleation of Ca and Ox crystals, and induces alkaluria due to production of bicarbonate.[33] Although reduction of urine Ca and oxalic acid concentrations by restriction of dietary Ca and oxalic acid appears logical, it is not without risk. Reducing consumption of only one of these constituents may increase urinary excretion. Because epidemiologic data in cats suggest marked dietary Ca restriction increases urolith risk, moderate levels of dietary Ca are advised in non-hypercalcemic cats.[33] Low dietary phosphorus is a risk factor for CaOx urolith formation in cats.[34] Reduction in dietary phosphorus may be associated with activation of vitamin D, which in turn promotes intestinal Ca absorption and hypercalciuria. Additionally, phosphate status determines pyrophosphate urinary concentrations, an inhibitor of CaOx urolith formation. Urinary magnesium forms complexes with oxalic acid, reducing the amount of oxalic acid available to form CaOx. Studies in cats associate low dietary magnesium with CaOx risk.[3,32,34-37] Increasing dietary protein from 35% to 57% (DM) increased urine Ca by 35% and decreased urine citrate by 45% in cats and was associated with increased urine saturation for CaOx.[16] However, a case control study showed that higher protein content in cat foods appeared protective against CaOx uroliths.[34] The diet should be adequately fortified with vitamin B_6 because vitamin B_6 deficiency promotes endogenous production and subsequent urinary excretion of oxalic acid.[38] There is no evidence that providing increased vitamin B_6 beyond meeting the nutritional requirement provides a benefit in cats. Certain types of fiber (soy or rice bran) decrease Ca absorption from the gastrointestinal tract, which may decrease urinary Ca excretion. Also, higher fiber diets tend to be less acidifying. In five cats with idiopathic hypercalcemia and CaOx uroliths, feeding a high fiber diet with supplemental potassium citrate resulted in normalization of serum Ca concentrations.[39] Increasing urine volume is a mainstay of preventative therapy for CaOx urolithiasis. By increasing water intake, urinary concentrations of calculogenic minerals are reduced. In addition, larger urine volumes typically hastens urine transit time and voiding frequency, thereby reducing retention time for crystal formation and growth. Feeding a canned food containing at least 70% moisture[40,41] is the most practical means of increasing water intake and lowering CaOx urine saturation. The goal is to dilute urine to a specific gravity of <1.040 in cats.[21,33] Flavoring water, enhancing water access and adding water to dry foods may be used in patients that refuse to eat canned foods. NaCl may be used to increase water intake and several "CaOx preventative" diets contain >1% NaCl (DM basis).[42,43] Increased dietary Na may increase urinary Ca excretion, however, and can contribute to ongoing renal damage in cats with marginal renal function[33] although this has not been a consistent finding.[44,45]

In cats, available diets are avail formulated to reduce Ca and oxalic acid concentrations in urine, promote high concentration and activity of inhibitors of CaOx crystal growth and aggregation in urine, and maintain dilute urine. Consumption of these diets by healthy cats results in production of urine that is undersaturated with CaOx.[37,46] In one study of cats with naturally-occurring CaOx uroliths, consumption of one "Ox preventative diet" resulted in a decrease of urine saturation from the oversaturated state to a metastable state; cats did not reform uroliths.[31] In cats with hypercalcemia and CaOx uroliths, prevention appears to be more successful when feeding a higher fiber containing diet, and administering potassium citrate.[39] Other treatments that have been proposed include vitamin B_6 and hydrochlorothiazide. A 65% decrease in urinary Ca and Ox was reported in healthy cats receiving hydrochlorothiazide.[47] In cats with idiopathic hypercalcemia, alendronate may effectively lower serum Ca concentration to normal. There are, however, no studies evaluating the hard endpoint of urolith recurrence in cats. Serial monitoring of a cat with a history of CaOx urolithiasis should be part of the preventative protocol. Periodically, a complete urinalysis should be performed to monitor urine specific gravity, pH, and presence of CaOx crystalluria. Urine should be dilute, with neutral to alkaline range, and CaOx crystalluria should not be present. Survey abdominal radiography should be performed every 6 months to evaluate for recurrence.

REFERENCES

1. Bartges JW *et al. Vet Clin North Am Small Anim Pract*. 1999;29(1):45.
2. Kirk CA *et al. J Am Vet Med Assoc*. 1995;207(11):1429.
3. Thumchai R *et al. J Am Vet Med Assoc*. 1996;208(4):547.
4. Lekcharoensuk C *et al. J Am Vet Med Assoc*. 2000;217(4):515.
5. Balaji KC *et al. Urol Clinc North Am*. 1997;24(1):1.
6. Bartges JW *et al. Vet Clin North Am Small Anim Pract*. 2004;34(4):969.
7. Bartges JW. In: August JR, ed. 2001;352.
8. Midkiff AM *et al. J Vet Intern Med*. 2000;14(6):619.
9. Bai SC *et al. J Nutr*. 1989;119(7):1020.
10. McKerrell RE *et al. Vet Rec*. 1989;125(2):31.
11. Gnanandarajah JS *et al. Urol Res*. 2012;40(5):467.
12. Dijcker JC *et al. Vet Rec*. 2012.
13. Dijcker JC *et al. Nutr Res Rev*. 2011:1.
14. Ching SV *et al. J Nutr*. 1989;119(6):902.
15. Buffington CA *et al. Am J Vet Res*. 1990;51(12):2025.
16. Passlack N *et al. J Nutr Sci*. 2014;3(e51).
17. Bartges JW *et al. Am J Vet Res*. 2013;74(10):1347.
18. Bartges JW *et al. Vet Clin North Am Small Anim Pract*. 1996;26(2):255.
19. Lulich JP *et al. J Vet Intern Med*. 2016;30(5):1564.
20. Bartges JW. *J Feline Med Surg*. 2016;18(9):712.
21. Lulich JP *et al*. In: Bartges J, Polzin DJ, eds. 2011;687.
22. Defarges A *et al. Compend Contin Educ Vet*. 2013;35(1):E1.
23. Runge JJ *et al. J Am Vet Med Assoc*. 2011;239(3):344.
24. Bartges JW *et al*. In: Bonagura JD, Feldman EC, eds. 2014;905.
25. Berent A. *J Feline Med Surg*. 2014;16(1):51.
26. Defarges A *et al. Compend Contin Educ Vet*. 2013;35(2):E3.
27. Brace MA *et al. Vet Surg*. 2014;43(2):199.

28. Berent AC *et al. J Am Vet Med Assoc.* 2014;244(5):559.
29. Kyles AE *et al. J Am Vet Med Assoc.* 1998;213(8):1150.
30. Mehl ML *et al. Vet Surg.* 2005;34(2):114.
31. Lulich JP *et al. J Am Anim Hosp Assoc.* 2004;40(3):185.
32. Lekcharoensuk C *et al. J Am Vet Med Assoc.* 2000;217(4):520.
33. Kirk CA *et al.* 2003;10.
34. Lekcharoensuk C *et al. J Am Vet Med Assoc.* 2001;219(9):1228.
35. Robertson WG. In: Nordin BEC, Need AG, Morris HA, eds. 1993;249.
36. Lulich JP *et al.* In: Ettinger SJ, Feldman EC, eds. 1999;1747.
37. Smith BHE *et al. J Vet Intern Med.* 1997;11(2):127.
38. Bai SC *et al. J Nutr.* 1989;119(1020).
39. McClain HM *et al. J Am Anim Hosp Assoc.* 1999;35(4):297.
40. Buckley CM *et al. Br J Nutr.* 2011;106 Suppl 1(S128).
41. Deng P *et al. J Anim Sci.* 2014;92(3):1271.
42. Lulich JP *et al. Am J Vet Res.* 2005;66(2):319.
43. Stevenson AE *et al. Res Vet Sci.* 2003;74(2):145.
44. Xu H *et al. J Feline Med Surg.* 2009;11(6):435.
45. Chetboul V *et al. PLoS One.* 2014;9(6):e97862.
46. Smith BH *et al. J Nutr.* 1998;128(12 Suppl):2763S.
47. Hezel A *et al. J Vet Intern Med.* 2006;20(3):741.

acvim330

Significance of Unique and Rare Uroliths

Jody Lulich, DVM, PhD, DACVIM (Small Animal Internal Medicine)
St. Paul, MN, USA

Urolithiasis is a general term referring to the causes and effects of stones anywhere in the urinary tract, and should not be viewed conceptually as a single disease with a single cause but rather as the sequela to multiple interacting physiologic and pathologic processes. Thus, the syndrome of urolithiasis may be defined as the occurrence of familial, congenital, or acquired pathophysiological risk factors that, in combination, progressively increase precipitation of excretory metabolites in urine to form stones (i.e., uroliths). Knowing the composition of uroliths reveals how the body, particularly the kidneys and urinary bladder, maintains homeostasis in the face of disease and metabolic disturbances. Several types of uroliths are commonly recognized in dogs and cats. The following are uncommon uroliths that, once analyzed, provide clues to their potential causes and possible management solutions.

Urethral Plugs in Dogs

Crystalline-matrix plugs are a common cause of urethral obstruction in male cats. Although not considered a cause for urethral obstruction in the dog, the Minnesota Urolith Center has analyzed plugs from over 100 dogs in the last decade. In canine urethral plugs, the crystalline material was struvite in 86%, not struvite in 8%, and non-mineral in 6% in which it was composed of organic matrix (i.e., protein and inflammatory cells) in the absence of mineral. Unlike most struvite stones in dogs in which urinary tract infection with urease-elaborating bacteria is a prerequisite, of the dogs evaluated at our facility, urinary tract infection was absent. Although the Minnesota Urolith Center has diagnosed urethral plugs in over 10 breeds of dog, 70% of urethral plugs are from male Pugs. Close evaluation of Pugs with this disease has not revealed an underlying cause for plug or struvite's incorporation into the plug. The clinical significance of this disease is that preventative with struvitolytic diets should be considered in male Pug dogs with persistent struvite crystalluria to avoid life-threatening obstructions, and that urethral plugs should be included as a diagnostic possibility for dogs, especially male Pugs, with urethral obstruction.

Figure 1. Urethral plug protruding from penis of a 2-yr-old pug

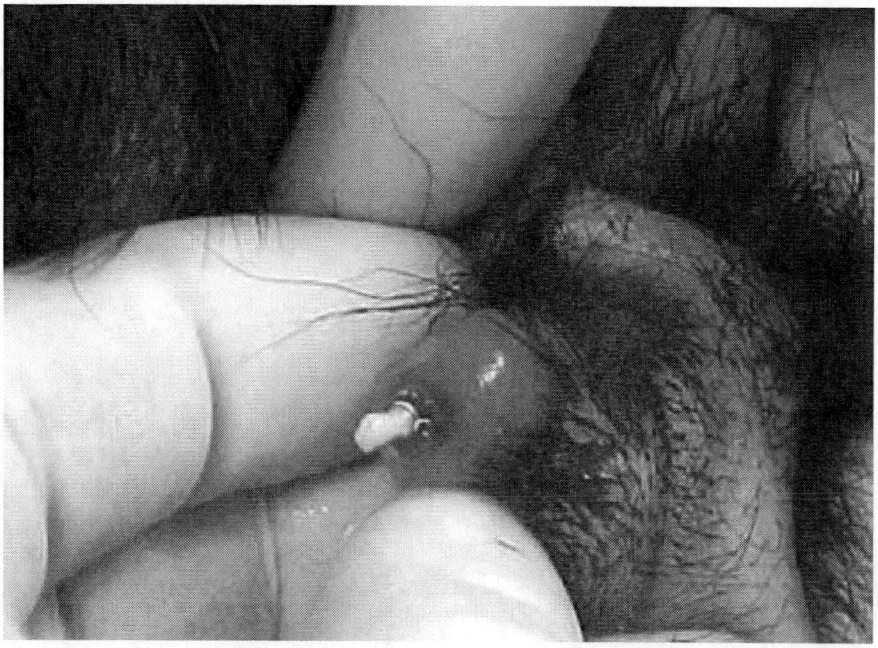

ECTOPIC BONE

Ectopic bone is not related to pathological calcification (a lesion in which calcium salts, usually in the form of calcium phosphate, deposit abnormally in soft tissues). Ectopic bone is bone in a tissue where it is not normally found. Ectopic bone can develop in two ways. Heterotopic bone is essentially normal osseous tissue that is believed to originate from embryonic cells that persist in tissues during development and maturation. Heterotopic ossification has been seen in several genetic disorders in humans caused by an activating mutation (617G>A; R206H) encoding activin receptor-like kinase 2, a bone morphogenic protein (BMP) type 1 receptor. The pathophysiology is not well understood, but excessive BMP signaling in an inflammatory environment together with increased expression of other osteogenic cytokines has been shown to result in heterotopic ossification in animal models. The mechanism underlying this mechanism in cats is not known.

The other mechanism for bone formation at abnormal sites is osseous metaplasia. Osseous metaplasia commonly develops from other types of connective tissue (i.e., fibrous connective tissue), at sites of chronic inflammation (trauma, neurogenic or surgical insults) and in florid reparative processes. Bone that dislodges from the urothelium looks and acts like a urolith. In most cases, however, the bone remains attached to the mucosal lining and serves as a nidus for mineral deposition.

Figure 2. Survey radiograph with mineralized plug in urethra (between arrows)

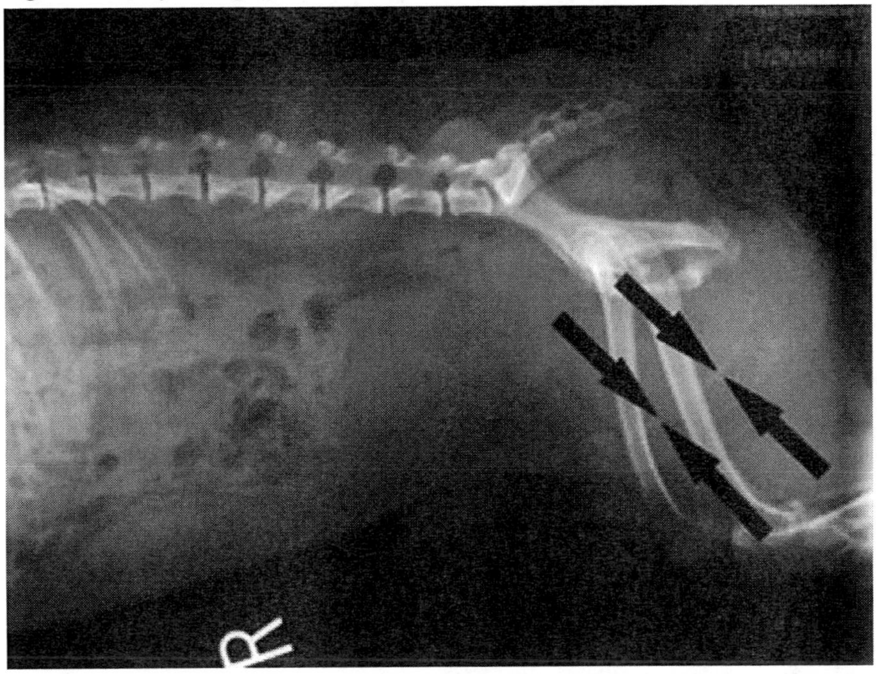

Figure 3. A calcium oxalate urolith from a 9-year-old domestic shorthair cat (A). The underside of the stone (B) has a flattened concave surface with a central indentation (umbilication; arrow) consistent with an area of attachment to the urothelium.

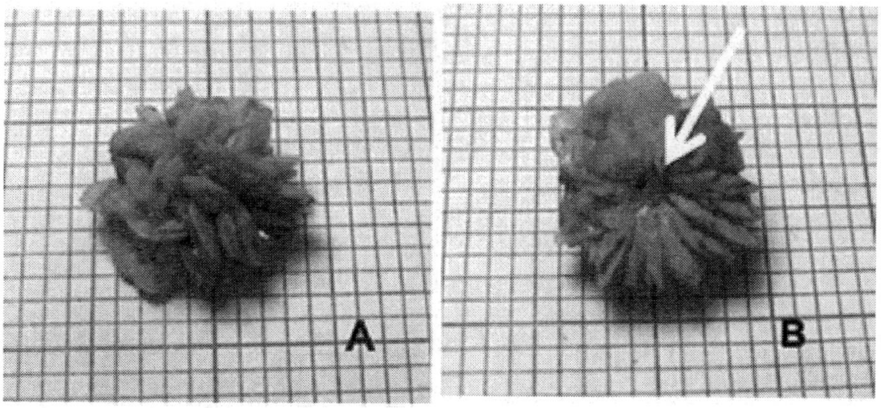

Figure 4. Two-dimensional slice (panel A) of the microcomputer tomographic reconstruction of the stone in Figure 1. The nidus of the stone (N) has a different density than the mineral layers surrounding it. Optical crystallographic analysis of the layers were calcium phosphate in the nidus (N), calcium oxalate monohydrate in the middle layer and calcium oxalate dihydrate in the outer stone. Panel B is an undecalcified section of the stone in panel A. Sections were stained with Stevenel's blue and Van Gieson's picro fuchsin;[8] vital bone stains red with variations in intensity depending on the maturity of the bone; non-vital bone and osteoid stain bright green. Panel C is a higher magnification of panel B showing the woven appearance of the nidus (N) that is consistent with bone, and the intersection (arrow head) of the nidus with the calcium oxalate (CaOx) deposited over the nidus. The mineral portion of the urolith is not vital and therefore does not retain vital stains.

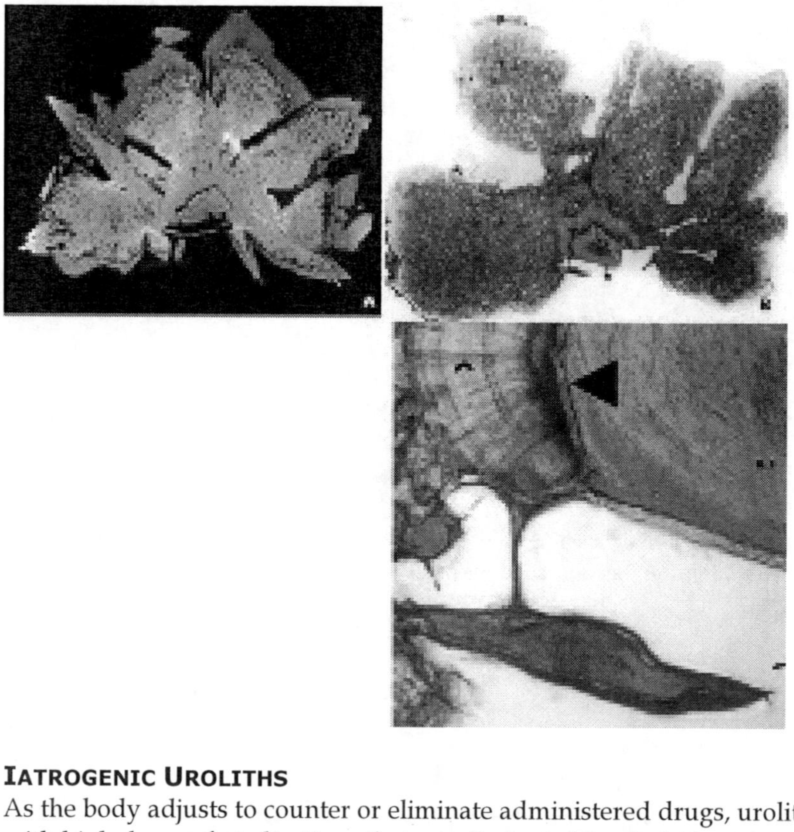

Iatrogenic Uroliths

As the body adjusts to counter or eliminate administered drugs, uroliths can form. This is commonly seen with high doses of medications that are eliminated by the kidneys or medications that alter metabolic pathways where end products are eliminated by the kidneys. Table 1 lists some of the iatrogenic uroliths evaluated at the Minnesota urolith Center.

Table 1. Iatrogenic uroliths: potential causes and potential solutions

Urolith type	Potential etiology	Management option
Ciprofloxacin	High-dose ciprofloxacin to counter variable intestinal absorption in the dog	Stop ciprofloxacin, increase fluid intake, & alkalinize urine. In some cases the stone may dissolve without surgery.
Sulfonamide	High-dose sulfa antibiotic administration	Reduce dose rate (especially in renal insufficiency), alkalinize urine, increase fluid intake, and use triple-sulfonamide (incorporating the law of independent solubility). Dissolution is unknown.
Xanthine associated with allopurinol administration	Allopurinol competitively inhibits xanthine oxidase. The disease is worse with high purine/protein diet consumption.	Stop allopurinol, increase fluid intake, alkalinize urine, reduce dietary purine/protein intake. Stones dissolve without surgery.
Suture nidus	Heterogeneous nucleation of minerals over suture. Suture exposed to urine lowers the supersaturation required for mineral precipitation.	Consider nonsurgical removal of uroliths. If not feasible, avoid suture entering the urinary space.

REFERENCES

1. Stiller AT, Lulich JP, Furrow E. Urethral plugs in dogs. *J Vet Intern Med*. 2014;28:324–330.
2. Tsuchida K, *et al*. Origin and therapeutic strategies for ectopic bone formation in skeletal muscle. *Human Genet Embryol*. 2013;3:110.

acvim331

USMI or Urethral Dyssynergia in Male Dogs: Are We Misdiagnosing Them?

Julie K. Byron, DVM, DACVIM
Columbus, OH, USA

Introduction

For the veterinary practitioner, the presentation of a male dog with urinary incontinence can be perplexing. The first inclination may be to use phenylpropanolamine to address urethral sphincter mechanism incompetence (USMI). However, response to treatment for USMI has been frustratingly poor and this may be due to misdiagnosis of a disorder of voiding with overflow incontinence rather than a disorder of storage as in classic USMI. Male dogs presenting for incontinence must be evaluated through history, physical examination, and observation of voiding in order to determine if outlet obstruction is the primary problem.

Functional urethral or bladder outlet obstruction has been recognized in people from a variety of causes including benign prostatic hyperplasia, pediatric voiding dysfunction, and detrusor hypocontractility, as well as a primary functional urethral obstruction in women. These disorders are treated with a variety of alpha-antagonists, PDE-5 inhibitors, skeletal muscle relaxants and biofeedback systems. Since the majority of veterinary patients with functional obstruction are males, and often neutered, direct extrapolation of underlying pathophysiology and treatment effect cannot be made from human medicine. This has complicated our ability to predict the efficacy or treatment response to various therapies in our patients and has led to challenges in diagnosis and management.

In veterinary patients, functional urethral obstruction, or detrusor-urethral dyssynergy (DUD), is suspected to arise from an abnormality in the reflex arc which allows the urethral sphincter to relax at the initiation of detrusor contraction and urination. The lesion is thought to be in the reticulospinal tract, Onuf's nucleus, or the caudal mesenteric ganglion, and it is possible that the lesion involves the loss of inhibitory signals to the pudendal and hypogastric nerves.[1] It is unknown if there is a more local lesion to the nerves, the neuromuscular junction, or the smooth or striated urethral sphincters[2] or if there is a dynamic outflow obstruction at the bladder neck, as has been described in people[3]. Unlike the 'upper motor neuron bladder' seen in animals with thoracolumbar intervertebral disc disease and other spinal cord lesions, these animals typically have an otherwise normal neurologic examination and have the ability to consciously initiate a detrusor contraction.

Diagnosis

The disorder affects primarily middle-aged, large and giant breed male dogs, although female dogs and cats can be affected. One case series of 22 dogs reported a mean age of 4.9 years.[4] Clinical signs are similar to those of mechanical obstruction. The animal often postures to urinate and is able to produce a urine stream that quickly becomes attenuated or stops completely. The animal may continue to posture to urinate or make several attempts without fully emptying the bladder. The presence of large amounts of residual urine typically leads to overflow incontinence and may be mistaken for USMI. This leakage can occur because the hypertonicity of the involved sphincter is likely dynamic and is triggered by the act of voiding. In chronic cases, bladder over-distension and subsequent detrusor hypofunction may develop. Unlike animals with mechanical obstruction, these dogs are typically easy to catheterize.

Presumptive diagnosis of DUD is often made by observing the dog urinate with a typical interrupted pattern, documentation of a large residual urine volume, easy passage of a urinary catheter, and ruling out of a mechanical obstruction. Normal residual urine volumes in 48 normal dogs were reported to be 0.1–3.4 ml/kg body weight with a mean of 0.2 ml/kg.[5] The author uses <1.0 ml/kg as a general guideline. Ultrasonography is recommended to assess the ureters and renal pelves for dilation secondary to chronic obstruction and ureterorenal reflux of urine. Additional diagnostics, including contrast urethrography, urethroscopy may be necessary to verify the diagnosis and rule out anatomic abnormalities in patients who fail to respond to medical therapy.

In people, voiding videofluoroscopy and measurement of bladder pressures with urine flow rates are considered standard diagnostic testing for bladder outlet obstruction. This can assist in determining if the obstruction is located at the bladder neck, internal, or external urethral sphincter. Due to the need for sedation in our patients when performing urodynamic studies, the utility of these studies has not been evaluated in dogs with signs of functional obstruction. In addition, the dynamic nature of this obstruction may lead to normal cystourethrography and normal urethral pressure profiles on urodynamic evaluation unless the animal is actively voiding during the study. Contrast urethrography may be normal or reveal areas of narrowing of the urethra (urethrospasm). Further investigation of the use of EMG studies and uroflow parameters needs to be performed to determine the feasibility and practicality of these measures in veterinary patients.

TREATMENT

At this time, very few studies have been performed evaluating the efficacy of any treatment for DUD in dogs. Treatment of the dyssynergic urethral sphincter mechanism (since we are unsure if it is the smooth muscle or striated muscle component that is a fault) generally consists of α-adrenergic blockade. Prazosin, an α_1-specific antagonist with demonstrated effects on both the internal and external urethral sphincter[6] and tamsulosin, which is specific for the α_{1A} subtype found in the internal urethral sphincter, have been successful in these dogs. Alpha blockade carries with it the side effect of systemic hypotension, particularly with the less subtype specific prazosin. Based on this risk, as well as the convenience of once a day vs. three times a day dosing, and a large safety margin, the author has begun to use tamsulosin as first-line treatment in these patients. Many dogs will require additional therapy to address striated muscle involvement. Benzodiazepines, such as diazepam, or other skeletal muscle relaxants, including acepromazine and methocarbamol, may be more effective if the external urethral sphincter is involved. Diazepam is typically administered 30 minutes before voiding to decrease external urethral sphincter pressure. Dantrolene and baclofen have been used in the past as skeletal muscle relaxants; however, the potential for adverse effects has decreased their use in veterinary patients.[6,7] In severe and refractory cases, intermittent sterile catheterization by the owner at home may be necessary. Medical therapy of associated bladder atony should only be started after adequate relief of the functional urethral obstruction has been reached (Table 1). Close monitoring of these patients for residual urine volume and UTI is needed to assess efficacy of treatment and prevent complications.

Table 1. Drugs frequently used to treat disorders of emptying[8]

Drug	Mechanism	Dose	Side effects/Caution
Prazosin	α_1-antagonism, smooth muscle relaxant	1 mg/animal < 15 kg 2 mg/animal > 15 kg q8–12h PO	Hypotension, weakness, syncope, GI upset/ renal disease, cardiac disease
Tamsulosin*	α_{1A}-antagonism, smooth muscle relaxant	0.01–0.2 mg/kg PO q24h	Hypotension
Phenoxybenzamine	Non-specific α-antagonism, smooth muscle relaxant	Dog: 0.25 mg/kg q8–12h Cat: 1.25–7.5 mg/cat q 8–12 h PO	Hypotension, tachycardia, miosis/ first dose hypotension
Acepromazine	Non-specific α-antagonism Skeletal muscle relaxant Anxiolytic	0.5–2.2 mg/k q6–8h PO	Sedation, hypotension
Diazepam	Skeletal muscle relaxant Anxiolytic	Dog: 0.5–2 mg/kg q8h PO or 30 min prior to voiding	Sedation, ataxia/liver dysfunction Do not use in cats

Methocarbamol	Skeletal muscle relaxant	22–44 mg/kg q8h PO	CNS sedation, weakness, hypersalivation
Baclofen*	Skeletal muscle relaxant	Dog: 1–2 mg/kg q8h PO	Weakness, GI upset/Do not use in cats
Dantrolene	Skeletal muscle relaxant	Dog: 1–5 mg/kg q8h PO Cat: 1–2 mg/kg q8h PO	Weakness/liver disease
Bethanechol	Parasympathomimetic	Dog: 2.5–15 mg/dog q8h PO Cat: 1.25–5 mg/cat q8h PO	Diarrhea/GI or urethral obstruction
Cisapride	Prokinetic	Dog:0.1–0.5 mg/kg q8–12h PO Cat: 2.5–5 mg/cat q8–12h PO	Ataxia, GI upset/GI obstruction

*From Lane IF, Westropp JL, Urinary incontinence and micturition disorders: pharmacologic management. In: *Kirk's Current Veterinary Therapy.* 14th ed., Elsevier:St. Louis, MO. 2009:955–959.

Anecdotally, some dogs have shown worsening of signs or relapse of previous response when stressed. The relationship of stress and voiding complications has not been investigated in these dogs, and may provide some assistance for sever cases. For these patients, the addition of an anxiolytic may be necessary to maintain normal voiding.

Prognosis for recovery of normal voiding is good, based on the limited available information, but most dogs will require life-long therapy for DUD. Attempts to taper medications to the lowest effective dose may be hampered by relapse of clinical signs after months of normal voiding.[4] Prognosis appears to be worse in patients with bladder atony or UTI secondary to urine retention. Anecdotally, urethral stenting may improve clinical signs; however, this is considered a salvage procedure with several potential complications, and indicated only in the most refractory of cases.

References

1. de Groat WC, Fraser MO, Yoshiyama M *et al*. Neural control of the urethra. *Scand J Urol Nephrol Suppl*. 2001:35–43; discussion 106–125.
2. Gookin JL, Bunch SE. Detrusor-striated sphincter dyssynergia in a dog. *J Vet Intern Med*. 1996;10:339–344.
3. Lin CD, Kuo HC, Yang SS. Diagnosis and management of bladder outlet obstruction in women. *Low Urin Tract Symptoms*. 2016;8:30–37.
4. Espineira MMD, Viehoff FW, Nickel RF. Idiopathic detrusor urethral dyssynergia in dogs: a retrospective analysis of 22 cases. *J Small Anim Pract*. 1998;39:264–270.
5. Atalan G, Barr FJ, Holt PE. Frequency of urination and ultrasonographic estimation of residual urine in normal and dysuric dogs. *Res Vet Sci*. 1999;67:295–299.
6. Fischer JR, Lane IF, Cribb AE. Urethral pressure profile and hemodynamic effects of phenoxybenzamine and prazosin in non-sedated male beagle dogs. *Can J Vet Res*. 2003;67:30–38.
7. Khorzad R, Lee JA, Whelan M *et al*. Baclofen toxicosis in dogs and cats: 145 cases (2004–2010). *J Am Vet Med Assoc*. 2012;241:1059–1064.
8. Papich MG. *Saunders Handbook of Veterinary Drugs, Small and Large Animal*. 3rd ed. Elsevier. 2010.

acvim332

Diagnosis of Canine Transitional Cell Carcinoma: A New Era

Shelly Vaden[1], DVM, PhD, DACVIM; Michael Nolan[2], DVM, PhD, DACVR; Claire Wiley[3], VMD
Raleigh, NC, USA

INTRODUCTION

Estimates suggest that bladder cancer will newly develop in over 20,000 dogs in the United States every year, and the prevalence appears to be on the rise.[1,2] When traditional methods are used to evaluate dogs with lower urinary tract signs, 90% of the tumors that are diagnosed are intermediate- to high-grade invasive transitional cell/urothelial carcinomas (TCC/UC); superficial, low-grade tumors are uncommon. Most TCC/UCs are located in the trigonal region of the bladder; urethral or prostatic involvement is common. By the time a diagnosis is established, 16% of dogs already have nodal or distant metastasis, a number that increases to 50% by the time of death.[2,3] Regional nodes and lung are the most common site of metastasis but the tumor can spread to a number of sites including bone, liver, kidney, spleen and skin.[2]

The mean age at diagnosis is 9 to 11 years but TCC/UC can develop at an earlier age.[2] Females are reported to be at 1.71 to 1.95 greater risk of TCC/UC than males, although this sex predilection may be less pronounced in high-risk breeds.[2] Several breed are at increased risk for TCC/UC, including Scottish terriers, Shetland sheepdogs, West Highland white terriers, Beagles, and American Eskimo dogs.[2] Other risk factors include obesity, urban dwelling, and exposure to lawn chemicals or older-generation flea control products.[4,5] Scottish terriers that consumed vegetables at least three times a week had a 70% reduction in risk of developing TCC/UC.[6]

Clinical signs are non-specific and indicative of lower urinary tract disease (hematuria, dysuria and stranguria, urinary incontinence). Late-stage signs may be attributed to urethral or ureteral obstruction. Urinary tract infection (UTI) is common in dogs with TCC/UC, with 55% having at least one positive culture.[7] TCC/UC should be suspected in dogs that fail to have complete resolution of clinical signs of UTI during treatment with appropriate antibiotic therapy, particularly if they are middle aged or older and belong to an at-risk breed.

CONVENTIONAL DIAGNOSIS OF TCC/UC

A common route to diagnosis of a TCC/UC is one in which the dog is first taken to a veterinarian for evaluation of lower urinary tract signs. In many cases the dog is treated based, on the assumption that there is a non-malignant cause, often with repeated courses of antibiotics for presumptive UTI. Sometimes anti-inflammatory medications are also administered. While this treatment approach may provide temporary relief of the symptoms, the underlying TCC/UC is progressing. The delay in diagnosis also increases the possibility that the tumor will have metastasized by the time of diagnosis. Eventually the dog is specifically evaluated for the presence of a TCC/UC via urine cytology, abdominal ultrasound, and/or cystoscopy. When a mass is detected, a biopsy serves to confirm the diagnosis of TCC/UC. Further imaging and evaluation of local lymph nodes may be performed to assess the spread of the disease. A delay in diagnosis is likely correlated with advancing clinical and pathologic stage of disease. The high predominance of advanced tumors detected by conventional means in many dogs may reflect the prolonged time taken to diagnose the tumors.

TYPICAL APPROACH TO TREATMENT

Treatment of canine TCC/UC most commonly includes the use of chemotherapy, NSAIDs, or combinations of these drugs. Where single agent therapy is used, the proportion of dogs entering remission is generally low (<20%), although this is increased to 35–50% with combined chemotherapy and cyclooxygenase inhibitors. While less common than drug-based intervention, surgery and radiation therapy are also used where appropriate.[3] Dogs treated with single agent NSAID therapy have a median survival time of approximately 6–7 months, whereas a combination of cytotoxic chemotherapy (typically mitoxantrone) and an NSAID yields median survival times closer to 10 months. Recent data suggest that the addition of full-course intensity-modulated and image-guided radiation therapy is associated with a 60% response rate, and a median survival time of >21 months.[8]

A Novel Test for Diagnosis of TCC/UC

In two recent independent studies, performed by research teams at North Carolina State University (NC State)[9] and the National Institutes of Health (NIH),[10] a single mutation (T>A) (generally heterozygous) in exon 15 of the canine BRAF gene was detected in pathology verified tumor biopsy specimens of canine TCC/UC. The discovery of the same mutation independently by two groups using two different approaches provides cross validation of the data. The result of this single mutation is one amino change (valine to glutamic acid) in the *BRAF* protein, resulting in a mutated protein with increased kinase activity. This change signals the cells to proliferate, leading to the development of a tumor. Over the following two years we showed that the *BRAF* mutation was not present in non-neoplastic bladder tissues, including inflammatory bladder tissue and polyps, or in urine from dogs with numerous other canine cancers. In a follow up study a rapid and highly sensitive test to detect the presence of this mutation in cells shed into the urine was developed,[11] which led to the launch of the CADET^{SM} *BRAF* Mutation Detection Assay. Importantly, the technology behind the new test means that it is not adversely affected by the presence of bacteria or blood in the urine and thus provides a highly effective means to detect the presence of malignant TCC/UC cells even in such specimens. Assessment of over 300 pathology verified specimens of TCC showed that 85% are associated with a BRAF mutation. Analysis of urine samples from over 100 biopsy verified specimens from the caseload at NC State and surrounding practices indicated a 100% concordance between the presence of a *BRAF* mutation detected in free-catch urine and the mutation in the mass. Presence of a *BRAF* mutation has not been detected in hundreds of biopsy and urine specimens from dogs that were shown to not have a TCC/UC (e.g., polyps, inflammation etc.). These data show that the presence of a BRAF mutation, detected with the CADET^{SM} assay is >99% is specific to TCC/UC. The technical sensitivity of the assay is such that it is also quantitative, with the level of mutation generally correlating with the extent of disease. With advanced disease the higher levels of *BRAF* mutated cells detected are associated with visible presence of a mass, or bladder wall thickening on ultrasound. Preliminary use of the test for monitoring of *BRAF*-positive tumors during treatment suggests that some TCC/UCs respond quickly and effectively to specific therapeutic approaches while others do not. Further work is needed to inform why there is such variation.

When the CADET^{SM} *BRAF* result is positive, indicating the presence of a TCC in the urinary tract, there are additional considerations to be made, depending on the clinical presentation of the dog. In all cases the dog should have imaging studies and/or cystoscopy to determine the location and extent of any visible disease. Dogs that are symptomatic and have a higher level of *BRAF* mutation generally have visible disease. Dogs with a low level of mutation may or may not have visible disease, depending on the size and location of an emerging mass, and the resolution of imaging equipment used. Treatment should be considered in symptomatic dogs, given the high likelihood that the mass/lesion is a TCC/UC and not merely a polyp or benign inflammation.

When a CADET^{SM} *BRAF* mutation is undetected, it could mean one of three situations: 1) The patient has a *BRAF* mutation associated TCC/UC, but the level of mutation is below the threshold of detection due to a low number of cells shed into the urine; 2) the patient does not have a TCC/UC; or 3) the patient is one of the 15% of dogs that has a TCC/UC not associated with a *BRAF* mutation. The threshold of detection is based on the number of cells recovered from the urine. In over 96% of cases analyzed to date the threshold of detection has been <1%, meaning that the cell number in the urine was sufficient for the assay to detect the presence of the mutation in a least 1% of the BRAF alleles. The few cases with a threshold of detection of >1% are those that presented with no symptoms and shed very few cells into the urine. All these cases were subsequently shown to be non-TCCs. To address all three options, the NC State team has been developing a parallel assay based on a secondary signature specific to TCC/UC. The sensitivity of the combined assay is expected to be greater than 85%.

Has the Game Changed?

The CADET[SM] *BRAF* Mutation Detection Assay is able to detect the presence of a TCC/UC in urine, even if the urine contains blood, bacteria, inflammatory cells etc. In addition, due to the high level of technical sensitivity to detection of a mutation, the assay is able to detect very early disease, before clinical signs are present. Ongoing random screening of dogs age 6 years and over from some of the higher risk breeds is revealing the presence of the *BRAF* mutation in cases 3–4 months (and sometimes longer) before clinical signs are subsequently reported. This provides exciting new opportunities for earlier detection that will most likely allow us to better characterize the disease. The test also gives the internal medicine/oncology community the opportunity to consider how to treat such dogs, or if they should be treated at all during these early phases. Imaging of these asymptomatic dogs has sometimes indicated the presence of very small lesions. Studies are needed to determine if these small lesions could be effectively treated using modalities other than what is considered standard today or if standard treatment is even more effective against these small lesions. These are all areas that will need to be evaluated in this new era of early detection.

References

1. Fulkerson CM *et al. Vet J*. 2015;205:217.
2. Knapp DW *et al. ILAR J*. 2014;55:100.
3. Knapp D *et al. Urol Oncol*. 2000;5:47–59.
4. Glickman LT *et al. J Toxicol Environ Health*. 1989;28:407.
5. Glickman LT *et al. J Am Vet Med Assoc*. 2004. 224(8):1290–1297.
6. Raghavan M *et al. J Am Vet Med Assoc*. 2004;225:389.
7. Budreckis DM *et al. J Vet Intern Med*. 2015; 29:828.
8. Nolan MW *et al. J Vet Intern Med*. 2012; 26:987.
9. Mochizuki H *et al. PLoS One*. 2015. 10(6): e0129534.
10. Decker B *et al. Mol Cancer Res*. 2015. 13(6):993–1002.
11. Mochizuki H *et al. PLoS One*. 2015. 10(12):e0144170.

acvim333

Fluid Therapy for Acute Kidney Injury: Lessons from Animal Models and Humans

J.D. Foster, VMD, DACVIM (SAIM)
Washington, DC, USA

Introduction

Derangements of sodium and water balance are commonly observed consequences of kidney failure which must be considered when creating the prescription for intravenous fluid therapy. Common consequences of disordered renal sodium and water homeostasis include dysnatremias, dehydration, and hypervolemia. These complications may be exacerbated by inappropriate type of fluid administered, or if an excessive amount is delivered.

Goal of Intravenous Fluid Therapy

Fluid therapy is often used to correct dehydration, hypovolemia, acid-base and electrolyte disturbances while it is generally assumed that nutritional requirements will be provided separately.[1] It has been commonly believed that fluid therapy is essential to maintaining cardiac output, renal perfusion, and systemic oxygen delivery as well as promoting diuresis.[2] Situations where cardiac preload is reduced often result in a "prerenal" azotemia, characterized by decreased renal plasma flow and glomerular filtration rate (GFR). Intravenous fluids can increase cardiac preload and stroke volume, thus improving systemic blood pressure and rapidly correcting the azotemia. The kidney functions to conserve >97% of the filtered sodium load; altered sodium reabsorption occurs in acute kidney injury (AKI) and the kidneys can easily be overwhelmed with chronic administration of high sodium fluids, eventually leading to salt and water accumulation.[3]

Choosing an Appropriate Fluid

Isotonic replacement crystalloids are the most commonly used intravenous fluid for volume replacement as well as maintenance of water balance in hospitalized patients. Several studies in humans have evaluated the chloride content of these fluids - comparing chloride-liberal fluids such as 0.9% NaCl to "balanced" or chloride-restricted fluids such as LRS, Plasmalyte, etc. Experimental studies have suggested that administration of chloride-rich solutions results in a high intratubular chloride concentration, which stimulates tubuloglomerular feedback. Subsequently, decreased renal blood flow and glomerular filtration rate may occur, an effect that has been documented in healthy humans receiving 0.9% NaCl.[4] Although there are some conflicting studies, numerous randomized prospective studies of hospitalized humans have shown chloride-liberal fluids to be associated with an increased risk of AKI and need for renal replacement therapy compared to those who received chloride-restricted fluids.[5-7] When administering both isotonic and hypotonic crystalloids to meet the patient's water requirements, sodium and chloride may be given in excess of their daily requirement.[1] The initial fluid choice for most veterinary patients with AKI should be an isotonic crystalloid. Correction of hypovolemia should be performed rapidly and dehydration should be corrected over the first 6–24 hours.[8] After fluid deficits have been replaced, the administered fluid should be transitioned to a maintenance fluid with a lower concentration of NaCl.

There is much controversy surrounding the administration of synthetic colloids in humans. The effects of these fluids on coagulation are well known and should not be discounted. However, recent meta-analysis of numerous studies has shown hydroxyethylstarch (HES) administration in critically ill humans is associated with increased risk of mortality, increased risk of AKI, and increased need of dialysis when compared to crystalloid or albumin administration.[9] Recently, there have been questions regarding the safety of HES in veterinary patients.[10,11] A retrospective cohort study found that Pentaspan (10% 250/0.5/5:1) administration to critically ill dogs was associated with a 4.87 odds ratio for developing AKI and a 1.69 odds ratio for death.[12] This relationship was shown to be dose dependent, where an increase of each ml/kg/h of HES increased the odds of developing AKI or death by 12.15%. Another retrospective study evaluating a less concentrated form of HES (Vetstarch 6% 130/0.4/6:1) did not find its administration increased SCr in dogs.[13] A third retrospective study of the same type of HES (Voluven) did not find its administration resulted in increased incidence of AKI compared to other fluids; however, the number of days of HES administration was associated with increasing risk of worsening grade of AKI.[14] It should be noted that all synthetic starch solutions have been found to increase the risk of AKI in people. Until data show HES administration resulting in improved survival in humans and animals with AKI, its administration should be performed cautiously, or avoided if possible.

Delivering an Appropriate Rate of Fluids

The rate of intravenous fluid administration should be sufficient to provide maintenance water requirement to meet metabolic water demands and insensible losses, match ongoing sensible losses, and correct volume deficits that may be present. Most formulae for canine and feline maintenance fluid requirements are based on studies in healthy animals that are kept in a cage. These formula produce varying results; one author has proposed daily water requirement of $97*(\text{body weight [kg]})^{0.655}$ for sedentary or sick animals.[1] However, providing this volume of fluid may result in delivery of electrolytes (particularly sodium and chloride) 15–25 fold greater than daily requirement recommended by the National Research Council. The kidneys must respond to this salt load by increasing the rate of sodium excretion. Patients with AKI may be unable to excrete the excessive sodium to maintain a neutral balance and may therefore develop hypervolemia, hypertension, and edema secondary to sodium retention.

Oliguria may be an appropriate response to decreased renal perfusion ("prerenal" azotemia) and may be treated with a bolus of intravenous fluids. However, oliguria may also occur regardless of hydration and renal perfusion status in patients with intrinsic renal injury. A study in oliguric humans showed that a bolus of intravenous fluids failed to increase cardiac stroke volume in most (73%) patients and only half demonstrated increased urine output.[15] Decreased urine output (UOP) should not indicate a need for increasing the rate of intravenous fluid delivery, unless the patient has developed hypovolemia or insufficient renal perfusion.

Prescribing fluid rates exceeding the patient's water requirement is commonly performed as part of "aggressive fluid therapy" or to induce diuresis in patients with kidney failure. This clinical approach is not evidence based, nor is it supported by renal physiology. Diuresis, defined as the promotion of urine production, does not necessarily involve improvement in GFR; evidenced by most diuretics causing increased urine production via decreased renal tubular sodium reabsorption and a concurrent decrease in glomerular filtration.[16] Fluids should be administered to maintain renal perfusion; which may increase urine output through decreased activation of the renin-angiotensin-aldosterone system. There is no evidence to suggest administration of fluids above this volume causes any beneficial effects in renal function. A study of anesthetized healthy dogs showed no improvement in GFR with the administration of 10 ml/kg/h of lactated Ringer's solution.[17] Similarly, 6 ml/kg/h of intravenous 0.9% NaCl produced no change in GFR in healthy cats.[18] Pharmacologically inducing polyuria, through intravenous fluids or diuretics, has not shown to improve mortality and need for dialysis in humans and dogs with AKI.[19,20] Actually, conservative rates of fluid administration have been associated with lower incidence of AKI in people compared to those who received liberal volumes.[21]

Potential Consequences of Intravenous Fluid Therapy

Experimental studies in dogs have demonstrated that increased venous pressure results in decreased renal tubular sodium and water excretion and therefore decreased urine production.[22,23] The high hydrostatic pressure of the renal vein is actually transmitted back into the distal tubule and will result in approximately 25% decreased renal plasma flow and GFR.[23,24] Mouse models have shown that renal venous hypertension (caused by renal vein clamping) produces renal congestion and more severe tubular injury than clamping of the renal artery.[25] The renal capsule prevents expansion of the interstitium during hypervolemic states, such that venous congestion increases intracapsular pressure and reduces venous, lymphatic, and arterial blood flow.[26]

Therefore, administering fluids in excess of the patient's need may result in hypervolemia and the sequelae of increased renal venous pressure, renal congestion, decreased renal plasma flow and GFR, and decreased UOP resulting from decreased renal excretion of sodium and water. Such a patient would likely develop oligoanuria and worsening renal failure. While many clinicians may inappropriately respond to this scenario by increasing the rate and volume of intravenous fluid administration, the most appropriate action would be to correct the excess volume, which would help normalize the increased renal venous pressure. This may be accomplished by discontinuation of intravenous fluids until normovolemia is achieved, or the cautious use of diuretic therapy. Injured kidneys may not have the capacity to respond normally to diuretics.

Aggressive fluid therapy in people has been demonstrated to be associated with increased risk of development of AKI, less recovery of renal recovery following AKI, increased need for dialysis, increased risk of multiple organ dysfunction, and higher mortality rates.[27-31] Recently, a study of critically ill dogs demonstrated an increased risk of death associated with fluid overload.[32] An appropriate fluid prescription considers the patient's water and electrolyte needs, and a fluid is chosen which minimizes excessive NaCl administration, while being delivered at a rate that meets the patient's water demand.

References

1. Hansen B, Vigani A. Maintenance fluid therapy: isotonic versus hypotonic solutions. *Vet Clin North Am Small Anim Pract*. 2017;47(2):383–95.
2. Prowle JR, Bellomo R. Fluid administration and the kidney. *Curr Opin Crit Care*. 2013;19(4):308–14.
3. Prowle JR, Echeverri JE, Ligabo EV, Ronco C, Bellomo R. Fluid balance and acute kidney injury. *Nat Rev Nephrol*. 2010;6(2):107–15.
4. Chowdhury AH, Cox EF, Francis ST, Lobo DN. A randomized, controlled, double-blind crossover study on the effects of 2-L infusions of 0.9% saline and plasma-lyte® 148 on renal blood flow velocity and renal cortical tissue perfusion in healthy volunteers. *Ann Surg*. 2012;256(1):18–24.
5. Yunos NM, Bellomo R, Hegarty C, Story D, Ho L, Bailey M. Association between a chloride-liberal vs. chloride-restrictive intravenous fluid administration strategy and kidney injury in critically ill adults. *JAMA*. 2012;308(15):1566–72.
6. Shaw AD, Bagshaw SM, Goldstein SL, Scherer LA, Duan M, Schermer CR, *et al*. Major complications, mortality, and resource utilization after open abdominal surgery: 0.9% saline compared to Plasma-Lyte. *Ann Surg*. 2012;255(5):821–9.
7. Yunos NM, Bellomo R, Glassford N, Sutcliffe H, Lam Q, Bailey M. Chloride-liberal vs. chloride-restrictive intravenous fluid administration and acute kidney injury: an extended analysis. *Intensive Care Med*. 2015;41(2):257–64.
8. Langston C. Managing fluid and electrolyte disorders in kidney disease. *Vet Clin North Am Small Anim Pract*. 2017;47(2):471–90.
9. Zarychanski R, Abou-Setta AM, Turgeon AF, Houston BL, McIntyre L, Marshall JC, *et al*. Association of hydroxyethyl starch administration with mortality and acute kidney injury in critically ill patients requiring volume resuscitation: a systematic review and meta-analysis. *JAMA*. 2013;309(7):678–88.
10. Glover PA, Rudloff E, Kirby R. Hydroxyethyl starch: a review of pharmacokinetics, pharmacodynamics, current products, and potential clinical risks, benefits, and use. *J Vet Emerg Crit Care (San Antonio)*. 2014;24(6):642–61.
11. Adamik KN, Yozova ID, Regenscheit N. Controversies in the use of hydroxyethyl starch solutions in small animal emergency and critical care. *J Vet Emerg Crit Care (San Antonio)*. 2015;25(1):20–47.
12. Hayes G, Benedicenti L, Mathews K. Retrospective cohort study on the incidence of acute kidney injury and death following hydroxyethyl starch (HES 10% 250/0.5/5:1) administration in dogs (2007–2010). *J Vet Emerg Crit Care (San Antonio)*. 2016;26(1):35–40.
13. Yozova ID, Howard J, Adamik K-N. Retrospective evaluation of the effects of administration of tetrastarch (hydroxyethyl starch 130/0.4) on plasma creatinine concentration in dogs (2010–2013): 201 dogs. *J Vet Emerg Crit Care (San Antonio)*. 2016;26(4):568–77.
14. Sigrist NE, Kälin N, Dreyfus A. Changes in serum creatinine concentration and acute kidney injury (AKI) grade in dogs treated with hydroxyethyl starch 130/0.4 from 2013 to 2015. *J Vet Intern Med*. 2017;31(2):434–41.
15. Legrand M, Le Cam B, Perbet S, Roger C, Darmon M, Guerci P, *et al*. Urine sodium concentration to predict fluid responsiveness in oliguric ICU patients: a prospective multicenter observational study. *Crit Care*. 2016;20(1):165.

16. Swärd K, Valsson F, Sellgren J, Ricksten S-E. Differential effects of human atrial natriuretic peptide and furosemide on glomerular filtration rate and renal oxygen consumption in humans. *Intensive Care Med.* 2005;31(1):79–85.
17. Boscan P, Pypendop BH, Siao KT, Francey T, Dowers K, Cowgill L, *et al.* Fluid balance, glomerular filtration rate, and urine output in dogs anesthetized for an orthopedic surgical procedure. *Am J Vet Res.* 2010;71(5):501–7.
18. McClellan JM, Goldstein RE, Erb HN, Dykes NL, Cowgill LD. Effects of administration of fluids and diuretics on glomerular filtration rate, renal blood flow, and urine output in healthy awake cats. *Am J Vet Res.* 2006;67(4):715–22.
19. Shilliday IR, Quinn KJ, Allison ME. Loop diuretics in the management of acute renal failure: a prospective, double-blind, placebo-controlled, randomized study. *Nephrol Dial Transplant.* 1997;12(12):2592–6.
20. Nielsen LK, Bracker K, Price LL. Administration of fenoldopam in critically ill small animal patients with acute kidney injury: 28 dogs and 34 cats (2008–2012). *J Vet Emerg Crit Care (San Antonio).* 2015;25(3):396–404.
21. Liu KD, Thompson BT, Ancukiewicz M, Steingrub JS, Douglas IS, Matthay MA, *et al.* Acute kidney injury in patients with acute lung injury: impact of fluid accumulation on classification of acute kidney injury and associated outcomes. *Crit Care Med.* 2011;39(12):2665–71.
22. Blake WD, Wégria R. Effect of increased renal venous pressure on renal function. *Am J Physiol.* 1949;157(1):1–13.
23. Winton FR. The influence of venous pressure on the isolated mammalian kidney. *J Physiol.* 1931;72(1):49–61.
24. Bradley SE, Bradley GP. The effect of increased intra-abdominal pressure on renal function in man. *J Clin Invest.* 1947;26(5):1010–22.
25. Li X, Liu M, Bedja D, Thoburn C, Gabrielson K, Racusen L, *et al.* Acute renal venous obstruction is more detrimental to the kidney than arterial occlusion: implication for murine models of acute kidney injury. *Am J Physiol Renal Physiol.* 2012;302(5):F519–25.
26. Ding X, Cheng Z, Qian Q. Intravenous fluids and acute kidney injury. *Blood Purif.* 2017;43(1–3):163–72.
27. Prowle JR, Kirwan CJ, Bellomo R. Fluid management for the prevention and attenuation of acute kidney injury. *Nat Rev Nephrol.* 2014;10(1):37–47.
28. Raimundo M, Crichton S, Martin JR, Syed Y, Varrier M, Wyncoll D, *et al.* Increased fluid administration after early acute kidney injury is associated with less renal recovery. *Shock.* 2015;44(5):431–7.
29. Wang N, Jiang L, Zhu B, Wen Y, Xi X-M, Beijing Acute Kidney Injury Trial (BAKIT) Workgroup. Fluid balance and mortality in critically ill patients with acute kidney injury: a multicenter prospective epidemiological study. *Crit Care.* 2015;19(1):371.
30. Han MJ, Park KH, Shin J-H, Kim SH. Influence of daily fluid balance prior to continuous renal replacement therapy on outcomes in critically ill patients. *J Korean Med Sci.* 2016;31(8):1337–44.
31. Kim IY, Kim JH, Lee DW, Lee SB, Rhee H, Seong EY, *et al.* Fluid overload and survival in critically ill patients with acute kidney injury receiving continuous renal replacement therapy. *PLoS One.* 2017;12(2):e0172137.
32. Cavanagh AA, Sullivan LA, Hansen BD. Retrospective evaluation of fluid overload and relationship to outcome in critically ill dogs. *J Vet Emerg Crit Care (San Antonio).* 2016;26(4):578–86.

1.
acvim334

Biofilms in the Urinary Tract; Implications for Implants

Ewan D.S. Wolff, PhD, DVM; Mohamed Seleem, DVM, PhD, MS; Lynn Guptill, DVM, PhD, DACVIM; Larry Adams, DVM, PhD, DACVIM
West Lafayette, IN, USA

BIOFILMS

Biofilms have been defined as an "assemblage of surface-associated microbial cells that are enclosed in an extracellular polymeric substance matrix."[1] Biofilms do not form with every urinary tract infection (UTI). Predisposing conditions include overdistension of the urinary bladder, pre-existing UTI, damage to mucosal surfaces, immunocompromise, and the presence of *in situ* devices such as urinary catheters or implants.[2,3] To start a biofilm, planktonic bacteria attach to a conditioning film composed of a "deposition of urinary components."[4] This film then allows for weak adsorption of bacteria that form a linking film bonded to the mucosal surface. A bacteria-rich base film then forms as the next layer. Planktonic bacteria are released from the top layer of the biofilm and may create other biofilm sites. Biofilms are associated with chronic infection, relapses, and treatment failure.[3,4] Functional zones also exist within the biofilm with the superficial layer decreasing introduction of antimicrobial agents, the center of the biofilm being an area of resistance gene exchange and the base of the biofilm serving as a breakdown byproduct layer where antibiotics do not function appropriately.[5] These layers are able to communicate via molecular crosstalk called **quorum sensing** that endows them with the ability to respond colonially to changing environments.[6]

URINARY CATHETERS AND IMPLANTS

Catheter associated UTIs (CAUTI) are an important potential cause of biofilms in the urinary tract.[7] Bacteria enter the urinary tract via either an intra- or extra-luminal route and attach to implants such as catheters, or damaged urinary tract mucosa. The CDC defines CAUTI in humans based on presence of a catheter for more than two days, a list of criteria of which one or more must be met (including fever >38°C, urinary urgency, urinary frequency, dysuria) and a urine culture with no more than two species of organism with a cfu/mL count of $>10^5$.[8] When urinary catheters are left in long-term, mineral encrustation may form from calcium precipitates. If catheters then become colonized with a biofilm there may be no way of resolving the infection without removal of the catheter.[3] In dogs there was a 27% increase in the likelihood of urinary tract infection per day of catheterization in hospital.[9] In a study of 39 hospitalized dogs with indwelling Foley catheters, 10% developed UTI.[10] Neither of these studies of dogs mentioned the presence of biofilms, however, the development of biofilms on indwelling urinary catheters has been demonstrated in dogs.[11]

Biofilms have also been reported from ureteral stents. In one study of 40 people, 23 (58%) stents were encrusted with 11 having confirmed biofilms, although encrustation and biofilm formation were not related in this study.[12] Synthetic urease raises pH and allows for encrustation *in vitro*.[13] *In vivo* studies have indicated a mix of mineral compositions occur in stent encrustation including calcium oxalate and apatite.[14,15] One study of ureteral stents in humans showed 68–90% of encrustations were colonized with urease-producing bacteria and only 27–30% of patients with encrusted stents had bacteriuria.[4] Other studies showed a range of 35% to >50% of stents showing biofilms.[16-18] In another study, 54% of 148 stents in 146 people were colonized when stent material was cultured at 6 weeks.[19] *Escherichia coli*, *Enterococcus*, *Enterobacter*, *Pseudomonas*, coagulase negative *Staphylococcus*. and *Candida* sp. were the most common bacterial and fungal species noted in six studies of stent biofilms in humans.[12,16-20] In cats, a urinary tract infection rate of up to 50% has been reported in the presence of ureteral stents. Five studies have reported stent encrustation, although the mineral analysis was not reported.[21-25] Furthermore, in the small animal literature, stent biofilm colonization has not been specifically researched, although a recent study lists biofilm as a specific concern.[21]

The subcutaneous ureteral bypass (SUB) was developed in veterinary medicine in 2009, and the subcutaneous nephrovesicular bypass has existed in human medicine since 1998.[26] In humans with nephrovesicular bypasses, 5 patients with metastatic disease developed encrustation with no biofilm. Two people with nephrovesicular bypasses and renal transplants in another study had *Candida* infections (one patient grew a fungal ball in the bypass system), and the other patient had a chronic bacterial infection.[27,28] In cats, a 21% UTI rate was reported in a study of stents and SUB systems in 106 patients.[29] Another recent study reported UTIs in 4 of 11 cats followed after SUB placement, with two patients having infections that did not resolve with antibiotic therapy.[30] The authors expressed the concern, based on resistant infections, that biofilms exist which cannot be eradicated in these implants. In a recent abstract concerning 174 SUB systems in cats, 25% were encrusted at a median of approximately 1 year but mineral analysis is not yet published.[31]

Biofilm Prevention and Treatment

Developing a strategy for biofilm prevention prior to placement of catheters, stents, and SUB systems could avoid serious complications long run. General recommendations and standard precautions for UTI prevention should be followed for all patients. Judicious use of perioperative antibiotics, limiting use of indwelling urinary catheters, and wearing gloves during handling are some general guidelines for patients with urinary tract implants.[32] For SUB systems, minimization of repeated port access, following strict sterile technique whenever handling ports, and placing a sterile cover on the ultrasound probe when imaging patients for SUB flush procedures are additional considerations suggested to minimize the risk of iatrogenic UTI. Suggestions also include screening for previous UTIs/biofilms (with *in vitro* biofilm culture) prior to placing any urinary tract implant, monitoring of urine pH as a screen for urease-producing organisms and more regular indirect monitoring for encrustation of implanted materials with ultrasound. One approach to preventing biofilm formation is materials modification such as integrating antimicrobial compounds into catheter and stent material. Coatings for stents and catheters have been shown to prevent biofilm growth including salicylic acid coatings for gram-negative rods and triclosan-eluting stents for *Pseudomonas* and *Enterococcus*.[33,34] Non-specific antimicrobial coatings such as chlorhexidine and silver have been shown to decrease biofilm and CAUTI respectively.[11,35]

Curative treatment once biofilms have formed may not be possible. EDTA, mandelic acid and antibiotics have been used as bladder infusions or SUB infusions.[36-40] Systemic antibiotics must be active within biofilms (e.g., macrolides, lincosamides, doxycycline, fluoroquinolones, sulfonamides.[41] Combination antibiotic therapy has proved more effective than use of individual agents).[41] Although the types of biofilms in cat and dog implants are not well described, common resistant urinary tract pathogens include uropathogenic *E. coli*, *Proteus*, *Klebsiella*, *Pseudomonas*, *Corynebacterium*, *Staphylococcus*, *Streptococcus* and *Enterococcus*.[32,42-45] Consistently testing multi-drug resistant/recurrent isolates for the ability to form biofilms *in vitro* and antibiotic susceptibility of the biofilm phase are possible directions for future research on catheter and implant infections.

Conclusions

Biofilms cannot be described as an emerging problem given that they have been a significant medical conundrum since the 1980s. In small animal medicine, there is enough knowledge concerning resistant infection, biofilm formation, and encrustation in catheters, stents, and SUB systems to suggest that more consideration of the occurrence of biofilms, and means of preventing and treating biofilms, is needed. Prevention of infection through strategies like screening for UTI, careful adherence to standard precautions and aseptic technique, material modification, and culture-guided therapy should be a priority.

References

1. Donlan RM. *Emerg Infect Dis.* 2002;8:(9):881–890.
2. Vasudeva P. *Neurourol & Microbiol.* 2014;33:(1):95–100.
3. Choong SKS, *et al. BJUI Int.* 2000;86:(8):935–941.
4. Tenke P. *World J Urol.* 2012;30:(1):51–57.

5. Stewart PS. *Lancet.* 2001;358:(9276):135–138.
6. Williams P, *et al. Philos Trans R Soc Lond B Biol Sci.* 2000;355(1397):667–680.
7. Choong S, *et al. Intl J Antimicrob Agents.* 2001;17(4):305–310.
8. CDC Website. www.cdc.gov/hai/ca_uti/uti.html. Accessed 2/15/17.
9. Bubenik LJ, *et al. J Am Vet Med Assoc.* 2007;231(6):893–899.
10. Smarick SD, *et al. J Am Vet Med Assoc.* 2004;224(12):1936–1940.
11. Segev G, *et al. J Am Vet Med Assoc.* 2013;27(1):39–46.
12. Keane PF, *et al. Br J Urol.* 1994;73(6):687–691.
13. Desgrandchamps F, *et al. Br J Urol.* 1997;79(1):24–27.
14. Robert M, *et al. Urol Int.* 1997;58(2):100–1004.
15. Schulze KA, *et al. Urology.* 1985;25(6):616–619.
16. Bonkat G, *et al. World J Urol.* 2013;31(3):541–546.
17. Bonkat G, *et al. Transpl Infect Dis.* 2012;14(1):57–63.
18. Uvin P, *et al. Urology.* 2011;78(1):154–158.
19. Lojanapiwat B. *World J Urol.* 2006;24(6):681–683.
20. Yeniyol CO, *et al. Int Urol Nephrol.* 2002;34(2):199–202.
21. Wormser C. *J Am Vet Med Assoc.* 2016;248(5):518–525.
22. Kuntz JA, *et al. J Am Vet Med Assoc.* 2015;246(2):216–225.
23. Kulendra LJ, *et al. J Feline Med Surg.* 2014;16(12):985–991.
24. Manassero M, *et al. J Feline Med Surg.* 2014;16(8):623–630.
25. Nicoli J, *et al. Vet J.* 2012;194(1):60–65.
26. Desgrandchamps F, *et al. Transpl Int.* 1998;11 Suppl 1:S150–151.
27. Haddad N, *et al. Can Urol Assoc J.* 2013;7(5–6):E417–E420.
28. Schmidbauer J, *et al. Eur Urol.* 2006;50(5):1073–1078.
29. Steinhaus J, *et al. J Vet Intern Med.* 2015;29(1):63–70.
30. Livet J, *et al. J. Feline Med Surg.* 2016;(09):Early View.
31. Berent A, *et al. J Vet Intern Med.* 2016;31(1):1539.
32. Olin S, *et al. Vet Clin North Am Small Anim Pract.* 2015;45:721–746.
33. Farber BF, *et al. J Biomed Mater Res.* 1993;27(5):599–602.
34. Cadieux PA, *et al. J Endourol.* 2009;23(7):1187–1194.
35. Lederer JW, *et al. J Wound Ostomy Continence Nurs.* 2014;41(5):473–480.
36. Torres, *et al. J Am Vet Med Assoc.* 2014;1:245(7):809–811.
37. Hannan TJ, *et al. FEMS Microbiol Rev.* 2012;36(3):616–648.
38. Percival SL, *et al. J Hosp Infect.* 2005;60(1):1–7.
39. Getliffe KA, *Br J Urol.* 1994;73(6):696–700.
40. Adams LG, *et al.* 2017 (unpublished data).
41. Wu H, *et al. Int J of Oral Sci.* 2015;7(1):1–7.
42. Alvarez-Perez S, *et al. Med Mycol Case Rep.* 2016;11:9–12.
43. Gulhan T. *Iran J Vet Res.* 2015;16(3):261–266.
44. Freitag T. *Vet Microbiol.* 2005;106(1–2):79–86.
45. Wolff EDS. *J Vet Intern Med.* 2016;30:1489.

acvim335

Novel Oral Anticoagulants: Human Medicine Perspective

Stephan Moll, MD
Chapel Hill, NC, USA

INTRODUCTION

Four direct oral anticoagulants (DOACs) are presently FDA approved for use in humans: apixaban, dabigatran, edoxaban, and rivaroxaban. This article summarizes key aspects of the drugs' characteristics and clinical use in humans.

BASELINE CHARACTERISTICS

The DOACs are small synthetic molecules, which directly inhibit specific coagulation factors in in an antithrombin-independent manner. The two classes of FDA approved DAOCs are the anti-IIa (thrombin) drug dabigatran and the three anti-Xa drugs apixaban, edoxaban, and rivaroxaban. All four drugs are FDA approved for the prevention of arterial thromboembolism in atrial fibrillation (AF) and the treatment of venous thromboembolism (VTE). Apixaban, dabigatran, and rivaroxaban are also approved for the prevention of VTE after knee and hip replacement surgery.

The advantages of the DOACs over the vitamin K antagonists are

1. Their wide therapeutic window, allowing fixed dosing without the need for anticoagulant level monitoring.
2. The lack of influence of oral vitamin K intake on the DOACs' anticoagulant effect.
3. A short time to full anticoagulant effect after oral intake of 1.5–3 hours compared to warfarin's 5 or more days, allowing DOACs to be started without the need for a parenteral anticoagulant.
4. A short half-life (in the range of 7–14 hours), making preoperative interruption of anticoagulation easier than with warfarin.
5. Less drug interactions.

These factors often make the DOACs easier and more convenient to use than warfarin. Key pharmacokinetic and pharmacodynamics characteristics of the DOACs are listed in Table 1.

Table 1. Key information about the various FDA-approved DOACs

	Dabigatran	Apixaban	Rivaroxaban	Edoxaban
Baseline characteristics				
Mechanism	Anti-IIa	Anti-Xa		
Tmax	1½–3 hours	1-3 hours	2–4 hours	1–1½ hours
Half-life	12–14 hours	8–15 hours	9–13 hours	9–11 hours
Renal excretion	80%	25%	33%	35%
Reversal agent	Idarucizumab[1]	Andexanet[2]		
FDA approval				
VTE prevention[3]	Yes			No
VTE treatment	Yes			
Atrial fibrillation	Yes			
Dosing				

VTE prevention[3]	110 mg first day, then 220 mg q.d. if CrCl >30[4]	2.5 mg b.i.d.	10 mg q.d.	Does not apply
VTE treatment	150 mg b.i.d.	10 mg b.i.d. for 1 week, then 5 mg b.i.d. After 6 months: 2.5 mg b.i.d.	15 mg b.i.d. for 3 weeks, then 20 mg q.d.	60 mg q.d. if CrCL 50–95 30 mg q.d. if CrCL 15–50 or weight ≤60 kg or use of certain P-gp inhibitors Do not use if CrCL >95
Atrial fibrillation	150 mg b.i.d. if CrCl >30 75 mg b.i.d. if CrCl 15–30	5 mg b.i.d. 2.5 mg b.i.d. if 2 of these criteria apply: - Age ≥80 years - Weight ≤60 kg - Creatinine ≥1.5 mg/dL	20 mg q.d. if CrCl >50: 15 mg q.d. if CrCl 15–50	60 mg q.d. if CrCL 50–95 30 mg q.d. if CrCL 15–50 Do not use if CrCL >95

1 FDA approved; 2 not FDA approved as of Feb 24th, 2017; 3 VTE prevention after hip and knee replacement surgery; 4 CrCl measured in mL/min.
Abbreviations: VTE = venous thromboembolism; b.i.d. = twice daily; q.d. = once daily; CrCl = creatinine clearance.

Dosing, Renal Clearance, Half-Life

All four DOACs are, to some degree, renally cleared (Table 1), making renal function, as assessed by serum creatinine and glomerular filtration rate (GFR), an important parameter to observe during clinical management. Different dosing regimens exist for the various DOACs depending on renal function and indication. While the DOACs' half-life is around 7–14 hours, there is considerable inter-individual variation in achieved plasma drug levels and drug half-life.[1] Significant intra-individual variation of plasma levels has also been demonstrated for dabigatran, but has not yet been studied for the other DOACs.[2] The DOACs' half-life is longer in patients with renal impairment and this is particularly relevant for dabigatran which is predominately (to 80%) cleared by the kidneys. Apixaban is FDA approved for use in renal failure and dialysis, but given limited published data, caution is to be exercised.[3,4]

Efficacy and Safety

Meta-analyses and systematic reviews have shown that the DOACs are at least as effective as warfarin in the prevention of stroke and systemic thromboembolism in AF, and in the prevention of recurrent VTE when used for the treatment of VTE. DOACs also lead to less major, intracranial and clinically relevant major bleeding than warfarin. Gastrointestinal and menstrual bleeding may be more with some of the DOACs than with warfarin. Based on efficacy and safety profile, prominent clinical guidelines now broadly endorse use of DOACs in atrial fibrillation and, for the treatment of VTE suggest DOACs over warfarin.[5,6] DOACs are a good treatment option in many patients with VTE and atrial fibrillation, including the elderly. While they are FDA approved to be used in obese patients, limited data exist on their use in patients with extreme body. This has led to a guidance document that suggests the use warfarin instead of a DOAC in patients with a body weight above 120 kg or a body mass index of >40 kg/m^2.[7]

Coagulation Testing

The DOACs influence many coagulation assays. The prothrombin time (PT) and activated partial thromboplastin time (aPTT) can be elevated, sometimes markedly. The degree of elevation depends on (a) the proximity of blood draw to the last DOAC drug intake, (b) the individual DOAC drug, and (c) the sensitivity of the PT and aPTT reagents. Individual procoagulant factor activity levels (e.g., factor VIII, IX and other) may be falsely decreased, anticoagulant factor activity levels (i.e., protein C, protein S, and antithrombin) falsely elevated; lupus anticoagulant results can be false positive.

While routine coagulation testing of DOACs in the majority of patients is not necessary in selected patients testing may be clinically helpful, such as in patients

1. With extremely low or high body weights
2. After bariatric surgery
3. With significant renal impairment
4. With recurrent thrombosis or bleeding in spite of being on a DOAC
5. Started on a drug that may majorly interfere with the metabolism of the DOAC
6. In whom non-adherence to therapy needs to be assessed

Assays to assess plasma level of dabigatran are the ecarin clotting time (ECT; either as a chromogenic or clotting assay), the dilute thrombin time (dTT), and the thrombin time (TT), as well as mass spectrometry testing.[1] Assays to assess levels of the anti-Xa drugs are the anti-Xa assay with drug-specific calibrators or mass spectrometry.[1] Expected peak and trough levels for the four drugs have been published.[1] DOAC dose adjustment based on plasma drug level results is not recommended, but results are rather used to understand why a patient on a DOAC had a recurrent thrombosis or bleed, is adherent to therapy, or should - in case of levels above or below expected range - rather be treated with warfarin.

For dabigatran and edoxaban correlation between drug levels and bleeding (at higher levels) and thrombosis (at lower levels) has been documented, but similar data have not been published for apixaban and rivaroxaban.[8,9]

Management of Bleeding

For moderate bleeding no specific treatment is typically needed beyond local hemostatic measures and discontinuation of the drug for some period of time. Given the relatively short half-life of the DOACs - unless in the presence of renal impairment - the DOACs are relatively quickly out of the system. For major bleeding on dabigatran the intravenous drug Idarucizumab is FDA approved and available, a humanized antibody directed against dabigatran. A reversal agent for the anti-Xa anticoagulants, Andexanet, is not FDA-approved at this point, but an FDA decision is expected in the second part of 2017.

Summary

Given that DOACs are at least equivalent to warfarin in respect to efficacy and safety and given that they are more convenient to take for the patient - no drug level monitoring, not vitamin K intake restrictions, less drug interactions, faster on-off mechanism, - the DOACS are now being widely used in the treatment of atrial fibrillation and venous thromboembolism.

Disclosures

Dr. Moll has consulted for Boehringer-Ingelheim, Janssen Pharmaceuticals, and Portola.

References

1. Samuelson BT, Cuker A, Siegal DM, Crowther M, Garcia DA. Laboratory assessment of the anticoagulant activity of direct oral anticoagulants: a systematic review. *Chest.* 2017;151:127–138.
2. Chan NC, Coppens M, Hirsh J, *et al.* Real-world variability in dabigatran levels in patients with atrial fibrillation. *J Thromb Haemost.* 2015;13:353–359.
3. Wang X, Tirucherai G, Marbury TC, *et al.* Pharmacokinetics, pharmacodynamics, and safety of apixaban in subjects with end-stage renal disease on hemodialysis. *J Clin Pharmacol.* 2016;56:628–636.
4. Chang M, Yu Z, Shenker A, *et al.* Effect of renal impairment on the pharmacokinetics, pharmacodynamics, and safety of apixaban. *J Clin Pharmacol.* 2016;56:637–645.
5. Kearon C, Akl EA, Ornelas J, *et al.* Antithrombotic therapy for VTE disease: CHEST guideline and expert panel report. *Chest.* 2016;149:315–352.
6. Moss JD, Cifu AS. Management of anticoagulation in patients with atrial fibrillation. *J Am Med Assoc.* 2015;314:291–292.
7. Martin K, Beyer-Westendorf J, Davidson BL, Huisman MV, Sandset PM, Moll S. Use of the direct oral anticoagulants in obese patients: guidance from the SSC of the ISTH. *J Thromb Haemost.* 2016;14:1308–1313.

8. Reilly PA, Lehr T, Haertter S, *et al*. The effect of dabigatran plasma concentrations and patient characteristics on the frequency of ischemic stroke and major bleeding in atrial fibrillation patients: the RE-LY trial (randomized evaluation of long-term anticoagulation therapy). *J Am Coll Cardiol*. 2014;63:321–328.
9. Ruff CT, Giugliano RP, Braunwald E, *et al*. Association between edoxaban dose, concentration, anti-Factor Xa activity, and outcomes: an analysis of data from the randomised, double-blind ENGAGE AF-TIMI 48 trial. *Lancet*. 2015;385:2288–2295.

1.
acvim336

Novel Oral Anticoagulants: Veterinary Medicine Perspective

Armelle deLaforcade, DVM, DACVECC
North Grafton, MA, USA

INTRODUCTION

Thrombotic complications of naturally occurring disease processes are being recognized with increasing frequency in veterinary medicine. The elements of Virchow's triad (hypercoagulability, endothelial damage, and altered blood flow) are accepted as predisposing factors to clot formation, and it is thought that changes in 2 of the 3 categories represent a clinically significant increase in risk for developing a thrombus. Retrospective studies of dogs with confirmed thrombosis consistently identify certain underlying diseases associated with thrombus formation. These include immune-mediated diseases, protein-losing nephropathy and enteropathy, various forms of neoplasia, sepsis, and cardiac disease. Common features frequently documented include corticosteroid administration and the presence of central venous catheters.

Hypercoagulability describes a state of altered coagulation predisposing to the formation of a thrombus. These states are characterized by increased platelet activation, increased activation of coagulation factors, a reduction in endogenous anticoagulants, or reduced fibrinolysis. Although states of factor deficiency seem to correlate nicely with bleeding tendencies, the same is not true of hypercoagulability. Studies of deep vein thrombosis (DVT) in people have shown that while DVT occurs more commonly among people with thrombophilia, only about 30% of people experiencing DVT have a defined condition associated with hypercoagulability, and many people with thrombophilia do not go on to develop a DVT. Therefore, it can be surmised that hypercoagulability in people is a risk factor for DVT formation but not a true cause. The degree to which this can be extrapolated to animals is unclear, since animals with the disease **states** described above are more likely to experience venous and arterial thrombosis rather than true DVT in response to disease states.

The concept of **blood stasis** (also referred to as 'impaired circulation') is accepted as a risk factor for thromboembolism. Some believe that blood lacking pulsatile flow is more likely to be affected by thrombosis. Impaired circulation is thought to play an important role in the development of DVT, most importantly reduced venous blood velocity and/or reduced venous return (congestion). Valves play an important role in promoting venous circulation, and the venous valvular sinus has been identified as a common source of thrombosis in people. Impaired flow in this region has been associated with hypoxia, potentially creating a microenvironment more susceptible to thrombosis. Diseases associated with aberrant blood flow include hypovolemia, vascular anomalies, cardiomyopathy, hyperviscosity disorders, and neoplasia.

The literature regarding **endothelial injury** can be confusing, as it is difficult to differentiate between endothelial injury as the cause or effect of thrombus formation. The contribution of endothelial injury to thrombotic states has traditionally centered on cytokine-mediated tissue factor expression triggering secondary hemostasis and fibrin production. It is currently believed that endothelial injury resulting from hypoxic events leads to increased platelet and leukocyte recruitment and increased concentration of activated clotting factors.

CHALLENGES

Routine empirical therapy with anticoagulants is complicated by lack of evidence in veterinary studies of naturally occurring diseases that are large enough to truly prove efficacy in preventing thrombosis or in impacting survival. Reliable tests for early identification of hypercoagulability are also not readily available. The use of viscoelastic testing for investigation of hemostatic changes in veterinary diseases have been helpful in identifying animals at risk of thrombosis. The decision to initiate anticoagulant therapy requires a risk/benefit assessment that is hard to do without supporting large-scale studies. Thus anticoagulation remains somewhat controversial, often driven by personal experience, and not practiced according to any published guidelines. The selection of a platelet blocker over an injectable anticoagulant is often driven more by cost and the owner's willingness/ability to administer injections at home, rather than evidence for one drug over another. Most commonly administered drugs to prevent thrombosis include platelet blockers (aspirin, clopidogrel) and heparin (unfractionated vs. low molecular weight). While there are several publications describing the use of warfarin in dogs following valve replacement, this medication requires frequent visits and is awkward to administer long term.

The ideal anticoagulant would be orally administered, administered once daily, have few drug interactions, have a wide therapeutic margin, and require no monitoring. In the past few years, several novel oral anticoagulants have been approved that place the future of routine warfarin therapy in people in question. Given their ability to be administered orally, these are of particular interest in veterinary medicine.

Novel Oral Anticoagulants

The two main types of oral anticoagulants that have gained popularity in people include direct thrombin inhibitors and factor Xa inhibitors. As thrombin and factor Xa are both key points of coagulation, it makes sense that they would be desirable targets for anticoagulation.

Dabigatran

Dabigatran functions as a direct thrombin inhibitor, with a half-life in people of 12–17 hours. This drug has been approved for the treatment of human non-valvular atrial fibrillation, acute DVT, and pulmonary embolism. Two large landmark studies compared dabigatran to warfarin in patients with acute venous thromboembolism (therapeutic anticoagulation). In these studies, dabigatran was found to be non-inferior to warfarin with recurrence of VTE as the outcome. Both trials of dabigatran involved administration of parenteral anticoagulants for 5–10 days prior to treatment. When compared to enoxaparin, dabigatran was also found to be non-inferior, with a similar safety profile as enoxaparin.

Rivaroxaban

Rivaroxaban acts by competitively inhibiting both free and clot-bound Xa, reducing thrombin formation. It has been approved for the treatment and prevention of DVT and pulmonary embolism. The EINSTEIN study compared rivaroxaban to traditional protocol consisting of enoxaparin followed by warfarin in people with DVT. The EINSTEIN-PE study focused on treatment of pulmonary embolism. Both studies suggested that rivaroxaban is similar in efficacy to the traditional protocol in preventing recurrence of DVT and PE, with a similar bleeding risk. Rivaroxaban has also been approved for the prevention of DVT in patients undergoing orthopedic surgery. Additional studies are emerging exploring the safety and efficacy of rivaroxaban in cancer-associated thrombosis.

Of the oral anticoagulants, rivaroxaban has gained the most attention in veterinary medicine. An *in vitro* study showed that rivaroxaban had a concentration-dependent anticoagulant effect in canine plasma, and confirmed the concentration at which rivaroxaban achieved 50% inhibition of thrombin propagation. Currently two small canine studies suggest that rivaroxaban is safe when administered to dogs. In one study, either rivaroxaban or a combination of clopidogrel and aspirin were administered to dogs with IMHA. No drug reactions or bleeding complications were noted. A case series of four dogs described the use of rivaroxaban in dogs with thrombotic complications.

Apixaban

This direct Xa inhibitor is also approved in people for treatment of DVT and PE as well as for prophylaxis against DVT. A series of studies called ADVANCE (1–3) were conducted to compare apixaban to enoxaparin for the prevention of DVT in patients undergoing knee replacement. In the ADVANCE-3 study of over 5,000 patients, apixaban was superior to enoxaparin with equal bleeding risks. For the treatment of PE and DVT, the AMPLIFY study indicated that apixaban was non-inferior standard protocol of enoxaparin followed by warfarin, with the advantage of a reduced bleeding risk.

One veterinary study investigated the pharmacokinetics and pharmacodynamics of apixaban administered orally and IV in purpose-bred cats. In this study, factor Xa was inhibited after a single dose of oral or IV apixaban, and inhibition was inversely correlated to plasma concentration of the drug, suggesting a potential role for prevention of thrombosis in cats.

Edoxaban

Another factor Xa inhibitor, edoxaban has been approved for treatment of DVT and PE, but not yet for prophylaxis for orthopedic surgery.

The major disadvantage of these oral anticoagulants has been the lack of available reversal agents when bleeding complications are encountered. Treatment is largely supportive, with prothrombin complex concentration and factor VIIa having varied success. New compounds for reversal are under investigation.

Lack of reversal agents is a real consideration for the utility of these drugs in veterinary medicine, given overall anxiety in using anticoagulants. Currently, treatment for any bleeding consists of discontinuation of the drug and support with fluids/blood products as needed.

References

1. Madan S, *et al. Cardiovas Diagn Ther*. 2016;6(6):570.
2. Mantha S, *et al. J Thromb Thrombolysis*. 2016;EPub.
3. Yang VK, *et al. J Vet Emerg Crit Care*. 2016;26(5):729.
4. Conversy B. *Vet J*. 2013;198(2):437.
5. Morassi A, *et al. J Vet Emerg Crit Care*. 2016;26(4):488.
6. Myers JA, *et al. Am J Vet Res*. 2015;76(8):732.

1.

acvim337

Year in Review: Transfusion Medicine

Marie Holowaychuk, DVM, DACVECC
Calgary, AB, Canada

INTRODUCTION

Transfusion medicine is a rapidly growing field with numerous articles being published in the veterinary literature each month. During the last year, more than 100 articles pertaining to veterinary transfusion medicine were published. The canine and feline studies related to blood product storage, blood types, compatibility testing, and alternative transfusion methods will be outlined below in the context of other pertinent literature.

RED CELL STORAGE

Interest in storage lesions and the impact of the age of blood products on patient morbidity and mortality has recently increased. Storage lesions are defined as the metabolic, biochemical, and morphological changes that occur *in vitro* in stored packed red blood cells (PRBCs), whole blood (WB), and platelets. *In vivo*, these changes affect the viability and efficacy of the product when it is transfused. The cumulative effect of storage lesions is thought to be responsible for harm associated with transfusion of older stored blood to human and veterinary patients. Biochemical changes alter the RBC and membrane structure, resulting in compromised microvascular blood flow after transfusion and increased hemolysis during storage, respectively. Some RBC changes are immediate, such as a decrease in 2, 3-DPG, whereas alterations in lactate, pH, or adenosine triphosphate (ATP) occur over days.[1]

In order to investigate the clinical consequences of administering stored RBC-containing blood products, a retrospective study described an association between product age and hemolysis as a transfusion-related complication in 210 dogs receiving RBC-containing transfusions. Analysis revealed that for every additional day of product storage, the odds of hemolysis increased significantly (OR = 1.11).[2] This study follows a large retrospective study investigating the consequences of administration of older (stored >14 days) PRBC units in more than 3,000 dogs that was published in 2014 and found that while the age of the unit administered was not associated with mortality when considering transfusions in all dogs, dogs with hemolysis, in particular those with immune mediated hemolytic anemia (IMHA), had increased mortality when PRBC stored for longer were administered.[3] Nevertheless, a meta-analysis of the most recently published randomized controlled trials investigating the impact of age of RBCs and survival in human patients shows no clinical benefit to using fresher RBCs when compared with older or standard-issue RBC-containing products.[4]

Leukoreduction

Leukocytes are metabolically active and produce cytokines that accumulate during storage and can contribute to storage lesions and possibly transfusion reactions in dogs. Leukocyte depletion or leukoreduction is the reduction in leukocyte numbers within a unit of blood to less than $5x10^6$ WBCs per unit. Leukoreduction is primarily achieved by passing donated blood through a filter prior to blood product storage. Previous veterinary studies have demonstrated a decrease in RBC hemolysis, loss of 2, 3-DPG, and accumulation of vascular endothelial growth factor and interleukin-8 (IL-8) in stored RBC products that have been leukoreduced when compared to non-leukoreduced products.

Recently, a group of investigators found no change in phosphatidylserine expression or eicosanoid concentrations when comparing leukoreduced and non-leukoreduced canine PRBCs stored for up to 21 days.[5] This follows a study by the same group, who found that concentrations of several pro-inflammatory eicosanoids increased in canine PRBC units during storage.[6] Likewise, another group demonstrated that leukoreduction decreased IL-8 concentrations that accumulated during storage of PRBC units collected from Greyhounds during hemorrhagic shock.[7] Unfortunately, there remains a paucity of research investigating the *in vivo* effects of transfusion of leukoreduced versus non-leukoreduced blood in veterinary patients, making clinical conclusions difficult to draw from these studies. Likewise, the evidence supporting the routine use of leukoreduction in human blood banks remains low to very low quality.[8]

Feline Stored Whole Blood

There are very few studies investigating storage lesions in feline blood or the impact of storage on RBC transfusions in feline patients. A recent study measured concentrations of ammonia, IL-6, IL-10, and CXCL-8 (previously IL-8), in stored feline whole blood units. Results showed that ammonia concentrations increased with storage time and that CXCL-8 was also detectable.[9] Further research investigating these markers in blood products transfused to feline patients is required in order to determine their clinical significance.

Microparticles

Microparticles are submicron phosphatidylserine-bearing lipid vesicles that accumulate in PRBCs during storage and might contribute to transfusion-related complications in blood product recipients. A study investigating methods to detect microparticles in stored canine PRBCs using different techniques found that three flow cytometric methods were comparable, although further investigation is needed to determine their accuracy.[10]

Bacterial Contamination

While the risk of bacterial contamination in stored canine and feline blood units is mitigated by the use of strict sterile technique during blood collection and processing, there are few studies investigating the incidence of contamination of stored units. One research group published a description of the detection and quantification of bacterial contamination in four cases of canine and feline blood unit contamination.[11] Subsequent to that, the same group published a prospective study to detect bacterial contamination in 49 stored canine WB units using bacterial culture and PCR. Two units had positive bacterial cultures, whereas 26 units had positive qPCR assays but at very low quantities. The authors concluded that the bacterial contamination detected was at an acceptably low level and unlikely to cause clinically-relevant sepsis in recipients.[12]

Canine Blood Types

Traditionally, dogs have been cited to have low numbers of naturally-occurring alloantibodies, which makes them more suitable candidates for a first transfusion without pre-transfusion compatibility testing. However, naturally-occurring anti-DEA 7 antibodies have been identified in up to 50% of DEA-7 negative transfusion-naïve dogs and have been implicated in causing delayed transfusion reactions such as decreased RBC lifespan post-transfusion. An Italian study measured the titers of anti-DEA 7 antibodies in 317 DEA-7 negative dogs and found that 23% of dogs had low titers of weakly agglutinating mostly IgM antibodies.[13] The same group measured anti-DEA 7 antibodies in 252 DEA-7 negative dogs in Spain and Italy and found that 38% of dogs had anti-DEA-7 antibodies. The authors estimated a 6% chance of developing a hemolytic reaction after transfusion with blood not crossmatched or typed for DEA 7.[14]

Previously, two monoclonal antibodies against two different canine RBC antigens were recently discovered in South Korea and named anti-Kai 1 and anti-Kai 2. A recent study was conducted in the USA to determine whether Kai 1 and Kai 2 exist in North America and are related to the other known canine RBC antigens. Blood typing was performed in 503 dogs and revealed that 60% were DEA-1 positive and 94% were Kai 1 positive and Kai 2 negative. No relationship was found between Kai 1 and Kai 2 at the other recognized canine blood types, nor did any DEA-1 or Kai-negative dogs exhibit naturally-occurring alloantibodies.[15]

Compatibility Testing

In dogs, the low incidence of clinically significant naturally-occurring antibodies makes the risk of a serious hemolytic reaction small during a first transfusion. As such, crossmatching has traditionally been recommended in previously transfused dogs, once sensitization to red cell antigens has occurred. However, in cats clinical incompatibilities are documented infrequently when AB-type-specific blood is administered. Therefore, recommendations for crossmatching vary among institutions ranging from always to never crossmatching.

A recent study investigated the incidence of incompatible crossmatch results in transfusion-naïve dogs and found that 17% of 149 dogs undergoing major crossmatch testing were incompatible with at least 1 potential donor. The authors suggested that consideration be given to performing crossmatch testing for all canine patients regardless of the transfusion history.[16] Additionally, a recent study measuring serial crossmatch results in hospitalized cats found that 25% of feline recipients had incompatible major crossmatch results as early as 2-days after the first whole blood transfusion, suggesting that crossmatch testing be performed for serial transfusions in cats, even when AB-type-specific blood is transfused.[17]

Recent studies investigated the accuracy of a commercially available gel column crossmatch kits in dogs. While the commercial available gel column kit was 3-times faster (15 minutes) than the manual tube crossmatch technique (45–50 minutes), 35% of dogs had an incompatible gold standard laboratory crossmatch result, even though the commercial crossmatch kit indicated compatibility.[18] This was attributed to microagglutination and low-grade hemolysis, which are difficult to identify using gel-based kits. Additionally, sample spreading and negative control tubes occurred in up to 50% of patients, making accurate test interpretation difficult. These factors might be exacerbated by weak agglutination reactions or less than ideal centrifuge settings.[19]

Autologous Transfusions

Given the growing concerns regarding storage lesions and transfusion-related complications in veterinary patients, efforts are being made to limit allogenic blood product transfusion whenever possible. As such, autologous transfusion using cell salvage devices or direct re-infusion are being described more frequently in the literature. A recent prospective *in vitro* experimental study investigated the impact of processing fresh canine WB using a cell salvage device. The authors concluded that while there were slight changes in some RBC properties such as mean corpuscular volume, erythrocyte osmotic resistance, RBC deformability, RBC aggregation, and the activity of lactate dehydrogenase, the changes were not clinically relevant and were comparable to that seen in human blood after cell salvage processing.[20]

Additionally, a novel technique for autologous transfusion using direct re-infusion of blood obtained during surgery for hemiabdomen was described. Four dogs underwent autologous transfusion using the 2-syringe technique and survived with no complications.[21]

Xenotransfusion

Xenotransfusion describes transfusion of blood between different species and is most commonly reported in veterinary medicine between dogs and cats. A recent report described the effects of xenotransfusion in 2 anemic cats that received canine blood. While both cats transiently improved after transfusion with canine blood, acute intravascular hemolysis occurred and the PCV declined rapidly. Blood typing post-xenotransfusion with DEA-1 immunochromatographic strips revealed a positive band consistent with canine RBCs, which was no longer detected 4 days after xenotransfusion. Major cross-matching resulted in both positive and negative reactions, whereas minor cross-matching showed incompatibility.[22] These results are consistent with previous work that demonstrates the average life span of xenotransfused RBCs is short compared to allogenic transfusions in cats (3.6 days vs. 30 days). If a dog-to-cat xenotransfusion is considered the only option, doses and rates should be similar to those used during allogenic transfusions and under no circumstances should the xenotransfusion be repeated.

REFERENCES

1. Obrador R, *et al. J Vet Emerg Crit Care*. 2015;25(2):187.
2. Maglaras C, *et al. J Vet Emerg Crit Care*. 2017;27(1):108.
3. Hann L, *et al. J Vet Intern Med*. 2014;28(6):1830.
4. Remy K, *et al. Vox Sang*. 2016;111(1):43.
5. Muro S, *et al. J Vet Intern Med*. 2017 [epub ahead of print].
6. Blake R, *et al. J Am Vet Med Assoc*. 2017;250(2):191.
7. Purcell S, *et al. Aust Vet J*. 2017;95(1–2):26.

8. Simancas-Racines D, *et al. Cochrane Database Syst Rev*. 2015;Dec 3(12):CD009745.
9. Cummings K, *et al. J Vet Emerg Crit Care*. 2016;26(5):639.
10. McEntire M, *et al. Vet Clin Pathol*. 2016;Dec 9 [epub ahead of print].
11. Stefanetti V, *et al. Vet Clin Pathol*. 2016;45(3):406.
12. Miglio A, *et al. J Vet Intern Med*. 2016;30(6):1830.
13. Spada E, *et al. J Vet Diagn Invest*. 2016;28(6):705.
14. Spada E, *et al. Am J Vet Res*. 2016;77(8):877.
15. Euler C, *et al. J Vet Intern Med*. 2016;30(5):1642.
16. Odynayo A, *et al. J Am Vet Med Assoc*. 2017;250(3):303.
17. Hourani L, *et al. J Feline Med Surg*. 2017;Jan 1 [epub ahead of print].
18. Guzman L, *et al. J Vet Emerg Crit Care*. 2016;26(2):262.
19. Villarnovo D, *et al. Vet Clin Pathol*. 2016;45(3):411.
20. Hofbauer N, *et al. J Vet Emerg Crit Care*. 2016;26(3):373.
21. Robinson D, *et al. J Vet Emerg Crit Care*. 2016;26(6):766.
22. Euler C, *et al. Vet Clin Pathol*. 2016;45(2):244.

1.
acvim338

Controversies in Veterinary Blood Banking

Sarah E. Musulin, DVM, DACVECC
Raleigh, NC, USA

INTRODUCTION

Medical and surgical advances in veterinary medicine continue to create a demand for readily available, high-quality blood transfusion products. Currently there are neither official guidelines nor governing bodies overseeing the quality of veterinary blood products or blood donor programs. In order to ensure the wellbeing of our blood donors as well as safe and efficacious blood products, responsible practices in veterinary blood banks are paramount. The careful extrapolation of human blood banking and transfusion research to the veterinary field has some value. Additionally, new research is being generated in veterinary medicine, allowing for some evidence-based practices in blood banking and transfusion medicine. This lecture serves to highlight some of these key controversies in veterinary blood banking with a review of the available research and literature. In 2016, Jagodich and Holowaychuk published their summarized results of an Internet-based survey investigating current canine and feline blood banking and transfusion practices at private referral hospitals (n=53) and veterinary teaching hospitals (n=20).[1] Results from this survey are incorporated throughout the lecture.

DONOR CONSIDERATIONS

Blood Donor Screening

- Donor pool - Should blood donors be privately owned or kept as part of an in-house donor colony? Which circumstance creates the least amount of stress for canine and feline donors? Is a closed colony required for minimization of infectious disease exposure?
- Donor health screening - What is the appropriate minimum database for donor health screening?
- Donor infectious disease screening - What are the optimal and minimal standards for blood donor screening of bloodborne pathogens? Should we be testing the donors or the collected blood for these pathogens?
- Donor blood type - Is there such thing as a universal canine blood donor? What is the ideal canine donor blood type? Is DEA 1 compatibility sufficient in dogs?

When selecting blood donors, the health and safety of both the donor and the recipient must be equally considered. A blood donor pool may utilize privately owned dogs and cats or animals that are kept as part of an in-house donor colony. For in-house donor colonies, the housing environment must be clean, comfortable (as dictated by Institutional Animal Use and Care Committee standards), and provide ample enrichment opportunities. Comfortably sized accommodations and adequate exercise and socialization are challenging to provide for in-house canine donors due to their size and energy. Cats can do well when housed in groups with creative options to expand their space vertically with climbing shelves, platforms, cat trees, and nooks for exercise and personal space. The length of donor participation in in-house programs should be less than four years, so that donors can be adopted out to permanent homes after their tenure in a program. Volunteer community donors must be at ease with transportation and the donation environment. Educating community donor owners on the importance of avoiding exposure of their pets to infectious diseases and complying with ectoparasite prevention is important. Pre-donation questionnaires can aid in the identification of potential exposures or risk behaviors. Both in-house and community donors should be observed for stress behaviors such as trembling, hiding, hypersalivation, inappropriate voiding, and be retired from the program when concerns arise.

Donors should be young adults with no history of prior transfusion, infectious disease, or chronic/active health conditions. The screening process should begin with temperament testing in animals that have historically demonstrated comfort in unfamiliar environments and minimal stress with handling. Temperament testing should include mock donations that evaluate ease with restraint and venipuncture. Donor selection can be divided into multiple visits to allow for continued temperament evaluation. Initial health screening may include physical examination and blood work. Standard laboratory screening to ensure the health of the donor may include complete blood count, chemistry panel, urinalysis, and fecal. The goal of donor health screening is to minimize the risk that the donor's health is compromised by regular blood donation. At a minimum, a hemoglobin or PCV to rule out anemia and completion of a health history questionnaire prior to each donation may be deemed acceptable for donor health screening. More extensive donor screening for whole blood, platelet, or plasma donations may include platelet count, clotting times, and/or coagulation factor quantification and plasma vWF concentration. The primary goal of coagulation testing in donors for plasma and platelet products is for product quality assurance. Some donor programs may rely solely on history (i.e., lack of bleeding events, uneventful surgical procedures) to assume adequate coagulation status in donors.

Infectious disease testing should be guided by knowledge of bloodborne pathogens endemic to the donor's geographical area. The American College of Veterinary Internal Medicine published updated consensus guidelines in 2016 for infectious disease testing for canine and feline blood donors in North America.[2] The recommendations are organized into optimal and minimal standards for bloodborne pathogen screening. The panel acknowledges the limitations of testing for infectious agents where no test has 100% sensitivity and specificity. Compared to the previously published 2005 guidelines,[3] the 2016 optimal standards include extensive PCR and IFA testing for many pathogens. Health screening and infectious disease testing should be completed annually or more frequently in donors that become ill, have high vector exposure, and/or live in bloodborne pathogen endemic areas. The overhead expense associated with bloodborne pathogen testing must be considered. Ideally, individual units of blood product would be screened for bloodborne pathogens prior to administration. This practice is currently limited in veterinary medicine due to turnaround time and costs of testing. Blood bank surveillance programs may include saving an aliquot from each donation for random or circumstantial testing for bloodborne pathogens or contaminants. Optimal preventative healthcare should be ensured in all donors, including vaccinations, deworming, and flea, tick, and heartworm prevention. Some blood donor programs assume the costs of screening and preventative healthcare as an incentive for healthy donor participation.

Blood type must be determined in all donors and may be complemented with plasma antibody screening for anti-DEA antibodies in donors with unknown histories. Dogs do not appear to have naturally occurring alloantibodies that cause acute hemolysis with first-time allogenic transfusions, although sensitization may occur. Our understanding of canine blood types is ever evolving beyond the DEA (dog erythrocyte antigen) classification with discoveries in the last decade of novel canine RBC antigens, such as Dal,[4] Kai 1 and Kai 2[5]. To date, the only Dal (-) dog breeds identified have been Dalmatians, Doberman Pinschers, and the Beagle.[6] In Dal (-) dogs, anti-Dal antibodies can be produced via isoimmunization.[6] In a 2017 paper characterizing anti-Dal alloantibody formation following sensitization of two Dal (-) dogs, anti-Dal alloantibodies were detected as early as 4 days posttransfusion, and a shortened RBC lifespan of 21 days was documented in one of the dogs.[6] The newly identified Kai 1 and Kai 2 blood types have unknown clinical significance and warrant further study. The DEA 1 blood type is considered the most clinically relevant blood type due to prevalence and immunogenicity. Naturally occurring alloantibodies to DEA 1 have not been documented in dogs; therefore, an AHTR will not occur with first-time transfusions. The transfusion of DEA 1 (+) RBCs to a sensitized DEA 1 (-) recipient will cause an acute hemolytic transfusion reaction (AHTR) and removal of the transfused RBCs within 12 hours. The presence of alloantibodies to DEA 3, 5, and 7 has been documented in dogs,[7] and transfusion of sensitized DEA 3, 5, or 7 (-) dogs with their respective DEA 3, 5, or 7 (+) RBCs will show sequestration and loss of RBCs within 3–5 days[7]. Given the reportedly high prevalence (up to 50%) of naturally occurring anti-DEA 7 alloantibodies in DEA 7 (-) dogs, Spada *et al.* further characterized the immunogenicity of these anti-DEA 7 antibodies.[8] They found low titers of warm, weakly agglutinating, naturally occurring IgM anti-DEA 7 antibodies in 23% of DEA 7 (-) dogs.[8] Finally, >98% of dogs are DEA 4.[9] There is a single case report of a serious hemolytic transfusion reaction occurring in a previously sensitized DEA 4 (-) dog from exposure to a DEA 4 (+) donor.[10]

In absolute terms, a universal canine blood donor would be DEA 1, 3, 5, 7 and Dal negative. Given that Dal (-) dogs have only been identified in Dalmatians, Doberman Pinschers, and the Beagle, it is unrealistic to populate a blood donor colony with only these breeds. Point-of-care canine blood type testing is only available for DEA 1, which allows expansion of blood donor programs to include both DEA 1 (-) and DEA 1 (+) dogs while minimizing sensitization in recipients. Because naturally occurring alloantibodies are present in DEA 3, 5, and 7 (-) dogs that may cause delayed loss of RBCs in sensitized dogs, DEA 3, 5, and 7 (+) dogs should ideally not be donors. These ideal considerations in donor blood types and antibody screening must be balanced with costs and feasibility.

Donation Frequency and Volume

- How often and how much can canine and feline donors safely donate?
- What is the appropriate screening protocol for iatrogenic iron deficiency in regular donors?
- Should iron supplementation be recommended for canine and feline blood donors?

Regular blood donation can lead to iron deficiency. Iron deficiency can progress to deficient erythropoiesis and anemia when iron stores become diminished. Hemoglobin concentration, hematocrit, and packed cell volume are not sensitive indicators of iron stores. In other words, iron deficiency can be present without anemia. Previous laboratory animal research found that canine donors can donate ~18 mL/kg (~20% of total blood volume) every 3–4 weeks for 1 year without evidence of anemia.[11] More recent canine studies have focused on measuring iron status parameters to evaluate if iron deficiency is a consequence of frequent or chronic blood donation. Mild iron deficiency has been documented in experienced canine donors (~14 consecutive blood donations every ~8 weeks), although in this study the indicators of iron status remained within normal canine reference intervals.[12] A 2015 study measuring RBC and reticulocyte indices in canine donors supports the presence of iron-deficient erythropoiesis in dogs donating ≥6 times annually.[13] Although further research investigating the consequences of chronic donation on RBC regeneration in dogs and cats is warranted, current considerations may include selecting for larger donors, increasing donation interval, providing iron supplementation, and monitoring donors for iron deficiency (rather than just anemia). According to the 2016 Jagodich and Holwaychuk Internet-based survey, the minimum accepted bodyweight for a canine blood donor was 25 kg in 68% (n=34/50) of hospitals; the majority of the hospitals (83%, n=43/52) reported collecting blood from canine donors less often than every 3 months; and none of the respondents reported supplementation of canine donors with oral ferrous sulfate.[1] For feline donors, the minimum accepted bodyweight was 5 kg in 47% (n=26/55) of respondents; donation frequency was less often than every 2 months at 45% (n=25/56) of hospitals; and supplementation with oral ferrous sulfate was only reported in 5% (n=3/55) of hospitals.[1]

Blood Product Considerations

Storage - RBC Components

- What is the clinical significance of RBC storage lesion in veterinary patients?
- Should we tailor the age of stored units to specific patients, conditions, or illness severity?
- How do hospitals ensure product availability while balancing length of storage and wastage of valuable expired product?

The storage of RBC transfusion products allows for immediate access to a lifesaving product. During storage of RBCs, biomechanical, biochemical, and immunologic changes occur that may adversely affect the recipient. These changes may affect the viability, deformability, and oxygen-carrying capacity of transfused RBCs as well as the microcirculatory flow and immune response in the recipient. The human experimental and clinical research on the significance of storage lesion is vast, with evidence both supporting and negating these potential adverse consequences. A 2017 systematic review and meta-analysis that includes the most recent randomized controlled trials (RCTs) concluded that the transfusion of fresher RBCs does not reduce mortality when compared to older RBCs.[14] These findings fall in line with the American Association of Blood Banks' revised 2016 RBC transfusion guidelines that state that patients "should receive red blood cell units selected at any point within their licensed dating period (standard issue) rather than limiting patients to transfusion of only fresher units."[15] Many of the previous human studies evaluating the effects of RBC storage on patient morbidity and mortality were observational and vulnerable to bias, including the often-referenced 2012 Wang *et al.* meta-analysis that included 18 observational studies (12 retrospective, 6 prospective) and 3 RCTs.[16] The authors of this systematic review found that the transfusion of older stored RBCs was associated with a significantly increased risk of death (odds ratio 1.16; 95% confidence interval, 1.07–1.24).[16]

Much of the veterinary literature on the clinical consequences of RBC storage is retrospective. Although inflammatory changes such as increased cytokine concentrations have been shown to be progressive with storage,[17] the duration of RBC storage has not been shown to be a major contributing factor on outcome (discharge, death, euthanasia, or 30-day survival) in an overall canine population.[18] In a subset of patients with hemolysis in the Hann *et al.* study, a longer duration of RBC storage was a negative risk factor for survival.[18] In a 2017 retrospective study, the age of stored RBC products was found to be associated with an increased risk of transfusion-related hemolysis, but no association was found with acute mortality.[19] Pre-administration analysis of donor units for hemolysis was not evaluated, making it unclear if the hemolysis noted was *in vitro* versus *in vivo* and associated with an immunologic reaction.[19]

The changes that occur within the RBC and storage media over time may be mitigated by preservative solutions, such as the addition of buffers or lower salt concentrations. In a 2014 experimental study, four additive preservative solutions were compared for their effect on the *in vitro* quality of canine leukoreduced RBCs stored for 41 days.[20] The pH maintained by the most alkaline solution that contained no chloride, phosphate-adenine-glucose-guanosine-gluconate-mannitol (PAGGGM), was significantly higher than that seen with SAGM or Adsol.[20] All of the solutions gave satisfactory results for % hemolysis and decreases in ATP and 2,3-DPG concentrations during the first 21 days, but by day 31 significant changes were seen in these variables.[20]

It is prudent for hospitals to manage their RBC product inventory with the goal to reduce storage time while meeting hospital needs. According to the 2016 Internet-based survey, canine and feline pRBCs were stored for ≤30 days by 73% (n=46/63) and 76% (n=47/62) of all hospitals, respectively, whereas 16% (n=10/63) and 11% (n=7/62) reported storage for 42 days, respectively.[1] The storage media used by the surveyed hospitals was not reported. The majority (92%, n=67/73) of all hospitals stored blood according to age with 75% (n=55/73) of hospitals storing the most recently collected units at the back of the refrigerator.[1] Almost 60% (n=41/73) of all hospitals reported always using the oldest blood products first; 32% (n=23/73) chose age of RBC products based on patient's condition; 11% (n=8/73) reported using blood products based on size and availability; and one hospital reported using the newest blood products first.[1] In a study comparing inventory management strategies at two veterinary hospitals, a standardized protocol for the priority administration of the oldest stored pRBCs resulted in the administration of significantly older RBCs, but fewer units were discarded due to expiration.[21]

Leukoreduction

- Does leukoreduction have clinically significant benefits in transfusion recipients?
- What are the disadvantages of leukoreduction?

Leukoreduction is the use of an inline filter to remove white blood cells (WBCs) ± platelets depending on the filter. Prestorage leukoreduction removes WBCs and platelets before they undergo apoptosis or necrosis and release potentially undesirable cytokines or breakdown products during storage. The purported benefits are numerous, including decreased febrile nonhemolytic transfusion reactions, decreased inflammatory mediators[17,22] and response in recipients,[23] decreased alloimmunization against leukocyte and platelet antigens, decreased microaggregation, prevention of transfusion-related immunomodulation, decreased postoperative infections, reduced procoagulant properties (decreased microparticle formation),[24,25] decreased transmission of infectious diseases or contaminating bacteria, and improved RBC function and survival[26]. Universal leukoreduction is performed in many countries with the notable exception of the United States. Leukoreduction is an uncommon practice in veterinary medicine. According to the 2016 Internet-based survey, leukocyte reduction filters were used by only 4% (n=2/53) of all hospitals, including one veterinary teaching hospital and one private referral hospital.[1] The major disadvantages of leukoreduction are costs and logistics. Additionally, not all studies support a substantive or clinical benefit of leukoreduction.[27,28]

References

1. Jagodich TA, Holowaychuk MK. *J Vet Emerg Crit Care*. 2016;26(3):360.
2. Wardrop KJ, *et al. J Vet Intern Med*. 2016;30:15–35.
3. Wardrop KJ, *et al. J Vet Intern Med*. 2005;19:135–142.
4. Blais M, *et al. J Vet Intern Med*. 2007;21:281.
5. Euler CC, *et al. J Vet Intern Med*. 2016;30:1642.
6. Goulet S, Blais MC. *Vet Pathol*. 2017;Epub.
7. Weiss DJ, Wardrop KJ, eds. *Schalm's Veterinary Hematology*. Wiley-Blackwell; 2011.
8. Spada E, *et al. Am J Vet Res*. 2016;77:877.
9. Hale A, *et al. J Vet Intern Med*. 2008;22(3):740.
10. Melzer KJ, *et al. J Vet Intern Med*. 2003;17:931–933.
11. Potkay S, *et al. Lab Anim Care*. 1969;19(2):192.
12. Zaldivar-Lopez S, *et al. J Vet Intern Med*. 2014;28:211.
13. Foy DS, *et al. J Vet Intern Med*. 2015;29:1376.
14. Chai-Adisaksopha C, *et al. Vox Sang*. 2017;Epub.
15. Carson JL, *et al. J Am Med Assoc*. 2016;316:2025.
16. Wang D, *et al. Transfusion*. 2012;52:1184.
17. Corsi R, *et al. J Vet Emerg Crit Care*. 2014;24(3):259.
18. Hann L, *et al. J Vet Intern Med*. 2014;28:1830.
19. Maglaras CH, *et al. J Vet Emerg Crit Care*. 2017;27(1):108.
20. Lacerda LA, *et al. Vet Clin Pathol*. 2014;43:362.
21. Holowaychuk MK, Musulin SE. *J Vet Emerg Crit Care*. 2015;25(5):679.
22. Purcell SL, *et al. Am J Vet Res*. 2015;76:969.
23. McMichael MA, *et al. J Vet Intern Med*. 2010;24:1131.
24. Herring JM, *et al. Vet Clin Pathol*. 2013;42(2):163.
25. Smith SA, *et al. Am J Vet Res*. 2015;76:35.
26. Ergul Ekiz E, *et al. Turk J Vet Anim Sci*. 2012;36(6):711.
27. Callan MB, *et al. Vox Sang*. 2013;105:319.
28. Muro SM, *et al. J Vet Intern Med*. 2017;31:410.

acvim144

Everyone's a Critic: Dealing with Negative Reviews and Cyberbullying

Kimberly May, DVM, MS

The Dark Side of Social Media

There are many wonderful things about social media, but it's a double-edged sword. When people can hide behind a keyboard, they're often more reactive and less civil than they would be in a face-to-face conversation. Social media amplifies your message, but it also amplifies the angry critics and provides a megaphone for outcry. Information, misinformation, and allegations can now be propagated faster than ever before, and crises can flare up in minutes or hours instead of days or weeks.

When animals are involved, emotions run high and can quickly spark passionate reactions that rapidly escalate via social media.

Review Sites: Friend, Foe...or Frenemy?

Like it or not, rating and review sites are here to stay, and they're only growing in use. The economic effect of these sites on veterinary income isn't (as far as I know) known, but data on restaurants on Yelp has shown that a one-star increase in Yelp ratings results in a 5–9% boost in revenue.[2]

Yelp.com, as of September 30, 2016

- 25 million monthly mobile unique users
- 72 million monthly mobile web unique users
- 77 million monthly desktop unique users
- 115 million reviews

Some veterinarians have discovered the hard way that you don't have to authorize or consent to a page being set up for your practice. Anyone can start a Yelp page (or even a Facebook community/page) for your practice.

And that's one reason why you should make the effort to own your presence on review sites and social media. Daniel Lemin, author of Manipurated[3], states,

"By neglecting to take control of your online rating and reviews, you're creating barriers for customers to find you and learn about you…Ignoring rating and review sites is as harmful to your business as shutting off your phone or taking down your store sign."

Common complaints about review platforms include 1) weak reviewer verification; 2) lack of support for business owners with grievances about reviews; and 3) lack of transparency of the algorithm that ranks the reviews. Add the three together, and it feels like a perfect storm of small business disasters. The pay-to-play aspect of solicitation calls often set off extortion alarms for veterinary practices.

It's easy to view review sites as foes, but considering reframing your approach and viewing these sites as, according to Lemin, "gold mines for free feedback and advice from your customers."

Getting Positive Reviews

Soliciting positive reviews puts you on the fast track to trouble on review sites (particularly on Yelp), and bribing clients for positive reviews will get you there even faster.

Here's the not-so-secret recipe to getting positive reviews:

1. Provide quality, compassionate medicine for your patients.
2. Effectively communicate with your clients.
3. Tell your clients (through all of your communication channels) what review sites you're on, and that you always appreciate their feedback and help.

DEALING WITH CRITICAL COMMENTS AND NEGATIVE REVIEWS

Common sense dictates that you're not going to satisfy 100% of the people 100% of the time. Negative reviews are a part of doing business, but we often take them personally and become immediately angry or defensive - this is a dangerous reaction when it comes to your online reputation. When has lashing back at an upset person actually helped the situation? (Hint: rarely, if ever.)

When you become aware of a negative comment or review, follow these steps:

1. Take the comment seriously, but take a step back and get a grip on your emotions. This is business, it's not personal.
2. Keep calm. Angry, hostile, or vengeful responses will only hurt you.
3. Reframe your view: this is an opportunity to learn, demonstrate your eagerness to listen, and respond.
4. Remember that there is a person on the other end of the comment/review, and they may have lashed out from fear, frustration, anger, or grief. Don't make assumptions about the person based on your emotional pain from the comment/review.
5. Read the comment carefully
6. Investigate the reviewer.
   a. Do you recognize the name as someone who's been a client?
   b. Do they have other reviews that suggest that all they do is complain about businesses and never seem happy?
7. Read the review itself with an open mind.
   a. Is there a valid concern or complaint in the review that you can address?
   b. Does it express a concern that's been included in other reviews, indicating a problem that you need to address?
   c. Is it full of **all caps** phrases and hyperbole (which most readers will dismiss as unreliable)?
8. Be proactive in your response, and acknowledge their feelings. In many cases, the commenter/reviewer just needs to feel heard. If there's a matter to address, do so in your response.
   a. Remember that your response will not only be read by the commenter/reviewer, but by anyone else who visits the page.

COMMON MISTAKES TO AVOID IN REPLYING TO CRITICISM

- Being **defensive or showing overt denial**; a response of this type makes you look unsympathetic and not open to feedback.
- Responding in anger and/or attacking the commenter.
- Formality and jargon; this makes you seem distant and unsympathetic.

The "Secret Sauce" for Responding to Critics and Negative Reviews

- **Compassion**
- **Confidence**
- **Competence**

HOW BAD REVIEWS CAN HELP YOU

The occasional bad review can actually help your practice. It seems counterintuitive, right? But they can. Panagiotis Ipeiratus, a researcher at NYU Stern School of Business, found that the presence of negative reviews actually helps BOOST the validity of the positive reviews around it. In other words, negative reviews reinforce the power of the positive review. He describes what he calls the "Perfect Review Sandwich:"[5]

- Great review
- Great review
- Amazing or okay review
- Bad review

- Okay review
- Great review
- Great review

Although you (as the business owner) focus on the negative review, consumers see the occasional bad review as a fact of life in the **age of the internet**. Your business actually seems more legitimate and authentic to them because of the occasional negative review. However, if you are getting many negative reviews, there's either 1) something you can do better, and you need to read them with an open mind; or 2) you're being cyberbullied.

FREEDOM OF SPEECH VS. DEFAMATION

I commonly hear from veterinarians who are irate over bad reviews or comments that they feel are defamatory. They're often made more irate when the review site refuses to remove the comment.

Defamation is a statement that injures a person's reputation. It's a pretty broad term, and it's easy to morph it into a definition supporting a claim against every negative comment or review. However, "truth" is always a defense to any claim of defamation; if the commenter/reviewer can provide any evidence that there is some degree of truth to their comment, the defamation claim may be refused. In order to win a defamation case, the plaintiff must prove four things: 1) a false statement represented as fact; 2) publication or communication to a third person; 3) fault/negligence in posting the comment/review; and 4) damage or harm was done.[6]

In general, the law is going to err on the side of favoring free speech and the expression of opinion. If you feel a remark is truly defamatory, seek legal counsel. But first, consider if it's worth the time and expense to pursue legal action. Second, note that even a court ruling in your favor isn't going to undo the reputational damage that's been done. In many cases, your money would be better spent on consultation to mitigate the situation.

CYBERBULLYING: WHEN NEGATIVITY GOES TOO FAR

A survey of AVMA members in 2014[7] revealed that one in five AVMA members had either been a victim of cyberbullying or knew a colleague who'd been a victim. This isn't likely to decrease, either.

In 2014, one of our colleagues (Dr. Shirley Koshi) was driven to suicide by cyberbullying attacks coupled with in-person attacks.[8] The 2015 AVMF "America's Favorite Veterinarian" contest was canceled after vicious cyberattacks were mounted against several of the finalists.[9] One finalist's website was hacked and redirected to a porn site; others were threatened with physical violence and profane diatribes via phone, social media, and/or email; one finalist's charitable organizations were also attacked and vilified; and several were subjected to language that most would consider defamatory.

WHAT CYBERBULLYING IS - AND IS NOT

Definitions of cyberbullying vary, but cyberbullying is not one or a few negative, but constructive, reviews or comments. It is not civil disagreement. In veterinary medicine, cyberattacks are often relentless attacks that may even be organized by a person or group.

The majority of cyberattacks reported in our 2014 survey were due to disputes over care, charges for care/services, diagnosis, or treatment. Seven out of 10 cyberattacks were initiated by former clients or staff. Staff infighting also was a notable source of cyberattacks, including an incident that resulted in a $200,000 hostile workplace lawsuit.

Cyberbullying can result in significant workplace tension and threatened well-being. Fortunately, the economic effects appear to be minor and short-lived in the majority of cases.

The single most important piece of advice I can give you when you're being cyberbullied is that you will survive it. It may be a rough few days, but you will survive. First and foremost, take care of yourself and your team.

How do you deal with it? Don't be afraid to ask for assistance. AVMA members can call the cyberbullying hotline (626-531-1140) to receive expert advice from Bernstein Crisis Management, Inc.

Responding to Cyberbullies

- Inform your team and provide them with messages to respond when appropriate.
- Remember the 3Cs (compassion, confidence, and competence) and use them every time.
- Disable auto responses, and check all scheduled posts to make sure their content won't fan the flames.
- Contact the authorities if you feel that you, your family, or your team is at risk of physical harm.
- Document the evidence, just in case you need it later. This includes taking screen grabs (with notations of time and date) before deleting any posts or comments.
- Seek professional crisis management help if the interactions escalate into an attack; if the media become involved; or if your search results are becoming strongly negative. This cannot be resolved by search engine optimization (SEO) or an SEO consultant, and trying to simply "bury" bad search results will not work. The solution requires active crisis management.

The TL; DR (Too Long; Didn't Read) Version

- If you incorporate the 3Cs (compassion, confidence, and competence), you won't go wrong.
- If you're attacked, always remember that you'll make it through the crisis.

References

1. AVMA's online reputation management website resources. https://www.avma.org/onlinereputation.
2. Luca M. Reviews, reputation and revenue: The case of Yelp.com. Harvard Business School working paper. http://www.hbs.edu/faculty/Publication.pdf. 2016.
3. Lemin D. Manipurated. Available at major booksellers and through Manipurated.com.
4. Bernstein Crisis Management, Inc. www.bernsteincrisismanagement.com.
5. Baer J. How to fight back against negative customer ratings and reviews. http://www.convinceandconvert.com.com/social-media-strategy/how-to-fight-back-against-negative-customer-ratings-and-reviews/.
6. Cornell University Law School. Defamation. Available at https://www.law.cornell.edu/wex/defamation.
7. Mandlik R, Larkin M. Fighting the cyberbully: how harassment can affect your practice. JAVMA News. www.avma.org/News/JAVMANews/Pages. November 15, 2014.
8. Outrage over ownership dispute, veterinarian's death plays out over internet. DVM360. http://veterinarynews.dvm360.com/outrage-over-ownership-dispute-veterinarians-death-plays-out-internet. March 2014.
9. Cyberbullies derail favorite veterinarian contest. JAVMA News. www.avma.org/News/JAVMANews/Pages. September 30, 2015.

acvim145

Preventive Care for Your Online Reputation

Kimberly May, DVM, MS

Why does your online reputation matter?

Think about the last time you visited a restaurant or purchased something online. What influenced your decision to choose that particular restaurant, brand, or item? Odds are, your perception of their reputation was a primary driver, if not the defining criterion that determined your choice. The same applies when prospective clients are searching for a veterinarian. When they go online, what impression do they get of your practice? It can make - or break - your opportunity to reach them.

Your online reputation is an extension of your overall reputation.

First contact

Think of your online reputation as your first point of contact with prospective clients – even before the in-person or phone contact takes place.

Why do you need to monitor your online reputation?

If you're not a participant in a conversation about you, does that mean the conversation isn't taking place, or that someone will defend your honor if misinformation is being provided?

Taking ownership and maintaining your online reputation provides a number of benefits:

- Online visibility - your online presence affects your search engine ranking and visibility to those searching for services
- Economic benefits - from increased visibility online through search engines
- Customer education and client relations - build client loyalty by providing an online community where you are a respected expert
- Environmental scanning and early detection system - know what's being said about you, so you can address problems to prevent crises
- Damage control - the ability to get involved right away to mitigate problems
- Lower stress - when you are aware and can respond, preventing or mitigating problems, there's less stress placed on your team

Online Presence vs. Advertising

For some, an online presence is "advertising" and "marketing," which is unappealing to those who don't feel veterinarians should advertise themselves. Yes, an online presence provides excellent opportunities to market your practice. But what is just as, if not more, important is the **community** aspect. By establishing connections with your clients (and prospective clients), you create opportunities to build bonds and improve compliance. You have an opportunity to increase your value proposition for your clients.

Reframe your perspective

View your online presence, particularly on social media, as community building. Focus on relationship building, education, and engagement with clients.

GETTING STARTED

There's a learning curve, but it is entirely possible for a veterinary practice to establish and maintain their online reputation by themselves - see the checklist in these notes. But if you don't have the interest or the resources to do it yourself, there are services available. For example, AVMA members receive a discount on DVM ReputationGuard (www.bernsteincrisismanagement.com/dvm-reputation-guard) through an agreement with Bernstein Crisis Management, Inc.

BEST PRACTICES

Utilizing best practices in medicine and surgery ensures you're providing quality care for your patients; similarly, utilizing best practices to protect your online reputation provides you with assurance that you're positioned to protect yourself. Many of these best practices provide benefits far beyond online reputation.

- Client- and patient-focused service to build a cushion of client loyalty
- Effective communication, especially about findings, options, and costs - including written estimates
- Staff training on conflict resolution, communication
- Support to prevent/reduce compassion fatigue
- HR policies and approaches to resolve conflict and prevent workplace bullying
- Security measures to protect access to your clinic's communications channels
- Clear policy (and training) regarding online and social media use (personal and for practice)
- Own your website domain and have a mobile-friendly site
- Own your review site presence
- Encourage reviews
- Read and respond (as appropriate) to reviews and comments

AVMA
Our Passion. Our Profession.

Checklist for Monitoring Your Online Reputation

Startup

- ☐ Set up your website; ensure that it is mobile-friendly and contains necessary and relevant information about your practice.
- ☐ Set up website analytics such as Google Analytics.
- ☐ Claim your business' presence on Facebook, Yelp, and other relevant sites (based on surveys/conversations to determine what review sites are used by local clients and friends, and website analytics showing which sites drive traffic to your website).
- ☐ Claim your business on Google My Business.
- ☐ Set a social media policy for staff and train staff.
  - Template available
- ☐ Establish community guidelines/moderation policy for social media and post them on each social media platform you use.
  - Sample
- ☐ Identify potential concerns/issues and write prepared responses to address concerns/criticism that may be expressed on social media, via telephone, via email and face-to-face.
- ☐ Designate staff members responsible for monitoring your social media pages and responding on behalf of your practice.
- ☐ Set up Google alerts for your clinic name, veterinarians' names, possibly other staff members.
- ☐ Set up social media tracking – either automated (e.g., Trackur, Mention, Sendible or Rankur) or manually (e.g. SocialMention) according to a schedule.
- ☐ Post signage announcing your presence on social media and review sites. Create your own, or use the samples below. You can find links to social media logos and usage guidelines here.
  - Sample: "Find us on..."
    - Request from Yelp
  - Sample: "We're on..."
  - Sample: "We're listening and want your feedback. Can you help us out?"
- ☐ Use your other client communication channels – e.g., newsletters – to let clients know how to find you on social media. Get more ideas in Marketing Your Social Media.

Every 6-12 months

- ☐ Re-evaluate, revise (if indicated) social media policy for staff.
- ☐ Re-evaluate, revise (if indicated) community guidelines/moderation policy for social media.
- ☐ Update Google alerts list if needed.
- ☐ Ensure that your clinic's information on social media and review sites is still accurate; update if indicated.

Every 2-3 months

- ☐ Re-evaluate template responses. Edit as needed; add more as needed.
- ☐ Review your website analytics to determine which sites are driving traffic to your website. (For Google Analytics, look at the Acquisition Source and Medium.)

Weekly

- ☐ Perform an incognito search* of your clinic's name to see what comes up in search engines (*this prevents filtering of results).

- ☐ Check social media attitude monitoring/mentions (if social media monitoring is performed manually, such as through SocialMention)

Daily

- ☐ Ask new clients how they found your clinic.
- ☐ Ask clients which review sites and social media channels they use.
- ☐ Promote your presence on review sites and social media (but do not solicit reviews).
- ☐ Check notifications on social media and review sites.
- ☐ Check and respond to new reviews.
    - Template responses available for both positive and negative reviews.
- ☐ Check new comments/engagement on social media.
- ☐ Check emails, other channels for client feedback.

Ongoing

- ☐ Monitor social media notifications.

During potential (or developing) crises

- ☐ Provide your team with consistent messaging and guidance on responding to criticism/comments provided via phone, in person, email and online.
- ☐ Review any scheduled posts to make sure they won't inflame the problem; edit, delete or unscheduled them if you have concerns.
- ☐ Make edits as needed to template responses; develop new messages as needed, and post when appropriate and timely.
- ☐ Monitor responses to your content and your responses to comments.
- ☐ Monitor social media mentions/attitude, and review pages more frequently.
- ☐ Regularly inform and de-brief team.

After problems

- ☐ Hold de-briefing to determine outcomes of actions and identify opportunities to improve future responses.
- ☐ Re-evaluate messaging templates; revise as needed.
- ☐ Re-evaluate website; revise as needed.
- ☐ Re-evaluate social media policy and community guidelines/moderation policy; update as needed.

Source: www.avma.org/onlinereputation

BEST PRACTICES: THE DIFFERENCE BETWEEN GOOD REVIEWS AND BAD REVIEWS

During a pilot review of more than 150 Yelp reviews from veterinary clinics across the U.S. (as found via CustomScoop alerts), I found some interesting things: First, veterinary clinics are reviewed as either 1-star reviews (poor) or 5-star reviews (excellent), with very few 2-, 3-, and 4-star reviews. The pie chart below shows the distribution of the reviews on the 5-star scale:

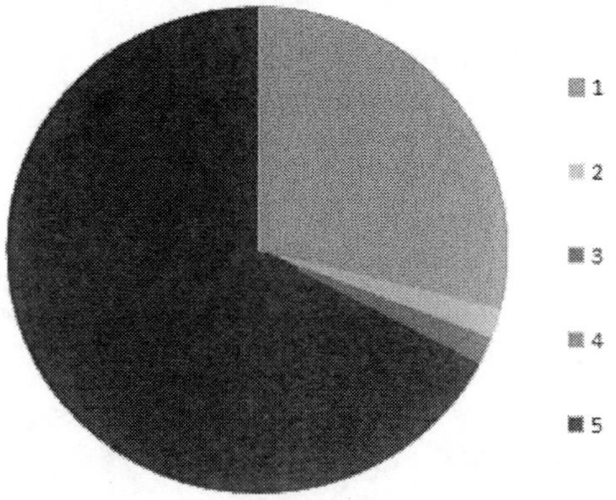

When examining the 5-star reviews, common threads emerged. Compare the word clouds below: the first is from the 5-star reviews, and the second is from the 1-star reviews.

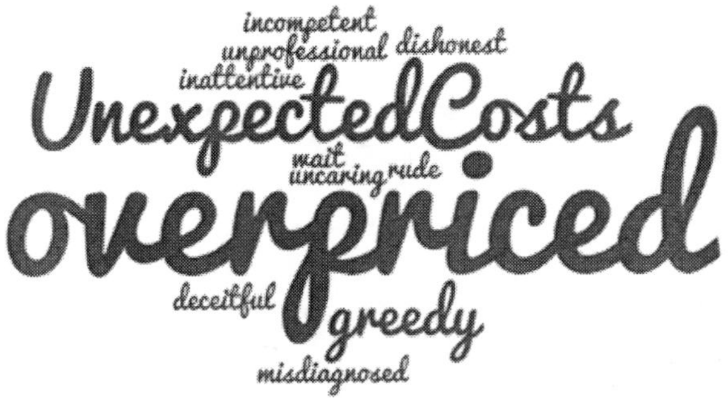

The TL;DR (Too Long; Didn't Read) Version

- Owning your online presence is critical to maintaining your overall reputation and preventing and mitigating crises that could otherwise cause severe damage.
- Utilize best practices, including policies and training, and look at social media and review sites as opportunities to collect feedback and engage with your community and clients.

References

1. AVMA's Online Reputation Management website resources: www.avma.org (VIN editor: Original link was modified as of 4-17-17.)
2. Luca M. *Reviews, Reputation, and Revenue: The Case of Yelp.com*. Harvard Business School working paper, 2016. Available at www.hbs.edu/faculty/Publication%20Files/12-016_a7e4a5a2-03f9-490d-b093-8f951238dba2.pdf.
3. Lemin D. *Manipurated*. Available at major booksellers and through Manipurated.com.
4. Bernstein Crisis Management, Inc. www.bernsteincrisismanagement.com.
5. Baer J. How to fight back against negative customer ratings and reviews. Available at www.convinceandconvert.com/social-media-strategy/how-to-fight-back-against-negative-customer-ratings-and-reviews
6. Cornell University Law School. Defamation. Available at www.law.cornell.edu/wex/defamation.
7. Mandlik R, Larkin M. Fighting the cyberbully: how harassment can affect your practice. *JAVMA News*, November 15, 2014. Available at www.avma.org/News/JAVMANews/Pages/141115a.aspx.
8. Outrage over ownership dispute, veterinarian's death plays out over Internet. DVM360, March 2014. Available at http://veterinarynews.dvm360.com/outrage-over-ownership-dispute-veterinarians-death-plays-out-internet.

9. Cyberbullies derail favorite veterinarian contest. *JAVMA News*, September 30, 2015. Available at www.avma.org/News/JAVMANews/Pages/151015u.aspx.

acvim147

Primary Care Veterinary-Specialist Collaborative Care: Research Update, VetSOAP

Saundra E. Willis, DVM, MvSc, DACVIM (SAIM)
Mukilteo, WA, USA

Introduction

Veterinary Specialist Outreach and Awareness Project (VetSOAP) embarked upon a 3-phase research project to demonstrate that through collaborative care between a primary care veterinarian (pcDVM) and specialist, patients will live longer and receive more positive healthy outcomes/improved quality of life, the pcDVM will receive a positive revenue benefit, and the relationship between primary care and specialty veterinarians will be enriched. This session will present the work that has been completed to date in order to demonstrate the financial benefit of referral/collaborative care to pcDVM practices. Discussion will review the VetSOAP literature review (Phase 1), key findings of the qualitative phases of the research (Phase 2), and research design and possible preliminary results of a KPI Collaborative Care Impact Study (Phase 3). The overall objective of VetSOAP's research is to provide basis for igniting dialog in the profession to embrace collaborative care between pcDVM and specialist and show it has positive revenue, patient retention, and treatment outcomes for the pcDVM.

VetSOAP Vision and Mission

VetSOAP is an initiative committed to the investigation of collaborative care, relationships, outcome, and success factors. The vision is to achieve optimal health care for companion animals and enhance the relationship between primary care and specialty veterinarians. The VetSOAP mission is to demonstrate how companion animals, their owners, and the profession benefit from collaborative care. The strategic goals of VetSOAP are to foster and promote optimal collaboration between primary care and specialty veterinarians, drive research regarding the benefits of a collaborative approach to patient care, establish strategic, sustaining partnerships with members of the companion animal health care community, increase companion owner awareness of the benefits of collaboration between primary care and specialty veterinarians, and maintain a strong financial base through sponsorships from all facets of the veterinary profession. All stakeholders are represented including pcDVMs, specialty veterinarians, corporate practices, independent practices, academicians, industry, specialty colleges, veterinary associations, and universities.

The overall plan for VetSOAP's research is similar to a subjective/objective/assessment/plan or SOAP for a diseased patient. We can't accurately and effectively form a specific treatment plan without diagnosing the disease. An initiative such as VetSOAP cannot create a plan to improve collaborative care between the pcDVM and specialist without knowing the current status of the relationship. Subjective is Phase 1 comprehensive literature review; what do we know (history and physical exam)? Objective consists of the Phase 2 qualitative research/discussion boards and Phase 3 KPI data collection; these are the "diagnostic tests." Assessment, the differential list and prognosis for a patient. For VetSOAP, this is the phase of data interpretation and collation of observations and key findings. Plan, the treatment. For VetSOAP, the results of our work will be a set of initiatives involving all facets of the veterinary industry to promote/foster seamless comprehensive, collaborative care for animals.

Background Research

Pet owner awareness of specialists is improving. In the 2013 ACVIM Member Engagement and Brand Assessment Survey, awareness of specialty medicine for pet owners who visit their veterinarian is 66% of cat owners, and 72% of dog owners.[1] We don't have data for awareness among general pet owners but it is believed to be significantly lower. Of those aware of specialty services, according to the ACVIM survey, 40% of companion animal owners and 35% of equine owners were made aware of these services by their primary care veterinarian. 61% of companion animal owners and 50% of equine owners visit the specialist as a result of a referral from their primary care veterinarian.

A retrospective study by Lefbom *et al.* was undertaken to assess the effects of in-person collaborative care by pcDVMs and board-certified veterinary cardiologists on survival time in small breed dogs after onset of congestive heart failure due to myxomatous mitral value disease, compared to care received solely by primary care veterinarians.[2] Dogs lived 74% longer when they received collaborative care, median survival time 254 days, compared with those who were not referred, median survival 146 days. Although not statistically significant due to the small sample size of the study, the longer the dogs survived, the great revenue reported by their primary care veterinarians. That revenue increased by 22% when the pcDVMs collaborated with a cardiologist. The study showed that collaborative care can provide a positive outcome for both patients and their pcDVM.

VetSOAP Phase 1 Results

The VetSOAP Specialty Referral Literature Review was the first effort of its kind to synthesize research and views among hundreds of published works.[3] Understanding the historical perspective helped to find gaps and contributed to forming questions and hypothesis for Phases 2 and 3. Gaps identified included: Financial impact of client referral on the pcDVM: Does robust referral dynamic result in a healthier more successful veterinary practice and by what measures? PcDVM referral dynamics: What are the characteristics of a successful relationship between primary care and specialty veterinarians? What is the current magnitude of the differential between a pcDVM's perception of a client's willingness to pay and a client's actual willingness to pay? How does the use of pet insurance and/or financing influence the perception of the referral? Client satisfaction and patient outcomes: What is the pet owner's awareness of, understanding of, and expectations related to comprehensive veterinary care? The results of Phase 1 implied that market dynamics, pcDVM referral dynamics, and pet owner dynamics are all critical factors in the process. Further understanding of the financial impact to the pcDVM, the relationship dynamics of the referral, and client satisfaction and patient outcomes of the referral warrant additional qualitative and quantitative study.

Phase 2 Qualitative Study Results

In late May 2016, VetSOAP completed Phase 2, a qualitative diagnostic conversation with 33 pcDVMs through online discussion boards. The participants joined three days of guided online discussions. Through an unbiased exploration, the purpose of this phase was to surface the unknown in preparation for a quantitative questionnaire development, discover and investigate attitudes, perceptions and behaviors related to referral behavior, and surface drives of these attitudes, perceptions, and behaviors. The results are not available publicly. These results do not allow conclusions, but create some validity to the mission of VetSOAP and need for quantitative research to create a plan.

Initial findings of the Phase 2 discussion boards were as follows: Initiating referral: Respondents exhausted their own diagnostics to draw their own diagnoses/conclusions prior to making a referral. Reasons included: ability to work on complex cases increased job satisfaction for pcDVMs, desire of pcDVMs to make sure they referred to the correct specialty, and their desire to maintain their expert status with their clients. Value: pcDVMs see referral as costing the client money and running in-clinic diagnostic tests, as opposed to using an outside veterinary diagnostic laboratory or specialty center, was seen as cost saving. Referral Communications: Respondents indicated improvement in recent years in communications with specialists. However, pre, during and post-referral communication preferences with specialists varied among pcDVMs. Early specialist dialogue with the pcDVM once the patient was seen, was a critical communication. Inclusion in discussion of cases and opportunity for the pcDVM to learn from the specialist appear as key elements in strengthening the collaborative relationship. PCDVM assumptions regarding pet owners; Respondents showed a strong emotional connection with and responsibility to their clients and patients. Responses reinforced the pcDVM's personal perspective and historical relationship with the client can influence their course of action. PCDVMs believed that treatment recommendations were not altered based on their perception of a client's ability to pay.

Phase 3 KPI Collaborative Care Impact Study

VetSOAP is currently in the final phase of a 3-phase research project. In October 2016, after consideration of a proposed behavioral quantitative survey of veterinarians, the VetSOAP Board of Directors instead saw the need to perform a systematic data collection (non-survey methodology) of financial key performance indicators (KPIs) and patient outcomes. A subteam "KPI and Data Strategy Team" met to establish goals and evaluate multiple data collection and analysis opportunities with various technology companies/experts. In October 2016, the VetSOAP BOD redirected Phase 3 of the collaborative care research from a behavioral quantitative survey of veterinarians to a data driven "key performance indicator" study. VetSOAP is in the process of seeking proposals from reputable research organizations to execute this multisite, in-hospital patient study. The study will capture patient outcomes and pcDVM financial impact for 5 specific disease types which may include lymphoma, seizures, cranial cruciate disease, equine inflammatory airway disease, and a dermatologic disease. This next phase of the research will focus on two primary hypotheses: 1) Referred patients have better life span/outcomes (patients with same condition) and generate better patient retention and average annual revenue for pcDVM and 2) Referring practices have better financial profile (revenue & profits), stronger client loyalty, etc. Cause and effect may be more difficult to prove with the second hypothesis but analysis of the results may be informative.

References

1. American College of Veterinary Internal Medicine. June, 2013. Available at www.acvim.org
2. Lefbom B, *et al. J Am Vet Med Assoc.* 2016;249(1):72.
3. Lefbom B, *et al.* In: Conference Proceedings, ACVIM Forum, 2016. Full Report available at www.vetsoap.org.

acvim148

Growing Leaders in Your Practice: What is a Leader?

Wendy Hauser, DVM
Parker, CO, USA

INTRODUCTION

While facilitating a leadership workshop, I asked "Who are the leaders in your practice?" This question stimulated some robust discussion. While some groups discussed the positions in their organizational leadership chart, one thoughtful response was "everyone is a leader in my practice." I asked this practice manager to describe to the group what leadership looked like in her hospital. She explained that every employee was invested in the success of the practice. Rather than defined leaders, the management team in the hospital believed every team member was a stakeholder and had an equal voice in the crafting of policies and procedures. There was a culture of trust within the practice which encouraged constructive conflict. Because every individual participated, the discussions incorporated many different perspectives which allowed for better decisions to be made. This attitude helped to drive a culture of accountability within her practice. These factors helped to shape a healthy and happy work environment where employee turnover was low, there was less burnout, better teamwork and high workplace satisfaction. How did the management team in this hospital create the environment described above?

QUALITIES OF A LEADER

The qualities of effective leaders have been described in many ways. Simply defined, a leader is "a person who motivates a group of people toward achieving a common goal."[1] I believe that this definition is incomplete and that effective leadership requires agility. We live in a world that is rapidly evolving, with a business climate that is uncertain and complex. Our workforces are shaped by the events occurring in the world as our employees reached maturity. As such, we have very diverse generational workforce with diverging ideas about what leadership is and who leaders are. Today's successful leaders are those that possess the following characteristics.

EMOTIONAL INTELLIGENCE

Identified in 1995 by Daniel Goleman, he applied this term to business in 1998, when studies indicated highly developed emotional intelligence was more than twice as important than IQ or experience in predicting success in truly effective leaders.[2,3] Emotional intelligence, or EQ, is defined as the "ability to identify and manage your own emotions and the emotions of others."[4] According to Goleman's research,[5] emotional intelligence is comprised of five separate skill sets. The first three represent self-management skills and are self-awareness, self-regulation, and motivation. The final two are relationship management skills: empathy and social skills. While there is a genetic as well as a nurture component to emotional intelligence, it is widely recognized that EQ increases with maturity. Furthermore, emotional intelligence can be learned. It requires that individuals break old habits and form new ones, focusing on the five EQ skills.

VULNERABILITY

In the not-so-recent past, leaders eschewed any sign of weakness. Admitting to being vulnerable would have been a career-limiting action. While often viewed as synonymous with weakness, vulnerability is a paradox in that it takes great courage to allow oneself to be vulnerable. Brene' Brown, a social scientist who has performed extensive research on wholehearted living and vulnerability, defines vulnerability as "uncertainty, risk and emotional exposure."[6] She further states that "vulnerability is the catalyst for courage, compassion and connection." Great leaders understand that when they allow themselves to be vulnerable, they invite others to be vulnerable as well. These actions require humility, transparency, and trust; they serve as the basis of relational interactions and as an inspiration for loyalty.

COMMUNICATION

Effective leaders are great communicators. They have learned the value in listening with the intent to understand what the speaker is trying to convey. They ask curious questions that "promote discovery and insight."[7] Leaders that build organizations like the one described in the opening paragraph understand the value of encouraging a culture of feedback to and between team members. They don't avoid critical conversations, understanding that unresolved crucial conversations become conflicts. They have learned the value of understanding others' perspectives.

Successful leaders utilize good communication tools such as open-ended questions, reflective listening, and allowing others to complete their thoughts without interrupting. They understand that 93% of communication is non-verbal; they seek to understand the non-verbal messages of those with whom they are conversing and are willing to explore those messages verbally. Exceptional leaders understand the value in being fully present in the moment. As described by James Autry in *The Servant Leader*,[8] being present means "having your whole self available at all times - available to yourself as you try to bring your values to bear on the work at hand, and available to others as you respond to the problems and issues and challenges of team members..."

These leaders know that it is impossible to multi-task and be fully present. They make a conscious decision to apply laser-like focus toward the situation, be it an employee's needs, a patient's care or a client's wants.

AUTHENTICITY

Remarkable leaders know who they are and in what they believe. They uphold their personal core beliefs and align their actions with their core values, regardless of how tempting it might be to compromise themselves for personal, professional, or financial gain. They are true to themselves. In doing so, they set the standard within their organizations for others to live and honor their own values.

SEEK CHANGE

The world around us is constantly evolving. Many leaders may embrace this change in their personal lives, but are reluctant to challenge the status quo within their own hospitals. Outstanding leaders understand the value of being forward-focused. While they are managing the present, they are actively planning for the future. These leaders understand the importance of maintaining a broad professional network, so that they may gain information and clarity around changes that are likely to impact their businesses. They are curious, seeking new ways of doing things. By challenging the status quo, they help to build a culture of innovation and possibility in their hospitals. Finally, these leaders are not afraid of failure. They establish a culture where employees are encouraged to try new ideas. When these ideas fail, the processes are dissected so that the failure becomes a valuable learning experience. They ask "what went well?" and "what could be better yet?" Team members are encouraged to apply these lessons and try again.

FOLLOWERSHIP

As our understanding of leadership has evolved, so too has an awareness of followers. Historically, leaders have been individuals that use influence to persuade followers to achieve pre-determined goals. What role do followers play in the leadership process? This question has formed the basis for a new field of research, followership. This area of study has accelerated in the past 9 years, when it became apparent that "leadership cannot be fully understood without considering the role of the follower in the leadership process."[9] It is important to recognize that leadership is impacted not by the role of follower, but as result of 'following behaviors.' Without this, leadership does not occur. As discussed by Mary Uhl-Bien,[9] "it is probably easier to recognize leadership in following behaviors than it is in leadership behaviors, as individuals attempting to be leaders are only legitimized in the responses and reception of those willing to follow them."

Currently, followership is viewed in two ways. The first is role theory,[9] seen as the rank or position the follower occupies in the traditional hierarchical relationship. Role theory reverses the focus of the interaction from a leader-centric position to consider "how followers influence leader attitudes, behaviors and outcomes."[9] Followership characteristics form the followership behaviors, which in turn impact leader perceptions and behaviors. The result is followership and leadership co-created outcomes.

The second way to consider followership is from a 'constructionist' view, in which followership is a "relational interaction through which leadership is co-created through combined acts of leading and following."[9] This approach considers how individuals and groups relate and connect to develop leadership and followership. Based on relational interactions, this model acknowledges a mutual influence process between leaders and followers, where the roles may shift as circumstances dictate.

MILLENNIALS IN THE WORKPLACE

An understanding of the followership concept is important in developing awareness around evolving leadership models. The significance of this concept becomes even more critical with the presence of the millennial generation in our hospitals. Millennials, as a group, share some unique characteristics that will impact their preferences for leadership style. They are the most diverse generation, socially and racially. Due to their acceptance of variety, this generation views problems broadly, applying concepts that work in one area to problems that arise in others. They have also been raised with an 'elevated status,' where their opinions have been solicited and considered in decision-making paradigms. Therefore, hierarchal boundaries are less clear. They will seek to collaborate with whomever can help in the attainment of their goals.

A commonly used model of leadership in many hospitals is transactional leadership, when followers behave in ways defined by their leaders. This leadership style is incompatible with the needs of the millennial generation. To succeed, leaders will be required to flex and adapt, adopting leadership styles that are more transformational in nature. Transformational leadership is described as leadership that occurs "when one or more persons engage with others in such a way that leaders and followers raise one another to higher levels of motivation and morality."[10] Culturally, hospitals will need to adapt in ways that support the core values defined by this generation, such as collaboration, teamwork, transparency, and relational interactions.

CONCLUSION

Through carefully selecting and cultivating their hospital leaders, the organization described in the first paragraph nurtured a culture where followers and leaders worked hand in hand, where all team members were empowered and felt valued. Their ethos was attractive to millennials, encouraging this promising generation to bring added value and innovation to the workplace. What changes can you make to emulate their success?

REFERENCES

1. Ward S. Leadership, about money. https://www.thebalance.com/leadership-definition-2948275.
2. Golemen D, Boyatzis R, McKee A. *Primal Leadership: Learning to Lead with Emotional Intelligence*. Boston, MA: Harvard Business School Press; 2002.
3. Fernandez-Araoz C. *Great People Decisions*. Hoboken, NJ: Wiley; 2007.
4. Emotional intelligence. https://www.psychologytoday.com/basics/emotional-intelligence.
5. What makes a leader. https://hbr.org/2004/01/what-makes-a-leader.
6. Brown B. *Daring Greatly*. 2012.
7. What great listeners actually do. https://hbr.org/2016/07/what-great-listeners-actually-do.
8. Autry J. *The Servant Leader*. 2001
9. Uhl-Bien M, Riggio R, Lowe K, Carsten M. Followership theory: a review and research agenda. *The Leadership Quarterly*. 2014;25(1):83–104.
10. Balda J, Mora F. Adapting leadership theory and practice for the networked, millennial generation. *Journal of Leadership Studies*. 2011;5(3):13–24.

acvim149

Growing Leaders in Your Practice: How to Cultivate Leadership?

Wendy Hauser, DVM
Parker, CO, USA

Introduction

As veterinary professionals, we are trained to practice preventive care. When our patients become ill, we are well-prepared to diagnose and treat illnesses. What if the "patient" was not furred or scaled, but composed instead of brick and mortar? What if the patient's ailment is a lack of direction fueled by miscommunication and an absence of shared purpose? What steps do you take to resuscitate an "ADR" hospital?

The importance of "organizational health" is underestimated in many veterinary hospitals. Organizational health refers to the way management, operations, culture, and strategy integrate form a cohesive organization. While proficient leadership is the cornerstone of organizational health, this critical component is often ignored or allowed to occur in happenstance fashion, resulting in aimlessness and missed opportunities.

Who Needs a Practice Manager?

Consider the question above: "Who needs a practice manager?" I believe that **every** veterinary hospital would benefit from hiring a professional manager. Veterinarians are rarely trained in business and infrequently possess keen interest in this aspect of hospital ownership. Oftentimes, the veterinarian relegates management responsibilities to a low priority item, giving attention to them after completing their primary roles as doctor in the practice. Professional managers are the key to developing a successful veterinary hospital. Their efforts directly impact the ability of the hospital to provide updated facilities, new equipment and continued staff training. This impacts the client experience, increased adherence to recommendations and ultimately, the health of our patients. Using the example of the practice manager position in a hospital, consider the following regarding recruitment and training for **all** hospital leadership roles.

How Managers are Selected

Often, managers are promoted from within as a "reward" for longevity or because it is "easy" and not carefully selected for skills, competencies, and potential. The outcome is a mismatch in the needs of the position and the ability of the selected individual to fill those needs. As leaders, we make these decisions based on an innate psychological bias that we as humans possess, which is over-rating capability. As described in his book *Great People Decisions*,[1] author Claudio Fernandez Araoz explains that over-rating capability is "based on two incorrect assumptions. The first is that people can change more quickly, and to a greater degree, than they actually can" and secondly "that a high correlation exists between the **motivation** to perform and the actual **capacity** to do so." He goes on to state that a strong interest in the position and high motivation is not enough to overcome a basic lack of key competencies, attributes, and experience.

Managers should be selected based on a variety of factors, including emotional intelligence, their potential (can they continue to grow and adapt in an ever-changing business climate, taking on larger challenges with an eye toward planning for the future?) and the alignment of their personal values with the hospital's core values. When selecting managers, the talents they possess should be carefully compared to the key competencies needed in the position.

The Role of the Manager in the Veterinary Hospital

The lack of clearly defined managerial priorities can sabotage the ability of a manager to perform even the most basic of responsibilities. For a manager to bring value to the role, there must be a clear understanding of the responsibilities the position assumes. Ideally, these competencies are defined prior to undertaking the search for the manager.

Once the manager is hired, it is critical that the veterinarians and owners **get out of the manager's way**. It is difficult for veterinarian/owners to "let go" and allow the manager to perform their job. I believe there are many reasons for this, including the difficulty in trusting the person selected with a business tied so intimately to the veterinarian's personal and professional well-being. Simply stated, veterinarians need to work to develop vulnerability-based trust in their managers, viewing them as partners in working toward the continued success of the hospital. When this trust is developed, the tendency of the veterinarian to micro-manage dissipates.

Mentor Your Manager

Mentoring is defined as an "employee training system under which a senior or more experienced individual (the mentor) is assigned to act as an advisor, counselor, or guide to a junior or trainee. The mentor is responsible for providing support to, and feedback on, the individual in his or her charge."[2]

This is very different from micromanagement, defined by businessdictionary.com as "close, detailed, and often de-motivating scrutiny of employees' work on a continuing basis."

Create an agreement between the manager and the veterinarian that holds each person accountable to the other. Take the time to understand what the manager needs to be successful in their position, such as regular meetings with the leadership team to discuss business metrics, team performance, and new opportunities. Create a system so that the veterinarian/owner knows where the manager is focusing his/her time and provide on-going feedback to the manager. The manager should be involved in the daily operations of the hospital; the veterinarian should be observing from a 30,000-ft. level.

Provide Educational Opportunities for Your Manager

Seek out ways for managers to grow in their competencies and skills. This investment is one of the most valuable a veterinarian will make for their business and in fulfilling a very important mission of a committed leader, to help followers achieve personal and professional development and fulfillment. There are many educational resources available managers; see the resources list at the end of this article for ideas.

Encourage Networking and Collaboration

Support managers in developing a strong professional network. Having other managers and leaders as peer resources provides the opportunity for collective and creative problem solving and insight. Ways to achieve this are to encourage participation in the Veterinary Hospital Manager's Association. Encourage managers to participate in a local manager's group. Consider joining a business group, such as VMG, that encourages veterinarians and their managers to work together to achieve specific business goals and attend meetings as a team (this is a critical component to realizing goals; if both the veterinarian and the manager are at the table together, they can hold each other accountable for reaching these benchmarks).

Conclusion

By approaching each formal leadership role in the hospital with forethought, clear expectations, and a healthy system of feedback owners will capitalize on the talents that their teams can contribute to their hospitals. Doesn't a veterinary hospital deserve a dedicated caregiver too?

Resources

There are many great learning resources for developing managers. Below are a few of my favorites:

- American Animal Hospital Association (www.aaha.org/professional/education/default.aspx)
    - Veterinary Management School (VMS)
    - Veterinary Management Institute (VMI)
    - AAHA Yearly Conference
    - **Coming soon: Peak Executive Transformations** (perfect for advanced level managers)
- Veterinary Hospital Managers Association (www.vhma.org)
    - Webinars
    - Certified Veterinary Practice Manager Certification program

   - Regional and National Conferences
- Patterson University (www.pattersonvet.com/SignatureSolutions/Education/Patterson-Veterinary-University)
   - Virtual (on-line learning) and regional classes

REFERENCES

1. Fernandez Araoz, C. *Great People Decisions*. 2007.
2. Mentoring. www.businessdictionary.com/definition/mentoring.html.
3. Today's Veterinary Practice. http://todaysveterinarypractice.epubxp.com/i/473047-mar-apr-2015.

acvim150

You Are a Skilled Veterinarian...What's Next?

Wendy Hauser, DVM
Parker, CO, USA

INTRODUCTION

I vividly remember the conclusion of my academic career. I was in a room, participating in a rounds discussion with nine classmates when we were dismissed. As my classmates jumped and hooted with glee, fleeing the room to meet other classmates exhibiting the same amount of enthusiasm, I stayed seated. I recall thinking that I was going to miss the academic rigor and structure in which I had been immersed for the past four years. I also recall being scared. I had the book knowledge to be a veterinarian, but I didn't feel like one.

Retrospectively, this was the first of many career transitions that I would undertake. The first years of my career were spent learning how to be a proficient veterinarian, how to communicate effectively with teams and clients and gaining confidence in my abilities. As I became a self-assured doctor, I found that I wasn't satisfied with my role as an associate veterinarian. I discovered a new passion, veterinary business, and with determination began to learn how hospitals become successful. These lessons, often painful, continued as I transitioned from associate to hospital owner. As I pushed myself to become a knowledgeable owner, I maintained my clinical role. As the years passed, I learned new skills, such as advanced dentistry, to continue feeling challenged as a doctor. After a successful hospital sale, I have appreciated the flexibility that a DVM degree provides. I have been fortunate to have worked in industry as a technical services veterinarian, to be an entrepreneur again as a veterinary business consultant, and most recently to re-enter industry as a veterinary relations and business development executive.

How did I decide what path my career would take? What factors determined in which roles I would flourish and grow? Was it blind luck, or calculated risks? As I made career transitions, I found the following to be most helpful in making 'jumps to the next lily pad.'

WHO ARE YOU?

To determine the types of positions that will be most satisfying, it is important to understand yourself. There are many tools to aid in identifying these criteria. Below are suggestions to help you begin.

PERSONALITY PREFERENCE ASSESSMENTS

As individuals, each of us have personality preferences that influence our thought processes, emotions, and behaviors. These psychological characteristics also influence how we relate to others, in how we act and what we say. Self-understanding and acknowledgement of these preferences will help us identify our work preferences. Simply defined, work preferences are "the outcomes individuals desire from their engagement in paid work."[1] When considering work preferences, those positions with the best alignment between personal characteristics and work environment produce the most positive outcomes to both the individual and the organization.

How do you learn what your personality preferences are? There are many on-line assessment tools to provide this insight, such as Myers-Briggs Personality Type Indicator, DiSC, and Insights Discovery. Gaining this knowledge will help you begin to identify reasons that your current work environment is unsatisfactory. Is it the particular work setting or is it the type of work that is inadequate in meeting your needs?

STRENGTHSFINDER

Strengthsfinder is a different type of assessment tool. Rather than focusing on personality traits, Strengthsfinder assesses what you do best. Many veterinary professionals view their capabilities from the other side of the lens; what do they need to do better? As such, we miss the opportunity to capitalize on our innate skills and knowledge. A cost-effective on-line assessment tool, Strengthsfinder determines your top five strengths and provides resources to help you understand and build on these competencies.

What's Your Why?

Simon Sinek, a social scientist, popularized the concept of the 'golden circle' in books[2] and a Ted-X talk[3]. He postulated that individuals and organizations don't understand 'why' they do what they do, because they start with the 'what' and the 'how.'

He explains that 'what' and 'how' are functions of the neocortex. This part of the brain is responsible for processing facts, figures, details, and logic; all important functions but ones that rarely move a person to action. Rather, our behaviors are best motivated through our limbic system, which process feelings and emotions and corresponds to the 'why.' As he states, the 'why' is why we get out of bed in the morning and why anyone should care? It is the essence of who we are, our primary driving force behind our actions.

When considering career moves, a clear understanding of your 'why' is essential. With this information, you will be able to determine which positions and organizations offer the best alignment to and for you. I have engaged in some healthy debates around the evolution of a personal 'why.' I believe that because your 'why' represents what you hold sacrosanct, one's 'why' is a constant throughout life.

To illustrate this, consider my 'why' as a practicing DVM: "I am a veterinarian because I love helping people by keeping their beloved pets healthy. By forming strong partnerships with my clients, my patients benefit."

Now consider how my statement changed as I entered the various permutations of my professional career: "I am a veterinarian because I love helping people by keeping their beloved pets healthy. By forming strong partnerships with my clients, their clients and patients benefit."

Despite many role changes, my fundamental 'why' has remained untouched. The 'what,' or details and facts, changed.

Core Values

Core values "are the guiding principles that dictate behavior and action."[4] Individuals have core values, as do organizations.

It is necessary to understand your core values before actively pursuing a career move. With this knowledge, you can assess which career changes and organizations have the highest likelihood of creating an impactful and long-lasting match.

What's Next?

Once you have an understanding and can define what is important to you, it is time to think about "what's next?" Consider the following viewpoints as you contemplate options.

The Starting Point

Why are you contemplating a career move? Dissecting what is going well with your current position and what could be better will help provide insight into possibilities for your future. Are the core values of your current organization in alignment with your own values? Does your personal 'why' fit with the focus of the organizational 'why'? Are you still learning and being challenged in your current role?

New Skills Versus a New Job?

If you are feeling stagnant in your position, what would make it more fulfilling? Do opportunities exist to develop new competencies? Considerations include advanced communication skills, dentistry, behavior, new surgical techniques, rehabilitation, business management skills, and leadership. When negotiating with a hospital owner or manager to add new or enhanced profit centers and services, be prepared to discuss the return on investment to the patient, the team, your growth, and the business. Provide hard facts: what will additional training and equipment cost? How long will it take to implement? How will it be marketed to clients? How sustainable is the investment? How will you measure outcomes?

As a practice owner, I had three associates, at separate times, which believed in the benefits of acupuncture and wanted to offer those services in our hospital. I agreed to pay for these associates to attend classes and become certified. As a profit center, acupuncture represented less than 1% of our total revenues. Sometimes return on investment is measured tangentially. Patients benefitted, clients were appreciative and my associates developed new skills and interests.

If your work preferences have changed, options include practice ownership, self-employment, employment in industry, a different discipline within veterinary medicine, or exploring new career fields outside of veterinary medicine. As you evaluate these possibilities examine what the benefits of such a move would be personally, professionally, and financially. What commitments are necessary to make this move? For example, pursuing a different discipline within veterinary medicine might require additional education. Pursuing residency training often require decreases in household income, increased debt, and physical relocation. How do the benefits of such a decision outweigh the costs? Practice ownership often requires huge time commitments, undertaking of large sums of debt, and the need to perform many different job roles within the hospital. The rewards can be huge, in terms of financial performance, personal and professional growth, flexibility, and the satisfaction of working with a team to bring your visions to life.

Networking

A critical component to answering the question "what's next" includes "who can help?" In all aspects of these mid-career and late-career roles, trusted advisors offer differing perspectives and valuable life lessons that impact the decision-making paradigm. Many career positions are more readily accessed through personal connections. A robust network of acquaintances is invaluable as you position yourself to move toward new career paths.

Networking can be a challenging skill to master. It can feel disingenuous; for introverts, making new connections can be uncomfortable. When approached with the right mindset, networking is liberating. By meeting new people, you are exposed to new ideas and have opportunities to broaden your perspectives. Networking allows you to identify common interests and form collaborative relationships. By focusing on what you offer, rather than take, networking feels more authentic and is infinitely more rewarding.

How and where can you network? Everywhere! Opportunities to form strong networking groups exist within your community, your local and state Veterinary Medical Associations, other organized veterinary medical groups, veterinary meetings, industry sales representatives, your veterinary school classmates, former employers, LinkedIn, and professional groups in the community such as Chamber of Commerce and dedicated networking groups.

Assessing the Fit

What methods and tools will provide confidence that the career move you are deliberating is a positive one? This multifactorial question is answered by evaluating the data that you have collected from the "who are you" section above, in addition to the exploratory actions undertaken in "what's next." I have found a particularly helpful tool to assess the fit of a new opportunity is to perform a SWOT analysis. This analysis evaluates the strengths, weaknesses, opportunities and threats of the contemplated change. When combined with the strengths from Strengthsfinder, this assessment helps bring clarity to appraise the wisdom of the career move. An example may be found at the conclusion of this article.

In assessing the fit, the relative risk of the move and available resources should be taken into account. If the decision is made to become a practice owner, particularly if the hospital will be a start-up with no cash flow, or if you plan to establish another form of self-employment such as consulting, mobile ultrasound, or rehabilitation services, do you have enough income to meet your expense obligations for the next 6–12 months?

If your chosen career path will be in industry, will you be able to relocate within a timely fashion? To what parts of the country would you be willing to move? Are you willing to spend a percentage of your time away from your home and family, traveling? Are you willing to work some weekends?

Finally, when would you like to make the career change? As you can determine from this article, wise career changes require effort, time, patience, and perseverance. In my experience, the journey is worthwhile!

SWOT Analysis Example

Hypothetical position: Technical Services Veterinarian (TSV).

Use the competency descriptions from the job posting to initiate the SWOT analysis, complete with information about your core values, 'Why,' Strengthsfinder, and personality preferences. For this example, we will use the strengths: learner, achiever, input, individualization, relational, indicated in the chart in brackets.

Strengths	Weaknesses
Relational job [relator, input, individualization] Educational; work with vets , AHC teams and students [relator, input, individualization] Business development [achiever] Coaches/rides with field team [relator, input, individualization, achiever, learner] Good income Great benefits Company Car	Overnight travel Accountable for sales targets, regardless of environmental impacts such as recession or manufacturing problems Corporate Setting/less personal?
Opportunities	**Threats**
Provides ability to develop higher level networks within profession [relator, individualization, input] Will require development of new skills [learner, input] Will function on different teams such as territory sales team, TSV Team [relator, input, individualization] Will have access to advanced training opportunities such as leadership, coaching [learner, relator, input] Will work with R&D to understand new product development [learner, input, relator]	Highly competitve market Impact of economic downturn on position? Possibility of continued consolidation within Industry and loss of job Not a challenging enough role ? Loss of clinical skills if not practicing medicine

SWOT

References

1. Gilbert RG, Sohi SR, McEachern GA. Measuring work preferences. A multidimensional tool to enhance career self-management. *Career Development International*. 2008:13(1):56–78.
2. Sinek S. *Start With Why*. 2009.
3. Sinek S. Ted Talk: Start with why. https://www.youtube.com/watch?v=sioZd3AxmnE.
4. Examples of core values. http://examples.yourdictionary.com/examples-of-core-values.html.

acvim151

What Motivates You?

Wendy Hauser, DVM
Parker, CO, USA

Introduction

I was facilitating an interactive workshop for leaders when I asked "How do you motivate your employees?" This seemingly simple question resulted in silence as each participant considered the question, uncertain how to best respond. What would you have said?

A trait highly desirable in leaders and followers alike, motivation is defined as "internal and external factors that stimulate desire and energy in people to be continually interested and committed to a job, role or subject, or to make an effort to attain a goal."[1] As authors Tomas Chamorro-Premuzic and Lewis Garrad note "the science of motivation is robust and well-established, it is rarely applied to real-world management practices, which tend to be based on managers' intuition and subjective experience."[2] Is it any wonder that the question posed above caused confusion? How does a leader or an individual learn to apply motivation to life situations?

Types of Motivation

Fred Lee, in *If Disney Ran Your Hospital,*[3] describes four basic levels of motivation.

The first, or level one, centers around 'compliance,' which is described as "a disposition to yield to others."[4] This type of motivation is often referred to as the carrot and stick model. It depends on extrinsic motivators to achieve the completion of a task by a person. Once the external factor is removed or is no longer desirable, the motivation dissipates. As such, this is considered the weakest of the motivational types.

Level two is imposed motivation, termed by Fred Lee as 'willpower.' Categorized as an extrinsic motivator, 'willpower' occurs when the individual completes a task not because they enjoy it. Instead, they feel motivated out of a sense of duty, obligation, or fear of disappointment.

The third level in Mr. Lee's motivational diagram is 'imagination.' As he explains, this is a stronger level of motivation because the person is moved to do something that they enjoy and find rewarding. The purpose to attain goals comes from within, so this level is classified as intrinsic motivation. It is at this level that motivational forces are sustainable.

The final level, level four, is 'habit.' This is such a strong intrinsic motivator that it would be unquestionable not to perform the associated tasks as the unfinished tasks would the individual feeling incomplete.

Other authors focus on defining motivation by examining our basic needs as humans. Maslow, in the mid-1950s, suggested that the fulfillment of certain psychological needs is as essential to humans as vitamins. In the decades that followed, many additional researchers built on these concepts. Deci and Ryan proposed the foundations of what has become self determination theory (SDT). SDT states that there are three basic psychological human needs: autonomy, relatedness, and competence. These needs are uniform for all people, despite cultural, geographic, and genetic differences. This needs-as-requirements perspective is often used to support the premise that "satisfying these needs is leads to positive outcomes, such as increases in well-being or performance towards goals, whereas thwarting the need leads to ill-being or performance decrements."[5]

In her book, *Why Motivating People Doesn't Work…and What Does*[6] author Susan Fowler describes autonomy as "the human need to perceive we have choices…It is our perception that **we are the source of our actions**." In her explanation of relatedness, she states it "is our need to care about and be cared about by others…it is our need to feel that we are contributing to something greater than ourselves." She says "competence is our need to feel effective at meeting every day challenges and opportunities. It is demonstrating skill over time. It is a sense of growth and flourishing." She further explains that these three needs are inter-related. When basic psychological needs are being satisfied, the outcomes are "positive energy, vitality and a sense of well-being." However, if one of the needs is unfulfilled, the advantages rendered by meeting the remaining needs is diminished.

To more fully explain how the SDT impacts our motivational state, Fowler developed the *Spectrum of Motivation*[6] model.

It is based on the observation that individuals are **always** motivated; it is the degree of motivation that varies. This model describes six motivational attitudes. These viewpoints are driven both by self-regulation, which she defines as "mindfully managing feelings, thoughts, values, and purpose for immediate and sustained positive effort" as well the psychological needs previously discussed. She has further identified two separate motivation states: suboptimal and optimal. Suboptimal, or low quality motivation, occurs when self-regulation and fulfillment of the psychological needs are depressed. Needs go unmet and the motivation is not constructive nor sustainable. Optimal, a high quality motivational outlook, ranks highly in both satisfying psychological needs and self-regulation. This category promotes positive outcomes that are energizing and rewarding. Because the mental state experienced is so satisfying, we seek additional opportunities that would promote similar feelings.

As defined by Fowler, the six motivational attitudes are, from lowest quality (low self-regulation, low psychological needs) to highest quality (fully self-regulated and meeting all three psychological needs) are disinterested, external motivational outlook, imposed motivational outlook, aligned motivational outlook, integrated motivational outlook and inherent motivational outlook. You might appreciate some similarities between some of these attitudes and the four basic levels of motivation as described by Lee. Alignments are seen when comparing 'compliance' and external motivational outlook, 'willpower' and imposed motivational outlook, 'imagination' and aligned motivational outlook and 'habit and integrated motivational outlook.

A defining feature between the two authors' depictions of motivation is best appreciated by recognizing that Lee's model portrays highly categorized, stationary motivational states. By contrast, Fowler's model is predicated on the theory that motivational states are constantly in flux. During the completion of one activity, it is possible to be in all six states at different stages.

An example of this fluidity might be illustrated by examining the process of writing this proceedings article. While I very much enjoy the process of writing, there are some topics that are less challenging to write about than others. The topic of motivation fascinates me; however, explaining the psychological aspects of motivation required hours of research. I started this project from the unhealthy starting point of an imposed motivational state: I believe it is important to provide attendees of the ACVIM conference with information that will aid in the understanding and application of the topics on which I will be presenting. Therefore, the option of not writing an article about motivation was untenable. We don't live in a vacuum; as such, multiple activities and obligations competed for my time. This article, the fourth of four, was authored as I had speaking engagements and the initiation of a new employment situation. My psychological needs of competence and autonomy were not balanced, which led to a sub-optimal motivation quality. As I delved further into the writing project and the research, I shifted to an aligned motivational outlook. This optimal motivation state is characterized by a higher self-regulation state in that my work had purpose and satisfied my psychological needs. Writing this article shifted from being a chore to an activity that I enjoyed.

To further underscore Fowler's concept of fluctuating motivational states, remember that as humans we rarely ever perform one task exclusively. As I mentioned in the paragraph above, within the time frame of authoring this article, I was also preparing for a three-hour workshop on communications. My motivational state for that activity ranged between imposed (updating power points and practicing the presentation) to aligned as I added new material, and finally integrated as I presented the workshop. Assessing my motivational state after the meeting, I was in an inherent state, the highest and most healthy of all the six motivational attitudes. Similarly, I was experiencing a range of motivational states as I prepared to embark on a new career path.

Applying Motivation to the Workplace

Understanding Fowler's spectrum of motivation is key in recognizing that individuals are always motivated. It is the quality of the motivation that varies. The reason that the participants in the workshop had such a difficult time answering the question "How do you motivate your employees?" is that on a foundational level they understood that their employees were already motivated. The more valuable approach is to ask each team member "What motivates you?" When managers tailor an individual's job expectations with the qualities of self-regulation[6] (mindfulness, values, and purpose), a shift in the motivational outlook occurs. The employee experiences an optimal motivational state, finding meaning and fulfillment in their work.

Fowler ascertains that motivation is a skill and as such, can be coached. To effectively teach this skill, leaders must first master their own motivational states. This internal dialog necessitates that you break old habits and learn new habits by retraining the limbic system of the brain, responsible for feelings and impulses.

Some steps that are useful in reprogramming this system include:

1. Where are you now on the "Spectrum of Motivation"?
   Consider which activities result in your optimal motivation states and "why" this happens. Ask trusted peers, mentors and family members if they have observed endeavors in which you have been fully engaged. Which ones have been harder for you to accomplish, as seen by procrastination, sub-standard performance, lack of enthusiasm, or manifested emotionally, in responses that seem out of proportion to the situation? Do your tasks align with your values and purpose? Are your basic psychological needs of autonomy, relatedness, and competency being satisfied?
2. Where do you want to be?
   Are you in a healthy motivational state? How can you reframe necessary activities in which you find yourself in a perpetually sub-optimal motivation state so that the activities align to your values and purpose? Are there some non-critical activities that need to be purposely abandoned? What do you need to do to consciously shift to a healthier motivational outlook?
3. How will you get there?
   Recognizing your motivational states involves self-reflection. Taking time during the activity to ask "Where am I on the Spectrum of Motivation and why am I there?" This self-awareness is critical in honestly managing your self-appraisal process. This is the first step in learning to consistently stay in the optimal motivational positions.

To help others understand their unique answers to "what motivates" the individual, leaders must be willing to engage in conversations with each team member. It is helpful for the team to first undergo some core values work, as well as understanding their own "why." Please reference the information contained in You Are a Skilled Veterinarian...What's Next?[7] Observations that might encourage these conversations are when leaders see behaviors that suggest sub-optimal motivation, as discussed in #1 above. Other factors that might result in these conversations are when there are team members that have unrealized, recognizable potential or when workplace situations occur that are unhealthy to the leader, an individual team member, or the entire team. These are often recognized by feelings of resentment, frustration, tension, stress, or dissatisfaction.

When engaging in these conversations, your goal is to help facilitate a move to an optimal motivational outlook. I use the word 'facilitate,' because the only person who can make this shift is the individual with whom you are conferring. This is a coaching conversation, best approached by asking questions to help the employee find their own answers. As such, leaders should refrain from problem solving and imposing their needs in the conversation. Leaders should identify what questions could be asked to help the team member understand their motivational state and find ways to shift their current outlook. It is important to be fully present for these conversations; conduct them at a time when there will be no distractions. Finish clearly with the team member, and schedule times for follow-up conversations and coaching.

CONCLUSION

When contemplating motivation, remember that it is not a leader's job to motivate their teams! They are already motivated. It is a leader's job to help themselves, as well as their team members, understand 'why' they are motivated and to help them learn how to manage their motivational outlooks. When your team members are in an optimal motivation state, their purpose and values align with their daily activities. Their psychological needs are met. These factors help them to be fully functional, thriving contributing members to your workplace culture. As Susan Fowler concludes "People can flourish as they succeed. This is the promise of optimal motivation."[6]

References

1. Motivation. http://www.businessdictionary.com/definition/motivation.html.
2. If you want to motivate employees, stop trusting your instincts. https://hbr.org/2017/02/if-you-want-to-motivate-employees-stop-trusting-your-instincts.
3. Lee F. *If Disney Ran Your Hospital*. 2004.
4. Compliance. www.merriam-webster.com/dictionary/compliance.
5. Prentice M, Halusic M, Sheldon K. Integrating theories of psychological needs-as-requirements and psychological needs-as-motives: a two process model. *Social and Personality Psychology Compass*. 2014;8(2):73–85.
6. Fowler S. *Why Motivating People Doesn't Work…and What Does*. 2014
7. Hauser W. You are a skilled veterinarian…what's next? In: Proceedings from the American College of Veterinary Internal Medicine Proceedings; 2017.

1.
acvim340

Giardia, Hookworms, Pythium, and Other Small Bowel Parasites of the Dog and Cat

Dwight D. Bowman, MS, PhD, DACVM (Parasit-Hon); Kenneth W. Simpson, BVM&S, PhD, DECVIM, DACVIM

INTRODUCTION

There are a number of common and not-so-common parasitic infections of the small intestine of dogs and cats that can be frustrating, dangerous, or sometimes deadly, that remain in spite of continued improvements in anthelmintic preventive therapy and care by gastroenterologists. The parasites covered in this discussion will include two protist species of the genera *Giardia* and *Cryptosporidium*, two nematodes *Ancylostoma caninum* and *Strongyloides stercoralis*, a trematode *Heterobilharzia americana*, and two Stramenopiles of the genera *Blastocystis* and *Pythium*.

GIARDIA

Giardia is a common occurrence in the intestines of dogs and cats around the world and is perhaps the most commonly occurring protist occurring in dogs and it is also common in cats. Beginning shortly maps showing the prevalence of *Giardia* positive fecal samples in dogs in the United States will appear on the maps presented monthly on maps of the Companion Animal Parasite Council [CAPC]. Currently, these maps that combine fecal flotation with zinc sulfate and antigen testing are showing national prevalence levels of around 7% based on samples of 10^6 to 10^7 dogs. There are not going to be maps for cats initially on the updated presentation, but it is expected that the number might be similar to that reported for dogs. Unfortunately, the maps will not include signalment, and most of the samples are likely to represent routine annuals screens for parasites. However, overall it does show the ubiquitous nature of this protist in samples of feces from canine (and probably also feline) pets.

When molecular methods have been used, the majority of *Giardia* isolated from dogs has been identified as being of the assemblages C and D, the dog adapted assemblages, and the majority of *Giardia* isolated from cats has been identified as being assemblage F, the cat adapted assemblage;[1] however, there is a report that suggests that dogs may commonly be shedding much higher percentages of other assemblages including A&B that are commonly considered to be zoonotic or human adapted isolates.[2] The C, D, and F assemblages are of zoonotic significance only in the immunosuppressed or immunocompromised, have been recognized in only a small subset of infections of humans with this immunologic status, and overall, it would appear that the preponderance of evidence suggests that dogs and cats are not a major reservoir for human infections. This raises the question as to how critical is the need to treat dogs that do not have clinical signs.[3]

Diagnosis of *Giardia* infection is by fecal flotation, fecal ELISA or IFA.[4] At high prevalence flotation, ELISA and IFA perform relatively similarly. However, the IFA appears more sensitive at lower prevalence, and is recommend for testing patients who have failed treatment where cryptosporidium is a possibility. Treatment for dogs is fenbendazole (50 mg/kg SID) for 5 days. Alternatively, it may be combined with metronidazole (25 mg/kg BID) for 5 days. If treatment combined with bathing does not eliminate infection (as evidenced by testing feces for persistence of cysts), treatment with either fenbendazole alone or in combination with metronidazole may be extended for another 10 days. In cats fenbendazole (50 mg/kg SID) for 5 days or metronidazole (25 mg/kg BID) for 5 days, or a combination of the two as described for dogs. Repeated treatment is not required if clinical signs are absent, and dogs and cats not treated for zoonotic reasons. The question with *Giardia* is why in some cases although treatment appears successful, the signs never stop; this is also the case with human giardiasis following clinical outbreaks of disease where symptoms persist for years after successful treatment.[5] This may reflect underlying disease that is not related to the presence of *Giardia*. The other conundrum is that many dogs will become clinically normal after treatment despite continuing to shed organisms or antigens in their feces.

CRYPTOSPORIDIUM CANIS AND *CRYPTOSPORIDIUM FELIS*

The dog and cat have their own species of *Cryptosporidium*, and dogs and cats appear refractory to infection with the *Cryptosporidium* parvum from calves that is cross-transmissible to people and appears to also be refractory to infections with *Cryptosporidium* hominis, the human adapted species. Also, as with assemblages C, D, and F of *Giardia*, the canine and feline *Cryptosporidium* species are only found in people who are immunosuppressed or immunocompromised. As with people, naïve dogs and cats and dogs and cats that are immunosuppressed can develop clinical cryptosporidiosis.[6] Thus, as dogs and cats are now being treated immunologically for cancer and are being immunosuppressed for transplantation and other surgical reasons, it needs to be remembered that it was an unsuspected outcome in persons who had been medically immunosuppressed for reasons such as transplants where this disease was first recognized.[7] Most *Cryptosporidium* infections in dogs and cats are subclinical. Diarrhea and dehydration are the primary clinical signs. Diagnosis is by detecting oocysts in feces using sucrose flotation, which is complicated by their very small size, or IFA/ELISA. Persistent diarrhea with oocyst shedding may be treated with tylosin: 10 to 15 mg/kg TID for 14 to 21 days (cats), azithromycin: 5 to 10 mg/kg BID for 5 to 7 days (dogs); 7 to 15 mg/kg for 5 to 7 days (cats) or paromomycin: 150 mg/kg SID for 5 days (dogs and cats) (toxicity when absorbed through inflamed GIT) (www.capcvet.org/capc-recommendations/cryptosporidia/).

ANCYLOSTOMA CANINUM

The canine hookworm due to its ability to infect dogs by transmammary transmission and because it routinely repopulates the intestine from larvae sequestered in the tissues continues to occasionally kill puppies and to confound diagnostics, because the worms can and do often reappear after successful adulticide treatment. The disease is one of anemia due to blood loss into the bowel, and the amount of loss per worm has been determined to be easily sufficient to kill puppies and adult dogs. This was carefully explained by Doctor Georgi who studied the blood loss due to *A. caninum* in dogs: "The amount of blood removed is directly proportional to the weight of adult hookworms in the small intestine and amounts to a daily loss of about 30 μL of erythrocytes per mg of worm. An adult female *A. caninum* weighs 2.0–2.6 mg and a male, about 0.8 mg. Therefore, a population of only 50 females and 25 males would exact a toll of about 4 ml of erythrocytes each day, a substantial drain on the hemopoietic resources of any pup. The imposition of such a burden during the nursing period could prove fatal. To make matters worse, peak blood loss is experienced even before eggs appear in the feces. This not only renders diagnosis difficult but constitutes a sudden hemorrhagic shock that very young pups are ill-equipped to compensate. The other species of *Ancylostoma* remove less blood than *A. caninum*, and *Uncinaria stenocephala* least of all."[8] We have recently captured images endoscopically of both the marked bleeding caused within the small intestine by young hookworms and the presence of repopulating hookworms in a dog who was taken off its routine monthly heartworm and internal parasite treatment. Diagnosis is by fecal flotation with centrifugation. In dogs fenbendazole, milbemycin oxime, moxidectin, and pyrantel pamoate are approved for adult *A. caninum*; pyrantel pamoate, fenbendazole, and moxidectin are approved for adult *U. stenocephala*; and pyrantel pamoate is approved for adult *A. braziliense*. (www.capcvet.org/resource-library/parasite-product-applications-for-dogs). In cats Emodepside, ivermectin, milbemycin oxime, moxidectin, pyrantel, and selamectin are approved for adult *A. tubaeforme*. (www.capcvet.org/resource-library/parasite-product-applications-for-cats)

STRONGYLOIDES STERCORALIS

This is a parasite that like *Cryptosporidium* is liable to become problematic in dogs undergoing medicinal immunosuppression as has also occurred in many humans. We now know based on cases related to both Allied and Japanese soldiers from WWII that these infections can persist undetected in people for greater than 60–70 years.[9,10] Dogs were shown experimentally to develop autoinfection and hyperinfection like humans.[11] Recently a dog with inflammatory bowel diseases was treated with immunosuppressives and developed strongyloidiasis.[12] Fortunately, such cases are preventable due to the high efficacy of elevated doses of ivermectin, however, it has to be considered if treatment is to be administered.

HETEROBILHARZIA AMERICANA

This is basically schistosomiasis of dogs in the Americas, the reservoir hosts are mainly opossums and raccoons. Dogs are infected by cercariae penetrating the skin, and the infection is confined to the southeastern United States which is likely due to a good part by the fastidiousness of these parasites for their snail intermediate host. Due to the calcification associated with the eggs of these vein-dwelling schistosomatids, these parasites can produce marked radiographic and ultrasound images.[13] Unfortunately, as with human schistosomiasis, these infections can be associated with lymphosarcoma.[14] Young, large breed dogs, Southern USA appear predisposed. Swimming and raccoon exposure are considered risk factors for infection which can be associated with lethargy, weight loss, anorexia, vomiting, diarrhea and PU/PD. Poor body condition and ascites may be present. Anemia, hyperglobulinemia, azotemia, and hypercalcemia (ionized) (59% of cases) may be present. Enteropathy associated with granulomatous inflammation and congestion of submucosa. Flukes may be visible in vessels. Diagnosis is by fecal sedimentation with saline or fecal PCR or histopathology. Praziquantel (10 mg/kg PO TID for 2 days) and fenbendazole (24 mg/kg PO SID for 7 days) may result in resolution of clinical signs in some infected dogs. Hypercalcemia did not resolve without definitive treatment with praziquantel 12/17 were alive for 6 months to 3 years after diagnosis.[15]

BLASTOCYSTIS HOMINIS

There is disagreement as to whether or not there are truly signs associated with infections with this parasite and the life cycle is not really known; both gaps in our knowledge can be easily demonstrated if one visits the CDC-DPDx website. Diagnosis with molecular methods has made it easier to find *Blastocystis*, and there have now been many surveys of animals in which the organism has been identified around the world in all sorts of hosts. A recent survey in the northwestern USA found it in about 10% of canine and feline samples from a shelter in Portland, OR, but in none of the 50 pet dog and pet cat samples examined.[16] Like assemblages in *Giardia*, there are sub-types of *Blastocystis*, and the number of these subtypes has not yet stabilized while anywhere from 6 to 14 or more have been recognized by different authors.[17,18]

At this time, it remains unclear whether this infection actually causes disease, and it is unclear whether or not it is transmitted between people and animals. There are papers that suggest that pets and their care-givers, and also, zoo-workers and zoo animals may share the same subtypes.[19,20] It will likely be several years before this is all worked out and whether it is determined that *Blastocystis hominis* is a parasite or simply a commensal, and whether it is shared between people and pets.

PYTHIUM INSIDIOSUM

This parasite is much better known to most veterinarians than is *Blastocystis hominis*. It was for a long time considered to be a fungus because they have structures like hyphae. Also, they are markedly different from fungi in that they have biflagellate zoospores that swim to infect the next host. There are many *Pythium* species, and almost all of them are parasites of plants. They are basically omnipresent in the environment, and this just makes it hard to understand why the organism will penetrate the tissues of dogs and cause such horrible lesions. The disease, pythiosis, also occurs in people, mainly in Thailand, and in people, like dogs, is sometimes fatal.

Pythiosis has been considered to be restricted to the hot and humid portions of the world, buts division may no longer be as seemingly strict as it once was with cases having occurred in areas of California and Arizona which do not fit the general pattern.[21] Also, recently were reports of infected dogs from Wisconsin and the upper peninsula of Michigan.[22]

The disease is due to a slowly growing difficult-to-treat fungal-like infection or cancer-like growth that will not respond to antibiotics and to many antifungals.[23-25] Dogs present with two forms of infection, granulomatous skin lesions and the gastrointestinal form. Young male large breed dogs are predisposed. The gastrointestinal form is associated with vomiting, anorexia, diarrhea, weight loss, and palpable abdominal masses by the gastric outflow tract and ileocolic junction. Clinicopathological abnormalities may include anemia, hypoalbuminemia (9/9), hyperglobulinemia (8/9), and eosinophilia (7/9). Focal thickening and lymphadenopathy can be detected with abdominal sonography. Diagnosis is made by finding the agents in histological sections in biopsy (GMS, immunocytochemistry), along with serology for antibodies, or testing of the biopsy for *Pythium* DNA by PCR. Treatment is surgical resection combined with itraconazole (10 mg/kg PO SID) and terbinafine (5 to 10 mg/kg PO SID) for 2 to 3 months post-surgery. Adjunct therapy may include immunotherapy with *Pythium* vaccine and the fungicide mefenoxam. ELISA titers before and after TX may help to monitor remission. Prognosis is generally poor.

References

1. Bowman DD, *et al. Exp Parasitol.* 2010;124(1):121.
2. Covacin C, *et al. Vet Parasitol.* 2011;177(1/2):28.
3. Tysnes KR, *et al. Trends Parasitol.* 2014;30(11):520.
4. Simpson KW, *et al. J Vet Intern Med.* 2009;23:476.
5. Robertson LJ, *et al. Trends Parasitol.* 2009;26(2):75.
6. Willard MD, *et al. J Am Anim Hosp Assoc.* 1999;35(5):405.
7. Beaver PC, *et al. Clinical Parasitology.* 1984;150.
8. Georgi JR, *et al. Clinical Canine Parasitology.* 1991;155.
9. Robson D, *et al. Ann Trop Med Parasitol.* 2009;103(2):145.
10. Suzuki T, *et al. Jpn J Med.* 1989;28:96.
11. Schad GA, *et al. Exp Parasitol.* 1984;57:287.
12. Cervone M, *et al. Rev Vet Clin.* 2016;51:55.
13. Kvitko-White HL, *et al. Vet Radiol Ultrasound.* 2011;52(5):538.
14. Stone RH, *et al. J Vet Diagn Invest.* 2011;23:1065.
15. Fabrick C, *et al. J Vet Intern Med.* 2010;24(1):140.
16. Ruaux GC, *et al. PLoS One.* 2014;9:e107496.
17. Mattiucci S, *et al. Epidemiol Infect.* 2016;144:635.
18. Alfellani MA, *et al. Parasitology.* 2013;140:966.
19. Nagel R, *et al. Intern Med J.* 2012;42:1187.
20. Parkar U, *et al. Vet Parasitol.* 2010;169:8.
21. Berryessa NA, *et al. J Vet Intern Med.* 2008;22:1065.
22. Oldenhoff W, *et al. Vet Dermatol.* 2014;25(52):e21.
23. Schmiedt CW, *et al. J Am Vet Med Assoc.* 2012;241(3):358.
24. Hummel J, *et al. Med Mycol.* 2011;49(5):539.
25. Pereira DI, *et al. Mycopathologia.* 2013;176(3–4):309.

acvim341

Tritrichomonas, Whipworms, and Other Large Bowel Associated Parasites of Dogs and Cats

Kenneth W. Simpson, BVM&S, PhD, DECVIM, DACVIM (SAIM); Dwight D. Bowman, MS, PhD, DACVM (Parasit-Hon)
Ithaca, NY, USA

INTRODUCTION

The parasites of the large bowel of dogs and cats that regularly cause any amount of disease are much rarer than are the parasites of the small intestine. However, two, infections with whipworms and trichomonads, are quite common and cause a good deal of disease or, at least, concern and consternation on the part of owners and veterinarians.

TRITRICHOMONAS

The *Tritrichomonas* of cats goes by two names: *Tritrichomonas blagburni* (Stockdale Walden *et al.* 2013) and *Tritrichomonas foetus* (Riedmuller 1928; Wenrich, Emmerson 1933). Also, there is still debate as to whether or not the one in cattle is the one that was originally described from the nasal passages, stomach, and cecum of pigs *Tritrichomonas suis* (Gruby, Delafond 1843; Levine 1961).[1] On some occasions, dogs with diarrhea may also be infected with the same tritrichomonad found in cats. Whatever the name, this parasite is associated with large bowel disease in the cat with associated chronic soft or mucoid stools.

It is assumed that cats are infected by the oral ingestion of trophozoites, very likely by direct fecal oral contact. It has been shown that the trichomonad trophozoites from cat feces can survive transit through the alimentary tract of slugs suggesting that they have some resistance to environmental extremes. A survey of 117 cats from 89 catteries revealed the prevalence of *T. foetus* at 31% among cats (36 out of 117) and catteries (28 out of 89) based on results of fecal smear examination, fecal culture, or PCR diagnosis. Among catteries where *T. foetus* was identified, they were more likely to have had a recent history of diarrhea, an historical diagnosis of coccidia infection in adult cats, and less square feet of housing per cat. Infections in cats are felt to be most common in young purebred domestic show cats that are group housed and in other cats housed in high-density populations. There is evidence that the organisms are found not uncommonly in healthy show cats, and that they may proliferate in cats with other causes of bowel inflammation.[2]

The infection is associated with diarrhea - frequently contains blood and mucus and may be foul smelling. Affected cats frequently appear in good health, apart from the diarrhea. Some kittens have rectal prolapse associated with infection. In rare cases, weight loss and intestinal thickening, lymphadenopathy are appreciated. Concurrent *Cryptosporidium* infection is associated with more severe disease.

Diagnosis can be made by PCR of feces or using the InPouch TF transport and culture kit combined with PCR for species confirmation. The InPouch TF-Feline medium is not specific for this feline organism since *Pentatrichomonas* will also thrive in the pouch medium;[3] this means that if a culture is positive that PCR is required for validation before subjecting the cat to treatment. The current treatment of choice to clear these cats of their infection is the oral administration of ronidazole (30 mg/kg, SID, for 14 days). Post-treatment testing to verify clearance; monitor for neurotoxicity (e.g., lethargy, inappetence, ataxia, or seizures). These can resolve if the drug is withdrawn immediately. Pregnant, nursing, systemically ill and kittens <12 wks should not be treated. Dr. Gookin at NC State (www.JodyGookin.com) continues to lead in the diagnosis and treatment of *T. foetus* (=*T. blagburni*). It remains a mystery as to when it will cause disease and when it will resolve without treatment.[4]

ENTAMOEBA HISTOLYTICA

Almost all reports of entamoebiasis in dogs are quite old. Levine (1961)[5] notes the following cases: "Kartulis (1891, 1913) found *E. histolytica* causing dysentery in 3 dogs in Egypt; in one of these, a liver abscess was also present. Darling (1915) reported a fatal infection in a dog in Panama. Ware (1916) reported an outbreak in a pack of foxhounds in the Nilgiri Hills of India. Boyd (1931) reported an outbreak in another pack of hounds in India, and more recently Ganapathy and Alwar (1957) reported 2 cases of amoebic dysentery in dogs in India. Fischer (1918) reported a case of amoebic dysentery in a dog in China, Bauche and Motais (1920) reported one in Indochina, and Morcos (1936) found 5 cases in Egypt. In the United States, Faust (1930) found 2 dogs in New Orleans with amoebic dysentery, Andrews (1932) found *E. histolytica* in the feces of a diarrheic dog in Baltimore, and Thorson, Seibold and Bailey (1956) reported a case of systemic amoebosis in a puppy which also had distemper. *E. histolytica* was found in large numbers in the lungs and amoebae were also seen in the liver, kidneys and spleen." Since this, things have basically gone silent with a few reports from India and China.

In the 1930s and 1940s, dogs were experimentally infected with *E. histolytica* and developed disease,[6] they did then develop significant disease. Also, it was through the experimental infection of cats that the initial description of the non-pathogenic *Entamoeba dispar* was recognized that is now distinguished from *E. histolytica* by immunologic and morphologic methods.[7,8] Thus, after some 100 years it has become possible to distinguish the pathogenic from the non-pathogenic forms, and the majority of people are now readily identified as having the non-pathogenic species.

Reports of *Entamoeba histolytica* are scant in the USA. Clinical signs are bloody mucoid large-bowel diarrhea. Diagnosis is by analysis of feces for trophozoites, PAS staining of colonic mucosa, and immunocytochemistry and PCR of infected tissues. Treatment is metronidazole (25 mg/kg BID) for 5 days.

WHIPWORMS

Canids around the world share infections with *Trichuris vulpis*. This is a common parasite of the colonic mucosa. It appears more regionally confined than *Toxocara* and hookworms, perhaps because it has a greater requirement for specific soil types. Historically, the United States has been without feline whipworms, but it seems that now feline whipworms are beginning to appear on fecal examinations in the southeastern United States.

Trichuris vulpis

The life cycle of *T. vulpis* is direct, but poorly described. Parasitologists have not been able to routinely infect dogs with adult worms to test drug efficacy, and also, they have never been able to get more than a few worms to mature in dogs. In natural infections, dogs can sometimes have thousands of worms lining their colonic mucosa. The new anthelmintics are excellent, but because there is so little known about larval development of these worms, it is not known if larvae can be killed or if they are at what stage of development they might have reached at the time they are killed. Thus, this very common worm remains a parasitic enigma.

The egg is readily diagnosed by the easily recognizable football-shaped (and colored) egg. The size is larger than that of the human or pig whipworm, and there should not be any cause for misidentification; however, the three species of eggs from dogs (*T. vulpis*), people (*Trichuris trichiura*), and pigs (*Trichuris suis*), are sometimes produced by parents who make eggs that are much larger in size.[9] This can lead people to suspect that there are infections of *T. vulpis* in humans, which has been reported, but the reports are not very convincing when given careful scrutiny. More recently, an immunodiagnostic test has been developed that detects *Trichuris* antigen in canine feces.[10] This test can detect prepatent whipworm infections in dogs.

Disease due to whipworms is typically manifest as bloody mucoid diarrhea. Severe infections may be accompanied by signs of systemic involvement and clinicopathological analysis may reveal anemia, hypoproteinemia, eosinophilia, and a Na:K ratio <27:1, raising suspicion of hypoadrenocorticism. However, these patients are normoresponsive to ACTH and are categorized as "pseudo Addison's" disease. Diagnosis of *T. vulpis* is by fecal flotation, which can be complicated by episodic shedding. Antigen testing may also be employed. Treatment is with fenbendazole 50 mg/kg for 3d. Prevention: milbemycin oxime (0.5 mg/kg), moxidectin 2.5mg/kg, monthly. Prognosis is typically excellent.

Trichuris serrata

Cats in St. Kitts have been shown to be infected with *Trichuris serrata*.[11,12] There have been no signs described associated with these infections; however, one of the 12 cats in a study in St. Kitts had 117 worms (mean 17.4/cat).[11] Treatment in cats would probably be best with Interceptor, Advantage Multi, or Profender. Drontal does not contain fenbendazole, which is present in the canine product Drontal Plus.

PROTOTHECA

Prototheca sp. are unicellular, achlorophyllic, aerobic algae that are abundant in the environment. They can also be found in cow's milk and are associated with bovine mastitis. Disease associated with *Prototheca* in dogs has been infrequently reported since the first description by Van Kruningen.[13] Dogs most commonly have involvement of the gastrointestinal tract (colon) and ocular and/or neurologic symptoms. Large-bowel diarrhea, blindness (due to chorioretinitis and retinal detachment), deafness, seizures, and ataxia may be present.[14] Colonic and rectal involvement is a consistent feature, even in the absence of overt signs of colitis. Signs of large-bowel disease can be present for some months before evidence of dissemination. Boxer dogs and Collies appear to be overrepresented. The presence of concurrent ocular/neurological and GI signs should prompt consideration of protothecosis and systemic fungal disease (*Histoplasma*). There are typically no specific clinicopathological or imaging findings associated with protothecosis. Diagnosis can be made with cytology (e.g., aspirate from eye, other lesion rectal scrape), biopsy (GMS/PAS-stained colonic biopsies/other tissues) and culture of urine. Histopathology of the colon typically reveals granulomatous colitis. FISH analysis for *E. coli* has been negative in Boxer dogs with *Prototheca*-associated colitis. Dogs with cutaneous restricted disease may be successfully treated. Disseminated disease is typically fatal despite treatment with combinations of amphotericin and itraconazole.[15]

REFERENCES

1. Morin-Adeline V, *et al. Vet Parasitol.* 2015;212:111.
2. Tysnes KR, *et al. Acta Vet Scand.* 2011;53:39.
3. Ceplecha V, *et al. Vet Parasitol.* 2013;196:503.
4. Gookin JL, *et al. J Clin Microbiol.* 2004;42:2707.
5. Levine ND. *Protozoan Parasites of Domestic Animals and of Man.* 1961:136.
6. Swartzwelder JC. *Am J Hyg.* 1939;29:89.
7. Brumpt E. *Bull Acad Med (Paris).* 1925;94:943.
8. Diamond LS, *et al. J Eukaryot Microbiol.* 1993;40:340.
9. Yoshikawa H, *et al. Parasitol Res.* 1989;75:649.
10. Elsemore DA, *et al. J Vet Diagn Invest.* 2014;26:404.
11. Ketzis JK, *et al. Springer Plus.* 2015;4:115.
12. Ketzis JK, *et al. J Infect Dev Ctries.* 2015;9:111.
13. Van Kruiningen HJ, *et al. Pathol Vet.* 1969;6(4):348.
14. Van Kruiningen HJ. *J Am Vet Med Assoc.* 1970;157(1):56.
15. Stenner VJ, *et al. Med Mycol.* 2007;45(3):249.

acvim343

Nutritional Management of Chronic Enteropathies

Aarti Kathrani, BVetMed (Hons), PhD, DACVIM (SAIM), DACVN, MRCVS
Bristol, UK

Definition of Food-Responsive Enteropathy

Adverse reaction to food in humans encompasses many conditions with different etiologies (Figure 1).[1] The same is likely true of adverse reaction to food in companion animals. The term food-responsive enteropathy describes a group of diseases resulting in chronic gastrointestinal signs that is responsive solely to dietary management; the underlying etiology of these diseases is unknown but likely encompasses immune-mediated and non-immune-mediated causes.

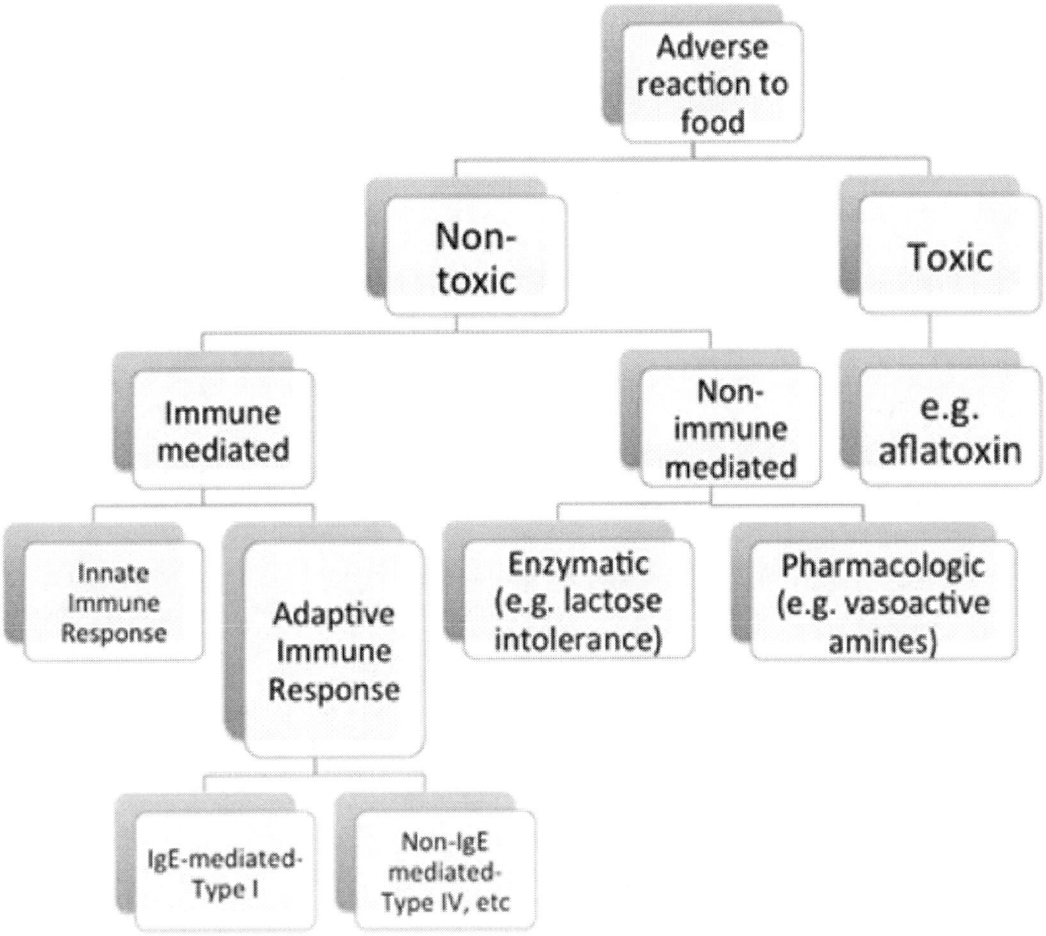

Figure 1. Sub-categorization of adverse reaction to food in humans

Diagnosis of Food-Responsive Enteropathy

The index of suspicion for a food-responsive enteropathy may be increased if the dog is younger and has predominantly large intestinal diarrhea, as one study documented that such dogs were more likely to respond to diet.[2] In this study, only 5% of dogs with exclusively small intestinal diarrhea had food-responsive enteropathy. Similarly, cats with food-responsive enteropathy also present at a relatively younger age (5 years) and more than half of cats in one study had large intestinal diarrhea.[3]

Diagnostic work-up of dogs and cats with suspected food-responsive enteropathy includes ruling out secondary causes of chronic gastrointestinal signs with blood work, urinalysis, and imaging and depending on the results of work-up, an empirical diet trial (see Treatment of Food-Responsive Enteropathy in Companion Animals below) or endoscopy with intestinal biopsies can be performed; the latter may be chosen if the animal is inappetent, lethargic, losing weight, hypoalbuminemic, hypocobalaminemic or has ultrasonographic changes of the gastrointestinal tract.[4] However, it is important to note that dogs and cats with food-responsive enteropathy cannot be differentiated from those diseases requiring steroid treatment based on the severity of endoscopic or histologic lesions and therefore food trial and challenge is currently the gold standard in the diagnosis of food-responsive enteropathy in companion animals. Intradermal skin testing, skin patch testing and measuring circulating food allergen-specific serum IgE are of no diagnostic value because of their low sensitivity and specificity.[5,6]

Although there is no standard acceptable criterion, the author considers companion animals that fail to respond to at least two appropriate commercial therapeutic diet trials and a limited-ingredient home-cooked diet as likely not having a food-responsive enteropathy.

Treatment of Food-Responsive Enteropathy in Companion Animals

There are three categories of commercial therapeutic diets that are available for the diagnosis and management of chronic enteropathy in dogs and cats; hydrolyzed diets, limited-ingredient novel protein diets, and gastrointestinal diets. In addition, a limited-ingredient, complete and balanced home-cooked diet, formulated by a board certified veterinary nutritionist could also be used. When selecting one of these diets for the diagnosis and treatment of food-responsive enteropathy, the diet should be the sole source of food for at least 6 weeks and no additional foods, treats, snacks, flavored medication, or toothpaste should be fed during this time. Although at least 6 weeks of diet trial is recommended, most animals with food-responsive enteropathy respond relatively quickly.[7] In humans, dietary therapy reduced inflammation and led to changes in the microbiome within one week.[8]

Commercial Therapeutic Hydrolyzed Diets

Commercial therapeutic hydrolyzed diets contain proteins that have been hydrolyzed to a size that theoretically evades a type 1 hypersensitivity response. However, it is currently unknown what type of immunological disturbance is present in canine and feline chronic enteropathy. In addition, some animals that have been sensitized to the intact protein can still react to the hydrolyzed protein.[9] This may be due to the size of the hydrolyzed protein that is present in the current commercial therapeutic hydrolyzed diets; as there is evidence that the size needed to evade a type 1 hypersensitivity response may actually be smaller than 3–5 kDa.[10] All of the available commercial therapeutic hydrolyzed diets, apart from Royal Canin Veterinary Diet Canine Ultamino dry food contain proteins that have an average size of approximately 10–12 kDa. In addition, the range of size of proteins in these diets has not been reported. Therefore, any remaining intact protein in these diets may continue to elicit an immune response. However, there is anecdotal and scientific evidence that commercial therapeutic hydrolyzed diets help companion animals with chronic enteropathies attain remission.[11-14] However, it is unknown which attributes of these diets are helping to resolve the clinical signs, especially as the controlled diet that was used differed in many parameters compared to the test diet.[15] In addition, these studies had a relatively small sample size,[12,14,15] were unblinded[15] or uncontrolled[12-14]. Other attributes of commercial therapeutic hydrolyzed diets that may help animals attain remission include higher digestibility, lower fat content, differing fiber content, omega 3 fatty acid, soy ingredients which are known to be immunomodulatory and vegetarian diets such as Purina Pro Plan Veterinary Diets Canine HA. Therefore, further studies are needed to determine which attributes of the commercial therapeutic hydrolyzed diets are helping these animals attain remission. This is important as if the hydrolyzed approach is not necessary, these diets may be more palatable and economically feasible.

Commercial Therapeutic Limited-Ingredient Novel Protein Diets

Although commercial therapeutic limited-ingredient novel protein diets are more palatable than hydrolyzed diets, determining what is novel for the animal may be difficult if a complete diet history is unavailable from birth. In addition, some over-the-counter commercial diets contain novel ingredients that are more unusual such as tilapia fish or bison and therefore selecting a protein that is truly novel for the animal may be difficult to impossible. Another disadvantage of novel protein diets is the concern regarding relapse with these diets as the immune system may become sensitized to the dietary antigens resulting in a recurrence of clinical signs. Although this is a theoretical concern, this has been anecdotally recognized. In addition, commercial therapeutic limited-ingredient novel protein diets may be higher in fat due to protein sources such as venison, duck, and salmon and therefore this strategy may exacerbate diarrhea or be contraindicated in dogs with concurrent pancreatitis or lymphangiectasia. However, the potential lower cost and higher palatability of these diets makes them a desirable option over hydrolyzed diets.

One study demonstrated the presence of other common food antigens in over the counter limited ingredient diets[16] and therefore these diets should be avoided for the diagnosis and treatment of suspected food-responsive enteropathy.

Commercial Therapeutic Gastrointestinal Diets

Some cases of food-responsive enteropathy may respond to a commercial therapeutic gastrointestinal diet as these diets have higher digestibility and therefore less intact protein is presented to the gastrointestinal associated lymphoid tissue. These diets are also lower in cost compared to hydrolyzed diets. Although, one study demonstrated that such diets were able to induce remission in dogs with food-responsive enteropathy; they were less likely to remain asymptomatic at subsequent rechecks when compared to dogs managed with a hydrolyzed diet.[15] Therefore, the author prefers to use hydrolyzed diets in the first instance if the animal's appetite is good and there are no financial constraints. However, the option of low fat formulas in this category of diets makes them particularly desirable for the management of lymphangiectasia.

Limited-Ingredient Novel Protein Home-Cooked Diets

There is a concern that some animals with food-responsive enteropathy may be reacting to antigens that have either been unmasked or developed during to the commercial processing of food and therefore these animals may not be able to tolerate any commercial diet. These animals can usually be diagnosed by their positive response to a limited ingredient novel protein complete and balanced home-cooked diet. There are several advantages of a home-cooked diet for these cases, including higher palatability, higher digestibility, ability to precisely set macronutrient profile, use of very limited ingredients (two to three plus vitamin/mineral supplements) and the ability to choose ingredients that are known to be novel or palatable and tolerated by the animal. However, home-cooked diets may not be feasible for many owners due to financial and time constraints. In addition, animals may relapse on home-cooked diets, especially if they become sensitized to the novel proteins and there may also be recipe drifts over time potentially resulting in an unbalanced diet. Also, as the formulation of the recipe is a computer-driven process, the bioavailability of the nutrients has not been tested and therefore any animal on a home-cooked diet should be regularly monitored for any signs of a nutrient deficiency. If an animal's clinical signs fail to respond to a home-cooked diet then it is unlikely that a food-responsive enteropathy is the underlying cause and further investigation or other treatment modality should be pursued.

Studies of Individual Nutrients in Human and Animal Models of Inflammatory Bowel Disease

Epidemiological studies support the role of diet in the etiopathogenesis of human IBD. In addition, many human patients with IBD report diet as a trigger factor in the relapse of their disease.[17] A western diet, which is high in fat and red meat and low in fruits and vegetables is associated with an increased risk of IBD.[18] A high-fat diet increased mRNA levels of the pattern recognition receptors NOD2 and TLR5 in an experimental Crohn's disease model.[19] High fat diets have also been shown to increase taurocholic acid production, which can promote changes in colonic microbiota that exacerbate colitis in IL10 knockout mice.[20] Dimerization of TLR4, a pattern recognition receptor for LPS, is facilitated by saturated fatty acids, while unsaturated fatty acids such as DHA inhibit its dimerization.[21] Interestingly, 21 days of fish oil supplementation downregulated mRNA levels of NOD2 and TLR4 in LPS treated pigs.[22] However, current evidence for use of omega 3 fatty acid in human IBD is weak.[23] Amino acids can also target the gastrointestinal immune system by several mechanisms; inhibition of pattern recognition receptors (arginine, glutamine), antimicrobial production (arginine, isoleucine) and by influencing the microbiota, however, further studies determining efficacy and therapeutic dose are need in human and companion animal IBD.[24] Vitamin D also plays a role in the innate immune system and low levels have been shown to be a risk factor for IBD.[25] Similarly, retinoic acid, a derivative of vitamin A plays a crucial role in the innate immune system by maintaining homeostasis of dendritic cell subsets[26] and has been shown to attenuate experimental colitis via inhibition of NF-kB signaling[27]. Curcumin was shown to down-regulate colonic TLR and MyD88 expression in rats with TNBS induced colitis.[28] Although curcumin shows promise in patients with ulcerative colitis, further larger scale studies are needed.[23] Oral administration of iron sulfate or dietary iron present in red meat increased the severity of chemically induced colitis in animal models.[29] Dietary intake of zinc has been shown to be a protective factor for Crohn's disease.[30] A high-fiber diet decreases the risk of developing IBD in humans,[31] although intervention studies do not support a role for increased intake of dietary fiber in the management of Crohn's disease; however, further studies are needed to assess the individual role of soluble and insoluble fiber.[23] Clinical studies assessing the effects of vitamin C and E supplementation in Crohn's disease have shown effects on biomarkers of oxidative stress, but have so far not been shown to have significant clinical efficacy.[32]

Dietary Management of Inflammatory Bowel Disease in Humans

Exclusive enteral elemental nutritional therapy is the only dietary intervention that has been rigorously tested and shown to induce remission of Crohn's disease.[33-35] Elemental formulas contain individual amino acids, glucose polymers, and are low in fat with only about 2 to 3% of calories coming from long-chain triglycerides. Semi-elemental formulas contain peptides of varying chain length, simple sugars, glucose polymers, or starch and fat that is primarily medium chain triglycerides. Polymeric formulas contain intact proteins, complex carbohydrates, and mainly long chain-triglycerides. Elemental diets are used for the induction of remission, which is generally achieved over a period of 6 to 8 weeks, during this time the liquid feed is fed exclusively after which time a low-residue diet is slowly introduced. Continued intermittent feeding of the elemental diet can prolong the length of remission.[36] Several studies and a meta-analysis compared elemental diets with corticosteroids in pediatric patients with active Crohn's disease and found no significant difference in remission rates between the two forms of therapy.[35,37,38] Similar remission rates have also been reported in adults with newly diagnosed Crohn's disease.[39] The mechanisms of action of elemental diets include improved mucosal healing[40] (74% versus 33% seen with the use of corticosteroids[33]), decreased inflammatory cytokines,[41] alterations in gut microbiota[42] and improved nutritional status. Interestingly, there have been no differences seen in remission rates between elemental, semi-elemental and polymeric formulas.[43] Also, no differences in remission rates have been seen between different fat content of diets.[44]

A specific carbohydrate diet, which restricts refined sugar and complex carbohydrate consumption helped to improve clinical signs in pediatric patients with IBD.[45,46] A FODMAP diet is recognized to be helpful in patients with IBD and concurrent functional intestinal signs.[47]

References

1. Valenta R, Hochwallner H, Linhart B, Pahr S. Food allergies: the basics. *Gastroenterology.* 2015;148(6):1120–1131 e4.
2. Allenspach K, Wieland B, Gröne A, Gaschen F. Chronic enteropathies in dogs: evaluation of risk factors for negative outcome. *J Vet Intern Med.* 2007;21(4):700–708.
3. Guilford WG, Jones BR, Markwell PJ, Arthur DG, Collett MG, Harte JG. Food sensitivity in cats with chronic idiopathic gastrointestinal problems. *J Vet Intern Med.* 2001;15(1):7–13.

4. Day MJ, Bilzer T, Mansell J, *et al.* Histopathological standards for the diagnosis of gastrointestinal inflammation in endoscopic biopsy samples from the dog and cat: a report from the World Small Animal Veterinary Association Gastrointestinal Standardization Group. *J Comp Pathol.* 2008;138(Suppl 1):S1–43.
5. Ishida R, Masuda K, Kurata K, Ohno K, Tsujimoto H. Lymphocyte blastogenic responses to inciting food allergens in dogs with food hypersensitivity. *J Vet Intern Med.* 2004;18(1):25–30.
6. Kunkle G, Horner S. Validity of skin testing for diagnosis of food allergy in dogs. *J Am Vet Med Assoc.* 1992;200(5):677–680.
7. Schmitz S, Glanemann B, Garden OA, *et al.* A prospective, randomized, blinded, placebo-controlled pilot study on the effect of *Enterococcus faecium* on clinical activity and intestinal gene expression in canine food-responsive chronic enteropathy. *J Vet Intern Med.* 2015;29(2):533–543.
8. Lewis JD, Chen EZ, Baldassano RN, *et al.* Inflammation, antibiotics, and diet as environmental stressors of the gut microbiome in pediatric Crohn's disease. *Cell Host Microbe.* 2015;18(4):489–500.
9. Jackson HA, Jackson MW, Coblentz L, Hammerberg B. Evaluation of the clinical and allergen specific serum immunoglobulin E responses to oral challenge with cornstarch, corn, soy and a soy hydrolysate diet in dogs with spontaneous food allergy. *Vet Dermatol.* 2003;14(4):181–187.
10. van Beresteijn EC, Meijer RJ, Schmidt DG. Residual antigenicity of hypoallergenic infant formulas and the occurrence of milk-specific IgE antibodies in patients with clinical allergy. *J Allergy Clin Immunol.* 1995;96(3):365–374.
11. Mandigers PJ, Biourge V, van den Ingh TS, Ankringa N, German AJ. A randomized, open-label, positively-controlled field trial of a hydrolyzed protein diet in dogs with chronic small bowel enteropathy. *J Vet Intern Med.* 2010;24(6):1350–1357.
12. Mandigers PJ, Biourge V, German AJ. Efficacy of a commercial hydrolysate diet in eight cats suffering from inflammatory bowel disease or adverse reaction to food. *Tijdschr Diergeneeskd.* 2010;135(18):668–672.
13. Walker D, Knuchel-Takano A, McCutchan A, *et al.* A comprehensive pathological survey of duodenal biopsies from dogs with diet-responsive chronic enteropathy. *J Vet Intern Med.* 2013;27(4):862–874.
14. Marks SL, Laflamme DP, McAloose D. Dietary trial using a commercial hypoallergenic diet containing hydrolyzed protein for dogs with inflammatory bowel disease. *Vet Ther.* 2002;3(2):109–118.
15. Mandigers PJ, Biourge V, van den Ingh TS, Ankringa N, German AJ. A randomized, open-label, positively-controlled field trial of a hydrolyzed protein diet in dogs with chronic small bowel enteropathy. *J Vet Internal Med.* 2010;24(6):1350–1357.
16. Raditic DM, Remillard RL, Tater KC. ELISA testing for common food antigens in four dry dog foods used in dietary elimination trials. *J Anim Physiol Anim Nutr (Berl).* 2011;95(1):90–97.
17. Limdi JK, Aggarwal D, McLaughlin JT. Dietary practices and beliefs in patients with inflammatory bowel disease. *Inflamm Bowel Dis.* 2016;22(1):164–170.
18. Shoda R, Matsueda K, Yamato S, Umeda N. Epidemiologic analysis of Crohn disease in Japan: increased dietary intake of n-6 polyunsaturated fatty acids and animal protein relates to the increased incidence of Crohn disease in Japan. *Am J Clin Nutr.* 1996;63(5):741–745.
19. Martinez-Medina M, Denizot J, Dreux N, *et al.* Western diet induces dysbiosis with increased *E. coli* in CEABAC10 mice, alters host barrier function favouring AIEC colonisation. *Gut.* 2014;63(1):116–124.
20. Devkota S, Wang Y, Musch MW, *et al.* Dietary-fat-induced taurocholic acid promotes pathobiont expansion and colitis in Il10-/- mice. *Nature.* 2012;487(7405):104–108.
21. Lee JY,Plakidas A, Lee WH, *et al.* Differential modulation of Toll-like receptors by fatty acids: preferential inhibition by n-3 polyunsaturated fatty acids. *J Lipid Res.* 2003;44(3):479–486.
22. Liu Y, Chen F, Odle J, *et al.* Fish oil enhances intestinal integrity and inhibits TLR4 and NOD2 signaling pathways in weaned pigs after LPS challenge. *J Nutr.* 2012;142(11):2017–2024.
23. Richman E, Rhodes JM. Review article: evidence-based dietary advice for patients with inflammatory bowel disease. *Aliment Pharmacol Ther.* 2013;38(10):1156–1171.
24. Marion-Letellier R, Savoye G, Ghosh S. IBD: in food we trust. *J Crohns Colitis.* 2016;10(11):1351–1361.
25. Raman M, Milestone AN, Walters JR, Hart AL, Ghosh S. Vitamin D and gastrointestinal diseases: inflammatory bowel disease and colorectal cancer. *Therap Adv Gastroenterol.* 2011;4(1):49–62.
26. Klebanoff CA, Spencer SP, Torabi-Parizi P, *et al.* Retinoic acid controls the homeostasis of pre-cDC-derived splenic and intestinal dendritic cells. *J Exp Med.* 2013;210(10):1961–1976.
27. Hong K, Zhang Y, Guo Y, *et al.* All-trans retinoic acid attenuates experimental colitis through inhibition of NF-kappaB signaling. *Immunol Lett.* 2014;162(1 Pt A):34–40.
28. Lubbad A, Oriowo MA, Khan I. Curcumin attenuates inflammation through inhibition of TLR-4 receptor in experimental colitis. *Mol Cell Biochem.* 2009;322(1–2):127–135.

29. van der Logt EM, Blokzijl T, van der Meer R, Faber KN, Dijkstra G. Westernized high-fat diet accelerates weight loss in dextran sulfate sodium-induced colitis in mice, which is further aggravated by supplementation of heme. *J Nutr Biochem*. 2013;24(6):1159–1165.
30. Ananthakrishnan AN, Khalili H, Song M, *et al*. High school diet and risk of Crohn's disease and ulcerative colitis. *Inflamm Bowel Dis*. 2015;21(10):2311–2319.
31. Reif S, Klein I, Lubin F, Farbstein M, Hallak A, Gilat T. Pre-illness dietary factors in inflammatory bowel disease. *Gut*. 1997;40(6):754–760.
32. Aghdassi E, Wendland BE, Steinhart AH, Wolman SL, Jeejeebhoy K, Allard JP. Antioxidant vitamin supplementation in Crohn's disease decreases oxidative stress. a randomized controlled trial. *Am J Gastroenterol*. 2003;98(2):348–353.
33. Borrelli O, Cordischi L, Cirulli M, *et al*. Polymeric diet alone versus corticosteroids in the treatment of active pediatric Crohn's disease: a randomized controlled open-label trial. *Clin Gastroenterol Hepatol*. 2006;4(6):744–753.
34. Day AS, Whitten KE, Lemberg DA, *et al*. Exclusive enteral feeding as primary therapy for Crohn's disease in Australian children and adolescents: a feasible and effective approach. *J Gastroenterol Hepatol*. 2006;21(10):1609–1614.
35. Heuschkel RB, Menache CC, Megerian JT, Baird AE. Enteral nutrition and corticosteroids in the treatment of acute Crohn's disease in children. *J Pediatr Gastroenterol Nutr*. 2000;31(1):8–15.
36. Esaki M, Matsumoto T, Nakamura S, *et al*. Factors affecting recurrence in patients with Crohn's disease under nutritional therapy. *Dis Colon Rectum*. 2006;49(10 Suppl):S68–74.
37. Dziechciarz P, Horvath A, Shamir R, Szajewska H. Meta-analysis: enteral nutrition in active Crohn's disease in children. *Aliment Pharmacol Ther*. 2007;26(6):795–806.
38. Heuschkel RB. Enteral nutrition in children with Crohn's disease. *J Pediatr Gastroenterol Nutr*. 2000;31(5):575.
39. Gonzalez-Huix F, de León R, Fernández-Banares F, *et al*. Polymeric enteral diets as primary treatment of active Crohn's disease: a prospective steroid controlled trial. *Gut*. 1993;34(6):778–782.
40. Grover Z, Muir R, Lewindon P. Exclusive enteral nutrition induces early clinical, mucosal and transmural remission in paediatric Crohn's disease. *J Gastroenterol*. 2014;49(4):638–645.
41. Nahidi L, Day AS, Lemberg DA, Leach ST. Differential effects of nutritional and non-nutritional therapies on intestinal barrier function in an *in vitro* model. *J Gastroenterol*. 2012;47(2):107–117.
42. Leach ST, Mitchell HM, Eng WR, Zhang L, Day AS. Sustained modulation of intestinal bacteria by exclusive enteral nutrition used to treat children with Crohn's disease. *Aliment Pharmacol Ther*. 2008;28(6):724–733.
43. Ludvigsson JF, Krantz M, Bodin L, Stenhammar L, Lindquist B. Elemental versus polymeric enteral nutrition in paediatric Crohn's disease: a multicentre randomized controlled trial. *Acta Paediatr*. 2004;93(3):327–335.
44. Zachos M, Tondeur M, Griffiths AM. Enteral nutritional therapy for induction of remission in Crohn's disease. *Cochrane Database Syst Rev*. 2007(1):CD000542.
45. Cohen SA, Gold BD, Olivia S, *et al*. Clinical and mucosal improvement with specific carbohydrate diet in pediatric Crohn disease. *J Pediatr Gastroenterol Nutr*. 2014;59(4):516–521.
46. Suskind DL, Wahbeh G, Gregory N, Vendettuoli H, Christie D. Nutritional therapy in pediatric Crohn disease: the specific carbohydrate diet. *J Pediatr Gastroenterol Nutr*. 2014;58(1):87–91.
47. Gearry RB, Irving PM, Barrett JS, Nathan DM, Shepherd SJ, Gibson PR. Reduction of dietary poorly absorbed short-chain carbohydrates (FODMAPs) improves abdominal symptoms in patients with inflammatory bowel disease-a pilot study. *J Crohns Colitis*. 2009;3(1):8–14.

acvim344

Adopting Best Practices: Rational Therapy for Gastrointestinal Ulceration and Erosion

M. Katherine Tolbert, DVM, PhD, DACVIM (SAIM)
Knoxville, TN, USA

Introduction

The gastric mucosa is repeatedly exposed to noxious substances including an acidic pH, mechanical and chemical irritants, and digestive enzymes. The gastric mucosal barrier (GMB) is comprised of defense mechanisms that protect the gastric mucosa against these noxious substances. The GMB is comprised of a single layer of highly resistant epithelial cells that repel gastric acid and are capable of rapid repair, a double mucus layer that is rich in bicarbonate ions, the local production of prostaglandins that serve to regulate blood flow and stimulate the secretion of mucus and bicarbonate ions, and a rich mucosal blood supply. Injury to the upper GI mucosa develops when there is an imbalance between injurious agents and protective mechanisms. Gastric hyperacidity disorders, such as mastocytosis and gastrinoma, and diseases exacerbated by exposure to gastric acid, such as peptic ulceration and stress-related mucosal bleeding, are increasingly identified in canine and feline patients. The rational and appropriate use of gastroprotectants for the prevention and treatment of upper GI ulceration and erosion (GUE) is a matter of great concern. Herein, the objective of these proceedings is to review the evidence for the use of gastroprotectants as prophylaxis and treatment for GUE in dogs and cats.

Acid Suppressants

Review and Indications

Healing of upper GI ulceration and erosion (GUE) is partially dependent on achieving an increased gastric pH. For this reason, acid suppressants are widely used in both human and veterinary medicine and are covered in greater detail in this review. Prior to their discovery, treatment of GUE and reflux esophagitis often required highly invasive surgery. Thus, the efficacy and role of acid suppressant therapy in the adjunctive treatment of upper GUE is not up for debate. Despite the limitations described below, **acid suppressants should not be withheld from patients with clear indications for their use**. These indications, based on studies in human or veterinary patients, include upper GUE or upper GI bleeding resulting from GI or pancreatic neoplasia, administration of ulcerogenic drugs (e.g., NSAIDs, methylprednisone, toceranib phosphate), gastroesophageal varices, inflammatory bowel disease (IBD), stress-related mucosal disease (SRMD), and GI foreign bodies.[1] Other diseases where acid suppressants may have a benefit include upper GUE or upper GI bleeding resulting from IVDD, hemostatic disorders, and intrahepatic shunt, although bleeding from the latter is proposed to be due to concurrent IBD.[2]

Choices for commercially available acid suppression include histamine-2 receptors antagonists (H2RAs) and proton pump inhibitors (PPIs). The H2RAs, which target the histamine-2 receptor on the acid-producing gastric parietal cells, have some advantages. The H2RAs are maximally effective within hours of administration, have a good safety profile, and, unlike the PPIs, have a bioavailability that is unaffected by food. Unfortunately, prolonged use of famotidine leads to tachyphylaxis, which may occur as soon as the second day of therapy.[3,4] These effects appear to be independent of dose or form of administration. Evidence suggests that prolonged famotidine may provide continued symptom relief because of mucosal desensitization but mucosal injury may be ongoing.[3] Thus, H2RAs may be a good treatment for as-needed symptom control or as short-term therapy (e.g., preventative therapy for reflux esophagitis) but are not the preferred treatment for upper GUE. If H2RAs are the only option, a dose of 1 mg/kg BID is suggested.

Proton pump inhibitors, being superior to H2RAs, revolutionized the medical management of GUE and upper GI tract bleeding. The PPIs alone are ninth among the highest selling drugs in the U.S., accounting for over $9 billion in sales in 2012. Although PPIs can take up to four days to reach peak effect, evidence suggests they are likely as effective as H2RAs on day one of administration. For most causes of GUE, dogs and cats should be treated twice-daily with a PPI (e.g., omeprazole, esomeprazole, pantoprazole) at 1 mg/kg.[5-11] However, once-daily treatment may be effective for certain causes of GUE such as SRMD.[12] A variety of omeprazole formulations (tablet, capsule, reformulated paste, suspension) have been evaluated and demonstrated to be effective in raising the intragastric pH in healthy dogs and cats.[6-8,13,14] For the reasons discussed below, long-term use of PPIs is discouraged for the majority of the indications listed above.[15,16]

Prophylaxis

Guidelines for acid suppressant prophylaxis in dogs and cats to reduce risk for GUE have not been established; however, guidelines are available for humans at high risk for GUE. Direct implementation of these guidelines for treatment of dogs and cats is not possible because of the confounding effect of *H. pylori* infection on the development of upper GUE in people. In addition, interventional procedures used in people for refractory or severe GUE, such as endoscopic hemostasis, are not common practice in veterinary medicine. However, some recommendations are sensible and can be adopted until appropriate guidelines have been established for veterinary patients. Most guidelines recommend reserving the prophylactic use of acid suppressants unless patients have multiple risk factors for the development of clinically important GUE or have a previous history of GI bleeding.[17] For example, the Surviving Sepsis Campaign guidelines suggests the use of acid suppressant prophylaxis in critically ill patients that also have high-risk factors for GUE such as mechanical ventilation or coagulopathies.[18] Other important risk factors for the development of GI bleeding in critically ill patients include acute renal injury or hepatic failure, shock, severe head or spinal cord injury, acute lung injury, or major surgery lasting more than 4 hours.[17] Patients being treated with warfarin that are concurrently receiving NSAIDS or anti-platelet drugs also benefit from acid suppressant therapy to reduce the risk of warfarin-induced GI bleeding.[19] For the majority of these prophylactic indications, PPIs are superior to H2RAs. Prophylactic use of acid suppressants in patients taking NSAIDs, whilst effectivee,[20] is not recommended unless the patient has a previous history of GUE or has an additional risk factor such as critical illness or the concurrent use of other ulcerogenic drugs. Prophylaxis for patients receiving corticosteroid therapy is also not recommended unless other risk factors are present. There is also no evidence to suggest the prophylactic use of acid suppressants in patients with cirrhotic liver, chronic kidney disease, or acute pancreatitis unless other risk factors for GUE are present.[17] Indeed, prophylactic PPI therapy in these patients may do more harm than provide benefit. For example, PPI therapy in patients with cirrhosis is associated with the development of spontaneous bacterial peritonitis.[16] The degree of acid suppression needed to promote platelet aggregation and clot formation in patients with thrombocytopenia-induced bleeding, intragastric pH > 6.8, does not occur with standard doses of acid suppressants. Therefore, prophylaxis is also not recommended for ITP patients without other risk factors. Some studies have identified a beneficial effect of PPIs for protection of pancreatic enzyme replacement therapy (ERT) in human patients with refractory chronic pancreatitis or exocrine pancreatic insufficiency. Despite the enzymes being enteric-coated, addition of PPIs has been shown to improve steatorrhea in this subset of patients. Thus, PPI administration should be considered for dogs and cats that do not initially respond to ERT.[17] The appropriate dose and frequency of PPIs for the prevention of GUE for the aforementioned conditions in dogs and cats is unknown.

Adverse Effects

The misuse of PPIs is pervasive.[21] PPI prescriptions for inappropriate indications are common. Often, patients continue to take these drugs long-term without re-evaluation.[22] PPIs are also available over-the-counter. Thus, many patients take them without clinician consent. Although PPIs appear to be well-tolerated during short-term administration in dogs and cats, the consequences of inappropriate and prolonged acid suppressant use may be grave. Studies in human patients suggest causation or association of PPI use with chronic kidney disease, acute kidney injury, cobalamin deficiency, osteoporosis and pathologic fractures, community-acquired pneumonia, hypomagnesemia, cardiovascular events, *Clostridium difficile*-associated diarrhea, dementia, and drug interactions. It is important to emphasize that many of these studies suggest an association and cannot clearly implicate PPIs in the induction of these disorders. As practitioners, we must be judicious in our use of PPIs and critically evaluate the risks versus benefits of PPI use on a case-by-case basis. The implementation of a standardized guideline may help to limit the abuse of this class of drugs. Introduction of a standardized guideline describing appropriate indications for PPI use significantly lowered rates of PPI use at a tertiary center among inpatients that were not receiving PPIs at the time of hospital admission.[23] Practitioners should also consider other options including short-term H2RA use, nutritional intervention, and stress reduction when GUE is not documented or suspected.

Coating Agents

Coating agents include sucralfate and barium. **Sucralfate**, a polyaluminum sucrose sulfate, forms a protective layer on the proximal GI mucosa. Sucralfate is activated in the presence of an acid to form a gel-like substance that covers areas of denuded epithelium. Sucralfate may also stimulate the production of protective prostaglandins. The results of human studies suggest that sucralfate may be more effective in the adjunctive treatment of duodenal ulcers compared to gastric ulcers. There are very few studies that have evaluated the efficacy of sucralfate in the treatment of GUE in dogs and cats. However, studies investigating the use of sucralfate in the polypharmacy treatment of GUE and mucositis in humans suggest there is a benefit to this practice. Moreover, sucralfate is associated with very few adverse effects aside from constipation. Its use may be discouraged in cats with CKD.[24] Sucralfate does change the pH of the stomach and therefore may interfere with the metabolism of drugs that are dependent on an acidic gastric pH (e.g., PPIs). It also interferes with drugs affected by the aluminum component of sucralfate (e.g., tetracyclines, ciprofloxacin). Therefore, these drugs should be administered at least two hours before or after sucralfate administration.[4]

Barium, like sucralfate, is proposed to have mucosal protecting effects and hemostatic properties. Barium enemas are effective for treatment of lower GI bleeding in people. To the author's knowledge, no published studies have evaluated the efficacy of barium in the treatment of dogs and cats with GUE, however, studies in the treatment of rectal bleeding in people and anecdotal information in dogs and cats suggests that it may be an effective adjunct treatment of GUE. The dose recommended for mucosal hemostasis can be occasionally challenging to administer especially in a patient with a history of inappetence or vomiting. Although barium is inert, aspiration of barium with gastric fluid contents can be fatal. Discontinue barium for at least 24 hours prior to gastrointestinal endoscopy and do not use in animals where GI perforation is suspected.

Prostaglandin Agonists

The most commonly used prostaglandin agonist in veterinary medicine is misoprostol, a PGE1 analog. By simulating endogenous eicosanoids, **misoprostal** increases mucosal blood flow and epithelial repair and stimulates mucus and bicarbonate secretion. Despite its mechanism of action, misoprostol is only effective for NSAID-induced injury and has no effect on steroid-associated ulceration. As mentioned for PPIs, prophylactic use of misoprostol for NSAID-induced GUE is effective but not recommended unless other risk factors are identified.[20] Misoprostol may increase GI and urogenital smooth muscle contractions leading to side effects of cramping, diarrhea, and abortions.

Other Therapies

Other therapies including **tranexamic acid** (TXA), **yunnan baiyou**, and **somatostatin analogs** may also have a place in the treatment of GI hemorrhage. Of these, TXA shows the most promise for severe GUE. Tranexamic acid, a lysine derivative, inhibits fibrinolysis. Because of the concern for thromboembolic events, TXA is generally reserved for life-threatening GUE.[25] Topical delivery to the site of bleeding may reduce the risk for thromboembolic disease. Pharmacologic and non-pharmacologic endoscopic techniques are effective for the treatment of GUE in humans but are not widely available to veterinary practitioners and, therefore, are not discussed here. The reader is referred to the following reference for a review of those therapies.[26]

References

1. Swann JW *et al. J Vet Emerg Crit Care* (San Antonio). 2016;26:419–427.

2. Weisse C *et al. J Am Vet Med Assoc.* 2014;244:78–94.
3. McRorie JW *et al. World J Gastrointest Pharmacol Ther.* 2014;5:57–62.
4. Tolbert MK *et al. J Vet Intern Med.* 2016.
5. Jenkins CC *et al. Am J Vet Res.* 1991;52:658–661.
6. Parkinson S *et al. J Vet Intern Med.* 2014.
7. Bersenas AM *et al. Am J Vet Res.* 2005;66:425–431.
8. Tolbert K *et al. J Vet Intern Med.* 2011;25:47–54.
9. Lane M *et al. J Feline Med Surg Open Reports.* 2016;2.
10. Neiger R *et al. J Vet Intern Med.* 2000;14:33–36.
11. Lane MB *et al. J Am Vet Med Assoc.* 2017;250.
12. Williamson KK *et al. J Vet Intern Med.* 2010;24:285–288.
13. Sutalo S *et al. J Vet Intern Med.* 2015;29:840–846.
14. Tolbert MK *et al. J Vet Intern Med.* 2015;29:556–560.
15. Khuroo MS *et al. J Gastroenterol Hepatol.* 2005;20:11–25.
16. Lo EA *et al. Ann Pharmacother.* 2015;49:207–219.
17. Scarpignato C *et al. BMC Med.* 2016;14:179.
18. Alshamsi F *et al. Crit Care.* 2016;20:120.
19. Ray WA *et al. Gastroenterology.* 2016;151:1105–1112 e1110.
20. Rostom A *et al. Cochrane Database Syst Rev.* 2002:CD002296.
21. Patel HR *et al. J Clin Diagn Res.* 2016;10:FC12–15.
22. Ahrens D *et al. Eur J Clin Pharmacol.* 2010;66:1265–1271.
23. Yachimski PS *et al. Arch Intern Med.* 2010;170:779–783.
24. Quimby J *et al. J Am Anim Hosp Assoc.* 2016;52:8–12.
25. Jiang M *et al. Cell Physiol Biochem.* 2016;39:2477–2491.
26. Garber A *et al. Clin Endosc.* 2016;49:421–424.

acvim346

Indications and Emerging Therapies for Use of Antiemetics and Appetite Stimulants

David Twedt, DVM, DACVIM (SAIM); Chad M. Johannes, DVM, DACVIM (SAIM, Oncology)
Fort Collins, CO, USA

PART A - ANTIEMETIC THERAPY

Antiemetics

Nausea and vomiting are common clinical conditions and antiemetics play an important role in the management of many of these cases.[1] There is also increasing evidence that some antiemetics (in particular NK1 and 5HT3 receptor antagonists) have other therapeutic applications beyond that of antiemetic actions and will be highlighted below.

Vomiting Mechanism and Antiemetic Receptors

Antiemetics work either peripherally, centrally or both.[2,3] In the brain stem the dorsal motor nucleus of the vagus with the nucleus tractus solitarius (NTS) make up what is referred to as the emetic center (EC) and the area postrema (AP), an area lacking a blood-brain barrier that detects humoral emetogens, is referred to as the chemoreceptor trigger zone (CRTZ). These areas all function in integrating vomiting. Neural stimulation of the EC arises through either afferent vagal, sympathetic, vestibular and cerebrocortical pathways. There is also evidence in the dog that vestibular stimulation passes through the CRTZ before activating the EC but not in the cat. The EC contains $5HT_1$, NK_1 and α_2 receptors.[2] The CRTZ contain dopamine (D_2) histamine (H_1), muscarinic cholinergic (M_1), alpha adrenergic (α_2), serotonin ($5HT_3$), enkephalin ($ENK_{\mu,\hat{e}}$) and neurokinin 1 (NK_1) receptors that are activated by blood borne substances.[2] Vestibular activation is associated with M_1, H_1 and NK_1 receptors. The $5HT_3$ and NK_1 receptors appear to have a pivotal role in the process of vomiting being abundant on vagal afferent neurons, other neurons and smooth muscle within the abdominal cavity and presynaptically on vagal afferent terminals in the medulla.[1] The role of $ENK_{\mu,\hat{e}}$ receptors within the cerebral cortex has been implicated but not yet been established to be present in dogs and cats.

Metoclopramide

Metoclopramide (Reglan™) has antiemetic effects through several mechanisms. At low doses it inhibits dopamine (D_2) receptors and at higher doses inhibits serotonin ($5HT_3$) receptors in the CRTZ.[1] Metoclopramide is useful for the control of CRTZ mediated vomiting. Cats, however, are reported to have few, if any, CRTZ dopamine receptors and consequently a poor antiemetic choice. Metoclopramide is rapidly metabolized through the liver and the best results are observed when given at continuous rate infusion. Metoclopramide also has gastric prokinetic effects and is also used for the management or prevention of reflux esophagitis. There are conflicting reports of the gastric prokinetic effects of metoclopramide in dogs. The general consensus is metoclopramide likely has weak if any effects in increasing gastric motility.[4] Studies using high resolution manometry found metoclopramide did not increase pressure of the lower esophageal sphincter nor did it prevent gastric reflux episodes in anesthetized dogs.[5] In another study, metoclopramide treatment did not change gastric myoelectric and motor activities in a way that would promote increased gastric emptying in dogs recovering from GDV.[6] We found using a wireless motility capsule that metoclopramide also did not significantly alter gastrointestinal motility.[7] Consequently, metoclopramide is not considered to be a potent prokinetic agent.

Ondansetron

The serotonin antagonists ondansetron (Zofran™) or dolasetron (Anzemet™) were developed as antiemetics primarily for the control of vomiting and nausea in humans receiving chemotherapy. They are effective at blocking $5HT_3$ receptors found peripherally, in the CRTZ and EC but not very effective in relieving motion sickness.[1] Ondansetron can be given PO or IV and is frequently given to animals vomiting from parvovirus, pancreatitis or azotemia. Pharmacokinetic studies in cats show the oral bioavailability of ondansetron to be poor (35%) and the half-life very short (1 hour) but SQ ondansetron had a slightly longer half-life of 3 hours.[8] A pharmacokinetic study in dogs found that the oral bioavailability of ondansetron was very poor (<10%) indicating this may not be an appropriate route of administration in dogs although injectable administration has been found to be clinically effective in prevention of vomiting.[9] We recently performed a blinded study comparing ondansetron to maropitant in parvovirus and found maropitant to likely be a better choice for managing parvovirus (see below).[10]

Ondansetron has effects on gastrointestinal motility in the dog by virtue of $5HT_3$ receptor antagonism. We found ondansetron slowed intestinal transit using a wireless motility capsule and others found it slowed colonic motility in a colitis model and reduced migrating motor complexes (MMC) during the interdigestive phase of GI motility.[7,11,12] Consequently, ondansetron may slow gastrointestinal motility and could prove useful in patients having both vomiting and diarrhea.

Maropitant

Maropitant (Cerenia™) is the first FDA approved antiemetic for the control of vomiting and motion sickness in the dog and cat. Numerous reports find it to be an excellent broad-spectrum antiemetic with a very high margin of safety.[3,13-15] Maropitant is approved for dogs over 8 weeks of age and more recently for IV administration and for prolonged use with blood levels remaining stable after 4 days of administration. Because the drug requires hepatic metabolism, it should be used with caution in patients that have hepatic dysfunction. I tend to decrease the dose by approximately 25–50% in cases having advanced hepatic disease.

Our experience finds maropitant very effective in preventing vomiting and nausea associated with chemotherapy, in management of parvovirus cases and many other causes of vomiting such as pancreatitis.[1,16] We recently compared maropitant to ondansetron in the therapy of parvovirus. We observed no difference between the two drugs in hospitalization time but found the maropitant group gained weight during hospitalization while the ondansetron group lost weight ($p<0.05$).[17] Although the sample size was small (n=22) our impression was that the maropitant group may have had less nausea and ate and drank more that the ondansetron group. In another CSU study evaluating maropitant for the control of vomiting and inappetence in chronic kidney disease cats found a significant reduction in vomiting in the maropitant treatment but not in their appetite scores compared to placebo.[18]

Maropitant is one drug that appears to exert many therapeutic effects in addition to being an antiemetic. A large area of research and emerging therapies involve maropitant's role in blocking visceral pain and its use as a preanesthetic agent. Maropitant blocks the effects of the inflammatory neuropeptide substance P preventing binding to NK_1 receptors. NK_1 receptors are abundant found in both vomiting but also visceral pain pathways. We developed a model for visceral pain involving stretching the ovarian ligament in anesthetized dogs and cats.[8,9] Using this model we found IV (but not epidural) maropitant to significantly lower the anesthetic minimum alveolar concentration (MAC) by 30% following visceral pain stimulation when compared to placebo.[19] The amount of visceral pain lowering of MAC was similar to that of morphine reported in another visceral pain model. There are also numerous studies showing maropitant decreased nausea and vomiting when given as preanesthetic prior to opioid administration in both dogs and cats.[20-24] Opioids also exert emetogenic effects via direct stimulation of the chemoreceptor trigger zone, inhibition of gut motility, and stimulation of the vestibular apparatus by activation of mu, kappa and delta opioid receptors. We compared preanesthetic effects of morphine or maropitant in an OHE model.[25] We found significant a difference at surgery with lower heart rates, blood pressure and isoflurane MAC concentration in the maropitant group and the impression of a more stable the plane of anesthesia, significant lower postoperative pain scores at the time of extubation, less nausea and significantly greater return to eating (65% vs. 15%) during the three-hour postoperative recovery period. A similar OHE model found maropitant cases had a smoother postoperative recovery with 91% of dogs returning to full appetite at 20 hours compared to 42% in placebo the group.[26]

Substance P and its NK_1 receptor are considered to play a role in modulating gastrointestinal motility. Although unpublished we found in our OHE study comparing maropitant to morphine there was less gastroesophageal reflux based on esophageal pH measurement but did not determine if it had any effect on esophageal pressures.[25] In an earlier study we found maropitant to have no effect on GI motility when compared to saline placebo using a wireless motility capsule.[7] Another study evaluating the prokinetic effects of maropitant on gastric motility found no effect compared to placebo based on radioscintigraphy and breath testing.[27] However, in a mouse model maropitant given at very high doses (8 mg/kg) increased intestinal transit of a luminal dye.[28] At this point, there is little compelling evidence that maropitant clinically modulates GI motility.

Since substance P is an inflammatory neuropeptide there is considerable interest in its effect on modulating inflammation for conditions such as inflammatory bowel disease, atopy, chronic rhinitis, chronic tracheitis or bronchitis, feline asthma and feline lower urinary tract disease, to name a few examples. Some of these theories are purely anecdotal impressions or come from extrapolation of laboratory animal models or *in vitro* testing. There is yet no convincing studies to show maropitant's ability to reduce inflammation. In a mouse model of IBD there was no improvement but in a mouse ulcerative dermatitis study there was improvement but likely this effect was on NK_1 receptors in sensory neurons in the skin reducing the itch.[28,29] Maropitant did reduce coughing in canine bronchitis (blocking sensory pathways?) but failed to reduce the inflammatory component.[30] Maropitant also failed to ameliorate clinical signs, airway hyper-responsiveness or airway eosinophilia in an experimental model of feline asthma.[31] We have used maropitant in the management of visceral pain associated with feline interstitial cystitis however any impression of improvement is only anecdotal.

Mirtazapine

Mirtazapine (Remeron™) is serotonin antagonist often used in cats as an appetite stimulant. This drug works primarily by blocking $5HT_3$ receptors, similar to the action of ondansetron. In a CSU placebo-controlled crossover study in cats with CKD, mirtazapine-treated cats had significantly increased appetite ($p=0.02$) and decreased vomiting ($p=0.04$).[32] Due to similar actions as ondansetron in blocking $5HT_3$ receptors is unlikely that there would be additive effects in antiemetic action if given together.

Part B - Appetite Stimulant Therapy

Chronic Causes of Inappetence in Dogs and Cats

Appetite is often a key determinant in a pet owners' perceived quality of life evaluation for their dog or cat with cancer or other chronic disease process. Lack of appetite is often the first recognizable sign by pet owners that their dog or cat is not feeling well and often triggers inquiry to their veterinarian's office. Inappetence in pets with chronic disease is often multifactorial with potential contributing factors including: the tumor or disease process itself (e.g., dysphagia, pain, ascites), increased inflammatory cytokines, chemotherapy or other drug side effects (e.g., dysmotility, nausea) and effects of concomitant medications.

While inappetence is a commonly described side effect of many chemotherapeutic agents in dogs and cats, the true incidence is not well documented. Previous studies (Table 1) focusing on the incidence of inappetence in dogs and cats receiving chemotherapy are limited and methods of measuring inappetence are variable and inconsistent. In addition to cancer,[41] other chronic conditions in dogs and cats associated with inappetence and related weight loss/cachexia include chronic kidney disease (CKD),[42-45] congestive heart failure (CHF),[46] and inflammatory bowel disease.[47]

Table 1. Reported incidence of inappetence in dogs and cats receiving chemotherapy

Chemotherapeutic	Species	Inappetence incidence	Reference
Vinorelbine	Dogs	17% (4/24)	Grant (2008)1
Epirubicin	Dogs	19% (27/139)	Marrington (2012)2
Carboplatin	Dogs and cats	25% (7/28)	Bowles (2010)3
Doxorubicin	Dogs	35-51% (n=49)*#	Rau (2010)4
Cyclophosphamide	Dogs	36% (15/42)*	Mason (2014)5
Toceranib phosphate	Dogs	39% (34/87)	London (2009)6

Vincristine	Dogs	43% (25/57)*	Mason (2014)5
Epirubicin	Dogs	44% (8/18)	Kim (2007)7
Lomustine	Dogs	48% (39/81)	Vail (2012)8
Paclitaxel	Dogs	76% (128/168)	Vail (2012)8

*Subset of dogs treated with maropitant following chemotherapy administration; #Crossover study

Clinical Impact of Inappetence

Weight loss frequently occurs with several types of human cancer and is often associated with reduced survival and quality of life in cancer patients.[48] There are multiple causes of weight loss in cancer patients, key among them inappetence along with numerous catabolic drivers including direct tumor metabolism, systemic inflammation, and insulin resistance.[49] This phenomenon is referred to as cancer-related anorexia/cachexia syndrome (CACS), a metabolic, paraneoplastic syndrome characterized by decreased food intake, involuntary weight loss, and altered body composition (loss of fat and muscle).[49] Cachexia is reported in 15–40% of human cancer patients, approaching 80% in those with advanced illness.[49]

For decades, it has been debated whether or not veterinary patients develop CACS. This may be attributed in part to veterinarians not recognizing the syndrome due to lack of awareness and/or absence of established diagnostic criteria. While some veterinary patients may be humanely euthanized before the syndrome becomes clinically evident, the reality is there is limited information on the incidence of CACS in pets. A study by Michel *et al.* reported that 37% of dogs with cancer had lost >5% of their body weight at diagnosis.[50] A number of studies in veterinary medicine have indicated dogs with cancer experience metabolic alterations similar to those observed in human patients with cachexia: increased resting energy expenditure and decreased rates of protein synthesis in dogs with osteosarcoma;[51] alterations in carbohydrate metabolism in dogs with solid tumors;[52] and differences in serum cytokine profiles in dogs with lymphoma compared to healthy controls.[53] Dogs underweight at the time of lymphoma diagnosis have been shown to experience significantly shorter survival times.[54]

Beyond cancer, studies in veterinary medicine have further indicated that weight loss is not only an early indication or predictor of underlying CKD in cats[11-13] but is associated with survival.13 Dogs with CHF that gained body weight have been shown to have longer survival times compared to those losing or maintaining weight.[43-45] Similarly, cats with CHF having low body weights experienced shorter survival times compared to cats with moderate or high body weights.[56]

Emerging Treatment Options for Inappetence in Dogs and Cats

If weight loss/cachexia impacts canine and feline survival as it does in humans with cancer, CHF and other chronic illnesses, management of this syndrome may improve survival times and quality of life for pets diagnosed with similar conditions. The challenge for veterinarians lies in having effective, proven therapies available for clinical use. The list of drugs utilized in an attempt to stimulate appetite in dogs and cats is quite extensive with common choices including mirtazapine, cyproheptadine, corticosteroids and antiemetics. The challenge with these drugs is none of them has stimulation of appetite as a primary mechanism. The appetite stimulation component is a secondary or side effect for these drugs, explaining the lack of reliability from patient to patient. Except for mirtazapine in cats, additional challenges arise from limited data on dosing and pharmacokinetics and sparse to non-existent data on efficacy for appetite stimulation in dogs and cats.

ENTYCE®/Capromorelin/AT-002 (Aratana Therapeutics, Inc.)

Given the impact of CACS on quality of life and survival in humans with cancer, a number of therapeutic approaches are being explored to positively treat this syndrome. A ghrelin agonist, anamorelin, is one example and has demonstrated positive effects on body weight and appetite in humans with non-small cell lung cancer (a cancer with an extremely high incidence of CACS) and may also impact survival.[57]

Capromorelin is an orally active small molecule in this same category of therapeutic that mimics the action of ghrelin causing growth hormone secretion and appetite stimulation. Capromorelin oral solution dosed at 3 mg/kg PO q24h has been shown to cause increased food intake and weight gain in both healthy laboratory and inappetent client-owned dogs.[58-60] Capromorelin oral solution has also demonstrated a wide margin of safety, being well-tolerated at daily doses up to 40 mg/kg once daily for 12 consecutive months.[61] When administered to healthy dogs for 7 days, capromorelin treatment produced increased insulin-like growth factor 1 (IGF-1) concentrations on Day 1 and this increase was sustained through Day 7.[58] Additionally, capromorelin has been demonstrated to cause increases in IGF-1 and increased food intake and body weight in cats.[62]

ENTYCE was granted FDA approval in May 2016. Aratana plans to make it commercially available to veterinarians in late 2017.[63] This therapeutic has significant potential to positively impact the treatment of inappetence related to several chronic conditions in our canine patients. Aratana is also developing a cat-specific formulation with additional studies pending.

Melanocortin-4 Receptor (MC-4R) Antagonist

Another mechanism to directly impact the appetite pathway is via the MC-4 receptor which plays an important role within the hypothalamic appetite center, affecting meal size and choice. MC-4R antagonism is associated with increased food intake. MC-4R antagonists have been evaluated and shown promise in mouse models for the treatment of cancer cachexia.[64]

A veterinary MC-4R antagonist product is currently being investigated in dogs with cachexia at the University of Missouri-Columbia. The product is an MC-4R antagonist peptide designed to minimize the potential cardiovascular side effects seen in this category of drug. Rodent models have shown increased quality of life despite tumor growth. Safety has been demonstrated in healthy dog studies. The ongoing funded clinical trial is enrolling dogs with cachexia (5% or greater weight loss in 3 months) of any cause without severe underlying cardiac, renal or hepatic disease (Personal communication with Sandra Bechtel, DVM, DACVIM).

Mirtazapine

Mirtazapine is a 5-HT3 receptor antagonist with appetite stimulant properties documented in healthy young cats at a dose of 1.88 mg PO q24h.[65] Additional studies have shown that dosing of 1.88mg PO q48h is more appropriate in cats with CKD.[66] A double-masked, placebo-controlled crossover clinical trial in cats with CKD demonstrated that mirtazapine administration at 1.88 mg PO q48h for 3 weeks resulted in a significant increase in appetite and activity and significant decrease in vomiting compared to placebo. Mirtazapine-treated cats also experienced a significant gain in bodyweight (median gain 0.18 kg) compared to placebo.[67] Used as an anti-depressant in human medicine with the observed side effect of increased appetite and weight gain, the specific method of appetite stimulation is not well documented. The antiemetic effect noted in cats with CKD is interesting as mirtazapine is sometimes used for refractory chemotherapy-induced nausea and vomiting in humans.

Clinical experience would indicate that mirtazapine is a less reliable/predictable appetite stimulant in dogs. A small pilot study evaluating the pharmacokinetics of mirtazapine in healthy Beagle dogs has been performed.[68] Results of this study indicate that mirtazapine appears to be metabolized much more quickly in dogs and twice daily dosing may be more appropriate in this species. As the appetite stimulation component is essentially a side effect of mirtazapine rather than a primary mechanism, it is not surprising that its clinical effect is variable and rather unpredictable.

References

1. Elwood C *et al. J Small Anim Pract.* 2010;51(1):4.
2. Boothe DM. *Vet Clin North Am Small Anim Pract.* 1999;29(2):343.
3. Benchaoui HA *et al. J Vet Pharmacol Ther.* 2007; 30(4):336.
4. Burger DM *et al. J Vet Med A Physiol Pathol Clin Med.* 2006;53:97.
5. Kempf J *et al. Am J Vet Res.* 2014;75:361.
6. Hall JA *et al. Am J Vet Res.* 1996;57:1616.
7. McCord KW *et al. J Vet Intern Med.* 2009; 23:735.
8. Quimby JM *et al. J Vet Pharmacol Ther.* 2014;37:348.
9. Saynor DA *et al. Eur J Cancer Clin Oncol.* 1989;25:S75.
10. Venn EC *et al. J Vet Emerg Crit Care.* 2017;27:52.
11. Hayashi K *et al. Surgery.* 2010;147:789.
12. Nakajima H *et al. J Gastroenterol.* 2010;45:506.

13. Benchaoui HA *et al. Vet Rec.* 2007;29;161(13):444.
14. Lesman SP *et al. J Vet Pharmacol Ther.* 2013;36(5):462.
15. Ramsey DS *et al. J Vet Pharmacol Ther.* 2008;31(6):538.
16. Mason SL *et al. J Small Anim Pract.* 2014;55(8):391.
17. Lenberg *et al. J Vet Intern Med.* 2012;26(3),795.
18. Boscan P *et al. Am J Vet Res.* 2011;72:1576.
19. Boscan P *et al. Am J Vet Res.* 2011;72(12):1576.
20. Niyom S *et al. Vet Anaesth Analg.* 2013;40(4):425.
21. Lorenzutti AM *et al. Can Vet J.* 2017;58:35.
22. Martin-Flores M *et al. J Feline Med Surg.* 2016; Epub ahead of print.
23. Martin-Flores M *et al. J Feline Med Surg.* 2015; Epub ahead of print.
24. Lorenzutti AM *et al. Vet Anaesth Analg.* 2016;43:195.
25. Marquez M *et al. PLoS One.* 2015;10:e0140734.
26. Ramsey, D *et al.* Int J Applied Res Vet Med. 2015;12:229.
27. Schmitz S *et al. Tierarztl Prax Ausg K Kleintiere Heimtiere.* 2016;44:163.
28. Mikawa S *et al. J Vet Med Sci.* 2015;77:1195.
29. Williams-Fritze MJ *et al. J Am Assoc Lab Anim Sci.* 2011;50(2):221.
30. Grobman M *et al. J Vet Intern Med* 2016;30:847.
31. Graham A *et al. ACVIM Forum* (abstract). 2013; Seattle.
32. Quimby JM *et al. Vet J.* 2013;197:651.
33. 1Grant IA *et al. J Vet Intern Med.* 2008;22:388–393.
34. Marrington AM *et al. Vet Comp Oncol.* 2011;10(2):113–123.
35. Bowles DB *et al. J Small Anim Pract.* 2010;51:104–112.
36. Rau SE *et al. J Vet Intern Med.* 2010;24:1452–1457.
37. Mason SL *et al. J Small Anim Pract.* 2014;55:391–398.
38. London CA *et al. Clin Cancer Res.* 2009;15(11):3856–3865.
39. Kim SE *et al. J Am Vet Med Assoc.* 2007;231:1550–1557.
40. Vail DM *et al. J Vet Intern Med.* 2012;26(3).
41. Rajagopaul S *et al. J Small Anim Pract.* 2016;57:484–490.
42. O'Neill DG *et al. J Vet Intern Med.* 2013;27(4):814–821.
43. Greene JP *et al. J Am Vet Med Assoc.* 2014;244(3):320–327.
44. Markovich JE *et al. J Feline Med Surg.* 2015;17(12):979–983.
45. Freeman L *et al.* 2015 ACVIM Forum; Indianapolis, IN. Abstract.
46. Freeman LM *et al. J Vet Intern Med.* 1998;12(6):440–448.
47. Craven M *et al. J Small Anim Pract.* 2004;45(7):336–342.
48. Mondello P *et al. BMC Cancer.* 2014;14:282.
49. Fearon K *et al. Lancet Oncol.* 2011;12:489–495.
50. Michel KE *et al. J Vet Intern Med.* 2004;18(5):692–695.
51. Mazzaferro EM *et al. Am J Vet Res.* 2001;62(8):1234–1239.
52. Ogilvie GK *et al. Am J Vet Res.* 1997;58(3):277–281.
53. Calvalido J *et al.* 2016 ACVIM Forum; Denver, CO. Abstract.
54. Romano FR *et al. J Vet Intern Med.* 2016;epub Jun 8.
55. Slupe JL *et al. J Vet Intern Med.* 2008;22:561-565.
56. Finn E *et al. J Vet Intern Med.* 2010;24:1369-1374.
57. Currow DC *et al.* J Clin Oncol. 2015;33(29-suppl):175.
58. Zollers B *et al.* J Vet Pharmacol Ther. 2016;doi:10.1111/jvp.12344.
59. Zollers B *et al. J Vet Intern Med.* 2016;30(6):1851–1857.
60. Zollers B *et al. BMC Vet Res.* 2017;13(1):10.
61. Zollers B *et al. J Vet Pharmacol Ther.* 2016;doi:10.1111/jvp.12358.
62. Zollers B *et al.* 2015 ACVIM Forum; Indianapolis, IN. Abstract.
63. Aratana Therapeutics. http://aratana.investorroom.com/2017-02-03-Aratana-Therapeutics-to-Attend-North-American-Veterinary-Conference.
64. Dallmann R *et al. J Cachexia Sarcopenia Muscle.* 2011;2(3):163–174.
65. Quimby JM *et al. J Vet Pharmacol Ther.* 2011;34(4):388–396.
66. Quimby JM *et al. J Vet Intern Med.* 2011;25:985–989.
67. Quimby JM, Lunn KF. *Vet J.* 2013;197(3):651–655.
68. Giorgi M, Yun H. *Vet J.* 2012;192(2):239–241.

acvim347

Animal Models: The Increasing Role of the Microbiome

James G. Fox, DVM, DACLAM
Cambridge, MA, USA

Defining the Microbiome

The host microbiome is routinely defined as all the microorganisms inhabiting a specific environment. The term **microbiome** was coined in the "-ome" and "-omics" era to signify the ecological community of commensal, symbiotic, and pathogenic microorganisms that share a host's body.[1] The mammalian gastrointestinal tract is colonized with a wide array of microbes from diverse phyla including viruses, prokaryotic bacteria, and eukaryotic microbes. The latter, referred to broadly as the eukaryome, is comprised of a myriad of fungi, helminths, and protists, which include protozoa. As an example, in a recent paper, authors have shown the protozoan *Tritrichomonas musculis* activates the host epithelial inflammasome to induce IL-18 release, which promotes dendritic cell-driven Th1 and Th17 immunity and confers dramatic protection from mucosal *Salmonella* infections. Along with its role as a protistic antibiotic, colonization with *T. musculis* exacerbates the development of T-cell-driven colitis and sporadic colorectal tumors. This finding demonstrates a novel mutualistic host-protozoan interaction that increases mucosal host defenses at the cost of an increased risk of inflammatory disease.[2]

Research in the microbiome field has stimulated increasing interest in the scientific community over the past 16 years (Figure 1). There is an increasing realization that the microbiota has an indispensable impact on the phenotype of various animal models. Molecular methods have become available, allowing the analysis of complex microbial communities to a degree that was not realized using traditional culture methods.

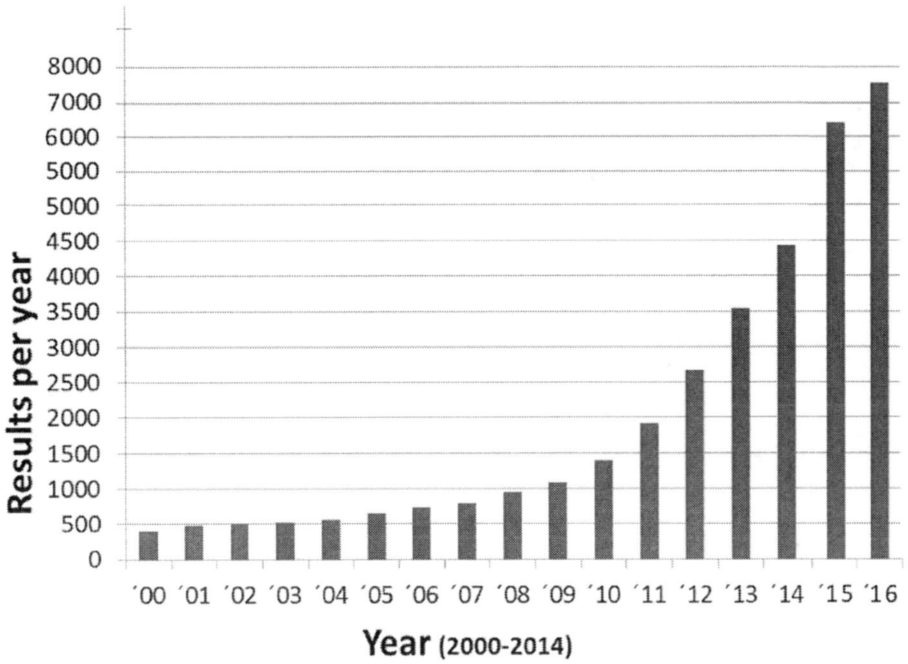

Figure 1. PubMed Search using terms microbiome or microflora, total hits per year (Updated from[15])

The Impact of the Microbiota on Animal Models

The impact of the microbiota on host physiology is readily observed when comparing conventional to germfree mice. A variety of mucosal parameters differ in germfree mice, such as decreased epithelial renewal, enzyme production, and mucosa thickness. Microbiota are important in bile salt metabolism, production of short-chain fatty acids, or vitamin K and B-complex vitamins in the large intestine. For germfree animals, there are reduced populations of various innate and adaptive immune cells and reduced or altered cell-specific activities. While effects on gut inflammatory phenotypes and intestinal disease models could be expected from absence of gut microbiota,[3] other conditions such as diabetes, obesity, arthritis, allergy, inflammatory pain, or atopic dermatitis can be affect as well[4]. Changes in microbiota more subtle than presence or absence lead to phenotypic variation. It is possible that a limited diversity could lead to artificial or even loss of phenotypes in widely used animal models (Figure 2). Variation of complex microbiota might lead to phenotypic alterations. For example, diabetes incidence was reduced in non-obese diabetic (NOD) mice after they received acidified water which altered the intestinal microflora.[5] In principle, all of these models are influenced by variation of the microflora, and the number of published models being affected by microbial variation continues to grow (Figure 1).

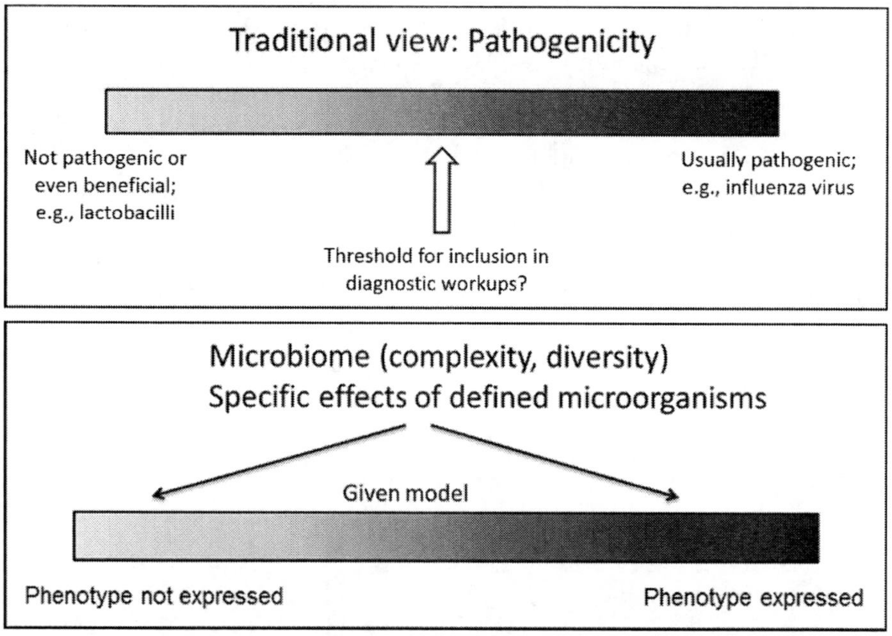

Figure 2. Concepts for microbial inclusion in animal models. (A) Traditional and current focus on health monitoring is based on virulence of microorganisms (symbiotic, opportunistic, pathogenic) and aims to exclude or report pathogens (and potential pathogens), depending on the barrier level and specified excluded agents. (B) Microbial relevance (other than pathogenicity) to the host is increasingly recognized as is the complexity and diversity of the microbiota (Modified from[15])

Microbiota and Their Effect on the Host: The IBD Example

Different approaches are used to identify relevant microbes as well as mechanisms of microbiota-related phenotype variation. Approaches range from analyzing the effect of single species, simplified communities, or even complex microbiota. For example, using models of intestinal inflammation, protective effects have been observed using *Lactobacillus* spp., which are considered to be probiotic bacteria, whereas proinflammatory effects were observed using *Enterococcus faecalis, E. coli, Bacteroides vulgatus, Helicobacter* spp., and murine norovirus. The interleukin-10-deficient mouse model of inflammatory bowel disease (IBD), when maintained at different institutions, develops either spontaneous or *Helicobacter*-induced disease at differing rates and severity.[6-8]

Inflammatory bowel disease in dogs is a common, chronic, inflammatory, spontaneous enteropathy; it has a multifactorial aetiology consisting of an aberrant host immune system, genetics, environmental factors, and gut microbiome. A recent study indicates that microbiome dysbiosis network underlying IBD in dogs differs from that in humans. Select bacteria, such as *Fusobacterium* spp., switch roles between dogs and humans. A dog-specific dysbiosis index achieved a high correlation with the overall dog microbial community diversity patterns. These relationships were not detected in previously published dog IBD data sets due to their limited sample size and statistical power.[9,10]

Another example of the importance of a single commensal bacterium are segmented filamentous bacteria (SFB), known to induce Th17 cells in the intestine; importantly, this effect is transferrable with transfer of SFB.[11,12] Indeed, simplified bacterial communities or complex microbiota administered to germfree animals revealed pathophysiological effects that vary with the microbial communities.[13,14] Molecular analysis using sequencing approaches is increasingly used to elucidate the effect of complex microbial communities on the phenotype expressed in different animal models.[8]

Microflora and Metabolic Disorders

Many studies have found correlations between intestinal composition and the incidence of metabolic disorders, including obesity, type 2 diabetes mellitus, metabolic syndrome, and other related diseases. Probiotics, prebiotics, and antibiotics have been used to modulate gut microbiota as a possible treatment for metabolic disorders. Metabolites generated by gut bacteria appear to also alter host gene expression by dysregulating energy intake through insulin signaling and fat deposition through adipogenesis. Other associated host gene expression changes disrupt normal intestinal wall permeability causing metabolic bacteremia or endotoxemia. Intestinal or systemic inflammation is often associated with and believed to contribute to metabolic diseases. Modulation of bile acid metabolism and bacterial hepatotoxic bioproducts also provides clues to the connection between gut microbiota, nonalcoholic steatohepatitis, and fatty liver disease. Indeed, commercial entities are now promoting real-time PCR-based, application-specific microbial identification or profiling to identify taxon or genera that are incriminated in these metabolic disorders increasingly common in humans. Further studies on this topic are needed in animals, including pet dogs and cats.

References

1. Lederberg J, McCray AT. 'Ome sweet 'omics - a genealogical treasury of words. *Scientist*. 2001;15:8.
2. Chudnovskiy A, Mortha A, Kana V, Kennard A, Ramirez JD, *et al*. Host-protozoan interactions protect from mucosal infections through activation of the inflammasome. *Cell*. 2016;167:444–456.
3. Bleich A, Mahler M. Environment as a critical factor for the pathogenesis and outcome of gastrointestinal disease: experimental and human inflammatory bowel disease and *Helicobacter*-induced gastritis. *Pathobiology*. 2005;72:293–307.
4. Bleich A, Hansen AK. Time to include the gut microbiota in the hygienic standardisation of laboratory rodents. *Comp Immunol Microbiol Infect Dis*. 2012;35:81–92.
5. Wolf KJ, Daft JG, Tanner SM, Hartmann R, Khafipour E, *et al*. Consumption of acidic water alters the gut microbiome and decreases the risk of diabetes in NOD mice. *J Histochem Cytochem*. 2014;62:237–250.
6. Keubler LM, Buettner M, Hager C, Bleich A. A multihit model: colitis lessons from the interleukin-10-deficient mouse. *Inflamm Bowel Dis*. 2015;21:1967–1975.
7. Mahler M, Leiter EH. Genetic and environmental context determines the course of colitis developing in IL-10-deficient mice. *Inflamm Bowel Dis*. 2002;8:347–355.
8. Yang I, Eibach D, Kops F, Brenneke B, Woltemate S, *et al*. Intestinal microbiota composition of interleukin-10 deficient C57BL/6J mice and susceptibility to *Helicobacter hepaticus*-induced colitis. *PLoS One*. 2013;8:e70783.
9. Honneffer JB, Minamoto Y, Suchodolski JS. Microbiota alterations in acute and chronic gastrointestinal inflammation of cats and dogs. *World J Gastroenterol*. 2014;20:16489–16497.
10. Vazquez-Baeza Y, Hyde ER, Suchodolski JS, Knight R. Dog and human inflammatory bowel disease rely on overlapping yet distinct dysbiosis networks. *Nature Microbiology*. 2016;1:16177.
11. Ivanov II, Atarashi K, Manel N, Brodie EL, Shima T, *et al*. Induction of intestinal Th17 cells by segmented filamentous bacteria. *Cell*. 2009;139:485–498.
12. Gaboriau-Routhiau V, Rakotobe S, Lecuyer E, Mulder I, Lan A, *et al*. The key role of segmented filamentous bacteria in the coordinated maturation of gut helper T cell responses. *Immunity*. 2009;31:677–689.
13. Faith JJ, Ahern PP, Ridaura VK, Cheng JY, Gordon JI. Identifying gut microbe-host phenotype relationships using combinatorial communities in gnotobiotic mice. *Sci Transl Med*. 2014;6:220ra211.
14. Slezak K, Krupova Z, Rabot S, Loh G, Levenez F, *et al*. Association of germ-free mice with a simplified human intestinal microbiota results in a shortened intestine. *Gut Microbes*. 2014;5:176–182.
15. Bleich A, Fox JG. The mammalian microbiome and its importance in laboratory animal research. *ILAR J*. 2015;56:153–158.

acvim348

Overdiagnosis: Don't Just Do Something, Stand There!

Brennen McKenzie, MA, MSc, VMD
Los Altos, CA, USA

WHAT IS OVERDIAGNOSIS?

A 5-year-old Labrador Retriever presents for an acute cranial cruciate ligament rupture. Otherwise, the dog is healthy in every way with no clinical symptoms other than lameness. However, pre-operative blood work shows moderate elevations in ALT, and an ultrasound exam shows some indistinct, mildly hypoechoic nodular lesions in the liver. An ultrasound-guided needle biopsy is performed which ultimately shows benign nodular hyperplasia. Unfortunately, the dog dies of complications from the biopsy procedure. What this dog really died from was overdiagnosis.

Veterinarians are often fearful of misdiagnosis; incorrectly identifying disorders in their patients or diagnosing diseases the patients do not actually have. However, few worry about the dangers of overdiagnosis; the correct diagnosis and treatment of disorders patients do have but which will never cause clinical symptoms or mortality.

Overdiagnosis is now recognized as a common and serious problem in human medicine which causes significant harm in terms of cost, wasted resources, and suffering for patients and their caregivers. A consortium of seventy specialty groups has created the online resource Choosing Wisely to help physicians and patients make better decisions and reduce overdiagnosis and treatment. Major medical journals have ongoing features devoted to overdiagnosis, such as the Journal of the American Medical Association's Less is More feature, and the British Medical Journal's Too Much Medicine. Changes in clinical practice guidelines for many conditions, including highly publicized changes in practices such as breast cancer and prostate cancer screening programs, have resulted from the recognition that overdiagnosis harms patients. Yet the subject of overdiagnosis is virtually unknown in veterinary medicine.[1]

HOW COMMON IS OVERDIAGNOSIS

A challenge in controlling overdiagnosis is that there is no absolute way to know in advance if a finding in a particular patient is going to prove clinically important or not. This can only be known after sufficient time has passed to evaluate whether or not the finding has resulted in disease or death. However, it is possible to evaluate the frequency of overdiagnosis in a population based on the frequency of diagnosis and mortality data for specific conditions and populations. As an example, CT imaging of clinically healthy people frequently leads to diagnosis of cancers which, based on mortality figures, are never going to lead to early mortality (Table 1).

Table 1. Detection and risk of mortality from cancer using CT imaging in asymptomatic humans (from Welch, 2012)

Organ	% with lesion detected by CT (a)	10-year risk of cancer mortality (b)	Chance lesion is lethal cancer (c=b/a)	Chance lesion is not lethal cancer (d=1–c)
Lung (smokers)	50	1.8	3.6	96.4
Lung (never smoked)	15	0.1	0.7	99.3
Kidney	23	0.05	0.2	99.8
Liver	15	0.08	0.5	99.5
Thyroid (US)	67	0.005	<0.01	>99.99

Similar data is available for many other diseases and diagnostic tests, suggesting that overdiagnosis is extremely common in human medicine. There is no published data specifically addressing the frequency of overdiagnosis in veterinary medicine.

What is the Harm of Overdiagnosis?

The diagnosis of diseases that are unlikely to ever cause significant illness or mortality causes harm in several ways. The testing leading to diagnosis, and the treatment often offered once such a diagnosis is made, have financial costs. It is estimated, for example, that overdiagnosis and overtreatment of clinically irrelevant lesions detected through mammography costs $4 billion annually in the United States alone.[3] Another study suggests that unnecessary treatment of people with mild hypertension in the U.S., with no benefit in terms of reducing symptoms or early mortality, may cost $32 billion annually.[4] These are only estimates for the costs of overdiagnosis for two diseases in one country, so the global financial cost is undoubtedly much greater. And despite the impulse to feel that no price should be put on efforts to improve health and treat disease, it is undeniable that such waste raises the overall costs of healthcare and reduces access for some people, all without any benefit to patients.

There are no estimates of the costs of overdiagnosis in veterinary medicine. The economic model of the veterinary profession is quite different from human medicine, and the financial costs of overdiagnosis may not impact the overall cost of veterinary care or the availability of care as dramatically. However, these costs are still a waste of client resources, and they can reduce the ability of some clients to pay for subsequent care that may be necessary or beneficial for their animals.

The financial costs of overdiagnosis, however, have not been the major drivers of change in clinical practice in human medicine. This has been the physical and emotional harm to patients. Diagnostic testing and treatment for conditions not destined to cause illness or death can cause both physical injury an psychological distress.

It has been estimated that, for example, that prostate specific antigen (PSA) screening for prostate cancer in men will identify cancer in 30–100 patients who would never have been clinically effected for every death prevented.[2] And for these men who are overdiagnosed and go on to have biopsies or treatment, up to 50% will experience sexual dysfunction, 30% will have urination difficulties, and 1–2 per thousand screened will die as a result of unnecessary treatment. Research has also shown that quality of life diminishes after a diagnosis of prostate cancer, and that the risk of suicide and cardiovascular death increases immediately following such a diagnosis.[5,6]

Similar evidence of physical and psychological harm from overdiagnosis is available for many other conditions in human medicine. There has been, however, no published research on the risks of overdiagnosis in the veterinary field.

What Causes Overdiagnosis?

Overdiagnosis is driven by numerous factors. Screening tests, imaging, and other diagnostics employed without specific clinical justification frequently lead to the detection of abnormalities. Such abnormalities are far less likely to be clinically important than those which cause symptoms, and therefore they often represent overdiagnosis. However, once an abnormality is detected, some of the psychological harm to patients or caregivers has already occurred. And because of the anxiety induced by the finding and the desire of both clients and doctors to take some action, even when it is unclear this will benefit the patient (a phenomenon known as "commission bias"[6]), further testing and even therapy often results from an initial overdiagnosis.

Overdiagnosis also stems from the development of more sensitive tests, which identify conditions earlier and prior to the onset of clinical symptoms. Expanded definitions of disorders which encompass patients previously not consider to have these disorders can also lead to overdiagnosis and overtreatment. Psychologically, doctors are prone to overdiagnosis because they are likely to be punished, in the form of blame or even litigation, for failing to diagnose a medical condition if it does eventually lead to symptoms or death, but there are no sanction for unnecessarily diagnosing and treating conditions which would never have cause any harm if undiagnosed since such an outcome is rarely detected.

The inappropriate reliance on anecdotal evidence and clinical experience to guide diagnostic practices also contributes to overdiagnosis. Even in the face of strong evidence, for example, that mammography of women under 50 years of age leads to significantly more harm from overdiagnosis than benefit from earlier diagnosis and treatment, there has been significant resistance to the change in screening guidelines implemented to reduce this harm. Much of that resistance is justified with the use of anecdotes of women who had been diagnosed and treated for breast cancer because they and their doctors believed, rightly or wrongly, that this intervention had saved their lives. Any doctor who believes their use of screening or other diagnostic interventions in asymptomatic patients has saved someone's life will be very reluctant to stop using those interventions regardless of the evidence that they do more harm than good. Individual stories are always more psychologically compelling than statistically data. But acting on our emotional response to such stories and ignoring the evidence for regarding overdiagnosis ultimately causes more unnecessary suffering for real patients.

Patients are similarly inclined to seek diagnosis and take action on it even if the statistical evidence suggests it is in their best interests not to. Doing so gives people a sense of control over their fate, or that of their animals. Even if this sense of control is an illusion, it tends to outweigh rational considerations. One survey found that 98% of people mistakenly diagnosed with cancer through screening were still glad they had had the test once the follow-up evaluation showed they actually did not have cancer.[8] Like doctors, patients are inclined to take action rather than choosing inaction, even when inaction is demonstrably the better choice.

Finally, we cannot ignore the potential influence of financial interests on overdiagnosis. Companies selling diagnostic tools and veterinarians using them receive income from the use of these tools. And the follow-up testing and treatment of diseases, even when they are overdiagnosed, also generates revenue. While doctors are unlikely to intentionally pursue unnecessary testing and treatment purely for financial gain, it would be naive to imagine such revenue has no impact on doctors' decision making. Federal law prohibits doctors from referring patients to diagnostic facilities in which they have a financial interest because research has shown such interests increase the number of tests done and the costs to patients.[9,10] There is no reason to believe veterinarians to be exempt from the same potential for financial self-interest to influence clinical decisions.

How Do We Reduce Overdiagnosis & Overtreatment?

The first step in reducing the harms from overdiagnosis is to understand the phenomenon and its causes. This includes developing the data to identify overdiagnosis of specific conditions. Because overdiagnosis can only be identified in retrospect in individual patients, we must gather and analyze epidemiologic data to recognize the level of risk for overdiagnosis of particular diseases using specific tests. We cannot safely rely on anecdotes and uncontrolled clinical experience alone to drive our diagnostic and therapeutic practices. We need data.

In the meantime, in the absence of such data, the best strategy is to understand the limitations of our diagnostic tests, including important measures such as their positive and negative predictive value, which help us to appreciate the likely significance and reliability of test results in particular patient populations. We should also ensure that we have an appropriate clinical index of suspicion for any condition before we begin testing for it. "Fishing expeditions," "shotgun diagnostics," indiscriminate imaging, and other such irrational diagnostic practices raise the risk of overdiagnosis.

We must also learn to accept the inevitable uncertainty in medicine and be honest with clients about our ability to predict and control all patient outcomes. We need to recognize and disclose that testing and treatment have costs and risks as well as benefits, especially in patients without significant clinical symptoms associated with the disorders we are trying to diagnose and treat. Though it is psychologically more difficult for us, it is often wiser to avoid action when there is not good evidence to show that our actions will truly benefit our patients. Don't just do something, stand there!

References

1. McKenzie BA. Overdiagnosis. *J Am Vet Med Assoc.* 2016:249(8);884–889.

2. Welch HG. Schwartz L. Woloshin S. *Overdiagnosed: Making People Sick in the Pursuit Of Health*. Boston, MA: Beacon Press; 2012.
3. Ong M, Mandl KD. National expenditure for false-positive mammograms and breast cancer overdiagnosis estimated at $4 billion a year. *Health Aff*. 2015;34(4):576–583.
4. Martin SA, Boucher M, Wright JM, *et al*. Mild hypertension in people at low risk. *Br Med J*. 2014;349:g5432.
5. Heijnsdijk EA, Wever EM, Auvinen A, *et al*. Quality-of-life effects of prostate-specific antigen screening. *N Eng J Med*. 2012;367:595.
6. Fang F, *et al*. Immediate risk of suicide and cardiovascular death after a prostate cancer diagnosis: cohort study in the United States. *J Natl Cancer Inst*. 2010;102:307.
7. McKenzie BA. Veterinary clinical decision-making: cognitive biases, external constraints, and strategies for improvement. *J Am Vet Med Assoc*. 2014;44(3):271–276.
8. Schwartz LM, Woloshin S, Fowler FJ, *et al*. Enthusiasm for cancer screening in the United States. *J Am Med Assoc*. 2004;291(1):71–78.
9. Levin DC, Rao VM. Turf wars in radiology: updated evidence on the relationship between self-referral and the overutilization of imaging. *J Am Coll Radiol*. 2008;5(7):806–810.
10. Gazelle GS, Halpern EF, Ryan HS, Tramontano AC. Utilization of diagnostic medical imaging: comparison of radiologist referral versus same-specialty referral. *Radiology*. 2007;245(2):517–522.

acvim349

Why Use Scientific Literature to Answer Clinical Questions?

Robert (Bob) L. Larson, DVM, PhD, ACT, ACAN, ACVPM (Epi)
Manhattan, KS, USA

INTRODUCTION

While clinical experience provides important observations about disease onset risk factors, therapeutic intervention efficacy, and patient prognosis, clinical observations lack controls for bias and confounding and as such must be tempered with data from controlled experiments. Clinical experience plays a critical role in veterinary decision making and these experiences form the framework through which all literature will be interpreted. While our own experience and the experience of colleagues that is passed through communications can be useful, the human mind has significant limitations managing normal variation in results, dealing with bias, and overcoming the difficulty of interpreting the complicated biologic systems that make up the animals and populations with which we contend on a daily basis. While every research study has important limitations in either internal validity, external validity or both, well-designed studies have incorporated at least some level of control for bias and confounding and use statistical tests to distinguish treatment effects from random variation.

USING SCIENTIFIC LITERATURE TO ANSWER CLINICAL QUESTIONS

Limitations of Clinical Experience

Biologic systems are inherently more complex than other natural systems. Only biologic systems have complex homeostatic controls as well as the capability to remove insults, self-repair, and substitute for deficient factors. In addition, the hierarchy of: tissues, animals, herds, populations, and ecosystems adds complexity that is absent in other natural systems. The relative simplicity of chemical, geological, and mechanical systems allowed for thousands of years of sustained advancements and innovations in metallurgy, horology, and engineering. These advancements in other sciences were occurring even as basic understanding of disease processes was stagnant. Simple experiments with small sample size and only rudimentary controls for bias and confounding and no statistical tests to differentiate between random chance and treatment effects were used to gain great insight into the sciences of chemistry, astronomy, and physics. In contrast, until the development of specific strategies in the first half of the twentieth century to design experiments with the purpose of controlling bias and to rigorously analyze hypotheses with statistical tests, advancements in disease prevention and treatment were slow or non-existent.[1,2,3] The reasons that for hundreds of years medical sciences lagged behind mechanical and other sciences are the same reasons that veterinarians today must embrace the scientific method and the results of rigorous, well-designed studies that control for variation, minimize bias, and effectively explore interactions.

Random Variation, Biased and Confounded Observations, and Complex Interactions

Veterinarians continually witness the phenomena of random variation of animals and biologic systems. We recognize that animals treated identically exhibit a range of average daily weight gain, pain tolerance, response to anesthesia, immunologic response to vaccination, and countless other examples. Differentiating the expected variation in response between animals treated identically from the effects of an intervention requires careful collection of data and probability calculations (e.g., statistical tests) because most diagnostic, therapeutic, and prognostic questions concern subtle differences that are impossible to identify without careful data collection and statistical tests.

In addition to accounting for the expected range of outcomes that animals exposed to the same factors will exhibit, well-designed studies also differ from clinical experience because explicit restraints are implemented to control for sources of bias. Clinical experience is particularly prone to bias because the same person provides and then evaluates interventions. Biases that occur commonly, and often inadvertently, when relying on clinical experience are typically grouped into categories of selection bias, information bias, and confounding.

Observations in clinical settings are plagued by selection bias because animals with certain signalment, history, or physical examination findings are treated differently than animals without those characteristics. While clinically reasonable, this practice prevents accurate comparisons of alternative risks or treatments. Similarly, information bias is very common because we intentionally gather different types and amounts of information about different animals. For example, it is reasonable to observe some animals more closely, under different circumstances, or for longer periods of time than other animals. However, this bias can lead to incorrect associations with either disease-causation or treatment factors.

Confounding can occur when two potential explanatory factors are correlated but not evenly distributed and it is not clear which factor is causing the outcome of interest. Because many risk factors for disease tend to occur clustered together, confounding is a common problem when using our clinical experience to make disease causation or treatment efficacy associations. Without careful statistical control, it is impossible to determine which of two confounding factors is the most important factor associated with a disease outbreak or therapy success. Controlled, well-designed experiments use a number of tools including random allocation of animals, blinding (aka masking) of observers, and extensive data collection to combat bias and confounding.

Another limitation of clinical experience compared to well-designed experiments is the difficulty for the human mind to detect and quantify multiple relationships between two or more factors that affect clinically important outcomes. Unfortunately, because of the complexity of disease avoidance and the multi-faceted nature of most diseases, health outcomes are rarely influenced by a single factor. Factors that may interact with each other to affect disease onset or recovery include age; sex; breed; concurrent exposure to other infectious, toxic, or metabolic insults; nutritional status; and stress; to list a few. The interpretative challenge is that the effect of each factor may not be constant and the interrelationships between factors may alter disease risk. Interactions usually involve many variables, making it essentially impossible to accurately determine the entire causal web involved with disease onset or recovery.

Evaluating the Body of Evidence

Although all or most veterinary clinical decisions are based on evidence of some type, some evidence is very strong (rigorously tested in the target-species under natural conditions in experiments designed to prove a theory to be false) and some is very weak (not tested) and some is intermediate.[4-6] The body of evidence relevant to answering clinical questions is the sum of multiple studies investigating the same area of interest and each study can be ranked on a scale from weak evidence to very strong evidence. The first consideration is the internal validity of the research - which is determined by the study method and appropriate use of controls for bias. Research reports with good internal validity provide assurance that the results represent an unbiased estimate of the true direction and magnitude of the treatment effect. The second consideration is the population used in the research and its appropriateness as a model for the population that generated the clinical question. Generally, the target species in similar housing and husbandry environments provides stronger evidence than the target species in significantly different housing and husbandry environments, related species, unrelated species, or *in vitro* methods. And thirdly, the clinical relevance of the outcomes of the research should be considered with patient-/herd-oriented outcomes providing more direct evidence than disease-oriented outcome measurements.

These three considerations can be used to evaluate the strength and quality of evidence in a published study, by asking the following questions: First, are the results of this individual study valid? This can be addressed by determining if the assignment of patients to treatments was randomized and if the caregivers and clinicians were blind to those assignments. Studies in which treatment is allocated by any method other than randomization tend to show larger and frequently false-positive treatment effects than do randomized trials.[7-9] It is also important that all animals in a group or treatment were evaluated (even drop-outs) and that all groups were treated equally except for the experimental therapy. In addition, it is important that the trial last long enough to recognize both potential positive and negative outcomes.

If it seems that the study results are valid, the second question one needs to ask is, are the valid results of this individual study important? Or asked another way, how large was the treatment effect? This question may be addressed by a report of the magnitude of treatment effect, such as relative risk reduction, absolute risk reduction, or number needed to treat. To determine if a particular magnitude of effect is important, the effect produced by the therapy in question should be compared to the magnitude of effect for other interventions and durations of therapy. The true magnitude of effect of a therapy can never be known, but a valid trial provides a point estimate around which the true effect is expected to lie. The 95% confidence interval of the point estimate of the treatment effect is commonly reported to communicate the precision of the estimate of the treatment effect, and indicates the range that includes the true treatment effect 95% of the time. The larger the sample size the greater is one's confidence that the true treatment effect is near the point estimate (narrow 95% confidence interval). To determine the importance of a reported positive treatment effect, one should look at the lower boundary of the confidence interval. If this lower boundary is important to you as a clinician, then you are confident that the treatment in question only has a 2.5% chance of failing to reach this lower boundary and the treatment can be deemed important. In contrast, if the treatment effect indicated by the lower boundary is not clinically important to you, the study cannot definitively indicate an important treatment effect even if the results are statistically significant. Similarly, to determine the risk of a reported neutral or negative treatment effect, one should look at the upper boundary of the confidence interval. If the upper boundary is clinically important, you cannot exclude the possibility that an important treatment effect is possible even though the trial indicated either no difference between treatment and controls or a negative treatment effect.

And finally, if a study appears to be valid and important, are the valid, important results of this study applicable to my client's patient or herd?[10] If the study animals or treatment setting is sufficiently different from your patient/herd that initiated the question, or if the treatment is either not feasible or not consistent with the client's/manager's values and expectations, a valid and important therapy would not be applicable in your current situation. For a reported trial to influence treatment strategies, one should ask whether all clinically important outcomes were considered. And most importantly, were the outcomes of most importance to you, the clinician, and the owner/manager reported? Trials that report substitute endpoints such as blood cell counts, blood chemistry, or body temperature are not as valuable for evaluating treatment options as indications of reduced mortality risk or improved health or production (weight gain, conception, etc.). Even when a report indicates one or more positive outcomes associated with a treatment, one should consider whether other negative outcomes are also associated with that treatment. This is particularly important if the negative outcomes are more important to the owner/manager than are the positive outcomes.

The Role of Scientific Literature in Critical Thinking

Critical thinking is a process, and as such is never completed. Critically thinking about a clinical problem involves repeating the steps of gathering information, evaluating that information, reflecting on the information and coming to tentative conclusions. Because no single study can fully address most clinical questions and because every study has limitations either in internal and/or external validity, using scientific studies to enhance clinical decision making requires combining different pieces of evidence of varying strengths. While gaining competence in each step of the critical thinking process requires education, skill, and experience, probably the most difficult to master is to reflect deeply about what is known and what is unknown and how to tie multiple pieces of information and evidence together.

Essentially, good clinical decisions are based on good critical thinking skills. Because the quality of thinking is limited by the amount and accuracy of pertinent information available, gaining competence to access, evaluate, and integrate the scientific literature is critical to develop good critical thinking skills. However, good information literacy skills does not guarantee good critical thinking skills in that a person with poor critical thinking skills may accrue very limited benefit from gaining access to valid scientific studies.

Limitations in the Use of Scientific Literature in Clinical Decision Making

One of the greatest challenges to increased use of scientific literature in clinical decision making is that by knowing the strength of evidence for potential interventions, veterinary practitioners will often be less certain of his/her diagnostic, therapeutic, and prognostic pronouncements compared to client expectations for greater certainty. This puts the expectations and values of the client at odds with the expectations and values of the practitioner. It appears that many people gravitate toward the clarity of an untested (or under-tested), but strongly stated pronouncement in preference to the skeptical, methodical, testing-rejecting-retesting pace of the scientific method.

SUMMARY

Practicing high-quality medicine by incorporating scientific literature into veterinary practices leads one to establish a problem-solving process that recognizes the important limitations that bias and complex interactions places on our ability to gain knowledge about animal health through clinical observations, and the value that well-designed studies provide to reduce bias and explore some aspects of complex interactions. However, understanding how to use the scientific literature in clinical practice also requires that one recognizes that even well-designed studies have incomplete control of bias and have limited ability to explore complex interactions. And, finally, integrating scientific literature into high-quality veterinary practice causes one to realize that gaining accurate knowledge is a very slow, iterative process with no short-cuts for the complex problems commonly encountered by veterinary practitioners.

REFERENCES

1. Fisher RA. *Statistical Methods for Research Workers*. Edinburgh, UK: Oliver and Boyd; 1925.
2. Fisher RA. *The Design of Experiments*. Edinburgh and London, UK: Oliver and Boyd; 1935.
3. Mann HB. *Analysis and Design of Experiments*. New York, NY: Dover Publications; 1949.
4. Sackett DL. *Can J Cardiol*. 1993;9:487.
5. Dans AL, *et al*. *J Am Med Assoc*. 1998;279:545.
6. Berg AO. In: Geyman JP, Deyo RA, Ramsey SD, eds. *Evidence-Based Clinical Practice: Concepts and Approaches*. Boston, MA: Buterworth-Heinemann; 2000.
7. Sacks HS, *et al*. *Am J Med*. 1982;72:233.
8. Chalmers TC, *et al*. *N Eng J Med*. 1983;309:1358.
9. Colditz GA, *et al*. *Stat Med*. 1989;8:441.
10. Sackett DL, *et al*. In: Sackett DL, Straus SE, Richardson WS, Rosenberg WA, Haynes RB, eds. *Evidence-Based Medicine: How to Practice and Teach EBM*. 2nd ed. Philadelphia, PA: Churchill-Livingstone; 2000.

acvim350

Reporting Quality and Risk of Bias in Veterinary Clinical Trials

Brennen McKenzie, MA, MSc, VMD
Los Altos, CA, USA

INTRODUCTION

Evidence-based medicine (EBM) is a set of principles and techniques intended to improve the quality of patient care through the explicit integration of scientific research evidence into clinical decision making. A fundamental assumption underlying this approach is that properly designed and conducted scientific research can identify the true effects of medical interventions more accurately than clinical experience and trial-and-error.

However, the utility of research evidence in informing more effective patient care is proportional to the reliability of that evidence. Poor quality evidence will not support practices that improve patient outcomes. Even though controlled research provides more reliable information than uncontrolled observation, it is still subject to residual bias and error. The process of integrating research evidence into clinical decision making, therefore, must incorporate structured assessment of the limitations of relevant research evidence.

Critical appraisal is the process of evaluating published research in order to establish the reliability of the results and the extent to which these should influence clinical practice. This process uses the information about research studies available in published study reports to evaluate the reliability of the research. Therefore, assessing the quality of the information reported is an important element of critical appraisal.

The type of study most directly relevant to clinical decision making is the clinical trial. Much of the emphasis in EBM is on encouraging clinicians to incorporate the results of clinical trials into their decision-making process. Therefore, it is vital that we understand the strengths and limitations of the clinical trial literature and continuously seek to improve the quality and reliability of clinical trial evidence. While there are established mechanisms for this and an extensive body of literature assessing the clinical trial literature in human medicine, there have been few studies evaluating the reliability of veterinary clinical trial evidence. The purpose of this recent review was to add to the limited evidence available assessing the reporting quality and risk of bias of published veterinary clinical trials.

STUDIES EVALUATING CLINICAL TRIAL REPORTS

Table 1 lists the extant studies evaluating reporting quality and risk of bias in veterinary clinical trials, including a recent unpublished review.

Table 1. Characteristics of previous reviews of reporting and bias in veterinary clinical trials with information for the current study for comparison

First author & year	Study focus	# articles	# journals	Date range
Elbers, 19951	Clinical field trials	147	1	1988–1992
Lund, 19982	Clinical trials	49	11	1989–1990
Brown, 20063	Parallel group controlled clinical trial	97	12	2000–2005
Brown, 20074	Parallel group controlled clinical trial	63	12	2000–2005
Sargeant, 20095	Clinical trial	100	50	2006–2008
Sargeant, 20106	Clinical trial	100	32	2006–

				2008
Arlt, 20107	Meta-analyses, clinical trials, case reports, observational study	287	65 (1 conference proceedings)	1996–2006
Simoneit, 2011[8]	Meta-analyses, clinical trials, case reports, observational study	268	Not described	1998–2008
Giuffrida, 2012[9]	Parallel group randomized clinical trials	195	11	2004–2010
Giuffrida, 2014[10]	Parallel group randomized clinical trials	240	11	2005–2012
McKenzie, unpublished[11]	Clinical trial	300	26	2004–2013

Each study looks at reports in different journals, different time periods, and involving a wide variety of research subjects and species. They also utilize a variety of different methods and criteria to assess reporting and risk of bias. While in human medicine there are several validated and widely used instruments, such as the CONSORT reporting checklist[12] and the Cochrane Collaboration risk of bias tool,[12] there are no such generally accepted tools for appraising published veterinary trials. Recently, however, attempts have been made to adapt some of these instruments for veterinary use.[11,15]

Because of these significant methodological differences, the data from these reviews cannot be directly compared. However, they provide a general overview of the veterinary clinical literature in which some patterns can be discerned that are likely significant and which suggest areas in which there is room for improving the reliability of the literature.

The general themes that emerge from these reviews are:

1. Reporting quality in the veterinary clinical literature is poor. Published reports often fail to include even the most basic and crucial information necessary for readers to assess the reliability of the studies reported, and very few reports achieve high-quality, comprehensive reporting.
2. Because of the limitations in reporting quality, it is very difficult to assess the reliability of the evidence base. In the human literature, there are well-established associations between study design criteria and reporting characteristics and the risk of bias in clinical trial reports. Some similar associations appear to be present in the veterinary literature, but there are few studies evaluating this and often the poor quality of reporting precludes effective assessment of the risk of bias.
3. Statistical methods employed in veterinary trials often lower the reliability of conclusions. Small samples with low power, failure to specify outcomes in advance and to properly account for multiple *post hoc* comparisons, and other deficiencies in analytic methods are common.

In terms of reporting quality, Table 2 lists the percentage of studies evaluated in various reviews that properly reported key methodological features. Even core design features, such as randomization method and extent of blinding, were not reported by a majority of studies according to most of these reviews.

Table 2. Percentage of studies fully describing key reporting criteria (nr=not reported)

	Elbers 1995[1]	Lund 1998[2]	Brown 2006[3]	Sargeant 2009[5]	Sargeant 2010[6]	Arlt 2010[7]	Simoneit 2011[8]	Giffrida 2012[9]	Giuffrida 2014[10]	McKenzie unpublished[11]
Randomization sequence	47.6	22.0	35.0	20.3	15.7	39.6	12.0–34.0	nr	nr	19.5
Allocation concealment	nr	nr	11.3	3.0	2.9	nr	nr	nr	nr	15.4

Blinding of investigators	6.0	61.0	nr	26.0	76.0	13.8	4.0–7.0	58.0	nr	40.3
Eligibility, inclusion & exclusion criteria	nr	35.0	nr	39.0	43.5	nr	nr	nr	nr	64.0
Baseline comparison of groups	nr	nr	84.0	14.0	34.1	nr	nr	nr	nr	62.0
Discussion of results, including assessment of study limitations	nr	nr	nr	33.0	57.6	44.2	nr	nr	nr	45.3
Primary outcome specified	nr	83.0	nr	13.0	8.2	nr	nr	nr	32.4	11.7
Sample size justified	nr	0.0	nr	5.1	1.2	nr	nr	nr	nr	13.7
Power calculation	0.0	0.0	nr	nr	nr	nr	nr	nr	22.0	12.0

Incomplete reporting is certainly associated with methodological quality and risk of bias in human clinical trials. It is very difficult to evaluate this for veterinary trials. There are no established and widely used bias assessment tools for veterinary studies. There is also no clinical trial registration for the vast majority of veterinary trials, so study protocols are not published in advance and the extent of adherence to protocol, including the *post hoc* alteration of outcomes assessed, cannot be effectively evaluated. However, some reviews have found associations between the reporting of key methodological features and the likelihood of statistically significant outcomes.[5,6,11] In other words, there is some evidence that studies with incomplete reporting are more likely to have positive or statistically significant results than studies with more complete reporting. This might suggest that incomplete reporting represents deficiencies in study design or conduct that increase risk of bias and false positive reports. Such associations have been found repeatedly in the human clinical trial literature. However, the existing evidence on this question involving veterinary trials is scant and inconsistent, so no definitive conclusion can be made.

CONCLUSIONS

There are a number of possible reasons for the incomplete reporting in published veterinary clinical trials. The principles and practices of evidence-based medicine are not as widely utilized in veterinary medicine as in human medicine, and the importance of implementing and reporting bias control methods may not be as widely appreciated among veterinary researchers. Veterinary medicine is also subject to less regulatory oversight than human medicine, and there is no clinical trial registry in which protocols and data are available for outside review. This may reduce the perceived need for full reporting of trial procedures. Finally, reporting guidelines are not widely available, and journals do not require adherence to such guidelines as a condition for publication. The promulgation and use of such reporting guidelines could improve the quality and utility of the veterinary clinical trial literature.

The failure of most authors to specify primary outcomes for statistical analysis or to report methods of correcting for multiple statistical comparisons suggests that there is a significant risk of false positive findings in published veterinary clinical trials. This is a risk that could be reduced by adoption of practices widely used in human clinical research, such as registering clinical trials and publishing protocols, including primary and secondary outcome measures, in advance and allowing outside access to all data once trials are complete.

Future research is needed to more clearly identify the reasons for poor reporting of veterinary clinical trials and potential approaches to improve this, to better assess the risk of bias in the veterinary clinical trial literature, to further clarify the role of funding bias and publication bias, and to evaluate the efficacy of measures to improve the reliability of the literature, such as adoption of mandatory reporting guidelines and pre-registration of clinical trials. Clinical trial reports, and synthesis of these in systematic reviews and meta-analyses, are the gold standard evidence upon which clinical practice should rely. It is critical to ensure that the clinical trial literature in veterinary medicine is reliable and of high quality so that evidence-based practice can be more widely employed.

REFERENCES

1. Elbers A, Schukken Y. Critical features of veterinary field trials. *Vet Rec.* 1995;136(8):187–192.
2. Lund EM, James KM, Neaton JD. Veterinary randomized clinical trial reporting: a review of the small animal literature. *J Vet Intern Med.* 1998;12(2):57–60.
3. Brown DC. Control of selection bias in parallel-group controlled clinical trials in dogs and cats: 97 trials (2000–2005). *J Am Vet Med Assoc.* 2006;229(6):990–993.
4. Brown DC. Sources and handling of losses to follow-up in parallel-group randomized clinical trials in dogs and cats: 63 trials (2000–2005). *Am J Vet Res.* 2007;68(7):694–698.
5. Sargeant JM, Thompson A, Valcour J, Elgie R, Saint-Onge J, Marcynuk P, *et al.* Quality of reporting of clinical trials of dogs and cats and associations with treatment effects. *J Vet Intern Med.* 2010;24(1):44–50.
6. Sargeant JM, Elgie R, Valcour J, Saint-Onge J, Thompson A, Marcynuk P, *et al.* Methodological quality and completeness of reporting in clinical trials conducted in livestock species. *Prev Vet Med.* 2009;91(2–4):107–115.
7. Arlt S, Dicty V, Heuwieser W. Evidence-based medicine in canine reproduction: quality of current available literature. *Reprod Domest Anim.* 2010;45(6):1052–1058.
8. Simoneit C, Heuwieser W, Arlt S. Evidence-based medicine in bovine, equine and canine reproduction: quality of current literature. *Theriogenology.* 2011;76(6):1042–1050.
9. Giuffrida MA, Agnello KA, Brown DC. Blinding terminology used in reports of randomized controlled trials involving dogs and cats. *J Am Vet Med Assoc.* 2012;241(9):1221–1226.
10. Giuffrida MA. Type II error and statistical power in reports of small animal clinical trials. *J Am Vet Med Assoc.* 2014;244(9):1075–1080.
11. McKenzie BA. *Reporting, Risk of Bias, and Statistically Assessed Outcomes in Published Veterinary Clinical Trials: A Systematic Review* [Unpublished master's thesis]. London, England: London School of Hygiene and Tropical Medicine; 2015.
12. Higgins JPT, Green S, eds. Handbook for systematic reviews of interventions version 5.1.0. The Cochrane Collaboration. http://handbook.cochrane.org/. Updated March 2011.
13. Moher D, Hopewell S, Schulz KF, Montori V, Gøtzsche PC, Devereaux PJ, *et al.* CONSORT 2010 explanation and elaboration: Updated guidelines for reporting parallel group randomised trials. *Int J Surg.* 2012;10(1):28–55.
14. Sargeant JM, O'Connor AM, Gardner IA, *et al.* The reflect statement: Reporting guidelines for randomized controlled trials in livestock and food safety: explanation and elaboration. *Zoonoses Public Health.* 2010;57(2):105–136.

acvim351

From Evidence-based Practice to Practice-based Evidence: The Electronic Health Record as a Platform for Clinical Decision Support

Stuart W. Turner, DVM, MS (Informatics)
Davis, CA, USA

INTRODUCTION

Evidence-based medicine (EBM) or evidence-based practice (EBP) (also **evidence-informed medicine**) has evolved from its inception around 1990 (Sur RL, *et al.*) as a methodology to train interns and residents in the skills required to meet their clinical information needs within a more rigorous and reliable framework. This was typically done while caring for patients during rounds. Despite its science-based foundation, medical training is foundationally vocational. The degree of scientific rigor as "practiced" by a clinician can vary depending on the extent of their post-graduate education or even their own personal belief systems. EBM was designed to engender behaviors such as good lifelong learning patterns after being untethered from academia or to realize some of Donabedian's principles from his classic work *Evaluating the Quality of Care* (Donabedian *et al.*) such as the gap between "presumptive evidence" and "substantial evidence" or to help a clinician understand and subsequently recognize inherent cognitive errors or biases and how best to mitigate them at the point-of-care.

Historically, the canonical steps to EBM include (1) develop a well-formed and focused clinical question using comparators (e.g., comparing a new drug or procedure to placebo); (2) use attributes from that question to identify appropriate information resources (e.g., PubMed, CABI or the Cochrane Library) and query them effectively (good precision and recall); (3) use clinical epidemiology skills (biostatistics) to appraise the biomedical literature and evaluate the fundamental quality of the various research methodologies and their publication types - from curbside consultations (e.g., professional forums) to case reports to well-executed randomized controlled trials to systematic reviews; (4) apply the evidence from population-based research to your specific patient or patient population within its proper context (e.g., questions about etiology, diagnosis, prognosis or therapy or the use of quality-adjusted life years or client values or cost concerns); and (5) evaluate the outcome of this approach and communicate this new knowledge effectively with colleagues and clients.

While most of the principles and tools of EBM have been developed within human medicine, much of it is crosscutting with veterinary medicine and these techniques have mostly been wholly adopted. The interest in developing evidence-based **veterinary** medicine (EBVM) resources is growing.

There are several organizations devoted to EBVM, such as the Evidence-based Veterinary Medicine Association (EBVMA, International, 2004), the Centre for Evidence-Based Medicine at Nottingham University (CEVM, UK, 2009) and the Royal College of Veterinary Surgeons *RCVS Knowledge Network* and *Veterinary Evidence* journal (UK, 2012). There are recurring EBVM conferences and workshops, textbooks, published EBVM toolkits and special online resources such as *VetSRev,* a database of veterinary systematic reviews published by the CEVM. Librarians are providing specialized training and support to veterinary researchers who wish to properly conduct systematic reviews using established reporting guidelines (e.g. PRISMA). Journals have published special editions dedicated to systematic reviews (O'Connor *et al.*). ROBIS, developed at the University of Bristol, is a new tool for assessing the risk of bias in systematic reviews. Existing reporting standards have been specialized for veterinary medicine, such as the REFLECT Statement for interventions in livestock and food safety and the ARRIVE guidelines for animal studies focused on human disease.

Clinical Knowledge Summaries (CKS) are brief structured summaries of the best available evidence in response to specific clinical questions with relatively broad appeal. They are designed as both an EBM teaching tool and a method to share this knowledge for others to learn from or to modify. They are being developed, some with peer review, by several institutions in various flavors, such as Critically Appraised Topics (CATs) or BestBETs for Vets (Best Evidence Topics) or simply as **evidence summaries**. This author has developed a CKS schema, an extension of an existing open-structured data model and vocabulary developed by a collaboration between major search engine providers (e.g., Google, Microsoft) that provides semantic **microdata** bound to these resources in order to improve their indexing and discovery online.

Despite this interest, when making comparisons between human and veterinary medicine, a persistent negative critique has been the relative paucity in the quantity and quality of biomedical literature between the two. Case reports continue to dominate over publications of higher quality such as prospective studies (e.g., RCTs). The gap is especially true for systematic reviews and clinical guidelines. A simple PubMed query for all publications indexed as a clinical guideline (MeSH: "Practice Guideline Publication Type OR Guideline Publication Type") constrained to veterinary medicine yields approximately 549 results (April 2017). There is some redundancy (e.g., recurring guideline publications regarding vaccines or pain management) and only a subset covers small companion animals. The US AHRQ Guidelines Clearinghouse (guidelines.gov) lists approximately 1,700 published between 2000–2017. The same query in PubMed yields approximately 29,000 results. VetSRev has cataloged approximately 500 systematic reviews in veterinary medicine, much of them developed over the last 5 years. In humans, the Cochrane library is approaching 10,000 published systematic reviews or protocols or "seventy-five trials and eleven systematic reviews a day" in 2010 (Bastian *et al.*).

While it's important to note this disparity and to continue to move forward to develop EBM training programs and promote the publication of high-quality reviews and prospective research, including the outcome of **all** clinical trials, there is the ethical and pragmatic obligation to use the "evidence you've got." The challenge here is not the paucity of good literature, but the skills, time and cognitive burden needed to discover what's available and to execute the synthesis, implementation and the evaluation of this effort (the 5–7 EBM steps listed previously) during patient care.

Major trends in the evolution of EBM today are the integration of knowledge resources such as clinical prediction rules more directly into the workflow by using patient data directly to support knowledge-based **clinical decision support systems** (CDSS), the use of **patient (client) reported outcomes**, the training of medical students about the nature and use of federated queries of clinical data inside institutional repositories as part of emerging **Learning Health System** programs (Altman M) and the practice of **precision medicine** (Frey L) using genomic and other data. Therefore, the "evidence you've got" extends to the clinical data you've already collected, or have the potential to collect, about your patients.

Clinical Decision Support Systems

Clinical Decision Support (CDS) Systems depend on compiled clinical knowledge. Their sophistication runs along a continuum, from simple published guidelines to patient data-derived triggers of rules or algorithms to remind the clinician of actions they intend to do, or where they are unsure of what to do (e.g., help provide an accurate diagnosis) or to correct errors or invoke a change of plans (e.g., drug interactions or adverse events based on pharmacogenomic data). CDS capabilities can be segmented into four axes: **triggers** (e.g., updating a problem list or ordering a drug), **input data** (e.g., an abnormal laboratory observation), **interventions** (suggested actions), and more directly, **offered choices** (e.g., an alert to change an order due to a drug interaction) (Wright *et al.*).

When compiled best evidence is matched to patient data, there is the opportunity to fulfill what (Osheroff *et al.*) described as the five rights of clinical decision support: "CDS should be designed to provide the right information to the right person in the right format through the right channel at the right time." Research has shown that CDSS are more effective than manual processes (e.g., guideline review or calculations), when presented within the clinical workflow and when clinical decisions are made. It's also more effective for those that make explicit recommendations to direct the clinician (veterinarian or nurse) vs. passive evaluation summaries.

Electronic Health Records

The foundation for this type of integration is the Electronic Health Record (EHR). In larger environments this is part of a clinical information system (CIS) integrated with an EHR, as well as other components (e.g., a laboratory information system). For example, the UC Davis Medical Center has approximately 60 standards-based interfaces sharing data between various systems.

The design of EHRs today allows for a broad array of detailed patient data to be collected. There is a considerable body of research; masters, doctoral and post-doctoral programs in informatics; a new board-certified subspecialty (clinical informatics under the American Board of Preventive Medicine), extensive international normative clinical architecture and terminology standards and implementation experiences published in dedicated peer-review journals. There have also been federal mandates for the design of EHRs (under the US Meaningful Use Program), a federal EHR certification body and financial incentives for physicians to implement EHRs. Certification standards in some cases specifically require EBM components such as decision support systems, computerized provider order entry systems that use knowledge bases to identify potential drug interactions, adverse events and other errors or context-aware knowledge retrieval or "Infobuttons" (Cook DA *et al.*); all of which depend upon the proper utilization of the current best evidence. There is considerable emphasis on **working semantic interoperability** between disparate systems. This depends on the use of international reference standards comprised of software-independent information and behavioral (workflow) models, specifications, profiles and archetypes and the extensive use of biomedical terminologies.

In some larger institutions or regional health information networks, the EHR has become an important clinical repository to conduct evidence-based cohort research. They are used directly for patient care when published evidence was either not available, not applicable or of low quality (e.g., levels III through V on the evidence hierarchy) - an environment especially true in veterinary medicine. They support clinical trial research and translational medicine programs or are used to perform more sophisticated activities such as data mining or machine learning. Leading examples include SHRINE (the Shared Health Research Information Network) clinical repository developed at the Center for Biomedical Informatics at Harvard (Weber GM *et al.*) and STRIDE (Translational Research Integrated Database Environment) at Stanford University (Frankovich J *et al.*).

Practice-Based Evidence Capabilities in Veterinary Medicine

The essence of this paper is to highlight a few of the significant gaps in incentives, sophistication, and resources between human and veterinary medicine regarding the potential to use practice-based evidence and to begin to describe pragmatic next steps.

Computerized patient systems in veterinary medicine have historically been weighted considerably on "practice management," not "patient management." They are designed primarily to capture procedural information for invoicing, to manage client reminders (e.g., vaccinations) and with some integration of radiology and laboratory information systems. Connections between systems are mostly proprietary "integrations." The patient record components of these systems are mostly free-text with little coded entry and not based on published standards for clinical or document profiles or based on reference models used for interoperability between systems.

Perhaps the most commonly used standard in veterinary medicine is DICOM (Digital Communications in Medicine) as used by radiological systems, including PACs, because of this natural standards inheritance from human systems retrofitted for veterinary medicine. We know of no major commercial laboratory vendor regularly using LOINC (Logical Identifier Names and Codes), the putative international standard for laboratory and clinical observations (McDonald CJ *et al.*). In veterinary medicine, LOINC is used primarily in the public health space (Laboratory Response Network). SNOMED CT (Systemized Nomenclature of Medicine Clinical Terms) is an international standard and the most comprehensive healthcare terminology in existence with over 300,000 concepts and over 1 million relationships. Its use of concept inheritance and description logics allows for inheritance relationships to link concepts together as well as post-coordination of atomic concepts into more complex expressions. SNOMED CT has been relatively well supported in veterinary medicine with a veterinary extension developed at the Veterinary Medical Informatics Program at the Virginia-Maryland College of Veterinary Medicine. That group has developed a subset, sponsored by the American Animal Hospital Organization (AAHA). The Veterinary Medical Database (VMDB) is a database limited to SNOMED CT encoded discharge diagnoses collected from some veterinary medical teaching hospitals in the United States. It's principally a limited index to snapshots of confirmed or presumptive diagnoses for a subset of patients at local institutions. It's not a clinical repository, nor could it serve as one, since it has a limited data element set, has not published or is based on any reference model or has an API to integrate this knowledge with other systems. In the UK, a proprietary veterinary terminology, VeNOM is available and a recent effort is underway there to capture structured patient data from health record systems for clinical trial research (Doit H *et al.*). There is a large number of task or domain-oriented terminologies or ontologies, too many to mention here, such as RxNorm for drugs or RadLex for imaging or UCUM for units of measure, with no known implementations of significance in veterinary medicine.

Working semantic interoperability requires shared information models. Structural vocabularies are bound to them to unambiguously describe those models to both humans and computers. Patient data are further annotated with the types of vocabularies mentioned above. While there has been significant work on several terminologies in veterinary medicine, the development of information models is relatively nonexistent.

If we are to leverage patient data to perform any notion of practice-based evidence, the community needs to develop those shared models and build terminology "value sets" that carry vocabulary agnostic metadata to describe and mix relevant concepts from one or more terminologies for example. We must leverage existing standards for things like healthcare data types that enforce engineering principles and archival quality and safety by clearly describing what resources are being used and how semantic drift may impact patient data when encoded to terms of questionable semantic equivalence. The EHR is ostensibly the ideal EBM platform.

The following are a just a few approaches to work toward these goals. More detailed examples and analysis (links and tables) are available in the companion online document listed below.

Problem Lists

In medical cognition research, the **active problem** and **health concern** fit within an epistemology (as a facet in computer science parlance) that are most effective in segmenting the problem space for clinical decision support, queries, patient interventions and so on (see https://pico.vet/help.html). They are aggregations of findings and observations that may not be directly actionable. Diagnoses are a higher-level abstraction that defines a patient's condition overall and when known (confirmed), but not all diabetes patients are alike. A patient's unique list of active problems and comorbidities are more finely granular and appropriate (in addition to a diagnosis list) to trigger CDS interventions or build focused clinical questions. However, problem lists must change often to be current and representative and are difficult if not impossible to maintain manually. They can be invoked using background agents, such as laboratory results that activate a problem of "hyperkalemia" or "anemia" automatically. The problem list is an ideal source for building focused questions and integration with existing EBM-based standards, such as the Context-Aware Knowledge Retrieval ("Infobuttons") that link out to patient-centered information resources such as a clinical knowledge summary or a systematic review.

Continuity of Care Record

The Continuity of Care Document (CCD) is an international standard (Clinical Document Architecture under Health Level Seven) that captures a summary of patient health. It embodies the essence of relevant and current patient information, especially for the transfer of care between providers. In the United States, it's also a required component within the EHR Meaningful Use program and the past EHR certification program under the Certification Commission for Healthcare Information Technology (CCHIT) and now the Healthcare Information and Management Systems Society (HIMSS). It carries patient demographics (signalment), active problems, medications, allergies and intolerances, vital signs, clinical and anatomic pathology results, procedures, providers, and care plans and instructions. It's the putative knowledge source, built using semantic standards, to serve as an EBM platform. Because it's primarily designed for patient care, its adoption is more likely. EBM implementations (and other uses, such as biosurveillance or public health reporting) therefore become part of the continuous use case environment.

Nurses, Librarians, Scribes and Chief Medical Information Officers

It may seem odd that a relatively technical discussion mentions the important human actors needed to implement practice-based evidence. Some CDS systems are better targeted at those directly providing patient care (nurses). While it's uncertain if veterinary medicine has the financial resources to support the growing cottage industry of scribes, who shadow physicians and perform much of the necessary documentation required today, it emphasizes that these colleagues serve an important role. The curation of knowledge resources requires the expertise of librarians. The Infobutton standard has an authoring environment where health science librarians use their expertise to curate knowledge resources for their institution. It's reasonable to expect larger private group practices have a librarian on staff to support this as well. Lastly, the clinical and research enterprise requires informatics leadership and expertise. There is an emerging role in human medicine of the informatics/data-science trained physician or nurse who function as a Chief Medical Information Officer (CMIO). These are not traditional IT roles. These are clinicians with graduate training (certificate, masters, doctoral, residency or fellowship) in informatics, data or computer science, library science or education that help guide larger organizations to reach these goals.

References

NB: This paper discusses several evolving normative technology reference standards and research developments with online resources that cannot be effectively described, updated or hyperlinked within these proceedings. Therefore, an openly licensed companion online resource will be available for a period of time by the author at https://stuartturner.org/acvim (VIN editor: Link not accessible as of 5/1/17).

1. Sur RL, Dahm P. History of evidence-based medicine. *Indian J Urol.* 2011;27(4):487–489.doi: 10.4103/0970-1591.91438
2. Donabedian A. Evaluating the quality of medical care. *The Milbank Quarterly.* 2005;83(4):691–729. doi:10.1111/j.1468-0009.2005.00397.x.
3. VetSRev Database. https://webapps.nottingham.ac.uk/refbase/. Accessed April 26, 2017.
4. The PRISMA Statement. www.prisma-statement.org/. Accessed April 26, 2017.
5. O'Connor AM, Sargeant J, *et al.* Special issue: systematic reviews and meta-analysis in animal agriculture and veterinary medicine. *Zoonoses and Public Health.* June 2014:1–63.
6. ROBIS: www.bristol.ac.uk/social-community-medicine/projects/robis/robis-tool/. Accessed April 26, 2017.
7. Bastian H, Glasziou P, Chalmers I. Seventy-five trials and eleven systematic reviews a day: how will we ever keep up? *PLoS Med.* 2010;7(9):e1000326. https://doi.org/10.1371/journal.pmed.1000326 Kalra AM-CTAJM-CCLP-CD.
8. Osheroff JA. Improving medication use and outcomes with clinical decision support: a step-by-step guide. Chicago, IL: The Healthcare Information and Management Systems Society (HIMSS); 2009.
9. Cook DA, *et al.* Information needs of generalists and specialists using online best-practice algorithms to answer clinical questions. *J Am Med Inform Assoc.* February 2017.
10. Cook DA, Teixeira MT, Heale BS, Cimino JJ, Del Fiol G. Context-sensitive decision support (infobuttons) in electronic health records: a systematic review. *J Am Med Inform Assoc.* 2017;24(2):460. doi:10.1093/jamia/ocw104.
11. Evaluation of clinical information modeling tools. *J Am Med Inform Assoc.* 2016;23(6):1127–1135.
12. Schema.org, health and medical types. https://schema.org/docs/meddocs.html. Accessed April 26, 2017.
13. Frey LJ, Bernstam EV, Denny JC. Precision medicine informatics. *J Am Med Inform Assoc.* 2016;23(4):668. doi:10.1093/jamia/ocw053.
14. Harle CA, Listhaus A, Covarrubias CM, *et al.* Overcoming barriers to implementing patient-reported outcomes in an electronic health record: a case report. *J Am Med Inform Assoc.* 2016;23(1):74. doi:10.1093/jamia/ocv085.
15. Ford E, Carroll JA, Smith HE, Scott D, Cassell JA. Extracting information from the text of electronic medical records to improve case detection: a systematic review. *J Am Med Inform Assoc.* 2016;23(5):1007. doi:10.1093/jamia/ocv180.
16. Canadian Medical Association. The Appraisal of Guidelines for Research and Evaluation (AGREE) Reporting Checklist. The AGREE Trust; 2016.
17. Mandel JC, Kreda DA, Mandl KD, Kohane IS, Ramoni RB. SMART on FHIR: a standards-based, interoperable apps platform for electronic health records. *J Am Med Inform Assoc.* 2016;23(5):899. doi:10.1093/jamia/ocv189.
18. Wu DT, Hanauer DA, Mei Q, *et al.* Assessing the readability of ClinicalTrials.gov. *J Am Med Inform Assoc.* 2016;23(2):269. doi:10.1093/jamia/ocv062.

19. Marcos C, González-Ferrer A, Peleg M, Cavero C. Solving the interoperability challenge of a distributed complex patient guidance system: a data integrator based on HL7's Virtual Medical Record standard. *J Am Med Inform Assoc.* 2015;22(3):587. doi:10.1093/jamia/ocv003.
20. Klann JG, Mendis M, Phillips LC, *et al.* Taking advantage of continuity of care documents to populate a research repository. *J Am Med Inform Assoc.* 2015;22(2):370. doi:10.1136/amiajnl-2014-003040.
21. McCoy AB, Wright A, Sittig DF. Cross-vendor evaluation of key user-defined clinical decision support capabilities: a scenario-based assessment of certified electronic health records with guidelines for future development. *J Am Med Inform Assoc.* 2015;22(5):1081. doi:10.1093/jamia/ocv073.
22. Lee D, de Keizer N, Lau F, Cornet R. Literature review of SNOMED CT use. *J Am Med Inform Assoc.* 2014;21(e1):e11. doi:10.1136/amiajnl-2013-001636.
23. Hruby GW, McKiernan J, Bakken S, Weng C. A centralized research data repository enhances retrospective outcomes research capacity: a case report. *J Am Med Inform Assoc.* 2013;20(3):563. doi:10.1136/amiajnl-2012-001302.
24. Weber GM. Federated queries of clinical data repositories: the sum of the parts does not equal the whole. *J Am Med Inform Assoc.* 2013;20(e1):e155. doi:10.1136/amiajnl-2012-001299.
25. Frankovich J, Longhurst CA, Sutherland SM. Evidence-Based Medicine in the EMR Era. *New England Journal of Medicine.* 2011;365(19):1758–1759. doi:10.1056/NEJMp1108726.
26. Brouwers M KMBGCFFGFBHSMJO. AGREE II: Advancing guideline development, reporting and evaluation in healthcare. *Canadian Medical Association Journal.* 2010.
27. U.S. Department of Health and Human Services. Common Terminology Criteria for Adverse Events (CTCAE) Common Terminology Criteria for Adverse Events (CTCAE) Version 4.0. National Institutes of Health, National Cancer Institute (United States); 2009.
28. Weber GM, Murphy SN, McMurry AJ, *et al.* The Shared Health Research Information Network (SHRINE): a prototype federated query tool for clinical data repositories. *J Am Med Inform Assoc.* 2009;16(5):624. doi:10.1197/jamia.M3191.
29. Weiner JP, Kfuri T, Chan K, Fowles JB. "e-Iatrogenesis": the most critical unintended consequence of CPOE and other HIT. *J Am Med Inform Assoc.* 2007;14(3):387–388. doi:10.1197/jamia.M2338.
30. Altman M. The Clinical Data Repository: A Challenge to Medical Student Education. *J Am Med Inform Assoc.* 2007;14(6):697. doi:10.1197/jamia.M2483.
31. Maviglia SM, Yoon CS, Bates DW, Kuperman G. KnowledgeLink: Impact of context-sensitive information retrieval on clinicians' information needs. *J Am Med Inform Assoc.* 2006;13(1):67. doi:10.1197/jamia.M1861.
32. Liu S, Ma W, Moore R, Ganesan V, Nelson S. RxNorm: prescription for electronic drug information exchange. *IT Professional.* 2005;7(5):17–23. doi:10.1109/MITP.2005.122.
33. Kawamoto K, Houlihan CA, Balas EA, Lobach DF. Improving clinical practice using clinical decision support systems: a systematic review of trials to identify features critical to success. *BMJ.* 2005;330(7494):765. doi:10.1136/bmj.38398.500764.8F.
34. McDonald CJ, Huff SM, Suico JG, *et al.* LOINC, a universal standard for identifying laboratory observations: a 5-year update. *Clinical Chemistry.* 2003;49:624–633.
35. Einbinder JS, Scully K. Using a clinical data repository to estimate the frequency and costs of adverse drug events. *J Am Med Inform Assoc.* 2002;9(Supplement_6):S34. doi:10.1197/jamia.M1222.
36. Stein HD, Nadkarni P, Erdos J, Miller PL. Exploring the degree of concordance of coded and textual data in answering clinical queries from a clinical data repository. *J Am Med Inform Assoc.* 2000;7(1):42. doi:10.1136/jamia.2000.0070042.
37. BestBets for Vets. https://bestbetsforvets.org. Accessed April 26, 2017.
38. Hannah Doit. The Veterinary Clinical Trials Network - a Pragmatic Approach to Filling the Evidence Gaps for Veterinary Practice. RCVS Knowledge, Veterinary Evidence Online. Published 26 April 2017. Presentation (podcast).

acvim352

Best of 2016 ISCAID Symposium, Bristol, UK

Annette L. Litster, BVSc, PhD, FANZCVSc (Feline Medicine), MMedSci (Clinical Epidemiology)
Chicago, IL, USA

This is a selection of presentations from the International Society for Companion Animal Infectious Diseases 4th Biennial Symposium, Bristol, October 16–19, 2016. Presentations from the 13th International Feline Retrovirus Research Symposium and the Clinician Day, held in conjunction with the ISCAID Symposium, were not eligible for this selection.

The Comparative Value of Feline Virology Research: Can Findings From the Feline Lentiviral Vaccine Be Translated to Humans?

Prof. Margaret Hosie, University of Glasgow[1]

Recent studies in our laboratory demonstrated that the interaction between FIV and its primary receptor CD134 changes as disease progresses, reminiscent of the R5-X4 receptor switch observed as disease progresses in HIV infected individuals. Our findings suggested that the switch in receptor usage is a consequence and not the cause of disease progression in lentivirus infection. It has become evident that the virus-receptor interaction that occurs with disease progression has implications for lentivirus vaccine design. The whole inactivated virus, dual-subtype FIV vaccine confers immunity to a range of heterologous strains, but does not protect cats against challenge with the filed isolate GL8. GL8 has the phenotype of an "early" strain of FIV, being CRD2-dependent and requiring the CRD2 domain of CD134 for infection. Such "early" isolates are typical of the viruses that are transmitted in the field, since epidemiological evidence suggests that transmission is most likely to occur from healthy cats with high viral burdens (i.e., before the switch to the "late") CRD2-independent phenotype has occurred. Therefore, the next generation of FIV vaccines should be designed to boost immunity to CRD2-dependent strains. If our hypothesis that vaccinal immunity is related to receptor phenotype is confirmed, perhaps the affinity of the FIV-CD134 interaction might dictate the ability to escape from neutralizing antibody and to turn a protective response into an enhancing one. Perhaps by understanding the limitations of the current FIV vaccine, it will be possible to design the next generation of FIV vaccines that will protect cats from challenge with CRD2-dependent strains. Improving the prospects for vaccination of cats against FIV will encourage new strategies for the development vaccines against other lentiviral infections, such as human immunodeficiency virus.

Feline Leukemia Virus Infection: New Advances In Diagnosis.

Prof. Regina Hofmann-Lehmann, University of Zurich[2]

The outcome of FeLV infection can be characterized using the results from virus isolation, immunofluorescence assays, detection of FeLV p27 capsid antigen in serum or plasma, and detection of FeLV provirus or viral RNA in the blood, saliva or some tissues by molecular assays. Antigenemia (presence of p27 capsid antigen) is a marker of infection and in most, but not all, cats a parameter for viremia (presence of replication-competent virus). Detection of viremia using virus isolation is not widely available, laborious and time consuming. In contrast, antigenemia can be easily detected using ELISA in a specialized laboratory or by rapid point-of-care tests. Most infected cats test antigen positive within 3–6 weeks after FeLV exposure. The detection of FeLV provirus in the peripheral blood or bone marrow using real-time PCR (RT-PCR) is more sensitive than the detection of antigenemia or viremia to demonstrate previous FeLV exposure. Provirus PCR from blood tests positive sooner after FeLV exposure than antigen detection; in experimental studies, cats were provirus positive 1–2 weeks after FeLV infection. Finally, some specialized laboratories offer RT-PCR for the detection of viral RNA, for example in the saliva of infected cats. The detection of viral RNA in the saliva generally correlates well with the detection of antigen in the blood of infected cats. FeLV viral RNA in saliva and blood of experimentally infected cats was found as early as 1 week after FeLV exposure (i.e., at least 2 weeks prior to antigen detection in the blood).

The FeLV status of every cat should be established. Importantly, testing should be performed before FeLV vaccination or entry into a new home/shelter, etc. Any positive or questionable initial antigen test result must be confirmed, especially, if local FeLV prevalence is low or the expected FeLV exposure risk is low. With all testing platforms, low disease prevalence reduces the predictive value of a positive result; this increases the number of false positive results. Confirmed positive test results can be the result of the following infection classifications - regressive infection during transient viremia or progressive infection. Retesting should be performed 1–2 months after an initial positive result to determine if regressive (second test negative) or progressive (second test positive) infection status is present.

Negative initial test results in an exposed cat could be the result of abortive infection, or the specimen having been collected during the incubation period. Testing during the incubation period is an important cause of false negative results that is often forgotten in clinical practice. For this reason, an initial negative result should always be followed by retesting if a risk of exposure within 6 weeks prior to testing is possible. If RT-PCR is used, retesting should be performed in a FeLV-negative exposed cat within 1–2 weeks of the first test. For exposed cats that test negative using point-of-care or laboratory-based ELISA testing, a second test should be performed 3–6 weeks after the first test.

ANGIOSTRONGYLUS - WHERE ARE WE NOW?

Dr. Eric Morgan, University of Bristol[3]

Angiostrongylus vasorum causes a variety of respiratory and other disease in dogs, and appears to have expanded its geographic range in recent years. Surveys in foxes and - using new diagnostic modalities - in dogs confirm that this spread is real and not explained by increased awareness alone. Understanding of the underlying causes of this emergence remains limited, but recent studies suggest that the transmission pathways and ecological context of maintenance in reservoir hosts is more complex than previously imagined. New diagnostic tests and drugs to prevent patent infection provide better support than ever to veterinary practitioners managing the disease. Nevertheless, key gaps in knowledge remain, including the pathogenesis of associated bleeding disorders, quantification of climatic and ecological influences on the life cycle, and the elucidation and management of risk factors for infection in dogs. The rat lungworm *A. cantonensis*, a cause of eosinophilic meningitis in tropical and sub-tropical areas, also appears to be emerging and similar uncertainty exists on its basic biology and ecology. Increasing researcher interest and improving tools available to study these knowledge gaps should lead to further advances in understanding and control in future.

TOXOPLASMA GONDII - LATENT INFECTION AND CLINICAL TOXOPLASMOSIS

Prof. Vanessa Barrs, University of Sydney[4]

Until recently the latent form of toxoplasmosis was thought to be benign. However, *T. gondii* induces behavioral changes in infected rodents, such that they become attracted to cat odors that usually incite an aversive response, increasing the chance of predation by the cat and completion of the parasite life-cycle. Similarly, *T. gondii* is implicated in certain neuropsychiatric disorders in humans, including schizophrenia. Key to the ability of *T. gondii* to change behavior of infected hosts is its ability to cross the blood-brain barrier in latent infection and to alter host-cell signaling and neurogenic pathways involving dopamine tryptophan, GABA, and vasopressin. The recognition that latent infection is not always benign further increases the significance of this common parasitic zoonosis to human health.

Few studies have examined associations between *T. gondii* strain and clinical disease in cats. However, genotype has a significant effect on outcome in rodent infections. Clonal type I lineages are lethal (LD_{100}=1), while clonal types II & III lineages are much less virulent ($LD_{100} \geq 10^3$). A recent study compared genotypes of *T. gondii* in 4 Australian cats with naturally occurring systemic toxoplasmosis and 4 cats with latent infection using multilocus PCR-RFLP. The infecting genotype was the same in all cats and was identified as clonal type II variant (ToxoDB#3). Three of the four cats with toxoplasmosis had co-morbidities including FIV-infection, lymphoma or treatment with cyclosporine and prednisolone, while one was immunocompetent. The infecting strain of *T. gondii* has been genotyped in only eight other cases of naturally occurring toxoplasmosis in cats. A Type II variant (ToxoDB genotype#3) was detected in a cat with multivisceral toxoplasmosis in Switzerland. Isolates from another 6 cats with multivisceral toxoplasmosis from Switzerland were genotyped using microsatellite markers and found to be clonal type II (ToxoDB genotype #1), while one cat from the US was infected with a common endemic strain Type 12 (ToxoDB genotype #4).[15]

More recently, it has been shown that rhoptry organelle proteins (ROPS) are critical to host-cell invasion, replication and resistance of IFN-δ stimulated innate immunity. ROP5 pseudokinase and ROP18 kinase are key determinants of virulence in mice. The combination of ROP18 and ROP5 allele types appears to be highly predictive of parasite virulence, not just among archetypal clonal lineages of *T. gondii* (I-III) but among diverse genotypes from Central/South America and Asia. Genotyping of clinical isolates of *T. gondii* from cats using ROP18 and ROP5 is warranted to determine if these are useful virulence markers.

FIP: An Update on the Pathogenesis of an Enigmatic Disease

Dr. Anja Kipar, University of Zurich[5]

FIP is an immune-mediated disease and there are several key processes that are required for the development of FIP lesions.

1. **Systemic infection with virulent FIPV.**
   FIPV mainly arise *in vivo* in infected animals, from mutations of the horizontally transmitted, largely avirulent FECV; this explains the sporadic nature of the disease. Molecular studies indicate that the ability of a FCoV to maintain a systemic infection (i.e., in monocytes/macrophages) is mediated by a mutation (M1058L) in the fusion peptide of the S gene, possibly alongside other, so far not defined mutations.
2. **Effective, sustainable viral replication in monocytes.**
   This is a feature of FIPV: in addition to replicating in isolated feline monocyte/macrophage cultures (FECV are also capable of this), FIPV spread the infection in the cultures. In line with, FIP is associated with significantly higher viral loads in tissues and significantly higher viral replication in the blood (i.e., in monocytes) than FCoV infection in healthy animals.
3. **Activation of FIPV-infected monocytes.**
   This is the main process ultimately responsible for the disease, as it is a prerequisite for the development of granulomatous phlebitis and periphlebitis which is the earliest lesion and histological hallmark of FIP. It is most likely during a phase of high-level monocyte-mediated viraemia with substantial viral replication that monocytes are sufficiently activated to initiate the process; this is represented by their strong expression of cytokines (TNF-α, IL-1β etc.), adhesion molecules (CD18 etc.) and enzymes, such as MMP-9 which can destroy the vascular basement membrane at sites of monocyte attachment and emigration. At this stage, vascular endothelial cells are systemically activated as well, likely as a consequence of systemic cytokine release. Peripheral blood leukocytes were also found to be activated; this could initiate emigration of neutrophils and lymphocytes and the formation of granulomatous lesions outside vessels. *In vitro* studies have shown that monocyte activation is a direct effect of FIPV. Increased VEGF transcription in monocytes of cats with FIP and increased VEGF levels in their blood could explain the development of effusions, as VEGF is a strong mediator of vascular permeability.
4. **Viral persistence and recrudescence.**
   This is likely the source of replicating virus and offers the opportunity for viral mutation from FECV to FIPV potentially at any time point after (initial) infection of the cat. The main site of viral persistence is the colon, but virus can also be detected in other tissues (mainly in liver, lung, and lymph nodes) in the absence of viraemia, in particular in tissue macrophages. This indicates that viraemia can also occur in the absence of persistent enteric infection. Recent studies suggest that at

least part of the viral FECV-FIPV "transition" occurs at the enterocyte-macrophage/monocyte interface. Considering that the viral mutation rate appears to be higher in macrophages than in enterocytes and that FECV can infect monocytes without intestinal replication, it seems possible that monocytes or tissue macrophages rather than enterocytes are the site of definite FIPV generation.

5. **Lack of effective immune response**.
   This may allow more pronounced virus production and/or a quiescent infection state and is indicated by the fact that FCoV-infected isolated feline monocytes rapidly internalize viral glycoprotein that expressed on their surface as antigen-antibody complexes, and that FIPV inhibits the complement-mediated lysis of infected cells, even if they express viral antigen on their surface. Interestingly, however, FCoV infection leads to macrophage proliferation in hemolymphatic tissues regardless of FIP. This likely ensuring the supply of viral target cells (i.e., mature circulating monocytes or tissue macrophages) but is generally more pronounced and associated with macrophage activation in cats with FIP.

REFERENCES

1. Hosie MJ, *et al.* ISCAID Bristol Conference Proceedings No. iscd002; 2016.
2. Hofmann-Lehmann R. In: ISCAID Bristol Conference Proceedings No. iscd014; 2016.
3. Morgan E. In: ISCAID Bristol Conference Proceedings No. iscd038; 2016.
4. Barrs V. In: ISCAID Bristol Conference Proceedings No. iscd017; 2016.
5. Kipar A. In: ISCAID Bristol Conference Proceedings No. iscd029; 2016.

1.
acvim354

Diagnosing FIP: Has Recent Research Made It Any Easier?

Emi Barker, BSc, BVSc, PhD, DECVIM-CA (Internal Medicine); Séverine Tasker, BSc, BVSc, PhD, DSAM, DECVIM-CA (Internal Medicine), PGCertHE, FHEA, MRCVS
Bristol, UK

INTRODUCTION

Feline infectious peritonitis (FIP) is a fatal, systemic, infectious disease of cats. Its cause, feline coronavirus (FCoV), is globally distributed, and highly contagious, with high levels of exposure within multi-cat households (~87% seropositive),[1] in particular rescue shelters and breeding catteries. However, it is estimated that only 1–5% of cats infected with FCoV will go on to develop FIP.[2,3] Disease manifestations typically comprise a vasculopathy resulting in ("wet") effusions, granuloma formation resulting in ("dry") mass lesions, or a combination of the two.

A single minimally invasive FIP test that can definitively diagnose all FIP cases simply does not exist. Although a high index of clinical suspicion may be obtained from history, physical examination, and routine clinicopathological results, confirmation of diagnosis may prove elusive, invasive, expensive, and delayed. Rapid confirmation of disease is becoming more pressing, as potential treatments may become available.

FELINE CORONAVIRUS

Avirulent FCoVs, and those that cause self-limiting gastrointestinal signs (vomiting, and/or diarrhea), are those that are usually transmitted between cats. Avirulent or self-limiting FCoV infection was believed to be confined to the intestinal tract, but we now know that healthy FCoV-infected cats can have systemic FCoV infection, albeit with lower FCoV viral loads than cats with FIP.[4-7] Variants of FCoVs, arising through mutation of FCoVs during replication in infected cats, occur and it is these viruses that can cause clinical FIP, likely influenced by host factors. Thus, pathogenic variants usually arise by mutation during an existing infection with an avirulent FCoV,[8] and so these viruses are not commonly transmitted and usually die out within the host cat. However, occasional reports of FIP outbreaks in colonies suggest that direct transmission of pathogenic FIP-causing FCoVs is occasionally possible.

FCoV is a large enveloped RNA virus. As the genome is RNA based it is prone to errors during replication, resulting in a large number of quasispecies within infected hosts. Mutations allow replication outside the intestinal tract, within monocytes and macrophages. The greater the frequency of viral replication the greater the likelihood of a FCoV arising with the potential to cause FIP. The viral envelope is derived from components of the host cell membrane (proteins and phospholipids), in combination with viral glycoproteins (Figure 1). One of these viral glycoproteins, spike (S) protein, is a transmembrane protein that mediates host cell binding and cell entry. The S protein contains a fusion peptide that enables fusion of the FCoV envelope with the host cell membrane. Recent studies have identified mutations in the S protein fusion peptide, implicated in monocyte/macrophage tropism, that were thought to be markers of FIP.[9] Two alternative amino acid differences, **M1058L** and/or **S1060A**, together distinguished FCoVs found in the tissues of FIP cats from those found in the feces of healthy non-FIP cats in >95% of cases. More recently we have shown that these mutations are likely to be markers of systemic FCoV infection, which can occur in both FIP and non-FIP cats, rather than FIP *per se*.[7] These mutations are still important as it is probably via these, and/or perhaps other mutations, that the FCoV acquires its monocyte tropism to allow it to spread systemically and contribute to the development of FIP.

Figure 1. FCoV virion structure

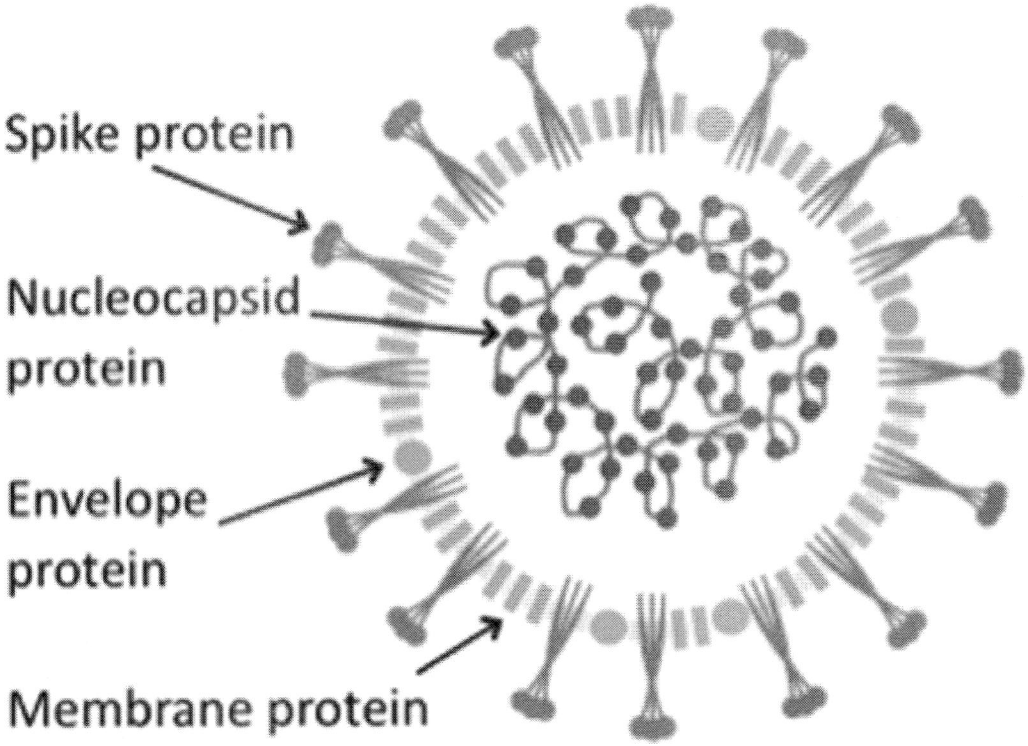

Other viral genomic mutations have been associated with FIP. Amino acid differences in the furin cleavage motif[10] and the heptad HR1 region of the S2 subunit[11] of the S protein have been correlated with disease. Truncations of the accessory protein 3c gene have been associated with loss of ability to replicate within enterocytes, and as such have only been reported in FIP-associated FCoVs.[12] Other viral and host factors are likely to be important in determining whether FIP results.

DIAGNOSING FIP

Evaluation and comparison of the diagnostic accuracy of available tests is complicated by the use of different criteria to definitively diagnose FIP in publications, (e.g., routine histopathology vs. immunohistochemistry [IHC], fluid analysis, PCR, or diagnostic algorithms).

Signalment/Clinical History

Male cats and younger cats are consistently over-represented among FIP cases, although older age and female status does not exclude FIP as a diagnosis. Although residence in multi-cat households increases the likelihood of FCoV seropositivity, a recent large study found that the majority of cats presented to a university hospital with FIP were from small households;[13] however, this may not reflect origin, and study recruitment may have been biased as pursuing a definitive diagnosis in cats from large households such as shelters may not be possible due to financial limitations. Whether pure-bred status is a risk factor remains unclear, with pedigree cats over-represented in some studies,[14,15] but not in others[13]. However, experimental loss of genetic heterozygosity appears to increase risk of FIP development following challenge.[16] A recent history of stress is common.[13]

Commonly reported features of cats with FIP include lethargy and weight loss typical of the "unwell" cat. Other clinical signs may pertain to the affected organs and can be highly variable (e.g., progressive abdominal distension with peritonitis, tachypnea with pleural effusion, seizures with CNS involvement).

Physical Examination

Physical examination findings of FIP are extremely variable. Clinical signs can change over time so **repeated** examinations are important to detect newly apparent signs. Although the majority present with evidence of effusive disease (e.g., dyspnea, ascites), these may be preceded by a period of non-specific findings, including fever, presenting the clinician with more of a challenge. A significant percentage of cats with dry FIP present with exclusively neurological or ocular signs.[17] Rarely cats may present with dermatological changes.[18,19]

A thorough physical examination, including retinal visualization, is mandatory, and will provide information to support a diagnosis of FIP (e.g., presence of uveitis) as well as regarding potential sites for diagnostic sampling (e.g., ascites, renal masses). If there is any query regarding neurological status, a full neurological examination should be performed as any region of the CNS may be involved (e.g., multi-focal localization). The presence of fever (80% cases >39.5°C; 39% cases >40°C) and/or icterus in particular are supportive of FIP, especially the wet form, but their absence does not exclude FIP.[13]

Blood Analysis

Routine hematology can support a diagnosis of FIP but changes are non-specific. **Lymphopenia** is particularly common (55–77% of cases; although a recent study found only 49.5% of FIP cases to be lymphopenic[13]), with **neutrophilia** (39–57%), a **left shift**, and **mild-moderate normocytic, normochromic anemia** (37–54%) also reported.[13,20-22] An association between FIP and microcytosis (with or without anemia) was recently reported.[13] Severe immune-mediated hemolytic anemia can occasionally occur with FIP.[20]

Serum biochemistry often reveals a **hyperproteinemia** (up to 60% cases, especially the dry form) due to a **hyperglobulinemia** (89% cases; typically, a polyclonal gammopathy on serum protein electrophoresis), usually with a low or low-normal serum albumin (64.5%). Consequently the **albumin: globulin (A:G) ratio is low** (<0.4 makes FIP very likely, >0.8 makes FIP very unlikely; reference range 0.45–1.2).[20-22] A study[23] in a population of cats with a low prevalence of FIP (akin to the situation usually encountered in practice) found that an A:G ratio >0.6 was useful to **rule out** FIP, but that lower values were not helpful in ruling in FIP. However, the frequency and extent of hypoalbuminemia, hyperglobulinemia, low A:G ratio, and frequency of electrophoretic abnormalities, in FIP cases have decreased recently,[13,24] with more recent cases trending towards elevated α2-globulins in absence of elevated γ-globulins, possibly due to earlier diagnosis of disease.[24] **Alpha1-acid glycoprotein** (AGP), an acute phase protein, is elevated in many inflammatory and non-inflammatory diseases (>0.48 mg/ml), but serum AGP levels in FIP are often markedly elevated (>1.5 mg/ml), so the magnitude of increase may be helpful in aiding diagnosis of FIP.[25-28] However, markedly elevated AGP is not completely specific with overlap between FIP and non-FIP cases. Hyperbilirubinemia can occur (21–63% cases), especially in wet FIP, often without marked elevations in hepatic enzyme activities; however, it should be noted that lymphocytic cholangitis (a differential for ascites with icterus) cannot be excluded on this basis. Worsening and increased prevalence of hyperbilirubinemia are reported during FIP disease progression.[22]

Interpretation of **FCoV serology** is complex. Breed-related differences in median FCoV antibody titre have been detected, and may reflect differences in breed response to FCoV infection.[29,30] Although FIP cases tend to have higher FCoV antibody titers than non-FIP cats, there is much overlap, so the value in an individual cat is limited,[30] and clinicians vary as to whether they perform serology or not in suspected cases. Many clinically healthy cats (especially those in multi-cat households) have positive, often very high, titers, whilst ~10% of cats with FIP are seronegative (sometimes due to FCoV complexing antibody in the sample, thus rendering it unavailable to the serological test).[31]

Tissue/Fluid Analysis

Analysis of effusion samples is very helpful in the diagnosis of FIP; so, obtaining samples should be prioritized in investigations (e.g., repeat ultrasonography to detect fluid formation). FIP effusions are usually clear, viscous, straw-yellow, protein-rich (thick proteinaceous backgrounds are often described on cytology) with a total protein concentration of >35 g/l. They usually have similar low A:G ratios and raised AGP concentrations to those in serum. A recent study found that effusion AGP concentrations (of

>1.55 mg/ml) were more useful in differentiating FIP and non-FIP cases than levels in the serum (although diagnosis of FIP was not always based on histopathology and immunostaining in this study).[26] FIP effusions are poorly cellular (usually <$5x10^3/\mu L$ cells), typically with non-degenerate neutrophils and macrophages. **Immunocytochemistry** (ICC) for FCoV antigen and reverse-transcriptase (RT-) polymerase chain reaction (PCR) for FCoV on effusion samples can also be performed (see later).

Tissue samples (e.g., liver, kidney, mesenteric lymph nodes) collected ante-mortem (by ultrasound-guided percutaneous TruCut biopsy, laparoscopy, or laparotomy) or at postmortem examination, can be evaluated for **characteristic histopathological changes of FIP**, which are considered reliable for diagnosis.[32] However, a lack of pathological lesions is more difficult to interpret as lesions may be missed in small samples (e.g., TruCut biopsies) due to their multifocal distribution, or due to non-affected organs being sampled.[33] Biopsy of affected tissue may not be possible without significant surgical intervention and/or significant financial implications in what is currently a fatal disease. Additionally, typical histopathological changes consistent with FIP may not be present, and **IHC** for FCoV antigen may be required for diagnosis.[25]

Gold standard confirmation of diagnosis is by the use of **IHC** to demonstrate the presence of the virus within macrophages associated with histopathological changes. **ICC** may also be used to demonstrate virus laden macrophages within effusions typical of FIP, if present. However, these methods are not without their limitations. Both methodologies require submission to specialist laboratories and therefore may be subject to time delays, they also have the potential for false positive staining and should be interpreted in the light of appropriate controls. ICC on effusion samples has shown variable sensitivity (ranging from 57 to 100%),[34-39] attributable in part due to the relatively low cellularity of typical FIP-associated effusions. Although initially thought to be very specific, false positive ICC have been reported for non-FIP effusions and CSF,[34,36,40] although methodology may play a role in this.

Molecular Diagnostics

RT-PCR targeting FCoV can detect cell-associated transcriptional mRNA and genomic viral RNA as well as cell-free, virion-associated, genomic viral RNA in fluid, tissue, and feces. Due to the presence of cell-associated mRNA, even where a quantitative PCR is used, template copy number does not necessarily reflect viral load. Quantitative RT-PCRs (RT-qPCRs) for FCoV may be used in combination with control PCR assays e.g., to confirm the absence of PCR inhibitors.

PCRs are reliant on binding of short "primer" sequences to the target viral sequence. TaqMan qPCRs further rely on binding of a "probe" sequence to the amplified target viral sequence. Mutations at the site of primer and/or probe binding may result in loss of assay efficiency, and ultimately sensitivity. PCR conditions may be altered to tolerate such mutations, but this can result in a loss of specificity and potential primer-dimer formation.

Loop mediated isothermal amplification (LAMP) has been suggested as a more economical alternative to PCR for screening purposes,[41,42] but more data regarding sensitivity and specificity are required before making recommendations.

FCoV copy number tends to correlate with histopathological findings suggestive of FIP.[43] Recently an abstract[44] suggested that ultrasound-guided fine needle aspirates could be a good alternative to tissue samples for FCoV RT-PCR. FCoV RNA has also been detected in most (72–89%) FIP effusion samples but not in any non-FIP effusion samples,[45-47] showing good sensitivity and excellent specificity. Blood/plasma/serum samples are not suitable for FCoV RT-PCR due to very poor sensitivity.[45,47] FCoV RT-PCR on CSF is a useful tool for diagnosis of FIP in cats with neurological signs,[48] with excellent specificity and moderate to good sensitivity, so a negative result does not rule out FIP.

As described above, M1058L and/or S1060A mutations in the FCoV S protein have been suggested as markers of FCoVs associated with FIP,[9] but our research suggested they were more markers of systemic FCoV infection c.f. FIP *per se* because, these mutations were present in tissues from both FIP and non-FIP cats.[7] Therefore, any assay targeting these S protein mutations in tissues cannot be used to definitively diagnose FIP.

Studies evaluating these S protein mutations in effusions have found that the majority of FCoVs in the effusions of FIP cats do indeed have M1058L and/or S1060A mutations.[45,46] However, it is not always possible to determine whether mutations are present as FCoV sequencing is not possible in all FCoV RT-PCR positive samples due to, for example, low levels of FCoV RNA present, alternative sequences being present, which limits the number of samples that yield results for S protein mutation analysis. This limitation, together with the finding of mutations in non-FIP cases questions the benefits of looking for mutations subsequent to finding FCoV in the samples of FIP cats by RT-PCR.

So Has Recent Research Made Diagnosis of FIP Any Easier?

Recent studies have shown that cats with FIP are now less likely to demonstrate the "classical" signalment or clinicopathological data (e.g., pedigree status, hyperglobulinemia with polyclonal gammopathy) that in previous years used to maintain a diagnosis of FIP high on the differential list. In addition, advances in molecular diagnostics have provided us with more techniques to rapidly identify both the presence of virus in fluid or tissue samples and to characterize the virus in terms of S protein mutations. However, no tools are 100% sensitive, particularly with clinicians suspecting FIP earlier in the course of disease, such that a negative result should not exclude a diagnosis of FIP. Furthermore, demonstration of systemic FCoV infection, albeit highly supportive of, does not necessarily equate to FIP development.

References

1. Drechsler Y, *et al. Vet Clin North Am Small Anim Pract.* 2011;41:1133.
2. Addie DD, Jarrett O. *Vet Rec* 1992;130:133–137.
3. Addie DD, *et al. Am J Vet Res.* 1995;56:429.
4. Kipar A, *et al. J Feline Med Surg.* 2006;8:69.
5. Kipar A, *et al. J Gen Virol.* 2010;91:1698.
6. Meli M, *et al. J Feline Med Surg.* 2004;6:69.
7. Porter E, *et al. Vet Res.* 2014;45.
8. Vennema H, *et al. Virology.* 1998;243:150.
9. Chang HW, *et al. Emerg Infect Dis.* 2012;18:1089.
10. Licitra BN, *et al. Emerg Infect Dis.* 2013;19:1066.
11. Lewis CS, *et al. J Gen Virol.* 2015;96:1358.
12. Pedersen NC, *et al. Virus Res.* 2012;165:17.
13. Riemer F, *et al. J Feline Med Surg.* 2016;18:348.
14. Worthing KA, *et al. J Feline Med Surg.* 2012;14:405.
15. Pesteanu-Somogyi LD, *et al. J Feline Med Surg.* 2006;8:1.
16. Pedersen NC, *et al. Vet Immunol Immunopathol.* 2016;171:17.
17. Pedersen NC. *J Feline Med Surg.* 2009;11:225.
18. Bauer BS, *et al. Vet Ophthalmol.* 2013;16(Suppl 1):160.
19. Cannon MJ, *et al. J Feline Med Surg.* 2005;7:233.
20. Norris JM, *et al. Aust Vet J.* 2005;83:666.
21. Sparkes AH, *et al. Vet Rec.* 1991;129:209.
22. Tsai HY, *et al. J Feline Med Surg.* 2011;13:74.
23. Jeffery U, *et al. J Feline Med Surg.* 2012;14:903.
24. Stranieri A, *et al. J Feline Med Surg.* 2016 [Epub ahead of print].
25. Giori L, *et al. J Small Anim Pract.* 2011;52:152.
26. Duthie S, *et al. Vet Rec.* 1997;141:299.
27. Hazuchova K, *et al. J Feline Med Surg.* 2016 [Epub ahead of print].
28. Paltrinieri S, *et al. J Vet Diagn Invest.* 2007;19:266.
29. Bell ET, *et al. Aust Vet J.* 2006;84:2.
30. Bell ET, *et al. Aust Vet J.* 2006;84:74.
31. Meli ML, *et al. J Feline Med Surg.* 2013;15:295.
32. Kipar A, Meli ML. *Vet Pathol.* 2014;51:505.
33. Giordano A, *et al. Vet Clin Pathol.* 2005;34:368.
34. Felten S, *et al. J Feline Med Surg.* 2016 [Epub ahead of print].
35. Hartmann K, *et al. J Vet Intern Med.* 2003;17:781.
36. Litster AL, *et al. Vet J.* 2013;198:362.
37. Paltrinieri S, *et al. J Vet Diagn Invest.* 1999;11:358.

38. Cammarata Parodi M, *et al. J Small Anim Pract.* 1993;34:609.
39. Hirschberger J, *et al. Tierarztl Prax.* 1995;23:92.
40. Gruendl S, *et al. J Feline Med Surg.* 2016 [Epub ahead of print].
41. Stranieri A, *et al.* In: Proceedings of ISCAID. Bristol, UK. 2016.
42. Weber K, *et al.* In: Proceedings of ISCAID. Bristol, UK. 2016.
43. Pedersen NC, *et al. Vet Microbiol.* 2015;175:157.
44. Freiche GM, *et al. J Vet Int Med.* 2016:411.
45. Felten S, *et al. J Feline Med Surg.* 2015 [Epub ahead of print].
46. Longstaff L, *et al. J Feline Med Surg.* 2017;19:240.
47. Doenges SJ, *et al. J Feline Med Surg.* 2016 [Epub ahead of print].
48. Doenges SJ, *et al. J Feline Med Surg.* 2016;18:104.

1.
acvim355

Chagas Disease at the Canine-Wildlife-Human-Vector Interface

Sarah A Hamer, PhD, DVM, DACVPM; Rachel Curtis-Robles, PhD
College Station, TX, USA

Introduction

The hemoflagellate protozoan *Trypanosoma cruzi* is the etiologic agent of Chagas disease, and infects humans, nonhuman primates, dogs, and other wild and domestic mammalian species. Active transmission of *T. cruzi* occurs throughout the Americas, including the southern USA, where infected hematophagous triatomine "kissing bug" vectors (Fig. 1) and wildlife reservoirs co-occur.[1] Infection can occur through the introduction of infectious triatomine feces to skin lesions during or shortly after blood feeding. Oral transmission to animals may result from consumption of infected bugs or infected prey species.[1] Congenital transmission has been documented in humans, canines, and other animals.[1] *T. cruzi* genetic strain differences occur throughout the Americas,[2] where parasite discrete typing units (DTUs) TcI-TcVI circulate and may be associated with different reservoir hosts, vector species, and variation in disease outcome. Infected individuals do not always develop clinical disease;[1] and prognosis in initially asymptomatic individuals is largely unknown. In the USA, clinical *T. cruzi* infection in dogs is most commonly associated with cardiac disease.[1,3] Dogs are important hosts throughout the Americas, and in some cases serve as sentinels of human disease risk across a landscape.[4] Diagnosis of Chagas disease is most commonly based on clinical suspicion and antibody detection, but diagnostic tests are imperfect and confirmation of antibody status is best derived from the results of more than one independent assay, which is usually impracticable. Chagas disease is a notifiable infectious disease for humans in four states, including Texas, Tennessee, Massachusetts, and Arizona. Using a One Health approach, we have conducted epidemiological studies of *T. cruzi* in vectors, dogs, wildlife, and humans, with a focus on the high-risk state of Texas in order to characterize transmission cycles, identify risk factors, and identify points for management interventions to reduce disease risk; highlights are reviewed below.

Figure 1. Three species of kissing bugs commonly found in Texas. L to R: *Triatoma protracta, T. gerstaeckeri, T. sanguisuga*. Bar represents 1 in (~25 mm). Photo credit: Gabriel Hamer. Originally published in (5).

Research Highlights

Triatomine Vectors

T. cruzi transmission risk is affected by the density of infected triatomine ("kissing bug") vectors (Fig. 1). Eleven species of triatomine insects have been recorded across the southern USA (Fig. 2), where they are colloquially known as "kissing bugs" or "conenose bugs."[1] These nocturnal Reduviidae insects are blood feeding throughout all five nymphal stages and as adults. They are easily confused with other look-alike species that do not feed on blood and do not pose risk of *T. cruzi* transmission. First records of triatomines in the USA date back to the 1850s - they are not new to the USA. With seven documented species, Texas has the highest triatomine species richness of any state.[1] In contrast to areas of Central and South America where domestic and peridomestic populations of triatomines exist, the species in the USA are generally sylvatic.[1] Nymphs are rarely documented in human domiciles (presence of wingless, flightless nymphs in houses is a hallmark of a domestic infestation), and triatomines found in peridomestic and domestic settings are usually dispersing adults that are attracted to lights.[1] Blood meal analyses of triatomines have revealed a generalist feeding strategy appearing to reflect local, easily-accessible blood sources, including nidiculous mammals such as woodrats.[1] Since infection prevalence[5] and sylvatic host associations[1] are known to vary among triatomine species, risk of parasite transmission to humans and domestic animals likely varies according to the vector species found in a local area. Despite the recognized ecological importance of *T. cruzi* genetic variation, few studies have examined the DTUs found in triatomines in the USA: TcI has been detected in limited sample sizes from several states, and TcIV has rarely been documented.[6] Knowledge of the prevalence and distribution of *T. cruzi* distribution in USA triatomine vectors - particularly those encountered by humans in peridomestic settings - will afford a greater understanding of disease risk and potential consequences. Our research efforts aim to: determine *T. cruzi* infection prevalence and DTUs in vector species, map occurrences of *T. cruzi* DTUs, determine temporal and spatial variation of triatomine activity, and define contemporary and unique distributions of *Triatoma* spp. across Texas.

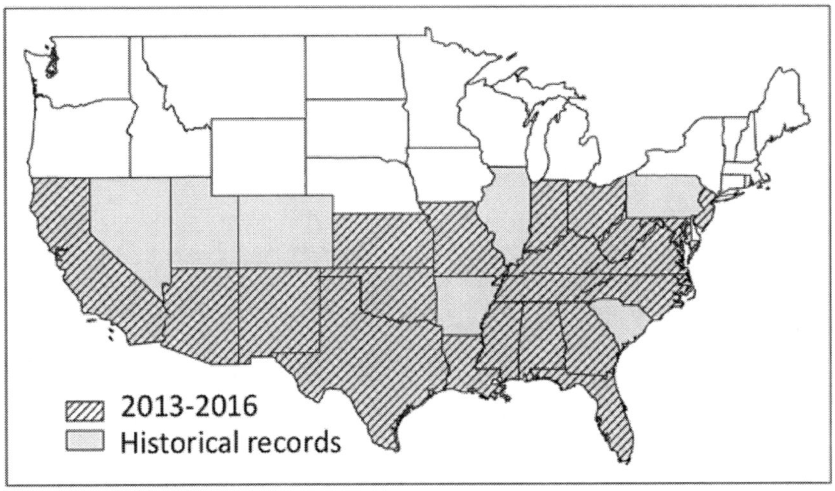

Figure 2. Records of triatomine insects in the USA. Historical records date back to the 1850s, while submission to our citizen science program are from 2013–2016.

Through a citizen science program and field collections from 2013–2016, we collected over 4000 kissing bugs of 7 species from diverse ecological regions, mainly from Texas, but including over 150 bugs from 20 additional USA states. The two most common species, *T. gerstaeckeri* and *T. sanguisuga*, exhibited activity peaks in mid-summer and early fall, respectively. We tested over 1200 triatomines and found 59% *T. cruzi* infection prevalence. PCR revealed two *T. cruzi* DTUs: TcI and TcIV. *T. gerstaeckeri* was significantly more likely to be infected with TcI than *T. sanguisuga*. Relative abundance of parasite DTUs varied spatially, with both TcI and TcIV co-circulating nearly equally in central Texas, while TcIV predominated in northern Texas. Given prior findings implicating TcI and TcIV in infections in USA humans and animals, knowledge of *T. cruzi* infection and DTUs in vectors is important to understanding public health risk of *T. cruzi* infection. Consideration of local temporal and spatial heterogeneity of *Triatoma* species and *T. cruzi* strains will allow targeting of vector control and medical/veterinary outreach initiatives to reduce human and animal vector contact.

Mammalian Wildlife

Diverse mammalian wildlife species are known reservoirs of *T. cruzi* in the USA, including raccoons, opossums, armadillos, skunks, and many rodent species.[1] Wildlife are important sources of blood - and therefore, infection - for triatomine vectors. Studies of infected USA wildlife populations have revealed association of TcI with opossums and TcIV with raccoons.[7] Recently, TcII was found in small rodents from Louisiana.[8] Our research efforts aim to document seroprevalence, parasitemia, and *T. cruzi* strains in USA wildlife, to better understand the important role wildlife species play in maintaining *T. cruzi* and infection risk to humans and domestic dogs.

In a study of cardiac tissue and blood samples from central Texas wildlife, we found 70% of raccoons were infected, whereas 12–14% of bobcats, coyotes, and foxes were infected. Despite PCR evidence of *T. cruzi* in cardiac samples, few infected individuals exhibited histological evidence of infection, and no amastigotes were observed, suggesting that the infected raccoons either did not experience severe cardiac changes, or, perhaps those with advanced cardiac disease were not represented in our sample. Strain typing of the infected raccoons revealed mainly TcIV, indicating that raccoons could be a possible sylvatic source of TcIV. In contrast, our investigations of feral swine and bats in Texas have identified very few *T. cruzi*-infected individuals (6% and 0.2%, respectively).[9,10] Relative to other wildlife species we have tested, our data suggest that raccoons may be important reservoirs of TcIV in Texas and a source of infection for indigenous triatomine bugs. Although the relationship between the sylvatic cycle of *T. cruzi* transmission and canine or human disease risk in the United States has yet to be defined, our data show robust sylvatic transmission with the potential for spillover to canine and human hosts.

Other Domestic Animals

T. cruzi-infected cats have been reported in other regions of the Americas,[4,11] and we have a growing dataset of apparently serologically positive adult cats from south Texas. However, the degree to which *T. cruzi* causes clinical disease in cats is unknown. A major limitation in our understanding of feline Chagas disease is the lack of validated feline diagnostic tests available to practitioners. Infections in domesticated livestock (cattle, horses, goats, sheep, pigs, etc.) have been occasionally documented in other areas,[4] but there has been limited study of *T. cruzi* in domesticated livestock in the USA, in which cattle, horses, and sheep were identified as potentially infected.[12] Additionally, *T. cruzi* was recently implicated in a neurological infection in a Texas horse.[13]

Shelter Dogs and Working Dogs

Cases of *T. cruzi* infection in dogs in the USA are widespread, especially in Texas.[3] There appears to be broad variation in clinical presentation in infected dogs, ranging from asymptomatic to acute death or chronic heart disease,[14] with little prognostic ability. Acute onset disease followed sudden death is more commonly reported in infected puppies than older dogs.[3,15] Clinical case management typically focuses on treatment of cardiac complications, and case management is complicated by the absence of a vaccination or approved antiparasitic treatment against *T. cruzi*. *T. cruzi* transmission to dogs can occur via infected triatomine feces, consumption of infected bugs or infected prey species, and congenitally, although the frequency with which congenital transmission occurs is unknown. Owners of seropositive breeding bitches may consider removing positive females from breeding roles.[14] Parasitemia has been shown to occur within days to four weeks after infection, during which time a dog would likely be most infectious to a vector, but becomes undetectable after a short period of time.[14] Anti-*T. cruzi* antibodies are first detected an average of 3 weeks post infection.[14] Serologic testing, namely, indirect florescent antibody assay (IFA), is a common tool for detecting anti-*T. cruzi* antibodies in dogs. In the USA, mainly TcIV has been found in canine infections, with TcI less commonly documented;[7,15,16] however, small sample sizes have restricted ability to determine clinical effects of TcI and TcIV in USA canines. Our research efforts aim to document seroprevalence and parasitemia in dogs, and examine risk factors for infection. We are interested in characterizing the clinical signs associated with *T. cruzi* strain infections, to determine whether TcI and TcIV result in different disease outcomes.

A study of 205 dogs from a network of 7 shelters across different ecological regions in Texas revealed 8.8% seroprevalence across the state, with infected dogs in all regions sampled.[17] While stray and shelter dogs may be associated with an elevated risk of Chagas disease due to increased exposure to vectors in outdoor settings, infections have been diagnosed across many types of dogs, including household pets, hunting dogs, and working dogs. A cross-sectional analysis of working hound dogs in south central Texas documented markedly high seroprevalence (58%), in which many dogs were also parasitemic, and parasite DTUs TcI and TcIV infected cardiac, uterine, and other canine tissues.[16] Over 80% of the triatomines collected from the working dog kennels were infected with *T. cruzi* and had molecular evidence of having fed on dogs, leading to the conclusion that multi-dog kennels can be locations of particularly high transmission risk.[16] Vector control and screening of bitches prior to breeding are the most common methods owners/breeders use to reduce transmission risk in their kennels.

Humans and Their Pet Dogs

Throughout Latin America, an estimated 6 million people are infected with *T. cruzi*;[18] and estimates of infected immigrants living in the USA range from approximately 72,000 to over 300,000.[19,20] Locally-acquired human cases have been recognized in the southern USA, where Texas has documented transmission,[1] and the small number of typed human cases in the USA were TcI.[7] Along the USA-Mexico border, approximately 1.7 million individuals inhabit "colonias" - economically distressed and unincorporated border communities with inadequate sewer, water, and/or electric services.[21] Such populations may be at higher risk for Chagas disease if they have emigrated from countries with high disease burdens;[19] and health disparities related to limited health education, poor access to health care, low incomes, and housing structures that can support vector populations may be important risk factors for locally-acquired infections. Additionally, domestic dogs may serve as important reservoirs and sentinels of *T. cruzi*. Our research efforts aim to determine Chagas disease prevalence in sympatric human and dog populations of rural colonias in south Texas.

We conducted a cross-sectional prevalence study in seven south Texas border colonias.[22] We found 1.3% seroprevalence in 233 humans, including one child born in the USA with only short duration travel to Mexico. Among 209 dogs, seroprevalence was 19.6%, but adjusted to 31.6% when including those dogs positive on only one (rather than two) serological test and extrapolating potential false negatives. Parasite DNA was detected in 5 dogs, indicating potential parasitemia. Seropositive dogs lived in all sampled colonias with no difference in odds of positivity across age, sex, or breed. Colonia residents collected 2 adult *Triatoma gerstaeckeri* and 1 nymph triatomine from around their homes; 1 of 3 bugs was infected with *T. cruzi*, and blood meal hosts were molecularly determined to include dog, human, and raccoon. Dogs and the infected vector all harbored *T. cruzi* TcI. Such active *T. cruzi* transmission cycles in impoverished communities should be a priority in outreach and vector control initiatives.

SUMMARY

Veterinarians and dog owners are faced with increasing diagnoses of canine *T. cruzi* infections, with little information available as a basis for understanding the implications to veterinary and public health. The lack of antiparasitic veterinary treatments complicates clinician recommendations to clients. Major recommendations to reducing risk in canines generally focus on integrated vector control methods, as well as careful attention to infection status of breeding bitches. Vector control recommendations include insecticides, screening windows and doors, reducing use of night lighting (which may attract flying adult *triatomines*), removing brush piles and other microhabitats conducive to wildlife nests that serve as sources of blood meals for kissing bugs, and sealing crevices that may serve as hiding places for *triatomines*.

As an additional resource, we initiated a citizen science program widely used by members of the public, veterinarians, and other public health professionals; the website for the program is: http://kissingbug.tamu.edu/. Through the website, we provide outreach materials about *triatomine* vectors and Chagas disease and identify vectors submitted to the program. Anyone is able to access the website to learn about *triatomines*, Chagas disease, and vector control, view interactive maps of county-level locations of insects submitted to the program, and submit individual inquiries answered by a member of our research team.

References

1. Bern C, *et al. Clin Microbiol Rev.* 2011;24:655.
2. Zingales B, *et al. Infect Genet Evol.* 2012;12:240.
3. Kjos S, *et al. Vet Parasitol.* 2008;152:249.
4. Gürtler RE, *et al. Acta Trop.* 2015;151:32.
5. Curtis-Robles R, *et al. PLoS Negl Trop Dis.* 2015;9(12):e0004235.
6. Brenière SF, *et al. PLoS Negl Trop Dis.* 2016;10:e0004792.
7. Roellig DM, *et al. Emerg Infect Dis.* 2008;14:1123.
8. Herrera CP, *et al. Parasit Vectors.* 2015;8:1.
9. Comeaux JM, *et al. J Wildl Dis.* 2016;52(3):627.
10. Hodo CL, *et al. Acta Trop.* 2016;164:259.
11. Jiménez-Coello M, *et al. Zoonoses Public Health.* 2012;9(Suppl 2):102.
12. Hanford EJ, *et al. Soc Sci Med.* 2007;65:60.
13. Bryan LK, *et al. Vet Parasitol.* 2016;216:13.
14. Snowden KF, *et al. Chapter 72. In: Greene, ed. Infectious Diseases of the Dog and Cat.* 4th ed. 2012:722.
15. Vitt JP, *et al. J Vet Intern Med.* 2016;30:1210.
16. Curtis-Robles R, *et al. PLoS Negl Trop Dis.* 2017;11(1):e0005298.
17. Tenney TD, *et al. Emerg Infect Dis.* 2014;20:1323.
18. World Health Organization. *Wkly Epidemiol Rec.* 2015;6(90):33.
19. Bern C, *et al. Clin Infect Dis.* 2009;49:e52.
20. Schmunis GA, *Mem Inst Oswaldo Cruz.* 2007;102:75.
21. Housing Assistance Council. *Housing in the Border Colonias.* 2013.
22. Curtis-Robles R, *et al. Am J Trop Med Hyg.* 2017;in press.

acvim356

Diagnostics and Management of Pyelonephritis

Jodi L. Westropp, DVM, PhD, DACVIM; Jane E. Sykes, DVM, PhD, DACVIM
Davis, CA, USA

INTRODUCTION

Pyelonephritis can be an acute life-threatening condition or a chronic condition, which results from bacterial or fungal invasion of the renal parenchyma. The remainder of this article refers specifically to bacterial pyelonephritis. The true prevalence of bacterial pyelonephritis in dogs and cats has not been reported. Bacterial pyelonephritis usually results from ascending infection by fecal commensals in the face of inadequate host defense mechanisms, but less commonly can follow hematogenous spread of bacteria to the renal parenchyma and pelvis. Clinical signs and physical examination findings of pyelonephritis can include lethargy, hyporexia, anorexia, polyuria, polydipsia, pain upon kidney palpation, and other signs associated with acute kidney injury (AKI). Furthermore, fever may be documented in the patient. A diagnosis of pyelonephritis can be suspected based on positive aerobic bacterial urine culture when accompanied by these clinical signs, as well as clinical laboratory findings of azotemia, cylindruria, and peripheral neutrophilia with or without bandemia (in the absence of another identifiable cause); and supportive ultrasonographic findings such as renal pelvic dilation or blunting of the renal papilla.[1] These ultrasonographic findings are not specific for pyelonephritis and renal pelvic dilation can be noted in patients with clinically normal renal function with diuresis and early ureteral obstructions. Pyonephrosis, a collection of sloughed urothelium and inflammatory cells in the dilated renal collecting system is a complication that can be noted in severe prolonged urinary stasis and secondary infection. It was reported that in the majority of 18 dogs with suspected pyonephrosis, the renal pelvis was distended with hyperechoic contents; hyperechoic retroperitoneal mesentery and effusion was noted in 11 dogs suggestive of retroperitoneal inflammation.[2] This complication appears to be more common in dogs with ureteral obstruction.[3]

In many cases, the diagnosis of pyelonephritis, particularly if the animal is not presenting with AKI, can be a diagnostic challenge. Some animals with pyelonephritis have negative bacterial urine cultures due to intermittent shedding of bacteria in the urine, systemic signs and laboratory abnormalities may be absent, and ultrasonographic findings suggestive of pyelonephritis may be absent. While nephropyelocentesis can be performed, this may not be possible in the absence of significant pyelectasia. Data evaluating the sensitivity and specificity of this procedure for pyelonephritis have not been published. There have been many publications evaluating the diagnostic utility of various urinary and serum biomarkers for evaluation of AKI in dogs and cats. For example, spot urinary N-acetyl B D-glucosaminidase (NAG) and γ-glutamyl transferase (GGT)/creatinine were reported to be helpful in the diagnosis of early AKI in experimental dog models.[4] More recently, we have reported that neutrophil gelatinase-associated lipocalin (NGAL), a protein induced marker during AKI, may permit detection of AKI sooner than creatinine. NGAL/creatinine was an early marker of gentamicin-induced AKI; further its decrease documented the onset of recovery.[5] Moreover, NGAL/creatinine was found to be specific for naturally occurring AKI and can be used to screen patients that are at risk for AKI.[6] Serum symmetric dimethylarginine (SDMA) can also support the presence of chronic kidney injury in association with bacteriuria. Unfortunately none of these biomarkers are specific for pyelonephritis and only SDMA is currently clinically available for use.

In humans, increased concentrations of urine NGAL, urine kidney injury molecule-1 (uKIM-1), serum cystatin C (sCysC), plasma C-reactive protein, procalcitonin, erythrocyte sedimentation rate among others have been used to support a diagnosis of pyelonephritis in patients with bacteriuria, although they lack high sensitivity and specificity.[7-10] Other biomarkers that may have value for diagnosis of pyelonephritis in humans include urine heparin-binding protein (HBP) and urine interleukin-6 (IL-6), the concentrations of which are significantly higher in elderly humans with suspected pyelonephritis when compared with those that have asymptomatic bacteriuria.[11] Clearly, there is a need for a sensitive and specific marker in humans and veterinary medicine to confirm the diagnosis of pyelonephritis.

DIAGNOSTICS

In all dogs and cats with suspected pyelonephritis, urine collected by cystocentesis should be submitted to the laboratory for aerobic bacterial urine culture and susceptibility testing. If the patient is severely thrombocytopenic or has another coagulopathy, a urine sample can be obtained by midstream free-catch collection or aseptic urinary catheterization, although these methods are not preferred because of increased likelihood of contamination with commensal organisms. Results need to be interpreted with all other clinical signs and using quantitative bacterial counts (CFU/mL) if urine is collected in this manner. As mentioned above, if the urine culture is negative, urine collected by pyelocentesis could be submitted if renal pelvic dilation is present and the suspicion for pyelonephritis remains high. Blood cultures are recommended at the same time as urine cultures in immunosuppressed patients. Blood culture should also be considered for diagnosis of pyelonephritis if fever and azotemia are present but urine sediment examination is unremarkable and/or the urine culture is negative. Other general principles of diagnosis for simple bacterial cystitis apply for pyelonephritis. Interpretation of susceptibility data should be based on antimicrobial breakpoints for serum rather than urine drug concentrations because high tissue drug concentrations are needed to treat pyelonephritis effectively. If multiple organisms are isolated, the suspected relative relevance of these should be considered, especially for urine specimens collected by free-catch or catheterization.

TREATMENT

The treatment for patients suspected of pyelonephritis should begin prior to the availability of urine culture results, particularly in those patients that are azotemic, febrile and/or systemically unwell. Initial treatment should involve antimicrobial drugs known to have local or regional efficacy against Enterobacteriaceae, based on the predominance of those organisms in pyelonephritis. If regional data are supportive, treatment with a fluoroquinolone excreted in urine in the active form (i.e., not difloxacin) is a reasonable first choice. If ascending infection is suspected, any recently obtained urine culture results might be the basis for initial therapy. If the upper UTI results from hematogenous spread, initial therapy should be based on cultures of blood or the infected site, whenever available. Oral antimicrobial therapy may be sufficient in animals that otherwise appear systemically well and have normal appetite. Intravenous therapy is recommended for animals that are dehydrated, hyporexic, or lethargic.

Once culture and susceptibility results are available, treatment should be revised if necessary. If combination therapy was initiated and the isolate is susceptible to both drugs, one should be discontinued. If resistance is reported to one of the drugs, that antimicrobial should be discontinued. Depending on the time elapsed, consideration should be given to substituting a drug to which the isolate is susceptible if the patient has not responded sufficiently; substitution is not necessary if patient response has been sufficient. If resistance is reported to both antimicrobials and clinical evidence of improvement is not evident, antimicrobial treatment should be changed to a drug to which the offending organism is susceptible *in vitro*. Consultation with a specialist with an interest in microbiology, urinary or clinical infectious diseases is recommended for treatment of multidrug-resistant organisms. If resistance to the drug(s) that are used is reported but there has been good clinical response, continuation with the initial therapy could be considered, provided there are no other reasons (such as fluid therapy) that might explain clinical improvement. The clinician can confirm eradication of the bacteria by submitting another aerobic bacterial urine culture 5–7 days after antimicrobial therapy in such a scenario to confirm microbiological cure if clinically indicated. Failure of treatment response in the face of bacterial susceptibility to the initially chosen drug should also result in re-evaluation of the diagnosis and other potential reasons for treatment failure (e.g., non-infectious cause of azotemia such as ureteroliths and neoplasia accompanied by bacteriuria). At that time, consideration should be given to discontinuing antimicrobial therapy.

Treatment for 4–6 weeks has previously been recommended for veterinary patients with pyelonephritis. However, as with simple (uncomplicated/sporadic) urinary tract infections, shorter duration therapy has been investigated in humans with pyelonephritis and should be considered in veterinary patients. A recent study in humans evaluated short-duration IV levofloxacin or conventional IV and oral levofloxacin in adults with complicated UTI and acute pyelonephritis. Treatment with IV levofloxacin for 5 days was not found to be inferior to treatment with conventional therapy (IV/oral regimen of levofloxacin for 7–14 days).[12] While even the "conventional" therapy may seem short to most veterinary clinicians, the recommended duration of therapy for acute bacterial pyelonephritis in children is 7–14 days[13] and the recommendation for adults is 10–14 days for beta-lactams or trimethoprim-sulfamethoxazole and 7 days for ciprofloxacin[14]. There is no reason to suspect that a longer duration would be necessary for dogs and cats, and a recommendation for 10 to 14 days of treatment is currently under consideration by the Antimicrobial Use Working Group of the International Society for Companion Animal Infectious Disease as part of revised guidelines for treatment of urinary tract infections in dogs and cats.[15] Antimicrobial prophylaxis in preventing recurrence is still debated in humans and children with pyelonephritis and due to the risk of antimicrobial resistance the authors do not recommend it.

Monitoring the Patient

A recheck examination that includes physical examination, serum creatinine, urinalysis, and aerobic bacterial urine culture is recommended 1 to 2 weeks after discontinuation of treatment. However, if clinical signs and azotemia have resolved, consideration has to be given to the clinical relevance of a positive urine culture, as it may represent subclinical bacteriuria. Re-isolation of the same bacterial species as that identified initially should stimulate an investigation for reasons for treatment failure. If clinical signs persist, blood values do not return to within the reference range, or subclinical bacteriuria is present, further diagnostics should be considered to investigate the patient for bacterial resistance, urolithiasis, anatomic defects or immune deficiency.

References

1. Anjou M, *et al. Vet Radiol Ultrasound*. 2011;52(1):88.
2. Choi J, *et al. Vet Radiol Ultrasound*. 2010;51(5):548.
3. Lulich J, *et al. J Vet Intern Med*. 2016;30(10):564.
4. Grauer G, *et al. J Vet Intern Med*. 1995;9(5):324.
5. Palm C, *et al. J Vet Intern Med*. 2016;30(1):200.
6. Segev G, *et al. J Vet Intern Med*. 2013;27(6):1326.
7. Sim J, *et al. Pediatr Res*. 2015;78(1):48.
8. Yim H, *et al. Pediatr Nephrol*. 2014;29(11):2181.
9. Shaikh N, *et al. Cochrane Database Syst Rev*. 2015;20(1) DOI:10.1002.
10. van der Starre Q, *et al. Clin Microbiol Infect*. 2014;20(10):1048.
11. Kjolvmark C, *et al. Diagn Microbiol Infect Dis*. 2016;85(2):243.
12. Ren H, *et al. Int Urol Nephrol*. 2017;20(10): DOI 10. 1007.
13. Morello W, *et al. Pediatr Nephrol*. 2016;331(8):1253.
14. Gupta K, *et al. Clin Infect Dis*. 2011;1(52):103.
15. Weese S, *et al. Vet Med Int*. 2011;2011:263768.

acvim358

Ductal Plate Malformations (DPM)

S.A. Center, DVM, DACVIM
Ithaca, NY, USA

Bile ducts and hepatocytes derive from a common hepatoblast precursor. Differentiation into biliary epithelium involves development of the ductal plate, a composite formed by a double layer of embryonic epithelium surrounding a developing portal vein. Remodeling by tubulogenesis (involution & proliferation of differentiating cholangiocytes) and integration of the evolving structure with portal mesenchyme, eventually defines & matures biliary epithelium into intrahepatic bile ducts. This is a complex process orchestrated by regulatory factors interacting in a time- and context-dependent manner. Involved are numerous transcription factors and genes influencing epithelial, mesenchymal, and vascular development, regulating apoptosis, adhesion molecules, and cell orientation or polarity. Tubulogenesis requires orchestrated involution of non-essential cells, proliferation of other cells, and epithelial to mesenchymal transformation that contributes to deposition of substructural extracellular matrix (ECM).

Essential to the process of tubulogenesis is cell orientation & polarity dependent on function of **primary cilia**. These are solitary nonmotile organelles located on the apical surface of most mammalian cells that function as mechano-, osmo-, and chemoreceptors, designation of planar cell polarity, influence cell cycle control, and numerous additional signaling pathways. Pathomechanisms specific to ductal plate malformations (DPM) remain elusive owing to the complexity & number of interactions associated with tubulogenesis. Several gene mutations affecting structure or functional interaction of primary cilia causal to DPM syndromes have been identified (humans & knock-out mouse models). Examples include genes coding for polycystin 1 [PC1], polycystin 2 [PC2], hepatocystin & SEC63, and fibrocystin/polyductin. During remodeling of the ductal plate, brisk mesenchymal proliferation that spatially separates bile ducts and liver parenchyma, is balanced by a high rate of hepatocyte and cholangiocyte apoptosis. Disruption of primary ciliary function attributable to mutations involving PC1 or PC2 leads to anomalous signaling between extracellular and intracellular compartments of cholangiocytes. This leads to imbalanced proliferation & apoptosis during the remodeling process leaving remnant malformative bile duct structure enveloped within abundant dense ECM. When severe, malformative intra- & interlobular bile ducts entrapped within exuberant ECM forming fibroductal partitions culminates in **congenital hepatic fibrosis** (CHF). Because primary ciliary dysfunction is a common factor linking various DPM syndromes, these are now considered congenital or inherited **cholangiociliopathies**. One gene defect in veterinary medicine is polycystic kidney disease in Persian cats; involving **polycystin-1 precursor gene** - influences structure of **primary cilia** -~15% of affected Persians develop DPM.

Defective remodeling, excessive proliferative or involution responses, or errant epithelial to mesenchymal transformation, evolves malformative ductal structures with embryonic phenotypes & variable accumulation of ECM. DPM have geometrically expanded portal tracts but vary in the severity of the fibroductal phenotype. In severe malformations, bridging fibroductal partitions extend deep into the hepatic parenchyma and segregate islands of hepatic parenchyma. These are not regenerative nodules as these islands retain normal hepatic cord organization and substructural reticulin structure. The hepatic tissue is subdivided by overgrowth of portal fibrous tissue with no parenchymal regenerative activity - allowing differentiation from cirrhotic regenerative nodules - the usual misdiagnosis in cases we have examined as a second opinion. DPM morphologies encompass a diversity of phenotypes that we have simplified into 6 subtype classifications - these vary in severity, clinical presentation, & prognoses. CHF encompasses two discrete phenotype - Caroli and small proliferative-like DPM.

Ductal Plate Malformations

1) **Simple Biliary Cysts:** single, few
2) **von Meyenberg Complexes**
   - microscopic, NO signs
3) **Biliary Cystadenoma**
4) **Choledochal Cysts**

5a) **Caroli's Malformation**
   large ducts
   - may be mineralized
   - cholangitis, cholelithiasis

5b) **Caroli Syndrome w/ CHF**

6a) **Proliferative-like DPM**

6b) **Proliferative-like DPM w/CHF**

Congenital Hepatic Fibrosis (CHF)
- portal hypertension
- APSS, Ascites, ± HE

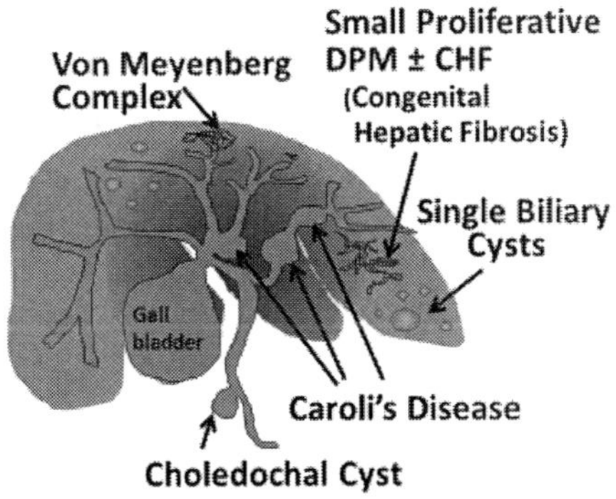

Stereotypic Feature of Portal Venous Hypoperfusion Accompany Most DPM Malformations in Dogs

Desmet, the pioneering pathologist that first recognized DPM, discerned aberrant remodeling of the ductal plate associated with aberrant branching of portal veins, resulting in what he termed as a "pollard willow" tree-like branching - that is - numerous too small tiny malformed closely spaced vascular structures - the equivalent of hypoplastic portal veins. Transverse sectioning through a pollard willow pattern appears as an expanded portal region corresponding to conjoining of several small portal tracts - each with its own malformed ductal plate, hypoplastic portal vein or lack of portal vein silhouette. However, arterial cross sections are plentiful & prominent. It is unclear if diminished portal vein profiles are developmental or reflect the portal accumulation of non-compliant ECM causing enough resistance to thwart forward flow in the portal vasculature. The portal vascular system is a capacitance bed with low splanchnic vascular pressure (5–8 mm Hg) lacking valves. Thus, it circulates blood into the pathway of least resistance. In this circumstance, this may evolve development of acquired portosystemic shunts (APSS) in the most severe phenotype - **congenital hepatic fibrosis**. In less severe forms of DPM, reduced portal flow is inconsistent in severity across liver lobules and there is reapportioning of splanchnic circulation within the liver. Reduced portal venous perfusion provokes a number of stereotypic changes reflecting **lobular atrophy** and the arterial changes associated with the **hepatic arterial buffer response**. Thus, most DPM liver biopsies disclose small hepatocytes, close spatial proximity of **lobular elements** (portal & centrilobular regions), increased numbers of miniaturized portal tracts, and prominent thick-walled arterials in cross section in large portal tracts and increased numbers of plexiform/serpiginous arterials entangled within the small malformative bile duct profiles. Whether portal veins are developmentally hypoplastic or merely reflect under-perfusion and collapse as a result of the noncompliant exuberant extracellular matrix in the portal tract remains unresolved.

Other Malformative Features

In animals with severe malformations, all centrilobular regions are anatomically conjoined with or "fused" with portal tracts, merging within expansive fibroductal partitions. In animals with less severe forms, fusion complexes are inconsistent and scattered or may not be observed. **Fusion complexes** also variably are observed in dogs with microvascular dysplasia (MVD) and dogs with portosystemic vascular anomalies (PSVA) not associated with DPM. Other malformative features have included: agenesis or atresia of the GB, asymmetric atresia of entire liver lobes - left sided only, right sided only, quadrate lobe only, concurrent single extrahepatic PSVA, and concurrent large unusual intrahepatic vascular malformations, and congenital diaphragmatic hernia. As I predominantly see patients with liver disease and biopsies of liver, only a subset seen at Cornell have also been evaluated other than by US imaging for renal involvement - indeed, polycystic renal disease is observed in a subset of dogs and more often in cats. Cystic ductal malformations also may involve the pancreas.

So Why Have DPM Suddenly Emerged?

They have not suddenly emerged, rather, diagnosis has been "demystified" with characterization of unique phenotypes that distinctly segregate these disorders. Until recently, dogs & cats with these malformations had been merged with animals thought to have portal vein hypoplasia or grouped into small case series as subsets with hepatoportal fibrosis in the veterinary literature. While a diverse spectrum of DPM has been described in various animal species, only a few appear to be familial. We recently published a case series of 30 non-related boxer dogs to help clarify syndrome presentation, phenotype variation, and survival in a single breed to exclude complications introduced by cross-breed studies. In this study we morphometrically demonstrated significantly increased area of hepatic parenchyma comprised of portal tracts & quantitatively compared the area of collagen deposition (digitized scanning of biopsy sections and positive pixel proprietary algorithm using Masson's trichrome staining).

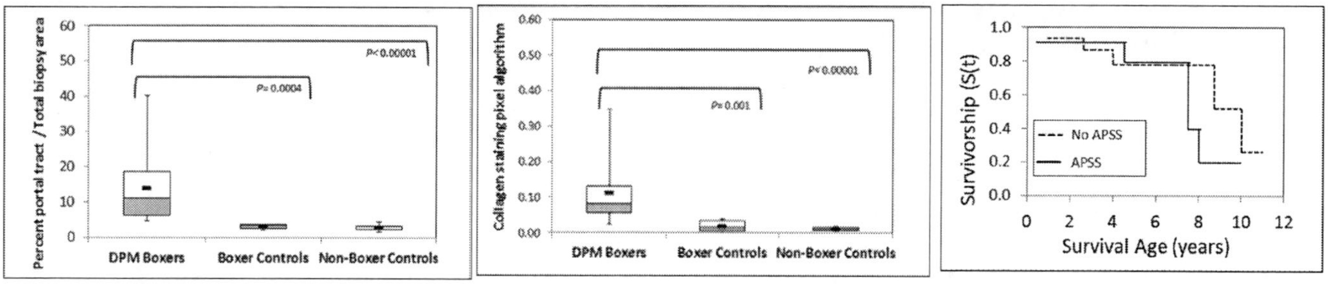

Ductal Plate Malformation Phenotype

Diverse phenotypes are defined by involved structures and variable development of hepatic fibrosis, portal hypertension, and acquired portosystemic shunts (APSS). Propensity for septic suppurative cholangitis, choledochitis, cholelithiasis, and extrahepatic bile duct obstruction (EHBDO) reflect the diversity of structural malformations and propensity for stasis of bile flow in sacculated large ducts.

1. **Simple biliary cysts** - usually asymptomatic unless these enlarge causing pressure on adjacent visceral structures (stomach, abdominal distention, portal hepatis). These typically display an attenuated ductal epithelium (flattened) and are filled with serous thinly mucinous material lacking bile (no bilirubin, no bile acids). These can be extracorporeally drained, marsupialized, removed, or ablated with alcohol. However, often these are detected serendipitously on US imaging and are asymptomatic.
2. **von Meyenburg complexes** - isolated single microscopic DPM comprised of clusters of proliferative intralobular bile ductule profiles embedded in an expanded ECM profile; also called bile duct hamartomas. These are inconsequential and typically identified near the liver capsule.
3. **Biliary cystadenomas** - in cats may represent biliary malformations that enlarge with aging; I see these more commonly in geriatric cats with hyperthyroidism. This either reflects their macroscopic expansion with aging or perhaps increased ductal secretion under the influence due to hypermetabolism. These often develop adjacent to the gallbladder and, when expanded, may dip into the portal hepatis. These are multi-partitioned cystic cavities septated and enclosed by flattened ductal epithelium embedded in dense ECM with arterial perfusion. Attempts for complete resection may result in critical perioperative arterial hemorrhage. This is a rare diagnosis in the dog. These often are asymptomatic in cats and discovered serendipitously during US abdominal imaging.
4. **Choledochal cyst** - these are diverticula protruding from or within the structure of the common bile duct (CBD) or cystic duct. These have propensity for infection - similar to appendicitis in humans causing sepsis, extrahepatic bile duct obstruction (EHBDO), and a cranial abdominal mass effect. These are a feline DPM malformation. During US imaging these may be confused with the GB, a distended CBD secondary to suspected EHBDO. These may be discovered at the sphincter of Oddi

involving the duodenal wall, develop in the CBD ampulla where they directly impact on the pancreatic duct, be located near or fused with the GB, or run elongate adjacent to the CBD. These generate considerable confusion at surgery with the conundrum of interventional options. Depending on the anatomical malformation, these may be 1) excised, 2) marsupialized to the intestines, or 3) used as an enteric-biliary bypass when the GB is infected and needs removal. These must be drained as often they are infected in the symptomatic cat. Cultures may be negative from GB bile & liver and only positive from the choledochal cyst, so it is essential to collect this fluid and potentially a portion of wall for cytologic examination and culture, along with liver & bile. These may present as suppurative cholangitis or because of EHBDO due to cholelithiasis.

5. **(a & b) Caroli malformations** - characterized by progressive irregular segmental saccular or cystic dilation of hepatic, interlobular and large intralobular bile ducts. This malformation associates with degradation of the laminin & type IV collagen (major basal laminar components) that normally provides structural support for larger bile duct structures. This may associate with diffuse fibroductal partitions causing severe hepatic fibrosis and development of APSS. If there is no severe fibrosis - classify as Caroli disease; if severe fibrosis causing portal hypertension and APSS - classify as Caroli syndrome with congenital hepatic fibrosis (CHF). Histologically, ducts have variable morphology appearing elongate, irregular, lanceolate, sacculated, or may have numerous budding evaginations/diverticula and sometimes numerous small satellite bile duct profiles surround a larger duct structure. Often the biliary epithelium is irregularly attenuated, may appear corrugated, and there often is intraluminal debris in large ducts. In animal with bacterial infection (suppurative cholangitis), bacteria are sometimes visualized in these ducts. Some of these patients develop intrahepatic microcholeliths - these may contain intra-structural bacteria that will sustain chronic bile borne infection. In a well-studied knockout mouse model of DPM forming a Caroli phenotype (ARPKD-PKHD1 mutation), dilated saccular ducts can progress to dissociate with the biliary tree forming stand-alone isolated biliary cysts. US imaging is essential in this disorder to rule out bile duct obstruction as a causal factor in duct dilation or sacculation. Saccular distended bile ducts may or may not be identified during US imaging. If these are histologically diagnosed and no distended bile ducts observed on images, it is prudent to retrospectively re-examine that patient for sacculated/distended intra- and extra-hepatic ductal structures branching into & within different liver lobes, intraductal debris, mineralized densities, & GB mineralization. When diagnosed in a symptomatic animal, these associated with infection, cholelithiasis, duct mineralization (dystrophic due to chronic inflammation/infection). These are commonly infected. If not done already, you need to scrutinize US images for cystic malformations involving large ductal structures. If there are no US Images, you need to retroactively collect these images for baseline as this animal will require chronic monitoring that will involve sequential US imaging - as Caroli phenotype worsens with aging. Occasionally, Caroli phenotype transforms into a cholangiocarcinoma in feline patients (I have seen 4 at this point).
6. **Diffuse intralobular small bile duct malformation (proliferative-like DPM) without (5a) OR with (5b) diffuse portal-to-portal bridging fibrosis** - If fibrosis=small proliferative-like CHF - Abnormal tubulogenesis of small intralobular bile ducts reflect abnormal resorption of extraneous ductule anlage, resulting in the retention of numerous small malformed proliferative ductules lacking luminal apertures and variously embedded in excessive extracellular matrix (ECM) within expanded portal tracts. The amalgamation of proliferative ductules, numerous stout arterials & ECM that dimensionally expands portal tracts forms variable fibroductal partitions. These may be severe enough to bridge most or all portal tracts with merging of centrilobular regions. Portal veins are usually inconspicuous, arterials prominent with thick muscular walls & the entwining plexiform/serpiginous phenotype. Dense fibrosis & severe portal venous hypoperfusion may cause presinusoidal portal hypertension causing development of APSS in the CHF phenotype.

Clinical Signs

Ductal plate malformations represent non-inflammatory hepatopathies, typically recognized in juvenile and young adult dogs or cats. These have been diagnosed in dogs as old as 13 yrs and cats as old as 17 yrs. As in humans, these are clinically heterogenous with some animals manifesting illness shortly after birth whereas other are recognized later in life. Late-onset diagnosis is either serendipitous or associated with acquired complications. Co-morbidities reflect hepatic presinusoidal portal hypertension, APSS, predisposition for ascites, increased risk for bacterial infections of the biliary tree, and cholelith formation. Secondary conditions (infections, cholelithiasis, portosystemic shunting) may lead to episodic fever, lethargy, increased liver enzyme activity (ALT, ALP, GGT), hyporexia, weight loss, jaundice, and rarely, abdominal distention or abdominal pain, and ammonium biurate urolithiasis. These features motivate clinicopathologic assessments, imaging studies, and liver biopsy. In cats with DPM I have observed survivals as long as 19 yrs but calculate a median survival of 9.1 years in 36 feline cases. Dogs with small duct phenotypes may have a normal life expectancy based on our Boxer study - with or without acquired shunts and ascites.

A substantial number of dogs (~40–50%) with DPM also have pathologic copper accumulation. Depending on quantification, some of these require D-penicillamine chelation, antioxidants, and lifetime dietary copper restriction.

Approximately 15–20% of dogs & cats with DPM also display arterial thrombi on liver biopsy. None have been symptomatic for this phenomenon. I suspect it reflects the rheology associated with coiled thick-walled arterials in the non-compliant ECM expanding portal tracts. We check d-dimer & fibrinogen concentrations, antithrombin & protein C activities, examine blood smears for RBC shearing effects, and systemically for other signs of small arterial occlusion whenever these thrombi are encountered. Typically they are historic & not active and have not comprised the vasa vasorum of the portal tract.

Treatment

- **1. Simple biliary cyst** - If frequent enlargement causing abdominal distress, US-guided aspiration (these may quickly re-accumulate fluid), surgical marsupialization, or alcohol ablation.
- **2. von Meyenburg plexus phenotype** - no treatment needed, dissociated biliary malformation as a "histologic curiosity" most of the time.
- **3. Biliary cystadenoma** - Do not aggressively pursue with surgical debulking unless large and causing compression-associated effects on adjacent viscera. Arterial perfusion and multiseptated compartmentalization and mucinous serous contents complicate debulking.
- **4. Choledochal cyst phenotype** - If identified by US imaging & suspected infected, start trimodal antibiotic therapy (metronidazole, enrofloxacin, Clavamox or spectrum-like Clavamox) for several days before surgical intervention for 1) removal, 2) marsupialization to intestines, 3) enteric-biliary bypass if GB diseased and removed. Choleretic therapy recommended - ursodiol 10–15 mg/kg PO BID given with food for best bioavailability and high dose Denosyl for GSH-mediated choleresis (40 mg/kg PO SID on empty stomach).
- **5. Caroli DPM phenotype** - High risk for bile-borne bacterial infection & suppurative bacterial cholangitis, cholelithiasis, complicating gallbladder sepsis. Collection of bile (large volume with particulate debris which contains bacterial biofilm) combined with liver for aerobic & anaerobic cultures. If suspected before surgery and procedure targeting cholelith removal - start trimodal antimicrobials before surgery. Include a crushed lith in bacterial culture. May require lifelong intermittent culture-tailored antimicrobial therapy.
- **6. Small proliferative bile duct phenotype DPM** - High risk for chronic suppurative bacterial cholangitis. Choleresis may not have impact on assisted clearance of infection, inflammatory cytokines, debris, microscopic choleliths as the deformed bile duct structures may not have continuity with the biliary tree or even luminal apertures. Chronic tailored antimicrobial therapy based on bile cultures. If no GB because of atresia, culture via Tru cut needle biopsy if during post-diagnostic interval for follow-up cultures.
- **5b & 6b. Congenital hepatic fibrosis phenotypes** - Experimental work has shown that fibrosis relates to bacterial cholangitis causing inflammatory provocation of macrophage TGFβ, a potent inducer of epithelial to mesenchymal transition prompting cholangiocytes transformation to activated mesenchymal cells (fibroblasts) producing collagen. Other work confirms activation of the renin-angiotensin system as the mechanism stimulating TGFβ.

Recent experimental studies in the Caroli CHF rodent model demonstrated positive response to telmisartan in reducing fibrogenesis and cystic expansion of malformed ducts. In humans, diagnosis of DPM with CHF often culminates in liver transplantation. Ascites is managed with sodium restriction, furosemide but beware that concurrent spironolactone & telmisartan may result in critical hypotension, collapse, and acute renal failure at conventional dosing. Nutritional therapy should be optimized to protein tolerance to avoid unnecessary protein restriction.

References

1. Brown DL, Van Winkle T, Cecere T, *et al.* Congenital hepatic fibrosis in 5 dogs. *Vet Pathol.* 2010;47(1):102–107.
2. Desmet VJ. Congenital diseases of intrahepatic bile ducts: variations on the theme ''ductal plate malformation.' *Hepatology.*1992;16:1069–1083.
3. Desmet VJ. What is congenital hepatic fibrosis? *Histopathology.* 1992;20:465–477.
4. Wills ES, Roepman R, Drenth JPH. Polycystic liver disease: ductal plate malformation and the primary cilium. *Trends Mol Med.* 2014;20(5):261–270.
5. Drenth JP, Chrispijn M, Bergmann C. Congenital fibrocystic liver disease. *Best Pract Res Clin Gastroenterol.* 2010;24(5):573–584.
6. Drögemüller M, Jagannathan V, Welle MM, *et al.* Congenital hepatic fibrosis in the Franches-Montagnes horse is associated with the polycystic kidney and hepatic disease 1 (*PKHD1*) gene. *PLoS One.* 2014;9(10):e110125.
7. Eaton KA, Biller DS, DiBartola SP, *et al.* Autosomal dominant polycystic kidney disease in Persian and Persian-cross cats. *Vet Pathol.*1997;34(2):117–126.
8. Gunay-Aygun M, Gahl WA, Heller T. Congenital hepatic fibrosis overview. *GeneReviews.* April 24, 2014.
9. Kerkar N, Norton K, Suchy FJ. The hepatic fibrocystic diseases. *Clin Liver Dis.* 2006;10(1):55–71.
10. Masyuk T, Masyuk A, LaRusso N. Cholangiociliopathies: genetics, molecular mechanisms, and potential therapies. *Curr Opin Gastroenterol.* 2009;25(3):265–271.
11. Pillai S, Center SA, McDonough SP, *et al.* Ductal plate malformation in the liver of Boxer dogs: clinical and histological features. *Vet Pathol.* 2016;53(3):602–613.
12. Görlinger S, Rothuizen J, Bunch S, *et al.* Congenital dilatation of the bile ducts (Caroli's disease) in young dogs. *J Vet Intern Med.* 2003;17(1):28–32.
13. Haechler S, van den Ingh TS, Rogivue C, *et al.* Congenital hepatic fibrosis and cystic bile duct formation in Swiss Freiberger horses. *Vet Pathol.* 2000;37(6):669–671.
14. Kawasaki T, Carmichael FJ, Saldivia V, *et al.* Relationship between portal venous and hepatic arterial blood flows: spectrum of response. *Am J Physiol Gastrointest Liver Physiol.* 1990;259(6, pt 1):G1010–G1018.
15. Lanier VC Jr., Buchanan RD, Foster JH. Hepatic morphologic changes following end-to-side portocaval shunt in dogs. *Am Surg.* 1968;34(3):185–195.
16. Lautt WW. Mechanism and role of intrinsic regulation of hepatic arterial blood flow: hepatic arterial buffer response. *Am J Physiol Gastrointest Liver Physiol.* 1985;249(5, pt 1):G549–G556.
17. McAloose D, Casal M, Patterson DF, *et al.* Polycystic kidney and liver disease in two related West Highland White Terrier litters. *Vet Pathol.* 1998;35(1):77–81.
18. McKenna SC, Carpenter JL. Polycystic disease of the kidney and liver in the Cairn terrier. *Vet Pathol.* 1980;17(4):436–442.
19. Sato Y, Ren XS, Nakanuma Y. Caroli's disease: current knowledge of its biliary pathogenesis obtained from an orthologous rat model. *Int J Hepatol.* 2012;2012:ID 107945, 10 pages, doi:10.1155/2012/107945.
20. Terada T, Nakanuma Y. Detection of apoptosis and expression of apoptosis-related proteins during human intrahepatic bile duct development. *Am J Pathol.*1995;146:67–74.
21. van den Ingh TS, Rothuizen J. Hepatoportal fibrosis in three young dogs. *Vet Rec.* 1982;110(25):575–577.
22. van den Ingh TSGAM, Rothuizen J. Congenital cystic disease of the liver in seven dogs. *J Comp Pathol.* 1985;95(3):405–414.
23. van den Ingh TS, Rothuizen J, Meyer HP. Portal hypertension associated with primary hypoplasia of the hepatic portal vein in dogs. *Vet Rec.* 1995;137(17):424–427.
24. Wills ES, Roepman R, Drenth JPH. Polycystic liver disease: ductal plate malformation and the primary cilium. *Trends Mol Med.* 2014;20(5):261–270.

acvim359

Understanding Biliary Atresia: A One Health Story

Rebecca G. Wells, MD
Philadelphia, PA, USA

Introduction

Biliary atresia (BA) is a major liver disease in children, occurring worldwide with an incidence of 1:5000–1:19,000.[1,2] The disease presents exclusively in neonates and never in mothers or in older children; affected patients are typically diagnosed between the age of 1 and 3 months after appearing healthy at birth. BA is characterized by fibro-obliteration and obstruction of the extrahepatic bile ducts and variable, although lesser, damage to intrahepatic ducts. It is always fatal without treatment. The primary initial treatment is performance of a Kasai portoenterostomy to enhance drainage; however, the disease rapidly progresses to cirrhosis, even with successful drainage. Half of affected babies will require transplant by age 2, and most of the remaining patients will need a transplant before reaching adulthood. [1,2] BA is the major indication for liver transplant in children.

The Etiology of Biliary Atresia Is Unknown

Neither the etiology nor the early course of BA is understood. While 10–20% of BA patients have associated anomalies, including polysplenia and situs inversus, most patients have isolated (non-syndromic) BA. BA is not a genetic disease; there is no evidence for Mendelian inheritance and most pairs of twins (including monozygotic twins) are discordant for the disease. GWAS studies have identified susceptibility loci, but none are fully characterized as yet.[3] Epidemiological data suggest an environmental cause, either a toxin or infection, although no specific insult has been identified in spite of extensive work searching for viral associations.[2,4,5] Recent data suggest that the etiological injury in BA occurs before birth: in two Texas studies, 66 newborns who went on to develop BA all had abnormalities in conjugated bilirubin within several days of birth (including many within the first 24 hours).[6-9] The disease clearly progresses after birth - although most babies with BA are diagnosed after they develop significant biliary obstruction and liver damage, a small percentage without overt disease come to medical attention earlier. The Texas newborns with BA who were identified retrospectively as having elevated conjugated bilirubin shortly after birth also clearly underwent postnatal progression,[6,9] which appears to be driven by inflammation and immune-mediated damage[10-13]. The initial injury is primarily extrahepatic, with epithelial damage, fibrosis, and obstruction. Based on the observation that patients with successfully draining livers after Kasai have decreased cholestasis but progressive ductal damage and fibrosis,[14] it is assumed that intrahepatic bile duct (IHBD) damage also occurs, but this has not been possible to study directly. In a livestock model of BA (see below), we have shown via histology that there is minimal intrahepatic damage at early stages, even with complete extrahepatic duct obstruction and fibrosis,[15] and that intrahepatic damage and cirrhosis are later events.

The major model used to study BA today is the mouse rhesus rotavirus (RRV) model, in which BALB/c mice are infected with RRV within 24 hours of birth. This model is not perfect - there is no definitive evidence for RRV or any other virus as a cause of human BA[16] - but it does mimic human BA in its specificity for young neonates and patterns of extrahepatic bile duct and liver damage[17-19] and has yielded important information about inflammatory and immune mechanisms of bile duct damage.

Biliary Atresia in Australian Livestock

There have been four large outbreaks of a BA-like disease in genetically mixed lambs (and in one case calves) in Australia, all associated with severe drought and grazing in atypical pastures.[20,21] Each outbreak was associated with a specific pasture; between 14 and 100% of newborns were affected, with totals in some cases numbering in the hundreds. Mothers were spared and signs of disease in the young were observed shortly after birth, with acholic diarrheal stools and poor growth (**Figure 1**). Younger lambs showed extrahepatic bile duct obliteration but minimal intrahepatic damage, while older animals had cirrhosis, consistent with the course of human disease.[15,20,21] Although a variety of hypotheses were considered by veterinarians at the time, consumption of a *Dysphania* species plants by pregnant animals was considered the most likely cause, with either a plant toxin or a plant-associated fungus being the injury agent.

Figure 1. Australian sheep from outbreak. The two lambs in the front have BA; the rear-most lamb and ewes are unaffected.

Isolation of a Plant Toxin

We isolated a novel isoflavonoid, biliatresone, from *Dysphania glomulifera* ssp. *glomulifera* and *D. littoralis* (which are distinguishable only microscopically) collected from a pasture associated with one outbreak. We carried out 10 rounds of sequential fractionation of the plants using a zebrafish bioassay. Larval fish ingested fluorescent lipids which, under normal circumstances, collected in and highlighted the gallbladder; in the setting of duct obstruction, the gallbladder signal was not visible. We isolated a never-described isoflavonoid we called biliatresone (**Figure 2**). Immuno- and histochemical staining showed that biliatresone causes atresia specific to the extrahepatic bile ducts in larval zebrafish.[15,22] The immune response does not appear to play a role in the fish disease.[22] Biliatresone is likely excreted by the liver into bile ducts; data from mutant zebrafish lacking IHBD suggest that secretion into the biliary system is required for biliatresone toxicity.[22] An extensive literature on structurally similar plant isoflavones supports a mammalian model whereby a small amount of biliatresone, after ingestion by the mother, crosses the placenta and reaches the fetal liver in either free or glucuronide-conjugated form.[23-29]

Synthesis of biliatresone confirmed that the compound we isolated from plants was responsible for the biliary effects in fish and mammalian systems (below). The final plant fraction examined included several structurally similar compounds which had no effect on zebrafish and which notably lacked the highly reactive alpha-methylene ketone moiety of the C ring; the C ring was closed. One of these compounds, betavulgarin, is found widely in foods consumed by humans around the world, although there is no evidence that it causes disease.

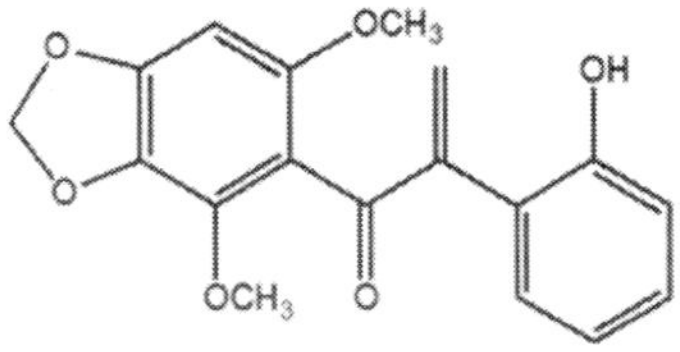

Figure 2. Structure of biliatresone. Note alpha-methylene ketone moiety (free double bond at C ring).

Biliatresone Causes Biliary Epithelial Cell Damage in Mammalian Cell Systems

Biliatresone causes loss of cellular tubulin staining (likely due to destabilization) in neonatal and adult mouse biliary epithelial cells (BECs) in 2D culture, with loss of primary cilia. There is no effect of similar doses on either neonatal or adult hepatocytes.[22] In 3D spheroid culture, biliatresone causes a rapid (beginning at 1 hour), but reversible, loss of BEC polarity, with occlusion of the spheroid lumen, loss of cellular tubulin staining, and loss and mislocalization of apical polarity markers such as ZO-1 and E-cadherin. Spheroid permeability is dramatically increased; whereas rhodamine is secreted into the spheroid lumen and remains there without leakage for greater than 12 hours, preloaded spheroids treated with biliatresone demonstrated a >50% loss of luminal rhodamine in 4.4 ± 0.6 hours.[15]

We also tested biliatresone in neonatal bile duct explant cultures. Isolated ducts from 0–3 day-old BALB/c mice can be cultured in high oxygenation incubators for up to 3 days without evidence of damage. When these explants were treated with biliatresone for as short as 1 hour, they demonstrated a delayed (12 hours later) but reversible BEC monolayer disruption with luminal occlusion, and, remarkably, myofibroblast activation and collagen and fibronectin deposition in the submucosa.[15]

We are currently treating pregnant mice with biliatresone by gavage, with preliminary results showing significant effects of maternal treatment during late gestation on the pups.

MECHANISMS OF BEC DAMAGE

Biliatresone reacts rapidly and spontaneously with glutathione, suggesting this as a potential mechanism of damage. We examined glutathione levels in the extrahepatic bile ducts compared with levels in the liver (which includes intrahepatic bile ducts) and observed that, while there was no difference between levels in neonates versus adults, levels in the extrahepatic ducts were significantly lower than in the liver. This was also observed in zebrafish.[30]

When we treated BECs in culture with biliatresone, we observed a marked drop in levels of reduced glutathione (GSH) within 1 hour, with rebounding to reach or exceed the baseline by 12 hours.[15] This drop in GSH is key to the mechanism of biliatresone toxicity: for cholangiocyte organoids and neonatal bile duct explants, N-acetyl-L-cysteine (L-NAC), which raises intracellular levels of GSH, prevents damage.[15] Although L-NAC both binds biliatresone and increases cellular GSH, the latter is most important to its action since the D-stereoisomer of NAC, which binds biliatresone equally well, does not prevent damage. Conversely, buthionine sulfoximine (BSO), which decreases cellular GSH, mimics the effects of biliatresone in both organoids and explants, protecting against BEC damage and fibrosis. Similar to biliatresone, even 1 hour of exposure to BSO results in damage and, in explants, fibrosis.[15] Thus, biliatresone causes cell damage by decreasing GSH. This is interesting given that decreases in GSH in some cell types destabilize tubulin.[31-33]

In an effort to determine the downstream signaling pathways targeted by biliatresone, we carried out a microarray comparing biliatresone-treated primary neonatal mouse BECs with cells treated with vehicle, betavulgarin, or another biliatresone-like compound with a closed C-ring. The three conditions without biliatresone were remarkably similar, while there were significant differences between these and biliatresone. Further investigation of differences demonstrated that biliatresone caused a significant decrease in the transcription factors Sox17 and Hes1, but increases in Hey2 and the small GTPase RhoU; artificial knockdown of Sox17 or overexpression of Hey2 or RhoU in spheroids mimicked the effects of biliatresone.

Biliatresone treatment of BECs also caused significant increases in their expression of chemokines including fractalkine (CX3CL1), which has been previously linked to inflammation in human BA.

RELEVANCE TO HUMAN BA: THE POWER OF ONE HEALTH

Pregnant women are highly unlikely to ingest biliatresone. Nonetheless, the livestock model and, where applicable, the zebrafish model, mimic human BA in crucial features, including prenatal susceptibility, maternal sparing, primary extrahepatic damage, postnatal progression, and non-genetic etiology. This suggests that biliatresone is a tool that can provide important new information on mechanisms of bile duct injury, repair, and fibrosis at the tissue level. We have already generated data about the mechanism of action of biliatresone, and this has led to important observations about the human bile duct, some of which may have therapeutic implications for affected children. This work, which began with observations by field veterinarians in 1964, illustrates the power of One Health concepts in studying human and veterinary diseases.

References

1. Haber BA, *et al. Gastroenterol Clin North Am.* 2003;32:891.
2. Hartley JL, *et al. Lancet.* 2009;374:1704.
3. Mezina A, *et al. Dig Dis.* 2015;33:408.
4. Moreira RK, *et al. Arch Pathol Lab Med.* 2012;136:746.
5. Rauschenfels S, *et al. Eur J Pediatr.* 2009;168:469.
6. Harpavat S, *et al. Pediatrics.* 2011;128:e1428-33.
7. Zhou K, *et al. PLoS One.* 2012;7:e49270.
8. Wang KS. *Pediatrics.* 2015;136:e1663.
9. Harpavat S, *et al. J Pediatr Gastroenterol Nutr.* 2016;62:799.
10. Mack CL, *et al. Semin Liver Dis.* 2012;32:307.
11. Mack CL. *Cell Mol Gastroenterol Hepatol.* 2015;1:267.
12. Lu BR, *et al. Gastroenterology.* 2010;139:1753.
13. Bessho K, *et al. Annu Rev Med.* 2011;62:171.
14. Lampela H, *et al. J Clin Gastroenterol.* 2014;48:721.
15. Waisbourd-Zinman O, *et al. Hepatology.* 2016;64:880.
16. Rauschenfels S, *et al. Eur J Pediatr.* 2009;168:469.
17. Petersen C, *et al. J Pediatr Surg.* 1997;32:1190.
18. Keyzer-Dekker CM, *et al. J Pediatr Surg.* 2015;50:1304.
19. Karjoo S, *et al. J Pediatr Gastroenterol Nutr.* 2013;57:96.
20. Harper P, *et al. Aust Vet J.* 1990;67:18.
21. Robson S. New South Wales Animal Health Surveillance. 2007;2007/2:2.
22. Lorent K, *et al. Sci Transl Med.* 2015;7:286ra67.
23. Balakrishnan B, *et al. Placenta.* 2010;31:506.
24. Doerge DR, *et al. Reprod Toxicol.* 2001;15:105.
25. Maruti SS, *et al. Nutr Cancer.* 2010;62:208.
26. Nielsen IL, *et al. Nutr Cancer.* 2007;57:1.
27. Prasain JK, *et al. J Nutr.* 2006;136:2975.
28. Soucy NV, *et al. Toxicol Sci.* 2006;90:230.
29. Wahajuddin, *et al. Curr Drug Metab.* 2013;14:369.
30. Zhao X, *et al. Hepatology.* 2016;64:894.
31. Chou IN, *et al. Cell Biol Int Rep.* 1984;8:441.
32. Oliver JM, *et al. J Cell Biol.* 1976;71:921.
33. Urani C, *et al. Toxicol Lett.* 1995;76:135.

acvim360

Update on Pathogenesis of Gallbladder Mucoceles

Jody L. Gookin, DVM, PhD, DACVIM (Small Animal Internal Medicine)
Raleigh, NC, USA

Introduction

Gallbladder mucocele formation in dogs is characterized by an unrelenting accumulation of thick, immobile, and rubbery mucus within the gallbladder until the contents eventually obstruct bile flow or rupture into the abdomen. The syndrome was rarely reported[1,2] prior to 15 years ago and has emerged as one of the most common, deadliest and poorly understood causes of gallbladder disease in the dog. Mucocele formation afflicts older dogs (mean, 9–10 years of age) and has a strong breed predisposition. It is particularly common in Shetland sheepdogs, Cocker Spaniels, Pomeranians, Miniature Schnauzers, and Chihuahuas as well as others.[3-13] Gallbladder mucocele formation is uncommon in mixed breed dogs. There does not appear to be a sex-predilection; however, the impact of gonadectomy remains unclear.

Diagnosis

Clinical Signs

Clinical signs of mucocele formation range from asymptomatic (occult disease), to nonspecific gastrointestinal distress, to progressive cholestasis, or acute bile peritonitis. The more common clinical signs and physical examination abnormalities in dogs with gallbladder mucocele formation include vomiting, anorexia, lethargy, abdominal pain, icterus, and fever.[3,4] These signs are presumably secondary to gallbladder pain, gallbladder rupture, or common bile duct obstruction caused by mucus accumulation. Clinical signs are frequently vague and could easily be attributed to more common differential diagnoses such as pancreatitis or acute or chronic gastroenteritis. This is confounded by the fact that many dogs with mucocele formation have concurrent pancreatitis. Many dogs with mucocele formation are asymptomatic and diagnosed incidentally at the time of abdominal ultrasound for a different disease indication.

Clinical Pathology Abnormalities

The most common complete blood cell count and serum biochemistry profile abnormalities in dogs with mucocele formation include an inflammatory leukogram, and elevations in ALP, ALT, and GGT. Approximately 50% of dogs have an elevated serum bilirubin at the time of diagnosis of mucocele formation.[14] Importantly, many dogs with gallbladder rupture do not have hyperbilirubinemia.[10] Due to the rubbery nature of the gallbladder contents, rupture often exudes solid mucus, rather than liquid bile, into the abdomen. Many dogs will have evidence at surgery of repeated loss of gallbladder wall integrity ameliorated by adhesions to the ruptured regions. Studies have shown that a ruptured mucocele is more likely to be associated with a higher venous lactate,[4] higher liver enzyme activities,[4] and higher white blood cell/neutrophil counts[4,9].

Abdominal Imaging

Abdominal ultrasound is the keystone to diagnosis of mucocele formation. Currently the only alternative ways to diagnose mucocele formation are by means of exploratory laparotomy or at autopsy. The extent to which burgeoning use of abdominal ultrasonography has contributed to increasing diagnosis of gallbladder mucocele formation is unknown. We are likely becoming more vigilant in our ultrasonographic examination of gallbladder contents in dogs; however, this cannot be the only explanation.

The classic ultrasonographic appearance of mucocele formation is the finding of an enlarged gallbladder containing immobile/non-gravity-dependent contents with a finely striated or stellate pattern reminiscent of a slice of kiwi.[6] In reality, the appearance of a mucocele can show considerable variation. Attempts to correlate clinical disease status to sonographic appearance of the gallbladder mucocele have been unsuccessful.[10] Sensitivity of ultrasound for gallbladder rupture is reported to be only 78.6% with a specificity 100%.[10] Ultrasonographic findings considered suggestive of gallbladder rupture include a discontinuous gallbladder wall, pericholecystic hyperechoic fat or fluid[6] or the presence of echogenic peritoneal fluid in the abdomen[9]. Recent descriptions of contrast-enhanced ultrasonography hold promise in increasing the sensitivity of ultrasound for diagnosis of gallbladder necrosis or rupture in dogs but it is currently limited by the availability of the contrast agent.[15] Abdominocentesis may reveal cytological evidence of long, dense, fibrillary strands of bright blue to blue gray material that represents gallbladder mucus.[16]

MicroRNA

Recent studies have characterized the expression of microRNA in 5 dogs with gallbladder mucocele formation. Compared to control dogs, dogs with mucocele formation have a 267-fold increase in serum concentrations of miR-122 (95% CI = 40–1,768). Among dogs categorized as having biliary disease, miR-21 and miR-222 were only increased in dogs with mucoceles.[17] The authors speculate that measurement of miRNA may prove useful as a future diagnostic marker for mucocele formation in dogs.

TREATMENT

Surgical

Cholecystectomy is currently considered the treatment of choice for dogs with symptomatic gallbladder mucocele formation. Many dogs are diagnosed when the disease is advanced and either rupture has occurred or is impending or bile flow has become obstructed by mucus, and surgery to remove the gallbladder is required to save their lives. After surgery, a median of 27% of dogs (range, 7–45%) don't survive long enough to be discharged from the hospital.[3-7,9,12] For dogs that do survive, surgical removal of the gallbladder appears to carry a good long-term prognosis. Reports of rupture or evidence of previous bile leakage are seen in approximately 50% of dogs at the time of surgery.[4,9] There does not appear to be a significant difference in mortality based on presence/absence of gallbladder rupture.[4] Factors associated with non-survival in dogs with mucocele formation include increases in serum lactate concentration (OR 1.3, 95% CI = 1.1–1.6) and packed cell volume,[3] post-operative hypotension (hazard ratio 20.4, 95% CI: 2.7–154.4),[3] higher total WBC and neutrophil counts,[7,12] and lower serum concentrations of potassium[7]. Laparoscopic cholecystectomy has been described in 6 dogs with mucocele formation.[18] Where offered, laparoscopy is generally limited to uncomplicated/unobstructed dogs without evidence of mucus extension into the cystic, common hepatic, or common bile ducts.

Medical

Medical management of gallbladder mucocele formation is somewhat controversial. It is unclear to what extent the process can be ameliorated or reversed. Total or partial resolution of mucocele formation has been reported in only a handful of dogs.[7,10,19] Treatments administered to these dogs have consisted of various combinations of fat-restricted diet, ursodeoxycholic acid, S-adenosylmethionine, silymarin, omega-3 fatty acids, and antibiotics. In 2 dogs, a concurrent diagnosis of hypothyroidism was treated with levothyroxine.[19]

Histopathology

A histologic diagnosis of mucocele formation is based on the presence of a large viscous accumulation of mucus that fills and distends the gallbladder lumen. The gallbladder mucosa is characterized by formation of long, thin and branching fronds of well-differentiated gallbladder epithelial cells containing modestly distended clear cytoplasm filled with mucus and supported by a scant amount of submucosa that extends into the mucus. Inflammatory infiltrates are uncommon except in cases of overt bacterial infection or gallbladder necrosis or rupture. A common finding in dogs with mucocele formation is the presence of gallbladder arterial thrombosis.[14] Gallbladder infarction in dogs with mucocele formation is reminiscent of a case series of dogs in which thrombi and atheromatous vascular changes were accompanied by transmural coagulative necrosis and suggests the possibility of a shared mechanism between the two conditions.[20]

The earliest reported description of what is believed to be an ancestor of the modern day mucocele was published in 1965 as "cystic mucinous hypertrophy of the mucosa of the gallbladder in the dog."[2]

Despite many descriptions in people of proliferative changes in the gallbladder epithelium,[21] there exist no diseases of the gallbladder in people that closely mirror the histological appearance of gallbladder mucocele formation in dogs. The only descriptions of gallbladder mucosa that are similar to gallbladder mucocele formation are in animals experimentally treated with progestins[22-24] or lacking functional cystic fibrosis transmembrane regulatory (CFTR) protein expression[25,26].

Recent studies have identified that mucocele formation in dogs involves excess secretion by the entire gallbladder epithelium of gel-forming mucins with abnormal properties and characterized by a disproportionally significant increase in Muc5ac relative to Muc5b, defective mucin un-packaging, and the presence of a unique composition of mucin-interacting proteins that are capable of dramatically altering the physical and functional properties of mucus. These findings are sufficient to explain the thick and rubbery properties of the mucus; however, the inciting cause for the mucus hypersecretion remains unknown.[14]

Concurrent Diseases

Biliary Infection

Positive results of aerobic and/or anaerobic culture of gallbladder contents in dogs with mucocele formation range from 2.7%[3] to 67%[6] (median, 12.5%; mean, 20%)[4,5,7,9,11,12,18]. Identities of the reported bacteria include alpha and beta hemolytic *Streptococcus* spp., *Streptococcus pyogenes*, other non-hemolytic *Streptococcus* sp., *Staphylococcus* sp., *Staphylococcus intermedius*, *Enterococcus* sp., *Micrococcus* sp., *Propionibacterium* sp., *Pseudomonas* sp., *Escherichia coli*, *Proteus mirabilis*, and *Enterobacter* sp. The role of bacterial infection in pathogenesis of mucocele formation is unclear.

Endocrinopathy

There appears to be a significant association between gallbladder mucocele formation and diagnosis of hypothyroidism[14,27] or hyperadrenocorticism[4,14,27] in dogs. Whether these endocrinopathies are a contributing cause or consequence of mucocele formation is unclear. It is important to consider that these endocrinopathies have existed long before emergence of the gallbladder mucocele as a clinical entity and are not found in every dog diagnosed with the disease. Accordingly, these are unlikely to be a direct cause, but rather an exacerbating factor to disease pathogenesis.

Dogs with mucocele formation are 3 times as likely to be diagnosed with hypothyroidism compared to dogs without mucocele formation (95% CI = 1.2–7.7).[27] Dogs with mucocele formation are 29 times as likely to be diagnosed with hyperadrenocorticism compared to dogs without mucocele formation (95% CI = 3.8–220).[27] Dogs with mucocele formation and pituitary-dependent hyperadrenocorticism (PDH) have higher post-ACTH serum cortisol levels than dogs with normal gallbladders, PDH alone, or other cholestatic disease and PDH.[28] Dogs with mucocele formation and PDH required higher doses of trilostane.[28] Other parallels have been made between mucocele formation and exposure to exogenous steroids, atypical adrenal hyperactivity, diabetes mellitus, and pancreatitis.[3,4,7,8]

Hyperlipidemia

Hypertriglyceridemia (OR 3.55, 95% CI = 1.12–15.91)[8] and hypercholesterolemia (OR 2.92, 95% CI = 1.02–8.36)[8] are significantly associated with gallbladder mucocele formation in dogs. These elevations frequently occur in dogs despite concurrent inappetence[7] and are commonly seen in dogs having no serum biochemical abnormalities indicative of cholestasis. There is also no significant correlation between cholesterol and triglyceride levels in dogs with mucocele formation,[8] suggesting that they may not share the same pathogenesis. Several breeds of dog that are predisposed to mucocele formation are also genetically predisposed to either hypercholesterolemia (Shetland sheepdog)[29] or hypertriglyceridemia (Miniature Schnauzer).

Proteinuria

Several studies have noted the presence of proteinuria in dogs with gallbladder mucocele formation. Whether there is a clear association between the two conditions is unknown. Concurrent diseases in dogs with mucocele formation can be associated with proteinuria including hyperadrenocorticism, pancreatitis, and hyperlipidemia.[3,4,7,8] A recent study of proteinuric Miniature Schnauzers demonstrated that hypertriglyceridemic dogs had ultrastructural evidence of osmophilic globules consistent with lipid in glomerular tufts. Some dogs had lipid thromboemboli in glomerular capillary loops.[30] This observation is of interest, because Miniature Schnauzers are predisposed to both hyperlipidemia and gallbladder mucocele formation. Dogs with mucocele formation frequently have concurrent gallbladder infarction and gallbladder infarction in dogs has been described as associated with atherosclerotic lesions.[20] These findings suggest that gallbladder infarction in dogs with mucocele formation may be related to hyperlipidemia.

CURRENT THEORIES ON GALLBLADDER MUCOCELE FORMATION

Change in Gallbladder Composition of Bile Acids

Hydrophobic bile acids stimulate mucin secretion by dog gallbladder epithelial cells *in vitro*[31] and long exposures of the gallbladder epithelium to hydrophobic bile acids *in vivo* have been speculated to be a contributing factor to gallbladder mucocele formation in dogs[32]. In an effort to link hyperadrenocorticism with a change in bile acid composition as a mechanism for promoting mucocele formation, studies were undertaken to treat dogs with 8.5 mg/kg of hydrocortisone orally, twice a day for 84 days, and evaluate the resulting change in gallbladder bile acid composition. This treatment resulted in an increase in total unconjugated bile acids and a decrease in concentration of total taurine-conjugated bile acids in bile.[32] Recent studies, however, have demonstrated that the bile acid composition of bile from the gallbladder contents of dogs with mucocele formation actually has lower total bile acid concentrations and lower amounts of secondary/hydrophobic bile acids compared to normal bile.[33]

Poor Gallbladder Motility and the Role of "Sludge"

Several authors have noted an association between the presence of a forming gallbladder mucocele and an apparent decrease in gallbladder emptying. Using ultrasonography as an index of gallbladder emptying, minimal to no post-prandial contraction of the gallbladder was described in 3 dogs with non-gravitationally dependent gallbladder debris.[7] A second study described the presence of decreased gallbladder ejection fraction in dogs with mobile sludge, immobile sludge, or mucocele compared to normal dogs.[34] It is important to recognize that poor gallbladder motility as a cause for gallbladder mucocele formation is difficult to prove once the gallbladder is filled with mucus. Moreover, gallbladder paresis does not result in mucocele formation in people.

Recent studies indirectly support the idea that gallbladder motility may be impaired in dogs with mucocele formation. Serum concentrations of leptin are higher in dogs with mucocele compared to healthy dogs and their gallbladder tissue has increased mRNA expression of leptin and leptin receptor in relation to normal dog gallbladders.[35] Leptin plays an important role in cholecystokinin-mediated gallbladder contraction. Therefore the authors speculate that increased secretion of leptin may represent a compensatory response to poor gallbladder contractility.

There is considerable interest and speculation on whether or not the presence of gallbladder "sludge" is indicative of poor gallbladder motility or a precursor to mucocele formation. Gallbladder sludge is generally considered to be an incidental finding in dogs and is more commonly found in older dogs.[36,37] Dogs with sludge tend toward larger gallbladder dimensions than dogs without sludge,[36,37] which may reflect greater gallbladder flaccidity. Other studies have observed no association between gallbladder volume and the presence of sludge.[38] Dogs with hyperadrenocorticism or hypothyroidism are more likely to have more severe sludge,[37] which may indirectly suggests a mechanism by which these endocrinopathies could promote mucocele formation. Importantly however, treatment of dogs with exogenous steroids did not increase the incidence of ultrasonographically detectable gallbladder sludge nor to an increase in bile constituents that are commonly implicated in sludge formation in humans.[39] Efforts to follow the spontaneous appearance of biliary sludge in 42 dogs over the course of a year did not disclose evidence for progressive worsening of biliary sludge or an association with mucocele formation.[38]

Genetics

A strong breed predisposition to gallbladder mucocele formation suggests a significant contribution of genetics. Efforts to link a genetic defect in ABCB4,[40] a hepatocyte canalicular membrane phosphatidylcholine floppase, in the Shetland sheepdog was initially promising but later disproven[41]. Whether or not the mutation could be a contributing factor to mucocele formation cannot be entirely ruled out. Genetics are unlikely to be a direct cause of mucocele formation in dogs, because the breeds associated with gallbladder mucocele formation have existed long before emergence of the disease.

To date, gallbladder mucocele formation has only been reported in a single cat. The cat was a 3-year-old domestic shorthair with concurrent hepatic lipidosis and was treated by means of a cholecystojejunostomy. While the description of the gallbladder content was consistent with that of mucocele-type mucus, the gallbladder histology was quite different from that described in dogs and consisted of papillary and adenomatous hyperplasia that greatly thickened the gallbladder mucosa.[42]

Environmental Exposures

A matched case-control study investigating the association between histologic diagnosis of mucocele formation and reported use of drugs for flea and tick infestation, heartworm prophylaxis, or degenerative joint disease found that Shetland sheepdogs with mucocele formation were 9 times as likely to have had reported use of imidacloprid as were control Shetland sheepdogs (95% CI = 1.103–78.239). This association was not observed in non-Shetland sheepdog breeds.[13] The results of this study do not suggest that imidacloprid is a primary cause of GBM formation in dogs, but possibly a contributing or exacerbating factor in Shetland sheepdogs.

REFERENCES

1. Newell SM, *et al. J Am Anim Hosp Assoc*. 1995;31:467–472.
2. Kovatch RM, *et al. Pathologia veterinaria*. 1965;2:574–584.
3. Malek S, *et al. Vet Surg*. 2013;42:418–426.
4. Pike FS, *et al. J Am Vet Med Assoc*. 2004;224:1615–1622.
5. Worley DR, *et al. J Am Vet Med Assoc*. 2004;225:1418–1422.
6. Besso JG, *et al. Vet Radiol Ultrasound*. 2000;41:261–271.
7. Aguirre AL, *et al. J Am Vet Med Assoc*. 2007;231:79–88.
8. Kutsunai M, *et al. Vet J*. 2014;199:76–79.
9. Crews LJ, *et al. J Am Vet Med Assoc*. 2009;234:359–366.
10. Guess SC, *et al. J Am Vet Med Assoc*. 2015;247:1412–1414.
11. Mizutani S, *et al. J Vet Med Sci*. 2016.
12. Uno T, *et al. J Vet Med Sci*. 2009;71:1295–1300.
13. Gookin JL, *et al. J Vet Intern Med*. 2015;29:1464–1472.
14. Kesimer M, *et al. PLoS One*. 2015;10:e0138988.
15. Bargellini P, *et al. Vet Radiol Ultrasound*. 2016;57:611–620.
16. Escobar MC, Neel JA. *J Am Vet Med Assoc*. 2011;239:65–67.
17. Dirksen K, *et al. J Vet Intern Med*. 2016;30:1816–1823.
18. Mayhew PD, *et al. Vet Surg*. 2008;37:625–630.
19. Walter R, *et al. J Am Vet Med Assoc*. 2008;232:1688–1693.
20. Holt DE, *et al. Vet Pathol*. 2004;41:416–418.
21. Scott AJ. *Gut*. 1978;19:558–562.
22. Selman PJ, *et al. Vet Q*. 1995;17:128–133.
23. Mawdesley-Thomas LE, Noel PR. *Vet Rec*. 1967;80:658–659.
24. Geil RG, Lamar JK. *J Toxicol Environ Health*. 1977;3:179–193.
25. Meyerholz DK, *et al. Am J Pathol*. 2010;176:1377–1389.
26. Sun X, *et al. Am J Pathol*. 2014;184:1309–1322.
27. Mesich ML, *et al. J Small Anim Pract*. 2009;50:630–635.
28. Kim KH, *et al. J Vet Intern Med*. 2017 Jan 8. doi: 10.1111/jvim.14608.
29. Sato K, *et al. J Vet Med Sci*. 2000;62:1297–1301.
30. Furrow E, *et al. Vet Pathol*. 2016:300985816681412.
31. Klinkspoor JH, *et al. Gastroenterology*. 1995;109:264–274.
32. Kook PH, *et al. Am J Vet Res*. 2011;72:1607–1612.
33. Kakimoto T, *et al. Am J Vet Res*. 2017;78:223–229.
34. Tsukagoshi T, *et al. Vet Radiol Ultrasound*. 2012;53:84–91.
35. Lee S, *et al. J Vet Intern Med*. 2017;31:36–42.
36. Bromel C, *et al. Vet Radiol Ultrasound*. 1998;39:206–210.

37. Cook AK, *et al. J Am Anim Hosp Assoc.* 2016;52:125–131.
38. DeMonaco SM, *et al. J Vet Intern Med.* 2016;30:771–778.
39. Kook PH, *et al. Vet J.* 2012;191:225–230.
40. Mealey KL, *et al. Comp Hepatol.* 2010;9:6.
41. Cullen JM, *et al. J Vet Diagn Invest.* 2014;26:434–436.
42. Bennett SL, *et al. Aust Vet J.* 2007;85:397–400.

acvim306

One Medicine in Action: Climate Changes, Vector Biology and Arthropod-Borne Viruses

Sanjay Kapil, DVM, PhD, DACVM
Stillwater, OK, USA

Per European Environmental Agency, global surface temperature has increased 0.74°C in the 20th century. Further, increased air travel and spread of mosquito vectors have propelled the migration of vector-borne viruses (such as ZIKV). Due to increase in ambient temperature, glaciers are melting in Iceland leading to changes in geographic spread of some of the marine animals that were restricted to smaller area. Due to the increase in air travel by humans and concomitant expansion of the geographic range of arthropod-borne vectors, such as mosquitoes and ticks, the arthropod-borne viruses have crossed into continents with naïve human populations that did not have background immunity to these exotic viruses that were endemic in some parts of the world.

Zika virus (ZIKV) offers an excellent example of newly emerged and recently introduced zoonotic virus in the USA that is spread by *Aedes* mosquitoes.

For 60 years, ZIKV was restricted to Africa and no human clinical cases were reported. In 2007, ZIKV spread to Yap islands in Pacific and then to Samoa, Puerto Rico, US Virgin Islands. In 2016, ZIKV has caused human cases in Florida and Texas, USA. Several of these cases were imported to continental USA by travel to ZIKV endemic areas in Caribbean and South America. In 2015, Brazil reported widespread outbreak of ZIKV. ZIKV was detected in fetal brain and amniotic fluids. Factors such as large population mobility, availability of transmitting vector, favorable climatic conditions, lead to about 1.3 million human cases of ZIKV in Brazil. Based on molecular epidemiology, there was a single introduction of ZIKV in Brazil. The virus was phylogenically interspersed with those from South American and Caribbean countries.

ZIKV has been experimentally reproduced in monkeys and mice. Mice develop neuronal degeneration and cellular infiltrate. Monkeys develop transient pyrexia. In humans, ZIKV can be asymptomatic to self-limiting acute febrile illness (influenza virus-like syndrome). Clinical symptoms include headache, rash, myalgia, arthralgia of hands and feet joints, and conjunctivitis. ZIKV is transmitted by *Aedes aegypti* and *A. albopictus*. ZIKV can pass vertically from mother to fetus. Only people with symptoms need ZIKV testing and pregnant women that are increased risk of ZIKV infection of developing fetus. Risk factors include travel to ZIKV areas. Health department should be contacted for collection of blood, saliva and urine samples. Testing can be done at health department or CDC in the USA.

Reference

1. Faria *et al*. 2016. *Science*. 2016;352:245–249.

acvim307

Climate Change and One Health: What About Tuberculosis in Populations?

D.F. Lawler, DVM; B.P. Tangredi, DVM; R.H. Evans, DVM, MS
Laguna Beach, CA, USA

INTRODUCTION

Tuberculosis (*Mycobacterium tuberculosis*, MTB) remains one of the top ten causes of human mortality.[1] The World Health Organization noted that during 2015 alone, 10.4 million new MTB cases were recorded world-wide; 1.8 million recorded human deaths reflected poorer regions of the world.[1] These data are consistent with previous reports estimating that around one-third of all humans harbor causative organisms, with an average 9 million new reported cases and 1.5 million deaths each year.[2] However, the **total** number of human deaths world-wide, from complicated and uncomplicated MTB, would be quite difficult to estimate more precisely because of varying accessibility to health care and inconsistent record-keeping.

Many world-wide domestic and wildlife species also are infected with tuberculosis-causing mycobacteria, with host and microorganism tropisms that are recognized for some, but not all, susceptible species. Thus, the actual prevalence of animal tuberculosis world-wide is unknown. Considering both human and animal aspects of tuberculosis in a One Health context, in a world of climate change, is a pathway to more comprehensive understanding.

Mathematical modeling is a well-established approach to evaluating and predicting complex biomedical problems. However, too often, the mathematical approaches themselves are valid but they are compromised because the parameters on which they operate are incomplete and do not reflect the relevant population biology. Frequent causes for this problem include inadequate knowledge of applicable scientific literature, poorly done epidemiology, and failure to understand the biological ranges for quantifiable parameters. An incompletely informed model, especially a predictive model, is less likely to produce biologically correct and applicable outcomes.

PREVIEW TO TUBERCULOSIS MODELS

Numerous mathematical modeling studies of MTB have been published, along with a very large volume of related genomic, microbiological, biological, clinical, and population literature. Almost certainly, climate-related changes in tuberculosis demographics will necessitate revisiting some modeling studies and creating some new ones, because many human cultures and animal populations likely will be altered by forced migrations and remixing in different environmental circumstances. This level of predictive modeling demands broadly interdisciplinary biomedicine. Thus, a One Health approach is critical.

In advance of new modeling studies, special preparatory consideration needs to be given to the multilevel complexity of tuberculosis. Several studies of bovine tuberculosis (*Mycobacterium bovis*, BTB) in Ethiopia (east-central Africa) provide an instructive example, with respect to influential factors in naturally-occurring circumstances.

A 2002 report from southern Ethiopia indicated PCR-recognized presence of MTB (83%) and BTB (17%) infection in humans, occasionally occurring simultaneously; the latter was more frequent in HIV-positive patients.[3]

Ethiopia presently has the largest cattle population in Africa,[4] historically primarily Zebu (*Bos indicus*), reared and maintained pastorally. However, recent increased consumer desire for milk consumption has resulted in more intensive dairying near population centers, often with other cattle breeds or breed crosses.[5] A study of dairy farms within 50 km of Addis Ababa (central Ethiopia) revealed that 30% of nearly 3000 tested cattle were positive for BTB, with about 50% of herds affected at some level.[5] The investigators used the comparative intradermal tuberculin test (CIDT), necropsy findings from infected cattle, bacteriological culture, and molecular typing, to establish a thorough database. From the data, the authors recommended new disease control measures on farms and in dairies, as well as new policies for local and broader BTB control.[5]

The high endemic prevalence of *M. bovis* (1–10%) in Ethiopia's 51 million cattle[6] raises questions about organism interactions in human disease. In a study of 964 tuberculosis-positive humans, culture isolates and genotyping from suspect TB patients revealed only 0.4% *M. bovis* prevalence. Since the patients presented to health centers, a less rural demography was implied.[6] Recent demography suggests increased tuberculosis risk to humans and cattle, secondary to increasingly intensive urban-proximate milk production that frequently includes imported cattle.

Further, various observations suggest significant human-directed cattle movement in Ethiopia today, possibly increasing future dissemination risk for MTN and BTB. Given the observations from these studies, what is in play for a climate-related predictive modeling effort? There exist many moving parts. The following parameters list can be derived from the studies described above. Many of these parameters can be quantified or estimated from existing literature. Thus, they can be included in models, as appropriate to biomedical and ecological circumstances.

From the studies reviewed here, the following parameters were recognized.

- Model BTB and MTB together?
- Largest cattle population in Africa.
- Most cattle have been rural; increasing intensive rearing nearer population centers.
- BTB geographic prevalence.
- Method of screening cattle for BTB and effectiveness.
- Method of molecular BTB typing and effectiveness.
- BTB and MTB type diversity.
- Breed differences in BTB susceptibility.
- Current (1–10%) and future endemic BTB.
- BTB varies 0.4–17% depending on geography.
- Patterns and pathways of cattle movements.
- BTB effect on milk or meat yield and carcass value.
- *M. bovis*, *M. avium*, and differential bovid population effects.
- Geographic regions and varying cattle population density.
- Reasons for cattle sale and associated numbers, and their effects.
- On-farm consumption of animal products.
- Educational programs in-industry and on-farm (costs, effects) .
- Post-mortem data and slaughter MTB data.
- Herd, region, country certification programs and effects (and costs).
- Age susceptibility of cattle to BTB.
- Cattle access to health care.
- Varying human-cattle direct and indirect contact.
- On-farm sanitation; dairy sanitation.
- Endemic MTB; 261 cases/100,000 population in Ethiopia.
- Long pre-clinical time period; long disease course.
- Increasing human demand for milk
- Pasteurized vs. unpasteurized milk consumption by humans.
- Human migrations because of climate or geopolitical conflict.
- Clean food and water resource availability.
- Human sensitization to environmental mycobacteria.
- Human and cattle morbidity and mortality, from increasing antibiotic resistance.
- Humans access to health care.
- Health care cost impacts on human and cattle numbers.

Depending on geography and human culture, metrics of human co-morbid disease effects that also need to be considered should include HIV, silicosis, diabetes mellitus, drug abuse, immunosuppressive drugs, gastrectomy, jejunoileal bypass, malnutrition, severe weight loss, and post-transplant health.[7] Other reported co-morbid diseases include smoking, alcoholism, chronic lung diseases such as COPD.[8]

Should BTB and MTB be included in the same model structures? In tuberculosis models that are intended for application in co-endemic regions, especially where climate change and disease x environment effects are expected to exert yet-unknown influences, more complete model parameterization and greater flexibility likely are important. Certain closely related parameters may need to be synthesized to a single numerical mean and range to reduce the parameter load in the model.

This discussion can serve as an example of the levels of complexity that need to be sought out and evaluated for inclusion in new tuberculosis modeling studies. Veterinarians who are experienced with animal populations are a valuable resource for identifying, acquiring, and quantifying needed information. New tuberculosis models should be done with recognition of the need to inform their structure with a range of location-specific biological, biomedical, and ecological data. Comprehensive integration of influential organism, environmental, and biological factors represents a pathway to more dependable predictive models that involve effects of climate change on animal and human disease. Thus, the need for interdisciplinary participation is obvious.

Note: The oral presentation also will contain a short section discussing a simple way to conceptualize the mathematics of biomedical modeling.

References

1. World Health Organization Global Tuberculosis Report. 2016.
2. American Lung Association. *Trends in Tuberculosis Morbidity and Mortality*. 2013.
3. Kidane B, *et al. J Clin Microbiol*. 2002;40(11):4230.
4. Vordermeier M, *et al. Comp Immunol Microbiol Infect Dis*. 2012;35:227.
5. Firdessa R, *et al. PLoS One*. 2012;7(12):e52851.
6. Firdessa R, *et al. Emerg Inf Dis*. 2013;19(3):460.
7. Raviglione, *et al. Harrison's Principles of Internal Medicine*. 2005:953–966.
8. Marais, *et al. Lancet Infect Dis*. 2013;13:446–448.

1.
acvim308

Real-Time One Medicine Experience: Emerging Novel Viruses

Sanjay Kapil, DVM, MS, PhD, DACVM (Virology and Immunology)
Stillwater, OK, USA

Emerging novel viruses can be defined as viruses that were never reported before, because they emerged recently; became obvious to scientists and health professionals by causing recent outbreaks of diseases in humans and animals; classified as genetically novel by modern molecular methods (PCR and sequencing); and by causing more severe disease or different symptoms in fully vaccinated animals and humans. Recent developments in metagenomics and sequencing technologies have generated large amounts of sequence data on unknown viruses of animals and humans. Developments in sequence analysis and assignments to known virus families based on sequence homology to known ORFs have contributed to identification of new members of virus families from humans, animals, and wildlife. By shotgun sequencing and improvements in sequencing technology, novel ORFs have been identified.

However, there are bottlenecks in our understanding of these novel viruses, because many of them can also be detected in animals and humans with no clinical diseases or symptoms. Further, shotgun sequencing does not discriminate between source of the sequence that could originate from environmental sources, including food and feed elements, insects, and plants that may have been ingested by the humans and animals.

Information generated by studying novel viruses (such as influenza A) could provide better tools for prevention and control of emerging viruses. Influenza A viruses evolved by exchange of RNA segments or recombination between human and animal influenza A viruses in parts of the world where birds, pigs, and humans have closer contact. The talk covers improvements that occurred in the last 100 years in the influenza A research.[1]

Despite tremendous progress in sequencing methods and diagnostic methods, some important limitations remain. These limitations include inability to propagate novel viruses; inability to assign all the sequences to their viral origin, especially contaminating sequences; and our inability to understand the clinical significance of these sequences generated by high-throughput sequencing methods. These limitations hamper our ability to correctly diagnose and control novel viruses (such as development of vaccines) that may be part of the clinical syndromes (including cancers).

References

1. Saunders-Hastings, Kweski. *Pathogens*. 2016.

1.
acvim309

Pathobiome of Blood and Complex Disease Expression

Edward B. Breitschwerdt, DVM, DACVIM
Raleigh, NC, USA

Introduction

For over two decades, our research group has contributed to the development of diagnostic, therapeutic, and preventive strategies for the management of infections caused by vector-transmitted intracellular organisms, which often induce persistent intravascular infections in reservoir and at times non-reservoir-adapted hosts. As a result of these research efforts, and those of many other investigators throughout the world, clinician scientists have collectively begun to gain a unique perspective on the immunopathological consequences of chronic intravascular infections. More recently, we have used a novel bacterial isolation approach to culture *Bartonella* spp. and other fastidious bacteria from the blood of sick dogs and human patients, suggesting that concurrent infections with vector-borne and potentially non-vector-borne organisms occur concurrently.

In recent years, researchers have focused their efforts on the characterization of the "normal" microbiome of various tissue sites within the body, including documentation of the presence of bacterial DNA in blood of healthy individuals, DNA incorporated into various types of tissue or metallic implants, and the presence of biofilms within blood vessels and other anatomical sites. Characterization of "persister" bacteria is a current research focus that potentially explains antibiotic treatment failures and recrudescent infectious disease processes. On an evolutionary basis, there has been extensive and variable microbiological adaptation within healthy individuals, resulting in complex microbiomes that are critical for the overall wellbeing of the individual. In addition to other causes, a pathobiome of blood develops when arthropod vectors transmit bacteria, protozoa, and viruses that alter the "normal" and as-yet incompletely characterized microbiome of blood.

Bacterial Adaptation and Complex Disease Expression

It is important for the clinician to remember that vector-transmitted bacteria and protozoa have a very long evolutionary history, which spans millions of years. Many vector-borne organisms, such as *Anaplasma, Babesia, Bartonella, Ehrlichia, Leishmania*, and *Rickettsia* species have both an intracellular and intravascular lifestyle. Certain organisms are obligate intracellular pathogens.[1] Obviously this means that the organism must reside within the cell of an arthropod (such as a louse, flea, or tick) or within the cell of a mammalian host. As one example, *Rickettsia* has lost several critically important genes that are required for extracellular survival.[2] In the absence of these genes, *Rickettsia* species have developed mechanisms that allow intracellular survival by manipulating selected biochemical pathways within the mammalian cell. For the clinician, it is critical to recognize that evolutionary adaptation has facilitated complex interactions between arthropod vectors, bacteria and protozoa and the blood microbiome of mammalian reservoir hosts.

In some instances, disease expression occurs when a vector-borne bacteria is transmitted from a reservoir-adapted host such as a rodent to a non-reservoir host, such as dog or man.[3,4] *Anaplasma phagocytophilum, Borrelia burgdorferi*, and *Bartonella washoensis* are examples of organisms for which small mammals and rodents are seemingly healthy reservoirs, whereas transmission of these same organisms to dogs or humans can induce disease.[3,4] In the case of *A. phagocytophilum*, cats, dogs, horses, and human beings can all be infected following transmission by *Ixodes scapularis* or *Ixodes pacificus* in the northeastern, northcentral, and western United States.[5,6] Numerous wild and domestic animals also serve as seemingly healthy reservoirs for specific host-adapted intravascular *Bartonella* species, supporting the evolutionary adaptation of these bacteria as a component of the blood microbiome. Transmission to cats, dogs, or humans can occur when a reservoir-adapted *Bartonella* species is introduced by a scratch, bite, needle stick, blood transfusion, or arthropod vector into a non-reservoir host species.[4,7] *Ehrlichia* and *Anaplasma* spp. inhibit apoptosis of monocytes and neutrophils respectively, so as to selectively prolong the natural lifespan of organism-infected cells.[8,9] Obviously, this ability to influence the rate of apoptosis would infer substantial benefit to a group of organisms that have the ability to persist within the host's immune cells for a prolonged period of time. Although as clinicians we perceive these organisms as pathogens that induce disease, bacteria and protozoa are merely attempting to protect and perpetuate their species through highly sophisticated adaptive mechanisms and intravascular persistence. As another example of these evolutionary relationships, mammalian hosts serve as blood reservoirs for future generations of *Ehrlichia* spp.; thus it is not in the best interest of vector-transmitted organisms to induce immune recognition and disease (ehrlichiosis) in the host animal. In this context, disease develops when the organism, the host, or perhaps the veterinarian (by providing an antigenic challenge via vaccination or suppressing the immune system by administering corticosteroids) induces a state of immunological imbalance in the cat or dog that is persistently infected with one or more bloodborne organisms.

IMMUNE RECOGNITION AND DISEASE EXPRESSION

The development of disease in an animal infected with an obligate intracellular organism would seem to represent a strategic error on the part of the organism or the host. Excessive immune recognition of an organism that can persist within erythrocytes (*B. canis* or *B. gibsoni*), the vascular endothelium (i.e., *Bartonella henselae*), or within circulating immune cells (i.e., *E. canis* in monocytes or *Anaplasma phagocytophilum* in granulocytes) would result in cellular damage, induction of autoantibodies, excessive generation of immune complexes, or elimination of the intravascular niche for the organism's perpetuation. In the context of infectious diseases, it has been suggested that circulating macrophages represent the "Trojan horse" for a variety of intracellular pathogens.[10,11] Once infected, monocytes disseminate organisms throughout the body. Differences in immune reactivity to various organisms also influence disease expression and severity.[12] Regardless of the factors that induce alterations in immune recognition, the consequences in a chronically infected animal could include immune-mediated hemolytic anemia (babesiosis, bartonellosis), immune-mediated thrombocytopenia (anaplasmosis, babesiosis, bartonellosis, ehrlichiosis), vasoproliferative lesions such as peliosis hepatis secondary to *B. henselae* or glomerulonephritis secondary to immune complex deposition (borreliosis [Lyme nephropathy], ehrlichiosis, leishmaniasis).[13-17]

BIOFILMS, PERSISTERS, AND DISEASE RECRUDESCENCE

Biofilms are most often comprised of polymicrobial communities and are commonly associated with chronic infections. By including multiple bacterial and/or fungal species in a single community, biofilms obtain numerous advantages, such as passive resistance, metabolic cooperation, byproduct influence, quorum sensing systems, an enlarged gene pool with more efficient DNA sharing, and many other synergies, which give them a competitive advantage, particularly in adverse environments.[18] Routine clinical cultures are ill-suited for evaluating biofilms that consist of polymicrobial infections composed of a mixture of viable, dormant, and "persister" or "viable but non-culturable" (VBNC) bacteria.[19,20] DNA methods utilizing PCR, PCR/mass spectroscopy, and sequencing have demonstrated their ability to identify microorganisms and quantitate their contribution to biofilms in clinical infections. Many bacteria, including a variety of important animal and human pathogens, are known to respond to various environmental stresses by entry into a novel physiological state, where the cells remain viable but are no longer culturable on standard laboratory media. These bacteria have been referred to as dormant, "persisters," or "viable but non-culturable" (VBNC).[18-20] Upon resuscitation from this VBNC state, bacterial cells regain culturability and the renewed ability to cause infection, inflammation, and recrudescent disease. It is likely that persisters or the VBNC state represents a bacterial survival strategy that allows a small subset of bacteria to survive, despite immunological or therapeutic elimination of their viable counterparts.

An important example of biofilm and chronic infection relates to the putative role of *Chlamydia pneumoniae* in the development of human atherosclerosis. *Chlamydia pneumoniae* was first described as a human pulmonary pathogen in 1987. During subsequent decades, numerous studies have attempted to determine if there is a causative relationship between infection with *C. pneumoniae* and atherosclerosis. Although *C. pneumoniae* has been detected in atherosclerotic lesions by electron microscopy, immunohistochemistry, *in vitro* cultivation, as well as PCR or *in situ* hybridization, the controversy regarding causation continues unabated. It is possible that a complex pathobiome consisting of biofilm formation, persister bacteria, and chronic inflammation contributes to the ultimate development of atherosclerosis; thus, previous studies that focused on a single etiology, such as *C. pneumoniae*, generated inconsistent and conflicting results.

Pathobiome of Blood and Complex Disease Expression

In most instances, the influence of concurrent or sequential intravascular infection with multiple vector-borne organisms on the host immune response is unknown. Experimental infection of rodents with *B. burgdorferi* and *A. phagocytophilum* induces more severe immunopathology than infection with either organism alone,[21,22] and a similar association appears to exist in dogs co-infected with *B. burgdorferi* and *A. phagocytophilum*. Under natural exposure conditions, polymicrobial infection may be much more complex than an experimental infection in which two organisms are inoculated at a specific dose, to a specific inbreed strain of mouse and administered at a single point in time. For example, while attempting to clarify an increase in unexplained deaths in a Walker hound kennel, we were able to amplify DNA of up to six different organisms (i.e., 6 different species from 4 different genera) from a single EDTA-anticoagulated blood sample, obtained at a single point in time.[23] PCR data from several dogs in this kennel identified complex patterns of co-infection with multiple bloodborne pathogens. The clinical implications of this study suggested that large-breed hunting dogs with extensive tick, flea, and louse exposure can be simultaneously infected (based upon detection of DNA) with a complex pathobiome. Secondly, working dogs appear to be "genetically" capable of compensating to a substantial degree for simultaneous bloodborne infections with bacteria, protozoa, rickettsiae, and potentially other organisms including viruses that were not specifically targeted. However, over half of the dogs in the kennel had ocular abnormalities including anterior uveitis, ocular hemorrhage, or retinitis, generally accompanied by thrombocytopenia, mild anemia, and hyperglobulinemia.[23] Finally, age, sex, breed, nutritional status, and variation in the pathobiome could influence the pattern of disease expression for each dog in this kennel. Presumably, the pathobiome of blood complicates the clinical evaluation of individual cats and dogs presented to veterinarians on a daily basis.[15,24-26] One possible conclusion, derived from recent research, is that the mammalian body is an ecosystem comprised of numerous microbiomes, which evolved in conjunction with a highly developed immune system that interacts constantly with microorganisms on the skin, mucosal surfaces, and within blood and other tissues. Disease occurs when the microbiome evolves into a pathobiome.[27]

BAPGM Enrichment Blood Culture and the Pathobiome of Blood

Based upon phylogenetic analysis of DNA sequences deposited in GenBank, these are estimated to be >36,000 unique bacterial species inhabiting earth, but the majority of these organisms have never been successfully isolated using contemporary bacteriologic methods.[28,29] Clinical and microbiological utilization of molecular-based diagnostic tests such as PCR, in conjunction with conventional microbiological approaches, support a much more complex role for normal bacterial flora, as well as fastidious, intracellular, or uncultivable bacteria in the spectrum of animal and human diseases.[27-29] In 2007, Belen Cadenas from our research group used 16S rDNA PCR and DNA sequencing to identify non-*Bartonella* BAPGM subculture isolates obtained from human blood, fluid, and tissue samples.[29] Subsequently, Amy Davenport used the same approach to identify non-*Bartonella* bacteria isolated by means of BAPGM culture from sick dogs. In addition, she compared BAPGM isolate identities with isolates obtained from a cohort of sick dogs that had conventional blood cultures performed during the same time interval.[30] Between 2004 and 2008, 695 specimens from 513 dogs were tested by the NCSU-IPRL using the *Bartonella* alpha-Proteobacteria growth medium (BAPGM) enrichment blood culture.[30] Over the same period of time, blood samples from 270 dogs were cultured by the NCSU-CML using Bactec-Plus Aerobic/F media. BAPGM isolates were characterized using *Bartonella* genus primers and 16S rDNA primers followed by DNA sequencing. NCSU medical records were retrospectively reviewed. Blood culture results from the NCSU-CML were compared with BAPGM blood culture results. Seventy-nine non-*Bartonella* isolates were obtained from 69/513 dogs. The most commonly isolated phylum was Proteobacteria (48.1%), with alpha-Proteobacteria being the most commonly isolated class. The majority of the isolates were bacteria that are rarely or have never been isolated from canine culture samples. Comparison of NCSU-CML and IPRL (BAPGM) blood culture isolates showed alpha-Proteobacteria were isolated more often from BAPGM. Future studies are needed to elucidate the utility of BAPGM and other "nonconventional" growth media and methods for isolation of fastidious organisms that may constitute the blood microbiome or alternatively play a causal role in the pathobiome of disease.[31-33]

Chronic Inflammation, Pathobiome, and Disease Correlations

Clinicians frequently make a distinction between infectious, non-infectious inflammatory, autoimmune, immune-mediated, and neoplastic diseases. From a clinical perspective, dogs with endocarditis, ehrlichiosis, systemic lupus erythematosus, and cancer can share similar or identical disease features and laboratory test results. As is the case for any clinical data generated for a patient, there are multiple interpretations for each diagnostic parameter in a sick individual. However, the overlapping occurrence of these parameters in association with a bacterial infectious process (endocarditis), a systemic vector-borne infection (ehrlichiosis), an idiopathic immune-mediated disease (SLE), and a plasma cell cancer illustrates the potential diagnostic complexity that might be encountered in the clinical setting and in the context of causation, divisions among disease categories are less than clear. It is now well recognized that babesiosis should be a differential diagnosis for dogs with immune-mediated hemolytic anemia.[34-36] It also appears that *B. henselae* and *B. vinsonii* (*berkhoffii*) can be infectious causes of IMHA and ITP in dogs.[37,38] Although considered a hallmark for SLE, antinuclear antibodies have been described in dogs with endocarditis, ehrlichiosis, leishmaniasis, bartonellosis, and heartworm disease.[39] Antiplatelet antibodies have also been demonstrated in association with both acute (gram-negative sepsis and Rocky Mountain spotted fever) and chronic (ehrlichiosis and babesiosis) infectious diseases.[17] "Idiopathic immune-mediated" polyarthritis can be documented in dogs with chronic infections such as *B. burgdorferi, B. henselae, B. vinsonii* subspecies *berkhoffii*, *E. ewingii*, or *L. infantum*.[4-42] Due to the relatively low percentage of dogs that develop polyarthritis or glomerulonephritis, when infected with any of these organisms, one might suspect that a more complex pathobiome should be considered in this subdivision of patients.[27,40-42]

Immunosuppressive Drug Administration and the Blood Pathobiome

Occult bloodborne infection can persist in cats or dogs for prolonged periods in a state of immunological balance. A variety of factors including antibiotic administration, vaccination, and immunosuppressive drugs can disrupt this state of balance and ultimately result in disease expression in an individual animal. As an academic internist, I now more clearly understand the words of Voltaire, a famous French philosopher, when he stated: "Doctors put drugs of which they know little, into our bodies of which they know less, to treat disease of which they know nothing at all."

References

1. Walker DH, *et al. Emerg Infect Dis*. 1996;2:18.
2. Parola P, *et al. Clin Microbiol Rev*. 2005;18:719.
3. Daszak P, *et al. Science*. 2000;287:443.
4. Chomel BB, *et al. Emerg Infect Dis*. 2006;12:389.
5. Adelson ME, *et al. J Clin Microbiol*. 2004;42:2799.
6. Holden K, *et al. Vector-Borne Zoonotic Dis*. 2006;6:99.
7. Chomel BB, *et al. Ann N Y Acad Sci*. 2003;990:267.
8. Ge Y, *et al. Cell Microbiol*. 2006;8:1406.
9. Rikihisa Y. *Curr Opin Microbiol*. 2006;9:95.
10. Nguyen L, *et al. Trends Cell Biol*. 2005;15:269.
11. Verani A, *et al. Mol Immunol*. 2005:42:195.
12. Hossain H, *et al. Curr Opin Immunol*. 2006;18:422.
13. Scorpio DG, *et al. Clin Vac Immunol*. 2006;13:806.
14. Dumler JS, *et al. Clin Infect Dis*. 2007;45:199.
15. Tuttle AD, *et al. J Am Vet Med Assoc*. 2003;233:1306.
16. Kitchell BE, *et al. J Am Vet Med Assoc*. 2000;216:519.

17. Grindem CB, *et al. J Am Anim Hosp Assoc*. 1999;35:55.
18. Gupta P, *et al. Arch Microbiol*. 2016;198:1–15.
19. Oliver JD. *FEMS Microbiol Rev*. 2010;34:415–425.
20. Lewis K. *Ann Rev Microbiol*. 2010;64:357–372.
21. Zeidner NS, *et al. Parasite Immunol*. 2000;22:581.
22. Holden K, *et al. Infect Immun*. 2005;73:3440.
23. Kordick SK, *et al. J Clin Microbiol*. 1999;37:2631.
24. Breitschwerdt EB, *et al. J Clin Microbiol*. 1999;37:3618.
25. Breitschwerdt EB, *et al. J Clin Microbiol*. 1998;36:2645.
26. Mexas AM, *et al. J Clin Microbiol*. 2002;40:4670.
27. Vayssier-Taussat M, *et al. Cellular Infect Microbiol*. 2014;4:e29.
28. Keller M, Zengler K. *Nat Rev Microbiol*. 2004;2:141–150.
29. Cadenas MB, *et al. J Mirobiol Meth*. 2007;71:147–155.
30. Davenport AC, *et al. J Vet Intern Med*. 2013;27:854–861.
31. Potgieter M, *et al. FEMS Microbiol Rev*. 2015;39:567–591.
32. Rowe SE, *et al. Methods Mol Biol*. 2016;1333:17–28.
33. Paisse S, *et al. Transfusion*. 2016;56:1138–1147.
34. Matthewman LA, *et al. Vet Rec*. 1993;133:344.
35. Birkenheuer AJ, *et al. J Clin Microbiol*. 2003;41:4172.
36. Birkenheuer AJ, *et al. J Am Vet Med Assoc*. 2005;227:942.
37. Breitschwerdt EB, *et al. J Am Anim Hosp Assoc*. 2004;40:92.
38. Goodman RA, *et al. Am J Vet Res*. 2005;66:2060.
39. Smith BE, *et al. J Vet Intern Med*. 2004;18:47.
40. Henn JB, *et al. J Vet Res*. 2005;66:688.
41. Goodman RA, *et al. J Am Vet Med Assoc*. 2003;222:1102.
42. Gaskin AA, *et al. J Vet Intern Med*. 2002;16:34.

AUTHOR INDEX

AUTHOR INDEX

AUTHOR INDEX

AUTHOR INDEX

AUTHOR INDEX

American College of Veterinary Internal Medicine (ACVIM)
8301 E. Prentice Avenue, Suite 300
Greenwood Village, CO 80111
www.acvim.org

CURRAN ASSOCIATES INC.
proceedings.com

ISBN 978-1-5108-4279-3
9 781510 842793

American College of Veterinary Internal Medicine (ACVIM) Forum 2017

National Harbor, Maryland, USA
8 - 10 June 2017

Volume 2 of 2

ACVIM
American College of Veterinary Internal Medicine

American College of Veterinary Internal Medicine (ACVIM) Forum 2017

National Harbor, Maryland, USA
8 - 10 June 2017

Volume 2 of 2

ISBN: 978-1-5108-4279-3

Printed from e-media with permission by:
Curran Associates, Inc.
57 Morehouse Lane
Red Hook, NY 12571

Some format issues inherent in the e-media version may also appear in this print version.

Printed by Curran Associates, Inc. (2017)

Address correspondence to:
American College of Veterinary Internal Medicine (ACVIM)
8301 E. Prentice Avenue, Suite 300
Greenwood Village, CO 80111
www.acvim.org

Additional copies of this publication are available from:

Curran Associates, Inc.
57 Morehouse Lane
Red Hook, NY 12571 USA
Phone: 845-758-0400
Fax: 845-758-2633
Email: curran@proceedings.com
Web: www.proceedings.com

TABLE OF CONTENTS

VOLUME 1

CARDIOLOGY

LARGE ANIMAL INTERNAL MEDICINE

NEUROLOGY

NUTRITION

ONCOLOGY

SMALL ANIMAL INTERNAL MEDICINE

SMALL ANIMAL INTERNAL MEDICINE & LARGE ANIMAL INTERNAL MEDICINE

TRANSFUSION

RESEARCH ABSTRACTS – ORAL PRESENTATIONS

RESEARCH ABSTRACTS – POSTER PRESENTATIONS

RESEARCH ABSTRACTS – ORAL PRESENTATIONS

RESEARCH ABSTRACTS – POSTER PRESENTATIONS

RESEARCH ABSTRACTS – ORAL PRESENTATIONS

RESEARCH ABSTRACTS – POSTER PRESENTATIONS

RESEARCH ABSTRACTS – ORAL PRESENTATIONS

RESEARCH ABSTRACTS – POSTER PRESENTATIONS

RESEARCH ABSTRACTS – ORAL PRESENTATIONS

RESEARCH ABSTRACTS – POSTER PRESENTATIONS

RESEARCH ABSTRACTS – ORAL PRESENTATIONS

RESEARCH ABSTRACTS – POSTER PRESENTATIONS

RESEARCH ABSTRACTS – ORAL PRESENTATIONS

RESEARCH ABSTRACTS – POSTER PRESENTATIONS

RESEARCH ABSTRACTS – ORAL PRESENTATIONS

RESEARCH ABSTRACTS – POSTER PRESENTATIONS

RESEARCH ABSTRACTS – ORAL PRESENTATIONS

RESEARCH ABSTRACTS – POSTER PRESENTATIONS

RESEARCH ABSTRACTS – ORAL PRESENTATIONS

RESEARCH ABSTRACTS – POSTER PRESENTATIONS

RESEARCH ABSTRACTS – ORAL PRESENTATIONS

RESEARCH ABSTRACTS – POSTER PRESENTATIONS

RESEARCH ABSTRACTS – ORAL PRESENTATIONS

RESEARCH ABSTRACTS – POSTER PRESENTATIONS

RESEARCH ABSTRACTS – ORAL PRESENTATIONS

RESEARCH ABSTRACTS – POSTER PRESENTATIONS

RESEARCH ABSTRACTS – ORAL PRESENTATIONS

RESEARCH ABSTRACTS – POSTER PRESENTATIONS

RESEARCH ABSTRACTS – ORAL PRESENTATIONS

RESEARCH ABSTRACTS – POSTER PRESENTATIONS

RESEARCH ABSTRACTS – ORAL PRESENTATIONS

RESEARCH ABSTRACTS – POSTER PRESENTATIONS

AMERICAN ASSOCIATION OF VETERINARY NUTRITION/AMERICAN COLLEGE OF VETERINARY NUTRITION

AMERICAN COLLEGE OF VETERINARY CLINICAL PHARMACOLOGY/AMERICAN ACADEMY OF VETERINARY PHARMACOLOGY AND THERAPEUTICS

AMERICAN SOCIETY FOR CLINICAL PATHOLOGY

AMERICAN SOCIETY OF VETERINARY NEPHROLOGY & UROLOGY

ASSOCIATION OF VETERINARY HEMATOLOGY AND TRANSFUSION MEDICINE

BUSINESS MANAGEMENT

COMPANION ANIMAL PARASITE COUNCIL

COMPARATIVE GASTROENTEROLOGY SOCIETY

EVIDENCE BASED VETERINARY MEDICINE ASSOCIATION

INTERNATIONAL SOCIETY FOR COMPANION ANIMAL INFECTIOUS DISEASES

LIVER STUDY GROUP

VOLUME 2

ONE MEDICINE

SOCIETY FOR COMPARATIVE ENDOCRINOLOGY

VETERINARY COMPARATIVE RESPIRATORY SOCIETY (LARGE ANIMAL)

VETERINARY COMPARATIVE RESPIRATORY SOCIETY (SMALL ANIMAL)

VETERINARY EMERGENCY AND CRITICAL CARE SOCIETY

CARDIOLOGY

NEUROLOGY

ONCOLOGY

SMALL ANIMAL INTERNAL MEDICINE

EQUINE

acvim361

Endocrine Regulation of Appetite

Audrey K. Cook, BVM&S, DACVIM (SAIM), DECVIM (CA), DABVP (Feline)
College Station, TX, USA

Introduction

The regulation of appetite is a complex process, involving integrated endocrine and neurological responses from the brain, pituitary, gastrointestinal (GI) tract, pancreas and adipose tissue. The goal of this lecture is to review the physiology of appetite regulation with a focus on both familiar hormones and newly recognized endocrine influences.

Background

The hypothalamus plays a key role in food intake, with two distinct neuronal subpopulations driving either food-seeking behaviors or promoting a feeling of satiety.[1] Within the ventrolateral arcuate nucleus are the anorexigenic neurons; these are pro-opiomelanocortin (POMC) neurons and produce cocaine and amphetamine-related transcript (CART) and α-melanocyte stimulating hormone (α-MSH). These bind to melanocortin receptors and activate the satiety center in the hypothalamus. They also increase energy expenditure. Within the ventromedial arcuate nucleus are the orexigenic neurons; these release neuropeptide Y (NPY) and agouti-related peptide (AgRP) which activate the hypothalamic hunger center and reduce energy expenditure. Both groups also release their mediators within other areas of the brain, brainstem and spinal cord.

The vagus nerve, via the general visceral axons, controls digestive and propulsive events in the GI tract. Activation of afferent fibers by gastric distention inhibits appetite and diminishes food-seeking behavior. The vagus also contains receptors for various hormones and can therefore rapidly relay signals to the brain regarding digestive events, fullness, etc. Transection of all gut sensory fibers results in increased meal size and meal duration.[2,3]

Endocrine Events

Numerous hormones affect food intake and interact with the satiety and/or hunger centers in the brain. Although some of these players (e.g., cortisol, thyroxine and insulin) have well-established effects on food intake, the actual mechanisms by which they modulate hunger and satiety are often still unclear. In addition, numerous hormones secreted by the GI tract, pancreas, and adipose tissue directly and indirectly influence food intake, and have both long- and short-term influences on energy homeostasis.

Glucocorticoids

Supraphysiologic doses of glucocorticoids predictably increase food intake in dogs; the effect in cats is often more subtle. Similarly, dogs with spontaneous hyperadrenocorticism (HAC) are often markedly polyphagic. Acutely, glucocorticoids increase utilizable energy by promoting glycogen and protein metabolism in the liver, along with enhancing catecholamine-induced lipolysis of adipose tissue. However, persistently high concentrations have different metabolic effects, and are linked to responses designed to mitigate chronic stress. For example, pleasurable activities, including the ingestion of 'comfort foods' are enhanced in people with hyperadrenocorticism, possibly through the action of glucocorticoids on central dopamine secretion. Although the exact mechanism for appetite stimulation is still unclear, studies in rodent models of HAC suggest that glucocorticoid-induced hyperphagia is associated with increased hypothalamic AgRP.[4]

Corticotrophin Releasing Factor (CRF)

Although we traditionally think of this hormone as simply triggering the release of adrenocorticotrophic hormone (ACTH) from the pituitary, it actually has a substantial role in the regulation of energy balance, food intake and satiety, with receptors expressed in many areas of the brain.[5] Based on pharmacological studies, CRF causes an acute, dose-dependent, inhibition of food intake, and attenuates the expected stimulatory effect of NPY. In addition to limiting energy intake, CRF also increases energy expenditure via the activation of the sympathetic nervous system and brown fat thermogenesis.

Thyroid Hormones
Thyroid hormones (primarily triiodothyronine [T3]) are crucial for the maintenance of energy equilibrium through their stimulatory effects on basal metabolic rate (BMR), oxygen consumption and heat production. The primary product of the thyroid glands is T4, which is peripherally deiodinated by various tissues to T3. This process is directly influenced by food intake, as nutrients entering the portal circulation trigger the conversion of T4 to T3 by the liver. This drives up BMR and promotes thermogenesis. In addition, these enhanced T3 levels (higher than basal, but still within the normal range) promote further food intake, most likely through the activation of AgRP and NPY neurons in the hypothalamic ventromedial nucleus. Starvation therefore has a predictable negative effect on circulating T3 levels, which results in a drop in BMR and conservation of resources.

Central regulation of thyroid hormone production is modulated by the release of thyrotropin releasing hormone (TRH) and thyroid stimulating hormone (TSH); these are inhibited by thyroid hormone in a classic negative feedback loop. However, activity in this system is also influenced by nutritional signals such as AgRP (which inhibits TRH release) and leptin (which stimulates TRH secreting neurons).[6] In obese patients with high leptin levels, central leptin resistance prevents hyperthyroidism.

Insulin
There is substantial evidence to suggest that insulin plays a key role in the hypothalamic feedback loop of energy balance. Circulating insulin concentrations depend on food consumption, diet type and adiposity. Insulin crosses the blood-brain barrier (BBB) via a receptor-mediated transport system, and stimulates receptors in the anorexigenic pathways, leading to decreased food intake. Postprandial inhalation of insulin in people limits food intake, and simultaneously reduces the desirability of previously palatable items.[7] Insulin likely also interacts with orexigenic neurons and inhibits expression of NPY. This hypothesis is supported by studies in rats with type I diabetes mellitus (DM), as intracerebroventricular administration of insulin mitigates hyperphagia.

Cholecystokinin (CCK)
This peptide is released by I-cells in the proximal small intestine, in response to the arrival of chyme from the stomach, and was the first GI hormone to be implicated in the control of food intake.[3] Proteins and fat are particularly potent secretagogues, with plasma levels rising within 15 minutes of the start of a meal. In addition to triggering pancreatic acinar secretion and gallbladder contraction, CCK suppresses food intake via activation of anorexigenic neurons within the hypothalamus. As CCK does not cross the BBB, it is likely that vagal afferent fibers transmit the peripheral CCK signal to the brain to signal satiation and limit further ingestion.[8] CCK receptor antagonists result in substantial increases in meal size. Interestingly, the appetite-limiting effects of CCK are mitigated by chronic consumption of a high fat diet.[9]

The Incretins
The incretins (glucose-dependent insulinotrophic peptide [GIP]; glucagon-like peptide 1 [GLP-1]) are also released by specialized cells in the small intestine in response to the arrival of nutrients and suppress the further intake of food. GLP-1 release may also be triggered by the anticipation of feeding via vagal pathways and appears to synergize with CCK to potentiate its satiety signaling effect.[10] GLP-1 is also released within the brain; this central source appears to be an important step in the anorexigenic process and is blocked by central injection of GLP-1 receptor antagonists. Receptors for GLP-1 are present on both orexigenic and anorexigenic neurons. GLP-1 analogues are routinely used in people for the management of type 2 DM and have been shown to delay gastric emptying and abate feelings of hunger. Meal size is also reduced, suggesting impacts on satiation.[11,12]

Ghrelin: 'The Hunger Hormone'
Ghrelin is a relative endocrine newcomer, as it was first reported in 1999.[13] It is essential for survival as it plays a pivotal role in the generation of appetite. It is a 28-amino-acid protein, produced primarily by X/A cells of the gastric mucosa (>75%), but also from the small and large intestine, pancreas, and hypothalamus. There are two forms: acyl (10–30%) and des-acyl (70–90%) ghrelin. The acyl form has a half-life of about 30 minutes and is able to attach to the ghrelin receptor. Plasma ghrelin levels rise before

eating. Secretion is quickly suppressed by the arrival of food within the gastric lumen, although it has been suggested that absorbed nutrients within local vessels may also inhibit ghrelin release.[8]

Interestingly, its receptor (the growth hormone secretagogue receptor [GHS-R]) was discovered first and the subsequent hunt for its endogenous ligand led to the discovery of ghrelin. Receptors for ghrelin have been identified in numerous peripheral tissues (pituitary, stomach, intestine, pancreas, adipose tissue, adrenal gland, thyroid gland, myocardium, spleen, immune cells, and vagal afferents) and within the brain (arcuate nucleus, ventromedial nucleus, hippocampus, and other sites). Ghrelin directly activates the orexigenic neurons in the hypothalamus, and indirectly inhibits the anorexigenic neurons via gabaminergic inputs. These effects trigger the sensation of hunger and the search for food. In addition, ghrelin affects homeostasis by slowing down energy expenditure and supporting the conservation of resources. It also activates mesolimbic cholinergic-dopaminergic reward systems in the brain and is therefore linked to perceptions of pleasure.

Activation of GHS-Rs within the hypothalamus results in the secretion of GH-releasing hormone (GHRH) and the subsequent secretion of GH by the somatotrophs of the pituitary.[14] Ghrelin also directly stimulates GH release from the anterior pituitary. GH and its product, insulin-like growth factor 1 (IGF-1), promote weight gain and an increase in muscle mass. In addition to its appetite-promoting and anabolic effects, ghrelin is also an immunomodulator, with an overall anti-inflammatory action. It increases the synthesis of anti-inflammatory interleukin (IL) 10, whilst decreasing the production of various proinflammatory cytokines (e.g., IL-1β, IL-6 and TNF-α). It also inhibits the activation of nuclear factor-κB, a substance that stimulates proinflammatory cytokine synthesis and protein degradation.

Ghrelin also impacts GI motility and gastric acid secretion, likely via the vagus. Much like its close family member, motilin, ghrelin induces hunger contractions in the fasted state and speeds gastric emptying post-prandially.[15] In patients with idiopathic gastroparesis, ghrelin has been shown to hasten the exit of both fluids and semisolid material from the stomach.[16]

Peptide YY (PYY)

This hormone (also referred to as peptide tyrosine-tyrosine) is co-secreted from intestinal L cells with GLP-1 and was identified in 1982.[17] It was viewed initially as a local GI hormone influencing pancreatic and gastric secretions, and GI motility. However, its role in energy regulation was subsequently identified, and it was shown to reduce food intake via a central action in the arcuate nucleus. PPY is now known to be a link between nutrient ingestion and satiety, and a key postprandial regulator of energy homeostasis. Levels are low in the fasted state but rise in proportion to caloric content of the meal, with differential responses based on the specific macronutrient composition of the ingesta. Two forms of PYY are recognized, PYY1-36 and PYY3-36, with divergent actions on energy and glucose homeostasis. PYY3-36 appears to affect directly brain regions that control energy balance and modulate the reward value of food.

Dieting is accompanied by a drop in PPY levels; interestingly, these stay low even when caloric restriction has ended, thereby contributing to rebound weight gain. Gastric bypass likely has its positive results because of modulation of PPY secretion.

PPY is also expressed in the pancreas and may play a role in the paracrine regulation of insulin secretion. It may also have some glucomodulatory effects via modulation of GLP-1 secretion.

Orexin/Hypocretin

This is a neuropeptide involved in arousal, wakefulness and appetite. It is produced by cells in the lateral hypothalamus, with axons extending throughout the brain. Orexin-producing cells are triggered by ghrelin. Secretion increases food craving and promotes the development of brown adipose tissue. Lack of orexin results in a common form of narcolepsy.

Oxyntomodulin

This peptide is released from the GI tract in the post-prandial state, primarily from the L-cells in conjunction with GLP-1.[18] Although oxyntomodulin has many similar effects to GLP-1, it may also inhibit ghrelin secretion by the stomach.

Leptin: 'The Satiety Hormone'

Leptin is secreted by adipocytes, primarily those in white adipose tissue, and circulating levels are exponentially proportional to fat mass.[2] This hormone is primarily involved in the long-term regulation of food intake and energy balance; low levels act as a starvation signal to promote food-seeking behaviors and the conservation of resources, whereas adequate leptin levels communicate repletion of fat stores to the brain and facilitate non-food-seeking activities. Leptin is moved across the blood-brain barrier by a saturable transport system and exerts its anorectic effect in the arcuate nucleus. Interestingly, receptors are present on both the anorexigenic and orexigenic nuclei, which are stimulated or inhibited, respectively. Leptin receptors have also been identified on the vagus and may play a role in satiety (i.e., the cessation of food ingestion), possibly in synergy with CCK.[3] Ironically, leptin levels are persistently elevated in obese individuals but no longer inhibit food intake. This phenomenon, known as leptin resistance, may be due to changes in the transportation across the blood-brain barrier or the reduced receptor signal transduction. Hyperleptinemia limits ghrelin secretion in response to short-term energy restriction. Obese people often have lower than normal levels of ghrelin.[19]

SUMMARY

The regulation of appetite and metabolism is extremely complex, with an elaborate interplay between numerous hormones and the peripheral and central nervous systems. A better understanding of this process may pave the way for successful pharmacological manipulations in hyporexic or obese patients and improve both short- and long-term outcomes.

REFERENCES

1. Meister B. *Physiol Behav*. 2007;92:263.
2. Suzuki K, *et al. Endocr J*. 2010;57:359.
3. Owyang C, *et al. J Neurogastroenterol Motil*. 2011;17:338.
4. Nakayama S, *et al. Endocr J*. 2011;58:279.
5. Diz-Chaves Y. *Int J Pept*. 2011;doi:10.1155/2011/898450
6. Muller R, *et al. Physiol Rev*. 2014;94:355.
7. Hallschmid M, *et al. Diabetes*. 2012;61:782
8. Stengel A, *et al. Curr Protein Pept Sci*. 2011;12:293.
9. Lo C, *et al. Endocrinology*. 2014;155:1700.
10. Dailey MJ, *et al. Trends Endocrinol Metab*. 2013;24:85.
11. Horowitz M, *et al. Diabetes Res Clin Pract*. 2012;97:258.
12. Naslund E, *et al. Int J Obes Relat Metab Disord*. 1999;23:304.
13. Kojima M. *Nature*. 1999;402:656.
14. Broglio F, *et al. J Clin Endocrinol Metab*. 2001;86:5083.
15. De Smet B, *et al. Pharmacol Ther*. 2009;123:207.
16. Tack J, *et al. Aliment Pharmacol Ther*. 2005;22:847.
17. Manning S, Batterham RL. *Annu Rev Physiol*. 2004;76:585.
18. Pocai A. *Mol Metab*. 2014;3:241.
19. Somogyi V, *et al. Nutr Res Rev*. 2011;24:132.

acvim362

Endocrine Aspects of Obesity

Thomas A. Lutz, DMV, PhD
Zurich, Switzerland

Introduction

The incidence of obesity and other eating disorders continues to climb worldwide. Eating is controlled by a plethora of factors; most importantly peripheral gastrointestinal hormones that act directly or indirectly on the central nervous system play a major role. The increase in the incidence of obesity seems to be largely due to increased energy intake, combined with a reduction in energy expenditure. For larger animals (including humans), it seems to be easier to influence overall energy balance via a control of energy intake rather than energy expenditure (the situation may be opposite in small animals, e.g., mice). Hence, we must understand controls of eating to understand their potential contribution to the development of obesity. It is generally accepted that factors that influence food intake and consequently body fat are conceptualized as those that influence when individuals start eating and those that influence when eating, once begun, will end; i.e., factors that stimulate appetite or eating *per se* and those that stimulate fullness or satiation. Except in rare circumstances, eating is not initiated by physiological factors associated with energy needs but by factors such as habit, time of day, the social situation, food availability, and so on. The amount eaten (i.e., meal size), on the other hand, is determined by satiation factors generated by the gastrointestinal system interacting with ingested food. The best known satiation factor is the intestinal peptide, cholecystokinin (CCK). Satiation factors interact in the brain with signals emanating from adipose tissue and other organs indicating how lean or fat the individual is. These adiposity signals, such as leptin and insulin, signal the amount of adipose tissue to the brain and interact with receptors in the hypothalamus; they have potent effects on food intake and energy expenditure. It is important to realize, though, that both groups of signals interact, and that the concentrations of the adiposity signals insulin and leptin also fluctuate in relation to food intake. Apart from the distinction of short-term meal-associated signals versus long-term adiposity signals, eating controls can be classified into homeostatic controls of eating versus hedonic controls; the latter influence reward driven processes. Such processes, which are also influenced by gastrointestinal hormones, can overcome the homeostatic controls of eating under certain conditions. Finally, sex differences in the incidence and in particular the extent of obesity may depend on estrogen's modulation of eating controls.

Association Versus Causation

While it is well established that the concentration and action of many hormones is altered in obesity, their causal contribution to the development of obesity is much less clear. To name just one example, leptin is an important control of eating and energy metabolism (in particular when leptin or leptin signaling is reduced), and baseline leptin levels are elevated in most forms of obesity. However, it seems unlikely that a lack of leptin or leptin action contributes to the development of common obesity; rather, it seems that obesity is associated with leptin resistance and high leptin levels which may explain why the attempts to use leptin as an anti-obesity treatment were largely unsuccessful.

Leptin

Since its discovery in 1994, leptin is considered the most important adiposity signal and since then, research in leptin action has clearly dominated over insulin action which was the first adiposity signal discovered. Circulating leptin levels are higher in most cases of obesity, with the rare exception of genetic leptin deficiency in some spontaneously obese mouse models (the *ob/ob* mouse) and rare cases of human obesity. Most cases of obesity are therefore characterized by high circulating leptin levels and prevailing leptin resistance. Leptin resistance in obesity may be due to reduced leptin transport across the blood brain barrier and reduced leptin action in its hypothalamic target neurons. Hyperleptinemia itself has also been claimed to contribute to leptin resistance in obesity; this means that elevated circulating levels of leptin (and insulin) in obesity eventually result in the decreased peripheral and central sensitivity to these signals which only further potentiates the obese state.

Downstream of the Leptin Receptor

Leptin's effects are probably mediated by a coordinated network of action including several hypothalamic nuclei. Proopiomelanocortin (POMC) expressing neurons in the hypothalamic arcuate nucleus are direct targets of leptin. POMC is the precursor of several biologically active peptides including α-melanocyte-stimulating hormone (αMSH). In the brain, αMSH is a potent anorexigenic neuropeptide that reduces eating and increases energy expenditure by activating melanocortin (MC) 3 and 4 receptors in the paraventricular nucleus of the hypothalamus and elsewhere. This pathway seems to integrate many factors controlling energy balance and its function is critical to the maintenance of metabolic control. Several mutations in this pathway are known to be associated with genetic causes of obesity in a number of species, including POMC in dogs (in particular Labrador retrievers) and the MC4 receptor in cats.

Gut Hormones and Their Modulation

Gut hormones are among the best studied meal-associated satiation signals. Their role in the physiological control of eating is clearly established but it is less clear whether dysfunction of these systems contributes to the development of obesity. In most instances, this may actually not be the case. Obesity can affect the sensitivity to satiation hormones such as glucagon-like peptide-1 (GLP-1) and cholecystokinin (CCK) but overeating in obesity does not seem to be due to reduced secretion of these gut hormones.

CCK, GLP-1, Amylin

CCK was the first hormone whose physiological role in satiation was demonstrated in a number of species. At least in humans, allele variants of the CCK receptor has been found to correlate with average meal size which may contribute to overeating, and hence the development of obesity. CCK or CCK agonists have been explored for their potential therapeutic role as anti-obesity drugs. Interestingly, chronic administration often failed to reduce caloric intake and body weight under *ad libitum* feeding conditions (because smaller meal sizes were compensated by increased meal frequency), but CCK has been shown to be effective if the number of meals is controlled. GLP-1 is well known for its incretin effect and is successfully used to treat type 2 diabetic individuals. GLP-1 also reduces eating and chronic administration has a weight-lowering effect, even though GLP-1 sensitivity seems to be reduced in obesity. Nonetheless, GLP-1 analogues are now among the most potent non-surgical means to treat obesity in humans. The pancreatic hormone amylin controls eating and energy expenditure; even though baseline amylin concentrations are elevated in obesity, this does not seem to be associated with amylin resistance.

Ghrelin

Ghrelin is one of the few peripheral orexigenic peptide hormones known thus far. Ghrelin secretion increases before meals and behavioural and electrophysiological evidence shows that ghrelin acts in the hypothalamus via homeostatic pathways to signal hunger, and increase food intake and adiposity. Most obese individuals have normal or reduced ghrelin levels so that it seems unlikely that overeating which drives obesity is caused by excessive ghrelin secretion.

Modulation of Gut Hormones by Estradiol

Gonadectomy in females leads to an increase in body weight associated with an increase in eating. This effect can largely be explained by the lack of estradiol which leads to reduced effectiveness of meal controllers such as CCK, GLP-1, glucagon and others.

Transgenerational Effects

The intrauterine and perinatal environment plays an important role for the life-long control of energy balance. Considerable evidence indicates that early programming can influence metabolism and various aspects of the control of energy intake and expenditure. Feeding pregnant mouse or rat dams a high-fat

(HF) diet has a marked effect on their offspring. Further, offspring of obesity-prone dams are heavier and more obese than offspring of diet-resistant animals (i.e., animals that usually do not become fat despite exposure to HF diets), and the effect is greater in obesity-prone dams exposed to HF diet. The obesity in this offspring extends into adult life. Similarly, maternal HF feeding in humans leads to offspring that have a markedly increased risk of obesity in adult life; this is paralleled by hyperphagia and a higher preference for fat and sweet foods.

Anti-Obesity Surgery

Today, by far the most effective treatment option that leads to a rapid and in particular sustained loss in body weight is bariatric surgery, such as the Roux-en-Y gastric bypass (RYGB). Interestingly, the treatment success of RYGB appears to be associated with a change in the hormonal profile, in particular of gastrointestinal hormones that control meal size, and of bile acids. In addition to weight loss, RYGB also leads to pronounced changes in the metabolic situation, e.g., rapid resolution of type 2 diabetes and other obesity-related comorbidities. The causal role of these gut hormones in the treatment success of RYGB is still debated but there is strong evidence that GLP-1 and bile acids play important roles for some of the observed effects.

References

1. Asarian L, Geary N. *Am J Physiol Regul Integr Comp Physiol.* 2013;305:R1215–1267.
2. Bates SH, Stearns WH, Dundon TA, Schubert M, Tso AW, Wang Y, Banks AS, Lavery H, Haq AK, Maratos-Flier E, Neel BG, Schwartz MW, Myers MG, Jr. *Nature.* 2003;421:856–859.
3. Bray GA, York DA. *Physiol Rev.* 1979;59:719–809.
4. Campfield LA, Smith FJ, Guisez Y, Devos R, Burn P. *Science.* 1995;269:546–549.
5. Clegg DJ, Gotoh K, Kemp C, Wortman MD, Benoit SC, Brown LM, D'Alessio D, Tso P, Seeley RJ, Woods SC. *Physiol Behav.* 2011;103:10–16.
6. Halaas JL, Gajiwala KS, Maffei M, Cohen SL, Chait BT, Rabinowitz D, Lallone RL, Burley SK, Friedman JM. *Science.* 1995;269:543–546.
7. Leibel RL. *Int J Obes (Lond).* 2008;32 Suppl 7:S98–108.
8. Levin BE, Govek E. *Am J Physiol.* 1998;275:R1374–1379.
9. Levin BE, Lutz TA. *Trends in Endocrinology and Metabolism.* 28:153–164.
10. Lutz TA, Woods SC. *Curr Protoc Pharmacol.* 2012;chapter 5:unit 5.61.
11. Lutz TA, Bueter M. *Am J Physiol.* 2014;307:R1275–R1291.
12. O'Rahilly S. *Nature.* 2009;462:307–314.
13. Osto M, Lutz TA. *Eur J Pharmacol.* 2015;759:240–252.
14. Schwartz MW, Woods SC, Seeley RJ, Barsh GS, Baskin DG, Leibel RL. *Diabetes.* 2003;52:232–238.
15. Tamashiro KL, Terrillion CE, Hyun J, Koenig JI, Moran TH. *Diabetes.* 2009;58:1116–1125.
16. Woods SC. *Psychol Rev.* 1991;98:488–505.

acvim363

Endocrine Causes of Hypertension in Cats

Tim L. Williams, MA, VetMB, PhD, FRCPath, DECVCP, MRCVS
Cambridge, UK

INTRODUCTION

Hypertension is defined as an elevation in systolic or diastolic blood pressure; however, the risks of target organ damage are thought to most closely correlate with systolic blood pressure. In dogs and cats, generally veterinarians focus on measurement of systolic blood pressure (SBP) for risk classification with respect to target organ damage (TOD). Cats with a SBP <150 mm Hg are considered at minimal risk of TOD, whereas risk of TOD is considered severe when the SBP is >180 mm Hg.[1]

Elevated SBP can occur secondary to stress (white coat hypertension), other disease processes that increase SBP (secondary hypertension), or it can occur in the absence of other potential causes (idiopathic or primary hypertension). Primary hypertension is thought to occur in approximately 20% of cases, therefore secondary hypertension is more prevalent. Secondary hypertension in turn is subclassified into renal and endocrine hypertension, although in some disease states for which renal disease and endocrinopathy can occur concurrently, such as hyperthyroidism, there may be multiple etiologies to the observed hypertension. The most common endocrinopathies that are associated with elevated SBP and/or hypertension in cats are hyperthyroidism, primary hyperaldosteronism, and diabetes mellitus. Hyperadrenocorticism and pheochromocytoma may also cause hypertension but are very uncommon diagnoses in the cat.

HYPERTHYROIDISM AND HYPERTENSION

The diagnosis of hypertension in cats can be problematic due to the 'white coat' effect; therefore it is important that an experienced clinician takes blood pressure measurements following a short period of acclimatization and before any physical examination. In hyperthyroid cats, it has been suggested that there is a more pronounced white coat effect, if care is not taken to minimise stress;[2] therefore it is possible that the prevalence of true systolic hypertension in this group is overestimated.

The relationship between hyperthyroidism and hypertension is supported by a number of association studies in humans and cats. Hypertension was reported in 26% of human hyperthyroid patients,[3] and the prevalence of hypertension was higher in human hyperthyroid patients than euthyroid controls[4]. Early studies in hyperthyroid cats also identified a high prevalence of hypertension.[5,6] Kobayashi reported that hypertension was present in 87% of hyperthyroid cats; however, hypertension was defined as a SBP of >140 mm Hg (comparatively low, particularly given the exaggerated white coat effect reported in hyperthyroid cats[2]), and no fundic examination was performed to confirm if hypertensive retinopathy was present[5]. A subsequent study by Stiles and others, in which hypertension was diagnosed as a SBP >160 mm Hg, reported the prevalence of hypertension to be lower (23%).[6] Furthermore, studies which reported the presence of ocular lesions associated with hypertension in cats, identified concurrent hyperthyroidism in only 6–7% of cases,[7,8] and most of these cases also had concurrent chronic kidney disease (CKD). Interestingly, one study of ocular abnormalities in hyperthyroid cats reported hypertensive retinopathy in only 2 out of 100 hyperthyroid cats that were assessed.[9] This could indicate that the prevalence of hypertension in hyperthyroidism is not as high as earlier studies had suggested, or could indicate that hypertensive hyperthyroid cats are more resistant to TOD, since 41% of hypertensive cats with CKD have ocular abnormalities.[10]

More recent epidemiological studies have evaluated the prevalence of hypertension (defined as either persistent SBP >170 mm Hg or SBP >160 mm Hg with concurrent hypertensive retinopathy) in larger numbers of hyperthyroid cats treated in first-opinion practice. Ten per cent of hyperthyroid cats had concurrent hypertension at the time of diagnosis and a further 4% of hyperthyroid cats had previously been diagnosed as hypertensive.[11] In addition, 23% of initially normotensive cats developed hypertension after treatment of hyperthyroidism.[12] Therefore, although hypertension is present in around

14% of hyperthyroid cats at the time of diagnosis, it would appear that hypertension occurs in a greater number of cats following treatment of hyperthyroidism. Overall, the prevalence of hypertension in hyperthyroid cats (summating pre- and post-treatment hypertension) is around 30–35%, which is probably higher than expected for the geriatric population as a whole. This would support the hypothesis that hyperthyroidism can cause hypertension in cats.

Pathophysiological Mechanisms for the Development of Hypertension in Hyperthyroidism
Hypertension could occur in hyperthyroidism secondary to increased sympathetic nervous system (SNS) activity, increased activity of the renin-angiotensin-aldosterone system (RAAS), the presence of concurrent chronic kidney disease (CKD) or altered vascular tone. Increased activity of the SNS could contribute to hypertension because this will lead to enhanced renal sodium retention, activation of the RAAS, and increased cardiac output. Treatment of hyperthyroid cats with atenolol did reduce heart rate; however, the SBP normalised (<160 mm Hg) in only 30% of cats,[13] suggesting that SNS activation is not the sole cause of hypertension in the majority of hyperthyroid cats.

Renin-angiotensin-aldosterone system (RAAS) activation in hyperthyroidism could contribute to the development of hypertension since this would be associated with increased sodium and water retention, expansion of plasma volume, and peripheral vasoconstriction. Elevated plasma renin activity (PRA) and plasma aldosterone concentration have been observed in human hyperthyroid patients in several studies, and RAAS activation occurs in feline hyperthyroidism.[14] RAAS blockade markedly decreases arterial pressure in rats with experimentally induced hyperthyroidism (following parenteral injection of thyroxine),[16] and long-term parenteral administration of captopril prevents the development of hypertension in rats with experimentally induced hyperthyroidism[17]. Both of these studies would support a role for RAAS activation in the development of hypertension in hyperthyroid rats; however, such studies have not been performed in cats. However, decreased PRA has been reported in one study of human patients with isolated systolic hypertension related to hyperthyroidism,[15] which could suggest that the development of hypertension in human hyperthyroid patients is independent of RAAS activation; therefore the role of RAAS activation in the pathogenesis of hypertension in hyperthyroidism remains controversial.

In a cross-sectional study of hyperthyroid cats,[14] PRA and plasma aldosterone concentrations were not significantly different between hyperthyroid cats with and without concurrent hypertension, suggesting that RAAS activation does not contribute to the development of hypertension in these patients. Longitudinal analysis indicated that PRA and plasma aldosterone concentrations both decrease following treatment of normotensive hyperthyroid cats; however, in hyperthyroid cats which developed hypertension following treatment, only PRA decreased following treatment, whereas plasma aldosterone concentrations did not change significantly.[14] This suggests that changes in plasma aldosterone concentration could be independent of PRA in hyperthyroid cats which developed hypertension following treatment. Similarly, plasma aldosterone concentrations appear to be independent of PRA in hypertensive cats with CKD.[18] One interpretation of these data could be that aldosterone secretion is autonomous of control by PRA (and hence angiotensin II) in hypertensive cats, perhaps reflecting the presence of adrenal hyperplasia, which has also been observed in high numbers of geriatric cats.[19] Dynamic testing of adrenal gland function (such as the captopril response test or fludrocortisone suppression test) could be performed to determine if aldosterone secretion was autonomous of control of PRA in hypertensive cats. It could be speculated that autonomous aldosterone secretion might contribute to the development of hypertension in hyperthyroid cats, and cats with CKD, because the RAAS is an important determinant of blood pressure.

It is also possible that the presence of hypertension in hyperthyroid cats is associated with concurrent CKD, since hypertension is present in at least 19% of cats with azotemic CKD.[10] Ten per cent of hyperthyroid cats have concurrent azotemic CKD at the time of diagnosis,[11] and a further 15–49% of initially non-azotemic hyperthyroid cats develop azotemic CKD within 6 months of treatment for hyperthyroidism;[11,20] therefore CKD and secondary renal hypertension could be present in a large number of cats with hyperthyroidism. However, concurrent azotemic CKD was only present in 33% of hyperthyroid cats with hypertension at the time of diagnosis (T. Williams, unpublished data), and 36% of hyperthyroid cats which developed hypertension following treatment of hyperthyroidism.[12] Additionally, no significant difference in the prevalence of hypertension in cats with hyperthyroidism and CKD, and cats with CKD alone, has been identified,[10] and the presence of hypertension at the time of diagnosis of hyperthyroidism is not a predictor of the development of azotemia following treatment[11]. These data suggest that hypertension occurs in hyperthyroid cats with and without concurrent azotemic CKD, and therefore it could be speculated that hyperthyroidism itself contributes to the development of hypertension in some cats. Alternatively, hypertension and hyperthyroidism may be unrelated co-morbid conditions in geriatric cats.

Altered vascular responsiveness is reported in experimental models of hyperthyroidism, although the results are conflicting. Nitric oxide synthase (NOS), which regulates vascular tone and sodium retention, is upregulated in hyperthyroid rats;[21] however, such studies have not been performed in hyperthyroid cats. Hypertension is also associated with increased urinary cadmium excretion in euthyroid cats,[22] and it is postulated that the pathophysiological mechanism for hypertension could involve altered vascular tone; therefore evaluation of the association between cadmium exposure and hypertension in hyperthyroid cats could be an interesting avenue of investigation.

Overall, the pathophysiology of hypertension in hyperthyroidism is likely to be multifactorial and thus complex, which means that different drugs may be required to treat hypertension effectively in individual hyperthyroid cats.

Primary Hyperaldosteronism and Hypertension

Primary hyperaldosteronism (PHA) will cause hypertension because aldosterone stimulates sodium and water retention by the kidney, leading to expansion of plasma volume, and it induces vasoconstriction, which will increase systemic vascular resistance. PHA is generally regarded as an uncommon disease in cats, although the frequency of diagnosis is increasing, most likely as a result of increased awareness. In humans, around 5–10% of all hypertensive patients have PHA, although large studies evaluating the prevalence of PHA in cats with hypertension have not been performed. The majority of cats with PHA are hypertensive, hypokalemic and either non-azotemic or mildly azotemic.[23] It has been postulated that PHA could contribute to renal injury, because aldosterone can promote fibrosis, and serum urea and creatinine increased over time in cats with PHA in a single observational study;[23] however, further controlled studies are needed to confirm these findings. Treatment of PHA with spironolactone and amlodipine in combination, or adrenalectomy lead to normalisation of the SBP in most cases.

It could be speculated that the prevalence of PHA is underestimated in hypertensive cats, given that azotemic hypertensive cats have higher plasma aldosterone concentrations than normotensive azotemic cats.[18,24] However, the majority of these cats only have mildly increased plasma aldosterone concentrations, which were substantially lower than those reported in feline PHA. In addition, the response of azotemic hypertensive cats to treatment with angiotensin-converting enzyme inhibitors (ACE-I) is generally poor, which suggests that primary hyperaldosteronism is not the sole contributor to hypertension in most hypertensive cats with CKD.

Diabetes Mellitus and Hypertension

Both type 1 and type 2 diabetes mellitus are associated with hypertension in humans; however, the relationship in dogs and cats is less clearly defined. Dogs with recently diagnosed diabetes mellitus have higher blood pressures than healthy dogs,[25] although most diabetic dogs are not hypertensive[26]. It is possible that hypertension develops in chronic diabetes, given that blood pressure is correlated with the duration of diabetes, and 46% of dogs with long-standing diabetes (median 6 months) were hypertensive.[27] In cats, most studies have found no association between diabetes mellitus and hypertension, and the prevalence of hypertension in diabetics was similar to healthy control cats.[28,29] One study did report a higher SBP in diabetic cats compared to healthy controls, although the prevalence of hypertension (SBP >160 mm Hg) was not different between the groups.[30]

Hyperadrenocorticism and Hypertension

Hyperadrenocorticism (HAC) is rare in cats, but relatively common in dogs. In canine patients with HAC, the prevalence of hypertension is high (37–81%);[31,32] however, in one case series of cats with HAC, the prevalence of hypertension was lower (19%)[33]. The pathophysiological mechanism for hypertension in HAC is unclear; a role of hyperaldosteronism in the pathogenesis of hypertension in dogs has been investigated with conflicting results; however, altered vascular reactivity to norepinephrine has been observed in experimentally induced hypercortisolism in dogs.[34]

References

1. Brown S, *et al. J Vet Intern Med.* 2007;21(3):542–58.
2. Stepien R, *et al. J Vet Intern Med.* 2003;17:754 (abstract).
3. Hurxthal L. *Arch Intern Med.* 1931:167–181.
4. Saito I, Saruta T. *Endocrinol Metab Clin North Am.* 1994;23(2):379–86.
5. Kobayashi DL, *et al. J Vet Intern Med.* 1990;4(2):58–62.
6. Stiles J, *et al. J Am Anim Hosp Assoc.* 1994;26:647–651.
7. Sansom J, *et al. J Small Anim Pract.* 1994;35(12):604–611.
8. Maggio F, *et al. J Am Vet Med Assoc.* 2000;217(5):695–702.
9. van der Woerdt A, Peterson ME. *J Vet Intern Med.* 2000;14(2):202–203.
10. Syme HM, *et al. J Am Vet Med Assoc.* 2002;220(12):1799–804.
11. Williams TL, *et al. J Vet Intern Med.* 2010;24(4):863–869.
12. Morrow L, *et al. J Vet Intern Med.* 2009;23:700 (abstract).
13. Henik RA, *et al. J Feline Med Surg.* 2008;10(6):577–82.
14. Williams TL, *et al. J Vet Intern Med.* 2013;27(3):522–9.
15. Marcisz C, *et al. Neuro Endocrinol Lett.* 2012;33(1):81–9.
16. Garcia-Estan J, *et al. Am J Physiol.* 1995;268(5 Pt 1):E897–901.
17. Garcia del Rio C, *et al. Eur J Endocrinol.* 1997;136(6):656–60.
18. Jepson RE, Syme HM, Elliott J. *J Vet Intern Med.* 2014;28(1):144–53.
19. Keele S, *et al. J Vet Intern Med.* 2009;23:1328 (abstract).
20. Boag AK, *et al. Vet Rec.* 2007;161(21):711–5.
21. Quesada A, *et al. Eur J Endocrinol.* 2002;147(1):117–22.
22. Finch NC, *et al. Vet Rec.* 2012;170(5):125.
23. Javadi S, *et al. Domest Anim Endocrinol.* 2005;28(1):85–104.
24. Jensen J, *et al. Am J Vet Res.* 1997;58(5):535–40.
25. Bodey AR, Michell AR. *J Small Anim Pract.* 1996;37(3):116–25.
26. Reusch CE, *et al. Vet Clin North Am Small Anim Pract.* 2010;40(2):335–52.
27. Struble AL, *et al. J Am Vet Med Assoc.* 1998;213(6):822–5.
28. Sennello KA, *et al. J Am Vet Med Assoc.* 2003;223(2):198–201.
29. Norris CR, *et al. J Am Vet Med Assoc.* 1999;215(10):1455–9.
30. Al-Ghazlat SA, *et al. Top Companion Anim Med.* 2011;26(3):154–7.
31. Smets PM, *et al. J Vet Intern Med.* 2012;26(3):565–74.
32. Ortega TM, *et al. J Am Vet Med Assoc.* 1996. 209(10):1724–9.
33. Valentin SY, *et al. J Vet Intern Med.* 2014;28(2):481–7.
34. Martínez NI, *et al. Res Vet Sci.* 2005;78(1):25–31.

acvim364

Feline Hyperaldosteronism

Hans S. Kooistra, DVM, PhD, DECVIM-CA (Internal Medicine)
Utrecht, The Netherlands

SUMMARY

Primary hyperaldosteronism is the most common disorder of the adrenal gland in cats. Cats affected with primary hyperaldosteronism are typically presented with muscle weakness due to hypokalemic myopathy and/or complications of arterial hypertension, such as acute blindness. The plasma concentration of aldosterone is relatively high and plasma renin activity is low in cats with primary hyperaldosteronism. Unilateral adrenalectomy is the treatment of choice for confirmed unilateral primary hyperaldosteronism. If surgery is not possible or if the adrenocortical disease is bilateral, cats can be treated with the mineralocorticoid-receptor antagonist spironolactone.

INTRODUCTION

Primary hyperaldosteronism, also referred to as primary aldosteronism, low-renin hyperaldosteronism and Conn's syndrome, is an adrenocortical disorder which is characterized by autonomous aldosterone secretion originating from either neoplastic or hyperplastic zona glomerulosa tissue.[1] Primary hyperaldosteronism was first described in a cat in 1983, and was initially regarded a rare disease in felines. However, the number of case reports has risen considerably in the last two decades. Increased awareness of the disease will probably lead to a further increase in detected cases.

The inappropriately high aldosterone secretion in primary hyperaldosteronism has a number of serious implications. The most acute repercussions are associated with the typical mineralocorticoid actions of aldosterone on the kidneys. Excessive mineralocorticoid receptor activation in the epithelial cells of the distal nephron results in potassium depletion and sodium and water retention, and thereby leads to hypokalemia, systemic arterial hypertension, or both. Hence, cats affected with primary hyperaldosteronism are typically presented with muscle weakness due to hypokalemic myopathy and/or complications of arterial hypertension, such as acute blindness associated with retinal detachment and/or intraocular hemorrhage.[2] In addition, primary hyperaldosteronism has been associated with progressive renal dysfunction in cats.[3]

DIAGNOSIS

In primary mineralocorticoid excess, the plasma concentration of aldosterone is characteristically high and plasma renin activity (PRA) is immeasurably low. In hyperaldosteronism due to adrenocortical tumor plasma aldosterone concentration (PAC) is usually highly elevated. In idiopathic hyperaldosteronism PAC is usually only slightly elevated or within the upper limit of the reference range. As hypokalemia is a predominant factor in lowering PAC, in the presence of hypokalemia moderately elevated aldosterone values can be regarded as inappropriately high. The PRA must also be taken into account. The combination of a high-normal or elevated PAC and low PRA indicates persistent aldosterone synthesis in the presence of little or no stimulation by the renin-angiotensin system. In humans the PAC:PRA ratio (ARR) is considered to be a very useful aid in diagnosing primary hyperaldosteronism. This also seems to be true for cats with idiopathic primary hyperaldosteronism.

The ARR is elevated in about 10% of human patients with arterial hypertension and most of these have excess aldosterone production from both adrenal cortices. The diagnostic value of the ARR is principally determined by the sensitivity of the renin assay and interpretation should rest upon comparison with an appropriate control population. The ARR is currently regarded as the most reliable means of detecting primary hyperaldosteronism, but the measurements should be repeated if the initial result is inconclusive or difficult to interpret because of suboptimal sampling conditions.

An alternative diagnostic approach may be measurement of the urinary aldosterone:creatinine ratio (UACR). Cats excrete smaller quantities of aldosterone and its 18-glucuronidated metabolite in urine than do humans or dogs, but nevertheless the UACR can be determined. This would allow the development of a dynamic test, such as employing a suppressive agent that would reduce the UACR in healthy individuals but have little or no effect in those with primary hyperaldosteronism. In 42 healthy cats the upper limit for the UACR was 46.5×10^{-9}. The administration of sodium chloride did not significantly lower the UACR but administration of fludrocortisone (0.05 mg/kg body weight) reduced it by 44–97% (median 78%). In a cat with an aldosterone-producing adrenocortical carcinoma the UACR was within the reference range and was not lowered by fludrocortisone administration.[4] In a group of nineteen client-owned cats with arterial hypertension, nine had primary hyperaldosteronism and ten had another cause for the arterial hypertension. All PHA cats had basal UACRs $>7.5 \times 10^{-9}$. In all non-PHA cats with a basal UACR $>7.5 \times 10^{-9}$, fludrocortisone administration induced >50% suppression. In contrast, fludrocortisone administration resulted in <50% suppression in six of the nine PHA cats.[5] This test may prove to be a practical noninvasive diagnostic tool, but further evaluation is required, particularly with regard to its discriminatory power in diagnosing idiopathic hyperaldosteronism.

Subtype classification - differentiating between tumorous and non-tumorous mineralocorticoid excess - requires diagnostic imaging. Ultrasonography and computed tomography have been used in cats to identify and characterize adrenal tumors.[6] As in humans the findings are not always immediately conclusive. The visualization of a small aldosteronoma may pose problems while nodular hyperplasia might be interpreted as microadenoma.

Treatment

Unilateral adrenalectomy is the treatment of choice for confirmed unilateral primary hyperaldosteronism. Preoperatively and perioperatively hypokalemia should be controlled as well as possible, by oral and/or intravenous supplementation. Analogous to the postoperative management of hypercortisolemia due to adrenocortical tumor, temporary fludrocortisone therapy could be considered. However, in the reported cases such postsurgical measures have not been necessary and their omission does not seem to have had deleterious effects.

If surgery is not possible or if the adrenocortical disease is bilateral, medical treatment is possible with the mineralocorticoid-receptor antagonist spironolactone and oral supplementation with potassium gluconate. The initial doses are 2 mg spironolactone/kg and 0.5 mmol potassium gluconate/kg, twice daily. Persistent arterial hypertension can be treated with the calcium blocker amlodipine (0.1–0.2 mg/kg). In cases of adrenocortical tumor medical treatment may lead to resolution of symptoms and signs such as the myopathy in cats, but complete normalization may not be achieved. Particularly plasma potassium tends to remain below the reference range, despite increasing doses of both spironolactone and potassium. Doses of spironolactone >4 mg/kg may cause anorexia, diarrhea, and vomiting. These side effects may be due to interference by spironolactone with aldosterone action on transepithelial electrolyte transport in the distal colon.

Experience is very limited, but medical treatment appears to be preferable in cats with hyperaldosteronism due to bilateral adrenocortical hyperplasia. The hyperaldosteronism is usually somewhat milder than in cases due to tumor and normokalemia may be maintained for a long period with spironolactone alone or together with low doses of potassium.

Prognosis

After complete removal of a unilateral non-metastasized mineralocorticoid-producing tumor, the prognosis can be excellent, without any medication. In both forms the disease may be associated with renal insufficiency. Successful removal of the tumor will probably prevent further progression of aldosterone-induced arteriolar sclerosis and interstitial fibrosis in the kidneys. The prognosis may not be as favorable in cats with idiopathic hyperaldosteronism treated with spironolactone, for this treatment will not abolish the mineralocorticoid excess as definitely as surgery may do.

References

1. Galac S, *et al.* In: *Clinical Endocrinology of Dogs and Cats*. 2nd ed. 2010:93.
2. Djajadiningrat-Laanen SC, *et al. J Feline Med Surg*. 2011;13:641.

3. Javadi S, *et al. Domest Anim Endocrinol.* 2005;28:85.
4. Djajadiningrat-Laanen SC, *et al. J Vet Intern Med.* 2008;22:1283.
5. Djajadiningrat-Laanen SC, *et al. J Vet Intern Med.* 2013;27:1493.
6. Ash RA, *et al. J Feline Med Surg.* 2005;7:173.

acvim219

Evidence-Based Regenerative Medicine for the Internist

Andrew M. Hoffman, DVM, DVSc, DACVIM (LAIM)
North Grafton, MA, USA

Introduction

Evidence is emerging that stem cells have the potential to uplift endogenous mechanisms of repair and regeneration.[1,2] Why are stem cells viewed as the major player in regenerative medicine? What 'stem cells' have been evaluated for safety and activity in dogs and cats? What diseases were investigated and what is the quality of the evidence? Were stem cell transplantations safe? Based on endpoints, what benefits were noted? What does the collective analysis of all trials say about stem cell applications in dogs and cats? What is the regulatory environment for stem cell-based therapies in veterinary medicine? The intent of this discussion is to address these questions.

Why Are Stem Cells Viewed as the Major Player in Regenerative Medicine?

Classically, regeneration has been viewed as the **spontaneous return of form and function of an organ**[3] that has been lost through voluntary shedding ('autotomy'), traumatic injury, or experimental excision. The capacity for regeneration is a feature of many animals (e.g., Amphibia, Reptilia, Arthropods, Cnidaria/Ctenophora, Echinodermata) which have provided important cellular-molecular clues about this phenomenon. Examples include regeneration of limbs in salamanders and newts, tails of certain fish, and the claws of crustaceans. In contrast, mammals have limited capacity for 'true' regeneration, restricted to the tips of digits (young children), and the antlers of Cervidae. Hepatectomy or nephrectomy evokes **compensatory regrowth** at a distant site from excision, although this is often misrepresented as regeneration *per se*. Apart from hematopoiesis and above examples, what is limiting regeneration in mammals? Animals capable of regeneration are capable of (1) rapid initiation of anti-inflammatory mechanisms, (2) wound closure with epithelium, (3) dedifferentiation of cells (e.g., connective tissue fibroblasts) to a pool of regenerative stem cells, (4) positional memory of those stem cells, (5) nerve dependence of patterned growth, and (6) and species-specific proteins (e.g., Prod1 in salamander limbs) that support branching patterns of organs. Regeneration may also involve epigenetic controls such as inhibition of miRNA that suppress dedifferentiation of mature cardiomyocytes to stem cells enabling cardiac regeneration in some species such as zebra fish, a process that can be experimentally induced in mammals.[4] Undeniably, **stem cells through differentiation to specialized cells are paramount to classical regeneration**. However, attempts to recapitulate this process with exogenous stem cells has consistently failed **due to our current inability to engraft stem cells**. This has led to a shift in regenerative medicine toward exploiting stem cells for their **paracrine, drug-like activities rather than stable engraftment**.

What 'Stem Cells' Have Been Evaluated for Safety and Activity in Dogs and Cats?

Endogenous stromal cells sustain the niche of resident cells, support angiogenesis, extracellular matrix formation, limit fibrosis, and contributing to barrier function. Stromal cells of this nature include interstitial and perivascular fibroblasts and macrophages (e.g., M2 polarized macrophages). Mesenchymal stem-stromal cells (MSC) are a subset of these stromal cells in all tissues and are readily isolated from fat, bone marrow, or discarded placental-umbilical tissues for therapeutic studies. MSC lack MHCII and co-stimulatory molecules that promote antigen presentation. The MSC were used in the vast majority (21/23) of 'stem cell' trials reported in the veterinary literature; these included MSC from adipose tissue (AD-MSC, 14/23 trials), bone marrow (BM-MSC, 6/23 trials), dental pulp (1/23), fetal bone marrow (1/23), or embryonic stem cells (1/23). Most stem cell trials have employed **autologous** MSC (15/23 trials). More recently in 2016, 3/6 trials reported using allogeneic MSC, suggesting a trend toward 'off-the-shelf' allogeneic sources, consistent with trends in industry to comply with new FDA guidelines (see below).

WHAT DISEASES WERE INVESTIGATED AND WHAT IS THE QUALITY OF THE EVIDENCE?

MSC were employed for anti-inflammatory/immunomodulatory action in several diseases including osteoarthritis,[5-9] cruciate rupture,[10] meningoencephalitis of unknown origin (MUO),[11] perianal furunculosis (fistulitis),[12] inflammatory bowel diseases,[13,14] keratoconjunctivitis sicca, and chronic gingivostomatitis (in atopic dermatitis, no benefit was noted)[15]. Mitigation of hypoxic-ischemic injury and associated inflammation was sought by testing the effects of MSC in intervertebral disc degeneration (i.e., spinal disc herniation).[16-20] In degenerative valve disease[21] and chronic kidney disease,[22,23] the main focus was abrogating remodeling and fibrosis. Evidence supporting potency or specific mechanisms of action of MSC (e.g., immunomodulation) as delivered was either not shown[9] or lacking with few exceptions[12,15]. Hence, it is not entirely clear if the cells employed for clinical trials had consistent potency for the intended application. The quality of the evidence and ability to compare between studies was also reduced by wide variations in cell manufacturing process (e.g., cell passage, culture medium, storage), donor age, donor cell adventitious agent screening, and cell dosages, or a complete lack of sufficient documentation.[1] Studies employing an open-label study design were useful mainly to establish feasibility of the protocol and safety of the cell lines, dosages, and route of administration, as well as endpoints.

WERE STEM CELL TRANSPLANTATIONS SAFE?

No adverse effects were observed in 21/23 trials employing MSC implying a **high degree of safety** was employed. No tumors, ectopic tissue, transfer of adventitious infectious agents, immune reactions, or other adverse or unanticipated effects were observed in dogs treated with MSC, including intraspinal, intraarticular, and intralesional (perianal fistulas) routes of administration at varying dosages. However, Quimby *et al.*[22] noted that allogeneic AD-MSC that were cryopreserved using standard DMSO formulations and thawed immediately prior intravenous injection (implying off-the-shelf application) evoked vomiting and respiratory embarrassment in 4/5 cats treated in this way, but not by freshly harvested cells. In cats with gingiva-stomatitis, autologous AD-MSC infusion induced transfusion rate-related reactions (vomiting, tachypnea, urination, apathy) in 2/7 subjects, which spontaneously reversed upon slowing of the transfusion rate.[15] The exact molecular basis and role for disease diagnosis in these adverse reactions is unknown. However, these data alert us to the sensitivity of feline patients to novel cell-based therapies indicating that dose-transfusion conditions need to be well understood.

BASED ON ENDPOINTS, WHAT BENEFITS WERE NOTED?

Acknowledging the caveats described above, 19/23 studies (**Table 1**) reported clinically relevant benefits. In four double-blind randomized controlled clinical trials (RCT) in the canine species highlighted here, MSC transplantations produced statistically significant clinical benefits, based on owner and/or veterinary metrics. In the first published study employing FDA-INAD filing, Harman *et al.* (2016) on behalf of VetStem Biopharma, Inc. conducted a double-blind RCT to evaluate safety and clinical effects of a single intraarticular injection of AD-MSC in dogs with OA. AD-MSC were isolated and expanded from a single laboratory canine, and grown and assayed for potency using proprietary methods. Based on both veterinary and owner observations, dogs improved significantly in the AD-MSC (~92% success) vs. vehicle (saline, ~50% success) treated groups at 60 days post-injection. In a double-blind RCT, dogs with chronic (3 mo) disc herniation and no deep pain, were treated intra-spinal via percutaneous route with autologous olfactory epithelial cells (OEC, $p75^{pos}$).[17] This resulted in significant improvements in forelimb-hindlimb coordination at 6 mos post-injection. In a double-blind RCT in dogs at the time of diagnosis of heart failure secondary to degenerative mitral valve disease (ACVIM class C), dogs received intravenous allogeneic dental pulp MSC (1*10e6/kg bwt) on days 0 and 14, resulting in significant improvements in ejection fraction and quality of life scores, although not in fractional shortening or left ventricular internal diameters.[21] Survival benefits were not evaluated.

TABLE 1. **Stem cell trials conducted in dogs or cats between 2008 and 2017 excluding retrospective studies, case reports, experimental injury models or those involving laboratory animals, or related to cancer (modified from: Hoffman, Dow. *Stem Cells* 2016;34:1709–1729).**

Disease (Canine unless indicated)	Cells	Study design	Adverse Events	Benefit	Ref
Osteoarthritis (hip joint)	Auto AD-MSC	Blinded, baseline controlled study	none	Y	Vilar 2014
Osteoarthritis (hip joint)	Auto AD-MSC	Blinded, baseline controlled study	none	Y	Vilar 2013
Osteoarthritis (hip joint)	Auto AD-MSC	**Double blind randomized comparative study**	none	Y	Cuervo 2014
Hip dysplasia (partially refractory)	Auto AD-MSC	Open label baseline-controlled study	none	Y	Marx 2014
Osteoarthritis (hip joint)	Allo AD-MSC	**Double Blind RCT**	none	Y	Harman 2016
Osteoarthritis humero-radial joint (refractory)	Auto AD-MSC	Open label baseline-controlled study	none	Y	Guercio 2012
Partial cruciate rupture (w/TPLO)	Auto BM-MSC	Open label baseline controlled study	none	Y	Muir 2016
Intervertebral disc degeneration (IVDD) >6 mo	Auto BM-MSC	Open label baseline-controlled study	none	Y	Penha 2014
IVDD, chronic (> 3 mo), no deep pain	Auto-OEC	**Double blind RCT**	none	Y	Granger 2012
IVDD, chronic (>60 d), no deep pain	Allo fetal BM-MSC	Open label baseline controlled study	none	Y	Sarmento 2014
IVDD, no deep pain, surgery within 7 d	Allo BM-MSC	Open label randomized controlled	none	Y	Kim 2016
IVDD, no deep pain >42d post-op	Neural lineage BM-MSC	Open label baseline-controlled	none	Y	Besalti 2015
Meningoencephalitis of unknown origin (MUO)	Auto BM-MSC (IT, IA)	Open label baseline controlled study	none	Y	Zeira 2015
Dilated cardiomyopathy	AAV2-SDF1 AD-MSC	Open label, historical controls	none	N	Pogue 2013
Degenerative mitral valve disease	Dental pulp MSC	**Double blind RCT**	none	Y	Petchdee 2016
Atopic dermatitis (non-seasonal)	Auto AD-MSC	Open label, baseline controlled study	none	N	Hall 2010
Perianal fistulas (refractory)	xeno hESC-MSC	Open label, baseline controlled study	none	Y	Ferrer 2016
Inflammatory bowel disease (partial refractory)	Auto AD-MSC	Open label, baseline controlled study	none	Y	Perez-Merino 2015-6
Inflammatory bowel disease (feline)	Allo AD-MSC	**Single (owner) blind RCT**	none	Y	Webb 2015
Chronic kidney disease (feline)	Allo AD-MSC	Open label, baseline-controlled study	Cryo only (4/5)	N	Quimby 2013
Chronic kidney disease (feline)	Allo AD-MSC	**Double blind RCT**	none	N	Quimby 2015
Keratoconjunctivitis sicca (refractory)	Allo AD-MSC	Open label, baseline controlled study	none	Y	Villatoro 2015
Chronic Gingivostomatitis (feline)	Auto AD-MSC	Open label, baseline controlled study	yes (2/7)	Y	Arzi 2016

What Does the Collective Analysis of All Trials Say About Stem Cell Applications in Dogs and Cats?

At the present time it is still very difficult to interpret the broad significance of MSC therapies. First, there is a strong effect, as much as 50% improvement, of participation in a clinical trial as noted in the double-blind RCT by Harman *et al.*[9] thus obscuring the actual benefits from test articles in open-label studies. There are also lingering questions about the benefits versus the potential costs of using stem cells - Are the magnitude and duration of effects sufficient to warrant the costs to consumers? Studies with longer endpoints (1 yr) are warranted. Furthermore, studies employing allogeneic MSC or other cells (e.g., OEC) generally involved a single injection, whereas it is plausible that multiple treatments will be needed to sustain the benefits. However, the safety and efficacy of multiple injections of allogeneic MSC are largely unknown although numerous human studies support their strong safety profile. It is clear from the 23 trials submitted that OA and IVDD were top priorities for investigators, perhaps related to the prevalence of OA and the catastrophic and translational relevance of spinal cord injury to humans. Interestingly, open-label studies have not led to fully powered studies in veterinary medicine, suggesting reluctance, lack of resources, or response to the shifting regulatory environment.

What Is the Regulatory Environment for Stem Cell-Based Therapies in Veterinary Medicine?

New FDA guidelines were issued concerning stem cell trials in client-owned animals (GFI 218 June 4 2015), largely based on previously issued human guidelines. The essence of premarket evaluation in the guidance was stated as evaluations of stem cell tumorigenicity, formation of ectopic tissues, immunogenicity, donor selection criteria, transmission of adventitious agents, survival endpoints, toxicity, and biodistribution (where relevant),[1] reviewed recently by Devireddy *et al.*[24]. It is clear that the focus is on allogeneic and non-homologous applications of stem cells. The implication is that veterinarians cannot market stem cells that are not under investigation employing FDA guidelines. The implications of these guidelines on the trajectory of veterinary stem cell trials will be discussed.

References

1. Hoffman AM, Dow SW. Concise review: stem cell trials using companion animal disease models. *Stem Cells.* 2016;34:1709–1729.
2. Kol A, *et al.* Companion animals: translational scientist's new best friends. *Sci Transl Med.* 2015;7(308):308ps21.
3. Mason C, Dunnill P. A brief definition of regenerative medicine. *Regen Med.* 2008;3:1–5.

4. Aguirre A, *et al.* *In vivo* activation of a conserved microRNA program induces mammalian heart regeneration. *Cell Stem Cell.* 2014;15:589–604.
5. Cuervo B, *et al.* Hip osteoarthritis in dogs: a randomized study using mesenchymal stem cells from adipose tissue and plasma rich in growth factors. *Int J Mol Sci.* 2014;15:13437–13460.
6. Vilar JM, *et al.* Assessment of the effect of intraarticular injection of autologous adipose-derived mesenchymal stem cells in osteoarthritic dogs using a double blinded force platform analysis. *BMC Vet Res.* 2014;10:143.
7. Vilar JM, *et al.* Controlled, blinded force platform analysis of the effect of intraarticular injection of autologous adipose-derived mesenchymal stem cells associated to PRGF-Endoret in osteoarthritic dogs. *BMC Vet Res.* 2013;9:131.
8. Marx C, *et al.* Acupoint injection of autologous stromal vascular fraction and allogeneic adipose-derived stem cells to treat hip dysplasia in dogs. *Stem Cells Int.* 2014;2014:391274.
9. Harman R, *et al.* A prospective, randomized, masked, and placebo-controlled efficacy study of intraarticular allogeneic adipose stem cells for the treatment of osteoarthritis in dogs. *Front Vet Sci.* 2016;3:81.
10. Muir P, *et al.* Autologous bone marrow-derived mesenchymal stem cells modulate molecular markers of inflammation in dogs with cruciate ligament rupture. *PLoS One.* 2016;11:e0159095.
11. Zeira O, *et al.* Adult autologous mesenchymal stem cells for the treatment of suspected non-infectious inflammatory diseases of the canine central nervous system: safety, feasibility and preliminary clinical findings. *J Neuroinflammation.* 2015;12:181.
12. Ferrer L, *et al.* Treatment of perianal fistulas with human embryonic stem cell-derived mesenchymal stem cells: a canine model of human fistulizing Crohn's disease. *Regen Med.* 2016;11:33–43.
13. Perez-Merino EM, *et al.* Safety and efficacy of allogeneic adipose tissue-derived mesenchymal stem cells for treatment of dogs with inflammatory bowel disease: clinical and laboratory outcomes. *Vet J.* 2015;206:385–90.
14. Webb TL, Webb CB. Stem cell therapy in cats with chronic enteropathy: a proof-of-concept study. *J Feline Med Surg.* 2015;17:901–908.
15. Arzi B, *et al.* Therapeutic efficacy of fresh, autologous mesenchymal stem cells for severe refractory gingivostomatitis in cats. *Stem Cells Transl Med.* 2016;5:75–86.
16. Penha EM, *et al.* Use of autologous mesenchymal stem cells derived from bone marrow for the treatment of naturally injured spinal cord in dogs. *Stem Cells Int.* 2014;2014:437521.
17. Granger N, Blamires H, Franklin RJ, Jeffery ND. Autologous olfactory mucosal cell transplants in clinical spinal cord injury: a randomized double-blinded trial in a canine translational model. *Brain.* 2012;135:3227–3237.
18. Sarmento CA, Rodrigues MN, Bocabello RZ, Mess AM, Miglino MA. Pilot study: bone marrow stem cells as a treatment for dogs with chronic spinal cord injury. *Regen Med Res.* 2014;2:9.
19. Kim Y, Lee SH, Kim WH, Kweon OK. Transplantation of adipose derived mesenchymal stem cells for acute thoracolumbar disc disease with no deep pain perception in dogs. *J Vet Sci.* 2016;17:123–126.
20. Besalti O, *et al.* Intraspinal transplantation of autologous neurogenically-induced bone marrow-derived mesenchymal stem cells in the treatment of paraplegic dogs without deep pain perception secondary to intervertebral disk disease. *Turk Neurosurg.* 2015;25:625–632.
21. Petchdee S, Sompeewong S. Intravenous administration of puppy deciduous teeth stem cells in degenerative valve disease. *Vet World.* 2016;9:1429–1434.
22. Quimby JM, Webb TL, Habenicht LM, Dow SW. Safety and efficacy of intravenous infusion of allogeneic cryopreserved mesenchymal stem cells for treatment of chronic kidney disease in cats: results of three sequential pilot studies. *Stem Cell Res Ther.* 2013;4:48.
23. Quimby JM, *et al.* Assessment of intravenous adipose-derived allogeneic mesenchymal stem cells for the treatment of feline chronic kidney disease: a randomized, placebo-controlled clinical trial in eight cats. *J Feline Med Surg.* 2016;18:165–171.
24. Devireddy L, Boxer L, Myers M, Skasko M, Screven R. Questions and challenges in the development of mesenchymal stromal/stem cell-based therapies in veterinary medicine. *Tissue Eng Part B Rev.* 2017.

acvim220

Next-Generation Regenerative Medicine (Beyond Stem Cells): Exosomes and Extracellular RNA

Andrew M. Hoffman, DVM, DVSc, DACVIM (LAIM)
North Grafton, MA, USA

INTRODUCTION

The overwhelming emphasis in regenerative medicine at this moment is the use of stem cells, with the goal to reach a broad range of pro-regenerative molecular targets through paracrine signals. However, the regenerative medicine 'sandbox' is increasingly multimodal, including new technologies such as gene therapy, oligonucleotides (RNAi, antisense), nano-pharmaceuticals, biomimetic scaffolds, and stem cell products (extracellular vesicles). There is also a large push toward the application of induced pluripotent stem cells which have the capacity to differentiate into multiple lineages and thus any cell type for replacement therapy. As engraftment is still challenging, and pluripotent stem cells have a propensity for errors resulting from reprogramming and differentiation, their application is a long way off in veterinary medicine.

With respect to stem cells, there is increasing evidence that mesenchymal stem-stromal cells (MSC) commonly used for this purpose employ both soluble mediators (e.g., PGE2, NO, TGFβ1, IL-10, adenosine) and those which are shed in the form of extracellular vesicles after transplantation. These intriguing vesicles may redefine how we go about designing therapies based on stem cells. This discussion will define the biological characteristics of extracellular vesicles (EV), provide context for this emerging area in regenerative medicine, and discuss their potential contributions to new diagnostic and therapeutic strategies.

WHAT ARE EXTRACELLULAR VESICLES, MICROVESICLES, AND EXOSOMES?

Vesicles were first identified as procoagulant platelet-derived particles, later termed 'platelet dust.'[1] It was decades later when it was recognized that **all cells (including RBCs) and platelets produce EV**. Extracellular vesicles are bilipid membrane structures (50–1000 nm) with two known biogenesis pathways: (1) endosome invagination into multivesicular bodies (MVB) for packaging of cargo, modification, and eventual fusion of the MVB for release of smaller '**exosomes**' (50–200 nm) or (2) direct blebbing of the outer cell membrane to produce '**microvesicles**' (~200–500 nm),[2,3] as shown in **Figure 1**. In addition, cancer cells release oncosomes (1000–5000 nm) by a mechanism akin to microvesicles, which appear in tumor tissue and biofluids (e.g., blood, CSF); these oncosomes contain molecular signatures of the parent cancer cells, including mutated sequences in ss/dsDNA and mRNA.[4] Larger apoptotic bodies are often included in the definition of EV, but they clearly differ in genesis as they represent the fragmentation of the cell and chromatin bodies prior to phagocytosis and destruction.

Fig 1. Extracellular vesicle biogenesis, release, and transfer to intercellular spaces, other cells, biofluids *in vivo*, or culture supernatants *in vitro*, leading to diagnostic and therapeutic applications

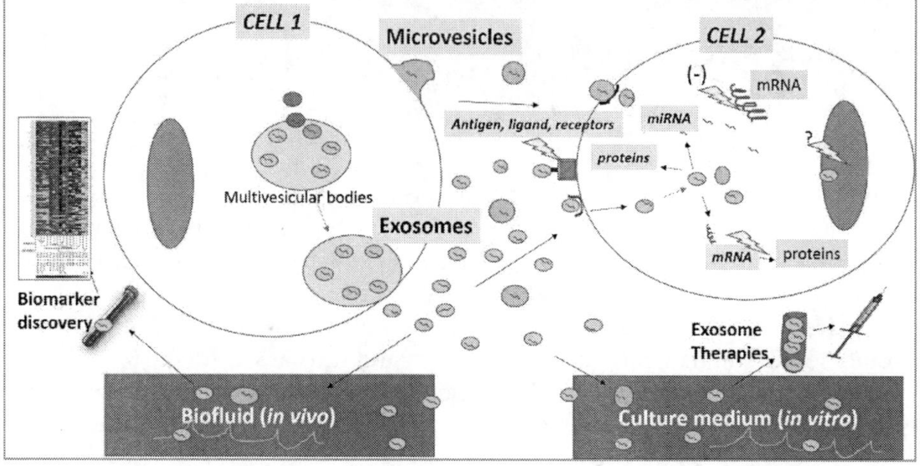

While EV are nanoparticles, it is important to make the distinction between these naturally shed structures and artificial nanoparticles such as liposomes, lipid-based products, PEGylated polymers, protein-drug conjugates, virosomes, called nanopharmaceuticals.[5] Natural EV are more efficiently internalized by cells than nano-pharmaceuticals, which has attracted interest in EV application as vectors for drugs and biologics. However, they are produced by cells, a process which introduces many biological variables and costs absent in pharmaceuticals. It is noteworthy that gene therapy vectors (adeno associated virus or AAV) are shed by cells in exosomes to some extent, and this has led to the notion that exosomes may be a powerful vector to increase efficiency of AAV transfection and transduction into recipient cells.[6]

What Is Extracellular RNA?

Until recently, it was thought that all RNA was restricted to the intracellular environment, and all other RNA represented degraded fragments mRNA emanating from damaged cells.[7] Two major forms of extracellular RNA (exRNA) exist in nature, including (1) RNA found within EV (mainly mRNA and miRNA but also tRNA, Y-RNA, lncRNA, circRNA), and (2) miRNA which is protected from RNase degradation by complexing with Argonaute 2. There are significant technical hurdles to isolate exRNA and determine whether RNA originates from within vs. outside EV in samples. Methods include degradation of samples with proteinases and RNases to remove non-EV miRNA, or various means of physical separation (differential ultracentrifugation, density gradient centrifugation, size exclusion chromatography, spin columns, precipitation methods). The two major forms of exRNA (EV associated, non-EV) are biologically active but their role in health and disease is a nascent field.

How Are Exosomes and Extracellular RNA Involved in Cell Physiology and Pathology?

Based on their molecular structure, exosomes are thought to be critical to intercellular communication.[1] EV are rich in heat shock proteins, tetraspanins, integrins, and cytokines suggesting respective roles in downstream signaling of recipient cells. For example, cytokines that may be released in association with EV include IL-1β, IL-6, IL-18, TNFα, and TGFβ1, although soluble forms of each are also released. While EV and soluble forms of mediators overlap in downstream signaling, major differences have been revealed.[8,9] EV may also present antigen in a cell-free manner, contributing to the adaptive immune response, a phenomenon leveraged for dendritic cell tumor vaccines.

More intensely studied is the role for EV in pathology. EV are intimately involved in several important disease problems in humans, e.g., (1) Tau protein transfer in Alzheimer's disease, (2) prion transmission, (3) transfer of misfolded proteins (e.g., SOD1 in ALS), (4) retroviral packaging, (5) immunopathogenesis of infectious agents (e.g., *Cryptococcus*, *Leishmania*, and HIV, (6) propagation of primary and secondary pulmonary hypertension, (7) initiation of procoagulant states (e.g., pre-eclampsia), and (8) cancer cell evasion of the immune system, recruitment of vessels, transformation of neighboring cells, and induction of cancer-associated fibroblasts.

What Are Diagnostic Applications of Exosomes and Extracellular RNA?

Because EV payloads reflect molecular pathology in many disease states, there is mounting interest in the diagnostic value of EV. Analysis of EV molecular signatures or contents (e.g., infectious agents/components) fulfill the goals of precision medicine, to harmonize molecular pathology, diagnostic biomarkers, and therapeutic targets. Hence, EV may transform the way we screen, diagnose, prognosticate, and monitor patients as the EV signatures are more disease relevant. The first FDA-approved exosome-based product was a diagnostic kit to detect mRNA mutations in patient's urine with suspected prostate cancer produced by Exosome Diagnostics (Cambridge, MA). It is expected to markedly reduce false positives from the current application of PSA in humans. Similar diagnostic tests could be developed for the veterinary market, and some of these are currently in progress.

Therapeutic Application of MSC Exosomes?

Like all cells, MSC shed exosomes, a subgroup of nanoscale (50–200 nm) extracellular vesicles that are rich in RNA (in particular miRNA), proteins, and lipids.[1] There is emerging evidence that **MSC exosomes** have unique potential to be exploited therapeutically. It has been observed that MSC exosomes promote neurite growth and survival, angiogenesis, recovery from ischemia (e.g., stroke, limb ischemia), wound healing, and mitigate neurodegeneration, acute kidney failure, and cirrhosis for example.[10-22] MSC exosomes also appear to be strongly anti-inflammatory although the mechanism is currently unclear (i.e., whether canonical or non-canonical pathways are involved).

The conceptual and practical advantage of advancing MSC exosomes for therapeutic applications include: (1) cells are taken out of the equation, thus reducing manufacturing problems associated with variations in donors, cell quality, cell storage and shipping (viability), and cell transformation in recipients; (2) biomanufacturing of exosomes will result in a more consistent product; (3) exosomes can be engineered as vectors independent of parent cells; (4) exosomes are the core mechanism by which MSC exert biological activity so this component can be concentrated enhancing outcome; and (5) exosome biodistribution is significantly better for intravenously transplanted agents. The disadvantages include (1) rapid uptake, so short half-life, (2) complicated process to isolate, concentrate, and purify, and (3) potential for entry into off-target sites including the brain. Furthermore, there is currently no FDA guidance on extracellular vesicles in veterinary or human medicine.

The implications for a future shift from stem cells to exosomes and multimodal therapies in companion animals will be discussed in the context of the internal medicine.

References

1. Yanez-Mo M, *et al*. Biological properties of extracellular vesicles and their physiological functions. *J Extracell Vesicles*. 2015;4:27066.
2. Kowal J, Tkach M, Thery C. Biogenesis and secretion of exosomes. *Curr Opin Cell Biol*. 2014;29:116–125.
3. Urbanelli L, *et al*. Signaling pathways in exosomes biogenesis, secretion and fate. *Genes (Basel)*. 2013;4:152–170.
4. Peterson MF, Otoc N, Sethi JK, Gupta A, Antes TJ. Integrated systems for exosome investigation. *Methods*. 2015;87:31–45.
5. Weissig V, Pettinger TK, Murdock N. Nanopharmaceuticals (part 1): products on the market. *Int J Nanomedicine*. 2014;9:4357–4373.
6. Hudry E, *et al*. Exosome-associated AAV vector as a robust and convenient neuroscience tool. *Gene Ther*. 2016;23:819.
7. Valadi H, *et al*. Exosome-mediated transfer of mRNAs and microRNAs is a novel mechanism of genetic exchange between cells. *Nat Cell Biol*. 2007;9:654–659.
8. Webber JP, *et al*. Differentiation of tumour-promoting stromal myofibroblasts by cancer exosomes. *Oncogene*. 2015;34:290–302.
9. Shimoda M, Khokha R. Proteolytic factors in exosomes. *Proteomics*. 2013;13:1624–1636.

10. Lopez-Verrilli MA, *et al*. Mesenchymal stem cell-derived exosomes from different sources selectively promote neuritic outgrowth. *Neuroscience*. 2016;320:129–139.
11. Sun L, *et al*. Safety evaluation of exosomes derived from human umbilical cord mesenchymal stromal cell. *Cytotherapy*. 2016;18:413–422.
12. Zhang B, Yeo RW, Tan KH, Lim SK. Focus on extracellular vesicles: therapeutic potential of stem cell-derived extracellular vesicles. *Int J Mol Sci*. 2016;17:174.
13. Aliotta JM, *et al*. Exosomes induce and reverse monocrotaline-induced pulmonary hypertension in mice. *Cardiovasc Res*. 2016;110:319–330.
14. Teng X, *et al*. Mesenchymal stem cell-derived exosomes improve the microenvironment of infarcted myocardium contributing to angiogenesis and anti-inflammation. *Cell Physiol Biochem*. 2015;37:2415–2424.
15. Ko SF, *et al*. Adipose-derived mesenchymal stem cell exosomes suppress hepatocellular carcinoma growth in a rat model: apparent diffusion coefficient, natural killer T-cell responses, and histopathological features. *Stem Cells Int*. 2015;2015:853506.
16. Hu GW, *et al*. Exosomes secreted by human-induced pluripotent stem cell-derived mesenchymal stem cells attenuate limb ischemia by promoting angiogenesis in mice. *Stem Cell Res Ther*. 2015;6:10.
17. Shabbir A, Cox A, Rodriguez-Menocal L, Salgado M, Van Badiavas E. Mesenchymal stem cell exosomes induce proliferation and migration of normal and chronic wound fibroblasts, and enhance angiogenesis *in vitro*. *Stem Cells Dev*. 2015;24:1635–1647.
18. Nakamura Y, *et al*. Mesenchymal-stem-cell-derived exosomes accelerate skeletal muscle regeneration. *FEBS Lett*. 2015;589:1257–1265.
19. Zhang B, *et al*. Human umbilical cord mesenchymal stem cell exosomes enhance angiogenesis through the Wnt4/beta-catenin pathway. *Stem Cells Transl Med*. 2015;4:513–522.
20. Lai RC, Yeo RW, Lim SK. Mesenchymal stem cell exosomes. *Semin Cell Dev Biol*. 2015;40:82–88.
21. Lener T, *et al*. Applying extracellular vesicles based therapeutics in clinical trials - an ISEV position paper. *J Extracell Vesicles*. 2015;4:30087.
22. Yu B, Zhang X, Li X. Exosomes derived from mesenchymal stem cells. *Int J Mol Sci*. 2014;15:4142–4157.

acvim222

Respiratory Distress in Camelids: A Case Based Discussion

Daniela Bedenice, DMV, DACVIM, DACVECC
North Grafton, MA, USA

Introduction

The evaluation of respiratory disease in South American camelids (SAC) may pose a particular challenge to veterinarians, as llamas and alpacas may merely display subtle clinical abnormalities even in the face of progressive or severe respiratory disease. In part, this relates to evolutionary adaptations to environmental hypoxia. For example, the small elliptical red blood cells of camelids have a greater affinity for oxygen, a high hemoglobin concentration and high oxygen transfer conductance (facilitated by the small RBC size) (Jurgens *et al.* 1988). The oxygen dissociation curves of the blood of lamoids are thus shifted to the left, even in camelids adapted to low altitudes. Furthermore, llamas were identified to carry a higher blood oxygen content than, for example, sheep (Moraga *et al.* 1996). A high muscle myoglobin concentration, more efficient oxygen extraction at the tissue level (Benavides *et al.* 1989) and high lactate dehydrogenase activity (Llanos *et al.* 2003) further improve the camelid's resistance to hypoxemia. Camelids are primary nasal breathers rather than obligate nasal breathers, and commonly display evidence of open mouth breathing in response to marked hypoxemia. This has to be clinically differentiated from behavioral responses to stress (handling) and systemic compromise. The resting respiratory rate of adult llamas or alpacas varies from 10–30 bpm, where normal lung sounds are often muted in heavily fleeced animals. Camelids have 12 ribs, and lamoid lungs are most similar to those of horses. A cardiac notch separates the apical portion of the lung, but there are no lobes, except for a small accessory lobe of the right lung. The mediastinum is complete (Viera *et al.* 1968).

Emergency Procedures

The unique anatomical and physiological characteristics of SAC will impact both the assessment and management of respiratory distress in llamas and alpacas. Most emergency procedures, such as tracheostomy, thoracostomy, and orotracheal intubation are performed using similar techniques to those established in companion animal species. **Endotracheal intubation** is necessary for ventilation (manual/mechanical) during resuscitation or anesthesia, and prevents aspiration if the patient should regurgitate. Most commonly, size 12–16 mm inner diameter (ID) endotracheal tubes (55 cm length) are appropriate for adult llamas and size 7–11 mm ID for adult alpacas. It is important to use a blunt ending wire or polyethylene catheter to serve as a guide during orotracheal intubation, while the animal's head is maintained in maximal extension. If the epiglottic cartilage is not visible, the soft palate may be lifted from beneath the epiglottis with the tip a long laryngoscope blade. The polyethylene catheter/guide wire is then inserted into the trachea and the laryngoscope blade removed. The endotracheal tube should be threaded over the catheter and pushed gently between the cheek teeth into the trachea. Retrograde intubation as an alternative to traditional normograde intubation in camelids, was recently described in llamas by Byers *et al.* (2009). The technique involves introducing a trocar and stylet into the trachea and advancing the stylet to exit the mouth. The stylet is then used to guide the endotracheal tube into the correct position. The latter study found no statistical difference in time to place the stylet or endotracheal tube.

An **emergency tracheotomy** may be indicated for camelids with obstruction of the nasal, nasopharyngeal, or upper trachea. The procedure is usually performed just proximal to the junction of the upper and middle third of the cervical region, using aseptic techniques, if possible. Following a linear (midline) incision, the paired sternohyoideus muscles are bluntly separated to expose the fascia over the trachea. A horizontal incision (from lateral to midline) is then performed between two tracheal rings of less than half (50%) of the tracheal circumference, to prevent tracheal instability (Huber 2012). Commercial (human) tracheostomy tubes (size 7–11 ID for adult alpacas) with a removable inner cannula are preferred to maintain airway patency. Placement of a stay suture in both the upper and lower tracheal rings will facilitate replacement of tracheostomy tubes as needed, especially in small-sized animals (such as crias).

An **arterial blood sample** can most easily be obtained from the saphenous artery of the medial thigh (alternatively the auricular artery may be used). Normal arterial blood gas result have been documented in 14 adult, healthy male and non-pregnant female alpacas (mean weight: 62 ± 21 kg, mean age: 4.5 years) as well as 12 healthy neonatal alpaca crias (2–7days old, mean 7.8 kg body weight). Data showed a median (95% CI interval) PaO_2 of 104.6 (100.7–108.4) mm Hg in adults and 92 (87.7–96.4) mm Hg in crias. The reported $PaCO_2$ was notably lower in both adult (27.7 mm Hg, 26.4–28.9) and neonatal alpacas (30.5 mm Hg, 27.2–33.8) compared to domestic ruminants.[13] If arterial blood gas analysis is delayed, samples should be stored on ice in glass syringes, to reduce variability in PaO_2. Storage related changes in other blood gas parameters may be clinically insignificant for the first 2 hours following sample collection (Buckley *et al.* 2014).

Sampling of the Respiratory Tract

The preferred site of **pleurocentesis** is the 6th or 7th intercostal space, 10–15 cm dorsal to the ventrum of the sternum or 2–4 cm dorsal to the costochondral junction of the ribs in normal llamas. Penetration can be made from either side and may utilize a 14–16 gauge needle or cannula (2 inch length) inserted near the cranial border of the rib to avoid intercostal vessels. In normal llamas fluid should flow from the needle at an approximate depth of 2–3 cm (1–1.5 inches). The pleural fluid of 17 clinically healthy adult llamas identified a normal total nucleated cell count of 200–1500 cells/mL with a mean of 576 ± 361 cells/microliter. In these fluids, small lymphocyte, mononuclear cell, and neutrophil percentages ranged from 80%–100%, 0%–20%, and 0%–10%, respectively. Pleural fluid total protein concentrations, determined by refractometry, ranged from <2.5 to 3.5 g/dL. The pleural fluid glucose and lactate concentrations were very similar to the plasma concentrations in normal llamas, with a pleural fluid specific gravity of 1.013 ± 0.002, glucose concentration of 135.1 ± 9.02 mg/dL and lactate of 2.95 ± 1.34 mg/dL (Gerros, Andreasen 1999; Tornquist 2009).

The cytological characteristics of tracheal fluid obtained by trans-tracheal aspiration have also been described in 17 clinically healthy adult llamas. The technique for **transtracheal aspiration** was similar to that used in horses. A 10 cm^2 area over the ventral mid-cervical region was clipped, aseptically prepared and locally infiltrated with Lidocaine HCl (2% solution). A stab incision was made through the skin with a #15 scalpel blade to allow insertion of a 10-gauge, through-the-needle Winch catheter between the tracheal rings. Approximately 20 mL of sterile saline was injected through the catheter into the trachea, then immediately aspirated back into the syringe. Cytologic evaluation of TTA fluid revealed that the majority of cells were vacuolated macrophages (60–100%), with 0–40% neutrophils, and fewer lymphocytes (0–2%), eosinophils (0–3%), and ciliated respiratory epithelial cells (0–10%) (Gerros, Andreasen 1999).

Upper Airway Dysfunction

Choanal atresia is considered the most common congenital upper airway disease in SAC, occurring in 0.48% and 0.75% of live births in alpacas and llamas, respectively (Johnson *et al.* 2008). The choanae are the caudal openings of the nasal cavities into the nasopharynx (Nykamp *et al.* 2003). Failure in the embryologic development of these openings may thus lead to unilateral or bilateral, membranous or osseous, and complete or partial choanal atresia. Since camelids are primary nasal breathers rather than obligate nasal breathers, complete choanal atresia is not immediately fatal, but often presents with congenital respiratory distress. Partial choanal atresia can be particularly difficult to diagnose due to minimal or complete lack of clinical signs. Choanal atresia is considered a heritable disorder with a complex mode of inheritance (Hathcock *et al.* 1995). A recent evaluation of CHD7 as a candidate gene for choanal atresia in alpacas showed that, in contrast to humans, CHD7 mutations do not appear to be the primary cause of choanal atresia in alpacas (Reed *et al.* 2013). Positive contrast rhinography, nasal catheter placement, endoscopy, and computed tomography (CT) have all been used to diagnose choanal atresia. If surgical intervention is considered, CT can provide the spatial resolution necessary to differentiate the types of choanal atresia and estimate the length of the obstructive lesion (Nykamp *et al.* 2003). Surgery may be beneficial in the membranous forms of choanal atresia, and has been successfully performed using endoscopically guided fenestration of choanal membranes in juvenile animals. Fulton *et al.* described the surgical perforation of choanal membranes in a 2-day-old llama cria, using an endotracheal tube guarded Steinman pin. It is recommended to maintain nasal stent tubes in place for 2–3 weeks, to prevent reobstruction following surgical repair (Fulton *et al.* 1989).

Snake envenomation, bilateral laryngeal paralysis, laryngeal swelling, traumatic head injury, drug hypersensitivity, nasopharyngeal feed impaction, and aberrant nasopharyngeal bots may lead to upper airway obstruction in both juvenile and adult camelids. A diagnosis is commonly based on rhinoscopy, radiography, or computed tomography. In these situations, a tracheotomy is often required to provide a functional, low-resistance temporary airway (Huber 2012).

Diaphragmatic Paralysis

Diaphragmatic paralysis may induce a varying degree of respiratory failure in camelids, since the diaphragm normally functions as the main inspiratory muscle of mammals. In healthy camelids, the chest and abdominal movements should be synchronous throughout the respiratory cycle. However, in animals with diaphragmatic dysfunction or failure, a characteristic pattern of "paradoxical breathing" is evident. In this case, chest expansion via intercostal muscle contraction initiates or leads inspiration. A negative inspiratory, intrathoracic pressure is thus generated, and the inactive diaphragm is passively deflected into the chest, resulting in a concurrent inward shift of the abdominal wall. As a consequence, the abdominal and thoracic respiratory movements are 180 degrees out of phase in patients with complete diaphragmatic failure. Complete thoraco-abdominal asynchrony has been associated with airway obstruction, pneumothorax, diaphragmatic rupture and bilateral diaphragmatic paralysis. The etiologic diagnosis of diaphragmatic paralysis remains challenging. However, genetic, metabolic (oxidative stress due to hyperglycemia in camelids), infectious (aberrant parasite migration, tick paralysis, EHV-1), malignant, traumatic, nutritional (vitamin E deficiency), and toxicologic causes (heavy metals and several organic compounds) have been proposed. Diaphragmatic paralysis due to phrenic nerve degeneration was originally reported in a llama (Bedenice *et al.* 2002), and was associated with multifocal diaphragmatic myofiber degeneration. The clinical disease in the latter report led to thoraco-abdominal asynchrony and was characterized by non-invasive pulmonary function testing. Similarly, diaphragmatic paralysis in 11 juvenile alpacas (Byers *et al.* 2011), with 3 alpacas presenting from a single farm within 6 months, was diagnosed antemortem via fluoroscopic examination in 7/11 animals. Phrenic nerve degeneration was identified in 6/6 alpacas submitted for necropsy.

Pulmonary Aspergillosis

Pulmonary disease associated with *Aspergillus fumigatus* infection may cause a wide spectrum of clinical syndromes, including a variety of infectious and allergic diseases, depending on the host's immune status or pulmonary structure (Kousha *et al.* 2011). Inhalation and colonization of *Aspergillus* spores may be inconsequential in the normal, immune competent host, while chronic necrotizing aspergillosis, invasive pulmonary aspergillosis and aspergilloma development have been specifically associated with the presence of underlying chronic lung disease (or mildly immunocompromised hosts), immune deficiency and cavitary lung disease, respectively. Allergic bronchopulmonary aspergillosis is most likely observed in humans with a history of asthma, cystic fibrosis, or atopy (Soubani *et al.* 2002). In camelids, disseminated aspergillosis has been previously described in association with forestomach ulceration. More recently, cavitary pulmonary aspergillosis was identified in an adult alpaca with type II diabetes, showing large, rounded, cavitated lesions on thoracic radiographs. Trans-tracheal aspirate and intralesional, ultrasound-guided fluid aspiration both identified neutrophilic inflammation with evidence of cellular necrosis and fungal hyphae. The numbers and location of these hyphae in abnormal cystic structures within the pulmonary parenchyma was consistent with a primary fungal pathogen. Subsequent culture confirmed *Aspergillus fumigatus* infection. The alpaca failed to respond to both systemic (Voriconazole, 250 mg IV once daily) and intralesional (Amphotericin B) antifungal therapy long-term. Post-mortem examination confirmed marked, multifocal and locally extensive, suppurative, interstitial pneumonia with extensive cavitary pulmonary aspergillosis or chronic necrotizing aspergillosis (CNA). Chronic necrotizing aspergillosis (CNA) is an indolent, cavitary, and infectious process of the lung parenchyma secondary to local invasion by *Aspergillus* species, which is usually slowly progressive over weeks to months. This syndrome is rare in both humans and animals and the available literature is based on individual case reports and small case series. In humans, CNA most commonly affects middle-aged and elderly patients with altered local defenses, secondary to diabetes mellitus, alcoholism, chronic liver disease, low-dose corticosteroid therapy, malnutrition, and connective tissue diseases such as rheumatoid arthritis. CNA is generally associated with underlying chronic lung diseases such as COPD, previous pulmonary tuberculosis, thoracic surgery, radiation therapy, pneumoconiosis, cystic fibrosis, lung infarction, or (less commonly) sarcoidosis (Zmeili *et al.* 2007). Risk factors in the affected alpaca may have included age, chronic type II diabetes, and underlying pulmonary parenchymal disease.

Acute Lung Injury (ALI) and Acute Respiratory Distress Syndrome (ARDS) in Camelids

The clinical syndrome of ALI and ARDS generally reflects a state of acute, diffuse alveolar damage that is characterized by increased capillary permeability (inflammation), pulmonary edema and refractory hypoxemia, following a variety of direct and indirect insults to the parenchyma or vasculature of the lung. A retrospective cohort analysis recently applied the diagnostic criteria of neonatal ALI and ARDS to 237 crias (1999–2014) admitted to a university hospital under 4 weeks of age. All patients with lateral thoracic radiographs and arterial blood gas results obtained on room air within 24 hours of each other were included, while eliminating any neonatal patients born following C-section, induction, or with a diagnosis of prematurity. The diagnostic PaO_2/FiO_2 threshold was adjusted for age as previously described (Wilkins *et al.* 2007). Overall, merely 10 of 237 (4.2%) crias met the diagnostic criteria of neonatal ALI (non-survivors: 4/7) and ARDS (non-survivors: 1/3). Pulmonary histopathology was available in 3 animals, showing diffuse alveolar damage in one cria with a clinical diagnosis of ALI, and in one cria with ARDS. Severe diffuse alveolar edema and atelectasis were noted in the third (Bedenice *et al.* 2014). In contrast to ALI and ARDS, neonatal respiratory distress syndrome (RDS) requires a primary developmental surfactant dysfunction and lack of an inflammatory component. Successful medical management of RDS has been recently described in a neonatal alpaca cria, using equine surfactant and positive pressure ventilation (Tinkler *et al.* 2015).

References

1. Bedenice D, *et al.* In: Proceedings from the Veterinary Comparative Respiratory Society; 2014.
2. Bedenice D, *et al. J Vet Intern Med.* 2002;16:603.
3. Benavides CE, *et al. Respir Physiol.* 1989;75:327.
4. Buckley GJ, *et al.* WJVSC 2, eCollection 2014.
5. Byers S, *et al. J Vet Intern Med.* 2011;25:380.
5. Fulton C, *et al. J Zoo Wildl Med.* 1989;20:488.
6. Gerros TC, *et al. Vet Clin Pathol.* 1999;28:29.
7. Hathcock J, *et al. Vet Radiol US.* 1995;36:290.
8. Huber M. ACVS. 2012.
9. Jurgens KD, *et al. J Comp Physiol.* 1988;B158:469.
10. Llanos AJ, *et al. High Alt Med Biol.* 2003;4:193.
11. Moraga F, *et al. Comp Biochem Physiol.* 1996;115:111.
12. Nykamp SG, *et al. Vet Radiol US.* 2003;44:534.
13. Reed KM, *et al. The Vet J.* 2013;198:295.
14. Tinkler S, *et al. Can Vet J.* 2015;56:370.
15. Tornquist SJ. *Vet Clin North Am Food Anim Pract.* 2009;25:311.
16. Viera RF, *et al.* Rev Fac Med Vet Univ Nac Mayor San Marcos 1968;22:54.
17. Wilkins PA, *et al. JVECC.* 2007;17:333.

1.
acvim365

Feline Bronchiolar Disease: Clinical Features and Ct/Histopathologic Correlates

Kurt Williams[1], DVM, DACVP, PhD; Carol Reinero[2], DVM, DACVIM, PhD
[1]East Lansing, MI, USA; [2]Columbia, MO, USA

Overview

In veterinary medicine, with increasing use of sophisticated diagnostic imaging tests such as thoracic computed tomography (CT), veterinary specialists are faced with the need to "re-learn" or perhaps learn for the first time how to interpret results for diagnosis and classification of certain respiratory diseases. We believe this is particularly true for a variety of pulmonary diseases including vascular disease, fibrotic lung disease, other interstitial lung diseases and, the subject of this lecture, bronchiolar disease. Having histopathologic correlates to information obtained through CT is critical to confirm and describe the underlying disorder. While in widespread use and relatively inexpensive, thoracic radiographs are comparatively crude means of assessing thoracic disease. It is becoming increasingly recognized that bronchial, interstitial, and alveolar patterns do not necessarily localize microscopic disease to the airways, interstitium, and alveoli. Additionally, there are many lesions completely missed on thoracic radiography. Lesions associated with the small airways (bronchioles), are a classic example of this. Bronchioles are defined as airways less than 2 mm in diameter that lack cartilage in their walls inclusive of terminal, respiratory and membranous bronchioles. They are accompanied by branches of arteries (arterioles) as they head to gas exchange units of the lung. While in rare circumstances they have been described on histopathologic examination as part of another disease process, they may also be the primary site of pathology. While not yet described in the veterinary literature, cats develop bronchiolar disease that has important implications in terms of morbidity and mortality.

Why Are Bronchiolar Diseases New in the Cat?

It is highly likely that feline bronchiolar diseases are not emerging disorders, but disorders that have been missed for a wide variety of reasons. Bronchiolar diseases represent a spectrum or group of disorders sharing in common involvement of the small airways; this spectrum of disorders shares many clinicopathologic features among itself and in common with other respiratory disease mimics (see Respiratory Disease in the Cat: Typical Considerations, below). Importantly, since bronchiolar diseases have not yet been described as distinct clinical entities, they are not on the radar of clinicians. Because the bronchioles are very small, bronchiolar disorders have no pathognomonic radiographic features and in fact, are not visible; thus, they are often overlooked by radiologists viewing plain survey thoracic radiographs. Similarly, bronchiolar disorders are not well recognized by most pathologists, especially in chronic end-stage lesions where bronchioles may have been largely destroyed. So how have clinicians been tipped off to the existence of bronchiolar disorders? The answer lies in multidisciplinary collaboration between clinicians, radiologists, and pathologists with clinical cases that just did not make "sense" (i.e., a lack of correlation between clinical findings, imaging results, and histopathologic features).

The recent use of HRCT high resolution computed tomography (HRCT) has revolutionized the antemortem detection of lesions that are compatible with bronchiolar disease. The presence of these direct and indirect lesions, especially with dynamic studies including inspiratory and expiratory breath hold series, has been instrumental in localizing disease to the small airways. Ultimately, histopathologic examination is critical for diagnosis and classification of bronchiolar disease. This is not commonly performed antemortem in cats, again leading to a missed opportunity for disease identification. Because few veterinary pathologists are aware of bronchiolar disease as a primary disease entity, (despite descriptions of small airway involvement in infectious, neoplastic, and interstitial lung diseases included on a final pathology report), clinicians and radiologists will need to discuss having their pathologist take another look at the small airways. "End stage" airways that have been completely obliterated are often just absent from biopsies, and their absence without other dramatic nearby lesions (e.g., inflammation) is very easy to miss. Special stains highlighting elastic remnants of small airways adjacent to arterioles are required to highlight these lesions. End stage airways fail to provide information on what may have caused the fibrotic lesions, similar to how identification of cirrhosis on a liver biopsy reflects end stage damage without a specific cause. This makes it especially critical to have samples from multiple regions of the lung or multiple lungs to look for active, ongoing pathology. Even knowing that bronchiolar lesions are present in histologic samples, to have meaningful clinical interpretation it will be necessary to understand more about the natural history of these diseases, distinct clinical features, and provide histopathologic correlates to lesions found in minimally invasive diagnostics such as HRCT. In human medicine where bronchiolar disease has been recognized for decades, there is still no consensus terminology used to classify the disorders with terms such as bronchiolitis obliterans meaning different things to different people. Another challenge both in human and feline medicine will be to identify exposures or triggers causing bronchiolar disorders. With the marked variability in presentation and uncommon recognition of bronchiolar disorders, this is likely to remain an obstacle.

RESPIRATORY DISEASE IN THE CAT: TYPICAL CONSIDERATIONS

Classically, major categories of respiratory disease in the cat affecting the airways and pulmonary parenchyma include infectious disorders (bacterial, fungal, viral, protozoal, parasitic), inflammatory processes (e.g., chronic bronchitis, asthma, aspiration pneumonia, foreign body pneumonia, eosinophilic pneumonia, etc.), neoplasia (primary or metastatic), pulmonary edema (non-cardiogenic or cardiogenic), pulmonary hemorrhage, interstitial lung diseases, and structural or functional disorders (bronchiectasis, bronchopulmonary dysplasia, bullae or cavitary lesions). Obtaining a thorough history and completing both a distance and hands on physical examination can provide important clues to help narrow the list of differentials, but ultimately other diagnostic tests need to be performed. Thoracic radiography is a common subsequent diagnostic test that has value in discriminating cardiac from respiratory causes of disease and in determining structural disorders. Radiography can also help determine if disease is focal, multifocal, or diffuse and can provide a general idea of anatomic severity (i.e., how much of the pulmonary parenchyma appears to be involved and if the changes are mild, moderate, or severe). However, there are important limitations to thoracic radiography that makes computed tomography appealing. Computed tomography allows for three-dimensional reconstruction of anatomic structures with very good discrimination of tissue types and is capable of detecting subtle lesions which are missed on thoracic radiography. Normal bronchioles are not visualized on CT scans, but when diseased, there are signs that indicate their presence.

CT changes in humans compatible with bronchiolar disease are categorized into direct and indirect signs. Direct signs are noted by changes to the bronchioles themselves and include a centrilobular pattern of nodules, tree-in-bud pattern, ground glass opacity surrounding the bronchioles and bronchiolectasis. Indirect signs pertain to changes induced to the lung tissue distal to the bronchioles including a mosaic pattern of attenuation and air trapping. In cats, for optimal interpretation of thoracic HRCT, general anesthesia to allow for both inspiratory and expiratory holds will allow detection of lesions of air trapping. In veterinary medicine, efforts to interpret and correlate thoracic HRCT scans to the underlying histopathologic features is in an early stage. To maximize the value of such efforts histopathologic specimens must be obtained and processed appropriately and pathologists must be included in multidisciplinary discussions.

RESPIRATORY DISEASE IN THE CAT: PATHOLOGIC CONSIDERATIONS

In human medicine, the use of open lung biopsies began in the 1960s; it is now a routine diagnostic procedure widely used in a variety of lung diseases. Lung biopsy has been particularly important in facilitating the classification of human interstitial lung diseases and has also been key for diagnosis and classification of bronchiolar disorders. From the perspective of the pathologist in veterinary medicine, lung biopsy is woefully underutilized. Lung biopsy promises to alter our understanding of the pathogenesis, and even identification, of respiratory disease in our profession. To be able to recognize, classify, and correlate the histologic appearance of bronchiolar lesions in cats with clinical outcomes and response to therapy, early and more aggressive antemortem biopsies will be needed. Additionally, while lesions are more likely to be advanced, post-mortem histopathologic examination also has value in identification of feline bronchiolar disease.

How to Obtain and Handle Lung Biopsies: A Pathologist's Perspective

Surgical lung biopsies in veterinary medicine will likely only be performed once in the lifetime of the patient so it is incumbent on the clinicians and pathologists to maximize the amount of information that can be obtained from each biopsy. Site selection for the biopsy is of paramount importance, thus communication between the clinician, surgeon, radiologist, and pathologist is important (though rarely done). If a lesion is determined clinically to be solitary then biopsy of the single site is appropriate, though it is important to include the border between normal lung and lesion in the biopsy sample. In cases with more widespread disease selection of the site(s) to be biopsied may determine the value of the information obtained from the biopsy.

In general, the following recommendations should be considered:

1. Biopsy more than one lobe if possible.
2. If radiographic or CT information suggests heterogeneity of the process, then multiple sites reflecting the various stages of progression of disease should be biopsied.
3. **Do not** only biopsy lesions considered to be the oldest/most well developed: also sample lesions deemed intermediate in progression as well as lung considered to be normal.
4. Avoid obtaining the apex of lung lobes as these may have incidental lesions, i.e., sub-pleural fibrosis, that may be unrelated to the disease process of interest.
5. Keep a section of fresh unfixed lung for potential microbiologic examination.
6. If an entire lung lobe is removed, it is advisable to inflate the lobe with fixative (usually 10% neutral buffered formalin) via the lobar bronchus using a 60-cc syringe until the pleura is uniformly expanded and smooth.
7. Thoracoscopically-obtained biopsies can be gently inflated with fixative using a needle and syringe inserted through the pleura.
8. It may be valuable to consult with the pathologist prior sectioning the biopsy material to discuss proper regions to be processed for histologic evaluation.

Bronchiolar Disease in the Cat: Case Examples

In this lecture, a series of clinical cases will be presented in cats with bronchiolar disease highlighting the clinical presentation, CT scan results and their histopathologic correlates. Awareness of the existence of these disorders will help in their classification, which in turn may provide useful for future investigation of optimal management and treatment strategies.

References

References are available upon request.

1.
acvim367

How Dysphagia Hurts the Lung: An Overview of Cough-Swallow Disorders

Megan Grobman, DVM, DACVIM (SAIM), MS; Carol Reinero, DVM, DACVIM (SAIM), PhD
Columbia, MO, USA

Overview

A relationship between pathology of the gastrointestinal and respiratory tracts is linked to important clinical syndromes. Aspiration of gastric contents leads to direct chemical injury, and in some cases influx of bacteria leading to aspiration pneumonia/pneumonitis. However, this is only the most common condition in a large group of respiratory diseases associated with gastrointestinal dysfunction. Macro-aspiration and microaspiration have been associated with upper airway disorders, large airway obstruction, bronchiectasis, diffuse aspiration bronchiolitis, aspiration pneumonitis, and exogenous lipid pneumonia in addition to aspiration pneumonia.[1]

It is estimated that over 50% of healthy adult humans aspirate during sleep. Additionally, reflux is a common source of acute and chronic pulmonary disease mostly through chronic microaspiration. In human medicine, reflux is common with a prevalence of 50% in patients with chronic cough, asthma, and chronic obstructive pulmonary disease (COPD). While this has not been exhaustively evaluated in veterinary medicine, the association between reflux disease and airway dysfunction has been supported by a number of clinical case reports and experimental canine and feline models.[2-5]

In discussing the link between pulmonary and GI disease, it is important to recognize risk factors for aspiration-related respiratory syndromes. GI diseases leading to airway dysfunction include dysphagia, regurgitation, recurrent/chronic vomiting, gastroesophageal reflux disease, and depressed consciousness. Over time, these conditions can result in architectural changes and loss of normal airway protective mechanisms, leading to sequelae such as bronchiectasis and secondary infection. These changes predispose to a vicious cycle of airway inflammation, infection, structural damage, and decreasing respiratory function. Disorders impacting the oral cavity, oropharynx, and esophagus are common in dogs and cats and are typically associated with dysphagia, regurgitation, chronic vomiting, and gastroesophageal reflux disease. An understanding of the basics of the swallowing mechanism allows for improved clinical recognition and treatment of these conditions.

The Physiology of Swallowing

The swallow can be separated into voluntary and involuntary phases. The voluntary phase involves the prehension and mastication of food and propulsion of the food or liquid bolus to the oropharynx. The presence of a bolus in the proximal oropharynx initiates the pharyngeal phase of the swallow. This and all subsequent phases are involuntary. The pharyngeal phase is characterized by the initiation of airway protective behaviors (blockage of the nasopharynx and larynx) and passage of the bolus through the upper esophageal sphincter and into the esophagus. Passage into the esophagus marks the beginning of the esophageal phase of swallowing. Contraction of the UES and accompanying peristaltic contraction represents the primary wave. Luminal distention triggers progressive contractions (secondary waves). The esophageal phase ends with the opening of the lower esophageal sphincter to allow the passage of food into the stomach and closure to prevent reflux. Disorders of swallowing and regurgitation represent defects in one of these processes.

Dysphagia, Regurgitation, and Vomiting

Dysphagia refers to difficulty swallowing. This term is typically used to describe defects in the oral and oropharyngeal phases; however, in some cases esophageal diseases such as esophagitis can also present with signs of dysphagia. Dysphagia may be classified based on location/swallowing phase (oral vs. oropharyngeal vs. esophageal) and/or mechanism. Dysphagia may be secondary to functional disturbances of the swallowing reflex (e.g., myasthenia gravis and cricopharyngeal achalasia) or secondary to structural abnormalities (e.g., severe dental disease, foreign bodies, strictures, neoplasia, and trauma). Key clinical features associated with dysphagia include difficulty with prehension, gagging, retching, and repetitive swallowing behaviors. Though these features may be consistent across location/swallowing phase, treatment is not. While some conditions can be managed medically (pharyngeal esophagitis, myasthenia gravis), disorders of motility (cricopharyngeal achalasia) require targeted intervention. Incorrect localization may lead to significant worsening of clinical signs, making correct identification critically important.

The clinical feature most commonly associated with defects in the esophageal phase of swallowing is regurgitation. Regurgitation refers to the **passive** retrograde expulsion of esophageal contents. Regurgitation may be classified as congenital vs. acquired and, like dysphagia, may be secondary to anatomic/obstructive lesions or diseases of motility. Obstructive disorders include vascular ring anomalies, strictures, and esophageal/paraesophageal tumors. Motility disorders include defects of the lower esophageal sphincter (e.g., LES achalasia), muscle (e.g., esophageal hypomotility), inflammatory disease (e.g., esophagitis), neuromuscular dysfunction (e.g., botulism, myasthenia gravis, and polyneuropathy/polymyopathy), and toxins (e.g., lead and organophosphates).

In contrast to regurgitation, vomiting is a centrally mediated process that involves the coordinated closure of the glottis and nasopharynx in order to protect the airways. Vomiting is the **active**, conscious, retrograde expulsion of GI contents. However, despite this attempt to protect the airways, chronic vomiting is considered a major risk factor for respiratory disease by overwhelming these protective mechanisms.[2,4,6,14]

Reflux and Reflux Disease

A less well-known link between respiratory and GI disease is gastroesophageal reflux (GER). Gastroesophageal reflux refers to the backflow of stomach contents into the esophagus. Extraesophageal manifestations of GER can also occur once the gastric contents leave the esophagus. Gastroesophageal reflux is often an occult process in veterinary species, in part because they cannot communicate clinical signs such as heartburn. It is important to note that GER can occur in healthy, asymptomatic humans, and there is evidence of this in dogs.[6] However, reflux can also result in clinically relevant syndromes in some individuals, such as esophagitis, laryngeal dysfunction, regurgitation, and secondary respiratory disease.

The combination of GER and clinical signs and gross pathology is then termed gastroesophageal reflux disease (GERD) or extraesophageal reflux disease (EERD). Clinical signs are dependent on frequency, volume, and duration of contact between the refluxate and the esophagus, larynx, pharynx, airways, and pulmonary parenchyma. Causes of reflux include spontaneous relaxation of the LES (most common), diminished basal LES pressure, straining (coughing, vomiting, increased intraabdominal pressures), and hiatal hernia. Pathology-associated wither GERD can occur by direct contact of low pH refluxate and digestive enzymes on the tissue, leading to tissue damage and stimulation of regional nerve terminals, and by macro-/microaspiration. Indirect mechanisms include stimulation of esophageal bronchial reflex whereby stimulation of the esophagus leads to distant changes in airway resistance.

Clinical Approach

The assessment of patients presenting with features consistent with swallowing disorders requires a thorough history and review of the patients' signalment. Puppies can be diagnosed with a number of congenital predisposing abnormalities, including persistent right aortic arch (PRAA), cleft palate, and cricopharyngeal achalasia. Breeds that have been associated with a high incidence of dysphagia include German Shepherd dogs (vascular ring anomaly), large-breed dogs (masticatory muscle disorders), and Golden retrievers (cricopharyngeal achalasia). Additionally, drug and anesthetic history may increase suspicion for esophageal strictures and/or esophagitis. A thorough physical examination, evaluation for comorbidities, and evaluation of swallowing are also critically important.

Specific imaging modalities can be used to assess for both swallowing dysfunction and respiratory disease. These include survey radiography, computed tomography, and videofluoroscopic swallow studies (VFSS).[6] Survey radiographs are often the first line for evaluation of an animal with respiratory disease, dysphagia, or both. Positioning is critical, as rotation can lead to summation artifact and difficulty in interpreting images, particularly in the cervical region. Radiographs of the thorax and the neck (3-view thorax and at least 1 lateral projection of the neck) are intended to rule out anatomic and obstruction lesions that result in dysphagia, and to detect the presence of comorbidities such as aspiration pneumonia. Aspiration pneumonia is common in patients with dysphagia and is a documented negative prognostic indicator for patients with megaesophagus.[7] The lung lobes most commonly involved include the right middle and the right and left cranial lung lobes. As the right middle is considered the most common site of involvement, a left lateral thoracic radiograph is required for thorough assessment of aspiration pneumonia. The addition of contrast can be used to help better define the origin of the problems - for example, with vascular ring anomalies in young puppies.

Computed tomography (CT) is helpful for evaluation of thoracic masses, extraluminal/intraluminal esophageal lesions, pharyngeal/retropharyngeal masses, and thorough evaluation of pulmonary diseases secondary to chronic aspiration (aspiration bronchiolitis and bronchiectasis). CT is considered the most sensitive diagnostic modality for the identification of bronchiectasis. Survey radiography and CT scans are both limited, however, in that they can only reliably detect static anatomic and obstructive changes. As swallow disorders are by definition dynamic, a more sensitive imaging modality would be one that can evaluate a swallow in real time.

Videofluoroscopic swallow studies (VFSS) have the benefit of allowing real-time imaging. A recent study has standardized the VFSS to allow for natural positioning and bolus size as well as limiting the need for force feeding.[6] By avoiding forced feeding, there is reduced risk of aspiration. This is an excellent means of detecting subtle, segmental, and dynamic motility disorders that do not result in, or may underlie, visible megaesophagus. Reflux and sliding hiatal hernias may also be detected by VFSS.

Summary

In veterinary medicine, our understanding of the link between gastrointestinal and respiratory disorders is still in its infancy. This lecture will focus on the use of clinical case scenarios where veterinary patients with defects in swallow subsequently developed a wide variety of respiratory disorders. The diagnostic and therapeutic approach will be highlighted.

References

1. Nafe L, *et al. J Am Vet Med Assoc*. 2017; in press.
2. Loughlin CJ, *et al. The Laryngoscope*. 1996;106(12 Pt 1):1506–1509.
3. Lux CN, *et al. J Am Vet Med Assoc*. 2012;240(9):1100–1103.
4. Poncet CM, *et al. J Sm Anim Pract*. 2006;47(3):137–142.
5. Tuchman DN, *et al. Gastroenterology*. 1984;87(4):872–881.
6. Harris RA, *et al. J Vet Intern Med*. 2017, in press.
7. McBrearty AR, *et al. J Am Vet Med Assoc*. 2011;238(12):1622–1628.

acvim369

Bronchoalveolar Lavage: Picking Protocols to Match the Case

Eleanor C. Hawkins, DVM, DACVIM (SAIM)
Raleigh, NC, USA

Introduction

Bronchoalveolar lavage (BAL) can be successfully performed in many different ways. However, each variation in protocol has the potential to influence the quality of the specimen collected. This presentation will focus on the clinical implications of BAL protocol decisions, with an emphasis on maximizing diagnostic yield in individual patients. Such a goal is different than what is desired in a research setting where uniformity of methods is essential in an attempt to minimize variation in results between patients. The need for uniformity may also become greater in the future if the measurement of concentrations of specific enzymes and inflammatory mediators within the BALF itself becomes routine for clinical patients. Concentrations of such compounds are influenced by the dilution of epithelial lining fluid (ELF; already a very small percentage of return volume), and therefore by variables regarding collection technique, and also variation in anatomy and disease-related differences between individual patients. Ideally, there would be concurrent measurement of a normative compound in the ELF, such as urea, but even these methods are imperfect. In a review article of BAL in people[1] the following variables were identified as affecting BAL results: the disease process itself, suction pressure, handling of fluid (filtered or not; concentrated or not), volume instilled, handling of first aliquot recovered, number of aliquots, position of patient, area lavaged, number of areas lavaged, variability of lavage return, reporting of measurements of acellular components (per mL of recovered fluid vs. normalized to urea or another compound), and sample storage. In the research setting, these factors should be kept as consistent as possible and reported.

Bronchoalveolar lavage fluid (BALF) has the same benefits and shortcomings as other specimens that are not tissue. The collection of specimen generally carries less risk and expense than lung or airway biopsy. However, to provide an accurate and complete diagnosis, the specimen must be representative of the disease process or processes that are present. Failure to obtain a representative specimen can result from sampling an unaffected region of the lung, alteration in inflammatory or neoplastic cell populations or obscuring of infectious organisms as a result of recent treatments, secondary processes such as infection obscuring underlying disease, or relative adhesion of different cell populations to the lung.

Possible cytologic findings from BALF are limited primarily to classification of the inflammatory cell population and the identification of infectious agents and criteria of malignancy. Evaluation for more specific information is possible with BALF because of the relatively large specimen size. Fluid can be cultured for aerobic and anaerobic bacteria or fungi; PCR for a variety of infectious agents can be performed; and cell populations can be further evaluated with techniques such as flow cytometry or cell function testing (the latter limited primarily to research settings). Decision-making related to specific techniques for BAL begins with the top differential diagnoses for the particular patient and how to maximize the likelihood of confirming, or at least supporting, those diagnoses.

What Makes BAL Fluid Different From Other Airway Fluid Specimens

Fluid can be obtained from patient airways through direct collection of sputum, in patients considerate enough to actually cough out secretions and not immediately swallow them, tracheal wash (via transtracheal or endotracheal techniques), bronchial wash (a slightly deeper version of the tracheal wash), and BAL. Tracheal and bronchial wash are collected through relatively small-gauge catheters with their tip situated near the carina (proximal to the carina for tracheal wash; within the bronchi for bronchial wash). A relatively small volume of saline (usually boluses of 3–5 ml) is infused, then retrieved along with secretions within the large airways. Such specimens are often representative of the disease process in patients with diffuse bronchial disease or with bronchopneumonia. Exudate in patients with bronchopneumonia commonly extends into the large airways as a result of mucociliary clearance, cough, and/or extension of disease.

Bronchoalveolar lavage is the retrieval of fluid following the instillation of a sufficient volume of saline to fill the alveoli dependent on the bronchus being sampled. The bronchoscope or lavage catheter is lodged in an airway so to increase the volume of fluid retrieved from distal regions of the lung. Secretions from the bronchi are obtained, but the primary site of collection is the "deep lung" - smaller airways, alveoli, and in some diseases, the interstitium. In comparison to tracheal and bronchial washes, the sample volume and, importantly, the number of cells for cytologic examination is much greater. These qualities may make the identification of an etiologic diagnosis more likely. In some cases where a top differential diagnosis is the absence of disease, an assessment of "unremarkable" is more comfortable from BALF because even in health there is the expectation of several hundred cells/μl with a predominance of alveolar macrophages. With tracheal and bronchial washes, a healthy airway will yield few cells leaving open the possibility of an inadequate specimen. In comparison to lung aspiration cytology (another technique for sampling alveolar or interstitial infiltrates), a large volume of lung is sampled and a much large specimen volume is retrieved.

What Protocol Variations Are There and How Might they Affect the Specimen?

Protocol variations that will be discussed include blind vs. visualized techniques; diameter of scope or catheter; suction by syringe or via pump into a collection chamber; number of lobes sampled; number of boluses per site; and volume per bolus. Each of these factors may affect results of BALF analysis and diagnostic yield.

Blind vs. Visualized Techniques

Bronchoscopically guided BAL has several benefits over blind collection that must be weighed against the need for bronchoscopic equipment and expertise, expense, duration of anesthesia, and obstruction of airway. One obvious benefit is the added information from bronchoscopy itself. In focal or multifocal disease, bronchoscopy allows BAL to be performed in the most diseased portions of the lung based on lesions identified by radiography, CT, or bronchoscopy. Visualized guidance importantly allows for several different regions of the lung to be sampled (discussed later). Visualization also assists in lodging of the scope in an appropriately sized airway to allow for a seal to be created, maximizing return volume, and can be useful in observing airway collapse as an indication to decrease suction pressure.

The general region of lung sampled by blind BAL techniques can sometimes be predicted. Blind BAL performed on dogs in dorsal recumbency using a 16F stomach tube (size selected to mimic the diameter of a pediatric bronchoscope), reliably sample one of the caudal lung lobes.[2] Blind BAL performed in cats directly through an endotracheal tube presumably samples the dependent lung lobes.[3] The final lodging site of smaller catheters may be difficult to predict. It may also be possible to determine the site of blind BAL using fluoroscopic guidance, though the time and effort involved may not outweigh the added benefits of performing bronchoscopy.

Diameter of Scope or Catheter

The diameter of the scope or catheter dictates how deep in a bronchus that lodging will occur. Smaller scopes and catheters will reach deeper into the lung, will result in relatively less large airway representation, and will sample fewer alveoli compared with BAL collected through larger diameter scopes and catheters. A decrease in diagnostic yield is possible. In addition, the volume of saline needed to fill the dependent alveoli will be relatively less. In research settings, such as for the measurement of distribution of drugs, minimizing dilution of the ELF can be desirable. In these situations, small-diameter catheters have been used to reach far into the lung where a small volume, single bolus may be instilled and retrieved. There is no concern for heterogeneity in this research setting, as exists in clinical disease states. There is a wide range of choices when choosing a catheter for blind BAL, but patient size and scope availability often limit the size bronchoscope used for visually guided BAL. It may be prudent to use a larger number of smaller volume boluses when performing BAL through a small-diameter catheter or a particularly small scope.

Suction via Syringe vs. Suction Pump

Fluid can be retrieved manually by syringe or with a suction pump and trap. The main advantage of a suction pump and trap is that less coordination of the bronchoscopist is required, and there may be less chance for contamination of the specimen. Maximal return of BALF requires small adjustments in the position of the scope tip to maintain position within the bronchus where fluid was instilled and to avoid obstruction of the suction channel with mucosa. On the other hand, suction pressure can be readily adjusted by syringe, which is particularly helpful in retrieving fluid in patients with bronchomalacia. A study comparing results of suction by syringe, compared with a suction pump and trap, in healthy dogs found a higher return volume and total cell count when fluid retrieval was performed with a suction pump and a fluid trap compared with manually by syringe. Relative cell counts were not affected.[4] My slight preference is to use the syringe for suction because the majority of my patients are dogs with chronic cough and bronchomalacia. However, when I am using a small gastroscope for bronchoscopy in a large dog, I find it difficult to control the scope and get a good seal with a syringe. In this instance use a suction trap.

Number of Lobes Sampled

Several studies reporting BAL results in dogs and cats with respiratory disease have shown variation in results between lobes, and in some instances have documented greater diagnostic yield even in diseases considered generalized.[5-8] Statistical significance was not reached in all of these studies. Many lung diseases, including some bacterial infections, are multifocal with areas of unaffected lung adjacent to affected lung rather than homogeneously diffuse. However, if BAL is performed to sample a relatively large number of alveoli (i.e., with the catheter or bronchoscope that is not particularly small), affected areas in diffuse disease seem likely to be reached. In infectious disease processes, such as systemic fungal infections or toxoplasmosis, the number of organisms seen on a slide prepared through cytocentrifugation of 100–200 μl of fluid may only be 1 or 2 per slide.[7] Perhaps sampling multiple lobes simply increases the odds that the specimen actually processed contains the organism. In our hospital, visible mucus is also examined by direct smear, in the belief that some organisms are more likely to be seen trapped in the mucus.

Number of Boluses per Site/Total Volume Infused

In human medicine, the volume of BAL fluid instilled is thought to influence diagnostic yield and a total volume of at least 100 ml per site is recommended.[1,9] The number of aliquots instilled to achieve this total volume is not considered to be important, with the exception that the first aliquot is known to have greater representation from the larger airways than subsequent aliquots. This effect is sufficiently great that in people, some clinicians and investigators discard the return from the first aliquot when investigating interstitial lung disease. However, most feel that the large surface area sampled when large numbers of alveoli are bathed with lavage fluid comprise the majority of a full volume (100 ml or more) lavage.

The relative representation of large airways can be used to advantage in particular veterinary patients, depending on whether the disease being investigated is primarily airway centered or primarily involving the deeper lung. But it is critical to note that the reference ranges for BALF cytology may be different for relatively small volume, single aliquot lavage compared with those for BALF collected from dogs and cats using traditional total volumes of 50 ml/site in 1–5 aliquots. In contrast, relatively well adhered cells from the deep lung may be more likely retrieved when larger volumes of fluid are used and sampling error in heterogeneous diseases may be minimized.

Rather than aiming for a standard total volume of instilled fluid, use of weight-adjusted volumes has been recommended for BAL in dogs.[10] A comparison of BAL performed with 50 mL divided in 2 equal aliquots with BAL performed with 2 mL/kg body weight divided in 2 equal aliquots found that the recovery of ELF was more uniform using the weight adjusted technique (2.28% compared with 2.89%). The study was performed in healthy beagles weighing 9–20 kg. These results have strong implications for research settings and may become more relevant to clinical patients as measurements of markers of disease become more readily available.

One should also consider the body weight of the patient, along with the size scope/catheter being used when considering instillation volumes from several angles. The intention of the volume used is to fill the alveoli dependent on the bronchus where the scope or catheter is lodged. When BAL was nearly always performed (as in human medicine) through a bronchoscope, the tendency for the scope to lodge more proximally in a small dog than a large dog likely tended to equalize the volume below the scope. Now, with the availability of smaller scopes and the use of blind catheter techniques, the dependent volume may be much smaller. Another consideration if using weight-adjusted volumes of 2 ml/kg relates to particularly small dogs. The calculated volume may be quite small and could minimize the number of alveoli sampled. This would cause me greater concern if the disease was interstitial or patchy, compared with patient with diffuse airway disease.

PATIENT RISK CONSIDERATIONS

Healthy animals, including dogs and cats, are quite tolerant of even high volume lavages and the effect on pulmonary function and lung histology have been studied.[2,3,11] However, in the clinical setting lung function is often compromised and the risk of desaturation greater. Fortunately, supplemental oxygen for an hour or two following lavage will be effective in overcoming relative hypoxemia in most patients that are breathing comfortably in room air prior to BAL. Cats should also be pre-treated with bronchodilators. However, patients that are showing increased respiratory effort in room air may already have a need for supplemental oxygen before the insult of anesthesia and BAL. These patients should only undergo BAL if it is considered essential and the capability of ventilator support is at hand. For patients that are judged to be relatively fragile (perhaps having exercise intolerance, co-morbidities, or poor oxygenation as measured by arterial blood gas analysis), total lavage volume and number of sites sampled should be limited to the minimum that clinical judgement suggests should be sufficient to obtain a representative specimen. These patients are deserving of serious consideration of the variables described, but even so there is no objective way to quantitate relative risk.

EXAMPLES OF IMPLICATIONS FOR SPECIFIC CASES

Considering the variables discussed above, decisions can be made for performing BAL in individual patients depending upon prior test results and the top differential diagnoses for the patient. For an obvious example, a patient with disease limited to a small, focal region of the lung will more likely benefit from bronchoscopic BAL rather than a blind technique. The potential advantages of bronchoscopic BAL in dogs or cats with diffuse disease need to be weighed against the expense and increased anesthesia time, particularly in relatively fragile patients.

There are numerous considerations for patients, usually dogs, with presumed bacterial bronchopneumonia. In my experience the greatest challenge for getting an adequate specimen to confirm the diagnosis and obtain good culture results is having the patient off antibiotics for a prolonged period of time, rather than how the specimen is collected. Bronchoscopy can be used to guide the sample collection, but in patients with classic history and radiographic signs, the added anesthesia and expense may not be warranted. Blind BAL as described using a 16F stomach tube consistently sampled one of the caudal lung lobes,[2] while the disease in these patients is likely to be in the gravity dependent lung lobes. As such, tracheal or bronchial washings may be more appropriate. If foreign body is among the differentials for the signs, bronchoscopy may be indicated. Generally, any method that samples the exudate within the airway should be sufficient as long as a sufficient volume of fluid is recovered to perform the appropriate cytologic and microbiologic tests. For fragile patients with presumed bacterial bronchopneumonia, I may use a minimal volume of saline for infusion. In specific cases, one may theoretically find higher yield for organisms such as *Bordetella* or *Mycoplasma* from large airway washings or brushings, compared with deeper lung specimens, and there may rarely be an advantage to collecting both large airway specimens and BAL.

In patients with diffuse, primarily interstitial, lung disease and top differential diagnoses of fungal or protozoal infections, or neoplasia, I will try sample at least two or three lung lobes using a standard-diameter pediatric bronchoscope, and use a full 50 ml/site in hopes of increasing diagnostic yield. With a blind technique, I will use a larger-bore catheter to sample relatively more alveoli. If I have parasitic lung disease among my differential diagnoses, I will also avoid using a small catheter for specimen collection for the additional advantage of sampling relatively larger airways. Patient status must always be the first consideration (first do no harm!).

For patients with diffuse bronchitis, I am less concerned with the alveoli and interstitium and so will readily sample fewer sites and may use smaller volumes of fluid if the patient's status is concerning. I am particularly cautious if there are significant co-morbidities such as severe tracheobronchomalacia or heart disease. One must remember that if a smaller volume or a single aliquot is used, the expected differential cell count may be different than that expected from a full volume BAL. Importantly, the finding of infectious agents, eosinophils, or neoplastic cells will remain of great diagnostic value.

REFERENCES

1. Baughman RP. *Semin Respir Crit Care Med.* 2007;28:475.
2. Hawkins EC, *et al. J Am Vet Med Assoc.* 1999;215:1635.
3. Hawkins EC, *et al. Am J Vet Res.* 1989;50:855.
4. Woods KS, *et al. J Am Vet Med Assoc.* 2014;75:85–90.
5. Hawkins EC, *et al. J Vet Intern Med.* 1995;9:386.
6. Hawkins EC, *et al. J Am Vet Med Assoc.* 1990;197:79.
7. Hawkins EC, *et al. J Am Vet Med Assoc.* 1997;210:648.
8. Ybarra WL, *et al. J Vet Intern Med.* 2012;26:1281.
9. Klech H, *et al. Eur Respir J.* 1989;2:561.
10. Melamies MA, *et al. Am J Vet Res.* 2011;72:694.
11. Muggenburg BA, *et al. Am Rev Respir Dis.* 1972;106:219.

acvim370

Using Airway Pressures and Lung Function in Clinical Practice

Elizabeth Rozanski, DVM, DACVIM (SAIM), DACVECC
North Grafton, MA, USA

Pressure determinations are potentially useful in the ICU and ER, and may be considered as point-of-care evaluation of physiologic responses to critical illness or injury. Most critical care units have easy access to some sort of electronic pressure transducer for measurement of direct arterial pressure, or at least a water manometer. Measurements of any body cavity pressure may be obtained via manipulations of these devices. The utility of the measurement of any pressure specific gradient varies, but in most cases significant amount of knowledge can be gained by monitoring pressures. Additionally, for hospitals involved in teaching students or house officers, a more physiological understanding may help further cement knowledge of a specific fact. For example, visually observing a large volume pleural effusion drain spontaneously from the chest cavity will reinforce the idea that pleural pressure is positive to atmosphere when the effusion is large. Alternatively, measurement of airway pressures during ventilation is mandatory, and subsequent calculation of lung compliance may help determine if the lung function is improving separate from simple measurements of PaO_2:FiO_2 ratio. In fact, decreasing compliance over days 1–6 has been independently associated with death in people with ARDS (Seeley *et al.*). The oxygenation index, defined as = (FiO_2 x Mean Airway Pressure)/PaO_2 may also be used in people but has not been well described in dogs.

Pleural Pressure

Pleural pressure may be monitored during thoracocentesis for pleural effusion or pneumothorax or continuously if a thoracostomy tube is in place. As a brief review, pleural pressure in a healthy dog or cat is negative in relation to the atmosphere. This negative pressure gradient helps promote/maintain lung inflation and facilitate a lower work of breathing. A small amount (<5 ml) of effusion is present in the pleural space. However, the pleural space is dynamic with extravasated fluid being rapidly reabsorbed via lymphatic and capillaries. Pleural space disease may be acute or chronic, with most effusions being somewhat chronic. As pleural effusion forms, there is collapse of the lung parenchyma, and also an **increase** in intra-thoracic pressure. As respiratory distress develops, the intra-pleural pressure is less negative or positive to the atmosphere. While typically as clinicians we assume that removal of isolated pleural effusion will result in immediate improvement in lung function, this is not always the case. In people, two separate categories of non-recruitable lung associated with pleural effusion are recognized, the first termed **lung entrapment**, which is a condition, which develops in association with active pleural inflammation or neoplasia. Immature fibrin and overlying inflammation prevent re-expansion and contribute to the failure to recruit lung after thoracocentesis. During longer standing effusions, there is thickening and constriction of the visceral pleural and thickening of the parietal pleura. This may lead to the development of "**trapped lung**" or lung that is tightly constricted by overlying visceral pleura, and can not adequately re-expand, even in the presence of negative intra-thoracic pressure. Entrapped lung is likely more common in veterinary patients, as most effusions are associated with more active inflammation. However, the non-recruitable lung has been much less evaluated in dogs and cats than people.

Pleural manometry was designed to monitor the pleural cavity for signs of non-recruitable lung, as evidenced by large decreases in intra-pleural pressures, which may be subsequently associated with the development of pain or spontaneous pneumothorax. Spontaneous pneumothorax is of particular interest in veterinary medicine; iatrogenic pneumothorax associated with thoracocentesis is most commonly thought to be associated with inadvertent laceration of the lung. The concept of non-recruitable lung is compelling, as larger increases (e.g., more negative) in pleural pressures might be associated with spontaneous pneumothorax as well. Pleural manometry may easily be performed with an over-damped electronic manometer. Additionally, the use of pleural manometry is also able to answer the question as if the needle or catheter is still in the thoracic cavity following either re-positioning, patient movement, or apparent completion of the procedure which may be almost as helpful!

Pulmonary (Lung) Pressures

During spontaneous breathing, inspiration occurs due to a negative inspiratory pressure with an open glottis. Lung compliance determines how easy or how hard it is for the lungs to inflate, with lower compliance being associated with higher pressures for the same tidal volume. Compliance can be calculated by looking at the slope of the pressure-volume (PV) curve, or by measuring changes in volume/changes in pressure. A PV curve looks as below.

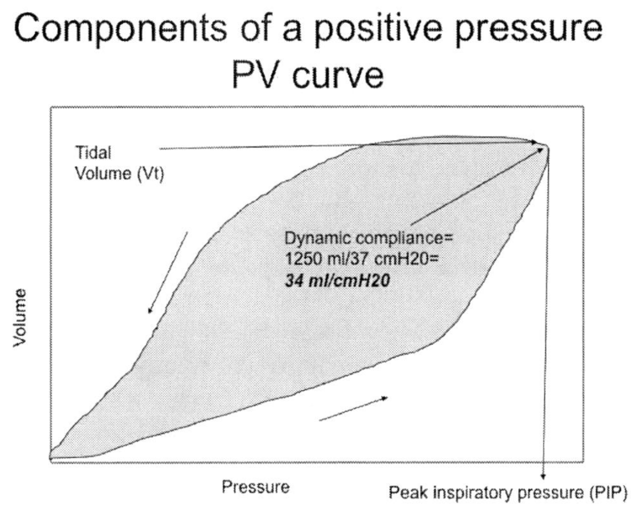

Thus when the lung starts to become stiffer, the curve becomes flatter. On newer critical care ventilators, algorithms built in the software can calculate compliance and resistance and help to better define improvement or decline of the clinical condition. Without a ventilator, King and colleagues described a simple technique for determining static lung compliance with an anesthesia machine with a pressure gauge and a handheld spirometer. Decreasing compliance is associated with progressive lung disease or atelectasis. A sigh or deep breathing can help to improve compliance but recruiting de-recruited airways.

Other pressures that may be evaluated during positive pressure ventilation include peak and plateau pressures. Peak airway pressure is the highest pressure that the lungs are exposed to during inspiration, and larger reflects airway flow rate and is a marker of airway resistance (larger airways). Plateau pressure is more reflective of pressure to the smaller airways and alveoli, and requires a moment of no flow.

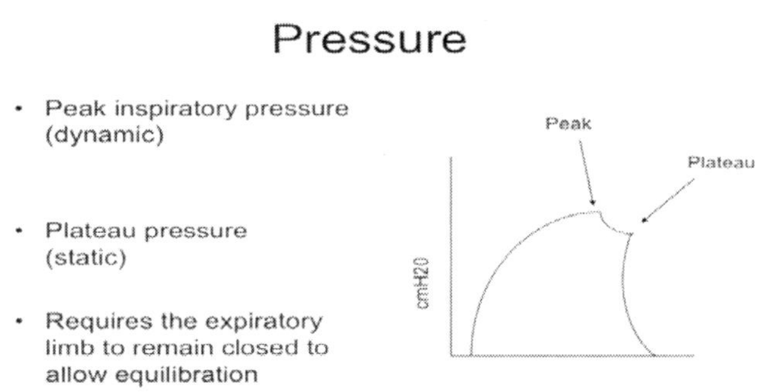

To better discuss airway pressure, it is important to recognize that mechanical ventilators work by delivering a breath to either a set volume (aka volume-limited ventilation) or to a set pressure (aka pressure-limited ventilation). The speed that the set volume or set pressure is reached is dependent on the inspiratory flow rate and the inspiratory time. For example, if your set tidal volume is 500 ml, and your inspiratory flow rate is 1000 ml/sec, it will take 0.5 second to reach 500 ml. However, if you set your patient's inspiratory flow at 5000 ml/sec, it will only take 0.1 second to reach 500 ml. In this case, the peak inspiratory pressure will be **much** higher in the patient the higher inspiratory flow rate, but plateau pressures will be similar. Pressure changes will also be seen if the cuff is leaking, tube is kinked or if there is a pneumothorax.

OTHER LUNG FUNCTION MEASUREMENTS (AWAKE)

6-Minute Walk Test

The 6-minute walk test (6MWT) is used in people as a functional assessment of the cardiopulmonary reserves. It is not designed to evaluate maximal effort, but instead to more closely evaluate the activities of daily living. The test is performed by measuring the distance an individual may walk comfortably over six minutes. The 6MWT is used in human medicine as a predictor of morbidity and mortality from heart or lung disease as well as a method of assessing an objective response to a therapeutic intervention (surgery/medical). Importantly, the 6MWT is not performed in those individuals that are short of breath, but rather in those that are potentially affected with a cardiopulmonary disease that might affect exercise tolerance. The 6MWT has recently been reported in healthy dogs and shown in a pilot study to be lower in dogs affected with pulmonary disease. Healthy dogs walked approximately 522 meters, while dogs with pulmonary disease walked about 384 meters. There were only mild to moderate correlations between body size characteristics and distance walked.

SUMMARY

Use pleural pressure to

1. Ensure placement of needle/catheter in the thoracic cavity
2. Watch for excessively negative pressure that could cause spontaneous pneumothorax

Use pulmonary pressures

1. To monitor for lung injury
2. As surrogate markers of resistance and compliance
3. To adjust ventilator settings
4. To identify air leak (pneumothorax or cuff leak)

REFERENCES

1. Feller-Kopman D, *et al. Chest.* 2009;135(1):201–9.
2. Huggins JT, *et al. Chest.* 2007;131(1):206–3.
3. Koo, *et al. Crit Care Med.* 2011;39(7):1613–8.

acvim371

Diagnosis and Management of Pulmonary Thromboembolism in Dogs

Robert Goggs, BVSc, DACVECC, DECVECC, PhD, MRCVS
Ithaca, NY, USA

Introduction

The presence of a thrombus in a pulmonary artery constitutes pulmonary thromboembolism (PTE), a term that encompasses both *in situ* thrombus formation and embolization from elsewhere in the vasculature. Splenic and portal vein thrombosis have been frequently reported, but few reports of limb deep vein thrombosis in small animals exist. As such, it is unclear if pulmonary thrombi in small animals are emboli or form directly in the right heart or pulmonary arteries. These are rarely differentiated clinically and share common pathophysiology, so the distinction is largely academic. PTE has been associated with numerous diseases and most patients have more than one risk factor. The incidence of PTE is unknown but it is likely more common than the literature suggests, likely because PTE remains a challenging diagnosis to confirm. A thorough, logical diagnostic approach should incorporate assessment of clinical probability, identification of risk factors and disease biomarkers, and diagnostic imaging including echocardiography and CT pulmonary angiography (CTPA). Since PTE carries a poor prognosis and is potentially fatal, the importance of early detection cannot be overstated. Consistent diagnosis will be required before we can evaluate thrombolytics, anticoagulants, and antiplatelet agents in canine PTE. In the interim, therapy for PTE should consist of close monitoring, good supportive therapy, and judicious, individualized use of thrombolytics, anticoagulants, and antiplatelet agents.

Diagnosis

Establishing a clear diagnosis of PTE is difficult and requires a logical approach. Since PTE likely carries a substantial mortality rate, early detection is vital. Risk stratification is the first step towards definitive diagnosis. Patients with known risk factors and compatible clinical signs should undergo expedited evaluation. History, clinical examination and initial diagnostic tests have limited sensitivity and specificity for PTE, but a combination of consistent findings significantly improves the pre-test probability that an individual dog has PTE. This is important because the positive predictive value of CTPA is enhanced by increasing the pre-test probability. If the likelihood of PTE is low, patients may be saved expensive and invasive diagnostic testing. The value of clinical assessment in human PTE diagnosis has been repeatedly demonstrated,[1] aided by the development of clinical prediction scores such as the Wells and Geneva rules[2]. The Wells rule incorporates historical information, intercurrent disease and risk factors, and the presence of appropriate clinical signs and physical examination findings. The revised Geneva score is similar but involves only objective criteria.[3] Both have been adequately validated in people.[4] No formalized scoring systems exist in veterinary medicine to aid the evaluation of patients with PTE. Until such a validated score is developed, patients at high risk must be identified based on clinical judgment. The clinical signs of PTE are variable, inconsistent, and nonspecific. The degree of physiological impairment and thus the severity of clinical signs reflect both the magnitude of the PTE and the patient's ability to compensate. The most common signs are dyspnea, tachypnea, and depression. Other signs include coughing, hemoptysis, cyanosis, syncope, collapse, and sudden death. In dyspneic patients, physical examination may reveal harsh lung sounds and crackles suggestive of interstitial or alveolar edema. Occasionally, lung and heart sounds may be muffled due to pleural effusion. In eupneic patients, lung field auscultation will likely be normal. On cardiac auscultation, tachycardia with a split, second heart sound may be noted, or more commonly a loud second heart sound associated with pulmonary hypertension. Signs compatible with backward or forward heart failure may be present. Complicating the picture may be clinical signs associated with the predisposing disease. PTE should be suspected in patients with PTE risk factors, but no history of cardiopulmonary disease, that develop respiratory distress acutely.

Thoracic Imaging

Most patients with new-onset respiratory distress and PTE risk factors will undergo thoracic radiography. Occasionally this will be diagnostic, but most patients will require additional imaging for a definitive diagnosis. This step should be omitted if the patient recently underwent thoracic radiography. Bedside lung ultrasound and echocardiography is a reasonable next step in high-risk patients, aiming to identify any evidence of pulmonary parenchymal disease denoted by the appearance of B-lines (comet tail artifacts) and to evaluate the size and function of the right heart. Given the precarious state of these patients, care should be taken to minimize stress. The presence of B-lines on lung ultrasound and/or identification of right heart dysfunction, dilation or tricuspid regurgitation should prompt the clinician to consider CTPA to obtain a definitive diagnosis. Up to 30% dogs with PTE have normal thoracic radiographs, but many will have non-specific abnormalities visible.[5-7] Pulmonary infiltrates are the most common finding and are typically alveolar or alveolar-interstitial in appearance, present focally or multifocally.[6] These infiltrates likely represent edema, atelectasis, hemorrhage or infarction and often have indistinct borders. Hypovascular lung areas (Westermark sign) are hyperlucent regions representing zones of reduced blood flow distal to sites of vascular occlusion, are rare but pathognomonic for PTE.[6,8] Pulmonary infarcts appear as distinct pleural-based, wedge-shaped densities. Other potential abnormalities include uneven vessel diameter, pulmonary arterial enlargement, lobar vessel attenuation, pleural effusion and right-sided cardiomegaly.[6,9,10]

Rapid, multislice spiral CTPA is key to diagnosis of PTE.[11,12] These studies are obtained by performing rapid, spiral thoracic CT, concurrent with bolus injection of contrast media. The contrast agent is infused through a peripheral catheter using a pressure injector, linked electronically to the scanner to ensure peak enhancement of the pulmonary arteries.[13] The critical technical aspect of these studies is the co-ordination of the contrast injection with the image acquisition to maximize visualization of the pulmonary vasculature. This can be done manually, by initiating the scan at a fixed time interval after the injection of the contrast material, although this may result in suboptimal opacification of the pulmonary arteries limiting diagnostic utility. CTPA is therefore best performed using a process of "bolus tracking," also called region of interest (ROI) scanning. In ROI scanning, a field of interest is predefined and a contrast detection threshold set (e.g., 100 HU). Once the set level of X-ray beam attenuation in that area is detected, the multislice scan is initiated. Depending on the capabilities of the scanner, the ROI can be defined anywhere between the vena cava and the proximal pulmonary artery. State of the art (256-slice) CT scanners enable acquisition of all of the necessary images within a 1 second X-ray exposure, thereby avoiding the problems of respiratory motion, cardiac motion, and contrast drainage. Currently, the fastest CT scanners installed in veterinary centers in the US are 128-slice machines. These multislice CT scanners permit the imaging of the whole thorax in sedated patients and can eliminate breathing artifact.

Diagnostic criteria for PTE using CTPA are failure to enhance the entire arterial lumen due to occlusive filling defects, partial filling defects surrounded by contrast material, producing the "polo mint" or "railway track" signs, and peripheral intraluminal filling defects that form acute angles with the arterial wall.[14] A normal CTPA study essentially rules out PTE as the cause of the patient's respiratory distress, unless the index of suspicion is very high.[15,16] In most cases where CTPA is negative, the CT scan will provide sufficient information to enable an alternative diagnosis to be made. Undoubtedly, the risk:benefit profile for CTPA in these patients must be considered and some clinicians will opt to manage these patients presumptively rather than risk deterioration in the CT scanner. The risk of CTPA is frequently overestimated however, and the potential benefit of a definitive diagnosis outweighs the small increase in risk associated with the procedure in most patients.

Multislice CTPA has been reported in dogs using 16-slice[13] and 64- slice CT scanners,[17] and protocols to maximize PA contrast enhancement published[18]. Although the majority of publications to date have focused on normal patients, multislice CT angiography has been successfully used to detect PTE in experimental dogs,[19] and to identify a descending aortic thrombus secondary to spirocercosis in a dog[20]. Promisingly, a recent study reported the successful use of a 16-slice scanner for identification of PTE in dogs with IMHA.[21] In this study, dogs with naturally occurring PTE were definitively identified using a sedation-only protocol. This confirms that CTPA can be readily applied to the diagnostic investigation of canine PTE and should enable more frequent identification of this disorder and improve our ability to investigate the predictive ability of other diagnostic test criteria. This suggests that with the now widespread availability of scanners with 32, 64 and 128 slice capabilities, definitive diagnosis of PTE with CTPA should become routine in veterinary medicine.

Baseline Bloodwork

Complete blood counts and chemistry panels are not discriminating for PTE, but may identify predisposing conditions such as hyperadrenocorticism, protein-losing nephropathy, diabetes mellitus, or hypothyroidism. Complete blood counts may identify a nonspecific inflammatory leukogram or myeloproliferative disorders such as polycythemia or essential thrombocytosis that can predispose to thrombosis.[22] Thrombocytopenia or schistocytosis, as markers of DIC, may increase the index of suspicion for PTE. Blood-gas abnormalities associated with PTE in dogs are hypoxemia, hypocapnia, and an increased A-a gradient.[5] In some cases, this hypoxemia may respond poorly to oxygen therapy due to intrapulmonary shunt.[5,23,24] These findings are not specific for PTE however, and should be interpreted in light of other data. Similarly, normal blood-gas values do not rule out PTE.

Coagulation Testing

In patients with appropriate clinical signs but without known risk factors, diagnostic testing should focus on identifying any underlying predisposing factors and on interrogation of the coagulation system. Coagulation assessment with measurement of D-dimer, AT activity, viscoelastic tests or thrombin generation assays may help identify patients with evidence of fibrin degradation, thrombin activation, and hypercoagulability. Irrespective of whether an underlying disease process is identified, recognition of a new and unexplained episode of respiratory distress in a dyspneic, hypercoagulable patient mandates imaging studies for PTE.

Clotting times and fibrinogen assays are of limited value in PTE because they may be normal, and any abnormalities are nonspecific.[5] In patients with active thrombin production, plasma antithrombin (AT) activity is reduced by consumption and urinary AT loss has been linked to the hypercoagulable states associated with protein-losing nephropathy,[25] and hypercortisolemia[26-28]. Measuring patient levels of AT activity may allow thrombosis risk stratification. Based on extrapolations from humans, reductions of AT activity between 50–75% moderately increase risk, while activities below 30–50% markedly increase thrombosis risk.[29] Increased plasma D-dimer concentrations indicate plasmin-mediated cleavage of FXIII crosslinked fibrin.[30] Since D-dimers require activation of both thrombin and plasmin for their formation, they are considered more specific for fibrinolysis following thrombosis than FDPs. Rapid, accurate, bedside D-dimer assays are integral to decision making in humans with possible PTE.[2] People with high clinical probability scores (previous TE disease, deep vein thrombosis, dyspnea) undergo thoracic imaging without a D-dimer test. For all other people presenting with compatible clinical signs, rapid, quantitative D-dimer ELISAs are performed prior to selection of further diagnostic testing.[31] Positive D-dimer tests prompt thoracic imaging, while negative tests suggest no further investigation for PTE is required. Similar guidelines for small animals cannot be formulated currently because we lack sufficient data and more fundamentally a convenient and highly accurate D-dimer assay is not available for small animals. A recent study evaluated the diagnostic utility of D-dimer concentrations in 10 dogs with necropsy-confirmed PTE compared to controls. Using 250 ng/mL as a cut-off D-dimer concentration, the sensitivity and specificity were 80 and 30%, for the diagnosis of PTE, while the negative predictive value was 60% and the positive predictive value was 53.0%. No dog with a D-dimer value <100 ng/mL had PTE. The authors concluded that PTE still can occur in dogs with normal D-dimer concentrations and that increased D-dimer concentrations were not specific for PTE.[32] Irrespective of the assay chosen, D-dimers should be evaluated within 1–2 hours of the suspected embolic event because in experimental canine PTE, D-dimers were increased by 30 minutes, peaked at 1–2 hours before returning to normal after 24 hours.[33]

Whole blood tests of coagulation such as thromboelastography (TEG), thromboelastometry (ROTEM), and the Sonoclot system, graphically represent the viscoelastic properties of blood during clot formation and lysis.[34] Using TEG/ROTEM, hypercoagulability has been identified in multiple settings in dogs.[35-39] Hypercoagulable patients produce characteristic TEG/ROTEM tracings. These patients typically have short reaction and clot formation times, steep α angles and large maximum amplitudes. In people, MA values derived from conventional TEG and from a rapid-TEG assay have been demonstrated to predict thrombotic risk in human surgical patients,[40] and risk of PTE following trauma[41]. It is unclear which (if any) of these parameters best relates to thrombotic risk in veterinary patients. It is also essential to recognize that preanalytical variables, such as patient hematocrit, fibrinogen concentration, and platelet count, in addition to sample handling and assay parameters contribute to the final TEG/ROTEM tracings.[42] A recent pilot study was unable to identify a correlation between TEG and CTPA findings in dogs with PTE, but this study was likely underpowered.[21] A separate study evaluated TEG in dogs with thrombosis in various anatomic locations, and found that although the TEG G-values of dogs with thrombosis were significantly greater than controls, half of the dogs had normal TEG tracings and suggests standard TEG/ROTEM assays may not be discriminating for PTE. Modified TEG/ROTEM assays may augment our ability to identify hypercoagulable patients. A recent study used t-PA modified TEG assays to assess 20 dogs with disease processes that predispose to thrombosis. Compared to healthy controls, the patients with thrombotic diseases were more "resistant" to the lytic effects of t-PA addition in the TEG assay. This suggests that occult hypofibrinolysis may contribute to thrombotic risk in some patients, particularly those with protein-losing disease or systemic inflammation.[43] These tests may help identify at-risk patients who require further investigation, but more work is needed to identify which tests and which parameters are most predictive of risk.

Cardiovascular Function Assessments

Electrocardiographic abnormalities are likely uncommon in acute PTE, with sinus tachycardia due to hypotension or hypoxemia the most likely alteration.[9] Echocardiography is useful for identification and quantification of the cardiovascular compromise that occurs secondary to acute PTE, which causes right ventricular (RV) pressure overload and dysfunction. Typical echocardiographic findings in humans with PTE include right ventricular dilatation and hypokinesis, septal flattening and paradoxical septal motion, diastolic left ventricular impairment, pulmonary arterial hypertension, right ventricular hypertrophy and patency of the foramen ovale. RV dilation is found in at least 25% of people with PTE and is associated with prognosis. The degree of RV dysfunction is related to the cross-sectional area of the pulmonary vasculature affected; occlusion of >30% is frequently associated with RV hypokinesis.[44] The absence of RV dysfunction does not preclude PTE however and RV distension and dysfunction may result from processes other than PTE. If a patient has hemodynamic compromise due to PTE, there will be echocardiographic changes present. In other words, if PTE is suspected and hemodynamic instability exists, a normal echocardiogram can exclude PTE as the cause of the shock and may provide an alternative mechanism such as pericardial tamponade, acute valvular dysfunction, cardiomyopathy or severe volume depletion. Where PTE is definitively diagnosed such as by CTPA, identification of hemodynamic compromise including RV pressure overload and dysfunction provide justification for thrombolysis.[45]

There are currently few data regarding echocardiography in canine PTE, but typical findings include RA and RV enlargement, paradoxical septal motion, pulmonary hypertension and thrombi in the heart or PA.[5,46-48] In a recent prospective study of at-risk dogs that underwent CTPA studies for PTE, 10/12 dogs had echocardiography performed. Nine dogs had evidence of pulmonic regurgitation and 6 had tricuspid regurgitation. Three additional measures of right heart function were evaluated: TAPSE - a marker of RV ejection fraction; Tei index - an index of myocardial performance; and PEP/RVET - a marker of PVR.[49,50] PH causes increased PEP/RVET and Tei index and decreased TAPSE. No association between these indices and the PTE was identified however.[21]

In humans, echocardiographic RV dysfunction secondary to PTE is associated with elevated cTn concentrations.[51] A recent meta-analysis of 20 studies suggested that humans with PTE and an increased cTn level are at high risk of short-term death and adverse events.[52] In dogs with PTE secondary to dirofilariasis, cTnI values were significantly higher than in controls,[53] and in experimental PTE, cTnI levels increase in proportion to the severity of occlusion as indicated by PVR and MPAP[54]. As yet, there are no reports from clinical PTE patients. cTn values are unlikely to enhance PTE diagnosis due to a lack of specificity, but they may be of value in assessing PTE severity and predicting complications.[55] cTnI levels peak at 8 hours post-embolism in people.[56] The availability of POC tests for cTn in dogs makes accurate timing of this assay a viable proposition.

Atrial and B-type natriuretic peptides are produced by cardiac muscle and released in response to volume overload, hypertrophy and hypoxia to elicit natriuresis and vasodilation.[57] In PTE, NT-proBNP concentrations are closely related to the degree of right ventricular dysfunction,[58,59] and this likely underpins their relationship with adverse events and outcome in people with PTE[60, 61]. In experimental PTE, both ANP and NT-proBNP increase significantly when severe pulmonary arterial hypertension is present,[62] but there are no clinical studies in dogs to date. The NT-proBNP test is limited by *ex vivo* stability, the lack of a canine POC assay, and interference from reduced kidney function.

MANAGEMENT

Therapy for patients with PTE aims to reduce the clot burden, improve pulmonary function and hemodynamics, identify and manage the underlying disorder, and provide thromboprophylaxis. Not all PTE patients are the same however and while some warrant aggressive interventions such as thrombolysis, in others these therapies do not have a favorable risk:benefit profile and a more conservative approach is appropriate. The optimal strategy for management of PTE in small animals is not known because of the scarcity of specific information in the veterinary literature. Tailored supportive treatment for PTE patients is easy to recommend and should be universal. When PTE is definitively diagnosed or the index of suspicion is sufficiently high, then antithrombotic therapy is also warranted. Where PTE is definitely diagnosed and the patient is hemodynamically compromised, thrombolysis should be considered. Successful treatment for acute PTE is typically followed by chronic antithrombotic therapy and should be accompanied by management of any underlying predisposing disorder.

Supportive Therapy

All dyspneic patients should receive oxygen therapy to improve oxygenation and to produce pulmonary vasodilation by combating hypoxic pulmonary vasoconstriction. If the patient's hypoxemia is severe and responds poorly to oxygen, then intubation and PPV may be necessary, but the decision to intubate a patient with PTE should not be taken lightly. If intubation and PPV is considered, then induction should be performed using cardiovascular-sparing drugs such as ketamine, ketamine-propofol, etomidate or fentanyl-midazolam. Mechanical ventilation has complicated but mostly deleterious effects on PVR. Low tidal-volume strategies are indicated in PTE patients, but can be challenging to achieve because of the conflicting demands of hypoxemia and hypercapnia. Low VT strategies are commonly coupled with high PEEP to maintain oxygenation, but high PEEP will further compromise RV function. If PPV is being considered for a patient with PTE, then thrombolysis is also likely to be necessary and appropriate.

For some patients, PTE is rapidly fatal irrespective of therapy. Normal dogs thrombi lyse rapidly though, so supportive therapy buys time for fibrinolysis to occur.[63] Death in PTE is typically due to hemodynamic collapse rather than hypoxemia or hypercapnia. As discussed above, PTE causes RV pressure overload and dilation with impairment of left ventricular filling, systemic hypotension and reduced perfusion of the RV myocardium. This downward spiral can result in PTE patients decompensating rapidly - once a tipping point is reached, ongoing deterioration occurs unless we intervene. Routine fluid administration in these patients should be avoided and if fluid administration is being considered for management of shock, then echocardiography to determine RV filling should be used as a guide.[64] Hypotension in PTE compromises perfusion of the brain, kidneys and the RV myocardium. Hypotension in the PTE patient should be identified early and addressed proactively with norepinephrine,[65] which will cause venoconstriction, thereby increasing preload, and arterioconstriction increasing LV afterload and augmenting MAP.

In PTE, much of the increase in PVR is secondary to pulmonary vasoconstriction resulting from the action of vasoactive mediators (thromboxane and serotonin) rather than mechanical obstruction.[66] Pulmonary vasodilators are therefore rational therapies to reduce PVR and improve hemodynamics. Inhaled nitric oxide (NO) is effective but prohibitively expensive. Epoprostenol (prostacyclin) is an alternative ($8–10/h) that can be delivered intravenously or by inhalation. Epoprostenol may reduce vascular leakage in canine models of lung microembolization,[67] and reduce RV afterload,[68] but its efficacy in canine experimental PTE is limited[69]. Theophylline may be beneficial as a bronchodilator and as pulmonary vasodilator and may also improve diaphragmatic contractility and reduce respiratory muscle fatigue. In dogs, sildenafil has beneficial effects in pulmonary hypertension,[70] and in canine PTE models it reduces MPAP through selective pulmonary arterial vasodilation[71]. In people sildenafil has beneficial effects in both acute PTE and in the treatment of chronic PTE-induced pulmonary hypertension.[64] Although sildenafil has not been reported in small animal PTE to date, it should be considered if pulmonary hypertension is documented echocardiographically.

Thrombolysis

Therapeutic thrombolysis is the administration of supraphysiologic doses of plasminogen activators to accelerate plasmin generation and induce thrombus dissolution. Thrombolysis aims to reverse right heart failure, improve systemic and pulmonary hemodynamics, limit neurohumoral activation and reduce recurrence.[72] Risk stratification of people with PTE underpins successful treatment and patients are divided into massive, high-risk submassive, low-risk submassive, and non-massive (stable segmental). Massive PTE is defined as systemic arterial hypotension due to PTE. High-risk submassive patients have pulmonary embolism with evidence of significant cardiopulmonary dysfunction, but not hypotension. Low-risk submassive patients have limited evidence of RV dysfunction and are not hypotensive, while non-massive PTE patients have no evidence of cardiovascular compromise. There is consensus on how to manage patients at the extremes of the spectrum. Anticoagulant therapy alone, typically unfractionated heparin (UFH) or low-molecular weight heparin (LMWH) is recommended for low-risk submassive patients or those with non-massive PTE. Patients with massive PTE are best treated with thrombolytics followed by anticoagulation,[73] with route and dose of thrombolytic delivery based on bleeding risk. There is equipoise regarding the optimal drug and dosing strategy for patients with submassive PTE, however, despite several large human interventional trials in this area.[74-76]

Thrombolytic drugs have distinct pharmacokinetics, thrombolytic activities, and fibrin specificities.[77] The t-PA products are fibrin specific, activating fibrin-bound plasminogen more rapidly and more effectively than plasminogen in circulation. At pharmacologic concentrations however, a systemic lytic state can still occur and all t-PA products are associated with a risk of hemorrhage in humans.[78] Alteplase has the shortest half-life of the available recombinant thrombolytics.[79] Alteplase has been successfully used in experimentally induced canine PTE.[80] In veterinary medicine, no consensus on their administration exists and clinicians must determine the potential benefits and risks of administration for individual patients.[81] Recent advances in small animal interventional radiology may provide novel methods or routes for thrombolysis in future.[82] The data from people argues that thrombolytic therapy be administered to small animals with confirmed hemodynamically unstable acute PTE, and where continuous hemodynamic monitoring is available.[83] In appropriate cases, fibrin-specific drugs with short half-lives like alteplase are preferred. The human data also argue that alteplase may be worthwhile in dogs with submassive PTE and RV dysfunction, albeit at the cost of an increased risk of bleeding. Patients with a high risk of bleeding are probably best managed with heparin therapy alone.

Antithrombotic Strategies

Antithrombotics are administered to minimize thrombus propagation[84] and to reduce risk of recurrence. However, the optimal antithrombotic strategy for small animals is not known. Some suggest that PTE should be managed with anticoagulant drugs and that antiplatelet agents are not appropriate.[85] Viewing arterial and venous thrombosis as separate pathophysiological entities is likely an oversimplification however,[86,87] and there is evidence of overlapping efficacy of antiplatelet and anticoagulant agents in the treatment of both venous and arterial thromboembolic conditions[88]. Platelets are integral to the cell-based model of hemostasis, and thrombocytosis is associated with an increased risk of PTE in people.[89] In venous thrombosis, tissue-factor-mediated thrombin generation likely precedes platelet activation, but both arms of the hemostatic system are activated.[90]

Anticoagulants remain the mainstay of antithrombotic therapy for people with PTE. Humans with massive PTE are treated with UFH before, during, and after thrombolysis. In people with submassive PTE, initial parenteral anticoagulation is initiated with low-molecular-weight heparin (LMWH) or fondaparinux.[83] Oral anticoagulation is instituted for PTE patients for at least 3 months to prevent recurrence, typically using a direct oral anticoagulant. To date there are no clinical trials of anticoagulants in PTE in small animals on which to base recommendations. Unfractionated or low-molecular-weight heparin are appropriate for in-hospital management of PTE,[85] but there is little consensus on dosage, frequency of administration or monitoring strategy. Empirically dosed rivaroxaban, apixaban or parenteral LMWHs are sensible options for long-term anticoagulation.

Unfractionated heparin, a heterogenous mixture of polysaccharides, complexes with and amplifies the inhibitory activity of AT against FIIa (thrombin) and FXa,[91] although factors IXa, XIa, XIIa and XIII are also inhibited. Thrombin inactivation prevents fibrin formation and reduces thrombin-induced platelet activation.[92] Variation in the size distribution of heparin molecules in UFH and unpredictable bioavailability in critical illness cause variation in heparin dose responses necessitating therapeutic monitoring,[93] since severe bleeding complications can result from exceeding the therapeutic range. Therapeutic monitoring can be performed using the ACT, aPTT,[94,95] thrombin generation,[96] viscoelastic testing,[97] or anti-FXa levels[98]. This need for therapeutic monitoring reduces the cost benefit of UFH, but individually adjusted dosing may offer benefits over fixed dosing.[99] LMWHs derive from depolymerization of UFH and were developed to provide more consistent PK-PD than UFH. The reduced size of the LMWH polysaccharides limits their ability to simultaneously bind AT and thrombin. Thus LMWHs are principally anti-FXa agents[100] and are probably best monitored with anti-Xa assays,[101] although TF-activated TEG can also be used[102]. LMWHs are less protein-bound than UFH, have more predictable PK profiles and better bioavailability after SC injection.[103] The PK profiles of distinct LMWHs preparations are not interchangeable however.[100] LMWH PKs have been evaluated in normal healthy dogs and therapeutic anti-Xa levels extrapolated from people are achievable.[104,105]

Antiplatelet agents are widely used in thromboprophylaxis for both arterial and venous thrombi, likely because of their low cost, ease of administration, and minimal monitoring requirements. In small animals, antiplatelet agents are principally used for long-term oral maintenance therapy or as thromboprophylaxis for at-risk patients. Common oral antiplatelet agents in small animals include aspirin,[106] and clopidogrel[107]. There is limited evidence of efficacy for either aspirin or clopidogrel in PTE in veterinary medicine however, which precludes formulation of evidence-based recommendations. The cyclooxygenase inhibitor aspirin irreversibly inhibits platelet thromboxane A2 (TXA2) synthesis, inhibiting platelet function.[106] To date, no studies have been performed in veterinary medicine evaluating the efficacy of aspirin against objective measures of thrombosis or comparing the efficacy of aspirin against other antithrombotic therapies, although one such study is reportedly underway.[85] Aspirin likely does inhibit platelet function in dogs, but its effects appear limited and inconsistent. A recent study suggested that 1 mg/kg/d aspirin consistently inhibited platelet function in only a third of dogs.[108] In humans, clopidogrel may be more effective than aspirin, while combining both agents may produce beneficial additive effects.[109] In dogs, clopidogrel may provide more consistent antiplatelet effects than aspirin.[107] A multicenter double-blinded prospective study comparing aspirin with clopidogrel in cats with ATE was recently completed.[110] The study enrolled 72 cats and demonstrated that relative to aspirin, clopidogrel significantly increased the time to ATE recurrence and significantly prolonged median survival time for cats with a history of thrombosis. It is unclear if this benefit would directly translate to canine PTE, but it does suggest that clopidogrel is superior.

References

1. Musset D, *et al. Lancet*. 2002;360(9349):1914–1920.
2. Wells PS, *et al. Thromb Haemost*. 2000;83(3):416–420.
3. Le Gal G, *et al. Ann Intern Med*. 2006;144(3):165–171.
4. Lucassen W, *et al. Ann Intern Med*. 2011;155(7):448–460.
5. Johnson LR, *et al. J Vet Intern M*ed. 1999;13(4):338–345.
6. Fluckiger MA, Gomez JA. *Vet Radiol*. 1984;25(3):124–131.
7. Dennis JS. *Compend Contin Educ Vet*. 1993;15(12):1595–1603.
8. Westermark N. *Acta Radiologica*. 1938;19:357–372.
9. Larue MJ, Murtaugh RJ. *J Am Vet Med Assoc*. 1990;197(10):1368–1372.
10. Mitchell CW. *Can Vet J*. 2009;50(2):199–202.
11. Fesmire FM, *et al. Ann Emerg Med*. 2011;57(6):628–652.e675.

12. Goggs R, *et al. J Vet Emerg Crit Care*. 2009;19(1):30–52.
13. Habing A, *et al. Vet Radiol Ultrasound*. 2011;52(2):173–178.
14. Wittram C, *et al. Radiographics*. 2004;24(5):1219–1238.
15. Quiroz R, *et al. JAMA*. 2005;293(16):2012–2017.
16. Torbicki A, *et al. Eur Heart J*. 2008;29(18):2276–2315.
17. Drees R, *et al. Vet Radiol Ultrasound*. 2011;52(4):362–367.
18. Makara M, *et al. Vet Radiol Ultrasound*. 2011;52(6):605–610.
19. Takahashi A, *et al. Vet Parasitol*. 2008;155(3–4):242–248.
20. Kirberger RM, Zambelli A. *Vet Radiol Ultrasound*. 2007;48(5):418–420.
21. Goggs R, *et al. J Small Anim Pract*. 2014;55(4):190–197.
22. Neel JA, *et al. Vet Clin Pathol*. 2012;41(2):216–222.
23. Dantzker DR, *et al. Circ Res*. 1978;42(1):92–103.
24. Dennis JS. *Compend Contin Educ Vet*. 1991;13(12):1811–1817.
25. Lennon EM, *et al. J Vet Intern Med*. 2013.
26. Green RA, Kabel AL. *J Am Vet Med Assoc*. 1982;181(9):914–917.
27. Jacoby RC, *et al. Arch Surg*. 2001;136(9):1003–1006.
28. Rose L, *et al. J Vet Intern Med*. 2013;27(1):207–211.
29. Feldman BF. Thrombosis - diagnosis and treatment. In: Kirk RW, editor. *CVT IX: Small Animal Practice*. 1986:505–509.
30. Stokol T. *Vet Clin North Am Small Anim Pract*. 2003;33(6):1419–+.
31. Stein PD, *et al. Radiology*. 2007;242(1):15–21.
32. Epstein SE, *et al. J Vet Intern Med*. 2013;27(6):1646–1649.
33. Ben S-q, *et al. Thromb Res*. 2007;120(4):575–583.
34. McMichael MA, Smith SA. *Vet Clin Pathol*. 2011;40(2):140–153.
35. Kristensen AT, *et al. J Vet Intern Med*. 2008;22(1):140–147.
36. Wiinberg B, *et al. J Vet Intern Med*. 2008;22(2):357–365.
37. Goodwin LV, *et al. J Vet Intern Med*. 2011;25(2):273–277.
38. Goggs R, Wiinberg B. *J Vet Emerg Crit Care*. 2012;22(2):145–147.
39. Bentley AM, *et al. J Vet Emerg Crit Care*. 2013;23(1):14–22.
40. McCrath DJ, *et al. Anesth Analg*. 2005;100(6):1576–1583.
41. Cotton BA, *et al. J Trauma Acute Care Surg*. 2012;73(2):365–370.
42. Smith SA, *et al. Am J Vet Res*. 2012;73(6):789–798.
43. Spodsberg EH, *et al. Vet Clin Pathol*. 2013;42(3):281–290.
44. Ribeiro A, *et al. Am Heart J*. 1998;135(5):868–874.
45. Kucher N, *et al. Eur Heart J*. 2003;24(4):366–376.
46. Bunch SE, *et al. J Am Vet Med Assoc*. 1989;195(12):1748–1753.
47. Palmer KG, *et al. J Am Vet Med Assoc*. 1998;213(2):220–+.
48. Venco L, *et al. Vet Radiol Ultrasound*. 1998;39(6):564–565.
49. Teshima K, *et al. J Vet Med Sci*. 2006;68(12):1307–1313.
50. Baumwart RD, *et al. J Vet Intern Med*. 2005;19(6):828–832.
51. Janata K, *et al. BMJ*. 2003;326(7384):312–313.
52. Becattini C, *et al. Circulation*. 2007;116(4):427–433.
53. Carreton E, *et al. Vet Parasitol*. 2011;176(4):313–316.
54. Uzuelli JA, *et al. Clin Chim Acta*. 2008;388(1–2):184–188.
55. Jung SK, *et al. Shock*. 2013;39(1):50–54.
56. Ferrari E, *et al. Heart*. 2012;98(9):732–735.
57. Oyama MA, Singletary GE. *Vet Clin North Am Small Anim Pract*. 2010;40(4):545–558.
58. Henzler T, *et al. Eur Respir J*. 2012;39(4):919–926.
59. Pasha SM, *et al. Thromb Haemost*. 2012;108(2):367–372.
60. Lega JC, *et al. Thorax*. 2009;64(10):869–875.
61. Sanchez O, *et al. Am J Respir Crit Care Med*. 2010;181(2):168–173.
62. Hori Y, *et al. Vet J*. 2012;194(2):215–221.
63. Moser KM, *et al. Circulation*. 1973;48(1):170–178.
64. Carlbom DJ, Davidson BL. *Chest*. 2007;132(1):313–324.
65. Ventetuolo CE, Klinger JR. *Ann Am Thorac Soc*. 2014;11(5):811–822.
66. Smulders YM. *Cardiovasc Res*. 2000;48(1):23–33.
67. Hirose T, *et al. Microvasc Res*. 1983;26(2):193–204.

68. Zwissler B, *et al. J Cardiothorac Vasc Anesth*. 1995;9(3):283–289.
69. Zwissler B, *et al. J Cardiothorac Vasc Anesth*. 1995;9(6):634–640.
70. Brown AJ, *et al. J Vet Intern Med*. 2010;24(4):850–854.
71. Dias-Junior CA, *et al. Pulm Pharmacol Ther*. 2005;18(3):181–186.
72. Tapson VF. *Semin Thromb Hemost*. 2013.
73. Esponda O, Tafur A. *Curr Treat Options Cardiovasc Med*. 2013;15(2):137–152.
74. Sharifi M, *et al. Am J Cardiol*. 2013;111(2):273–277.
75. Kline JA, *et al. J Thromb Haemost*. 2014;12(4):459–468.
76. Meyer G, *et al. N Engl J Med*. 2014;370(15):1402–1411.
77. Whelan MF, O'Toole TE. *Compend Contin Educ Vet*. 2007;29(8):476–482.
78. Tapson VF. *Curr Opin Cardiol*. 2012;27(6):585–591.
79. Patel SC, Mody A. *Prog Cardiovasc Dis*. 1999;42(3):217–233.
80. Yang W, *et al. Clin Appl Thromb Hemost*. 2010;16(5):537–542.
81. Lunsford KV, Mackin AJ. *Vet Clin North Am Small Anim Pract*. 2007;37(3):579.
82. Dunn ME. *J Vet Emerg Crit Care*. 2011;21(2):144–150.
83. Guyatt GH, *et al. Chest*. 2012;141(2 Suppl):7S–47S.
84. Morris TA, *et al. Am J Physiol Heart Circ Physiol*. 2004;287(2):H818–H822.
85. Smith SA. *Top Companion Anim Med*. 2012;27(2):88–94.
86. Sobieszczyk P, *et al. Circulation*. 2002;106(14):1748–1749.
87. Lowe GD. *J Thromb Haemost*. 2006;4(9):1882–1885.
88. Watson HG, Chee YL. *Blood Rev*. 2008;22(2):107–116.
89. Ho KM, Chavan S. *Thromb Haemost*. 2013;109(2):272–279.
90. Lopez JA, *et al. Hematology Am Soc Hematol Educ Program*. 2004:439–456.
91. Garcia DA, *et al. Chest*. 2012;141(2 Suppl):e24S–43S.
92. Beguin S, *et al. Thromb Haemost*. 1988;60(3):457–462.
93. Mischke RH, *et al. Am J Vet Res*. 2001;62(12):1887–1891.
94. Mischke R, Jacobs C. *Res Vet Sci*. 2001;70(2):101–108.
95. Mischke R. *J Vet Diagn Invest*. 2003;15(6):588–591.
96. Allegret V, *et al. Vet Clin Pathol*. 2011;40(1):24–31.
97. Babski DM, *et al. J Vet Intern Med*. 2012;26(3):631–638.
98. Brooks MB. *Vet Clin Pathol*. 2004;33(4):208–214.
99. Helmond SE, *et al. J Vet Intern Med*. 2010;24(3):597–605.
100. Bick RL, *et al. Hematol Oncol Clin North Am*. 2005;19(1):1–51, v.
101. Alwood AJ, *et al. J Vet Intern Med*. 2007;21(3):378–387.
102. Jessen LR, *et al. Vet Clin Pathol*. 2008;37(4):363–372.
103. Grebe S, *et al. Berl Munch Tierarztl Wochenschr*. 2000;113(3):103–107.
104. Mischke R, *et al. Am J Vet Res*. 2001;62(4):595–598.
105. Lunsford KV, *et al. J Am Anim Hosp Assoc*. 2009;45(6):261–267.
106. Wiinberg B, *et al. J Vet Emerg Crit Care*. 2012;22(1):42–58.
107. Brainard BM, *et al. Am J Vet Res*. 2010;71(7):822–830.
108. Dudley A, *et al. J Vet Intern Med*. 2013;27(1):141–149.
109. Eikelboom JW, *et al. Chest*. 2012;141(2 Suppl):e89S–119S.
110. Hogan DF, *et al. J Vet Cardiol*. 2015;17(Suppl 1):S306–317.

1.
acvim372

Heparin Monitoring in Critically Ill Dogs

Rita M. Hanel, DVM, DACVIM (SAIM), DACVECC
Raleigh, NC, USA

Introduction

Studies evaluating the efficacy of anticoagulants in veterinary medicine are lacking, and most involve the use of fixed dosages of unfractionated heparin (UFH) with variation in outcome measures. One prospective study of the use of UFH in 18 dogs with immune mediated hemolytic anemia (IMHA) evaluated the effect of fixed dosages of UFH (300 IU/kg subcutaneously every 6 hours) on activated partial thromboplastin time (aPTT) as well as factor Xa inhibition (e.g., anti-Xa or AXa activity).[1] During the initial 40 hours, only eight of 18 dogs attained therapeutic AXa activities. Fifteen out of 18 dogs survived to discharge and 11 were alive at one year. In a second prospective report of hemostatic abnormalities in 20 dogs with IMHA, 17 of 20 dogs were treated with UFH at a starting dose of 100 IU/kg body weight subcutaneously every 6 hours, for a range of 3 to 28 days.[2] The UFH dose was adjusted in order to prolong aPTT by 1.5 to 2.0 times above reference range. Four dogs died during hospitalization and two additional died during the first month after discharge. Thromboembolism was the most common finding on necropsy. Further information regarding UFH therapy, such as time to therapeutic concentration, was not mentioned. A recent randomized, prospective, controlled clinical trial in dogs with IMHA showed an 85% reduction in 180-day mortality when UFH anticoagulant therapy was adjusted based on AXa activity versus standardized or fixed dose administration.[3] Seven out of 8 dogs with individually adjusted UFH survived, whereas only one out of seven survived in the fixed dosage cohort.

Low molecular weight heparins (LMWH) are a preferred parenteral anticoagulant in human medicine, owing to their predictable pharmacokinetics and pharmacodynamics. Based on their predictability, LMWH is not typically titrated and monitored but administered as a fixed dosage. One retrospective study examined a LMWH, dalteparin, in 38 critically ill dogs, and found that fixed dosages were often not therapeutic and dose adjustment in response to AXa activity interpretation inconsistently resulted in subsequent attainment of the target AXa range.[4] Also, dogs that appeared more hypercoagulable on pre-treatment thromboelastography were significantly less likely to achieve the target AXa activity. Thus, once cost and the aforementioned results are taken into consideration, the benefits of LMWH in veterinary medicine compared to UFH are not as clear.

This session will focus on the use of UFH in dogs and will address 3 topics: 1) Perspective and pitfalls surrounding the monitoring of UFH, 2) utility of current methods of monitoring UFH in healthy dogs including recently published data by the author, 3) utility of accepted methods of monitoring UFH in critically ill dogs, including recently completed research on its use in critically ill dogs with immune mediated hemolytic anemia (IMHA).

Heparin

Heparin is a mixture of negatively charged mucopolysaccharides. UFH is comprised of molecules weighing 3,000 to 30,000 Daltons (Da), with an average molecular weight (MW) of 15,000–18,000 Da or 40 to 50 monosaccharides. There is a large amount of nonselective binding of UFH molecules to both cells and proteins, which creates individual variation in pharmacodynamics and dose-dependent variations in pharmacokinetics. LMWH is obtained by various methods of fractionation or depolymerization of UFH. The average MW of LMWHs are usually less than 8,000 Da, and LMWH molecules are renally excreted via a non-saturable route. This is in contrast to UFH, whereby a saturable mechanism is responsible for a large portion of UFH renal clearance and endothelial cells and macrophages mediate depolymerization. Thus, LMWHs have dose-independent clearance and theoretically more predictable pharmacokinetics, in the absence of renal insufficiency, than UFH.

Heparins bind to antithrombin via a specific pentasaccharide sequence, inducing a conformational/allosteric change, which enhances its ability to inactivate multiple serine proteases by up to 1000-fold. This conformational change facilitates the inhibition of Xa, but the inhibition of thrombin and the majority of other coagulation proteases requires a more extensive interaction between the molecules whereby heparin serves as a template for both proteins and tethers them together. This is typically facilitated by heparin molecules larger than 14 saccharide units. Because, ultimately, UFH promotes inhibition at multiple steps in the coagulation pathway, it can profoundly affect global hemostatic assays, such as the activated partial thromboplastin time, whereas LMWH will barely be able to affect a change.

HEPARIN MONITORING

There are three main categories of available monitoring: global clotting assays measuring primarily the intrinsic and common pathways of coagulation, global assays utilizing viscoelastic technology, and methods of measuring heparin concentration. These three approaches will be mentioned during the session, albeit basic familiarity is assumed.

Activated Partial Thromboplastin Time (aPTT)

This is a global clotting assay measuring primarily the intrinsic and common pathways of coagulation. It is performed by adding a surface activator and a diluted phospholipid to citrated plasma. Calcium is then added and the time to clot formation is measured. After a bolus and subsequent infusion of UFH, the dose of heparin is typically adjusted to achieve an aPTT of 1.5–2.5 times the control value. The landmark study upon which this protocol is based was published by Basu *et al.* in 1972 and demonstrated a decreased incidence of recurrence of thromboembolism in people.[5] This protocol was verified by this same group in a rabbit model of thrombus extension. However, it is important to note that this now widely discussed guideline of a therapeutic aPTT ratio between 1.5 and 2.5 was chosen based on correlation with protamine titration. The main limitation of the aPTT is the marked amount of variation in performance due to pre-analytical, analytical, and biological variables, which are summarized in Table 1. The marked variability in aPTT results from different reagent-instrument combinations has been demonstrated by multiple studies in humans as well as in dogs.[6] Recent results regarding analytical variability in point of care testing in dogs receiving UFH will also be presented and discussed. Regardless, the widespread use and acceptance of this technique continues based on cost effectiveness and ease of use.

Table 1

Preanalytical and analytical factors	APTT	AXa activity
Blood sampling in the morning/evening (diurnal variations)	D/I*	N
Underfilled tubes/too much citrate	I*	N
Sampling too close to site/improper flushing; heparin contamination	I*	I*
Delay in analysis/processing sample (>1 hour)	D*	D*
Inadequate centrifugation (platelet contamination; heparin neutralization by platelet factor 4)	D*	D*
Hemolysis	N	D*
Change in reagents; necessitates re-calibration	I or D	I or D
Change in machine/coagulometer; necessitates re-calibration	I or D	I or D
Biological factors	**APTT**	**AXa activity**
Hyperbilirubinemia (>6.6 mg/dl in humans)	N	D*
Hyperlipidemia	N	I*
Antithrombin deficiency; congenital or acquired form of "heparin resistance"	D	D or N (1 or 2 stage)
Increased acute phase reactants (fVIII, fibrinogen, etc.); mimics "heparin resistance" if using aPTT	D*	N
Increased heparin binding proteins (infection, inflammation, malignancy); common cause of "heparin resistance" requiring higher heparin dosages	D	D

Obesity (increased volume of distribution)	D	D
Impaired renal function (decreased UFH elimination)	I	I
Liver dysfunction/disease or Vitamin K antagonism (decreased factor production)	I*	N
Consumptive coagulopathy	I*	N
Clotting factor deficiencies (fXII, VIII, IX, etc.)	I*	N

Factors known to influence activated partial thromboplastin time (APTT) and anti-Factor Xa (AXa) activity in humans when used in the monitoring of unfractionated heparin (UFH). I=increase in laboratory result; D=decrease in laboratory result; N=minimal effect on laboratory result. * Indicates that dose adjustment based on result may be incorrect.

Measures of Heparin Concentration

There are three ways to measure heparin concentration. The first is an assay that directly measures the concentration. These are not favored as they are difficult to perform and do not reflect the pharmacodynamics of the drug; as not all measured molecules have the pentasaccharide sequence needed for biologic activity. The second is estimation of heparin concentration via quantification of the lowest amount of protamine needed to neutralize the effects of the heparin-induced prolongation of the thrombin time. The recommended therapeutic concentration of heparin based on protamine titration is 0.2–0.4 unit/ml. This is an accepted method, just not as widely used due to difficulties with automation and lack of clinical availability. The final method is via measurement of the functional activity, or specifically the functional activity of factor Xa. This is a chromogenic assay that has been automated and is able to be used for UFH, LMWH, and fondaparinux. In this assay, the heparin-antithrombin complex is allowed to bind to available proteases in the plasma. A chromogenic factor Xa substrate is then added, which binds to any residual unbound factor Xa. This produces a color, which can be read optically and plotted against a standard curve. The amount of color produced is inversely proportional to the activity of the heparin. The recommended UFH therapeutic range using the AXa concentration is 0.3–0.7 unit/ml. Based on the colorimetric nature of this assay, it has some analytical and biological variables that can skew the results. These and other factors affecting the assay are also summarized in Table 1.

Viscoelastic Monitoring

Attempts have been made to utilize viscoelastic techniques, such as thromboelastography (TEG), to monitor heparin therapy. The methodology of these tests makes them similar to AXa concentrations, since they are measuring the functional activity of the drug, but the methods of activation used are comparable to global clotting assays. However, TEG is very sensitive to heparin, which has previously limited its application. Activation of TEG by re-calcification alone was proven to be ineffective and does not correlate with therapeutic AXa concentrations.[7] However, further investigation into the use of variable strong activators in the author's laboratory showed that the use of RapidTEG® is feasible, and a significant correlation ($R=0.90$, $p<0.005$) was found between the TEG R value and factor Xa inhibition, which was comparable to aPTT ($R=0.84$ and 0.86 for two different coagulometers, $p<0.001$) in healthy dogs.[8] Conclusions from this study and its usage to develop a heparin nomogram will be discussed.

Nomograms

How to interpret this information in veterinary medicine is still challenging due to a lack of clinical trials with sufficient patients to evaluate their impact on morbidity and mortality. Regardless, it is important to note that the measurement of the functional activity of heparin, via anti-Factor Xa, is considered the optimal method for heparin monitoring in people and is recommended by the American College of Chest Physicians (ACCP) and the College of American Pathologists (CAP) as the gold standard.[9,10] If this is not readily available, it is recommended to establish institution specific references and nomograms for therapeutic aPTT ranges corresponding to heparin activity, based on anti-Xa concentrations or protamine titration. A nomogram is a prediction tool, whereby the relations between variables are arranged so that the value of one variable can be predicted by another. In this scenario, the linear relationship between the available assay (TEG, aPTT, and aPTT ratios) and AXa concentrations can be plotted using the least-squares regression technique with a line of best fit, according to recommendations for establishing therapeutic ranges for heparin therapy in people.[11] The therapeutic range can then be determined using the ordinate values on the least squares regression line that corresponds to plasma AXa activity of 0.3 to 0.7 IU/mL. This technique was used to create a nomogram using the data acquired from McLaughlin *et al.*[8] which is presented in Table 2.

Table 2

Predicted AXa value	Value of R (min) Rapid TEG® (CWB)	Value of clinical pathology PTT ratio* (CWB) (STA Satellite®, Stago)	Value of POC PTT ratio* (WB) (Coag DX™, IDEXX)	Heparin multiplier IV CRI§	Heparin multiplier SQ‖
0.0-0.1	0.0-0.45	0.0-0.93	0.0-1.03	3.0	3.0
0.11-0.15	0.46-0.56	0.94-1.07	1.04-1.08	2.4	2.4
0.16-0.2	0.57-0.67	1.08-1.20	1.09-1.12	1.9	1.9
0.21-0.25	0.68-0.78	1.21-1.34	1.13-1.17	1.7	1.7
0.26-0.3	0.79-0.89	1.35-1.48	1.18-1.21	1.5	1.5
0.31-0.35	**0.90-1.00**	**1.49-1.61**	**1.22-1.26**	1.3	**1.3**
0.36-0.4	**1.01-1.11**	**1.62-1.75**	**1.27-1.31**	1.2	**1.2**
0.41-0.45	**1.12-1.22**	**1.76-1.88**	**1.32-1.35**	1.1	**1.2**
0.46-0.5	**1.23-1.33**	**1.89-2.02**	**1.36-1.40**	1.1	**1.1**
0.51-0.55	**1.34-1.43**	**2.03-2.16**	**1.41-1.44**	1.0	**1.1**
0.56-0.6	**1.44-1.54**	**2.17-2.29**	**1.45-1.49**	1.0	**1.0**
0.61-0.65	**1.55-1.65**	**2.30-2.43**	**1.50-1.54**	1.0	**1.0**
0.66-0.7	**1.66-1.76**	**2.44-2.56**	**1.55-1.58**	1.0	**1.0**
0.71-0.75	1.77-1.87	2.57-2.70	1.59-1.63	0.9	0.9
0.76-0.8	1.88-1.98	2.71-2.84	1.64-1.68	0.8	0.8
0.81-0.85	1.99-2.09	2.85-2.97	1.69-1.72	0.7	0.7
0.86-0.9	2.10-2.20	2.98-3.11	1.73-1.77	0.5	0.5
0.91-0.95	2.21-2.30	3.12-3.24	1.78-1.81	0.4	0.4
0.96-1	2.31-2.42	3.25-3.38	1.82-1.86	0.3	0.3
>1.00	>2.42	>3.38	>1.86	0.2	0.2

Unfractionated Heparin (UFH) Nomogram. The UFH dosage is adjusted using the multiplier supplied for either intravenous continuous rate infusion (IV CRI) or subcutaneous (SQ) administration. Citrated whole blood (CWB) or whole blood (WB) should be obtained 3 hours post UFH dose adjustment§ or administration‖. The thromboelastography derived R value can be directly applied. *The activated partial thromboplastin (aPTT) ratio is obtained by dividing the current aPTT by the baseline (pre-UFH) aPTT obtained by same method. Values for predicted Anti-Xa (AXa) were derived using linear regression and hemostatic assay results obtained from 8 healthy male dogs.[8]

Heparin in the Critically Ill

Factors affecting the usage of aPTT and AXa activity, the most commonly used methods of monitoring in veterinary medicine, have been summarized in Table 1. Biologic factors of significant concern in our critically ill dogs include: antithrombin deficiency, increased acute phase reactant and inflammatory proteins, liver dysfunction or a consumptive coagulopathy. Antithrombin, as the target of heparin activity, may blunt the functional effect of heparin if it becomes decreased, which has been demonstrated in dogs with sepsis and IMHA.[12,13,1] Increased acute phase reactant proteins, such as fibrinogen or factor VIII, can shorten aPTT but may or may not affect the capacity for thrombin generation. Elevations in these factors have been demonstrated in dogs in the setting of progressive heart failure,[14] and increases in fibrinogen and C-reactive protein, along with a decrease in von Willebrand Factor, have been documented in dogs with inflammatory conditions such as babesios.[15] Consumptive coagulopathies are an unfortunate and possible consequence of most multisystemic disease states, which can have a myriad of effects on both aPTT and thrombin generation that may be divergent depending upon the stage of disease. In a previously mentioned manuscript involving hemostatic abnormalities in dogs with IMHA, the authors used the presence or absence of seven criteria (thrombocytopenia, prolonged PT or aPTT, low antithrombin, low fibrinogen, increased fibrin degradation products, and increased D-Dimer concentrations) to diagnose disseminated intravascular coagulation (DIC).[2] Based on the defined criteria, DIC was diagnosed in nine of the 20 (45%) dogs in this study and was suspected in an additional four (20%). Taking into consideration the aforementioned factors and their impact on hemostatic monitoring tools, the utility of commonly used methods, such as aPTT in a critically ill population of animals requires scrutiny. Final data from the **TOP HATT** trial (**T**eg **or P**TT for **h**eparin **a**nticoagulation to prevent **t**hrombosis) will be presented comparing aPTT and viscoelastic methods to AXa concentrations in a critically ill cohort of dogs with IMHA. Trends in fibrinogen, antithrombin, and thrombin generation will also be presented.

SUMMARY

UFH was deemed likely to fall out of favor as newer drugs have been introduced. However, in the setting of the critically ill dog, it will probably continue to be used due to its short half-life, ability to be reversed, and partially non-renal route of elimination. As such, it continues to be prudent to be familiar with the currently available methods of monitoring and their targets. Based on available evidence, fixed weight based dosage strategies do not appear adequate for the prevention of thrombosis in dogs, and therapeutic monitoring should be considered. It is this author's recommendation to use AXa concentrations, when available, for UFH monitoring in healthy and critically ill dogs. Although a canine specific target therapeutic concentration for AXa has not been established, it appears reasonable to initially adopt human standards based on the results from Helmond *et al.*[3] When AXa is not readily available, the use of RapidTEG® activated TEG is superior to aPTT for monitoring UFH in critically ill dogs, albeit both assays perform equally well in healthy dogs. Institution specific nomograms should be developed for the use of TEG or aPTT in either setting.

REFERENCES

1. Bruehl EL, *et al. J Am Anim Hosp Assoc.* 2009;45:125.
2. Scott-Moncreiff JC, *et al. J Am Anim Hosp Assoc.* 2001;37(3):220.
3. Helmond SE, *et al. J Vet Intern Med.* 2010;24(3):597.
4. Lynch A, *et al. J Vet Emerg Crit Care.* 2014;24(4):421.
5. Basu D, *et al. N Engl J Med.* 1972;287:324.
6. Mischke RH, *et al. Am J Vet Res.* 2001;62(12):1887.
7. Pittman JR, *et al. J Vet Emerg Crit Care.* 2010;20(2):216.
8. McLaughlin, *et al. J Vet Emerg Crit Care.* 2017;27(1):71.
9. Guyatt GH, *et al. Chest.* 2012 Feb. 141(2 Suppl):7S.
10. Olson JD, *et al. Arch Pathol Lab Med.* 1998;122:782.
11. Brill-Edwards P, *et al. Ann Intern Med.* 1993;119:104.
12. Bentley AM, *et al. J Vet Emerg Crit Care.* 2013;23(1):14.
13. deLaforcade AM, *et al. J Vet Intern Med.* 2008 Jan-Feb;22(1):26.
14. Prihirunkit K, *et al. J Vet Med.* Epub 2014 Oct 16.
15. Barić Rafaj R, *et al. J Vet Intern Med.* 2013;27(5):1172.

acvim373

Prothrombotic Mechanisms in Critical Illness

Alex Lynch, BVSc (Hons), DACVECC, MRCVS
Gainesville, FL, USA

In health, hemostasis simplistically can be considered the balance between the antithrombotic and prothrombotic states. The antithrombotic state is necessary for maintenance of blood in the fluid phase within the closed, high pressured circulatory system. In the event of this closed circulatory system being breached through injury, the prothrombotic state ensues. This enables temporary blood stasis through thrombus formation limiting excessive extravascular blood loss. When the injured blood vessel has been repaired, the thrombus is no longer necessary and is removed. This restores blood vessel patency and enables blood to circulate in the fluid phase once again. Our knowledge of hemostasis has evolved considerably over time, and likely will continue to do so, with applications for health, disease, and therapeutics. Conventionally, we have considered hemostasis in three distinct phases following vascular injury.[1] Primary hemostasis described the chain of events leading to the formation of a temporary, yet somewhat unstable, platelet plug. The cementing of this platelet plug is achieved via fibrin, which is the end-product of the sequential activation of a series of serine proteases, collectively referred to as the clotting factors. The sequential activation of these clotting factors is termed secondary hemostasis. Secondary hemostasis is traditionally further subdivided into the extrinsic, intrinsic, and common pathways. Tertiary hemostasis is more commonly described as fibrinolysis, which involves the degradation and subsequent removal of the fibrin cemented platelet plug. Fibrinolysis restores vessel patency once the injured vessel has been repaired. This understanding of hemostasis stood the test of time for many years enabling the development of commonly utilized diagnostic tests (e.g., prothrombin time, activated partial thromboplastin time). Since the early 21st century, however, increasing focus has been placed on the cell based model of hemostasis.[2] This model likely reflects *in vivo* hemostasis more accurately than the discrete phases of the traditional theory. The cell-based model aims to take into account the complex interaction between vascular tone, the endothelium, and the cellular and soluble components of the blood. This cell based model aims to better describe the experimental data illustrating the concomitant deposition of platelet thrombus and fibrin in rodent models of vascular injury.[3] Some key points to recall with respect to the cell-based model of hemostasis are described below.

Under normal conditions, the intact endothelium promotes a naturally antithrombotic phenotype. This is achieved in four major ways: (1) The intact endothelium separates the components of blood from the prothrombotic constituents of the subendothelial layer (e.g., collagen); (2) the endothelium produces several substances that inhibit platelet function, specifically prostacyclin, nitric oxide, and ADPase; (3) some anticoagulant substances are also expressed by the endothelium including antithrombin, thrombomodulin, and tissue factor pathway inhibitor; and (4) a profibrinolytic state is maintained, enabling the breakdown of any clots that may form, via the production of tissue plasminogen activator. Upon injury of a vessel wall, rapid reflex vasoconstriction occurs allowing the injured vessel wall to come into closer contact with the constituents of the blood. The exposed subendothelium is naturally highly prothrombotic but the most potent stimulator of thrombosis is tissue factor. Tissue factor (TF) is a glycoprotein usually only contained on the extravascularly situated fibroblasts and macrophages. Once injury has occurred to the vessel wall, exposed tissue factor on these cells is able to contact and complex with factor VII in the blood. A small amount of FVII is circulating in the blood in the activated form, denoted as FVIIa, enabling rapid initiation of hemostasis upon contact with these TF bearing cells. The complexing of TF with FVIIa is the key step in initiating hemostasis and produces a small amount of FXa. In turn, FXa combines with FIXa, forming a prothrombinase complex. The prothrombinase complex catalyzes the activation of inactive prothrombin (FII) to thrombin (FIIa). This series of steps is termed the initiation phase in the cell based model of hemostasis. The thrombin generated from the initiation phase is a key step in the activation, aggregation, and recruitment of platelets to the site of injury. Thrombin interacts with surface receptors on the platelet leading to several changes in platelet shape, changes in the prothrombotic tendency of the platelet surface, and also release of several vasoactive substances from intracellular granules. These substances include other potent activators of platelets including collagen and ADP, leading to an overall activation and aggregation of platelets local to the vessel wall injury. This process has been described as the amplification phase. The activated platelets described in the amplification phase also serve an important function as a phospholipid bilayer upon which other soluble clotting factors within the blood can interact. These clotting factors result in the formation of two major enzyme complexes, the tenase complex (constituting FVIIIa and FIXa) and the prothrombinase complex (composed of FVa and FXa). The tenase complex catalyzes the activation of inactive FX to active FXa, necessary for this prothrombinase complex. Prothrombinase catalyzes the conversion of further prothrombin to thrombin, with a resultant large thrombin burst. Thrombin is a key player with several important roles within hemostasis as alluded to already. Specifically thrombin is necessary for (1) platelet activation, (2) conversion of fibrinogen to fibrin, (3) negative feedback systems for natural anticoagulants (e.g., antithrombin, protein C, protein S, thrombomodulin), and (4) crosstalk with fibrinolysis via thrombin activatable fibrinolysis inhibitor (TAFI). This final process of the cell based model of hemostasis, characterized by these key events leading to massive thrombin generation, is described as the propagation phase.

In health, the complex process of hemostasis described above is tightly controlled so that there is balance between blood fluidity and clot formation. In the critically ill patient, however, abnormalities in inappropriate thrombosis and hemorrhage may be encountered. This lecture will focus on the prothrombotic tendency in the critically ill rather than hypocoagulable states. Many of our traditional tests of hemostasis, however, are actually better suited to the identification of hypocoagulability. This has made accurate characterization of prothrombotic states historically more challenging. While the hypercoagulable nature of blood is an important aspect of the prothrombotic state, other features of the critically ill patient should be remembered. Another key point to recall is that the prothrombotic tendency of the critically ill usually occurs independent of apparent blood vessel wall injury in contrast to the normal process of hemostasis described above. Conveniently the distinct, yet interacting, components of this prothrombotic tendency can be described by Virchow's triad.[4,5] This triad was first described in the 1850s, yet took another century before really coming to the forefront of hemostasis theory. The triad describes that the prothrombotic state in illness is influenced by (1) hypercoagulability within the blood, (2) blood stasis, and (3) vessel wall or endothelial dysfunction. It is generally considered necessary for at least two of these components to be disordered in order for thrombosis to occur.

Hypercoagulability can be due to both plasma and cellular procoagulant activity. This is in fact the most straightforward, and therefore best, characterized aspect of Virchow's triad. With respect to plasma components, there can simplistically either be an excess of prothrombotic factors (e.g., fibrinogen, prothrombin) or a relative deficiency of antithrombotic factors (e.g., antithrombin, protein C). The cellular contribution to a hypercoagulable state includes hyperactivity of platelets and increased platelet number. Much like TF bearing cells are the key initiator of hemostasis under normal circumstances, the role of TF bearing cells in pathological conditions is also pivotal.[5] In some disease states including neoplasia and sepsis, abnormal expression of TF on endothelial cell surfaces and circulating monocytes is thought to increase the risk of a prothrombotic tendency even with an intact endothelium.[8] The role of microparticles is also an important, emerging area of research relevant to hypercoagulable states.[9,10] Microparticles are small vesicles derived from the plasma membranes of eukaryotic cells that are found circulating in the blood. These microparticles express a variety of surface antigens depending on their original source (e.g., platelets, white blood cells) and are important in the cross talk between different pathological process (e.g., inflammation and thrombosis). A key contribution of microparticles to a prothrombotic state is the surface expression of phosphatidylserine (PS) that promotes the assembly of vitamin K dependent enzyme complexes of the coagulation cascade.[10] Moreover, microparticles are the major reservoir of circulating TF in the blood, and therefore are an important contributor to *de novo* thrombin generation in the critically ill patient.[11] The presence of hypercoagulability in conjunction with another derangement of Virchow's triad is generally considered necessary for pathological thrombosis to occur.

Alterations in blood flow are more difficult to measure but likely have a considerable influence on the nature of thrombosis. The histopathology of blood clots typically varies between arterial and venous clots. Arterial clots are usually platelet rich, while venous clots are typically more densely packed with erythrocytes. Two important variables are considered in this context, specifically shear stress and shear rate.[5] In clinical cases, shear stress is really referring to blood vessel wall shear stress, and in venous low- or no-flow states, wall shear stress is decreased. In people, the best characterized example of this would be hospitalized or travel-associated immobility increasing the risk of deep vein thrombosis.[12] In veterinary patients equivalent immobility potentially could be noted in large, sick, down dogs or in immobilized patients receiving mechanical ventilation. The Krüppel-like transfusion factors are implicated in the expression of the antithrombotic endothelial surface proteins (e.g., thrombomodulin).[5] Shear stress increases the expression of Krüppel-like transfusion factors. In low shear states there may therefore be a transition to a prothrombotic endothelial surface, which could promote local thrombosis.[5,13] Shear rates refer to how rapidly blood constituents (e.g., platelets) flow downstream through a vessel, as well as in a short radial distance from the vessel wall.[5] Rheological changes may prematurely augment platelet activation and promote a prothrombotic state. This may be most relevant for arterial thrombi where rheology can be considerably altered due to the presence of atherosclerotic plaques.[14] Interestingly, concurrent venous and arterial thrombi are infrequently identified in the critically ill suggesting distinct risk factors for their development likely occur. In people, smoking is a considerable risk factor for peripheral arterial disease, for example, yet has only a modest impact on venous thrombosis.[5,15] The most common and best described form of arterial thrombosis in veterinary species is, of course, feline arterial thromboembolism. Underlying and frequently clinically silent cardiomyopathy underpins the majority of these cats. An important contributor to their pathology is the altered flow of blood within the enlarged left atrium of affected cats. There is as an association between the visible spontaneous echocardiographic contrast ('smoke') visualized in cats with cardiomyopathy and diminished left atrial appendage flow velocity.[16]

Blood vessel or endothelial damage can occur in several ways in the critically ill. The placement of indwelling catheters, especially long stay catheters in large veins (e.g., jugular catheters, saphenous catheters), are a risk factor for thrombosis. The nature of the thrombosis seen with catheters is variable but can include partially occlusive intraluminal clots, partially occlusive intramural clots adjacent to the catheter attached to the vessel wall, and completely occlusive thrombi within vessels.[17] Vessel wall damage can occur with several other disease states including sepsis, vasculitis, and severe trauma. Endothelial dysfunction may be difficult to definitively prove in clinical cases but several studies have made strides towards documenting such changes. In cases of arterial thromboembolism in cats, data has identified increased von Willebrand factor antigen concentrations in affected cats compared to controls, as a downstream indicator of endothelial dysfunction.[18] An alternative to measuring circulating biomarkers of endothelial damage has been investigated in dogs with congestive heart failure.[19] This novel technique involves using a sphygmomanometric cuff to occlude flow through the brachial artery for 3 minutes. Ultrasonography is used to assess the pulse wave brachial artery velocity before and after occlusion. This technique assesses flow mediated vasodilation as a correlate to the vascular endothelial capacity to release vasodilatory substances like nitric oxide in response to occlusion induced hypoxia (reactive hyperemia). The magnitude of the reactive hyperemia response is used as an indicator of endothelial function. In the cohort of dogs with heart failure, an attenuated reactive hyperemic response was noted. This technique could be applied in the future to other diseases where endothelial dysfunction may occur. Growing interest in the relevance of dysfunctions of the glycocalyx may also shed further light on endothelial dysfunction in critical illness.[20,21] The glycocalyx contributes to the maintenance of vascular permeability, vascular tone, and prevents leukocyte adhesions.[20] Plausibly derangements of the glycocalyx could therefore have negative downstream effects with respect to thrombosis risk, including the risk of platelet hyperaggregation and local hypercoagulability (e.g., loss of antithrombin and thrombomodulin).[21]

When faced with a critical patient, it can be difficult to know which components of the triad may be contributing to the overall prothrombotic tendency. Philosophically, the end product of increased thrombotic risk may be globally more helpful to know for the clinician. Improved appreciation of the individual components, however, could hypothetically individualize management of these cases. Several common disease states are recognized as having increased prothrombotic risk. In the septic patient, for example, there is a strong body of evidence to suggest important cross talk between inflammation and coagulation.[22] Cytokines mediate the link between these two systems and several of them have a well-established role in this process (e.g., TNF-α, IL-1β).[5] Venous thromboembolic risk predominates in septic patients associated with an inappropriate intravascular procoagulant environment. Tissue factor mediated initiation of coagulation appears most important in this setting.[22] In sepsis, natural endogenous anticoagulants (e.g., antithrombin, protein C)[22,23] plus inhibitors of fibrinolysis (e.g., tissue factor pathway inhibitor) are decreased[22], favoring a prothrombotic tendency. Inflammatory states may also induce increase platelet aggregation[24,25], as well as increase TF expression on endothelial cells[24]. Sterile causes of severe systemic inflammation (e.g., severe acute necrotizing pancreatitis) also have increased thrombotic risk and likely shares similarities with sepsis. Venous thromboembolic complications are also an important consideration for patients with neoplasia.[26] Neoplasia may induce a prothrombotic state through several mechanisms including surface expression of TF, downregulation of endogenous anticoagulants, platelet hyperactivity, endothelial dysfunction, and as sequelae to chemotherapy.[26] Venous thromboembolic complications of immune mediated hemolytic anemia (IMHA) are also significant cause for concern in dogs with respect to morbidity and mortality risk. In particular, pulmonary thromboembolism is an important and potentially life threatening consequence of IMHA. Several important factors likely contribute to the prothrombotic tendency in these cases, including hyperfibrinogenemia, the nitric oxide scavenging effects of free hemoglobin, TF bearing microparticles, platelet hyperactivity, and overall increased TF expression.[27] Multiple mechanisms may exist in other disease states, highlighted by recent research into the prothrombotic tendencies of some dogs with protein losing nephropathy. Glomerular loss of antithrombin was conventionally considered the explanation for increased thrombotic risk in these dogs. While antithrombin deficiency is a characteristic of this process, other components of hemostasis may also be dysregulated including platelet hyperreactivity and increased prothrombotic factors (e.g., fibrinogen).[28]

Laboratory assessment of the prothrombotic state can be achieved through a number of means. Viscoelastic monitoring with thromboelastography or thromboelastometry has been a busy area of research in the veterinary community. These tests can provide information suggestive of a hypercoagulable tendency but fails to pinpoint the trigger for such changes (e.g., prothrombotic excess versus antithrombotic deficiency) in individual patients. Exogenous factors including fibrinogen concentration and red cell mass also influence the data obtained through these tests and, therefore, should be concurrently interpreted in clinical cases29. Guidelines to appropriate usage of viscoelastic testing in small animals has been published through the American College of Veterinary Emergency and Critical Care.[30] Individual contributions to hemostasis including fibrinogen, antithrombin, protein C, von Willebrand factor antigen, thrombin-antithrombin complexes, and individual clotting factor concentrations may provide useful information in select cases. Moving forward, further evaluation of thrombin generation as a simple global downstream indicator of a prothrombotic tendency likely deserves further attention. A small number of veterinary publications have utilized this technique to date and further work is required to assess the full potential of thrombin generation as an assessor of hemostatic dysfunction in the critically ill patient.[31-33]

References

1. 1. Hall JE. *Guyton and Hall Textbook of Medial Physiology*. 12th Ed. Ch 36, pp. 451.
2. 2. Hoffman M, Monroe DM. *Thromb Hemost*. 2001;85(6):958–65.
3. 3. Furie B, Furie BC. *N Eng J Med*. 2008;359:358–49.
4. 4. Kumar DR *et al. Clin Med Res*. 2009;8:168–172.
5. 5. Wolberg AS *et al. Anesth Analg*. 2012;114(2):275–285.
6. 6. Machlus KR *et al. Blood*. 2011;117:4953–63.
7. 7. Geenen *et al. Thromb Hemostasis*. 2012;108:863–871.
8. 8. Hopper K, Bateman S. *J Vet Emerg Crit Care*. 2005;15(2):83–91.
9. 9. Morel O *et al. Semin Immunopathol*. 2011;33:469–486.
10. 10. Herring JM *et al. J Vet Intern Med*. 2013;27:1020–1033.
11. 11. Giesen PL *et al. Proc Natil Acad Sci USA*. 1999;96:2311–2315.
12. 12. Kesieme E *et al. J Blood Med*. 2011;2:59–69.
13. 13. Parmar KM *et al. J Clin Invest*. 2006;116:49–58.

14. 14. Nesbitt WS *et al. Nat Med.* 2009;15:665–73.
15. 15. Brogneaux C *et al. Rev Med Liege.* 2012;67:560–5.
16. 16. Schober KE, Maerz I. *J Vet Intern Med.* 2006;20:120–130.
17. 17. Baskin JL *et al. Lancet.* 2009;374:159.
18. 18. Stokol T *et al. J Vet Intern Med.* 2008;22(3):546–52.
19. 19. Cunningham SM *et al. J Vet Intern Med.* 2012;26:547-557.
20. 20. Kolarova H *et al. Mediators Inflam.* 2014;2014:694312.
21. 21. Tarbell JM, Pahakis MY. *J Int Med.* 2006;259:339–350.
22. 22. Levi M *et al. Semin Thromb Hemost.* 2013:39:559–566.
23. 23. De Laforcade AM *et al. J Vet Intern Med.* 2003;17(5):674–9.
24. 24. Oikonomopoulou K *et al. Semin Immunopathol.* 2012;34:151–165.
25. 25. Major SB *et al. Am J Vet Res.* 2015;76:328–37.
26. 26. De Cicco M. *Crit Review Onc Hem.* 2004;50:187–196.
27. 27. Kidd L *et al. J Vet Emerg Crit Care.* 2013;23(1):3–13.
28. 28. Donahue *et al. J Vet Emerg Crit Care.* 2011;21(4):346–355.
29. 29. Smith SA *et al. Am J Vet Res.* 2012;73:789–798.
30. 30. Goggs R *et al. J Vet Emerg Crit Care.* 2014;24(1):1–22.
31. 31. McMichael MA *et al. J Vet Intern Med.* 2015;29:499–504.
32. 32. Dengate AL *et al. J Vet Emerg Crit Care.* 2016;26:446–452.
33. 33. Gruber EJ *et al. Am J Vet Res.* 2016;77(4):404–12.

1.
acvim374

Thromboelastography at the Point of Care in ICU: Research Update

Rita M. Hanel, DVM, DACVIM (SAIM), DACVECC
Raleigh, NC, USA

Introduction

The condition of a critically ill (CI) patient can significantly change on a daily basis, making serial assessment a crucial part of patient management. Cage-side diagnostics can be performed in addition to a daily physical examination to make this assessment more objective. With the increase in available choices, this can become costly to the client. Thus, learning to maximize tests that are readily accessible and relatively inexpensive can make a large difference in your quality of care. One of the most difficult systems to predict and monitor is the hemostatic system. For years, we have been searching for a tool that will predict if and when the patient will bleed or form a pathologic thrombus. In this author's opinion, the available information is sometimes difficult to interpret as the "event" is already happening in the several of manuscripts at the time of the assay. As such, consumption has occurred and in the case of a thrombus, fibrinolysis may already be ongoing, thereby skewing commentary and conclusions on its predictive power. That being said, whole blood thromboelastography (TEG) appears to be the best available tool to evaluate the hemostatic phenotype in a critically ill patient at a single point in time. However, there are some quandaries and obstacles to consider before making it a cage-side tool. This session will address 3 main concerns relative to TEG: 1) prediction of bleeding or pathologic thrombus formation, 2) analytical concerns with respect to activation, and 3) the impact of anemia on the TEG tracing.

Prediction of Thrombosis or Hemorrhage

With the advent of the cell-based model of coagulation, it has become increasingly apparent that the hemostatic system is reliant on both plasmatic and cell based factors, as well as the endothelium and the microenvironment in which it lives. Any test interrogating just one portion of this system at a single point in time is likely to fail in its ability to predict how the system will behave as a whole in all disease states. That being said, the recent exploitation of viscoelastic monitors as a means of global hemostatic evaluation has enabled veterinarians to document hypercoagulable states in patients with variable diseases, including: protein losing nephropathy, protein losing enteropathy, neoplasia, disseminated intravascular coagulation (DIC), parvovirus, hepatic disease, immune mediated hemolytic anemia (IMHA), thrombocytopenia, etc. This signifies a risk for thrombosis, but still represents only one arm of the so-called Virchow's triad, which includes venous stasis, vascular endothelial injury, and a hypercoagulable state. However, the fact that diseases known to frequently result in thromboembolic complications, such as canine IMHA,[1,2] have significant correlations with hypercoagulable tracings by thromboelastography does lend some potential to the assay.

Unfortunately, the veterinary literature is scant with regard to clinical outcome assessment for hemostatic tests, and most manuscripts reflect assays performed at the time of the event. In a report of portal vein thrombosis in 33 dogs, D-dimers were elevated in two out of three dogs tested.[3] In the seven dogs in which TEG was performed, three were hypercoagulable and four were hypocoagulable. A retrospective case series of aortic thrombosis in 31 dogs reported increased D-dimer concentrations in nine out of eleven dogs tested.[4] In the five dogs in which TEG was performed, four were hypercoagulable and one was normal based on the G value. In a prospective evaluation of hemostatic changes in seven dogs with documented thrombosis, both normo- and hypercoagulable states were documented.[5] Four out of seven dogs had elevated D-dimers, and three out of six dogs were hypercoagulable based on TEG results, with the remaining three being normocoagulable. Thus, on face value, some may conclude that TEG is not a sensitive screening tool. However, these documented changes are more likely attributed to multitude of ongoing factors including chronicity of the thrombus, consumption of coagulation factors, localization of the thrombus due to endothelial factors, or venous stasis without systemic effects, and not truly representative of predictive performance.

With respect to hemorrhage, there is also no perfect test. In people, it is well documented that platelet count, bleeding time (BT), and global assays such as prothrombin time (PT) fail to predict clinical bleeding.[6-11] In veterinary medicine, there are two studies of low grade evidence that attempt to address this question. In a retrospective study correlating coagulation profile findings and bleeding complications after ultrasound-guided biopsies in 434 dogs and cats, patients were found to be more likely to bleed if they were thrombocytopenic (<80,000/uL), had a prolonged PT if canine, a prolonged PTT if feline, and if the biopsy was of the kidney.[12] In another retrospective study by Vaden *et al.* documenting the complications of renal biopsies, findings included an increased number of patients that bled with pre-procedural abnormal coagulation profiles.[13] Albeit the number was reportedly too small for statistical analysis. In a prospective study by the author involving canine bone marrow transplant patients, predictable thrombocytopenia occurred post radiation and hemostatic indices as well as clinical bleeding events were documented.[14] Thromboelastography, BT, PT, aPTT, fibrinogen, D-dimer, and platelet counts were collected at predetermined time points and analyzed for correlation with mucosal or clinically significant bleeding events. The cohort was made up of 10 dogs with 43 time points. Only the TEG derived maximum amplitude was found to have correlation with mucosal or clinical bleeding. However, its predictive performance lied in its ability to discern who would not bleed versus identifying patients with an increased risk of bleeding. A second prospective study involving 27 dogs, which were retrospectively matched with controls, recorded and correlated clinical signs of bleeding with coagulation profiles versus TEG.[15] Thromboelastography correctly identified dogs with clinical signs of bleeding with a positive predictive value (PPV) of 89% and a negative predictive value (NPV) of 98% based on G alone. In comparison, the coagulation profile had a PPV between 50–81% and a NPV between 92–93% for detection of bleeding, depending on the observer.

Quandaries of Viscoelastic Testing

Analytical Variability

The question of variability and which activator to choose to obtain the most consistent and reliable result is always a concern with thromboelastography. In evaluating the available literature and guidelines supplied through the Partnership on Rotational ViscoElastic Test Standardization (PROVETS), the use of an activator is recommended with citrated whole blood samples and stronger activators lend to less variation and a more robust assay for use in a cage-side scenario.[16,17] There are significant differences in results between activators and, as such, results are not comparable.[18-20] It is recommended that reference intervals be established at each institution or hospital, and human and veterinary guidelines suggest that 40 samples (animals) are needed for accurate reference intervals, albeit the use of robust methodology may enable the usage of 20 as the minimum.[21,22] Although multiple combinations of activators have been reported in aforementioned studies involving dogs and cats of variable numbers, no one study has performed a head to head evaluation of all commonly used activators in 20 or more animals. Data from 48 healthy dogs will be presented by the author evaluating inter- and intraassay variability using proprietary kaolin, standard tissue factor dilutions, and the RapidTEG® assay, a proprietary mixture of both kaolin and tissue factor.

Impact of Anemia

One question that remains is the effect of anemia on TEG. Publications have shown that anemia is associated with TEG changes consistent with hypercoagulability.[23,24] This led many individuals to question if anemia is in fact creating an artefactual result, as patients with anemia often have increased bleeding times more typical of hypocoagulable states.[25] This author urges caution in this interpretation as these hypocoagulable effects may be predicated on interactions between the blood and the endothelium. Thromboelastography is not meant to evaluate this interaction, which is evidenced by the fact that dogs with Von Willebrand disease have normal TEG tracings. It is unknown what is actually happening *in vivo* in this scenario, and it is possible that the microenvironment created in the anemic state is a hypercoagulable one. Another recent publication on this topic involving an *ex vivo* model showed that a decrease in viscosity independent of red cells will cause hypercoagulability, and hypocoagulability is present with anemia once viscosity is normalized.[26] Regardless, it is a fact that thromboembolic disease, consistent with a hypercoagulable state, has been repeatedly documented in certain anemic conditions, such as dogs with IMHA, and recent studies have documented the presence of a hypercoagulable TEG tracing in these same cohorts.[1,2] In this session, the author will present data from recently completed research refuting red cell mass as the main contributor to hypercoagulable TEG tracings in the face of anemia.

Summary

Whole blood thromboelastography is likely the best available tool to evaluate the global hemostatic phenotype in a critically ill patient at a single point in time. As such, its routine use in the intensive care unit would be of benefit. Based on available evidence, TEG appears to be of benefit in the determination of hyper- and hypocoagulability in dogs. From a methodological standpoint, proprietary kaolin is the recommended activator to be used in dogs in a clinical setting, and the presence of anemia does not create an artefact in the tracing.

References

1. Goggs R, *et al. Vet J*. 2012;191:347.
2. Sinnott VB, *et al. J Vet Emerg Crit Care*. 2009;19:484.
3. Respess M, *et al. J Vet Intern Med*. 2012;26:230.
4. Lake-Bakaar G, *et al. J Am Vet Med Assoc*. 2012;241:910.
5. Bauer N, *et al. J Small Anim Pract*. 2013;54:129.
6. vanVeen JJ, *et al. Br J Anaesth*. 2011;106:1.
7. Manno C, *et al. Kidney Int*. 2004;66:1570.
8. Ewe K. *Dig Dis Sci*. 1981;26:388.
9. Segal JB, *et al. Transfusion*. 2005;45:1413.
10. Chee YL, *et al. Br J Haematol*. 2008;140:496.
11. Friedmann AM, *et al. Transfus Med Rev*. 2002;16:34.
12. Bigge LA, *et al. J Am Anim Hosp Assc*. 2001;37:228.
13. Vaden SL, *et al. J Vet Intern Med*. 2205;19:794.
14. Bucknoff MC, *et al. Am J Vet Res*. 2014;75:425.
15. Wiinberg B, *et al. Vet J*. 2009;179:121.
16. Goggs R, *et al. J Vet Emerg Crit Care*. 2014;0:1.
17. Smith SA, *et al. Blood Coagul Fibrinolysis*. 2010;21:692.
18. Banerjee A, *et al. J Vet Diagn Invest*. 2011;23:1109.
19. Flint SK, *et al. Vet Clin Pathol*. 2012;41:249.
20. Marschner CB, *et al. Acta Vet Scand*. 2010;52:38.
21. Solberg HE, *et al. Clin Chem Lab Med*. 2004;42:710.
22. American Society for Veterinary Clinical Pathology(ASVCP). Guidelines. asvcp.org. 2011.
23. Smith SA, *et al. Am J Vet Res*. 2012;73:789.
24. McMichael M, *et al. Blood Coagul Fibrinolysis*. 2011;22:424.
25. Valeri CR, *et al. Transfusion*. 2001;41:977.
26. Brooks AC, *et al. Transfusion*. 2014;54:727.

acvim375

NETs, Cell-Free DNA and Biomarker-Guided Therapy in Sepsis

Robert Goggs, BVSc, DACVECC, DECVECC, PhD, MRCVS
Ithaca, NY, USA

Introduction

Sepsis is common in dogs and is associated with substantial morbidity and a high risk of death, with reported mortality rates of 20% to 68%.[1-3] Sepsis is a clinical syndrome defined as the systemic inflammatory response to infection that is diagnosed by documenting the hallmarks of systemic inflammatory response syndrome (SIRS) (fever, tachycardia, tachypnea, and leukocytosis) and identifying an infective agent. Commonly used SIRS criteria are sensitive for sepsis diagnosis,[2] but lack specificity and have limited prognostic ability[4]. Illness severity in septic dogs can be assessed using scoring systems such as SPI2,[5] APPLE,[6] and the SOFA score,[7] but they are insensitive to the underlying immune dysfunction that characterizes sepsis. In humans, biomarkers have been widely evaluated in sepsis in an attempt to provide accurate diagnosis and prognostication.[8] The ideal biomarker provides biologic insight into the disease process and improves our understanding of pathogenesis. Recently, there has been a growing interest in identifying diagnostic and prognostic biomarkers in canine sepsis. Acute phase reactants such as serum amyloid A and C-reactive protein (CRP) are increased in canine sepsis but are not prognostic.[9,10] Increased procalcitonin (PCT) concentrations are prognostic in humans with septic shock,[11,12] and PCT may be superior to CRP in this regard, but efforts to measure it in dogs have been unsuccessful[13]. Identification of a biomarker that can be rapidly measured patient-side and that correlates with illness severity and prognosis in canine sepsis would advance the pursuit of individualized medicine. To this end, several groups have begun investigating the potential of novel biomarkers that relate to a recently identified neutrophil function - the formation of neutrophil extracellular traps (NETs).[14,15] In addition, related biomarkers that act as endogenous damage-associated molecular patterns (DAMPs) are also the focus of attention.[16] Here we review some of the recent data in this field.

NETs and NETosis

Neutrophils are the principal effector cells of the innate immune response to bacterial infection. Neutrophils have two well-recognized roles in combating bacterial pathogens, namely phagocytosis and degranulation, but it was recently recognized that neutrophils can also release extracellular traps (NETs).[17] This process has been termed NETosis,[18] and is an orchestrated process involving intracellular signaling, de-condensation of chromatin, disintegration of the nuclear envelope, and culminating in controlled release of DNA and granular contents into the extracellular space[19]. NETosis requires recognition of specific pathogens via pathogen-associated molecular patterns (PAMPs).[20] A recent study in dogs employed a PKC activator (phorbol myristate acetate) to activate isolated neutrophils and demonstrated that canine neutrophils can produce NETs.[15]

NETs are extracellular DNA decorated with citrullinated histones and antimicrobial proteins including myeloperoxidase (MPO) and neutrophil elastase (NE).[17] *In vivo* microscopy experiments in septic mice have confirmed that these structures can trap *E. coli* within liver sinusoids and that these NETs facilitate bacterial clearance by the resident Kupffer cells.[21] In addition to trapping and prevention of bacterial dissemination, NETs may also engage in direct microbial killing, although the antimicrobial properties of NETs are still not completely understood. Neutrophil granules contain MPO, NE, and cathepsin G, which are released into the extracellular space via NETosis,[22] and NE has been shown to neutralize certain virulence factors in bacteria[23]. It is theorized that the trapped bacteria can be killed by high local concentration of NE, MPO and histones.[24] NETs clearly play a role in host defense, but their role is complex and may be temporally regulated. For instance treatment of *Staphylococcus aureus* skin infections with DNase, an enzyme that digests DNA, led to systemic bacterial dissemination in mice.[25] Similarly early treatment of septic mice with DNase led to a deterioration in the condition, but late treatment reduced mortality.[26] A similar study also demonstrated that systemic treatment with recombinant DNase attenuated sepsis-induced organ dysfunction and improved survival rate in septic mice.[27]

Current evidence suggests that NETs are an essential component of innate immunity, but excessive NETosis, especially during sepsis, can be detrimental to the host. In an LPS-induced endotoxemia model in mice, presence of NETs was associated with multiple organ dysfunction,[27] although the cause of this relationship is unclear. Histones within NETs may act as DAMPs to initiate apoptosis, cytotoxicity, endothelial cell death, and systemic inflammation via toll-like receptor-mediated signaling pathways.[28] Certainly, the histones appear to play a key role in the host response to NET formation. Citrullination of histone H3 by peptidylarginine deminase-4 (PAD-4) leads to chromatin decondensation, a process that precedes NETosis.[29] Systemic treatment of mice with inhibitors of PAD improve survival in sepsis models.[29] Extracellular histones released during NETosis can also be scavenged by heparin - noteworthy, because it has also been demonstrated in mouse sepsis models that treatment with heparin improves survival.[30] This has potential clinical relevance in both humans and dogs.[31,32]

Dysregulation and derangements of the coagulation system are well-recognized abnormalities in sepsis.[3,33-36] There is now a body of evidence that suggests NETs play a key role in mediating thrombosis during sepsis, a process known as immunothrombosis.[37] These derangements can lead to thrombosis, microvascular dysfunction, tissue necrosis, multiple organ failure, disseminated intravascular coagulation, and death.[38] NETs are able to activate the contact pathway, potentially through intravascular expression of a polyanionic surface.[39,40] Histones also facilitate thrombus formation in platelet-dependent and independent manners.[41] Nucleosomes (DNA-histone complexes) may also play physical roles in promoting and propagating clot formation, since they contribute to the physical clot structure and impair clot dissolution by inhibiting fibrinolysis.[42] Cell-free DNA (cfDNA) competes with plasmin on fibrin, which results in decreased clot lysis.[43] Cell-free DNA can also suppress fibrinolysis by accelerating the inactivation of tissue plasminogen activator (tPA) by plasminogen-activator inhibitor-1 (PAI-1).[40] Platelets are also involved in immunothrombosis. Platelets facilitate neutrophil activation and the release of NETs,[41,44] further emphasizing the links between coagulation and inflammation[45]. It was recently demonstrated that binding of platelet P-selectin to neutrophil P-selectin glycoprotein ligand-1 is required for NETosis in mice.[46]

Cell-Free DNA and Nucleosomes

Circulating cfDNA, histones and nucleosomes have been studied as surrogate markers for NETs. In people, increased plasma concentrations of cfDNA have been demonstrated in burns, cancer, trauma, and myocardial infarction. Elevated levels of cfDNA have also been detected in the blood of critically ill people and in those with sepsis. In febrile human ICU patients, cfDNA measurement had high negative and positive predictive values, similar to PCT and superior to CRP, whereas normal cfDNA concentrations excluded infection, while high concentrations predicted mortality. Due to their cytotoxic and proinflammatory effects, extracellular histones have been shown to lead to excessive cell damage and death, thus contributing to the pathogenesis of sepsis and multiple organ failure. In septic human patients, high histone levels were significantly associated with new onset of left ventricular dysfunction. In people with non-thoracic blunt trauma, circulating histones were shown to rise to toxic levels quickly after the trauma and to contribute to distant organ damage, with the lungs being the most susceptible. In a study of critically ill people, plasma levels of circulating nucleosomes on admission correctly distinguish septic patients from non-septic patients. Furthermore, concentrations of nucleosomes were correlated with the severity of the immunosuppressive response and organ dysfunction in septic patients. Components of NET (e.g., cfDNA, histones, and nucleosomes) appear to be potential new biomarkers of disease and possible therapeutic targets in sepsis and other inflammatory diseases.

NETosis has now been described in a number of diseases and syndromes in people including sepsis, and acute respiratory distress syndrome.[41,47] Elevated cfDNA has been shown to have prognostic significance in human patients with severe sepsis.[48] Nucleosomes, complex of DNA and histones, are also markedly elevated in human patients with severe sepsis, suggesting that excessive NET release may be detrimental to the host. In humans, increased plasma concentrations of cfDNA have been demonstrated in burns, cancer and trauma.[49-51] This cfDNA may originate from apoptotic cells, necrotic tissue or from

NETs.[52-54] Where the rate of NETosis, apoptosis or necrosis is high, and the mechanisms responsible for removal become saturated, the plasma concentrations increase.[55] As such, cfDNA has the potential to provide biologic insight into the extent and nature of the innate immune response to infection and inflammation. In addition to being an indicator of nature of the disease process and its severity, it has been demonstrated that NET formation may directly contribute to pathogenesis in sepsis through the phenomenon of immunothrombosis,[37] while cfDNA can inhibit plasmin-mediated fibrinolysis[40]. Increased cfDNA concentrations have also been detected in the blood of critically ill humans and in those with sepsis,[56,57] and increased cfDNA concentrations are prognostic in humans with bacteremia[58,59]. In febrile human ICU patients, cfDNA measurement has high negative and positive predictive values, wherein normal cfDNA concentrations exclude infection, while high concentrations predict mortality.[60]

Nucleosomes are complexes formed by DNA and histone proteins that are released into circulation during cell death and cellular damage such as apoptosis and necrosis.[61] They can also be due to the process of NET formation. As such, cfDNA and nucleosomes share potential origins, but are distinct entities,[62] with differential potential for immune cell activation through pattern recognition receptors[61]. Measuring both biomarkers may provide better insights into the disease process than either alone.[62] Humans with septic shock have significantly higher nucleosome concentrations than patients with sepsis, non-infectious SIRS, or fever.[61] In humans, plasma nucleosome concentrations distinguish septic from non-septic critically ill patients and correlate with the severity of immunosuppression and with organ dysfunction.[55] Nucleosome concentrations are increased in horses with colic,[63] and have recently been reported in dogs with sepsis[64].

Biomarker-Guided Therapy in Sepsis

Many clinical trials in sepsis have tested various interventions to modulate the immune system, but have not resulted in the integration of new drugs into clinical practice. This suggests novel strategies to treat sepsis should be considered. Our current understanding of the pathophysiology of sepsis may be inadequate to design such interventions, particularly when faced with changing patient populations, the effects of therapeutic interventions and the ever-changing dynamic interaction between host and pathogen. The heterogeneity of sepsis and of the patients we manage is also part of the hurdle any novel therapy must overcome. Given the complexity of the host response to sepsis, it may be impossible for a single drug to benefit all patients with sepsis, but biomarkers may help us to overcome some of these issues. Indeed, a recent reanalysis of a trial evaluating an IL-1R antagonist identified benefit in a subgroup of patients with signs of a macrophage activation syndrome,[65] despite the overall trial being negative. A good biomarker provides biologic insight as well as sensitivity, specificity, diagnosis and prognostication. Using sets of biomarkers that provide insight into biologic activity and pathophysiology can be used to target specific patients for intervention, to stratify risk and to monitor response to therapy. Veterinary medicine is not that far behind and personalized medicine for our patients may yet be within reach.[66]

Despite therapeutic advances over the last decade, published survival rates for sepsis have not significantly improved, and range from 30–64%.[67,68] This high mortality rate is attributed to the dysregulated host response to the infection that manifests as cardiovascular instability, hypotension, and organ dysfunction.[35,69] Early recognition and aggressive management comprising cardiovascular stabilization, administration of broad-spectrum antimicrobials and surgical source control is essential to maximizing survival.[70,71] Dogs with bacterial pneumonia or intra-abdominal infections are typically prescribed antimicrobials for 3 to 6 weeks, to limit potential complications such as recurrence, novel infection development and death.[72] Studies in people with intra-abdominal infections show similar outcomes when treated for four days compared to eight days.[73] Similar findings are described in patients with ventilator-associated pneumonia with no change in outcome when treated for 8 or 15 days.[74] Thus, human guidelines for treatment of intra-abdominal infection following surgery and for ventilator-associated pneumonia now recommend shorter courses of antibiotics.[75] Despite this, the average duration of antimicrobials in people remains longer than recommended,[76] likely due to concerns of increased complications for individual patients with premature antimicrobial discontinuation. A recent study in dogs suggested that for urinary tract infections, a 3-day, high-dose course of antimicrobials was non-inferior to a standard 14-day course.[77] This suggests that, as in people, dogs may not require extended antimicrobial courses for some disease processes.

Inflammatory biomarkers are now widely used in people to guide antimicrobial administration in sepsis,[78] and to support decisions to discontinue antimicrobial therapy while minimizing risk[79-82]. Various inflammatory biomarkers have been evaluated to aid identification of the appropriate time to safely discontinue antimicrobials. Procalcitonin is a hormokine associated with the calcitonin gene family whose expression is upregulated in response to endotoxins or cytokines. Retrospective and prospective studies using absolute or relative changes in PCT concentrations have shown that patients in whom antimicrobials were discontinued based on PCT values received shorter courses of treatment with no difference in outcome.[81] The acute phase reactant C-reactive protein has also been evaluated as part of a decision-making algorithm for antimicrobial therapy in patients suffering from sepsis.[83] CRP-guided antimicrobial therapy in people is associated with shorter antimicrobial administration without changes in patient outcome,[84] while a pilot study in canine pyometra suggests that novel post-operative increases in CRP may help identify post-operative complications such as surgical wound infections[85]. The optimal duration of antimicrobial in dogs with sepsis is unknown, but data from people suggest we may be able to safely reduce the duration of antimicrobials for those diseases. Veterinarians typically rely on clinical, clinicopathologic and radiographic criteria to discontinue therapy. For instance, current recommendations for canine pneumonia are to continue antimicrobials for a least 1-week post-radiographic resolution. Biomarkers of the inflammatory response might provide superior guidance.[86]

REFERENCES

1. King LG. *J Am Vet Med Assoc*. 1994;204(3):407–414.
2. Hauptman JG, *et al. Vet Surg*. 1997;26(5):393–397.
3. de Laforcade AM, *et al. J Vet Intern Med*. 2003;17(5):674–679.
4. Okano S, *et al. Vet Rec*. 2002;150(8):245–246.
5. King LG, *et al. Am J Vet Res*. 2001;62(6):948–954.
6. Hayes G, *et al. J Vet Intern Med*. 2010;24(5):1034–1047.
7. Ripanti D, *et al. Schweiz Arch Tierheilkd*. 2012;154(8):325–330.
8. Pierrakos C, Vincent JL. *Crit Care*. 2010;14(1):R15.
9. Jitpean S, *et al. BMC Vet Res*. 2014;10:273.
10. Gebhardt C, *et al. J Vet Emerg Crit Care*. 2009;19(5):450–458.
11. Jekarl DW, *et al. Diagn Microbiol Infect Dis*. 2013;75(4):342–347.
12. Georgopoulou AP, *et al. J Crit Care*. 2011;26(3):331 e331–337.
13. Floras AN, *et al. J Vet Intern Med*. 2014;28(2):599–602.
14. Letendre J-A, Goggs R. *J Vet Emerg Crit Care*. 2017; In Press.
15. Jeffery U, *et al. Vet Immunol Immunopathol*. 2015;168(3–4):262–268.
16. Hatada T, *et al. Thromb Haemost*. 2005;94(5):975–979.
17. Brinkmann V, *et al. Science*. 2004;303(5663):1532–1535.
18. Steinberg BE, Grinstein S. *Sci STKE*. 2007;2007(379):pe11.
19. Fuchs TA, *et al. J Cell Biol*. 2007;176(2):231–241.
20. Pisetsky DS. *Clin Immunol*. 2012;144(1):32–40.
21. McDonald B, *et al. Cell Host Microbe*. 2012;12(3):324–333.
22. Egesten A, *et al. Blood*. 1994;83(10):2985–2994.
23. Weinrauch Y, *et al. Nature*. 2002;417(6884):91–94.
24. Medina E. *Methods Mol Biol*. 2010;602:411–415.
25. Yipp BG, Kubes P. *Blood*. 2013;122(16):2784–2794.
26. Mai SH, *et al. Shock*. 2015;44(2):166–172.
27. Czaikoski PG, *et al. PLoS One*. 2016;11(2):e0148142.
28. Chen R, *et al. Cell Death Dis*. 2014;5:e1370.
29. Li Y, *et al. Surgery*. 2014;156(2):229–234.
30. Wildhagen KC, *et al. Blood*. 2014;123(7):1098–1101.
31. Wada H, *et al. Thromb Res*. 2010;125(1):6–11.
32. Bentley AM, *et al. J Vet Emerg Crit Care (San Antonio)*. 2013;23(1):14–22.
33. de Laforcade A. *Top Companion Anim Med*. 2012;27(2):59–64.
34. de Laforcade AM, *et al. J Vet Intern Med*. 2008;22(1):26–30.

35. Kenney EM, *et al. J Am Vet Med Assoc*. 2010;236(1):83–87.
36. Hoppensteadt D, *et al. Clin Appl Thromb Hemost*. 2015;21(2):120–127.
37. Gould TJ, *et al. J Thromb Haemost*. 2015;13(Suppl 1):S82–91.
38. Fourrier F, *et al. Chest*. 1992;101(3):816–823.
39. Martinod K, Wagner DD. *Blood*. 2014;123(18):2768–2776.
40. Liaw PC, *et al. Blood Rev*. 2015.
41. Gould TJ, *et al. Arterioscler Thromb Vasc Biol*. 2014;34(9):1977–1984.
42. Schulz C, *et al. J Thromb Haemost*. 2013;11(Suppl 1):233–241.
43. Esmon CT. *Thromb Haemost*. 2013;109(3):416–420.
44. Clark SR, *et al. Nat Med*. 2007;13(4):463–469.
45. Sreeramkumar V, *et al. Science*. 2014;346(6214):1234–1238.
46. Etulain J, *et al. Blood*. 2015;126(2):242–246.
47. Pigozzi L, *et al. Intensive Care Med*. 2016;42(4):583–586.
48. Dwivedi DJ, *et al. Crit Care*. 2012;16(4):R151.
49. Chiu TW, *et al. Clin Chem Lab Med*. 2006;44(1):13–17.
50. Johnson PJ, Lo YMD. *Clin Chem*. 2002;48(8):1186–1193.
51. Lo YMD, *et al. Clin Chem*. 2000;46(3):319–323.
52. Lichtenstein AV, *et al. Ann N Y Acad Sci*. 2001;945(1):239–249.
53. Hamaguchi S, *et al. Mediators Inflamm*. 2015;2015:614518.
54. Margraf S, *et al. Shock*. 2008;30(4):352–358.
55. Chen Q, *et al. Int J Infect Dis*. 2012;16(7):e558–564.
56. Saukkonen K, *et al. Intensive Care Med*. 2007;33(9):1624–1627.
57. Saukkonen K, *et al. Clin Chem*. 2008;54(6):1000–1007.
58. Forsblom E, *et al. PLoS One*. 2014;9(2):e87741.
59. Huttunen R, *et al. PLoS One*. 2011;6(7):e21700.
60. Moreira VG, *et al. Ann Clin Biochem*. 2010;47(Pt 3):253–258.
61. Zeerleder S, *et al. Crit Care Med*. 2003;31(7):1947–1951.
62. Marsman G, *et al. Cell Death Dis*. 2016;7(12):e2518.
63. Bauquier JR, *et al. J Vet Intern Med*. 2016;30(1):260–268.
64. Letendre J-A, *et al. J Vet Emerg Crit Care*. 2016;26(S1):S7.
65. Shakoory B, *et al. Crit Care Med*. 2016;44(2):275–281.
66. Davenport EE, *et al. Lancet Respir Med*. 2016;4(4):259–271.
67. Bentley AM, *et al. J Vet Emerg Crit Care*. 2007;17(4):391–398.
68. Dayer T, *et al. J Small Anim Pract*. 2013;54(12):625–629.
69. Singer M, *et al. JAMA*. 2016;315(8):801–810.
70. Abelson AL, *et al. J Vet Emerg Crit Care*. 2013;23(5):551–556.
71. Dellinger RP, *et al. Intensive Care Med*. 2008;34(1):17–60.
72. Dear JD. *Vet Clin North Am Small Anim Pract*. 2014;44(1):143–159.
73. Sawyer RG, *et al. N Engl J Med*. 2015;372(21):1996–2005.
74. Chastre J, *et al. JAMA*. 2003;290(19):2588–2598.
75. Sartelli M, *et al. World J Emerg Surg*. 2013;8(1):3.
76. Sartelli M, *et al. World J Emerg Surg*. 2013;8(1):1.
77. Westropp JL, *et al. J Vet Intern Med*. 2012;26(3):506–512.
78. Quenot JP, *et al. Ann Intensive Care*. 2013;3(1):21.
79. Bouadma L, *et al. Lancet*. 2010;375(9713):463–474.
80. Albrich WC, *et al. Arch Intern Med*. 2012;172(9):715–722.
81. Maseda E, *et al. J Crit Care*. 2015;30(3):537–542.
82. Hochreiter M, *et al. Crit Care*. 2009;13(3):R83.
83. Han JH, *et al. Antimicrob Agents Chemother*. 2015;59(10):6494–6500.
84. Oliveira CF, *et al.* Crit Care Med. 2013; 41(10): 2336–2343.
85. Dabrowski R, *et al. Theriogenology*. 2009;72(4):471–476.
86. Viitanen SJ, *et al. J Vet Intern Med*. 2017;31(1):124–133.

acvim376

SIRS, MODS and Sepsis 3.0: Where Are We Now?

Deborah C. Silverstein, DVM, DACVECC
Philadelphia, PA, USA

SIRS and MODS

The systemic inflammatory response syndrome (SIRS) describes the clinical condition seen with widespread activation of an inflammatory system secondary to an infectious or non-infectious insult. It is well known that a localized inflammatory response normally occurs at a site of tissue damage or injury in order to protect the host. In patients with SIRS, however, severe inflammation leads to an "overflow" of mediators into the systemic circulation and results in numerous global abnormalities that may have life-threatening consequences. If not treated appropriately, SIRS often results in multiple organ dysfunction syndrome (MODS) and death. The distinction between SIRS and sepsis is based upon the presence or absence of infection as described below.

Definitions

- **Systemic inflammatory response (SIRS)**: the clinical syndrome of systemic inflammation that results from either an infectious or non-infectious insult (i.e., septic or nonseptic SIRS). Dogs must have 2/4 and cats 3 of the 4 criteria listed below to have SIRS:

	Dogs	Cats
Temperature (°F)	<100.6 or >102.6	<100 or >104
Heart rate (beats/minute)	>120	<140 or >225
Respiratory rate (breaths/minute)	>20	>40
WBC ($x10^3$); % bands	<6 or >16; >3%	>19 or <5

- **Sepsis**: the clinical syndrome of systemic inflammation that results from infection (bacterial, viral, protozoal or fungal).
- **Severe sepsis**: sepsis complicated by dysfunction in one or more organs.
- **Septic shock**: acute circulatory failure and persistent arterial hypotension despite adequate volume resuscitation associated with sepsis.
- **Multiple organ dysfunction syndrome (MODS)**: physiologic derangements in endothelial, coagulation, cardiopulmonary, renal, nervous, endocrine, and gastrointestinal systems associated with the systemic inflammatory response syndrome.

Pathogenesis of SIRS & MODS

The pathophysiologic mechanisms responsible for the generation of the SIRS are complex and incompletely understood. The initial insult that stimulates SIRS can come from a variety of sources such as trauma, burns, aspiration, pancreatitis, or infectious agents (e.g., products of gram-positive or gram-negative bacteria). The induction of systemic inflammation may initially start locally (i.e., an abscess localized to a limb) but can lead to systemic signs when "mediators" enter the circulation causing global activation of inflammation.[1] Although mediators such as platelets, polymorphonuclear leukocytes, and the endothelium also play a role, it appears that the stimulation of macrophages and the release of inflammatory cytokines are pivotal in the generation of SIRS. While cytokines trigger a beneficial inflammatory response that promotes local coagulation to confine tissue damage, the excessive production of these proinflammatory cytokines can be even more dangerous than the original stimulus, overcoming the normal regulation of the immune response and producing the clinical signs notably seen in patients with SIRS. When the immune system is fighting pathogens, these cytokines signal immune cells, such as T-cells and macrophages, to travel to the site of infection.

In addition, cytokines activate these recruited immune cells, stimulating them to produce even more cytokines. This production of a "cytokine storm" and global activation of white blood cells ultimately overwhelms the compensatory anti-inflammatory response and is a key component in the pathogenesis of SIRS.

In addition to systemic activation of white blood cells, other pathologic effects of inflammatory mediators include increased capillary permeability, vasodilation, activation of coagulation, myocardial dysfunction, and mitochondrial dysfunction. These pathologic effects of systemic inflammation can lead to the syndrome of multiple organ dysfunction or MODS. MODS is characterized by abnormalities in organs that were not affected by the original insult and is associated with a high morbidity and mortality rate. Organs at risk for dysfunction include lungs, cardiovascular (heart and vasculature), kidneys, gastrointestinal tract, endocrine, nervous, and coagulation systems.

Implications

Sepsis is one of the leading cause of human deaths worldwide; it is also commonly fatal in dogs and cats. Our understanding of the concept of sepsis has come a long way since the time of the Ancient Greeks, when sepsis was referred to as rot, decay or putrefaction. More recently, much effort has been invested into studying prevention, early recognition and the best treatment approach to maximize survival. The initial sepsis consensus definitions for humans were published in 1991, and many of the current definitions used in veterinary medicine are still based on this consensus. However, updated definitions were published in 2001, and more recently in 2016. These three generations of definitions are now referred to as Sepsis-1, Sepsis-2, and Sepsis-3, in chronological order.

The 1991 consensus for the definition of sepsis revolved around the notion that sepsis was the host's systemic inflammatory response syndrome (SIRS) to invading organisms. SIRS was diagnosed if the patient had 2 or more abnormalities with regards to temperature, heart rate, respiratory rate/$PaCO_2$ or white blood cell count. These rather simple criteria enabled researchers to identify clinical patients for enrollment in sepsis trials. Subsequently, the veterinary SIRS criteria were developed and have been utilized in countless clinical animal studies. Although the specific values for the various criteria differed between the humans and small animals, the physiology behind them is nearly identical.

The main shortcomings of the SIRS criteria are threefold: 1) SIRS criteria are so sensitive that up to 90% of human patients admitted to an intensive care unit (ICU) meet the criteria. Similar problems likely exist in veterinary ICUs, although data are not available to confirm this. Since sepsis was defined as the presence of SIRS plus an infection and a large number of patients meet the SIRS criteria, sepsis effectively equals infection. Yet, it is important to recognize that not all patients with an infection have sepsis, but they are almost equivalent since so many critically ill patients meet the SIRS criteria and have an infection. 2) A variable degree of host response is inherent to any infection; this actually differentiates between infection and mere colonization. It is the body's response to even minor infections that keeps people and animals healthy. 3) Determining the role of infection in the pathogenesis of SIRS has been challenging because non-septic SIRS states, as occurs with pancreatitis, trauma, heatstroke, immune-mediated diseases, etc., can elicit similar clinical changes due to acute systemic inflammation. Actually, sepsis differs from sterile inflammation only due to the presence of an underlying infectious process, not the nature of the activated host response pathways or the type of organ dysfunction. It is therefore a diagnostic priority to identify any focus of invasive infection when managing a patient with sepsis, and therefore it is desirable to have screening parameters for sepsis that can be easily applied.

A second consensus conference aimed at revisiting the SIRS criteria convened in 2001, but did not come up with a user-friendly list of variables to define sepsis. In an attempt to expand the list of potential clinical criteria, the definition of sepsis actually became less specific. However, the concept of staging patients using a PIRO scheme, which examines the predisposition, infection (or insult), response and organ dysfunction, was created. This approach takes into account the fact that each individual patient behaves differently in response to the same illness/insult/injury, as well as the response to treatment. Of equal importance is the understanding that not all infections are similar and the prognosis differs according to the infection site, type of organism, extent of infection, and response of the patient to both the infection as well as the treatment. The PIRO score helped to predict response to treatment using characterization and classification of a patient's phenotype. A recent study by Kumar *et al.* examining the clinical usefulness of PIRO staging system found that it was an important tool for assessing risk stratification and prognosis of hospitalized patients with sepsis. However, it did not advance the definition of sepsis per se, therefore the previous criteria from 1991 have remained largely unchanged for more than 2 decades.

The impetus for a new definition therefore lies in the desire to differentiate sepsis from an uncomplicated infection and to find new definitions for sepsis and septic shock that reflect the improved understanding of their pathophysiology. Clinicians want to be able to identify patients with sepsis, a condition whereby the host's deleterious, non-resolving inflammatory response to infection often leads to organ dysfunction. At a Merinoff Symposium, the International Sepsis Forum wrote "sepsis is a life-threatening condition that arises when the body's response to an infection injures its own tissues and organs." It is well recognized that sepsis is a syndrome, but there is no validated criterion or standard diagnostic test. In addition, there are no clear criteria to identify the elements of sepsis (i.e., the infection, host response, and organ dysfunction). In addition, the definition should be fairly simple to obtain promptly and without a substantial cost. If the new criteria could be used in all veterinary hospitals and emergency rooms, earlier identification and therefore treatment might be possible.

CHALLENGES

Since sepsis is more of a syndrome than a specific disorder, it likely requires a constellation of clinical abnormalities in patients with suspected infection. Sepsis involves a multifaceted host response to the infecting agent and this response is dependent on endogenous factors. The original definition focused on the proinflammatory state associated with sepsis, but did not take into account the anti-inflammatory response that occurs concurrently, along with changes in non-immunologic pathways that include the cardiovascular, autonomic, hormonal, neuronal, bioenergetics, metabolic, and coagulation. All of these changes likely have prognostic significance, as does even mild organ dysfunction that could lead to the ultimate demise of the patient. SIRS is identified in many hospitalized patients, including those who never develop infection or suffer from adverse outcomes. The SIRS criteria, thus, do not indicate a dysregulated, life-threatening response. Interestingly, 1 in 8 patients admitted to critical care units in Australia and New Zealand with infection and new organ failure did not have 2/4 SIRS criteria necessary to fulfill the definition of sepsis, but did have protracted courses of illness with significant morbidity and mortality.

Since the host response to an infection is based on both biological and clinical differences in affected individuals, taking into account the age, co-morbidities, concurrent injury or surgery, medications and source of the infection are also important, although difficult to study. The new sepsis definitions of 2016, also referred to as "Sepsis-3" have attempted to help clinicians identify patients that may be septic in an attempt to treat them promptly and appropriately.

SEPSIS-3

Between January 2014 and January 2015, a task force comprising 19 critical care, infectious disease, surgical and pulmonary specialists who were selected by the European Society of Intensive Care Medicine and the Society of Critical Care Medicine, engaged in 4 face-to-face meetings. Their primary goal was to re-examine the current sepsis definitions and differentiate sepsis from uncomplicated infection. Unrestricted funding was provided by the societies and the task force retained complete autonomy. Drs. Deutschman and Singer were nominated as co-chairs and they selected members according to their scientific expertise in sepsis epidemiology, clinical trials, and basic or translational research. Sepsis is now defined as "a life-threatening organ dysfunction caused by a dysregulated host response to infection" and is essentially the same as what was previously referred to as severe sepsis. Organ dysfunction corresponds to an acute change in total SOFA score of ≥2 points as a result of infection.

Multiple large electronic health record databases were evaluated to create a simplified quick SOFA (sequential organ failure assessment) score, also known as qSOFA. A clinical model was developed with multivariable logistic regression and found that any 2 of 3 clinical variables - a Glasgow Coma Scale score of 13 or less, systolic blood pressure of 100 mm Hg or less, and a respiratory rate of 22 breaths/min or greater - offered predictive validity (AUROC=0.81; 95% CI, 0.80–0.82) similar to that of the full SOFA score outside of the ICU. This model was strong in the face of multiple sensitivity analyses including a more simple assessment of altered mentation (Glasgow Coma Score <15) and in the out-of-hospital, emergency department, and ward settings within the external US and non-US data sets.

The qSOFA allows for the rapid identification of patients with suspected infection by both physicians and nurses. In summary, the criteria are as follows:

- Respiratory rate >22 breaths/min
- Altered mentation measured as any GCS<15
- Systolic blood pressure <100 mm Hg (this was included because it was the most widely recorded criterion in the electronic health record data sets)

Bedside qSOFA allows for rapid identification of patients with suspected infection that are likely to have a prolonged ICU stay or to die while in the hospital. It can easily be performed by clinicians and nurses and does not require laboratory tests. It is quick and repeatable over time and should prompt clinicians to further investigate for organ dysfunction, to initiate or change therapy as appropriate, and to consider referral to a specialty critical care unit or increase the frequency of monitoring. It should also prompt a search for infection in patients not previously recognized as infected. The full SOFA score requires more values, such as platelet count, bilirubin, Glasgow Coma Scale, creatinine, PaO_2/FiO_2, and mechanical ventilation status. An increase in the SOFA score of more than two points also indicates that sepsis is likely and reflects an overall mortality risk of ~10%.

Rather than using the terms sepsis, severe sepsis, and septic shock, severe sepsis has been removed from the terminology of these patients and only sepsis and septic shock remain. Septic shock is a subset of sepsis in which underlying circulatory and cellular/metabolic abnormalities are profound enough to substantially increase mortality. In order to identify patients with septic shock, the following parameters should be present: hypotension requiring vasopressor therapy to maintain a MAP >65 mm Hg and a lactate >2 mmol/L despite adequate volume resuscitation. Humans that meet these criteria have a mortality greater than 40%. Although it is likely that veterinarians will continue to use the older, species-specific SIRS and sepsis definitions in the near future, sepsis-3 certainly gives much food for thought and the need for updated veterinary definitions.

Updated Surviving Sepsis Campaign Group Recommendations for Screening and Treatment Using Sepsis-3

Screening patients for infection is the first step towards identifying those at risk. Early identification is key. Patients that are identified as having an infection should have blood and other cultures performed, as indicated, and antimicrobial therapy initiated. A complete bloodwork evaluation will help identify patients with infection-related organ dysfunction. Patients with sepsis should still be identified using the same organ dysfunction criteria (including lactate >2 mmol/L), and the qSOFA should serve as a secondary screening tool for patients at risk for deteriorating in the hospital. Two out of three qSOFA criteria (altered mental status, RR ≥20 bpm, or systolic BP ≤100 mm Hg) is therefore a flag that these patients should be monitored more closely. For clarification, the qSOFA is a tool for identifying patients at risk of sepsis with a higher risk of hospital death or prolonged ICU stay, both inside and outside of the ICU. qSOFA does not define sepsis, but is only a predictor of increased mortality and ICU stay.

If organ dysfunction is present, a 3-hour bundle approach to treatment should be a priority. Patients with infection and hypotension and lactate level ≥4 mmol/L should be given a 30 mL/kg isotonic crystalloid bolus followed by reassessment of volume responsiveness or tissue perfusion. Within the six-hour bundle, a repeat lactate measurement should be obtained if the initial lactate was >2 mmol/L.

It is important for all clinicians to note that the new definitions do not change the primary focus of early sepsis identification and initiation of rapid treatment for the management of septic patients.

New Challenges

Despite the updated terminology, sepsis remains a broad term that is applied to an incompletely understood process, therefore pragmatic compromises were required with emphasis on generalizations. An initial retrospective evaluation of qSOFA revealed that it could be a useful tool for screening of sepsis in clinical patients. Future prospective validation studies will help to confirm its accuracy. It is important to note that neither qSOFA nor SOFA criteria is a stand-alone definition of sepsis and failure to meet the new recommended values for sepsis should not defer investigation or treatment of infection or delay any part of hospital care. There was disagreement among the task force regarding the use of lactate as an important biochemical identifier of sepsis in infected patients. Lactate was therefore not listed in the simple bedside criteria, but its use should not be limited as it may be a useful guide for therapeutic response or indicator of response to therapy in many patients.

The new criteria are focused on adult human patients, but further updates for pediatrics and veterinary populations are needed. There are also some societies that did not endorse Sepsis-3, including the Latin American Sepsis Institute, the American College of Chest Physicians, and the American College of Emergency Physicians. It is also interesting to note that there were no members from the low- and middle-income countries on the task force. Patients in these areas, similar to veterinary patients, often have more variable care available/affordable and therefore likely have important differences in assessment, diagnostics and treatment approaches.

Although the SIRS criteria are now recognized as non-specific and will no longer be used for sepsis recognition or screening in adult human patients, the presence of SIRS criteria, especially in small animals, indicates that the patient may be suffering from severe illness. Further updates to human and veterinary medicine are imperative as additional evidence-based medicine is available. One of the first studies to evaluate the prognostic accuracy of the SOFA score, SIRS criteria, and qSOFA score in adults with suspected infection that were admitted to the ICU found that only an increase in SOFA score of 2 or more had greater prognostic accuracy for in-hospital mortality than the SIRS criteria or the qSOFA score. However, when examining patients presenting to the emergency department with suspected infection, the qSOFA had better prognostic accuracy for in-hospital mortality than either SIRS or severe sepsis.

Overall Summary of New Definitions

Sepsis → life-threatening organ dysfunction caused by a dysregulated host response to infection
Septic shock → need for vasopressors AND lactate >2 mmol/L
Severe Sepsis → OUT
SIRS is OUT and qSOFA and SOFA are IN

References

1. DeClue AE, Sharp CR, Harmon M. Plasma inflammatory mediator concentrations at ICU admission in dogs with naturally developing sepsis. *J Vet Intern Med.* 2012;26(3):624–630.
2. Greiner M, Wolf G, Hartmann K. A retrospective study of the clinical presentation of 140 dogs and 39 cats with bacteraemia. *J Small Anim Pract.* 2008;49(8):378–383.
3. Brady CA, Otto CM, Van Winkle TJ, King LG. Severe sepsis in cats: 29 cases (1986–1998). *J Am Vet Med Assoc.* 2000;217(4):531–535.
4. Kenney EM, Rozanski EA, Rush JE, *et al.* Association between outcome and organ system dysfunction in dogs with sepsis: 114 cases (2003–2007). *J Am Vet Med Assoc.* 2010;236(1):83–87.
5. Singer M, Deutschman CS, Seymour CW, Shankar-Hari W, *et al.* The Third International Consensus Definitions for Sepsis and Septic Shock (Sepsis-3). *JAMA.* 2016;315(8):801–810.
6. Rathour S, Kumar S, Hadda V, Bhalla A, Sharma N, Varma S. PIRO concept: staging of sepsis. *J Postgrad Med.* 2015;61:235–42.
7. Copotoiu SM, Copotoiu R. Much ado about the new definitions of sepsis. *J Crit Care Med.* 2016;2(2) :67–72.
8. Vincent JL, Opal SM, Marshall JC, Tracey KJ. Sepsis definitions: time for a change. *Lancet.* 2013;381:774–775.
9. Dellinger RP, Levy MM, Rhodes A, *et al.* Surviving sepsis campaign: international guidelines for management of severe sepsis and septic shock: 2012. *Crit Care Med.* 2013;41(2):580–637.
10. Seymour CW, Liu VX, Iwashyna TJ, *et al.* Assessment of clinical criteria for sepsis: for the third international consensus definitions for sepsis and septic shock. *JAMA.* 2016;315(8):762–774.

11. Rhee C, Gohil S, Klompas M. Regulatory mandates for sepsis care - reasons for caution. *N Engl J Med*. 2014;370(18):1673–1676.
12. Singer W, De Santis V, Vitale D, Jeffcoate W. Multiorgan failure is an adaptive, endocrine-mediated, metabolic response to overwhelming systemic inflammation. *Lancet*. 2004;364(9433):545–548.
13. Bone RC, Balk RA, Cerra FB, *et al*. Definitions for sepsis and organ failure and guidelines for the use of innovative therapies in sepsis. The ACCP/SCCM Consensus Conference Committee. American College of Chest Physicians/Society of Critical Care Medicine. *Chest*. 1992;101(6):1644–1655.
14. Brady CA, Otto CM, Van Winkle TJ, *et al*. Severe sepsis in cats: 29 cases (1996–1998). *J Am Vet Med Assoc*. 2000;217(4):531–535.
15. Hauptman JG, Walshaw R, Olivier NB. Evaluation of the sensitivity and specificity of diagnostic criteria for sepsis in dogs. *Vet Surg*. 1997;26(5):393–397.
16. Raith EP, Udy AA, Bailey M, *et al*. Prognostic accuracy of the SOFA score, SIRS criteria, and qSOFA score for in-hospital mortality among adults with suspected infection admitted to the intensive care unit. *JAMA*. 2017;317(3):290–300.
17. Freund Y, Lemachatti N, Krastinova E, *et al*. Prognostic accuracy of sepsis-3 criteria for in-hospital mortality among patients with suspected infection presenting to the emergency department. *JAMA*. 2017;317(3):301–8.

acvim377

Antimicrobial Decision Making in Sepsis

Alex Lynch, BVSc (Hons), DACVECC, MRCVS
Gainesville, FL, USA

Antimicrobial prescription is undoubtedly a mainstay of our treatment for sepsis. While implicitly the decision to provide antimicrobials in the septic pet is not questioned, the actual practicalities of choosing an appropriate drug and dosing scheme, along with accurate identification of sepsis may prove more challenging. Up until recently the term sepsis was used to describe the systemic inflammatory response to infection.[1,2] The criteria for the systemic inflammatory response is not standardized in veterinary medicine and differs between dogs and cats.[3-5] Alterations in heart rate, respiratory rate, temperature, and the leukon characteristics are utilized to clue the clinician in to the likely presence of SIRS. Sepsis was used to describe the scenario where an infectious trigger initiated a cytokine driven syndrome of systemic inflammation. Potent non-infectious stimuli like acute necrotizing pancreatitis or severe polytrauma could also produce a similar, yet fundamentally non-infectious, clinical situation. When the patient with sepsis has evidence of organ dysfunction (e.g., azotemia, coagulopathy), their syndrome was classified as **severe sepsis**. Those septic patients with persistent hypotension refractory to fluid resuscitation were described as having **septic shock**. Clinically, animals with infection related illness therefore presented on a spectrum with respect to severity of illness and reversibility of their disease. Last year updated sepsis and septic shock definitions were introduced in people.[6] Sepsis was reclassified as a **life-threatening organ dysfunction caused by a dysregulated host response to infection**, similar to the previous definition of severe sepsis. Septic shock was further defined as **a subset of sepsis in which profound circulatory, cellular, and metabolic abnormalities are associated with a greater risk of mortality than with sepsis alone**. These definitions highlight a few points relevant to our treatment of animals with infection. Firstly, not all animals have the same disease course and consequently no one-size-fits-all approach to antimicrobials likely exists for them. Secondly, the dysregulated host response to infection will not be corrected or reversed through the use of antimicrobials alone.

The first challenge for the clinician is answering the question of how likely is infection in this patient. Several different pieces of data can prove helpful. A comprehensive history is important, including reproductive status, travel history, and relevant antecedent events (e.g., vomiting prior to onset of respiratory distress, coughing, and pyrexia in the dog with aspiration pneumonia). An appreciation of common pathologies in specific breeds and species is also helpful. For instance, cats with a radiographic pulmonary pattern are more likely to have a non-infectious etiology, yet pneumonia will be seen from time to time. The classical SIRS criteria used for dogs and cats are highly sensitive but not specific. Consequently, the clinician may be faced with a sick pet that has many compelling characteristics of sepsis that ultimately proves to have non-infectious disease. Additional pieces of clinical data may improve our confidence that we are on the right lines (e.g., hypoglycemia, mild to moderate hyperbilirubinemia, compelling diagnostic imaging findings), while confirmation of infection usually necessitates a cytological or culture based diagnosis. There is growing interest in novel biomarkers that could be applicable to the septic patient (e.g., NT-proCNP[7], CXCL-8[8], procalcitonin[9]). Identification of suitable candidate biomarkers is a busy ongoing area of research to help characterize which tests have helpful diagnostic utility for sepsis in the future.

In reality the accumulation of data necessary to confidently rule in or rule out sepsis may take some time. While this data is being gathered the clinician ought to start treatment. The decision whether or not to prescribe antimicrobials is a judgement call typically on a case by case basis. The Surviving Sepsis Campaign strongly recommends that intravenous antimicrobials are provided as soon as possible after recognition, and certainly within an hour of identification, of both sepsis and septic shock.[10] The clock certainly appears to be ticking therefore for the veterinarian potentially dealing with a septic pet and erring on the side of caution fits with the Surviving Sepsis guidelines. The seminal study cited as the evidence base for these recommendations identified an association between improved outcome and prompt (<60 minute) administration of antimicrobials in septic people with hypotension.[11] The longer the delay in providing antimicrobials, the increased chance of both morbidity (e.g., subsequent organ failure)[12-14] and mortality.[11,15] Interestingly a meta-analysis on this topic did not find early antimicrobial administration in people positively impacted outcome.[16] A veterinary study examined the impact of an emergency room antimicrobial protocol for dogs with septic peritonitis.[17] The median time to antimicrobial administration was significantly decreased after implementation of the protocol, yet overall survival to discharge was not different. The study, however, does provide an insight into the potential help of clinical guides in the emergency room. This is especially true in those cases that deviate from the textbook expectation of the disease and may accelerate the correct care being provided. Other practical considerations for this topic are hospital specific delays, outside of disease recognition, in providing antimicrobials that may encompass communication issues within the team or challenges with the pharmacy.[18] The balance of this accelerated approach to antimicrobial decision making, however, is that the call we make could prove to be an inappropriate one. Studies have described that an accelerated protocolized approach for elderly people with suspected pneumonia inadvertently resulted in inappropriate antimicrobial usage, actual misdiagnosis, delays in eventually obtaining the correct diagnosis, and unforeseen adverse effects (e.g., *Clostridium difficile* outbreaks).[19-21]

The impact of appropriate or inappropriate antimicrobial usage is also relevant, since empirical choices are made definitive confirmation via culture and susceptibility are available. Inappropriate antimicrobial usage, either from choosing an ineffective drug or providing antimicrobials in sterile disease states, risks increasing antimicrobial resistance. Culture-guided antimicrobial usage is supported by the Surviving Sepsis guidelines. Specifically, the guidelines recommend ideally obtaining cultures prior to the initiation of antimicrobials provided this does not delay antimicrobial administration by more than hour.[10] Sterilization of cultures can occur rapidly after antibiotic administration.[22,23] This appears a reasonable guide to follow in veterinary patients also but there may be some situations where rapid culture is not always especially practical (e.g., airway washing in unstable respiratory case). In people, blood cultures are recommended as part of the standard work up of sepsis, but this is uncommonly performed in dogs unless there is significant concern for bacteremia (e.g., infective endocarditis, polyarthropathy). A recent clinical study in dogs with septic peritonitis evaluated the effect of intraoperative peritoneal lavage on bacterial culture.[24] Pre-operative and postoperative bacterial cultures were compared and lavage did yield less bacteria. There was no significant difference on the ability to select an appropriate antimicrobial based on the post-lavage culture, however. In these situations, it therefore appears reasonable to initiate empirical antimicrobials while performing initial resuscitation and then utilize either pre- or post-lavage samples obtained intra-operatively.

In general and in accordance with Surviving Sepsis guidelines, broad spectrum empirical antimicrobial coverage is recommended. Broad spectrum classically addresses the 'four quadrant' approach of gram positives, gram negatives, aerobes, and anaerobes. In practice this may involve utilizing more than one class of antimicrobial, especially if this improves spectrum of coverage or if the selected drugs have synergism with each other. Ideally, once cultures are obtained, the antimicrobial plan can be de-escalated, allowing earlier discontinuation of one of the drugs. One veterinary study investigated antimicrobial therapy and aerobic culture patterns in an academic ICU.[25] They described that empirical choices were appropriate in their hospital 75% of the time, but that antimicrobials were often still administered even in cases of negative culture rather than being discontinued. Empirical antimicrobial choices should represent an educated best guess for the given patient in question based on key factors including the anatomical site affected, the most likely pathogens involved, resistance patterns in the hospital, and confounding patient factors (e.g., prior antimicrobial exposure, immunosuppression).[10] Prior antimicrobial exposure is an important piece of clinical history to clarify. Recent veterinary studies have described a high prevalence of recent antibiotic usage in dogs with bacterial pneumonia and septic peritonitis.[26,28] In both studies, prior exposure increased the likelihood of bacterial resistance to the same class of antimicrobial, suggesting an alternative class of drug would be a fair empirical choice in such circumstances. Appreciation for differences in resistance patterns for community and hospital acquired infections may also influence the empirical antimicrobial choices made. The likelihood of multidrug resistance characteristics of bacterial isolates was increased for dogs that has been hospitalized for 48 hours or more.[25] The use of hospital-specific antibiograms, based upon specific culture and susceptibility results derived from your own patient population, aids in improving likelihood of making appropriate initial antimicrobial choices.[28]

An appreciation of the pharmacokinetic and pharmacodynamic profiles of antimicrobials should also be considered. Optimal therapeutic targets differ between antimicrobials, for example concentration-dependent drugs like aminoglycosides, require achievement of peak plasma targets.[29] Time-dependent drugs like beta lactams benefit from prolonged therapeutic concentrations in excess of the pathogen minimum inhibitory concentration.[29] To that effect, there is some evidence to suggest a paradigm shift of delivering beta-lactam antibiotics via infusion to people with sepsis rather than via intermittent bolus may improve outcome.[30] Polypharmacy for the purposes of broad spectrum coverage is also recommended, at least in the short term, by the Surviving Sepsis guidelines. This approach of using multiple drugs has improved outcome in certain disease entities in people (e.g., severe pneumonia[31], toxic shock syndrome[32]) but is a fairly controversial recommendation in people with gram negative sepsis. A Cochrane Review examining monotherapy in septic people with beta lactams compared to those treated with beta lactams and aminoglycosides, found a trend towards improved outcome for the monotherapy group and increased adverse effects in the combination therapy group.[33] Combination therapy could also increase the risk of resistance within hospitals as has been described in people receiving fluoroquinolones as part of combination therapy.[34]

A meta-analysis in septic people reported significantly higher mortality when inappropriate empirical choices were made but patient illness severity contributed to the overall risk of death (i.e., people with septic shock had increased mortality regardless).[35] Not all data necessarily supports the idea that empirical antimicrobial choices are the make or break for sepsis however. One systematic review examined the impact of the appropriateness of antimicrobials in people with bacteremia and found no positive impact of making correct empirical choices.[36] A retrospective study in dogs with septic peritonitis failed to find an association between worse outcome and inappropriate antimicrobial choices.[27] Interestingly a veterinary review on the appropriate usage of antimicrobials in the septic patient reminds us of the risks of producing meta-analyses.[37] Meta-analyses are influenced by the design and execution of its composite studies and therefore can provide misleading global interpretations of topics. In the previously mentioned meta-analysis, a high number of cases failing to identify positive outcome of empirical therapy were excluded due to insufficient data availability. Another difference to recall between human and veterinary antimicrobial recommendations for sepsis is the recommended duration of therapy. The Surviving Sepsis guidelines suggest, albeit as a weak recommendation, that a 7 to 10 day course may be sufficient even for serious infections.[10] By convention, certain bacterial diseases in veterinary medicine are typically treated for longer than this, with protracted courses being recommended for a variety of conditions including pyothorax and bacterial cholangitis/cholangiohepatitis. In people, procalcitonin can be used as an objective biomarker to help guide antimicrobial discontinuation[38] but is not helpful in the dog.[9] Consequently, we may lack the confidence as clinicians to provided shorter courses of antimicrobials in septic pets when we lack a good objective method of confirming clearance of infection.

Much in the same way that our definitions for sepsis change over time, theories relating to the pathophysiology of sepsis also adapt. Traditionally, sepsis was considered the systemic inflammatory response to infection as mentioned earlier. Infection triggers a host response, mediated by endogenous inflammatory cytokines, leading to cellular dysfunction and risk of morbidity and mortality. The clinical presentation varies in severity with some cases having non-survivable sequelae to this infection. The immunological response to infection is initially pro-inflammatory (the systemic inflammatory response or SIRS) but could progress to (or potentially concurrently develops) a state of immunoparalysis characterized by the compensated anti-inflammatory response syndrome (CARS).[39] If early antimicrobials are administered, infection can be cleared quickly, which should attenuate the downstream effects of SIRS and CARS. Cases where antimicrobial administration seemed not to positively impact outcome could represent individuals in which the dysregulated host response is so extreme that conventional therapy was 'not enough'. The patient with advanced disease refractory to traditional therapies is of course sometimes encountered in practice. An updated theory for these cases has been suggested, which places a far greater emphasis on the role of the initial bacterial load.[40] The updated theory consolidates the recommendations to ensure early appropriate antimicrobial administration in cases of sepsis. It suggests that bacteria initiate infection but that the persistence of bacterial antigens (e.g., contained within bacterial toxins) continue to drive the host's immune response leading to a shock syndrome characterized by cellular and tissue dysfunction. At a critical point, termed the shock threshold, shock itself becomes the central driver of the pathophysiology, and failure to rapidly reverse shock may lead to terminal disease. This theory therefore reinforces traditional clinical concepts that prompt antimicrobial administration, effective source control, and targeted resuscitation of patients is the key to success. The theory also suggests minimizing the time that patients spend above this critical shock threshold is the key to a positive outcome. This theory also may explain why infection related disease seemingly occurs on a spectrum of very treatable and reversible conditions right through to true septic shock with a high likelihood of mortality. Regardless of the theory of sepsis we read or the results of meta-analyses, prompt and academically sound empirical antimicrobial choices continue to be at the forefront of our management of sepsis in people and animals alike.

References

1. Bone RC *et al. Chest*. 1992;101(6):16455-16455.
2. Levy MM *et al. Crit Care Med*. 2003;31:1250-1256.
3. Hauptman JG *et al. Vet Surg*. 1997;26(5):393-397.
4. De Laforcade AM *et al. J Vet Intern Med*. 2003;17:674-679.
5. Brady *et al. J Am Vet Med Assoc*. 2000;217:531-535.
6. Singer M *et al. JAMA*. 2016;315:801-810.
7. DeClue AE *et al. J Vet Intern Med*. 2011;25:453-459.
8. DeClue AE *et al. J Vet Intern Med*. 2012;26(3):624-630.
9. Floras AN *et al. J Vet Intern Med*. 2014;28:599-602.
10. Rhodes A *et al. Intensive Care Med*. 2017;Epub ahead of print.
11. Kumar A *et al. Crit Care Med*. 2006;34:1589-1596.
12. Bagshaw *et al. Intensive Care Med*. 2009;35:871-881.
13. Iscimen R *et al. Crit Care Med*. 2008;36:1518-1522.
14. Garnacho-Montero J *et al. Crit Care*. 2006;10:R111.
15. Ferrer R *et al. Crit Care Med*. 2014;42(8):1749-1755.
16. Sterline SA *et al. Crit Care Med*. 2015:43(9):1907-1915.
17. Abelson AL *et al. J Vet Emerg Crit Care*. 2013;23:551-556.
18. Funk DJ, Kumar A. *Crit Care Clin*. 2011;27:53-76.
19. Welker JA *et al. Arch Intern Med*. 2008;168:351-356.
20. Kanwar M *et al. Chest*. 2007;131:1865-1869.
21. Polygreen PM *et al. Infect Control Hosp Epidemiol*. 2007;28:212-214.
22. Zadroga R *et al. Clin Infect Dis*. 2013;56:790-797.
23. Kanegaye JT *et al. Pediatrics*. 2001;108:1169-1174.
24. Swayne SL *et al. Can Vet J*. 2012;53:971-977.
25. Black DM *et al. J Vet Emerg Crit Care*. 2009;19(5):489-495.
26. Proulx A *et al. J Vet Emerg Crit Care*. 2014;24(2):194-200.
27. Dickinson AE *et al. J Vet Emerg Crit Care*. 2015;25(1):152-159.
28. Green DL. *J Intensive Care Med*. 2005;20(5):296-301.
29. Papich MG. In: *Small Animal Critical Care Medicine*. 2nd ed. 949-955.
30. Roberts JA *et al. Am J Respir Crit Care Med*. 2016;194(6):681-691.
31. Shorr AF *et al. BMJ Open*. 2013;3(6).
32. Linner A *et al. Clin Infect Dis*. 2014;59(6):851-857.
33. Paul M *et al. Cochrane Database Syst Rev*. 2014;(1):CD003344.
34. Charbonneau P *et al. Clin Infect Dis*. 2006;42(6):778-784.
35. Niederman MS. *Clin Infect Dis*. 2006;42Suppl2:S72-81.
36. Paul M *et al. Antimicrob Agents Chemother*. 2010;54(11):4851-4863.
37. McGregor JC *et al. Clin Infect Dis*. 2007;45(3):329-337.
38. Keir I, Dickinson AE. *J Vet Emerg Crit Care*. 2015:25:55-62.
39. Matthaiou DK *et al. Intensive Care Med*. 2012;38:940-949.
40. Ward NS *et al. Clin Chest Med*. 2008;29(4):617.
41. Kumar *et al. Virulence*. 2014;5(1):80-97.

acvim378

Resuscitation of the Patient with Septic Shock

Deborah C. Silverstein, DVM, DACVECC
Philadelphia, PA, USA

INTRODUCTION

Sepsis is a life-threatening syndrome caused by a dysregulated host response to infection. If sepsis is not recognized early and rapidly treated, patients can quickly deteriorate and suffer from fatal organ dysfunction. Resuscitation of the patient with sepsis and septic shock should aim to return the patient to a physiologic state that enables adequate organ perfusion and restores the balance between metabolic supply and demand. Endpoints for resuscitation are commonly used to determine success and need for additional treatment. Protocolized resuscitation of people with sepsis or septic shock has been recently popularized and is recommended for use in small animal patients suffering from sepsis or septic shock. This protocol-driven resuscitation is best performed within 6 hours of recognition, although the evidence is mixed regarding the best monitoring goals. Research has shown that intravenous volume replacement for the treatment of patients with septic shock helps restore tissue perfusion, attenuate the cytokine response, and reverse the cellular swelling and organ injury that commonly occurs in these animals. The ideal resuscitation fluid would be a small-volume cocktail that improves cellular perfusion, enhances oxygen delivery and diffusion, provides adequate metabolic substrates, neutralizes toxic molecules released by inflamed tissues, provides antimicrobial activity, renders the recipient less susceptible to the effects of shock, and has sustained beneficial effects. In addition, the fluid would be stable for prolonged time periods at a variety of temperatures, easy to prepare and administer, and reasonably priced. This perfect fluid does not exist, but science has progressed in helping clinicians to better understand the use and abuse of fluid therapy for the treatment of septic shock and the future holds continued excitement and promise.

Approximately 50% of hypotensive, septic humans will have normalization of cardiovascular hemodynamics with fluid therapy alone. This number may be as high as 66% in dogs. However, the patients that do not respond will require additional monitoring and treatment in order to maximize the chances for survival. This lecture will describe major pathologic players in patients with sepsis, as well as resuscitation strategies and endpoints that are currently recommended.

HYPOVOLEMIA IN SEPSIS

Animals with sepsis frequently suffer from a true hypovolemia secondary to vomiting, diarrhea, panting, body cavity effusion, and other exogenous losses without adequate fluid intake. In addition, there is a relative hypovolemia from maldistribution of blood within the vasculature that frequently occurs due to arterio- and venodilation, peripheral blood pooling, microcirculatory dysfunction, and transvascular fluid extravasation due to increased capillary endothelial permeability.

Subsequently, there is a decrease in blood volume and therefore preload, diastolic ventricular pressure, stroke volume, cardiac output and systemic oxygen delivery. Compensatory changes within the body help restore tissue perfusion by increasing circulating catecholamines to vasoconstrict vascular smooth muscle and shunt blood to the vital organ beds, increase heart rate and augment contractility; cortisol, aldosterone, renin-angiotensin-aldosterone and vasopressin are also released to increase intravascular volume and oxygen delivery.

Compared to patients with classic hypovolemic shock, those with sepsis or septic shock are not always responsive to fluid administration. Nonetheless, aggressive fluid resuscitation to achieve a central venous pressure greater than 8 mm Hg was the standard of care in human patients within the last decade, based on the early goal-directed therapy (EGDT) guidelines originally popularized by Rivers *et al.* in 2001. Recent multicenter trials, including ProCESS, ARISE and PROMISE, as well as a meta-analysis of EGDT have questioned these guidelines, but it is hard to argue that prompt recognition and treatment of septic shock patients, regardless of the specific goals set forth, is not a necessary goal in order to maximize the outcome of these critically ill people and animals. In addition, the most recent Surviving Sepsis Guidelines recommend early goal-directed therapy. Preliminary research in dogs suggests that goal-directed therapy is beneficial in dogs with sepsis and septic shock.

Vascular Dysfunction in Sepsis

As blood vessels fail to constrict in patients with septic shock, subsequent vasoplegia occurs. This is most likely due to increased nitric oxide (NO), activation of KATP channels causing hyperpolarization of the smooth muscle, increased natriuretic peptides (that function synergistically with NO) and a relative vasopressin deficiency. In addition to the resultant arterial hypotension, profound venodilation of the splanchnic and cutaneous vascular beds leads to a marked decrease in venous return and cardiac output. Since ~70% of blood volume is within the venous system, these alterations play a major role in determining venous return. Interestingly, crystalloid fluid resuscitation in human patients with septic shock has been shown to cause a decrease in systemic vascular resistance, thus serving as another cause of vasodilation. This is why the early use of vasopressors has gained popularity in recent years.

In addition to the vascular tone, septic patients have an increased expression and activation of endothelial adhesion molecules leading to adherence and activation of platelets, neutrophils and mononuclear cells and activation of the coagulation cascade. Subsequently, there is diffuse endothelial damage, microvascular thrombosis, endothelial intercellular gaps, and shedding of the endothelial glycocalyx layer. The end result is a reduced functional capillary density, heterogeneous capillary blood flow, and increased permeability of the microcirculation. These changes are more severe in non-survivors, and lack of improvement with treatment has been associated with organ failure and death. Interestingly, intravenous fluids have been shown to improve the microcirculation if given in the early phase of sepsis, but have no effect in patients with late-phase sepsis. This again stresses the importance of early fluid resuscitation combined with appropriate vasopressor administration in the treatment of these patients.

Cardiac Dysfunction in Sepsis

Diastolic and/or systolic dysfunction is present in up to 50% of human patients with severe sepsis and septic shock, and there is likely a similar incidence in small animals. Diastolic function, in particular, has been found to correlate not only to prognosis, but also to fluid responsiveness. A landmark study by Ognibene and colleagues found that there was only a mild increase in left ventricular stroke work index and end-diastolic volume index in patients with septic shock following administration of a fluid challenge. There was an increase in cardiac filling pressures, increase in venous and pulmonary hydrostatic pressures, and increased release of natriuretic peptides with little (if any) increase in stroke volume. This explains the lack of fluid responsiveness in many patients with septic shock; their hearts are functioning on the flat portion of the Frank-Starling curve where fluid administration only serves to increase atrial, venous and pulmonary hydrostatic pressure, disrupt the endothelial glycocalyx and subsequently enable the movement of intravascular fluid to the interstitial space. This tissue edema is deleterious as it impairs oxygen and metabolite diffusion, distorts tissue architecture, prevents capillary blood flow and lymphatic drainage, and interferes with cell-to-cell interactions. Vital organ edema also affects function of the organ, especially in the kidneys. The use of positive inotropes in patients with systolic function is often beneficial.

The Resuscitation Bundle Concept

The resuscitation bundle is not actually a novel strategy for resuscitation. In 1976, Wilson *et al.* wrote several expert opinions that included the basic tenets of early sepsis management (i.e., early recognition, source control, antibiotic therapy, and early hemodynamic optimization of oxygen delivery). In the popular publication of 2001, similar "early goal-directed therapy" (EGDT) was tested against standard care in a randomized control trial and resulted in a mortality benefit greater than 16%. Over the last 15 years, the components of EGDT and the "resuscitation bundle" have been examined, validated and incorporated into evidence-based human guidelines. These guidelines include the following:

Within 3 hours of presentation, do the following:

1. Measure lactate level
2. Obtain blood cultures prior to administration of antibiotics
3. Administer broad-spectrum antibiotics
4. Administer 30 ml/kg crystalloid for hypotension or lactate ≥4 mmol/L

And within 6 hours of presentation, do the following:

5. Apply vasopressors (for hypotension that does not respond to initial fluid resuscitation) to maintain a mean arterial pressure (MAP) ≥65 mm Hg
6. In the event of persistent hypotension after initial fluid administration (MAP <65 mm Hg) or if initial lactate was ≥4 mmol/L, re-assess volume status and tissue perfusion
7. Re-measure lactate if initial lactate elevated

FLUID ADMINISTRATION

The detrimental effects of overly aggressive fluid therapy in septic patients have been well documented in affected humans. Humans have evolved to handle hypovolemia much better than hypervolemia. Thus, excessive amounts of intravenous fluids can counter vital life-preserving homeostatic mechanisms in critically ill patients, leading to death. Since patients with septic shock do not always show normalization of hemodynamic stability following limited fluid resuscitation, fluids alone are likely to worsen the vasodilatory shock and increase vascular leak syndromes and tissue edema. Despite the Surviving Sepsis guidelines above, there are researchers in the human medical field that argue for administering smaller initial volumes, with a maximum of 20 ml/kg. Since 0.9% sodium chloride is not a "balanced" crystalloid and causes a hyperchloremic metabolic acidosis following infusion of large volumes, it is not currently the fluid of choice in most patients. The increase in chloride can lead to a decrease in renal blood flow, thereby increasing the risk of acute kidney injury. In addition, the use of normal saline in septic human patients has been associated with an increased risk of death when compared to the more physiologic, balanced isotonic crystalloid solutions. Any patients that remain hypotensive despite fluid administration should be treated with a vasopressor agent. Both dopamine and norepinephrine are commonly used in small animals, although dopamine has fallen out of favor in human medicine. Low-dose epinephrine and vasopressin infusions have received more attention in recent years. Early vasopressor therapy in patients with septic shock will likely decrease the incidence of fluid-related complications while maintaining perfusion to the vital organs. However, it is important that the clinician document a fluid-refractory vasodilatory state prior to inducing vasoconstriction. Otherwise, further decreases in vital organ perfusion may result.

EARLY ANTIMICROBIAL THERAPY

Animals with confirmed or highly suspected sepsis should receive appropriate antimicrobial therapy as soon as possible. Mortality rates escalate for every hour of delay in administration of antimicrobials in human patients with septic shock. Ideally, samples for culture and sensitivity testing should be obtained prior to starting antimicrobial therapy, they should not delay treatment. Intravenous bactericidal antimicrobial drugs that target the organisms and site of infection should be used; inappropriate antimicrobial selection is associated with a 5-fold increase in death in humans with septic shock. Results of culture and susceptibility testing should be used when changing therapy in order to de-escalate the spectrum to decrease the incidence of resistant organisms.

MEAN ARTERIAL PRESSURE

The target mean arterial pressure (MAP) is somewhat controversial; one human study found that a MAP<60–65 mm Hg during the first 24–48 hours was most predictive of 30-day mortality and organ function. This endpoint should be achieved using fluid and vasopressor therapy, as appropriate, although fluid administration is preferred to vasopressor use within the first six hours, especially since hypotension tends to be more refractory to fluid therapy in the later stages of disease presentation. Another human study showed that targeting a MAP >70 mm Hg using vasopressor therapy may increase mortality. Although Rivers *et al.* used a target MAP >65 mm Hg, most patients exceeded this value greatly. A retrospective feline study found that a Doppler blood pressure <90 mm Hg was associated with decreased survival.

Red Blood Cell Transfusions

Animals suffering from hypoperfusion secondary to SIRS or sepsis commonly present with or develop anemia secondary to the pre-existing disease, volume resuscitation and dilution, impaired bone marrow response, and possible decrease in sensitivity of the erythropoietin receptors. A compensatory increase in systemic oxygen extraction occurs to maintain oxygen needs, but may not be adequate. Red blood cell transfusions may therefore be warranted during this delivery-dependent state if the lactate is elevated and the $S_{cv}O_2/S_vO_2$ decreased. One human study found that a $S_{cv}O_2$ of 69.5% was a useful transfusion trigger in order to maximize survival. Although the use of red blood cells in these patients has been controversial over the years, a large observational human study recently found that red blood cell transfusions were associated with decreased mortality rates. Although 10 mg/dL of hemoglobin is the current transfusion trigger for people with septic shock, the best trigger for small animals remains unknown.

Inotrope Therapy

Myocardial dysfunction in people and animals with SIRS or sepsis is common. Patients treated with EGDT had 12.9% greater frequency of (early recognition of) myocardial dysfunction requiring inotropic therapy. Research has shown that neither the physical examination nor the $S_{cv}O_2$ can predict systolic dysfunction. The surviving sepsis campaign guidelines recommend the use of dobutamine in human patients that have ongoing evidence of hypoperfusion despite adequate intravascular volume and mean arterial pressure.

Lactate and Lactate Normalization

Type A lactic acidosis is associated with tissue hypoxia and anaerobic energy production. The clearance of elevated lactate seems to be more important than a single value; the use of lactate clearance as a marker of global hypoxia dates back to the 1960s. However, a normal lactate level does not necessarily mean that oxygen delivery is normal. Type B lactic acidosis may also be present in septic patients due to microvascular shunting, mitochondrial dysfunction, and/or dysregulation of pyruvate dehydrogenase activity. However, the recent study in dogs with sepsis due to pyometra found a significant decrease in lactate over time in survivors. Another study evaluating EGDT and lactate in 8 critically ill dogs showed that death was more likely if the animal's lactate failed to decrease by at least 50% during the first 6 hours of aggressive treatment. Human studies have found a correlation between early lactate clearance and a decrease in pro- and anti-inflammatory biomarkers, improved organ function and reduced mortality. In some studies, resuscitation based on CVP, MAP, and lactate clearance of 10% was as effective as resuscitation based on CVP, MAP, and $S_{cv}O_2/S_vO_2$ >70%. This remains controversial, however, since lactate exceeds clearance only when the critical oxygen extraction ratio has been reached, a less sensitive indicator of impending tissue hypoxia compared to $S_{cv}O_2/S_vO_2$. However, studies examining the use of lactate normalization as the primary resuscitation endpoint have shown not only a mortality benefit, but also a decrease in organ failure, duration of catecholamine therapy, and length of mechanical ventilation. However, it is important to remember that an elevated lactate in septic patients may indicate more than just poor tissue perfusion (e.g., cytopathic hypoxia, microcirculatory dysfunction), and conversely a normal lactate does not exclude the possibility of tissue hypoperfusion. Therefore, a bundle approach is most logical in order to most effectively treat patients based on early indices of perfusion abnormalities.

Monitoring Resuscitation in the Septic Shock Patient

In addition to the goals discussed above, there are many other possible monitoring tools. The value of frequent physical reassessments cannot be overemphasized. Monitoring patients with sepsis or septic shock can be difficult; microcirculatory distress can be present despite macrocirculatory hemodynamic variables. In addition to mean arterial blood pressure and lactate, urine output, central venous pressure, mixed venous oxygen saturation, arterial-venous CO_2 gradient and base excess may also be valuable. Although hyperlactatemia in these cases is most frequently attributed to hypoperfusion, metabolic failure in septic patients can also affect this value. Nevertheless, persistently elevated blood lactate levels appear to have good prognostic value in both humans and dogs. Close monitoring of patients during fluid resuscitation using the physical examination is of utmost importance. Although we still do not know all of the answers regarding fluid resuscitation in septic patients, we do know that early and adequate fluid therapy improves outcomes. Future studies may help answer many of the questions that still remain.

REFERENCES

1. Rivers E, Nguyen B, Havstad S, Ressler J, Muzzin A, Knoblich B, *et al.* Early goal-directed therapy in the treatment of severe sepsis and septic shock. *N Engl J Med.* 2001;345(19):1368–77.
2. Mouncey PR, Osborn TM, Power GS, Harrison DA, Sadique MZ, Grieve RD, *et al.* Trial of early, goal-directed resuscitation for septic shock. *N Engl J Med.* 2015;372(14):1301–11.
3. Investigators A, Group ACT, Peake SL, Delaney A, Bailey M, Bellomo R, *et al.* Goal-directed resuscitation for patients with early septic shock. *N Engl J Med.* 2014;371(16):1496–506.
4. Investigators TP. A randomized trial of protocol-based care for early septic shock. *N Engl J Med.* 2014;370(18):1683–93.
5. Ognibene FP. Hemodynamic support during sepsis. *Clin Chest Med.* 1996;17(2):279–87.
6. Conti-Patara A, de Araujo Caldeira J, de Mattos-Junior E, de Carvalho Hda S, Reinoldes A, Pedron BG, *et al.* Changes in tissue perfusion parameters in dogs with severe sepsis/septic shock in response to goal-directed hemodynamic optimization at admission to ICU and the relation to outcome. *J Vet Emerg Crit Care (San Antonio).* 2012;22(4):409–18.
7. Butler AL. Goal-directed therapy in small animal critical illness. *Vet Clin North Am Small Anim Pract.* 2011;41(4):817–38.
8. Rhodes A, Phillips G, Beale R, Cecconi M, Chiche JD, De Backer D, *et al.* The Surviving Sepsis Campaign bundles and outcome: results from the International Multicentre Prevalence Study on Sepsis (the IMPreSS study). *Intensive Care Med.* 2015;41(9):1620–8.
9. Gaieski DF, Mikkelsen ME, Band RA, Pines JM, Massone R, Furia FF, *et al.* Impact of time to antibiotics on survival in patients with severe sepsis or septic shock in whom early goal-directed therapy was initiated in the emergency department. *Crit Care Med.* 2010;38(4):1045–53.
10. Silverstein DC, Wininger FA, Shofer FS, King LG. Relationship between Doppler blood pressure and survival or response to treatment in critically ill cats: 83 cases (2003–2004). *J Am Vet Med Assoc.* 2008;232(6):893–7.

acvim065

Development of a Standardized, Individual, Submaximal Treadmill Exercise Test in Healthy and CHIEF B1/B2 Dogs

Leona F. Wall; Sephan Oscar Hungerbuehler; Annika Mohr; Florenza Lüder Ripoli; Nayeli Schulze; Camila Penter; Ingo Nolte

Small Animal Clinic, University of Veterinary Medicine Hanover, Foundation, Hannover, Niedersachsen, Germany

Cardiovascular examinations in veterinary practice lack an objective method for assessing exercise capacity. Therefore, the present study aimed to provide a submaximal exercise test (SET) for canine patients.

To investigate the underlying parameters for exercise capacity, 12 healthy (H) and 12 CHIEF B1/B2 dogs (P) underwent a six-stage SET on a motorized treadmill. They trotted in their individual speed for three minutes per stage and slope was increased by 4% per stage. Venous blood samples were taken at rest, within two 3-minute pauses, after the SET, and after a 3-hour recovery period. Various parameters including lactate, N-terminal B-type natriuretic peptide (NT-proBNP) and cardiac troponin I (cTnI) were measured.

Lactate decreased from 9.8 to 8.3 mg/dl (H) and from 12.4 to 9.8 (P) after the second stage. It increased with further workload to 10.13 (H) and 14.2 (P) and 11.5 (H) and 14.6 mg/dl (P). NT-proBNP increased from 435 to 523 pmol/l (H) and from 700 to 803 pmol/l (P). cTnI elevated from 0.02 to 0.024 ng/ml in H and from 0.05 to 0.06 ng/ml (P). One-way ANOVA revealed a significant ($p<0.05$) influence of the different stages on the parameters, but not for every time point.

The SET in the present study yielded workload-dependent alterations in various parameters such as lactate, NT-proBNP and cTnI. All of the parameters showed higher baseline values in P than in H and showed a more pronounced increase after exercise. Thus, SET could be useful to supplement cardiovascular examinations in canine heart patients.

acvim066

Implantable Cardiac Loop Recorders: A Novel Diagnostic Tool for Atrial Fibrillation Detection in Horses

Rikke Buhl[1]; Eva Z. Hesselkilde[1]; Helena Carstensen[1]; Merle Frederike Fenner[1]; Thomas Jespersen[2]; Jacob Tfelt-Hansen[3]; Stefan M. Sattler[3]

[1]Department of Large Animal Sciences, Faculty of Health and Medical Sciences, University of Copenhagen, Taastrup, Hovedstaden, Denmark; [2]Department of Biomedical Sciences, Faculty of Health and Medical Sciences, University of Copenhagen, Copenhagen, Hovedstaden, Denmark; [3]Copenhagen University Hospital, Copenhagen, Hovedstaden, Denmark

Atrial fibrillation (AF) is the most common pathological cardiac arrhythmia affecting performance in horses. The prevalence of paroxysmal AF is unknown as the disease may go unnoticed by the trainer or rider due to the short duration of the AF episodes. Implantable loop recorders (ILRs) are small ECG devices placed subcutaneously, capable of automatically detecting and storing predetermined arrhythmic events. ILRs are used in human patients to detect AF and other arrhythmias and have shown superior diagnostic yield compared to surface ECGs for long term monitoring, although the function is highly dependent on the quality of the recorded signals.

The purpose of the present study was to evaluate the best anatomical localization for the ILR in horses in order to obtain a good quality ECG recording. By comparing three different anatomical positions in a horse developing AF, the ability of this new diagnostic technique to detect AF was described and compared to Holter surface ECG.

Three ILRs (Reveal Linq, Medtronic) were placed subcutaneously in local anesthesia at three different locations in a 13 year old Standardbred mare. One recorder was placed in the superficial pectoral region (Front), one was placed in the left sixth intercostal space at the level of the shoulder joint (Left) and finally the third recorder was placed ventrally at the level of the xiphoid process in a cranio-caudal direction (Ventral). The horse was tachypaced into AF by pacemaker stimulation and arrhythmic events were registered for a two month period by the ILRs. During the period, a Holter ECG was placed on the horse at selected time points.

All ILRs showed stable R wave amplitude and were able to detect the occurring AF episodes, with the ILR at the Left position showing the highest R wave amplitudes compared to the other two positions ($p<0.001$). P waves were only visible at the Left position. During AF, f waves were constantly visible (100%) prior to all R waves for the Left position compared to 80% and 13% for the Front and Ventral position, respectively. Movement artifacts for all three ILRs were rare and none of them led to a false positive AF diagnosis.

This study shows that ILRs can detect AF occurrence in horses, but the anatomical location is essential for optimal ECG quality with the Left position superior compared to the two other locations. ILRs may add diagnostic value in horses presented with poor performance or for detection of AF recurrence after cardioversion.

acvim067

Epigallocatechin-3-gallate Recouples Myofilament Calcium Sensitivity to Troponin I Phosphorylation in Feline Hypertrophic Cardiomyopathy

David J. Connolly[1]; Andrew E. Messer[2]; Jasmine Chan[1]; Alex Daley[1]; Steven B. Marston[2]

[1]Royal Veterinary College, Hatfield, England, UK; [2]Imperial College London, London, England, UK

Development of left ventricular hypertrophy is a late stage event in hypertrophic cardiomyopathy (HCM). Myofilament mutations generally enhance calcium sensitivity, maximal force production, and ATPase activity leading to energy deficiency and altered calcium handling as major common pathways promoting HCM irrespective of specific genotype.

In the normal myocardium, the contraction relaxation cycle is regulated by calcium and modulated by β-adrenergic stimulation, leading to the phosphorylation of several key proteins which increase inotropy. In addition, phosphorylation of cardiac troponin I (cTnI) results in a faster dissociation of calcium from cardiac troponin C (cTnC) and decreases the sensitivity of the cardiac myofilament for calcium (lusitropy). Mutations in sarcomeric proteins that cause HCM in humans and engineered rodents abolish the relationship between cTnI phosphorylation and increase calcium sensitivity even if the mutation is not in the troponin complex. This phenomenon has been termed uncoupling. Recently it has been shown that calcium desensitisers epigallocatechin-3-gallate (EGCG) can restore the calcium sensitivity of the myofilament (at least in part) by "recoupling" the relationship between calcium sensitivity and cTnI phosphorylation.

The purpose of this study was to investigate if uncoupling was common in cats with HCM and whether it could be reversed using EGCG.

In vitro flow motility assays were performed with troponin extracted from 7 cats diagnosed with HCM, including a ragdoll homozygous for the MYBPC R820W mutation and 5 wild type (WT) unaffected cats. All HCM cats had higher calcium sensitivity than the WT cats with an average increase of 1.8 times, which is equivalent to the finding in human HCM patients. Uncoupling of calcium sensitivity from cTnI phosphorylation was observed in all 7 HCM cats examined, whereas in all 5 WT cats coupling was normal. Furthermore, EGCG was tested in 1 WT cat and 3 HCM cats and was found to completely reverse the uncoupling in all HCM cats and reduce the calcium sensitivity to a level similar to WT cats. We performed protein transfer experiments which indicated that the uncoupling in HCM cats is linked to cTnT, similar to what has previously been reported for HCM in humans.

A

HCM mutations including the Ragdoll MYBPC/R820W mutation caused an increase in Ca^{2+} sensitivity

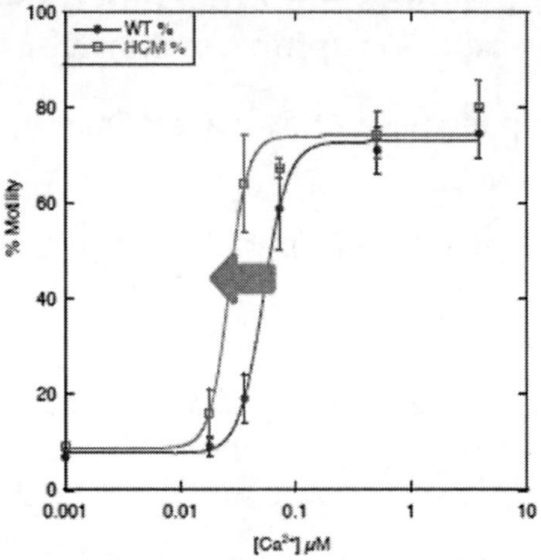

A

EGCG restored the difference between phosphorylated and dephosphorylated troponin with Ca^{2+} sensitivities being restored to wild type.

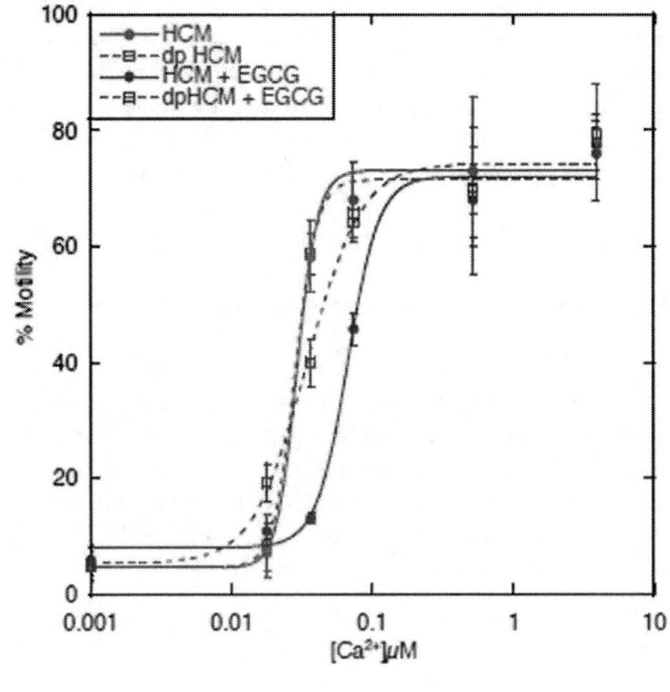

acvim068

Early Experiences with 3D Modeling and Printing of Hearts with Complex Cardiac Disease

Susanne M. Stieger-Vanegas

College of Veterinary Medicine, Oregon State University, Corvallis, OR, USA

Three-dimensional (3D) models of computed tomography angiography (CTA) images provide the benefit of a haptic experience, allow manipulation and enhance the understanding of the cardiovascular anatomy and underlying pathologies.

The aim of this study was to create anatomically correct and precise 3D printed models of the heart of veterinary patients with complex cardiac disease to facilitate a better understanding of the complex cardiac abnormality.

Based on CTA images, regions of interest were segmented using the Mimics software (Materialise, Leuven, Belgium). The segmented areas were blood volume and cardiac surface or area of major abnormality. After generating a stereolithography (STL) file out of the patient's data set, the 3D printer created a 3D model. The patient's individual 3D printed models were evaluated for geometric accuracy by comparing measurements of the cardiac structures in the CTA images and 3D printed model.

3D models of hearts of veterinary patients with complex cardiac abnormalities can be created and printed. These 3D models were accurate and precise. The higher the quality and degree of contrast enhancement in the CTA images of the cardiac structures, the easier and faster the segmentation of the heart could be performed. The CTA images and measurements of the 3D models were similar.

3D modeling and printing is a novel technique, which allows to obtain accurate 3D models of hearts with complex cardiac disease. Once a 3D model is created, it can be printed multiple times and made available for teaching. These 3D printed hearts or files could easily be shared between facilities.

acvim094

Use of Moxidectin/Imidacloprid and Doxycycline for Non-Arsenical Heartworm Adulticidal Therapy

Marisa K. Ames[1]; Philip VanVranken[2]; Chelsea Plath[2]; Clarke Atkins[3]

[1]Colorado State University, Fort Collins, CO, USA; [2]Dickman Road Veterinary Clinic, Battle Creek, MI, USA; [3]North Carolina State University, Cary, NC, USA

Arsenical therapy has long been the basis for heartworm adulticidal therapies and is safe and effective when used as directed. Arsenical therapy, however, has limited availability periodically and is unavailable in many countries. Topical moxidectin/imidacloprid (M/I) is different than other macrocyclic lactone preventives, in that higher blood and tissue concentrations are reached and maintained for a longer period. M/I has also been shown to be safe and well tolerated at accelerated dosing protocols in dogs with demodectic mange.

Eighteen dogs positive for HW by antigen testing received M/I at the preventive dosage (2.5–7.4 mg/kg), yet given every 15 days for 90 days, then every 30 days thereafter. Concurrently, each dog received doxycycline for 15 days. Additional courses of doxycycline or minocycline could be given at the discretion of the clinician. Dogs were also tested for microfilaria on the day of diagnosis. Heartworm antigen tests were repeated on Day 90, 120, and 150, then every 60 days thereafter. When dogs became antigen-negative (by the conventional test), heat-treated serum was also tested. For dogs remaining HW antigen-positive after day 210, the owner was offered melarsomine therapy. If the owner elected to continue with the study protocol, then M/I was continued monthly at the preventive dosage. Dogs were not strictly exercise restricted.

At the time of this writing, 18 dogs have been entered into the prospective study, with 13 having completed the study. The median (interquartile range; IQR) age and weight of the 18 dogs was 4.9 (3.0–8.8) years and 24.8 (17.1–29.7) kilograms, respectively. The median (IQR) doxycycline dosage was 12.6 (12.0–16.1) mg/kg/day. Three dogs received additional courses of doxycycline or minocycline at the attending clinician's discretion. Eight dogs developed a cough during M/I treatment. Treatment for cough included prednisone (n=3), dexamethasone (n=1), and dextromethorphan hydrobromide and guaifenesin (n=1). Seventeen of the 18 dogs were tested for microfilaria upon study enrollment and 11 were microfilaremic. Eight of these dogs had additional microfilaria tests. This study was not designed to test the efficacy and speed of the microfilaricidal effect of M/I, yet the median (IQR) time to negative was 85 (59–111) days. For the 13 dogs that have finished, the median (IQR) time to the first negative on the (conventional) antigen test was 208 (162–282) days. Six dogs were also antigen-negative when the serum was heat-treated. Five dogs had a negative heat-treated test 33, 38, 41, 67 and 107 days after their conventional test was negative, and 2 dogs have not yet been re-tested. Five dogs remain antigen positive and the median (IQR) time that these dogs have been in the study is 340 (340–396) days. All five of these dogs are past study day 210 and when their owners were offered melarsomine therapy, they declined and elected to continue giving M/I once monthly.

The combination of M/I and doxycycline provides a potential alternative when/where arsenicals are unavailable. In some cases, the conversion to antigen-negative status using heat-treated serum will lag behind the conventional test. The clinical importance of this is not known, but at present, we do not advocate heat-treating serum to confirm negative status. Exercise restriction would likely have reduced the frequency of cough seen in this study. Though preliminary, these data suggest that, based on conventional antigen testing, the efficacy, safety, and convenience of this protocol is acceptable, when melarsomine cannot be employed.

acvim097

Validation of a Quality of Life Survey for Owners of Dogs with Heart Disease

Lisa M. Freeman[1]; John E. Rush[2]; Audry Helburn[1]

[1]Cummings School of Veterinary Medicine, Tufts University, North Grafton, MA, USA; [2]Tufts University, North Grafton, MA, USA

Perceived QoL of a pet with heart disease is an important factor in owner decision-making. However, the owner's QoL may be an equally important factor in the care of a pet with heart disease. Studies have assessed the QoL of caregivers for people with heart disease, but no studies have been reported on QoL for owners of pets with heart disease. Therefore, the purpose of this study was to develop and validate a questionnaire to assess self-reported QoL of owners of dogs with heart disease. A 7-item questionnaire was developed based on areas of a person's life that could be impacted by caring for a sick animal, with a possible range of 0–35, with higher scores indicating worse owner QoL. The survey was self-administered to owners of dogs with heart disease presented to the Cardiology Service. The owner questionnaire was evaluated for readability, face validity, criterion-related validity, and test-retest reliability. The International Small Animal Cardiac Health Council (ISACHC) classification of heart failure was used as an independent measure of the severity of dogs' disease and the Functional Evaluation of Cardiac Health (FETCH) questionnaire was used to assess the owners' perception of their dogs' QoL. The owner QoL questionnaire was completed by 217 owners of dogs with heart disease and repeated after one to three weeks by 11 owners. Underlying diseases included degenerative mitral valve disease (n=164) and cardiomyopathy (n=53). Dogs' ISACHC stages were 1a (n=38), 1b (n=108), 2 (n=44), 3a (n=25), and 3b (n=2). Median FETCH QoL score was 7 (range, 0–70). Median owner QoL score was 6 (range, 0–31). The owner QoL score was correlated with the FETCH score ($r=0.76$, $p<0.001$) and the ISACHC stage ($r=0.53$, $p<0.001$). Test-retest reliability was high ($r=0.83$, $p<0.001$). These results suggest that the owner QoL questionnaire has good reliability and validity, and that further testing of this survey is warranted to optimize communication with owners regarding the care of dogs with heart disease.

acvim106

Influence of Chronic Kidney Disease on Measurements of Cardiac Structure and Function in Dogs

Melanie J. Hezzell[1]; J.D. Foster[2]; Mark Oyama; Jesse Buch[3]; Giosi Farace[3]; John Quinn[3]; Mahalakshimi Yerramilli[3]

[1]University of Bristol, Bristol, England, UK; [2]Friendship Hospital for Animals, Washington, DC, USA; [3]IDEXX Laboratories, Inc., Westbrook, ME, USA

The interplay between cardiac and renal disease (cardiorenal syndrome [CRS]), is poorly understood in veterinary medicine. Type IV CRS describes cardiac dysfunction secondary to primary renal disease; however, there is little data regarding the baseline cardiac function in dogs with chronic kidney disease (CKD). The study objective was to describe renal and cardiac function and biomarkers in a cohort of dogs with CKD.

Dogs with IRIS stage 2 and 3 CKD and healthy dogs were studied. Dogs were excluded if they were <10 kg, hypothyroid, had a heart murmur, evidence of primary cardiac disease, or active urine sediment, or were receiving glucocorticoid or diuretic therapy. Dogs with CKD were examined 6 months apart. Standard echocardiography and blood and urine tests, as well as symmetric dimethylarginine (SDMA), glomerular filtration rate (GFR) by iohexol clearance, plasma cardiac troponin I and N-terminal-pro-B-type natriuretic peptide (NT-proBNP), and urinary clusterin, cystatin B, and neutrophil gelatinase-associated lipocalin (NGAL) were measured. Groupwise comparisons were made using one-way ANOVAs or Kruskal-Wallis tests, as appropriate.

Twenty-three dogs were recruited (12 healthy controls, 5 IRIS stage 2, and 6 IRIS stage 3). Significant differences among groups were detected for body condition score (p=0.044), PCV (p=0.005), sodium (p=0.005), BUN (p=0.002), creatinine (p<0.001), SDMA (p<0.001), GFR (p<0.001), NT-proBNP (p=0.005), USG (p=0.001), UPC (p=0.001), cystatin B (p=0.006), NGAL (p=0.010), and E wave velocity (p=0.047). No variables changed significantly between visits.

In conclusion, E wave velocity and various biomarkers increased with increasing CKD severity, suggesting that cardiac function is affected by declining renal function.

acvim117

Accuracy of Angiographic Measurements During Transcatheter Procedures: Comparison of Two Calibration Methods in 25 Dogs

Lauren E. Markovic[1]; Brian A. Scansen[2]

[1]Colorado State University Veterinary Teaching Hospital, Fort Collins, CO, USA; [2]Colorado State University, Fort Collins, CO, USA

Obtaining an accurate measurement of an angiographic structure is critical for successful intervention. The accuracy of a measurement depends on both precise measurement of the lesion and an accurate reference calibration to compensate for image magnification. The aim of this study was to evaluate angiographic measurements using two commonly employed reference calibration methods, an esophageal marker catheter (C) versus a table ruler (R).

Twenty-five dogs undergoing transcatheter intervention were prospectively enrolled. Interventions included balloon pulmonary valvuloplasty (PS, n=14), balloon aortic valvuloplasty (n=1), ductal occlusion (PDA, n=8), or cor triatriatum membranostomy (CTD, n=3) from May 2016 to February 2017. All angiograms were reviewed and relevant measurements performed in triplicate and averaged when calibrated to either C or R. Comparisons between methods were made by paired t-test and the method of Bland and Altman.

Measurements were larger when calibrated to C versus R ($p<0.0001$). The mean bias was 0.8 mm having 95% limits of agreement from 0.1 mm to 1.7 mm, with greater bias for larger measurements. These differences persisted when PDA and PS cases were analyzed individually ($p=0.0003$ and $p<0.0001$, respectively), but were not significant for CTD cases alone ($p=0.051$).

Calibration of an angiographic image using a ruler placed on the fluoroscopy table during transcatheter procedures underestimates the actual size of a structure, when compared to a calibrated esophageal marker catheter placed at the same level as the heart. This effect is greatest when measuring larger structures such as the pulmonary valve annulus.

acvim137

INSPECT Survey: An Initiative for Simple Pragmatic End-Point Clinical Trials in Veterinary Cardiology

Brian A. Scansen[1]; Sonya Gordon[2]; Mark Rishniw[3]

[1]Colorado State University, Fort Collins, CO, USA; [2]College of Veterinary Medicine & Biomedical Sciences, Texas A&M University, College Station, TX, USA; [3]Cornell University, Ithaca, NY, USA

Randomized controlled clinical trials (RCT) offer the best approach for questions with clinical equipoise. However, funding sources for RCT in veterinary cardiology are sparse. Therefore, we utilized a short survey to assess the possibility of developing a cost-effective infrastructure for multicenter RCT specific to veterinary cardiology to answer simple, relevant, clinical questions with pragmatic end-points.
We distributed the survey to 605 members of a cardiology listserve of diplomates, residents, certificate holders, and clinicians interested in cardiology. The survey was completed by 155 respondents (32 provided partial responses) composed of 81 cardiology diplomates, 12 certificate holders, 7 residents, 6 other, and 49 undeclared. A majority of respondents (n=133, 86%) declared an interest in contributing cases, 11 (7%) supported the concept but lacked time to be involved, and 2 (1%) felt the concept would not succeed. Of the 3 conditions proposed as initial studies, 75% of respondents considered the highest priority to be beta-blocker treatment of feline subclinical hypertrophic obstructive cardiomyopathy, followed by canine subaortic stenosis (15% of respondents) and canine pulmonary valve stenosis (10% of respondents). Estimated case numbers by respondents suggested that robust sample sizes could be anticipated for these studies, resulting in high power for addressing these questions.

Our results suggest widespread interest in developing an infrastructure for multicenter RCT in veterinary cardiology, as has been done successfully by other specialties (e.g., veterinary oncology). Further analyses of the survey will be presented and the INSPECT framework will be proposed for discussion.

acvim153

Unusual Causes and Complicated Clinical Courses in Congestive Heart Failure Patients

Rebecca L. Stepien, DVM, DACVIM (Cardiology); Heidi B. Kellihan, DVM, DACVIM (Cardiology)
Madison, WI, USA

Introduction

The straightforward, stepwise march from heart disease to congestive heart failure (CHF) to successful therapy of heart failure is not followed by all patients. In a subset of clinical patients, this orderly progression is disrupted. In these patients, standard evaluation protocols and therapy for CHF may not be fully effective due to the presence of non-cardiac disease. Recognizing these patients early provides the chance for successful therapy.

Diagnostic Testing

Most "non-standard" causes of cardiac dysfunction and/or congestive heart failure will be identified with typical cardiac testing, including history and physical examination, thoracic radiography, echocardiography, electrocardiography (ECG), and blood pressure (BP) measurement. In addition, targeted laboratory assessment can provide information leading to the diagnosis of systemic disease having an impact on the heart. **Key to recognizing unusual circumstances is realizing that they may exist, and keeping an open mind when interpreting the results of diagnostic testing**.

History

A thorough history may uncover evidence of recent changes in the patient's life that may have altered cardiac status. Most importantly, even though the patient may be presented in obvious congestive heart failure, the history should be reviewed with attention to evidence of non-cardiac diseases that may have had an impact on the heart. These may include renal disease, endocrine diseases like hypothyroidism, hyperthyroidism or hyperadrenocorticism, and history of exposure to infectious diseases (including heartworm or myocardial parasites). In patients who have developed CHF after anesthetic events or fluid administration, special attention should be directed at screening for systemic hypertension, including a retinal examination.

Physical Examination

Physical examination of the patient may reveal evidence of concurrent systemic illness (e.g., hair loss or weight gain in endocrine patients, retinal detachment in patients with systemic hypertension). Do the signs of CHF match the patient's auscultation results, or are the auscultation results and clinical signs discordant? Is an arrhythmia present or has the patient's arrhythmia changed?

Thoracic Radiography

Thoracic radiography is one of the most important imaging modalities in patients with confusing clinical signs. Thoracic radiographs allow the clinician to evaluate the size and shape of the heart as well as to evaluate the interaction between the heart and the airways, pulmonary parenchyma, and pulmonary vasculature. In cases where external forces are acting on the heart (e.g., pericardial effusion or mass lesions), these may be suspected or identified with certainty based on thoracic radiographs. In dogs, the orderly and predictable progression of pulmonary edema (from pulmonary venous engorgement, to infiltrates in the caudal dorsal lung fields, to spread of the infiltrates ventrally then to the cranial ventral lung fields) allows differentiation of pneumonia and CHF.

Echocardiography

Echocardiography, whether a complete examination with Doppler evaluation or a brief ultrasound screen of an emergency patient, can provide a rapid answer to the question of "Is heart disease the cause of these clinical signs?" in some cases. Echocardiography is best suited to answering questions about the internal anatomy of the heart and heart function, but may also identify mass lesions and cavitary effusions. In the case of unexpected CHF, echocardiographic imaging may allow the clinician to differentiate CHF from other causes of pulmonary infiltrates or pleural effusions by screening for the presence of the atrial enlargement that supports a diagnosis of CHF.

ECG and Blood Pressure Assessment

These two cardiac assessment modalities share a common characteristic - each is the **only** way to assess "their" type of abnormality. An arrhythmia may be suspected based on clinical examination, but can only be diagnosed with an ECG. In the case of systemic hypertension, if severe clinical signs (typically ocular signs of retinal detachment or hyphema), or a suggestive clinical course (CHF develops after fluid therapy) are present, systemic hypertension may be strongly suspected but cannot be quantified or followed during therapy without BP assessment.

Situation 1: The Patient Doesn't Tolerate Fluid Administration

Patients that develop signs of CHF after fluid administration typically have a systemic abnormality with cardiac repercussions (e.g., systemic hypertension, hyperthyroidism), have primary cardiac abnormalities that may have been underestimated, or develop arrhythmias that limit the heart's ability to deal with additional volume. Feline patients with renal disease or canine patients with serious systemic illnesses like pancreatitis may develop CHF after fluid administration for anesthetic events or fluid resuscitation. Prior to fluid administration, especially in patients with known systemic disease or those middle-aged or older, careful auscultation should be performed, with special attention paid to all patients with preexisting arrhythmias or murmurs, and cats with gallop rhythms (which may indicate increased cardiac stiffness). Baseline cardiac testing (which may include radiographs, echocardiography, ECG, and BP) may be chosen on an individual basis to investigate the degree of cardiac compromise that may be present prior to fluid administration. All cats with known renal disease should have a BP measured prior to fluid therapy. If the fluid administration has already occurred and the patient has signs of CHF, the first priority is stabilization (typically with oxygen, furosemide, amlodipine +/- pimobendan therapy). When the patient is stable, the underlying disease or cardiac abnormality can be investigated and addressed. In patients with no preexisting CHF, resolution of the CHF incident and restabilization of underlying disease may allow discontinuation of some (i.e., furosemide) or all cardiac medications.

Situation 2: The Patient Looks Like CHF, But May Not Be....

The presence of known cardiac abnormalities predisposes the clinician to assume that any clinical signs similar to those of CHF likely represent CHF (think horses, not zebras). However, older patients often bring more to the table than disease of a single body system, or can develop non-cardiac complications when being treated acutely or chronically for heart disease or CHF. The two-step process that allows diagnosis of these problems involves both recognizing that a second disease is possible and allowing the diagnostic testing to show you the problem (i.e., keeping an open mind when interpreting test results). This is particularly important when a patient is generally following an expected clinical course, then seems to lag in recovery or develop new signs despite stable treatment. The general timeline for recovery from CHF signs when treated with IV furosemide is typically 2–3 days. Patients that take longer may have arrhythmias (e.g., atrial fibrillation) preventing full recovery, may have developed another problem (e.g., pulmonary hypertension or pneumonia). Detection and diagnosis of these problems depends on close physical examination for new or changing findings and confirmation with appropriate testing.

Situation 3: We Have a Diagnosis, But This Finding Just Doesn't Fit....

Not all patients have just one disease, and this is especially true of middle-aged to older cardiac patients, who may be juggling several related or unrelated disease conditions. During the initial evaluation for cardiac abnormalities, the diagnosis may appear to be obvious (e.g., arrhythmia, valve disease), but careful scrutiny of the laboratory test findings may reveal blood chemistry abnormalities (e.g., hypercholesterolemia), urinalysis abnormalities (e.g., proteinuria) or CBC abnormalities (e.g., anemia) that may have an impact on therapeutic choices and signal the need for additional testing. Similarly, when there are unexpected radiographic findings (e.g., organ displacement) or echocardiographic findings (e.g., abnormal muscle function or mass lesions) that are not consistent with the primary diagnosis, further imaging and evaluation may be required. Commonly available imaging modalities are often all that is needed to identify unusual findings or complications, but additional, specific testing (e.g., blood cultures, titers, biopsies or aspirates) may be required to finalize a diagnosis.

Summary

A good clinical history, physical examination, cardiac and systemic testing (often including laboratory assessment) will help the clinician differentiate common from uncommon causes of CHF or identify unusual developments. Therapy is targeted based on (often temporary) stabilization of CHF signs while systemic diseases are addressed, or direct therapy of non-cardiac diseases (e.g., mass lesions, pericardial lesions), which may have surgical or chemotherapeutic cures.

Identifying unusual causes or clinical courses of CHF in dogs and cats involves recognizing what is "usual" (i.e., known presence of heart disease prior to development of CHF signs or expected response to therapeutic interventions), so that the unusual can be suspected early. Judicious choices of cardiac and laboratory testing and thoughtful evaluation of results lead to successful therapeutic outcomes.

References

1. Sampedrano C, *et al. J Vet Intern Med.* 2006;20(5):1106.
2. Meurs KM, *et al. J Am Vet Med Assoc.* 1998;213:497.
3. Gerritsen RJ, *et al. Vet Q.* 1996;18(2):49.

acvim154

The Great Pretender: Pulmonary Hypertension - Unusual Cases and Complicated Clinical Courses in Pulmonary Hypertension Patients

Heidi B. Kellihan, DVM, DACVIM (Cardiology); Rebecca L. Stepien, DVM, DACVIM (Cardiology)
Madison, WI, USA

INTRODUCTION

In veterinary medicine, pulmonary hypertension (PH) is typically a sequelae of a primary underlying disease process (i.e., left-sided heart disease, heartworm disease, pulmonary disease, congenital heart disease, protein-losing kidney or gastrointestinal disease resulting in thromboembolic disease). These patients are often smaller-breeds and middle- to older-aged. This distribution can represent the population of dogs that acquire myxomatous mitral valve disease and chronic pulmonary disease. In addition, these dogs can have more than one disease process occurring at the same time, complicating the clinical picture even more.

Clinical Approach Overview

The clinical approach to PH is based on recognizing that PH can be the cause of clinical signs, diagnosing PH, identifying the etiology of PH and determining an optimal treatment plan. **Key to recognizing unusual circumstances is realizing that they may exist, and keeping an open mind when interpreting the results of diagnostic testing.**

History

Pulmonary hypertension can be present in many dogs with underlying diseases that result in PH, yet typically the dogs with moderate and severe PH are displaying clinical signs associated with this phenomenon. The common clinical picture often reads like this: a small-breed, older dog is collapsing after excitement. The history of a dog with clinically significant PH is important because the collapse episodes may be described by different terms, such as: seizures, slipping, wobbliness. The timing of the collapse episodes are also critical since collapse from PH typically occurs after excitement or exercise. This is in contrast to the dog having a seizure, at night while sleeping or the dog that has intermittent sinus arrest and is syncopal while not doing anything.

Signalment

There are some more "classic" signalments for dogs that present with clinically significant PH. They are the small-breed, older dogs that develop myxomatous mitral valve disease and congestive heart failure. They are the older, terrier-breeds that develop interstitial pulmonary fibrosis (IPF). They are the young dogs in the south or rescue dogs from the south that have heartworm disease.

Clinical Signs

Clinical signs associated with PH are non-specific and typically represent signs attributable to the underlying disease. Cough, respiratory distress, exercise intolerance, and lethargy are common clinical signs seen in dogs with PH and could be secondary to: left atrial enlargement, pulmonary edema, IPF, pulmonary thromboembolism, heartworm disease, or congenital reversed shunts. The presence of collapse after excitement or exercise, or the presence of ascites in dogs with the above clinical signs is helpful in leading the clinician to a suspicion of PH.

Physical Examination

The physical examination of a dog with PH is again, often non-specific and abnormalities are usually attributed to the underlying disease process. Mitral and tricuspid murmurs are often present, there may be a split secondary heart sound, and pulmonary crackles are frequent.

Thoracic Radiography

Thoracic radiography is an important imaging modality in patients with clinical signs associated with PH. Thoracic radiographs allow the clinician to evaluate the size and shape of the heart as well as to evaluate the interaction between the heart and the airways, pulmonary parenchyma, and pulmonary vasculature. Thoracic radiography can help determine the **etiology** of PH, such as: left-sided heart disease, pulmonary neoplasia, congenital heart disease with a ductus bump in the descending aorta, and IPF in a dog with pulmonary crackles and no pulmonary infiltrates. Thoracic radiography can **support** a diagnosis of PH when there is: right heart enlargement, main pulmonary artery enlargement, enlarged branch pulmonary arteries, and tortuous and blunted pulmonary arteries.

Echocardiography

Echocardiography is crucial in the diagnosis of PH. Doppler echocardiography and the interrogation of tricuspid regurgitation is the only way to non-invasively estimate the systolic pulmonary artery pressure. There are many other echocardiographic assessments that can be used to support a diagnosis and estimate the severity of PH. Echocardiography is also helpful in determining if left-sided heart disease or congenital heart disease is potentially the etiology of the PH.

ECG and Other Diagnostic Tools

Electrocardiography is helpful in the workup of patients with clinical signs associated with PH because arrhythmias should be excluded as contributory to clinical signs such as syncope. An ECG can also show evidence of right-heart enlargement with a right axis shift, which could be suggestive of cardiac enlargement seen with PH. A cardiac troponin I can also be used to support a diagnosis of clinically significant PH if there is a high-suspicion of PH in the absence of left-heart disease and the absence of tricuspid regurgitation.

SITUATION 1: THE PATIENT DOES NOT FIT THE PROFILE

As previously described, there is a "typical" profile of patients that are presenting with PH. Sometimes they do not follow these rules. The history may be unclear and ambiguous. The signalment may also be misleading. Large breed dogs and young dogs can present with clinically significant PH. Ultimately, PH should always be on a clinician's radar and therapy based solely on clinical signs that could be consistent with heart disease and arrhythmias could result in worsening clinical signs and deleterious results.

SITUATION 2: STABLE MITRAL VALVE DISEASE THAT DEVELOPS ASCITES

Pulmonary hypertension secondary to myxomatous mitral valve disease is a common finding. Ultimately, the severity of PH will determine how the patient responds, clinically. Pulmonary hypertension secondary to elevated left atrial and pulmonary venous pressures will develop as a compensatory mechanism to keep forward blood flow through the pulmonary vasculature. Disproportionate PH can develop in patients with left-heart disease and this occurs secondary to remodeling of the pulmonary arterial vasculature. Unfortunately, many of our veterinary patients have more than one disease that could contribute to the development of PH. A dog with mitral valve disease may also have IPF or protein-losing kidney disease that complicate the picture by elevating the pulmonary artery pressure above what would compensatorily be need in a dog with high left atrial pressures. Attempts to tease out the underlying cause of PH will help guide therapy and potentially prognostication.

SITUATION 3: DOESN'T PULMONARY EDEMA HAVE TO MEAN HEART FAILURE?

As dogs with PH are often complicated at baseline, the presence of pulmonary edema can add another piece of confusion. Non-cardiogenic pulmonary edema can result from PH directly. These dogs are, again, typically small-breed, older dogs with murmurs, cough, respiratory distress, and collapse. They present with diffuse, patchy alveolar infiltrates. These pulmonary infiltrates are thought to be a direct result of PH where there are regions of high pulmonary blood flow (no vasoconstriction) and low pulmonary blood flow (vasoconstriction), resulting in leakage of fluid out of the capillary beds. These pulmonary infiltrates (non-cardiogenic edema) are a direct result of the PH and treatment of PH can lead to the resolution of the pulmonary edema. Again, these are very difficult clinical decisions to make in a patient that could be presenting for other causes of these clinical signs. Collapse and the distribution of pulmonary edema are helpful hints in suspecting PH in this clinical scenario.

SUMMARY

A good clinical history and physical examination in combination with thoracic radiography and echocardiography can identify patients with PH and subsequently leading to the most appropriate treatment and prognostication. While much has been learned about PH in veterinary medicine, the key is to recognize that PH may be contributing to abnormal clinical signs.

REFERENCES

1. Kellum HB, *et al. J Vet Intern Med.* 2007;21(6):1258.
2. Kellihan HB, *et al. VCNA.* 2010;40(4):623.
3. Kellihan HB, *et al. JVC.* 2012;14(1):149.
4. Kellihan HB, *et al. JVC.* 2011;13(3):171.
5. Kellihan HB, *et al. JVC.* 2015;17(3):152.

acviml56

Pathophysiology of Congestive Heart Failure - Beyond the Heart

Yolanda Martínez Pereira, LdaVet, CertVC, DECVIM-CA (Cardiology), MRCVS
Roslin, Scotland, UK

Pathophysiology of Heart Failure, From Pump Failure to Neurohormonal Theories

Heart failure (HF) is a clinical, progressive syndrome caused by the inability of the cardiovascular system to meet the physiological demands of the body, despite adequate ventricular filling.

Historically, HF was explained using a simple cardiorenal model,[1] based on the clinical observation of fluid retention, causing the clinical signs of HF. This model of HF was predominant during the 1940s and until the 1960s. At this time, the concepts of "forward failure" and "backward failure" were proposed, where the assumption was that an insult to the heart would impair its ability to either "pump" blood to the kidney (therefore interfering with the renal excretion of sodium and water), or its ability to "receive" the blood returning from the veins (therefore leading to increased ventricular filling pressures and altering hydrostatic pressures within the microcirculation).

During the 1960s–1980s, the chronic nature of HF was recognised. At this time, haemodynamic measurements were available and confirmed the presence of a reduced cardiac output and excessive peripheral vasoconstriction, leading to the cardiocirculatory or haemodynamic model of HF.[2] Based on this model, the peripheral vasoconstriction would increase afterload leading to increased ventricular wall stress and therefore decreasing the systolic function. In addition, peripheral vasoconstriction would decrease perfusion to organs such as the kidneys (leading to decreased renal blood flow and therefore causing sodium retention) and the skeletal muscle (leading to decreased exercise tolerance).

It was only from the 1980s to the 1990s that clinicians recognised the detrimental effects of HF in survival time. Further advances in the understanding of the progressive nature of HF led to the development of the neurohormonal model.[3] Based on this model, an initial damage to the heart triggered the activation of compensatory systems that exerted a deleterious effect on the circulation.

Unfortunately, the neurohormonal model of HF does not completely address the gradual progression and eventual resistance to treatment observed in HF patients. It is recognised that alterations in myocyte biology, myocardial changes and changes in left ventricular chamber geometry (dilatation, increased sphericity) occur during HF.[3] These new insights into the pathophysiology of HF has led to the introduction of the biomechanical model,[3] in which HF develops and progresses as a result of the chronic neurohormonal activation, that leads to alterations in cardiac function and cardiac remodelling.

The Neurohumoral Responses in HF

Progression from Compensatory Cardiac Remodelling to Failure

The pathophysiology of the neurohormonal activation in HF starts with an initial cardiac insult of some sort. This would include conditions such as those that lead to volume overload (e.g., myxomatous mitral valve disease), pressure overload (e.g., subaortic stenosis), impaired left ventricular filling (e.g., hypertrophic cardiomyopathy) and primary myocardial disease (e.g., dilated cardiomyopathy). Cardiac compensatory mechanisms occur that aim at restoring the left ventricular wall stress, either by concentric or eccentric hypertrophy. The molecular mechanisms that lead to compensatory hypertrophy are complex and beyond the scope of this lecture, but involve the cytoskeleton sensing the abnormal load and triggering local angiotensin II production that will activate growth pathways (such as mitogen-activated protein kinase, MAPK; nuclear factor κβ; Janus kinase-signal transducer and activator or transcription, JAK-STAT), increased intracellular calcium entry, which will also stimulate growth via calcineurin, and activation of cytokine pathways involving tumour necrosis factor-α (TNF-α). Unfortunately, these compensatory mechanisms often fail to normalise the left ventricular wall stress and can trigger undesirable effects such as fibrosis, matrix remodelling, apoptosis/necrosis, and abnormalities of the calcium cycling. These mechanisms further exacerbate the left ventricular function, precipitating the syndrome of HF.

Activation of the Sympathetic Nervous System in HF

It is believed that sympathetic activation results from relative hypotension caused by impaired myocardial function, which activates the baroreceptor reflex. Stretch receptors located in the aortic arch and carotid bodies, carotid and peripheral chemoreceptors and low-pressure cardiopulmonary stretch receptors send afferent signals (via the vagus and the glossopharyngeal nerves) to the *tractus solitarius* in the brain stem. This reflex stimulates the vasomotor centre, which mediates an efferent neuronal sympathetic response and the liberation of catecholamines to the blood stream, alongside inhibition of the parasympathetic nervous signals to the heart. The cardiovascular effects include heart rate acceleration, increase in cardiac contractility, reduction of venous capacitance and constriction of resistance vessels. In HF, it has been demonstrated that a dysfunction of the autonomic nervous activation exists[4,5] with a reduced sensitivity of various sympatoinhibitory reflexes (arterial baroreflex and cardiopulmonary reflexes) combined with additional modulation of the sympathetic response by other neurohormonal components (angiotensin II, nitric oxide, reactive oxygen species, vasopressin, endothelin, aldosterone, etc.) and pro-inflammatory cytokines. The sympathetic activation also mediates renal renin release, further contributing to the positive feedback between the sympathetic and the renin-angiotensin-aldosterone system (RAAS).

Chronic stimulation of the sympathetic system in HF leads to down-regulation of β1-adrenoreceptors, and alterations in the cardiac beta-adrenergic signal transduction. In addition, the chronic effects of sympathetic stimulation of the failing heart can lead to calcium overload, increased oxygen wastage and contribute further to the progression of disease.

Activation of the Renin-Angiotensin-Aldosterone System (RAAS) in HF

The RAAS, together with the sympathetic nervous system and arginine vasopressin, is a defense mechanism designed to preserve arterial volume and circulatory homeostasis during periods of low cardiac output. However, in HF the activation is sustained and chronic, contributing to the progressive ventricular remodelling and worsening HF. The trigger for renin release from the yuxtaglomerulal apparatus in the macula densa in the kidney is a decrease in renal perfusion pressure, reduced sodium absorption by the distal renal tubules and sympathetic or vasopressin activation. Renin mediates the conversion of the precursor angiotensinogen to angiotensin I, which in turn is converted to angiotensin II by the angiotensin-converting enzyme (ACE). Angiotensin II causes vasoconstriction and mediates stimulation of aldosterone from the adrenal cortex, which increases sodium ion reabsorption by the distal renal tubules. Angiotensin II also stimulates the thirst centre. Vasopressin release by the pituitary gland is stimulated by osmoreceptors in the hypothalamus (in response to increase osmolality of the extracellular fluid component) and by baroreceptors (left ventricle, aortic arch and carotid artery, renal afferent arteriole). Vasopressin mediates vasoconstriction and myocardial hypertrophy (receptors V1) and free-water reabsorption in the renal collecting duct (receptors V2). It is believed that in HF, excessive vasopressin release occurs in response to low arterial pressure and not by osmotic stimuli.[6]

The chronic, sustained activation of the RAAS and vasopressin lead to excessive fluid retention and eventually precipitates oedema formation. Other neurohormones that contribute to the progressive deterioration of cardiac function are endothelin, a vasoconstrictor that can exert toxic effects in the heart, and cytokines such as TNF-α, which also promotes cardiac remodelling. Compensatory neurohormones such as atrial and brain natriuretic peptides exert vasodilatory and diuretic actions and inhibit aldosterone secretion; the effects, however, are overcome by the effects of the RAAS.

Effects of HF on Other Body Systems

The Cardiorenal Syndrome

The interrelationship between the heart and the kidney depends on the RAAS, the sympathetic nervous system and the atrial natriuretic peptides.[7] Haemodynamic changes in the heart affect the renal function and vice-versa. Whereas the kidneys help to maintain fluid and electrolyte homeostasis under normal condition, they contribute to progression and worsening in HF. In addition, renal venous congestion combined with poor perfusion can contribute to deterioration of the renal function. Vascular complications can arise as a result of systemic hypertension, as well as anaemia associated to a reduced production of renal erythropoietin. In humans, abnormal calcium and phosphorus metabolism can contribute to the acceleration of coronary atherosclerosis, and azotaemia can contribute to the progression of cardiomyopathy.[7] A genetic predisposition and presence of underlying diseases (diabetes, obesity) have been proposed in humans as a risk factor for cardiorenal syndrome.

Interactions Between RAAS and Parathyroid Hormone in HF

Parathyroid hormone (PTH) is involved in calcium homeostasis. Its production by the parathyroid glands is induced by a decrease in the extracellular concentration of calcium in the extracellular fluid and it mediates calcium absorption in the small intestines (mediated by vitamin D), renal reabsorption and bone resorption. There is evidence from human studies that increased PTH levels in patients with HF is an independent predictor of all-cause and cardiovascular mortality.[8] It is believed that in patients with advanced HF, urinary and fecal excretion of calcium induced by hyperaldosteronism, together with calcium excretion caused by high dosages of loop diuretics, results in secondary hyperparathyroidism. Beyond the calcium homeostasis function, PTH receptors have been identified within the cardiovascular system[9] (cardiomyocytes, vascular smooth muscle and endothelial cells). Equally, mineralocorticoid receptors have been identified in non-epithelial cells (vascular smooth muscle and cardiomyocytes) and in the nuclei of parathyroid cells, which can be activated by aldosterone. It is believed that PTH stimulates aldosterone secretion by direct and indirect methods[9] and therefore the interplay between aldosterone and PTH can contribute to the progression of HF.

In our institution, we analysed PTH and aldosterone levels in a group of 15 control dogs, 12 dogs with asymptomatic heart disease (ISACHC Ib) and 16 dogs with advanced HF (ISACHC III). PTH levels were increased in dogs with heart disease compared to control dogs, and higher PTH levels were found in patients with advance HF, compared to asymptomatic heart disease. Furthermore, PTH levels showed a moderate, positive correlation with aldosterone and a moderate, negative correlation with ionised calcium. These results suggest that secondary hyperparathyroidism may occur in association to hyperaldosteronism in canine patients with heart disease.

Vitamin D deficiency is associated with heart failure events in human medicine[10] and probably in dogs[11]. Supplementation with Vitamin D down-regulates renin-angiotensin-aldosterone system hormones, and leads to reduced blood pressure and myocardial hypertrophy in experimental models.[12] It is therefore plausible to hypothesize that supplementation with Vitamin D in dogs with CHF may decrease hyperaldosteronism and therefore hyperparathyroidism, although clinical studies in this field in dogs are lacking.

Heart-Gut Interactions in HF

In humans, it is believed that gut-heart interactions are contributing to the pro-inflammatory status found in HF through morphological and functional alterations leading to bacterial translocation.[13] Studies have demonstrated increased bacterial endotoxin concentration,[14] altered intestinal permeability associated to increased intestinal thickness and augmented bacterial biofilm,[15] reduced active carrier-mediated intestinal transport and increased lipopolysaccharide, which related to inflammation[16] and reduced intestinal blood flow in patients with congestive heart failure[13]. It is believed that the intestinal dysfunction contributes to the development of cardiac cachexia, a well-recognised predictor of poor survival.[13] In canine cardiac patients, cardiac cachexia has also been associated with decreased survival and perception of poor quality of live by owners, contributing to the decision of euthanasia.[17] Omega-3 fatty acids are recommended for their anti-inflammatory and antiarrhythmic properties, although longitudinal clinical studies assessing its beneficial effects are lacking.

References

1. Packer M. *Am J Cardiol.* 1993;71:3C.
2. Mann DL. *Circ.* 1999;100:999.
3. Mann DL, *et al. Circ.* 2005;111:2837.
4. Kishi TJ. *Card.* 2012;59:117.
5. Floras JS. *J Am Col Card.* 2009;54:375.
6. Chatterjee K. *Am J Cardiol.* 2005;95:8B.

7. Boudoulas KD, *et al. Prog Cardiovasc Dis*. 2017;32:203.
8. Zhang S, *et al. Arch Gerontol Geriat*. 2015;60:359.
9. Tomaschitz A, *et al. Metab Clin Exp*. 2014;63:20.
10. Boxer RS, *et al. J Card Fail*. 2014;20:334.
11. Kraus MS, *et al. J Vet Int Med*. 2014;28:109.
12. Li YC, *et al. J Clin Inv*. 2003;88:327.
13. Sandek A, *et al. J Am Col Cardiol*. 2014;64:1092.
14. Niebauer J, *et al. Lancet*. 1999;353:1838.
15. Sandek A, *et al. J Am Col Cardiol*. 2007;50:1561.
16. Sandek A, *et al. Int J Cardiol*. 2012:157:80.
17. Freeman LM, *et al*. 2012;26:3.

acvim157

Syncope: A Review

Fiona E. Meyers, BVSc (Hons), PhD, MANZCVS, DACVIM
Queensland, Australia

Syncope is defined as sudden transient loss of consciousness and postural tone with spontaneous and complete recovery that occurs as a result of reduced cerebral perfusion. Presyncope is the condition described in human patients where cerebral hypoperfusion produces signs of episodic dizziness, weakness, or light-headedness without loss of consciousness. In cats and dogs, this may be seen as transient weakness or appendicular instability without loss of consciousness.

Syncope is a distressing clinical event for most pet owners and their concern is not misplaced. While some syncopal etiologies, such as reflex syncope, are very benign, others such as ventricular tachycardia are malignant; failure of sustained ventricular tachycardia to spontaneously revert to sinus rhythm will precipitate sudden death.

Assessment of the syncopal patient presents a diagnostic challenge for the veterinarian since recovery is typically quick and complete prior to presentation and physical examination may reveal no abnormality.

Syncopal Pathophysiology & Etiology

Cessation of cerebral blood flow for 6–8 seconds producing a drop in cerebral arterial pressure to less than 25 mm Hg is necessary to provoke a complete loss of consciousness.[1] Reduced cerebral blood flow may be a result of reduced cardiac output or reduced peripheral vascular resistance or a variable combination of both.

In human patients, syncope is classified by pathophysiology into three categories: reflex (neutrally-mediated) syncope, orthostatic syncope, and cardiovascular syncope.[1] Reflex syncope is subdivided into the categories of vasovagal (i.e., mediated by emotional distress or orthostatic stress), situational (i.e., cough, micturition, defecation, exercise, others) and carotid sinus syncope. Orthostatic syncope describes the positional hypotension exacerbated by primary autonomic failure (e.g., Parkinson's disease), secondary autonomic failure (e.g., diabetes), drugs (e.g., vasodilators, diuretics) or volume depletion (with hemorrhage, diarrhoea, vomiting). Cardiovascular syncope includes syncope elicited by cardiovascular pathology including arrhythmias (i.e., tachycardia, bradycardia), acquired structural cardiac disease (e.g., cardiomyopathies, cardiac masses, pulmonary thromboembolus, pulmonary hypertension, pericardial disease, others) and congenital structural disease (e.g., subaortic/pulmonic stenosis, mitral stenosis, others).

Reflex Syncope

This form of syncope refers to a heterogenous group of conditions in which normal physiologic reflexes aimed at preservation of cardiac output and tissue perfusion are executed inappropriately in response to a specific trigger to precipitate vasodilation and bradycardia sufficient to compromise cerebral perfusion and result in loss of consciousness. The afferent limb of the reflex is poorly understood, but essentially, the nucleus tractus solitarii of the brainstem is either directly or indirectly stimulated by the inciting trigger. The efferent limb of the reflex is characterized by augmented parasympathetic tone (producing a cardioinhibitory response of negative chronotropy and inotropy) and withdrawal of sympathetic tone (producing a vasodepressor response of peripheral vasodilation). Either the cardioinhibitory or the vasodepressor response may predominate the efferent limb of the reflex, but often there a variable contribution of these two mechanisms. For veterinary patients, triggers for reflex syncope include exercise/excitement,[2,3] cough, micturition, defecation and advanced myxomatous mitral valve degeneration.

Cardiovascular Syncope

Congenital (e.g., subaortic stenosis and pulmonic stenosis) and acquired (e.g., pulmonary hypertension, pulmonary thromboembolism, cardiac neoplasia) cardiovascular diseases that increase the pressure against which ventricular contraction occurs are associated with risk for syncope. When sympathetically stimulated, augmented inotropy and chronotropy provoke an abrupt rise in intraventricular pressure stimulating ventricular mechanoreceptors to initiate vagal afferent pathways to the brainstem where the efferent response is increased vagal tone and withdrawal of sympathetic activity to effect a reduction in cardiac output coupled with peripheral vasodilation which can precipitously drop cerebral perfusion pressure to trigger syncope.

Primary cardiomyopathic diseases and cardiac diseases with secondary cardiac remodelling (e.g., subaortic stenosis) histologically feature necrosis with replacement fibrous or fibrous/fibro-fatty infiltrates which provide a known substrate for ventricular arrhythmias. Rapid and sustained supraventricular tachycardia, sufficient to provoke syncope, can occur with congenital accessory pathways and may also arise secondary to atrial remodelling or primary myopathic processes involving the atria. The abrupt reduction in diastolic filling associated with sustained supraventricular or ventricular tachycardia in these patients can confer a sufficient reduction in cardiac output, particularly if coupled with systolic myocardial dysfunction, to provoke syncope.

Conduction system diseases (e.g., high grade AV block and sick sinus syndrome) can precipitate syncope when a prolonged sinus pause/arrest is not supported by cardiac output via subsidiary pacemakers in a diseased conduction system. For high grade AV block, syncope can result if the escape focus is unreliable resulting in period of complete ventricular quiescence of if a patient's activity level cannot be matched by the rate-limited ventricular output.

Diagnostic Tests

Minimum Screening Tests

A complete blood count, biochemistry, and urinalysis is likely to be low-yield in most dogs and cats with syncope but is necessary to rule out metabolic causes of collapse (e.g., seizures associated with hypoglycaemia, hepatic insufficiency) and concurrent or precipitating disease (e.g., protein loosing nephropathy and low ATIII levels may theoretically increase risk of pulmonary thromboembolism).

Echocardiography

Echocardiography is indicated for any patient with syncope to identify structural or mechanical heart disease which is either primarily implicated (e.g., outflow tract obstruction, pulmonary hypertension) or associated (e.g., dilated cardiomyopathy) with syncopal etiology. It is also necessary to indirectly measure pulmonary arterial pressures via determination of tricuspid and/or pulmonic regurgitant velocity to identify pulmonary hypertension.

Electrocardiography

A resting ECG is incapable of definitively identifying a causative arrhythmia unless it is recorded at the time of the syncopal event. However, a resting ECG may provide supportive evidence for arrhythmia-associated syncope in some dogs and cats. For example, the identification of pre-excitation (short PR interval and delta waves) on sinus complexes is indicative of a congenital accessory pathway with risk for orthodromic AV reciprocating tachycardia. Similarly, the identification of ventricular premature complexes of left bundle branch block morphology on a 2-minute resting ECG of a Boxer is poorly sensitive but highly specific for a diagnosis of arrhythmogenic right ventricular cardiomyopathy.[4]

Ambulatory 24-hour Holter ECG

Ambulatory 24-hour Holter ECG monitoring is indicated whenever an arrhythmogenic cause for syncope is suspected, to screen for occult dilated cardiomyopathy or arrhythmogenic right ventricular cardiomyopathy in at-risk dogs, and to confirm efficacy of antiarrhythmic medication following commencement. It has a far greater sensitivity at detection of arrhythmias than the resting ECG which records <0.1% of the patient's daily cardiac electrical activity and which may miss intermittent or transient arrhythmias.

The presence of >50–100 VPCs during a 24-hour Holter ECG recording is diagnostic for occult dilated cardiomyopathy[5] or arrhythmogenic right ventricular cardiomyopathy[6]. Documentation of cardiac rhythm during a syncopal event is unnecessary and 24-hour Holter ECG diagnosis of occult dilated cardiomyopathy or arrhythmogenic right ventricular cardiomyopathy is typically definitive for syncopal etiology.

Event Recorders

Event recorders are portable ECG recording devices that record a 10–15 minute loop of ECG by continually over-writing previously acquired data. When the device is manually triggered (by the owner upon witnessing the collapse event), the ECG from a few minutes prior to the trigger, and a few minutes after the trigger, is transferred from the buffer into memory, and can be subsequently downloaded and analyzed.

Event recorders are used to assess the cardiac electrical activity at the time of a witnessed event to identify an arrhythmogenic syncopal etiology. They are not used for patients suspected of having potentially fatal collapse etiology, such as a Boxer with suspected arrhythmogenic right ventricular cardiomyopathy where syncopal events may precipitate sudden death. Instead an event recorder is valuable at excluding an arrhythmogenic etiology in patients with collapse or identifying an arrhythmogenic cause in patients with disease unlikely to confer a risk of sudden death (e.g., sinus pauses/profound bradycardias in patients with cardioinhibitory reflex syncope or sick sinus syndrome).[7]

Event monitors require a patient to experience an event while the monitor is worn. Typically, event monitors are attached for 7 days and require that collapse frequency to be sufficient to facilitate capture of an event during this period.

Implantable Loop Recorders

Implanted subcutaneously over the left hemithorax, these recording devices have a battery life of up to 36 months and have recently been validated for use in canine patients.[8,9] These devices have a solid state loop memory that stores retrospective ECG recordings, when activated by an owner witnessing a syncopal episode or automatically activated if predefined arrhythmias are recognized. Although cost may be prohibitive for some patients, this technology may facilitate accurate arrhythmia etiology for a subset of patients where more routine diagnostics have failed to identify an elusive arrhythmia or in patients with infrequent and unpredictably inducible reflex-syncope.

Tilt Table Testing

For human patients with presumed vasovagal syncope, tilt table testing may facilitate definitive diagnosis. Essentially, tilt table testing aims to provoke syncope during simultaneous documentation of heart rate and blood pressure to demonstrate a vasodepressor and/or cardioinhibitory response as the syncopal etiology. Patients are positioned supine on the table and tilted upright to 60–80 degrees for 20–45 minutes. Various pharmacological agents may be employed to increase test sensitivity but obviously do so at the expense of reduced specificity. Clearly, this diagnostic methodology is not applicable to veterinary patients, where reflex-syncope may remain a diagnosis of exclusion, particularly for patients with a predominant vasodepressor form where profound bradycardias need not contribute to syncope and definitive diagnosis is not possible even for patients with ambulatory ECG recording during an event.

TREATMENT

Ventricular Tachycardia

Large multi-centre placebo-controlled, double-blinded trials in human cardiology (notably the CAST trials) have shown that although antiarrhythmic drugs can suppress ventricular arrhythmias, this does not correspond to a reduction in risk for sudden cardiac death.[10] Several studies in canines have similarly documented the efficacy of antiarrhythmics to minimize frequency and severity of ventricular arrhythmias but clinical translation to reduced risk for syncope and sudden death is tenuous.[11]

Several antiarrhythmics have been used in veterinary patients including: mexiletine, flecainide, procainamide, atenolol, propanolol, tocainide, sotalol, amiodarone.

Combination therapy with sotalol and mexiletine (but not either agent alone) has demonstrated efficacy at suppression of arrhythmias in German shepherd dogs with inherited ventricular arrhythmias.[12] In Boxers with arrhythmogenic right ventricular cardiomyopathy, sotalol and combined mexiletine/atenolol (but not atenolol or procainamide) effectively reduced the frequency and severity of ventricular arrhythmias without altering the frequency of syncopal events.[13] Similarly, sotalol and combination sotalol/mexiletine therapy has demonstrated efficacy at reduction of arrhythmia frequency and grade in Boxers with arrhythmogenic right ventricular cardiomyopathy, however, a commensurate reduction in syncopal frequency and reduction in risk for arrhythmogenic sudden death has not be documented.[14]

Supraventricular Tachycardia
Supraventricular tachycardia sufficient to result in syncope, is often associated with a congenital accessory pathway and rapid rate orthodromic AV reciprocating tachycardia (OAVRT). For these patients, Class IA, IC, and Class III agents are typically indicated to retard the conduction properties of the accessory pathway and reduce the propensity for OAVRT. Radiofrequency ablation is a curative option for these patients.[15,16]

Bradycardias
Positive chronotropes such as sympathomimetics (e.g., terbutaline, phenylpropanolamine), anticholinergics, and PDEII-inhibitors (e.g., theophylline) have been utilized in syncopal patients with bradycardic conduction system disease. However, response to medical therapy is typically partial and temporary at best and permanent pacemaker implantation is typically necessary. Permanent pacing is curative for these patients and various pacing modalities are well-described in the veterinary literature.

Outflow Tract Obstructions
Beta-blockers are widely administered to patients with severe subaortic stenosis with aims to attenuate sympathetically-stimulated rise in left ventricular pressure against the fixed outflow obstruction and minimize risk for syncope/sudden death. However, available data suggests that beta-blockade does not affect survival time of severely affected individuals.[17] Surgical excision of the obstructive lesion and balloon valvuloplasty have similarly failed to extend survival times when compared to long-term atenolol treatment, even when the obstruction is significantly attenuated.[18,19] Combined cutting balloon and high pressure balloon valvuloplasty has recently been described for dogs with subaortic stenosis but it's clinical merit is yet to be determined.[20]

Conversely, the treatment of choice for syncopal patients with severe valvular pulmonic stenosis is balloon valvuloplasty which has been demonstrated to reduce minimize clinical signs and extend survival time when compared to untreated dogs.[21,22]

Reflex Syncope
Effective treatment of reflex syncope is predominantly dependent on the avoidance of known triggers for an individual. For affected dogs, this may mean avoiding abrupt onset or extremes of activity, particularly in high ambient temperatures where peripheral vasodilation may increase risk for an event. Medical cough suppression may also be sufficient to attenuate syncopal frequency in patients with cough-syncope.

For human patients when trigger-avoidance is not possible, there is no convincing data to support the use of one therapy over another as a first line therapy. Beta-blocker therapy was once the most commonly utilized medication but this has recently been demonstrated to be ineffective and according to the guidelines of the European Society of Cardiology, beta-blockers should no longer be used for this indication. Medications which may prove effective for affected people may include: fludrocortisone, midodrine (an alpha-agonist vasoconstrictor) or fluoxetine or paroxetine (serotonin reuptake inhibitors), but experience with these agents in veterinary patients is lacking. For people with the cardioinhibitory form of vasovagal syncope, pacemaker implantation may be effective.

References

1. Moya A, *et al. Euro Heart J.* 2009;30:2631–2671.
2. Thomason JD, *et al. J Vet Int Med.* 2008;22:931–936.
3. Calvert CA, *et al. J Vet Int Med.* 1996;10(2):88–93.
4. Meurs KM, *et al. J Am Vet Med Assoc.* 2001;218(2):222–224.
5. Calvert CA, *et al. J Am Vet Med Assoc.* 2000;216(1):34–39.

6. Meurs KM. ACVIM proceedings; 2011.
7. Bright JM, *et al. J Am Vet Med Assoc.* 2000;216:1110–1114.
8. MacKie BA, *et al. J Vet Cardiol.* 2010;12(1):25–33.
9. Santilli RA, *et al. J Am Vet Med Assoc.* 2010;236(1):78–82.
10. Echt DS, *et al. New Eng J Med.* 1991;324(12);781–788.
11. Clavert CA, *et al. J Am Anim Hosp Assoc.* 2004;40;24–28.
12. Gelzer ARM. *J Vet Cardiol.* 2010;12:93–106.
13. Meurs KM, *et al. J Am Vet Med Assoc.* 2002;221(4):522–527.
14. Prosek R, *et al.* ACVIM. 2006.
15. Palermo V. ACVIM proceedings; 2013.
16. Webber D. ACVIM proceedings; 2013.
17. Eason B. ACVIM proceedings; 2013.
18. French A, *et al. J Small Anim Pract.* 2000;41(10):451-456.
19. Monnet E, *et al. J Am Vet Med Assoc.* 1996;209(7):1255.
20. Kleeman. ACVIM. 2013.
21. Stafford-Johnson MS. *J Vet Intern Med.* 2004;18(5):656–662.
22. Locatelli C. *J Small Anim Pract.* 2013;54(9):445–452.

acvim158

Feeding for Two: Nutrition for the Cardiac Patient with Concurrent Diseases

Lisa M. Freeman, DVM, PhD, DACVN
North Grafton, MA, USA

INTRODUCTION

More and more, nutrition is becoming an integral component of medical therapy. In many diseases, including cardiac disease, nutrition can not only make the patient feel better but also can help to modulate the disease. Therefore, it is incumbent on the veterinary healthcare team to make specific nutritional recommendations for all animals with cardiac disease. Although there are diets marketed for dogs with cardiac disease (none for cats), there is not a single diet that is appropriate for **all** dogs with cardiac disease. Instead, we must view animals as individuals and to feed them appropriately according to the underlying disease **and** individual patient characteristics. Not all patients with congestive heart failure (CHF), for example, are the same or require the same diet. Some dogs with CHF will be overweight and some will have lost weight and muscle, some will be hypokalemic while others will be hyperkalemic, and a small number will be taurine deficient while most are not. Therefore, the same diet will not meet the requirements of all dogs with CHF. Cats with cardiac disease can be even more challenging because of their unique nutritional needs.

In addition to the need to individualize the diet for animals with cardiac disease, many animals with cardiac disease have concurrent diseases, which makes selection of an optimal diet even more challenging. Our research has shown that 56% of cats and 61% of dogs with cardiac disease have concurrent diseases, which complicates the process of diet selection. This talk will focus on how to optimize diets for patients with cardiac disease, as well as a concurrent disease that would benefit from nutritional modulation.

GENERAL APPROACH TO SELECTING A DIET

The process of choosing an appropriate diet for an animal with cardiac disease and concurrent medical conditions involves examining the patient, the diet, and the owner's feeding practices. There is not a single "best" diet for all animals with a particular disease. What might be the best diet for one animal may be contraindicated in another. So, it is important to assess a number of different factors to determine which diet or diets would best suit an individual patient. And when more than one disease is present, it becomes even more important to carefully consider all the issues at hand before making a nutritional recommendation.

Nutritional Assessment

To be able to make an optimal dietary recommendation, one must first perform a brief nutritional assessment, which has four key components:

1. Body weight: Use a consistent scale to assess weight at each visit and record it in the medical record to identify subtle changes or trends.
2. Body condition score (BCS): Use a consistent method to determine BCS to assess current status and changes over time (either a 9-point scale or a 5-point scale). The BCS evaluates body fat, with a goal for healthy dogs and cats of 4–5 on a 9-point scale. However, in dogs and cats with CHF, a BCS of 5–7 may be more desirable.
3. Muscle condition score (MCS): The MCS differs from the BCS in that the MCS evaluates muscle mass. Dogs and cats can be very obese and yet have severe muscle loss and conversely, cats can be thin but have normal muscle mass. Evaluation of muscle mass includes visual examination and palpation over the temporal bones, scapulae, lumbar vertebrae, and pelvic bones. Assessing MCS is important as dogs and cats with CHF lose primarily muscle, compared to a healthy animal that would lose fat. Early identification of muscle loss, at the "mild muscle loss" stage is valuable for

successful intervention. Muscle condition scoring systems have been developed for both dogs and cats (see the WSAVA Nutrition Toolkit).

4. Diet history: A complete diet history includes the pet food (commercial or home-prepared, including the exact brand, product, and flavor for commercial foods, and exact recipe for home-prepared foods), treats, table food, rawhides/bully sticks, dental products, dietary supplements, and foods used for medication administration. Tools have been developed to assist in this process (e.g., WSAVA Nutrition Toolkit). The nutritional assessment can help to determine if the diet is contributing to the underlying disease(s) or to suboptimal body weight or body composition. It also can identify what subsequent nutritional modifications will be needed:
   a. Overall nutritional balance of the diet: Unless formulated by a veterinary nutritionist, nearly all home-prepared diets are nutritionally unbalanced. Even commercial diets, if made by a small company with inadequate nutritional knowledge and/or quality control, can have nutritional deficiencies or excesses. Over-the-counter diets labeled as "for intermittent or supplemental use" will be deficient in one or more nutrients and should not be fed as a significant portion of the diet.
   b. Calories: The diet(s), treats, and other foods, may provide too many or not enough calories to maintain ideal body weight, BCS, and MCS in an individual animal.
   c. Protein: Diets vary tremendously in their protein content so the diet history can identify whether protein intake is optimized. Ensuring adequate protein intake is critical in dogs and cats with heart disease to minimize cachexia. Dietary protein also may require modification depending upon concurrent diseases.
   d. Sodium: The sodium content of the diet, treats, table food, and foods used for medication administration should be addressed for the individual animal, as all contribute to the total dietary intake of sodium.
   e. Other nutrients: Deficiencies or excesses of other nutrients also can occur, particularly if the owner is feeding an unbalanced diet, giving large amounts of table food or treats (or foods used to administer medications), or dietary supplements. In addition, various other nutrients may require modification depending on concurrent diseases (e.g., phosphorus in animals with chronic kidney disease).

Selecting Optimal Diet(s)

Patient's Nutrients of Concern

In general, the nutrients of concern in patients with cardiac disease are calories, sodium, protein, potassium, and magnesium. However, patients with cardiac disease vary tremendously in terms of their clinical signs, laboratory parameters, and food preferences and these will all affect diet selection. For example, dogs with asymptomatic cardiac disease require less severe sodium restriction than those with CHF. Obese cats require a less calorically dense diet than would a thin cat. Laboratory results (e.g., hypo- vs. hyperkalemia) also influence diet choice. A list of the nutrients of concern and the desired modification for each patient should be made.

Example: 10-year-old castrated male domestic shorthair cat with hypertrophic cardiomyopathy and CHF (and no other medical conditions). He weighs 5.2 kg, has a body condition score of 5/9, moderate muscle loss, and mild hypokalemia, with other electrolytes within the reference range.

For this cat, the nutrients of concern would be:

1. Sodium (moderate restriction)
2. Calories (adequate intake to maintain ideal body weight and body condition score)
3. Potassium (increased - if hypokalemia is present)
4. Protein (increased)

When animals have more than one medical condition, the nutrients of concern for the second (and potential other) conditions should be added to the above list. Therefore, if the same cat above also had gastrointestinal disease, the additional nutrients of concern for that cat would be:

5. Consistent diet
6. Fat (mild to moderate restriction)
7. Fiber (increased if no response to initial dietary modifications)

Owner Issues

In addition to the nutrients of concern for a patient, the owner also is an important determinant of a patient's diet. Therefore, meeting the owner's expectations in terms of diet is important for animals with cardiac disease. Their pet's quality of life is of tremendous importance to owners so providing diets that are palatable and readily eaten is critical (and this is obviously important for maintaining body weight). Be sure to address **all** components of the diet. Most dogs (and many cats) receive treats and the majority of dog owners give the medications with "people food." Dietary supplements, foods used to administer medications, rawhides, dental products, etc., also may be fed. Including this information in the overall diet plan is important to achieve success with nutritional modification. Also, cost preferences should be considered as veterinary diets may be out of the price range for long-term use by some owners; in this case, lower-priced, over-the-counter options should be offered.

The author also tries to identify "non-negotiables" for the owner (e.g., they wish to continue giving treats, they want to feed a home-cooked diet, or they have to give medications in foods). This can also help to identify diet options that will meet the owner's expectations and other areas to address, which can be helpful in terms of adherence. If appropriate diets cannot be identified that meet all the owner's expectations, this should be explicitly discussed.

Selecting the Optimal Diet(s)

Based on these parameters, a diet (and ideally, multiple diets) can be identified that meet these goals for the individual patient. In most cases, one or more commercial diets can be identified that meet all of the goals. When all the nutritional goals cannot be met with a single diet (e.g., when the goals contradict one another), the goals should be prioritized based on what medical issues are the most serious/life-threatening. In some patients, a home-cooked diet can provide more flexibility and can be used to meet the nutritional goals for animals with multiple conditions.

Make Specific Recommendations

Always make a specific recommendation for diet - this may be only to say that the diet the animal is currently eating is a good choice and that the owner is feeding an appropriate amount to keep the dog or cat in ideal body condition. It is important to reinforce good nutritional choices that owners are making, since there are so many myths about pet nutrition. The author typically determines several diets that would be appropriate for an individual patient based on the patient, diet, and feeding practices. These diets are offered as choices for the owner and for the pet. Once they determine which diet the pet prefers, this diet can be used. For animals with CHF, having multiple dietary options is particularly important since dysrexia, a cyclical or selective loss of appetite, is common. In complicated cases, especially those with concurrent diseases, consider referring the patient to a board-certified veterinary nutritionist (www.acvn.org or www.esvcn.eu/college).

Also, be sure to discuss treats, table food, rawhides, dental products, appropriate methods for administering medications, and dietary supplements. Omega-3 fatty acid supplementation may provide benefits in dogs and cats with heart disease. The omega-3 fatty acids, eicosapentaenoic acid (EPA) and docosahexaenoic acid (DHA), are normally are found in very low concentrations in cell membranes compared to the omega-6 fatty acids, but they can be increased by a food or supplement enriched in omega-3 fatty acids. Omega-3 fatty acids reduce the production of the more inflammatory eicosanoids and of the inflammatory cytokines, TNF and IL-1. The anti-inflammatory, antiarrhythmic, and antithrombotic effects of omega-3 fatty acids have many potential benefits in dogs and cats with heart disease (e.g., reduced muscle loss, improved appetite, association with longer survival; Slupe *et al*, 2008; Freeman *et al*, 2010). However, many owners of animals with cardiac disease have "researched" supplements on the internet so they may be receiving a variety of supplements with questionable safety, efficacy, and quality control.

Communicate with the Owner

Discuss the nutritional plan (for all components of the diet) and answer questions. Consider referring the owner to a board-certified veterinary nutritionist for complicated cases or when owners have a large number of questions about nutrition. Warn owners about the potential for hyporexia and dysrexia in animals with CHF. Reduced food intake can herald a change in CHF control or a need to change medications, but fluctuation in diet does happen even when CHF is well controlled. Owners often describe this as the animal preferring different foods than it used to or the animal eating one food well for a week and then not wanting it anymore (but that they'll eat a different food). It is helpful to warn the owner about appetite and to provide multiple appropriate diet options for the owner to feed (i.e., they can feed one diet for several days until the animal no longer has interest for that food and then switch to a different food. They'll often be willing to eat the original food again so owners can use a "rotation" of foods to keep the cat eating). It is important to advise the owner that if the animal also is acting lethargic, is tachypneic, won't eat any foods, or if it has been eating very well and then suddenly stops eating, this usually indicates the need for reassessment and adjustment of medications.

Concurrent Diseases

When concurrent diseases are present in patients with cardiac disease, selecting optimal diet(s) can become more challenging. The most common concurrent diseases include chronic kidney disease, feline and canine lower urinary tract disease, gastrointestinal disease, liver disease, and diabetes. Specific examples will be discussed to illustrate the process of selecting the best diet for a patient, as well as updates on nutritional management for these various concurrent diseases.

References

1. Fascetti AJ, Delaney SJ. *Applied Veterinary Clinical Nutrition.* West Sussex: Wiley-Blackwell, 2012.
2. Freeman LM. Beneficial effects of omega-3 fatty acids in cardiovascular disease. *J Small Anim Pract.* 2010;51:462–470.
3. Freeman LM. Cachexia and sarcopenia: Emerging syndromes of importance in dogs and cats. *J Vet Intern Med.* 2012;26:3–17.
4. Freeman L, Becvarova I, Cave N, *et al.* WSAVA nutritional assessment guidelines taskforce: 2011 nutritional assessment guidelines. *J Small Anim Pract.* 2011;52:385–396.
5. American College of Veterinary Nutrition. Board-certified veterinary nutritionists, many of which perform appointments and consultations. www.acvn.org In Europe, the European College of Veterinary Comparative Nutrition can be contacted: www.esvcn.eu/college.
6. Pet Nutrition Alliance. Nutritional resources for the veterinary healthcare team. http://petnutritionalliance.org/.
7. Tufts HeartSmart website (information on heart disease for pet owners, including nutrition [e.g., low sodium diets for dogs and cats, treats, omega-3 fatty acid supplements]). www.tufts.edu/vet/heartsmart.
8. Tufts Petfoodology website (information on pet food and pet nutrition for pet owners). www.petfoodology.org.
9. World Small Animal Veterinary Association Global Nutrition Committee. Muscle condition score chart, body condition score chart, diet history form, nutrition on the internet, recommendations for selecting a pet food, and other nutritional tools. www.wsava.org/nutrition-toolkit.

acvim159

Muscle Matters: Cardiac Cachexia in Dogs and Cats

Lisa M. Freeman, DVM, PhD, DACVN
North Grafton, MA, USA

INTRODUCTION

Traditionally, nutritional modifications have been considered only in dogs and cats with congestive heart failure (CHF) and few, if any nutritional recommendations have been made for animals in earlier stages of heart disease. However, if one waits to address nutrition until this later stage of disease, the benefits of optimizing nutrition during the longer, and more common, earlier stages are lost. In addition, there is some evidence for a role of diet in the development of heart disease. Therefore, attention to body composition appears to be increasingly important in both the development and progression of heart disease in dogs and cats.

CACHEXIA

Cachexia, a loss of muscle, occurs in a variety of acute and chronic diseases in people and companion animals, including cancer (cancer cachexia), heart failure (cardiac cachexia), chronic kidney disease (renal cachexia), chronic obstructive pulmonary disease, and a variety of acute illnesses and injuries seen in the intensive care unit (Freeman 2012). The weight loss that occurs in cachexia is unlike that seen in a healthy dog, cat, or person that loses weight. In a healthy animal that is receiving insufficient calories to meet requirements, metabolic adaptations allow fat to be used as the primary fuel source, thus preserving lean body mass. Conversely, the primary fuel source in animal with chronic disease is amino acids from muscle so that these animals quickly catabolize muscle and lean body mass. Therefore, loss of lean body mass is a hallmark of cachexia. Fat and bone also are lost to lesser degrees, although in advanced cachexia, depletion of all compartments is apparent. In addition to quantitative changes in muscle, there are also qualitative changes in muscle function. Loss of lean body mass and functional muscle changes have direct and deleterious effects on strength, immune function, and survival. In people, cachexia is an independent predictor of survival.

Severe weight loss in a patient with advanced CHF or cancer represents a classical picture of cachexia. However, cachexia actually begins gradually and is initially more subtle. It is at these early stages, when muscle loss is only mild, when identification and treatment is most effective. Therefore, developing clinically applicable techniques to identify lean body mass loss at an early stage is important because it is at this stage that treatments are much more likely to be successful. Although numerous research techniques are available for assessing lean body mass (e.g., dual-energy X-ray absorptiometry, MRI), they have inherent assumptions that affect their interpretation and are not readily available for clinical patients. The author has validated an ultrasound-based technique for quantifying muscle loss in dogs and cats that can be used for research. However, an easy and clinically applicable method of assessing muscle is the muscle condition score (MCS), which should be evaluated in every patient at every visit, in addition to body weight, body condition score (BCS), and diet history (WSAVA Nutritional Assessment Guidelines and WSAVA Nutrition Toolkit).

One of the forms of cachexia that has been studied in dogs and cats is cardiac cachexia (Freeman *et al.* 2012). In one study of dogs with CHF, 54% of patients had cachexia (Freeman *et al.* 1998). Cardiac cachexia is a multifactorial process caused by the reduced calorie intake, increased energy requirements, and metabolic alterations. Appetite is commonly affected in animals with CHF and can present as anorexia (complete loss of appetite), hyporexia (reduced appetite), or, most commonly, dysrexia (changes in appetite and food preferences). However, it is inflammatory cytokines, such as tumor necrosis factor (TNF) and interleukin-1 (IL-1) that are primary mediators of cachexia (Freeman *et al.* 1998; Freeman *et al.* 2012). These inflammatory cytokines are known to directly decrease appetite, to increase energy requirements, and to increase the catabolism of lean body mass. Of particular pertinence to heart disease, TNF and IL-1 also cause cardiac myocyte hypertrophy and fibrosis and have negative inotropic effects.

Sarcopenia

Sarcopenia is similar to cachexia in that it is characterized by muscle loss, but sarcopenia is a syndrome seen during aging in the **absence** of disease (although cachexia and sarcopenia can occur concurrently). In sarcopenia, the loss of lean body mass often is accompanied by an increase in fat mass so the total weight may not change (or may even increase), thus masking the sarcopenia. Like cachexia, sarcopenia has important clinical implications because it is associated with increased mortality. The mechanisms of sarcopenia involve physical inactivity, elevated cytokine production, decreased concentrations of growth hormone and testosterone, changes in type II muscle fibers, insulin resistance, and decreased protein synthesis. Dogs and cats also lose lean body mass during aging. Because heart disease typically affects older dogs and cats, sarcopenia (and cachexia) can complicate the treatment of patients with heart disease.

Obesity Paradox

While obesity is a well-known risk factor for the development of heart disease in people, once CHF is present, obesity may actually be associated with a protective effect - this is known as the obesity paradox and has been documented in numerous human studies. Given the adverse effects of cachexia, the association between obesity and improved survival in heart failure appears to be due to a **lack** of cachexia, rather than to the obesity *per se*. This is due to the increased reserve of lean body mass in overweight and obese people (when one gains weight, the weight gained is typically composed of 75% fat and 25% lean). The obesity paradox also has been demonstrated in dogs and cats with CHF (Slupe *et al.* 2008; Finn *et al.* 2010). These data emphasize the importance of avoiding weight (and muscle) loss in the patient with CHF.

Practical Implications for Dogs and Cats

Key nutritional issues that are important for optimal care of dogs and cats with heart disease include:

Assess Diet History

A complete diet history includes the pet food (commercial or home-prepared, including the exact brand, product, and flavor for commercial foods and exact recipe for home-prepared foods), treats, table food, dietary supplements, and foods used for medication administration. This can help to determine if the diet is contributing to suboptimal body weight or body composition:

- Overall nutritional balance of the diet: Unless formulated by a veterinary nutritionist, nearly all home-prepared diets are nutritionally unbalanced. Even commercial diets, if made by a small company with inadequate nutritional knowledge and/or quality control, can have nutritional deficiencies or excesses. Over-the-counter diets labeled as "for intermittent or supplemental use" will be deficient in one or more nutrients and should not be fed as a significant portion of the diet.
- Calories: The diet(s), treats, and other foods, may provide too many or not enough calories to maintain ideal body weight and BCS in an individual animal.
- Protein: Diets vary tremendously in their protein content so the diet history can identify whether protein intake is appropriate. However, ensuring adequate protein intake is critical in dogs and cats with heart disease to minimize cachexia.
- Sodium: The sodium content of the diet, treats, table food, and foods used for medication administration should be addressed for the individual animal.
- Other nutrients: Deficiencies or excesses of other nutrients also can occur, particularly if the owner is feeding an unbalanced diet, giving large amounts of table food or treats (or foods used to administer medications), or dietary supplements.

Physical Examination

Alterations in body composition are common and important issues in dogs and cats with heart disease or CHF and can negatively affect the outcome and quality of life.

- Weight loss: Use a consistent scale to assess weight at each visit to identify subtle changes or trends.
- Body condition score: Use a consistent method to determine BCS to assess current status and changes over time (either a 9-point scale or a 5-point scale). The BCS evaluates body fat, with a goal

for healthy dogs and cats of 4–5 on a 9-point scale. However, in dogs and cats with CHF, a BCS of 5–7 may be more desirable.

- Muscle loss: The MCS differs from the BCS in that the MCS evaluates muscle mass. Dogs and cats can be very obese and yet have severe muscle loss and conversely, cats can be thin but have normal muscle mass. Evaluation of muscle mass includes visual examination and palpation over the temporal bones, scapulae, lumbar vertebrae, and pelvic bones. Assessing MCS is important as dogs and cats with CHF lose primarily muscle, compared to a healthy animal that would lose fat. Early identification of muscle loss, at the "mild muscle loss" stage is valuable for successful intervention. Muscle condition scoring systems have been developed for both dogs and cats (see the WSAVA Nutrition Toolkit).

Make Specific and Individualized Nutritional Recommendations

Always make a specific recommendation for diet - this may be only to say that the diet the animal is currently eating is a good choice and that the owner is feeding an appropriate amount to keep the dog or cat in ideal body condition. It is important to reinforce good nutritional choices that owners are making, since there are so many myths about pet nutrition. It is also important to make recommendations specific for an individual animal based not only on the stage of disease, but also on the animal's nutritional assessment, physical examination, laboratory results, individual food preferences, and concurrent diseases. Ensuring adequate intake of protein and other nutrients is critical to maintain lean body mass. Avoid renal diets unless advanced chronic kidney disease or proteinuria are present. Even certain maintenance or senior diets with protein levels at or just above the Association of American Feed Control Officials minimums may not have enough protein for animals with heart disease, especially if they have cachexia. In complicated cases, especially those with concurrent diseases, consider referring the patient to a board-certified veterinary nutritionist (www.acvn.org or www.esvcn.eu/college).

Omega-3 fatty acid supplementation may provide benefits in dogs and cats with heart disease. The omega-3 fatty acids, eicosapentaenoic acid (EPA) and docosahexaenoic acid (DHA), are normally found in very low concentrations in cell membranes compared to the omega-6 fatty acids, but they can be increased by a food or supplement enriched in omega-3 fatty acids. Omega-3 fatty acids reduce the production of the more inflammatory eicosanoids and of the inflammatory cytokines, TNF and IL-1. The anti-inflammatory, antiarrhythmic, and antithrombotic effects of omega-3 fatty acids have many potential benefits in dogs and cats with heart disease (e.g., reduced muscle loss, improved appetite, association with longer survival; Slupe *et al.* 2008; Freeman *et al.* 2010).

Websites

1. Tufts HeartSmart website [information on heart disease for pet owners, including nutrition (e.g., low-sodium diets for dogs and cats, treats, omega-3 fatty acid supplements)]. Available at: www.tufts.edu/vet/heartsmart
2. Tufts Petfoodology website (information on pet food and pet nutrition for pet owners). Available at: www.petfoodology.org
3. World Small Animal Veterinary Association Global Nutrition Committee. Muscle condition score chart, body condition score chart, diet history form, Nutrition on the Internet, Recommendations for Selecting a Pet Food, and other nutritional tools. www.wsava.org/nutrition-toolkit.

References

1. Finn E, Freeman LM, Rush JE, *et al.* The relationship between body weight, body condition, and survival in cats with heart failure. *J Vet Intern Med.* 2010;24:1369–1374.
2. Freeman LM, Rush JE, Kehayias JJ, *et al.* Nutritional alterations and the effect of fish oil supplementation in dogs with heart failure. *J Vet Intern Med.* 1998;12:440–448.
3. Freeman LM. Beneficial effects of omega-3 fatty acids in cardiovascular disease. *J Small Anim Pract.* 2010;51:462–470.
4. Freeman LM. Cachexia and sarcopenia: Emerging syndromes of importance in dogs and cats. *J Vet Intern Med.* 2012;26:3–17.

5. Freeman LM, Rush JE, Meurs KM, *et al*. Body size and metabolic differences in Maine Coon cats with and without hypertrophic cardiomyopathy. *J Feline Med Surg*. 2013;15:74–80.
6. Slupe JL, Freeman LM, Rush JE. The relationship between body weight, body condition, and survival in dogs with heart failure. *J Vet Intern Med*. 2008;22:561–565.
7. WSAVA Nutritional Assessment Guidelines Taskforce: Freeman L, Becvarova I, Cave N, *et al*. 2011 Nutritional Assessment Guidelines. *J Small Anim Pract*. 2011;52:385–396.

acvim162

Advanced ECG Interpretation

N. Sydney Moise, DVM, MS, DACVIM (Cardiology, Small Animal Internal Medicine); Roberto A. Santilli, DVM, PhD, DECVIM (Cardiology); Romain Pariaut, DVM, DACVIM (Cardiology), DECVIM (Cardiology)
Ithaca, NY, USA

A variety of electrocardiograms will be discussed during this session. Particular points will be made for each image. In this proceedings some of the terminology which may be used has been defined. This is not an exhaustive list of all the topics that will be discussed.

Aberration/Aberrancy

This describes temporary bundle branch block that is frequently the result of impulses arriving during a time when full repolarization has not occurred. The availability of sodium channels is critical to a rapid upstroke of the action potential during phase 0. The more sodium channels available, the steeper phase 0 and the more rapid conduction. If the stimulus occurs before full repolarization such that the resting membrane potential is not -90 mV (which is when the majority of sodium channels are available) conduction will be slowed. Aberrancy can be tachycardia-dependent, long-short cycle dependent or bradycardia-dependent.

Aberrancy can occur during rapid rates and is known as tachycardia-dependent aberrancy or phase 3 aberrancy. This type of aberrancy can occur in normal fibers with a very premature complex. In pathology the action potential duration may be prolonged (even beyond the duration of the action potential) and phase 3 aberrancy may be more likely. Block in the right bundle branch is more likely because the action potential duration and refractory period are longer than the left bundle branch. The right bundle branch has a refractory period that is even longer than the atrioventricular node and his bundle. At rapid rates both the left and right bundle branch shorten; however, the right bundle shortens to a greater degree in this phenomenon results in the effective refractory period of the left bundle now exceeding that of the right. Consequently, during sinus tachycardia or atrial fibrillation premature beats or short cycles result in left bundle branch block or aberration while at normal rates right bundle branch block occurs. Another type of aberration that can be caused with an increase in heart rate is one which occurs gradually rather than with abrupt premature beats. This form occurs when there is a failure of the normal shortening of the refractory period with increased heart rate. As this form of aberrancy can occur with very small changes in rate long recordings may be required.

Ashman phenomenon is a commonly discussed form of aberrancy. This develops when there is a long-short cycle. This is common with atrial fibrillation. Therefore, right bundle branch block or left bundle branch block/aberrancy may be seen with atrial fibrillation depending upon the cycle lengths. Normally, as the heart rate increases the refractory period of the His-Purkinje system shortens (known as a **peeling back** of the refractoriness). Conversely, as the heart rate decreases the refractory period of the His-Purkinje system lengthens. With Ashman phenomenon a long cycle increases the refractory period and if the next cycle is very short it will hit during the refractory period and result in aberrancy which is most commonly a right bundle branch block. This same phenomenon can occur with atrial depolarization and this is termed **Chung's phenomenon**. Aberrancy can be distinguished from a ventricular premature beat by assessing the coupling interval. If over several cycles the coupling interval is constant it is more likely to be a ventricular premature beat then supraventricular with aberrancy. The aberrancy of Ashman's phenomenon can go over several cycles and has also been related to concealed transseptal activation.

Bradycardia-dependent or phase 4 aberrancy occurs long after the effective refractory period. This more complicated form of aberrancy to understand occurs with pathology that has resulted in some depolarization of the myocardial tissue. This depolarization results in a less negative membrane potential which then results in fewer sodium channels available for depolarization. Again with fewer sodium channels available, phase 0 is less steep and conduction is slowed. Thus the longer after a beat that the next stimulus comes the resting membrane potential has become less negative and aberrancy is more likely. Left bundle branch block is more likely.

Cavotricuspid Isthmus

This region of the right atrium is located between the caudal vena cava and tricuspid valve leaflet. It is a region of slowed conduction in the macroreentrant loop of atrial flutter. For counterclockwise or clockwise typical atrial flutter the cavotricuspid isthmus is the target for ablation.

Crista Terminalis/Sulcus Terminalis

The sulcus terminalis is the name given to the epicardial landmark that corresponds with the endocardial landmark of the crista terminalis. These 2 landmarks (epicardial view versus endocardial view) which extend between the cavae (intercaval strip) represent the junction of the embryonic sinus venosus and the pectinate muscles of the right auricle. This line of demarcation, which is composed of fibrous tissue, is important in the function of the sinus node. This is an important structure because it identifies the location of the sinus node and is a common site of origin for focal atrial tachycardia in the dog.

Concealed Conduction

This is a phenomenon whereby the mechanism for alteration in conduction (most commonly but not limited to the AV node) is not appreciated from the surface ECG. Thus, the adjective 'concealed' is used; however, with electrophysiologic studies the mechanisms for alteration due to premature beats antegrade or retrograde is discovered. On the surface ECG concealed conduction due to the effects of incomplete penetration of an impulse are recognized by an unexpected prolongation of conduction, unexpected failure for the impulse to occur, unexpected facilitation, or unexpected initiation of a spontaneous pacemaker. The most common recognition of concealed conduction in dogs occurs during atrial fibrillation or following ventricular premature complexes that affect AV nodal conduction. The former due to antegrade conduction in the latter due to retrograde conduction. Concealed conduction is linked to the inhibitory effects of subthreshold electrotonic potentials.

Concordance/Discordance

In wide complex tachycardia it can be difficult to differentiate ventricular tachycardia from supraventricular tachycardia. One means to assist in this differentiation is to determine if there is concordance or discordance between the leads in the frontal plane and the precordial leads. With supraventricular tachycardia if the QRS has a left (right) bundle branch block morphology in the limb leads it will also have a left (right) bundle branch block morphology in the precordial leads and this is known as concordance. With ventricular tachycardia this consistent relationship with regards to the bundle branch block pattern may not exist and this is known as discordance.

Decremental Conduction

This is a property almost exclusively of the atrioventricular node whereby as the frequency of impulses increase conduction slows. The slowed atrioventricular conduction is appreciated by a prolongation of the PQ/PR interval. It can be appreciated with spontaneous atrial tachyarrhythmias or during rapid atrial pacing. Some accessory pathways may exhibit the property of decremental conduction. This property can also be appreciated during Wenckebach periodicity whereby faster stimulation rates result in a gradual prolongation of AV nodal conduction (increase in PQ/PR interval) before AV block.

Electrotonic Potentials

These potentials which are considered passive **may alter conduction and impulse initiation.** Experimentally subthreshold impulses can have a temporary effect on excitability or may even inhibit subsequent excitation. The effects of electrotonic potentials may be seen in concealed conduction or parasystolic rhythms. In the latter these electrotonic potentials modulate the intervals of a parasystolic rhythm. When they occur early during phase 4 there is a delay in the next discharge, while if they occur late there is an acceleration. The alteration in the spontaneous cycle can be expressed in a phase-response curve which plots the percent of the spontaneous cycle against the percent change in the basic cycle length of the parasystolic rhythm.

Focal Atrial Tachycardia - Coronary Sinus

The P′ is negative in the inferior leads (II, aVF, III) and isoelectric or slightly positive in lead I. Focal atrial tachycardia can originate from the coronary sinus ostium (coronary sinus ostial tachycardia) or deep within the coronary sinus/great cardiac vein (the latter described in humans) Mechanisms described in humans for this arrhythmia most commonly include automaticity or triggered activity, but this cannot be determined from the surface ECG alone. The coronary sinus originates from the sinus venosus with a remnant of myocardium surrounding the ostium which is continuous with the right atrial myocardium. In this area the orientation of muscle fibers changes causing variable anisotropic conduction. Often the negative P waves associated with coronary sinus ostial tachycardia are wide because of the slowed conduction in this region of the right atrium. Focal atrial tachycardia from the coronary sinus musculature (a distance away from the ostium) has not yet been documented in the dog. In humans the P′ differentiation of coronary sinus muscular tachycardia from the coronary sinus ostial tachycardia is that the P wave polarity in V1 is positive and negative in the inferior leads.

Gap Phenomenon and Supernormal Conduction

These two electrophysiologic properties may be confused, although the mechanisms prescribed to each of these phenomena are completely different. In fact, when reading the literature over the years it is clear that the classifications may overlap. At times the gap phenomenon is considered a mechanism behind supernormal conduction but in fact more recently the terminology has been clarified. Often what was believed to be supernormal conduction is actually now known to be described as pseudo-supernormal conduction with the observations of improved conduction due to one of the following mechanisms: gap phenomenon, rate related changes in refractoriness, dual AV node physiology, concealed conduction, peeling back of the refractory period. The gap phenomenon occurs when block more proximal permits improved conduction more distal. Supernormal conduction occurs when conduction is better-than-expected, although it does not have to be normalized.

An understanding of the **effective refractory period** and **functional refractory period** is required to understand the gap phenomenon. The effective refractory period is the longest premature coupling interval after a fixed pulse train that fails to result in capture. The effective refractory period is equal to the absolute refractory period (the latter is a period during which no sodium channels are available and no matter how strong the stimulus no depolarization will occur) plus a short period during phase 3 of the action potential during which a stimulus may cause some cellular depolarization. The functional refractory period is the shortest premature coupling interval after a fixed pulse train that results in capture. The gap phenomenon occurs when the functional refractory period of the proximal tissue is shorter than the effective refractory period of the more distal tissue. That is, at a given stimulus frequency, block exists in the more distal tissue (longer relative refractory period), but as the stimulus frequency increases a point is reached whereby the proximal tissue has block or conduction delay (shorter functional refractory period) that then permits the more distal tissue to have recovery so that conduction is improved. Again, the conduction delay in the proximal tissue must exceed the refractory period of the distal tissue. The gap phenomenon can occur when there is proximal delay the atrioventricular node, proximal His-Purkinje system, His bundle, or atrium which then causes delay respectively in the His-Purkinje system, distal His-Purkinje system, or proximal or distal atrioventricular node. Thus, although 6 different types of gap phenomenon have been described, the most common is a delay in the AV node that then permits conduction in the His-Purkinje system.

An example of a gap phenomenon in dogs with an accessory pathway can be seen with variation in preexcitation due to antegrade conduction across the accessory pathway with certain PP intervals (sinus rates) versus conduction through the AV node due to block in the accessory pathway. This happens with right-sided accessory pathways in humans as well. This is been identified in dogs that have varying preexcitation with sinus arrhythmia and may be particularly noticeable when treated with drugs such as sotalol. The accessory pathway having a longer effective refractory period.

During supernormal conduction there is a brief period after repolarization following phase 3 of the action potential during which depolarization can occur. It is commonly associated with ischemic myocardium. Supernormal conduction has been associated with the influence of increased sympathetic tone. Purkinje fibers exhibit supernormal conduction, but the His bundle and ventricular myocardium do not. The supernormal conduction requires a relatively long refractory period and occurs during a relatively consistent time period during repolarization although it will occur earlier with faster rates and later at slower rates.

P Wave Morphology/Polarity/Location

During the sinus rhythm of a dog or during an arrhythmia, the P wave morphology, duration and mean vector (polarity) should be assessed. Because the majority of dogs have a wandering pacemaker in conjunction with a sinus arrhythmia the morphology/polarity changes normally. The P waves are taller with a mean vector directed towards lead II with short PP intervals (faster sinus rates) and the P waves are shorter with a mean vector directed more leftward with longer PP intervals (slower sinus rates). Experimental studies in the dog have shown that P′ wave morphology is primarily controlled by the direction of left atrial and septal depolarization. Assessment of the P′ wave can assist in determining the type of supraventricular tachycardia. Conceptually, such rhythms originate from the **roof of the right atrium** (e.g., focal atrial tachycardia-crista terminalis or cranial vena cava: P′ wave similar to sinus P wave), **floor of the right atrium** (e.g., focal atrial tachycardia-coronary sinus ostial tachycardia: P′ wave negative in leads II, III, and aVF; slightly positive in lead I; lead aVL slightly more positive than lead aVR; wide duration of P′ wave), **junction of the atrioventricular node** (e.g., focal junctional tachycardia or non-paroxysmal junctional tachycardia: P′ wave negative in leads II, III, and aVF; isoelectric in lead I; lead aVL and aVR equally positive), **roof of the left atrium** (e.g., focal atrial tachycardia - left superior pulmonary vein: P′ wave positive in leads II, III and aVR; negative in lead I and aVL; positive in lead aVR), **floor of the left atrium** (e.g., focal atrial tachycardia - ligament of Marshall: P′ wave negative in leads II, III and aVF; negative in lead I and aVL. It is very important to understand that these P wave vectors are reflecting atrial depolarization only in the frontal plane. Properly positioned precordial leads may be of assistance. In humans the mean vector of the P wave in V1 is considered important. It is also important to realize that the confirmation of the thorax and phenotype of the dog may have some alteration in the exact P wave vector.

The location of the P′ wave with focal atrial tachycardia may be in the first portion of the ST segment, ascending portion of the T wave, or descending portion of the T wave. The location of the P′ wave with junctional tachycardia may be immediately following the QRS complex at the onset of the ST segment which because of the negative P′ wave can sometimes give the appearance of an S wave (pseudo-S wave). With orthodromic atrioventricular reciprocating tachycardia a negative P wave is seen during the first portion of the ST segment. The F wave of atrial flutter is located just before or immediately after the QRS complex or during the first portion of the ST segment.

QT:TQ Restitution

Importantly, formulas developed in humans to correct for QT interval relationships to heart rate are inaccurate in the dog. The relationship of the QT to RR interval is nonlinear. Some formulas over-correct while others under-correct at varying heart rates. Moreover, there is a QT hysteresis that occurs in its relationship to the RR interval. Development of methods to relate the QT interval to heart rate are an ongoing struggle. Of notation is the fact that in the dog during expiration when the RR interval is greater than 500 ms the QT interval changes very little, but in contrast with inspiration and an increase in sympathetic tone the QT/RR interval relationship becomes more variable and steeper. A procedure used in the analysis of 24-hour ambulatory electrocardiographic recordings evaluates the stability of beat-to-beat QT interval with respect to the TQ interval (electrical diastole) this is called the ECG restitution. The restitution is the ability of the heart to recover from one beat to the next. During very rapid heart rates the TQ interval can approach zero (e.g., ventricular tachycardia). The QT/TQ ratio provides an assessment of the electrical systolic period to the electrical diastolic period. Restitution curves of this relationship have been associated with transition from ventricular tachycardia to ventricular fibrillation based on the steepness of the restitution relationship. The steeper the slope (>1) the more likely instability exists. The longer the QT interval relative to this TQ interval the more likely for fibrillation.

Triggered Activity

A mechanism responsible for atrial, junctional, and ventricular arrhythmias. Two types of triggered activity have been described based on the time of depolarization during the action potential. **Early afterdepolarizations** occur during phase 2 or phase 3 of the action potential and are frequently pause-dependent. They developed because of a decreased outward current (K^+) or an increased inward current (Na^+ or Ca^{++}). Phase 2 or plateau early afterdepolarizations result from L-type calcium current while phase 3 may be caused by activation of the Na/Ca exchanger. Early afterdepolarizations often originate from the Purkinje system. Early afterdepolarizations are known to be the trigger for the inherited polymorphic ventricular tachycardia identified in young German shepherds. **Delayed afterdepolarizations** occur during phase 4 of the action potential and are often associated with an acceleration in rate prior to their development. Delayed afterdepolarizations develop in situations with calcium overload (intracellular). These include digitalis toxicity which increases intracellular calcium because of the inhibition of the Na^+/K^+ pump or high catecholamine or cAMP both of which enhance calcium uptake by the sarcoplasmic reticulum. It should be noted that not all early- and delayed-afterdepolarizations result in propagation to the muscle with a premature beat. Drugs that prolong the action potential duration and thus the QT interval can predispose to arrhythmias caused by afterdepolarizations. Drugs that shorten the action potential duration such as lidocaine or mexiletine can decrease early after depolarizations. It should be noted that in German Shepherds with inherited ventricular arrhythmias lidocaine is very effective in eliminating the arrhythmias; however, mexiletine has limited effectiveness. The latter may be due to inadequate concentrations.

References

1. Okumura K, *et al. J Am Coll Cardiol.* 1991;17:509–518.
2. http://www.eplabdigest.com/article/7871.
3. http://cdn.intechopen.com/pdfs-wm/27014.pdf.
4. Fossa AA. *Pharmacology and Therapeutics.* 2008;118:231–238.
5. Issa ZF, Miller JM, Zipes DP. *Clinical Arrhythmology and Electrophysiology.* 2nd ed. 2012.

acvim163

Advanced Holter Analysis

N. Sydney Moïse, DVM, MS, DACVIM (Cardiology, Small Animal Internal Medicine)
Ithaca, NY, USA

As with any diagnostic test we use in cardiology its helpfulness and accuracy depends on the quality of the interpretation. The depth of the interpretation depends on the questions that are asked. The questions that are asked depend on the thoroughness required to meet the needs for treatment. To maximize the value that can be obtained from the analysis of a 24-hour Holter recording demands a meticulousness in the evaluation and the knowledge of the potential information. During this session the advanced interpretation of 24-hour electrocardiographic monitoring will be discussed and illustrated extensively.

Monitoring of the cardiac rate and rhythm for 24 hours is indicated in the assessment of syncope (transient loss of consciousness), near-syncope (partial transient loss of consciousness), collapse, arrhythmias (bradycardias and tachycardias), antiarrhythmic treatment effect, pacemaker malfunction, or screening for cardiomyopathies. Certainly other means of rhythm assessment outside of the hospital environment are of value in the appraisal of cardiac rate and rhythm. Some of these include implantable loop recorders or event recorders. However, the 24-hour Holter which can have extended monitoring time, provides a great deal of information that may not be obtained from other modalities. The technology is advancing to make application and use better in veterinary medicine. Analysis software has improved but still demands careful editing.

The evaluation of heart rate entails the examination of numerous parameters. **Average heart rate** is of course one of the initial parameters assessed. It is vitally important to understand and know for a given Holter analysis system what this average heart rate means. Is this based on the sinus rhythm? Is this based on the total heartbeats regardless of beat identification? Some systems will be able to give you both. The average heart rate is only the most basic assessment of rate. Obviously 2 dogs can have the same average heart rate and have vastly different distributions of R-R intervals. **Hourly average heart rate** can be assessed from the hourly table provided by most analysis programs. Such tables often give the minimum and maximum heart rate for any given hour as well. Again, it is vital to know if this heart rate is sinus or total R-R intervals. A **heart rate tachogram** provides a graphic display of **geometric heart rate variability** over the 24-hour time period. From this graph the heart rate distribution at certain points during the day may be correlated with the diary that is kept by the owner. The majority of dogs have a sleep cycle that is most consistently seen with a slower heart rate between midnight and 7 AM. Of course depending upon the household activities this will vary. It is also very common for some of the fastest heart rates to occur during the application or removal of the Holter monitor. It must be determined for any given analysis program how the average heart rate is being determined and displayed on this graph. Some use a rolling average of 8 beats with a display of the minimum and maximum heart rate displayed and others will use the standard deviation of the average heart rate. Quite often associated with the heart rate tachogram there is a graph indicating the timing and duration of pauses. What is defined as a pause can be selected but usually within a prescribed range. Some prefer to set this definition of a pause as greater than 2 seconds and others greater than 3 seconds. It is vitally important to recognize that if a pause is indicated in the narrative or graphic presentation this area needs to be visualized on the full disclosure adequately. Quite often pauses occur with an interruption by an escape beat and repeated pauses. Therefore, if an individual relies totally on the narrative report without investigating the existence of secondary pauses, they will not be identified and recognized as potential problems. Quite often the minimum and maximum heart rate are over emphasized as it must be remembered that these are singular points in time. A more important parameter to assess is the amount of time that a dog spends with a heart rate less than 50 bpm or greater than 120 bpm. These 2 parameters can be adjusted but in general they work pretty well in the assessment of what the heart rate distribution should be in a dog. In general the majority of dogs spend greater than 80% of their day with a heart rate between these 2 values. Most dogs spend less than one hour with a heart rate less than 50 and less than 3 hours with a heart rate greater than 120.

Heart rate variability can be determined from the 24-hour recording. Such assessment can be done for all or part of the recording. **Geometric**, **time domain**, and **frequency domain** can be used. It is important to remember that when heart rate variability is considered it is in the context of what the sinus node is doing; however, the PP interval is not what is being assessed. Consequently, if heart rate variability to is to be used accurately the actual P-QRS complexes throughout the 24-hour period must be assessed to ensure that there is an association between the P and the QRS for heart rate variability to be accurate for the assessment of the sinus rhythm and autonomics during the timeframe that is considered. Although heart rate variability is primarily considered in the assessment of the autonomic system and its effect on the sinus node, importantly, geometric heart rate variability serves a very valuable role in understanding rhythm disorders with specific patterns developing with certain types of arrhythmias. Geometric heart rate variability involves the use of graphs and other visual means in the assessment of R-R intervals. These graphs include heart rate tachogram, sinus interval histogram, ventricular interval histogram, supraventricular interval histogram, RR interval 24-hour tachogram, hourly RR interval tachogram, and Poincaré plots. It should be noted that the RR interval tachogram is a linear 2 dimensional representation over time that appears as a compressed image. This time series can be expanded if the R-R intervals are exported simply to files like Excel. In this manner you can evaluate shorter periods of time and appreciate the beat-to-beat intervals. In the standard 24-hour RR interval tachogram because of the compression repeated intervals overlay each other and therefore the appreciation of beat density is lost. This can be regained by the use of three-dimensional RR interval tachograms which incorporate a histogram with the tachogram. The Poincaré plot is also a 2 dimensional geometric representation of the heart rate variability; however, it is a nonlinear representation of the beat-to-beat intervals. A Poincaré plot is created by plotting a given RR interval on the X-axis against the next RR interval on the Y-axis (RR interval + 1). Importantly, the next interval that serves as the location on the X-axis overlaps by one and is the previous RR interval + 1. Thus the graph is RR interval versus RR interval + 1. In the Holter analysis packages these are often represented by the hour but if a particular algorithm permits a specific time interval to be assessed this may provide information as desired. Additionally, a full 24-hour image of the Poincaré plot can be obtained; however, the beat density is so great that real information is difficult to obtain. Importantly the still images cannot portray the exact dynamics that is in play. To assist in this, animated Poincaré plots can be used. Some specific Holter analysis programs such as that of Forest Medical provide this possibility. However, regardless of the particular Holter system used, if the cardiologist can export RR interval files they can be used in a variety of software programs such as Matlab. Also to obtain an animated Poincaré plot to assess the beat-to-beat variability in real time at specific time points, the exported R-R files (saving them as a plain text file) can be uploaded to the following site and visualized. https://wyattflanders.com/poincareplot/.

Time domain heart rate variability includes:

1. **SDNN: standard deviation of all NN intervals**
2. **SDANN: standard deviation of all 5-minute NN interval means**
3. **SDNN Index: mean of all 5-minute NN interval standard deviations**
4. **RMSSD: square root of the mean squared successive NN interval differences**

The **triangular index** is available in many software packages; however, this is likely not a usable parameter as it is determined from the total number of beat intervals divided by the height of the histogram. Because the dog does not have a singular distribution of beats represented by the histogram it would not be accurate. Time domain heart rate variability can be determined over the entire 24-hour period or by selecting time periods of interest during a designated period of time. Frequency domain heart rate variability is usually determined over shorter periods of time, although longer periods may also be done. For short-term frequency domain assessment time is usually chosen as a 5-minute interval, 256 beats or 512 beats. Additionally different filtering methods are selected (e.g., Hamming, Hanning). With frequency domain heart rate variability the values are reported as the power spectral density (sec^2/Hz) **on the Y-axis with frequency (Hz) on the X-axis. The power in msec^2 as the total <0.4 Hz, very low frequency <0.04 Hz, low frequency <0.04–0.15 hertz, and high-frequency <0.15–0.4 hertz.**

Additionally the low-frequency/high-frequency ratio is reported, and importantly the indexed power in units of n.u. is given.

Geometric heart rate variability has provided insight with regards to the unique patterning of the sinus beating in the dog. The dog does not have the same type of pattern of humans, cats, or horses. The canine distribution is unique with a bifurcating pattern associated with high vagal tone. This bifurcated pattern is recognized on the RR interval tachogram. It appears as a paucity of beat intervals (zone of avoidance) between approximately 700 and 1200 ms. Numerous bands or additional bifurcations are seen in dogs with sinus node dysfunction. The low number of R-R intervals or zone of avoidance, is also appreciated on the Poincaré plot with very few data points in the center of the graph. Dogs frequently will have a comet shaped shorter RR interval pattern indicating sympathetic tone that then bifurcates displaying 2 arms of long and short R-R intervals. Some dogs will have a separation of these arms whereby another region with a paucity of beats is seen. This is due to the exit block of sinus beats from the sinus node and the normal dog. Frequently dogs will have a scattered cloud of longer RR intervals above the zone of avoidance as seen in the Poincaré plot. On the sinus beat RR interval histogram there are frequently 2 to 3 distributions that approximate Gaussian distributions. That is, the sinus rhythm of the dog does not beat with a single Gaussian distribution but hypothesized as multiple. This is hypothesized to be the result of the sinus beat intervals corresponding to exit from the sinus node exit pathways under the control of the autonomic nervous system. Geometric heart rate variability can be of value not only in the assessment of normal and abnormal sinus node function, but can give evidence to the type of arrhythmia that is present. For example, atrial fibrillation has a distinctive pattern, as does 2:1, 3:1 etc. AV nodal heart block, atrial flutter, couplets, or abrupt changes in rhythm such as cardioversions. Each of these will be shown during the presentation. With atrial fibrillation the actual refractory period of the AV node can be calculated.

Of course one of the primary reasons for performing 24-hour Holter monitoring is to assess the presence and severity of arrhythmias. Each of the above assessments are made but to diagnose the specific arrhythmia the actual beats are examined. Examination of the P-QRS-T complexes is essential and can be done from looking at selected report strips, full disclosure at a speed that permits adequate assessment of coupling intervals and short strips whereby the intervals between beats are indicated. It is not possible to fully appreciate the results of a Holter recording by simply looking at a very compressed image of a full disclosure. This is particularly true if one is trying to assess the response to treatment with arrhythmias. In addition to knowing the frequency and complexity of arrhythmias, the distribution or triggers of these arrhythmias can lend important information that may have an impact on the type of treatment that is given. For example, do the ventricular arrhythmias occur more with excitement or more during the evening during sleep? The type of patterning over many beats may give evidence for particular mechanism for the arrhythmia. This may be assisted additionally with study of the animated Poincaré plots. It is also important to realize that some dogs may not have a large ventricular load, but they have dangerous ventricular tachycardia. As mentioned earlier the assessment of pauses in rhythm are vitally important. Careful evaluation with regards to the beat-to-beat patterning before the pause is also important. For example, evidence of a neurocardiogenic reflex would be a tachycardia followed by gradual slowing and then excessively long pauses. The word pauses is used because sometimes there will be a single pause that is then followed by several secondary pauses, such that although the first pause may be for 7 seconds an additional 3 pauses may be present that average 5 seconds each so that the period of time whereby the dog has extremely low cardiac output far exceeds that necessary to have a syncopal episode. Such careful examination of a 24-hour Holter will give deeper understanding to the mechanism behind syncope in dogs associated with arrhythmias.

As a general statement if the dog has sufficient cardiac arrhythmias (supraventricular or ventricular) to demand the institution of treatment that dog should have a baseline 24-hour Holter recording and follow-up recordings. If this type of monitoring is not done, then the response to medication will not be known. It is impossible to make a full and accurate appraisal based on a short electrocardiogram performed in the clinic. It is also important to realize and accept that in general the majority of dogs can tolerate a 24-hour period to obtain a baseline recording before antiarrhythmic medication is instituted. If a baseline is not obtained it is impossible to know if the medication is having an antiarrhythmic or a proarrhythmic effect or more commonly no effect at all. Of course, there are circumstances for which a baseline is not in the best interest of the patient, but these are the exception. In general one wants to see an 85% reduction in the frequency of an arrhythmia with treatment in order to be sure that it is just not day-to-day variability.

Commonly Holter analysis packages include graphs for the QT plotting and assessment. In addition they often offer one of several QT correction formulas to use in the assessment of the QT interval. Such formulas have been established to be flawed and of limited value in the assessment of canine heart rate and QT relationships. The relationship of the QT to RR interval is nonlinear. Some formulas over-correct while others under-correct at varying heart rates. Moreover, there is a QT hysteresis that occurs in its relationship to the RR interval. Development of methods to relate the QT interval to heart rate are an ongoing struggle. Of notation is the fact that in the dog during expiration when the RR interval is greater than 500 ms the QT interval changes very little, but in contrast with inspiration and an increase in sympathetic tone the QT/RR interval relationship becomes more variable and steeper. A procedure used in the analysis of 24-hour ambulatory electrocardiographic recordings evaluates the stability of beat-to-beat QT interval with respect to the TQ interval (electrical diastole) this is called the ECG restitution. The restitution is the ability of the heart to recover from one beat to the next. During very rapid heart rates the TQ interval can approach zero (e.g., ventricular tachycardia). The QT/TQ ratio provides an assessment of the electrical systolic period to the electrical diastolic period. Restitution curves of this relationship have been associated with transition from ventricular tachycardia to ventricular fibrillation based on the steepness of the restitution relationship. The steeper the slope (>1) the more likely instability exists. The longer the QT interval relative to this TQ interval the more likely for fibrillation. In the Forest Medical system analysis package for QT:TQ restitution is available for use in the dog.

Some miscellaneous comments with regards to evaluation of the Holter recording are warranted. Dogs change their body position often when wearing the monitor and this will result in dramatic alterations and changes in the P-QRS-T complex. So such changes must be understood and not misdiagnosed as a problem. Most often these changes are accompanied by motion artifact list the observer. Variability in the frequency of arrhythmias can occur with no medical intervention and this is why very strict criteria must be used to be sure that a drug has an antiarrhythmic effect or to blame a drug for a proarrhythmic effect. Moreover, with the analysis of many recordings in dogs it can be realized that the normal young adult dog has virtually no arrhythmias and that arbitrary numbers to define a dog as affected are not affected during screening based on some limit of premature complexes could be problematic. Repeated Holters are very important in the assessment of treatment and creating package deals for owners with regards to the finances can increase the yield for providing a better medical interpretation. Additional studies are required to truly know normal values. For example, this includes an evaluation of geriatric dogs and their heart rate patterns and frequency of arrhythmias. Treatment effects of medications demand study. The discoveries possible looking at more than 100,000 QRS complexes per day in a dog provides a wealth of information and an exciting place for discovery.

References

1. Moïse NS, *et al. J Vet Cardio*. 2010;12(3):191–196.
2. Gladuli A, Moïse NS, *et al. J Vet Cardio*. 2011;13(1):63–70.
3. Fedorov VV, *et al. Circulation*. 2010;122(8):782–789.
4. Boineau JP, *et al. Circulation*. 1978;58(6):1036–1048.

acvim164

Electrophysiology and Ablation of Atrial Flutter

Roberto Santilli, DVM, PhD, DECVIM-CA (Cardiology)
Samarate, Varese, Italy

Introduction

The generic term atrial tachycardia refers to a group of arrhythmias that originate in the atria and then propagate to the ventricles with variable atrioventricular (AV) conduction ratio depending on the electrophysiologic properties of the AV node. According to human guidelines two different groups of atrial tachycardias can be identified: focal atrial tachycardia (FAT) and macro reentrant atrial tachycardia or atrial flutter (AFL).[1] Focal atrial tachycardia is defined as an atrial activation starting rhythmically at a small area (focus) from which it spreads out centrifugally and without activation over a significant portion of the cycle length (CL).[2] Focal atrial tachycardias manifest an atrial rate between 100 and 250 bpm and rarely at 300 bpm.[2] Despite the fact that FATs exhibit three different electrogenic mechanisms (abnormal automaticity, micro-reentry, and triggered activity), all of them are grouped in the same category. According to the response of the electrophysiologic tests FATs caused by triggered activity and micro-reentry can also be called non-automatic FAT.[3-5]

Prevalence of FATs in people is low (0.34–0.46%), FATs account for 10–23% of supraventricular tachycardias (SVT) in children with spontaneous remission often reported (24–63%).[6] Automatic FATs tend to be more common in young people and they are present in 80% of patients with tachycardia-induced cardiomyopathy (TICM).[7] Sixty-three percent of patients with FAT have left ventricular dysfunction most commonly caused by a TICM.[7] Focal atrial tachycardias have been described also in dogs of different breeds, with a mean body weight of 33 ± 17.4 Kg and a median age of 51 months (range 5–156 months).[8] Dogs are usually presented for episodic weakness or for congestive heart failure compatible with TICM (19%).[8] Focal atrial tachycardia resulted incessant in 19% of dogs, paroxysmal in 56% of dogs and deteriorated into persistent atrial fibrillation in 25%.[8] Mean atrial CL reported in dogs is 238.2 ± 69.2 msec; it is conducted in 1/3 of the dogs with different degree of AV nodal block with a mean ventricular CL of 292.7 ± 72.5 msec.[8]

In people FATs have a characteristic anatomic distribution with typical clustering around the crista terminalis, coronary sinus, para-hisian region, tricuspid and mitral valve annulus, right and left interatrial septum, right and left appendages, pulmonary veins, superior vena cava and non-coronary aortic cusp.[9-17] In people 63% of FATs are right-sided, while 37% arise from the left atrial wall or pulmonary veins.[9] One-half to 2/3 of right atrial FATs and 21% of total FATs arise from the crista terminalis an area with marked anisotrophy and rich of automatic tissue, factors that promote, respectively, reentry and abnormal automaticity.[9] Coronary sinus ostium and body account for approximately 10% of FATs in people, the region of Thebesian valve presents an abrupt change in myocardial fiber orientation that facilitates reentry.[9] Coronary sinus musculature can be also a source of abnormal automatism foci.[18] Interatrial septum and perinodal area accounts for 34% of FATs while tricuspid valve annulus for 13–22%.[9,12] Pulmonary veins tachycardias comprise 46% of left atrial FATs and 24% of all FATs.[9,15,19] Left and right superior pulmonary veins are the ones most commonly affected.[15] Mitral valve annulus is the second most common location of left FATs and accounts for 28–36% of total FATs.[15] Four to 17% of human patients with FATs have more than one focus in the right free wall (29.5%) or in the right free wall and right septal area.[20] Patients with multiple ectopic foci presented a cardiovascular co-morbidity, shortest tachycardia CL, and lower radiofrequency catheter ablation (RFCA) success rate.[21] According to detailed endocardial mapping 63% of the dogs with FATs presented right sided foci (crista terminalis 41%; triangle of Koch 25%; tricuspid valve annulus 16%; right interatrial septum 9% and right auricle 9%) and 37% presented FATs arising from the pulmonary veins (right superior pulmonary veins 57%; left superior pulmonary veins 28%; right inferior pulmonary veins 15%). Twelve percent of dogs presented multiple ectopic foci.[8]

The reported success rate of RFCA of FAT in people ranges between 69 and 100%, with higher success rate in the right FATs compared to left FATs and septal tachycardias (80%, 72%, and 52%, respectively).[22-24] The recurrence rate ranges from 0–33%. Positive predictor for successful RFCA of atrial foci are right atrial location, while the negative ones are male gender, older age, multiple foci, and repetitive forms.[22-24] Recurrence rate reported is approximately 8%, complication rate reported is around 5% and the most common sites of damage are the AV node and the phrenic nerve during RF energy delivery to ablate the cristal tachycardias.[22-24]

Macro-reentrant atrial tachycardia is a reentrant activation around a large central obstacle (normal or abnormal structures, fixed, functional, or a combination) in at least one of its dimension.[25] There no part of atrial tissue outside the circuit that are activated. Atrial flutter is a macro reentrant atrial tachycardia characterized by an organized atrial rhythm with a rate typically between 250 and 350 bpm in people[1,25,26] and usually more than 300 bpm in dogs[27,28]. According to the location of protected isthmus, an area of slow conduction which is part of the anatomic circuit of the arrhythmia, AFL are classified in typical and atypical1.[25,26] The most common anatomic circuit of AFL includes the cavo-tricuspid isthmus (CTI, an area of the low posterior right atrium bounded by the inferior vena cava and the Eustachian ridge posteriorly and the tricuspid valve annulus anteriorly.[29] According to the rotation pattern around the tricuspid valve annulus, two forms of CTI-dependent AFL have been identified: typical and reverse typical.[26] During typical AFL the activation wavefront turns counterclockwise around the tricuspid valve annulus, while during reverse typical AFL it turns clockwise.[25,26] Other forms of CTI-dependent AFL reported in people include the double-wave and lower-loop reentry.[1] A further type of AFL, for some authors classifiable as atypical, is the left atrial macroreentrant atrial tachycardia, a stable arrhythmia that can originate in the left atrium.[25] A variety of left atrium circuits associated with large silent areas over the posterior left atrium has been described.[31] Last type of AFL reported in people is the left septal atrial flutter, an arrhythmia that shows a slowing of the electrical conduction in the left atrial septum caused by antiarrhythmic treatments and/or atrial myopathy.[32] We reported two cases of endocardial mapping and radiofrequency catheter ablation of CTI-dependent AFL typical and typical reverse in two dogs with episodic weakness and documented SVT.[28] Atrial CL found were respectively 165 msec and 130 msec. Electrocardiographic pattern of typical AFL was continued succession of depolarizing segments and electrical deflections that conferred a classical saw-tooth pattern with predominant negative polarity in inferior leads (F waves). Reverse typical atrial flutter waves appeared as positive deflections in inferior leads with isoelectric lines between them.[28] We also reported two atypical forms of AFL in dogs with the isthmus located at the level of right atrial free wall and right interatrial septum.[33] Average atrial rate during atypical atrial flutter was 280 ms, in all but one dog reported atrioventricular conduction ratio was 1:1.[33]

RFCA is considered the treatment of choice for AFL in people with a long-term success rate after induction of bidirectional block in the protected isthmus area of 90–100%, a recurrence rate of 7% and a complication rate of 1.34%.[1,34,35]

Endocardial Mapping and Radiofrequency Catheter Ablation Techniques

Radiofrequency catheter ablation of either FAT or AFL should be performed after a detailed endocardial mapping to assess the arrhythmogenic mechanism and the location of targets for ablation, respectively, ectopic foci and isthmus. In the dog electrophysiological studies are performed in general anesthesia as previously described, after discontinuing antiarrhythmic drugs for at least 5 drug half-lives. The animals are placed in dorsal recumbence. Two multipolar electrode catheters and one 7-F ablation catheter are inserted in the venous system, with a modified Seldinger technique, using fluoroscopic and intra-cardiac ECG guidance.[8-28] A decapolar catheter is inserted through the right external jugular vein into the coronary sinus, a second decapolar electrode catheter is inserted through the right femoral vein and placed around the tricuspid valve annulus to record with the proximal electrodes pairs the potentials of the antero-lateral aspect of the crista terminalis (HRA), with the middle pairs the potentials of the middle lateral right atrium (MRA) and with the distal pairs the potentials of the low postero-lateral right atrium (LRA). A 7-F ablation catheter, 8 mm, with deflectable curve is advanced through the right femoral vein and placed alternatively at the His bundle, right ventricle, different area of right and left atrial wall and

CTI to map endocavitary atrial electrograms and to perform programmed stimulation.[8,28] To map left atrial chamber or pulmonary veins ostium and bodies, in case a patent foramen ovale is not present, a transeptal puncture should be performed using the described technique.[36] From the right femoral vein, under fluoroscopic guidance, a specially designed guide wire is first inserted into a transeptal needle and pushed across the fossa ovalis into the left chamber. A jump of the tip of the needle from the thicker muscular septum to the thin wall of the fossa ovalis is used to localize fossa ovalis. Along the wire positioned into the left atrium the transeptal needle is pushed across the interatrial septum to enlarge the hole, and then an introducer is left in place and used to move the mapping and the ablation catheters. Lasso catheters are special electrodes positioned in a circular fashion that are used to map pulmonary vein ostium, during ablation procedures. In some cases left atrial tachycardias arising from right superior pulmonary vein, can be identified mapping high posterior right atrium.[37] The analysis of double potentials in this area can help to establish the location of the ectopic foci, if the amplitude of the first potential is greater than that of the second potential, this indicated a right posterior atrial focus, and in the opposite case a right superior pulmonary vein focus should be suspected. All intra-cardiac electrograms, 12-lead surface ECG are displayed and analyzed on an EP recorder.

Macro-Reentrant Atrial Tachycardia

During macro-reentrant atrial tachycardia the reentrant activation is continuous and should be present of the entire CL of the tachycardia, and therefore also during the isoelectric line of the ECG. In case of AFL line of block with double potentials can be recorded in the area of functional or anatomic barriers.[25] In the isthmus area or in sites with local conduction disturbances fractioned electrograms can be recorded.[25] Transient entrainment of the tachycardia is present in most cases and a return CL equal to 20 ms of the tachycardia CL at the site of pacing identifies that the area is part of the circuit. In case the return CL is present in at least 2 pacing sites separated by at least 2 cm a diagnosis of macro-reentry can be done.[38] CTI-atrial flutter is diagnosed when a concealed entrainment during LRA pacing can be shown.[39] Typical AFL presents a counterclockwise reentry descending along the anterior and lateral walls and ascending along septal and posterior wall of right atrium. A reverse direction of rotation is observed during typical reverse AFL with an ascending direction along the anterior wall and descending along the posterior and septal walls.[1,25,26] The circuits of left atrial flutter and of septal flutter are less commonly studied. The criteria to diagnose the macro-reentrant circuit and the mapping technique of the protected isthmus are although similar. In our EP laboratory we identified beyond, several cases of CTI-dependent atrial flutter, cases of atypical atrial flutter with the isthmus area located in the right interatrial septum, right atrial free wall and cranial vena cava.

Radiofrequency catheter ablation of AFL circuits is performed in controlled temperature with a thermocouple-tipped steerable 7-F catheter. Maximal catheter tip temperature and power output are usually set respectively at 65°C and 100 W. Radiofrequency delivery for each target lasts at least 45 seconds. Persistence of block should be verified for at least 20–30 minutes. The target sites for ablation to cure AFL are the protected isthmus. Radiofrequency catheter ablation of the CTI is performed using both anatomical and an electrophysiological approach using the right lateral fluoroscopic view and the presence of concealed entrainment at the target site. Radiofrequency delivery can be performed during tachycardia or sinus rhythm. The catheter should be then gradually withdrawn towards the caudal vena cava to produce a linear lesion. The endpoint of ablation is the induction of bidirectional conduction block at the CTI demonstrated by a collision of waveform, while pacing the coronary sinus, with a supero-inferior activation of the right free wall post-ablation, by the presence of a slow conduction from low right free wall and coronary sinus during pacing of contralateral site, and by recording double potential at the site of linear ablation at the CTI.[29,34,40] Reported success rate of RFCA of CTI-dependent AFL and our experience of RFCA of right septal, right atrial free wall and cranial vena cava flutters in dogs are promising.

References

1. Blomstrom-Lundqvist C, *et al.* ACC/AHA/ESC Guidelines. *J Am Coll Cardiol.* 2003;42:1493.
2. Saoudi N, *et al. Eur Heart J.* 2001;22:1162.

3. Chen SA, *et al. Circulation*. 1994;90:1262.
4. De Bakker JM, *et al. J Cardiovasc Electrophysiol*. 1994;5:335.
5. Kammeraad JA, *et al. Pacing Clin Electrophysiol*. 2003;26:736.
6. Poutiainen AM, *et al. Eur Heart J*. 1999;20:694.
7. Chen SA, *et al. J Cardiovasc Electrophysiol*. 1998;9:355.
8. Santilli RA, *et al. J Vet Intern Med*. 2010;24:539.
9. Kistler PM, *et al. J Am Coll Cardiol*. 2006;48:1010.
10. Badhwar N, *et al. J Am Coll Cardiol*. 2005;46:1921.
11. Chang KC, *et al. Jpn Circ J*. 2001;65:1034.
12. Chen CC, *et al. J Cardiovasc Electrophysiol*. 2000;11:744.
13. Das S, *et al. J Cardiovasc Electrophysiol*. 2008;19:231.
14. Freixa X, *et al. Europace*. 2008;10:105.
15. Hachiya H, *et al. Circ J*. 2005;69:205.
16. Kalman JM, *et al. J Am Coll Cardiol*. 1998;31:451.
17. Kistler PM, *et al. Circulation*. 2003:108:1968.
18. Tritto M, *et al. J Cardiovasc Electrophysiol*. 2001;12:1187.
19. Haisseguerre M, *et al. N Engl J Med*. 1998;339:659.
20. Hillock RJ, *et al. Heart Rhythm*. 2007;4:435.
21. Hu YF, *et al. Heart Rhythm*. 2009;6:198.
22. Anguera I, *et al. Am J Cardiol*. 2001;87:886.
23. Hindricks G. *Eur Heart J*. 1993;14:1644.
24. Kay GN, *et al. J Am Coll Cardiol*. 1993;21:481.
25. Saoudi N, *et al. Eur Heart J*. 2001;22:1162.
26. Cosio FG. *Card Electrophysiol Rev*. 2002;6:356.
27. Schoels W, *et al. Circ Res*. 1990;67:35.
28. Santilli RA, *et al. J Vet Cardiol*. 2010;12:59.
29. Olgin JE, *et al. J Cardiovasc Electrophysiol*. 1996;7:1112.
30. Morrouche F, *et al. Circulation*. 2004;109:2440.
31. Jais P, *et al. Circulation*. 2000;101:2928.
32. Marrouche NF, *et al. Circulation*. 2004;109:2440.
33. Santilli RA. *J Vet Cardiol*. 2014;16:9.
34. Chen SA, *et al. J Am Coll Cardiol*. 1996;27:860.
35. Kay GN, *et al. J Cardiovasc Electrophysiol*. 1993;4:371.
36. Brockenbrough E, *et al. Am J Cardiol*. 1960;6:219.
37. Yamada T, *et al. J Cardiovasc Electrophysiol*. 2004;15:745.
38. Lin YJ, *et al. J Cardiovasc Electrophysiol*. 2005;16:163.
39. Stevenson WG, *et al. J Cardiovasc Electrophysiol*. 2005;16:173.
40. Saoudi N, *et al. PACE*. 1998;21:2105.
41. Calkins H, *et al. Am J Cardiol*. 1994;73:353.

acvim165

Diagnosis and Treatment of Accessory Pathway002DMediated Arrhythmias

Roberto Santilli, DVM, PhD, DECVIM-CA (Cardiology)
Samarate, Varese, Italy

Introduction

Congenital AV accessory pathways are muscular tracts which connect directly atrial to ventricular myocardium, bypassing the AV node-His-Purkinje system.[1-5] In dogs APs can be single or multiple, they conduct in an all-or-none pattern, in 25% of the animals APs present bidirectional conduction with overt electrocardiographic signs of ventricular pre-excitation (VPE) during sinus rhythm.[6-11] In the remaining cases antegrade conduction is intermittent, unapparent or absent (concealed), depending on the electrophysiological properties of the AV bypass tract and normal AV-His-Purkinje system.[6-11] Differently from human APs in dogs are usually right-sided along the postero-septal and right posterior part of the tricuspid valve annulus, more rarely in the mid-septal or antero-septal area.

Accessory pathway-mediated tachycardias are classified according to the direction of the electrical impulse within the anatomical circuit, into orthodromic and antidromic AVRT. During orthodromic AVRT antegrade conduction proceeds down the AV node, while retrograde conduction occurs over an accessory pathway. During antidromic AVRT the re-entrant impulse travels in reverse direction.[1-4] Orthodromic AVRT is the most common arrhythmias correlated with an accessory pathways in dogs[6-11], while antidromic AVRT has not yet been reported. Others bypass tract-mediated tachyarrhythmias reported in dog are: permanent junctional reciprocating tachycardia (PJRT)[11] and pre-excited atrial fibrillation[12]. Permanent junctional reciprocating tachycardia or Coumel tachycardia is a form of circus movement tachycardia in which the anterograde limb of the circuit is the AV node and the retrograde limb is a concealed slowly conducting fiber with decremental properties usually located in the posteroseptal region.[11]

Accessory pathway are commonly present in young Labradors and Boxer with a common history of narrow QRS complex tachycardia causing weakness and heart failure due to arrhythmic-induced cardiomyopathy (AICM).[6-11]

Surface Electrocardiogram

In people several studies described the value of 12-lead surface ECG to discriminate SVT and particularly orthodromic AVRT from AVNRT and focal atrial tachycardia (FAT).[13-20] Also in dogs different electrocardiographic criteria such as ventricular rate, cycle length alternans (beat to beat oscillation of the cycle length of ≥20 ms) QRS alternans (beat to beat variation of QRS amplitude ≥1 mm in at least one lead), presence of ventricular pre-excitation, evaluation of atrial depolarization position and axis on frontal, RP and PR ratio, AV relation and repolarization abnormalities have been used to differentiate the most common types of SVT.

Dogs with FAT had faster heart rates (278±62 versus 229±42 bpm; P 5.049) and less QRS alternans (33 versus 86%; P 5.022). P waves appeared during tachycardia in 22 dogs, with a superior axis in 100% of OAVRT and 22% of FAT (P 0.001). OAVRT was characterized by a shorter RP interval (85.0±16.8 versus 157.1±37.3 ms; P 0.001) and smaller RP/PR ratio (0.60±0.18 versus 1.45±0.52; P 0.001). Repolarization anomalies were present in 64% of OAVRT and no FAT (P 0.001). Multivariate analysis identified QRS alternans and a positive P wave in aVR during tachycardia as independent predictors of arrhythmia type.

Reported electrocardiographic findings in dogs with PJRT are: narrow QRS complex (60 ms) tachycardia with a regular ventricular cycle length (RR interval) of 260 ms, negative P waves in the inferior leads (II, III and aVF) with a mean electrical axis in the frontal plane of -45°, a long RP' (140 ms) and RP'/P'R of 1.16.

Treatment of Accessory Pathways-Mediated Tachycardias

Recommendation for long-term therapy in human patients with Wolf-Parkinson-White (WPW) syndrome and well-tolerated arrhythmias is catheter ablation (class I-level of evidence B) and pharmacological therapy (IIa-C); while in case of WPW syndrome and atrial fibrillation or poorly tolerated arrhythmias only catheter ablation is indicated (I-B).[22] Although the safety and efficacy of catheter ablation of APs have not been tested by prospective, randomized clinical trials, large single-center and multicenter trials showed an initial success rate of 95% with a reoccurrence of conduction along the AP in approximately 5% of cases.[4-5,23-30] Different types of complications can occurs during catheter ablation of APs as reported by the Multicenter European Radiofrequency Survey (MERFS) with and overall complication rate of 4.4%, including 0.13% of procedure-related deaths. Others complications are related to radiation exposure, vascular access, catheter manipulation and radiofrequency delivery. The incidence of iatrogenic AV block ranges between 0.17 and 1%.[31-32]

In veterinary medicine no studies prospectively tested the efficacy of antiarrhythmic drugs on accessory pathway-mediated tachycardia. Most cases reports anecdotally suggested short-term efficacy of class I agents (lidocaine and quinidine), class III (sotalol) and class IV (diltiazem) in the control of OAVRT and PJRT in dogs.[6-12]

Radiofrequency Catheter Ablation Of Accessory Pathways

Radiofrequency catheter ablation of APs and consequent elimination of macro-reentrant circuit should be performed after a detail intracavitary mapping to study the exact location of the bypass tract, its electrophysiological properties and the underlining mechanism of documented and/or inducible tachyarrhythmias. In dog electrophysiological study are performed under general anesthesia with the animal in dorsal recumbence. Three multipolar electrode catheters are inserted in the venous system, with a modified Seldinger technique, using fluoroscopic and intra-cardiac ECG guidance: A decapolar catheter is inserted through the right external jugular vein into the coronary sinus, two quadripolar catheter through the right femoral vein, one to record the His bundle potential and one for programmed electrical stimulation, is placed, alternatively, in high right atrium and right ventricular apex. Twelve lead surface electrocardiogram and intra-cardiac signals are displayed and analysed with an electrophysiological recorder at paper speed of 100 or 200 mm/second. The intracardiac electrograms are recorded at filter settings of 50–500 Hz. Pacing is usually performed using stimuli that are twice the diastolic threshold and 2 ms in duration. The first part of the study consists in the evaluation of normal conduction intervals (A-H and HV), A-ERP, R-ERP and Wenckebach point of the normal AV node and their comparison with reference values for dog.[33] Effective refractory period of the AV node is defined as the longest S1-S2 interval at which S2 failed to reach the His bundle (antegrade) or ventral right atrium (retrograde). Forty-six percent of dogs present ventricular-atrial (VA) conduction during the basal state, and it is enhanced by high sympathetic tone.[33] Antegrade and retrograde AV nodal Wenckebach point is obtained with incremental atrial and ventricular pacing. Pacing cycle length at which type I atrioventricular or VA block occurs is considered the Wenckebach point.[33] During the study presence and conduction properties of AV bypass tract should be assessed. The anatomical distribution of the AP can be studied by adapted left anterior oblique fluoroscopic projection and intracardiac electrograms guidance. Antegrade conduction is considered present in case of ventricular pre-excitation (either abnormally short, or negative His to ventricular interval) during normal sinus rhythm or while atrial pacing at progressively shorter cycle lengths. The presence of retrograde conduction through the accessory pathways is demonstrated in most cases by the evidence of eccentric VA activation during ventricular programmed pacing or tachycardia, and retrograde non-decremental conduction by programmed ventricular stimulation.[9-10] In case of midseptal or anteroseptal APs concentric VA activation can be differentiated from retrograde atrial activation through the AV node with an accurate para-Hisian pacing or with tachycardia resetting.[9-10] A single ventricular extrastimulus during tachycardia is given, while His was refractory, looking for preexcitation of the atrium with identical activation sequence.[9-10] Programmed atrial stimulation is also used to induce tachycardia, and to study its location, type and mechanism. Tachycardia induction is considered reproducible if initiated with the same S1-S2 interval. Reentrant tachycardia can be initiated and terminated as soon as S1-S2 interval encounter the refractory period of one of the two limb of the circuit. Another electrophysiological test that is usually used to differentiate ectopic atrial tachycardia from reentrant tachycardia is the measure of first VA interval after cessation of overdrive atrial pacing during tachycardia. In case of reciprocating tachycardia, it is usually fixed to testify the presence of a predetermined circuit.[9-10] The sequence of atrial and ventricular activation during tachycardia is crucial to define the type and the location of tachycardia. Orthodromic AVRT usually present fixed ventricular-atrial interval, with a duration depending of R-ERP of APs. Retrograde atrial activation during orthodromic AVRT starts at atrial insertion of AP. It could be defined, either during tachycardia or ventricular pacing, as the point of earliest atrial activation of bipolar electrogram.[9-10] Concomitant unipolar recording are used to optimize the accuracy of accessory pathways localization. Sharp and negative unipolar waveform indicates the exact site of impulse spreading.[9-10]

During RFCA a thermocouple-tipped steerable 7F catheter connected to a RF generator is generally used. The radiofrequency delivery is performed in controlled temperature (65°, 75 W) and maximal ablation time for a particular target is 60 seconds. In case of antero-septal accessory pathways a lower temperature (50°, 50 W) or cryoablation.[34] For AVRT ablation catheter is placed at the atrial endocardial insertion of AP around the tricuspid valve annulus or within the coronary sinus. Ablation procedures is guided by intra-cardiac bipolar recording, searching for the shorter VA interval, or bypass tract potentials, during OAVRT or ventricular pacing, and with unipolar recording with sharp and negative waveform, which permit to assess the pole closest to the bypass tract. In case of antegrade conduction is present the ablation target is represented by the earliest site of ventricular activation during sinus rhythm or atrial pacing.[9-10]

Detailed intracavitary mapping of a total of 25 APs in 15 dogs have been reported.[6-11] Twenty-four APs were right-sided (11 postero-septal, 9 right atrial free wall, 3 midseptal), 2 were left-sided (left atrial free wall). The majority of bypass tracts had unidirectional retrograde conduction, and in 5 cases were multiple.[6-11] Correlated arrhythmias documented or inducible in these reports were OAVRT in 14 dogs, PJRT in 1 dog and atrial fibrillation in 2 dogs.[6-11] Radiofrequency catheter ablation (RFCA) of anomalous AV bypass tract was successfully performed in all the cases.[6-11]

In our electrophysiological laboratories RFCA acute success rate all APs is 98%, with 2% recurrence rate at 6–12 months follow-up. Complication rate ranges from 1 to 2% and include major complication (iatrogenic AV block) and minor complications (venous access bleeding).

References

1. Gallagher JJ, *et al. Prog Cardiovasc Dis.* 1978;20:285.
2. Ferrer MI. *Am J Med.* 1977;62:715.
3. de Chillou C, *et al. J Am Coll Cardiol.* 1992;20(3):666.
4. Swartz JF, *et al. Circulation.* 1993;87(2):487.
5. Lesh MD, *et al. J Am Coll Cardiol.* 1992;19:1303.
6. Atkins CE, *et al. J Vet Intern Med.* 1995;9:43.
7. Wright KN, *et al. J Vet Intern Med.* 1999;13:361.
8. Sherlag BJ, *et al. J Am Vet Med Assoc.* 1993;203:1147.
9. Santilli RA, *et al. J Vet Cardiol.* 2006;8:157.
10. Santilli RA, *et al. J Am Vet Med Assoc.* 2007;231:393.
11. Santilli RA, *et al. J Vet Cardiol.* 2013;15:225.
12. Santilli RA *et al. J Am Vet Med Assoc.* 2010;237:1142.
13. Kalbfleisch SJ, *et al. J Am Coll Cardiol.* 1993;21:85.
14. Morady F *et al. J Am Coll Cardiol.* 1987;9:489.
15. Erlinler I, *et al. A.N.E.* 2002;7:120.
16. Green M, *et al. Circulation.* 1983;68:368.
17. Kay GN, *et al. Am J Cardiol.* 1987;59:296.
18. Tai CT, *et al. J Am Coll Cardiol.* 1997;29:394.
19. Jaeggi ET,*et al. Am J Cardiol.* 2003;91:1084.
20. Riva SI, *et al. J Am Coll Cardiol.* 1996;27:1480.
21. Santilli RA. *J Vet Inter Med.* 2008;22:915.
22. Blomstrom-Lundqvist C, *et al. J Am Coll Cardiol.* 2003;42:1493.
23. Jackman WM, *et al. N Engl J Med.* 1991;324:1605.

24. de Chillou C, *et al. J Am Coll Cardiol.* 1992;20(3):666.
25. Calkins H *et al. Circulation.* 1992;85:1337.
26. Haïssaguerre M, *et al. J Cardiovasc Electrophysiol.* 1994;5:532.
27. Warin JF, *et al. Pacing Clin Electrophysiol.* 1990;13:1609.
28. Gaita F, *et al. G Ital Cardiol.* 1992;22:1245.
29. Xie B, Heald SC, *et al. Eur Heart J.* 1996;17:1072.
30. Huang JL, *et al. Am J Cardiol.* 1996;78:1375.
31. Hindricks G, *et al. Eur Heart J.* 1993;14:1644.
32. Calkins H, *et al. Circulation.* 1999;99:262.
33. Wright KN. *et al. Am J Vet Res.* 1996;57:1695.
34. Lanzotti ME. *Italian Heart J.* 2002;3:128.

acvim166

The Use of 3-Dimensional Electroanatomic Mapping Systems for the Treatment of Supraventricular Tachycardia

Sandra P. Tou, DVM, DACVIM (Cardiology, Small Animal Internal Medicine)
Raleigh, NC, USA

Introduction

Radiofrequency catheter ablation (RFCA) has largely replaced antiarrhythmic drug therapy for the treatment of various cardiac arrhythmias in human medicine.[1] Its high success rate against certain supraventricular tachyarrhythmias makes RFCA an attractive therapy option, also eliminating issues surrounding medication administration and untoward drug side effects. In veterinary medicine, paroxysmal atrioventricular reciprocating tachycardia attributed to an accessory atrioventricular pathway is a common indication for RFCA in dogs.[2-4] Although many affected dogs respond well to medical management, RFCA is a good option in the setting of poor owner compliance, inadequate arrhythmia control despite drug therapy, intolerable drug side effects, and pet owners looking to avoid lifelong drug administration. Although the use of RFCA for supraventricular tachycardia in clinical canine patients is well-documented,[3-6] this treatment remains uncommon in veterinary medicine due to cost and limited availability of electrophysiology (EP) equipment and expertise.

Conventional Mapping for RFCA

RFCA procedures are performed in the EP laboratory, requiring specialized equipment and fluoroscopy. A full EP study to determine the precise origin of activation and activation mapping is followed by catheter ablation. Resistive heating is then applied through RF electrode catheters to create thermal lesions within the myocardium to destroy arrhythmogenic tissue. Conventional RFCA is often associated with long procedure and fluoroscopy times, increasing the safety concern of radiation exposure to medical professionals.

Electroanatomic Mapping Systems

Non-fluoroscopic electroanatomic mapping systems such as the EnSite NavX™ system (St. Jude, St. Paul, MN, USA) use electrical impedance to generate a three-dimensional (3D) image of cardiac structures. Low-level current passes through 3 pairs of surface (skin) electrode patches. Standard EP catheters detect electrical signals (voltage and impedance) transmitted between surface electrode patches. The position of the electrode is then computed relative to a reference electrode to generate a 3D real-time geometry of the endocardial surface. These systems also calculate and display activation sequences in 3D views by superimposing color-coded EP data on the generated anatomic map. This technology has allowed for reduced fluoroscopy time and reduced radiation exposure.[7]

References

1. Page RL *et al. J Am Coll Cardiol.* 2015;67(13:e27–e115).
2. Wright K. *Vet Clin Small Anim.* 2004;34:1171–1185.
3. Wright KN *et al. J Vet Intern Med.* 1999;13:361–371.
4. Santilli RA *et al. J Vet Cardiol.* 2006;8:157–165.
5. Santilli RA *et al. J Vet Cardiol.* 2010;12:59–66.
6. Santilli RA *et al. J Vet Cardiol.* 2014;16:9–17.
7. Casella M *et al. J Interv Card Electrophysiol.* 2011;31:109–118.

acvim167

Classification of Congenital Heart Disease: Learning From Pathology

Diane Spicer, BS, PA (ASC)
Valrico, FL, USA

Abstract

Analysis of congenital heart disease in both humans and animals cannot be successful without first knowing the normal anatomy and morphology of the cardiac segments. In the human population, the morphologic method and sequential segmental analysis have long been used to successfully analyze and diagnose both simple and complex disease. The anatomy in animals with congenital heart disease can be very closely compared to those defects seen in the human population allowing for the classifications to be studied along the same line. This review will look at the value of the morphologic method and sequential segmental analysis as it relates to the diagnosis of congenital heart disease and the comparative anatomy of malformations in both the human and animal population.

Introduction

Utilization of the 'Morphologic Method' is the key to understanding and analyzing hearts with congenital cardiac malformations. The basis of this principle is that one anatomic feature which is variable should not be used to define another feature which is itself variable. This system allows for the morphologic distinction of the atriums, the ventricles, and the arterial trunks which make up the three segments of the heart. With sequential segmental analysis, the three cardiac segments can be analyzed for their intrinsic morphology. The cornerstone of sequential segmental analysis is establishing the arrangement of the atrial appendages. Once this is determined, the type of atrioventricular connection or junction can be assessed, with this being a separate feature from the morphology of the atrioventricular valve itself. In the normal heart there are two atrioventricular junctions, each guarded by its own atrioventricular valve. Ventricular morphology is determined by the apical trabecular component, allowing for adequate assessment of the atrioventricular junctions, ventricular topology, and the relationships of the chamber within the ventricular mass. The last of the three segments are the arterial trunks, which are identified by their branching pattern. Once identified, the description of the ventriculo-arterial junctions and the relationships of the arterial trunks to one another as they leave the ventricular mass can be achieved.

Morphology of the Atrial Appendages

The most constant morphologic feature of the atriums is the appendage and the extent of the pectinate muscles. The atriums have venous component, a vestibule, a septum, and an appendage. The right atrium on outward appearance has a blunt, often triangular tip and many think that this is all that makes up the appendage when in reality, it is the entire anterior aspect of the chamber. The right atrium has a broad attachment with the venous component, this junction is marked by the terminal groove. On the right, the superior and inferior caval veins serve as the systemic venous return with the venous return from the heart returning via the coronary sinus. As the superior caval vein joins the roof of the right atrium there is a prominent muscle bundle that crosses the junction anteriorly, known as the terminal crest. The terminal crest runs along the interior surface of the right atrium lying immediately adjacent to the previously mentioned terminal groove that marks the broad junction with the venous component. The pectinate muscles arise from the terminal crest and run in a parallel fashion, extending laterally into the appendage. The pectinate muscles extend around nearly the entire right atrioventricular junction and are the single most important and constant morphological determinate of a right atrial appendage. The pectinate muscles extend into the diverticulum inferior to the coronary sinus or the sub-Thebesian sinus. Taking their origin from the terminal crest are the Eustachian and Thebesian valves, which guard the openings of the inferior caval vein and the coronary sinus. The right atrial vestibule composes a small part of the right atrium and is the narrow, smooth-walled portion that inserts into the leaflets of the tricuspid valve.

The left atrium is the most posterior of the cardiac chambers and also has a venous component, a vestibule, a septum, and an appendage. The left atrial appendage outwardly has a tubular appearance, often times with a hooked tip and a scalloped outer border. Its attachment with the venous component is narrow and unlike the right atrium, this area is not marked by a terminal groove or terminal crest. The pectinate muscles are confined to the appendage itself with this being the most constant morphological feature of the left atrium. The vestibule is typically smooth, particularly the portion surrounding the posterior aspect of the left atrioventricular groove. The flap valve at the floor of the oval fossa forms the septal surface and it has a characteristic 'horseshoe' appearance where it overlaps the superior interatrial fold. The pulmonary veins enter the roof of the smooth-walled vestibule. The left atrium has a significant body which is the area joining the vestibule, the appendage and the septum with the body remaining prominent even in cases of anomalous pulmonary venous return.

Morphology of the Ventricles

The right and left ventricles should be analyzed in a tripartite fashion, both with an inlet, an apical trabecular, and outlet components. The apical trabecular component is the most constant feature and is the morphological determinate.

The majority of the anterior aspect of the ventricular mass is composed of the muscular walls of the right ventricle. The right ventricular inlet is guarded by the tricuspid valve, the inlet component extending from the hinge point of the valve to the attachments of the tendinous cords. The tricuspid valve possesses attachments to the septal surface and to papillary muscles. The apical trabeculations within the right ventricle are coarse with a trabeculated septal surface. There is a prominent trabeculation on the septal surface known as the septomarginal trabeculation or the septal band. The septal band divides into two limbs toward the base of the heart, oriented in a cephalic and caudal direction. The two limbs join the supraventricular crest with the medial papillary muscle arising from the caudal arm and the cephalic arm extending to the subpulmonary muscular infundibulum. The anterior papillary muscle extends from the body of the septal band towards the apex along with the moderator band which traverses the space between the septum and the right ventricular free wall. The moderator band is known as an echocardiographic marker of the right ventricle. Septoparietal trabeculations also arise from the anterior surface of the septal band and extend to the parietal wall along the right ventricular outflow tract. Within the right ventricular outlet, the pulmonary valve is supported by a free standing muscular sleeve or infundibulum. The sleeve of muscle lifts the pulmonary valve away from the ventricular mass.

The left ventricular inlet is guarded by the mitral valve and it consists of two leaflets supported by tendinous cords that extend to two papillary muscles. The inlet component extends from the hinge point of the valve to the insertion of the cords into the tips of the papillary muscles. The mitral valve has no attachments to the septum. The apical trabecular component of the left ventricle is fine and crisscrossing with a relatively smooth septal surface toward the outlet component. The aortic valve marks the outlet of the left ventricle and is supported in part by muscle and fibrous tissue. The fibrous component is made up of the fibrous continuity between the anterior or aortic leaflet of the mitral valve and the aortic valve. The muscular support is typically beneath the right coronary aortic sinus and a portion of the left coronary aortic sinus.

Morphology of the Arterial Trunks

The pulmonary trunk and the aorta make up the third segment and are identified morphologically by their branching pattern. In the normal heart, the aorta forms the centerpiece of the heart with the pulmonary trunk spiraling around it as the arterial trunks rise from the ventricular mass. Both arterial trunks have an intra- and extrapericardial component. The first arteries to branch from the aorta are the right and left coronary arteries. The intrapericardial portion of the aorta is typically referred to as the tubular portion or the ascending aorta and it extends from the sinotubular junction to the bifurcation of the brachiocephalic arteries in humans and the brachiocephalic trunk in animals. The brachiocephalic arteries branch from the aortic arch at or slightly distal to where the pericardial reflection crosses the anterior surface of the aorta. Normally, the aortic arch extends over the root of the left lung and along the spine. The pulmonary trunk has a short intrapericardial segment prior to bifurcating into the right and left pulmonary arteries.

ANALYZING THE HEART

The atrioventricular and ventriculo-arterial connections can be assessed once the morphology of the atrial and ventricular chambers and the arterial segments have been determined. The interarterial relationships and the relationships between the ventricular chambers within the ventricular mass deserve description as well.

As previously stated, atrial morphology is the cornerstone of sequential segmental analysis and is of particular importance along with the arrangement of the atriums. The classifications are as follows: usual atrial arrangement, mirror image atrial arrangement, and presence of bilateral morphologically right or morphologically left atrial appendages (isomerism).

There are five basic types of atrioventricular connection:

1. The normal arrangement exists when a morphologically right atrium is connected to a morphologically right ventricle and a morphologically left atrium is connected to a morphologically left ventricle. This can also be referred to as concordant atrioventricular connections and represents a biventricular connection.
2. A morphologically right atrium connected to a morphologically left ventricle and a morphologically left atrium connected to a morphologically right ventricle is known as a discordant atrioventricular connection. This also represents a biventricular connection.
3. A biventricular and mixed connection occurs when isomeric atrial chambers are connected to separate ventricular inlets. This type of atrioventricular connection renders the terms concordant and discordant inappropriate.
4. Double inlet connection exists when both atriums and both ventricular inlet components connect to the same ventricular trabecular component. The dominant ventricular chamber may be of right, left or indeterminate morphology. Whether the dominant ventricle is of right or left morphology, there will be an incomplete, rudimentary ventricular chamber that will lack an inlet component. If the rudimentary chamber is of right ventricular morphology, it is typically located in an antero-superior position and when of left morphology, the location is posterior or postero-inferior within the ventricular mass. This is also known as a univentricular connection.
5. A second type of univentricular connection exists when only one atrium is connected to a ventricular inlet, with absence of the other atrioventricular connection. A second ventricular chamber is typically present and is usually incomplete and rudimentary.

Ventriculo-Arterial Connections

There are three components that need attention when analyzing the ventriculo-arterial junctions, including how the arteries are connected to the ventricles, the relationship of the arteries to one another, and the morphology of the outlet component of the ventricle that supports the arterial valvar apparatus. Four possible types of ventriculo-arterial connection exist:

1. Concordant ventriculo-arterial connections are the normal relationship where the aorta arises from the morphologically left ventricle and the pulmonary trunk from the morphologically right ventricle.
2. Discordant ventriculo-arterial connections describe the situation typically referred to as transposition with the aorta arising from the morphologically right ventricle and the pulmonary trunk from the morphologically left ventricle.
3. A double outlet ventriculo-arterial connection describes the situation where both arterial trunks, either totally or in their greater part, arise from the same ventricular chamber. This chamber can be of right or left ventricular morphology or rarely a solitary and indeterminate ventricle.
4. A single outlet ventriculo-arterial connection can exist with a common arterial trunk, a solitary arterial trunk or when there is an absent connection between the ventricular mass and either the aorta or the pulmonary trunk. The common arterial trunk is usually shared to varying degrees

between the two ventricles and supplies the coronary, systemic, and pulmonary circulations. A solitary arterial trunk exists when only one intrapericardial trunk can be identified and it supplies the coronary and systemic circulations with the pulmonary circulation supplied by arteries arising along the descending portion of the aorta, distal to the brachiocephalic arteries. When two arterial trunks are identified with only one outlet, there is typically a hypoplastic arterial trunk with associated valvar atresia (aortic or pulmonary atresia).

Comparative Anatomy

When congenital cardiac malformations occur in both the human and animal populations, they exhibit many of the same characteristics and associations between the cardiac segments. They often present with similar clinical findings as well, this applying to the fetal, newborn, and adult groups. In fetal life, premature closure of the oval foramen, or a restrictive oval foramen are incompatible with life and this can often be associated with other congenital cardiac malformations. The anatomic considerations are very similar if the morphologic method and sequential segmental analysis are incorporated into the assessment of the heart. Several examples of pertinent comparative specimens with sequential segmental analysis will be shown to demonstrate the affinity and difference in comparative analysis between the domestic mammals and the humans.

acvim168

Vascular Ring Anomalies: Using Cross-Sectional Imaging to Guide Surgery

Brian A. Scansen[1], DVM, MS, DACVIM (Cardiology); Augusta Pelosi[2], DVM, DACVS, DACVIM (Cardiology)
[1]Fort Collins, CO, USA; [2]Palm Beach Gardens, FL, USA

Introduction

Several malformations of the arterial system have been described in animals; the most common clinical presentation is a pediatric animal with dysphagia. Indeed, the original human medical term for a vascular ring anomaly was **dysphagia lusoria**, which is from the Greek **dysphagia lusus natural** meaning difficulty eating and freak of nature.[1] The term "vascular ring" was used for the first time in 1945 by Robert Gross.[2] The basic pathophysiology of these malformations is a constriction of the esophagus at the level of the heart base, dilation of the cranial esophagus, and regurgitation. Due to its position in the chest, the trachea may also be entrapped within the ring.

Several classifications have been suggested; however, we are currently lacking a systematic organization of these conditions as they occur in veterinary species. The use of cross-sectional imaging has significantly enhanced our understanding of thoracic anatomy, which is particularly relevant when considering 3D structures such as vascular ring anomalies.

Embryology

The heart takes origin as paired endocardial heart tubes that arise from splanchnic mesoderm, just ventral to the pericardial coeloms on either side of the embryo. The heart tubes gradually fuse (from cranial to caudal) to form a single tube consisting of mesoderm, endoderm, and extracellular material (cardiac jelly). From cranial to caudal, these chambers are the aortic sac and truncus arteriosus, the bulbus cordis, the primitive ventricle, the primitive atrium, and the sinus venosus.

During cardiovascular development, this tube begins to loop or fold upon itself to form the fetal heart. The bulbus cordis folds and bends to the right, caudally, and ventrally; the primitive ventricle moves to the left; the sinus venosus and primitive atrium migrate cranially and dorsally. The truncus arteriosus gives origin to the first portion of the great vessels. The primitive ventricle will become the adult left ventricle, and the trabeculated portion of the proximal bulbus cordis will become the adult right ventricle. The middle third of the bulbus cordis, now called the conus cordis, will form the outflow portions of both ventricles, and the terminal third of the bulbus cordis will develop the proximal parts of the ascending aorta and main pulmonary artery.

Truncus Arteriosus and Conus Cordis

An invagination of the aortic sac between the 4th and the 6th aortic arch forms the truncus arteriosus. The truncal cushions form first and expand into the lumen of the truncus from the right and left sides. The two sides eventually fuse to form the truncal septum, which has a slight spiral. The bulbar cushions arise from the right dorsal and left ventral margins of the heart wall and are continuous with the truncal ridges. The spiral orientation of the bulbar cushions produces the spiral septum when they ultimately fuse with each other and with the truncal septum.

The Aortic Arches

The blood, pumped out of the cardiac ventricles into the embryonic circulatory system, reaches a pair of dorsal aortae via six paired aortic arches originating from the aortic sac. Not all the arches are present at the same time, but develop and atrophy at various stages of development.

In mammals, the first 2 arches atrophy and disappear. The 3rd arches form the common carotid arteries and proximal segments of the internal carotid artery. The left 4th arch develops into the aortic arch and the beginning portion of the left subclavian artery and brachiocephalic trunk. The right 4th arch forms the proximal segment of the right subclavian artery. The 5th arches persist in reptiles but not in mammals. Persistency of the 5th arch has been reported in humans, but not in veterinary medicine. The proximal segment of the left 6th arch completes the pulmonary trunk and partially the left pulmonary artery, and the distal segment forms the ductus arteriosus. The proximal segment of the right 6th arch forms the right pulmonary artery and the distal segment atrophies. The 7th intersegmental arteries arise from the dorsal aorta caudal to the aortic isthmus.

The final aortic arch anatomy is different among species. Briefly, horses and ruminants have only one vessel (the brachiocephalic trunk) branching cranially from the ascending aorta; while dogs, cats, pigs, rabbits, and mice have two cranial vessels (brachiocephalic trunk and left subclavian artery).[3] The European hamster, rat, and human have three cranial branches: the brachiocephalic trunk (often called the innominate artery), the right common carotid, and the left subclavian. Interestingly, 15–20% of people have only two great vessels off the ascending aorta, which is frequently (and incorrectly) termed a "bovine arch."[4]

Arterial Malformations

In a suggested classification, the vascular rings are divided based on the vessels constituting the ring and location of the different abnormal segments: 1) Persistent right aortic arch with left ligamentum arteriosus with or without an aberrant left subclavian artery; 2) Persistent left aortic arch with right ligamentum arteriosus with or without an aberrant left subclavian artery (mirror image of the previously described anomaly); 3) Aberrant subclavian; 4) Double aortic arch.

Persistent Right Aortic Arch with Left Ductus/Ligamentum Arteriosus

The most common form of vascular ring anomaly in dogs and cats is a persistent right aortic arch (commonly known as PRAA) with regression of the left arch and presence of a left-sided ligamentum or ductus arteriosus. The intersegmental arteries are pulled forward along the aorta, as usual, but the left arch, instead of the right arch, separates from the dorsal aorta. Consequently, the right 4th aortic arch matures and a left-sided ductus/ligamentum arteriosus and pulmonary artery form a vascular ring around the esophagus. Though this condition is well-known as PRAA, it is technically incorrect to define it as such since the presence of a stand-alone PRAA will not produce an obstruction without the concomitant presence of a left-sided ductus/ligamentum. There are numerous reports in dogs,[5-13] with the anomaly first described in 1925[14]. PRAA has also been reported in the cat,[15-21] horse,[22-25] cow,[26,27] lamb,[28] pig,[28] llama,[29,30] elephant seal,[31] and tiger[32]. It is believed to be familial, with an increased incidence in German Shepherd dogs and Irish Setters.[5] The Great Dane breed is often mentioned in case reports as well.

Persistent Left Aortic Arch with Right Ductus/Ligamentum Arteriosus

Mirror image of the previously described vascular ring is a ring formed by a normally developed left aortic arch from the left 4th arch and a persistent right 6th arch leading to formation of a right ductus/ligamentum arteriosus. This anomaly has been reported in the dog,[10,33,34] cat,[35] and alpaca[30,34]. Surgical reports[34] describe the origin of the right ductus/ligamentum as more caudal than would be expected. A right ductus/ligamentum arteriosus is a remnant of the distal right 6th aortic arch. It may coexist with any of the other arch abnormalities and cause independent vascular ring compression of the esophagus. Ventrally, the right ligamentum attaches to the right pulmonary artery. Dorsally, it may attach directly to the aorta, to an aberrant right subclavian artery, or to an atretic or functional right 4th arch. These can all be dissected from the dorsal aspect and separately ligated and divided if appropriate.

Aberrant (Retroesophageal) Subclavian Arteries

Aberrant origin of the left subclavian artery directly from a more dorsal location of the aortic arch/descending aorta, the descending aorta or a ductus represents a vascular variation causing incomplete, partial and dorsal esophageal obstruction. It develops when the left 7th intersegmental artery, being pulled forward along the dorsal aorta, fails to reach the left 4th arch before the latter separates from the dorsal aorta. This has been found in 1/3 of dogs with PRAA,[36] leading to a dorsal and cranial compression of the esophagus in addition to the constriction caused by the persistent right aortic arch and left-sided ductus/ligamentum arteriosus. A right subclavian artery, originating from an aberrant location, in the presence of a normal left arch, may also cause compression of the dorsal esophagus. This abnormality occurs when the right 7th intersegmental artery, being pulled forward along the dorsal aorta, fails to reach the right 4th arch before the latter separates from the dorsal aorta. The surgeon enters the thorax expecting to find a typical vascular ring and discovers the aortic arch on the left side along with a left ligamentum arteriosus. Further dissection reveals the right subclavian artery arising on the dorsal aorta and crossing over the esophagus. This anomaly has been reported in dogs.[5] Esophageal signs may still be seen with partial compression of the esophagus and surgical correction is advised. Unique variants of left subclavian anatomy have also been observed by one of the authors (BAS) - isolation and ductal origin. In one dog with a PRAA, the left subclavian artery was not attached to the aortic arch - rather, it had an atretic segment that resulted in esophageal compression and had retrograde flow from the left vertebral artery via the circle of Willis. Ductal origin of the left subclavian artery has been observed and also recently reported.[37]

Double Aortic Arch

The final form of vascular ring in the dog, often reported as the most common form in humans, is the double aortic arch. Both left and right 4th arches persist and connect to the dorsal aorta, leading to constriction of the esophagus and trachea in addition to the constriction caused by the ductus/ligamentum. This anomaly has been described in the dog,[38-42] cat,[43] and lion[44]. Frequently, one of the arches will be atretic or less well-developed than the contralateral arch. Tracheal stenosis and malformed tracheal rings with associated respiratory signs are commonly encountered in humans with double aortic arch and have been reported in dogs.[5]

DIAGNOSTICS

The clinical history and signalment are often highly suggestive of a vascular ring anomaly. A typical history includes a young animal with frequent regurgitation beginning at the time of solid food introduction. Physical examination may include poor body condition or abnormal growth pattern; a cardiac murmur is appreciated if the anomaly includes a patent ductus arteriosus.

Standard **chest radiographs** can only identify presence of aspiration pneumonia but provide limited information regarding the anatomy of the vascular ring. However, Buchanan[36] has suggested that a dorsoventral or ventrodorsal projection is sufficient for the diagnosis. In his study of 52 dogs with PRAA, all had leftward deviation of the trachea at the heart base, in contrast to 30 dogs with megaesophagus in which this finding was not present. Radiography also reveals dilation of the cranial esophagus, often air-filled or impacted with food. Contrast radiography provides confirmatory evidence of the location of the esophageal compression and barium swallows are frequently performed for this purpose.

Esophagoscopy may be considered to rule out intraluminal esophageal constriction and can support a vascular compression as a pulsatile vessel may be seen coursing adjacent to the level of constriction. However, esophagoscopy does not provide confirmatory evidence as to the anatomy of the vascular ring nor of concurrent arterial or venous anomalies.

Angiography either selective, non-selective, or **with contrast-guided cross-sectional imaging** (e.g., CTA or MRA) can confirm the diagnosis and fully delineate the venous and arterial anatomy. The simplest, though least precise, of these methodologies is a non-selective angiogram (injection of contrast into a peripheral vein). The best site for contrast injection is the left cephalic vein as this will allow delineation of a persistent left cranial vena cava, if present. Optimally, digital subtraction angiography should be employed, as a lower volume of contrast can be given for a diagnostic study. Following injection, fluoroscopic cine images are taken until the dye has circulated to the left heart and the aortic arch is apparent. The position of the ascending aorta as it crosses the tracheal bifurcation defines the arch - if the ascending aorta crosses the left bronchus, the animal has a left 4th arch; if it crosses the right, a diagnosis of PRAA can be made. Care should be taken to evaluate the cranial branches in addition to the arch laterality to screen for aberrant subclavian arteries or other malformations. Selective injection into the ascending aorta provides a clearer study, but is more costly and invasive. Advanced imaging techniques including CTA[45,46] have been described in animals and are commonly employed in people to define aortic arch laterality and associated anomalies.

Treatment

The treatment of choice for vascular ring anomalies is to surgically correct the vascular structure responsible for the compression. This can be achieved by isolating and dividing the ductus/ligamentum arteriosus either left or right, an aberrantly located subclavian artery, or one of the aortic arches (persistent double aortic arches) through a left or right intercostal thoracotomy. Thoracoscopic dissection of ligamentum in dogs with PRAA has also been described.[47]

This presentation will focus on CTA findings in all of the above anatomies, providing a framework for interpretation and highlighting the value of preoperative cross-sectional imaging. Cross-sectional imaging with CTA teaches us that more variations of vascular rings exist in veterinary species than was previously believed and obtaining these studies limits uncertainty for the surgeon once the thorax is opened and surgical intervention attempted.

References

1. Ciocca RG, *et al. Ann Vasc Surg*. 1995;9(2):204.
2. Gross RE. *N Engl J Med*. 1945;233(Journal Article):586.
3. Noden DM, *et al. The Embryology of Domestic Animals: Developmental Mechanisms and Malformations*. Williams & Wilkins; 1985.
4. Layton KF, *et al. Am J Neuroradiol*. 2006;27(7):1541.
5. Ellison GW. *Compendium*. 1980;2(9):693.
6. Imhoff RK, *et al. J Am Vet Med Assoc*. 1963;143:599.
7. Reed JH, *et al. Can Vet J/La Revue Veterinaire Canadienne*. 1964;5(9):240.
8. Buergelt CD, *et al. J Am Vet Med Assoc*. 1970;156(8):1026.
9. Lucker WF Jr., *et al. J Am Vet Med Assoc*. 1970;157(12):2090.
10. van den Ingh TS, *et al. J Am Vet Med Assoc*. 1974;164(9):939.
11. Fingeroth JM, *et al. J Am Vet Med Assoc*. 1987;191(8):981.
12. Ellison GW.
13. Gunby JM, *et al. J Am Vet Med Assoc*. 2004;224(7):1120.
14. Gorton B. *Vet Rec*. 1925;5:483.
15. Berry AP, *et al. Vet Rec*. 1984;114(14):336.
16. Core SH, *et al. Vet Med Small Anim Clin*. 1979;74(6):822.
17. Evans I, *et al. Vet Med Small Anim Clin*. 1971;66(11):1090.
18. Hathaway JE. *J Am Vet Med Assoc*. 1965;147:255.
19. KiChang L, *et al. J Vet Clin*. 2003;20(2):248.
20. Uzuka Y, *et al. Companion Anim Pract*. 1988;2(8):14.
21. Vögtli T, *et al. Schweiz Arch Tierheilkd*. 1994;136(10):335.
22. Bartels JE, *et al. J Am Vet Med Assoc*. 1969;154(4):406.
23. Petrick SW, *et al. J S Afr Vet Assoc*. 1978;49(4):355.
24. Mackey VS, *et al. Vet Surg*. 1986;15(4):325.
25. Butt TD, *et al. Can Vet J*. 1998;39(11):714.
26. Roberts SJ, *et al. Cornell Vet*. 1953;43(4):537.
27. Rooney JR 2nd, *et al. J Am Vet Med Assoc*. 1956;129(1):5.
28. van den Ingh TS, *et al. Vet Q*. 1986;8(1):37.
29. Butt TD, *et al. Vet Rec*. 2001;148(4):118.
30. Margiocco ML, *et al. Vet Clin North Am Food Anim Pract*. 2009;25(2):423.
31. Maclean RA, *et al. J Wildlife Dis*. 2008;44(2):499.
32. Ketz CJ, *et al. J Zoo Wildlife Med*. 2001;32(2):268.
33. Hurley K, *et al. J Am Vet Med Assoc*. 1993;203(3):410.
34. Holt D, *et al. Vet Surg*. 2000;29(3):264.
35. van der Linde-Sipman JS, *et al. Zentralbl Veterinarmed A*. 1981;28(7):569.
36. Buchanan JW. *J Vet Intern Med*. 2004;18(4):510.
37. Saunders AB, *et al. J Vet Cardiol*. 2013;15(2):153.
38. Martin DG, *et al. J Am Vet Med Assoc*. 1983;183(6):697.
39. Ferrigno CR, *et al. Anat Histol Embryol*. 2001;30(6):379.
40. Vianna ML, *et al. J Am Vet Med Assoc*. 2004;225(8):1222.
41. Du Plessis CJ, *et al. J Small Anim Pract*. 2006;47(1):31.
42. Moonan N, *et al. J Vet Cardiol*. 2007;9(1):59.

43. Yarim M, *et al. Vet Pathol.* 1999;36(4):340.
44. Goldin JP, *et al. J Zoo Wildlife Med.* 1999;30(1):145.
45. Joly H, *et al. Vet Radiol Ultrasound.* 2008;49(1):42.
46. Pownder S, *et al. J Vet Cardiol.* 2008;10(2):125.
47. Isakow K, *et al. J Am Vet Med Assoc.* 2000;217(9):1333.

acvim170

The VALVE Study - Is Triple Therapy Superior to Double Therapy for Heart Failure Treatment Due to Endocardiosis?

Vasotop® (Ramipril) in Addition to Lasix® and Vetmedin® in Canine Endocardiosis (VALVE Study)

Gerhard Wess (On Behalf of the VALVE Investigators)
Munich, Germany

Myxomatous mitral valve degeneration (MMVD) is the most common cardiac disease in dogs and the prevalence increases with age.[1,2] In the course of the disease some dogs develop severe clinical signs of CHF.[3] After having suffered from an episode of CHF, most dogs die within one year despite medical management.[4-6]

ACE-inhibitors used to be a standard treatment for the treatment of asymptomatic and symptomatic dogs with mitral valve disease. However, two prospective, randomized studies showed no benefit of ACE-I concerning the time to the onset of congestive heart failure given to asymptomatic dogs with MMVD.[7,8] The recommendation of giving an ACE-I in addition to diuretics to dogs with MMVD in CHF is based on two short term studies, COVE and IMPROVE as well as two long term studies (LIVE and BENCH). The COVE study (28 days) included dogs with dilated cardiomyopathy (DCM) and MMVD and it showed an improvement of the heart failure class, pulmonary edema scores and overall evaluation for both groups.[9] More pronounced short-term effects were seen for DCM. The IMPROVE (study 21 days) included DCM and MMVD and showed decreased pulmonary capillary wedge pressures in dogs with DCM.[10] The LIVE study followed a subpopulation of dogs from the two short-term studies DCM and MMVD (n=67) and showed an improved survival or longer in study than placebo 160 vs. 77.0 days for the dogs receiving an ACE-I.[6] The BENCH study included dogs with DCM and MMVD (n=125) in ISACHC class II and III and showed an improved survival or being longer in study than placebo 436 vs. 151 days, p=0.015 in favors for ACE-I.[11] However, not all dogs received diuretics, which makes it unlikely that those dogs were in CHF. Also, the survival times of the dogs receiving ACE-I were much longer compared to the dogs in the QUEST trial receiving an ACE-I.[4,11]

The guidelines for humans with mitral valve disease recommend valve repair as optimal surgical treatment in patients with severe MR. When repair is not feasible mitral valve replacement is recommended. Medical treatment consists of ACE-I in severe cases (no study cited).[12] The recent AHA/ACC guidelines for the management of patients with valvular disease recommend that "medical therapy is reasonable for systolic dysfunction in symptomatic patients with chronic primary MR (stage D) and LVEF less than 60% in whom surgery is not contemplated".[13] They state that only sparse data exist on the use of beta blockade, ACE-I, ARBs, aldosterone antagonists, as no convincing studies for ACE-I in CHF are conducted. Furthermore, a recent analysis of all mega trials using ACE-I in humans revealed, that ACE-I are beneficial in patients with reduced ejection fraction (DCM) and post myocardial infarcts.[14] The major effect of ACE-I in humans is being effective lowering the blood pressure and thus providing renal protective and additional cardiovascular protective effects, but doubts remain over the concept of blood-pressure-independent CV protection.[14]

Dogs with MMVD that have developed congestive heart failure have a clear benefit concerning survival times when they receive Pimobendan in combination with diuretic therapy compared with treatment consisting of an ACE-inhibitor plus diuretic therapy according to the QUEST trial (267 versus 140 days).[4] The ACVIM guidelines currently recommend the use of triple therapy including Pimobendan, ACEI and furosemide for the treatment of dogs in congestive heart failure, as there might be an additional beneficial effect of adding an ACEI to Pimobendan. However, there is currently no solid proof besides expert opinion that the triple therapy is superior compared to Pimobendan plus furosemide alone.

The objective of the VALVE study (comparative investigation into an extension of survival time benefits of **V**asotop® (Ramipril) in **A**ddition to **L**asix® and **V**etmedin® in canine **E**ndocardiosis) was to determine whether chronic oral administration of the ACEI ramipril (Vasotop®) in addition to pimobendan (Vetmedin®) and furosemide (Lasix®) would result in a survival time prolongation compared to a group of dogs treated with pimobendan and furosemide alone (standard therapy).

In order to be enrolled in the study dogs had to meet the following criteria:

- A presence of a moderate to high intensity heart murmur with maximal intensity over the mitral area.
- Echocardiographic evidence of advanced myxomatous mitral valve disease defined as characteristic valvular lesions of the mitral valve apparatus (leaflet thickening, valve prolapse, ruptured chordae tendineae).
- Presence of mitral regurgitation in colour Doppler echocardiography.
- Echocardiographic evidence of moderate to severe left atrial and/or left ventricular dilatation; i.e., LA/Ao ratio measured by 2D mode >1.5 and LV values above breed specific normal reference values.
- All animals must show or have shown radiographic evidence of pulmonary oedema and cardiomegaly (VHS score >10.5).
- Clinical signs of decompensated congestive heart failure must be present or have been resolved in the past with treatment (that is still being administered and in the view of the attending clinician necessary to stop the return of previous clinical signs).
- Animals must be five years or older.
- Animals must have a body weight under 20 kg.

Dogs were excluded if they had any of the following:

- Presence of other cardiac disease (congenital or acquired). Dog with tricuspid valvular insufficiency attributable to myxomatous mitral valve disease may be included if it is present together with significant mitral regurgitation where the latter is judged to be the major contributor to the presenting signs of disease.
- Evidence of other significant disease, systemic or organ related, such as liver disease, renal disease not consistent with pre-renal azotemia, etc., gastrointestinal disease that could interfere with drug absorption.
- Body weight >20 kg
- Female dog which is pregnant or lactating

The trial was a prospective single-blinded, randomized, international, multi-center clinical study. Dogs were enrolled by veterinary cardiologists at four trial centres in Germany and Switzerland.

158 canine patients were enrolled in the trial and were randomly allocated to either a pimobendan plus furosemide (standard therapy) or ACEI plus pimobendan and furosemide (triple therapy) group.

Scheduled reexaminations were at day 7, day 28, and 3 months after inclusion. Thereafter, the dogs were scheduled for reexamination every 3 months. On every occasion dogs were examined, the following occurred: a case history was obtained, a complete physical examination performed, a 3-minute ECG recorded, and blood was taken to measure creatinine, protein, sodium, and potassium concentrations. Echocardiography and thoracic radiographic examinations were scheduled every 6 months after inclusion. At the discretion of the investigator, additional testing and visits were permissible.

Recruitment to the study began in October 2005 and was finished in October 2015.

The final results of the study will be presented during this session. It is anticipated that the results will be published in 2017.

References

1. Detweiler DK, Patterson DF. The prevalence and types of cardiovascular disease in dogs. *Ann N Y Acad Sci*. 1965;127:481–516.
2. Buchanan JW. Prevalence of Cardiovascular Disorders, in Textbook of Canine and Feline Cardiology Principles and Clinical Practice. In: Fox PR, Sisson D, and Moise NS, eds. Saunders: Philadelphia. 1999;457–470.

3. Borgarelli M, Savarino P, Crosara S, Santilli RA, Chiavegato D, Poggi M, Bellino C, La Rosa G, Zanatta R, Haggstrom J, Tarducci A. Survival characteristics and prognostic variables of dogs with mitral regurgitation attributable to myxomatous valve disease. *J Vet Intern Med.* 2008;22:120–128.
4. Haggstrom J, Boswood A, O'Grady M, Jons O, Smith S, Swift S, Borgarelli M, Gavaghan B, Kresken JG, Patteson M, Ablad B, Bussadori CM, Glaus T, Kovacevic A, Rapp M, Santilli RA, Tidholm A, Eriksson A, Belanger MC, Deinert M, Little CJ, Kvart C, French A, Ronn-Landbo M, Wess G, Eggertsdottir AV, O'Sullivan ML, Schneider M, Lombard CW, Dukes-McEwan J, Willis R, Louvet A, DiFruscia R. Effect of pimobendan or benazepril hydrochloride on survival times in dogs with congestive heart failure caused by naturally occurring myxomatous mitral valve disease: the QUEST study. *J Vet Intern Med.* 2008;22:1124–1135.
5. BenchStudyGroup. The effect of benazepril on survival times and clinical signs of dogs with congestive heart failure: Results of a multicenter, prospective, randomized, double-blinded, placebo-controlled, long-term clinical trial. *J Vet Cardiol.* 1999;1:7–18.
6. Ettinger SJ, Benitz AM, Ericsson GF, Cifelli S, Jernigan AD, Longhofer SL, Trimboli W, Hanson PD. Effects of enalapril maleate on survival of dogs with naturally acquired heart failure. The Long-Term Investigation of Veterinary Enalapril (LIVE) Study Group. *J Am Vet Med Assoc.* 1998;213:1573–1577.
7. Kvart C, Haggstrom J, Pedersen HD, Hansson K, Eriksson A, Jarvinen AK, Tidholm A, Bsenko K, Ahlgren E, Ilves M, Ablad B, Falk T, Bjerkfas E, Gundler S, Lord P, Wegeland G, Adolfsson E, Corfitzen J. Efficacy of enalapril for prevention of congestive heart failure in dogs with myxomatous valve disease and asymptomatic mitral regurgitation. *J Vet Intern Med.* 2002;16:80–88.
8. Atkins CE, Keene BW, Brown WA, Coats JR, Crawford MA, DeFrancesco TC, Edwards NJ, Fox PR, Lehmkuhl LB, Luethy MW, Meurs KM, Petrie JP, Pipers FS, Rosenthal SL, Sidley JA, Straus JH. Results of the veterinary enalapril trial to prove reduction in onset of heart failure in dogs chronically treated with enalapril alone for compensated, naturally occurring mitral valve insufficiency. *J Am Vet Med Assoc.* 2007;1061–1069.
9. Group TCS. Controlled clinical evaluation of enalapril in dogs with heart failure: results of the Cooperative Veterinary Enalapril Study Group. The COVE Study Group. *J Vet Intern Med.* 1995;9:243–252.
10. Group TIS. Acute and short-term hemodynamic, echocardiographic, and clinical effects of enalapril maleate in dogs with naturally acquired heart failure: results of the Invasive Multicenter PROspective Veterinary Evaluation of Enalapril study. The IMPROVE Study Group. *J Vet Intern Med.* 1995;9:234–242.
11. Group BS. The effect of benazepril on survival times and clinical signs of dogs with congestive heart failure: Results of a multicenter, prospective, randomized, double-blinded, placebo-controlled, long-term clinical trial. *J Vet Cardiol.* 1999;1:7–18.
12. Vahanian A, Alfieri O, Andreotti F, Antunes MJ, Baron-Esquivias G, Baumgartner H, Borger MA, Carrel TP, De Bonis M, Evangelista A, Falk V, Lung B, Lancellotti P, Pierard L, Price S, Schafers HJ, Schuler G, Stepinska J, Swedberg K, Takkenberg J, Von Oppell UO, Windecker S, Zamorano JL, Zembala M. Guidelines on the management of valvular heart disease (version 2012): the Joint Task Force on the Management of Valvular Heart Disease of the European Society of Cardiology (ESC) and the European Association for Cardio-Thoracic Surgery (EACTS). *Eur J Cardiothorac Surg.* Germany. 2012;S1–44.
13. Nishimura RA, Otto CM, Bonow RO, Carabello BA, Erwin JP, 3rd, Guyton RA, O'Gara PT, Ruiz CE, Skubas NJ, Sorajja P, Sundt TM, 3rd, Thomas JD, 2014 AHA/ACC Guideline for the Management of Patients With Valvular Heart Disease: a report of the American College of Cardiology/American Heart Association Task Force on Practice Guidelines. *Circulation.* 2014;e521–643.
14. Dusing R. Mega clinical trials which have shaped the RAS intervention clinical practice. *Ther Adv Cardiovasc Dis.* 10:133–150.

acvim171

The Longitudinal Outcome of Mitral Valve Disease (LOOK-Mitral) Study: Preliminary Results

Michele Borgarelli, DVM, PhD, DECVIM-CA (Cardiology); LOOK-Mitral Investigators
Blacksburg, VA, USA

Introduction

Degenerative mitral valve disease due to myxomatous degeneration (MMVD), is the most common heart disease in dogs and the most frequent cause of congestive heart failure in this species. However, there are very few studies evaluating natural history and potential risk factors for progression of the disease in the canine.

Natural History

MMVD is characterized by a prolonged preclinical period and many affected dogs die from causes other than congestive heart failure. A study of 558 dogs affected by MMVD of differing clinical severity demonstrated that more than 70% of dogs were alive at the end of the follow-up period of 6.6 years. In another more recent study, 256 dogs affected with preclinical MMVD, mortality was 27% with cardiac death accounting for 11% during the 833 days observation period. The same study showed that 30/83 dogs, that were available for re-examination, progressed to a more advanced stage of MMVD within 1 year. A more recent study (VetCompass) assessed median survival time in a population of affected dogs in a primary-care veterinary practice in England. The reported median survival time was 25.4 months after disease detection. Another study aimed to assess the prognostic values of echocardiography in dogs affected with MMVD reported a median survival time of 297 days.

Prognostic Indicators

Reported risk factors associated with progression of the disease or death include: age, gender, intensity of heart murmur, degree of valve prolapse, severity of valve lesions, the degree of mitral valve regurgitation, degree of left atrial enlargement, severity of eccentric hypertrophy, rupture of chordae tendinae, and increased concentration of natriuretic peptides. Left atrial enlargement assessed by echocardiography appears to represent one of the most reliable independent predictors of death/risk factors. The risk of death from cardiac disease for dogs with a left atrium/aortic root ratio exceeding 1.7 was 2.1 times that of dogs with smaller atrial dimensions. E wave peak velocity of transmitral flow was also found to represent an independent risk factor for death or onset of congestive heart failure in multiple studies. We have also recently reported that the presence of pulmonary hypertension increases risk of death in dogs with MMVD.

The value of monitoring natriuretic peptides in affected dogs has been investigated in few studies. One study reported that plasma NT-proBNP concentration was an independent risk factor for development of first onset-congestive heart failure. In another study, NT-proBNP level <965 pmol/l was associated with a significant longer survival. A study conducted on Cavalier King Charles Spaniels, reported that NT-proANP >1000 pmol identifies dogs with an increased risk of developing heart failure. In humans, several studies have reported that combining multiple cardiac biomarkers may be advantageous to better stratify patients with cardiac diseases. In veterinary medicine, it has been reported that combining NT-proBNP with high sensitivity troponin I had a higher prognostic value than either single cardiac biomarker in identifying MMVD dogs with higher risk of death.

The presence of atrial fibrillation was also associated with an increased risk of death in dogs with MMVD.

The LOOK-Mitral Study

The Longitudinal Outcome of Mitral valve disease (LOOK-Mitral study) is a multi-center prospective longitudinal study intended to:

1. Assess the natural history of canine mitral valve disease (MMVD) in a large referred population.
2. Assess the prevalence and effects of co-morbidities on the natural history of the disease.
3. Identify risk factors for progression and death.

The study was initiated in October 2015 and represents a collaborative research project between the Virginia-Maryland College of Veterinary Medicine and the CVCA Cardiac Care for Pets group. Inclusion criteria for study enrollment is a diagnosis of MMVD confirmed by echocardiographic exam. Dogs are enrolled in the study at the time the diagnosis is confirmed by a cardiology specialist. Collected data include: physical exam, radiographic findings, ECG findings, echocardiography, co-morbidities, laboratory results and treatment. Data are collected at each reevaluation until the termination of the study (October 2020) or until the death of the dog.

To date, the database contains clinical information from 2,053 dogs of 104 different breeds. Follow-up examination data is available for approximately 30% of the dogs.

REFERENCES

1. Borgarelli M, *et al. J Vet Int Med.* 2015;29(2):564–574.
2. Borgarelli M, *et al. J Vet Int Med.* 2012;26(1):69–75.
3. Borgarelli M, *et al. J Vet Int Med.* 2008;22(1):120–128.
4. Hezzell MJ, *et al. J Vet Int Med.* 2012;26(20):302–311.
5. Jung SW, *et al. J Vet Int Med.* 2016;30:51–57.
6. Mattin MJ, *et al. Preventive Vet Med.* 2015;122:436–442.
7. Reynolds CA, *et al. J Vet Cardiol.* 2012;14(1):193–202.
8. Sargent J, *et al. J Vet Cardiol.* 2015;17(1):1–12.
9. Wolf J, *et al. J Vet Cardiol.* 2012;14:399–408.

acvim172

Feline Cardiomyopathies: New Perspectives

Philip R. Fox, DVM, MS, DACVIM (Cardiology), DECVIM-CA (Cardiology)
New York, NY, USA

Overview

By general consensus, myocardial disease accounts for 95% of cardiovascular disorders in the cat, and experience affirms that these diseases can levy considerable morbidity.[1] Thanks to widespread availability of echocardiography and diagnostic cardiac biomarkers, public and veterinary awareness of these diseases has increased remarkably. Indeed, familial disease patterns are increasingly reported for cardiomyopathies, most notably hypertrophic cardiomyopathy (HCM).[2-4] This has led to replacement of the now outdated impression that held these conditions to be rare entities, with more contemporary appreciation that considers their prevalence to be relatively common.

Epidemiology

The complications of HCM have been widely reported, but the natural history of preclinical ('occult') heart disease has not been elucidated. Moreover, there is very little outcome data for other forms of feline heart disease and overall, epidemiologic studies are wanting.

Disease Recognition

During early development of echocardiography, it was presumed that this technology would remove the diagnostic uncertainties surrounding disease phenotype. Often, echocardiography accomplishes this task.[5] In cats it seems, however, that more data do not always result in clarity. Enhancements in echo-Doppler imaging acuity - while providing finer distinctions in cardiac structure and myocardial function, may not necessarily improve clinical interpretation or lead with certainty, to what form of heart disease is actually present- or annoyingly in some cases, whether heart disease is even present or evident at all.

Clinical Diagnosis

Over the past 50 years, a number of cardiomyopathy classification schemes have been proposed for humans, intended to categorize and elucidate forms of myocardial disorders from each other. These diagnostic systems have been regularly updated and modified, accompanied by task force guidelines from American, European, and other working groups.[6] Increasingly, genetic screening is used to clarify diagnosis in human patients. This has contributed to disease recognition and evolution of disease classification, sometimes leading to newer, often molecular-based schemes. These schemes can also spin away from the simple, classic phenotypes that originally defined basic forms of feline cardiomyopathy, and which still constitute the basic templates for veterinary diagnoses.

Disease Heterogeneity and Variability Over Time

Virtually all cats with cardiomyopathy are essentially 'vulnerable' to changes in phenotype and function over time. Whether a consequence of structural and functional remodeling affecting myocytes, interstitium, and microcirculation; inflammation/infection; infarction; neuroendocrine modulation; or unknown but presumed genetic factors - shifting and vacillating phenotypes both affect the criteria and judgements used to compose clinical diagnoses. In addition, under-appreciated pathologic changes affecting the mitral and tricuspid valve apparatus, myxomatous valve degeneration, infarction, and other factors contribute to health outcomes, and collectively, influence the disease in ways not yet identified.

Despite these vagaries, there is still diagnostic value using, where appropriate, the four basic categories of feline cardiomyopathy. These include hypertrophic, restrictive, dilated, and arrhythmic forms. Ventricular noncompaction may represent a new, rare form of disease.

Collectively, they continue to represent the most reasonable heart disease 'snapshot' at any given time course, and are often embellished with slight adjustment.

Cardiovascular practice involving feline heart disease should continue to progress in concert with advances in imaging technology, biomarkers, and especially, evidence-based research. Additionally, clarification of natural history, risk factors, and management strategies will lead to improved diagnosis and therapy for these complex myocardial diseases.

References

1. Fox PR, *et al. J Vet Cardiol.* 2015;17(Suppl 1):S150.
2. Maron BJ, *et al. J Vet Cardiol.* 2015;17(Suppl 1):S6.
3. Trehiou-Sechi E, *et al. J Vet Intern Med.* 2012;26(3):532.
4. Wess G, *et al. J Vet Intern Med.* 2010;24(3):527.
5. Maron BJ, *et al. J Am Coll Cardiol Cardiovasc Imaging.* 2016;9(7):858.
6. Sen-Chowdhry S, *et al. Nat Rev Cardiol.* 2016;31(11):651.
7. Kittleson MD, *et al. J Vet Intern Med.* 2017;epub ahead of print.

1.
acvim173

An Update and Paradigm Shift on Heartworm Resistance to Macrocyclic Lactones: Where Have All the LOEs Gone?

Clarke Atkins, DVM, DACVIM (Small Animal Internal Medicine, Cardiology); Tobi Ku; Lee Payne, DVM
Raleigh, NC, USA

Introduction

Beginning in the mid-2000s, there has been a change in the experience concerning dogs infected with HW in the Mississippi Delta.[1] Specific concerns include heartworm infections in dogs purported to have been treated appropriately with HW preventives and increased difficulty in clearing HW microfilariae with macrocyclic lactone antibiotics (ML). This was perceived and widely accepted as a manifestation of resistance to ML, used as both heartworm preventives and, off-label, as microfilaricidal agents.

This is despite the fact that classical "rules" of resistance have been violated, including:

1. **Resistance is much less likely to occur in a region with large refugia**. The Delta has the largest refugia for HW in the USA.
2. **Resistance is characterized as rapidly spreading**. LOE reports have not occurred in numbers greater than expected outside of the MS Delta (bordered by lines drawn between Memphis, TN; Gulfport, LA; Lake Charles, LA; and Port Girardeau, MO).
3. **Resistance is forever**. The incidence of LOEs peaked in 2009 and had fallen by nearly half by 2011.

In addition, the validity of LOE reports has fallen under question, as 80–98% of 301 reimbursement claims have been invalidated with careful scrutiny of the medical records and product purchase history.[2] Nevertheless, resistant HW isolates exist and should be of great concern to our profession, pharmaceutical companies, the FDA-CVM, and the public.

Hypothesis

1. While resistant HW isolates have been discovered, this is not the cause of the MS Delta HW crisis.
2. Furthermore, the "epidemic of LOEs" is not due to resistance, but to an unrelated, multi-factorial heartworm epidemic that opened a vigorous and successful search for resistance.
3. A significant role in the HW epidemic was played by the spate of hurricanes hitting the gulf coast from 2002–2008.
4. The HW and LOE epidemic has declined to low levels or even disappeared in some regions, approximately a decade later.

Methods - Abbreviated

The 2000–2015 medical records of the Animal Clinic of Oxford, MS were scrutinized for numbers of LOEs, HW diagnoses, and HW tests. The number of hurricanes, rainfall, and lake levels were tabulated by year.

Results

To be presented.

Conclusion

The explosion of LOEs from ~2004–2010 that created such angst for Delta veterinarians and their clients, as well as for pharmaceutical companies, was in fact an explosion of heartworm cases, which resulted in an intense and successful search for resistant HW biotypes. The decline in HW cases and LOEs after 2009 was likely caused by multiple factors. These might include changes in preventive and housing practices, the use of doxycycline, application of mosquito blocking agents, and a renewed emphasis on HW prevention. But we believe the most important factor to be the almost complete silence of hurricane activity. Implicit in this conclusion is the belief that the LOE explosion did not cause the HW epidemic but was simply a by-product of the HW epidemic and the concern regarding resistance that it spawned. The MS Delta, as defined, appears to be a unique ecosystem(s), regarding HW infection, but is not the only source of resistant biotypes, as 50% of recognized ML-resistant biotypes have come from areas outside the MS Delta.

References

1. Hartogensis M. *FDA Veterinarian.* 2005;20:2–3.
2. Atkins C, *et al. Vet Parasitol.* 2014;206:106–113.

1.
acvim174

Cardiovascular Anesthesia

Luisito S. Pablo, DVM, MS, DACVAA
Gainesville, FL, USA

Introduction

Dogs and cats presented for general anesthesia may have preexisting cardiovascular disease. Depending upon the severity of the heart condition, oxygen delivery to the tissue may be reduced in these cases. The reduced cardiac output is more likely due to decreased stroke volume. The determinants of stroke volume are preload, afterload, and myocardial contractility. The hemodynamic goals for patients with cardiovascular diseases involve targeting these factors with the ultimate objective of providing adequate cardiac output and oxygen delivery to the anesthetized patients. Different heart conditions will have different hemodynamic goals. A drug may be a good choice for a certain cardiac disease and not for another heart condition.

The main objective of this presentation is to present the anesthetic management of specific cardiovascular diseases in dogs and cats. The cardiovascular conditions in this presentation are the more commonly encountered problems in practice.

Dogs with Mitral Regurgitation (MR)

The main cause of MR in dogs is endocardiosis. The clinical consequences of mitral valve endocardiosis are seen in elderly small-breed dogs. The anesthetic risk associated with MR depends on its severity. Mild MR does not have hemodynamic consequence. Dogs with untreated congestive heart failure will have the highest risk. Dogs being treated with an ACE inhibitor and diuretics for congestive heart failure can be anesthetized, but extreme care should be taken. Anesthesia should not be performed in dogs with untreated congestive heart failure.

With MR, part of the left ventricular stroke volume is ejected through the incompetent mitral valve into the left atrium. It creates a volume load on the left atrium and left ventricle, resulting in increased pressure in those chambers. With these increased pressures, the LA and LV dilate and develop hypertrophy. Later in the disease process, the increased pressures will be reflected back upon the pulmonary venous circulation and this ultimately results in pulmonary edema.

The basic hemodynamic problem in MR is the reduction in forward left ventricular stroke volume and cardiac output. To improve the hemodynamic state in MR, the following should be done: (1) reduce preload slightly to reduce regurgitant flow (preload); (2) avoid acute increases in afterload; (3) maintain contractility; (4) avoid bradycardia; (5) maintain sinus rhythm; and (6) do not increase myocardial oxygen requirement by avoiding severe tachycardia and hypotension.

To accomplish the above-mentioned goals, there are specific steps that can be done. Do not overload the patient with fluids. Do not use fluid boluses to treat hypotension during anesthesia. Fluid should be given at a lower rate of 3.0–5.0 ml/kg/hour. Those patients with compensated CHF should receive fluid with lower sodium load, like 2.5% dextrose and half-strength LRS. Sinus bradycardia should be avoided, because if there is prolonged systole, the regurgitant fraction is increased. Sinus bradycardia may also lead to left ventricular volume overload. It is best to maintain a normal heart rate; however, slight tachycardia is acceptable. Atropine (0.02 mg/kg IV) or glycopyrrolate (0.01 mg/kg IV) should be given to correct sinus bradycardia. Myocardial contractility should be maintained. If hypotension develops, beta-1 agonist like dobutamine or dopamine is administered as a constant-rate infusion. Dobutamine and dopamine can be given at 5.0–10.0 µg/kg/min. Increases in systemic vascular resistance should be avoided, because it can cause decompensation of the left ventricle. Slight vasodilation is acceptable. $Alpha_2$-agonists should not be used in dogs with MR. The patients should be handled gently, as excitement may cause vasoconstriction as a result of catecholamine release.

Anesthetic Considerations

Low dose of acepromazine is acceptable as a premedicant. This will reduce the afterload. It will provide sedation and minimize catecholamine-induced dysrhythmias. It can be administered at 0.01–0.02 mg/kg IM, SC. Opioids are indicated if there will be pain involved. Most opioids produce minimal myocardial depression and are good choices as premedicants. However, they may cause sinus bradycardia. An anticholinergic (atropine or glycopyrrolate) can be added to the preanesthetic protocol if the patient has preexisting bradycardia or if a relatively high dose of an opioid is used. The opioids that can be used for premedication are morphine, hydromorphone, methadone, butorphanol, and buprenorphine. These are the recommended dosages for the different opioids: (1) morphine and methadone: 0.2–0.5 mg/kg IM, SC; (2) hydromorphone: 0.05–0.1 mg/kg IM, SC; (3) butorphanol: 0.2–0.4 mg/kg IM, SC; and (4) buprenorphine: 0.01–0.02 mg/kg IM, SC. Benzodiazepines are acceptable premedicants for MR. They produce minimal cardiopulmonary depression. They can produce paradoxical agitation in dogs if given alone or with an opioid. Instead of sedation, the dog may show restlessness and dysphoria. It is best to avoid the use of benzodiazepines in dogs that are not depressed. Diazepam and midazolam are the two most commonly used benzodiazepines in dogs. Midazolam is water soluble and can be administered IM, SC, or IV. Diazepam is not water soluble and when given IM will result in unpredictable absorption. These are the dosages for the benzodiazepines: (1) diazepam: 0.2–0.4 mg/kg IV; and (2) midazolam: 0.2–0.4 mg/kg IM, SC, IV.

Propofol is an acceptable option for dogs with MR. It causes some degree of peripheral vasodilation, which will be beneficial for MR. When used in these cases, it should be given slowly following premedication. The dose of propofol is 4.0 mg/kg IV; half of the dose should be given over 40–60 seconds, and the remaining dose is given to effect. Propofol can depress myocardial contractility, which is considered a negative effect for MR. To minimize the amount of propofol used during induction, diazepam or midazolam (0.25 mg/kg IV) can be used with propofol. With this technique, propofol at 1.0 mg/kg is given first, followed by the whole dose of the benzodiazepine. The rest of the propofol is given to effect (1.0 mg/kg at a time) until the dog can be intubated.

A combination of diazepam-ketamine or midazolam-ketamine can be used in MR. The combination will result in transient sinus tachycardia. Ketamine by itself does not result in significant increase in peripheral vascular resistance. Diazepam is given at 0.25 mg/kg IV and ketamine at 5.0 mg/kg IV. Lower doses should be used in dogs that have profound sedation from the premedicants and are depressed. An alternative IV dosing for diazepam and ketamine can also be used, featuring a lower dose of ketamine (3.3 mg/kg) and a higher dose of diazepam (0.33 mg/kg).

Etomidate is the induction agent with best cardiovascular profile and an excellent option for MR. However, it is still expensive and not readily available. It is the drug of choice in patients with enlarged heart and manifesting signs of congestive heart failure. The dose of etomidate is 0.5–1.0 mg/kg IV. Benzodiazepine (0.25 mg/kg) can also be administered with etomidate to reduce its adverse effects (myoclonus, retching, sneezing) during induction.

A neuroleptanalgesic combination is another option for MR. This combination maintains myocardial function. Bradycardia may occur due to the opioid's effect on the vagal tone. Anticholinergic can be added to the protocol to prevent sinus bradycardia. The two most common combinations used in practice are diazepam-hydromorphone and diazepam-fentanyl. The doses for diazepam and hydromorphone are 0.25 mg/kg IV and 0.2 mg/kg IV, respectively. When fentanyl is chosen, instead of hydromorphone, the dose is 10 μg/kg IV. This combination should be reserved for very depressed patients with MR.

Anesthesia should be maintained with either isoflurane or sevoflurane. Dogs that develop hypotension and tend not to respond to the positive inotrope should be managed by "balancing" the anesthetic technique. A fentanyl CRI at 0.2–0.6 μg/kg/min can be added to the inhalant administration. This will allow the use of the lower concentration of the inhalant.

Drugs with vasoconstrictive effect should be avoided in dogs with MR, and they include xylazine, medetomidine, dexmedetomidine, ephedrine, and phenylephrine.

Body temperature of the patient should be supported using an external heat source. Hypothermia can result in increased oxygen demand. Oxygen should be provided until the patient is ventilating adequately. Hypoventilation will result in hypoxemia when breathing room air. It is important to provide optimal recovery conditions. Stress should be minimized. Analgesics should be administered to dogs perioperatively. It is not recommended to give NSAIDs in patients with compensated congestive heart failure because of the decrease in effective circulating blood volume.

Cats with Hypertrophic Cardiomyopathy (HCM)

Many cats may have hypertrophic cardiomyopathy without manifesting any signs. These cats can have mild to severe left ventricular thickening. Cats with hypertrophic cardiomyopathy can live longer than two years after a bout of pulmonary edema. It is prudent not to perform anesthesia in cats with signs of congestive heart failure due to hypertrophic cardiomyopathy.

Left ventricular myocardium thickens (concentric hypertrophy) in HCM. This results in stiff chamber. The basic problem in HCM is the diastolic dysfunction of the left ventricle. It is characterized by delayed or incomplete myocardial relaxation and reduced compliance of the left ventricle. The stiff chamber leads to increased diastolic intraventricular pressure and then left atrial enlargement. The increased pressure in left atrium will be transmitted to the pulmonary veins, and pulmonary edema will follow. Mitral regurgitation will develop as a result of the distortional change in the mitral valve apparatus caused by the hypertrophy of the left ventricle. In those cats with asymmetric hypertrophic cardiomyopathy, dynamic left ventricular outflow tract (LVOT) obstruction can develop when the anterior leaflet of the mitral valve moves to contact the septum during systole. Conditions that can lead to increased outflow obstruction include increased myocardial contractility, decreased preload (hypovolemia, tachycardia, and controlled ventilation), and decreased afterload (vasodilator).

Hemodynamic goals for HCM include: (1) maintain preload in cats without congestive heart failure; (2) avoid reduction of afterload; (3) do not increase myocardial contractility; (4) maintain normal heart rate and avoid sinus tachycardia; (5) maintain sinus rhythm; and (6) avoid increases in myocardial oxygen consumption. To accomplish these goals, there are specific steps that should be considered. Adequate diastolic filling can be accomplished by large ventricular volume and sinus rhythm. Increased contractility, reduced afterload and preload will decrease the ventricular volume. Ketamine should be avoided because it increases myocardial contractility and heart rate. It also increases the myocardial oxygen demand. Acepromazine at the higher dose will cause vasodilation resulting in augmentation of the outflow obstruction. Dysrhythmias should be controlled before anesthesia. Minimize stress and excitement because of the catecholamine release that will increase heart rate and contractility. The drugs (beta blocker or calcium channel blocker) being taken by the patient for HCM should be administered up to and including the day of anesthesia.

Anesthetic Considerations

Opioids are considered good choices because they have minimal effects on myocardial contractility, preload, and afterload. The opioids that are commonly used for premedication are hydromorphone, morphine, methadone, butorphanol, and buprenorphine. The premedicant doses for the opioids are as follows: hydromorphone (0.03–0.05 mg/kg IM, SC), morphine (0.1 mg/kg IM, SC), methadone (0.1 mg/kg IM, SC), butorphanol (0.2 mg/kg IM, SC), and buprenorphine (0.01–0.02 mg/kg IM, SC).

Benzodiazepines are also suitable for these cases. They produce minimal cardiopulmonary depression. However, they can produce paradoxical agitation and restlessness in cats. To minimize these side effects, midazolam or diazepam can be given with the IV induction agent. A small amount of the induction agent is given first. The dose of midazolam or diazepam is given next. The induction is completed with the primary induction agent given to effect. The dose for diazepam and midazolam is 0.2–0.4 mg/kg. Diazepam is preferably given IV while midazolam can be given by IM, SC, or IV route.

Medetomidine has been shown to eliminate outflow tract obstruction in cats with dynamic left ventricular outflow tract obstruction. The dose used in the study is 20 μg/kg IM. It appears that medetomidine or now dexmedetomidine may be used in cats with HCM that are not amenable to physical restraint during examination. Dexmedetomidine causes peripheral vasoconstriction and bradycardia. It may also decrease the myocardial contractility. These effects are known to decrease outflow obstruction.

The use of propofol in cats with hypertrophic cardiomyopathy is debatable. There is a study that showed that it impairs regional myocardial function in ischemic myocardium. It also causes peripheral vasodilation. If used, it should be given very slowly and with a benzodiazepine. The calculated dose is 4.0 mg/kg IV; half of the dose should be given over 40–60 seconds, and the remaining dose is given to effect.

Etomidate is an excellent option for HCM. However, its cost limits its use in veterinary practice. It is the author's drug of choice in cats with enlarged heart and have shown signs of congestive heart failure at presentation. The dose is 0.5–1.0 mg/kg IV. Half of the calculated dose is given as a slow bolus, and the rest of the dose is given to effect. A benzodiazepine can also be administered with etomidate.

Neuroleptanalgesic combination is another option in cats with HCM. The combination maintains myocardial function. Diazepam and hydromorphone are given at 0.25 mg/kg IV and 0.2 mg/kg IV, respectively. The fentanyl can replace the hydromorphone and is given at 10 µg/kg IV. This combination works well in depressed cats. When this combination is planned, a dose of propofol should be available so that in case the cat cannot be intubated because of swallowing, induction can be completed with a low dose of propofol.

Mask induction using sevoflurane or isoflurane should be avoided because of the stress and excitement associated with this induction method. The outflow obstruction will be worse because of the increased heart rate and myocardial contractility due to the sympathetic stimulation.

The drugs that should be avoided are high to moderate doses of acepromazine, ketamine, atropine, glycopyrrolate, and thiopental.

Anesthesia should be maintained with either isoflurane or sevoflurane. Cats that develop hypotension should be managed by administering phenylephrine CRI at 1–5 µg/kg/min. It is advisable to avoid dobutamine and dopamine because they will increase myocardial contractility. Positive pressure ventilation should be instituted with caution because it can decrease preload. A fentanyl CRI at 0.2–0.6 µg/kg/min can be added to the inhalant administration. This will allow the use of the lower concentration of the inhalant.

The patient should be monitored closely during anesthesia. The monitoring tools should include pulse oximetry, blood pressure measurement, capnography, ECG, and temperature probe.

The measures taken postoperatively will minimize mortality and morbidity in these cases. External heat support should be provided. Hypothermia will result in increased oxygen demand during recovery. Adequate ventilation and oxygenation should be ensured. High concentration of oxygen should be administered until the cat can show that it is ventilating well. Stress should be minimized to prevent unwanted catecholamine release. Pain control should be instituted throughout the perioperative period. It is advisable not to give NSAIDs in patients with compensated congestive heart failure because of the reduction in effective circulating blood volume.

Dilated Cardiomyopathy (DCM)

DCM is common in large-breed dogs. The main concerns in this condition are the reduction in myocardial contractility and dysrhythmias. The hemodynamic goals in these patients that need anesthesia include: (1) reduce preload, (2) avoid increases in afterload, (3) maintain or increase contractility, (4) maintain normal heart rate, (5) maintain sinus rhythm, and (6) do not increase myocardial oxygen requirement by avoiding severe tachycardia and hypotension. These goals are almost similar to those set for mitral regurgitation. Therefore, a detailed discussion of the management of DCM is not included in these notes.

References

1. Clutton RE. *Manual of Small Animal Anaesthesia and Analgesia.* 1999:155.
2. Skubas N, *et al. Clinical Anesthesia.* 5th ed. 2006:886.
3. Bready LM, *et al. Decision Making in Anesthesiology.* 4th ed. 2007:130.

acvim175

Echocardiography versus Radiography for the Diagnosis and Management of Heart Disease

Deborah M. Fine-Ferreira, DVM, MS, DACVIM (Cardiology)
Kailua-Kona, HI, USA

Introduction

Thoracic radiography and echocardiography are both invaluable tools in the diagnosis and management of heart disease. Although there is certainly overlap in the information obtained by these two modalities, they should be viewed as complementary, rather than competing tools.

Echocardiography provides two general categories of information about the heart: morphology and function. Morphology of the heart provides information such as chamber dimensions, appearance of the valves, evidence of effusions, and presence of tumors. Morphology is assessed using 2D and M-mode. Systolic and diastolic function is assessed using a combination of these modes and Doppler imaging.

Unfortunately, ultrasound has extremely limited utility for evaluation of air-filled structures, thus thoracic radiography is an invaluable diagnostic tool for evaluation of pulmonary parenchyma and vasculature. Thoracic radiographs also provide a good overall sense of the size of the cardiac silhouette relative to the thorax. Although viewers often use the terms "heart" and "cardiac silhouette" interchangeably, it is important to remember that this structure on a radiograph simply represents the pericardium and its contents. If the cardiac silhouette is interpreted as enlarged, then the next question becomes "what is causing the enlargement"?

Thoracic Radiography Reviewed

Alterations in the cardiac silhouette on thoracic radiographs are useful for suggesting specific chamber enlargements. On lateral views, the cranial aspect of the cardiac silhouette is composed of the right atrium and ventricle and the caudal aspect is composed of the left atrium and ventricle. The cranial dorsal region, the "cranial cardiac waist," is composed of three overlapping structures: the right atrium, pulmonary artery, and aorta. Loss of the cranial cardiac waist can be due to enlargement of any of these structures and evaluation of the orthogonal view is necessary to determine which is involved. The caudal dorsal region, the "caudal cardiac waist," is composed only of the left atrium. In dogs, serial radiographs to assess for progressive left atrial enlargement are very useful for monitoring for progression of degenerative mitral valve disease. In cats, left atrial enlargement is more easily appreciated on the DV/VD projections. Left atrial enlargement causes widening of the heart base, commonly referred to as a valentine-shaped or shield-shaped heart.

A clock-face analogy is often used to identify structures on the DV/VD view. The aortic arch is approximately at 11:00–1:00; the pulmonary artery is from 1:00–2:00; the left auricle is from 2:00–3:00; the left ventricle is 3:00–6:00; the right ventricle is 6:00–9:00; and the right atrium is 9:00–11:00. The left atrium is central and caudal to the bifurcation of the mainstem bronchi.

Various methods have been developed to quantify enlargement of the cardiac silhouette using thoracic radiographs. On lateral views, the cranial-to-caudal width of the heart should span less than 2.5 intercostal spaces for deep-chested dogs, and less than 3.5 intercostal spaces for barrel-chested dogs. The height of the heart from the carina to the apex should be ≤2/3 of the height of the thoracic cavity. On DV/VD views, the maximal width of the heart should be ≤1/2 the width of the hemi-thorax at the 9th rib. These are relatively subjective measures, and a vertebral heart scale (VHS) system was developed to try to provide a more objective determination of cardiac size. Using the lateral view, the long and short axes of the heart are measured. The measurements are then superimposed over the vertebral column, both starting at the cranial edge of the fourth thoracic vertebral body (T4), and the number of vertebrae covered is summed. Normal dogs have a VHS of 8.5–10.6 although there are some breed variations with Boxers, Dachshunds, Labradors and Cavaliers having normal VHS up to 12. Normal cats have a VHS of 7.2–7.8. I routinely use a combination of methods to interpret cardiac size.

Abnormal lung patterns are described as interstitial, alveolar, or bronchial. Frequently, a combination of patterns is present in the same patient. An interstitial pattern appears as an increase in the overall opacity of the lungs. All normal structures are still visible, but the margins are hazy. Interstitial patterns can be structured (e.g., nodular) or unstructured. An alveolar pattern is characterized by complete loss of margins of the vasculature, and the presence of air bronchograms. An air bronchogram is caused by fluid or cells filling the terminal alveoli, resulting in a uniform soft tissue opacity surrounding the radiolucent bronchus. Alveolar patterns occur with pulmonary edema, pneumonia, hemorrhage, atelectasis, lung lobe torsion, or neoplasia. A bronchial pattern is caused by thickening of the airways, either from inflammation or mineralization of the airways themselves (e.g., asthma, bronchitis) or interstitial fluid tracking along the airway. Radiographically, this appears as "donuts" and "tramlines."

Changes in the pulmonary vasculature are indicative of the underlying pathology. Enlargement of the pulmonary arteries is implicative of pulmonary hypertension. Enlargement of the veins is a hallmark feature of congestive heart failure, and is the result of increased cardiac preload. Symmetrical enlargement of both the arteries and veins indicates a combination of pulmonary hypertension and increased left atrial pressure. In a young animal, this is most likely due to a left-to-right shunting cardiac defect. In an older animal, this is most likely due to a combination of left heart failure and chronic pulmonary disease.

Early cardiogenic pulmonary edema appears as interstitial to bronchointerstitial infiltrate. As failure worsens, the interstitium becomes saturated and an alveolar pattern may be seen. Cardiogenic edema is typically bilaterally symmetrical. In the dog, it is most prominent in the caudal-dorsal and perihilar regions. In cats, it is typically found in the ventral regions, although a diffuse distribution can also be seen. Large volume pleural effusion as a manifestation of heart failure is much more common in cats than in dogs.

Generally, left-sided congestive heart failure is considered to be present and active by the finding of three radiographic changes: 1) left atrial enlargement, usually accompanied by left ventricular enlargement; 2) distention of the pulmonary veins (with or without pulmonary arterial distension); and 3) interstitial to alveolar pattern in a location consistent with heart failure for the species. The presence of these three signs indicates that, at very minimum, therapy with a loop diuretic such as furosemide, and an ACE-inhibitor should be initiated. Most dogs will also benefit from the addition of pimobendan to the above combination.

Echocardiography Reviewed

Echocardiography, in contrast to radiographs, allows very precisely measurements of chamber dimensions. Echocardiograms usually begin with views from the right hemithorax, and the first view commonly acquired is the right parasternal long-axis four-chamber view. The heart is positioned horizontally across the screen with the left atrium (LA) and ventricle (LV) in the far field and the right atrium (RA) and ventricle (RV) in the near field. By making a slight rotation of the transducer, the aorta is brought into view, the so-called "five" chamber view. Here the LA is much smaller, but the aorta is in full view and represents the fifth chamber. From this view the transducer is rotated 90 degrees and the heart is seen in the short axis. There are three standard short-axis views: 1) LV at the papillary muscles, 2) LV at the mitral valve (MV), and 3) the heart base showing the aortic valve, LA, RA, RV and RV outflow tract (RVOT).

To complement these 2D views, a series of M-Mode views are also acquired. M-Mode uses only one ultrasound beam, requiring less raw data to be processed. Consequently, M-Mode has greater image resolution which usually allows more precise measurements of wall thickness and chamber dimensions than 2D. To measure the LV chamber dimensions and wall measurements, a cursor is placed across the 2D image so that the cursor crosses the interventricular septum (IVS) and LV posterior (or free wall; LVPW) between the papillary muscles. The resulting image is a graphical representation of the heart with Time along the X axis and Depth along the Y axis. The RV and IVS are in the near field and the LV lumen appears in the middle, with the LVPW in the far field. From this image, measurement of the LV can be made in diastole and systole. From the mitral valve view, the motion of the MV can be recorded during the cardiac cycle. From the base of the heart, a measurement of the LA and aortic diameters is made. A ratio of LA to aorta (LA/Ao) can also be calculated.

Diastolic wall thicknesses obtained on M-mode are compared to weight-matched reference ranges. In cats with HCM or systemic hypertension, and dogs with sub-aortic stenosis, these measurements can be markedly increased. Diseases causing eccentric hypertrophy (e.g., mitral regurgitation, dilated cardiomyopathy, and left-to-right shunts) result in chamber dilation. An increase in the LV systolic dimension is implicative of systolic dysfunction. Only one M-Mode measurement is taken from the MV, the "E point" to septal separation (EPSS). The EPSS measures the distance from the MV at its maximal opening to the IVS. This distance increases in the face of systolic dysfunction. Multiple views are used to measure the LA. In addition to the short-axis M-mode, a 2D measurement of the LA can be measured from the same view: the measurement line through the LA extends from and parallel to the commissure of the aortic valve between the non-coronary and left-coronary cusps. An LA/Ao ratio can be calculated and in a healthy patient this ratio is usually <1.25. Increased preload causes the LA to increase in size; however, the aorta remains static. Patients in heart failure frequently have an LA/Ao ratio of >2 or more. Another common measure of 2D LA size is obtained from the long-axis view, at the maximal width of the LA, parallel to the mitral valve annulus.

No single measurement provides a fool-proof method of assessing systolic function. The most common measures used are percent fractional shortening (%FS) and ejection fraction (%EF). Each of these measurements determines the percent change in the LV from diastole to systole. Fractional shortening is a percent change in the linear measurement of the LV diameter from on M-Mode using the following formula: (LVIDd - LVIDs)/LVIDd*100 = %FS (where LVIDd=LV internal diastolic dimension and LVIDs=LV internal systolic dimension). %FS is the easiest measure of systolic function, but it has important limitations. The %FS measures the change in the LV lumen in a single line, but the heart moves in three dimensions. For a 3-D evaluation of LV systolic function, the %EF is used. To measure the LV volume requires some mathematical assumptions since the LV is not a perfect cylinder, but more of a truncated ellipse. Using a formula called the Modified Simpson's rule, the area of the LV is traced from the long-axis four-chamber view. The computer measures the maximal length and then fills this area with a series of disks approximating the area at various levels of the LV. The area of these disks is summed to generate an approximate volume. Another advantage of the %EF is that the difference between the LV volume in diastole and systole is equal to the stroke volume, which allows for calculations of cardiac output by multiplying stroke volume by heart rate. In most clinical situations, if the %FS is normal, then the patient is considered normal. If the %FS is not normal, then the other assessments of systolic function are utilized. However, keep in mind the most important limitation of all of these measures: they are simply derived from the ratio of the starting point in diastole to the ending point in systole; thus, abnormalities of either of these dimensions will alter the final product. You cannot assume because a patient has an ejection fraction below 50% that systolic dysfunction is present (i.e., the diastolic dimensions could be small due to volume depletion). Similarly, a normal ratio does not guarantee that systolic function is adequate: a large increase in diastole due to increased preload can result in this ratio being "normal" in a patient with decreased contractility. The bottom line: treat your patient, not the numbers.

In addition to these standard views and measurements, other image planes are often useful. Views of the main pulmonary artery and its branches may allow visualization of heartworms or a patent ductus arteriosus, if present. In the presence of pericardial effusion, a tipped short-axis view may bring a tumor on the tip of the right auricle into view. From the left hemithorax, the heart is seen in a four-chamber view with the atria in the far field and the apex of the ventricles in the near field. From this orientation, accurate Doppler velocity measurement across the valves may be acquired.

A complete discussion of Doppler is beyond the scope of this article. However, Doppler is used to assess direction and velocity of flow in order to determine if regurgitant or stenotic valves are present, identify and quantitate shunts, and assess tissue motion for diastolic abnormalities. Most commonly, Doppler is used to quantify the severity of obstruction at a stenotic valve by measuring the velocity of flow out the valve. The severity of obstruction is directly proportional to the velocity and is related by the formula: pressure gradient = $4 \times V^2$ where V is velocity in m/sec. Understanding normal cardiac physiology is crucial to interpret the findings. For example, normal velocity out of the aorta is about 1.0 to 1.7 m/sec. Using this equation, this indicates a LV to aortic PG of ~ 4 to 12 mm Hg. In subaortic stenosis, this velocity can be markedly increased. If the velocity recorded by CW Doppler measured 5.0 m/sec, then a PG of 100 mm Hg exists.

Which Test When?

The answer is based upon the following question: "**What kind of information are you trying obtain by performing the test?**" The test(s) listed below represents my opinion of the easiest way to answer the question in a routine clinical setting.

Information desired	Recommended initial diagnostic test
What is the cause of this patient's murmur?	Echocardiogram
Is this patient with a heart murmur safe for general anesthesia?	Thoracic radiographs ± echocardiogram
Why is this patient coughing?	Thoracic radiographs
Is this patient in heart failure? Should I start medications?	Thoracic radiographs
Did the heart failure medications work? Why is this patient still coughing?	Thoracic radiographs
Why is this patient having syncopal episodes?	Echocardiogram, ECG, ± thoracic radiographs
Why is this patient having labored breathing?	Thoracic radiographs
Does this patient have pericardial effusion?	Echocardiogram
Does this old dog with a murmur have severe heart disease?	Thoracic radiographs ± echocardiogram
Does this young dog with a murmur have severe heart disease?	Echocardiogram ± thoracic radiographs
Is this patient safe for doxorubicin?	Echocardiogram
Is this dog's mitral valve disease getting worse over time?	Thoracic radiographs ± echocardiogram
Does this large breed dog have dilated cardiomyopathy?	Echocardiogram
Does this cat have hypertrophic cardiomyopathy?	Echocardiogram
Does this patient have pulmonary hypertension?	Echocardiogram

In summary, echocardiography provides important information about cardiac morphology and function, whereas thoracic radiographs allow evaluation of pulmonary parenchyma and determination of the functional consequences of heart disease. Thoracic radiographs are irreplaceable for patients with respiratory signs. These two modalities are complementary and invaluable resources for the diagnosis and treatment of heart disease. Knowing what answers a clinician is attempting to learn, will guide the diagnostic approach.

References

References are available upon request.

acvim176

Point-of-care Thoracic Ultrasound in the Emergency Diagnosis of Congestive Heart Failure in Dogs and Cats

Jessica Ward, DVM, DACVIM (Cardiology); Teresa DeFrancesco, DVM, DACVECC, DACVIM (Cardiology)
Ames, IA, USA

INTRODUCTION

Emergency diagnosis of congestive heart failure (CHF) in dogs and cats can be challenging. Thoracic radiographs (TXR) are considered a relatively high-yield test for diagnosis of dyspneic patients, but images are sometimes equivocal, and positioning for TXR can exacerbate respiratory distress in a compromised patient. Point-of-care thoracic ultrasound has emerged as a rapid noninvasive screening test for dyspneic veterinary patients that can accelerate the diagnosis and treatment of CHF. Point-of-care thoracic ultrasound can function as a "visual stethoscope"[1] to supplement auscultation findings and improve diagnostic accuracy of the physical examination prior to acquisition of TXR.

FLUID DETECTION AND VOLUME STATUS: AFAST AND TFAST

Point-of-care ultrasound examinations become common practice as an extension of the physical exam in human emergency and urgent care facilities. The original codified ultrasound technique for emergency clinicians was the FAST examination (Focused Assessment with Sonography for Trauma), designed to detect free abdominal fluid in patients with blunt abdominal trauma.[2] Application of FAST quickly expanded to include fluid detection in the thorax (TFAST), allowing assessment of pleural and pericardial effusion.[3] More recently, the RUSH (Rapid Ultrasound in SHock) examination was developed to add information about volume status and cardiac function to the basic FAST scan.[4] RUSH incorporates information about vena cava size and distensibility, as well as basic echocardiographic views, often obviating the need for central venous pressure measurements in the intensive care unit.[5-7]

Protocols for point-of-care ultrasound to detect free fluid have also been established in veterinary medicine. A 4-quadrant AFAST procedure has been codified to detect and quantify abdominal fluid in dogs and cats.[8] Additionally, a 5-point TFAST examination has been described to detect pleural effusion, pericardial effusion, and pneumothorax in veterinary patients. The TFAST protocol involves imaging bilaterally at caudodorsal sites ("chest tube site," or CTS), bilaterally over the heart ("pericardial site," PCS) and subcostally ("diaphragmatico-hepatic site," DH).[8] Images of the caudal vena cava (CVC) and hepatic veins can also be obtained from the DH view and may be useful in estimating central venous pressure and right heart volume overload.[9] Basic 2-dimensional echocardiography images (short-axis views at the level of the left ventricle and heart base) can also be obtained from right parasternal views; these images can be used to assess left ventricular volume overload or underload, left ventricular systolic function, and left atrial enlargement.

The combination of AFAST, TFAST, vena cava imaging, and basic echocardiography views provides a wealth of information that can aid diagnosis and treatment of critical patients. First, this point-of-care imaging rapidly establishes the presence of cavitary effusions and can guide either diagnostic or therapeutic centesis. Such a rapid diagnosis is particularly critical in patients with pericardial effusion and cardiac tamponade causing right-sided CHF (ascites). Second, among patients with ascites or pleural effusion, point-of-care thoracic ultrasound can greatly increase or decrease clinical suspicion of CHF as the cause of the cavitary effusion. For example, in a dog with ascites, an enlarged right heart as well as distension and decreased distensibility of the CVC make right-sided CHF more likely, while normal right heart dimensions and CVC size and distensibility suggest a noncardiac cause of abdominal effusion. In a cat with pleural effusion, a left atrium-to-aorta ratio of >2.0 makes CHF more likely, while a normal left atrial size can rapidly rule out CHF. In this way, AFAST and TFAST can provide additive information to the clinical history and physical examination findings (e.g., heart rate and rhythm, presence of heart murmur, presence of jugular venous distension) to stratify index of suspicion of CHF and guide initial diagnostic testing and treatments.

Detection of Alveolar-Interstitial Disease: Lung Ultrasound

Another advancement in point-of-care ultrasound for emergency assessment of patients is lung ultrasound (LUS).[10-14] Although the presence of air in alveoli attenuates the ultrasound beam and impedes **direct** imaging of the lungs themselves, LUS can suggest the presence of lung disease by identification of certain ultrasound **artifacts** caused by pulmonary pathology. Specifically, LUS can suggest the presence of pulmonary edema by identifying ultrasound artifacts (B-lines, also called "ring-down artifacts," "comet-tails," "lung rockets," and others) caused by increased lung water.[10,15] B-lines (the terminology advocated by a recent international consensus statement)[10] are created when small fluid-filled alveoli, which are below the resolution threshold of the ultrasound beam, are surrounded by air, creating a high impedance gradient. These artifacts are identified ultrasonographically as discrete narrow-based vertical hyperechoic artifacts, which extend from the pleural-pulmonary interface to the far aspect of the ultrasound screen without fading, and which move synchronously with respiration.[10,13,16,17]

The most widely adopted LUS protocol in humans, termed bedside lung ultrasound in emergency (BLUE), codifies the examination of 4 sites on each side of the thorax. For each site, the examiner assesses presence and number of B-line artifacts within a single intercostal space. As more B-lines are seen, eventually the discrete linear artifacts become so numerous that they begin to coalesce and can no longer be individually enumerated. The number of B-lines per intercostal space is therefore scored as follows: 0, 1, 2, 3, or >3 (see Figure 1). The "cutoff" criterion for suspicion of CHF in a dyspneic person is the presence of multiple, bilateral B-lines: specifically, >3 B-lines at 2 or more locations on either side of the thorax. Using this criterion in human emergency settings, LUS can differentiate CHF from noncardiogenic causes of dyspnea with high sensitivity and specificity,[18-21] and similar or greater positive predictive value than NT-proBNP[13,17,18,22] or TXR[23-25]. Increased numbers of B-lines are also linearly correlated with the amount of lung water and the pulmonary capillary wedge pressure.[26,27] The diagnostic utility of LUS makes sense in this context, since the most relevant differential diagnoses for dyspnea in people are CHF (pulmonary edema; lung water and many B-lines) and chronic obstructive pulmonary disease (no lung water, no B-lines). A recent international consensus panel[10] and meta-analysis[14] confirm the utility of LUS in dyspneic people and the diagnostic accuracy for prediction of CHF.

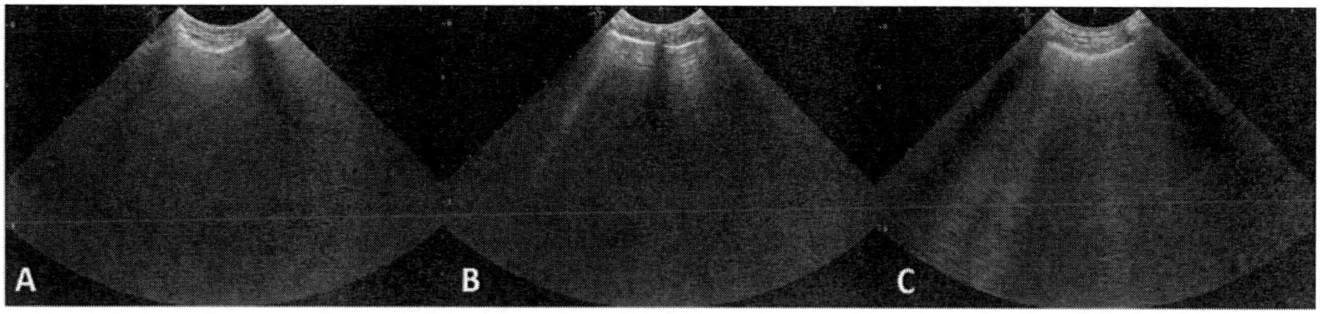

Figure 1. Example of lung ultrasound images demonstrating variable number of B-lines per intercostal space (rib shadows display in black). (A): Normal lung ultrasound. (B): Single B-line per intercostal space. (C): Multiple (>3) coalescing B-lines per intercostal space.

LUS is also emerging as an extension of point-of-care ultrasound in veterinary medicine. A protocol for veterinary LUS, modeled after the human protocol and termed Vet BLUE, has been published to standardize LUS examinations (see Figure 2).[28] Similar to the human protocol, Vet BLUE involves assessment of 4 sites on each hemithorax, with quantification of B-lines at a single intercostal space within each site. Initial studies demonstrated that B-lines are relatively rare in normal dogs and cats, but common and widely distributed in dogs diagnosed with CHF.[28,29] More recently, the diagnostic utility of

LUS was assessed in 100 dyspneic dogs and cats. Using the same criterion as in people (>3 B-lines in at least 2 sites on either side of the thorax), LUS had a sensitivity of 84% and specificity of 74% for identifying CHF (similar to the accuracy of TXR).[30] Diagnostic accuracy was even higher in cats, with sensitivity of 85% and specificity of 87%. In another study of 368 patients, absence of **any** B-lines on LUS ruled out CHF with specificity of 88% in dogs and 96% in cats.[31] Species differences in diagnostic utility of LUS are logical when considering the most common **noncardiac** differential diagnoses for respiratory distress in dogs versus cats. The most common noncardiac cause of dyspnea in cats is inflammatory airway disease ("feline asthma"); like human COPD, this disease lacks interstitial/alveolar infiltrates, and B-lines are absent. In contrast, noncardiac causes of dyspnea in dogs are more varied and include diffuse alveolar/interstitial diseases such as ARDS, noncardiogenic pulmonary edema, disseminated fungal disease, and pulmonary thromboemboli. Such diseases would result in multiple multifocal B-lines and would be miscategorized by LUS as suggestive of CHF. The diagnostic utility of LUS (particularly specificity) therefore depends on the most likely differential diagnoses for dyspnea in a given patient.

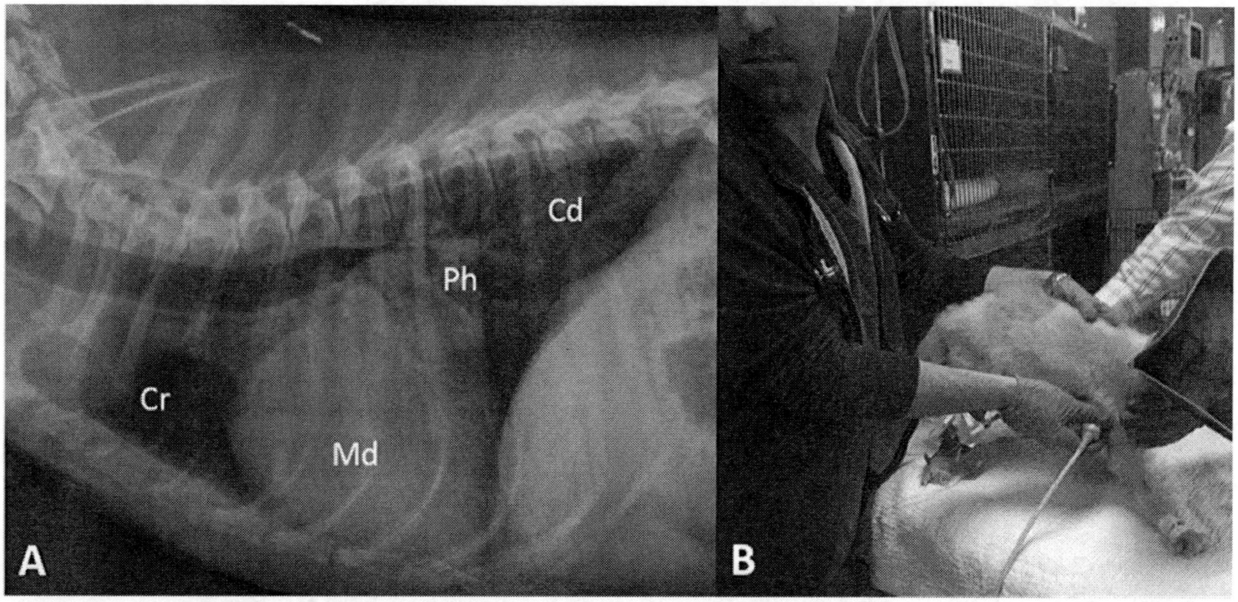

Figure 2. Vet BLUE technique for LUS. (A): The ultrasound probe is held horizontally at 4 sites on each hemithorax to record images (Cd = caudal, Ph = perihilar, Md = middle, Cr = cranial), corresponding to locations on a lateral TXR. (B) An emergency clinician performing a Vet BLUE examination in a dyspneic patient.

In addition to B-lines, other artifacts can be detected on lung ultrasound and may have diagnostic value. These findings include the "shred" sign (irregularity in the pulmonary-pleural interface, suggestive of lung consolidation or severe focal alveolar disease); the "tissue sign" (a triangular discontinuity between lung lobes at the pulmonary-pleural interface, also suggestive of consolidation or potential pulmonary thromboembolism); and the "nodule" sign (focal circular or ellipsoid structure with hyperechoic border, suggestive of a pulmonary nodule or mass).[28,32,33] When performing LUS, it is important to assess not only the presence of B-lines and other artifacts, but also their distribution. For example, multiple B-lines and a "shred" sign in the right middle lung lobe area (but not in any other location) may be suggestive of aspiration pneumonia; multifocal small "nodules" might suggest metastatic neoplasia or disseminated fungal disease. Future studies in veterinary medicine will assess the diagnostic utility of pattern-based approaches to analyzing LUS artifacts in patients with pulmonary disease.

Conclusion: From FAST to Focused, Toward a Global POCUS (Point-of-Care Ultrasound)?

Over the past 2 decades, point-of-care ultrasound has evolved from a simple "fluid check" to a sophisticated suite of protocols capable of evaluating volume status, cardiac function, and pulmonary pathology. These data have particular application for hastening the diagnosis and treatment of CHF. Despite the capacity to obtain ever more diagnostic information, point-of-care ultrasound remains rapid, noninvasive, and minimally stressful for compromised patients, making it an excellent first-line diagnostic screening technique in the emergency setting. Veterinarians have followed physician trends in incorporating point-of-care ultrasound into practice. Protocols will continue to be refined and additional diagnostic uses examined as veterinarians progress toward a comprehensive "global point-of-care ultrasound."

References

1. Kobal *et al. Am J Cardiol.* 2005;96:1002.
2. Boulanger *et al. J Trauma.* 1996;40:867.
3. Tayal *et al. J Ultrasound Med.* 2004;23:467.
4. Perera *et al. Emerg Med Clin North Am.* 2010;28:29.
5. Stawicki *et al. J Am Coll Surg.* 2009;209:55.
6. Chen *et al. Acad Emerg Med.* 2007;14:841.
7. Nagdev *et al. Ann Emerg Med.* 2010;55:290.
8. Lisciandro. *J Vet Emerg Crit Care.* 2011;21:104.
9. Nelson *et al. Vet Radiol Ultrasound.* 2010;51:313.
10. Volpicelli *et al. Intensive Care Med.* 2012;38:577.
11. Moore *et al. N Engl J Med.* 2011;364:749.
12. Lichtenstein, Meziere. *Chest.* 2008;134:117.
13. Liteplo *et al. Acad Emerg Med.* 2009;16:201.
14. Al Deeb *et al. Acad Emerg Med.* 2014;21:843.
15. Lichtenstein *et al. Am J Respir Crit Care Med.* 1997;156:1640.
16. Lichtenstein. *Minerva Anestesiol.* 2009;75:313.
17. Prosen *et al. Crit Care.* 2011;15:R114.
18. Gargani *et al. Eur J Heart Fail.* 2008;10:70.
19. Vitturi *et al. J Ultrasound.* 2011;14:147.
20. Hassdenteufel. *J Vet Emerg Crit Care.* 2013;23(4):416.
21. Cibinel *et al. Intern Emerg Med.* 2012;7:65.
22. Manson *et al. West J Emerg Med.* 2011;12:102.
23. Volpicelli *et al. Am J Emerg Med.* 2006;24:689.
24. Gargani *et al. Crit Care Med.* 2007;35:2769.
25. Peris *et al. Anesth Analg.* 2010;111:687.
26. Jambrik *et al. Am J Cardiol.* 2004;93:1265.
27. Agricola *et al. Chest.* 2005;127:1690.
28. Lisciandro. *Vet Radiol Ultrasound.* 2014;55:315.
29. Rademacher *et al. Vet Radiol Ultrasound.* 2014;55:447.
30. Ward *et al. J Am Vet Med Assoc.* In press.
31. Lisciandro *et al. J Vet Emerg Crit Care.* In press.
32. Lichtenstein. *Expert Rev Respir Med.* 2010;4:533.
33. Lichtenstein. *Intensive Care Med.* 2004;30:276.

acvim177

Update on Congenital Heart Disease: Nomenclature

Brian A. Scansen, DVM, MS, DACVIM (Cardiology)
Fort Collins, CO, USA

INTRODUCTION

Congenital heart disease (CHD) represents a spectrum of lesions, occurring in isolation or combination, with historically complex nomenclature. The complexity that can arise during maldevelopment of the mammalian heart has resulted in various classification schemes to characterize CHD. Pediatric cardiology has adopted a sequential segmental approach to analysis of CHD, providing a systematic template to follow during description of a congenitally malformed heart.[1-4] Each cardiac segment (atria, ventricles, great arteries) and connection (atrioventricular, ventriculoarterial) is analyzed and malformations noted. The segmental analysis evolved due to a need to describe increasingly complex lesions. Sequentially analyzing each segment avoids the risk of associating adjacent connections as hallmarks of the segment of interest. That is, while the left atrium may be expected to receive the pulmonary veins, the left atrium is not defined by this relationship. In an animal with complex CHD, the pulmonary veins may have an anomalous connection to the right atrium or systemic vasculature. Each cardiac segment is defined morphologically by features that are largely specific to that segment. These include the narrow-based attachment of the left atrial appendage vs. a broad-based attachment of the right atrium; coarse apical trabeculations with a prominent moderator band of the right ventricle compared to a more smooth-walled left ventricle; bifurcation and lung connection of the pulmonary trunk compared to cranial branches for the aorta.[4] While reports describing this type of analysis exist in veterinary medicine,[5-8] widespread adoption of sequential segmental analysis in the description of animals with CHD has not occurred. This may be due in part to high rates of perinatal mortality for complex CHD that limit the number of complex cases presented. However, complex lesions are occasionally encountered in veterinary cardiology and the stepwise approach of segmental analysis helps to organize the cardiologist's approach in the evaluation of such hearts. Although not all known forms of CHD have been fully described in animals, it is likely that all developmental anomalies that occur in humans, and perhaps even more forms, also occur in other mammalian species. A sequential categorization of CHD forms follows below and will be highlighted in the presentation.

ABNORMALITIES IN LATERALITY

Situs Inversus and Dextrocardia

Abnormalities in left-to-right laterality can occur during development, which may result in abnormal cardiac positioning within the thorax. If these structures are positioned opposite to normal, with the left atrium on the right and vice versa, the animal is classified as having atrial situs inversus. Atrial situs inversus is almost always seen as a component of the entire viscera (thoracic and abdominal) exhibiting a mirror-image arrangement of left-to-right laterality, termed situs inversus totalis. Mammals can develop with uncertain atrial arrangements, termed situs ambiguous. Typically, such cases are defined by the presence of 2 morphologically similar atria with absence of the other (right atrial isomerism = 2 right atria; left atrial isomerism = 2 left atria). Dextrocardia is a term used to describe a heart with a rightward apex-to-base axis, in contrast to the normal leftward axis.[9]

ABNORMALITIES IN VENOATRIAL CONNECTIONS

Anomalous Pulmonary Venous Connection

The primitive common pulmonary vein develops from an out-pouching of the left atrium and fuses with the primitive pulmonary veins; at this time, the connections from the pulmonary veins to the systemic venous system regress and go away. If the embryologic communications from pulmonary veins to systemic veins persist, either partially or completely, the animal is born with partial or total anomalous pulmonary venous connection (TAPVC), respectively. These defects can be considered comparable to an atrial septal defect (ASD) as they result in a left-to-right shunt at the atrial level. Classification of TAPVC is based on the site of abnormal connection and classified as supracardiac, cardiac, infracardiac, and mixed.

Cor Triatriatum Sinister

If the connection of the common pulmonary vein to the left atrium is stenotic, the left atrial chamber may appear as 2 discrete compartments with a dividing membrane. This clinical entity is termed cor triatriatum denoting 3 atria within a single heart; in veterinary medicine, the modifying term sinister is added to distinguish this condition from a divided right atrium (cor triatriatum dexter). Types of cor triatriatum sinister are categorized as an accessory atrial chamber that receives all pulmonary veins and communicates with the left atrium (type A), receives all pulmonary veins and does not communicate with the left atrium (type B), or receives some of the pulmonary veins with the others connecting normally or anomalously (type C). Among veterinary species, cor triatriatum sinister is seen most frequently in the cat.

Abnormalities in the Atria

Atrial Septal Defect and Patent Foramen Ovale

An atrial septal defect (ASD) is a hole in the intracardiac septum that divides the upper chambers of the heart, of which there are 4 principal forms. The ostium primum ASD is a defect in the ventral septum directly above the AV valves and is a form of atrioventricular septal defect (AVSD). Secundum ASDs are the most common form in veterinary species; secundum defects result from insufficiency of tissue at the fossa ovalis in the central interatrial septum. Sinus venosus defects occur in the dorsal interatrial septum and can be located either caudally or cranially. A rare form of ASD is the unroofed coronary sinus, in which there is a communication between the left atrial wall and the great cardiac vein, often associated with persistence of the left cranial vena cava, which allows for the shunting of blood from left atrium to right atrium through the coronary sinus.

Abnormalities in Atrioventricular Connection

Atrioventricular Septal Defect

Atrioventricular (AV) septal defects, formerly called endocardial cushion or atrioventricular canal defects, result from maldevelopment of the atrioventricular septum - that portion of the heart that normally exists due to an offset of the tricuspid annulus toward the cardiac apex relative to the mitral annulus. Defects of the AV septum result in the tricuspid and mitral annuli located along the same anatomical plane and variably result in maldevelopment of the ventral atrial septum, dorsal interventricular septum, and portions of the AV valves. These defects may be partial, transitional/intermediate, or complete.

Abnormalities in Atrioventricular Valves

Mitral Valve & Tricuspid Valve Dysplasia

Malformations of the mitral valve complex include shortened and thickened chordae, direct papillary to valve attachments, papillary muscle fusion, and altered leaflet geometry. These malformations often lead to mitral valve regurgitation; valvular stenosis is less common. The presence of a supravalvular ring of fibrous or fibromuscular tissue directly above the mitral annulus and resulting in impaired left ventricular inflow is considered within the spectrum of MVD.

Dysplasia of the tricuspid apparatus is similar to the mitral - thickened and elongated leaflets, shortened or absent chordae, direct papillary to leaflet attachment, papillary muscle fusion and hypertrophy, and variable degrees of stenosis or regurgitation. A common eponym of this condition is Ebstein's anomaly. This implies specific meaning from human medicine, namely that the leaflets of the tricuspid valve fail to delaminate from the ventricular myocardium resulting in the appearance of an apically displaced and cranially rotated tricuspid annulus. In the author's experience, true Ebstein-like malformations are exceptionally rare in animals.

Abnormalities in the Ventricles

Ventricular Septal Defect

Ventricular septal defect (VSD) classification is variable, but broadly attributed to which portions of the septum are deficient when viewed from the perspective of the right ventricle. Defects that result in fibrous continuity of the aortic and tricuspid valve are perimembranous, while defects that result in fibrous continuity between the aortic and pulmonary valves are juxtaarterial. Defects that are entirely surrounded by septum are muscular defects. Defects in the membranous septum are the most common in humans and animals, excluding small muscular defects seen in neonates.

Double-Chambered Right Ventricle and Primary Infundibular Stenosis

Double-chambered right ventricle (DCRV) refers to the presence of anomalous muscle bundles traversing the middle of the right ventricle, separating it into a hypertrophied proximal chamber and a distal, low-pressure chamber. This lesion is often associated with VSD. Primary infundibular stenosis is a similar condition characterized as a fibromuscular band of tissue at the junction of the right ventricular cavity and pulmonary infundibulum (proximal right ventricular outflow tract).

Abnormalities in Ventriculoarterial Connection

Tetralogy of Fallot

Although thought of as four lesions, the tetralogy of Fallot (TOF) may be better considered as a singular defect. There is malrotation and improper division (septation is displaced cranially) of the aortic and pulmonary outflow tracts, which leads to the four components: pulmonary stenosis, right ventricular hypertrophy, VSD, and dextroposition of the aorta. A characteristic feature of TOF is hypertrophy of the septoparietal trabeculation, which distinguishes this from pulmonary stenosis with VSD.

Pulmonary Atresia

Pulmonary atresia can be considered an extreme form of TOF in which the aorta receives nearly all the embryologic truncus and the pulmonary trunk persists as an atretic stalk. Severe cyanosis develops early in life and this lesion can be considered "ductal-dependent," meaning pulmonary blood flow, and therefore systemic oxygenation, relies upon patency of the ductus or other major aorticopulmonary collateral vessels. Formerly, this defect was termed pseudotruncus arteriosus given similar imaging findings with truncus arteriosus, though Tet-PA is preferred as it reflects the comparable etiopathogenesis between this defect and TOF.

Double Outlet Right Ventricle

Double outlet right ventricle (DORV) is a cardiac malformation in which both great vessels arise entirely or predominately from the right ventricle. The forms of DORV relate to the position of the interventricular communication relative to the aorta and pulmonary trunk - the defect may be subaortic, subpulmonary, doubly committed, or remote/non-committed.

Transposition of the Great Arteries

Transposition of the great arteries (TGA) refers to ventriculoarterial discordance, with the pulmonary trunk arising from the left ventricle and the aorta arising from the right, and results in severe cyanosis. Congenitally corrected TGA is a rare defect, in which ventriculoarterial discordance is present along with atrioventricular discordance. Aside from the ventricular 'inversion,' this combination maintains the normal series path of blood flow (RA to LV to pulmonary trunk to lungs to pulmonary veins to LA to RV to aorta).

Truncus Arteriosus Communis

Truncus arteriosus communis (common arterial trunk) is characterized by a single great vessel leaving the heart through a common valve to provide systemic, pulmonary, and coronary circulations. The malformation results from a lack of septation within the embryologic outflow tracts resulting in no division between the aortic and pulmonary trunks. Classification relates to aortic dominance or pulmonary dominance, relating to the morphologic features of the proximal trunk. Pulmonary circulation in truncus arteriosus is reduced if there is a restriction or stenosis at the origin of the pulmonary arteries or high-resistance pulmonary vascular disease, while it is increased if entry is unrestricted because the pulmonary circulation originates from a high-pressure vessel.

Abnormalities in the Semilunar Valves

Aortic Valve & Pulmonary Valve Stenosis

Stenoses of the semilunar valves are common defects in animals, particularly the dog. Obstruction to ventricular outflow results in concentric hypertrophy, myocardial ischemia, and clinical signs of heart failure or syncope.

Abnormalities in the Great Arteries

Patent Ductus Arteriosus

Persistent patency of the ductus arteriosus is commonly regarded as the most frequent congenital heart defect of dogs in North America and in Australia. Left-to-right shunt volume varies with ductal anatomy and pulmonary vascular resistance. The development of progressive pulmonary vascular disease is associated with reduced velocity of left-to-right shunting on Doppler echocardiography.

Aortopulmonary Window

Aortopulmonary window is a rare CHD that is pathophysiologically comparable to PDA. Rather than persistence of the fetal circulation, as occurs with PDA, aortopulmonary window results from failure to close the aortopulmonary foramen in the developing aortic sac. Classification of aortopulmonary window includes three types: small proximal defects (Type I), origin of the right pulmonary artery from the aortic aspect of the window (Type II), and a large defect that extends from the sinotubular junction to the origin of the left pulmonary artery (Type III).

Coarctation of the Aorta, Tubular Hypoplasia, Interrupted Aortic Arch

The coarctation is a ridge of tissue at the aortic isthmus - the junction of the arch and the descending aorta - adjacent to the site of the ductus or ligamentum attachment and immediately distal to the left subclavian artery. In addition to a focal constriction, narrowing of the aortic arch may occur over a longer segment and this is termed tubular hypoplasia. In the most severe case, the ascending aorta is separated from the descending aorta, a condition referred to as an interrupted aortic arch.

Pulmonary Arterial Stenosis

Stenosis of the pulmonary trunk or branch pulmonary arteries appears less frequently than stenosis of the pulmonary valve or infundibulum. Four forms exist: Type I lesions are confined to a single constriction of the pulmonary trunk or the right or left pulmonary arteries; Type II lesions begin at the bifurcation of the pulmonary trunk and extend into the ostia of the left and right pulmonary arteries; Type III lesions involve multiple segmental pulmonary arteries; and Type IV lesions involve multiple stenoses of both peripheral and central pulmonary arteries.

Unilateral Absence of a Pulmonary Artery

Unilateral absence of a pulmonary artery is a rare defect, in which one of the branch pulmonary arteries is absent from birth. Systemic-to-pulmonary collateral vessels are common and help to provide some perfusion to the intrapulmonary arteries in the lung ipsilateral to the absence.

References

1. Van Praagh R. *Circulation*. 1977;56(2):139.
2. Shinebourne EA, *et al. Br Heart J*. 1976;38(4):327.
3. Van Praagh R. *Cardiovasc Intervent Radiol*. 1984;7(3–4):115.
4. Anderson RH, *et al. Ann Pediatr Cardiol*. 2009;2(1):24.
5. Schwarzwald CC. *Equine Vet Educ*. 2008;20(6):305.
6. Abduch MC, *et al. J Small Anim Pract*. 2003;44(8):374.
7. Michaelsson M, *et al. Congenital Heart Malformations in Mammals: An Illustrated Text*. 2000:19.
8. Scansen BA, *et al. J Vet Cardiol*. 2015;17(S1):10.
9. Evans WN, *et al. Pediatr Cardiol*. 2010;31(1):1.

acvim178

Balloon Valvuloplasty for Subaortic Stenosis Update in Dogs: Who, How and Why?

Amara H. Estrada[1], DVM, DACVIM (Cardiology); Mandi Kleman[2], DVM, DACVIM (Cardiology)
[1]Gainesville, FL, USA; [2]Atlanta, GA, USA

While the diagnosis of severe forms of subaortic stenosis in dogs is straightforward, treatment and decisions on when to consider medical versus interventional therapy remain a challenge.[1-5] Combined cutting balloon and high-pressure balloon valvuloplasty (CB/HPBV) as a treatment for dogs with severe SAS has proven to be effective in reducing the systolic pressure gradient across the left ventricular outflow tract (LVOT) in both the short- and medium-term follow-up. However, long-term postoperative outcomes in dogs with severe SAS have varied.[3,4] While some dogs have maintained the reduction in LVOT gradient and alleviation of clinical signs, other dogs have re-stenosed, developed recurrence of syncopal events, sudden death or have progressed to the development of left ventricular myocardial failure and congestive heart failure.[3,4] Many of the dogs enrolled in our cohort of dogs for our pilot study have now been lost to follow-up and sufficient numbers of dogs are not available for long-term survival statistics. Thus, most of our information regarding the long-term efficacy of the procedure for our patient population remains anecdotal.

The cause of the patient variation in response to the procedure is unknown. Studies in humans and adult dogs have shown that SAS can be associated with a measurably abnormal aortoseptal angle (AoSa).[6-8] One hypothesis for this variation may be variable shear stress on the basal most interventricular septum immediately apical to the aortic valve, as a result of irregular blood flow caused by this abnormal aortoseptal angle AoSA. This sheer stress may lead to fibrosis and creation or exacerbation of an SAS lesion.[6-8] We have recently evaluated the AoSA in our cohort of dogs to assess whether the magnitude of the AoSA may serve as a useful criterion for determining which dogs will respond better to balloon valvuloplasty and may therefore be considered better candidates for the procedure.[9] The results of this study suggest that dogs with severe SAS (mean left ventricular to aortic PG = 143 mm Hg; range = 80–322 mm Hg) and an AoSA >160° as measured on right-sided parasternal long-axis LVOT echocardiographic view, respond to CB/HPBV better than dogs with an AoSA <160°, and that this difference is detectable 24 hours, 6 months, and 12 months postoperatively.[9] The difference in response to the same procedure suggests that the magnitude of the AoSA could help determine if a patient is a good candidate for treatment with CB/HPBV, especially if measurement of the AoSA can be helpful in the determination of the patient's long-term ability to maintain a persistent reduction in systolic left ventricular to aortic PG.

Currently, the authors' approach to the management of severe SAS is to first up-titrate dogs on atenolol therapy (or sotalol if needed for ventricular arrhythmias), starting at 1 mg/kg orally twice daily to achieve a heart rate that averages no greater than 120 bpm on re-evaluation. If following beta blocker therapy (up to 2 mg/kg), the LVOT velocity is still above 130–140 mm Hg, or in patients who are symptomatic (exercise intolerant, syncopal), balloon valvuloplasty should be considered. It is important that realistic expectations, prognosis, necessary expense and expertise, and the unknown long-term outcomes associated with CB/HPBV are explained to owners as they are evaluating treatment options.

References

1. Orton EC, *et al. J Am Vet Med Assoc.* 2000;216:364–367.
2. Meurs K, *et al. J Am Vet Med Assoc.* 2005;227:420–424.
3. Kleman ME, *et al. J Vet Cardiol.* 2012;14:351–361.
4. Kleman ME, *et al.* In: Proceedings ACVIM Forum; 2013 June 12–15; Seattle, WA.
5. Eason BD, *et al. J Vet Intern Med.* 2014;28:857–862.
6. Belanger MC, *et al. J Vet Intern Med.* 2014;28:1498–1503.
7. Quintavalla C, *et al. Vet J.* 2010;185:332–337.
8. Yap SC, *et al. Int J Cardiol.* 2008;126:138–139.

9. Shen L, *et al. J Vet Cardiol.* 2017 Jan 20.

acvim181

SAS Update in Dogs: Practical Updates on Palliative Balloon Valvuloplasty

Ashley Saunders, DVM, DACVIM (Cardiology); Sonya Gordon, DVM, DVSc, DACVIM (Cardiology)
College Station, TX, USA

Cutting balloon valvuloplasty (CBV) has been described as a treatment for subaortic stenosis (SAS) in the dog.[1] This session will report the outcome of a case series of six dogs that underwent palliative CBV for SAS and represent the initial experience with this procedure at Texas A&M University. Breeds included two each of Newfoundland, Golden Retriever, and German Shepherds. The average age at CBV was 21 months (range 5–72 months) and average weight was 26.5 kg. Dogs in this case series were characterized by clinical signs associated with myocardial failure, congestive heart failure, atrial fibrillation, ventricular arrhythmias or a combination that were poorly managed with medical therapy including inotropic support. Three dogs had at least one other congenital heart defect; two had pulmonary hypertension. Prior to CBV, dogs were receiving furosemide (4), pimobendan (4), atenolol (4), ACE inhibitor (4), amiodarone (4), spironolactone (2), diltiazem (1). The six dogs underwent successful palliative CBV. Success was determined based on each of the following: survived the procedure, reduction in number and dose of medications after CBV, reduction or resolution of owner-reported clinical signs including syncope/seizure, weakness/collapse, respiratory distress (intermittent despite treatment for congestive heart failure), and echocardiographic acute/chronic gradient reduction. At the time of this writing, one dog was known to have died suddenly 6 months after CBV and the remaining dogs are out an average of 13.5 months. Additional case and procedural-related details will be presented and will emphasize patient selection, procedural complications, and clinical outcome.

References

1. Kleman ME, *et al.* *J Vet Cardiol.* 2012;14:351–361.

acvim183

Cutting and High Pressure Balloon Valvuloplasty for Dysplastic Pulmonary Valve Stenosis in Dogs

Brian A. Scansen, DVM, MS, DACVIM (Cardiology); Lauren E. Markovic, DVM, DACVIM (Cardiology)
Fort Collins, CO, USA

INTRODUCTION

Congenital pulmonary valve stenosis (PS) is among the most commonly diagnosed congenital heart defect of dogs.[1,2] As in children,[3] PS in dogs is described in three principal forms: (1) a dome-shaped valve with commissural fusion, (2) a dysplastic and markedly thickened valve with or without hypoplasia of one or more leaflets, and (3) hypoplasia of the pulmonary annulus.[4,5] While the human form of PS is predominately that of commissural fusion, the canine form is more commonly characterized by thickened, dysplastic valves and variable hypoplasia of the pulmonary annulus.[4,6,7] Without therapy, dogs with valvar PS are at risk for symptoms including exercise intolerance, syncope, sudden cardiac death, congestive heart failure, and cyanosis from a right-to-left shunt.[5, 8-10]

CONVENTIONAL BALLOON PULMONARY VALVULOPLASTY

Balloon pulmonary valvuloplasty (BPV) was first performed in a dog in 1980 and reported in a child in 1982.[11,12] There is evidence that BPV improves the clinical outcome of human and canine patients with valvar PS, both with a reduction in clinical symptomatology and an improvement in survival.[6,9,13,14] The procedure is now routinely performed in clinical canine practice with low morbidity and mortality for those patients with a moderate to severe gradient or the presence of clinical signs referable to their disease.

Although effective for many patients, human and canine studies suggest patients with valve dysplasia and a hypoplastic annulus show less improvement following BPV, as well as a poorer prognosis with or without BPV, compared to those with purely valvar fusion.[7,15,16] The primary benefit of BPV is tearing of commissural fusion; it is unlikely to alleviate obstruction secondary to annular hypoplasia or thick, redundant valve tissue. In a study of 30 dogs with valvar PS, 94% of the 18 dogs with a normal pulmonary annular diameter and minimal valve thickening had resolution of their clinical signs and survived more than 1 year after the procedure; however, for the 12 dogs with marked leaflet thickening and a hypoplastic pulmonary annulus, only 66.6% had a favorable outcome and only 50% showed a resolution in their clinical signs.[7] Another study which retrospectively reviewed 172 dogs with moderate or severe PS, 107 of which underwent BPV, revealed that one of the factors negatively affecting survival in the overall population of dogs was dysplastic valve morphology with hypoplasia of the pulmonary annulus.[16]

NEW INTERVENTIONAL THERAPIES FOR PS

Newer treatment options have been developed in children and variably reported in dogs to address the frequently disappointing results for BPV in the setting of a highly dysplastic pulmonary valve and/or annular hypoplasia. These include cutting balloon dilation as well as intravascular stent placement.[17-19] Stenting of the right ventricular outflow tract in dogs with severe dysplastic PS and a right-to-left shunt has been shown to be a feasible palliation in dogs;[19] however, cutting balloon dilation for dysplastic PS has yet to be reported in veterinary medicine. The use of transcatheter cutting balloon technology to treat PS was first attempted over 20 years ago when a double-blade balloon was used in an attempt to open a stenotic pulmonary valve in a dog model, followed by its use in 3 children.[20] Modern cutting balloons incorporate small blades or atherotomes, which are bonded to the balloon and serve to make small incisions in a vessel resistant to standard balloon angioplasty. A small case series described the use of cutting balloons in 4 children with variable degrees of valvar PS and infundibular narrowing/hypertrophy in the setting of tetralogy of Fallot.[17] These authors found the technique to be safe and effective in this population, resulting in significant improvement in clinical status and obviating the need for prompt surgical intervention. More recently, 5 children with dysplastic PS who failed to respond to standard BPV underwent cutting balloon dilation followed by immediate conventional BPV.[21] This small case series revealed that the cutting balloon technique was safe and provided partial relief of resistant PS in children who failed standard BPV, avoiding the need for surgical intervention.

Cutting Balloon Dilation Followed by High-Pressure BPV in Dogs

We investigated the use of cutting and high-pressure BPV in a cohort of dogs with dysplastic PS - patients had severe valve thickening, subvalvar fibrous tissue, or a hypoplastic pulmonary valve annulus. The ideal criteria for cutting balloon selection in canine patients with dysplastic PS have yet to be determined. In our case series, a cutting balloon dilation catheter (Peripheral Cutting Balloon Dilatation Device, Boston Scientific, Natick, MA, USA) was selected and sized within approximately 1–2 mm of the smallest orifice measurement (valvar or subvalvar) based on both transthoracic echocardiography and right ventriculography of the pulmonary annular and subvalvar orifice dimensions. The cutting balloon selected should not be larger than the size of the pulmonary valve annulus to avoid disruption of the annulus, however should ideally be larger than the true orifice of the stenotic pulmonary valve to engage this tissue.[21] In the majority of dogs, high-pressure BPV was performed with an Atlas Gold PTA balloon dilatation catheter (ATLAS PTA Balloon Dilatation Catheter, Bard Peripheral Vascular, C.R. Bard, Inc., Tempe, AZ, USA), ranging from 12 to 22 mm in diameter and 2 to 4 cm in length, which was prepped to ensure all remained air was extracted and attached to an inflation device. Sizing of the high-pressure balloon was determined using both transthoracic echocardiographic and angiographic measurements of the pulmonary valve annulus. Although the ideal criteria for high-pressure balloon selection for dysplastic PS has yet to be determined, the authors currently advise a balloon-to-annulus ratio of approximately 1.2 to 1.4, as previously described for conventional BPV.[22]

Considerations When Performing Cutting Balloon Dilation and High-Pressure BPV

Cutting balloon dilation followed by high-pressure BPV may be used safely and effectively to achieve a reduction in transpulmonary pressure gradient in cases of dysplastic PS in dogs. This procedure may improve clinical status and avoid the need for surgical intervention; however, long-term follow-up of these patients is necessary along with future studies to validate initial findings. Another potential avenue for the use of this technique is in patients with valvar PS that develop restenosis after an initially successful BPV, requiring the need for a second procedure. Restenosis was recently described in a dog who had a doming pulmonary valve with commissural fusion prior to BPV, and years later developed a markedly thickened valve, confirmed on postmortem evaluation.[23] Potential complications with the cutting balloon dilation technique include disruption of the pulmonary annulus, a tear in the pulmonary artery, and damage to the right ventricular outflow tract, tricuspid valve apparatus, or other nearby cardiac structures or vasculature. Partial fracture of a microtome blade was noted upon removal of the cutting balloon in one dog from our series, without apparent consequence. It is preferable to advance the cutting balloon through a long guiding sheath placed within the proximal right ventricle to avoid damage to the above structures, though in dogs with very small right ventricular lumens this may not be possible. Disadvantages of cutting balloon dilation and high-pressure BPV include added procedural time/fluoroscopy time, as well as the additional cost and training required for the equipment involved.

The purpose of this presentation is to review the morphology of PS in dogs with particular emphasis on therapeutic decision-making in this disease. The indications for cutting balloon dilation and high-pressure BPV will be presented and discussed with clinical case examples. Technical aspects and clinical results of cutting balloon dilation followed by high-pressure BPV in a small series of clinical canine patients with dysplastic PS will be reported. Novel interventional techniques, which are currently being performed in pediatric interventional cardiology, will also be highlighted with a discussion of the potential to transfer these techniques to the care of canine patients. Our initial experience with this novel approach for the treatment of dysplastic PS, its potential for widespread application, and the currently unresolved issues will be offered for discussion.

References

1. Scansen BA, *et al. Current Veterinary Therapy XV*. 2014:756.
2. Oliveira P, *et al. J Vet Intern Med*. 2011;25(3):477.
3. Waller BF, *et al. Clin Cardiol*. 1995;18:45.
4. Patterson DF, *et al. Am J Cardiol*. 1981;47:631.
5. Fingland RB, *et al. J Am Vet Med Assoc*. 1986;189:218.
6. Ristic JM, *et al. J Vet Cardiol*. 2001;3:13.
7. Bussadori C, *et al. J Vet Intern Med*. 2001;15:553.
8. Lombard CW, *et al. J Am Vet Med Assoc*. 1989;194:71.
9. Johnson MS, *et al. J Vet Intern Med*. 2004;18:656.
10. Francis AJ, *et al. J Small Anim Pract*. 2011;52:282.
11. Buchanan JW, *et al. J Vet Intern Med*. 2002;16:116.
12. Kan JS, *et al. N Engl J Med*. 1982;307:540.
13. Fawzy ME, *et al. Am Heart J*. 2007;153:433.
14. Rao PS. *Catheter Cardiovasc Interv*. 2007;69:747.
15. McCrindle BW. *Circulation*. 1994;89:1751.
16. Locatelli C, *et al. J Small Anim Pract*. 2013;54:445.
17. Carlson KM, *et al. Catheter Cardiovasc Interv*. 2005;64:507.
18. Dohlen G, *et al. Heart*. 2009;95:142.
19. Scansen BA, *et al. J Vet Cardiol*. 2014;16:205.
20. Yang SY, *et al. Pediatr Cardiol*. 1991;12:224.
21. Gavri S, *et al. Interv Cardiol*. 2011;5:543.
22. Scansen BA. *Veterinary Image-Guided Interventions*. 2015:575.
23. Sunahara H, *et al. J Vet Med Sci*. 2015;77:95.

acvim184

SAS Panel Discussion

Amara H. Estrada[1], DVM, DACVIM (Cardiology); Mandi Kleman[2], DVM, DACVIM (Cardiology); Ashley Saunders[3], DVM, DACVIM (Cardiology); Sonya Gordon[3], DVM, DVSc, DACVIM (Cardiology)
[1]Gainesville, FL, USA; [2]Atlanta, GA, USA; [3]College Station, TX, USA

Cutting balloon valvuloplasty (CBV) has recently been described and reported as a treatment for subaortic stenosis (SAS) in the dog.[1] This session will use a series of cases to engage the audience in discussions surrounding CBV for the treatment of SAS in the dog. This discussion will build on the content presented in the earlier sessions and focus on practical concepts relevant to cardiologists including patient selection, procedural tips, and possible pitfalls.

Patient Selection

- Echocardiographic features of SAS: gradient, systolic function of left ventricle, anatomy of the obstruction, mitral regurgitation, co-morbidities and concurrent heart disease (patent ductus arteriosus, ventricular septal defect, mitral dysplasia)
- Clinical signs: congestive heart failure, exercise intolerance, collapse
- Arrhythmias: atrial fibrillation, ventricular arrhythmias, ST segment depression
- Medications: atenolol, pimobendan, furosemide, amiodarone, sotalol, diltiazem
    - Pre-procedural
    - Peri-procedural
    - Post-procedural
- Other considerations: cost, prognosis

Procedural Tips and Possible Pitfalls

- The Florida, Georgia experience
- The Texas experience
- This session will try to answer the common questions asked by cardiologists about this procedure before they perform one.

References

1. Kleman ME, *et al. J Vet Cardiol.* 2012;14:351-361.

1.
acvim188

Emerging Surgical and Hybrid Interventions for Canine Structural Heart Disease

Christopher Orton, DVM, PhD, DACVS
Fort Collins, CO, USA

INTRODUCTION

Congenital and acquired structural heart disease remains a significant cause of progressive heart failure and premature death in dogs. Several important structural heart conditions do not have established or proven treatment options in the dog. Among these are tricuspid valve dysplasia, aortic insufficiency, subvalvular aortic stenosis, and degenerative mitral regurgitation. Emerging surgical or hybrid interventional options for these and other conditions could become increasingly available for dogs but will likely require development of canine-specific techniques and devices. These new interventions will require development of canine-specific intra- and post-procedural antithrombotic management protocols.

TRICUSPID VALVE DYSPLASIA

Tricuspid valve dysplasia (TVD) accounts for approximately 5% of canine congenital cardiac malformations. Typical morphology consists of direct insertion of malformed papillary muscles to the mural leaflet and complete fusion or tethering of the septal leaflet to ventricular septum by shortened or absent chordae. These morphologic changes combine to severely restrict the closing motion of the valve leaflets and allow wide-open tricuspid regurgitation (TR). The resultant volume overload causes progressive, severe right heart dilation, which in turn dilates the valve annulus, further worsening the TR. Atrial fibrillation is a common late sequela. Most dogs with severe TR develop progressive right-sided congestive heart failure within the first 3 years of life.

The only reported intervention for TVD in dogs to date has been open valve replacement with a bioprosthetic valve.[1] This approach was associated with moderately good surgical survival (83%) in a small cohort of dogs and has been completely corrective for periods of up to 7 years in some dogs. However, complications associated with the valve prosthesis including thrombosis, endocarditis, and late inflammatory pannus has limited long-term outcomes to <5 years in about 70% of dogs. Valve repair for TVD would have several advantages over valve replacement, including the avoidance of prosthesis-associated complications, reduced need for antithrombotic therapies after surgery, and better right ventricular function. One of the major challenges for valve repair for TVD in dogs is that there is little reported precedent for such a repair in humans, as this defect is quite rare in humans.

Recently we have begun to explore open valve repair for canine TVD. The technique involves mobilizing the restrictive native leaflets and annuloplasty to correct secondary dilation of the valve annulus. Papillary muscles inserting directly onto the mural leaflet are divided close to the leaflet margin and lengthened with pledget-reinforced ePTFE mattress sutures. The tethered septal leaflets are completely freed from the septum by dividing shortened abnormal chordae and resuspended with pledget-reinforced ePTFE artificial chordae from the septum to the leaflet edge. Annuloplasty is performed to reduce the annulus by about 30% to 40%, equivalent to a 28-mm to 32-mm orifice diameter. At the time of this writing, 5 dogs with severe TR secondary to TVD had undergone valve repair. One dog died the day after surgery. The remaining dogs were alive and judged to have 80% or greater reduction in TR severity. Two dogs experienced pre-caval syndrome secondary to thrombosis of the cranial vena cava after surgery that required treatment with TPA and anticoagulation therapy. These results suggest that open valve repair is feasible in dogs with severe TVD, but the significant challenges of performing cardiopulmonary bypass in dogs remain. Emerging transcatheter interventions for severe TR may be an option in the future but will likely require development of canine-specific devices.

AORTIC INSUFFICIENCY

Aortic insufficiency (AI) is a common problem in dogs, but tends to have a rapidly progressive course when it becomes severe. The most common cause of acquired AI in dogs is endocarditis.[2] Orthotopic aortic valve replacement or repair has not been reported in dogs. Open orthotopic aortic valve replacement presents several challenges in dogs, including lack of availability of appropriate-sized valve prostheses for the dog. Heterotopic valve implantation in the descending aorta resulted in significant palliation in a dog with severe AI.[3] Heterotopic valve implantation (Hufnagel procedure) was first performed in humans for the treatment of severe AI in 1953 prior to availability of cardiopulmonary bypass.[4] The validity of the approach was subsequently reaffirmed for human patients who were considered high risk for orthotopic aortic valve surgery.[5] Heterotopic aortic valve implantation can be performed without cardiopulmonary bypass. Currently available valve bioprostheses limit the procedure to large- or giant-breed dogs.

Transcatheter aortic valve implantation (TAVI)/replacement (TAVR) has rapidly emerged as a viable alternative to open valve replacement for high-risk and intermediate-risk human patients with calcific aortic stenosis.[6,7] Clinical trials for TAVI in low-risk human patients are currently underway. Available devices of human patients are a balloon-expandable bioprosthetic valve (Sapien, Edwards) and a self-expanding nitinol bioprosthetic valve (CoreValve, Medtronic). These valves can be delivered via transfemoral or transapical approach. Expense and size limitations make these human devices an unrealistic option for dogs. Nevertheless, the success of TAVI in humans strongly supports development of a canine-specific device. The preferred device would be implanted orthotopically via either a transarterial or hybrid transapical approach. Because the aortic valve annulus of dogs with AI typically is not calcified, a canine orthotopic aortic valve would likely need to employ a novel anchoring mechanism. A short-term solution could be transcatheter implantation of a valved stent into the descending aorta, analogous to the Hufnagel procedure. Such a device could be delivered percutaneously via a transfemoral approach or by a hybrid transaortic approach. We have used the latter approach to implant a nitinol stent bioprosthetic valve in 1 dog with severe AI secondary to endocarditis.

Subvalvular Aortic Stenosis

Subvalvular aortic stenosis (SAS) is one of the most important congenital heart conditions in dogs, accounting for 15% to 25% of congenital heart defects in this species, depending on study population. Several studies have documented an increased risk for sudden cardiac death in dogs with moderate to severe SAS. Reported treatment options to date have included lifetime medical therapy with beta-blockers,[8] transarterial balloon valvuloplasty,[9] transapical aortic valve dilation,[10] and open resection under cardiopulmonary bypass.[11] To date, no therapy has been convincingly shown to decrease the risk of sudden cardiac death. Congestive heart failure is a less common manifestation in dogs with SAS. When it occurs, it is usually associated with concurrent primary or secondary MR and invariably has a rapidly progressive course. Currently few treatment options are available for dogs with severe SAS, particularly dogs with concurrent MR and congestive heart failure.

Aortic valve bypass (AVB) or apicoaortic conduit involves the placement of a valved conduit between the left ventricular apex and aorta for the relief of severe aortic stenosis. The surgery can be performed on the beating heart without the need for cardiopulmonary bypass. Off-pump AVB has been shown to be an effective procedure with favorable hemodynamics in human patients with severe aortic stenosis.[12,13] AVB was actually pioneered in the 1970s using dogs as an experimental model.[14] Some of these dogs included an NIH colony of Newfoundland dogs with severe SAS with survivals of over 1 year (personal communication with Dr. J.W. Brown). The procedure was reported in 8 clinical dogs with SAS by Breznock *et al.* in 1983.[15] Despite a high operative mortality rate, 4 dogs in this series demonstrated short- to intermediate-term survival. Currently valved conduits and LV connectors are not commercially available for humans or animals. Development of AVB for dogs would require availability of an appropriate-sized conduit and left ventricular connector (12 mm or 14 mm) suitable for large dogs. The most challenging aspect of the surgery is safe coring of the left ventricle and insertion of the LV connector. Techniques of LV coring and connector insertion in a single step have been developed and could be made available for dogs. Efforts to develop a canine-specific device and technique for AVB in dogs with severe SAS and concurrent CHF are in progress.

ANTITHROMBOTIC PROTOCOLS

Future development of surgical and interventional devices for structural heart disease in dogs will require development and validation of antithrombotic protocols that are effective and safe in dogs. Despite an abundance of pharmacokinetic and pharmacodynamic data for antithrombotic drugs in dogs, established antithrombotic protocols do not exist. Antithrombotic protocols include antiplatelet drugs, anticoagulant drugs, or combinations of these.

Antiplatelet drugs available for dogs include aspirin and clopidogrel. In human patients, aspirin is typically administered as a "low-dose" regimen (81 mg/day PO) to achieve long-term platelet suppression that minimizes the risk of side effects. In dogs, an equivalent "low-dose" aspirin regimen (1 mg/kg PO q 24 h) produces inconsistent effects on platelet aggregation in healthy dogs.[16,17] Administration of high-dose aspirin (5 mg/kg PO q 12 h) to dogs consistently inhibits platelet aggregation; however, the safety of long-term administration of aspirin to dogs at this dose has not been established. Pharmacodynamic and pharmacokinetic data suggest that clopidogrel administered at a dose of 1 mg/kg PO q 24 h produces consistent and sustained suppression of platelet aggregation in dogs. In human patients, antiplatelet strategies are typically employed for arterial thrombosis (e.g., coronary artery syndrome). Antiplatelet strategies are considered inadequate for prophylaxis of venous and cardiac (e.g., atrial fibrillation) thrombosis.

Anticoagulant drugs potentially available for use in dogs include traditional drugs such as unfractionated heparin, fractionated heparin, and warfarin, as well as the newer oral anticoagulants such as Xa inhibitors rivaroxaban and apixaban, and direct thrombin inhibitors such as dabigatran. The use of unfractionated heparin is largely restricted to intraoperative and immediate-postoperative anticoagulation. Intraoperative dose of heparin is typically 100 to 150 U/kg for cardiovascular device implantation not involving CPB. Continuous IV infusion of heparin at 20–60 U/kg/h with a therapeutic goal of 1.5x baseline aPTT can be employed as an effective antithrombotic strategy in dogs in the acute perioperative period. Long-term anticoagulation therapy in humans has historically been accomplished by administration of a vitamin K antagonist such as warfarin. The dose of warfarin is highly individualized, necessitating periodic testing. Warfarin is affected by changes in diet and has numerous drug interactions. The dose of warfarin is adjusted based on the international normalized ratio (INR) calculated from the prothrombin time (PT). Target INR is typically 2.5 to 3.0 depending on indication. Indications in humans have included treatment or prophylaxis for deep venous thrombosis, atrial fibrillation, and prosthetic heart valves. We have reported effective use of oral warfarin therapy in dogs for the long-term management of aortic thromboembolism[18] and prophylaxis after surgical valve replacement.[1] Several of the new oral anticoagulants have emerged as alternatives to warfarin therapy for its traditional indications. The oral Xa inhibitor apixaban in particular has been shown to be superior and have a better safety profile compared to warfarin or fractionated heparin in the management of atrial fibrillation[19] and venous thrombosis.[20] New-generation oral anticoagulants have not yet been approved for heart valve prophylaxis. Although the clinical dose, efficacy, and safety of apixaban in dogs have not been established, pharmacokinetic and pharmacodynamic data for apixaban dogs have been published.[21] Review of this data suggests that the dose of apixaban in dogs would be between 0.2 and 0.5 mg/kg q 12 h to accomplish a 50% decrease in factor Xa activity. This dose is several-fold higher than the equivalent effective dose in humans (0.07 mg/kg PO q 12 h).

Anecdotal experience suggests that dogs may be more prothrombotic than humans in response to implanted cardiovascular devices. One reason for this is the high levels of coagulation factors in canine plasma compared to humans. Factor Xa activity in canine plasma is approximately twice that of humans.[21] Standard prophylactic therapy for surgically implanted bioprosthetic heart valves in humans is warfarin for 3 months. Duel antiplatelet therapy (DAPT) with clopidogrel and aspirin is currently standard for humans after TAVI unless they are in atrial fibrillation.[22] If patients are in atrial fibrillation, then therapy consists of combined oral anticoagulant and antiplatelet drug. Based on currently available data and drugs, we are exploring combined therapy with apixaban and clopidogrel as a longer-term antithrombotic strategy for cardiovascular devices in dogs. This strategy will require validation in the clinical setting.

References

1. Arai S, Griffiths LG, Mama K, Hackett TB, Monnet E, Boon JA, Carter L, Orton EC. Bioprosthesis valve replacement in dogs with congenital tricuspid valve dysplasia: technique and outcome. *Journal of Veterinary Cardiology*. 2011;13:91–99.
2. Calvert CA. Valvular bacterial endocarditis in the dog. *Journal of the American Veterinary Medical Association*. 1982;180:1080–1084.
3. Arai S, Wright BD, Miyake Y, Boon JA, Orton EC. Heterotopic implantation of a porcine bioprosthetic heart valve in a dog with aortic valve endocarditis. *Journal of the American Veterinary Medical Association*. 2007;231:727–730.
4. Hufnagel CA, Harvey WP. The surgical correction of aortic regurgitation preliminary report. *Bull Georgetown Univ Med Cent*. 1953;6:60–61.
5. Cale AR, Sang CT, Campanella C, Cameron EW. Hufnagel revisited: a descending thoracic aortic valve to treat prosthetic valve insufficiency. *The Annals of Thoracic Surgery*. 1993;55:1218–1221.
6. Kumbhani DJ, Banerjee S. 3-year results of a TAVR trial in high surgical risk patients. *Journal of the American College of Cardiology*. 2016;67:2575–2577.
7. Miller DC, Blackstone EH, Mack MJ, Svensson LG, Kodali SK, Kapadia S, Rajeswaran J, Anderson WN, Moses JW, Tuzcu EM, Webb JG, Leon MB, Smith CR, Investigators PT, Patients, Group PSSW, Executive C. Transcatheter (TAVR) versus surgical (AVR) aortic valve replacement: occurrence, hazard, risk factors, and consequences of neurologic events in the PARTNER trial. *Journal of Thoracic and Cardiovascular Surgery*. 2012;143:832–843 e13.
8. Eason BD, Fine DM, Leeder D, Stauthammer C, Lamb K, Tobias AH. Influence of beta blockers on survival in dogs with severe subaortic stenosis. *Journal of Veterinary Internal Medicine*. 2014;28:857–862.
9. Meurs KM, Lehmkuhl LB, Bonagura JD. Survival times in dogs with severe subvalvular aortic stenosis treated with balloon valvuloplasty or atenolol. *Journal of the American Veterinary Medical Association*. 2005;227:420–424.
10. Linn K, Orton EC. Closed transventricular dilation of discrete subvalvular aortic stenosis in dogs. *Veterinary Surgery: VS*. 1992;21:441–445.
11. Orton EC, Herndon GD, Boon JA, Gaynor JS, Hackett TB, Monnet E. Influence of open surgical correction on intermediate-term outcome in dogs with subvalvular aortic stenosis: 44 cases (1991–1998). *Journal of the American Veterinary Medical Association*. 2000;216:364–367.
12. Thourani VH, Keeling WB, Guyton RA, Dara A, Hurst SD, Lattouf OM. Outcomes of off-pump aortic valve bypass surgery for the relief of aortic stenosis in adults. *The Annals of Thoracic Surgery*. 2011;91:131–136.
13. Vliek CJ, Balaras E, Li S, Lin JY, Young CA, DeFilippi CR, Griffith BP, Gammie JS. Early and midterm hemodynamics after aortic valve bypass (apicoaortic conduit) surgery. *The Annals of Thoracic Surgery*. 2010;90:136–143.
14. Brown JW, Myerowitz PD, Cann MS, McIntosh CL, Morrow AG. Apical-aortic anastomosis: a method for relief of diffuse left ventricular outflow obstruction. *Surgical Forum*. 1974;25:147–149.
15. Breznock EM, Whiting P, Pendray D, Thomas B, Strack D, Bauer T, Koblik P, Hornof W, Ludders JD. Valved apico-aortic conduit for relief of left ventricular hypertension caused by discrete subaortic stenosis in dogs. *Journal of the American Veterinary Medical Association*. 1983;182:51–56.
16. Haines JM, Thomason JM, Seage EC, Wills RW, Bulla C, Lunsford KV, Mackin AJ. *In vitro* and *in vivo* assessment of platelet function in healthy dogs during administration of a low-dose aspirin regimen. *American Journal of Veterinary Research*. 2016;77:174–185.
17. Hoh CM, Smith SA, McMichael MA, Byron JK. Evaluation of effects of low-dose aspirin administration on urinary thromboxane metabolites in healthy dogs. *American Journal of Veterinary Research*. 2011;72:1038–1045.
18. Winter RL, Sedacca CD, Adams A, Orton EC. Aortic thrombosis in dogs: presentation, therapy, and outcome in 26 cases. *Journal of Veterinary Cardiology*. 2012;14:333–342.
19. Granger CB, Alexander JH, McMurray JJ, Lopes RD, Hylek EM, Hanna M, Al-Khalidi HR, Ansell J, Atar D, Avezum A, Bahit MC, Diaz R, Easton JD, Ezekowitz JA, Flaker G, Garcia D, Geraldes M, Gersh BJ, Golitsyn S, Goto S, Hermosillo AG, Hohnloser SH, Horowitz J, Mohan P, Jansky P, Lewis BS, Lopez-Sendon JL, Pais P, Parkhomenko A, Verheugt FW, Zhu J, Wallentin L, Committees A and Investigators. Apixaban versus warfarin in patients with atrial fibrillation. *The New England Journal of Medicine*. 2011;365:981–992.
20. Greig SL, Garnock-Jones KP. Apixaban: a review in venous thromboembolism. *Drugs*. 2016;76:1493–1504.
21. He K, Luettgen JM, Zhang D, He B, Grace JE, Jr., Xin B, Pinto DJ, Wong PC, Knabb RM, Lam PY, Wexler RR, Grossman SJ. Preclinical pharmacokinetics and pharmacodynamics of apixaban, a potent and selective factor Xa inhibitor. *European Journal of Drug Metabolism and Pharmacokinetics*. 2011;36:129–139.
22. Sharma A, Goel S, Lavie CJ, Arbab-Zadeh A, Mukherjee D, Lazar J. Antithrombotic therapy before, during and after transcatheter aortic valve replacement (TAVR). *Journal of Thrombosis and Thrombolysis*. 2015;39:467–473.

acvim069

The Molecular Etiology of Hansen's Type I Disc Disease Provides a Framework for Disease Elimination

Danika L. Bannasch[1]; Emily A. Brown[1]; Peter J. Dickinson[1]; Beverly Sturges[2]; Tamer Mansour[1]; Jenna Lind[1]; Courtney Korff[1]; Miriam Aguilar[3]; Cassie Ettinger[1]; Amy Young[1]; Sam Varon[4]; Terje Raudsepp[5]; Titus Brown[1]

[1]University of California Davis, Davis, CA, USA; [2]Surgical and Radiological Sciences: Veterinary Medicine: Veterinary Medicine, Davis, CA, USA, [3]UC Davis Population Health and Reproduction: Veterinary Medicine, Davis, CA, USA; [4]Barney and Russum Animal Clinic, Fairfield, CA, USA; [5]Texas A&M University, College Station, TX, USA

Hansen first described type I intervertebral disc disease (IVDD) in the dog using five breeds, which he classified as chondrodysplastic: French bulldog, Dachshund, Pekingese, Dachsbrache and the Spaniel. Dogs from these breeds have abnormal long bone growth plates, abnormal vertebral growth, and abnormal intervertebral disc material, referred to as endochondrosis intervertebralis. Later, Braund performed systematic histopathological analysis of the Beagle breed and classified it as chondrodystrophoid-like with the same predisposition to type I disc herniation. While these breeds are not all severely disproportionately short legged, they all share the common features of abnormal endochondral ossification of long bones leading to shorter legs. The Nova Scotia Duck Tolling Retriever is a medium sized dog and the smallest of the retriever breeds. They have a form of skeletal dysplasia that in its severest form has been likened to chondrodysplasia. We performed a genome wide association study (GWAS) for this severe phenotype using the Illumina 174K single nucleotide polymorphism (SNP) array and obtained a significant association (p_{raw}=6.803 x 10^{-8}, $p_{Bonferroni}$=0.007232) on chromosome 12. Underlying this association was a large region of homozygosity. When sharing of this region was investigated with other breeds, portions of it were shared with Beagles, Dachshunds, American Cocker Spaniels, Pekingese and French Bulldogs, the same breeds predisposed to Type I IVDD. In order to investigate the association with diseased animals, samples were collected from 36 dogs with presence of mineralized intervertebral discs confirmed by vertebral column radiographs and/or magnetic resonance imaging followed by surgery. These dogs were selected from many different breeds and mixed breed dogs. A statistically significant (p_{raw}=3.191 x 10^{-15}, $p_{Bonferroni}$=9.438 x 10^{-10}) genome wide association to the same region on chromosome 12 was obtained, which narrowed the critical interval to ~1 MB. The most associated SNP from the GWAS was 32.67 (95% CI = (30.12, 35.44)) times more likely to be present in dogs with IVDD than in healthy dogs. Genes within the area were evaluated for expression in neonatal disc from predisposed animals and healthy animals. Whole genome sequence data was mined to identify the causative mutation, which has an odds ratio of 51.23 (95% CI = (46.69, 56.20)) for Hansen's Type I disc disease across all dog breeds. In addition, this mutation segregates in the majority of breeds where it is found allowing elimination of this disease from dogs.

acvim071

Serum Levetiracetam Concentrations After Chronic Administration of Once Daily Extended-Release Levetiracetam in Healthy Cats

Heidi L. Barnes Heller[1]; Martin Granick[1]; Dawn M. Boothe[2]

[1]University of Wisconsin-Madison, Madison, WI, USA; [2]Auburn University, Auburn, AL, USA

The purpose of this study was to determine if once daily administration of extended-release levetiracetam for 10 days in cats can achieve serum concentrations above the minimum human therapeutic interval (5 mcg/mL) and to evaluate side effects during the treatment course. Eight healthy staff- or student-owned cats weighing more than 5 kg body weight were enrolled. Extended-release levetiracetam (500 mg once daily) was administered for 10 days with food. On day 11, median serum levetiracetam concentrations at 0, 4, 6, and 8 hours post-pill were 7.0, 78.6, 92.7, and 72.2 mcg/mL, respectively. Mean estimated Tmax was 6.3 hours (range: 4–8+ hours). Samples for some time points were not collected from 2 cats owing to lack of cooperation. Six of 8 cats had serum levetiracetam concentrations above 5 mcg/ml time 0. The neurologic examination was normal for all cats on the day of sampling. No significant changes were noted on serum biochemical analysis on day 11 compared with the day of enrollment. Side effects were limited to mild ataxia on day 1 in one cat. The results from this study suggest that once daily administration of 500 mg extended-release levetiracetam in cats weighing more than 5 kg is well tolerated and able to provide serum concentrations above the minimum human therapeutic interval at time 0 in the majority of cats. Peak and trough samples and seizure monitoring should be considered on an individual basis for epileptic cats.

acvim074

Use of 3T Magnetic Resonance Imaging to Analyze Craniocervical Junction Cerebrospinal Fluid Flow in Cats

Sofia Cerda-Gonzalez[1]; Philippa J. Johnson[1]; Eve M. LoCastro[2]

[1]College of Veterinary Medicine, Cornell University, Ithaca, NY, USA; [2]Imaging Data Evaluation and Analytics Lab, Weill Cornell Medicine, New York, NY, USA

Abnormalities in cerebrospinal fluid (CSF) flow are described in dogs and humans, particularly in association with craniocervical junction (CCJ) anomalies, arachnoid scarring and syringomyelia. In cats, although subarachnoid and spinal cord pressure changes have been described caudal to focal arachnoid scarring,[1] and secondary syringomyelia has been suspected to result from intracranial/CCJ anomalies,[2] CSF flow has not yet been characterized in this species. We aimed to use phase-contrast MRI (PCMRI) to measure normal CSF flow parameters in cats and develop novel software to automate flow analyses. Ten healthy, neurologically normal, domestic shorthaired cats were imaged with a 3T MRI unit. T2-weighted sagittal sequences were used to rule out structural brain, CCJ and cervical spinal cord abnormalities. With the CCJ in extended (i.e., straightened) and flexed positions, CSF flow velocities, stroke volume, and flow rates were measured at the atlantooccipital, C1-C2 and C2-C3 junctions using PCMRI.

Flow could be assessed in all cats at all levels, although atlantooccipital junction flow was best assessed in flexion. Flow velocities were lower than those reported in dogs and humans. In preliminary analyses, peak systolic velocities were higher at the atlantooccipital and C12 junctions when positioned in extension; peak diastolic velocities were also higher in extension at the atlantooccipital junction. Surface area, stroke volume, and flow rates did not vary significantly with position.

In conclusion, PCMRI can be effectively used to measure CSF flow in cats. Feline CSF flow is slower than that of dogs and humans, and may be influenced by head and neck positioning.

References

1. Klekamp J, Volkel K, Bartels CJ, *et al.* Disturbances of cerebrospinal fluid flow attributable to arachnoid scarring cause interstitial edema of the cat spinal cord. *Neurosurgery.* 2001;48:174–186.
2. Okada M, Kitagawa M, Ito D, *et al.* MRI of secondary cervical syringomyelia in four cats. *J Vet Med Sci.* 2009;71:1069–1073.

acvim076

Biomechanical Properties of Virgin PEEK, Glass-Fiber PEEK, and Carbon-Fiber-Reinforced PEEK in the Adamo Spinal Disc™

Filippo Adamo[1]; Tanya Garcia-Nolen[2]; Susan Stover[2]; Robert Kroll[3]; Brian Hynes[4]

[1]East Bay Veterinary Specialists, Walnut Creek, CA, San Mateo, CA, USA; [2]Anatomy, Physiology and Cell Biology: Veterinary Medicine, Davis, CA, USA; [3]VCA Northwest Veterinary Specialists, West Linn, OR, USA; [4]Rogue Research, Inc., Montréal, QC, Canada

The Adamo Spinal Disc™ is an articulating surgical implant used to treat disc-associated-wobbler syndrome in dogs. Material for the articulating connection between the two vertebrae must resist shear, compressive forces, and be resistant to wear and fatigue failure.

An articulating ball was created using three polyether-ether-ketone (PEEK) composites [Virgin PEEK (V-PEEK), 30% glass-fiber-PEEK (GF-PEEK), 30% carbon-fiber-reinforced PEEK (CFR-PEEK)] to compare biomechanical properties. Three replicates were tested by aligning the two mating faces of the titanium discs with a PEEK ball and loading the construct in shear to failure at 0.01 mm/sec in a mechanical testing system. One construct with each PEEK ball was loaded in compression to 0.2 mm displacement at 0.01 mm/sec (i.e., until the space closed between the metal faces).

In shear, GF-PEEK was 40% stiffer and CFR-PEEK was 60% stiffer than V-PEEK. GF-PEEK was 20% stronger and CFR-PEEK was 30% stronger than V-PEEK. GF-PEEK had 370% less post-yield energy and CFR-PEEK had 170% less post-yield energy than V-PEEK. In compression, CFR-PEEK was 3 times as stiff and twice as strong as V-PEEK and GF-PEEK in the linear elastic region.

CFR-PEEK was stronger and stiffer than V-PEEK and GF-PEEK, making it more resistant to acute failure. Based on these findings, CFR-PEEK appears to be the best suited for articulating implants. However, V-PEEK was highly ductile and tough, properties that may contribute to longer fatigue life. Because fibers in CFR-PEEK also contribute to increasing fatigue life, further studies investigating fatigue life associated with chronic repetitive motion are warranted.

acvim077

Evaluation of Risk Factors in the Development of Inflammatory Brain Disease in Dogs

Heidi L. Barnes Heller[1]; Martin Granick[1]; Marie Pinkerton[1]

[1]University of Wisconsin-Madison, Madison, WI, USA

The purpose of this case-control study is to identify specific triggers significantly associated with development of granulomatous meningoencephalitis (GME), necrotizing leukoencephalitis (NLE), and necrotizing meningoencephalitis (NME) in this population of dogs. A retrospective review of the hospital records database identified 31 dogs with adequate historical information diagnosed with GME, NME, or NLE on histopathologic examination at necropsy from 2003–2015. Three matched (age-breed-year of presentation) control dogs were randomly selected from the hospital population for each case dog. One case had only one control match available. GME, NME and NLE are inflammatory diseases of the central nervous system in dogs that are suspected to be immune-mediated in at-risk individuals; however, a trigger has yet to be identified. Triggers, such as geographic location, temporal association with vaccination, sex, seasonal distribution, obesity, and population density, have been investigated for other animal and human immune-mediated diseases with variable results. No association between body condition score ($p=0.72$), weight ($p=0.28$), temporal relationship to vaccination ($p=0.40$), or population density ($p=0.45$) was found in this study. Male dogs were more likely to develop disease compared with female dogs ($p=0.02$). In summary, no significant association between select triggers and the development of GME, NLE, and NME was identified. The association with sex may suggest a hormonal influence on development of disease. Pursuit of other triggers, such as exposure to infectious agents, toxins, and animal crowding, could be investigated in future studies.

acvim079

Experimental Therapies for Canine Krabbe Disease

Allison Bradbury

University of Pennsylvania, Philadelphia, PA, USA

Globoid cell leukodystrophy (GLD), or Krabbe disease, results from a deficiency in the hydrolytic enzyme galactosylceramidase (GALC), which is responsible for degrading the central and peripheral nervous system (CNS, PNS) myelin lipids galactosylceramide and galactosylsphingosine (psychosine). Symptoms in infants include irritability and stiffness with disease progressing rapidly to a vegetative state and death typically resulting by 2 years of age. GLD is naturally occurring in dogs, and signs include ataxia, tremors, pelvic limb paralysis, loss of hearing, and blindness, with disease progression warranting euthanasia at 15.9±4.6 weeks. Motor and sensory conduction velocities in peripheral nerves of GLD dogs are significantly reduced compared to normal and brainstem auditory evoked response testing exhibits a significant increase in central conduction time, both consistent with demyelination. Magnetic resonance imaging of the brain shows T2-weighted hyperintensity of the white matter, widened sulci, and enlarged ventricles and magnetic resonance spectroscopy identified decreases in N-acetylaspartate and increases in choline. Diffusion tensor imaging established decreased fractional anisotropy and increased radial diffusivity in multiple white matter tracts including the corpus callosum and posterior internal capsule. Histologically, GLD dog brains show severe loss of myelin, a decrease in oligodendrocytes, astrogliosis, and microgliosis.

Gene therapy has been evaluated in GLD affected dogs using AAVrh10 encoding canine *GALC* (AAVrh10-cGALC). Two GLD dogs received a low dose of AAVrh10-cGALC, 1.2E12 total vector genomes divided evenly between combination intravenous (IV) and intracerebroventricular (ICV) injection routes and survived to 17.9 and 22.1 weeks of age. Two additional GLD dogs were treated with an increased dose, 1.9E13 total vector genomes, and survival was further increased to 30.3 and 43.1 weeks of age. Both low and high doses prevented the onset of tremors and delayed the onset of other neurological dysfunction. High dose AAVrh10-cGALC, but not low dose, resulted in normalization of peripheral nerve conduction velocity. After high dose treatment, GALC activity reached normal levels in the cerebellum and sciatic nerve and psychosine accumulation was reduced. High dose AAV resulted in CSF psychosine concentrations lower than untreated and low dose AAV-treated GLD dogs, despite being older. Histologically, the therapy largely attenuated the loss of myelin, accumulation of storage material, and onset of neuroinflammation. AAV gene therapy targeted to the CNS and PNS delayed the onset of neurological dysfunction, increased survival, and corrected pathology in the canine model of Krabbe disease.

acvim080

Gene Therapy in Animal Models of the Gangliosidoses: From Presentation to Human Clinical Trials

Heather L. Gray-Edwards

Auburn University, Auburn, AL, USA

GM1 and GM2 gangliosidosis (Tay-Sachs, Sandhoff disease, GM2 activator protein deficiency) are monogenetic diseases that result from deficiencies of β-galactosidase (β-gal), hexosaminidase (Hex), or GM2 activator protein, which are enzymes of the ganglioside degradation pathway. These are global diseases resulting in ganglioside buildup in all tissues, but have primary neurodegenerative presentation due to the inherent concentration of gangliosides within neurons. These diseases are clinically indistinguishable in humans and symptoms present in children with lack or loss of developmental milestones (sitting, crawling, etc.) and progress to inability to swallow, seizures and eventual semi-vegetative state. As with most human diseases, these diseases are also naturally occurring in animals (cats and sheep), and are maintained for study and testing of novel therapeutics.

Cats with GM2 gangliosidosis have a deficiency in the β-subunit of the isoenzyme hexosaminidase and have the genetic equivalent of Sandhoff disease in humans. Affected cats exhibit a profound cerebellar phenotype with resting and intention tremors that progress to severe ataxia and inability to stand at 4.5 +/- 0.5 months. GM2 cats also exhibit a mucopolysaccharidosis like phenotype with skeletal abnormalities and glycosaminoglycan storage. Cats with GM2 develop prolonged latencies in waves II, III and V on brainstem auditory evoked responses (BAER). 7T T2 weighed (T2w) MRI shows an intensity inversion, where white matter becomes hyperintense to surrounding gray matter correlating with cell body storage of GM2 and demyelination. MRS reveals severe cerebellar disease with a large increase in the toxic metabolite N-acetylhexosamine and elevations in demyelination markers glycerophosphocholine and phosphocholine (GPC+PCh) and the glial cell marker myoinositol (INS). Cerebrospinal fluid (CSF) levels of aspartate aminotransferase (AST), lactate dehydrogenase (LDH) increase with disease severity.

GM2 gangliosidosis sheep have a deficiency in α-subunit of hexosaminidase and have the genetic equivalent of Tay-Sachs disease in people. GM2 sheep present with proprioceptive deficits, also develop generalized proprioceptive ataxia and eventually lose the ability to ambulate at 9.4±0.8 months. Approximately 10% of GM2 sheep exhibit loss of consciousness episodes that are suspected epileptic events. Sheep exhibit continuous semi-rhythmical slowing of neural oscillations with a predomination of delta waves on electroencephalography (EEG), with superimposition of sharp spike discharges. On 7T T2 weighed (T2w) MRI, the brain develops areas of isointensity due to ganglioside storage and demyelination, which are most evident in the cingulate gyrus, parietal cortex and temporal lobes. MRS reveals the greatest disease severity in the thalamus with a reduction in neuronal markers N-acetylaspartate (NAA), glutamine + glutamate (GLX) and increases in GPC+PCh, taurine and INS. Elevations of AST, LDH, and GM2 ganglioside as well as other sphingolipids were noted in the CSF of TSD sheep.

Adeno-associated (AAV) gene therapy is ideal for treatment of monogenetic neurologic disorders. We performed intracranial injection of AAV gene therapy that has resulted in widespread distribution of the encoded therapeutic protein(s) and increased survival in in all three diseases. AAV treated GM1 cats are now living into the 7th year of life with mild neurologic disease and seizures that are largely controlled with anti-epileptics. AAV treated GM2 cats survival has increased to almost 3 years of age (a >four-fold increase in lifespan), with marked improvement in neurologic disease with the largest therapeutic limitation being peripheral dysfunction. AAV treated GM2 sheep are now living to almost 2 years of age with the largest therapeutic weakness being reduced distribution to the spinal cord. Success of these studies in animals has led to submission of pre-Investigational New Drug (IND) applications with the FDA and clinical trials for both GM1 and GM2 gangliosidosis are anticipated to start within the next year.

acvim081

AAV Gene Therapy Studies in the Feline Model of Niemann-Pick Type C

Brittney L. Gurda

University of Pennsylvania, Collegeville, PA, USA

Niemann-Pick type C1 (NPC1) disease is characterized by progressive cerebellar ataxia, dementia, and death in adolescence. It is caused by mutations that result in deficient function of lysosomal, membrane-bound NPC1, resulting in the intralysosomal accumulation of cholesterol and sphingolipids. There are no FDA-approved therapies for NPC1 disease; however, we have previously shown that administration of 2-hydroxypropyl-beta-cyclodextrin (HPβCD) via the cerebromedullary cistern (CBMC) in presymptomatic cats with NPC1 disease prevented the onset of cerebellar tremor and resulted in Purkinje cell (PC) survival and near normal concentrations of cerebral cortex and cerebellar cholesterol and sphingolipids. Remarkably, these cats remained alive for greater than three years of age, in contrast to untreated NPC1 cats that died before six months of age. This therapy has advanced to clinical trials where patients receive biweekly HPβCD.

In an attempt to both reduce the number of HPβCD injections and to treat uncorrected brain regions, we studied the feasibility of viral gene therapy, providing a one-time injection, in NPC1 cats. Preliminary studies utilized administration of an adeno-associated virus (serotype 9; AAV9) carrying the feline NPC1 transgene (AAV9.fNPC1) via the CBMC of pre-symptomatic NPC1 cats. These treated cats showed reduced ataxia compared to untreated cats. Histologic evaluation of the brains revealed that NPC1 expression had been restored to some PCs and that this was associated with increased PC survival. This initial success prompted us to further explore the potential of various promoters to express functional NPC1 protein in the brain. In this study we evaluated the ability of four different promoters, three ubiquitous (CB7, GUSB, and USP) and one neuron-specific (CamKII), to drive GFP expression in the central nervous system of normal cats.

Our preliminary therapeutic studies used the GUSB promoter (pGUSB) to mediate fNPC1 expression. Expression of both fNPC1 and a GFP reporter under pGUSB were low in PCs, and in cerebrocortical neurons and astrocytes. Overall, the CB7 promoter drove the highest level of expression in similar regions; however, this large promoter (~970 bp) limits transgene size (where NPC is ~4.3 kbp) and led us to examine a smaller exogenous promoter based on CB7 (termed USP; UNC via Dr. Steven Gray). This promoter was similar in expression levels to the pGUSB, leading to low-level expression at our limit of detection using DAB-based immunohistochemistry. Interestingly, a 6-month therapeutic study using the USP promoter had a similar outcome as pGUSB where treated cats had reduced clinical signs at the end of study. Expression driven by the neuron-specific promoter CamKII produced levels similar to the CB7 promoter, although PCs of the cerebellum had low to undetectable expression levels. These data and preliminary therapeutic data in AAV-treated NPC1 cats will be further explained in this presentation.

acvim082

Generating Novel Large Animal Models of Nervous System Disease for Therapeutic Study Using the CRISPR/Cas9 System

Brittney L. Gurda

University of Pennsylvania, Collegeville, PA, USA

Large animal models of disease continue to play a strong role in the advancement of novel therapies into the clinical setting. While numerous naturally occurring models of disease have been previously identified, the rates at which new mutations are found and characterized generate a significant bottleneck in the field. With advancements in cloning strategies we now harness a more direct way to target mutations in the genome and thus generate large animal models for study. The most recent technological advances in cloning are based on an immune defense mechanism identified in bacteria and archaea. This novel system uses clustered regularly interspaced short palindromic repeats (CRISPRs) and their associated proteins (Cas) to guide a small section of foreign DNA into the host genome that leads to immunity against the invader. Specifically, the CRISPR/Cas9 system has recently been repurposed to initiate genome engineering in cells, tissues, and animals as well as novel treatment strategies for genetic and infectious diseases.

In this study we sought to generate a novel feline model of neurological disease using the CRISPR/Cas9 system. Niemann-Pick disease, type C (NPC), is a recessively inherited neurodegenerative lysosomal storage disease characterized by the accumulation of low-density lipoprotein-derived unesterified cholesterol in the late endocytic organelles. Both NPC1 and NPC2 proteins participate in the egress of cholesterol from cells and disruptions in either of these genes have been associated with NPC disease. About 90–95% of mutations occur in the NPC1 gene while the remaining 5% are attributed to disruptions in NPC2. Since a NPC type C1 model already exists and has been well characterized in felines, we sought to create an NPC2 model that will complement our NPC1 feline model studies. Using a naturally occurring defect previously identified in a domestic cat and the CRISPR/Cas9 system we have successfully generated a single-nucleic acid mutation in the NPC2 gene of feline zygotes. Implantation of NPC2-defective zygotes into surrogate female cats is currently underway and methodical data leading up to and beyond the implantation stage will be discussed in this presentation.

acvim085

Volumetric Modulated Arc Radiotherapy of Canine Trigeminal Nerve Tumors

Mario Dolera[1]; Luca Malfassi[2]

[1]La Cittadina Fondazione Studi e Ricerche Veterinarie, Romanengo, Lombardia, Italy; [2] La Cittadina Fondazione Studi e Ricerche Veterinarie, Crema, Lombardia, Italy

The aim of this study was to evaluate the feasibility and effectiveness of curative high dose hypofractionated (HDH) frameless volumetric modulated arc radiotherapy (VMAT) in trigeminal peripheral nerve sheath tumours (PNST).

A prospective clinical trial was conducted from February 2010 to December 2013 on client-owned dogs with presumptive imaging-based diagnosis of trigeminal PNST. Seven dogs were enrolled and treated with 37 Gy in 5 fx using a 6 MV linear accelerator equipped with an external micro-multileaf beam collimator. The plans were computed using a Monte Carlo algorithm. Overall survival was estimated using a Kaplan-Meier curve analysis.

MRI follow-up examinations revealed complete response in one dog, partial response in four dogs, and stable disease in two dogs. Median overall survival was 952 days with a 95% confidence interval (CI) of [543–1361] days.

VMAT appears to be feasible and effective and offers the best median survival time in dogs suffering from trigeminal PNST compared with other radiotherapy treatment modalities, such as LINAC cone based radiotherapy or surgery.

Further information available on website www.lacittadina.org

acvim086

Frameless Stereotactic Volumetric Modulated Arc Radiotherapy of Brachial Plexus Tumors in Dogs: 10 Cases

Mario Dolera[1]; Luca Malfassi[2]

[1]La Cittadina Fondazione Studi e Ricerche Veterinarie, Romanengo, Lombardia, Italy; [2] La Cittadina Fondazione Studi e Ricerche Veterinarie, Crema, Lombardia, Italy

Treatment of canine peripheral nerve sheath tumors (PNSTs) is challenging, and prognosis after surgical resection is considered poor. The aim of this study was to evaluate the feasibility and the effectiveness of stereotactic radiotherapy of these tumors.

Dogs with clinical signs and MRI findings consistent with PNSTs of the brachial plexus, branches and nerve roots were treated with LINAC-based volumetric modulated arc radiotherapy (VMAT) with a dose of 35 Gy/5 fractions. Clinical and MRI follow-up examinations were planned and radiotoxicity and survival times investigated.

Ten dogs were enrolled. Tumors involved the plexus and proximal nerves in three dogs, the plexus, proximal nerves and nerve roots in five dogs, and the nerve roots and proximal nerves in two dogs. Partial or complete reductions of neurological deficits were observed in all treated dogs. Partial responses were observed in 10/10 patients. Local recurrence was observed in 9/10 treated dogs. Mean overall survival of 371 days and mean progression-free survival of 240 days are comparable to surgical literature data regarding the plexus and proximal nerve localization but are superior in comparison to nerve root localization.

VMAT can be a safe and viable alternative to surgery in cases of canine brachial plexus PNSTs involving the proximal nerves and nerve roots.

To our knowledge, this is the first prospective observational clinical study regarding VMAT stereotactic RT for canine brachial plexus PNSTs and suggests that VMAT may achieve at least similar clinical outcome as surgery in a safer way.

Further information available on website www.lacittadina.org

acvim087

Electrophysiologic Quantification of Motor Neuron Pool Excitability in Chronically Paralyzed Dogs and Association with Gait

Melissa Lewis[1]; James F. Howard[2]; Natasha Olby[1]

[1]College of Veterinary Medicine, North Carolina State University, Raleigh, NC, USA; [2]Department of Neurology, School of Medicine, University of North Carolina, Chapel Hill, NC, USA

Acute, severe thoracolumbar spinal cord injury (TLSCI) results in variable recovery. Plasticity of local spinal circuitry has been demonstrated in rodent SCI models. We hypothesized that reorganization of reflex circuitry including altered motor neuron excitability contributes to recovery of pelvic limb stepping in dogs with severe TLSCI. Our aim was to quantify motor neuron pool excitability electrophysiologically in normal dogs and dogs with chronic deficits due to acute TLSCI and correlate findings to gait.

Nineteen dogs without pain perception at least 3 months following acute TLSCI were prospectively enrolled and compared to 5 normal dogs. H reflex was evaluated. Gait was categorized as ambulatory (yes/no) and quantified using an ordinal scale (OFS) and treadmill-based stepping and coordination scores (SS, RI). H latency, H:M ratio, H threshold intensity were compared between cases and controls using student's t-test and investigated for associations with gait using linear regression.

H threshold in cases (mean: 2.87 mA±2.4) was significantly lower than controls (mean: 6.9 mA±4.3, $p = 0.016$). H threshold was not associated with walking, OFS ($p \geq 0.05$) but was inversely associated with SS ($p=0.031$), RI ($p=0.029$). No differences in H latency and H:M ratio ($p>0.05$) were identified between cases and controls and these were not associated with gait.

These findings establish reference values for H wave variables in dogs with permanent deficits after TLSCI. H threshold might be a useful measure of motor neuron pool excitability in this population. These findings suggest dogs with greater excitability manifest greater stepping after injury.

acvim100

Cochlear Changes in Dogs with Hereditary Deafness

Nevra Keskin[1,2]; Hasan Albasan[1,2]; **Irem Gul Sancak**[1,3]; Michael M. Paparella[1,4]; Sebahattin Cureoglu[1]

[1]Department of Otolaryngology-Head and Neck Surgery, University of Minnesota, Minneapolis, MN, USA; [2]Department of Internal Medicine, Faculty of Veterinary Medicine, University of Ankara, Diskapi, Ankara, Turkey; [3]Department of Surgery, Faculty of Veterinary Medicine, University of Ankara, Diskapi, Ankara, Turkey; [4]Paparella Ear Head and Neck Institute, Minneapolis, MN, USA

The hearing ability has a promising effect on human-animal bonding. The purpose of this study was to evaluate any otopathologic changes in temporal bone specimens of dogs with hereditary deafness. Specimens were obtained from the temporal bone collection of Otopathology Laboratory at the University of Minnesota. Temporal bone collection consisted of dogs clinically confirmed with hereditary deafness or hearing ability.

Morphometric analysis included the measurement of the stria vascularis and spiral ligament areas and the count of spiral ganglion cells. In addition, the presence of the organ of Corti and cochlear hair cells are noted. The location of Reissner's membrane was also assessed.

The results of this report indicated a generalized degeneration in the cochlea including deformed organ of Corti and tectorial membrane, collapsed Reissner's membrane and loss of spiral ganglion cells ($p<0.05$) in addition to lateral wall changes including decreased area ($p<0.05$) of stria vascularis, spiral ligament and spiral prominence in deaf dogs. Hereditary deafness is untreatable with current known therapeutic methods; therefore, these animals may not be considered for breeding and future therapeutic approaches should be designed to target all the elements of the cochlea.

acvim108

Diffusion Tensor Imaging Analysis of Severe Thoracolumbar Spinal Cord Injury in Chronically Paralyzed Dogs

Melissa Lewis[1]; Pew-Thian Yap[2]; Natasha Olby[1]

[1]College of Veterinary Medicine, North Carolina State University, Raleigh, NC, USA; [2]Department of Radiology, Biomedical Research Imaging Center, School of Medicine, University of North Carolina, Chapel Hill, NC, USA

The variability and complexity of lesions amongst chronically paralyzed dogs after acute, functionally complete thoracolumbar spinal cord injury (TLSCI) is poorly understood. We hypothesized that diffusion tensor imaging (DTI) would capture lesion heterogeneity in dogs with permanent deficits following TLSCI better than standard magnetic resonance imaging (MRI) sequences.

Twenty-one dogs without pain perception at least 3 months following acute TLSCI were enrolled and compared to 6 control dogs. All dogs underwent standard thoracolumbar MRI with DTI. Fractional anisotropy (FA) and mean diffusivity (MD) were calculated on regions of interest (ROI) at the lesion epicenter and cranial and caudal to the visible extent of the lesion in normal-appearing areas on conventional MRI. FA and MD were calculated on corresponding ROI in controls. Tractography was performed to detect trans-lesional fibers. FA and MD values were compared between cases and controls using Wilcoxon rank sum test.

FA at the lesion epicenter (median: 0.230, 0.107–0.320), cranial (median: 0.420, 0.391–0.516), and caudal to the lesion (median: 0.365, 0.265–0.513) were lower than corresponding ROI in controls (median: 0.611, 0.554–0.825, median: 0.605, 0.579–0.774, median: 0.589, 0.342–0.817, respectively, $p \leq 0.0043$). MD at the lesion epicenter (median: 2.09×10^{-3}, $1.33–2.96 \times 10^{-3}$) and cranial to the lesion (median: 1.52, $1.03–1.87 \times 10^{-3}$) were higher than corresponding ROI in controls (median: 1.29×10^{-3}, $0.82–1.35 \times 10^{-3}$, median: 1.31×10^{-3}, $0.98–1.41 \times 10^{-3}$, respectively, $p \leq 0.0162$). Tractography revealed absence of trans-lesional fibers in 3 cases and partial disruption in 18.

These findings suggest quantitative DTI can detect the cranial-caudal extent of severe TLSCI better than standard MRI and tractography may identify physical transection. DTI could allow for establishment of a continuum of severity amongst chronically paralyzed dogs following severe TLSCI.

acvim122

The Effect of Post-Operative Rehabilitation on Recovery from Acute Thoracolumbar Intervertebral Disc Herniations

Natasha Olby[1]; Cory Sims[1]; Joseph Fenn[2]

[1]College of Veterinary Medicine, North Carolina State University, Raleigh, NC, USA; [2]Royal Veterinary College, Hatfield, UK

Standard treatment of non-ambulatory dogs with acute intervertebral disc herniations (IVDH) is surgical decompression. Following surgery, recommendations for management range from a period of strict rest to early intensive rehabilitation and the most appropriate management is unclear. This randomized, blinded trial compared the effect of a conservative versus an intensive rehabilitation protocol on gait score and coordination.

Thirty non-ambulatory dogs treated surgically for acute IVDH were recruited and entered the trial on day 2 after surgery. Dogs had to be paraparetic or paraplegic with pain perception. They were randomized 1:1 to a conservative protocol of passive range of motion and sling walking or an intensive staged rehabilitation protocol. Dogs were treated for 2 weeks and had daily neurological examinations and videotaped gait acquisition on days 1, 14 and 42. The primary outcome was gait and coordination score at 14 days.

Thirty dogs completed the trial. All recovered independent ambulation and intensive rehabilitation was well tolerated with no adverse events recorded. Mean gait score at 2 weeks was 7.8/12 ±1.2 and coordination score was 53.4% ±35. There was no difference in gait score or coordination at 14 days between groups ($p>0.05$). Gait and coordination scores improved to 9.4/12 ±1.2 and 70.4% ± 24.8 by 42 days, again with no significant differences between groups ($p>0.05$).

We conclude that early intensive rehabilitation is well tolerated by postoperative dogs, but does not influence the final gait grade or coordination in dogs with incomplete spinal cord injuries.

1.
acvim250

Recognition and Treatment of Movement Disorders

Dennis P. O'Brien, DVM, PhD, DACVIM (Neurology)
Columbia, MO, USA

Movement disorders are a group of conditions characterized by abnormal movements. This can be either excessive, involuntary movements or difficulty initiating movements. Seizures would technically fall under this definition, but they are classified separately. Seizures are characterized by abnormal electrical activity in the cerebral cortex seen on EEG, and are typically accompanied by loss of consciousness and post-ictal behavior changes. Focal seizures, however, may not affect consciousness, and some movement disorders are also episodic. Thus the border between episodic movement disorders and seizure disorders is often unclear. Diseases that cause visible contractions of muscle fibers, such as fasciculations, but do not cause movement of the body part are not typically classified as movement disorders. In this presentation, we will review the characteristics of some of the movement disorders in animals and discuss what is known about the pathogenesis of these conditions and some of the therapeutic options available.

Myoclonus

Myoclonus is a very brief, sharp contraction of a muscle, group of muscles, or the entire body. It can be single or repeated rhythmically. Rhythmic myoclonus lacks the smooth, oscillatory character of a tremor. Focal myoclonus is a common sequelae to canine distemper infections. Generalized myoclonus is often a sign of a more generalized neurodegenerative disease.[1,2] Sometimes myoclonus can be elicited by stimulation. A hereditary myoclonus in Irish Wolfhounds called startle disease or hyperekplexia is caused by a dysfunction of the inhibitory neurotransmitter glycine.[5] Without this inhibition, the normal startle response is exaggerated. In other conditions, the jerks occur spontaneously. The latter is sometimes called myoclonic epilepsy, but it is often not clear if the cortical epileptic discharges that would classify it as a seizure disorder are present or not. Negative myoclonus looks similar clinically, but is actually caused by brief loss of tone in a muscle followed by a jerk as the animal catches itself.

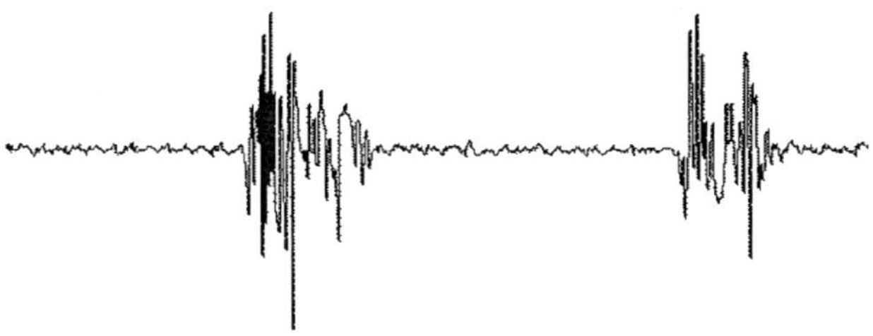

An EMG recording of muscle activity in a dog with distemper shows the repetitive, brief, "lightning-like" burst of firing that characterizes myoclonus.

Opsoclonus-myoclonus-ataxia is an unusual form of myoclonus affecting the eyes as well as the body. The body appears more like a generalized tremor, but the eyes make very rapid, uncontrollable darting movements. It is thought to be an autoimmune disease in humans.[3] Idiopathic tremors ("little white shakers") in dogs are also thought to be an immune-mediated disease with generalized tremors. While many breeds can be affected, it is most commonly reported in Maltese, West Highland White Terriers, and Jack Russell Terriers. Affected dogs generally respond to prednisolone therapy, though relapses can occur.[4]

Tremors

Tremor is defined as a rhythmic, oscillatory movement of body parts. They are further broken down by the frequency of the oscillations and when the tremors occur. Feedback systems have a natural tendency to oscillate, and in muscles such feedback systems may produce a physiologic tremor. Under most conditions, physiologic tremor is barely perceptible; but stress, fear, fatigue, weakness, and drugs such as caffeine can produce an exaggerated physiologic tremor. These tremors are typically fast and small amplitude and tend to occur when standing still, diminishing with movement. Disease processes such as peripheral nerve disease, pheochromocytoma, and hypoglycemia will exaggerate physiologic tremors. Tremor can also be incidental in aged dogs.

Orthostatic tremors occur when standing and affect primarily the hind limbs. They tend to be a little bit slower and coarser than physiologic tremor. The dog is often hesitant to sit or stand, and thus appears weak. When the dog walks, however, the movements show no evidence of weakness, and the tremors resolve. They also resolve when the dog is sitting or lying down. The cause is not known, but they have been reported most commonly in Great Danes.[6] Bulldogs, Boxers, Labrador Retrievers, and Doberman Pinschers show benign head tremors that appear to be hereditary.[7,8] These may be a variation on orthostatic tremors since the head must also be supported against gravity. Both appear to be benign conditions. Beta blockers and primidone are used as symptomatic therapy for tremors in humans but have not been extensively investigated in animals.

Intention tremors are a hallmark of cerebellar disease of any cause. The term 'intention tremor' may not be appropriate for animals since we cannot know what they intend to do, but it captures the key feature of the tremor, so the term is still used. These tremors only occur when the animal is making a goal-directed movement. They disappear completely at rest. They are most apparent in the head but can involve the entire body or limbs. The head tremors can become dramatic when the animal eats, and the limb tremors when they try to walk down stairs. The tremor is slow and coarse. Cerebellar ataxia is also a movement disorder, but beyond the scope of this discussion.

Parkinsonism

Parkinson's disease is a common neurodegenerative disease of older people. It is caused by degeneration of the substantia nigra. Neurons in this area that use dopamine as a neurotransmitter project forward to synapse on the basal nuclei as well as other forebrain areas. This system functions as the gate-keeper for movement. Dopamine is an important component of the "reward system" in the brain. Based on past experience, this system gives either a "go" or a "no-go" signal to the motor programs. Without the dopamine input, the system is stuck in "no-go" and movement cannot be initiated. Although the resting tremor is the sign of Parkinson's disease that everyone recognizes, it is the inability to initiate movement that devastates patients with Parkinson's disease. These signs of difficulty initiating movement and resting tremors are called Parkinsonism and can occur in other hereditary, toxic, and degenerative diseases in people.

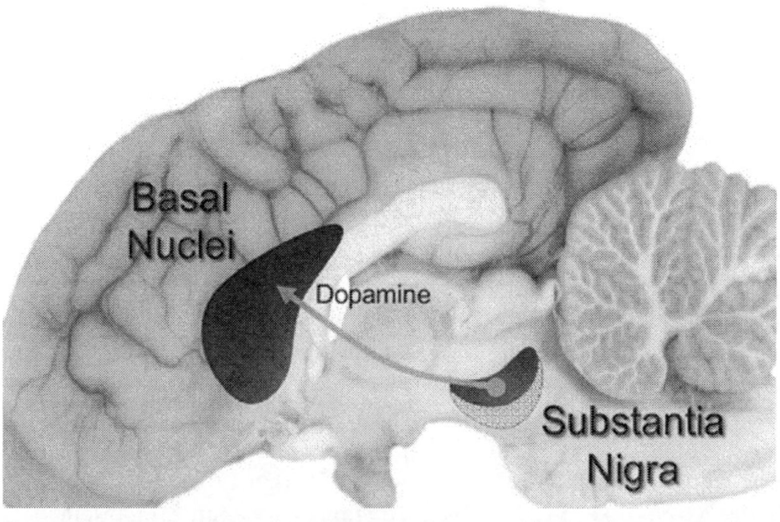

Dopamine neurons from the substantia nigra provide a go/no-go signal to motor programs from the basal nuclei.

Parkinson's disease per se has not been identified in dogs or cats, but the signs of Parkinsonism are seen in some hereditary movement disorders of dogs. Chinese Crested dogs and Kerry Blue Terriers have a juvenile-onset, hereditary disease called canine multiple system degeneration.[9,10] As the name implies, multiple parts of the motor system in the brain degenerate, beginning about 4 months of age. The cerebellum is the first to degenerate, and the dogs have classic cerebellar ataxia. Affected dogs have tremors, but they are intention tremors associated with movement, in contrast to the tremors of Parkinsonism, which occur at rest. Except for primates, none of the experimental or hereditary forms of Parkinsonism in animals show a resting tremor. Beginning at about 6–8 months of age, affected dogs begin having difficulty initiating movement. They shift their weight forward until they finally propel themselves forward in a gait called festination. With further progression of the disease, they are unable to initiate voluntary movements. They adopt a hunched-up posture when they try to walk, and they rock until they lose their balance and fall. In addition to degeneration of the cerebellum, affected dogs have degeneration of the substantia nigra and basal nuclei.

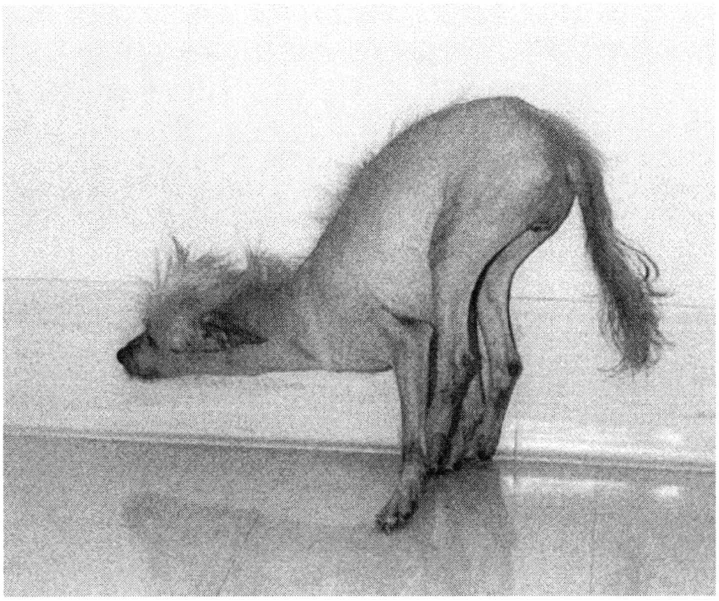

Parkinsonism is characterized by an inability to initiate movement. Affected dogs adopt a hunched posture and lean forward, but they are unable to step forward.

Dyskinesia and Dystonia

Parkinsonism is considered a hypokinetic disease of the basal nuclei, because there is difficulty initiating movement. Another group of diseases is called hyperkinetic movement disorders because the basal nuclei are unable to stop movements. Involuntary movements (dyskinesia) or abnormal muscle tone (dystonia) then occur. Dyskinesias can be focal or generalized. Focal dyskinesia can involve the facial muscles or single limbs. In the limbs, proximal or distal muscles may be preferentially affected. The movements can either be slow or rapid, and they may be continuous or episodic (paroxysmal). In humans, dyskinesias are divided into three general categories: athetosis, chorea, and ballism. Athetosis is a slow, continuous, writhing movement involving primarily the distal arm. The movements may be somewhat purposeful. Chorea (from the Greek "to dance") also affects predominantly distal limb muscles, but the movements are totally involuntary. They are more rapid and fragmented. Ballism is also a rapid movement, but it affects predominantly proximal muscles. The result is a flailing or flinging movement. Dystonia is a sustained contraction of muscles. The axial muscle or limb muscles can be affected. The limbs can either be held in flexion or extension, and sometimes a tremor of the affected muscles accompanies dystonia. As with dyskinesias, the episodes can last from minutes to hours but tend to be more infrequent. Often a mix of different dyskinesias and dystonia will occur in the same patient. Acquired dyskinesias can be drug induced or secondary to basal nuclei lesions from infections or stroke.

The best-characterized dyskinesias in veterinary medicine are paroxysmal and appear to be hereditary. Like seizures, the animals are normal between episodes. In contrast to generalized seizures, animals remain conscious during paroxysmal dyskinesias and they do not have the life-threatening complications sometimes seen in seizures. Some dyskinesias have clear triggers such as stress or excitement, while others do not. The episodes can occur infrequently to multiple times per day, and they can last minutes to hours. The episodes can vary from simple flexion of one limb while walking (which looks like a lameness), to frantic, alternating flexion and extension of multiple limbs.

Chinook dogs have a paroxysmal dyskinesia with primarily dystonia that appears to be inherited as an autosomal recessive trait.[11] Affected dogs have episodes of dystonia and tremors lasting up to an hour. They are conscious during the episodes and appear perfectly normal on recovery. Occasional flailing movements of a limb can occur also. In Cavalier King Charles Spaniels with episodic falling syndrome, a mutation has been identified in *BCAN*, a gene involved in connections between cells and fiber guidance in the brain development in dogs.[12] Although the dogs do collapse during these episodes, they do not lose consciousness, and they have increased tone in the limbs, resulting in them adopting a "deer-stalking" posture. A similar syndrome, Scottie cramp, has been recognized in Scottish Terriers since the 1940s, but sequencing the *BCAN* gene did not reveal any mutations.[13] Canine epileptoid cramping syndrome or Spike's disease is a movement disorder that affects Border Terriers. As the name implies, the episodes resemble seizures, but the dog remains conscious throughout. Most episodes last from 2–30 minutes but may last for hours. The episodes consist primarily of tremors and dystonia affecting the head and body. Excessive intestinal noise and vomiting or diarrhea may accompany the episodes, leading some owners to suspect a dietary link.[14] Recently, antibodies suggestive of gluten sensitivity have been found in affected dogs, and some have shown improvement on a gluten-free diet.[15] Finally, a paroxysmal dyskinesia has recently been recognized in Soft Coated Wheaten Terriers and "whoodles." Affected dogs have episodes of dyskinesia or dystonia, which last minutes to hours and may occur multiple times per day, and a mutation in *PIGN* - a gene involved in anchoring proteins to the cell surface - has been identified.[16] The main carbonic anhydrase found in brain is anchored through this mechanism, and some affected dogs have shown a dramatic response to acetazolamide therapy. Other breeds with reports of dyskinesia are Labrador Retrievers, Jack Russell Terriers, Bichon Frise, Boxers, and Springer Spaniels.[17-20]

Chinook dogs have a paroxysmal dyskinesia consisting of primarily dystonic episodes with flexion of the limbs and twisting of the torso. The episodes may last for up to an hour. The dogs remain conscious during the episodes and show no post-ictal behavior.

Mild episodes of paroxysmal dyskinesia in SCWT and whoodles may only affect one rear limb and can resemble a lameness.

REM Sleep Disorder

Abnormal movements can also occur during sleep. During the rapid eye movement (REM) phase of sleep, the cerebral cortex is as active as it is during wakefulness. This corresponds to the time when people report having dreams. A nucleus in the brainstem prevents any motor commands generated by this activity from being translated into movements, except for the eye movements in people and some small, distal limb movements in animals. When this system fails, larger, sometimes violent movements will occur during REM sleep. This is called REM behavior disorder or REM without atonia in older research literature. This sleep disorder is a prominent sign of a hereditary disorder in Nova Scotia Duck Tolling Retrievers associated with degeneration of the caudate nuclei.[20] In humans, REM behavior disorder can be seen in diseases of the basal ganglia, such as Parkinson's disease. It has been reported in dogs and cats, but aside from the NSDTR disease has not been associated with any particular cause. Benzodiazepines or bromide may decrease the frequency and severity of the episodes.

Conclusion

The increased availability of video capture with mobile devices has meant that movement disorders are being increasingly recognized in veterinary medicine. In the case of hereditary diseases, current gene discovery techniques permit us to identify the mutations responsible and eliminate the disease from affected breeds. Understanding the basis of these hereditary diseases can suggest new pathways to explore for effective therapies for acquired conditions. What we can learn from our patients can also shed light on the comparable human diseases.

References

1. Lohi H, *et al. Science*. 2005;307:81.
2. Awano T, *et al. Mol Gen Metab*. 2006;89:254–260.
3. Markakis I, *et al. Clin Neur Neurosurg*. 2008;110:619–621.
4. Hazell, *et al. Aust Vet Pract*. 2011;41:167–171.
5. Gill JL, *et al. Neurobiol Dis*. 2011;43:184–189.
6. Garosi LS, *et al. J Vet Intern Med*. 2005;19:606–609.
7. Guevar J, *et al. Mov Dis*. 2014;29:191–194.
8. Wolf M, *et al. Mov Dis*. 2011;26:2381–2386.
9. O'Brien DP, *et al. J Hered*. 2005;96:727–734.
10. deLahunta A, Averill DR. *J Am Vet Med Assoc*. 1976;168:1119–1124.
11. Packer RA, Patterson EE, Taylor JF, *et al*. Characterization and mode of inheritance of a paroxysmal dyskinesia in Chinook dogs. *J Vet Intern Med*. 2010;24:1305–1313.
12. Gill JL, *et al. Neurobiol Dis*. 2012;45:130–136.
13. Urkasemsin G, Olby N. *Vet Clin North Am (Small Anim Pract)*. 2014;44:1091–1102.
14. Black V, *et al. J Small Anim Pract*. 2014;55:102–107.

15. Lowrie M, *et al. J Vet Intern Med.* 2015;29:1564–1568.
16. Kolicheski AL, *et al. Neurogenetics.* 2017;18:39–47.
17. Lowrie M, Garosi L. *Vet J.* 1997:2016;213:33–37.
18. Penderis J, Franklin RJ. *J Small Anim Pract.* 2001;42:24–25.
19. Shelton GD. *Vet Clin North Am Small Anim Pract.* 2004;34:1483–1496.
20. Woods CB. *J Am Anim Hosp Assoc.* 1977;13:255–257.

acvim251

C-reactive Protein in the Monitoring and Diagnosis of Diskospondylitis

William Bush, VMD, DACVIM (Neurology)
Leesburg, VA, USA

INTRODUCTION

C-reactive protein (CRP) is an acute phase protein and serves as a biomarker in both human and veterinary medicine to diagnose and follow the treatment of inflammatory disease. CRP has a useful role in the diagnosis and treatment of vertebral osteomyelitis in humans and potentially the equivalent veterinary disease called diskospondylitis. This talk will describe the role of CRP as an acute phase protein (APP), describe its use as a biomarker in multiple veterinary diseases and in human vertebral osteomyelitis. Lastly, we will discuss diskospondylitis and CRP's potential role as a biomarker for this disease.

C-REACTIVE PROTEIN

CRP was the first described APP and named for its ability to bind the C-polysaccharide of *Pneumococcus pneumonia*. CRP binds to microorganisms and then to a receptor on macrophages, enhancing phagocytosis. CRP is a major, positive APP meaning that it increases quickly and dramatically in response to tissue damage. The APP response is considered part of the innate host defense system which is responsible for the survival of the host in the critical early stages of the attack and in evolutionary terms it predates the acquired immune response. Interleukin-6 and interleukin-1 drive the production of CRP from hepatocytes. In dogs the acute phase response is very fast and develops before the activation of a specific immune response and in many cases even before the development of clinical signs. As a major APP, CRP rises early and quickly and then rapidly declines, and is therefore a very useful test to indicate the clinical situation at the time of sample collection. Although sensitive, the APP response is non-specific and develops secondary to numerous conditions that can produce tissue injury like infection, immunologic, neoplastic and traumatic events. In addition to promoting opsonization, APP regulate the immune response controlling inflammation and promote the repair and recovery of damages tissue.

In this talk we will examine CRP's role as a biomarker in part because CRP has several advantages over monitoring a WBC count. Unlike the WBC count, CRP is not subject to extravasation or bone marrow suppression and is 8 times more sensitive than WBC for detecting inflammation. Additionally CRP is more stable and can be sampled as frozen serum or plasma.

CRP AND STEROID RESPONSIVE MENINGITIS ARTERITIS (SRMA)

SRMA is an inflammatory disease occurring in young (mostly <2 years) medium- to large-breed dogs. Clinical signs pursue a waxing and waning course and include neck pain (neck flexion and short-strided on all 4 limbs), fever and lethargy. Diagnosis is based on supportive clinical signs, history and abnormal CSF cell counts, along with response to therapy. In this disease CRP is most sensitive biomarker and one study 100% (20/20) dogs had elevated CRP compared to 85% (17/20) having elevated WBC count. CRP is specific for SRMA because it is significantly elevated compared to other neurological disease like meningoencephalitis, intervertebral disk disease, central nervous system tumors, idiopathic epilepsy and negative controls.

With treatment the CSF counts decline and this is paralleled by drops in the serum CRP concentrations making CRP a useful monitoring parameter in SRMA. At resolution of disease the CRP was normal in all 20 dogs. Relapse rates are stated at about 25% and tend to occur while dogs are still on immunosuppressive therapy. CRP sampling at 4 weeks into treatment demonstrates a trend towards relapse ($p=0.63$) in dogs with higher serum CRP concentrations. In 7 dogs with a putative relapse 100% (7/7) had very elevated CRP, 43% (3/7) had elevated WBC count and only 1/7 (14%) had elevated CSF cell count. CRP monitoring is therefore more sensitive and less invasive and expensive than a CSF analysis in assessing relapse.

CRP in Idiopathic Immune-Mediated Polyarthropathy (IMPA)

IMPA is the most common cause of relapsing fever in dogs and is typically seen in young adult dogs manifesting as pain, shifting leg lameness, reluctance to rise and anorexia. In dogs with IMPA, CRP is much higher at the time of diagnosis than in control dogs or dogs with osteoarthritis and is significantly lower at 2 and 4 weeks in dogs treated successfully with 50 mg/m^2 of prednisone and tramadol. CRP correlates significantly with synovial fluid cell counts, pain scores as performed by owners using the canine brief pain inventory and mobility by accelerometry collars. This work shows CRP is a sensitive biomarker for IMPA and because it reflects synovial fluid cell counts, CRP could substitute for joint taps in evaluating response to treatment.

CRP and Bacterial Pneumonia

CRP is significantly elevated over control dogs in bacterial pneumonia. Antimicrobial treatment is currently recommended for 3–6 weeks or 1–2 weeks beyond resolution of clinical signs. In one study, antibiotics were stopped 5–7 days after normalization of CRP in one group of 8 dogs and pursued for conventional timeframes in another group of 9 dogs. When CRP was used to guide duration of antibiotic treatment, treatment duration was significantly decreased without increasing number of relapses.

Diskospondylitis

Diskospondylitis is infection of the cartilaginous vertebral endplates and intervertebral disk. It is a well-recognized but uncommon disease of typically large, purebred dogs (Great Dane, Labrador, Boxer) with risk factors for having the disease being progressive age and male gender. Clinical presentation is variable and non-specific and may include spinal pain, fever, lethargy, decreased appetite, weight loss and with progression weakness, ataxia and paralysis secondary to spinal cord involvement. Hematologic (leukocytosis, neutrophilia) and serum chemistry (hyperglobulinemia) findings are reported to be nonspecific and unremarkable. Delayed recognition is common because nonspecific signs, insensitivity of bloodwork, limitations with radiographs (delayed appearance of lesions, difficult to distinguish chronic from healing lesions or spondylosis) and lack of referral for advanced imaging. A delayed diagnosis is thought to lead to more severe clinical signs and a worse outcome. Specific treatment is based on bacterial culture; however, dogs typically have only 30 to 35% of samples showing microbial growth in urine or blood samples. Mean duration of antibiotic therapy was 54 weeks in one study and generally it is recommended to follow the response to therapy with only radiographs about every 4–8 weeks until there is resolution of endplate lysis and vertebral fusion. Adjunctive surgery has been recommended when there is extradural compression, worsening of neurological signs or evidence of instability.

Vertebral Osteomyelitis in People

Vertebral osteomyelitis (VO) or spondylodiscitis is the same disease in people and shares many characteristics with the canine disease. VO is known to be an insidious condition with delays in diagnosis of 2 weeks to 4 months and a delayed diagnosis is thought to be associated with a worse outcome. Males are predisposed and patients are generally older; in one study mean age was 66 year (range 28–88). In this same study the initial clinical signs include back pain in 68% (70/105) and fever in 33% (35/105). WBC elevations were noted in 69% (72/105) hospitalized patients; however, in out-patients it is noted that the WBC is an unreliable laboratory marker and only slightly elevated and often normal in the elderly or immune compromised. In the in-patient study, blood cultures were positive in 49%, percutaneous CT guided biopsy in 58%, and surgical biopsy in 50%. Overall 56% (59/105) patients had a confirmed microbiology. MRI is the diagnostic of choice with a 95% sensitivity compared to 66% with CT and 83% with PET scans. In this tertiary facility study 12% (13/105) patients died on the hospital. Mortality was associated with comorbidity, advanced age and CRP protein levels at admission.

In another study of out-patients antibiotic therapy was on average 8 weeks (±4.1 weeks). Treatment failure was noted in 15% (6/39) where patients either had recurrence or disease progression. In this report they note that the evaluation of treatment effectiveness is not simple - imaging abnormalities can persist despite clinical and biological sure and they are not associated with greater risk of relapse. Clinical signs can improve but this does not preclude recurrence or treatment failure.

CRP and Vertebral Osteomyelitis

CRP serum concentrations have played a prominent role in the diagnosis and treatment of VO. CRP is most sensitive biomarker and elevated in about 84% of human VO cases but only 42% have elevated WBC count. Admission CRP predicts mortality in people where a higher mortality is noted when admission CRP>10 mg/dl compared to <10 mg/dl (p=0.028). In one study, predicting outcome is best done 4 weeks into therapy and a ROC curve showed patients with a CRP>2.75 mg/dl were at higher risk for treatment failure. In addition, it has been shown a CRP drop of 50% each week is indicative of a good clinical response. In terms of patient monitoring, it has been recommended that for 1 year after completion of antibiotics patients be followed with clinical exam and CRP and then imaging if either is abnormal.

Also of note is that admission CRP in humans is that it is significantly inversely related to duration of clinical signs and also when it its elevated there is a much higher chance of a positive culture from blood or a biopsy. Lastly, CRP significantly different when spinal implants fail from infection compared to aseptic failure (p=0.0011).

Bush Veterinary Neurology Multicenter Retrospective Study (2011–2016)

We examined records from our Richmond, Leesburg, Springfield, VA, and Rockville, MD, to identify dogs with a MRI diagnosis of diskospondylitis. MRI findings used to make an imaging diagnosis included T2w and STIR hyperintensity, with variable contrast enhancement of affected vertebral endplates. From this population we sought to answer many questions: (1) What was our population of dogs, clinical presentation, frequency of leukocytosis, neutrophilia, hyperglobulinemia, radiographic and CRP abnormality? (2) What was the outcome and does admission CRP predict outcome? (3) Were follow-up CRP serum concentrations useful in predicting outcome? (4) Did NSAID/steroid therapy have an influence on whether the CRP was positive or negative? (5) How frequent were microbiological cultures positive and did this correlate to CRP concentrations?

Signalment and Clinical Presentation

There were 18 dogs in the study 9 males (2 intact, 7 neutered) and 9 females (2 intact, 7 neutered), mean age was 5.3 years (range 1-12) and mean weight was 24.75 kg (range 4-79). Breeds included 2 Labrador Retrievers, 2 Rhodesian Ridgebacks, 2 German Shepherds, 2 Boxers, 1 Mastiff, 1 Newfoundland, 1 Doberman, 1 English Bulldog, 1 German Shorthair Pointer, 1 St. Bernard, 1 Beagle and 3 mixed breed dogs. Spinal pain noted in all dogs, weakness noted in 50% (9/18) and fever noted in 28% (5/18). Duration of signs prior to MRI diagnosis ranged had a median of 90 days (range 6-299). Duration of clinical signs was significantly shorter among dogs with elevated CRP than among dogs with normal CRP concentrations (p=.046).

Biomarker and Radiographs

Elevated CRP serum concentrations present in 61% (11/18), elevated WBC count 6% (1/18), neutrophilia 35% (6/17), elevated globulin 33% (6/18). Increased CRP was significantly greater (10 fold) than the sensitivity for leukocytosis (p=0.002). NSAID or steroid therapy was noted in 67% (12/18). Elevated CRP was in 33% of the cases treated with anti-inflammatories compared to 50% in untreated dogs (p=0.62). This is similar to SRMA studies where prednisone treated group did not have a significantly different CRP concentrations from untreated dogs. Radiographs were abnormal in only 20% (2/10), highlighting the need for a more sensitive diagnostic imaging.

Bacterial Culture and Admission CRP Concentrations

Bacterial culture was performed on blood 4/18 cases, urine in 17/18, and CSF in 5/18 cases. A positive microbiological culture was noted in 24% (4/17) of cases - 3/4 in the blood and 1/17 in the urine. CRP elevations were not significantly different among dogs with a positive and negative culture. The 75% frequency of positive culture in the blood was similar to a recent study by Harris *et al.* where 82% (14/17) were positive, however having just 6% of urine cultures as positive was much lower than in Harris' study where 45% (9/20) were positive. Regardless, these studies support the idea that blood culture are more fruitful than urine in identifying a causative bacterial agent.

Admission and Follow-up CRP concentrations and Outcome

A good outcome was defined as improved clinical signs and noted in 83% (15/18) whereas 17% had a poor outcome defined as static to worsening clinical signs or death. This is comparable to VO in people. Median follow-up time was 120 days (12 to 870) and a good outcome was reported in 83% (15/18) and a poor outcome in 17% (3/18) cases. All dogs with a poor outcome had an elevated CRP and there was a trend for the median CRP to be higher in the dogs with a poor outcome (p=0.17). When a ROC analysis was performed, a CRP concentration of >8.6 was 100% specific for predicting a poor outcome. In the 6 patients with a follow-up CRP, all 5 patients with a normal CRP had a good outcome and the 1 patient with abnormal CRP had a poor outcome - this trended towards being significant (p=0.17).

Conclusion

CRP is emerging as a useful biomarker in veterinary medicine. CRP has an established role in humans with VO where CRP assays aid in the diagnosis and assessment of therapy. Similarly, in dogs, CRP is the most sensitive biomarker for diskospondylitis and admission and follow-up CRP serum concentrations appear to be helpful in predicting outcome. Currently dogs are treated for about a year with antibiotics using the clinical exam and radiographs as endpoints for therapy. This could likely be shortened if dogs had a combination of advanced imaging and CRP concentrations after therapy was initiated and based on human literature a 4 week follow-up would be a good starting point to evaluate therapy.

References

1. Bathen-Noethan A *et al*. Concentrations of APP in dogs with SRMA. *J Vet Intern Med*. 2008;22:1149–1156.
2. Burkett BC, *et al*. Signalment and clinical features of diskospondylitis in dogs: 513 cases (1980–2001).
3. Ceron JJ *et al*. Acute phase proteins. *Vet Clin Path*. 2005;34(2):85–99.
4. Foster JD *et al*. Serum biomarkers in IMPA. *J Vet Intern Med*. 2014:28:905–911.
5. Gorgi AA *et al*. Diskospondylitis. *Standards of Care: Emer and Crit Care Medicine*. 2007;95:11–15
6. Harris JM *et al*. Clinical features and MRI of diskospondylitis in 23 dogs. (1997–2010). *J Am Vet Med Assoc*. 2013;242,359–365.
7. Loibl M *et al*. Outcome-related and co-factors in 105 cases of VO. *Infection*. 2014;42:503–510.
8. Lowrie M *et al*. APP in diagnosis and management of SRMA in dogs. *Vet J*. 2009;182:125–130.
9. Lowrie M *et al*. SRMA: A prospective study of potential disease markers. *J Vet Int Med*. 2009;23:862–870.
10. Viitanen SJ, *et al*. APP in assessment of treatment response in dogs with bacterial pneumonia. *J Vet Int Med*. 2016;1–11.
11. Yoon SH *et al*. VO: identification of microorganism, lab markers and outcome. *Eur Spine J*. 2010;19:575–582.

1.
acvim253

Comprehensive Epilepsy Management

Holger A. Volk, PhD, DECVN, MRCVS
London, UK

The most common chronic neurological brain disease in people and dogs is epilepsy, with an estimated prevalence of 0.4–1% (Cowan 2002; Sander and Shorvon 1996) and 0.6–2%, respectively (Schwartz-Porsche 1986; Kearsley-Fleet *et al.* 2013; Heske *et al.* 2014). A bit more than 20% of human patients with epilepsy will continue to have seizures despite chronic medical treatment with two adequate antiepileptic drugs (AEDs) (Picot *et al.* 2008). Around 20–80% of dogs with suspected idiopathic epilepsy (depending on the examined dog population [hospital vs. primary care]) will not get into remission and continue to have recurrent seizures (Heynold *et al.* 1997; Berendt *et al.* 2002; Arrol *et al.* 2012; Boothe *et al.* 2012), and around 20–30% will remain inadequately controlled (<50% reduction of seizure frequency) despite adequate treatment with the standard AED phenobarbital (PB) and/or potassium bromide (KBr) (Trepanier *et al.* 1998; Schwartz-Porsche *et al.* 1985; Podell and Fenner 1993).

Epilepsy is more than a just pure seizure disorder (De Risio *et al.* 2015). Dogs suffering from epilepsy, especially drug-refractory epilepsy frequently characterized by a high seizure frequency and cluster seizures, have an increased risk of developing behavior changes, premature death, and a reduced owner-perceived quality of life (Berendt *et al.* 2007; Shihab *et al.* 2011; Chang *et al.* 2006; Wessmann *et al.* 2012). Behavioural changes, the effects of recurrent seizure activity and anti-epileptic drugs can affect significantly the quality of life (QoL) of affected dogs (e.g., injury, disorientation, fear, anxiety, restlessness, polyphagia, and polydipsia/polyuria). There is also a potentially high burden upon the owners of dogs with epilepsy that may negatively affect their QoL, including unpredictable and uncontrollable seizure activity, medicating their pets several times a day, and emergency vet visits (Wessmann *et al.* 2012; Chang *et al.* 2006). Owner-perceived dog's QoL is especially affected by high seizure frequency and if the dog is on more than 2 drugs (Wessmann *et al.* 2016). The QoL of the owner is, on the other hand, mainly influenced by the sedation and ataxia level of their AED-treated dog with epilepsy. It is not surprising, but relevant, that reductions in the perceived dog's QoL scores is associated with reductions in owner's QoL, and vice versa.

Understanding what influences QoL is of uttermost importance. Podell *et al.* (2016) state in the ACVIM Small Animal Consensus statement on seizure management that the most important outcome measure is QoL, as this will ultimately affect the owner's decisionmaking in looking after a pet with epilepsy or ultimately can force a decision if the pet has a life worth living for. Owners will elect euthanasia if caring for a pet with epilepsy carries too much emotional stress, becomes financially challenging, and it has an impact on the owner's own psychosocial environment. As specialists, we often focus mainly on the question of how to improve seizure control better with current available AEDs, especially as specialists will usually see the more difficult-to-control patients. It is no doubt important to improve our understanding of which patient will respond to which pharmacological interventions and which patient might not - potentially requiring a more aggressive medical treatment approach early on in the disease process. However, it is equally as important to understand how different management options are tolerated and influence the QoL of the affected pet and owner.

Key for reducing stress and improving QoL for the owner and the dog is to consider the influence of the environment, diet, comorbidities, and antiepileptic treatment. Some reports have indicated the importance of diets, such as the ketogenic diet, hypoallergenic diet, and fatty-acid supplementation, as new or alternative treatment strategies for canine epilepsy. Food supplementation with omega-3 showed inconclusive results (Matthews *et al.* 2012). There is some anecdotal evidence that a hypoallergenic diet might improve seizure control in dogs with gastrointestinal hypersensitivity (Lujan *et al.* 2004). A traditional high-fat low-carb/protein ketogenic diet failed to improve seizure control (Patterson *et al.* 2005). A more promising ketogenic diet is based on medium-chain triglycerides, which improved seizure control in the majority of cases (Law *et al.* 2015).

Pet owner education is another key to successful management of the patient with epilepsy. The better the pet owner is educated about epilepsy, its comorbidities, and antiepileptic treatment (side effects, pharmacodynamics, and pharmacokinetic aspects), the more the owner will learn to live with the condition successfully and help you in the care of the patient (e.g., active monitoring of seizure frequency - e.g., paper or electronic seizure diary using an APP [Bhatti *et al.* 2015]). The owner can also help to reduce environmental stress factors and help with the individualization of the drug treatment depending on the seizure status of the dog (Packer *et al.* 2015). The principle of comprehensive epilepsy care is that 'every little bit helps' to increase the seizure threshold and therefore improve the management of epilepsy.

Conceptually, epilepsy is not one disease with one unique and simple pathophysiology. The term embraces a heterogeneous group of chronic conditions, with seizures being its clinical manifestation. Complex simplicity - the brain is composed by a complex structure; however, it can only function and dysfunction in certain limited ways. This explains why a wide variety of disturbances of brain structure and function can result in seizures and epilepsy. It is thought that similar pathophysiological pathways cannot only lead to epilepsy, but also to other neurodevelopmental and neurobehavioral disorders (Johnson and Shorvon 2011; Shihab *et al.* 2011). These disorders share pathophysiological commonalities at various levels. Understanding the common and discrete pathophysiological pathways can help improve our understanding of these diseases and how to diagnose, monitor, and treat them appropriately. Furthermore, pharmacologically characterizing these disorders and understanding why treatment does fail in some cases cannot only improve our knowledge of the mechanisms behind drug resistance and how to overcome them, but potentially even more importantly deepen our understanding of their pathophysiology and natural clinical course. Furthermore, understanding comorbidities of epilepsy, such as behavior abnormalities, might also give us novel treatment angles, which could improve QoL of the affected dog and potentially seizure control.

Shihab *et al.* (2011) first described that dogs with idiopathic epilepsy have an increase in fear/anxiety and defensive aggression behavior. In the Lagotto Romagnolo dog, a benign "childhood" or juvenile epilepsy form has been described with the first seizures occurring when they are around 6 weeks of age (Jokinen *et al.* 2007). Nearly all of the affected dogs come into remission at the age of 10 weeks, generally without receiving medical treatment. Jokinen and colleagues (2015) looked at long-term outcome of these dogs and found that despite these dogs remaining seizure free, they had an increase in inattention and excitability/impulsivity behavior compared to controls. ADHD-type behavior has also been described in a recent study from the UK (Packer *et al.* 2016), which could in part be improved when a medium-chain triglyceride diet was fed.

Drug resistance should not be thought of as a static state but rather as a dynamic process, which is dependent on the underlying pathophysiology (Kwan *et al.* 2010). Seizure severity and frequency can fluctuate over time (Berg *et al.* 2009; Muñana *et al.* 2010), and the 'state' of drug resistance is determined on the time point of evaluation (Kwan *et al.* 2010). Like in humans, some epilepsy syndromes might change even to the better. Epidemiological and basic science studies have investigated which clinical factors of the disease phenotype might predict poor or good response to AED(s). There continues to be a debate in human medicine, if seizures can be controlled or stopped when treated early, best after the first seizure. Epilepsy is thought to be a chronic condition, and, as aforementioned, drug responsiveness can be dependent on factors independent of the disease process itself.

Epidemiological studies in third-world countries, in which AEDs are not readily available or affordable, revealed remission rates being similar to those in industrial countries (Placencia *et al.* 1993). AEDs have been developed to suppress seizure activity and therefore were often referred to as 'anticonvulsant.' However, not every seizure will be associated with motor activity, and therefore the term 'antiepileptic' is commonly used. Nevertheless, the current AEDs have little or no effect on the disease process itself (epilepsy). Promising AEDs such as levetiracetam, which have been thought to have disease-modifying and/or antiepileptogenic effect, have failed in chronic epilepsy models and have not prevented the development of epilepsy (Brandt *et al.* 2007). In an epidemiological study from 1997, Labrador Retrievers, which stopped seizing, received medical treatment for longer after their first seizure than those epileptic patients, which continued to have epileptic activity (Heynold *et al.* 1997). This could mean that timing of treatment in relation to the first seizure might not be the most important factor for drug responsiveness. We have, however, good data from studies in dogs, rodents, and humans showing that a severe disease phenotype, such as prolonged seizure activity (status epilepticus), high density of seizures (cluster seizures), and a general high seizure frequency, are risk factors for the development of AED drug resistance (Kwan and Brodie 2000; Heynold *et al.* 1997; Hülsmeyer *et al.* 2010; Weissl *et al.* 2012; Löscher and Brandt 2010). The severity of epilepsy will determine how fast and aggressive a clinician will treat the disease (e.g., a dog presented with a high seizure frequency will be treated earlier and potentially more aggressively than a dog that has a very low seizure frequency). This needs to be considered when evaluating epidemiological datasets.

Drug-resistant epilepsy continues to be a major challenge for veterinary medicine, and further studies are needed to unravel its pathophysiology and determine clinical and behavior factors, which might provide new treatment angles so that new treatment strategies can be developed that are individually tailored for the patient to maximize responses to AEDs. AEDs will remain the mainstay for epilepsy management, but environmental factors, diet management, and treatment of comorbidities might also need to be considered to improve ultimately the QoL of the patient and the owner.

REFERENCES

1. Arrol L, *et al. Vet Rec.* 2012;170:335.
2. Berendt M, *et al. J Vet Intern Med.* 2002;16:262.
3. Berendt M, *et al. J Vet Intern Med.* 2007;21:754.
4. Berg AT, *et al. Ann Neurol.* 2009;65:510.
5. Boothe DM, *et al. J Am Vet Med Assoc.* 2012;240:1073.
6. Brandt C, *et al. Neuropharm.* 2007;53:207.
7. Chang Y, *et al. J Small Anim Pract.* 2006;47:574.
8. Cowan LD. *Mental Retard Dev Dis Res Rev.* 2002;8:171.
9. De Risio L, *et al. BMC Vet Res.* 2015.
10. Heske L, *et al. Vet J.* 2014;202:3.
11. Heynold V, *et al. J Small Anim Pract.* 1997;38:7.
12. Hülsmeyer V, *et al. J Vet Intern Med.* 2010;24:171.
13. Johnson MR, Shorvon SD. *Epilep Behav.* 2011;22:421.
14. Jokinen TS, *et al. J Vet Intern Med.* 2007;21:464.
15. Jokinen TS, *et al. J Vet Intern Med.* 2015;10:1111.
16. Kersley-Fleet L, *et al. Vet Rec.* 2013;171(13):338.
17. Kwan P, *et al. Epilepsia.* 2010;51:1069.
18. Kwan P, Brodie MJ. *New Engl J Med.* 2000;342:314.
19. Löscher W, Brandt C. *Epilepsia.* 2010;51:89.
20. Muñana KR, *et al. J Vet Intern Med.* 2010;24:166.
21. Picot MC, *et al. Epilepsia.* 2008;49:1230.
22. Packer RMA, *et al. PlosONE.* 2015.
23. Packer RMA, *et al. Epilepsy & Behavior.* 2016;55.
24. Placencia M, *et al. Epilep Res.* 1993;14:237.
25. Podell M, Fenner WR. *J Vet Intern Med.* 1993;7:318.
26. Podell, *et al. J Vet Intern Med.* 2016.
27. Sander JW, Shorvon. *J Neurol Neurosurg Psych.* 1996;61:433.
28. Schwartz-Porsche D, *et al. J Vet Pharm Therap.* 1985;8:113.
29. Schwartz-Porsche D. *Proc ACVIM.* 1986:1161.
30. Shihab N, *et al. Epilep Behav.* 2011;21:160.
31. Trepanier LA, *et al. J Am Vet Med Assoc.* 1998;213:1449.
32. Weissl J, *et al. J Vet Intern Med.* 2012;26:116.
33. Wessmann A, *et al. J Vet Intern Med.* 2012;26:1.
34. Wessmann A, *et al. Vet Rec.* 2016;10:1136.

acvim254

Impact of New Radiation Therapy Technologies on the Management of Brain and Spinal Tumors

Susan M. LaRue, DVM, PhD, DACVS, DACVR (Radiation Oncology)
Fort Collins, CO, USA

Introduction

Surgery and fractionated radiation therapy have been the most important tools for achieving local tumor control in patients with brain and spinal tumors. Both modalities are geographically oriented and abide by the principles that 1) microscopic tumor extension must be anticipated and addressed, 2) critical normal tissue structures must be spared, and 3) surgical decompression may be required on an emergency basis prior to more definitive procedures.

Meningiomas are the most common intracranial tumor in dogs. Surgery has long been considered the standard of care; however a subset of tumors are found in locations not amenable to surgery. For these, fractionated radiation therapy has been used. This session will discuss stereotactic radiation therapy, delivered in 1–3 fractions, as an alternative to fractionated radiation therapy, as well as new strategies for combining radiation therapy and surgery. The role of SRT for treatment of glial tumors and choroid plexus tumors will be discussed.

Vertebral tumors: These challenging tumors have been treated using a number of different surgical and radiation therapy procedures. Decompressive surgery with extensive cytoreduction followed by fractionated radiation therapy makes medical and biological sense, but outcome has not improved. In the last decade SRT and intensity modulated radiation therapy (IMRT) have become available in veterinary medicine. This lecture will focus on how these new modalities, such as separation surgery combined with SRT, may change the treatment paradigm.

Meningiomas

Meningiomas

Meningiomas are the most common intracranial tumor in dogs. Surgery is considered the standard of care, however, many tumors are not found in a location where surgical excision is possible. In addition, some of these tumors are friable and difficult to completely remove, so radiation therapy is often advocated following excision. Until recently, fractionated radiation therapy in veterinary medicine was administered using a technique called 3 dimensional conformal radiation therapy. For many brain tumors this technology, even though it is older, is still very appropriate. Reported median survival times for surgery followed by 3D-CRT range from 540–900 days. Radiation alone is associated with tumor control of up to 699 days.

A number of advances in radiation oncology have vastly improved the ability to focus radiation dose on the tumor while decreasing dose to surrounding tissues. Intensity modulated radiation therapy (IMRT) controls the intensity of each beam through motorized collimator leaves (MLC) that move during treatment, allowing the dose to be sculpted around important normal structures. This technique requires a specific treatment planning system that uses **inverse planning**. Inverse planning requires that the various tumor structures as well as critical normal tissue structures be identified. The gross tumor volume (GTV) is the macroscopic portion of the tumor. The clinical target volume is the area where microscopic disease may be present. The planning target volume (PTV) accounts for potential error in patient positioning. For inverse treatment planning these structures of interest are identified by the radiation oncologist and outlined (contoured) on each slice of the CT scan, which has been downloaded into the treatment planning computer. Dose objectives for each structure are entered, and a sophisticated algorithm attempts to meet all objectives.

The 3 major benefits associated with IMRT are: 1) dose to adjacent normal tissue structures can be minimized, dramatically reducing acute effects, 2) radiation dose to the tumor can be increased, improving the probability of tumor control and 3) IMRT is appropriate for use following surgery. Although most IMRT is most often administered using a linear accelerator, it can also be administered using a technology called Tomotherapy™. IMRT is now available at a number of veterinary radiation therapy centers.

Interestingly, for brain tumors IMRT may not always be clinical advantage over 3D-CRT, which has been the standard of care for many years in veterinary medicine. That is because, depending on location, many brain tumors do not have adjacent acutely responding normal tissues, like skin, that would benefit from decreased dose. Dogs with tumors near the eyes (olfactory tumors) or near the auditory apparatus experience fewer acute effects with IMRT. Perhaps more critical is fact that IMRT can decrease the dose to normal brain in the region. This is important because the dose that can be delivered to the tumor is constrained by the dose delivered to normal surrounding brain. If higher doses could be delivered to the tumor (dose escalation) using IMRT while maintaining appropriate dose constraints to normal tissues, then greater tumor control could be achieved. This has been demonstrated in a number of tumors in people treated with IMRT, including prostate tumors and some head and neck tumors. Immobilization devices, which allow greater confidence in treatment reproducibility, has also contributed to reduced radiation dose to normal tissues when using 3D-CRT or IMRT.

Stereotactic radiation therapy (SRT) is a new technology that has been used for the treatment of brain tumors in dogs.[5] SRT involves the use of high doses per fraction but overcomes the radiobiological limitations with stereotactically verified positioning and treatment delivery techniques that leave a minimal volume of normal tissue in the high dose area. Stereotactic radiation therapy, by definition, requires: 1) a tumor for targeting (not microscopic disease); 2) treatment planning and administration that will provide a dramatic dose drop off between the tumor and the surrounding normal tissue structures; and 3) a method of stereotactically verifying patient positioning. The result is that normal late responding tissue structures are spared through dose avoidance rather than by administering small doses per fraction. This technique is appealing because only 1–5 fractions of radiation are required and treatments are generally well tolerated by the patient. A number of different systems have been developed for SRT including the Body Gamma Knife™, Cyberknife™ and linear accelerator based systems, such as the Varian Trilogy™. Like IMRT, acute radiation effects are minimized using this technique. A series of 30 dogs with image-based diagnoses of meningiomas were treated at Colorado State University using SRT. The median survival time due to death from any cause was 561 days. The 1- and 2-year survival rates were 61 and 33%.

Vertebral Tumors

Combining surgery and radiation therapy can improve treatment outcome in tumors that are difficult to control with a single modality treatment. Surgery can remove bulky disease, but can fail at the periphery of the tumor if microscopic tumor remains. Fractionated radiation therapy is not effective for large tumors, but can spare surrounding normal tissue while killing residual disease. Combining the two allows the surgery to remove bulky disease while the radiation eliminates local microscopic disease and regional nodal disease. IMRT is now available at a number of veterinary radiation centers. Although a newer technology that can provide improved dose painting and decrease dose to normal tissue structures, it is still fractionated in much the same fashion as 3D-CRT. As such, the approach to combining surgery and IMRT are the same as for surgery and 3D-CRT. Depending on the location and size of the lesion, IMRT can provide improved sparing to normal tissue that hypothetically could result in improved outcomes. But the following principles are the same, so this discussion is appropriate for either IMRT or 3D-CRT.

Much has been written about the sequencing of surgery and radiation therapy. From a biological aspect, radiation before surgery requires a smaller radiation field and the vascular supply of the tumor is not compromised by surgery. Vascular compromise could increase hypoxia, which makes radiation treatment less effective. Also, radiation first can sometimes make a non-resectable tumor into a tumor with surgical options. Treating with radiation first is associated with an increase in wound healing, while post-operative radiation is associated with greater late fibrosis. Surgery before radiation therapy provides optimal tumor staging/grading and patients can physically and psychologically benefit from having reduced tumor burden. Surgery can delay the onset of radiation therapy, as delays (over 1 month post-surgery) have been associated with increased local recurrence for some tumor types. Surgery can also adversely impact radiation dosimetry and treatment planning. For instance, hardware can create severe artifacts that make anatomical identification of structures less precise, and divots and sharp angles created by tumor resection can make impact dose distribution making treatment planning goals difficult to achieve.

Since surgery often precedes fractionated radiation therapy for the treatment of spinal tumors, the surgeon needs to plan the procedure carefully to have minimal impact on the radiation planning and delivery. When using this sequence, tumor control will be enhanced by removing as much of the tumor burden as possible. Unfortunately this often requires more rigorous stabilization procedures. If possible, surgical implants for stabilization should be placed out of the radiation field. Disruption of blood supply to normal tissues should be minimized when possible. For tumors such as osteosarcoma that can literally wrap around the cord, multiple laminectomies may be required to decompress. Radiation therapy should begin as soon after wound healing as possible.

Tumors of the vertebra are rare in dogs and cats. Treatment of vertebral osteosarcoma with decompressive surgery plus radiation and/or chemotherapy is associated with a poor prognosis and median survival of about 4 months. Plasma cell tumors can often be treated successfully with fractionated radiation therapy alone; surgery for decompression or stabilization may or may not be needed.

Combining Surgery and Stereotactic Radiation Therapy for Vertebral Tumors: The New Paradigm

Stereotactic radiation therapy (SRT) is a new technology that has been used for the treatment of brain tumors in dogs[4,5] and its use is being explored for vertebral tumors in humans[6-10] and in dogs.(personal correspondence, Katie Swift) SRT involves the use of high doses per fraction but overcomes the radiobiological limitations with stereotactically verified positioning and treatment delivery techniques that leave a minimal volume of normal tissue in the high dose area. Stereotactic radiation therapy, by definition, requires: 1) a tumor for targeting (not microscopic disease); 2) treatment planning and administration that will provide a dramatic dose drop off between the tumor and the surrounding normal tissue structures; and 3) a method of stereotactically verifying patient positioning. The result is that normal late responding tissue structures are spared through dose avoidance rather than by administering small doses per fraction. This technique is appealing because only 1–5 fractions of radiation are required and treatments are generally well tolerated by the patient. A number of different systems have been developed for SRT including the Body Gamma Knife™, Cyberknife™ and linear accelerator based systems, such as the Varian Trilogy™. Acute radiation effects are minimized using this technique.

In human patients with primary or metastatic tumors of the vertebra, SRT is now being combined with surgery. This sounds counterintuitive since SRT requires a target. This new approach a much less aggressive surgery that focuses on spinal cord decompression and stabilization, if necessary. There is no attempt to significantly cytoreduce the tumor. If possible, hardware associated with stabilization should be kept out of the radiation field. A one week delay between surgery and SRT is recommended to decrease the probability of delayed wound healing or other surgical complications. SRT alone may be indicated for patients with sable neurological status and with minimal cord compression.

References

1. Axlund B, *et al. J Am Vet Med Assoc.* 2002;221:1597.
2. Uriarte A, *et al. Can Vet J.* 2011;52:748.
3. Theon P, *et al. J Am Vet Med Assoc.* 684;216:701
4. Bley C, *et al. J Vet Intern Med.* 2005;6:849–854.
5. Lester N, *et al. J Am Vet Med Assoc.* 2001;219:1562.
6. Griffin L, *et al. Vet Comp Oncol.* 2014.
7. Dunning M, *et al. J Vet Intern Med.* 2009;23:243–249.
8. O'Sullivan B, *et al. Lancet.* 2002;359:2235.
9. Davis A. *J Clin Oncol.* 2002;20:4472.
10. Davis A. *Radiother Oncol.* 2005;75:48.
11. Moussazadeh N, *et al. Cancer Control.* 2014;21:168.

12. Moussazadeh N, *et al. Int J Radiat Oncol Biol Phys*. 2015.
13. Chang U, *et al. Neurol Res*. 2014;36:597.
14. Chang U, *et al. J Neurosurg Sci*. 2013;57:87.
15. Chang U, *et al. J Neurooncol*. 2012;107:551.

acvim255

Hyperbaric Oxygen Therapy and Its Use in Neurology and Neurosurgery: Physiology and Applications with Examples

Anne E. Chauvet, DVM, DACVIM (Neurology)
Sarasota, FL, USA

History

The first recorded efforts of use of hyperbaric (HB) oxygen for medical treatment dates to 1662 when Henshaw used compressed air for medical purposes. Bellows system created a chamber for hyper- and hypoventilation for acute and chronic conditions, respectively. Note though that the oxygen molecule was not even discovered until 1773 and named in 1777.

By 1834, Junod (France) build a HB chamber for pulmonary affliction treatment. These chambers were promoted and used specifically for pulmonary diseases, including tuberculosis, laryngitis, tracheitis and pertussis, as well as apparently unrelated diseases such as deafness, cholera, rickets, menorrhagia and conjunctivitis. The first mobile Chamber came about in 1877 with Fontaine and later, in 1921, Cunningham built a six-story hotel with all amenities and funded by a grateful patient. It is also in 1937 that the first decompression hyperbaric oxygen (HBO) chamber was built. Because regulations were poor, and in an attempt to define the use of HBO, in 2001, the Undersea and Hyperbaric Medical Society established medical accreditation, and to this day continues to supervise the medicine, training, certification and art of HBO therapy (HBOT).

Rational Behind Oxygen Therapy - Physiology and Physics

The science of HBOT is based on a number of principles such as Boyle's law where the volume of gas is inversely proportional to the pressure under constant temperature. Dalton's law states that the sum of the partial pressures of gases in any mixture equals the total pressure and Charles law states that for a body of ideal gas at constant pressure, the volume is directly proportional to the absolute temperature. Hyperoxygenation is an application of Henry's law where partial pressure of a gas dissolved in a liquid is equal to the partial pressure of that gas exerted on the surface of the liquid at equilibrium. Oxygen is normally transported mostly in combination with hemoglobin (3% in plasma). Since hemoglobin is only 97.5% saturated, oxygen carried by 100 ml of blood is actually 19.5 ml and passing through tissue capillaries, this amount is reduced to 5 ml (paO_2 40 mm Hg and 75% oxygen saturation). With increased ATA, more oxygen concentrates in the plasma thus increasing delivery to the mitochondria and other cellular organelles. At 1 ATA, partial pressure is 760 mm Hg with the PaO_2 is 102 mm Hg with 673 mm Hg if breathing 100% oxygen. At 2 ATA, those values are 1520, 262 and 1433, respectively, showing linear increase of PaO_2 with the pressure rise.

A secondary effect of HBOT is vasoconstriction which reduces post traumatic edema and thus finding use for crush injuries, compartment syndromes and burns. Note that the vasoconstriction does not cause hypoxia and thus existing hypoxic areas benefit of the increased plasma oxygen and capillary pressure. On a cellular level, fibroblasts need high levels of oxygen for their effectiveness after elevated lactate triggers them to produce collagen. As well, hydroxylation of lysine and proline residues during collagen synthesis benefits from oxygen. Hypoxia promotes angiogenesis but proper capillary network development needs oxygen. Because HBOT increases the gradient of oxygen between the center and periphery of the wound, angiogenic stimulus is prominent and neovascularization.

Oxygen is also found to be bacteriostatic at 150 mm Hg and at higher levels may be more than static and has a great antibiotic synergy. HBOT is particularly effective against anaerobic bacteria because of lack superoxide dismutase. It also negates many toxins of certain bacteria while potentiating antibiotics that use oxygen for transport across cell membranes.

HBOT also has shown prominent effects on bone defects and repair, correcting non-unions, osteomyelitis and improving speed of healing of bone repair, stimulating osteoclasts. Autoimmune and immune responses are tamed by HBOT due to marked decrease in substance P and the role of TNF-alpha in mediating HBO effects. Red cells are deformed by HBOT and more able to pass in capillaries. Neutrophils consume high amounts of oxygen for phagocytosis (10–15 times normal) and have decreased adherence. HBO inhibits xanthine oxidase system and thus allows skin flaps to heal faster. Regarding neurology, HBOT has shown improvement in nerve regeneration, rescues fibers from ischemic fiber degeneration. The most recent and most significant documented advances with hyperbaric medicine have emerged with the utilization of magnetic resonance imaging (MRI) and single photon emission computed tomography.

Oxygen Toxicity and Negatives

Oxygen toxicity is noted with pallor, bradycardia, palpitations, apprehension, tinnitus, vertigo, seizures, syncope lip twitching and more. This occurs probably due to too high or too long exposure to oxygen at high pressures (ATA). As well there are studies showing possible pulmonary surfactant alterations with HBOT. Myopia and cataracts, the former being transient are ocular changes noted in some patients. Oxygen toxicity is usually transient and can be managed with adjustment of treatment protocol. Factors affecting tolerance to oxygen include such as adrenal hormones, certain pharmaceuticals, hypothyroidism starvation, hypothermia and more.

HBOT has been shown to induce some DNA damage at the first treatment, the opposite occurs in following treatments.

The generation of oxygen-free radicals with HBOT damage the DNA and inhibit bacterial metabolic functions. It is the superoxide dismutase, catalase, glutathione and glutathione reductase that keep the formation of free radicals in check until oxygen eventually overwhelms the enzymes.

The Central Nervous System and Oxygen Therapy

Decades of thinking the damaged brain cannot repair have been erased by HBOT findings. The holy triad in neuronal repair plays a role in that change: Mg, Li and Zn. Magnesium is effective in facilitation recovery of brain function, sciatic damage and depression in humans. Zinc and lithium prevents oxidative stress and allows better wound healing. Brain injury, whether acute or chronic leads to depletion of these minerals, oxygen, vitamins. Replenishment is essential to increase neural growth factors, stem cell mobilization and healing processes, recovery of function. We know that HBOT creates a subset of beneficial radicals that stimulate healing, neurogenesis, activate intracellular healing processes, modify gene expression and stem cell. It induces cerebral plasticity promoting repair of even chronically injured brain. HBOT induces both somatic and stem cell production as shown in numerous studies. When applied for spinal cord injury, increases in levels of vascular endothelium growth factor and connexin-43 are noted and inhibition of NO induced apoptosis. The upregulation of the anti- apoptotic protein Bcl-2 in IVDD in humans is noted and promotes cellular survival. The complete understanding of the mechanisms of HBOT on healing of neurological conditions remains to be explored. At this time, insurance and Medicare consider only the following conditions for HBOT to be covered for payment: air or gas embolism, carbon monoxide poisoning, compartment syndrome/crush injury/other traumatic ischemia, decompression sickness, diabetic and selected wounds, exceptional blood loss, gas gangrene, intracranial abscesses, necrotizing soft tissue infection, osteoradionecrosis and radiation tissue damage, osteomyelitis, skin grafts and flaps, thermal burns. However, the following conditions are off-label which may or may not be covered by insurance or Medicare: autism, cerebral palsy, Lyme disease, migraine, multiple sclerosis, near drowning, recovery from plastic surgery, sports injuries, stroke, traumatic brain injury and PTSD. In the veterinary field, HBOT has been used for all the above and snake/spider bites, pancreatitis, immune mediated disorders, wounds, post op recovery, and a myriad of other disorders.

There are contraindications which include pneumothorax, upper respiratory infection (hypothetical), emphysema, air cysts, thoracic surgery and ear surgery, uncontrolled fever, early pregnancy (all hypothetical now). Where neoplasia used to be a contraindication it is now on the forefront of HBOT research including brain tumors.

Examples in Veterinary Medicine

During the past years of use of hyperbaric oxygen therapy, our patients have educated us on the almost limitless use of this modality. The most common applications in veterinary field are post-surgical recovery and, when possible, pre-surgical treatments. Wounds, spider bites, snake envenomation no longer occupy the ICU must longer than a couple of days and return as outpatients for treatments. Older arthritic dogs walk again when treated with laser combination with polysulfated aminoglycans and HBOT. Pneumonias, pancreatitis, abscesses, anal gland disorders, otitis media, rhinitis, hepatitis, renal diseases, tick borne disorders, leptospirosis and the list goes on, have all be managed with great advantages of HBOT.

GME is a difficult disease to explain to clients when addressing survival. Since the use of HBOT, this author is able to combine CCNU and lower doses than recommended, sometimes without steroids, and treat every 4–6 weeks for a year and occasionally more and patients continue to improve, with some being over 4 year survival at the time of this writing, and normal. All cases of meningitis are also addressed with HBOT when possible financially and physically.

Hypoxic episodes are common in hospitals particularly with anesthesia and referrals for anesthetic accidents have been useful in both acute and chronic cases with recovery in days to weeks, respectively, to resume normal life with normal behaviors.

Post-operative management of spinal surgeries includes HBOT in all this author's patients and has shown faster healing, minimal to no scarring, decreased recovery time of neurological status. This is particularly true of craniotomy cases. Any and all head traumas and spinal cord injuries including atlanto axial subluxations are addressed with HBOT.

Seizure cases with status epilepticus and clusters also have benefited from HBOT for their recovery. Contrary to the concern of oxygen toxicity causing seizures, this author found that maintained at 1.75 ATA, the patients manage positively with no complication.

Perhaps the most interesting findings are those of intracranial cancer management and degenerative myelopathy. In the case of the former, multiple cases have been managed with debulking and CCNU wafer *in situ* followed by HBOT and repeat chemo and HBOT every 4–6 weeks. Cancers are managed long term even when not debulked (Miko and Samantha videos). This author is interested in pursuing further evaluation of management of intracranial tumors with chemotherapy and HBOT with even ketogenic diet management.

In the case of degenerative myelopathy, it is this author's early observation that patients improve with HBOT, particularly with multiple treatments. As well, stem cell therapy has been attempted in one patient with HBOT. Five sessions HBOT prior to stem cell collection yielded 40% more stem cells than the average patient VET STEM received. Two injections of stem cells were possible intrathecally a couple of days apart. The patient's condition, although not improving further is no longer deteriorating. There were some complications in injections which may have altered this initial pilot case but it seems that the combination of stem cell therapy and HBOT should be investigate for the management of DM.

FCE are by far the most rewarding cases as well as intracranial vascular accidents and idiopathy or old dog vestibular disease that respond rapidly, in days or less in most cases. The benefits to the client is so striking that they often request treatment longer than needed.

Some of the neurological conditions that the author does not feel have significantly benefited from HBOT include Chiari malformations, hydrocephalus, LS and SI subluxations in cats and dogs, MPNST. Many other diagnoses still need to be explored for treatment with HBOT and with more facilities using this technology results are likely to be more promising and encouraging.

References

1. Kindwall, EP, Whelan HT. *Hyperbaric Medicine Practice*. 3rd ed. 2008.
2. Zhang, J. *Hyperbaric Oxygen for Neurological Disorders*. 2008.
3. Jain, KK. *Textbook of Hyperbaric Medicine*. 2009.
4. Poff AM *et al. Comp Physiol*. 2017;7:213.
5. Bhutani S *et al. Indian J of Plastic Surg*. 2012;45(2):316.
6. Figueroa SA *et al. Neurology*. 2015.

acvim256

Innervation of the Respiratory System and Neurogenic Causes for Breathing Problems in our Patients

Devon Wallis Hague, DVM, DACVIM (Neurology)
Urbana, IL, USA

INTRODUCTION

Multiple areas of the nervous system are involved with normal respiration. The brainstem is influenced by afferent information from sensors and the cerebral cortex. This complex central respiratory center in the medulla and pons ultimately drives respiration by controlling the rate and rhythm of breath. The inspiratory and expiratory neurons project to the brainstem and spinal cord to innervate peripheral nerves and muscles involved in respiration. Disruption of these areas can result in abnormal or absence of respiratory function. In evaluating clinical cases with a combination of respiratory and neurologic symptoms, the ability to localize a lesion involves both an understanding of the neuroanatomy and accurate neurologic examination findings.

NEUROANATOMY

Afferent Input

The respiratory centers are stimulated by three receptors: central chemoreceptors, peripheral chemoreceptors, and stretch receptors. Central chemoreceptors are located in the central nervous system (medulla). When the cerebrospinal fluid becomes more acidic due to an increase in carbon dioxide, central chemoreceptors stimulate the respiratory center to increase inspiration. Peripheral chemoreceptors are located in peripheral nerves within the aortic arch and carotid bodies and behave similarly to the central chemoreceptors. The peripheral chemoreceptors stimulate the respiratory center to increase respiration when the pH or pCO_2 increases or pO_2 decreases. Stretch receptors are located within the smooth muscle of the lungs and respond to excessive stretching during inspiration. This is called Hering-Breuer inflation reflex and is a protective mechanism to prevent overinflation of the lungs.

The peripheral chemoreceptors and pulmonary mechanostretch receptors stimulate the vagus and glossopharyngeal nerves and mostly terminate on the nucleus of the solitary tract in the medulla. Breathing is also influenced by higher brain centers both by afferent and efferent pathways to and from the brainstem. In the awake animal, CO_2 affects the breathing pattern via suprapontine structures.

Inspiration

The dorsal respiratory group (DRG) and rostral ventral respiratory group (rVRG) contain primary inspiratory neurons. The apneustic center (located in the caudal pons) stimulates the DRG and prolongs inspiration. Conversely, the pneumotaxic center (also located in the caudal pons) inhibits respiration and prevents overinflation of the lungs.

The preBötzinger complex (located rostral to the rVRG) is the site of rhythm-generating inspiratory neurons. The preBötzinger neurons act as a central chemoreceptor to directly stimulate the nucleus of the solitary tract. They also are thought to be involved with sighs and gasps due to their function as an oxygen chemosensor.

The DRG and rVRG inspiratory neurons are excitatory (via glutamate and nitric oxide) and project to the phrenic nucleus in the C5–7 spinal cord segments to the phrenic nerve and ventral horn of the thoracic spinal cord. The phrenic nerve innervates the diaphragm. The thoracic motor neurons innervate the external intercostal muscles.

Expiration

The caudal ventral respiratory group (cVRG) contains primary expiratory neurons. They function to modulate the amplitude of respiration. The Bötzinger complex (located rostral to the VRG) is activated during expiration to inhibit inspiratory drive. The cVRG bulbospinal neurons project through the thoracic and lumbar spinal cord to innervate abdominal muscles and the expiratory intercostal muscles.

Expiration is primarily a passive process, but the cVRG can be activated during exercise or when active breathing is necessary.

Abnormal Respiratory Patterns

Abnormal respiratory patterns have been best described in people and can be useful with neurolocalization.

Respiratory pattern	Description	Neurolocalization
Cheyne Stokes respiration	Waxing and waning depth of respiration with regular periods of hyperpnea and apnea	Bilateral cerebral or diencephalic disease
Biot's breathing	Irregular periods of apnea alternating with periods of 4-5 breaths of identical depth	Increased intracranial pressure
Central neurogenic hyperventilation	Tachypnea to the point of panting	Mesencephalic (midbrain) lesions
Apneustic breathing	Short periods of breathing (1-2 breaths) and apnea	Pontine lesions
Ataxic breathing	Irregular rate and rhythm	Caudal brainstem (medulla)

Neurolocalization and Clinical Cases

Any lesion in the brainstem or cervicothoracic spinal cord has the potential to produce an upper motor neuron lesion affecting the respiratory drive. Patients with a brainstem lesion may show concurrent abnormalities of alerted level of consciousness, ataxia, proprioceptive abnormalities, and cranial nerve deficits. Cervicothoracic spinal cord lesions may manifest as gait abnormalities (proprioceptive ataxia, paresis or plegia), proprioceptive deficits, and changes in reflexes. Possible differentials include trauma, inflammatory, vascular (i.e., cerebrovascular infarction or fibrocartilaginous embolism), degenerative, neoplasia or malformations. Additionally, acute intervertebral disc disease could be a differential in patients with evidence of a myelopathy.

A lesion in the lower motor neuron pathways innervating respiratory function can also manifest in patients as the only symptom or as a conglomerate of multiple neurologic abnormalities. The lower motor neuron involves the ventral spinal cord, peripheral nerve, muscle and neuromuscular junction. Symptoms include weakness, decreased tone, decreased reflexes, and muscle atrophy.

References

1. Alheid GF, McCrimmon DR. The chemical neuroanatomy of breathing. *Res Physiol Neurobiol.* 2008;164:3–11.
2. Cohen MI. Central determinants of respiratory rhythm. *Ann Rev Physiol.* 1981:43:91–104.
3. DeLahunta A, Glass E, Kent M. *Veterinary Neuroanatomy and Clinical Neurology*. St. Louis, MO: Elsevier; 2014.
4. Doi A, Ramirez J. Neuromodulation and the orchestration of the respiratory rhythm. *Res Physiol Neurobiol.* 2008;164:96–104.
5. Donkelaar HJ, *et al.* The autonomic nervous system. In: Donkelaar HJ, ed. *Clinical Neuroanatomy Brain Circuitry and its Disorders*. New York: Springer; 2011:583–91.
6. Dowell AR, Buckley CE, *et al.* A review of clinical manifestations and critique of physiology mechanism. *Arch Intern Med.* 1971;127:712-726.
7. Evans HE, deLahunta A. *Miller's Anatomy of the Dog*. 4th ed. St. Louis, MO: Elsevier;2013.
8. Gautier H. Pattern of breathing during hypoxia or hypercapnia of the awake or anesthetized cat. *Respir Physiol.* 1976;27:193–206.
9. Guyton AC, Crowell JW, *et al.* Basic oscillating mechanisms of Cheyne-Stokes breathing. *Am J Physiol.* 1956;187:395–398.
10. Hendricks JC, Kline LR, *et al.* The English bulldog: a natural model of sleep-disordered breathing. *J Appl Physiol.* 1987;63:1344–1350.
11. Hoff HE, Breckenridge CG. The medullary origin of respiratory periodicity in the dog. *Am J Physiol.* 1949;158:157–172.
12. Kelly BJ, Luce JM. The diagnosis and management of neuromuscular disease causing respiratory failure. *Chest.*1991;99:1485–1494.
13. Stornetta RL. Identification of neurotransmitters and co-localization of transmitters in brainstem respiratory neurons. *Res Physiol Neurobiol* 2008;164: 18-27.

14. Walter MS, Glasser RL, *et al*. Apneustic breathing after vagotomy in cats with chronic pneumotaxic center lesions. *Respir Physiol*. 1971;12:239–250.
15. Weber CL, Speck DF. Experimental Biot periodic breathing in cats: effects of changes in PiO_2 and $PiCO_2$. *Respir Physiol* 1981;46:327–344.

acvim441

I Want to Be a Veterinary Neurologist/Neurosurgeon: An In-Depth Look at Working in Academia versus Private Practice

Belinda Comito, DVM, DACVIM (Neurology); Stephanie A. Thomovsky, DVM, MS, DACVIM (Neurology), CCRP
West Lafayette, IN, USA

Once a veterinary resident has finished his or her training, employment can be daunting decision. When it comes to being a specialist who wants to continue to regularly see clinical cases there are two main avenues of employment in the USA - working in a private specialty practice or working in an academic institution. When one graduates from a residency the decision between pursuing employment in the private sector versus an academic institution can be a difficult, challenging decision. The goal of this talk it to discuss what it is like to be a veterinary specialist in both the private sector and also in an academic environment. We will explore the differences while using the specialty of neurology as an example.

Private Practice

There are two different types of private practice opportunities available for clinicians who want to practice clinical medicine in the private sector: multispecialty clinics and neurology-only practices. Multispecialty hospitals vary in size, that being said they are typically larger hospitals each containing a full complement of veterinarians boarded in a variety of specialties. In this vein the multispecialty practice is a parallel to a teaching hospital setting. The advantages of multispecialty practices mainly lie in the ability to collaborate with other specialists, convenience for consultation with other specialists, and proximity of a busy, active emergency service. Being in the same building as an active emergency service improves neurology caseload and can take the pressure of a busy neurologist who does not have the time to fit emergencies within their busy appointment schedule.

Neurology-only practices offer the advantages of efficiency (you may be one of two or three doctors compared to one of ten) and often have multiple neurologists that allows for real time case collaboration. You are totally immersed in all things neurology and do not have the distractions that can come with multispecialty clinics including the equipment and staffing needs that are necessary for other specialists but not utilized by neurologists, for example. As these clinics are solely in existence to see neurologic animals it may be easier to purchase equipment that neurologists require such as an electrophysiology machine.

Whether you choose to work in a multi-specialty or neurology-only clinic a decision also needs to be made as to whether you want to be the sole neurologist or one of several neurologists. Your preference and comfort level of joining a practice with an established neurologist and neurology service versus joining a practice as the solo neurologist. A multi-neurology private practice allows advantages of joining an established service (technical help, equipment, referring veterinarian relationships, shared on call duties, and a set schedule). A solo neurology practice has the benefit of establishing the practice from the ground up in a way that is conducive for the individual neurologist. Often establishing a practice does present challenges in the early phases including training and establishing staff, purchasing equipment, and developing both a caseload and relationships with the referral community.

Another important consideration when choosing a private practice is the type of management for which you want to work. Types of management include: private versus cooperative ownership. Several years into employment you may want the opportunity to own a portion of the clinic. Each private practice is going to have its own unique management and ownership. Depending on the practice you join you may have opportunities to buy into the practice, become a medical director or chief of staff.

The main focus of a private practice neurologist is to see clinical cases. These cases are typically scheduled to maintain a four day work week. That being said, less commonly some private specialist do work a more traditional five day work week. It is not uncommon for a neurologist who has just completed a residency to work a 5-day workday initially; four of the days are days during which clinical cases are scheduled while the fifth day is reserved for paperwork and other administrative duties. Eventually after experience is garnered these young specialists then transition into a four-day work week.

Maintaining a high case load is very important as salary can often be dictated by the number of cases seen (if all or a portion of the salary is based on production). Thus, it is uncommon to have formal, "off clinics" time as is common in the university setting making it rare to develop or collaborate with other institutions on research. That being said, private practice neurologists are often encouraged to pursue projects of interest; these projects (publications, continuing education seminars, house officer learning materials) are developed outside of work hours.

Prior to seeing the first appointment of the day, ICU rounds are typically scheduled. These rounds allow for a collaborative approach to hospitalized cases and for formal case transfers between services. In a given day, the private neurologist will see scheduled appointments, emergencies and accept ICU transfers (if affiliated with an ER clinic or critical care service). Typically appointments are seen in the morning with diagnostic work ups in the afternoon; alternatively, some practices with multiple neurologists may have designated appointment and work up days.

As mentioned earlier case load is important in the private sector. Case load can have a direct relationship to salary. In general compensation for the private practice neurologist is higher than that of the average academic neurologist. Frequently compensation for the private neurologist will consist of a base salary or a base salary in addition to production. Neurologists on a base salary will earn a flat salary regardless of the revenue that is generated. The advantage of this predetermined base salary is a lack of fluctuation if case load reduces/it is a slow time of year. Many new, young neurologists will garner a base salary for the first 6 to 12 months of employment while they transition into practice - allowing the neurologist time to build an adequate case load. This is especially important for solo practicing neurologists who are establishing a brand new service from scratch. Neurologists on base salary **and production** generate income on top of the base salary (production). Production is made if enough revenue is generated to cover the base salary; the neurologist earns a percentage of the total neurology revenue. If you are unable to generate enough revenue to cover your base salary, some practices will continue to pay your base salary while other clinics will consider you to be in a short fall. A short fall is defined as the time during which the specialist falls below their base salary. During this time the practice continues to pay the base salary, however, future production will be taken away from the employee to make up for the lost salary.

Neurologists on a production-based income will generate a percentage of the incoming revenue; this type of contract carries the most risk as salary is completed dictated by case load. Production percentage is variable (typically 19–26%). When negotiating your contract is important to have an attorney review the contract. It is also important for your future employer to provide an accurate assessment the employee's earning potential. When joining a multi-neurologist practice it is important to consider the total case load and whether it can support an additional neurologist. Similarly when employed as the solo neurologist you must remember that your case load is very dependent on you fostering relationships with the local veterinary population. Neurologists that practice both medical and surgical neurology have more earning potential in private practice; you can be successful in private practice as a medical only neurologist especially in a practice where revenue can be generated off of the magnet if the employer owns the magnet. If the MRI is owned by an outside company and leased you may only generate revenue after a report is generated. Vacation time in private practice is variable but generally, if making production, you have more flexibility in the amount of time you take off.

Lastly, in the private sector the opportunity to teach does exist. It is becoming exceedingly more common for neurologists to work in practices with both residents and/or interns. These house officers shadow the neurologist all while learning on the job, clinical skills. In some situations the house officers not only shadow cases but may actually take on primary case responsibility. Because the expectations of clients who choose to bring their pets to a private specialty hospital can be different than those who bring their pets to a teaching hospital, it is not uncommon that the clients expect to see a board certified neurologist, rather than a resident or intern. This may mean more primary client contact for a specialty neurologist working in private practice as compared to academia. Veterinary student teaching can also be a reality for the private neurologist. Student externs are a reality. These students may spend time shadowing the neurology service for 1–2 weeks and enable clinical case discussion and learning. As compared to academia, often this teaching is performed following completion of the appointment rather than during, as time is less of a luxury in the private sector as compared to academia. In the vast majority of private specialty practices formal journal club and rounds sessions are regularly scheduled.

ACADEMIA

In most academic hospitals a faculty member's time is defined in three areas: learning, discovery and engagement. Learning encompasses didactic teaching (in the classroom and laboratory) including time spent giving the lectures but also composing the PowerPoint and notes for these lectures. Discovery encompasses publications, research and writing grants. It also includes things related to mentoring and supervising graduate students or residents such as resident rounds and assisting/editing/mentoring residency projects. Engagement can include clinical services related to a service but not involving veterinary students. So, if you change your discharge instruction template or work on your clinical services webpage, for example. It is time spent doing those things. Also included is continuing education, consulting in locations or with people or institutions unrelated to your employer. Engagement also involves administrative work including being chief of a section or being the supervisor of staff. Committee work is also encompassed in engagement. There are many committees in academia that one can join or be nominated to be a part of. These committees include those related to the veterinary school, but also main campus or national committees such as ACVIM work. What percent of the year a faculty member spends on learning, discovery and engagement is determined the type of faculty position. There are, typically, three main types of clinical faculty members in an academic institution: clinical instructor (CI), clinical track faculty (CTF) and tenure track faculty (TTF). Table 1 outlines the major differences in percent of time spent on teaching, research and committee work.

Table 1. Comparison between CI, CTP and TTP in academia*

Employment position	Clinical instructor	Clinical track faculty	Tenure track faculty
Percent on clinics (learning)	Often higher than 75%-80% or more of the year is spent on clinics	Often 65-75% of the year is spent on clinics	Often 50% or less of the year is spent on clinics
Research and discovery	Typically no mandatory research requirement.	This is loosely defined in many institutions, but often 5-10% of your total time.	Often 50% or higher
Engagement	Little to no time is spent on this	Roughly 10-15%	10-20%
Salary**	Lowest of the three	Middle of the three	Highest of the three
Other	Unable to become tenured. Most individuals spend 1-2 years maximum in this position. It is often used as an entry level faculty position only - not a long-term position. Your contract goes up for renewal yearly	Promotion from clinical assistant to associate professor is possible. Usually after about 5 years promotion is discussed. Tenure is often not possible in this position. In many institutions your contract goes up for renewal yearly.	Tenure is a reality. Typically a faculty member goes up for tenure after about 5 years. If you are not granted tenure, many times your position changes or your employment can be put in jeopardy. That being said, once tenured, it is your contract does not go up for yearly renewal.

*This table contains general concepts, specifics will vary between institutions

** Salary is often flat with limited upward mobility other than at the time of promotion or tenure.

When working in an academic environment the opportunity exists to work in a school wherein you would be the sole neurologist and the opposite is also true where you can work in an environment with 1-3 other neurologists. These are two different working environments and often times the neurology service is not running 52 weeks a year when there is only one neurologist. In these situations internal medicine may do neuromedical cases while the orthopedic surgery department performs neurosurgery when the neurology faculty member is off on research blocks, for example. Many times the neurologist in solitary neurologist schools is available for a consulting basis during his or her research blocks.

A typical academic work week consists of two different situations. The work week when on clinics versus the work week when off clinics. When on clinics the week very much mimics that of a private practice specialty neurologist with respect to seeing cases and emergencies, usually on a daily basis. The faculty clinician often works a 5-day work week but comes in on Saturdays and Sundays to work with in house patients. The faculty clinicians often splits neuromedical call with his or her fellow neurologist or internal medicine specialists while they share neurosurgery call with their fellow neurologist or surgeons. In addition to overseeing residents and interns with neurologic cases, the academic neurologist also teaches students on clinics. This teaching is performed in once or twice daily in patient rounds sessions in addition to didactic topic rounds. Teaching is also done at the level of the new patient. The student and neurologist discuss the physical and neurologic examinations, formulate a localization and then speak to the owners. To get a patient into advanced imaging often involves patience and time as often they are at the whim of the anesthesia and radiology services. In addition to student teaching resident and intern teaching is often performed daily with cases but also in structured journal club and resident class/rounds sessions. Teaching is also done daily via answering referring veterinarian phone calls, referral questions and emails. Many times the faculty member also has the opportunity to teach and work with externing students and visiting residents and interns. As a general rule, the teaching of veterinary students causes a ripple effect in the schedule wherein less appointments are scheduled in the books each day and academic institutions may carry a lower case load than specialty hospitals (especially those servicing larger populated areas).

When off clinics the faculty uses their time at their own discretion. This is often time reserved for the 'heavy duty/time consuming' research that they are performing. Faculty try to schedule meetings with main campus researchers at this time, as well. It is also a time when the faculty member tries to schedule meetings with students, and committees. It is also when the faculty member hopes to be able to do didactic classroom teaching and laboratories. It is a time to meet with underclassman and mentees (normally faculty members are mentees to interns, residents but also to undergraduate students and veterinary medicine graduate students). That being said, many times meetings and classroom teaching is interspersed with clinics and cannot be reserved for off clinics time. Many academic hospitals allow their faculty members to consult or perform locums; off clinics time is also reserved for these opportunities. It is also a time when many faculty members work on formulating and composing continuing education lectures, proceedings, write book chapters and submit and write publications. This is also a time when sabbaticals can be performed and vacations taken. Most clinicians in academia have roughly 2–4 weeks per year of vacation time (the amount of time is dependent on the type of faculty position held).

An academic neurologist has many opportunities to research, write and publish. This research could be clinical research, more common for CTF and TTF but also bench top research, more common for TTF. The opportunity exist to be director of and head up a laboratory - this is more common for those TTF who are on clinics much less than 50% of the year. With research comes the opportunity to write internal (grants granted by the University; i.e., your employer) or external grants (Morris animal grants or AKC grants, for example). That being said publications can also be in the form of review articles or summary articles and do not have to stem from primary research. Book chapters are also a reality for the academic clinician. Another aspect of grant writing is advising a graduate student or resident on his or her grant. In many academic institutions being second author to one of these students/residents counts as being primary author on the publication when it comes to promotion and tenure. Reviewing other internal and external grants in addition to reviewing peer reviewed publications for various journals is also something that is a reality for the academic clinician. Many times these reviews and the writing are times when one is off clinics, if possible.

An academic neurologist is very much involved in committee work. These committees may service the University as a whole, more commonly the veterinary school and professional groups such as ACVIM. ACVIM committees are usually on a volunteer basis. Committees are open to any board certified neurologist and include such things as: examination grade committee, credentials committee, residency training committee, etc. University committees can be quite varied and include terms on the university senate and with various on campus groups. Participation in veterinary school committees can mean anything from serving on the art in the veterinary library committee, to the grade appeals committee to the hospital improvements committee. There truly are a wide world of options that exist. There can also be time for being a locum at a private practice or separate university hospital. Often 1–2 week are spent at a hospital unaffiliated with the university during the clinician's off clinics block. The clinician is able to experience life working in a different academic or private hospital and the other hospital benefits from being able to keep their neurology service open while the normal staff neurologist is studying for board examinations, on pregnancy leave or ill, for example.

The salary of an academic neurologist is dependent on the type of faculty position the clinician holds. That being said, unlike private practice, case volume and production are rarely considered. While it is important to maintain an adequate caseload for monetary purposes and also student learning, there is not a pressure to see a certain specified number of cases or make above a certain amount of money, to make commission, for example. There is also no opportunity to become a business owner. The hospital is a part of either a state or privately run institution and therefore ownership is unlikely. That being said the ability to oversee and be "boss" to staff is a reality. There is typically a head of the neurology department and that person is in charge of the service. Or, in some universities, neurology is under the umbrella of either the medicine or surgery services, in which case the department head may be an internal medicine specialist or surgeon. The "boss" of the department head is both the veterinary hospital director and also the chair of veterinary clinical sciences (VCS). Upward movement for an academic clinician is normally into VCS chair or some sort of Deanship.

Exciting opportunities exist for veterinarians skilled in a specialty field. A veterinary neurologist can have a successful and rewarding career in both the private and academic sector. It is also important to remember that lateral movement is a viable option, as well. If you start in academia and decide that you would most enjoy working in private practice the switch can easily be made. The same is true for a private practitioner who yearns for an academic opportunity.

References

1. Packer RA *et al. Journal of Veterinary Medical Education*. 2012. Vol 39, 152.
https://www.thebalance.com/veterinary-specialist-salary-125665

acvim257

Preparing for the Feline Neurologic Examination

Amanda R. Taylor, DVM, DACVIM (Neurology); Sharon C. Kerwin, DVM, MS, DACVS, DACVIM (Neurology)
Auburn, AL, USA

Introduction

One could argue that one of the most difficult cases seen in practice is the cat with neurologic disease. As a result, the admission of a cat with neurologic disease can result in an apprehensive veterinarian. A veterinarian's level of stress in dealing with these patients, and the emotional environment they create for themselves, their staff, their patients and clients can set up failure or success. Cats have many unique behaviors and natural tendencies which make neurologic evaluation difficult when appropriate care is not taken to take these differences into consideration. We will review unique behavior and needs of cats and how to prepare for the neurologic evaluation of the cat.

Feline Behavior

Cats are fight or flight animals that are solitary hunters by nature. This results in a unique combination of predatory and prey behavior. A cat's first reaction to confrontation is to flee from the confrontation or hide from it and only fight when escape is not possible. Loud noises, bright lights, unexpected or sudden movement, aggressive handling and agitated handlers can all result in negative reactions from the cat either resulting in the cat attempting to hide or react aggressively. While these behaviors can be overtly seen, their initial signs may be subtle and go unnoticed by the unprepared observer.

Learning to recognize when a cat is fearful or combative can prepare the veterinary team to change their approach when these signs are first seen, rather than reacting once a cat has already acted aggressively in response to their experience in the clinic. One should pay particular attention to a cat's posture, facial expression, tail movement and body posture. When examined, most cats prefer to not be quickly approached face to face, and prefer physical contact over the back of the head or neck.

Postures that are displayed in stressed cats include arching of the back, pulling the feet close to the body, and lowering the head. Fearful cats may display mydriasis, whiskers pulled toward the face, wide opening of the eyes and a general tightening of the muscles of facial expression. The ears will change from their normal erect posture to a gradual flattening. The tail may begin to twitch or flick back and forth. Signs that an interaction has resulted in more aggressive behavior from the cat include further arching of the back, forward turning of ears, narrowing of the pupils, and finally vocalization or swatting of the paw toward the veterinarian or handler. Recognition of the initial signs of stress in a cat should prompt the veterinary team to take a break from the examination and allow the cat time free from handling.

Feline Clinic Admission

Admission to the clinic is one of the most stressful periods of a cat's visit to the hospital due to the noises, smells and other stimuli they may encounter in the lobby of the clinic. While a separate feline waiting area and entrance are ideal, this is not a realistic solution for many clinics that are limited in their space or flexibility. Prior to admission of a cat, client liaisons or front desk staff should inform the clients that all cats must be in their own carrier when visiting the hospital and cannot be removed from the carrier until they are transferred to the examination room. This is for the safety of their cat and the other cats and dogs in the waiting area. A carrier will limit the number of stimuli the cat is exposed to during admission. When it is not possible to admit cats in a separate area, expediting their transfer to a feline examination room as quickly as possible can minimize exposure to stressful situations.

Feline Examination Rooms

Examination rooms that are exclusively maintained for cats (even if it is just one room in the clinic) allow for control of all of the environmental factors that can either contribute to the stress of the cat or enhance the comfort of the cat. Strong smells can be a negative stimulus for cats as they have a very sensitive sense of smell and what they smell can cause them to react with fear or aggression. Elimination of odors or keeping odors in the room to a minimum can help to avoid this issue. Dogs should not be allowed in examination rooms used for cats, or if they are, all surfaces should be cleaned and the room opened to allow for dissipation of remaining odor from cleaning products. Some scents may be positive for cats, in particular the use of a synthetic feline facial pheromone analog. This analog has been shown to increase normal behavior in cats and reduce behavioral displays of anxiety, fear and combative behavior. Be very aware of visual and auditory input in the feline exam room. Bright lights and loud noises can be negative stimuli for cats and by secluding cat examination to a feline- only room you can have more control over the surrounding environment.

The room should be equipped with tools to make a cat more comfortable. This begins with covering the examination table with a thick towel so that the cat can rest on a soft surface that is not cold to the touch. Catnip, laser pointers, cat toys and cat treats may facilitate cat handling and should be stored in this room for easy access during examinations.

Feline Handling Techniques

The further you push a cat during examination, the more likely you are to elicit a negative response and be unable to complete the examination. Aggressive handling such as scruffing, muzzling, or verbally scolding the cat are likely to result in fear responses from the cat which, in addition to the obvious detriment to the patient, will prohibit the examiner from being able to assess the patient. Educating the staff of the veterinary clinic as to appropriate handling techniques is the basis for a successful neurological examination being completed.

Acceptable restraint techniques include the use of sturdy Elizabethan collars (the authors prefer reusable collars that snap in to place), thick towels and patience for the cat. The cat should never be forcefully removed from a carrier or forced to comply in general during the examination. If a cat does not tolerate examination, a hands-off approach consisting of observation is recommended.

Client Education

Cat owners are often astute and in tune with their pets. We have been most successful in performing feline examinations for neurologic disease when we explain to the client prior to performing the examination what will occur. Some of the tests performed during examination can appear strange or even scary to a client, not to mention the negative reactions these tests may elicit from the cat. We explain to the client that if their cat indicates that the examination is too stressful, we will cease the examination and continue to observe the patient. Open communication with the client prior to examination will create a partnership between the client and veterinarian that will facilitate the other aspects of the cat's treatment and diagnostics.

References

1. Frank D *et al. J Am Vet Med Assoc.* 2010;236:1308–16.
2. Gunn-Moore DA *et al. J Feline Med Surg.* 2004;6:133–138.
3. Garosi L. *J Feline Med Surg.* 2009;11:340–348.
4. Kronen PW *et al. Vet Anaesth Analg.* 2006; 33:258–265.
5. Rodan I *et al. J Feline Med Surg.* 2011;13:364–375.
6. Sundahl E *et al. Feline Behavioral Health and Welfare.* St. Louis, MO: W.B. Saunders; 2016.

acvim259

Performing the Feline Neurologic Examination

Amanda R. Taylor, DVM, DACVIM (Neurology); Sharon C. Kerwin, DVM, MS, DACVS, DACVIM (Neurology)
Auburn, AL, USA

Introduction

While any neurologic evaluation can be challenging due to the nature of the diseases encountered, cats with neurologic disease may present additional obstacles due to the relative lack of tolerance for restraint and handling typically used during neurologic evaluations. Key features of the authors' neurologic evaluation of the cat are: limiting repetition, utilizing tests that are reliable, and focusing the exam on what is most likely to render useful information that will aid in determining a neuroanatomical localization in the cat. Once this has been achieved one must determine a ranked list of diseases that occur in the cat and could account for the clinical signs seen. One of the key aspects in examining the cat is first determining whether a neurologic disease is present or whether another disease is affecting the cat in ways that cause signs that seem to be neurologic in origin. Some of these diseases include systemic hypertension, orthopedic disease, aortic thromboembolism, respiratory disease, cardiovascular disease and metabolic disease.

The order of the examination listed herein is how the authors proceed unless, as they begin to evaluate the case, the cat becomes intolerant of the examination. Should this occur, the order of the examination is changed to prioritize the most helpful tests or a break is taken. For example, a cat presenting for inability to walk (nonambulatory paraparetic) becomes difficult to evaluate after gait evaluation. In this case the authors would proceed to segmental reflexes prior to performing testing of the cranial nerves. While during neurologic examination of the dog we may repeat tests multiple times and come back to the same test more than once, this opportunity may not be granted by the cat. The authors attempt to move through the examination with efficiency and safety. If there is an abnormality that needs to be reassessed, and the cat is tolerating examination, we will revisit this portion of the exam after other tests are complete.

Components of the Examination

Observational

Video Recording

While the tests and observations listed below are ideal, there are some cats that do not tolerate nor participate in their evaluation to a point where it becomes necessary to base diagnostic and therapeutic decisions on historical findings. One of the most helpful ways to supplement the information gained from the history is to ask the owner to video record their observations. This can be done prior to the initial appointment, or if the examination is unsuccessful can be sent electronically. With the ready availability of smart phones, this is very practical for most clients. Video can be essential in cases where the cat does not allow thorough examination or where a cat does not display the abnormalities an owner has observed at home. A video can also be helpful when the primary veterinarian is seeking consultation with a neurologist or other specialist for help with a case.

Mentation

As discussed in the previous lecture, feline behavior in the clinic can be fearful or aggressive. Whenever possible, a cat's mentation should be assessed prior to removing it from the carrier. Once a cat has been taken to the examination area, the front of the cage should be opened to allow the cat to exit if it is willing. Alternatively, the top of the carrier can be removed and then the cat can explore its surroundings. In some cases, the cat will not participate in this way. One can gently lift the cat from the carrier and place it on the floor away from the carrier to observe its interactions with the environment.

One must also ask key history questions of the owner to ascertain what the cat's mentation is like at home. These include open ended questions such as: How does your cat's demeanor now compare to six months ago? What are your cat's favorite activities? How is your cat interacting with you?

Posture

Posture is concurrently assessed with gait and mentation. Common abnormal postures seen include head tilts and turns, ventroflexion of the neck and plantigrade stance.

Gait

Challenges with gait evaluation arise when the cat is fearful, as these cats will stay as low to the ground as possible when walking or will crouch in one place and be very reluctant to move. In cats with the pelvic limbs affected, one should always evaluate the femoral pulses (generally done prior to neurologic examination) and perform an orthopedic examination (generally following neurologic examination). When watching the gait of the cat, several questions must be asked: Is the cat off-balance (ataxic), does the cat appear weak in a limb? Is there lameness associated with the gait? What is the length of the steps the cat is taking? Listen to the gait, can you hear scuffing of the dorsal aspect of the paws? Several techniques can be utilized to encourage ambulation in the cat in a gentle manner. Gait should always be evaluated on a non-slick surface.

Laser pointers: Many cats are attracted to these devices to the point where a previously reluctant cat will follow this light on the floor. This can be an excellent time to observe the cat for gait, but also for intention tremors.

Movement of a carrier: Some cats identify their carrier as a safe place to hide and reside while in the clinic. One can slowly pick up the carrier and carry it across the room which may cause the cat to follow. Alternatively, the carrier can be placed far away from the cat on the ground and the cat may walk toward it.

Stepstool: A short step stool can be utilized to get the cat to move. The authors use a step stool shorter than two feet and place the cat on the top step. Most feline patients will jump down from the stool and walk away from the stool.

Direct Tests

Cranial Nerves

Cranial nerves are assessed prior to other tests as frightened or stressed cats may not respond appropriately to cranial nerve testing and false negative responses can occur. A cotton-tip applicator may be used instead of one's finger to perform many of the cranial nerve tests in the cat. Prior to performing any of the below tests the cat is first assessed for facial symmetry. We recommend performing the assessment without touching the cat when possible, as the facial expression may change significantly in response to touch.

Menace is assessed first as the more relaxed the cat is, the more likely they are to respond (in the authors' experience). One eye is covered while the other is assessed as tolerated by the patient. If no response is initially seen, a gentle tap at the medial and lateral canthus of the eye (which will also serve as the palpebral test) may cause a cat to respond to menace testing appropriately. Many cats seem to be less interested in cotton ball testing than canine patients and the authors have not found this to be as reliable as menace testing.

Palpebral is assessed after menace by gently tapping the medial and lateral canthus of each eye to elicit a blink. Physiologic nystagmus can be more difficult to elicit and assess in the cat by turning the head from side to side. It is most helpful to pick up the cat, and with it facing you twist your body from side to side at the waist, moving the cat with you. An assessment for positional strabismus is performed by quickly elevating the head and watching the globes for lack of central positioning.

Facial sensation is assessed by touching either side of the muzzle, temporal region, mandible and nasal mucosa. The cat should respond by blinking or consciously responding to the sensation.

Gag can be assessed directly by touching the caudal pharyngeal wall with a finger. In patients where you do not believe this will be tolerated, you can externally place pressure on the pharynx and observe the patient for swallowing and licking.

Pupillary light reflex is assessed by shining a strong light in one pupil and observing both irises for constriction. Fearful cats may have pupils that are mydriatic at rest and less responsive to PLR assessment due to increased sympathetic tone, however a direct and indirect response should still be present.

Cranial nerve test	Afferent arm	Efferent arm
Menace	CN II	CN VII
PLR	CN II	CN III
Palpebral	CN V	CN VII
Sensation, face	CN V	CN VII
Gag	CN IX	CN X
Physiologic nystagmus	Vestibular apparatus	CN III and VI

Postural Reactions

We begin the postural reaction testing utilizing the same step stool that was helpful in gait evaluation. When the cat jumps down from the stool, a cat with deficits in the proprioceptive system may stumble upon landing. The authors typically perform one or two methods of postural reaction testing in the cat: hopping (most reliable) or tactile placing where the patient is brought to the edge of the table and allowed to place one paw at a time on the surface of the table. Knuckling is typically not performed as the reaction in most cats is withdrawal of the limb and inability for the examiner to place the dorsal surface of the paw on the ground.

Segmental Reflexes

While we typically restrain dogs in lateral recumbency to perform segmental reflexes, this technique is often unsuccessful in the cat. The degree of restraint used to maintain the cat on its side may result in the cat no longer tolerating examination. Therefore, these reflexes should be performed with either the cat laying on its back in the lap of the examiner, or while the cat is standing. Only flexor withdrawal reflexes and patellar reflexes are typically performed in the limbs as the authors have found the other reflexes to be unreliable in performance and interpretation. Following these reflex tests, cutaneous trunci and perineal reflex are assessed in the standing cat. In some cases other reflexes are assessed, such as cranial tibial, gastrocnemius, triceps, extensor carpi and biceps reflexes. For example, a cat with monoparesis of a thoracic limb might have all the thoracic limb reflexes assessed if tolerated.

Reflex	Spinal cord segments
Flexor withdrawal, thoracic limb	C6-T2
Biceps	C6-8
Triceps	C7-T2
Extensor carpi radialis	C7-T1
Flexor withdrawal, pelvic limb	L4-S2
Patellar	L4-L6
Gastrocnemius	L6-S2
Cranial tibial	L6-L7
Cutaneous trunci	C8-T1 (efferent), T3-L3
Perineal	S1-S3

Nociception Testing

Nociception testing is only performed in cats with absent voluntary motor function. If voluntary motor function is present, the degree of dysfunction is not severe enough to result in loss of sensation. Superficial nociception is tested by compressing the interdigital webbing of the affected limb with fingers or hemostats. The cat is observed for a conscious response to the stimulus, such as turning of the head, vocalization or attempts to bite. If a response is not elicited with fingers, hemostats are utilized. If superficial nociception is not detected, deep nociception is tested by clamping the digit with hemostats.

Hyperesthesia Detection

The cat is gently palpated while it is standing for paraspinal discomfort and head pain. This is performed with mild pressure placed with the forefinger and middle finger of the dominant hand on the epaxial muscles on dorsal midline proceeding from caudal to cranial. The neck is put through a range of motion to detect for resistance or painful reactions from the patient. Gentle pressure is placed over the head using the thumb and middle finger in the temporal region, as if picking up a six-pack. If a response is not elicited with gentle pressure, more firm pressure is applied and the tests repeated.

While in the dog we can observe for subtle signs such as ceasing to pant as indications of pain, cats may display other signs. You may detect a sudden contraction of the epaxial muscles or lowering of the head in regions that are painful. This is in addition to typical responses such as vocalization, turning of the head or other behaviors.

References

1. De Lahunta A *et al. Veterinary Neuroanatomy and Clinical Neurology*. Elsevier Health Sciences; 2014.
2. Dewey CW *et al. Practical Guide to Canine and Feline Neurology*. John Wiley & Sons; 2015.
3. Garosi L. J Feline Med Surg. 2012;14:85-93.
4. Lorenz MD *et al. Handbook of Veterinary Neurology*. Elsevier Health Sciences; 2010.
5. Fingeroth JM *et al. Advances in Intervertebral Disc Disease in Dogs and Cats*. John Wiley & Sons; 2015.

acvim261

Feline Neurologic Cases of Intracranial Disease: Reviewing the Classic and Puzzling over Unusual Case Presentations

Amanda R. Taylor, DVM, DACVIM (Neurology); Sharon C. Kerwin, DVM, MS, DACVS, DACVIM (Neurology)
Auburn, AL, USA

INTRODUCTION

Cats presenting with signs of intracranial disease must be carefully considered when reviewing their history, signalment, physical examination and neurologic signs. Due to the greater emphasis placed on dogs in the neurology specialty, there is less awareness of some of the differentials for intracranial disease in the cat. Systemic diseases, such as hepatic encephalopathy, renal failure, electrolyte abnormalities, hyperadrenocorticism, hypoglycemia, hypertension and hyperthyroidism can cause signs of intracranial disease such as abnormal mentation, seizures, tremors and ataxia. Systemic causes of neurologic signs should be ruled out prior to the pursuit of advanced diagnostics such as magnetic resonance imaging and cerebrospinal fluid analysis.

Due to the prevalence of extraneural organ involvement with the some of the most prevalent diseases, retinal examination is recommended in all cats to attempt to detect ocular changes associated with LSA, FIP, cryptococcosis, toxoplasmosis. Prior to case discussion of unusual cases seen by the authors, common intracranial diseases will be reviewed. For the purposes of this discussion we will review neuroanatomical localizations of intracranial disease, common disease presentations, diagnostic testing, and prognosis associated with each disease.

NEUROANATOMICAL LOCALIZATION

	Forebrain	Brainstem	Cerebellum
Mentation	N, obtunded, demented	A (obtunded, stupor, coma) or N	N
Posture	N, ipsilateral head turn, head pressing	Head tilt, decerebrate rigidity	Intention tremors, head tilt
Gait	N, wide ipsilateral circling	Proprioceptive or vestibular ataxia	Cerebellar ataxia
Postural reactions	Contralateral deficits	Ipsilateral Deficits	A or N
CN deficits	Contralateral deficits in sight and perception of facial sensation; CN I and II	CN III-XII deficits possible	Ipsilateral menace deficits
Segmental reflexes	N	N or increased ipsilateral	N or increased ipsilateral

N=normal, A=abnormal

CONGENITAL

Cerebellar hypoplasia results from an *in utero* parvoviral infection that leads to the destruction of granule cells of the cerebellum. The clinical signs of cerebellar disease, intention tremors and cerebellar ataxia, are first identified when the cat starts walking. These signs are not progressive and are symmetrical. Diagnosis can be presumptively made based on signalment and history with supportive changes on MRI of a cerebellum that is hypoplastic or aplastic. No treatment is available for this disease. Prognosis for recovery of cerebellar function is abysmal, but for quality of life is good as these cats have normal mentation.

METABOLIC

Hepatic encephalopathy (HE) occurs when the liver is unable to clear the body of gut-derived toxins which results in altered function of the brain. Cats develop HE most commonly when a congenital anomaly of the portal vasculature is present. The clinical signs include abnormal behavior, seizures, ptyalism (occurs in up to 75% of cats), and abnormal mentation, in addition to signs of systemic illness due to liver failure (weight loss, decreased appetite, vomiting, etc.). Diagnosis of hepatic encephalopathy can be made with ammonia or bile acid levels (with concurrent response to therapy for HE), although this will not determine the underlying cause of liver disease. MRI abnormalities have been identified in the brains of animals with hepatic encephalopathy, although this test is typically not performed on these patients. Prognosis for recovery depends on the underlying cause of the liver dysfunction, however animals with high levels of ammonia tend to respond better to treatment.

Thiamine deficiency can cause an acute onset of central vestibular signs, ventroflexion of the neck, tremors, pupillary dilation, vision loss and mentation changes. Non-specific signs such as anorexia, vomiting and lethargy are often reported prior to the development of neurological dysfunction. Thiamine is an important cofactor in the Krebs cycle and insufficient amounts of it lead to energy deprivation in areas of the CNS because of dependence on glucose for energy. Cats are dependent upon dietary thiamine in relatively high amounts. Thiamine is high in diets with grains, meats, vegetables, and eggs, but it is also easily destroyed by inappropriate food processing (temperatures that are too high or inappropriate pH). Despite the addition of thiamine to commercial cat food, cats eating commercial diets still occasionally develop thiamine deficiency. Canned foods in particular may be deficient in thiamine due to the differences in how they are processed, although a recent outbreak in cats exclusively eating commercial dry food has been reported. Diagnosis is typically based on clinical signs, history, and response to thiamine supplementation. Low erythrocyte transketolase activity with elevations of pyruvate and lactate or measurement of low thiamine in the blood support the diagnosis. T2 hyperintense changes in the thalamus, vestibular nuclei and midbrain are identified on MRI. If treatment is initiated when neurologic signs begin prognosis is good.

NEOPLASTIC

Meningiomas are tumors that arise from the meningeal tissue that covers the brain (extra-axial tumors). Meningiomas are the most common brain tumor in the cat (60% of feline intracranial tumors). Middle aged to senior cats are typically affected. The clinical signs reflect the location of the mass. Commonly reported signs are both nonspecific: lethargy and inappetence, and specific to the brain: altered mentation, circling, ataxia, and behavior changes. MRI of meningiomas tend to display well-marginated masses that are supratentorial in location and that strongly contrast enhance. Up to ¼ of cats with meningiomas have multiple meningiomas. While treatment options will depend on the location of the tumor, those in a surgically accessible location can be completely excised. Cats can live for several years following surgical removal or radiation therapy of meningioma.

Lymphoma is the second most common brain tumor of the cat. Both primary CNS and secondary CNS forms of lymphoma occur, resulting in large masses or diffuse neoplastic lesions in the brain. In cats where palliative and definitive treatment has been attempted the response has varied from a very short survival times (9–30 days) to months.

INFLAMMATORY INFECTIOUS

FIP is caused by mutated feline coronavirus. Involvement of the central nervous system (CNS) occurs frequently in cats with dry FIP, although cats with effusive (wet) FIP may also be affected. While young cats with exposure to multi-cat households are typically affected, it can take years for the dry form of FIP to manifest. Therefore, FIP should still be considered a differential in middle-aged cats without recent exposure to other cats. The signs seen with FIP are a result of the immune system response to FIP infected macrophages. Cats with neurological manifestations of this disease often have the "dry" form of the disease where they develop lesions on the meningeal or ependymal surfaces that may extend deeper into the parenchyma. While a probable diagnosis can be made based on the clinical features of elevated serum globulins, elevated CSF titers, and MRI changes, definitive diagnosis requires immunohistochemistry. Prognosis is poor.

IDIOPATHIC

Feline idiopathic vestibular disease is a common cause of peripheral vestibular disease. Cats of any age may be affected by this disease. Bilateral vestibular signs may occur. Diagnosis can only be presumptively made by excluding other causes of vestibular disease, such as the most common cause of peripheral vestibular disease (otitis media/interna) and central vestibular disease (FIP). Animals should be treated with supportive care and rehabilitation to encourage balance. Recovery is typically complete within two weeks, although the head tilt may not resolve.

Idiopathic epilepsy is less common in cats than dogs, but is still relatively common. Cats were not thought to be affected often by idiopathic epilepsy, however recent studies report that 25–38% of cats with seizures have idiopathic epilepsy. Young cats (1–4 years of age) are more likely to have idiopathic epilepsy than epilepsy as symptom of a brain lesion (symptomatic epilepsy). It has not been determined in cats whether certain breeds are predisposed to idiopathic epilepsy. In the interictal period, these cats should be normal on neurologic examination; however, a normal examination does not eliminate other causes of seizures. While focal seizures were historically viewed as an indication of structural brain disease, such as a tumor, this has not been found to be the case in cats. Up to half of cats with idiopathic epilepsy have both focal and generalized seizures. Seizures in cats are often complex focal seizures characterized by facial twitching, mydriasis, hyper salivation, and tremoring. Due to the lower frequency of occurrence of this disease process in cats than dogs, a full work up of diagnostics including MRI and CSF analysis is recommended to rule out other causes of seizures.

TOXIN

Pyrethrins and permethrins facilitate excitation and block inhibition in the central nervous system. Within hours of application of topical permethrin product, cats can develop signs of toxicity including ear twitching, behavior changes (hyperexcitability), tremors, ataxia, salivation and seizures. Diagnosis is supported by history and clinical signs, although definitive diagnosis can be made with chemical analysis for the insecticide on the skin, hair or in the gastrointestinal tract. Treatment consists of decontamination of the skin or gastrointestinal tract, methocarbamol for tremors, anticonvulsants for seizures and supportive care. Most cats completely recover within three days of onset of signs.

VASCULAR

Feline ischemic encephalopathy causes severe and peracute neurological disease. While this disease is caused by aberrant larval migration of *Cuterebra* species through the brain, it is categorized under vascular because clinical signs develop due to the cerebral infarction caused by this migration. It has been proposed that the migrating parasite causes either a vasoconstriction or irritation of vessels resulting in infarction. Any cat can be affected by this disease. The clinical signs reflect the area of the brain that is damaged, the cerebrum supplied by the middle cerebral artery, and include: seizures, behavior changes, vision loss (unilateral), and wide circling. In some cases the neurologic signs occur after upper respiratory signs, which is thought to be due to parasitic migration through the nasal passages prior to entry of the calvaria through the cribriform plate. Cats are most often affected by this disease during the fall and late summer in the Northeastern part of the United States (U.S.). Diagnosis may be made if an MRI is performed that identifies a "worm tract" from migration of the parasite in combination with changes consistent with cerebral ischemia. Cats may recover with supportive care and anticonvulsants alone. More definitive treatment with ivermectin, glucocorticoids and diphenhydramine has been recommended by some authors.

REFERENCES

1. Chang Y *et al. J Feline Med Surg*. 2016:1098612X15625353.
2. De Lahunta A *et al. Veterinary Neuroanatomy and Clinical Neurology*. Elsevier Health Sciences; 2014.
3. Dewey CW *et al. Practical Guide to Canine and Feline Neurology*. John Wiley & Sons; 2015.

4. Fingeroth JM *et al. Advances in Intervertebral Disc Disease in Dogs and Cats*. John Wiley & Sons; 2015.
5. Garosi L. *J Feline Med Surg*. 2012;14:85–93.
6. Kent M. *J Feline Med Surg*. 2009;11:395–407.
7. Lorenz MD *et al. Handbook of Veterinary Neurology*. Elsevier Health Sciences; 2010.
8. Markovich JE *et al. J AM Vet Med Assoc*. 2013;243:649–656.
9. Pakozdy A *et al. J Feline Med Surg*. 2010;12:910–916.
10. Pakozdy A *et al. J Vet Intern Med*. 2014;28:255–263.
11. Pakozdy A *et al. J Feline Med Surg*. 2013;15:267–273.
12. Rossmeisl JH. *Vet Clin N Am Small*. 2010;40:81–100.
13. Schriefl S *et al. J AM Vet Med Assoc*. 2008;233:1591–1597.

acvim263

Feline Neurologic Cases of Spinal Cord Disease: Reviewing the Classic and Puzzling over Unusual Case Presentations

Amanda R. Taylor, DVM, DACVIM (Neurology); Sharon C. Kerwin, DVM, MS, DACVS, DACVIM (Neurology)
Auburn, AL, USA

Introduction

Cats presenting with signs of myelopathy require careful review of their history, signalment, physical examination and neurologic signs. Due to the greater emphasis placed on dogs in the neurology specialty, there is less awareness of some of the differentials for spinal cord disease in the cat. While the two most common spinal cord diseases in cats are lymphosarcoma (LSA) and feline infectious peritonitis (FIP), many other disease processes should be considered.

Metabolic diseases, such as renal failure, electrolyte abnormalities, diabetic neuropathy, hyperadrenocorticism, and hyperthyroidism can cause signs of weakness, plantigrade stance, difficulty jumping, and others that can be confused with animals with myelopathy. Metabolic causes of neurologic signs should be ruled out prior to the pursuit of advanced diagnostics such as magnetic resonance imaging and cerebrospinal fluid analysis.

Osteoarthritis occurs commonly in older cats and can result in lameness or weakness in the limbs. Aortic thromboembolism (ATE) can result in paraparesis secondary to ischemia of the nerves and muscles of the pelvic limbs. A complete orthopedic examination should be performed after neurologic evaluation when presented with cats that are having difficulty walking. Femoral pulses should always be palpated and compared to detect deficits consistent with ATE. Cardiomyopathy is often the underlying cause of ATE, therefore careful thoracic auscultation should also be performed in patients with paraparesis.

Due to the prevalence of extraneural organ involvement with the most common diseases, retinal examination is recommended in all cats to attempt to detect ocular changes associated with LSA, cryptococcosis, toxoplasmosis, and FIP. Prior to case discussion of unusual cases seen by the authors, common spinal cord diseases will be reviewed. For the purposes of this discussion we will review neuroanatomical localizations of spinal cord disease, common disease presentations and diagnostic testing associated with each disease.

Neuroanatomical Localization

Cranial nerves are normal in spinal cord diseases with the exception of Horner's syndrome that may be present in cervical myelopathy cases.

	C1-C6	**C6-T2**	**T3-L3**	**L4-Cd5**
Mentation	N	N	N	N
Gait	Tetraparesis Long-strides TL & PL	Tetraparesis Short strides TL, long strides PL	Paraparesis Long-strides PL	Paraparesis Short strides PL
Postural reactions Thoracic	D to A	D to A	N	N
Postural reactions Pelvic	D to A	D to A	D to A	D to A
Reflexes Thoracic	N to I	D to A	N	N
Reflexes Pelvic	N to I	N to I	N to I	D to A

Muscle tone	I in all limbs	D-TL I-PL	N-TL I-PL	N-TL D-PL
Posture			Schiff-Sherrington	
Cutaneous trunci	N	A on side of lesion with severe injury or brachial plexus disease	Cutoff 2 spaces cranial to stimulation point	N

N=normal, I=increased, D=decreased, A=absent, TL=thoracic limbs, PL=pelvic limbs

INFLAMMATORY/INFECTIOUS

The Most Likely Differential in Cats <2 Years Old

FIP is caused by mutated feline coronavirus. Involvement of the central nervous system (CNS) occurs frequently in cats with dry FIP. FIP most commonly affects the cervical spinal cord. The signs seen with FIP are a result of the immune system response to FIP infected macrophages. Cats with neurological manifestations of this disease have the "dry" form of the disease where they develop lesions on the meningeal or ependymal surfaces that may extend deeper into the parenchyma. While a probable diagnosis can be made based on the clinical features of elevated serum globulins, elevated CSF titers, and MRI changes, definitive diagnosis requires immunohistochemistry. Prognosis is poor.

Cryptococcosis is ubiquitous throughout the United State (U.S.), but is more common in California and the south. Cats are more frequently affected that dogs. Extra-neural signs such as uveitis, stertor, and upper respiratory tract disease may occur. There is a highly sensitive and specific test for *Cryptococcus neoformans* (latex agglutination) commercially available. Fluconazole (crosses blood brain barrier) is utilized for treatment. Prognosis is guarded.

Toxoplasmosis in the CNS in cats can result in neurologic signs localizing to the spinal cord and the brain. Cats that are immunosuppressed or have concurrent feline immunodeficiency virus (FIV) may have reactivation of previous infections or may be more likely to develop systemic infections. Almost all cats that have systemic infections of toxoplasmosis have involvement of the central nervous system. Antibody titers, with rising convalescent titers being strongly suggestive of an active infection, can support a diagnosis, but identification of *Toxoplasma gondii* organisms in tissue through biopsy of muscle or nerve is necessary for definitive diagnosis. Prognosis is poor when neurological signs are present.

Neoplasia - The Most Likely Differential in Cats >8 Years Old

While some tumors of the spinal cord do not result in signs of pain, the majority of cats with spinal cord or vertebral column neoplasia have detectable signs of pain.

Lymphosarcoma was traditionally thought of as a disease that was the most prevalent in young cats infected with feline leukemia virus (FeLV). More recent studies suggest that cats affected by this disease are have a bimodal age distribution (less than 4 years old or between 11–16 years old), and are not concurrently infected with FeLV. Lymphosarcoma can affect the spinal cord with an extradural location and spinal cord compression or an intradural location and spinal cord disruption. Diagnosis may be made if tumor cells are identified on cytology of the cerebrospinal fluid (CSF) in up to one third of cases. The most common localizations for this disease are L4-S3 and T3-L3, although it is not uncommon for LSA to extend to multiple spinal cord localizations and the brain. The majority of cats with spinal cord LSA have other organ systems affected by LSA (up to 84% of cats with LSA), often the kidney or bone marrow. This may explain why cats often display anorexia, weight loss and lethargy when affected by this disease. Prognosis in cats with concurrent FeLV is poor and is guarded to poor in FeLV negative cats.

Osteosarcoma is the most common primary vertebral column tumor in the cat, followed closely by other forms of sarcoma (fibrosarcoma, undifferentiated, chondrosarcoma). Cats with osteosarcoma are middle aged to senior. The majority of these cases have an identifiable lesion on radiographs, however CT may help to identify less aggressive/advanced lesions. Palliative treatment with cytoreductive surgery or radiation therapy is available. The prognosis is poor.

Meningioma is the second most common spinal cord neoplasia in the cat and is intradural extramedullary in its location. Clinical signs reflect the region of the spinal cord affected. Surgical resection is the therapy most often reported in the literature and has resulted in a median survival time of over a year. Prognosis is guarded.

Trauma

Trauma can result in vertebral fractures and luxations, secondary disc extrusion and in some cases, spinal cord ischemia. A disc may be extruded due to trauma in up to one-fifth of cases (see below). Traumatic injuries to the vertebral column/spinal cord account for 13% of spinal cord diseases in cats. Cats exposed to trauma should be thoroughly evaluated as up to 20% have a second luxation/subluxation or fracture of the vertebral column. If deep nociception is intact, cats have a favorable chance of recovery from traumatic injury. A full-body CT for the purposes of trauma assessment will decrease the amount of manipulation, and therefore increase safety, for trauma patients.

Acute non-compressive disc extrusion (previously identified as traumatic disc extrusion) should be a differential for cats that present with a history of trauma and a peracute onset of myelopathy. Presumptive diagnosis is made with MRI when T2 hyperintensity in the spinal cord overlying an intervertebral disc space is identified. The first study of this disease, retrospective in nature, indicates that the outcome of these cases is good with 90% returning to ambulatory status and continence (fecal and urinary) despite presenting as non-ambulatory.

Congenital/Inherited

Vertebral column abnormalities are present in up to 20% of cats (excluding the Manx breed). The most common abnormalities identified are block and transitional vertebrae. These occur without clinical signs frequently. Transitional vertebrae occur most commonly at the sacrocaudal and thoracolumbar junctions. In Manx cats sacrocaudal dysgenesis results in L4-S3 clinical signs. There are no treatments for this disease.

Vascular

Fibrocartilaginous emboli are one of the more common spinal cord diseases in cats. Cats with disease present similarly to dogs, with an acute occurrence of myelopathy signs which are asymmetric and may initially progress. The origin of the embolus is believed to be the intervertebral disc, and while there are several theories as to how this embolus finds its way in to the spinal cord vasculature, the route has not been identified. The C6-T2 spinal cord intumescence and lumbosacral intumescence are the most frequently reported sites affected. Care for cats with this disease is similar to that of dogs, with supportive care and rehabilitation therapy cats can make a full recovery in many cases.

Degenerative

Intervertebral disc disease is most common in the thoracolumbar (T13-L1) and lumbosacral (L4-L6) region. These cats are usually older (>8 years old). Clinical signs of this disease may be subtle. The authors have seen these cases presented for decreased tail tone, difficulty with urination or defecation due to inability to posture, and decreased willingness to jump. Purebred cats are overrepresented in this disease category but it is unknown whether they are predisposed to this disease. While both extrusive (Hansen Type I) and protrusive (Hansen Type II) disc herniations occur in the cat, cats affected by extrusive discs are more acutely and severely affected than cats with protrusive disc disease. Due to the common localization of L4-S3 for this disease, it is of high importance to rule out aortic thromboembolism and orthopedic disease as the underlying cause of signs.

Neuraxonal dystrophy is thought to be a hereditary disease that results is degeneration of axons characterized by swelling of the axons (spheroids) within the CNS> Clinical signs begin with pelvic limb proprioceptive ataxia and paresis and progress to signs of brainstem disease. Cats are affected at <1 year old. There is no treatment for this disease and prognosis is poor.

References

1. De Decker S *et al. J Feline Med Surg*. 2016;1098612X16630358.
2. Foley JE *et al. J Vet Intern Med*. 1998;12:415-23.
3. Garosi L. *J Feline Med Surg*. 2012;14:85-93.
4. Gonçalves R *et al. J Feline Med Surg*. 2009;11:53–59.
5. Kent M. *J Feline Med Surg*. 2009;11:395–407.

6. Marioni-Henry K *et al. J Vet Intern Med*. 2004;18:851–858.
7. Marioni-Henry K *et al. J Am Vet Med Assoc*. 2008;232:237–43.
8. Marioni-Henry K *et al. Vet Clin N Am Small Anim Pract*. 2010;40:1011–1028.
9. Negrin A *et al. J Feline Med Surg*. 2009;11:361–72.
10. Newitt A *et al. Vet Radiol Ultrasound*. 2008;49:35–41.
11. Pedersen NC. *J Feline Med Surg*. 2009;11:225–58.
12. Rayward RM. *Vet Comp Orthop Traumatol*. 2002;15:137–44.
13. Singh M. *J Feline Med Surg*. 2005;7:77–93.
14. Sykes JE *et al. J Vet Intern Med*. 2010;24:1427–38.
15. Taylor-Brown FE *et al. J Feline Med Surg*. 2015;19:1–6.

1.
acvim266

Electrodiagnostics: Review of Basic Concepts and Clinical Applications

Monica Aleman[1], MVZ Cert., PhD, DACVIM (LAIM, Neurology); Cody Alcott[2], DVM, DACVIM (LAIM, Neurology)
[1]Davis, CA, USA; [2]Tucson, AZ, USA

Introduction

Electrodiagnostics are an extension of the clinical neurologic examination.[1-3] These electrodiagnostics aids include electromyography (EMG), nerve conduction (NC) of motor and sensory nerves, repetitive nerve stimulation (RNS), single fiber EMG (SFEMG), electroencephalography (EEG), brainstem auditory evoked responses (BAER), visual evoked potentials (VEP), electroretinography (ERG), otoacoustic emissions (OE), and transcranial magnetic stimulation (TMS).[1-3] This session will review the principles of electrodiagnostics for its use in patients with neurological disease. Cases (small and large animals) will be provided for the illustration of various conditions and electrodiagnostic modalities that aid in the diagnosis of disease.

Electromyography

Electromyography (EMG) refers to the study of the electrical activity of the muscle.[2,3] The distribution of EMG abnormalities can assist in the localization of lesions to spinal cord segments, nerves, and/or muscles.[2,3] Quantitative EMG can further assist in the determination of disorders as myopathic, neurogenic or both.[2,3] EMG might not provide a definitive diagnosis but will aid to support or refute a suspected clinical diagnosis and further our understanding of disease processes. Electromyography evaluates muscle electrical activity during insertion of a needle and spontaneous activity (normal and abnormal). These activities are influenced by the integrity of the nerve innervating the muscle under study and the muscle itself. Insertional activity is observed in muscles following needle insertion due to muscle fiber mechanical disruption.[2,3] Spontaneous activity, asynchronous and spontaneous firing of muscle fibers, occurs subsequent to denervation injury or myopathy. The characteristics of this spontaneous activity are one of the most important features of the electromyographic examination. Some of these abnormalities include fibrillation potentials, positive sharp waves, complex repetitive discharges, and myotonic discharges, among others.

Nerve Conduction Studies

Nerve conduction velocity studies assess the motor and sensory evoked responses to stimulation of nerves.[1-4] Nerve conduction latency, amplitude and dispersion give clues to nerve health. Motor nerve conduction requires stimulation of a nerve while recording from a muscle innervated by that nerve. Motor nerve conduction (MNC) is the orthodromic conduction of nerve potentials (m-wave) between two recording locations, limiting NMJ influence. Sensory nerve conduction requires stimulation of a mixed nerve while recording from a cutaneous nerve.[2,3] Sensory nerve conduction (SNC) recordings can be collected along the course of the peripheral nerve, at the nerve spinal cord junction (cord dorsum potential), cranial projections within the spinal cord (sensory evoked potential) or at brain projections (somatosensory evoked potentials). NCV studies aid in the localization of nerve, neuromuscular junction and muscle disease. Furthermore, it aids in the distinction between axonal degeneration and demyelination.[2,3] These studies should be done under general anesthesia for safety reasons, patient cooperation, avoid pain to the patient, and obtain interpretable recordings.

Peripheral nerve stimulation results in orthodromic and antidromic salutatory nerve conduction. Strategic recording needle placement allows for examination of nerve conduction following stimulation via compound muscle action potential (CMAP) recordings. F-waves represent antidromic-neuron-orthodromic interactions following nerve stimulation providing evidence of proximal limb and motor neuron function. In contrast, motor evoked potentials can originate from the brain (transcranial motor evoked potentials) or along the spinal cord (motor evoked potentials) and recorded via CMAP waveforms in distal muscles.

Repetitive Nerve Stimulation

This is the most commonly diagnostic modality used for neuromuscular junction disorders in human and small animal medicine.[2,3,5] A compound muscle action potential (CMAP) or M wave reflects the number of muscle fibers activated by a train of stimuli (repetitive stimulation) applied to a single nerve. The test is considered abnormal if fewer muscle fibers are activated during repetitive stimulation (usually 10 consecutive stimuli).[2,3,5] This is called a decremental response as seen in cases of myasthenia gravis.[2,3] The opposite, an incremental response of the M waves in its amplitude and area under the curve (more muscle fibers activated) is seen in cases of botulism.[2,3] Multiple patterns of abnormalities have been described in acute OP toxicity in people.[2,3]

Single Fiber EMG

This is a selective EMG for the study of junctionopathies by allowing the measurement of neuromuscular transmission of single end-plates.[2,3] The motor unit action potentials from a single motor unit are recorded over a small area with a concentric EMG electrode. Variations in myofiber end-plate potential amplitude and firing threshold result in variations in latency of subsequent action potentials during nerve stimulation.[2,3] This phenomenon is termed neuromuscular jitter. Jitter is increased in diseases on which neuromuscular transmission is disturbed.[2,3]

Electroencephalography

Electroencephalography is the graphic representation of the difference in voltage in microvolts between two different locations within the cerebral cortex plotted over time. EEG recordings represent spontaneous electrical activity primarily synaptic, excitatory post synaptic potentials (EPSP), and inhibitory post synaptic potentials (IPSP) from cortical neurons. Pyramidal cells are the most important neuronal surface of EEG. It is estimated that a minimum of 108 neurons or an area of 6 square centimeters are required to produce a detectable change in voltage. EEG is a valuable tool for the investigation of brain disease particularly seizures and sleep disorders, localization of abnormal activity, and extent (focal or diffuse). Two EEG montages have been described: bipolar and referential.[6] Electrode placement could be rostral to caudal or transverse in a bipolar montage. EEG recordings are evaluated for symmetry, waveform, morphology, frequency, and amplitude. Waveform nomenclature is as follows: Delta 0–3 Hz, theta 4–7 Hz, alpha 8–12 Hz, beta 13–29 Hz, gamma >30 Hz. Paroxysmal discharges such as sharp waves and spike and wave have been also referred as epileptic discharges.

Brainstem Auditory Evoked Responses

Auditory stimulus applied in the ear canal generates an evoked potential that originates in the sensory receptors within the organ of Corti on basilar membrane in the cochlea of inner ear and depolarization of the cochlear nerve (wave I). Propagation of the nerve stimulus goes from the sensory receptors, cochlear nerve, cochlear nucleus in the medulla, olivary nucleus in the medulla, lateral lemniscus in the pons, caudal colliculus in the midbrain, medial geniculate in the thalamus, and auditory cortex in the cerebrum. Brainstem auditory evoked response testing evaluates the integrity of the auditory pathway from its peripheral part (cochlear nerve) to the brainstem. It is useful in the detection of unilateral and bilateral deafness, and differentiation between conductive and sensorineural hearing loss.[7] Five waves (in Roman numbers from I to V) are usually identified on a BAER in animals and represent the various regions of the auditory pathway from the cochlear nerve (wave I) to the brainstem (waves II through V). Absence or abnormalities in the various waveforms assist in neurolocalization. BAERs vary with age, gender, breed, head size, and temperature.

References

1. Aleman M. *Vet Clin North Am Equine Pract.* 2011;27:481.
2. Kimura J. *Electrodiagnostics In Diseases of Nerve and Muscle.* 3rd ed. Springer 2001.
3. Daube JR, *et al. Muscle Nerve.* 2009;39:224.
4. Aleman M, *et al. J Vet Intern Med.* 2013;27:1571.
5. Aleman M, *et al. J Vet Intern Med.* 2011;25:365.

6. Williams DC, *et al. J Vet Intern Med*. 2008;22:630.
7. Aleman M, *et al. J Vet Intern Med*. 2014;8:1310.

1.
acvim267

Autophagy in Neurologic Disease

Dennis P. O'Brien, DVM, PhD, DACVIM (Neurology)
Columbia, MO, USA

Autophagy (or more specifically macroautophagy) is one of the processes involved in the degradation of superfluous or damaged proteins, lipids, or organelles. It progresses in a stepwise fashion from initiation, through isolation of membrane from the endoplasmic reticulum to enclosure of the substrate targeted for degradation within a double-membrane vesicle, the autophagosome. The autophagosome then fuses with a lysosome, and the substrate is degraded.[1-6] The basic process was originally delineated in yeast by Yoshinori Ohsumi[7] who received the 2016 Nobel Prize in Medicine for his work.

When yeast are starved for nutrients, metabolism shifts from growth and reproduction to survival, and cytoplasmic components are recycled through autophagy. Autophagy can also be selectively directed toward degradation of specific components such as abnormally folded proteins, damaged organelles, or invasive organisms. Recently, disorders of autophagy have been associated with a number of neurodegenerative diseases in both humans[1,3-6,8-13] and dogs[14-19]. In this presentation, we will review the autophagy pathway and illustrate how this pathway is disrupted in neurodegenerative disease.

Regulation of Autophagy

Autophagy is regulated in response to the energy balance of the cell. In addition to opening glucose transporters, insulin and insulin-like growth factor act through a common second messenger, AKT, to activate mammalian target of rapamycin complex (mTORC1). This "life is good" signal stimulates the cell to store glycogen, grow, and reproduce. At the same time, it inhibits the initiator of autophagy, a kinase called Atg1 in yeast or ULK1 in mammals. In the face of nutritional deficits, the AKT-mTOR signal will cease while decreasing ATP and increasing AMP levels in the mitochondria will stimulate AMP-activated protein kinase (AMPK), which in turn activates ULK and initiates autophagy.

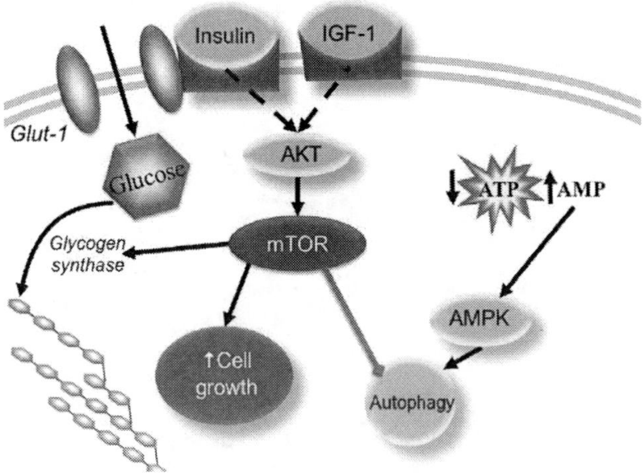

Initiation at the Endoplasmic Reticulum

Phosphorylation of ATG13 and RB1CC1 (FIP200) by ULK initiates autophagy by moving this complex from the cytosol to the endoplasmic reticulum. ULK also triggers another protein, AMBRA, which complexes with Beclin-1 and VPS (vacuolar protein sorting) molecules to form a phosphatidylinositol 3-kinase complex. PI3P mediates a protrusion of the endoplasmic reticulum membrane. From this base, WIPI proteins serve as a protein interaction platform for ATG9 to transport membrane from cytoplasmic membrane stores to the expanding isolation membrane (also called the phagophore) that will ultimately form the autophagosome.

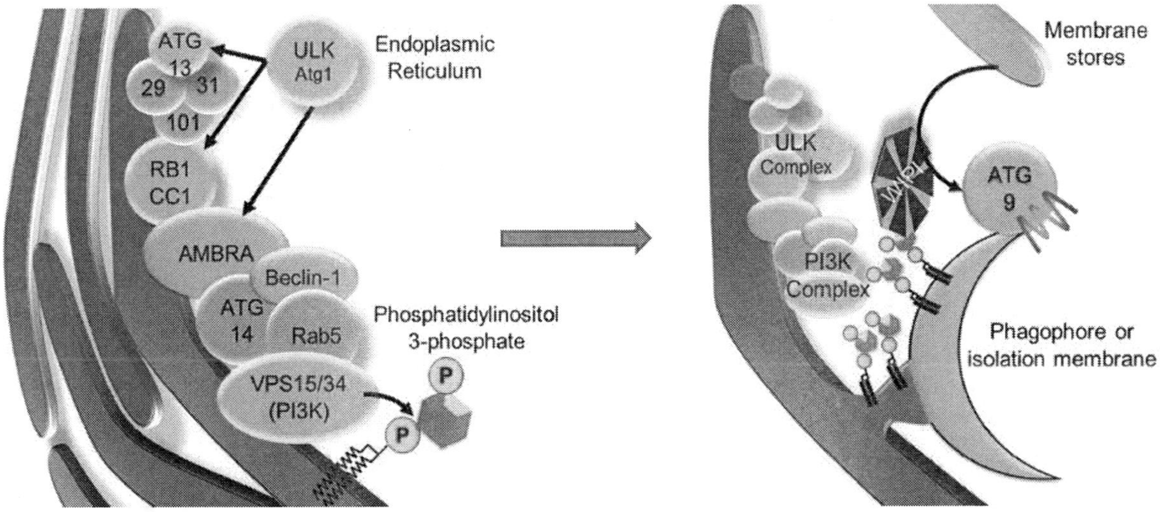

AUTOPHAGOSOME FORMATION

As the phagophore matures, light chain 3 (LC3) is activated through a series of ubiquitin-like conjugation steps. Like WIPI, TECPR2 is a beta propeller protein that serves as a platform for protein interactions, in this case activating LC3. Rab3GAP enhances the inherent GTPase activity of certain Rab proteins, which serve as molecular switches for processes such as membrane transport. In autophagy, Rab3GAP is essential to bind ethanolamine phosphate to LC3, which in turn facilitates transfer of the LC3 to the inside of the forming autophagosome. Molecular tags bind substrates to LC3. The phagophore closes around the substrate forming the double-membrane autophagosome. The autophagosome then fuses with the single membrane-bound lysosome. Another Rab protein, Rab24, binds to LC3 where it appears to play a role in fusion with the lysosome. Ultimately the acid hydrolases within the lysosome will degrade the autophagosome along with its contents.

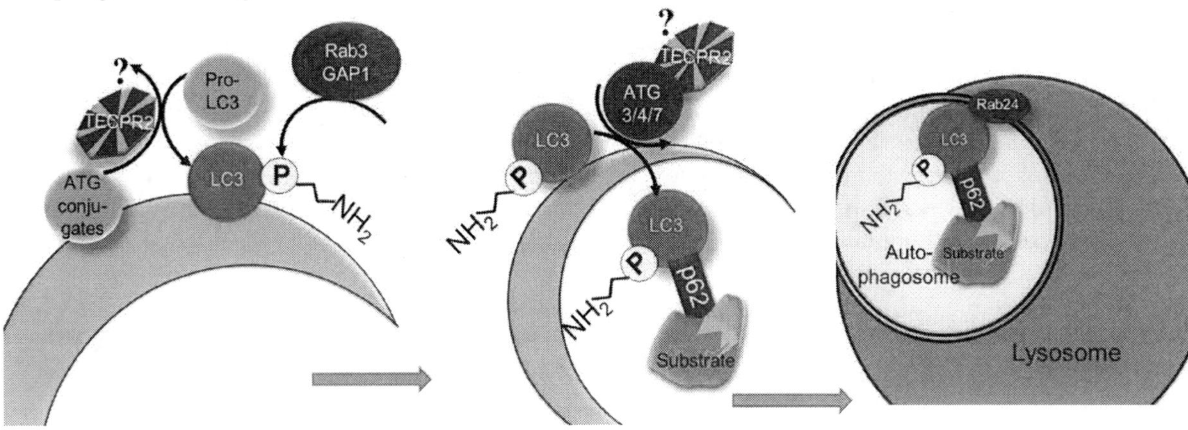

SUBSTRATE TARGETING

In some cases, specific components such as protein aggregates or unwanted or damaged organelles are targeted for autophagy. The protein p62 (also called sequestosome 1; SQSTM1) targets protein aggregates for degradation. When mitochondria lose their membrane potential, Pink1 ubiquitinates (Ub) surface proteins and recruits Parkin, a ubiquitin ligase that conjugates more ubiquitin tags to the mitochondrial surface. The ubiquitinated mitochondria can then bind to LC3 and be incorporated into an autophagosome. The seed bed of autophagy, the endoplasmic reticulum is not immune from damage and degeneration. The protein FAM134B targets damaged endoplasmic reticulum for LC3 binding and degradation.

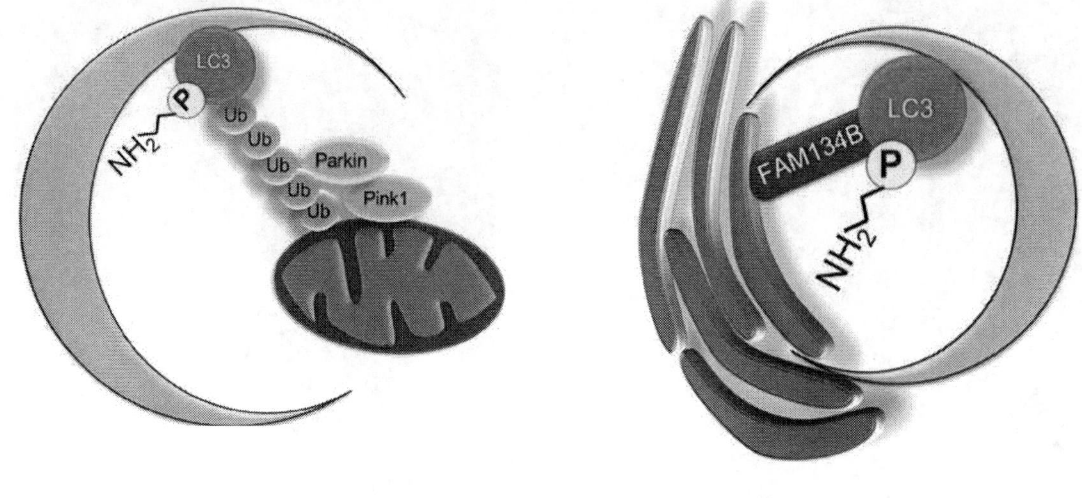

ROLE IN NEURODEGENERATIVE DISEASES

Disorders of the end stage of autophagy, the lysosomal storage diseases, have been well recognized in veterinary medicine, and many of the mutations leading to loss of enzyme function have been identified. The single-membrane-bound storage bodies are readily recognized on histopathology, and the staining or morphology of the bodies or specific enzyme assays point toward candidate genes. It is only recently that diseases associated with disruption of the earlier stages of autophagy have been recognized. Now next-generation gene discovery strategies have identified mutations associated with a number of neurodegenerative diseases in humans and animals that disrupt one of these earlier points in the signaling pathway.

A category of human hereditary neurodegenerative diseases are called neurodegeneration with brain iron accumulation (NBIA) and are characterized by iron deposits in the basal ganglia. One rare form of NBIA in children called beta-propeller protein-associated neurodegeneration (BPAN) is associated with mutations in one of the WIPI protein genes *WDR45*. Affected children show a static encephalopathy in childhood with neurodegeneration as adults (SENDA syndrome). Affected infants have developmental delay, seizures, and spastic paraplegia but then appear to stabilize. As young adults, they develop progressive dementia and movement disorders. Impaired autophagy is thought to lead to the iron accumulation which is apparent on MRI.[20] So far no brain iron accumulation has been identified in the canine diseases with mutations of genes in the autophagy pathway.

TECPR2 is another beta-propeller protein in autophagy. It is involved in the ATG conjugates that process LC3, though the exact role in that process is unclear. Mutations have been associated with hereditary spastic paraplegia in humans and neuroaxonal dystrophy (NAD) in Spanish Water Dogs.[16,20] Affected dogs show progressive neurologic signs beginning at 6–11 months of age. They show behavior changes and gait abnormalities with predominantly sensory deficits. On histopathology, neuronal loss with spheroid formation is seen primarily in the dorsal nuclei of the brainstem and spinal sensory pathways. The spheroids contain abundant, double-walled, small vacuoles which bind antibodies for LCB3 identifying them as autophagosomes whereas few mature lysosomes could be identified. The accumulation of autophagosomes within spheroids highlights the role of autophagy in the maintenance of axonal function.[10,12]

Black Russian Terriers, Alaskan Huskies, and Rottweilers develop a hereditary disease characterized by polyneuropathy with ocular abnormalities and neuronal vacuolation (POANV, also called juvenile-onset laryngeal paralysis/polyneuropathy or neuronal vacuolation and spinocerebellar degeneration).[17,18,21,22] As the names imply, affected dogs develop laryngeal paralysis, progressive weakness and sensory loss, and cataracts at a young age, while Rottweilers also show a prominent cerebellar ataxia. On histopathology, neuronal vacuolation in the brain (A) initially raised concern that this was a prion disease, but no abnormal prion proteins were identified, and sequencing the gene did not reveal any variants. Instead, mutations in the *RAB3GAP1* gene have been identified in the disease. In addition to its role in autophagy, *RAG3GAP1* is involved in other aspects of membrane processing. The large and small (B) vacuoles seen in the disease are single membrane bound as opposed to the double-membrane autophagosomes and contain only scant debris or a single dense body, respectively, and did not stain with oil red O for lipids. Thus they are unlikely to be lysosomes and are most likely dilated endoplasmic reticulum membranes. Spheroids (C) and membranous vesicular (D) structures within axons suggest that disruption of axonal membrane transport contributes to the pathogenesis of this disease also.

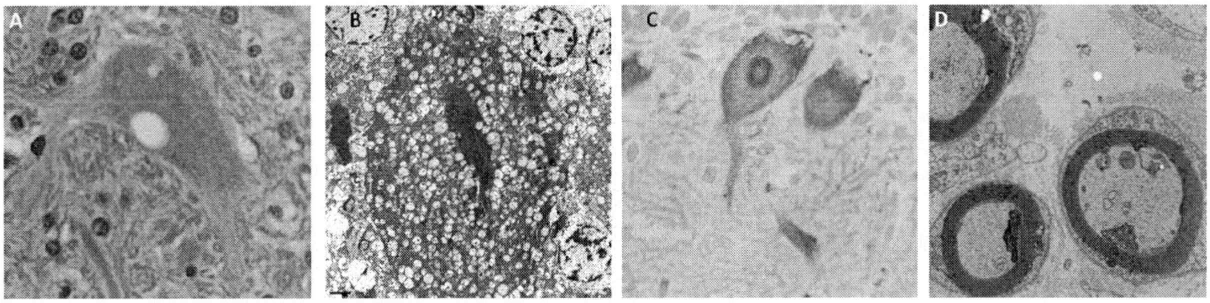

Small to large intraneuronal vacuolation and spheroids are also seen in Lagotto Romagnolo dogs with progressive cerebellar ataxia and behavioral changes.[14] The vacuoles are also single-membrane bound but contain membranous profiles or floccular material that did not stain with PAS. Normal autophagosomes were sometimes seen. A missense mutation was found in *ATG4D*, and spheroids showed immunoreactivity to autophagy proteins LC3 and p62. Some of the vacuoles were positive for lysosome markers.

Mutations in genes for the molecular tags that target proteins or organelles for autophagy also lead to neurodegenerative diseases. Mutations affecting either of the proteins involved in ubiquitinating nonfunctional mitochondria (pink1 and parkin) are associated with hereditary Parkinson's disease in humans.[13] Mutations in the *SQSTM1* gene (p62) have been associated with ALS and frontotemporal dementia and p62 immunoreactive aggregates have been found in ALS associated with *SOD1* mutations in humans.[23] Border Collies with hereditary sensory neuropathy show ataxia, loss of pain perception, and self-mutilation. An inversion that disrupted expression of the gene *FAM134B* was identified in dogs with this disease. *FAM134B* is highly expressed in sensory and autonomic ganglia where it mediates selective autophagy of the endoplasmic reticulum.[24]

Hereditary ataxia in Old English Sheepdogs and Gordon Setters is associated with a mutation in *RAB24*.[15] In contrast to other RAB proteins, *RAB24* has poor GTPase activity, and its function is unclear. It co-localizes with autophagosomal markers such as LC3 and is thought to play a role in fusion of autophagosome with the lysosome. The spheroids found in affected dogs contained double-walled vesicular structures, suggesting impaired autophagy.[25]

Conclusions

In dogs, mutations affecting autophagy pathway genes are associated with cerebellar degeneration, neuroaxonal dystrophy, neuronal vacuolation, and other neurodegenerative diseases. The neuronal vacuolation is particularly intriguing since it has not been described in human disease and is reminiscent of transmissible spongiform encephalopathies. Whether these vacuoles represent aberrant membrane trafficking, non-functional lysosomes, or expanded endoplasmic reticulum remains to be investigated. As more diseases of autophagy are identified, drugs that affect these pathways are being investigated as potential therapies. Rapamycin, which inhibits mTOR, is currently being investigated as an "anti-aging" therapy in elderly dogs. Blocking mTOR would enhance autophagy. This may not overcome downstream defects in the pathway, however, and potential deleterious effects on autophagy and the other metabolic processes regulated by mTOR would need to be considered.

Table 1. Some mutations in autophagy genes associated with neurologic diseases in humans and animals[4,14-20]

Gene	Human disease	Canine disease	Stage of autophagy and effect
WDR45	BPAN (SENDA)	-	Early: phagosome elongation
TECPR2	Hereditary spastic paraplegia 49	Neuroaxonal dystrophy	Early/mid: phagosome elongation and maturation?
RAB3GAP1	Warburg micro syndrome	POANV/JLPP	Mid: LC3 lipidation
ATG4	-	Hereditary cerebellar ataxia	Mid: LC3 activation
PINK1	Parkinson's	-	Mid: mitophagy induction
PARK2	Parkinson's	-	Mid: mitophagy induction
SQSTM1	ALS/FTD	-	Mid: selective autophagy
SPG11	Hereditary spastic paraplegia 11	-	Late: lysosome genesis
EPG5	Vici syndrome	-	Late: autophagosome/lysosome fusion
SNX14	Spinocerebellar ataxia recessive 20	-	Late: autophagosome/lysosome fusion?
RAB24	-	Hereditary cerebellar ataxia	Late: autophagosome/lysosome fusion
ZFYVE26	Hereditary spastic paraplegia 15	-	All stages

References

1. Damme M, *et al. Acta Neuropathol.* 2015;129:337–362.
2. Feng Y, *et al.* The machinery of macroautophagy. *Cell Res.* 2014;24:24–41.
3. Ghavami S, *et al.* Autophagy and apoptosis dysfunction in neurodegenerative disorders. *Prog Neurobiol.* 2014;112:24–49.
4. Jiang P, *et al. Cell Res.* 2014;24:69–79.
5. Komatsu M, *et al. Autophagy.* 2006;2:315–317.
6. Nixon RA. *Nat Med.* 2013;19:983–997.
7. Nakatogawa H, *et al. Nat Rev Mol Cell Biol.* 2009;10:458–467.
8. Eskelinen EL, *et al. Bioch Biophys Acta.* 2009;1793:664–673.
9. Martinez-Vicente M, *et al. Lancet Neur.* 2007;6:352–361.
10. Yang Y, *et al. Trends Neurosci.* 2013;36:418–428.
11. Webb JL, *et al. J Biol Chem.* 2003;278:25009–25013.
12. Maday S. *Brain Res.* 2016;1649(Pt B):142–150.
13. Yamano K, *et al. EMBO Reports.* 2016;17:300–316.
14. Kyostila K, *et al. PLoS Genet.* 2015;11:e1005169.
15. Agler C, *et al. PLoS Genet.* 2014;10:e1003991.
16. Hahn K, *et al. PLoS One.* 2015;10:e0141824.
17. Mhlanga-Mutangadura T, *et al. Neurobio Dis.* 2015;86:75–85.
18. Mhlanga-Mutangadura T, *et al. J Vet Intern Med.* 2016;30:813–818.
19. Forman OP, *et al. G3* (Bethesda, MD). 2016;6:2687–2692.
20. Ebrahimi-Fakhari D, *et al. Brain.* 2016;139:317–337.
21. Kortz GD, *et al. Vet Path.* 1997;34:296–302.
22. Wiedmer M, *et al. G3* (Bethesda, MD). 2016;6:255–262.
23. Gal J, *et al. J Neurochem.* 2009;111:1062–1073.
24. Khaminets, *et al. Nature.* 2015;522:354–358.
25. Ao X, *et al. Cell Death Dif.* 2014;21:348–358.

acvim268

Autophagy and Its Therapy in Lysosomal Storage Diseases

Charles H. Vite, DVM, PhD, DACVIM (Neurology)
Philadelphia, PA, USA

Introduction

Niemann-Pick disease type C (NPC) is a progressive disorder characterized by cerebellar ataxia, cognitive impairment, seizures, vertical supranuclear gaze palsy, dysphagia, hearing loss, hepatic disease, and death typically by 20 years of age.[1,2] Over 350 disease-causing mutations have been identified in either the *NPC1* or *NPC2* gene.[3] NPC1 and NPC2 proteins are critical for the movement of unesterified cholesterol and glycosphingolipids from the endosomal/lysosomal compartment to the Golgi apparatus, plasma membrane, and endoplasmic reticulum.[4,5] Brain histology includes widespread neuronal cytoplasmic vacuolization, neuronal loss most severely affecting the Purkinje cells, and axonal spheroid formation.[2,6]

The mechanism by which deficient activity of NPC1 or NPC2 results in neurodegeneration is not understood. Purkinje cell death in NPC has been described as: 1) cell autonomous, based on studies in chimeric mice expressing functional npc1 in a subset of cells, and, 2) due to autophagy, based on the ultrastructural observation of increased numbers of autophagic vacuoles containing cytoplasmic contents.[7] Lipidated microtubule-associated protein 1 light chain 3 (LC3-II), which associates with autophagosome membranes, has been widely used as an autophagosomal marker.[8] LC3-II accumulation occurs in the brains of npc1-/- mice,[7,9,10] with greater amounts occurring in the cerebellum and brainstem than in the cerebral cortex[9].

We investigated the distribution of LC3 accumulation in the CNS in the naturally occurring feline NPC1 model. Naturally occurring feline NPC1 closely resembles the biochemistry, pathology, and clinical presentation of juvenile-onset NPC1.[6,10-14,15] Our initial staining for LC3 in the NPC1 cat brain demonstrated specific staining patterns within olivocerebellar projections which could explain the anterior-to-posterior Purkinje cell loss that occurs over time in the mouse model.[16,17] Afferent inputs reach the cerebellar cortex via two excitatory systems - climbing fibers and mossy fibers. Climbing fibers originate from the inferior olivary nucleus and synapse on the cell bodies and dendrites of Purkinje cells. In contrast, mossy fibers arise from the spinocerebellar and cuneocerebellar tracts, the vestibular system, trigeminal pathways, reticular formation, and pontine nuclei and synapse on the dendrites of granule cells. The specific distribution of LC3 accumulation in the inferior olives of NPC1 cat brain can explain the spatial and temporal distribution of Purkinje cell loss and suggests that Purkinje cell loss may not be solely due to cell-autonomous autophagy.

Results

LC3 Expression in CNS Neurons of the Normal Cat

Little or no LC3 staining was seen in the majority of neurons in the CNS of 6-month-old normal cats (comparable in age to endpoint NPC1 cats). Those that were labeled tended to be medium or large in size (e.g., Purkinje cells, pyramidal neurons, brain stem neurons, spinal motor neurons) with expression present in perikarya and proximal neurites. In the cerebellar cortex, LC3 staining was present in a subpopulation of Purkinje cell perikarya and their proximal dendrites. In the cerebral cortex, staining was limited to the neurons of layers II, III and V. In the brain stem, nearly all neuronal perikarya in some nuclei were immunoreactive while adjacent nuclei were negative. In the spinal cord, diffuse LC3 immunoreactivity was seen in perikarya and proximal neurites of motor neurons. Most of the staining was diffuse and only rarely were well-defined punctae seen. White matter was largely unstained for LC3.

LC3 Expression in the CNS of the NPC1 Cat

In end-stage, 6-month-old NPC1 cats, LC3 appeared as variable-sized aggregates within distended axons and presynaptic neuropil; accumulation was not seen within neuronal perikarya or dendrites. LC3 co-localized with the axonal and presynaptic markers neurofilament-light (NFL), synapsin I, synaptophysin, and alpha synuclein. Within axonal spheroids, LC3 punctae were seen surrounded by myelin basic protein. In contrast, LC3 did not co-localize with the neuronal cell body/dendrite marker MAP2, neuronal nuclear marker NeuN, or the postsynaptic marker gephyrin. LC3 also did not co-localize with GFAP, oligodendrocyte-specific protein, S-100, Iba1 or LAMP1. Interestingly, LC3 did not co-localize with calbindin-labeled axonal swellings in the cerebellum.

The Feline NPC1 Cerebellar Cortex

Purkinje cell loss proceeded in a stereotypical manner with depletion beginning in the anterior vermis and then spreading posteriorly and laterally into the hemispheres. The nodulus (lobule X) remained generally unaffected at disease end-stage. Purkinje cell loss was readily visible by decreased numbers of calbindin-positive cells.

LC3 accumulation was concentrated in the regions of greatest Purkinje cell loss (lobules I–IV) and found predominantly in the granular cell layer and white matter. NFL-positive spheroids greatly exceeded the number of calbindin-positive spheroids, suggesting that the majority of the swollen axons did not arise from Purkinje cells and were likely afferent in origin. Only rarely did NFL and calbindin co-localize to the same axons. Most NFL-positive axonal swellings stained for LC3; however, calbindin-positive axonal swellings did not stain positive for LC3. This association of disrupted autophagy and axonal degeneration in NFL-positive but not in calbindin-positive axonal swellings suggests different degenerative mechanisms may be occurring in different neuronal populations.

Olivocerebellar Projections

Climbing fibers originate from the three divisions of the inferior olivary nucleus (principal, medial accessory, dorsal accessory), and project to specific regions of the cerebellum where they synapse on the cell bodies and dendrites of Purkinje cells. These divisions of the inferior olivary nucleus were regularly and differentially impacted by disrupted autophagy as evaluated by LC3 staining. Serial sections were taken through the medulla of two affected 6-month-old NPC1 cats and were stained for LC3. Pathology was seen throughout the caudal-rostral extent of the dorsal accessory olive (DAO). In the rostral two-thirds of the DAO, a sharply defined boundary between very large amounts of LC3 pathology in the ventrolateral and very small amounts in the dorsomedial portions were seen. LC3 pathology was also seen through much of the caudal-rostral extent of the medial accessory olive (MAO). The concentration of aggregates in the MAO was greater in the more caudal regions and also was predominately laterally located (although the lateral-medial boundary was less clearly defined than it was in the DAO). Significant pathology in the dorsal lamina of the principal olive (dlPO) was seen in its most medial portion around Brodal levels VIII-XI.[18] Pathology in the junction region of the ventral and dorsal lamina of the principal olive was seen in the rostral region. Moderate pathology was seen in the dorsomedial cell column while pathology in the dorsal cap and nucleus beta was insignificant.

A dense concentration of LC3 aggregates was present in the rostrolateral DAO as early as 11 weeks of age, an age that preceded significant Purkinje cell loss. It is noteworthy that the climbing fibers, which project to the Purkinje cells of the anterior cerebellar vermis, originate from the DAO, thus the areas first affected in the cerebellum and those first affected in the inferior olives are linked anatomically. Similarly, the PO projects to the cerebellar hemispheres, which were the last portion of the cerebellum to show Purkinje cell loss. The nodulus, which shows no Purkinje cell loss in NPC1 disease, is innervated by the dorsal cap of the PO and the rostral portion of the MAO, neither of which shows severe LC3 accumulation.

Next, we evaluated LC3 accumulation in the afferents that project to the inferior olivary nuclei. Sections of normal spinal cord stained for LC3 showed strong expression in perikarya and neurites of lamina IX motor neurons, less staining in perikarya in lamina VII and VIII, and the least staining in the dorsal horn. In contrast, NPC1 motor neuron perikarya did not show LC3 accretions, although LC3 accumulation was seen in neuropil at all spinal cord levels. Overall, the distribution and density of accretions was moderate in the dorsal horn and lamina IX motor neuron areas and somewhat heavier in lamina VII/VIII.

A semi-quantitative scoring system, employed throughout the CNS, showed that in the remaining afferents to the inferior olivary nuclei, pathologic LC3 staining was increased in the gracile, cuneate, and red nuclei. The pretectal nucleus, caudate nucleus, trigeminal nucleus, periaqueductal gray, and globus pallidus were only mildly affected. No LC3-indicated pathology was seen in NPC1 specimens at 2 or 4 weeks of age and only very subtle aggregates were detected at 6 weeks. Substantial LC3 pathology had appeared by 11 weeks (see above) and continued to increase until end-stage (average 24 weeks). An example of this temporal accumulation of LC3 pathology can be seen by comparing 17-week and 27-week-old NPC1 brainstems. An increase in the number and density of LC3 puncta and aggregates within the gracile and cuneate nuclei was evident. An apparent swelling or expansion of the gracile nucleus was also seen at 27 weeks.

Mossy fibers arise from the spinocerebellar and cuneocerebellar tracts, the vestibular system, trigeminal pathways, reticular formation, and the pontine nuclei and synapse on the dendrites of cerebellar granule cells. Moderate pathology was seen throughout the medullary reticular formation but was very heavy within the lateral reticular nucleus, particularly in the more caudal region. There was very little LC3 pathology seen in the pontine nuclei, with the exception of the dorsal lateral pontine nucleus where staining was densely concentrated. Heavy labeling of the tegmental pontine reticular nucleus was also seen.

Little LC3 staining was observed in the cerebellar and vestibular nuclei, which receive efferent axons from the Purkinje cells.

LC3 Accumulation in the CNS in Other Feline Storage Diseases

We evaluated the brains from other feline lysosomal storage diseases and examined LC3 pathology in order to determine what similarities may exist between them and feline NPC1. No LC3 pathology was noted in the brainstem, cerebellum or midbrain in the mucopolysaccharidosis type VI or VII (MPS VI, VII) cats and staining appeared identical to that of normal brain. In the mucolipidosis II (MLII cat), only a trivial number of LC3 aggregates were found located in the internal capsule. Cerebellum appeared as seen in normal brain. In the alpha-mannosidosis (AMD) cat, LC3 pathology was generally absent from most areas examined. Brainstem areas affected included moderate concentration of aggregates in the caudal-lateral portions of the medial and dorsal accessory olive nuclei and in the gracile nucleus. There were detectable small aggregates seen in the cerebellar white matter and granular cell layers but the amount was trivial when compared to NPC1 brain (~<5%). No pathology was seen in the thalamus with the exception of the ventral lateral geniculate body, the optic tract near its merger with the LGN and in the thalamic reticular formation adjacent to the LGN. No midbrain pathology was seen.

Discussion

The results of our study are the first to identify regional CNS autophagic abnormalities in any species affected with NPC1 and provide evidence of differential vulnerabilities between discrete cell populations and pathways. These results also provide an additional method for assessing neuropathology in NPC1. The pattern of LC3-labeled aggregates observed was not random but was concentrated within specific CNS nuclei and their subregions; the changes were specific to the NPC1 model when compared to MPS, AMD, and MLII models. The consistency of labeling seen in all end-stage NPC1 specimens examined (n=21) was particularly evident in the inferior olive where the boundaries between heavily and lightly affected subregions showed a degree of congruency with the pattern of Purkinje cell loss and with the afferent nuclei which project to each region of the olivary nuclei.

The hodology of the inferior olive was developed over 70 years and it is beyond the scope of this paper to recapitulate it;[19,20] interestingly, most of the data were acquired in the cat. The dorsal accessory (DAO), medial accessory (MAO), and principal (PO) olive are readily distinguished in the cat and their climbing fibers project to specific regions of the cerebellar cortex.[21] Climbing fibers originate from the inferior olive and form excitatory glutaminergic synapses on the primary dendrites of cerebellar Purkinje cells. Each Purkinje cell is innervated by a single olivary neuron and each olivary neuron innervates approximately ten Purkinje cells.[22] Specific subnuclei of the inferior olive project topographically to the cerebellar cortex and terminate in parasagittal zones.[22,23] We hypothesize an association between patterned Purkinje cell loss and LC3 accumulation in regions of the olive projecting to them.

The most clearly defined pathologic boundary in the inferior olive was seen in the DAO, where intense pathology of the ventrolateral region and an essential lack of pathology in the dorsomedial region were found. Afferents to the ventrolateral region are from (principally) the gracile,[24-26] but also from the cuneate and the lumbosacral spinal cord,[27] areas that show correspondingly high levels of LC3 pathology. Injection of tracer into the ventrolateral DAO (at approximately level XI shown in our figure) demonstrated climbing fiber termination in lobules I-IV of the anterior paravermis.[27] Similarly, the very limited pathology in the dorsomedial portion of the DAO (and its adjoining rostral medial part of the ventral lamina of the PO) matches that of its principal afferent, the spinal trigeminal nucleus.[28,29] These findings suggest that autophagic abnormalities in specific inferior olivary subnuclei pathways may explain the spatial pattern of Purkinje cell loss seen in NPC1.

Loss of autophagy results in neurodegeneration.[30,31] Since the establishment of this general principle, the impact that lysosomal storage diseases and, particularly relevant to the present study, cholesterol storage has on aberrant autophagy has become the focus of intensive research.[32-38] Cholesterol storage (as a primary or secondary substrate) and disrupted autophagy have been shown to be common features of many lysosomal storage diseases. Although fibroblasts are an imperfect model for neurons, their use in the elucidation of the role that excess cholesterol plays in impaired autophagic flux is of great importance. Abnormal cholesterol accumulation has been shown to impair endolysosome/autophagic vesicle fusion in mouse embryonic fibroblasts (MEFs) of the lysosomal storage diseases mucopolysaccharidosis type IIIa and multiple sulphatase deficiency.[39] In NPC1 patient fibroblasts, NPC1-/- MEFs, and Chinese hamster ovary-K1 NPC1 null cells, a significant reduction in the number of amphisomes formed due to the failure of the fusion of late endosomes and autophagosomes was seen.[40] Mechanistically, both of these studies concluded that excess membrane cholesterol disrupted the recruitment and correct localization of the N-ethylmaleimide-sensitive factor attachment protein (SNAP) receptors (SNARE) proteins in the membrane that are required for autophagosome fusion and maturation. Neuronal autophagosomes form primarily in distal axons, fuse with late endosomes and are transported to the soma for fusion with lysosomes.[41,42] The accumulation of LC3-positive punctae and aggregates in axons and synaptic neuropil, but not in neuronal perikarya or dendrites here reported, may be due, at least in part, to the failure of amphisome formation, resulting in the impaired retrograde axonal transport of autophagic cargo.[43,44]

REFERENCES

1. Pentchev PG, *et al. Proc Natl Acad Sci U S A*. 1985;82:8247–8251.
2. Vanier MT, Millat G. *Clin Genet*. 2003;64:269–281.
3. Runz H, *et al. Hum Mutat*. 2008;29:345–350.
4. Li H, *et al. J Neuropathol Exp Neurol*. 2005;64:323–333.
5. Infante RE, *et al. Proc Natl Acad Sci U S A*. 2008;105:15287–15292.
6. Walkley SU, Suzuki K. *Biochim Biophys Acta*. 2004;1685:48–62.
7. Ko DC, *et al. PLoS Genet*. 2005;1:81–95.
8. Kabeya Y, *et al. EMBO J*. 2000;19:5720–5728.
9. Liao G, *et al. Am J Pathol*. 2007;171:962–975.
10. Lowenthal AC, *et al. Acta Neuropathol*. 1990;81:189–197.
11. Brown DE, *et al. Am J Pathol*. 1994;144:1412–1415.
12. Munana KR, *et al. J Vet Intern Med*. 1994;8:117–121.
13. March PA, *et al. Acta Neuropathol*. 1997;94:164–172.
14. Somers KL, *et al. Mol Genet Metab*. 2003;79:99–103.
15. Vite CH, *et al. Pediatr Res*. 2008;64:544–549.
16. Tanaka J, *et al. J Neuropathol Exp Neurol*. 1988;47:291–300.
17. Higashi Y, *et al. Acta Neuropathol*. 1993;85:175–184.
18. Brodal A. *Z Neurol*. 1940;169:1053.
19. Voogd J. *Cerebellum*. 2011;10:334–350.
20. Voogd J, *et al.* Cerebellar nuclei and the inferior olivary nuclei: organization and connections. In: Manto M, Gruol DL, Schmahmann J, Koibuchi N, Rossi F, eds. *Handbook of the Cerebellum and Cerebellar Disorders*. Netherlands: Springer; 2013:377–436.
21. Ruigrok TJ. *Cerebellum*. 2011;10:464–474.
22. Mlonyeni M. *J Comp Neurol*. 1973;147:1–10.
23. Groenewegen HJ, *et al. J Comp Neurol*. 1979;183:551–601.

24. Boesten AJ, Voogd J. *J Comp Neurol*. 1975;161:215–237.
25. Berkley KJ, Hand PJ. *Brain Res*. 1978a;153:263–283.
26. Bull MS, *et al. Brain Res*. 1990;525:1–10.
27. Molinari HH. *J Comp Neurol*. 1984;223:110–123.
28. Berkley KJ, Hand PJ. *J Comp Neurol*. 1978b;180:253–264.
29. Huerta MF, *et al. J Comp Neurol*. 1985;241:180–190.
30. Hara T, *et al. Nature*. 2006;441:885–889.
31. Komatsu M, *et al. Nature*. 2006;441:880–884.
32. Ballabio A, Gieselmann V. *Biochim Biophys Acta*. 2009;1793:684–696.
33. Jaeger PA, Wyss-Coray T. *Mol Neurodegener*. 2009;4:16.
34. Yue Z, *et al. Biochim Biophys Acta*. 2009;1793:1496–1507.
35. Nixon RA. *Nat Med*. 2013;19:983–997.
36. Menzies FM, *et al. Nat Rev Neurosci*. 2015;16:345–357.
37. Fraldi A, *et al. Annu Rev Neurosci*. 2016;39:277–295.
38. Ward C, *et al. Biochim Biophys Acta*. 2016;1861:269–284.
39. Fraldi A, *et al. EMBO J*. 2010;29:3607–3620.
40. Sarkar S, *et al. Cell Rep*. 2013;5:1302–1315.
41. Maday S, Holzbaur EL. *Autophagy*. 2012;8:858–860.
42. Maday S, Holzbaur EL. *Dev Cell*. 2014;30:71–85.
43. Cheng XT, *et al. Cell Biol*. 2015a;209:377–386.
44. Cheng XT, *et al. Autophagy*. 2015b;11:1434–1436.

acvim091

Grenadian CTVT: Data From Mitochondrial Genetic Diversity, Selection and Recombination in a Canine Transmissible Cancer

Anne M. Corrigan[1]; Marta Lanza Perea[1]; Andrea Strakova[2]; Maire Ni Leathlobhair[2]; Elizabeth Murchison[4]

[1]St. George's University, Grenada, West Indies; [2]Wellcome Sanger Trust University of Cambridge, Cambridge, England, UK; [3]Wellcome Trust Investigator, University of Cambridge, Cambridge, England, UK

Canine transmissible venereal tumor (CTVT) is a naturally occurring clonally transmissible cancer prevalent worldwide and in Grenada. Clinical data and population characteristics including the signalment, clinical signs, island location, tumour size, number of treatments, speed of regression of tumours, and concurrent diseases in CTVT were collected. This is a portion of data collected in conjunction with a worldwide genetic study "Mitochondrial genetic diversity, selection and recombination in a canine transmissible cancer" (Strakova and Leathlobhair *et al.* 2016), where 449 complete mitochondrial genomes (mtDNAs) that track two millennia of dog global migration were sequenced. Five different tumour clades from 39 countries over 6 continents were identified. Mutations have occurred and captured in the canine infectious tumours highlighting the role of mtDNA in cancer evolution.

Fifty Grenadian CTVT dogs were selected through the Pothounds Against Pregnancy program, with IACUC and European ethical standards approval. Clinical and patient data were collected upon presentation to the St. George's University Junior Surgery and Anesthesia Laboratory and tissue samples for genetic analysis were obtained during ovariohysterectomy or castration. IDEXX 4DX SNAP was performed concurrently. Illumina Platform sequencing in the Wellcome Sanger Trust Cancer Genome Laboratory was utilized to perform low coverage whole genome sequencing, genome analysis and genome variant calling. With the aim of estimating the frequency and timing of mtDNA horizontal transfer, CTVT was confirmed with quantitative PCR to detect LINE-MYC genomic rearrangements.

All of the treated dogs currently living have undergone complete clinical cure. Three dogs were lost to follow-up, with one showing possible side effects of vincristine treatment. The mutations found include 1005 single point mutations and 27 indels revealed 5 distinct CTVT mtDNA clades with independent origins. The population descriptors of the 50 Grenadian TVT dogs and the worldwide mapping of these mutations will be presented.

There is variability in the population characteristics in our Grenada CTVT dogs. The number of treatments varied as well as the IDEXX 4DX SNAP test results. The genetic data are still under analysis and we hope to identify a correlation with the CTVT genomic changes and the different treatment responses. The Grenada CTVT mapped to both Clade 2 and Clade 3, revealing dynamic recent history of CTVT lineage.

acvim092

Lymphoma and Symmetric Dimethylarginine Concentration in Dogs: A Preliminary Study

Anthony Abrams-Ogg[1]; Bronwyn Rutland[2]; Phillippe Levis[2]; Vicky Sabine[3]; Allison Majeed[3]; Dorothee Bienzle[4]; Alex Zur Linden[3]; Danielle Richardson[5]; Anthony Mutsaers[6]; Paul Woods[3]

[1]University of Guelph, Guelph, ON, Canada; [2]VCA Canada 404 Veterinary Emergency and Referral Hospital, Newmarket, ON, Canada; [3]Department of Clinical Studies, Ontario Veterinary College, Guelph, ON, Canada; [4]Department of Pathobiology, University of Guelph, Guelph, ON, Canada; [5]Animal Cancer Centre, Ontario Veterinary College, Guelph, ON, Canada; [6]Department of Biomedical Sciences, Ontario Veterinary College, Guelph, ON, Canada

Symmetric dimethylarginine (SDMA) is a product of arginine metabolism that is cleared by the kidney. It is considered a more sensitive and specific biomarker than serum creatinine for the diagnosis and staging of chronic kidney disease in dogs. It is possible that serum SDMA levels may be increased in animals with malignancies because of increased protein metabolism. We observed that some dogs with untreated lymphoma appeared to have disproportionately increased serum SDMA levels compared to other biomarkers of kidney function, and that SDMA levels normalized when the dogs achieved remission. A previous study in humans reported that asymmetric dimethylarginine, but not SDMA levels, may be increased with hematologic malignancies, including non-Hodgkin's lymphoma (Szuba *et al. Leuk Lymphoma.* 2008;49:2316–2320). A preliminary study in dogs reported that SDMA levels may be increased in tumor-bearing dogs with normal creatinine levels, which was attributed to kidney disease or neoplastic infiltration (Yerramilli *et al. J Vet Intern Med.* 2017;31:251), but changes in SDMA levels after treatment were not reported. Therefore, the purpose of this study was to retrospectively and prospectively investigate the effect of lymphoma on SDMA levels in dogs. Cases were retrospectively identified in autumn 2016. Prospective case enrollment began December 1, 2016.

Inclusion criteria were: 1) cytologic ± histologic diagnosis of lymphoma by a pathologist; 2) measurement of serum SDMA level at any stage of disease; 3) routine abdominal ultrasound examination for cases prospectively studied, to identify evidence of renal lymphoma; 4) informed owner consent. Dogs with indolent lymphomas were not included. All treated cases initially received the 25-week Madison Wisconsin CHOP combination chemotherapy protocol using prednisone, vincristine, cyclophosphamide and doxorubicin. Complete remission (CR) was defined as resolution of lymphadenopathy ± previous imaging abnormalities attributed to lymphoma. Clinicians assessing remission status were not aware of SDMA levels. Cases include: 1) Dogs with a diagnosis of lymphoma and measurement of SDMA but no further assessment or treatment; 2) Dogs currently receiving CHOP in CR or partial remission (PR); 3) Dogs currently in CR after completing CHOP; 4) Dogs in relapse receiving rescue therapy; and 5) Dogs euthanized after relapse. Groups 2–5 include dogs with or without pre-treatment measurement of SDMA. Creatinine was measured on a cobas® analyzer (Roche Diagnostics). Samples for SDMA measurement were submitted to IDEXX Laboratories, Inc., Markham, Ontario, Canada. Data were summarized with descriptive statistics. Differences between means were analyzed by t-tests and paired t-tests. Differences between medians were analyzed by Wilcoxon rank-sum and signed-rank tests. Correlations were examined by Spearman Rank method.

Fifty-two dogs are included to date with mean, median and age range of 8, 7 and 3–13 years, respectively. Fifty dogs had WHO Stage III–V multicentric nodal lymphoma, one dog had primary mediastinal lymphoma, and one dog had primary bilateral renal lymphoma with sternal lymphadenopathy. No dog had clinical or clinicopathologic signs of pre-existing chronic kidney disease when diagnosed with lymphoma. Results for serum SDMA and creatinine levels are shown in Table 1.

Table 1. SDMA level, creatinine level, and SDMA-to-creatinine ratio (SDMA:Creatinine) for 52 dogs with lymphoma

Status	N	SDMA mean, median (range) [% > RI 0–14 μg/dL]	Creatinine mean, median (range) [% > RI 44–133 μmol/L]	SDMA:Creatinine mean, median (range)
Untreated	36	18, 16 (5–90) [72%]	97, 81 (47–313) [12%]	0.21, 0.19 (0.06–0.53)
CR	19	10, 11 (4–18) [11%]	90, 79 (24–169) [17%]	0.12, 0.11 (0.06–0.24)
PR	6	16, 16 (14–19) [83%]	90, 78 (65–152) [17%]	0.22, 0.20 (0.14–0.35)
Relapse	5	27, 21 (14–39) [80%]	129, 105 (71–216) [40%]	0.21, 0.20 (0.15–0.27)
B-cell	13	20, 15 (5–90)		
T-cell	6	18, 19 (10–24)		

Results are grouped by status of lymphoma burden and immunophenotype determined by flow cytometry. Groups are not mutually exclusive. (RI: reference interval).

Comparing all dogs with untreated lymphoma and all dogs with CR ranging from 1 week to >2 years, differences between mean and median SDMA levels and mean and median SDMA:Creatinine were significant ($p<0.001$), while differences between mean and median creatinine levels were not significant. Overall correlation between SDMA and creatinine was moderate (rho=0.29, $p=0.04$). In the one case where SDMA was measured twice prior to treatment, SDMA increased from 18 to 22 μg/dL over 37 days, while creatinine remained stable at 76–78 μmol/L (SDMA:Creatinine increased from 0.24 to 0.28). Fourteen dogs had serial SDMA and creatinine measurements beginning before treatment and then after achieving CR. Eleven of these 14 dogs (79%) had elevated SDMA before treatment, of which 3 of these 11 dogs (which included the dog with primary renal lymphoma) also had elevated creatinine. All 11 dogs with increased SDMA at diagnosis had normal SDMA levels in CR, and differences in the means and medians from baseline were significant ($p<0.05$). Ten of the 11 dogs also had normal creatinine levels when in CR. For the 2 dogs that did not have primary renal lymphoma but had increased pre-treatment creatinine levels, in one dog normalization of SDMA from 15 to 10 μg/dL may have been due to correction of pre-renal azotemia, as creatinine decreased from 149 to 109 μmol/L, and SDMA:Creatinine decreased from 0.10 to 0.09. However, in the other dog, the decrease in SDMA from 90 μg/dL to 12 μg/dL was disproportionately large compared to the decrease in creatinine from 227 mmol/L to 166 mmol/L, with SDMA:Creatinine decreasing from 0.47 to 0.07.

Five dogs had lymphoma in relapse. Results in these dogs were similar to those with untreated lymphoma, while results for dogs in PR were in between those of dogs with untreated lymphoma/lymphoma in relapse, and dogs in CR.

No relationships between SDMA level and signalment were identified. Nineteen (38%) dogs had immunophenotyping. Differences in SDMA levels between dogs with B-cell and T-cell lymphomas were not significant.

Twenty-nine (56%) dogs had abdominal ultrasound examinations performed as part of staging. For the dog with primary bilateral renal lymphoma, at diagnosis, SDMA was 15 μg/dL, creatinine was 149 μmol/L and SDMA:Creatinine was 0.10. In CR 4 months later, SDMA was 12 μg/dL, creatinine was 109 μmol/L, and SDMA:Creatinine was 0.11. The dog relapsed 1 month later with mediastinal involvement, and SDMA was 21 μg/dL, creatinine was 105 μmol/L, and SDMA:Creatinine was 0.20. Abdominal ultrasound examination was not repeated. Only one of the 27 dogs with multicentric nodal lymphoma that had abdominal ultrasound examinations had evidence of renal infiltration. Ultrasound examination of this dog was performed again 6 weeks and 6 months later, and on both examinations the lesion had resolved. At diagnosis SDMA was 22 μg/dL, creatinine was 92 μmol/L, and SDMA:Creatinine was 0.24, and in CR SDMA was 8 μg/dL, creatinine was 105 μmol/L, and SDMA:Creatinine was 0.08. The dog with the highest SDMA level (90 μg/dL), had no evidence of renal lymphoma and had normal kidney architecture on ultrasound examination. No dog examined by ultrasound had ultrasonographic evidence of chronic kidney disease at diagnosis.

Serum lactate dehydrogenase (LDH) activity was analyzed retrospectively in 29 available samples as a marker of lymphoma metabolic activity. There was no correlation between LDH activity and SDMA levels. Markers of spontaneous tumor lysis were analyzed to investigate acute kidney injury secondary to tumor lysis as a potential cause of increased SDMA. All dogs had normokalemia at all stages. Two dogs with untreated lymphoma had hyperphosphatemia and 5 dogs had mild hypocalcemia, while no dogs in CR had hyperphosphatemia and 2 had mild hypocalcemia; these differences were not significant. Uric acid was prospectively measured in 8 dogs. Prior to treatment uric acid levels were: mean=17 mmol/L, median=16 mmol/L, range=9–28 mmol/L. When these dogs achieved CR, results were mean=12 mmol/L, median=12 mmol/L, range=12–14 mmol/L. These results were significantly different (p=0.02). Uric acid levels were moderately correlated with SDMA levels (rho=0.26, p=0.04, N=59). Although uric acid is eliminated by the kidney, uric acid levels were not significantly correlated with creatinine levels (rho=-0.08, p=0.57) which may indicate that uric acidemia was caused more by increased release by lymphoma than by decreased renal excretion.

We conclude that some dogs with lymphoma and large tumor burden may have increased SDMA levels which may normalize if dogs achieve CR, and that SDMA should not be interpreted in these dogs as evidence of age-related chronic kidney disease. SDMA might have potential as a biomarker of treatment response in some dogs, although the data are insufficient to make firm conclusions in this regard. The data may be underpowered to detect relationships of SDMA to immunophenotype or other characteristics of the patients or malignancy. The pathophysiology of increased SDMA in dogs with lymphoma is unknown. If the assumption is correct that SDMA production is not increased by lymphoma, potential mechanisms include 1) reversible acute kidney injury, and 2) reversible decrease in SDMA excretion without compromise to other aspects of kidney function, with either of these mechanisms being secondary to renal infiltration below the threshold of routine sonographic detection, or kidney injury resulting from spontaneous tumor lysis.

acvim096

In Vitro Macrophage Polarization by Canine Osteosarcoma

Jonathan Fogle[1]; Joanne Tuohy[2]; Duncan Lascelles[1]; William Eward[4]; Jason Somarelli[4]

[1]North Carolina State University College of Veterinary Medicine, Raleigh, NC, USA; [2]North Carolina State University, College of Veterinary Medicine, CBS Immunology, Raleigh, NC, USA; [3]Duke University School of Medicine, Duke Cancer Institute - Comparative Oncology, Durham, NC, USA

Published reports describe the effect of human cancers on circulating monocyte populations and macrophage polarization within the tumor microenvironment. We have previously reported defects in monocyte chemotaxis from dogs with osteosarcoma. However, there is much less information regarding the effects of osteosarcoma on canine macrophage polarization. M1 type macrophages are typically associated with inflammation and heightened antigen processing and presentation. M2 type macrophages exhibit more heterogeneity, but in general are associated with wound healing and immunosuppressive cytokine production. We hypothesized that canine osteosarcoma induces a macrophage phenotype more closely related to the M2 phenotype, which may ultimately contribute to tumor progression and the development of metastases. The primary purpose of this study was to compare cytokine profiles of canine macrophages cultured with or without osteosarcoma. Macrophages were cultured from canine peripheral blood monocytes obtained from healthy control dogs, and either maintained as macrophages cultured alone, or cocultured with osteosarcoma cells (Abrams canine osteosarcoma cell line). Culture supernatants were analyzed using a canine-specific multiplex assay for IFN-γ, TNF-α, MCP-1, IL-10, and an ELISA for TGF-β. Macrophages cocultured with osteosarcoma secreted increased levels of both inflammatory (TNF-α) and anti-inflammatory (IL-10, TGF-β) cytokines compared to macrophages cultured alone. To further evaluate the phenotype of these cultured macrophages, we analyzed various macrophage markers (CD11b, MHCII, CD206, iNOS, and arginase) using flow cytometry. Our findings indicated that macrophages cocultured with osteosarcoma exhibited increased MHCII expression and suggested polarization to an M2-like phenotype.

acvim107

Serum Inflammatory Cytokine Values in Cachectic and Non-Cachectic Cats with Cancer and Normal Cats

Erika L. Krick; Kathy Michel; Molly Church

School of Veterinary Medicine, University of Pennsylvania, Philadelphia, PA, USA

Cancer cachexia is a complex syndrome in humans associated with poor tolerance and response to treatment. Patients often have elevated serum inflammatory cytokines. Some cancer-bearing cats exhibit clinical signs of cancer cachexia. The study purpose was to compare body weight, body condition, complete blood count, serum chemistry and inflammatory cytokine values in cachectic and non-cachectic cancer-bearing cats and healthy cats.

Client-owned cancer-bearing and healthy cats prospectively underwent weight measurement, body condition and muscle mass scoring, complete blood count and chemistry panel, and serum amyloid A and alpha1-acid glycoprotein measurement. Cachexia was defined as muscle mass score below 2 and/or body condition score below 4.

Seventy-two (17 cachectic, 27 non-cachectic, 30 healthy) cats have been enrolled. Mean body weight and body condition score were not significantly different among groups. Cachectic cats had significantly lower hematocrit and albumin compared to non-cachectic ($p<0.0001$, $p<0.0001$) and healthy cats ($p<0.0001$, $p=0.0003$). Cancer-bearing cats had significantly higher white blood cell counts compared to healthy cats ($p<0.0001$). Preliminary cytokine results showed that cachectic cancer-bearing cats had significantly higher serum amyloid A compared to non-cachectic and healthy cats ($p=0.0086$, $p=0.05$).

The similar body condition score and weight among all groups underscores the importance of specifically assessing muscle mass in cats with cancer. Differences in hematocrit, albumin, and serum amyloid A in cachectic versus non-cachectic cats suggest that therapies directed to the altered metabolic state of these cats combined with cytotoxic treatment is worth investigation.

acviml16

Efficacy and Safety of Electrochemotherapy in the Treatment of Soft Tissue Sarcomas in Dogs

Ron Lowe[1]; Filippo Torrigiani[2]; Alessio Pierini[2]; Joe Impellizeri[3]; **George Lubas**[2]

[1]Aisleigh Veterinary Clinic, Knaresborough, England, UK; [2]Dept. Veterinary Sciences, University of Pisa, San Piero a Grado, Pisa, Toscana, Italy; [3]Veterinary Oncology Services, PLLC Visiting Scholar-Vassar College, NY Director, Barrymore Center for Advanced Cancer Therapeutics, New York, NY, USA

Soft tissue sarcomas (STS) represent up to 15% of all skin and subcutaneous tumours in dogs. Local tumour control with wide surgical margins or a combination of surgery and radiation therapy is the standard of care in the management of STS; however, these approaches are not always possible due to anatomic location and/or expense of the treatment. Electrochemotherapy (ECT) is an emerging, locally ablative technique, now available in the United States. ECT has been available in Europe for over a decade and is a standard therapy for cancer treatment in people and animals.
The purpose of this study was to evaluate the safety and efficacy of ECT with intravenous bleomycin in the treatment of STS in dogs.

Fifty-three client owned dogs with 55 STS were enrolled (two dogs each had two STS in different anatomical sites). Four dogs with STS located orally were excluded according to Bray (2016). Tumour location, size, histologic grade, electroporation parameters, recurrence rate, recurrence time, disease free interval (DFI), and treatment toxicity were evaluated. The owners elected to pursue ECT as an alternative treatment to surgery and/or radiotherapy. For all dogs, the diagnosis of STS was based on histopathology. Most dogs were staged with physical examination, fine needle aspiration of regional lymph node (if palpable), and three view thoracic radiographs. Complete blood cell count, minimal serum biochemistry profile, and urinalysis were also obtained pre general anesthesia. For ECT treatment, dogs were premedicated with medetomidine (0.01 mg/kg, IM) and butorphanol (1 mg/kg, IM). Once sedated, patients were pre-oxygenated via face mask, and anesthesia was induced with intravenous alfaxalone (2 mg/kg). Dogs were intubated and anesthesia was maintained with oxygen, nitrous oxide and isoflurane. Atipamezole (0.05 mg/kg IM) was administered to reverse medetomidine at the conclusion of the procedure. Bleomycin was diluted in saline solution and administered IV at a dosage of 15,000 UI (Pharmacopeia Europe)/m^2 (equivalent to 15 U-US Pharmacopeia/m^2). Treatment areas were prepared for surgery, skin was clipped and aseptically prepared with povidone-iodine wash (10% iodine) and surgical spirit. All dogs received anti-inflammatory and pain medicine in the perioperatory period with meloxicam (0.2 mg/kg SC) and butorphanol. After ECT, all patients received antibiotic coverage (Amoxicillin + clavulanic acid 12.5 mg/kg, BID, PO or clindamycin 5.5 mg/kg BID PO) for at least one week. Patients were assessed every 3–10 days for the initial 4 weeks after ECT treatment, then monthly for three months. Thereafter, progress was monitored until at least 2 years either by examination of the patient or by telephone contact with the owner. Any progression in the treated tissue was recorded by digital camera images. ECT was performed using Cytopulse PA4000 in 9 cases or Cytopulse Oncovet in 46 cases (Cyto Pulse Sciences, Inc., now distributed by BTX Harvard Apparatus, MA, USA), with Gehl needle electrodes of 1–2.5 cm length from the same manufacturer. The pulse pattern employed was 8 monophasic square pulses of 100 microseconds each at a frequency of 1 Hz with the PA4000, or 1 Hz or 5 kHz with the Oncovet. The electrode pattern comprised two parallel rows of six 1–2.5 cm length needles with rows 6 mm apart. The pulse amplitude was 600–720 volts (1000–1200 volts/cm). The time between bleomycin injection and pulse delivery was 8 minutes, and maximum electroporation treatment time was 20 minutes (Tozon *et al.* 2016).

Patients were divided into three categories according to the ECT modality:

1. ECT alone, performed in smaller lesions, where STSs were clearly visible macroscopically and could be resected by conventional surgery but the owners elected to pursue ECT.
2. Intraoperative ECT, performed in larger lesions. In these cases ECT was performed at the same time of the surgical cyto-reduction of the tumour. During surgery the majority of the tumour was excised, with no attempt on obtaining complete margins (as standardly accepted of 2–3 cm according to Bray 2016), and ECT was applied to lateral and deep margins before wound closure; in some cases macroscopic tumour was still present. This method was used where radical excision would have resulted in functional or cosmetic compromise due to anatomic limitation.
3. Postoperative (adjuvant) ECT: this method was applied when the surgical resection of the tumour was initially performed attempting wide margins but where histological examination showed inadequate margins. Adjuvant ECT was applied after a period of 2–4 weeks from the surgical intervention (mean time 22.4 days). The reason for the delay was to perform ECT treatment as early as possible after adequate wound strength had developed following surgery and the final decision was left to the surgeon's judgement.

In dogs receiving ECT alone, the tumour plus a 1–2 cm margin in all planes into grossly normal tissue were accessed by penetration of the electrode needles. The treatment was started in the margins and progressed concentrically into the tumour mass, thus avoiding transference of tumour cells from the centre to the margins. Moreover, this pattern of pulse application is reported to achieve better reduction of the blood flow to the tumour, thus obtaining a better drug retention into the tumour stroma (Cemazar *et al.* 2016). Similarly, when ECT was used intraoperatively, a 1–2 cm margin in all directions was used. In dogs treated with ECT adjuvant to surgery, where there was no macroscopic disease, the treatment margin extended at least 1.5 cm from the healed surgical scar site in all planes. Local tumour response in dogs treated with ECT alone was evaluated according to the response evaluation criteria for solid tumors established by the veterinary cooperative oncology group consensus document (Nguyen *et al.* 2015). A complete remission (CR) was defined as total reduction of the tumour. A partial remission (PR) was defined as ≥30% reduction in tumour diameter. Stable disease (SD) was defined as <30% reduction in tumour diameter or <20% increase in tumour diameter, and progressive disease (PD) was defined as ≥20% increase in tumour diameter. The presence of new lesions near to the primary tumour was considered as PD. A minimum duration of two to three weeks was required for a response to qualify as positive. In order to assess the local treatment toxicity a grading score (5-point scale) for tissue necrosis was established by the authors and previously described (Lowe *et al.* 2016): 0=none, 1=slight swelling, 2=swelling/necrosis <1 cm, 3=severe swelling, 4=deep necrosis, and 5=severe swelling and tissue loss. DFI was calculated from the date of treatment with ECT to the first recurrence or death of the animal unrelated to the tumour; or alternatively when we censored the DFI at the date we prepared this paper. The statistical analysis was performed using Chi-squared test and Fisher's exact test.

Of the 53 dogs included in the study, 22 were females (16 neutered) and 31 were males (14 neutered). The majority of dogs were pure breeds (38/53, 71.6%); Labrador Retrievers (8) and Staffordshire Bull Terrier (5) were overrepresented. Patient age ranged from 4 to 15 years (median 9.1). Tumour grade was assessed in 46 cases, most tumours were grade I (21/46, 45.6%) and II (23/46, 50.0%), only two were grade III (4.4%). The large majority of tumours affected the limbs (42/55, 76.4%), other locations were trunk (8/55, 14.5%) and head (5/55, 9.1%). Tumour size ranged from 0.4 to 20 cm (median 4.5 cm). Median follow-up period was 482 days (range 8–2,483). The first ECT treatment was performed as follows: ECT alone in 5 cases, intraoperative ECT in 26, and postoperative (adjuvant) ECT in 24 cases. Five dogs underwent a second treatment due to tumour recurrence, in 3 of them a further recurrence occurred and a third ECT treatment was performed. The treatment response in the ECT alone group was as follows: two CR, one PR, and two SD. At the end of the observation period, twenty-two dogs had died; of these, 7 had tumour recurrence, which was the reason for euthanasia in two cases. Of the other 29 dogs, 8 had tumour recurrence and in three cases the affected limb was amputated as rescue treatment. Patient outcome was not available in four cases. Overall tumour recurrence rate (OTRR) was 27.2% (15/55). Recurrence rate was not associated with tumour grade, anatomical site, tumour size, pulse frequency, and pulse voltage ($p>0.05$). For all dogs, median DFI was not reached. For dogs with tumour recurrence, median DFI was 68 days (range 0–1,025). None of our patients developed systemic adverse effects related to the ECT treatment. Local toxicity was well tolerated in 65% of cases (36/55 had toxicity score ≤2 in a scale with 5 as maximum) and was not associated with tumour grade, tumour size and pulse frequency ($p>0.05$). Local toxicity was associated with higher pulse voltage (1000 vs. 1200 V), which was statistically significant ($p=0.003$).

The overall tumour recurrence rate (OTRR) of ECT alone or in combination with surgery (intra- or postoperative) was 27.2%.

Adjuvant ECT with local injection of bleomycin has been successfully applied in the treatment of canine sarcomas in a previous study (Spugnini *et al.* 2007). However, these results cannot be compared with ours because the tumour list included neoplastic entities that are no more considered as "true canine STS" (i.e., hemangiosarcoma, malignant fibrous histiocytoma, leiomyosarcoma and neurofibrosarcoma) (Liptak & Forrest 2013). Moreover, the criteria of tumour response (i.e., complete response and progressive disease) were not extensively illustrated in the paper, thus making a comparison impossible.

Other therapies including adjuvant radiation therapy and intralesional chemotherapy have been reported to have an OTRR of 18.0% and 16.6%, respectively (Demetriou *et al.* 2012; Havlicek *et al.* 2009). When possible, a wide margin excision still remains the best choice to treat canine STS (3.2%; Prpich *et al.* 2013). In this study, ECT with intravenous bleomycin was well tolerated with 65% showing mild toxicity. This finding is similar to the results previously reported by one of the authors using the same toxicity score scale (53%; Lowe *et al.* 2016). To our knowledge, no data has been reported regarding postsurgical time of healing or complication rate with intraoperative ECT. Local toxicity was similar in intraoperative and postoperative ECT groups, suggesting that treatment delay may be not necessary in a postoperative setting.

ECT with bleomycin should be considered an option to treat STS in dogs, especially when other treatment options are declined or not applicable. However, further investigation is necessary to identify ideal treatment parameters.

References

1. Bray JP. Soft tissue sarcoma in the dog - Part 1: a current review. *Journal of Small Animal Practice*. 2016;57:510–9.
2. Chase D, *et al.* Outcome following removal of canine spindle cell tumours in first opinion practice: 104 cases. *Journal of Small Animal Practice*. 2009;50:568–74.
3. Cemazar M, *et al.* Efficacy and safety of electrochemotherapy combined with peritumoral IL-12 gene electrotransfer of canine mast cell tumours. *Veterinary and Comparative Oncology*. 2016.
4. Demetriou JL, *et al.* Intentional marginal excision of canine limb STSs followed by RT. *Journal of Small Animal Practice*. 2012;53:174–81.
5. Havlicek M, *et al.* Intra-operative cisplatin for the treatment of canine extremity soft tissue sarcomas. *Veterinary and Comparative Oncology*. 2009;7:122–9.
6. Liptak & Forrest. In: Withrow SJ, Page R, Vail DM. *Withrow and MacEwen's Small Animal Clinical Oncology*. Elsevier Health Sciences; 2013.
7. Lowe R, *et al.* The treatment of canine mast cell tumours with electrochemotherapy with or without surgical excision. *Veterinary and Comparative Oncology*. 2016.
8. Nguyen SM, *et al.* Response evaluation criteria for solid tumours in dogs (v1. 0): a Veterinary Cooperative Oncology Group (VCOG) consensus document. *Veterinary and Comparative Oncology*. 2015;13:176–83.
9. Prpich CY, *et al.* Second intention healing after wide local excision of soft tissue sarcomas in the distal aspects of the limbs in dogs: 31 cases (2005–2012). *Journal of the American Veterinary Medical Association*. 2014;244:187–94.
10. Spugnini EP, *et al.* Adjuvant electrochemotherapy for the treatment of incompletely excised spontaneous canine sarcomas. *In Vivo*. 2007;21:819–22.
11. Tozon N, *et al.* Operating procedures of the electrochemotherapy for treatment of tumor in dogs and cats. *Journal of Visualized Experiments*. 2016;116.

acviml19

Evaluation of Chimeric Antigen Receptor T Cell Therapy in Dogs with B Cell Lymphoma

Nicola Mason; Kazim Panjwani; Martha MaloneyHuss

University of Pennsylvania, Philadelphia, PA, USA

Chimeric antigen receptors (CARs) are synthetic receptors that combine an antigen-specific antibody moiety with T cell intracellular signaling domains. T cells that have been engineered to express a CAR can therefore recognize a specific antigen in an MHC independent manner and subsequently mediate potent cytotoxic responses against cells that express the target antigen. In this study, we aimed to determine whether canine T cells could be genetically modified using a second generation, CD20 targeting CAR and whether these re-directed T cells could proliferate in response to CD20+ target cells and become CD20-specific serial killers. We used a third-generation self-inactivating lentiviral vector to successfully deliver a second generation CD20-CD28-CD3-zeta CAR to canine T cells isolated from healthy dogs and dogs with B cell lymphoma. Cells were first activated using artificial antigen presenting cells (aAPC) and then transduced with the lentiviral CAR. We demonstrate that canine T cells can be successfully re-directed and upon engagement with CD20, can undergo antigen-specific proliferation, cytokine production, and serial killing. Having demonstrated proof-of-concept *in vitro*, we treated 4 dogs with B cell lymphoma with autologous, second generation, CD20 targeted CAR T cells. Although numbers of adoptively transferred CAR T cells were limiting, and B cell aplasia was not induced in any dog, we did observe evidence of *in vivo* CAR T cell activity. Taken together, this early data provides proof-of-concept that autologous CD20-targeting CAR T cell products can be generated from canine B cell lymphoma patients and should be further explored for the treatment of CD20+ malignancies.

acvim138

Survivin Inhibition with EZN-3042 for Canine Lymphoma Therapy: Phase I Investigation

Douglas H. Thamm[1]; Jenette Joseph[2]; Travis Meuten[3]; Kristen Weishaar[3]

[1]Colorado State University Flint Animal Cancer Center, Fort Collins, CO, USA; [2]University of Colorado Comprehensive Cancer Center, Aurora, CO, USA; [3]Colorado State University, Fort Collins, CO, USA

Lymphoma is one of the most common cancers in dogs. While remission is experienced in most dogs receiving chemotherapy, very few are cured, with median survival times typically in the 12-month range. Novel treatment approaches are needed. The IAP family member survivin, one of the most commonly overexpressed proteins in human cancer, plays a critical role in apoptosis resistance, a major cause of drug-resistant treatment failure. Therapies targeting survivin have shown promise preclinically; however, none have been evaluated in dogs to date. We demonstrated previously that survivin is abundantly expressed in canine lymphoma and osteosarcoma (OSA), and that high expression correlates with inferior clinical outcome. Furthermore, survivin inhibition with siRNA or the locked nucleic acid antisense oligonucleotide EZN-3042 reduced proliferation, increased apoptosis and enhanced chemosensitivity in canine lymphoma and OSA cells *in vitro*, and increased chemosensitivity in a canine OSA xenograft. The goal of this study was to investigate the safety and biologic effect of systemic EZN-3042 delivery in dogs with lymphoma. The primary endpoint was modulation of survivin expression in lymphoma tissues.

This was a prospective, phase I dose-escalation trial in dogs with biopsy accessible multicentric lymphoma. Dogs received 3 biweekly doses of EZN-3042 as a 2-hour IV infusion. Starting dosage was 3.25 mg/kg. Escalation occurred at 1.25 mg/kg increments, according to a standard 3+3 cohort design. A CBC, serum biochemistry profile, coagulation profile, and urinalysis were performed before the first and at each subsequent treatment. Peripheral blood mononuclear cells (PBMC) and tumor tissue were collected prior to the first treatment and 24 hours after the 3rd treatment. Adverse events (AEs) were graded according to the VCOG-CTCAE v1.1. Survivin expression was assessed in pre- and post-treatment tumor tissue and PBMCs utilizing qRT-PCR and immunohistochemistry (IHC). Tumor cell proliferation and apoptosis were assessed using Ki67 and activated caspase-3 IHC.

Eighteen dogs were enrolled in 5 dose cohorts, ranging from 3.25 to 8.25 mg/kg twice weekly. One of 6 dogs treated at the highest dose cohort developed grade 3 thrombocytopenia. No other AEs were noted. Reduction in survivin mRNA was observed in PBMC and tumor tissue in 3 of 5 evaluable dogs treated at the highest dose cohort (8.25 mg/kg). Reduction in tumor survivin protein was observed in 3 of 6 cases at 8.25 mg/kg, but there was no correlation between protein and mRNA reduction. Reduction in Ki67 was observed in 2/5 cases, and increase in activated caspase was observed in 1/5 cases at 8.25 mg/kg. There was no correlation between survivin inhibition and changes in proliferation or apoptosis.

In conclusion, we have demonstrated reduction in survivin expression in lymphoma tissues and PBMC in the majority of dogs treated with EZN-3042 at a dose of 8.25 mg/kg twice weekly, a dose and schedule associated with acceptable toxicity. This biologically effective dose can be utilized in future studies evaluating EZN-3042/chemotherapy combinations in dogs with lymphoma and/or OSA.

acvim269

Early Phase Translational Studies in Canine Lymphoma

William C. Kisseberth, DVM, PhD, DACVIM (Oncology)
Columbus, Ohio, USA

Introduction

Although the occurrence of spontaneous cancers in dogs has been recognized as a potential model for human cancers for many years, it has only been much more recently that full potential of the model has been more widely appreciated, recognized, and utilized. The key drivers for this change have been the sequencing of the canine genome and resultant gradual understanding of the genomic landscape of different canine cancers, an increasing number of clinician scientists implementing clinical trials in pet dogs and the establishment of different networks of clinical investigators, both in the private sector and academia (e.g., National Cancer Institute Comparative Oncology Program Comparative Oncology Trials Consortium [COTC] and European Canine Lymphoma Network [ECLN] conducting clinical trials).[1,2]

Canine lymphoma in particular has been the focus of many of these trials. This is for a variety of reasons. First, the disease clinically, histopathologically, and molecularly shares many similarities with its human counterpart. It is a relatively common cancer, and as such, patients can accrue to trials rapidly. Significantly, the tumors (commonly lymph nodes) are readily accessible for assessing response (often caliper measurements) and obtaining biologic samples (e.g., fine needle aspirates, biopsies), often serially over time. Furthermore, dogs are more similar in size to humans than other animal models and can be treated and imaged in most cases with identical, or similar, technologies, increasing the clinical relevance and translation to humans. For all of these reasons, dogs with spontaneous lymphoma are commonly used in early phase studies, not only for investigating new therapies that are thought to be uniquely relevant to lymphoma (e.g., a CD20-targeted agent) but also for investigation of agents with potentially broader applications (e.g., SINE inhibitors).

Study Design

The spontaneous canine cancer model, lymphoma included, can be used to inform potentially every aspect of the traditional cancer drug development pathway; however, the model has been most commonly used by academic researchers in development of investigator developed therapies and by pharma in preclinical development, often to further enhance an investigational new drug (IND) application. Sometimes these canine clinical trials precede first-in-human trials and other times they may occur concurrently with a human phase I study. In either case, it is common for the dog study to help inform the conduct of the human trial. Early phase translational trials in dogs with lymphoma lend themselves to being able to be designed in a way to address many different drug development questions in a single study, including: pharmacokinetics, pharmacodynamics, toxicity, safety, dose, and scheduling. Furthermore, correlative molecular, genomic, and biomarker discovery often can be easily incorporated.

Recent Studies

Ibrutinib

Ibrutinib (PCI-32765) is a first-in-class small molecule Btk inhibitor. Although the trial evaluating ibrutinib in dogs with lymphoma was small (data from only 8 dogs shown), the impact of this work was significant.[3] First, the study demonstrated that occupancy of the Btk active site by ibrutinib could be monitored *in vivo* in dogs with lymphoma using an affinity probe for Btk and that active site occupancy of Btk was tightly correlated with the blockade of BCR signaling and *in vivo* efficacy. Most significantly, the study demonstrated proof of concept, that ibrutinib could induce objective clinical responses in dogs with spontaneous B-cell lymphoma.

Acalarutinib

Acalarutinib (ACP-196), a second-generation small molecule Btk inhibitor, has greater specificity, less toxicity, and greater potency than ibrutinib.[4] In a dose escalation study, we showed that acalarutinib was well-tolerated and that the objective response rate in dogs with lymphoma appeared to be similar to that of ibrutinib. Significantly, the response rate of the Btk inhibitors ibrutinib and acalarutinib in dogs with B-cell lymphoma appeared to be very similar to that observed in humans with diffuse large B-cell lymphoma (DLBCL), further supporting that canine B-cell lymphoma is a good model for preclinical evaluation of Btk inhibitors for people.

SINE Inhibitors

SINE (selective inhibitors of nuclear export) inhibitors are compounds that inhibit the function of the nuclear export protein Exportin 1 (XPO1/CRM1) that carry macromolecules, including several major tumor suppressor and growth regulatory proteins (e.g., p53, p75, Rb, p21, p27, STAT3, FOXO and IκB) from the nucleus to the cytoplasm. In a phase I clinical trial, the SINE inhibitor KPT-335 was evaluated in 17 dogs with spontaneous cancers with the majority of the cancers being lymphoma (n=14).[5] The MTD, which included dose and scheduling, was established. Clinical benefit was observed in 9/14 dogs (partial response, n=2; stable disease, n=7). Toxicities were primarily gastrointestinal consisting of anorexia, weight loss, vomiting, and diarrhea and were manageable with supportive care, dose modulation, and administration of low dose prednisone; hepatotoxicity, anorexia, and weight loss were the dose limiting toxicities.

Adoptive T-cell

The feasibility of adoptive T-cell therapy for canine B-cell lymphomas was demonstrated using autologous *ex vivo* expanded T-cells.[6] In this study, graded doses of *ex vivo* expanded autologous T-cells were infused after a CHOP-based chemotherapy regimen. *Ex vivo* expanded T-cells were found to be safe, persisted for up to 49 days, homed to the tumor and significantly improved survival. Significantly, the study demonstrated that clinical-grade canine and human T cells can be propagated using identical approaches.

CD40 Vaccine

A cell-based active immunotherapy approach was tested in dogs with spontaneous B-cell lymphoma.[7] In this study, dogs that achieved clinical remission with a CHOP-based chemotherapy protocol were vaccinated with CD40-B cells electroporated *ex vivo* with autologous tumor RNA. This CD40-activated B-cell vaccine stimulated immunity *in vivo* and although it did not improve time-to-progression or lymphoma-specific survival, it did appear to improve second remission and survival. Importantly, the authors point out that the study of dogs with spontaneous lymphoma provides a unique opportunity to evaluate the therapeutic efficacy of immune therapies in the setting of minimal residual disease.

cCD20 CAR T

An initial proof-of-principle study demonstrating the feasibility and safety of RNA-transfected CD20-specific chimeric antigen receptor T cells (CAR T cells) was performed in a single dog with spontaneous B-cell lymphoma.[8] In this study, an *ex vivo* autologous T-cell expansion methodology was developed and mRNA electroporation was used to express a canine CD20-specific CAR. The autologous CD20-specific CAR T cells were infused weekly, three times, to a dog with relapsed B-cell lymphoma. Treatment was generally well tolerated, other than transient fever. Modest antitumor activity (transient increase, then decrease in lymph node size) was seen. Although only an "n" of one, this study demonstrated the translatability and feasibility of using the spontaneous canine lymphoma model for the preclinical development of CAR T cell and other cell based therapies for people.

LLP2A Imaging

Dogs with spontaneous cancers, especially lymphoma, are excellent models for the development of radiopharmaceuticals and other imaging agents and modalities. The relatively large body size and ability to use exactly the same imaging instrumentation and other technologies greatly facilitates translation of findings from the preclinical model (dog) to humans. This was illustrated in a study evaluating the novel theranostic agents ^{99m}Tc-LLP2A-HYNIC-PEG and ^{99m}Tc-LLP2A-HYNIC in dogs with lymphoma.[9] LLP2A targets activated $\alpha 4\beta 1$ integrin. *In vivo* molecular targeted imaging and biodistribution was performed in 3 normal dogs and 5 dogs with lymphoma. In dogs with lymphoma, tumors showed moderately increased activity over background, with tumor activity decreasing after chemotherapy.

Conclusions

In many respects, the comparative oncology approach, particularly with respect to lymphoma, is coming of age. First driven by sequencing of the canine genome, the focusing of cancer drug development on targeted therapies and the recent explosion of immune-oncology drug development has helped spur on the search for more relevant and predictive models. This is further driven by the high and escalating costs of cancer drug development, which also can be addressed, in part, by better animal models. Notably, for lymphoma drug development, consideration of conducting a clinical trial in dogs as part of the preclinical and IND package is now commonplace, and for some drugs may almost be a necessity (e.g., Btk inhibitors). For the first time, NIH is explicitly investing in development of the spontaneous canine cancer model, soliciting research proposals directed at canine comparative genomics, model development, and immune-oncology. Important for our veterinary patients, early phase studies of novel agents in dogs is a win-win for both canine and human cancer patients. Our canine patients obtain access to new drugs even before they are available to people, while human cancer patients, hopefully, will collectively gain faster access to new drugs, as dogs help facilitate and accelerate the drug development process.

References

1. Gordon I, *et al. PLoS Med*. 2009;6(10):e1000161.
2. Comazzi S, *et al. Vet Comp Oncol*. 2015;13(4):494–497.
3. Honigberg LA, *et al. Proc Natl Acad Sci USA*. 2010;107(29):13075–13080.
4. Harrington BK, *et al. PLoS One*. 2016;11(7):e0159607.
5. London CA, *et al. PLoS One*. 2014;9(2):e87585.
6. O'Connor CM, *et al. Sci Rep*. 2012;2:249.
7. Sorenmo KU, *et al. PLoS One*. 2011;6(8):e24167.
8. Panjwani MK, *et al*. 2016;24(9):1640–1642.
9. Zwingenberger AL, *et al. PLoS One*. 2012;7(4):e34404.

acvim270

Progenitor Cell Transplantation for the Treatment of Canine Hematologic Malignancies

Steven E. Suter, VMD, MS, PhD, DACVIM (Oncology)
Raleigh, NC, USA

Introduction

For close to 6 decades, the domestic dog (*Canis familiaris*) has served as a preclinical model for hematopoietic cell transplantation (HCT).[1] Based on limited knowledge gained in early, uniformly unsuccessful, human clinical trials and later in murine studies, dogs were identified as a suitable species for study based on their large body size, longer life span, genetic diversity, random-bred nature, ease of handling, and relatively economical colony maintenance costs.[2] For these reasons, there is a tremendous body of literature documenting the utility of dogs for studies of HCT, total body irradiation (TBI), myeloablation, cell dose requirements, graft manipulation, engraftment, and GVHD. Therefore, the vast majority of clinical HCT problems encountered in people were solved using dogs. It is estimated that >95% of human transplant procedures were first perfected using canines.[1]

Hematopoietic cell transplantation was revolutionized by a number of important discoveries in both humans and dogs dating back to the 1990s. First, the cloning and large-scale production of recombinant hematopoietic cytokines, including granulocyte colony-stimulating factor (rhG-CSF, recombinant human G-CSF and rcG-CSF, recombinant canine G-CSF), stem cell factor (rhSCF and rcSCF), and granulocyte macrophage colony-stimulating factor (rhGM-CSF and rcGM-CSF), that could increase circulating hematopoietic progenitors and support their *in vivo* growth.[3,4] Second, the finding that peripheral blood contains small numbers of hematopoietic progenitor cells and that recombinant cytokines dramatically increase these numbers via recruitment from the bone marrow.[5,6] Third, the production and characterization of anti-CD34 monoclonal antibodies directed against human[7], mouse[8], primate[9], and canine mononuclear cells[10] that allowed the isolation of a population of cells that can reconstitute the peripheral blood cell lineages after lethal marrow ablation. And, finally, the development of sophisticated continuous-flow blood separators that are able to remove and separate blood components via centrifugation on the basis of specific cell densities.[11,12]

The NC State Canine Bone Marrow Transplant Unit currently utilizes refurbished TerumoBCT COBE® Spectra apheresis machines (www.terumobct.com/COBE-Spectra) to perform a number of apheresis procedures, including peripheral blood mononuclear cell collections for HCT. We are using HCT, in conjunction with chemotherapy, to perform both autologous and allogeneic HCT in dogs with hematologic malignancies, including high-grade B- and T-cell lymphomas and lymphoid leukemias.

Initial Studies

Apheresis Procedure, Product Analysis, and Protocol Refinement

Our initial data was generated using a Baxter-Fenwal CS-3000 Plus with Access Management System continuous blood separator (Fenwal, Baxter Healthcare Corporation, Deerfield, IL) that were kindly donated by the Mayo Clinic, Rochester, MN. While using a slightly different technology than the TerumoBCT machines, we were able to refine our apheresis procedure and show that the use of rhG-CSF (Neupogen, Amgen, USA: 5 μg/kg SQ BID x 5 days) clearly increased the peripheral white blood cell count and increased the number of CD34+ cells in the final harvest product.

Figure 1. Total white blood cell count of 6 normal dogs receiving rhG-CSF for 5 days. In all dogs, the target dose of 30,000/µl was reached. All dogs received a double dose of Neupogen (10 µg/kg) 2 hours before the apheresis procedure

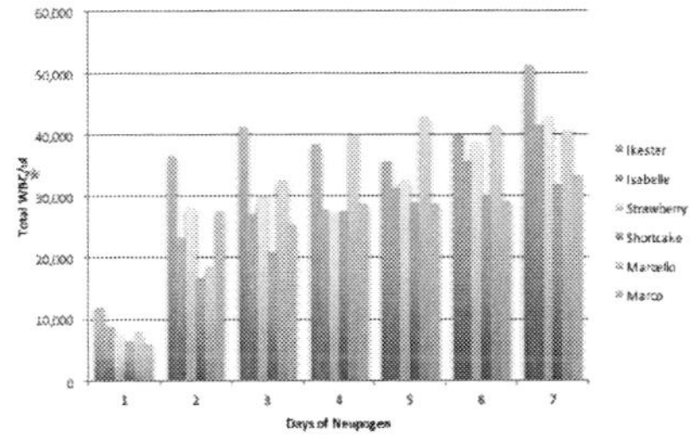

!!

Fig 2. Flow cytometric analysis of the apheresis product from a normal rhG-CSF treated dog. The left dot plot represents forward scatter (x-axis) vs side scatter (y-axis) of the harvested product with gating (circle) on the cells that are analyzed for CD34/CD14 content in the right dot plot. The CD34+ percentage was 0.35% (upper left quadrant)

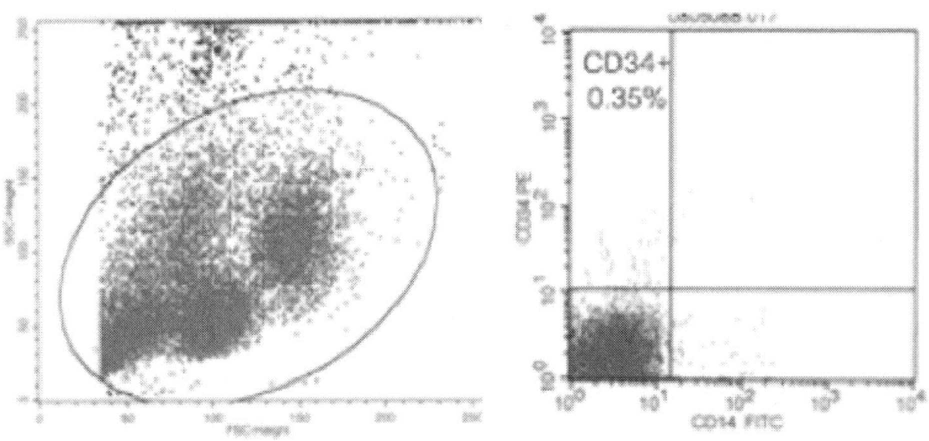

In all 6 dogs receiving rhG-CSF, the target dose of $2\text{x}10^6$/kg CD34+ cells/kg was easily achieved, with a range of $7.8\text{x}10^6$/kg to $3.3\text{x}10^8$/kg. In addition, all dogs tolerated the procedure extremely well, with no evidence of any clinical abnormalities with appropriate monitoring and support.[13,14]

CLIENT-OWNED DOGS WITH HEMATOLOGIC MALIGNANCIES

Based on our results and previous reports documenting the utility of this procedure when treating dogs with hematologic malignancies[14,15], we began performing HCT in conjunction with chemotherapy on client-owned dogs diagnosed with hematologic malignancies beginning in November, 2008. All dogs were diagnosed with a hematologic malignancy by their referring veterinarians and received standard multi-agent systemic CHOP-based chemotherapy (cyclophosphamide, Adriamycin, vincristine, prednisone).

Fig 3. Flow cytometric analysis of the apheresis product from an rhG-CSF primed treated dog. The CD34+% of this product was 3.5% (upper left hand quadrant). Based on the total number of cells harvested in 50 ml of product, ~1.94 x 10^{7} CD34+ cells/kg were harvested

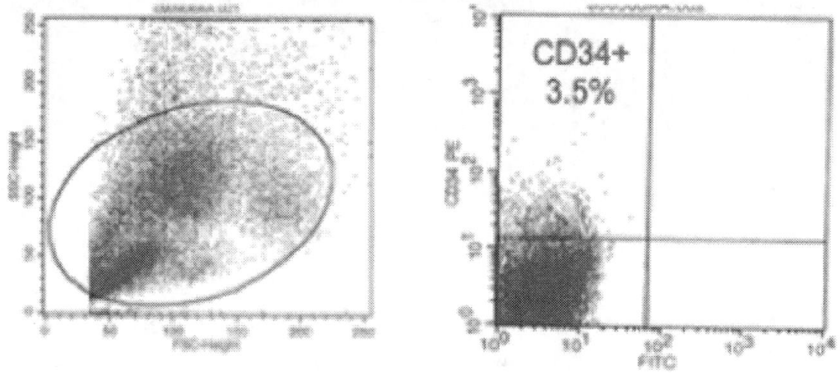

Autologous HCT

B-Cell Lymphoma

Most of the dogs with B-cell lymphoma (24 total) were in a complete clinical remission before HCT, although five of these dogs had either relapsed with disease and driven back into remission or were not in clinical remission prior to HCT. All dogs received high-dose cyclophosphamide (300–650 mg/m^2 IV) in an effort to clear their blood and bone marrow of all cancer cells two weeks before starting rhG-CSF for 6 days.[15] Blood and bone marrow PARR (PCR for antigen receptor rearrangements)[16] was performed upon presentation to NC State and all samples were negative using B cell primers. In all dogs, the target dose of >30,000 WBC/µl after six days of rhG-CSF was achieved (apheresis product analysis from 1 of these dogs are seen in Figure 4)

Fig 4. Flow cytometric analysis of the apheresis product from a dog in remission for B-cell lymphoma. A total of 4.726x10^{9} total cells were harvested. Based on 3% CD34+ cells, a total of 3.9x10^{6} CD34+ cells/kg were harvested, well above the target of 2x10^{6} CD34+ cells/kg

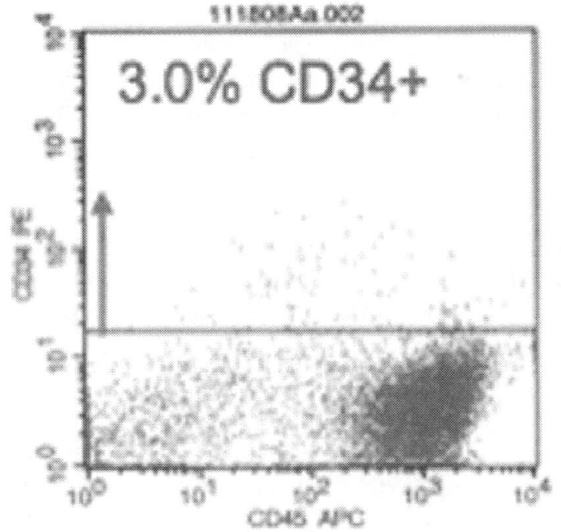

CD34+ cell enumeration, the apheresis products are refrigerated overnight. The following day, all patients underwent 10 Gy total body irradiation (TBI) using a previously described radiation protocol.[15] The harvested cells were then returned to the patient intravenously immediately after TBI. The patients remained in the hospital for 2 weeks post-TBI for supportive care relating to TBI toxicities, which included variable gastrointestinal signs and severe cytopenias. All patients showed complete hematologic recovery during this time (Figure 5).[17]

Figure 5. Post-TBI/HCT hematologic recovery in 10 dogs. The neutrophil nadir in most dogs occurred ~day +4–7, while the platelet nadir occurred ~day +9–11. All dogs cytopenias either normalized (neutrophils) or increased significantly (platelets) within 14 days post-TBI

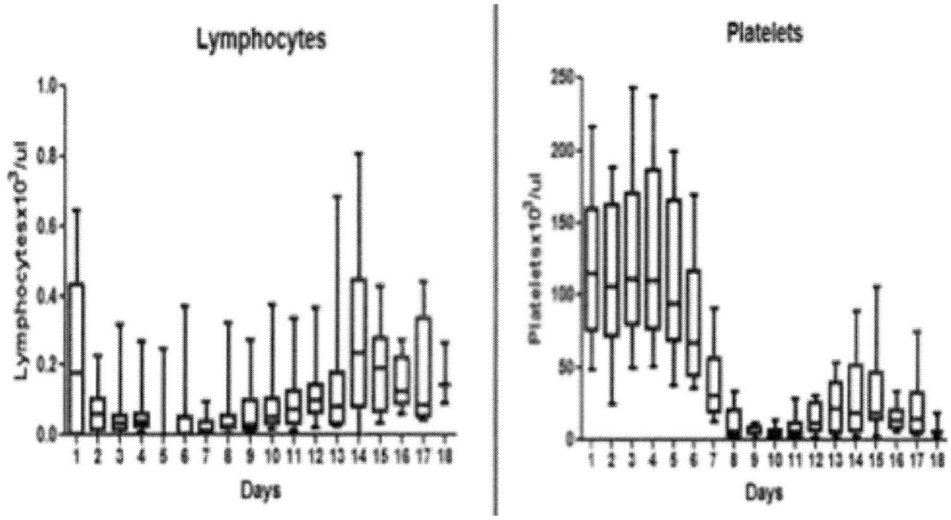

Figure 6. Disease-free interval and overall survival time based on relapse or no relapse before HCT

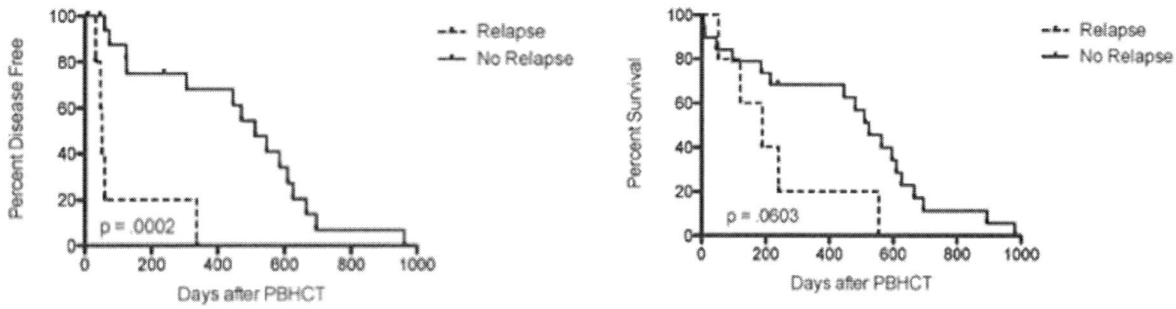

The median disease-free interval and overall survival (OS) of all dogs from the time of PBHCT was 271 and 463 days, respectively. In addition, 33% (5/15) dogs who underwent HCT before relapse remain in clinical remission.[18]

T-Cell Lymphoma

These 15 dogs received high-dose cyclophosphamide (500–750 mg/m^2 IV) as described previously[18], and underwent apheresis with either the Baxter-Fenwal CS-3000 Plus cell separator (1) or the Terumo/BCT cell separator (14). Greater than 2x10^6 CD34+ cells/kg were harvested from all dogs. TBI consisted of 10 Gy (3), 11 Gy (9 dogs), or 12 Gy (2 dogs). One dog received only 8 Gy due to a dose miscalculation. HCT and hematologic recovery were similar to that previously described in the cohort of B-cell lymphoma dogs.

Figure 7. Disease-free interval and overall survival time of dogs with TCL

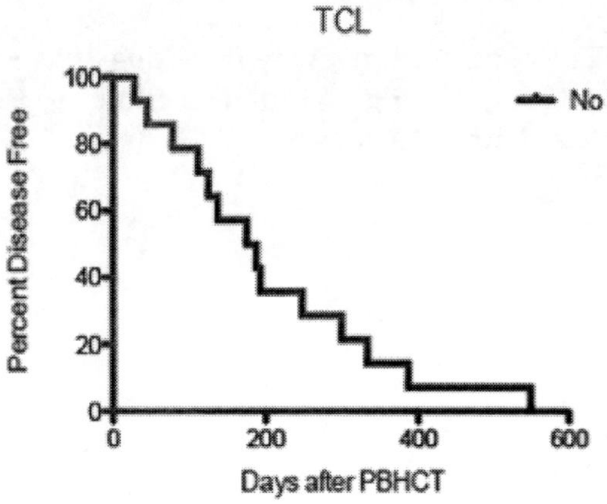

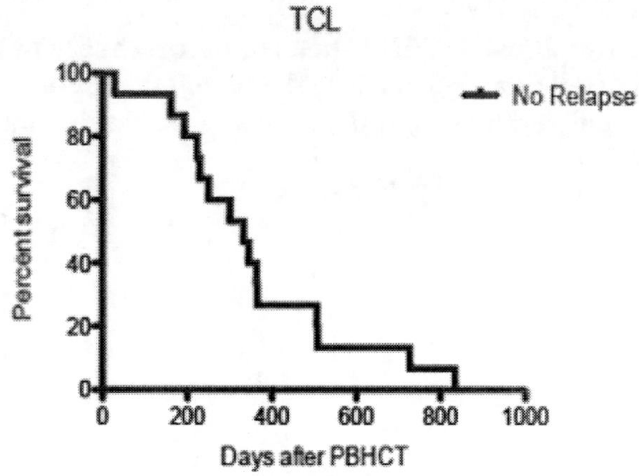

The median disease-free interval and overall survival (OS) of the 13 dogs transplanted (2 dogs died in hospital) in first remission from the time of HCT were 184 and 240 days, respectively. Two of 13 (15%) dogs are alive 4 yrs and 5 yrs post-HCT.[19]

Allogeneic HCT

We have performed 4 allogeneic HCTs where a DLA-matched donor was located who donated peripheral blood CD34+ cells after 5 days of rhG-CSF priming (many more allogeneic HCTs have been performed in private practices).

Zeke Ben-Dor: Dx Splenic LGL Leukemia via Flow Cytometry

Figure 8. Left: DLA-88 ad DRB1 matching of Zeke and donor (#3). Right: Zeke exhibited complete donor chimerism within 2 weeks after HCT

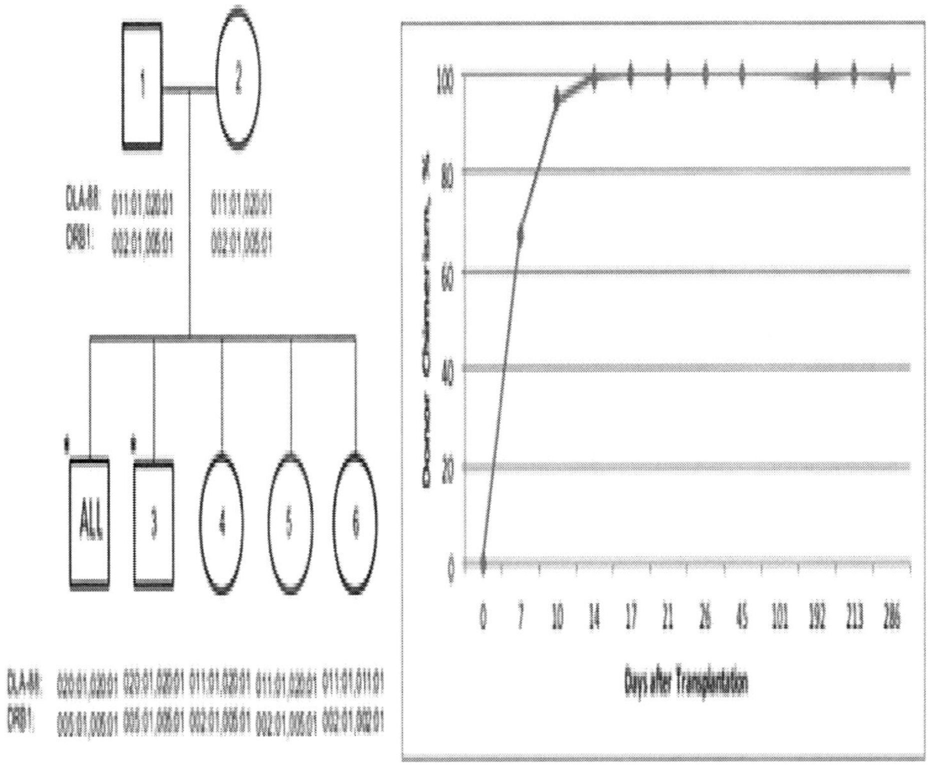

Zeke remains alive today (~5 yrs post-HCT) and is clinically normal.20 2/3 other allogeneic HCT dogs remain alive and in clinical remission >4 years post-HCT. One dog died of unrelated causes.

These initial results support the notion that canine patients with lymphoma can undergo autologous and allogeneic HCT in a clinical setting and become long-term survivors. Although we report an ~33% cure rate of canine B-cell patients (as defined as living >2 yrs post-HCT), we have much more to learn about the procedure itself and the long-term consequences of HCT. Therefore, more dogs need to be transplanted and more time needs to elapse before we can begin to determine the true cure rate (surviving >2 yrs) of lymphoma using this technique. High-grade T-cell lymphoma remains a challenging disease to cure, even with chemotherapy and HCT. Finally, we now know that dogs with long-term GI issues, such as food sensitivity secondary to chronic allergies, dogs with chronic pruritis, and dogs with long histories of vomiting/diarrhea before HCT are at an increased risk for transplant-related mortality.

Facilities that Perform HCT

North Carolina State University (Dr. Steven Suter)
VCA West LA, CA (Dr. Johnny Chretin)
Bellingham Veterinary Emergency & Critical Care, Bellingham, WA (Dr. Ed Sullivan)
MedVet, Worthington, Ohio (Dr. Lisa Fulton)

References

1. Lupu M *et al*. *Vet Comp Oncol*. 2007;5(1):14.

2. Ostrander EA *et al. Genome Res.* 2005;15:1706.
3. de Revel T *et al. Blood.* 1994;83:3795.
4. Sandmaier BM *et al. Blood.* 1996;87:3508.
5. Appelbaum FR. *Exp Hematol.* 1979;7:7.
6. Storb R *et al. Blood.* 1967;30:805.
7. Ratajczak MZ *et al. Leukemia.* 1998;12:942.
8. Morel F *et al. Blood.* 1996;88:3774.
9. Andrews RG *et al. Blood.* 1992;80:1693.
10. McSweeney PA *et al. Blood.* 1998;91:1977.
11. Buckner D *et al. Blood.* 1969;31:653.
12. Herbst EW *et al. Blut* [German]. 1975;30:265.
13. Storb R *et al. J Vet Intern Med.* 2008;22:74.
14. Suter SE. *J Vet Intern Med.* 2011;25:1406–1413.
15. Lupu M *et al. J Amer Vet Med Assoc.* 2006;228:728.
16. Burnett RC *et al. Vet Pathol.* 2003;40:32.
17. *Vet Pathol.* 2012 Mar;49(2):341–343.
18. Wilcox JL *et al. J Vet Intern Med.* 2012;26:1155–1163.
19. Warry EE *et al. J Vet Intern Med.* 2014;28:529–537.
20. Suter SE *et al. J Am Vet Med Assoc.* 2015;246:9.

acvim271

How Understanding DNA Methylation May Help Treat Lymphoma More Effectively

Jeffrey N. Bryan, DVM, MS PhD, DACVIM (Oncology)
Columbia, MO, USA

Introduction

First reported in 1949, methylcytosine in DNA sequences has become a central focus as an epigenetic mechanism of gene regulation in normal, chronic disease, and neoplastic disease processes. Methylcytosine is created post-replication by the addition of a methyl group to the 5 position of the cytosine ring when a cytosine is followed by a guanine (CpG) in sequence. The addition is made by DNA methyltransferases (DNMTs) using the methyl donor s-adenosyl methionine. DNMT1 is responsible for maintenance methylation by which parental strand methylation patterns are copied onto daughter strand CpGs post-replication. DNMT3a is responsible for placing methylation marks on CpGs under conditions of dynamic epigenetic change. Mutations of both proteins have been associated with alterations of both global genomic methylation and gene promoter-specific methylation associated with cancer. Across mammalian genomes, CpG dinucleotides are relatively rare sequential base pairs. Sporadically, CpG may be found in high density in regions called CpG islands that are frequently located in association with the 5′ control regions of genes. Hypermethylation of these islands was originally thought to result in persistent silencing of the associated gene. New work has demonstrated, however, that chromatin marks on histones are responsible for the active silencing of the gene, which can often be expressed even when methylated. These hypermethylated genes will revert to a silenced state in relatively short order when the signal for the histone changes recedes.[1] Recently, a third concept of CpGs has emerged. The CpGs residing in the lower density regions near CpG islands are considered to reside on the "shores" and "shelves" of those islands and appear to play a significant role in cell development and stem cell function. The first reported examination of DNA methylation in canine cancer was in 2003 when lymphomas were shown to be characterized by the same global loss of DNA methylation seen in human lymphomas.[2] Subsequently, the often hypermethylated *FHIT* gene was studied in canine lymphoma cell lines and found to be aberrantly expressed, but hypermethylation could not be documented to be the cause.[3] Conversely, the *DLC1* gene was shown to be significantly hypermethylated in canine lymphoma patient samples, but no effect on the expression of the gene could be identified.[4] Put most simply, epigenetic changes like DNA methylation are the mechanism by which cells establish persistent and stable phenotype while sharing a common genotype across all cells. Certain sets of genes in each cell are mostly permanently turned on or off to limit the set of proteins that are expressed. These mechanisms become disrupted in the course of carcinogenesis, changing the gene sets available to cancer cells and resulting in significant genetic and epigenetic plasticity.

Cancer has long been known to be a genetic disease. However, recent reports have identified at least as many genes (sometimes hundreds) that are dysregulated by gain or loss of DNA methylation as are mutated, deleted, translocated, or amplified.[5] In fact, some diseases, such as prostate cancer, are more frequently defined by characteristic hypermethylation in genes than by characteristic mutation. The degree to which epigenetic derangement characterizes some cancers is startlingly underscored by a series of experiments in which mouse melanoma nuclei were reprogrammed by nuclear transplantation into an oocyte and implantation for development or chimeric growth.[6] Cells with transplanted melanoma nuclei were able to participate as embryonic stem cells in growth and organ differentiation. Adult chimeras had a high rate of melanoma formation, but were otherwise normal, functional mice. These experiments demonstrate that epigenetic reprogramming of embryogenesis abrogates many of the changes that characterize cancer, raising the possibility that therapeutic induction of such reprogramming could lead to apoptotic death of the cells in patients with cancer. DNA methylation plays important roles in lymphocyte maturation, differentiation, and functional specialization. It is likely that derangement of these mechanisms are part of the epigenetic process of lymphomagenesis.

Over a thousand papers (1,311 reports in PubMed on 2/17/17) have now been published on DNA methylation in lymphoma. Of those, 11 are papers describing canine lymphoma. Significant progress has been made in understanding the role of DNA methylation in development, stem cell generation and reprogramming, and in gene expression in individual cancer types. However, these understandings have not yet translated into strong biomarker development for identification of risk, early diagnosis, or prognosis in many cancers. Nor have huge strides been made in the abrogation of cancer risk for individuals with known environmental and lifestyle risk factors. Because of the relatively shorter lifespan of dogs, the shared environment as human companions, and the relatively identifiable breed risk of many cancers, the comparative evaluation of DNA methylation changes in dog cancers would seem a fruitful pursuit to understand the interaction of environment, genes, epigenetic responses to environmental stresses, and the development of cancers. To date, hypermethylation has not been shown to regulate expression of the MGMT gene or the ABCB1 gene in canine lymphoma cell lines. DNA methylation changes could be identified and studies as markers of risk, early diagnosis, prediction of therapeutic response, prognosis, and testing of novel therapeutic interventions.

Epigenetic Progenitor Origin of Cancer

Recently, a model of carcinogenesis has been proposed that has been called the epigenetic progenitor origin of cancer.[7] Unlike the clonal genetic model of cancer development, this model suggests that progenitor cell populations in tissues are affected by environmental interactions that lead to an expanded pool of epigenetically altered progenitor cells with altered differentiation potential and polarity. On this background of expanded progenitor cells, a gatekeeper genetic mutation occurs characteristic of the particular type of tumor. This then becomes the primed source of a clonal cancer. Following this is a period of epigenetic and genetic plasticity that results in expansion of the clone and development of the hallmarks of cancer.

DNA Methylation Marks as Biomarkers

Because DNA methylation patterns are altered early in the process of carcinogenesis, it is likely that DNA methylation marks could serve as biomarkers of disease risk. In fact, recent work has demonstrated that differentially methylated regions of CpG island shores, CpG islands, and even intergenic regions reliably distinguish normal from neoplastic colon tissue.[8] Further, benign polyps had an intermediate frequency of methylation changes between normal and neoplastic epithelium. These same marks are reliably altered in several other types of cancer in a similar fashion to colon cancer and are enriched in regions differentially methylated during stem cell reprogramming during induced pluripotent stem cell formation, supporting the concept that it is a progenitor cell population that is initially affected in cancers.[8] If such regions are reliably affected, it stands to reason that their detection in tissue might serve as a harbinger of increased risk of full-blown carcinogenesis. Because of the difficulty and expense of longitudinal studies in human subjects, identification and validation of methylated DNA risk biomarkers in dogs could streamline the development of such assays in humans.

Biomarkers of Disease

Methylated DNA in tumor-associated genes can be detected in tissue samples as well as remote media such as plasma and urine. Gene-specific methylation has been identified to be strongly associated with large number of tumor types. *GSTP1* in prostate cancer. ADAM23 in breast cancer is frequently methylated and associated with a more aggressive, more highly metastatic form of the disease.[9] The company Exact Sciences has released a test of four methylation markers with a reported 100% sensitivity to detect colon cancer.[9] Methylated genes can be assayed in tissue biopsies to guide therapy decisions. When the drug metabolizing gene *MGMT* is methylated and silenced in glioblastomas and other brain tumors, human patients have a better response to temozolamide.[10,11] When unmethylated, the gene product is expressed which results in removal of DNA adducts and loss of drug efficacy. As mentioned earlier, this pattern of sensitivity and resistance is seen in canine lymphoma cell lines, but does not yet appear to be under control of methylation. Expansion of biomarker development in naturally occurring canine tumors will accelerate the application of such assays to human disease. The Comparative

Oncology Radiobiology and Epigenetics Laboratory at the University of Missouri has a project, funded by the Canine Health Foundation, identifying DNA methylation biomarkers in diffuse large B cell lymphoma of Golden Retrievers. We have identified distinct methylation marks that reliably separate lymphoma from normal lymphocytes. Further studies are underway to evaluate these marks in a population of dogs for screening and diagnostic application.

To date, methylation of a handful of genes has been studied in canine lymphoma cell lines and tissues. DNA methylation has been shown to regulate the expression of p16 in canine lymphoma cell lines and high-grade lymphoma samples.[12,13] The gene *TFPI-2* was identified to be hypermethylated in 19/23 canine DLBCL samples evaluated.(14) Since our initial report of hypermethylation of the *DLC1* gene, we have found that it is consistently hypermethylated in all canine DLBCL samples studied since (unpublished data). These represent the potential for biomarker discovery.

Epigenetically Active Therapy

DNA methylation and histone modifications can be modified with drug therapy, causing the re-expression of silenced genes and resulting in cell death and tumor regression *in vitro* and in mouse models of disease. However, these drugs have translated incompletely to the clinic. Currently, Vidaza (5-azacytidine) and decitabine (5-aza,2-deoxycitidine) (DNMT inhibitors) have been found to be useful in treating myelodysplastic syndrome in humans, resulting in transfusion independence.[9] Histone deacetylase (HDAC) inhibitors such as vorinostat (SAHA) have found application in a number of tumors with SAHA being approved for epitheliotropic lymphoma in humans. Clinical trials of DNMT inhibitors as well as HDAC inhibitors have been combined with conventional chemotherapy with demonstrated tolerability and better efficacy compared to chemotherapy alone.[9] Demethylating drugs active in human lymphoma cells have been shown to be active in canine cells as well (FLESNER). At the University of Missouri, we are studying demethylation therapy prior to doxorubicin-based therapy for canine lymphoma. The hypothesis of this study is that pre-treatment with demethylating drugs will sensitize lymphoma cells to cytotoxic chemotherapy and prolong remissions compared to doxorubicin alone. Given the delicate balance of gene control across many cell types within the body, it is likely that generalized inhibition of DNMTs or HDACs is not desirable, even for individuals with cancer, and particularly for a protracted course of therapy. Brief courses of inhibition to sensitize to chemotherapy are attractive for better management of disease. Ideally, inhibition of these molecules will be targeted selectively to cancers, avoiding these effects on the rest of the somatic cells. Animal models of naturally occurring cancers could model such targeted therapies and be used to identify potentially optimal drug combinations.

Identification of DNA Methylation

A multitude of techniques exist to query chromatin for epigenetic marks related to cancer.[15] Individual genes can be studied using bisulfite techniques to identify methylation changes that alter gene expression or serve as a mark typical of a particular histology. DNA methylation arrays exist for humans and have been designed for dogs that allow elucidation of DNA methylation patterns by isolating a pool of fragmented DNA enriched for methylcytosine-containing regions. Hybridization with a DNA methylation array identifies the genomic regions containing significant methylcytosine compared to normal tissue of the same origin. Next generation sequencing techniques now allow a pool of fragmented DNA enriched for methylcytosine to be sequenced. This DNA can be in native form, identifying gene regions with higher methylcytosine content, or can be bisulfite sequenced to determine the methylcytosine patterns at a single-base level. The advent of the sequencing of the canine genome provides the bioinformatic framework to identify the sequences of desired genes or to locate reads from large-scale sequencing experiments. The output from these experiments can then be used to compare methylation patterns across species, potentially identifying the unifying changes that are most at the root of these diseases. These more recent, high throughput, high output methods yield massive quantities of data. Such output almost invariably yields statistically significant results. The challenge of this field has been identifying clinically significant results. The comparative evaluation of similar cancers across species promises to identify the subset of information that represents the underlying malfunctions in gene regulation leading to cancer. Identifying such marks could lead to preventive interventions in high risk groups, earlier cancer diagnoses, markers of prognostic and therapeutic import, and sensitive markers of residual disease. Companion animals offer highly relevant models of naturally occurring cancers that can be evaluated at much less expense in a much shorter time-frame than humans with similar disease. Exploiting these models could lead to significant progress in understanding these important mechanisms of disease.

References

1. Raynal NJ *et al. Cancer Res*. 2012.
2. Pelham JT *et al. Res Vet Sci*. 2003;74(1):101–4.
3. Hiraoka H *et al. J Vet Med Sci*. 2009;71(6):769–77.
4. Bryan JN *et al. BMC Genet*. 2009;10(1):73.
5. Jones PA, Baylin SB. *Cell*. 2007;128(4):683–92.
6. Hochedlinger K *et al. Genes Dev*. 2004;18(15):1875–85.
7. Feinberg AP *et al. Nat Rev Genet*. 2006;7(1):21–33.
8. Hansen KD *et al. Nat Genet*. 2011;43(8):768–75.
9. Costa FF. *Cancer Manag Res*. 2010;2:255-65.:255–65.
10. Hegi ME *et al. N Engl J Med*. 2005;352(10):997-1003.
11. Donson AM *et al. Pediatr Blood Cancer*. 2007;48(4):403–7.
12. Fujiwara-Igarashi A *et al. Res Vet Sci*. 2014;97(1):60–3.
13. Fujiwara-Igarashi A *et al. Vet J*. 2014;199(2):236–44.
14. Ferraresso S *et al. PloS One*. 2014;9(4):e92707.
15. Bryan JN *et al. Cancer Ther*. 2008;6:137–48.

acvim272

Chemoimmunotherapy for Canine Lymphoma: Evidence for Efficacy 2017

Gregory K. Ogilvie, DVM, DACVIM (Small Animal Internal Medicine, Oncology), DECVIM-CA (Oncology)
Carlsbad, CA, USA

INTRODUCTION

Research to develop safe, effective, and affordable treatments for lymphoma continues. This is an ethical and reasonable thing to do; this research makes sense financially because lymphoma is the most common hematopoietic tumor in the dog. The nonprofit organization, Canine Lymphoma Education and Awareness and Research (CLEAR) estimates that only 3% of all dogs with lymphoma are treated. This is shocking to many because most agree that many types of lymphoma respond well to existing chemotherapy treatments resulting in an excellent quality of life and happy clients. Regardless, the duration of remission for most patients is limited to a year or less. Hence, the continuing effort to break this one year barrier continues. This is exactly the same problem physicians had as they treated human patients with lymphoma. The treatment of lymphoma is evolving due to recent research that affirms that, like in people, there are many types of lymphoma, many with a unique biological behavior, response to therapy and thus, prognosis. The disease is usually not painful. It responds very well to systemic chemotherapy yet only one cytotoxic agent has been approved for the treatment of lymphoma. The U.S. Food and Drug Administration announced recently that the lymphoma drug, Tanovea-CA1, is approved for use in canine patients and will be available to veterinarians nationwide in the very near future. Regardless, it is estimated that this new medication will likely only provide modest benefits. Something more must be done. If the experience in the treatment of people with lymphoma is a guide, only the use of immunotherapeutics to traditional chemotherapy will likely enhance quality of life while improving the duration of remission and the overall survival time. The purpose of this brief manuscript is to review some key points about canine lymphoma and its treatment, including the amazing new world of chemoimmunotherapy.

OVERVIEW

As stated previously, canine lymphoma is a common disease worthy of continued research, especially in the field of chemoimmunotherapy. Lymphoma accounts for approximately 90% of canine hematopoietic tumors in the dog. The development of a single immunotherapeutic effective for all dogs with lymphoma is nearly impossible given the multiple variables that coexist in this disease. Genetics and the environment have both been cited as factors that can influence the prevalence and possibly the type of this disease in the dog.[1-5] Some believe this disease is increasing in prevalence, with Boxers, Bulldogs, Bull Mastiff, Golden Retrievers and Rottweiler being over represented in several studies.[1-6] Boxers may have more T cell lymphoma than the general population.[7] Golden Retrievers have been reported to have three times and Rottweilers are 2.5 and 6 times more likely to develop lymphoma than other breeds.[8-13] C type virus particles have been observed in cases of canine lymphoma and leukemia although it is unknown if these viruses are innocent passengers or involved in causality.[14,15] Similarly, factors in the immediate environment have been implicated such as environmental factors, industrial paints, solvents, herbicides such as 2,4-D, exposure to electromagnetic fields, and diseases of the immune system such as immune thrombocytopenia.[14-24]

Lymphoma is uncommon in young dogs; affected dogs are typically middle-aged. One controversial study suggested gender and neutering is a predisposing factor for developing lymphoma, at least in Golden Retrievers. This is the reason for the Morris Animal Foundation Lifetime Study in this breed. In that study, the authors stated that nearly 10% of male dogs neutered early in their lives were diagnosed with lymphoma, 3 times more than intact males.

Diagnostics[1-24]

When evaluating an animal for treatment of lymphoma, it is important to not only obtain a definitive diagnosis via cytology, histopathology, and importantly by flow cytometry but also assess the general health of the patient by clinical examination and ancillary diagnostics. The clinical stage of a dog with lymphoma is determined in part by performing a physical examination. All lymph nodes should be palpated; accurate measurements should be recorded to help the clinician assess response to therapy. Careful attention should be directed to examine the eyes for any evidence of iritis, uveitis, or hyphemia. A delayed identification of this finding can result in blindness. Other, extranodal sites should be examined as well. Each dog is clinically staged based on the results of physical examination, clinical laboratory testing (i.e., complete blood count [CBC], biochemical profile, urinalysis, and bone marrow cytology), and imaging procedures (i.e., radiography and ultrasonography). In each case, a biopsy or cytology and flow cytometry of the affected lymphoid tissue is necessary to confirm the diagnosis and determine the cell type (histologic grading and immunophenotyped via flow cytometry). The purpose of doing these tests is to confirm the diagnosis and extent of lymphoma, the health of the patient, and to determine which therapy is likely to be most effective which should result in the best quality of life. By knowing where the lymphoma begins, it is possible to subsequently reassess the patient to determine the response to therapy.

TREATMENT[1-18]

Treatment for malignant lymphoma must be systemic in nature since it is a multi-system disease. Chemotherapy is most frequently utilized and it is essential to downsize the disease if immunotherapy is to be used as well. Since most immunotherapeutics require an intact immune system to be effective, chemotherapy must be given in such a way as to ensure that there is minimal disease and an intact, functional immune system. Untreated, dogs affected with malignant lymphoma live an average of only six weeks once a diagnosis has been made. With chemotherapy, dogs can survive for 6–10 months with an excellent quality of life. Since chemotherapy is a key component of the chemoimmunotherapy approach, a brief summary is outlined below.

Chemotherapy[15-32]

Malignant lymphoma is one of the most responsive forms of cancer presented to the practitioner. With appropriate chemotherapy or chemoimmunotherapy protocols, nearly 90% of animals placed on therapy should reach complete remission. Of those that reach remission, approximately 80–90% will maintain a reasonable (>6 months) time frame of excellent quality life. Cost is a concern, especially when insurance is not available or not to be used; however, it is within reasonable financial reach of many clients, thus allowing therapy to be possible.

Although marginally effective, prednisone is inexpensive and often used in combination with other drugs to treat lymphoma. With prednisone therapy, the average pet lives 2 months. One-third of the dogs and cats treated with prednisone will go into complete remission, one-third will go into partial remission, and one-third will not respond at all.

Doxorubicin is one of the most effective single agent treatments for lymphoma in dogs.[29] Of the dogs treated with doxorubicin, 70–80% developed a complete and partial remission. The duration of remission is approximately 9 months. Dogs treated with doxorubicin and then switched to COP (cyclophosphamide, Oncovin, prednisone) had a higher second remission rate compared to those started on COP and then switched to doxorubicin.

The COP protocol is effective for inducing a remission in 75% of dogs with lymphoma.[30] A median duration of remission of 6 months is commonly seen. Approximately 20% of the dogs treated with the COP regimen are in remission at 1 year. The addition of doxorubicin to the COP regiment resulted in longer remission time (7 months versus 6 months). A complete remission was attained in 84% of dogs treated with COPA. An additional 7% achieved a partial remission. Twenty-two percent of dogs treated with the COPA protocol were in remission at one year.

Garrett *et al.*[28] compared a maintenance-free chemotherapy protocol based on CHOP to a similar protocol with a maintenance phase for the treatment of canine lymphoma. Fifty-three dogs with multicentric lymphoma were treated with a 6-month modified version of the University of Wisconsin (UW)-Madison chemotherapy protocol. Disease-free interval (DFI) and survival were compared to a historical control group of 55 dogs treated with a similar protocol with a prolonged maintenance phase.

Remission rate for the study dogs was 94.2%. The remission and survival between the 2 groups did not differ significantly. Thus, we, and others, believe the 6-month chemotherapy protocol based on CHOP with no maintenance phase provided is equal to a similar protocol with a prolonged maintenance phase. CHOP is considered the standard of care for dogs with lymphoma. It is often used to precede either active or passive immunotherapy.

Chemoimmunotherapy[32-36]

One monoclonal antibody developed for the treatment of cancer in humans or animals is canine lymphoma/monoclonal antibody 231 (CL/MAb 231; Synbiotics Corporation, San Diego, California), which recognizes canine lymphoma cells.[38,39] While no longer commercially available, it was determined to mediate antibody-dependent cellular cytotoxicity against a canine lymphoma cell line and has been reported to prolong remission duration when used in combination with chemotherapy in dogs with lymphoma.[39] The median survival time of dogs treated with the monoclonal antibody and chemotherapy was 591 days, as compared to historical controls, which had a median survival time of 189 days.

Rituximab, a chimeric murine/human IgG monoclonal antibody that targets the B-cell antigen CD20 is now considered the standard of care for the management of human B-cell lymphoproliferative disorders including diffuse large B-cell lymphoma (DLBCL).[33,34] This aggressive form of DLBCL has many similarities to the most common form of lymphoma in dogs.[33-35] In both species, DLBCL generally responds well to combination chemotherapy, however, the duration of remission is often limited. The addition of the species-specific antibody rituximab in combination with chemotherapy to treat DLBCL in people is associated with little increase in toxicity, yet enhanced efficacy. The increased frequency in the use of rituximab in the management of lymphoid malignancies has grown to the point that it presented a huge logistical challenge for the human medical system because the antibody was originally given by slow infusion, a time intensive procedure. Therefore, more cost and time effective procedures have been developed to infuse the antibody safely and effectively.[36] Thus, the availability of a similar antibody that targets the canine CD20 antibody combined with the lessons learned on how to incorporate this type of therapy safely and effectively could have a powerful impact for the management of DLBCL.

The discovery, development, and commercial production of multiple rituximab-like chimeric murine/canine monoclonal antibodies to specifically and precisely target the B-cell antigen CD20 in dogs has gone on from several investigators. A group from Novartis Research Foundation and Elanco Animal Health[36] developed a rituximab-like anti-CD20 antibody intended as a candidate treatment for canine B-cell lymphoma. Their antibody 1E4-cIgGB was selected for further testing and development as an agent for the treatment of canine B-cell lymphoma. A group from the University of Edenborough and elsewhere[37] has developed a novel anti-CD20 monoclonal antibody (NCD1.2) that binds both human and canine CD20. They determined that their recombinant anti-CD20 scFv might form a useful tool for evaluation in bioconjugate-directed anti-CD20 immunotherapies in comparative medicine. Ito and Modiano's group[38] developed a novel anti-canine CD20 mAb designated as 6C8. 6C8 recognized the extracellular domain of canine CD20 and showed high-affinity binding to canine CD20 in solution, as well as in its native conformation on canine B-cells. Our group[39] developed a set of molecules including one that appeared initially to have a complex mechanism of action that may or may not also target B-cell antigen C20. This construct is in the early phases of being evaluated by us and others. If any of these constructs is confirmed to be safe and effective, then the potential to benefit dogs with a long list of benign or malignant diseases that are caused by B-cell lymphocytes that express CD20 is high. That is exactly what occurred with the development and subsequent use of rituximab. Our group[39] therefore embarked on this double blind, randomized, placebo controlled study to test the hypothesis that the rituximab-like chimeric murine/canine monoclonal antibody is safe and effective for the treatment of CD20 positive high or intermediate grade B cell lymphoma in the dog. The initial phase of our work showed that the antibody was safe and effective when used alone to treat dogs with B cell lymphoma. Remissions were short but efficacy was confirmed. In a subsequent study, we enrolled dogs into a prospective, randomized, double blind, placebo controlled clinical trial. All dogs were treated with one 4-week cycle of an L-CHOP chemotherapy protocol followed by either the canine lymphoma antibody or a placebo. Once remission was lost, a single dosage of doxorubicin was given followed by the antibody.

Assessment of antitumor efficacy was based on modified Response Evaluation Criteria in Solid Tumors (RECIST) guidelines and toxicity based on criteria delineated by Veterinary Cooperative Oncology Group's common terminology criteria for adverse events (VCOG-CTCAE). The median proliferation free interval and overall survival time respectively of dogs with high or intermediate grade B cell lymphoma treated with one cycle of the L-CHOP protocol followed by the antibody or the placebo was much longer in the antibody arm compared to the placebo arm. Toxicities that were observed in this protocol were mild and restricted to the L-CHOP cycle with no significant adverse events in the antibody or placebo arms. On the contrary, there was a significant improvement in client perceived quality of life parameters in the antibody, but not the placebo-treated dogs. Other studies have completed patient accrual by us and an Aratana-funded group patient assess this construct when used to treat canine B cell lymphoma using a two cycle CHOP protocol (mini-CHOMP study).[40] Another study that has completed patient accrual is a double-blind study to assess the B cell construct vs. a placebo during a full CHOP protocol. We and others were involved in the evaluation of a second monoclonal antibody has received the first conditional licensure for use as an aid to treating T cell lymphoma in dogs that express CD-52. These CD-52 antigens are expressed on most dogs with T cell lymphoma and thus, a monoclonal antibody directed to this epitope can bind and initiate a specific cytotoxic event. The initiation of remission prior to the administration of the antibody with combination chemotherapy will increase the probability of inducing a sustained response. This antibody was purchased by Aratana during the acquisition of Veterinary Therapeutics. This company will likely market the product. Two interesting studies were done by an Italian group headed by Dr. Maconato.[35] The aim of the first non-randomized controlled trial was to compare time to progression (TTP), lymphoma-specific survival (LSS), and safety of an autologous vaccine (consisting of hydroxyapatite ceramic powder and heat shock proteins purified from the dogs' tumors, HSPPCs-HA) plus chemotherapy versus chemotherapy alone in dogs with newly diagnosed, clinically advanced, histologically confirmed, multicentric indolent B-cell lymphoma. The vaccine was prepared from dogs' resected lymph nodes and administered as an intradermal injection. The results of the first study suggested that vaccination with HSPPCs-HA may produce clinical benefits with no increased toxicity, thereby providing a strategy for enhancing chemotherapy in dogs with advanced indolent lymphoma. The second study involved 19 dogs with naturally occurring diffuse large were entered into a prospective randomized placebo-controlled double-blinded trial of HSPPCs-hydroxylapatite plus chemotherapy versus chemotherapy alone. The results of this trial demonstrated that the autologous vaccine tested here is safe and efficacious in prolonging TTP and LSS in dogs with DLBCL when used in combination with dose-intense chemotherapy. Based on these results, additional evaluation of this novel therapeutic strategy is warranted in human DLBCL. A novel xenogeneic murine CD20 DNA therapeutic vaccine has been developed and marketed by Merial for use in dogs with B-cell lymphoma.[43] The vaccine was conditionally licensed by the USDA in September 2015. Utilizing a similar technology platform as ONCEPT® for canine melanoma, the anti-CD20 vaccine is administered using a needle-free transdermal device. A prospective, multicenter clinical trial in dogs with B-cell lymphoma receiving CHOP-based chemotherapy followed by 4 doses of anti-CD20 vaccine has been described, but not published.

References

1. Dorn CR, *et al. Am J Vet Res.* 1967;28:993–1001.
2. Schneider R. *J Natl Cancer Inst.* 1983;70:971–977.
3. Teske E. *Vet Q.* 1994;16:209–219.
4. Dobson JM, *et al. J Small Anim Pract.* 2002;43:240–246.
5. Clancy E, *et al.* In: Proceedings of the Veterinary Cancer Society 19th Annual Conference; 1999: 71.
6. Priester WA. *J Natl Cancer Inst.* 1967;39:833–845.
7. Lurie DM, *et al. Vet Comp Oncol.* 2004;2:171–175.
8. Onions DE. *J Natl Cancer Inst.* 1984;72:909–912.
9. Edwards DS, *et al. Vet Comp Oncol.* 2003;1:200–206.
10. Moore AS, *et al. J Vet Intern Med.* 2001;15:348–354.
11. Zemann BI, *et al. J Vet Intern Med.* 1998;12:465–470.
12. Jagielski D, *et al. J Vet Med A Physiol Pathol Clin Med.* 2002;49:419–424.

13. Teske E, *et al. Vet Q.* 1994;16:134–136.
14. Onions D. *Eur J Cancer.* 1980;16:345–350.
15. Tomley FM, *et al. Brit J Cancer.* 1982;45:644.
16. O'Brien DJ, *et al. Prev Vet Med.* 1999;42:1–15.
17. Gavazza A, *et al. J Vet Intern Med.* 2001;15:190–195.
18. Hayes HM, *et al. J Natl Cancer Inst.* 1991;83:1226–1231.
19. Reynolds PM, *et al. Cancer Epidemiol Biomarkers Prev.* 1994;3:233–237.
20. Carlo GL, *et al. Regul Toxicol Pharmacol.* 1992;16:245–252.
21. Kaneene JB, *et al. Vet Human Toxicol.* 1999;41:164–170.
22. Hayes HM, *et al. Environ Res.* 1995;70:119–125.
23. Reif JS, *et al. Am J Epidemiol.* 1995;141:352–359.
24. Keller ET. *Cancer.* 1992;70:2334–2337.
25. Torres de la Riva G, *et al. PLoS One.* 2013;8(2):e55937.
26. Burnett RC, *et al. Vet Pathol.* 2003;40:32–41.
27. Morrison-Collister KE, *et al. Vet Comp Oncol.* 2003;1:180–190.
28. Garrett LD, *et al. J Vet Intern Med.* 2002;16:704–709.
29. Valerius KD, *et al. J Am Vet Med Assoc.* 1997;210:512–516.
30. Cotter SM, *et al. J Am Anim Hosp Assoc.* 1983;19:159–165.
31. Rosales C, *et al. Cell Immunol.* 1988;115:420–428.
32. Jeglum KA. Monoclonal antibody treatment of canine lymphoma. In: Proceedings of the Eastern States Veterinary Conference; 1992: 222–223.
33. Vose JM, *et al. J Clin Oncol.* 2001;19(2):389–397.
34. Vose JM, *et al. Leuk Lymphoma.* 2005;46(11):1569–1573.
35. Marconato L, *et al. Hematol Oncol.* 2013;31(1):1–9.
36. Sehn LH, *et al. Blood.* 2007;109(10):417.
36. Rue SM, *et al. Vet Immunol Immunopathol.* 2015;164(3–4):148–59.
37. Jain S, *et al. PLoS One.* 2016;11(2):e0148366.
38. Ito D, *et al. Leuk Lymphoma.* 2015;56(1):219–25.
39. Ogilvie GK, *et al.* In: Proceedings of the Veterinary Cancer Society. St. Louis, MO; 2014 (abstract).
40. Rodriguez CR, *et al.* ACVIM Forum. Indianapolis, IN; 2015 (abstract).
41. http://aratana.investorroom.com/2015-09-24-Aratana-Therapeutics-Provides-Product-Updates.
42. Rue SM, *et al. Vet Immunol Immunopathol.* 2015;164(3–4):148–159.
43. http://c.ymcdn.com/sites/www.nysvms.org/resource/collection/824A08A0-F7AF-4D95-A850-C088F60C9C8E/NYSMVS_SeptOct_proof_06.pdf

acvim273

Diverse DNA Damage Response Mechanisms in Cancer: From Tumor Suppression to Therapeutic Targets

Kelly R. Hume, DVM, DACVIM (Oncology)
Ithaca, NY, USA

Cellular DNA damage response (DDR) mechanisms exist to protect against genomic instability and phenotypes such as aging and cancer. Sources of DNA damage include metabolic by-products like reactive oxygen species and estrogen and cholesterol metabolites, as well as environmental exposure to ultraviolet light. Also, spontaneous hydrolysis reactions can create abasic sites and deamination.[1] Molecular DDR networks first sense or recognize the damage, and then transduce signals to appropriate effector molecules. Possible outcomes for an executed response can include cell cycle arrest, DNA repair, and apoptosis, depending on the severity of the damage. Adult stem cells can also activate terminal differentiation in response to DNA damage.[2] Incorrectly repaired or unrepaired DNA, if allowed to persist, can result in DNA "scars" that can foster further mutations, chromosomal aberrations, cellular transformation and cancer development.[3] Somewhat paradoxically, cancer cells with DDR defects can have enhanced resistance or sensitivity to DNA-damaging ionizing radiation or cytotoxic drugs, depending on the particular defect.

The serine/threonine kinases ATM, ATR, and DNA-PKcs are the upstream signaling factors responsible for inducing the DDR. All three kinases are members of the PI3 (phosphatidylinositide3-kinases)-like kinases family. Successful activation of these kinases is dependent on the type of damage that occurs and on appropriate localization of specific protein complexes to the site of damage. ATM and DNA-PKcs respond to double strand breaks (DSBs), whereas ATR responds to single-stranded (ss) DNA breaks, replication errors (with exposed ssDNA) and DNA adducts. Cross talk occurs and it is not always an all or none phenomenon. Table 1 indicates proteins important in the recruitment of each kinase.

Table 1. Proteins important for recruitment of PI3-like kinases to sites of damaged DNA

PI3-like kinase	Proteins important for recruitment to damaged DNA
ATM	MRE11, RAD50, NBS1
ATR	RPA, HUS1, RAD1, RAD9, TOPBP1, ATRIP
DNA-PKcs	KU70, KU80

All three kinases can phosphorylate H2AX (γH2AX) at serine 139, a process long recognized as one of the earliest hallmarks of DNA damage.[4] Substrates of ATM include the cohesin SMC1 and the effector kinase CHK2. Other important targets include BRCA1 and 53BP1.[5] CHK1 is the effector kinase of ATR signaling. Substrates of DNA-PKcs include XRCC4, ARTEMIS, XLF, and DNA-LIGASE IV.[6] Members of the ATM pathway have long been recognized as tumor suppressors. Similarly, complete loss of DNA-PKcs is also associated with a high incidence of lymphoma.[7] How members of the ATR pathway influence tumorigenesis and cancer biology has been less clear, as these are essential genes that cannot be simply knocked out and studied in mouse models. Expression is upregulated in many cancers, and they are generally regarded as pro-tumorigenic, although gene dosage effects can occur.

Following DNA damage, cell cycle delay is necessary to facilitate DNA repair and cell survival. These stopping points are referred to as the G1 (G1/S), G2 (G2/M), and intra-S checkpoints.[5] P53 is important in the G1 checkpoint, whereas the G2 checkpoint relies on inhibition of the CDK1-CYCLINB complex. Fork stabilization and origin suppression are features of the intra-S checkpoint. ATM signaling has a role in all three checkpoints, whereas CHK1 is an effector kinase in the G2 and intra-S checkpoints.

Homologous recombination (HR) and non-homologous end joining (NHEJ) are the repair pathways activated in response to DSBs, whereas nucleotide excision repair (NER), mismatch repair (MMR), and base excision repair (BER), are pathways used to respond to other DNA defects. HR requires an intact

DNA template for repair, and thus is only possible during S and G2 phases of the cell cycle. ATM is the signaling kinase involved in this repair pathway. BRCA1 is important in DNA end resection at the site of the DSB. BRCA2 is necessary for localization of the RAD51 recombinase to the newly exposed ssDNA. RAD51 binding is needed for invasion of the homologous double helix. Once strand invasion occurs, DNA polymerases can synthesize new DNA based on the homologous template strand. Finally, DNA ligases and endonucleases are required in order to resolve the complex structures that arise.[8] In contrast to HR, NHEJ can occur throughout all phases of the cell cycle and is thus the prominent DSB repair pathway. Because DNA end groups at the break site may require processing prior to ligation, nucleotides can be lost from either side of the break, making NHEJ potentially error prone.[9] DNA-PKcs is critical to activation of this repair pathway. Enzymes that may be responsible for processing DNA ends include ARTEMIS, DNA polymerases μ and λ, PNK, and WRN. This processing is necessary to remove blocking end groups, fill in gaps, and/or remove damaged DNA or secondary structure elements surrounding the break so that the broken DNA strands can be pieced back together. Ligation of the processed DNA ends requires DNA LIGASE IV and XRCC4. Due to the nature of the initial DNA lesions, ATR signaling is involved in NER and MMR. Interestingly, the response to base lesions seems to involve ATM and not ATR signaling, although various pathway members are known to interact with BER proteins independent of checkpoint signaling.[10]

Apoptosis is a consequence of extensive DNA damage; this can occur through either the Fas pathway or the mitochondrial apoptosis pathway.[11] For this to happen, the DDR in the nucleus must activate the apoptosis machinery in the cytoplasm. ATM, ATR, CHK2 and CHK1 all have the ability to phosphorylate and stabilize P53, albeit at different residues. P53 has the ability to both upregulate pro-apoptotic genes such as *BAX* and *PUMA* as well as anti-apoptotic genes like *DDB2*, *MGMT*, and *MSH2* that encode DNA repair proteins. Inhibition of transcription by bulky DNA lesions can also trigger apoptosis. Another regulatory step exists in that ATM can complex with NEMO in the nucleus, and then passage to the cytoplasm to ultimately affect apoptosis signaling through interactions with the NF-κB pathway.[12]

Tumors exhibit extreme genomic instability, causes of which include replication stress, telomere shortening, oncogene activation, and increased reactive oxygen species. In contrast to normal cells, tumor cells may continue to replicate in the presence of significant DNA damage, and thus can be resistant to damage associated with cytotoxic chemotherapy and ionizing radiation. In prostate cancer, copy number aberrations in the DDR gene *NBN*, which encodes NBS1, are associated with relapse following radiation therapy.[13] Interestingly, the effect of genomic instability is magnified by intratumoral hypoxia, even in patients treated with radical prostatectomy.[14] High expression of γH2AX is associated with shorter overall survival in early non-small cell lung cancer.[15] In breast cancer, the presence of BRCA1-, RAD51, or γH2AX foci before treatment is inversely correlated to response to epirubicin chemotherapy regimens.[16] On the other hand, tumor cells that have specific DNA repair defects may be exquisitely sensitive to certain therapeutics. For example, cancers with *BRCA1/2* mutations and thus defective HR are very responsive to platinum agents and PARP (poly ADP ribose polymerase) inhibitors. (PARP is an enzyme involved in repair of ss breaks; it is necessary for the release of DNA repair proteins.) The synthetic lethality of HR defects and PARP inhibition has been particularly important in breast, ovarian, and prostate cancer. Deficiencies in ATM also seem to sensitize cancer cells to PARP inhibition.[17] Screening for mutational signatures, rather than specific defects may help identify tumors that have "BRCAness," meaning similar phenotypic defects in HR.[18] DNA repair defects do seem to segregate by cancer type; for example, HR defects in ovarian cancer, MMR defects in colon and endometrial cancer, and NER defects in bladder cancer.[19] Defects in HR and NER seem to sensitize to platinums, whereas topoisomerase II sensitivity is associated with HR and NHEJ defects. Inhibitors of various DDR components have been developed and are currently being evaluated in a variety of human clinical trials. These include inhibitors of ATM, ATR, DNA-PKcs, and CHK1. Inhibitors of MRE11, RAD51, and WEE1 (a kinase in the G2 checkpoint) are in preclinical development. A variety of CDK inhibitors are also in preclinical development. Some of these agents do have single agent activity, but there is also a lot of motivation to use them to enhance chemo- and radio-sensitivity. Synergism with mTOR inhibition and CHK1 inhibition has also been noted.[20] Successful use of these therapies and other strategies targeting DNA repair defects will ultimately require appropriate biomarkers of response. Investigation of these mechanisms in veterinary cancers is in its infancy.

References

1. Hoeijmakers JH. *NEJM*. 2009;361:1475–1485.
2. Oh J, *et al. Nat Med*. 2014;20:870–880.
3. Matt S, Hofmann TG. *Cell Mol Life Sci*. 2016;73:2829–2850.
4. Scully R, Xie A. *Mutation Res*. 2013;750:5–14.
5. Smith J, *et al. Adv Cancer Res*. 2010;108:73–112.
6. Jette N, Lees-Miller SP. *Prog Biophys Mol Biol*. 2015;117:194–205.
7. Reiling E, *et al. PLoS One*. 2014;9:e93568.
8. Lord CJ, Ashworth A. *Nature Rev Cancer*. 2016;16:110–120.
9. Mahaney BL, *et al. Biochem J*. 2009;417:639–650.
10. Chou W-C, *et al. EMBO J*. 2008;27:3140–3150.
11. Roos WP, Kaina B. *Cancer Letters*. 2013;332:237–148.
12. Brzoska K, Szumiel I. *Mutagenesis*. 2009;24:1–8.
13. Berlin A, *et al. Oncotarget*. 2014;5:11081–11090.
14. Lalonde E, *et al. Lancet Oncol*. 2014;15:1521–1532.
15. Matthaios D, *et al. Onco Targets Ther*. 2012;5:309–314.
16. Asakawa H, *et al. Breast Cancer Res*. 2010;12:R17.
17. Williamson CT, *et al. Mol Cancer Ther*. 2010;9:347–357.
18. Nik-Zainal S, *et al. Nature*. 2016;534:47–54.
19. Stover EH, *et al. Clin Cancer Res*. 2016;22:5651–5660.
20. Massey AJ, *et al. Mol Oncol*. 2016;10:101–112.

acvim274

Molecular Diagnostics in Veterinary Oncology

Chand Khanna, DVM, PhD, DACVIM (Oncology), DACVP (Hon)
Washington, DC, USA

Overview

Novel molecular diagnostics are rapidly being delivered to the field of veterinary medicine and veterinary oncology. These molecular diagnostics, most commonly applied to oncology and inflammatory conditions, have most recently emerged from the sub-discipline of pathology referred to as molecular pathology. This sub-discipline is based on the study of disease through the interrogation of molecules (often nucleic acids) within tissues or other biospecimens. These diagnostics rely less on conventional anatomic/clinical pathology and biochemistry, and are more closely aligned with the scientific platforms of molecular biology, and genomics. Nonetheless the integration of conventional anatomic/clinical pathology skills are necessary for the successful development of such novel diagnostics, as clinically informative tests. An over-riding expectation linked to this field of molecular diagnostics is that more accurate diagnosis is possible when the diagnosis is based on the detection of the true drivers of a disease rather than cellular or biochemical responses to a driver event. This connection to a disease driver may suggest a greater opportunity to design and deliver tests beyond simple diagnostics to now include predictors of treatment response and disease progression, the susceptibility of individuals to develop specific disorders and tolerate specific medicines (pharmacogenomics). The detection of such drivers often includes molecular techniques that allows detection of disease markers in progressively smaller sample volumes and from samples with limited prospective preservation (as in, forensic pathology).

Assay Platforms

The opportunity to develop and now deliver novel molecular diagnostics, comes from a convergence of the release of the first publicly available canine genome in 2004, improved annotation of both the canine and feline genomes, and the development of novel DNA sequencing (i.e., next-generation sequencing), which collectively now provide a broader spectrum of methodologies to be delivered on a reasonable cost basis in a clinical context.

These advances are not limited to DNA sequencing technologies, but have a carryover to numerous molecular approaches, including: polymerase chain reaction (qPCR), multiplex PCR; *in situ* hybridization, antibody based immunofluorescence tissue assays, and even proteomic strategies.

Biospecimens

In the setting of oncology, the most common biospecimen used for disease interrogation by molecular diagnostic continues to be the diseased tissue itself (i.e., tumor biopsy). Nonetheless, the recognized evolution of cancer, and the difficulty to rebiopsy metastatic lesions over the course of disease progression, has prioritized consideration of applying novel molecular diagnostics to blood or other biofluids (so-called liquid biopsy).

Diseased Tissue (Tissue Biopsy)

Advances in DNA sequencing techniques, and annotation for the dog genome, has now allowed the extraction and analysis of smaller fragments of DNA. This provides new opportunities to conduct DNA sequencing analysis of tumor samples that were previously not usable.

Most noteworthy from the analysis of such smaller fragments of DNA, is the opportunity to utilize tissues from formalin fixed and paraffin embedded tissues. There, however, continue to be a consequences to formalin fixation, which can limit viable DNA extraction. Strategies to limit the consequence of excessive DNA damage, from formalin induced cross-linking, can include short-term fixation in formalin, but subsequent transfer of tissues to ethanol, prior to paraffin embedding. Nonetheless for the most part many DNA sequencing techniques and associated diagnostic strategies can utilize FFPE; however, clinicians should seek evidence that a novel molecular diagnostic technique successfully applied to frozen or nucleic acid preserved materials can also be used in FFPE.

In many instances novel molecular diagnostics used in oncology will seek to identify prognostic and therapeutic signals from a tumor sample. The heterogeneity of cancer specimens and the biology of cancer predicts that most signals (i.e., mutations or other variants), present in the primary tumor will not persist or drive disease progression and metastasis. Accordingly, it is unclear how valuable extensive analysis of the primary tumor will be, when patient outcomes are most commonly defined by metastatic progression and not any change in the primary tumor (see below for liquid biopsy).

Blood (Liquid Biopsy)

As discussed above, the heterogeneity and genetic evolution of cancer from a primary tumor to metastatic progression, recognizes that many genetic alterations found in a primary tumor may not be present in the metastatic lesions. As such it is reasonable to ask how valuable the analysis of primary to material for prognostic or therapeutic information will be.

This challenge may be addressed through the use of novel noninvasive biopsy techniques to allow the sampling of metastatic lesions, as a more valuable sample for prognostic and therapeutically linked molecular diagnostics.

Alternatively the use of a liquid biopsy (i.e., the analysis of cancer associated genetic alterations in blood or other biofluids) maybe a more valuable starting place for use in novel molecular diagnostics. It is reasonable that true driver events for cancer will be present in the blood of patient during disease progression, and therefore more reflective of the prognosis of the patient.

In cancer, molecular materials that may be found in blood may come from a variety of sources.

Circulating Cell Free DNA

Given the fragility and proclivity for cell death in tumor cells, it is not surprising that cellular DNA from tumor cells can be detected in the blood of patients. Exemplified by the experience in human patients with BRAF mutated melanoma. BRAF mutated cell free DNA can be readily detected in patient blood, and furthermore the amount of mutated DNA correlates with the burden of disease in a patient. Since the mutation of DNA (in this case B-RAF), is a hallmark of this cancer, the detection of this mutated DNA in the blood, is easily understood to be derived from cancer cells, and is distinguished from non-tumor DNA that may also be circulating in the blood. Accordingly, molecular diagnostics that measure mutated or variant DNA maybe easily adapted to the study of such a liquid biopsy. Conversely, platforms that measure gene expression alone, cannot so easily distinguish tumor from non-tumor gene expression.

Circulating Tumor Cells

Both in the presence and absence of measurable disease it has long been recognized that high numbers of viable cancer cells may be present and detected in some cancer patients. The detection of cancer cells can involve tumor-specific antibodies and alternative approaches that detect subtle differences in the plasma membrane of tumor versus normal cells. Despite the promise, it is unclear if the detection of circulating tumor cells in the blood will be an informative approach to the liquid biopsy strategy in oncology.

Circulating Exosomal DNA

Both in cancer and systemic inflammatory conditions, exosomal DNA may be purified and used as input material for novel molecular diagnostics.

Exosomes are cell-derived vesicles that are present in many biofluids including blood, urine, saliva and others. Exosomes are lipid bilayer and actin vesicles that carry DNA.

In cancer exosomes are believed to be actively released by highly motile cancer cells and passively released during apoptosis or other events at the cell membrane. Such "passive release" is also predicted in the setting on inflammatory conditions from a variety of host cells including leukocytes, endothelial cells, and stem cells. Irrespective of the mechanism of release exosomes appear to be actively involved in disease pathobiology and may serve as a stable reservoir of DNA that can be detected as biomarkers for disease diagnosis, prognosis, and therapy selection.

Progressive Adoption of Diagnostic Innovation

Amid the excitement linked to the broader availability and use of genomic and molecular techniques, one approach to the delivery of novel molecular diagnostics could be to prioritize speed and a first to market strategy. Although attractive this approach carries a risk of providing diagnostic innovations that are not familiar to the community, with the potential for misuse or miscommunication.

An alternative approach would be the delivery of progressive innovations in molecular diagnostics so as to introduce products and services that are more familiar to the community as an initial step, and then progressively deliver innovations that may be better understood and effectively used in the clinical environment. Such an approach may additionally provide confidence in rigor and validation, of each progressive offering.

Molecular Pathology Meets the Reference Laboratory

In the setting of a rapidly changing portfolio of diagnostic offerings, as suggested above there is an expected tension between early use of novel diagnostic tools, and the depth and nature of validation of these methodologies.

As clinicians increasingly embrace these novel methodologies, it is important to recognize that most of these genomic and molecular platforms are fundamentally distinct from more conventional reference laboratory offerings from the biochemistry era. For example a conventional molecular assay used in oncology, involves PCR analysis of T cell receptor repertoire/clonality. Such an assay should be expected to be conducted in fundamentally distinct ways by different laboratories, and as such results and validation in one laboratory should not be viewed as sufficient confidence for the use of that same assay in a distinct laboratory.

This expected laboratory to laboratory difference is much greater then differences in the measurement of a specific analyte by a biochemical process (i.e., creatinine) which should be expected to have less laboratory to laboratory differences. Accordingly there is a greater obligation of laboratories to provide transparency into the conduct and the performance of a molecular diagnostic test, and to publish the performance of their specific laboratory assay in a relevant population of animals. Similarly there is an obligation of the clinician to recognize the likelihood of differences in the performance of an assay from laboratory to laboratory and to insist on the publication of data from a given laboratory as transparent evidence of the success of that assay. Ideally, laboratories should consider collaborative studies which would allow sharing of patient samples and co-publication of their assay performance across these shared samples.

What About My Cat?

For a number of reasons the dog is often prioritized over the cat for the development of novel molecular diagnostics and novel therapeutics. Accordingly there is a greater unmet need for novel molecular diagnostics and the cat.

The development of such feline specific diagnostics will be facilitated through maturation of the feline genome and its annotation, and the development of biospecimen repositories inclusive of feline tissues for assay development and validation.

acvim275

Clinical Impact and Management of Chemotherapy-Induced Vomiting and Inappetence

Chad M. Johannes, DVM, DACVIM (SAIM & Oncology)
Ames, IA, USA

Clinical Impact of Chemotherapy-Induced Nausea and Vomiting (CINV)

Chemotherapy has played a critical role in improving outcomes and increasing survival times for a number of types of cancer in recent decades, both in human and veterinary oncology. As more chemotherapeutic agents have become available and more intensive protocols have developed, one of the most common challenges encountered from an adverse effect aspect is CINV. Significantly impacting health-related quality of life in cancer patients, it is important to understand the incidence and pathophysiology of CINV and therapeutic interventions that can be made to reduce its effect on patient care and clinical outcomes.

As many as 80% of human cancer patients receiving chemotherapy experience some level of adverse effects, the most common and dreaded of which is CINV. The impact CINV can have on cancer patients is profound: adversely affecting quality of life and daily activities, including work productivity; giving rise to medical complications (poor nutrition, dehydration, electrolyte imbalances); causing treatment delays or refusal to pursue additional therapies; increasing hospital visits and subsequent cost of medical care.[1-3] In spite of the availability of more effective antiemetics (newer 5-HT_3 antagonists, NK1 antagonists) in recent years, control of CINV in human cancer patients remains a challenge. The role of healthcare practitioners is vital to improving the control of CINV. Studies indicate that underestimating the incidence, failing to understand the pathophysiology and ineffectively implementing antiemetic treatment guidelines contributes to the ongoing clinical burden of CINV.[1,4,5]

Nearly two-thirds of human patients receiving chemotherapy experienced one or more episodes of CINV in one prospective study. Delayed CINV (greater than 24 hours after administration) occurred nearly twice as often as acute CINV (within 24 hours of administration).[5] Also of note is that nausea occurred twice as often as vomiting. Patients indicated that nausea was more distressing than vomiting and had a greater impact on their quality of life.[5,6] This point has particular significance for veterinary health professionals where evaluation of nausea in patients is much more subjective but nonetheless important to monitor.

In human oncology, chemotherapeutic agents and radiation therapy locations/procedures have been categorized as to their emetic risk (high, moderate, low, minimal) and CINV treatment guidelines are published to serve as recommendations for effectively controlling signs in patients being treated.[7] Although studies have found that patients receiving antiemetic therapy according to these guidelines experience less frequent delayed CINV compared to those receiving inadequate therapy, findings also indicate that the guidelines are frequently not followed. Several studies have identified that healthcare providers greatly overestimate the control of delayed nausea and emesis following both highly and moderately emetogenic chemotherapeutics and that patients do not always adequately report symptoms.[4,5,8,9]

CINV Pathophysiology in Human Oncology

CINV is typically classified as acute or delayed. Acute CINV occurs within 24 hours of chemotherapy administration whereas delayed CINV is defined as nausea or vomiting occurring 24–120 hours after chemotherapy. The distinction between acute and delayed CINV implies more than timing. Physiologic differences in the pathways involved account for acute versus delayed CINV. Acute CINV appears to be primarily mediated by serotonin (5-HT^3) pathways while delayed CINV is primarily substance P (NK1) pathway mediated. This difference becomes important when preventative therapies are considered. Breakthrough CINV is nausea and vomiting that occurs in spite of antiemetic therapy and requires the use of rescue (additional antiemetic) treatment.[10-12]

Initiation of emesis (vomiting) requires activation of the vomiting center within the brainstem which has three main regional components (area postrema, nucleus tractus solitarius, dorsal vagal complex) that integrate the emetic responses. In general, activation of the vomiting center can occur as the result of stimuli from many sources including drugs (i.e., chemotherapeutics), gastrointestinal input/irritation/stretching, motion, smells, sights, or emotions.

CINV in particular is believed to result from the presence of chemotherapeutic agents or their metabolites within the blood stream and/or the cerebrospinal fluid acting on the chemoreceptor trigger zone (CRTZ) in the area postrema (lies outside the blood-brain barrier). Signals from the area postrema are relayed to the nucleus tractus solitarius and the emetic reflex is induced. Chemotherapeutic drugs may also cause the release of serotonin and substance P from the gastric mucosa (enterochromaffin cells) which triggers signals via vagal sensory fibers to the nucleus tractus solitarius, initiating the emetic reflex via another route. In addition to the substance P (NK_1) and serotonin ($5\text{-}HT_3$) pathways, cannabinoid and dopamine (D_2) pathways are believed to be involved in CINV. Alternative pathways involved in nausea and vomiting do not appear to be activated in CINV, including: acetylcholine/muscarinic (M), histamine (H_1), endorphin and gamma-aminobutyric acid pathways.[1,12]

These findings have practical implications as antiemetic therapeutic options for CINV are considered. Antiemetic therapy is administered prophylactically to minimize the clinical impact of CINV and the treatment utilized depends on the chemotherapy regimens employed. Patient-related risk factors and prior response to treatment regimens are also carefully considered.[7,12] It has been well-documented that patients experiencing acute CINV have a higher risk of developing delayed CINV compared to those who are completely protected during the acute phase. This recognized carry-over effect from the acute to the delayed phase is an important consideration and emphasizes the importance of preemptive emetic control strategies.[12]

Practical Management of CINV Signs in Dogs and Cats

Although vomiting is a commonly described side effect of many chemotherapeutic agents in dogs and cats, the actual incidence is not well documented. Published studies (Table 1) focusing on the incidence and/or treatment of vomiting in dogs and cats receiving chemotherapy are limited and methods of measuring vomiting inconsistent. Beyond medical therapy, fasting dogs prior to administration of doxorubicin has been shown to reduce the incidence of delayed-type vomiting.[13] Other non-medical preventative measures are not well documented. The most commonly utilized medical antiemetic treatment options in dogs and cats are summarized below.

Table 1. Reported incidence of vomiting and inappetence in dogs and cats receiving chemotherapy

Chemotherapeutic	Species	Vomiting incidence	Inappetence incidence	Reference
Vinorelbine	Dogs	21% (5/24)	17% (4/24)	Grant (2008)47
Epirubicin	Dogs	24% (34/139)	19% (27/139)	Marrington (2012)48
Carboplatin	Dogs and cats	11% (3/28)	25% (7/28)	Bowles (2010)49
Doxorubicin	Dogs	8-35% (n=4-17/49)*#	35-51% (n=17-25/49)*#	Rau (2010)16
Cyclophosphamide	Dogs	21% (9/42)*	36% (15/42)*	Mason (2014)17
Toceranib phosphate	Dogs	32% (28/87)	39% (34/87)	London (2009)50
Vincristine	Dogs	24% (14/58)*	43% (25/57)*	Mason (2014)17
Epirubicin	Dogs	72% (13/18)	44% (8/18)	Kim (2007)51
Lomustine	Dogs	51% (41/81)	48% (39/81)	Vail (2012)52
Paclitaxel	Dogs	79% (133/168)	76% (128/168)	Vail (2012)52

*Subset of dogs treated with maropitant following chemotherapy administration; #Crossover study.

Neurokinin (NK) Receptor Antagonists

Substance P (a tachykinin) has been recognized as a key component in the emetic reflex by stimulation of NK_1 receptors in the emetic center, CRTZ and abdomen/gastrointestinal tract. Development of effective NK_1 receptor antagonists which block the binding of Substance P to NK_1 receptors provided a significant breakthrough in the control of CINV in human oncology. As a result, this class of drug is part of the recommended protocol for patients receiving chemotherapeutics characterized by a high emetic risk.[7]

Maropitant (Cerenia®, Zoetis) was approved by the FDA for the prevention and treatment of acute vomiting in dogs in 2007. Two separate studies utilizing maropitant for the prevention of CINV in dogs receiving the highly emetogenic chemotherapy agent cisplatin showed that dogs treated with maropitant experienced significantly fewer subsequent emetic events compared to placebo-treated dogs.[14,15] When maropitant was administered prior to cisplatin treatment in a prevention regime, 94.9% did not vomit compared with only 4.9% of placebo-treated dogs.[14]

Maropitant, when administered orally for 5 days following doxorubicin chemotherapy, was effective in preventing delayed vomiting in a prospective, randomized trial. Interestingly, dogs receiving maropitant also experienced less frequent and severe diarrhea than placebo-treated dogs.[16] Another prospective study in dogs treated with vincristine and cyclophosphamide showed, however, that prophylactic administration of maropitant did not reduce the frequency of vomiting.[17] Additional studies in dogs indicate that maropitant may be beneficial in reducing visceral pain[18,19] which may be another mechanism by which it helps reduce the effects of CINV. Some dogs with cancer may benefit from chronic therapy with maropitant. The pharmacokinetics and clinical utility of longer-term maropitant use has been reported.[20,21] Approved for use in cats, published data on the safety and efficacy of maropitant as an antiemetic in cats is available.[22]

5-HT_3 Antagonists

Selective 5-HT_3 antagonists block the binding of serotonin to receptors in the gut and brain (nucleus tractus solitarius).[12] Newer generation drugs of this category (i.e., palonosetron) are recommended in the antiemetic protocols for human oncology patients receiving highly and moderately emetogenic chemotherapeutic drugs as well as for many patients at risk for radiation-induced nausea and vomiting.[7] While not approved for use in dogs and cats, ondansetron and dolasetron are utilized fairly commonly as antiemetic agents in dogs and cats[23] due to their generic availability, relatively low cost, and empirical efficacy. The oral bioavailability of ondansetron has been shown to be low in both dogs[24] and cats[25] with high variability between individuals,[24] indicating that oral administration may not be the most effective delivery route for this drug.

Metoclopramide

An antagonist of dopamine D_2 receptors (CRTZ) at lower dosages, metoclopramide may inhibit 5-HT_3 receptors in the gut and CRTZ at higher dosages. While human cancer patients receiving the high-dose metoclopramide experienced less CINV subsequent to cisplatin therapy, extrapyramidal reactions (involuntary movements of face, tongue extremities; trismus; tremors) were experienced by a number of patients. Diphenhydramine was often administered to alleviate these signs and studies showed that addition of dexamethasone improved the antiemetic effect compared to metoclopramide alone. As a result, the three-drug combination of high-dose metoclopramide, a corticosteroid and diphenhydramine was considered the standard antiemetic regimen for human chemotherapy patients until the development of 5-HT_3 and NK_1 antagonists.[12] Based on the findings that the dopaminergic D_2 pathway plays little role in the development of CINV and the side effects observed with high-dose therapy, metoclopramide is typically only utilized in combination therapy for patients experiencing signs in spite of optimal prophylaxis.[7]

Other Agents

Butorphanol was commonly used as part of standard antiemetic protocols in dogs receiving cisplatin prior to the availability of 5-HT_3 and NK_1 antagonists.[26] A number of studies have evaluated the use of **corticosteroids** as antiemetic agents in human oncology patients. Studies have indicated that, as single agents, corticosteroids have an antiemetic effect in patients receiving moderately emetogenic chemotherapies. Additionally, it has been shown that corticosteroids increase the effect of other antiemetics (i.e., low-dose and high-dose metoclopramide, 5-HT_3 antagonists, NK_1 antagonists).[12] As a result, dexamethasone is commonly utilized as a component of therapy for CINV associated in human cancer patients being treated with high, moderate, and low emetogenic chemotherapy agents as well as radiation patients at high or moderate risk of nausea or vomiting.[7] The potential role of corticosteroids such as dexamethasone as antiemetic agents in dogs and cats is not well documented and this category of drug is not routinely included in CINV prophylactic protocols for veterinary patients. **Mirtazapine**, often considered for its potential appetite stimulant effects (see below), is also a consideration for refractory CINV as is being recognized in human oncology.[27]

Clinical Impact of Inappetence in Dogs and Cats With Cancer

Appetite is often a key determinant in a pet owners' perceived quality of life evaluation for their dog or cat with cancer. Lack of appetite is often the first recognizable sign by pet owners that their dog or cat is not feeling well and triggers inquiry to their veterinarian's office. Inappetence in pets with cancer is often multifactorial with potential contributing factors including: the tumor itself (e.g., dysphagia, pain, ascites), increased inflammatory cytokines, chemotherapy or other drug side effects (e.g., dysmotility, nausea) and effects of concomitant medications. Similar to vomiting, inappetence is a commonly described side effect of many chemotherapeutic agents in dogs and cats. Published studies (Table 1) focusing on the incidence of inappetence in dogs and cats receiving chemotherapy are limited and methods of measuring inappetence are inconsistent.

Weight loss frequently occurs with several types of human cancer and is often associated with reduced survival and quality of life in cancer patients.[28] There are multiple causes of weight loss in cancer patients, key among them inappetence along with numerous catabolic drivers including direct tumor metabolism, systemic inflammation, and insulin resistance.[29] This phenomenon is referred to as cancer-related anorexia/cachexia syndrome (CACS), a metabolic, paraneoplastic syndrome characterized by decreased food intake, involuntary weight loss, and altered body composition (loss of fat and muscle).[29] Cachexia is reported in 15–40% of human cancer patients, approaching 80% in those with advanced illness.[29]

For decades, it has been debated whether or not veterinary patients develop CACS. This may be attributed in part to veterinarians not recognizing the syndrome due to lack of awareness and/or absence of established diagnostic criteria. While some veterinary patients may be humanely euthanized before the syndrome becomes clinically evident, the reality is there is limited information on the incidence of CACS in pets. A study by Michel *et al.* reported that 37% of dogs with cancer had lost >5% of their body weight at diagnosis.[30] A number of studies in veterinary medicine have indicated dogs with cancer experience metabolic alterations similar to those observed in human patients with cachexia: increased resting energy expenditure and decreased rates of protein synthesis in dogs with osteosarcoma;[31] alterations in carbohydrate metabolism in dogs with solid tumors;[32] and differences in serum cytokine profiles in dogs with lymphoma compared to healthy controls.[33] Dogs underweight at the time of lymphoma diagnosis have been shown to experience significantly shorter survival times.[34]

Emerging Treatment Options for Inappetence in Dogs & Cats

If weight loss/cachexia impacts canine and feline survival as it does in humans with cancer, management of this syndrome may improve survival times and quality of life for pets diagnosed with similar conditions. The challenge for veterinarians lies in having effective, proven therapies available for clinical use. The list of drugs utilized in an attempt to stimulate appetite in dogs and cats is quite extensive with common choices including mirtazapine, cyproheptadine, corticosteroids, and antiemetics. The challenge is that none of these drugs has stimulation of appetite as a primary mechanism. The appetite stimulation component is a secondary or side effect, explaining the lack of reliability from patient to patient. Except for mirtazapine in cats, additional challenges arise from limited data on dosing and pharmacokinetics and sparse to non-existent data on efficacy for appetite stimulation in dogs and cats.

ENTYCE®/Capromorelin/AT-002 (Aratana Therapeutics, Inc.)

Given the impact of CACS on quality of life and survival in humans with cancer, a number of therapeutic approaches are being explored to positively treat this syndrome. A ghrelin agonist, anamorelin, is one example and has demonstrated positive effects on body weight and appetite in humans with non-small cell lung cancer (a cancer with an extremely high incidence of CACS) and may also impact survival.[35]

Capromorelin is an orally active small molecule in this same category of therapeutic that mimics the action of ghrelin causing growth hormone secretion and appetite stimulation. Capromorelin oral solution dosed at 3 mg/kg PO q 24 h has been shown to cause increased food intake and weight gain in both healthy laboratory and inappetent client-owned dogs.[36-38] Capromorelin oral solution has also demonstrated a wide margin of safety, being well-tolerated at daily doses up to 40 mg/kg once daily for 12 consecutive months.[39] When administered to healthy dogs for 7 days, capromorelin treatment produced increased insulin-like growth factor 1 (IGF-1) concentrations on Day 1 and this increase was sustained through Day 7.[36] Additionally, capromorelin has been demonstrated to cause increases in IGF-1 and increased food intake and body weight in cats.[40]

ENTYCE was granted FDA approval in May 2016. Aratana plans to make it commercially available to veterinarians in late 2017.[41] This therapeutic has significant potential to positively impact the treatment of inappetence related to several chronic conditions in our canine patients. Aratana is also developing a cat-specific formulation with additional studies pending.

Melanocortin-4 Receptor (MC-4R) Antagonist

Another mechanism to directly impact the appetite pathway is via the MC-4 receptor which plays an important role within the hypothalamic appetite center, affecting meal size and choice. MC-4R antagonism is associated with increased food intake. MC-4R antagonists have been evaluated and shown promise in mouse models for the treatment of cancer cachexia.[42] A veterinary MC-4R antagonist product is currently being investigated in dogs with cachexia at the University of Missouri-Columbia. The product is an MC-4R antagonist peptide designed to minimize the potential cardiovascular side effects seen in this category of drug. Rodent models have shown increased quality of life despite tumor growth. Safety has been demonstrated in healthy dog studies. The ongoing funded clinical trial is enrolling dogs with cachexia (5% or greater weight loss in 3 months) of any cause without severe underlying cardiac, renal, or hepatic disease. (Personal communication with Sandra Bechtel, DVM, DACVIM.)

Mirtazapine

Mirtazapine is a 5-HT$_3$ receptor antagonist with appetite stimulant properties documented in healthy young cats at a dose of 1.88 mg PO q 24 h.[43] Additional studies have shown that dosing of 1.88 mg PO q 48 h is more appropriate in cats with CKD.[44] A double-masked, placebo-controlled crossover clinical trial in cats with CKD demonstrated that mirtazapine administration at 1.88 mg PO q 48 h for 3 weeks resulted in a significant increase in appetite and activity and significant decrease in vomiting compared to placebo. Mirtazapine-treated cats also experienced a significant gain in bodyweight (median gain 0.18 kg) compared to placebo.[45] Used as an anti-depressant in human medicine with the observed side effect of increased appetite and weight gain, the specific method of appetite stimulation is not well documented. The antiemetic effect noted in cats with CKD is interesting as mirtazapine is sometimes used for refractory chemotherapy-induced nausea and vomiting in humans, as noted previously.

Clinical experience would indicate that mirtazapine is a less reliable/predictable appetite stimulant in dogs. A small pilot study evaluating the pharmacokinetics of mirtazapine in healthy Beagle dogs has been performed.[46] Results of this study indicate that mirtazapine appears to be metabolized much more quickly in dogs and twice daily dosing may be more appropriate in this species. As the appetite stimulation component is essentially a side effect of mirtazapine rather than a primary mechanism, it is not surprising that its clinical effect is variable and rather unpredictable.

References

1. Hawkins R, *et al. Clin J Oncol Nurs.* 2009;13(1):54.
2. Hamadani M, *et al. J Oncol Pharm Pract.* 2007;13(2):69.
3. Shih YCT, *et al. Cancer.* 2007;110(3):678.

4. Grunberg SM, *et al. Cancer.* 2004;100(10):2261.
5. Ihbe-Heffinger A, *et al. Ann Oncol.* 2004;15(3):526.
6. Bloechl-Daum B, *et al. J Clin Oncol.* 2006;24(27):4472.
7. Basch E, *et al. J Clin Oncol.* 2011;29:4189.
8. Erazo Valle A, *et al. Curr Med Res Opin.* 2006;22:2403.
9. Majem M, *et al. Support Care Cancer.* 2011;19:1983.
10. Jordan K, *et al. Oncologist.* 2007;12(9):1143.
11. Schwartzberg L. *J Support Oncol.* 2006;4(2, Suppl. 1):3.
12. Herrstedt J. *Nat Clin Pract Oncol.* 2008;5(1):32.
13. Withers SS, *et al. Transl Oncol.* 2014;7(3):377–383.
14. Vail DM, *et al. Vet Comp Oncol.* 2007;5(1):38.
15. de la Puente-Redondo VA, *et al. Am J Vet Res.* 2007;68(1):48.
16. Rau SE, *et al. J Vet Intern Med.* 2010;24:1452.
17. Mason SL, *et al. J Small Anim Pract.* 2014;55:391–398.
18. Boscan P, *et al. Am J Vet Res.* 2011;72(12):1576.
19. Alvillar BM, *et al. Vet Anaesth Analg.* 2012;39(2):201–205.
20. Lesman SP, *et al. J Vet Pharmacol Ther.* 2013:36(5):462–470.
21. Clifford C, *et al.* Abstract, 2016 Veterinary Cancer Society Conference; Orlando, FL.
22. Hickman MA, *et al. J Vet Pharmacol Therap.* 2008;31:220.
23. Ogilvie GK. *J Am Anim Hosp Assoc.* 2000;36:481.
24. Baek IH, *et al. J Vet Pharmacol Ther.* 2015;38(2):199–202.
25. Quimby JM, *et al. J Vet Pharmacol Ther.* 2014;37(4):348–353.
26. Moore AS, *et al. J Am Vet Med Assoc.* 1994;205(3):441.
27. Dupuis LL, *et al. Paediatr Drugs.* 2010;12(1):51–61.
28. Mondello P, *et al. BMC Cancer.* 2014;14:282.
29. Fearon K, *et al. Lancet Oncol.* 2011;12:489–495.
30. Michel KE, *et al. J Vet Intern Med.* 2004;18(5):692–695.
31. Mazzaferro EM, *et al. Am J Vet Res.* 2001;62(8):1234–1239.
32. Ogilvie GK, *et al. Am J Vet Res.* 1997;58(3):277–281.
33. Calvalido J, *et al.* Abstract, 2016 ACVIM Forum; Denver, CO.
34. Romano FR, *et al. J Vet Intern Med.* 2016;epub Jun 8.
35. Currow DC, *et al. J Clin Oncol.* 2015;33(29_suppl):175.
36. Zollers B, *et al. J Vet Pharmacol Ther.* 2016;doi:10.1111/jvp.12344.
37. Zollers B, *et al. J Vet Intern Med.* 2016;30(6):1851–1857.
38. Zollers B, *et al. BMC Vet Res.* 2017;13(1):10.
39. Zollers B, *et al. J Vet Pharmacol Ther.* 2016;doi:10.1111/jvp.12358.
40. Zollers B, *et al.* Abstract, 2015 ACVIM Forum; Indianapolis, IN.
41. http://aratana.investorroom.com/2017-02-03-Aratana-Therapeutics-to-Attend-North-American-Veterinary-Conference.
42. Dallmann R, *et al. J Cachexia Sarcopenia Muscle.* 2011;2(3):163–174.
43. Quimby JM, *et al. J Vet Pharmacol Ther.* 2011;34(4):388–396.
44. Quimby JM, *et al. J Vet Intern Med.* 2011;25:985–989.
45. Quimby JM, Lunn KF. *Vet J.* 2013;197(3):651–655.
46. Giorgi M, Yun H. *Vet J.* 2012;192(2):239–241.
47. Grant IA, *et al. J Vet Intern Med.* 2008;22:388–393.
48. Marrington AM, *et al. Vet Comp Oncol.* 2011;10(2):113–123.
49. Bowles DB, *et al. J Small Anim Pract.* 2010;51:104–112.
50. London CA, *et al. Clin Cancer Res.* 2009;15(11):3856–3865.
51. Kim SE, *et al. J Am Vet Med Assoc.* 2007;231:1550–1557.
52. Vail DM, *et al. J Vet Intern Med.* 2012;26(3).

acvim278

Panel: The Economics of Veterinary Oncology: Opportunities and Challenges

Chad M. Johannes, DVM, DACVIM (SAIM, Oncology); Philip Bergman, DVM, MS, PhD, DACVIM (Oncology); Craig A. Clifford, DVM, MS, DACVIM (Oncology); G. Sylvester Price, DVM, PhD, DACVIM (SAIM, Oncology)
Ames, IA, USA

Veterinary Oncology Market Growth Trend

The oncology specialty has experienced exceptional growth over the past 25 years, with an estimated compound annual growth rate (CAGR) of boarded specialists of nearly 13% (Figure 1).[1,2] In addition to these oncologists that have completed the boarding process, 39 candidates have completed residencies but are not yet fully boarded and 86 more candidates are currently in residency training programs.[1] This growth rate experienced in medical oncology has outpaced the overall growth of boarded specialists (~8.5% over the past 40 years), where there are reported to be >11,000 boarded specialists in 22 recognized specialty boards.[2,3] The apparent market demand for board-certified oncologists remains robust with more than 50 advertised positions in the US (~90% private practice, ~10% academic) currently listed on the ACVIM Career Center.[5]

Factors within the market driving this growth of oncologists are varied but likely include: 1) Client expectations for the advanced medical care of their pets; 2) More veterinarians pursuing advanced training due to the desire to practice specialty medicine, the typically higher annual income for specialists compared to primary care and rising student debt; 3) Corporatization of practices. The result is expanding client/patient access to oncology care across regions (including previously untouched locales), advanced treatment/care for patients and new markets for veterinary pharmaceutical development.

Figure 1. Number of ACVIM-boarded oncologists[1]

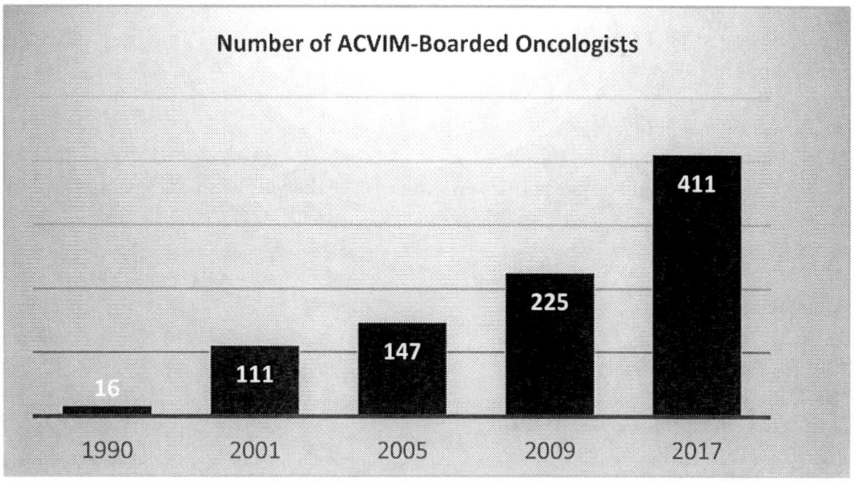

Economic Opportunities in Veterinary Oncology

Corporate Specialty Practice

- Massive number of jobs open for specialists currently, but this begs the question of which specialty hospitals will survive, especially in highly fractionated market regions?
- Increased ability to collaborate and interact with other specialists within the corporation.
- Increased ability to be able to afford the latest technology and innovation in corporate practices.
- Increased number of clinical studies available in many corporate practices due to increased infrastructure and breadth of corporate network (sponsored clinical studies as well as homegrown "organic" non-sponsored studies by clinicians, house officers, technicians, etc.).

- Increased development of SOPs, consensus statements, and evidence-based diagnostics/therapeutics due to aforementioned corporate infrastructure.
- Increased ability for other career pathways as One Health/Medicine expands.
- Increased penetration of companion animal insurance leading to increased ability to afford recommended care options.
- As the human-animal bond continues to evolve and grow, this will continually improve the role of pets in families leading to increased care.

Private Specialty Practice

- Staying competitive: With the rapidly changing landscape many of the larger referral clinics have become corporately owned; this represents a concern and an opportunity of veterinarian-owned clinic.
- Pros:
    - The ability to be "quick and nimble" and make relative rapid decisions without the need for oversight or multiple layers of management.
    - The ability to allow clinicians to build their service with its own identity (not follow a model).
    - The ability to modify invoices (discount owners)
        - We are allowed to discount but are required to list reasons (drop-down list - ethical, military, clinical interest, clinical study, etc.).
        - These go a long way with owners who feel you care and are working with them.
    - As a clinician, you have a say in associated costs of drugs, etc.
        - Not having steadfast 3x markup, being flexible in order to utilize newer agents even if this may decrease compensation.
        - Owners "price shop" and some will look for the clinic that is considered less expensive.
    - Ability to offer ownership for those who are inclined
        - Many of the new grads have no interest in owning, which is perpetuating the shift to corporate ownership.
    - Salary (Base + percentage): most privately owned clinics have advantage of not having a negative accrual to base pay if under set point.
    - There is a subset of veterinarians who are anti-corporate and more inclined to refer to a privately owned clinic.
    - Ability to participate in clinical trials as part of larger networks (ACI, One Health etc.).

Academic Specialty Practice

- Expansion in private/specialty oncology practice caseload provides growing caseload for cooperative clinical trials/research opportunities; increases speed at which trials can be completed.
- Increasing interest in veterinary oncology therapeutic development opens new research funding avenues.
- High competition to fill academic positions has resulted in increased base salary for faculty hires, although they remain well below starting salaries for oncologists in private/corporate practice.

Industry Specialty Market

- Advances in immuno-oncology and genome-driven oncology care represent exciting and promising approaches for new therapeutics.
- The use of dogs (and cats) as translational models for human oncology has introduced veterinary oncologist to exciting novel compounds.
- There are a large number of small veterinary pharmaceutical and biopharmaceutical companies exploring novel oncologic therapeutics.

ECONOMIC CHALLENGES IN VETERINARY ONCOLOGY

Corporate Specialty Practice

- Potentially fewer cases as more specialty clinics open and fractionate remaining areas.
- Potentially fewer cases from pcDVMs as the most recent studies in the literature suggest younger pcDVMs want to keep these cases and referral may be a last report option for some.
- As technology and innovation continue, the cost of care increases, which may price out many clients.
- Proving to pcDVMs that early referral can lead to longer & healthier patient outcomes, which benefits all parties (i.e., patient, owner, pcDVM, and specialist). We are currently working hard with rigorous scientific investigations to prove this with VetSOAP.org.
- Continued pervasiveness and use of Dr. Google resulting in a perceived decrease need in specialty care.
- Workups by pcDVMs appear to be on the rise prior to referral, which may leave owners with reduced financial abilities after meeting with the specialist.
- Overcoming perceptions about corporate veterinary medicine ruining the profession since this has been stated for over 25–30 years now.
- Fewer veterinary students are headed to internships since they are being taught this is financially not prudent by veterinary schools, which is then likely to decrease residency candidate pools which may lead to an incubator crisis.
- Increased use of unproven technologies and/or methods of treatment such as autologous cancer vaccines, etc.
- Production-based financial remuneration models which reward less time away from the clinic, but these are true whether corporate or not. Furthermore, these models also inflate the cost of care, especially when expensive diagnostics and/or therapeutics are utilized since these line items must build in costs that flow to production.

Private Specialty Practice

- Staying competitive in regards to marketing, purchase power, equipment, hiring, CE.
- Privately owned has a much smaller budget
  - Do new oncologists understand or care about the difference?
- Referrals:
  - Working with primary care doctors in an effort to promote "case sharing."
  - Not to be seen as a "black hole."
  - Promoting a partnership.
  - Economic downturn often leads to less referrals.
  - No way to change this…
- Millennials?
  - Many new grads have a different approach and are more concerned over improved quality of life for the clinician vs. the needs of the hospital.
  - Keeping clinicians happy
    - Recognizing this and working to help modify situations (work and home).
    - Creating a TEAM approach.
    - Worry about the team, not the individual.
    - Hiring another clinician will initially affect revenue but…will grow the service and practice.
    - Leading by example.
- Clinical trials: will corporate eventually take over all clinical trials in private practice by offering everything "in house" within their network?

Academic Specialty Practice

- Many academic centers are geographically isolated creating challenges in maintaining caseload and recruiting talent; exacerbated by extensive geographic expansion of private/corporate specialty practices in neighboring metropolitan areas.
- Reduced caseload makes it more difficult to enroll clinical trials and train students and house officers effectively.
- Inherent inefficiencies of process at many academic centers preclude ability to maximize caseload; clients often opt for regional private/corporate specialty practices, which may be more expensive but often are much more time efficient.
- Variability of funding (state, granting agencies) and subsequent impact on investment in the practice create difficulty in consistent growth and development.
- In spite of training the majority of medical oncology residents (20/21 medical oncology residencies in 2017 VIRMP were academic based),[6] it is significantly difficult to effectively recruit to academic practice, given the salary differential compared to private/corporate specialty practices.

Industry Specialty Market

- The number of clinical trials conducted on new therapeutics continues to increase, sometimes leading to (1) competition for cases of a given tumor type and/or (2) increased client awareness and requests for involvement but enrollment may be limited to a few centers.
- Trials involving compounds not intended for subsequent veterinary use may lead to a temporary sense of optimism and positive outcomes that may not be sustained.
- Clinical trials conducted to achieve regulatory approval and/or meet regulatory requirements do not necessarily provide clinical data oncologists need to recommend the use of a new therapeutic to their clients.
- Conditional license approval process allows a therapeutic to be commercially available, but the sample size and patient populations studied to achieve the conditional license may be less than ideal.
- After a product has been approved, non-randomized non-controlled studies conducted on a small patient sample size by independent clinician researchers sometimes yields results that conflict with those generated in the trials conducted to obtain regulatory approval.
- Despite the positive potential outcomes of targeted therapy and immuno-oncologic therapeutics, combination therapy is likely to be the mainstay; however, investigations on multimodal therapy efficacy and scheduling studies are not typically conducted by industry.

References

1. ACVIM Office, February 2017.
1. Data on file, Johannes CM.
2. www.petmd.com/dog/care/evr_dg_veterinary_specialists_who_are_they_really.
3. www.avma.org/ProfessionalDevelopment/Education/Specialties/Pages/default.aspx.
4. http://jobs.acvim.org/jobseeker/search/results/?str=51&max=25&t731=43304&vnet=0&long=1 (accessed 08FEB2017).
5. www.virmp.org/Program/List?categoryIDs=32&institution=&stateID= (accessed 08FEB2017).

acvim280

Just Because We Can Doesn't Mean We Should: Anecdote vs. Evidence in Current Clinical Oncology

Carolyn J. Henry, DVM, MS, DACVIM (Oncology)
Columbia, MO, USA

INTRODUCTION

All medical decisions come at a cost. Whether paid in a currency of time, emotion, money, quality of life, or some combination thereof, it is incumbent upon us to clearly, honestly, and realistically present treatment options so that our clients understand the costs and can make informed decisions appropriate to their personal situation. Somewhere along the course of our veterinary professional training, most of us were taught the axiom, "above all, do no harm." Despite a general acceptance of this axiom as a guiding principle of the profession, day-to-day clinical decision making often disregards the concept. Look no further than the latest online listserve case discussion to find a barrage of anecdotes offered with disclaimers such as "for what it's worth" (FWIW) or "my two cents' worth" (MTCW). The acronyms suggest that one should not place too much stock in the information, yet we commonly witness treatment decisions being made on the basis of wanting to do **something**, as opposed to nothing - even if that something is unproven and potentially harmful. But just because we can doesn't mean we should implement some therapeutic options. This session will address the propriety of "doing something" and will include a discussion of the levels of evidence that should be applied to veterinary clinical decisions. We will examine real-life examples of clear disregard for evidence-based medicine (EBM) and will also consider the effect of product and treatment availability, client financial constraints, and the option of euthanasia for their role in shaping pet healthcare decisions.

DON'T JUST DO SOMETHING, STAND THERE

The "need to do something" is a phenomenon not only amongst veterinarians, but also witnessed in other health-related professions. A classic example is the treatment of "the death rattle" that is routine practice by those caring for terminal human patients. Anticholinergic drugs are frequently administered to decrease the excessive respiratory secretions associated with the end of life in people in order to decrease caretaker anxiety and to fulfill the request to "do something." This is done in spite of the fact that systematic review of the clinical outcomes after administration of antimuscarinic drugs demonstrates no clinical benefit.[1-4]

In an effort to address such questionable practices, there is renewed advocacy for clinical audits in human medical centers. These audits rely upon a Plan-Do-Study-Act (PDSA) process to assess outcome with current practices and to make adjustments when these practices are suboptimal.[5] While audits have come to be considered a critical component of best practice implementation in human medicine, they are not routinely conducted in veterinary medicine.[5,6] In the spirit of an audit, veterinary medicine often relies upon retrospective studies to generate outcome data from current protocols/practices and use that information to identify those protocols in need of modification. In the ideal setting, this leads to subsequent prospective studies to assess outcomes with and without the protocol modification. Unfortunately, veterinary studies are sporadic and often underpowered or poorly designed.[5-7] Retrospective studies are commonly over-interpreted and substituted for controlled prospective studies to decide when protocol changes are warranted.

ANECDOTE VS. EVIDENCE-BASED MEDICINE (EBM)

Levels of Evidence in Veterinary Oncology Literature

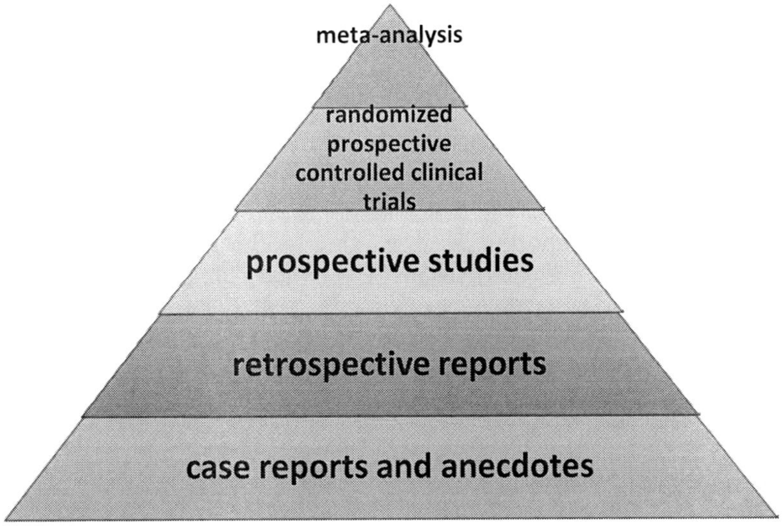

By definition, EBM is "the conscientious, explicit, and judicious use of the current best evidence in making decisions about the care of individual patients."[2,8] The general hierarchy of evidence is shown in the above diagram, with the highest level of evidence being meta-analysis of multiple randomized, prospective controlled, clinical trials. In order to apply EBM to clinical decision making, we must have high-quality, published clinical research findings to inform our choices. As addressed in the 2010 JVIM paper by Sahora *et al.* this prerequisite for EBM is not always available in the veterinary literature.[9]

Some have suggested that there are contexts within which the principles of EBM are not entirely appropriate. In the 2008 article, "Evidence-based medicine beyond the bedside: keeping an eye on context," Dr. Jon Tilburt of the Mayo Clinic asserts that additional factors including clinical circumstances and patient preferences should be considered when applying EBM to daily practice.[10] In fact, the aforementioned definition of EBM was subsequently updated to include the "integration of best research evidence with clinical expertise and patient values."[8] The latter definition is more in line with the principles we apply to veterinary medicine, yet does not entirely address the additional confounding factors we deal with in veterinary medicine - namely, treatment availability, client financial constraints, and the option of euthanasia.

Even when evidence to support or refute the use of a treatment intervention exists, there is still a certain reassurance that comes from hearing the anecdotal experience of our colleagues. This phenomenon occurs amongst physicians, as well. In their 2008 published survey, Hay *et al.* reported on responses of 29 individual physicians divided into three focus groups and asked about the sources of information they use to make clinical decisions.[11] They reported the following:

"We found that physicians tend to favour experience, either theirs or that of trusted colleagues, in making clinical decisions, referring to the EBM literature (largely through electronic clinical resources) either for general information about a condition or to double check that a therapy does not have a published negative outcome. Decisions are made with some reference to EBM, but experience weighs more heavily in clinical decision making about therapeutics. Experience certainly is built up with reference to EBM, but the learning of traditions of practice through apprenticeships and learning from one's particular cases seem to be at the heart of clinical practice."

Thus, the reliance upon our own clinical experience and that of our colleagues for clinical decision-making is apparently not unique to the veterinary oncology community. That said, we should not necessarily endorse such practice. For a broader discussion of this phenomenon and how it impacts our approach to cancer care, we will consider examples from the board-certified oncology veterinary listserve community and first-opinion practitioners who rely upon their recommendations. Time permitting, we will explore the evidence-based [or not] approach to the following topics:

- Metronomic therapy
- New and older tumor vaccines

- Complementary and alternative medicine
- Compounded formulations
- Chemotherapy protocols that have not been proven efficacious
- The ethics of clinical trials

Access to Therapy Options

The choice between applying evidence vs. anecdote to clinical decision-making is often influenced by the inaccessibility of certain treatment modalities to practicing veterinarians. The paucity of approved drugs in veterinary oncology has made off-label use of human oncology drugs commonplace. Drug dosages and indications are extrapolated from human oncology literature, sometimes without appropriate prospective evaluation of safety or efficacy in healthy companion animal species, much less in veterinary patients with cancer. The limited availability of approved veterinary cancer drugs does not, however, negate the need for clinical practices to be evidence-based. Anecdotal reports of drug efficacy in a handful of patients should not serve as the basis for justification of therapy decisions, but often do. This is in spite of a fundamental understanding amongst practitioners that one must objectively evaluate clinical responses in light of inherent and potential biases in patient selection or outcome assessment. The same is true of other cancer treatment interventions including surgery, radiation therapy, and complementary or alternative medical care.

The Impact of Financial Resources

It is estimated that <2% of dog and cat owners in the US have insurance policies for their pets.[12] Accordingly, finances often dictate treatment decisions and may cause pet owners to select cheaper alternatives over data-proven, effective therapy. This unfortunate reality means that we are obligated to advise clients about how to best meet their treatment goals within their financial limitations - even when we know more expensive alternatives would likely be more effective. With the ever-increasing number of funded clinical trials available, referral to clinical trial centers may provide a less expensive option for the [fully-informed, consenting] client needing to balance treatment costs with financial constraints. A searchable database of veterinary oncology clinical trials in the US can be found at www.VetCancerTrials.org (VIN editor: link is redirected to https://ebusiness.avma.org/aahsd/study_search.aspx 03/21/17). Established last year, the website: www.avma.org/FindVetStudies lists veterinary clinical trials in other specialties, as well.

At the other extreme of treatment decision-making concerns is the 'do everything' client without financial constraints. For these pet owners, practitioners must advocate only for therapy that is truly in the best interest of the pet. A variety of advanced care options including radiation therapy and nuclear medicine, tumor vaccines, multimodality therapy, and bone marrow transplantation have become available at a higher price for selected companion animal cases. Where financial constraints are not a concern, these options may offer the chance for a cure and, as such, may be considered reasonable pursuits. However, it is important that we counsel these clients who have greater financial resources about potential side effects and provide realistic expectations for the success of therapy.

Palliative Care, End-of-Life Care, and Euthanasia

Central to the evidence-based medicine debate in human medicine is the field of palliative care. In many regards, the same issues apply to veterinary oncology. Whereas the veterinary oath recognizes relief of suffering as a tenant of good veterinary practice, this is not part of the Hippocratic Oath for physicians.[13] The World Health Organization defines palliative care as "an approach that improves the quality of life of patients and their families facing the problem associated with life-threatening illness, through the prevention and relief of suffering by means of early identification and impeccable assessment and treatment of pain...."[10] The WHO definition falls noticeably short of advocating for euthanasia as a means of preventing and relieving suffering. The Hippocratic Oath, in fact, speaks against it: "I will give no deadly medicine to anyone if asked, nor suggest any such counsel."[13] As a result, physicians may be more apt to consider end-of-life care than veterinarians, who have patient euthanasia as an option.

Conclusions

It is time that we, as veterinarians, engage in a dialogue about the cost of "doing something." While we enjoy the flexibility our profession provides with regard to individualizing our patients' treatment plans beyond the confines of third-party payer issues or institutionally defined standard-of-care practice, we must commit to providing care that is in keeping with our medical ethics, the principles of the veterinary oath, and our professional obligation to serve as responsible stewards of our medical knowledge and expertise.

References

1. Wee B, Hillier R. *Cochrane Database Syst Rev.* 2008:CD005177.
2. Visser C, *et al. Cancer Biol Med.* 2015;12:193.
3. Wee B, *et al. Palliat Med.* 2008;22:173.
4. Hirsch CA, *et al. Palliat Med.* 2013;27:732.
5. Rose N, Toews L, Pang DS. *BMC Vet Res.* 2016;12:40.
6. Viner B. *J Feline Med Surg.* 2010;12:561.
7. Toews L. *J Vet Med Educ.* 2011;38(2):123.
8. Sackett DL, *et al. BMJ.* 1996;312:71.
9. Sahora A, Khanna C. *J Vet Intern Med.* 2010;24(1):51.
10. Tilburt JC. *J Eval Clin Pract.* 2008;14(5):721.
11. Sackett DL, Straus SE, Richardson WS, Rosenberg W, Haynes RB. *Evidence-Based Medicine: How to Practice and Teach EBM.* 2nd ed. Edinburgh: Churchill Livingstone; 2000.
12. Hay MC, *et al. J Eval Clin Pract.* 2008;14(5):707.
13. World Health Organisation. WHO Definition of Palliative Care 2015. www.who.int/cancer/palliative/definition/en/

1.
acvim281

Canine Non-Hematopoietic Gastrointestinal Tumors

Angela E. Frimberger, VMD, MANZCVS, DACVIM (Oncology); Ashley A. Smith, DVM
Wauchope, NSW, Australia

Canine gastrointestinal tumors are relatively uncommon tumors that can have marked impact on quality of life and cause significant morbidity and mortality of affected dogs. This lecture will exclude tumors arising from the hematopoietic system as they have been well reviewed elsewhere. Likewise, oral tumors have been extensively reviewed elsewhere and will not be discussed here. Carcinomas and sarcomas affecting the gastrointestinal tract are only moderately well characterized in the published literature.

GASTRIC TUMORS

The most common non-hematopoietic stomach tumor in dogs is adenocarcinoma. Most affected dogs are elderly, with a median age of 10 years. Belgian Shepherds, Rough Collies, and Chow-Chows, and some terriers, are considered as being at high risk.

Although gastric carcinomas can be experimentally induced in dogs by various compounds, there are no epidemiologic studies to support a role for these compounds in naturally occurring disease so a direct cause is uncertain. *Helicobacter* have been identified in canine gastric carcinomas, although an epidemiologic link to carcinogenesis has not been made.

In most affected dogs, ulceration is common and often deep and crater-like, causing inappetence and vomiting, hematemesis and/or melena. Clinical signs are often chronic with a median duration of two months. For symptomatic management, and a multimodality antiemetic approach may be most effective. In humans, the NK-1 inhibitors (maropitant in dogs) are best for treatment of delayed vomiting, while the serotonin inhibitors (ondansetron and dolasetron) are best for acute vomiting. We suggest using maropitant +/- ondansetron prophylactically in affected dogs rather than waiting until it is needed. Pain management is probably best done with opioids, rather than corticosteroids or NSAIDs which could exacerbate gastric signs. **Nutritional support is important for dogs with any tumor involving the gastrointestinal tract.**

Gastric adenocarcinomas metastasize in approximately 75% of affected dogs, particularly to perigastric/mesenteric lymph nodes and to other viscera. Tumors usually arise in the mucosa, but most extend to or through the serosa. Extension through the serosa creates an intense scirrhous reaction in the mesentery and omentum, and this may cause ascites.

Surgery is considered the primary treatment for gastric carcinoma; however, the advanced stage of gastric adenocarcinomas at diagnosis and their diffuse nature mean that complete resection is not often possible, and their high rate of metastasis means even complete resection is unlikely to be curative. Most occur in the lower two thirds of the stomach, particularly the pylorus, and often involve a large area of the stomach wall, making them unresectable. Even with complete resection, most dogs die within four months of surgery from local recurrence and metastases. The median survival in another study was 2 days. Wide resection often requires an aggressive gastroduodenostomy. In one study where dogs were treated with this surgery alone, recurrence of signs was delayed for 10 months in one dog, but most had much earlier recurrence. On the other hand, although anecdotal, we have seen one patient that had a complete resection of a small gastric carcinoma live more than 2 years. Poor quality of life, intractable vomiting, and hematemesis are the usual reasons for euthanasia.

There are few reports of chemotherapy for this disease. In humans, platinum and fluoropyrimidine chemotherapy remains the backbone of therapy with ongoing debate as to the benefit of triplet therapy with either an anthracycline or taxane. In case reports, dogs were treated with (1) 5-fluorouracil and cyclophosphamide and was euthanized due to progression after 9 weeks; (2) 5-fluorouracil, doxorubicin, and cyclophosphamide followed by cisplatin and lived 7 ½ months; (3) cisplatin and 5- fluorouracil with no response and lived for 4 months after incomplete resection of the tumor. Three dogs were treated with toceranib and had (1) response for 3 to 4 months, (2) progressive disease, and (3) stable disease (in combination with CCNU). In humans sunitinib (very similar to toceranib) results in a response rate of less than 3%. Anecdotally, we have seen responses to doxorubicin.

This is a difficult disease in all species. In human patients with unresectable gastric carcinoma treated with chemotherapy, the best response rates are around 40%, for a median of less than 4 months. Recently published trials of biological agents, especially targeting the Her2 receptor (not used in dogs because of biological incompatibility), have provided some signs of improvement.

Intestinal Carcinomas

In dogs, intestinal carcinomas are more common than sarcomas, and both can affect either the small or large intestine.

Small Intestinal Carcinomas

Although adenocarcinoma is the most common non-hematopoietic small intestinal tumor in dogs, it is still infrequently described. Small intestinal adenocarcinomas are usually located in the jejunum and occur in older dogs causing vomiting, anorexia, and weight loss in most patients. Anemia is common, secondary to chronic disease or gastrointestinal bleeding. Ultrasonography shows bowel wall thickening, loss of layering, decreased motility, and regional lymphadenopathy.

Metastasis to the regional lymph nodes is present at surgery in 60% of affected dogs, and involvement of the mesentery, peritoneum, and omentum (carcinomatosis) are also common. Pulmonary metastasis appears to be rare, although pleural spread and bone metastasis has been reported; but most metastases are intra-abdominal.

Wide resection and anastomosis is the basis of treatment for intestinal carcinoma, and lymph nodes should be biopsied at surgery. In published studies, overall survival for 38 dogs with small intestinal carcinomas treated by surgery alone was 8 to 10 months; 41% of patients were alive 1 year after surgery. Dogs without histologic evidence of lymph node metastasis had longer median survival (approximately 15 months compared with 3 months if metastases were detected). Age, weight, severity of clinical signs, laboratory testing, and location of tumor had no effect on survival. A minimum surgical margin of 5 cm on either side of the visible tumor is usually recommended, but has not been examined as a prognostic factor in studies. Although this may not seem as important when mesenteric spread is present, local recurrence is often a greater source of morbidity than mesenteric recurrence.

There is little published information regarding chemotherapy for dogs with intestinal adenocarcinoma. Case reports describe possibly extended survival in dogs treated postoperatively with doxorubicin, carboplatin, cisplatin, and 5- fluorouracil, including regression of mesenteric lymph node metastasis with doxorubicin. In cats with colonic carcinoma, both doxorubicin and carboplatin improve median survival rates. We have also seen anecdotal response to doxorubicin in dogs with metastatic or recurrent intestinal carcinoma, occasionally of long duration (greater than 18 months). In human patients with colonic carcinoma, postoperative survival times are improved by chemotherapy; however, small intestinal carcinoma is rare in people, and while chemotherapy is used when metastases are detected, there are insufficient data to know whether adjunctive therapy after surgery improves survival. The drugs used in people are 5-flurouracil, oxaliplatin, capecitabine, and irinotecan; of which only 5-fluorouracil has been used extensively in dogs.

Nearly half of the canine small intestinal carcinomas examined in one study showed marked expression of cyclo-oxygenase-2 (cox-2) in tumor cells. This has led to speculation that non-steroidal inhibitors of cox-2, such as piroxicam, may play a role in the treatment of intestinal carcinoma in dogs.

Large Intestinal Carcinomas

Canine colorectal carcinomas may extend to involve a few centimeters of the colon, and occasionally they can involve most of the large intestine. The size of the tumor does not necessarily indicate whether the tumor is benign or malignant; some rectal polyps can be 3 to 7 cm in diameter. Many colorectal carcinomas occur in the distal rectum, causing tenesmus and hematochezia in most patients. Distal rectal carcinomas are mostly palpable as a single pedunculated mass; whereas tumors located proximally are more likely to be multiple ("cobblestone appearance") or appear as an annular constriction. In one study of colorectal adenocarcinomas, dogs with annular tumors had the shortest average survival (1.6 months);

dogs with nodular or "cobblestone" lesions had an average survival of 12 months; and single pedunculated lesions were associated with the longest survival (32 months) after surgery - these survival times are probably related to the ease of surgical excision and imply that surgical margins are important.

Metastasis of colorectal adenocarcinoma is considered uncommon. In one large study, there was no evidence of metastasis in 78 dogs with this disease, and other work found a total metastatic rate of 23% to the sublumbar lymph nodes. When we have seen metastasis, it has also been to regional lymph nodes.

Aggressive surgery is the basis of treatment for rectal carcinoma. In earlier literature, recurrence after local excision was common however, in more recent studies that used rectal pull-through surgery, recurrence was seen in only 11% and 19% of dogs; with median control of 24 months, and median survival times 23 to 45 months reported. Reported postsurgical complications of rectal pull-through may be common and can included fecal incontinence, diarrhea, tenesmus, stricture formation, rectal bleeding, constipation, dehiscence, and infection. This is probably a surgery best performed by a specialist surgeon.

There is no published information regarding chemotherapy for dogs with rectal adenocarcinoma, mostly because the metastatic rate is low and local control is emphasized in treatment reports. Chemotherapy alone is unlikely to be as effective as surgery; and if the lesion is completely resected, and there is no vascular invasion on the excised sample, then chemotherapy after surgery may not be advantageous. If metastases are found on staging, or suspected on histopathology, then chemotherapy should be considered. As above, both doxorubicin and carboplatin have been shown to be beneficial in cats with colonic carcinoma; and anecdotally we have seen responses to doxorubicin.

Intestinal Mesenchymal Tumors

Most early literature classified all intestinal mesenchymal tumors as leiomyosarcomas. More recently some of these have been reclassified as gastrointestinal stromal tumors (GIST); therefore, some of the tumor behavior for GIST is found in the early literature for leiomyosarcomas.

Conventionally, GIST are distinguished by positive immunostaining for c-kit (CD-117), however, neither c-kit staining nor mutation are uniform. In one study, one-third of GIST stained with smooth muscle actin (as did nearly all leiomyomas), and all GIST were negative for desmin staining (compared with 62% desmin-positive leiomyomas). Mitotic index was 0 in all leiomyomas, and 0–19 per 10 hpf in GIST.

In humans, about 5–10% of GIST stain negative or only weakly positive for c-kit. In those cases, a further antibody (DOG1 [Discovered On GIST-1]) is considered highly sensitive and specific. A recent study evaluated a commercially available DOG1 antibody for use in diagnosis of canine GIST. Fifty-five primary mesenchymal gastrointestinal tumors were stained for c-kit, DOG1, and desmin and 33 (60%) were diagnosed as GIST; the great majority (97%) of these were positive for both c-kit and DOG1 but DOG1 expression was identified in 2 of those tumors that were negative for c-kit and desmin and our clinical experience has been similar to this.

Gastrointestinal sarcomas are less likely to metastasize than carcinomas; metastases occur in less than 30% of affected dogs, often long after definitive surgery. In 21 GIST patients with clinical follow-up; 6 (29%) developed metastases to liver or other sites within the abdominal cavity. In another study, 28% of GIST, but no leiomyosarcomas, metastasized. In one study the median survival for dogs that had complete excision of GIST was 36 months, which was longer (but not statistically significantly) than that for dogs with leiomyosarcoma (9 months).

Aggressive surgery is the basis of treatment for these tumors and many dogs with GIST have long survival with complete surgery alone. Overall the risk for metastasis from GIST in dogs is around 30%. For dogs with metastatic or invasive tumors or tumors that cannot be completely resected, or that have high risk characteristics, adjunctive therapy should be considered. Many dogs with non-GIST intestinal sarcomas have long survival with surgery alone, particularly if excision is complete and the overall metastatic rate is moderate.

In human patients with GIST without metastasis, low-risk status is assigned to patients with a tumor 2–5 cm in diameter, and with a mitotic count of <1 per 10 hpf. High-risk status is assigned to patients with a tumor >10 cm in diameter, mitotic count of >2 per 10 hpf, or that have ruptured, or are multifocal. The management of patients with "grey-zone" characteristics remains controversial. The same system appeared predictive for metastasis in a small preliminary study in dogs.

For soft tissue sarcomas in other locations, the most important prognostic feature is tumor grade, consisting of: mitotic index, percent necrosis, and differentiation. This system has not been applied to GI sarcomas, but some oncologists believe it may predict metastases.

There is very little published information regarding chemotherapy for dogs with GIST. In humans, sunitinib and imatinib are the clear drugs of choice, providing spectacular clinical responses and some cures in what was previously considered an otherwise untreatable cancer via both anti-tumor effects and antiangiogenic effects. Therefore, sunitinib's homologue, toceranib, is considered probably the drug of choice for dogs with GIST. Masitinib may also have utility but is not as closely homologues to sunitinib; and there are 2 case reports of imatinib use in dogs with GISTs.

For intestinal non-GIST sarcomas, there is no clear guidance in the literature as to the potential benefit of chemotherapy or which drug to choose. In the case of tumors that are not completely resectable, or where there is invasion or metastasis present, the best alternative may be to extrapolate from the more plentiful published data for spindle cell sarcomas in other anatomic locations. In dogs with metastatic or recurrent sarcomas doxorubicin is usually considered the drug of choice, but CCNU, ifosfamide, or possibly carboplatin can also be considered.

References

Gastric Carcinoma

1. Patnaik AK, *et al. Vet Pathol.* 1978;15:600.
2. Sullivan M, *et al. Vet Rec.* 1987;120:79.
3. Sellon RK, *et al. J Vet Intern Med.* 1996;10:333.
4. Gualtieri M, *et al. Vet Clin North Am Small Anim Pract.* 1999;29:415.
5. Swann HM, *et al. J Am Anim Hosp Assoc.* 2002;38:157.
6. Spangler WL, *et al. Vet Pathol.* 2006;43:136.
7. Eisele J, *et al. Vet Surg.* 2010;39:261.
8. von Babo V, *et al. Tierärztl Prax Kleintier.* 2012;40:243.
9. Seim-Wikse T, *et al. Acta Vet Scand.* 2013;55:25.
10. Lee HC, *et al. J Vet Med Sci.* 2014;76:1033.
11. Hugen S, *et al. Vet Comp Oncol.* 2016; epub ahead of print.

Small Intestinal Carcinoma

1. Birchard SJ, *et al. J Am Anim Hosp Assoc.* 1986;22:533.
2. Slawienski MJ, *et al. J Am Vet Med Assoc.* 1997;211:878.
3. Crawshaw J, *et al. J Am Anim Hosp Assoc.* 1998;34:45.
4. Esplin DG, *et al. J Am Anim Hosp Assoc.* 1998;34:287.
5. Prater MR, *et al. J Am Anim Hosp Assoc.* 2000;36:169.
6. Abrahams NA, *et al. Dis Colon Rectum.* 2002;45:1496.
7. Paoloni MC, *et al. Vet Radiol Ultrasound.* 2002;43:562.
8. McEntee MF, *et al. Vet Pathol.* 2002;39:428.
9. Penninck DG, *et al. Vet Radiol Ultrasound.* 2003;44:570.
10. Stanclift RM, *et al. J Am Vet Med Assoc.* 2004;225:1412.
11. Arteaga TA, *et al. J Am Anim Hosp Assoc.* 2012;48:399.
12. Pourmand K, *et al. Curr Gastroenterol Rep.* 2016;18:1.
13. Ecker BL, *et al. Cancer.* 2016;122:693.

Large Intestinal Carcinoma

1. Church EM, *et al. J Am Vet Med Assoc.* 1987;190:1427.
2. Danova NA, *et al. Vet Surg.* 2006;35:337.
3. Knottenbelt C, *et al. J Smal Anim Pract.* 2006;47:196.
4. Morello E, *et al. Vet Surg.* 2008;37:420.
5. Yoon HY, *et al. J Am Vet Med Assoc.* 2008;232:1016.
6. Sarathchandra SK, *et al. Aust Vet J.* 2009;87:356.
7. Nucci DJ, *et al. J Am Vet Med Assoc.* 2014;245:684.

Intestinal Mesenchymal Tumors
1. Frost D, *et al. Vet Pathol.* 2003;40:42.
2. Van der Zwan SM, *et al. Cancer.* 2005;104:1781.
3. Maas CPHJ, *et al. Vet Surg.* 2007;36:302.
4. Russell KN. *J Am Vet Med Assoc.* 2007;230:1329.
5. Gillespie V, *et al. Vet Pathol.* 2010;48:283.
6. Hanazono K, *et al. J Vet Med Sci.* 2012;74:1477.
7. Hayes S, *et al. Vet Pathol.* 2013;50:779.
8. Dailey DD, *et al. J Vet Diagn Invest.* 2015;27:268.
9. Jakhetiya A, *et al. World J Gastrointest Surg.* 2016;8:345.
10. Kiśluk J, *et al. Prz Gastroenterol.* 2016;11:115.
11. Takanosu M, *et al. Vet J.* 2016;207:118.

acvim283

The Node Knows: Importance of Lymph Node Status in Canine Mast Cell Tumors

Erika L. Krick, VMD, DACVIM (Oncology)
Philadelphia, PA, USA

Introduction

Cutaneous mast cell tumors are common skin tumors in dogs. Many cutaneous mast cell tumors can be cured with complete surgical excision, but some tumors may metastasize. The regional lymph nodes are the most common site for metastasis from canine mast cell tumors. Lymph node status is also predictive of distant metastasis (or lack thereof). Both low and high grade mast cell tumors can metastasize to regional lymph nodes. Historically such metastatic behavior was thought to indicate a high risk of systemic spread and a poor survival time. More recent literature has shown that with appropriate treatment, dogs with low grade mast cell tumors with lymph node metastasis (Stage 2 disease) can experience a good prognosis. In contrast, lymph node metastasis is associated with a worse prognosis in dogs with high grade mast cell tumors, despite appropriate treatment. It is clear that lymph node status is an important part of canine cutaneous mast cell tumor staging, as it informs the recommended treatment plan, and, in some cases, the prognosis. The purpose of this talk is to review the evidence on which the above conclusions are based and to highlight remaining gaps in our understanding of the relationship between canine cutaneous mast cell tumors and regional lymph nodes.

Sampling The Appropriate Lymph Node

Who Is at Risk for Lymph Node Metastasis?

Dogs with high grade mast cell tumors and tumors of the muzzle and oral cavity are most likely to have lymph node metastasis at diagnosis.[1-6] Lymph node metastasis is also associated with a shorter survival time for these patients, though treatment aimed at the primary tumor and metastatic lymph node can improve the prognosis.[2-6] Although these patients are at higher risk for lymph node metastasis, even dogs with low grade tumors can have lymph node metastasis at diagnosis.[1] Because this type of metastasis changes the treatment recommendations, regional lymph node evaluation is recommended for any dog with a mast cell tumor, ideally prior to surgical removal. If the lymph node status is known prior to surgery, then removing the lymph node can be done at the same time as the primary tumor resection. In addition, if a dog has a metastatic lymph node, then it is recommended to stage for distant metastasis. One study found that 30.9% of 220 dogs with mast cell tumors who underwent full staging had lymph node metastasis at diagnosis.[7] Of those dogs, 6.8% had distant metastasis as well, while none of the dogs without lymph node metastasis had distant metastasis.[7]

Which Node Is the "Right" One to Sample?

The typical recommendation, and what is described in most of the literature regarding mast cell tumor lymph node metastasis in dogs, is to sample the regional draining lymph nodes. Charts that describe the flow of lymph drainage in dogs are available and can be helpful in identifying the most appropriate node for evaluation.[8] The draining lymph node is not always the closest lymph node to the primary site; for example, the pinna drains to the prescapular, not mandibular, lymph node.[8] In order to improve the diagnostic accuracy of lymph node evaluation, methods for detecting sentinel lymph nodes (lymph nodes that drain a certain tumor as opposed to simply the closest anatomic lymph nodes) have been evaluated. Lymphoscintigraphy was reported in 20 mast cell tumors from 19 dogs.[9] Eight of the dogs had sentinel lymph nodes that were not the closest lymph node to the primary tumor.[9] Two of these dogs had pinna mast cell tumors for which the sentinel lymph node was the prescapular lymph node as opposed to the mandibular lymph node (as expected), and for the other dogs the sentinel lymph node was another lymph node further in the chain of the regional lymph nodes (for example, an inguinal lymph node as opposed to a popliteal lymph node for a dog with a stifle mast cell tumor).[9] Indirect lymphography (IL) using iodized oil has also been described to identify sentinel lymph nodes in 29 dogs with a variety of tumor types.[10] Mast cell tumor was the most common diagnosis, comprising 15 of the 30 tumors described. Iodized oil was injected intra- or peri-tumorally, then radiographs of the area were obtained 24 hours later to identify lymph node uptake.[10] At the time of surgical removal of the primary tumor, methylene blue (MB) was injected in the same area 15 minutes prior to the start of the surgical excision.[10] Of the 26 instances of both primary tumor and sentinel lymph node removal, 22 cases had agreement between the two methods (IL and MB).[10] Sentinel lymph nodes were identified via IL for 25 of 26 tumors, and only 7 of the identified lymph nodes were palpable.[10] Specific information about which sentinel lymph nodes were identified for dogs with mast cell tumors was not provided, but 7 of the 15 identified lymph nodes were metastatic.[10] These methods are associated with a need for specialized equipment (scintigraphy),[9] additional procedures prior to surgery,[10] longer anesthesia time and increased expense. Although there is some use for them, the literature that describes evaluating the lymph node in closest proximity following the canine lymph drainage pattern has given consistent results regarding the likelihood of metastasis and the prognosis. If a significant number of metastatic sentinel lymph nodes were being undetected and thus untreated, one would expect reports of dogs with mast cell tumors having disease progression in unexpected lymph node locations. Currently these reports are rare to non-existent, indicating that current methods of determining which lymph nodes to evaluate in dogs with mast cell tumors are appropriate.

Is Lymph Node Palpation Enough? No!

When the regional lymph node is a normally palpable lymph node, such as the mandibular, prescapular, or popliteal lymph node, clinicians may be tempted to go by the feel of the lymph node instead of performing an aspirate and cytology prior to surgical removal of the cutaneous mast cell tumor. When results of palpation, cytology, and histopathology of regional lymph nodes were compared in 44 dogs and cats with a variety of tumors, 1 in 5 palpably normal lymph nodes was metastatic on excisional lymph node biopsy.[11] Interestingly, only 36% of palpably enlarged lymph nodes were metastatic.[11] The sensitivity of lymph node palpation to detect metastatic disease was 93%, but the specificity was only 14%.[11] When patients with normal and mildly enlarged lymph nodes were grouped together compared to patients with moderately or severely enlarged lymph nodes, the sensitivity decreased to 60%, but the specificity increased to 72%.[11] Overall, patients with normal feeling lymph nodes had a low chance of metastasis, but, perhaps surprisingly, only 53% of patients with moderately or severely enlarged lymph nodes were found to have lymph node metastasis.[11] Seven dogs with mast cell tumors of varying grades were included in this study, and six of them had lymph node metastasis.[11] Specific information about the results of palpation of those lymph nodes was not provided in this study; however, the general findings indicate that lymph node sampling (as opposed to palpation only) is prudent as part of staging for dogs with cutaneous mast cell tumors.

Although palpation is not necessarily an accurate predictor of lymph node metastasis, fine needle aspiration and cytology does seem to be an accurate predictor of histopathology, which is considered the gold standard test for determining lymph node metastasis. Fine needle aspiration and cytology had a sensitivity of 100% and specificity of 96%.[11] Specifically for the seven dogs in that study with mast cell tumors, 6 of 7 dogs had metastatic lymph nodes on cytology and histopathology, while one dog had metastasis on cytology but not on histopathology.[11] A more recent study comparing the results of cytology and histopathology of lymph nodes in dogs found that the sensitivity and specificity of cytology for diagnosing neoplasia in the lymph node was 90% and 73%, respectively, which contrasts to the earlier reported findings.[12] Of the 48 lymph nodes with histopathologically confirmed metastatic mast cell tumor, cytology matched that diagnosis in 33 patients.[12] False negatives occurred in 15 lymph nodes, and false positives occurred in 7 lymph nodes when cytologic diagnosis was compared to histopathologic diagnosis.[12] Based on this information, it may be reasonable to consider lymph node histopathology in dogs with mast cell tumors when the regional lymph nodes are accessible for excision. This study did not report on lymph node palpation results, however, so it is unknown if the false positive or negative rate changed with lymph node palpation. Abnormal feeling lymph nodes are more likely to be sampled, so this study may be biased towards patients with palpably abnormal lymph nodes and may not represent the general group of dogs with mast cell tumors and lymph nodes of varied sizes.

In practicality, it may not be feasible to obtain a diagnostic sample of the regional lymph node - it may be not palpable or it may be too small to obtain a diagnostic aspirate. If the regional lymph node is an internal lymph node, such as the medial iliac lymph node, the author recommends diagnostic imaging of the lymph node and image-guided sampling if the lymph node is enlarged and sampling is feasible. If the non-palpable lymph node is normal on diagnostic imaging, or if an external lymph node is not palpable or a diagnostic cytological sample cannot be obtained from it, it is reasonable to conclude that the risk of metastasis to that lymph node is low. According to the study cited above, the risk of lymph node metastasis in that scenario is 20%.[11] If the lymph node is accessible for excisional biopsy at the time of primary tumor removal, then it is reasonable to remove it at the time of surgery for histopathological evaluation.

Interpreting Lymph Node Cytology And Histopathology

The next challenge after determining which lymph node to sample and obtaining a cytologic and/or histopathologic sample is to correctly interpret the pathologist's evaluation of the sample. Until recently, no guidelines for diagnosis lymph node metastasis via cytology or histopathology in dogs with mast cell tumors were available, so methods of diagnosis such metastasis (or lack thereof) were not standardized among research reports. Many neoplastic cells are not normally found in lymph nodes, so their presence alone is enough to make a diagnosis of metastasis; however, because mast cells are present in normal canine lymph nodes, simply the presence of mast cells is not sufficient to diagnosis metastatic disease. Cytological indicators of lymph node metastasis from mast cell tumors include effacement of lymphoid tissue by mast cells, aggregates of poorly differentiated mast cells, and/or greater than 5 aggregates of at least 3 mast cells.[5] This definition of "certain" metastasis was associated with a shorter survival time compared to lymph nodes categorized as "probable" or "possible" metastasis, reactive, or normal.[5] In that study, Grade 3 tumors were significantly more likely to have a metastatic lymph node compared to lower grade tumors.[5] Histopathologic criteria of mast cell tumor lymph node metastasis include aggregates of at least 3 mast cells in the sinuses or sheets of mast cells in the parenchyma or sinuses (early metastasis) and disruption or effacement of the normal lymph node anatomy by mast cells (metastasis).[13] Having standardized criteria for diagnosing lymph node metastasis in dogs with mast cell tumors allows better comparison of study results from different populations of dogs.

Outcome For Stage 2 Low Grade Mast Cell Tumors

The typical definitive treatment recommendation for a dog with a low grade mast cell tumor with lymph node metastasis is adequate local treatment of the primary tumor and metastatic lymph node followed by systemic chemotherapy. However, one prior study indicates that dogs with Stage 2 mast cell tumors can have an extended survival time following adequate loco-regional treatment and prednisone, without chemotherapy.[14] In that study, 19 dogs with Stage 2 mast cell tumors (2 Grade 3, 16 Grade 2, 1 Grade 1) received surgical removal of the primary tumor with or without the lymph node, followed by radiation therapy to the primary tumor site and regional lymph node and prednisone.[14] The median disease free survival was 1240 days, with 9 of the dogs alive with no mast cell disease progression at the time of data analysis.[14]

More recent studies have evaluated the outcome in dogs with Stage 2 low grade mast cell tumors following adequate loco-regional treatment and systemic chemotherapy.[15,16] One study specifically compares outcome between dogs with Grade 2 mast cell tumors with and without lymph node metastasis (55 dogs with lymph node metastasis compared to 35 dogs without lymph node metastasis).[15] Dogs received a variety of treatments in addition to removal of the primary tumor; 83% of dogs with lymph node metastasis received chemotherapy in addition to loco-regional treatments, while 51% of dogs in the no metastasis group received chemotherapy.[15] Lymph node metastasis did not significantly affect survival (median survival time not reached in either group because more than half of the dogs were still alive for at least 4 years after diagnosis); however, among dogs with lymph node metastasis, removal of the lymph node was associated with a longer survival time (median not reached versus 43 months).[15] A second study again focused on Grade 2 mast cell tumors and reported the outcome of 21 dogs with Stage 2 tumors who received adequate loco-regional treatment and systemic chemotherapy.[16] Median disease free interval for these patients was 2120 days, and median survival time was 1359 days.[16] Local recurrence within the radiation therapy field occurred in 2 of the 21 dogs, and distant metastasis was not noted in any dog.[16]

The prognosis for dogs with Stage 2 low grade mast cell tumors is considered good with appropriate definitive treatment. Such treatment means complete excision of the primary tumor and lymph node or incomplete excision of the primary tumor followed by radiation therapy of the primary tumor and surgical excision of or radiation therapy to the lymph node, followed by systemic chemotherapy. Based on these studies, the next question to ask is how important is chemotherapy for achieving such positive outcomes - is chemotherapy necessary, or is adequate loco-regional treatment sufficient?

OUTCOME FOR STAGE 2 HIGH GRADE MAST CELL TUMORS

In contrast, lymph node status affects prognosis as well as treatment recommendations for dogs with Grade 3 mast cell tumors. Dogs with high grade mast cell tumors with no signs of metastasis at diagnosis treated with surgery and chemotherapy have a median progression free survival time of 349 days, compared to 77 days for dogs with lymph node metastasis (reported median overall survival was 540 versus 194 days, respectively).[6] Treatment of the lymph node in addition to adequate local control of the primary tumor and systemic chemotherapy was associated with a median survival time of 240 days compared to 42 days for dogs that did not have treatment specifically directed to the lymph node, but only lymph node status remained a significant factor in multi-variate analysis.[6] That study also found that recurrence of the primary tumor and lymph node metastasis were the two most common locations for tumor progression (occurred in 27 dogs compared to 5 dogs who went on to develop distant metastasis), underscoring the importance of lymph node evaluation and treatment in dogs with mast cell tumors.[6] An older study describes the outcome of full course radiation therapy to the primary tumor site and regional lymph node for dogs with incompletely excised Grade 3 mast cell tumors.[17] Records of dogs that received chemotherapy were excluded from this retrospective study to more accurately assess the efficacy of radiation therapy alone following surgery for these patients.[17] The median remission duration was 27.7 months, and the median survival time was 28 months.[17] Local recurrence and lymph node metastasis were also the more common locations for progressive disease (16 dogs progressed at the primary or lymph node site, while 1 dog developed distant metastasis).[17] Interestingly, both of these studies found that a size cut-off of 3 cm was significantly associated with prognosis, with dogs with tumors >3 cm diameter having a shorter survival time.[6,17]

Based on these studies, surgical removal of or radiation therapy to the regional lymph node in addition to adequate local control and systemic chemotherapy is often recommended for patients with Grade 3/high grade mast cell tumors at the author's institution.

CONCLUSION

Regional lymph nodes are the most common site for metastatic spread for canine mast cell tumors and are also an indicator of the risk of distant metastasis. Reliable methods of lymph node evaluation include fine needle aspiration for cytological evaluation and biopsy for histopathological evaluation. Regional lymph node status affects treatment recommendations for low grade mast cell tumors but does not significantly affect prognosis if appropriate treatment is pursued. On the other hand, lymph node status affects both treatment recommendations and prognosis for dogs with high grade/Grade 3 mast cell tumors. Surgery or irradiation of regional lymph nodes regardless of lymph node status may be recommended for dogs with high grade mast cell tumors due to the high risk of lymph node metastasis and the significant effect on prognosis of lymph node metastasis. We are currently unable to predict which dogs with low grade mast cell tumors are at risk for lymph node metastasis, and the added benefit of chemotherapy in addition to adequate loco-regional therapy for dogs with stage 2 low grade mast cell tumors is unknown. Essentially, when you think cutaneous mast cell tumor, think lymph node next!

References

1. Stefanello, *et al. J Am Vet Med Assoc*. 2015;246:765.
2. Hillman, *et al. J Am Vet Med Assoc*. 2010;237:936.
3. Gieger, *et al. J Vet Med Intern Med*. 2003;17:687.
4. Elliott, *et al. Vet Comp Oncol*. 2013;14:101.
5. Krick, *et al. Vet Comp Oncol*. 2009;7:130.
6. Hume, *et al. J Am Anim Hosp Assoc*. 2011;47:37.
7. Warland, *et al. Vet Comp Oncol*. 2014;12:287.
8. Rogers, *et al. Compendium*. 1993;15:397.
9. Worley, *Vet Comp Oncol*. 2014;12:215.
10. Brissot, *et al. Vet Comp Oncol*. 2016;DOI:10.1111/vco.12214.
11. Langenbach, *et al. J Am Vet Med Assoc*. 2001;218:1424.
12. Ku, *et al. Vet Comp Oncol*. 2016;DOI:10.1111/vco.12256.
13. Weishaar, *et al. J Comp Path*. 2014;151:329.
14. Chaffin, *et al. Vet Radiol Ultrasound*. 2002;43:392.
15. Baginski, *et al. J Am Anim Hosp Assoc*. 2014;50:89.
16. Lejeune, *et al. Vet Comp Oncol*. 2015;13:267.
17. Hahn, *et al. J Am Vet Med Assoc*. 2004;224:79.

acvim284

The Role of Radiation Therapy in the Management of Cardiac Tumors

Susan M. LaRue, DVM, PhD, DACVS, DACVR (Radiation Oncology)
Fort Collins, CO, USA

Introduction and Background

Cardiac tumors in any species present a therapeutic challenge. Surgical options are limited, and chemotherapy is generally not effective. New technologies in imaging, radiation treatment planning, and treatment delivery have elevated the role of radiation therapy for both palliative and definitive treatment of these tumors. The lecture associated with these proceedings will focus on specific strategies for the treatment of cardiac tumors in dogs.

Hemangiosarcoma (HSA), a highly malignant tumor of the endothelium, is by far the most common canine cardiac tumor. Located in the right atrium and/or auricle, hemangiosarcomas are the most common cause of pericardial effusion. Golden Retrievers may be over represented. Treatment of HSA in any location is challenging because this tumor is highly metastatic, and even aggressive chemotherapeutic regimens do not prevent metastatic spread. The goal of treating cardiac HSA is to palliate clinical signs and improve quality of life for a finite period of time. Surgery, chemotherapy, and radiation can all be used palliatively for these patients.

The term **heart base tumor** refers to tumors found in that location. Aortic body tumors are the most common heart base tumor (85%) but ectopic thyroid tumors can also occur in that location. Aortic body tumors are neuroendocrine tumors, and are also referred to as chemodectomas and paragangliomas, depending on the cell of origin. They originate from chemoreceptors that are sensitive to changes in blood carbon dioxide, oxygen, and acidity. Chemodectomas have a breed predilection for brachycephalic breeds. It is hypothesized that low oxygen concentrations due to their unique anatomy combined with genetic susceptibility may cause these cancers. Most chemodectomas are slow growing and may not impact cardiac function. However, a subset can be more aggressive, and metastatic rates may be as high as 30%. Surgery is rarely an option for these tumors; however, a pericardial window can improve survival. These tumors are responsive to radiation therapy.

Other primary tumors, such as fibrosarcoma, chondrosarcoma, rhabdomyosarcoma, and myxosarcoma as well as benign tumors such as fibromas and myxomas have been reported. Treatment options for these unusual treatment are based on tumor location. Lymphoma is infrequently reported, but chemotherapy and/or radiation are treatment options.

Diagnosis

Dogs with cardiac tumors can present with pericardial effusion, blood flow obstruction within the heart, and/or abnormalities of the electromechanical function of the heart. Clinical signs will be based on which manifestations are present. Clinical signs of pericardial effusion may be vague and may develop acutely. Common signs of pericardial effusion include lethargy, dyspnea, collapse, anorexia, and abdominal distension. Physical exam findings include muffled heart sounds, a weak pulse, and jugular distension. If ascites is present there may be an abdominal fluid wave. Signs of blood flow obstruction may include syncope. Heart failure can also be a sequela to heart tumors. Radiographs are not always diagnostic, but are an important part of the evaluation process for pericardial effusion, the presence of a heart base tumor, and concomitant metastasis to the lungs. Electrocardiography may be normal, although 50% of dogs with effusion have low voltage QRS complexes. Echocardiography is the most sensitive and specific non-invasive evaluation for any of the cardiac tumors. While image guided aspirates or minimally invasive endoscopic biopsies provide histological confirmation, the correspondence between imaging presumed diagnosis and cytology/histology is better than for tumors in other body locations.

Management Before Treatment

Pericardiocentesis to remove fluid is necessary for patient stabilization. Keep in mind that pericardial fluid is advantageous for echocardiography, so image before tapping if possible. Patients in heart failure should be appropriately medicated to potentially improve outcome for patients undergoing surgery or radiation therapy. Blood flow obstructions that are life threatening should be ameliorated using non-invasive stenting before embarking on definitive radiation therapy treatments.

TREATMENT

Heart Base Tumors and Ectopic Thyroid Tumors

While it may be surprising to young oncologists, veterinary radiation oncologists have been treating heart base tumors for decades, using very basic technology. Diagnosis was based on thoracic radiography, the only imaging available at the time. Treatment was administered using fractionated radiation therapy, prescribed at a dose/volume that was considered safe for the normal tissues in the irradiated field. Many of these tumors were responsive to treatment, however, the tumor shrinkage occurs slowly, over a period of months. Radiation therapy relies on mitotic death to kill tumor cells and decrease tumor size. For dogs with advanced cardiac dysfunction who needed immediate relief, radiation therapy was often too little too late.

As echocardiography became the standard of imaging in veterinary cardiology, heart based tumors were identified more frequently. Many aortic body tumors are considered slow growing and need only active surveillance; if the tumor starts growing or impacts function, radiation therapy should be considered. However, some aortic body tumors are aggressive and can rapidly lead to serious cardiac dysfunction. These dogs can present with life-threatening dysfunction and need emergent stenting procedures prior to radiation therapy. Heart base tumors of any size are generally not amenable to surgical excision. Pericardectomies and pericardial windows have been shown to increase survival (median 730 days) compared to dogs that did not have the surgical procedure (median 42 days).

Prior to radiation therapy, the patient must undergo a CT scan optimized for radiation treatment planning. The patient is positioned using a variety of positioning devices to help accurately replicate the positioning at the time of treatment. In addition to standard non-contrast imaging, an arteriogram and venogram may more clearly identify tumor and normal structures. The images are networked to the planning system, and the tumor and all normal structures in the radiation field are identified and contoured. The pre-contrast study can be fused with the arteriogram and/or venogram to help this identification process. Additionally, assistance from the cardiology team is required to incorporate echo information into the anatomically based CT scans. A radiation therapy plan is created and approved, as is a quality assurance plan to verify dose administration. The patient can then be scheduled for treatment. Three radiation therapy technologies have vastly improved the ability to safely and effectively treat cardiac tumors.

Intensity Modulated Radiation Therapy (IMRT)

IMRT controls the intensity of each beam through motorized collimator leaves (MLC) that move during treatment, allowing the dose to be sculpted around important normal structures. This technique requires a specific treatment planning system that uses **inverse planning**. Inverse planning requires that the various tumor structures (gross tumor volume [GTV], clinical target volume [CTV], and planning target volume [PTV] as well as critical normal tissue structures (lung, esophagus, trachea) be identified so dose to those structures can be minimized. This is achieved by outlining (contouring) each structure of interest on each slice of the CT scan. Dose objectives for each structure are entered, and a sophisticated algorithm attempts to meet all objectives. The 2 major benefits associated with IMRT are: 1) dose to adjacent normal tissue structures can be minimized, dramatically reducing radiation effects and 2) radiation dose to the tumor can be increased, improving the probability of tumor control. For heart base tumors, this means that the tumor dose can be focused on the tumor and not on the normal portion of the heart. Dose to the lungs, esophagus, and trachea can be minimized. Inverse planning using intensity modulation can be used to develop plans for both finely fractionated radiation therapy (generally 15–20 fractions in veterinary medicine) and for stereotactic radiation therapy (1–5 treatments).

Image-Guided Radiation Therapy (IGRT)

This is another key component in the treatment of cardiac tumors in dogs. IGRT utilizes imaging systems built into the therapeutic accelerator to confirm the location of the target immediately prior to treatment.

For tumors associated with soft tissue structures (such as heart, lung, esophagus, and trachea), a cone beam CT (CBCT) is required. The patient is set-up for treatment using the customized positioning devices. Prior to treatment a CBCT is acquired and compared in 3 dimensions, to the CT used for the radiation treatment planning. IGRT allows the oncologist to confirm the location of the treatment target. The importance of this is that it allows a substantial decrease in the size of the PTV expansion, drastically decreasing the amount of normal tissues in the field.

Stereotactic Radiation Therapy (SRT)

Stereotactic radiation therapy is a new technology that has been applied to a wide variety of canine and feline tumor sites. SRT involves the use of high doses per fraction but overcomes the radiobiological limitations of high doses/fraction by avoiding dose to normal structures within the field. Stereotactic radiation therapy, by definition, requires: 1) a tumor for targeting (not microscopic disease); 2) treatment planning and administration that will provide a dramatic dose drop off between the tumor and the surrounding normal tissue structures (like IMRT); and 3) a method of stereotactically verifying patient positioning. The result is that normal late-responding tissue structures are spared through dose avoidance rather than by administering small doses per fraction. The normal tissue structures still receive dose, and the dose per fraction is higher than for traditional radiation therapy. This technique is appealing because only 1–5 fractions of radiation are required and treatments are generally well tolerated by the patient. Late effects of radiation are an extremely important issue in human patient treatment sites such as breast cancer and Hodgkin's disease. These late effects may take years to develop in humans, and at a somewhat shorter timespan in veterinary patients. Radiation therapy may therefore be the best acceptable therapy for canine cardiac tumors.

Clinical Outcome

In an ongoing study, twenty-nine patients with presumed chemodectomas or ectopic thyroid tumors were treated with finely fractionated radiation therapy or SRT. Median survival was over 400 days.

Cardiac Hemangiosarcoma

HSA is a highly metastatic disease. Palliative local treatment ± chemotherapy (to delay onset of metastatic disease) is a reasonable and achievable treatment option. Pericardiocentesis temporarily improves quality of life. Tumors located in the right auricle can be excised, and in one report, this surgery combined with chemotherapy resulted in a survival of 175 days. Less invasive options are preferred by some clients. Unlike heart base tumors, my belief is that pericardial resection or windows are not indicated for this condition. Pericardial taps are more effective than pleural taps, and dogs can potential exsanguinate if cardiac bleeding is left unabated. While pericardial fluid can cause tamponade, informed owners will return for additional pericardiocentesis. A recent publication indicated that palliative radiation therapy significantly decreased the frequency of pericardiocentesis compared to pre-irradiation. Median overall survival was 79 days.

Conclusions

The new renaissance in radiation oncology has introduced exciting technologies that may benefit canine and presumably feline patients with cardiac tumors. It is important that progress in this area be documented. Determining normal tissue tolerance of organs at risk is crucial to the development of optimal radiation therapy protocols. Minimally invasive procedures such as appropriate stenting of cardiac vasculature is important to stabilize patients prior to treatment and allow the radiation therapy to gradually reduce tumor size. As local tumor control improves, it is also critical that chemotherapy protocols are carefully evaluated for efficacy. Radiation therapy also offers a non-invasive treatment for cardiac HSA.

References

1. Janus I. *et al. Acto Veterinaria Hungarica*. 2016;64(1):90.
2. Rajagopalan V, *et al. Journal of Veterinary Internal Medicine*. 2013;27:1092.
3. McChesney Gillette S, *et al. Radiation Research*. 1991;125(1):34.
4. McChesney Gillette S, *et al. Radiation Therapy and Oncology*. 1992;23:41.

5. Hillers K, *et al. Journal of the American Animal Hospital Association*. 2007;43:187.
6. Weisse C, *et al. Journal of American Veterinary Medical Association*. 2005;226(4):575.
7. Nolan M, *et al. Journal of Veterinary Cardiology*. 2017; in press.

acvim286

Medical, Surgical & Radiation Oncology: Practical Approach to Diagnosis & Management of Osteosarcoma in Dogs and Cats

Sarah Boston[1], DVM, DVSc, DACVS, CVS Founding Fellow of Surgical Oncology; Michelle Turek[2], DVM, DACVIM (Oncology), DACVR (Radiation Oncology); Paul Woods[3], DVM, MS, DACVIM (Small Animal Internal Medicine & Oncology)
[1]Gainesville, FL, USA; [2]Madison, WI, USA; [3]Guelph, ON, Canada

INTRODUCTION

Appendicular osteosarcoma is the most common form of primary bone cancer in dogs. It primarily affects middle-aged to older dogs of large and giant breeds. It is important to note that the median age at presentation is 7 years, but there are two peaks in age at presentation. The large peak is in dogs 7–9 years of age, but there is also a small peak at 18–24 months of age. The presentation will vary with the progression of disease and the timing with which the owner elects to bring their dog for evaluation. The lameness can be mild to severe, acute to chronic. It is possible that mild cases that are early in their progression will respond to rest and NSAID therapy. If this treatment is elected in large-breed dogs, it is important to recheck the patient after 5–7 days to evaluate whether or not the pain or lameness persists and further workup should be recommended in these cases. It is also possible that this disease will present as a pathological fracture. This is generally associated with low-impact trauma such as playing in the park or jumping out of the car or off furniture.

Typically, osteosarcoma (OSA) is metaphyseal and endosteal in origin. The most common sites are the distal radius and proximal humerus, followed by the distal femur and proximal tibia. This has led to the adage, "away from the elbow, towards the knee." Although this holds fairly true in the forelimb, the tibia and femur can present in proximal or distal locations and it is important to remember that although these are the most common sites, other sites can be affected.

DIAGNOSTIC WORKUP

The decision to biopsy a suspect primary aggressive bone lesion will depend on the clinician's preference and the characteristics of the patient and lesion. Some clinicians will biopsy every bone site and others will only biopsy if there is an atypical signalment, location or radiographic appearance in cases that are consistent with osteosarcoma. It is important to consider fungal osteomyelitis in endemic areas. Potential risks of bone biopsy include increased lameness after the procedure, pathological fracture, a non-diagnostic sample and increased complications in limb-spare candidates. The risk of a non-diagnostic sample is decreased by taking the biopsy from the center of the lesion. For distal radial lesions where limb spare may be an option, the biopsy must be taken from one small incision as the biopsy tract must be removed with resection of the distal radius. The risk of pathological fracture after treatment with stereotactic radiosurgery is also a concern for patients that may receive this treatment modality. Because of the risks, time and expense involved in bone biopsy, many clinicians opt out of this diagnostic test. A fine needle aspirate (FNA) is an option with less risks and the benefit of a diagnosis of sarcoma or osteosarcoma in most cases. The diagnostic yield may be increased using ultrasound guidance. This technique will be painful and heavy sedation or general anesthesia is required. I use a large-gauge needle (18–20G) to improve the diagnostic yield. ALP staining may help to differentiate an osteosarcoma from a sarcoma cytologically. However, in most cases, the treatment of all primary sarcomas of bone will be similar.

Most cases (>90%) of canine osteosarcoma have micrometastasis at the time of presentation. Staging is done to determine if there is evidence of gross metastasis. Current recommendations for staging of osteosarcoma include evaluation of the lungs with three-view thoracic radiographs and a bone scan to assess for bone metastasis. The rate of finding gross lung metastasis is around 10%. This is slightly increased by thoracic CT and this is becoming a more common method of staging the lungs. Bone staging is more challenging at centers where scintigraphy is not readily available. Long-bone survey radiography has been reported as a method of long-bone staging, with up to 7.8% of dogs being positive for bone metastasis. However, in the author's experience, the rate of detecting bone metastasis with long-bone survey radiography is much lower than this. Whole-body CT scan is used at some centers for long-bone staging. However, a recent study indicated that this method is an insensitive test for long-bone staging and should be reserved for specific assessment of lesions that were previously detected on scintigraphy. Draining lymph nodes should be assessed with palpation, FNA if accessible and histopathology if the limb is amputated. The rate of metastasis to the lymph nodes is low, but carries a worse prognosis. Full blood work and urinalysis should be performed to ensure that the patient is overall healthy and can undergo treatment. The ALP has been shown to be a prognostic indicator in dogs and humans with osteosarcoma.

Treatment

The most common treatment of dogs with osteosarcoma is limb amputation. Most dogs tolerate this procedure extremely well and will adapt quickly to being on three legs. It is important that a full orthopedic examination is performed prior to amputation to assess for potential contraindications for limb amputation and to assess for pain at other bone sites that may be consistent with metastasis. Mild or moderate DJD in other joints is not considered a contraindication for amputation. Most dogs with osteosarcoma are not using the affected limb well and this will give the owner and attending clinician an idea of how well they will do on three legs.

For patients that will not tolerate amputation, limb-salvage procedures are available at select centers. A surgical limb spare is most often performed for lesions of the distal radius and usually involves resection of the affected distal radius with an adequate margin and the use of a metal endoprothesis spacer to fill the defect. The endoprothesis is secured using a specialized bone plate that spans from proximal radius to metacarpals. Stereotactic radiotherapy (SRT) is an option for lesions at other sites. This involves the administration of a high dose of radiation to the bone tumor, with little to no radiation given to the surrounding normal structures. This is a relatively new procedure but it holds promise for treatment of osteosarcoma in dogs. It is important to remember that although it is rare, it is possible to perform a limb-salvage procedure for primary bone tumors of the ulna by ulnectomy or partial ulnectomy or of the scapula by complete or partial scapulectomy. Other than cost, the biggest downside to SRT is the high risk of pathological fracture, with rates of up to 60% reported.

After aggressive local therapy, systemic chemotherapy is recommended as adjuvant therapy for osteosarcoma. This has been shown to significantly increase survival time. Dogs that are treated with amputation alone will succumb to disease in 3–5 months, whereas dogs treated with amputation or limb spare and chemotherapy will live for 9–12 months. Palliative amputation may still be appropriate in cases where the owner is not willing to pursue chemotherapy because the quality of life will be much better with removal of the primary tumor.

Palliative therapy may also be elected in the form of palliative radiation therapy and analgesics. Of all of the palliative approaches, palliative radiation is thought to have the most significant impact on pain control. Analgesia provided should be a combination of NSAIDs, bisphosphonates, opioids, and gabapentin. Chemotherapy may also help to treat pain and slow the progression of disease. These patients must be monitored carefully to assess their quality of life and euthanasia should be recommended in cases where the patient remains painful or becomes painful again. Many dogs with chronic pain will not cry out or change their behavior much and it is very important to educate owners that the best way to assess pain due to osteosarcoma is to evaluate limb use. Other factors that should be assessed include activity, appetite, willingness to participate in normal activities with the owners, and ability to sleep soundly and dream.

Radiotherapy is an important tool in the armamentarium of treatment options for OSA. As radiation-delivery technology has advanced, our uses of radiotherapy have evolved. We will discuss the tried-and-true approach of palliative radiotherapy for OSA, as well as new cutting-edge techniques that may have the potential to offer effective, non-surgical, limb-sparing therapy.

Generally speaking, radiation therapy can be used for two distinct purposes to treat cancer: **palliation** and **definitive tumor control**. The goal of the **palliative** approach is to reduce clinical signs, which in the case of OSA means to reduce tumor-associated bone pain. The goal with this approach is not to optimize tumor control, nor to extend patient longevity (i.e., cure), but rather to reduce pain associated with the tumor so that the patient may be more comfortable as it lives with the disease. In contrast, **definitive therapy** (also called curative-intent or full-course radiation) is designed to sterilize as many tumor cells as possible, achieve long-term tumor control and extend patient survival. In order to achieve tumor sterilization, high doses of radiation must be delivered to kill the malignant cells. To prevent severe, non-healing late-radiation effects, the radiation dose delivered per fraction has to be low. So, when a high total radiation dose is delivered with the intent of sterilizing a maximum number of tumor cells, the dose is divided into many daily low-dose treatment sessions delivered over a period of weeks in order to maintain the integrity of the normal tissues. This conventional approach to definitive radiotherapy is used for most tumor types and has been investigated for appendicular OSA in dogs. The results did not show a substantial improvement over palliative protocols and no survival benefit was observed, even in combination with chemotherapy. Thus, conventionally fractionated definitive radiotherapy (daily low-dose per fraction with a definitive intent) is generally not recommended for OSA.

Palliative radiotherapy is an important option for pet owners who may not have the financial resources, time or desire to pursue standard-of-care amputation and chemotherapy. For these pet owners, palliative radiation provides a less time-intensive and less expensive treatment option and it affords the patient an improved quality of life with reduced tumor-associated pain. Animals with advanced disease (metastases) or those that are poor candidates for amputation because of other mobility issues can also benefit from palliative radiation as an alternative to euthanasia. Palliative protocols are designed to reduce pain without causing significant side effects, since side effects would defeat the purpose of palliation. Common protocols include weekly hypofractionated regimes (8 Gy x 4 weekly treatments or 8 Gy x 3 treatments on days 0, 7 and 21) and two consecutive-day protocols of 8–10 Gy/fraction. The delay between treatments in the weekly protocols, and the short duration of the two-day protocol, result in a low risk of acute skin side effects (sunburn). The higher dose per fraction (8–10 Gy) increases the risk of late effects; however, this is considered acceptable given the poor long-term prognosis and given that late effects take months to years to develop. There is controversy over the optimal palliative treatment schedule. The studies suggest that the rate of achieving pain relief is 75–95% for these protocols and the median duration of effect is about 2–4 months. The time to pain relief may be shortest for the two-day protocols (median 2 days compared to 1–2 weeks). In people with painful bone cancer (most commonly bone metastasis), multiple studies have shown no additional benefit in palliation when using multifraction protocols compared to a single fraction (8 Gy/fraction). Single-fraction protocols are recommended as standard of care in people with bone pain due to cancer. The advantages of the shorter protocols in both species are decreased hospitalization time, owner/patient convenience, lower cost, and the option for re-treatment upon recurrence of pain. The median survival time associated with most palliative regimes is approximately 3–5 months, with one report suggesting as long as 10 months. Although the goal of palliative radiation is not to extend life expectancy, the act of improving quality of life may in fact result in a survival advantage by delaying euthanasia.

The benefit of adding chemotherapy to a palliative radiation protocol is unclear. Some oncologists feel that adjuvant chemotherapy is worthwhile given its known benefit in the curative-intent setting. Others feel that the added expense and risk of toxicity defeats the conservative approach of palliation. The use of bisphosphonates has been shown to be safe and to improve pain scores. While studies in dogs with OSA have suggested that there is no increase in pain relief or patient survival when radiation is combined with bisphosphonates, methodological difficulties in studying these therapies limit the interpretation of the reports. In people, pain control associated with bone cancer is undisputedly improved when a bisphosphonate, particularly zoledronate, is added to radiotherapy. Studies in people show a reduction in skeletal-related events and a trend towards increased survival in patients with bone metastasis. Additional, well-designed studies are necessary to confirm this effect in dogs. Based on the clear benefit in people, it is standard practice to combine palliative radiotherapy with bisphosphonates, preferably zoledronate, in dogs with OSA.

The evolution of radiotherapy technology has allowed for greater conformal dose delivery to tumors using intensity-modulated therapy (IMRT). IMRT techniques have made it possible to deliver high doses of radiation to the tumor target while reducing irradiation of critical, surrounding normal tissues. Since IMRT allows for reduced normal tissue side effects, it follows that the next opportunity for improvement in cancer therapy is to deliver higher doses of radiation for more effective tumor killing and in fewer treatments for greater patient convenience. This approach is what is referred to in veterinary medicine as stereotactic radiosurgery (SRS) or stereotactic body radiotherapy (SBRT). The first term is a misnomer; there is no surgery involved. SRS refers to the delivery of a single ablative radiation dose to a defined tumor target whose location is precisely defined. SBRT is similar, except that it involves the use of 3 to 5 high-dose fractions (the dose/fraction is lower than with SRS). Radiation is delivered in high doses with high precision to the tumor with steep dose gradients at the periphery of the tumor. Thus, adjacent normal tissues receive low doses of radiation that do not cause side effects. Stereotactic radiotherapy requires the tumor target to be very well defined. Tendrils of microscopic disease extending into normal tissue cannot be targeted effectively because the tumor-ablative dose of radiation needed to kill them would damage the healthy tissues infiltrated by the cancer cells. Thus, the high therapeutic ratio of stereotactic radiation cannot be achieved for invasive tumors with microscopic extension into normal tissues. OSA is believed to be more-or-less of a well-defined cancer that is minimally invasive into surrounding normal tissues, making it theoretically amenable to SRS or SBRT.

Due to the promise of effective tumor-ablative treatment without amputation, SRS/SBRT has become a highly desirable and sought-after approach. Despite the popularity and increased use of SRS/SBRT, there is still a lot to learn about how to optimize clinical outcome and safety. Based on limited studies of dogs that received ablative radiation doses of approximately 20–30 Gy in a single fraction, the median survival times are promising at 10–12 months. The high rate of 100% tumor necrosis following irradiation is also encouraging. It is important to note that the treatment has not yet been proven to be safe in terms of clinically important side effects. The rate of pathologic fracture associated with SRS (20–30 Gy in a single fraction) may be >50% with an average time to fracture of 5–6 months. The greatest risk of fracture appears to be with subchondral bone involvement. Other potential risks include skin necrosis and non-healing wounds in dogs where the tumor abuts the skin. All of these complications limit the feasibility of the approach. Additional studies are needed to investigate the safety and efficacy of fractionated stereotactic approaches (SBRT). When combined with chemotherapy, it is possible that this approach will provide a non-surgical, limb-sparing, definitive therapy for dogs with OSA with comparable tumor control to amputation and chemotherapy.

So, what should we offer as radiation treatment for our canine OSA patients? For dogs that are poor candidates for amputation or when pet owners decline amputation, and the option of definitive-approach SRS/SBRT is not interesting to the owner, then a palliative approach with the aim of reducing pain and improving quality of life is appropriate. This should also be considered for dogs with advanced disease (metastasis) that are otherwise feeling well. Any of the palliative radiation treatment schedules listed above is suitable. The addition of a bisphosphonate, preferentially zoledronate, is recommended. Concurrent use of chemotherapy is controversial. For pet owners interested in pursuing a definitive-intent treatment to maximize life expectancy, amputation and adjuvant chemotherapy currently represent the only validated approach that optimizes clinical outcome in dogs without gross metastatic disease. For pet owners interested in a definitive-intent approach using a non-surgical limb-sparing option, SRS/SBRT could be offered. In these cases, it should be emphasized to pet owners that these radiation techniques are new and safety and efficacy investigations are ongoing. Notwithstanding these unknowns, SRS/SBRT is an exciting advancement in veterinary cancer therapy that has the potential to impact the treatment paradigm for canine OSA.

Radiotherapy for OSA in cats has been investigated. Theoretically, the same radiation treatment options that exist for dogs are available for cats; however, clinical outcomes have not been evaluated.

References

1. Talbott JL, *et al. Vet Surg.* 2017;46(1):75.
2. Pagano C, *et al. Vet Radiol Ultrasound.* 2016;57(5):551.

3. Khanna C. *Vet Comp Oncol.* 2016;14(2):e1.
4. Kubicek L, *et al. Vet Radiol Ultrasound.* 2016;57(3):321.
5. Schmidt AF, *et al. Prev Vet Med.* 2016;125:116.
6. Mitchell KE, *et al. Vet Surg.* 2016;45(1):36.
7. Selmic LE, *et al. J Vet Intern Med.* 2014;28(2):554.
8. Schmidt AF, *et al. Prev Vet Med.* 2013;112(3-4):414.
9. Oblak ML, *et al. Vet Comp Oncol.* 2015;13(1):28.
10. Ryseff JK, *et al. Vet Clin Pathol.* 2012;41(3):391.
11. Armbrust LJ, *et al. J Am Vet Med Assoc.* 2012;240(9):1088.
12. Neihaus SA, *et al. J Am Anim Hosp Assoc.* 2011;47(5):317.
13. Eberle N, *et al. Vet Comp Oncol.* 2011;9(2):131.
14. McMahon M, *et al. J Vet Intern Med.* 2011;25(3):511.
15. Walter CU, *et al. Vet Comp Oncol.* 2005;3:1.
16. Green EM, *et al. J Am Anim Hosp Assoc.* 2002;38:445.
17. Ramirez O III, *et al. Vet Radiol Ultrasound.* 1999;40:517.
18. Mueller F, *et al. In Vivo.* 2005;19:713.
19. Knapp-Hoch, *et al. J Am Anim Hosp Assoc.* 2009;45:24.
20. Dennis K, *et al. Radiother Oncol.* 2013;106:5.
21. Fan TM, *et al. J Vet Intern Med.* 2008;22:380.
22. Lopez-Olivo MA, *et al. Support Care Cancer.* 2012;20:2985.
23. Farese JP, *et al. J Am Vet Med Assoc.* 2004;225:1567.
24. Kubicek L, *et al. Vet Radiol Ultrasound.* 2016;57(3):321.
25. Pagano C, *et al. Vet Radiol Ultrasound.* 2016;57(5):551.
26. Covey JL, *et al. Vet Surg.* 2014;43:174.

acvim288

Medical, Surgical & Radiation Oncology: Approach to Diagnosis & Management of Feline Injection-Site Sarcoma

Michelle Turek[1], DVM, DACVIM (Oncology), DACVR (Radiation Oncology); Sarah Boston, DVM, DVSc, DACVS, CVS Founding Fellow of Surgical Oncology; Paul Woods, DVM, MS, DACVIM (Small Animal Internal Medicine & Oncology)
[1]Madison, WI, USA; [2]Gainesville, FL, USA; [3]Guelph, ON, Canada

The association between vaccination and development of feline sarcoma was first suggested in 1991 following three vaccination-related events that occurred in the 80s: 1) a shift to production of a killed, adjuvanted rabies vaccine, 2) an increase in perceived threat of rabies in some areas that led to widespread vaccination of cats, and 3) introduction of the first FeLV vaccines. Since then, vaccine-associated sarcoma (VAS) has become a well-recognized phenomenon, and numerous studies have been performed to elucidate the pathogenesis of the disease and to identify effective therapies. Although the true incidence of feline injection-site sarcoma is unknown, it is reported to be approximately 1–100 per 10,000 cats. The term feline injection-site sarcoma (ISS) is preferred by some because tumors have reportedly developed after administration of other injectable medications such as methyl prednisolone acetate and certain antibiotics.

To address this complex and ethical issue, the Vaccine-Associated Feline Sarcoma Task Force (VAFSTF) was formed in 1996. The objectives of the task force were to present a unified perspective on behalf of the veterinary profession, make recommendations for diagnosis and management, and further the understanding of the disease by funding and directing research.

The VAFSTF made the following vaccination recommendations to reduce the possibility of sarcoma development and to improve chances of successful treatment in affected cats:

1. **Avoid over-vaccination**. The risk of tumor development has been shown to increase with the number of vaccines administered. Indiscriminate use of the FeLV vaccine should be avoided. Older exposed animals are less likely to develop persistent FeLV infection, so the risk:benefit ratio should be considered prior to vaccination of these animals.
2. **Single-dose administrations** should be utilized.
3. **Thoughtful vaccination protocols:** Administration of any vaccine should proceed only after strong consideration of the patient's exposure risk, the zoonotic potential, and the overall medical significance.
4. **Standardization of vaccination protocols and documentation** of injection site, type of vaccine or other injectable medication, manufacturer and serial number.
5. **Location of vaccine administration:** Vaccines containing rabies antigen should be given as **distally** as possible in the right hind limb; vaccines containing feline leukemia virus (unless also containing rabies antigen) should be given as **distally** as possible in the left hind limb; vaccines containing other antigens should be given on the right shoulder, avoiding the midline or interscapular space. These recommendations are to facilitate aggressive surgery and increase the likelihood of a complete excision in the case of tumor development. Locations where management is particularly difficult, such as the interscapular space, should be avoided.
6. **Vaccines should be administered subcutaneously.** This allows for easier and earlier detection of tumor growth (vs. intramuscular administration).

When considering vaccination sites, vaccines should be administered below the elbow or stifle. The routine administration of vaccinations over a hind limb has resulted in a shift of the location of ISS from the intrascapular region to the hip/flank area. If the cat is in a hunched position when it is vaccinated, this results in a vaccine that is administered in the flank region. The resulting tumor will usually be in a location that requires a surgical excision with a hemipelvectomy and body wall resection. Distal vaccination would convert this to a limb amputation. One pilot study has investigated tail vaccination as a potential vaccination site with promising results.

Tumors that develop at injection sites are typically mesenchymal in origin, and fibrosarcoma is most common. Other histologies include malignant fibrous histiocytoma, osteosarcoma, chondrosarcoma, undifferentiated sarcoma, and rhabdomyosarcoma. Compared to non-vaccination-related soft tissue sarcoma, ISS tends to have a more aggressive biological behavior and, histologically, it is distinguished from non-vaccine-related tumors by the presence of an inflammatory cell infiltrate, more areas of necrosis, increased mitotic activity, variation in cellular size and shape, and the presence of foreign material (adjuvant). A grading scheme has been proposed; however, unlike in dogs with soft tissue sarcomas, tumor grade has not been shown consistently to correlate with clinical outcome. The latency period of tumor development is variable, ranging from several weeks to years (median 1 year). To date, no single vaccine brand or manufacturer has been associated with VAS development.

The exact pathogenesis of ISS has not been elucidated. The most accepted hypothesis suggests that inflammation induced by injection, particularly that caused by vaccine components, promotes neoplastic transformation of fibroblasts and myofibroblasts in some cats leading to tumor development. Genetic changes may be present in affected cats that promote neoplastic transformation. Genetic screening for susceptible cats is an intriguing idea but is not feasible at this time as a causal genetic link to ISS development has not been found. Viral etiologies (FeLV, FIV, feline sarcoma virus, papillomavirus and polyomavirus) do not seem to play a role in the pathogenesis of this tumor.

At the time of vaccination, cat owners should be informed about the risks of ISS development. Aggressive intervention early in the course of disease is critical to effective treatment. Cat owners should be advised to evaluate the vaccination site regularly. Local inflammatory reactions (granulomas) are common and can form in the weeks after vaccination. These will regress.

The VAFSTF has made the following recommendations for diagnosis of lumps appearing at the vaccination site to differentiate postvaccinal inflammatory reactions from ISS:

1. Tissue biopsy should be performed if the mass (3-2-1 rule):
   a. Persists more than **3** months post-vaccination
   b. Is larger than **2** cm in diameter
   c. Is increasing in size after **1** month post-injection
2. Record anatomic location, size and shape.

Tissue biopsy is required for definitive diagnosis and for differentiation of postvaccinal inflammatory reaction versus ISS. Cytology is unreliable since pleomorphic reactive fibroblasts cannot be easily distinguished from malignant mesenchymal cells. Biopsy should be performed **prior** to surgical excision so as not to render a second surgery more difficult and extensive. The biopsy should be performed in such a way that subsequent surgery or radiation therapy can readily include the biopsy tract without having to enlarge the field of treatment. Principles of biopsy should be maintained so as to avoid contamination of unaffected tissues since tumor cells can potentially track along fascial planes and seed tumor at a distance from the primary mass.

Important information to glean from the biopsy report includes: 1) confirmation of the diagnosis, and 2) margins of excision. The diagnosis of an injection-related sarcoma, as opposed to the less aggressive non-injection-related soft tissue sarcoma, is supported by the presence of inflammation and/or foreign material representing adjuvant. The margins of excision inform the decision for adjuvant therapy, especially radiation. Despite the prognostic value of clean margins, most studies show that the recurrence rate following an excision deemed complete on histopathology is high. Histopathologic tumor grading has not been consistently shown to have prognostic value in ISS.

If a diagnosis of ISS is made, tumor staging should be performed prior to aggressive therapy. The metastatic rate at the time of initial presentation is relatively low, ranging from 6% to 23%. Metastasis occurs primarily in the lungs; however, other sites include regional lymph nodes, skin, and visceral organs. Cats diagnosed with ISS should undergo 3-view thoracic radiography, lymph node palpation with cytologic evaluation if indicated, and abdominal ultrasound in selected cases. The gross appearance of ISS and apparent size can be misleading. Advanced imaging using computed tomography or MRI is essential for accurate assessment of tumor extension in soft tissues. Determination of the extent of tumor is critical for surgical and radiotherapy planning. CT is a convenient method of combining local and distant staging and also combining a pretreatment CT to plan surgery and/or radiation therapy. This is particularly important if the definitive therapy will be determined based on CT scan. However, a recent study showed the challenges in differentiating neoplastic from non-neoplastic peritumoral changes based on imaging alone.

Due to the relatively low metastatic rate, treatment is aimed primarily at achieving local tumor control. If gross and microscopic disease is excised, then surgery in theory can be curative. Simple (marginal) excision inevitably leads to local recurrence, which is often more difficult to control than the original tumor. This is due to the invasive nature of this disease, which cannot be overstated. Median time to tumor recurrence after a conservative surgery alone is 3 months. Recurrence rates are as high as 70%. The most important prognostic indicator appears to be the completeness of surgical excision. It follows that the most favorable clinical outcome after surgery alone is achieved in cats treated at referral institutions where wide or radical surgical excision is more common. When a more aggressive surgery is performed, the median time to tumor recurrence is extended. Radical surgery has been associated with the lowest recurrence rates and favorable survival times. Radical surgery is defined as removal of 5 cm margins and 2 fascial planes deep often involving body wall resection, ostectomy and/or amputation. A recent study of 91 cats treated with radical surgery reported a recurrence rate of 14% and a median survival time of 2.5 years. Unfortunately, a histopathological report that indicates a complete excision may still be associated with a risk of recurrence of up to 50%. Feline ISS located distally on a limb and treated with amputation or hemipelvectomy are reported to have the longest time to tumor progression, which is likely attributable to the efficacy of the procedure at removing gross and microscopic disease. A recent study showed that since the VAFSTF was formed, there has been a shift in the location of tumor development in correlation with vaccination recommendations (i.e., there has been an increase in tumor development on the limbs). However, due to the challenges of vaccinating distally on the limb, many tumors occur proximally near the body wall where local control is less attainable with surgery and radiation. It is important to keep in mind that in order to optimize the chances for local control with amputation in the event of tumor formation, vaccines should be administered distally on the limb, or on the tail.

If a radical excision is not performed (5 cm margins and 2 fascial planes), it is unlikely that a lesser surgery will be of benefit unless combined with radiation therapy. Since surgery alone may not be curative with large tumors, multimodality therapy has been investigated. Radiation therapy can be used successfully in both the preoperative and postoperative settings to increase the tumor control rate when microscopic disease remains after a surgical excision. When used in this setting, a treatment of radiation is delivered daily for about 4 weeks in order to deliver a total dose in the range of 50–60 Gy. The most favorable clinical outcome (as measured by time to tumor progression) is reported for combination therapy of preoperative radiotherapy and surgery. The aim of preoperative radiotherapy is not necessarily to reduce the size of palpable tumor, but rather to "sterilize" any residual microscopic disease that might be left after radical surgical excision. Among the benefits of this approach is the fact that the radiation treatment field is smaller than it would be in the postoperative setting, because no surgery has been performed and no surgical scar is present. This limits the amount of normal tissue that is exposed to radiation. The smaller radiation treatment field makes it more possible to deliver an effective, tumoricidal radiation dose with low morbidity. The primary disadvantage of this approach is the possibility of impaired wound healing after surgery in tissues previously exposed to radiation, which is rare.

In a series of 92 cats with ISS treated with preoperative radiotherapy followed by surgical excision, median time to tumor progression was 1.5 years. The completeness of surgical excision (i.e., dirty vs. clean margins) was shown to be important for prognosis. In cats with a complete surgical excision (i.e., cancer cells did not extend up to the margin of excision), median time to tumor progression was 32 months versus 10 months for cats with an incomplete excision following RT. In the author's practice (MT), preoperative radiotherapy followed by aggressive or, if possible, radical surgical excision is the approach of choice when curative treatment is the goal. This requires a collaborative effort between the radiation oncologist and the surgeon. Many cats with ISS are now referred prior to any therapeutic intervention so that a multimodality team approach may be taken from the onset of treatment to optimize the likelihood of a successful outcome.

Radiotherapy also plays an important role in the palliation of cats with ISS. For cats with ISS-related clinical signs or a growing tumor when curative-intent treatment is not possible or not desired by the client, hypofractionated radiotherapy can result in favorable tumor responses and short-term palliation. Palliative radiation usually involves weekly hypofractionated radiation treatments (e.g., 8 Gy x 4 weekly treatments). Alternatively, 4 Gy x 5 daily fractions is also an option. A tumor response rate of 50%, including complete responses, has been reported with a median progression free interval of about 4 months. Pet owners of cats with newly diagnosed ISS for whom a curative-intent approach is not feasible or desirable, should be made aware that palliative radiation is a future treatment option to extend good quality of life as clinical signs begin to develop.

Stereotactic radiotherapy (SRT) is a new radiotherapy delivery approach available at some veterinary facilities with advanced radiotherapy equipment. It delivers precisely targeted radiation to the gross tumor mass in fewer high-dose treatments than traditional therapy, while preserving tissue outside of the gross tumor mass. The aim is to deliver tumoricidal doses of radiation to the tumor in 1–3 days and to spare normal surrounding tissues. ISS is a highly invasive tumor with extensions of microscopic cells into the surrounding normal tissues. Microscopic disease is not targetable with SRT. Thus, SRT is not a curative treatment option for ISS. A recent study evaluating SRT in cats with ISS showed strong tumor responses including complete and partial remissions of the primary tumor and a median survival time of 10 months. SRT is a reasonable and convenient treatment option for palliation of ISS.

The role of chemotherapy has not been robustly demonstrated but chemotherapy that has been employed after surgery ± radiation therapy includes doxorubicin, carboplatin, cyclophosphamide, lomustine (CCNU), and toceranib.

References

1. Nolan M, *et al. J Am Vet Med Assoc.* 2013;243(4):526.
2. Eckstein C, *et al. Vet Comp Oncol.* 2009;7(1):54.
3. Kobayashi T, *et al. Vet Radiol Ultrasound.* 2002;43(5):473.
4. Phelps H, *et al. J Am Vet Med Assoc.* 2011 239(1):97.
5. Hendricks C, *et al. J Feline Med Surg.* 2014;16(4):275.
6. Shaw SC, *et al. J Am Vet Med Assoc.* 2009 Feb 1;234(3):376-80.
7. Barber LG, *et al. J Am Anim Hosp Assoc.* 2000;36:416.
8. Cohen M, *et al. J Am Vet Med Assoc.* 2001;219:1582.
9. Poirier VJ, *et al. J Vet Intern Med.* 2002;16:726.
10. Bregazzi VS, *et al. J Am Vet Med Assoc.* 2001;218:547.
11. Katayama R, *et al. Cancer Chemother Pharmacol.* 2004;54:25.
12. Martano M, *et al. Vet J.* 2005;170:84.
13. Hahn KA, *et al. J Am Vet Med Assoc.* 2007;231:742.
14. Hampel V, *et al. J Vet Intern Med.* 2007;21:1340.
15. Kisseberth WC, *et al. J Vet Intern Med.* 2008;22:83.
16. VAFST *J Am Vet Med Assoc.* 2005;226:1821.
17. Holterman N, *et al. Vet Comp Oncol.* 2016 doi:10.1111/vco.12207.

acvim292

Medical, Surgical & Radiation Oncology: Practical Approach to Diagnosis and Management of Transitional Cell Carcinomas in Dogs & Cats

J. Paul Woods[1], DVM, MS, DACVIM (Small Animal Internal Medicine & Oncology); Sarah Boston[2], DVM, DVSc, DACVS; Michelle Turek[3], DVM, DACVIM (Oncology), DACVR (Radiation Oncology)
[1]Guelph, ON, Canada; [2]Gainesville, FL, USA; [3]Madison, WI, USA

Introduction[1-6]

Urinary bladder cancer accounts for ~2% of all cancers in dogs. Transitional cell carcinoma (TCC), also known as urothelial carcinoma, is the most common cancer of the urinary bladder in dogs (75–90%) (other primary bladder cancers are squamous cell carcinoma, adenocarcinoma [ACA], undifferentiated carcinoma, fibrosarcoma, leiomyosarcoma, rhabdomyosarcoma). TCC is much less common in cats but is still the most common urinary bladder cancer (others are SCC, ACA, leiomyosarcoma).

Risk factors for TCC include obesity, sex (female dogs and male cats at higher risk), breed (Scottish terriers, WHWT, Sheltie), and chemical carcinogen exposure (bladder storage site) such as insecticides, pesticides, herbicides, 2nd-hand cigarette smoke and cyclophosphamide. TCC affects older dogs (average 11 years at diagnosis). The clinical signs associated with TCC are most commonly due to lower urinary tract signs: dysuria (84%), hematuria (50%), pollakiuria (37%), stranguria, urinary incontinence, chronic urinary tract infections, and signs of partial or total obstruction of the urethra. Other signs include lameness with bone metastasis (or HO secondary to pulmonary metastases), respiratory difficulty with pulmonary metastases, lethargy, and weight loss. Patients with chronic urinary tract infections that do not resolve with an appropriate course of antibiotics should have further investigation of their lower urinary tract with ultrasound or cystoscopy. The limiting quality of life factor is usually due to clinical signs of urinary bladder discomfort.

If a TCC is suspected, both ultrasound and cystoscopy are useful for local staging. In general, ultrasound-guided aspirate is not recommended if it can be avoided because of the high risk of tumor seeding with TCC. Cystoscopy or traumatic catheterization (with or without ultrasound guidance) are recommended to obtain a diagnosis. The most common sites for metastases are the iliac and other abdominal lymph nodes, liver, and lung (10–30% have metastases at the time of diagnosis). Abdominal ultrasound can also be used to evaluate the local lymph nodes (iliac) for evidence of metastasis. Three-view thoracic radiographs should be taken to evaluate for pulmonary metastasis. Complete staging, also CBC, biochemical profile, urinalyses (up to 30% have clumps of epithelial cells in the urine sediment suggestive of neoplasia), urine bacterial culture, and a rectal exam may reveal thickened urethra and enlarged sublumbar lymph nodes. Physical exam may be normal in 30% of dogs with TCC. Histologic diagnosis can be obtained via a biopsy obtained via surgery, ultrasound-guided cystoscopy, or traumatic catheterization (80% accuracy of surgical biopsy). TCC is both aggressive locally and systemically, so therapy should be directed at both local and distant disease, even in cases with no gross metastasis.

Therapy

Surgery

Surgical options depend on location, size, and extent of TCC tumor. Clean margins are difficult to achieve with surgery; therefore, surgery often only provides local control of tumor growth and curative surgical removal is rare. Tumors located at the apex or body of the bladder offer the best chance for surgical resection. Partial cystectomy can be performed in these cases with excision of up to 70% of the bladder wall without the need for urinary diversion (if ureters not involved).[7] Local or distant recurrence of tumor growth occurs in 80% of patients that undergo this surgery. This is likely due to multiple foci of TCC within the bladder due to a "field effect" in which the entire bladder epithelium is suggested to undergo a malignant change in response to exposure to carcinogens in the urine.[6]

Transitional cell carcinoma located at the trigone (66% in dogs) is difficult to remove surgically and there is likelihood of urethral (56%), ureteral or prostatic involvement. Hence, surgical removal is only possible by performing a total cystectomy and ureteral diversion.[8] Total cystectomy is an uncommon procedure for treatment of bladder neoplasia in dogs. The technique is rarely used because of complications associated with ureteral reimplantation. Kadosawa *et al.* reported 14 cases of total cystectomy for treatment of cancer. The method used was creation of a single lumen with the remaining ureters and anastomosis of the distal ureteral lumen with the distal urethra, vaginal, or preputial mucosa. The results of Kadosawa were promising, with long-term survival in some cases and an overall median survival time of 6 months without adjunctive chemotherapy. All cases managed with total cystectomy and ureteral reimplantation into the urethra, prepuce or vagina will obviously be incontinent and will need to be managed with a diaper and considerable care on the part of the owners to ensure that these dogs do not develop urine scald. The acceptance of incontinence and a diaper by an owner will vary. However, many of these dogs are partially or completely incontinent and the owners are already managing this at home. Total cystectomy is a technically feasible procedure that serves two important purposes in the treatment of lower urinary tract TCC: to remove the source of pain due to the tumor, inflammation and urinary tract obstruction and to remove the primary tumor, which may limit or prevent metastatic spread of disease. Major disadvantages of this procedure are incontinence and the potential for ascending infection. Often a total cystectomy is not offered during consultation for a bladder TCC because it is considered too radical, because it is assumed that the owners will not manage urinary incontinence, or simply because it is below our collective consciousness as a treatment option.

Emergency palliative surgery has been used to debulk nonresectable tumors to relieve obstruction. Other procedures to bypass urinary obstruction include ureteral and urethral stents and cystostomy tubes.[9,10] Surgical debulking has been helpful with prolonged survival in some (but not other) studies as adjunctive therapy along with chemotherapy.[11,12] The use of urethroscopy and a laser to ablate tumors of the urethra is a new technique and also requires specialized equipment, but may hold promise as a palliative technique for urethral tumors.

Radiation Therapy[13-15]

With the increasing availability of specialized radiation equipment capable of conformal radiation delivery and on-board 3D imaging (image-guided intensity modulated radiotherapy), the use of radiotherapy in the definitive or 'curative-intent' treatment of transitional cell carcinoma has increased. The close proximity of radiation-sensitive critical normal structures, such as the large bowel, is a challenge that can be better managed with conformal dose delivery using intensity-modulated radiotherapy (IMRT) compared to older techniques. The ability to image the bladder daily in three dimensions using on-board CT allows for adaptation of the treatment plan to account for inconsistencies in the position/size of the bladder. These advances have resulted in renewed interest in the use of radiation as a tool for primary treatment of TCC. A recent study reported an encouraging median survival time of 654 days in 21 dogs with bladder or prostatic carcinoma treated with IMRT to 54–58 Gy in 20 daily fractions. Median time to first event was over 300 days. Acute radiation toxicity included low-grade effects that were manageable and self-limiting. Clinically important late complications were high at 19% and included stricture of the rectum, urethra or ureter. Radiation-induced fibrosis was the suspected cause in many cases, although tumor-related obstructions were difficult to rule out. Sixty percent of pet owners reported improvement in the quality of life of their pets after radiotherapy. In the author's practice, definitive radiotherapy is offered in combination with standard medical treatment for resectable tumors treated in the microscopic disease setting, as well as for nonresectable tumors without evidence of distant metastasis. Further studies are needed to corroborate the favorable outcomes mentioned above, and daily radiation doses should be low to minimize the risk of late complications. Due to internal motion of the bladder and the close proximity of the large bowel, stereotactic radiotherapy using ablative radiation doses is probably not suitable for this tumor type.

Radiotherapy can also be used in the palliative setting if the definitive approach is not indicated or desired. Multiple protocols exist, such as weekly regimes and daily treatments for 1–2 weeks. Studies are limited but suggest that amelioration of urinary signs and quality of life is achieved in approximately

75% of cases, including some complete responses. Despite the benefits of palliation, median survival times are generally no better than with medical therapy alone.

Chemotherapy

Medical management of TCC is indicated in nonresectable or metastatic tumors. Rather than cure, remission or stable disease is the goal and most treatments are well tolerated. Chemotherapy and nonsteroidal anti-inflammatory drugs (NSAIDs), cyclooxygenase (COX) inhibitors have been used alone and in combination. Cisplatin, carboplatin, doxorubicin, mitoxantrone, cyclophosphamide, vinorelbine, gemcitabine, toceranib, vinblastine, metronomic chlorambucil, intravesical mitomycin C, and intravesical thiotepa chemotherapy protocols have been reported with measurable tumor response rates of 0–38% and median survival times between 4–11 months.[16-24]

Piroxicam (Feldene™), a non-selective COX inhibitor, has been used alone as a single agent (median survival 195 days). Gastrointestinal toxicity (vomiting, melena, anorexia) is the dose limiting effect of piroxicam. If toxicity occurs, then the drug is stopped and supportive care is given as needed. Misoprostol may be given if piroxicam is reinstituted. It is not known if selective COX-2 inhibitors have the same antitumor effect as nonselective piroxicam. However, COX-2 is overexpressed in most canine TCC and selective COX-2 inhibitors (deracoxib, firocoxib) have shown antitumor activity with fewer gastrointestinal side effects (Knapp). Several mechanisms of antitumor activity of COX inhibitors have been proposed including direct induction of apoptosis, immunomodulatory effects, and antiangiogenic activity by blocking the effects of chronic inflammation. Other possible mechanisms include restoring chemosensitization through modulation of microRNAs expression, inhibiting prostaglandin E_2-induced repopulation of cancer stem cells, and blocking multidrug resistance drug transporters. Sequential rather than simultaneous combination therapy of chemotherapy and a COX inhibitor may give a survival advantage, which may be due to development of resistance to both drugs simultaneously when the drugs are given concurrently.[25-30]

In the future, more individualized and targeted therapies (e.g., intra-arterial vs. intravenous, folate-targeted [with high folate uptake in TCC], bladder cancer specific peptide [PLZ4]), epigenetic modifying drugs (5-azacitidine, zebularine), and nanoparticle conjugates may result in more antitumor effects with less risk of toxicity to the patient.

Urinary tract infections (UTIs) are common with TCC, due to the higher risk of secondary bacterial infections due to cancer-induced abnormal urination, urine retention, acquired anatomical defects in the bladder and urethra, altered uroepithelium, and possibly compromised immune function.[4] UTIs can lead to a worsening of lower urinary tract clinical signs, which leads to a decreased quality of life and falsely suggests the cancer is progressing. In addition, bacteria are becoming more resistant to antibiotics, which may result in a bacterial infection that only responds to toxic, expensive antibiotics or in some cases is not sensitive to antibiotics. Hence, euthanasia may be due to noncontrollable UTI, even if the TCC is controlled. The effect of UTI-induced inflammation to possibly enhance tumor progression and hinder treatment response is another concern. Therefore, periodic urinalyses are indicated with urine culture and sensitivity tests when infection is suspected.[31]

Prognosis

Without treatment, dogs with TCC typically survive 2–3 months with death due to progression of clinical signs or urinary outflow obstruction. In contrast, treatment for TCC, although not curable at present, can result in good quality of life. At present, objective response rates to most chemotherapy protocols <20% so better ways and drugs are needed to treat (or prevent) TCC.

Chemotherapeutic Protocols for Canine TCC

- Piroxicam 0.3 mg/kg/day PO[25]
- Mitoxantrone 5 mg/m^2 IV q 3 weeks and piroxicam 0.3 mg/kg/day PO[18]
- Vinblastine 2–2.5 mg/m^2 IV q 2 weeks and piroxicam 0.3 mg/kg/day PO simultaneously or sequentially[30]
- Chlorambucil 4 mg/m^2 PO q 24 h[24]

References

VIN editor: references 1 and 5 are the same.

1. Knapp DW. *Urol Oncol*. 2000;5:47.
2. Knapp DW, *et al*. In: Withrow SJ, *et al*., eds. *Withrow and MacEwen's Small Animal Clinical Oncology*. 5th ed. St. Louis, MO: Saunders; 2013:572.
3. Mutsaers AJ, *et al*. *J Vet Intern Med*. 2003;17:136.
4. Fulkerson CM, *et al*. *Vet J*. 2015;205(2):217.
5. Knapp DW, *et al*. *Urol Oncol*. 2000;5(2):47.
6. Knapp DW, *et al*. *ILAR J*. 2014;55(1):100.
7. Bacon N. In: Chew DJ, *et al*., eds. *Canine and Feline Nephrology and Urology*. 2nd ed. St. Louis, MO: Saunders; 2010.
8. Kadosawa T, *et al*. VCS 26th Annual meeting; 2006:66.
9. Weisse C, *et al*. *J Am Vet Med Assoc*. 229;226:2006.
10. Blackburn AL, *et al*. *J Am Vet Med Assoc*. 2013;242:59.
11. Molnar T, *et al*. *Acta Vet Hung*. 2012;60:55.
12. Robat C, *et al*. *J Small Anim Pract*. 2013;54:67.
13. Poirier VJ, *et al*. *J Am Vet Med Assoc*. 2004;40:131.
14. Nolan MW, *et al*. *J Vet Intern Med*. 2012;26:987.
15. Choy K, *et al*. *Vet Radiol Ultrasound*. 2016;57:341.
16. Chun R, *et al*. *J Vet Intern Med*. 1997;11:279.
17. Ogilvie GK, *et al*. *J Am Vet Med Assoc*. 1991;198:1618.
18. Henry CJ, *et al*. *Clin Cancer Res*. 2003;9:906.
19. Boria PA, *et al*. *Vet Comp Oncol*. 2005;3:73.
20. Greene SN, *et al*. *J Am Vet Med Assoc*. 2007;231:1056.
21. Abbo AH, *et al*. *J Vet Intern Med*. 2010;24:1124.
22. Arnold EJ, *et al*. *J Vet Intern Med*. 2011;25:1385.
23. Knapp DW, *et al*. *J Vet Intern Med*. 2013;27:126.
24. Schrempp DR, *et al*. *J Am Vet Med Assoc*. 2013:242,1534.
25. Knapp DW, *et al*. *J Vet Intern Med*. 1994;8:273.
26. Mohammed SI, *et al*. *Cancer Res*. 2002;62:356.
27. Mohammed SI, *et al*. *Mol Cancer Ther*. 2003;2:183.
28. McMillan SK, *et al*. *J Am Vet Med Assoc*. 2011;239:1084.
29. Khan KN, *et al*. *Am J Vet Res*. 2000;61:478.
30. Knapp DW, *et al*. *Bladder Cancer*. 2016;2:241.
31. Budreckis DM, *et al*. *J Vet Intern Med*. 2015;29:828.

acvim295

Medical, Surgical & Radiation Oncology: Practical Approach to Diagnosis and Management of Squamous Cell Carcinomas in Dogs & Cats

J. Paul Woods[1], DVM, MS, DACVIM (Small Animal Internal Medicine & Oncology); Sarah Boston[2], DVM, DVSc, DACVS; Michelle Turek[3], DVM, DACVIM (Oncology), DACVR (Radiation Oncology)
[1]Guelph, ON, Canada; [2]Gainesville, FL, USA; [3]Madison, WI, USA

Introduction

In dogs, squamous cell carcinoma (SCC) is the second most common oral cancer 20–30% (following melanoma).[1] Non-tonsillar SCC is more common with less invasion and rare metastasis.[2] Approximately 80% of dogs with non-tonsillar SCC have a favourable prognosis and only 19% develop metastasis.[3] Gingiva is the most common site for non-tonsillar SCC, while cancers on the tongue occur less frequently. Tonsillar SCC is rare and has aggressive local invasion with metastasis. Most dogs with non-tonsillar SCC have a good prognosis after surgical or radiation therapy.[4]

Cutaneous SCC accounts for 5% of cutaneous tumors in dogs and it most often occurs in thinly haired, nonpigmented, sun-damaged skin and may be preceded by actinic (solar) keratosis. Sunlight-induced tumors on the flank and ventrum occur most commonly in lightly pigmented dog breeds such as Dalmatians, Beagles, Whippets, and White English Bull terriers. The incidence of solar-induced squamous cell carcinoma is highest in geographic areas with intense sunlight. Squamous cell carcinoma appears usually as single, but possibly multiple, proliferative or ulcerative lesions on the trunk, legs, digits, scrotum, nose, and lips. Nail bed tumors usually involve one digit, but multiple digits may be involved, especially in large, black-coated dogs such as black Labradors, German shepherds, and Standard Poodles.

In cats, SCC is the most common oral cancer in cats. Ten percent of all tumors diagnosed in cats are oral tumors and 60–75% of feline oral cancers are SCC.[5,6] Gingiva followed by sublingual location and tongue (commonly in the ventral part of the tongue) are reported as the most commonly affected sites for feline SCC. SCC accounts for 15% of cutaneous tumors in cats with highest incidence reported in older white cats most commonly involving nonpigmented ear pinnae, nose, and eyelids. Similar to dogs, the incidence of solar-induced squamous cell carcinoma is highest in geographic areas with intense sunlight.

Squamous cell carcinoma frequently occurs in aged cats with no apparent breed or sex predisposition.[6] Cats are usually presented with SCC at a late stage of disease and the advanced state of disease in many of these patients makes treatment with surgery, chemotherapy, and radiation therapy challenging.

Squamous cell carcinoma in dogs, unlike cats, varies markedly in behaviour and response to treatment. In dogs, surgery with wide margins and adjuvant radiation therapy or chemotherapy gives a good-to-guarded prognosis for partial or complete remission. However, in cats with oral SCC, the cancer often invades into the bone or under the tongue, which results in a grave prognosis.

Therapy

Surgery

In dogs with oral SCC, the goal is clean surgical margins (in all directions) of 2 cm. If not possible, then palliative surgery can achieve temporary local control or be combined with radiation therapy. The prognosis for lingual SCC in dogs is poor. Partial glossectomy up to 40–60% of the tongue can be tolerated but aggressive excision can result in a dog that will not have the ability to eat. For cutaneous SCC, the treatment of choice is early, complete surgical excision that includes amputation of digital tumors. Excised tissues should be examined histopathologically for completeness of excision (margin assessment). Laser ablation may be appropriate for superficial lesions. Cryotherapy or laser ablation may be appropriate for small, superficial lesions. For nonresectable or partially resectable lesions, adjuvant radiotherapy (especially electron beam radiation) or strontium-90 (a form of superficial radiotherapy) may be effective.[7,8]

In cats with oral SCC, surgery alone is generally unrewarding with a median survival of 6-9 months. Tumor location dictates the extent of resection possible. Tumors of the mandible and maxilla may be removed with mandibulectomy or maxillectomy aiming for 1-cm margins. Partial glossectomy is feasible; however, many cats are not candidates for surgery because their tumors are sublingual and nonresectable because acceptable surgical margins are not achievable.

Cats with cutaneous SCC are treated with surgery unless there are too many lesions or lesions are in a difficult location, in which case imiquimod, an immunomodulatory with antitumor and antiviral effects, can be used (5% cream daily or 3/week).[9] Site, stage of disease, and available facilities dictate if surgery, photodynamic therapy (PDT), cryotherapy, RT, etc., will be employed.[10]

Radiation Therapy[11-20]

Radiotherapy in canine oral squamous cell carcinoma can be offered in the definitive or 'curative-intent' setting, or for palliation of gross tumors causing clinical signs. Tumor control with definitive radiation is highest when used after surgery in the microscopic disease setting. Treatments are delivered daily for 3–4 weeks. If surgical excision is not possible due to the location and/or size of the tumor, then definitive radiotherapy can still be offered in the gross disease setting. Based on limited clinical studies, clinical outcomes can be favorable even when bulky tumors are treated. Tumor size is an important prognostic indicator, with small tumors doing better. Median progression free survival for T1 and T2 tumors is reported to be 1–2 years or longer with a reported high rate of durable complete or partial responses. Larger tumors or those with locoregional metastasis may not fare as well. Another study suggests that young dogs may have a more favorable prognosis. The role of concurrent chemotherapy, usually carboplatin, is unclear. Some oncologists recommend chemotherapy in the face of gross disease for its potential additive local effects, even though the metastatic rate is low. Palliative radiotherapy is a good option for dogs with macroscopic tumors causing clinical signs, when a more definitive approach is not indicated or desired. Palliative protocols include weekly regimes, or daily treatments for one week. While durable tumor responses are not necessarily expected, clinical signs may be successfully palliated. Stereotactic radiotherapy has not been evaluated for canine oral squamous cell carcinoma.

Canine tonsillar squamous cell carcinoma is less commonly treated with definitive-intent radiation in light of the tumor's high metastatic rate. Survival is often compromised by progressive distant disease. That said, older studies suggest that outcomes are optimized when full-course radiotherapy is used in conjunction with chemotherapy. Palliative protocols can successfully ameliorate clinical signs of tonsillar tumors or lymph node metastases, and are usually the radiotherapy treatment of choice for this form of the disease. In the author's practice (MT), definitive therapy is offered in cases without metastasis in which primary tumors are excised when pet owners are interested in a tolerable, but aggressive approach to treatment. Concurrent chemotherapy is indicated in light of the high risk of metastasis.

Feline oral squamous cell carcinoma is an aggressive disease for which radiation is available, but not always beneficial. When tumors are small, located on the mandible and resectable, then adjuvant definitive radiotherapy may play a role in slowing tumor recurrence. A small study of seven cats suggested a median survival time of 14 months in this setting. Since most cats present with nonresectable tumors and since definitive therapy (daily treatments for 3–4 weeks) has not been shown to improve outcomes in the face of bulky disease, palliative protocols are often recommended. Large tumors can respond favorably to palliative radiation, but responses are often short-lived. Many protocols have been evaluated including weekly regimes, daily treatments, and treatments using chemotherapy radiosensitizers. Median survival times remain dismal at 2–3 months due to tumor regrowth. Accelerated schedules requiring two treatments per day for 9–10 days have been associated with the longest median survival times of 5–6 months. Occasionally, more durable tumor responses can be observed when small tumors (T1) are treated. Concurrent chemotherapy has not been shown to improve outcomes.

Chemotherapy

For nonresectable lesions in dogs, intralesional chemotherapy (cisplatin, carboplatin, 5-fluorouracil), local hyperthermia, or photodynamic therapy (using HPPH, Photochlor) may be effective in some cases.

Systemic chemotherapy is less consistently effective for treatment of squamous cell carcinoma. Cisplatin or carboplatin in combination with piroxicam has anti-tumorigenic effects with acceptable toxicity levels in dogs with SCC.[2,21,22] Toceranib demonstrated clinical benefit against canine SCC in 8 dogs with 1 complete response (CR), 5 PR and 1 SD.[23] Piroxicam (0.3 mg/kg/day), a NSAID, has been used for palliation.

For cats' SCC, chemotherapy (doxorubicin, cyclophosphamide, mitoxantrone, carboplatin, bleomycin) has demonstrated minimal efficacy.[24] NSAIDs (piroxicam, meloxicam) have been utilized; however, there have been variable results (9–67%) on the expression of COX-2 by feline SCC. Also, serial monitoring for toxicity is required. Others have used bisphosphonates (zoledronate, pamidronate), tyrosine kinase inhibitors (toceranib), and polyamine inhibitors.[25] Epidermal growth factor receptor (EGFR) overexpression has been found in 70% of feline SCC; therefore, tyrosine kinases that block EGFR may be used in the future.[26]

A recent study in 6 cats had encouraging results (5 CR and 1 PR with long survival times) utilizing multimodality therapy with neoadjuvant chemotherapy (bleomycin 10 UI/m^2 q weekly, piroxicam 0.3 mg/kg PO q 48 h and thalidomide 2 mg/kg q 24 h) for at least 2 weeks duration followed by marginal surgical excision, and hypofractionated RT (4.8 Gy twice per day for 5 days [48 Gy total]) followed by adjuvant medical therapy.[27]

Buprenorphine (0.01–0.03 mg/kg PO q 8 h) has been used for pain management. Oral antibiotics that penetrate bone and/or have anaerobic spectrum activity (doxycycline, amoxicillin/clavulanate, clindamycin) are used for infections of cancer tissue secondary to normal flora.

Prognosis

The prognosis for dogs is variable, depending on the degree of differentiation and the site of the lesion. Most tumors are locally invasive and slow to metastasize, although squamous cell carcinoma of the digit tends to be more aggressive and may metastasize more readily.

The prognosis in cats depends on the size and degree of differentiation, with smaller, well-differentiated tumors having a better prognosis than large or poorly differentiated ones. Most cases have a low risk of metastasis but often need to be diagnosed early because site and stage determine best chance for surgical cure. To prevent new solar-induced lesions from developing, future ultraviolet light exposure should be avoided.

References

1. Hayes AM, *et al. J Small Anim Pract.* 2007;48:394.
2. de Vos JP, *et al. Vet Comp Oncol.* 2005;3:16.
3. Fulton AJ, *et al. J Am Vet Med Assoc.* 2013;243:696.
4. Grier CK, *et al. Can Vet J.* 2007;48:1189.
5. Supsavhad W, *et al. Vet J.* 2016;210:7.
6. Soltero-Rivera MM, *et al. J Feline Med Surg.* 2014;16:164.
7. Liptak JM, *et al.* In: Withrow SJ, *et al.*, eds. *Withrow & MacEwen's Small Animal Clinical Oncology.* 5th ed. St. Louis, MO: Saunders/Elsevier; 2013:381.
8. Northrup NC, *et al. J Am Vet Med Assoc.* 2006;42:350.
9. Gill VL, *et al. Vet Comp Oncol.* 2008;6:55.
10. Bexfield NH, *et al. J Vet Intern Med.* 2008;22:1385.
11. Rejec A, *et al. J Vet Dent.* 2015;32(4):212.
12. LaDue-Miller T, *et al. Vet Radiol Ultrasound.* 1996;37(1):74.
13. Theon A, *et al. J Am Vet Med Assoc.* 1997;210(6):778.
14. Brooks M, *et al. J Vet Intern Med* 1988;2:206.
15. MacMillan R, *et al. J Am Anim Hosp Assoc.* 1982;18:311.
16. McDonald C, *et al. Vet Radiol Ultrasound.* 2012;53(3):358.
17. Fidel J, *et al. J Vet Intern Med* 2011;25:504.
18. Fidel J, *et al. Vet Radiol Ultrasound.* 2007;48(5):482.
19. Hutson C, *et al. J Am Vet Med Assoc.* 1992;201(5):777.
20. Poirier V, *et al. Vet Radiol Ultrasound.* 2013;54(1):81.
21. Boria PA, *et al. J Am Vet Med Assoc.* 2004;224:388.

22. Ogilvie GK, *et al. J Am Vet Med Assoc.* 1993;202:1839.
23. London C, *et al. Vet Comp Oncol.* 2012;10:194.
24. Hayes A, *et al. J Comp Pathol.* 2006;135:93.
25. Wiles V, *et al. J Feline Med Surg.* 2017;19:185.
26. Looper JS, *et al. Vet Comp Oncol.* 2006;4:33.
27. Marconato L, *et al. Vet Comp Oncol.* 2013;11:101.

1.
acvim297

The Principles of Lymphocyte Development and Lymphocyte Trafficking Support the Diagnosis of Lymphoma

Peter F. Moore, BVSc, PhD, DACVP
Davis, CA, USA

Introduction

A diagnosis of lymphoma is currently made on the basis of clinical, histologic, and immunophenotypic criteria. Histologic evidence of lymphoma includes the presence of dense cellular infiltration, a monomorphic appearance of infiltrating lymphocytes, cytological immaturity of the lymphoid infiltrate, and disruption, effacement, and replacement of the normal structures of the involved tissue by infiltrating cells. Immunophenotypic assessment is a useful aid in lymphoma diagnosis, and it may increase the index of suspicion of lymphoma when the majority of cells express the same phenotype. Difficulties arise when lymphomas arise in sites of chronic inflammation, especially in skin and gut. In these instances, it may be difficult to identify lymphoma in a background of inflammatory infiltration. Molecular clonality assessment is used to demonstrate clonal expansion of lymphocytes consistent with lymphoma in these instances.

Molecular Clonality Assessment in Lymphoproliferative Disease

Lymphocyte antigen-receptor gene rearrangement analysis is a methodology used to detect clonality in B and T cell populations. While it is generally accepted that clonality is a property of neoplasia, it is not synonymous with malignancy. Benign clonal expansion is most often discovered in T cell infiltrates in a variety of clinical settings, ranging from the response to infection to the response to neoplasia.[8,18] The detection of a clonal population in an equivocal lymphoproliferative lesion can be used as an important adjunctive diagnostic tool to determine the likelihood of neoplasia if interpreted in conjunction with clinical, morphological, and immunophenotypic findings.[6] Polymerase chain reaction (PCR)-based methods are most frequently used for molecular clonality assessment (commonly referred to as "PARR"). This methodology provides the advantage of allowing the detection of B and T cell clonality in formalin-fixed paraffin-embedded (FFPE) tissues, which are routinely used for diagnosis.[17]

Specific antigen recognition is essential for adaptive immunity. Lymphocytes of B cell and T cell lineage have antigen receptors, which consist of multiple protein subunits/chains. T cells have either alpha and beta chains (alpha beta T cells) or gamma and delta chains (gamma delta T cells). B cells have immunoglobulin heavy chains (IGH) and light chains, either kappa (IGK) or lambda (IGL). These proteins have variable (V) and constant (C) domains, which are encoded by antigen-receptor genes. Antigen binding occurs in the V domain. The V domain is encoded by 2 or 3 genes - the variable, diversity (D), and joining (J) genes. Although multiple V, D, and J genes exist in antigen-receptor loci, they are relatively limited in number and diversity. Gene rearrangement during lymphocyte development leads to random joining of V, D (optionally), and J genes (most commonly one of each), leading to formation of a complete V domain exon. The diversification of the antigen-receptor repertoire is markedly enhanced by modification of the junctions between the V, D, and J segments. For instance, random nucleotides (N or non-template encoded) are added by the enzyme terminal transferase. This creates the highly diverse third hypervariable region of the V domain, also known as the complementarity-determining region 3 (CDR3). The CDR3 region is at the center of the antigen-binding site and hence is the major contributor to antigen specificity. The CDR3 is unique for each antigen-receptor subunit and creates a "molecular fingerprint" unique to each lymphocyte.[14]

During T cell development in the thymus, T cells rearrange their antigen-receptor genes in the following order: TRD (delta), TRG (gamma), TRB (beta), and TRA (alpha), and in the process create 2 lineages of T cells, alpha beta and gamma delta T cells. Rearrangement of TRA results in deletion of TRD on at least one chromosome, since the TRD locus occurs within the TRA locus. A TRG gene rearrangement remains in the majority of T cells regardless of the surface T cell receptor (TCR) phenotype, and hence TRG is the target of choice for T cell clonality.[15]

B cell development occurs in bone marrow. B cells first rearrange IGH followed by IGK. The B cell either expresses a kappa light chain or subsequently rearranges IGL and expresses a lambda light chain. If IGL is rearranged, IGK is inactivated by rearrangement of the kappa-deleting element (KDE). The IGH locus is rearranged in all B cells and is the preferred target for molecular clonality. Unlike T cells, B cells can modify the sequence of IGH (and IGK or IGL) by somatic hypermutation in germinal centers of secondary lymphoid tissues (spleen, lymph nodes, and Peyer's patches). This is an important property of humoral immunity and is responsible for affinity maturation of the immunoglobulin response to antigen, which occurs in germinal centers. Somatic hypermutation is a necessary, albeit dangerous process, and targeting of genes other than immunoglobulin V, D, and J genes can precipitate lymphoma. In this regard, most B cell lymphomas have follicular origins. Another consideration is that the sensitivity of IGH clonality assays is reduced due to somatic hypermutation of primer binding sequences in the affected V and J genes. Lambda light chains are dominant in cats, dogs, and horses. Hence, IGK is frequently inactivated by rearrangement of KDE. This occurs during B cell development in the bone marrow prior to antigen encounter in germinal centers. Therefore, KDE rearrangements are not subject to somatic hypermutation and are an attractive supplemental target for molecular clonality.

Lymphocyte antigen-receptor gene rearrangements of necessity are lineage associated, but cross-lineage rearrangements are possible. In these instances, B cells rearrange T cell loci and T cells rearrange B cell loci. Also, rearrangement of B or T cell loci has been observed in myeloid leukemia. Because of these events, it is not recommended to assign cell lineage solely based on clonality results. Cell lineage assignment by immunophenotyping is more accurate.[17]

In principle, clonality assays assess the diversity of antigen-receptor gene rearrangements within a population of lesional lymphocytes by PCR-based amplification of CDR3. The products (amplicons) are size-separated by gel electrophoresis. In order to amplify CDR3 in rearranged genes, forward primers are designed for V genes and reverse primers are designed for J genes. Hence, the primers flank the CDR3. In reactive/inflammatory lesions, amplicons are derived from multiple unique cells and differ in size and sequence; this is a polyclonal result. In lymphoma, amplicons are derived from multiple identical cells and are homogeneous in size and sequence; this is a clonal result.

Clonality assays in veterinary medicine use heterogenous primer sets that differ by institutions conducting the assays. Ideally, these assays should cover all appreciably rearranged gene segments in a locus. The reality is markedly different, and the sensitivities of clonality assays are quite variable.[6] Antigen-receptor genes can be identified within high-quality genome assemblies of species of interest (dog, cat, and horse); these are becoming available but are not annotated to indicate the location (topology) of the genes on a chromosome. Antigen-receptor genes can also be identified by high-throughput sequencing of the expressed V/J repertoire in mRNA. This approach exposes V/J usage but not locus topology. By using both approaches, the primer design process can be optimized to provide the best sensitivity.[6] Initial primer sets used for IGH and TRG clonality assessment were based on limited sequence data derived from mRNA prior to the genome era.[1,13,16,22] Modifications have occurred through limited genomic search or expanded mRNA sequence data.[2,3,10,11,19-21,23] The release of genome assemblies for dog, cat, and horse has facilitated detailed description of antigen-receptor loci and design of clonality assays that cover all genes in a locus. For instance, a complete description of the canine TRG locus revealed a more complicated topology than was previously anticipated. The canine TRG locus consisted of 7 complete V-J-C clusters/cassettes compared to one in humans.[9] Using a new multiplex assay that covered all potentially rearranged genes, Keller and Moore demonstrated that canine T cells rearrange multiple V-J-C cassettes resulting in >1 rearrangement per chromosome.[4] This novel finding changed the premise that more than 2 clonal rearrangements in a sample indicated oligoclonality. Indeed, knowledge of the locus topology is important in the design and interpretation of clonality assays. Preliminary indications are that cat and horse also possess multi-cassette TRG loci. So in the future, existing clonality assays in veterinary medicine, which are currently prone to false-negative results, will be modified to consider locus topology and high-resolution V/J usage.

Immunophenotypic Assessment in Lymphoproliferative Disease

To classify lymphomas by the WHO system, it is important to have access to markers for immunohistochemical analysis. Table 1 lists the markers of value for determining cell lineages in leukocytic proliferations in dogs and cats (with the exception of CD11d and CD45). These markers are suitable for use in formalin-fixed paraffin-embedded tissues with appropriate antigen-retrieval protocols. Markers are available for the detection of B and T cells. However, markers for the unequivocal detection of NK cells are not available, so the existence of NK cell lymphomas in dogs and cats is not easily assessable. Determination of T cell-receptor usage and major subsets of T cells (CD4 or CD8), as well as dissecting the lineages of histiocytes (macrophages, interstitial type DC and Langerhans cells) is best done in unfixed cell smears or snap frozen tissues or by flow cytometry.

Table 1. Markers of diagnostic importance to leukocytic disease investigation - formalin-fixed tissue

CD3ε	Signaling component of the T cell antigen receptor. Expressed by αβ T cells and γδ T cells. Cytoplasmic expression by NK cells is possible - especially if activated.
CD79a	Signaling component of the B cell antigen receptor. Expressed by all stages of B cell differentiation. Expression is less in plasma cells.
CD20	Surface molecule expressed at all stages of B cell differentiation except for plasma cells. CD20 plays a role in regulation of B cell activation and proliferation. CD20 is not lineage specific and has been observed uncommonly in T cell lymphomas. Caution is advised in interpretation of diffuse cytoplasmic expression, which can occur in several cell types.
Pax5	Transcription factor essential for maintenance of B cell differentiation. Useful B cell marker.
MUM1/IRF4	Transcription factor essential for plasma cell differentiation. Useful plasma cell marker.
CD11d	αD subunit of β2 integrin (CD18) family. Expressed by macrophages and T cells in hematopoietic environments, especially splenic red pulp, bone marrow, and lymph-node medullary sinuses. CD11d is consistently expressed in diseases emanating from splenic red pulp (LGL form of chronic lymphocytic leukemia; hepatosplenic lymphoma, and hemophagocytic histiocytic sarcoma).
CD18	β subunit of the β2 integrin family of leukocyte adhesion molecules. Expressed as a heterodimer of CD11a, CD11b, CD11c, or CD11d with CD18. Leukocytes express at least one form of the heterodimer. Hence, CD18 is expressed on all leukocytes - the expression level on myeloid cells is especially high compared to normal lymphocytes. CD18 has been used as a marker of histiocytes, but this is dependent upon exclusion of lymphocyte differentiation by the use of other markers (CD3 and CD79a).
CD45	Surface molecule expressed by most leukocytes - formerly known as "leukocyte common antigen." Antibodies to CD45 bind to the extracellular domain outside of the 3 variably spliced exons (A, B, and C). A useful marker of leukocytic origin. A subset of peripheral T cell lymphomas lacks expression of CD45 - this is encountered in T zone lymphomas of lymph nodes - these are indolent in behavior.
CD204	Class A scavenger receptor. Expressed by macrophages in tissues. Not expressed by dendritic cells in lymphoid tissues or Langerhans cells in the epidermis. Expressed by neoplastic cells in canine histiocytic sarcomas, but not in histiocytomas (LC origin).
c-Kit	Surface molecule and member of the receptor tyrosine kinase family (type III). Expressed by most hematopoietic progenitor cells and by mast cells. Intense expression in high-grade mast cell tumors.
E-cadherin	Adhesion molecule expressed by epithelia and by some leukocytes. Especially useful in cutaneous round cell tumors to identify Langerhans cells indicative of cutaneous histiocytoma.
Granzyme B	Serine protease located in the granules of cytotoxic T cells (CD8+) and NK cells. GrB is expressed at high levels in activated cells and leads to rapid target cell death by apoptosis. Mast cells may also express GrB.
Myeloperoxidase	Myeloperoxidase (MPO) is a lysosomal protein stored in the azurophilic granules of neutrophils (and monocytes). MPO is an important marker of myeloid differentiation.

Ki-67	Cell proliferation marker (nuclear) - expressed in all phases of the cell cycle except G0 and early G1. Excellent marker for determining the growth fraction of a cell population. Indolent lymphomas such as marginal zone B cell lymphoma and T zone lymphoma have proliferative fractions <5–10%.

Antigen loss in lymphoma is an important occurrence, because it increases the diagnostic difficulty. Partial or complete loss of CD3 expression in T cell lymphomas has been recognized for decades.[7] A similar phenomenon exists in high-grade B cell lymphomas with CD20, CD79a, and Pax-5 expression. This necessitates the assessment of more than one of these markers in many instances. Co-expression of CD20 and CD3 occurs in some T cell lymphomas; this underscores the importance of recognizing the limitations of lineage-associated molecules (CD20), which may be expressed more broadly than lineage-specific molecules (CD79a). High-grade, large granular T cell lymphomas (LGL lymphomas) are readily recognized in aspirate biopsies with Wright-Giemsa staining, but the recognition of cytoplasmic granules in routine HE sections is often difficult. Granzyme B is a frequent component of these granules, and staining for this molecule facilitates the recognition of T-LGL lymphoma.[12]

The beta-2 integrin CD11d is dominantly expressed in splenic red pulp. It is also expressed in the neoplastic counterparts of normal CD11d expressing red pulp inhabitants. For instance, CD11d+ gamma delta T cells are enriched in splenic red pulp compared to peripheral blood. Hepatosplenic T cell lymphoma is a high-grade lymphoma of gamma delta T cells in which CD11d and granzyme B are consistently expressed.[5] Knowledge of this immunophenotype is useful in establishing the diagnosis.

High-grade lymphomas have an elevated mitotic index, which correlates well with the proliferative fraction assessed by Ki-67 staining. There is marked interobserver variability in determining the mitotic index, and mitotic figures can be difficult to discern in some lesions. Ki-67 staining adds cost but avoids some of these problems.

References

1. Burnett RC, *et al. Vet Pathol.* 2003;40(1):32–41.
2. Gentilini F, *et al. Vet Immunol Immunopathol.* 2009;127(1–2):47–56.
3. Henrich M, *et al. Vet Immunol Immunopathol.* 2009;130(1–2):59–69.
4. Keller SM, *et al. Vet Immunol Immunopathol.* 2012;145(1–2):410–419.
5. Keller SM, *et al. Vet Pathol.* 2013;50(2):281–290.
6. Keller SM, *et al. Vet Pathol.* 2016;53(4):711–725.
7. Knowles DM. *Am J Pathol.* 1989;134:761–785.
8. Malik UR, *et al. Med Oncol.* 1996;13(4):207–213.
9. Massari S, *et al. Mol Immunol.* 2009;46(13):2728–2736.
10. Mochizuki H, *et al. Vet Immunol Immunopathol.* 2011;143(1–2):38–45.
11. Mochizuki H, *et al. Vet Immunol Immunopathol.* 2012;145(1–2):402–409.
12. Moore PF, *et al. Vet Pathol.* 2012;49(4):658–668.
13. Moore PF, *et al. Vet Immunol Immunopathol.* 2005;106(3–4):167–178.
14. Murphy K. *Janeway's Immunobiology.* New York: Garland Science; 2012.
15. Theodorou I, *et al. J Pathol.* 1994;174(4):233–242.
16. Valli VE, *et al. Vet Pathol.* 2006;43(3):241–256.
17. van Dongen JJ, *et al. Leukemia.* 2003;17(12):2257–2317.
18. Wedderburn LR, *et al. Immunology.* 2001;102(3):301–309.
19. Weiss AT, *et al. Vet Immunol Immunopathol.* 2008;124(1–2):63–74.
20. Weiss AT, *et al. Vet Med Int.* 2010;2010:953272.
21. Weiss AT, *et al. J Comp Pathol.* 2011;144(2–3):123–134.
22. Werner JA, *et al. Vet Pathol.* 2005;42(5):596–607.
23. Yagihara H, *et al. Vet Immunol Immunopathol.* 2007;115(3–4):375–382.

1.
acvim298

The Principles of Lymphocyte Development and Lymphocyte Trafficking Support the Classification of Lymphoma

Peter F. Moore, BVSc, PhD, DACVP
Davis, CA, USA

INTRODUCTION

Lymphomas have been classified in many ways; initially the classification distinguished follicular from diffuse lymphomas. Diffuse lymphomas are the most common form of lymphoma in animals, and they may be of B or T cell derivation. In most species, B cell lymphomas account for about 80% of all lymphomas. Most B cell lymphomas originate in follicular structures, and most commonly from germinal centers. Immunoglobulin (Ig) V segment somatic hypermutation is known to target genes other than Ig genes. This probably explains the much greater incidence of B cell lymphomas compared to T cell lymphomas, because T cell antigen-receptor genes are not subject to somatic hypermutation.

Lymphomas may arise in lymphoid tissues or in non-lymphoid tissues such as intestinal mucosa and skin due to the extended trafficking range of memory B and T cells. Hence, lymphomas often infiltrate multiple sites as a consequence of cell trafficking. Consequently, the important lymphoma types will be discussed as clinicopathological entities, and the focus of this presentation will largely be on gut and skin lymphomas. The new approach to classification has been pioneered by the lymphoma study group of the ACVP, and it is based on the **World Health Organization (WHO) classification of tumors of hematopoietic and lymphoid tissues**, which is widely accepted in human medicine.[29,30] It is clear from this work that there are distinctive subtypes of canine lymphoma that most often correspond to equivalent human entities as defined by the WHO classification scheme. An important issue that has emerged is that lymphoma is an encompassing term for a number of discrete clinical and morphological entities with quite different outcomes (Figures 1–3). Yet treatment strategies often do not take this heterogeneity into account. In applying the WHO classification scheme to canine lymphomas, it was apparent that morphological assessments required immunohistochemical stains in many instances to reinforce the diagnosis. In some instances, molecular clonality determination by lymphocyte antigen-receptor gene rearrangement analysis was also needed to confidently diagnose lymphoma versus lymphoid hyperplasia.

WHO CLASSIFICATION OF TUMORS OF HEMATOPOIETIC AND LYMPHOID TISSUES

The following figures list the main entities that have been identified in domestic species.

Figure 1. WHO precursor lymphoid neoplasia

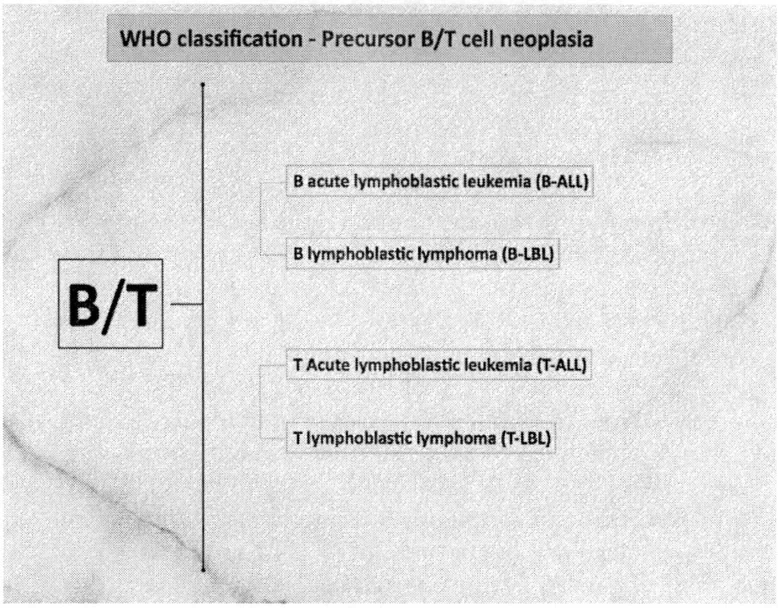

Figure 2. WHO mature B cell neoplasia - major types

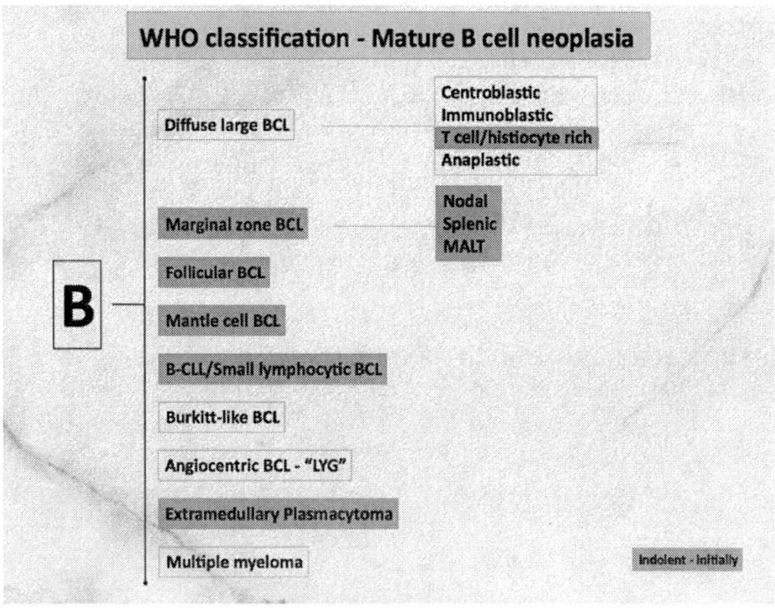

Figure 3. WHO mature T cell neoplasia - major types

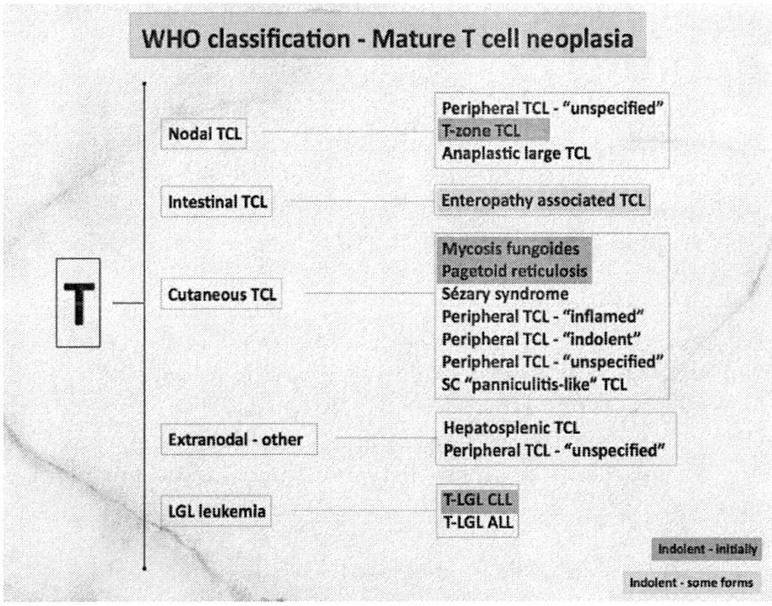

BCL = B cell lymphoma; TCL = T cell lymphoma; MALT = mucosal associated lymphoid tissue; B-CLL = B cell chronic lymphocytic leukemia; T-CLL = T cell chronic lymphocytic leukemia; LYG = "lymphomatoid granulomatosis"; SC = subcutaneous; LGL = large granular lymphocyte

Lymphocyte Trafficking to Secondary and Tertiary Lymphoid Organs

Naïve lymphocytes (i.e., antigen inexperienced) leave thymus and bone marrow where they acquired specific molecules, which allow them to traffic to secondary lymphoid organs represented by lymph nodes and spleen.[18] T cells recirculate from blood to lymph nodes where they enter via high endothelial venules (HEV) in the paracortex. Memory T cells develop the capacity to exit peripheral blood through capillaries in sites rich in antigen deposition (lamina propria of gut, and skin). B cell trafficking follows similar rules. These events are mediated by specific homing receptor/ligand pairs (chemokine/chemokine receptors and adhesion molecules), which are present either on lymphocytes or on vascular endothelium. T cells acquire their homing receptors during differentiation in the thymus to naive peripheral T cells. Dendritic cells (DCs) are instructive to T cells in the conversion from a naïve to memory cells and influence the acquisition of site-specific surface homing molecules by memory T cells. Skin and gut homing are mediated by distinctive T cell surface molecules, which interact with ligands entrapped or expressed on the surface of vascular endothelial cells in the skin or gut.

Lymphocytes that enter lymph nodes via HEV express the trafficking molecules L-selectin and CCR7. The HEVs express the peripheral lymph node addressin (PNAD) and chemokine (C-C) ligand 21 (CCL21), which are the ligands for the lymphocyte trafficking molecules.[18]

The requirements for lymphocyte intestinal trafficking have been elucidated in humans and rodents.[4,18,24] Essentially, beta 7 integrins (alpha 4 beta 7) and specific chemokines and chemokine receptors (CCL25 and CCR9) induce lymphocytes to enter the intestinal lamina propria via the endothelium of small vessels. The endothelial cells express mucosal addressin cell adhesion molecule (MAdCAM), which is the ligand for alpha 4 beta 7 expressed by mucosal homing lymphocytes. Intestinal epithelial cells (especially crypt epithelium) produce the chemokine CCL25, which is displayed on the glycocalyx of mucosal vascular endothelial cells. The receptor for CCL25 is CCR9, which is expressed by mucosal homing lymphocytes. Once lymphocytes enter the lamina propria, epithelial-derived TGF beta induces some to express the integrin alpha E beta 7, which is highly expressed by intra-epithelial lymphocytes (IEL). Feline IEL have also been shown to express alpha E beta 7, and canine lamina propria blood vasculature expresses MAdCAM. Hence the principles of intestinal mucosal lymphocyte trafficking established for humans and rodents likely applies to cats and dogs.[12,33]

Lymphocyte trafficking to skin is facilitated by DCs, which migrate from skin and present antigens to naïve T cells in skin-draining lymph nodes. These T cells are converted to effector-memory T cells, which express cutaneous lymphocyte antigen (CLA). CLA is a carbohydrate modification of P-selectin glycoprotein ligand-1 (PSGL-1). PSGL-1 is expressed constitutively on human T cells. The modification of PSGL-1 by fucosyl transferase VII adds CLA and confers the ability to also bind E-selectin, which is expressed on the endothelium of the dermal vasculature.[11] Skin-homing T cells also express chemokine receptors (especially CCR4 and CCR10), which are involved in skin homing. Dermal endothelium presents chemokine ligands to activate these receptors (CCL17 and CCL27) in the process of migration into skin.[17]

Memory T cells in skin exist as resident cells and recirculating cells.[32] Resident memory T cells are generated in local skin immune responses and globally protect skin. They provide rapid local immune protection against infection or other antigen exposure and do not readily leave the skin. Recirculating memory T cells are generated in local skin immune responses and recirculate between blood, lymph nodes, and skin. They are much less effective in local skin immune protection. Recirculating memory T cells occur as 2 populations: central memory T cells and migratory memory T cells. Central memory T cells express CCR7 and L-selectin, and migrate more rapidly than migratory memory T cells.[32] The patterns of skin disease involving activation or proliferation of memory T cells can be traced to one or other of these skin memory T cell subsets. For instance, in humans, cutaneous epitheliotropic T cell lymphoma (mycosis fungoides) is a disorder of resident memory T cells, while Sezary syndrome is a disorder of central memory T cells.[6,32] The nature and pattern of lesion development differs markedly between these 2 diseases and largely reflects the migratory capability of the underlying cell of origin.[6,32] There is mounting evidence that the concept of resident and recirculating memory T cells applies to other barrier tissues such as gut.[5]

The immunological concepts underlying memory T cell trafficking are important in understanding the topographical patterns of lymphomas occurring in tertiary lymphoid tissues such as skin and gut.

Gastrointestinal Lymphoma

Feline intestinal lymphomas are most frequently of T cell origin and originate predominately in the diffuse MALT of the small intestine (jejunum > duodenum > ileum), and may remain in the mucosal environment for extended periods. Progression to involvement of mesenteric lymph nodes occurs with or without transmural migration. Extension to the stomach and large bowel occurs with much lower frequency; this is likely due to the different lymphocyte trafficking requirements in these sites. T cell lymphomas occur as mucosal and transmural lesions.[23] Canine gastrointestinal lymphomas are also most frequently of T cell origin, and broadly follow these topographic patterns.[3,7,10,25]

Feline B cell lymphomas originate in the organized lymphoid tissues, such as the Peyer's patches and mucosal lymphoid nodules, which are concentrated in the distal small intestine, cecum, and colon. Intestinal B cell lymphomas are most frequently transmural lesions. Feline B cell lymphomas also originate in the stomach, but it is not clear whether they arise from existing gastric mucosal lymphoid nodules or diffuse gastric mucosal lymphoid infiltrates, which are found in association with chronic gastritis. B cell lymphomas are often multiple and occur simultaneously in several locations; for example, in the stomach and at the ileo-cecal-colic junction. Feline B cell lymphomas have not been observed in the diffuse MALT of the small intestine.[23] Data for canine gastrointestinal B cell lymphomas with respect to precise location in the gastrointestinal tract are less clear.[10]

Mucosal T cell lymphoma of small cell type, defined by expression of CD3, is the dominant gastrointestinal lymphoma in aged cats (67% of all lymphomas).[23] This lymphoma closely resembles human EATCL type II in the WHO classification.[29] Small T cell lymphomas are associated with prolonged survival with appropriate treatment (median survival - 29 months).[23] Small T cell lymphomas have been increasingly recognized in dogs since 2010 due to the increased utilization of TRG molecular clonality in the assessment of canine inflammatory bowel disease (IBD) (Moore, unpublished data). Application of the WSAVA guidelines for the assessment of gastrointestinal inflammation without concomitant molecular clonality assessment leads to underdiagnosis of small T cell lymphoma, because some of the morphological features of lymphoma and severe IBD are shared (particularly lymphocyte epitheliotropism).[31] Enhanced lymphocyte epitheliotropism (villous and/or crypt epithelium) is an important diagnostic feature and occurs in about 60% of cases of feline small T cell lymphoma.[23] Enhanced lymphocyte epitheliotropism is also an important feature of canine small T cell lymphoma (Moore, unpublished data).[7] Distinctive patterns of T cell infiltration of the intestinal lamina propria are encountered with disease progression, and are also diagnostically important.[23] A diagnosis of mucosal T cell lymphoma is best confirmed by CD3 immunohistochemistry and T cell antigen-receptor gene rearrangement analysis ("PARR").[23] The sensitivity of feline TRG molecular clonality assessment is about 90% in cats with morphological features of T cell lymphoma.[23] The revised canine TRG molecular clonality assay is about 94% sensitive in diverse T cell neoplasia (Keller and Moore, manuscript in preparation).

Mucosal T cell lymphomas are readily diagnosed in endoscopic biopsy specimens utilizing CD3 immunohistochemistry and TRG molecular clonality assessment. Laparoscopic (full-thickness) biopsy specimens are superior to endoscopic specimens because the entire small intestine can be sampled (not just duodenum or ileum) and transmural progression is assessable.[9,23]

Transmural extension of T cell lymphoma occurs focally or multifocally. Mass formation, intestinal obstruction, and even perforation with peritonitis are commonly encountered.[23] Transmural T cell lymphoma consists most frequently of large lymphocytes (nuclear diameter >2 red blood cell diameters), although transmural lesions involving small- to intermediate-sized lymphocytes occur.[23] Feline transmural large-cell lymphomas have a markedly worse prognosis than mucosal small-cell lymphomas.[23,26] T cell lymphomas of granular lymphocytes (LGL) occur most frequently as transmural large-cell lymphomas, but they are recognizable as mucosal lesions at an earlier stage. Feline transmural LGL T cell lymphomas have the worst prognosis.[8,15,23,27] Transmural large T cell lymphomas (including LGL T cell lymphomas) are the most frequently encountered lymphomas in the canine small intestine; these are high-grade lymphomas with short median survival times (Moore, unpublished data).[7,10,25] Fine-needle aspirate cytology readily identifies cytoplasmic granules in LGL lymphomas. Immunostaining for the cytoplasmic granule protein, granzyme B, is useful in FFPE tissue sections.[23] LGL lymphomas routinely invade mesenteric lymph nodes and spread rapidly to liver, spleen, and kidney.[23,27] LGL intestinal T cell lymphomas may invade peripheral blood; this may be sufficient to cause secondary leukemia. Although well described, the frequency of this event has not been quantified.[27]

Cutaneous T Cell Lymphoma

Cutaneous T cell lymphomas (CTCL) are a heterogeneous group of disorders with a variety of clinical presentations and morphological features in humans and dogs.[14,16,21,22] Epitheliotropic CTCL is also known as mycosis fungoides (MF), which accounts for the majority of cases of CTCL. It is distinguished from other peripheral T cell lymphomas that affect skin by virtue of the prominent epitheliotropism of the T cell infiltrates. Pagetoid reticulosis and Sezary syndrome are rare forms of epitheliotropic T cell lymphoma that in the past have been considered variants of MF. Pagetoid reticulosis is a highly epitheliotropic lymphoma in which the lymphoid infiltrate is almost entirely confined to the epidermis in patch-plaque stage lesions. Sezary syndrome is a generalized epitheliotropic T cell lymphoma with erythroderma, lymphadenopathy, and neoplastic T cells in the blood. Recent data in humans indicate that Sezary syndrome is a unique disease, which originates in skin recirculating effector memory T cells, while MF is a disease of skin resident effector memory T cells.[6] Human MF is predominantly a lymphoma of CD4+ alpha beta T cells; expression of CD8 occurs less commonly.

Epitheliotropic CTCL is an uncommon, but well-recognized entity in dogs, and diseases similar to MF, Sezary syndrome, and pagetoid reticulosis have been described.[22] Lesions closely resemble the equivalent human diseases except for the marked tropism of the lymphoid infiltrate for the hair follicles and apocrine sweat glands in canine epitheliotropic CTCL.[22] This tropism is maintained even in tumor stage lesions. The clinical disease course of canine epitheliotropic CTCL ranges from 3 months to 4 years from initial diagnosis. Progression from erythematous patches and plaques to nodules and tumors, which are initially localized to the skin and mucous membranes, but eventually metastasize to lymph nodes and beyond, is observed in canine epitheliotropic CTCL.[22] The immunophenotype of T cells in canine epitheliotropic CTCL is markedly different from its human counterpart. In canine CTCL, epitheliotropic T cells most commonly express CD8 (about 80% of dogs) or lack expression of CD4 and CD8.[22] T cell expression of CD4 has not been observed. T cells in human MF dominantly express TCR alpha beta; expression of TCR gamma delta occurs in about 10–15% of cases.[29] Whereas in canine epitheliotropic CTCL, T cells more frequently express TCR gamma delta (60%). In the case of the highly epitheliotropic pagetoid form of canine MF, T cells exclusively express TCR gamma delta.[19]

The topographical localization of canine epitheliotropic CTCL lesions provides insight into the trafficking patterns of skin homing memory T cells in the dog. Lesions occur at mucocutaneous junctions or in tissues of the oral cavity with high frequency. The feet are also frequently involved. Facial mucocutaneous junctions and feet are common sites of involvement in canine atopic dermatitis. Furthermore, atopic dermatitis has been shown to be a risk factor for later development of epitheliotropic CTCL in dogs.[28] Indications that inflammatory conditions can precede and merge into cutaneous lymphoma in humans are increasingly apparent.[2]

Non-epitheliotropic CTCLs also occur in dogs. They are variants of the WHO peripheral T cell lymphoma entities. They were thought to be relatively less common than epitheliotropic CTCL.[21] However, with the advent of molecular clonality testing, recognition of other disease syndromes occurred (Figure 3). Indolent forms of non-epitheliotropic CTCL have been described in dogs and cats.[1,13] Inflamed non-epitheliotropic CTCL is a recently described entity, which bears considerable resemblance to cutaneous reactive histiocytosis.[20] The distinction of the 2 diseases relies heavily upon immunohistochemistry and molecular clonality assessment.

CONCLUSION

Mucosal T cell lymphoma of small to intermediate cell type (WHO EATCL type II) is the dominant lymphoma in the gastrointestinal tract of cats and has a relatively long median survival time. Mucosal T cell lymphoma is characterized by distinctive patterns of infiltration of the villous and/or crypt epithelium and lamina propria. These patterns are readily recognizable in either endoscopic or laparoscopic (full-thickness) biopsy specimens, and they are highly correlated with clonal TRG gene rearrangement. However, despite the success in recognition of mucosal T cell lymphoma in endoscopic biopsy specimens (largely from duodenum), it should be noted that the highest incidence of lymphoma is in the jejunum. Hence, an endoscopic biopsy procedure, despite its convenience, would potentially miss some mucosal T cell lymphomas. Mucosal small T cell lymphoma has been increasingly recognized in dogs due to the increasing utilization of molecular clonality in diagnostic investigation. Transmural T cell and B cell lymphomas are readily diagnosed by morphological assessment and immunohistochemistry for CD3/CD79a/CD20/Pax5, usually without the need for molecular clonality analysis. Transmural T cell lymphoma of large-cell type (WHO EATCL type I) is most often an LGL lymphoma; this is readily demonstrable with immunohistochemistry to detect the cytotoxic granule protein, granzyme B. Transmural T cell lymphoma of large-cell type has a relatively short median survival time, but is the least common T cell lymphoma in the gastrointestinal tract of cats. Transmural large T cell lymphoma was the dominant form of intestinal lymphoma in dogs, but this may change with the relatively recent recognition of small T cell lymphoma. However, the overall incidence of gastrointestinal lymphoma in dogs is much lower than in cats.

Cutaneous T cell lymphomas occur as epitheliotropic or non-epitheliotropic forms. The topographical localization of cutaneous lymphomas corresponds to sites commonly afflicted by allergic skin diseases (especially atopic dermatitis). It is also apparent that epitheliotropic CTCLs also commonly involve tissues of the oral cavity. This has revealed the ability of skin homing memory T cells to also traffic to the oral cavity. The immunophenotype of canine CTCLs is markedly different from human CTCLs. In dogs, there is a high incidence of gamma delta T cell lymphomas compared to humans. Also, the existence of CD4+ epitheliotropic CTCLs has not been recognized, and CD4+ non-epitheliotropic CTCLs are uncommon in dogs. The role for gamma delta T cells as a responsive element in the canine skin-associated immune system has yet to be defined. Furthermore, there are likely to be significant operational differences between the canine and human skin immune systems.

REFERENCES

1. Affolter VK, *et al. Vet Dermatol*. 2009;20(5-6):577-585.
2. Burg G, *et al. Arch Dermatol*. 2001;137(7):949-952.
3. Carrasco V, *et al. Vet Pathol*. 2015;52(4):668-675.
4. Cheroutre H, *et al. Nat Rev Immunol*. 2004;4(4):290-300.
5. Clark RA. *Sci Transl Med*. 2015;7(269):269rv261.
6. Clark RA, *et al. Sci Transl Med*. 2012;4(117):117ra117.
7. Coyle KA, *et al. Vet Pathol*. 2004;41(2):141-146.
8. Darbes J, *et al. Vet Pathol*. 1998;35:323-460.
9. Evans SE, *et al. J Am Vet Med Assoc*. 2006;229(9):1447-1450.
10. Frank JD, *et al. J Am Anim Hosp Assoc*. 2007;43(6):313-321.
11. Fuhlbrigge RC, *et al. Nature*. 1997;389(6654):978-981.
12. German AJ, *et al. J Comp Pathol*. 1999;121(3):249-263.
13. Gilbert S, *et al. Vet Dermatol*. 2004;15(1):3-12.
14. Girardi M, *et al. N Engl J Med*. 2004;350(19):1978-1988.
15. Kariya K, *et al. Vet Pathol*. 1997;34(2):156-159.
16. Kerl H, *et al. Semin Diagn Pathol*. 1991;8(2):55-61.
17. Kupper TS, *et al. Nat Rev Immunol*. 2004;4(3):211-222.
18. Miyasaka M, *et al. Nat Rev Immunol*. 2004;4(5):360-370.
19. Moore PF, *et al. Vet Dermatol*. 2009;20(5-6):569-576.
20. Moore PF, *et al. Vet Dermatol*. 2013;24(1):204-211.
21. Moore PF, *et al. Clin Dermatol*. 1994;12:499-505.
22. Moore PF, *et al. Am J Pathol*. 1994;144:421-429.
23. Moore PF, *et al. Vet Pathol*. 2012;49(4):658-668.
24. Nagler-Anderson C. *Nat Rev Immunol*. 2001;1(1):59-67.
25. Ozaki K, *et al. Vet Pathol*. 2006;43(3):339-344.
26. Richter KP. *Vet Clin North Am Small Anim Pract*. 2003;33(5):1083-1098, vii.
27. Roccabianca P, *et al. Vet Pathol*. 2006;43(1):15-28.
28. Santoro D, *et al. Vet Dermatol*. 2007;18(2):101-106.
29. Swerdlow SH, *et al. WHO Classification of Tumors of the Haematopoietic and Lymphoid Tissues*. 4th. ed. Lyon: International Agency for Research on Cancer (IARC); 2008.
30. Valli VE, *et al. Vet Pathol*. 2011;48(1):198-211.
31. Washabau RJ, *et al. J Vet Intern Med*. 2010;24(1):10-26.
32. Watanabe R, *et al. Sci Transl Med*. 2015;7(279):279ra239.
33. Woo JC, *et al. Vet Immunol Immunopathol*. 2002;85(1-2):9-22.

acvim282

Epigenetically Active Therapies for Osteosarcoma

Jeffrey N. Bryan, DVM, MS, PhD, DACVIM (Oncology)
University of Missouri, Columbia, MO, USA

Introduction

The methylation of cytosine bases occurs post-replication with the addition of a methyl group to the 5 position of the cytosine ring when a cytosine is followed by a guanine (CpG) in sequence. The addition is made by DNA methyltransferases (DNMTs) using the methyl donor s-adenosyl methionine. DNMT1 is responsible for maintenance methylation by which parental strand methylation patterns are copied onto daughter strand CpGs post-replication. DNMT3a is responsible for placing methylation marks on CpGs under conditions of dynamic epigenetic change. Mutations of both proteins have been shown to cause alterations of both global genomic methylation and gene promoter-specific methylation associated with cancer. Methyl groups are removed primarily by removal of the methylated base and repair to cytosine. Oxidation of the methylcytosine to hydroxymethylcytosine, formylcytosine, or carboxylcytosine can result in removal of part or all of the base and repair to unmethylated cytosine. The AID or APOBEC enzymes catalyze deamination of the base. TET enzymes that catalyze conversion to formyl- or carboxylcytosine lead to removal of the adduct and direct restoration of unmethylated cytosine. Across mammalian genomes, CpG dinucleotides are relatively rare sequential base pairs. Sporadically, CpG may be found in high density in regions called CpG islands that are frequently located in association with the 5′ control regions of genes. Hypermethylation of these islands was originally thought to result in persistent silencing of the associated gene. New work has demonstrated, however, that chromatin marks on histones are responsible for the active silencing of the gene, which can often be expressed even when methylated. These hypermethylated genes will revert to a silenced state in relatively short order when the signal for the histone changes recedes.[1] Recently, a third concept of CpGs has emerged. The CpGs residing in the lower-density regions near CpG islands are considered to reside on the "shores" and "shelves" of those islands and appear to play a significant role in cell development and stem cell function. Put most simply, epigenetic changes like DNA methylation are the mechanism by which cells establish persistent and stable phenotype while sharing a common genotype across all cells. Certain sets of genes in each cell are mostly permanently turned on or off to limit the set of proteins that are expressed. These mechanisms become labile in the course of carcinogenesis, changing the gene sets available to cancer cells and resulting in significant genetic and epigenetic plasticity. Further, labile somatic cells, including important immune-regulating cells, utilize DNA methylation as a persistent epigenetic mark to modify phenotype during immune reactions and immune-repressed states. Each of these manifestations of DNA methylation represents a therapeutic target. The approach of using epigenetically active therapy to modify anti-tumor therapy sensitivity, immune response, and modify the metastatic cascade has been minimally explored, but may represent a large, untapped potential in cancer treatment.

Epigenetic Progenitor Origin of Cancer

Recently, a model of carcinogenesis has been proposed that has been called the epigenetic progenitor origin of cancer.[2] Unlike the clonal genetic model of cancer development, this model suggests that progenitor cell populations in tissues are affected by environmental interactions that lead to an expanded pool of epigenetically altered progenitor cells with altered differentiation potential and polarity. On this background of expanded progenitor cells, a gatekeeper genetic mutation occurs characteristic of the particular type of tumor. This then becomes the primed source of a clonal cancer. Following this is a period of epigenetic and genetic plasticity that results in expansion of the clone and development of the hallmarks of cancer. It is unclear to which cancers this mechanism applies and whether it applies to osteosarcoma. However, if it is shown to be part of the mechanism of carcinogenesis, the theory would suggest a role for demethylator therapy in osteosarcoma.

EPIGENETICALLY ACTIVE THERAPY

DNA methylation and histone modifications can be modified with drug therapy, causing the re-expression of silenced genes and resulting in cell death and tumor regression *in vitro* and in mouse models of disease. However, these drugs have translated incompletely to the clinic. Currently, Vidaza (5-azacytidine) and decitabine (5-aza,2-deoxycitidine) (DNMT inhibitors) have been found to be useful in treating myelodysplastic syndrome in humans, resulting in transfusion independence.[3] Demethylating drugs function similarly to antimetabolite chemotherapy drugs, being inserted into the growing DNA strand, leading to an inability to extend the strand and arresting DNA polymerase while covalently bound to the base. The entire protein/base molecule is then excised from the strand to remove the methylated base, replacing it with an unmethylated cytosine. Histone deacetylase (HDAC) inhibitors such as vorinostat (SAHA) have found application in a number of tumors with SAHA being approved for epitheliotropic lymphoma in humans. Clinical trials of DNMT inhibitors as well as HDAC inhibitors have been combined with conventional chemotherapy with demonstrated tolerability and better efficacy compared to chemotherapy alone.[3] Demethylating drugs active in human lymphoma cells have been shown to be active in canine cells as well.[4] Oral demethylators have been developed, including Zebularine and guadecitabine. Zebularine was abandoned in preclinical evaluation because of poor reactions in monkeys and lack of metabolism in certain dog breeds. Given the delicate balance of gene control across many cell types within the body, it is likely that generalized inhibition of DNMTs or HDACs is not desirable, even for individuals with cancer, and particularly for a protracted course of therapy. Brief courses of inhibition to sensitize to chemotherapy are attractive for better management of disease. Interestingly, the effects of demethylation therapy on gene expression and radiosensitization appear to be much less in normal cells than in cancer cells. Ideally, inhibition of these molecules will be targeted selectively to cancers, avoiding these effects on the rest of the somatic cells. New generations are already in development that inhibit particular methyltransferases or deacetylases very specifically. Even this level of specificity may not be optimal and may become so selective as to cross species poorly. The results of gene activation can be varied. Classically, reactivating tumor suppressor pathways is a desired outcome. Recent work has suggested that modulation of miRNA levels may have a profound role on cell survival following demethylating or deacetylating treatments and in combination with cytotoxic therapies. Animal models of naturally occurring cancers could model such targeted therapies and be used to identify potentially optimal drug combinations.

DEMETHYLATION AS A CHEMOSENSITIZING STRATEGY

Little work has been published evaluating the interaction between demethylation therapy and cytotoxic chemotherapy. However, much work has been done in ovarian cancer to study the interaction, particularly with platinum-based therapy. Established chemoresistant cell lines of ovarian cancer cells were demonstrated to have very low expression levels of p27, which was restored to normal following decitabine exposure.[5] This also restored sensitivity to cisplatin. In women with recurrent, platinum-resistant ovarian cancer, treatment with decitabine resulted in multifold changes in 10 circulating miRNAs that correlated with restoration of sensitivity to carboplatin chemotherapy.[6] Further, serial biopsies of ovarian cancer patients treated with decitabine identified reactivated pathways associated with subsequent response to platinum-based therapy.[7] However, a randomized clinical trial of decitabine followed by carboplatin compared to carboplatin alone found no survival benefit for the combination in ovarian cancer.[8] In the patients receiving combination, neutropenia and carboplatin hypersensitivity were of higher grade compared to the single-agent treated women. While there is theoretical benefit to combining demethylation with cytotoxic chemotherapy, the current application has not yielded clinical improvement across the board.

DEMETHYLATION AS A RADIOSENSITIZING STRATEGY

Many studies have shown *in vitro* evidence of the radiosensitizing properties of demethylating drugs. Tantalizingly, demethylating drugs have had much less of a radiosensitizing effect on normal stromal cells, and gene expression changes have been of much lower magnitude in response to this class of drug. The therapeutic implication of this opposite effect is that demethylating drugs may exert some selectivity in their radiosensitization effect. This may allow for enhanced tumor response with minimal increase in normal tissue acute and late effects. The mechanisms of radiosensitization appear to be multifactorial. Decitabine has been reported to activate the G2/M checkpoint and promote apoptosis in human osteosarcoma cells. Treated cells showed cell cycle synchrony along with increased apoptotic fraction and upregulated expression of 14-3-3σ, CHK2, and DAPK-1 genes in concert with demethylation of the promoter CpG island of those genes.[9] Similarly, decitabine was shown to modulate expression of a number of pro-apoptotic genes, including GADD45A, HSPA9B, PAWR, PDCD5, NGKBIA, and TNFAIP3.[10] In further evaluation, decitabine was shown to demethylate and upregulate the GADD45A gene, promoting apoptosis spontaneously in human osteosarcoma cells.[11] The enhancement of radiosensitivity could be blocked by anti-sense siRNA treatment of cells reducing GADD45A expression. While not shown to directly contribute to radiosensitivity, such enhancement of apoptosis would likely contribute to therapeutic responsiveness to radiotherapy. Finally, decitabine, through immune mechanisms which will be discussed below, has been shown to upregulate anti-osteosarcoma T cell responses when tumors are treated with ionizing radiation.[12] This dual effect of pro-apoptosis and anti-tumor immunostimulatory property of demethylating drugs in conjunction with ionizing radiation deserves further exploration.

Demethylating Agents as Immunomodulatory Drugs

Systemic demethylation therapy has the potential to affect any cell population in the body that experiences effective levels of drug. This includes the cells of the immune system which use the stable but modifiable mechanism of DNA methylation to establish phenotype and function of immune effectors and regulators. Multiple mechanisms of action can contribute to improved immune recognition, breaking of immune tolerance, and immune reaction to the tumor cells to contribute to the therapeutic effect. As mentioned above, T cell responses to solid-tumor cell lines have been shown to be upregulated when demethylation is combined with ionizing radiation therapy. In an *ex vivo* model, cancer cells exposed to decitabine demonstrated upregulated MHCI and costimulatory molecules, making them more attractive targets for the T cells.[12] NK cell function was not shown to be modified in this model. The effects of demethylating drugs on NK activity have been reported to be conflicting. While demethylation appears to upregulate killer immunoglobulin-like receptors that enhance recognition of MHCI, the cytotoxicity of NK appears to be inhibited by azacytidine exposure.[13] The net benefit to this arm of anti-tumor immunity is unclear. In human T CD8+ cells, transcription of IL-2 is modified by methylation of a promoter-enhancer region that increases transcription in the unmethylated state.[13] Mouse T cells cultured in the presence of azacytidine also express greater levels of IFN-γ. *Ex vivo* evaluations of human and mouse T cells have suggested that they are more reactive when cultured in the presence of demethylating drugs.

It appears that demethylator exposure can also modify the function of helper T cells. Human CD4+ T cells cultured in the presence of decitabine produce greater levels of IL-17A, a pro-inflammatory cytokine. Human patients treated with azacytidine also have higher numbers of circulating IL-17A-secreting T cells than prior to treatment.[13] Finally, decitabine treatment resulted in polarization of naïve T cells toward a Th1 phenotype and in depolarization of Th2 cells to produce IFN-γ.[13]

Conclusions

Treatment with demethylating agents before cytotoxic chemotherapy treatment has many theoretical benefits. It may alter the epigenetic set of cancer cells to sensitize those cells to chemotherapy or radiotherapy, resulting in greater cell kill. There is some evidence that this effect is greater for cancer cells than normal cells, making this strategy relatively tissue sparing. The interaction between demethylation and radiation may enhance immune responses against tumor neoantigens. Demethylating drugs may enhance anti-tumor immune responses by other mechanisms as well. Given the current high rate of failure of platinum-based chemotherapy against canine osteosarcoma, epigenetically active therapy should be investigated in these patients as comparative models for human osteosarcoma.

REFERENCES

1. Raynal NJ, Si J, Taby RF, *et al.* DNA methylation does not stably lock gene expression but instead serves as a molecular mark for gene silencing memory. *Cancer Res.* 2012.
2. Feinberg AP, Ohlsson R, Henikoff S. The epigenetic progenitor origin of human cancer. *Nat Rev Genet.* 2006;7:21-33.
3. Costa FF. Epigenomics in cancer management. *Cancer Manag Res.* 2010;2:255-265.
4. Flesner BK, Kumar SR, Bryan JN. 6-Thioguanine and zebularine down-regulate DNMT1 and globally demethylate canine malignant lymphoid cells. *BMC Vet Res.* 2015;10:290.
5. Zhao Y, Li Q, Wu X, *et al.* Upregulation of p27Kip1 by demethylation sensitizes cisplatin-resistant human ovarian cancer SKOV3 cells. *Mol Med Rep.* 2016;14:1659-1666.
6. Benson EA, Skaar TC, Liu Y, *et al.* Carboplatin with decitabine therapy, in recurrent platinum resistant ovarian cancer, alters circulating miRNAs concentrations: a pilot study. *PloS One.* 2015;10:e0141279.
7. Fang F, Zuo Q, Pilrose J, *et al.* Decitabine reactivated pathways in platinum resistant ovarian cancer. *Oncotarget.* 2014;5:3579-3589.
8. Glasspool RM, Brown R, Gore ME, *et al.* A randomised, phase II trial of the DNA-hypomethylating agent 5-aza-2'-deoxycytidine (decitabine) in combination with carboplatin vs carboplatin alone in patients with recurrent, partially platinum-sensitive ovarian cancer. *Br J Cancer.* 2014;110:1923-1929.
9. Li CH, Lim SH, Ryu HH, *et al.* Enhancement of radiosensitivity by inhibition of c-Jun N-terminal kinase activity in a Lewis lung carcinoma-bearing subcutaneous tumor mouse model. *Oncol Rep.* 2016;36:3397-3404.
10. Al-Romaih K, Somers GR, Bayani J, *et al.* Modulation by decitabine of gene expression and growth of osteosarcoma U2OS cells *in vitro* and in xenografts: identification of apoptotic genes as targets for demethylation. *Cancer Cell Int.* 2007;7:14.
11. Al-Romaih K, Sadikovic B, Yoshimoto M, *et al.* Decitabine-induced demethylation of 5' CpG island in GADD45A leads to apoptosis in osteosarcoma cells. *Neoplasia.* 2008;10:471-480.
12. Son CH, Lee HR, Koh EK, *et al.* Combination treatment with decitabine and ionizing radiation enhances tumor cells susceptibility of T cells. *Sci Rep.* 2016;6:32470.
13. Saleh MH, Wang L, Goldberg MS. Improving cancer immunotherapy with DNA methyltransferase inhibitors. *Cancer Immunol Immunother.* 2016;65:787-796.

acvim299

The Future of Small Animal Oncology Therapeutics

Chad M. Johannes, DVM, DACVIM (SAIM, Oncology)
Ames, IA, USA

Veterinary Oncology Therapeutic Market

The diverse expanse of specialized oncology therapeutics on the horizon for dogs and cats makes remaining current on the latest advances challenging. Providing updates since ACVIM Forum 2016, this session reviews the status, science, available supporting data and clinical use of several new canine and feline cancer therapeutics. Although many of these therapeutics are in the earlier stages of development, they hold strong potential and are beginning to become clinically available for patient care. Those therapeutics, regulated by either the Food and Drug Administration (FDA) or Unites States Department of Agriculture (USDA) and currently commercially available to veterinarians in the US, are summarized in Table 1.

Immunotherapy Developments

Anti-Cancer Monoclonal Antibodies (MAbs)

1. **BLONTRESS® - Canine lymphoma monoclonal antibody, B-cell (Aratana Therapeutics, Inc.)**
    a. Caninized MAb fully licensed by the USDA in 2015 as an aid in the treatment of canine B-cell lymphoma.
    b. No peer-reviewed data is available to date. A clinical study (Mini B-CHOMP) has been conducted to further evaluate use of this MAb in conjunction with cytotoxic chemotherapy, however results have not been released.
    c. Aratana has stated that BLONTRESS is not as specific to the CD20 target as expected.[1]
2. **TACTRESS™ - Canine lymphoma monoclonal antibody, T-cell (Aratana Therapeutics, Inc.)**
    a. Caninized MAb received full licensure by the USDA in 2016 as an aid in the treatment of canine T-cell lymphoma.
    b. No peer-reviewed data is available to date. Two prospective studies have been conducted to further evaluate use of this MAb in conjunction with cytotoxic chemotherapy (CHOP-based protocol and lomustine). Initial analysis indicates TACTRESS does not add significant progression free survival in these two studies.[1]
    c. Aratana has stated that TACTRESS is not as specific to the CD52 target as expected.[1]
3. **Anti-CD20 monoclonal antibody for canine lymphoma (Elanco)**
    a. The generation and characterization of a rituximab-like canine-specific anti-CD20 antibody (1E4-cIgGB) intended as a candidate treatment for canine B-cell lymphoma has been described.[2]
    b. 1E4-cIgGB elicited a therapeutic response against canine diffuse large B-cell lymphoma *in vivo* (murine xenograft model) as a single agent. Augmented responses were noted when combined with CD47 blocking therapies.[3]
    c. Development status for commercial use in dogs is unknown at this time.
4. **PD-1/PD-L1 - Likely future immunotherapeutic targets**
    a. Inhibition of T-cell checkpoint molecules such as programmed death 1 (PD-1) or its ligand (PD-L1) has produced significant responses in advanced human tumors (melanoma, renal cell carcinoma, and non-small cell lung cancer).[4]
    b. Recent studies have reported expression of PD-L1 in several canine malignant tumors including oral melanoma, osteosarcoma, hemangiosarcoma, mast cell tumor, mammary and prostate adenocarcinoma and lymphoma.[5,6]
    c. It is highly probable that new immunotherapeutics in veterinary oncology will target this pathway given the success noted by such therapies for various human cancers.

Therapeutic Anti-Cancer Vaccines

1. Canine Lymphoma Vaccine, DNA (Merial/Boehringer Ingelheim)

a. Xenogeneic murine CD20 DNA therapeutic vaccine for use in dogs with B-cell lymphoma was conditionally licensed by the USDA in 2015.
b. No peer-reviewed data is available to date. The initial trial reported with this therapeutic described vaccine administration (four doses at two-week intervals) after completion of CHOP-based cytotoxic chemotherapy.[7,8]
c. Additional prospective, multi-center clinical trials in dogs with B-cell lymphoma have been described but results have not yet been published.

2. **Canine Osteosarcoma Vaccine/AT-014 (Aratana Therapeutics, Inc.)**
   a. Recombinant HER2/neu expressing *Listeria* therapeutic vaccine (AT-014) is being investigated as an aid in the treatment of canine osteosarcoma.
   b. AT-014 was administered to dogs with appendicular osteosarcoma following amputation (or limb salvage surgery) and chemotherapy (4 doses of carboplatin). In 18 dogs with no evidence of metastatic disease at enrollment and treated with AT-014, the median disease free interval (DFI) was 615 days and median survival time (MST) was 956 days. Adverse events were mild to moderate and primarily consisted of fever, lethargy, and nausea/vomiting.[9]
   c. As of the third quarter of 2016, Aratana was continuing to enroll dogs in the pivotal field safety study for AT-014. Conditional licensure by USDA is anticipated in the first half of 2017.[10]
3. **Feline IL-2 Immunomodulator (Merial/Boehringer Ingelheim)**
   a. Recombinant canarypox virus (ALVAC) expressing feline interleukin-2 (IL-2).
   b. Approved by EMA in 2013 (Oncept IL-2); conditional license granted by USDA in 2015 (indicated to delay postsurgical recurrence of feline fibrosarcoma in adult cats with stage 1 disease).[11]
   c. Adjuvant treatment of feline injection-site sarcomas with this immunotherapy in complement to surgery and brachytherapy has been described.[12]
   d. Studies further demonstrating the efficacy and safety of this vaccine as an adjunct to surgical resection of primary (first occurrence) stage I feline fibrosarcoma are being pursued.[13]
4. **Telomerase Vaccine (Takis)**
   a. A DNA telomerase therapeutic vaccine targeting dog TERT (dTERT) has been described.[14] TERT activity is largely confined to tumor tissues making it may a valid target for immunotherapy.
   b. Two clinical field studies in dogs with lymphoma have evaluated this therapeutic in combination with cytotoxic chemotherapy [cyclophosphamide, vincristine, prednisone (COP)].[15,16]
   c. Reported therapeutic targets include lymphoma and hemangiosarcoma. Veterinary sites have been identified for additional studies.[14]
5. **ImmuneFx™ Cancer Vaccine (Morphogenesis Inc./Veterinary Oncology Services, Inc.)**
   a. A reportedly multi-indication cancer vaccine, IFx-VET, is available. As a cell therapy, this product is reported to break tolerance to multiple tumor associated antigens via inter-antigenic epitope spreading.[17]
   b. Clinical benefit of this therapeutic is not well defined as no peer-reviewed data is available to date.
   c. As an autologous vaccine, this therapeutic is not currently approved or licensed by a regulatory agency.
   d. The USDA has authorized the company to provide the personalized (autologous) form of the vaccine to veterinarians; reportedly utilized in 37 different types of cancer in over 113 breed of dogs, cats and horses.[18]

Other Immunotherapies

1. **ELIAS Cancer Immunotherapy [ECI™] (Elias Animal Health)**

   a. Adoptive T-cell therapy. Treatment protocol vaccinates the dog with its own cancer cells to produce an immune response. Personalized T-cells are then obtained via apheresis, activated *ex vivo* and reinfused into the dog.[19]
   b. ELIAS is currently conducting studies utilizing this therapy in dogs with osteosarcoma and B-cell lymphoma.[19]
   c. Additional trials for other types of cancer are in development.[19]
2. **EBC-46 (QBiotics)**
   a. Novel natural product small molecule PKC activator with reported anticancer activity in mouse models.[20]
   b. Utilized for treatment of solid tumors in dogs, cats and horses via intra-tumor injection and topical application.[21]
   c. No peer-reviewed data in dogs/cats is available to date.
   d. Not yet registered as a veterinary medicine and thus not available; not actively recruiting patients for veterinary clinical trials. Intend to pursue registration in US, EU, and Australia.[21]

Small Molecule Inhibitor Developments

1. **Verdinexor/KPT-335 for Canine Lymphoma (Karyopharm Therapeutics)**
   a. Orally bioavailable XPO1 inhibitor. Exportin 1 (XPO1) is the sole nuclear exporter of several major tumor suppressor and growth regulatory proteins (p53, p75, Rb, p21, p27, STAT3, FOXO, IkB). Expression is known to be upregulated in a variety of both hematologic malignancies and solid tumors, correlating with a poor prognosis.[22]
   b. Early phase clinical trials of Verdinexor in dogs with lymphoma have been conducted.[22,23]
   c. Verdinexor has received a minor use designation from the Center for Veterinary Medicine (CVM) of the FDA for the treatment of lymphomas in dogs. The effectiveness and safety technical sections are complete to support conditional approval (manufacturing section not complete).[24]
2. **RV-1001 for Canine Lymphoma (Rhizen Pharmaceuticals)**
   a. Orally bioavailable inhibitor of PI3 kinase (PI3K), an important cell signaling pathway. In lymphomas and leukemias, PI3K isoforms γ and δ are noted to be particularly important in maintaining cancer cell growth.[25]
   b. A phase I clinical trial was performed in dogs with naïve or relapsed B or T-cell lymphoma.[26,27]
   c. RV-1001 was granted minor use designation in July 2015 by FDA-CVM and a phase II study has been initiated for treatment of lymphoma in dogs.[27]
3. **VDC-597 (VetDC)**
   a. Orally bioavailable dual inhibitor of PI3Kα and mTOR pathways, both of which are implicated as playing key roles in many types of cancer, including lymphoma, osteosarcoma, hemangiosarcoma, and many solid tumors.[28]
   b. In preclinical models, VDC-597 achieved good distribution into tumors, target inhibition and suppression of tumor growth. VDC-597 demonstrated significant blood concentrations and bioactivity in dogs at well-tolerated doses.[28]

Cytotoxic Chemotherapeutic Developments

1. **Paccal Vet®-CA1 (Oasmia Pharmaceutical)**
   a. The active ingredient is paclitaxel, an antimicrotubule agent that works by stabilizing microtubules.
   b. Conditional approval by FDA; labeled to treat: 1) nonresectable stage III, IV, or V mammary carcinoma in dogs that have not received previous chemotherapy or radiotherapy; and 2) resectable and nonresectable squamous cell carcinoma in dogs that have not received previous chemotherapy or radiotherapy.[29]

c. Phase III clinical trials for dogs with mammary carcinoma (ongoing) and dogs with squamous cell carcinoma (planned) are being pursued for full approvals.[30]

2. **Rabacfosadine/Tanovea™-CA1 (VetDC)**
   a. Prodrug of the nucleotide analogue 9-(2-phosphonylmethoxyethyl) guanine (PMEG); effectively loads lymphoid cells while reducing levels of PMEG in plasma and target organs of toxicity.
   b. A phase I/II trial was conducted in 38 dogs with non-Hodgkin lymphoma (NHL) using different dose schedules. Thirty (79%) dogs receiving Tanovea monotherapy achieved clinical remission [23 (61%) CR and 7 (18%) PR]. Median first remission duration for all dogs was 128 days.[31]
   c. A number of abstracts at recent conferences have reported on treating dogs with lymphoma:
      i. Tanovea-CA1 with prednisone in naïve and relapsed cases (n=74).[32]
      ii. Tanovea-CA1 at different doses in naïve and relapsed cases (n=62).[33]
      iii. Tanovea-CA1 alternating with doxorubicin in naïve cases (n=54).[34]
      iv. Tanovea-CA1 at different doses in relapsed cases receiving one prior doxorubicin-containing regimen (n=52).[35]
   d. Efficacy of Tanovea-CA1 has also been evaluated in dogs with cutaneous T-cell lymphoma (n=11)[36] and multiple myeloma (n=14).[37]
   e. Tanovea-CA1 received conditional approval from FDA in January 2017 for the treatment of lymphoma in dogs. VetDC anticipates that Tanovea-CA1 will be available to veterinarians in the spring of 2017.[38]

ONCOLOGY SUPPORTIVE THERAPEUTIC DEVELOPMENTS

1. **Canalevia™ (Jaguar Animal Health)**
   a. Canine-specific formulation of crofelemer, an active pharmaceutical ingredient isolated and purified from the *Croton lechleri* tree, that contains anti-secretory properties.[39]
   b. Canalevia is an oral, enteric-coated, twice daily formulation, being developed for the treatment of chemotherapy-induced diarrhea (CID) in dogs. The product is not absorbed systemically at the therapeutic dose, but acts locally in the gastrointestinal tract.[39]
   c. Results of the proof-of-concept study for secretory diarrhea were announced in October 2016. At that time, it was stated that the pivotal trial to evaluate the safety and efficacy of Canalevia for treatment of acute diarrhea in dogs was ongoing. An estimated 200 dogs will be enrolled with estimated completion near the end of 2016.[40]
   d. Jaguar expects to file the new animal drug application (NADA) for Canalevia in 2017 for the indication of acute diarrhea in dogs.[40]
2. **Capromorelin Oral Solution/ENTYCE® (Aratana Therapeutics, Inc.)**
   a. Capromorelin is an orally active small molecule that mimics the action of ghrelin which causes growth hormone secretion and appetite stimulation.
   b. Capromorelin oral solution dosed at 3 mg/kg PO q 24 h has been shown to cause increased food intake and weight gain in both healthy laboratory and inappetent client-owned dogs.[41-43]
   c. Capromorelin oral solution has also demonstrated a wide margin of safety, being well-tolerated at daily doses up to 40 mg/kg once daily for 12 consecutive months.[44]
   d. When administered to healthy dogs for 7 days, capromorelin treatment produced increased insulin-like growth factor 1 (IGF-1) concentrations on Day 1 and this increase was sustained through Day 7.[41] IGF-1 is believed to play a role in cellular proliferation and increased metastatic ability in a number of human cancers.[45,46] It is unclear what role, if any, the increased IGF-1 levels produced by capromorelin could impact the clinical progression of various canine cancers. Anamorelin, a therapeutic with similar mechanism, is however currently being studied in treatment of cancer-related cachexia in human non-small cell lung cancer.[47,48]

e. Additionally, capromorelin has been demonstrated to cause increases in IGF-1 and increased food intake and body weight in cats.[49]
f. ENTYCE was granted FDA approval in May 2016. Aratana continues to interact with the FDA regarding the manufacturing of ENTYCE and intends to make it commercially available to veterinarians in late 2017.[50]
g. This therapeutic has significant potential to positively impact the treatment of inappetence related to chemotherapy and/or underlying cancer in our canine patients.
h. Aratana is also developing a cat-specific formulation with additional studies pending.

Table 1. Commercially available, regulated, veterinary oncology therapeutics (USA)

Trade name	Compound name	Company	Indication	Regulatory status (USA)	Species
Blontress®	Canine lymphoma MAb, B-cell	Aratana	B-cell lymphoma	USDA Licensed (2015)	Canine
NA	Canine lymphoma vaccine, DNA	Merial/BI	B-cell lymphoma	USDA Conditional license (2015)	Canine
NA	Feline interleukin-2 immunomodulator	Merial/BI	Primary stage I fibrosarcoma	USDA Conditional license (2015)	Feline
Immunocidin®	Mycobacterium cell wall fraction	NovaVive	Mammary tumors	USDA Licensed (2009)	Canine
Oncept®	Canine melanoma vaccine, DNA	Merial/BI	Melanoma	USDA Licensed (2010)	Canine
Paccal Vet®-CA1	Paclitaxel for injection	Oasmia	Mammary CA SCC	FDA Conditional approval (2014)	Canine
Palladia®	Toceranib phosphate	Zoetis	Grade II/III mast cell tumor	FDA Approved (2009)	Canine
Tactress™	Canine lymphoma MAb, T-cell	Aratana	T-cell lymphoma	USDA Licensed (2016)	Canine
Tanovea™-CA1	Rabacfosadine for injection	VetDC	Lymphoma	FDA Conditional approval (2017)	Canine

References

1. http://aratana.investorroom.com/2015-09-24-Aratana-Therapeutics-Provides-Product-Updates.
2. Rue SM, *et al. Vet Immunol Immunopathol.* 2015;164(3–4):148–159.
3. Weiskopf K, *et al. Cancer Immunol Res.* 2016;4(12):1072–1087.
4. Regan D, *et al. Vet J.* 2016;207:20–28.
5. Maekawa N, *et al. PLoS One.* 2016;11(6):e0157176.
6. Shosu K, *et al. In Vivo.* 2016;30(3):195–204.
7. Merial LYM15DETAILER (10/15).
8. Bergman PJ, *et al.* In: Proceedings of the 28th Annual Conference of the Veterinary Cancer Society. Seattle, WA. 2008;55.
9. Mason NJ, *et al. Clin Cancer Res.* 2016;22(17).
10. http://aratana.investorroom.com/2016-11-03-Aratana-Therapeutics-Reports-Third-Quarter-2016-Financial-Results.
11. http://todaysveterinarypractice.epubxp.com/i/504690-may-jun-2015/18.
12. Jas D, *et al. Trials in Vaccinology.* 2015;4:1–8.
13. Merial IL215NA1PGDETAILER (04/16).

14. http://www.takisbiotech.it/index.php/takis-biotech/tk-004.
15. Peruzzi D, *et al. Mol Ther.* 2010;18(8):1559–1567.
16. Gavazza A, *et al. Hum Gene Ther.* 2013;24(8):728–738.
17. http://www.morphogenesis-inc.com/rd/technologies/immunefx/.
18. http://www.morphogenesis-inc.com/veterinary-oncology/.
19. http://eliasanimalhealth.com/.
20. Boyle GM, *et al. PLoS One.* 2014;9(10):e108887.
21. https://qbiotics.com/index.php/vet-pipeline/ebc-46-animals.
22. London CA, *et al. PLOS One.* 2014;9(2):e87585.
23. http://karyopharm.com/pipeline/oral-verdinexor-kpt-335/.
24. http://investors.karyopharm.com/releasedetail.cfm?releaseid=854785.
25. http://www.rhizen.com/vetresearch.html.
26. London CA, *et al.* Abstract, 2015 American Association for Cancer Research; Philadelphia, PA.
27. https://globenewswire.com/news-release/2015/09/22/769986/0/en/Rhizen-Pharmaceuticals-S-A-announces-initiation-of-Phase-2-study-of-RV1001-in-dogs-with-T-or-B-cell-Lymphoma.html.
28. http://vet-dc.com/products/vdc-597-for-various-cancers/.
29. http://www.fda.gov/AnimalVeterinary/ResourcesforYou/ucm402476.htm.
30. http://oasmia.com/en/research-development/.
31. Vail DM, *et al. Clin Cancer Res.* 2009;15(10):3503–3510.
32. Thamm D, *et al.* Abstract, 2015 ACVIM Forum; Indianapolis, IN.
33. Thamm D, *et al.* Abstract, 2015 Veterinary Cancer Society Conference; Tysons, VA;73.
34. Thamm D, *et al.* Abstract, 2016 ACVIM Forum; Denver, CO.
35. Vickery K, *et al.* Abstract, 2016 Veterinary Cancer Society Conference; Orlando, FL;46.
36. Morges MA, *et al. J Vet Intern Med.* 2014;28(5):1569–1574.
37. Thamm DH, *et al. BMC Vet Res.* 2014 Jan 25;10:30.
38. http://vet-dc.com/news/vetdc-receives-fda-conditional-approval-tanovea-ca1-first-new-animal-drug-treating-lymphoma-dogs/.
39. http://www.businesswire.com/news/home/20150701005278/en/Jaguar-Completes-Study-Dogs-Chemotherapy-Induced-Diarrhea-Commercial#.Vc4B8ZNViko.
40. http://phx.corporate-ir.net/phoenix.zhtml?c=253723&p=irol-newsArticle&ID=2212567.
41. Zollers B, *et al. J Vet Pharmacol Ther.* 2016;doi:10.1111/jvp.12344.
42. Zollers B, *et al. J Vet Intern Med.* 2016;30(6):1851–1857.
43. Zollers B, *et al. BMC Vet Res.* 2017;13(1):10.
44. Zollers B, *et al. J Vet Pharmacol Ther.* 2016;doi:10.1111/jvp.12358.
45. Burtscher I, Christofori G. *Drug Resist Updat.* 1999;2:3–8.
46. Armakolas N, *et al. Crit Rev Oncol Hematol.* 2016;108:137–145.
47. Currow DC, *et al. J Clin Oncol.* 2015;33(29_suppl):175.
48. Temel JS, *et al. J Clin Oncol.* 2016;34(26_suppl):203.
49. Zollers B, *et al.* Abstract, 2015 ACVIM Forum; Indianapolis, IN.
50. http://aratana.investorroom.com/2017-02-03-Aratana-Therapeutics-to-Attend-North-American-Veterinary-Conference.

acvim300

Advances in Our Understanding of Metastasis Biology and Therapy

Chand Khanna, DVM, PhD, DACVIM (Oncology), DACVP (hon)

The Problem of Cancer Metastasis

The development of metastasis is a universally grave development for cancer patients irrespective of specific cancer histology. Emblematic of this clinical problem is the biology associated with progression in osteosarcoma. Indeed, for both human and canine cancer patients afflicted with osteosarcoma, the most common primary tumor of bone, metastasis to the lungs is the most common cause of death. Lung metastases develop in these patients despite highly effective treatment of the primary tumor. Improving our understanding of the biology of metastasis is needed to improve outcomes for these patients. It is important to emphasize that the study of metastasis must include two distinct biologic entities that are both referred to as metastasis. First metastasis is a verb that describes the process of cancer progression from a primary tumor site to a distant secondary site. Second metastasis also refers to the actual metastatic lesion that exists at the secondary site and is responsible for patient mortality. Progress towards understanding metastatic disease (both the verb and noun) and its inherent resistance to conventional treatments is limited by many factors. First, the process of metastasis is believed to begin very early in the course of disease progression and has occurred in most patients at the time of their initial presentation. This provides limited opportunities to study these events in human patients alone. Second, the genetic aberrations that are responsible for the development of metastasis are complex, heterogeneous, and difficult to distinguish from the events responsible for the actual development of cancer. Finally, the access to well described patient samples is limited and often is available only after treatment with cytotoxic chemotherapy. For relatively rare human cancers such as osteosarcoma, these problems become quickly amplified.

Genes and Pathways Driving Osteosarcoma Development

A consistent genomic hallmark of osteosarcoma and its metastasis is a highly complex and significantly and aneuploid karyotype. It is reasonable that this complex karyotype, has limited the ability of multi-platform genomic studies in both the human and the dog to define recurrent alterations in the genome that explain the development or metastatic behavior of this cancer. Furthermore, this complex karyotype has prompted hypotheses surrounding the metastatic biology of this cancer. One hypothesis argues that a genetic alteration in DNA recognition and repair pathways may allow the progressive acquisition of this genetic complexity in the absence of the expected induction of programmed cell death. Recurrent alterations in cell cycle associated genes have been identified in both human and canine osteosarcoma, but are unlikely sufficient to explain the unique karyotypic alterations associated with this disease. An alternative and non-mutually exclusive hypothesis, may suggest a genetic alteration that result from an immediate and sudden acquisition of multiple genomic alterations (i.e., chromothripsis). Indeed, structural events in the osteosarcoma genome have been identified in support of chromothripsis.

Comparative Approach to Osteosarcoma Metastasis: Stress Adaptation and Ezrin-Associated Pathways

In the absence of simple genetic variants, in a single species, that may explain the metastatic phenotype of osteosarcoma, we have utilized a cross-species comparative approach that includes murine, canine, and human systems for gene and pathway identification and evaluation. This approach has leveraged the availability of murine, canine, and human genomes, to survey the expression of genes in normal and diseased tissues and then identify either patterns of gene expression of individual genes and pathways responsible for or associated with metastasis. As an example, we used this comparative approach to identify and then associate the cytoskeleton linker protein, ezrin, with metastasis. Ezrin is the best characterized of the ERM (ezrin-radixin-moesin) family. ERM proteins exist in the cytoplasm in an inactive "closed conformation" though N-terminal to C-terminal associations within the protein or with other ERM members. The stability and consistency of our findings involving ezrin across species lines (mouse, dog, human) has strengthened our belief in this cross-species approach. Ezrin's linkage of the cell membrane to the actin cytoskeleton directly allows cells to interact with its micro-environment and functionally provides an "intracellular scaffolding" that facilitates signal transduction through a number of growth factor receptors and adhesion molecules. We have demonstrated that ezrin is necessary for metastasis in murine transplantable osteosarcoma and genetically engineered murine models, that it is relevant in human-murine sarcoma xenograft models, and that its expression is associated with metastatic progression in pet dogs with naturally occurring osteosarcoma, finally we have found an association between ezrin expression and risk of metastatic relapse in pediatric osteosarcoma patients. The value of this comparative approach has been to define biologically relevant motifs that have sustained importance as we cross species lines.

Evaluating Antimetastatic Therapies Based Stress Adaptation and Ezrin Biology

The comparative approach is uniquely suited to rapidly translate findings from the laboratory to the clinic. With this interest in mind we have recently linked ezrin expression and the mTOR (mammalian target of rapamycin) pathway in the translation of proteins by metastatic cells. We have begun to validate these associations in murine, canine, and human cancer cells and have conducted preclinical studies in mice that support the therapeutic role of inhibiting mTOR using rapamycin (and novel analogs). Indeed novel inhibitors of the mTOR pathway have entered early clinical trials for human osteosarcoma and sarcoma patients. There are many questions that remain unanswered regarding the optimal use of these agents in human patients. Indeed, many of the unanswered questions that exist in the development path of new drugs may be effectively answered by integrating studies that include pet dogs with cancer. To complete the translational effort based on ezrin and mTOR inhibition, we have now completed dose and regimen finding studies in pet dogs with osteosarcoma that will allow the evaluation of the inhibition of translation initiation (by rapamycin) in pet dogs with osteosarcoma. Ongoing studies will now define optimal treatment schedules for the use of rapamycin in children and dogs with osteosarcoma. Efforts to validate reagents and further characterize these models using more sophisticated techniques has been ongoing within several comparative oncology laboratories around the world. Contributing to this effort is the Comparative Oncology Program of the US National Cancer Institute's Center for Cancer Research and the not for profit, Canine Comparative Oncology and Genomics Consortium.

Based on the opportunities provided by the completion of the canine genome, the identical biological behavior of canine and human osteosarcoma, and the increased prevalence and aggressiveness of this disease in dogs we recently used our cross-species comparative gene expression approach to uncover specific genes, gene families/functions, or pathways that were conserved across the dog and human and are commonly linked to metastasis. Collectively, our comparative approach has provided a novel, necessary, and informative perspective for the study of cancer biology. The approach exemplifies the values of One Medicine in the study of complex biomedical problems. Ongoing work, within this paradigm will be essential in the translation of our biological understanding of osteosarcoma metastasis towards therapeutic opportunities.

References

1. Paoloni M, Khanna C. Translation of new cancer treatments from pet dogs to humans. *Nat Rev Cancer*. 2008;8:147–156.
2. Khanna C. Novel targets with potential therapeutic applications in osteosarcoma. *Curr Oncol Rep*. 2008;10:350–358.
3. Khanna C, Wan X, Bose S, *et al*. The membrane-cytoskeleton linker ezrin is necessary for osteosarcoma metastasis. *Nat Med*. 2004;10:182–186.
4. Wan X, Mendoza A, Khanna C, Helman LJ. Rapamycin inhibits ezrin-mediated metastatic behavior in a murine model of osteosarcoma. *Cancer Res*. 2005;65:2406–2411.

1.
acvim301

Novel Osteosarcoma Therapies

Douglas H. Thamm, VMD, DACVIM (Oncology)
Fort Collins, CO, USA

Introduction

There have been no major breakthroughs in adjuvant therapy for canine (or human) osteosarcoma (OSA) in the last 20 years. However, there are a variety of drugs, targets, and approaches that have been evaluated in preclinical or early clinical studies in dogs, which may have efficacy in canine OSA. The hour will discuss some targets that have been explored recently and could be candidates for future study.

Histone Deacetylase/Valproic Acid

A number of compounds that inhibit HDACs have progressed rapidly through clinical trials.[1] For example, suberoylanilide hydroxamic acid (SAHA, also known as Vorinostat and marketed as Zolinza®; Merck & Co., Inc., NJ, USA) was approved recently for treatment of cutaneous T cell lymphoma. Approval was granted as part of the FDA orphan drug program, following a priority review. A variety of other natural, semi-synthetic, and synthetic inhibitors of HDAC are in various stages of clinical development. As with most novel, targeted therapies, they may find their greatest utility when combined with standard forms of therapy such as chemotherapy, radiation therapy, and surgery. Preclinical data support the concept that HDAC inhibition can potentiate the efficacy of both chemotherapy and radiation therapy.

The HDAC inhibitors phenylbutyrate and sodium valproate (VPA) are FDA approved for other indications, and are comparatively inexpensive versus SAHA. VPA is marketed as an anticonvulsant and is available in both oral and injectable forms, and preliminary pharmacokinetic information in dogs is available from studies performed in the 1980s. VPA is rarely used as an anticonvulsant in dogs due to its very short half-life. Sustained-release oral formulations of VPA are now available. Human clinical trials have evaluated VPA as a single-agent, as well as in combination with demethylating agents, retinoids, or conventional cytotoxic agents. It appears that VPA concentrations capable of modulating HDAC activity are achievable *in vivo* in humans.

Our laboratory demonstrated that pretreatment of canine and human OSA cells with clinically relevant concentrations of the HDAC inhibitor VPA enhances the antiproliferative and pro-apoptotic effects of doxorubicin and significantly inhibits tumor growth in a murine subcutaneous xenograft model. This appears to be due, at least in part, to increased nuclear accumulation of doxorubicin by the tumor cells following VPA exposure.[2] We completed accrual to a phase I clinical trial in tumor-bearing dogs of escalating VPA followed by a standard dose of doxorubicin; no toxicity was observed at the highest doses evaluated, and increased histone acetylation in peripheral blood mononuclear cells increased linearly with dose. Increased histone acetylation was documented in tumor tissue from 4 of 8 evaluable patients.[3] We have performed gene expression profiling to interrogate pathways modulated by VPA pretreatment in OSA cells. Oxidative phosphorylation, cytoskeleton remodeling, cell cycle, and ubiquitin-proteasome pathways were among those pathways most affected by VPA treatment, and synergy with proteasome inhibition was demonstrated *in vitro*.[4]

Survivin/EZN3042

Resistance to apoptosis (programmed cell death) is one of the hallmarks of cancer. The inhibitors of apoptosis proteins (IAP) constitute a family of highly conserved inhibitors of cell death that have been identified in yeast, invertebrates, and vertebrates. Survivin, a product of the *Birc5* gene, is an important anti-apoptotic IAP family member that is unique in that it exhibits peak expression during mitosis and is critical for normal cell division. Although survivin is highly expressed in fetal tissues, expression is nearly undetectable in most terminally differentiated adult cells. Importantly, an analysis of 3.5 million human transcripts from 19 normal and diseased tissue types identified survivin among the most common transcripts uniformly upregulated in cancer but not in normal tissues.

Several survivin-targeted therapeutics have been or are currently undergoing human clinical evaluation. These include the small molecule YM155, and the oligo-based therapeutics LY2181308, and EZN-3042, a LNA antisense oligonucleotide that downregulates survivin mRNA and protein. Importantly, the sequence of EZN-3042 has 100% homology with the canine survivin sequence. We recently demonstrated that elevated survivin expression in primary canine OSA tissue correlated with increased histologic grade and mitotic index and a decreased disease-free interval (DFI). Survivin attenuation using either siRNA or EZN-3042 in canine OSA cells inhibited cell-cycle progression, increased apoptosis, mitotic arrest, and chemosensitivity, and cooperated with chemotherapy to significantly improve *in vivo* tumor control.[5,6]

We recently completed a phase I clinical trial of single-agent EZN-3042 in dogs with lymphoma. No dose-limiting toxicities were observed at any dose level, and significant downregulation of survivin expression in tumor tissues was observed in dogs treated at the highest dose level (8.25 mg/kg twice weekly).

Thioredoxin Reductase/Auranofin

Gold compounds have been used in medicine dating as far back as 2,500 BCE. Their primary use has been as immunomodulatory agents for the treatment of inflammatory diseases such as rheumatoid arthritis. The oral gold compound 2,3,4,6-tetra-o-acetyl-1-thio-β-D-glucopyranosato-S-(triethyl-phosphine) gold, manufactured under the drug name auranofin (AF, Ridaura®), was developed in the 1980s as a potential antirheumatic agent. More recently, potent antineoplastic activity has been observed *in vitro* and *in vivo*. Potential mechanisms of action include modulation of secretion of tumor-promoting cytokines such as IL-6 and IL-8 by monocytes and macrophages, modulation of intracellular signaling and survival pathways such as MAPK, NF-κB, STAT3 and telomerase activity, and intracellular generation of reactive oxygen species (ROS) through inhibition of thiol-redox enzymes such as thioredoxin reductase (TxR) and thioredoxin glutathione reductase.

A recent gene expression-based study compared transcripts from poorly metastatic and highly metastatic human OSA cell lines and identified upregulation of TxR. AF inhibited TxR activity in human OSA cells at low doses and inhibited metastases *in vivo*.[7] A second recent study demonstrated *in vitro* and *in vivo* inhibition of canine OSA growth with the related gold salt aurathiomalate.[8] We recently completed an unbiased screen of FDA-approved compounds that were selected for growth inhibition of canine **and** human OSA cells but not normal human osteoblasts or fibroblasts. Interestingly, AF was one of the top hits. Synergy was observed with the mTOR inhibitor rapamycin and the HDAC inhibitor SAHA both *in vitro* and *in vivo*.[9]

Multiple studies have evaluated the bioavailability, tolerability, pharmacokinetics, and efficacy of AF in normal dogs and dogs with immune-mediated disease. The oral bioavailability of AF ranges from 15 to 38%. Diarrhea is the most common side effect observed in dogs. Rapid reversal of the diarrhea is seen with drug withdrawal and lowering the administered dose. Proteinuria and thrombocytopenia/anemia, thought to be immune-mediated, are less common adverse effects and tend to occur after prolonged treatment (48 to 72 months) at high doses (2.4 to 3.6 mg/kg/d). Corticosteroid administration may be necessary to treat these side effects. There is evidence of efficacy in multiple canine inflammatory diseases at tolerated concentrations, including immune-mediated polyarthritis and pemphigus complex diseases refractory to corticosteroids alone.

A phase I clinical trial of AF is currently ongoing to determine a dose that is associated with modulation of TxR and/or generation of reactive oxygen species in canine cancer patients.

Autophagy/Hydroxychloroquine

Autophagy is the process by which proteins, organelles, and other cellular components are delivered to the lysosome for degradation and recycling. Mammalian autophagy is controlled by a complicated network of signaling pathways, most of which feed into the PI3K/mTORC1 pathway. These signaling pathways regulate the evolutionarily conserved core autophagy machinery that involves over 30 autophagy regulators encoded by ATG genes. There have been numerous studies addressing how autophagy affects cancer therapeutics. Many recent studies have concluded that the role of autophagy is primarily protective and have suggested that autophagy induction is a mechanism of chemoresistance. However, in some cases it has also been concluded that the opposite holds (i.e., that autophagy is not protective but is actually necessary for the antitumor effect of the drug).

One of the drugs that appears to have autophagy-inhibitory activity is the antiparasitic drug hydroxychloroquine (HCQ). Recent *in vitro* studies have demonstrated high basal autophagic flux in a panel of canine OSA cell lines. When compared with most other tumor cells, canine OSA cells appeared to be quite sensitive to HCQ, and HCQ and doxorubicin appeared to have additive antiproliferative effects *in vitro*.[10]

We recently completed a phase I clinical trial of combined HCQ and doxorubicin in dogs with spontaneous tumors. HCQ was administered daily by mouth throughout the trial, beginning 72 h prior to doxorubicin, which was given intravenously on a 21-day cycle. Pharmacokinetic analysis revealed a 100-fold increase in HCQ in tumors compared with plasma. There was a trend that supported therapy-induced increase in LC3-II (the cleaved and lipidated form of microtubule-associated protein 1 light chain 3/LC3, which serves as a maker for autophagosomes) and SQSTM1/p62 (sequestosome 1) after treatment.[11]

Personalized Medicine/The COXEN Approach

Recent breakthroughs in cancer genomics have made the emerging field of personalized medicine an attractive alternative to traditional approaches to cancer treatment. Although success of single biomarkers in predicting response to molecularly targeted agents has been observed, this is not the case with traditional chemotherapeutics, which are still heavily used in human and veterinary medicine. The use of gene expression prediction signatures has been used for the accurate prediction of traditional chemotherapy response in many cancers. One such strategy developed by Lee and Theodorescu was termed the Co-expression Extrapolation (COXEN) method. It combines microarray gene expression and drug sensitivity data from a reference set to identify gene signatures that are then extrapolated to a different dataset for drug sensitivity prediction. This method has been used to successfully predict *in vitro* sensitivity as well as clinical outcome in human bladder, breast, and non-small cell lung cancers.

The Gustafson Lab at CSU evaluated the use of gene expression-based models built in an intra- or interspecies manner to predict chemosensitivity and treatment outcome in canine OSA. Models were built and evaluated using microarray gene expression and drug sensitivity data from human and canine cancer cell lines, and retrospective canine OSA tumor datasets. The "COXEN" method was utilized to filter gene signatures between human and dog datasets based on strong co-expression patterns. Models for doxorubicin and carboplatin sensitivity were built using linear discriminant analysis via the misclassification penalized posterior algorithm. The best doxorubicin model involved genes identified in human lines that were co-expressed and trained on canine OSA tumor data, which accurately predicted clinical outcome in 73% of dogs. The best carboplatin model utilized canine lines for gene identification and model training, with canine OSA tumor data for co-expression. Dogs whose treatment matched predictions had significantly better clinical outcomes than those that didn't ($p=0.0006$), and this predictor significantly associated with longer disease-free intervals in a Cox multivariate analysis (hazard ratio=0.3102, $p=0.0124$).[12]

An ongoing prospective trial is subjecting tumors from dogs with appendicular OSA undergoing amputation to Affymetrix-based gene-expression profiling, and then using the COXEN algorithms developed for doxorubicin and carboplatin sensitivity to inform choice of postoperative chemotherapy (single-agent carboplatin, single-agent doxorubicin, alternating carboplatin/doxorubicin).

Conclusions

Multiple novel strategies could hold promise for canine OSA treatment in the future, not limited to those discussed above. The recent Morris Animal Foundation canine OSA initiative has provided a remarkably important vehicle for the large-scale evaluation of novel antimetastatic therapies in dogs with OSA, the first of which is the mTOR inhibitor rapamycin. Prioritizing additional agents for study will be key moving forward.

References

1. Xu WS, *et al. Oncogene*. 2007;26:5541.
2. Wittenburg LA, *et al. Cancer Chemother Pharmacol*. 2011;67(1):83.
3. Wittenburg LA, *et al. Clin Cancer Res*. 2010;16(19):4832.
4. Wittenburg LA, *et al. J Cell Biochem*. 2012;113:773.
5. Shoeneman JK, *et al. Cancer Res*. 2012;72:249.
6. Shoeneman JK, *et al. Vet Comp Oncol*. 2016;42:e45.
7. Topkas E, *et al. Oncotarget*. 2016;7:831.
8. Scharf VF, *et al. Anticancer Drugs*. 2014;25:332.
9. Fulbright JM, *et al*. Proc AACR Annu Meeting. 2015;Abstr #3689.
10. Carnicello C, *et al*. Proc AACR EORTC NCI Molecular Targets and Cancer Therapeutics. 2016;Abstr #328.
11. Barnard RA, *et al. Autophagy*. 2014;10:1415.
12. Fowles JS, *et al. BMC Bioinformatics*. 2016;17:93.

acvim303

Biology, Molecular Targets, and Clinical Interventions for Canine Osteosarcoma

Timothy M. Fan, DVM, PhD, DACVIM (Oncology, Internal Medicine)
Urbana, IL, USA

Introduction

Canine osteosarcoma (OS) remains a deadly disease for the majority of dogs affected with only a fraction of patients achieving long-term survival. Incremental knowledge has been gained in the molecular pathways which might participate in disease genesis and progression; however, it remains ill-defined if any unifying and conserved driver aberrations are shared across the majority of dogs with OS. Given the complexity of OS pathogenesis, it would be unexpected that a singular treatment intervention could serve as a "silver bullet," and there has been a dearth of transformative therapeutic breakthroughs over the past 20 decades. However, with continued research efforts, new information is being discovered which should be incorporated into multimodality therapeutic approaches to improve the management of canine OS. The hour will discuss some aspects of OS biology, molecular targets, and potential future clinical interventions for dogs diagnosed with OS.

Canine Osteosarcoma

Osteosarcoma is a malignant tumor derived from primitive mesenchymal stem cells with the capacity to produce osteoid matrix. Consistent with its stem cell derivation, OS can adopt a spectrum of histomorphologic characteristics, permitting its categorization into chondroblastic, fibroblastic, osteoblastic, and telangiectatic subtypes. Clinically, OS develops predominantly in the metaphyseal regions of weight-bearing bones, most commonly the distal radius, proximal humerus, distal femur, proximal tibia, and distal tibia.[1,2] The incidence of OS is bimodal with the majority of cases arising in geriatric patients (greater than 7 years old) and a much smaller peak of OS observed in younger animals (2–3 years old).

Biologically, OS behaves as a high-grade sarcoma possessing aggressive osteolytic properties and the strong predilection for distant metastases. Typically, OS originates within the intramedullary cavity of metaphyseal bone. The growth and progression of appendicular OS result in effacement and erosion of the immediate bone microenvironment, including the marrow cavity and circumferential cortical and trabecular bone. In addition to localized skeletal perturbations, OS metastasizes to distant visceral organs, most commonly the pulmonary parenchyma via hematogenous dissemination. Effective treatment of appendicular OS necessitates the management of both local disease and distant metastases; however, even with the institution of multimodality therapy, the majority of dogs ultimately succumb to metastatic progression.

Despite the commonality of OS, a unifying etiopathogenesis for its development and progression remains incompletely elucidated. However, based upon molecular and preclinical investigations, OS genesis likely involves the complex interplay of aberrant molecular signals associated with environmental, bone metabolic, and genetic factors. Unlike the orderly multistep progression associated with epithelial malignancy development, considerable evidence supports a chromothripsis genomic event for the development of OS. Given the complexity of OS etiopathogenesis, there is a clear and justified need to intensively investigate putative molecular pathways which participate in the development and progression of OS in dogs. It would be anticipated that with the identification of druggable targets would lead to future therapeutic strategies that improve the long-term management of OS.

Molecular Pathogenesis

Unlike certain cancers which are characterized by recurrent chromosomal aberrations, the genetic derangements associated with OS tend to heterogeneous and chaotic. As such, the molecular origins of

OS have been difficult to ascribe to specific intracellular signaling anomalies. Nonetheless, substantive evidence derived from experimental, preclinical, and comparative investigations have underscored the potential importance of key dysregulated intracellular signaling and cell survival pathways likely involved in OS development and progression. Many of these molecular pathways have been identified in not only immortalized cell lines, but also spontaneously arising OS samples; which further substantiates the importance of specific signaling pathways in naturally-occurring canine OS. It is the hope of cancer biologists and clinicians alike that the identification of druggable targets will promote the development and validation of novel treatment strategies for improving the long-term outcome in dogs diagnosed with OS.

MOLECULAR TARGETS

Mammalian Target of Rapamycin (mTOR)

The mammalian target of rapamycin (mTOR) pathway critically regulates cellular signals induced by growth factors, nutrients, energy, and stress for the purposes of regulating cell cycle progression and growth. Excessive or aberrant signaling through the mTOR pathway contributes to growth, survival, and chemotherapy resistance in multiple tumor types, and might participate in canine OS progression. Based upon work conducted in canine OS cell lines, the operative functionality of the mTOR pathway was supported by basal phosphorylation of mTOR and p70S6K, a downstream effector protein.[3] More importantly, phosphorylation of mTOR and could be inhibited by rapamycin; resulting in apoptosis and reduced growth of OS cells *in vitro*, suggesting that mTOR inhibition might serve as a therapeutic target. Recently, mouse preclinical models have demonstrated that mTOR inhibition with rapamycin is capable of delaying the onset of both experimental and spontaneous lung metastases.[4] Through the incorporation of elegant reporter probes, it was demonstrated that rapamycin perturbs mRNA translation efficiency during times of cellular stress with consequent reduced survivability of early metastatic cells as they reach a new microenvironment such as the lung parenchyma. Extending upon these promising murine preclinical studies of rapamycin, a translational study was conducted in dogs with OS, testing the feasibility of mTOR inhibition as a treatment strategy was supported in a phase I trial, which showed that biologically effective concentrations of rapamycin were safely obtainable in dogs, and that modulation of mTOR target proteins by rapamycin was achievable within the bone tumor microenvironment.[5] Most recently, the derivation of a pharmacokinetically-driven dosing regimen for oral rapamycin was completed in dogs with OS, which has led to the launch of an ongoing randomized study evaluating the potential antimetastatic activities of adjuvant open rapamycin therapy in dogs following treatment with standard of care protocols (amputation + carboplatin).

ErbB-2 Proto-Oncogene

The proto-oncogene *erbB-2* encodes human epidermal growth factor receptor 2 (HER-2), which is a tyrosine kinase receptor capable of promoting cell transformation and growth. The overexpression of HER2 protein as a result of gene amplification is a negative prognostic factor in mammary carcinoma in dogs; however less clarity exists for the role of HER2 overexpression in the pathogenesis of OS.[6-7] Given that HER2 is a druggable target in people (Herceptin), there remains potential interest in HER2 as a novel therapeutic target for canine OS, however, recent findings with combining Herceptin with standard-of-care therapy (limb salvage + methotrexate/doxorubicin/cisplatin) for the management of metastatic pediatric OS patients failed to demonstrate any improvement in progression-free or survival times.[8] The overexpression of HER2 at the mRNA and protein levels have been demonstrated in canine OS cell lines, as well as spontaneously-arising OS tumor samples.[9] Most recently, the expression of HER2 immune dominant peptides in context with major histocompatibility complexes as a target for canine OS has been described with highly promising results.[10] This investigation demonstrated that the adjuvant administration of a modified HER2-expressing *Listeria monocytogenes* vaccine to dogs could produce specific immune responses and impressive prolongation in progression-free and overall survival times. These highly promising findings in dogs with OS treated with a vaccine strategy has fueled further enthusiasm for developing additional immunotherapeutic strategies for the management of OS in dogs and people alike, including toll-like receptor agonists and checkpoint inhibitors.

Aminobisphosphonate - Zoledronate

Aminobisphosphonates specifically inhibit osteoclastic activity and are used for the treatment of pathologic bone resorption disorders in people. Aminobisphosphonates directly inhibit bone resorption by binding and stabilizing hydroxyapatite crystals. Once incorporated into bone matrix, aminobisphosphonates inhibit further calcium and phosphorous mineral dissolution. Most importantly, aminobisphosphonates interfere with the geranylgeranylation of certain proteins involved in cell signaling, which ultimately impedes osteoclast activity and induces osteoclast apoptosis.[11-13]

While there is general consensus and agreement that aminobisphosphonates are indicated for the palliative bone pain management of diverse malignant skeletal malignancies such as OS, considerable research has identified direct anticancer activities of aminobisphosphonates which could delay the onset of metastatic progression. With their capacity to inhibit geranylgeranylation of key signaling proteins involved in cytoskeletal reorganization and chemotaxis, aminobisphosphonates have been investigated in canine OS cell lines for their ability to alter cell migration through the perturbation of a specific chemokine receptor, CXCR4.[14] This investigation identified the ability of zoledronate to downregulate and inhibits signaling mediated through CXCR4 in a subset of cell lines, and more importantly showed that the primary OS tumor of dogs receiving zoledronate therapy prior to amputation had reduced CXCR4 expressions in comparison with dogs not receiving aminobisphosphonate therapies. Based upon these observations, perturbation of lung metastases was speculated as a potential therapeutic effect of zoledronate. Despite the findings reduced CXCR4 expressions in dog OS tumor samples following zoledronate exposure, the inclusion of adjuvant zoledronate to standard backbone therapy in pediatric patients failed to demonstrate any improvement in progression-free or overall survival times.[15]

CONCLUSIONS

Canine osteosarcoma remains a challenging tumor to treat based upon its complex and incompletely understood pathogenesis. Multiple molecular targets have been identified and clinical trials testing the feasibility and anticancer activities of novel therapeutics are ongoing. Currently, the most promising avenues of pursuit for OS management is likely through immune enhancement strategies, however, it should be recognized that rational design of combinatorial strategies inclusive of immune system manipulation will require additional future studies to demonstrate definitive activities.

REFERENCES

1. Brodey RS, *et al. J Am Vet Med Assoc.* 1963;143:471.
2. Knecht CD, *et al. J Am Vet Med Assoc.* 1978;172:72.
3. Gordon IK, *et al. Am J Vet Res.* 2008;69:1079.
4. Morrow JJ, *et al. Clin Cancer Res.* 2016;22(24):6129.
5. Paoloni MC, *et al. PLoS One.* 2010;5:e11013.
6. Gorlick R, *et al. J Clin Oncol.* 1999;17:2781.
7. Scotlandi K, *et al. Eur J Cancer.* 2005;41:1361.
8. Ebb D, *et al. J Clin Oncol.* 2012;30(20):2545.
9. Flint AF, *et al. Vet Pathol.* 2004;41:291.
10. Mason NJ. *Clin Cancer Res.* 2016;22(17):4380.
11. Rogers MF, *et al. Cancer.* 1997; 80(8):1652.
12. Clezardin P, *et al. Joint Bone Spine.* 200;67(1):22.
13. Rogers MJ, *et al. Cancer.* 2000;88(12 Suppl):2961.
14. Byrum ML, *et al. J Vet Intern Med.* 2016;30(4):1187.
15. Piperno-Neumann S, *et al. Lancet Oncol.* 2016;17(8):1070.

acvim093

Selective Osmotic Shock for Canine Pancreatic Islet Isolation

Christopher A. Adin[1]; Elizabeth Thompson[1]; Jennifer Sollinger[1]; Emmanuel Opara[2]

[1]College of Veterinary Medicine, North Carolina State University, Raleigh, NC, USA; [2]Wake Forest Center for Regenerative Medicine, Winston Salem, NC, USA

Pancreatic islet transplantation could provide a cure for diabetes mellitus in dogs. Currently islet isolation is performed using harsh collagenases that cause non-specific injury to both islets and exocrine tissue, negatively affecting the outcome of cell transplantation. We evaluated a novel islet isolation protocol utilizing high concentrations of glucose to cause selective osmotic shock. Islets have a membrane glucose transporter (GLUT2) that allows adaptation to changes in glucose concentrations while exocrine tissue can be selectively destroyed by these osmolar shifts. Canine pancreata were obtained within 15 minutes after euthanasia from animals (n=6) euthanized for reasons unrelated to this study. Each pancreas was divided into 4 segments that were randomized to receive 300 mOsm glucose for 20 min (group 1); 600 mOsm for 20 minutes (group 2); 300 mOsm for 40 minutes (group 3); or 600 mOsm for 40 minutes (group 4). Islet yield, purity and viability were compared between groups. Mean±SEM islet yield for groups 1-4 was 533±445, 616±445, 1237±445, and 1494±445 islet equivalents per gram, respectively. Purity ranged from 37-45% and was not significantly different between groups. Islet cell viability was significantly affected by treatment group (p<0.001) and was highest in groups 1 (90%) and 2 (89%), which experienced shorter durations of exposure to hyperosmolar solutions. Exocrine cell viability for these same groups was 36% and 45% showing selective cell destruction when compared to tissue. Selective osmotic shock provides a widely available means for researchers to isolate canine islets for use in islet transplantation or in studies of canine islet physiology.

acvim098

Novel Medical Approach to Canine Hyperadrenocorticism

Karin Sanders[1]; Jan Mol[1]; Hans Kooistra[1]; Jan Willem Hesselink[1]; **Sara Galac**[1]

[1]Faculty of Veterinary Medicine, Utrecht, Netherlands

Current medical therapy options for canine hypercortisolism consist of mitotane and trilostane. While effective in controlling the clinical signs, both have disadvantages. Another approach in dogs with pituitary-dependent hyperadrenocorticism could be to block the glucocorticoid pathway by a melanocortin 2 receptor (MC2R) antagonist (i.e., an ACTH receptor antagonist). In cortisol-secreting adrenocortical tumors (ATs), suppression of steroidogenic factor 1 (SF-1) activity by SF-1 inverse agonists could be beneficial, as SF-1 is a critical factor associated with malignancy, but also steroidogenesis.

This study was performed *in vitro* in cultured canine primary adrenocortical cells. Cell cultures were prepared from normal adrenals (NA, n=8) and cortisol-secreting ATs (n=6). Hypercortisolism was mimicked by incubation with synthetic ACTH. Cultures were treated with various concentrations of two different MC2R antagonist compounds [BIM-22776 (776) and BIM22A299 (299)], and three SF-1 inverse agonist [IsoQ A (SID7969543 (#IsoqA)), compound 31 (F1808-0154, (#31)) and compound 32 (F1808-0165, (#32)). Effects on steroidogenesis were determined by cortisol measurements in the culture media, and mRNA relative expression of steroidogenic enzymes with RT-qPCR analysis. The effect of incubations on cell proliferation was evaluated by EdU assay.

Compound 299 significantly inhibited ACTH-stimulated cortisol production with a mean drop of cortisol production of 79% at 500 nM ($p=0.004$) and 91% at 5000 nM ($p<0.0001$). Compound 776 was less effective. Among SF-1 inverse agonists, #31 significantly ($p<0.0001$ and $p=0.003$) inhibited cortisol concentrations with a mean of 39% and 46% at 10 µM in ACTH stimulated and non-stimulated NA, respectively. In ATs, #31 was the most effective with a mean suppression of cortisol to 48% at 10 µM ($p=0.021$). RT-qPCR data showed significant downregulation of steroidogenic enzymes in ACTH-stimulated cells treated with MC2R antagonist 299. Among SF-1 inverse agonists, #31 significantly reduced mRNA expression of CYP17 and CYP21. Effects of compounds on cell proliferation were only detectable in ATs and not in NAs. Incubations with #31 resulted in 45% suppression of proliferation.

We conclude that MC2R antagonist compound 299 is a potent MC2R antagonist *in vitro* and could become a promising new treatment option for pituitary-dependent hypercortisolism. SF-1 inverse agonist #31 has a great potential in treatment of cortisol-secreting ATs, because it suppresses not only cortisol synthesis, but may also affect proliferation.

This *in vitro* study is a revolutionary approach to select the most potential treatment for canine hypercortisolism. It can lead to the development of new drugs and improvement of the current medical management.

acvim123

Assessment of Transdermal (Lipoderm) Mirtazapine as an Appetite Stimulant in Cats with Chronic Kidney Disease

Jessica Quimby[1]; Stacie Summers[2]; Kellyi Benson[3]; Andrea Herndon[3]; Dan Gustafson[3]

[1]The Ohio State University, Columbus, OH, USA; [2]College of Veterinary Medicine, Colorado State University, Fort Collins, CO, USA; [3]Colorado State University, Fort Collins, CO, USA

Transdermal mirtazapine (Lipoderm gel) effectively increases appetite in normal cats. The purpose of this study was to assess the appetite stimulation properties of the transdermal mirtazapine in cats with chronic kidney disease (CKD).

Nine client-owned cats with stable Stage II or III CKD and history of decreased appetite completed a double blind placebo-controlled crossover prospective study. Required diagnostics included serum biochemistry profile, complete blood count, urinalysis, urine culture, blood pressure, and total thyroxine measurement. Exclusion criteria included concurrent systemic illness, pyelonephritis, ureteral obstruction, or decompensation of CKD. Mirtazapine tablets were compounded into a 3.75 mg/0.1 ml dose transdermal Lipoderm gel by the CSU Veterinary Medical Center pharmacy along with an identical placebo. Predetermined randomization (AB or BA) was performed and as cats were enrolled they were assigned consecutively to a treatment regime. The randomly predetermined gel (A or B) was administered to the inner ear pinna every other day for three weeks and then crossed over after a 4 day washout. Owners documented appetite, rate of food ingestion, begging, activity and vocalization daily in the home environment. Physical exam, weight, body condition score, muscle mass score and serum biochemistry were performed at the end of each treatment phase. Change in weight, number of vomiting episodes over 21 days, and summed clinical scores for appetite, begging, vocalization and activity were compared between placebo and mirtazapine treatment phases using a Wilcoxon matched pairs test. Mirtazapine concentrations in compounded gel were measured using liquid chromatography/tandem mass spectrometry.

When CKD cats (six IRIS Stage 2 cats and three IRIS Stage 3 cats) received 3.75 mg transdermal mirtazapine every other day for three weeks, a statistically significant change in weight ($p=0.008$), increase in appetite ($p=0.05$), and increase in rate of food consumption ($p=0.02$) was seen. A statistical trend in increase in begging behaviors was seen ($p=0.06$). No significant difference in activity or vocalization was seen. Median weight change after transdermal mirtazapine administration was 0.22 kg (range 0.04 to 0.44 kg), while median weight change during the placebo period was -0.04 (range -0.6 to 0.08). Improvement in body condition score was seen in 5/9 cats during mirtazapine administration and improvement in muscle mass score was seen in 3/9 cats. Gel concentrations varied by 77–114% (median 88%) of target dose.

Transdermal mirtazapine increased appetite and resulted in weight gain in CKD despite inconsistencies in compounding and may be helpful with the management of CKD-associated dysrexia.

acvim124

Relationship Between Fibroblast Growth Factor-23 and Survival in Dogs with Chronic Kidney Disease

Adam Rudinsky[1]; Laura Harjes[1]; Ramiro Toribio[1]; Dennis Chew[2]; Valerie Parker[1]

[1]The Ohio State University Veterinary Clinical Sciences, Columbus, OH, USA; [2]The Ohio State University, Columbus, OH, USA

Fibroblast growth factor-23 (FGF23) and parathyroid hormone (PTH) concentrations increase with advancing stage of canine chronic kidney disease (CKD), while 25-hydroxyvitamin D (25[OH]D) and 1,25-dihydroxyvitamin D (1,25[OH]2D) concentrations decrease. Plasma FGF23 has previously been reported to be an independent predictor of CKD progression and survival in cats and people with CKD. The objective of this study was to investigate the relationship between FGF23, PTH, and vitamin D metabolites with survival time in dogs with naturally occurring CKD.

Twenty-seven dogs with azotemic CKD were included. Based on the International Renal Interest Society (IRIS) staging guidelines, nine dogs had IRIS stage 2 CKD, twelve dogs had stage 3 CKD, and six dogs had stage 4 CKD. At the time of study enrolment, a complete blood count, biochemical profile, urinalysis and urine protein:creatinine (UPC) ratio were performed. Concentrations of serum PTH, 25(OH)D and 1,25(OH)2D and plasma FGF23 were measured.

Survival times were calculated from the date of study inclusion to either death (all-cause) or right censoring. Cox regression was used to assess variables associated with survival including hematocrit, creatinine, CKD stage, serum phosphorus, UPC, PTH, 25(OH)D, 1,25(OH)2D and FGF23.

Significant hazard ratios (hazard ratio; 95% confidence interval; p value) were as follows: hyperphosphatemia (2.4; 1.2–8.8; 0.02), elevated UPC (3.2; 1.4–7.7; 0.01), and increased FGF-23 (2.7; 1.1–6.7; 0.03). CKD stage was also significantly associated with survival at all stages (p=0.01).

In conclusion, in addition to other factors previously identified to be associated with survival time in dogs with CKD (e.g., CKD stage, UPC, serum phosphorus), FGF23 was also correlated with survival.

acvim125

Evaluation of Serum Thyroxine, Free Thyroxine, and Thyrotropin Concentrations for Diagnosis of Feline Iatrogenic Hypothyroidism

Mark E. Peterson[1]; Rhett Nichols[2]; Mark Rishniw[3]

[1]Animal Endocrine Clinic, New York, NY, USA; [2]Antech Diagnostics, Lake Success, NY, USA; [3]Cornell University, Ithaca, NY, USA

Radioiodine is generally considered the treatment of choice for feline hyperthyroidism but can cause iatrogenic hypothyroidism in up to 75% of cats. However, these hypothyroid cats rarely develop overt clinical signs of the disease (hair-coat changes), and there are no published guidelines for how to diagnose or manage iatrogenic hypothyroidism in cats. Treating hyperthyroidism can also unmask subclinical chronic kidney disease (CKD), with up to 49% of cats developing azotemia by 3–6 months of treatment. Cats with iatrogenic hypothyroidism are more likely to develop CKD than treated cats that remain euthyroid; iatrogenic hypothyroidism can also reduce survival time, especially in cats with concurrent azotemia. Therefore, it is important to diagnose hypothyroidism in these cats, since thyroid hormone replacement might preserve kidney function and improve survival.

In this study, we sought to determine which serum thyroid hormone or thyrotropin test best identifies iatrogenic hypothyroidism in cats that first develop azotemia after radioiodine treatment. Further, we sought to determine which test would best differentiate these azotemic, hypothyroid cats from azotemic sick euthyroid cats following radioiodine treatment, as well as from azotemic cats with CKD but no history of thyroid disease.

Five hundred and twenty-four hyperthyroid cats treated with radioiodine at the Animal Endocrine Clinic from June 2013 to June 2015 were evaluated for inclusion in this study. None of these cats had pre-existent azotemia (defined as serum creatinine >1.9 µg/dL), either when hyperthyroid or controlled with methimazole. Following radioiodine, 42 cats that developed post-treatment azotemia (creatinine >2.5 µg/dL) had serum concentrations of thyroxine (T4), free T4 by dialysis (fT4), and thyrotropin (TSH) measured at 3, 6, and 12 months. Iatrogenic hypothyroidism was confirmed (n=28) or excluded (n=14) in these 42 cats on the basis of thyroid scintigraphy. Fourteen cats with CKD and 166 clinically normal cats underwent similar serum thyroid testing and scintigraphy. Nineteen of the 28 cats with documented hypothyroidism had T4, fT4, and TSH reevaluated after 1–3 months of levothyroxine (L-T4) treatment to determine the influence of replacement therapy on these analytes.

Serum concentrations of T4 and fT4 were lower and TSH higher in hypothyroid cats than in all three groups of euthyroid cats ($p<0.0001$). Of the hypothyroid cats, T4 and fT4 concentrations were low in 15 (53.6%) and 7 (25%), respectively. Low serum T4 and fT4 concentrations were also measured in 7 (50%) and 2 (14.3%) of the cats with CKD. TSH concentrations were clearly high in all hypothyroid cats, whereas TSH remained within the reference interval in all three groups of euthyroid cats. After levothyroxine supplementation, serum T4 and fT4 increased and TSH concentrations decreased in all 19 cats treated, normalizing in 18. Serum creatinine also decreased significantly after L-T4 supplementation, with 4/19 (21.1%) cats becoming non-azotemic (creatinine ≤1.9 µg/dL).

Our results indicate that measurement of serum TSH concentration is a very sensitive and specific diagnostic test for iatrogenic hypothyroidism in cats that develop azotemic CKD after treatment. The finding of high serum TSH concentrations best identifies feline iatrogenic hypothyroidism and differentiates it from nonthyroidal illness syndrome in cats that develop azotemia following treatment. After L-T4 supplementation of cats with hypothyroidism, high serum TSH concentrations decreased to within the reference interval, indicating that serum TSH measurement can also aid in post-pill monitoring. L-T4 replacement in our hypothyroid cats also appeared to improve renal function, as indicated by the decrease in serum creatinine concentrations.

acvim126

Evaluation of Fecal Canine Calprotectin as a Biomarker in Dogs with Chronic Inflammatory Enteropathies

Romy M. Heilmann[1]; Nolie Parnell[2]; Nora Berghoff[3]; Joanne Mansell[4]; Corinne Gurtner[5]; Niels Grutzner[5]; Jan S. Suchodolski[6]; Joerg M. Steiner[6]

[1]College of Veterinary Medicine, University of Leipzig, Leipzig, Sachsen, Germany; [2]College of Veterinary Medicine, Purdue University, Lafayette, IN, USA; [3]College of Veterinary Medicine, Michigan State University, East Lansing, MI, USA; [4]College of Veterinary Medicine and Biomedical Sciences, Texas A&M University, College Station, TX, USA; [5]Vetsuisse Faculty, University of Bern, Bern, Switzerland; [6]Gastrointestinal Laboratory, College of Veterinary Medicine, Texas A&M University, College Station, Texas, USA

Canine calprotectin (CP) and S100A12 have potential as markers of inflammation in dogs. Fecal CP and S100A12 concentrations were associated with clinical disease activity, and the latter also correlated with the severity of endoscopic lesions and the response to treatment in dogs with chronic enteropathies (CE). Serum C-reactive protein (CRP) concentration has been suggested as a monitoring tool for canine CE. The association between fecal CP and microscopic lesions or response to treatment has not been evaluated in dogs with CE. The aim of this study was to evaluate fecal CP concentrations in dogs with chronic inflammatory enteropathies (CIE) in relation to the severity of clinical signs, histologic lesions, fecal S100A12 and serum CRP concentrations, disease classification, and response to treatment.

Serum and fecal samples were collected from dogs with CIE. Fecal CP and S100A12 were measured using species-specific in-house ELISA assays. Serum CRP was determined by a commercial ELISA test. Each dog was assigned a canine chronic enteropathy clinical activity index (CCECAI) score, and the severity of histologic lesions was assessed using the WSAVA grading system. Clinical response to treatment was either complete remission (CR), partial response (PR), or no response (NR). Cases of food-responsive diarrhea (FRD), antibiotic-responsive diarrhea (ARD), and steroid-responsive or -refractory disease (SRD) were distinguished. Statistical analysis was performed using non-parametric 2-group comparisons and a Spearman rho calculated for correlation analyses.

A total of 127 dogs diagnosed with CIE (median age: 7 years; 64 males/63 females) were included in the study. Median CCECAI score was 7 (moderate disease) and median histologic score was 2 (moderate lesions). Fecal CP concentrations ranged from 0.04–1,659.5 µg/g, and were correlated with CCECAI scores ($p=0.0111$) and fecal S100A12 concentrations ($p<0.0001$) but not with serum CRP concentrations ($p=0.1184$) nor the overall severity of histologic lesions ($p=0.1705$). However, fecal CP correlated with the increase in intraepithelial lymphocytes in the duodenum ($p=0.0059$) and lymphoplasmacytic cells in the ileal lamina propria ($p=0.0282$). Significantly higher fecal CP concentrations were detected in dogs with SRD than in those dogs with FRD or ARD ($p=0.0353$). Within the SRD group of dogs, PR or NR was associated with higher fecal CP concentrations than CR, but the difference did not reach statistical significance ($p=0.0535$).

We conclude that, similar to fecal S100A12 and in line with our previous findings, fecal CP may be a useful surrogate marker of disease severity and may have utility in predicting the response to treatment in dogs with CIE. Further research is needed to evaluate the utility of serial fecal CP measurements in dogs with CE and to compare the clinical utility of CP with that of S100A12.

acvim127

Expanded Panel of Cytometric Markers for Canine Immune Thrombocytopenia

Marjory Brooks[1]; Dana N. LeVine[2]; Robert Goggs[3]; John Randolph[3]; John Lucy[4]; Michael Koch[5]; Kelly M. Makielski[6]; Marnin Forman[7]; Haruhiko Maruyama[8]

[1]Animal Health Diagnostic Center/Cornell University, Ithaca, NY, USA; [2]Iowa State University, Ames, IA, USA; [3]College of Veterinary Medicine, Cornell University, Ithaca, NY, USA; [4]Oradell Animal Hospital, Hoboken, NJ, USA; [5]Veterinary Specialists and Emergency Services, Rochester, NY, USA; [6]Dept. Veterinary Clinical Sciences/University of MN, Minneapolis, MN, USA; [7]Cornell University Veterinary Specialists, Stamford, CT, USA; [8]Laboratory of Veterinary Clinical Pathology, Department of Veterinary Medicine, Nihon University, Fujisawa, Kanagawa, Japan

Platelet count alone is not predictive of bleeding severity or response to treatment for dogs with ITP. We evaluated dogs with severe thrombocytopenia to identify platelet abnormalities with potential diagnostic or prognostic relevance.

Flow cytometric assays detected platelet membrane-bound IgG and IgM (PsAIg), fibrinogen receptor (CD61) expression, and phosphatidylserine (PS) externalization. Microspheres conjugated with defined concentrations of canine immunoglobulin were used as assay controls. To assess sample stability, replicate analyses were performed on EDTA blood samples from healthy dogs (n=7) through 72 hours of storage. Pre-treatment samples were also analyzed within 48 hours of collection from 21 dogs with severe thrombocytopenia (<50,000 platelets/µL) in 3 categories: primary/autoimmune (PI, n=7); infectious/lymphoproliferative/drug-associated secondary immune (SI, n=7); non-immune (NI, n=7) due to consumption or bone marrow suppression.

During storage, healthy dog platelets had stable CD61 expression but progressively increased PsAIg and PS externalization, with significant differences from baseline by Day 3 (Table 1). While thrombocytopenic dogs had significantly higher median PsAIg than controls (IgG^{pos}=22% vs. 6.3%; IgM^{pos}=23.6% vs. 7.3%; p<0.01), the differences among disease groups were not significant. Disease groups varied, however, in CD61 expression (significant decreases for PI and SI) and significant increases in PS externalization (SI) vs. controls (Figure 1).

Table 1. Healthy Dog Platelets: Stability of Cytometric Assay Parameters During 3 Days of Storage

Storage Times	$CD61^{pos}$ (%)	PS^{pos} (%)	IgG^{pos} (%)	IgG^{pos} (MFI)	IgM^{pos} (%)	IgM^{pos} (MFI)
Day 0	99	1.7	1.1	541	0.4	180
Day 1	97	3.4	3.8	347	4.7	248
Day 2	95	6.2	6.3	479	7.3	271
Day 3	94	**11**	**6.8**	533	**10.4**	474

Median values displayed; **Bold** = significant difference from Day 0 (P < 0.01, Dunn's multiple comparison test)

Figure 1. Platelet CD61 expression and PS externalization for healthy dogs (Controls-Day 2), dogs with non-immune (NI), primary immune (PI), or secondary immune (SI) thrombocytopenia. Horizontal bars = median and interquartile ranges, *** = significant difference from controls ($P < 0.001$, Dunn's multiple comparison test)

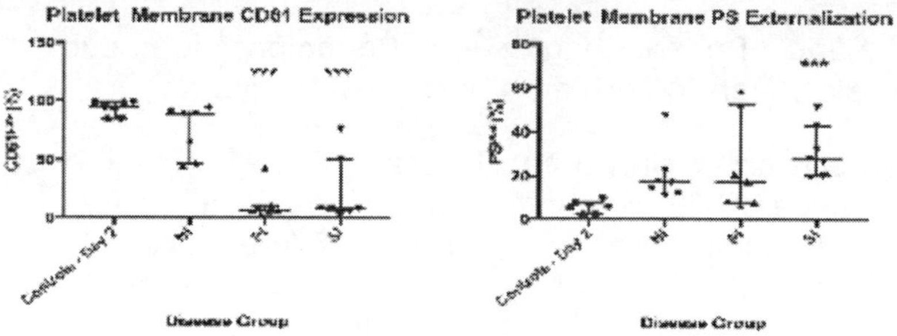

Platelets from canine ITP cases typically lack the CD61 antigen, with an apparent increase in PS externalization for SI-triggered disease. These abnormalities may reflect membrane injury caused by immune targeting and activation or apoptosis induced by an underlying disease process. Together with PsAIg, these parameters may serve as biomarkers of active platelet destruction and metrics to guide treatment intensity.

acvim128

Cell-free DNA, a Potential Thrombosis Trigger, Is Increased in Canine Immune-Mediated Hemolytic Anemia

Dana N. LeVine[1]; Unity Jeffery[2]; Rita Hanel[3]; Laura Ruterbories[4]

[1]Iowa State University, Ames, IA, USA; [2]Texas A&M University, College Station, TX, USA; [3]BluePearl Veterinary Partners, Raleigh, NC, USA; [4]College of Veterinary Medicine, North Carolina State University, Raleigh, NC, USA

Immune-mediated hemolytic anemia (IMHA) has a high mortality rate due to thrombosis. Cell-free DNA from hypoxia-induced cell death and neutrophil extracellular trap release may contribute to IMHA hypercoagulability because it activates clot formation and impairs fibrinolysis. Reduced endogenous DNase activity reported in human autoimmune diseases may impair cell-free DNA clearance. We hypothesized that cell-free DNA would be increased in dogs with IMHA, in part due to reduced DNase activity.

DNA extracted from plasma from primary IMHA dogs (n=28) and healthy controls (n=20) was quantified by DNA dye fluorescence. Serum DNase activity was measured using a DNA lysis radial diffusion assay. Thrombotic outcomes and death/euthanasia were recorded for 14 days.

Cell-free DNA was increased in dogs with IMHA compared with healthy controls (Figure 1A). DNA lysis area was not reduced in IMHA dogs. Four dogs died or were euthanized with thrombosis confirmed post-mortem. Surviving dogs had one confirmed and one suspected thrombotic event. Dogs with cell-free DNA in the upper quartile were at increased risk of death or euthanasia (odds ratio 15, 95% confidence interval 1.62–201; p=0.04); there was no association between cell-free DNA and thrombosis (odds ratio 2.4, 95% confidence interval 0.34–14.3, p=0.57). Area under the receiver operator curve for cell-free DNA prediction of death or euthanasia was 0.86 (95% confidence interval 0.69–1.0; p=0.02, Figure 1B).

Figure 1A. Cell-free DNA is increased in dogs with IMHA (median, range; Mann-Whitney test)

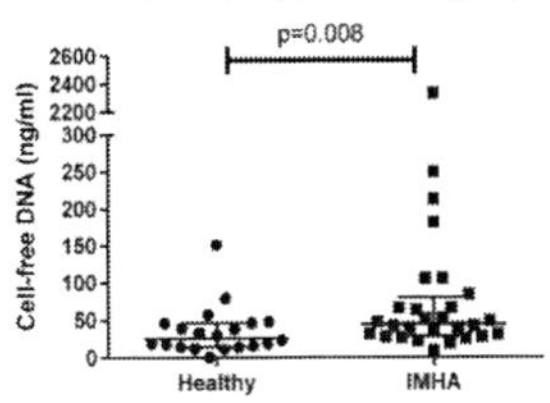

Figure 1B. ROC curve for cell-free DNA as a predictor of death/euthanasia for dogs with IMHA

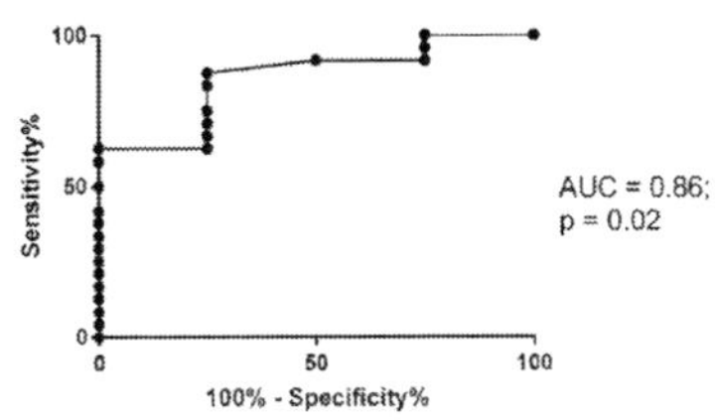

Cell-free DNA is increased in IMHA dogs due to increased DNA release, rather than impaired DNA degradation. Cell-free DNA could have prognostic value in IMHA, but evaluation is needed in a larger cohort.

acvim129

Mutations Causing Feline Factor XII Deficiency in the United States

Haruhiko Maruyama[1]; Amelia Frye[2]; Marjory Brooks[3]

[1]Laboratory of Veterinary Clinical Pathology, Department of Veterinary Medicine, Nihon University, Fujisawa, Kanagawa, Japan; [2]Comparative Coagulation Section, College of Veterinary Medicine, Cornell University, Ithaca, NY, USA; [3]Animal Health Diagnostic Center, Cornell University, Ithaca, NY, USA

Coagulation Factor XII (FXII) biologic activities include initiation of the intrinsic pathway of blood coagulation, activation of the kallikrein-kinin system and regulation of vascular permeability. FXII deficiency is hypothesized to prevent pathologic thrombosis. Unlike other species, FXII deficiency is a common hereditary factor deficiency in cats. To date, two causative feline F12 mutations have been described: p.L441CfsX119, found in a feline research colony in the United States and p.G544A, identified in 2 Japanese DSH siblings. The aim of this study was to identify F12 mutations responsible for the high prevalence of feline FXII deficiency in the United States.

Study subjects included 17 unrelated pet cats (n=14 Domestic Shorthair; n=2 Domestic longhair; n=1 Siamese) with Factor XII activity (FXII:C) ≤10% of normal. Genomic DNA was extracted from EDTA blood for amplification of all F12 exons, followed by direct amplicon sequencing.

Nineteen nucleotide differences were identified in the FXII deficient cats compared to the published feline reference sequence (GQ_981174). After exclusion of synonymous base changes and non-synonymous changes present in cats with normal FXII:C, 3 mutations unique to FXII deficient cats remained (Table 1). These mutations included the 2 previously described mutations and a novel mutation, p.G517X. The 17 FXII deficient cats were classified based on their combination of mutation-types: 1) p.L441CfsX119: Homozygous (HOM) and p.G544A: HOM (n=9), 2) p.L441CfsX119: Heterozygous (HET) and p.G544A: HOM (n=6), 3) p.G544A: HOM (n=1), and 4) p.L441CfsX119: HET, p.G544A: HET, and p.G517X: HET (n=1). F12 genotype was related to residual FXII:C. Cats classified as p.L441CfsX119 HOM (type 1) had ≤1 FXII:C%, whereas cats heterozygous or lacking this mutation (type 2 and 3) had moderate FXII deficiency (FXII:C = 3 to 10%).

Table 1. Feline Factor XII Deficiency Demographic and Mutation Summary

F12 mutation-types	Breeds (number of cats)	APTT in seconds (number of cats)	Factor XII:C % (number of cats)	US Census Region (number of cats)
p.L441CfsX119: HOM p.G544A: HOM	DSH (7), DLH (1), Siamese (1)	>120 (8), 100 to 120 (1)	≤1 (9)	Midwest (4), Northeast (3), South (2)
p.L441CfsX119: HET p.G544A: HOM	DHS (5), DLH (1)	100 to 120 (4), 50 to 99 (2)	3 to 10 (6)	Midwest (3), South (2), Northeast (1)
p.L441CfsX119: WILD p.G544A: HOM	DSH (1)	89.4	10	South
p.L441CfsX119: HET p.G544A: HET p.G517X: HET	DSH (1)	>120	1	South

APTT = activated partial thromboplastin time; feline reference range =14.5 - 19.0 seconds. Factor XII:C = Factor XII coagulant activity; feline reference range = 50 to 200%. HOM = homozygous for mutation type. HET = heterozygous for mutation type. WILD = wild type sequence at mutation site.

The different mutation types were geographically widespread (Table 1) and none of the cats had experienced abnormal bleeding. Future studies are warranted to identify potential beneficial effects of FXII deficiency in modulating the inflammation/coagulation axis that might explain the widespread propagation of inactivating F12 mutations in cats.

acvim130

Performance of Four In-Clinic Tests for *Giardia* Antigen Compared with Direct Immunofluorescence Assay

Dwight D. Bowman[1]; Janice L. Liotta[1]; Jennifer S. Mizhquiri-Barbecho[1]; Mark Rishniw[1]; Kenneth Simpson[1]

[1]College of Veterinary Medicine, Cornell University, Ithaca, NY, USA

Giardia are protozoan parasites capable of infecting a variety of mammals including both dogs and cats. Diarrhea is a common clinical manifestation of dogs presented to veterinary clinics and determination of *Giardia* infection status is a key consideration in the treatment of these patients. Fecal floatation using $ZnSO_4$ followed by microscopy is commonly used to identify *Giardia* cysts in-clinic. However, this method lacks sensitivity and specificity because cysts can be few in number, are shed intermittently, and are commonly not present in diarrheic feces. Direct immunofluorescence assay (DFA) using fluorescent antibodies directed against *Giardia* cysts has been shown to be more sensitive and specific than flotation methods and has been adapted as the gold standard for identifying fecal *Giardia* cysts. The DFA can only be performed in a specialized laboratory which limits its usefulness in-clinic. More recently, *Giardia* antigen detection assays that detect a soluble *Giardia*-specific cyst wall antigen have been developed and are available from several sources for use in-clinic. Comparison studies for the in-clinic assays are for the most part lacking. The purpose of this study was to compare the performance of in-clinic antigen assays using results of DFA as the gold standard. Four in-clinic antigen test kits were evaluated: VetScan® Canine *Giardia* Rapid Test (Abaxis), Witness *Giardia* Test (Zoetis), Anigen® Rapid CPV/CCV/*Giardia* Antigen Test (BioNote), and SNAP® *Giardia* Test (IDEXX).

Archived canine fecal specimens originally submitted by veterinarians to a commercial reference laboratory for general ova and parasite testing were used. These specimens were first examined by DFA for presence or absence of *Giardia* cysts using the Meriflour IFA detection reagent. This evaluation identified 101 cyst positive and 76 cyst negative samples. Fecal samples were tested on each of the 4 in-clinic assays following instructions supplied in the package insert for each kit.

Compared to the DFA results, IDEXX SNAP was the most sensitive test, followed by the BioNote Rapid CPV/CCV/*Giardia* Antigen Test, then Zoetis Witness and Abaxis VetScan; specificity (highest to lowest) was IDEXX SNAP, Abaxis VetScan, BioNote Rapid CPV/CCV/*Giardia* Antigen Test and Zoetis Witness.

A limitation of this study is that the clinical status and environmental history of the dogs were not available for the samples evaluated. Notwithstanding this limitation, the results demonstrate that the SNAP *Giardia* test is the most accurate test among the 4 in-clinic tests evaluated.

In-clinic Test	Sensitivity (%)	Specificity (%)
SNAP *Giardia* Test	87.1	93.4
Rapid CPV/CCV/*Giardia* Antigen Test	80.2	80.3
Witness *Giardia* Test	73.3	71.1
VetScan Canine *Giardia* Rapid Test	70	85.5

acvim131

Worldwide SNAP® 4Dx® and SNAP® 4Dx® Plus Clinic-Based Serologic Survey of Tick-Borne Diseases in Dogs

Jesse Buch[1]; Guillermo Couto[2]; Guadalupe Miro[3]; Leif Lorentzen[1]; Ramaswamy Chandrashekar[1]

[1]IDEXX Laboratories, Inc., Westbrook, ME, USA; [2]Couto Veterinary Consultants, Hilliard, OH, USA; [3]Department of Animal Health, Veterinary Clinic Hospital, Faculty of Veterinary Medicine, Universidad Complutense de Madrid, Madrid, Spain

Tick-borne diseases in dogs are a growing issue worldwide and increasing emphasis is being placed on studying canine populations as sentinels for tick-borne diseases in people. The aim of this study was to evaluate macrogeographic trends in major tick-borne antibody test results in pet dogs over a five-year period (2011 to 2015). The data were obtained from a comprehensive international database of SNAP 4Dx and SNAP 4Dx Plus field results and used to assess the frequency of infection with or exposure to three vector-borne disease agents in dogs: *Anaplasma* spp. (AP); *Ehrlichia* spp. (EC); and *Borrelia burgdorferi* (BB). Within the United States, data were collected from 2,526,712 samples distributed across 10,722 unique postal codes. Outside of the US (OUS), data were derived from over 585,000 samples distributed across 3,122 unique postal codes from 66 countries spanning 7 regions of the world: North America (2 countries, n=345,116); Caribbean (8 countries, n=28,008); Latin America (13 countries; n=17,587); Northern Europe (15 countries, n=51,357); Southern Europe (13 countries, n=57,446); Middle East-Africa (5 countries, n=2,255 samples); and Asia-Pacific (9 countries, n=83,720).

Latin America had the highest overall positive results at 39.6%, followed by the Caribbean (28.7%), Middle East-Africa (25.9%), Northern Europe (25.1%), Asia-Pacific (12.9%), United States (11.6%), Southern Europe (10.6%), and OUS North America (3.3%). Seropositive rates by region for AP, EC, BB, along with a breakdown of the overall major co-infection (Co+) rates are summarized in Table 1.

Table 1. Frequency of seropositive result types by region.

Region	AP	EC	BB	Overall Co+	AP-EC Co+	AP-BB Co+
United States	3.0%	2.1%	8.0%	1.4%	0.1%	1.1%
North America (OUS)	0.5%	0.5%	2.4%	0.2%	0.1%	0.1%
Caribbean	9.4%	25.2%	0.3%	6.1%	6.0%	0.0%
Latin America	13.9%	33.5%	0.6%	8.3%	8.0%	0.1%
Northern Europe	15.0%	1.3%	11.8%	3.0%	0.3%	2.4%
Southern Europe	4.8%	6.1%	1.1%	1.4%	1.2%	0.2%
Middle East-Africa	11.2%	19.2%	0.6%	5.1%	4.7%	0.0%
Asia-Pacific	4.6%	10.0%	0.3%	2.0%	1.9%	0.0%

This study represents the broadest worldwide canine tick-borne disease survey reported to date utilizing data from a single test platform. Co-infection data from this multiplex test may have use as an indicator of the predominant tick vectors to which dogs are being exposed in specific regions. Overall, the global seropositive rates documented in this study support the need for continued awareness of tick-borne diseases and monitoring the risks they pose to animal and public health.

acvim132

Prevalence of *Borrelia Burgdorferi, Anaplasma* spp., *Ehrlichia* spp. and *Dirofilaria immitis* in Canadian Dogs, 2007–2016

Michelle Evason[1]; Jason Stull[2]; Jesse Buch[3]; Zachary Lailer[3]; Tom O'Connor[3]; Ramaswamy Chandrashekar[3]; Claire Jardine[4]; David Pearl[4]; Andrew Peregrine[4]; J. Scott Weese[4]

[1]University of Guelph, Guelph, ON, Canada; [2]The Ohio State University, Columbus, OH, USA; [3]IDEXX Laboratories, Inc., Westbrook, ME, USA; [4]University of Guelph, Guelph, ON, Canada

Vector-borne pathogens are emerging concerns in multiple regions of Canada, for both animals and humans. This has created knowledge gaps in awareness of shifting prevalence, disease risk, and information regarding true geographic distribution of these pathogens. A dataset of test results for *Dirofilaria immitis* (HW) antigen (ag) and *Borrelia burgdorferi* (BB), *Anaplasma phagocytophilum/platys* (AP), and *Ehrlichia canis/ewingii* (E) serology (SNAP® 4Dx® and SNAP® 4Dx® Plus) performed on dogs residing in Canada at a diagnostic laboratory or reported by veterinary clinics was analysed. Data were available from March 2007 until July 2016, along with canine signalment and geographic distribution. The objectives were to investigate overall (country-wide) and individual province vector-borne pathogen positive status, determine prevalence of co-infection of BB with other tested pathogens, evaluate factors associated with positivity and assess changes (and emergence) in annual prevalence over time (2008–2015), both overall and within individual provinces in Canada. Descriptive statistics and chi-squared tests were used.

A total of 937,004 tests were identified during the study period. Most tests were from Ontario (ON) 76.9%, Québec (QC) 11.7% and Manitoba (MB) 8.9%; the remaining provinces each accounted for <1% of total tests. Test submissions increased annually (2008–2015) from 8,082 to 181,205. Overall positive tests (2007–2016) were 0.2% for HW Ag and 2.0% BB, 0.4% AP and 0.5% E for serology.

Seropositivity for BB (Table) varied among provinces over the study period (2007-2016). Seroprevalence was highest in ON, QC, MB and within the maritime provinces of Nova Scotia (NS), New Brunswick (NB) and Newfoundland (NL), and lowest in the western Canadian provinces of Alberta (AB), Saskatchewan (SK) and British Columbia (BC). There were significant differences in prevalence among some of the provinces ($p<0.001$). Assessment of Canadian complete year data (2008–2015) showed an annual increase in BB seroprevalence ($p<0.001$). The magnitude of this shift was greatest within the provinces with the highest BB seroprevalence (ON, MB, QC and NS), and was most dramatic in NS.

Table: Percent positive seroprevalence for *Borrelia burgdorferi* from 2007-2016, and change between 2008 and 2015. *

	2007-2016 (total)	% Change (2008 to 2015)
Canada*	2.0%	90.0%
ON	1.8%	216.7%
QC	2.4%	233.3%
MB	2.4%	146.2%
NS	9.5%	441.7%
NB	7.0%	31.3%
NL	2.8%	**
AB	0.8%	25.0%
SK	0.7%	**
BC	0.4%	**

*As only three tests were reported for Prince Edward Island and 12 tests for Yukon, these are not included in the Table.
**Unable to compute.

Co-infections with BB were identified in 0.2% of tests (2007–2016), the most common being the combination of BB and AP (0.06%). Significant associations between BB seroprevalence and co-infection with AP, E and HW Ag were also noted (all $p<0.001$). Further analyses of Canadian complete year data (2008–2015) revealed a positive association between increasing dog age and BB seropositivity; $p<0.001$. Additionally, increased seroprevalence for both E (0.2 to 0.5%; $p<0.001$) and AP (0.2 to 0.4%; $p=0.006$) were observed between 2008–2015. A corresponding prevalence increase was not found for HW Ag. The increase in prevalence detected for most of these pathogens over time is consistent with anecdotal information on heightened seropositivity and disease in some regions of Canada; specifically a rise in Lyme disease in parts of ON, MB and NS. This improved understanding of the temporal and spatial changes in these vector-borne pathogens, together with the information on co-infections and additional study findings will serve as the base for future research efforts. The study highlights needs for targeted education and emphasis on preventive tick and mosquito health care across Canada, particularly within specific high-risk regions.

acvim133

Occurrence of Renal Azotemia and Associated Clinicopathological Findings in Canine Babesiosis

Pieter Defauw[1]; Sylvie Daminet[1]; Andrew L. Leisewitz[2]; Amelia Goddard[2]; Dominique Paepe[1]; Luc Duchateau[3]; Johan P. Schoeman[2]

[1]Department of Medicine and Clinical Biology of Small Animals, Faculty of Veterinary Medicine, Ghent University, Merelbeke, Oost-Vlaanderen, Belgium; [2]Department of Companion Animal Clinical Studies, Faculty of Veterinary Science, University of Pretoria, Pretoria, South Africa; [3]Department of Comparative Physiology and Biometrics, Faculty of Veterinary Medicine, Ghent University, Merelbeke, Oost-Vlaanderen, Belgium

The reported occurrence of acute kidney injury and overt acute renal failure in canine babesiosis is not well documented, mostly due to the use of different study populations and variable criteria to diagnose renal azotemia (RA). Furthermore, interpretation of urine specific gravity (USG) to assess renal concentrating ability is hampered by the presence of hemoglobinuria in babesiosis. The objectives of this study were to determine the occurrence of RA in dogs with babesiosis, using the gold standard to measure urinary concentration (i.e., urine osmolality), and to examine possible associations between the presence of RA and selected clinicopathological findings at presentation.

Data of 296 dogs, presented with babesiosis and included in 3 previously performed prospective studies, were reviewed retrospectively. Diagnosis of *Babesia rossi* infection was confirmed by PCR in all dogs. Cases were included in this study if a complete blood count, biochemistry profile, and urinalysis were performed at admission. All samples were collected before initiating fluid therapy. Urine osmolality was measured by freezing point depression in all azotemic dogs. Renal azotemia was defined as the presence of azotemia combined with urine osmolality <1,110 mOsmol/kg (i.e., the equivalent of USG<1.030). Differences between dogs with RA and dogs without RA were assessed by the Wilcoxon rank-sum test for continuous and ordinal categorical variables, while the Fisher's exact test was used for binary variables using a significance level of 5%.

A total of 152 dogs were included, of which 26 (17%) were azotemic at admission. The occurrence of RA was 14% (21/152), hence it was present in 21/26 (81%) of all azotemic dogs. When RA was defined as the presence of azotemia combined with USG<1.030, only 6/26 (23%) of the azotemic dogs were considered to have RA.

Dogs with RA were significantly older than dogs without RA (p=0.016). Dogs aged 7 years or older were 3.8 times more likely to present with RA compared to dogs younger than 7 years of age. Dogs with RA had a significantly higher body weight (p<0.001) and lower body temperature (p=0.036). Presence of collapse was significantly associated with the presence of RA (p=0.007), while gender (p=1.000), duration of illness (p=0.137) and mean arterial pressure (p=0.595) were not. Most hematological parameters were not different between both groups, except for band neutrophil count, which was significantly higher in dogs with RA (p=0.013). Blood urea nitrogen (p<0.001), total bilirubin (p=0.039), serum hemoglobin (p=0.003) and lactate concentrations (p=0.026) were significantly higher in dogs with RA. Hyperphosphatemia (p<0.001) and hypoglycemia (p=0.002) were significantly more common in dogs presenting with RA, while the presence of systemic inflammatory response syndrome was not (p=0.439). Visually, urine was significantly more turbid (p<0.001) and darker (p=0.009) in dogs with RA, while urinary protein (p=0.781), bilirubin (p=0.809) and hemoglobin concentrations (p=0.052), assessed by dipstick analysis, were not different between both groups.

Our results clarify that RA is a common complication of babesiosis caused by *B. rossi*. This study also demonstrates that USG determination is not a reliable method to evaluate renal concentrating ability in azotemic dogs with babesiosis. As a consequence, 15/21 (71%) of the dogs with RA were misclassified as having prerenal azotemia when USG was used to classify azotemia. If urine osmolality cannot be measured, presence of several factors found to be associated with RA in this study could help to identify azotemic dogs that are more likely to have RA. Further work is needed to investigate the significance of these associations in relation to the pathogenesis of acute kidney injury caused by *B. rossi*.

acvim134

P16Ink4a-Mediated Cellular Senescence Is Increased in Geriatric Cats and Cats with Chronic Kidney Disease

Jessica Quimby[1]; Andrea Herndon[2]; Shannon M. McLeland[3]; David Maranon[2]; Katharine Lunn[4]; Jonathan Elliott[5]; Rachel Cianciolo[6]; Susan Bailey[2]

[1]The Ohio State University, Columbus, OH, USA; [2]Colorado State University, Fort Collins, CO, USA; [3]College of Veterinary Medicine, Iowa State University, Ames, IA, USA; [4]North Carolina State University, Raleigh, NC, USA; [5]Royal Veterinary College, London, England, UK; [6]Ohio State University, Columbus, OH, USA

P16^{Ink4a} is a molecular mediator of cellular senescence that has been shown to increase with age in the rat kidney. The purpose of this study was to explore p16^{Ink4a} as a possible senescence pathway for renal aging and disease in cats.

P16^{Ink4a} was assessed in kidney tissue from 29 cats with chronic kidney disease (CKD; 10 cats IRIS Stage 2, 7 cats IRIS Stage 3, 13 cats IRIS Stage 4), 15 young non-CKD controls (10 years). Kidney sections were dual stained with p16^{Ink4a} and aquaporin-1 (AQP) immunohistochemistry to identify a proximal tubular epithelial cell population for analysis. Ten consecutive high power fields (60x) were evaluated from both cortical and corticomedullary junction regions by a blinded pathologist. Within each field tubules were designated either AQP+ or AQP- and the number of P16^{Ink4a} positive tubules counted. Semi quantification of tubular staining was also performed by assigning a score of the distribution of staining (none, focal, segmental, or diffuse). Sum of percentage of positive tubules and mean semiquantification scores were compared between groups using Kruskal-Wallis with Dunn's post hoc analysis.

CKD cats had significantly increased p16^{Ink4a} expression in comparison to young cats in all analysis categories: percentage p16^{Ink4a} cortex ($p=0.0003$), percentage p16^{Ink4a} corticomedullary junction ($p=0.001$), semiquantification cortex ($p=0.0002$) and semiquantification corticomedullary junction ($p=0.0009$). Geriatric cats had significantly increased p16^{Ink4a} expression in comparison to young cats in two analysis categories: percentage p16^{Ink4a} corticomedullary junction ($p=0.03$) and semiquantification corticomedullary junction ($p=0.006$). There was no statistically significant difference between CKD cats and geriatric cats for any of the analysis categories.

Results support p16^{Ink4a} as a mediator of senescence in geriatric cats and cats with CKD. Further work will be needed to determine if senescence in geriatric cats predisposes to development of CKD.

acvim135

Raw Food Diets Alter the Microbiome and Metabolome of Dogs Compared to Commercial

Jan S. Suchodolski[1]; Milena Schmidt[2]; Julia B. Honneffer[1]; Blake C. Guard[1]; Jonathan A. Lidbury[1]; Joerg M. Steiner[1]; Stefan Unterer[3]

[1]Gastrointestinal Laboratory, College of Veterinary Medicine, Texas A&M University, College Station, TX, USA; [2]Ludwig Maximilian University of Munich, Munich, Bayern, Germany; [3]Clinic of Small Animal Medicine, LMU Munich, Bayern, Germany

Feeding raw food diets are gaining in popularity among dog owners. Feeding raw food has been associated with increased risk for transmission of bacterial enteropathogens. No detailed studies have been reported that evaluate the effect of raw food diets on the fecal microbiome and metabolome of healthy dogs. The aim of this study was to evaluate potential differences in microbial populations as well as functional metabolites in the intestine of healthy dogs fed either a raw food diet or commercially available wet or dry food diets.

Fecal samples were collected from healthy dogs fed a commercial diet (n = 20) and healthy dogs fed a raw food diet (n = 27). The microbiota was analyzed by Illumina sequencing of the 16S rRNA gene and by qPCR for selected bacterial groups and calculation of the dysbiosis index. The metabolome was analyzed using an untargeted approach using a gas chromatography/mass spectrometry (GC/MS) platform and by targeted assays for fecal bile acid concentrations and sterols using GC/MS. Univariate statistics (Mann-Whitney test) were used to compare bacterial taxa and metabolic compounds between diet groups. P-values were adjusted for multiple comparisons using the Benjamini and Hochberg False Discovery Rate and significance was set at $p<0.05$.

Significant differences in microbiota structure were observed between the diet groups (ANOSIM; $p<0.05$), but microbiota richness was not significantly altered ($p=0.200$). The dysbiosis index was significantly increased in dogs fed a raw food diet ($p<0.001$), and this increase was driven by an increase in *E. coli* ($p<0.001$) and *C. perfringens* ($p<0.001$). In contrast, *Faecalibacterium* was significantly lower in dogs fed a raw food diet. Several metabolites were altered between diet groups (e.g., 4-hydroxybutyric acid, tocopherol). Fecal cholesterol was significantly increased in the dogs fed a raw food diet, but no significant changes were observed for fecal bile acid concentrations.

In conclusion, dogs fed a raw food diet have an altered microbiome and metabolome when compared to dogs on a commercial diet.

acvim136

Ciprofloxacin Population Pharmacokinetics in Dogs After Oral Administration

Mark G. Papich

North Carolina State University, Raleigh, NC, USA

Ciprofloxacin human generic tablets are frequently administered to dogs for treatment of bacterial infections because they are inexpensive and easily available. However, previous work by our laboratory indicated low and variable oral absorption in healthy research dogs. The aim of the study reported here was to examine a larger group of diverse patients from the North Carolina State University veterinary hospital. Ciprofloxacin intact generic human-label tablets (250 mg, 500 mg, or 750 mg) were administered to 34 client-owned clinical patients. The mean dog weight was 22.95 kg (range 4.6–57 kg). The mean age was 5.9 years (range 1–16 years). The mean dose was 23.5 mg/kg (±4.75). We used a sparse sampling design so that blood samples were collected from each dog four times covering a time range of 0 to 24 hours. Population pharmacokinetic results were obtained using nonlinear mixed effects modeling and the Phoenix® NLME™ software (Certara, St. Louis, MO). The data was parameterized as a one-compartment model with first order input and elimination. The true values (theta) for volume of distribution and elimination rate were 10.7 L/kg (11.7%) and 0.16/hour (7.62%), respectively. (All values shown as the estimate and CV%.) Secondary parameters were: T_{MAX} 3.9 hours (10.1%), C_{MAX} 1.9 µg/mL (11.66%), absorption half-life 1.78 hours (20.28%), and elimination half-life 4.35 hours (7.62%). Various covariates were examined to determine the source of variation. The size of the dog was an important covariate with larger dogs showing less oral absorption than smaller dogs. These data were used to estimate a breakpoint for susceptibility testing. Monte Carlo simulations were used to determine the probability of target attainment (PTA) for an AUC/MIC ratio of ≥100, which is the accepted ratio for fluoroquinolones. Using this analysis, a breakpoint (susceptible) of ≤0.06 µg/mL should be considered when ciprofloxacin tablets are administered to dogs at a dose of 25 mg/kg once daily. This is much lower than the current human breakpoint of ≤1 µg/mL.

CIPROFLOXACIN PHARMACOKINETICS IN CLINICAL CANINE PATIENTS

Mark G. Papich, DVM, North Carolina State University, College of Veterinary Medicine, Raleigh, North Carolina, USA

ABSTRACT

Background – Ciprofloxacin human generic tablets are frequently administered to dogs for treatment of bacterial infections because they are inexpensive and easily available. However, previous work indicated low and variable oral absorption in healthy research dogs.
Objective – To examine oral ciprofloxacin in a group of clinical canine patients using population pharmacokinetics. The data was used to derive minimum inhibitory concentrations (MIC) that could potentially be treated with oral ciprofloxacin in dogs.
Animals – 34 clinical canine patients; mean weight 22.95 kg (range 4.6-57 kg).
Methods – Ciprofloxacin intact generic human tablets were administered to dogs in a prospective study (mean dose 23.5 mg/kg). A sparse blood sampling was used to obtain population pharmacokinetic results with nonlinear mixed-effects modeling (NLME). These data were used to estimate a breakpoint for susceptible bacteria. Monte Carlo simulations were used to determine the probability of target attainment (PTA) for an AUC/MIC ratio of ≥ 100, the pharmacokinetic-pharmacodynamic (PK-PD) target for fluoroquinolones.
Results – The values for volume of distribution, peak concentration, and half-life were 10.7 L/kg (11.7%), 1.9 μg/mL (11.66%), and 4.35 hours (7.62%), respectively (mean, % CV). The size of the dog was an important covariate with larger dogs showing lower plasma drug concentrations than smaller dogs. 90% PTA was obtained for a MIC ≤ 0.06 μg/mL.
Conclusions and clinical importance – A breakpoint (susceptible) of ≤ 0.06 μg/mL should be considered when ciprofloxacin tablets are administered to dogs at a dose of 25 mg/kg once daily which is much lower than the human breakpoint of ≤ 1 μg/mL.

---

acvim379

The Gallbladder Halo Sign: Update on the Diagnosis & Complications of Canine Anaphylaxis

Gregory R. Lisciandro, DVM, DABVP, DACVECC
Spicewood, TX, USA

Introduction

The lecture will not cover detail regarding the pathophysiology of anaphylaxis of which there exist many excellent published descriptions. Rather, the lecture will focus on how to rapidly gain supportive evidence for the diagnosis of canine anaphylaxis, including the sonographic marker of gallbladder wall edema; and important rule outs for sonographically-detected gallbladder wall edema that can confound the diagnosis at the expense of the patient; and a newly described fascinating complication referred to by the author as canine "anaphylaxis-related, heparin-induced hemoabdomen" that is medically-treated.

The major reason for this lecture is that many veterinarians are now using point-of-care ultrasound, specifically FAST exams, as a life-saving, first-line, screening test. As a result, ultrasound detects free-fluid and other soft tissue changes that are not recognized without ultrasound; and many triaged collapsed dogs with anaphylaxis have no cutaneous signs nor witnessed inciting envenomation. The focus of the source of anaphylaxis in this lecture disregards the obvious canine with classic cutaneous signs of angioedema, pruritus, and urticaria, but focuses on the single Hymenoptera envenomation that is often unwitnessed. All veterinarians need to recognize the strengths and limitations of sonographically-detected gallbladder wall edema since its presence is not pathognomonic; and that AX-related, heparin-induced hemoabdomen is a medically treated canine complication, because without this knowledge gallbladder wall edema will be misinterpreted and surgical intervention will likely lead to a fatal exploratory surgery with the findings of hepatic swelling, intra-abdominal blood, and non-specific histopathology.

The FAST Diaphragmatico-Hepatic (DH) View

Imaging the Gallbladder and Detecting Intramural Edema

In normalcy, the gallbladder sonographically is generally oval in longitudinal (sagittal) orientation with a lumen that is homogeneously anechoic (black). The gallbladder wall is quite reliably and seen sonographically as a thin hyperechoic (white) line in both canines and felines despite normal thickness reported to a thickness of 2–3 mm. In summary, the sonographic features of the canine and feline gallbladder are easy to appreciate by non-radiologist sonographers when imaging the FAST DH view. In a 2009 study, Quantz *et al.* published in JVECC a brilliant study that correlated the presence of a thickened, edematous, striated gallbladder wall, referred to as the gallbladder halo sign, with canine anaphylaxis (AX). Their study design was a result of recognizing that in a canine AX research model, gallbladder wall edema was commonly present.

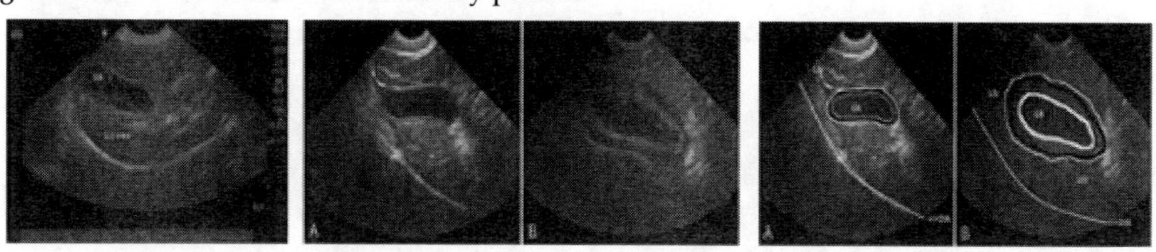

Figure 1. Normal expected sonographic appearance of the gallbladder wall in canines. Unlabeled middle figure with A and B shows more subtle gallbladder wall striation/intramural edema in A and more obvious gallbladder wall striation/intramural edema in B. Far right image is now outlined. This material is reproduced and modified with permission of John Wiley & Sons, Inc, *Focused Ultrasound Techniques for the Small Animal Practitioner*, Wiley ©2014.

Gallbladder Wall Edema as a Sonographic Marker for Canine Anaphylaxis (AX)
AX-related gallbladder edema is specific to canines because their shock organ, where the highest concentration of mast cells are located, is their liver and gastro-intestinal tract. In contrast, the shock organ of felines and humans is the lung, thus gallbladder wall edema is not a hallmark of AX in these species. The cause of gallbladder wall edema is the result of massive histamine release causing hepatic venous sphincter constriction and massive generalized hepatic venous congestion. Simply put, when the liver swells, so does the gallbladder wall. This is important to remember when considering other rule outs for canine gallbladder wall edema. The AX-related intramural gallbladder edema is recognized sonographically as sonographic striations. These striations appear as a hyperechoic (white) lines representing the inner and outer aspects of the gallbladder wall, and a sonolucent anechoic (black) line striation representing the intramural edema. In other words, the gallbladder wall becomes layered as white, black, and white, (sometimes it is white, gray, white), and thus has been dubbed the "gallbladder halo sign." In the Quantz *et al.* study, it was found that AX-induced gallbladder edema is an immediate occurrence within seconds/minutes that lasts up to 24–48 hours post-insult.

Serum Alanine Transaminase as a Serum Marker for Canine Anaphylaxis (AX)
Because the liver and gastro-intestinal tract are the shock organ for the canine species, traditionally serum alanine transaminase (ALT) has been used as supportive evidence for canine AX. However, in the Quantz *et al.* study, it was stated that serum ALT marker was not as immediate as gallbladder wall edema; and that the ALT may not spike in value for up to 2–4 hours post-insult. Quantz *et al.* documented a mean ALT of ~400 IU/L for their case study population of canine AX.

The Classic Constellations of Signs for Canine Anaphylaxis
Keeping in mind the canine shock organ, traditional means of diagnosing anaphylaxis have relied on a history of acute collapse often associated with gastro-intestinal signs in a previously healthy dog. The collapse is often associated with gastro-intestinal signs (i.e., vomiting and defecation). The great majority of these dogs have no obvious cutaneous signs. Upon presentation due to massive fluid shifts caused by histamine release and likely other factors that increase vascular permeability, dogs with AX are commonly hemoconcentrated with packed cell volumes in the 50% range and even higher (in contrast dogs with a hemoabdomen from a bleeding tumor are not hemoconcentrated); and as previously mentioned, the serum ALT is likewise a marker because of the hepatic insult. The weather should also be considered because many warm days and cool nights make Hymenoptera species lethargic and less likely to move away from the dog walking and sniffing in the grass in the cool evenings and mornings (author's experience).

Gallbladder Wall Edema is Not Pathognomonic for Canine Anaphylaxis - It's also a "Cardiac Gallbladder"
In the collapsed or acutely weak hypotensive canine triaged with the finding of gallbladder wall striation/edema, other important rule outs include pericardial effusion, right-sided heart and generalized systolic dysfunction (DCM). The pathophysiology of gallbladder wall edema in these cases is mechanical obstruction of blood, in which backflow of blood leads to a distended caudal vena cava (CVC), and hepatic venous congestion. Simply put, when the liver swells, so does the gallbladder. These rule outs are addressed by looking past the diaphragm for pericardial effusion, the classic racetrack sign of pericardial effusion rounding the muscular apex of the heart; and adding the right TFAST pericardial (PCS) view to assess contractility at the left ventricular short-axis view. Moreover, the really savvy sonographer always, always looks at the caudal vena cava (CVC) where it traverses the diaphragm at the FAST DH view. The CVC is a marker for central venous pressure (CVP) and its human counterpart, the inferior vena cava at the analogous location, is being taught to medical doctors. The CVC in AX is flat (no volume, low CVP) with no variation in its diameter vs. the CVC characterization in pericardial effusion and cardiac cases is diametrically opposed as FAT (too much volume, high CVP), or distended with minimal variation in it diameter. When the CVC is FAT from a high CVP, hepatic veins, not normally obvious in lateral or standing/sternal recumbency, are obvious branching structures from venous downstream obstruction. The upshot is that gallbladder wall edema is not pathognomonic for canine anaphylaxis in the collapsed or weak canine; that veterinarians must resist satisfaction of search error and minimally look past the

diaphragm for pericardial effusion and the TFAST right PeriCardial View to evaluate contractility before administering large volumes of crystalloids.

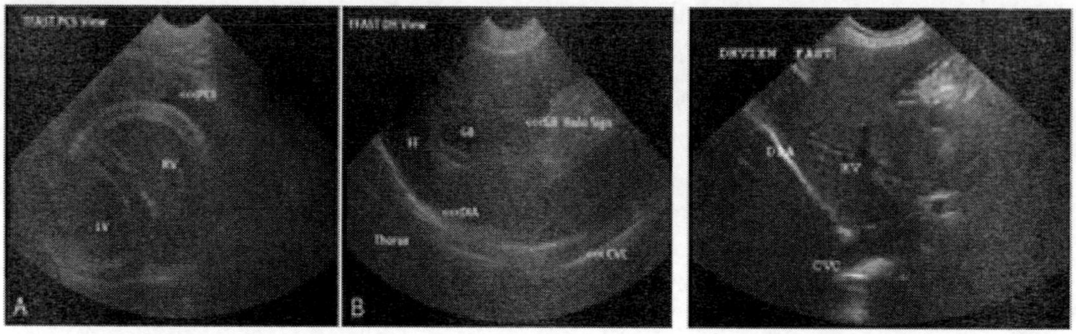

Figure 2. The gallbladder halo sign and FAT (distended) CVC in a dog with pericardial effusion (PCE) in A and B. PCE should be ruled out in collapsed, weak dogs suspected of having AX. The single figure on the right showing the FAT (distended) caudal vena cava (CVC) as it traverses the diaphragm (DIA); and the associated distended branching hepatic veins appearing as tree trunks (referred to as the tree trunk sign). The character of the CVC is completely different between canines with AX (flat CVC) and canines with pericardial effusion or right-sided heart failure/DCM (FAT CVC). LV: left ventricle; RV: right ventricle; PCE: pericardial effusion; DIA: diaphragm; GB: gallbladder; CVC: caudal vena cava; FF: free abdominal fluid. This material is reproduced with permission of John Wiley & Sons, Inc, *Focused Ultrasound Techniques for the Small Animal Practitioner*, Wiley ©2014.

Causes of gallbladder wall edema (the gallbladder halo sign)

Anaphylaxis (acute collapse, flat caudal vena cava) - massive histamine release results in hepatic venous congestion
Right-sided heart failure/dysfunction (collapse, weakness, FAT caudal vena cava) - backflow of blood to the right heart results in hepatic venous congestion
Pericardial effusion (acute collapse, weakness, FAT caudal vena cava) - obstruction of blood flow to the right heart results in hepatic venous congestion
Cholecystitis
Pancreatitis
Hypoalbuminemia, 3rd spacing
Right-sided volume overload (iatrogenic), from intravenous fluid therapy
Immune-mediated hemolytic anemia (IMHA), unknown cause, speculate immune-mediated
Post-blood transfusion, unknown pathogenesis, speculate immune-mediated, volume overload

Other potential causes for gallbladder wall edema that are generally present in non-collapsed dogs include 3rd spacing from hypoalbuminemia and vasculitis, primary gallbladder disease including cholecystitis, pancreatitis, and iatrogenic right-sided volume overload. Gallbladder wall edema is often observed in dogs with in immune-mediated hemolytic anemia and post-blood transfusion. Its presence in these subsets of patients does not necessarily indicate canine AX, so it is important to look at the complete clinical profile of these canine patients.

Canine Anaphylaxis-Related, Heparin-Induced Medically-Treated Hemoabdomen

Dogs with anaphylaxis commonly develop abdominal effusion often scored as an abdominal fluid score (AFS) 1 or 2 using the AFAST-applied fluid scoring system; and most commonly positive at the FAST DH view. These low-scoring effusions are often self-resolving, the canine patient is non-coagulopathic, and the volume too small for safely performing abdominocentesis. Serial AFAST with AFS, minimally one repeat AFAST 4-hours post-admission, is justified to detect worsening (increasing score) or resolution (decreasing score) of the AX-related effusion; and depending on clinical course, a repeat PCV/TS and coagulation profile; and AFAST with AFS again after daily patient rounds until the attending veterinarian is certain that AX has resolved. In fact, Global FAST - AFAST with AFS, TFAST, and Vet BLUE - is an even better format over AFAST alone because Global FAST provides information on volume status, lung status, and other potential complications occult by physical exam, blood and urine testing, radiography, and vital signs.

Other cases of canine AX have large volume effusions of AFS 3 and 4. Even in these large volume effusions, the coagulation profile may be close to normal and stay close to normal (<25% above baseline) on repeat PCV/TS and coagulation profile 4-hours post-admission along with AFAST with AFS. These large volume effusions will likewise generally self-resolve within 24-hours if the patient responds favorably to initial resuscitation and therapy for AX including fluid resuscitation ± epinephrine, histamine 1 blocker, histamine 2 blocker, and glucocorticoids. Abdominocentesis should be performed if the free fluid is safely accessible, generally at the most gravity-dependent regions of the abdominal cavity, when your patient is in lateral recumbency. In our experience, these effusions are hemorrhagic with a comparative abdominal PCV of minimally ≥50% of the peripheral PCV. In canine AX cases with abnormal coagulation profiles, clotting factors should be replaced as soon as possible (e.g., fresh frozen plasma [FFP]). As a crude guideline, 1 in 5–7 canine AX cases require FFP, and 1 in 20 canine AX cases require pRBCs. Some of the coagulopathic canines require repeated FFP over several days. The author treats all canine AX cases with famotidine for several days and a 7-day tapering regimen of anti-inflammatory dosing of prednisone to attenuate the second wave of AX-related inflammation.

Pathophysiology of Canine AX-Related Heparin-Induced Medically-Treated Hemoabdomen

In theory, the aPPT is more affected by heparin, a natural component of the mast cell granule; thus, PT and aPTT times should be discordant with the aPTT far more prolonged than the PT (opposite of the warfarin or Coumadin effect). So when the PT is near normal or mildly elevated with an out of range aPTT, a flag should be raised that the coagulopathy is a result of AX and heparin and tryptase release by mast cells. However, the discordance seems unreliable and the entire clinical picture must be considered. In our case series of 11 dogs from nearly 4-years ago, all survived without surgery with complete resolution of their hemorrhagic effusion; and of the clients that responded to long-term follow-up, all 7 said AX had not recurred in their dog. The data from this case series is available off our FASTVet Facebook page or may be requested.

Conclusion

It is important to recognize the limitations and additional rule outs for the sonographic finding of a striated gallbladder wall, the so-called gallbladder halo sign; and that dogs have a unique AX-related, heparin-induced medically-treated hemoabdomen complication, and not over-react to stable patients with normal to relatively normal clotting times (<25% over baseline) since many will self-resolve with standard AX therapy. In other words, many dogs with mildly abnormal coagulation profiles will resolve without transfusion products. Equally important is to know not to take an AX-related canine hemoabdomen to surgery, which could be catastrophic for the dog likely resulting in death. Larger case studies and more sophisticated coagulation assessment are needed to fully understand this perplexing canine AX-related phenomenon.

References

1. Quantz JE, Miles MS, Reed AL, White GA. Elevation of alanine transaminase and gallbladder wall abnormalities as biomarkers of anaphylaxis in canine hypersensitivity patients. *J Vet Emerg Crit Care*. 2009;19(6):536–544.
2. Lisciandro GR. Abdominal FAST (AFAST)-detected hemorrhagic effusion in 11 dogs with acute collapse and gallbladder wall edema (halo sign) with presumed anaphylaxis (abstract). *J Vet Emerg Crit Care*. 2016.
3. Lisciandro GR. Focused abdominal (AFAST) and thoracic (TFAST) focused assessment with sonography for trauma, triage and monitoring in small animals. *J Vet Emerg Crit Care*. 2011;20(2):104–122.
4. Lisciandro GR. Case series of dogs with the gallbladder halo sign associated with pericardial effusion and right-sided heart failure during FAST exams. Unpublished, 2014.
5. World allergy organization guidelines for the assessment and management of anaphylaxis. *World Allergy Organization Journal*. 2010;4(2):13–37.

acvim380

AFAST, TFAST, and Vet BLUE Point-of-Care Ultrasound for Traditionally Difficult Conditions

Gregory R. Lisciandro, DVM, DABVP, DACVECC
Spicewood, TX, USA

INTRODUCTION

We will discuss traditionally difficult conditions, treated more often based on suspicion rather than evidence-based information. The Global FAST ultrasound approach of using AFAST, TFAST and Vet BLUE in combination can tackle many of these traditional conundrums.

TRADITIONALLY DIFFICULT CONDITIONS ADDRESSED BY AFAST, TFAST, AND VET BLUE

Where Is My Patient Bleeding?

Global FAST - AFAST, TFAST, and Vet BLUE - survey 4 spaces and lung. Global FAST should be applied to all coagulopathic and at risk for bleeding cases. The abdominal cavity, retroperitoneal space, pleural space, pericardial sac, and lung are surveyed and interventions may take place to better treat the patient prior to overt signs.

Is It Feline Asthma or Left-Sided Congestive Heart Failure?

The use of Vet BLUE lung ultrasound rapidly answers this historically difficult question in <90 seconds because the Vet BLUE profiles are radically different.

Does My Patient Have Pulmonary Thrombo-Embolism?

Better evidence may be gained for the presence of PTE by using the TFAST long-axis 4-chamber view and evaluating the right ventricular to left ventricular ratio (RV:LV); and through the use of Vet BLUE and the finding of the wedge sign representative of small infarcts on the lung surface.

Does My Patient Have Pulmonary Hypertension?

The respiratory case being treated for pneumonia that in fact has pulmonary hypertension. Using the TFAST long-axis 4-chamber view and evaluating the right ventricular to left ventricular ratio (RV:LV); and characterizing the caudal vena cava are means to support the presence of PHT, explaining why your patient is not improving on antibiotics and triggering an echo in an otherwise occult condition.

Left Atrial Tears in Mitral Valve Disease More Common Than You Think?

Using Vet BLUE and TFAST first line and during hospitalized care as part of your cardiac evaluation helps detect this otherwise missed condition.

Does Your Collapsed Dog Have Anaphylaxis?

Gallbladder wall striation due to gallbladder intramural edema is a marker for canine anaphylaxis. How to recognize the sonographic finding and why you should look beyond the diaphragm for other important rule outs will be discussed; and briefly discuss the canine anaphylaxis-related heparin-induced hemoabdomen.

How to Better Interpret the Nebulous Thoracic Radiograph?

You are missing pulmonary disease by not using proactive Vet BLUE lung ultrasound. Use Vet BLUE to complement thoracic radiography and better help pick your next best respiratory test to more accurately and expediently diagnose and treat.

Why You Should Do Vet BLUE in All Vomiting Patients?

Aspiration pneumonia may be detected point-of-care in your hospitalized patients using Vet BLUE.

How to Monitor Pneumothorax by Using the Lung Point?

Pneumothorax can not only be diagnosed point-of-care using TFAST but may also be monitored through the use of the lung point.

References

1. Lisciandro GR, Fosgate GT, Fulton RM. Frequency and number of ultrasound lung rockets (B-lines) using a regionally based lung ultrasound examination named Vet BLUE (veterinary bedside lung ultrasound exam) in dogs with radiographically normal lung findings. *Vet Radiol and Ultrasound*. 2014;55(3):315–322.
2. Lisciandro GR, *et al*. Frequency and number of ultrasound lung rockets (B-lines) using a regionally based lung ultrasound examination named Vet BLUE (veterinary bedside lung ultrasound exam) in cats with radiographically normal lung findings. *J Vet Emerg Crit Care*. 2016; In press.
3. Ward JL, Lisciandro GR, Tou SP, Keene BW, DeFrancesco TC. Evaluation of point-of-care lung ultrasound (Vet BLUE protocol) for the diagnosis of cardiogenic pulmonary edema in dogs and cats with acute dyspnea. *J Am Vet Med Assoc*. 2015; In press.
4. Lisciandro GR, *et al*. Absence of B-lines on lung ultrasound (Vet BLUE protocol) to rule out left-sided congestive heart failure in 368 cats and dogs (abstract). *J Vet Emerg Crit Care*. 2016.
5. Ward JL, Lisciandro GR, DeFrancesco TC. Distribution of alveolar-interstitial syndrome in dyspneic veterinary patients assessed by lung ultrasound versus thoracic radiography. *J Vet Emerg Crit Care*. 2016; In press.
6. Kulhavy DA, Lisciandro GR. Use of a lung ultrasound examination called Vet BLUE to screen for metastatic lung nodules in the emergency room (abstract). *J Vet Emerg Crit Care*. 2015.
7. Lisciandro GR. Abdominal FAST (AFAST)-detected hemorrhagic effusion in 11 dogs with acute collapse and gallbladder wall edema (halo sign) with presumed anaphylaxis (abstract). *J Vet Emerg Crit Care*. 2016.
8. Lisciandro GR. Focused abdominal (AFAST) and thoracic (TFAST) focused assessment with sonography for trauma, triage and monitoring in small animals. *J Vet Emerg Crit Care*. 2011;20(2):104–122.
9. Lisciandro GR, Lagutchik MS, Mann KA, *et al*. Evaluation of an abdominal fluid scoring system determined using abdominal focused assessment with sonography for trauma (AFAST) in 101 dogs with motor vehicle trauma. *J Vet Emerg Crit Care*. 2009;19(5):426–437.
10. Lisciandro GR. Case series of dogs with the gallbladder halo sign associated with pericardial effusion and right-sided heart failure during FAST exams. Unpublished, 2014.

acvim381

Proactive Vet BLUE® Lung Ultrasound in Feline Respiratory Distress

Gregory R. Lisciandro, DVM, DABVP, DACVECC
Spicewood, TX, USA

Introduction

The reluctance to pro-actively apply lung ultrasound (LUS) to small animals with respiratory signs or distress is irrational in many respects. The overriding belief that air-filled lung creates insurmountable sonographic obstacles, and that sonographic imaging of lung is difficult, perpetuates these myths, and delays the widespread acceptance of what is true and reality, that the ultrasound probe is your new stethoscope. TFAST (thoracic FAST) was the first standardized abbreviated veterinary ultrasound exam of the thorax and the proactive survey of lung via the TFAST chest tube site (CTS) view for the diagnosis of pneumothorax.

However, because of the finding of lung pathology during TFAST, extended lung surveillance from the TFAST CTS with the addition of 6 more lung views and its applications beyond trauma to non-trauma subsets of small animal patients was logically the next step. The name of this novel pattern-based, regional LUS exam is Vet BLUE (veterinary **b**edside **l**ung **u**ltrasound exam); and BLUE also for cyanosis implying respiratory conditions. The Vet BLUE's 8 regional sites applied bilaterally include the caudo-dorsal (Cd) lung region, the peri-hilar (Ph) lung region, the middle (Md) lung region, and the cranial (Cr) lung region. It is important to recognize that the views are not anatomical per se, and that different parts or different lobes may dynamically enter and exit the intercostal spaces of each of these 8 views. The regional concept is important to appreciate because 2 parts of a lung lobe or 2 different lung lobes may be coming into view over the same intercostal space during Vet BLUE. As an example, dry lung, then wet lung, then dry lung, then wet lung may interchange during phases of respiration; or dry lung, and a shred sign, and dry lung then a shred sign, as the patient inspires and expires, This phenomenon in fairly common.

The next logical step is developing a lung language that is easy to understand. We have sided with the visual and explanative terms of dry lung, wet lung, shred sign, wedge sign, tissue sign, and nodule sign, while also acknowledging the lung ultrasound consensus statement of 2012. The terms have been adopted and modified primarily from Volpicelli and Lichtenstein. In 2017, many medical doctors default to these easily understandable terms over less descriptive B-lines, A, B, and C profiles. Having an easily understandable lung language allows for easy communication regarding findings, and thus, LUS use can accelerate over the next several years bucking the lag time between human and veterinary medicine. Finally, having an easily understandable standardized method of recording findings is mandatory for interpreting LUS, and establishing patient profiles that may be compared on serial exams. Having applied Vet BLUE to >1,000 small animal patients since 2010, with >400 cases compared to thoracic radiography, we have learned expected patterns, perfected a lung language, and a perfected a simple procedure for recording findings. Most recently, we have been determined to define the Vet BLUE acoustic windows more accurately by embracing the pitfalls and peri-thoracic anatomy to locate the Vet BLUE sites more consistently for better inter-observer performance.

How to Perform Vet BLUE

Patient Positioning & Preparation

Although Vet BLUE may be performed in lateral recumbency, Vet BLUE is generally performed in standing or sternal recumbency, which is safer for dogs and cats in respiratory distress, or those that are hemo-dynamically or respiratory fragile. For ventral views, a roll of towels or paper towels under the forelegs of a sternally recumbent feline or canine is an easy tolerated maneuver to gain access to the lower ventral Vet BLUE and TFAST pericardial site views.

Generally no Vet BLUE sites are shaved. All images in the lecture were from unshaved sites. The fur is parted and alcohol is applied to the skin with or without a small amount of acoustic coupling gel or alcohol-based hand sanitizer on the probe head. Keep in mind that most ultrasound manufacturers warn against placing alcohol directly on the probe head because of alcohol's potentially damaging effects. To maximize the image quality, the probe head should be applied as directly as possible to the skin surface without fur in between the probe head and the patient's skin.

Basic Vet BLUE Orientation

The Gator Sign

The gator sign typically gets some chuckles, however, the concept is important to understand to ensure that the pulmonary-pleural interface, also referred to as the lung line, is properly imaged. By mistaking air reverberation artifacts or another structure for the lung line, makes your LUS ineffective. The probe is held stationary perpendicular to the long-axis of the ribs.

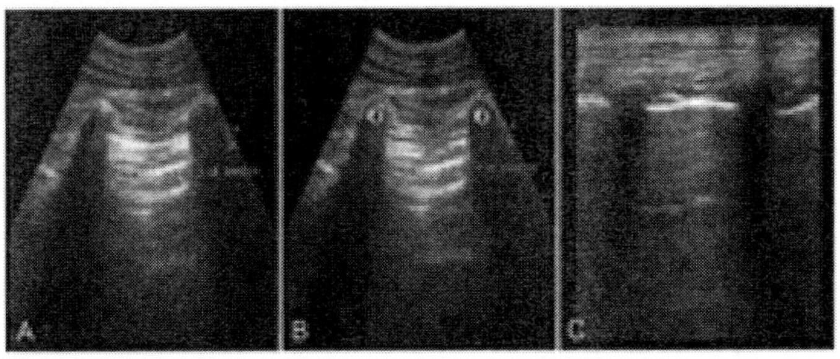

Figure 1. The probe is positioned over lung perpendicular to the long-axis of the ribs. The rounded rib heads are likened to the eyes, and the pulmonary-pleural (PP-line) interface, also called the lung line, to the bridge of its nose, as a partially submerged gator (alligator) peers at the sonographer (middle image, B). The proximal PP-line white line is the focus of all LUS. The major orientation error is looking beyond the PP-line and mistaking A-line artifacts for the PP-line, or being over the abdomen and mistaking liver, stomach (especially when air-filled), or the gallbladder for lung pathology, or being over peri-thoracic soft tissue such as the thoracic inlet, the pectoral or scapular muscles. This material is reproduced with permission of John Wiley & Sons, Inc, *Focused Ultrasound Techniques for the Small Animal Practitioner*, Wiley ©2014. Reproduced with permission. Lisciandro, *JVECC* 2011;20(2):1104–122.

Probe Orientation, Type, and Depth Settings

LUS orientation always begins with establishing the gator sign so that the sonographer has properly identified the pulmonary-pleural interface or lung line for confident imaging and assessment of the surface of the lung. The depth is generally set between 4–8 cm depending on the size of the patient. In felines, depth is generally set between 4–6 cm. Being in closer than 4 cm lends itself to missing pathology that extends from the lung surface deeper than your depth setting. Frequency is generally set between 5–10 MHz. A microconvex probe is preferred over a linear probe because the microconvex probe is acceptable for Global FAST - combining AFAST, TFAST, and Vet BLUE - as a whole body surveillance of the patient, a concept and strategy gaining momentum on the human side. A phase-array or sector probe is generally not recommended because its focal point is too small, although usefulness of a phase array probe is unknown. A linear probe will provide superior imaging of the lung surface, however, how much is gained over a microconvex probe is unknown, and the linear probe is not ideal for the AFAST and TFAST portions of Global FAST. Either the abdominal preset or cardiac preset may be used for Vet BLUE. The preference is machine and user dependent. We prefer the abdominal preset for the entire Global FAST exam in the great majority of cases.

FELINE VET BLUE EXAM

How to Perform

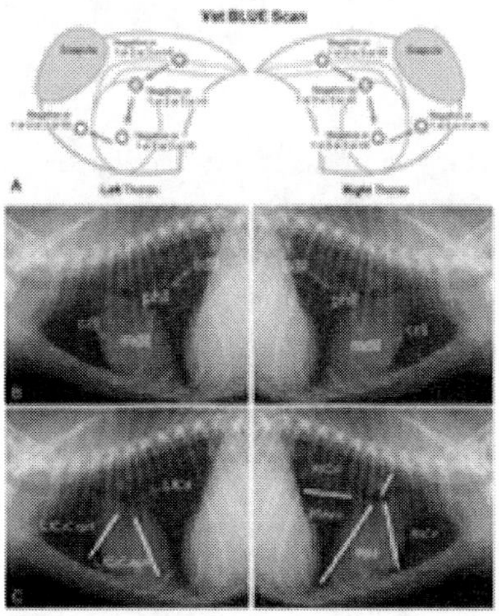

Figure 2. The Vet BLUE lung examination is a screening test performed identically as the probe is positioned at the CTS view of TFAST. The probe is then moved through regional locations that are bilaterally applied as follows: caudo-dorsal lung region (Cd - same as the TFAST CTS view, upper third, 8–9th intercostal space), peri-hilar lung region (Ph - 6–7th intercostal space, middle third), middle lung region (Md - 4–5th intercostal space, lower third), and cranial lung region (Cr - 2nd–3rd intercostal space, lower third). This material is reproduced with permission of John Wiley & Sons, Inc, *Focused Ultrasound Techniques for the Small Animal Practitioner*, Wiley© 2014 and FASTVet.com© 2014.

The Best Way to Perform Vet BLUE Accurately

Locate the left TFAST chest tube site directly above the xiphoid in the area of the 8–10th intercostal space in the upper 1/3rd of the thorax, finding the transition zone of lung and abdominal cavity, and then moving 1 or 2 intercostal spaces cranially to make sure the probe is over lung/pleural space and not over liver/stomach/abdominal contents. From the left TFAST CTS view which is the same as the Vet BLUE caudo-dorsal (Cd) view, draw an imaginary line or an actual line with your alcohol or acoustic coupling gel to the elbow. Halfway to the elbow is the Vet BLUE peri-hilar (Ph) view, and near the elbow is the Vet BLUE middle (Md) view. If the heart is in view at the Vet BLUE Md view, slide above the heart until you see the lung line. In most cats and dogs, sliding caudally at the Md view often images the transition zone of lung and abdominal cavity, confounding those unfamiliar or inexperienced with the pitfall. The final site is the Vet BLUE cranial (Cr) view, which requires gently extending the foreleg cranially to get the probe placed in the 2nd–3rd intercostal space. If too ventral at the Cr view, you will see the striations of the pectoral muscles; and if too high and cranial at the Cr view, you will be in the thoracic inlet with soft tissue and vessels. We use the thoracic inlet as a landmark for then sliding over the first 2 intercostal spaces for the Cr view. Our preference is to start high (dorsal) on the left hemithorax moving from Cd to Cr, and then performing the same routine on the right hemithorax. By always performing Vet BLUE in the same manner, findings are better remembered.

THE 5 BASIC VET BLUE FINDINGS

Wet vs. Dry Lung - Basic Lung Ultrasound

Basic easily recognizable LUS findings are categorized into the wet lung vs. dry lung concept. A glide sign with A-lines (reverberation artifact) at the lung line is considered "Dry Lung" only to be confounded with PTX (A-lines and no glide sign). However, many patients in which the probability of PTX is very low, then spending additional time finding the glide sign becomes less important and A-lines alone suffice. Ultrasound lung rockets (ULRs) also called B-lines are considered "Wet Lung" and oscillate to and fro with inspiration and expiration and must extend without fading to the far field obliterating A-lines.

Shred Sign, Tissue Sign, & Nodule Sign (plus Wedge Sign) - Advanced Lung Ultrasound
These are the 3 more advanced LUS signs we have described in progressive order of increasing consolidation/infiltration. The shred sign is similar to an air bronchogram on TXR or rather consolidation with aeration of the lung; the tissue sign is similar to hepatization of lung or rather consolidation with**out** aeration; and the nodule sign or rather consolidation/infiltration in discreet nodules. The wedge sign is a subset of the shred sign and represents pulmonary thrombo-embolism (PTE) or rather infarcts at the lung periphery.

Regionally-Based Respiratory Pattern Approach Using Vet BLUE
The clinical relevancy of Vet BLUE's patterned-based regional approach includes the following:

1. Dry lung all fields rules out clinically relevant left-sided congestive heart failure, and suggests upper airway obstruction, feline asthma, COPD, PTE, and non-respiratory look-a-likes.
2. Wet lung or ULRs in high numbers (>3 or infinity) in > than 2 sites bilaterally is good evidence for the likely diagnosis of cardiogenic lung edema in non-trauma even more reliably in felines than canines, especially when involving upper Vet BLUE views.
3. Wet lung in ventral fields with or without signs of consolidation (shred sign/tissue sign) suggest pneumonia.
4. Multiple nodules suggest metastatic disease or granulomatous disease.
5. Wedge sign suggests PTE.

THE TALE OF 4 FELINES
The use of thoracic auscultation and breathing patterns for respiratory distress is insensitive and prone to error, coupled with the dangers of transport and restraint in radiology and thoracic radiographic interpretation, making the proactive use of Vet BLUE lung ultrasound incredibly powerful for differentiating respiratory distress in cats. Vet BLUE is a pattern-based regional approach of using our 5 described basic lung ultrasound findings - dry lung (glide with A-lines), wet lung (ultrasound lung rockets also called B-lines), shred sign (wedge sign, PTE), tissue sign and nodule sign.

A Veterinary Cardiologist Once Said - "I Can't Tell the Difference Between Feline Asthma and CHF" - However, You Can Using Point-of-Care Vet BLUE in <60–90 Seconds
In the Tale of 4 Cats, we have Cat #1 with left-sided CHF that is wet all fields; Cat #2 with feline asthma that is dry all fields; Cat #3 with pleural effusion, but don't stop at the pleural effusion, look through the lung line; and Cat #4 with metastatic disease that has the nodule sign with various-sized nodules at nearly every Vet BLUE view. On physical examination, considering auscultation and breathing patterns, all 4 cats look exactly the same with nostril flaring, abdominal breathing, and harsh lung sounds. We will work through these 4 cats emphasizing the evidence-based power of Vet BLUE that better directs care and diagnostic testing.

REFERENCES

1. Lisciandro GR, Fosgate GT, Fulton RM. Frequency and number of ultrasound lung rockets (B-lines) using a regionally based lung ultrasound examination named Vet BLUE (veterinary bedside lung ultrasound exam) in dogs with radiographically normal lung findings. *Vet Radiol Ultrasound.* 2014;55(3):315-322.
2. Lisciandro GR, *et al.* Frequency and number of ultrasound lung rockets (B-lines) using a regionally based lung ultrasound examination named Vet BLUE (veterinary bedside lung ultrasound exam) in cats with radiographically normal lung findings. *J Vet Emerg Crit Care.* 2016; In press.
3. Ward JL, Lisciandro GR, Tou SP, Keene BW, DeFrancesco TC. Evaluation of point-of-care lung ultrasound (Vet BLUE protocol) for the diagnosis of cardiogenic pulmonary edema in dogs and cats with acute dyspnea. *J Am Vet Med Assoc.* 2015; In press.
4. Lisciandro GR, *et al.* Absence of B-lines on lung ultrasound (Vet BLUE protocol) to rule out left-sided congestive heart failure in 368 cats and dogs (abstract). *J Vet Emerg Crit Care.* 2016.
5. Ward JL, Lisciandro GR, DeFrancesco TC. Distribution of alveolar-interstitial syndrome in dyspneic veterinary patients assessed by lung ultrasound versus thoracic radiography. *J Vet Emerg Crit Care.* 2016; In press.
6. Kulhavy DA, Lisciandro GR. Use of a lung ultrasound examination called Vet BLUE to screen for metastatic lung nodules in the emergency room (abstract). *J Vet Emerg Crit Care.* 2015.

acvim382

Global FAST℠: AFAST®, TFAST®, and Vet BLUE® for Patient Monitoring

Gregory R. Lisciandro, DVM, DABVP, DACVECC
Spicewood, TX, USA

Introduction

The evolution of veterinary abbreviated ultrasound formats has extended beyond the abdominal format as a simple "flash exam" of fluid positive or fluid negative since the landmark publication by Boysen *et al.* in 2004. In 2008, the thoracic FAST format was developed by Lisciandro *et al.* and referred to as TFAST for the rapid diagnosis of pneumothorax and other thorax-related injury. In 2009, the abdominal FAST referred to as AFAST was a modification from the original FAST study. AFAST renamed the views with a target-organ approach rather than naming of external sites so that sonographer would be more aware anatomically about the actual organs and structures at each of the AFAST views; and AFAST directed the probe more strategically into the gravity-dependent regions of each view; and AFAST has its applied fluid scoring system to make more sense of a positive scan; and the same study advocated for 4-hour serial exams with repeat scoring for all hospitalized patients. In 2014, a 3rd abbreviated lung ultrasound format was published by Lisciandro *et al.* named Vet BLUE to complement AFAST and TFAST. In combination, these 3 formats are called Global FAST. Global FAST provides a huge amount of clinical information (<6–8 minutes) regarding your patient by taking advantage of basic echo views of the heart, and non-echo views that reflect left- and right-sided cardiac status; and determining if your patient is losing volume internally because the abdominal cavity, retroperitoneal space, pleural cavity, pericardial sac, and lung are also surveyed for free fluid and edema, respectively; and urinary bladder volume may be estimated and serial exams using our AFAST cysto-colic formula can non-invasively estimate urine output. A similar strategy has evolved referred to as the rapid ultrasound in shock exam (RUSh) in emergent patients, although the RUSh exam does not have a fluid scoring system, lacks a lung screening component other than ruling in or out pneumothorax, and does not evaluate urinary bladder volume. Most recently, the analogous global approach as a screening test is gaining some momentum on the human side because focused exams are dangerous, and traditional complete abdominal ultrasound and complete echocardiography are often not in the right cavity. We advocate for a baseline Global FAST recorded on goal-directed templates for all admitted patients prior to intervention; and with proper training Global FAST takes <6–8 minutes with no shaving, minimal restraint.

Global FAST for Patient Monitoring

The Abdominal Fluid Scoring System

Small animals are placed in preferably right lateral recumbency because it facilitates the right TFAST pericardial views for TFAST basic echo views; however, either lateral recumbency is validated for the AFAST abdominal fluid scoring system. The abdominal fluid score (AFS) of 1 is given to any positive AFAST views so the scoring system ranges from 0–4. The use of the abdominal fluid scoring system gives more value to effusions over mild, moderate and severe, and allows for better tracking of resolution or worsening or effusions. In cases of hemorrhage, the fluid scoring system helps categorize intra-abdominal bleeding as small volume bleeding, AFS 1 and 2, and large volume bleeding, AFS 3 and 4. Small volume AFS 1 and 2 dogs and cats do not have enough intra-abdominal hemorrhage for anemia, so if they are or become anemic then pre-existing anemia existed or they are losing blood somewhere else (i.e., retroperitoneal, pleural, pericardial, lung, intrapelvic, fracture sites, gastro-intestinal tract, or externally, respectively). On the other hand, large volume bleeders of AFS 3 and 4, considered life-threatening, have enough intra-abdominal hemorrhage to become anemic and many require blood transfusions dependent on the subset of patient and the degree of fluid resuscitation. The use of the AFAST-applied AFS is also a monitoring tool for all at-risk for bleeding, post-interventional cases, including those with percutaneous needle and Tru-cut biopsies. Patient AFS helps better make decisions regarding ongoing bleeding, resolving bleeding, and need for blood transfusion and/or exploratory laparotomy.

TFAST Right Pericardial Echo Views

Left Ventricular Short-Axis View for Volume and Contractility

The left ventricular short-axis view (LVSA) is acquired just below the mitral valves at the level where the chordae tendinae come off the left papillary muscles referred to as the LV short-axis "mushroom" view. The filling and size of the "mushroom" is a reflection of patient volume status as long as the sonographer is aware of how to locate the proper level on short-axis. Contractility is also assessed subjectively using the eyeball approach. It does not take a whole lot of training to be able to screen for poor filling and poor contractility. Poor filling indicating poor volume can be supported or refuted by assessing the caudal vena cava; and contractility by triggering a complete echocardiography. In the meantime though, a patient thought to have poor contractility (i.e., dilated cardiomyopathy) may be treated and better stabilized during the delay of acquiring complete echocardiography.

Long-Axis 4-Chamber View for the Right Ventricular to Left Ventricular Ratio (RV:LV)

The normal RV:LV ratio is 1:3–4 with the RV being a small triangle when compared to the LV. When the RV is nearly the same size of the LV then right heart problems and pulmonary hypertension should be suspected, and complete echocardiography is indicated until proven otherwise. However, by recognizing the abnormality, patient therapy may be adjusted to better head off complications. In an acutely respiratory distressed cat or dog that develops acute RV dilation, massive PTE has likely occurred. Your non-echo fallback view for right-sided heart problems is the caudal vena cava at the FAST DH View (see below).

Left Ventricular Short-Axis View for the Left Atrial to Aortic Ratio (LA:Ao)

The normal LA:Ao ratio is <1.3 in dogs and <1.6 in cats. This is the most challenging view to obtain. Your non-echo fallback strategy is performing the easier, less stressful, Vet BLUE lung exam. Absent B-lines in all views (ABAV) is an effective means to rapidly rule out left-sided congestive heart failure (see below).

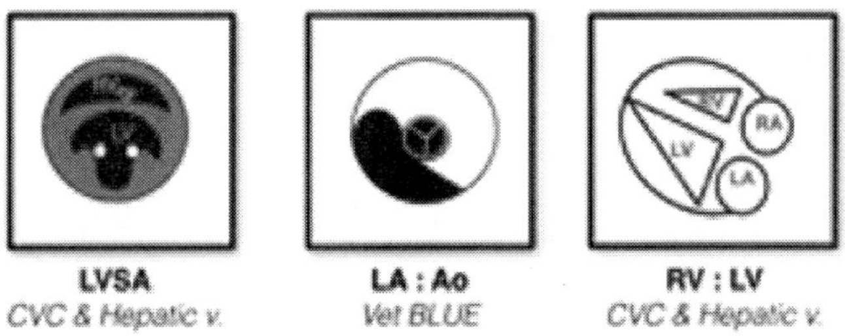

Figure 1

The Non-Echo Fallback Views for Left- and Right-Sided Cardiac Problems

Characterizing the Caudal Vena Cava and Hepatic Veins

The caudal vena cava (CVC) where it traverses the diaphragm rapidly reflects preload and is your new non-invasive central venous pressure (CVP). In fact, central lines have been debunked in human medicine since 2013; and it is common practice to use the analogous view of the inferior vena cava (IVC). We simply eyeball and characterize the CVC as being 1) FAT or distended with <10% change in diameter (high CVP), or 2) flat (collapsed with <10% change in diameter, low CVP), or 3) having a bounce (~50% change in diameter, in the ballpark of normal). The normal "bounce" reflects the dynamic changes in CVC diameter during inspiration and expiration as blood is drawn/squeezed into the heart in spontaneously ventilating dogs and cats. Measuring the CVC using M-mode can be challenging and difficult with a lot of patient movement. However, by visually characterizing the CVC at the FAST DH view, called the "eyeball approach," and correlating with clinical impression and other findings (blood pressure, physical exam findings, blood lactate), the clinician has a much better idea of patient preload (CVP) and right-sided cardiac status. Moreover, if the sonographer wants a numeric value, then imaging the CVC in B-mode and freezing and rolling the cine ball to get minimal and maximal diameter is another approach. These measurements can then be used to calculate its distensibility index. Because of the great differences in sizes in dogs absolute measurements are less likely to be as reliable. The hepatic veins are not normally seen in lateral or sternal/standing, so their distension is another clue that CVP is high.

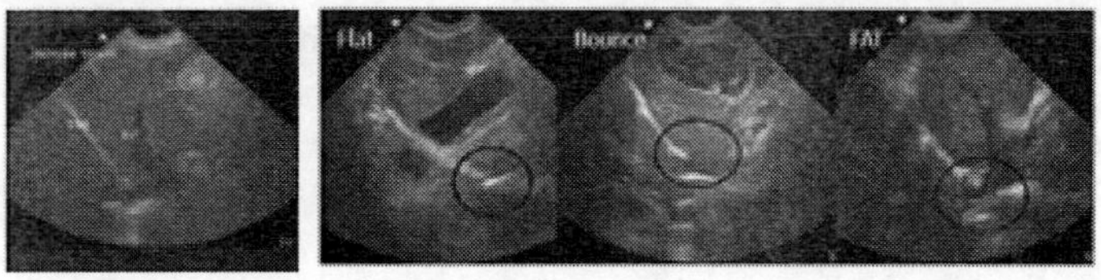

Figure 2. Far left image shows the classic FAT or distended CVC as it traverses the diaphragm with associated hepatic venous distension likened to tree trunks and branching referred to as the "Tree Trunk Sign." The 3 images labeled FAT, Bounce, and Flat represent a high CVP, a ballpark normal CVP, and a low CVP, respectively. This material is reproduced with permission of John Wiley & Sons, Inc, *Focused Ultrasound Techniques for the Small Animal Practitioner*, Wiley ©2014.

Use of Vet BLUE - "Wet Lung" vs. "Dry Lung"

Because the "wet lung" vs. "dry lung" concept is easily recognized during Vet BLUE, the presence or absence of ultrasound lung rockets (ULRs), also called B-lines, provides important clinical information regarding left-sided cardiac status and left-sided volume overload. Moreover, Volpicelli *et al.* showed that numbers of ULRs correlate with the degree of alveolar-interstitial edema when compared to CT; and Vet BLUE requires minimal patient restraint, is rapid (<60–90 seconds), safe and point-of-care. Thus, acquire a baseline Vet BLUE prior to fluid therapy on all hospitalized dogs and cats. ULRs have been shown to correlate with extravascular lung water, and thus are sentinels for worsening respiratory status and pulmonary failure. If treatment strategy is not adjusted, then alveolar-interstitial edema may progress to alveolar flooding, which is much more difficult to treat. Using the regionally-based Vet BLUE patterned approach, other causes of wet lung artifacts, such as pneumonia, can often be discriminated. Moreover, the use of Vet BLUE potentially triggers additional testing and imaging.

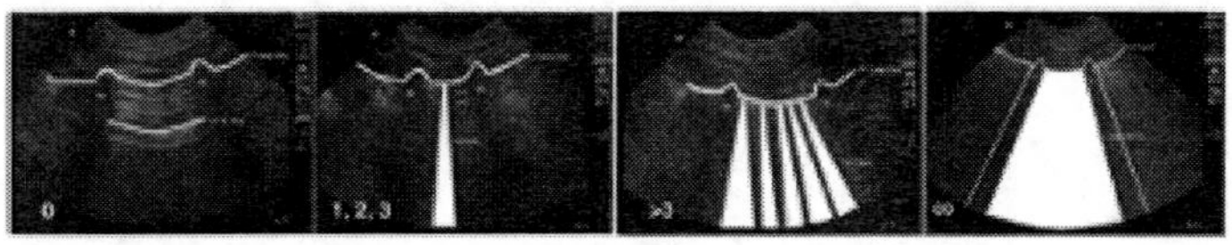

Figure 3. Showing the counting scheme published by the author for counting ULRs as 1, 2, 3 or >3 and infinity ∞.

The Urinary Bladder Volume Formula

At the AFAST cysto-colic view the urinary bladder is imaged in longitudinal (sagittal) and the best oval is acquired in this plane and measured followed by transverse orientation and acquiring the largest oval and measured. Measurements in cm will give you an estimation of urinary bladder volume in ml by using Length x Width x Height x 0.625. With measurements over time, urine output can be non-invasively estimated.

Use of the Lung Point for Monitoring Pneumothorax (PTX)

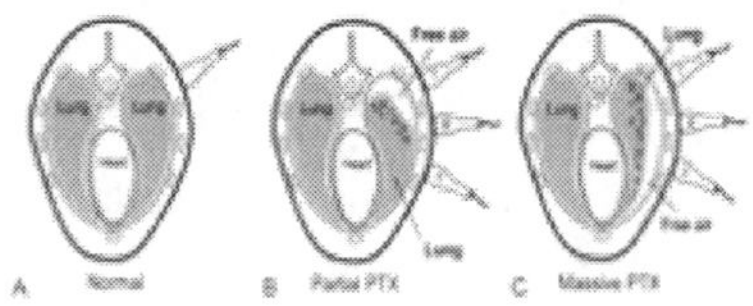

Figure 4

The use of the lung point, the transition zone of where there is pneumothorax (PTX) and lung re-contacting the thoracic wall, is a means to increase the sensitivity for the diagnosis of PTX and to track worsening, or resolving, PTX. Post lung lobe aspirate, chest tube placement/removal, or other invasive thoracic procedures, the lung point qualifies the PTX, point-of-care with minimal patient restraint and stress, as follows: 1) upper 1/3 of the thorax, trivial or 2) middle 1/3 of thorax moderate and concerning warranting thoracocentesis or 3) lower 1/3 of thorax as severe/massive warranting thoracocentesis (author's experience). Clinical judgment is required to maximize decision-making.

Baseline Admission Global FAST and Serial Exams are Key

The repeating of Global FAST exams, serial exams, cannot be overemphasized. Minimally a 4-hour post-admission Global FAST exam should be performed in all admitted patients; and the author incorporates Global FAST as part of daily rounds immediately after a complete physical exam.

REFERENCES

1. Lisciandro GR. Focused abdominal (AFAST) and thoracic (TFAST) focused assessment with sonography for trauma, triage and monitoring in small animals. *J Vet Emerg Crit Care*. 2011;20(2):104–122.
2. Lisciandro GR, Fosgate GT. Use of AFAST cysto-colic view urinary bladder measurements to estimate urinary bladder volume in dogs and cats. *J Vet Emerg Crit Care*. 2016; In press.
3. Ward JL, Lisciandro GR, Tou SP, Keene BW, DeFrancesco TC. Evaluation of point-of-care lung ultrasound (Vet BLUE protocol) for the diagnosis of cardiogenic pulmonary edema in dogs and cats with acute dyspnea. *J Am Vet Med Assoc*. 2015; In press.
4. Lisciandro GR, *et al*. Absence of B-lines on lung ultrasound (Vet BLUE protocol) to rule out left-sided congestive heart failure in 368 cats and dogs (abstract). *J Vet Emerg Crit Care*. 2016.
5. Perera P, Mailhot T, Riley D, Mandavia D. The RUSH exam: rapid ultrasound in shock in the evaluation of the critically ill. *Emerg Med Clin North Am*. 2010;28(1):29–56.

acvim436

Selection, Delivery, and Monitoring of Transfusions in Cats and Dogs

Julie Walker[1], DVM, DACVECC; Alex Lynch[2], BVSc (Hons), DACVECC, MRCVS; John Thomason[3], DVM, MS, DACVIM
[1]Madison, WI, USA; [2]Gainesville, FL, USA; [3]Starkville, MS, USA

INTRODUCTION

This interactive session is arranged in three segments. The background section will present concepts of blood component therapy and whole blood transfusion, with a goal of assisting veterinary professionals in selecting blood products for individual patients. Concepts of delivering and monitoring transfusions will be reviewed, with a focus on identifying and treating/preventing transfusion-associated complications. The second segment involves the audience in assessing patients and making treatment decisions. The third segment will allow time for discussion of concepts among presenters and to answer questions from attendees.

OVERVIEW OF BLOOD COMPONENT PRODUCTION

After collecting blood from a donor, the unit can be transfused immediately, stored, or processed into components (then stored or administered). At the time a unit of blood is collected, it is called fresh whole blood (FWB). If the unit of blood will not be used immediately, it should be refrigerated (4°C) or separated into components then stored properly. Once a whole blood unit is refrigerated, it is deemed stored whole blood (SWB). If a unit of FWB is centrifuged, it can then be separated into packed red blood cells (PRBC) and plasma. Plasma that is separated and frozen (≤-18°C) within 8 hours of collection is called fresh frozen plasma (FFP), and plasma frozen >8 hours after collection is called frozen plasma (FP). Other products such as platelet-rich plasma (PRP), platelet concentrate (PC) and cryoprecipitate are prepared using special techniques. The production, contents, storage, and indications for the most commonly used blood components are described in further detail below.

COMMONLY USED BLOOD COMPONENTS

Packed Red Blood Cells (PRBC)

Following donation, a unit of whole blood is centrifuged at 4°C for 5–7 minutes at 5000xg. Plasma is expressed into an attached bag and frozen. A RBC additive solution, containing ingredients which support RBC osmotic balance (mannitol, normal saline) and ATP production (dextrose, adenine), is combined with the concentrated RBC prior to storage as a unit of PRBC (Wardrop 1995; Fung *et al.* 2014). PRBC contain red blood cells, anticoagulant, remnant plasma, and RBC preservatives. Unless leukoreduction is performed, white blood cells and inactivated platelets are also present within the PRBC unit. PRBC units can be stored in a refrigerator at 4°C for 35 to 42 days, depending on the additive solution used. The administration of PRBC is appropriate for treatment of clinically significant anemia of any underlying cause. Since units contain a minimal volume of plasma, administering PRBC is advantageous in normovolemic animals with anemia due to hemolytic or non-regenerative etiologies. Alternatives to PRBC transfusion includes administration of FWB or SWB.

Fresh Frozen Plasma (FFP)

Within 8 hours of collection, whole blood is centrifuged at 4°C for 5–7 minutes at 5000x*g*. The plasma supernatant is expressed into an attached blood bag and frozen at <-18°C. This product is known as FFP, which can be stored in a freezer for up to one year. For the first year of storage, FFP contains all circulating procoagulant and anticoagulant proteins, albumin, globulin, anticoagulant, electrolytes, and water. If FFP is not used within one year, it is re-labeled as FP. Administration of FFP is indicated in patients with hemorrhage due to acquired or congenital coagulopathies. Examples of congenital coagulopathies include von Willebrand disease, hemophilia A (factor VIII deficiency), hemophilia B (factor IX deficiency), and factor VII deficiency. Acquired coagulopathies are commonly due to anticoagulant rodenticide intoxication (deficiency of activated factors II, VII, IX, X) and the hemorrhagic phase of disseminated intravascular coagulation. FFP transfusion is also indicated for resuscitation of patients with severe blood loss requiring massive transfusion. More controversial uses for FFP include treatment of patients with severe hypoalbuminemia and pancreatitis.

Frozen Plasma (FP)

Plasma is designated as FP if it is separated from whole blood more than 8 hours after blood collection. Also, units of FFP stored in a freezer for more than one year are re-labeled as FP. A unit of FP can be stored in a freezer for up to 5 years from the date of initial blood collection (Walker 1993). Plasma loses potency of labile clotting factors (factors V, VIII) and von Willebrand factor over time, which are no longer present in high concentrations in FP (Benjamin, McLaughlin 2012; Allain *et al.* 1983). FP contains albumin, globulin, and all clotting factors except for factors V, VIII, and von Willebrand factor. Therefore, indications for FP include treatment of patients with hemophilia B (Factor IX deficiency), factor VII deficiency, and anticoagulant rodenticide intoxication. FP should not be administered for von Willebrand disease or hemophilia A (Factor VIII deficiency).

Cryoprecipitate

To produce cryoprecipitate, a unit of FFP is slowly thawed under refrigerated conditions (4°C). When the product is fully thawed, a translucent, white precipitate can be identified in the bag. The bag of plasma is then centrifuged and most of the supernatant plasma is expressed into a separate bag. The remaining precipitate and small volume of plasma that remains is referred to as cryoprecipitate. Cryoprecipitate is rich in von Willebrand factor, factors VIII and XIII, fibrinogen, and fibronectin. Cryoprecipitate can be re-frozen and stored for up to 12 months. A lyophilized (freeze dried) form of cryoprecipitate is also available through a commercial blood bank, which can be held in refrigerated storage for up to 18 months. Cryoprecipitate is most commonly used to treat von Willebrand disease and hemophilia A (factor VIII deficiency).

Cryoprecipitate-Poor Plasma (CPP)

In the preparation of cryoprecipitate described above, the supernatant plasma that is expressed off of the concentrated precipitate is known as CPP (or cryosupernatant). CPP is re-frozen and stored for up to 12 months. CPP contains albumin, globulin, and all clotting factors except for factors VIII, XIII, von Willebrand factor, fibrinogen, and fibronectin. Therefore, CPP can be used to treat patients with anticoagulant rodenticide intoxication and hemophilia B (factor IX deficiency).

Platelet-Rich Plasma (PRP) and Platelet Concentrate (PC) – Whole Blood Derived

To prepare PRP, a unit of FWB is centrifuged at a slower speed (1000xg for 4 minutes) and a higher temperature (20°C) to preserve platelet function. The supernatant (PRP) is expressed into an attached bag, which includes plasma, platelets, and white blood cells. When a unit of PRP is re-centrifuged at 2000xg for 10 minutes and supernatant plasma is removed, the remaining platelet product is called PC (Abrams-Ogg *et al.* 1993; Fung *et al.* 2014). PRP and PC must be stored at room temperature (20–24°C) under constant agitation to maintain viability of platelets. The products may be stored for up to 5 days. Platelet transfusions are indicated in patients with life-threatening hemorrhage due to severe thrombocytopenia or thrombopathia. Platelet transfusions are also indicated in the treatment of patients with severe blood loss requiring massive transfusion, and in patients with severe thrombocytopenia or thrombopathia in which an invasive procedure must be performed.

Other Platelet Products

Short platelet storage times limit veterinary practices to producing PRP and PC at the time the product is needed. The preparation of two platelet products has extended storage times for platelets: DMSO-frozen platelet concentrate (cryopreserved platelets) and lyophilized platelets. Cryopreserved platelets are produced by collecting platelets from donors via apheresis and adding 6% DMSO prior to freezing the unit for up to 6 months. Lyophilized (freeze dried) platelets have been produced from canine platelets by a commercial blood bank and offer an extended storage time of up to 2 years. Unfortunately, lyophilized platelets are not yet available for clinical use.

Storage of Blood Components

A refrigerator suitable for blood product storage should have a temperature monitoring system and be dedicated solely to the storage of blood products. Opening and closing the refrigerator door should be minimized to avoid fluctuations in storage temperatures. Refrigerated blood products should be maintained in a temperature range between 1–6°C (ideally 4°C). Colder storage predisposes red blood cell units to freezing, which induces hemolysis and subsequent recipient morbidity (Patterson *et al.* 2011). Frozen blood products should be stored at or below -18°C (Fung 2014). An appropriate freezer should also have a temperature monitoring system, be dedicated to the storage of blood products, and should **not** have an auto-defrost function. Also, ensuring that the freezer has a secure closure and/or lock to avoid unnecessary opening and closing of the door will help prevent temperature fluctuations. If plasma is stored above -18°C, blood product quality and shelf-life could be compromised.

Advantages and Disadvantages of Component Therapy

The use of blood components can be advantageous to the veterinary hospital, to the individual patient, and from the perspective of blood product conservation. By separating a unit of whole blood into its constituents, a patient can receive only the portion(s) of blood needed (cells, plasma, platelets), limiting unnecessary volume administration and antigen exposure. Component production also prolongs storage times (e.g., plasma stored for 1–5 years), thus reducing product wastage and enabling veterinary practices to store product for use on-demand even if transfusions are administered infrequently. However, the in-house production of blood components requires access to a centrifuge capable of maintaining set temperatures (4–25°C) and large enough to safely accommodate blood bags. Blood components are frequently purchased from commercial animal blood banks and shipped to an individual practice for storage or use. Since most practices are not in close proximity to these blood banks, planning is required to have product on hand at the time of need. Patients requiring the transfusion of red blood cells, plasma, and platelets often receive two or three different blood components (e.g., PRBC, FFP, ± PC) to equate to one unit of fresh whole blood. In the process of providing these components, the recipient is often exposed to antigens from multiple blood donors. Component production also induces some loss of platelets (Abrams-Ogg *et al.* 1993) in the preparation process, which might reduce the efficacy of the blood product.

Ideally, veterinarians would have access to both whole blood and blood components so that critically ill patients receive the most appropriate and most immediately available products. While access to all blood products might be feasible in large private or academic veterinary hospitals, it is not practical for most hospitals. The decision to transfuse stored blood components or fresh whole blood must be made by individual veterinary practices based on patient need and on these logistical considerations.

Fresh Whole Blood Transfusion

Fresh Whole Blood - When to Consider It?

As mentioned above, there are many benefits to advocating component therapy for veterinary patients. Many animals will fall into discrete categories with respect to their transfusion needs. Component therapy typically enables the clinician to address specific issues while preserving blood bank supplies. A dog with a euvolemic anemia associated with IMHA, for example, would benefit from red cell administration but does not need volume expansion with plasma. In contrast, a dog with a congenital bleeding disorder would benefit from preemptive clotting factor administration before an invasive procedure, yet likely does not have a clinical indication for red cells. Since canine and feline blood products are a precious resource, splitting one unit of whole blood into at least two additional component units is logical. There may be some select patients where fresh whole blood offers clinical advantages over components however. Simplistically, it can be helpful to consider when 'fresher' blood is more ideal, and when the 'wholeness' of blood may offer advantages.

Does the 'Freshness' Matter?

When a unit of blood is collected from a donor, it can be stored at room temperature for 8 hours and still be considered fresh. This fresh blood contains red blood cells, white blood cells (of a variable number depending on whether leukoreduction has been performed), platelets, and plasma. The plasma contains all clotting factors, coagulation proteins, albumin, and globulins. This unit of fresh whole blood can be split into further components or left whole for immediate usage. For hospitals that use blood products infrequently, there is a real risk of wasting stored units. In this scenario, calling in donors for bleeding on an 'as needed' basis may work well. For other hospitals where blood product usage is more common, relying on donors in this manner is somewhat impractical, inefficient, and non-sustainable. Beyond 8 hours, the 'freshness' of the blood will begin to wane with loss of platelet viability (Tsuchiya *et al.* 2003), and is then termed stored whole blood. Clotting factor and hemostatic protein concentrations are acceptable in whole units for up to 24 hours (Walton *et al.* 2014), meaning separation of whole units into packed red cells and fresh frozen plasma may still be performed within this time frame. Stored whole blood maintained in refrigerators (e.g., 1–6°C) can be used for up to 35 days depending on the anticoagulant preservative utilized (Walker 2016). These stored units have a progressive decline in the concentration of the labile clotting factors (FVa and VIIIa) over the course of their storage (Jobes *et al.* 2011).

Over time morphological and biochemical changes in packed red blood cells, termed the storage lesion, also occur. The clinical consequences of the storage lesions raise concern for recipients. Two major questions arise when reflecting on the storage lesion. Firstly, might older stored red blood cells be less effective in patients compared to younger stored cells? Secondly, could older blood products increase morbidity or mortality in recipients? One change that has been characterized is the rapid depletion of 2,3-diphosphoglycerate (2,3-DPG) in canine red cells. Hemoglobin affinity for oxygen is modified by 2,3-DPG and concerns therefore arise regarding oxygen unloading potential for transfused red cells. The role of 2,3-DPG in cats is less important, with a greater reliance on chloride for hemoglobin-oxygen affinity in this species (Herrmann, Haskins 2005; Wong, Haskins 2007). There is actually little compelling evidence to suggest that 2,3-DPG depletion in stored red cells negatively impacts oxygen delivery in recipients however (Raat *et al.* 2005). The concentration of ATP in red cells progressively declines with storage also, but at a variable rate depending on the preservative utilized (Price *et al.* 1998; Wardrop *et al.* 1994). Red cells lack mitochondria and depletion of energy sources will lead to Na^+ K^+ ATPase membrane pump dysfunction. The downstream effects of this are impaired cellular deformability, which limits red cell movement through the microcirculation, as well as accelerating clearance from the body via the erythrophagocytic system (Baskurt 1999). Energy depletion also impairs red cell resistance to oxidative changes with time (Tavazzi *et al.* 2000). Erythrocyte membrane stability will change associated with energy depletion. The classic biconcave disc shape of a normal erythrocyte will change (Obrador *et al.* 2015). The abnormally shaped red cells risks the formation of microparticles. Microparticles are small vesicles derived from the plasma membranes of various cells, including red cells, which may contribute to local thrombosis (Herring *et al.* 2013). Microparticles contribute to a local prothrombotic state by surface expression of various products including phosphatidylserine and tissue factor.

Hemolysis that may develop secondary to red cell membrane changes releases free hemoglobin that scavenges local nitric oxide. Nitric oxide is an important local vasodilator, meaning reduced concentrations of nitric oxide promote regional vasoconstriction, endothelial dysfunction, and microthrombosis (Gladwin *et al.* 2009). Biochemical changes noted in stored units include increased ammonia concentration in canine (Waddell *et al.* 2001) and feline (Cummings *et al.* 2016) stored units. Animals with normal hepatic reserve likely would not experience adverse effects associated with ammonia-rich products. A subset of patients however reasonably could be at higher risk of hyperammonemia and may benefit from younger blood (e.g., encephalopathic lipidotic cat). Another consideration for stored units is the immunological impact of older blood. The presence of white cells within units may lead to increased cytokine concentrations over time that could contribute to transfusion reactions. This aspect of the storage lesion is explored in greater depth below, with specific discussion of the impact of leukoreduction.

Based on the information available, it would appear logical that older red cells could be less functional than fresher cells, and consequently risk impairing oxygen delivery to tissues. The evidence to suggest this occurs *in vivo* after receiving older units requires careful scrutiny however. Species variation means that rodent models cannot necessarily be extrapolated directly to people (Obrador *et al.* 2015).

Meta-analyses have reported a high risk of death in people receiving older stored red cells (Wang *et al.* 2012) but may depend upon patient underlying disease (e.g., septic or not septic). A study performed in a canine model of sepsis, utilizing *Staphylococcus aureus* pneumonia, found that a large volume 42-day-old allogenic blood transfusion resulted in both higher morbidity (worse lung injury) and mortality compared to 7-day-old blood (Solomon *et al.* 2003). Based upon on the data, it may be reasonable to consider specific patient populations whereby fresher blood might be advantageous if available. Reasonably, this might reflect patients with higher baseline illness severity or specific disease processes (e.g., sepsis). Further work is required to characterize the impact of the age of cells on recipient outcome in veterinary patients.

Does the 'Wholeness' Matter?

As described above, fresh whole blood contains a source of platelets for up to 8 hours after collection. Alternative sources of platelets may be harder to access in many hospitals. Platelet-rich plasma, for example, derived from a soft spin of fresh whole blood has a shelf life of 5 days if agitated frequently. In emergent situations where severe, life-threatening hemorrhage has occurred due to severe thrombocytopenia, the extra step necessary to prepare the platelet-rich plasma from whole blood may seem unappealing. Long-term stored platelet units (e.g., lyophilized platelets) are infrequently available, limiting their applicability to emergency cases as well. The clinician should consider the likely benefit of platelet administration in individual cases. For example, dogs with immune-mediated thrombocytopenia will rapidly destroy transfused platelets, limiting platelet transfusion as a useful therapeutic strategy in these patients. The exception to this statement ,however, is when severe bleeding associated with thrombocytopenia has occurred, to the extent that death may ensue without an intervention (e.g., pulmonary hemorrhage, intracranial hemorrhage), in which case platelet administration may be reasonable even if an immune-mediated condition exists. From a clinical perspective, certain other situations may arise when whole blood may seem advantageous. A good example would be severe acute hemorrhage resulting in both clinically relevant anemia and hypovolemia. A classic example would be the dog that has developed hemostatic dysfunction following ingestion of an anticoagulant rodenticide. These dogs benefit from vitamin K-dependent clotting factor replacement that can be achieved in several ways (e.g., fresh frozen plasma, frozen plasma, cryo-poor plasma). These dogs also typically have moderate anemia and hypovolemia. The administration of plasma products will typically volume expand effectively, yet risks inadvertent dilution and worsening of their anemia. It is therefore quite common for dogs to ultimately also receive red cells in this scenario. Acute and severe traumatic hemorrhage may also be a situation where fresh whole blood would be considered reasonable. Similar to the anticoagulant rodenticide situation, both anemia and hypovolemia may be present, that could be resuscitated effectively with whole blood. Component therapy would not necessarily fare any better or worse in these scenarios but an appreciation of replacing 'like with like' appears intuitive. When a massive transfusion of blood products is administered to people (e.g., after trauma), there are recommendations to administer ratios of products that replicate the constituents of whole blood (e.g. packed cells:plasma:platelets of 1:1:1) (Miller 2013). In veterinary practice, alternative strategies to allogenic transfusion (e.g., transfusion of shed cavitary blood) may also be helpful. Preemptive autologous transfusion prior to elective surgeries with anticipated high risk of hemorrhage (e.g., adrenalectomy) has also been utilized by the author.

Component therapy will still remain the backbone of transfusion medicine in small animal patients. There may be select instances, however, when either fresher or more whole blood may be preferable in some patients. Consequently, fresh whole blood can be a helpful additional tool in the transfusion armamentarium for small animal veterinarians.

Transfusion-Associated Complications

Red blood cell transfusions are commonly used for the treatment of life-threatening anemia in critical veterinary patients. From major blood loss and hemolytic diseases to severe erythropoietic failure, there is a wide range of common clinical conditions that require blood transfusions. Although transfusion is a necessary and frequently life-saving therapy, all transfusion recipients, regardless of blood product used, remain at risk for the development of adverse events and severe, potentially life-threatening reactions. In fact, in veterinary medicine, over 10% of all blood transfusion recipients develop a transfusion reaction (Hahn *et al.* 2014; Callan *et al* 1996; Kerl *et al.* 1993). There are several pre-transfusion tests, such as blood typing and cross-matching, which are designed to assess the compatibility and safety of blood products prior to administration to the recipient. However, despite these pre-transfusion tests, some recipients still develop reactions during and after a blood transfusion.

There are multiple causes of transfusion reactions, including the transfusion of a hemolyzed blood product, and the presence of naturally occurring or induced antibodies in the recipient's plasma directed against antigens located on transfused red blood cells, leukocytes and platelets (Patterson *et al.* 2011). Antibody-antigen interactions can result in an increase in systemic levels of cytokines and other proinflammatory mediators. Antibody binding to transfused mast cells and basophils can lead to degranulation and secretion of proinflammatory and vasoactive substances such as histamine, cytokines, and eicosanoids (Tocci 2010; Blake *et al.* 2017; Muro *et al.* 2017). With all of these potential mechanisms for transfusion reactions, multiple predisposing factors and conditions may interact to increase the likelihood of a reaction.

TRANSFUSION REACTIONS

Transfusion reactions can be categorized into acute or delayed reactions and immunologic or non-immunologic reactions.

Acute Immunologic Reactions

Acute Hemolytic Transfusion Reactions (AHTR)

Acute hemolytic transfusion reactions (AHTRs) can cause several significant abnormalities in the recipient. This type of reaction occurs when the transfused erythrocytes interact with circulating antibodies, naturally occurring or acquired, already present in the recipient. After this interaction takes place, complement and cytokines are released, leading to a systemic inflammatory response, fever, restlessness, salivation, and shock. However, acute intravascular hemolysis, renal failure, disseminated intravascular coagulopathy, and death are serious complications associated with AHTRs. The severity of AHTRs is directly related to the volume of blood transfused and destroyed. Treatment of AHTRs consists of stopping the transfusion and providing adequate fluid therapy to reduce hypotension and maintain renal blood flow (Bracker *et al.* 2005).

Recently, there was a published report of the death of three of four dogs that developed an acute hemolytic reaction due to inappropriate storage of red blood cell products. In humans, it has been shown that, as well as contributing to red cell fragility, the simple act of storing blood products for later use also leads to the accumulation of several vasoactive and proinflammatory molecules, including cytokines (IL-1β, IL-6, and IL-8) and arachidonic acid-derived eicosanoids. Accumulation of these molecules is believed to significantly increase the risk of transfusion reactions (Blake *et al.* 2017; Muro *et al.* 2017; Bracker *et al.* 2005).

Non-Hemolytic Transfusion Reactions (FNHTR)

The presence of platelets and leukocytes in blood products has historically been associated with transfusion reactions. Febrile non-hemolytic transfusion reactions (FNHTR) are the most common adverse transfusion reactions in both human and veterinary medicine. This type of transfusion reaction is often associated with an increase in body temperature by more than 1°C without any other explanation. These reactions are most likely associated with leukocyte-derived cytokine and/or circulating anti-leukocyte antibodies. Most of these reactions are self-limiting and patients may not demonstrate significant clinical signs. Treatment of FNHTRs consists of decreasing the transfusion administration rate or stopping the transfusion altogether. Also, adding an anti-inflammatory medication to try and decrease the fever. Platelet transfusions are most commonly associated with FNHTR; however, any blood product that contains platelets or leukocytes can cause this type of reaction. In humans, prestorage leukoreduction, removal of leukocytes and platelets via specialized filters, has been associated with a decrease in frequency of post-transfusion FNHTRs. However, in dogs, it is unknown if the removal of platelets and leukocytes via leukoreduction filters would influence the incidence of FNHTRs (Blake *et al.* 2017; Muro *et al.* 2017; Bracker *et al.* 2005).

Allergic Reactions

Acute allergic reactions can develop following transfusions when the sensitized recipient is exposed to substances (allergen, protein) in donor plasma. Allergic reactions are type 1 hypersensitivity responses that develop when preformed IgE binds to an antigen, elicits an immune response and degranulates mast cells and basophils. Most allergic reactions are mild or self-limiting and manifested as cutaneous abnormalities (erythema, urticaria, and pruritus). Even though uncommon, more clinically significant allergic reactions, anaphylactic/anaphylactoid reactions can occur, and patients may develop hypotension, tachycardia, bronchoconstriction, vomiting, and diarrhea. Anaphylactic reactions will commonly present with mild cutaneous or mild gastrointestinal upset first, and special attention should be given to these transfusion recipients. Anaphylactic reactions in dogs manifested primarily with gastrointestinal signs, but cats will demonstrate respiratory distress, along with pruritus and gastrointestinal signs. If an allergic reaction is suspected, the transfusion should be slowed or stopped and diphenhydramine is administered. If the patient does not improve, a short-acting steroid, oxygen supplementation, intravenous fluid therapy, and supportive care can be administered (Bracker *et al.* 2005).

Acute Non-Immunologic Reactions

Volume Overload

Volume overload can occur if the rate of infusion is too rapid or if the volume of blood is excessive. Blood products are colloidal suspensions and do not rapidly diffuse into the interstitial space. Risk factors for volume overload include underlying structural heart disease and renal insufficiency. Neonates and cats do not tolerate intravascular volume expansion as well as adult patients, potentially contributing to volume overload. Euvolemic patients that have a chronic anemia may have a compensated euvolemic state, increasing the risk of volume overload. Volume overload may cause pulmonary edema, increased respiratory rate, cough, and dyspnea. Supplemental oxygen, stopping IV fluids, and furosemide may improve patients that are volume overloaded (Bracker *et al.* 2005).

Citrate Toxicity

Citrate, the anticoagulant most commonly used in blood products, chelates calcium and prevents clot formation. During large-volume or rapid transfusions, normal citrate metabolism can be overwhelmed leading to calcium chelation in the recipient by unmetabolized citrate. Citrate toxicity can cause hypocalcemia and hypomagnesemia and will commonly present with clinical signs of hypocalcemia. Therapy includes supportive care and calcium supplementation. Recipients with normal hepatic function will metabolize citrate in a few hours after administration. If calcium supplementation is required, a second intravenous line might be required to avoid the administration of calcium gluconate and anticoagulated blood products in the same line because infusion in the same line may precipitate clot formation (Bracker *et al.* 2005).

Hypothermia

The infusion of large volumes of refrigerated or frozen blood products, especially in smaller animals, can precipitate clinically signs associated with hypothermia, such as arrhythmias and hypothermia-induced coagulopathy. If administered through a central line, small volumes of cold blood products can be detrimental because cold transfused blood does not mix with enough warm blood before reaching the heart. To minimize the effects of hypothermia, cold blood products should be warmed to body temperature before administration (Bracker *et al.* 2005).

Bacterial Contamination

Patients receiving bacterial-contaminated blood products can develop a fever, signs of systemic inflammation, severe sepsis, and potentially death. Contamination of blood products usually develops from occult bacteremia of the blood donor, skin contamination at collection, or contaminated collection devices (Bracker *et al.* 2005).

Delayed Immunologic Reactions

Delayed Hemolytic Transfusion Reaction (DHTR)

Delayed hemolytic transfusion reactions develop several days after blood administration and occurs when antibodies in the recipient identify, adhere, and prematurely remove the transfused cells. Antibodies could be present in the recipient at birth (anti-DEA 7 antibodies) or the recipient could have developed antibodies following a previous transfusion that contains a foreign antigen (usually 4–14 days post-transfusion). DHTRs usually cause extravascular hemolysis and patients can develop hyperbilirubinemia or bilirubinuria with an unexpected drop in the PCV. Clinical signs are usually mild with a transient fever, but signs could be clinically undetectable. With an expected decrease in the PCV, it might be difficult to differentiate a DHTR from erythrocyte destruction due to the underlying cause of anemia. Therapy is not generally indicated (Bracker *et al.* 2005).

Delayed Non-Immunologic Reactions

Infectious Disease Transmission

Transmission of an infectious agent to the recipient can occur if the blood donor is carrying an infectious disease. Screening blood donors prior to blood donation is essential in the prevention of delayed infectious disease transfusion reactions. Recently, the ACVIM published a consensus statement on screening canine and feline blood donors for infectious agents (Wardrop *et al.* 2016). For dogs, the infectious agents that can be transmitted by transfusion include babesiosis, ehrlichiosis, leishmaniasis, and brucellosis. Although heartworm disease (*Dirofilaria immitis)* is common in some geographic regions of the United States and may be included in the screening of donors, heartworm disease cannot be transmitted via transfusion (Wardrop *et al.* 2016).

Conclusion

Numerous factors influence what blood products are available to veterinarians across the world, including proximity to commercial blood banks and access to blood donors, centrifugation equipment and blood storage equipment. By having a wide knowledge of commonly used blood products and their indications, the veterinary professional can select products that are most suitable to the patient and most practical for the veterinary practice. Administering blood products requires close inspection of blood products and serial examination of transfusion recipients. Pre-transfusion testing and early recognition of a transfusion-related complication are important to minimize risks to our critically ill blood product recipients.

References

1. Abrams-Ogg ACG, *et al. Am J Vet Res.* 1993;54:635.
2. Allain JP, *et al. Vox Sanguinis.* 1983;44:246.
3. Baskurt OK. *Clin Hemorheol Microcirc.* 1999;20:181.
4. Benjamin RJ, McLaughlin LS. *Transfusion.* 2012;52:9SS.
5. Blake RR, *et al. J Am Vet Med Assoc.* 2017;250:191.
6. Bracker KE, *et al. Compendium.* 2005;500.
7. Callan MB, *et al. J Am Anim Hosp Assoc.* 1996;32:303.
8. Cummings KA, *et al. J Vet Emerg Crit Care.* 2016;26:639.
9. Fung MK, *et al. AABB Technical Manual.* 18th ed. Bethesda, MD: AABB; 2014.
10. Gladwin MT, *et al. Curr Opin Hematol.* 2009;16(6):515.
11. Hann L, *et al. J Vet Intern Med.* 2014;28:1830.
12. Herring JM, *et al. J Vet Intern Med.* 2013;27:1020.
13. Herrmann K, Haskins S. *J Vet Emerg Crit Care.* 2005;15(1):26.
14. Jobes D, *et al. Transfusion.* 2011;51:43.
15. Kerl ME, *et al. J Am Anim Hosp Assoc.* 1993;202:1495.
16. Miller TE. *Periop Med.* 2013;2:13.
17. Muro SM, *et al. J Vet Intern Med.* 2017;31:410.
18. Obrador R, *et al. J Vet Emerg Crit Care.* 2015;25:187.
19. Patterson J, *et al. J Vet Intern Med.* 2011;25:927.

20. Price GS, *et al. J Vet Intern Med.* 1988;2:126.
21. Raat NJ, *et al. Crit Care Med.* 2005;33:39.
22. Solomon SB, *et al. Blood.* 2013;121:1663.
23. Tavazzi B, *et al. Eur J Biochem.* 2000;267:684.
24. Tocci LJ. *Vet Clin North Am Small Anim Pract.* 2010;40:485.
25. Tsuchyia R, *et al. J Vet Med Sci.* 2003;65:825.
26. Waddell L, *et al. J Vet Emerg Crit Care.* 2001;11:23.
27. Walker JM. Component therapy. In: Yagi K, Holowaychuk MK, eds. *Manual of Veterinary Transfusion Medicine and Blood Banking.* Ames, IA: Wiley-Blackwell; 2016:13–26.
28. Walker RH, ed. *AABB Technical Manual.* 11th ed. Bethesda, MD: AABB; 1993:61.
29. Walton JE, *et al. J Vet Intern Med.* 2014;28:571.
30. Wang D, *et al. Transfusion.* 2012;52(6):1184.
31. Wardrop KJ. *Vet Clin North Am Small Anim Pract.* 1995;25:1263.
32. Wardrop KJ, *et al. J Vet Intern Med.* 1994;8:253.
33. Wardrop KJ, *et al. J Vet Intern Med.* 2016;30:15.
34. Wong C, Haskins SC. *J Vet Emerg Crit Care.* 2007;17:32.

acvim383

Anti-Platelet Therapy: What is New?

John M. Thomason, DVM, MS, DACVIM
Starkville, MS, USA

INTRODUCTION

Thromboembolism is a significant complication in many commonly encountered diseases, and can be a devastating sequel to otherwise treatable conditions. Platelets play an essential role in the hemostatic process and, consequently, are associated with thrombus formation. Platelets adhere to denuded vascular subendothelium, recruit additional platelets and cells, aggregate, and provide the catalytic surface for thrombin production and fibrin formation. Therapy to prevent unwanted thrombus formation and thromboembolic crises is essential in the management of hypercoagulable patients. Unfortunately, many of the medications used in veterinary medicine that inhibit or modulate coagulation factors, such as the heparins, are cost prohibitive, only effective when administered by injection or require frequent drug monitoring, and are therefore poor choices for long term at home therapy. Veterinarians often resort to anti-platelet therapy in the management of patients at risk for thromboembolic complications, because many anti-platelet medications are inexpensive, require minimal drug monitoring, and can be given orally. The aim of this discussion is to review the anti-platelet therapies that are currently being used or being considered for use to inhibit platelet function and reduce thromboembolic complications in hypercoagulable dogs and cats.

ASPIRIN

One of the most commonly used anti-platelet medications, aspirin (acetylsalicylic acid) is highly effective at inhibiting both human and canine platelet function at high enough doses.[1,2] Aspirin blocks one of the key pathways of platelet activation, thromboxane A2 synthesis. Thromboxane A2 is synthesized from arachidonic acid that has been released from phospholipids in the platelet membrane.[3,4] Once mobilized, arachidonic acid is converted into prostaglandin H2 by cyclooxygenase (COX)-1 and COX–2 and subsequent production of biologically active prostaglandins, particularly thromboxane A2 and prostacyclin.[5] Unlike thromboxane A2, which is primarily produced by platelets to stimulate vasoconstriction and enhance aggregation, prostacyclin (prostaglandin I2) originates from the vascular endothelium and causes vasodilation and inhibition of platelet aggregation.[3] These two prostaglandins work in concert to maintain a balanced, normal hemostatic system.

Aspirin inhibits platelet function by irreversibly acetylating the hydroxyl group on the COX enzyme.[4,5] This acetylation will permanently block arachidonic acid from reaching the active site of the COX enzyme, and prevent platelet thromboxane A2 synthesis.[2,4,5] Since platelets do not contain a nucleus, and are unable to synthesize new and functioning COX enzymes, aspirin-induced platelet dysfunction will persist for the lifespan of the platelet.[6] Thromboxane A2 produced by platelets via the actions of COX-1 is sensitive to inhibition by aspirin. Prostacyclin produced by the vascular can be derived through the actions of both COX-1 and COX-2,[2,3] and the actions of COX-2 of the vascular endothelium are less sensitive to inhibitory actions of aspirin at anti-platelet doses.[7] Potentially allowing continued COX-2 dependent prostacyclin synthesis in the face of anti-platelet doses of aspirin.[2]

Aspirin-induced inhibition of the COX enzyme not only inhibits the prostaglandins involved in platelet function, but also inhibits all prostaglandins that play key roles in mediating inflammation, renal blood flow, and gastrointestinal wall protection. Aspirin's ability to inhibit the COX enzyme will also inhibit pro-inflammatory prostaglandin synthesis, making this medication an effective anti-inflammatory therapy. Unfortunately, in both humans and dogs, by inhibiting the COX enzyme and decreasing prostaglandin synthesis, aspirin can also cause several life-threatening side effects that include gastrointestinal hemorrhage and renal failure.[8-10] A 'high' dose of aspirin will consistently inhibit platelet function in dogs, but can lead to multiple gastrointestinal and renal side effects, which makes this aspirin dose undesirable as a preventative anti-platelet therapy.

A high dose of aspirin permanently inhibits COX function in all cells that express this enzyme, including vascular endothelial cells. However, unlike the higher, anti-inflammatory dose, it has been suggested that a 'low' dose of aspirin (0.5-1 mg/kg once daily) may potentially inhibit platelet thromboxane synthesis[11] without permanently inhibiting COX function of the vascular endothelium, allowing prostacyclin synthesis to continue.[1] Lower doses will also minimize the side effects that occur with high dose aspirin. In fact, based on endoscopic evaluation, when administered concurrently with an immunosuppressive dose of prednisone, there was no difference in gastrointestinal lesions between dogs receiving low dose aspirin (liquid suspension) and dogs receiving a placebo, although the concurrent administration of prednisone and low dose aspirin did increase the frequency of mild, self-limiting diarrhea.[12] Aspirin is however unstable in liquid suspension, and becomes inactivated within minutes of suspension,[13] so the systemic and antiplatelet effects of low dose aspirin in liquid form are uncertain.

Unfortunately, despite preventative therapy, some dogs treated with aspirin will continue to develop thromboembolic complications.[14] A similar phenomenon occurs in humans that are receiving low doses of aspirin, suggesting that these patients are poorly responsive to the anti-platelet effects of aspirin (so-called "aspirin resistant"). Aspirin resistance is a poorly defined term that describes the inability of aspirin to protect patients from thrombotic complications. The true incidence of aspirin resistance in humans is unknown, but the published incidence ranges from 8–45% of human patients.[2] In dogs, it has been shown that only 33% of healthy dogs treated with low dose oral aspirin (1 mg/kg once daily) consistently had impaired platelet function as measured using platelet function analyzer (PFA-100®) closure times; however, when groups were separated based on the presence or absence of inhibition of platelet function, there was no difference in thromboxane synthesis between groups.[11] However, again, thromboxane synthesis in dogs treated with low dose aspirin appears to be independent of whether or not aspirin impaired platelet aggregometry.[15] Since canine patients inconsistently respond to low dose aspirin, it is likely that, similar to humans, dogs also experience aspirin resistance. The mechanism and true incidence of aspirin resistance in dogs is, however, unknown.

Although there are many proposed mechanisms of aspirin resistance, one potential mechanism is aspirin underdosage. Previous studies have reported that aspirin-associated platelet dysfunction in dogs appears to be highly dose-dependent.[15,16] While low-dose aspirin has inconsistent and variable effects in individual dogs, platelet function in dogs has been shown to be consistently inhibited by high doses of aspirin.[16,17] Using both PFA-100® and optical aggregometry, previous studies have demonstrated that 28% to 57% of dogs treated with low-dose aspirin (1 mg/kg, q24h) were considered non-responders, while all dogs were considered aspirin responders when treated with a much higher dose (10 mg/kg, q12h).[17] Furthermore, when canine platelets are exposed to aspirin *in vitro*, at a concentration that consistently inhibits activation, platelet function is consistently markedly inhibited.[15] Recently a study suggests that, in healthy dogs, an oral aspirin dosage of 2 mg/kg once daily will consistently and adequately inhibit platelet function without decreasing the degree of prostacyclin synthesis significantly more than lower aspirin dosages.[18] Additional studies will be required to determine if this higher aspirin dosage is associated with an increased risk of adverse drug effects.

ADP Receptor Antagonists

There are several pathways of platelet activation that are independent of arachidonic acid, including pathways mediated by ADP. Following release from platelet granules, ADP interacts with two platelet G-protein-linked purinergic receptors, $P2Y_1$ and $P2Y_{12}$. The $P2Y_1$ receptor is widely distributed in many cells, including the platelet.[19] Activation of this receptor initiates influx of calcium and subsequent actin polymerization, causing weak and transient platelet activation.[19] The $P2Y_{12}$ receptor is primarily located on the platelet, and amplifies the platelet activation effects of ADP initiated by the $P2Y_1$ receptor. Additionally, the $P2Y_{12}$ receptor is responsible for ADP-dependent amplification of other agonists and full activation of the GPIIb/IIIa complex.[20,21] Normal ADP-induced platelet aggregation is dependent on the co-activation of both $P2Y_1$ and $P2Y_{12}$, and inhibition of either receptor will decrease platelet activation.[19] Due to its platelet specific location and strong response to ADP, the $P2Y_{12}$ receptor has become the primary molecular target of anti-platelet therapy, but inhibition of the $P2Y_1$ receptor may be a target for future anti-platelet drugs.[19]

Ticlopidine was one of the first ADP receptor antagonists to be used in both human and veterinary medicine. In cats, only a high dose of ticlopidine consistently inhibits platelet function, but associated dose dependent gastrointestinal side effects preclude the use of this medication as a reliable, safe anti-platelet therapy.[22] Due to several severe side effects identified in people on ticlopidine, including neutropenia and hepatotoxicity, clopidogrel has largely replaced ticlopidine.[21]

Clopidogrel is an irreversible ADP receptor antagonist that permanently inhibits the function of circulating platelets. Similar to humans, clopidogrel in dogs requires hepatic conversion into the active metabolite. When measured by whole blood aggregation, platelet function is inhibited within 3 hours of clopidogrel administration in most dogs.[23] Recently, a small prospective study evaluated the safety of clopidogrel (single oral loading dose of 10 mg/kg, followed by a maintenance dose of 2–3 mg/kg once daily) in canine IMHA patients, and compared the clinical efficacy of single agent clopidogrel to single agent low-dose aspirin (0.5 mg/kg orally once daily) or a combination of low-dose aspirin and clopidogrel. When compared to low-dose aspirin, clopidogrel appeared to be an equally safe treatment option, with no difference in the incidence of gastrointestinal side effects and no evidence of excessive hemorrhage. There were no significant differences in survival rate, duration of hospitalization, transfusion requirements, thromboembolism, or rate of relapse between the three treatment groups, the study had insufficient power (only 8 cases per group) to truly determine whether each treatment option actually provided a comparable outcome.[14]

Clopidogrel has also been used for the prevention and treatment of canine aortic thromboembolism. In one case report, clopidogrel was used as the sole therapy for 6 months in one dog with ATE, but the patient showed progressive ambulatory dysfunction.[24] In a second small case series, three dogs with ATE were treated with clopidogrel and aspirin, and clinical signs improved in one dog, slowly progressed in the second, and acutely reoccurred in the third. None of these 3 dogs were treated with clopidogrel as the sole anti-platelet therapy.[25] Given the small number of cases published to date, a true determination of the value of clopidogrel for the treatment and prevention of canine ATE cannot be determined.

Clopidogrel is a prodrug that require metabolism by hepatic cytochrome p450 enzymes to form active metabolites that irreversibly block the platelet $P2Y_{12}$ receptor. Concurrent administration of medications requiring metabolism via the cytochrome p450 pathway could also inhibit clopidogrel metabolism to its active form. In humans, there is controversy regarding the risk of cardiovascular adverse events during the co-administration of clopidogrel and proton pump inhibitors, specifically omeprazole. Some, but not all, studies have demonstrated that concurrent use of proton pump inhibitors will reduce the anti-platelet effects of clopidogrel.[26] In contrast, a recent study found no evidence that the concurrent use of clopidogrel and omeprazole significantly alters the anti-platelet effects of clopidogrel in healthy dogs. Although blood concentrations of an inactive clopidogrel metabolite increased when clopidogrel was administered concurrently with omeprazole.[27]

Because of the many potential mechanisms of resistance to the anti-platelet effects of both drugs, either aspirin or clopidogrel alone may not be sufficient to reliably prevent thromboembolic complications in all at-risk patients. Combined aspirin and clopidogrel 'dual' anti-platelet therapy may therefore be more predictably reliable at decreasing platelet function. When compared to single agent therapy (aspirin or clopidogrel alone), dual anti-platelet therapy in humans often results in a reduction in the risk for major cardiovascular events, myocardial infarction, and stroke.[28] Unfortunately, dual anti-platelet therapy in people also significantly increases the risk for major bleeding events when compared with single agent therapy, potentially outweighing the preventative benefits of combination therapy.[28] In veterinary medicine, when comparing dual anti-platelet therapy with single agent therapy in canine IMHA patients, there were no significant differences in hemorrhagic events between the treatment regimens; however, this study did not have ample power to detect significant clinical difference in hemorrhagic events.[14]

GPIIB/IIIA Receptor Antagonists

Regardless of the inciting agonist, the expression and transformation of functionally active platelet GPIIb/IIIa and subsequent fibrinogen binding is the final common pathway of platelet aggregation, and the GPIIb/IIIa receptor is, therefore, a promising target for anti-platelet mediations.[20] In comparison to aspirin, GPIIb/IIIa receptor antagonists are platelet specific, and are also more potent inhibitors of platelet function. GPIIb/IIIa has become the target of several anti-platelet drugs including monoclonal antibodies against the receptor.[29] The first GPIIb/IIIa receptor antagonist used in human medicine was abciximab. Following intravenous administration, abciximab rapidly binds to platelets and inhibits platelet function in a dose-dependent manner. Despite the potential benefits of abciximab in dogs, there have been no studies in naturally occurring diseases that have evaluated the clinical efficacy or safety of this medication. Additionally, the high cost of the medication probably contributes to the lack of clinical veterinary usage. Eptifibatide and tirofiban are more recently developed GPIIb/IIIa receptor antagonists, and have been evaluated in dogs as models for coronary artery disease, and are effective at inhibiting platelet function and thrombus formation in a dose-dependent manner.[30] However, there have been no studies evaluating the clinical efficacy and safety of tirofiban and eptifibatide in dogs with naturally occurring disease.

Conclusion

Anti-platelet medications are a common treatment option in patients that are at risk for thromboembolic complications. Unfortunately, there is currently not a single superior anti-platelet medication that effectively and predictably inhibits platelet function in all patients at risk for thromboembolism, suggesting a potential role for dual or combination anti-platelet therapy, or the concurrent use of other anticoagulant drugs. Low-dose aspirin has been the most common anti-platelet therapy administered in both dogs and cats, but more recently ADP-receptor antagonists such as clopidogrel have also attained popular usage. With a better understanding of platelet function and structure, more effective and safer therapies can be initiated to improve the survival of patients at risk of thromboembolic complications.

References

1. Rackear D *et al. J Am Anim Hosp Assoc.* 1988;24(1):23.
2. Patrono C *et al. Chest.* 2001;119:39S.
3. McAdam B *et al. Proc Natl Acad Sci USA.* 1999;96:272.
4. Bjorkman D. *Am J Med.* 1998;105(1B):8S.
5. Guthikonda S *et al. J Thromb Haemost.* 2007;5:490.
6. Pedersen AK *et al. N Engl J Med.* 1984;311:1206.
7. Topper JN *et al. Proc Natl Acad Sci USA.* 1996;93:10417.
8. Data J *et al. Am J Physiol.* 1976;230:940.
9. Johnston S *et al. J Vet Intern Med.* 1995;9:32.
10. Shaw N *et al. J Am Anim Hosp Assoc.* 1997;33:215.
11. Dudley A *et al. J Vet Intern Med.* 2013;27:141.
12. Graham AH *et al. J Vet Intern Med.* 2009;23:482.
13. Bica K *et al. Phys Chem Chem Phys.* 2010;12:2011.
14. Mellett A *et al. J Vet Intern Med.* 2011;25:71.
15. Haines J *et al. Am J Vet Res.* 2016;77:174.
16. Thomason J *et al. J Vet Intern Med.* 2011;25:1106.
17. Thomason J *et al. J Vet Intern Med.* 2016;30:1022.
18. McLewee N *et al. J Vet Intern Med.* 2016;30:1465.
19. Hechler B *et al. Purinergic Signal.* 2011;7:293.
20. Ji X *et al. J Hematol Oncol.* 2011;4:44.
21. McNicol A *et al. J Pharmacol Sci.* 2003;93(4):381.
22. Hogan D *et al. Am J Vet Res.* 2004;65:327.
23. Brainard B *et al. Am J Vet Res.* 2010;71:822.
24. Winter R *et al. J Vet Cardiol.* 2012;14:333.
25. Lake-Bakaar G *et al. J Am Vet Med Assoc.* 2012;241:910.
26. Bhatt D *et al. N Engl J Med.* 2010;363:1909.
27. Thames B *et al. J Vet Pharmacol Therap.* 2016; doi: 10.1111/jvp.12340.
28. Zhou Y *et al. PLoS ONE.* 2012;7(2):e31642.
29. Scarborough R *et al. Circulation.* 1999;100:437.
30. Rukshin V *et al. Circulation.* 2002;105:1970.

acvim384

Altered Fibrinolysis in the Clinical Setting

Armelle deLaforcade, DVM, DACVECC
North Grafton, MA, USA

Introduction

Fibrinolysis is a process that describes clot degradation and is triggered by the production of plasmin from the inactive zymogen plasminogen. Plasmin generation on the surface of the fibrin clot results from generation of plasminogen activators. Tissue plasminogen activator (released by endothelial cells) and urokinase (from monocytes, macrophages and urinary epithelium) lead to degradation of the fibrin clot once endothelial injury has resolved. As with most aspects of hemostasis, fibrinolysis is tightly controlled, and checks and balances exist to balance fibrin formation with fibrin degradation. Inhibitors of fibrinolysis are responsible for preventing excessive fibrin deposition and include plasminogen activator inhibitors 1 & 2 (PAI-1, PAI-2), alpha 2-antiplasmin, and thrombin-activatable fibrinolysis inhibitor (TAFI). These work primarily by targeting both plasmin and the plasminogen activators, thus controlling dissolution of the fibrin clot.

As is often the case, pro- and anti-fibrinolytic factors exert some effect beyond clot degradation, highlighting the complexity of hemostasis. For example, plasmin reduces fibrinogen, and factors Va and VIIIa, while fibrin degradation products have both pro- and anti-inflammatory effects. Dysfibrinogenemias and afibrinogenemia have been described in people and are rarely reported in animals; however, acquired disorders of fibrinolysis are gaining attention for their potential role in morbidity and mortality in a variety of disease processes.

It has long been observed that dogs are hyperfibrinolytic compared to people. Older experimental models indicate that experimentally induced thromboemboli resolve more rapidly in dogs compared to people, and other studies have shown a greater concentration of circulating tPA in dogs compared to people. Given this information, it isn't clear that the dose for effective inhibition of fibrinolysis in dogs can be directly extrapolated from the human literature. In a recent study where varying concentrations of aminocaproic acid were added to both canine and human plasma, considerable higher concentrations of aminocaproic acid were required for complete inhibition of fibrinolysis (assessed using thromboelastography) in dogs than were required for the same effect in human plasma.

Fibrinolysis in Greyhounds

In considering how to manage fibrinolytic alterations in dogs with varying diseases, it is helpful to evaluate what is already known in the greyhound. Altered hemostasis and a greater tendency for delayed bleeding after provocative procedures have been clearly demonstrated in the greyhound, and accelerated fibrinolysis is a significant contributor to this phenomenon. Based on a study showing reduced postoperative bleeding compared to placebo in greyhounds administered aminocaproic acid, many hospitals have adopted preoperative protocols for this breed of dog to minimize the likelihood of postoperative complications. Additional studies are needed to confirm the dosing needed for optimized inhibition of fibrinolysis in non-greyhound dogs.

Disseminated Intravascular Coagulation (DIC)

The traditional example of a hyperfibrinolytic state occurs as a result of severe systemic inflammation, triggering tissue factor exposure and excessive microvascular fibrin formation. Consumption of inhibitors leading to a consumptive coagulopathy leads to concurrent microvascular fibrin deposition and clinical evidence of bleeding. DIC as a complication of severe illness contributes to reduced tissue oxygen delivery and the development of multiple organ failure. Markers of fibrinolysis such as fibrin degradation products or D-dimer are incorporated into a scoring system for identification of DIC in people, and are expected in animals with a diagnosis of DIC.

Fibrinolysis & Cancer

Hemostatic alterations in people with cancer have been an active area of interest for several decades. Fibrinolysis, in particular, has been investigated for its potential contribution to metastasis and as a prognostic marker in several forms of cancer, including prostate cancer. The role of angiogenesis in providing nutrients and oxygen to promote tumor cell growth is widely accepted, and inhibition of angiogenesis is considered an important aspect of anti-cancer therapy. In addition to cleaving fibrin to split products, plasmin can exert its effect on other proteins such as metalloproteinases and extracellular matrix proteins, and these actions stimulate cell invasion and migration. The active form of TAFI downregulates fibrinolysis through inhibition of plasmin activation from plasminogen, but also has anti-angiogenic and anti-metastatic properties including reduced endothelial proliferation, reduced cell migration and invasion.

In two studies of canine intracranial neoplasia (meningioma and glioma), D-dimer deposition was among the factors investigated using immunolabelling to assess the role of fibrin deposition and breakdown. These studies both noted extravascular D-dimer deposits providing evidence for extravascular activation of coagulation and intra-tumoral fibrinolysis. Another study of 71 dogs with malignant neoplasia confirmed hypercoagulability as the most common hemostatic dysfunction identified, but also noted that dogs with metastatic disease had significantly increased fibrinogen and D-dimer compared to dogs with local invasive and local non-invasive cancers. Finally, it has recently been shown that dogs with spontaneous hemoperitoneum have increased fibrinolysis and that this hemostatic alteration correlates to plasma lactate concentration (and shock). While the assumption that hemorrhage in these dogs is due to a bleeding mass, the potential contribution of accelerated fibrinolysis offers potential therapeutic options that could impact morbidity and transfusion needs in this patient population.

Fibrinolysis & Trauma

Fibrinolysis and reduced fibrinogen are important contributors to the phenomenon referred to as acute traumatic coagulopathy (ATC). The pathogenesis of ATC is complex and multifactorial. As it relates to fibrinolysis, fibrinolysis is thought to occur both due to catecholamine-induced tissue plasminogen activator (t-PA) release, and depletion of antifibrinolytic factors. Some studies suggest an active role for protein C in trauma-associated hyperfibrinolysis. Therapy with anti-fibrinolytics within 3 hours of trauma significantly reduced bleeding-related mortality in people and has led to widespread use of antifibrinolytics (tranexamic acid or aminocaproic acid) in the treatment of human trauma. Despite the findings of the CRASH 2 trial, some remain skeptical about the widespread use of fibrinolysis inhibitors, given that mortality increases when administered greater than 3 hours of trauma. Another argument for more targeted use of antifibrinolytics is the finding of fibrinolysis shutdown, describing a state of deficient clot breakdown characterized by LY30 less than the normal range. Both hyperfibrinolysis and fibrinolysis shutdown are associated with increased mortality in severe trauma.

Several studies have investigated fibrinolysis in veterinary trauma. In a multicenter study of canine trauma, plasmin inhibition as assessed by measurement of antiplasmin and plasminogen activities was shown to be reduced in dogs following severe acute trauma, suggesting a potential role for fibrinolysis in traumatic hemorrhage. Another study of dogs experiencing blunt trauma did not find hemostatic changes consistent with ATC using thromboelastography. One case report of a dog with suspected acute traumatic coagulopathy described hyperfibrinolytic changes using thromboelastography along with clinical bleeding that was not responsive to blood product administration. Both resolved with aminocaproic acid administration.

Fibrinolysis & Hepatic Disease

Extensive investigation of hemostatic alterations in hepatic disease have led to the concept of 'rebalanced hemostasis' where alterations in hemostatic proteins having a procoagulant effect are balanced by alterations in those with an anticoagulant effect. Clinically, spontaneous bleeding is rare in patients with liver disease, but more common following provocative procedures such as liver biopsy. Veterinary studies using thromboelastography to characterize the hemostatic state of dogs with liver disease have identified some hypercoagulable, some normocoagulable and a contingent of dogs with hyperfibrinolysis. Given the inability to accurately predict the bleeding risk in these dogs undergoing hepatic biopsy using conventional coagulation testing, the potential role of fibrinolysis in post-biopsy bleeding is of great interest and currently being studied. The mechanism through which fibrinolysis is upregulated in dogs with hepatic disease is unknown. In people, the reduced synthesis of alpha 2-antiplasmin has been described in chronic hepatic disease.

While viscoelastic testing has heightened the awareness of fibrinolysis in dogs with neoplasia, trauma and hepatic disease, it may be that manipulation of the assay can increase the sensitivity of TEG for the detection of accelerated fibrinolysis. In the Fletcher study of dogs with spontaneous hemoperitoneum, addition of t-PA to the standard TEG assay resulted in profound changes that highlighted increased fibrinolysis in this patient population. Additional studies investigating fibrinolysis in canine diseases using this methodology may be helpful to identify more subtle fibrinolytic change that may contribute to clinical bleeding.

Fibrinogen

Interestingly, coagulation factors do not decrease uniformly during bleeding and fibrinogen is the first to fall to a critically low level. Because of this, it is considered one of the more vulnerable coagulation factors in surgical and trauma patients. Decreased fibrinogen levels have been reported in feline patients with HL (12/19 patients in one study) and in dogs with CPSS, acute liver disease and CH/cirrhosis. Whether these changes in fibrinogen correlate with fibrinolytic changes is not known.

Hypofibrinolysis

While excessive fibrinolysis and its role in hemostatic alterations in veterinary diseases is currently of great interest, the role of hypofibrinolysis and resulting impaired clot dissolution is unknown. In people, hypofibrinolysis is recognized as a cause of major thrombosis in antiphospholipid syndrome, alcoholic liver disease, hypothyroidism, and multiple myeloma. Additionally, in studies of human trauma, a phenomenon called fibrinolysis shutdown has been described, where a LY30 below the reference range results in increased mortality in this patient population. While acquired hypofibrinolysis has not been widely investigated in veterinary medicine, a few studies are worthy of mention. In a study of 35 dogs with various diseases, dogs with malignant neoplasia had the highest activities of TAFI, and dogs with bacterial sepsis as a group had significantly increased activity of TAFI compared to controls. This suggests one mechanism through which hypofibrinolysis may contribute to an overall hypercoagulable state in sepsis.

Therapeutic Implications

Heightened awareness of increased fibrinolysis in disease states as well as the findings of the CRASH-2 trial in people have led many to consider the incorporation of fibrinolysis inhibitors in treatment of clinical bleeding. The wide safety margin and relative inexpense of lysine analogs (aminocaproic acid and tranexamic acid) have led to its empirical use in veterinary medicine. Presurgical treatment with aminocaproic acid has been recommended in greyhounds to prevent delayed postoperative bleeding. Isolated case reports have described cases where aminocaproic acid administration has both corrected the thromboelastographic abnormalities and reduced bleeding in dogs. Further studies will be needed before the effect of these drugs on morbidity and mortality can be established.

References

1. Bazzi ZA, *et al. Clin Exp Metastasis*. 2017;34(2):155. DOI 10.1007/s10585-017-9837-y.
2. CRASH-2 trial collaborators. *Lancet*. 2010;376(9734):23.
3. Brown JC, *et al. Am J Vet Res*. 2016;77(11):1258.
4. Fletcher DJ. *J Vet Emerg Crit Care*. 2016;26(1):41.
5. Leeper CM. *J Trauma Acute Care Surg*. 2017;82(1):27.
6. Chapin JC, *et al. Blood Rev*. 2015;29(1):17.
7. Fletcher DJ, *et al. J Vet Emerg Crit Care*. 2012;22(3):289.

8. Holowaychuk MK, *et al. J Vet Emerg Crit Care*. 2014;24(1):93.
9. Fletcher DJ, *et al. J Vet Emerg Crit Care*. 2016;26(1):41.
10. Font C, *et al. Vet J*. 2015;206(3):426.
11. Kelly D, *et al. J Vet Intern Med*. 2015;29(4):1053.
12. Fletcher JD, *et al. Am J Vet Res*. 2014;75(8):731.

acvim385

CT Angiography of Portal Venous Anomalies in Dogs & Cats

Brian A. Scansen, DVM, MS, DACVIM (Cardiology)
Fort Collins, CO, USA

Introduction

Angiography refers to the administration of contrast agents into the vascular system. This has traditionally been performed for fluoroscopic imaging, but can also be utilized for cross-sectional imaging (computed tomographic angiography [CTA] or magnetic resonance angiography). The vasculature of the liver is complex and 3-dimensional, yet traditional imaging of the liver and vasculature has relied on 2-dimensional imaging techniques such as radiography, fluoroscopy, and ultrasonography. The adoption of CT as an advanced diagnostic tool has been substantial over the last 30 years, with increased use of CTA in the last decade. The benefit of cross-sectional imaging, such as CTA or MRA, is the avoidance of selective central catheterization, as only peripheral venous access is required for these techniques. Further, the ability to generate 3-dimensional models from high-resolution CTA has allowed a level of anatomic understanding not previously available. Within the vasculature, cross-sectional imaging has been reported for detailed characterization of vascular malformations, such as portocaval communications,[1] vascular ring anomalies,[2] and arterial-venous fistulae[3]. The focus of this lecture is on the technical aspects of CTA in the hepatic vasculature, providing a framework for interpretation of hepatic vascular CTA, with case-based examples of anomalies encountered in dogs and cats.

CT Technology

Early-generation CT scanners lacked sufficient spatial and temporal resolution to characterize vascular anatomy. The development of multi-detector CT (MDCT), also known as multi-slice CT, has improved both spatial and temporal resolution, such that highly detailed visualization of vascular structures is now clinically feasible. A detector is the portion of the scanner that receives the photons generated by the X-rays passing through the patient. A single detector is comprised of many elements along its circumference, but is limited to a narrow width in order to resolve structures of interest. Multiple detectors allow the radiation effect to be collected along a broader slab width, known as the z-axis, which increases the slab thickness that can be scanned at one time. The first-generation MDCT scanners used 4 detectors to improve z-axis thickness, with subsequent generations increasing to 8, 16, 64, 128, 320, and even 600 detectors. A MDCT with 320 detectors can typically cover the z-axis of a human liver allowing the entire organ to be evaluated in one spin of the CT gantry. It should be noted that the length of the z-axis that can be imaged by a given scanner is dependent on both the number of detectors as well as the width of each detector such that two different 64-slice scanners may not have the same z-axis coverage. As an example, a 64-slice scanner with detectors of 1.25 mm width could cover a z-axis thickness of 80 mm while a 64-slice scanner from another manufacturer with detectors of 1.0 mm width would yield a z-axis thickness of 60 mm with each gantry rotation. Both the width in the z-axis that can be covered by each rotation, as well as the speed of gantry rotation, impact the final image quality of the scanner, as will be discussed below.

Image Resolution in CTA

Spatial resolution is defined as the ability to resolve structures that lie next to one another on an image. Spatial resolution is predominately affected by the size of the detector and the spacing between detectors; the detector must be smaller than an object to resolve it and detectors must be sufficiently close together to resolve two objects that lie close to one another in the patient. For CT of a large organ, the z-axis spatial resolution is equally important to allow reconstruction of the image in all planes without distortion and for this reason MDCT scanners with a high number of detectors are desired to optimize imaging. Spatial resolution is also improved by advancements in image detector technology and reconstruction software. New high-definition scanners are available that can achieve spatial resolution in the x- and y-axes to 0.23

mm using novel detector material and advanced reconstruction algorithms. As dogs and cats are smaller in size than humans, spatial resolution of small structures (e.g., hepatic arterial branching) remains a challenge in these species with all but the newest generations of scanners.

Temporal resolution describes the ability of the scanner to resolve a structure moving in time and is best measured as the time required to acquire successive images. Temporal resolution is dependent on gantry rotation speed, mode of image acquisition, gating, and pitch. The fastest scanners currently on the market can perform a full gantry rotation in ~250 ms, which means that they can resolve an image in roughly 125 ms since a CT image is created/extrapolated from ½ of a gantry rotation, as the X-ray source moves from 0 degrees to 180 degrees. Adding an additional X-ray source placed 90 degrees to the other is one strategy to minimize the time required to circumferentially rotate around the patient and improve temporal resolution. These dual-source CT scanners have two unique X-ray sources (often of different energy spectra) as well as 2 sets of detectors and need to only rotate 90 degrees in order to acquire a half gantry rotation, reducing temporal resolution down to 66 ms. Pitch is defined as the distance the table advances per gantry rotation divided by the total thickness of the acquired slab (in the z-axis). With MDCT, the thickness of the slab is the sum of the width of all detectors activated during acquisition. A pitch greater than 1 implies gaps in the z-axis coverage as the table moved too rapidly to allow the scanner to image every slab in the field of view. Some degree of overlap in the z-axis is preferred to optimally reconstruct the full path of a blood vessel images of the heart. Fast scanners are able to image the organ of interest rapidly enough to avoid artifacts of respiration, patient movement, etc.

Timing of Contrast Administration

Timing of contrast administration relative to scanning the structure of interest can be challenging, but is paramount to obtaining a useful imaging study. CTA is achieved by delivering a bolus of contrast into a peripheral vein (typically cephalic in the dog or cat) and then initiating the CT scan when the contrast reaches the vascular structure of interest. Multiple phases are performed to highlight both the arterial and venous anatomy. A delayed phase is also commonly obtained roughly 3 minutes after the arterial and venous phases. Various protocols for contrast timing are available, with the two primary methods being test scan and bolus-tracking. A test scan involves monitoring a single slice that includes the vessels of interest (typically around the porta hepatis for liver scans) and then measuring the time it takes for each vascular structure to reach peak opacification. The true scans are then planned based on these times so that the areas of interest are scanned when the contrast reaches the desired vasculature (arterial and venous). Alternatively, bolus-tracking techniques employ a machine-based monitoring scan that automatically initiates the full scan once an increase in radiodensity (operator determined) is noted in the structure being monitored. For the liver, bolus-tracking is often performed to initiate the arterial phase, with the scan triggered when the contrast bolus reaches the descending aorta. A time delay is then employed after the arterial scan to allow sufficient time for perfusion of the abdominal organs and anticipated venous return (often 20 to 30 seconds in dogs); this second scan is the venous phase of the study and highlights both portal and systemic venous anatomy.

Gated techniques are used to time the acquisition sequences to the cardiac cycle during cardiac CT, thereby freezing cardiac motion in the acquired image series; gating is not needed for vascular CTA of the liver. Respiratory gating is seldom employed, but for slower scanners subjecting the animal to a breath-hold or apneic period during the scan is occasionally needed to avoid movement during scan acquisition (**Figure 1**).

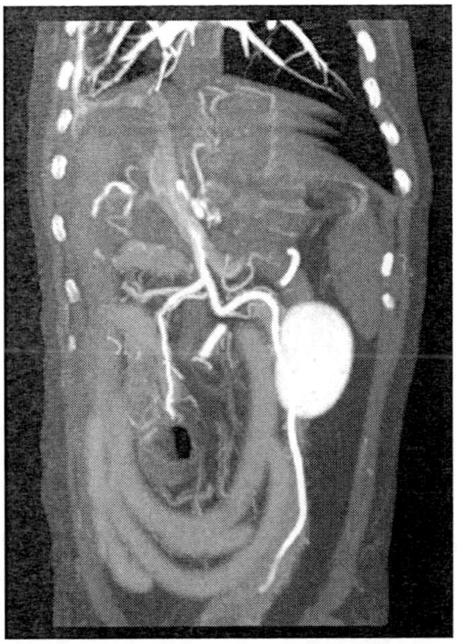

Figure 1. An example of respiratory artifact from a dog during CTA acquisition. Note the stair-step appearance of the diaphragm which occurs with respiration and which prevents visualization of the path of the hepatic artery.

IMAGE RECONSTRUCTION

Viewers that support maximum intensity projections (MIP), minimum intensity projections (MinIP), and 3-dimensional volume rendering with the ability to stack images can improve image interpretation significantly - in particular for vascular studies. Most scanners allow reformatting of the axial images into sagittal or dorsal planar reformats. MIP is a technique whereby the image viewing software creates a single image by projecting the voxel with the highest attenuation value on every view throughout the volume onto a 2-dimensional image. This is particularly useful for vascular contrast studies as the vessels (and bones) are highlighted. When stacks of images are processed in this manner, the entire course of a vessel and its relationship to surrounding structures can be more clearly appreciated. The comparable (though opposite) technique of MinIP maximizes detection of low-density structures from a study by projecting the voxel with the lowest attenuation value on every view onto a 2-dimensional image. This technique is not used much for CTA of the liver, but is useful for evaluating subtle attenuation patterns in the lungs. 3-dimensional volume rendering of a CTA allows for the full spatial relationship of the vasculature to be realized and can greatly enhance surgical or interventional planning. **Figure 2** displays the differences between these techniques from the same arterial phase CTA of a dog. As can be seen, the arterial anatomy is best appreciated in the stacked MIP and 3-dimensional images.

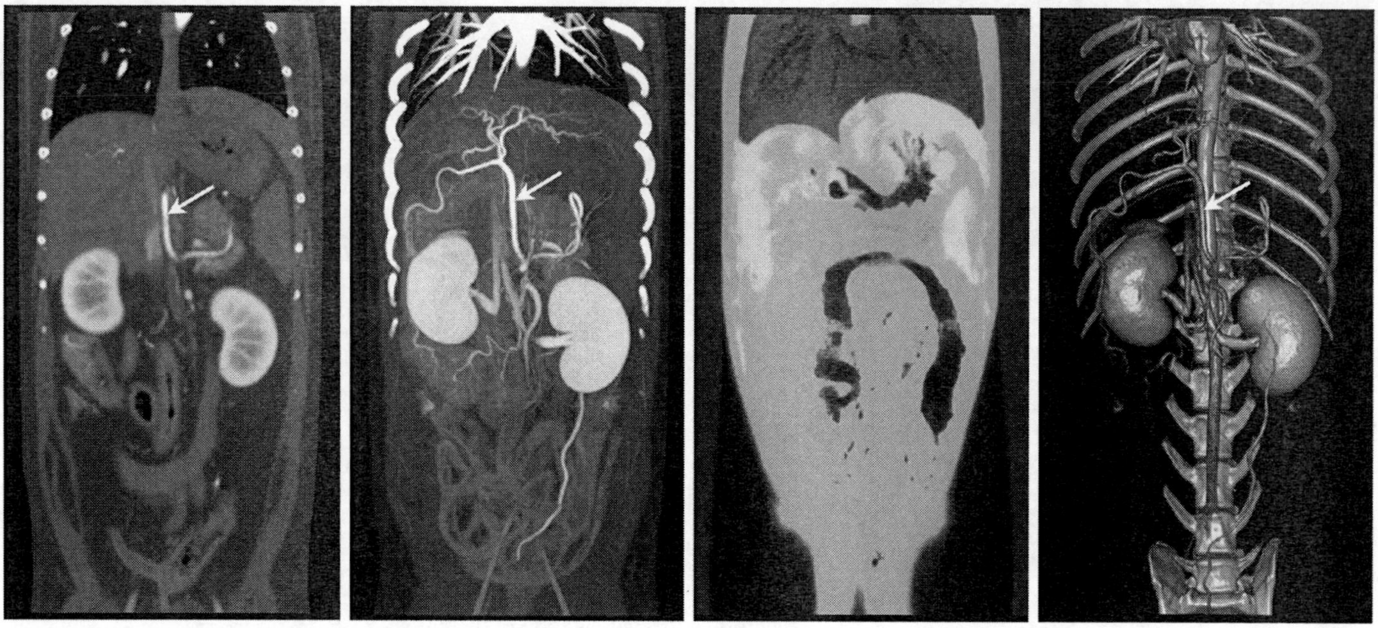

Figure 2. CTA arterial phase planar reformats from the same dog using various image processing techniques. The left most panel shows a thin dorsal reformat which only allows visualization of the proximal common hepatic artery (arrow). Panel B shows a stacked MIP that permits visualization of the full hepatic arterial course. Panel C shows a MinIP, which highlights poorly attenuating structures. Panel D is a volume-rendered 3D reformat of the same study showing the arterial anatomy of this dog.

NORMAL PORTAL VASCULATURE

When evaluating a hepatic CTA, an understanding of the normal anatomy is important to aid interpretation. The portal vein begins as a confluence of the cranial and caudal mesenteric veins before coursing craniad. The gastrosplenic (also termed splenic) vein enters at the left lateral aspect of the portal vein in the mid-abdomen as short gastrosplenic trunk that quickly divides into the left gastric vein, which runs cranial toward the inner curvature of the stomach, and the splenic vein, which branches before entering the spleen. The gastroduodenal vein is the last vein to enter the portal system prior to the liver and it typically enters on the right lateral aspect of the portal vein just before the liver and is composed of the cranial pancreatoduodenal vein and the right gastric and right gastroepiploic veins. Once in the liver, the portal branching begins with a small right divisional branch that enters the right lateral lobe as well as the caudate process, and a left branch that courses cranial to give a tributary to the right medial lobe and a larger left branch that crosses midline and supplies the quadrate process, left medial lobe, left lateral lobe, and papillary process. The arterial supply to the liver begins in the first branch of the abdominal aorta, the celiac artery, which trifurcates into the common hepatic artery, splenic artery, and left gastric artery. The common hepatic artery enters the liver and gives three to five cranial branches to supply the liver lobes before continuing to the right and dividing into the gastroduodenal artery and right gastric artery.[4]

ANOMALIES OF THE PORTAL VASCULATURE

Both CTA and MRA have been employed to evaluate hepatic vascular structures in animals.[1,5-8] Most of the published literature on these modalities has focused on anatomic delineation of portosystemic shunts, though they can be used to investigate any vascular anomaly. After reviewing an interpretative approach for evaluation of the hepatic vasculature by CTA, this lecture will provide examples of and discuss the CTA findings in intra- and extra-hepatic portosystemic shunts of variable anatomy, arterial portal malformations.

REFERENCES

1. Zwingenberger A. *Vet Clin North Am Small Anim Pract*. 2009;39(4):783.
2. Henjes CR, *et al*. *BMC Vet Res*. 2011;7:57.
3. Zwingenberger AL, *et al*. *Vet Radiol Ultrasound*. 2005;46(6):472.
4. Bezuidenhout AJ. *Miller's Anatomy of the Dog*. 4th ed. 2013:428.
5. Frank P, *et al*. *Vet Radiol Ultrasound*. 2003;44(4):392.
6. Zwingenberger AL, *et al*. *Vet Radiol Ultrasound*. 2005;46(1):27.
7. Bruehschwein A, *et al*. *Vet Radiol Ultrasound*. 2010;51(2):116.
8. Nelson NC, *et al*. *Vet Radiol Ultrasound*. 2011;52(5):498.

acvim386

High Resolution Computed Tomography in the Diagnosis of Idiopathic Pulmonary Fibrosis

Brendan M. Corcoran, MVB, PhD, MRCVS
Edinburgh, Scotland, UK

INTRODUCTION

Idiopathic pulmonary fibrosis (IPF) is recognised as the most common idiopathic interstitial pneumonia (IIP) in humans and dogs, and for both species carries a poor to grave prognosis.[1,2] For both dog and human there are shared clinical similarities, which can be used as diagnostic criteria, such as, no other identifiable cause, older than 50 years (>9 in dog), insidious onset of dyspnoea on exertion, duration of symptoms greater than 3 months, inspiratory crackles, normal bronchoalveolar lavage, and characteristic abnormalities on high resolution computed tomography (HRCT).[1,3,4] Furthermore, in the dog there is a particularly high prevalence in the West Highland white terrier (WHWT), but the disease has been reported in other breeds.[1,3,5] There are familial forms of human IPF, but invariably these are rare and clinical signs appear much earlier in life.

While much more is known about HIPF, only recently have we begun to understand in detail the clinical features, diagnosis and treatment of CIPF and as of yet still have limited information on pathology in the dog.[6,7] In human medicine IPF is defined as having the pathology of usual interstitial pneumonitis (UIP), which can be differentiated from a range of IIPs including non-specific interstitial pneumonitis (NSIP) and connective tissue diseases. A definitive set of pathological correlates for CIPF are not yet available, but the majority of changes seen might be more reminiscent of NSIP. If strict definitions are applied that would mean CIPF and HIPF are not the same disease.[3,6]

UIP in humans is characterised at low magnification by areas of fibrosis and honeycombing (cystic fibrotic airspaces) alternating with areas of normal lung, and on histopathology, by the presence of subpleural distribution, marked fibrosis, myofibroblastic foci and mild inflammation.[2] However, there is a reasonably wide spectrum of pathology seen with the IIPs, confounded by patchy distribution, which can make confirmatory pathology, particularly on surgical lung biopsy (SLB), problematic. In dogs, the main pathological findings, from a limited number of dogs, are mature interstitial and alveolar fibrosis with type II pneumocyte hyperplasia, subpleural and peribronchial fibrosis, but in a small number of more severely affected individuals myofibroblastic foci have been seen.[3,6] This suggests a mixed pathology picture of NSIP and UIP in the dog, but without the extensive fibrosis often reported in human UIP IPF. All these pathological changes likely reflect what is seen on HRCT in the dog.

Currently, diagnosis of IPF in humans relies on exclusion of other IIPs, the presence of a UIP pattern on HRCT and specific combinations of HRCT and SLB findings.[2] For the dog, it is the first two that are now readily available diagnostic criteria, but surgical biopsy is rarely, if ever, undertaken.

COMPUTED TOMOGRAPHY (CT) FINDINGS IN CANINE IPF

There have been limited reports on CT and HRCT findings in canine IPF.[1,6,8] This will likely change as more HRCT is performed on veterinary patients. The changes reported include ground glass attenuation, peribronchovascular interstitial changes, bronchial wall thickening, parenchymal bands, subpleural thickening, the interface sign, honeycombing and traction bronchiectasis, with ground glass opacity being the most common finding (Table 1; Figure 1).[1,6,8] Honeycombing and traction bronchiectasis are relatively rare and are more readily seen in severely affected dogs.[8] Applying a CT score, with measure of lung attenuation in Hounsfield units, has identified a moderate positive correlation between CT findings with clinical stage, and negative correlation with survival time.[6] Furthermore, applying densitometry to measure lung attenuation can help with diagnosis of IPF, with a HU value of -702 taken as the cut-off for a likely diagnosis for IPF.[6]

Table 1. Severity grading computed tomographic lung patterns in IPF dogs (adapted from Thierry *et al.* In press)

Grade 0 (normal)	Normal lung
Grade 1 (mild severity)	Ground glass attenuation
Grade 2 (moderate severity)	Mosaic ground glass pattern, focal reticular pattern, pulmonary consolidation or bronchiectasis
Grade 3 (marked severity)	Generalized reticular pattern, honeycombing, traction bronchiectasis or nodular pattern

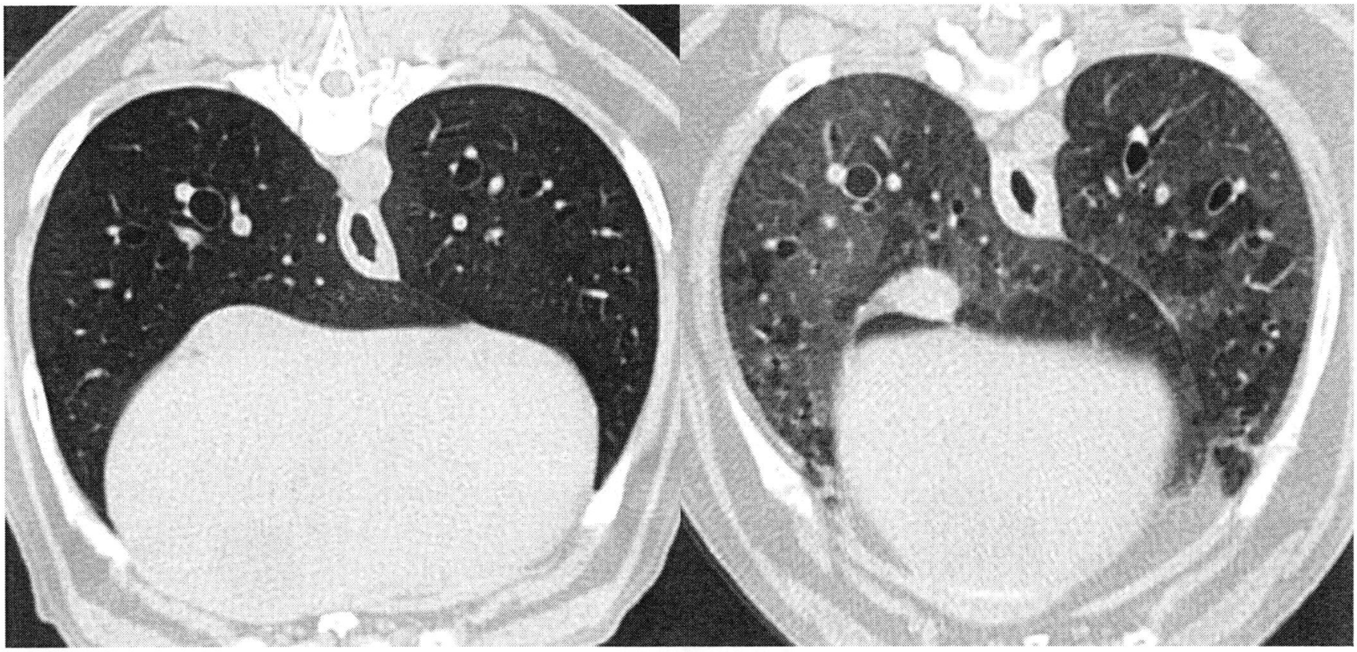

Figure 1. Cross-sectional HRCT canine images. Left image, normal. Right image, generalised ground glass attenuation, with patches of normal lung, in a dog with canine IPF.

Bronchoscopy with collection of BALF samples are readily undertaken in dogs being examined for IPF because of the common co-morbidity of chronic bronchitis (CB). While there is some disagreement it appears that dogs with consistent HRTC findings of IPF and no evidence of CB have relatively acellular BALF samples.[1,5] What is not known is to what extent raised BALF cell numbers might equate with a NSIP-like disease and ground glass attenuation.

Computed Tomography Findings in Human IPF

Consensus documents have been prepared and updated several times by the American Thoracic Society and the European Respiratory Society covering all aspects of the disease definition, diagnosis, treatment and management.[2,9] In the most recent consensus statements and updates (2011, ATS/ERS/JRS/ALAT and 2013, ATS/ERS) on diagnostic criteria, HRCT has been given a central role, with specific changes associated with UIP identified.[2,9] While this has reduced (not removed) the need for lung biopsy to achieve a definitive diagnosis, other confounding factors have arisen, not least problems of interpretation by inexperienced radiologists and disagreement on the diagnostic value of bronchoalveolar lavage. To a certain extent IPF in human patients is a diagnosis of exclusion, where all other possible explanations (environmental and occupational lung diseases being of particular importance) have been accounted for, and is supported by HRCT.

The HRCT findings of UIP IPF can be divided into probable and definitive categories. In the former, the changes are bilateral, predominantly subpleural with basilar reticular abnormalities and absence of features incompatible with IPF (Table 2; Figure 2). For the latter, there must be honeycombing, which has a positive predictive value greater than 90% for UIP pathology. Where the HRCT findings are not definitive, particularly if there is absence of honeycombing, then surgical lung biopsy should be considered or at least is necessary to make a definitive diagnosis. HRCT is often used in that instance to direct surgical lung biopsy by identifying appropriate regions. Lastly, there are situations where, despite HRCT not being consistent with UIP, UIP has been found to be the case on SLB. However, one study from Japan found discordance between HRCT and UIP on SLB and actually better concordance in identifying NSIP.[10] A potential problem with that study is that only one of the UIP cases had honeycombing on HRCT.

Table 2. High-resolution computed tomography criteria for human UIP pattern (adapted from Raghu *et al.* 2011)

UIP pattern (All four features)	Possible UIP pattern (All three features)	Inconsistent with UIP pattern (Any of the seven features)
Subpleural, basal predominance	Subpleural, basal predominance	Upper or mid-lung predominance
Reticular abnormality	Reticular abnormality	Peribronchovascular predominance
Honeycombing with or without traction bronchiectasis	Absence of features listed as inconsistent with UIP pattern	Extensive ground glass abnormality (extent >reticular abnormality)
Absence of features listed as inconsistent with UP		Profuse micronodules (bilateral, predominantly upper lobes)
		Discrete cysts (multiple, bilateral, away from areas of honeycombing
		Diffuse mosaic attenuation/air-trapping (bilateral in three or more lobes)
		Consolidation in bronchopulmonary segment(s)/lobe(s)

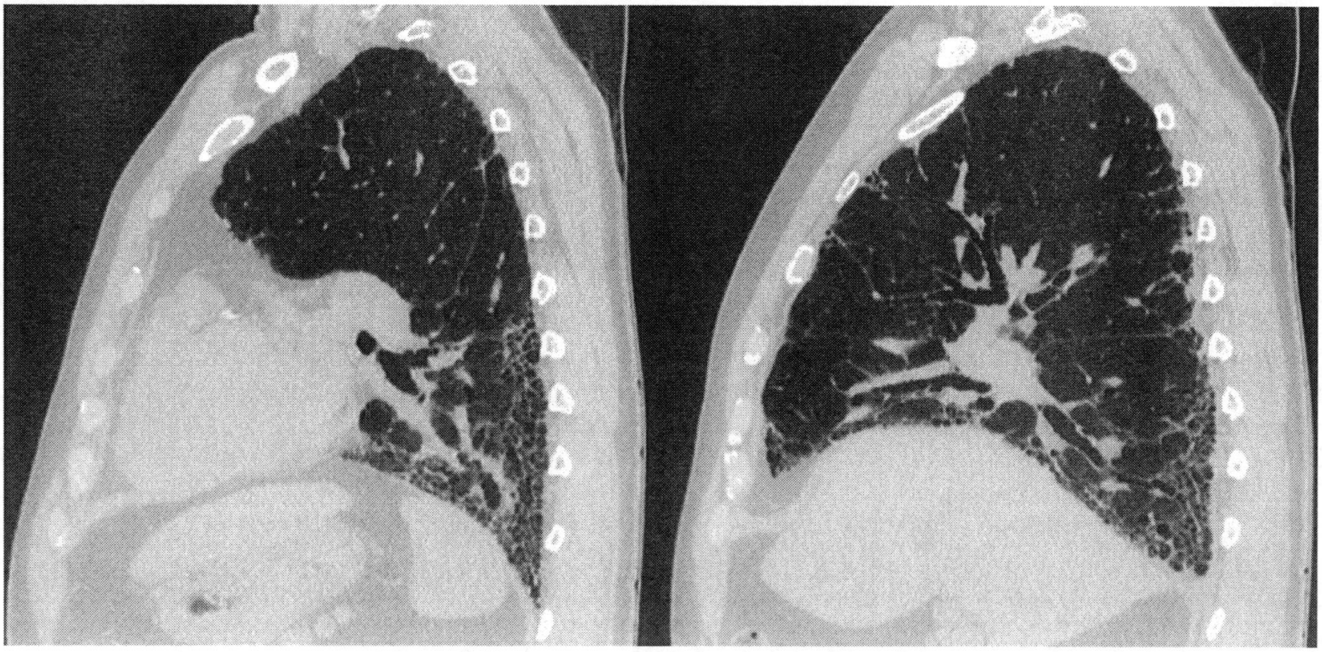

Figure 2. Axial and coronal HRCT images of a human patient with IPF. There is septal thickening, some ground glass attenuation, traction bronchiectasis and honeycombing all consistent with a diagnosis of UIP (with kind permission from Professor John Murchison, NHS Lothian, Scotland).

To date a similar set of criteria has not been determined for CIPF, but it is not unreasonable to assume a dog with honeycombing on HRCT likely has UIP IPF. One contentious issue is how to interpret the presence of ground glass opacity on HRCT, which is usually taken to be representative of NSIP, and which itself is the most common HRCT finding in dogs. However, there is concern that the over-proscriptive application of HRCT criteria in human IIPs might be excluding NSIP-like changes as part of the spectrum of changes seen with IPF. One approach is to assess BALF results in human patients, which is much more widely used by European clinicians compared to the USA, particularly where definitive diagnosis of UIP IPF is absent. BALF lymphocytosis >30% is taken as suggestive of a diagnosis other than UIP, but only if the UIP HRCT findings are equivocal.

Conclusion

HRCT has become the diagnostic test of choice in both canine and human IPF. While correlation between HRCT findings and pathology have not been determined yet in dogs, the close correlation between HRCT and pathology in human IPF has resulted in significant decline in the use of surgical lung biopsy in diagnosis, and the data can be extrapolated to the dog with reasonable confidence. For a diagnosis of UIP IPF it is preferable that honeycombing is present on HRCT. This is a rare finding in dogs. For NSIP ground glass attenuation is likely the best HRCT correlate and this is more readily seen in canine IPF. However, there is undoubtedly a spectrum of pathological changes found in both dogs and humans and this is reflected in the range of HRCT findings for both species, all of which eventually lead to extensive and irreversible fibrosis.

References

1. Corcoran *et al. Vet Rec.* 2011;168:355.
2. Ragu *et al. Am J Respir Crit Care Med.* 2011;183:788.
3. Corcoran *et al. Vet Rec.* 1999;144:611.
4. Thierry *et al. Vet Radiol Ultrasound.* 2017 (in press).
5. Heikkila *et al. J Vet Intern Med.* 2011;25:433.
6. Syrja *et al. J Comp Pathol.* 2013;149:303.
7. Norris *et al. Vet Pathol.* 2005;42:35.
8. Johnson *et al. J Small Anim Pract.* 2005;46:381.
9. Travis *et al. Am J Respir Crit Care Med.* 2013;188;733.
10. Sumikawa *et al. Eur J Radiol.* 2012;81:2919.

1.
acvim387

When to Consider Abdominal Computed Tomography (CT) Over Abdominal Ultrasound?

Angela J. Marolf, DVM, DACVR
Fort Collins, CO, USA

INTRODUCTION

Ultrasound has been performed in dogs and cats for evaluation of abdominal abnormalities for many years in veterinary medicine. Advantages to abdominal ultrasound include ability to perform the study on awake or sedated patients, use of nonionizing soundwaves, and ultrasound's ability to distinguish fluid and soft tissue changes. Disadvantages to sonographically evaluating the abdomen include incomplete assessment of the organs due to overlying stomach and bowel gas, food within the gastrointestinal tract and/or fecal material in the colon that interferes with the ultrasound wave propagation, and patient discomfort or poor compliance during the study. Additionally, ultrasound is highly operator dependent, and thorough evaluation of abdominal organs can be limited due to experience level of the sonographer.

With the increased availability of CT scanners in veterinary hospitals, this imaging modality is more readily available for use in dogs and cats for diagnosis of abdominal diseases. The ability to image veterinary patients without superimposition of other organs or bowel gas and perform a complete assessment of the multiple abdominal organs offers distinct advantages. With the advent and increased availability of multislice CT scanners, the speed at which CT studies can be performed has increased, and the ability to sedate animals for some studies is possible.

CT imaging utilizes intravenous iodinated contrast to better evaluate the blood flow to organs and adjacent tissues. To specifically evaluate vascularity, CT angiography must be performed. This type of study uses a timed bolus of intravenous contrast followed by multiple, specifically timed scans to capture the arterial, venous, and delayed phases of contrast passage through the organ (3-phase angiogram).[14,15] It is important to remember that iodinated contrast can be problematic for dehydrated patients or patients with renal compromise. CT imaging provides excellent contrast resolution, and multiple planes can be reconstructed with CT imaging, allowing for visualization of all anatomic structures in the abdomen via unlimited angles for thorough assessment. Disadvantages to CT imaging include ionizing radiation, added expense, and decreased availability compared to ultrasound.

Due to familiarity of the uses of abdominal ultrasound, the focus of these proceedings will be on conditions under which CT may be the preferred imaging modality over ultrasound. Patient size and specific conditions will be discussed.

PATIENT SIZE CONSIDERATION

Large and deep-chested dogs can present a challenge to the sonographer. These challenges include limited field of view, imaging the cranial abdomen through the ribs, and interference/lack of visibility of organs of interest due to overlying bowel gas. Additionally, ultrasound examinations in these types of patients may take more time as the sonographer attempts to identify and assess specific organs. Direct comparisons between sedated abdominal CT imaging with contrast and abdominal ultrasound have been performed.[1,2] In one study, for patients over 25 kg, abdominal CT had better overall lesion detection than abdominal ultrasound including clinical relevant lesions such as masses, portosystemic shunts, and vascular invasion by adrenal masses.[1] There was also a lower percentage of agreement between CT and ultrasound in dogs over 25 kg with better agreement between the two modalities in smaller dogs (less than 25 kg). This study also compared time to complete studies and found that average CT scan time was approximately 8 minutes compared to 45 minutes for the abdominal ultrasound.[1] However, when adding in the time to interpret the CT scan, the total time for evaluation of the CT study would increase. A second study also compared CT and abdominal ultrasound but did not discriminate dogs by weight.[2] This study found better agreement between ultrasound and CT imaging; however, in about half of the cases, the ultrasonographic evaluation of the abdomen was considered incomplete due to inability to visualize specific organs that would normally be included in a complete study.[2] All organs were identified in all patients with CT abdominal imaging.[2] Based on these findings, patient size should be a consideration when choosing abdominal ultrasound or CT imaging.

PREFERRED CT CONDITIONS

Causes of Portal Hypertension

Portal hypertension is often associated with several conditions including hepatic and extrahepatic causes and arterioportal shunts. Hepatic causes of portal hypertension include chronic hepatitis, cirrhosis, neoplasia, and primary hypoplasia of the portal vein. Extrahepatic causes include portal vein thrombosis or stenosis. Arterioportal shunts are congenital communications between the hepatic artery and intrahepatic portal branches. Ultrasound and computed tomography (CT) can be used to assist in the diagnosis of portal hypertension.

Sonographic identification of changes associated with portal hypertension can be difficult to find, and many abnormalities can be missed. The ultrasound changes associated with portal hypertension include splenomegaly, free fluid (ascites), identifying multiple acquired shunts, and enlargement of the diameter of the portal vein.[3-6] Portal vein velocities can decrease to less than 10 cm/s in dogs with portal hypertension.[7] Reversal of blood flow within the portal vein (hepatofugal) or bidirectional blood flow can also be identified.[3-5] Portal vein thromboses can be identified with ultrasound imaging but are often missed due to free fluid in the abdomen or poor patient compliance decreasing the ability of the sonographer to apply pressure to visualize the entire portal vein.

CT angiography has been used to identify multiple acquired shunts and portal vein thromboses and is often definitive where ultrasound is not, as the entire portal system can be evaluated. The use of a 3-phase CT angiogram captures the arterial, venous, and delayed phases after intravenous contrast administration. Multiple, small tortuous vessels are often identified between the portal vein and left renal vein with drainage into an enlarged caudal venal cava.[8] However, multiple small shunting vessels are often identified throughout the abdomen. Other sites for acquired shunts include mesenteric, gastroesophageal, paraesophageal, and choledochal varices.[9] These have been described based on CT angiographic studies.[9] Portal vein thromboses are best identified in CT angiographic studies. The extent and number are reliably noted with this imaging modality and are often missed on ultrasound examination.

Arterioportal fistulas should be suspected in young dogs with severe portal hypertension. Sonographic changes include dilation of multiple portal vein tributaries, reversal of portal flow (hepatofugal), identification of multiple acquired shunts, with or without free fluid.[5,10] CT imaging requires a minimum of an arterial- and venous-phase angiogram. During the arterial phase, there is immediate filling of portal veins, often via a large plexus of vessels.[36] The intrahepatic portal veins are typically dilated, and multiple acquired shunts are identified.[11] The aorta may decrease in diameter caudal to the celiac artery due to "stealing of blood" into the portal system via the fistula.[11]

Pancreatitis

Pancreatitis is the most common exocrine pancreatic disorder. The pancreas is a challenging organ to image due to its anatomic location, thin angular shape, and fat/soft tissue composition. Ultrasound imaging has been the mainstay in diagnosing pancreatic disease in dogs and cats, but it has its limitations, including false-negative sonographic findings, overlying gastric and bowel gas, and poor patient cooperation/discomfort. As discussed earlier, patient size can contribute to incomplete evaluation of this organ. Imaging of the pancreas with CT is commonly performed in people and allows for the entire pancreas to be evaluated without superimposition of surrounding organs. The ability to use intravenous contrast allows evaluation of perfusion changes to the pancreas and evaluation of adjacent vascular structures for thrombosis.

In diagnosing acute pancreatitis, the gland typically is enlarged with or without peripancreatic changes. Sonographically, the pancreas is hypoechoic with or without hyperechoic mesentery and adjacent free fluid.[12] The stomach and duodenum may become thickened and dilated, and the common bile duct may dilate as well.[12] In diagnosing chronic pancreatitis, the pancreas may be more normal in size and of normal to increased echogenicity due to fibrosis.

In veterinary patients, CT imaging of pancreatitis demonstrated a hypoattenuating parenchyma, homogeneous to heterogeneous contrast enhancement and peripancreatic changes.[13] The ability to image the entire pancreas in dogs was an advantage compared to ultrasound imaging.[13] The heterogeneous contrast enhancement of the pancreas identified in 3 of the dogs in a recent study may have represented necrosis or decreased vascularity, which appeared to affect long-term outcome in these dogs.[13,14] Identification of the sequelae of pancreatitis is also an advantage of CT imaging. Venous and arterial thromboses can occur in patients with pancreatitis and carries a poorer prognosis in people.[14] In a recent study evaluating acute pancreatitis in dogs, 3 of 10 dogs had CT evidence of portal vein thrombosis, none of which were diagnosed with ultrasound.[13] The additional imaging information that CT provides may be a distinct advantage of CT imaging of pancreatitis in dogs and cats in the future.

Congenital Portosystemic Shunts

While ultrasound can be very valuable in diagnosing portosystemic shunts, inherent physical limitations exist, and the quality of the study is dependent on the skill level of the sonographer.[15,16] The small field of view, deep location of portal vasculature, and overlying bowel and stomach gas can limit the view of the sonographer. CT angiography is well suited for diagnosing congenital portosystemic shunts due to contrast within the vasculature, lack of superimposition, determination of location, and course of the anomalous vessel. This includes identification of the origin and insertion of the anomalous vessel. Additionally, CT information is obtained in a volumetric form, so three-dimensional formatting is possible. Clinicians and surgeons are better able to visualize the vessel anatomy and define anatomic relationships. CT angiography has become the imaging gold standard for evaluation of portosystemic shunts, with 100% sensitivity and 100% specificity.[17-20]

Liver and Adrenal Masses

Ultrasound is usually performed for initial imaging of the liver; however, there is overlap in the appearance of different neoplastic and inflammatory processes.[21,22] Additional information regarding vascular invasion of the mass at the porta hepatis can be challenging to diagnose with ultrasound. Several studies in dogs have looked at the CT characteristics of liver masses.[23-26] Two of these studies utilized CT angiographic methods to evaluate the arterial, venous, and delayed phases to assess tumor type.[24,26] Both studies showed that hepatocellular carcinomas had characteristic enhancement patterns: hypoattenuation in the later phases and heterogeneous, marginal, or central early (arterial) contrast enhancement. Nodular hyperplasia had a diffuse enhancement pattern in the arterial phase and was isoattenuating to normal liver in the delayed phase. Liver metastatic lesions were also reviewed in one of the studies and were found to be hypoattenuating in both arterial and delayed phases.[26] These differences can potentially be used to distinguish liver masses and metastases noninvasively and assist in therapeutic planning. A more recent retrospective study evaluated dual-phase CT imaging in dogs with liver and splenic masses, which acquired images in early (<30 seconds) and delayed (>60 seconds) post contrast administration.[23] This study did not find any significant differences in enhancement patterns of malignant and benign liver lesions with this imaging technique.

Adrenal masses have been evaluated with ultrasound imaging with good success. Size, shape, and echogenicity can be determined.[27,28] Assessment of vascular invasion can be more challenging. Caudal vena caval invasion detection by ultrasound is 100% sensitive and 96% specific; however, the sensitivity and specificity drop to 76% and 96%, respectively, when all types of vascular invasion are taken into account.[28] Recently, CT has been used more frequently to evaluate adrenal masses prior to surgical removal to better assess tumoral contrast enhancement patterns and evidence of vascular invasion. In multiple studies, there is excellent agreement between CT's assessment of vascular invasion compared to surgery or necropsy, with one study indicating 92% sensitivity and 100% specificity.[29-31] This accuracy is due to the better anatomical detail possible with CT imaging. The smaller phrenicoabdominal and renal veins are able to be assessed for tumor thromboses. Additionally, with 3-phase angiography, differentiation of adenoma, adenocarcinoma, and pheochromocytoma is potentially possible.[29]

Ectopic Ureters

Ultrasound can be utilized for evaluation of ectopic ureters; changes most often identified include hydronephrosis and hydroureter which extends beyond the ureterovesical junction.[32] Not all dogs with ureteral ectopia have these imaging characteristics, and CT excretory urography has become the more accurate imaging choice for diagnosis of ureteral ectopia.[33-35] With CT excretory urography, the course of the ureters can be followed to the point of termination, and diagnosis of potential type of ectopia including intramural or extramural origin is often determined.[34,35]

References

1. Fields E, *et al. Vet Radiol Ultrasound.* 2012;53(5):513.
2. Shanaman M, *et al. Vet Radiol Ultrasound.* 2013;54(6):604.
3. d'Anjou M, *et al. Vet Radiol Ultrasound.* 2004;45(5):424.
4. d'Anjou M. *Clin Tech Small Anim Pract.* 2007;22:104.
5. Szatmari V, *et al. J Am Vet Med Assoc.* 2004;224:717.
6. Lamb C. *J Vet Intern Med.* 1999;13:498.
7. Nyland T, *et al. Vet Radiol Ulrasound.* 1990;31(4):189.
8. Zwingenberger A. *Vet Clin Small Anim.* 299;39:783.
9. Bertolini G. *Vet Radiol Ultrasound.* 2010;51(1):25.
10. Szatmari V, *et al. Vet Radiol Ultrasound.* 2000;41(3):284.
11. Zwingenberger A, *et al. Vet Radiol Ultrasound.* 2005;46(6):472.
12. Hess R, *et al. J Am Vet Med Assoc.* 1998;213(5):665.
13. Adrian A, *et al. J Vet Intern Med.* 2015;29:97.
14. Scaglione M. *Semin Ultrasound CT MR.* 2008;29(5):322.
15. d'Anjou M. *Clin Tech Sm Anim Pract.* 2007;22:104.
16. d'Anjou M, *et al. Vet Radiol Ultrasound.* 2004;45(5):424.
17. Zwingenberger A, *et al. Vet Radiol Ultrasound.* 2005;46(1):27.
18. Zwingenberger A. *Vet Clin Small Anim.* 2009;39:783.
19. Bertolini G, *et al. Vet Radiol Ultrasound.* 2006;47(5):439.
20. Nelson N, *et al. Vet Radiol Ultrasound.* 2011;52(5):498.
21. Feeney D, *et al. Am J Vet Res.* 2008;69(2):212.
22. Murakami T, *et al. Am J Vet Res.* 2012;73(6):821.
23. Jones I, *et al. Vet Radiol Ultrasound.* 2016;57(2):144.
24. Fukushima K, *et al. Vet Radiol Ultrasound.* 2012;53(3):252.
25. Phillips K, *et al. Vet Radiol Ultrasound.* 2012:103.
26. Kutara K, *et al. Vet Radiol Ultrasound.* 2014;55(1):7.
27. Besso J, *et al. Vet Radiol Ultrasound.* 1997;38(6):448.
28. Davis M, *et al. Vet Radiol Ultrasound.* 2012;53(4):442.
29. Yoshida O, *et al. Vet Surg.* 2016;45:427.
30. Gregori T, *et al. Vet Radiol Ultrasound.* 2015;56(2):153.
31. Schultz R, *et al. Vet Radiol Ultrasound.* 2009;50(6):625.
32. Lamb C, *et al. Vet Radiol Ultrasound.* 1998;39(3):218.
33. Samii V, *et al. J Vet Intern Med.* 2004;18:271.
34. Anders K, *et al. J Am Anim Hosp Assoc.* 2012;48:390.
35. Fox A, *et al. J Sm Anim Pract.* 2016;57:210.

1.
acvim388

Imaging in GI Disease: Beyond Radiographs and Ultrasound

Jill S. Pomrantz, DVM, DACVIM (SAIM)
Menlo Park, CA, USA

Introduction

GI signs are one of the most common reasons for dogs to be presented to both general and specialty practices. Because GI signs are often nonspecific, and much of the GI tract is not accessible for direct examination, many clinicians rely on a variety of imaging and laboratory tests to determine a diagnosis and guide therapy. Although technological advancements have greatly improved the quality of GI imaging, no single test is a panacea. The ideal modality would be one that is both highly sensitive and specific, inexpensive, easy to use, and safe for the patient. This session will explore the attributes of commonly utilized GI imaging modalities and introduce capsule endoscopy as a relatively new technique in veterinary medicine. Capsule endoscopy's role as a complement to other types of imaging will be discussed.

Radiography and Ultrasonography

Imaging for GI disease typically starts with survey radiographs and/or ultrasound, which have the benefits of being noninvasive, readily available, and diagnostic in some cases (such as mechanical obstruction). These modalities, however, can have poor sensitivity and specificity, especially in the setting of inflammatory disease, GI bleeding, or other processes that may exhibit primarily mucosal abnormalities.[1,2] Additionally, ultrasonography can be time consuming and operator dependent, based on the user's skill and experience. Patient size, compliance, and the presence of gas in the bowel can also affect image quality. Because of these limitations, studies may be normal or inconclusive, leading to a choice between empiric therapy and additional testing. However, the appropriate subsequent diagnostic test may not always be obvious, such as in cases of GI bleeding where the source can be difficult to localize with traditional imaging studies. Consequently, treatment in the absence of a specific diagnosis may lead to inappropriate therapy or a delay in diagnosis.

Computed Tomography (CT)

Historically, CT studies required general anesthesia to restrict patient movement. The advent of helical scanners has reduced imaging time and allowed conscious sedation to replace general anesthesia in some situations, making CT a more practical diagnostic test. One study that compared CT to abdominal ultrasound showed that in dogs weighing greater than 25 kg, more abdominal lesions were detected overall with CT than with ultrasound. However, regardless of weight, ultrasound was more sensitive in identifying abnormalities specifically involving the bowel (duodenal and colonic wall thickening, mucosal speckles and striations, fluid-filled duodenum), suggesting that CT may be less useful than abdominal ultrasound in the evaluation of dogs with GI disease.[3] Other disadvantages of CT include its cost, the physical space, and radiation shielding requirements needed to house the scanner, as well as the need for a trained technician to perform scans and radiologists to interpret them. These factors continue to limit widespread availability of CT, often necessitating a referral to a specialty practice.

Traditional Endoscopy

Endoscopy is routinely performed in specialty practices and in some general practices. The benefits of endoscopy include the ability to directly visualize the mucosa and obtain biopsies. One study comparing endoscopy to ultrasound in dogs and cats with histologically confirmed gastric neoplasia found endoscopy to be more accurate in identifying neoplastic lesions.[4] Disadvantages of endoscopy are the need for special equipment that must be properly cleaned and maintained and the need for general anesthesia, which may not be ideal for older or compromised patients. Like ultrasonography, endoscopy is also highly operator dependent and requires a certain skill level to be performed proficiently. Another disadvantage of endoscopy is that the length of the endoscope limits the area that can be examined. Even when upper and lower endoscopy are performed, the jejunum (which accounts for the largest surface area of the small bowel) cannot be visualized.

Capsule Endoscopy (CE)

CE offers an additional noninvasive means of imaging the GI tract. Because it is also less expensive than CT or traditional endoscopy, and simple to use, the technology is an appealing option for the evaluation of GI disease. The technique involves patients swallowing a camera system that is contained within a capsule. The capsule is propelled by peristalsis and captures images of the mucosal surface of the entire GI tract. CE has been approved for use in human medicine since the year 2000 and has become the imaging modality of choice in many situations, such as cases of iron deficiency anemia when bidirectional endoscopy has failed to identify a source of bleeding.[5] CE has been reported to be diagnostically superior to barium and push enteroscopy in humans[6] and has also been shown to be useful in the management of inflammatory conditions, such as Crohn's disease[7].

CE was introduced into veterinary medicine in the last 5 years and became widely available in 2015. The capsules available to veterinarians vary in the number and location of cameras, ranging from 1 front-view camera to 4 side-view cameras. The other main difference between CE systems is the accessory equipment needed to perform and interpret studies. Front-view cameras require a dedicated workstation that is used to view images and generate reports. They also utilize external recording equipment that is attached to the patient for the duration of the study. Although the capsule does not need to be recovered, the recorders are bulky, expensive, and could be subject to damage or loss during the study. The high cost of these CE systems may limit the practicality in most veterinary practices. The most common front-view camera used in dogs is EndoCapsule, made by Olympus. The capsule measures 26 x 11 mm, provides a 160° wide-angle field of view, and has a battery life of 12 hours.

The ALICAM capsule, made by Infiniti Medical, has the benefit of being an all-in-one system, with no equipment needed other than the capsule itself, which greatly reduces costs. The product is also bundled with an optional image interpretation/consultation service to assist the clinician in the care of the patient. It has 4 side-view cameras, which provide a 360° panoramic view of the GI tract. The other main components of the capsule are a light source, memory, clock, and batteries. The capsule measures 31 x 11 mm. Each camera captures up to 5 images/sec and the capsule can store up to 50,000 images, with a battery life of up to 18 hours. The average study yields 20,000 images over 15 hours, which is equivalent to 1–2 hours of video in real-time. The images can be viewed using special software, which allows for the playback speed to be adjusted, such that the video can be viewed in under 30 minutes. The program is capable of capturing video clips and still images. Transit times can be calculated by correlating the internal clock with visual assessment of the capsule passing through landmarks, including the lower esophageal sphincter, pylorus, and ileocolic sphincter.

Although CE is a relatively new technology and its role in the evaluation of dogs with GI signs has yet to be fully determined, the evidence continues to grow demonstrating its utility in a broadening array of clinical scenarios. CE is a valuable tool in dogs with anemia or overt GI hemorrhage, because it can detect bleeding mucosal lesions.[8,9] One advantage over endoscopy is its ability to image the sections of the GI tract that lie beyond the reach of the endoscope. When a lesion is found, the patient can be appropriately triaged to exploratory surgery or endoscopy based on its location. There is also evidence that CE may be more sensitive than abdominal ultrasound for GI mucosal lesions. One study looked at the utility of CE in dogs with GI signs and normal abdominal ultrasound examination, and demonstrated the ability of CE to detect lesions not seen on ultrasound, such as jejunal masses and dilated lacteals.[10]

Cases

Several cases will be discussed that highlight the utility of capsule endoscopy, how it can be incorporated into the diagnostic workup of dogs with GI disease, and its ability to favorably impact patient care. The cases will also demonstrate how CE may add additional information when ultrasound and endoscopy are negative.

Conclusion

CE is an attractive, evolving imaging modality in veterinary medicine. It overcomes many of the pitfalls of traditional imaging and has the potential to positively impact the care of patients in a variety of settings.

Disclosure

Dr. Pomrantz is a full-time employee of Infiniti Medical.

References

1. Pastore GE, *et al. Journal of the American Animal Hospital Association*. 2007;43:264.
2. Rudorf H, *et al. Journal of the American Animal Hospital Association*. 2005;46:322.
3. Fields EL, *et al. Veterinary Radiology & Ultrasound*. 2012;53:513.
4. Marolf AJ, *et al. Journal of Small Animal Practice*. 2015;56:339.
5. Rondonotti E, *et al. Techniques in Gastrointestinal Endoscopy*. 2015;17:12.
6. Hartmann D, *et al. Zeitschrift für Gastroenterologie*. 2003;41:377.
7. Dussault C, *et al. Digestive and Liver Disease*. 2013;45:558.
8. Pomrantz JS, *et al. Journal of Veterinary Internal Medicine*. 2016;30:1540 [abstract].
9. Davignon DL, *et al. Journal of Small Animal Practice*. 2016;57:148.
10. Pomrantz JS, *et al. Journal of Veterinary Internal Medicine*. 2016;30:1540 [abstract].

acvim389

Shedding Light on Canine Pituitary Dwarfism

Annemarie M.W.Y. Voorbij, DVM, PhD; Peter A.J. Leegwater, PhD; Hans S. Kooistra, DVM, PhD, DECVIM
Sydney, NSW, Australia

INTRODUCTION

The development of the adenohypophysis is a highly differentiated process that is tightly regulated by the coordinated actions of numerous transcription factors. During embryogenesis, the adenohypophysis develops from Rathke's pouch, which arises from the oral ectoderm. The individual hormone-secreting cells emerge from this pouch in a sequential order.

Of the adenohypophyseal endocrine cells, the corticotropic cells are the first to develop. Any defect in the development of the pituitary gland may result in a form of isolated or combined pituitary hormone deficiency. In dogs, congenital growth hormone (GH) deficiency or pituitary dwarfism is the most striking example of pituitary hormone deficiency. Pituitary dwarfism is an autosomal, recessively inherited disorder in shepherd dogs.[1] The condition is encountered most often in German shepherd dogs. German shepherd dwarfs have a combined deficiency of GH, thyroid-stimulating hormone (TSH), prolactin (PRL), and the gonadotropins. In contrast, adrenocorticotropic hormone (ACTH) secretion is preserved in these animals.[2-4]

Clinical Manifestations

Pituitary dwarfism can lead to a wide range of clinical manifestations and not all dwarfs display the same clinical signs. The most common clinical manifestations are marked growth retardation, retention of secondary hairs (puppy hair coat) with concurrent lack of primary hairs, and bilateral symmetrical alopecia. Other often encountered manifestations are skin problems such as hyperpigmentation, scales, and bacterial infections. In addition, a pituitary cyst can be identified in most dwarfs.[5]

Diagnosis

Although the physical features of pituitary dwarfism may seem obvious, the definite diagnosis of pituitary dwarfism should rely on evaluation of pituitary responsiveness to provocative testing, i.e., challenging the adenohypophyseal cell types by stimulation with four releasing hormones: GH-releasing hormone (GHRH), corticotropin-releasing hormone (CRH), thyrotropin-releasing hormone (TRH), and gonadotropin-releasing hormone (GnRH).[2,4,6]

Another test to determine the responsiveness of the somatotropic cells is the ghrelin stimulation test. Ghrelin is a potent stimulator of GH release in dogs. In young dogs it is an even more potent stimulator than GHRH.[7]

The morphology of the pituitary may be investigated with computed tomography or magnetic resonance imaging. In most German shepherd dwarfs, an intrapituitary cyst can be identified at a young age, and the size of this cyst gradually enlarges during life.[5] Because healthy dogs may have pituitary cysts as well, a definitive diagnosis of pituitary dwarfism cannot be based solely upon the presence of pituitary cysts.[4]

Treatment

The most logical option would be to treat the dwarfs with canine GH. Unfortunately, this is not possible, since canine GH is not available for therapeutic use. Another option is the use of heterologous GH. In the past, there have been attempts to treat pituitary dwarfs with human GH. Not only is this a very expensive therapy, formation of antibodies directed against human GH also precludes its use. A good option is the use of porcine GH. Administration of porcine GH will not result in the formation of antibodies, because the amino acid sequence of porcine GH is identical to that of canine GH.[4]

Whether therapy will lead to linear growth of the dwarf is dependent on the status of the growth plates at the time that treatment is started. A beneficial response in the skin and hair coat usually occurs within 6 to 8 weeks of the start of therapy.[4]

Alternatively, long-term treatment with medroxyprogesterone acetate (MPA) or proligestone can be used, since progestins are able to induce the expression of the GH gene in the mammary gland of dogs.[5,8]

The secondary hypothyroidism should be treated with synthetic levothyroxine.

Prognosis

Without proper treatment, the long-term prognosis is poor. By the age of 3 to 5 years the animal has usually become a bald, thin, and dull dog. At this stage, owners usually request euthanasia for their dog. Although the prognosis improves significantly when dwarfs are properly treated with levothyroxine and either porcine GH or progestins, their prognosis still remains guarded.[4]

The good news is that pituitary dwarfism is easily preventable. As long as mating between 2 carriers of the mutation that leads to pituitary dwarfism is prevented, no dwarfs will be born. Unfortunately, carriers of the mutation associated with pituitary dwarfism cannot be distinguished easily from dogs free of the mutation. The availability of a diagnostic DNA test, however, would enable breeders to prevent dwarfs from being born by testing the carrier status of potential breeding animals and applying a correct breeding policy. But before such a test could be developed, the mutated gene had to be identified first.

A Contracted DNA Repeat in *LHX3* Intron 5 Is Associated with Aberrant Splicing and Pituitary Dwarfism in German Shepherd Dogs

LHX3 is a member of the LIM homeodomain protein family of DNA-binding transcription factors. These factors regulate the expression of genes that pattern the body and are critical for cell specialization during embryonic development. Molecular defects in the *LHX3* gene are associated with the combined pituitary hormone deficiency (CPHD) syndrome in humans.[9-14] Most human patients display a complete deficit of all pituitary anterior lobe hormones, except for adrenocorticotropic hormone. Also in mice, LHX3 is essential for differentiation and proliferation of pituitary cell lineages. Homozygous *LHX3*-knockout mice display a complete absence of the differentiated hormone-secreting cells, except for some corticotropes.[15,16]

Because the endocrinological phenotype of humans with *LHX3* mutations and *LHX3*-knockout mice is in accordance with the phenotype of the German shepherd dog dwarfs, *LHX3* was considered an excellent candidate gene for involvement in pituitary dwarfism associated with GH deficiency in this dog breed. Analysis of intron 5 revealed that dwarfs have a deletion of 1 of 6 imperfect 7-base pair (bp) repeats. This deletion reduces the intron size to 68 bp and is associated with defective splicing of a proportion of the transcripts *in vivo* and *in vitro*. The aberrant splicing products result from skipping of exon 5 or retention of intron 5. Skipping of exon 5 results in a frame shift; the translation product will lack the homeodomain and will therefore probably not be functional. Retention of the mutant intron also leads to a frame shift in the part of the mRNA that codes for the homeodomain.[17]

Pituitary Dwarfism in Saarloos and Czechoslovakian Wolfdogs Is Associated with the Same *LHX3* Mutation as in German Shepherd Dogs

Congenital dwarfism associated with GH deficiency is also known in Saarloos wolfdogs and Czechoslovakian wolfdogs. Both are German shepherd dog-wolf cross-breeds.

Two Saarloos wolfdogs and 4 Czechoslovakian wolfdogs with pituitary dwarfism were subjected to genetic testing. All dwarfs were found to be homozygous for the same 7-bp deletion in intron 5 of *LHX3*. This mutation is identical to the one described in German shepherd dogs.

A large group of clinically healthy Saarloos and Czechoslovakian wolfdogs were screened for the mutation associated with pituitary dwarfism. The percentage of carriers of the mutated allele was 31% and 21%, respectively.[18]

These results clearly demonstrate that pituitary dwarfism is a relevant disorder and emphasize the need for screening.

Atlanto-Axial Malformation and Instability in Dogs with Pituitary Dwarfism Due to an *LHX3* Mutation

In humans with pituitary dwarfism due to an *LHX3* mutation, anatomical abnormalities in the occipito-atlantoaxial joint have been reported.[13,14]

We found similar anatomical malformations, leading to instability and dynamic compression of the cervical spinal cord in 3 Czechoslovakian wolfdogs and a German shepherd dog with pituitary dwarfism due to an *LHX3* mutation.[19] All dogs displayed neurological signs indicative of a cervical spinal cord disorder. Magnetic resonance imaging (MRI) and computed tomography (CT) images revealed incomplete ossification of the atlas with 1 or more of the 3 suture lines between the 3 ossification centres of C1 still open in all dogs. The incomplete ossification of the bony elements of C1 is expected to have resulted in instability of C1. This instability and resulting movements between the bony elements of C1 caused excessive soft tissue formation in the open sutures in C1. Additionally, flexion of the C1-C2 junction resulted in an increased dorsal displacement of the dens axis, causing compression of the cervical spinal cord.

These findings suggest that the anatomical abnormalities of the atlanto-axial joint are associated with canine pituitary dwarfism due to an *LHX3* mutation. Consequently, pituitary dwarfs should be monitored closely for neurological signs.

Central Hypothyroidism in Miniature Schnauzers

The most important endocrine differential diagnosis of pituitary dwarfism due to GH deficiency is juvenile hypothyroidism. Defects at any level of the hypothalamus-pituitary-thyroid axis can lead to deficient secretion of thyroid hormones. Hypothyroidism can be classified as primary or central, and both forms can be congenital or acquired. In central hypothyroidism, the thyroids are not affected primarily but are deprived of stimulation by TSH. Primary hypothyroidism is a common endocrinopathy in dogs. In contrast, isolated TSH deficiency has only been reported in a family of Giant Schnauzers,[20] in a young Boxer,[21] and in a 2-week-old Portuguese water dog[22].

We diagnosed central hypothyroidism in 7 Miniature Schnauzers. Three of these dogs had disproportionate dwarfism and at least 1 of them had a combined deficiency of TSH and PRL.[23]

Central hypothyroidism could be an underdiagnosed disorder in Miniature Schnauzers and should be considered in dogs of this breed with symptoms indicative of deficient thyroid hormone secretion. Primary hypothyroidism, on the other hand, is diagnosed regularly in Miniature Schnauzers, based either on thyroid scintigraphy or a TSH-stimulation test. However, due to secondary atrophy of thyroid tissue, these tests produce the same results in dogs with primary hypothyroidism as in dogs with central hypothyroidism.

The fact that 7 dogs of the same breed are affected by the same rare disorder strongly suggests that central hypothyroidism has a genetic background in Miniature Schnauzers.

References

1. Andresen E, *et al. Nord Vet Med.* 1976;28:481.
2. Hamann F, *et al. Vet Rec.* 1999;144:646.
3. Kooistra HS, *et al. Domest Anim Endocrinol.* 2000;19:177.
4. Voorbij AMWY, *et al. J Vet Clin Sci.* 2009;2:4.
5. Kooistra HS, *et al. Domest Anim Endocrinol.* 1998;15:93.
6. Meij BP, *et al. Domest Anim Endocrinol.* 1996;13:161.
7. Bhatti SFM, *et al. Res Vet Sci.* 2006;81:24.
8. Knottenbelt CM, *et al. J Small Anim Pract.* 2002;43:164.
9. Netchine I, *et al. Nat Genet.* 2000;25:182.
10. Bhangoo AP, *et al. J Clin Endocrinol Metab.* 2006;91:747.
11. Pfaeffle RW, *et al. J Clin Endocrinol Metab.* 2007;92:1909.
12. Rajab A, *et al. Hum Mol Genet.* 2008;17:2150.
13. Kriström B, *et al. J Clin Endocrinol Metab.* 2009;94:1154.
14. Sobrier ML, *et al. J Clin Endocrinol Metab.* 2012;97:E503.
15. Sheng HZ, *et al. Science.* 1996;272:1004.

16. Sheng HZ, *et al. Science*. 1997;278:1809.
17. Voorbij AMWY, *et al. PLoS One*. 2011;6:e27940. doi:10.1371/journal. pone.0027940.
18. Voorbij AMWY, *et al. J Vet Intern Med*. 2014;28:1770.
19. Voorbij AMWY, *et al. J Vet Intern Med*. 2015;29:207.
20. Greco DS, *et al. J Vet Intern Med*. 1991;5;57.
21. Mooney CT, *et al. J Small Anim Pract*. 1993;34:31.
22. Gal A, *et al. Can Vet J*. 2012;53:659.
23. Voorbij AMWY, *et al. J Vet Intern Med*. 2016;30:85.

acvim390

New Approaches to Diagnosis and Interpretation of Intestinal Dysbiosis

Jan S. Suchodolski, DrMedVet, PhD, AGAF, DACVM (Immunology)
College Station, TX, USA

Introduction

The intestinal microbiota is defined as the collection of all living microorganisms (i.e., bacteria, fungi, protozoa, and viruses) that inhabit the gastrointestinal (GI) tract. With the advance of molecular techniques, we are beginning to understand how intestinal bacteria interact with the host. Studies that used sequencing of bacterial 16S rRNA gene have reported that the GI microbiota of dogs and cats is highly diverse, comprising several hundred bacterial species.[1,2] The intestine harbors an estimated number of 10^{10}–10^{14} microbial cells, approximately 10 times more than the number of host cells. This highly complex microbiota has a crucial impact on host health. Gut bacteria and their metabolites protect the host by fending off transient enteropathogens, aid in digestion of dietary components, provide nutritional support for enterocytes, and stimulate the host's immune system. On the other side, alterations in gut microbiota (dysbiosis) have also been linked to chronic GI diseases of dogs and cats.[3] More and more evidence also implicates dysbiosis to extraintestinal disorders such as obesity, diabetes mellitus, autism, and others. Therefore, proper recognition of alteration in microbiota (i.e., dysbiosis) in disease and the reestablishment of normobiosis would be a desired treatment outcome. Molecular approaches, such as PCR-based assays for specific bacterial groups, the calculations of microbial indices (i.e., dysbiosis index) to assess normal vs. dysbiotic microbiota, and the use of fluorescence *in situ* hybridization (FISH) to visualize the translocation of bacteria into the mucosal epithelium are now routinely available. Measurement of serum cobalamin and folate concentrations can help with assessment of dysbiosis in the small intestine. The assessment of functional dysbiosis is an emerging area that allows to link changes in bacterial groups to a specific function in the GI tract, and development of such assays for microbial-derived functional metabolites are currently underway.

Gut Bacteria in Healthy Dogs and Cats

Due to differences in anatomical and physiological factors along the GI tract, the microbial composition differs along the length of the GI tract. Also, differences exist between luminal and mucosa-adherent microbiota. The stomach harbors between 10^1 and 10^6 cfu/g of contents. Bacterial counts in the duodenum and jejunum of dogs and cats can range from 10^2 to 10^9 cfu/mL of contents. The distal small intestine (i.e., ileum) contains a more diverse microbiota and higher bacterial numbers (10^7 cfu/mL of contents) than the proximal small intestine. Bacterial counts in the colon range from 10^9 and 10^{11} cfu/ml of intestinal content. In addition to bacteria, the mammalian GI tract harbors various fungi, archaea, protozoa, and viruses. While recent molecular studies have provided more information about the diversity of these microorganisms in healthy animals, their role in health and disease remains unclear.

Characterization of Microbiota Dysbiosis

Microbiota dysbiosis is an altered composition of the bacterial microbiota. There are various reasons for the development of dysbiosis. It should be noted that some disease processes may be associated with changes in microbiota function (e.g., reduced production of SCFA and other metabolites) rather shifts in microbiota composition. These functional or immunological alterations are often not readily detected due to the current lack of assays to comprehensively capture the microbiota along the entire GI tract. Furthermore, it is currently not possible to assess interactions between microbiota and host immune system. Because all these inaccessible factors play a crucial role in the intricate communication between bacteria and the host immune system, crude assessment of bacterial changes in intestinal samples often does not explain the entire disease process. Nevertheless, much progress has been made in characterizing intestinal dysbiosis in GI diseases, and recent metabolomics studies have also provided initial insights into the functional consequences of dysbiosis, and its role in the pathophysiology of some GI disorders in dogs.[4]

There is currently no single best method for assessing GI microbiota and dysbiosis. Because the gut harbors a complex ecosystem, the best diagnostic approach for dysbiosis would be a combination of molecular tools that include next generation (NGS) sequencing, direct quantification of specific bacterial taxa by quantitative PCR (qPCR), and FISH to visualize the translocation of bacteria into the mucosal epithelium. However, such an extensive approach is currently available only in research settings. More recent studies are attempting to study the functional properties of the intestinal microbiota. This is important because it remains challenging to correlate the presence of specific bacterial groups with GI health and disease. For example, administration of antibiotics to healthy dogs leads to dysbiosis, but only a proportion of animals develop clinical signs such as diarrhea and vomiting.[5] This is likely due to a functional redundancy that exists in the GI tract, with several members of the bacterial community performing similar functions, and if one group is displaced because of perturbations (e.g., antibiotic therapy), other members of the community appear to maintain a stable ecosystem function. Therefore, it seems important to evaluate the intestinal microbiome as an entity, including phylogenetic relationships and metabolic functions. Novel metabolomics approaches using multiple mass spectrometry platforms allow assessing changes in metabolite profiles, either produced by the host or microbiota, and therefore both together yield a better understanding of the pathophysiology of GI disease. In human medicine, such assays are becoming increasingly available. For example, measurement of fecal bile acids can help diagnosing dysbiosis associated with bile acid diarrhea, and treatment with bile acid sequestrants (e.g., cholestyramine) may lead to improvement of diarrhea.[6]

Bacterial culture has not enough resolution to characterize the many different bacterial species within the complex GI ecosystem. The gut harbors predominantly anaerobic bacteria, and these are not cultivable using standard methods used in reference laboratories. Therefore, a culture-based approach underestimates total bacterial numbers and also does not allow identifying the majority of bacterial groups present in the GI tract. Bacterial culture may be useful for the detection of specific enteropathogens in a targeted approach, such as culture for *Salmonella* or *Campylobacter jejuni*. However, this would need to be specifically requested.

Similarly, fecal smears are of no utility to assess shifts in intestinal microbiota. The most abundant groups in the intestine are Bacteroidetes, Fusobacteria, and various Clostridiales (i.e., *Ruminococcus* spp., *Faecalibacterium* spp., *Dorea* spp.). Species like *C. perfringens* are a minor component of the intestinal microbiota. Of importance is that every animal has a very unique microbiota composition, and therefore, some healthy animals can have a mostly gram-negative bacterial flora, whereas other healthy animals may have a predominantly gram-positive flora, containing many commensal clostridial spores.[2] Therefore, associating a gram-negative or gram-positive shift on a fecal smear with GI disease may be misleading.

Studies using sequencing of 16S rRNA genes have described dysbiosis in dogs and cats with chronic enteropathies (CE) and acute diarrhea.[3] In CE there is an increase in proportions of bacterial genera belonging to Proteobacteria (e.g., *E. coli*) and decreases in Fusobacteria, Bacteroidetes, and members of Firmicutes (i.e., *Faecalibacterium*, Ruminococcaceae, *Turicibacter*, *Blautia*). The dysbiosis in the duodenum tended to correlate with the severity of histopathological scores, but not with clinical disease severity (i.e., clinical inflammatory bowel disease activity index [CIBDAI]).[7] In fecal samples of dogs with acute hemorrhagic diarrhea syndrome (AHDS), an increase in *C. perfringens* has been reported.[8] Recent (unpublished data) have linked this increase to *C. perfringens* strains that express the gene encoding NetF, a recently novel identified pore-forming toxin. In our studies we have observed the gene for NetF in 19/45 (42%) dogs with AHDS, 0/36 (0%) of dogs with chronic diarrhea, and 4/104 (4%) of healthy dogs (unpublished data). The dysbiosis pattern improves and also *C. perfringens* harboring the gene clear very rapidly in these dogs within 3–7 days, even without the use of antibiotics.

The canine microbiota dysbiosis index (DI) is a recently developed rapid PCR-based assay that quantifies the abundances of 8 bacterial groups. These bacterial groups (e.g., *Faecalibacterium, E. coli, Blautia, Streptococcus, Turicibacter)* have been reported to be commonly altered in dogs with chronic enteropathies.[3] The advantage of the dysbiosis index is that it summarizes the results in one single number and therefore allows to define a reference interval. A DI below 0 indicates normal fecal microbiota, while a DI of 0 or above indicates fecal dysbiosis. An increase in the dysbiosis index is observed in dogs with chronic enteropathy (food-responsive and antibiotic-responsive diarrhea, idiopathic IBD). Of importance is that dysbiosis is now recognized to be a component of chronic enteropathies, and the presence of dysbiosis does not predict to which therapy a dog will respond best (i.e., also dogs with food-responsive diarrhea may have an increased dysbiosis index). The dysbiosis patterns in CE do not resolve for at least several months, even when dogs improve clinically.[9] This is likely due to the remaining underlying inflammation due to residual histological inflammation. A small subset of dogs with acute diarrhea may have a DI >0, but the microbiota normalizes typically within 7–14 days (DI <0). A DI >0 is also observed in dogs with exocrine pancreatic insufficiency (EPI), and also antibiotics induce an increase in the DI. In most dogs, the microbiota recovers within a couple of weeks after cessation of antibiotic administration, but some dogs may have prolonged microbiota dysbiosis, potentially causing signs of intestinal disease. In our recent studies, we have observed that up to 15% of clinically healthy dogs may have an increased DI, with most dogs falling in the equivocal range between 0 and 2. The use of the dysbiosis index has been reported in a recent study evaluating the impact of fecal microbiota transplantation (FMT) on the microbiota in dogs with chronic enteropathies.[10] Weekly measurements of the dysbiosis index allowed monitoring the improvement of dysbiosis one week after FMT, but the dysbiosis index increased in one dog approx. 3 weeks after FMT.

As mentioned above, in dogs with chronic enteropathies, the dysbiosis can be a component of the multifactorial disease process. While in some dogs the dysbiosis is the main driver of clinical signs, in other dogs that dysbiosis is more of an effect of the underlying histological inflammatory process. Therefore, fecal dysbiosis is not a predictor whether a dog will benefit from antibiotics. There are, however, conditions where antibiotics are warranted to clear an infection or bacterial translocation. Fluorescence *in situ* hybridization (FISH) allows visualizing the translocation of bacteria into the mucosal epithelium and can be a particular useful assay for diseases like granulomatous colitis of Boxer dogs, which is one specific form of chronic enteropathy that responds to antibiotics. It has been associated with mucosal infiltration of invasive and adherent *E. coli* in the colon.

A subset of animals with chronic enteropathies has a dysbiosis in the small intestine (SID), and these dogs are typically responding well to antibiotic administration. Therefore, this disease syndrome is called by some authors as antibiotic-responsive diarrhea (ARD). It is difficult to definitively diagnose SID/ARD, as there is no routinely available duodenal culture or molecular assay that would allow diagnosing small intestinal dysbiosis or a bacterial overgrowth. A tentative diagnosis can be made by evaluating clinical signs, altered serum cobalamin and folate concentrations, and by an antibiotic therapeutic trial. However, since also diseases due to undetected intestinal pathogens may respond to antibiotic therapy, a positive response to therapy does not necessarily confirm the presence of small intestinal dysbiosis. Serum cobalamin concentrations may be decreased and serum folate concentrations may be increased in dogs with SID/ARD. Changes in small intestinal microbiota may lead to an increased bacterial utilization for cobalamin, resulting in decreased absorption of cobalamin in the ileum. Bacteria in the distal small and large intestine produce folic acid, but folate absorption via carriers takes place in the proximal small intestine. When folate-producing bacteria accumulate in the proximal small intestine, an increased amount of bacterial folate will be absorbed by the host, resulting in increased serum folate concentration. However, cobalamin and folate uptake from the small intestine is highly complex and can be affected by several mechanisms and it is, therefore, not highly specific for SID/ARD. A diet high in folate may lead to falsely increased serum folate concentrations. Inflammation of the ileum may damage cobalamin receptors and thus may lead to cobalamin malabsorption. Dogs with exocrine pancreatic insufficiency (EPI) have a decreased secretion of antibacterial products with subsequent small intestinal bacterial overgrowth. As a consequence, dogs with EPI often have increased serum folate concentrations. Thus, in dogs with an abnormal serum concentration of cobalamin and/or folate, serum trypsin-like immunoreactivity (TLI) should be evaluated to rule out EPI. It has been demonstrated that administration of tylosin does not lead, as would be expected, to a decrease in serum folate and an increase in serum cobalamin concentration. Therefore, serum folate concentrations may not reflect therapeutic success and serum folate concentrations should always be evaluated together with the clinical picture. When both, serum cobalamin and folate concentrations, are altered this is highly suggestive of SID. However, both have a rather poor sensitivity and specificity for the diagnosis of SID. The reported sensitivity of serum cobalamin concentration for a diagnosis ranges from 25% to 55% and from 50% to 66% for serum folate concentration (German *et al.* 2003).

Conclusions

Dysbiosis has a major impact on the host's immune system and metabolism. However, dysbiosis is now recognized as part of a multifactorial disease process, and at this time there are limited data available to make treatment decisions based on the identification of dysbiosis. Patients need to be treated based on the entire clinical picture, with improvement of clinical signs as the most useful outcome measure. While in some animals the use of antimicrobials is useful (e.g., animals with ARD), their use may exacerbate dysbiosis in other GI diseases that do not respond to antibiotics.[9,11] Therefore, more clinical studies are required to understand how to best therapeutically approach dysbiosis as part of chronic enteropathies.

References

1. Swanson KS, Dowd SE, Suchodolski JS, *et al.* Phylogenetic and gene-centric metagenomics of the canine intestinal microbiome reveals similarities with humans and mice. *ISME J.* 2011;5:639–649.
2. Handl S, Dowd SE, Garcia-Mazcorro JF, *et al.* Massive parallel 16S rRNA gene pyrosequencing reveals highly diverse fecal bacterial and fungal communities in healthy dogs and cats. *FEMS Microbiol Ecol.* 2011;76:301–310.
3. Honneffer JB, Minamoto Y, Suchodolski JS. Microbiota alterations in acute and chronic gastrointestinal inflammation of cats and dogs. *World J Gastroenterol.* 2014;20:16489–16497.
4. Honneffer J, Guard B, Steiner JM, *et al.* Mo1805: Untargeted metabolomics reveals disruption within bile acid, cholesterol, and tryptophan metabolic pathways in dogs with idiopathic inflammatory bowel disease. *Gastroenterology.* 2015;148:S-715.
5. Suchodolski JS, Olson E, Honneffer J, Guard B, Blake A, Alshawaqfeh M, Steiner JM, Barr J, Gaschen F. Effects of a hydrolyzed protein diet and metronidazole on the fecal microbiome and metabolome in healthy dogs. *J Vet Intern Med.* 2016;30:1455.
6. Lin S, Sanders DS, Gleeson JT, *et al.* Long-term outcomes in patients diagnosed with bile-acid diarrhoea. *Eur J Gastroenterol Hepatol.* 2016;28:240–245.
7. Suchodolski JS, Dowd SE, Wilke V, *et al.* 16S rRNA gene pyrosequencing reveals bacterial dysbiosis in the duodenum of dogs with idiopathic inflammatory bowel disease. *PLoS One.* 2012;7:e39333.
8. Minamoto Y, Dhanani N, Markel ME, *et al.* Prevalence of *Clostridium perfringens*, *Clostridium perfringens* enterotoxin and dysbiosis in fecal samples of dogs with diarrhea. *Vet Microbiol.* 2014;174:463–473.
9. Minamoto Y, Otoni CC, Steelman SM, *et al.* Alteration of the fecal microbiota and serum metabolite profiles in dogs with idiopathic inflammatory bowel disease. *Gut Microbes.* 2015:6(1):33–47.
10. Gerbec Z. Evaluation of therapeutic potential of restoring gastrointestinal homeostasis by a fecal microbiota transplant in dogs. Master's thesis. Pharmacy. University of Ljubljana, Slovenia, 2016.
11. Gevers D, Kugathasan S, Denson LA, *et al.* The treatment-naive microbiome in new-onset Crohn's disease. *Cell Host Microbe.* 2014;15:382–392.

acvim391

Feline Hypersomatotropism

Christopher John Scudder, MRCVS, DACVIM (SAIM)
London, UK

INTRODUCTION

The pituitary develops as a blend of two different ectoderms: one from the developing oropharynx and one from the developing brain. These tissues become adherent as they mature but remain functionally and histologically distinct. The oropharyngeal ectoderm develops into anterior pituitary which contains the following five predominant cell types: the Pit-1 lineage of cells are growth hormone (GH)-secreting somatotropes, prolactin-secreting lactotropes and TSH-secreting thyrotropes; the SF-1 lineage LH- and FSH-secreting gonadotropes and the Tpit lineage ACTH-secreting corticotropes. The cell lineage is important as it may explain why pituitary cells may co-express hormones such as TSH and GH and why primary hypothyroid dogs secrete excessive GH.[1,2]

Approximately 50% of the anterior pituitary in humans are somatotropes, and this is also the situation in the feline pituitary gland (unpublished). Somatotropes can be evenly dispersed amongst other pituitary cells or form dense clusters within a normal pituitary gland. Despite the numbers of other cell types in the pituitary being 'plastic,' and can change depending on the physiological status (i.e., thyrotrope hyperplasia in induced hypothyroidism or corticotrope hyperplasia in mitotane-treated dogs), physiological somatotrope hyperplasia has not been documented.[2-4]

Growth hormone stimulates hepatic synthesis and secretion of IGF-1. These hormones exert effects on almost all tissues in the body: the predominant anabolic effects are to enhance protein synthesis and bone growth and the main catabolic effect is to increase mobilisation of fat from adipose tissue. The insulin antagonistic effects are brought about by GH-induced hyperglycaemia at the level of the liver.[5] Growth hormone exhibits a circadian rhythm in humans, mice (and even plants!), peaking during times of rest, but whether this occurs in cats is unknown.[6]

Acromegaly is the term used in human medicine to describe the condition resulting from chronic excessive GH secretion. Acromegaly results in gigantism if the condition occurs before growth plate closure, and causes asymmetrical increased soft tissue and bone growth without gigantism when occurring following growth plate closure. Gigantism has not yet been recognised in cats and almost all feline acromegalics are diabetic middle-aged to older cats. The most recent prevalence data suggest one in four diabetic cats in the UK have concurrent acromegaly.[7]

We have undertaken preliminary investigations to identify a possible genetic cause of acromegaly in cats. This has been a candidate gene approach assessing the possible impact of variations within two genes: AIP and GNAS. The AIP gene was of interest because it is a tumour suppressor and many acromegalic cats have similar characteristics to humans who have acromegaly secondary to AIP mutations (large pituitary tumours and poor responsiveness to octreotide treatment). We identified a possible association between the heterozygous single nucleotide polymorphism (SNP) AIP:c.9T>G and development of acromegaly in 2/10 tested acromegalic cats and one pair of three sibling pairs tested.[8] Three of four of these cats had large pituitary tumours (8.3 mm, 8.4 mm and 16 mm). Further SNP analysis of 36 acromegalic cats and 81 control cats revealed a SNP incidence of 5% in both groups and the acromegalic cats who had the AIP:c.9T>G SNP in the second study had pituitaries measuring 6.1 mm and 10.0 mm. Therefore it remains possible this SNP increases the risk of large tumour development (the median height of the pituitary gland in acromegalic cats was 6.1 mm in 68 acromegalic cats[9]). The GNAS gene was of interest because 40% of human acromegalics have an oncogene promoting SNP affecting gene and activating mutations in the GNAS gene have been described in canine cortisol-secreting adrenal tumours.[10] Analysis of seven control cats and 17 acromegalic cats did not identify any somatic pituitary activating mutations in this gene.

DIAGNOSIS OF ACROMEGALY

The recommended investigation for diagnosis of acromegaly in humans is as follows:

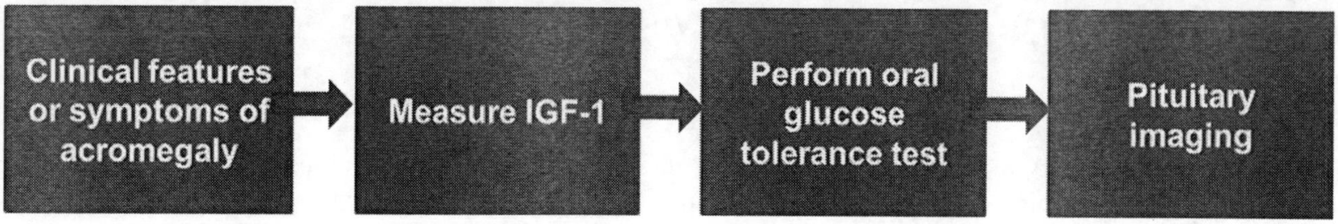

Acromegalic cats are typically heavier, have poorer diabetic control and receive greater insulin doses than 'routine' diabetic cats.[7] There is no known at-risk breed and some reports describe males being over-represented. We recognise a spectrum of disease severity with some acromegalic cats presenting as difficult to control diabetics without requiring high doses of insulin while others become non-responsive to exogenous insulin administration.

Clinical signs include those associated with diabetes mellitus although severe polyphagia is more commonly recognised, respiratory stridor and broad facial features. We have also seen cats that present with neurological signs associated with the presence of a very large pituitary tumour. There are also recent reports of non-diabetic acromegalic cats who have presented for investigation of upper respiratory signs, severe polyphagia or unexpected weight gain alone.[11,12]

Blood Tests

Not knowing whether there is a circadian rhythm or what the half-life of GH is in cats may limit the utility of single GH measurements for diagnostic purposes (the half-life of GH is between 10 to 25 minutes in humans[13]). A single measurement >8 μg/L was consistent with acromegaly in one study.[14] The measurement of GH is required for a glucose tolerance test in the context of testing for acromegaly. Hyperglycaemia should suppress GH and patients who fail to achieve GH suppression likely have acromegaly. However, unfortunately there is currently no commercially available GH assay.

IGF-1 has a longer half-life than GH (15 hours in humans)[15] and is not released in an episodic manner. Recent data suggest IGF-1>1000 ng/mL (131 nmol/L) as measured by the radioimmunoassay performed by Nationwide Specialist Laboratories, Cambridge, UK which measures IGF-1 binding protein 3 blocked total IGF-1 has a 95% positive predictive value for the diagnosis of acromegaly.[7] Non-acromegalic cats may have an IGF-1 above this threshold because insulin therapy increases IGF-1 concentration in cats (and dogs), and length of time receiving insulin therapy might affect endogenous IGF-1 levels.[16,17] What we don't know is whether there are breed, sex or age differences of IGF-1 in healthy cats as described in humans.[18,19]

Diagnostic Imaging

Acromegalic cats have been reported to have an increased organ size, and that this may be useful to increase the suspicion for acromegaly when performing diagnostic imaging in diabetic cats.[20] We have assessed kidney length in acromegalic, diabetic and control cats (unpublished data). Kidney length is greater in acromegalic cats compared to diabetics and control cats; however, kidney length correlated with body weight and the absolute measurement of kidney length did not help to differentiate acromegalic cats from diabetic cats.

The incidence of feline intra-cranial neoplasms was 2.2% in one study and pituitary tumors are the third most common type of intra-cranial neoplasia in cats.[21,22] A CT imaging study reported 1/62 cats without suspicion of pituitary disease having pituitary enlargement (i.e., pituitary height >4 mm[7,23]). We are confident in a diagnosis of acromegaly if a patient has the appropriate clinical signs, increased IGF-1 (>1000 ng/mL) and pituitary enlargement. However, recently we identified several patients that have appropriate clinical signs of acromegaly and increased IGF-1 concentration without pituitary enlargement identified on CT imaging. Some of these patients have undergone hypophysectomy with reduction of IGF-1 concentration and achieved diabetic remission. Although intra-cranial MRI would provide a higher resolution image of the pituitary gland, we are becoming increasingly confident that a diagnosis of acromegaly can be determined using clinical signs and IGF-1 testing alone. The clinical signs aid differentiation of these cats from cats with a false positive increased IGF-1 concentration.

Pituitary Histology

The World Health Organization classifies pituitary tumours into typical adenomas, atypical adenomas and carcinomas.[24] Adenomas are thought to be slow growing and may cause compression against the adjacent neuropil. Atypical adenomas infiltrate the adjacent neuropil and have increased Ki-67 labelling (>3%). Pituitary carcinomas, as defined by the identification of metastasis, are rare. All of the GH-secreting pituitary tumours we have seen are either typical or atypical adenomas. Routine histology using H&E staining reveals GH-secreting tumours are acidophilic, and the more acidophilic the staining the greater secretory capacity of that individual cell.[25] Pituitary GH immunohistopathology has shown a spectrum of pathology within the pituitary. We have observed a range of GH immunoreactivity from somatotrope hyperplasia, to small densely staining GH adenomas within a normal pituitary to large diffusely weakly GH-positive adenomas.

Treatment

Almost all acromegalic cats have concurrent diabetes mellitus. Their diabetic management should be optimised to the individual (our preference is to use long-acting insulin, low-carbohydrate diet if possible, weight loss programme if indicated, and effective management of concurrent conditions).

Medical Management Options

We have been able to demonstrate that the feline pituitary gland expresses the typical inhibitory receptors for GH release (somatostatin receptors and dopamine receptors). The only drug currently proven to induce a clinically relevant reduction of IGF-1 and improve diabetic control is the somatostatin agonist, pasireotide.[26,27] This is available as a short-acting drug given SC q12 h to q24 h, or long-acting drug given SC once monthly. The main adverse effects are gastrointestinal upset such as soft voluminous stools (affecting ¾ of treated cats) and gaseous distention of the intestines. These adverse effects often reduce with time. The main factor limiting the use of this drug is cost.

Hypophysectomy

We have performed over 40 hypophysectomy procedures in cats. The mortality rate within the first three weeks of this procedure at our hospital was 10%. Cats can achieve long-term survival following hypophysectomy (10 cats were alive after two years, 21 have died with an average time of death after surgery of 15 months). The known causes of death in the non-survivors were seizures, congestive heart failure and acute kidney injury. The predictors of response to treatment are neurosurgeon experience and pituitary tumour morphology.

Radiotherapy

Pituitary radiotherapy is particularly useful at improving neurological signs that occur as a consequence of large pituitary tumours and radiotherapy can improve diabetic control.[28-30] However, radiotherapy does not result in normalisation of IGF-1 concentrations. Additionally, hypofractionated radiotherapy treatment requires 8 to 10 doses given with the patient under general anaesthesia each time. This is a disadvantage because acromegalic cats appear to have an increased risk of congestive heart failure compared to normal diabetic cats (diastolic dysfunction and increased left atrial size compared to diabetics[31]). The time to respond to radiotherapy therapy is variable and those that do respond may relapse months later. Radiosurgical techniques using an adapted linear accelerator (LINAC) or gamma knife may prove to be useful therapeutic modalities in the future. A LINAC technique has recently been reported to successfully treat cat pituitary tumours.[32]

References

1. Diaz-Espiñeira MM, *et al. J Vet Intern Med.* 2009;23(1):100–7.
2. Diaz-Espiñeira MM, *et al. Domest Anim Endocrinol.* 2008;35(1):98–111.
3. Taoda T, *et al. Am J Vet Res.* 2006;67(8):1385–94.
4. Al-Gahtany M, *et al. Hormones.* 2003;2(3):149–58.
5. Clemmons DR. *J Clin Invest.* 2004;113(1):25–7.

6. Morris CJ, *et al. Mol Cell Endocrinol.* 2012;349(1):91–104.
7. Niessen SJM, *et al. PLoS One.* 2015;10(5):e0127794.
8. Scudder CJ, *et al. Domest Anim Endocrinol.* 2017;59:134–9.
9. Lamb CR, *et al. J Feline Med Surg.* 2014;16(2):99–108.
10. Kool MMJ, *et al. J Vet Intern Med.* 2013;27(6):1486–92.
11. Fracassi F, *et al. J Feline Med Surg Open Reports.* 2016;2(1):2055116916646585.
12. Fletcher JM, *et al. J Vet Intern Med.* 2016;30(4):1216–21.
13. Shah N, Aloi J, *et al. J Clin Endocrinol Metab.* 1999;84(8):2862–9.
14. Niessen SJM, *et al. Vet Rec.* 2007;160(1):902–7.
15. Rajaram S, *et al. Endocr Rev.* 1997;18(6):801–31.
16. Ciftci G, Yarim GF. *J Vet Sci.* 2011;12(4):325.
17. Starkey SR, *et al. J Feline Med Surg.* 2004;6(3):149–55.
18. Engström BE, *et al. Clin Chem.* 1998;44(6):1289–95.
19. Lombardi G, *et al. J Endocrinol Invest.* 2005;28(3 Suppl):99–108.
20. Lourenço BN, *et al. J Feline Med Surg.* 2015;17(8):698–703.
21. Zaki FA, Hurvitz AI. *J Small Anim Pract.* 1976;17(12):773–82.
22. Troxel MT, *et al. J Vet Intern Med.* 2003;17(6):850–9.
23. Tyson R, *et al. Vet Radiol Ultrasound.* 2005;46(1):33–8.
24. Lloyd RV, *et al. Pathology and Genetics. Tumours of Endocrine Organs.* Third ed. IARC Press; 2004:10–13.
25. Saeger W, *et al. Endocr Pathol.* 2016;27(2):104–14.
26. Scudder CJ, *et al. J Vet Intern Med.* 2015;29(4):1074–80.
27. Gostelow R, *et al. J Vet Intern Med.* 2017;[Epub ahead of print].
28. Littler RM, *et al. J Small Anim Pract.* 2006;47(7):392–5.
29. Dunning MD, *et al. J Vet Intern Med.* 2009;23(2):243–9.
30. Mayer MN. *J Vet Intern Med.* 2006;20(5):1151–4.
31. Borgeat K. *J Vet Intern Med.* 2015;29(4):1263.
32. Sellon RK, *et al. J Vet Intern Med.* 2009;23(5):1038–44.

acvim392

Standards Revised: Time for New Markers in Canine Hepatobiliary Diseases?

Iwan A. Burgener, DMV; Dr. Habil, PhD, DACVIM (SAIM), DECVIM-CA
Vienna, Austria

INTRODUCTION

Hepatobiliary diseases are commonly encountered in dogs and can be divided into 4 main groups: parenchymal, biliary, vascular, or neoplastic diseases.[1] In many cases, clinical signs of liver disease are not very specific, except in end-stage liver disease, where icteric mucous membranes and presence of ascites may indicate advanced liver disease. However, especially in cases of chronic hepatitis (CH), there is a long subclinical phase in which dogs do not show clinical signs but hepatitis is already present. Biochemical parameters including alanine aminotransferase (ALT), alkaline phosphatase (AP), and bile acids (BA) are often useful to make a presumptive diagnosis of liver disease in dogs with clinical signs that may be related to liver dysfunction. Although these biochemical indicators with high sensitivity for liver disease (i.e., ALT and AP) are essential to identify affected dogs in an early stage of disease, robust data considering sensitivity and specificity of ALT, AP, and BA in dogs with subclinical hepatitis is currently missing. Furthermore, current biochemical indicators cannot discriminate between parenchymal, biliary, vascular, or neoplastic hepatobiliary disease.

DIAGNOSTIC WORKUP OF HEPATOBILIARY DISEASES

To establish a tentative diagnosis, biochemical blood parameters are used as a first step in most cases. Several laboratory tests can be used to evaluate hepatocellular damage and hepatic function. The most commonly used biochemical indicators of hepatobiliary injury are ALT, AP, aspartate amino-transferase (AST), and gamma-glutamyltransferase (GGT).[2] To confirm the existence of significant hepatic impairment, liver function tests can be performed, including measurements of bilirubin, BA (often 12h preprandial and 2h postprandial), and ammonia levels in blood.[3] With the exception of some vascular hepatic diseases (i.e., congenital portosystemic shunts, cPSS),[4] biochemical indicators function at best establishing the presence of hepatobiliary disease but are usually not sufficient to specify the underlying disease.

A thorough and extensive diagnostic workup including imaging techniques, cytology and culture of bile, and cytologic and histopathologic evaluation of liver biopsies is usually necessary to establish a definitive diagnosis. A relatively noninvasive sensitive and specific blood-based biomarker profile that can diff erentiate between various types of hepatobiliary disease can potentially restrict or specify follow-up tests and be a valuable addition in current diagnostic workup protocols.

PRIMARY HEPATITIS (PH)

Primary hepatitis (PH) is one of the most frequently occurring group of parenchymal liver diseases in the dog.[5] Examples of PH include acute hepatitis (AH), CH, lobular dissecting hepatitis, and granulomatous hepatitis.[1] CH is by far the most common observed form of PH. In the majority of dogs with CH, the etiology remains undetermined and is considered to be of idiopathic origin. AH can be induced by a variety of stimuli, including toxins, adverse drug reactions, infectious disease (i.e., canine adenovirus-1 infection, leptospirosis), or is considered idiopathic.[1,5,6] In the last few decades, copper toxicosis is recognized with increased frequency as an etiologic factor in the development of both AH and CH in several dog breeds, including the Labrador retriever.[7,8] In the Labrador retriever[8] and the Bedlington terrier, a hereditary background has been established, while in other dog breeds familial predisposition for copper toxicosis is suspected.

Unfortunately, cases of CH are often only recognized in an advanced stage when there is extensive hepatocellular injury and/or loss of liver function. At this stage, treatment is less likely to be effective and prognosis guarded.[5] For successful management of the disease, early identification of (subclinical) dogs with PH is of great importance.

Sensitivity and Specificity of ALT, AP and BA in PH

The most commonly used indicators of hepatobiliary disease are liver enzymes and BA. ALT is a liver specific cytosolic enzyme and a sensitive indicator for hepatocellular necrosis and inflammation. AP is believed to have a good sensitivity for liver diseases, with higher activities in cholestatic disease and chronic hepatitis/cirrhosis, although being less liver specific.[2] Elevations in liver enzymes are, however, also possible findings in dogs wherein the liver parenchyma reacts to a variety of extra-hepatic diseases (i.e., inflammatory bowel disease, pancreatitis) or other stimuli. This results in a non-specific reactive hepatitis (RH). The concomitant increase in ALT and/or AP can make it difficult to distinguish RH from true primary liver disease.

Whereas liver enzymes establish the presence of liver injury, BA are often used in the identification of hepatobiliary dysfunction due to cholestasis or portosystemic shunting.[4,9] Measuring BA is a sensitive method to establish liver dysfunction,[3,4] but similar to liver enzymes these measurements are not sufficient in specifying the underlying liver disease. Therefore, definitive diagnosis relies on an extensive clinical work-up, usually including histopathologic evaluation of a liver biopsy specimen.

We performed a study with 191 clinically healthy and 51 clinically ill Labrador retrievers with known hepatic copper concentrations and hepatic histopathology with the aim to determine (1) sensitivity and specificity of ALT, AP and BA for detecting PH in clinically healthy Labrador retrievers, and (2) if ALT and AP can discriminate between dogs with a PH and non-specific RH.[10] In 64% (122/191) of the clinically healthy Labrador retrievers hepatic histology revealed abnormalities. Sensitivity of ALT, AP, and BA in this population in detecting AH was only 45%, 15%, and 15%, respectively. For CH sensitivity was 71%, 35%, and 13%, respectively. In Labradors with increased liver enzymes, median ALT was significantly higher (312 U/L, range 38–1,369) in dogs with PH compared to dogs with RH (91 U/L, range 39–139). There was no difference in AP between dogs with PH and RH.

In the present study, the specificity of ALT and AP in clinically healthy Labrador retrievers was found to be 93% and 99%, respectively. However, Center *et al.*[2] reported much lower specificities for both enzymes, especially AP. This difference can be explained because our study did not include dogs with known extra-hepatic disease, and therefore specificity may be overestimated compared to the general population.

We showed that the sensitivity of AP and BA for primary inflammatory lesions in these clinically healthy dogs is low and these parameters are therefore not appropriate to use as screening parameters to detect early primary hepatic disease. ALT, on the other hand, has shown to have a reasonable sensitivity to identify dogs with CH. As expected, ALT, AP and BA are also not able to detect hepatic injury in the face of excess copper accumulation. We also showed that, in Labrador retrievers in which ALT or AP is increased, the magnitude of ALT increase can aid in the discrimination between dogs with PH and RH. Nevertheless, ALT, AP, and BA have a low sensitivity for detecting hepatitis and new more sensitive biomarkers for inflammatory liver disease are needed.

Micro RNAs

Mature microRNAs are a class of small noncoding RNAs that are important regulators of posttranscriptional gene expression.[11] With critical functions in the regulation of multiple aspects of hepatic development, microRNAs have recently emerged as promising and stable candidate biomarkers for a variety of hepatobiliary diseases in humans.[12] Several hepatocyte-derived microRNAs (HDmiRs) and cholangiocyte-derived microRNAs (CDmiRs) have been shown to be sensitive and stable candidate biomarkers in human patients with acute or chronic liver injury caused by various etiologies, for example drug- or hepatitis C virus-induced liver injury. Studies focusing on neoplastic diseases demonstrated diff erent serum concentrations of several microRNAs in human patients with intra-hepatic cholangiocarcinoma or hepatocellular carcinoma versus normal patients.[13,14]

Hepatocyte-Derived MicroRNAs as Sensitive Serum Biomarkers of Hepatocellular Injury[15]

Common parenchymal liver diseases in dogs include RH and PH (acute or chronic). In CH, there is usually a long subclinical phase. Specific clinical signs become overt only when liver damage is severe and in this phase, treatment is usually less effective. Limited data are available regarding the sensitivity of liver enzyme activity or biomarkers for early detection of sub-clinical hepatitis. HDmiRs were recently identified as promising biomarkers for hepatocellular injury in multiple species. Here, the potential of the HDmiRs miR-122 and miR-148a as sensitive diagnostic biomarkers for hepatocellular injury in Labrador retrievers were investigated.

Samples from 66 Labrador retrievers with histologically normal livers, high hepatic copper, and with various forms of liver injury were evaluated for serum ALT activity and microRNA values. Median values of HDmiR-122 were 34.6 times higher in dogs with liver injury and high ALT than in normal dogs (95% confidence intervals [CI], 13–95; $p<0.001$). HDmiR-122 values were significantly increased in dogs with liver injury and normal ALT (4.2 times; 95% CI, 2–12; $p<0.01$) and in dogs with high hepatic copper concentrations and unremarkable histopathology (2.9 times; 95% CI, 1.1–8.0; $p<0.05$). Logistic regression analyses demonstrated that miR-122 and miR-148a were both predictors of hepatocellular injury. The sensitivity of miR-122 was 84% (95% CI, 73–93%), making it superior to ALT (55%; 95%CI, 41–68%) for the detection of hepatocellular injury in Labrador retrievers ($p<0.001$).

This study demonstrated that serum HDmiR, particularly miR-122, is a highly sensitive marker for the detection of hepatocellular injury in Labrador retrievers and is a promising new biomarker that may be used for early detection of subclinical hepatitis in dogs.

MicroRNAs as Biomarker for Hepatobiliary Diseases[16]

As already mentioned before, biochemical parameters including ALT, AP, and BA are often useful to make a presumptive diagnosis of liver disease in dogs, but they cannot discriminate between parenchymal, biliary, vascular, or neoplastic hepatobiliary disease. Therefore, the first aim of this exploratory study was to evaluate whether the concentrations of a selected group of hepatocyte-derived, cholangiocyte-derived, and oncogenic microRNAs, based on the current human literature, were measurable in serum of dogs with parenchymal, biliary, vascular, or neoplastic hepatobiliary diseases. Our second aim was to explore the possibility of using a serum microRNA profile to differentiate between parenchymal, biliary, vascular, or neoplastic hepatobiliary diseases.

Forty-six client-owned dogs with an established diagnosis of hepatobiliary disease and stored serum samples and eleven client-owned healthy control Labrador Retrievers were entered in this retrospective study. Medical records of dogs with parenchymal, biliary, vascular, or neoplastic hepatobiliary diseases and control dogs were reviewed. Afterwards, the concentrations of miR-21, miR-122, miR-126, miR-148a, miR-200c, and miR-222 were quantified in serum by real-time polymerase chain reaction.

No diff erent microRNA concentrations were found in the adenoma and cPSS groups. In all other diseases, miR-122 concentrations were elevated with the highest concentration in the mucocele group (267-fold, CI: 40–1,768, $p<0.001$). In dogs with biliary diseases, miR-21 and miR-222 were only increased in dogs with mucoceles (26-fold, CI: 5–141, $p=0.005$ and 13-fold, CI: 2–70, $p=0.025$, respectively). Uniquely increased microRNAs were found in the hepatocellular carcinoma group (miR-200c, 35-fold increase, CI: 3–382, $p=0.035$) and the chronic hepatitis group (miR-126, 22-fold increase, CI: 5–91, $p=0.002$).

This study demonstrated that a microRNA panel consisting of miR-21, miR-122, miR-126, miR-200c, and miR-222 can distinguish between parenchymal, biliary, and neoplastic hepatobiliary diseases. Serum microRNA profiling is a promising new tool that might be a valuable addition to conventional diagnostics to help diagnose various hepatobiliary diseases in dogs.

Potential Biomarkers for Hepatic Copper Concentrations[17]

Copper-associated hepatitis is one of the most common forms of PH in dogs as shown by a study conducted in 101 purebred and crossbred dogs in The Netherlands, of which up to 36% of the CH and 24% of the AH cases were copper associated.[5] Dog breeds known to develop copper-associated hepatitis are the Bedlington terrier, West Highland white terrier, Dalmatian, Doberman, and the Labrador retriever.[8] In the Bedlington terrier hereditary copper-associated hepatitis is due to deletion of exon 2 of the *COMMD1* gene and hepatic copper concentrations can easily reach 5,000 mg/kg dwl or higher.[18] Copper-associated hepatitis in the Labrador retriever is one of the best-studied examples of copper toxicosis with a complex molecular background. Besides the involvement of mutations in the copper transporters ATP7A and ATP7B8 the expression of the disease phenotype also relies on environmental factors such as dietary copper intake.[19,20]

Hereditary hepatic copper accumulation in Labrador retrievers leads to hepatitis with fibrosis and eventually cirrhosis. The development of a non-invasive blood-based biomarker for copper status in dogs could be helpful in identifying dogs at risk and to monitor copper concentrations during treatment. In this study, two cellular copper metabolism proteins, Cu/Zn superoxide dismutase (SOD1) and its chaperone (copper chaperone for SOD1, CCS) were measured in erythrocytes and tested for association with hepatic copper concentrations in 15 Labrador retrievers with normal or increased hepatic copper concentrations.

Antibodies against CCS and SOD1 were applicable for use in canine specimens. This was demonstrated by the loss of immune-reactive bands for CCS and SOD1 in siRNA treated canine bile duct epithelial cells. Erythrocyte CCS and CCS/SOD1 ratios were decreased 2.37 ($p<0.001$) and 3.29 ($p<0.001$) fold in the high copper group compared to the normal copper group. In conclusion, erythrocyte CCS and CCS/SOD1 ratio are potential new biomarkers for hepatic copper concentrations in Labrador retrievers and could facilitate early diagnosis and treatment monitoring for copper-associated hepatitis in dogs.

Conclusion

Clinical signs of liver disease are not very specific except in end-stage liver disease. Especially in cases of CH, there is a long subclinical phase in which dogs do not show clinical signs but hepatitis is already present. Biochemical parameters including ALT, AP, and BA are often useful to make a presumptive diagnosis of liver disease, but their sensitivity for liver disease is lower than expected. Results of these studies indicate that AP and BA are not sensitive enough to detect hepatitis in clinically healthy Labrador retrievers with histopathologic changes. Although ALT is not sensitive in detecting dogs with AH, sensitivity for detecting dogs with CH is reasonable. Even though hepatic copper accumulation is a well-described problem in de Labrador retriever, it is not correlated to ALT, AP, BA or to the existence of hepatocellular injury.

MicroRNAs have been identified as promising new biomarkers for hepatobiliary disease in humans and dogs. With a new microRNA panel, it was possible to distinguish between parenchymal, biliary, and neoplastic hepatobiliary diseases. Within these classes, a differentiation between the subclasses acute/chronic hepatitis, adenomas/carcinomas/malignant lymphomas, and mucoceles/other biliary diseases could be made.

References

1. *The WSAVA Liver Standardization Group*. Saunders Elsevier 2006:144.
2. Center SA. *Vet Clin North Am Small Anim Pract*. 2007;37:297–333.
3. Center SA *et al. J Am Vet Med Assoc*. 1991;199:217–226.
4. van Straten G *et al. Vet J*. 2015;204:282–286.
5. Poldervaart JH *et al. J Vet Intern Med*. 2009;23:72–80.
6. Lester C *et al. J Vet Emerg Crit Care*. 2016;26(4):559–67.
7. Hoffmann G *et al. J Vet Intern Med*. 2006;20:856–861.
8. Fieten H *et al. Dis Model Mech*. 2016;9(1):25–38.
9. Gerritzen- Bruning M *et al. J Vet Intern Med*. 2006;20:13–19.
10. Dirksen K *et al. J Vet Intern Med*. 2017; accepted for publication.
11. Krol J *et al. Nat Rev Genet*. 2010;11:597–610.
12. Chen Y *et al. Liver Int*. 2014;34:976–99.
13. Karakatsanis A *et al. Mol Carcinog*. 2013;52:297–303.
14. Borel F *et al. J Hepatol*. 2012;56:1371–1383.
15. Dirksen K *et al. Vet J*. 2016;211:75–81.
16. Dirksen K *et al. J Vet Intern Med*. 2016;30:1816–1823.

17. Dirksen K *et al. Vet.* 2016;218:1–6.
18. Twedt DC *et al. J Am Vet Med Assoc.* 1979;175:269–275.
19. Johnston AN *et al. J Am Vet Med Assoc.* 2013;242:372–380.
20. Fieten H *et al. J Vet Intern Med.* 2012;26:1274–1280.

acvim393

Unraveling the Role of Histamine Receptors H1, H2, H3, and H4 in the Canine Gastrointestinal Tract

Alyssa M. Sullivant, DVM, MS, DACVIM
Starkville, MS, USA

Introduction

The role of histamine in chronic gastrointestinal diseases has been increasingly recognized in humans, but the role of histamine in the canine gastrointestinal tract has not been thoroughly investigated. Histamine is a primary amine synthesized intracellularly through oxidative decarboxylation of the amino acid histidine and stored in granules within mast cells and basophils.[1] Histamine is released following immunological and non-immunological stimulation of histamine-producing cells and exerts its diverse biological actions through binding of four different G-protein coupled histamine receptors (H1, H2, H3, and H4) located throughout the body.[2] Each histamine receptor varies in its affinity for histamine and its level of expression throughout the body.[3] In the following sections, location of histamine receptors is based on research in humans and laboratory animals primarily, unless otherwise noted.

Histamine Receptors

H1 histamine receptor was the first of the four histamine receptors to be discovered.[4] The H1 receptor is ubiquitously expressed in vascular endothelium, vascular and airway smooth muscle, hepatocytes, chondrocytes, epithelial cells, neurons, glial cells and immune cells, including dendritic cells, monocytes, neutrophils, T cells and B cells.[2,5] It has also been identified in the gastrointestinal tract, genitourinary tract, heart, skin and adrenal medulla.[6] Common pathological processes mediated by the H1 receptor include allergic rhinitis, conjunctivitis, asthma, urticaria, atopic or allergic dermatitis, and anaphylaxis.[1,5]

Like the H1 receptor, the H2 receptor is widespread throughout the body. It is abundant on the parietal cells of the gastric mucosa and on the endothelial cells, astrocytes, and neurons in the brain.[7] It is also located in smooth muscle cells, myocytes, enteric neurons, neutrophils, eosinophils, monocytes, macrophages, dendritic cells, and T and B lymphocytes.[1,6,8] The H2 receptor is best known for its role in controlling gastric acid secretion. Unlike the H1 receptor, which has a pro-inflammatory role, the H2 receptor has an anti-inflammatory role through the inhibition of cytokine production and chemotaxis and dampening of T cell responses.[1,9] Therefore, H2 antagonism may predispose to or exacerbate allergy, inflammation, or infection, particularly in the gastrointestinal (GI) tract.[9,10]

Primary locations of the H3 receptor include the central and peripheral nervous system (CNS), including the enteric nervous system.[2] The H3 receptor is involved primarily in negative feedback-mediated inhibition of the synthesis and release of not only histamine but also other neurotransmitters such as dopamine, serotonin, glutamate, acetylcholine, and norepinephrine.[7]

In 2000, the H4 receptor was discovered.[7] Expression of H4 receptors is highest in the bone marrow and cells of the immune system.[5] H4 receptors have been identified in the spleen, thymus, brain, liver, lung, heart, epidermis, peripheral nerves, enteric nerves, and the large and small intestines.[7] The primary function of the H4 receptor is mediation of inflammation through induction of cytokine release and chemotaxis of inflammatory cells, especially eosinophils and mast cells.[2,4] Overall, the H4 receptor appears to exert a pro-inflammatory state, and evidence for its role in chronic inflammatory diseases and autoimmunity is abundant.[3,7]

Histamine in Gastrointestinal Inflammation

Histamine's contribution to gastrointestinal pathology has been a productive focus of research in humans, and all 4 histamine receptors have been identified in the human GI tract.[6,11] In addition to its central role in gastric acid secretion, histamine has an important role in many functions of the gastrointestinal tract, including neurotransmission, visceral nociception, and mucosal immunity.[1,12] Because of histamine's far-reaching effects on numerous physiological processes, an increase in histamine or a dysregulation of its receptors has a significant impact on normal organ structure and function. Levels of histamine in the gut lumen of human patients with inflammatory bowel disease are elevated, likely because gastrointestinal lesions in these patients have dense accumulations of histamine-producing mast cells.[13] Intestinal tissue samples and secretions of patients with chronic enteropathies such as Crohn's disease have elevated histamine levels, and the histamine levels correlate with disease severity.[14-17] Gastrointestinal mucosal IgE and mast cells, as well as urinary N-methylhistamine levels, are elevated some human and canine inflammatory bowel disease (IBD) patients.[14,18-20]

The H1 receptor is involved in intestinal smooth muscle contraction, mucosal ion transport, and nociception.[21] A recent study showed that visceral hypersensitivity in human irritable bowel patients is mediated by the H1 receptor, and that administration of the H1 receptor antagonist, ebastine, substantially reduces abdominal pain by blocking a common nociceptor involved in visceral hypersensitivity.[21] In addition to its previously described role in gastric acid secretion, the H2 receptor downregulates the immune response.[9] It promotes an anti-inflammatory cytokine profile and, therefore, is protective in the realm of gastrointestinal immunity.[9] Indeed, antagonism of the H2 receptor has led to exacerbations of clinical signs of Crohn's disease, has been linked to worsening of intestinal bacterial inflammation, and has been associated with susceptibility to immune-mediated inflammation.[9,10] Regulation of histamine and intestinal neurotransmitter release is mediated in part by the H3 receptors, which are densely concentrated in the enteric nervous system.[22] Gastrointestinal motility is altered by the H3 receptor due to influence on the release various neurotransmitters.[1,11] The H3 receptor also appears to have a protective role against the formation of gastrointestinal ulcers.[23] Many of histamine's effects on GI mucosal immunity are mediated by the H4 receptor. For example, the H4 receptor is a strong mediator of eosinophil chemotaxis, and eosinophilic enteritis is a frequent finding in gastrointestinal tissue of both human and canine IBD patients.[24,25] Treatment of experimentally induced peritonitis with an H4 antagonist blocks TLR-mediated inflammation.[13] The trinitrobenzene sulfonic acid model of colonic inflammation, which mimics Crohn's disease, involves TLR, T cell and dendritic cell pathology that is effectively reduced by H4 antagonism.[13] H4 antagonists also inhibit TNF-α and IL-6, cytokines involved in the pathogenesis colonic inflammation.[12]

Histamine Receptors in the Canine Gastrointestinal Tract

Certainly, histamine receptors have a significant role in human gastrointestinal immunology, and scientific evidence supports their role in gastrointestinal pathology. Histamine receptor distribution within the human GI tract has been extensively studied over the past decade, and alterations of histamine levels and histamine expression have been described in human patients with chronic enteropathies such as ulcerative colitis and Crohn's.[26,27] Human patients with irritable bowel syndrome and food allergy have higher expression of H1 and H2 mRNA within the GI tract compared to unaffected individuals.[6] Rodent models of inflammation have also been used to study GI tract histamine receptors. Studies using a common mouse model of GI inflammation that mimics Crohn's disease in humans have demonstrated increased expression of the H4 receptor in the GI tract, and a correlation of colitis progression with receptor density.[27,28] Additionally, a novel H4 receptor antagonist, JNJ7777120, has been shown to reduce histological inflammation and clinical signs in experimentally-induced colitis.[29]

The role of histamine in canine inflammatory bowel disease (IBD) has not been thoroughly investigated, but dogs with IBD have been shown to have increased GI tract mast cell density, and some have high levels of urine N-methylhistamine, a metabolite of histamine.[26] Limited data about the location and role of histamine receptors in the canine gastrointestinal tract exists. In 2008, the canine H4 receptor was cloned, and PCR techniques were used to identify H4 receptor expression in the canine small intestine.[30] Recently, H4 receptor mRNA was identified in the canine colon.[31] In 2016, the author published the presence and location of all four histamine receptors in the normal canine GI tract using a common immunohistochemical technique.

All 4 histamine receptors were identified in the canine gastrointestinal tract, and differed in location and density within sections (stomach, duodenum, jejunum, ileum, and colon) and tissues/layers (superficial mucosa, deep mucosa, lymphoid tissue, smooth muscle, submucosal ganglia and myenteric ganglia) of the canine gastrointestinal tract. Overall, the distribution of H1, H2, and H3 receptors was greater than that of H4 receptors throughout the entire GI tract. The H1 receptor was located in all sections of the GI tract. Mucosal staining was limited to the stomach. Staining in the stomach was intermittent and included staining of parietal cell, chief cells and superficial epithelial cells. Lymphoid follicles in the ileum and colon were positive for the H1 receptor, with the strongest staining in lymphoblasts and germinal centers. The H1 receptor was identified in all smooth muscle and ganglia. Distribution of the H2 receptor in the GI tract was widespread in the mucosa of the stomach, duodenum, jejunum, ileum, and colon. Gastric mucosal staining was strong, especially in the parietal and chief cells. Moderate to strong staining of lymphoid tissue was present in the duodenum, jejunum, ileum and colon. H2 receptors were not identified in smooth muscle but were present in all ganglia. The H3 receptor was most frequently located in the gastric mucosa and all ganglia. Mucosal staining was present in the stomach and colon, but not in the small intestine. Gastric mucosal staining was typically greater in the superficial mucosal enterocytes. Lymphoid follicles of the ileum and colon were positive for the H3 receptor. The H3 receptor was identified in all smooth muscle and ganglia. Immunostaining for the H4 receptor was identified in gastric and colonic mucosa, but not in the small intestinal mucosa. Lymphoid follicles in the ileum were weakly positive. Smooth muscle and ganglia staining for the H4 receptor was mild to absent throughout the GI tract.

Relevance of Histamine Receptors in Canine Gastroenterology

There is considerable variation in clinical signs and histologic disease severity amongst canine IBD patients. Treatment for canine IBD typically involves specific dietary changes and modulation of the immune system. Unfortunately, not all patients respond to standard treatment modalities. Individual-to-individual differences in disease severity and treatment responsiveness in our canine IBD patients highlight the pressing need to better understand the pathogenesis of this important condition. Clinically, patients who have comparable histological findings on GI biopsy can respond very differently to similar treatment regimens: some dogs have minimal response to aggressive therapy, while other respond well. Commonly used canine IBD protocols have been shown to lead to only partial or no treatment responses in up to 50% of patients.[32] In one large retrospective study of canine IBD, 13% of affected dogs had intractable disease and were euthanized, and a further 50% still had intermittent clinical signs despite treatment with a variety of immunosuppressants and antibiotics.[33] In another large prospective study in dogs with chronic enteropathies, 18% of dogs were euthanized due to refractory disease.[34] This lack of predictability in response to therapy serves to underscore our limited understanding of certain aspects of IBD in dogs, including the role of histamine receptors. Further investigation of the distribution of histamine receptors in canine enteropathies using our newly published technique is therefore warranted.

Conclusion

The role of histamine in allergic responses, anaphylaxis, and gastric acid secretion has been well described in both human and veterinary medicine. New roles for histamine and its receptors, however, have been elucidated over the last two decades, and novel histamine receptor antagonists have also recently been developed. Histamine has been shown to have an important role in many functions of the gastrointestinal (GI) tract, including neurotransmission, visceral nociception, and mucosal immunity.[2,5]

Whether histamine receptor distribution is altered in canine GI diseases remains unknown, but human research has documented altered histamine receptor density in certain GI diseases. Our study has established immunohistochemical techniques for the staining of H1, H2, H3, and H4 receptors in canine GI tissue and described the distribution of these receptors in normal dogs. The results of our study provide the foundation for similar work in dogs with GI disease. Of particular interest is the role of these receptors in canine inflammatory bowel disease, which shares similarities with human inflammatory bowel diseases such as Crohn's disease and ulcerative colitis. Future research in the role of histamine receptors in canine IBD may provide evidence to suggest additional treatment options for a common, yet sometimes refractory, disease.

References

1. Peters LJ *et al. J Vet Emerg Crit Care.* 2009;19(4):311.

2. Shahid M *et al. The Open Immunol.* 2009;2:9.
3. Migolavich-Sheikhet H *et al. Front Immunol.* 2012;3:238.
4. Panula P *et al. Pharmacol Rev.* 2015;67(3):601.
5. Jutel M *et al. Clin Exp Aller.* 2009;39(12):1786.
6. Sander LE *et al. Gut.* 2006;55(4):498.
7. Thurmond RL. *Histamine in Inflammation.* Landes Bioscience Springer Science. 2010.
8. Akdis CA *et al. Eur J Immunol.* 2006;553(1–3):69.
9. Gao C *et al. M Biol.* 2015;6(6):e01358.
10. Teuscher C *et al. Am J Pathol.* 2004;164(3):883.
11. Breunig E *et al. J Physiol.* 2007;583(Pt 2):731.
12. Deireren A *et al. B J Pharmacol.* 2014;172(5):1165.
13. Zhang M *et al. Pharmacol Therapeut.* 2007;113(3):594.
14. Nolte H *et al. Gut.* 1990;31(7):791.
15. Raithel M *et al. Inflamm Res.* 2010;59(suppl 2):S257.
16. Baenkler HW *et al. Hepatogastroenterol.* 1987;34(6):289.
17. KimpelS *et al. Inflamm Res.* 2007;56(suppl 1): S61.
18. Bischoff SC *et al. Semin Immunopathol.* 2009;31(2):185.
19. Bischoff SC *et al. Histopathol.* 1996:28(1):1.
20. Prinz C *et al. Am J Physiol.* 1999;227(5 pt 1):C845.
21. Wouters MM *et al. Gastroenterol.* 2016;150(4):875.
22. Coruzzi G *et al. J Physiol.* 2001;52(4 pt 1):539.
23. Adami M *et al. Pharmacol.* 2010;86(5-6):259.
24. Ling P *et al. B J Pharmacol.* 2004;142(1);161.
25. Frosali S *et al. Immunol Res.* 2015
26. Berghoff N *et al. Vet J.* 2014;201(3):289.
27. Coruzzi G *et al. Front Biosc.* 2012;Jan 1(4):226.
28. Schirmer B *et al. Biochem Pharmacol.* 2015;98(1):102.
29. Parsons ME *et al. B J Pharmacol.* 2006;147(suppl 1):S127.
30. Jiang W *et al. Eur J Pharmacol.* 2008;592(1–3):26.
31. Eisenschenk MNC *et al. Vet Derm.* 2010;22(5):297.
32. Jergens A *et al. J Vet Intern Med* 2010;24(2):269.
33. Craven M *et al. J Small Anim Pract.* 2004;45(7):336.
34. Allenspach K *et al. J Vet Intern Med.* 2007;21(4):700.

acvim394

Inflammatory Bowel Disease Revisited: Role of the Eosinophil

Idil Bastan; Savitha Rao; Srirama Rao; Aaron Rendhal; Robert Washabau
St. Paul, MN, USA

Chronic inflammation of the GI tract is referred to as inflammatory bowel disease (IBD). Dogs and cats affected with IBD often have moderate to severe clinical signs such as anorexia, vomiting, diarrhea, and weight loss. Factors that play a role in the pathogenesis of IBD include dietary antigens, environmental factors, immunologic factors, infectious agents, genetic susceptibility, and the microbiome. Eosinophilic enteritis is one of the variants of IBD and is recognized as a diffuse infiltration of the small intestinal mucosa with eosinophils (primarily), lymphocytes, and plasma cells. While eosinophils are present in the normal GI tract, clinical studies associate increased numbers of eosinophils with inflammatory GI diseases raising concerns that eosinophils may have a more harmful effect on the gut. Our studies are designed to determine the prevalence, location, clinical characteristics, and activation status of eosinophilic inflammation in canine inflammatory bowel disease.

Background

Eosinophils are non-dividing, fully differentiated, granulocytic leukocytes that express a wide array of proinflammatory mediators to initiate and sustain an inflammatory response. These cells serve as the histological hallmark of allergic diseases such as asthma, atopic dermatitis, allergic rhinitis, food allergy, and the emerging group of diseases termed eosinophilic gastrointestinal (GI) diseases. Dietary antigens/allergens can lead to abnormal infiltration of the GI tract with eosinophils resulting in chronic inflammation of the GI tract.

Eosinophil Biology

Eosinophils have numerous cellular functions. They are involved in inflammatory reactions associated with the innate immune defense against parasitic infection and allergy. It has been recently shown that eosinophils also play crucial non-inflammatory roles in the generation and maintenance of adaptive immune responses. Eosinophils are a major source of the plasma cell survival factor APRIL (activation and proliferation-induced ligand) and are essential for the long-term survival of plasma cells in the bone marrow and for the maintenance of these cells in the lamina propria.[1]

Eosinophil Pathobiology

Eosinophils have a critical role in allergen-induced inflammatory responses, such as asthmatic lung disease, atopic dermatitis, and GI food allergy. Eosinophils are regarded as end-stage effector cells, which, when activated, degranulate and release highly cytotoxic substances, such as the major basic protein or the eosinophilic cationic protein. In the case of an infection, these granule proteins may act directly against parasites; however, in an allergic reaction they contribute to tissue destruction. In patients with asthma, the damage is correlated with the number of influxing eosinophils.[1]

Eosinophil-Related Disorders

Feline Eosinophilic Gastrointestinal Fibrosis

Feline eosinophilic gastrointestinal fibrosis (FEGF) is an emerging clinical entity that is still poorly defined.[2] The underlying pathogenesis is incompletely understood although the eosinophil has clearly been impugned in the pathobiology. Many cases remain undiagnosed with varying degrees of morbidity and mortality. To further complicate matters, a related disorder (feline hypereosinophilic syndrome) has been reported that bears some resemblance to FEGF.[3]

Canine Eosinophilic Gastric Mass

Several cases of canine eosinophilic gastric mass (CEGM) have been characterized in the veterinary literature.[4] These are generally focal lesions involving eosinophil activation, but are inherently different from the more diffuse lesions of FEGF.

Eosinophilic Enteritis of Rottweiler Dogs[4,5]

Three Rottweiler dogs with marked peripheral eosinophilia and infiltration of the liver, spleen, lungs, and bone marrow with eosinophils were diagnosed with idiopathic hypereosinophilic syndrome (IHES). Mean serum IgE were markedly elevated in all three dogs. Pulmonary involvement was present in two of the three dogs.

Inflammatory Bowel Disease

Inflammatory bowel disease (IBD) broadly refers to a group of idiopathic, chronic gastrointestinal (GI) disorders characterized by mucosal inflammation. Although the prevalence of IBD is unknown, it is arguably the most common histopathologic diagnosis in dogs and cats affected with chronic vomiting or diarrhea. All breeds may be affected but several breeds that are uniquely at increased risk are the German shepherd, Rottweiler, Yorkshire terrier, soft-coated Wheaten terrier, basenji, and Irish setter. Despite the perceived prevalence, there has been no internationally accepted standard for making this diagnosis. With that recognition, the WSAVA Gastrointestinal Standard Group was convened in 2004 for the purpose of creating standards in the diagnosis and therapy of IBD. The group consisted of 10 experts in internal medicine or pathology. Three pivotal publications[6-8] were produced from these efforts: Another four papers were developed as a consequence of the study. The WSAVA standards are now recognized worldwide, and many publications have been produced using these standards.

Problems with the Model

In a set of experiments that our group published in 2017,[9] we showed that H&E staining insufficiently stained intact and degranulated eosinophils. We reported that a monoclonal antibody (EPXmAb) was much more effective in staining both types of eosinophils. This was an important finding in that it suggested that H&E staining had previously underestimated the role of eosinophils in the pathogenesis of IBD, and it is on the basis of this finding that we have developed our hypotheses and specific objectives. We carried out a set of experiments using EPXmAb to selectively stain intact and degranulated eosinophils in a control group of dogs, dogs with IBD, and dogs with IBD which had been treated with anti-inflammatory doses of prednisone. We found that untreated IBD dogs had a significantly higher number or intact and degranulated eosinophils compared with control and treated IBD dogs. We concluded that EPXmAb would facilitate a better understanding of the role of the eosinophil in IBD.[9]

The primary purpose of our study is to determine the role of the eosinophil in the pathogenesis of canine inflammatory bowel disease (IBD). Prior studies of canine IBD have focused on the involvement of lymphocytes and plasma cells mostly to the exclusion of the eosinophil. Standards for the reporting of tissue eosinophils are available, but it has now been shown that H&E staining dramatically underestimates intact and degranulated eosinophils. New strategies will be needed to more accurately assess the involvement of eosinophils in the pathogenesis of IBD.

Plans

1. Examine the prevalence, location, and activation state of eosinophils during inflammation of the GI tract using an eosinophil-specific marker (monoclonal antibodies against eosinophil peroxidase [EPXmAb]).
2. Relate severity of eosinophilic inflammation to severity of clinical signs using an established clinical scoring system.[10]
3. Relate severity of eosinophilic inflammation to serum and fecal biomarker activity.[11]
4. Relate severity of eosinophilic inflammation to changes in the intestinal microbiome using an intestinal dysbiosis index.[12]
5. Investigate the role of eosinophils at a cellular level with the aim of developing methods to reliably detect activated eosinophils for prevention, diagnosis, and development of treatment strategies.

REFERENCES

1. Berek C. Eosinophil: important players in humoral immunity. *Clin Exp Immunol*. 2015;183:57–64.
2. Linton M, *et al*. Feline gastrointestinal eosinophilic sclerosing fibroplasia. *J Feline Med Surg*. 2013;17:392–404.
3. Hendrick M. A spectrum of hypereosinophilic syndromes exemplified by six cats with eosinophilic enteritis. *Vet Pathol*. 1981;18:188–200.
4. Lyles SE, *et al*. Idiopathic eosinophilic masses of the gastrointestinal tract in dogs. *J Vet Intern Med*. 2009;23:818–823.
5. Sykes J, *et al*. Idiopathic hypereosinophilic syndrome in Rottweilers. *J Vet Intern Med*. 2001;15:162–166.
6. Day MJ, Washabau RJ. International standards for the histopathological diagnosis of gastrointestinal inflammation. *J Comp Pathol*. 2008;138:S1–S43.
7. Washabau RJ, Day MJ, Willard MD, Hall EJ, Jergens AE, Mansell J, Minami T, Bilzer T. ACVIM Consensus Statement - Endoscopic, biopsy, and histopathologic guidelines for the evaluation of gastrointestinal inflammation in companion animals. *J Vet Intern Med*. 2010;24:10–26.
8. Jergens AE, Willard MD, Hall EJ, Bilzer T, Minami T, Day MJ, Washabau RJ. Development of a modified histopathologic scoring system for the characterization of inflammatory bowel disease. *Vet Pathol*. 2014;51:946–950.
9. Bastan I, Robinson N, Rao *et al*. Assessment of eosinophil peroxidase as a potential diagnostic and prognostic marker in dogs with inflammatory bowel disease. *Am J Vet Res*. 2017;78:1–6.
10. Jergens AE, *et al*. A scoring index for disease activity in canine inflammatory bowel disease. *J Vet Intern Med*. 2003;17;291–297.
11. Collins MT. Canine inflammatory bowel disease: current and prospective biomarkers for diagnosis and management. *Compend Contin Educ Vet*. 2013;35:E1–E7.
12. Suchodolski JS, *et al*. 16S rRNA gene pyrosequencing reveals bacterial dysbiosis in the duodenum of dogs with idiopathic inflammatory bowel disease. *PLoS One*. 2012, http://dx.doi.org/10.1371/journal.pone.0039333.

1.
acvim395

The Somogyi Effect: Is It Clinically Significant?

Chen Gilor[1], DVM, PhD, DACVIM; Linda Fleeman[2], BVSc, PhD, MANZCVS
[1]Davis, CA, USA; [2]Melbourne, VIC, Australia

The "Somogyi Effect" Defined

In 1949, Dr. Michael Somogyi, a chemist, first presented the idea that people who had been given too large doses of insulin became "actually victims of chronic insulin poisoning."[1] In 1959 Somogyi published a paper titled "Exacerbation of diabetes by excess insulin action," in which he concluded that hypoglycemia, even mild and subclinical, leads to hyperglycemia and worsening of signs of diabetes. He hypothesized that this occurs as a result of activation of stress hormones.[2] Later, this became the "Somogyi effect" and the definition was reduced to "Early morning hyperglycemia due to treatment with excessive amount of exogenous insulin."[3] This definition is problematic because it can be easily confused with the dawn effect. The dawn effect is a well-established phenomenon in which morning hyperglycemia occurs as a result of increased growth hormone (GH) concentrations overnight, not as a result of insulin administration.[3]

There is early evidence for the existence of the Somogyi effect in people but in recent years overwhelming evidence accumulated against its existence. Simply demonstrating hyperglycemia that follows hypoglycemia does not prove the hypoglycemia is the cause of the hyperglycemia or that the hyperglycemia was caused by excess of counter regulatory hormones. The hyperglycemia could be the result of insulin simply wearing off or the dawn effect. Early studies, such as Somogyi's from 1959,[2] demonstrated association but not causality. Later studies on very small groups of patients demonstrated causality. For example, in a study of 10 patients, morning glycemia was compared between nights in which insulin-induced hypoglycemia was enabled using a CRI of insulin, vs. nights in which hypoglycemia was prevented by infusing dextrose while maintaining same rates of insulin CRI. In this study, nocturnal hypoglycemia was associated with increased epinephrine, cortisol, GH as well as morning fasting hyperglycemia.[4] Another study found similar results in 5 patients.[5] While these results supported Somogyi's hypothesis, multiple studies contradict it. For example, in a study of diabetic children, the counter-regulatory response to hypoglycemia was shown to be defective, resulting in prolonged nocturnal hypoglycemia.[6] While epinephrine increased as a result of hypoglycemia, the threshold for its response was remarkably low (34 mg/dL [1.9 mmol/L]). In the same study, nocturnal hypoglycemia did not result in an increase in norepinephrine, glucagon, and cortisol. GH did increase after nocturnal hypoglycemia but this could have been a normal nocturnal response.[6] More recent studies using continuous glucose monitoring (CGM) in hundreds of patients demonstrate that nocturnal hypoglycemia is not associated with morning hyperglycemia but rather with morning hypoglycemia.[7-9]

In the 5th edition of Ellenberg and Rifkin's Diabetes Mellitus textbook (1997) the Somogyi effect was still presented as a controversial cause of brittle diabetes (but not at all as a cause of insulin resistance).[10] In the next edition of this textbook (2002), the Somogyi effect was simply not discussed anymore.[11] Even if the Somogyi effect does exist in people, its frequency and clinical importance are low.

"Somogyi Effect" in Veterinary Medicine

Insulin-induced hyperglycemia was reported in dogs and cats by Nelson and Feldman.[12,13] These papers presented small series of clinical cases in which a high dose of insulin was administered, hypoglycemia was recorded, followed by hyperglycemia. However, a cause and effect between hypoglycemia and hyperglycemia was not clearly demonstrated. For example, hypoglycemia was often prolonged (sometimes lasting hours) without evidence of an acute counter-regulatory response. Also, hyperglycemia was often demonstrated only after feeding. Most importantly, no hormones were measured in these studies: The possibility that hyperglycemia was caused simply by insulin wearing off was not ruled out and the plausibility that hyperglycemia was caused by increased counter-regulatory hormones was not demonstrated. The assertion that these cases represented insulin-induced hyperglycemia was supported by improvement in clinical signs (and in some cases by lesser fluctuations in blood glucose based on glucose curves) after insulin dose was decreased. Indeed, the main importance of these studies is in describing clinical cases that presented for signs that were consistent with insulin deficiency but benefited from a reduction in insulin dose. Most of these cases were diagnosed with insulin-induced hyperglycemia when their dose of insulin was quite high (median [range] in dogs = 1.8 U/Kg [0.6–4.4] and in cats = 1.6 U/Kg [0.6–6.1]). In contrast, in cats treated with twice-daily administration of a long-acting insulin at lower doses (mean 3–4 U/cat, range 1–9 U/cat) that were managed with intensive glucose monitoring and the intention of maintaining blood glucose between 50–100 mg/dL (2.8–5.6 mmol/L), Roomp and Rand showed no correlation between hypoglycemic events and subsequent hyperglycemia.[14] Hyperglycemia was as likely to occur after an event of hypoglycemia as it was after euglycemia. However, a limitation of this study was that the majority of cases recruited were lost to follow-up before study completion and so data from cats with less stable diabetic control might not have been included.

It has been stated that "secretion of diabetogenic hormones during the Somogyi response may induce insulin resistance which can last 24 to 72 hours after the hypoglycemic episode."[15] This induction of insulin resistance is probably the most important reason to consider the diagnosis of the Somogyi effect because it is the only differential diagnosis for insulin resistance that requires the clinician to reduce the dose of insulin in the face of hyperglycemia. However, is there evidence that this induction of insulin resistance actually occurs?

The Importance of Using the Correct Terminology

The key observation of the commonly accepted veterinary definition of "Somogyi effect" is improvement of diabetic control when the insulin dose is decreased. As there is little evidence to support Somogyi's original hypothesis, a more accurate and clinically relevant term is "insulin-induced hyperglycemia."

The "Somogyi effect" is also often used to explain periods when both hypoglycemia and hyperglycemia occur. A more appropriate term which has been adopted in human medicine is "glycemic variability."[16] The term "brittle" diabetes has also been used, although this is now less preferred.

Physiological Responses to Hypoglycemia

The most serious and therefore the most well recognized consequence of insulin-induced hypoglycemia is neuroglycopenia. This can present as the patient sleeping or hiding more than normal, altered behavior, dull mentation, pacing/restlessness, weakness, star gazing, ataxia, collapse, seizure, vomiting, and/or bed-wetting. In addition, yowling may occur in cats or trembling in dogs. However, neuroglycopenia occurs only when the animal's counter-regulatory mechanisms fail.

In a healthy animal, the release of multiple counter-regulatory hormones is triggered whenever blood glucose concentration decreases towards the hypoglycemic range. These include glucagon, epinephrine, and cortisol, which together increase insulin resistance and so prevent further lowering of blood glucose. The counter-regulatory response is often deficient in diabetics. If the counter-regulatory response is appropriate, euglycemia is maintained during the period of excess insulin action. There are anecdotal reports of diabetic dogs and cats that remain euglycemic for periods up to several days following an insulin overdose.

Neuroglycopenia results if the counter-regulatory response to insulin overdose is inadequate. In contrast, the Somogyi effect is described as an excessive counter-regulatory response to hypoglycemia resulting in hyperglycemia that may persist for hours to days. Despite the lack of evidence for its existence, the concept that excess exogenous insulin is sometimes associated with unstable diabetic control (as suggested by Somogyi) is still important to explore because of the clinical observation that a reduction of insulin dose sometimes helps reducing the clinical signs of diabetes. Therefore, rather than refuting the Somogyi effect because of lack of evidence, more sophisticated hypotheses are warranted for the clinical observations made first by Doctor Somogyi and then backed by clinicians and diabetic persons over the subsequent decades.

Glycemic Variability and Hypoglycemia

Availability of CGM systems has allowed better understanding of the ongoing upward and downward fluctuations of interstitial glucose and provided insight that was not possible from intermittent blood glucose measurements. Glycemic variability becomes more pronounced as glucose intolerance progresses to overt diabetes.[17] Hypoglycemic events are more common than previously realized; in one study patients were unaware of 75% of the hypoglycemic events recorded by CGM.[18] Importantly, higher glucose variability indices including higher **mean absolute glucose** values have recently been shown to predict episodes of nocturnal hypoglycemia in insulin-treated people.[19] Individuals with increased glycemic variability therefore may present with poor diabetic control due to the occurrence of frequent hyperglycemia, whereas most of the hypoglycemic events are associated with no clinical signs and would be missed by standard intermittent blood glucose monitoring. Nevertheless, the frequent hypoglycemia can induce physiologic unawareness of hypoglycemia and impaired glucose counter-regulation, which in turn predisposes to increased risk of neuroglycopenia.[20]

Glycemic Variability and Insulin Resistance

Whereas the periods of hypoglycemia in patients with glycemic variability are likely caused by excess exogenous insulin, the mechanism(s) for the periods of hyperglycemia are poorly understood and probably relate to the multiple causes of hyperglycemia and insulin resistance in diabetes.

Diabetic cats and dogs that present with signs of poor glycemic control when receiving insulin doses >1.5 U/kg are typically assessed to have insulin resistance.

The main differential diagnoses to consider are

a. Error in insulin handling or administration
b. Obesity
c. Concurrent disease or drug therapy
d. Insulin-induced hyperglycemia (previously "Somogyi effect")

Clinical experience supports the continued inclusion of the final point on this list despite lack of strong supporting evidence because long-term improvement of clinical signs is sometimes observed following reduction of the insulin dose. The main value of refuting the "Somogyi effect" in veterinary medicine might be that there will no longer be requirement to document a hypoglycemic event coupled with rebound hyperglycemia to justify trialing insulin dose reduction. Instead, the animal's response to a decreased insulin dose can be assessed once the possibilities of dosing error, obesity, drug therapy, and the more common concurrent diseases have been ruled out. Additional strategies such as changing to a less potent, longer-acting insulin, or restriction of dietary carbohydrate can also be considered.

Close attention to the animal's clinical signs may allow a clinical suspicion that glycemic variability associated with insulin overdose is occurring. There is often a history of good diabetic control for a variable period, which might be very brief or may not have been noted in the patient's records if the owners did not contact their veterinarian when their pet was "going well." It is important to check if the insulin dose was the same or lower than the current dose during that period. Deterioration of diabetic control often begins with intermittent polydipsia, especially overnight, which sometimes follows a 3-day cycle. The owners might also report unusual behaviors, pacing/restlessness, yowling, vomiting, urinating, etc., especially in the early hours of the morning. In some cases, hypoglycemia might have been recorded or suspected, and then disregarded as an error. Affected animals might present with very unpredictable blood glucose results that vary anywhere from 30–650 mg/dL (1.5–35 mmol/L) despite consistent insulin dosing. Key factors include failure to improve with increases of the insulin dose, and in some cases weight gain despite otherwise suboptimal diabetic control.

Substantial improvement can often be achieved with small adjustments of the insulin dose; often with a decrease of just 0.5–1.0 U. The improvement can be immediate or may be gradual over 1–2 weeks. In contrast, dogs and cats with poor diabetic control due to an inadequate insulin dose will typically become much more polydipsic and polyuric within 12 hours of such a trial decrease of their insulin dose.

This immediate worsening of clinical signs will confirm that insulin overdose was not the reason for poor diabetic control in those cases, and the previous insulin dose can be promptly resumed and the investigation for another cause continued.

How to Manage and Avoid Glycemic Variability Due to Insulin Overdose

Rapid escalation of insulin dose in an effort to "stabilize" the diabetes appears to be one of the most common causes of glycemic variability in veterinary patients. It is usually recommended that frequent dose adjustment and "micro-management" of blood glucose is avoided. The animal's response to each insulin dose adjustment should ideally be monitored for 1–2 weeks before another increase of the dose is considered. The goal is for step-wise improvement of diabetic control with each increase of the insulin dose. If there is no apparent improvement with increasing insulin doses, it is important to re-evaluate the patient's treatment regimen.

It is therefore recommended that the insulin dose is increased only if all of the following conditions are met:

a. Some improvement of clinical signs (i.e., decreased water intake, decreased urine output, weight gain, decreased polyphagia, and/or increased activity/interaction) was clearly demonstrated with lower insulin doses.
b. Some clinical signs of diabetes are still present (e.g., polyuria/polydipsia, lethargy, polyphagia).
c. There is evidence that blood glucose is often above the renal threshold (e.g., based on blood or interstitial glucose measurements in the clinic or at home, or based on consistently positive urine glucose measurements at home).

If there is clinical suspicion that glycemic variability might be occurring, the insulin dose should be decreased, usually by 0.5–1.0 U, and the animal's response to the change closely monitored. If clinical signs were previously well controlled, a reasonable approach is to resume the insulin dose that the dog or cat was receiving when there was good diabetic control and monitor the response for a minimum of 2 weeks. This information will then provide baseline data for ongoing monitoring and treatment adjustment. Small dogs and cats likely have increased risk for glycemic variability because dose adjustments are typically larger in relation to their body weight. For these patients, insulin dosing pens provide more accurate and precise dosing than syringes, resulting in less day-to-day variability of insulin dose and lower risk of insulin-induced hypoglycemia.[21]

Although clinical signs are a reliable means of identifying periods of poor diabetic control, they are very unreliable at detecting when there is an increased risk of hypoglycemia. Regular urine dipstick testing for glucose at home is a practical method of monitoring for periods when the blood glucose is below the renal threshold. It is recommended that the insulin dose is reduced and monitoring continued whenever there is persistent negative glycosuria for 1–2 weeks.

Conclusion

The key observation with insulin-induced hyperglycemia is documenting improvement of diabetic control when the insulin dose is decreased. Although the mechanism(s) for insulin-induced hyperglycemia (or insulin-induced glycemic variability) are unproven and supporting evidence is mostly anecdotal, the concept has great clinical relevance. There is danger in refuting this phenomenon due to lack of evidence because insulin dose reduction might improve glycemic control and even be life-saving for many diabetic patients.

References

1. Walker H Jr. Michael Somogyi, Ph.D. (1883–1971). *Metabolism.* 1972;21:589–590.
2. Somogyi M. Exacerbation of diabetes by excess insulin action. *Am J Med.* 1959;26:169–191.
3. Rybicka M, Krysiak R, Okopien B. The dawn phenomenon and the Somogyi effect - two phenomena of morning hyperglycaemia. *Endokrynol Pol.* 2011;62:276–284.

4. Perriello G, De Feo P, Torlone E, *et al*. The effect of asymptomatic nocturnal hypoglycemia on glycemic control in diabetes mellitus. *N Engl J Med*. 1988;319:1233–1239.
5. Bolli GB, Gottesman IS, Campbell PJ, *et al*. Glucose counterregulation and waning of insulin in the Somogyi phenomenon (posthypoglycemic hyperglycemia). *N Engl J Med*. 1984;311:1214–1219.
6. Matyka KA, Crowne EC, Havel PJ, *et al*. Counterregulation during spontaneous nocturnal hypoglycemia in prepubertal children with type 1 diabetes. *Diabetes Care*. 1999;22:1144–1150.
7. Guillod L, Comte-Perret S, Monbaron D, *et al*. Nocturnal hypoglycaemias in type 1 diabetic patients: what can we learn with continuous glucose monitoring? *Diabetes Metab*. 2007;33:360–365.
8. Hoi-Hansen T, Pedersen-Bjergaard U, Thorsteinsson B. The Somogyi phenomenon revisited using continuous glucose monitoring in daily life. *Diabetologia*. 2005;48:2437–2438.
9. Choudhary P, Davies C, Emery CJ, *et al*. Do high fasting glucose levels suggest nocturnal hypoglycaemia? The Somogyi effect-more fiction than fact? *Diabet Med*. 2013;30:914–917.
10. *Ellenberg & Rifkin's Diabetes Mellitus*, 5th ed. McGraw-Hill; 1997.
11. *Ellenberg & Rifkin's Diabetes Mellitus*, 6th ed. McGraw-Hill; 2002.
12. Feldman EC, Nelson RW. Insulin-induced hyperglycemia in diabetic dogs. *J Am Vet Med Assoc*. 1982;180:1432–1437.
13. McMillan FD, Feldman EC. Rebound hyperglycemia following overdosing of insulin in cats with diabetes mellitus. *J Am Vet Med Assoc*. 1986;188:1426–1431.
14. Roomp K, Rand J. Rebound hyperglycaemia in diabetic cats. *J Feline Med Surg*. 2016;18:587–596.
15. Nelson RW. Canine Diabetes Mellitus. In: Feldman E, Nelson R, Reusch C, eds. Canine and feline endocrinology, 4th ed. Saunders; 2015.
16. Service FJ. Glucose variability. *Diabetes*. 2013;62:1398–1404.
17. Chen T, Xu F, Su JB, *et al*. Glycemic variability in relation to oral disposition index in the subjects with different stages of glucose tolerance. *Diabetol Metab Syndr*. 2013;5:38.
18. Gehlaut RR, Dogbey GY, Schwartz FL, *et al*. Hypoglycemia in type 2 diabetes - more common than you think: a continuous glucose monitoring study. *J Diabetes Sci Technol*. 2015;9:999–1005.
19. Klimontov VV, Myakina NE. Glucose variability indices predict the episodes of nocturnal hypoglycemia in elderly type 2 diabetic patients treated with insulin. *Diabetes Metab Syndr*. 2016;16:30182–30185.
20. Fanelli CG, Porcellati F, Pampanelli S, *et al*. Insulin therapy and hypoglycaemia: the size of the problem. *Diabetes Metab Res Rev*. 2004;20:S32–S42.
21. Thompson AL, Lathan P, Fleeman LM. Update on insulin treatment for dogs and cats: insulin dosing pens and more. *Veterinary Medicine: Research and Repo*rts. 2015;6:129–142.

acvim397

A Novel HbA1c Assay for the Monitoring of Canine Diabetes

Ian K. Ramsey, BVSc, PhD, DSAM, DECVIM, FHEA, FRCVS; Anne F. Goemans, BVMS, MSc, MRCVS; Susanna J. Spence, BVMS, MRCVS
Glasgow, UK

INTRODUCTION

Glucose chemically reacts with many proteins in the body. In the veterinary context, this is used diagnostically in the measurement of fructosamines, which provide an estimate of the average glucose concentration over the preceding 2 to 3 weeks.[1] However, several factors, other than plasma glucose concentration, affect fructosamine concentration including hypoproteinemia, hyperlipidemia, and azotemia.[2] Moreover, few diabetic patients visit their veterinarian every 2 to 3 weeks - many diabetic patients visit every 3 months or less.

Glucose also reacts with hemoglobin forming glycosylated hemoglobin. Glycosylation at one specific site on the hemoglobin molecule forms hemoglobin A1c (HbA1c). In people HbA1c has been established as a highly specific and reliable biomarker for the long-term control of diabetes mellitus.[3,4] Moreover, HbA1c concentrations are associated with outcome in human diabetes, whereas fructosamine values are not.[3] Compared to fructosamine measurements, HbA1c is less affected by pathologic conditions other than diabetes mellitus.[5] In dogs, there have been no studies looking at the relationship between fructosamine or HbA1c and outcome in diabetes mellitus.

The measurement of HbA1c in dogs has been previously described.[6-9] Several methods developed for human use have been evaluated for use in canine diabetes mellitus but none have been adopted into standard clinical practice, either due to technical difficulties or lack of commercial availability. All published studies have, however, shown that canine diabetics have higher average ratios of glycosylated hemoglobin to total hemoglobin compared to non-diabetic controls.[6-9]

An immunoturbidimetric assay, previously validated for use in humans, has been assessed over the last 2 years for use in canine diabetes mellitus. The DCA™ Vantage (Siemens, Surrey, UK) uses murine anti-human HbA1c monoclonal antibodies specific to the last few amino acids on the N-terminal of the β-chain.[10] HbA1c in the sample competitively inhibits latex agglutination by binding to anti-HbA1c antibodies. Total hemoglobin is measured concurrently and the result is expressed as millimoles of HbA1c per mole of hemoglobin.

VALIDATION

We have demonstrated using pooled hemolyzed erythrocytes from non-diabetic dogs that *in vitro* incubation with increasing glucose concentrations increases the proportion of HbA1c measured.[11] The linearity was assessed by mixing known samples in differing proportions and was found to be satisfactory within the expected working range.[11]

Repeatability

As each of the reagents for each assay are contained within a single use cartridge it was not possible to assess the intra-assay imprecision. The inter-assay precision when measured using 5 repeats of 6 samples in the same day was good (mean CV=4.08%, range 1.16%–6.10%), with a maximum difference of 2 mmol/mol between the highest and lowest measurement within any sample.[11] The inter-assay imprecision was also assessed over longer term using the quality control material (stabilized human hemoglobin) supplied by the manufacturer over the course of the study.[11] This was also acceptable for normal HbA1c concentrations with a mean CV of 2.97%. The inter-assay imprecision with the high HbA1c commercial control had a higher CV of 8.81%. These figures are comparable to CVs obtained by several independent studies validating the DCA Vantage for use in human specimens with a CV <3.1%.[12,13]

Storage and Interference

The effect of 5 days of storage at 4°C on measured HbA1c was minimal.[11] Samples could also be frozen for up to 4 weeks without significant variation but after 8 weeks of freezing small changes (maximum 14% from baseline) appeared.

The interference by lipemia was tested samples with gross hyperlipemia. There was a trend for increasing interference from lipids as the triglyceride and cholesterol concentrations increased. The largest difference was 6 mmol/mol. Severe hemolysis also increased measured HbA1c slightly. No significant difference was found between measurements obtained from blood anticoagulated with EDTA or heparin before for analysis which suggests that EDTA and heparin anticoagulants can be used interchangeably in the measurement of canine HbA1c.[11]

Reference Interval

The reference interval was **9–18.5 mmol/mol** for canine HbA1c in a hospital population of 60 dogs, which is lower compared to the human RI of 20–42 mmol/mol.[11] This partially explainable by the lower glucose permeability of canine erythrocytes compared to human erythrocytes.

Clinical Use

HbA1c is likely to provide a better estimate of long-term control than fructosamine because it assesses average glucose concentration over a longer period. A high HbA1c concentration should trigger an investigation into possible factors that might cause diabetic instability. This would include an assessment of the owner's storage of insulin, syringe selection, injection technique, concurrent diseases as well as specific tests for insulin dose such as a blood glucose curve.

To date we have studied three groups of dogs: an uncontrolled diabetic group which included a few patients with concurrent hyperadrenocorticism, a normoglycemic control group, and a normoglycemic group with hyperadrenocorticism.[14] We have found that diabetic dogs have statistically higher HbA1c concentrations than the other 2 groups that are sufficiently large to be clinically relevant. There is no statistical correlation with this group's fructosamine measurements. Several patients have had sequential HbA1c and fructosamine measured over at least a 1 year period. In these patients, HbA1c and fructosamine measurements have agreed with the clinical course.[14]

Limitations

The major limitation of this assay is that it does not work in cats. Feline hemoglobin has a different amino acid sequence replacing the N terminal valine with a glycine at its N - terminal end.[9]

Human	MET	**VAL**	**HIS**	LEU	THR	*PRO*	GLU
Dog	MET	**VAL**	**HIS**	LEU	THR	**ALA**	GLU
CAT	MET	*GLY*	*PHE*	LEU	THR	**ALA**	GLU

The second limitation is that - like fructosamines - a high HbA1c does not indicate the cause of the unstable diabetes mellitus. Dogs that are given too much insulin and develop insulin-induced hyperglycemia ("Somogyi" overswing) are likely to have a high HbA1c, similar to dogs that have been given too little insulin.

It is also worth noting that the HbA1c assay works on erythrocytes - so submitting a serum sample is likely to be a fruitless exercise!

Summary

Currently it is unclear which are the most important laboratory parameters that veterinarians should be monitoring in diabetic dogs in order to reduce complications (such as cataracts, hypoglycemic episodes, ketoacidosis, pancreatitis, urinary tract infections) and increase survival time. In humans, HbA1c has been widely accepted in this role but it does not indicate how to change an individual's dose, but rather HbA1c does indicate (far more clearly that fructosamines) which individuals are more likely to suffer from complications of their diabetes and therefore may need a change in insulin dose, timing, or nutrition. The immunoagglutination assay for HbA1c provides a reliable method of canine HbA1c measurement with good analytical performance. HbA1c measurements are being investigated for their correlation with long term glycemic control and as markers of outcome in recently stabilized diabetic dogs.

References

1. Reusch CE. *et al. J Vet Intern Med.* 1993;7:177.
2. Reusch CE, Haberer B. *Vet Rec.* 2001;148:370.
3. The Diabetes Control and Complications Trial Research Group. *N Engl J Med.* 1993;329:977.
4. Stratton IM, *et al. BMJ.* 2000;321:405.
5. Little RR, Rohlfing CL. *Clin Chim Acta.* 2013;418:63.
6. Davison LJ, *et al. J Small Anim Pract.* 2002;43:526.
7. Haberer B, Reusch CE. *J Small Anim Pract.* 1998;39:510.
8. Marca MC, Loste A. *J Small Anim Pract.* 2000;41:189.
9. Catchpole B, *et al. Vet Rec.* 2008;162:47.
10. Szymezak J, *et al. Clin Chem Lab Med.* 2008;46:1195.
11. Goemans AF, *et al. Vet Clin Path.* (accepted for publication) 2017.
12. Lenters-Westra E, Slingerland RJ. *Clin Chem.* 2014;60:1062.
13. Lenters-Westra E, Slingerland RJ. *J Diabetes Sci Technol.* 2009;3:418.
14. Spence S, *et al.* In: Proceedings of the British Small Animal Veterinary Association; 2016. http://www.vin.com/members/cms/project/defaultadv1.aspx?id=7227699&pid=14095.

acvim398

Monitoring Canine Cushing's Disease Without ACTH

Ian K. Ramsey, BVSc, PhD, DSAM, DECVIM, FHEA, FRCVS; Laura Macfarlane, BVMS, DECVIM, MRCVS
Glasgow, UK

Introduction

Hyperadrenocorticism (HAC) is commonly treated with trilostane (Vetoryl; Dechra, UK).[1-3] This treatment is usually monitored using a combination of clinical signs and ACTH stimulation tests (ACTH-STs), despite this test having never being validated for this purpose. The use of ACTH-STs has only been previously validated for mitotane (o,p'-DDD), a long-acting cytotoxic drug which causes persistent adrenal suppression.[4] As trilostane is relatively short-acting, there are concerns regarding the variation in the results from ACTH-STs, depending on the time since the last dose of trilostane and whether the results reflect clinical control.[5-7] In addition, in some countries synthetic ACTH is expensive or hard to obtain. In 2013, the supply of ACTH became inconsistent in Europe and so efforts were started to find an alternative to ACTH-STs. The result is pre-pill cortisol monitoring.[8] This technique appears to be cheaper, easier, and better correlated with clinical control than performing ACTH-STs. These notes will describe the research behind this technique and provide recommendations on how to interpret the results.

Research Behind Pre-Pill Monitoring

To provide an objective assessment of the clinical control of dogs with HAC receiving trilostane, a questionnaire for owners and an associated scoring system was developed using a panel of 14 independent veterinarians. Using this scored questionnaire, dogs were first classified as clinically unwell or well. The unwell dogs were further divided into 2 groups - those with low cortisol and those that had normal or high cortisol concentrations. The well dogs were further classified as under-controlled or well controlled according to the scored questionnaires. At the same time a blood sample was taken before trilostane administration (pre-trilostane). The trilostane dose was then administered, along with the dog's normal meal. Three hours after the trilostane had been given, an ACTH-ST was performed by taking a blood sample (3-hour post-trilostane) and then administering 0.005 mg/kg of tetracosactide intravenously. This time point was chosen, as in most dogs it is when the lowest cortisol concentration is found following trilostane administration.[9] A third blood sample was taken 1 hour after tetracosactide had been administered (post-ACTH). Cortisol concentration was measured concurrently on all three samples using a competitive chemiluminescent enzyme immunoassay (Immulite® 2000 cortisol; Siemens Healthcare, UK) that has been validated for dogs.[10] The ability of the 3 cortisol results to discriminate between dogs that were clinically well controlled or clinically under-controlled was then examined.

In total, 110 tests from 67 individual dogs were included in this study.[8] The inclusion of multiple visits from some dogs was shown to not introduce a significant bias into the statistical modelling. A total of 37 tests were performed at the University of Glasgow, and 73 tests were performed at external practices. In total, 94 tests were performed on dogs receiving once-daily trilostane and 16 on dogs receiving twice-daily trilostane. There were 4 dogs diagnosed with adrenal-dependent and 23 dogs with pituitary-dependent HAC. In 40 dogs, the distinction was not made. Seventeen of the tests from 11 dogs were assessed, from the owners' questionnaires, to have been unwell at the time of sampling and were removed from statistical analysis. None of the tests on unwell dogs had a pre-trilostane or post-ACTH cortisol concentration <40 nmol/l (1.4 μg/dl), thereby excluding iatrogenic hypocortisolism.

All three cortisol results were significantly correlated to the total owner score (pre-trilostane cortisol $r=0.38$ $p<0.000$; 3-hour post-trilostane cortisol $r=0.32$ $p<0.002$; and post-ACTH stimulation cortisol $r=0.27$ $p<0.010$). All three cortisol results were also significantly correlated to each other (pre- and 3-hour post-trilostane to post-ACTH $r=0.56$ $p<0.001$, $r=0.68$ $p<0.001$, respectively, and pre- to 3-hour post-trilostane $r=0.77$ $p<0.000$). All three cortisol results were significantly lower in the dogs with excellent control compared to those that were under-controlled. ROC curve analysis of pre-trilostane, 3-hour post-trilostane, and post-ACTH cortisol concentrations showed areas under the curves of 0.73, 0.73, and 0.64, respectively, suggesting that the first two measures were superior to the post-ACTH cortisol concentration. The optimal cutoff for pre-trilostane cortisol was found to be 138 nmol/l.

The 3-hour post-trilostane cortisol was <40 nmol/l (1.4 µg/dl) in 31 tests from well dogs, of which 23 had >40 nmol/l (1.4 µg/dl) pre-trilostane or post-ACTH, suggesting that 3-hour post-trilostane cortisol was not specific for over-suppression. There were a total of 14 tests that had a cortisol concentration <40 nmol/l (1.4 µg/dl) either pre-trilostane or post-ACTH or at both time points. All these tests were classified as coming from well dogs based on the results of the owner questionnaire. Many of the dogs had their doses of trilostane reduced after the veterinarians received the results (even though there were no clinical signs suggesting that this was necessary). This included the dogs that just had a low post-ACTH cortisol (i.e., pre-trilostane cortisol was >40 nmol/l (1.4 µg/dl). Too few dogs were maintained on their previous dose to draw meaningful conclusions; however, it is worth noting that two dogs with pre-trilostane cortisol <40 nmol/l (1.4 µg/dl) and post-ACTH stimulation cortisol >40 nmol/l (1.4 µg/dl) were identified. One dog responded positively to trilostane withdrawal, and a further case that was maintained on the same dose had signs consistent with hypoadrenocorticism 3 months later. These results suggest that dogs with a pre-trilostane cortisol <40 nmol/l (1.4 µg/dl) are at risk of hypocortisolism.

Therefore, using a **target range of 40 to 138 nmol/l (1.4 to 5.0 µg/dl),** the pre-trilostane cortisol concentrations were better than the post-ACTH cortisol concentrations at discriminating between dogs whose HAC was well controlled and those that were under-controlled with a **sensitivity of 55.4% and a specificity of 86.5%.**[8] In addition, pre-trilostane monitoring may be better able to identify over-suppression when the result is <40 nmol/l (1.4 µg/dl), but further case examples and research are needed. Post-trilostane (pre-ACTH) cortisol was also able to discriminate between dogs whose HAC was well controlled and those that were under-controlled; however, it was not able to distinguish between those dogs that were well controlled and those that were over-suppressed.[8] **No cortisol result correlated well enough with the clinical score to be used as standalone monitoring test.**

Interpretation of Pre-Trilostane Monitoring

The aim of trilostane therapy should be to satisfactorily control clinical signs without resulting in iatrogenic hypocortisolism. The studies that have been performed to date question the existing recommendations that ACTH-STs should be used to ensure either adequate control or to detect dogs with over-suppression. Pre-trilostane cortisol appears to be an objective measurement that has the potential to balance safety with effective therapy; however, further studies are needed, and the importance of owners' and clinicians' observations is paramount. It appears that this monitoring technique can also be used in dogs that receive twice-daily trilostane or that have adrenal tumors, but too few dogs have been monitored for long enough to be sure. The effect of stress on pre-pill cortisol has not been investigated yet, so it is recommended that efforts should be made to reduce stress at the time of sampling.

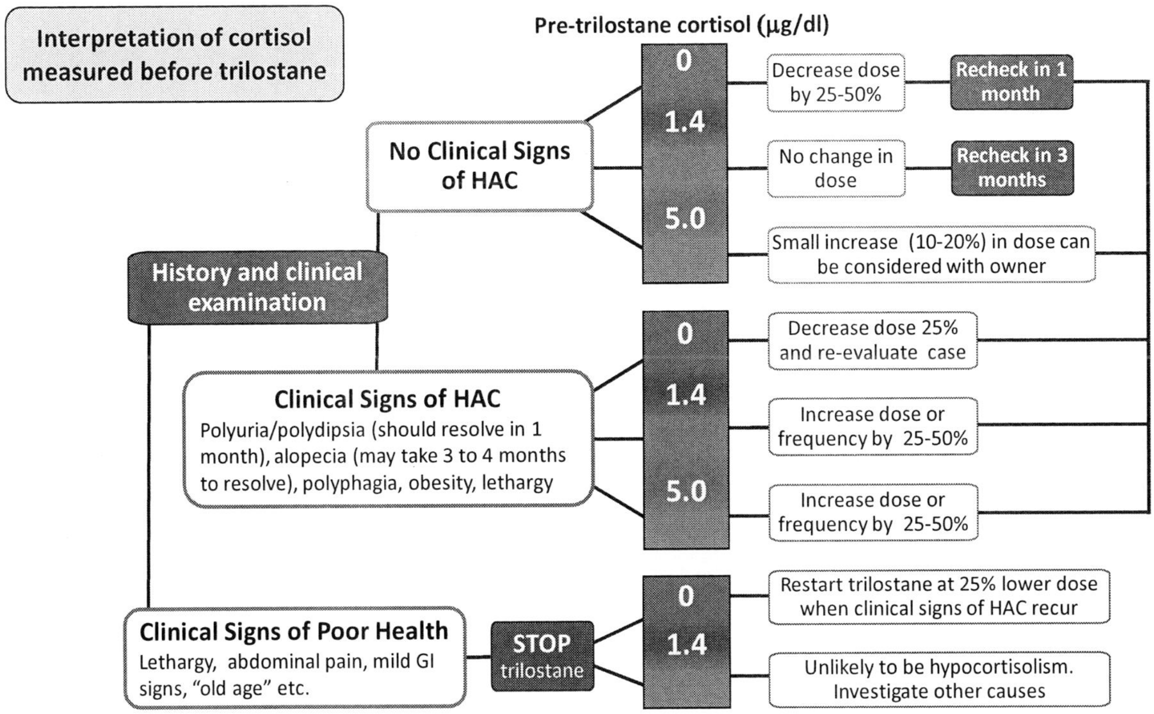

*Note: Dose increments are suggestions, and available capsules sizes will influence final selection.

Declaration of Conflict of Interest

The study was funded by Dechra® Veterinary Products.

References

1. Neiger R, *et al. Veterinary Record.* 2002;150:799.
2. Ruckstuhl NS, *et al. American Journal of Veterinary Research.* 2002;63:506.
3. Ramsey IK. *Veterinary Clinics of North America: Small Animal Practice.* 2010;40:269.
4. Dunn KJ, *et al. Veterinary Record.* 1995;137:161.
5. Bell R, *et al. Veterinary Record.* 2006;159:277.
6. Midence JN, *et al. Journal of Veterinary Internal Medicine.* 2015;29:1529.
7. Wehner A, *et al.* Proceedings of the 23rd ECVIM-CA Congress. *Journal of Veterinary Internal Medicine.* 2014;28:743.
8. McFarlane L, *et al. Veterinary Record.* 2016;179:597–605.
9. Griebsch C, *et al. Journal of Veterinary Internal Medicine.* 2014;28:160.
10. Singh AK, *et al. Journal of Veterinary Diagnostic Investigation.* 1997;9:261.

acvim400

Subclinical Bacteriuria in Dogs and Cats

Mary F. Thompson[1], BVSc (Hons), PhD, DACVIM (SAIM), MANZCVS (Assoc); Joanna D. White[2], BVSc (Hons), MVS (Epi), DACVIM (SAIM), PhD
[1]Murdoch, WA, Australia; [2]North Ryde, NSW, Australia

Abstract

There is increasing awareness that subclinical bacteriuria (SB), defined as the presence of a positive urine culture without clinical signs of lower urinary tract disease, occurs in dogs and cats. No clear differences in urinalysis or bacteria (species or colony count) have been identified between animals with urinary tract infections and SB. Currently, the treatment of SB in dogs and cats is not recommended for the majority of animals. While withholding treatment of SB in non-azotemic, older cats has been demonstrated to not adversely affect survival, the necessity of screening for or treating SB in animals with concurrent immunosuppression or urogenital disease is unknown.

Background

There has been increasing recognition that not all bacterial infections of the lower urinary tract in dogs and cats are associated with clinical signs. In people, asymptomatic bacteriuria (ASB) has been clearly defined. Observations that treatment of ASB is not associated with a reduction in the frequency of future episodes symptomatic urinary tract infections (UTI), pyelonephritis, the development of renal impairment or survival has informed the current recommendations that ASB not be treated in the vast majority of patients.[1]

Knowledge of subclinical bacteriuria (SB) in dogs and cats is evolving. SB is currently defined as the presence of bacteria in the urine as determined by positive bacterial culture, in the absence of clinical evidence of UTI.[2]

Prevalence

Dogs: Subclinical bacteriuria has been reported in a number of canine patient groups with varying prevalence. These include dogs with diabetes mellitus, hyperadrenocorticism, recurrent UTIs, obesity and dogs treated with long term cyclosporine or glucocorticoids for management of inflammatory skin disease.[3-8] Subclinical bacteriuria has also been detected in dogs presenting for elective orthopaedic (non-neurologic) and soft tissue surgery procedures and on screening of healthy female dogs.[9,10]

Cats: Determining the true prevalence of SB in cats is hampered by studies where UTIs and SB are not separately defined. Positive urine cultures are identified in 12% cats with hyperthyroidism, 13% cats with diabetes mellitus, and 22% cats with chronic kidney disease (CKD).[11,12] Subclinical bacteriuria specifically was identified in 29% of cats with a range of concurrent disorders,[13] 21% of cats with CKD[14] and 13% of healthy, older (>7 years) non azotaemic cats.[15]

Urinalysis Findings in SB

In people with ASB, pyuria is identified in between 30% (healthy adult women) and 90% (elderly people) of people with SB.[16,17] Similarly, pyuria is frequently identified in dogs and cats with SB. Pyuria has been detected in 19% of healthy dogs, 17% of morbidly obese dogs and 75% of dogs treated with long-term cyclosporine and SB. Pyuria is present in between 18 and 65% of cats with SB and concurrent disease (hyperthyroidism and CKD respectively), in 87% cats with SB and CKD and almost 50% of older, healthy cats with SB.[11,15,18]

Microbial Pathogens in SB

It is not yet evident that a UTI is associated with different bacterial species, different phylogenetic groups or different microbial loads when compared with SB or ASB. Escherichia coli is the most common pathogen isolated from the urine of humans, dogs and cats regardless of their clinical presentation but it is not the only pathogen associated with asymptomatic bacteriuria in humans. Asymptomatic bacteriuria has been reported in association with *Proteus mirabilis, Klebsiella pneumoniae, Enterobacter* spp., *Pseudomonas aeruginosa, Enterococcus* spp., *Staphylococcus aureus, Streptococcus agalactiae,* and others.[19] The prevalence of the different species varies markedly according to patient group; for example, *E. coli* and *Enterococcus* spp. each represent approximately 25% of isolates in healthy men with ASB, but *E. coli* is isolated in 86% of ASB episodes in pregnant women, with *Enterococcus* isolated in just 4%. The reasons for these predilections is unknown.

The numbers of reported cases of SB in veterinary patients are too small to draw such comparisons but a number of bacterial species have been implicated including *E. coli, Enterococcus faecalis, Enterococcus faecium,* beta-haemolytic *Streptococcus, Staphylococcus pseudintermedius, Proteus mirabilis, Klebsiella* spp., *Streptococcus canis, Enterobacter* spp., and *Pseudomonas aeruginosa*.[3,5,7,10]

It has been suggested that SB in dogs and cats is associated with low colony counts when cystocentesis samples are cultured. However, quantitative culture frequently reveals colony-forming unit numbers in excess of 10^5 ml.[5,10,15] Collection of voided urine is not recommended for diagnosis of subclinical bacteriuria, as false positive cultures are more likely due to the risk of external contamination and are especially difficult to interpret in the absence of clinical signs of LUTD.[20,21]

Treatment

Treatment of subclinical bacteriuria in humans is reserved for a small number of patient categories. It is recommended that pregnant women be routinely screened via urine culture and treated with antimicrobials if this test is positive, largely due to an increased risk of pyelonephritis (thought to be related to changes in ureteral pressure due to the enlarging uterus).[1] Likewise, individuals with ABU undergoing invasive urologic surgery; e.g., trans-urethral resection of the prostate, have an increased risk of post-operative symptomatic UTI and sepsis and antimicrobial treatment is recommended.[1]

In epidemiologic studies, SB was not associated with creatinine concentration in cats with CKD.[14] In non-azotemic older cats, the presence of SB was not associated with body weight or creatinine concentration. In this cohort, there was no association between SB and survival even though treatment was withheld.[15]

Current veterinary recommendations are that dogs and cats with SB not be treated with antimicrobials.[2] Pyuria is not an indication for antimicrobial treatment in people with ASB or dogs and cats with SB.[1,2] Treatment of SB in dogs and cats is not without potential adverse consequences. Antimicrobial treatment can result in the bacterial acquisition of antibiotic resistance[22,23] and uropathogenic *E. coli* are a potential zoonosis, capable of being transmitted between people and animals living in the same household.[24]

There are several categories of disease in dogs and cats that are associated with an increased risk of SB where clinical treatment trials of antimicrobials would be useful. In patients with CKD, there is an increased risk that if bacteriuria actually represents pyelonephritis/renal isolate, a failure to recognise and treat the infection could lead to progressive kidney disease. In dogs and cats with severe systemic disease and SB, the significance of the positive urine culture can be hard to determine. It would also be helpful to know more about the risks of untreated SB in animals with urinary tract implants, such as stents, and iatrogenically immunosuppressed animals such as those being treated with potent immunosuppressive or chemotherapy drugs.

Conclusion

SB is common in dogs and cats and is usually associated with cytological evidence of inflammation in the form of pyuria. Despite this, most dogs and cats with SB do not require treatment. However, there is a need for further studies to identify the clinical characteristics of any dogs or cats at risk of urosepsis or progressive renal damage from untreated SB.

References

1. Nicolle LE. *et al. Clin Infect Dis.* 2005;40(5):643-654.
2. Weese JS *et al.* In: ACVIM Forum. Denver, CO. 2016;420-424.
3. McGuire *et al. J Am Anim Hosp Assoc.* 2002;38:541-544.

4. Seguin MA *et al. J Vet Intern Med.* 2003;17:622-631.
5. Wynn SG *et al. PeerJ.* 2016;4:e1711.
6. Torres SM *et al. J Am Vet Med Assoc.* 2005;227(2):239–243.
7. Peterson AL *et al. Vet Dermatol.* 2012;23(3):201-e43.
8. Ihrke PJ *et al. J Am Vet Med Assoc.* 1985;186(1):43–46.
9. McGhie JA *et al. Aust Vet J.* 2014;92(1–2):33–37.
10. Wan SY *et al. J Am Vet Med Assoc.* 2014;245:106–112.
11. Mayer-Roenne B *et al. J Feline Med Surg.* 2007;9(2):124–132.
12. Bailiff NL *et al. J Vet Intern Med.* 2006;20(4):850–855.
13. Litster A *et al. Vet Micro.* 2009;136(1–2):130–134.
14. White JD *et al. J Feline Med Surg.* 2013;15(6):459–465.
15. White JD *et al. J Vet Intern Med.* 2016;30(6):1824–1829.
16. Nicolle LE. *Clin Geriatr Med.* 2009;25(3):423–436.
17. Hooton TM *et al. N Engl J Med.* 2000;342:992–997.
18. White JD *et al. J Feline Med Surg.* 2013;15(6):459–465.
19. Ipe DS *et al. FEMS Microbiol Lett.* 2013;346:1–10.
20. Weese JS *et al. J Vet Med Intern.* 2011;article ID:263768;9.
21. Sorensen TM *et al. Vet J.* 2016;216:168–173.
22. Drazenovich N *et al. J Vet Intern Med.* 2004;18:301–306.
23. Freitag T *et al. J Vet Intern Med.* 2006;20:245–249.
24. Johnson JR, Clabots C. *Clin Infect Dis.* 2006;43(10):e101–108.

acvim401

UTI Treatment and Prevention: What You Need to Know to Help

Gary P. Oswald, DVM, DACVIM (Small Animal Internal Medicine)
Clearwater, FL, USA

Introduction

Bacterial urinary tract infection (UTI) is a leading cause of urinary disease in dogs. Antibiotic therapy is routinely prescribed for dogs diagnosed with UTI. UTI treatment is a leading contributor to overall antibiotic use in companion animals. There is a genuine concern about the use, misuse and overuse of antibiotic therapy in our patients, especially in regard to the development of antibiotic resistance. There have been numerous individual studies and reviews regarding the diagnosis, treatment and prevention of canine UTI. In 2011 the International Society for Companion Animal Infection Diseases (ISCAID) published comprehensive guidelines to provide consensus guidelines to assist in the diagnosis and management of upper and lower urinary tract infections in dogs and cats. These guidelines were recently updated in 2016. The purpose of these publications is to improve antibiotic prescribing practices in UTI patients as part of a broader antibiotic stewardship program.

Diagnosis and Treatment of Uncomplicated Urinary Tract Infection

An uncomplicated UTI is an occasional bacterial infection of the urinary bladder in an otherwise healthy individual with likely normal urinary tract anatomy and function. Typical clinical signs of a lower UTI are present, characterized by dysuria, pollakiuria, and/or increased urgency of urination. Whenever possible, urine should be collected via cystocentesis when a UTI is suspected. Urinalysis generally reveals the presence of pyuria and bacteriuria, which supports evidence of a UTI; hematuria and proteinuria are also often present. Bacteria can be present in the urine in the absence of clinical signs (covert bacteriuria/subclinical bacteriuria) and is not always associated with an active UTI. Therefore, the clinician must interpret the clinical evaluation, gross and cytological appearance of the urine in parallel to determine the likelihood of a clinically significant UTI. Urine culture should be considered to confirm the presence of bacterial infection, identify the presence of resistant bacteria that may not respond to initial antibiotic therapy, and to help differentiate reinfection from relapse, should a UTI return following initial therapy.

Antimicrobial therapy is recommended and initial therapy with amoxicillin (11–15 mg/kg PO q 8h) or trimethoprim-sulfonamide (15 mg/kg PO q12 h) is recommended to provide a narrow antibiotic spectrum while maintaining optimal efficacy. Uncomplicated UTIs are generally treated for 7–14 days. Providing the full course of an appropriate antibiotic has been administered correctly by the owner, then there is no strong indication that measures beyond monitoring of clinical signs is necessary to determine the efficacy of treatment. If culture and susceptibility testing is performed and demonstrates an isolate that is resistant *in vitro* to initial antibiotic therapy but there has been a positive clinical response, then maintaining the current antibiotic is acceptable and follow-up urinalysis, including culture, is indicated after treatment has been completed to verify resolution of infection. If culture and susceptibility results indicate that an isolate is not susceptible to the chosen antimicrobial and there is a lack of clinical response, then therapy with the original antibiotic should be discontinued and treatment with an alternative drug begun based on the culture and susceptibility result.

Diagnosis and Treatment of Complicated Urinary Tract Infection

A complicated UTI is a bacterial infection that occurs in association with an anatomic or functional urinary tract abnormality or a comorbidity that predisposes the patient to persistent infection, recurrent infection, or treatment failure. An identifiable abnormality is not always confirmed because of the difficulty diagnosing some anatomical, functional, metabolic, or other abnormalities. Comorbid medical conditions such as urinary calculi, urinary neoplasia, prostatitis, neurogenic bladder, diabetes mellitus, and immunocompromising disorders (hyperadrenocorticism, immunosuppressive drug therapies) often are associated with recurrent UTIs. Recurrent UTIs often occur 3 or more times during a 12-month period.

Recurrent UTIs can be defined as bacterial reinfection or relapse. **Reinfection** is recurrence of a UTI within 6 months of completing apparently successful antibiotic treatment and isolation of a different bacterial microorganism. **Relapse** is recurrence of a UTI within 6 months of completing apparently successful treatment and isolation of an indistinguishable bacterial organism from the one that was present previously; presumably relapse occurs due to failure to completely eliminate the pathogen with prior treatment. Relapses tend to occur earlier than reinfections (i.e., within weeks rather than months). **Refractory** infection is similar to a relapse, except that it is characterized by persistently positive culture result during treatment (despite *in vitro* susceptibility to the antimicrobial), with no period of eliminated bacteriuria during or after treatment.

A thorough investigation is necessary in most cases to determine the presence of underlying factors that could be associated with recurrence or relapse. All drugs or supplements that are administered should be documented. A thorough physical examination, including prostatic examination via rectal palpation and examination of the vulva, is required. A complete blood cell count, serum biochemical profile, urinalysis, urine culture, radiographic and ultrasound imaging and, if appropriate, endocrine testing should be performed. Lower urinary endoscopic exam (vaginoscopy, urethrocystoscopy) or advanced imaging (contrast-enhanced CT), particularly in females, should be considered to further investigate underlying causes. Any underlying concurrent causes identified on examination or diagnostic testing should be managed appropriately, whenever possible. If an underlying cause cannot be found and corrected, it is possible therapy will ultimately be unsuccessful. Client compliance with previous antibiotic treatment should be determined; this is particularly important in cases where relapse is suspected.

Consideration should be given to waiting on culture results before starting antibiotic therapy. If treatment must be initiated immediately, a narrow antibiotic spectrum drug should be selected as recommended for initial treatment of uncomplicated UTI. The drug class used should be different from that used to treat prior UTI(s) (i.e., if amoxicillin was used initially, start treatment with trimethoprim-sulfa drug). Continued antibiotic treatment should be amended as indicated based on the results of culture and susceptibility testing. Preference should be given to drugs that are excreted in urine predominantly in an active form (individual drug descriptions and doses are presented in lecture).

There is no supporting evidence for administration of additional drugs for the purpose of breaking down bacterial biofilm. There is no supporting evidence that direct instillation of antimicrobials, antiseptics, or DMSO directly into the bladder via a urinary catheter is effective for treatment of recurrent UTIs; these compounds are quickly flushed out of the bladder when the animal urinates and may be locally irritating.

Antimicrobial therapy should be directed against all pathogenic organisms, when possible. If more than one bacterial species is identified on culture, the relevance of the each organism should be considered, based on the bacterial counts and the pathogenicity of the organisms. Certain bacterial species, such as *Enterococcus*, generally do not require specific treatment in mixed infections. A single effective antibiotic may not be available. Reasonable combination therapy that would be potentially effective against all organisms based on susceptibility testing should be employed when available.

Evidence supporting the duration of therapy for complicated UTI does not exist, but typically 4 weeks of appropriate antibiotic treatment is a reasonable recommendation. In patients with a non-recurrent but complicated UTI (e.g., first instance of UTI in a diabetic or cushinoid patient), a shorter-term treatment course may be considered.

A urine culture is recommended 5–7 days after initiation of antibiotic therapy to assess the efficacy of the particular antibiotic, especially in patients with previous relapsing or refractory infection, or those at high risk for ascending or systemic infection. Any bacterial growth during treatment indicates treatment failure and should prompt immediate re-evaluation. A second urine culture is recommended 7–14 days after completing antibiotic treatment. If a positive urine culture is obtained after treatment, more in-depth investigation of predisposing factors for relapse or reinfection should be performed. Unless there is clear evidence for the reason for failure, retreatment without any other investigation is not recommended. If no clinical signs of lower urinary tract disease are present, then the patient should be managed as described for subclinical bacteriuria.

Upper Urinary Tract Infection (Pyelonephritis)

Urine culture and susceptibility testing should always be performed; urine sampling should be performed by cystocentesis (or ultrasound-guided pyelocentesis). Treatment should be initiated immediately, while awaiting culture and susceptibility results. Initial treatment should involve antimicrobial drugs known to have efficacy against gram-negative Enterobacteriaceae, based on the predominance of those organisms in canine and feline pyelonephritis. Treatment with a fluoroquinolone is an acceptable first choice. The initial antibiotic selection should be reviewed when results are received. If resistance is reported and clinical evidence of improvement is not evident, the antibiotic selection should be changed to a drug to which the offending organism is susceptible. Antibiotic treatment for 4–6 weeks is generally recommended. Treatment efficacy and monitoring is generally the same as for a complicated UTI (i.e., multiple cultures).

Multidrug-Resistant Infections

There are individual patient and public health concerns with regard to resistant pathogens. Multidrug-resistant bacterial pathogens, including various Enterobacteriaceae, staphylococci, and enterococci, are increasingly problematic. These pathogens are often harder to treat because of limited drug choices. Because of the high incidence of antimicrobial use in UTIs of dogs and cats, veterinarians must be aware of the role of inappropriate treatment in the emergence and dissemination of multidrug-resistant pathogens. Use of antibiotics in the treatment of canine and feline UTIs can be justified as long as their use is prudent and proper, based on culture and susceptibility data. Infection must be documented based on clinical, cytological, and culture abnormalities. Antibiotic use in multidrug-resistant infections for the treatment of subclinical infection is not recommended as organisms may be replaced with susceptible organisms which can allow for self-resolution or practical treatment at a later time.

Subclinical Bacteriuria

Subclinical bacteriuria is the presence of bacteria in the urine as determined by urinalysis and confirmed positive by bacterial culture in the absence of clinical and cytological evidence of UTI. Quantitative culture result does not differentiate subclinical bacteriuria vs. UTI. Subclinical bacteriuria may be present in healthy dogs and cats but is more commonly identified in patients with obesity, diabetes mellitus, Cushing's disease, and immunosuppressive drug treatment. Subclinical bacteriuria has no association with subsequent UTI development. Antibiotic treatment may not be necessary in patients with no clinical signs of UTI and no evidence of UTI (pyuria) on urine sediment exam. In fact, a higher bacterium recurrence rate may be seen following antibiotic therapy.

Antibiotic treatment of subclinical bacteriuria may be considered if there is concern that there is a particularly high risk of ascending or systemic infection (e.g., immunocompromised patients, patients with underlying renal disease) or in patients that are unable to display clinical signs of UTI (e.g., spinal injury). The presence of multidrug-resistant bacterium does not represent an absolute indication for treatment. Multidrug-resistant organisms may be replaced with susceptible organisms if treatment is withheld, and subsequent treatment with routine antimicrobials may be more practical if bacterial decolonization is desired or if clinical disease develops. Treatment of subclinical *Corynebacterium urealyticum* should be considered because of its association with encrusting cystitis.

Prevention of Recurrent Urinary Tract Infections

Patients that are predisposed to UTI or have experienced recurrent infection may benefit from prevention strategies to reduce the likelihood of future infection. A variety of non-antibiotic drug treatment, supplement (nutraceutical) treatments, and elective surgery can be considered in individual patients.

A thorough examination of the vulva should be completed in all female dogs. Particular attention should be directed to determining if a "hooded" (juvenile, inverted) vulvar confirmation or excessive vulvar folds are present. Superficial fold pyoderma or abnormal waxy exudate may be present. All of these issues can promote superficial bacterial colonization with easier access to the lower urinary tract. Weight loss, corrective surgery (i.e., vulvoplasty), and superficial cleansing of the perivulvar area are all critical considerations in recurrent UTI prevention. The client should be questioned and the perivulvar hair and skin should be examined for evidence of moisture that might suggest mild involuntary urinary incontinence. Mild urethral hypotonus is associated with incontinence but also allows bacterial translocation and an opportunity for bacteria to gain easier access to the urinary bladder.

Castration should be considered in intact male dogs to reduce the likelihood of recurrent bacterial prostatitis development and subsequent UTI.

Phenylpropanolamine (PPA) is approved for the control of urinary incontinence due to urethral sphincter hypotonus. This drug acts via sympathomimetic agonist activity which results in an increase in urethral sphincter tone and closure of the bladder neck. PPA treatment trial (1.25 mg/kg PO q 8–12 h) should be considered in any individual that has recurrent UTI and clinical evidence of even subtle involuntary urinary incontinence. Promoting enhanced urethral tone helps restore an effective urethral defense mechanism to prevent ascending bacterial translocation. Long-term therapy is generally safe, so if a decreased incidence of UTI results with PPA treatment, then continued indefinite use should be considered. PPA stimulation of alpha- and beta-adrenergic receptors can result in increased vasoconstriction, heart rate, coronary blood flow, blood pressure, mild CNS stimulation, and decreased nasal congestion and appetite. Oral estrogen replacement therapy can also be considered in younger females that develop recurrent UTI following ovariectomy.

Cranberry extract supplementation has been suggested for UTI prevention. Initially it was thought that this extract produced an inhospitable acidic urine environment. However, it has now been shown that the American cranberry (*Vaccinium macrocarpon*) contains a natural bioactive tannin (proanthocyanidin, PAC-A) which inhibits *E. coli* fimbriae adhesion to the uroepithelium. This activity results in reduced bacterial numbers via bacterial elimination through urinary wash-out and reduced pathogenic colonization and infection. A similar activity has been shown against *Enterococcus faecalis.* Pharmaceutical cranberry extract with PAC-A is available in concentrated formulation in veterinary medicine. Recent *in vitro* and *in vivo* studies in dogs have demonstrated efficacy and safety.

D-mannose is a sugar moiety with antibacterial properties. Its presence in urine causes inhibition of bacterial adherence to urothelial cells. *In vitro* experiments have shown that D-mannose binds and blocks FimH adhesin, which is positioned at the tip of the type 1 fimbria of enteric bacteria. During bacterial colonization, FimH binds to carbohydrate-containing glycoprotein receptors on the epithelium of the urinary tract. D-mannose is similar in structure to the binding site of urothelial glycoprotein receptors, and acts as a competitive inhibitor of bacterial adherence; in sufficient concentration in urine D-mannose saturates FimH adhesins and prevents the bacteria from binding to urothelial receptors. *Escherichia coli, Pseudomonas aeruginosa* and *Streptococcus zooepidemicus* bacterial species have been shown to be effected by D-mannose.

Methenamine mandelate | Methenamine hippurate is used as an antimicrobial agent for prophylaxis of recurrent urinary tract infection. Following oral administration, plasma concentrations of methenamine are very low and have negligible systemic antibacterial activity. 70–90% of each dose is excreted unchanged into the urine. In an acidic urinary environment (pH <6.5), methenamine is converted to formaldehyde. Formaldehyde is a non-specific antibacterial agent that exerts a bactericidal effect. Some urea-splitting bacteria (e.g., *Proteus* and some strains of staphylococci, *Enterobacter* and *Pseudomonas*) may increase urine pH. The addition of a urinary acidifier may be required using dietary modification and acidifying drugs. Hippuric acid is added primarily to acidify urine, but it also has some non-specific antibacterial activity. Bacterial resistance to formaldehyde or hippuric acid does not usually occur. Methenamine also has reported activity against fungal urinary tract infections. It is not commonly used in veterinary medicine and little good evidence is available to confirm its efficacy in dogs or cats. Adverse effects are related to gastrointestinal upset, with nausea, vomiting, and anorexia noted; the drug can be given with food to prevent stomach upset. Tablets are very large, but can be split. Recommended anecdotal doses : Methenamine hippurate 500 mg PO q 12 h; methenamine mandelate usually range from 10–20 mg/kg PO q 8–12h (practically, this is rounded off to the nearest 250 mg as only available in 1-gram tablets).

Probiotics (oral, vaginal suppository) have been postulated to prevent recurrent UTI by increasing the number of lactic acid commensal bacterial flora present in the vagina (or presumably the prepuce) of affected dogs. Human studies are mixed as to the ability of probiotics to prevent recurrent infections. There are currently no evidence-based veterinary studies that provide data as to whether this therapy is effective. Probiotic treatment is not associated with any significant side effects so an empirical trial may be considered.

References

1. Weese J S, *et al. Vet Med Int.* 2011;Article ID 263768:9 pages..
2. Weese J S, *et al.* In: Proceedings of the American College of Veterinary Internal Medicine Forum: ACVIM 2016.
3. Byron J, *et al. J Vet Intern Med.* 2007;21(1):47–53.
4. Hsin-I C, *et al. Am J Vet Res.* 2016;77(4):421–7.

acvim402

Urinary Incontinence and OHE

Julie K. Byron, DVM, DACVIM
Columbus, OH, USA

Introduction

Urinary incontinence is a common problem affecting 4.0% to 20% of spayed female dogs and 30% of those weighing more than 20 kg. In most of these patients it occurs within 3 years of ovariohysterectomy (OHE), although in many it may not become a major problem until later in life when it can be complicated by diseases that increase water intake and urine production. There are several reasons for urinary incontinence, including ureteral ectopia and detrusor hyperreflexia (over-active bladder), but the most common in the dog is urethral sphincter mechanism incompetence (USMI). There is much discussion regarding the role of OHE in the risk of USMI. While the association is no longer in debate, the relationship of the timing of the procedure to incontinence is controversial. Here we will review the current literature and potential mechanisms as well as consideration of the timing of OHE in dogs at risk.

Role of Estrogen

Several factors are important for maintenance of urinary continence. These include an adequate 'vascular cushion' surrounding the urethra, adequate strength of the muscles of the pelvic floor and urethral sphincter, and the integrity of pubourethral ligaments and periurethral connective tissue.[1] In women, the strength and flexibility of the connective tissue components of the pelvic floor have been implicated in several conditions related to SUI including pelvic organ prolapse, urethrocele, and urethral hypermobility. Although these particular conditions have not been distinctly recognized in dogs with urinary incontinence, their increased prevalence in women with the decreased serum estrogen concentrations found after menopause suggests that the integrity of pelvic and periurethral tissues may play a role in urinary incontinence in spayed female dogs.

The relationship of hormone status to incontinence has long been recognized. Estrogen receptors are found in the urethral mucosa, vasculature, and the smooth and striated muscles of the urethra. The supporting ligaments and vaginal walls also have estrogen receptors. Estrogen appears to have a trophic effect on all of these areas. Decreased estrogen levels are associated with loss of urethral muscle tone, urethral vascular atrophy, and decreased glandular secretions, affecting major components of the continence mechanism. It is expected that decreased estrogen levels are a factor in the development of incontinence in dogs, however the lack of clinically recognized incontinence in anestrus intact females points to a more complex mechanism than simple lack of trophic effect. Some evidence suggests the concomitant increase in LH and FSH associated with estrogen (and testosterone in males after neutering) decrease may also play a role in the development of incontinence in dogs, although the mechanism has yet to be determined.

Age at Ovariohysterectomy

There is much discussion regarding the role of spay in the risk of incontinence. While the association is no longer in debate, the relationship of the timing of the procedure to incontinence is controversial. Urinary incontinence has also been reported to develop most often within 2–4 years of ovariohysterectomy (OHE).[2-4] Several studies have investigated the relationship of age at OHE and development of USMI, and the results have often appeared to be contradictory.

Several studies have implied a decreased risk of incontinence if neuter occurred prior to the first estrus. In a study of 206 bitches that were spayed before their first estrus (average age 7.1 months), one study found 9.7% developed urinary incontinence.[3] Interestingly, this study noted that the incidence was higher in dogs weighing >20 kg than those <20 kg (12.5% vs. 5.1%). A large evaluation (~1,000 females) of the effects of early spaying at a shelter looked at the hazard of urinary incontinence with respect to age at spay.[5] The authors found a hazard ratio of 1.2 with every month earlier OHE was performed. The risk increased to 3.46 for those dogs spayed at <3 months old. Another study of 38 dogs with USMI noted that the time to urinary incontinence was shorter in dogs spayed when they were older vs. those spayed when younger.[6] The authors concluded that this supported a lower risk of USMI in earlier spayed dogs, although that specific analysis was not performed. It was noted that most of the dogs were of large and giant breeds, but unfortunately no correlation between age at spay and body weight was investigated. Other studies have found little correlation between the dog's age at OHE or whether it had gone through estrus and the risk of incontinence.[2,7] A recent study of 566 dogs in the United States showed no relationship to age at OHE[8] and a systematic analysis of the available English language literature found no consistent evidence that supported a relationship between the age at OHE and development of urinary incontinence in female dogs.[7]

Based on these previous studies, it appears that neutering after 3 months of age and before the first estrus may be least associated with the development of incontinence; however, this does not prevent its occurrence. No study has looked at the interaction of age at neuter and adult body weight as continuous variables and only one has attempted to evaluate them together.[8] The small number of age and weight categories in that study may have masked their impact.

We have recently completed a retrospective study of 163 dogs with USMI and 193 control dogs in which the age at OHE was evaluated with respect to adult body weight. We found that the relationship between weight at diagnosis and age at OHE was impacted the hazard of USMI. A decrease in the hazard of USMI was found in dogs with an adult weight of >25 kg for every month delay of OHE in the first year. The hazard did not change for dogs who had an adult weight of <15 kg (*J Vet Intern Med*, in press) (See Figure 1).

Figure 1. Contour plot of the hazard of incontinence by OVH age and weight at presentation. This figure is based on 178 observations where the OVH age was ≤12 months

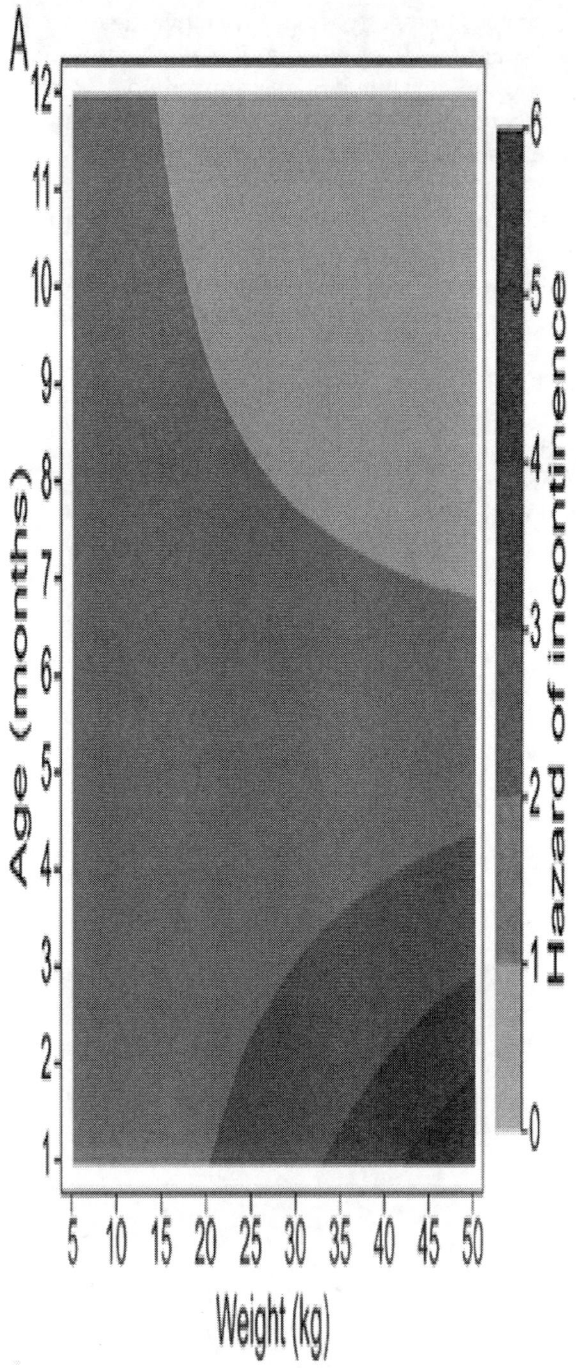

The most compelling clinical impact of the results of this study is the need to make age of neuter recommendations based on the projected adult weight of the bitch. In the past, recommendations to neuter females before the first estrus were based on reduction of unwanted litters and the decreased risk of mammary neoplasia.[9] This has been challenged by the increased risk of USMI, particularly in cases of early (<3 months) OHE.[6] The data presented here shows that the risk of USMI is not significantly reduced in smaller breed dogs (<15 kg) by delaying OHE, thus making the decision to neuter prior to initiation of estrus more appropriate. However, in dogs with projected adult weights over 25 kg, the timing of neuter and its associated risks and benefits, should be carefully considered and discussed with pet owners. Special consideration should be taken when making recommendations for females of breeds that have been identified as at increased risk of USMI such as Boxers, Old English sheep dogs, and Doberman pinschers. Simply put, larger female dogs may benefit from a delay in neuter through a reduction in USMI risk.

Previous studies have shown an increased risk of developing USMI in dogs where OHE was performed after the pubertal estrus.[3,4,10] We performed our regression model on dogs neutered within the first 12 months of life since the majority of OHE procedures in the United States occur during that time and most bitches experience pubertal estrus during the first year. After the dog reaches sexual maturity, it could be assumed that the impact of estrogen cessation on the adult tissues, and risk of USMI, has a different physiology than its lack during the developmental stages. Information on whether the dogs had OHE before or after pubertal estrus was not available in the current study. The wide variability in age of first estrus among and even within breeds[11] prevents the authors from making a recommendation for neutering related to sexual maturity.

Conclusion

There is still controversy about the optimal time to perform OHE on female dogs to reduce the risk of urinary incontinence; however, new research has shown that the projected adult body weight of the dog, as well as other risk factors, should be considered when discussing this with owners. As supported by previous studies, dogs with smaller adult body weights are less likely to be impacted by the age at OHE, however, larger breed dogs may benefit from delayed spaying by reducing the hazard of urinary incontinence.

References

1. Goepel C, Hefler L, Methfessel HD *et al.* Periurethral connective tissue status of postmenopausal women with genital prolapse with and without stress incontinence. *Acta Obstetricia et Gynecologica Scandinavica.* 2003;82:659–664.
2. de Bleser B, Brodbelt DC, Gregory NG *et al.* The association between acquired urinary sphincter mechanism incompetence in bitches and early spaying: a case-control study. *Vet J.* 2011;187:42–47.
3. Stocklin-Gautschi NM, Hassig M, Reichler IM *et al.* The relationship of urinary incontinence to early spaying in bitches. *J Reprod Fertil Suppl.* 2001;57:233–236.
4. Thrusfield MV, Holt PE, Muirhead RH. Acquired urinary incontinence in bitches: its incidence and relationship to neutering practices. *J Small Anim Pract.* 1998;39:559–566.
5. Spain CV, Scarlett JM, Houpt KA. Long-term risks and benefits of early-age gonadectomy in dogs. *J Am Vet Med Assoc.* 2004;224:380–387.
6. Veronesi MC, Rota A, Battocchio M *et al.* Spaying-related urinary incontinence and oestrogen therapy in the bitch. *Acta Vet Hung.* 2009;57:171–182.
7. Beauvais W, Cardwell JM, Brodbelt DC. The effect of neutering on the risk of urinary incontinence in bitches - a systematic review. *J Small Anim Pract.* 2012;53:198–204.
8. Forsee KM, Davis GJ, Mouat EE *et al.* Evaluation of the prevalence of urinary incontinence in spayed female dogs: 566 cases (2003–2008). *J Am Vet Med Assoc.* 2013;242:959–962.
9. Schneider R, Dorn CR, Taylor DO. Factors influencing canine mammary cancer development and postsurgical survival. *J Natl Cancer Inst.* 1969;43:1249–1261.
10. Arnold S, Jager P, DiBartola SP *et al.* Treatment of urinary incontinence in dogs by endoscopic injection of Teflon. *J Am Vet Med Assoc.* 1989;195:1369–1374.
11. Schaefers-Okkens A. Estrous cycle and breeding management of the healthy bitch. In: Ettinger SF, EC, ed. *Textbook of Veterinary Internal Medicine.* 7 ed. St. Louis, MO: Saunders Elsevier. 2010;1873–1883.

acvim403

CKD: Updates on Diagnosis, Treatment and Prevention of Progression

Dennis J. Chew, DVM, DACVIM (Small Animal Internal Medicine)
Columbus, OH, USA

Measurement of serum creatinine concentration is the most commonly used surrogate to estimate GFR in the clinics, as it is not technically feasible or practical to measure GFR directly. Serum creatinine is generally preferred over BUN for evaluation of renal function since creatinine has fewer non-renal variables. Symmetric dimethylarginine (SDMA) is a molecule that is also a surrogate for the evaluation of GFR in the dog[1] and the cat[2] that has some advantages over measurement of serum creatinine. SDMA results from methylation of arginine that occurs in all nucleated cells. SDMA is excreted exclusively into urine and consequently increases in blood when GFR decreases. SDMA is not influenced by lean muscle mass whereas serum creatinine is related to lean muscle mass. Since creatinine arises from muscles, animals with decreased muscle mass will have lower serum creatinine concentrations than would otherwise develop. During progression of CKD, loss of lean muscle mass can parallel loss of renal function which results in little change in serum creatinine concentration. This phenomenon reduces the ability for serum creatinine to detect ongoing CKD progression early. It is important to remember than any surrogate for GFR will be influenced by pre-renal, primary renal, and post renal factors. It is also important to remember that the finding of normal serum creatinine, BUN, and/or SDMA concentrations does **not** exclude the presence of primary renal disease. In a patient with primary renal disease, a serum creatinine concentration above the reference range (>2.0 mg/dl) is often interpreted to be associated with a loss of >75% of renal mass and GFR. When a lower value for the upper limit of serum creatinine such as 1.4, 1.5, or 1.6 mg/dl is used, an increased serum creatinine concentration is associated with a loss of about 50% renal mass and GFR. It appears that on average SDMA increases at a time when about 40% of renal mass and GFR has been lost. SDMA can increase in CKD patients when there is as little as 25% loss of renal mass and GFR at times.

Concentrations of SDMA have been shown to increase above the reference range before serum creatinine increased in multiple studies of dogs and cats eventually diagnosed with azotemic CKD.[1,3-5] SDMA increased many months before serum creatinine when relatively high values for the upper limit of creatinine reference (above 2.0 mg/dl) ranges were used to diagnose the onset of CKD. When lower creatinine values were used for the upper limit as the comparator, SDMA still increased before creatinine but by a shorter time. Initial studies were reported in dogs and cats living in colonies at a pet food company and in a colony of dogs with hereditary nephritis. Recent reports in client owned dogs and cats show that SDMA is a more sensitive diagnostic indicator for the development of azotemic CKD than is serum creatinine. LC/MS has been used as the gold standard to measure SDMA in most studies in dogs and cats but an automated clinical biochemistry method offered by a commercial veterinary laboratory is now available for use by practicing veterinarians. SDMA is increased much more frequently than serum creatinine in both dogs and cats when measured on the same sample (diagnostic discordance),[6] but it is not yet clear how many of these patients will progress to azotemic CKD.

IRIS CKD Staging now includes an "At Risk" category. This category is based on increased risk for a future diagnosis of CKD based on history and signalment that includes exposure to nephrotoxicants, breed, high prevalence of infectious disease in the area, and old age. IRIS CKS Stage 1 is the most difficult to assign with certainty. Since these patients are nonazotemic, some other index of renal disease must be present. This could include some combination of sub-maximally concentrated urine without identification of a non-renal cause, abnormal renal palpation or renal imaging, renal proteinuria, trend of serum creatinine to increase within the reference range, cylindruria and decreased or decreasing GFR (iohexol clearances). CKD Stages 2-4 are easily assigned, based on the magnitude of increased serum creatinine in a stable hydrated patient. Sub-staging is then based on the magnitude of UPC and systemic blood pressure.[7] SDMA has been added to the IRIS staging system to further characterize CKD in patients in which serum creatinine concentration is likely to underestimate a more appropriate stage assignment. The finding of a serum creatinine concentration <1.6 mg/dl for the cat and <1.4 mg/dl for the dog and an

SDMA that is increased (>14 ug/dl) provides one entry point into IRIS Stage 1 CKD and should trigger further diagnostic evaluation of the patient to verify or exclude the presence of primary renal disease (urinalysis, systemic blood pressure, UPC, renal imaging).[7] An SDMA ≥25 μg/dl in patients with IRIS CKD Stage 2 and low body condition scores may indicate underestimation of renal dysfunction; these patients may be more appropriately staged as Stage 3. Similarly, an SDMA ≥45 μg/dl in patients characterized as IRIS Stage 3 with low body condition scores may be more appropriately staged as Stage 4.[7]

Progression of CKD refers to further loss of functional renal mass and accumulation of primary chronic renal lesions (tubulo-interstitial nephritis, fibrosis, and nephron drop-out) associated with decreasing GFR. When enough nephron mass occurs, increases in BUN, creatinine, SDMA, and phosphorus will be detected. Progressive loss of excretory renal function will occur if the cause of the initial renal injury is not detected and stopped, as can occur with various types of immune-complex glomerulonephritis, familial glomerular disease, and renal amyloidosis. The "inexorable progression" of CKD occurs after a substantial amount of renal mass has already occurred. This type of progression can continue even when the original cause of the renal injury has been stopped. Based on experimental data, this type of self-progressive destruction of the kidney likely occurs in the dog and cat when there is loss of at least 75% of the original renal mass. Glomerular hyperfiltration and glomerular hypertension are considered to be important players that exist in the surviving single nephrons in advanced CKD that contribute to this type of progression. Systemic hypertension, PTH, CKD- BMD (bone mineral disease associated with phosphorus, FGF-23, Klotho, ionized calcium) and oxidative stress are also likely to contribute to progression.

Treatment

The feeding of a renal diet is usually the first step in treatment of stable CKD dogs and cats that are in IRIS CKD Stage 2–4. Studies involving the feeding of renal diets in dogs and cats with CKD have included mostly overtly azotemic patients. Increased survival and decreased uremic crises in general occur when renal diets are fed to these CKD patients. The feeding of lower protein diets to cats with CKD is considered controversial by some but is still considered the standard of care by most when the degree of protein restriction is modest.[8,9] Recently it has been suggested that the feeding of a renal diet to both geriatric dogs and cats in IRIS stage 1 could exert benefits. Serum creatinine and BUN decreased while SDMA did not increase in cats following the feeding of a renal diet for 6 months. In dogs fed a renal diet for 3 or 6 months, creatinine and SDMA decreased. Whether this was the result of improved excretory renal function or other factors was not determined.[10,11] It remains to be determined if the feeding of any renal diet can prevent progression of IRIS stage 1 CKD to higher stages.

A renal diet confers major benefits to decrease progression of CKD and increase survival in large part from its degree of phosphorus restriction for those in IRIS Stage 2 to 4. Little is known about the possible benefits of feeding a phosphorus restricted diet to patients in IRIS CKD stage 1. The feeding of a moderate protein and phosphate restricted renal diet to non-azotemic geriatric cats had some benefits to limit PTH and fractional urinary excretion of phosphorus but did not impact how many cats converted to azotemic CKD.[12] Phosphorus retention in the body occurs during progressive CKD with serum phosphorus initially maintained within the reference range, likely due to the correcting effects of PTH and FGF-23 that increase phosphaturia. Hyperphosphatemia develops as CKD advances and overcomes these correcting forces.

It is important to evaluate phosphorus content on an energy content basis (mg/100 kcal) in order to determine the current phosphorus intake and what diets will provide considerably less phosphorus intake. Other factors must also be considered such as the biological availability of the phosphorus in these diets, which can vary despite similar concentration in mg/100 kcal. Animal sourced proteins are usually higher in phosphorus content than proteins from vegetable sources.[13] Phosphorus restriction to a targeted level of serum phosphorus is achieved through feeding a veterinary diet restricted in phosphorus and intestinal phosphate binders. Renal diets may provide sufficient dietary phosphate restriction during early stages of CKD but often the addition of dietary phosphate binders will be needed to reach targeted control of serum phosphorus. An initial goal is to maintain serum phosphorus concentration within the reference range but a more optimal goal is to achieve a circulating concentration of phosphorus that is in the lower half of the reference range to achieve more salutary effects on ionized calcium, PTH, and FGF-23.[14-17]

A total calcium-phosphate product >70 mg^2/dl^2 was associated with a higher mortality rate compared to those with a lower product in a study of dogs with naturally occurring CKD; this finding appeared to be independent of serum creatinine.[18] As single variables, serum phosphorus, total calcium, or ionized calcium at the time of diagnosis did not predict survival in these dogs. The risk of death was 4.2 times higher for CKD dogs with a calcium-phosphate product >70 mg^2dl^2 compared to a lower product. IRIS Stage 3 and 4 CKD dogs with a >70 mg^2/dl^2 calcium phosphate produce survived a median of 30 days. The median survival for IRIS Stage 1,2 and 3 CKD dogs with a calcium-phosphate product <70 mg^2/dl^2 could not be determined, as over 50% of these dogs were still alive at 450 days. A targeted calcium-phosphate product <55 mg^2/dl^2 has been recommended in human nephrology. This study of CKD dogs did not evaluate outcomes for those with <55 mg^2/dl^2 or from 55 to 70 mg^2/dl^2 calcium-phosphate products.[18] Indexing serum FGF23 to urinary phosphorus excretion (fractional or absolute mg/day) has recently been reported in human nephrology as a method to assess the efficiency or appropriateness of urinary phosphorus excretion in relation to nephron mass and signal for phosphaturia. Decreases in this ratio reflect a declining number of nephrons and are associated with vascular mineralization.[19,20] Survival time in CKD dogs was adversely affected by a decreased ratio of urinary fractional phosphorus to serum FGF23 as reported at this Forum.[21]

Intestinal phosphate binders are mixed in the food. The goal is to bind or decrease the absorption of phosphates in the food so that they are not available to be absorbed across the intestine into the circulation. Aluminum salts are the most commonly prescribed intestinal phosphate binders in veterinary medicine since they are inexpensive and bind phosphate well. Aluminum retention is known to be toxic in humans with CKD and evidence is emerging that there is toxicity in dogs with CKD as well.[22] Calcium salts (carbonate and acetate) are alternative phosphate binders to aluminum, but they are not as good at binding phosphates and there is the potential for ionized hypercalcemia to develop. Calcium salts should be avoided when calcitriol treatment will be prescribed in the same patient. Sevelamer is an organic polymer that contains no aluminum or calcium and is used extensively in people, but reports are lacking as to its efficacy and safety in veterinary CKD patients. Lanthanum carbonate is an excellent phosphate binder with little known toxicity but it is expensive. A previously available affordable veterinary formulation of lanthanum carbonate is no longer manufactured.[13] A new oral phosphate binder based on iron oxide has been developed for use in cats but is not yet marketed.[23]

Calcitriol is either absolutely or relatively deficient in dogs and cats with azotemic CKD. Calcitriol treatments decrease PTH or prevent its increase during CKD. This occurs mostly by genomic effects to block PTH synthesis in addition to a mild calcemic effect, and antiproliferative effect that prevents parathyroid gland hyperplasia. During treatment of CKD patients with calcitriol, simultaneous monitoring of serum ionized calcium, serum phosphorus and PTH concentrations is the ideal way to document successful and safe control of renal secondary hyperparathyroidism. Calcitriol should not be administered until hyperphosphatemia has been controlled. If the Ca x P solubility product exceeds 60–70, calcitriol should be avoided because of the risk of soft-tissue mineralization. The beneficial effects of calcitriol are also lessened within the parathyroid gland when ionized calcium remains low. Phosphorus restriction relieves phosphate-mediated inhibition of the renal 1α-hydroxylase system, resulting in enhanced endogenous synthesis of calcitriol and subsequent inhibition of PTH synthesis. Though the PTH lowering effect following calcitriol administration has been emphasized as a central benefit from this treatment, many other effects can follow binding of calcitriol to the VDR (vitamin D receptor) in various tissues including the kidney. Calcitriol binding to VDR in the renal tissue exerts anti-inflammatory and anti-fibrotic effects as well as reducing the generation of intra-renal RAAS activity all of which are independent of any effect to decrease circulating PTH.[14]

Intermittent rather than daily dosing calcitriol treatment protocols have emerged as the standard of care since less hypercalcemia occurs. The equivalent daily dose of 2.5 to 3.5 ng/kg daily is given instead every 3.5 days at 9 to 12 ng/kg. This is the longest time in between dosing that will still suppress PTH synthesis within the parathyroid gland. This method of dosing is especially attractive for cat owners since medication will only be given twice weekly but this is also offered as the first option for dogs. Calcitriol treatment for azotemic CKD dogs increased survival to a median of 365 days compared to 250 days in dogs treated with placebo in one study.[24] A similar study was done by the same investigators who concluded that there is no advantage to calcitriol treatments in cats with azotemic CKD, but this study followed cats for just one year which is likely not enough time to demonstrate a treatment effect due to the inherently slow nature of CKD progression in cats.

It is common in some human nephrology practices to treat CKD patients with BOTH activated vitamin D metabolites like calcitriol and parent vitamin D (cholecalciferol).[25] Survival of human CKD patients correlates better with 25(OH)-vitamin D concentrations than to that of calcitriol, likely due to vitamin D receptor activation in local cells that then generate intracellular 1,25 $(OH)_2$-vitamin D. Low circulating 25(OH)-vitamin D is common in humans with CKD and this has also been observed in dogs with azotemic CKD.[26] In a recent study, 25(OH)-vitamin D, 1,25$(OH)_2$-vitamin D and 24,25$(OH)_2$-vitamin D concentrations were low in CKD dogs, with deficits more severe with advancing IRIS stage.[27] Low 25(OH)-vitamin D could be due to decreased dietary intake, decreased intestinal absorption, decreased hepatic hydroxylation of vitamin D to 25(OH)-vitamin D, or to increased loss in urine. 25(OH)-vitamin D and 1,25$(OH)_2$-vitamin D are reabsorbed along the proximal tubule following glomerular filtration - this process is mediated by megalin receptors that are upregulated by calcitriol-VDR interactions.[14] Regimens to best replete 25(OH)-vitamin D status have not yet been well developed in veterinary medicine for animals with or without CKD. Oral administration of 25(OH)-vitamin D achieved FDA approval in 2016 for repletion of 25(OH)-vitamin D status and control of secondary hyperparathyroidism in humans with CKD. Repletion of 25(OH)-vitamin D status in experimental dogs without CKD was more efficient when supplemented with oral 25(OH)-vitamin D compared to supplementation with parent vitamin D (cholecalciferol).[28] Vitamin D status based on 25(OH)-vitamin D concentrations can vary widely in normal dogs eating the same diet, so it is likely that repletion status will depend on measurement of 25(OH)-vitamin D post treatment and titration of dose to the targeted level of 25(OH)-vitamin D desired.[29] Whether baseline 25(OH)-vitamin D status predicts survival for CKD dogs and cats is not yet known. Also yet to be determined is the effect of 25(OH)-vitamin D repletion on survival time and stability of GFR in dogs and cats with CKD.

Inactivation of the RAAS (renin-angiotensin-aldosterone-system) can provide renoprotection to dogs and cats with CKD. This protection is provided by some combination of reducing intraglomerular hypertension and single nephron hyperfiltration, reducing systemic blood pressure, and by reducing damaging signals from angiotensin-II and aldosterone that promote renal inflammation and fibrosis. Most nephrologists advocate RAAS reduction in patients with renal proteinuria 30, but it is possible that benefits could be achieved in those without renal proteinuria also, but this is controversial. Most canine and feline CKD patients undergo reduction in UPC during treatment with enalapril or benazepril and some maintain higher GFR in the long term, but this does not necessarily translate into increased survival time. The angiotensin receptor blocker (ARB) telmisartan (Semintra® Boehringer Ingelheim) was approved by the European Commission in 2013 for use in the European Union as a drug for use in cats with CKD and is available for use in Canada but not yet in the USA. Semintra was found to be at least as effective as benazepril in reducing proteinuria in cats with CKD and was well tolerated.[31,32] It is not clear when or if an ARB should be chosen to reduce RAAS activity instead of an ACE-Inhibitor for treatment of CKD in veterinary patients to reduce proteinuria, systemic blood pressure, or intra-renal inflammation.

Most animals with azotemic CKD have normal serum potassium concentrations. Hyperkalemia more readily develops when the animal becomes oliguric or anuric. Hyperkalemia in CKD dogs with normal urine production or polyuria may, however, occur more commonly than generally appreciated. In one study 47% of CKD dogs experienced 1 or more episodes of hyperkalemia, with 16% experiencing 1 or more episodes of severe hyperkalemia (>6.5 mEq/L). Some of the development of hyperkalemia in these CKD dogs was attributed to the potassium content of commercial renal diets, as serum potassium declined in dogs fed diets specifically formulated to be lower in potassium content.[33]

Hypokalemia may occur in 10–30% of dogs and cats with azotemic CKD due to some combination of anorexia, loss of muscle mass, vomiting, and polyuria. A significant body deficit of potassium can exist without the development of hypokalemia (kaliopenia). Serum potassium levels can underestimate total body potassium since most of the body's potassium lies within the cells. As potassium is lost excessively from the plasma, potassium leaches out of the cells down a concentration gradient masking the initial loss of potassium. One study of cats with azotemic CKD and normokalemia revealed decreased potassium content of muscles.[34] Hypokalemic (kaliopenic) nephropathy has been described as a specific syndrome in clinical cats either as a result of CKD or resulting in CKD. Whether kaliopenia without hypokalemia can result in CKD in cats has not been established. In one study of experimental cats, the development of azotemic CKD and hypokalemia was discovered at the same time in cats fed a marginally replete urinary acidifying and magnesium restricted diet.[35] Hypokalemia can contribute to anorexia, depression, and weakness independent of the effects of CKD. Severe muscle weakness (the "hanging-head" posture, truncal ataxia) is seen in many cats with serum potassium less than 2.5 mEq/L and in some with less than 3.0 mEq/L. The fractional excretion of potassium into urine is high despite the low serum potassium. Cats with normokalemic CKD also have high fractional excretion for potassium as part of the adaptive nephron response to maintain potassium balance.

Hypokalemia can be associated with functional and/or structural changes in the kidneys. These changes include a reduced ability to concentrate the urine - initially functional, structural later. Decreased GFR initially can be a result of hemodynamic changes, but later due to structural changes. Structural lesions of tubulointerstitial inflammation and fibrosis associated with azotemic CKD have been reported to develop in cats fed marginally potassium replete diets that are highly acidifying and magnesium restricted. Some cats with azotemic CKD and hypokalemia will improve their excretory renal function following correction of the hypokalemia. Whether this improvement is due to correction of pre-renal factors, intrarenal hemodynamics, or resolution of some intrarenal lesions is not clear. Some cats with azotemic CKD periodically develop hypokalemia during the course of their disease. Hypokalemia in dogs and cats with azotemic CKD may be treated with oral potassium gluconate or citrate. The dosage of potassium for cats usually is 2–4 mEq/day, whereas dogs may require as little as 2 or as much as 40 mEq/day depending on their body size.

References

1. Nabity MB *et al. J Vet Intern Med.* 2015;29:1036.
2. Braff J *et al. J Vet Intern Med.* 2014;28:1699.
3. Relford R *et al. Vet Clin North Am Small Anim Pract.* 2016;46:941.
4. Hall JA *et al. Vet J.* 2014;202:588.
5. Hall JA *et al. J Vet Intern Med.* 2014;28:1676.
6. IDEXX. White Paper. https://wwwidexxcom/files/small-animal-health/solutions/articles/sdma-data-white-paper.pdf. 2016.
7. IRIS Straging of CKD. http://iris-kidneycom/pdf/staging-of-ckdpdf. 2015;6.
8. Scherk MA, Laflamme DP. View. *Vet Clin North Am Small Anim Pract.* 2016;46:1067.
9. Polzin DJ, Churchill JA. *Vet Clin North Am Small Anim Pract.* 2016;46:1049.
10. Hall JA *et al. PLoS One.* 2016;11:e0153654.
11. Hall JA *et al. PLoS One.* 2016;11:e0153653.
12. Geddes RF *et al. J Vet Intern Med.* 2016; 30:1690.
13. Chew DJ, Parker VJ. 2015 Ohio State & IAMS Symposium Proceedings: Small Animal Renal & Urinary Health. 2015;40.
14. de Brito Galvao JF *et al.. J Vet Emerg Crit Care.* 2013;23:134.
15. Foster JD. *Vet Clin North Am Small Anim Pract.* 2016;46:1131.
16. Cortadellas O *et al. J Vet Intern Med.* 2010;24:73.
17. Geddes RF *et al. J Vet Intern Med.* 2013;27:1354.
18. Lippi I *et al. J Am Vet Med Assoc.* 2014;245:1135.
19. Yamada H *et al. PLoS One.* 2016;11:e0160782.
20. Craver L *et al. BMC Nephrol.* 2013;14:221.
21. Martorelli CR *et al. J Vet Int Med.* 2017; in press.
22. Segev G *et al. Israel Journal of Veterinary Medicine.* 2016;71:31.
23. King JN *et al. J Vet Pharmacol Ther.* 2014;38:278.
24. Polzin D *et al. J Vet Int Med.* 2005;19:433.
25. Morrone LF *et al. J Nephrol.* 2016;29:305.

26. Galler A *et al. Vet J*. 2012;192:226.
27. Parker VJ *et al. J Vet Intern Med*. 2017; in press.
28. Young L, Backus RL. Waltham International Nutritional Sciences Symposium 2016;70.
29. Parker VJ *et al. J Am Vet Med Assoc*. 2017; in press.
30. Pressler B. *Vet Med*. 2013;108:392.
31. Sent U *et al. J Vet Intern Med*. 2015;29:1479.
32. Sent U *et al. JFMS*. 2013;15:824.
33. Segev G *et al. J Vet Intern Med*. 2010;24:546.
34. Theisen SK *et al. J Vet Intern Med*. 1997;11:212.
35. Buffington CA *et al. J Nutr*. 1991;121:S91.

acvim404

Evidence for Nutritional Management of Dogs & Cats with Stage 1 Chronic Kidney Disease

Jean A. Hall, DVM, PhD, DACVIM (Small Animal Internal Medicine)
Corvallis, OR, USA

INTRODUCTION

Chronic kidney disease (CKD) is a major cause of morbidity in dogs and cats, especially as they age. Regardless of the initiating cause, CKD tends to be progressive.[1] Progression occurs because of persistence of primary disease, or addition of other renal insults or complications. Early indicators of kidney damage include a reduction in glomerular filtration rate (GFR) and impaired urine concentrating ability. Later indicators include azotemia, with increased serum creatinine (Cr) and urea nitrogen (BUN) concentrations. Once GFR has decreased to 30 to 50% of normal, progression to end-stage renal failure tends to be inevitable. Glomerular hyperperfusion leads to glomerulosclerosis and proteinuria. Compensatory hyperfunctioning renal tubules progress to tubulointerstitial lesions, destructive fibrosis, and loss of nephrons. Even in laboratory dogs maintained under optimal environmental conditions, the development of renal lesions is progressive over their lifetime.[2] Progressive kidney disease is likely present when kidney disease is first diagnosed.[3] Thus, early recognition of CKD is desirable in order to test for renoprotective interventions, such as dietary modifications that might slow its progression.[4-6]

Once dogs and cats are diagnosed with CKD, they are staged according to guidelines developed by the International Renal Interest Society (IRIS) and accepted by the American and European Societies of Veterinary Nephrology and Urology. The four-tier IRIS CKD staging system is based on serum symmetric dimethylarginine (SDMA) and serum Cr concentrations, the magnitude of proteinuria as measured by the urine protein:creatinine ratio, and systolic blood pressure.

Currently, nutritional management of CKD is recommended when serum SDMA exceeds 14 μg/dL or Cr exceeds 1.4 mg/dL (dogs) or 1.6 mg/dL (cats) with IRIS-Stage 2 CKD , or when pets with IRIS Stage 1 CKD have a UPC ratio >0.5 (dogs) or >0.4 (cats). Dietary interventions that might benefit renal function in nonazotemic pets with moderately reduced GFR, yet persistently impaired urine concentrating ability or persistent proteinuria (e.g., IRIS-Stage 1 CKD), have not been well studied. Typical therapeutic renal foods have controlled protein, phosphorus, and sodium content, increased caloric density, increased buffering capacity, and increased soluble fiber, B-complex vitamins, antioxidants, and omega-3 fatty acids when compared with industry averages of foods for healthy adult pets.[7] The goals of nutritional therapy are to slow progressive loss of kidney function, reduce clinical and biochemical consequences of CKD, and maintain adequate nutrition.[7,8]

Five prospective studies are summarized (4 in dogs; 1 in cats) to determine if renal function is stabilized (based on serum renal biomarker concentrations) with nutritional intervention. Two studies were in geriatric dogs and three studies were in nonazotemic pets with IRIS-Stage 1 CKD treated with early dietary intervention. All studies show decreased serum SDMA concentrations across time with nutritional interventions. All test foods were dry foods that contained 7 to 9% moisture.

FIRST GERIATRIC DOG STUDY[9]

Forty-one healthy Beagles, mean age 9.9 years (range 3.1 to 14.8 years), were divided into three groups and studied over a 6-month period. Canine Prescription Diet® k/d® (Hill's Pet Nutrition, Inc.) was used as the control food as well as the base formula for two test foods. Both test foods had added L-carnitine (300 mg/kg, as fed) and fish oil (0.6 or 1.5%, as fed). The test food with 1.5% fish oil also had added medium-chain triglycerides (MCT) from coconut oil, but reduced arachidonic acid because of a reduction in animal fat in the formulation. The reduced animal fat was replaced with plant oils (2% coconut oil and 7% corn oil). Prescription Diet® k/d® contains protein and phosphorus concentrations that meet the nutrient requirements of dogs based on recommended allowances by the National Research Council (NRC). In addition, animal feeding tests performed by Hill's Pet Nutrition, Inc. using Association of

American Feed Control Officials (AAFCO) procedures substantiate that Prescription Diet® k/d® provides complete and balanced nutrition for maintenance of adult dogs. It is an antioxidant-enriched food that has been used to manage CKD in dogs for more than 60 years (although revisions of the original food have occurred). As fed, dry food contains 13.5% protein, 18.0% fat, 56.9% carbohydrate, and 0.28% phosphorus. Phosphorus concentrations of test foods were 0.31 and 0.29%. Serum biomarkers of renal function were measured prospectively at baseline, and 1, 3, and 6 months. SDMA concentrations, but not Cr concentrations were influenced by time on food. SDMA concentrations decreased across time ($p<0.001$). SDMA concentrations decreased significantly from baseline to 3 months, and continued to decrease at 6 months. Mean ± SD SDMA concentrations at baseline were 13.3 ± 5.2, at 1 month 12.6 ± 2.7, at 3 months 8.1 ± 2.1, and at 6 months 6.5 ± 1.7 μg/dL. Time on food was the only significant explanatory variable for SDMA concentration ($R^2=0.49$). In this study, we did not report GFR using iohexol clearance. However, we have previously shown that serum SDMA is significantly correlated with GFR in dogs.[10] Thus, the decrease in serum concentrations of SDMA in this study was consistent with an improvement in GFR. We speculated that this represents a dietary effect on renal function.

Second Geriatric Dog Study[11]

The purpose of this study was to determine the effects of feeding traditional and renal protective foods supplemented with functional food bioactives on GFR, lean body percent (LB%), and selected circulating biomarker and metabolite concentrations in a geriatric dog model. Eighty-one geriatric dogs, mean age 10.4 years (range 7.9 to14.2 years) and 30 mature, adult dogs, mean age 5.0 years (range 3.3 to 6.9 years) were fed one of three foods (n=27 per group) for 6 months: a traditional renal protective food (control), or control food supplemented with increasing amounts of functional food bioactives.

Pre-trial food was a dry food with 20.1% protein and 0.80% P, as fed. Control food was a dry food, Hill's Science Diet Mature Adult, with 19.3% protein and 0.70% P, as fed. Test foods were dry foods with 18.0 to 18.8% protein, and 0.60% P, as fed. Test foods differed from control food in degree of supplementation with functional lipids, antioxidants, and botanicals (fish oil, lipoic acid, fruits and vegetables), as well as egg and chicken protein. Both functional foods contained 0.5% fish oil and 100 mg/kg lipoic acid. Functional food 1 (FF1) contained unique protein sources compared with control food in that egg protein (7%), corn gluten meal (6.4%), and wet meat chicken (7.5%) replaced other carbohydrate and protein sources in control food. In addition, there were increased amounts of fruits and vegetables [i.e., beet pulp, citrus pulp, carrot granules, dried spinach, and tomato pomace (each present at 0.55% of formula; total 2.75%)] in FF1. FF2 differed from control and FF1 in that it contained additional amounts of fruits and vegetables [i.e., beet pulp, citrus pulp, carrot granules, dried spinach, and tomato pomace (each present at 1.50% of formula; total 7.5%)], more α-tocopherol acetate (≥260 IU/kg, as fed), and increased egg protein (14%) and wet meat chicken (10%), but no corn gluten meal. Fat content was increased from 8.9%, as fed, in pre-trial food to 14 to 15% in all three trial foods. In addition, all three foods contained at least 300 mg/kg L-carnitine, 213 mg/kg ascorbyl monophosphate, and 1147 IU/kg α-tocopherol acetate.

Geriatric dogs were compared before and after the feeding trial with mature adult dogs. Renal function was assessed by GFR, LB% was determined by dual energy X-ray absorptiometry, and circulating biomarkers and metabolites were measured in blood. Before the feeding trial, GFR (+28.2%), LB% (+18.6%), and serum total protein (+10.0%) were higher in mature versus healthy geriatric dogs (all $p<0.001$). Geriatric dogs consuming all three foods increased ($p<0.001$) GFR over time; group averages ranged from 13.0 to 16.9%. The increases in GFR associated with feeding FF2 for 6 months were large enough, such that when geriatric dogs were compared with mature adult dogs, GFR were no longer significantly different from mature adult dogs ($p=0.20$). Dogs fed the highest supplemented level of bioactives (FF2) had lower ($p<0.001$) SDMA concentrations (baseline, 8.5 ± 0.2 μg/dL; 6 months, 7.3 ± 0.4 μg/dL; -14.3%). Feeding FF2 for 6 months decreased SDMA concentrations (-1.2 ± 0.3 μg/dL) to the point that SDMA was significantly ($p< 0.001$) lower in dogs fed FF2 compared with mature dogs. In conclusion, supplementation of mildly protein-restricted, energy-dense currently available renal protective food with functional lipids (fish oil), antioxidants (lipoic acid), carnitine, increasing concentrations of botanicals (fruits and vegetables), and more bioavailable protein sources reversed the age-associated decline in GFR in healthy geriatric dogs. The greatest improvements were observed in dogs fed FF2, which had egg and wet meat chicken as protein sources and higher concentrations of fruits and vegetables.

Positive Impact of Nutritional Interventions on Serum SDMA and Creatinine Concentrations in Client-Owned Geriatric Cats[12]

A prospective study was conducted in client-owned geriatric cats to evaluate the short-term effects of a test food on serum SDMA and Cr concentrations. Test food (dry food; protein 32.94%; phosphorus 0.76%, as fed) contained functional lipids (fish oil), antioxidants (vitamins C and E), L-carnitine, botanicals (vegetables including oat groats, beet pulp, tomato pomace, broccoli powder), highly bioavailable protein (wet meat chicken), and amino acid supplements. Cats (n=80) were fed either test food or owner's-choice foods (non-nutritionally controlled cohort). Cats were included based on age (≥9 years), indoor only, neutered, and free of chronic disease. At baseline, all cats had serum Cr concentrations within the reference interval. Renal function biomarkers and urinalysis results at baseline and after consuming test food or owner's-choice foods for 3 and 6 months were evaluated. Cats consuming test food showed significant decreases in serum Cr and BUN concentrations across time. Overall, cats consuming owner's-choice foods showed significant increases in serum SDMA concentrations at 3 and 6 months compared with baseline ($p \leq 0.05$; baseline SDMA, 10.3 ± 0.4; 3 months, 11.2 ± 0.5; 6 months, 11.4 ± 0.5 µg/dL), whereas in cats consuming test food serum SDMA concentrations did not change (baseline SDMA, 10.3 ± 0.5; 3 months, 10.0 ± 0.6; 6 months, 9.7 ± 0.6 µg/dL). At baseline or during the 6-month feeding trial, 23 (28.8%) cats had increased serum SDMA, but normal serum Cr consistent with IRIS Stage 1 chronic kidney disease. This included 6 cats fed test food (baseline SDMA, 14.2 ± 1.6; 3 months, 13.2 ± 1.6; 6 months, 14.4 ± 1.6 µg/dL) and 17 cats fed owner's-choice foods (baseline SDMA, 12.3 ± 0.9; 3 months 15.1 ± 1.0; 6 months, 14.5 ± 0.9 µg/dL). In the 6 cats fed test food, serum SDMA decreased in 3 cats and remained stable in 1 cat, whereas in the 17 cats fed owner's-choice foods, serum SDMA increased in 13 cats and decreased or remained stable in 4 cats. The increase in serum SDMA concentration was significant ($p=0.02$) only for cats fed owner's-choice foods. These results suggest that nonazotemic cats with increased serum SDMA (early CKD) when fed a food designed to promote healthy aging are more likely to demonstrate stable renal function compared with cats fed owner's-choice foods. Cats fed owner's-choice foods were more likely to demonstrate progressive kidney disease based on increasing SDMA concentrations.

Positive Impact of Nutritional Interventions on Serum SDMA and Creatinine Concentrations in Client-Owned Geriatric Dogs[13]

A prospective study was conducted in client-owned geriatric dogs to evaluate the short-term effects of a test food on serum SDMA and Cr concentrations. Test food (dry food; protein 21.75%; phosphorus 0.68%, as fed) contained functional lipids (fish oil), antioxidants (lipoic acid, vitamins C and E), L-carnitine, botanicals (fruits and vegetables), controlled sodium concentration, and high-quality protein sources (high bioavailability egg and wet meat chicken and an ideal amino acid composition). Dogs (n=210) were fed either test food or owner's-choice foods (non-nutritionally controlled cohort). Dogs were included based on age and body weight: small (6.8 to 11.4 kg) and medium dogs (11.5 to 22.7 kg) were ≥9 years, whereas dogs >22.7 kg were ≥7 years at baseline. At baseline, all dogs had to have serum Cr concentrations within the reference interval and be free of chronic disease. Renal function biomarkers and urinalysis results at baseline, and after consuming test food or owner's-choice foods for 3 and 6 months, were evaluated. Only dogs consuming test food showed significant decreases in serum SDMA and Cr concentrations (both $p \leq 0.05$) across time (Test food baseline SDMA, 9.7 ± 0.2; 3 months, 8.7 ± 0.2; 6 months, 8.8 ± 0.2 µg/dL; Owner's-Choice foods baseline SDMA, 9.0 ± 0.2; 3 months, 8.8 ± 0.2; 6 months, 9.2 ± 0.2 µg/dL). At baseline or during the 6-month feeding trial, 18 dogs (8.6%) had increased serum SDMA, but normal serum Cr, consistent with IRIS Stage 1 chronic kidney disease. This included 9 dogs fed test food (baseline SDMA, 13.7 ± 0.8; 3 months, 12.2 ± 0.9; 6 months, 11.6 ± 0.8 µg/dL) and 9 dogs fed owner's-choice foods (baseline SDMA, 12.9 ± 1.2; 3 months, 12.5 ± 0.8; 6 months, 13.4 ± 0.9 µg/dL). Compared with baseline, after feeding 9 dogs test food for 6 months, serum SDMA decreased in 8 dogs and increased in 1 dog. After feeding 9 dogs owner's-choice foods for 6 months, serum SDMA decreased in 4 dogs and increased in 4 dogs (remained stable in 1 dog). The decreases in serum SDMA and Cr concentrations were significant (both p=0.03) only for dogs fed test food. These results suggest that nonazotemic dogs with increased serum SDMA (early CKD) when fed a test food designed to promote healthy aging are more likely to demonstrate improved renal function based on decreasing SDMA concentrations compared with dogs fed owner's-choice foods.

A Longitudinal Study on the Acceptance and Effects of a Therapeutic Renal Food in Pet Dogs with IRIS-Stage 1 Chronic Kidney Disease[14]

A prospective 12-month feeding trial was performed in client-owned dogs with IRIS-Stage 1 CKD to assess ease of transition to a therapeutic renal food and its effects on renal biomarkers and quality of life attributes. Dogs (n=36) were transitioned over 1 week from grocery brand foods to renal food. At 0, 3, 6, 9, and 12 months owners completed a questionnaire to assess their pet's acceptance of renal food and quality of life. Renal biomarkers, including serum Cr, BUN, and SDMA were measured. Of 36 dogs initially enrolled, 35 (97%) transitioned to renal food. Dogs moderately or extremely liked the food 88% of the time, ate most or all of the food 84% of the time, and were moderately or extremely enthusiastic while eating 76% of the time. All renal biomarkers (Cr, BUN, SDMA) were decreased ($p \leq 0.05$) from baseline at 3 months, and remained decreased from baseline at 12 months in dogs completing the study (n=20). Owners reported improvement in overall health and quality of life attributes, and hair and coat quality (all $p<0.01$). In summary, most dogs with IRIS-Stage 1 CKD readily transition to renal food. Decreasing serum biomarker concentrations suggest stabilization of renal function.

This was a "pre-and-post" study design. Participating clinics (44 sites) were in Kansas, Oklahoma, California, New York, Pennsylvania, Missouri, Tennessee, Georgia, Colorado, and Arizona. There were a total of 43 investigators that participated in recruitment of dogs between August 14, 2009 and August 26, 2011. Two hundred and nine dogs were screened for all stages of CKD (physical examination, CBC, serum biochemical profile, urinalysis) of which 36 were classified as IRIS-Stage 1 CKD. Prescription Diet® k/d® (dry food; 13.66% protein, 0.28% P, as fed) was the renal food. Of the 36 dogs categorized as IRIS-Stage 1 CKD, all had serum Cr concentration <1.4 mg/dL, 20 had persistently dilute urine (USG <1.020) without identifiable nonrenal cause, 7 had persistent proteinuria of renal origin (UPC ratio >0.5), and 9 had both persistently dilute urine and proteinuria. Dogs consuming renal food showed decreases in serum Cr and BUN concentrations across time (both $p \leq 0.01$) and a decreasing trend in serum SDMA concentration (p=0.09). All serum renal biomarkers (Cr, BUN, SDMA) were decreased ($p \leq 0.05$) from baseline by 3 months, and remained decreased from baseline at 12 months in dogs completing the study (n=20). These conclusions were also true when considering only the 20 dogs that completed the study. In this study, decreasing serum SDMA concentrations suggested improvement in GFR, reflecting stabilization and improvement in compensatory renal function.

References

1. Maxie MG, Newman SJ. Urinary system. In: Maxie MG, ed. *Jubb, Kennedy, and Palmer's Pathology of Domestic Animals.* 5th ed. Philadelphia, PA: Elsevier Saunders; 2007:425–521.
2. Pomeroy MJ, Robertson JL. The relationship of age, sex, and glomerular location to the development of spontaneous lesions in the canine kidney: analysis of a life-span study. *Toxicol Pathol.* 2004;32:237–242.
3. Finco DR, Brown SA, Brown CA, Crowell WA, Cooper TA, Barsanti JA. Progression of chronic renal disease in the dog. *J Vet Intern Med.* 1999;13:516–528.
4. Brown SA, Brown CA, Crowell WA, Barsanti JA, Allen T, Cowell C, Finco DR. Beneficial effects of chronic administration of dietary omega-3 polyunsaturated fatty acids in dogs with renal insufficiency. *J Lab Clin Med.* 1998;131:447–455.
5. Brown SA, Brown CA, Crowell WA, Barsanti JA, Kang CW, Allen T, Cowell C, Finco DR. Effects of dietary polyunsaturated fatty acid supplementation in early renal insufficiency in dogs. *J Lab Clin Med.* 2000;135:275–286.
6. Jacob F, Polzin DJ, Osborne CA, Allen TA, Kirk CA, Neaton JD, Lekcharoensuk C, Swanson LL. Clinical evaluation of dietary modification for treatment of spontaneous chronic renal failure in dogs. *J Am Vet Med Assoc.* 2002;220:1163–1170.
7. Polzin DJ. Evidence-based step-wise approach to managing chronic kidney disease in dogs and cats. *J Vet Emerg Crit Care (San Antonio).* 2013;23:205–215.

8. Brown S, Elliott J, Francey T, Polzin D, Vaden S, IRIS Canine GN Study Group Standard Therapy Subgroup. Consensus recommendations for standard therapy of glomerular disease in dogs. *J Vet Intern Med*. 2013;27:S27–S43.
9. Hall JA, Yerramilli M, Obare E, Yerramilli M, Melendez LD, Jewell DE. Relationship between lean body mass and serum renal biomarkers in healthy dogs. *J Vet Intern Med*. 2015;29:808–814.
10. Hall JA, Yerramilli M, Obare E, Yerramilli M, Almes K, Jewell DE. Serum concentrations of symmetric dimethylarginine and creatinine in dogs with naturally occurring chronic kidney disease. *J Vet Intern Med*. 2016;30:794–802.
11. Hall JA, Yerramilli M, Obare E, Yerramilli M, Panickar K, Bobe G, Jewell DE. Nutritional interventions that slow the age-associated decline in renal function in a canine geriatric model for elderly humans. *J Nutr Health Aging*. 2016;20(10):1010–1023.
12. Hall JA, MacLeay J, Yerramilli M, Obare E, Yerramilli M, Schiefelbein H, Paetau-Robinson I, Jewell DE. Positive impact of nutritional interventions on serum symmetric dimethylarginine and creatinine concentrations in client-owned geriatric cats. *PLoS One*. 2016;11(4):e0153654.
13. Hall JA, MacLeay J, Yerramilli M, Obare E, Yerramilli M, Schiefelbein H, Paetau-Robinson I, Jewell DE. Positive impact of nutritional interventions on serum symmetric dimethylarginine and creatinine concentrations in client-owned geriatric dogs. *PLoS One*. 2016;11(4):e0153653.
14. Hall JA, Fritsch DA, Yerramilli M, Obare E, Yerramilli M, Jewell DE. Acceptance and positive effects of renal protective food in pet dogs with IRIS-Stage 1 CKD. *J Anim Physiol Anim Nutr*. 2017.

acvim439

Canine Copper-Associated Hepatopathy- Where Are We Now?

S.A. Center, DVM, DACVIM
Ithaca, NY, USA

Primary & Secondary Copper (Cu)-Associated Hepatopathies (Cu-AH)

Copper (Cu) is a trace element essential for a wide range of biochemical processes with homeostatic balance dependent on Cu transporters, chaperones, and binding proteins. In this classification scheme, **primary copper-associated hepatopathy** indicates Cu is histologically associated with tissue injury or secondary reflects the consequence of liver injury, typically associated with cholestasis as shown in human and cats. Dogs are seemingly more resistant than other species to Cu accumulation secondary to cholestatic liver injury (e.g., bile duct obstructed cholestasis) or cirrhosis, based observationally on large numbers of examined liver biopsies. So take note - Primary does not indicate a genetic cause in this paradigm - it regards whether Cu accumulation is the **cause** of liver injury.

Enterocyte copper uptake

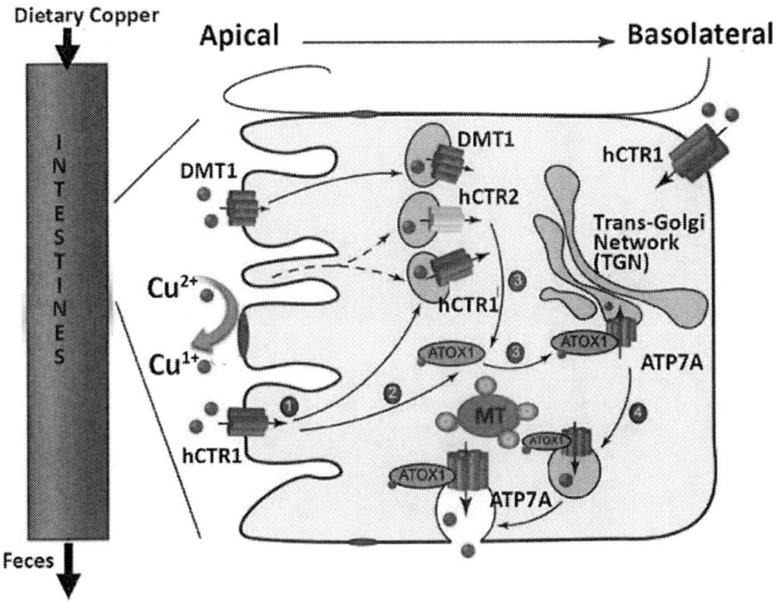

Enterocyte diagram, modified from: van den Berghe PV, Klomp LW. New developments in the regulation of intestinal copper absorption. *Nutrition Reviews*. 2009;67:658–72.
hCTR1= copper transport protein, ATOX1= antioxidant 1 copper transport chaperone. ATP7A= ATPase7A, TGN= Trans Golgi Network, MT= metallothionein. Metalloreductase reduction to Cu^{1+} before uptake. Cu CTR1 uptake at the apical surface at the alimentary lumen. Large quantity Cu uptake: internalizes bound to CTR1. The divalent metal transporter (DMT1) & CTR2 also may facilitate Cu uptake - dashed line indicating that possible contribution. MT buffers/scavenges excess cytosolic copper in enterocytes and liver along with GSH. ATOX1 binds to and transports Cu to ATP7A in enterocytes and also to ATP7B in liver. These ATPases (ATP7A & ATP7B) are essential for copper homeostasis and have essential biosynthetic roles in the trans-Golgi network (TGN) where they facilitate copper incorporation into proteins. This facilitates absorption and also guards against copper accumulation - redistributing copper in metal binding domains or cytoplasmic domains of certain proteins.

Dietary copper homeostasis should achieve neutral balance to ensure copper availability for a number of proteins (enzymes) with excess copper excreted in bile. Mutations causing dysfunction of ATP7A lead to **Menkes disease** or copper deficiency state whereas mutation in ATP7B causes copper accumulation and toxicity - demonstrated in **Wilson's disease in humans**. COMMD1 is a protein that facilitates ATPase copper trafficking. Mutation in COMMD1 is the cause of copper storage hepatopathy in Bedlington terriers. Recent genome-wide mapping of Labrador retrievers has implicated mutations in both ATPases that may have a complex effect on copper balance; estimated to contribute ~12% total heritability of the propensity for copper-associated hepatopathy in this breed.

Hepatocyte copper trafficking

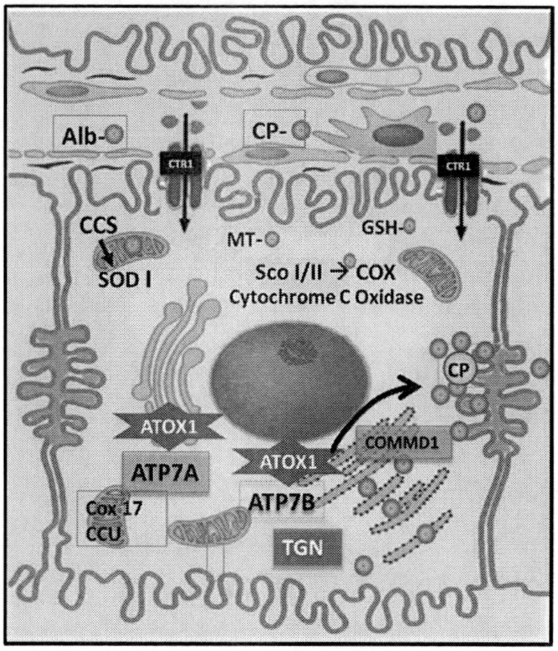

CTR1= High Affinity Cu transporter

Metallothionein & GSH – first interception

COX17= Cyclooxygenase chaperones cytosolic Cu to mitochondria for Cytochrome C Oxidase.

Sco 1 /II- may assist Cu → **COX-Cytochrome C Oxidase**- mitochondrial respiration.

CCS = transfers Cu → mitochondrial **SOD 1**

Antioxidant 1-Cu Chaperone (ATOX1)- transfers Cu ATPase transporters. Cu moved to Golgi secretory pathway-.Trans Golgi Network (TGN).

ATP7B: loads 6 Cu atoms on CP → circulation. When cellular Cu is high, loads lysosomes.

ATP7A: iCu uptake in gut & redistributes Cu systemically & within hepatocyte

COMMD1- partners with ATP7A & ATP7B navigating TGN.

Primary Cu Storage Hepatopathy

Prototypical Cu-AH is the COMMD1 mutation in Bedlington terriers (BT), a 13.6 kb deletion in exon 2 of this gene. Pathologic Cu accumulation correlates with functional and morphologic liver injury in this breed. A small number of BT families with Cu-AH lack this specific mutation. Considering that in humans with Wilson's disease, there are >500 mutations, this does not eliminate a gene-related mutation in those dogs. A recent genome-wide association study in Labrador retrievers identified missense mutations in metal-binding domains of ATP7A & ATP7B. As mutation in ATP7A is the cause of Menkes disease - (causing copper deficiency) the combined findings were interpreted as evidence of a complex trait involving the same ATB7B gene as in Wilson's disease. This complex & interesting study proposes one cause of the propensity of Labrador retrievers for copper-associated hepatopathy but it is far from clear cut. There are numerous dogs of many other breeds and mixed bred dogs with pathologic hepatic copper accumulation which suggests that dietary copper ingestion is an important contributing factor.

CLINICAL & CLINICOPATHOLOGIC FEATURES

Physical features & clinical signs vary depending on the stage of injury.

Preclinical: asymptomatic, serendipitous discovery of increased ALT activity. **Chronic**: vague chronic illness, variable appetite, weight loss, lethargy, with increase ALT +/- ALP, AST activity. **Peracute**: sudden severe illness with clinicopathologic parameters consistent with severe liver injury, small subset with euglycemic glucosuria (acquired Fanconi syndrome), rare hemolysis. All breeds are susceptible with highest prevalence in Labrador retrievers. Early recognition of Cu-AH is based on finding persistent unexplained increases in serum ALT +/- AST usually exceeding fold increases in ALP in the absence of liver function abnormalities. Suspected progressive hepatopathy leads to US imaging which may be normal or disclose homogenous or coarse hyperechoic hepatic parenchyma (multifocal glycogen or lipid vacuolation, inflammation).

Liver Fine Needle Aspiration Cytology

Liver aspirates may or may not disclose hepatocellular Cu and can be broadly misinterpreted, depending on sample cellularity. Estimating severity of Cu accumulation requires assessment of zonal distribution - that cannot be completed with aspiration samples. Glycogen vacuolation is common, induced by inflammatory cytokines. Inflammatory infiltrates largely focused in centrilobular regions and may not be represented or may be confused by blood contamination. Wrights Giemsa staining (not Diff Quik) will disclose cytosolic turquoise refractile granules in hepatocytes that must be confirmed with rhodanine staining as Cu; interpretation requires estimation of % affected sampled cells but can overestimate the severity of tissue Cu accumulation. Similar issues with scant needle biopsy samples.

ARE GENETIC OR NUTRITIONAL FACTORS CAUSAL TO CU-ASSOCIATED HEPATOPATHY?

A breed-specific genetic basis for Cu-AH remains unclear with exception of the BT and now the complex findings in Labrador retrievers. Small case series of individual dogs (pure breeds, e.g., Dalmatian, Corgi, Labradors, others), and mixed breeds with hepatopathies associated with Cu (≥1,000 μg/g dry matter basis [DMB])) are abundant. A compelling report of Dalmatians with lethal severe necrotizing Cu hepatopathy similar to BT circumstantially suggested a heritable disorder in some kindreds. A few Dalmatians seen at Cornell have peracute illness & prodigious Cu concentrations. West Highland White Terriers (WHWT) commonly have increased hepatic Cu that does not correlate with liver injury or illness. Common diagnosis of Cu-AH in Labrador retrievers (LR) and Doberman Pinchers (DP) has generated concern for a breed-related propensity in each breed. In the last 18 years, the Pathology Service at Cornell routinely observes excessive hepatic Cu concentrations in large and small pure breed and mix-bred dogs with >30% sample submissions. We document a significant increase in hepatic Cu in LR & DP with and without hepatitis since Cu chelates replaced copper oxide in premix supplements used for maintenance foods in 1993 (study of >500 dogs). It is relevant that the reference range for canine hepatic Cu concentrations has progressively increased since 1940 when commercial pet foods were introduced. From that time, hepatic Cu concentrations have progressively increased from 50–70 μg/g DWB to the current level of 400–500 μg/g DWB.

So why are some breeds seemingly predisposed? The recent GWA in Labrador retrievers explains propensity for copper accumulation. This may reflect selection of "hitchhiker" genes - genes regulating one or more of the copper chaperones or transporters, binding proteins, or storage proteins that influence Cu trafficking - naively selected because they are closely positioned to genes determining breed standards/favored characteristics under positive selection pressure.

Of interest is that using a liver-specific COMMD1 knockout mouse model, there is evidence for a biological role of COMMD1 in hepatic copper homeostasis. However, affected dogs progressively accumulate copper in their livers whereas mice only display hepatic copper accumulation when fed a copper-rich diet.

Defining Hepatic Copper Status

Qualitative Cu Assessment

On routine light microscopy with H&E staining, Cu granules appear as pink-brown or pink-grey intracytoplasmic refractile inclusions in the cytosol of hepatocytes, macrophages, and small pyogranulomatous foci - we designate as **copper granulomas**. That is of course, if tissue sections are cut thin enough to appreciate this change. We often encounter sections too thick for this assessment on second-opinion cases where tissue has been sectioned elsewhere. With rhodanine staining, Cu-protein aggregates can be observed when Cu >179 µg/g DWB. Excessive Cu must be confirmed with Cu-specific stains (e.g., rhodanine or rubeanic acid most commonly used) to allow qualitative grading. Qualitative grading considers zonal Cu distribution and lesion severity but are subjective and influenced by inter-individual differences and inconsistencies within a single interpreter. An objectified grading scheme is more valuable than an opinion biased +1 to +4 method. However, histologic assessments should always be reconciled with measured Cu concentrations (DWB).

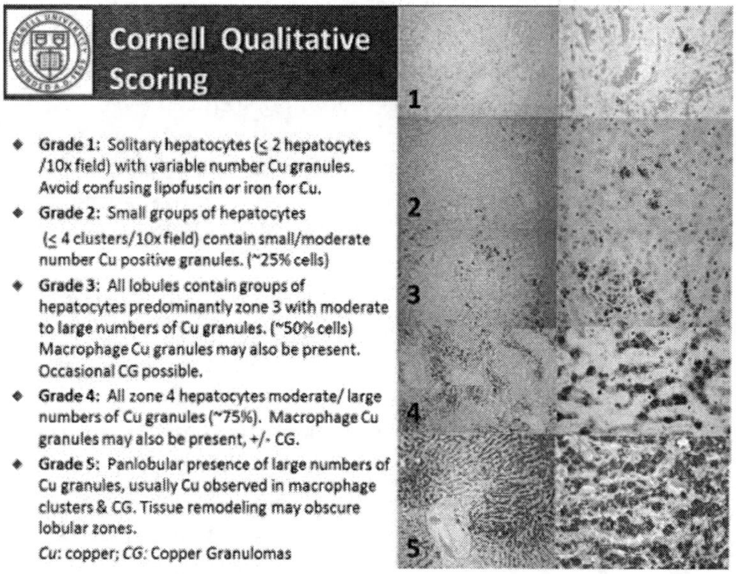

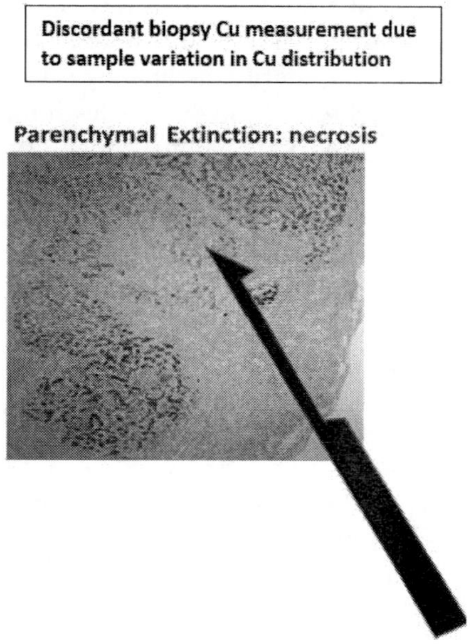

Copper Quantification

Copper quantification guides propriety & duration of chelation therapy and used to assess treatment response in some cases. Liver Cu ≥600 µg/g DWB is common in dogs with hepatitis, with values >1,000 µg/g DWB considered severe. Frozen tissue, formalin-fixed tissue, or paraffin-embedded tissue can be used for Cu quantification. However, if Cu is routinely measured after histopathologic assessment using formalin-fixed specimens, Cu content of formalin should be ascertained. Measurement in paraffin-embedded tissue requires xylene extraction of paraffin and wastes tissue that may be needed for additional histologic assessments. Variation in Cu distribution within a biopsy specimen associated with architectural remodeling can lead to erroneously low measurement when a separate biopsy sample is submitted for bench analysis. This occurs even when done by atomic absorption spectroscopy (AAS) considered the gold standard method. We also have encountered numerous erroneous Cu concentration measured by mass spectroscopy (high & low values) as used in several large diagnostic facilities - these may relate to incomplete sample desiccation but the matter is under investigation. Therefore, single measured samples external to the biopsy specimen histologically examined **may not** reconcile with visual biopsy assessment. Quantification methods must be validated for intra- and inter-assay performance, recovery, with quality control samples routinely measured, and results expressed as DMB.

Digital Scanning for Cu Quantification

To overcome the issue of marked discordance between visual qualitative & quantitative Cu assessments, we developed a digital scanning method to quantify hepatic Cu using the entire biopsy specimen that can be routinely completed. A digital scan (40x, Aperio-Leica Scanner) of the rhodanine-stained biopsy section is analyzed using a proprietary positive pixel algorithm that was validated against AAT replicate samples from the same tissue ($R^2 = 0.97$ linear regression) to calculate hepatic Cu DMB. This can accommodate even small fragmented needle core biopsies.

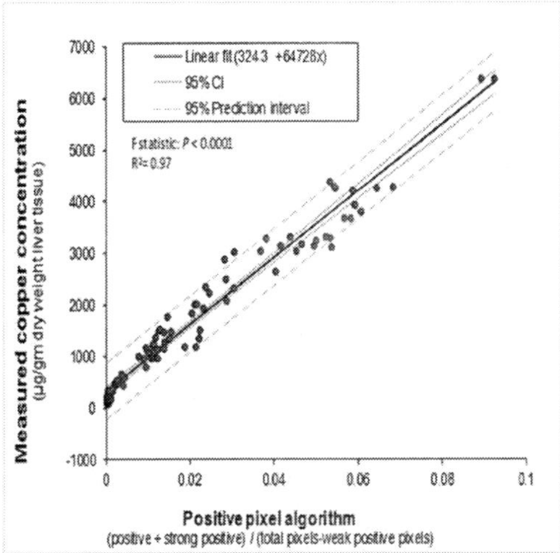

In human medicine, it is known that hepatic Cu concentration can increase before evident on light microscopy, where values >50 µg/g DMB are considered indicative of Wilson's disease - thus, bench measurement is needed to achieve early diagnosis of this genetically based disease before tissue injury is realized. Humans have low tolerance to hepatocellular Cu accumulation compared to dogs. This may reconcile with higher metallothionein concentrations and different response of ceruloplasmin in dogs. Preemptive treatment is necessary to avoid development of irreversible neurologic injury.

Histologic Lesions of Cu-Associated Hepatopathy

Variable with H&E staining ranging from finding eosinophilic Cu granules with no lesions to rare Zone 3 single cell necrosis, to numerous **copper granulomas** - mixed cellular granulomas surrounding foci of necrotic hepatocytes, Zone 3 parenchymal collapse, bridging/progressive tissue remodeling (regenerative nodules, dissecting fibrillar collagen (pericellular), parenchymal extinction). Accordingly, clinicopathologic features are variable. Immunohistochemistry for oxidative injury and TNF-α show positive signals in damaged hepatic tissues. Finding Cu granules in hepatocytes marginating regenerative nodules represents a causal role in liver injury. Large amounts of nitrotyrosine adducts may initiate immunoinjury. Some of these dog display **lymphocyte satellitosis** (aggregation of lymphocytes surrounding single necrotic hepatocytes) implicating some role for immunoinjury that may require addition of immunomodulatory intervention after initial response to D-penicillamine is observed, as this has immunomodulatory influences.

Pathogenesis of Cu-Associated Hepatopathy

Cu, like iron (Fe), is a transition metal (transition metals "bounce" electrons) that can initiate and augment oxidative injury. Each can play a pathologic role in necroinflammatory liver lesions. Fe usually accumulates in macrophages where it remains in flux with the redox reactive free Fe pool. Both Fe & Cu can catalyze the Haber-Weiss & Fenton reactions that generate injurious hydroxyl & superoxide radicals. Cu accumulates in hepatocyte lysosomes, leading to organelle membrane injury, rupture & release of autodigestive enzymes. Cytosolic release of Cu leads to nuclear injury (impairing gene transcription), oxidative mitochondrial & membrane damage, & enzyme dysfunction leading to cell death (cytolytic, apoptotic). Focal or massive hepatocellular necrosis may ensue. Hepatocellular injury is initially evident in Zone 3 (around central veins). Rarely, when large amounts of Cu are released into the circulation, hemolytic anemia may develop as well as a transient acquired Fanconi's syndrome. Routine measurement of plasma or serum Cu are poor tests for detection of Cu-AH. Measurement of ceruloplasmin in dogs has no relevance although is useful but not diagnostic in humans with Wilson's disease.

$$Fe^{2+} + H_2O_2 \xrightarrow{\text{Fenton Rxn}} Fe^{3+} + HO^- + O^\bullet$$
$$HO^\bullet + H_2O_2 \xrightarrow{\text{Haber-Weiss Rxn}} H_2O + O_2^{\bullet -} + H^+$$
$$O_2^{\bullet -} + H^+ + H_2O_2 \rightarrow O_2 + HO^\bullet + H_2O$$

$$O_2^{\bullet -} + Cu^{2+} \rightarrow O_2 + Cu+$$
$$Cu^+ + H_2O_2 \rightarrow Cu^{2+} + OH^\bullet + OH^-$$

Acquired Fanconi Syndrome

Severe Cu-induced proximal renal tubular injury causes glucosuria in the absence of hyperglycemia and often is accompanied by granular casts. Other toxins or infectious agents considered include aminoglycosides, NSAIDs, xylitol, chicken jerky treats, leptospirosis, or causes of proximal renal tubular injury. Treatment requires judicious fluid therapy, supplemental B vitamins (proximal tubular loss), and antioxidants (IV N-acetylcysteine dosed as for Tylenol toxicity: 140 mg/kg IV, given over 20 min, followed by 70 mg/kg q6–12 h; SAMe: 20 mg/kg/q24 PO, vitamin E: 10 U/kg/q24 h). Supplemental bicarbonate is used for metabolic acidosis (pH<7.2). Dogs can completely recover from Cu-Fanconi injury.

Rare Hemolytic Anemia

Typically, few spherocytes, Coombs negative, non-regenerative because of acute nature, & methemoglobinemia may develop. The cause is suspected to involve oxidative injury, damage to membrane transporters, depletion of GSH, and altered membrane phosphatidylserine. Supportive care requires acute institution of chelation; in humans intervention by plasmapheresis and administration of plasma & packed red cells have salvaged severely affected individual. Hemodialysis has removed less free Cu. These procedures in humans are used to bridge to liver transplant.

Urinary Copper Elimination - Before & After Chelation Therapy

In humans with Wilson's disease, an important test is 24 h urinary Cu-elimination before and after initiation of chelation. This has not been rigorously investigated in dogs. One study of 22 dogs, 7 normal beagles, 4 homozygous ATP7B mutation mixed breed dogs (Beagltons), and 11 Labrador retrievers D-penicillamine induced urinary copper excretion and copper/creatinine ratios did not correlate with hepatic copper concentrations. Basal urinary copper/zinc ratios correlated with hepatic copper concentrations in Labradors - this has not been further evaluated.

Treatment Considerations for Cu-AH

Treatment recommendations consider hepatic Cu concentration, histologic distribution of Cu, severity of tissue injury, clinical signs & clinicopathologic features. Quantitative Cu >600 but <800 µg/g DWB, Grade 1–2, undergo dietary Cu restriction, 2–4 months of d-penicillamine chelation with supplemental pyridoxine, antioxidants (bioavailable thiol donors, low dose mixed tocopherols [vitamin E; 10 u/kg PO SID with food], and lifelong restriction of Cu but not protein intake. Longer term chelation (~6 months) is recommended for dogs with tissue Cu ≥800 µg/g DMB, ≥Grade 3, and high ALT activity, that have no other discerning pathology. Dogs with Cu ≥800 µg/g DWB, ≥Grade 3 with histologic injury are chelated and given antioxidants for ≥6 mths, then either prescribed a 50% dose reduction of the chelator given twice weekly for life Or daily zinc supplementation, lifelong restriction of Cu intake, & antioxidants while ALT activity remains increased. Only liver biopsy confirms the extent of tissue Cu removal after chelation. However, ALT functions as a surrogate marker, notably declining after the first month and usually normalizing by 3 to 6 months. In dogs with hepatic Cu (>3,000 µg/g) longer term chelation may be needed to reduce tissue Cu and normalize enzyme activity up to 1 yr or sometimes longer when there is prodigious Cu accumulation.

REMOVING HEPATIC & SYSTEMIC COPPER

Anyone faced with a difficult treat dog will start combing PubMed for current recommendations for humans with Wilson's disease. Here are a few things to consider. Humans first diagnosed with Wilson's disease may have neurologic signs with risk for worsening with d-penicillamine and even with trientine initiation. In fact, some patients never recover pre-illness cognitive functions completely after this adverse response. Thus, there are numerous case series investigating response & survival of patients with neurologic signs treated with **zinc, trientine, D-penicillamine**, and **ammonium tetrathiomolybdate**. Zinc can initiate control of Cu status **only very slowly**, achieving a modest negative Cu-balance. Neurologic signs in humans can progress over this lag time. Chelators more rapidly remove Cu from tissue and blood; d-penicillamine being considered the gold standard first-line drug, trientine the alternative. Historically, BAL was initially used to treat humans with Wilson's disease before d-penicillamine treatment was developed. Zinc therapy should not be combined with chelation. The following are treatment pearls that may be useful in particular cases.

D-Penicillamine is a reductive chelator - meaning it reduces Cu, thereby diminishing its affinity for proteins, augmenting d-penicillamine-Cu binding. It was identified first as a breakdown product of penicillin but is a sulfhydryl-bearing amino acid cysteine doubly-substituted with methyl groups. Like dimercaptopropanol (or BAL - British anti-Lewisite - historic heavy metal chelator) it has sulfydryl groups that function as Cu chelators. A racemic mixture - no longer used, interferes with pyridoxine action (treatment with d-pen vs. DL-pen was done in humans, pyridoxine change was more substantial with DL form but were still seen with D-pen). Supplemental pyridoxine is still used in humans with Wilson's disease. D-penicillamine induces metallothionein in Wilson's patients which contributes to copper attenuation. Once Cu is bound to d-Pen, it is rapidly eliminated in urine. In human Wilson's patients, 200–300 μg/Cu/day may be eliminated in urine before d-Pen is administered; this escalates up to 1.0–1.5 mg/day on drug initiation, rapidly achieving a negative Cu balance. Dosing should be given 30–60 min before meal ingestion or at least 2 h after food ingestion - taking the drug with a meal reduces absorption by 50%. In man, therapy is followed by measuring plasma Cu and calculating free or unbound Cu concentration from the ceruloplasmin concentration (ceruloplasmin binds 3 μg of Cu per mg). Calculation of free Cu concentration between 10–15 μg/dL indicating good Cu control. This has not been used as a management strategy in dogs.

Dosing d-Penicillamine: 10–15 mg/kg PO BID 30 minutes before feeding or 2 h after feeding. A glucocorticoid dose may increase appetite if hyporexia occurs or a drug hypersensitivity response is suspected. In man, excess Cu is mobilized during the initial treatment year with improved health status realized **before** all Cu is mobilized from the liver, suggesting that Cu detoxification precedes tissue Cu chelation. D-penicillamine sequesters Cu in an innocuous state either through direct binding or by induction of Cu-binding proteins (e.g., metallothionein). However, in patients with prodigious hepatic copper concentrations, D-Pen may initially increase the flux of free copper causing illness (in humans associates wit neurologic decompensation). Urine is the major route got elimination of chelated copper. In addition to Cu chelation, d-Penicillamine provides antiinflammatory effects through suppression of leukotrienes & prostaglandin-associated inflammation, immunomodulation, has antifibrotic effects via inhibition of collagen cross-linking & stimulated collagenase activity, assists with cellular GSH restoration, & improves efficacy of mononuclear-macrophage surveillance against antigens and/or endotoxins. D-Penicillamine also may chelate zinc & other essential metals. Chronic dosing at initial therapeutic levels can rarely lead to Fe deficiency anemia. Co-treatment for these possible scenarios reduces its Cu chelating efficacy. In fact, co-treatment with zinc is strictly contraindicated as this will negate benefits of each therapy. If chelator and zinc are administered in combination, the interval between dosing must be as wide as possible.

Side effects of d-Penicillamine: tolerability of this drug may be enhanced by starting with incremental doses (1/3 calculated dose per day increasing dose q 4–7 days until the full intended dose is achieved). The most common side effects in dogs are hyporexia (may relate to dysgeusia), nausea, and vomiting. This may be ameliorated by drug administration with a small piece of meat. If signs are persistent, the drug should be discontinued and re-initiated at a lower dose, then gradually titrated upwards over several weeks. In some cases, a small dose of glucocorticoids moderates this response.

Acute hypersensitivity drug reactions are also possible- I see this in <5% of dogs; in humans this is associated with fever, malaise, rash, pruritus, and lymphadenopathy. In dogs this associates with unexplained lethargy, +/- fever, & vague illness. This may be modulated by a drug holiday, restarting a lower dose, again with low dose glucocorticoids. In some dogs, signs are only controlled if a lower drug dose is chronically used (50% dose reduction), increasing time for systemic copper chelation. If this is recurrent - change treatment to a different drug (i.e., trientine, zinc, or ammonia tetrathiomolybdate). Proteinuria, immune-mediated renal disease sometimes progressing to nephrotic syndrome is rarely observed; a urine P/C should be determined before starting treatment with d-penicillamine. A lupus-like syndrome is rare - this may induce a positive ANA, associated with polyarthritis or polymyositis, hematopoietic dyscrasias (leukopenia-granulocytopenia, thrombocytopenia, eosinophilia, aplastic anemia), peripheral neuritis, optic neuritis, a myasthenia-like syndrome, or dermatitis with cutaneous eruptions, and/or hair loss. D-penicillamine also may induce ALP activity associated with induction of a mild glycogen-type vacuolar hepatopathy in dogs - resolving with drug discontinuation, - unless there is an additional inflammatory process or adrenal hyperplasia syndrome inducing this change. Unnecessary dietary protein restriction may amplify this response (unknown mechanism). Because d-penicillamine mixtures imparts an anti-pyridoxine effect (vitamin B_6 inactivation), pyridoxine supplementation was and still is advised for humans; we use 25 mg pyridoxine/day without issue.

2-3-2 or 2-2-2 Tetramine (trientine, triethylene-tetramine dihydrochloride): The alternative chelator (Syprine, 250 mg capsules, Merck Sharp & Dohme; Cuprid) in one of a family of chelators with a polyamine-like structure chemically distinct from penicillamine; it lacks sulfhydryl groups. It is poorly absorbed from the alimentary canal; what is absorbed is metabolized and inactivated. Copper is chelated by forming a stable complex with four constituent nitrogen in a planar ring. It is used at an initial dose of 5–7 mg/kg PO BID, 1–2 hours before meals (this a lower dose than published that we use based on two dogs that developed acute renal failure after starting chelation. This drug is a preferred primary chelator by some physicians, touted as a more effective cupruretic/chelating agent compared to d-penicillamine but this remains controversial. It has been suggested to more effectively mobilize Cu from different compartments. Opinions vary regarding efficacy & safety of trientine in humans with neurologic signs of Wilson's disease caused by brain Cu toxicity. Case-based observations (humans) verify that trientine also may increase free circulating Cu to an extent that can precipitate or worsen neurologic signs as also observed with D-Pen chelation. This can lead to permanent disability or death. **Side effects** (humans): similar to d-Pen causing gastritis, nausea, and iron-deficiency anemia. Both drugs can increase free circulating Cu. While there were no severe side effects in a limited clinical trial in healthy dogs, I have observed acute renal injury in two dogs with severe hepatic Cu accumulation treated with the initially recommended 10–15 mg/kg/ day PO BID dose. This may have been due to increased free Cu uptake into proximal renal tubules. Trientine tablets are not stable for prolonged periods at high ambient temperature, problematic for warm climates.

Ammonium tetrathiomolybdate (Coprexa, Pipex Pharmaceuticals, not yet FDA approved, now being developed by Wilson Therapeutics as Decuprate - WTX 101-202, bis-choline tetrathiomolybdate, has orphan drug status) has been used in humans with neurologic signs, where d-Pen or trientine may induce critical decompensation. This application was first reported in 1991 - Brewer *et al*. Ammonium tetrathiomolybdate structurally has a molybdenum molecule shared between 4 sulfhydryl groups. Biochemically it forms a tripartite complex with protein (many proteins) & Cu and can attenuate systemic Cu accrual in two ways. First, it complexes Cu with ingested food protein, irreversibly binding Cu within the alimentary canal and preventing absorption. Drug administration with meals is necessary to achieve this effect. It also binds Cu in oral secretions & gastric juices. This treatment mechanism immediately imparts a negative copper balance as complexes are not digested and eventually eliminated in feces. Secondly, it complexes free Cu (Cu bound to proteins & albumin in the circulation) forming a tripartite complex - making Cu unavailable for cellular uptake. These complexes slowly undergo hepatic metabolism and are subsequently eliminated in bile. These mechanisms permit acute manipulation of free copper in the systemic circulation that can rapidly detoxify a patient with crisis toxicity within 2–3 weeks. The FDA published an admonishing letter for lack of FDA protocol compliance in the human clinical molybdate study, so approval of this drug was denied. Another company, Wilson Therapeutics in

Stockholm, Sweden states that the drug has achieved orphan disease status in Europe & USA, Decuprate or WTX 101-202, bis-choline tetrathiomolybdate.

Initial studies of ammonium tetrathiomolybdate were completed in 6 Wilson's patients with neurologic signs that were treated with non-pharmaceutical grade chemical purchased from Aldrich Chemical Co. Initial dosing ranged from 1.6 to 2.9 mg/kg/day - divided into 6 doses divided between meal-associated dosing and dosing during the interdigestive interval. Six doses were divided - 3 times per day **with meals** & 3 times per day **between meals** and given from 8 to 29 days based on estimated copper removal. This dosing strategy capitalized on the mechanistic drug impact on alimentary & systemic copper. Four patients were treated from 11 to 17 days to achieve rescue from critical Cu toxicity. Duration of treatment was based on sequential calculations estimating "closure of the Cu-Molybdenum gap" determined from total plasma Cu concentration & ceruloplasmin concentration (estimating amount of Cu bound to ceruloplasmin as 3 µg Cu/mg ceruloplasmin) to determine free Cu concentration (ceruloplasmin bound Cu being nontoxic), and plasma molybdenum concentration. A stoichiometric estimate of 1:1 molecular binding with Cu, estimated the amount of Cu needing removal. Enteric binding efficacy of tetrathiomolybdate was determined using oral Cu^{64} - verifying efficacy of fecal Cu elimination. Stability of raw chemical stored anaerobically at room temperature and stored in therapeutic capsules in closed vials in a pharmacy environment, were determined. Raw chemical was stable & capsules were stable with daily dose withdrawal from a capped vial - over 8 weeks. The eventual protocol determined was a total daily dose of 2–3 mg/kg with 1/3 given at meals times & 2/3 given between meals to optimize Cu mobilization. A 2 wk overlap treatment with zinc & molybdate were given during the transition to chronic zinc management. By 2009, the most studied dosing scheme has been 120 mg/day, with 20 mg given 3 times per day with food and 20 mg given 3 times per day during the interdigestive interval - for a total of 8 weeks.

Limited toxicity has been encountered but has included 1) development of anemia +/- leukopenia (10–15% of patients) thought to reflect bone marrow Cu depletion. This has been encountered in patients with the most rapid Cu removal and is ameliorated by a temporary drug holiday. On recovery, treatment is resumed with a 50% dose reduction. The other side effect (10–15% of treated humans) is an acute increase in ALT activity after ~2 weeks of drug administration suspected to reflect Cu mobilization causing transient aggravation of Cu-mediated liver injury. Because of this side effect, standard dosing has been modified to 120 mg/day x 2 weeks with reduction to 60 mg/day for the remainder of the treatment interval (~6 weeks). This drug has been used in other disorders without observation of ALT increase and thus is assumed to reflect a unique Cu shift and liver injury in Wilson's patients. A strong proponent of this initial therapy (Dr. Brewer) proposes that chelators (d-penicillamine & trientine) are contraindicated in initial treatment of Wilson's disease in humans with neurologic signs because of the risk of causing disastrous neurologic worsening. Zinc is too slow in action for these patients, taking 9–12 months to control copper toxicity - during which time the disease-associated neural injury can progress. Ammonium tetrathiomolybdate has also been shown to lead to neurological dysfunction with overly aggressive copper removal.

Zinc therapy was initially predicated based on recognition that it made humans with sickle cell anemia Cu deficient. After 20 yrs of animal and human trials it was approved by the FDA for chronic management of Wilson's disease (Galzin™ in the U.S. TEVA pharmaceuticals, Wilzin® in Europe - zinc acetate). As most know, zinc induces enterocyte metallothionein that avidly binds ingested Cu. Metallothionein has a greater avidity for Cu vs. zinc, thus preferentially binds Cu. Zinc also may induce hepatocellular metallothionein contributing to Cu removal from lysosomes & mitochondria. Because induced enterocyte metallothionein also binds copper in saliva and gastric juice, zinc treatment can generate a negative Cu balance. With a ~6 day turnover interval it disposes of bound Cu with effete enterocytes. In man, it takes ~2 wks of a minimal dose of 37.5 mg elemental zinc PO BID or 25 mg TID to induce metallothionein to obstruct Cu ingress. Loss of metallothionein induction takes a similar interval - meaning that an occasional missed dose will not substantially increase patient risk. A single large dose of zinc is not effective, rather, it must be given separated from food by 1 hour before or at least 2 h after meal intake for efficacy. Dose regimen studies and efficacy studies using Cu^{64} were done in humans **AND 6 dogs** - 3 Bedlington terriers with CuAH & 3 WH White Terriers. **In humans**, 50 mg elemental zinc PO

TID is the standard adult dose approved by the FDA based on Dr. Brewer's work. **In dogs**, based on detailed work in these 6 patients, zinc effects took ~3–6 mths to achieve a therapeutic plasma zinc response and suppressed Cu^{64} absorption. Dosing initially at 200 mg/day of elemental zinc and then lowered to 100 mg/day of elemental zinc (~Bedlington weight=8–10 kg so this is initially 16–20 mg elemental zinc/kg lowered to 8–10 mg/kg/day and then ½ of this, ~WHW Terrier weight=6–10 kg so this was initially 20–32 mg/kg/day subsequently lowered to 10–16 mg/kg/day and ½ of this). Dose adjustment was predicated on baseline zinc compared to post-dosing zinc with a target of at least doubled baseline value (>200 µg/dL) without reaching >1,000 µg/dL associated with hemolysis. The 200 mg/day dose more than doubled plasma post-dosing zinc concentrations. A 100 mg/day dose was adopted, with a further lowered dose to 50 mg/day given to some dogs (dosing always BID, 30 min before feeding). One dog of each breed was treated for 1 h and 2 dogs of each breed for 2 yrs. Follow-up liver biopsies at 2 yrs confirmed **variable de-coppering** (BT: 5,400 → 2,740 µg/g, 5,700 → 1,500 µg/g, WHWT: 1,300 → 432 µg/g, 1,890 → 400 µg/g DWB). Notably, BT were not free of pathologic Cu - so this would not be considered effective intervention. Encountered vomiting (AM dosing usually) was controlled by administration of zinc with a small piece of meat (meat is less interfering with zinc binding vs. vegetable phytates). Management solely with zinc should be reserved for asymptomatic patients with little evidence of liver injury.

Considerations regarding zinc therapy: While there are numerous peer-reviewed published manuscripts describing the response of humans with Wilson's disease to monotherapy treatment with zinc, I am particularly impressed by the data presented in a 2011 paper published in *Gastroenterology* - n=288; median follow-up time, 17.1 years (*Gastroenterology*. 2011;140:1189–1198) where poor response to zinc therapy vs. chelation with D-penicillamine or trientine is demonstrated - patients treated with zinc went to liver transplant or death significantly sooner than those treated with chelators or a combined complex chelator/zinc protocol.

Unfortunately, zinc administration to dogs often leads to gastritis, vomiting, and inappetence. Zinc acetate is the least noxious form of zinc for oral medication of dogs. If you try zinc, use 5–7 mg elemental zinc/kg body weight divided BID (dose thus depends on the molecular wt of the product and contribution of zinc to the total molecular weight). Measure plasma zinc at baseline and then at 2–6 wks of treatment to document an at least doubled plasma zinc concentration - if no increase is observed, then the zinc dose should be increased. Beware of zinc plasma concentrations approaching 800 µg/dL - a danger zone for hemolysis. Efficacy of long-term zinc therapy for complete control of Cu accumulation and **copper-associated hepatopathy** remains questionable based on the study summarized above - the only study of elemental zinc dosing. What dose does a Labrador retriever need? In dogs, dose optimization is hard to prove - in man the dose can be confirmed with radioisotope studies & urinary Cu excretion that we have not investigated as routine assessments.

RESTRICTING SYSTEMIC CU ALLOWANCE

Diet alone helps control Cu-**associated liver injury** & is a mandatory husbandry adjustment that must be accommodated. Humans with Wilson's disease consuming a severely Cu-restricted lactovegetarian diet, almost totally non-compliant with anti-Cu therapy (chelation), are able to avoid Cu accumulation. Prescription canine liver diets (L/d, [Hills], Hepatic Support, [Royal Canin]) provide ~4.9 mg/kg (ppm, DMB, ~0.1 mg Cu/100 kcal) predicated on achieving a neutral Cu-balance in an affected Bedlington Terrier; Purina NF delivers a Cu content close to these. These Cu-limited diets provide a base diet that delivers ~2.2 to 2.5 g protein/kg body wt/day when consumed for RER. Considering that most dogs with Cu-AH will not require protein restriction, additional 0.5 to 1.5 g protein/kg body weight/day of low Cu containing foods are supplemented. The USDA food tables are used to sort food Cu content to select a useful protein supplement. I recommend empowering owners to navigate this website and sort foods based on mineral - Cu content and select low Cu containing foods to keep their dog's diet engaging. The Super Tracker freeware web link at the USDA website can be used to determine protein content per food portion of the supplements. This process empowers an owner to adjust their dog's diet with foods within safe Cu limits. Or alternatively, engage assistance of a Clinical Nutritionist.

USDA Food Tables

http://www.nal.usda.gov/fnic/foodcomp/Data/ then Display Nutrient Lists VIN editor: link could not be accessed as of 5/11/17; the home page is www.nal.usda.gov/fnic). Scroll down and select Cu and sort by nutrient content. This shows Cu content per unit measured in all foods. SuperTracker is also available at this site. This is webware constructed for human diet tracking - but can be employed for a dog.

Limiting Water Cu

<0.1 ppm; if Cu plumbing, flush line for 5 min to remove Cu contamination. Determining Cu contamination in home water can eliminate this concern - of importance usually only if use well water. If an owner purchases water for home use at the grocery, distilled water is recommended.

Other Considerations

Glucocorticoids

Reserved for hepatocellular necrolytic or hemolytic crises; low dose stimulates appetite in dogs hyporexic from d-Penicillamine or with an initial hypersensitivity response.

Avoid Vitamin C Supplementation

May augment transition metal induced tissue injury. Use vitamin supplements specifically produced without copper. BalanceIT is one source of supplemental vitamins needed for a homemade diet.

Antioxidants

Vitamin E 10 IU/kg/day PO - used mixed tocopherols - predicated on determinations made in plasma & liver of humans with Wilson's disease and cell culture assessments and liver tissue collected from Bedlington terriers with Cu AH.
Bioavailable SAMe (Denosyl™): 20 mg/kg PO/day (empty stomach).

Ursodeoxycholic Acid

If cholestatic injury (high bile acids) due to liver injury: 10–15 mg/kg/day, divided BID give with food.

Immunomodulatory Drugs

Immunomodulatory therapy is **not routinely** used in most Cu-AH and is guided by findings on liver biopsy. Some dogs with CuAH develop inflammatory T-cell infiltrates (esp., Labrador Retrievers, Doberman Pinchers, others) reflecting immunoinjury - this may be instigated against epitopes formed by oxidative injury. These require immunomodulatory therapy as prescribed for other immune-mediated necroinflammatory liver disorders (e.g., azathioprine or mycophenolate or cyclosporin, with low-dose titrated glucocorticoids to an every-other day basis). Whether these are separate and discrete disease processes or reflect self-sensitization subsequent to Cu-related hepatocyte injury remains unclear.

Antifibrotics

Vitamin E (10 IU/kg/24h PO), SAMe (20 mg/kg/24h PO), and polyunsaturated phosphatidylcholine (dilinolylphosphatidylcholine: 20–50 mg/kg PO per day) seemingly provide benefit with this treatment protocol based on follow-up liver biopsies examined on dogs after appropriate treatment intervals. It is salient to point out that the best antifibrotic is control of the instigating inflammatory process. I do not often use colchicine for control of fibrosis because it can inhibit neutrophil function, cell replication important for hepatocellular regeneration, and cause adverse drug interactions. The role of Telmisartan as an antifibrotic in this process is not resolved. The treatment protocol I describe herein usually provides remarkable response without adding an ARB-2 blocker.

Role of milk thistle (silybin/silibinin) remains unresolved because of the very small doses absorbed with oral administration. There still remains a dearth of convincing data that this remedy can modify injury additional to restoration of GSH provided by thiol donors when not given by endovenous administration (high dose IV therapy used for death cap mushroom toxicity).

Judging Therapeutic Response

While only histology can confirm the extent of Cu removal. However, surveillance via sequential ALT determinations is used as a surrogate marker. ALT activity should decline over 2–6 months of treatment.

Needle aspirates cannot be used to judge response to therapy. Small needle biopsies also are unreliable for assessing success of Cu mobilization or in comparing histologic change/fibrosis between biopsies because of the differences between lobular regions when only tiny samples are evaluated - this is a problem that has plagued some of the research in CuAH in Labrador retrievers - and is a well-acknowledged problem in clinical practice in veterinary & human medicine. Recent investigation of erythrocyte SOD and the SOD Cu chaperone (CCS) in blood as a marker biomarker of hepatic copper concentration in Labrador retrievers suggests there may be potential utility for this marker in diagnosis and monitoring of hepatic copper accumulation. Diagnosis still will require liver biopsy to determine if copper accumulation associates with other disease processes.

Chronic Management

Lifelong mandatory dietary Cu restriction is recommended **unless** there was an identified unusual causal exposure to Cu ingestion. Chronic intermittent (2 x per week) BID low dose d-penicillamine or chronic zinc therapy are options. If chronic D-penicillamine is used, remain alert for iatrogenic iron deficiency - I have personally never encountered this with this approach. Chronic chelation dosing has rarely provoked iron deficiency when given for years.

References

1. Dirksen K, Roelen YS, van Wolferen, *et al.* Erythrocyte copper chaperone for superoxide dismutase and superoxide dismutase as biomarkers for hepatic copper concentrations in Labrador retrievers. *Vet J.* 2016;218:1–6.
2. Fedoseienko A, Bartuzi P, van de Sluis B. Functional understanding of the versatile protein copper metabolism MURR1 domain 1 (COMMD1) in copper homeostasis. *Ann NY Acad Sci.* 2014;1314:6–14.
3. Fieten H, Gill Y, Martin AJ, *et al.* The Menkes and Wilson disease genes counteract in copper toxicosis in Labrador retrievers: a new canine model for copper metabolism disorders. *Dis Model Mech.* 2016;9(1):25–38. doi: 10.1242/dmm.020263.
4. Fieten H, Biourge VC, Watson AL, *et al.* Nutritional management of inherited copper-associated hepatitis in the Labrador retriever. *Vet J.* 2014;199:429–433.
5. Forman OP, Boursnell ME, Dunmore BJ, *et al.* Characterization of the COMMD1 (MURR1) mutation causing copper toxicosis in Bedlington terriers. *Anim Genet.* 2005;36(6):497–501.
6. Haywood S, Boursnell M, Loughran MJ, *et al.* Copper toxicosis in non-COMMD1 Bedlington terriers is associated with metal transport gene ABCA12. *J Trace Elem Med Biol.* 2016;35:83–9. doi: 10.1016/j.jtemb.2016.01.015. Epub 2016 Feb 4.
7. Hoffmann G, van den Ingh TS, Bode P, Rothuizen J. Copper-associated chronic hepatitis in Labrador Retrievers. *J Vet Intern Med.* 2006;20(4):856–61.
8. Johnston AN, Center SA, McDonough SP, *et al.* Hepatic copper concentrations in Labrador Retrievers with and without chronic hepatitis: 72 cases (1980–2010). *J Am Vet Med Assoc.* 2013;242(3):372–80. doi: 10.2460/javma.242.3.372.
9. Langlois DK, Smedley RC, Schall WD, Kruger JM. Acquired proximal renal tubular dysfunction in 9 Labrador Retrievers with copper-associated hepatitis (2006–2012). *J Vet Intern Med.* 2013;27(3):491–9. doi: 10.1111/jvim.12065.
10. Shih JL, Keating JH, Freeman LM, Webster CR. Chronic hepatitis in Labrador Retrievers: clinical presentation and prognostic factors. *J Vet Intern Med.* 2007;21(1):33–9.
11. Vonk, WIP, Bartuzi P, de Bie, *et al.* Liver-specific Commd1 knockout mice are susceptible to hepatic copper accumulation. *PLos One.* 2011;6:e29183.

acvim419

Vector-Borne Diseases of the Cat

Remo Lobetti, BVSc (Hons), MMV (Med), PhD, DECVIM
Bryanston, South Africa

Introduction

Ticks on domestic cats are often overlooked as parasites and vectors of infectious disease agents. This oversight may result from the belief that cats are fastidious groomers and remove ticks before attachment takes place and thus insufficient time for transmission of infectious agents to occur. As some cats spend a majority of their time indoors, owners may be led to believe that their cats are not exposed to ticks. Feline vector-borne diseases (FVBD) have, however, emerged in recent years, showing a wider geographic distribution and increased global prevalence. In addition to their veterinary importance, domestic cats play a central role in the transmission cycles of some FVBD agents by acting as reservoirs and sentinels, an occurrence that requires a One Health approach.

Cytauxzoonosis

Cytauxzoon felis infection is transmitted by *Amblyomma americanum* (lone star tick) and *Dermacentor variabilis* (American dog tick) and is considered the most severe disease transmitted by ticks to cats. Infected bobcats act as the wild animal reservoir and chronically infected domestic cats are the urban reservoirs. Cytauxzoonosis is seen in domestic cats throughout the central, southeastern, and south central USA.

Seasonal activity of ticks is a key component in the transmission of *C. felis* as the number of cases increase with increased tick activity. In addition, outdoor cats living in wooded habitats are at higher risk for infection. Neither age nor breed predisposition for infection has been reported.

Typical clinical signs are pyrexia, inappetence, anorexia, dyspnea, and icterus with the disease having a severe acute nature, which rapidly progresses with death occurring 1–7 days after onset of clinical signs.

Diagnosis of cytauxzoonosis was historically based on finding the parasites in blood smears by light microscopy; however, the current gold standard is the use of PCR, which also allows the detection of infection prior to onset of clinical signs.

Recommended treatment for cytauxzoonosis is atovaquone (15 mg/kg PO TID) together with azithromycin (10 mg/kg PO SID). Imidocarb dipropionate has been used but not recommended due to its lack of efficacy and side effects. In the debilitated cat, administration of azithromycin intravenously as a slow infusion can be considered.

Babesiosis

Babesiosis is a tick-borne protozoan disease caused by parasites of the genus *Babesia* belonging to the Piroplasmida group. Multiple species of *Babesia* organisms have been documented in cats including *Babesia felis, B. herpailuri, B. cati, B. vogeli, B. canis* subsp. *presentii, B. canis* subsp. *canis, Theileria annae, B. pantherae, B. microti*-like, and *B. leo*. An unidentified *Babesia* sp. has been reported in Florida in panthers and Texas cougars with PCR indicating a prevalence of up to 90% and rRNA sequencing showing it to be genetically distinct from other feline babesial species, and most closely related to the canine *Babesia* spp. found in Japan (*B. odocoilei* and *B. divergens*). In addition, new species such as *B. venatorum* and *B. hongkongensis* have recently been discovered. The level of genetic diversity within *Babesia* species suggests that new subspecies can develop easily.

The pathogenicity of *Babesia* spp. is dependent on many variables including species, strain, age of the host, and immunologic response. The species and strain appears to be the most important factors in determining pathogenicity. The four babesial species that cause clinical disease are *B. felis, B. cati, B. leo,* and *B. microti*-like spp. The most and least virulent species in felines are *B. felis* and *B. cati*, respectively.

Common clinical signs are anorexia, lethargy, weakness, and rough hair coat. Pyrexia and icterus are uncommon. Both the clinical signs and laboratory changes can often be attributed to the hemolytic anemia that occurs secondary to the disease.

Babesiosis can be diagnosed on peripheral blood smear examination. However, when the level of parasitemia is low, detection of the organism in blood smears can be difficult. In addition, the different *Babesia* species and some other blood parasites, including *Cytauxzoon felis*, are morphologically indistinguishable. The best method for definitive diagnosis is PCR, which also allows differentiation between species and determination of co-infections.

The drug of choice for *B. felis* is primaquine at 0.5 mg/kg PO SID for 1–3 days, then repeated every 7 days for four doses. Although primaquine is capable of reducing parasitemia, it does not eliminate the parasite. Rifampicin and sulfadiazine-trimethoprim have shown to have some anti-parasitic effect but less effective than primaquine. Diminazene aceturate and imidocarb diproprionate have shown promising results in the treatment of the large babesial species of cats.

Unfortunately, there are no controlled clinical trials evaluating therapeutic regimens in feline babesiosis and such trials are desperately needed. Combination therapy with azithromycin and atovaquone or clindamycin and quinine, doxycycline, clindamycin, and metronidazole may all have some efficacy against the *Babesial* spp. of the cat.

Prognosis depends on the severity of disease, which in turn depends on both the organism and host factors. Mortality rates of 15–20% have been reported. Concurrent infection with *Mycoplasma haemofelis*, FeLV or FIV has a negative impact on the response to treatment and outcome of disease.

Feline Granulocytotropic Anaplasmosis

Feline granulocytotropic anaplasmosis is caused by *Anaplasma phagocytophilum* and is transmitted by *Ixodes scapularis* and *Ixodes pacificus* tick. Although the disease is common in both dogs and humans in the northeastern part of the USA, it has been poorly described in cats.

Clinical signs of feline anaplasmosis include pyrexia, lethargy, anorexia, arthralgia, lameness, lymphadenomegaly, weight loss, and neurologic signs.

Diagnosis can be based on clinical signs, history of tick exposure, and the microscopic detection of morulae within the neutrophils in acute infection. However, morulae are often difficult to find, as the parasitemia can be very low. Although, serology and PCR are the preferred methods confirming the diagnosis, studies have reported negative PCR results from cats with clinical illness and positive on serological tests. The latter, however, may merely indicate prior exposure to the parasite. Possible explanations for PCR negative cats would be that *A. phagocytophilum* is sequestered in low numbers in tissues, or that infected cats quickly clear the parasite.

Recommended therapy is doxycycline at 10 mg/kg PO SID for 28 days.

Feline Monocytotropic Ehrlichiosis

Feline monocytotropic ehrlichiosis is caused by an *Ehrlichia canis*-like agent, however, even though a large number of suspected feline ehrlichiosis cases have been reported, *E. canis*-like DNA has been demonstrated from only a few infected cats.

Clinical signs include pyrexia, lethargy, inappetence, weight loss, and polyarthritis.

Diagnostic criteria for feline ehrlichiosis have not yet been clearly established, although positive serology and organism detection in the blood supports the diagnosis.

For cats with suspected clinical ehrlichiosis, administration of doxycycline at 10 mg/kg PO SID for a minimum of 28 days is recommended.

Tularemia

Tularemia is caused by *Francisella tularensis*, which is a non-motile, aerobic, encapsulated gram-negative coccobacillus. The disease is a notifiable zoonotic disease. Cats can be exposed to *F. tularensis* through direct contact with infected wildlife (cottontail rabbits, hares, and rodents), vectors, or contaminated environment and are more susceptible to the disease than dogs. Vector transmission is via *Amblyomma americanum*, *Dermacentor variabilis*, and *D. occidentalis*.

Clinical signs include pyrexia, anorexia, marked depression, dehydration, lymphadenopathy, splenomegaly, hepatomegaly, icterus, and acute oral or lingual ulcers.

Diagnosis is most commonly based on clinical signs and confirmed by paired serum titers.

Therapy of choice is gentamicin at 5 mg/kg SC, IV, or IM SID for 7–14 days. Although doxycycline and fluoroquinolones may be effective, relapses are common.

Lyme Borreliosis

Lyme borreliosis is caused by *Borellia burgdorferi* and despite antibodies to *B. burgdorferi* being found in cats from endemic areas, a distinct clinical entity has to date not yet been described. Cats are hosts for the tick vectors, *Ixodes scapularis* and *Ixodes pacificus*, and a prevalence of antibodies to *B. burgdorferi* (up to 47%) has been reported in endemic areas of the northeastern USA.

Cats experimentally infected with *B. burgdorferi* show signs of arthritis and meningitis. Recommended treatment for suspected Lyme borreliosis in cats is doxycycline at 10 mg/kg PO SID for 28–30 days.

Hemoplasmosis

Hemoplasmosis is caused by *Mycoplasma*, of which currently there are 3 species, namely *M. haemofelis*, *M. haemominutum*, and *M. turicensis*. The incubation period of the disease varies between 6–17 days and has a range of 20–34 days. The disease can be divided into the preparasitemic, acute, recovery, and carrier phase.

Infection with *M. haemofelis* often causes a severe hemolytic anemia, whereas *M. haemominutum* infections do not usually induce clinical signs, although some reports have documented a mild or moderate anemia. It has also been suggested that co-infection with *M. haemominutum* and feline retroviruses may result in anemia. The pathogenic potential of *M. turicensis* seems to depend on cofactors but has been reported to result in hemolytic anemia.

Clinical signs are generally non-specific. Infected animals may exhibit depression, lethargy, anorexia, weight loss, weakness, and pale mucous membranes. Icterus is variably present. In the acute phase, pyrexia may be apparent whereas chronically infected cats tend to be normo-thermic. Hepato-splenomegaly and generalized lymphadenopathy may be present.

Hemoplasmosis can be diagnosed on a peripheral blood smear examination. However, when the level of parasitemia is low, detection of the organism in a blood smear can be difficult and often it is difficult to distinguish the organisms from Howell-Jolly bodies, basophilic stippling, and/or stain deposit. In addition, *Babesia felis* and *Cytauxzoon felis* create additional confusion. The best method for definitive diagnosis is PCR, which also allows differentiation between species and determination of co-infections.

Therapy is doxycycline, together with corticosteroids; the latter as the anemia associated with hemoplasmosis has an immune-mediated pathogenesis.

References

1. Day MJ. *Parasit Vectors*. 2011;4:48.
2. Lappin MR, *et al*. In: Green CE, ed. *Infectious Diseases of the Dog and Cat*. 2012;238.
3. Little SE. *Vet Clin North Am Small Anim Pract*. 2010;40:1121.
4. Meinkoth JH, *et al. Vet Clin North Am Sm Anim Pract*. 2005;35:89.
5. Otranto D, *et al. Parasit Vectors*. 2010;3:2.
6. Persichetti MF, *et al. Parasit Vectors*. 2016;9:247.
7. Yoko Nagamor, *et al. Vet Pract*. 2015;69.

acvim420

Canine Leptospirosis Prime Suspect

Annette L. Litster, BVSc, PhD, FANZCVSc (Feline Medicine), MMedSci (Clinical Epidemiology)
Chicago, IL, USA

Is Leptospirosis a Re-Emerging Disease?

Over recent years, there has been published evidence that leptospirosis is an increasing threat across the USA. Serological surveys have documented that the most commonly identified antibodies are against serovars Autumnalis, Grippotyphosa, Pomona, and Bratislava (Gautam *et al.* 2010), although this does not necessarily reflect the true prevalence of these serovars because of cross-reactivity between serovars. One study reported that 24.9% of healthy dogs from Michigan had significant *Leptospira* titers indicating exposure, with serovar Grippotyphosa being the most common. Interestingly, there was no breed predilection for seropositivity to *Leptospira* in that study and breeds ranging from German shepherds (12.5%) to Yorkshire terriers (11.0%) had positive MAT results (Gautam *et al.* 2010).

There was also a significant increase in the proportion of positive *Leptospira* microscopic agglutination tests (MATs) in a prevalence study of 23,005 US dogs over 2002–2004 (Moore *et al.* 2006). Additionally, a survey of US and Canadian veterinary teaching hospitals between 1983 and 1998 reported a significant increase in canine leptospirosis prevalence (Ward *et al.* 2002).

There have been various hypotheses proposed to explain the recent increase in prevalence. More frequent testing could explain some of the increase, but the main reason is likely to be urban sprawl resulting in increased exposure to wildlife reservoir hosts (Ward *et al.* 2004). One Connecticut study of peridomestic wildlife showed exposure to leptospirosis in 36% of raccoons (Icterohaemorrhagiae) and 13% of skunks (Grippotyphosa; Richardson 2003), while another Illinois study reported that 222/459 (48%) of raccoons were seropositive for leptospirosis (Mitchell *et al.* 1999). Wildlife studies have revealed that nearly 50% of rats, coypus (nutria) and muskrats are seropositive and almost 30% of rats and muskrats are renal carriers for serovars Icterohaemorrhagiae and Grippotyphosa, respectively. Additionally, the initial 2-serovar bacterin vaccine was not broadly protective, so vaccinated dogs were still at risk of infection.

Leptospira Serovars

Serovar classification is based on the expression of lipopolysaccharide (LPS) surface epitope antigens, which is the main antigen for *Leptospira*. Serovars are differentiated by biochemical tests. Antigenically related serovars are grouped into serogroups. Information on serovars is useful for epidemiological understanding, but relationships between serovars and clinical presentation or severity of disease are not well characterized in the veterinary literature. However, one recent study reported significant associations between serovar Grippotyphosa and renal disease and serovar Icterohaemorrhagiae and hepatic involvement (Tangeman and Littmann 2013). Additionally, another study reported that dogs infected with serovar Pomona had more severe renal disease and were statistically less likely to be discharged from hospital alive (Goldstein *et al.* 2006). Serovar Canicola is highly adapted to dogs, but it is thought from experimental studies that, given time, kidney disease and chronic shedding eventually develop.

Epidemiology of *Leptospira*

Leptospires are thin, flexible, spiral bacteria that are directly transmitted by contact infected urine and intact mucous membranes, abraded skin, or skin that has been immersed in water for a prolonged time. Indirect transmission can also occur via contaminated water sources, soil, food, or contaminated bedding. Stagnant or slow-moving warm water provides ideal environmental conditions for *Leptospira*, and increased incidence has been reported after high rainfall or flooding. There is increased seasonal incidence during summer/late fall. During periods of drought, infection is more common around water sources. Leptospires remain viable for several months in moist soil and temperatures of 0–25°C favor survival; freezing markedly decreases survival.

Pathogenesis of Leptospirosis

Organism numbers are highest in blood for the first 10 days after infection. Subsequently, the highest concentration of organisms is found in the urine. The incubation period is typically 7 days, but can range from 2 to 30 days. The severity of clinical disease is dependent on the infecting strain, number of organisms, and host age and susceptibility. The initial IgM antibody response develops within 1 week, increasing for up to 14 days before decreasing in favor of a rising IgG antibody-mediated response.

Clinical Signs of Leptospirosis

While many practitioners consider leptospirosis in dogs with acute renal failure and/or signs of hepatopathy, leptospirosis has also been responsible for pulmonary and hemorrhagic disease. Fever, uveitis, meningitis, and immune-mediated disease have also been associated with leptospirosis. Laboratory abnormalities can include proteinuria, thrombocytopenia, coagulopathy, hypoalbuminemia, and glucosuria (Tangeman and Littman 2013).

There are four main forms of clinical leptospirosis:

1. **Peracute** - Affected animals can experience massive leptospiremia and death with few preliminary signs.
2. **Acute** - This is characterized by fever, muscle pain, vomiting, dehydration, vascular collapse, tachypnea, hematemesis, hematochezia, melena, epistaxis, petechiae, icterus, and/or pain on renal palpation.
3. **Subacute** - In subacute infection, there can be fever, decreased appetite, vomiting, dehydration, polydipsia/polyuria, paraspinal hyperesthesia, petechial or ecchymotic hemorrhages, uveitis, icterus, coughing, and/or dyspnea.
4. **Chronic** - Chronic infection can be associated with uveitis, fever, progressive deterioration in renal or hepatic function, PD/PU, weight loss, and/or anorexia.

Diagnosis of Leptospirosis

Serology - Antibody Testing

Microscopic Agglutination Test (MAT) Titers

The MAT is the most commonly used test to diagnose canine leptospirosis. The test detects antibodies against specific serovars, by mixing serial dilutions of sera with cultured *Leptospira* organisms of different serovars representing different serogroups. The resulting titer is the highest dilution of the sera that causes 50% agglutination. Agglutination is visualized using dark field microscopy. Antibody titers become positive in about 1 week, peak in 3–4 weeks and remain elevated for months. If the titer is <1:800, retesting in 2–3 weeks is recommended; a four-fold increase in MAT titer is generally demonstrated in infected dogs. Low titers can be the result of an acute infection (in the first 7–9 days after initial infection); a less pathogenic strain; or antibiotic treatment. Generally, a single titer >1:800 in an unvaccinated dog with compatible clinical signs is adequate for the diagnosis of leptospirosis, although some workers feel that titers >1:800 or ≥3200 represent a more reliable threshold for a positive test. One recent study demonstrated that: (1) dogs vaccinated with *Leptospira* vaccines have variable MAT titers over time; (2) antibodies should not be used to predict resistance to *Leptospira* infection; and (3) MAT titers ≥1:800 can develop after *Leptospira* spp. vaccination (Martin *et al.* 2014).

There are a number of challenges with the interpretation of MATs. A positive *Leptospira* antibody titer does not necessarily mean that the dog has clinical leptospirosis, since many canine *Leptospira* infections are subclinical; therefore, seroprevalence studies that only include dogs with clinical disease might underestimate exposure rates in the canine population as a whole. Additionally, vaccination can also produce detectable antibody titers. Cross-reaction between serovars in infected dogs can make identification of the infecting serovar difficult. For example, in one study of naïve puppies vaccinated against four serovars, the MAT titers were often highest to the non vaccinal serovar Autumnalis. Therefore, high titers to a specific serovar may not definitively identify the causative serovar, because of cross-reactive antibodies. In a human study where urine cultures and MAT results were compared, the MAT accurately predicted serovar in only 46% of cases. Information on the infecting serovar is not crucial, as the disease appears to be similar and the treatment is identical, regardless of serovar. Serovar information is important mainly from an epidemiologic perspective. Additionally, intra- and inter-laboratory variability can be problematic, since there is no standardization for MAT methodology.

Polymerase Chain Reaction (PCR) Testing

Kansas State University Veterinary Diagnostic Laboratory offers PCR testing on urine and IDEXX Laboratories offer PCR testing on urine and blood. PCR allows identification of early infections prior to seroconversion and helps evaluate dogs for urinary shedding. PCR testing is significantly more sensitive than culture and MAT for identifying dogs shedding leptospires and for diagnosing infection, but it does not identify *Leptospira* to a serovar level. PCR testing has no crossreactivity with commercially available vaccines.

Visualization of *Leptospira* Organisms

Bacterial Culture

If bacterial culture is to be attempted, suitable specimens need to be submitted early after infection, since leptospiremia occurs early, often before clinical signs appear, and leptospires have moved out of the blood and into the urine by the end of the first week of acute illness. Special techniques are required for culture and blood is inoculated a semi-solid medium containing 5-fluorouracil. Multiple cultures, and using media dilutions may increase recovery.

Microscopic Demonstration

Dark-field microscopic examination is required to visualize leptospires. Suitable specimens are infected blood, urine, CSF or peritoneal dialysate fluid, but sensitivity and specificity are low.

Histopathology

Leptospires can be visualized in infected tissue sections using silver staining or immunohistochemical stains.

Ancillary Testing - Diagnostic Imaging

Pulmonary hemorrhage in infected dogs can result in an interstitial pattern on thoracic radiographs, especially in the caudal dorsal lung fields, which can progress to a nodular alveolar pattern. On abdominal radiology, there can be renomegaly, hepatomegaly, and/or abdominal effusion. Abdominal ultrasound can reveal renomegaly, perinephric effusion, increased cortical echogenicity, and/or a 'medullary rim sign.'

TREATMENT OF LEPTOSPIROSIS

The treatment of choice for leptospirosis is oral doxycycline (5 mg/kg PO q12h or 10 mg/kg q24h for 2 weeks), as only doxycycline will clear leptospires from the kidneys. However, some dogs cannot tolerate oral doxycycline, especially in the early stages of infection, so IV ampicillin (22 mg/kg q8h) or oral amoxicillin (22 mg/kg q12h) can be used initially. Therapy should switch to oral doxycycline as soon as the dog can tolerate it.

Supportive care is vital and general management of acute renal failure or liver failure should be instituted promptly. Management for acute renal failure includes IV fluid diuresis (hemodialysis is an option for some oliguric or anuric dogs); close monitoring of volume of fluids administered and urine output; electrolyte supplementation to correct any electrolyte or acid-base abnormalities; mannitol, furosemide, dopamine can be considered for oliguric or anuric dogs; blood pressure monitoring; use of H2 blockers and anti-emetics if there is vomiting; nutritional support if it does not result in vomiting, possibly with concurrent use of phosphate binders.

Prognosis

Published data are limited and survival rates are variable, depending on the manifestations of infection, severity of disease and general health status of the dog at the time of infection. Dogs with renal and/or liver failure can have high mortality rates and chronic organ disease is possible in dogs surviving the infection. One study reported that the prognosis for dogs with mild to moderate azotemia was good with conservative therapy and that treatment with hemodialysis appeared to improve prognosis for dogs with severe azotemia (Adin *et al.* 2000).

Zoonotic Potential

Leptospirosis is a zoonotic disease and it can be an occupational hazard for people who work outdoors or with animals, campers and people who participate in outdoor sports. Since there is not a vaccine available to prevent disease in people, avoiding contact with potentially infected material is crucial. Particular care should be exercised when handling urine from dogs with renal, liver or pulmonary disease, or those with coagulopathies. Any suspect case should be handled as if it had leptospirosis until proven otherwise and communication of any potential risks to all personnel involved is paramount. One study of 500 urine samples from dogs of any health status submitted to Kansas State University Veterinary Diagnostic Laboratory reported that 8.2% of specimens contained leptospires (Harkin *et al.* 2003). Interestingly, a CDC survey of 3,000 veterinarians indicated a large proportion utilize 'very poor practices when handling urine.'

Proper protection when handling potentially infected material includes wearing disposable gloves and a gown. Eye protection and a mask or full-face shield should be worn if there could be potential aerosolization of leptospires (e.g., a struggling, urine-soaked dog). Hands should be washed or hand sanitizer used after contact with a potentially infected dog, the dog's environment, and after glove removal. Special care is required for immunocompromised people or pregnant women as leptospirosis can cause miscarriage. If there has been potential exposure to infectious material, the person at risk should be monitored for early signs of infection (e.g., fever, flu-like signs) and should consult their physician. The incubation period for leptospirosis is 2–20 days.

Prevention - Vaccination

Immunization is effective in reducing the prevalence and severity of canine leptospirosis by preventing renal colonization and shedding (Wilson *et al.* 2013). With increasing prevalence of leptospirosis, there is significant risk of morbidity and mortality among unvaccinated dogs. Additionally, since leptospirosis is a zoonotic disease, there is potential risk to the owners of unvaccinated dogs, and to veterinary hospital and laboratory staff. Human exposure to leptospires from vaccinated dogs has never been documented. Annual vaccination is recommended against the four most common *Leptospira* serovars currently diagnosed in dogs - Icterohaemorrhagiae, Canicola, Grippotyphosa, and Pomona

References

1. Adin CA, *et al. J Am Vet Med Assoc.* 2000;216(3):371–5.
2. Gautam R, *et al. J Am Vet Med Assoc.* 2010;237(3):293–298.
3. Goldstein RE, *et al. J Vet Intern Med.* 2006;20(3):489–94.
4. Harkin, *et al. J Am Vet Med Assoc.* 2003;222:1230–1233.
5. Martin, *et al. J Vet Intern Med.* 2014;28(3):789–92.
6. Mitchell MA, *et al. J Wildl Dis.* 1999;35(2):347–55.
7. Moore GE, *et al. Emerging Infectious Diseases.* 2006;12:501–3.
8. Richardson DJ. *Vector Borne Zoonotic Dis.* 2003;3(4):187–93.
9. Tangeman and Littman. *J Am Vet Med Assoc.* 2013;243(9):1316–1322.
10. Ward MP, *et al. J Am Vet Med Assoc.* 2002;220(1):53–8.
11. Ward MP, *et al. J Am Vet Med Assoc.* 2004;225(1):72–7.
12. Wilson, *et al. Vaccine.* 2013;31:3131–3134.

acvim421

Non-Antibiotic Therapies for Infection Treatment and Prevention

Jennifer Granick, DVM, MS, PhD, DACVIM (SAIM)
St. Paul, MN, USA

The World Health Organization (WHO) has warned recently of a looming "post-antibiotic era" in which previously treatable infections will be fatal.[1] Antimicrobial resistance "occurs when bacteria change in response to the use of antibiotics to treat bacterial infections (such as urinary tract infections, pneumonia, bloodstream infections) making [antibiotic treatment] ineffective," according to WHO. The Centers for Disease Control and Prevention, the National Institutes of Health, the WHO and other local and global entities have sounded the alarm regarding emerging antimicrobial resistance and have made antimicrobial stewardship and finding alternatives methods to treat antibiotic resistant infections top priorities.

Non-Antibiotic Therapies Used in Clinical Practice

Several non-antibiotic therapies are already frequently used, either alone or as adjunctive therapy with antibiotics. Veterinary dermatologists utilize many topical antimicrobial therapies for the treatment of superficial bacterial folliculitis.[2] These include antiseptics like benzoyl peroxide and chlorhexidine. Clinical use of a variety of non-antibiotic agents for chronic wound therapy include silver and iodophor compounds, which have broad antimicrobial spectra and minimal resistance.[3] Additionally, honey is active against a large number of bacterial species and exhibits antimicrobial effects via its osmotic properties, acidic pH, and antioxidants, among other benefits. Indeed, honey-laden wound dressings are commercially available. Despite proven antimicrobial performance, no studies have validated use of honey as a systemic antibiotic substitute. However, recently honey has been investigated for prevention of IV catheter site infections in pediatric patients undergoing peritoneal dialysis and as a potential coating for urinary catheters.[4,5] Even select phenolic compounds extracted from honey show *in vitro* bactericidal effects.

Cranberry extract has gained popularity as a preventative for urinary tract infections. Results of efficacy studies have been mixed, though this is perhaps related to the formulation, dose, and frequency prescribed. A recent human clinical trial showed promise for prevention of UTIs in women in a nursing home setting, though ultimately was underpowered for drawing definitive conclusions.[6] Using the same commercial formulation, another study showed a significant reduction in the mean number of UTIs in juveniles between treatment and controls groups.[7]

Fecal microbiota transplantation (FMT) has become a key strategy for treatment of intractable *Clostridium difficile* infections in people. FMT is a relatively new therapy in companion animals and thus lacks peer-reviewed evidence in these species, though clinical use is increasing, with anecdotal successes. Better studied in canines are probiotics, though most studies are not focused on eradication or control of pathogens. A probiotic and prebiotic combination given to healthy dogs upon entry to a shelter environment significantly reduced incidence and duration of diarrhea compared to untreated controls.[8] Similarly, a double-blinded placebo controlled study of dogs with acute diarrhea showed probiotics improved stool consistency and decreased the concentrations of potential enteric pathogens.[9]

Therapies on the Horizon

There are a variety of promising new antibiotic alternatives on the horizon, some in preclinical and early phase clinical trials. These include therapies directed against bacteria such as bacteriophages, antimicrobial peptides, inhibition of bacterial secretion systems and virulence factors, and quorum sensing inhibitors; and those that alter host immune responses like vaccines and immune stimulants.

Bacteriophages are viruses that infect bacteria and utilize bacterial machinery for their own replication. Lytic bacteriophages are attractive as non-antibiotic therapeutics because they are extremely selective for bacterial species and even strains. This effectively eliminates collateral damage of the non-pathogenic microbiota while targeting specific pathogens. This selectivity is also a limitation of bacteriophage therapy, as a single bacteriophage is not sufficient to treat multi-species or multi-strain infections. Successful treatment of chronic *Pseudomonas aeruginosa* otitis using a six-phage cocktail has recently been demonstrated in humans.[10] Efforts to improve phage therapy by engineering bacteriophages with enhanced host range and infectivity are ongoing.

Antimicrobial peptides are small molecules produced by many organisms from innate immune cells of mammals to bacteria themselves as defense against other bacterial species. These peptides are thought to have great potential as alternative therapy because of their broad spectrum of activity. Clinical trials evaluating antimicrobial peptides have used topical administration, and despite preclinical promise, have not shown expected efficacy.[11] Further studies are needed to improve efficacy as well as to find ways to safely use these molecules systemically.

Quorum sensing is a used by bacteria to sense bacterial density and alter gene expression in response. Quorum sensing is important in the transition to biofilm formation as well as production of virulence factors. Quorum sensing inhibitors have the potential to block the formation of antimicrobial resistant biofilms and are unlikely to create selective pressure, which could be especially valuable against multi-drug resistant biofilm formers such as *Pseudomonas aeruginosa*.[12]

Many bacteria use secretion systems to inject toxins that aid in bacterial invasion of the host. Specifically targeting the secretion apparatus or the toxins themselves is another novel antibacterial approach. Small molecules targeting secretion system components as well as anti-toxin antibodies are already in various phases of preclinical and clinical trials.[13]

There is significant interest in vaccines for sepsis, including those directed against *Staphylococcus aureus*. A commercially available vaccine against components of *S. aureus* and *E. coli* has been shown to decrease the incidence of mastitis in the bovine species.[14] Immune stimulants to bolster host immune responses are also potential therapies. Similarly, inhibition of immune responses can sometimes improve healing, as over-exuberant innate immune system responses can create host tissue damage that paradoxically prolongs infection.

As long as antibiotics have been utilized in medicine, antibiotic resistance has soon developed, thus the problem of antimicrobial resistance is not going away. The development of new antibiotics is slow and expensive, thus turning efforts to investigation of non-antibiotic therapies, whether to replace or enhance the effects of current antibiotics is needed. As this field matures, we will likely see more non-antibiotic therapies come to market for clinical use.

References

1. *Antimicrobial Resistance: Global Report on Surveillance*. World Health Organization; 2014.
2. Hillier A, Lloyd DH, Weese JS, *et al*. Guidelines for the diagnosis and antimicrobial therapy of canine superficial bacterial folliculitis (Antimicrobial Guidelines Working Group of the International Society for Companion Animal Infectious Diseases). *Vet Dermatol*. 2014;25:163–175, e142–163.
3. Lipsky BA, Hoey C. Topical antimicrobial therapy for treating chronic wounds. *Clin Infect Dis*. 2009;49:1541–1549.
4. Emineke S, Cooper AJ, Fouch S, Birch BR, Lwaleed BA. Diluted honey inhibits biofilm formation: potential application in urinary catheter management? *J Clin Pathol*. 2017;70(2):140–144.
5. Forbes TA, Shaw L, Quinlan C. Topical honey in the management of pediatric peritoneal dialysis exit sites. *Perit Dial Int*. 2016:36(6):684–687.
6. Juthani-Mehta M, Van Ness PH, Bianco L, *et al*. Effect of cranberry capsules on bacteriuria plus pyuria among older women in nursing homes: A randomized clinical trial. *J Am Med Assoc*. 2016;316:1879–1887.
7. Ledda A, Belcaro G, Dugall M, *et al*. Highly standardized cranberry extract supplementation (Anthocran®) as prophylaxis in young healthy subjects with recurrent urinary tract infections. *Eur Rev Med Pharmacol Sci*. 2017;21(2):389–393.
8. Rose L, Rose J, Gosling S, Holmes M. Efficacy of a probiotic-prebiotic supplement on incidence of diarrhea in a dog shelter: a randomized, double-blind, placebo-controlled trial. *J Vet Intern Med*. 2017;31(2):377–382. LID - 10.1111/jvim.14666 [doi].
9. Gomez-Gallego C, Junnila J, Männikkö S, *et al*. A canine-specific probiotic product in treating acute or intermittent diarrhea in dogs: A double-blind placebo-controlled efficacy study. *Vet Microbiol*. 2016;197:122–128.

10. Wright A, Hawkins CH, Anggård EE, Harper DR. A controlled clinical trial of a therapeutic bacteriophage preparation in chronic otitis due to antibiotic-resistant *Pseudomonas aeruginosa*; a preliminary report of efficacy. *Clin Otolaryngol.* 2009;34:349–357.
11. Czaplewski L, Bax R, Clokie M, *et al.* Alternatives to antibiotics - a pipeline portfolio review. *Lancet Infect Dis.* 2016;16:239–251.
12. Rasmussen TB, Givskov M. Quorum sensing inhibitors: a bargain of effects. *Microbiology.* 2006;152:895–904.
13. Hauser AR, Mecsas J, Moir DT. Beyond antibiotics: new therapeutic approaches for bacterial infections. *Clin Infect Dis.* 2016;63:89–95.
14. Guccione J, Pesce A, Pascale M, *et al.* Efficacy of a polyvalent mastitis vaccine against *Staphylococcus aureus* on a dairy Mediterranean buffalo farm: results of two clinical field trials. *BMC Vet Res.* 2016;13:29.

acvim422

Enterococcal Infections in Dogs and Cats

Mary F. Thompson, BVSc (Hons), PhD, DACVIM (SAIM), MANZCVS (Assoc)
Murdoch, WA, Australia

Introduction

Enterococci have been described as "a second-rate pathogen that can become a first-rate clinical problem".[1] Indeed, isolation of *Enterococcus* spp. from the body fluids or tissues of dogs and cats can be frustrating for a veterinarian to manage for a number of reasons. For one, the significance of an enterococcal isolate in causing a disease process can sometimes be difficult to ascertain, especially when multiple bacterial species are isolated from the same site, and secondly, enterococci are well known for both intrinsic and acquired resistance to antimicrobials, thus can prove extremely difficult to eradicate with drugs.

Enterococci are non-sporulating gram-positive bacteria that were initially grouped within the genus Streptococcus but were officially grouped as Enterococcus in 1984. Enterococcal infections have been a growing concern in human medicine over the past 20 years, which have seen this ancient and ubiquitous intestinal commensal species evolve to become an important cause of opportunistic infection and antimicrobial resistance.[2] In humans, enterococci are most frequently associated with complicated urinary tract infection but they are also an important cause of bacteriaemia (accounting for approximately 10% of all bacteriaemia episodes globally), endocarditis, and meningeal, wound and biliary tract infections.[3] The two major species of enterococci that cause disease in humans and in dogs and cats are *Enterococcus faecalis* and *Enterococcus faecium*. In humans, a shift towards *E. faecium* colonization and infection has become apparent over time. In general, *E. faecalis* is more virulent and less antimicrobial resistant and *E. faecium* more antimicrobial resistant. The most feared *Enterococcus* subgroup is arguably vancomycin-resistant enterococci (VRE), and the majority of such strains are *E. faecium*.

Enterococci form part of the normal gastrointestinal flora in humans, animals and insects, with *E. faecalis* typically outnumbering *E. faecium* in healthy animals, and are also found in the mouth and genital tract.[4] The source of infection in individual patients has been traditionally considered to be their endogenous flora but the rise in health-care associated multidrug-resistant (MDR) clonal strains has made it apparent that strains can move from patient to patient, likely on the hands of healthcare workers, who can then directly inoculate catheters and wounds or facilitate establishment of the strain in the patients GI flora.[4]

Ampicillin-resistant *Enterococcus faecium* (ARE) belonging to human-associated clonal groups have been isolated from the feces of normal dogs and cats, albeit with some differences in antimicrobial susceptibility and/or virulence gene profiles, with the same strain isolated from a dog and a close contact human.[5,6] Multidrug-resistant enterococci have also been isolated from the feces of dogs leaving a veterinary ICU, the animals having been treated with antimicrobials, and these strains, many in possession of key virulence genes and biofilm-forming capacity, were also closely related or identical to strains associated with human hospital outbreaks.[7]

Antimicrobial Resistance in Enterococci

Enterococci are intrinsically resistant to a number of antimicrobials including penicillin (presence of low-affinity penicillin binding proteins), clinically achievable levels of aminoglycosides (so-called low-level gentamicin resistance, which precludes entry of the drug into the bacterial cell), clindamycin, and trimethoprim-sulfamethoxazole (ability to absorb folate from the environment).[8] Fluoroquinolones are not considered uniformly effective *in vivo*.[9] Regardless of *in vitro* susceptibility, enterococci are also described as "tolerant" to the activity of β-lactams, generally being inhibited rather than killed by them. For this reason β-lactams are not suitable for use as monotherapy in serious infections such as endocarditis or meningitis, rather synergism between a β-lactam or glycopeptide (e.g., vancomycin) and an aminoglycoside, the combination of which is bacteriocidal, is standard for treatment of severe infections.[1] Synergy between the cell wall active agents and aminoglycosides involves the cell wall synthesis inhibitor facilitating aminoglycoside entry into the bacterial cell.[8]

Enterococci are also capable of acquiring resistance to antimicrobials via mutation, transfer of genetic material through plasmids, and movement of transposable genetic elements (transposons).[8] Resistances conferred include increased resistance to penicillins, high-level gentamicin resistance (HLGR), and resistance to important anti-enterococcal antimicrobials including vancomycin, quinupristin-dalfopristin, linezolid and daptomycin. Beta-lactamase production is relatively uncommon in enterococci and described almost exclusively for *E. faecalis*. If high-level-aminoglycoside resistance is present then the synergistic bacterial killing resulting from combined use with β-lactams is lost.

Enterococcal Infections in Dogs and Cats

Enterococci have been most commonly isolated from the urine in dogs and cats and within the urinary tract have been associated with bacterial cystitis, pyelonephritis, and subclinical bacteriuria, including monomicrobial and polymicrobial infections.[10-15] One relatively recent study with identification of isolates to a species level determined that *E. faecalis* and *E. faecium* are most commonly implicated in canine urinary tract infections (only *E. faecalis* was isolated from cats), most of which were considered complicated, with multidrug-resistance detected in some isolates, particularly *E. faecium*.[16] An earlier study reported on higher numbers of canine urinary *E. faecium* in comparison to *E. faecalis* isolates; antimicrobial non-susceptibility was frequently observed and one isolate, in possession of a key human-associated resistance-conferring transposable genetic element (*Tn*1546), displayed high-level resistance to both vancomycin and gentamicin.[17]

Enterococci have also been isolated from bile in cats and dogs with cholecystitis and/or cholangitis. The majority of isolates identified to a species level were determined to be *E. faecalis*, with *E. faecium* not reported; some were polymicrobial infections.[18] In one study, 100% of Enterococcus isolates were susceptible to ampicillin but in another, all were MDR, isolated from dogs with a history of prior antimicrobial therapy.[19,20] In a final study, all enterococcal isolates from bile were resistant to multiple antimicrobials (4/8 identified as *E. faecalis*); 4/6 were initially fluoroquinolone resistant and two were resistant on repeat isolation.[20]

Enterococcus has also been isolated from the mouth of dogs with periodontal disease; the majority of isolates were *E. faecalis*, ampicillin-susceptible and capable of biofilm production.[22] Further to this, *Enterococcus* spp. has been isolated from canine endocarditis and in some cases the implicated strain appeared identical those that obtained from the gums.[23]

Treatment

As is appropriate for all bacterial infections and in light of concern around selection of a resistant population of enterococci (or other species) that could prove dangerous to animals and humans, a decision needs to first be taken as to whether treatment of an enterococcal infection in a given dog or cat is warranted. Factors to consider include clinical signs in the animal (e.g., subclinical bacteriuria is not associated with clinical signs and in most, if not all, cases antimicrobial therapy is not recommended), presence in polymicrobial infections (treating a more virulent coinfecting organism may be sufficient), evidence of inflammation suggestive of bacterial infection in cytology or tissue biopsy, response to treatment to date, and surgical options for infection control. Detection of multidrug-resistance is not an indication for treatment.

For treatment of monomicrobial infections such as uncomplicated urinary tract infections then an aminopenicillin can be used as a sole agent. In more serious infections involving enterococci susceptible to ampicillin; e.g., endocarditis then this drug, in combination with an aminoglycoside for effective killing, is the recommended treatment regimen in human medicine. For *E. faecalis* strains with β-lactamase production capability then treatment with amoxicillin-clavulanic acid or similar is warranted in addition to gentamicin. If ampicillin resistance is not associated with β-lactamase production (as for the great majority of *E. faecium* isolates) then vancomycin in addition to gentamicin is recommended. When HLGR is present and thus synergistic therapy with ampicillin or vancomycin cannot be exploited then therapeutic options are limited; ampicillin and ceftriaxone may prove synergistic and effective for *E.*

faecalis but not *E. faecium*.[4] Most commonly in human medicine monotherapy with ampicillin or vancomycin is subsequently employed but relapse or primary failure are major concerns.[24] Nitrofurantoin or fosfomycin may also be options for simple uncomplicated urinary tract infections but such a scenario is uncommon.[16]

References

1. Murray BE. *Am J Med.* 1997;102:284.
2. Van Tyne D, Gilmore MS. *Annu Rev Microbiol.* 2014;68:337.
3. Murray BE. *Clin Microbiol Rev.* 1990;3;46.
4. Agudelo Higuita NI, Huycke NM. Enterococcal disease, epidemiology, and implications for treatment. In Gilmore MS *et al*, eds. *Enterococci: From Commensals to Leading Causes of Drug Resistant Infection.* Boston, USA; 2014:1.
5. Damborg P *et al. Appl Environ Microbiol.* 2009;75;2360.
6. de Regt MJ *et al. PLoS One.* 2012;7;e30319.
7. Ghosh A *et al. PLoS One.* 2011;6;e22451.
8. Hollenbeck BL, Rice LB. *Virulence.* 2012;3;421.
9. Zervos MJ *et al. J Antimicrob Chemother.* 1988;21;1;113.
10. Martinez-Ruzafa I *et al. J Feline Med Surg.* 2012;14;729.
11. Wan SY *et al. J Am Vet Med Assoc.* 2014;245;106–112.
12. Puchot ML *et al. J Feline Med Surg.* 2017;1098612X16688806.
13. Wong C *et al. J Vet Intern Med.* 2015;29;1045.
14. Litster A *et al. Vet Microbiol.* 2009;136;130.
15. Litster A *et al. Vet Microbiol.* 2007;121;182.
16. KuKanich KS *et al. J Am Anim Hosp Assoc.* 2015;51;148.
17. Simjee S *et al. J Clin Microbiol.* 2002;40;4659.
18. Tamborini A *et al. J Vet Intern Med.* 2016;30;1046.
19. Wagner KA *et al. J Vet Intern Med.* 2007;21;417.
20. Lawrence YA *et al. J Am Vet Med Assoc.* 2015;246;982.
21. Peters LM *et al. J Vet Intern Med.* 2016;30;123.
22. Oliveira M *et al. Comp Immunol Microbiol Infect Dis.* 2016;46;27.
23. Semedo-Lemsaddek T *et al. PLoS One.* 2016;11;e0146860.
24. Marothi YA *et al. Indian J Med Microbiol.* 2005;23;214.

acvim423

Hyperadrenocorticism: Diagnosis and Management

Jo R. Smith, MA, VetMB, PhD, DACVIM (SAIM)
Athens, GA, USA

Introduction

Approximately 85% of spontaneous canine hyperadrenocorticism (HAC) cases are caused by pituitary-dependent hyperadrenocorticism (PDH). Hypersecretion of ACTH from an adenohypophyseal micro- or macroadenoma results in bilateral adrenocortical hyperplasia and subsequent hypercortisolemia. The remaining ~15% of spontaneous HAC cases are caused by a functional adrenal tumor (FAT), an autonomously cortisol-secreting adenoma or adenocarcinoma of the adrenal cortex. Diagnostic tests for HAC can be separated into screening (confirming the presence of HAC) and differentiation (PDH or FAT) tests.

The prevalence of disease will influence the predictive values of a test. The positive predictive value of a test (a positive test result truly identifies a patient with the disease) increases as the disease prevalence increases. Hence, the diagnostic accuracy of these tests is improved when they are used in a discriminating manner in patients with a history and physical examination findings consistent with HAC.

		Disease		
		Present	Absent	
Test	Positive	True positive **TP**	False positive **FP**	Positive predictive value **PPV = TP/(TP + FP)**
	Negative	False negative **FN**	True negative **TN**	Negative predictive value **NPV = TN/(TN + FN)**
		Sensitivity **TP/(TP+FN)**	Specificity **TN/(TN+FP)**	

Diagnosis

Screening Tests

The utility of the different diagnostic tests depends on their relative sensitivity and specificity, the index of suspicion that the patient has the disease, and presence of comorbid disease.

Diagnostic sensitivity is the ability to detect true positives. A highly sensitive test will minimize false negatives and is a good screening test; **SNOUT** - if a **S**ensitive test is **N**egative, the disease can be ruled **OUT**.

Diagnostic specificity is the ability to detect true negatives. A highly specific test will minimize false positives and is a good confirmatory test; **SPIN** - if a **S**pecific test is **P**ositive, the disease can be ruled **IN**.

Diagnostic test[1]	**Sensitivity (%)**	**Specificity (%)**
Urine cortisol:creatinine cortisol	75–100; single sample obtained at clinic 94–100; two samples obtained at home	20–25; single sample obtained at clinic 64–87; two samples obtained at home
Low-dose dexamethasone suppression test	85–100	44–73
ACTH stimulation test	57–95 PDH = 80–83; AT = 57–63	59–93

It has been recommended that the current reference intervals and cut-offs for HAC screening tests are reevaluated because the assay methodologies have changed over the intervening ~30–40 years.[1] Furthermore, the original evaluation of screening tests occurred in referral centers with a high disease prevalence; today, the tests are often performed in first-opinion practice where the disease prevalence is lower. Additionally, increased client awareness means dogs with "early" HAC may test negative initially; in dogs where there is a high index of suspicion of HAC, screening tests should be repeated in 2–3 months.

Urinary Cortisol:Creatinine Ratio (UCCR)

This test is a highly sensitive test and, counterintuitively, is used to rule out HAC when there is a low clinical index of suspicion.

- **Protocol**
  Stress or non-adrenal illness can cause false-positive results, so urine should not be collected in the hospital or within 48 hours of a clinic/hospital visit. The client should collect early-morning urine samples on 3 consecutive days and refrigerate them. An equal aliquot can be taken from each sample, pooled and a single UCCR test performed.
- **Interpretation**
  Normal ratio - HAC is very unlikely.
  Increased ratio - another screening test (LDDST or ACT stim) should be used to confirm HAC.

Low-Dose Dexamethasone Suppression Test (LDDST)

This test is more sensitive than specific; non-adrenal disease may result in false-positive test results/failure to suppress. It is used as a screening test to rule out HAC. However, it can also be a differentiating test if the result is positive and specific criteria are met.

- **Protocol**
  Collect baseline blood sample; give 0.01–0.015 mg/kg dexamethasone intravenously (preferred) or intramuscularly; collect blood samples 4 and 8 hours later.
- **Interpretation**
  Suppression below cut off at both 4 and 8 hours - HAC is unlikely.
  Lack of suppression at 4 and/or 8 hours - consistent with HAC (PDH or FAT).
  Suppression at 4 and "escape" 8 hours; or suppression at 4 and/or 8 hours to <50% baseline cortisol concentration – PDH.

ACTH Stimulation Test (ACTH Stim)

This test is more specific than sensitive; there is less chance of a false-positive result with non-adrenal illness. It is the only test that can confirm iatrogenic HAC.

- **Protocol**
  Collect baseline blood sample; administer synthetic (cosyntropin or tetrasactrin) intravenously at a dose of 5 mcg/kg (to a maximum dose of 250 mcg/dog); collect a blood sample 1 hour later.
- **Interpretation**
  One-hour sample cortisol concentration is greater than laboratory reference interval - consistent with spontaneous HAC.
  One-hour sample is below reference interval, or shows minimal response to ACTH - consistent with iatrogenic HAC.

Noncortisol Steroid Hormone Panel (aka Tennessee Adrenal Panel)

Used when the history, physical examination and clinicopathologic data support a diagnosis of HAC, but the UCCR, LDDST and/or ACTH stimulation test fail to confirm hypercortisolemia.

- **Protocol**
  Same as ACTH stimulation test; need to submit a minimum of 2 ml of serum.
- **Interpretation**
  An increased hormone 2–3 times the upper reference interval, ideally in >1 hormone - consistent with occult HAC.

Differentiating Tests

If only medical management is feasible, the merit of differentiating between PDH and FAT will be to provide prognostic information. Otherwise, if medical management with trilostane or mitotane, or surgery is feasible, then differentiation will allow selection of the optimal therapeutic regimen for the patient-client-clinician combination. It is worth noting that differentiation between PDH and FAT may be very difficult once therapy has been initiated.

High-Dose Dexamethasone Suppression Test (HDDST)

This test must be completed after a positive result in a screening test. A patient without HAC will demonstrate suppression at 4 and 8 hours, a result that is also seen in dogs with PDH. This test also cannot confirm a FAT.

- **Protocol**
  Same as LDDST, except administer 0.1 mg/kg dexamethasone.
- **Interpretation**
  Suppression at 8 hours - PDH
  Lack of suppression at 4 and/or 8 hours - consistent with PDH or FAT
  Suppression at 4 and "escape" 8 hours; or suppression at 4 and/or 8 hours to <50% baseline cortisol concentration - PDH.

Endogenous ACTH Concentration (eACTH)

This test must also be completed after a positive result in a screening test because eACTH concentrations cannot differentiate between normal dogs and those with PDH. This hormone is very labile and the plasma blood sample requires specialized and rapid handling.

- **Protocol**
  Collect sufficient blood into a pre-chilled plastic EDTA tube to provide 0.5–1 ml of plasma. Centrifuge immediately and separate plasma using a plastic pipette; store frozen in a plastic tube until shipping with frozen gel packs.
- **Interpretation**
  Definitively "low" eACTH concentration - adrenal tumor, or poor sample handling.
  Definitively "high eACTH concentration - PDH.

MANAGEMENT

Adrenalectomy is the treatment of choice for non-metastatic FAT. A recent case series has also investigated the use of stereotactic radiation therapy for adrenocortical tumors with vascular invasion.[2] Laparoscopy helps decrease surgical wound complications, reduce morbidity, and the faster recovery reduces likelihood of thromboembolic complications secondary to prolonged inactivity.[3,4] Due to atrophy of the contralateral adrenal cortex, prednisolone is typically administered (1 mg/kg) immediately postoperatively; this dose is gradually tapered over 8–12 weeks to eventual alternate-day therapy. The steroids are then discontinued for 2–3 days prior to performing an ACTH stimulation test to evaluate the function of the remaining adrenal gland. Alternatively, an ACTH stimulation test can be performed 6–8 hours postoperatively to assess the need for glucocorticoid supplementation. Mineralocorticoid supplementation is not always required; however, because of minimal adverse effects, a single dose of perioperative deoxycorticosterone pivalate is often administered.

Medical management is typically recommended for symptomatic PDH cases.[5] Its utility in dogs that only demonstrate clinicopathologic abnormalities is questionable, although some clinicians advocate for treating these cases to reduce the potential morbidity associated with progressive disease (chronic urinary tract infections and pyoderma, calcinosis cutis and calcium urolithiasis, development of insulin-resistant diabetes mellitus, and thromboembolic disease). The availability of transsphenoidal hypophysectomy is geographically limited; the optimal protocol for radiation therapy to control the clinical signs associated with pituitary macroadenomas has yet to be determined.[6] Medical management is also recommended for dogs with FAT when unilateral adrenalectomy is not feasible, and in dogs with metastatic FAT.[7]

Trilostane

Trilostane is the only FDA-approved treatment for canine HAC in the United States. Due to variations in trilostane content in capsules from compounding pharmacies, especially liquid formulations, use of the brand name product (Vetoryl, Dechra Veterinary Products) only is recommended.[8] Although a component of this drug's popularity with first-opinion practitioners is its relatively greater therapeutic index in comparison to mitotane, it can still cause adrenal necrosis and prolonged or permanent cortisol deficiency, with or without aldosterone deficiency.

The therapeutic regimen for PDH or FAT is the same. Doses of 0.5–1.5 mg/kg twice daily are recommended; dogs >25 kg may need lower doses on a per kilogram basis.[9,10] Dose adjustments are often needed and are based on clinical signs and the results of ACTH stimulation tests. The accuracy of baseline cortisol to predict the adrenal reserve in dogs being treated with trilostane for HAC has not been consistent.[11-13] The ACTH stimulation tests should be performed at a consistent time post-pill, whether the 2–4 or 4–6 hour time frame is adopted.[14] A dose of 1 mcg/kg cosyntropin can be for ACTH stims used for therapeutic monitoring.[15] The first ACTH stim should be performed after 10–14 days, unless a decreased appetite, vomiting, diarrhea, listlessness or normalization of water intake occurs before then. It may take 30 days for the full effect of trilostane to manifest, so the dose should not be increased after this visit; it should be decreased if the cortisol concentrations are too low. If the post-ACTH stim cortisol concentration is <1 mcg/dL (30 nmol/l), then trilostane should be discontinued and not reinstated until a recover of the adrenocortical function has been demonstrated by another ACTH stim. The next test is performed at day 30 of treatment, and the dose adjusted according to the clinical signs and post-stimulation cortisol concentration.[16] And dosage increase or decrease is likely to be dictated by the available capsule size; typically it should not exceed 25%.

Mitotane

Mitotane (o,p'-DDD; Lysodren, Bristol-Myers Squibb) is adrenocorticolytic and causes selective necrosis of the zone fasciculata and reticularis, decreasing secretion of cortisol and sex hormones, respectively. Although the zona glomerulosa is relatively resistant to mitotane, life-threatening aldosterone deficiency can occur, and complete adrenocortical insufficiency can occur in 6–10% dogs receiving mitotane.[17] Deliberate nonselective adrenocorticolysis has been attempted. The disadvantages of this approach are that the mortality can be ≤12%;[18] treatment of a hypoadrenal dog may be expensive; and failure to give medication to a hypoadrenal dog may be fatal, whereas missing a dose of mitotane will not have life-threatening consequences.

Selective adrenocorticolysis with mitotane occurs in two phases: induction (loading) and maintenance. Dogs with FAT are often more resistant to the effects of mitotane than those with PSH, and may require a longer induction period and/or higher mitotane doses during the induction and maintenance phases.

Additional Medical Therapy

Systemic hypertension and proteinuria are frequent complications in dogs with Cushing's syndrome and do not always resolve after treatment of hypercortisolism.[19] The hypertension is typically mild to moderate, and rarely associated with significant target organ damage (e.g., retinal hemorrhage or detachment). However, the hypertension can exacerbate left ventricular hypertrophy and congestive heart failure and may lead to glomerular damage and proteinuria. If hypertension or proteinuria (or both) persist, one should institute treatment with angiotensin-converting enzyme inhibitors and add amlodipine, if needed and monitor renal function.[20]

References

1. Behrend E, *et al. J Vet Intern Med.* 2013;27:1291.
2. Dolera M, *et al. J Small Anim Pract.* 2016;57:710.
3. Jiménez Peláez M, *et al. Vet Surg.* 2008;37:444.
4. Naan E, *et al. Vet Surg.* 2013;42:710.
5. Nagata N, *et al. J Vet Intern Med.* 2017;31:22.
6. Marcinowska A, *et al. Vet Radiol Ultrasound.* 2015;56:554.
7. Arenas C, *et al. J Vet Intern Med.* 2014;28:473.

8. Cook A, *et al. J Am Anim Hosp Assoc.* 2012;48:228.
9. Galac S, *et al. Vet J.* 2010;183(1):75.
10. Feldman EC. *J Am Vet Med Assoc.* 2011;238(11):1441.
11. Cook A, Bond K. *J Am Vet Med Assoc.* 2010;237:801.
12. Burkhardt W, *et al. J Vet Intern Med.* 2013;27:919.
13. Woolcock A, *et al. J Am Vet Med Assoc.* 2016;248:814.
14. Bonadio C, *et al. J Vet Intern Med* .2014;28:1239.
15. Aldridge C, *et al. J Vet Intern Med.* 2016;30:1637.
16. Dechra. Diagnosis, treatment and monitoring of hyperadrenocorticism. www.dechra-us.com/products/details/vetoryl%C2%AE-capsules-(trilostane)-5-mg.
17. Kintzer P, Peterson M. *J Vet Intern Med.* 1991;5:182.
18. den Hertog E, *et al. Vet Rec.* 1999;144:12.
19. Smets P, *et al. J Vet Intern Med.* 2012;26:565.
20. Brown S, *et al. J Vet Intern Med.* 2013;27;S27.

acvim424

Hyperadrenocorticism: Management with Diabetes Mellitus

Cynthia R. Ward, VMD, PhD, DACVIM (SAIM)
Athens, GA, USA

Hyperadrenocorticism (HAC) is a common endocrinopathy in middle to older aged dogs. Clinical signs to make the clinician suspicious of hyperadrenocorticism include many "Ps": polyphagia, polyuria/polydipsia, pyoderma, pot-bellied appearance, and persistent urinary tract infections. Physical examination findings include hepatomegaly, truncal obesity, muscle wasting, panting, bilaterally symmetrical alopecia, thin hyperpigmented skin, calcinosis cutis, and lipid deposits.

Hyperadrenocorticism and diabetes mellitus often occur concurrently and these patients can be a challenge to diagnose and manage.[1] They have similar clinical signs and the diagnostic tests are affected by the clinical diseases. Additionally, treatment of one disease will affect the treatment of the other. So, how to proceed with such challenging cases. Diabetes mellitus is usually diagnosed first, as it is the easiest to recognize. Clinical signs include polyuria/polydipsia, weight loss, persistent or recurrent urinary tract infections, weakness and muscle wasting, cataracts (usually dogs), and peripheral neuropathies (usually cats). Diagnosis can be made by recognition of appropriate clinical signs, and demonstration of persistent hyperglycemia and glucosuria. One confounding factor to this diagnosis is stress. Stress, alone, can cause hyperglycemia that can be high enough to cause spill over into the urine and glucosuria. Should the clinician have any doubt of whether hyperglycemia and glucosuria are due to diabetes mellitus, s/he should check a serum fructosamine level. This value gives the average of the blood glucose over the preceding 2–3 weeks. If elevated, then diabetes mellitus can be diagnosed.

After diagnosis, stable diabetics may be sent home with insulin to begin therapy. Optimal insulins for dogs include NPH, Lente (Vetsulin®), detemir (Levemir®). Of these, Vetsulin® is the only one approved for veterinary use by the FDA. NPH insulin is an inexpensive option for owners, although the duration of action may not be optimal in some dogs. Vetsulin® and detemir are more appropriate choices if insulin can only be given once per day. The optimal diet for diabetic dogs is one high in insoluble fiber. This diet slows glucose absorption from the gut and postprandial hyperglycemia. The patient should be re-examined in 7–10 days to determine insulin effect. Optimal monitoring includes evaluation of a glucose curve.[2-4]

Concurrent HAC in the diabetic patient is usually suspected if the diabetes mellitus is not controlled with insulin doses climbing over 1.5 U/kg (insulin resistance) or if clinical signs consistent with HAC persist with insulin therapy.[5] Other reasons for insulin resistance should be considered to include improper insulin injection technique, outdated insulin, or incorrect syringes. The practitioner should also check for other concurrent diseases such as urinary tract infections. If HAC is still suspected, adrenal function testing should be initiated. Do not test for HAC if the dog is clinically unstable. There should be loose control of diabetes mellitus with the blood sugar ideally ranging from 100–400 mg/dl for most of the day. Do not test for HAC if the dog is ketotic, hypoglycemic, or hyperosmolar.

Dogs get HAC in 2 ways: pituitary tumors (85%) and adrenal tumors (15%). Both produce an excess of cortisol that causes the disease entity. Three screening tests are available.[6] The urine cortisol:creatinine ratio is very sensitive (100%) although not specific. Therefore if the urine cortisol:creatinine ratio is positive, another screening test will be necessary in order to diagnose the disease. A second screening test for HAC is the low-dose dexamethasone suppression test. This should be the go to test unless there is significant concurrent disease such as diabetes mellitus. It is sensitive (89–92%) but with a decrease in specificity (75%). It is least specific in animals with concurrent disease. A third screening test for HAC is the ACTH stimulation test. It is less sensitive (75–83%) than the low-dose dexamethasone test but it is more specific (85%). It is more specific in animals with concurrent disease and recommended for use in dogs with concurrent diabetes mellitus to decrease the possibility of a false-positive response. The practitioner measures cortisol before and 60 min after injection of 5 μg/kg synthetic ACTH (Cortrosyn). In diabetic patients, the owner should feed and administer insulin as usual. A simultaneous glucose curve should be performed in conjunction with the ACTH stimulation test to ensure that the animal has blood glucose within the optimal range during testing. If ACTH stimulation testing supports HAC, then treatment should be initiated. If HAC is still suspected but is not supported by the ACTH stimulation test, then the test can be repeated 2–4 weeks later.

The practitioner should determine whether the patient has pituitary- or adrenal-dependent disease. This will direct therapy choice as well as help predict cost of treatment, explain potential neurologic disease, predict time of resolution of signs, and mitigate owner frustration with therapy (hopefully). An endogenous ACTH concentration can be evaluated. Dogs with adrenal tumors have low endogenous ACTH levels (<10 pg/ml). Mishandling of the plasma sample will result in artificially low ACTH levels, leading the practitioner to erroneously conclude the presence of an adrenal tumor. Abdominal ultrasound can be very helpful in identification of an adrenal tumor or bilateral adrenal hyperplasia. The practitioner should be cautious in interpreting results of an abdominal ultrasound in that adrenal masses may be nonfunctional, may be a pheochromocytoma, or the animal may have pituitary-dependent disease as well as an adrenal tumor. Bilateral adrenal hyperplasia may also occur in animals with concurrent non-adrenal disease.

Medical treatment of HAC includes trilostane (Vetoryl®) and mitotane (Lysodren®). Ketoconazole and deprenyl (Anipryl®) have also been advocated, but are substantially less effective than trilostane or Lysodren. Surgical options include adrenalectomy and hypophysectomy.

Mitotane is a DDT derivative. It kills adrenocortical cells with a preference for the cortisol-producing ones (zonae fasciculata and reticularis). The induction phase consists of administering 35 (big dog)–50 (little dog) mg/kg PO q.d. x 5–7 d. The maintenance phase is 35–50 mg/kg PO divided twice per week. The practitioner may work with the 500-mg tablet size or may have it compounded. Owners should be discharged with instructions to call the veterinarian if any of the danger signs including loss of appetite, lethargy, vomiting, diarrhea, or "just aint right" occur. The practitioner should also send home at least 2 doses of prednisone at 0.5 mg/kg to be given if the animal becomes ill from the mitotane. Side effects of mitotane include adrenal cortical destruction producing a glucocorticoid-deficient addisonian or a full-blown addisonian. Often the adrenocortical deficiency is reversible. Another side effect is liver toxicity. This often resolves once off Lysodren, but it cannot be used again.

Monitoring mitotane therapy should occur after the induction phase. An ACTH stimulation test should be performed the morning after the last dose of mitotane during the induction phase. Ideally, the cortisol should be within the normal range pre- and post-ACTH injection. ACTH should cause a minimal increase in the cortisol level. The practitioner should also perform bloodwork to evaluate liver health. Clinical signs of the animal should be discussed with the owner. In cases where the ACTH stimulation test results do not agree with the resolution (or non resolution) of clinical signs reported by the owner, the clinical signs should be believed.

Trilostane is a competitive 3B-OH steroid dehydrogenase inhibitor that inhibits the conversion of pregnenolone to progesterone in the adrenal gland. This blocks the formation of the end products of progesterone including cortisol and aldosterone. Trilostane can be compounded but a manufactured product is available (Vetoryl®, Dechra Pharmaceuticals). Vetoryl® is FDA approved for veterinary use so that the veterinarian is legally protected. Vetoryl® is much more consistent in active ingredients than compounded trilostane and should be used when possible. The drug must be given daily. The recommended dose is 1.5–3 mg/kg, with the lower dose given BID considered safer and more efficacious.[7] Side effects of trilostane are similar to those of mitotane and prednisone should be sent home with the owner similarly to that done with mitotane therapy.

Side effects of trilostane include glucocorticoid, mineralocorticoid or glucocorticoid + mineralocorticoid deficiency. Complete adrenal necrosis has been reported, although this a rare and idiosyncratic response. Over time adrenal glands will get enlarged and irregular, but no problems with this have been noted. Investigators have reported elevations of 17OH progesterone indicating that trilostane has other actions than those expected based on just its pharmacology.

Trilostane therapy should be monitored 10–14 days after initiation. An ACTH stimulation test measuring cortisol is performed 4–6 hours after the morning dose of trilostane is given. Results of the ACTH stimulation are ideally a cortisol in the normal range pre- and post-ACTH with little stimulation, similar to what is desired with Lysodren therapy. A sodium/potassium level should be measured as well. This often decreases with trilostane treatment because of increasing K^+ due to loss of aldosterone function.[8]

Which therapy should the practitioner choose to treat HAC? In comparing trilostane and mitotane, both are as efficacious in controlling clinical signs and in the longevity of the treated animals.[9,10] Both are sold commercially and cost is approximately the same. Some considerations are that trilostane must be given every day while mitotane can be given 2–3 times per week in a stabilized animal. Mitotane should not be given to animals with liver disease. Animals treated with trilostane should be monitored carefully since they can become glucocorticoid/mineralocorticoid deficient even after long-term stable treatment, although more consistent results are found with the use of Vetoryl®. The effects of adrenal gland hyperplasia/metaplasia by trilostane have not been fully investigated but appear to be unproblematic.

Treatment for HAC in the patient with concurrent diabetes mellitus should be approached cautiously.[11] Removal of excess glucocorticoids will lessen the concurrent insulin resistance and result in more profound glucose-lowering activity by the usual insulin dose. The current insulin dose should be decreased by 25% upon initiation of treatment of HAC. Owners should be sent home with Keto-Diastix® and instructed to measure urine at least once per day approximately 6 hours after insulin is given (if given twice per day). If ketones are detected or the absence of urine glucose is detected more than twice in a row, the owner should be instructed to inform the veterinarian. Lack of urine glucose may indicate hypoglycemia that may necessitate therapy. Alternately, if owners are able to collect blood glucose at home, they should perform spot checks 4–8 hours after the insulin is given. Severe hyperglycemia or hypoglycemia should be reported to the veterinarian.

A glucose curve should be performed at every recheck to monitor HAC treatment. After the HAC is controlled, tight control of glucose with insulin can be attempted.

References

1. Hess RS, Ward CR. Effect of insulin dosage on glycemic response in dogs with diabetes mellitus: 221 cases (1993–1998). *J Am Vet Med Assoc*. 2000;216:217–221.
2. Fracassi F, Corradini S, Hafner M, *et al*. Detemir insulin for the treatment of diabetes mellitus in dogs. *J Am Vet Med Assoc*. 2015;247:73–78.
3. Gilor C, Graves TK. Synthetic insulin analogs and their use in dogs and cats. *Vet Clin North Am Small Anim Pract*. 2010;40:297–307.
4. Rucinsky R, Cook A, Haley S, *et al*. AAHA diabetes management guidelines. *J Am Anim Hosp Assoc*. 2010;46:215–224.
5. Hess RS. Insulin resistance in dogs. *Vet Clin North Am Small Anim Pract*. 2010;40:309–316.
6. Behrend EN, Kooistra HS, Nelson R, *et al*. Diagnosis of spontaneous canine hyperadrenocorticism: 2012 ACVIM consensus statement (small animal). *J Vet Intern Med*. 2013;27:1292–1304.
7. Vaughan MA, Feldman EC, Hoar BR, *et al*. Evaluation of twice-daily, low-dose trilostane treatment administered orally in dogs with naturally occurring hyperadrenocorticism. *J Am Vet Med Assoc*. 2008;232:1321–1328.
8. Reid LE, Behrend EN, Martin LG, *et al*. Effect of trilostane and mitotane on aldosterone secretory reserve in dogs with pituitary-dependent hyperadrenocorticism. *J Vet Intern Med*. 2014;28:443–450.
9. Helm JR, McLauchlan G, Boden LA, *et al*. A comparison of factors that influence survival in dogs with adrenal-dependent hyperadrenocorticism treated with mitotane or trilostane. *J Vet Intern Med*. 2011;25:251–260.
10. Clemente M, De Andres PJ, Arenas C, *et al*. Comparison of non-selective adrenocorticolysis with mitotane or trilostane for the treatment of dogs with pituitary-dependent hyperadrenocorticism. *Vet Rec*. 2007;161:805–809.
11. McLauchlan G, Knottenbelt C, Augusto M, *et al*. Retrospective evaluation of the effect of trilostane on insulin requirement and fructosamine concentration in eight diabetic dogs with hyperadrenocorticism. *J Small Anim Pract*. 2010;51:642–648.

acvim425

Hyperadrenocorticism: Management with Urinary Disease

Joe Bartges, DVM, PhD, DACVIM (SAIM), DACVN
Athens, GA, USA

Introduction

Hyperadrenocorticism (HAC) occurs in middle-aged to old dogs with 80–85% being ACTH dependent and 15–20% ACTH independent. HAC is associated with many changes in physical examination findings, laboratory values, and imaging studies. HAC may be associated with altered renal function, systemic arterial hypertension (SAHT), proteinuria, urinary tract infection (UTI), and urolithiasis.

HAC Effects on Renal Function

Dogs with HAC have PU/PD. Urine specific gravity (USG) <1.020 occurs in up to 85% of dogs with HAC.[1-4] Cortisol maintains normal renal blood flow (RBF) and glomerular filtration rate (GFR). Excess cortisol increases GFR and RBF with increased renal tubular flow rates.[5] In dogs with induced HAC, GFR was increased at 16 weeks, but was not different at 24 weeks when compared with controls.[6] Hypercortisolemia disturbs osmoregulation of vasopressin (ADH) release, and causes renal resistance to ADH.[7,8] There is decreased inner medullary urea accumulation and interstitial osmolality[9] despite ADH increased expression of aquaporins (AQ). Decreased medullary tonicity results in a decreased concentration gradient. In dogs with spontaneous HAC, treatment was associated with increased USG and decreased PU.[10] SAHT occurs in 35% dogs with HAC.[11] In dogs with naturally occurring HAC, systolic BP was not significantly different prior to treatment and after treatment.[10] SAHT results in increased capillary hydrostatic pressure causing hyperfiltration and resulting in proteinuria and glomerulosclerosis. Glomerulosclerosis has been documented in dogs with HAC.[12-14] Dogs with induced HAC have light and electron microscopic glomerular changes unassociated with immune complex or antibody deposition[6,14] and tubular lesions.[6] Proteinuria, occurring in 44–75% of dogs with HAC[13,15,16], occurs because of increased intraglomerular pressure, damage to glomerular barrier, and decreased tubular absorption.[17] There is a direct correlation between BP and severity of proteinuria in dogs with HAC.[13,18] Most dogs have UPC <5; however, some dogs may have UPC >10.[19] In dogs with induced HAC, UPC increases within 1 to 2 months.[4,14] In dogs with HAC, treatment resulted in a significant decrease in proteinuria.[10] Cortisol excess saturates 11β-hydroxysteroid dehydrogenase enzyme activity and binds to mineralocorticoid receptors in renal distal convoluted tubule and collecting duct cells. Mineralocorticoid receptors regulate expression of epithelial sodium channels (ENaC) at luminal side of cells and Na^+K^+-ATPase at basolateral side causing increased sodium reabsorption.[20] Calcium (Ca) reabsorption is decreased leading to increased fractional excretion of Ca.[21] NAG is a lysosomal enzyme leaks into tubular fluid from proximal tubular cells when injured; however, urinary NAG (uNAG) levels may indicate a functional tubular disorder rather than damage. uNAG were approximately 4 times higher in dogs with untreated HAC compared with controls.[19] In dogs with HAC, uNAG was elevated prior to treatment and did not change with treatment. uNAG correlates with proteinuria and remains high in dogs with persistent proteinuria because of increased tubular processing.[10] The net effect of HAC is increased GFR, proteinuria, and sodium reabsorption with decreased calcium, phosphate, and urea reabsorption and urine concentrating ability.[2] Despite an increased urea excretion, serum urea concentrations were not different between dogs with induced HAC and control dogs.[6]

HAC and Chronic Kidney Disease

In human beings, CKD is considered rare with HAC despite decreases in GFR over time, presence of SAHT, increased risk of nephrolithiasis, and glomerular changes with proteinuria. An association in dogs has not been reported and no differences in serum creatinine or urea concentrations were found between dogs with HAC and controls dogs.[6] In a study of dogs with HAC, GFR decreased significantly with treatment although serum creatinine and urea concentrations did not increase significantly.[10]

HAC and Urinary Tract Infection

UTI refers to colonization of one or more parts of the urinary tract with an infectious agent. The urinary tract is in contact with external environment and bacteria normally reside in the distal urogenital tract. The development of a UTI depends upon the balance between infectious agents and host resistance, and a UTI occurs when an urovirulent organism invades the urinary tract and there is a break in host defenses.[22] Dogs and cats with UTI may be symptomatic or asymptomatic. Decision to treat is based on whether there are symptoms or other evidence of UTI (e.g., pyuria, hematuria). *Escherichia coli* is the most common isolate accounting for 1/3 to 1/2 of infections. Gram-positive organisms are the second most common cause with *Staphylococci* accounting for ¼ to 1/3 of infections. Other bacteria account for the remaining.[22] In a study of 19 dogs with HAC, only 2 dogs had UTI[10]; however, in a study of 119 dogs, approximately 50% had UTI.[23] In a retrospective study of 101 dog with HAC, DM, or both, 42% had a positive urine culture but only approximately 2% had signs of lower urinary tract disease. A single organism was cultured from 83% of dogs whereas 17% had 2 or more organisms. *E. coli* occurred most commonly (69%) while gram-positive cocci accounted for approximately 25%, and other organisms accounted for the remaining.[3] Interestingly, UTI did not occur in dogs with experimentally induced hyperadrenocorticism over 49 days.[4] Treatment of bacterial UTI is based on whether it is symptomatic and whether the breech in host defenses is temporary or permanent. Simple or uncomplicated UTI are treated either conventionally with a 10- to 14-day course of an antimicrobial agent based on culture and sensitivity or empirically or with a 3-day course of enrofloxacin (20 mg/kg given PO q24h) if the organism is susceptible. Dogs with HAC, however, should be considered to have a complicated UTI even if well controlled. Conventionally, the choice of antimicrobial agent is based on results of culture and sensitivity testing and is administered for a minimum of 3 weeks. Bacteriuria unassociated with clinical signs or changes on urinalysis are considered asymptomatic. This is not uncommon in dogs with HAC due to the absence of pyuria with bacteriuria. In studies, more than 80% of asymptomatic bacteriuria will resolve without treatment. The incidence of asymptomatic bacteriuria is up to 25% in morbidly obese dogs.[24] Asymptomatic bacteriuria does not warrant treatment; however, the decision to treat or not should be based on the individual patient and risks associated with treatment or not. Prophylactically, low dose or pulse antimicrobial therapy, methenamine, nitrofurantoin, estrogens, ecotherapeutics, cranberry extracts, and D-mannose may be tried.

HAC and Urolithiasis

Urolith formation occurs with sustained alterations in urine composition that promotes supersaturation of one or more substances in urine resulting in precipitation and subsequent organization and growth into uroliths.[25] Approximately 80% of canine uroliths are either struvite or calcium oxalate. Calcium oxalate and struvite occur at approximately even frequency although struvite occurs more commonly now. In dogs, struvite urolith formation primarily occurs because of an initiating UTI with an organism that produces the enzyme, urease.[26] Typically, this is *Staphylococcus*; however, *Proteus* and *Enterococcus* may also occur. HAC predisposes to both calcium oxalate and struvite formation. Because dogs with HAC are prone to UTI, if a UTI with a urease-producing organism occurs, then struvite uroliths may form.[26] Struvite uroliths are radiodense and often large and numerous. Infection-induced struvite uroliths typically form in young to middle-aged adult female dogs. They may be removed surgically or through minimally invasive procedures, but are amenable to medical dissolution. Struvite uroliths may be dissolved by inducing an acidic urine pH, increasing urine volume (decreased urine specific gravity), and feeding a diet that is restricted in magnesium and phosphorous. With infection induced struvite uroliths, an antimicrobial agent must also be administered. The average dissolution time is 8 weeks for infection induced struvite.[26] Once uroliths are dissolved radiographically, dissolution therapy should continue for an additional 2 to 4 weeks. An alternative dissolution protocol has been shown to be effective in greater than 80% of dogs. In this protocol, the diet is not changed; instead a urinary acidifier, d,l-methionine, is administered in combination with an appropriate antibiotic for the organism responsible for struvite formation (typically *Staphylococcus*).[27] With infection induced struvite uroliths, prevention and treatment of UTI will prevent recurrence because it is the UTI that causes these stones to form. Calcium oxalate urolith formation is not a specific disease, but results from underlying disorders that promote precipitation of calcium oxalate in urine. It occurs when urine is oversaturated with calcium and oxalate. Additional risk factors for urolith formation include breed, gender, age, and diet. Alterations in balance between urine concentrations of calculogenic substances (calcium and oxalic acid), and crystallization inhibitors (including citrate, phosphorus, magnesium, sodium, and/or potassium) have been associated with initiation and growth of calcium oxalate uroliths. In addition to these alterations in activities of ions, large molecular weight proteins occurring in urine, such as nephrocalcin, uropontin, and Tamm-Horsfall mucoprotein, have an influence on calcium oxalate formation.[25] Dogs with HAC are ten times more likely to have calcium-containing uroliths.[28] In humans, hypercortisolemia is associated with hypercalciuria, hypocitraturia, and increased urinary excretions of phosphorous, oxalate, potassium, cystine, and uric acid.[29] In human beings, nephrolithiasis has a prevalence of 50% in patients with active disease and 17–27% in treated patients.[29] Many of these patients have additional metabolic derangements despite normalization of cortisol levels; therefore, Cushing's syndrome is a risk for calcium oxalate nephrolithiasis in human beings but other factors exist. At present, there are no medical dissolution protocols for dogs and cats with calcium oxalate uroliths. If calcium oxalate uroliths are causing clinical signs, then they must be physically removed. Options include cystostomy, catheter-assisted retrieval, voiding urohydropropulsion, cystoscopy and retrieval, cystoscopy and laser lithotripsy, and percutaneous cystolithotomy.[30] Calcium oxalate uroliths are recurrent; therefore, preventative measures are warranted. There is an approximate 10% recurrence at 6 months and 35% recurrence at 12 months. The goal of prevention is lower the amount of calcium oxalate by decreasing urinary levels of calcium and oxalate and by increasing urine volume in order to dilute the minerals. Management of HAC may decrease recurrence; however, in an unpublished study of dogs with naturally occurring HAC, urinary calcium excretion decreased with treatment using Lysodren in 5 of 6 dogs.[31] Additional measures, therefore, may be required including a diet that is mineral restricted, diuresing, and alkalinizing. Alternatively, a higher fiber diet with supplemental potassium citrate may be used. Hydrochlorothiazide administration decreases urinary calcium excretion in dogs, but may induce hypercalcemia.[32]

References

1. Feldman EC, Nelson RW, eds. *Canine and Feline Endocrinology and Reproduction*. St. Louis: Elsevier Saunders; 2004;252.
2. Smets P *et al. Gen Comp Endocrinol*. 2010;169(1):1.
3. Forrester SD *et al. J Vet Intern Med*. 1999;13(6):557.
4. Forrester SD *et al. Res Vet Sci*. 2003;74(2):179.
5. Li C *et al. J Am Soc Nephrol*. 2008;19(10):1975.
6. Smets PM *et al. PLoS One*. 2012;7(2):e31702.
7. Biewenga WJ *et al. Acta Endocrinol (Copenh)*. 1991;124(5):583.
8. Knoepfelmacher M *et al. Eur J Endocrinol*. 1997;137(2):162.
9. Baas JJ *et al. Vet Q*. 1984;6(1):17.
10. Smets PM *et al. J Vet Intern Med*. 2012;26(3):565.
11. Chen HY *et al. Int J Endocrinol*. 2016;2016(3814034.
12. Littman MP *et al. J Am Vet Med Assoc*. 1988;193(4):486.
13. Ortega TM *et al. J Am Vet Med Assoc*. 1996;209(10):1724.
14. Waters CB *et al. J Vet Intern Med*. 1997;11(3):172.
15. Hurley KJ *et al. J Am Vet Med Assoc*. 1998;212(3):369.
16. Mazzi A *et al. Vet Res Commun*. 2008;32 Suppl 1(S299).
17. Lees GE *et al. J Vet Intern Med*. 2005;19(3):377.
18. Schellenberg S *et al. J Vet Intern Med*. 2008;22(2):273.
19. Smets PM *et al. Vet J*. 2012;192(3):532.
20. Bailey MA *et al. Hypertension*. 2009;54(4):890.
21. Ferrari P *et al. J Am Soc Nephrol*. 2002;13(10):2540.
22. Olin SJ *et al. Vet Clin North Am Small Anim Pract*. 2015;45(4):721.
23. Ling GV *et al. J Am Vet Med Assoc*. 1979;174(11):1211.
24. Wynn SG *et al. PeerJ*. 2016;4(e1711).
25. Bartges JW *et al. Vet Clin North Am Small Anim Pract*. 1999;29(1):45.
26. Seaman R *et al. Compend Contin Educ Pract Vet*. 2001;23(407).

27. Bartges J *et al. ACVIM Forum*; 2010.
28. Hess RS *et al. J Am Vet Med Assoc*. 1998;212(12):1889.
29. Faggiano A *et al. J Clin Endocrinol Metab*. 2003;88(5):2076.
30. Lulich JP *et al. J Vet Intern Med*. 2016;30(5):1564.
31. Kraje AC *et al. J Vet Intern Med*. 2000;14(384).
32. Lulich JP *et al. J Am Vet Med Assoc*. 2001;218(10):1583.

acvim426

Hyperadrenocorticism: Management with Pancreatitis and GI Disease

Tracy Hill, DVM, DACVIM, PhD, DECVIM-CA
Athens, GA, USA

Introduction

Hyperadrenocorticism (HAC) is one of the most commonly diagnosed endocrinopathies in dogs. Pancreatitis and various forms of GI disease are also common, particularly chronic enteropathies and gastroduodenal ulceration. Management of these diseases concurrently is often complicated by the presence of HAC and there is some suggestion of causality in some cases. In this lecture, we will discuss the evidence regarding hyperadrenocorticism and pancreatic/GI diseases and discuss optimizing management to maximize patient outcomes.

Hyperadrenocorticism and Pancreatitis

Hyperadrenocorticism has been identified to introduce changes in the canine pancreas that alters diagnostic modalities used in pancreatitis, cPLI and abdominal ultrasound. HAC, in dogs without clinical signs of pancreatitis, is associated with increased SPEC cPLI and more frequent positive SNAP cPLI.[1] Thirty-five percent of dogs with HAC without clinical signs of pancreatitis had cPLI over 400 µg/L and 55% had a positive SNAP (versus 6% in control dogs). It is unknown what effect controlled HAC would have on cPLI. Though it is possible some of these dogs had subclinical pancreatitis, it is unlikely that that was true in all cases. HAC is also associated with ultrasonographic changes to the pancreas. The pancreas of a dog with HAC with a normal cPLI is more heterogenous and hyperechoic than that of a control dog.[2] In this study, the proportion of uncontrolled HAC dogs was not reported; as such, the effect of treatment of HAC on pancreatic appearance remains undetermined. The morphologic change leading to this finding is unknown but is hypothesized to be related to increased pancreatic fat deposits due to altered fat metabolism and mineralization associated with elevated cortisol. Collectively, these studies demonstrate that the definitive diagnosis of pancreatitis would be complicated by HAC. The effect of controlled HAC on either cPLI or ultrasonographic appearance of the pancreas is unknown.

Hyperadrenocorticism is an associated co-morbidity with pancreatitis in dogs, though causality has not been determined. HAC was diagnosed in 12% of dogs with acute pancreatitis and was the second most common endocrinopathy (after diabetes mellitus); 75% of the HAC dogs were not currently receiving treatment.[3] Of course, this finding does not indicate causality and should be interpreted with caution. In dogs with fatal acute pancreatitis, HAC was identified as a risk factor for the development of fatal acute pancreatitis (OR, 4.3; 95% CI, 0.2 to 87.6) when compared to dogs dying from a traumatic event.[4] It was not reported whether dogs were controlled for HAC. Though HAC was more common in dogs that died from acute pancreatitis than dogs that died from trauma, causality cannot be extrapolated. The reasons behind this association are unknown. It could be that HAC is a predisposing factor for acute pancreatitis. Equally, HAC could increase mortality with acute pancreatitis by predisposing to complications known to be associated with both diseases, such as thromboembolic disease and/or sepsis.

In people, there is little known association with hypercortisolism and pancreatitis. There are case reports of HAC associated with acute necrotizing pancreatitis though HAC is not a recognized risk factor for pancreatitis in people.[5] People have similar echogenic changes when receiving low-dose exogenous corticosteroid therapy for over 6 months.[6] This does not, however, seem to manifest in increased susceptibility to pancreatitis or increased risk of complications from pancreatitis in most cases.

It appears clear that HAC does have effects on pancreatic appearance and function, as evidenced by altered ultrasonographic appearance and elevated cPLI. It is less clear, however, that these changes increase the risk of pancreatitis and what, if any, changes would persist in dogs where HAC is controlled. The diagnosis of pancreatitis in dogs may be complicated by the presence of HAC, especially in dogs with uncontrolled HAC. If history and clinical signs are supportive, however, these diagnostic modalities still have merit, even in uncontrolled HAC. Once both HAC and pancreatitis are presumptively or definitively diagnosed, it is likely prudent to more aggressively manage dogs for complications including sepsis and thromboembolic disease. As such, broad-spectrum antibiotics and anticoagulant therapy with clopidogrel may be warranted in the author's opinion.

Hyperadrenocorticism and Gastroduodenal Ulceration

Hypercortisolism, either exogenous or endogenous, has long been assumed to be a risk factor for gastroduodenal ulceration in people and dogs. A recent study in people examined the prevalence of gastroduodenal ulcers in healthy people versus people with endogenous hypercortisolism (HAC) and people receiving chronic corticosteroids.[7] There was no difference in the detection of ulcers among the three groups. Pangastritis, detected endoscopically, was more prevalent in both HAC and exogenous corticosteroid users versus the healthy controls, while antral gastritis was less common. Histopathologic findings were not different among the three groups for gastritis or *Helicobacter pylori* infection. The authors' conclusions were that there was no evidence that hypercortisolism, either endogenous or exogenous, increased the risk of ulceration. The remainder of the evidence for the role of corticosteroids and ulceration is largely limited to studies involving exogenous steroid use. In people receiving low-dose aspirin for secondary prevention of cardiovascular events, the concurrent use of oral steroids increased the risk of uncomplicated peptic ulcer disease (OR 1.88; 95% CI, 1.09–3.25).[8] In a 2014 metanalysis examining the risk of adverse GI events with corticosteroids, exogenous corticosteroids were associated with a 40% increased OR of experiencing GI bleeding or perforation compared to placebo (OR 1.43, 95% CI, 1.22–1.66).[9] Diseases for which steroids were being administered included traumatic injury, meningitis, sepsis, bronchopulmonary dysplasia, liver disease, rheumatoid arthritis, and others. Because these patients had preexisting conditions, it is difficult to determine causality from the steroids solely and not as a contributing factor to the underlying disease in most cases.

In dogs, exogenous corticosteroids have been a proposed risk factor for gastroduodenal ulceration in most retrospective studies. Corticosteroid use was associated with gastroduodenal ulceration in 9/43 dogs; corticosteroids were reported to be the single risk factor in 2/43 dogs (4.7%).[10] Other concurrent factors included hepatic disease, coagulopathy/shock, neurologic disease, and mastocytosis. In a separate study, corticosteroids were listed as risk factors in 3/16 dogs, all of which had secondary factors (stress, concurrent NSAID).[11] In the two available retrospective studies of all-cause gastroduodenal ulceration, 2/59 (3.3%) dogs had corticosteroids as the only risk factor. There are no prospective studies examining the incidence of adverse GI events in dogs with exogenous or endogenous hypercortisolism and no statistical analysis establishing that corticosteroids are a true risk factor for ulceration. Similarly, there are no reports of dogs with HAC that developed gastroduodenal ulceration without secondary risk factors.

Overall, there is very little evidence that corticosteroids, endogenous or exogenous, increase the risk of gastroduodenal ulceration without some concurrent predisposing factor in people or dogs. As such, it would not be expected that HAC alone would increase the risk of gastroduodenal ulceration. If a dog with HAC develops a secondary risk factor, including NSAID use, liver disease, sepsis/SIRS, renal disease, it would be prudent to administer appropriate prophylactic therapy with omeprazole at 1 mg/kg twice daily and attempt to treat or correct the concurrent disease. It is currently unknown whether dogs with controlled HAC would have the same increased risk of ulceration with secondary risk factors though it would likely be prudent to treat any HAC dog with prophylactic therapy as long as a secondary risk factor is present.

Hyperadrenocorticism and Chronic Enteropathy

One of the common treatments for canine chronic enteropathy (CE) that is non-responsive to dietary therapy or antibiotics (metronidazole, tylosin) is corticosteroid therapy. In dogs with concurrent chronic enteropathy and HAC, the management of both is compromised by the presence of the other. Successful control of HAC will reduce the level of endogenous cortisol, which may indeed be exerting some anti-inflammatory effect on the CE. Effective treatment of HAC may necessitate administration of oral immunosuppressants for management of the CE (again, usually with corticosteroids).

Monitoring for clinical signs during treatment of HAC, particularly with mitotane, is complicated by the presence of chronic GI signs in dogs with a concurrent CE. If a dog developed GI signs during the induction period or later in the maintenance phase, ACTH stimulation tests may be required with increased frequency to confirm that the GI signs are attributable to the underlying CE and not to iatrogenic hypoadrenocorticism. Though the same is true for trilostane, it is the author's preference to use trilostane in the cases where treatment of the HAC is necessary.

There is no evidence-based medicine to guide treatment in these difficult cases. Ultimately, treatment decisions must be made on a case-by-case basis. If a dog is more significantly clinical for its HAC, it would be appropriate to treat with trilostane. The goal in that case may not be completely controlling HAC but instead treating to control clinical signs of HAC without exacerbating the clinical signs of the CE. If the clinical signs of the CE are more significant, it may be more appropriate to withhold treatment of the HAC. If treatment of both is required simultaneously (significant chronic GI signs including uncontrolled vomiting and diarrhea, hypoalbuminemia, and weight loss along with PU/PD, pyoderma, urinary tract infections, and thromboemboli), one could consider treatment of HAC with trilostane along with management of the CE with an alternate immunosuppressive agent, such as budesonide, cyclosporine, or chlorambucil. VSL#3 may also be of benefit. Though still experimental, fecal microbiome transplants (FMT) have shown some promise anecdotally in the management of refractory CE.

References

1. Mawby DI, Whittemore JC, Fecteau KA. *J Vet Intern Med.* 2014;28(4):1244–50.
2. Granger LA, Hilferty M, Francis T, *et al. Vet Radiol Ultrasound.* 2015;56(5):540–8.
3. Cook AK, Breitschwerdt EB, Levine JF, Bunch SE, Linn LO. *J Am Vet Med Assoc.* 1993;203(5):673–9.
4. Hess RS, Kass PH, Shofer FS, Van Winkle TJ, Washabau RJ. *J Am Vet Med Assoc.* 1999;214(1):46–51.
5. Hiramatsu K, Moriuchi J, Arimori S, Ide A, *et al. Tokai J Exp Clin Med.* 1991;16(2):153–6.
6. Gupta AK, Arenson A-M, McKee JD. *J Clin Ultrasound.* 1987;15(3):171–4.
7. Hatipoglu E, Caglar AS, Caglar E, *et al. Endocrine.* 2015;50:489–495.
8. Ruigomez A, Johansson S, Nagy P, *et al. BMC Gastroenterol.* 2014;14:205.
9. Narum S, Westergren T, Klemp M. *BMJ Open.* 2014;4(5):e004587.
10. Stanton ME, Bright RM. *J Vet Intern Med.* 1989;3:238–244.
11. Hinton LE, McLoughlin MA, Johnson SE, *et al. J Am Vet Med Assoc.* 2002;38:176–187.

acvim427

Monitoring of Diabetic Patients: Could We Do Better?

Matthew Krecic, DVM, MS, MBA, DACVIM (SAIM)
Parsippany, NJ, USA

INTRODUCTION

Adequate monitoring of diabetic pets is critical for achieving short- and long-term success. The word "adequate," however, connotes different meanings to different stakeholders - pet owners and veterinary personnel. Likewise, "monitoring" may suggest many possibilities. Reviewing short- and long-term goals of monitoring and available options for monitoring will hopefully help with selection of the best options for diabetic pets and their owners.

Short- and Long-Term Goals

Pet owner short-term goals of monitoring of their diabetic pets likely differ from veterinary short-term goals. Consider the following short-term, pet owner goals.

1. Understand diabetes mellitus: Pet owners may not realize that dogs and cats can become diabetic. Additionally, they may need clarity on the need for insulin for their pets, given now the seemingly plethora of new pharmacological drugs for management of type 2 diabetes in humans.
2. Gain confidence: Pet owners may be timid with their abilities to successfully manage their pets' diabetes. They may believe that they cannot prepare and administer insulin properly and that they will harm their pets. They may have fear of overdosing insulin, leading to a hypoglycemic event.
3. Little to no negative effect on their relationships with their pets: Concern regarding their relationships with their pets - the human-animal bond - is likely based on fear that their pets will no longer seek them because of their association with a potential negative experience, namely insulin administration. I think we can all empathize with having to "chase" our own pets to administer a medication or parasiticide.

Now, consider the following short-term, veterinary goals.

1. Ensure correct (optimal) insulin type and dose: We desire comfort knowing blood glucose concentrations will be within an acceptable/tolerable range such that clinical signs do not arise with continued hyperglycemia (e.g., ketoacidosis) or with new hypoglycemia.
2. Ensure appropriate body weight and appetite: We desire planned weight loss for those diabetic pets that are overweight to obese; yet, we want weight stability or gain for those with reduced body weight. Although appetite is generally increased with diabetic pets, reduction of appetite is a warning sign for possible ketoacidosis or a concurrent illness.
3. Identify and treat/manage concurrent illnesses: To satisfy goals numbers one and two, we attempt to identify and treat or manage those concurrent illnesses (e.g., pancreatitis, urinary tract infection, hyperadrenocorticism). By doing so, we hope that the dose of insulin we have prescribed will lead to positive outcomes (i.e., none or few adverse consequences of diabetes).

Long-term goals for pet owners still include consideration of the human-animal bond; yet, not only on their relationships at home with their pets but also with their relationships with their pets in travel to and from the veterinary hospital. Pet owners have concern about the well-being of their pets with frequent, often day-long visits. Associated with the frequency of visits is the expense for such - as well as the expense of insulin, insulin supplies, etc. An added long-term goal for veterinary personnel is ensuring pet owner compliance with our instructions. We have this fear of pet owners "going rogue" - managing their diabetic pets without our medical consultation.

With an understanding of respective goals and an attempt at alignment of these goals, pet owners and veterinary personnel can, however, develop an optimal monitoring protocol for their pets/patients.

Monitoring Options

Based on our mutual decisions with our pet owners, we have many monitoring options: history, physical examination, routine laboratory evaluation, urine glucose/ketones, serum fructosamine, hemoglobin A1c, and blood glucose. Monitoring methods are not isolated; rather, they are used in different combinations to assess pet status/glycemic control and the need for any therapeutic adjustments.

History

Ask about appetite, water consumption, and urination habits. The latter is especially important for dog owners because nocturia and urinary accidents within the house are likely tolerable for only a short period of time once insulin has been started. Less or resolution of nocturia suggests improved glycemic control. Owner reported observations of activity are also important because reduced activity, increased sleeping, etc. may suggest periods of hypoglycemia.

Physical Examination

Assess body condition and body weight. Sometimes recording of body weight at each visit is missing and if recorded, trends are seemingly overlooked at subsequent appointments. Although we often think about weight loss for our diabetic patients, unplanned steady declines of body weights suggest suboptimal glycemic control.

Routine Laboratory Evaluation

At the initial visit, assess a complete blood count and a serum biochemistry profile and perform a urinalysis. Consider additional laboratory evaluations based on history, physical examination, and aforementioned blood and urine test results. If abnormal results are identified, best to re-evaluate the abnormal analytes, etc. to ensure normalization or trend to normalization. Given our short-term concern for concurrent illnesses interfering with glycemic control, re-evaluation of abnormal routine laboratory results may help us identify and treat or manage such, hopefully leading to optimal glycemic control.

Urine Glucose/Ketones

Some veterinarians request owners of diabetic pets to collect urine at home and test with prescribed urine dipsticks for glucose and/or ketones. Urine collection may be especially difficult, however, for owners of diabetic cats. Hydrophobic litter may help with retrieving a urine sample from a litter box. The presence of glucose - glucosuria - may be acceptable depending on the amount. Trace or 1+ glucose is often insignificant. Glucose amounts of greater than or equal to 2+ warrant patient re-evaluation. The absence of glucose likewise warrants patient re-evaluation; for some cats, lack of glucosuria may indicate diabetic remission. The presence of ketones - ketonuria - suggests insulin deficiency or insulin resistance, indicating the need for re-evaluation.

Because glucose and ketone concentrations within the urine may differ throughout the day, depending on their respective concentrations within the blood and length of time urine has accumulated within the urinary bladder prior to ketone and glucose testing, look for trends in their presence or absence. A single measure of 2+ glucosuria or any amount of ketonuria, for example, may not be significant as multiple measures of 2+ glucosuria or any amount of ketonuria.

Additionally, have owners for whom you have directed to evaluate their pets' urine for glucose and/or ketones periodically provide you with results of their testing and instruct them to notify you of abnormal results (e.g., persistent 2+ glucosuria). Ignoring or forgetting about this data you have requested will likely have pet owners questioning the reason for them to collect this. Therefore, consider the usefulness of this information - presence or absence of glucosuria or ketonuria - before instructing pet owners to collect it.

Serum Fructosamine

Fructosamine is a measure of glycosylated proteins, primarily albumin, within the circulation. Fructosamine concentration will increase with prolonged hyperglycemia. It is generally considered a measure of glycemic control over the previous 2–3 weeks. For cats, it may be helpful to distinguish between stress hyperglycemia and diabetic hyperglycemia. As a monitoring option for confirmed diabetic pets, high and low concentrations of fructosamine suggest poor glycemic control. Both may be attributed to insulin overdose, with the concentration of fructosamine dependent on the length of time a pet spends as hyper- and hypoglycemic. For this reason, its measurement may not be as helpful as it was once believed.

Hemoglobin A1c

Interest in measurement of hemoglobin A1c (HbA1c) has seemingly been renewed, possibly given the significance of type 2 diabetes in humans and much commercial advertising of new pharmaceuticals intended to lower the HbA1c of affected humans. Glycosylated hemoglobin is a hemoglobin product with a glucose attached to its N-terminal amino acid valine. Glycosylation is irreversible and concentration of HbA1c within the circulation is approximately 2–3 months, the lifespan of the red blood cell. A dried blood spot test is currently offered (http://baycomdiagnostics.com/); yet, published data on its performance remains unavailable.

Blood Glucose

Blood glucose concentrations (BG) are best for adjusting insulin dose. Yet, we do not want to ignore available information our other monitoring tools provide - i.e., do not want to simply judge glycemic control on BG ("the numbers"). Rather, knowledge of BG in conjunction with the results of other monitoring tools is important to optimally manage these diabetic pets. Ideally, we want to check BG over the duration of 10–12 hours with q 12 h insulin dosing. Deciding on a dose of insulin based on a single BG may be dangerous. "Spot checking" could lead to an erroneous decision to increase the amount of insulin, for example. Likely, we aim to obtain a blood sample for BG determination at the expected nadir - time of lowest BG/maximal effectiveness of insulin. However, the nadir will vary within the patient dose to dose and day to day - and likely dog to dog administered the same type of insulin. Therefore, we may not be coincidentally sampling the blood for BG determination at the actual nadir. One caveat regarding adjusting insulin amount based on a BG obtained at a single time point, however; consider decreasing the amount of insulin with confirmed (i.e., repeatable) BG less than 100 mg/dl because a clinical hypoglycemic event is possible.

Given the challenge of interpreting a single BG, checking multiple BG throughout an 8–12 h period - the "blood glucose curve" (BGC) - has become standard. The goals of the BGC are to determine: 1) duration of insulin; 2) glucose nadir; and 3) average BG throughout the day. Ideally, we aim for duration of insulin effect to be 10–12 h, BG nadir 100–150 mg/dl for the long-term diabetic pet, and average BG less than 250 mg/dl over that 10–12 h. The above parameters are ideal; yet, as previously mentioned, we should not assess glycemic control simply based on "the numbers." The BGC provides important and useful data to be considered alongside results from our other monitoring tools. An interactive algorithm available online may help with assessing the BGC and determining changes, if any, to insulin dose (http://www.veterinaryteambrief.com/article/blood-glucose-curves-whys-hows).

If we elect to perform a BGC, we must decide whether to perform it within the hospital or have the pet owner perform it within the home. We likely recognize the limitations of a BGC performed within the hospital; limitations include stress and abnormal pet routine (e.g., poor appetite, lack of play or rest), with both influencing the BG. Sometimes we have collected this BG data and cannot assess pet glycemic control because of these limitations. Therefore, collection of BG at home by the pet owner may minimally mitigate these limitations. Self-monitoring of BG among people with diabetes is the standard and although physicians were initially apprehensive of allowing their patients to self-monitor, physicians have uniformly adopted and have realized the benefits of their patients doing so. We are likewise concerned about empowering pet owners with collecting BG at home - "going rogue."

The benefits of home blood glucose monitoring (HBGM) have been published in a few studies, and HBGM has been advocated through organizations such as AAHA and ISFM/AAFP. Blood glucose concentrations are believed to be more accurate because pets remain within their homes, experiencing their normal routines. With more accurate BGC, we may better decide on optimal management of the pet; we may be better able to lessen the probability of a clinical hypoglycemic event; of ketoacidosis, considering 35% of dogs with diabetic ketoacidosis are already receiving insulin; of cataracts with diabetic dogs; and of peripheral neuropathy with diabetic cats.

Home blood glucose monitoring, like the other monitoring methods, may not be suitable for all pets and their owners. As mentioned above, pet owners may lack the confidence for which to obtain a blood sample. (Although we may want to refer to it as a capillary sample, such that pet owners do not incorrectly believe they have to phlebotomize their pets.) Recall, their minimum goal is to simply become confident at administering insulin. They may also have fear of disrupting their bond with their pets by adding acquisition of BG. Yet, in several studies to date, pet owners have not reportedly experienced a negative effect on their relationships with their pets with introduction of HBGM. In fact, of two-thirds of diabetic cat owners satisfied with the glycemic control of their cats, almost 75% reportedly monitored their cats' BG at home. Additionally, our fear of our pet owners not complying to our directions - one of our long-term fears - is often mitigated with repeated communication that the BGC will be used to inform us - the veterinarians - the need, if any, to alter management, namely insulin dose. Inform pet owners that frequent adjustments of doses of long-acting insulin will negate its effectiveness. Also consider contrasting this with physicians' management of their human patients, in which they direct them to only adjust their short-acting insulin dose based on meals and activities; their patients do not self-adjust their long-acting insulin dose. In veterinary medicine, we often only prescribe long-acting insulin.

Consider introducing online or printed resources to diabetic pet owners at the initial visits such that they can review at their leisure and focus on confidently administering insulin. Once they have mastered insulin administration, then our staff and us can instruct them on HBGM. For those stable long-term diabetic pets, consider having their owners perform BGC every 3–4 months and share the collected data with us. We can also instruct them to check BG at home if their pets are not behaving normally, to ensure lack of clinical hypoglycemia. Owners of ketoacidotic diabetic pets at the time of hospital discharge of their pets may be best served to begin HBGM to lessen the probability of recurrence of ketoacidosis.

REFERENCES

1. Rucinsky R, *et al. J Am Anim Hosp Assoc.* 2010;46:215.
2. Sparkes AH, *et al. J Fel Med Surg.* 2015;17:235.
3. Briggs CE, *et al. J Am Vet Med Assoc.* 2000;217:48.
4. Cook AK. *J Diabetes Sci Technol.* 2012;6:491.
5. Ford SL, *et al. Vet Clin Small Anim.* 2013;43:283.
6. Lin C, *et al. J Diabetes Metab.* 2014;5:1.
7. Rand JS, *et al. J Vet Intern Med.* 2002;16:123.
8. Hume DZ, *et al. J Vet Intern Med.* 2006;20:547.
9. Kley S, *et al. J Am Vet Med Assoc.* 2004;225:261.
10. Niessen SJM, *et al. J Vet Intern Med.* 2010;24:1098.
11. Niessen SJM, *et al. J Vet Intern Med.* 2012;26:953.
12. Aptekmann KP, *et al. J Am Anim Hosp Assoc.* 2014;50:247.

acvim428

Classification of Diabetes: Type I, Type II, and... Does It Matter?

Chen Gilor, DVM, PhD, DACVIM
Davis, CA, USA

Introduction: What Is Diabetes Mellitus?

Diabetes mellitus (DM) is not a single disease, but a syndrome characterized by hyperglycemia that can be the result of defects in insulin secretion, insulin sensitivity (in target tissues), or both.[1] There is a wide range of pathogenic processes that can lead to the development of DM, from autoimmune destruction of the b-cells of the pancreas with consequent insulin deficiency to abnormalities that result in resistance to insulin action. **Regardless of the pathogenic cause, deficiency in insulin action on target tissues leads to a myriad of abnormalities in carbohydrate, fat, and protein metabolism**. Abnormalities in insulin secretion and action frequently coexist in the same individual, and it is often unclear which abnormality, if either alone, is the primary cause of the hyperglycemia. The degree of hyperglycemia (if any) may change over time, depending on the extent of the underlying disease process and comorbidities. A disease process may be present but may not have progressed far enough to cause hyperglycemia. The same disease process can cause prediabetes [impaired fasting glucose (IFG) and/or impaired glucose tolerance (IGT)] without fulfilling the criteria for the diagnosis of diabetes.[1]

Thus, the clinical and clinicopathological manifestations of DM are not indicative of the cause/causes of the disease but might be indicative of the stage and severity of the process.

Classification of Diabetes Mellitus: Historical Perspective

Ideally, classification of diabetes should be based on etiology and pathogenesis. However, in the 20th century, the classification of diabetes was based on clinical manifestations such as age of diagnosis and insulin requirement (in terms of preventing ketosis). Based on these, human diabetics were divided into 3 categories: Insulin-dependent DM (IDDM, AKA juvenile, prone to ketosis), non-insulin-independent [NIDDM, AKA mature onset, including MODY (Mature-Onset Diabetes of the Young), not prone to ketosis] and others (DM secondary to pancreatitis, endocrinopathies, drugs, etc.). Impaired glucose tolerance and gestational diabetes were categorized as separate disease entities (causing abnormal glucose intolerance but not DM).[2] It was recognized however that an individual can change from one class to another and that insulin requirements vary and depend on the stage of the disease. With increased understanding of the pathophysiology, the terminology of IDDM and NIDDM was slowly replaced by type 1 and type 2. Initially the clinically based categorization overlapped completely with etiologic type (i.e., IDDM was termed type 1 and NIDDM was termed type 2) but on the turn of the century an etiologic-based classification became consensus and IDDM and NIDDM were abandoned.[3] An up-to-date classification of diabetes from the American Diabetes Association is presented in Table 1.[1] Key features of each type of DM and prediabetes based on this ADA classification[1] are presented in the next section and then discussed in relation to DM in cats and dogs below.

Table 1—*Etiologic classification of diabetes mellitus*

- I. Type 1 diabetes (β-cell destruction, usually leading to absolute insulin deficiency)
  - A. Immune mediated
  - B. Idiopathic
- II. Type 2 diabetes (may range from predominantly insulin resistance with relative insulin deficiency to a predominantly secretory defect with insulin resistance)
- III. Other specific types
  - A. Genetic defects of β-cell function
    1. MODY 3 (Chromosome 12, HNF-1α)
    2. MODY 1 (Chromosome 20, HNF-4α)
    3. MODY 2 (Chromosome 7, glucokinase)
    4. Other very rare forms of MODY (e.g., MODY 4: Chromosome 13, insulin promoter factor-1; MODY 6: Chromosome 2, *NeuroD1*; MODY 7: Chromosome 9, carboxyl ester lipase)
    5. Transient neonatal diabetes (most commonly ZAC/HYAMI imprinting defect on 6q24)
    6. Permanent neonatal diabetes (most commonly KCNJ11 gene encoding Kir6.2 subunit of β-cell K_{ATP} channel)
    7. Mitochondrial DNA
    8. Others
  - B. Genetic defects in insulin action
    1. Type A insulin resistance
    2. Leprechaunism
    3. Rabson-Mendenhall syndrome
    4. Lipoatrophic diabetes
    5. Others
  - C. Diseases of the exocrine pancreas
    1. Pancreatitis
    2. Trauma/pancreatectomy
    3. Neoplasia
    4. Cystic fibrosis
    5. Hemochromatosis
    6. Fibrocalculous pancreatopathy
    7. Others
  - D. Endocrinopathies
    1. Acromegaly
    2. Cushing's syndrome
    3. Glucagonoma
    4. Pheochromocytoma
    5. Hyperthyroidism
    6. Somatostatinoma
    7. Aldosteronoma
    8. Others
  - E. Drug or chemical induced
    1. Vacor
    2. Pentamidine
    3. Nicotinic acid
    4. Glucocorticoids
    5. Thyroid hormone
    6. Diazoxide
    7. β-Adrenergic agonists
    8. Thiazides
    9. Dilantin
    10. γ-Interferon
    11. Others
  - F. Infections
    1. Congenital rubella
    2. Cytomegalovirus
    3. Others
  - G. Uncommon forms of immune-mediated diabetes
    1. "Stiff-man" syndrome
    2. Anti-insulin receptor antibodies
    3. Others
  - H. Other genetic syndromes sometimes associated with diabetes
    1. Down syndrome
    2. Klinefelter syndrome
    3. Turner syndrome
    4. Wolfram syndrome
    5. Friedreich ataxia
    6. Huntington chorea
    7. Laurence-Moon-Biedl syndrome
    8. Myotonic dystrophy
    9. Porphyria
    10. Prader-Willi syndrome
    11. Others
- IV. Gestational diabetes mellitus

Patients with any form of diabetes may require insulin treatment at some stage of their disease. Such use of insulin does not, of itself, classify the patient.

Classification of Diabetes Mellitus in Human Medicine

I. Type 1 DM

Key Features

Beta cell destruction (without destruction of other cell types in the islet) usually leading to absolute insulin deficiency.

Immune-Mediated Diabetes

This form of DM was previously encompassed by IDDM (juvenile-onset DM). It results from a **cellular-mediated autoimmune destruction of the b-cells of the pancreas.** In people, markers of the immune destruction of the b-cell include islet cell autoantibodies, autoantibodies to insulin, autoantibodies to GAD (GAD65), and autoantibodies to the tyrosine phosphatases IA-2 and IA-2b. One or more (usually more than one) of these autoantibodies are present in 85–90% of human type I diabetics when fasting hyperglycemia is initially detected. Also, the disease has strong HLA associations, with linkage to the DQA and DQB genes, and it is influenced by the DRB genes. These HLA-DR/DQ alleles can be either predisposing or protective. In this form of DM, the rate of b-cell destruction is quite variable. While rapid progression is seen in juveniles, in adults there is often slow progression with residual b-cell function that may be retained for many years. Because b-cell function is often sufficient to prevent ketoacidosis in these adults, they can be easily confused with type 2 DM (see below). Eventually, these individuals become dependent on insulin. Although patients are rarely obese when they present with type 1 DM, the presence of obesity is not incompatible with the diagnosis. Type 1 diabetics are also prone to other autoimmune disorders (endocrine as well as others including myasthenia gravis, autoimmune hepatitis, etc.).

Idiopathic Diabetes

No evidence of autoimmunity. An absolute requirement for insulin replacement therapy in affected patients may come and go. In people, only a minority of patients with type 1 diabetes fall into this category and most of these are of African or Asian ancestry. This form of DM is strongly inherited, lacks immunological evidence for b-cell autoimmunity, and is not HLA associated. It has been recently suggested that this subtype should be reclassified under type 2 DM (see below).[4]

II. Type 2 DM

Key Features

Unknown etiology (and probably more than one). A diagnosis of exclusion. Pathogenesis: A combination of an insulin secretory defect with insulin resistance. Relative insulin deficiency.

Type 2 DM was previously encompassed by non-insulin dependent DM (or adult-onset DM). It is characterized by the combination of insulin resistance and impaired insulin secretion. Impaired secretion results in insulin deficiency that is usually relative (rather than absolute) to the overall increased insulin requirement that is the result of insulin resistance. At least initially, and often throughout their lifetime, these individuals do not need insulin treatment to survive. There are probably many different causes of this form of diabetes. Although the specific etiologies are not known, autoimmune destruction of b-cells does not occur, and patients do not have any of the other causes of diabetes listed above or below (i.e., it is a diagnosis of exclusion). Most patients with this form of diabetes are obese. Patients who are not obese by traditional weight criteria may have an increased percentage of body fat distributed predominantly in the abdominal region. Although obesity itself causes some degree of insulin resistance it is not the cause of type 2 DM. This is because for type 2 DM to develop a component of impaired insulin secretion and inability to compensate over insulin resistance are necessary.

Whereas patients with type 2 DM may have insulin levels that appear normal or elevated, the higher blood glucose levels in these patients would be expected to result in even higher insulin values had their b-cell function been normal. Thus, insulin secretion is defective in these patients and insufficient to compensate for insulin resistance. Insulin resistance may improve with weight reduction and/or pharmacological treatment of hyperglycemia but is seldom restored to normal. The risk of developing type 2 DM increases with age, obesity, and lack of physical activity. It occurs more frequently in women with prior GDM and in individuals with hypertension or dyslipidemia, and its frequency varies in different racial/ethnic subgroups. **It is often associated with a strong genetic predisposition, more so than is the autoimmune form of type 1 DM.** However, the genetics of this form of diabetes is complex and not fully defined.

III. Other Types of DM ("Type 3"?)

- **A. Genetic defects of the b-cell.** Several forms of diabetes are associated with **monogenetic** defects in b-cell function. They are referred to as **maturity onset diabetes of the young (MODY)** and are characterized by impaired insulin secretion with minimal or no defects in insulin action. They are inherited in an autosomal dominant pattern. Abnormalities at six genetic loci on different chromosomes have been identified to date. Previously classified under NIDDM.
- B. Genetic defects in insulin action. Rare in people.
- **C. Diseases of the exocrine pancreas**. Any process that **diffusely** injures the pancreas can cause diabetes. Acquired processes include pancreatitis, trauma, infection, pancreatectomy, and pancreatic carcinoma. Damage to the pancreas must be extensive for diabetes to occur.
- **D. Endocrinopathies**:
  - **Insulin resistance**: Several hormones (e.g., growth hormone, cortisol, glucagon, epinephrine) antagonize insulin action. Excess amounts of these hormones (e.g., acromegaly, Cushing's syndrome, glucagonoma, pheochromocytoma, respectively) can cause diabetes. This generally occurs in individuals with preexisting defects in insulin secretion, and hyperglycemia typically resolves when the hormone excess is resolved.
  - **Decreased insulin secretion**: Somatostatinomas, and aldosteronoma-induced hypokalemia, can cause diabetes, at least in part, by inhibiting insulin secretion. Hyperglycemia generally resolves after successful removal of the tumor.
- E. Drug- or chemical-induced diabetes.
  - Impairing insulin secretion or destroying beta cell (e.g., diazoxide, thiazides, Vacor [a rat poison], intravenous pentamidine).
  - Impairing insulin action: Thyroid hormones, glucocorticoids, nicotinic acid.
- F. Infections
- G. Other others…

IV. Gestational DM (GDM)

For many years, GDM has been defined as any degree of glucose intolerance with onset or first recognition during pregnancy, at any stage of pregnancy. Although most cases resolve with delivery, the definition applied whether or not the condition persisted after pregnancy and did not exclude the possibility that unrecognized glucose intolerance may have antedated or begun concomitantly with the pregnancy. After deliberations in 2008–2009, the International Association of Diabetes and Pregnancy Study Groups (IADPSG) recommended that high-risk women found to have diabetes at their initial prenatal visit, using standard criteria receive a diagnosis of overt, **not** gestational, diabetes. Thus, the diagnosis of GDM is made in those that were screened for DM and were negative early in their pregnancy and developed DM during pregnancy.

V. Prediabetes and Diabetes Risk

Prediabetes is defined as a condition in which abnormal hyperglycemia is present but does not meet criteria for diabetes. Impaired fasting glucose (IFG): fasting plasma glucose (FPG) levels = 100–125 mg/dl (5.6–6.9 mmol/l). Impaired glucose tolerance (IGT): 2h values in the oral glucose tolerance test (OGTT) of 140–199 mg/dl (7.8–11.0 mmol/l). Prediabetes (IFG and IGT) indicates the relatively high risk for the future development of diabetes. IFG and IGT **should not be viewed as clinical entities in their own right but rather risk factors for diabetes as well as cardiovascular disease. They can be observed as intermediate stages in the disease processes of any type of diabetes (1, 2 or others).** IFG and IGT are associated with obesity (especially abdominal or visceral obesity), with high triglycerides and/or low HDL cholesterol, and hypertension. Structured lifestyle intervention, aimed at increasing physical activity and producing 5–10% loss of body weight, and certain pharmacological agents have been demonstrated to prevent or delay the development of diabetes in people with IGT; the potential impact of such interventions to reduce mortality or the incidence of cardiovascular disease has not been demonstrated to date.

Several prospective studies demonstrate a strong, **continuous** association between glucose measures (FPG, 2h OGTT, hemoglobin A1c) and subsequent diabetes (i.e., the cutoffs for IFG and IGT are arbitrary). When recommending the use of the A1C to diagnose diabetes in its 2009 report, the International Expert Committee stressed **the continuum of risk for diabetes with all glycemic measures** and did not formally identify intermediate category for A1C that is equivalent to IFG or to IGT. However, they note that people with A1C at the high end of the normal range (5.5–6.0%) have a 5-year cumulative incidence of diabetes that is x3–x8 higher than the population as a whole. Above the laboratory "normal" range but below the diagnostic cut point for diabetes (6.0 to 6.5%) people are at very high risk of developing diabetes (x10 times that of people with lower levels). However, the 6.0 to 6.5% range fails to identify a substantial number of patients who have IFG and/or IGT. The cutoff above which preventive interventions are recommended is 5.5% (within the "normal range").

Can/should we attempt to translate this to fructosamine measurements? Or should A1C be reintroduced to veterinary medicine?

Classification of Diabetes Mellitus in Veterinary Medicine

The classification system of IDDM and NIDDM was adopted by veterinary medicine in the 20th century and with time, type 1 and type 2 were adopted as well, replacing IDDM and NIDDM. However, type 1 and type 2 are essentially still used in veterinary medicine as equivalent for IDDM and NIDDM, with disregard to etiopathogenesis.

What Do We Know About the Etiology and Pathogenesis of DM in Cats and Dogs and How Does It Fit with the Current Classification of DM in People?

Applying 20th century classification, practically 100% of dogs are IDDM, the majority of cats are IDDM, and the minority of cats is NIDDM. However, what are the causes of DM in dogs and cats and how does it affect the way we treat them? Would better understanding of the etiopathogenesis of DM in dogs and cats change the way we treat them in the near future?

DM in Dogs: Type 1?

Diabetes mellitus in dogs is most commonly associated with permanent hypoinsulinemia, no increase in C-peptide in response to insulin secretagogues, and an absolute necessity for exogenous insulin administration to avoid ketoacidosis.[5] This presentation is consistent with type 1 DM but it is also consistent with most other types of DM depending on stage of disease and depending on the degree of glucose toxicity. For example, this presentation is consistent with DM secondary to pancreatic disease, especially if the initial injury for beta cells is aggravated by inability to completely control hyperglycemia and resulting glucotoxicity. Dogs are sensitive to glucose toxicity and in its presence can become diabetic despite having a beta cell mass that was previously sufficient to maintain euglycemia.[6,7] **Thus, the importance of searching for a specific diagnosis of DM in dogs is in that it obviates the assumption that upon diagnosis of DM all dogs are at the end stages of an irreversible process (as they would be if they suffer from auto-immune type 1 DM).** Importantly, the hallmarks of type 1 DM in people are NOT present in the majority of dogs with diabetes. **There is evidence of cellular-mediated autoimmune destruction of the b-cells in about 50% of diabetic dogs or less, depending on the study.**[8-11] In a recent study there was lack of evidence for a role of islet autoimmunity in 121 dogs of 40 different breeds.[12] In that study, histopathological evaluation was performed in only 5 dogs but none of these had lymphocytic (or other) inflammation in pancreatic islets. Previously, infiltrating mononuclear cells, predominantly lymphocytes, were observed in 6 of 13 dogs (46%) with diabetes while in 5 of 18 dogs (28%), extensive pancreatic damage appeared to be responsible for the development of DM.[11] Based on these studies, it seems like **at least 50% of diabetic dogs have etiologies other than with immune-mediated type 1. What is the most likely etiopathogenesis of DM in the other 50%?**

As in people, there is an association between pancreatitis (acute and chronic) and DM in dogs.[13-17] In dogs and in people, it is difficult to determine a cause and effect relationship between the two and it is also possible that both result from the same primary disease process. Importantly, as mentioned above, concurrent or resulting pancreatitis can be the perpetuating process that causes an otherwise transient insult to beta cells to become permanent. If this is the case in a proportion of diabetic dogs, it would be possible to try and prevent establishment of a diabetic state by intensive protocol for glycemic normalization around the time of diagnosis. Another inconsistency between canine DM and type 1 DM is the association between canine DM and being overweight.[16] In an earlier study only 28 of 118 dogs presented for DKA were underweight and only 45 were treated with insulin for more than 1 week (i.e., insulin therapy could have explained normal-increased body weight in no more than half of the dogs).[13] While it is unlikely that insulin resistance *per se* (caused by being overweight) is the cause of DM in dogs, it is likely that in overweight dogs, especially those with acute pancreatitis and no historical manifestations of DM (no chronic PU/PD, etc.), an etiology other than auto-immune type 1 DM is present.

It has been previously suggested that canine DM is similar to type 1 DM in the presence of susceptibility and protective major histocompatibility complex haplotypes in canine diabetes mellitus.[18] This candidate gene approach has been criticized however because it runs a high risk of false positives in dogs, and the DLA locus may be particularly susceptible to false associations created by popular sires, inbreeding, or genetic drift.[12] Further studies are necessary to confirm these findings.

Precise diagnosis early in the course of a disease might be important in slowing down or preventing progression. Immune modulation prior to extensive loss of beta cell mass could become a viable preventive strategy in the near future and will require accurate and early diagnosis.[19] However, in dogs, precise diagnosis might have crucial implications also on treatment strategies in the established diabetic. Most importantly, the risk of life-threatening hypoglycemia might be very different in animals suffering from an isolated disease of pancreatic beta cells vs. animals suffering from a disease process involving other cell types in the islet of Langerhans. Decreased glucagon secretion, for example, could lead to increased susceptibility to relative overdosing with exogenous insulin and may necessitate a more conservative approach to treatment. Alternatively, pancreatic polypeptide deficiency might affect the sensitivity of the liver to insulin. In dogs, the liver is a primal site for developing insulin resistance[20] and the response to insulin might differ significantly between dogs with intact secretion of pancreatic polypeptide and dogs with dysfunctional secretion of this hormone.

DM in Cats: Type 2?

Based on clinical presentation, the epidemiology of the disease and the association with amyloid deposition in the islet of Langerhans, it has been previously suggested that the majority of diabetic cats suffer from a type-2-like disease.[5,21] As in type 2 DM in people, obesity and physical inactivity are major risk factors for feline DM. Also similarly, diabetic cats are often older, insulin-independence is not uncommon and genetic predisposition has been demonstrated in cats as in people.[22] Based on these and considering that type 2 DM is used in people as an umbrella diagnosis for unknown etiologies of DM, type 2 might be an appropriate category for most diabetic cats. It is important to recognize however that when a disease like hypersomatotropism is diagnosed, type 2 is, by definition, excluded. Correct classification in those cases can have important therapeutic, prognostic, and epidemiological consequences. When it was first suggested that feline DM is mostly similar to human type 2 DM, hypersomatotropism was considered a rare disease in diabetic cats.[21] However, recent studies suggest that hypersomatotropism is much more frequent than previously thought.[23] In contrast to type 2 DM, it is questionable whether a primary defect in beta cell function is present in hypersomatotropism. From a clinical perspective, drugs that improve beta cell function such as GLP-1 receptor agonists might be irrelevant in DM secondary to hypersomatotropism. From a research perspective, differentiating between these two types might be crucial for future studies that attempt to clarify the genetic predisposition and etiology of DM in cats. As mentioned above, precise diagnosis early in the course of a disease might also be important in slowing down or preventing progression.

References

1. American Diabetes Association. *Diabetes Care*. 2013;36(Suppl 1):S67–74.

2. Porte D Jr., Sherwin RS, Rifkin H, eds. *Ellenberg and Rifkin's Diabetes Mellitus*. 5th ed. 1997.
3. Porte D Jr., Sherwin RS, Baron A, eds. *Ellenberg and Rifkin's Diabetes Mellitus*. 6th ed. 2003.
4. Hameed S, *et al. Pediatr Diabetes*. 2011;12(3 Pt 1):142–9.
5. Nelson RW, Reusch CE. *J Endocrinol*. 2014;222(3):T1–9.
6. Imamura T, *et al. Diabetes*. 1988;37(5):600–9.
7. Fall T, *et al. J Vet Intern Med*. 2010;24(6):1322–8.
8. Hoenig M, Dawe DL. *Vet Immunol Immunopathol*. 1992;32(3–4):195–203.
9. Davison LJ, *et al. J Vet Intern Med*. 2008;22(6):1317–25.
10. Davison LJ, *et al. Res Vet Sci*. 2011;91(1):58–63.
11. Alejandro R, *et al. J Am Vet Med Assoc*. 1988;193(9):1050–5.
12. Ahlgren KM, *et al. PLoS One*. 2014;9(8):e105473.
13. Hume DZ, *et al. J Vet Intern Med*. 2006;20(3):547–55.
14. Watson PJ, *et al. Vet Rec*. 2010;167(25):968–76.
15. Papa K, *et al. Acta Vet Hung*. 2011;59(1):37–52.
16. Mattin M, *et al. Vet Rec*. 2014;174(14):349.
17. Bostrom BM, *et al. Vet J*. 2013;195(1):73–9.
18. Kennedy LJ, *et al. Tissue Antigens*. 2006;68(6):467–76.
19. van Belle TL, *et al. Physiol Rev*. 2011;91(1):79–118.
20. Kim SP, *et al. Diabetes*. 2003;52(10):2453–60.
21. Rand J. *J Feline Med Surg*. 1999;1(3):143–53.
22. Forcada Y, *et al. J Vet Intern Med*. 2014;28(2):458–64.
23. Niessen SJ. *J Feline Med Surg*. 2010;12(1):15–23.

acvim429

How Pain & Stress Impact the Hospitalized Patient

Sara Bennett, DVM, MS, DACVB, DACVN; Tammy Grubb, DVM, PhD, DACVAA;
Jennifer E. Slovak, DVM, MS, DACVIM
Pullman, WA, USA

Introduction

Currently in human medicine, a more patient-centric approach to patient care and comfort is becoming more intensely studied. This has been done by redesigning the environment of the patient including; hospital waiting, examination and treatment rooms, and by thoughtfully implementing pain assessments during physical examinations for both verbal and non-verbal people.

In veterinary medicine, there is evidence-based literature to support the statement that the stress and pain of an animal affects their physiologic parameters. Fortunately, just as in human medicine, there has been efforts into improving patient comfort and assessment during veterinary visits and hospitalization. However, additional work and improvement in assessing our veterinary patients still needs to be done.

Most veterinarians would agree that being pain and fear free is a fundamental measurement of quality of life, not only for ourselves, but for all living things. However, many of us feel inadequate when making these assessments for our patients. We are taught only the mere basics of pain and behavior in veterinary school. From there, an individual typically enters private practice or continues with a selected avenue of veterinary training to become specialists. Unfortunately, although many of us become experts in our respective fields, few of us re-evaluate our entire approach to a patient. This includes not only taking a thorough history and performing a complete physical examination, but the recognition and documentation of our patients' anxiety and pain levels. We become so fixed on the presenting complaint and the pace/cadence of our workday, that the patients' behavioral and pain assessment seem to become secondary or tertiary measures.

Discussion

We are challenged as veterinarians and veterinary caregivers in that our patients do not precisely articulate their stress and discomfort, this is not unlike pediatricians assessing their patients, or from physicians interacting with nonverbal adult patients. Therefore, certain objective and subjective assessment tools and parameters have been largely adopted from human medicine. The challenge lies in discerning pain from stress or anxiety. Unfortunately, many of the signs are similar, as they are sympathetically driven, affecting heart rate, temperature, and blood pressure to name a few.

Behavior

Therefore, initially creating a relaxed and calm environment for our patients is paramount to our assessment of them. The implementation of "seeing" things from the animal's perspective is now widely accepted, especially after Dr. Temple Grandin's groundbreaking research and publication in feedlot animals. We have adopted several of these principles in small animal medicine by implementing separate waiting and exam room spaces for dogs and cats and with additional work done by Dr. Sophia Yin and others. They have encouraged the veterinary community to approach and treat our small animal patients safely and efficiently. Goals which we all share as practitioners or specialists.

Research in sheltered or cage-housed cats has begun to show evidence of the association between risk of illness, strength of the immune system, and level of our feline patient's anxiety and frustration. For example, Stella and Buffington demonstrated an association between sickness behaviors and change in routine or interactions that would increase stress in cats. Additionally, studies from Gourkow have demonstrated associations between the level of secretory IgA measured and the stress level of cats. Cats that were more anxious and/or frustrated had lower IgA levels and were at an increased risk of developing upper respiratory infections. By going back and looking at these studies further, some of the conclusions can shed light on how we can treat and interact with our feline patients to help reduce their stress as much as possible. It can be extrapolated that our hospitalized patients that are stressed are less likely to have optimal immune strength, and are at increased risk for sickness behaviors or respiratory disease that could prolong their time to recovery.

Canine, and even more so feline, body language is an ever-changing continuum and evokes many subtle messages that are easily missed to the inexperienced observer. Additional interest has been paid to tools that have been used to measure stress levels in cats through behavioral observation. Though helpful in many situations, each still have drawbacks. However, the overarching goal of understanding of patient body language and the emotional state underlying different body postures and behaviors, should not be lost. Tools that use easily observable behaviors or body postures can be used to help identify patient stress and a negative emotional state. The first step in patient wellbeing is changing the environment and our own behavior to reduce stress in name of overall patient wellbeing.

Pain

As previously stated, recognizing an unpleasant sensory, emotional, and psychological experience in our veterinary patients is confusing at best. As pain is best described as multidimensional, we must recognize that simple everyday practices, or "unintentional pain" may be overlooked. Dogs and especially cats are notorious for masking their pain responses, especially in a stressful environment. Therefore, careful and deliberate individual pain assessment in our patients is absolutely necessary.

Pain recognition is important to the overall health and healing of the patient. It has been designated as the 'fifth vital sign' in human medicine. Pain initiates a profound stress response and a sympathetic overdrive. Stress and autonomic imbalance are not benign and the cascade of side effects include gastrointestinal (GI) ileus, GI ulceration, clotting dysfunction, hypertension, tachycardia, tachyarrhythmias, and many others. Furthermore, stress and pain cause a marked increase in cortisol release and a substantial increase in energy requirements, the latter of which may lead to a negative nitrogen balance, both of which impair healing. Clearly, pain compounds the negative effects of the initial disease, trauma, or surgical condition and treatment of pain should be as integral to therapeutic protocols as any other disease modifying drug.

Conclusion

In the preceding decades, we have made profound advancements in technology and therapeutics in veterinary medicine. Perhaps, it is time to reconsider the patient as an individual with unique and specific needs for their wellbeing versus a "case." We as clinicians should make every effort to discern between the signs of pain, stress, and anxiety and try to minimize those variables in order to better care for our patients.

References

1. Berry SH. Analgesia in the perioperative period. *Vet Clin North Am Small Anim Pract*. 2015:45(5):1013–1027.
2. Bragg R, Bennett J, Cummings A, Quimby JM. Evaluation of the effects of hospital visit stress on physiologic variables in dogs. *J Am Vet Med Assoc*. 2015;246:212–215.
3. Closs SJ, Dowding D, Allcock N, *et al*. Toward improved decision support in the assessment and management of pain for people with dementia in hospital: a systematic meta-review and observational study. *Health Services and Delivery Research*. 2016;4:1–192.
4. Ellis S, Rodan I, Carney H, *et al*. AAFP and ISFM feline environmental needs guidelines. *J Feline Med Surg*. 2013;15:219–230.
5. Gourkow N, Hamon SC, Phillips CJ. Effect of gentle stroking and vocalization on behavior, mucosal immunity and upper respiratory disease in anxious shelter cats. *Prev Vet Med*. 2014;117:266–275. http://dx.doi.org/10.1016/j.prevetmed.2014.06.005
6. Hammerle M, Horst C, Levine E, *et al*. 2015 AAHA canine and feline behavior management guidelines. *J Am Anim Hosp Assoc*. 2015;51:205–221.
7. Kessler MR, Turner DC. Stress and adaptation of cats (*Felis silvestris catus*) housed singly, in pairs and in groups in boarding catteries. *Anim Welf*. 1997;6:243–254.
8. Manworren R, Stinson J. Pediatric pain measurement, assessment, and evaluation. *Semin Pediatr Neurol*. 2016;23:189–200.
9. Merola I, Mills D. Systematic review of the behavioral assessment of pain in cats. *J Feline Med Surg*. 2016;2:60–76.

10. Minton C, Batten L. Rethinking the intensive care environment: considering nature in nursing practice. *J Clin. Nurs*. 2015;25:269–277.
11. Murrell JC, Psatha EP, Scott EM, Reid J, Hellebrekers LJ. Application of a modified form of the Glasgow pain scale in a veterinary teaching centre in the Netherlands. *Vet Rec*. 2008;162:403–408.
12. Stella JL, Lord LK, Buffinton CAT. Sickness behaviors in response to unusual external events in healthy cats and cats with feline interstitial cystitis. *J Am Vet Med Assoc*. 2011;238:67–73.

acvim433

Diagnosing & Managing the Painful Medical Patient

Tammy Grubb, DVM, PhD, DACVAA; Jennifer E. Slovak, DVM, MS, DACVIM
Pullman, WA, USA

Introduction

Veterinary patients are brought to a hospital for a variety reasons including; annual healthy pet examinations, a detected physical abnormality, surgery, or for suspicious clinical signs observed by their caregiver. Our initial challenge as a veterinarian is to assimilate the following; historical and current patient history, data from the physical examination and our own clinical observations, to then address and treat the pet. This can be daunting. Simply stated, our goal is to improve the health and well-being of our patients to the best of our abilities. As we practice veterinary medicine in the 21st century, most of us agree that pain management is an important aspect of patient care. However anticipating, recognizing, and implementing appropriate pain management strategies for our patients is challenging.

Unfortunately, pain and its interpretation is subjective. The gold standard of pain assessment in human medicine, is to self-report the extent, quantity, and degree of pain experienced by the patient. However, self-reporting is not possible in our veterinary patients. Instead, we rely on the observed patient pain assessment by the caregiver, technician, and/or doctor. Fortunately, pain assessment scales have existed in human medicine for decades, and have also been recently developed and validated in veterinary medicine to help us better evaluate and treat pain in our patients.

Discussion

Pain can be easily masked in animals. Hiding pain is a protective mechanism for many species. However, it is important to appreciate that lack of overt outward expression of pain does not negate it. Observing and documenting an animal's normal behavior may be more helpful in pain recognition in certain circumstances than a patient's expression. Some of these signs that can be a significant indicator of discomfort include; appetite, grooming habits, and mobility. These behaviors are complex and are patient dependent. In addition to a sensory component, pain has an emotional impact on the patient. Unmanaged pain can lead to weight loss, ileus, an altered immune status and changes in physiologic parameters.

Pain can present itself in many forms; somatic, visceral, neuropathic, acute, recurrent, chronic, or even acute on chronic. We have undertreated pain in our pets for decades, and although pain awareness is becoming more prevalent, we still fall short. This is likely the result of lack of realization and recognition of the different manifestations of pain in pets, and lack of a universally implemented and validated pain scale. Initially, the pain scales used in animals were simple and descriptive. However, those scales were deemed unreliable as they relied heavily on only one pain dimension, intensity. Currently, multidimensional or composite rating pain scales are favored in order to assess certain spontaneous/evoked behaviors, interactions with the animal, and clinical observations. This type of pain scale allows a more complete assessment of pain including sensory and affective qualities.

It is important to note that pain scoring should be done not only in association with painful procedures but should also be a part of the initial physical exam as many patients present with either pain from their disease or underlying pain due to conditions like osteoarthritis. When pain has been identified, or anticipated prior to painful procedures, appropriate therapeutics should be implemented. Because the source/type/location of pain can vary greatly, specific recommendations are difficult to make in an overview but generalizations regarding principles of pain management and suggested analgesic drug classes are included.

For most pain syndromes, especially if pain is moderate to severe, a variety of analgesic drugs, techniques and routes of administration (multimodal or balanced analgesia) should be utilized. This capitalizes on the additive or synergistic effects of analgesic drugs and allows us to provide analgesia that is more intense and/or of longer duration than analgesia provided with any one drug used alone. Typical treatment protocols for patients presented to the internal medicine service could include opioids, local anesthetic drugs, NSAIDs (if applicable), adjunctive drugs like ketamine, maropitant, gabapentin, and varied delivery techniques like local/regional blockade and constant rate infusions (CRIs).

Pain should be addressed not only while a patient is being hospitalized for their medical condition, but even after the patient has been discharged from the hospital. Many veterinarians feel that animals do not need analgesic drugs once they have left the hospital because the patients tend not to exhibit pain at home. However, we know that animals instinctively hide pain and that pain, even from elective procedures, does not go away once the animal is no longer in the hospital. Instead, the pain dissipates gradually over a period of days to weeks (depending on the severity of the disease, injury, or surgery) and the pain that the animal experiences in that time should be addressed. Gabapentin, NSAIDs, and long-acting opioids like Simbadol® (buprenorphine) may be good options, depending on the source of pain and the underlying health conditions of the patient.

Conclusion

Pain is a multi-dimensional and complex phenomenon. Many animals mask their pain, especially in stressful environments. Fortunately we have many options for pain relief for patients, which allows us to individualize their treatment. We, as veterinary care providers must be conscientious and diligent in the pain assessments of our patients. This can be accomplished by adequately assessing and documenting our patient's pain scores, providing training for our staff and owners in making pain assessments, frequently re-evaluating the patient's pain status and implementing the appropriate pain intervention for the duration of the pain and patient's needs.

References

1. Barratt L. Feline pain assessment and scoring systems. *The Vet Nurse*. 2013;(4):471–477.
2. Berry SH. Analgesia in the perioperative period. *Vet Clin North Am Small Anim Pract*. 2015;45(5):1013–1027.
3. Boscan P, Monnet E, Mama K, Twedt DC, Congdon J, Steffey EP. Effect of maropitant, a neurokinin 1 receptor antagonist, on anesthetic requirements during noxious visceral stimulation of the ovary in dogs. *Am J Vet Res*. 2011:72(12):1576–1579.
4. Brondani JK, Mama KR, Luna SP, *et al*. Validation of the English version of the UNESP-Botucatu multidimensional composite pain scale for assessing postoperative pain in cats. *BMC Vet Research*. 2013;(9):1–15.
5. Closs SJ, Dowding D, Allcock N, *et al*. Toward improved decision support in the assessment and management of pain for people with dementia in hospital: a systematic meta-review and observational study. *Health Services and Delivery Research*. 2016;(4):1–192.
6. Felix RH, Almeida CB, Cremaschi RC, *et al*. Pain-induced depression is related to overestimation of sleep quality in a very elderly population with pain. *Arq Neuropsiquiatr*. 2017;(1):25–29.
7. Mansfield C, Beths T. Management of acute pancreatitis in dogs: a critical appraisal with focus on feeding and analgesia. *J Small Anim Pract*. 2015;56(1):27–39.
8. Manworren R, Stinson J. Pediatric pain measurement, assessment, and evaluation. *Semin Pediatr Neurol*. 2016;23:189–200.
9. Merola I, Mills D. Systematic review of the behavioral assessment of pain in cats. *J Feline Med Surg*. 2016;18(2):60–76.
10. Murrell JC, Psatha EP, Scott EM, Reid J, Hellebrekers LJ. Application of a modified form of the Glasgow pain scale in a veterinary teaching centre in the Netherlands. *Vet Rec*. 2008;162:403–408.
11. Niyom S, Boscan P, Twedt DC, Monnet E, Eickhoff JC. Effect of maropitant, a neurokinin-1 receptor antagonist, on the minimum alveolar concentration of sevoflurane during stimulation of the ovarian ligament in cats. *Vet Anaesth Analg*. 2013;40(4):425–431.
12. Pieper MJ, van der Steen JT, Francke AL, Scherder EJ, Twisk JW, Achterberg WP. Effects on pain of a stepwise multidisciplinary intervention (STA OP!) that targets pain and behavior in advanced dementia: A cluster randomized controlled trial. *Palliat Med*. 2017;1–11.

acvim434

Pharmacology for Non-pharmacologists

Claire L. Fellman, DVM, PhD, DACVIM, DACVCP
North Grafton, MA, USA

Introduction

Pharmacology is an important branch of medicine with the potential to improve patient care through rational drug selection and recognition of drug interactions. This presentation will review important concepts in pharmacology including applying information from pharmacokinetic studies and principles of drug interactions, and will introduce Freedom of Information summaries as an additional resource for more detail on studies performed during drug approval.

Principles

Pharmacokinetics (PK) evaluates the concentration of drugs in the body, and is an effort to mathematically describe the physiological processes involved in drug absorption, distribution, metabolism, and excretion (ADME).[1] A general understanding of pharmacology principles is required to best use information presented in pharmacokinetic studies. The first important concept is that of first-order versus zero-order processes. In the context of elimination, zero-order elimination implies that a constant **amount** of the administered substance is eliminated in a given time period, while first-order elimination implies that a constant **proportion** of the substance is eliminated during that time. This is the difference between 20 mg/dL being removed (zero order) versus 20% of the total amount (first order) in a given time interval. Almost all commonly used medications are absorbed, distributed, and eliminated by first-order processes, meaning that these processes occur in proportion with the concentration gradient present.[1] In contrast, zero-order processes most often occur due to saturation. Examples of zero-order processes include the elimination of large amounts of alcohol, and other drugs once they have reached saturation of the metabolizing enzyme or transporter. Once below the saturation point, zero-order processes can then become first order.

The next concept is dose proportionality, which states that administering a multiple of the drug dose will result in a multiple of that drug blood concentration. This is the reason proportional dose adjustments can be performed during therapeutic drug monitoring (TDM) for many agents, and relies on first-order processes. Similarly, the elimination half-life is also due to first-order metabolism and excretion and is the time for the blood drug concentration to decrease by 50%. Half-life is not valid if processes are not first order. Five half-lives are required for drug levels to reach steady state, including when starting a medication and after making a dose adjustment.

The therapeutic index is a value that reflects the safety of a drug. It is calculated by dividing the concentration to produce an **adverse effect** in 50% of the population by the concentration to produce the **desired effect** in 50% of the population,[2] and is also considered qualitatively as a margin of safety. The larger this value, the less likely individual variations in PK are to cause an adverse event from a higher blood concentration than anticipated. This is distinct from the therapeutic range, however, which states the target blood concentrations for a drug to have a reasonable expectation of efficacy. Within the therapeutic range patient response can still be variable, and some animals will require blood concentrations at the high end of the range to achieve a clinical response, while other patients will experience toxicity at higher levels and require dose reduction to the lower end of the range. A low therapeutic index or narrow therapeutic range are two indications that TDM may be needed to ensure appropriate dosing. Drugs with variable bioavailability can also be candidates for TDM, especially if either of the other two factors are present. The therapeutic index and range are also considered when determining whether dose adjustments are indicated when systemic disease is present, most commonly with decreased hepatic or renal function.

PK DEFINITIONS & APPLICATION

It is important to remember that there are two types of modeling that can be performed when evaluating a pharmacokinetic study: compartmental and noncompartmental. Compartmental modeling refers to creating an equation to best fit the blood drug concentration curve, and can have multiple "compartments" depending on the distribution of the drug. Compartmental modeling is required to predict blood concentrations at times other than those sampled, and results in an exponential equation. An example is the generic two-compartment intravenous administration model equation:

$$C_p = A_1 e^{-\lambda_1 t} + A_z e^{-\lambda_z t}$$

Noncompartmental modeling still assumes first-order kinetics, but determines the area under the curve mathematically by adding the area of numerous small trapezoids. The elimination half-life can also be directly calculated from the graph.

When evaluating PK studies, parameters of greatest interest are often the peak blood concentration (Cmax), time to peak blood concentration (Tmax), area under the drug time-concentration curve which reflects drug exposure (AUC), and drug half-life ($T_{1/2}$). All of these parameters can be determined using only noncompartmental modeling. These values can have direct clinical application if a pharmacodynamic outcome is associated with a given PK parameter in that species. Pharmacodynamic (PD) outcomes reflect the biological response to a given drug concentration, and are often, at least initially, extrapolated from results in other species. Caution must be exercised, however, since responses can vary substantially by species. An example is the marked sensitivity of cattle to xylazine.[3]

Target blood concentrations for many agents are known, and using the information from a PK study, proportional dose adjustments can be performed (assuming first-order processes, as described above), time to drug washout can be evaluated (recognizing 97% of a drug is eliminated by 5 half-lives), and dosing regimens can be evaluated. Drugs with shorter half-lives and a narrow therapeutic range will require more frequent administration, while drugs with a long biological effect (such as benazepril, which inhibits angiotensin-converting enzyme for significantly longer than its PK half-life)[4] can be administered less frequently. Information from PK studies is particularly helpful in designing antimicrobial treatment regimens, where PD targets are better characterized and involve maintaining time above the bacterial minimum inhibitory concentration (MIC) for beta-lactams, obtaining Cmax as a multiple of the MIC for aminoglycosides, and having the AUC be a multiple of the MIC for fluoroquinolones.[5,6]

Understanding a few more PK concepts will allow calculation of loading doses and constant rate infusions. The first of these is volume of distribution (V). The volume of distribution is a theoretical value that reflects the apparent volume the drug distributes into to achieve the plasma concentration measured, and is calculated as the amount of drug in the body divided by the plasma concentration at that time. Simplistically, if the drug has extensive tissue binding, the drug concentration in the plasma will be lower than if the same dose of drug is limited to the vascular space. In the former case, the drug concentration would be low with a high V, while the latter case would have a high plasma concentration but low V. There are several volumes of distribution depending whether the calculation is evaluated immediately after administration when the drug is in the central compartment (V_c), at steady state (V_{ss}), or in the terminal phase (V_{area}). Values for V can range from being limited to the blood volume (0.05 L/kg), distributing to the extracellular fluid or all body water (0.3–0.6 L/kg), to distributing to all tissues and having extensive tissue binding (1 to >250 L/kg). It is important to remember, however, that merely having a high volume of distribution does not mean greater tissue concentrations at the site of action. It is important to evaluate whether the drug concentrates in the tissue of interest if the target is intracellular. If the target is extracellular, only the plasma free drug concentration is relevant as this is what equilibrates with interstitial fluid.

Total body clearance (Cl) represents the amount of plasma cleared of a drug per unit time, and reflects a combination of metabolism and elimination by the liver, kidney, and other routes (e.g., saliva, sweat). Clearance is challenging to directly measure, however, and is more commonly calculated using the results of other parameters.

Bioavailability (F) reflects the amount of the drug available to the body if given by a route other than intravenously. Bioavailability calculations are made by evaluating the AUC achieved for a given oral dose, compared to a given intravenous dose. Drugs with high first-pass hepatic metabolism will have low F values, as will drugs that have poor absorption.

There are many equations to describe the interrelationships between PK parameters. To add confusion, often the same parameter can be calculated using information from compartmental or noncompartmental modeling. Parameter values are often shown in pharmacokinetic papers, and can be applied to calculate loading doses and constant rate infusions (CRIs):

$$Loading\ dose = \frac{Desired\ plasma\ concentration * V_{ss}}{F}$$

$$Infusion\ rate = Desired\ plasma\ concentration * Cl$$

For agents with particularly long half-lives or where rapidly achieving target blood concentrations is therapeutically advantageous (such as with the antiepileptic medications potassium bromide and phenobarbital), loading doses allow patient blood concentrations to reach the therapeutic range rapidly. When it is desirable to maintain that blood concentration continuously, a loading dose can then be followed by the above calculated infusion rate, only replacing as much drug as is removed by clearance. CRIs are useful for sedation, pain control, and for some antibiotic regimens, including beta-lactams where the goal is to maintain concentrations above the MIC for an organism.[7]

EFFECTS OF DISEASE

The impact of present disease processes and concurrent medications should be considered carefully for each drug prescribed, especially those with narrow therapeutic indexes and ranges. The presence of liver or kidney failure can either prolong drug action if the drug is primarily metabolized or excreted through one of these organs, or alternatively, decrease drug effect if, for example, hepatic metabolism is needed to produce an active metabolite. If an agent undergoes both hepatic and renal clearance, often the unaffected organ is able to maintain drug elimination. Unfortunately, metabolism across species can differ significantly, so the ramifications of changes in organ function are often incompletely understood, and the clearest recommendations for alterations in dosing strategies are drawn from human medicine.

The World Health Organization Model Formulary (last updated in 2008) has a list of medications to consider dose adjustments for if hepatic or renal impairment is present.[8] For drugs with primarily renal elimination, either dose reduction or interval extension can be considered, depending whether smaller variations in blood drug concentrations are preferred, or if achieving a peak concentration is more important, respectively. Doses are often adjusted using either creatinine clearance, or serum creatinine as a surrogate. For dose adjustment, a standard first dose is administered to reach target blood concentrations, and subsequent doses are reduced. For interval adjustment, the standard interval is used for the first two doses, with subsequent intervals adjusted.

Kf (dose fraction) = patient GFR/normal GFR OR normal creatinine/patient creatinine
Dose adjustment = normal dose * Kf
Interval adjustment = normal interval/Kf

If renal excretion is less than 100%, Kf in the equations above is more correctly replaced by [f(Kf-1)+1], where f = fraction renally excreted.[9]

Liver disease poses a particular challenge because there is not a universal indicator of liver dysfunction that equates to impaired drug metabolism. Thus, even drugs with primarily hepatic metabolism can often be used successfully with hepatic disease, as demonstrated by the effective conversion of prednisone to prednisolone in humans with chronic active hepatitis.[10] Recommendations for hepatic disease are, therefore, somewhat arbitrary and require close evaluation of the role of the liver in drug metabolism and the margin of safety of the drug. For medications with majority hepatic metabolism, drug avoidance or dose reduction is generally recommended, with a 25–50% dose reduction suggested by one veterinary author.[8,11]

Adverse Drug Effects and Interactions

Drug interactions and adverse effects are frequently the result of competition for, or modification of, the function of drug metabolizing enzymes such as the cytochrome P450 (CYP) system, or transporters such as P-glycoprotein. Several medications are known inducers or inhibitors of the different enzymes of the CYP system and have the potential to affect other medications metabolized by these enzymes. Breed differences in CYP enzymes are also beginning to be identified.[12] Recognized CYP inducers in dogs include phenobarbital, rifampin, and omeprazole. CYP inhibitors in dogs include the azole antifungals (especially ketoconazole), cimetidine, chloramphenicol, and fluoroquinolones.[13] Less is known about CYP enzymes in cats, but differences in the function of other enzyme systems, such as those involved in glucuronidation and glycine conjugation are recognized to cause the differences in feline metabolism of acetaminophen, aspirin, and other compounds like benzoic acid, not similarly toxic to dogs.[14]

P-glycoprotein is the best studied drug transporter, involved in maintenance of the blood-brain barrier and in biliary excretion. Collies and herding breeds with the ABCB1 mutation have allowed study of the effects of P-glycoprotein deficiency, which allows substances like ivermectin and loperamide to concentrate in the central nervous system and impairs biliary excretion of substrates like doxorubicin and vincristine. Similar problems can arise with co-administration of P-glycoprotein inhibitors and P-glycoprotein substrates. Spinosad and ketoconazole have the most evidence supporting the potential for drug-drug interaction due to P-glycoprotein inhibition in dogs, though other drugs including cyclosporine, erythromycin, verapamil, and fluoxetine have also been shown to be P-glycoprotein inhibitors in humans or rats.[15] Less well known differences in the transporter ABCG2 (breast cancer resistance protein) in cats are suggested to be responsible for the retinotoxicity caused by fluoroquinolones in this species.[16] Organic anion transporters (OATs) in the renal tubular epithelium are also an increasingly recognized cause of drug interactions. OATs allow uptake of substrates from the blood into renal epithelial cells to facilitate the clearance of many compounds including steroid hormones, angiotensin-converting enzyme inhibitors, and loop and thiazide diuretics. OAT inhibitors include some antibiotics and non-steroidal anti-inflammatory drugs and can decrease the secretion of OAT substrates.[17] For example, NSAID co-administration with methotrexate, an OAT substrate, causes increased myelosuppression and nephrotoxicity from methotrexate in humans.[18,19] More study is needed to better understand drug interactions and species differences due to alterations in enzyme and transporter function.

Bioequivalence

According to the Food and Drug Administration, "Two products are considered to be bioequivalent when they are equally bioavailable; that is equal in the rate and extent to which the active ingredient(s) or therapeutic ingredient(s) is (are) absorbed and become(s) available at the site(s) of drug action."[20] Bioequivalence is most often assessed during generic drug approvals, though it is also important with changes in formulation. If a product is considered bioequivalent, then safety and effectiveness data for the original "pioneer" product are able to be applied to the generic or new formulation. To assess bioequivalence, Cmax and AUC are considered the pivotal parameters, and the 90% confidence interval for both parameters must fall within +/- 20% of the pioneer product (-20% to +25% if logarithmically transformed data are used). Bioequivalence is most commonly assessed using blood concentrations, but can also be less ideally evaluated using pharmacologic or clinical endpoints.[20]

Freedom of Information Summaries

Freedom of Information (FOI) summaries are prepared during every new and generic drug approval and include additional information on the safety and effectiveness studies mentioned in the product information sheet. These summaries can be very helpful when evaluating potential adverse events or accidental overdoses and are a valuable resource available to the public.

References

1. Riviere JE, Papich MG, eds. *Veterinary Pharmacology and Therapeutics*. 9th ed. 2009.
2. Boothe DM. *Small Animal Clinical Pharmacology and Therapeutics*. 2nd ed. 2012.
3. Hopkins TJ. *Aust Vet J*. 1972;48(3):109.
4. King JN, *et al. Am J Vet Res*. 1995;56(12):1620.
5. Drusano GL. *Clin Infect Dis*. 2007;45(Suppl 1):S89.
6. Llanos-Paez CC, *et al. J Antimicrob Chemother*. 2017;72:639.
7. Moore KW, *et al. Am J Vet Res*. 2000;61(10):1204.
8. WHO Model Formulary, 2008, www.who.int/selection_medicines/list/en/.
9. Shargel L, Yu A. *Applied Biopharmaceutics and Pharmacokinetics*. 3rd ed. 1993.
10. Uribe M, *et al. Clin Pharmacokinet*. 1982;7(5):452.
11. Trepanier LA. *Vet Clin North Am Small Anim Pract*. 2013;43(5):1013.
12. Court MH. *Vet Clin North Am Small Anim Pract*. 2013;43(5):1027.
13. Trepanier LA. *Vet Clin North Am Small Anim Pract*. 2006;36(5):975.
14. Court MH. *Vet Clin North Am Small Anim Pract*. 2013;43(5):1039.
15. Mealey KL, Fidel J. *J Vet Intern Med*. 2015;29(1):1.
16. Ramirez CJ, *et al. Pharmacogenet Genomics*. 2011;21(2):66.
17. Mealey KL. *Vet Clin North Am Small Anim Pract*. 2013;43(5):1067.
18. Furst DE. *Br J Rheumatol*. 1995;34(Suppl 2):20.
19. Masuda S, *et al. J Pharmacol Exp Ther*. 1997;283(3):1039.
20. FDA CVM GFI 35. *Bioequivalence Guidance*; 2006.

acvim435

The Role of Culture and Susceptibility Data in Minimizing Antimicrobial Resistance

Dawn Merton Boothe, DVM, PhD, DACVIM (Small Animal Internal Medicine), DACVCP
Auburn, AL, USA

INTRODUCTION

The Center for Disease Control indicates that antimicrobial use is the greatest risk factor for the advent of antimicrobial resistance. Although the most important approach minimizing resistance is de-escalation of use, once the decision is made to systemically treat and infection with systemic antimicrobials, the second most important approach is design of the dosing regimen. This includes both choosing the most appropriate drug (e.g., most effective, yet lowest tier), but also designing the dosing regimen such that resistance is minimized. The more "at risk" a patient is for therapeutic failure, the more important it is to remove the entire infecting inoculum ("dead bugs don't mutate"). Choosing the drug begins with identifying the organism, a step that is often made empirically. Yet, very few organs are infected consistently with a predictable organism. The exception might be skin and possibly UTI, but even for the latter, *E. coli* is the causative organism in only 50–65% of UTIs. The Target® Antimicrobial Handbook provides information regarding the most common **organisms** cultured from various tissues, but relevance to the cause of infection is not clear. As such, for complicated infections, or in patients at risk for therapeutic failure or developing resistance, identification of the infecting organism most appropriately should be based on **appropriately collected** cultures. Collection should avoid contamination (e.g., cystocentesis; deep tissue sample from wounds, etc.) and allow for sufficient tissue, ideally to allow for a CFU quantity count. C&S testing should be implemented by laboratories that follow the guidelines promulgated by the Clinical Laboratory Standards Institute (CLSI). CLSI determines which drugs to which an isolate will be tested, the range of concentrations tested and the interpretation of the MIC as to S, I or R. Determining to which drugs of interest an organisms is susceptible might be supported by population or patient C&S data. It is important to remember, however, that C&S is one of many tools that should guide therapy.

CULTURE AND SUSCEPTIBILITY TESTING

Two types of data may be available: agar gel diffusion data which provides simply a susceptible (S), (occasionally) intermediate (I) or resistant (R) or tube dilution data. The latter provides a minimum inhibitory concentration (MIC). The MIC is determined by exposing a standard amount of the isolated inoculum to wells containing twofold increasing concentrations of the drug of interest (i.e., 0.06, 0.125, 0.25, 0.5, 1, 2, 4, 8, 16, 32, etc.) mcg/ml. After a standard amount of time, the well that contains no visible growth and the lowest concentration of the drug contains the MIC (mcg/ml). If the drug is "bactericidal", the MIC may also be the minimum bactericidal concentration (MBC). For bacteriostatic drugs, the MBC will be several tube dilutions above the MIC. Presumably, the MIC is the concentration of drug that should be targeted in the patient (what you need). For agar gel diffusion procedures, the zone of inhibition that surrounds each drug containing disc correlates with MIC determined from tube dilution.

The concentrations tested in tube dilution procedures for each drug are determined by the CLSI and will be the same throughout the U.S. (Europe and other countries have similar standards setting organizations). Although the dilutions will always be (half) fractions or doublings of 1 mcg/ml, the actual range of concentrations tested vary with the drug and generally approach to the maximum drug concentration achieved in the plasma of the target species when treated with the drug at the recommended dose. For example, if the C_{max} of a drug is 24 mcg/ml, the highest concentration tested might be 16 mcg/ml and the lowest concentration might be 3 or 4 dilutions below (8, 4, 2 and perhaps 1). The number may be limited by the number of wells available on the "test card". Results often include entries such as ≤2 (for example for amikacin). This indicates that there was no growth in the well containing 2 mcg/ml and that was the lowest concentration tested; as such the MIC might be 2, but it could be lower. An entry of ≥ (e.g., ≥256 for ticarcillin) indicates that growth occurred in all wells and the highest concentration tested was 128 mcg/ml. The lowest the MIC could be is the next tube dilution which would be 256, but it may be higher.

E. coli UTI	MIC	Interp
Amikacin	≤2	S
Amoxi-Clav	≤8	S
Ampicillin	>0.5	R
Carbenicillin	≥512	R
Cephalothin	8	I
Ceftiofur	≤1	S
Chloramphenicol	≥32	R
Ciprofloxacin	<0.5	S
Enrofloxacin	≤0.25	S
Gentamicin	<0.5	S
Nitrofurantoin	≤32	S
Piperacillin	≥256	R
Tetracycline	≤	S
Ticarcillin	≥128	R
Tobramycin	<0.5	S
Trim-Sulfa	10	S

Susceptible vs. Resistant

The MIC alone does not provide an indication of susceptibility. For each drug on report, CLSI will have provide "breakpoints", the MIC above which an isolate is considered resistant and below which an isolate is considered susceptible. Breakpoints are based on population susceptibility (pharmacodynamic or MIC) data, pharmacokinetic data in the target species (i.e., C_{max}, AUC, half-life), and evidence that the breakpoints are clinically relevant. The resistant breakpoint (which must be a tube dilution) often approximates the C_{max} of the drug when given at the recommended dose. For example, the C_{max} of amikacin at the recommended dose of 22 mg/kg is 65 mcg/ml; any MIC above 32 mcg/ml is considered resistant because the next tube dilution is 64 mcg/ml and this is too close to the C_{max} for CLSI to consider the isolate susceptible. The resistant breakpoint for enrofloxacin is > 2 mcg/ml (or ≥4 mcg/ml); the C_{max} of enrofloxacin at 20 mg/kg is 4 mcg/ml. The further an isolate MIC is from the breakpoint (the further below), the more susceptible that isolate is to that drug.

Population Data

Both agar gel diffusion and tube dilution data collected from populations can be helpful. An antibiogram provides the % of isolates that are susceptible versus resistant during a given time period. The Target® Handbook is an example of national data (from one commercial laboratory); many academic environments generate local antibiograms which often deviate from national antibiograms. The more at risk a patient is to resistance, the more likely that patient will not be represented in the "susceptible" population. Tube dilution data is more useful. Population pharmacodynamics statistically represent the MIC of isolates collected from large (ideally > 100) sample populations. The low end of the range may reflect an isolate that has developed no resistance ("wild type"), but the high end (MIC 100) may represent outliers. The median and 75th percentile are useful to assess the distribution of the population but among the most useful parameters is the 90th percentile (MIC 90) which might be a reasonable surrogate MIC for a patient if the organism is known. MIC population statistics are available on many antimicrobial package inserts (see marbofloxacin data) and in literature; a study describing the pharmacokinetics of an antimicrobial in a sample target population is of little clinical relevance unless related to the MIC population statistics of potential pathogens towards which the drug is effective.

Table 4: MIC Values* (µg/mL) of marbofloxacin against pathogens isolated from skin, soft tissue and urinary tract infections in dogs enrolled in clinical studies conducted during 1994–1996.

Organism	No. of Isolates	MIC_{50}	MIC_{90}	MIC Range
Staphylococcus intermedius	135	0.25	0.25	0.125–2
Escherichia coli	61	0.03	0.06	0.015–2
Proteus mirabilis	35	0.06	0.125	0.03–0.25
Beta-hemolytic *Streptococcus*, (not Group A or Group B)	25	1	2	0.5–16
Streptococcus, Group D enterococcus	16	1	4	0.008–4
Pasteurella multocida	13	0.015	0.06	≤0.008–0.5
Staphylococcus aureus	12	0.25	0.25	0.25–0.5
Enterococcus faecalis	11	2	2	1–4
Klebsiella pneumoniae	11	0.06	0.06	0.01–0.06
Pseudomonas spp.	9	**	**	0.06–1
Pseudomonas aeruginosa	7	**	**	0.25–1

Interpreting C&S Data

To Which Drug Is the Isolate Most Susceptible?

C&S data accompanied by MIC (e.g., data from cystocentesis collected from 3 yr old [MN] Weimeraner) may have multiple drugs to which the isolate is most susceptible. Because different drugs achieve different C_{max}, one cannot compare the *E. coli* MIC; for example, of 0.25 mcg/ml for enrofloxacin to an MIC of 2 mcg/ml for amikacin and decide that the organism is more susceptible to enrofloxacin because the MIC is lower. In fact, the isolate is more susceptible to amikacin: the C_{max} at the recommended dose of 22 mg/kg is 65 mcg/ml, resulting in a C_{max}:MIC ratio of 65/2 or 32, compared to a C_{max} of 4.0 mcg/ml for enrofloxacin, resulting in a C_{max}:MIC ratio of 4/0.25, or 16. The organism is more susceptible to the drug for which the MIC is furthest from the C_{max}. Selecting a drug can also be based on a package insert using population data. Comparison of MIC 90 among organisms on the same package insert will reveal those organisms most and least susceptible to the drug. Comparing the MIC90 of an organism to the C_{max} (determining the ratio) for each drug can identify to which drug the isolate is most susceptible. For time-dependent drugs (see below), the duration at which drug concentrations remain above the MIC will also be important.

Designing a Dosing Regimen: The MIC Is KEY

Although a reasonable starting goal is to achieve the MIC of the infecting microbe in plasma, achieving well beyond that is important for several reasons: 1. **Bactericidal versus bacteriostatic** effects often are achieved only at concentrations well above the MIC. 2. **The site of infection**: C&S is based on plasma drug concentrations (PDC) yet infection is generally in tissue. The more difficult the tissue is to penetrate with the drug of choice, the higher the C_{max} must be compared to the MIC to compensate for limitations in drug movement, including into the cell for intracellular pathogens. 3. **The mutant prevention concentration (MPC)** is a more appropriate target than the MIC. The MIC is based on 2 to 3 colony forming units (CFU), yet clinical signs of infection are often associated with ≥10(5) CFU. The MIC on C&S most likely (statistically) is the median or 50th percentile. The larger the infecting inoculum, the more likely that at least on CFU has spontaneously mutated and has a much higher MIC that than provided on the report. The mutant prevention concentration is the highest MIC that occurs in the original (prior to exposure to antimicrobials) infecting inoculum. It is this MIC, that is the MPC, that should be targeted with the dosing regimen. However, the relationship between MIC and MPC is not known and can vary from 10 fold (common for many gram negative organisms) to over 125 (common for gram positive organisms). 4. **Concentration versus time-dependent drugs**. Concentration-dependent drugs - aminoglycosides and fluoroquinolones - are more effective when drug concentrations at the site of infection achieve 10x the MIC of the infecting organism. Thus, an MIC of 0.25 mcg/ml for enrofloxacin should lead to a dose that achieve 2.5 mcg/ml in the plasma. For amikacin, an MIC of 16 mcg/ml for *Pseudomonas aeruginosa* (which would be considered "S") would necessitate a PDC of 160 mcg/ml, more than twofold higher than what is achieved at the recommended dose. Such drugs are generally dosed once a day since the duration that drug concentrations persist is not relevant. For such drugs, if the target C_{max} cannot be achieved with a single dose, giving a second dose within the 24h period (dosing q12h) can be implemented to enhance efficacy (such an approach is often taken in humans when treating gram positive with aminoglycosides). For time-dependent drugs (generally acting reversibly), drug concentrations should be above the MIC of the infecting microbe for most or all (lower tier drugs) of the dosing interval. As such, the dose should be designed to be at least 4 times the MIC such that the dosing interval can be two drug half-lives. Penicillins are difficult to dose conveniently and effectively: The half-life is generally 1h and, as such, close to 90% of drug has left the blood stream by 3h. Amoxicillin (with or without clavulanic acid; AMX) ideally is a q6h drug unless MIC are extremely low: The C_{max} at 13 mg/kg approximates 5 mcg/ml. With an MIC for AMX for 0.25 mcg/ml, PDC concentrations will be above the MIC for only 4 half-lives or approximately 4h. For AMX, the MIC 90 for *E. coli* is 8 mcg/ml. Note that this is two-fold higher than PDC at the recommended dose. It is for this reason that CLSI has recently determined that AMX is no longer useful for treatment of soft tissue infections caused by gram negative organisms; it should no longer be included on C&S reports. However, if cultured from the urine, CLSI has determined that *E. coli* can be tested because it has provide urine specific breakpoints for it (and amoxicillin alone). Urine-specific breakpoints reflect renal concentration of the drug such that high concentrations are achieved in the urine. Urine-specific breakpoints also have been promulgated for cephalosporins. However, such breakpoints are only relevant in uncomplicated UTI, including patients for which renal function is normal.

Table 2. Selected Pharmacokinetic Data for Antimicrobials

Drug	HL	Dose	Route	Cmax
	hr	mg/kg		mcg/ml
Amikacin (C)	1	20	IV	65
Amoxicillin (T)	1	12.5	PO	4.5-6
Amoxi-clav (T)	1-1.5	12.5	PO	4.5-6
Ampicillin (T)+A29	0.5-1.5	30	PO	10
Cefotaxime	0.8	50	IV,IM	41
Cefpodoxime	5.6	5	PO	8.2
Ceftazidime (T)[3rd]	0.82	30	SC	42
Cephalexin (T)[1st]	1.3-2.5	20	PO	20
Chloramphenicol (T)	1.2-2.7	20	PO	23
Ciprofloxacin (C)	5.3	20	PO	2.8
Doxycycline (T)	4.6	2.5	PO	3
Enrofloxacin (C)	4.1	20	PO	4.2
Gentamicin (C)	1	10	IV	28
Imipenem/cilastin (T)	1	5	IM	13
Marbofloxacin (C)	11	2	PO	1.4
Orbifloxacin (C)	5	2.5	PO	2
Sulfadimethoxine (T)	13	55	PO	67
Tetracycline (T)	10	13.75	IV	7
Ticarcillin (T)	1	40	IV	80
Ticar -clav (T)	1	40	IV	80

C=concentration dependent

T=time dependent

To Wait or Not to Wait?

Frequently, antimicrobial therapy is begun before cultures are collected. This is particularly true and appropriate in critical patients; therapy also must begin in patients for which clinical signs are evident and are detracting from quality of life (patient or parent). However, should therapy begin and the choice prove to be wrong once C&S data is received, the original C&S data collected before the drug was begun may no longer be relevant to the patient. The use of the drug may change the pattern of resistance versus susceptibility, or may result in higher MIC (see mutant prevention concentration. A reculture may be indicated at that time, if possible. Certainly, dosing regimens with the appropriate drug should take into account the possibility that some level of resistance has developed toward the indicated drug. Note that increasing MIC on subsequent C&S are indicative of increasing resistance.

CULTURE AND SUSCEPTIBILITY TESTING: A TOOL TO GUIDE THERAPY

Pitfalls of Susceptibility Testing

Although culture and susceptibility data (C&S) can be a powerful tool to guide selection, it nonetheless is an *in vitro* test applied to *in vivo* conditions; over-reliance on the information can contribute to therapeutic failure. As early as the collection process, C&S data can be misleading.

To Treat or Not to Treat

Just as absence of growth does not indicate absence of infection, isolation of an organism is not necessarily evidence of infection, nor even if infection is present, does the isolated organism represent the infecting organism. Clearly, (properly collected) culture of an organism from a tissue that is normally sterile indicates infection. However, discriminating between normal and infecting flora can be difficult. Pure, vibrant (meaning special media was not needed to coax the growth of the organism) of a large number of colonies are indicators of infection. The isolation of three or more different organisms (including more than one strain of the same organism) may indicated contamination and reculture should be considered. The extent of growth should be strongly considered when deciding to treat. Generally, infection is considered to be present if $>10^7$ CFU/ml at the site. For C&S purposes, quantitative cultures can be helpful: the urinary tract is not considered infected until $>10^5$ CFU/ml are present whereas only $>10^3$ is indicative of infection in the respiratory tract. Laboratories may indicate "heavy, moderate or light" growth. For multiple organism, that with the greatest growth should be the primary focus of therapy. A call to the diagnostic lab might be prudent before marked financial commitment is put into treating an organism that is not causing infection. This is particularly important if the organisms' presence is unusual (e.g., *Lactobacillus* sp in urine)

To Trust or Not to Trust

The C&S procedures themselves are fraught with potential errors. For practices that provide in-house susceptibility testing, care must be taken to follow guidelines established and published by (or comparable to) the Clinical and Laboratory Standards Institute (CLASI) or comparable standard-setting agency. Materials, including interpretive standards, should be validated by the appropriate agency. Minor changes in pH, temperature, humidity, etc., can profoundly affect results. Personnel should be trained specifically in culture techniques and hospitals that provide this service (as do diagnostic labs) should maintain well-designed and adequately collected quality control data to validate their procedures (CLASI indicates control organisms). Pitfalls of susceptibility testing also reflect the drugs selected for testing. Not all companies are interested in establishing interpretive criteria and as such, not all drugs are available for testing. Because automated systems cannot accommodate and laboratories (nor clients) cannot afford to test all potential drugs used to treat an infection, one drug often is tested as a model for other drugs in the class. For example, oxacillin models methicillin, ampicillin models amoxicillin and amoxiclav models ampicillin sulbactam (but the former is more effective than the latter). Cephalothin models first generation cephalosporins, even though it is no longer used clinically. Recently, CLSI has indicated that, like cefazolin, cephalexin should be used rather than cephalothin. Enrofloxacin often represents the fluoroquinolones. In general, cross resistance can be expected among the FQs, although differences in potency do exist (for example, ciprofloxacin is more potent toward *Pseudomonas* or *E. coli,* but less to gram positives compared to enrofloxacin). Culture does not take into account active metabolites of some drugs (e.g., enrofloxacin converted to ciprofloxacin). An "I", regardless of the drug, might be interpreted as "R": used should occur only if the infection is at a site that the drug concentrations (e.g., urine, liver, white blood cells), the dose must be increased and/or combination therapy should be implemented. Some drugs should not be used as sole therapy even in the face of "S" and a good laboratory will warn the clinician that *in vitro* results may not reflect *in vivo*. For example, gentamicin is often more effect toward *Staphylococcus* sp compared to amikacin, aminoglycosides should not be used by themselves to treat staphylococci. Note that CLSI interpretive criteria are generated for specific species, and often for specific organisms and specific infections. Human laboratories will use human interpretive criteria, which often are not relevant to animals (e.g., ciprofloxacin). Fewer interpretive criteria are available for antibiotics used in animals.

acvim405

Indications and Techniques for Upper/Lower GI Endoscopy in Dogs and Cats

Albert E. Jergens, DVM, PhD, MS, DACVIM (SAIM)
Ames, IA, USA

INTRODUCTION

Flexible endoscopy provides non-invasive assessment of the GI mucosa and allows targeted collection of tissues, cells, and/or fluids for analysis. Tissue samples can be helpful in establishing a definitive diagnosis, prognosis, and therapeutic approach to many infiltrative chronic enteropathies (CE) in dogs and cats. This presentation will review current diagnostic utility of flexible GI endoscopy in regards to practical and technical considerations for endoscopic biopsy, optimal instrumentation for mucosal specimen collection, correlation of endoscopic indices to clinical activity and to histopathologic findings, and new developments in the endoscopic diagnosis of GI disease.

PRACTICAL/TECHNICAL CONSIDERATIONS FOR ENDOSCOPIC BIOPSY

Endoscopic biopsy of the GI tract has advantages and disadvantages. First, direct visualization of mucosal lesions undetectable from the serosal surface allows targeted biopsy. Second, being able to obtain numerous tissue specimens over a 10- to 20-cm length of intestine is more likely to detect mucosal lesions that can be regionally patchy in distribution (e.g., lymphoma, histoplasmosis, pythiosis, granulomatous colitis). However, while useful for detecting morphologic or infiltrative disease, endoscopy cannot detect functional disorders (e.g., abnormal motility, GI secretory disturbances, biochemical injury, brush border enzyme deficiencies). Furthermore, histopathology is of minimal use in diagnosing some forms of CE (e.g., food-responsive and antibiotic-responsive disorders).

The controversy between endoscopic versus surgical biopsy centers around the acquisition of quality mucosal samples from different sections of the GI tract. Endoscopy biopsy is a less invasive procedure, takes less time to perform in critical patients, and allows visualization of the mucosa to identify the best biopsy sites. The results obtained from endoscopic biopsy are correlated with clinical experience, operator expertise, and the acquisition of diagnostic biopsy specimens for histopathologic review. This is especially true with duodenal biopsy specimens where **marginal** specimens were defined as having at least one villus plus subvillus lamina propria, and **adequate** samples contained at least 3 villi with subvillus lamina propria that extended down to the muscularis mucosa. A correct diagnosis was most likely to be obtained in one study when 6 marginal or adequate biopsies of the feline stomach or duodenum were obtained by the endoscopist. The disadvantage is that only gastric, duodenal, and colonic mucosa can routinely be biopsied. High jejunal tissue can sometimes be obtained, and ileal tissue can be obtained in most patients if the operator is experienced. However, even competent endoscopists cannot always sample the ileum. There are surprisingly few data comparing the accuracy of full-thickness versus endoscopic biopsy for diagnosis of canine and feline GI disease.

GI ENDOSCOPIC INSTRUMENTATION

Endoscopic instruments for sampling GI mucosa include pinch biopsy forceps and guarded cytology brushes. Flexible pinch forceps are used to obtain mucosal specimens from the GI tract. These small flexible forceps with opposing 2- to 3-mm cups on their distal end are manufactured with numerous configurations. Cups may be smooth or serrated, standard or fenestrated, with or without a central needle; forceps may be multiple use or disposable, and some are designed to allow multiple samples to be taken before withdrawing the instrument. There is difference of opinion between endoscopists as to the best configuration for flexible endoscopic forceps. Fenestrated forceps may cause fewer crush artifacts and yield larger biopsy specimens than non-fenestrated models. Biopsy forceps with a central needle can help stabilize forceps on the mucosa and are useful for some endoscopists but may yield inferior tissue samples associated with puncture artifacts in the hands of other endoscopists. Others have shown that forceps cup size is what matters most by demonstrating that large capacity forceps (with or without a central spike) provided the highest quality of duodenal samples obtained from healthy dogs. The use of forceps that can be passed through smaller diameter endoscopes have also been shown to yield excellent quality endoscopic specimens of the ileum or in animals <10 kg (cats) because the mucosa is relatively thin.

Tissue samples should be as large and free of artifact as possible because the diagnostic quality of the endoscopic sample influences the ability of the pathologist to detect and define mucosal lesions. In general, six marginal or adequate samples should be collected for abnormalities to be detected in the feline gastric and duodenal mucosa, whereas six adequate or 10 to 15 marginal samples should be collected from the canine stomach and duodenum, depending on the lesion being sought. To be considered **adequate**, a biopsy sample should contain the full thickness of the mucosa and be wide enough to have at least three to four intact and preferably contiguous villi.

Post-Biopsy Considerations For Enhancing Histopathologic Results

Tissue specimens should be gently teased from the forceps with a needle and placed on lens paper, cucumber slice, or specially designed biopsy sponges. Commercial cassettes with precut "sponges" can be used for the submission of endoscopic biopsy specimens. Multiple biopsy specimens can be arranged on a sponge and the sponge and the closed cassette is then immersed in 10% neutral buffered formalin and submitted to the laboratory. Cucumber slices can be substituted for plastic sponges and are an excellent medium for the submission of endoscopic biopsy specimens. Biopsy samples are arranged in parallel rows on the thin slices of cucumber (preserved in alcohol), which are then deposited in formalin and submitted for routine processing. At the laboratory, the cucumber cassettes are removed from the formalin and the cucumber sections are reoriented 90 degrees on their side in the cassette (e.g., perpendicular to the cassette surface to optimize tissue orientation after sectioning). The tissues are then embedded in paraffin wax. The specimens do not have to be removed from the cucumber slices prior to embedding because the microtome can readily cut through the vegetable material. This technique minimizes specimen handling at the laboratory and consistently yields well-oriented tissues of high diagnostic quality. Attempts to reorient specimens on biopsy sponges or cucumber slices prior to formalin fixation should be avoided. Specimens should not be allowed to dry out on cucumber or sponge surfaces. Overly dried samples may adhere tightly to the sponge or cucumber and be damaged when removed by the pathology service.

Controversies in Histopathologic Interpretation of Endoscopic Specimens

Histopathologic interpretation may vary according to the quality of tissue specimens submitted (related to operator experience and processing artifacts) and inconsistent histopathologic criteria for defining GI inflammation in dogs and cats. Uniform grading criteria for defining GI inflammation in endoscopic specimens remain controversial. Numerous grading schemes for characterizing the nature and severity GI inflammation have been designed with most model systems emphasizing the type and degree of lamina propria cellular infiltrate. The emphasis on mucosal cellularity has meant that abnormalities in mucosal architecture have been overlooked, even though they may correlate more with inflammatory markers and clinical severity. Finally, mean cell populations (e.g., CD3+ T cells) do not differ in IBD dogs at diagnosis versus when in clinical remission and cats with and without signs of intestinal disease may have similar numbers of lymphocytes and plasma cells.

Inconsistency between pathologists, questionable quality of tissues submitted to the laboratory, and controversy as to which grading scheme to utilize are factors that have made it difficult to accurately correlate histopathologic findings to clinical disease activity. Different trials have attempted to correlate histopathologic changes in endoscopic specimens to disease severity at diagnosis or in response to treatment. For dogs with CE (predominantly IBD), the collective results indicated that there was no significant association between histopathologic findings and clinical signs, serum biomarkers, or responses to different treatments. Detection of the extent and severity of mucosal disease is aided by use of endoscopic indices in humans with IBD. There are relatively few reports in dogs where endoscopic lesions were associated with clinical severity and histopathologic lesions. Separate studies in dogs with small intestinal IBD have yielded conflicting results on the utility in endoscopic scoring as a measure of disease activity. These discordant results can be explained by differences in operator experience for detecting mucosal lesions of inflammation. Use of a simple endoscopic activity score based on qualitative criteria (e.g., friability, granularity, erosions, and lymphatic dilatation) has been recently validated in dogs with histopathologic IBD.

GIT endoscopic biopsy sample recommendations

Species	GI organ	# endoscopic specimens	Comments
Dog	Stomach	6 adequate	Biopsy gastric body unless focal lesions present
Dog	Duodenum	10–15 adequate	Up to 15 marginal samples may be required
Dog/cat	Ileum	3–5 adequate	Exact number unknown; blind forceps biopsies are OK
Dog/cat	Colon	9–12 adequate	Obtain 3–4 biopsies from each colonic region
Cat	Stomach	6 adequate	Six mucosal samples generally diagnostic
Cat	Duodenum	6 adequate	Six mucosal samples generally diagnostic

REFERENCES

1. Procoli F, Motskula PF, Keyte SV, Priestnall S, Allenspach K. Comparison of histopathologic findings in duodenal and ileal endoscopic biopsies in dogs with chronic small intestinal enteropathies. *J Vet Intern Med.* 2013;27:268–274.
2. Scott KD, Zoran DL, Mansell J, Norby B, Willard MD. Utility of endoscopic biopsies of the duodenum and ileum for diagnosis of inflammatory bowel disease and small cell lymphoma in cats. *J Vet Intern Med.* 2011;25:1253–1257.
3. Washabau RJ, Day MJ, Willard MD, Hall EJ, Jergens AE, Mansell J, Minami T, Bilzer TW. Endoscopic, biopsy, and histopathologic guidelines for the evaluation of gastrointestinal inflammation in companion animals. *J Vet Intern Med.* 2010;24:10–26.
4. Willard MD, Jergens AE, Duncan RB, Leib MS, McCracken MD, DeNovo RC, Helman RG, Slater MR, Harbison JL. Interobserver variation among histopathologic evaluations of intestinal tissues from dogs and cats. *J Am Vet Med Assoc.* 2002;220:1177–1182.
5. Willard MD, Lovering SL, Cohen ND, Weeks BR. Quality of tissue specimens obtained endoscopically from the duodenum of dogs and cats. *J Am Vet Med Assoc.* 2001;219:474–479.
6. Goutal-Landry CM, Mansell J, Ryan KA, Gaschen FP. Effect of endoscopic forceps on quality of duodenal mucosal biopsy in healthy dogs. *J Vet Intern Med.* 2013;27:456–461.
7. Jergens AE, Willard MD, Allenspach K. Maximizing the diagnostic utility of endoscopic biopsy in dogs and cats with gastrointestinal disease. *Vet J.* 2016;214:50–60.

acvim406

Recent Advances Into the (Patho)Physiology of Cortisol and Bile Acid Metabolism in Critical Illness

Pauline M. Jamieson, BVM&S, PhD, DECVIM-CA, MRCVS
Edinburgh, Scotland, UK

Introduction to Critical Illness-Related Corticosteroid Insufficiency

Acute critical illness activates the hypothalamic-pituitary-adrenal axis (HPAA) resulting in increased secretion of cortisol from the adrenal cortex which plays an essential role in allowing patients to preserve normal homeostasis in the face of the pathophysiological challenge. Crucially this includes the maintenance of blood pressure. However, in a subset of critically ill human and veterinary patients, profound hypotension that is poorly responsive to medical management with intravenous fluids and vasopressor medications is observed. A syndrome characterised by adrenal dysfunction in sepsis has been recognised for over a century, and is proposed as the reason for this refractory hypotension which is associated with a critical reduction in tissue perfusion and progression to multiple organ failure. This has been interpreted as a relative adrenal insufficiency, where despite normal to increased circulating cortisol concentrations the cellular cortisol levels are insufficient for the increased physiological demand. It is termed critical illness-related corticosteroid insufficiency (CIRCI) and it is argued that this subset of patients may benefit from the administration of physiological doses of corticosteroids.[1,2] However, despite decades of clinical research, the diagnosis, management, and even existence of CIRCI are matters of much debate, or in the words of Dr. Brian Fuller, 'a tour de force in evidence-based uncertainty.'[3]

Identifying individuals with CIRCI is notoriously difficult, as the relative as opposed to absolute nature of the cortisol deficiency means that baseline plasma cortisol levels are a poor indicator. In the past, recommendations for diagnosis of CIRCI have relied on identifying hypotensive patients with a blunted cortisol response to exogenous ACTH administration, but this criterion has also proved unreliable, and the recommendation was retracted in the Surviving Sepsis Campaign 2008 and subsequent consensus statements.[1,2] Diagnosis of CIRCI is now imprecisely based on failure of hypotension to respond to vasopressor and fluid therapy. This inability to identify which patients, if any, truly have a relative glucocorticoid deficiency clearly means that any intervention with glucocorticoids is likely to be suboptimal for the individual patient, and it probably accounts for the inconsistent conclusions from studies examining benefit of glucocorticoid administration in critical illness. High dose glucocorticoids have been shown to be detrimental to outcome, while smaller studies showing lower mortality with administration of lower physiological doses have not been validated by multicenter randomized controlled trials.[4] Consequently, there is currently continuing controversy regarding the value of, or protocols for, glucocorticoid administration to critically ill patients. The results of the ADjunctive coRticosteroid trEatment iN criticAlly ilL Patients With Septic Shock (ADRENAL) multicenter clinical trial which is recruiting 3800 human septic shock patients in 74 centres in 6 countries are awaited.[5]

In veterinary medicine, there is a paucity of evidence-based studies into CIRCI. Martin[6] and Creedon[7] provide inclusive reviews. Studies demonstrating that critically ill dogs with diminished cortisol responses to ACTH stimulation are more likely to be hypotensive and have decreased survival,[8] or are more likely to be pressor dependent,[9] suggest that adrenal responsiveness may be a marker of the severity of illness but stop short of demonstrating functional adrenal insufficiency when viewed in the light of the recent observations regarding cortisol metabolism in critically ill humans described below. Critically ill cats have been reported have higher baseline cortisol concentrations than controls, as would be expected,[10] but neither basal cortisol levels nor the cortisol response to ACTH stimulation have predicted survival[10,11]. Responses to treatment with glucocorticoids are limited to descriptive reports, and consequently recommendations in veterinary sources are not based on good evidence from clinical trial data.

Cortisol Physiology and Metabolism in Critical Illness

Abnormalities in HPAA function, cortisol binding to corticosteroid binding globulin (CBG), cellular metabolism of cortisol and tissue responsiveness to cortisol, have all been suggested to underlie the pathophysiology of CIRCI. In a subset of critically ill patients, a paradoxical dissociation between cortisol and ACTH levels is observed whereby ACTH levels are only transiently elevated while cortisol levels remain high. Recently, it has been suggested based on novel findings in critically ill human patients that reduced metabolism and clearance of cortisol may be responsible for this.[12] Tissue levels of cortisol are determined by the activity of the 11ß-hydroxysteroid dehydrogenase (11ßHSD) enzyme system, which has been shown to be altered in patients with sepsis. 11ßHSD1 and 11ßHSD2 convert inactive cortisone to active cortisol and *vice versa* respectively. The balance in activity between these enzymes will determine the intracellular levels of active cortisol. Meanwhile, renal 11ßHSD2 and hepatic 5α- and 5ß-reductases are then responsible for clearance of cortisol. The activities of these enzymes can be estimated non-invasively from the overall ratios of metabolites of cortisol and cortisone in urine.

The study by Boonen *et al.*[12] examined cortisol metabolism in human patients in the intensive care unit. They observed that plasma cortisol was consistently higher, but that ACTH levels were lower, in patients compared with controls. A 50% decrease in the clearance of cortisol tracer, and also of hydrocortisone administered at the clinically recommend dose for CIRCI, was observed, and this correlated with a lower adrenal cortisol response to ACTH stimulation. This reduced cortisol metabolism was associated with decreased 11ßHSD2 and 5α- and 5ß-reductase tissue expression and activities. They concluded that reduced tissue cortisol metabolism contributed to hypercortisolemia which in turn suppressed the HPAA and ACTH levels. On this basis they proposed that impaired cortisol metabolism and clearance during critical illness is potentially a beneficial adaptive response designed to maintain adequate tissue levels of glucocorticoid, and that therefore a reduced cortisol response to exogenous ACTH in these patients is physiologically appropriate. The enhanced negative feedback on the HPAA which is responsible for reduced adrenal cortisol production is therefore offered as explanation for the many conflicting data on the relationships between presence of CIRCI or glucocorticoid administration and patient survival, and particularly for the observation that high cortisol levels are associated with poor outcomes in critically ill children.[13] In accordance with these findings, recent data from our laboratory showed that both serum cortisol and total urinary cortisol metabolite excretion were higher in dogs with sepsis or septic shock (defined as systolic blood pressure <90 mm Hg unresponsive to fluid resuscitation) than in control dogs, and additionally that total cortisol metabolite excretion was greater in dogs with septic shock than those with sepsis alone. The bound serum cortisol fraction was also higher (with a trend toward higher total and free cortisol levels) in dogs with septic shock than with sepsis, and both groups had significantly lower renal 11ßHSD2 activity than controls.

No specific conclusions regarding therapeutic implications were drawn by Boonen *et al.*, but the point was made that with reduced cortisol clearance, the doses of hydrocortisone currently advocated in presumed CIRCI patients are almost certainly too high for the role they are proposed to fulfil (i.e., physiological replacement of the deficient cortisol production to 'stress levels'). It should, however, be noted that this study did not examine cortisol metabolism in the specific subset of critically ill patients with refractory hypotension, and so the findings do not preclude that those patients may have cortisol deficiency at the cellular level. Currently, the principal debate regarding CIRCI in human medicine is whether to institute glucocorticoid treatment at all. Meta-analyses indicate that while higher doses are detrimental, lower doses may or may not be beneficial, and could still potentially be harmful. In addition to the myriad of difficulties in diagnosing the presence CIRCI, the optimal dose of glucocorticoid for any individual patient cannot be determined, and these recent data draw into question whether it exists at all. While there is no doubt that adrenal and cortisol physiology are significantly altered in critical illness, the more controversial view that this represents beneficial physiological adaptation and is thus an indicator of illness rather than a disorder requiring specific medical intervention deserves consideration.

Bile Acids in Critical Illness

An additional observation in Boonen's study was that there was a significant negative correlation between serum bile acid (BA) levels and hepatic cortisol metabolizing 5ß-reductase enzyme activity in their critically ill patients. BAs are known inhibitors of cortisol-metabolizing enzymes including 11ßHSD2 and 5ß-reductase, and this has been observed in experimental and clinical models of cholestasis.[14,15] Clinical biochemistry commonly yields altered hepatic parameters in critically ill patients, and therefore the suggestion that BAs are at least in part responsible for the decreased cortisol metabolism in CIRCI does not seem improbable. What is more provocative is the idea that this may be part of a beneficial adaptive physiological response to critical illness,[16] particularly in light of the fact that hepatopathy is associated with negative outcomes in these patients.

It is important here to make the distinction between hypoxic hepatitis in critical illness, which is characterised by significantly quantitative elevations in hepatocellular enzymes in comparison to those seen in bilirubin and BAs as a result of hepatocellular necrosis, and the 'cholestatic liver dysfunction (CLD)' observed in a subset of critical patients where the elevations in bilirubin and BAs are the more salient feature of the serum biochemistry profile. The CLD associated with critical illness is not due to mechanical obstruction of bile flow, but is the result of complex biochemical alterations in the intrahepatic metabolism and transport of bilirubin and BAs. Sepsis and systemic inflammatory response syndrome (SIRS) are risk factors for CLD in critical illness, and animal models confirm that endotoxins do indeed precipitate this, but CLD can also occur in non-septic critical patients and is more likely to be induced by pro-inflammatory cytokines.

The primary mechanism behind the CLD in critical illness is an apparent reversal of the normal BA and bilirubin transport direction such that they are moved from the hepatocyte back into the blood circulation.[16] This altered hepatobiliary transport is mediated by downregulation of the Na^+ dependent bile acid transporter (NTCP) and organic anion transporting polypeptide (OATP) transporters, which are the primary molecules responsible of uptake of circulating BAs and bilirubin by hepatocytes. Coupled with upregulation of the multidrug resistance-associated proteins MRP3 and MRP4 which return bile salts directly from hepatocytes to the portal circulation, and downregulation of the bile salt export pump (BSEP) which secretes BAs from the hepatocyte to the bile canaliculi, the net result is an increase in circulating BAs. Furthermore, the normal negative feedback of BAs on the rate limiting enzyme in their own production (cholesterol 7-α-hydroxylase) appears to be suppressed despite their increased bioavailability. In addition to that from experimental models, there is some good evidence from human studies that these changes occur in naturally occurring critical illness.

Given that in severe or prolonged illness, hepatic cholestasis is associated with hepatocyte apoptosis and ultimately necrosis of the liver parenchyma and thus severe morbidity, what could be the basis for proposing that the changes in hepatobiliary transport system could be in any way a beneficial response? The simple explanation is that the transport of bile salts back into the circulation is a protective response secondary to pathological failure of the hepatocytes to excrete them to the biliary system. However, while severe CLD likely represents decompensation, the suggestion is that **mild** elevations in bilirubin and bile acids may in fact be beneficial for survival. It is well recognised that parenteral nutrition is associated with biochemical markers of hepatopathy, generally interpreted as CLD.[17] However, one large clinical study demonstrated that while enzymatic markers of cholestasis (ALP, GGT) were higher in patients given parenteral nutrition, serum bilirubin was in fact lower.[18] When parenteral nutrition was subsequently administered to the patients who had not received it, bilirubin levels decreased, suggesting that hyperbilirubinemia in critical illness may not reflect true cholestasis. Furthermore, the patients with higher bilirubin had more positive clinical outcomes. This concurs with reports of increased survival in animal models of sepsis following administration of bilirubin, and decreased survival in its absence. Beneficial effects of bilirubin on endothelial and hepatocyte functions, and antioxidant and cytoprotective effects are reported.[16] There is evidence this may be via inhibition of hepatic iNOS expression or improved redox cycling. Bilirubin has received most of the attention in respect of this, as it is the parameter commonly used to assess hepatic function or the presence of CLD in critical illness. However, reported anti-hypermetabolic and insulin-sensitizing effects of BAs as well as their effects on cortisol physiology described above may also be beneficial.

Conclusions

Recent advances into the (patho)physiology of cortisol and bile acid metabolism in critical illness have raised interesting questions and prompted provocative theories regarding interpretation of their function in this setting. Whether the hypothesis that the increased bile acids/bilirubin of CLD and their effects on cortisol physiology are indeed adaptive and beneficial rather than maladaptive is correct cannot be determined from the currently available evidence. However, given the therapeutic implications if it are, it is worthy of consideration and should be further investigated. In the meantime, cautious interpretation of the significance of biochemical evidence of cholestasis in critical illness and following the current recommendations for judicious use of glucocorticoids for refractory hypotension would appear justifiable.

References

1. Dellinger RP, Levy MM, Rhodes A, *et al.* Surviving sepsis campaign: international guidelines for management of severe sepsis and septic shock: 2012. *Critical Care Medicine.* 2013;41(2):580–637. PMID: 23353941.
2. Marik PE, Pastores SM, Annane D, *et al.* Recommendations for the diagnosis and management of corticosteroid insufficiency in critically ill adult patients: consensus statements from an international task force by the American College of Critical Care Medicine. *Critical Care Medicine.* 2008;36(6):1937–1949. PMID: 18496365.
3. Fuller BM. The adrenal gland and corticosteroid therapy in sepsis: I certainly remain uncertain. *Critical Care Medicine.* 2015;43(3):702–03. PMID: 25700055.
4. Annane D, Bellissant E, Bollaert PE, Briegel J, Keh D, Kupfer Y. Corticosteroids for treating sepsis. *Cochrane Database of Systematic Reviews.* 2015;(12):CD002243. PMID: 26633262.
5. ADjunctive coRticosteroid trEatment iN criticAlly ilL Patients With Septic Shock (ADRENAL) clinical trial. https://clinicaltrials.gov/show/NCT01448109.
6. Creedon JM. Controversies surrounding critical illness-related corticosteroid insufficiency in animals. *Journal of Veterinary Emergency and Critical Care.* 2015;25(1):107–112. PMID: 25516097.
7. Martin LG. Critical illness-related corticosteroid insufficiency in small animals. *Veterinary Clinics of North America Small Animal Practice.* 2011;41(4):767–782. PMID: 21757092.
8. Burkitt JM, Haskins SC, Nelson RW, Kass PH. Relative adrenal insufficiency in dogs with sepsis. *Journal of Veterinary Internal Medicine.* 2007;21(2):226–31. PMID: 17427381.
9. Martin LG, Groman RP, Fletcher DJ, *et al.* Pituitary-adrenal function in dogs with acute critical illness. *Journal of the American Veterinary Medical Association.* 2008;233(1):87–95. PMID: 18593315.
10. Prittie JE, *et al.* Hypothalamo-pituitary-adrenal (HPA) axis function in critically ill cats. *Journal of Veterinary Emergency and Critical Care.* 2003;13:165 (abstract).
11. Costello MF, *et al.* Adrenal insufficiency in feline sepsis. In: Proceedings of the ACVECC Postgraduate Course, Sepsis in Veterinary Medicine. 2006.
12. Boonen E, Vervenne H, Meersseman P, *et al.* Reduced cortisol metabolism during critical illness. *New England Journal of Medicine.* 2013;368(16):1477–1488. PMID: 23506003.
13. Levy-Shraga Y, Pinhas-Hamiel O, Molina-Hazan V, *et al.* Elevated baseline cortisol levels are predictive of bad outcomes in critically ill children. *Pediatric Emergency Care.* 2016. [Epub ahead of print] PMID: 27383405.
14. McNeilly AD, Macfarlane DP, O'Flaherty E, *et al.* Bile acids modulate glucocorticoid metabolism and the hypothalamic-pituitary-adrenal axis in obstructive jaundice. *Journal of Hepatology.* 2010;52(5):705–711. PMID: 2034717.
15. Quattropani C, Vogt B, Odermatt A, Dick B, Frey BM, Frey FJ. Reduced activity of 11 beta-hydroxysteroid dehydrogenase in patients with cholestasis. *Journal of Clinical Investigation.* 2001;108(9):1299–305. PMID: 11696574.
16. Jenniskens M, Langouche L, Vanwijngaerden YM, Mesotten D, Van den Berghe G. Cholestatic liver (dys)function during sepsis and other critical illnesses. *Intensive Care Medicine.* 2016;42(1):16–27. PMID: 26392257.
17. Carter BA, Shulman RJ. Mechanisms of disease: update on the molecular etiology and fundamentals of parenteral nutrition associated cholestasis. *Nature Clinical Practice Gastroenterology & Hepatology.* 2007;4(5):277–287. PMID: 17476210.
18. Vanwijngaerden YM, Langouche L, Brunner R, *et al.* Withholding parenteral nutrition during critical illness increases plasma bilirubin but lowers the incidence of biliary sludge. *Hepatology.* 2014;60(1):202–210. PMID: 24213952.

acvim407

The Emerging Role of Bile Acids in Chronic Enteropathies

Jan S. Suchodolski, DrMedVet, PhD, AGAF, DACVM (Immunology)
College Station, TX, USA

Introduction

Chronic enteropathies encompass several disease manifestations, including diet-responsive diarrhea, antibiotic-responsive diarrhea, and idiopathic inflammatory bowel disease. Chronic enteropathies (CE) are considered of multifactorial origin, caused by a combination of an underlying genetic susceptibility, an aberrant immune response, and microbiota dysbiosis. Over the last years several studies have evaluated the microbiota dysbiosis that occurs as part of chronic enteropathies. Studies using sequencing of 16S rRNA genes have described dysbiosis in dogs and cats with chronic enteropathies (CE) and acute diarrhea.[1] In CE there is an increase in proportions of bacterial genera belonging to Proteobacteria (e.g., *E. coli*) and decreases in Fusobacteria, Bacteroidetes, and members of Firmicutes (i.e., *Faecalibacterium*, Ruminococcaceae, *Turicibacter*, *Blautia*). While molecular approaches have solidified our knowledge about microbial alterations that occur during GI disease in dogs, relative limited information is available about the functional consequences of the dysbiosis.

Recently, initial studies in dogs have described metagenomics and metabolomic changes in the GI tract of dogs linked to dysbiosis.[2,3] Due to the important role of microbial-derived metabolites for host health, these alterations in bacterial groups are also associated with metabolic changes within the intestinal lumen. Some of these major metabolic pathways that are altered include bile acids, lactate, and tryptophan metabolism. These changes have an effect on the entire host metabolism. For example, an increase in serum D-lactate in cats with gastrointestinal disease is possibly due to disarrangements of the intestinal microbiota and increased bacterial production of D-lactate, and D-lactate has been shown to lead to neurological signs in some cats.[4,5] Bile acid dysmetabolism, leading to bile acid diarrhea (BAD), appears to be a common feature of GI disease in humans and dogs, as is either due to a dysbiosis or abnormal enterohepatic circulation of bile acids in the small intestine.[6-8] New diagnostic tests, such as for example measurement of fecal bile acids, can help diagnose dysbiosis associated with bile acid diarrhea, and treatment with bile acid sequestrants (e.g., cholestyramine) may lead to improvement of diarrhea.[6]

Bile Acid Metabolism

Bile acid metabolism involves multiple steps. At first, bile acids are synthesized in the liver from cholesterol via 7α-hydroxylase (CYP7A1) with two primary bile acids formed, cholic acid and chenodeoxycholic acid. Before excretion into the hepatobiliary ducts, primary bile acids in dogs are conjugated with taurine primarily and to a lesser extent glycine in the liver, and these are then stored in the gallbladder.[9] Digestion triggers the release of conjugated bile acids into the intestinal lumen. There, as a first reaction, the deconjugation of taurine or glycine by bile acid hydrolase occurs. In gut homeostasis, active reabsorption of conjugated bile acids takes place in the terminal ileum leading to the enterohepatic cycle. The reabsorption occurs through the apical sodium-bile acid transporter (ASBT). A small fraction of bile acids will reach the colon, where a second step in the transformation from primary to secondary bile acids occurs. Some bacterial species within the Firmicutes (e.g., *Clostridium scindens, C. hiranonis*) have 7α-dehydroxylation activity and are able to convert primary to secondary bile acids (e.g., deoxycholic acid and lithocholic acid). This reaction highlights the close interplay between intestinal microbiota and the host.

Bile acids are now recognized as important regulator of host health, as they are agonists for various receptors across the body. For example, activation of the farnesoid X receptor (FXR) and the transmembrane G protein-coupled receptor TGR5 activate a crucial feedback mechanism involved in bile acid synthesis. Furthermore, their binding activates various transcriptional factors, which lead to the expression of genes involved in cholesterol and carbohydrate metabolism. Bile acids also activate additional nuclear receptors such as pregnane X receptor and the vitamin D receptor. Activation of TGR5 results in activation of transcription factors that downregulate inflammation. Also, the activation of TGR5 in enteroendocrine cells through secondary bile acids promotes the secretion of glucagon-like peptide 1 (GLP-1), which helps in the regulation of insulin. Mounting evidence has suggested that chronic liver disease, hepatocellular cancer, and extrahepatic inflammatory and metabolic diseases may stand to benefit from targeting of these bile acid receptors, highlighting the important link between gut dysbiosis, dysregulation of bile acid metabolism, and various metabolic disorders.[10]

BILE ACIDS AND INTESTINAL DISEASE

There is emerging evidence that altered bile acid metabolism leads to bile acid malabsorption and therefore is a frequent cause of diarrhea in humans.[11] Bile acid diarrhea (BAD) can be due to various causes. Bile acid diarrhea Type 1 is caused by bile acid malabsorption that is secondary to ileal resection or inflammation. Type 2 is idiopathic bile acid malabsorption, likely due to a defective feedback inhibition of hepatic bile acid synthesis due to dysfunction of the ileal hormone fibroblast growth factor 19 (FGF19). Type 3 bile acid diarrhea occurs secondary to various GI diseases including small intestinal dysbiosis, radiation enteropathy, celiac disease, and chronic pancreatitis. It has been suggested that approx. 30% of human patients with irritable bowel syndrome (IBS) and 40% of patients with Crohn's disease have bile acid diarrhea. In humans with IBD, the destruction of apical sodium-bile acid transporter (ASBT) in the ileum leads to decreased reabsorption of bile acids in the small intestine, leading to increases in primary bile acids in the colon that cause secretory diarrhea. Interestingly, the use of glucocorticoids induces the expression of ASBT in the small intestine, potentially reducing bile acid malabsorption and improving diarrhea. Another cause of BAD could be due to dysbiosis. In a recent study in humans, inflammatory bowel disease (IBD) was associated with bile acid malabsorption. Fecal primary bile acids were significantly increased in active IBD, whereas fecal secondary bile acid concentrations were significantly decreased in IBD compared to healthy controls.[7] This change in bile acid metabolism was linked to the dysbiosis that was present in patients with IBD, suggesting that dysbiosis leads to a decrease in bacterial species with 7α-dehydroxylation activity.

Also, recent studies have suggested that altered bile acid metabolism is also present in dogs with chronic enteropathies. Our group evaluated fecal samples from dogs with histopathological evidence of gastrointestinal inflammation (n=15) and from healthy control dogs (n=15). The fecal metabolome, comprised of metabolic products of both the host and microbiota, was assessed using an untargeted approach combining LC-MS and GC-MS platforms. Primary bile acids were generally increased in dogs with disease, while secondary bile acids were generally decreased in dogs with chronic enteropathy, suggesting impaired bacterial conversion from primary to secondary bile acids.[3] A follow-up study used a targeted metabolic approach using gas chromatography coupled with mass spectrometry (GC-MS) platform to directly measure bile acids in fecal samples of dogs with chronic enteropathies. Fecal samples were collected from healthy dogs (n=13) and dogs with chronic enteropathy (n=13), with biopsy-confirmed inflammation. Significant decreases were observed in the secondary bile acids lithocholic acid and deoxycholic acid, with both being significantly decreased in dogs with chronic enteropathy compared to healthy dogs. A significant correlation was observed between an increased dysbiosis index and bile acid dysmetabolism. This would suggest that the dysbiosis observed in chronic enteropathies also leads to a decrease in bacterial species with 7α-dehydroxylation activity, as seen in humans.

Another recent study evaluated bile acid malabsorption in dogs with chronic diarrhea.[8] The authors measured serum concentrations of 7α-hydroxy-4-cholesten-3-one (C4), as a potential marker of bile acid malabsorption. C4 is an intermediate in the biochemical synthesis of bile acids from cholesterol, and serum C4 concentrations reflect the activity of the bile acid synthetic pathway. An increase in serum C4 concentrations is indicative of bile acid malabsorption in humans with chronic diarrhea, and these patients typically respond well to bile acid sequestrants (e.g., cholestyramine).[6] In the study by Kent *et al.*,[8] serum C4 was measured in 17 dogs with chronic diarrhea and 20 control dogs. Eleven of these cases were diagnosed with IBD. While there was no significant difference in the serum C4 concentration between both groups, 3 dogs in the diseased group were above the cut-off value established in the 20 healthy dogs. One of the dogs was partially responsive to a trial with the bile acid sequestrant cholestyramine in combination with a dietary trial. The authors suggested that bile acid malabsorption may be a clinically relevant disorder in dogs with chronic diarrhea.

The treatment of choice in humans with BAD is administration of bile acid sequestrants. In a recent study, 58 patients with diagnosed bile acid diarrhea were followed up.[6] Patients had a median time since diagnosis of 6 years. Of these, 38% were using bile acid sequestrants long-term, and their stool frequency decreased from seven stools per day to three stools per day. In contrast, those patients that ceased treatment with bile acid sequestrants had no change in their stool frequency.

Conclusions

There is emerging evidence in humans and also dogs that bile acid malabsorption is a component of chronic GI disease and could be partially responsible for clinical signs of chronic diarrhea. There are currently anecdotal reports about the use of cholestyramine in dogs with chronic diarrhea that are refractory to standard therapy. Clinical studies to evaluate the use of cholestyramine or other bile acid sequestrants as a tool for managing diarrhea are warranted.

References

1. Honneffer JB, Minamoto Y, Suchodolski JS. Microbiota alterations in acute and chronic gastrointestinal inflammation of cats and dogs. *World J Gastroenterol.* 2014;20:16489–16497.
2. Minamoto Y, Otoni CC, Steelman SM, *et al.* Alteration of the fecal microbiota and serum metabolite profiles in dogs with idiopathic inflammatory bowel disease. *Gut Microbes.* 2015:6(1):33–47.
3. Honneffer J, Guard B, Steiner JM, *et al.* Mo1805: Untargeted metabolomics reveals disruption within bile acid, cholesterol, and tryptophan metabolic pathways in dogs with idiopathic inflammatory bowel disease. *Gastroenterology.* 2015;148:S-715.
4. Packer RA, Moore GE, Chang CY, Zello GA, Abeysekara S, Naylor JA, Steiner JM, Suchodolski JS, O'Brien DP. Serum D-lactate concentrations in cats with gastrointestinal disease. *J Vet Intern Med.* 2012;26:905–10.
5. Packer RA, Cohn LA, Wohlstadter DR, *et al.* D-lactic acidosis secondary to exocrine pancreatic insufficiency in a cat. *J Vet Intern Med.* 2005;19:106–110.
6. Lin S, Sanders DS, Gleeson JT, *et al.* Long-term outcomes in patients diagnosed with bile-acid diarrhoea. *Eur J Gastroenterol Hepatol.* 2016;28:240–245.
7. Duboc H, Rajca S, Rainteau D, *et al.* Connecting dysbiosis, bile-acid dysmetabolism and gut inflammation in inflammatory bowel diseases. *Gut.* 2013;62:531–539.
8. Kent AC, Cross G, Taylor DR, *et al.* Measurement of serum 7alpha-hydroxy-4-cholesten-3-one as a marker of bile acid malabsorption in dogs with chronic diarrhoea: a pilot study. *Vet Rec Open.* 2016;3:e000163.
9. Washizu T, Tomoda I, Kaneko JJ. Serum bile acid composition of the dog, cow, horse and human. *J Vet Med Sci.* 1991;53:81–86.
10. Pavlidis P, Powell N, Vincent RP, *et al.* Systematic review: bile acids and intestinal inflammation-luminal aggressors or regulators of mucosal defence? *Aliment Pharmacol Ther.* 2015;42:802–817.
11. Johnston IM, Nolan JD, Pattni SS, *et al.* Characterizing factors associated with differences in FGF19 blood levels and synthesis in patients with primary bile acid diarrhea. *Am J Gastroenterol.* 2016;111:423–432.

acvim408

The Placebo Effect in Veterinary Medicine: Interpreting Clinical Study Results in Placebo-Controlled Studies

Linda Rhodes, VMD, PhD
Durham, NH, USA

The word placebo is derived from the Latin "I want to please" and was originally defined as a medicine that was prescribed to please a patient, not for its proven effectiveness. Commonly, placebos are called 'sugar pills' - that is, a pill that might look like a 'real' drug, but only be a dose of sugar, to which the patient responds by feeling better, in the context of some relevant belief or learning.[1] Naively, one could think of this as fooling the patient with a drug or therapy that didn't have a physiological therapeutic effect, but because the patient believed it was a "real" drug that would benefit their condition, they actually felt better. A placebo has its effect because it is expected to have that effect.[1] However, there is significantly more to the placebo effect than merely fooling the patient.

In human medicine, it has been shown that the placebo effect represents "complex neurobiological phenomena"[2] that are triggered by a variety of circumstances, including the form of the placebo administered, the context in which it is given, and the interaction of the therapist and the patient. One famous study showed that the placebo effect in pain studies could be reversed by the administration of naloxone,[3] demonstrating that the placebo treatment likely stimulated the release of endogenous endorphins. Even when patients were told that they were being given a placebo, and the pill bottle clearly was labeled with the word "Placebo," the patients still reported that the pill made them feel better.[4]

But how might this work in studies of animals? For example, when my dog is given a treatment that is hypothesized to reduce her pain from osteoarthritis (OA), she may have some pain relief, and one could expect that a dog's response would not be due to a placebo effect. After all, the dog does not anticipate that the drug will make her feel better; for all she knows, she is getting a treat. How, then, can a dog manifest a placebo effect?

I use the example of dog OA pain because in the veterinary literature, that is the condition that has been most studied using masked, placebo-controlled clinical studies, because those studies are what the FDA requires for approval of pain drugs. Dog OA pain studies show the percentage of placebo-treated dogs that are classified as a treatment success range from 25% to 50% or more. Measuring pain is difficult, and because pain is a subjective sensation, and the dog cannot self report her pain, there are generally three ways that pain is measured in these studies: owner assessments, veterinarian assessments, and force plate analysis. All three of these measurements are subject to being affected by the placebo effect, but in most cases, it is likely the humans participating in the trial that influence or experience the placebo effect; some call this the caregiver placebo effect.

How might this happen? The caregiver thinks the dog's pain is reduced or should be reduced by the treatment, and therefore they imagine they see a pain reduction when none is there. Another explanation is called placebo by proxy - for example, the owner, during the trial, takes better care of the dog, perhaps the dog gets more exercise and attention which may ameliorate pain, and the dog actually does feel better, resulting in the owner and veterinarian scoring the dog as less painful. The placebo effect has also been seen in dogs treated for seizures, where up to 29% of placebo-treated dogs were classified as responders, defined as 50% or more reduction in seizure frequency.[5]

Another possibility might be explained by conditioning: some dogs that are enrolled in a new dog OA study have been treated chronically with an effective pain drug. A dog that gets her pill each morning has come to expect to feel less pain after an hour or so. Now this dog is withdrawn from her pain medication, in anticipation of enrolling in a clinical trial of a new pain drug. When that daily treatment resumes after a few weeks, even if that treatment is a placebo, the dog is conditioned to feel better and does, in fact, experience less pain. In a study of the conditioned placebo effect in dogs, it was shown that separation-related behaviors could be affected by conditioning to a placebo.[6] Whether or not this is true in pain studies could be investigated by comparing the extent of the placebo effect in clinical studies that enrolled dogs that were naïve to treatment of OA, compared to dogs that had been on treatment and washed out before starting the trial, but in many cases, this detail of the enrollment criteria may not be publically available.

An important technique that is used in clinical studies to minimize the placebo effect is blinding or masking. In human clinical trials, the standard is "double blind" - that is neither the patient nor the clinician knows to which group the patient has been assigned. Veterinary clinical trials should be "triple blinded" in a sense. The animal subject, by definition, doesn't know the treatment group, the owner should not know the treatment group, and the clinician should also be blinded. But is that enough? For sufficient blinding, anyone participating in the trial - veterinary technicians, investigators, even scheduling staff, should not know the treatment group of each animal enrolled. Each person that interacts with the animal and the owner has an opportunity to influence, however subtly, the interaction of the owner with her animal, which could have an impact on the data.

It seems intuitive that if objective endpoints, such as body weight or body temperature, are used to measure treatment effects, the placebo effect will be minimized. For example, a large randomized, blinded, multi-site study was conducted in client-owned dogs, where two measurements were used to evaluate the action of capromorelin, an appetite-stimulating drug that has been shown to increase food intake in laboratory dogs.[7] The first endpoint was a subjective questionnaire about the dog's appetite over 4 days of treatment and the other was a measurement in the change in body weight over those 4 days. Treatment success was defined as the owner rating the appetite increasing over the four days of treatment by answering a simple question - "Did your dog's appetite increase, decrease or stay the same?" and 44.6% of dogs treated with placebo showed an increased appetite as reported by the owner, while 68.6% of dogs treated with capromorelin had increased appetite. If the dogs were truly eating more, one would expect body weights to increase, so measuring body weight should be a better measure of whether or not the placebo-treated dogs actually ate more, or if their owners just mistakenly thought they had a better appetite. But examining the body weights in the study, if you classify any increase in body weight as a treatment success (realizing the extent of weight increase over 4 days may only reflect filling up an empty gut), we see 76.0% of the dogs in the capromorelin group gaining some weight (defined as weight gain >0%) while 44.6% of the placebo dogs also showed some weigh gain. Notice that the placebo effect on weight gain is exactly the same as the placebo effect for the subjective measurement of owner rating of appetite!

This example illustrates a few interesting things about the placebo effect. Some dogs in any study, irrespective of treatment group, will 'get 'better' on their own. In the case of appetite, dogs that entered the study with owner's reporting inappetance of a couple of days duration may just have had, for example, a mild GI irritation that self-resolved. If these dogs were allocated to the placebo group, their improvement would contribute to the overall placebo effect. If they were in the treatment group, their improvement would be falsely attributed to a positive effect of the drug. Conducting a clinical trial that is randomized, and of a large enough size, should correct for this by randomly allocating dogs and assuming it is unlikely that all the self-cure dogs would be in one group. Nevertheless, these dogs will contribute to the measured magnitude of both the placebo and treatment group success.

Some clinical studies use comparator groups instead of placebo-treated groups. An example is again in studies of pain-relieving drugs in dogs with OA. Carprofen, for example, was tested against placebo, and shown to be effective in 57.3% of treated dogs on an overall veterinary assessment (improvement of 1 grade) compared to 41.3% of placebo dogs.[8] One would expect that in a study comparing a new veterinary drug for OA (tepoxalin) to carprofen in a comparator study, if the drugs had comparable effectiveness, then both carprofen and tepoxalin would have about 55–60% success rate in overall veterinary effectiveness, even if different questionnaires were used and studies had slightly different study designs. Instead what we see is that in overall veterinary assessment, carprofen-treated dogs had a 90% success rate, compared to 95% for tepoxalin-treated dogs.[9]

We see a similar effect in the clinical study of firocoxib,[10] another comparator study where dogs were treated with either firocoxib or etodolac. At day 29 of the study, veterinarians assessed 87.6% and 83.1% of the firocoxib- and etodolac-treated dogs, respectively, as having overall clinical improvement. This points out a flaw in using comparator studies for assessment of pain in dogs with OA. In a study where the investigator knows that all dogs are getting what is supposed to be an effective treatment, their expectation is that all dogs will improve, and in fact, this is what they see. A study design such as this makes it difficult to impossible to assess the effectiveness of a drug.

In some clinical studies, the high placebo effect also makes it difficult to measure a drug effect; if most of the subjects in the trial get better, then how does one distinguish a drug effect above the background of the placebo effect? In animal health, this is particularly true in studies of pain in cats. In a recent study of the effect of meloxicam in cats with OA,[11] at the study start, owner and investigators knew that all cats would be on placebo treatment for the first two weeks - this was a "practice" part of the study, to get the owners and the cats used to daily treatment. Then half the cats, in a triple-blind fashion, were switched to daily meloxicam treatment. Using an increase of ≥2 on the score of a Client Selected Outcome Measure (CSOM) scoring system as a measure of improvement, over three weeks of treatment, 65% of cats treated with placebo showed improvement, compared to 72% of cats treated with meloxicam, making it impossible to distinguish the effect, if any, of the meloxicam treatment. However, the study had one more arm - after three weeks of treatment, all cats were switched back to placebo treatment, but in a blinded fashion - owners did not know that their cats were all now receiving placebo. More of the owners of the cats that had been on meloxicam for three weeks reported their cat's pain got worse, compared to those who remained on placebo for the entire study. So to summarize this finding, almost all the treated cats got better no matter what their treatment, and while the placebo cats continued to be observed as doing well, only the cats withdrawn from meloxicam were observed to be getting worse, showing that meloxicam could be detected to have had an effect only after it was withdrawn.

In human clinical trials, this study design is called a placebo run-in study - and it has been used effectively to demonstrate a treatment effect in conditions where there is a high placebo effect such as benign prostatic hypertrophy in men,[12] but unlike the study in cats cited above, patients at study start are not informed of their treatment group. I am unaware of any placebo run-in studies that have been submitted to the FDA in support of regulatory review of a veterinary drug.

In order to have confidence that the drug or treatment effect we observe in a clinical study is due to the treatment and not a placebo effect, it is important to strive to minimize the potential biases that can creep into a study. The basics are randomization, to attempt to make groups as equivalent as possible, and blinding (also known as masking) of everyone possible at the clinical site as discussed above. The study also needs to have adequate power to statistically evaluate the potential differences between groups. Best practice is that the placebo should look, taste, smell and have the same texture as the active drug. When this is not possible, then a drug administrator should dispense drug to the owner, with a label that only gives a code, and the owner should not show the drug to the investigator or other people at the clinical site that may be involved in the study. No one conducting any patient evaluations, even simple things such as body weight, should know the treatment group of the animal. Enough patients should enroll to compensate for the percentage of patients in the study that may improve without treatment. When reading published clinical studies, look for a description of these details in the materials and methods, and use them to evaluate whether or not you are convinced that blinding and study power were adequate.

Other things to consider are how owners are informed of the study and the attributes of a new drug; enthusiastic investigators excited about the potential results can affect owner assessments. Anecdotally, we are aware of differences in the placebo effect at different clinical sites, but this has not been studied systematically. It may be easier to enroll studies where the drug: placebo ratio is 2:1, but when this is communicated to the owners, they may have increased expectations that their animal is getting active drug, and skew their evaluations accordingly. Other factors that might affect the placebo effect include the number of measurements taken in the study, the number of clinic visits required, whether or not veterinarians and owners are aware of each other's scores, the length of the study, and even the order in which study procedures are done. There are, to my knowledge, no published data that examine veterinary clinical trials to assess if these factors create bias or changes in the placebo effect, but it seems common sense that standardizing these variables across clinical sites might help the integrity of the data generated.

And finally, how should we think about setting success criteria for veterinary clinical studies? If the success criteria are set so as to make them difficult to meet, for example in pain studies, scores must improve by 50% or more in several different scores, then for each dog it would be harder to be classified as a treatment success. If the placebo effect is just the owner "imagining" that the dog in the pain study is getting better, perhaps these more stringent criteria will be harder to meet for placebo-treated dogs, and therefore the difference between placebo and treatment effect will be magnified. In a study of dogs treated with either placebo or carprofen, where the statistical power of setting various success criteria was examined, this does not appear to be the case.[13]

It is not surprising that we observe a placebo effect in veterinary clinical studies, and we are left with more questions than answers regarding its interpretation. Some end points that may appear objective, such as force plate measurements in pain studies, body weight measurements in appetite studies, or even seizure frequency in studies of epilepsy, may be influenced by subtle effects, such as the attitude of the person performing the test or measurement, the order in which dogs are tested and what is happening in the room. At least part of the placebo effect is what we learned early on in veterinary school - leave animals alone and do nothing, and many will get better on their own, or at least, their condition may wax and wane over time, making interpretation of therapeutic interventions difficult. In some cases, subtle conditioned responses may be involved. Because the clinical study design, including the setting of success criteria, and the details of study conduct differ so markedly across studies, be wary of comparing placebo response rates among different studies, even studies of the same treatment. Knowing the veterinary placebo effect is present in all clinical studies, be cautious of any open label or comparator study that presumes to show a treatment effect.

References

1. Stewart-Williams S, *et al. Psychol Bull*. 2004;130(2):324.
2. Bingel U, *et al. Gerontology*. 2011;57:354.
3. Benedetti F, *et al. Prog Neurobiol*. 1997;52:109.
4. Kaptchuk TJ, *et al. PLoS One*. 2010; doi:10.1371/journal.pone.0015591.
5. Munana KR, *et al. J Vet Intern Med*. 2010;24:166.
6. Sumegi Z, *et al. Applied Anim Behav Sci*. 2014;159:90.
7. Zollers B, *et al. J Vet Intern Med*. 2016;30:1851.
8. FOI summary, NADA 141-053, carprofen (Rimadyl®), 1996.
9. FOI summary, NADA 141-193, tepoxalin (Zubrin®), 2003.
10. FOI summary NADA 141-230, firocoxib (Previcox®), 2004.
11. Gruen ME, *et al. J Vet Intern Med*. 2014; doi: 10.1111/jvim.12312.
12. Nickel JC, *et al. BJU Int*. 2004;93:991.
13. Brown DC, *et al. Am J Vet Res*. 2013;74(12):1467.

acvim409

Changing Perspective on the Pathobiology of Bile Acids - Background Physiology and Diagnostic Use

S.A. Center, DVM, DACVIM
Ithaca, NY, USA

BACKGROUND

Our understanding of bile acid physiology including synthesis, interplay with metabolism, hepatic and enteric pathobiology, and utility in diagnostic testing has undergone a renaissance since recognition that bile acids are natural ligands for the intracellular nuclear farnesoid receptor (*FXR*, bile acid receptor NR1H4 restricted to cells involved with bile acid enterohepatic cycling) and the cell surface receptor TGR5 (G protein coupled bile acid receptor - located in enterocytes, Kupffer cells, sinusoidal endothelium, cholangiocytes, and tissues not participating in the enterohepatic circulation). For more than 5 decades, bile acids have been of interest to gastroenterologists and hepatologists in regard to their physiologic role in fat digestion/assimilation and pathology associated with cholestatic liver injury. There is escalating interest in the signaling capacity of bile acids as emerging disease mechanisms that are likely exploitable therapeutically for disease intervention. Of course of interest to me are emerging strategies for management of liver disease - but the spectrum of potential applications is far wider, impacting numerous pathologic processes external to the liver including the metabolic syndrome NASH/NAFLD so common in humans and in disorders involving the alimentary canal. That TGR5 cell receptor is widely expressed beyond tissues/cells involved with classic bile acid cycling, may add to the growing armamentarium of therapeutic targets applicable to diverse tissues and disease processes. This introductory discussion will lay the foundation for following lectures that will more discretely address emerging roles of bile acids in disease pathobiology.

HEPATIC SYNTHESIS AND MOLECULAR CONFIGURATIONS OF BILE ACIDS

Bile acids (salts) are amphipathic molecules - that is, they possess both hydrophobic and hydrophilic elements (e.g., detergent-like, see diagram below). These molecules uniquely self-associate in aqueous solutions forming simple micelles. Transport of bile salts into the canaliculus provides the major osmotic driving force for bile flow (bile salt dependent bile flow, BSDF). However, bile acid independent bile flow (BSIF) also contributes to bile flow, major contributor being GSH (reduced GSH, GSSG, GSH conjugates). Upon excretion of bile acids across the canalicular membrane, bile salts are rapidly incorporated into micelles and mixed micelles (bile salts, phospholipid, and cholesterol). Phosphatidylcholine (PC) is the major phospholipid in bile whereas cholesterol is the predominant sterol in bile. Bile being the major route of cholesterol elimination. Micelle formation maintains cholesterol solubility and minimizes free bile salt concentrations thereby protecting biliary epithelium from their detergent effect. Formation of mixed micelles is essential for assimilation and digestion of fat-soluble nutrients (fat-soluble vitamins, essential fatty acids). Integration in mixed micelles makes fats available to digestive lipases which otherwise cannot access the inside of lipid droplets (micelles increase surface area enabling digestive contact).

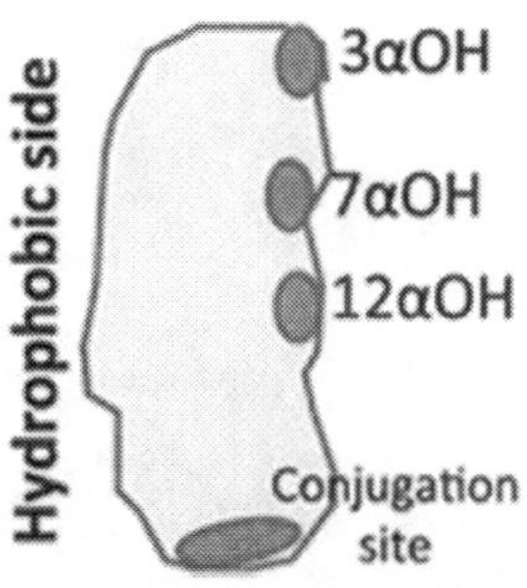

The primary bile acids, **cholic acid** (CA) and **chenodeoxycholic acid** (CDCA), are synthesized in hepatocytes in multistep processes involving the **neutral** (classic) and **acidic** (alternative) **bile acid synthetic pathways**. In health, the neutral pathway is the major synthetic source of CA and CDCA and is principally controlled by *7α-hydroxylase* (*CYP7A1*, a cytochrome p450 monooxygenase) located in the **endoplasmic reticulum**. *Sterol 12α-hydroxylase* (*CYP8B1*) is required for CA synthesis; lesser amounts of CDCA are synthesized by *CYP2A1* in the neutral pathway.

The "acidic" or alternative pathway is controlled by *steroid 27-hydroxylase* (*CYP27A1*) associated with the **mitochondria** (inner mitochondrial membrane) where the rate limiting factor is cholesterol availability. The "neutral" pathway proceeds with initial modification of the sterol ring of cholesterol (neutral products) while the "acidic" pathway initiates cholesterol side chain oxidation before steroid ring modification-forming acidic metabolites. The first step in the acidic pathway involves oxidation of cholesterol to *27-hydroxycholesterol* by sterol *27-hydroxylase* (*CYP27A1*). This is followed by conversion to *7α, 27-dihydroxycholesterol* by *oxysterol 7α-hydroxylase* (*CYP7B1*) - a microsomal enzyme specific to the acidic pathway. CDCA is the main bile acid synthesized by this pathway. This is the bile acid that predominates in the bile acid pool in cirrhotic livers- now understood to be in response to *FXR* modulation that controls enzyme activity in these synthetic pathways.

VARIABLES/TRANSCRIPTION FACTORS INFLUENCING BILE ACID SYNTHESIS

The gene encoding *CYP7A1* is highly regulated by the farnesoid receptor (*FXR*) that imparts feedback repression through an endocrine like response associated with fibroblast growth factor 15/19 (*FGF-15/19*, *FGF-15* **in rodents** and *FGF-19* **in humans**) initiated by enteric bile acid exposure in the ileum. *FGF-19* is released from ileal enterocytes into the portal circulation. It subsequently binds activates *fibroblast growth factor receptor-4* (*FGFR-4/β-Klotho complex*) which in hepatocytes activates the *JNK1/2* and *ERK1/2* signaling cascades. Activation of the *JNK1/2* pathway down regulates hepatocellular *CYP7A1* mRNA-effecting general suppression of bile acid synthesis and specifically, repression of CA synthesis. The important point here is that enteric *FGF-15/19* - synthesized in the intestines under activation by *FXR* has important negative regulatory feedback impacting hepatic bile acid synthetic pathways. This can have important consequences on each organ system. In humans, *FGF-19* displays a diurnal flux coordinated with postprandial gallbladder contraction and subsequent bile salt stimulated *FXR* expression in the ileum. Other regulatory factors that impact on the neutral synthetic pathway include down regulation of *CYP7A1* by glucagon and pro-inflammatory cytokines and its upregulation by glucose and insulin during the postprandial interval.

In the 1990s, discovery of a number of ligand-activated transcription factors of the (NHR) family led to the recognition that bile acids were endogenous ligands- introducing consideration that bile acids in μM concentration might act as signaling molecule. At this point, while it was realized that bile acids somehow modulated their own synthesis, mechanism(s) remained elusive. *Nuclear receptors* (*NR*s) are now realized as the most abundant class of transcriptional regulatory factors (a superfamily). These control numerous aspects of mammalian development and physiology. *NR* are ligand-activated "hormone receptors" functioning as a direct link between signaling and transcriptional responses. These are divided into 4 groups based on ligand-binding capacity, dimerization properties, and cellular localization (simplified version). *NR* lacking an identified ligand have been called "orphan receptors" (usually Class III or Class IV groups). In some cases, after initial discover, ligands have been subsequently discovered - the case for *FXR*. There are some *NR* that appear to activate gene expression without recognized ligand binding. Bile acid synthesis and enterohepatic circulation of bile acids are tightly controlled processes be a cadre of *NR*, the most notable being *FXR*. This receptor was originally classified as an orphan *NR* until investigators noted a ligand-type interaction with farnesol and other molecules. Soon thereafter broad ligand binding-specificity was recognized, notably and importantly including bile acids, along with numerous synthetic ligands, steroids, aromatics, terpenoids, alkaloids and fatty acids (there are more). This broad repertoire of ligand binding specificity, is typical for receptor diversity shown by *NR*s as these interact with diverse metabolic pathways. Of recognized ligands, primary and secondary bile acids are **the most physiologically relevant** with potency to activate *FXR*. The stimulating potency among different bile acid moieties from highest to least includes: CDCA>DCA>LCA>CA, refined based on study of the *FXR* null mouse. The amphophilic bile acid structure, its concave hydrophilic "face" with projecting OH groups and the its convex hydrophobic surface allow bile acids to snuggly fit against the hydrophobic pocket of the *FXR* ligand binding domain. Current understanding of the physiologic impact of *FXR* includes a breadth of metabolic processes including: hepatic bile acid transport, metabolism and synthesis, and influences affecting lipid and glucose metabolism, drug disposition, liver regeneration, hepatic and enteric inflammation, fibrosis, cell differentiation, and neoplastic transformation. It also imparts anti-inflammatory and immunomodulatory effects, modulates enteric integrity, permeability, and the enteric biome. Many of these interactions are now being exploited as possible mechanisms of disease pathogenesis and intervention - beyond the scope of this presentation.

NUCLEAR RECEPTORS

Nuclear receptors modulate the neutral and acidic bile acid synthetic pathways. The ratio of CA:CDCA is predominantly determined by activity of sterol *12a-hydroxylase* (*CYP8B1*). This enzyme is regulated by an orphan *NR* (**small heterodimer partner=SHP**) activated by FXR upon stimulation by enteric bile acids. Through interaction with several additional transcription factors (*hepatocyte nuclear factor-4* [*HNF-4a*] and *liver-related-homolog-1* [*LRH-1*]) *CYP8BI* activity is repressed favoring CDCA production. Expression of the gene encoding *steroidogenic acute regulatory protein* (*STARD1*) also can effect a 5-fold increase in synthesis of "acidic" bile acids (mitochondrial pathway). This transcriptional response can suddenly increase cholesterol availability, the rate limiting factor for the acidic synthetic pathway. This effect far exceeds the impact of *CP27A1* upregulation on CDCA production that induces a mere < 2-fold increase. *Liver X receptors* (*LXRs*) function as *NRs* sensing oxysterols. These regulatory genes also influence cholesterol and lipid metabolism. *LXRs* are classified as *LXR-a* (*NR1H3*) or *LXR-β* (*NR1H2*) based on their coding genes. *LXR-α* is highly expressed in liver and to a lesser extent in kidney, small intestine, spleen, and adrenals whereas *LXR-β* is ubiquitous. The "acidic" or adaptive pathway of bile acid synthesis is an important source of regulatory oxysterols - e.g., *25-hydroxycholesterol* and *27-hydroxycholesterol*. Thus, through regulation of this pathway, the natural ligands of *LXRs* influence cholesterol and lipid metabolism. *25-hydroxycholesterol* (formed via *CYP27A1*) can be sulfated in the liver and is now recognized to influence inflammation, lipid metabolism, as well as cell proliferation. It also functions as an activation ligand for the peroxisome proliferator receptor-γ (PPAR-γ) - linked with inflammation and lipid metabolism.

The membrane bound bile acid receptor - *TGR5* (*GPBAR1* or *M-BAR/BG37*), is a G-protein coupled bile acid receptor that does not depend on transport systems for cell ligand interaction. *FXR* is largely restricted to organs directly interacting with the enterohepatic circulation (liver, intestines, also defined in kidney, adrenals) whereas *TGR5* is systemically broadly expressed where it is thought to mediate bile acid initiated responses. It also is considered that this receptor provides an avenue for systemic effects and a diversity of cell types to influence bile acid signaling that is likely associated with disease related processes, perhaps even hepatic encephalopathy.

BILE ACID CONJUGATION

Once formed, primary bile acids undergo enzyme catalyzed conjugation with an amino acid (amidation human this can be either glycine or taurine [75%:25%, respectively], it is predominantly taurine in dogs, and exclusively taurine in cats. Amidation involves amino acid conjugation to the terminal side-chain carboxylic acid. Amino acid conjugation is orchestrated by the enzyme *bile acid CoA: amino acid N-acyltransferase* (*BAAT*). This enzyme is predominantly localized in peroxisomes (microbodies) with smaller amounts measurable in the cytosol of hepatocytes. There is some consideration that the cytosolic enzyme reflects enzymes "spilled" from the peroxisomal activity. Primary peroxisomal location of BAAT suggests it conjugates newly synthesized primary bile acids predominantly. The cytosolic BAAT may function for re-conjugating recycled primary and secondary bile acids, de-conjugated by enteric bacteria during enterohepatic circulation (see below). Conjugation of bile acids increases their water solubility and miscibility in the alimentary canal and also reduces potential adverse effects. Despite conjugation, bile aids are negatively charged over most of the physiological pH range- the term bile salt is more appropriate terminology. The chemical characteristics of conjugated bile acids (salts) necessitate specific transport proteins/receptors for efficient cell permeation- these are largely confined to intracellular signaling molecules and transporters confined to the small intestines and liver.

Hepatic Bile Acid Transport

Once conjugated to an amino acid, bile acids are actively transported from liver into bile. In health, this process renders a highly concentrated bile acid pool in the gallbladder (~100 mM total bile acids) compared to liver tissue (~20 μM total bile acids), small intestines (~2–10 mM total bile acids), serum (~2–15 μM total bile acids) and urine (~1 μM sulfated bile acids). Bile acids synthesized in hepatocytes cannot diffuse across membranes into circulation. They must have facilitated transport into bile (membrane transporter proteins). The most important transporter for bile acid movement into canalicular bile is the bile salt export pump (*BSEP, ABCB11*). This works in conjunction with the phospholipid transporter (*ABCB4/MDR2*-rodents/*MDR4* humans) and cholesterol transporter (*ABCG5/ABCG8*). On transport of phospholipids across the canaliculus, they facilitate formation of mixed micelles containing cholesterol, bile acids, and phospholipids. These effectively increase solubility of bile acids and minimize toxic impact of bile acids on biliary epithelium. Both *BSEP* and *ABCB4* are under control of *FXR* that also critically regulates bile acid flow by mediating expression of hepatic *BSEP* and *ABCB4*. Canalicular bile acid secretion is the main force driving bile flow - bile acid dependent predominated by *BSEP* and non-bile acid dependent predominated by MRP-2 (GSH transport), **see diagram below**.

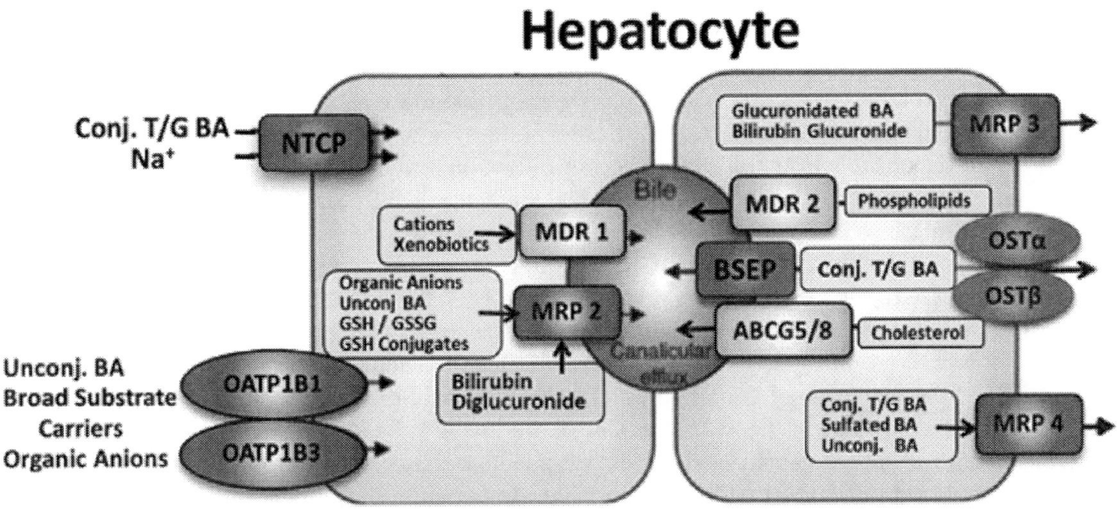

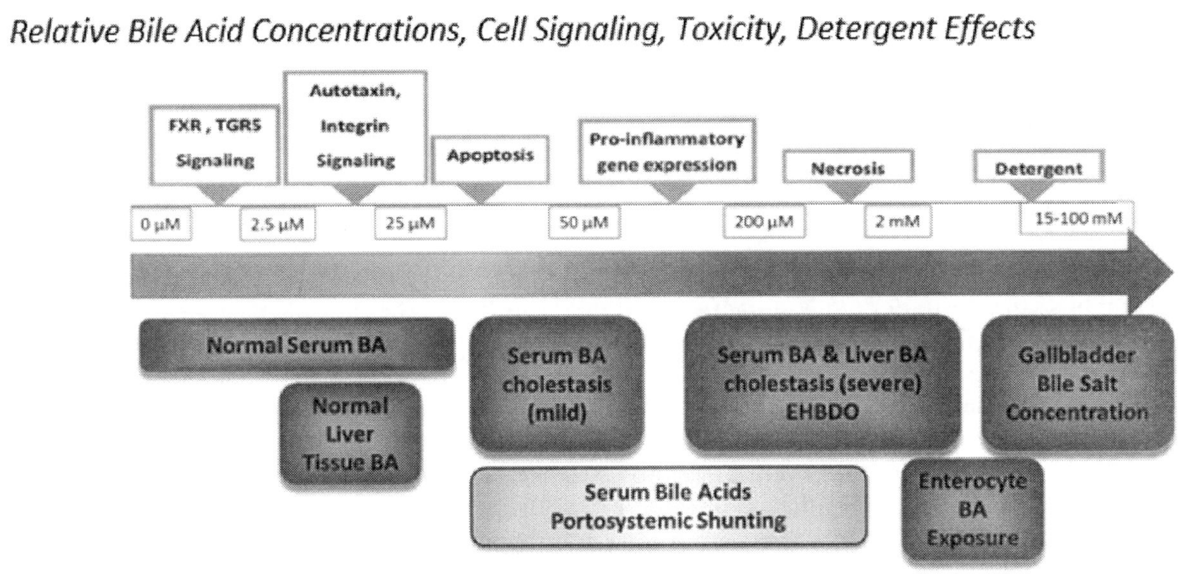

ENTEROHEPATIC CIRCULATION AND DIGESTIVE FUNCTIONS OF BILE ACIDS

Bile acids are concentrated in the gall bladder before release into the duodenum; concentrations can reach 100–300 mM/L. In response to meal ingestion, the neural enterogastric reflex and hormonal enterogastrone mechanisms stimulate gastric emptying. This provokes cholecystokinin that along with the enteric migrating motor complex orchestrates coordinated GB contraction synchronized with relaxation of the sphincter of Oddi. Ejection of concentrated bile into the duodenum (mM concentrations) concurrent with release of pancreatic juices initiates lipid digestion and mechanically cleanses the biliary tree of potential opportunistic bacteria. During the interdigestive interval, most of the bile acid pool collects within the GB or bile ducts. A bellows-like "flutter" of the GB during the interdigestive interval allows for small volume bile flow at all times. This associates with phasic contractions of the sphincter of Oddi and incremental rise in GB pressure. This maintains bile acid enterohepatic circulation in the interdigestive period. Meal composition (fat and protein content) and gastric acidity provoke cholecystokinin and secretin release, and along with neurohumoral stimuli mediate meal associated increased bile and HCO_3 - flow. The coordinated flux of bile acids and HCO_3 - rich biliary secretion assists with activation and function of enteric pancreatic digestive enzymes. Enteric bile acids also exert an early negative feedback control on postprandial cholecystokinin release, limiting the number of enterohepatic bile acid cycles per meal.

Bile acids are independent structures until a threshold concentration is reached that exceeds their *critical micellar concentration* (CMC). This is the point at which they spontaneously form micelles. Mixed micelles incorporate numerous polar bile acid molecules surrounding hydrophobic or insoluble fat particles. The detergent-like of bile acids enables pancreatic lipase to digest emulsified fat into monoglycerides and fatty acids. Bile acids facilitate enteric assimilation of these and fat-soluble vitamins. Bile acids also are indispensable for cholesterol bile solubilization and enteric uptake. Below the molecule specific CMC, bile salts behave as 1:1 strong electrolytes. Molecular configuration of hydrophobic and hydrophilic surfaces differ among the various bile salt structures. Physicostructural diversity influences solubilization of phospholipids, cholesterol and other lipids into mixed micelles. Hydrophilicity of common free & conjugated bile salts decrease in the order of: UDCA>CA>CDCA>deoxycholic acid (DCA)>lithocholic acid (LCA) with taurine conjugated>glycine conjugated>free species. These differences reflect orientation of the hydrophobic core and planar structure in relationship to configuration of the hydroxyl groups. Maximal polarity and hydrophilicity is reached when hydroxyl groups align along the same side of the molecule. Consequently, CA, DCA, and CDCA with α-OH groups, are the most efficient tensioactive bile acids (tensioactive material is one that changes the value of surface tension) compared to UDCA that has hydroxyl groups in both α- and β-configurations. One-sided single directed OH groups affords greater miscibility with other amphipathic configurations (i.e., biliary phospholipids). Having one hydrophilic face and one hydrophobic face is the most useful configuration for mixed micelle formation.

Modest enteric bile acid uptake can occur by passive uptake of conjugated bile acids along the small and large intestines. However, the majority of bile acid uptake is under *FXR* control. This occurs by regulation of four essential membrane transporters: 1) *enterocyte apical sodium-depended transporter* (*ASBT, SLC1042*-solute carrier family 10 member), 2) *bile acid-binding protein subclass 6* (*FABP6*, ***intestinal bile acid-binding protein***) and transporters essential for moving bile acids from the enterocyte into the portal circulation-3) *organic solute transporter alpha* (*OSTα*), and 4) *OSTβ*. The enterocyte ASBT apical Na-dependent bile salt transporter (*ileal BA transporter* [*IBAT; SLC10A2*]) determines active uptake of bile acids into the terminal ileum. This highly efficient transporter reabsorbs ≥90% of enteric bile acids into the portal circulation. The efficiency of the enterohepatic circulation recycles bile acids ~5–12 times per day. It has been estimated that a single bile acid molecule may undergo an average of 20 enterohepatic circulations before being lost with feces. In humans, <1 g of bile acids may reach the colon per day- this is < 5% of the BA pool. So, in addition to the other processes discussed to this point, the *enterohepatic bile acid circulation* **also** is under tight regulation by the *FXR*. Bile acids exert a negative feedback on *ASBT* expression via *FXR-SHP* induction - *SHP* repressing transcriptional activation properties of *LRH-1*. Once inside the ileal enterocyte, bile acids reversibly bind with *FABP6* (expression increased by bile acid binding to *FXR*); this shuttles bile acids from the apical to basolateral membrane where, *OSTα* and *OSTβ* bind to and transport them into the portal vein. Bile acids passed into the colon may be passively absorbed (deconjugation facilitated), molecularly transformed, or eliminated with feces. The high efficiency of the enterohepatic circulation limits bile acid exposure to tissues external to the liver. This tight system was medically exploited as an early method for lowering serum cholesterol in humans. Oral administration of cholestyramine (Questran®, a positively charged bile acid sequestrant, or colestipol or colesevelam) were used to bind negatively charged bile acids, limiting their enteric uptake. This provoked fecal bile acid and cholesterol wasting. With chronic use, substantial bile acid loss lowered hepatic bile acid concentrations stimulating enhanced bile acid synthesis. This process consumed cholesterol by creation of a fecal bile acid and cholesterol "sink". Complications included depletion of fat soluble vitamins and essential fatty acids causing serious nutritional deficiencies. Response to this process involved *FXR*.

The type of bile acids, the quantity resorbed vs. the quantity passed unabsorbed into the colon are inconstant and reflect physiologic variables impacting the enterohepatic bile acid circulation. These variables include: variation in dietary composition, enteric transit time, drug therapies, or enteric disease. The mixture of bile acids in bile typically represents a balance between primary and secondary bile acids. Secondary bile acids are formed during enteric transit by bacterial-deconjugation and hydroxyl group oxidation. In humans, a small subset of anaerobic colonic bacteria orchestrate 7α-dehydroxylation, deconjugation, and oxidation/epimerization of hydroxyl groups at C3, C7, and C12. The most common biotransformation involves dehydroxylation at C7 by bacterial dehydratases. Removal of this OH group yields a 7-deoxy bile acid; i.e., cholic acid → dehydrocholic acid (DCA) and chenodeoxycholic acid → lithocholic acid (LCA). Deconjugation by bacterial bile-salt hydrolases and removal of the C7 hydroxyl group (7α-dehydroxylation) increases molecular pKa and bile acid hydrophobicity, allowing for greater passive uptake of secondary bile acids across the colonic mucosa. In humans and rodent disease models, concentrations of secondary bile acids in feces, blood, and bile are linked to pathogenesis of cholesterol gallstone disease, enteric inflammation, and colon cancer. Bile acid modification by enteric bacteria can have several important consequences. First, it can favor passive colonic bile acid absorption by those bile acid moieties escaping ileal uptake - as noted above. Second, it increases potential for noxious enteric effects - can cause chronic severe diarrhea. Third, the bile acid pool varies when conditions of biotransformation are altered- as for example with altered enteric transit time or shifts in the typical enteric biome - effects introduced by drugs, diet composition, and disease processes. In humans and rodents, consumption of a high-fat diet and slowed enteric transit favors generation of DCA from CA and greater DCA absorption, increasing risk for cholesterol gallstones and neoplastic transformation. A high-fat, low-fiber diet can lead to a >10-fold increase in colonic delivery of taurine conjugated bile acids - conjugation impairing ileal absorption. Being overweight with accelerated intestinal transit leads to "idiopathic bile acid malabsorption". Because C7-dehydroxlation cannot be reversed by host enzymes, chronically increased LCA and DCA lead to increased accumulation of secondary bile acids within the bile acid pool - these have greater risk for hepatotoxicity.

HEPATIC UPTAKE OF BILE ACID

Major bile acid transporters orchestrating basolateral hepatocyte membrane uptake of bile acids are 1) the Na^+-taurocholate co-transporting polypeptide (*NTCP; SLC10A1*) and the organic anion-transporting peptides (*OATPs*). *NTCP* facilitates uptake of conjugated bile acids, while *OATPs* control uptake of unconjugated bile acids. The expression and activity of these transporters are influenced by numerous factors among which is negative feedback by decreased bile acid concentrations and *FXR* on *NTCP*. endotoxin and proinflammatory interleukin IL-1b also are known to reduce *NTCP* expression (see diagram above).

Major bile acids in human bile are CA, CDCA, DCA with minor amounts of UDCA and LTA. Feces contain mainly DCA, LCA, and minor amounts of CDCA, CA, and UDCA and a variety of bacterial transformed derivatives.

Major bile acids in plasma of healthy cats include: TCA>TCDCA>TDCA; <1% TLCA. Major bile acids in plasma of healthy dogs include: TCA>TDCA>TCDCA; <3.5% TLCA. Plasma and gallbladder bile from 6 studies examining bile acid profiles in healthy dogs and cats, animals with extrahepatic bile duct obstruction, plasma of cats with hepatic lipidosis, and plasma of germ-free dogs - are summarized in the table below. Fractionation done by HPLC or HPLC-Mass Spec or in germ-free dogs GLC. Fraction of plasma TDCA (secondary bile acid) is low in germ-free bile - reflecting absence of secondary bile acid formation in the alimentary canal (conjugated forms not differentiated in this study, measured by gas liquid chromatography. Control dog data entered for comparison). Not shown in this table are details derived from an investigation of bile acid profiles in dogs on 8.5 mg/kg PO prednisone for 84 days (6 treated, 6 placebo dogs). Daily high dose prednisone led to reversible reduction in taurine conjugated bile acids and a reversible increase in non-conjugated bile acids (more hydrophobic). This shift might contribute to the tenacious nature of mucin in mature GB mucocele aggregates.

Percentage of bile acid composition: plasma

	TCA	TDCA	TCDCA	CA	CCA	TLCA	DCA	GLCA
Dog normal	55.1%	26.2%	11.3%	2.9%	1.2%	None	None	None
Dog EHBDO	77.4%	3.1%	11.6%	None	None	None	None	1.9%
Cat normal	30.0%	13.3%	15.1%	30.1%	1.6%	1.1%	2.6%	None
Cat EHBDO	41.0%	1.0%	3.0%	2.0%	0.3%	None	None	None
Cat HL	22.0%	0.8%	2.1%	0.5%	0.3%	None	None	None

Percentage of bile acid composition: gallbladder bile

	TCA	TDCA	TCDCA	CA	CDCA	TLCA	DCA	GLCA
Dog normal	73.2%	20.9%	6.0%	0.1%	None	0.1%	None	None
Dog EHBDO	80.2%	3.5%	16.3%	None	None	None	None	None
Dog germ free	ND	ND	ND	94.9	4.6	ND	None	ND
Controls for germ-free dogs	ND	ND	ND	81.5	3.7	ND	12.1	ND
Cat normal	74.33%	13.5%	10.3%	0.3%	None	0.1%	None	None

Germ-free dogs no conjugation types differentiated, reported as non-conjugated

Diagnostic Application of Bile Acids in Dogs and Cats

Total serum bile acid (TSBA) measured enzymatically are routinely used in veterinary medicine. This test has been rigorously evaluated and proven as a diagnostic test for liver disease and detection of portosystemic shunting. While bile acids do not directly indicate presence of hepatic encephalopathy (HE), they sensitively detect causal hepatic dysfunction or perfusion abnormalities. TSBA supersedes utility of ammonia owing to their stability and convenience for detection of portosystemic vascular anomalies (PSVA). Best utility occurs with paired samples, one collected before meal consumption and the other ~2-h postprandial (PSBA). Either serum or heparinized plasma can be used; the latter permits submission of smaller blood volumes and faster sample processing. There is **no need to fast** for this test- the point of the test is to assess the dynamic enterohepatic BA challenge imposed by meal consumption that can reflect abnormalities in liver perfusion, hepatocyte function, and cholestasis. The test is best used adjunctively with routine biochemical parameters. For cholestatic disorders, it has higher sensitivity than bilirubin such that suspected hepatic hyperbilirubinemia relegates the test as superfluous.

Problems with the TSBA test involve interference by hemolysis and lipemia, the complexity/variable influencing dynamics of the enterohepatic "challenge", and potential for overlooking liver disease when only random or fasting SBA values are considered. Hemolysis interferes with enzymatic colorimetric testing because hemoglobin shares an absorption spectra with the endpoint dye. Lipemia can substantially interfere in endpoint detection causing spuriously high or low values -

with enzymatic colorimetric testing (two reactions run, one with 3α-OH steroid dehydrogenase that dehydroxylates the 3α position, the other without enzyme, color difference quantifies bile acid values). Lipemic samples should be clarified (chylomicrons removed by sample chilling and high speed centrifugation) or just ultracentrifugation. Understanding the complexity of the enterohepatic endogenous "challenge" is important in trouble shooting unusual test results: 1) 15–20% of dogs tested, 10–15% of cats tested, demonstrate "higher" fasting than postprandial SBA concentrations; 2) the need to feed a patient enough food to stimulate gastric motility and neurohumoral signals provoking gallbladder contraction, bile flow, and enteric motility; 3) test vulnerability to enteric disease causing malabsorption or too rapid enteric transit that may limit endogenous bile acid challenge; 4) some animals have a later optimal postprandial collection interval; and 5) physiologic variables cause TSBA value vacillation in dogs with microvascular dysplasia (MVD) and congenital macroscopic portosystemic vascular anomalies (PSVA) that some breeders insist is meaningful if sequential values are lower or higher than before. FSBA>PSBA reflect differences in gastric emptying and enteric motility, bile bypassing the gallbladder as occurs normally during the interdigestive interval (bellows like fluttering of the gallbladder). Unconjugated SBA quantitatively increase during the late postprandial interval owing to enhanced formation, higher rate of passive enteric uptake, and lower hepatic uptake compared to conjugated forms, however, only very small concentrations accumulate with intestinal bacterial overgrowth. That said, with the new signalling paradigm, what was once considered trivial is now recognized as reactive with NR & FXR signaling.

While some clinicians use TSBA for serial patient appraisal, there may be limited merit to this practice. In most chronic necroinflammatory liver disorders, architectural change and disturbed perfusion are unlikely to improve. In PSVA, permanent intrahepatic microvascular malformations (MVD). A quantitative association between absolute TSBA concentrations and the "degree", "severity" or nature of liver disease has never been proven (except that dogs with PSVA usually have markedly increased TSBA values, there may be a "shunting pattern" with low or normal fasting values that markedly increase at the 2-h postprandial test. Marked differences in TSBA values on enterohepatic bile acid challenge can appear within an individual at different sampling intervals or on different days at the same sampling interval. For example, there is no quantitative value in comparing a one day test of 90 µmol/L against a value of 45 µmol/L or 140 µmol/L - and especially common conundrum in dogs with MVD. Normalization of values would be the exception (unless there is enteric bile acid malabsorption). The need to withdraw treatment with ursodeoxycholic acid (UDCA) prior to submission of a TSBA test is often recommended by diagnostic laboratories. A study we conducted of UDCA (15 mg/kg PO) on serial TSBA concentrations in normal dogs (n=14) with blood collected q2h over a 12-h interval confirmed that 40% develop abnormally increased TSBA values. While median TSBA values remained in reference range, higher values developed (on inspection) at all intervals following oral UDCA with significant increases at the 2-, 3-, and 4-h intervals. Erroneous diagnosis of hepatobiliary "dysfunction" based on TSBA could therefore occur in dogs with normal liver function given UDCA.

Changes in serum bile acids with variable physiologic conditions (literature referenced)

	Total Serum Bile Acid Concentrations ***Fasting***	***Postprandial***	***Expected Qualitative Profile***
Normal Subjects			Standard for comparison: trihydroxy > dihydroxy bile acids
Ileal Dysfunction / Ileal Malabsorption	Subnormal	Subnormal, ↓ postprandial rise	↑↑ Unconjugated moieties
Fat Malabsorption / Maldigestion	Subnormal	Subnormal, ↓ postprandial rise	↑Unconjugated moieties
Bacterial Overgrowth	Variable	Variable	↑↑↑Unconjugated moieties, BUT does not pathologically increase total bile acid concentrations
Cholecystectomy Biliary Diversion	Normal to Slightly ↑	Reduced rise, early postprandial rise, postprandial may be equivalent to pre-meal	↑ Unconjugated moieties ↑ DCA vs CA
Delayed Gastric Emptying	Slightly ↑, exceeds postprandial	Expected increase not realized	Normal
Rapid Intestinal Transit	Subnormal, variable	Subnormal	↓ Unconjugated
Delayed Intestinal Transit	May be increased, exceeds postprandial	Postprandial may be less than fasting value	↑ Unconjugated
Portosystemic Shunting (congenital & acquired)	Normal after prolonged fast or ↑	Profound ↑↑↑, may see an acute change between fasting & postprandial values	↑ trihydr:dihydroxy ratio
Reduced Hepatic Mass	May be normal or↑ after prolonged fast	an acute change between fasting & postprandial	↑ trihydroxy:dihydroxy ratio, sulfated moieties (cat, not dog)
Hepatic Failure	Usually ↑↑	an acute change between fasting & postprandial	↓ trihydroxy:dihydroxy ratio ↑sulfated moieties (cat, not dog)
Cholestasis	Profoundly ↑↑↑	Profoundly ↑↑↑, fasting & postprandial have similar values	May have ↓ unconjugated, ↑↑ sulfated (cat), ↑↑ glucuronidated, ↓ trihydroxy: dihydroxy ratio

There is considerable interaction between bile acids and enteric flora during enterohepatic transit of bile acids. It has long been acknowledged that bile acids impose important regulatory influences on the enteric biome by regulating the bacterial population, suppressing growth of opportunistic pathogens, protecting against paracellular translocation, and improving intestinal integrity. Potential for cross talk between enterocytes, bacteria and bile acids in preventing disease and also in having pathologic consequence - is already recognized. Relevant to cases I often attend, absence of an enteric bile acid flux in patients with biliary tree obstruction is a recognized hazard permissive to complicating bile-borne bacterial infections that increases surgical and anesthetic risk.

ADDITIONAL BILE ACID RECEPTORS

Additional bile acid receptors discussed above are *FXR* and *TGR5*, but additional receptors may be relevant to disease processes or hepatic histologic responses in companion animals. Bile acids are relevant ligands for the *pregnane X receptor* (*PXR, NR1I2*), the *constitutive androstane receptor* (*CAR, NR1I3*), and vitamin D_3 *receptor* (*DR, NR1I1*). Although bile acids are one of many ligands for *PXR* and *CAR* they also respond to numerous unrelated xenobiotics. This repertoire of ligand interactions might relate to the common observation of glycogen-type vacuolar hepatopathy in dogs with diverse systemic disease processes. One study in rodents demonstrated that glucocorticoids (dexamethasone) blocked transcriptional activity of *FXR* contributing to hepatic cholestasis.

Novel interventional medications in clinical trials for human liver disease are predicated on the discoveries of bile acid signaling over the last decade. Briefly here is food for thought. See recent Trauner references.

Obeticholic acid (OTA) is the first in class agent of *FXR* agonists. With CDCA considered its natural ligand - *FXR* has a key role in bile acid homeostasis. OTA is a synthetic derivative of CDCA (6a-ethyl-CDCA) with remarkably higher antagonism compared to CDCA for FXR (~100 fold). OTA directly engages with FXR to influence bile acid synthesis, secretion, enteric transportation, absorption, and detoxification. Studies in models of liver disease demonstrate that OTA has potent anti-cholestatic effect. Trials in humans with primary biliary cirrhosis (PBC) dosed at: 10, 25, and 50 mg/day demonstrate increased serum FG19 concentration and reduced C4 precursor of bile acids- these two findings support FXR mediated repression of bile acid synthesis. Other clinicopathologic and serologic markers of PBC improved beyond effects achieved with UDCA. OTA side effects in humans may include debilitating pruritus (traditionally attributed to high bile acid concentrations - now likely *FXR* related). It also associates with unexplained significant decrease in high-density lipoprotein cholesterol with unchanged total cholesterol or with increased total cholesterol & increased low density lipoprotein cholesterol (observed in humans with NASH/NAFLD). This is undergoing NIH sponsored trials in combination with UDCA for PBC and NASH/NAFLD.

Nor-ursodeoxycholic acid (*NorUDCA*) is a side chain shortened UDCA derivative lacking a methylene group. It is thus resistant to amidation with taurine or glycine and can be passively absorbed from cholangiocytes. It undergoes "cholehepatic" shunting instead of complete enterohepatic cycling. The cholehepatic shunting at the level of the cholangiole generates and sustains a HCO_3^- anion resulting in induction of a HCO_3^- hypercholeresis. It is the strongest secretagogue of HCO_3^- identified to date. It is thought to protect cholangiocytes from hydrophobic bile acid induced injury, an important mechanism contributing to small duct injury cholangitis (immune-mediated destructive cholangiopathy). Neither Ursodeoxycholic acid or *NorUDCA* display relevant pathologic affinities for the *FXR* or other receptors. *NorUDCA* also has anti-lipotoxic, antiproliferative, antifibrotic and anti-inflammatory effects. This is undergoing clinical trials for PBC and is near FDA approval.

See ref: **Corpechot** C: Primary biliary cirrhosis beyond ursodeoxycholic acid. *Semin Liver Dis*. 2016;36:15–26.

FGF19 analogs: Proposed targeted outcome-hepatocyte *FGFR4* activation, *CYP7A1* repression, inhibition of bile acid synthesis, reduced circulation of the bile acid pool.

Second generation FXR agonists: Proposed targeted outcome - non-steroid, achieve *FXR* activation without bile acid specific effects with the goal of limiting human complication of pruritus.

TGR5 agonists: Proposed targeted outcome - activation of specific bile acid membrane receptors on cholangiocytes, achieve enhanced HCO_3^- secretion to improve choleresis, thwart bile acid "overload".

ASBT inhibitors: Proposed targeted outcome - interrupt enterohepatic circulation of bile acids, reduce circulating bile acid pool to reduce receptor interactions, reduce pruritus related to bile acids and cholestasis in humans.

UDCA: Effects are mediated at a post-translational level influencing *FXR*.

References

1. Alrefai WA, Gill RK. Bile acid transporters: structure, function, regulation and pathophysiological implications. *Pharm Res*. 2007 Oct;24(10):1803–23.
2. Beaver MH, Wostmann BS, Madsen DC: Bile acids in bile of germfree and conventional dogs. *Proc Soc Exp Biol Med*. 1978 Mar;157(3):386–9.
3. Beuers U, Trauner M, Jansen P, Poupon R. New paradigms in the treatment of hepatic cholestasis: from UDCA to FXR, PXR and beyond. *J Hepatol*. 2015 Apr;62(1 Suppl):S25–37. doi: 10.1016/j.jhep.2015.02.023.
4. Boyer JL: Bile formation & secretion. *Compr Physiol*. 2013;3:1035–78.
5. Camilleri M1, Klee EW, Shin A, *et al*. Irritable bowel syndrome-diarrhea: characterization of genotype by exome sequencing, and phenotypes of bile acid synthesis and colonic transit. *Am J Physiol Gastrointest Liver Physiol*. 2014 Jan 1;306(1):G13-26. doi: 10.1152/ajpgi.00294.2013.
6. Corpechot C. Primary biliary cirrhosis beyond ursodeoxycholic acid. *Semin Liver Dis*. 2016;36:15–26.
7. Gonzalez FJ. Nuclear receptor control of enterohepatic circulation. *Compr Physiol*. 2012; 2: 2811–28.

8. Hofmann AF. The enterohepatic circulation of bile acids in mammals: form and functions. *Front Biosci*. 2009;14: 2584–98.
9. Hofmann AF. Biliary secretion and excretion in health and disease: current concepts. *Ann Hepatol*. 2007;6:15–27.
10. Jansen PL, Ghallab A, Vartak N, *et al*: The ascending pathophysiology of cholestatic liver disease. *Hepatology*. 2017; Feb;65(2):722–38. doi: 10.1002/hep.28965.
11. Kook PH, Schellenberg S, Rentsch KM, *et al*. Effect of twice daily oral administration of hydrocortisone on the bile acids composition of gallbladder bile in dogs. *Am J Vet Res*. 2011;72:1607–12.
12. Komati R1, Spadoni D2, Zheng S3, Sridhar J4. *Ligands of Therapeutic Utility for the Liver X Receptors. Molecules*. 2017; Jan 5;22(1). pii: E88. doi: 10.3390/molecules22010088.
13. Lu Y, Zhang Z, Xiong, *et al*. Glucocorticoids promote hepatic cholestasis in mice by inhibiting the transcriptional activity of the Farnesoid X Receptor. *Gastroenterology*. 2012;143:1630–1640.
14. McMillin M, DeMorrow S. Effects of bile acids on neurological function and disease. *FASEB J*. 2016Nov;30(11):3658–68.
15. Schaap FG, Trauner M, Jansen PLM. Bile acid receptors as targets for drug development. *Nat Rev Gastroenterol Hepatol*. 2014;11(1):55–67.
16. Taylor W. The bile acid composition of rabbit and cat gall-bladder bile. *J Steroid Biochem*. 1977Oct;8(10):1077–84.
17. Trauner M, Fuchs CD, Halilbasic E, Paumgartner G. New therapeutic concepts in bile acid transport and signaling for management of cholestasis. *Hepatology*. 2016Dec20. doi: 10.1002/hep.28991.
18. Trivedi PJ, Hirschfield GM, Gershwin ME. Obeticholic acid for treatment of primary biliary cirrhosis. *Expert Rev Clin Pharmacol*. 2016;9:13–26.
19. Washizu T, Ishida T, Washizu M *et al*. Changes in bile acid composition of serum and gallbladder bile in bile duct ligated dogs. *J Vet Med Sci*. 1994;56:299–303.
20. Washizu T, Ikenaga H, Washizu M, Ishida T, Tomoda I, Kaneko JJ. Bile acid composition of dog and cat gall-bladder bile. *Nihon Juigaku Zasshi*. 1990Apr;52(2):423–5.
21. Zhou H, Hylemon PB. Bile acids are nutrient signaling hormones. *Steroids*. 2014Aug;86:62–8. doi: 10.1016/j.steroids.2014.04.016.

acvim410

Diagnosis of Immune-Mediated Hemolytic Anemia

Urs Giger, DACVIM (Small Animal Internal Medicine), ECVIM (Internal Medicine), DECVCP (Clinical Pathology)
Veterinary Hospital, University of Pennsylvania, Philadelphia, PA, USA

Introduction

Immune-mediated hemolytic anemia (IMHA) is one of the most common and serious hemolytic anemias in dogs, but occurs rarely in other animal species. In IMHA an immune response, including anti-erythrocytic antibodies, complement, and macrophages, targets directly or indirectly erythrocytes and a hemolytic anemia ensues. There are many triggers for IMHA such as infections, drugs and other agents, and cancer leading to secondary IMHA, but in many dogs no cause is identified (so-called idiopathic, autoimmune, or primary IMHA) or a genetic predisposition has been observed (Cocker Spaniels). Furthermore, alloimmune hemolytic anemias, such as hemolytic transfusion reactions, both acute and delayed, and neonatal isoerythrolysis (only litters from transfused bitches), are caused by specific anti-erythrocytic alloantibodies. In contrast to other species, dogs with IMHA also develop an often overwhelming inflammatory response resulting in thrombosis and necrosis of various organs. And while the anemia can be corrected with transfusions, these complications in dogs are causing severe morbidity and mortality despite aggressive immunosuppression and antithrombotic interventions.

This session will review the diagnostic approach to IMHA with special emphasis on autoagglutination, spherocytosis, and the Coombs' test. Other hemolytic anemias such as infectious, inherited, and toxic causes need to be carefully excluded.

Immune-Mediated Destruction of Erythrocytes

Regardless of the underlying cause, IMHA results from a breakdown in immune self-tolerance or from a deficit in the control mechanism that regulates B and T lymphocyte as well as macrophage reactivities. Immune destruction of erythrocytes is initiated by the binding of IgG or IgM antibodies to the surface of erythrocytes. Under most clinical circumstances, immune destruction is an extravascular process that depends on recognition of erythrocytes opsonized with IgG, IgM, and/or complement by specific receptors on reticuloendothelial cells. Macrophages with engulfed erythrocytes may be noted on cytological examination of blood and tissue aspirates as erythrophagocytosis, but this is not definitive proof of an immune-mediated process. Antibody-coated erythrocytes may also be lysed by complement fixation and the membrane attack complex, which is clinically noted as intravascular hemolysis.

A diagnosis of IMHA must demonstrate accelerated immune destruction of erythrocytes. Evidence of a hemolytic anemia is suggested clinically by icterus and a regenerative anemia with hyperbilirubinuria, and hemoglobinemia and hemoglobinuria refers to an intravascular process. However, the erythroid response in the bone marrow may be blunted by the immune, inflammatory, and necrotic process or the underlying disease thereby leading to non-regenerative anemias. It is difficult to diagnose a primary immune destruction of erythrocyte precursors and these anemias are not hemolytic, but rather a marrow disorder. Besides documenting a hemolytic anemia for IMHA, one or more of the following 3 hallmarks must be present to support a diagnosis of immune-mediated hemolysis: persistent autoagglutination, marked spherocytosis, and a positive direct Coombs' test result. As in human medicine, the Coombs' test should be considered the best test to definitively diagnose IMHA, although marked spherocytosis and persistent/true autoagglutination (after 3x washing of EDTA blood with saline) are other important parameters indicating immune-destruction of erythrocytes.

Autoagglutination

Anti-erythrocytic IgM and in large quantities IgG antibodies may cause direct erythrocyte autoagglutination. The autoagglutination may be seen by the naked eye in an EDTA tube or on a glass slide or may become apparent as small clumps of erythrocytes on blood smears. For yet unexplained reasons, canine erythrocytes have a tendency to unspecifically agglutinate in the presence of plasma and colder temperatures as well as possibly with excessive EDTA anticoagulant.

Mixing blood with one drop of saline may break up rouleaux formation but not other forms of unspecific red cell agglutination. It is, therefore, important to determine whether the agglutination persists after "saline washing," which has been coined persistent or true autoagglutination. This is accomplished by adding excessive physiologic saline to the tube containing a small amount of EDTA-anticoagulated blood (>4:1), mixing, centrifuging, and removing the supernatant including the plasma and repeating this saline washing 3 times. True or persistent autoagglutination is indicative of an immune process, but precludes the performance of Coombs' test and crossmatching procedures which are based upon an agglutination reaction as result. Typing kits based upon chromatographic strip techniques do not seem to be affected by autoagglutination (unless severe) as free red blood cells can move along the strip. In any case it is advisable to check for autoagglutination before performing typing and crossmatch kit tests and when agglutination is observed, washing should be first performed. If the agglutination breaks up after washing, the Coombs' test is expected to be positive, if it is a case of IMHA. There is no evidence for washing away red blood cell bound antibodies in dogs.

Spherocytosis

If erythrocytes are only partially phagocytized or lysed by complement in circulation, erythrocytes with reduced surface area to volume ratio, known as spherocytes, are formed. They appear spherical and microcytic with no central pallor and are considered fragile. Note proper areas on the blood smear need to be reviewed in order to find spherocytes in between single regular discoid red blood cells. Large numbers of spherocytes (>20/microscopic high power field) are nearly diagnostic for IMHA, whereas small numbers may be seen with other conditions including DIC, endotoxemia, and zinc intoxication. In the author's experience, all dogs with marked spherocytosis and suspected to have IMHA also had a positive Coombs' test. However, only 60–80% of dogs with a positive direct Coombs' test or clinically diagnosed with IMHA had marked spherocytosis. Hereditary spherocytosis due to genetic membrane defects has rarely been seen in dogs, but should be considered as a differential diagnosis in dogs with negative Coombs' test results.

Because of the difficulties with the Coombs' test (see below), Slappendale had proposed to use the erythrocytic osmotic fragility test at specific saline concentrations as a mean to diagnose IMHA and this test is currently used in various clinics in Europe. However, there are many other reasons for increased fragility of erythrocytes beside IMHA, including hereditary red blood cell defects. This test is not used in human medicine and has not been shown to be superior to determination of marked spherocytosis and a positive direct Coombs' test in dogs with IMHA. The osmotic fragility test is also a cumbersome and not well standardized technique further limiting its usefulness.

Positive Direct Coombs' Test

The direct Coombs' test is also known as direct antiglobulin test (DAT) and is used to detect antibodies and complement on the surface of erythrocytes when the anti-erythrocyte antibody strength or concentration is too low to cause spontaneous agglutination (sub-agglutinating titer). Separate canine-specific IgG, IgM, and C3b antibodies as well as polyvalent antiglobulin reagents are available. They are added at various concentrations after washing the patient's erythrocytes free of plasma (3x as shown above) and mixtures are generally incubated at room temperature or 37°C (cold agglutinins appear to be rarely of clinical importance and rarely cause hemolysis). The strength of the Coombs' reaction does not necessarily predict the severity of hemolysis, but reaction changes are useful in monitoring the disease progression and response to therapy.

Typically, tube or microtiter methods have been used exclusively in the reference or teaching laboratory setting, but a flow cytometric method has also been introduced in a couple of places. A standardized, sensitive, and simple gel column method was available by DiaMed (Switzerland), but unfortunately the company was sold to another company which decided to not pursue the veterinary market. A novel standardized antiglobulin test method has recently been developed by Alvedia (France) similar to the immunochromatographic strip technique for blood typing and crossmatching of dogs and cats. Although many commercial veterinary laboratories offer Coombs' testing for dogs, clinicians have questioned the test's sensitivity and specificity and often forgo the test and/or use response to therapy as a diagnostic. However, negative Coombs' test results may potentially be seen because of technical reasons, insufficient quantities of bound antibodies, the presence of weakly bound antibodies (no data available), or the disease in remission. The Coombs' test stays positive for days to months after initiating treatment. A few days of immunosuppressive therapy will not reverse the Coombs' test result, as unlikely a transfusion would cause a positive Coombs' test result. Also a persistently anemic dog with suspected IMHA even when treated for weeks to months with immunosuppressive therapy should have a positive Coombs' test. Thus, anemic dogs with negative Coombs' test results should be reevaluated for other causes of hemolytic anemia. Conversely, dogs with primary IMHA or with secondary IMHA, when the trigger has not been found and removed, may well have a positive Coombs' test for months to years.

In a recent prospective study of anemic and non-anemic dogs comparing various direct Coombs' test methods including microtiter plate assays, gel column, capillary, and immunochromatographic techniques using polyvalent antiglobulins in a laboratory setting, found excellent correlations between different Coombs' tests and with spherocytosis and without noticeable interference by immunosuppressive or transfusion therapy in anemic dogs.

In conclusion, a diagnosis of IMHA requires the documentation of red blood cell destruction and an immune process. While regenerative anemia, icterus, and hyperbilirubinuria are suggesting a hemolytic anemia, evidence of true autoagglutination, spherocytosis, and/or a positive direct Coombs' test are required to document immune destruction. The authors also recommend monitoring IMHA patients for the disappearance of these immunological parameters to adjust and taper therapy.

Acknowledgements

Author's studies were supported in part by a grant from the National Institutes of Health (OD 010939). Alvedia, Lyon, France provided the immunochromatographic strips for the studies. The author is the director of the non-for-profit PennGen Laboratory offering genetic and hematological testing.

Author's address: Urs Giger, Veterinary Hospital of the University of Pennsylvania, 3900 Delancey Street, Philadelphia, PA 19104-6010 USA. Phone: 215-898-8830; Fax: 215-573-2162; giger@vet.upenn.edu; http://research.vet.upenn.edu/penngen.

References

1. Caviezel LL, Raj K, Giger U. Comparison of 4 direct Coombs' test methods with polyclonal antiglobulins in anemic and non-anemic dogs for in-clinic or laboratory use. *J Vet Intern Med.* 2014;28:583–591 (and references therein).

acvim411

Individualizing Drug Therapy: Advances in the Pharmacodynamic Monitoring of Immunosuppressive Agents

Andrew Mackin, BSc, BVMS, MVS, DVSc, FANZCVSc, DACVIM (SAIM)
Starkville, MS, USA

For as long as veterinarians have been giving drugs to animals, it has been recognized that individual patients can respond to therapy in unpredictable ways. Most animals will respond as expected, but in some patients the drug will appear to be completely ineffective, and in others the drug will cause unacceptable side effects. While, in many patients, simple dose adjustments can be used to address the problem, sometimes a particular drug simply cannot be used in an individual patient. When dealing with serious diseases that require aggressive therapy, where a drug that is ineffective can lead to death from the undertreated disease, and a drug that is not tolerated can lead to death from adverse effects, there is a natural tendency for veterinarians to never want to again use a particular drug with which they have had an untoward experience. This is particularly true with the immunosuppressive agents, where the immune-mediated diseases being treated can often be severe and life-threatening, and the therapeutic window that provides immunosuppression without adverse effects or excessive predisposition to infection is narrow. With these drugs, there is a tendency for veterinarians to play "musical medications," abandoning a drug that they have a bad experience with forever, and switching to a newer immunosuppressive agent, often one they have never used before. The reality, however, is that there is no single immunosuppressive agent that is perfect for treating every immune-mediated condition in every individual patient. Moreover, even in drugs that are effective and well-tolerated in an individual animal, there is no single dose rate that is suited to all patients.

Understanding that there is no one drug nor single dose rate that suits all patients and all diseases is valuable to veterinarians, as it serves to dampen unrealistic expectations, and to encourage use of a problem-solving approach when determining the best medication and dosing regime for each patient. Unfortunately, however, even if clinicians recognize that a "one size suits all" approach to immunosuppressive therapy is not going to succeed with all patients, this recognition alone does not provide them with all of the tools needed to address these problem. The appropriate use of immunosuppressive agents can be complex, and there is no one single approach that works in all situations. Over that past decade, as our understanding of the use of immunosuppressive agents has grown, a range of pharmacokinetic, pharmacodynamic, and molecular tests have become available, and more are in the process of being developed, that will allow veterinarians to more accurately tailor therapy to the needs of the individual patient. These evolving and expanding tools will help veterinarians to embrace the concept of "individualized drug therapy."

To better understand individualized drug therapy, it is helpful to focus initially on cyclosporine, a drug that has been recently studied much more extensively in animals than the other immunosuppressive agents.

CYCLOSPORINE

Cyclosporine is a potent immunosuppressive agent, one which served to revolutionize transplant medicine. Cyclosporine was originally derived from a soil fungus, *Beauveria nivea* (*Tolypocladium inflatum*). Interestingly, many other important drugs can be derived from so-called "medicinal fungi," including the immunosuppressive agents mycophenolic acid and vincristine, although the vinca alkaloids such as vincristine can also be extracted from plants. The primary immunosuppressive mechanism of action of cyclosporine is inhibition of T cell function. Cyclosporine inhibits calcineurin, an intracellular protein phosphatase that activates gene transcription factors. Activation of T cells via the T cell receptor (TCR), with resultant intracellular efflux of calcium, results in activation of calcineurin, an enzyme which dephosphorylates inactive nuclear factor (NFAT). Dephosphorylation allows NFAT to translocate into the T cell nucleus, where it upregulates transcription of genes coding for a number of cytokines that play important roles in the immune response, including IL-2, IL-4, TNF-α, and INF-γ. Calcineurin inhibitors, including cyclosporine, act by binding to intracellular cyclophilins, which are proteins that facilitate protein folding. Binding of intracellular cyclosporine to cyclophilins, primarily cyclophilin A, creates a complex with high affinity for calcineurin. Through inhibition of calcineurin, cyclosporine specifically inhibits T cell function and thus, cell-mediated immunity, but has little immediate impact on humoral immunity. Decreased IL-2 expression leads to inhibition of proliferation and activation of both T-helper and T-cytotoxic lymphocytes, and blunting of the immune response. Cyclosporine also has many local anti-inflammatory and immunosuppressive effects, especially in the skin.[1]

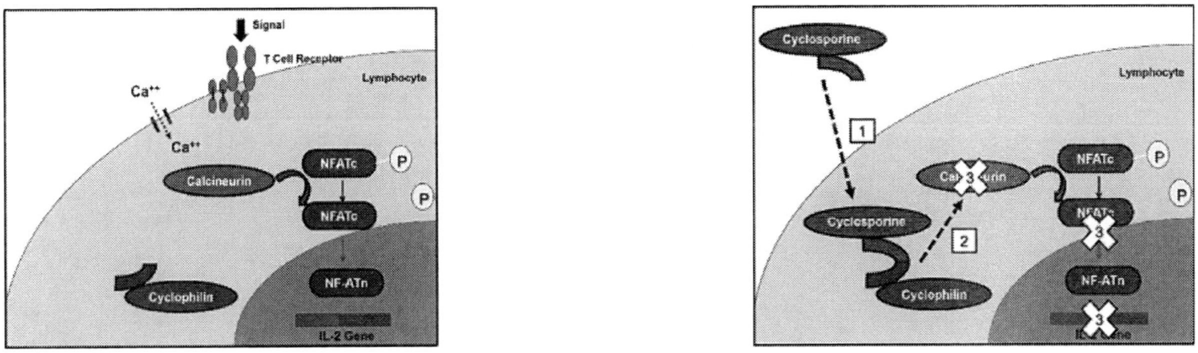

In a normal T cell (left), activation of TCR and influx of Ca^{2+} stimulates calcineurin and eventual production of cytokines such as IL-2. Cyclosporine disrupts T cell production of IL-2 (right) via: 1)Drug entrance into the cell and binding to cyclophilin, 2)Cyclosporine/cyclophilin complex binding to calcineurin, 3)Inhibition of calcineurin, and inhibition of dephosphorylation and transfer of NFAT into nucleus.

Cyclosporine is a large lipophilic molecule which must be solubilized prior to intestinal absorption. The commercial veterinary cyclosporine product, and the human proprietary and generic equivalents, are ultramicronized into a microemulsion in order to improve absorption, but even in this modified form bioavailability can vary greatly from patient to patient, and will be altered by fasting or feeding with a meal. The amount of active cyclosporine actually reaching the portal circulation will be influenced by enterocyte P-glycoprotein and cytochrome P-450 activity. P-glycoprotein (permeability glycoprotein), also known as multidrug resistance protein 1 (MDR1) or ATP-binding cassette sub-family B member 1 (ABCB1), is an efflux pump that pumps xenobiotics (drugs and toxins) back into the intestinal lumen, thereby reducing drug bioavailability. Enterocyte cytochrome P-450 (CYP) enzymatic metabolism of cyclosporine yields a number different metabolites, such as AM1, AM4N, and AM9, some of which may retain therapeutic efficacy, although much of the cyclosporine makes it to the portal circulation unmetabolized. Individual patient P-glycoprotein and CYP activity are both affected by a combination of genetic factors and environmental factors such as exposure to other drugs.[2,3] Although clinicians consider cyclosporine to be a "drug," to mammalian tissues cyclosporine is essentially treated as a fungal-derived toxin, and a range of efflux pumps and metabolizing enzymes exist to prevent the toxin from entering the body and susceptible tissues, and to metabolize it to non-active forms if it does gain access to the body. While these protective mechanisms serve to protect the body from toxins, unfortunately they can also "protect" the body from the desirable effects of xenobiotic drugs such as cyclosporine. Additionally, since every animal may have different levels of pump and enzymatic activity, these mechanisms serve to make cyclosporine bioavailability highly unpredictable from patient-to-patient.

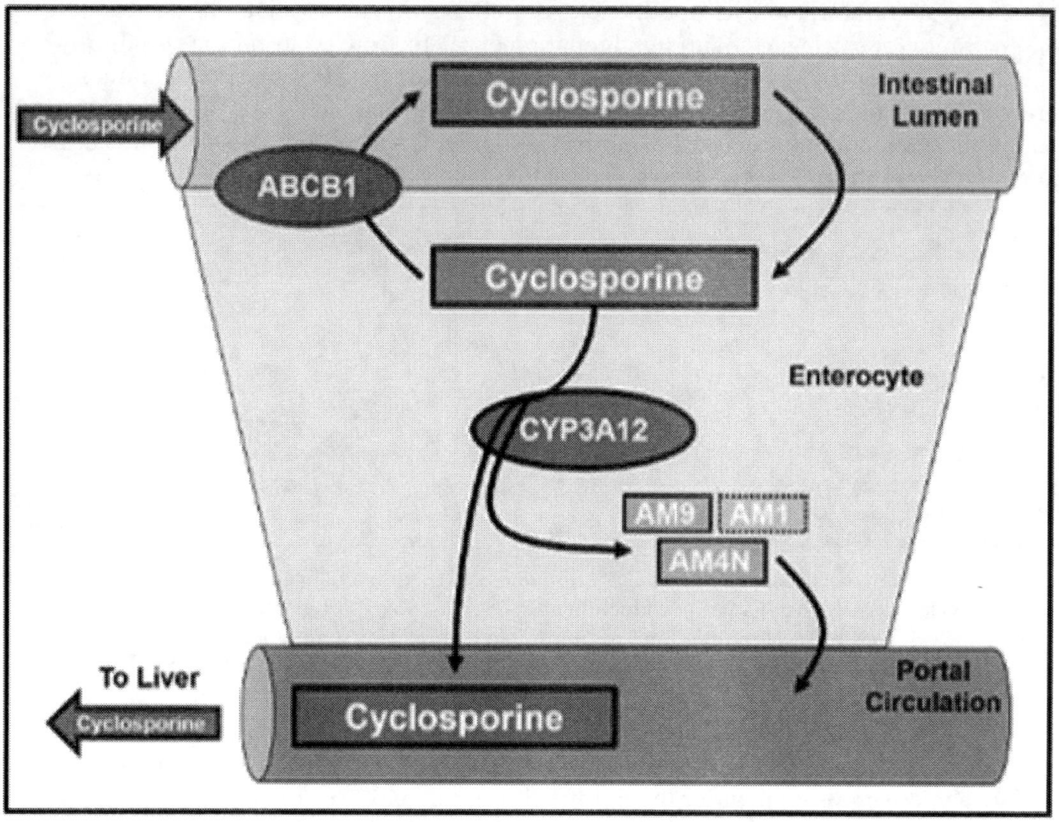

Once cyclosporine enters the portal circulation, it is subject to hepatic metabolism. Metabolism of cyclosporine by hepatic CYP enzymes yields further drug metabolites and, as in the enterocytes, much of the cyclosporine exits the hepatocytes unmetabolized. Drugs that inhibit or enhance hepatic CYP enzyme activity can markedly affect cyclosporine metabolism and, subsequently, the amount of cyclosporine entering the systemic circulation. In fact, drugs that inhibit CYP enzymes can be given concurrently with cyclosporine in order to reduce drug costs by decreasing the dose needed to maintain adequate blood drug concentrations. Ketoconazole, in particular, has been used to decrease in oral cyclosporine dosages in dogs by as much as 75%.[4] Interestingly, in people, even drinking grapefruit juice can alter blood cyclosporine levels, because grapefruit inhibits CYP enzyme activity.[5] In people, inherited variations in the activity of the CYP enzymes that metabolize cyclosporine have also been shown to have a significant effect on cyclosporine blood levels.[6] Even if the effects of variability in enterocyte handling of cyclosporine could be controlled for such that drug levels in the portal circulation could be standardized, variable hepatocyte drug metabolism would still lead to highly variable systemic cyclosporine concentrations.

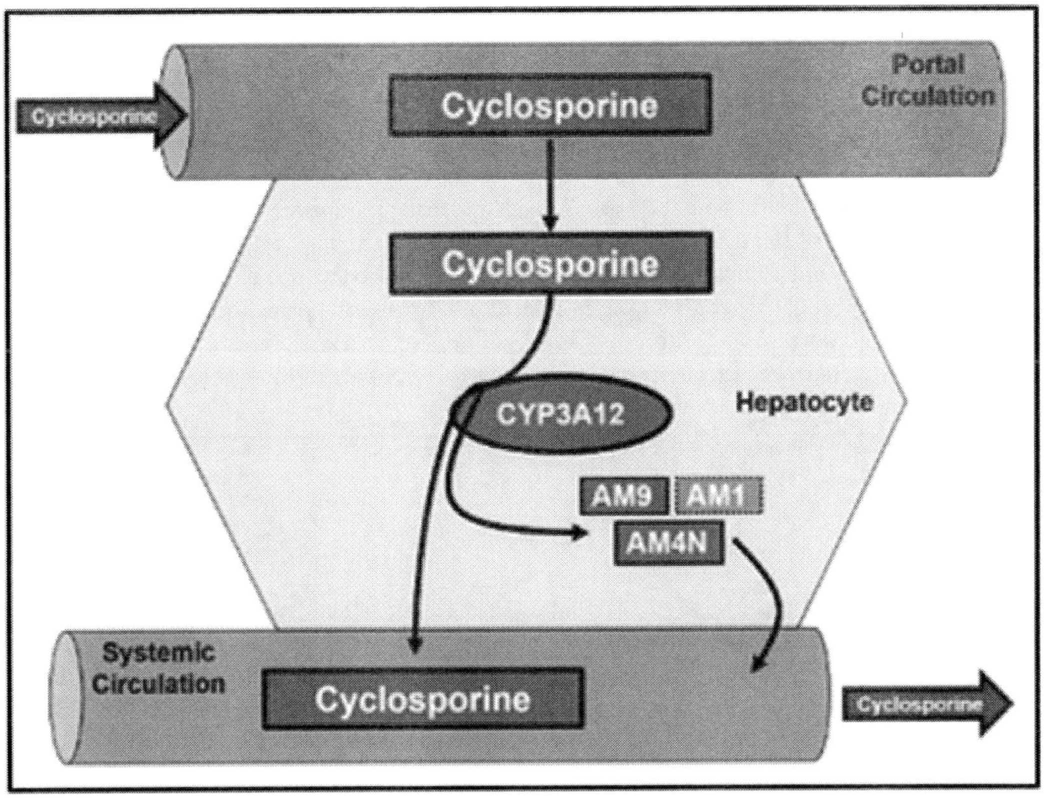

The net effect of highly variable individual-to-individual enterocyte and hepatocyte handling of cyclosporine is that, for any given oral dose, eventual blood drug concentrations will also be highly variable, and clinical effects will as a result be unpredictable. Since the initial major indication for cyclosporine in people was to prevent transplant rejection, "unpredictable clinical effects" essentially translated into the highly undesirable outcomes of either loss of transplanted organs or septic complications. In human transplant medicine, it was soon recognized that therapeutic drug monitoring (TDM) through measurement of blood cyclosporine concentrations, with subsequent adjustment of oral doses to achieve "target" blood levels, would eliminate the problems associated with unpredictable enterocyte and hepatocyte drug handling. Cyclosporine was the first immunosuppressive agent where TDM, and assessment of individual patient pharmacokinetic parameters, became a part of the standard clinical usage of the drug, with an associated decrease in the rates of transplant rejection and sepsis.

Pharmacokinetic Monitoring of Cyclosporine

Pharmacokinetics, simply defined, is the study of "what the body does to a drug." For cyclosporine, when looking at systemic immunosuppressive effects, the key element of pharmacokinetic TDM is measurement of blood drug concentrations. Peak blood drug concentrations occur approximately 2 hours after oral administration of cyclosporine, and then rapidly decline. Once in the circulation, cyclosporine distributes widely, accumulating in tissues such as the skin, liver, and fat. In those tissues where cyclosporine accumulates, such as the skin and, arguably, the liver and gastrointestinal tract, blood drug levels are not likely to predict local immunosuppressive effects. Additionally, in these tissues, once daily dosing of cyclosporine is often sufficient, whereas for systemic immunosuppression twice daily dosing is usually required. In the skin, where cyclosporine tends to gradually accumulate over time, sometimes dosing every second or third day will eventually be effective.[4]

In human transplant medicine, trough blood levels (typically, 12 hours post-dosing when the drug was given twice daily) were the initial basis for adjustment of drug dosages. Many studies in people have since suggested that the area under the plasma drug concentration time curve (AUC) or 2 hour peak drug concentrations are preferred. With peak cyclosporine concentrations requiring only a single blood sample, as opposed to the multiple samples needed for AUC calculations, adjusting drug doses to attain target peak drug levels has become the most common measurement used for human organ transplantation. In veterinary medicine, measurement of trough cyclosporine concentrations prevailed for many years based on initial work done in canine and feline renal transplant studies. Recommendations from veterinary laboratories offering TDM, particularly the Auburn Clinical Pharmacology Laboratory, now often involve measurement of both peak and trough cyclosporine blood levels, especially in dogs, although ideal target concentrations have not yet been well established. The real determinant of ideal target peak and trough drug concentrations is correlation with clinical efficacy, but clinical studies correlating TDM and efficacy can be very challenging to coordinate for many naturally-occurring diseases. For this reason, most laboratories conducting TDM actually appreciate clinical feedback regarding efficacy and side effects from veterinarians using the service, in order to develop a "feel" for ideal target drug concentrations.

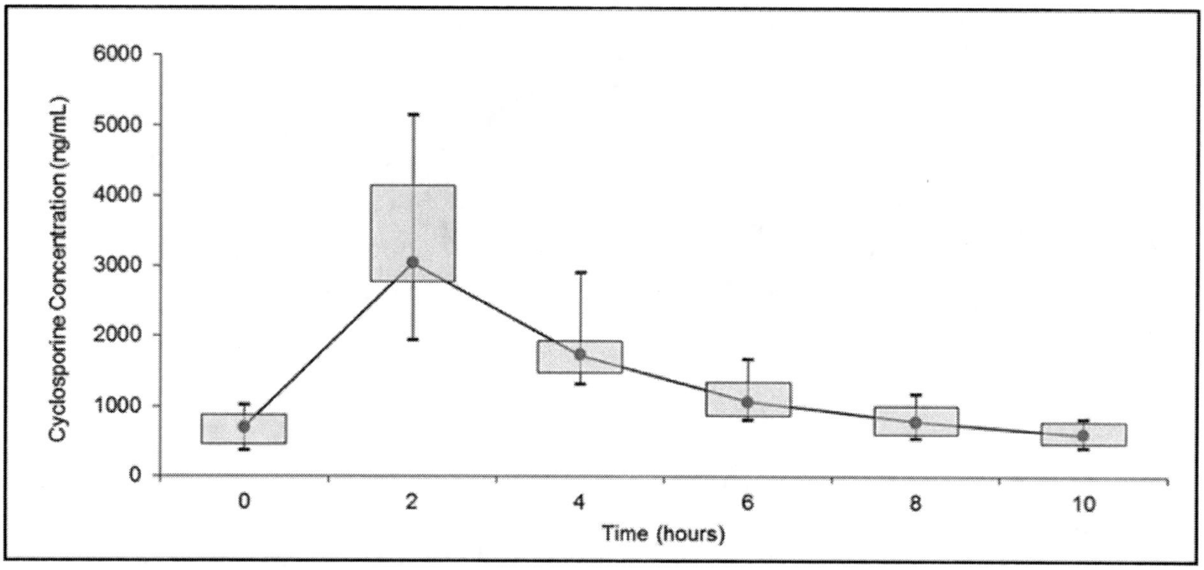

Whole blood drug concentrations in 8 research dogs receiving oral cyclosporine twice daily at relatively high doses.[7] Even with oral dose rates adjusted to achieve target concentrations of at least 600 ng/mL, subsequent trough and especially peak drug levels are highly variable, with an almost three-fold variation in peak concentrations. Trough drug concentrations therefore do not accurately predict peak concentrations, and cannot be used as a "surrogate marker" for peaks. Peak drug concentrations are very transient, and timing of sample collection must therefore be very accurate.

In human transplant medicine, despite the standard practice of TDM of blood cyclosporine levels, it became apparent over time that modifying doses based on target trough, peak, or AUC drug concentrations, although associated with improved response rates, still did not predict undertreatment or overtreatment with complete clinical accuracy. Individual patients with almost identical blood cyclosporine concentrations were still observed to have very different responses in terms of transplant tolerance or rejection, and predisposition to infection. Variations in drug handling by the target cell, the T lymphocyte, explain this variability in response rates despite comparable blood cyclosporine levels. Once cyclosporine enters the T cell cytoplasm, effects on cell are influenced by both lymphocyte efflux pump activity, and intracellular cyclophilin levels. P-glycoprotein pumps on the surface of lymphocytes serve to pump cytoplasmic cyclosporine out of the cell, essentially "protecting" the T cell from the effects of this xenobiotic "toxin." Lymphocytes with high levels of pump activity would be expected to be less affected by cyclosporine than lymphocytes with low levels of activity. In fact, drug-associated induction of P-glycoprotein in malignant lymphocytes is a key mechanism of resistance to chemotherapeutic agents, hence the alternate efflux pump name of multidrug resistance protein 1 (MDR1, also known as ABCB1). In healthy lymphocytes, exposure to other drugs can also potentially increase or decrease P-glycoprotein pump activity, and reduced pump activity due to the inherited MDR1 gene mutation also has the potential to increase T cell responsiveness to cyclosporine. Additionally, inherited variations in T lymphocyte cyclophilin levels may alter responsiveness to cyclosporine. T cells in mutant mice with low cyclophilin levels, for example, are highly resistant to the effects of cyclosporine.[8] The end result of these individual-to-individual variations in T cell handling of cyclosporine is that, even at comparable blood drug concentrations, the effect of the drug on T cell function can still be highly unpredictable. In response to this challenging situation, human transplant medicine evolved from TDM based solely on blood cyclosporine concentrations to TDM that also evaluated T cell responses to the drug, so-called "pharmacodynamic" monitoring.

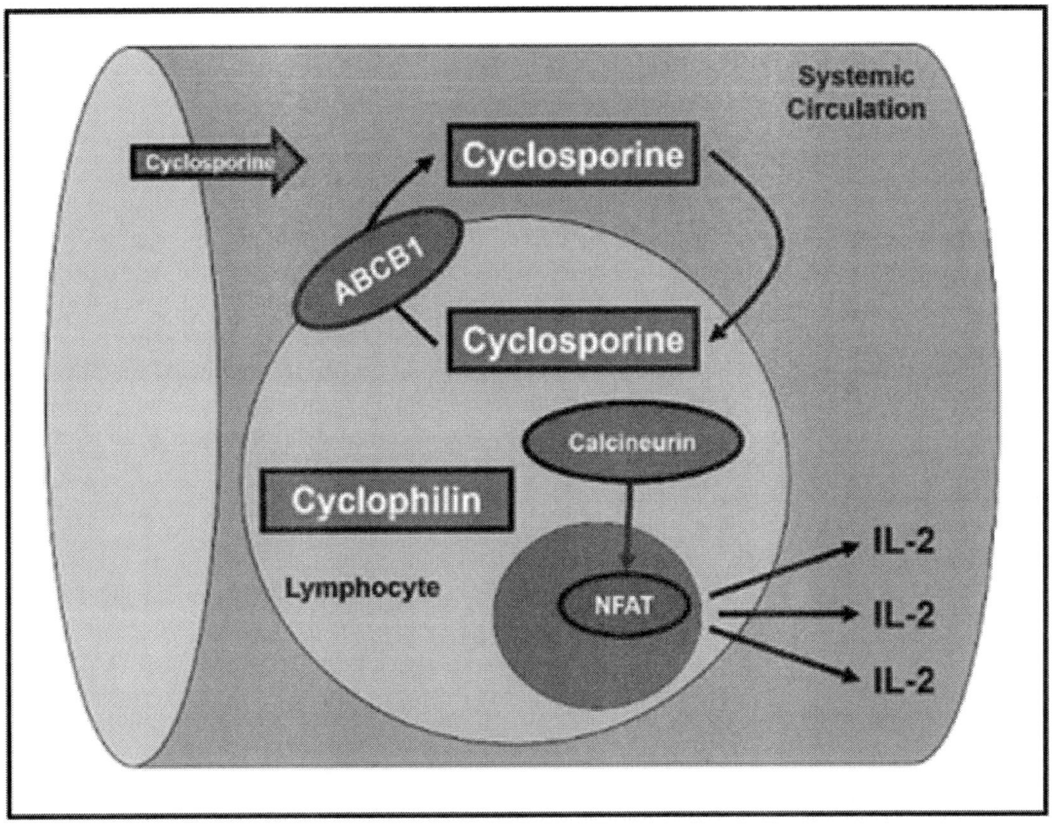

Pharmacodynamic Monitoring of Cyclosporine
Pharmacodynamics is defined as the study of the biochemical and physiologic effects of drugs with a particular emphasis on dose-response relationships: which, more simply put, is the study of "what a drug does to the body." Pharmacodynamics is just a fancy word for something that good veterinarians have been doing since drugs were invented: giving medications, and then adjusting doses on a trial-and-error basis based on clinical signs, apparent response to therapy, and side-effects. The best "marker" of the effectiveness of a drug at treating a disease is the disease itself and, in its most elementary form, pharmacodynamics is centered on simply observing the disease. For cyclosporine, the best pharmacodynamic indicator of correct dosage for treatment of atopy is resolution of pruritus, for anal furunculosis is resolution of ulcerative lesions, and for inflammatory bowel disease is resolution of vomiting, diarrhea or hypoalbuminemia.

The problem with this simple "observational" trial-and-error pharmacodynamic approach occurs when clinicians are dealing with life-threatening conditions where the end result of undertreatment is significant cost or morbidity, or even death, and the end result of overtreatment is life-threatening sepsis. Loss of an organ is not an acceptable indicator of inadequate treatment after transplantation, and death due to anemia, or transfusion dependency, are not desirable indicators of inadequate treatment of immune-mediated hemolytic anemia. With these more severe conditions, it is preferable to find a "surrogate marker" that serves as a reliable indicator of the effects the drug is having on the target organ or disease, and that may serve as an early warning to undertreatment or overtreatment that allows dose adjustment before complications such as organ failure, sepsis, or death occur. Such surrogate markers are the cornerstone of modern pharmacodynamic testing, and are centered on the known mechanism of action of the drug. For cyclosporine, pharmacodynamic testing is mostly centered on assays that either measure inhibition of T cell calcineurin enzyme activity, or inhibition of T cell NFAT-regulated cytokine production. Most of these assays are not convenient or robust enough for routine clinical use and therefore, although they can provide useful information in an experimental setting about median or average effective drug dosage, dosing interval, drug mechanism of action, time to onset of effects, duration of those effects, and likelihood and magnitude of variations in individual responsiveness to the drug, they cannot be used to fine tune individualized therapy in specific patients. Calcineurin enzyme assays, for example, although providing very useful general information, including that inhibition of calcineurin is very rapid (within 2 hours of drug administration) and very transient in that it is closely but inversely correlated with peak cyclosporine drug levels, are not practical to offer to veterinarians as a diagnostic test.[9] Similarly, flow cytometric measurement of activated T cell expression of cytokines such as IL-2, Il-4, and IFN-γ, although it is also not adaptable for routine clinical use, has yielded highly valuable clinical information in dogs, including the observation that, although inhibition of calcineurin is known to only last a few hours, resultant inhibition of cytokine production can last for 12 hours or more.[7] These flow cytometry results suggest that twice daily dosing is appropriate for attaining systemic immunosuppressive effects with oral cyclosporine in dogs. Flow cytometry has also revealed that, even at the low dose of cyclosporine approved for use to treat canine atopy (5 mg/kg orally once a day), a dose which has been considered to be low enough to avoid predisposing to immunosuppression and infection, expression of IFN-γ and IL-2 is still markedly suppressed in some dogs.[10]

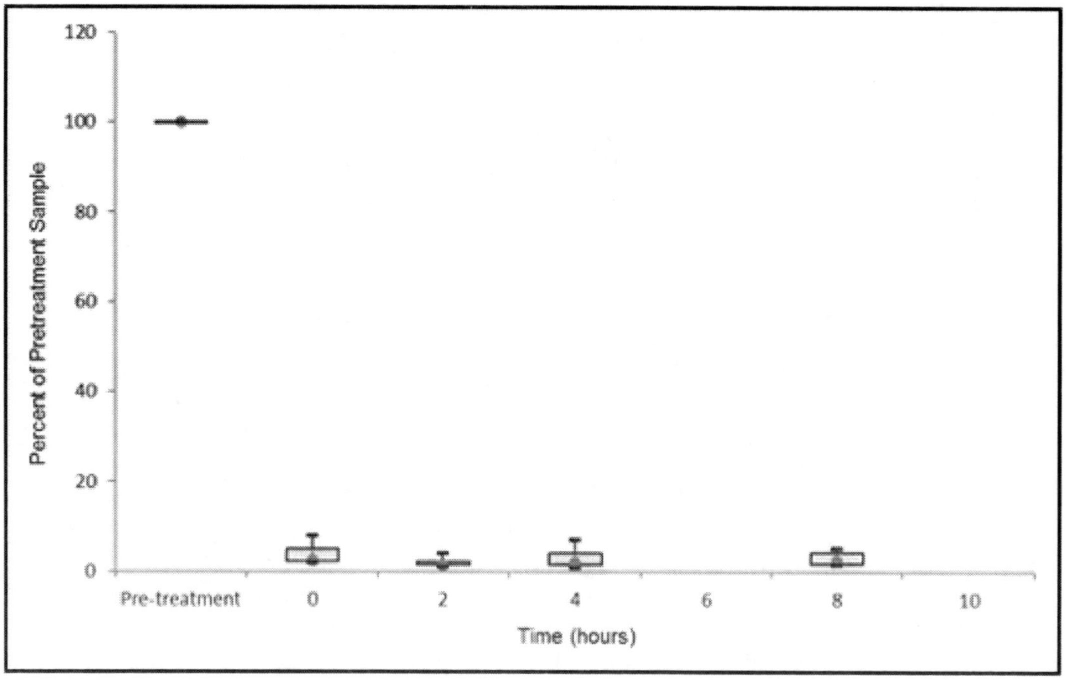

Activated T cell expression of IL-2, as measured using RT-qPCR in 8 dogs receiving oral cyclosporine twice daily at high doses.[7] In all dogs, lymphocyte IL-2 expression is markedly suppressed, to less than 5% of pre-treatment values, over an entire 12 hour dosing interval.

Subsequent studies at the Mississippi State University Pharmacodynamic Laboratory have determined that measuring activated T cell mRNA IL-2 and IFN-γ expression using a reverse transcriptase quantitative polymerase chain reaction (RT-qPCR) assay yields results comparable to flow cytometry.[9]

One advantage of the RT-qPCR assay is that it is robust enough to be performed on blood samples mailed in by clinicians, even from as far away as the UK. The PCR-based assay measures the mRNA coding for the cytokines, and results essentially parallel measurement of the cytokines themselves. The T cells must first be activated using a known activator such as PMA/ionomycin, because the effects of inhibition of calcineurin are only detectable in a cell that is activated.

The RT-qPCR assay is based on similar assays used in human transplant medicine, and has now been offered by MSU for several years. "Target" levels of T cell suppression were originally modeled on targets used in humans to prevent transplant rejection whilst avoiding sepsis, and have since been adjusted based on clinical experience. The assay is a classic example of a laboratory-based pharmacodynamic assay, in that it measures what the drug does to the body (or, more specifically, what it does to a T cell), with a focus on a very specific mechanism of action. This mechanism-targeted approach means that the assay is not likely to be markedly affected by the concurrent presence of other immunosuppressive drugs, but does unfortunately limit assay usefulness to assessment of only those drugs that inhibit calcineurin (cyclosporine and tacrolimus). Utilization of this assay in research dogs has revealed that inhibition of IL-2 expression is maximal within 3 days of commencing cyclosporine, and that this inhibition wears off within 3 days of discontinuing the drug. One enormous advantage of utilizing an assay that focuses on the specific target end result (IL-2 production) in the specific cell of interest (activated T cell) is that it effectively bypasses all of the unknown enterocyte and hepatocyte variables that affect blood drug concentrations, and all of the unknown lymphocyte variables that affect T cell responses. Clinical experience with this assay has revealed marked patient-to-patient variability in T cell responsiveness to cyclosporine: our laboratory has identified some dogs that were dangerously immunosuppressed (with resultant drug-associated infection) on doses as low as 3 mg/kg twice daily, and other dogs that were minimally immunosuppressed (with an associated lack of disease remission) on doses of over 10 mg/kg twice daily. The median immunosuppressive oral cyclosporine dose in dogs appears to be around 7 mg/kg twice daily. In our opinion, it is impossible to guarantee adequate but not excessive systemic immunosuppressive effects unless activated T cell responses are measured after commencement of cyclosporine therapy, and doses adjusted accordingly. This process is, however, cumbersome, because dogs must be on therapy for at least 3 days (ideally, 5–7 days) before RT-qPCR testing, and because activation and subsequent RNA processing steps can delay results for up to a week. The assay is not yet available for cats.

The fact that the effects of cyclosporine on the tissues are unpredictable and vary from animal-to-animal is hardly surprising, since this is true for most xenobiotics. In fact, in toxicology, the concept of median lethal dose (LD50), the dose required to kill half the members of a test population, is centered on the understanding that, at a given dose, some animals are more susceptible to the effects of a toxin than others. Since the same principles that apply to toxicology also apply to drug pharmacology, it should be expected that there will be a comparable individual-to-individual variably in drug effects (and adverse effects).

Our experimental and clinical work with cyclosporine has emphatically confirmed that cyclosporine is a complex drug, with effects that are so unpredictable and variable that the drug should probably not be used for systemic immunosuppression unless some form of pharmacokinetic or pharmacodynamics (or, ideally, both in combination) TDM is utilized. The complexity of the cyclosporine usage, the need for expensive TDM, and the high cost of the drug has probably contributed to a decreased usage of cyclosporine over the past few years, and an increased usage of other less expensive drugs such as azathioprine, mycophenolate mofetil, and leflunomide. Clinicians are probably also hopeful that the use of these other drugs is less complicated than the use of cyclosporine, and that there will be less patient-to-patient variability in response to standard dose rates. The reality, however, is that the pharmacokinetic and pharmacodynamic properties of these other immunosuppressive agents have been much less extensively studied in small animals, and that the "lack of complexity" may just reflect a general lack of knowledge about these drugs in the veterinary species. The PharmGKB Pharmacogenomics Database (www.pharmgkb.org/) web resource developed with NIH funding through Stanford University, for example, reveals stunningly complex metabolic pathways for the common immunosuppressive agents, with involvement of multiple different efflux pumps and metabolizing enzymes whose gene expression and subsequent activity is known to be highly variable from individual to individual. Azathioprine handling by the body, for example, involves the complex interplay of at least 6–8 intracellular metabolic enzymes, and at least 2 different efflux pumps. In my opinion, given this information, it is highly optimistic, to the point of being completely unrealistic, to expect that a single oral dose rate of azathioprine (or leflunomide, or mycophenolate mofetil) will have entirely predictable results, without undertreatment or overtreatment, in all treated animals. In fact, a strong species difference in azathioprine metabolism has already been identified in our small animal patients. Compared to dogs, cats have very low levels of thiopurine methyltransferase (TPMT), an enzyme involved in azathioprine inactivation and, as a result, are highly susceptible to azathioprine toxicity at standard canine drug doses.[11]

While it is important for veterinarians to be aware that the potential for variable individual responsiveness to the immunosuppressive agents is probably extremely high, this awareness is of little clinical value unless strategies are developed to account for this variability. While, with the exception of cyclosporine, the range of pharmacodynamic options that are available is still relatively limited, there are some strategies that can be used by veterinarians now, or hopefully, in the near future, to adjust for patient-to-patient variability in drug responsiveness.

Therapeutic Drug Monitoring

Monitoring of blood drug concentrations, when available, is always recommended for the immunosuppressive agents, particularly when pharmacodynamic TDM is not available. Measurement of drug concentration, while it does not allow for variable effects on target cells and tissues, does help the clinician adjust for variations in drug bioavailability and metabolism. Measurement of cyclosporine blood concentrations is available through a number of human and veterinary commercial laboratories, and through the Auburn University Clinical Pharmacology Laboratory. Auburn also offers measurement of teriflunomide, the active metabolite of leflunomide. Both Auburn and the Pharmacokinetics Laboratory at Washington State University have also developed assays for mycophenolic acid (MPA), the active metabolite of mycophenolate mofetil (and mycophenolate sodium, which is rarely used in veterinary medicine), although this assay is not yet commercially available. Ideal target ranges for these newer drugs in dogs and cats have not yet been well established based on either routine clinical usage or pharmacodynamic testing, and are mostly developed by extrapolation from human medicine. Measurement of blood drug concentrations is not necessarily as straightforward as it seems, because interpretation of results for clinical use will depend on such variables as the type of assay used, whether the test is performed on whole blood, serum or plasma, the degree of drug protein binding that occurs in each species and in each individual, and what drug metabolites are measured. The species that is being tested can also make a difference on what drug metabolite or metabolites should be measured. For cyclosporine and leflunomide, the same assay is typically used in both dogs and cats. Mycophenolic acid in humans is primarily excreted conjugated to glucuronide and, since cats lack the glucuronyl transferases responsible for glucuronidation of drugs such as MPA, concern has been expressed that the drug should be used with caution, if at all, in this species. Recent work at WSU, which measured various MPA metabolites, reassuringly, has found that cats seem to be able to metabolize mycophenolate rapidly, by actually utilizing glucosidation rather than glucuronidation.[12] Other metabolites of drugs such as mycophenolate and azathioprine may actually be responsible for many adverse effects, but may be too labile to allow development of routine clinical assays.

Genetic Testing

For the past few decades, the WSU Veterinary Clinical Pharmacology Laboratory has been extensively studying the effects of genetic variations in activity of the P-glycoprotein efflux pump, coded for by the MDR1 (or ABCB1) gene. Deficient P-glycoprotein pump activity has the potential to increase oral drug bioavailability, to decrease biliary drug excretion, and reduce to the protective effects of the blood-brain barrier, thereby increasing the likelihood of drug toxicity (so-called "multidrug sensitivity"). Certain dog breeds, such as the Collie, Australian Shepherd and Long-Haired Whippet, are known to have a high incidence of MDR1 gene mutations that lead to low efflux pump activity. WSU offers genetic testing on cheek swabs or whole blood samples to test for the incidence of homozygous (MDR1 mutant/mutant) and heterozygous (MDR1 mutant/normal) affected dogs in at-risk breeds. Several immunosuppressive agents used in dogs, particularly vincristine and cyclosporine, are handled by the P-glycoprotein pump. Decreased biliary excretion of vincristine is believed to contribute to a higher rate of drug toxicity in MDR1 deficient dogs.[13] For this reason, pre-treatment testing for the MDR1 gene mutation is recommended in at-risk breeds prior to receiving vincristine chemotherapy, with a 50% dose reduction in homozygous affected dogs, and a 25% reduction in heterozygotes. The use of vincristine as an immunosuppressive agent, typically reserved for "Day 1" therapy in dogs with immune-mediated thrombocytopenia (IMT), presents more of a problem, since the main indication for vincristine in these patients is to hasten recovery of platelet counts by several days, and the canine IMT patient is likely to be either dead or better by the time results of genetic testing return. One possible approach would be just to move ahead with giving vincristine in severe IMT patients, since the dose of vincristine is lower than that typically used for chemotherapy, and the risk of toxicity are probably low. Since, however, severe toxicity with single IMT bolus doses (0.05 mg/kg) of vincristine in at-risk breeds has been anecdotally reported, I prefer not to treat affected Collies and Australian Shepherds with vincristine at all, or to at least use the dose reductions recommended for known MDR1 deficient dogs.

Cyclosporine bioavailability and metabolism and, consequently, blood drug concentrations, are not at this time believed to be markedly affected by MDR1 deficiency. This does not necessarily mean, however, that the drug is safe in MDR1 deficient dogs. Recently, our laboratory identified a Border Collie that had profound T cell suppression and multiple abscesses despite a very low oral cyclosporine dose (less than 3 mg/kg twice daily) and blood drug concentrations at the low end of the target range, and subsequent genetic testing at WSU confirmed that this dog was MDR1 mutant/normal (heterozygous). Since P-glycoprotein is responsible for pumping cyclosporine out of lymphocytes, we suspect that MDR1 deficiency, even at "acceptable" blood drug concentrations, may predispose T cells to oversuppression with cyclosporine. In suspected or known MDR1 deficient dogs, we recommend monitoring of T cell suppression 5–7 days after commencing cyclosporine, with dose reductions if oversuppression is observed. Neurologists are ambivalent about using cyclosporine for inflammatory brain disease, because an intact blood-brain barrier has the potential to limit access of cyclosporine to the central nervous system. Clinical experience in our laboratory, however, suggests that adequate T cell suppression is associated with a higher likelihood of remission in inflammatory brain diseases. Since both effectiveness of the blood brain-barrier and T cell responses to cyclosporine are affected by P-glycoprotein pump activity, it is possible that effective T cell suppression may serve as a surrogate marker for adequate brain penetrance of the drug.

Individual immunosuppressive drugs are known to be handled by a wide array of different CYP and other metabolic enzymes, and by a wide array of efflux pumps beyond P-glycoprotein. Genetic polymorphisms in metabolic enzyme and efflux pump activity are known to occur in dogs and cats, and it is entirely likely that recognition of these polymorphisms prior to treatment would affect immunosuppressive drug choices and dose rates. Genetic testing, for example, can detect TPMT deficient people that should receive a much lower dose of azathioprine, or another drug entirely. Beyond the MDR1 gene mutation, genetic testing for metabolic enzyme and efflux pump polymorphisms is not clinically available and, until it is, pharmacodynamic testing such as the activated T cell suppression assay can be used to detect the effects of possible gene mutations.

Enzyme Assays

Given that the primary mechanism of action of many immunosuppressive agents is inhibition of a particular intracellular enzyme, it would seem make intuitive sense to utilize tests of enzyme activity as a surrogate pharmacodynamic marker. Certainly, experimentally, enzyme assays provide useful information regarding mechanism of action. Mycophenolic acid, for example, is a selective inhibitor of inosine monophosphate dehydrogenase (IMPDH), an enzyme that controls the rate of synthesis of guanine monophosphate in the *de novo* pathway of purine synthesis, and IMPDH assays in research dogs have demonstrated inhibition of this enzyme within 4 hours of oral administration of mycophenolate mofetil.[14] Most enzyme assays, however, are either too complex or insufficiently robust for routine clinical use, and it is likely that these assays with remain confined to a controlled research setting.

Lymphocyte Proliferation

The key final effect of many immunosuppressive agents is inhibition of lymphocyte proliferation. Since B and T cell proliferation is an important component of immune responses, drugs that inhibit lymphocyte proliferation would be expected to have significant immunosuppressive effects. Most immunosuppressive agents inhibit lymphocyte proliferation by directly disrupting the nucleotide "building blocks" needed for cell division: alkylating agents such as cyclophosphamide and chlorambucil by cross-linking DNA, azathioprine and mycophenolate by inhibiting purine synthesis, and leflunomide by inhibiting pyrimidine synthesis. Cyclosporine also inhibits lymphocyte proliferation, via a mechanism that appears to be independent of suppression of IL-2 production. Incubation studies with isolated dog lymphocytes have confirmed that cyclosporine and the active metabolites of azathioprine, mycophenolate, and leflunomide all inhibit lymphocyte proliferation in a concentration-dependent fashion.[15] Interestingly, recent studies in our laboratory have revealed that, although MPA rapidly inhibits canine B and T cell proliferation in a concentration-dependent fashion, at concentrations that are readily achievable at standard oral mycophenolate mofetil doses, it takes a full two weeks of oral drug administration to see the same effect in healthy dogs. Lymphocyte proliferation is not affected at all after a week of mycophenolate and, even after 2 weeks, proliferation is only inhibited to about 50% of pre-treatment values. These results illustrate two important concepts. Firstly, rapid onset of the initial mechanism of action does not predict speed of onset of immunosuppression. In the case of mycophenolate, inhibition of IMPDH enzyme activity, which is known to occur within 4 hours of drug commencement in dogs, does not predict eventual immunosuppressive consequences like inhibition of lymphocyte proliferation, which lags behind IMPDH inhibition by almost 2 weeks. The reason for this lag is not known, but may be because affected lymphocytes can initially find alternative sources of purine for DNA synthesis, despite rapid loss of *de novo* synthesis pathways. Crudely comparable mechanisms of action suggest that the same lag may occur with azathioprine and, potentially, leflunomide. Even with cyclosporine, where maximal suppression of T cell IL-2 production occurs within a few days of drug commencement, "downstream" effects that cascade through the remainder of the immune system could possibly take much longer to cause maximal immunosuppression. Secondly, although by 2 weeks mycophenolate had met one of the standard pharmacodynamic benchmarks for significant effects, that is a 50% suppression of lymphocyte proliferation values (in studies evaluating the concentration-dependent effects of drugs *in vitro*, a drug concentration that causes an effect of this magnitude is often termed the half maximal inhibitory concentration, or IC50), it is uncertain whether a 50% suppression is sufficient to effectively treat a patient with immune disease. With cyclosporine, for example, this degree of suppression of normal function would often not be sufficient for clinical effect. Ideal "target" levels of suppression can only be determined from published clinical studies in humans, from clinical trials in diseased dog and cats, or from clinical experience using the assay (which entails offering the assay to veterinarians before ideal target results are fully established).

Although various lymphocyte proliferation assays are available for use in dogs and cats, and are relatively simple to establish in an experimental setting, none are currently practical for routine clinical use for samples submitted from veterinarians. While experimental studies using lymphocyte proliferation can provide very useful clinical information, such as, for example, demonstrating that mycophenolate is almost certainly not suitable for emergency situations where rapid immunosuppressive effects are needed, the lack of a convenient clinically available lymphocyte proliferation assay unfortunately means that this methodology cannot be used to "individualize" therapy by adjusting doses in individual patients.

"Wholistic" Assays of Immunosuppressive Effects

In human medicine, there has long been a recognized need for a "wholistic" assay of the overall effect of immunosuppressive therapy, one that could be offered commercially nation-wide for all patients with all diseases and on all immunosuppressive drugs, including combination therapy with multiple drugs. One such assay that has become commercially available in the US over the past decade is the Immuknow®, which is reported to be robust enough for routine clinical submission, and sensitive enough to detect immunosuppression in patients on a variety of immunosuppressive agents. Essentially, the Immuknow® test measures ATP production in activated and magnetically sorted T cells using a fluorescent assay. Recent publications describing correlation between results of this assay and clinical outcomes in people, however, have not established clinical usefulness for many disease and drug combinations.[16] Furthermore, a version of this assay developed in our laboratory unfortunately failed to detect significant suppression of T cell ATP production in normal dogs on standard doses of common immunosuppressive drugs. Such simple "wholistic" assays of immunosuppression that can actually provide consistently usable clinical information, while attractive in principal, are in practice probably a long way from availability for routine clinical use.

Proteomics/Transcriptomics

Currently, pharmacodynamic assays concentrate almost exclusively on known drug mechanisms of action and expected consequences in the immune system. This approach is narrowly focused and, in my opinion, puts the cart before the horse, in that, because a mechanism of action must be established before an assay can be developed, only a very limited range of assays will even be considered for development. In reality, it is highly likely that immunosuppressive agents will have multiple different effects on cell function, and that not all of these effects have been discovered. Discovery of extra unexpected mechanisms of action may facilitate exploration of new creative uses for these drugs, explain some drug toxicities, and suggest novel pharmacodynamic markers for development. Proteomic and transcriptomic methods have recently become much more user friendly and affordable, and offer a creative way for researchers to thoroughly explore all of the effects of immunosuppressive drugs on target tissues. Proteomics is the large-scale study of the complete set of proteins produced by a cell or tissue, and transcriptomics is a comparable study focusing instead on the complete set of RNA transcripts that produce these proteins. High-throughput methods are now available for both proteomic and transcriptomic analysis. Simple "before and after" drug exposure proteomic and transcriptomic studies of target cells, such as lymphocytes, may provide a wealth of information that will facilitate development of future pharmacodynamic assays.

References

1. Archer T, *et al. J Vet Intern Med.* 2014;28:1.
2. Court MH. *Vet Clin North Am Small Anim Pract.* 2014;43:1027.
3. Mealey KL. *Vet Clin North Am Small Anim Pract.* 2014;43:1067.
4. Gray L, *et al. Vet Dermatol.* 2013;24:118.
5. Sridharan K, *et al. Eur J Drug Metab Pharmacokinet.* 2016;41:665.
6. Tao X, *et al. Eur J Pharm Sci.* 2015;76:238.
7. Fellman C, *et al. J Vet Pharmacol Ther.* 2015;39:237.
8. Colgan J, *et al. J Immunol.* 2005;174:6030.
9. Karamperis N, *et al. Basic Clin Pharmacol Toxicol.* 2006;98:569.
10. Archer TM, *et al. J Vet Intern Med.* 2011;25:1391.
11. White S, *et al. J Vet Intern Med.* 2000;14:499.
12. Rivera SM, *et al. Biomed Chromatogr.* 2017;DOI:10.1002/bmc.3942.
13. Mealey K, *et al. J Vet Intern Med.* 2008;22:996.
14. Langman LJ, *et al. Transplantation.* 1996;61:87.
15. Nafe LA, *et al. Can J Vet Res.* 2014;78:168.
16. Wang Z, *et al. Transplant P.* 2014;46:3343.

acvim412

Big Data Research: What Is It, and What Can It Teach Us?

Kate E. Creevy, DVM, MS, DACVIM (SAIM)
College Station, TX, USA

INTRODUCTION

The first hit on the phrase "big data" appears in PubMed in 2003; at the time of this writing there are 2876 PubMed hits for that phrase in unrestricted search. What is Big Data, and where does it come from? How did it get so big? What is it good for, what is it not, and how can we incorporate it into our practice? This overview presentation will begin to address these questions.

Most authors describe three key features of Big Data, or the "3 Vs": **volume, variety**, and **velocity**.[1]

- Regarding volume, Big Data datasets are commonly measured in terabytes (10^{12} bytes), petabytes (10^{15}), or exabytes (10^{18}).
  One definition of Big Data which is gaining some momentum is that data qualify as "big" if
  $\text{Log}(n*p) \geq 7$
  where n is the number of individuals in the dataset, and p is the number of variables captured for each individual. Quick math reveals that such an equation yields a dataset with a minimum of ten million points. It is also possible to define Big Data in more relative terms, as being data of a scale and scope that it is difficult to sort or analyze with existing computing technologies. The problem with this approach is that it suggests data called "Big" now may not be "Big" at some future point because advances in technology will facilitate analysis, enabling us to "get ahead" of the data acquisition. "However, the exponential growth of data suggests this is unlikely to occur, a point that further reinforces the view that Big Data describes a break with prior practice."[2] Thus, while size is intrinsic to the concept of Big Data, it is important to recognize that other features distinguish it from prior types of research.
- Regarding variety, these data may come from non-traditional sources. Examples include the metadata (timestamp, point of entry, author, etc.) associated with medical records entries rather than the records themselves; unprocessed images as data; or free-text communication logs. The notion of variety also commonly includes the fact that Big Data sets are constructed from several such sources (the medical record, **and** its metadata, **and** its raw images, **and** its client communication notes.) "Aggregation of data in this way is justified by the idea that things can be learned from a large body of data that cannot be comprehended from smaller amounts, revealing the implicit link between Big Data and complexity."[2]
- Velocity refers to the speed with which information is generated and captured in the modern world of smartphones, activity monitors, social media, and the internet of things. All such captured information can become Big Data. "Electronic traces of human behavioural patterns have accumulated at a breathtaking pace over the last two decades, along with the spread of the social web and of mobile applications running on smartphones."[3] Such velocity is possible partly because these "electronic traces" are often collected passively, or with minimal analytic intent or design, and are often not subject to the stringent collection and storage requirements typical of the intentional research setting. Use of these vast depots of unexamined data is too tempting for the hungry researcher to easily ignore; however, this opportunistic data collection carries issues such as lack of controls, unidentified bias, missing or irregular data and abundant confounders.

Thus, some authors propose consideration of two further conceptual components: **veracity** and **value.** By contrast to the characteristic 3 Vs, these are cautionary concepts. Due to some of the common means of its collection, certain aspects of Big Data may be of questionable veracity, and its value may derive from its sheer size (or inexpensiveness) rather than its quality.

Whether one adheres to 3 Vs or 5 Vs, these features of Big Data challenge all aspects of our customary workflow, from data acquisition, to storage, processing, and harmonization of results, to computational analysis, and finally to verification of our own or others' findings. Such bulky, complex, diverse data also challenge the traditional peer-review process (more than one veterinary peer reviewer has broken out in a cold sweat when faced with thousands of cases of **anything**), frustrate customary choices for graphical representation of our data (pie charts and bar graphs prove wildly inadequate), and require sophisticated statistical analysis (simple chi-squares and uncorrected p-values won't cut it). Despite these challenges, Big Data research is here to stay. To benefit from its power in our labs and clinics, we need to navigate its novel features and understand its strengths.

RELATED DEFINITIONS

Big Data often derives from unconventional sources and is being used to envision our world - not just biomedical sciences, but also business, finance and human behavior - in new ways. Familiarity with these concepts is required to enter this space.

Sources of Data

The internet of things. Smart devices from our phones, to our activity trackers, to our cars, thermostats and ovens, collect data about our use of them intended to enhance our experience, or provide us with feedback. Most of these devices share their data over the internet with manufactures (extraction of time settings on home thermostats to enable development of product enhancements), with user-selected communities (use of an activity monitor to engage in a fitness challenge with co-workers), or with other devices (a phone's map app interacts with data from traffic cameras to provide to-the-minute travel recommendations). Data collected for one task can be used for another by the product maker or marketer - the log of an exerciser's peak activity times might suggest whether sale to that person of a reflective vest or a sun visor is more likely. Similarly, data collected by certain devices could be used for research - a municipality could purchase records of travel times of users of a certain map app in its ZIP code to plan construction of new roadways, or schedules for public transit. By 2030 it is estimated that there will be 50 billion smart objects worldwide, tracking, recording and storing very Big Data about our every move, choice, and behavior.

Unstructured data. In contrast to carefully designed surveys, or clinical trial examination scoresheets, unstructured data do not correspond to an *a priori* specification. Historically, data such as free text fields in medical records, social media or other public opinion posts, or raw images have been challenging or impossible to analyze. The advent of computer programs to analyze text or speech, and even crowd-sourced citizen-science efforts to interpret or score images, now enables us to include these components in our analyses. While messy, such unstructured data offer the possibility of a greater nuanced understanding of a situation or outcome than might be possible from a forced-choice survey containing limited options.

Quantified self. This grassroots movement, also known as "lifelogging" is the deliberate incorporation of data acquisition technology into multiple aspects of a person's daily life - food consumption, physical activity, local air or water quality, mood or alertness, performance on daily tasks, etc. Devices range from the common (diet or exercise apps on smartphones) to the sophisticated (ECGs, EEGs, and insulin or cortisol monitors). Enthusiasts of this movement are primarily interested in improved self-understanding, and engage in user communities to share best practices. However, the growth of such user communities and conferences is also driving the integration of self-monitoring devices with one another, and sometimes user-accumulated data can spur targeted biomedical research.

Next-generation sequencing. This obscurely named process, also called high-throughput sequencing, is a set of technologies which parallelize the DNA or RNA sequencing process, producing thousands or millions of sequences concurrently, which are then assembled into a final sequence. These technologies have lowered cost of sequencing and increased the accessibility and speed of obtaining DNA or RNA sequences for all types, and all funding strata of biomedical research.

Analytic Concepts

Machine learning. Within computer science, this approach enables computers to learn through data acquisition, without specific programming for each decision. This is accomplished by creation of a basic algorithm or pattern-recognition system, which is expanded over time as it is used. In this way, a computer initially given only a simple framework can construct an increasingly nuanced model or sorting system for a particular kind of data, and can even grow to relate the initial data to new forms or sources of data.

Signal-to-noise ratio. This concept derives from engineering where it first literally applied to a sound, or electrical signal. The louder the background noise, the more difficult to hear, detect or measure the targeted signal becomes. In Big Data there is always tremendous "noise" because of the nature, sources, and size of the data collected - it would always be possible to inadvertently analyze the noise (e.g., coincidences, spurious relationships, indirect or confounded associations). Therefore, most Big Data enthusiasts believe that for a "signal" to be meaningful it needs to be very strong, in order to be certain that it's not just an over-interpretation of background noise. This leads to deliberate adjustments of p-values used in Big Data analytics to more stringent criteria.

Data overfitting. This outcome is a risk of any complex mathematical model applied to data. The more variables are included, the greater the chance of identifying a coincidental relationship within the dataset on which the model is developed, that is to say, the model could be describing the "noise" not the "signal." A model that matches its development dataset too precisely (i.e., a model that "overfits") will not be generalizable to future sets of data from similar sources. In this way modeling with Big Data runs the risk of developing an excellent description of particular past events, which is useless to predict future events.

Areas of Use

Precision medicine. This approach aims to optimize medical decision-making or medical outcomes for highly specific groups of patients, by combining conventional medical information with racial or ethnic origin, genetic, environmental, dietary, or other lifestyle data. This quintessential Big Data endeavor would focus not, for example, on treatment of heart failure, but on treatment of heart failure in African American women with a particular HLA haplotype who live in the Southeast US.

GWAS (Genome-wide association studies). This is one of the more commonly seen Big Data applications in veterinary medicine and relies upon fast and cheap genome mapping or sequencing. Given a large population of organisms with a described phenotype (snout length, height, diagnosis/non-diagnosis of disease X, responsiveness to therapy Y), and genetic information from each individual (whether a SNP map or a full genome), a machine learning system can identify correlations between certain phenotypes and certain genetic elements. Correlations discovered in this way become targets for future research.

STRENGTHS OF THE BIG DATA APPROACH

The analytical power of Big Data creates the possibility that small patterns and relationships can be detected in highly complex datasets, in a manner that would not be possible with traditional, or manual, techniques. This includes discovery of relationships that only hold for certain subsets of individuals within a large diverse population. In this way, Big Data approaches are particularly suited to exploratory analyses. Detected correlations can subsequently be interrogated in a more focused, intentional way. "Given the hypothesis-free nature of data mining techniques, big data can be used to obtain a global perspective that complements the focused mechanistic studies typical of experimental biology and enables the detection of high-level information patterns that would otherwise be impossible to perceive."[4] Multiple contributory causes are present in most examples of clinical illness - few diseases occur solely as a result of a single exposure or risk factor. With this in mind, Big Data offers the ability to detect multiple concurrent correlations that may indicate several components within a network of factors contributing to disease. For example, single point mutations cause a minority of genetic diseases; the more typical scenario is that multiple genetic variations sum to a risk state. "When it became possible to identify orders of magnitude more human polymorphisms using single-nucleotide polymorphisms (SNPs), and the coverage of the genome allowed definition of common allelic variants, it was then possible to begin to map genes that contributed to a phenotype, without being the sole cause."[5]

Big Data techniques can also unfold from more classic clinical practice. Consider the following example:

- Individual cardiologists prescribe Holter monitors to specifically selected patients with clinical or genetic suspicion of dangerous arrhythmias. Each doctor interprets her patient's recording in the context of her information about that patient as a whole. (conventional)
- Over time, the bank of Holter recordings grows and grows. Machine learning techniques are applied to banked recordings, and are able to identify novel risk factors for relatively uncommon, but dangerous, arrhythmias. (Big Data)
- These new findings enable clinicians to modify their prescription of Holter monitors to additional specific patients who fall into newly detected risk groups. This remains an individualized, patient-centered, prospective and deliberate choice in each instance. (conventional)
- Over time, the bank of Holter recordings includes a larger number of repeated samples from dogs before and after therapy. This time-based, or therapy-based, information is incorporated into the machine learning algorithm and begins to predict which types of patients will or won't respond to particular therapies. (Big Data)

Another example might include the use of Big Data techniques to analyze outcomes from patients all being treated with the same drug. Rather than examining response to therapy for the indication for which the drug was prescribed, relationships may be detected between the drug and non-target organ system changes or benefits. The most well-known example of such an occurrence - the discovery of sildenafil's effects on penile erectile tissue - was a serendipitous, not a Big Data, discovery. But the economic value of that observation is hard to overstate. More deliberate screenings of large patient databases for unexpected desirable effects of commonly used medications could allow for quicker release of drugs for novel indications through repurposing, than the time required to develop and obtain regulatory approval for a novel compound or agent.

Big Data may provide an unprecedented power to **detect** a phenomenon, pattern or relationship that would otherwise have been missed, but it remains incumbent on the investigator to then determine the mechanism, meaning, or significance of that finding - even the most powerful computers and modeling simulations can only interpolate, not extrapolate. Human interpretation, often in a much smaller and more tightly controlled environment, remains necessary to get at the "why" of any finding. The ability to generate knowledge, or to make decisions, through use of a tremendous amount of raw information hinges on the ability to make reliable inferences, both in the logical sense and in the strict statistical sense. Thus, Big Data are not a replacement for the traditional scientific method!

POTENTIAL PITFALLS OF BIG DATA

Identification of (sometimes relatively weak) patterns within huge data sets runs the risk of false correlations. Furthermore, global assessment of Big Data aggregated from diverse sources after collection in order to identify novel or unanticipated relationships is by definition both a retrospective and post-hoc analysis. By its retrospective nature it will be more useful for detection of correlation, but harder to deploy for questions of causation. Additionally, results of post-hoc analysis may "overfit" the data, meaning that they describe a target dataset with such a high degree of precision that it is not likely to accurately predict future results of similar studies.

Partly because we often obtain Big Data from sources collected for reasons other than our research question, it is possible that the collection system did not accurately capture the phenomenon of greatest interest, for example, "...one can intuitively understand how a random sample of measurements of height above sea level across a golf course is unlikely to reveal the all-important locations of the holes..."[6] Knowledge of the study subject is another possible origin of inaccurate data capture, for example, whether the registered owner of the fitness tracking device is actually the one wearing it when a particular dataset was downloaded. Changes in relevant conditions may not have been captured during non-research data collection. For example, increase in after-work exercise recorded on a fitness tracker might be erroneously attributed to a workplace wellness promotion program, when in fact the owner of the fitness tracker previously had child-care obligations immediately after work, and now no longer does. When data are **gathered** for analysis, rather than intentionally **collected** for analysis, it is important to recognize the sharp boundaries between what was obtained and what cannot be known.

Finally, because of the complexity of much Big Data, the algorithms used to analyze it may be incomprehensible to most people, including even the scientists interpreting the analyses. "As a result, questioning the validity of relationships and findings based upon analysis of Big Data becomes increasingly difficult not just for the general public but also for experts, whose critical investigations may become comparable to questioning the outputs of a 'black box'."[2]

Conclusions

Big Data presents us with challenges and opportunities and it is obviously here to stay. Recognition of its best uses, and biggest pitfalls is essential to the incorporation of Big Data into our research and clinical practice. Replacement of conventional research paradigms is not the goal; rather, we should seek an approach that blends thoughtful Big Data analysis and interpretation with careful testing of mechanisms and attribution of meaning.

References

1. Baro E, Degoul S, Beuscart R, Chazard E. Toward a literature-driven definition of big data in healthcare. *BioMed Research International*. 2015:2015:639021.
2. Mittelstadt BD, Floridi L. The ethics of big data: current and foreseeable issues in biomedical contexts. *Science and Engineering Ethics*. 2016;22(2):303–341.
3. Grossglauser M, Saner H. Data-driven healthcare: from patterns to actions. *European Journal of Preventive Cardiology*. 2014;21(2 Suppl):14–17.
4. Wooden B, Goossens N, Hoshida Y, Friedman SL. Using big data to discover diagnostics and therapeutics for gastrointestinal and liver diseases. *Gastroenterology*. 2017;152(1):53–67.e53.
5. Frelinger JA. Big data, big opportunities, and big challenges. *The Journal of Investigative Dermatology. Symposium Proceedings*. 2015;17(2):33–35.
6. Coveney PV, Dougherty ER, Highfield RR. Big data need big theory too. *Philosophical Transactions of the Royal Society A. Mathematical, Physical, and Engineering Sciences*. 2016;374(2080):20160153.

acvim413

Immunoglobulin Fc Receptors: Linking Humoral Immunity with Immune Cell Function

Jonathan E. Fogle, DVM, PhD, DACVIM (Small Animal Internal Medicine)
Raleigh, NC, USA

ANTIBODY STRUCTURE

An antibody molecule is composed of two heavy chains and two light chains joined together by a series of disulfide bonds. The antibody molecule is quite flexible and can change its shape due to the flexibility generated around these disulfide bonds. Together, the heavy and light chains form the classic "Y" shape of an antibody molecule. Functionally, the antibody molecule is composed of an antigen-binding region, also known as the variable region, which has two identical antigen-binding sites (each arm of the "Y"). The other end of the antibody molecule contains the Fc region (the biochemical designation of the crystallizable fragment of the molecule). Unlike the antigen-binding regions, each of which are unique to the plasma cell producing them, the Fc region of the antibody molecule is relatively constant. The five classes of immunoglobulins, IgA, IgD, IgE, IgG, and IgM are defined by their constant region. Specialized receptors, known as Fc receptors (FcRs), interact with the Fc portion of antibody molecules to coordinate the immune response.

FCR NOMENCLATURE

In general receptors are designated with the Greek letter for the antibody class that they interact with: IgE can interact with Fc epsilon receptors (FcεRs) and IgG interacts with Fc gamma receptors (FcγRs) and so on. Within each category of FcR, there are multiple FcR types with various binding affinities and either stimulatory or inhibitory cell signaling motifs. For example, there are at least three (or more depending upon the species) FcγRs: denoted FcγRI (CD64), FcγRII (CD32), and FcγRIII (CD16). Further, each receptor may have multiple isoforms such as FcγRIIa and FcγRIIb.

FCε RECEPTORS (FCεRS)

Fcε receptors are found on a variety of immune cells including mast cells, basophils, eosinophils, and dendritic cells. The surface of mast cells is covered with the high-affinity FcεRI which binds the Fc portion of IgE in large quantities. The antibody-binding region of the IgE is now exposed and ready to contact antigen. The intracellular "tail" of the FcεRI is associated with an immune-activating tyrosine motif. Multivalent antigen can bind multiple IgE molecules, leading to cross-linking of surface IgE (bound to FcεRIs). Cross-linking of the IgE propagates a strong intracellular signal via immunoreceptor tyrosine-based activation motifs (ITAMs for short), leading to mast cell degranulation. This process is central to the immune response to helminths and to Type I hypersensitivity reactions such as allergy, atopy and anaphylaxis. Helminth parasites or other antigens (such as certain allergens) that are coated by IgE are targeted by eosinophils. Eosinophils engage IgE/antigen via FcεRI, and upon activation release their contents to mediate elimination of the parasite or in the case of certain allergens, tissue damage. Dendritic cells and macrophages also express FcεRI which efficiently uptake antigen/IgE complexes for antigen processing and presentation. This phenomenon has been implicated in the propagation of allergic responses but may also be important for development of immune tolerance in the neonatal animal. FcεRII (CD23) exhibits an intermediate affinity for IgE. It is expressed on eosinophils, B cells and intestinal epithelium. On eosinophils, FcεRII usually contributes to mast cell degranulation. On B cells, FcεRII likely plays a role in regulating (positively and negatively) differentiation and antibody production. In the intestine, FcεRII facilitates the transport of IgE from the lamina propria into the intestinal lumen. Most recently, IgE fragments that engage FcεRI are being explored for therapeutic use in IgE-mediated diseases.

Fcγ Receptors (FcγRs)

Often, antibody, such as IgG, binds antigen first and immune complexes are formed. Neutrophils and macrophages can migrate to the sites of immune complex deposition or the immune complexes can be carried to lymphoid tissues, such as the spleen, for processing. IgG bound to antigen is recognized by FcγRs on the surface of macrophages and neutrophils. Engagement of the FcγR stimulates the immune cell to perform its immune function such as the release of toxic granules or phagocytosis of the immune complex. In general, FcγR activation stimulates immune cell activity; however, some subclasses of FcγRs can inhibit the immune response, particularly when inflammation is resolving. For example, FcγRs on B cell surfaces can inhibit further maturation and antibody production. FcγRI is a high-affinity binding receptor that is expressed predominantly on macrophages and is saturated with IgG at concentrations present in normal plasma. FcγRII and FcγRIII generally function to bind immune complexes. Two forms of FcγRII (FcγRIIa and b) have been described in people and mice. FcγRIIa is an activating receptor while FcγRIIb is inhibitory. In humans, FcγRII isoforms are expressed on B lymphocytes, macrophages, neutrophils, platelets, and monocytes. In humans, FcγRIIIa and b have been characterized. FcγRIIIa is expressed on T cells and NK cells. FcγRIIIb is expressed on human neutrophils as a truncated version and its signaling mechanism (if any) is unclear. To date FcγRI, FcγRIIb, and FcγRIII have been identified for the dog. Although there is only a handful of studies describing canine FcγRs, the distribution among cell types and immune functions seem to be similar to those of humans. Even less is known about FcγR distribution and function in the cat.

IgA Fc Receptors

FcαRs engage IgA molecules and are of particular importance to initiate phagocytosis of opsonized bacteria at mucosal surfaces. FcαRI (CD89) is expressed by macrophages, eosinophils, neutrophils, B cells and T cells. Fcα/μR is a recently discovered receptor that can engage either IgA or IgM. The Fcα/μR is expressed mainly on macrophages and B cells within the intestinal lamina propria and in lymph node germinal centers. This receptor likely plays an important role in coordinating the opsonization and elimination of pathogens at mucosal surfaces.

Neonatal Fc Receptors

Another category of FcRs is responsible for the transfer of maternal immunity. These receptors, known as Fc receptor neonatal or FcRn, have been identified in most mammalian species and are likely present in most, if not all, mammals. During the last part of pregnancy, FcRns are upregulated in the mammary gland. The FcRns facilitate transfer of maternal antibody into the colostrum. Following parturition, the neonate ingests the antibody-rich colostrum. Along the gut of the newborn the same type of FcRns are present in abundance. These FcRns facilitate the transfer of maternal IgG from the gut of the newborn into circulation for redistribution to all mucosal sites, providing protection from opportunistic infection. In general, neonatal FcRn expression in the gut decreases to low levels rapidly during the first 24 hours after birth. Uptake of antibody or antigen/antibody complexes by neonatal FcRns may help shape the developing immune response. For example, FcRn knockout (FcRn-/-) mice developed more allergen-induced airway inflammation early in life than mice with functional FcRns, suggesting that ingestion of colostrum is one important factor in preventing allergic airway disease. Later in life, intestinal epithelial cells transport antigen to underlying antigen-presenting cells using FcRn-dependent transport.

Conclusion

FcRs play a unique role in linking the antigen specificity of the humoral immune response with various immune effector mechanisms. FcRs are central in the immune response to parasites and the development of allergic disease. FcRs can bind antigen-antibody complexes, which typically facilitate uptake by phagocytic cells. Finally, FcRs are essential for neonatal immunity and the shaping of the neonatal immune response.

References

1. Murphy K. *Janeway's Immunobiology*. 8th ed. Garland Science; 2012.
2. Bergeron LM, *et al. Vet Immunol Immunopathol*. 2014;157:31.
3. Sibley T, Fogle J. *Vet Immunol Immunopathol*. 2013;151:229.
4. Abramson J, Pecht I. *Immunol Rev*. 2007;217:231.
5. Nakata K, *et al. Biochem Biophys Res Commun*. 2010;395:238.

acvim414

Reviewing the Evidence: How to be a Discriminating Reader

Annette L. Litster, BVSc, PhD, FANZCVSc (Feline Medicine), MMedSci (Clinical Epidemiology)
Chicago, IL, USA

Study Design

The first step in reviewing a journal article is to determine the study design type. Use the following algorithm.

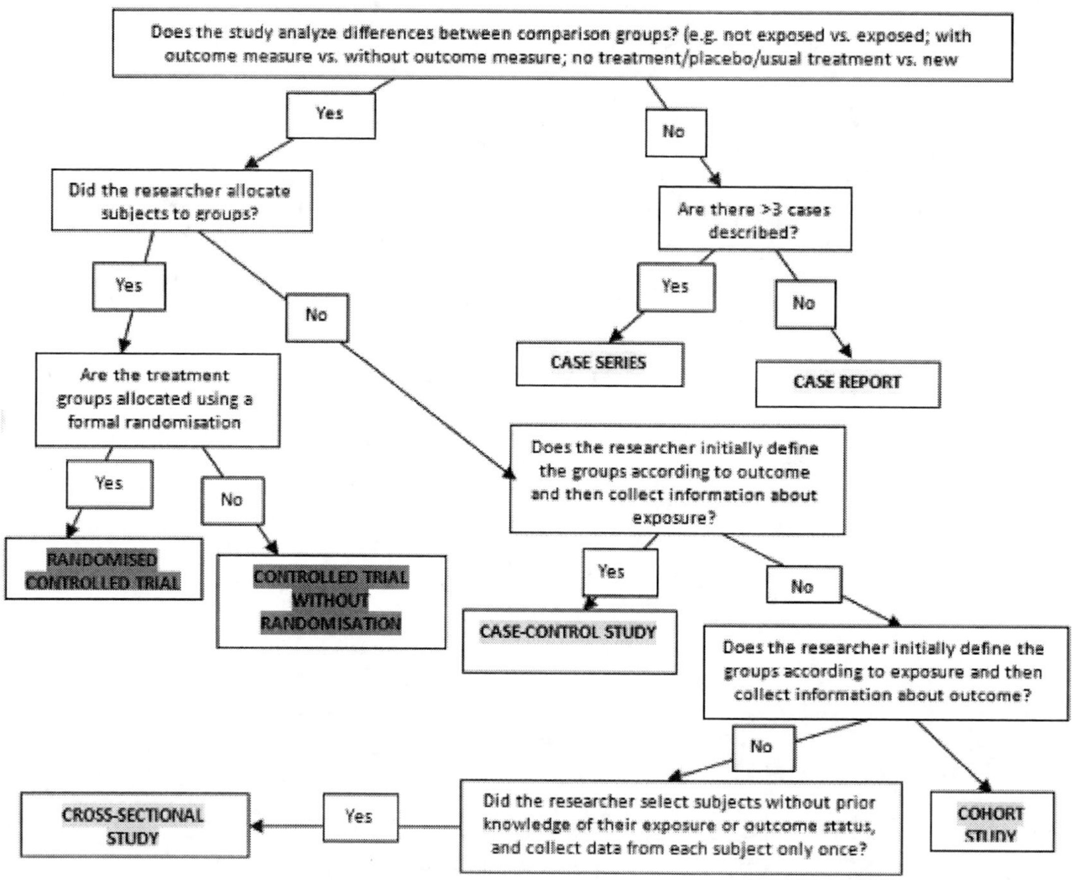

Figure 1. Study design algorithm

Once the design is correctly identified, you have a framework for understanding the most likely strengths and limitations of the study and the conclusions that can be accurately drawn from the results (Table 1).

Table 1. Comparison of observational study designs

	Case reports	Case series	Cross sectional study	Case control study	Cohort study
Subjects	Single case up to 3 cases	>3; variable number	Several subjects assessed individually	Two groups of subjects	Two groups of subjects

Characteristics	Describes unusual feature of a case	A group of cases; no control group	Picture of disease status at a single point in time for each animal in the study	One group has disease of interest and other is disease-free	One group is exposed to risk factor of interest and other is non-exposed
Time orientation	Present	Present	Present	Past	Future
Time factor	Quick	Quick	Quick	Quick	Time consuming
Cost	Inexpensive	Inexpensive	Inexpensive	Inexpensive	Expensive
Follow up period	None	None	None	None	Often prolonged
Purpose	Generates a hypothesis	Generates a hypothesis	Generates a hypothesis	Tests a hypothesis	Tests a hypothesis.
Cause and effect relationship	Cannot be interpreted since there is no control group	Cannot be interpreted since there is no control group	Cannot be interpreted	Can be suggested but not proven as the temporal sequence between exposure and outcome not established; good for studying diseases with long latencies between exposure and outcome.	Can be interpreted as the temporal sequence between exposure and outcome can be established
Measure of disease frequency and risk	None	None	Prevalence Association; several factors or outcomes may be measured simultaneously Incidence cannot be measured as new cases and pre-existing cases measured together	Odds ratio can be calculated.	Prevalence Incidence Relative risk Attributable risk Exposure is measured without bias since outcomes are unknown when study commences; known confounders can be measured. Suitable for studying the effects of rare exposures as these animals are preferentially recruited at commencement of study.
Potential problems	May lead to speculation about spurious associations	May lead to speculation about spurious associations	Cannot assess seasonal or other time-related variation; potential bias in sampling and measuring exposure; not practical for rare conditions	Prone to recall bias regarding exposure and misclassification and selection bias when cases and controls are selected; inefficient study design for studying rare exposures.	Attrition, loss to follow-up etc.; if period between exposure and outcome is prolonged, study may not be cost- or time-effective
Remarks	Descriptive	Descriptive	Observational	Observational study -	Observational

	study	study	study	Can study multiple causes of disease; useful for studying rare diseases; only one outcome can be studied per study.	study - Can study multiple outcomes of a single exposure

HIERARCHY OF EVIDENCE

Table 2. Designation of levels of evidence (Merlin *et al.* 2009)

Level of evidence	Study design
I	Evidence obtained from a systematic review of all relevant randomized controlled trials
II	Evidence obtained from at least one properly-designed randomized controlled trial
III-1	Evidence obtained from well-designed pseudo-randomized controlled trials (alternate allocation or some other method)
III-2	Evidence obtained from comparative studies (including systematic reviews of such studies) with concurrent controls and allocation not randomized, cohort studies, case-control studies, or interrupted time series with a control group
III-3	Evidence obtained from comparative studies with historical control, two or more single arm studies, or interrupted time series without a parallel control group
IV	Evidence obtained from case series, either post-test or pre-test/post-test

JOURNAL IMPACT FACTORS AND RANKINGS

The Journal Impact Factor is published each year by Thomson Reuters. It measures the number of times an average paper in a particular journal has been referred to. The impact factor of journal in a calendar year is the number of citations received by the journal in that year (any item), divided by the number of items published in the journal. Journal rankings are ordered by impact factor.

NINE STEPS - HOW TO READ A JOURNAL ARTICLE (MODIFIED FROM DARZINS *ET AL.* 1992)

1. What was the research question asked?
2. What was the study design?
3. What was the group size?
4. Was there a control group? Was it appropriate?
5. Treatment/exposure
    a. What was the treatment/exposure and how was it measured?
    b. How were animals assigned to treatment groups?
6. Outcomes
    a. What outcomes were measured and how (e.g., disease)?
    b. How long was the follow up? Were important sources of potential bias and/or confounding considered and eliminated?
7. Were appropriate statistical tests performed?
8. Are the results clinically relevant?
9. What were the conclusions drawn about the research question?

REFERENCES

1. Merlin, *et al. BMC Medical Research Methodology*. 2009;9:34.
2. Darzins, *et al. The Medical Journal of Australia*. 1992;157(6):389–394.

acvim415

Primary Immunodeficiencies and Infections

Urs Giger, DACVIM & ECVIM (Internal Medicine), DECVCP (Clinical Pathology)
Philadelphia, PA, USA

INTRODUCTION

Animals with recurring or persistent, antimicrobial-unresponsive and unusual infections likely suffer from a hereditary (primary) immunodeficiency disorder. Immunodeficiencies represent a large heterogeneous group of dysfunctions of host immunity increasing the risk for infections. They can arise through disturbances in antigen-specific defense mechanisms mediated by lymphocytes, the nonspecific defense system (which includes phagocytes, plasma proteins, and physical barriers), or both. Many genetically determined immune defects have been described in the dog, whereas only a few are known in cats. A definitive diagnosis often requires specific immune testing in addition to routine laboratory tests, and therapeutic interventions are limited. The molecular defects for several primary immunodeficiencies have been elucidated allowing for DNA screening. A few hereditary immunodeficiency disorders are prevalent within certain breeds of dogs, whereas others occur in isolated families/cases.

The nonspecific immune system, also known as **innate** or **natural immunity**, should be functional at birth and available on short notice to protect the host from invasion by all sorts of organisms. It includes physicochemical barriers, phagocytes, complement and other plasma proteins, and natural killer cells. Congenital barrier defects particularly involve the skin and mucous membrane surfaces and are associated with infections of particular organs. A variety of hereditary skin diseases are being further defined, and barrier, but also other immunodeficiencies, are being recognized. The Ehlers-Danlos syndrome, causing fragile, hyperextendable skin in many dogs and cats as well as the myxedematous skin and immunodeficiencies of Shar-peis, predisposes the animals to pyoderma, whereas ciliary dyskinesia in dogs increases the susceptibility to rhinosinusitis and pneumonia. Similarly, X-chromosomal ectodermal dysplasia in German shepherds is associated with skin disease as well as other immunodeficiencies.

Disorders of the phagocytic system involve defects of neutrophils and monocytes as well as the complement system and can lead to pyogenic and granulomatous infections. The granulomatous reaction can occur when neutrophils malfunction and mononuclear cells are recruited. A wide variety of pyogenic bacteria (e.g., staphylococci, *Escherichia coli*, *Klebsiella*, *Enterobacter*) are usually involved, most of which represent normal microflora or pathogens of relatively low virulence. Recurrent infections of the skin, respiratory tract, and oral cavity are common, and intermittent bacteremia and overwhelming sepsis are also seen. Multisystemic amyloidosis, vasculitis, and immune complex disease are complications that can occur as a result of chronic recurrent or persistent infection. Cyclic hematopoiesis and leukocyte adhesion deficiency (LAD) are examples of serious quantitative and qualitative phagocytic defects, respectively. A unique immunodeficiency causing a predisposition to avian tuberculosis in miniature schnauzers (and also bassets) has recently been elucidated at the molecular level.

The specific immune system can be divided into humoral and cell-mediated immune systems and includes B and T lymphocytes, immunoglobulins, and cytokines. Deficiencies of B lymphocytes or humoral immunity affect the production of immunoglobulins and lead to increased susceptibility to pyogenic bacterial infections. Deficiencies of T lymphocytes or cell-mediated immunity (CMI) are associated with viral and fungal infections, but intracellular bacterial infections may also occur. Animals with cellular immunodeficiencies may have smaller thymic and tonsillar tissues as well as intestinal and peripheral lymph nodes and decreased numbers of circulating lymphocytes.

The degree of immunodeficiency varies greatly between defects. Infections may be systemic or restricted to a particular organ system like the skin or respiratory tract. Some immunodeficiencies lead to overwhelming infections and death within the first few days to weeks of life, whereas others, such as morphologic leukocyte changes, are not consistently associated with any noticeable predisposition to infection. Chédiak-Higashi syndrome in smoke-colored Persian cats is characterized by abnormally large eosinophilic granules in polymorphonuclear leukocytes. It causes no immunodeficiencies but does cause a bleeding tendency resulting from a platelet storage pool disease. Similarly, Birman cats with acidophilic granulation of neutrophils and dogs and cats with various lysosomal storage diseases (e.g., mucopolysaccharidosis, gangliosidosis, mannosidosis) have granulation or vacuolation of leukocytes without being immunocompromised; they also exhibit frequently a lymphocytosis. The Pelger-Huët anomaly, which is characterized by hyposegmentation of granulocytes, causes no immunodeficiency in animals, despite the fact that the leukograms of affected dogs and cats reveal the most severe left shift with a normal leukocyte count.

Although an increased susceptibility to opportunistic infections develops, the type of infection varies depending on the type of defect within the immune system. A few immunodeficiency disorders predispose animals to a restricted group of unusual infectious agents. Some male dachshunds appear predisposed to *Pneumocystis* pneumonia, and German shepherd dogs may be prone to systemic aspergillosis or rickettsiosis. Doberman pinschers and Rottweiler dogs are more likely to develop parvoviral disease. Golden and Labrador retrievers and Bernese mountain dogs that had high serum antibody titers to *Borrelia burgdorferi* were more likely to have glomerulonephritis. Basset hounds and miniature schnauzers have an increased susceptibility to systemic avian mycobacteriosis, and possibly toxoplasmosis, and neosporosis. American and English foxhounds appear to be predisposed to developing leishmaniasis. Great Danes and Dobermans may be more susceptible to cryptococcal infections. A genetic predisposition to demodicosis has been proposed in various canine breeds and families. Feline infectious peritonitis has also been suggested to have a genetic basis. The mechanisms predisposing particular animals to specific infections remain unknown in many breeds but was recently discovered in miniature schnauzers with increased susceptibility to systemic fatal avian tuberculosis.

Major Clinical Signs of Primary Immunodeficiency Disorders

1. Recurrent infections, chronic and protracted course of infection, or both
2. Infection with common nonpathogenic (opportunistic) or aberrant infectious agents
3. Severe and often atypical infectious disease manifestations
4. Delayed, incomplete, or lack of response to antimicrobial therapy
5. Adverse reactions to modified-live virus vaccines

The above-mentioned key signs of infection develop in animals with a primary immunodeficiency generally early in life. Despite receiving colostrum, clinically affected animals may have illness during the neonatal to juvenile period and may develop recurrent and overwhelming infections that lead to severe debilitation and death before 1 year of age. Several animals, but typically not all, in a litter may be affected, whereas the parents are usually healthy. A genetic predisposition to infection is rarely noted after 1 year of age (e.g., avian tuberculosis in miniature schnauzers). Furthermore, animals with primary immunodeficiencies may have other special clinical manifestations. Hypersensitivity reactions may occur and reflect an overall dysregulation of the immune system caused by a lack of one or more components or a chronic antigen stimulation from inadequate clearance of infections. Chronic systemic infections may also hamper the growth rate. Characteristic coat color dilutions and increased tendency for surface bleeding are seen (e.g., in collies with cyclic hematopoiesis, Persian cats with Chédiak-Higashi syndrome, and Weimaraners with an incompletely defined immunodeficiency). Nude Birman kittens and ectodermal dysplasia are associated with a complete lack or loss of hairs.

The mode of inheritance of primary immunodeficiencies has not yet been determined in all cases. Autosomal recessive transmission, with affected males and females born to healthy parents, is usual, but a few exceptions exist. The Pelger-Huët anomaly is inherited as an autosomal dominant trait. Severe combined immunodeficiency caused by two different mutations in the common γ-chain interleukin-2 (IL-2) receptor in basset hounds and Cardigan Welsh corgis are X-chromosomal recessive disorder, so only males are affected, and the dams and half of her female littermates are carriers. Thus, the breed, gender, age of onset, type of infections, and other special characteristics may suggest a particular immunodeficiency. Furthermore, it follows that within a breed the immunodeficiency is typically caused by the same defect and mutation while different breeds may have mutations in the same or different genes.

Diagnostic Studies

Although an immunodeficiency may be suspected on the basis of clinical evidence, specific laboratory tests are generally required to reach a definitive diagnosis. A minimum database of information, including results of a complete blood count, serum chemistry screen, and urinalysis, should always be obtained and may suggest a specific disorder. The differential leukocyte count and microscopic evaluation of a blood smear are the most important test results. Leukopenia in the presence of an active bacterial infection is by far the most feared condition. It should be noted that generally, some breeds, such as greyhounds, have normally low white blood cell counts. Neutropenia may be transient, as it occurs with cyclic hematopoiesis every 12 to 14 days or parvovirus infection, or persistent, as it is seen in animals with cobalamin malabsorption or overwhelming infections (sepsis). Lymphopenia may be observed in dogs with a T-cell or severe combined immunodeficiency. Although leukocytosis is expected during periods of infection, defects in leukocyte adhesion and egress from blood circulation at sites of infection may be associated with disproportionately high leukocytosis for the degree of infection as seen with hereditary LAD and glucocorticoid usage. Dachshunds with *Pneumocystis* pneumonia also have very marked leukocytosis. Anemia of chronic disease is often observed in infected animals caused by several factors, but the erythrocyte count may be in the normal range even if the animals have active infections and during periods of treatment and remission. Careful review of a blood smear may reveal leukocyte abnormalities such as granulation and vacuolation resulting from lysosomal storage diseases or Chédiak-Higashi syndrome, acidophilic granulation of leukocytes in Birmans, phagocytized microorganisms, or toxic leukocyte changes that suggest overwhelming bacterial infections.

Serum globulin concentrations are generally higher during chronic infections. Low or normal globulin levels in infected animals may suggest major external losses or diminished production from a humoral (B-cell) immune defect. Indeed, specific immunoglobulin deficiencies have been recognized in dogs. Serum protein electrophoresis may identify a γ-globulin deficiency, but immunoelectrophoresis is required to detect the class and degree of immunoglobulin deficiency. Maternal immunoglobulins can only be absorbed during the first day of life and influence the values during the first few weeks. IgM can be synthesized very early in life, whereas the development of IgA may be delayed for months. Thus it is important to compare values with data from age-matched controls. Titers against specific antigens can be measured, followed by evaluation of the antibody response to vaccination against particular agents. T-cell or combined immunodeficiencies cause defective CMI responses. The animal may have prolonged allograft rejection times and decreased delayed-type hypersensitivity to skin testing with viral vaccines, tuberculin, or dinitrochlorobenzene (DNCB). Reduced *in vitro* lymphocyte stimulation results may also be caused by a primary lymphocyte defect or the infection.

The identification of the agents infecting an animal is important for diagnostic as well as therapeutic reasons. Appropriate cultures of tissues, body fluids, and excretions for microorganisms and antigen and serologic blood tests are addressed in the chapters on specific infectious agents. Antibody titers may also be used to assess a response to vaccines and humoral immunity.

Gross and microscopic histopathology and cytology may reveal certain microorganisms but are most helpful in characterizing the architecture, morphology, maturation, and function of the immune system, such as of the leukocytes, bone marrow, lymph nodes, thymus, and spleen, as well as other barrier systems. In ciliary dyskinesia, morphologic abnormalities of cilia may be identified by electron microscopy, but functional studies by imaging techniques or on respiratory epithelial biopsy specimens are also indicated.

For additional characterization of the immunodeficiencies, special leukocyte studies are often required. Surface marker studies by fluorescent-assisted cell sorters or flow cytometers can differentiate between T- and B-cells, determine T- and B-cell ratios, and determine the presence or absence of leukocyte adhesion proteins (CD11/18) or IL-2 receptors. Lymphocyte function studies include lymphocyte stimulation and plaque-forming assays for *in vitro* immunoglobulin production. Phagocyte function studies assess leukocyte adhesion, migration, chemotaxis, phagocytosis, "respiratory burst," and bactericidal activity. All functional assays should be performed on fresh blood cells (<1 day) and compared simultaneously with an age- and breed-matched control. Furthermore, *in vitro* lymphocyte functions are generally impaired and phagocyte functions are enhanced during periods of active infection. Whenever possible, it is advisable to control the infection before studying leukocyte function.

Treatment and Prevention

Successful control of infection in immunodeficient animals depends on the underlying disease as well as the type and severity of the immune defect. In immunocompromised patients, early and aggressive antimicrobial therapy is indicated even for mild infections with nonpathogenic agents. Because of the immunodeficient host's potential inability to kill bacteria, bactericidal antibiotics are recommended until bacterial infections are controlled.

No practical treatments for primary immunodeficiencies exist (except for parenteral cobalamin administration to animals with cobalamin malabsorption). Immunocompromised animals with infection generally have a guarded to poor prognosis. Despite aggressive antimicrobial therapy, their infections are difficult to control, leading to overwhelming infections, protracted courses, and recurrences. Some leukocyte defects cause death before 1 year of age, whereas others may not lead to a markedly increased predisposition to infection. In experimental studies, bone marrow transplantation and gene therapy corrected several canine leukocyte defects. Indeed, dogs with hereditary immunodeficiencies and other genetic defects have served as intermediate between experiments in murine models and its application in humans to test safety and efficacy of novel therapies.

Owners must consider the potential zoonotic risks involved with keeping an immunodeficient animal with infections that may be contagious to humans, particularly immunosuppressed humans exposed to foxhounds with leishmaniasis and miniature schnauzers and basset hounds with avian mycobacteriosis.

Examples of Primary Immunodeficiencies

Disease (synonyms)	Inheritance	Breeds	Characterization
Ciliary dyskinesia (immotile cilia syndrome)	AR	Many dog breeds	Rhinosinusitis, bronchopneumonia with bronchiectasis, situs inversus
Complement component 3 (C3 deficiency)	AR	Brittany spaniel	Pyogenic infections, lack of complement-mediated phagocytosis
Bactericidal neutrophil defect	U	Doberman pinscher	Upper respiratory infections, reduced bactericidal activity
Cyclic hematopoiesis (cyclic neutropenia)	AR	Collie (gray)	Severe neutropenia every 12–14 days, reactive amyloidosis
Leukocyte adhesion deficiency (LAD or CD18 deficiency)	AR	Irish setter, Red and white setter, DLH cat	Severe leukocytosis, infection with limited pus formation, lack of neutrophil adhesion
Pelger-Huët anomaly	AD	Australian shepherd, foxhound, others, cats	Hyposegmented granulocytes, no immunodeficiency
Selective cobalamin malabsorption (cubilin or amnionless deficiency)	AR	Giant schnauzer, border collie, beagle, Australian shepherd	Weight loss, inappetence, leukopenia with hypersegmentation megaloblastic bone marrow, methylmalonic aciduria
Increased susceptibility to avian mycobacteriosis	U	Miniature schnauzers, basset hound	Systemic avian tuberculosis and few other unusual infections
Increased susceptibility to *Pneumocystis* pneumonia	AR	Dachshund	*Pneumocystis* pneumonia
Susceptibility to fungal and rickettsial infections; pyoderma	U	German shepherd	Severe ehrlichiosis, Rocky Mountain spotted fever, disseminated aspergillosis, pyoderma
X-linked severe combined	XR	Basset hound,	Severe bacterial and viral infections, no

immunodeficiency (X-SCID)		Cardigan Welsh corgi	IgG and IgA, deficient lymphocyte blastogenesis
Severe combined immunodeficiency (SCID)	AR	Jack Russell terrier, Frisian water dog	Severe serum immunoglobulin deficiency, hypoplasia of lymphoid tissues
Thymic abnormalities and dwarfism	U	Weimaraner	Reduced growth, thymosin responsive
Recurrent infections/inflammation	U	Weimaraner	Pyoderma, severe abscess, bleeding tendency
Selective IgA deficiency	U	Beagle, Shar-Pei, German shepherd	Respiratory and GI infections
Hypotrichosis congenital and thymic atrophy	AR	Birman	Nude kittens, neonatal death, no thymus
Chédiak-Higashi syndrome	AR	Persian	No immunodeficiency, large granules in phagocytes, bleeding tendency

Author's studies were supported in part by grants from the National Institutes of Health (OD 010939) and AKC Canine Health Foundation. The author is the director of the non-for-profit PennGen Laboratory offering genetic and hematological testing.

Author's address: Urs Giger, Veterinary Hospital of the University of Pennsylvania, 3900 Delancey Street, Philadelphia, PA 19104-6010 USA. Phone: 215-898-8830; Fax: 215-573-2162; giger@vet.upenn.edu; http://research.vet.upenn.edu/penngen.

References

1. Giger U, Smith J. Immunodeficiencies and infectious diseases. In: Greene C, ed. *Infectious Diseases of the Dog and Cat*. 4th ed. Elsevier; 2012:1101–1123, and references therein.

acvim416

Canine Splenic Disease: An Update

Alex Lynch, BVSc (Hons), DACVECC, MRCVS
Gainesville, FL, USA

Splenic disease is frequently encountered in the dog and can pose considerable clinical challenges for the small animal veterinarian. Splenic masses are identified most commonly with the associated concern for poor long-term prognosis. Benign lesions are seen with some regularity, but malignant neoplasia usually features highly on the differential diagnosis list until convincing evidence of the contrary is confirmed. Some important non-neoplastic diseases are seen quite infrequently, however, including splenic torsion,[1] traumatic injury to the spleen, and splenitis.[2] Dogs with splenic disease may be presented in two major ways. Firstly, many dogs will have acute, and frequently profound, hemodynamic compromise associated with acute hypovolemia, hemorrhage, and anemia, secondary to splenic mass rupture. These dogs have impaired systemic oxygen delivery resulting in a syndrome of collapse, tachycardia, tachypnea, and pallor. These dogs may have had only minor antecedent events prior to their collapse, the significance of which may be overlooked or even confused with natural aging in older dogs. The second group of dogs seen in small animal practice have their splenic disease somewhat serendipitously identified during examination of the pet for an unrelated reason (e.g., annual examination, work up of unrelated clinical signs). In both circumstances, but especially in the emergent case, rapid and accurate identification of malignancy would be ideal. Hemangiosarcoma is of course the malignancy of greatest concern in these dogs, accounting for about 20% of all sarcomas in the species,[3] and about 4% of all tumors in general.[4] Its aggressive metastatic nature means the long term prognosis remains poor, even when treatment is sought.[5] For the dogs with acute hemodynamic compromise, resuscitation with fluids and potentially blood products,[6] along with emergent abdominal surgery to achieve hemostasis is usually necessary. Owners are left with a difficult decision. Their dog requires urgent, potentially costly, care given their life threatening condition. A definitive diagnosis for their splenic mass has likely yet to be determined and in a worst case scenario could result in a survival of just a few months.[5,7-8] In cases of benign lesions, however, splenectomy may prove curative and be associated with a good long-term prognosis. The small animal veterinarian therefore has to use any available clues to the likely final diagnosis in that patient to guide owners. This lecture will review recent literature relevant to splenic disease in dogs to build upon the existing evidence base for this disease.

How Likely Is Malignancy in a Dog With a Splenic Mass?

As mentioned above, the holy grail of canine splenic disease would be the rapid, accurate pre-operative identification of malignancy. Traditionally about half of splenic masses in the dog were expected to be benign and therefore potentially definitively cured with splenectomy. For the remaining half, malignancy would be expected with a variable outcome depending on the tumor type present. These original tumor prevalence studies were derived from general submissions to veterinary pathology laboratories and therefore gave little indication of the associated clinical picture.[9-10] This approach inevitably misses some dogs, specifically those that die without a definitive confirmation of splenic neoplasia either through lack of submission or lack of biopsy. More recent studies have examined subgroups of dogs with splenic disease that may prove clinically more helpful in practice.[6,11-13] While overall numbers generally reflect a similar 50:50 rule for benign:malignant, with a greater tendency towards more cases being malignant than not, there are certain cohorts of dogs that likely have a far greater prevalence of malignancy. Certain breeds of dog, for instance, have a well-publicized increased likelihood of hemangiosarcoma including retrievers, shepherds, and the molossoid breeds. Genomic studies in the USA have described particularly high lifetime risk of hemangiosarcoma in Golden Retrievers (e.g., greater than 1 in 5).[14-15] Recent research examined the genomic profile of 75 clinical cases and found great heterogeneity in somatic DNA copy aberrations fitting with the somewhat heterogenous anatomical and histopathological nature of hemangiosarcoma in the dog.[16] Overall, no difference in the prevalence of benign or malignant lesions were identified in dogs based on their body weight in a recent study. Despite the similar likelihood of malignancy in small and large dogs, with small dogs being defined as a body weight <27.8 kg, hemangiosarcoma was significantly less likely in these smaller dogs.[17] Other characteristics that have been recognized more commonly with malignancy include increasing patient age,[18] the presence of hemoperitoneum,[11] lower mass-to-spleen ratio,[19] lower splenic weight,[19] lower serum total protein,[18] lower platelet count,[18] and the requirement for transfusion.[6] No cytological or biochemical data on peripheral blood or abdominal effusion samples reliably differentiates between causes of hemoperitoneum in dogs, unfortunately.[20]

The clinical utility of pre-operative imaging in cases of splenic masses is also contentious. In the incidentally identified case, where the pressures of active hemorrhage and ensuing emergency surgery are not present, abdominal imaging is generally recommended. In the more emergent case, a balance between the data obtained from imaging studies and the delay of definitive treatment must be made. The ultrasonographic appearance of splenic masses rarely differentiates benign from malignant disease with any certainty. The number of lesions visualized with ultrasound was also not predictive of tumor type in cases of hemoperitoneum.[21] Ideally, ultrasonography would accurately identify metastases and aid with surgical planning. The completeness of an abdominal ultrasound examination can be compromised in larger, deep-chested dogs, however, especially for accurate characterization of more cranial abdominal organs. Computed tomography may enable improved visualization of abdominal organs (e.g., dogs >25 kg).[22] The discriminatory ability of CT scanning for differentiating the likely underlying pathology present in dogs is debatable and may depend on the nature of the CT study performed. One study described the use of single phase contrast enhanced CT in dogs and found predictable appearances of pre- and postcontrast images in malignant splenic disease.[23] More recently, a study evaluating dual phase CT found no significant differences in any quantitative or categorical CT data for the discrimination of benign or malignant disease.[24] In this study, there was overlap between all histological types, especially between hemangiosarcoma and hematoma that were seen to commonly have contrast accumulation associated with active hemorrhage. In reality, the availability of CT, plus the associated time for interpreting CT images, may considerably lessen the likelihood of recommending this modality in the emergent case. The most promising imaging modality may actually prove to be MRI with better contrast resolution compared to CT and has been described as a helpful modality in characterizing splenic and liver nodules.[25] The same argument applies, however, to emergent cases, where the requirement for a lengthy anesthetized period is not appealing in the actively bleeding dog. Survey thoracic radiographs assessing metastases are sensible, along with checking for pericardial effusion. A study described that 8.7% of dogs that underwent splenectomy had concurrent right atrial masses, although overt pericardial effusion was rarely seen.[26] In the incidentally identified splenic mass, pre-operative aspiration and/or biopsy of splenic masses can absolutely be considered. Combined ultrasound-guided fine-needle aspiration and needle core biopsy has been described as safe and effective means of collecting further information pre-operatively.[27]

In some cases, given the patient's instability or the attending surgeon's preference, imaging may be omitted altogether and the patient taken for exploratory laparotomy. A difficult intraoperative decision can be interpreting the significance of hepatic lesions in dogs undergoing splenectomy. In this context, hepatic lesions could represent advanced malignancy or incidental inconsequential benign nodular change. A study examined 79 dogs undergoing splenectomy and described 58 of the 79 having grossly abnormal liver appearance at surgery. Biopsies were obtained in the 21 dogs with normal liver appearance and no occult neoplasia was recognized. In the dogs with abnormal liver appearance, there was a 50:50 split with respect to benign and malignant disease. This study did describe an increased association with metastatic neoplasia if multiple, dark colored and actively bleeding lesions were visualized.[28]

What Prognostic Indicators Exist in Dogs With Splenic Masses?

The focus of most of the discussion so far has been on long-term outcome for dogs with splenic masses, which is largely determined by the presence of benign or malignant disease. In the short term, however, several factors have also been associated with a poorer outcome including preoperative thrombocytopenia, anemia, intraoperative ventricular arrhythmias and blood product administration.[6,12-

[13] Ventricular arrhythmia occurs in these dogs independent of heart disease and may occur secondary to hypovolemia or anemia, metastasis to myocardium, or catecholamine excess.[29] Hemostatic abnormalities may be noted in dogs with splenic masses and spontaneous hemoperitoneum, including a moderate consumptive thrombocytopenia, prolongation of prothrombin and activated partial thromboplastin times, and a hypocoagulable-hyperfibrinolytic tendency on viscoelastic monitoring.[30] Hyperlactatemia, as an indicator of systemic hypoperfusion, was associated with the extent of hyperfibrinolysis in this study. The most common reasons for short-term death in another study of dogs undergoing splenectomy for splenic masses were uncontrolled hemorrhage and thrombotic complications, specifically portal system thrombosis (PST) and pulmonary thromboembolism.[13] Acute PST is described as an uncommon complication of splenectomy in dogs[31] and people.[32-34] In people, a hypercoagulable tendency and ligation of the splenic vein creating a low flow 'cul de sac' are believed to contribute to postoperative PST.[34] Pre-operative viscoelastic evidence of hypercoagulability was common in hemodynamically stable, non-anemic dogs undergoing splenectomy for splenic masses.[35] The specific circumstances necessary for PST to occur in individual dogs is currently unknown and likely is multifactorial in nature. Hemangiosarcoma cell lines are also known to constitutively produce tissue factor associated mRNA and express TF antigens on the cellular membrane surfaces.[36] Consequently, these tissue factor bearing hemangiosarcoma cells line are associated with proportionally greater thrombin generation, compared to other neoplastic cell lines. Tissue factor bearing neoplastic cells therefore may play a pivotal role in hemostatic abnormalities recognized in clinically case. The potential role of thromboprophylaxis in some dogs with splenic masses is as yet uncertain. Dogs that develop PST following splenectomy tend to develop an acute onset of abdominal pain, cardiovascular collapse associated with massive fluid losses through the GI tract secondary to mesenteric ischemia, tachypnea and abdominal effusion. The thrombus may be identified with ultrasound and demarcation of normal and abnormal tissues may be visualized at necropsy. The prognosis for these dogs is generally poor in the authors' experience although more chronic portal system thrombosis may be associated with improved outcome.[31]

Another important scenario encountered in practice is the anticipated outcome for dogs with incidentally identified, but not yet ruptured, splenic masses. This was investigated in a study of 105 dogs and found a higher than usual prevalence of benign lesions (70.5%) compared to malignant (29.5%).[37] This study corroborates one of the criteria of malignancy described earlier, where the lack of hemoperitoneum is more commonly recognized in those dogs with benign disease. These dogs are therefore expected to have no long-term sequelae of their splenic pathology. A specific examination of the clinical course for dogs with histopathological evidence of splenic hematoma has also been described.[38] As anticipated, the median survival time for most of these dogs was generally excellent (MST 647 days, range 0–3,287 days). Interestingly, 4 of the 35 dogs did reportedly develop metastatic disease. This highlights that truly malignant lesions can be overlooked (e.g., benign spleen with malignant liver nodules) or the histopathological assessment of splenic masses be inaccurate. The histological patterns identified in cases of hemangiosarcoma can be variable, ranging from irregular vessels (capillary growth pattern) through to expanded caverns (cavernous growth patterns) all lined by variably differentiated endothelial cells.[39] This apparent heterogeneity means multiple sections of masses should be examined to gain as representative a sample as possible. No specific histological pattern was significantly associated with a better or worse outcome in dogs, but there was a trend towards a shorter median survival time with the cavernous growth pattern.[39] In cases of equivocal histopathology, immunohistochemistry panels including vWF-FVIII and CD31 may prove to be diagnostically helpful.[40-41]

What Is the Expected Outcome for Dogs with Splenic Malignancy Undergoing Advanced Care?

The aggressive nature of splenic malignancy, especially when hemangiosarcoma is present, is traditionally associated with a poor outcome. Staging of hemangiosarcoma gives a better idea of anticipated prognosis, with the modified WHO staging scheme being most commonly utilized. Stage I cases (non-ruptured splenic hemangiosarcoma without metastases) have accordingly longer survival time than stage II (ruptured splenic hemangiosarcoma or nodal metastases) and stage III (metastatic hemangiosarcoma).[42] A recent study identified that clinical stage was the only prognostic factor significantly associated with survival in dogs with splenic hemangiosarcoma that had undergone splenectomy now receiving chemotherapy.[5] Currently doxorubicin or epirubicin based chemotherapy is most commonly recommended for cases of hemangiosarcoma.[42-44] Recent data suggests that the addition of cyclophosphamide-based metronomic protocols to doxorubicin may only be met with modest prolongation of survival.[5] Splenectomy alone for splenic hemangiosarcoma was associated with a median survival time of 1.8 months (range 1.3–2.9 months), while splenectomy with chemotherapy generally extended the median survival time to 3.4 months (range 2.8–4.3) depending on the protocols employed and stage of the disease.[5]

Are There Any Promising Novel Biomarkers for Canine Splenic Disease?

From a diagnostic perspective, it would be advantageous to have an accurate non-invasive means of identifying malignancy in cases of splenic disease prior to committing to treatment. The best investigated biomarker for this purpose is thymidine kinase.[45] Thymidine kinase is an enzyme produced by proliferating cells (e.g., neoplastic cells) and has been specifically assessed in canine splenic hemangiosarcoma. Thymidine kinase was significantly higher in dogs with splenic hemangiosarcoma and hemoperitoneum compared to controls. The biomarkers also showed some promise in differentiating benign from malignant disease. Currently, this assay is not available as a point-of-care test limiting its relevance in the current clinical practice setting. Several other potential markers of malignancy have been investigated in canine hemangiosarcoma cell lines, including vascular endothelial growth factors (e.g., VEGF2),[4] tyrosine kinases (e.g., C-kit),[46] sphingosine-1-phosphate pathway,[47] and the endothelins.[48-50] Another potential application of these proteins would be as molecular targets for therapy. Inhibition of C-kit via the use of dasatinib and imatinib reduced the viability of hemangiosarcoma cell line viability and therefore could be considered as an additional chemotherapy option in clinical cases.[51] Similarly, since the endothelin axis is upregulated in cases of canine hemangiosarcoma, the use of endothelin receptor antagonists (e.g., ambristen, bosentan) has been shown to limit growth of cell cultures, with potential applications for clinical cases.[52] Interestingly, the Chinese herb Yunnan Baiyao that is used as non-specific hemostatic agent typically during the palliation of hemangiosarcoma cases, was described as exerting a time and concentration dependent effect on the death of canine hemangiosarcoma cell lines *in vitro*.[53]

References

1. DeGroot W *et al. J Am Vet Med Assoc.* 2016;248(6):661–668.
2. Ferri F *et al. Vet Pathol.* 2017;54(1):147–154.
3. Smith AN. *Vet Clin North Am Sm Anim.* 2003;33:533–552.
4. Goritz M *et al. J Comp Path.* 2013;149:30–39.
5. Wendelburg KM *et al. J Am Vet Med Assoc.* 2015;247(4):393–403.
6. Lynch AM *et al. J Am Vet Med Assoc.* 2015;247(6):636–642.
7. Hamer AS *et al. J Vet Intern Med.* 1991;5:160–166.
8. Sorenmo K *et al. J Vet Intern Med.* 2000;14:395–398.
9. Spanger WL *et al. J Am Vet Med Assoc.* 1992;200:829–834.
10. Day M *et al. J Small Anim Pract.* 1995;36:426–433.
11. Hammond TN *et al. J Am Vet Med Assoc.* 2008;232(4):553–558.
12. Lux CN *et al. J Am Vet Med Assoc.* 2013;242:1385–1391.
13. Wendelburg KM *et al. J Am Vet Med Assoc.* 2014;245:1382–1390.
14. Schulthiess PC. *J Vet Diagn Investig.* 2004;16:522–526.
15. Shearin AL, Ostrander EA. *Dis Model Mech.* 2010;3:27–34.
16. Thomas *et al. Chromosome Res.* 2014;22:305–319.
17. Sherwood MJ *et al. J Am Anim Hosp Assoc.* 2016;52:220–226.
18. Prymark C *et al. J Am Vet Med Assoc.* 1988;193:706–712.
19. Mallinckrodt MJ, Gottfried SD. *J Am Vet Med Assoc.* 2011;239(10):1325–1327.
20. Wong *et al. Vet Surg.* 2015;44:852–857.
21. Levinson JG *et al. J Vet Emerg Crit Care.* 2009;19:496–500.
22. Fields EL *et al. Vet Radiol Ultrasound.* 2012;53:513–517.
23. Fife WD *et al. Vet Radiol Ultrasound.* 2004;45:289–297.
24. Jones I *et al. Vet Radiol Ultrasound.* 2016:57(2):144–153.

25. Clifford CA *et al. J Vet Intern Med.* 2004;18:330–338.
26. Boston SE *et al. J Am Anim Hosp Assoc.* 2011;47:336–341
27. Watson AT *et al. Vet Radiol Ultrasound.* 2011;52:317–322.
28. Clendaniel DC *et al. J Am Anim Hosp. Assoc.* 2014;500(4):e6–10.
29. Keyes ML *et al. J Vet Emerg Crit Care.* 2007;3:33–38.
30. Fletcher DJ *et al. J Vet Emerg Crit Care.* 2016;26(1):41–51.
31. Respess M *et al. J Vet Int Med.* 2012 Mar–Apr;26(2):230–237.
32. Rattner DW *et al. Arch Surg.* 1993;128:565–569.
33. Pietrabissa A *et al. Surg Endosc.* 2004;18(7):1140–1143.
34. Winslow ER *et al. Ann Surg.* 2005;242:745.
35. Lynch AM *et al. J Vet Intern Med.* 2015;29:1193–1194.
36. Witter LE *et al. Am J Vet Res.* 2017:78:69–79.
37. Cleveland MJ, Casale S. *J Am Vet Med Assoc.* 2016;248(11):1267–1273.
38. Patten SG *et al. Can Vet J.* 2016;57:842–846.
39. Goritz M *et al. J Comp Path.* 2013;149:30–39.
40. Von Beust BR *et al. Vet Pathol.* 1988;25:251–255.
41. Sabattini S, Bettini G. *J Comp Path.* 2009;140:158–168.
42. Kim SE *et al. J Am Vet Med Assoc.* 2007;231:1500–1557.
43. Kahn SA *et al. Can Vet J.* 2013;54:237–242.
44. Wiley JL *et al. Vet Comp Oncol.* 2010;8:221–233.
45. Thamm DH *et al. Vet Comp Oncol.* 2012; 10:292–302.
46. Adachi M *et al. J Vet Med Sci.* 2016;78(4):649–656.
47. Rodriguez AM *et al. J Vet Int Med.* 2015;29:1088–1097.
48. Yanagisawa M *et al. Nature.* 1988;332:411–415.
49. Fukomoto S *et al. Life Sciences.* 2014;118:329–332.
50. Fukomoto S *et al. Vet J.* 2015;204:269–274.
51. Dickerson EB *et al. Transl Oncol.* 2013;6(2):158–168.
52. Fukomoto S *et al. Life Sciences.* 2016;159:55–60.
53. Wirth KA *et al. Vet Comp Oncol.* 2014;14(3):281–294.

acvim417

Limitations and Dangers in Analysis of Retrospective Study Data

Nick Jeffery, BVSc, PhD, MSc, DECVN, DECVS, FRCVS
College Station, TX, USA

INTRODUCTION

In veterinary medicine, reports on retrospective studies are far more numerous than those describing prospective studies. Whilst there is nothing inherently 'wrong' with describing the presentation, treatment and outcome of a series of cases, the manner in which retrospective case material is reported may lead to unwarranted conclusions that, in turn, may harm our animal patients in future. In the past, particularly in the 1970s and 1980s, case series were usually reported in simple descriptive terms together with some summary statistics. In contrast, nowadays, probably because of the increasing accessibility of high power software and perhaps because of the pressure to publish 'impactful' studies, similar case series material is often presented with extensive statistical analysis, which can be problematical. The multitude of pitfalls that can arise through erroneous interpretation of retrospective data have been associated with numerous early deaths and unnecessary costs in human medicine [e.g.,[1]] and it would be good if veterinary medicine can avoid making the same mistakes.

WHAT DOES THE P-VALUE MEAN?

Many published studies focus on inferential statistical analysis and highlight determination of p-values for various analyses; often a p-value of a specific magnitude is used to imply that a specific 'truth' about the data has been uncovered. This in itself is problematical. p-values were originally applied to experimental data and low values used to alert researchers to an outcome that was noteworthy and required further investigation. The p-value estimates the probability, if the same experiment were to be repeated an infinite number of times, of acquiring data as extreme as those observed (or more so) if the random play of chance was the only influence on the outcome. This then implies that a low p-value means that the probability of obtaining the observed results is low if random events are the only influence on those data.

In contrast, more commonly veterinary researchers wish to know 'how likely is my hypothesis to be true?' - which is a very different question! Just because a p-value is low it does not imply that a specific alternative hypothesis is true, it merely tells you that it is unlikely that this spread of data would occur by chance if the experiment were to be duplicated again and again. Similarly, not finding a 'significant' difference between two groups does not imply that they are not different - it might be that there is insufficient study power to detect the difference.

The main differences in retrospective clinical data (as compared with a formal prospective experiment) are that there are usually many reasons to suppose that it is not only chance that is playing on the data (see below) and also that the 'experiment' cannot often realistically be duplicated because of the specific (and heterogeneous) constituents of the population sample.

When evaluating the impact of an intervention, it is also important to emphasize the size of effect and p-values do not help with this. A low p-value does not imply a large effect but may simply reflect the amount of data available for analysis. Large sample sizes may be able to detect small effects with associated low p-values but this does not imply clinical relevance.

PROVENANCE OF THE DATA

The most reliable statistical comparisons between two groups are attained by eliminating all sources of difference between the groups apart from a single factor of interest and, typically, this is how scientific experiments are designed. In retrospective studies on clinical data this can rarely be achieved. It is not possible to eliminate the many sources of bias that may play a role in generating differences, most notably including clinician influence on owner decisions, poorly defined diagnostic criteria for diagnosis or treatment (and which may both also change over time) and loss to follow-up (often of a biased subpopulation of patients). This means that it will be difficult to make fair 'apples versus apples' comparisons.

More recently, multivariable analyses have increasingly been applied to retrospective datasets with the aim of accounting for these confounding factors. However, this does not eliminate the problem of systematic bias introduced by (often intangible) influences on decision-making at the time of initial diagnosis and commencement of treatment. A further problem with this approach is the risk of fragmenting the data into small groups - each of which has increasing likelihood of generating extreme results (see below).

Provenance of the Hypotheses That Are Tested

To have the best chance of achieving reliable results it is important that the questions that are asked with the dataset - i.e., the hypotheses that are tested - should be pre-defined - that is, before the data are collected or analyzed. This means that the research question can be focused and will not be biased by the way the data look after collection. It is also important that the questions are framed in a fair way - for instance that they do not result from a recent run of a specific outcome - which, of course, may well bias the results of the whole study.

The contrasting method - in which the data are examined, summarized and then tested for a variety of hypotheses - is well-known and disparaged as 'data dredging.' Any 'significant' results achieved this way will be highly suspect because of risks inherent in multiplicity - where the chances of a false positive are inflated by a factor relating to the number of tests performed (error rate = $1 - [1 - \alpha]^k$). In addition, there is a problem that analyzing data in this way becomes similar to analyzing the results of a horse race after it has been run - the results risk turning into a 'just so' story, where the analysis fits the fact that a specific outcome occurred.

Instead, it should be clear why a specific hypothesis is being tested - the provenance of this hypothesis should be based on some previous data or deduction from known biological mechanisms. On the other hand, the investigation of 'straw man' hypotheses, such as a significant difference between an animal that has been in a road traffic accident and a normal animal may not be of much practical value - we already know these animals are different. Similarly, some hypotheses cannot be considered to be clinically useful - an example would be a comparison of outcomes of an intervention in cats versus dogs. Not only does the null hypothesis (that there will not be a difference between the results in dogs and cats) not really make sense (because we know they are different anyway), there is also nothing useful to be gained from this comparison - we cannot change a cat into a dog.

A particularly pernicious aspect of retrospective data analysis has recently been termed HARKing - (hypothesizing after the data are known)[2] in which a hypothesis is retrospectively fitted to the data. Retrospective analyses and perceived pressure to achieve p-values of <0.05 encourage this activity and it is considered by some to be a form of research fraud.

Small Sample Size

Although the rationale for retrospective data analysis in veterinary medicine is often that it will (likely) aid in acquiring larger datasets, many of those analyzed and published are not really sizeable enough to avoid the problems of small sample size. There are many problems with analysis of small datasets, such as how representative they are of the population from which they are derived, but the most problematical is that analysis of small datasets is more likely to yield extreme results. For instance, it is intuitive that a series of six coin tosses is more likely to yield 100% heads than a series of 60 coin tosses. This means that the conclusions drawn from small samples are likely to include inflated estimates of effect size.

It is well-recognized that small sample sizes have a reduced chance of detecting a true effect - and that any such effect is likely to be large to be able to be detected. Similarly, it is well-known that large sample sizes will be able to detect small effect sizes - sometimes too small to be clinically relevant. However, what is less well-known is that, because of the low probability of there actually being a true effect in any individual investigation, small sample size also implies a reduced likelihood that an observed, apparently 'statistically significant' effect, is real.[3]

So - Is There a Better Way to Handle Retrospective Data?

The risk of drawing unwarranted conclusions may perhaps constitute a persuasive argument to avoid extensive statistical analysis and, instead, return to the previous practice of reporting raw data with summary statistics. Examination of the means or odds ratios plus summaries of the data spread will allow estimation of the likely clinical importance of an intervention. For instance, a difference of 20 between means of PCV in two groups is likely to be clinically meaningful and might imply that further investigation is warranted, even if a formal statistical test doesn't 'reach significance,' whereas an odds ratio of 1.1 (with CI of 1.01–1.19) that is found to be 'statistically significant' may have very little clinical importance.

Although it might be thought that avoiding statistical analysis might mean that possible new insights might be overlooked, there is little reason in practice to fear that this will occur. For instance, an article might report that 6/27 dogs treated by corticosteroids developed anemia versus 18/25 treated with azathioprine. This information suggests that anemia might be a risk of azathioprine in this specific type of patient but, presented in this form alone, would also suggest that the reader should examine the data more carefully to determine if there might be another reason to account for this pattern of results. Perhaps more cases treated with azathioprine had cancer or advanced renal failure and so were more likely to develop anemia anyway. It may be difficult to come to a definite conclusion, but this would also be true even if a statistical test had been applied: such a test will not explain why the groups differ, because animals were not randomly allocated to the two treatment groups there are many effects that might cause the differences apart from simple chance. Instead, the introduction of the statistical test might be detrimental in providing a mental short-cut that tends to preclude more thought about causes. The conclusion, either with or without the statistical test, is that a prospective study will be required in order to confirm the impact of the treatment.

Nowadays it is possible for journals to make full raw datasets available to all readers because there is almost unlimited 'space' available for it to be published online. This, together with access to summary statistics, means that it is possible to provide readers with access to all the material that is required for them to form their own conclusions, without the need to be told what to think by the authors!

References

1. Peters WP, *et al. J Clin Oncol.* 2005;23:2191.
2. Kerr NL. *Pers Soc Psychol Rev.* 1998;2:196.
3. Button KS, *et al. Nat Rev Neurosci.* 2013;14:365.

acvim070

Clinical and Endocrine Parameters in Horses with Pituitary Pars Intermedia Dysfunction Treated with Pergolide

Heidi E. Banse[1]; Dianne McFarlane[2]; Prasanth Chelikani[1]

[1]University of Calgary, Calgary, AB, Canada; [2]Oklahoma State University, Stillwater, OK, USA

Pergolide is commonly used in treatment of pituitary pars intermedia dysfunction. However, there have been limited placebo-controlled trials evaluating efficacy of treatment. Furthermore, seasonal variation exists in both clinical signs and endocrine testing results, which may impact interpretation of treatment efficacy. The purpose of this study was to evaluate the impact of pergolide on endocrine test results and intramuscular adiposity from spring (May) to early fall (July).

Thirteen equids with PPID (based upon resting ACTH or ACTH response to TRH) were randomly assigned to treatment (n=6; PERG) or placebo (n=7; CON) groups. There were no equids with insulin dysregulation at study initiation. At enrollment, median body condition score was 4.5 (range 3-6; PERG) and 4 (range, 3-6; CON). Mean body weight was 513±34 kg (PERG) and 458±59 kg (CON). Horses were treated with Prascend® (1 mg/horse) or placebo for twelve weeks starting in May. Twelve out of thirteen horses had a controlled diet for the duration of the study. Horses were weighed, and oral sugar tests, body condition scoring, and muscle biopsies performed prior to treatment (week 0) and following 12 weeks of treatment (week 12). Muscle biopsies were assessed for intramuscular fat using NMR. TRH stimulation tests were performed at weeks 0, 4, 8, and 12. Dose of pergolide was increased by 1 mg in treated horses at week 5 (n=4) and/or week 9 (n=3) if ACTH concentration did not normalize or decrease by >75% from week 0. Results were analyzed using two-way analysis of variance with repeated measures.

There was an effect of treatment on fasting insulin concentrations, with fasted insulin decreasing in PERG (p=0.03) but not CON (p>0.99) horses. There was an effect of time (p=0.04) but not treatment (p=0.17) on post-OST insulin. There was a significant time x treatment interaction for body weight, with PERG horses losing weight over time (p=0.04). There was no effect of treatment or time on body condition score or intramuscular fat percent (p>0.20). There was an effect of time on TRH stimulation results (p=0.03) and a time x treatment interaction for resting ACTH (p=0.001) and response to TRH (p=0.005). The findings of this study suggest that treatment with pergolide in horses with PPID may decrease body weight and fasting insulin concentrations, without affecting intramuscular adiposity. Furthermore, implementing pergolide treatment in the spring may help limit the seasonal increase in ACTH.

acvim072

Validation of Stall-Side Strangles Diagnosis Using Lamp

Ashley G. Boyle[1]; Shelley Rankin[2]; Kathleen O'Shea[2]; Darko Stefanovski[1]; Jinzhao Song[2]; Haim Bau[2]

[1]University of Pennsylvania New Bolton Center, Kennett Square, PA, USA; [2]University of Pennsylvania, Philadelphia, PA, USA

A low-cost point-of-care "stall-side" diagnosis of strangles would greatly reduce the spread of this infectious disease by identifying *S. equi* subsp. *equi* in index or carrier horses. We previously demonstrated that the *eqbE* loop-mediated isothermal nucleic acid amplification (LAMP) assay was comparable to the *seeI* PCR in guttural pouch lavage samples. We implemented LAMP in a minimally instrumented microfluidic device that can use a cellular phone, offering point-of-care detection. We hypothesized that the point-of-care *eqbE* LAMP device was comparable to the bench-top *eqbE* LAMP assay and the triplex quantitative PCR[1] (qPCR). Stored guttural pouch samples from strangles convalescent horses were analyzed using a benchtop *eqbE* LAMP, *eqbE* LAMP on the microfluidic device, and the triplex qPCR[1] assay targeting *S. equi* specific genes *eqbE* and *SEQ2190*. Samples tested positive for *S. equi* were as follows: 31/64 on point-of-care *eqbE* LAMP with median results at 15 minutes (range 11 to 38 minutes); 27/68 on benchtop *eqbE* LAMP; and 12/67 on *eqbE SEQ2190* triplex qPCR.[1] In comparison to triplex PCR[1] as a reference variable, benchtop *eqbE* LAMP showed excellent discrimination (ROC Area=0.813, 95% CI: 0.711–0.915). The newly developed point-of-care LAMP also showed excellent discrimination power when compared to triplex PCR (ROC Area=0.811, 95% CI: 0.529–0.782). There was no significant difference between benchtop and point-of-care LAMP PCR (p=0.974). Our point-of-care device and method could be used for fast, sensitive, accurate, and cost efficient diagnosis of strangles in the field.

Reference

1. Webb K, Barker C, Harrison T, Heather Z, Steward KF, Robinson C, *et al*. Detection of *Streptococcus equi* subspecies *equi* using a triplex qPCR assay. *Vet J*. 2013;195:300–304.

acvim073

Evaluation of a Rapid Test Strip for Detection of *Salmonella enterica* in Equine Fecal Samples

Brandy A. Burgess[1]; Kristy Pabilonia[2]; Rebecca McConnico[3]; Helen Aceto[4]; Nathan Slovis[5]; Paul S. Morley[2]

[1]University of Georgia, Athens, GA, USA; [2]Colorado State University, Fort Collins, CO, USA; [3]Louisiana State University, Baton Rouge, LA, USA; [4]University of Pennsylvania, Kennett Square, PA, USA; [5]Hagyard Equine Medical Institute, Lexington, KY, USA

Salmonella enterica is one of the most commonly reported organisms associated with epidemic disease at veterinary hospitals, significantly impacting animal morbidity and mortality and adversely affecting the public's perception of these hospitals. Comprehensive screening and rapid detection of *S. enterica* in relevant samples is essential for effective infection control in high-risk populations. The objective of this project was to estimate the sensitivity and specificity of a commercially available rapid test strip for detection of *Salmonella* in comparison to optimized enriched aerobic cultures and a commercial PCR kit, when used to test fecal samples via state-of-the-art, latent class ("no-gold-standard") analytical methods.

One-gram fecal samples (n=569) from equine patients were tested using 6 techniques: 1) tetrathionate broth (43C, 18 hrs) onto XLT-4 agar (43C, 18 hrs); 2) tetrathionate broth (43C, 18 hrs) onto Hektoen enteric agar (36C, 18 hrs); 3) selenite broth (36C, 18 hrs) onto Hektoen enteric agar (36C, 18 hrs); 4) selenite broth (36C, 18 hrs) onto XLT-4 agar (43C, 18 hrs), 5) rapid test strip, and 6) PCR. Data were stratified by expected prevalence (high versus low) for analysis based on previous surveillance data.

In general, the proportion of positive samples was greater when testing with the commercially available rapid test than with any other method evaluated, irrespective of prevalence group. Among low prevalence group - the proportion positive were similar for all culture methods and PCR testing; whereas, within the high prevalence group, the proportion positive was higher when using PCR testing as compared to the culture methods evaluated.

The findings of this project demonstrate the role test selection may have on the ability to detect *Salmonella* in equine fecal samples obtained from hospitalized patients. Having knowledge of a test's epidemiological sensitivity and specificity can greatly impact test interpretation in clinical practice. This study shows that commercially available rapid detection systems may be valuable tools when managing animal facilities; however, users should be prepared to manage false positive test results.

acvim078

Pharmacokinetics, Pharmacodynamics and Clinical Use of Trazodone in the Horse

Jennifer L. Davis[1]; Jenna Schirmer[2]; Emily Medlin[2]

[1]VA-MD College of Veterinary Medicine, Blacksburg, VA, USA; [2]NCSU-CVM, Raleigh, NC, USA

The purpose of this study was to evaluate the pharmacokinetics of oral trazodone and determine its effects on experimental and clinical horses for use as a long-term sedative. Six healthy adult horses were administered trazodone at 7.5 and 10 mg/kg in a crossover fashion. Plasma samples were collected for determination of trazodone and its active metabolite (m-CPP) via UPLC-MS/MS. Noncompartmental pharmacokinetic analyses were performed. Physical examinations, sedation and ataxia scores were determined following administration. Medical records of clinical cases treated with trazodone for behavioral disorders while on prescribed stall rest were reviewed and data collected. Trazodone was rapidly absorbed after oral administration with a maximum concentration of 2.5–4.1 μg/mL and an elimination half-life of approximately 7 hours. The metabolite was detected, although it only represented 2.5% of the total area under the curve. In experimental horses, concentration dependent sedation and ataxia were noted, beginning within 20 minutes of administration and lasting up to 12 hours. Trazodone was successful in modifying behavioral vices to some degree in 17 of 18 clinical cases at doses of 2.5–10 mg/kg once or twice daily. Tolerance developed in 2 of 18 clinical cases following 14 or 21 days of use. Adverse effects attributed to trazodone included oversedation, muscle fasciculations, and transient arrhythmias. Six horses received trazodone concurrently with reserpine with no apparent increase in adverse effects. In conclusion, oral trazodone can be used to cause sedation and facilitate stall rest in horses. Dose adjustments should be made based on individual response to the drug.

acvim083

Insulin and Glucose Responses to Pasture in Insulin-Dysregulated and Normal Ponies

Melody A. de Laat[1]; Danielle M. Fitzgerald[1]; Martin N. Sillence[1]; Christopher C. Pollitt[2]; Donald M. Walsh[3]

[1]Queensland University of Technology, Brisbane, QLD, Australia; [2]The University of Queensland, Gatton, QLD, Australia; [3]Animal Health Foundation, St. Louis, MO, USA

Pasture consumption is a risk factor for equine laminitis. However, the factors affecting insulin responses to grazing are poorly characterized. This study investigated whether pasture fertilization affected insulin and glucose responses to grazing in insulin-dysregulated (ID) and normal (IS) ponies (n=16). Adjacent fertilized and unfertilized pasture was grazed for five consecutive days (for 4 h), in a randomized, cross-over design with an interim washout phase (10 days; lucerne hay). Blood samples were collected (0 to 8 h) for the first three days. Pasture intake was estimated using total fecal DM output and forage digestibility. Ponies were classified as IS or ID using oral glucose tests, but were otherwise healthy. The fertilized pasture had increased ($p<0.05$) sward height and density but lower ($p=0.03$) non-structural carbohydrate (NSC) content than unfertilized, which resulted in similar areas under the insulin curve ($AUC_{insulin}$) during grazing for both pastures. The $AUC_{insulin}$ and $AUC_{glucose}$ did not differ between sampling days and the mean values were positively correlated ($p=0.002$). The NSC intake/cm height (wither) was positively correlated ($p=0.02$) with $AUC_{insulin}$, but not $AUC_{glucose}$. Insulin Cmax and $AUC_{insulin}$, but not $AUC_{glucose}$, were greater ($p<0.05$) in ID ponies for both pasture types, without a difference in pasture intake between groups. Thus, NSC intake, which is affected by pasture density and NSC content, is a more important predictor of pasture-associated insulin responses and laminitis risk than pasture fertilization status. Ponies with ID exhibit greater hyperinsulinism during grazing than IS ponies.

acvim084

Myelographic Findings at the Cervicothoracic Junction in Non-Neurologic Horses

Krista Estell[1]; Mathieu Spriet[2]; Kathryn Phillips[2]; Monica Aleman[2]; Carrie Finno[2]

[1]Marion duPont Scott EMC, Leesburg, VA, USA; [2]University of California-Davis, Davis, CA, USA

Previously published myelographic studies do not report findings at the junction between the seventh cervical and first thoracic vertebrae (C7-T1). Current state-of-the-art radiographic equipment allows visualization of this area. Based on clinical experience, we hypothesized that a 50% reduction of the height of the dorsal myelographic column (DMC) as a supportive finding for spinal cord compression does not apply at C7-T1.

Twelve healthy, neurologically normal Thoroughbred and Warmblood horses were included in the study. A myelographic study was performed under general anesthesia. Four independent observers assessed the myelographic studies for 50% reduction of the DMC at C6-C7 and C7-T1. A site was considered narrow if at least ¾ observers were in agreement. The dorsal myelographic ratios were calculated as the intervertebral DMC divided by the intravertebral DMC.

Narrowing of the DMC was observed at C6-7 and C7-T1 in three and six horses, respectively. The DMC ratio at C7-T1 was 52±12%, while the C6-C7 DMC ratio was 67±17% (mean±sd). Based on these findings, 25% of horses would have been diagnosed with myelographic evidence of spinal cord compression at C6-7 and 50% of horses would have been similarly diagnosed at C7-T1.

Narrowing of the height of the DMC is observed at C6-C7 and C7-T1 in horses without clinical evidence of spinal cord compression. This should be taken into consideration when assessing clinical cases. To avoid high numbers of false positive myelographic cases, applying a threshold of 60% narrowing of the height of the DMC (DMC ratio=40%) at C7-T1 is advisable.

acvim090

A 2.7 kb Deletion near HOXD3 is Associated with Occipitoatlantoaxial Malformation in an Arabian Horse

Carrie Finno[1]; Matthew Bordbari[1]; M. Cecilia Penedo[1]; Monica Aleman[1]; Stephanie Valberg[2]; James Mickelson[3]

[1]University of California-Davis, Davis, CA, USA; [2]Michigan State University, East Lansing, MI, USA, [3]University of Minnesota College of Veterinary Medicine, St Paul, MN, USA

The term occipitoatlantoaxial malformation (OAAM) is used to describe a developmental defect in which the first cervical vertebra (atlas) resembles the base of the skull (occiput) and the second cervical vertebra (axis) resembles the atlas. Affected individuals demonstrate an abnormal posture, with the neck extended, and varying degrees of ataxia. OAAM is presumed to be inherited as a monogenic autosomal recessive trait in Arabian horses. The homeobox (HOX) gene cluster is involved in the development of both the axial and appendicular skeleton. *Hoxd3*-null mice demonstrate a strikingly similar phenotype to Arabian foals with OAAM. We initially studied a radiographically confirmed OAAM affected horse, its sire and dam, as well as seven unaffected Arabian horses. Whole-genome sequencing was performed at 12x coverage of the OAAM-affected horse. Visual inspection of the raw reads within the region of *HOXD3* identified a 2.7 kb deletion located 4.4 kb downstream of *HOXD4* and 8.2 kb upstream of *HOXD3*. The sire and dam were heterozygous for the deletion and it was not present in controls. Additional genotyping identified 2 other heterozygous Arabians and the deletion was not present in 159 other Arabians as well as 371 horses of other breeds. The 2.7 kb deletion was not detected in two additional Arabian foals diagnosed with OAAM. Closer examination of these cases revealed notable phenotypic variation, including facial malformations and a patent ductus arteriosus. Comparative genomic studies reveal that the region containing the deletion is highly conserved across species and that the entire genomic region between *Hoxd4* and *Hoxd3* is transcribed in mice. We conclude that a deletion between *Hoxd4* and *Hoxd3* is a rare occurrence in Arabians and, when present as a recessive trait, causes one but not all forms of OAAM. Genetic heterogeneity therefore exists across the *HOXD* locus in Arabian foals with OAAM.

acvim099

Latency in Equid Herpesvirus-1 Purpose-Infected Horses

Kim Giessler[1]; Susanna Samoilowa[1]; Carlos E. Medina Torres[2]; Gisela Soboll Hussey[3]; **Lutz S. Goehring**[1]

[1]Ludwig-Maximilians University Munich, Munich, Bayern, Germany; [2]The University of Queensland, Brisbane, QLD, Australia; [3]Michigan State University, East Lansing, MI, USA

One of the hallmarks of infection with *Herpesvirinae*, including equid herpesvirus -1 (EHV-1), is the ability of the virus to establish latent infection during or shortly after lytic infection of the upper respiratory tract (URT) epithelium. As there is a connection between the URT and respiratory tract associated lymphoid tissue (RALT), EHV-1 latency is typically localized in respiratory associated lymphoid tissues (mandibular, retropharyngeal and bronchial lymph nodes), as well as the trigeminal ganglia (TG). Latent cell infection is characterized as a non-immunogenic infection with few viral genome copies per single cell, limited transcription of the immediate early-gene (IE) only, and, although not conclusively confirmed, the presence of additional (anti-sense) latency associated transcripts (LAT). The aim of this study was to determine whether latency is limited to RALT and/or TG in infection studies using EHV-1 wild-type (wt) or mutant strains.

Three groups of yearling horses were infected intranasally with wt EHV-1, strain Ab4 (group I, n=6), an EHV-1 polymerase mutant at position 752 (D>N) (group II, n=9), or an EHV-1 glycoprotein D (gD) where gD of EHV-1 was replaced with gD of EHV-4 (group III, n=8). All animals were euthanized 70 days post-infection (p.i.). During post-mortem examination, TG, and RALT including tonsillar pharyngeal roof tissue, mesenteric lymph nodes, and spleen were collected. Viral genomic DNA and RNA were extracted, and RNA was converted to cDNA. Established primers and probes were used to detect viral DNA and cDNA transcripts of the immediate early (IE) gene and late (L) gene using quantitative PCR. Results are reported as either positive (pos) or negative (neg) for genomic DNA, IE-mRNA, and L-mRNA. A tissue was considered latent-infected when IE-mRNA was detected while L-mRNA was not. Genomic DNA and IE-mRNA were detected in animals of all three groups, while L-mRNA was not. IE-mRNA copy numbers typically outnumbered genomic DNA copy numbers. IE-mRNA was detected in TG, as well as in RALT in all 3 groups, although with a different distribution between groups. In group III, for example, TG contained higher copy numbers of IE-mRNA than RALT, while in groups I & II copy numbers were higher in TG. Surprisingly, genomic DNA and IE-mRNA were also found regularly in mesenteric lymph nodes and spleen.

In conclusion, the number of IE-mRNA copies outnumbers viral genome copies during latency, which has been described for human herpes simplex 1 latency. We confirmed previously identified latency locations for EHV-1: RALT and TG. However, distribution preferences could be strain dependent. Moreover, EHV-1 latent infection of lymphoid tissue was not limited to RALT.

acviml09

Capsule Endoscopy (ALICAM) in Horses: Assessing Gastrointestinal Tract Transit Time, Image Quality and Lesions

Renaud Leguillette[1]; Mei Steinmann[1]; Rebecca Bezugley[2]; Jill S. Pomrantz[3]

[1]Faculty of Veterinary Medicine, University of Calgary, Calgary, AB, Canada; [2]Faculty of Veterinary Medicine, University of Calgary/ Moore Equine Veterinary Centre, Calgary, AB, Canada; [3]Infiniti Medical, Menlo Park, CA, USA

ALICAM is a veterinary-specific capsule endoscopy system that could be a breakthrough technology for diagnosing gastrointestinal (GI) pathologies in equine medicine.

The purposes of this study were to 1) Test the feasibility of ALICAM in horses; 2) Optimize protocols for shortest transit time and best visualization of the mucosa; 3) Measure transit time in healthy horses; and 4) Describe intestinal mucosa lesions in healthy and GI pathology cases.

Five healthy horses were used in a randomized crossover design to test 3 protocols. Three clinical cases with confirmed protein-losing enteropathies were also used.

Capsule recovery was tested using manure screening and radiography. Protocols were tested for three variables: fasting time, fluid volume administered, amount of exercise. Measured outcomes were: 1) Transit time, 2) Image quality score (image fields showing mucosa assessed by 2 observers), 3) Presence of GI mucosal lesions.

No side effects were noted. 3/20 capsules were not recovered. Radiography of manure was effective to recover the capsules. Battery life was 13.9±3.5 hours. There was no significant difference in transit time between the 3 protocols. Transit time was 5.8±2.9 days in healthy horses. The inter-observer agreement on image quality scores was excellent. Detailed images of the pylorus, duodenal papilla, villi, and ileocecal junction, but not the colon mucosa, were obtained. Lesions seen in healthy horses included fissures, ulcers, and submucosal hemorrhage. In addition, dilated lacteals were observed in clinical cases.

ALICAM allows measurement of transit time and detailed visualization of the pylorus and small intestinal mucosa in horses.

acvim110

A Pilot Investigation into the Safety and Efficacy of a Long-acting, Injectable Formulation of Omeprazole

Ben Sykes[1]; Krishna Kathawala[2]; May Song[3]; Sanjay Garg[3]; Paul Mills[4]

[1]The University of Queensland, Caringbah, NSW, Australia; [2]University of South Australia, Adelaide, SA, Australia; [3]University of South Australia, Adelaide, NSW, Australia; [4]The University of Queensland, Gatton, QLD, Australia

This study investigated the safety and efficacy of a novel, long-acting, injectable formulation of omeprazole in the treatment of naturally occurring equine gastric ulcer syndrome (EGUS).

Gastroscopy was performed on Thoroughbred racehorses in training. The squamous mucosa was graded using the EGUS council grading system. The glandular mucosa was graded using a system as follows; grade 0/2 - normal mucosa with no evidence of loss of mucosal integrity, grade 1/2 - mild-moderate lesion(s) present with evidence of loss of mucosal integrity, and grade 2/2 - severe lesion(s) present with evidence of loss of mucosal integrity.

Horses with grade ≥2/4 squamous or ≥1/2 glandular disease, or a combination of both, were considered eligible for enrollment. Once enrolled horses received 2.0 grams of the 100 mg/ml formulation of omeprazole by intramuscular injection on days 0 and 7.

Twenty-six horses met the inclusion criteria. Two were lost to follow-up. Repeat gastroscopy was undertaken on day 14 (23 horses) or 16 (1 horse). Squamous and glandular gastric disease were present in 22/24 and 12/24 of horses, respectively. Healing, defined as a return to grade 0, occurred in 22/22 (100%) horses with squamous disease and 9/12 (75%) horses with glandular disease. Improvement, defined as a reduction in grade by at least 1, occurred in 12/12 (100%) horses with glandular disease. No side effects were observed.

The results of the present study compare favorably with published reports for oral omeprazole. Further investigation into the efficacy of the formulation in larger scale clinical trials is warranted.

acvim111

Electroanatomical Cardiac Mapping in an Adult Horse in Sinus Rhythm

Gunther van Loon[1]; Tim Boussy[2]; Lisse Vera[1]; Dominique De Clercq[1]; Stijn Schauvliege[1]; Tim Vandecasteele[1]; Annelies Decloedt[1]; Glenn Van Steenkiste[1]; Glenn Van Langenhove[3]

[1]Faculty of Veterinary Medicine, Ghent University, Belgium, Merelbeke, Oost-Vlaanderen, Belgium; [2]AZ Groeninge/Middelares, Kortrijk, Oost-Vlaanderen, Belgium; [3]Fulgur Medical, Merelbeke, Belgium, Merelbeke, Oost-Vlaanderen, Belgium

Arrhythmias are very common in horses. Nevertheless, invasive electrophysiological techniques have not fully been explored.

The aim of this study was to establish whether an automatic, high-resolution mapping system (Rhythmia, Boston Scientific) would be applicable to identify 3D cardiac geometry and electrical activation patterns from the right and left heart in an adult horse.

In 2 horses, a ten-pole steerable catheter was placed in the coronary sinus. Under general anaesthesia, in dorsal recumbency, an 8.5 F bidirectional steerable basket mapping catheter with 8 splines and 64 low-noise electrodes (Intellamap Orion, Boston Scientific) was inserted through the jugular vein into the right atrium. The magnetic field generator for 3D localisation was placed over the left thorax. The right atrium and right ventricle were mapped using a stable coronary sinus electrogram and the surface ECG as timing reference, respectively. Subsequently, through a surgically inserted sheath in the right common carotid artery, the mapping catheter was advanced into the left ventricle under ultrasound guidance, and left ventricular mapping was performed. The left atrium was only mapped in one horse.

Three-dimensional geometry of the heart corresponded with ultrasound findings. Activation maps from right atrium, right ventricle, and left ventricle could be generated and showed earliest atrial activation in the high right atrium. His bundle recordings were successfully recorded from the right atrium.

We concluded that high resolution 3D electroanatomical mapping is feasible in adult horses and provides a promising and powerful tool for electrophysiological research and for examination of complex arrhythmias.

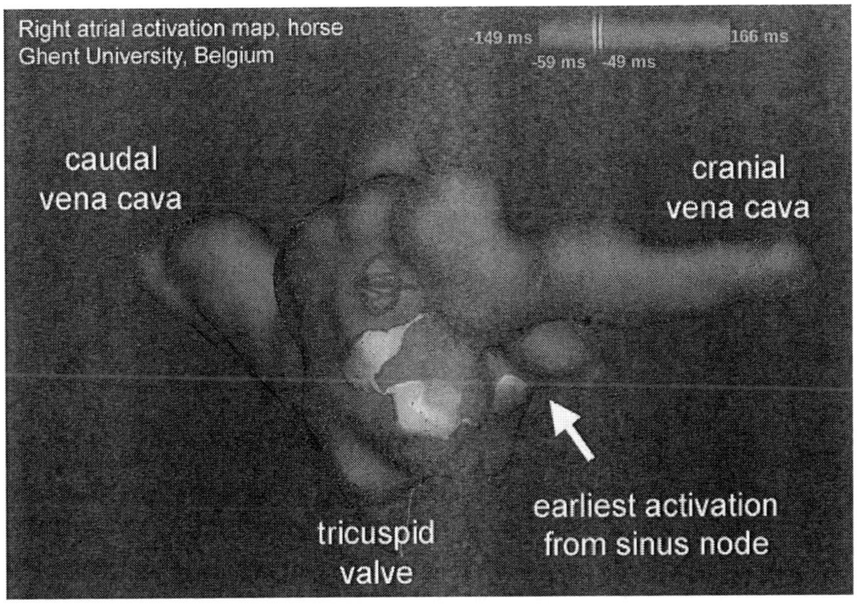
Right atrial activation map, horse
Ghent University, Belgium
-149 ms
166 ms
-59 ms
-49 ms
caudal
vena cava
cranial
vena cava
earliest activation
from sinus node
tricuspid
valve

acvim112

Heart Rate Monitor Derived Heart Rate Variability to Diagnose Atrial Fibrillation in Horses

Gunther van Loon[1]; Barbara Broux[1]; Dominique De Clercq[1]; Sofie Ven[1]; Lisse Vera[1]; Glenn Van Steenkiste[1]; Annelies Decloedt[1]

[1]Faculty of Veterinary Medicine, Ghent University, Belgium, Merelbeke, Oost-Vlaanderen, Belgium

Atrial fibrillation (AF) is the most common clinically important arrhythmia in horses. Monitoring cardiac rhythm in successfully treated patients is important because recurrence may occur, but it might be challenging to the owner. Heart rate monitors are user-friendly, can be applied by owners and may offer basic heart rate variability (HRV) parameters such as root mean square of the successive differences (RMSSD) that might differentiate AF from sinus rhythm. The purpose of this study was to assess if RMSSD, automatically generated by a heart rate monitor, can distinguish between AF and sinus rhythm (SR), based upon a short recording. In 14 horses a 2-minute recording with a heart rate monitor (Polar V800) was made at rest, walk and trot, both in AF and after treatment when back in SR. RMSSD was obtained from the heart monitor's software after importing the recordings. Statistical analysis was performed using ANOVA and receiver operating curves. At all recording points, RMSSD was significantly higher in AF compared to SR ($p<0.001$). Area under the curves were high (>0.9) and cut-off values at rest, walk and trot were set at 134 ms (100% sensitivity, 93% specificity), 55.5 ms (94% and 73%) and 14 ms (100% and 87%), respectively. We concluded that RMSSD obtained from a 2-minute heart rate monitor recording is an easy-to-use tool to monitor horses at risk for developing AF. Recordings at rest revealed best results. Loss of electrode contact and movement artifacts are possible confounders that increase RMSSD and may lead to false-positive results.

acvim113

The Effect of p38 MAPK Inhibition on the Inflammatory Response to Bacterial Toxins in Horses

Jennifer R. Bauquier[1]; Brett Tennent-Brown[1]; Elizabeth Tudor[2]; Simon Bailey[2]

[1]University of Melbourne Faculty of Veterinary and Agricultural Sciences, Werribee, VIC, Australia; [2]University of Melbourne Faculty of Veterinary and Agricultural Sciences, Parkville, VIC, Australia

The systemic inflammatory response occurs commonly in horses with gastrointestinal disease. Current treatments can be inadequate, therefore there is a need for novel anti-inflammatory treatments. p38 MAPK is an important intracellular signalling pathway in inflammatory cells leading to the transcription of inflammatory cytokines. In this study, a p38 MAPK inhibitor drug (BIRB-796) was used to determine the effect of p38 MAPK inhibition on the inflammatory response to bacterial toxins *in vitro* and *in vivo*.

The p38 MAPK inhibitor was investigated *in vitro* using equine whole blood assays. Increasing concentrations were added to equine whole blood diluted 1:1 with cell culture medium, and stimulated with either lipopolysaccharide (LPS), lipoteichoic acid (LTA) or peptidoglycan (PGN). Supernatant concentrations of TNF-α were measured using a murine L929 bioassay. BIRB-796 was then administered to 6 healthy horses at 0.5 mg/kg IV for safety analysis. Following this, the same formulation or placebo was administered IV at the same dose to 6 horses in a cross-over design, immediately prior to the administration of LPS (30 ng/kg) IV over 30 minutes. Clinical parameters were measured and blood samples taken for leukocyte count and measurement of TNF-α at baseline and then every 30 minutes for 6 hours.

BIRB-796 showed high potency *in vitro* with IC_{50} values for TNF-α inhibition of (geometric mean (95% CI)) of 0.07 (-0.18–0.25) μM for LPS, 0.03 (-0.21–0.27) μM for LTA and 0.46 (-0.69–1.60) μM for PGN. Maximum inhibition values (mean±SEM) were 66.7 (±9.2)%, 71.4 (±8.5)%, and 73.1 (±12.2)% for LPS, LTA, and PGN, respectively. IL-1beta was also inhibited. No clinical or clinicopathological abnormalities were found following the administration of this drug to healthy horses.

When pre-treated with BIRB-796, horses given IV LPS had lower peak heart rates (mean±SEM, 47±6 beats/min) compared to controls (57±5 beats/min) which approached significance; p=0.06. The AUC for heart rate in treated horses (mean±SEM of 1050±336.4 arbitrary units) was significantly lower than controls (mean±SEM of 1862±323.4); p=0.03. The AUC for rectal temperature of treated horses (mean±SEM of 175.6±42.92) was significantly lower than controls (mean±SEM of 344.2±67.49); p=0.03. Systolic and diastolic blood pressure increased significantly in the control group (mean (±SEM) systolic increase of 17.3±3.2 mm Hg; mean (±SEM) diastolic increase of 11±1.9 mm Hg) but not the treated group (mean (±SEM) systolic increase of 4.3±3.1 mm Hg, mean (±SEM) diastolic increase of 6.2±3.0 mm Hg) (p=0.03 [systolic]; p=0.03 [diastolic]).

A significantly smaller decrease in leukocyte count from baseline occurred in treated horses (mean ±SEM = 1.296±0.36 x 10^9/L) compared to controls (mean±SEM = 3.391±0.666 x 10^9/L); p=0.03. AUC for TNF-α concentration was significantly lower for treated horses (mean±SEM = 43469±11474) than controls mean±SEM = 189119±51413); p=0.03. AUC for IL-1beta concentration was also significantly lower for treated horses than controls (p=0.03).

The inhibition of p38 MAPK by BIRB-796 significantly reduced the systemic effects of LPS. This compound might be a useful treatment in equine clinical cases of SIRS. A clinical trial is needed to confirm its value in this setting.

acviml14

Next Generation DNA Sequencing to Identify Bacteria in Equine Transtracheal Aspirate Samples

Bonnie Barr; Stephen Reed

Rood and Riddle Equine Hospital, Lexington, KY, USA

Efficient detection of respiratory tract pathogens is crucial in the management of horses with lower respiratory tract infections. Next generation DNA sequencing allows identification of bacteria that may not grow by traditional culture methods due to small sample size, prior treatment, anaerobic growth, or formation of biofilm. The purpose of this study was to compare results of transtracheal aspirate samples identified by next generation DNA sequencing and traditional bacterial culture.

Transtracheal aspirate samples obtained from equine patients with suspected lower respiratory tract disease were submitted for next generation DNA sequencing (16S rRNA gene sequencing) and traditional bacterial culture. Samples were grouped according to whether prior antibiotic usage had been administered.

Fifty-five samples were collected from 55 horses. Thirty of the patients received antibiotic therapy prior. Next generation DNA sequencing identified 103 bacterial species whereas traditional culture identified 25 bacterial species.

Next generation DNA sequencing resulted in identification of more bacterial species than traditional bacterial culture and many of the bacteria identified by next generation DNA sequencing were not identified by bacterial culture. There were notable differences between the results of the next generation DNA sequencing when patients were grouped based on antibiotic usage. Increased knowledge of the microbial ecology provides veterinarians with the capacity to manage their patients more effectively. Next generation DNA sequencing is an important and valuable tool for practicing veterinarians because it allows them to survey the full bacterial landscape, not merely the culturable members of the microbial community.

acvim115

Changes in the Hematological and Biochemical Profile in Foals during the First Year of Life

Babak Faramarzi[1]; Lon Rich[2]; Sara Carman[3]

[1]College of Veterinary Medicine, Western University of Health Sciences, Pomona, CA, USA; [2]Antech Diagnostics, Irvine, CA, USA; [3]Western University of Health Sciences, Pomona, CA, USA

PART A: HEMATOLOGICAL PROFILE CHANGES IN FOALS DURING THE FIRST YEAR OF LIFE

Foals' hematological values change constantly during the first year of life. The use of updated age-based reference intervals is imperative for providing accurate diagnosis and optimum care for sick foals. Our objective was to provide updated reference intervals for 13 hematological values in 2, 7, 14, 30, 90, 180, and 365 day-old foals and to investigate statistical differences in each of the measured values between those ages. Venous blood was collected at 2, 7, 14, 30, 90, 180, and 365 days of age from 81 clinically healthy foals. Following American College of Veterinary Clinical Pathologists guideline, 13 hematological values were analyzed. The normal distribution of data was examined; reference intervals were determined using bootstrap analysis and over-time differences were examined using Friedman test followed by post-hoc analysis and adjusted p values were considered. Significant over time statistical changes were found between several values including WBC ($p=0.011$) and MCH ($p=0.011$) between days 2 and 90 and lymphocyte counts between days 2 and 180 ($p=0.033$). Our results revealed over time differences for several hematological values, specifically at 2, 90, and 180 days of age. High WBC, lymphocyte, and monocyte counts after birth may be attributed to developing lymphoid systems while decreases in MCH may be a result of microcytosis secondary to decreased serum iron from an increased growth demand. The use of age-based reference ranges is critical for providing the most accurate diagnosis and optimum care in young foals.

PART B: THE CHANGES IN BIOCHEMICAL PROFILE IN FOALS DURING THE FIRST YEAR OF LIFE

Biochemical profiles of the equine neonate and foals significantly change throughout the first year of life. It is critical that clinicians use updated age-based reference intervals. Our objective was to provide updated reference intervals for biochemical profile in foals at 2, 7, 14, 30, 90, 180 and 365 days of age and to examine statistical differences among those values among those ages. Left jugular venous blood was collected from 81 clinically healthy foals at 2, 7, 14, 30, 90, 180, and 365 days of age; submitted to and analyzed by a reference laboratory. Using routine laboratory techniques and following American College of Veterinary Clinical Pathologists guideline, 25 biochemical profiles were measured. The data normality was examined; using Bootstrapping analyses reference intervals determined and over time differences were examined using Friedman test following post-hoc analysis when appropriate and adjusted p values were considered. Median, standard deviation and reference intervals for each of the measured variables were calculated. Several variables significantly differed including calcium ($p=0.033$) between days 2 and 180; chloride ($p=0.033$) between days 7 and 365, as well as alkaline phosphatase ($p=0.01964$) and magnesium ($p=0.014$) between days 2 and 365. Observed physiological changes are related to growth and development of the animals which necessitate the use of age-based reference intervals. Our results revealed significant over time differences for several variables, more specifically at 30, 90, 180 and 365 days of age. The use of age-based reference ranges allows clinicians to provide most accurate diagnosis and optimum care in young foals.

acvim139

Prevalence Factors Associated with EHV-2/5 Among Equines with Signs of Upper Respiratory Infection in US

Kaitlyn James[1]; **Wendy E. Vaala**[2]; Duane Chappell[3]; Craig Barnett[3]; Earl Gaughan[3]; Bryant Craig[3]; Fairfield Bain[3]; Nicola Pusterla[4]

[1]UC Davis Graduate Group in Epidemiology, Boston, MA, USA; [2]Merck Animal Health, Alma, WI, USA; [3]Merck Animal Health, Lexington, KY, USA; [4]Department of Medicine and Epidemiology, School of Veterinary Medicine, University of California, Davis, Davis, CA, USA

The clinical relevance of the two equine gammaherpesviruses EHV-2 and EHV-5 has remained elusive, mainly because these viruses can be found in horses with respiratory signs as well as healthy horses. The purpose of this study was to characterize associations between demographic factors, clinical signs, and potential coinfections with EHV-2/-5 infection amongst a group of 3,030 horses that presented with acute onset of fever and/or respiratory signs from 2012–2016. Nasal secretions were tested for common and less characterized respiratory pathogens by qPCR. Out of the 3,030 horses in the study, 11% tested EHV-2 qPCR-positive (n=353), 20% tested EHV-5 qPCR-positive (n=630), 28% tested dually qPCR-positive for EHV-2 and -5 (n=858), and 36% tested qPCR-negative for both viruses (n=1,189).

While less severe clinical signs were associated with singular EHV-2 and EHV-5 infection in this study sample, fever and coinfection with ERBV and *S. equi* subspecies *equi* were associated with dual EHV-2/-5 infection in a multivariate model. The low pathogenicity observed with EHV-2 and -5 could be due to strain differences or age distributions that differ from the general equine population, and the effect of these singular infections should be explored further. In this equine sample population, dual infection leads to increased susceptibility to common respiratory pathogens and more prominent respiratory signs. Future studies should examine and further characterize the equine EHV-2/-5 dual infection model as a potential respiratory pathogen complex different from EHV-2 and -5 infection alone.

Table 1
Demographic factors associated with EHV-2 and EHV-5 qPCR-defined positivity amongst 3,030 United States horses, from 2012-2016.

	EHV-2 positive (only) (n=353)	**EHV-5 positive (only) (n=630)**	**Positive for both EHV-2 and EHV-5 (n=858)**	**Negative for both viruses (n=1,189)**
Breed				
Quarter Horse (n=1122)	124 (11%)	247 (22%)	334 (30%)	417 (37%)
Warmblood (n=306)	37 (12%)	76 (24%)	37 (12%)	156 (51%)
Thoroughbred (n=416)	46 (11%)	82 (19%)	198 (48%)	90 (21%)
Paint (n=151)	15 (10%)	33 (22%)	28 (19%)	75 (50%)
Arabian (n=268)	30 (11%)	42 (16%)	96 (36%)	100 (37%)
Draft Horse (n=82)	5 (6%)	18 (22%)	12 (15%)	47 (57%)
Pony/Miniature (n=124)	23 (19%)	18 (15%)	26 (21%)	57 (46%)
Other Breed (n=382)	50 (13%)	70 (18%)	85 (22%)	177 (46%)
Unknown (n=179)	23 (13%)	44 (25%)	42 (24%)	70 (40%)
Use				
Competition (n=1259)	128 (39%)	256 (43%)	463 (46%)	412 (38%)
Pleasure horse (n=1110)	125 (38%)	210 (35%)	308 (31%)	467 (43%)
Breeding (n=180)	20 (6%)	45 (8%)	52 (5%)	63 (6%)
Other Use (n=62)	8 (2%)	15 (2%)	20 (2%)	19 (2%)
Unknown (n=419)	51 (15%)	72 (12%)	165 (17%)	131 (12%)
Sex				
Female (n=1346)	158 (12%)	279 (21%)	408 (30%)	501 (37%)
Male[a] (n=1498)	173 (12%)	303 (20%)	397 (27%)	625 (43%)
Unknown (n=186)	22 (12%)	48 (26%)	53 (28%)	63 (34%)
Age[b]	6	7	3	8
Less than 1 (n=308)	59 (20%)	26 (8%)	178 (58%)	45 (15%)
1-5 (n=943)	91 (10%)	221 (23%)	328 (35%)	303 (32%)
6-10 (n=683)	82 (12%)	144 (21%)	115 (17%)	342 (50%)
11-15 (n=444)	44 (10%)	101 (23%)	80 (18%)	219 (48%)
16-20 (n=240)	27 (10%)	48 (20%)	50 (21%)	115 (48%)
Over 20 (n=181)	24 (12%)	35 (19%)	45 (25%)	77 (43%)
Unknown (n=231)	26 (11%)	55 (24%)	62 (26%)	91 (39%)
History of transport				
Yes (n=818)	101 (12%)	188 (23%)	213 (26%)	316 (39%)
No (n=1896)	215 (11%)	368 (20%)	548 (29%)	765 (40%)
Unknown (n=316)	37 (12%)	74 (23%)_	97 (31%)	108 (34%)
Affected horses on property				
Single (n=1749)	205 (12%)	380 (22%)	425 (24%)	739 (42%)
Multiple (n=949)	113 (12%)	181 (20%)	332 (35%)	323 (34%)
Unknown (n=332)	35 (11%)	69 (21%)	101 (30%)	127 (38%)

[a] "Male" includes male gelding and male stallion
[b] Median

1.
acvim194

Report of the Fourth Havemeyer Workshop on Infection Control in Equine Populations

Brandy A. Burgess, DVM, MSc, PhD, DACVIM (LAIM), DACVPM; Paul S. Morley, DVM, PhD, DACVIM (LAIM)
Athens, GA, USA

Background

Disease epidemics associated with healthcare-associated infections (HAIs) are commonly reported in veterinary medicine.[1-4] While these outbreaks garner much attention, there is minimal awareness regarding the occurrence of sporadic HAIs (i.e., the endemic rate). As such, we are likely underestimating the amount of harm done to patients. We have an ethical responsibility to take all reasonable actions to prevent foreseeable risks to our patients - including the occurrence of HAIs. Unfortunately, while there have been great strides to understanding the epidemiology of HAIs in human medicine, there is a very limited understanding in veterinary medicine. Most veterinary hospitals do not use a systematic approach for documenting or investigating the occurrence of HAIs, whether they are considered epidemic or endemic, thus there is little evidence upon which to base current recommendations.

In human medicine, the Study on the Efficacy of Nosocomial Infection Control (SENIC) demonstrated the profound impact an infection control program may have on the occurrence of HAIs if there are trained personnel on staff, routine hospital surveillance, and reporting of findings back to stakeholders.[5] The key here is that we cannot manage what we do not measure. This is not to say that we need to measure all adverse outcomes or monitor all processes. Rather, we should focus on the preventable fraction - that proportion of events above an expected baseline rate that reasonably could have been prevented through the use of risk-reduction strategies.

In human healthcare, surveillance is conducted prospectively by trained personnel using very specific definitions. Typically, one person per department will be dedicated to this task rather than a primary healthcare provider, as is more common in veterinary medicine. Unfortunately, primary healthcare providers are often too busy and either do not complete surveillance tasks or rely on retrospective medical records searches, which are of limited value without the use of standardized forms and denominator data. The National Healthcare Safety Network (NHSN) provides a framework in human medicine for developing a deeper understanding of the 'how' and 'why' of HAIs - specifically the idea that the occurrence of HAIs is tied to patient risk status.[6] The NHSN contains multiple modules - including 1) procedural-related infections, 2) device-associated events (e.g., CLABSI, CAUTI, VAP), 3) etiology-based events (e.g., MRSA, *C. difficile*, other MDR organisms), 4) antimicrobial use and resistance, and 5) compliance-associated events - which can motivate module development for use in veterinary medicine.

With this in mind, the 4th Havemeyer Workshop on Infection Control in Equine Populations took place in April 2016, in Duck, North Carolina - bringing together a diverse group of recognized experts in fields related to infection control. The overarching goal for this collaborative working meeting was to develop a standardized surveillance system for HAIs in veterinary hospitals - the Veterinary Health Surveillance Network (VHSN) - which will promote the continued systematic improvement of patient care through the prevention of HAIs, and to develop benchmarking resources to allow individual hospitals to assess the efficacy of their infection prevention efforts.

Surveillance Network Framework

Long-term surveillance should focus on the greatest burden - events that are significant with respect to health impact and number of patients affected (i.e., magnitude and frequency) - rather than low-risk, low-impact events. Therefore, we elected to focus on three specific issues: 1) *Salmonella* surveillance, 2) surgical site infections, and 3) catheter-associated inflammation. By developing 3 modules focusing on common issues, we can more quickly collect data upon which to base initial benchmarking activities and create templates for future modules. Framework development focused on equine hospitals - with the intention of expanding to include multiple species in the future - and concentrated on five main areas: 1) criteria for surveillance systems, 2) data collection mechanisms, 3) standardized definitions, 4) reporting mechanisms, and 5) data sharing and security. The idea here is to define what is being measured and to capture information on the population at risk (i.e., denominator data) to gain an understanding of the burden associated with an outcome or process of interest. De-identified data will be reported in a secure, Web-based, centralized system that facilitates data compilation for individualized benchmarking against standard baseline rates.

CONCLUSION

As the University Veterinary Teaching Hospital model is becoming obsolete, we need to consider the bigger market of animal healthcare (e.g., large referral hospitals) and how this surveillance system will be used down the road. This proposed surveillance network, the Veterinary Health Surveillance Network, is a progressive approach to infection control and ultimately may inform or become standard of practice.

REFERENCES

1. Steneroden KK, *et al. J Vet Intern Med.* 2010;24(3):606–616.
2. Goehring LS, *et al. J Vet Intern Med.* 2010;24(5):1176–1183.
3. Burgess BA, *et al. J Vet Intern Med.* 2012;26(2):384–392.
4. Dallap Schaer BL, *et al. J Vet Intern Med.* 2010;24(5):1138–1146.
5. Haley RW, *et al. Am J Epidemiol.* 1985;121(2):182–205.
6. NHSN-CDC: www.cdc.gov/nhsn/index.html.

acvim195

Nutritional Management of Equine Gastric Ulcers: What Do We Really Know?

Frank M. Andrews, DVM, MS, DACVIM (LAIM)
Baton Rouge, LA, USA

Introduction

Ulcers found in the terminal esophagus, proximal nonglandular (squamous) stomach, distal (glandular) stomach and proximal duodenum are collectively referred to as the equine gastric ulcer syndrome (EGUS).[1] All ages and breeds of horses are susceptible to EGUS and diet and management factors play an important role in the formation and maintenance of ulcers.[2-7] Pharmacologic agents are the mainstay of treatment; however, diet and nutritional management provide long-term solutions to reduce the risk of recurrence and complications associated with chronic ulceration. This proceedings focuses on the currently knowledge of nutritional and dietary factors important in the cause of squamous and glandular stomach ulcers in mature horses and how the horse's diet can be managed to lessen ulcer severity and help prevent recurrence and chronic ulceration.

Anatomy and Pathogenesis

Squamous Ulcers

Ulcers occur primarily in the squamous region of the stomach predominantly near the Margo plicatus, which is lined by nonglandular stratified squamous epithelia. The squamous region of the equine stomach has an osmiophilic phospholipid surfactant-like layer that contributes to the mucosal barrier, but it lacks a significant mucus layer, has poor blood supply and has a variable ability to heal spontaneously once injured.[8,9] Spontaneous healing of squamous ulcers varied from 4 to 8% in untreated stall-confined horses maintained in race training, and from 9 to 33% in horses under field conditions and 33% of horses when they were moved from pasture to stalls.[10-12] In addition, in one study, when horses were housed in stalls and moved to pasture, spontaneous healing of lesions occurred in 6/11 horses (55%) and new lesions occurred in 5/12 horses (42%), during a 6-week pasture recovery period.[13] Thus, the actual spontaneous healing rate and formation of new squamous ulcers is variable.

Glandular Ulcers

The distal two-thirds of the stomach consists of glandular mucosa, which secretes a thick protective mucus and bicarbonate layer, as well as hydrochloric acid (HCl), pepsinogen (converted to pepsin at pH <4.0) and gastric lipase, which are all involved in digestion (although the importance of gastric lipase and pepsin is unknown).[14-16] The glandular region is also where gastrin, histamine and somatostatin are synthesized. The glandular mucosa has an extensive capillary network and, as extrapolated from human studies, undergoes rapid restitution once injured.[17] It was once thought that ulcers in this region were infrequent; however, a recent review reported that glandular gastric ulcers varied in prevalence from 8 to 63% in certain populations of horses.[18,19]

Risk Factors

Although acid injury has been implicated as the cause of squamous ulcers, the cause of glandular ulcers is less well defined. Several risk factors were identified and include exercise intensity, feeding practices, stall confinement, diet type (high concentrate feeding), intermittent feeding, and forage type. Nutritional risk factors are intimately tied to management factors, such as exercise and housing and these factors should be taken into account when making dietary management changes. The methods for dietary management are listed below.

Environmental, Nutritional and Dietary Management

Without alterations in management or initiation of preventative therapy, squamous ulcers will quickly return if horses are maintained in training.[5] Unfortunately, glandular ulcer recurrence has not been evaluated. Reducing exercise intensity, increasing pasture turnout and amending dietary risk factors might help decrease ulcer severity and the risk of recurrence in squamous ulcers, but it is unknown if this will prevent recurrence of glandular ulcers. The following recommendations are primarily for squamous ulcers, but might have a positive effect on glandular ulcers as well.

Modification of Exercise Intensity and Duration

Intense exercise, racing, and race training have been shown to contribute to worsening of squamous ulcers in horses and pasture turnout or change in management might improve the condition.[20-22]

Pasture Turnout vs. Stall Confinement

Horses housed in pastures and constantly grazing, overall, seem less likely to have squamous ulcers.

Stall-confinement generally has been associated with an increased risk of squamous ulcers, which therefore might not improve in stall-confined horses even when horses are fed grass hay *ad libitum*. While pasture turnout may be helpful in controlling squamous ulcers, the presence of certain other 'stressors' and whether the horse continues to be fed high NSC-containing feedstuffs may be greater determining factors.[7,18] In some situations, changes such as moving horses from pasture to stall housing, might be effective in reducing squamous ulcer scores in some horses.[12]

Diet

Pasture grazing, high forage (dry matter [DM] intake, ≥1.5% bodyweight) and low concentrate (≤0.5% bodyweight) diets, a diet low in NSC, feeding smaller more frequent meals and providing *ad libitum* forages might reduce the risk of squamous ulceration as discussed below.

Eliminate Bolus Feeding and Increase Forage/Fiber Intake

One management practice that may help decrease squamous ulcers is to feed a forage based diet ad libitum or at least ensure that forage is provided every few hours and before exercise (even in small amounts). Along with hay, stabled horses should be fed (if required) 3 or 4 smaller meals (not to exceed 0.5 kg/100 kg bodyweight every 6 h and containing 20% NSC or lower) per day. Smaller low NSC meals will decrease intragastric fermentation, reduce production of VFAs and improve gastric emptying rate.

Providing alfalfa hay and good quality grass hay as the forage and/or adding alfalfa chaff to any complementary feed may be advantageous, although the total ration needs to be balanced to account for the alfalfa inclusion rate. However, straw as the only or main forage should be avoided as it was found to be at least 4.4 times more likely to increase squamous gastric ulcer severity scores.[7] Although absolute requirements have not been determined for horses, current recommended levels of forage (grass and preserved forages) are ideally ≥1.5 kg DM/100 kg bwt for all horses including those with high energy requirements (in which case young, less mature, high energy forages should be considered).[23] A target minimum, even for race horses, is 1.25 kg DM/100 kg bwt. Lower levels may be required for those horses on a strict veterinary monitored weight loss program (but not less than 1 kg DM/100 kg bwt) in which case appropriate measures need to be put into place to maximize the time spent chewing the restricted forage intake.[24,25] Appropriate protein, vitamin and mineral balancers will be required to nutritionally balance the ration especially if fortified concentrate feeds are not required to be fed.

Reduce the Intake of Nonstructural Carbohydrate

Smaller grain diets should be fed. Larger grain meals, 700 g/100 kg bwt, resulted in slower gastric emptying compared to a smaller, 300 g/100 kg bwt grain meal.[26] Increased gastric retention time will increase the fermentation by resident bacteria, resulting in higher VFA production and a greater potential for squamous injury. The intake of non-starch carbohydrates and particularly grains should, therefore, be restricted through the use of lower NSC complementary feeds. Based on previous studies, feeding <0.5 kg grain-based ('sweet feed')/100 kg bwt (NSC = 20%) should help keep stomach VFA concentration of acetic acid below threshold (20 mmol/l) and minimize the effect on squamous ulcers.[5] Feeding <2 g NSC/day or <1 g NSC/meal has also been recommended.[7] This can be a challenge for some trainers of high intensity exercising horses to achieve without an apparent (real or perceived) loss in performance, often resulting in the inclusion of various supplements as discussed below.

SUMMARY

There is clearly a role of nutrition in formation and maintenance of squamous gastric ulcers; however, the role of nutrition in glandular ulcer disease is yet to be defined. Current evidence supports feeding a diet high in forage and low in nonstructural carbohydrates and feed being provided throughout the day and night to reduce the risk of squamous gastric ulcers. In contrast, the role of diet in management of glandular gastric ulcers remains largely unknown. However, currently the authors recommend that the principles for feeding and management are the same for horses with squamous and/or glandular EGUS. The authors' current principles for diet and management for EGUS are:

- Horses in general should be fed as much fiber/forage as possible. Many horses and ponies do not require additional cereal-based feed but rather just an appropriate vitamin/mineral/protein forage balancer. Avoid prolonged periods without forage provision.
- Horses in light or no work, or that are extremely 'good doers', may benefit from being fed lower energy fiber sources so that time spent chewing is maximized. If the amount (not less than 1.5% bodyweight on DM basis without nutritional advice) needs to be restricted, still try to maximize the time taken to ingest (i.e., use double haylage nets/appropriately chopped chaff). Avoid feeding straw as the only or predominant forage source (other than for donkeys), as it has been shown to increase risk of gastric ulcers.
- Ideally feed <1 g/kg bwt NSC/meal and preferably <2 g/kg bwt NSC per day.
- Consider adding some chaff, which includes some alfalfa, into the meal.
- If additional energy is required, consider the gradual introduction of supplemental vegetable oil (up to 1 ml/kg bwt) but check the vitamin, mineral balance of the resultant diet or choose commercial, balanced high oil, high fiber feed.
- Provide pasture turnout whenever possible (although it is important to note that ulcers can be found in pasture managed horses especially if they are still being fed high levels of starch and sugar feeds and/or exercised intensively).
- Provide water continuously even when out at pasture.
- Whenever possible and especially in a horse with known predisposition to this problem, avoid stressful situations such as travelling long distances, changing environment and long periods of confinement where the horse cannot freely move around. Consider forage provision during transportation, and immediately on arrival.
- Medical management may be required especially in the initial stages and in particular for those with severe ulceration and those that do not respond to management changes.
- If horses need to be treated with nonsteroidal anti-inflammatory drugs (such as phenylbutazone or flunixin meglumine), a therapeutic dose should be used and limited to one and not 'stacked'. The authors recommend that horses predisposed to glandular ulcers be treated with preventive or treatment doses of antiulcer pharmaceutical agents while on NSAIDs.

REFERENCES

1. Andrews, F.M. *et al.* (1999a) Recommendations for the diagnosis and treatment of equine gastric ulcer syndrome (EGUS). *Equine Vet. Educ.* 11, 262–272.
2. Nadeau, J.A. *et al.* (2000) Evaluation of diet as a cause of gastric ulcers in horses. *Am. J. Vet. Res.* 61, 784–790.
3. Nadeau, J.A. *et al.* (2003a) Effects of hydrochloric, acetic, butyric and propionic acids on pathogenesis of ulcers in the nonglandular portion of the stomach of horses. *Am. J. Vet. Res.* 64, 404–412.
4. Nadeau, J.A. *et al.* (2003b) Effects of hydrochloric, valeric and other volatile fatty acids on pathogenesis of ulcers in the non-glandular portion of the stomach of horses. *Am. J. Vet. Res.* 64, 413–417.
5. Andrews, F.M. *et al.* (2006) *In vitro* effects of hydrochloric acid and various concentrations of acetic, propionic, butyric, or valeric acids on bioelectric properties of equine gastric squamous mucosa. *Am. J. Vet. Res.* 67, 1873–1882.
6. Lybbert, T. *et al.* (2007) Feeding alfalfa hay to exercising horses reduces the severity of gastric squamous mucosal ulceration. *Proc. Am. Assoc. Equine Pract.* 53, 525–526.
7. Luthersson, N. *et al.* (2009) Risk factors associated with equine gastric ulceration syndrome in 201 horses in Denmark. *Equine Vet. J.* 41, 625–630.

8. Ethell, M. *et al.* (2000) Evidence of surfactant contributing to the gastric mucosa barrier of the horse. *Equine Vet. J.* 32, 470–474.
9. Bullimore, S.R. *et al.* (2001) Surface mucus in the non-glandular region of the equine stomach. *Res. Vet. Sci.* 70, 149–155.
10. Andrews, F.M. *et al.* (1999) Efficacy of omeprazole paste in the treatment and prevention of gastric ulcers in horses. *Equine Vet. J.* 31, Suppl. 29, 81–86.
11. MacAllister, C.G. *et al.* (1999) Effects of omeprazole paste on healing of spontaneous gastric ulcers in horses and foals: a field trial. *Equine Vet. J.* 31, Suppl. 29, 77–80.
12. Woodward, M.C. *et al.* (2014) Effect of pectin, lecithin, and antacid feed supplements (Egusin) on gastric ulcer scores, gastric fluid pH and blood gas values in horses. *BMC Vet. Res.* 10, Suppl. 1, 54–62.
13. McGowan, C.M. *et al.* (2007) Induction and recovery of dietary induced gastric ulcers in horses. *J. Vet. Int. Med.* 21, 603.
14. Argenzio, R.A. (1999) Comparative pathophysiology of non-glandular ulcer disease: a review of experimental studies. *Equine Vet. J.* 31, Suppl. 29, 19–23.
15. Murray, M.J. (1999) Pathophysiology of peptic disorders in foals and horses: a review. *Equine Vet. J.* 31, Suppl. 29, 14–18.
16. Merritt, A. and Julliand, V. (2013) Equine GI physiology - Some species specific features. In: *Proceedings of the European Equine Health and Nutrition Congress.* pp 94–107.
17. Silen, W. (1987) Gastric mucosal defense and repair. In: *Physiology of the Gastrointestinal Tract.* Ed: L.R. Johnson. Raven Press, New York, NY. pp 1055–1069.
18. Malmkvist, J. *et al.* (2012) Behavior and stress responses in horses with gastric ulceration. *Appl. Anim. Behav. Sci.* 142, 160–167.
19. Sykes, B. and Jokisalo, J.M. (2015) Rethinking equine gastric ulcer syndrome: Part 3 - equine glandular gastric ulcer syndrome (EGGUS). *Equine Vet. Educ.* 27, 372–375.
20. Vatistas, N.J. *et al.* (1999a) Induction and maintenance of gastric ulceration in horses in simulated race training. *Equine Vet. J.* 31, Suppl. 29, 40–44.
21. Vatistas, N.J. *et al.* (1999b) Cross-sectional study of gastric ulcers of the squamous mucosa in Thoroughbred racehorses. *Equine Vet. J.* 31, Suppl. 29, 34–39.
22. Nieto, J.E. *et al.* (2009) Effect of gastric ulceration on physiologic responses to exercise in horses. *Am. J. Vet. Res.* 70, 787–795.
23. Harris, P.A. *et al.* (2013) Controversial areas in equine nutrition and feeding management: the editors' views. In: *Equine Applied and Clinical Nutrition: Health, Welfare and Performance.* Eds: R.J. Geor, P.A. Harris and M. Coenen, Elsevier Health Sciences, Waltham. pp 455–468.
24. Bruynsteen, L. *et al.* (2014) Changes in oxidative stress in response to different levels of energy restriction in obese ponies. *Br. J. Nutr.* 112, 1402–1411.
25. Ellis, A.D. *et al.* (2015) The effect of presenting forage in multi-layered haynets and at multiple sites on night time budgets of stabled horses. *Appl. Anim. Behav. Sci.* 165, 88–94.
26. Metayer, N. *et al.* (2004) Meal size and starch content affect gastric emptying in horses. *Equine Vet. J.* 36, 436–440.

acvim196

The Equine Gut as a Sensory Organ: Entero-Endocrine Cells and Incretin Hormones - A Review of Current Understanding

Philip Johnson, BVSc (Hons), MS, DACVIM (LAIM), DECEIM, MRCVS
Columbia, MO, USA

Introduction

The observation that an orally-administered dose of glucose would elicit a greater secreted insulin response than that measured following the intravenous administration of an isoglycemic dose led to the discovery of the gut-derived hormones during the early part of the twentieth century.[1] This intensification of the pancreatic response to absorbed glucose by gut-derived hormones, subsequently referred to as incretins, led to recognition of an enteroinsular axis (EIA), which is now known to consist of both endocrine and neuronal components.[2] Using a small number of animals, the existence of an EIA in equids was first reported in 2001 based on, among other things, the observation that plasma GIP concentrations increased significantly following oral glucose administration, but that intravenous glucose infusion had no effect on GIP.[3] In that study, elevated plasma GIP concentration following glucose administration in one pony, characterized as being affected with insulin dysregulation, was elevated to an exceptional extent. Those authors commented that abnormalities in the EIA might play a role in the equine insulin dysregulation (equine metabolic syndrome) and hyperlipemia syndromes.[3] The extent to which diet and postprandial incretin and insulin hormone responses might be involved in the equine metabolic syndrome was recently reviewed.[4]

The fact that hyperinsulinemia is a risk factor for laminitis was recently documented;[5] factors that promote hyperinsulinemia are likely important in terms of risk for endocrinopathic laminitis. The importance of a role for elevated incretin response (augmenting hyperinsulinemia) in some breeds (ponies and Andalusian horses developed higher insulin concentrations compared to Standardbred horses could have been attributed to either or both relatively lower insulin sensitivity or increased GLP-1 release) was recently documented.[6] Certainly, risk of insulin-associated endocrinopathic laminitis appears to be more likely in some breeds.[7] The importance of insulin dysregulation in both human and animal (including equine) health is emerging and represents an area of increasingly intense research scrutiny. The fact that some horses/ponies develop substantially greater post-prandial hyperinsulinemia responses when provided the same food is likely key to understanding how these individuals are predisposed to laminitis. In the past, hyperinsulinemia associated with the equine metabolic syndrome was largely attributed to a state of peripheral insulin resistance.[7] More recently, it has been postulated that the primary pathophysiological disturbance is post-prandial hyperinsulinemia and that insulin resistance develops as a consequence of hyperinsulinemia (as a result of peripheral insulin receptor down-regulation).[8]

A recently-reported detailed investigation that compared glucose, insulin, and incretin responses to orally- and parenterally-administered D-glucose between normal and insulin-dysregulated ponies demonstrated that, although all ponies had similar IV responses, insulin-dysregulated were markedly hyper-responsive to oral D-glucose.[8] In that study, insulin responses to fed D-glucose and grain were positively associated with the active amide of the incretin, GLP-1.[8] These results further affirmed the existence of a functional enteroinsular axis in ponies and that it might contribute to insulin dysregulation, predisposing affected individuals to endocrinopathic laminitis.[8] An additional very interesting finding in that study was the observation that affected ponies (insulin dysregulated) either absorbed more glucose from the diet or metabolized less glucose through the liver than did the normal ponies. In those ponies, it was glucose (not incretin hormones) that was the principal driver of insulin secretion.[8] Only approximately 23% of the variation in insulin response was explained by variation in active-GLP-1 levels.[8]

INCRETINS

Incretin hormones are secreted by entero-endocrine cells in the gastrointestinal tract as a result of nutrient ingestion. These neuropeptide hormones are named for the fact that they augment or amplify the extent to which insulin is released in response to orally-administered glucose when compared to an equivalent (isoglycemic) intravenous administration of glucose. The two principal incretins are glucagon-like peptide-1 (GLP-1) and glucose-dependent insulinotropic polypeptide (formerly known as gastric inhibitory peptide, GIP). Intestinally-derived incretin hormone secretion amplifies the glucose-stimulated insulin secretion in response to ingested carbohydrates, acting to tune insulin output to the feeding state of the host. On the other hand, well-documented, counter-regulatory mechanisms that are inhibitory to insulin secretion during or in the context of starvation have been recognized but are incompletely understood. It has been suggested that intestinally-derived (enteroendocrine) decretin hormones may be inhibitory to insulin release to prevent hypoglycemia under starvation conditions.[9] Results of recent work has pointed to an ancient decretin signaling pathway that inhibits insulin output in Drosophila.[10]

In the human medical context, GLP-1 and GIP are very important for normal function of beta cells in the islets of Langerhans and for blood glucose regulation in both health and type-2 diabetes mellitus. For example, GLP-1 and GIP may account for approximately 70% of the insulin response to ingested glucose in people,[11] approximately 23% of the variation in insulin response in ponies,[8] and a virtually negligible extent in some carnivorous species[12]. Not only do incretin hormones stimulate insulin secretion by pancreatic islet cells in response to ingested glucose, they also exert numerous other actions both within the islet of Langerhans and throughout the body ("extra-glycemic actions").[13] For example, the incretins retard nutrient absorption into the circulation through an inhibitory effect of the rate of gastric emptying and may directly reduce the intake of food. These truly pleiotropic hormones are considered to constitute an "incretin axis." Both incretins are secreted by entero-endocrine cells that are primarily located along the gastrointestinal epithelial lining. Specifically, GLP-1 is secreted by L-cells and GIP is secreted by K-cells.[14] Although L- and K-cells can be found throughout the length of the intestine, in most species, L-cells are located mostly in the ileum and the colon and K-cells are located predominantly in the duodenum and jejunum.[15] Interestingly, some so-called "L/K cells" may also exist in the proximal aspect of the small intestine, from which GLP-1 and GIP are co-secreted.[16] Therefore, following a sufficient stimulus (entry of ingesta into the small intestine or an intra-duodenal infusion of glucose), plasma GIP concentration is elevated initially and elevated GLP-1 levels are relatively delayed. A higher threshold of glucose delivery into the small intestine is needed to provoke a GLP-1 response compared to that for GIP.[17]

Both GLP-1 and GIP are rapidly inactivated by the enzyme dipeptidyl peptidase-4 (DPP-4), which is located on the surface of endothelial cells near to the L- and K-cells, the liver, and free in the circulation.[18] In recent years, there has been substantial interest in the role of incretins in health and disease, especially - from the human perspective - type-2 diabetes. A range of new drugs that influence the incretin axis has been made available to practitioners by the pharmaceutical industry. For example, incretin mimetics, such as exenatide and lixisenatide are GLP-1 analogues, are being employed for the management of type-2 diabetes. On the other hand, endogenous incretin effectiveness can be increased by using drugs (such as sitagliptin) that inhibit DPP-4. DPP-4 inhibitors are also known as 'gliptins' and have the advantage of being administered orally (incretin analogues must be injected).[19]

In the resting, fasted state, basal incretin secretion is regarded as low. It has been reported that fasting plasma GLP-1 levels are increased by exercise (influence of the myokine, IL-6) and inflammation (LPS, IL-1, and IL-6).[20] Substantially increased incretin secretion occurs following the ingestion of food. Although carbohydrate and fat components of food yield the highest incretin responses, both incretins are released in response to all principal nutrient components (carbohydrates, fat, protein). More maximal incretin release, ingested nutrients must first be digested (for example, orally-supplemented pancreatic enzymes cause greater levels of post-prandial plasma incretin concentrations in patients affected with pancreatic exocrine insufficiency). Nutrient-induced incretin secretion results significantly from the interplay between food components and an array of diverse chemo-sensors at the luminal aspect of entero-endocrine cells. Subsequent patterns of GLP-1 and GIP secretion are dependent on the rates of gastric emptying and small intestinal transit as well as the speed of digestion and absorption of nutrients.

Control of incretin secretion is incompletely understood and other "proximal-distal" endocrine loops (e.g., cholecystokinin released at the duodenum stimulates GLP-1 release more distally) likely exist.[21] When released into the circulation, both incretins exert diverse and pleiotropic glycemic and extra-glycemic effects.[13]

In healthy people, GLP-1 is reported to contribute to an insulinotropic effect when the plasma glucose concentration exceeds a threshold of 77.4 mg/dl.[22] GLP-1 contributes to the regulation of blood glucose through several direct and indirect mechanisms: direct stimulation of beta-cell exocytosis (enhanced insulin secretion); increased peripheral insulin sensitivity; inhibition of glucagon secretion; decreased gastric emptying; lowered intake of food. GLP-1 is also needed for maintenance of beta-cell sensitivity to ambient plasma glucose concentrations. Furthermore, GLP-1 is stimulatory to the biosynthesis of insulin by beta-cells, beta-cell survival, and beta-cell proliferation. To an undetermined extent, the insulinotropic effects of GLP-1 are also mediated through stimulation of neurons in the CNS (via vagal afferent innervation). In light of the fact that, following secretion by L-cells, GLP-1 is so quickly metabolized (by DPP-4), it has been suggested that GLP-1-stimulated insulin release may be further provoked through GLP-1 sensory neurons in both the intestinal tract and the hepatic portal veins (via hepatic vagal afferent nerves).[18]

Interesting recent work points to the existence of a GLP-1-dependent neural connection between the peripheral and central nervous systems. Numerous "extra-glycemic" actions of GLP-1 have been reported: promotion of lipolysis (white adipose) and thermogenesis (brown adipose); offsetting of hepatic lipid accumulation; numerous effects on the cardiovascular system (beneficial to endothelial cell function, improves myocardial function). Whereas GLP-1 exerts its effects through both endocrine and neurocrine mechanisms, GIP acts principally through simple endocrine effects.[23] Although binding of GIP to its receptor on beta-cells results in the same effects as described for GLP-1 (similarly in glucose dependent fashion), GIP stimulates alpha-cells to release glucagon (glucagonotropic effect). The resulting effect of GIP in terms of glycemic control depends on the extent to which alpha- and beta-cells are activated overall. It has been suggested that, in some species, the glucagonotropic action of GIP (stimulation of endogenous glucose production) may be important for maintenance of glucose homeostasis during episodes of hypoglycemia.

The extent to which both GLP-1 and GIP play a role in other tissues with importance regarding glucose homeostasis is being increasingly recognized (extra-glycemic effects). The two incretins can be distinguished by their action on adipose tissue. Some work has demonstrated that activation of GLP-1 receptors in adipose tissue may lead to activated lipolysis. Other work demonstrated that activation of GLP-1 receptors within the CNS caused reduced adiposity through activation of the sympathetic nervous system innervation of white adipose tissue and to stimulate thermogenic activity of brown adipose tissue.[24] On the other hand, GIP acts differently and stimulates lipid uptake by liver and muscle cells, promotes lipogenesis, and increases lipoprotein lipase activity. Unlike GLP-1, GIP is regarded as anabolic in its extra-glycemic actions (promotes weight gain). Elevated plasma GIP levels are associated with lowering of the plasma triglyceride concentration (plasma triglyceride concentration tends to increase when the effect of GIP is diminished). Pharmaceutical inhibition of GIP acts to prevent diet-associated obesity.[25]

It has been suggested that diets characterized by minimal effects on the secretion of GIP might be recommended for the prevention and management of obesity, fatty liver disease, and insulin resistance.[26] Whereas GLP-1 signaling at the liver (through actions on hepatocyte lipid transport, lipid oxidation, lipid synthesis and secretion) has been linked to reductions hepatic lipid content and has attracted interest regarding pharmaceutical agents being developed for the purpose of preventing and treating (non-alcoholic) fatty liver disease, GIP signaling is associated with promotion of fatty liver disease. Although substantially less is known about the role of GIP in this regard, there is also evidence that GLP-1 is important from the context of the regulation of myocardial and endothelial functions. Both incretins may exert actions on bone metabolism,[13] but the relative importance of this observation differs considerably between species. In rodents, GLP-1 acts by stimulating thyrocalcitonin release by C-cells in the thyroid gland, which is inhibitory to osteoclasts.[27] Use of GLP-1 receptor agonist drugs in human patients has not been linked to observed effects on bone strength, bone mineral density, or increased plasma thyrocalcitonin concentrations (human C-cells have significantly fewer GLP-1 receptors compared with rodents). Both osteoblasts and osteoclasts possess receptors for GIP which appears to stimulate and inhibit these cells, respectively, in rodents. As for GLP-1, the role of GIP in human bone metabolism appears to be less significant than that for rodents.

Pharmaceutical strategies based on GLP-1 and GIP biology have been broadly introduced for the treatment and management of type-2 diabetes mellitus in human patients. Although the secretion of either incretin does not appear to be abnormal in either obese or diabetic states, impaired glucose regulating and insulinotropic effectiveness of GIP has been reported. Other areas of future interest include the facts that GIP receptor activation (via stimulated glucagon secretion) could be useful for both diabetes-associated hypoglycemic complications and control of diet-induce obesity, fatty liver disease, and insulin resistance. The extent to which gastric surgical by-pass has demonstrated high efficacy for the management of type-2 diabetes mellitus in human patients is partly explained by the fact that GLP-1 secretion by L cells in the more distal aspect of the bowel is enhanced as a result of the provided 'short cut' that results in digesta arriving at this location more quickly.[28]

Emerging realization that endocrinopathic laminitis associated with the equine metabolic syndrome results from an equine hyperinsulinemia syndrome, distinct from insulin resistance has led to recognition and broadening interest in the equine entero-insular axis. The lining of the gastrointestinal tract is exposed to and, through multiple discrete mechanisms, monitors and responds to a remarkable and diverse array of molecular signals in the luminal content. Entero-endocrine cells represent only one component of this complex sensory array. Understanding the integrated response of the gut to all of the sensory information it receives represents a significant challenge.

References

1. Bayliss W, *et al. J Physiol (London)*. 1902;28(5):325.
2. Marks V, *et al. Proc Nutr Soc.* 1991;50:545.
3. Dühlmeier R, *et al. Comp Biochem Physiol A Mol Integr Physiol*. 2001;129(2–3):563.
4. de Graaf-Roelfsema E. *Eq Vet J*. 2014;199(1);11.
5. Asplin KE, *et al. Vet J*. 2007;174(3):530.
6. Bamford NJ, *et al. J Animal Sci*. 2015;93(7):3377.
7. Frank N, *et al. J Vet Intern Med*. 2010;24(3):467.
8. de Laat MA, *et al. Am J Physiol Endocrinol Metab*. 2016;310:E61.
9. Ensinck JW, *et al. J Clin Invest*. 1997;100:2295.
10. Alfa RW, *et al. Cell Metab*. 2015;21(2):323.
11. Vilsboll T, *et al. Diabetologia*. 2004;47:357.
12. Gilor C, *et al. Domest Anim Endocrinol*. 2011;40:205.
13. Baggio LL, *et al. Gastroenterol*. 2007;132:2131.
14. Furness JB, *et al. J Gastroenterol Hepatol*. 2013;10:729.
15. Gunawardene AR, *et al. Int J Exp Pathol*. 2011;92:219.
16. Theodorakis MJ, *et al. Am J Physiol Endocrinol Metab*. 2006;290:E550.
17. Ma J, *et al. Diabet Med*. 2012;29:604.
18. Holst JJ. *Physiol Rev*. 2007;87:1409.
19. Tasyurek HM, *et al. Diabetes Metab Res Rev*. 2014;30(5):354.
20. Kahles F, *et al. Diabetes*. 2014;63(10):3221.
21. Beglinger S, *et al. J Clin Endocrinol Metab*. 2010;95:879.
22. Nauck MA, *et al. J Clin Endocrinol Metab*. 2002; 87:1239.
23. Meier JJ, *et al. Am J Physiol Endocrinol Metab*. 2004;286:E621.
24. Lockie SH, *et al. Diabetes*. 2012;61:2753.
25. Pathak V, *et al. Mol Cell Endocrinol*. 2014;401:120.
26. Keyhani-Nejad F, *et al. Diabetologia*. 2015;58:374.
27. Yamada C, *et al. Endocrinol*. 2008;149:574.
28. Laferrere B, *et al. J Clin Endocrinol Metab*. 2008;93:2479.

acvim197

Biochemical Analysis of Peritoneal Fluid from Horses with Colic

Christopher K. Cebra, VMD, MA, MS, DACVIM (Large Animal)
Corvallis, OR, USA

INTRODUCTION

Blood and peritoneal fluid analysis have been used for many years to assess horses with colic signs. Gross interpretations of color, clarity, and general appearance could be made by visually examining peritoneal fluid in a tube, but most of this analysis was not "field-friendly": It involved microscopes and large analyzers, which made it most suitable for larger hospitals, or for field cases only when urgent results were unnecessary. These analyses focused mainly on analysis of cell populations and protein concentration, and were based on the concept that a normally-functioning capillary membrane should restrict the movement of both cells and protein. Increased hydrostatic pressure or inflammation increased protein loss and white blood cells exuded through damaged capillary walls or were recruited to inflamed intraabdominal structures. Red blood cells could similarly exude through damaged capillary walls or out of congested vessels.

Biochemical analysis of peritoneal fluid has also been reported. This has mainly focused on measurement of lactate and glucose concentrations. This was initially cumbersome and required major lab equipment, but the relatively recent proliferation of portable body fluid analyzers has put some of this analytical capacity into the hands of the field practitioner.[1-5] Most point-of-care analyzers are designed for use on blood, but may yield satisfactory results from other body fluids, such as peritoneal fluid. The great advantage of peritoneal biochemical analysis is that the results are available in minutes, which allows timely treatment decisions to be made. The goal of this presentation is to review some of the capabilities for biochemical analysis of blood and peritoneal fluid, and to give an overview of abnormalities which might be encountered.

PERITONEAL FLUID FORMATION

Normal peritoneal fluid arises from three principal sources: ultrafiltration from hepatic and gastrointestinal capillary beds, secretion from mesothelial cells,[6] and in cycling females, exudation of ovarian fluid via the oviducts.[7] Ultrafiltration from capillary beds is the most variable with pathologic conditions in the horse. The hepatic beds may be the most contributory under normal conditions, because their sinusoidal membranes are considered far more permeable (1,000x) than in the visceral peritoneum. Mesothelial cells are known to secrete glycolipids, glycoproteins and glycosaminoglycans, including hyaluronan, to provide for frictionless movement of organs with the peritoneal cavity. They can also secrete or absorb fluid and particulates, depending on surrounding conditions. Ovarian exudation increases during the follicular phase of the menstrual cycle to peak during the early luteal phase.[8,9] This is a major cause of peritoneal fluid volume variation in human females, but has not been investigated extensively in horses.

Fluid secreted or exuded into the peritoneal cavity should not remain there long. In humans, initial downward flow is followed by movement toward the diaphragm. There, the right lymphatic duct and to a lesser degree the thoracic duct, transport the fluid into the thorax and back into the vascular compartment. Additional non-lymphatic interstitial absorption has been documented in mice.[10] In horses, the main role in abdominal lymph drainage has been attributed to the thoracic duct. The rate of fluid removal frequently matches or exceeds the rate of fluid formation, meaning the peritoneal cavity is primarily a potential space with minimal local pockets of accumulation. Because of this overcapacity for removal, increased formation **and** impaired removal may both be required to allow persistence of excess peritoneal fluid.

Ultrafiltration is driven by pressure and concentration gradients. The hydrostatic pressure of the capillary bed is usually higher than of the interstitium, promoting the net outflow of water. Electrolytes follow to maintain concentration gradients. Proteins exude, but in a restricted fashion, leaving a higher protein concentration and oncotic draw in the plasma, and since most circulating proteins are anionic, also a net negative charge. This affects the equilibration of electrolytes - the Gibbs-Donnan Effect; sodium and potassium shift from the interstitium into the plasma to compensate for the anionic proteins, and chloride follows to maintain electrical neutrality. Thus, normal peritoneal fluid has lower concentrations of sodium, potassium, chloride and protein than plasma and a higher pH. (Table 1).

Table 1. Plasma and fluid peritoneal biochemical properties from healthy horses

	Peritoneal fluid	Plasma
Na^+	129±4 mEq/L	135±4 mEq/L
K^+	3.3±0.5 mEq/L	3.4±0.5 mEq/L
pH	7.64±0.17	7.42±0.06

Since ultrafiltration provides proportionally more fluid than protein, the protein secreted or exuded from mesothelial cells and the ovaries is diluted in peritoneal fluid. The overall result is a fairly low peritoneal fluid protein concentration in healthy animal.

Abnormal Peritoneal Fluid

Disorders of ultrafiltration can occur for a variety of reasons, which all affect either the comparative hydrostatic or oncotic pressures in the vascular and peritoneal compartments, or both. Factors which increase the hydrostatic pressure or decrease the oncotic pressure in the vascular space promote the movement of fluid into the peritoneal cavity, and the opposite factors draw fluid in the other direction. These affects may be local or global. Hypodynamic shock or dehydration with an increase in blood protein tend to cause water to be drawn into the vascular compartment from surrounding tissues and cavities, whereas vasodilation, local or global vascular congestion, or protein leakage into the interstitium promote the outflow of water. Thus, increases in peritoneal protein concentration can happen either with loss into the cavity through the capillary and peritoneal membranes, or through simple concentration with decreased fluid entry. This makes protein concentration an unreliable indicator of inflammation.

With an increase in peritoneal protein, the Gibbs-Donnan Effect is diminished. Peritoneal electrolyte concentrations climb closer to systemic levels and the alkalinity of peritoneal fluid is reduced.

Lactate and Glucose

In addition to protein and the main electrolytes, peritoneal fluid contains a number of other biochemical of diagnostic interest. Some of these are fairly specialized, such as the hormonal or inflammatory markers of endometriosis or certain types of cancer released from the ovaries or mesothelial cells. Others are more general markers of homeostasis and are useful in diagnosing gastrointestinal disorders. The most commonly used and easily measured of these are glucose and lactate concentrations. Interpretation of these concentrations is more complicated than analysis of elements, because they may be consumed (glucose) or generated (lactate) locally, and thus reflect local metabolism in addition to ultrafiltration.

Glucose enters the peritoneal fluid through ultrafiltration. Plasma changes in glucose are thought to equilibrate throughout the peritoneal fluid over 20 to 60 minutes, so there may be temporary physiologic differences. This is clinically most notable in stressed horses, where the acute mobilization of glucose leads to a rapid increase in plasma glucose and a lagging increase in peritoneal glucose. Glucose can also be catabolized by cells within the peritoneal cavity. The major variable users include microbes and inflammatory cells. The association between lower peritoneal glucose concentrations and septic peritonitis has been investigated,[2] but the effects of large cellular response have not.

Lactate tends to be higher in blood than normal peritoneal fluid, because a considerable amount of body lactate is produced in the blood by erythrocytes. It also enters peritoneal fluid through ultrafiltration. This can be a global effect, with relatively homogeneous contributions from various different capillary beds throughout the body, or a local affect, where specific capillary beds drain tissues with increased anerobic glycolysis. Examples include tissue beds whose vascular supply has been compromised by thrombosis, vasoconstriction, twisting of the vessels, mass lesions, luminal pressure, or poor arterial pressure. These focal compromised tissue beds may be in the abdomen or somewhere else in the body; in the former case, they are more likely to have a direct effect on peritoneal fluid.

Additionally, cells within the peritoneal cavity can produce lactate. This appears to be more important when peritoneal cell counts climb precipitously, but it may also be related to the volume of peritoneal fluid: when the cavity is a potential space, sufficient oxygen may diffuse across the surface membranes to adequately supply cells for aerobic glycolysis, but as the volume increases and the distance between peritoneal linings also increases, this limited diffusion may be insufficient. Hemoperitoneum may also measurably increase peritoneal lactate, because erythrocytes lack mitochondria and rely on anerobic glycolysis to produce ATP.

INTERPRETATION

A number of different biochemical abnormalities have been reported from horses with gastrointestinal lesions. Most of these reports are on small numbers of horses with specific types of lesions, and in some cases, lesions were induced. Reported abnormalities include increases in L-lactate, D-lactate, and phosphorus and decreases in pH, pO_2, pCO_2, and chloride with strangulated bowel; increases in L-lactate, protein, and glucose and decreases in chloride, sodium, bicarbonate, and pH in horses with undifferentiated colic and decreases in pH and glucose with septic peritonitis or rupture.[1-5,11-19] Comparative studies on horses with different kinds of lesions are rare, and a general interpretation could be that, unlike in ruminants, where specific blood electrolyte and acid-base changes are strongly associated with specific types of gastrointestinal lesions, many of the blood changes seen in horses with colic speak more to the general condition of the horse than the specific nature of the lesion. Horses with severe or advanced disease develop evidence of stress, dehydration, and shock, eventually progressing to evidence of multi-organ failure or possibly sepsis. These abnormalities develop at different rates in horses with different types of gut disorder, with the general consensus that horses with gut strangulations or ruptured bowel deteriorate faster than those with displacements or impactions. Generally speaking, the less serious the disorder, the more "normal" the blood work, but even with serious disorders, it is preferable to address them before blood work becomes too abnormal. Thus, the described blood abnormalities are not necessarily good predictors of disease or outcome early in the course, when critical treatment decisions (referral or field treatment, surgery or non-surgical treatment) may be required.

Of the reported abnormalities, systemic and peritoneal lactate concentrations have drawn the most attention. In early studies, high blood concentrations (>8 and later >4 mmol/L) were associated with a poor prognosis, and similar conclusions were reached with peritoneal fluid.[15,16] In more recent work, peritoneal lactate >4 mmol/L was associated with strangulating lesions. The cut-off of >4 mmol/L of lactate is a common poor prognostic indicator in medicine. In blood, it is considered to reflect the level at which hepatic clearance begins to fail.[20] More recently, a cut-off of 5 mmol/L has been proposed in other species.[21] These values relate to systemic lactate and generally assume global production related to general underperfusion. The significance of local production, such as from a discrete segment of strangulated bowel, is harder to interpret, but very high peritoneal lactate concentrations have been associated with a poor prognosis. Conversely, low blood lactate concentrations have been associated with a successful outcome.

In our clinic, horses with strangulations or ruptured viscera, or those with surgical lesions overall, have an overall trend towards hyperglycemia, hypochloremia, hyperlactemia and metabolic acidosis. Peritoneal fluid shows similar lactate and acid-base changes and an increase in sodium concentration. Horses with non-surgical, survivable lesions generally have more normal blood and peritoneal fluid, with the exceptions of those which are likely to be pre-diarrhea colitis cases and those with more chronic impactions. Both these groups of horses often have systemic evidence of shock, in some cases severe. In the case of impactions, peritoneal lactate concentrations can also become quite high, presumably related to mucosal ischemia brought on by high intraluminal pressures. In contrast, some horses diagnosed with colon volvulus often have mild to no evidence of shock. These presumably signify horse where the twist has not progressed to strangulation.

A small number of investigators have compared blood and peritoneal values of some analytes, most commonly glucose[2] and lactate.[3,5,12,15] Possible advantages to this relate to the potential for local consumption (glucose) or production (lactate). Horses could have global disturbances in these analytes, such as hyperglycemia, hypoglycemia or hyperlactemia, which could confound interpretation of low or high concentrations of them in peritoneal fluid. Methods of comparison include differences and ratios. With lactate, the value of comparisons appears to diminish with rising concentrations, and this confounded some early attempts to demonstrate their usefulness. It is likely that horses with severe progressive intestinal lesions fairly quickly go into shock and recruit additional and physically large extraabdominal foci of lactate production, and that these quickly overshadow the original abdominal source in magnitude. However, a comparison appears to offer a fairly sensitive and early indicator of strangulation, and therefore appears to be most useful in cases that have not yet developed severe hyperlactemia or high peritoneal lactate concentrations. It is also useful in distinguishing horses whose colic originates from infectious/inflammatory lesions, which often have evidence of shock without specifically having local abdominal ischemia.

A number of researchers, including our group, have attempted to create diagnostic or prognostic models based on blood, peritoneal, or physical parameters. Most conclude that multivariate analysis is superior to any univariate attempt.[3,5,22] We would also conclude that clinical judgment plays an important role in any decision making. Even though certain biochemical values, such as a high lactate in peritoneal fluid or a higher peritoneal than systemic value cast strong suspicions on specific diagnoses in certain cases, there are always exceptions. The clinician discovering an impaction on rectal exam, having a strong suspicion of impending diarrhea, or dealing with relentless pain in a horse with an early colon volvulus should factor these findings into the decision-making process.

References

1. Peloso JG *et al. J Am Vet Med Assoc.* 2012;240:1208.
2. Van Hoogmoed L *et al. J Am Vet Med Assoc.* 1999;214:1032.
3. Delesalle C *et al. J Vet Intern Med.* 2007;21:293.
4. Saulez MN *et al. Vet Rec.* 2005;157:217.
5. Latson KM *et al. Equine Vet J.* 2005;37:342.
6. Mutsaers SE. *Respirology.* 2002;7:171.
7. Koninckx PR *et al. Br J Obstet Gynaecol.* 1980;87:177.
8. Rock JA *et al. Contrib Gynecol Obstet.* 1987;16:138.
9. Maathuis JB *et al. J Endocrinol.* 1978;76:123.
10. Nagy JA. *Blood Purif.* 1992;10:148.
11. Ruggles AJ *et al. Am J Vet Res.* 1993;54:333.
12. Hjortkjaer RK *et al. Nord Vet Med.* 1979;31:466.
13. Yamout SZ *et al. Vet Surg.* 2011;40:817.
14. Moore JN *et al. Res Vet Sci.* 1977;23:117.
15. Parry BW *et al. Equine Vet J.* 1983;15:337.
16. Genn HJ *et al. Dtsch Tierarztl Wochenschr.* 1982;89:295.
17. Svendsen CK *et al. Nord Vet Med.* 1979;31:Suppl 1.
18. Freden GO *et al. J Am Vet Med Assoc.* 1998;213:1012.
19. Parry BW. *Vet Clin North Am Equine Pract.* 1987;3:529.
20. Tashkin DP *et al. Am J Physiol.* 1972;223:968.
21. Stacpoole PW *et al. Am J Med.* 1994;97:47.
22. Furr MO *et al. Vet Surg.* 1995;24:97.

1.
acvim200

Approaches to Investigating a *Salmonella* Outbreak on a Breeding Farm

Ann Chapman[1], DVM, MS, DACVIM; Brandy Burgess[2], DVM, MSc, PhD, DACVIM, DACVPM; Paul Morley[3], DVM, PhD, DACVIM
[1]Baton Rouge, LA, USA; [2]Athens, GA, USA; [3]Fort Collins, CO, USA

INTRODUCTION

Salmonellosis is concerning to equine breeding facilities owing to the potentially high morbidity and mortality of this disease, particularly among neonates.[1] Consequently, biosecurity measures on equine breeding farms should be considered essential elements of daily operations.[2] This report will describe how to recognize a salmonellosis outbreak, develop a diagnostic plan to identify the magnitude and scope of the disease, implement measures to reduce the likelihood of neonatal transmission, and design a biosecurity plan to minimize the spread of the disease. A recent *Salmonella* outbreak which occurred during the 2016 foaling season on a Thoroughbred breeding farm in Louisiana will be described.

EQUINE NEONATAL SALMONELLOSIS

The equine neonate is uniquely vulnerable to a host of environmental pathogens including *Salmonella* owing to the fragile nature of the developing immune system. The immediate postpartum period is a critical window when exposure to organisms is most likely to result in disease especially if passive transfer of maternal antibodies is deficient.

Diarrhea is a common clinical problem in foals with some estimates suggesting 70–80% of foals develop diarrhea during the first 6 months of life.[3] Differentiating between non-infectious and infectious causes of diarrhea remains a challenge. Approximately 55% of foals with diarrhea may have at least one infectious pathogen identified during hospitalization with *Salmonella* isolation being more likely in foals >4 weeks of age.[4] Additionally, 14% of foals with gastrointestinal disease, aged 2 days to 17 weeks, may have *Salmonella* identified by fecal PCR, and co-infection with other gastrointestinal pathogens is common (57%).[5]

Neonatal salmonellosis is characterized by different forms. The most serious form is peracute to acute septic shock in which affected foals develop altered mentation, fever, tachycardia, tachypnea, leukopenia, and inappetence. Septicemia accompanied by enterocolitis with or without diarrhea can occur in foals with neonatal salmonellosis. Both blood and fecal cultures should be collected when there is a suspicion of *Salmonella* septicemia/bacteremia since not all foals will be positive on blood culture. Chronic diarrhea may be a noted among foals that survive the acute form of the disease.

RECOGNIZING AN OUTBREAK

Since diarrhea is a common problem among suckling foals, recognizing the presence of an outbreak is challenging. Early recognition is vital to initiating measures to limit the spread.[6] A sudden rise in the number and severity of diarrhea cases among foals should raise an index of suspicion regarding *Salmonella*, particularly if it is accompanied by clinical evidence of colitis/enterocolitis in other age groups within the population. Highly virulent multidrug resistant strains of *Salmonella* have been identified in recent years in both human and veterinary medicine.[7,8] These strains are linked with increased morbidity and mortality, and incidentally are more likely to lead to outbreak situations. In addition to septicemia/enterocolitis described in foals, preterm infection of pregnant mares is possible. Increased incidence of embryonic loss, fetal abortion, and premature delivery may be related to exposure to highly pathogenic strains of *Salmonella*.

In the 2016 farm outbreak reported here, initial concerns began between January 9 and January 11 when two neonatal foals developed enteritis requiring hospitalization. Subsequently, both of these foals developed systemic complications despite conventional therapy. The index case was a 1-day-old foal that was admitted to the veterinary teaching hospital for enteritis and suspected sepsis. Despite conventional broad spectrum antibiotic therapy and supportive care, the foal developed intussusception and subsequent peritonitis and was euthanized on January 13. A blood culture sample obtained at admission was negative for bacterial growth; however, laboratory results for daily fecal cultures eventually revealed *Salmonella*. A postmortem examination identified *Salmonella* in multiple tissues including the spleen and liver. The second foal was admitted 2 days after the index case for enteritis at 6 days of age. This foal subsequently developed pneumonia and septic arthritis. On January 19, the second foal was euthanized due to lack of clinical response and poor prognosis. Likewise, blood culture upon admission was negative for bacterial growth, but fecal cultures confirmed *Salmonella* with antibiotic susceptibility pattern similar to the index case. On postmortem examination, *Salmonella* was isolated from the lung, joint fluid, lymph node, and small intestine of this foal. Eventual serotyping at a reference laboratory identified the isolates as *Salmonella enterica* subsp. *enterica* serovar Javiana, (O:9 [D_1] serogroup.) All the *Salmonella* isolates from both foals had identical antibiotic susceptibility patterns: resistant to ampicillin, cephalothin, chloramphenicol, doxycycline, gentamicin, sulfisoxazole, tetracycline, and trimethoprim/sulfamethoxazole; intermediate sensitivity to cefotaxime, ceftiofur, ceftriaxone, cefuroxime, and ticarcillin/clavulanic acid; and susceptible to amikacin, and nitrofurantoin.

Diagnostic Investigation

On January 22, a preliminary field investigation was performed due to a high clinical suspicion of community acquired MDR salmonellosis in the two initial enteritis cases. Initial field investigation and management of an outbreak are aimed at 1) defining the problem, 2) treating those affected, 3) prevention further disease transmission, 4) identifying the source(s), and 5) preventing its reoccurrence.[9]

To further define the problem - in this particular case a suspected outbreak of salmonellosis - active surveillance of the resident populations on the affected farm was undertaken of all foals and their mares, resident preterm mares and stallions. Fecal cultures were collected daily for 5 consecutive days. Expedited sampling can be performed in cases where clinical priority necessitates. In these circumstances, fecal samples can be obtained every 12 hours until 5 sample are collected. Although PCR is a highly sensitive and rapid method for detecting *Salmonella*, it does not provide phenotypic or genotypic information which is necessary for case mapping. Therefore, bacterial culture is necessary in epidemiologic investigations and should not be replaced with PCR. Additional data that is useful in identifying clinical disease in the neonatal foal population include physical examination, hematology, plasma biochemistry, thoracic ultrasound, abdominal ultrasound, and blood culture sampling.

During the initial field investigation, 7 additional neonatal foals (age range 2–10 days) underwent physical examination, thoracic and abdominal ultrasonography, hematology, and fecal sampling. Multidrug resistant *Salmonella* Javiana was isolated on fecal culture of all 7 foals. One of the foals evaluated was referred to the veterinary teaching hospital due to fever, tachypnea, and ultrasonographic evidence of pneumonia. It was suspected that post-partum mares were a source of infection for the neonatal foals; however, *Salmonella* was only identified in one mare of the initial group investigated. On January 24, a fourth foal was referred to a veterinary teaching hospital at 2 days of age for sepsis and tendon contracture. Similar to all previous cases, *Salmonella* Javiana was eventually isolated from fecal samples from this foal.

Infection Control Measures

As a result of the initial field investigation, biosecurity zones were established to lessen the spread of *Salmonella* on the farm. An existing barn was designated as an isolation area and all known *Salmonella*-positive animals were moved to that location. *Salmonella*-positive foals and their mares were strictly housed in stalls for ease of treating sick foals and for *Salmonella* containment. This procedure, which will limit paddock/pasture contamination, has the potential to create high contamination areas within each stall. Cleaning and disinfection procedures are essential to reducing sources of infection and minimizing transmission of pathogenic agents. As part of mitigation efforts to improve the environmental hygiene, the original neonatal barn underwent extensive renovations to improve the cleanability and disinfection of surfaces. Wooden materials were removed and replaced with industrial plastic polymer panels. Dirt flooring was removed 8–10 inches below grade and replaced with solid concrete. Cleaning protocols were established which included removal of organic debris with detergent scrubbing, followed by rinsing and disinfection with a peroxygen based product.

Foaling procedures were modified to reduce the potential exposure to *Salmonella* in the immediate post-partum period. Although preterm mares were maintained in their established groups, a new foaling location was established away from the previous foaling yard. Based on procedures utilized in other neonatal *Salmonella* outbreaks, mares were bathed with dish detergent immediately prior to parturition, paying particular attention to the udder and perineum.[1] Personnel assisting with foal delivery wore personnel protective equipment to include exam gloves and rubber footwear. After delivery, foals were permitted to nurse and afterward the mare/foal pair was transported to a designated "clean" neonatal barn. Patient handling procedures were implemented to include the routine use of latex exam gloves when handling animals, daily rectal temperature monitoring on all foals, and footbaths placed at the entrance to every stall. Additional renovations were recommended to other barns beyond the index barn. These included sealing of porous surfaces such as cement block walls with a concrete-sealing paint and eliminating openings between stalls where stall-to-stall horse contact could occur.

Active fecal surveillance for *Salmonella* implemented in which all neonatal foals were sampled within 24 hours of birth and every 2–3 days thereafter until 5 fecal samples were collected. Additional targeted surveillance was performed in foals that developed diarrhea or other gastrointestinal disease. Fecal samples were collected from preterm and postnatal mares at the time of rectal ultrasound during reproductive exams. Therefore, sample collection from mares was more sporadic compared to neonatal foals. Environmental surveillance was performed farm-wide to identify potential sources of infection. Surfaces on stalls (floors and walls), aisles, cleaning implements, wheelbarrows, buckets, feeders, fence railings, wash areas, feed rooms, trailers, breeding shed, and offices were sampled using disposable electrostatic cloths. Bedding and soil samples were also collected. This database was useful in determining locations for targeted disinfection.

RE-ASSESSMENT AND REVISION

Despite these efforts, additional *Salmonella*-positive foals were identified over the next 30 days. A team of infection control epidemiology experts were invited for a site visit to assess the infection control practices and provide guidance. Prior to such a visit, data should be compiled outlining the results of surveillance sampling efforts so a thorough appraisal of the scope of the problem can be determined. Relevant diagnostics should be completed so that groups can be identified including *Salmonella*-positive sick animals, *Salmonella*-positive healthy animals, *Salmonella*-negative animals, non-resident (transient) animals. A property map describing animal numbers and location is a useful tool to gain perspective regarding animal distribution and movement.

After a complete assessment, a number of revisions were made to the infectious disease procedures. Cleaning protocols were revised to exclude the use of household bleach as a sole surface disinfectant in areas where high traffic flow can lead to pathogen dispersion (e.g., barn aisles) as bleach is rapidly inactivated in the presence of organic debris. Instead, a quaternary ammonium or peroxide disinfectant was applied to these areas since they are more effective in the presence of organic debris. Additionally, improved contact time can be achieved by the use of metered foam spray systems. The stall cleaning procedure was revised so that scrubbing with a commercial dish detergent was replaced with a commercial laundry detergent for improved removal of organic debris. Similarly, the stall disinfection procedure was modified so that stalls housing a known *Salmonella*-positive horse underwent the cleaning procedure twice prior to housing a new animal, in order to permit increased contact time on all surfaces.

Animal management procedures were modified so that healthy foals shedding *Salmonella* were no longer confined strictly to stalls, rather were given access to a designated "dirty" pasture. While the period of time which these pastures should be considered contaminated is difficult to determine, the idea is aimed at decreasing the likelihood of a foal coming into contact with an infectious dose of *Salmonella*. Experimentally, dairy pastures contaminated with MDR *Salmonella* Newport remained infective for up to 1 year,[10] although routine removal of fecal pats, dragging to disperse any remaining fecal debris, and short mowing of contaminated pastures will likely decrease its survival time in this environment.

Fecal culturing methods were also changed to develop a complete understanding of the scope of the situation. Systemic sampling of foals continued, however, the sampling time was shortened to daily collection for 5 days or collection every 12-hours for 5 samples. This allowed for more rapid assessment of foal shedding status and improved the decision-making process. Among the foals shedding *Salmonella*, the median interval between the first positive samples until the last positive sampling was 21 days (range 1–101 days). The sampling of mares was also modified by collection of composite samples from mare groups. This was achieved through pasture/paddock sampling of fresh fecal piles. One fecal ball per horse housed in the paddock/pasture was collected from a given group to create a composite sample. Individual sampling was performed to determine the status of each mare if composite samples were positive for *Salmonella*. Horse management changes were also implemented based on this visit which included formal numbering of stalls and associated feed/water buckets in order to reduce cross-contamination between stalls.

A reduction in the number of *Salmonella*-positive foals was noted after the modified infection control protocol was implemented. During the remainder of the foaling season only one new case was identified on a single fecal sample (1 of 9 samples collected). This was thought to represent transient shedding rather than disease.

CONCLUSIONS

Managing a *Salmonella* outbreak on an equine breeding farm, particularly involving a MDR serovar, requires a concerted effort between the veterinarian, owners, and animal caretakers. Rapid recognition should prompt a systematic sampling protocol to identify scope of the situation. Mitigation efforts require "buy-in" and compliance by all personnel involved in animal care.

REFERENCES

1. Walker RL, *et al. J Vet Diagn Invest.* 1991;3:223–227.
2. Burgess BA, Morley PS. *Vet Clin North Am Equine Pract.* 2014;30:623–640.
3. Urquhart K. *In Pract.* 1981;3:22–23, 25, 27, 29.
4. Frederick J, *et al. J Vet Intern Med.* 2009;23:1254–1260.
5. Slovis NM, *et al. Equine Vet J.* 2014;46:311–316.
6. Steneroden KK, *et al. J Vet Intern Med.* 2010;24:606–616.
7. Center for Disease Control. www.cdc.gov/narms/reports/interactive-data-displays.html. 2013:1–114.
8. Hong S, *et al. PLoS One.* 2016;11:20.
9. Kane AMP. In: Annual Convention American Association of Equine Practitioners. 1999;137–141.
10. Toth JD, *et al. J Dairy Sci.* 2011;94:5238–5246.

acvim201

Pasture-Associated Asthma

Cyprianna E. Swiderski[1], DVM, PhD, DACVIM (Large Animal Internal Medicine); Lais Costa[2], DVM, MS, PhD, DACVIM (Large Animal Internal Medicine), DABVP
[1]Mississippi State, MS, USA; [2]Davis, CA, USA

Introduction

Equine pasture asthma is a seasonal remitting and recurring respiratory disease that is characterized by episodic dyspnea in horses grazing pasture during the summer in hot humid climates.[1,2] Inhaled particulates present in pasture during conditions of high heat and humidity are associated with inflammation, mucus hyper-secretion, and bronchoconstriction that obstruct the airways and impair ventilation, most severely in small caliber distal airways.[3,4] Seasonally recurring cough, wheeze, and dyspnea that are clinically reversible, as well as non-septic purulent inflammation are diagnostic features.[1-3] Equine pasture asthma is chronic and progressive, with increased disease severity during successive summers in horses that remain in the offending environment.[4] Available treatments are palliative and removing the horse from the offending environment is necessary.[5] This disease has been previously referred to as summer-pasture-associated pulmonary disease (SPAOPD) and summer pasture-associated recurrent airway obstruction (SPARAO). Equine pasture asthma shares clinical characteristics with severe equine asthma that is associated with stabling and inhalation of moldy hay dust (a.k.a, heaves), but differs in its specific association with pasture during warm months of the year. Horses with pasture asthma demonstrate marked improvement when they are effectively isolated from pasture aeroallergens and particulates.[5]

Epidemiology

Equine Pasture Asthma was first described in horses residing in Louisiana, and is commonly diagnosed in North American states with a similar subtropical climate, including Mississippi, Alabama, and Florida.[1,2,5] The exact prevalence is unknown; but has been estimated to be 5%.[6] The condition reliably demonstrates an adult onset (mean of 12±6 years).[2] No sex predilection was identified and Quarter Horse breeds (e.g., Quarter Horses, Paint Horses) and ponies were over-represented, perhaps due to over-representation in the senior population.[2] Pasture asthma has also been reported in horses grazing pastures in Scotland and Great Britain.[7,8]

Exact particulates responsible for eliciting pasture asthma are not identified, a temporal association with pasture-associated particulates is extrapolated because disease is clearly associated with housing on pasture and clinical improvement occurs when horses with pasture asthma are adequately isolated from inhalation of pasture particulates.[1,5] Mair reported that only 2 of 16 horses with equine pasture asthma improved when housed in stalls.[8] Similarly, we have identified advanced cases of pasture that necessitated aggressive isolation from pasture particulates in a climate controlled ventilated barn in order to elicit improvement. Hypoxic vasoconstriction should be ruled out in severe cases that are respond poorly to β_2-adrenergic agonist therapy when moved to stall (see note regarding severe hypoxemia).

Costa demonstrated that the probability of clinical exacerbation of equine pasture asthma in subtropical Louisiana was positively correlated to high temperature, humidity, dew point temperature, grass pollen and mold spores.[1-3] Molds and barn dust particulates are well delineated as relevant to exacerbations of Barn Dust Asthma, and moldy hay is reliably used to elicit clinical disease.[9] However, similar to findings by Costa, Bullone recently correlated severity of barn dust asthma to increased temperatures and pollen in a humid continental climate.[10]

Clinical Signs

Clinical exacerbations of pasture asthma are characterized by cough, increased expiratory effort, flared nostrils, tachypnea, and dyspnea that occur predictably during warm months of the year in horses maintained on pasture.[1-4] Early signs of exercise intolerance and occasional cough may be dismissed by owners or managed with anti-histamines and corticosteroids without definitive diagnosis.[2] With more severe respiratory impairment, horses may stand with head and neck extended, become dehydrated and anorexic, and lose weight. Mildly increased (<102.5°F) body temperature is not uncommon, but higher temperatures should lead to the consideration of other disease processes. In mild cases, thoracic auscultation reveals increased bronchovesicular sounds at rest, end-expiratory wheezes and crackles are evident during forced breathing (rebreathing bag or exercise). In severely affected horses, auscultation without a rebreathing bag reveals wheezes, generally expiratory (sometimes inspiratory or continuous crackles). Wheezes may be audible without a stethoscope. External abdominal oblique muscle hypertrophy ("heave line") occurs with disease progression. Disease progression can be expected if exposure to the inciting environment is not controlled. The authors have identified airway hyper-responsiveness to methacholine bronchoprovocation during seasonal disease exacerbation.[11] In a cohort of research horses with pasture asthma that were maintained for 1–5 years using environmental management, without corticosteroid administration, hyper-responsiveness to inhaled methacholine persisted during seasonal disease remission.

PATHOPHYSIOLOGY

Human asthma has been traditionally founded on three key characteristics: 1) episodic airway obstruction that is reversed by the administration of bronchodilatory agents, 2) airway constriction to agents that do not cause this effect in non-diseased individuals (a.k.a, airway hyperresponsiveness), and 3) chronic airway inflammation. Horses with pasture asthma similarly exhibit airway obstruction that can be reversed with β_2-adrenoceptor agonist and/or parasympatholytic therapy, excessive constriction responses to inhaled spasmogens of a magnitude noted in severe human asthma, and chronic airway inflammation that is predominately neutrophilic.[4,5,11]

Homology between defining human asthma criteria and characteristics of recurrent airway obstruction (RAO) and inflammatory airway disease (IAD) have led to advancement of the name 'equine asthma' as an umbrella term to include both RAO and IAD.[12-14] Within this terminology, severe equine asthma is applied to horses with lower airway inflammation and lower airway obstruction that is of a magnitude to increase respiratory effort at rest, presuming significant improvement with bronchodilator administration (i.e., RAO).[12,14] Mild/moderate asthma is reserved for horses, also with evidence of lower airway inflammation or increased mucus, that appear clinically normal at rest yet demonstrate chronic exercise intolerance or cough not attributable to other causes (i.e., IAD).[13] Some differences between the equine nomenclature and evolving nomenclature in human asthma are noteworthy.[15] Ventilation impairment observed during an isolated asthma episode is reflective of the magnitude of airway obstruction, and is scored along a continuum of mild to severe using measures of pulmonary function. Severe asthma in human medicine is a term that links to more longitudinal measures of asthma severity, and is now specifically distinguished from less severe disease based upon poor response to inhaled corticosteroids when used with appropriate compliance.

Central to the definition of asthma is an abnormal response to inhaled particulates that manifests as reversible and variable respiratory impairment and does not occur in healthy individuals exposed to identical challenge conditions.[16] This phenomenon of airway hyperresponsiveness (AHR) is significant in the diagnosis and management of human asthma, where longitudinal measures of asthma severity including risk of exacerbation, declining lung function, and a need for increased treatment to control symptoms, all correlate to AHR severity.[17-20] AHR is measured by inducing controlled bronchoconstriction with serial doses of well characterized airway spasmogens by nebulization, in order to achieve a threshold decrease in pulmonary function (bronchoprovocation).[21] Several agents are employed for bronchoprovocation and can elicit bronchoconstriction via direct effects on the airways (methacholine and histamine), or indirectly (antigen, mannitol, hypertonic saline) by inducing release and synthesis of contractile inflammatory mediators.[22] AHR can also be classified as specific, or non-specific, referring to whether the process of eliciting AHR is, or is not, antigen-specific.[23] Accordingly, the relationship between the specific spasmogen and outcomes of associated bronchoprovocation yield important pathophysiologic insight that have improved asthma recognition, selection of appropriate diagnostic testing modalities, and management strategies. Variability in responses to direct, indirect and antigen-induced bronchoconstriction agents have been documented in horses with barn dust asthma.[24-28] Employing protocols previously used for conventional mechanics testing (esophageal balloon) in non-compliant humans, we have identified non-specific AHR to methacholine in horses with pasture asthma.[11]

Decades of bronchoprovocation testing in human asthmatics have yielded an understanding that AHR is not stable. Measurement of AHR must be interpreted in the clinical context of the patient.[29-31] Bronchoprovocation has been demonstrated to be particularly useful in human asthmatics that report symptoms, yet have normal baseline lung function.[32] Congruent with this finding, we have identified, using methacholine bronchoprovocation, that AHR persists during seasonal disease remission in horses known to have pasture asthma. The diagnostic utility of bronchoprovocation has been similarly well documented in horses with mild/moderate asthma (previously IAD), that appear clinically normal at rest.[33,34] Corticosteroids as well as withdrawal of the inciting etiological agents can decrease the magnitude of AHR, particularly to indirect bronchoprovocation.[30,31] In horses with barn dust asthma, antigen withdrawal has been demonstrated to decrease measures of AHR to those of control horses within 4 days to 1 month.[25,36] Intercurrent infectious processes worsen AHR (measured using spasmogens that act indirectly) and it is likely that such infectious processes may not be detectable with conventional bacteriologic methods.[36,37] Viral infection is documented as an important factor that increases AHR and asthma exacerbation in both humans and horses with asthma.[38-40] AHR is therefore a diagnostic consideration in human asthma diagnosis, such that negative bronchoprovocation in the presence of active asthma symptoms in human medicine is an indication that diagnoses other than asthma should be considered.[15,29]

Congruent with this understanding, AHR persists in horses with pasture asthma during seasonal disease remission with exacerbations elicited by exposure to particulates that are inhaled while grazing pasture, especially during extremes of heat and humidity.[3] Increases in mold spores and grass pollen correlate to disease exacerbations and airway obstruction reflects bronchoconstriction, mucus hypersecretion, decreased mucociliary clearance, and airway inflammatory exudate.[3,4] In the southeastern U.S., affected horses demonstrate marked improvement when adequately isolated from pasture aeroallergens.[5] Persistence of signs in the face of moving horses to an indoor environment should prompt questions about the adequacy of isolation from the offending agents, is likely indicative of the magnitude of AHR, and is influenced by severity of airway structural pathology that mirrors human asthma and limits reversibility of airway obstruction and response to therapy.[4,5] Mildly affected horses may be identified in the summer months based upon complaints of exercise intolerance and coughing. Profound ventilation-perfusion inequalities and hypoxemia are common.[5]

It is clear that human asthma is much more complicated than a TH2-dominated inflammatory disease. Similarly, there is little evidence that equine pasture asthma is a Th2-driven phenomenon. TH17 cells are responsive to IL1, IL23 and IL-6, and producing IL-17A, IL-17F, IL-17AF, IL-21, IL-22, IL-26, GM-CSF, MIP-3α, and TNFα. Airway inflammation in pasture asthma is neutrophilic, not classically TH2-mediated eosinophilic inflammation. Seahorn also failed to identify a relationship between TH2-mediated antigen-specific IgE concentrations (in tracheal lavage fluids) and pasture asthma.[42] In fact, control horses had greater antigen-specific-IgE concentrations. Differential gene expression analysis in bronchoalveolar lavage (BAL) cells, BAL fluid, and peripheral blood mononuclear cells from horses with pasture asthma identified increases in IL-4 mRNA (a classical TH2 cytokine), in association with an increase in interferon-γ (IFN- γ, a classical TH1 cytokine), while IL-5 was not increased (TH2 cytokine).[43,44] Increased IL-13 mRNA was also demonstrated in BAL cells.[44] This 'mixed' cytokine response is also problematic for a TH17 regulated process at first glance, for which IL-4 is inhibitory. These conflicts have been attributed to differences clinical stage of disease. However, the collective cytokine findings point to likelihood that antigens capable of polarizing a TH17 response in a manner that includes IFN-γ production are relevant to pasture asthma, and are of sufficiently chronic exposure to render TH17 cells resilient to the regulatory effects of IL-4.[45,46]

Clearly, a body of knowledge from human asthma indicates that specific characteristics of AHR in equine disease reveals distinctive aspects of pathophysiology that will improve recognition, diagnostic testing modalities and management of these conditions.

Diagnosis

The primary differential diagnoses for pasture asthma are anhydrosis and infectious lower airway disease. Anhydrosis is typically diagnosed by inappropriate sweating, pyrexia, and tachypnea, as well as by confirming sweating in response to serial dilutions of epinephrine or terbutaline that are injected intradermally. Pharyngeal disease may cause chronic coughing and increased respiratory effort. Clinical examination reveals normal lung sounds and absence of typical increased end-expiratory effort.

Pasture asthma is considered a non-infectious inflammatory lower airway disease. Bacterial isolation from the lower respiratory tract of horses with pasture asthma, using standard bacteriological isolation techniques, is uncommon.[4] However, the absolute exclusion of infectious processes from the diagnosis of equine asthma requires further investigation. Horses with pasture asthma have decreased mucociliary clearance and opportunistic bacterial infections can occur that worsen clinical signs. Intracellular bacteria identified from tracheal aspirates or bronchoalveolar lavage, as well as positive bacteriologic cultures of tracheal aspirates, are generally considered evidence of colonization of the lower airway. Differentiation from bacterial pneumonia is aided by historical, and epidemiologic factors. Extremely high neutrophil percentages should be suspect. Viral agents are important precipitating factors for clinical asthma exacerbation in humans and horses.[38-40] Other agents that are not identified by conventional bacterial cultures are known precipitating factors in certain human asthma patients including *Mycoplasma* and *Chlamydophila* pneumonia, and dysbiosis of the lung microbiome.[36-40] Fungal sensitization has also been demonstrated to be particularly relevant to severe asthma in human patients.[47] The role of infectious agents in pasture asthma is an area of current investigation.

The CBC of horses with pasture asthma is generally unremarkable, but may demonstrate mild increases in WBC, segmented neutrophils, and fibrinogen. Cytologic examination of tracheal aspirates reliably yields mucopurulent inflammatory exudate with <90% nondegenerate neutrophils.[4] Endoscopy reveals mucopurulent exudate in the trachea that can be copious. Chronic cases will have thickening of the carina. Bronchoscopy may cause severe bronchospasm that can be minimized by the administration of butorphanol and intratracheal installation of small volumes of sterile lidocaine ahead of the advancing endoscope. Thoracic radiographs may fail to reveal abnormalities or increased bronchointerstitial pattern.

Diagnosis of pasture asthma relies on seasonal onset of dyspnea, cough, and wheeze in association with exposure to pasture in adult horses. Bronchoalveolar lavage is the diagnostic of choice, because it samples the distal airways. Horses with pasture asthma have increased nondegenerate neutrophils in the BAL fluid (>10–20%). Arterial blood gas sampling confirms hypercapnia ($PaCO_2$>40 mm Hg.) with hypoxemia (PaO_2 often <80 mm Hg) that can become profound (PaO_2 ~40 mm Hg). Horses with pasture asthma exhibit excessive airway responsiveness to methacholine bronchoprovocation (≤2 mg/ml). Lung biopsy carries an increased risk of rare but potentially fatal hemorrhage that is minimized by using a rapid fire biopsy instrument, for example a Bard® Monopty® while occluding the nostrils to temporarily halt respiratory movement and minimize tissue shearing. At this time, the diagnostic value of a peripheral lung biopsy in cases of equine pasture asthma does not clearly justify the associated risk. Lung biopsy should follow other diagnostic procedures.

Pathologic Findings

Horses euthanized during disease exacerbation have grossly overinflated lungs due to air trapping in the alveoli, not true emphasema.[4] Airway structural changes that are collectively termed airway remodeling in human asthma are identified in horses with pasture asthma, including increased airway smooth muscle, goblet cell hyperplasia/metaplasia, mild peribronchiolar fibrosis, and increases in fibers of the peribronchiolar elastic network.[4,41]

Treatment

Treatment is directed at restoring ventilation. Recommendations for therapeutic management are extrapolated from investigations in the horses with barn dust asthma as well as human asthma. It is important to recognize that horses with Pasture Asthma develop hypoxemia that can be quite profound (PaO_2<60 mm Hg). Inhaled oxygen supplementation is recommended, particularly when blood gas monitoring is not available. Oral β_2-adrenergic agonist bronchodilators including clenbuterol, albuterol, and terbutaline sulfate have low efficacy, the latter because of its poor bioavailability.

β_2-Adrenergic Receptor Agonists (β_2-Agonists)

Horses with significant respiratory impairment should receive inhaled selective β_2-adrenergic receptor agonist (β_2-agonists) for their bronchodilatory effect. During asthma exacerbation, nebulized β_2-agonists provide optimal penetration to the distal airways that are responsible for the majority of airway obstruction. Mirroring standards of practice in emergency human asthma management, the authors routinely deliver levalbuterol (1.25 mg in a total volume of 2 ml of sterile saline) by nebulization either in a fitted equine mask, or at the nostril to horses experiencing exacerbations of pasture asthma, every 4 hours, until signs abate. Nebulization is rapid and effective when delivered using the PARI® Vios PRONEB Nebulizer Compressor System paired with the LC Sprint Pari-UltraNeb, achieving 3.5 µM droplets and a delivery time for a 2 ml volume of under 6 minutes.[a] Other options for nebulization delivery include the Flexineb® (Haygain; Lambourn, Berkshire, UK). The short-acting β_2-agonist albuterol, is available as pressurized metered-dose-inhalers (pMDI) (1–2 µg/kg) and is commonly used, also at 4-hour intervals when treating severe disease exacerbation. Appropriate delivery devices such as the Aerohippus® (Trudell Medical) and Equinehaler® must be used with pMDI to assure adequate delivery to the lung. Long-term use of β_2-adrenoceptor agonists may result in tachyphylaxis.[48] In cases of severe hypoxemia, bronchodilatory agents can exacerbate ventilation-perfusion inequalities, temporarily worsening hypoxemia, supporting the recommendation for oxygen therapy in severe cases. In addition to their bronchodilatory effect, β_2-agonists have mucokinetic properties including decreased viscosity of the respiratory secretions, and increases in mucociliary beating that are beneficial to resolving airway obstruction.

Corticosteroids

Corticosteroid administration along with strict environmental control should be a mainstay of therapy for horses with is pasture asthma. Corticosteroid therapy is critical to controlling airway inflammation and remodeling that are significant factors in the perpetuation of airway hyper-responsiveness.[49,50] In human asthma management, regular administration of inhaled corticosteroids are a mainstay of therapy that maximizes delivery to the respiratory tract while minimizing systemic side effects. In horses moved from pasture to an stall environment, fluticasone propionate (2,000 µg, q12h, decreasing to 1,000 µg, q48h; Flovent HFA; GlaxoSmithKline), and beclomethasone dipropionate (240 µg inhaled q12h, decreasing to 120 µg q48h; QVAR HFA; IVAX LLC TEVA) can be recommended.[50,51] However, corticosteroid inhalers are quite expensive (Flovent: $400–500 per inhaler, $33–50/2,000 µg; QVAR: $200/inhaler, <$10 per dose). Oral and parenteral corticosteroids are significantly less expensive, and arguably more effective. Dexamethasone powder (0.05 mg/kg q24h) as well as prednisolone (2 mg/kg) have both been demonstrated to improve pulmonary function following 3 and 7 day of treatment, respectively.[52] Dexamethasone provides superior clinical efficacy.[52] The response to inhaled corticosteroids takes longer than that to systemic steroids, and combination of inhaled bronchodilators and corticosteroids may be warranted.

Parasympatholytic (Anticholinergic)

Parenteral administration of the parasympatholytic (anticholinergic) bronchodilator N-butylhyoscinium bromide (Buscopan®, 0.3 mg/kg) provides an alternative to administration of atropine and glycopyrrolate, which are both of concern for their adverse effects on gastrointestinal motility. The onset of bronchilation with N-butylhyoscinium is approximately 10 minutes following injection, and wanes by 1 hour post treatment.[53]

The inhaled muscarinic receptor antagonist bronchodilator, ipratropium bromide pMDI (2–3 µg/kg), is effective in horses. Despite minimal systemic side effects, it potently increases the viscosity of respiratory secretions and is therefore not used by the authors.

Magnesium Sulfate

Noteworthy is the phenomenon of hypoxic vasoconstriction, a normal physiologic response to bronchoconstriction that minimizes ventilation to perfusion mismatching in regions of bronchoconstriction. This can lead to pulmonary hypertension and right heart failure. Thoracic auscultation reveals lung fields that are inappropriately quiet. Prolonged hypoxic vasoconstriction leads to a refractory response to bronchodilator treatment which could account for particularly difficult cases of pasture asthma that fail to respond to initial bronchodilatory therapy. Such horses necessitate oxygen therapy and may benefit from intravenous magnesium sulfate therapy (40 mg/kg given over a 20 minute period, this is the equivalent of a 500 ml bag - 40 mg/ml - for a 450 kg horse). IV magnesium sulfate has been demonstrated to improve outcomes in severe human asthma.[54] The bronchodilatory effect of magnesium sulfate is limited to the duration of the infusion. Concurrent doses have not been evaluated and have the potential for toxicity. **Clinicians note**: Failure to respond to bronchodilator therapy, a diagnostic criteria for equine asthma, should be judged cautiously in horses with clinically severe pasture asthma until adjunctive therapies including oxygen and magnesium have been used to augment therapy.

Environmental Management

Regardless of disease severity, removing horses with pasture asthma from the inciting pasture to an indoor environment is very important to successful management and longevity of affected horses. In horses with confirmed pasture asthma, this can be initiated prior to the onset of clinical disease during warm months. In the authors' experience, most clients find this aspect of disease management difficult and impractical because it is completely contrary to equine husbandry to maintain horses without pasture access. While corticosteroids suppress inflammation and clinical disease, preventing exposure to the inciting environmental triggers is also warranted to limit inflammation and progression of airway pathology that leads to irreversible airflow obstruction.[49,50,55] In some cases, the response to appropriate environmental control is striking and no drug therapy is required. Cases that fail to improve or improve slowly despite strict environmental management warrant therapeutic intervention. Efforts to limit dust are quite beneficial, including the use of stalls with rubber mats (without bedding), cardboard or pelleted bedding.

Many horses can be successfully managed on good quality (low dust) wood shavings. The clinical picture of some horses with pasture asthma, particularly those with more long-standing disease, indicates that these horses highly responsive to pasture-associated particulates. Since complete isolation from pasture particulates is difficult, these cases may have inadequate resolution of signs when moved to a barn environment. However, in our experience, even advanced cases of pasture asthma become asymptomatic when maintained in a climate-controlled barn for a period of days to weeks. Though it may be possible to maintain less severely affected animals on pasture, the clear association between increasing pasture exposure and disease progression must be recognized. Clients can be taught to monitor clinical scores of respiratory effort, which correlate to ventilation impairment, removing affected horses from pasture to prevent progression to an 'asthma attack'.[3] However, respiratory signs usually recur within days of re-exposure to pasture during warm months of the year.[3] The pasture should be cut very short (only a few inches high), and the horse offered a complete diet in an effort to decrease grazing and inhalation of particulates. Attention to minimizing particulates during cooler months of clinical remission may be warranted by keeping pasture relatively short.

The diet should be consistent throughout the year, consisting of pelleted feed and cubed forage. Round hay bales should not be offered to affected horses, even during clinical remission. Hay should be fully submersed in water (not sprayed) immediately prior to feeding. This is easily accomplished by submersing a full hay net in a clean muck bucket filled with water.

Prognosis

Acute disease is clinically reversible with adequate environmental control coupled with medical therapy. The onset of clinical exacerbation is fairly predictable within the same 2-week period from year to year, indicating that environmental control and prevention are critical.[3] Inadequate environmental control is associated with worsening of the condition (increase severity of signs and shorten length of clinical remission) from year to year. Due to difficulties in isolating horses from pasture associated particulates and allergens, pasture asthma should be considered chronic and progressive. Treatments are palliative and their duration of effect is shortened when environmental management is inadequate. Though affected horses can typically be ridden for many years during seasonal disease remission, most riding is done during the season of disease exacerbation rendering affected horses unsuitable for athletic performance. Pasture asthma is not uncommon in the southeastern states. The guarded to grave prognosis for athletic performance and longevity dictates that veterinarians examining horses for purchase during the seasons of disease remission should be vigilant for historical evidence of warm season respiratory disease in horses greater than 6 years of age. While AHR that characterizes pasture asthma suggests the possibility of risk for barn dust asthma, it is important to realize that horses that fail to improve when moved to a barn environment may not have adequate isolation from the offending particulates/aeroallergens. Regional differences in improvement when moved to a barn environment have been reported and could also reflect regional differences in barn dust particulates.

REFERENCES

1. Seahorn TL, Beadle RE. *J Am Vet Med Assoc*. 1993 Mar 1;202(5):779.
2. Seahorn TL. *J Am Vet Med Assoc*. 1996 Jan 15;208(2):248.
3. Costa LR *et al. Am J Vet Res*. 2006;67(9):1635–1642.
4. Costa LR *et al. Am J Vet Res*. 2000;61(2):167.
5. Costa LR *et al.* Equine pasture asthma. In: Lavoie JP, Kinchcliff K, eds. *Five-Minute Veterinary Consult*. Wiley. (In press).
6. Beadle RE. In: Robinson NE, ed. *Current Therapy in Equine Medicine*. Philadelphia, PA: WB Saunders Co. 1983;512–516.
7. Dixon PM. *Vet Rec*. 1990;126:9.
8. Mair TS. *Equine Vet J*. 1995;27:402.
9. Leclere M. *Respirology*. 2011;16(7):1027.
10. Bullone M *et al. Equine Vet J*. 2016;48(4):479.
11. Hunter C *et al. Am J Respir Crit Care Med*. 2013;187:A6080.
12. Lavoie JP. *Equine Vet Educ*. 2015;27(5):2256.
13. Couëtil LL *et al. J Vet Intern Med*. 2016;30(2):503–515.
14. Pirie RS *et al. Equine Vet J*. 2016;48(4):403.
15. Global Initiative for Asthma. Global Strategy for Asthma Management and Prevention (2017 Update). www.ginasthma.org.
16. O'Byrne PM, Inman MD. *Chest*. 2003;123(3 Suppl):411S.
17. Brutsche MH *et al. Thorax*. 2006;61(8):671.
18. Parker DR *et al. Am Rev Respir Dis*. 1990;141(3):589.
19. Sears MR *et al. N Engl J Med*. 2003; 349(15):1414.
20. Leuppi JD *et al. Am J Respir Crit Care Med*. 2001;163(2):406.
21. Crapo RO *et al. Am J Respir Crit Care Med*. 2000 Jan;161(1):309.
22. Leuppi JD. *Curr Opin Pulm Med*. 2014 Jan;20(1):31.
23. Cockcroft DW *et al. Clin Allergy*. 1977;7(6):503.
24. Armstrong PJ *et al. Am Rev Respir Dis*. 1986;133(3):357.
25. Derksen FJ *et al. J Applied Physiol*. 1985;58(2):598.
26. Murphy JR *et al. Vet Immunol Immunopathol*. 1979;1(1):89.
27. McPherson EA *et al. Equine Vet J*. 1979;11(3):159.
28. Beeler-Marfisi J *et al. Am J Vet Res*. 2010;71(6):682.
29. Chapman DG *et al. Clin Exp Allergy*. 2015 Apr;45(4):706–719.
30. Boulet LP *et al. J Allergy Clin Immunol*. 1983;71(4):399.
31. Sumino K *et al. Ann Allergy Asthma Immunol*. 2014;112(4):354.
32. Sont JK *et al. Am J Respir Crit Care Med*. 1999;159(4 Pt 1):1043.
33. Wichtel M *et al. Equine Vet J*. 2016 Jul;48(4):466.
34. Hoffman AM *et al. Am J Vet Res*. 1998 Feb;59(2):176.
35. Thompson JR *et al. Equine Vet J*. 1984;16(1):35–38.
36. Rollins DR *et al. J Allergy Clin Immunol Pract*. 2014 Sep–Oct;2(5):511.
37. Sandrock CE, Norris A. *Clin Rev Allergy Immunol*. 2015 Feb;48(1):104.
38. Sterk PJ. *Eur Respir J*. 1993;6(6):894.
39. Matera MG *et al. Pulm Pharmacol Ther*. 2008;21(4):624.
40. Houtsma A *et al. Multidiscip Respir Med*. 2015 Nov 3;10:33. eCollection.
41. Ferrari CF *et al. Am J Respir Crit Care Med*. 191;2015:A3539.

42. Seahorn TL. *Am J Vet Res*. 1997 Dec;58(12):1408–1411.
43. Beadle RE. *Equine Vet J*. 2002 Jul;34(4):389.
44. Horohov DW. *Vet Immunol Immunopathol*. 2005 Oct 18;108(1–2):237.
45. Zielinski CE. *Nature*. 2012 Apr 26;484(7395):514.
46. Cooney LA. *J Immunol*. 2011 Nov 1;187(9):4440.
47. Agarwal R, Gupta D. *Med Mycol*. 2011 Apr;49 Suppl 1:S150.
48. Read JR *et al*. *Am J Vet Res*. 2012 Jan;73(1):140.
49. Leclere M *et al*. *Am J Respir Cell Mol Biol*. 2012 Nov;47(5):589.
50. Couëtil LL *et al*. *Am J Vet Res*. 2005 Oct;66(10):1665.
51. Couteil LL *et al*. *J Vet Intern Med*. 2006 Mar–Apr;20(2):399.
52. Leclere M *et al*. *Equine Vet J*. 2010 May;42(4):316.
53. de Lagarde M *et al*. *Equine Vet J*. 2014 Jul;46(4):474.
54. Blitz M *et al*. *Cochrane Database Syst Rev*. 2005 Oct 19;(4):CD003898.
55. Johnson JR *et al*. *Am J Physiol Lung Cell Mol Physiol*. 2008 Nov;295(5):L780.
56. Product literature. Pari Respiratory Equipment. Midlothian, VA. www.pari.com.

1.
acvim202

Wearable Devices for Gait Analysis in Ataxic Horses

Yvette S. Nout-Lomas, DVM, PhD, DACVIM, DACVECC
Fort Collins, CO, USA

Introduction

In equine neurology, evaluation of gait is necessary for diagnostic and prognostic purposes and for determination of efficacy of therapeutic regimens. Currently, the severity of gait deficits assessed during a neurological examination is graded using a modified ordinal grading scale (modified Mayhew scale) with grades 0–5.[1] It has been shown, however, that there is poor agreement, even between skilled observers of equine gait abnormalities when performing a neurological examination, especially when clinical signs are subtle.[2] This has been shown to be the case not only for evaluation of ataxic horses, but also for evaluation of horses with lameness.[3,4] In addition, bias is likely to affect outcomes of subjective gait assessment.[5]

Although watching the horse in motion is standard practice for both neurological and lameness examinations, it has thus become evident that subjective scoring systems do not suffice for reliable and accurate assessment of horses with gait deficits, in particular when deficits are mild. The field of quantitative gait analysis is rapidly developing, and numerous studies have now been performed to evaluate the use of movement tracking devices to supplement subjective gait analysis in lame horses[6,7] and to characterize gait patterns in normal horses with the goals of improving training programs and methods for judging and selection of horses.[8,9] There have only been few studies performed to assess use of these techniques for evaluation of ataxic horses.[10]

Wearable Devices

Where traditional equine gait analysis requires large laboratories to house equipment and to bring horses in for instrumentation and measurements, wearable devices are designed to be directly attached to the horse and work under field conditions. Gait analysis is typically divided into kinematic and kinetic studies, and although there are some products available for kinetic studies, most wearable devices are designed to be used for kinematic studies, which we will here focus on. Such systems use multiple microelectromechanical components (e.g., accelerometers, gyroscopes, pedometers, global positioning system, and inertial measurement units) and have wireless and telemetric components for data transmission.

Development of these devices for use by people has been mainly for health monitoring purposes, whereas developers of wearable devices for equine use are mainly interested in providing objective, quantitative data for fitness training purposes and for early detection of lameness. The main advantage of using wearable devices is that one is no longer reliant on hardwired technology but can take the horse away from laboratory constraints and evaluate gait in the field.

Movement Tracking Options

- Pedometer: This system either has a mechanical or digital operating system. The mechanical system consists of a pendulum that swings back and forth with each step; a switch then moves up and down, a dial turns, and the dial advances. Interestingly, Leonardo Da Vinci had early sketches of this device! The digital system uses current whereby the pendulum completes a circuit when swung, and allows for current to flow and steps to be added to the count.
- Global positioning system (GPS): These units calculate the two-dimensional position of an object or person using three satellites to triangulate the location based on latitude and longitude. When a fourth satellite is added, this technology allows for tracking altitude, thus giving three-dimensional positioning. These units use radio signals that can be received by any GPS device. Signaling can occur passively or actively, which means units can receive signals or send out a location.
- Accelerometer: These are electromechanical devices that measure acceleration forces, which can be static (such as with gravity) or dynamic (such as with movement). Accelerometers can provide

measures of tilt and acceleration and work through use of the piezoelectric effect (stressed microcrystals generate voltage) or through change in capacitance. In the latter, built-in circuitry converts capacitance to voltage. Accelerometers are available in analog or digital versions and can provide two-dimensional or three-dimensional data (along three axes). Maximum swing (i.e., ±2 g, 5 g, 16 g) can be selected based on device chip, and bandwidth (i.e., 50 Hz, 100 Hz, 200 Hz) can be selected based on chip and desired use or accuracy of movement desired. Accelerometer output is affected by gravity forces, and units provided are typically m/s^2; however, they can also be provided in "g," which is a measure of acceleration relative to gravity at Earth's surface (9.81 m/s^2).

- Gyroscope: This is a device that measures rotation about an axis. Units are provided in degrees/second or radians/second (velocity). Gyroscopes allow for measurement of angular or rotational movement (i.e., circumduction) and not just linear two-dimensional or three-dimensional movement.
- Inertial measurement unit (IMU): IMUs are units that contain a combination of previously described devices or they contain a set of chips that combine multiple movement measurements, usually an accelerometer and a gyroscope. A 6-degree of freedom IMU is an IMU that combines a 3-axis or 3 degrees of freedom accelerometer with a gyroscope (3-axis). IMUs frequently also include a magnetometer which when used in 3-axis chips will allow the IMU to go up to a 9-axis system. An IMU with 9 degrees of freedom is capable of measuring 9 independent readings, which correspond to 3 from the accelerometer which give linear acceleration along the 3 principal axes, 3 from the gyroscope which give the orientation, and 3 from the magnetometer which give the strength of the magnetic field. The onboard processor in an IMU is equipped with sensor fusion algorithms which take the readings of the surrounding magnetic field and correct the readings in acceleration and gyroscope, providing readings of higher accuracy. Importantly, IMUs may be sensitive to other external magnetic forces, and they are not designed to track position.

Examples of Wearable Devices

Devices that contain movement tracking options described above are commercially available for consumer use, and examples are listed below. These commercially available devices contain data analysis software and algorithms and may be paired with applications that can be used from computers, smartphones, or smartwatches for easy consumer use. The devices vary in what they can measure and how the data that is measured is used to assess gait. It is possible to place individual units or chips into a casing and use these devices to collect raw data from gait or other movements in humans and animals. These devices are more commonly used for research purposes, because raw data is often difficult to retrieve from commercial units. Aspects that should be taken into consideration when choosing a device should also include field conditions like water, time between motion and when human interaction with device is needed, distance traveled by animal, body shape of animal, movements expected, dirt and debris expected, and other conditions.

The file size generated from data can also constrain the amount of and type of data collected. Examples of wearable devices are:

- Fitness bands: These include units such as Fitbit® or Jawbone® that are attached on the wrist, waist, or chest and measure steps and some also measure heart rate. These units are limited in movement-measuring options since they typically contain only a pedometer; however, advanced models may also include GPS or accelerometer. Advantages of these units include that they last a long time, data can be stored in the devices and easily retrieved using standard connections.
- Smartphones: Nearly all modern smartphones contain IMU chip sets, a GPS, and a pedometer. These are typically expensive units and limited in options for data retrieval. Smartphones have been used for several studies in gait, but the devices are large and are either difficult to attach or may restrict natural movement.
- Smartwatches: These units typically contain an accelerometer, GPS, pedometer, and sometimes a gyroscope. A significant limitation to use of these devices is the access to the data. One needs to use an application with developer-level options to gain access to raw data files.

- IMU specific devices: These are available for many purposes; for example, to measure motion of cars, people, phones, computers, and other gadgets. The chips used in most devices are very small and can be placed into many types of casings including flexible material for clothing. Extreme public examples include Google Glass and Oculus. A selection of wearables developed for the equestrian market is shown in Table 1.

Table 1. Selection of commercially available wearable devices for use in horses

System	Country	Advertised use	Parameters	Sensors and placements	Purpose and users
GaitSmart Pegasus ETB-Pegasus Limb Phasing System	UK	Monitors the degree and timing of cannon bone swing for each leg within a stride; assists in identification of more effective movements	Stride duration; stride length; speed; gait; GPS position; temporal phasing of all limbs (% of stride); measures hock and cannon angles	Four sensors on distal limb; one sensor on rider or saddle pad	Training; gait symmetry For use by owners, trainers, riders, veterinarians, and researchers
E-Trakka	Australia	Monitor horse fitness through physiological readings	Speed; position; stride length; heart rate	Saddle pad with GPS and heart rate monitor	Fitness; gait For use by owners, trainers
EquiGait	UK	Gait asymmetry; back movement; rider - horse interaction	Acceleration; angular rotation; steps taken	1–8 sensors (poll, withers, tuber coxae, tuber sacrale); smartphone	Gait symmetry, riding pattern For use by owners, trainers, riders, veterinarians, and researchers
EquiMetrix	France	Training, lameness	Symmetry; cadence	One sensor girth; accelerometer	Training; lameness; selection For use by owners, trainers, riders, veterinarians
EquiMoves	The Netherlands	Monitor gait characteristics	Stride duration; speed; stride length	Eight IMUs (four on distal limbs; one at poll, sternum, withers, sacrum)	Gait symmetry For use by veterinarians and researchers
Equinosis - Lameness Locator	USA	Assess for lameness	Symmetry provided in indices; stride rate; stride number	Three sensors (two uniaxial accelerometers, one on head, one on sacrum; one gyroscope on distal limb)	Lameness For use by veterinarians and researchers
Equisense	France	Monitor work intensity and gait Early detection of gait anomalies	Symmetry; time of each lead/gait; elevation; cadence; number jumps; speed; line; work intensity	One sensor on girth; application on phone	Training; gait symmetry For use by owners, trainers, veterinarians
SeeHorse	Canada	Scan and display real-time data for	Vital signs (TPR); movement	One sensor anywhere	Long-term health monitoring

		horse's vital signs			For use by owners, trainers, riders, breeders, veterinarians, and researchers
Tekscan Hoof System	USA	Provides pressure data for analysis, interpretation, and treatment of large animal conditions	Pressure data; contact area; average and peak pressure; center of pressure	Hoof pressure system; one on each leg	Isolate and analyze regions of the hoof for gait studies and imbalance measurements For use by veterinarians and researchers

Gait Analysis in Equine Lameness and Ataxia

Traditional kinetic and optical kinematic analyses have been shown to be effective for detection of mild lameness in horses;[11-13] however, because of the shortcomings of this equipment, wearable devices have been designed and evaluated to objectively characterize lameness in both research and clinical practice settings. Several of these systems are sensitive enough for detection of mild single forelimb or hindlimb lameness at a trot.[6,14] Current techniques for lameness evaluation use sensors placed on head, trunk, and sacrum and take advantage of the two most common movement characteristics of a lame horse, which are the head nod and hip hike. Using these characteristics, there have been many studies evaluating different wearable systems and comparing those to optical systems; however, recent studies indicate that including kinematics of the distal limb is likely to improve accuracy and precision of gait analysis.[15-17]

There have only been few investigations that evaluated gait in clinically ataxic horses using traditional kinematic[18,19] and kinetic[20] studies, and fewer using accelerometers in clinically ataxic horses.[21] Up until now, studies into quantitative gait analysis of ataxic horses have used accelerometers in a sedative-induced ataxia model.[22-24] Although these horses display ataxia, it has been shown that movements are different from those seen in clinical ataxia.[10,19]

While objective assessment of gait has a place in lameness diagnostics and has the potential to improve clinicians' subjective evaluation of gait, use of these systems carries limitations, which were recently highlighted.[25] Such limitations include factors such as what are considered normal horses for generation of normal IMU data and the fact that it remains important for clinicians to develop an eye for the detection of unsoundness and compensations that horses adopt in the face thereof.

References

1. Mayhew IG, *et al. Cornell Vet.* 1978;68, Suppl.6:1–207.
2. Olsen E, *et al. J Vet Intern Med.* 2014;28:630–638.
3. Hewetson M, *et al. Vet Rec.* 2006;158:852–858.
4. Keegan K, *et al. Equine Vet J.* 2010;42:92–97.
5. Arkell M, *et al. Vet Rec.* 2006;159:346–349.
6. Keegan K, *et al. Am J Vet Res.* 2011;72:1156–1163.
7. Keegan K, *et al. Am J Vet Res.* 2013;74(1):17–24.
8. Barrey E, *et al. Equine Vet J.* 1994;Suppl. 17:7–12.
9. Back W, *et al. Vet Q.* 1994;16:Suppl. 2:S91–96.
10. Nout-Lomas YS, *et al. Equine Vet J.* 2016;Epub ahead of print.
11. Ishihara A, *et al. Am J Vet Res.* 2005;66:1805–1815.
12. Keegan KG, *et al. Am J Vet Res.* 2000;61:612–619.
13. Moorman VJ, *et al. Am J Vet Res.* 2013;74:1183–1191.
14. Pfau T, *et al. Equine Vet J.* 2007;39:407–413.
15. Moorman VJ, *et al. Am J Vet Res.* 2012;73:1160–1170.
16. Olsen E, *et al. J Biomech.* 2013;46:2320–2325.
17. Bragança FM, *et al. Equine Vet J.* 2016;Epub ahead of print.

18. Keegan KG, *et al. Equine Vet J*. 2004;36(8):712–717.
19. Strobach A, *et al. Equine Vet J*. 2006;Suppl. 36:423–426.
20. Ishihara A, *et al. J Am Vet Med Assoc*. 2009;234:644–651.
21. Nout-Lomas, *et al*. ACVIM. 2016;abstract.
22. López-Sanromán FJ, *et al. Vet J*. 2012;193:212–216.
23. López-Sanromán FJ, *et al. Am J Vet Res*. 2013;74:828–834.
24. López-Sanromán FJ, *et al. Res Vet Sci*. 2014;97:391–396.
25. Dyson S. *Vet J*. 2014;201:245–248.

acvim205

The Use of Angiotensin-Converting Enzyme Inhibitors for the Management of Heart Failure in Horses

Tiago Afonso, DVM, PhD, DACVIM (Large Animal Internal Medicine)
Adelaide, South Australia

Introduction

Congestive heart failure (CHF) occurs when left, right, or biventricular dysfunction leads to venous congestion. The term is also appropriate when a horse presents decreased cardiac output due to ventricular failure without venous congestion. This leads to the activation of compensatory mechanisms such as neurohormonal and cytokine systems that maintain cardiovascular homeostasis.

The Renin-Angiotensin-Aldosterone System

An important compensatory adaptation is the activation of the renin-angiotensin-aldosterone system (RAAS). The release of renin from the juxtaglomerular cells initiate this pathway when stimulated by the decreased renal perfusion, lower concentration of sodium presented to the macula densa in the distal tubule, and increased sympathetic drive to the kidneys. Renin interacts with angiotensinogen to form angiotensin-I (ANG-I), which is then cleaved by angiotensin-converting enzyme (ACE) to produce angiotensin-II (ANG-II). The latter is a potent vasoconstrictor and induces the adrenal cortex to secrete aldosterone, ultimately leading to the retention of sodium and water by the kidneys. This is a critical short-term compensatory mechanism; however, the prolonged action of ANG-II and aldosterone results in the progression of the cardiac disease, hypertrophy and fibrosis of vessels and cardiac muscle, as well as endothelial cell dysfunction, baroreceptor dysfunction, and inhibition of norepinephrine uptake.

Heart Failure in Horses

Conditions such as valvular insufficiency, pericarditis, congenital defects, aorto-cardiac fistulas, and myocardial disease have been reported to cause CHF in horses. Among these, mitral valve insufficiency (MI) is the most common underlying disease leading to left-sided volume overload and subsequent heart failure.[1-3] Albeit aortic valve insufficiency (AI) is more common than MI in the horse, it is less likely to lead to CHF. The exact proportion of horses with AI or MI that eventually develop CHF is unknown. Both AI and MI may lead to cardiac arrhythmias, which can be associated with a worse prognosis.[3]

Medical Management of Heart Failure

The ultimate goals of treatment are to improve cardiac output and decrease venous congestion. Diuretics (furosemide and spironolactone) and digoxin have long been used in horses. However, digoxin therapy is now controversial in the management of CHF, and mainly used by some to manage supraventricular tachycardia.

Vasodilators are critical in the control of CHF. Due to the neurohormonal activation, arterial and venous vasoconstriction causes an excessive increase in preload and afterload, leading to an increased demand from an already failing heart. Venodilation is therefore important to decrease ventricular filling pressures, wall stress, and preload. Arterial vasodilatation will increase forward stroke volume through decreased afterload. This has long been addressed by the use of ACE inhibitors, which represent a major advance in the medical management of cardiovascular diseases in people and dogs. In dogs with CHF due to myxomatous mitral valve disease or dilated cardiomyopathy, ACE inhibitors have beneficial hemodynamic and clinical effects, improving quality of life and extending survival times by 2- to 3-fold.[4] In the equine patient, ACE inhibitors have been widely recommended for the treatment of severe valvular diseases and CHF despite the paucity of available data regarding their efficacy. Nevertheless, studies have shown the activation of both, sympathetic nervous system and RAAS in horses with heart failure.[5,6]

ANGIOTENSIN-CONVERTING ENZYME INHIBITORS

Angiotensin-converting enzyme inhibitors prevent the conversion of ANG-I, a relatively inactive peptide, into ANG-II. Additionally, ACE is identical to kininase II which breaks down bradykinin. Therefore, blocking the action of ACE increases the concentration of bradykinin, further contributing to vasodilation.

The ACE inhibitor **enalapril** has been widely used in horses. However, recent studies have demonstrated that it has very low oral bioavailability and does not inhibit ACE in horses when administered via this route.[7-9] Administration of **ramipril** at 0.2 mg/kg to healthy horse's results in a significant decrease in blood pressure.[10]

In horses with MI, daily oral administration of **quinapril** at a dose of 120 mg increased stroke volume and cardiac output, and decreased mitral regurgitation velocity time integral.[11] However, the study lacked an untreated control group; therefore, it is difficult to differentiate a true effect of quinapril from that of changes in hemodynamics between time points. A more recent study on **quinapril** revealed a significant inhibition of plasma ACE activity with single doses of 120 mg IV, 120 mg PO or 240 mg PO in healthy horses. Mean (±SD) maximal ACE inhibition was 72.51±7.17%, 47.11±10.21% and 53.43±12.98%, respectively. Furthermore, no statistically significant increases in plasma renin were noted.[12] before publication, this study had been presented as a research abstract and was followed by another, from a different group, comparing the ability of different ACE inhibitors to reduce circulating ACE activity in healthy horses. A considerable variation in the relative potency of orally administered **benazepril**, **perindopril**, **quinapril**, and **ramipril** was documented. Of the 4 drugs tested, benazepril at a dose of 0.5 mg/kg was the most effective at inhibiting serum ACE activity, achieving a mean (±SD) maximal percentage of inhibition of 86.9±7%. Moreover, feeding did not affect the pharmacodynamics of benazepril negatively. Finally, daily administration of benazepril for 7 days did not result in a cumulative effect of the drug on serum ACE inhibition.[13] Based on these results, benazepril was selected for subsequent studies.

Given that circulating ACE comprises only 5–30% of the total body ACE pool, measurement of serum ACE activity is not the most accurate method to determine the lowest maximally effective dose of an ACE inhibitor in healthy, normotensive individuals. In normotensive people, measurement of the percentage attenuation of the pressure response to a dose of exogenous ANG-I that increases systolic blood pressure by 20–30 mm Hg is considered the best method for assessing the predicted efficacy of an ACE inhibitor. In these studies, doses resulting in >75% attenuation of the pressor response were considered likely to be clinically efficacious.[14] This idea lead to the investigation of the lowest dose of **benazepril** that results in optimal attenuation of the arterial blood pressure response to exogenous ANG-I administration. Administration of benazepril at a dose of 0.5 mg/kg resulted in a significantly lower mean (±SD) maximal attenuation (I_{max}) (19±14%) in response to ANG-I than dosages of 1 mg/kg (47±9%), 2 mg/kg (43±11%), and 4 mg/kg (56±17%). The target of ≥75% inhibition was not achieved despite increasing dosages. This is unlikely to be simply the result of lower oral bioavailability in horses compared to other species because maximal serum ACE inhibition was greater than 90% at a benazepril dose of 1 mg/kg and nearly 100% with benazepril doses of 2 mg/kg and 4 mg/kg. Enalapril, benazepril and lisinopril have been shown to be clinically effective at the lowest labeled dose in people with CHF despite I_{max} of only 45% to 60%.[14] Therefore, the maximal attenuation of 47±9% observed at a dose of 1 mg/kg in this study might be sufficient for clinical efficacy in horses with CHF. Overall, this study documented a modest attenuation of the rise in systolic blood pressure in response to ANG-I after administration of benazepril, despite adequate serum ACE inhibition. Based on these results, a dose of 1.0 mg/kg orally every 12 h were recommended for future investigations of benazepril for the management of cardiovascular diseases in horses.[15] Such recommendations lead to a preliminary randomized placebo-controlled study (unpublished data by the author). The objective was to assess the cardiovascular and hormonal effects of **benazepril** (1 mg/kg, orally, every 12 h) in 11 horses with left-sided valvular insufficiency and mild to moderate volume overload. This was the first randomized placebo-controlled study assessing the efficacy ACE inhibitors in horses. Despite considerable serum ACE inhibition in horses receiving benazepril, significant differences were not found between groups in any of the other RAAS constituents. These results must be interpreted with caution due to the small sample size and resulting low statistical power. Nevertheless, horses treated with benazepril had a significantly higher reduction in the left ventricular internal diameter in systole and severity of the aortic regurgitant jet, and a significantly greater increase in cardiac output and fractional shortening compared to horses receiving a placebo. These differences suggest a decrease in afterload, which is a major determinant of myocardial oxygen consumption. Additional studies with a much larger sample size will be necessary to determine if administration of benazepril delays progression of valvular regurgitation and prolongs survival times relative to horses receiving a placebo.

A different group of researchers, from the University of Córdoba, Spain, evaluated the effects of oral administration of **ramipril** in 6 healthy horses at incremental dosages. The highest dose tested, 0.8 mg/kg, resulted in the highest mean (±SD) circulating ACE inhibition of 84.27±4%.[16] Comparison of ACE inhibitions between studies must be done carefully. This should only be done when the assay used to measure ACE inhibition is the same,[17] which is true for this study and the previous studies on benazepril.[13,15] Such fact suggests that higher doses of benazepril (1 to 4 mg/kg) are likely to result in higher ACE inhibition than that achieved with ramipril at 0.8 mg/kg.

Finally, the same group evaluated the ability of 4 different orally administered ACE inhibitors in attenuating the increase in blood pressure in response to a standardized exercise test, 2 hours after drug administration.[18] The increase in indirect systolic blood pressure in horses receiving benazepril (0.5 mg/kg) was significantly lower than horses receiving **ramipril** (0.2 mg/kg). The latter was also significantly lower compared to the horses receiving **quinapril** (1 mg/kg) or **enalapril** (2 mg/kg). Interestingly, this was the highest reported dose of quinapril administered in horses and the mean ACE inhibition achieved (46.4%) was very similar to a previously reported mean (±SD) ACE inhibition with a dose of 0.25 mg/kg (40.6±12%).[13] This evidences the fact that orally administered ACE inhibitors reach a plateau in the ability to block ACE. Both enalapril and quinapril, have been demonstrated to reach a plateau at a circulating ACE inhibition of lower than 50%, which is considerably lower than the inhibitions achieved with both ramipril and benazepril.[9,13,15,16,18]

Other ACE Inhibitors and the Blockade of RAAS at a Different Level

Angiotensin-converting enzyme inhibitors can be divided in different groups according to their chemical structure. All aforementioned ACE inhibitors are classified as dicarboxyl-containing ACE inhibitors. Fosinopril, however, is a phosphorus-containing ACE inhibitor. This may result in different bioavailability in horses, and further research on this drug may be warranted.

Angiotensin (II) receptor blockers have been shown to attenuate the increase in blood pressure response to exogenous administration of ANG-I in both dogs and cats.[19,20] Such drugs have never been reported in horses and studies are needed.

Conclusion

The activation of the RAAS and its deleterious effects are well recognized in the setting of heart failure. Of the ACE inhibitors studied in horses, benazepril at a dose of 1 mg/kg, orally, twice daily, presents the best evidence for the management of CHF. Additional studies with a much larger sample size are still necessary to determine if administration of benazepril delays progression of valvular regurgitation and prolongs survival times relative to horses receiving placebo.

Further research on higher doses of oral ramipril are needed (>0.8 mg/kg). Blockade of RAAS at the ANG-II receptor level is warranted.

References

1. Davis JL *et al. J Am Vet Med Assoc.* 2002;220(10):1512.
2. Leroux AA *et al. J Vet Intern Med.* 2013;27(6):1563.
3. Marr CM. In: *Cardiology of the Horse*, 2nd ed. Saunders; 2010:207.
4. Lefebvre HP *et al. Curr Pharm Des.* 2007;13(13):1347.
5. Horn JNR. Sympathetic nervous control of cardiac function and its role in equine heart disease. University of London; 2002.
6. Gehlen H *et al. Res Vet Sci.* 2008;85(2):340.

7. Gardner SY *et al. J Vet Intern Med.* 2004;18(2):231.
8. Sleeper MM *et al. J Vet Cardiol.* 2008;10(2):111.
9. Gomez-Diez M *et al. Res Vet Sci.* 2014;97(1):105.
10. Luciani A *et al. Vet Res Commun.* 2007;31 Suppl 1:297.
11. Gehlen H *et al. J Vet Med A Physiol Pathol Clin Med.* 2003;50(9):460.
12. Davis JL *et al. Equine Vet J.* 2014;46(6):729.
13. Afonso T *et al. J Vet Intern Med.* 2013;27(5):1185.
14. Buchwalder-Csajka C *et al. Br J Clin Pharmacol.* 1999;48(4):594.
15. Afonso T *et al. Equine Vet J.* doi: 10.1111/evj.12593.
16. Serrano-Rodriguez JM *et al. Vet J.* 2016;208:38.
17. Costa MF *et al. J Vet Sci.* 2011;12(1):21.
18. Munoz A *et al. Vet J.* 2016;208:33.
19. Jenkins TL *et al. Am J Vet Res.* 2015;76(9):807.
20. Coleman AE *et al. J Vet Intern Med.* 2014;28(3):1002.

acvim206

Effect of Atrial Fibrillation on Exercise Performance in Horses

Dagmar S. Trachsel, DMV, PhD, DECEIM; Helena Carstensen, DVM; Eva Z. Hesselkilde, DVM; Merle F. Fenner, DVM; Rikke Buhl, DVM, PhD
Copenhagen, Denmark

Introduction

Atrial fibrillation (AF) is the most relevant performance limiting arrhythmia in horses.[1-3] This is especially true in horses performing at high intensity levels. During AF the coordinated contraction of the atria is absent and without cardioversion to sinus rhythm (SR), the patient in AF has to perform without the atrial systole contributing significantly to ventricular filling. Therefore, a conversion to SR is advised if performance level should be maintained.[4] However, among veterinarians, debates exist whether to treat horses used at lower level of performance as they are expected to perform satisfactory despite the lack of atrial systole. This review will highlight the role of the atria to the heart's overall performance both during SR and AF.

Atrial Function in Atrial Fibrillation

The active atrial contraction contributes to the late diastolic ventricular filling, and thus, through the Frank-Starling mechanism, it influences the contractility and the stroke volume of the ventricles. In horses and in man, it has been estimated that at rest the active contribution of the atria accounts for approximately 10–20% of the ventricular stroke volume.[5-7] Aside from this pumping function, the atria have a reservoir function supporting a fast filling of the ventricles in the early diastole and a conduit function taking place during the time frame between opening and closing of the atrio-venticular (AV) valves.[8] These two less known atrial functions also contribute to the overall performance of the ventricles.[8]

In healthy hearts in SR the reservoir function permits an almost continuous flow of blood from the pulmonary veins and the vena cava via the atria into the ventricles avoiding congestion of the upstream structures.[9] This is achieved through storage of the blood during the ventricular systole and relies on the compliance of the atrial myocardium. The expanding atrial volumes building up during ventricular systole generate a raise in atrial pressure. During the early diastole, the atrial-ventricular pressure difference and the elastic energy stored in the atrial wall allow a rapid filling of the ventricles. This represents the conduit phase.[8,10] The described reservoir and conduit function of the atria play a significant role in the heart's performance not only at rest, but also during exercise. In man, at rest approximately 63% of the left atrial (LA) volume is flowing passively in to the left ventricle (LV) during early diastole. Similarly, at light intensity exercise the LA reservoir function is the main contributor to LV filling. With increasing intensity of exercise the LA pumping function gains more and more importance and the relative part of passive filling is reduced but still contributes to 42% of the overall filling.[11]

During AF, in addition to the obvious loss of pumping function, the reservoir and to a lesser extent the conduit function of the atria are also affected.[12] In a goat animal model, long lasting AF induced remodeling of the atrial tissue and reduced atrial contractile function. The loss in contractile function led to an increased compliance of the atrial tissue, especially at low filling pressures. The increased compliance was also associated with increased dimensions of the atria.[13,14] The altered dimensions of the atria further modify the structural characteristics of the atria worsening the atrial function even more.[13] Similar modifications have been reported in horses with AF and a decrease in atrial contractile function has been documented to last up to 7 days after conversion.[15,16] Further, the reduction in contractile function has been associated with echocardiographic increased LA size.[15]

Heart Rate in Atrial Fibrillation

The heart rate (HR) determines the time frame during which filling of the ventricles takes place. The AV node plays an important role in regulating the electrical impulse conduction to the ventricles and thus influences the filling time.[9,17] In AF, the electrical disturbances in the atria have profound effects on the electrophysiological function of the AV node.[18] At normal physiological HR - like in SR without AV-blocks - there is a one-to-one coupling between atrial and ventricular depolarization. At very high atrial rates - like in AF - there will be an alteration in the electrical conduction trough the AV node resulting in only some of the atrial impulses passing to the ventricles and thereby protecting the ventricles towards very high pacing rates, known to be related to tachycardia-induced cardiomyopathy.[19] The inhomogeneous impulses from the atria, hitting randomly the AV node, modify the conduction properties of the AV node, resulting in a typical irregularly irregular rhythm in AF.[18] The mechanisms, responsible for transmission of the electrical impulses from the atria to the ventricles, include the refractory period of the cardiomyocytes in the AV node, but also the phenomena called concealed conduction and electronic modulation of propagation.[18,20] However, the conduction properties of the AV node depend also on the autonomous innervation.[17,21] Parasympathetic stimulation reduces the conduction through the AV node and thus in AF, the ventricular rate decreases.[22] In horses at rest the parasympathetic tone predominates and therefore, the conduction in the AV node is slow[9] and leads to normal resting HR in most horses with AF.[1,3,23-25] During exercise, the increased adrenergic stimulation increases the conduction in the AV node[20,21] and therefore, the electrical activity can spread to the ventricles more rapidly and induce disproportional high HR as seen in some horses with AF.[26] At such high ventricular rates, the ventricular function can be severely affected with reduced cardiac output, elevated ventricular filling pressure, increased vascular resistance and intense neurohumoral stimulation, mimicking ventricular systolic failure.[19] In addition to the reduction in the ventricular systolic function, the increased ventricular filling pressure impairs the reservoir and the conduit function of the atria leading in some cases to congestion and dilatation in the atria and in the upstream structures.[27] An increased LV filling pressure was also associated with reduced exercise capacity in humans with AF.[28]

ATRIAL FIBRILLATION AND PERFORMANCE IN HORSES

At rest the HR is often normal in horses with AF.[1,3,23-25] Several studies have reported that during exercise the HR reaches significant higher values in horses with AF than in healthy horses, even at relatively low intensity exercise.[23,25,26] In almost all horses with naturally occurring AF, participating in a lounging exercise, the HR was above the expected HR and reached values as high as 248–492/min.[26] Further, a marked reduction in exercise tolerance has recently been documented in horses after induction of AF.[29] Reduced cardiac function, as associated with tachycardia-induced cardiomyopathy, has been described in a horse with naturally occurring ventricular tachycardia.[30] Additionally, in horses with AF an increased pulmonary capillary wedged pressure has been documented during exercise.[23] The pulmonary capillary wedged pressure is a surrogate of LA pressure[31] and the reported increase with AF could be an indication for the increased upstream filling due to the affected function of the atrium and reduced systolic function of the heart.[7,19] Moreover, broad QRS complexes and R-on-T episodes, signs of aberrant conduction or ventricular ectopic activation, are frequent in horses with AF during exercise.[26]

The studies mentioned above would therefore rather support the statement that the altered atrial functions as seen in AF and the related changes in hemodynamics affect the performance of the heart in horses, especially at higher HR. Hence, poor performance is reported as the major complain in high performing horses with AF.1-3 Reduced performance is however not reported in 10–22% of horses with well tolerated AF and where the diagnosis is an incidental finding.[1-3]

Comparison with human studies concerning recommendations for exercise is difficult as most studies in man with AF concern elderly persons often showing comorbidities.[32] Further, the expected exercise capacity in this population is very different from that expected in horses, even if compared to leisure horses. However, guidelines used for human athletes allow a certain amount of exercise with AF but under strict HR control and in absence of hemodynamic impairment and structural heart diseases.[33] Further, athletes are instructed to stop activity when they experience palpitations. To which extend these guidelines can be applied in equine medicine remains to be established. In the recent published consensus statement it has been suggested that horses in AF can undergo exercise as long as HR during exercise is acceptable and there is no evidence of severe ventricular arrhythmias.[4] However, still a large amount of work is needed to define the range of HRs at exercise that represents minimal risk for the horse and the rider and how to detect reliably and promptly sudden occurring life-threatening ventricular arrhythmias during exercise.

References

1. Deem DA *et al. J Am Vet Med Assoc.* 1982;180:261–265.
2. Reef VB *et al. J Vet Intern Med.* 1995;9:57–67.
3. Reef VB *et al. J Vet Intern Med.* 1988;2:1–6.
4. Reef VB *et al. J Vet Intern Med.* 2014;28:749–761.
5. Gehlen H *et al. Vet Res Commun.* 2004;28:317–329.
6. Kubo K *et al. Exp Rep Equine Health Lab.* 1975;12:101–106.
7. Mitchell JH *et al. Am J Cardiol.* 1969;23:556–567.
8. Pagel PS *et al. Anesthesiology.* 2003;98:975–994.
9. Marr CM *et al. Cardiology of the Horse.* 2010.
10. Barbier P *et al. Circulation.* 1999;100:427–436.
11. Wright S *et al. J Appl Physiol* (1985). 2015;119:328–333.
12. Kazui T *et al. J Thorac Cardiovasc Surg.* 2015;150:1602–1608.
13. Schotten U *et al. Prog Biophys Mol Biol.* 2003;82:151–162.
14. Schotten U *et al. Am J Physiol Heart Circ Physiol.* 2004;287:H2324–2331.
15. Decloedt A *et al. Vet J.* 2013;197:583–588.
16. Schwarzwald CC *et al. J Vet Intern Med.* 2007;21:820–827.
17. James TN *et al. Prog Cardiovasc Dis.* 2003;45:327–360.
18. Meijler FL *et al. J Cardiovasc Electrophysiol.* 1996;7:843–861.
19. Umana E *et al. Am J Med.* 2003;114:51–55.
20. Toivonen L *et al. J Am Coll Cardiol.* 1990;16:1194–1200.
21. Gordan R *et al. Circulation.* 1999;99:2806–2814.
22. Gehlen H *et al. Res Vet Sci.* 2006;81:134–139.
23. Marr CM *et al. J Vet Intern Med.* 1995;9:336–340.
24. Buntenkotter S *et al. Equine Vet J.* 1976;8:26–29.
25. Verheyen T *et al. Equine Vet J.* 2013;45:309–314.
26. Heywood JT *et al. J Atr Fibrillation.* 2014;7:1117.
27. Chen SM *et al. J Geriatr Cardiol.* 2016;13:546–551.
28. Carstensen H *et al.* Proceedings from the ACVIM Forum. 2017.
29. Stern JA *et al. J Vet Cardiol.* 2012;14:445–451.
30. Walston A *et al. Am Heart J.* 1973;86:159–164.
31. Kirchhof P *et al. Europace.* 2016;18:1609–1678.
32. Heidbuchel H *et al. Eur J Cardiovasc Prev Rehabil.* 2006;13:475–484.

acvim207

Overreaching, Overtraining and Peripheral Fatigue

Modest Vengust[1], DVM, DVSc, PhD, DACVIM (LAIM), DACVSMR (Equine), ECEIM;
George Heigenhauser[2], BA (hons), MSc, PhD, FACSM
[1]Ljubljana, Slovenia; [2]Hamilton, ON, Canada

Introduction

Physical training stimulates adaptive changes, which improve physical performance. Inadequate intensity of exercise will not lead to improvement in performance capacity. Only during an overreaching (OR) phase of overtraining (overload training) the induced physical stress will generate adaptations, which, when combined with an appropriate recovery, will eventually improve performance. Overtraining always causes a variable level of performance decrement. The improvement of performance after a short-term performance decrement is called supercompensation and is the result of functional overreaching (FOR). Therefore, the target supercompensation due to short-term overtraining is an acceptable part of the athletic training. The phase of training when the accumulation of physical stress (overtraining) and inadequate recovery cause a longer-term decrement in performance is defined as nonfunctional overreaching (NFOR). Recovery from such physical stress can last more than two weeks. Nonfunctional overreaching phase of (over)training can, if not managed appropriately, lead to the overtraining syndrome (OTS), which can negatively affect performance for several months.[1,2]

Because there are no reliable and universal diagnostic tools to identify NFOR and OTS, and NFOR data are often skewed by OTS data or vice versa, it is also hard to gather reliable epidemiological data. However, NFOR and OTS have been recognized as a significant problem in athletic humans, horses, and dogs.[1-4] The terms often used in the equine industry, such as staleness, sourness, burnout, do not strictly apply to NFOR and/or OTS.[3] Therefore, terminology used in this paper is adopted from the European College of Sports Science and the American College of Sports Medicine joint consensus statement on OTS.[1]

Exercise fatigue is a psychophysical quantity, which includes central and muscular (peripheral) fatigue. Only peripheral/muscular fatigue will be discussed herein; some information on the central effect on fatigue is given in the next chapter. Muscle fatigue is a series of deteriorations related to the working muscle, which are expressed as an acute failure to maintain power output. Unlike NFOR/OTS, it is not a pathological entity. Muscle fatigue is directly related to the duration and intensity of exercise, and further denoted by the power and capacity of the athlete.[5,6]

Etiopathogenesis and Diagnosis of NFOR/OTS

It is important to realize that quantitative estimation (physiological responses) of overtraining is not the sole mechanism for OTS. Several psychological or qualitative (mal)adaptations are also evident just before or/and during OTS. Both qualitative (i.e., behavior changes) and quantitative (i.e., subclinical diseases of the respiratory and/or gastrointestinal systems, low-grade orthopedic pain, muscular pain) factors, often recognized in OTS, are the focus of intensive investigations, which primarily aim at defining sensitive and specific diagnostic tools for the diagnosis of NFOR and OTS.[2,3] To date, no reliable, minimally invasive or noninvasive, and technically accessible "bedside" diagnostic markers are identified. Consequently, some scientists have suggested that the OTS be renamed as the unexplained underperformance syndrome (UUS), and focus rather on key symptom of the syndrome rather than on the physiological (mal)adaptations. The UUS terminology, however, has not been universally adopted.[1,3]

Because of the nature of physical training, athletes competing in endurance sports are expected to have higher incidence of NFOR/OTS.[1] NFOR/OTS is currently a diagnosis of exclusion. Thorough history, including chief complaint, training program, diet, medications, illness, body weight fluctuations, assessment of training/competition goals, transportation, herd status, mood changes, and mares' estrous cycle, should be documented.[1,3] A thorough evaluation of the diet is necessary, because unbalanced macro- and micronutrient intake can lead to NFOR/OTS. In young animals, physiologic stress also includes growth.[1] Horses that have lost their level of performance for longer than two weeks, and may or may not show changes in their demeanor, have evident weight loss, have persistent low-grade infections, and/or have increased incidence of musculoskeletal accidents during faster and/or complex elements of movement, should be considered to have NFOR/OTS.[3] Some authors consider disturbances in mood state as an early sign of NFOR/OTS in horses. Personal experience of author (MV) is that the loss of appetite after increased training loads, with or without mood alterations, is the earliest complaint in NFOR/OTS horses, which is in contrast to the published evidence.[7] A routine and advanced clinical and laboratory tests should be performed and other common reasons for poor performance in horses excluded. Prescribing absolute rest for two weeks is necessary before considering the diagnosis of NFOR/OTS.[1,3] Even though some other diseases may coincide with NFOR/OTS (i.e., low-grade respiratory disease, EGUS,…),[8] if no performance improvement is noted after two weeks of rest, adjustments to training regimen are necessary to return the horse to normal level of performance. This may take several weeks to months.[1,3]

NFOR/OTS share some clinical signs with the inflammatory airway disease (IAD).[1,3,9] Because of the complexity and inconsistency of diagnostic and therapeutic procedures for NFOR/OTS, owners/trainers and veterinarians are more likely to ignore NFOR/OTS and implement treatment for IAD, which is better defined with regards to epidemiology, pathophysiology, diagnostic, and therapeutic procedures (personal experience MV).

Several quantitative and qualitative markers to provide early and unequivocal diagnosis of NFOR/OTS have been investigated in horses and humans.[1,3] Inconsistent research findings in patients or research subjects are the major obstacle for identifying specific marker(s) of NFOR/OTS.[1] A reliable controlled longitudinal research model for induction of the OTS in horses was published by Tyler *et al.* (1996), which serves as the foundation for further research of adaptations related to OTS.[10] Unfortunately, a number of equine studies failed to prove that horses in their NFOR/OTS research protocols reached the state of OTS.[3] Below is a list of most often investigated NFOR/OTS markers in the equine athlete, which also have a strong connection to the etiopathogenesis of the disease:

Complete blood count: No population-specific alterations in CBC were reported, except for eosinopenia, which was then proposed as an indicator for training stress in horses.[3,7]

Serum biochemistry: Creatine kinase (CK) and aspartate aminotransferase (AST) are often elevated in horses with OTS. Increasing training load also often results in a linear increase in gamma glutamyl transferase (GGT). Increased activity of CK, AST, and GGT cannot be considered as a specific indicator for NFOR/OTS.[3,4,7]

Energy metabolism: Prolonged training causes depletion of glycogen stores; however, muscle glycogen concentration is typically normal in horses induced with OTS.[11] Downregulation of glycogenolysis and glucose transport is associated with intense and prolonged physical training in horses and is not specific to NFOR/OTS. Blood lactate concentration is also not a marker for NFOR/OTS.[6,11]

Hormonal responses: The endocrine system has a specific response to acute and chronic stress.[12] Anabolic adaptation in the exercising muscle in most instances causes downregulation of cortisol, linked with an increase in free testosterone. However, free testosterone concentration and its ratio with cortisone have not been proven to be the reflection of NFOR/OTS. Free testosterone concentrations cannot be used in gelded horses to assess adaptations to overtraining.[3] In contrast, there seems to be a general agreement in human and equine medicine that maladaptation of hypothalamic axes coincide with the development of NFOR/OTS.[12] Reports from equine studies are inconsistent with regards to the behavior of cortisol.[13-15] Resting concentration of cortisol does not change in NFOR/OTS patients. It decreases in response to near- or maximal exercise, which also is not specifically related to NFOR/OTS. NFOR/OTS cases also did not show any reduction in cortisol response to administration of ACTH. Other pituitary hormones also respond to the stress of exercise.[3,14,16] An exercise test was designed in human sports medicine which can detect and distinguish between NFO and OTS patients using the concentration of ACTH, prolactin, and growth hormone.[1] Several confounders and analytical/laboratory considerations with applying such tests at the "patients' bedside" make these diagnostic tools inappropriate for the routine clinical use. Evidence for maladaptation of hypothalamic axes in overtraining is often considered as the Addison type (parasympathetic) overtraining syndrome.[17]

Cytokines and the immune system: Excessive exercise, NFOR and OTS cause an increase in pro- and anti-inflammatory cytokines. Proinflammatory cytokines (IL-1β, IL-6, TNF-α, ...) centrally induce "sickness behavior," which is common in NFOR/OTS. A similar cascade of central events is evident in animals/humans suffering from other inflammatory conditions. IL-6 (proinflammatory cytokine and an anti-inflammatory myokine) increases several-fold during exercise. It stimulates the immune system response, cortisol production, and glucose metabolism. Dysregulated production of IL-6 during NFOR/OTS has a negative effect on the immune system, fatigue sensation, and energy metabolism. Despite such evidence, depressed immune system with variable alteration in circulating cell numbers is still only the reflection of overtraining and cannot distinguish between the beneficial adaptation to overtraining and NFOR/OTS.[18-20]

Physiological markers: Besides biochemical/hormonal responses to overtraining, the identification of physiological changes is equally important to understand the etiopathogenesis of NFOR/OTS. Simple measurements/markers can be investigated in relation to NFOR/OTS (resting pulse, pulse recovery, time to fatigue at or near VO_{2max}, heart rate variability, specific submaximal time trials such as velocity at a predetermined heart rate or lactate thresholds).[1,3] Equine treadmill studies reported a decrease in V200; however, later reviews identified several possible confounders affecting the heart rate in overtrained equine (and human) athletes.[1,3] Heart rate variability (HRV) is easy to measure and can be used to monitor the cardiovascular responses during exercise in horses[21] and humans[1]. In some studies in OTS subjects, HRV decreased, which indicates increased sympathetic stimulus (Basedow type OTS).[17] However, evidence from human and recent equine studies indicate that HRV does not provide consistent results.[1,22,23] Similarly, inconsistent results were reported for resting, exercise, and recovery heart rates.[1]

VO_{2max} is also a relatively easy variable to measure. It is directly related to muscular work. If muscular work is decreased in normal, fatigued or NFOR/OTS athletes, VO_{2max} will decrease as the reflection of the decreased level of exertion and not the state of overtraining.[24]

Physiology of Fatigue

Muscular fatigue is likely the interaction of several intra- and extramuscular metabolic events. Intramuscular energy (ATP) depletion was often implicated as a direct cause for fatigue. However, the reduction in muscle ATP concentration is not significant enough to be considered as an independent and/or direct mechanism for fatigue. It is more likely that the products of ATP hydrolysis (ADP and Pi) impair muscle contractility.[5] Action potential is affected by electrolyte disturbances (K^+, Na^+), which are counteracted by the increased concentration of H^+.[25] Acidosis seems to have several beneficial effects on the energy/electrolyte metabolism.[5] Action potential stimulates the release of Ca^{2+}. Exercise-derived increase and/or decrease in intracellular and extracellular [Ca^{2+}] and [K^+], and increased [Pi] may have a suppressive effect on excitation-contraction coupling. Reduced sensitivity to and changes in [Ca^{2+}] do not only cause the reduction in muscular force, but also reduce shortening velocity and slow muscular relaxation.[5,26]

Muscular fatigue can be identified by more reliable clinical and laboratory markers. However, physiological markers and adaptations related to peripheral fatigue are mostly investigated in an "*in vivo*" setting.[5]

Prevention and Treatment

Periodization of training, appropriate diet, and hydration reduce the incidence of NFOR/OTS.[1,3,18] The only treatment is rest and/or very light exercise. The diet and hydration depend on the type of exercise.[19] Muscular fatigue should not be subjected to treatment. Recovery from fatigue should be optimized with appropriate diet and rest. Appropriate training regimens enhance fatigue resistance.[5,19]

Conclusion

A cascade of metabolic events leading to NFOR/OTS, and also peripheral fatigue, are complex and specific to the individual athlete and physical activity. Multiple clinical signs arise from training overload, inadequate rest, nutritional imbalance, psychological effectors, and mild/occult diseases. The syndromes are so complex that the proposed classical division on the parasympathetic (Addison) and the sympathetic (Basedow) types of overtraining are inadequate to explain the etiopathogenesis of NFOR/OTS. It is also important to realize that peripheral fatigue is more of a protective cascade of events than a physiological problem to protect the integrity of the muscle metabolism.

REFERENCES

1. Meeusen R, *et al. Med Sci Sports Exerc.* 2013;45:186.
2. Carfagno DG, Hendrix JC. *Curr Sports Med Rep.* 2014;13:45.
3. McGowan CM, Whitworth DJ. *Comp Exerc Phys.* 2008;5:57.
4. Snow DH, Harris RC. In: Saltin B. *Biochemistry of Exercise.* 1986:447.
5. Sahlinn K. In: Hargreaves N, Spriet L. *Exercise Metabolism.* 2006:163.
6. Jones NL, Killian KJ. *N Engl J Med.* 2000;343:632.
7. Tyler-McGowan CM, *et al. Equine Vet J Suppl.* 1999;30:621.
8. de Graaf-Roelfsema E, *et al. Equine Vet J Suppl.* 2010;38:58.
9. Couëtil LL, *et al. J Vet Intern Med.* 2016;30:53.
10. Tyler CM, *et al. J Appl Physiol.* 1996;81:2244.
11. McGowan CM, *et al. Equine Vet J Suppl.* 2002;34:257.
12. McEwen BS. *N Engl J Med.* 1998;338:171.
13. de Graaf-Roelfsema E, *et al. Vet Q.* 2007;29:82.
14. Golland LC, *et al. Pflügers Arch.* 1999;439:11.
15. Bruin G, *et al. J Appl Physiol.* 1994;76:1908.
16. de Graaf-Roelfsema E, *et al. Am J Physiol.* 2009;297:R403.
17. Brooks KA, Carter JG. *J Nov Physiother.* 2013;3.
18. Carfagno DG, Hendrix JC 3rd. *Curr Sports Med Rep.* 2014;13:45.
19. Robson-Ansley P, Costa R. *The Encyclopaedia of Sports Medicine.* An IOC Medical Commission Publication; 2013;19:404.
20. Smith LL. *Med. Sci. Sports Exerc.* 2000;32:317.
21. Kinnunen S, *et al. Equine Vet J Suppl.* 2006;36:214.
22. Becker-Birck M, *et al. J Anim Physiol Anim Nutr (Berl).* 2013;97:322.
23. Eggensperger BH, Schwarzwald CC. *J Vet Cardiol.* 2017 doi: 10.1016/j.jvc.2016.10.006.
24. Vengust M, *et al. Equine Vet J Suppl.* 2010;38:40.
25. Vengust M. *Slo Vet Res.* 2005:42:59.
26. McKenna MJ, Hargreaves M. *J Appl Physiol.* 2008;104:286.

acvim208

The Internist's Approach to Equine Dynamic Upper Airway Obstructions

Emmanuelle van Erck, DVM, PhD, DECEIM
Waterloo, Belgium

INTRODUCTION

Whatever the equestrian discipline, age or breed, the respiratory system represents the main limiting factor to exercise in the healthy horse. Respiratory diseases are the second most common cause of wastage in sport horses and racehorses. Dynamic upper airway (UA) obstructions are duly represented in the high prevalence of respiratory problems causing poor performances in equine athletes. In addition to being functionally disabling, some upper airway obstructions can generate respiratory noise, which can be judged as deleterious in competition and decrease the commercial value of the horse. The advent of overground endoscopes has revolutionized our diagnostic workup of UA disorders in practice: "real-life" exercising conditions can be recreated and UA function analyzed in situations similar to work or competition. This novel approach has highlighted that UA disorders can be a succession of complex events leading to impaired biomechanics. These disorders can occur without associated noise and can present with atypical clinical signs, such as behavioral issues or cough. With new diagnostic tools such as overground endoscopy, we need to revise and adapt our diagnostic methodology. Essential factors interfering with UA mechanics, such as rider intervention, equitation or the presence of lower airway disease, need to be accounted for.

Any UA dysfunction observed during exercise is linked to a primary cause: identifying that underlying cause is essential to devise an adapted treatment, be it medical or surgical. The traditional approach to resolving dynamic obstruction of the upper airways has mainly been surgical, with variable success rates reported. However studying cases with poor surgical success rates and carrying out a more in-depth and systematic analysis of UA biomechanics enables to propose therapeutic alternatives and improve outcome.

UNDERSTANDING UPPER AIRWAY PHYSIOLOGY

Horses are obligate nasal breathers and thus are particularly prone to developing dynamic upper airway collapse. They cannot switch to oral breathing and alleviate the high negative pressures associated with nasal breathing, as other species do during exercise. The airways, from the nostrils to the bronchi, represent an important dead space. The diameter of the larynx and naso-pharyngeal cavity is small in comparison to the horse's size and, consequently, the resistance of the upper airways to airflow can become very high during exertion. The nasopharynx is a long sleeve of muscular mucosa, which is not supported by osseous, or cartilaginous structures. The equine UA are thus highly collapsible, especially in the nasopharyngeal region, and they rely on a rapid, coordinated and sustained neuromuscular response to maintain its stability and patency. The drastic variations in airway flow and transmural pressures that are encountered during intense exertion promote instability and potential secondary dynamic obstruction of the UA.

Not all UA obstructive conditions have the same impact on respiratory function but they commonly create an increase in respiratory resistance, which may result in either reduced airflow or an increase in the trans-upper airway pressures required to maintain airflow. This increase in airway resistance will lead to an increase in respiratory workload, and where airflow is reduced, the resulting hypoventilation may lead to decreased oxygen consumption, increased blood lactate concentration and exacerbation of arterial hypoxaemia and hypercapnia.[1]

The equine UA are long, narrowed in some critical areas such as the larynx, bended and compliant in others such as the pharynx, as a consequence, they account for 80% of total respiratory resistance to airflow at rest, and over 90% during exercise. Any abnormality that reduces the airway section is a potential factor of impaired ventilation and consequently impaired performance. Complex obstructions, where more than one structure collapses into the airway, or combinations of upper and lower airway disorders occur frequently. Multiple abnormalities will more commonly be associated with significant gas exchange impairment than single disorders.[1] It is important to remember that, whilst videoendoscopy enables visualisation of any dynamic airway collapse, it does not enable to directly quantify the functional effects of an obstruction. Inclusion of a video-endoscopy to a simple standardized exercise test can be a means of evaluating the impact of the UA problem.

Leading an Appropriate Diagnostic Test

Observation of the horse in usual working conditions is of paramount importance if the horse is involved in racing or sports. This allows to elicit symptoms that would remain unobservable at rest. In some cases of UA obstruction, the horse can fight against the rider, become nervous and difficult to handle and these can be the only apparent symptoms, as UA is not necessarily associated to noise. The clinical exam of a horse suspected with UA requires a thorough and global evaluation of the patient. Any UA disorder can be associated to other clinical problems such as orthopedic issues, lower airway diseases or even cardiac problems. Conditions resulting in hypoxia, such as cardiac or lower airway disorders, will increase ventilatory drive and potentially accentuate UA instability.

Exercising videoendoscopy has become the "gold standard" for making a definitive diagnosis of dynamic upper airway collapse in horses, where findings at rest are often unreliable or absent. Cases relating to palatal instability, dorsal displacement of the soft palate or pharyngeal collapse do not correlate with resting observations. Exercising endoscopy was traditionally performed on a high-speed treadmill but overground endoscopes should be preferred as they allow to assess UA mechanical behavior in normal exercising conditions, both in race and sport horses. Prior to undertaking the exercise test, thorough questioning of the rider should be done to determine the circumstances in which the horse shows his symptoms. This allows selecting the appropriate tack and recreating the right level of exertion or type of exercise. This may involve asking the usual rider to perform the test, make the horse go over jumps or run with other horses on a racetrack. Certainly a number of factors are implicated in the development of dynamic UA collapse in an individual horse. Some horses will only display UA collapse during strenuous work: inspiratory pressures become more negative as speed increases, fatigue of the respiratory musculature may also play a role.[2] Hence the type of exercise test performed will have an impact on the ability to make a definitive diagnosis.

Equitation and Tack

Riding (or equitation) is implicated in the development of dynamic UA collapse. These factors are particularly important in pleasure or sport horses, where dynamic airway collapse appears to commonly occur at lower exercise intensities than in racehorses. We found that pharyngeal collapse was more readily diagnosed in warmbloods during overground endoscopy compared with treadmill endoscopy. This is likely due to the fact that riding factors, including increased tension in the reins and head flexion, enhance of dynamic airway collapse these horses and changes in poll flexion are easier to recreate during ridden exercise.

Ridden examination is of particular importance, as equitation-related maneuvers or movements will affect UA dynamics. Poll flexion has been recognized as a factor contributing to the decrease of laryngeal diameter and the subsequent increase in UA resistance. Rider equitation and head flexion have a combined effects of on upper airway morphology and aggravate UA dynamic obstruction.[4] The combined observation of the horse's attitude, gait, relative head-neck position and rider interventions during the overground endoscopic examination must be integrated in the interpretation of the behavior and stability of the upper airways. In a recent study looking at the effects of riding on upper airway function, 90% and 81% of the horses developed or showed an aggravation in the severity of upper airway obstructive disorders with poll flexion and rider intervention, respectively. In dressage horses, usually worked with a more acute head-neck angle, head flexion and riding had a more significant influence on the development of upper airway obstruction than in showjumpers (positive predictive value (PPV) for head flexion =86% vs. 65%; PPV for other riding maneuvers =83% vs. 57%). There was a significant association between rider intervention and exercising with increased head flexion for the detection of all dynamic upper airway obstructive conditions, except dorsal displacement of the soft palate (DDSP).[4]

Some horses will display DDSP only when the head is in hyperflexion, others only when the neck is in an extended position. The underlying pathological mechanisms of DDSP are markedly different in both cases and require different rehabilitation exercises.

Changing tack or exercise during the examination and evaluating potential changes in the UA biomechanics gives us insight on the underlying pathophysiology of instability or collapse. As an example, a horse that is comfortable and stable when exercised in a hackamore but shows exacerbated palatal instability with a Pelham bit, can be mispositioning his tongue. Rather than invasive surgery, these cases can be resolved with progressive rehabilitation exercises.

Pain and Inflammation

The pathophysiology of DDSP, palatal instability and of other forms of nasopharyngeal collapse, although not fully elucidated, involves neuromuscular dysfunction of the UA. The integrity of the nerves and muscles involved in maintaining UA stability should be systematically assessed. Underlying airway inflammation has been suggested as a possible etiology or predisposing factor to UA obstruction. Pharyngeal instability, under the form of palatal instability or nasopharyngeal collapse, was significantly affected by the presence of upper (pharyngeal) or lower airway inflammation.[5] Inflammatory conditions such as pharyngeal lymphoid hyperplasia can result in nasopharyngeal instability and possibly DDSP.

Expectoration of mucoid secretions and frequency of swallowing should be monitored throughout the exercising endoscopy as these signs can be indicative of lower airway inflammation. Presence of infection or inflammation, including guttural pouches and lower airways, should be carefully assessed, even if it requires post-exercise endoscopy of the lower airways to check for mucus or blood or to proceed to sampling material for cytology and culture.

Using topical nasopharyngeal anaesthesia in the last phase of the overground endoscopy yields interesting information on the horse's local sensitivity and proprioception: we have found that topical anaesthesia resolved instability or DDSP in cases of horses suffering with "painful throats" whereas other had more frequent or prolonged DDSP when anaesthetised probably because it altered regional proprioception. These findings are also essential to devise an appropriate medical therapeutic approach.

References

1. Durando M *et al. Equine Vet J.* Suppl. 2006;36:523.
2. Allen K, Franklin S. *Equine Vet J.* Suppl. 2010;38:587.
3. van Erck E. *Equine Vet J.* Suppl. 2011;(40):18.
4. van Erck-Westergren E *et al. Equine Vet J.* 2013;45(3):376.

acvim209

Designing EMS Treatment Plans Based on Phenotype and Concurrent Disorders

Elizabeth M. Tadros, DVM, PhD, DACVIM (LAIM)
East Lansing, MI, USA

Objectives

Upon completion, participants will be able to design customized dietary plans for horses with equine metabolic syndrome (EMS) by factoring in individual phenotype, degree of obesity, and comorbidities such as pituitary pars intermedia dysfunction (PPID), geriatric concerns, systemic illness, and laminitis.

Introduction

The classic triad of abnormalities defining EMS includes: 1) generalized obesity and/or regional adiposity, 2) insulin dysregulation (ID; hyperinsulinemia, insulin resistance and/or exaggerated insulin responses to oral/intravenous carbohydrates), and 3) clinical or subclinical endocrinopathic laminitis.[1] While the majority of affected animals are obese, there is recognition of a lean phenotype; such animals appear physically normal, but might have increased adiposity of less visible internal fat depots that adversely affects metabolism. Although abnormal adiposity and ID are associated, their relationships to one another are complex and incompletely understood in equids. The degree to which obesity induces or exacerbates ID likely depends on its interactions with each individual's underlying genetic predispositions, as well as concurrent environmental and physiological factors.[2]

Animals with EMS can have comorbidities that complicate treatment. Concurrent PPID is commonly encountered in middle-aged and older equids, while severe metabolic derangements, including type 2 diabetes mellitus and dyslipidemias, occur occasionally. Many EMS-affected animals have painful orthopedic conditions such as laminitis. Individualized nutritional plans, husbandry modifications, and pharmacological therapies should be customized around these factors to effectively manage EMS.

The Role of Obesity in Equine Metabolic Syndrome

Obesity is a growing problem in equine health, with prevalence estimates of 50% or more in some populations.[3] In other species, obesity-induced adipocyte dysfunction creates a systemic proinflammatory state, with resultant disruption of post-receptor insulin signaling, insulin resistance, and compensatory hyperinsulinemia.[4] The role of inflammation in equine obesity remains controversial, as abnormal basal inflammatory states are inconsistently identified; however, EMS-affected animals experience exaggerated systemic inflammatory responses to immunological challenges[5] and oral carbohydrate administration[6]. Because inflammation is a variable finding, it is possible that other mechanisms, such as abnormal adipokine release, play a more important role in obesity-related ID in this species. In the obese state, secretion of the insulin-sensitizing adipokine adiponectin is reduced; some studies suggest that hypoadiponectinemia is more closely associated with insulin resistance than the degree of obesity, *per se*.[7] Mild hepatic dysfunction, possibly from intrahepatic fat accumulation, is thought to decrease insulin clearance and contribute to hyperinsulinemia in obese horses.[8] That said, liver function in EMS-affected equids has not yet been thoroughly evaluated. Should it prove to be an important factor in ID, even modest weight loss of 5–10% could improve the hepatic lipid profile, as demonstrated in humans.[9]

Animals that become obese do not invariably develop ID.[7,10] The reason why some individuals tolerate being overweight better than others likely depends on whether they have underlying genetic predispositions towards metabolic derangements[11] or concurrent exposure to additional risk factors, such as a high-sugar diet[7]. Improved understanding of these relationships will enable better customization of nutritional plans. Because weight loss is difficult to achieve and requires a substantial commitment on the part of owners and caretakers, it may be helpful to identify those animals that would benefit most from losing weight in addition to reducing dietary nonstructural carbohydrate (NSC) intake. While it is possible for horses managed in their home environments to lose weight and experience improvement in ID, poor owner compliance is a major factor in treatment failure[12] and success often relies on sufficient veterinary guidance and support[13]. Furthermore, some animals are weight-loss resistant and require severe caloric restriction before losing body condition.[14] In certain situations, emphasizing owner compliance with dietary recommendations and eliminating excess carbohydrates might still improve ID, even in the absence of substantial weight loss.

While obesity and/or regional adiposity is the most prevalent phenotype, lean animals can also suffer from ID. These metabolic abnormalities are sometimes related to PPID, but ID can occur as an isolated abnormality in phenotypically normal equids. It can also persist in formerly obese equids that have lost weight.[15] Carbohydrate restriction is the focus of managing such cases and calorie requirements must be met using low-glycemic forage diets; if needed, diets can be supplemented with vegetable oils (1.7 Mcal/cup; up to 2 cups daily for a 500-kg horse), beet pulp, rice bran, and commercial fat supplements.

INSULIN DYSREGULATION

Obesity was once viewed as central to EMS pathophysiology, but a recent paradigm shift has focused attention on ID as the fundamental abnormality.[2] Development of ID is the culmination of interactions between genetic, physiological, and environmental risk factors. A horse with underlying genetic risks may avoid developing ID if its environment is optimal. However, decompensation occurs when at-risk individuals experience multiple factors that negatively affect metabolism. The genetics of EMS is currently under investigation. Predisposing traits likely include "thrifty" genes that promote metabolic efficiency.[11] Being able to survive on limited nutritional resources was historically advantageous. However, modern management conditions, where food is plentiful and exercise is in short supply, put thrifty individuals at risk of becoming obese. Breed differences in innate insulin sensitivity have been demonstrated. For example, insulin sensitivity is lower and insulinemic responses to carbohydrate challenge (oral and intravenous) are higher in ponies and Andalusians, compared to Standardbreds.[16] Insulin resistance and compensatory hyperinsulinemia are more easily triggered in these animals. Genetic differences may also promote heightened incretin responses to oral carbohydrates. Incretin hormones, such as glucagon-like peptide-1 (GLP-1), are released by the intestine after carbohydrate ingestion and augment postprandial insulin responses by stimulating pancreatic β-cells. Breeds such as ponies and Andalusians secrete more GLP-1 after oral carbohydrate ingestion than Standardbreds and this correlates positively with higher postprandial insulin concentrations.[17] There is evidence that abnormally exaggerated GLP-1 responses might be responsible for postprandial hyperinsulinemia in ponies with ID.[18] Innate function of the enteroinsular axis may therefore predispose at-risk individuals to hyperinsulinemia and laminitis, particularly if fed high-NSC diets. In these cases, hyperinsulinemia can develop in the absence of tissue insulin resistance.[18] Hyperinsulinemia is believed to be the most detrimental component of ID because it damages laminar tissues and disrupts cell cytoskeletal integrity through aberrant activation of the insulin-like growth factor-1 receptor.[19]

Because each individual's genetic predispositions and threshold for developing ID are unknown, it is prudent to preemptively eliminate controllable factors that exacerbate ID, particularly in high-risk breeds. Obesity is better avoided through good husbandry than treated after it develops. Nonstructural carbohydrate content should routinely be minimized in the diets of animals at risk of developing hyperinsulinemia. Certain situations are unavoidable, including development of concurrent illnesses that disrupt glucose and insulin metabolism; the most common of these are PPID and systemic inflammation/sepsis. However, awareness that these conditions exacerbate ID is important because measures such as routine PPID screening and close attention to energy homeostasis during systemic illness can help mitigate the development of severe metabolic derangements.

PITUITARY PARS INTERMEDIA DYSFUNCTION AND INSULIN DYSREGULATION

Aged horses increasingly comprise a larger segment of the equine population as medical advances maintain their athletic abilities and quality of life well into their 20s and beyond. Up to 30% of them develop PPID, which generally manifests in animals older than 10 years of age.[20] Equids with PPID range in body condition from obese to underweight; however, regional adiposity is common, even in lean animals. When formulating feeding plans, it is important to consider whether ID is a component of an individual's clinical presentation. While common, ID is not universally present in all PPID-affected animals; its presence may therefore have prognostic and therapeutic implications, as equids with concurrent PPID and ID (specifically, hyperinsulinemia) appear to be at the highest risk of developing laminitis.[21] Knowledge of insulin status is particularly important when faced with an insulin-dysregulated animal that cannot maintain weight, where pasture grazing and concentrate feeds must be eliminated due to carbohydrate intolerance. Even with proper dental care, such animals are often unable to meet their caloric needs if consuming only long-stem hay. For animals with severe ID, many commercial pelleted senior feeds are too high in NSC and attention should be paid to selecting complete feeds specifically formulated to be "low-sugar/low-starch" (NSC <10%). When needed, additional calories should be provided in the form of fat supplementation.

Whether PPID directly causes ID, or identifying them concurrently represents the coexistence of 2 common endocrinopathies in the same animal, is unclear. Pathologically high pituitary hormone concentrations likely induce or exacerbate ID, with antagonistic effects of hypercortisolemia on insulin sensitivity playing a potential role in some cases. Corticotropin-like intermediate lobe peptide[22] and other adrenocorticotropic hormone fragments[23] act as insulin secretagogues in pancreatic β-cells. Chronic overexposure to α-melanocyte stimulating hormone and resultant downregulation of the hypothalamic melanocortin system might also promote increased adiposity and hyperinsulinemia.[2] Medical management of PPID consists of inhibiting hormonal secretion from the pars intermedia with the dopamine agonist pergolide mesylate. There is conflicting evidence regarding pergolide's efficacy in improving ID,[24,25] but clinical experience suggests that it helps normalize metabolism in some animals. It is conceivable that even mild ongoing pituitary dysfunction could perpetuate ID, suggesting that incorrect pergolide dosing might be responsible for the apparent lack of effect in some cases. Because pituitary lesions and function are heterogenous in PPID, accompanying ID might improve more with pergolide therapy in certain animals than others. When ID persists despite dietary modification and pergolide therapy, metformin hydrochloride can be considered to help block enterocyte glucose absorption and blunt postprandial glucose and insulin spikes.

Ageing

Insulin sensitivity naturally decreases with age and insulin responses to oral and intravenous glucose challenges are exaggerated.[26-28] Age-related insulin resistance occurs due to inflammation ("inflammaging"), oxidative stress, mitochondrial dysfunction, increased visceral adiposity, decreased lean muscle mass, and reduced physical activity.[29] Because geriatric horses are already prone to heightened insulin responses, pathological conditions that exacerbate ID might greatly increase their risk of developing hyperinsulinemia and endocrinopathic laminitis. Aged horses may also be particularly sensitive to the metabolic effects of excessive dietary NSC. In 1 study, pancreatic insulin responses to intravenous glucose administration were 36% higher in healthy aged horses receiving a high-sugar diet for 4 weeks than in those receiving only forage (hay); however, the high-sugar diet did not alter pancreatic responses in younger adult animals.[27]

Systemic Illness

Animals with ID are at risk of decompensating metabolically when stressed by severe systemic illness. Should equids with ID become systemically ill or anorectic, meeting nutritional requirements must remain a priority during treatment. If allowed to lapse into negative energy balance, insulin-dysregulated animals can develop life-threatening hyperlipemia and hepatic lipidosis. Transient type 2 diabetes mellitus can also occur during systemic illness and sometimes requires exogenous insulin therapy; diabetes often resolves with recovery and might be related to transient suppression of pancreatic function by systemic inflammation.[2] As in humans, where severe hyperglycemia during sepsis is sometimes the first indication of occult diabetes mellitus, critical illness in horses can unmask subtle ID that has gone unnoticed. Because endocrinopathies complicate case management, clinicians should consider the possibility of preexisting ID in critically ill equids and closely monitor metabolic status for signs of decompensation.

References

1. Frank N, *et al. J Vet Intern Med*. 2010;24:467.
2. Frank N, *et al. Equine Vet J*. 2014;46:103.
3. Thatcher CD, *et al. J Vet Intern Med*. 2012;26:1413.
4. Shoelson SE, *et al. Gastroenterology*. 2007;132:2169.
5. Tadros EM, *et al. Am J Vet Res*. 2013;74:1010.
6. Elzinga SE, *et al. Domest Anim Endocrinol*. 2017;60:1.
7. Bamford NJ, *et al. Vet J*. 2016;214:14.
8. Toth F, *et al. Equine Vet J*. 2010;42:149.
9. Hsu CC, *et al. Adv Nutr*. 2017;8:253.
10. Ungru J, *et al. Vet Rec*. 2012;171:528.
11. Lewis SL, *et al. J Anim Sci*. 2017;95:1071.
12. Gill JC, *et al. J Equine Vet Sci*. 2016;39:80.
13. Morgan RA, *et al. Equine Vet J*. 2016;48:422.
14. Argo CM, *et al. Vet J*. 2012;194:179.
15. Tadros EM, *et al. Am J Vet Res*. 2013;74:1020.
16. Bamford NJ, *et al. Domest Anim Endocrinol*. 2014;47:101.
17. Bamford NJ, *et al. J Anim Sci*. 2015;93:3377.
18. de Laat MA, *et al. Am J Physiol Endocrinol Metab*. 2016;310:E61.
19. Lane HE, *et al. Equine Vet J*. 2017.
20. McFarlane D. *Vet Clin North Am Equine Pract*. 2011;27:93.
21. Karikoski NP, *et al. Equine Vet J*. 2016;48:472.
22. Marshall JB, *et al. J Clin Invest*. 1984;74:1886.
23. Beloff-Chain A, *et al. Nature*. 1983;301:255.
24. Andrews FM, *et al*. 2011:141 (www.fda.gov/downloads/AnimalV/UCM280354.pdf).
25. Gehlen H, *et al. J Equine Vet Sci*. 2014;34:508.
26. Liburt NR, *et al. Comparative Exercise Physiology*. 2012;8:173.
27. Rapson JL. Michigan State University (MS thesis); 2015.
28. Malinowski K, *et al*. Equine Vet J Suppl 2002:147.
29. De Tata V. *Front Endocrinol (Lausanne)*. 2014;5:138.

acvim210

Part 1: The Old Horse: Medical Management with a Focus on PPID

Harold C. Schott II, DVM, PhD, DACVIM (LAIM)
East Lansing, MI, USA

Objectives

Upon completion, participants will be able to:

1. Describe demographics and reasons for death/euthanasia of the ageing equine population
2. Understand the prevalence and impact of PPID on health of ageing equids
3. Understand current recommendations for diagnosis of PPID
4. Develop individualized treatment plans for PPID-affected equids

Ageing Equine Population

The equine population is ageing and this demographic change can be attributed to less breeding during the recent economic downturn in combination with the transition to a companion animal species that has been accompanied by improved health care and nutrition of older equids. So what is an "old horse"? In human medicine, "geriatric" is defined as a loss of physical condition, organ function, and immunity.[1] However, the rate of physiologic decline varies between individuals within a species making it difficult to link physical ageing with chronological ageing. Nevertheless, chronological age remains the most practical means of categorizing horse life stages but to date the descriptors "old" and "aged" have been inconsistently used.[2] For example, both 15 and 20 years of age have been used to describe "old" and "aged" horses,[3-5] while the descriptor "very old" has been more consistently used for horses 30 years of age and older[3,4]. In the United States, data from the USDA NAHMS equine studies revealed an increase in equids aged 20–29 years from 6.9 to 9.9% of the total estimated equine population from 2005 to 2015. Although fewer in number, horses aged >30 years increased from 0.7 to 1.5%.[6,7] Similarly, owner surveys in the United Kingdom and Australia revealed that 29 and 33% of equids were 15 years or older, respectively, with 2.2% of the population >30 years old in the UK.[3,5] One factor that likely contributes to a greater number of older equids in these countries may be a higher proportion of ponies, as compared to the US, as ponies are recognized to live longer than horses. For more accurate comparison of study results of old and aged equids, it is imperative that the research community comes to an agreement for describing both type of equids used as research subjects as well as age categories.

Mortality in Aged Horses

Mortality rate of ageing horses increases from 5–6% per year for horses 15–20 years of age to 35–40% per year for horses >30 years of age.[8] Causes of death or reasons for euthanasia vary between reports, depending on whether information is collected from horse owners or from veterinary pathology labs.[9] For example, in the NAHMS 2015 study owner-reported reasons for euthanasia of equids 20 years of age or older included colic (13%), cancer (13%), lameness/injury (12%), and neurological (12%) or respiratory disorders (4%). However, non-specific complaints of weight loss (12%) and other (26%, largely "old age") were additional reasons.[7] In contrast, necropsy data of old horses revealed that the most common system affected was the gastrointestinal system, followed by the musculoskeletal system.[10] Lameness due to arthritis and other injuries is a serious problem in older horses and may lead to difficulty laying down or an inability to rise, prompting owners to elect euthanasia of their ageing companions.

The NAHMS 2015 study also revealed that about 60% of equid operations had end-of-life plans for their equids, with 86% of owners utilizing guidance by veterinarians as the primary source for developing these plans.[7] Assessment of quality of life (QOL), utilizing a standardized set of questions, can assist animal owners facing end-of-life decisions. However, tools for objective assessment of QOL have not been well developed for ageing equids and end-of-life decisions remain largely subjective.[9]

PPID in Ageing Equids

The prevalence of PPID in the ageing equine population also varies with chronological age and whether or not the diagnosis is based on clinical signs, primarily regional or generalized hypertrichosis and/or laminitis, or supportive endocrine test results.[11] Most reports support a PPID prevalence of 15–20% in horses over 15 years of age, increasing to 30% in horses aged >30 years.[12,13] As such, PPID is perhaps the most common medical problem of ageing equids. However, owner recognition of signs of PPID is considerably lower.[13] Although client education and screening of ageing equids for PPID by veterinarians has increased dramatically over the past decade, it is likely that a large population of equids with PPID remain unrecognized. Although recognition and effective management of PPID likely improves QOL in most affected equids, it remains unclear whether or not treatment of PPID prolongs life. This somewhat paradoxical statement is due to the fact that decisions for euthanasia of most PPID-affected equids are also based on other "old age" problems including colic and lameness, although refractory laminitis and other complications of PPID (e.g., chronic infections) also contribute to a decision for euthanasia in a subset of PPID-affected equids.[14]

Diagnosis of PPID

Presence of clinical signs, coupled with measurement of an increased plasma adrenocorticotropin (ACTH) concentration, has become the most common diagnostic approach for PPID.[11] For advanced cases of PPID, clinical signs alone (usually generalized hypertrichosis) are essentially pathognomonic for PPID, such that laboratory confirmation may not be necessary. However, recent work has shown that laboratory data, including measurement of elevated plasma ACTH or melanocyte-stimulating hormone concentrations or failure to suppress endogenous cortisol concentration during an overnight dexamethasone suppression test, are usually only supportive with more advanced PPID (histologic grade 5/5; D. McFarlane, personal communication 2016). In contrast, endocrine test results are frequently normal in the earlier stages of PPID (histologic grades 2–4). Thus, diagnosis of "early PPID" (a poorly defined term itself) remains challenging because clinical signs may be nonspecific (e.g., laminitis, loss of topline musculature). Although veterinarians were initially deterred from testing for PPID in fall months, due to a late summer/early fall rise in ACTH in normal equids, it has now been demonstrated that PPID-affected equids have a more dramatic seasonal increase in plasma ACTH concentration, peaking in the last 2 weeks of September and the first 2 weeks of October in the northern hemisphere.[15] Thus, measuring ACTH in the fall months in "early PPID" cases is now a recommended screening test. Another test that is currently advocated to screen for "early PPID" is the thyrotropin releasing hormone stimulation test, with a plasma ACTH concentration >110 pg/mL 10 min after IV administration of 1 mg of TRH supportive of PPID during non-fall months (reference ranges for fall months are not yet validated for this test).[16] The reader is referred to the Equine Endocrinology Group's "Recommendations for the Diagnosis and Treatment of Pituitary Pars Intermedia Dysfunction" (revised August 2015, http://sites.tufts.edu/equineendogroup/) for algorithms for diagnosis of PPID.

Despite recent advances in knowledge about PPID, novel laboratory tests are urgently needed to assist veterinarians in diagnosis of "early PPID." Further, all equids with PPID are currently lumped together with a single diagnosis; however, varying clinical syndromes are recognized. For example, some equids may simply have persistent hypertrichosis while others develop severe muscle wasting. Others may also show intermittent seizure-like neurological signs. Additionally, why some PPID-affected equids have evidence of insulin dysregulation and/or laminitis while others do not remains a perplexing question (addressed in Part 2 of this presentation). In the future, novel diagnostic tests may allow subtypes of PPID to be recognized and allow better prognostication of response to management and medical treatment.

Management of PPID

Management of equids with PPID consists of improved husbandry, including proper nutrition and limiting competition for feed, body-clipping, preventive health care, dentistry, and appropriate treatment of concurrent medical problems.[11] In addition, specific treatment with the dopamine agonist pergolide can improve QOL and reverse many clinical signs of PPID. Combination treatment with both pergolide and cyproheptadine, in the author's experience, may also prove beneficial in more advanced cases. Assessment for concurrent glucose and insulin dysregulation, especially in patients with laminitis, is also warranted (see Part 2). Due to the costs of lifelong management and medication(s), the approach to treatment of PPID-affected equids should be made on a case-by-case basis, in consideration of the owner's goals.

Findings during long-term treatment of a cohort of PPID-affected equids have provided additional useful information about management of PPID.[14] The cohort included 30 equids (28 horses and two ponies) that completed the 6-month clinical efficacy study for Prascend™ at Michigan State University and were subsequently enrolled in an extended-use study in August 2009. Most enrolled horses were re-examined after 24 and 36 months of pergolide treatment and these equids had continued to receive the same dose (either 2 or 4 µg/kg, PO, q 24 h) that they had been receiving at the 6-month endpoint of the initial study. After 2.5 years of treatment (Spring 2011), all owners reported ongoing clinical improvement and overnight dexamethasone suppression test (ODST) results remained normal in 79% of tested equids (19 of 24 equids). Four horses were not tested at this time and two horses were lost due to causes unrelated to PPID (one sudden death and one euthanasia for neurological disease). Of interest, six horses with abnormal ODST results after 3 and 6 months of treatment had normal ODST results after 2.5 years of treatment with no increase in pergolide dose. After more than 5 years of treatment (May 2014), owners were satisfied with the remaining equids' attitude and condition, although patients had clearly aged during the extended-use study. Of the 12 remaining equids tested in Spring 2014, plasma ACTH concentration was <50 pg/mL in 9 animals (75%) and ODST results also remained normal in 75% of horses (9 of 12 equids). At present (Spring 2017), 6 equids remain in the extended-use study and of the 24 that died, four died and 20 were euthanized for age-related disorders and only four were euthanized for recurrent laminitis, a problem attributable to PPID. To date, the results of the extended-use study support that long-term treatment of equids with PPID with pergolide results in clinical improvement, normalization of endocrine test results, and owner satisfaction in a high percentage of cases. Further, equids with PPID can be maintained for several years on pergolide therapy without a need for a progressive increase in drug dosage.

References

1. Traub-Dargatz JL, *et al.* In: *Equine Geriatric Medicine and Surgery*. Elsevier; 2006:1–4.
2. Ireland JL, *et al. Vet Clin North Am Equine Pract.* 2016;32:195.
3. Ireland JL, *et al. Equine Vet J.* 2011;43:30.
4. Mellor DJ, *et al. Vet Rec.* 1999;145:299.
5. McGowan TW, *et al. Aust Vet J.* 2010;88:420.
6. Anon. USDA/APHIS National Animal Health Monitoring System; 2005.
7. Anon. USDA/APHIS National Animal Health Monitoring System; 2015.
8. Ireland JL, *et al. Prev Vet Med.* 2011;101:204.
9. McGowan CM, *et al. Vet Clin North Am Equine Pract.* 2016;32:355.
10. Williams N. *Equine Pract.* 2000;22:32.
11. Durham AE. *Vet Clin North Am Equine Pract.* 2016;32:301.
12. McGowan TW, *et al. Aust Vet J.* 2010;88:465.
13. McGowan TW, *et al. Equine Vet J.* 2013;45:74.
14. Schott HC, *et al.* Proc AAEP. 2014;60:329.
15. Copas VEN, *et al.* Equine Vet J 2012;44:440.
16. Beech J, *et al. J Am Vet Med Assoc.* 2011;238:1305.

acvim211

Part 2: The Old Horse: Medical Management with a Focus on PPID and Insulin Dysregulation

Elizabeth M. Tadros, DVM, PhD, DACVIM (LAIM)
East Lansing, MI, USA

OBJECTIVES

Upon completion, participants will be able to:

1. Describe pathophysiological mechanisms linking pituitary pars intermedia dysfunction (PPID) and insulin dysregulation (ID)
2. Develop treatment plans for PPID-affected equids with ID

INTRODUCTION

When formulating management plans for equids with PPID, it is important to consider whether ID is a component of an individual's clinical presentation. While common, ID is not universally present in all PPID-affected animals; its presence may therefore have prognostic and therapeutic implications, as equids with concurrent PPID and ID (specifically, hyperinsulinemia) appear to be at the highest risk of developing laminitis.[1] Thus, controlling ID when it is present should be an explicit goal of PPID therapy. Knowledge of insulin status is also important when designing appropriate diets for PPID-affected equids that have difficulty maintaining weight; in animals with ID, sufficient calories must be provided for weight maintenance or gain, while still limiting dietary nonstructural carbohydrates (NSC) that worsen insulin resistance.

One question that arises frequently is why only one third of PPID-affected horses and ponies show evidence of ID.[2,3] There are currently 2 schools of thought. In the first, PPID and ID are independent conditions that can occur either individually or concurrently. Alternatively, PPID may induce or exacerbate ID, particularly in animals that already have predisposing genetic or environmental risk factors. Heterogeneity in the severity of pituitary lesions, as well as individual variability in the specific proopiomelanocortin-derived peptide profile secreted by the diseased pars intermedia, might explain why PPID only induces ID in a subset of affected animals. Additionally, equids with preexisting ID might be at an increased risk of developing PPID. In this presentation, we will review proposed physiological mechanisms linking PPID and ID. We will then discuss therapeutic strategies to manage ID in PPID-affected animals.

MECHANISMS OF INSULIN DYSREGULATION IN EQUIDS WITH PITUITARY PARS INTERMEDIA DYSFUNCTION

Because PPID and ID are both risk factors for laminitis, it is prudent to aggressively address ID to reduce an individual's cumulative risk. If PPID and ID are entirely separate conditions, then identifying them concurrently may simply represent the coexistence of 2 common endocrinopathies in the same animal. The prevalence of obesity, which is linked closely with ID, is increasing in equine populations. Inevitably, some of these metabolically unhealthy animals also develop PPID with age. Equids with PPID range in body condition from obese to underweight; however, regional adiposity is common, even in lean animals. An external phenotype that appears normal does not preclude ID, in part because increased adiposity of less visible internal fat depots can adversely affect metabolism. Despite their close association, the mechanisms by which obesity and regional adiposity induce ID are complex and incompletely understood. In other species, obesity-induced adipocyte dysfunction creates a systemic proinflammatory state, with resultant disruption of post-receptor insulin signaling, insulin resistance, and compensatory hyperinsulinemia.[4] While the presence of a basal proinflammatory state in obese horses remains controversial, these animals experience exaggerated inflammatory responses to immunological challenges, suggesting an analogous role for systemic inflammation in this species.[5] Furthermore, in the obese state, secretion of the insulin-sensitizing adipokine adiponectin is reduced.[6] Finally, mild hepatic dysfunction associated with obesity could decrease insulin clearance and contribute to hyperinsulinemia.[7]

Beyond simply co-occurring, PPID might directly cause ID and hyperinsulinemia. In the past, hyperinsulinemia in equids with PPID was attributed to the antagonistic effects of hypercortisolemia on insulin sensitivity. However, improved understanding of PPID led to the conclusion that cortisol excess is not the dominant feature of the disease in most animals.[8] While hypercortisolemia might play a role in some cases, pathologically high pituitary hormone concentrations are likely the more important drivers of ID. Corticotropin-like intermediate lobe peptide[9] and other adrenocorticotropic hormone fragments[10] act as insulin secretagogues in pancreatic β-cells. Downregulation of the hypothalamic melanocortin system in response to chronic α-melanocyte stimulating hormone overexposure might also promote increased adiposity and hyperinsulinemia.[11]

Although research has not yet been undertaken to answer this question, younger animals with insulin dysregulation are suspected to be at increased risk of developing PPID. One proposed mechanism is that chronic obesity-related inflammation and oxidative stress hastens the dopaminergic degeneration that underlies PPID. It is also possible that both conditions share common genetic or environmental predispositions. In support of the latter, there is evidence that horses with a "thrifty" (metabolically efficient) phenotype have chronic hyperactivity of the hypothalamic-pituitary-adrenal (HPA) axis. The magnitude of adrenocorticotropic hormone secretion in response to administration of thyrotropin-releasing hormone is greater in thrifty horses compared to normal animals.[12] Likewise, mares with ID display exaggerated α-melanocyte stimulating hormone responses to thyrotropin-releasing hormone and sulpiride administration.[13] It is conceivable that chronic hyperactivity might eventually injure dopaminergic neurons of the hypothalamus and lead to PPID. Lastly, recent findings that free cortisol fraction is higher in insulin-dysregulated horses with obesity or PPID raises the possibility of a vicious cycle, where cortisol excess promotes the development of adiposity and obesity overactivates the HPA axis or increase cortisol bioavailablity.[14] Genetic predispositions could potentially underlie either HPA axis dysfunction or obesity, thereby initiating the cycle.

An important dimension that has recently gained attention is considering what happens when pathological conditions that induce hyperinsulinemia are superimposed on the normal age-related decline in insulin sensitivity. As in other species, insulin sensitivity decreases with age in horses and insulin responses to oral and intravenous glucose challenges are exaggerated.[15-17] Mechanisms implicated in causing age-related insulin resistance include inflammation ("inflammaging"), oxidative stress, mitochondrial dysfunction, increased visceral adiposity, decreased lean muscle mass, and reduced physical activity.[18] Because geriatric horses are already prone to heightened insulin responses, pathological conditions that exacerbate ID might greatly increase their risk of developing hyperinsulinemia and endocrinopathic laminitis. Aged horses may also be particularly sensitive to the metabolic effects of excessive dietary NSC. In 1 study, pancreatic insulin responses to intravenous glucose administration were 36% higher in healthy aged horses receiving a high-sugar diet for 4 weeks than in those receiving only forage (hay); however, the high-sugar diet did not alter pancreatic responses in younger adult animals.[16]

Management

The approach to managing ID in PPID-affected equids is 2-fold, involving both dietary modification and drug therapy to control PPID. An individual animal's feeding plan should address both body condition and ID status. This becomes particularly important when faced with an insulin-dysregulated animal that cannot maintain weight, where pasture grazing and concentrate feeds must be eliminated due to carbohydrate intolerance. Even with proper dental care, such animals are often unable to meet their caloric needs if consuming only long-stem hay. For animals with severe ID, many commercial pelleted senior feeds are too high in NSC and attention should be paid to selecting complete feeds specifically formulated to be "low-sugar/low-starch" (NSC <10%). Additional calories can be provided using vegetable oils (1.7 Mcal/cup; up to 2 cups daily for a 500-kg horse), beet pulp, rice bran, and commercial fat supplements. Should geriatric equids with ID become systemically ill or anorectic, meeting nutritional requirements must remain a priority during treatment. If allowed to lapse into negative energy balance, insulin-dysregulated animals can develop life-threatening hyperlipemia and hepatic lipidosis. Transient type 2 diabetes mellitus can also occur during systemic illness and sometimes requires exogenous insulin therapy.

Medical management of PPID consists of inhibiting hormonal secretion from the pars intermedia with the dopamine agonist pergolide mesylate. There is conflicting evidence regarding pergolide's efficacy in improving ID,[19,20] leading to speculation that PPID *per se* may not be the underlying cause of ID in equids with both endocrinopathies. However, the single published study directly examining this question did not assess individual patients' ID before and after pergolide therapy to evaluate improvement, nor were pergolide doses titrated to normalize pituitary function tests in all animals.[20] It is conceivable that even mild ongoing pituitary dysfunction could perpetuate ID. Additionally, ID was measured only by means of an intravenous glucose/insulin challenge (quantifies tissue insulin sensitivity) and response to oral glucose, which also integrates enteroinsular axis function, was not tested. Clinical experience suggests that pergolide therapy can improve ID in some PPID-affected horses. Because pituitary lesions and function are heterogenous in PPID, accompanying ID might improve more with pergolide therapy in certain animals than others. When ID persists despite dietary modification and pergolide therapy, metformin hydrochloride can be considered to help block enterocyte glucose absorption and blunt postprandial glucose and insulin spikes.

REFERENCES

1. Karikoski NP, *et al. Equine Vet J*. 2016;48:472.
2. McGowan TW, *et al. Equine Vet J*. 2013;45:74.
3. Mastro LM, *et al. Domest Anim Endocrinol*. 2015;50:14.
4. Shoelson SE, *et al. Gastroenterology*. 2007;132:2169.
5. Tadros EM, *et al. Am J Vet Res*. 2013;74:1010.
6. Bamford NJ, *et al. Vet J*. 2016;214:14.
7. Toth F, *et al. Equine Vet J*. 2010;42:149.
8. McFarlane D. *Vet Clin North Am Equine Pract*. 2011;27:93.
9. Marshall JB, *et al. J Clin Invest*. 1984;74:1886.
10. Beloff-Chain A, *et al. Nature*. 1983;301:255.
11. Frank N, *et al. Equine Vet J*. 2014;46:103.
12. Fredrick J, *et al*. In: Proceedings of the 3rd Equine Endocrine Summit (abstract). 2014;4.
13. Valencia NA, *et al. Domest Anim Endocrinol*. 2013;44:204.
14. Hart KA, *et al. J Vet Intern Med*. 2016;30:653.
15. Liburt NR, *et al. Comparative Exercise Physiology*. 2012;8:173.
16. Rapson JL. Michigan State University (MS thesis); 2015.
17. Malinowski K, *et al. Equine Vet J Suppl*. 2002:147.
18. De Tata V. *Front Endocrinol (Lausanne)*. 2014;5:138.
19. Andrews FM, *et al*. 2011:141 (www.fda.gov/downloads/AnimalV/UCM280354.pdf).
20. Gehlen H, *et al. J Equine Vet Sci*. 2014;34:508.

acvim212

The Old Horse: Life Stage Nutrition to Maximize Health

Pat Harris, MA, PhD, DECVCN, VetMB, MRCVS
Waltham-on-the Wolds, UK

Introduction

Aging is usually associated with changes in body composition, physical strength, and physiological function. Intrinsic factors, associated with the ageing process itself, contribute to this process as well as factors indirectly associated with age: in particular, degenerative diseases secondary to environmental or physical insults that progress with time and typically the increased physical inactivity associated with ageing. However, the physical age that an individual horse starts to show signs of ageing is very variable and many horses can be maintained in good body condition and continue to be used for athletic endeavours well into their 20s and even 30s.[1,2] Therefore, we should have a similar goal of optimal nutrition and physical activity for horses, as recommended for ourselves, based around 'a healthy lifestyle, incorporating a well-balanced diet and physical activity, cannot stop the years advancing, but may play a useful part in a healthful, active and independent old age.'[3]

The two main clinical situations, (which are often associated with difficulty in maintaining weight), loss of teeth plus abnormal wear patterns and pituitary pars intermedia dysfunction (PPID), are common in the older horse. However, obesity may be as much an issue with older equines[4] as it is in the general horse population, with the associated complications of increased risk of lameness, laminitis and insulin resistance/dysregulation. Obesity may also play an important role in the apparent age-related dysregulation of inflammatory cytokine production.[5] Weight loss or difficulty in maintaining adequate body condition is, however, not uncommon in aged horses and in one survey[4] 17% of owners reported weight loss in their elderly horse within the last 12 months. Clinical evaluation of weight loss in an old horse should also rule out other common causes such as renal and hepatic disease as well as pain associated with arthritis (especially of the neck, restricting ability to reach grass and/or ground-based feed bowls, etc.), acute or chronic diarrhoea, and inflammation associated with infections. The risk of weight loss may be higher in cold winter or very hot summer months due to the type of forage available and potentially increased difficulties with thermoregulation.[6] Therefore, ration and general management may be contributory factors in the failure of old horses to maintain acceptable body condition, especially in the winter months.

It has been hypothesized that chronic parasitic scarring of the large intestine and abnormal dentition may have contributed to the apparent malabsorption and/or maldigestion (reduced apparent digestion of fibre, protein and phosphorus) reported in some earlier studies of digestion in aged horses.[7] Decreased diversity of the gastrointestinal microflora was found in healthy, older horses regardless of diet,[8] but in non-obese healthy mares of similar stock-type breeding, all of which had received regular anthelmintic treatment and had normal dentition, no decrease in the digestibility of key macro- and micro-nutrients was found[9,10]. Sarcopenia (decrease in skeletal muscle mass) is often noted in older horses (especially those with PPID[11]) although it has been suggested[12] that whilst there may be lower rate of muscle protein synthesis in aged horses, 'age had no effect on any other indicators of whole body or muscle protein synthesis or on measures of systemic or muscle inflammation.' These authors suggested, therefore, that protein requirements may not differ according to age in healthy individuals. However, previously in another study of both older (>20 years) and younger horses on a light exercise programme, improved maintenance of muscle mass was shown, regardless of age, when supplementary lysine and threonine were fed,[13] suggesting that protein quality continues to be important in older animals. Interestingly, the presence of certain common dental abnormalities, such as points or hooks <3 mm in size, did not adversely affect digestion of nutrients in middle-aged (10 to 15 years old) horses[14] and may not affect fecal particle size or apparent digestibility of nutrients[15]. However, these researchers noted that extremely poor dentition might adversely impact feed intake and may contribute to weight loss through reduced feed intake. In addition, there is evidence that age is associated with increasing severity of degenerative changes in the equine temporohyoid joint[16] which may predispose horses to temporohyoid osteoarthropathy, which in turn may cause difficulty in chewing. Even if significant dental abnormalities are not present, periodontal disease (inflamed gums and buccal/lingual lesions) can be painful[14] and affected horses may demonstrate a reduced rate of feed intake (an issue with group feeding in particular) with partial or even complete anorexia[17]. Further work is still needed to evaluate the nutrient requirements of compromised older horses and those with certain dental disorders.

In humans, there is a progressive decrease in insulin sensitivity with age. Work in horses[18] has shown that in response to a sweet feed meal, healthy aged horses had a greater peak insulin concentration and area under the curve for insulin, than adult horses, regardless of the background diet they had been fed for 5 weeks prior to the meal challenge and our recent work has suggested that dietary effects on insulin dynamics need to be considered with respect to the optimal diet for older horses[19,20].

MANAGEMENT CONSIDERATIONS

General management considerations in the elderly horse include:

- Provision of adequate shelter but avoidance of prolonged confinement to stalls if possible.
- Protection from extremes of environmental conditions.
- Thorough, regular dental care and strict attention to control of internal/external parasites.
- Clipping long hair coats in hot weather if necessary. If the horse has hypertrichosis, it will have increased sweat losses in warm weather (unless clipped) and should have free access to both water and salt.
- Active strategies to ensure all year round adequate water intake.
- Water and feed containers at an appropriate location and height to optimize access - especially if any arthritis in the neck, etc.
- Regular monitoring of bodyweight/condition - the amount of feed needs to be adjusted according to whether body condition loss or gain is required - and particular attention paid during extreme environmental conditions.
- Choosing field companions carefully to avoid bullying.
- Regular foot trimming and ideally choosing a flat paddock, free from poaching and ruts, to decrease the strain on joints.
- Regular blood work/veterinary evaluations to detect onset of common disease/ageing infirmities.

NUTRITIONAL CONSIDERATIONS

The author's current practical recommendation is to maintain the vitamin and micronutrient intake at least at recommended maintenance levels and perhaps more optimally at the levels typically recommended for exercising animals. Vitamin C supplementation may be beneficial, in that aged horses have been reported to have lower plasma ascorbic acid concentrations than younger healthy horses with an apparently beneficial effect of supplemental Vitamin C on the antibody response to vaccination.[21] In addition, the incidence of RAO increases with age in horses and the concentration of ascorbic acid in plasma and pulmonary epithelial lining is significantly reduced in RAO-affected horses.[22]

With respect to their other nutritional needs, old horses can perhaps be divided into four groups as discussed below (for more information see[1,2]). It is important to realise that certain products marketed for 'the senior horse' may not be appropriate for all individuals and equally products not specifically marketed for the senior horse may be suitable for a particular individual. The potential benefits or otherwise of nutraceuticals will not be discussed:

Those that Are Maintaining Good Body Condition, Clinically Normal and "In Use"

This is the horse that is old in years but has no apparent age-related or clinically important disorders and is still being used for structured activity and/or reproduction. Such horses can maintain optimal bodyweight and condition on a standard "adult" ration formulated for their level of activity and reproductive status. For these animals, the nutritional advice would be to continue on the current management regimen, assuming it is balanced for their needs, but continue to monitor weight and to maintain routine vaccination, dental and parasite control programmes.

Those that Are Overweight/Obese But Clinically Normal

This is the horse that is old in years and still functioning normally, it may or may not have a reduced level of physical activity but does have a tendency to gain weight. For these animals, nutritionally it would seem, potentially, even more important to prevent them from being overweight compared to a younger animal – but, practically, this has to be balanced with the often increased difficulties of maintaining the body weight of older horses during the winter - especially when they have dental issues. For many older horses, especially if not prone to laminitis, having them in a slightly higher body condition (perhaps 6–7/9) just prior to entering the winter period may be the preferred option, although this is currently not proven. However, especially if overweight/obese coming out of winter, consider implementing appropriate managemental practices for weight loss. Continue to monitor weight and to maintain routine vaccination, dental and parasite control programmes.

Those with Difficultly to Maintain Weight But Clinically Normal

This is the horse that is old in years, clinically normal but no longer able to maintain weight on a standard adult diet. For these animals it is important to ensure that teeth maintenance and parasite control programmes are adequate and that hepatic and renal functions are within normal limits. Gradually change to a more calorie- and nutrient-rich diet. However, the author's current advice would be to try and obtain any increase in body condition score, through the use of meals that promote low (or low-moderate) insulin responses, i.e.,:

- Consider the use of more digestible forage (i.e., less mature grass hay).
- Consider including highly digestible fibre sources such as unmolassed sugar beet pulp/soya hulls as a means to increase the energy intake if required without using starch or sugar. Especially for those prone to laminitis, the author tends to recommend soaking and throwing away the water even from unmolassed sugar beet pulp to reduce as far as possible the WSC (and fructose) intake.
- If there are no contraindications, consider using oil as an energy source rather than cereal starch - especially for those animals not being exercised (remember to add gradually, balance the overall diet and add additional vitamin E at 1–1.5 IU/1 ml of oil). Not more than 1 ml oil/kg BW should be added unless advice has been obtained from an appropriately qualified nutritionist/veterinarian.
- If any cereals, other than oats, are fed make sure they are processed by cooking (e.g., steam flaking, micronizing) to make the starch more easily digested, reducing the risk of starch overload.
- Avoid feeding large grain-based meals: restrict meal sizes to <0.3 kg/100 kg BW of a cereal-based feed. For all horses ensure overall starch intake is <1 g starch/kg BW per meal and ideally <2 g/kg BW/day.
- Ideally ensure any meal/forage fed is sufficiently small in size and low in starch/sugar content so that it only produces a low/low-moderate insulin response.
- However, again practically for many older horses with poor dentition - access to grazing seems to be the most consistent way to promote a good body condition, especially during the late summer/autumn in preparation for winter.

The Truly Geriatric Horse

For these animals which have one or more physical or metabolic abnormalities as well as old age the nutritional advice would be to provide treatment and support for the specific underlying conditions. However, it may not be possible to address all of the issues in such horses adequately. It is better to feed something that reasonably meets the horse's nutritional needs and that it will consume voluntarily than to feed a ration it will not consume. Although some PPID animals may appear to be obese in certain areas (i.e., have abnormal fat pads), they may actually be in energy deficit.

- If a horse has clinical evidence of PPID but has not had laminitic episodes, it may be possible to maintain on good quality forage (pasture and/or hays) with the addition of a reduced starch/sugar adult or "senior" product with additional digestible fibre or oil supplementation if necessary to maintain or increase body weight/condition (see above).
- Especially for horses or ponies that have experienced repeated episodes of laminitis, and potentially for all PPID horses, it is recommended to analyse the forage and to feed a forage with <12% (DM basis) non-structural carbohydrate (NSC: starch, sugar and fructan and <10% WSC) content. Many

hays will be higher than this. Soaking grass hay in clean water (>8°C: winter tap water and ideally around 16°C: summer tap water) for around 90 min to 3 h may help to reduce the water-soluble carbohydrate (WSC) content but there will also be loss of DM and other nutrients. It is important to note that any residual water has to be treated as effluent and soaking for prolonged periods especially in warm conditions is not currently recommended. As the results from soaking are variable, it is advisable, if concerned, to ensure the original forage has a low NSC content or feed an appropriate forage replacer. Obviously reducing the WSC content will reduce the energy content of the hay and this, coupled with the loss of DM content, needs to be taken into consideration, as many PPID horses show weight loss.

- Feed a broad-spectrum vitamin, amino acid and mineral supplement/balancer, especially if reduced levels of complementary feed are given (or forage is the main component of the diet and especially if soaked). Ensure an adequate and balanced intake of magnesium. The author is not aware of any published scientific evidence, however, to suggest that high levels of magnesium will be protective and reduce the risk of laminitis or PPID.
- Old horses with PPID have been documented to have higher fecal gastrointestinal parasite egg counts than healthy age-matched controls or younger animals. Heavy intestinal parasite burdens may therefore contribute to weight loss in these animals and the implementation of a more rigorous anthelmintic treatment regimen may be indicated in these circumstances.

HORSES WITH INADEQUATE DENTITION

For all old horses, even those with severely compromised dentition, the ration should be based on forage/high-fibre feeds (as for all horses). Unless choke is a problem, hay, preferably high-quality grass or grass/legume mix, can be offered free choice. However, these animals may not be able to adequately masticate long stem, especially mature hays. When available, and if laminitis is not an issue, pasture turnout may be the most appropriate forage source, as grass appears to require less mastication than dried long-stem forages and is usually an excellent source of most of the nutrients required by healthy aged horses not in active work (an appropriate 'balancer' should be fed to ensure overall nutrition). Alternatively chopped hays or haylages, soaked beet pulp or hay/grass cubes divided into at least two or, preferably, three or more feedings per day (offering at least 1.5 to 2.0% BW in dry feed weight overall) can be used. Processed fibre and nutrient sources, especially the latter two, can be soaked in water to reduce the risk of choke and increase water intake. Pelleted or extruded "complete" feeds may also be used, especially for horses that have little or no ability to masticate even chopped hays adequately, etc. Again these should also be soaked in water to reduce the risk of choke, gradually introduced where possible, and ideally offered in limited amounts (0.1–0.5% BW) per feeding to prevent spoilage and maximize intake. Ensure good protein quality as well as adequate total intake. Additional vegetable oil (see above) can be added if required (and clinically appropriate) to help maintain weight. Whole grains and even processed grain mixes may not be suitable for horses with marked dental abnormalities.

Ageing horses should be treated as individuals, and by providing targeted nutrition alongside appropriate management and veterinary attention, we can help to ensure they maintain as healthy and active a life as possible.

REFERENCES

1. Harris PA, Ralston SL. In: *Equine Clinical and Applied Nutrition* Elsevier;2013:289.
2. Jarvis, *et al. Equine Vet Educ.* 2017 (in press).
3. Phillips F. *Nutrition Foundation.* 2003;28:253.
4. Ireland JL, *et al. Equine Vet J.* 2011;43(1):37.
5. Adams AA, *et al. Vet Immunol Immunopathol.* 2009;127:286.
6. McKeever KH, *et al. Med Sci Sports Exercise.* 2001;32:s156.
7. Ralston SL, *et al. J Equine Vet Sci.* 2001;21(7):310.
8. Dougal K, *et al. PLoS One.* 2014;9(2):e87424.doi:10.1371 /journal.pone.0087424.
9. Elzinga, *et al. J Equine Vet Sci.* 2014;34(10):1164.
10. Elzinga, *et al. J Equine Vet Sci.* submitted.

11. McFarlane D. *Vet Clin North Am Equine Pract*. 2011;27:93.
12. Wagner AL, *et al*. *Am J Vet Res*. 2013;74(11):1433.
13. Graham-Thiers PM, Kronfield DS. *J Anim Sci*. 2005;83:2783.
14. Ralston Sl, *et al*. *Equine Vet J*. 2001;33(4):390.
15. Carmalt JL, Allen A. In: AAEP proceedings; 2008:386.
16. Naylor, *et al*. *Equine Vet J*. 2010;42(5):425.
17. Knottenbelt DC. In: *Equine Dentistry*. Philadelphia, PA: WB Saunders; 2003:127.
18. Rapson JL. PhD thesis. Michigan University; 2015.
19. Jacob S, *et al*. 2017 ACVIM abstract.
20. Jacob S, *et al*. *Equine Vet J*. 2017 Submitted.
21. Ralston SL, *et al*. *Am J Vet Res*. 1988;49:1387.
22. Deaton, *et al*. *Exp Lung Res*. 2005;31:653.

acvim213

Vascular Function, Insulin Dysregulation, and Endocrinopathic Laminitis

Anne A. Wooldridge, DVM, MS, PhD, DACVIM (Large Animal Internal Medicine)
Auburn, AL, USA

Introduction

Laminitis is a clinical syndrome resulting from disruption of the lamellar attachments that suspend the horse's skeleton within the hoof. This leads to lameness and in severe cases, catastrophic displacement of the third phalanx. Broad categories of causes of laminitis include inflammatory events, endocrinopathic (hormonal) influences, and mechanical overload. The diseases that commonly put a horse at risk of laminitis due to inflammation include those that lead to severe compromise of the gastrointestinal wall (i.e., intestinal ischemia, colitis, anterior enteritis) and septicemia (i.e., acute metritis, coliform pleuropneumonia). In recent years, it has become evident that endocrinopathic causes of laminitis associated with insulin resistance and dysregulation may be the most common.

Insulin Resistance/Dysregulation

Fasting hyperinsulinemia, excess insulin response to carbohydrate feeding, and decreased tissue response to physiologic concentrations of insulin (insulin resistance) collectively are referred to as insulin dysregulation in horses. A key feature of equine metabolic syndrome is insulin dysregulation. Insulin dysregulation is evaluated through measurement of serum insulin concentrations after fasting, post-prandially, after a measured oral carbohydrate load, or after intravenous administration of both insulin and glucose. Many studies have shown direct and indirect links between hyperinsulinemia and laminitis; however, the exact mechanism is still unclear. In a landmark study (Asplin *et al.* 2007), laminitis was induced experimentally in healthy ponies with no history of laminitis by maintaining prolonged hyperinsulinemia with euglycemia, but the mechanism by which insulin induces laminitis has not been determined.

Endocrinopathic/Hyperinsulinemic Laminitis

Endocrinopathic laminitis due to hormonal dysfunction is most commonly seen in association with pituitary pars intermedia dysfunction (PPID), pasture grass consumption, corticosteroid use, and equine metabolic syndrome. Insulin resistance/dysregulation is common to these causes. The pathology of endocrinopathic laminitis from hyperinsulinemia models demonstrates architectural distortion that is characterized by lengthening and attenuation of the secondary epidermal lamellae (SEL) and rounding of basal cell nuclei, epidermal apoptosis and mitosis. Basement membrane separation is present but less acutely and less severely than inflammatory causes of laminitis (Karikoski *et al.* 2014). Lesions in naturally occurring cases of endocrinopathic laminitis (with hyperinsulinemia) have features of both inflammatory and hyperinsulinemic models of laminitis with apoptosis, hyperplasia, aberrant keratin, tapering of epidermal lamella and no extensive basement membrane failure (Karikoski *et al.* 2015). Analysis of laminar pathological change in horses with PPID showed that only horses with both PPID and insulin dysregulation had clinical or pathological changes of laminitis and lesions were almost identical to those found in naturally occurring hyperinsulinemic cases (Karikoski *et al.* 2016). The evidence for insulin affecting the lamina is strong; the mechanism is still elusive.

Insulin Effects on Endothelial and Vascular Function

The only lamellar tissues with the tyrosine kinase insulin receptor (IR) are the vascular endothelium and interstitial cells, so these may be target tissues for insulin to directly assert effects (Burns *et al.* 2013). Another putative target of insulin's effects on laminar tissue is signaling through the insulin growth factor receptor (IGF-1R), which is highly expressed in laminar epithelial cells (Burns *et al.* 2013). This receptor also requires higher concentrations of insulin to activate, which would be consistent with pathology induced by hyperinsulinemia. The IR is not expressed on laminar epithelial cells (Burns *et al.* 2013). The lack of cell culture models for laminar epithelial cells or laminar endothelial cells (or the combination) in horses has made research on this type of signaling challenging. Research in humans, animal models, and cell culture models has shown that endothelial function and repair mechanisms are impaired by insulin resistance.

A key feature of insulin resistance is specific impairment of phosphatidyl inositol 3 kinase (PI3K)-dependent insulin signaling pathways, whereas other insulin-signaling branches including Ras/mitogen-activated protein kinase (MAPK)-dependent pathways are unaffected (Muniyappa *et al.* 2007). The PI3K branch of insulin signaling regulates glucose metabolism in skeletal muscle, adipose tissue, and liver while stimulating nitric oxide (NO) production and vasodilation in vascular endothelium; while the MAPK branch regulates growth and mitogenesis and controls secretion of endothelin-1 (ET-1), a potent vasoconstrictor. Insulin-resistant animals often have compensatory hyperinsulinemia to maintain euglycemia. As previously mentioned, laminitis was induced experimentally in healthy ponies with no prior history of laminitis by maintaining prolonged hyperinsulinemia with euglycemia (Asplin *et al.* 2007). Another study showed that glucose uptake to hoof lamellae is insulin-independent; lamellar epidermal basal cells are rich in glucose transporter (GLUT)-1 proteins, but lack insulin receptors (Asplin *et al.* 2007). In the vasculature, hyperinsulinemia will overdrive unaffected MAPK-dependent pathways leading to an imbalance between PI3K- and MAPK-dependent functions of insulin. Upregulation of the MAPK-dependent pathways increases endothelin-1 (ET-1) gene expression and ET-1 release from endothelial cells. This promotes vasoconstrictor tone and contributes to abnormal vascular function. In addition, increased stimulation of MAPK-dependent pathways has been implicated in disruption of endothelial integrity as well as increased vascular permeability and microvascular leaking (Muniyappa *et al.* 2007).

Evidence for Vascular Changes in Equine Laminitis

The main physiologic observation in animals with endothelial dysfunction is deficient endothelium-dependent vasorelaxation, and insulin resistance is a major cause of this in humans. It has been shown that ET-1 expression in the lamellae is increased during both acute and chronic laminitis and likely plays a role in the pathophysiology of laminitis via its vasoconstrictive effects (Katwa *et al.* 1999). Dysfunction of the laminar vasculature at both the level of the digital arteries and laminar microvasculature during the prodromal phases of experimentally induced laminitis have been shown (Adair *et al.* 2000; Eades *et al.* 2007). The laminar venules are predisposed to venoconstriction (Peroni *et al.* 2006), and the contractile responses of laminar vessels are altered by incubation with insulin (Keen *et al.* 2012).

At physiologic concentrations, insulin actually causes endothelial-dependent vasorelaxation due to PI3K-activated release of NO (Figure 1). Laminar veins and arteries have a differential relaxation response to insulin *ex vivo*. Laminar veins do not demonstrate insulin-induced relaxation *ex vivo*, but laminar arteries do relax in response to insulin (Wooldridge *et al.* 2014). Vascular dysfunction in response to increased insulin concentrations can be induced in digital vessels *ex vivo* and may be useful for studying various insulin signaling pathways and interventions (Venugopal *et al.* 2011). Intact laminar veins, arteries and small facial arteries from horses with endocrinopathic laminitis in a recent *ex vivo* vessel contractility study demonstrated deficient endothelial-dependent vasorelaxation (Morgan *et al.* 2016). All affected horses in that study had fasting hyperinsulinemia. Since all vessels were affected, this suggested systemic endothelial dysfunction supporting an underlying endocrinopathy in these cases. Contractile responses of the laminar veins from affected horses were also increased, further supporting endothelial dysfunction (Morgan *et al.* 2016). Failure of blood vessels in the hoof to adequately dilate could result in overall reduction of blood flow and venous return and lead to tissue damage.

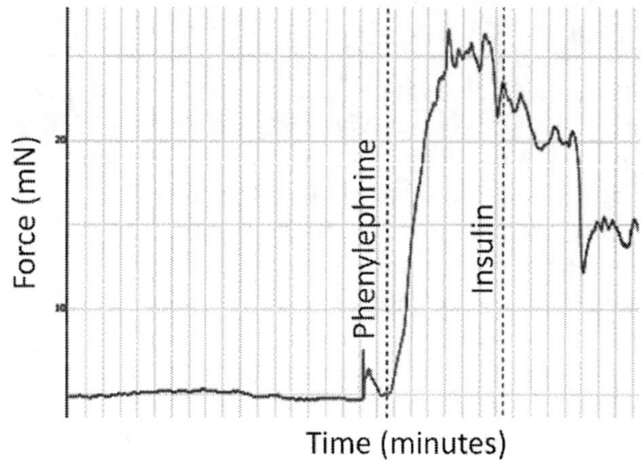

Figure 1. Example of insulin-induced vasorelaxation in *ex vivo* laminar arteries

Prolonged hyperinsulinemia did not increase expression of insulin signaling proteins in the lamina, but did in other tissues. There was a trend toward downregulation of intracellular insulin signaling molecules in the lamina, but changes did not reach significance in that small sample set (Campolo *et al.* 2016). Gene expression of both the IR and IGF-1R decreased with hyperinsulinemia at 24 h and 46 h after beginning insulin infusion in another study, which may indicate that both receptors were activated and then downregulated (de Laat *et al.* 2013). The question remains, though, what happens after activation of those receptors to induce laminitis? It may be that both receptors are activated and there is subsequent cross-talk between the endothelial cells and the epithelial cells. A combination of effects on epithelial cell proliferation and phenotype and endothelial dysfunction/blood flow may be occurring.

The progression to chronic laminitis, once the original insult has been leveled, may also include disrupted blood flow and impaired vascular function. Endothelial dysfunction is likely persistent in naturally occurring cases of endocrinopathic laminitis and may contribute to repeated episodes of laminitis. Vascular filling defects and tissue distortion are evident even 7 days after experimental laminitis induction using venography (Baldwin, Pollitt 2010). More displacement of the distal phalanx results in compression of vessels and ischemia to affected areas. Therapeutic interventions to restore vascularization and help reverse pathology are needed in chronic cases.

ENDOTHELIAL REPAIR MECHANISMS

A new area of endothelial dysfunction associated with insulin resistance that needs exploration in the horse is repair of damaged endothelium. Endothelial progenitor cells (EPCs) function in *de novo* neovascularization and in repair of damaged vasculature. EPCs are mobilized from the bone marrow or from the vessel wall in response to endothelial damage. They home to sites of injury, roll along the vessel wall, and firmly adhere (Seeto *et al.* 2013). After adherence, they proliferate and differentiate into mature endothelial cells to reconstruct areas of damage. A decrease in the number and functional ability of circulating EPCs has been shown in people with atherosclerosis, type 2 diabetes, obesity, and metabolic syndrome (Bakogiannis *et al.* 2012). Endothelial progenitor cells have been used therapeutically to treat non-healing wounds, limb ischemia, and cardiovascular ischemia, and they are important for the vascularization of engineered tissues (Chavakis, Dimmeler 2011). EPCs have recently been isolated and characterized in horses (Figure 2) (Salter *et al.* 2015; Sharpe *et al.* 2016). Consistent yield of EPCs from peripheral blood will allow future studies comparing number and function in various disease states and age ranges in the horse. Horses with metabolic disease, obesity, and insulin dysregulation may have similar dysfunctions in vascular homeostasis and repair mechanisms mediated by EPCs, as do people. Optimization of isolation and characterization of EPCs from horses is needed to determine if these cells will be therapeutically or diagnostically useful.

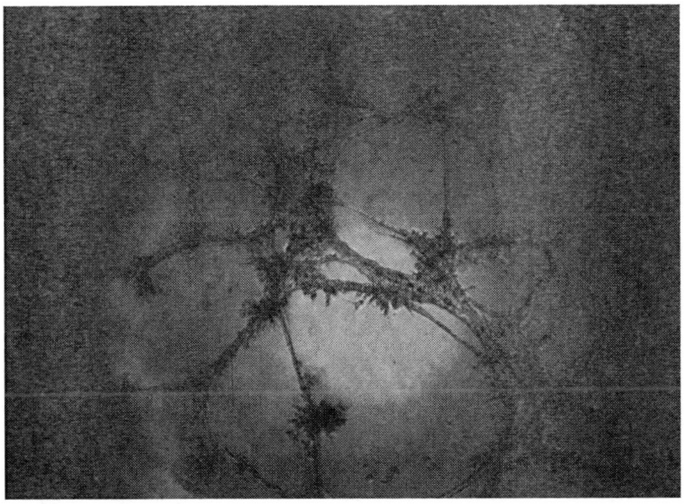

Figure 2. Endothelial progenitor cells from a horse forming vascular tubules *in vitro*

CONCLUSIONS

Insulin dysregulation has a primary role in the development of the most common type of laminitis, endocrinopathic laminitis. Although the mechanism is still unknown, large strides are being made in our understanding of the role of hyperinsulinemia in the pathogenesis of laminitis. There is evidence for endothelial dysfunction and vascular damage in endocrinopathic laminitis, as well as changes in laminar epithelial cell proliferation and phenotype. Specific identification of the relationship between insulin and laminitis will lead to new therapeutic interventions.

REFERENCES

1. Adair HS 3rd, *et al. Am J Vet Res.* 2000;61:862-868.
2. Asplin KE, *et al. Vet J.* 2007;174:530-535.
3. Bakogiannis C, *et al. Curr Med Chem.* 2012;19:2597-2604.
4. Baldwin GI, Pollitt CC. *Vet Clin North Am Equine Pract.* 2010;26:135-140.
5. Burns TA, *et al. Equine Vet J.* 2013;45:326-332.
6. Campolo A, *et al. Domest Anim Endocrinol.* 2016;55:41-45.
7. Chavakis E, Dimmeler S. *Antioxid Redox Signal.* 2011;15:967-980.
8. de Laat MA, *et al. Vet J.* 2013;197:302-306.
9. Eades SC, *et al. Am J Vet Res.* 2007;68:87-94.
10. Karikoski NP, *et al. Vet Pathol.* 2015;52:945-956.
11. Karikoski NP, *et al. Am J Vet Res.* 2014;75:161-168.
12. Karikoski NP, *et al. Equine Vet J.* 2016;48:472-478.
13. Katwa LC, *et al. Equine Vet J.* 1999;31:243-247.
14. Keen JA, *et al. J Vet Pharmacol Ther.* 2012.
15. Morgan RA, *et al. PLoS One.* 2016;11:e0163815.
16. Muniyappa R, *et al. Endocrine Reviews.* 2007;28:463-491.
17. Peroni JF, *et al. J Appl Physiol.* 2006;100:759-763.
18. Salter MM, *et al. Am J Vet Res.* 2015;76:174-187.
19. Seeto WJ, *et al. Acta Biomater.* 2013;9:8279-8289.
20. Sharpe AN, *et al. Am J Vet Res.* 2016;77:1157-1165.
21. Venugopal CS, *et al. Equine Vet J.* 2011;43:744-749.
22. Wooldridge AA, *et al. Vet J.* 2014;202:83-88.

acvim214

Adipokines and Equine Metabolic Function

Anne A. Wooldridge, DVM, MS, PhD, DACVIM (Large Animal Internal Medicine)
Auburn, AL, USA

INTRODUCTION

Insulin resistance and dysregulation in the horse is primarily associated with conditions such as the equine metabolic syndrome and pituitary pars intermedia dysfunction (PPID). A driving force behind much of the interest and research in this area in horses is the relationship between laminitis and insulin dysregulation. Research into the relationship between obesity and insulin resistance has resulted in identification of a class of signaling molecules from white adipose tissue called adipokines or adipocytokines (Gualillo *et al.* 2007). Adipose tissue was originally thought to be only an energy storage depot but is now recognized as an endocrine organ with an essential role in the maintenance of whole body homeostasis (Scherer 2006). The adipocytokines include leptin, adiponectin, resistan, visfatin, tumor necrosis factor (TNF)-alpha, plasminogen activator inhibitor (PAI)-1, interleukin (IL)-1, IL-6, and others. Adipokine release from the adipocyte changes with obesity and inflammatory stimuli, and obesity is now known to lead to a chronic inflammatory state in humans.

Equine metabolic syndrome (EMS) is a metabolic/endocrine disorder that commonly affects ponies and certain breeds of horses. The clinical definition of EMS is regional or generalized obesity, insulin dysregulation (fasting hyperinsulinemia, excessive insulin responses to carbohydrate feeding, and insulin resistance), and subclinical or clinical laminitis. Other abnormalities that may be caused by or linked to EMS are dyslipidemia, hyperleptinemia, hypoadiponectinemia, arterial hypertension, altered reproductive cycling in mares, and increased systemic markers of inflammation in association with obesity. Insulin dysregulation is an extremely important part of the pathophysiology of EMS as evidenced by the finding that severe hyperinsulinemia has been used to induce laminitis in otherwise clinically normal horses (Asplin *et al.* 2007). Research into causes of insulin dysregulation in horses has led to investigation into components of a metabolic profile which includes adipokines. The two most well-characterized adipokines in horses are leptin and adiponectin.

LEPTIN

Leptin is an adipokine that controls appetite and metabolism through its effects on specific receptors in the brain, and circulating leptin concentrations are tightly correlated with body fat mass. Adipocytes are the primary sources of leptin which is encoded by the *ob* gene. Leptin transcription increases with inflammatory stimulation and obesity and is inversely correlated with testosterone concentrations. Two well-known mouse models for obesity and diabetes research are the leptin deficient mouse (*ob/ob*) and the leptin receptor deficient mouse (*db/db*). These mutations are rare in humans and have not been identified in horses.

Leptin Function

The primary function of leptin is to suppress appetite and increase thermogenesis. Appetite suppression is centrally mediated and is due to stimulation of catabolic and anorexigenic neurons and suppression of anabolic and orexigenic neurons. Leptin also functions as a growth factor and mitogen. When leptin is increased in the obese state or in an aging animal, partial leptin resistance may occur. This could be due to saturation of leptin receptors, receptor down regulation, or modification of downstream signaling pathways. Leptin resistance results in alteration of metabolic function and contributes to the development of cardiac, renal, and vascular dysfunction seen in humans with obesity and type 2 diabetes. Excess leptin is pro-inflammatory and pro-thrombotic.

Leptin Signaling

The principal leptin receptor is the long isoform which is primarily expressed in the satiety center of the hypothalamus. Both the long and short isoforms of the leptin receptor are found in other tissues. There is also a circulating leptin receptor which acts as a binding protein. The long form isomer signals via the janus kinase (JAK) family with subsequent activation of the STAT transcription factors (JAK/STAT pathway). The shorter leptin receptor isoforms cannot activate STAT. The various STATS promote transcription of genes mediating the various metabolic activities of leptin. STAT signaling also induces suppressors of cytokine signaling (SOCS) that work in negative feedback loops to suppress leptin signaling. Other signaling targets of the leptin receptor are extracellular signal related kinases (ERK) and insulin receptor substrate 1 (IRS-1) (Radin *et al.* 2009).

Adiponectin

Adiponectin is an adipokine of particular interest because of its profound insulin sensitizing and anti-inflammatory effects and its inverse relationship with insulin and fat mass. Adipocytes produce adiponectin as 30 kDa monomers, which are assembled into higher-order multimers within the adipocyte for secretion. The majority of circulating adiponectin secreted by adipocytes consists of trimers (low molecular weight, LMW), middle molecular weight hexamers (MMW), and high molecular weight multimers (HMW), each of which has different biological activities that contribute to insulin-sensitizing and anti-inflammatory functions. A proteolytic fragment of adiponectin called globular adiponectin has also been identified in human plasma. Adiponectin circulates at high concentrations (2–20 μM) in plasma, and decreased total and HMW adiponectin concentrations have been identified in humans with cardiovascular disease, obesity, metabolic syndrome, and type 2 diabetes. There is not much fluctuation in adiponectin concentrations during the day and the half-life is short (Ruan, Dong 2016).

Adiponectin Function

The different forms of adiponectin have distinct target tissues and functions which typically increase insulin sensitivity and signaling and decrease inflammation. In liver, adiponectin decreases gluconeogenesis and lipolysis. In skeletal muscle it increases fatty acid oxidation and insulin-stimulated glucose uptake and decreases triglyceride accumulation. In adipose tissue, adiponectin stimulates proliferation and storage of fat in small adipocytes and increases insulin-stimulated glucose uptake. Adiponectin stimulates insulin secretion in the pancreas and supports beta cell survival. Endothelial oxidative stress is decreased and endothelial nitric oxide synthase activity is increased by adiponectin which support endothelial function. Importantly adiponectin also mitigates inflammation in adipose tissue and promotes the anti-inflammatory macrophage phenotype (Ruan, Dong 2016).

Adiponectin Signaling

There are currently 3 receptors that have been identified for adiponectin: AdipoR1, AdipoR2, and T-cadherin. All forms of adiponectin signal through AdipoR1 and AdipoR2, but only the HMW multimers bind the T-cadherin. The function of the T-cadherin adiponectin receptor is currently unknown but it may be a binding protein that modulates HMW adiponectin signaling. HMW adiponectin is the most active and insulin sensitizing form of adiponectin, presumably from its greater affinity for adiponectin receptors. AdipoR1 is ubiquitously expressed but highest concentrations are found in skeletal muscle. AdipoR2 is found mainly in the liver. Both are essential for insulin sensitivity in target tissue and expression of metabolic genes. Downstream targets of adiponectin signaling include adenosine monophosphate kinase (AMPK, an important sensor for metabolic changes), peroxisome proliferator receptor (PPAR)α, PPARγ, mitochondrial biogenesis, ceramide, and crosstalk with the canonical insulin signaling pathway (Ruan, Dong 2016).

Adipokine Evaluation in Horses as Part of the Metabolic Profile

Leptin

Measurement of leptin is commercially available in the USA through Cornell University Animal Health Diagnostic Center and normal ranges with the current test are considered to be less than 10 ng/ml. Leptin concentrations are consistently elevated in horses with obesity as a component of part of equine metabolic syndrome (Frank, Tadros 2014). Induced obesity in two EMS-prone breeds and one low-risk breed of horses showed no correlation between loss of insulin sensitivity and leptin concentrations (Bamford *et al.* 2016). A large ongoing study looking at metabolic derangements associated with the laminitis-prone phenotype has shown association of leptin with obesity but no association of leptin with laminitis (McCue *et al.* 2015).

Adiponectin

Adiponectin has been measured by radioimmunoassay in several studies but concentrations were lower than what was reported for other species. Validation of a HMW adiponectin ELISA showed concentrations comparable to other species, but commercially available ELISAs for total adiponectin have not been successful in the horse. Concentrations of HMW adiponectin in healthy, lean mares and geldings were 8.0 ± 4.6 μg/mL versus 3.7 ± 3.8 μg/mL in obese horses (Wooldridge *et al.* 2012). Note the microgram concentrations of adiponectin versus nanogram concentrations of leptin. Measurement of total or HMW adiponectin is not commercially available for horses in the USA at this time. Induced obesity in two EMS-prone breeds and one low-risk breed of horses showed strong correlation between loss of insulin sensitivity and decreased HMW adiponectin concentrations (Bamford *et al.* 2016). Total adiponectin concentrations were significantly decreased in ponies with pasture-associated laminitis in a group of 446 ponies (Menzies-Gow *et al.* 2016). Decreased HMW adiponectin concentrations were observed in the laminitis-prone phenotype in initial reports from a large ongoing study (McCue *et al.* 2015). A genome-wide association study in a cohort of 232 Welsh ponies showed that low adiponectin concentrations were associated with loci on equine chromosome 1 suggesting a genetic potential for hypoadiponectinemia that could be linked with EMS-prone breeds (Norton *et al.* 2016).

ADIPOKINES AS DRUG TARGETS

Current treatment for equine metabolic syndrome focuses on management changes (low carbohydrate diet and consistent exercise) and therapeutic hoof care. However, laminitis risk is still very high in horses with persistent hyperinsulinemia, and management changes are often difficult to consistently implement. Pharmacologic interventions are used to more effectively regulate hormonal functions, and more research is needed to find appropriate interventions in the horse.

The importance of adiponectin in metabolic disease and the possible link between hypoadiponectinemia and the laminitis prone phenotype in horses with EMS makes adiponectin a powerful potential therapeutic target. However, its short half-life, high circulating concentrations, and mixture of complexes that contribute to overall function make exogenous adiponectin administration difficult. Utilizing pharmaceutical interventions to increase endogenous production of biologically active fractions of adiponectin may provide a new avenue for management of EMS.

The most commonly prescribed medication in humans and horses for insulin resistance is metformin, an orally administered biguanide agent, which is primarily utilized in humans to stabilize blood glucose concentrations and improve insulin sensitivity by its action on insulin receptor expression in target tissues. Metformin has shown inconsistent effects on circulating adiponectin levels in humans. Use of metformin in horses has shown inconsistent results and challenges in appropriate dosing. Pharmacokinetic studies using a dose of 15 mg/kg orally showed limited oral bioavailability and fast elimination (Hustace *et al.* 2009). Some work in horses using metformin at 15 mg/kg has shown improved insulin sensitivity in the short term (Durham *et al.* 2008); however, other studies were unable to show an effect (Tinworth *et al.* 2012). Further research has shown metformin (30 mg/kg before a meal) resulted in improved insulin response and lower glucose peak following an oral sugar test in normal horses and those with induced insulin resistance (Rendle *et al.* 2013). No studies have examined metformin's effects on adipokine secretion in horses.

Another class of insulin sensitizing drugs is the thiazolidinedione class, which contains the commonly-prescribed drug pioglitazone, which is considered very safe in humans. Thiazolidinediones bind to PPARγ, which results in proliferation of smaller, more insulin-sensitive adipocytes and increased production of adiponectin, both of which have powerful effects on improving insulin sensitivity (Otto *et al.* 2006; Rasouli *et al.* 2006). Pioglitazone is also capable of consistently increasing production of total and HMW adiponectin in healthy and insulin-resistant humans (Sharma *et al.* 2006; Ikeda *et al.* 2007). The pharmacokinetics of pioglitazone in horses show it is bioavailable with 1 mg/kg oral dosing every 24 hours (Wearn *et al.* 2011). Initial research for pioglitazone in a lipopolysaccharide-induced model of insulin resistance did not show clinical benefit in terms of improving insulin sensitivity or reducing inflammatory markers (Suagee *et al.* 2011; Wearn *et al.* 2012). Triglyceride concentrations and leptin transcript abundance in adipose tissue did not change after 12 days of pioglitazone treatment in healthy horses (Suagee *et al.* 2011; Wearn *et al.* 2012). The potential ability of pioglitazone to increase circulating adiponectin concentrations has not been investigated in horses.

CONCLUSIONS

Adipokines are essential signaling molecules involved in metabolic regulation. Leptin and adiponectin are the most well characterized in horses, and evidence suggests that high leptin concentrations are associated with obesity, but low adiponectin concentrations are associated with predisposition to laminitis and insulin resistance but not necessarily obesity. Therapies that change adipokines, particularly adiponectin, may be useful drug targets for control and treatment of insulin dysregulation in horses.

REFERENCES

1. Asplin KE, *et al. Vet J.* 2007;174:530–535.
2. Bamford NJ, *et al. Vet J.* 2016;214:14–20.
3. Durham AE, *et al. Equine Vet J.* 2008;40:493–500.
4. Frank N, Tadros EM. *Equine Vet J.* 2014;46:103–112.
5. Gualillo O, *et al. Trends Cardiovasc Med.* 2007;17:275–283.
6. Hustace JL, *et al. Am J Vet Res.* 2009;70:665–668.
7. Ikeda Y, *et al. Diabetes Care.* 2007;30:e48.
8. McCue ME, *et al. J Equine Vet Sci.* 2015;35:367–375.
9. Menzies-Gow NJ, *et al. Equine Vet J.* 2016.
10. Norton EA, *et al. J Vet Intern Med.* 2016;30:1504.
11. Otto C, *et al. J Endocrinol Invest.* 2006;29:231–236.
12. Radin MJ, *et al. Vet Clin Pathol.* 2009;38:136–156.
13. Rasouli N, *et al. Am J Physiol Endocrinol Metab.* 2006;290:E42–e46.
14. Rendle DI, *et al. Equine Vet J.* 2013;45:751–754.
15. Ruan H, Dong LQ. *J Mol Cell Biol.* 2016;8:101–109.
16. Scherer PE. *Diabetes.* 2006;55:1537–1545.
17. Sharma PK, *et al. Clin Endocrinol (Oxf).* 2006;65:722–728.
18. Suagee JK, *et al. J Vet Intern Med.* 2011;25:356–364.
19. Tinworth KD, *et al. Vet J.* 2012;191:79–84.
20. Wearn JG, *et al. Vet Immunol Immunopathol.* 2012;145:42–49.
21. Wearn JM, *et al. J Vet Pharmacol Ther.* 2011;34:252–258.
22. Wooldridge AA, *et al. Am J Vet Res.* 2012;73:1230–1240.

acvim215

Ataxia Is Not the Only Clinical Manifestation of Neck Injuries

Jean-Marie Denoix, DVM, PhD, agrégé, HDR, Cert.ISELP, LAIA-ECVDI, DACVSMR
Goustranville, France

Abstract - Take Home Message

The neck is an essential part of the horse locomotor system. Movements of the neck are required for the horse's longitudinal and lateral balance, as well as acceleration and deceleration. Lesions of the cervical spine can be responsible for neck pain, neck stiffness, back stiffness, poor performance, forelimb lameness and, in some cases, for peripheral or central neurological manifestations. In most cases, a careful clinical examination is required to identify subtle manifestations which can be correlated to bone or soft-tissue injuries of the cervical vertebral column. The routine diagnostic imaging approach of these injuries consists in the combination of radiography and ultrasonography. Cross-sectional imaging, such as CT scan or MRI, improves 3D representation of the anatomical relationships between bone and intervertebral joints, soft-tissue structures and the central and peripheral nervous system. The presentation will consist in a review of anatomy and biomechanics of the neck and a selection of didactic clinical cases illustrated by videos of the dynamic examination. Diagnosis will be supported by radiographs and ultrasound pictures as well as outcome after specific treatments.

Introduction

A large variety of clinical manifestations are related to neck problems. They include neck stiffness and pain, back stiffness, forelimb lameness, and athletic manifestations such as poor performance and left-to-right asymmetrical restriction of movements. The diagnosis of the cause(s) of these manifestations requires a careful clinical examination followed by conventional investigations of locomotion asymmetry ending in adequate imaging of the cervical region. Combination of radiography and ultrasonography has improved the understanding of the correlation between neck injuries and different complaints and clinical presentations. Advanced imaging techniques, such as CT scan and magnetic resonance imaging, will contribute to a better knowledge of this field of pathology of the locomotor system.

Functional Anatomy of the Cervical and Thoracolumbar Spines

The cervical intervertebral joints present thick intervertebral discs and flat articular processes allowing a wider intervertebral mobility than in the thoracolumbar spine.[1,2] The centre of rotation of the intervertebral joints is located within the vertebral head; therefore, during flexion there is a dorsal sliding of the vertebral head in the vertebral fossa of the previous vertebra, opening of the intervertebral foramens and widening between the articular processes. During extension there is a dorsal displacement of the vertebral fossa compared to the vertebral head of the following vertebra, narrowing of the intervertebral foramens, and increased pressure between the articular processes.

Flexion of the cervical spine (lowering of the neck) induces tension of the nuchal ligament and elongation of the dorsal cervical muscles. Maximum tension is observed in the nuchal ligament (especially the funicular, dorsal part) when lowering of the neck is associated with flexion of the head.[3] Because of the tension applied by the nuchal ligament over the top of the spinous processes of the withers, flexion of the neck is accompanied with flexion of the thoracic spine. Extension of the cervical spine (elevation of the neck) induces relaxation of the nuchal ligament, as well as relaxation of the supraspinal ligament in the thoracic area.

Lateroflexion is a wide movement in the cervical spine. Like in the thoracolumbar spine, lateroflexion on one side is associated with rotation on the opposite side between C3 and T1.[1,2] In the nuchal area, lateroflexion on one side is associated with rotation on the same side.

Clinical Manifestations Associated with Cervical Vertebral Column Lesions

Ataxia has often been considered as the main manifestation correlated to cervical spine abnormalities and little attention has been paid to other clinical manifestations.[4,5] Lesions of the cervical spine can be responsible for neck pain, neck stiffness, poor performance, forelimb lameness, and, in some cases, for peripheral or central neurological manifestations.

Physical Examination

Muscle atrophy of the dorsal cervical muscles is commonly associated with chronic neck pain, even at a low level. This is mainly visible over the splenius and semispinalis capitis location, between the vertebral axis and rhomboideus cervicis muscle (acting mainly on the thoracic limb). Limited or asymmetric active neck lateroflexion (performed using grass or a carrot) can be observed. In a limited number of cases palpation can reveal a thickening of one or several articular process joint(s). When a forelimb lameness is present, foot atrophy with high heels can be seen on the side of the lame limb which is often positioned in retraction while standing.

Dynamic Examination

At the walk on a 8 figure several manifestations can be observed, isolated or in variable combinations:

- Neck stiffness, visible mainly on the circles; the horse reduces neck lateroflexion and dorsoventral flexibility, especially at the base of the neck
- Reduction of the cranial phase of the stride of one or both forelimb(s)
- Neurological manifestations of dyskinesia (such as hypermetria or hypometria) on one or both forelimb(s)
- Neurological manifestations of ataxia or dyskinesia on one or both hind limb(s)

At the trot on a hard surface in straight line, head and neck motion may be reduced (again, especially at the base of the neck) and this induces also some thoracic stiffness because of the anatomo-functional relationships between these vertebral areas. A forelimb (and sometimes an ipsilateral hind limb) lameness can be seen; it typically shows up as a reduction of the cranial phase of the stride and defect of fetlock extension during the stance phase.

Neck stiffness can be observed in most of the clinical situations including the physical examination and examination of the horse at different gaits on hard and soft surfaces, on straight line and circles.

Some horses can demonstrate episode of **neck distress**, being unable to raise the neck even when stimulated. Because of the biomechanical relationships between the neck and thoracolumbar spine, **back stiffness** has been mentioned as a complaint and seen in relation with cervical injuries.

Forelimb lameness without ataxia is typically characterized by a reduction of the cranial phase of the stride (a manifestation not specific to neck problems), especially during the swing phase at walking (reduction of protraction). At trot, asymmetrical oscillations of the withers are clearer than those of the head. The lameness is often worse on the opposite circle. Reduction of fetlock extension of the lame limb during the stance phase (a manifestation not specific to neck problems) is also observed. As concomitantly to reduction of cranial phase of the stride, the horse shows an extension of the caudal phase putting stresses on the podotrochlear apparatus, diagnostic analgesia of the foot can change ('improve') the gaits. A bilateral reduction of protraction with short gaits is seen if both forelimbs are affected by neck lesions. How and why neck injuries can cause forelimb lamenesses? The reason is alteration of nerve conduction in the roots of the brachial plexus (ventral rami of the last cervical nerves and first thoracic nerves), responsible of muscle weakness and sensory deficits involving proprioception and muscle coordination. Forelimb dyskinesia during the wing phase and alteration of the carpus stability can be discrete clinical manifestations of neurological deficits affecting the forelimbs.

Hindlimb lameness or deficits (e.g., lack of abduction in some exercises) without ataxia can be seen in horses having cervical lesions and no injuries of the hindlimb joints and tendons nor any response to diagnostic analgesia. These manifestations can be correlated to weakness of specific muscle groups.

Any **kind of combination** between neck manifestations, forelimb and hindlimb lameness or dyskinesia can be seen on a single horse.

Athletic manifestations when the horse is at training or competing are sometimes difficult to reproduce during the classical clinical examination and require a careful listening to the complaints and examination of the horse at work to be identified. They include abnormal neck carriage, asymmetry on turns (left/right), difficulties in changing leads and manifestations of back pain (thoracolumbar stiffness, bucking, kicking…).

Differential Diagnosis Before Considering Further Investigations

Different pain locations can induce forelimb lamenesses with reduction of the cranial phase of the stride; the most common include suspensory ligament pain (especially proximal suspensory disease), shoulder joint pain, fetlock joint pain, distal and proximal interphalangeal joint pain, and elbow joint pain.

The differential diagnosis must also be done with thoracolumbar pain which may be responsible for reduction of motion in the neck.

Diagnostic Imaging of Cervical Lesions

Diagnostic imaging of the equine neck and back lesions is routinely based on 3 complementary techniques: radiography, ultrasonography, and nuclear scintigraphy. Some MRI machines enable imaging of the neck and CT scan machines with wide opening permit examination of most of the cervical spine in adult horses.[6-8] Our routine imaging approach of problems arising from the cervical spine consists in a combination of plain radiography and ultrasonography. Less than 20% of our patients are also examined using nuclear scintigraphy.

1 - Radiography

Radiography remains the basic imaging technique for the evaluation of the cervical spine. Our routine approach consists in making 3 laterolateral projections including the cranial part from the external occipital protuberance to the third cervical vertebra (C3), the intermediate part from C3 to C6, and the caudal part from C5 to the first (T1) and second thoracic vertebra (T2). Good superimposition of the left and right sides of the vertebra is crucial. Any rotation or oblique projection alters the evaluation of the size and shape of the articular processes, symphyses and vertebral canal. The caudal projection (C5-T2) must be exposed enough to allow good visualisation of the last cervical disc (between C7 and T1). Forty-degree oblique dorsolateral-ventrolateral projections are useful to evaluate separately the left and right articular processes. They provide a nice representation of the joint space between the articular processes on the side of the generator and a good imaging of the articular processes on the opposite side. A good knowledge of radiographic anatomy and its variations is essential to assess carefully the presence of findings.[9-11]

The radiographic criteria in relation with the presence of neck lesions are well known; they include the vertebral alignment, the size and shape of the vertebral canal,[12] the size, shape and architecture of the articular processes, the density of the vertebral pedicle, the size of the intervertebral disc, as well as shape of the vertebral fossa and corresponding vertebral head.

2 - Ultrasonography

Ultrasonography is now systematically used in association with radiography in the routine evaluation of neck problems in horses.[13] It is performed with 7.5 or 5 MHz convex probes. On a transverse section made at an intervertebral joint space, different anatomical structures can be seen in a dorsoventral direction: the caudal articular process of the cranial vertebra, the intervertebral joint space between the articular processes, the cranial articular process of the caudal vertebra, the vertebral artery and vein, the transverse process of the caudal vertebra.

Ultrasonography is very sensitive to any bone surface abnormalities such as hypertrophy or fragmentation of the articular processes. With this procedure, soft-tissue injuries of the synovial intervertebral joint space between the caudal and cranial articular processes (synovial fluid effusion, synovial membrane proliferation, capsulitis…) can be identified.

3 - Nuclear Scintigraphy

Nuclear scintigraphic examination of the neck is performed on both left and right sides from the cranium to the scapulae. Additional ventral views are made to establish significance or location of findings over the vertebral body axis.

FINDINGS AND LESIONS

Several findings and lesions can be found in horses presenting the clinical manifestations described above (ataxia not considered).

In the nuchal and cranial cervical area (cranium to axis) several lesions may have incidence on the horse behaviour and ability to perform. Dorsally to the vertebral axis, they include nuchal desmopathy and enthesopathy, semispinalis capitis tendinopathy and enthesopathy, nuchal subligamentous bursitis,[14] and obliquus or rectus capitis muscle dystrophy. At the ventral aspect of the vertebral axis, they include atlantooccipital membrane dystrophic mineralisation, longus capitis as well as longus colli enthesopathies.

In the mid-cervical region (between C2 and C5), abnormalities such as malalignment, stenosis of the vertebral canal, and fractures are more often associated with central neurologic manifestations.[15]

In the caudal cervical region and cervicothoracic junction (C5 to T1), a variety of injuries can be responsible for neck and back manifestations, lameness, and athletic manifestations.

- **Malalignment** between vertebrae is often associated with hypoplasia of the corresponding (cranial) vertebral fossa, and relative caudal elongation of the vertebral arch, a ventral flattening of the corresponding vertebral head (dysplasia) as well as a dorsal subluxation of the vertebral head inducing narrowing of the median diameter of the vertebral canal.
- **Articular process joints conditions** include juvenile osteochondral fragmentation, dysplasia and hypertrophy (often asymmetric), and primary or secondary osteoarthrosis. Synovitis and capsulitis are often associated with bony changes. Acquired fractures after fall are most commonly seen in this segment of the cervical spine.
- **Intervertebral symphyses lesions** include collapse and prolapse of the intervertebral discs (segment C6-T1), ventral enthesophyte proliferation, and fractures. Septic diskospondylitis may induce severe neck pain and forelimb lameness.
- **Congenital abnormalities of the first rib(s)** such as synostosis or hypoplasia can be seen in horses presenting forelimb lameness or dyskinesia.

ATAXIA IS NOT THE ONLY CLINICAL MANIFESTATION OF NECK INJURIES

Upon completion, participants will be aware of the variety of clinical manifestations induced by neck injuries. Based on the physical and dynamic examination, they will be able to suspect neck problems as a part of the differential diagnosis between neck, back, and forelimb conditions. Combining radiography and ultrasonography, they will be able to diagnose most of the bone and soft-tissue injuries responsible for the different neck conditions.

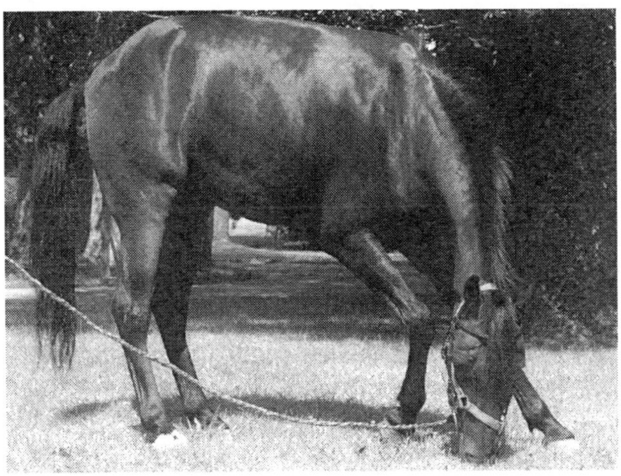

Figure 1. Neck stiffness and pain. The horse is positioning the forelimbs in an attempt to reduce caudal neck flexion while grazing.

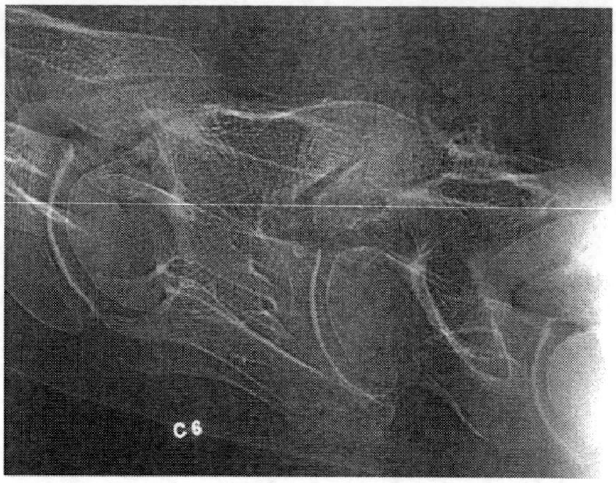

Figure 2. Severe lesions of the C6-C7 joint space with hypertrophy and fragmentation of the articular processes, stenosis of the C7 vertebral canal and dorsal subluxation of the C7 vertebral head.

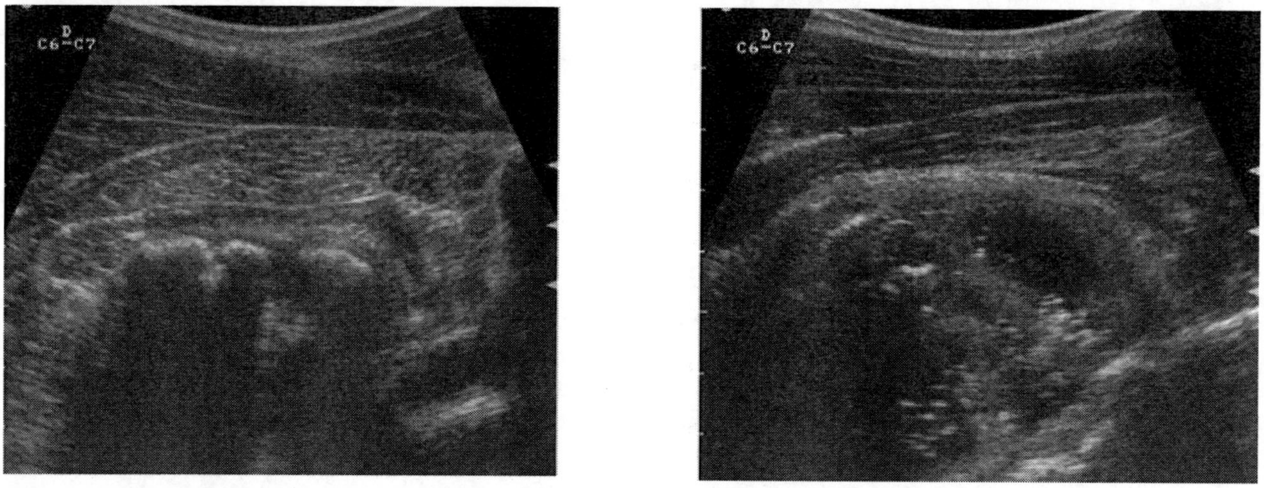

Figure 3. Ultrasonographic images of the right C6-C7 synovial intervertebral joint between articular processes showing fragmentation of the articular processes (left scan) as well as synovitis and capsulitis (right scan).

REFERENCES

1. Denoix J-M: Spinal biomechanics and functional anatomy. *Vet Clin North Am Equine Pract.* 1999;15(1):27–60.
2. Denoix J-M, Audigié F. The neck and back. In: Back W & Clayton H, eds. *Equine Locomotion.* London, Edinburgh, New York: Saunders;2000:167–192.
3. Denoix J-M. *Biomechanics and Physical Training of the Horse.* Boca Raton, FL: CRC Press; 2014:191pp.
4. Ricardi G, Dyson SJ. Forelimb lameness associated with radiographic abnormalities of the cervical vertebrae. *Equine Vet J.* 1993;25(5):422–6.
5. Dyson SJ. Lesions of the equine neck resulting in lameness or poor performance. *Vet Clin North Am Equine Pract.* 2011;27(3):417–37.
6. Mitchell CW, Nykamp SG, Foster R, Cruz R, Montieth G. The use of magnetic resonance imaging in evaluating horses with spinal ataxia. *Vet Radiol Ultrasound.* 2012;53(6):613–20.
7. Janes JG, Garrett KS, McQuerry KJ, Pease AP, Williams NM, Reed SM, MacLeod JN. Comparison of magnetic resonance imaging with standing cervical radiographs for evaluation of vertebral canal stenosis in equine cervical stenotic myelopathy. *Equine Vet J.* 2014;46(6):681–6.
8. Veraa S, Bergman W, van den Belt AJ, Wijnberg I, Back W. *Ex vivo* computed tomographic evaluation of morphology variations in equine cervical vertebrae. *Vet Radiol Ultrasound.* 2016;57(5):482–8.

9. Demondion X, Lefebvre G, Fisch O, Vandenbussche L, Cepparo J, Balbi V. Radiographic anatomy of the intervertebral cervical and lumbar foramina (vessels and variants). *Diagn Interv Imaging*. 2012;93(9):690–697.
10. Santinelli I, Beccati F, Arcelli R, Pepe M. Anatomical variation of the spinous and transverse processes in the caudal cervical vertebrae and the first thoracic vertebra in horses. *Equine Vet J*. 2016;48(1):45–9.
11. DeRouen A, Spriet M, Aleman M. Prevalence of anatomical variation of the sixth cervical vertebra and association with vertebral canal stenosis and articular process osteoarthritis in the horse. *Vet Radiol Ultrasound*. 2016;57(3):253–8.
12. Hahn CN, Handel I, Green SL, Bronsvoort MB, Mayhew IG. Assessment of the utility of using intra- and intervertebral minimum sagittal diameter ratios in the diagnosis of cervical vertebral malformation in horses. *Vet Radiol Ultrasound*. 2008;49(1):1–6.
13. Berg LC, Nielsen JV, Thoefner MB, Thomsen PD. Ultrasonography of the equine cervical region: a descriptive study in eight horses. *Equine Vet J*. 2003;35(7):647–55.
14. García-López JM, Jenei T, Chope K, Bubeck KA. Diagnosis and management of cranial and caudal nuchal bursitis in four horses. *J Am Vet Med Assoc*. 2010;237(7):823–9.
15. Janes JG, Garrett KS, McQuerry KJ, Waddell S, Voor MJ, Reed SM, Williams NM, MacLeod JN. Cervical vertebral lesions in equine stenotic myelopathy. *Vet Pathol*. 2015;52(5):919–27.

acvim216

Diagnosis of Low Back (Lumbar and Lumbosacral) Pain in Equine Athletes

Jean-Marie Denoix, DVM, PhD, agrégé, HDR, Cert.ISELP, LAIA-ECVDI, DACVSMR
Goustranville, France

ABSTRACT

Low back pain is suspected in horses presenting poor performance with hindlimbs stiffness, lack of propulsion, and discomfort during sport exercises such as short turns, jumping and dressage exercises. In race horses, low back pain is a cause of poor propulsion and irregular gait at high speed in trotters. Lesions of the thoracic vertebral column, such as kissing spines and ventral spondylosis, are well known and easy to image using radiography. For anatomical reasons, investigation of the lumbar and lumbosacral areas is more difficult in horses. A complete radiographic evaluation of the lumbar spine requires powerful X-ray generators and adequate image acquisition systems. An additional assessment of the articular processes and spinal processes can be made with ultrasonography. Transrectal ultrasonography is the key procedure for the routine evaluation of lumbosacral injuries in equine athletes, as diagnosis and documentation of several types of injuries can be made as well as differential diagnosis with sacroiliac lesions. Nuclear scintigraphy helps in determining the clinical significance of radiographic or ultrasonographic findings and in identifying osteoarticular injuries in the equine lumbar spine but has limitations to pick up deep lesions in the lumbosacral junction.

The objectives of this paper are to attract attention on the clinical incidence of lumbar and lumbosacral injuries in sport and race horses and to highlight the diagnostic value of transrectal ultrasonographic examination of this area.

INTRODUCTION

An objective assessment of back pain and identification of the cause(s) of the pain are not easy in horses. Osteoarticular thoracolumbar lesions have been well documented in horses.[1,2] Several papers have been published on the clinical and nuclear scintigraphic evaluation of the sacroiliac joint.[3-5] But little attention has been focused on the lumbar spine and the lumbosacral junction.[6,7] Our routine diagnostic approach of low back pain in horses includes a complete radiographic examination of the lumbar spine down to the fourth lumbar vertebra (L4) and a transrectal ultrasonographic evaluation of the caudal part of the lumbar spine (from L4 to L6), lumbosacral junction, and sacroiliac joint.[6-8] In a lot of sport and race horses, nuclear scintigraphy is also used to help localize bone pathology in these areas.

The objectives of this paper are to attract attention on the clinical incidence of lumbar and lumbosacral injuries in sport and race horses, to highlight the diagnostic value of combining standing radiographic examination of the lumbar spine and transrectal ultrasonographic examination of the lumbosacral junction.

COMPLAINTS - ANAMNESIS

Low back pain is a common complaint in sport and race horses. A number of different problems can be related to lumbar and lumbosacral and sacroiliac pain. Generally speaking, affected horses present athletic problems inducing performance limitations such as lack of engagement, lack of propulsion, hindlimb stiffness. In sport horses, asymmetrical lateral bending, defenses, discomfort during sport exercises such as short turns, jumping, and dressage exercises are reported. In race horses, low back pain is a cause of poor propulsion or asymmetrical gait at high speed for trotters without hindlimb lameness in trotters.

CLINICAL EXAMINATION

At physical examination, few horses exhibit postural abnormalities. Rarely owners report that the horse likes to place the caudal aspect of the pelvis on a support in the stall (to be seated on the hindquarters). At inspection, reduction of muscle volume (muscle hypotrophy) of the erector spinae and gluteal muscle can be seen in the thoracolumbar and lumbosacral area; this finding is usually symmetrical (Figure 1). But a number of athletes with lumbar and lumbosacral lesions do not present any muscle hypotrophy. At palpation, muscle contractions (spasms) are occasionally found, but this finding is lacking in sensitivity in our hands. Some horses with lumbosacral (LS) pain show a reduction of flexion and extension movements during active mobilisation induced by digital stimulation. A reluctance to lateroflexion is not specific to pain or ankylosis in the lumbar area; there is little or no lateroflexion at the lumbosacral junction. Pressure over the tuber sacrale for inducing lumbosacral extension provokes more often sinking (flexion) of the hindlimbs in trotters than in sport horses. Retraction of the hindlimbs (to be done after careful examination of the horse behavior) is another technique to assess the horse capacity and tolerance to lumbosacral extension.

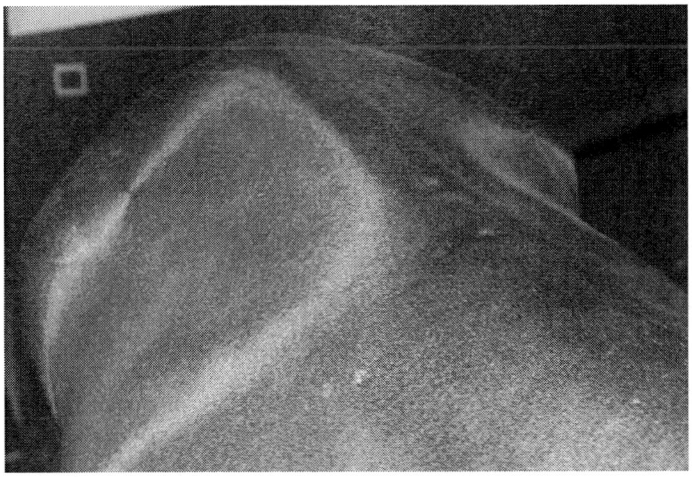

Figure 1. Lumbar muscle hypotrophy in a 12-year-old female show jumper presenting osteoarthrosis of the lumbosacral intertransverse joint.

Dynamic evaluation is first made during walking. Horses with lumbar and/or lumbosacral pain may exhibit a restricted gait with short strides, especially on short circles. At examination from the side at trot, the passive flexion and extension movement of the thoracolumbar and lumbosacral junctions can be assessed as well as the gait amplitude, engagement and propulsion of the hind limbs. If lameness is present, it will be investigated using a conventional approach, but horses with low back pain often do not present asymmetrical gait parameters at the walk and trot. Cantering on lunge in a deep footing is useful to assess the horse propulsion as well as power and strength on the hind legs. Defect of lateroflexion with dissociation or lack of engagement of the hindlimbs or wrong placement of the inside hindlimb is often seen in horses with lumbar and lumbosacral problems. Examination of the horse being ridden or on the track (for Standardbred trotters) may help to identify the gait abnormalities in training or competing situations. Assessment of the horse capacity to flex and extend the low back in show jumpers can be done examining the horse jumping on lunge a low fence (70–100 cm).

Diagnostic Imaging

As most of the clinical or athletic manifestations are not specific of the lumbar, lumbosacral or sacroiliac areas, a systematic investigation of the lumbar spine and lumbosacroiliac area is performed in an attempt to identify the most significant lesions correlated to the clinical or athletic manifestations.

Lumbar Spine

Lesions of the thoracic vertebral column, such as kissing spines and ventral spondylosis, are well known and easy to image using **radiography**. For anatomical reasons, investigation of the lumbar and lumbosacral areas is more difficult in horses. A complete radiographic evaluation of the lumbar spine (from the thoracolumbar junction to L4) on the standing horse requires powerful X-ray generators and adequate image acquisition systems. Strict lateral views are performed, avoiding any rotation. On good radiographs, the articular processes and transverse processes of the lumbar vertebra are superimposed in the middle of the image and slightly separated on both cranial and caudal directions. On normal images, the spinal processes are wider than in the thoracic spine, oblique craniodorsally and separated. The articular process joints (APJ) show a thin, linear, radiolucent joint space, oblique between 40 to 60°, with the vertebral axis separating dense subchondral bone of the articular processes. The caudal articular process of one vertebra is superimposed to the cranial articular process of the following one. Kissing spines can be seen in the cranial lumbar area (T18-L2); they are less frequently identified in the middle lumbar region, but are frequently seen between L4-L5 and L5-L6. Several types of findings involving the APJ have been identified; they include left-to-right asymmetry, sclerosis, lysis or decreased density, ventral periarticular proliferation, dorsal periarticular proliferation (the most common findings with different grades), intervertebral ankylosis created by a dorsal bridge or complete lysis and fusion of the articular processes (Figure 2a), and fractures (fatigue or traumatic).[9,10] Ventrolateral spondylosis on the vertebral bodies is much rarer in the lumbar spine than in the mid-thoracic region.

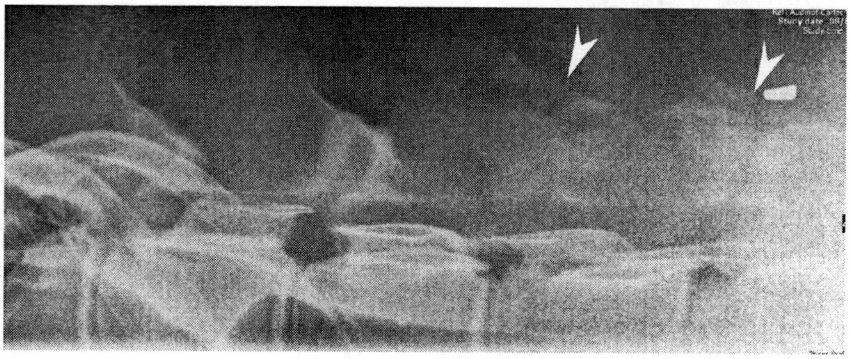

Figure 2a. Lateral radiographic projection of the lumbar spine of a 10-year-old international three-day eventer female. Severe osteoarthrosis of the articular process joints L2-L3 and L3-L4 intervertebral joints.

An additional assessment of the articular processes and spinal processes can be made with **ultrasonography**.[7,8] A low-frequency convex probe is required for adequate imaging of the articular processes of the lumbar spine This can also be done with a linear probe on thin and small horses. The technique consists in placing the probe on a transverse plane over the APJ on side and symmetrically on the other side to improve sensitivity and specificity in detecting findings. On both sides, the process is performed from T18-L1 to L5-L6. T18-L1 is identified using the convex profile of the last rib as a landmark. On normal images, the joint space of the APJ is thin and separates the caudal articular process of one vertebra medially and the cranial articular process of the following vertebra laterally. The most common finding is asymmetric periarticular bone proliferation which elevates the bone surface and hides the joint space. Dysplasia, modelling or lysis are less commonly seen (Figure 2b). Complete regular bone bridge between the articular processes is correlated to intervertebral ankylosis. Synovial fluid effusion and capsulitis have been identified. Asymmetric hypotrophy of the multifidus muscle has been seen on the side of the most affected APJ.

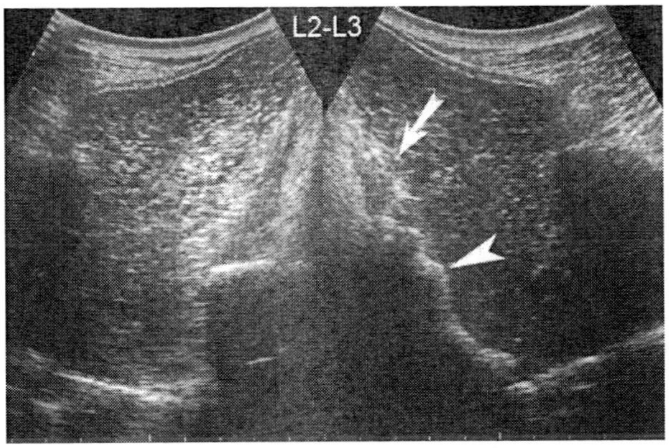

Figure 2b. Transverse ultrasound scan of the L2-L3 intervertebral joint of the same three-day eventer female. Severe deformation of the right articular process joints (arrowhead) and hypotrophy of the right multifidus muscle.

Nuclear scintigraphy helps in determining the clinical significance of radiographic or ultrasonographic findings and in identifying osteoarticular injuries in the equine lumbar spine. Three views are systematically performed: a left dorsolateral oblique, a right dorsolateral oblique, and a dorsal view.

Lumbosacral Junction

A ventrodorsal radiographic projection of the lumbosacral joint can be performed under general anaesthesia, but this view cannot be done in routine practice and has limitations because of superimposition of abdominal viscera. Nuclear scintigraphy has limitations to pick up deep lesions in the lumbosacral junction.

Transrectal ultrasonographic evaluation of the lumbosacral junction, the caudal part of the lumbar spine (from L4 to L6), and the sacroiliac joints is an invaluable technique to assess the presence of osteoarticular and soft-tissue injuries in sport and race horses.[8,11,12] The procedure is performed with a linear or microconvex rectal probe. Care should be taken to avoid any injury of the rectal wall and mucosa. Nervous horses should be sedated.

Several findings can be detected with ultrasonography at the lumbosacral junction. Lumbosacral ankylosis (or sacralization of L6), a congenital abnormality, is less common than the intervertebral ankylosis between L5 and L6.[8,9] Lumbosacral disc degenerative lesions are common; they include fissuration, cavitation (Figure 3), ventral herniation and dystrophic mineralization. Intervertebral malalignment (spondylolisthesis) of the lumbosacral joint or the joint between L5 and L6 has been identified on rare cases. Lumbosacral intertransverse osteoarthrosis with periarticular osteophytes or remodeling of the joint margins are common on sport and race horses. These findings have been documented at postmortem examination on 13 horses examined *in vivo* and subsequently dissected.

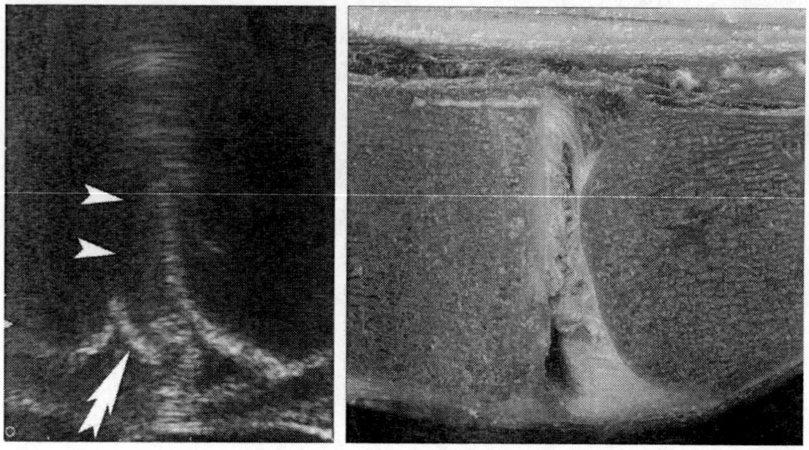

Figure 3. *In vivo* median ultrasound scan (left) and median postmortem section (right) of the lumbosacral disc in a 6-year-old Standardbred trotter female. The disc degenerative cavitation shows up as an anechogenic zone (arrowheads) limited by an echogenic line (arrow).

Nuclear scintigraphic examination of the pelvis is performed using a dorsal approach (camera horizontal), a dorsocaudal approach (camera parallel to the sacrum) and 2 symmetrical dorsolateral approaches (camera parallel to the coxal bone). As the ilium is superimposed to the lumbosacral joint, the location of increased radiopharmaceutical uptake (IRU) must be established carefully. Increased capture of radiation (ICR) over the auricular surfaces of the sacroiliac joint is in relation with osteoarticular disease of this joint. ICR can be seen between the auricular surfaces and tuber sacrale, where it is correlated either to the insertion of the interosseus sacroiliac ligament or the lumbosacral joint. ICR detected cranially to the auricular surfaces corresponds to the transverse processes of the last lumbar vertebra and the adjacent intertransverse joints.

REFERENCES

1. Jeffcott LB. Disorders of the thoracolumbar spine of the horse - a survey of 443 cases. *Equine Vet J*. 1980;12(4):197–210.
2. Denoix J-M, Dyson SJ. Thoracolumbar spine. In: Ross MW, Dyson SJ, eds. *Diagnosis and Management of Lameness in horses*. Philadelphia, PA: Saunders;2003:509–521.
3. Dyson S, Murray R, Branch M, *et al*. The sacroiliac joints: evaluation using nuclear scintigraphy. Part I: the normal horse. *Equine Vet J*. 2003;35(3):226–232.
4. Dyson S, Murray R, Branch M, *et al*. The sacroiliac joints: evaluation using nuclear scintigraphy. Part II: lame horses. *Equine Vet J*. 2003;35(3):233–239.
5. Dyson S, Murray R. Pain associated with the sacroiliac joint region: a clinical study of 74 horses. *Equine Vet J*. 2003;35(3):240–245.
6. Denoix J-M, Audigié F, Coudry V. A review of diagnosis and treatment of lumbosacral pain in sport and race horses. In: Proceedings of the 51st Annual Convention of the American Association of Equine Practitioners;2005:366–373.
7. Denoix J-M, Dyson S. Thoracolumbar spine. In: Ross M, Dyson S, eds. *Equine Lameness*. 2nd edition. Philadelphia, PA: Saunders;2011:592–605.
8. Denoix J-M. Ultrasonographic evaluation of back lesions. *Vet Clin North Am Equine Pract*. 1999;15(1):131–159.
9. Haussler KK, Stover SM, Willits NH. Pathologic changes in the lumbosacral vertebrae and pelvis in Thoroughbred racehorses. *Am J Vet Res*. 1999;60(2):143–53.
10. Collar EM, Zavodovskaya R, Spriet M, Hitchens PL, Wisner T, Uzal FA, Stover SM. Caudal lumbar vertebral fractures in California Quarter Horse and Thoroughbred racehorses. *Equine Vet J*. 2015;47(5):573–9.
11. Nagy A, Dyson S, Barr A. Ultrasonographic findings in the lumbosacral joint of 43 horses with no clinical signs of back pain or hindlimb lameness. *Vet Radiol Ultrasound*. 2010;51(5):533–9.
12. Espinosa P, Benoit P, Salazar I, de la Fuente J, Heiles P. Transrectal ultrasonography of equine lumbosacral nerves: pilot study in 28 healthy warmblood horses. *Vet Radiol Ultrasound*. 2017;58(2):228–236.

acvim075

Maternally-Derived BVDV and BHV-1 Antibodies in Calves Born to Dams Vaccinated or Not During Gestation

Emily J. Reppert; Lacey Robinson; Joyce Wick; **Manuel F. Chamorro**

Kansas State University College of Veterinary Medicine, Manhattan, KS, USA

Respiratory disease in nursing beef calves often occurs during the first 30 days of life or around 90–150 days of age. Failure in the transfer of passive immunity or rapid decay of maternally-derived antibodies to respiratory viruses, such as BVDV and IBR, have been associated with this condition. The objective of this study was to determine the effect of vaccination with two doses of a killed virus (KV) multivalent vaccine containing BVDV1, BVDV2, and BHV-1 during late gestation on the level of total serum IgG and neutralizing antibodies to BVDV1, BVDV2, and BHV-1 in newborn calves. 47 pregnant beef heifers, previously vaccinated with a modified live virus (MLV) multivalent vaccine containing BVDV1, BVDV2, and BHV-1 before breeding, were allocated into treatment groups. Group A (n=20) received two doses of a KV vaccine 21 days apart between 6.5 to 8 months of gestation. Group C (n=27) received 5 mL of saline in lieu of vaccination. Heifers were closely monitored for vaccine reaction, abortion, and normal parturition by personnel blinded to treatment allocation. Newborn calves were allowed to nurse colostrum naturally without assistance. Serum samples were collected from dams before vaccination and at calving and from newborn calves at 24 hours of age for neutralizing antibodies to BVDV1, BVDV2, and BHV-1. Additional serum was collected from calves at 24 hours to determine total IgG level. With respect to heifers, before vaccination the mean Log2 serum antibody titers to BVDV1, BVDV2, and BHV-1 were similar among groups A and C ($p>0.05$). At calving, the mean Log2 serum antibody titers to BVDV1, BVDV2, and BHV-1 were higher in group A compared with group C ($p>0.05$; Figure 1). With respect to newborn calves, the mean serum IgG levels at 24 hours were higher in calves born to heifers from group C compared with calves from heifers of group A (C=31.7 g/L; A=28.9 g/L; $p<0.05$). The mean Log2 serum antibody titers to BVDV1, BVDV2, and BHV-1 at 24 hours was higher in calves born to heifers in group A compared with calves born to heifers in group C (Figure 2). Vaccination of pregnant beef heifers during late gestation with two doses of a KV multivalent vaccine 21 days apart has been demonstrated to increase serum antibody titers to BVDV1, BVDV2, and BHV-1 at calving and at 24 hours in newborn calves. Higher serum antibody titers to BVDV 1, BVDV 2, and BHV-1 observed in calves born to vaccinated heifers might result in prolonged duration of maternally-derived immunity and better protection against respiratory disease during the pre-weaning period. Serum IgG at 24 hours in calves from this study could have been influenced by other factors different from the dam's vaccination protocol during gestation.

Figure 1. Mean Log2 +/- SEM Antibody titers to BVDV and BHV-1 at calving of heifers vaccinated or not during gestation

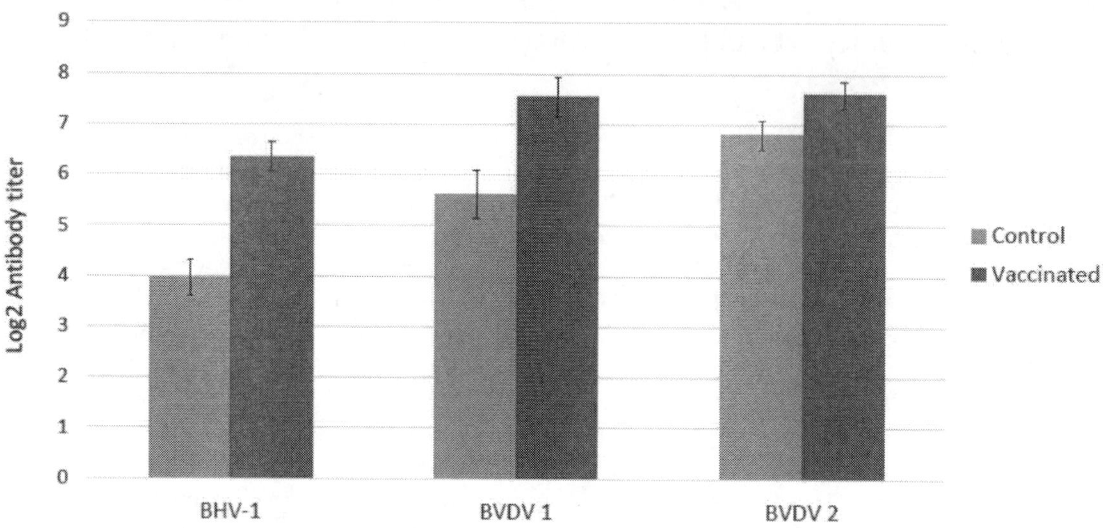

Figure 2. Mean Log2 +/- SEM Antibody titers to BVDV and BHV-1 of calves born to vaccinated or non-vaccinated heifers

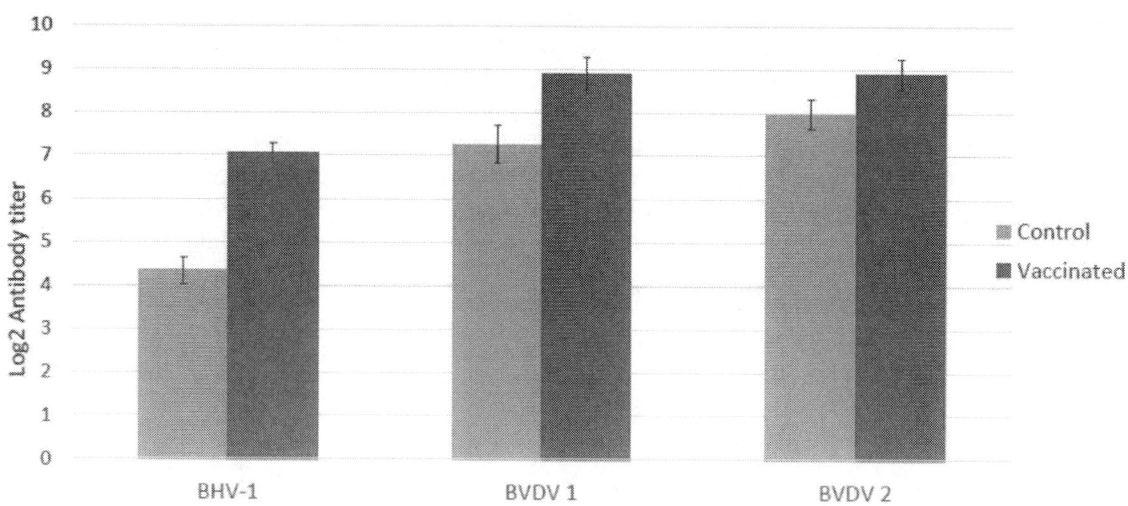

ac

acvim089

Changes in the Fecal Microbiota of Calves Experimentally Infected with *Mycobacterium avium* subsp. *paratuberculosis*

Marie-Eve Fecteau[1]; Daniel P. Beiting[2]; Raymond W. Sweeney[1]

[1]Department of Clinical Studies-New Bolton Center, School of Veterinary Medicine, University of Pennsylvania, Kennett Square, PA, USA; [2]Department of Pathobiology, School of Veterinary Medicine, University of Pennsylvania, Philadelphia, PA, USA

Johne's disease (JD) is a widespread chronic gastrointestinal disease of cattle caused by *Mycobacterium avium* subsp. *paratuberculosis* (MAP). The mammalian gastrointestinal tract is colonized by hundreds of different species of bacteria, living in a delicate balance in a symbiotic relationship with the host. Dysbiosis, or disruption of this balance, can result in inflammation of the intestines, and has been reported in human patients suffering from various infectious and inflammatory disorders of the gastrointestinal tract, and more recently, in cows naturally infected with MAP. It is however unknown if the dysbiosis observed in MAP-infected cattle is a consequence of MAP infection or a preexisting condition.

The purpose of this study was to investigate the impact of MAP infection on the fecal bacterial populations in calves prior to, and following experimental infection, when compared to age-matched uninfected calves, using metagenomic analysis.

Sixteen Holstein calves were used in this study. At time 0 (approximately 4 weeks of age), fecal samples were collected on all calves (preinfection samples). Twelve of the calves were then orally challenged with 5×10^9 CFU MAP on 2 consecutive days. The remaining 4 calves were used as negative controls and housed separately from the infected calves. At 16 weeks post-infection, fecal samples were collected again on all 16 calves. All fecal samples were kept frozen at -70°C until analysis. The fecal samples were processed for DNA extraction, purification, and amplification using specific primers. Genomic sequencing was performed using NextSeq Illumina platform.

Although all 16 calves had similar fecal microbiota prior to infection, there is a statistically significant difference in microbial diversity between infected and uninfected calves at 16 weeks post-infection (Figure 1). A more detailed analysis will be presented.

The results of this study suggest that the dysbiosis observed in MAP-infected cattle is a consequence of MAP infection, and this dysbiosis may play a role in the intestinal inflammation observed in cattle infected with MAP.

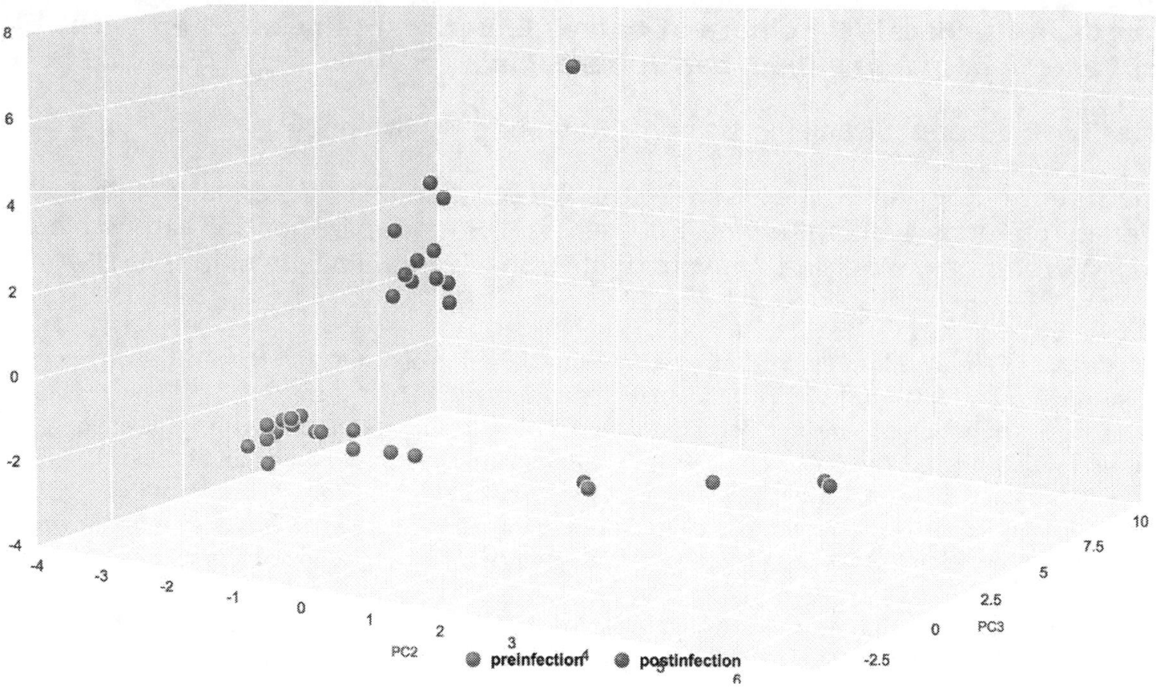

Figure 1. Principal component analysis (PCoA) of 12 infected and 4 uninfected calves prior to (orange dots) and following (green dots) experimental infection with MAP

acvim101

Algorithm-based Antimicrobial Therapy in Diarrheic Calves: Impact on Antimicrobial Usage, Health and Fecal Microbiota

Diego Gomez-Nieto[1]; Luis Arroyo[2]; Zvonimir Poljak[3]; Laurent Viel[2]; J. Scott Weese[4]

[1]Department of Pathobiology, Ontario Veterinary College, University of Guelph, Guelph, ON, Canada; [2]Department of Clinical Studies, University of Guelph, Guelph, ON, Canada; [3]Department of Population Medicine, University of Guelph, Guelph, ON, Canada; [4]University of Guelph, Guelph, ON, Canada

There is a strong campaign from the medical and veterinary communities and general public to limit antimicrobial use in food-producing animals. This has motivated the development of farm management practices to refine and reduce antimicrobial usage. The objective of this study was to evaluate the impact of an algorithm targeting antimicrobial therapy to systemically ill diarrheic calves on four outcomes (incidence of diarrhea, antimicrobial usage rates, overall mortality, and mortality in diarrheic calves). Additionally, the fecal microbiota of healthy, non-antimicrobial-treated calves before and after implementation of algorithm was compared.

A combination of retrospective (before-algorithm period [BAP], 12 months) and prospective (after-algorithm period [AAP], 12 months) cohorts was used. Two large dairy farms (**F1** and **F2**) were enrolled. The algorithm was designed to target antimicrobial therapy in calves with diarrhea, decreased milk intake and fever. Difference in the cumulative incidence risk (CIR) of developing each of the 4 outcomes was evaluated using Mantel-Haenszel approach with time BAP and AAP as the main exposure of interest, and farm as the potential confounder. Measures of associations were expressed as incidence risk ratio (IRR).The fecal microbiota of 15 healthy age-matched calves (<4 wks of age) from the BAP and 12 months after algorithm implementation (from each farm) was characterized using the Illumina MiSeq platform targeting the V4 region of the 16S rRNA gene.

A total of 529 and 639 calves were enrolled in the BAP from **F1** and **F2**, respectively, and 768 and 842 in the AAP. The CIR of using antimicrobials for treatment of diarrhea in the AAP was 81% lower than in the BAP, and these estimates were similar between farms (IRR: 0.19, 95%CI: 0.17 to 0.21). On **F1**, CIR of diarrhea was 19% lower in the AAP in **F1** (IRR: 0.81; 95%CI: 0.77 to 0.85), but there were no differences in **F2** (IRR: 1, 95%CI: 0.97 to 1.04). No difference in cumulative overall mortality could be detected in this study between BAP and AAP, in either the overall population (IRR: 0.97; 95%CI: 0.67 to 1.40) or in diarrheic calves (IRR: 0.93; 95%CI: 0.62 to 1.37).

There were pronounced differences in the fecal microbiota of non-antibiotic treated calves between BAP and AAP. Both farms had a significant higher richness in the AAP (Chao-1, p<0.001). **F1** had higher diversity in the AAP (Inverse Simpson, p=0.01), but no difference was observed in **F2** (p=0.2). There were significant differences in community membership (Jaccard index) and structure (Yue & Clayton index) between BAP and AAP in both farms (Parsimony and AMOVA, p<0.001). **LEfSe** analysis showed that among others, the genera *Faecalibacterium*, and unclassified genus of the family Ruminococcaceae were enriched in the AAP in **F1**, whereas *Bacteroides* and unclassified genera from the families Lachnospiraceae and Ruminococcaceae were enriched in the AAP in **F2** (p<0.05 and LDA score >3).

A simple farm-based algorithm for treatment of diarrhea drastically decreased the use of antimicrobials without impacting negatively calf health. It appears that decreasing the usage of antimicrobial on the farm can have an impact on the fecal microbiota of untreated calves, suggesting that antimicrobial use practices on farms can influence the microbiota of both treated and untreated calves.

acvim102

Pharmacokinetics of Fentanyl and Fentanyl Transdermal Patches in Healthy and Hospitalized Calves

Joe Smith[1]; Jonathon Paul Mochel[1]; David Borts[1]; Larry Wulf[2]; Kristen Hayman[1]; Jennifer Schleining[1]; Hans Coetzee[3]

[1]College of Veterinary Medicine, Iowa State University, Ames, IA, USA; [2]Veterinary Diagnostic Laboratory, Iowa State University, Ames, IA, USA; [3]College of Veterinary Medicine, Kansas State University, Manhattan, KS, USA

Fentanyl is a mu opioid receptor agonist that has been used for analgesia in multiple species. One of the more common routes of administration of fentanyl is the transdermal route. There have been multiple reports of the use of fentanyl transdermal patches (FTP) in large animal practice as a longer-term method of analgesia. The use of the patches has been described for analgesia in horses, sheep, goats, pigs, and llamas, but there are no data available for the use of this extended-release formulation in cattle.

The primary objectives of this study were to determine the pharmacokinetics (PK) of injectable fentanyl citrate, as well as the PK and effect of fentanyl transdermal patches in healthy bovine calves. An additional objective was to observe the pharmacokinetics and effect of fentanyl transdermal patches on healthy calves. The secondary objectives were to report the PK, clinical and analgesic effects of fentanyl transdermal patches on hospitalized calves.

Eight commercial female Holstein calves were administered a 2.5 μg/kg intravenous bolus dose of fentanyl citrate. Plasma concentrations of fentanyl and two of its metabolites, norfentanyl & despropionyl fentanyl, were determined by LC-MS. Initial non-compartmental analysis yields a clearance of 2.55 mL/h/kg, and a half-life of 1.73 h and a maximum concentration of about 1 μg/mL.

Four patient calves and four healthy calves had fentanyl patches applied as described for sheep. Six out of the eight calves exhibited episodes of tachycardia, tachypnea, bellowing, and ataxia. These signs resolved within 2–6 hours in 6 calves when the patches were removed and when the patches were removed and a commercial opioid antagonist was administered in 2 calves. Maximum concentration ranged from 0.726 to 3.292 μg/mL, and a time to maximum concentration ranged from 4–10 hours at the time of patch removal due to clinical presentation.

These data indicate that fentanyl can be administered intravenously and achieve plasma levels that are considered analgesic in other species. The adverse effects of the administration of fentanyl patches in calves require more research to determine an appropriate clinical regimen and environment for this sustained-delivery therapy.

acvim103

Accuracy of Refractometry for Diagnosis Failure of Passive Transfer in Calves: Systematic Review and Meta-analysis

Sébastien Buczinski[1]; Elise Gicquel[1]; Gilles Fecteau[1]; Yemisi Takwoingi[2]; Munashe Chigerwe[3]; Jean-Michel Vandeweerd[4]

[1]Faculté de Médecine Vétérinaire, Université de Montréal, St-Hyacinthe, QC, Canada; [2]Institute of Applied Health Research, University of Birmingham, Birmingham, England, UK; [3]Department of Medicine and Epidemiology, University of California-Davis, Davis, CA, USA; [4]Integrated Veterinary Research Unit, Namur Research Institute for Life Sciences, Department of Veterinary Medicine, Faculty of Sciences, University of Namur, Namur, Belgium

Failure of passive transfer of immunity (FPT) is traditionally diagnosed in calves based on serum immunoglobulin G (IgG) concentrations below a specific threshold during the first week of life. On farm direct determination of IgG concentration is not easily performed and serum refractance (converted in g/L or Brix % scale) is frequently used in practice. Our objective was to systematically review the diagnostic accuracy (sensitivity and specificity of refractometer against direct IgG concentrations (reference standard) determined by radial immunodiffusion or turbidimetric immunoassay.

To identify relevant studies, we searched the following databases from June 1986 to June 2016: Medline, CAB-abstract, Searchable Proceedings of Animal Conferences, and Google scholar. To be eligible, a study had to include calves sampled between 1 and 10 days of age, and data allowing recording the number of true positives, false negatives, false positives, and true negatives. Data extraction was performed by 2 different authors and any disagreement was resolved by a third author. For each threshold reported in 5 or more studies, pooled estimates of sensitivity and specificity were obtained using hierarchical models.

We identified 13 studies that all defined FPT as a calf with serum IgG.

In summary, when using refractometry to rule out FPT problem in herds, the 55 g/L cut-off could be used, whereas for ruling in FPT problem in herds, the 52 g/L cut-off could be used.

acvim105

A Comparative Evaluation of Diagnostic Findings and Mortality Risks in Hospitalized Neonatal Camelids and Foals

Daniela Bedenice[1]; Bailey Avila[2]; Mary Rose Paradis[1]

[1]Tufts Cummings School of Veterinary Medicine, North Grafton, MA, USA; [2]Wentworth Institute of Technology, Boston, MA, USA

Species-specific differences in disease prevalence, manifestation and outcome may impact client investments into neonatal intensive care. Our retrospective cohort study compared the presenting diagnostic evaluation, and mortality risks between 246 critically-ill (CI) neonatal crias (1/1999–5/2016) and 356 CI foals (2/2001–5/2016) admitted to a university hospital under 4-weeks old. Critically ill patients required hospitalization with intensive care management >4 times daily. All data were presented descriptively, and compared between groups using univariate and multivariate analyses (SPSS-22).

Female crias (142/246, 57.7%) were significantly overrepresented in comparison to fillies (132/352, 37.5%). Congenital defects, failure of passive transfer, and clinical signs of prematurity were more often observed in neonatal crias, while colic, diarrhea, patent urachus, septic arthritis and omphalitis were significantly more common in CI foals. Foals also remained hospitalized for longer than crias (mean 6 versus 4 days; $p<0.001$). Overall survival to discharge (excluding fatal congenital defects) was comparable between crias (174/224; 77.8%) and foals (287/347, 82.1%), while crias (26/48; 54.2%) were more likely than foals (21/60; 35%) to die naturally than undergo euthanasia. Lower respiratory disease and indications for oxygen or intravenous glucose support increased mortality in multivariate outcome models of both species. The latter analyses also associated ceftiofur treatment with better survival of crias and foals. A species-specific adaptation of pediatric diagnostic criteria of SIRS and sepsis was significantly associated with mortality in both species, including all animals with complete data sets.

Despite species-specific differences, the success of intensive care management is comparable between neonatal camelids and foals.

acvim118

Mycobacterium Cell Wall Fraction as an Aid in Reduction of Antibiotic Use in Young Calves

Frédéric Beaulac[1]; Annie Dubuc[2]; Stanley Alkemade[3]; **Aleksandar Masic**[3,4]

[1]Services Vétérinaires Dr Frédéric Beaulac-Veterinary Hospital, St-Hyacinthe, QC, Canada; [2]Délimax, St-Hyacinthe, QC, Canada; [3]NovaVive Inc., Belleville, ON, Canada; [4]Ontario Veterinary College, University of Guelph, Belleville, ON, Canada

High levels of mortality in dairy calves represent an important health issue and a major source of economic losses. Survey from NAHMS in the United States showed that 9.1% of female heifer calves and 7.3% of male dairy calves die during the post-calving or growing period. The majority (42%) of the mortality occurred in the first 21 days following arrival at veal farms. Consequently, economic costs are high and losses associated with early calf mortality were estimated to be USD$ 30–70 million in North America alone. The most common preventative measure used to reduce mortality in veal and dairy heifer raising operations is antimicrobials incorporated into the milk or milk replacer fed to the calves. Approximately 47.6% of dairy heifer producers fed an antibiotic in pre-weaned calves' milk. The majority of producers introduce antimicrobials prophylactically to calves immediately on arrival, or metaphylactically within the first week following arrival. As a result of the abundant use of antimicrobials in the veal industry, high levels of antimicrobial-resistant bacteria have been detected, as compared to the conventional cattle population. With antimicrobial resistance becoming of greater concern from both an animal and public health perspective, it is imperative to determine strategies to reduce antimicrobial use (and resistance) without compromising animal welfare.

Mycobacterium cell wall fraction (MCWF), Amplimune™ (NovaVive Inc., Canada), is the first immunostimulant licensed registered for bovine infectious disease therapy. Amplimune™ is indicated as an immunotherapeutic treatment in neonatal calves for the reduction of clinical signs and mortality associated with infection caused by K99+ enterotoxigenic *Escherichia coli* (ETEC), a major cause of diarrhea in neonatal calves. Amplimune™ has been shown to induce a non-specific immune response and participate in the development of both cell-mediated and humoral immunity. It could, therefore, be considered as either a stand-alone or adjunctive metaphylactic therapy to reduce the incidence of neonatal diarrhea and other neonatal diseases (e.g., bovine respiratory disease complex [BRDC]) and, consequently, the use of antibiotics in young calves.

The main objective of the study was to demonstrate the efficacy of MCWF in reducing the incidence of respiratory and gastrointestinal clinical conditions and reducing the use of antimicrobials administered as adjunct metaphylactic therapy to neonatal calves in a veal operation. A total of 699 calves, 1–10 days of age were randomly assigned to two experimental groups (N=354 MCWF; N=345 Control) and housed in four barns (Délimax, Québec). Animals were group weighed on arrival and prior to slaughter. In addition to standard farm procedures and metaphylactic therapies (electrolytes and antibiotics) which were administered to all animals upon arrival, 1 mL of MCWF was administered subcutaneously to the treatment group (N=345). This was repeated 7 days later. Following MCWF administration, study calves were monitored daily for adverse events and all calves were treated if detected as being ill. Data were collected for the incidence of BRDC, diarrhea and other conditions for the duration of the study (154 days). Data for additional antibiotic usage, supplemental treatments (duration of use and costs) and weight of animals were also collected and analyzed.

The post-study analysis revealed that the incidence of clinical conditions that required treatment was significantly higher in the control group compared to the MCWF group ($p<0.0001$; 123 vs. 41). In addition, the volume of antimicrobials and supportive therapies administered in the control group was significantly higher than in the MCWF group ($p<0.0001$; 1,090 mL vs. 298.5 mL). Furthermore, the total number of calves requiring treatments was significantly higher in the control group than in the MCWF group (p<0.01; 62 vs. 42 calves). The total number of treatment days was 2.9 times higher (191 vs. 65 days) in the control group than in the MCWF group. The average number of treatment days per calf was 1.6 days in the MCWF group versus 3.8 days in the control group (p<0.01).

Based on a preliminary analysis of the data, MCWF (Amplimune™) can be used as an aid or an antimicrobial alternative to not only reduce morbidity and mortality, but also reduce the number and volume of antimicrobials used in veal production. Further studies are underway to demonstrate MCWF efficacy as a stand-alone metaphylactic treatment in both dairy and beef calf operations.

Figure 1. Incidence and type of clinical conditions observed in MCWF and control groups

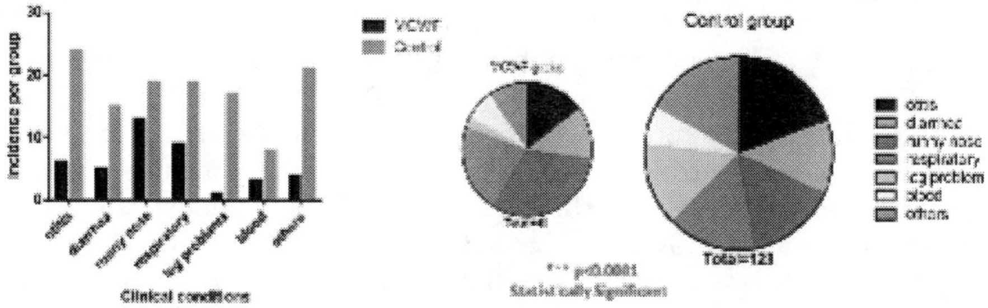

Figure 2. Supportive antibiotic and fluid therapy (mL) administered in both groups following MCWF treatment

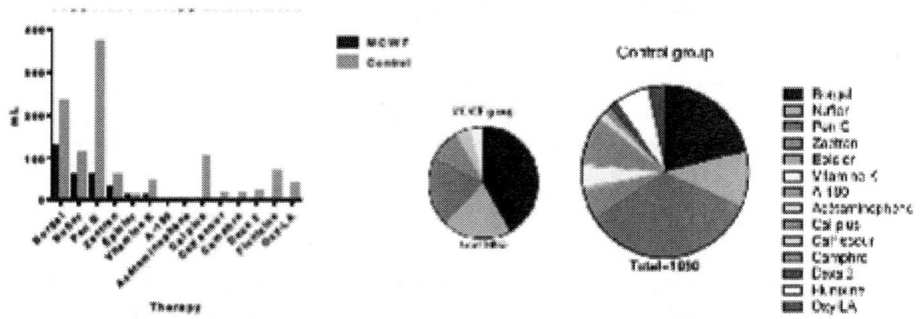

Figure 3. Total number of treated calves (A), total number of treatment days (B) and average number of treatment days per calf (C)

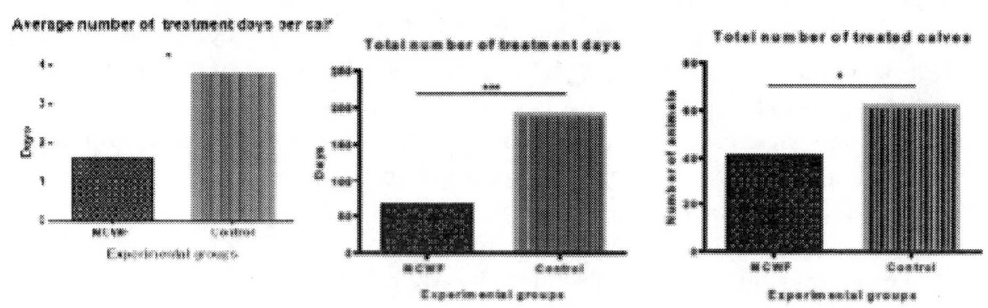

acvim121

Comparing *Salmonella enterica* Detection in Feces using Aerobic Culture, Targeted Molecular Tests, and Shotgun Metagenomics

Enrique Doster[1]; Pablo Rovira[1]; Noelle R. Noyes[1]; Brandy A. Burgess[2]; Xiang Yang[3]; Maggie Weinroth[1]; Lyndsey Linke[1]; Roberta Magnuson[1]; Kristy Pabilonia[1]; Christina Boucher[4]; Jaime Ruiz[4]; Keith E. Belk[1]; **Paul S. Morley**[1]

[1]Colorado State University, Fort Collins, CO, USA; [2]University of Georgia, Athens, GA, USA; [3]University of California-Davis, Davis, CA, USA; [4]University of Florida, Gainesville, FL, USA

Salmonella enterica is a common enteric pathogen of domestic animals, especially in livestock such as horses and cattle, and has important implications for public health. Diagnostic testing is an important tool in the management of this agent, but there is no ideal or perfect test available. Poor sensitivity (detection of *Salmonella* in truly infected animals) is perhaps the biggest problem with available diagnostic tests. Aerobic culture is the most common traditional test that is used, and enrichment is considered essential to improve sensitivity. While sensitivity of aerobic culture is variable, it is recognized as generally having fairly poor sensitivity. Polymerase chain reaction (PCR) is the other common method used for detection of *S. enterica*, but enrichment is still required when testing feces or other types of samples that contain PCR inhibitors or have quantities of related bacteria in the samples, as is true of feces. Another type of molecular test that is commercially available for targeted detection of *Salmonella* are lateral flow immunoassays (LFIA) that use monoclonal antibodies for specific detection of *Salmonella*-specific surface antigens; these tests are designed for use in food safety applications, but have been explored for use in detection of *S. enterica* in feces.[1,2] Recently, high-throughput genetic sequencing technologies have become highly affordable and widely available such that their use in research and clinical application is becoming increasingly common. Shotgun sequencing of metagenomic samples has the potential to open new avenues of research and provide improved understanding of the ecology of various microbes such as *S. enterica*, especially when our understanding is undoubtedly colored by imperfect diagnostic tools.

The objective of this study was to compare aerobic culture, PCR, LFIA, and shotgun metagenomic sequencing as tools for detection of *Salmonella enterica* in fecal samples obtained from animals that had a high probability of infection and shedding. This study was conducted as part of a trial evaluating the impact of metaphylactic antimicrobial treatment on the occurrence of antimicrobial resistance in fecal flora. Individual fecal samples were collected per rectum from 30 cattle upon arrival at a commercial feedlot, and again 11 days later. Each of the 60 samples was processed for testing in the 4 different tests. For aerobic culture, PCR, and LFIA testing, 4 grams of feces were enriched in tetrathionate broth at 42°C for 24 hrs before further processing. Samples of enrichment broth were plated to XLT4 agar for aerobic culture, and were tested in the LFIA. DNA was extracted from the broth for PCR testing using commercially available regents. For shotgun sequencing, metagenomic (whole microbial community) DNA was extracted from feces, and after library preparation was shotgun sequenced using the Illumina HiSeq 2000 platform. After sequencing, metagenomic reads were trimmed for increased quality and taxonomically classified with the Kraken software and the National Center for Biotechnology Information's (NCBI) RefSeq genome database.

Culture and PCR results showed 100% agreement, detecting *S. enterica* in only 0.05% (3/60) of samples. In contrast, there were positive LFIA tests in 22% (13/60) of samples, but none of these results agreed with culture or PCR results. Further testing of samples with the LFIA assay suggested that cattle eating high concentrate rations had a high prevalence of false-positive tests, which may explain these findings. In contrast with the results of culture and PCR which indicated a low prevalence of *S. enterica* shedding, initial analysis of shotgun metagenomic sequencing identified *S. enterica* in 60% (36/60) of samples. However, through further investigation it was recognized that the RefSeq genome database contains non-specific sequences found in individual isolates of one specific bacterial species that might be also found in isolates of other species. This is particularly true for mobile elements such as plasmids. As such, plasmid sequences are included within the reference genome files in the NCBI database, and as such plasmid sequences are considered to be specific to whichever bacterial species carried the plasmid when it was sequenced and submitted to NCBI. To investigate the impact of incorrectly classifying these plasmid sequences as being associated with a specific bacterial species, all of the plasmid sequences included with RefSeq genomes were identified and annotated to form one unique taxa representing all plasmids. Using this updated database, further analysis of the metagenomic sequencing data found that only 35% (21/60) of samples were classified as *S. enterica,* suggesting increased relative specificity. Not surprisingly, sequencing reads that were classified as belonging to plasmids in this subsequent analysis represented a small minority of all reads classified as bacterial. However, these plasmid reads were originally misclassified as representing sequences of bacteria belonging to 22 different phyla.

This study shows that the traditional techniques of aerobic culture and PCR provide similar results for *Salmonella enterica* identification in cattle feces, but that shotgun metagenomic sequencing provides new opportunities for novel investigations of microbial ecology of this important pathogen. However, classification of bacterial species must be considered cautiously as the results are largely influenced by the reference database employed in the bioinformatic analyses.

REFERENCES

1. Burgess BA, Noyes NR, Bolte DS, Hyatt DR, van Metre DC, Morley PS. Rapid *Salmonella* detection in experimentally inoculated equine faecal and veterinary hospital environmental samples using commercially available lateral flow immunoassays. *Equine Vet J*. 2015;47:119–122. [doi: 10.1111/evj.12234].
2. Burgess BA, Weller CB, Pabilonia KL, Bolte DS, Van Metre DC, Morley PS. Detection of different serotypes of *Salmonella enterica* in experimentally inoculated equine fecal samples by commercially available rapid tests. *J Vet Intern Med*. 2014;28:1853–1859 [doi:10.1111/jvim.12440].

acvim223

Surgical Management of Urinary Tract Obstructions: A Surgeon's Perspective

Pierre-Yves Mulon, DMV, DACVS
Knoxville, TN, USA

Surgical management of complete urinary tract obstruction in ruminants is often the primary treatment approach. Successful medical management of certain types of struvite obstructions in small ruminant has been described using percutaneous intravesicular infusion of Walpole's solution; however, if urinary outflow fails to be re-established quickly, surgery is required to avoid further trauma to the bladder or urethra. Several options can be offered to the clients with different complications and success rates. For all cases, medical treatment is needed to counter-balance the metabolic and electrolyte changes and is paramount to offer the best prognosis prior to the surgery. This presentation will focus on the surgical procedures for breeding-intended animals as well as the long-term management of permanently obstructed males.

The Challenge of Intact Males Intended for Breeding

Integrity of the urethra is extremely important for all intact males intended to be used as breeders. For males in which the owner desires to maintain breeding service soundness and when the glans penis and urethra appear to be normal on physical examination, tube cystotomy is offered to provide immediate relief of the urinary obstruction and to reduce the urethral spasm. Combined with anti-inflammatory medication, the aim is to dislodge the obstructive stones, minimizing the urethral mucosal damage.

For adult bulls, the surgery is performed standing via a left caudal paralumbar fossa laparotomy. If the bladder is already ruptured, repair should be attempted. Bladder repair is done blindly because in nearly all cases the bladder cannot be exteriorized. If the defect is located on the dorsal aspect of the bladder, it could be left to heal by second intention in some cases. A 22 to 24 French Foley catheter is inserted in the abdomen through a second independent incision in the flank determined to minimize the tension on the bladder after the completion of the procedure. It is then introduced into the bladder via a stab incision and is secured in the bladder by inflating the balloon and, when possible, installing a purse string suture around it. The Foley catheter is then pulled out gently; apposition of the bladder to the body wall often is not achievable. The Foley catheter should be secured to the skin using a finger trap suture pattern to avoid back and forth movement. The tube is left patent and protected by making a one-way Heimlich-type valve in order to avoid air ingress into the bladder. Instillation of acidifying solutions, such as Renacidin or Walpole's, can be initiated after a few days. Instillation is followed by temporary challenge of the urethra by occlusion of the catheter provided that the urinary stones or crystals have been identified as magnesium ammonium phosphate. Longer periods of challenge are started 8 to 10 days after surgery if urine has not begun flowing spontaneously.

In case of unsuccessful alleviation of the urethral obstruction, an ischial urethrotomy can be attempted to allow more time for the urethra to heal after completion of retrograde and normograde urohydropulsion. This increases the risk of further exacerbating the existing urethritis. Ischial urethrotomy is performed in standing position under epidural anesthesia and pudendal regional anesthesia. A 10–15 cm long incision is created on the perineal midline, starting just proximal to the ischium. The retractor penis muscles are identified and bluntly separated to reveal the bulbospongiosus muscle. The urethra is invaded by longitudinally incising the bulbospongiosus muscle in the median groove. Providing that the obstruction is resolved, a long sterile polyethylene tube first is inserted into the bladder and then redirected in a normograde fashion down the urethra. This can serve as a stent traversing the surgical incision along the perineal urethra.

If the obstruction is not resolved, this procedure can be converted into a permanent ischial urethrostomy. Although semen can be collected via ischial urethrostomy, breeding soundness prognosis is guarded.

For steers and bulls no longer intended to be used for breeding, penectomy is a cost-effective salvage procedure. Using a perineal approach, a median 10–15 cm incision is made ventral to the ischial arch and extended through the subcutaneous tissues to the level of the retractor penis muscles which are separated and retracted laterally. Blunt dissection is used to identify the penile body, which is freed from the surrounding tissue. The penile body is pulled caudally and dorsally and transected proximal to the sigmoid flexure at the ventral aspect of the surgical wound. The proximal portion of the penile body is pulled further caudally and dorsally, exteriorized through the wound and oriented ventral and caudal. The penis is secured to the skin using large monofilament suture with horizontal mattress pattern. The urethra and corpus spongiosum are incised longitudinally and sutured together on each side with 2 simple continuous patterns to limit hemorrhage and spatulate the distal opening.

Urolithiasis in Small Ruminants, Decisions About Urethrostomy?

Complications of both antepubic and perineal urethrostomy are common, with contracture and recurrence of obstruction being the most frequent. Both surgical procedures temporarily relieve urethral obstruction and should be considered as salvage procedures. A narrow vertical elliptical skin incision on both side of the PU can be performed in order to maintain the opening of the mucocutaneous sutures and eventually prevent contracture. Recently, Tobias and van Amstel reported on a modified perineal urethrostomy procedure to improve longer-term outcomes. This modification of the standard perineal urethrostomy technique includes transecting the ischiocavernosus muscles at their attachment to the pelvis. This modification improves the mobilization of the penis and reduces the tension on the mucocutaneous wound. In a report of 11 cases, 10 survived to discharge and 9 were still alive after 12 months.

Perineal urethrostomies do not allow complete lavage of the bladder or removal of the remaining stones. Surgical tube cystotomy remains the preferred surgical procedure to allow complete removal of the bladder stones, as well as attempting to relieve the urethral obstruction while preserving the integrity of the anatomy. With the patient anesthetized and positioned in dorsal recumbency, a caudal paramedian approach to the abdomen is created to expose the urinary bladder. After adequate protection of the abdomen and positioning of stay suture on the bladder, an apex cystotomy is performed to remove the remaining uroliths. A 12 F to 24 F Foley catheter is then passed through the body wall via a separate stab incision, inserted into the bladder and secured to it with a pure string suture after inflation of the cuff. Abdominal closure is routine and then the Foley catheter is secured to the body wall using a butterfly retainer technique or finger trap pattern. When compared to percutaneous tube cystotomy, surgical tube cystotomy was associated with a reduced chance of requiring a second intervention.

Long-Term Management of Permanent Urinary Tract Obstruction

When all procedures have failed to restore patency within the distal urinary tract, owners should decide whether or not they want to pursue alternative procedures for permanent deviation of urine outflow. Options would be long-term management of the tube cystostomy, bladder marsupialization, or vesicopreputial anastomosis.

Permanent Tube Cystotomy

Long-term management of tube cystotomy seems to be well tolerated by pet goats. The tube needs to be replaced on a regular basis because the tube degrades and becomes brittle or the animals pull them out leading to rapid contraction of the wound. Changing the tube involves sedating the animal followed by clipping and aseptically preparing the site of the stoma. By using a set of increasing size dilatators, the stoma is restored to accommodate the appropriate size Foley catheter that is replaced in the bladder and secured to the abdomen using standard techniques. It is our experience that ascending urinary tract infections occur infrequently and most often are readily managed.

Bladder Marsupialization

Marsupialization of the bladder is performed under general anesthesia or deep sedation and local anesthesia. A 10-cm paramedian approach to the abdomen (exploratory laparotomy and cystotomy approach) is performed approximately 3 cm lateral to the preputial sheath. Upon identification of the urinary bladder, a cystocentesis is performed to reduce the tension and facilitate exteriorization of the bladder. The bladder is held by stay suture and a ventral longitudinal cystotomy allows removal of the remaining urine and scavenging of the bladder stones.

A second, 4-cm long paramedian incision (marsupialization site) is performed on the opposite side of the prepuce. Stay sutures facilitate adequate positioning of the bladder and the bladder is held in place with several simple interrupted sutures between the external sheath of the rectus abdominis and the seromuscular layer of the bladder, after which the edges of the cystotomy are sutured to the skin with small-diameter monofilament absorbable suture material (2-0 polydioxanone).

Urine scalding is inevitable due to the constant dripping of urine on the skin. Daily application of petroleum-based ointment may limit the intensity of the inflammation. Ascending urinary tract infection occurs more commonly with this technique.

Vesicopreputial Anastomosis (VPA)

The VPA technique has been recently reported to prevent urine scalding following bladder marsupialization and the recurrent replacement of Foley catheters. This surgery is performed under general anesthesia. Location of the preputial fornix is mandatory prior to performing the surgical approach, as it has to be done centered to this landmark. A 4- to 5-cm long paramedian incision is made 3 cm away and parallel to midline to enter the abdomen. The bladder is located and cystocentesis done to facilitate manipulation. The bladder is exteriorized and secured to the internal sheath of the rectus abdominis. After dissection toward the craniomedial aspect of the wound, the preputial fornix is located and a circular incision is performed around the penis prior to its retraction and caudal amputation. A longitudinal cystotomy is performed and the edges are sutured to the circumferential preputiotomy. A 20 to 24 F Foley catheter is inserted from the preputial orifice to the bladder and the cuff inflated within the bladder lumen prior to the conclusion of the vesicopreputial anastomosis with simple interrupted sutures. Closure of the body wall is then performed as routine. The Foley catheter is secured to the prepuce and the distal extremity protected with one-way valve system. Complications related to this surgery are ascending cystitis and stricture of the anastomosis.

References

1. Cypher EE, Van Amstel SR, Videla R, *et al*. Vesicopreputial anastomosis for treatment of obstructive urolithiasis in goats. *Vet Surg*. 2017;46:281–288.
2. Byers SR. Urolithiasis. In: Smith BP, ed. *Large Animal Internal Medicine*. 5th ed. St. Louis, MO: Elsevier; 2015:897–903.
3. Videla R, van Amstel S. Urolithiasis. *Vet Clin North Am Food Anim Pract*. 2016;32:687–700.
4. Tobias KM, van Amstel SR. Modified proximal perineal urethrostomy technique for treatment of urethral stricture in goats. *Vet Surg*. 2013;42:455–462.
5. Rakestraw PC, Fubini SL, Gilbert RO, *et al*. Tube cystostomy for treatment of obstructive urolithiasis in small ruminants. *Vet Surg*. 1995;24:498–505.
6. Chigerwe M, Heller MC, Balcomb CC, *et al*. Use of a percutaneous transabdominal catheter for management of obstructive urolithiasis in goats, sheep, and potbellied pigs: 69 cases (2000–2014). *J Am Vet Med Assoc*. 2016;248:1287–1290.
7. May KA, Moll HD, Wallace LM, *et al*. Urinary bladder marsupialization for treatment of obstructive urolithiasis in male goats. *Vet Surg*. 1998;27:583–588.
8. Fortier LA, Gregg AJ, Erb HN, *et al*. Caprine obstructive urolithiasis: requirement for 2nd surgical intervention and mortality after percutaneous tube cystostomy, surgical tube cystostomy, or urinary bladder marsupialization. *Vet Surg*. 2004;33:661–667.
9. Van Metre DC, Ortved K. Urolithiasis. In: Fubini SL, Ducharme NG, eds. *Farm Animal Surgery*. 2nd edition. St. Louis, MO: Elsevier; 2017:584–595.

acvim224

Surgical Management of Urolithiasis - An Internist's Perspective

Robert J. Callan, DVM, MS, PhD, DACVIM
Fort Collins, CO, USA

INTRODUCTION

Obstructive urolithiasis continues to be a significant medical problem in livestock species, particularly in small ruminants. Medical and surgical management of this problem in ruminants, camelids, and swine is well described in many texts and publications.[1,2] However, overall successful management of individual patients is quite variable and complications with establishing short- or long-term urine excretion remain common. This presentation will discuss some of the challenges observed and concepts developed over the last 20 years of treating patients with urethral obstruction due to urolithiasis at Colorado State University.

KNOW YOUR STONE

The type of urinary calculi present is often unknown at the time of presentation. However, the ability of the stone to be dissolved through acidification of the urine has significant impact on surgical approach and outcome if urethral stones are present. Magnesium ammonium phosphate (struvite) uroliths are generally soluble in aqueous solution with a pH of 5.5 to 6.5. Other urolith types including calcium carbonate, calcium oxalate, and silica are essentially non-dissolvable in physiological aqueous solutions. In review of the literature and discussion with clinicians across the country, the types of urinary calculi observed, particularly in small ruminant patients, seems to have changed over time and may be influenced by the geographical location, management and nutrition of the patient. At CSU, we see an overwhelming prevalence of non-struvite uroliths in obstructed small ruminants based on analysis of stones recovered at surgery or necropsy. This is similar to that reported by the Minnesota Urolith Center.[3] Magnesium ammonium phosphate (struvite) was identified as the most common urolith type in Boer goats in a different study.[4] In potbellied pigs, the most common stone reported was magnesium calcium phosphate.[5]

Table 1. Distribution of major uroliths in livestock identified at the Minnesota Urolith Center (1981–2007)[3]

Urolith	Cattle #, (%)	Goat #, (%)	Llama #, (%)	Sheep #, (%)	Domestic Pigs #, (%)
Calcium carbonate	8 (3.7)	224 (42.6)	1 (4.2)	33 (29.7)	3 (3.2)
Calcium oxalate	4 (1.8)	4 (0.8)	0	1 (0.9)	2 (2.1)
Calcium phosphate (apatite)	17 (7.8	16 (3)	1 (4.2)	14 (12.6)	33 (34.7)
Magnesium ammonium phosphate (struvite)	73 (33.6)	19 (3.6)	2 (8.3)	16 (14.4)	18 (18.9)
Magnesium calcium phosphate	8 (3.7)	24 (4.6)	0	0	3 (3.2)
Magnesium calcium phosphate Carbonate	28 (12.9)	101 (19.2)	1 (4.2)	22 (19.8)	9 (9.5)
Silica	43 (19.8)	76 (14.5)	18 (75)	12 (10.8)	3 (3.2)

True commercial meat animals still seem to have a higher prevalence of magnesium ammonium phosphate (struvite) urolithiasis due to the high concentrate diet. Nutritional prevention of struvite urolithiasis in livestock can be successfully managed by reducing phosphorus (<0.25–0.30% DMI) and magnesium (<0.12–0.18% DMI) intake, and keeping the Ca:P ratio near 2:1. The importance of knowing if struvite urolithiasis is involved is because this is the only stone type that is readily dissolved by acidification of the urine. Thus, medical management is possible, and urinary acidification following surgical procedures can often successfully restore urethral patency. Struvite should be suspected if fine gritty material is palpable on the hairs at the end of the prepuce. If calculi are collected from the distal urethra during examination or from the urinary tract at surgery, a sample of the stones can be placed in acetic acid (vinegar) to quickly evaluate whether or not they are likely to dissolve with urine acidification.

Other urinary calculi including calcium carbonate, calcium oxalate, silica, or calcium phosphate are not likely to dissolve as readily with urinary acidification and must be small enough to be flushed through the urethra or they must be removed surgically in order to restore urethral patency. If this is not accomplished at the time of initial surgical intervention such as tube cystotomy, then persistence of urethral obstruction is common. This concept cannot be under stated as patients presented with struvite urolithiasis seems to be decreasing in frequency while the other stone types, particularly calcium carbonate in pet small ruminant populations, seems to be increasing. While struvite urolithiasis can be controlled from a nutritional standpoint, the risk factors and nutritional causes of calcium carbonate urolithiasis are not well understood. Excessive calcium urinary excretion likely plays a role; however, the factors underlying calciuresis and calculogenesis are not fully understood and calcium intake is only one of many factors involved.

Stabilize or Immediate Surgery? What Options Are There?

Many patients with urethral obstruction present with marked discomfort and are uremic with multiple electrolyte and metabolic disturbances. For these reasons stabilization of the patient is often indicated but must be balanced with the need to restore urine excretion. Since stabilization in part will rely on correction of azotemia and electrolyte imbalances, fluid therapy is often indicated. However, this results in obvious challenges if relief of urine accumulation and restoration of urine excretion is not achieved.

Patients that are significantly compromised and have marked or complete urethral obstruction may not be considered sufficiently stable for general anesthesia or surgery. This results in the challenging decision of how to relieve urine accumulation in a safe and effective manner while providing fluid and electrolyte support to stabilize the patient. If the bladder has already ruptured, uroperitoneum can be relieved by abdominocentesis or placement of an abdominal Foley catheter for drainage and peritoneal lavage. If the bladder is intact, temporary relief can be provided by cystocentesis. However, the cystocentesis site is likely to leak as soon as the bladder becomes distended again. Percutaneous transabdominal catheterization of the bladder is described as a temporary procedure to provide bladder drainage and allow time to stabilize the patient or treat the urethral obstruction medically.[6-9] Methods include placement of a coiled or other self-retaining catheter (Bonanno Catheter, Becton Dickinson, Dewan Suprapubic Urodynamics Catheter, Cook Medical, Stamey Percutaneous Malecot Suprapubic Catheter, Cook Medical); Foley catheter through a percutaneous insertion sleeve (One-Step Suprapubic Introducer, Cook Medical, or Supra-Foley Introducer, Utah Medical Products, Inc.); Foley catheter with a needle stylet (Rutner Suprapubic Balloon Catheter, Cook Medical); or using a minimally invasive paralumbar approach[10] to place an infusion set tube into the bladder. The most common complications with these procedures are obstruction of the catheter, loss of placement within the bladder, leakage around the catheter, and intestinal perforation. These catheters will generally remain patent for up to 3 days allowing time for stabilization and the opportunity to perform surgery under more elective conditions.

Don't Neglect Diagnostic Imaging

Diagnostic imaging consisting of both ultrasound evaluation of the bladder and abdomen as well as radiographs of the bladder and urethra is helpful in confirming a diagnosis, establishing potential sites of obstruction, directing surgical intervention, and refining a prognosis. In particular, radiographic imaging is extremely helpful in identifying the location of calculi within the urethra and provides useful information to plan for surgery or to intervene post-operatively if urethral obstruction is not resolved.[11] Unless the urinary calculi are known to be struvite and will dissolve with acidification, the surgical procedure must plan on removal of all stones within the bladder and urethra in order to have the greatest likelihood restoring and maintaining urethral patency.

CHOICE OF SURGICAL PROCEDURE

Several surgical procedures for treatment are described.[1,2,12,13] Salvage procedures are often recommended for commercial animals (Table 2). These procedures are intended to provide temporary relief of urethral obstruction with low cost. However, there is a high incidence of recurrence of urethral obstruction within 12 months. The two main causes for the recurrence of urethral obstruction are stricture at the surgical site and re-obstruction by stones still present in the bladder. Presurgical radiographic imaging can help identify the extent of urinary calculi and may aid in determining the suitability of these procedures.

Table 2. Salvage and long-term surgical procedures for urinary obstruction caused by urolithiasis

Salvage procedures	Long-term procedures
Removal of the urethral process	Tube cystostomy
Perineal urethrostomy	Tube cystostomy with urethrotomy
Urethrotomy	Modified perineal urethrostomy
Penile amputation	Bladder marsupialization
	Vesicopreputial anastomosis

Tube cystotomy is generally recommended for pet or breeding animals to restore normal urinary function. Failures with this procedure are observed and may in part be related to the type of urolith and the ability for obstructing uroliths to be passed, surgically removed, or dissolved. In early descriptions of tube cystostomy, overall success rates were reported near 70%.[14] However, in our population at CSU, the success of tube cystostomy without surgical removal or urethral stones is less. The lower success rate appears to be related to the high prevalence of non-struvite urolithiasis and the difficulty to remove these uroliths from the urethra by flushing or dissolution with a low pH.

LEAVE NO STONE BEHIND

The observed failure to restore urethral patency following tube cystotomy has led to a more aggressive surgical goal of removing all bladder and urethral stones during surgery unless there is strong confidence that the uroliths are struvite. Accomplishing this goal is greatly aided by pre-surgical radiographs of the bladder and urinary tract. At CSU, we have combined a tube cystotomy and urethrotomy surgical approach in order to achieve the goal of removing all stones from the bladder and urethra in order to surgically restore urethral patency. Our current approach is to perform a cystotomy to remove uroliths from the bladder. A second perineal surgical approach is performed to assist in passing a normograde urethral catheter to locate and remove uroliths through one or more urethrotomies. Once the catheter is successfully passed to the end of the penis, a second catheter is attached and then pulled retrograde into the bladder. Urethrotomy incisions are closed with 4-0 or 5-0 absorbable suture. The cystotomy is closed and a routine tube cystotomy is completed. Key points that we have learned with this technique will be discussed.

TUBE CYSTOTOMY COMPLICATIONS

Post-operative complications are often encountered during the management of tube cystotomies. What appears to be persistent urinary discomfort is relatively common and may be partly related to both bacterial cystitis and mechanical irritation from the cystotomy tube. Bacterial cystitis is expected and managed by appropriate antibiotics. This bacterial cystitis will persist until the cystotomy tube is removed. The balloon of the cystotomy tube should not be over-inflated and 5-ml balloon catheters are adequate in most cases. However, partial filling of a 5-ml balloon increases the risk of collapse of the balloon and retraction of the catheter from the urinary bladder. The cystotomy tube should exit the body wall in a natural position for a normal sized bladder in the caudal abdomen. There is a tendency to exit the cystotomy tube more anteriorly at the time of surgery when the bladder is very distended. This should be avoided to minimize stretching of the bladder when it returns to a normal size. Care should be taken if the catheter is tunneled in the subcutaneous tissues because this can increase the risk of kinking and obstruction of the tube at the point where it exits the external rectus sheath.

Rescue Procedures

Surgical intervention may fail due to the inability to resolve urethral obstruction or development of urethral stricture resulting in re-obstruction, particularly in non-struvite urolithiasis. In these cases rescue procedures such as the modified proximal perineal urethrostomy technique[12] or vesicopreputial anastomosis[13] may provide adequate long term urine excretion and comfort for the patient.

Don't Cut and Run

In order to better evaluate the success and the causes of failures of surgical procedures, it is important for the surgeons performing the procedures to remain highly involved in the post-surgical management and aftercare of these patients. The more that surgeons are aware of and involved in managing the post-operative challenges, the more likely that we will investigate adaptations and alternative surgical procedures that may improve success.

References

1. Byers SR. Urolithiasis. In: Smith BP, ed. *Large Animal Internal Medicine*. 5th ed. St. Louis, MO: Elsevier; 2015:897–903.
2. Videla R, van Amstel S. Urolithiasis. *Vet Clin North Am Food Anim Pract*. 2016;32:687–700.
3. Osborne CA, Albasan H, Lulich JP, *et al*. Quantitative analysis of 4468 uroliths retrieved from farm animals, exotic species, and wildlife submitted to the Minnesota Urolith Center: 1981 to 2007. *Vet Clin North Am Small Anim Pract*. 2009;39:65–78.
4. Sun WD, Zhang KC, Wang JY, *et al*. The chemical composition and ultrastructure of uroliths in Boer goats. *Vet J*. 2010;186:70–75.
5. Chigerwe M, Shiraki R, Olstad EC, *et al*. Mineral composition of urinary calculi from potbellied pigs with urolithiasis: 50 cases (1982–2012). *J Am Vet Med Assoc*. 2013;243:389–393.
6. Streeter RN, Washburn KE, McCauley CT. Percutaneous tube cystostomy and vesicular irrigation for treatment of obstructive urolithiasis in a goat. *J Am Vet Med Assoc*. 2002;221:501,546–549.
7. Janke JJ, Osterstock JB, Washburn KE, *et al*. Use of Walpole's solution for treatment of goats with urolithiasis: 25 cases (2001–2006). *J Am Vet Med Assoc*. 2009;234:249–252.
8. Fortier LA, Gregg AJ, Erb HN, *et al*. Caprine obstructive urolithiasis: requirement for 2nd surgical intervention and mortality after percutaneous tube cystostomy, surgical tube cystostomy, or urinary bladder marsupialization. *Vet Surg*. 2004;33:661–667.
9. Chigerwe M, Heller MC, Balcomb CC, *et al*. Use of a percutaneous transabdominal catheter for management of obstructive urolithiasis in goats, sheep, and potbellied pigs: 69 cases (2000–2014). *J Am Vet Med Assoc*. 2016;248:1287–1290.
10. Fazili MR, Malik HU, Bhattacharyya HK, *et al*. Minimally invasive surgical tube cystotomy for treating obstructive urolithiasis in small ruminants with an intact urinary bladder. *Vet Rec*. 2010;166:528–531.
11. Kinsley MA, Semevolos S, Parker JE, *et al*. Use of plain radiography in the diagnosis, surgical management, and postoperative treatment of obstructive urolithiasis in 25 goats and 2 sheep. *Vet Surg*. 2013;42:663–668.
12. Tobias KM, van Amstel SR. Modified proximal perineal urethrostomy technique for treatment of urethral stricture in goats. *Vet Surg*. 2013;42:455–462.
13. Cypher EE, van Amstel SR, Videla R, *et al*. Vesicopreputial anastomosis for the treatment of obstructive urolithiasis in goats. *Vet Surg*. 2017;46:281–288.
14. Rakestraw PC, Fubini SL, Gilbert RO, *et al*. Tube cystostomy for treatment of obstructive urolithiasis in small ruminants. *Vet Surg*. 1995;24:498–505.

acvim225

Population Medicine Using a New Lens: Investigating Veterinary and Producer Mental Health and Wellbeing

Joanne Hewson, DVM, PhD, DACVIM (Large Animal Internal Medicine)
Guelph, ON, Canada

INTRODUCTION

Veterinarians show elevated rates of psychological distress when compared to other professions or to the general population, and death by suicide is proportionally increased in this profession. Veterinary schools have similarly seen a dramatic increase in mental health challenges in their student population. We have an urgent need to elucidate and address the factors that contribute to this increased mental health risk. Like veterinarians, the farming population shares an increased rate of mental health distress and elevated suicide rates compared to the general population. Understanding mental health and wellbeing in this population ties closely with our roles in the veterinary profession. This review will present emerging research by the AWAR^{2}E (Advancement of Wellness and Resilience via Research and Education) team at the University of Guelph investigating mental health in veterinarians, veterinary students, producers, and associated animal health workers. Interventions to not only raise awareness of mental health in the veterinary curriculum, but also to facilitate student learning of self-care and resilience, will be introduced.

UNDERSTANDING THE PROBLEM

i) Towards a Better Understanding of Mental Health in Veterinarians

Surveys of veterinarians in the United Kingdom, USA, and Australia have consistently demonstrated an increased rate of death by suicide in veterinarians. Suicide rates are approximately four times higher in veterinarians than seen in the general population,[1] with more than one in five veterinarians reporting having "seriously considered suicide"[2]. Suicidal ideation was similarly reported by just under 20% of veterinarians in a Canada-wide survey by the Canadian Veterinary Medical Association.[3] Despite a high prevalence of suicidal ideation within the profession, only 11% of veterinarians surveyed perceived suicide to be an issue for the profession, indicating a general lack of awareness of this mental health crisis pervasive throughout the veterinary profession.[2]

Anxiety and depression are common precursors to suicidal ideation, and increased rates of anxiety and depression are frequently reported for veterinarians. Veterinarians in the UK demonstrated high levels of anxiety (26%) and depression (6%),[4] and an Australian study reported that more than a quarter of veterinarians experienced depression[5].

In order to better understand the complexities of mental health within the veterinary profession, our AWAR^{2}E team has undertaken both quantitative and qualitative approaches to investigating this crisis. A cross-sectional study, first conducted within Ontario, but more recently expanded across Canada, surveyed veterinarians looking at prevalence of depression, anxiety, compassion fatigue, and burnout using validated self-reporting mental health scales. As a protective factor in mental health, resilience of veterinarians was also assessed. Preliminary results from the Ontario data indicated that one-third of veterinarians reported overwhelming anxiety, and depression was reported for roughly one of every ten veterinarians surveyed. Of concern, reduced resilience in Ontario veterinarians was identified compared to reported norms for the general population.

A qualitative investigation into veterinary mental health was also done, using key informant interviews and thematic analysis, to seek the lived experiences of mental health issues in veterinarians and the perceived impact on patient care, the human-animal bond, client satisfaction, and the wellbeing of the veterinarian's practice. Data analysis is ongoing for this component of the veterinary mental health research.

ii) Towards a Better Understanding of Mental Health in Veterinary Students

It is unrealistic to believe that all issues relating to anxiety and depression in veterinarians developed post-graduation. Mental health illness in adults often links back to learned behaviours and coping mechanisms adopted in earlier years. There is increasing recognition that a high prevalence of mental health challenges already exists within students engaged in postsecondary education, and that mental health may be further impacted by lifestyle choices. Results of early surveys at two Canadian universities serve as examples of the mental health environment in postsecondary education; about 50% of students reported feelings of hopelessness and/or overwhelming anxiety within the year prior to the survey, and one third reported feeling depressed.[6,7] The Ontario University and College Health Association (OUCHA) performed a more widespread survey of over 2500 students attending Ontario colleges and universities in the spring of 2016.[8] Overall, students reported high rates of overwhelming anxiety (65%) and depression impacting functionality (46%) within the past year. Suicidal ideation was reported for 13% of students, with 9.3% reporting self-harm activities (intentional cutting, bruising, burning or other self-injury) and 2.2% of students having attempted suicide within the past year (9% having attempted suicide sometime in their lifetime). Self-care habits pertaining to sleep, nutrition, and exercise were significantly below recommended guidelines, and use of alcohol, drugs, stimulants, and prescription medications belonging to other individuals were frequently reported.

Within veterinary training, even higher levels of anxiety and depression than the population norms have been reported.[9-13] A recent study of 1245 students across 33 North American veterinary schools showed clear evidence that veterinary students function under a high level of stress, with a strong correlation between stress and depression symptomology.[14]

iii) Towards a Better Understanding of Mental Health in Agricultural Producers

While large scale disasters such as drought, flooding, or animal disease outbreaks present as extreme adversity to overcome for farmers, more common day-to-day threats to mental health include elevated expectations for hard work and productivity despite lack of control over external factors such as government oversight, market fluctuations, weather conditions, and physical injury or illness. Many farmers associate their sense of identity closely with the land, and may also experience a loss of sense of self during periods of poor farm productivity.

Prior studies have identified similar trends in poor mental health of farmers as for veterinarians. Evidence of poor self-care was demonstrated by higher prevalence of short-term, high-risk alcohol consumption, obesity, abdominal adiposity, overweight, hypertension, and risk for diabetes or cardiovascular disease in farmers compared to the national average.[15,16] Higher rates of psychological distress and suicide were also reported in farmers compared with non-farming occupations.[17,18]

Members of our AWAR^{2}E team conducted a recent online survey of more than 1100 producers across a range of industry sectors in animal agriculture in Canada. Results demonstrated that a significant number of producers experience high stress, anxiety, depression, emotional exhaustion and cynicism. Of similar concern as with veterinarians, producers also demonstrated reduced resilience compared to the general population in two-thirds of the survey respondents. While many producers expressed a positive attitude toward the merits of help from a mental health professional, less than half of producers expressed a perception that adequate resources existed in the industry to support the mental health needs of the farming community.

A significant proportion of the survey respondents indicated reluctance to seek such professional help due to the perceived stigma of "what people might think." While this is generally true of the farming community, farmers that die by suicide presented to their general practitioners more commonly with physical symptoms rather than identifying psychological problems - general practitioners should consider depression as a potential cause of unexplained conditions in this at-risk group.[19]

Effecting Change Using an Evidence-Based Approach

A first step to effecting change in the mental health crisis of our profession and those engaged in related occupations is to raise awareness. Leading this charge is a growing number of veterinary wellness summits, retreats, and conference tracts that are focusing on both mental illness as well as means to improve resilience. The recent publication of a special issue on mental health and well-being by the *Journal of Veterinary Medical Education* (Volume 44, Issue 1, Spring 2017) attests to the growing recognition of the importance mental health holds in our profession. Just like veterinarians and general medical practitioners serve as the points of contact to first detect mental ill health in producers, faculty within veterinary colleges are often the first to become aware of students showing compromised mental wellness. Mental health literacy and a working knowledge of available resources is essential for veterinary faculty, who must be ready to listen and support students with mental illness but must also be able to effectively recognize the boundaries of their expertise in this area and instead aim to serve as a conduit to more appropriate long-term resources for students.

Past literature on mental health in the veterinary profession has focused on states of illness - anxiety, depression, suicidal ideation - with little evidence of positive psychology to focus instead on personal skills and resources that support coping strategies and an ability to thrive.[20] The elements of psychological capital (self-esteem, optimism, hope and resilience) were positively associated with an adaptive response (improving depressive symptomatology over time) in veterinary students, and support the need to help train students in these important life skills.[13]

This shift to positive psychology focuses on building competence rather than correcting weakness. Students already learn other non-cognitive, behavioural ("transferable") skills such as responsibility, organization, independent work, collaboration, initiative and self-regulation in their earlier education. Resilience, as a transferable skill, can also be taught and practiced. Delany and colleagues demonstrated that teaching university students resilience strategies, such as positive coping skills, can lead to improved self-efficacy, cognitive control, and self awareness.[21] Teaching resilience has been a recent charge to colleges and universities - the Okanagan Charter, arising from the 2015 International Conference on Health Promoting Universities and Colleges/VII International Congress, represented an urgent call to action for post-secondary institutions internationally to directly integrate health and personal development into campus policies and services, in order to create environments that facilitate a culture of well-being. Similarly, the Higher Education Quality Council of Ontario (HEQCO) has identified resilience as a desired education outcome and is actively seeking optimal approaches to both teach **and** assess this transferable skill in higher education.[22]

The definition of resilience has evolved over time from an innate personal character trait to more recent views that resilience is a dynamic state impacted by self, coping strategies, as well as a wide spectrum of external factors contributing together towards a person's ability to bounce back. From this viewpoint, it is proposed that some elements of resilience can be learned and others fostered. External protective factors, such as support from family and friends, financial stability, and the care and consideration by faculty, can serve to strengthen the resilience of students. Recognizing the importance of these protective factors and their impact can help students to value these external resources, and serves to remind faculty of the importance of treating students respectfully and with empathy during all interactions.

Several veterinary schools, including Guelph, have introduced training in wellness into the core veterinary curriculum. Wellness across the domains of emotional, environmental, financial, and social wellbeing is taught and practiced in the first year of the program, with additional wellness domains of intellectual, occupational, physical, and spiritual wellbeing integrated across subsequent years of the program. In addition to clinical rotations in their final year, students also have the opportunity to train in mental health and resilience as an elective rotation before they graduate. Fostering nonjudgmental, nonreactive mindfulness and self-compassion, as predictors of resilience, has been proposed as a mechanism by which to arm veterinary students with the tools to maintain their mental wellbeing despite moments of adversity throughout their future careers.[23] These skills require regular practice and application in order to show a positive impact on lowering signs of anxiety and depression.[24] By integrating core training in resilience and self-care across the program, we hope to foster a culture of attentiveness to mental wellbeing and support regular practice of skills such as mindfulness over time to promote adoption and habituation of these practices. Qualitative and quantitative study of the impact of these changes in the curriculum to veterinary student mental health is a priority by the AWAR^{2}E group, in order to take an evidence-based approach to this training.

The AWAR[2]E group directly supports the activities of our student partner, OVC THRIVE (Teaching Healthy Resilience in the Veterinary Environment) to actively engage the student population in raising awareness, reducing stigma and connecting veterinary students with resources to support their mental health needs. Wellness Wagons, that provide recreational activities to reduce stress and connect with colleagues in a social context, have been situated at strategic and accessible sites throughout the College. Through OVC-THRIVE, students and faculty have participated in panel discussions on mental health both internally as well as across the campus as a way to connect with the wider university community. Most recently, the first annual "Let's THRIVE Week" was launched, delivering a mix of information sessions, sharing of lived experiences, social events, and providing each student with a Mental Health Resource Kit. Based on participant survey feedback following this event, powerful statements of improved sense of community, normalizing mental health struggles, and reduced stigma resulting from this event were evident. Encouraging students and faculty to share their lived experiences with mental illness has openly raised awareness, improved understanding of each other, and provided an opportunity for struggling students to speak and be heard. In general, being able to talk about stressors or events perceived as traumatic is associated with improved mental health and physical health outcomes,[25] whereas the prior culture of avoidance, denial, and a lack of ability to talk about stressors or events perceived as traumatic is associated with worsening signs of secondary traumatic stress. Future activities to build resilience through the curriculum include the implementation of a camp/retreat for first-year veterinary students, with the goal being to further build a sense of belonging and connection to their peers, establish friendship bonds that will serve as a support system, and to allow incoming students to get to know each other outside of the boundaries defined by the term "veterinary" or "student" and see each other as people.

While the majority of discussion on effecting change has focused here on methods to help veterinary students thrive throughout their training and to be prepared for a productive career, the same tenets of breaking down stigma, starting the conversation about mental health, and building skills for resilience are also beginning to be applied to agricultural producers. A coordinated effort to identify and improve mental health support resources across industry sectors is in progress, and training in mental health first aid has been made available as a pilot to Ontario producers.

Summary

Veterinarians and producers are populations uniquely vulnerable to mental illness, and while much research has been invested towards quantifying this risk, more is needed to understand and mitigate this risk. A paradigm shift towards positive psychology, with an emphasis on building resilience as a life skill, is essential as we work towards improved mental health within veterinarians, veterinary students, and farmers.

References

1. Mellanby RJ. *Vet Rec.* 2005;157(14):415.
2. Skipper GE, Williams JB. *Vet Med Educ.* 2012;39(1):79.
3. 2012 Veterinary Summit - Report of the CVMA Task Force on Member Wellness, www.canadianveterinarians.net
4. Bartram DJ, *et al. Occup Med.* 2009;59(5):334.
5. Hatch PH, *et al. Aust Vet J.* 2011;89(11):460.
6. Craggs S. CBC News 2012.
7. Lunau K. Macleans 2012.
8. American College Health Association; 2016.
9. Kogan LR, McConnell SL. *J Vet Met Educ.* 2005;33(2)193.
10. Hafen M, *et al. J Vet Met Educ.* 2006;33(3):432.
11. Hafen M, *et al. J Vet Met Educ.* 2008;35(1):102.
12. Reisbig AM, *et al. J Vet Met Educ.*` 2012;39(4):341.
13. Bakker DJ, *et al. J Vet Met Educ.* 2017;44(1):50.
14. Killinger SL, *et al. J Vet Met Educ.* 2017;44(1):3.
15. Brumby S, *et al. Aust J Rural Health.* 2012;20(3):131.

16. Brumby S, *et al. J Vet Met Educ.* 2013;29(3)311.
17. Fraser CE, *et al. Int J Soc Psychiatry.* 2005;51(4):340.
18. Hounsome B, *et al. Community Ment Health J.* 2012;48:503.
19. Booth N, *et al. Occup Environ Med.* 2000;57(9):642.
20. Cake MA, *et al. J Vet Med Educ.* 2017;44(1):95.
21. Delany C, *et al. Adv Health Sci Educ Theory Pract.* 2015;20(5):1303.
22. Patry D, Ford R. Toronto: HEQCO 2016.
23. McArthur ML, *et al. J Vet Med Educ.* 2017;44(1):106.
24. Correia HM, *et al. J Vet Med Educ.* 2017;44(1):125.
25. Bartram DJ, *et al. Vet Rec.* 2010;166(17):518.

acvim226

Clinical Listeriosis in Ruminants

Keith P. Poulsen, DVM, PhD, DACVIM (LAIM)
Madison, WI, USA

Listeria monocytogenes is a gram-positive rod shaped bacterium that is normally found in soil and is frequently found in the fecal flora of clinically normal domestic animals and humans. Recognition of human listeriosis is relatively new, starting in the 1980s, compared to clinical disease in ruminants, which has been recognized for decades and is typically referred to as "circling disease" or one of the differential diagnoses for an abortion storm. Listeriosis in ruminants in current practice is usually sporadic but remains relevant because it is a zoonotic foodborne disease. This comprehensive review will focus on the pathophysiology of listeriosis, which is unique as an intracellular microbe. The review will then shift to diagnosis, management, and treatment of three main clinical syndromes that veterinarians manage in practice including neonatal septicemia, meningoencephalitis, and abortion.

Listeria monocytogenes (most commonly the 4b and 1/2a strains) became a major interest in human medicine as a foodborne infectious pathogen in 1985 when contaminated hotdogs caused a widespread outbreak in the eastern part of the USA. We now recognize that ready to eat (RTE) foods and raw dairy products are a risk factor for human disease that the Centers for Disease Control estimate to cause 3,500 infections that result in 500 deaths annually in the United States. This number is small relative to salmonellosis and norovirus infections that cause estimated 2–3 and 100 million human infections, respectively, in the USA. However, the mortality rate associated with listeriosis is significantly higher and high risk groups are children under the age of 5, pregnant women, immunosuppressed individuals, and the elderly. Pregnancy is associated with a 20 times greater risk of clinical listeriosis and clinical attack rates are high with ingestion of contaminated RTE foods.

Listeria has a host of virulence factors to survive within the cells of the innate immune system and to replicate in endothelial tissues. Perf A is responsible for mediating expression of all of *Listeria*'s virulence mechanisms. Internalin A (InlA) allows for non-professional phagocytes to uptake *Listeria* and binds E-cadherin. Internalin B (InlB) is important for phagocytosis as well, but the exact mechanisms are not as well-known. It is generally accepted that non-virulent strains lack the ability to express InlA and InlB. Listeriolysin (LLO) is the major virulence factor and mediates intracellular survival by binding cytolysin and allows escape from the phagosome. Listeriolysin functions the best in acidic environments (like in the phagosome) and may explain why *Listeria* survives so well in acidic feedstuffs, such as inadequately prepared ensiled feeds. Listeria is quite mobile in the cytoplasm with expression of ActA that polymerizes proteins behind the cell to push it to the neighboring cell to start the process over again. This cycle requires about 60 minutes to complete.

Clinical listeriosis can manifest in several distinct syndromes. Circling disease from meningoencephalitis and micro-abscesses in the brainstem is typically seen in adults and is thought to occur from retrograde movement of the pathogen up the trigeminal nerve to the brainstem. A separate clinical syndrome that usually does not occur concurrently with circling disease is abortion and/or abortion storms. Both are thought to be from *Listeria* in feeds, particularly ensiled feeds, but is sporadic in focal meningoencephalitis and potentially wide spread in the herd with abortion. *Listeria* is a common Gram-positive cause of neonatal sepsis and meningitis in ruminant and pseuodoruminant species. Infection sources for neonates are not fully understood with potential *in utero* or inoculation via the birth canal as sources.

There is not much new information on diagnostics or treatment for veterinary cases. Diagnostics for neurological deficits include CSF cytology and advanced imaging. Definitive diagnosis of CSF with PCR and/or bacterial culture rarely yields a positive result. Cytology findings of a monocytic pleocytosis ± hyperproteinemia is accepted as diagnostic. For neonatal sepsis, blood culture is needed ante mortem. Post-mortem diagnosis may include gross visualization of micro abscesses in the brainstem with or without immunohistochemical staining of histopathological specimens. There are few reports of antimicrobial resistance of Listeria isolates and the mainstay of therapy in veterinary medicine has been penicillin or oxytetracycline. Because of this, at the Wisconsin Veterinary Diagnostic Laboratory, we do not run antimicrobial sensitivity testing. When considering treatment options, bioavailability and of the drug (and interactions of drug combinations) at the site of infection is more likely the limiting factor.

Listeria infections in large animal practice will likely continue to be sporadic and dependent on quality of ensiled feed preparation. Because Listeria is considered part of the normal fecal and environmental flora, vaccine development is very unlikely. Interestingly, due to the microbe's unique and efficient methods of intracellular invasion and survival, it is being developed as a potential vehicle for vaccine development in human medicine. The public health and food safety aspect is also important for veterinarians to know and be able to communicate to our human medical colleagues and the general public.

acvim227

Exploiting *Salmonella* Biology to Reduce Disease Burden

Johanna R. Elfenbein, DVM, PhD, DACVIM (LAIM)
Raleigh, NC, USA

Introduction

Salmonella enterica is zoonotic pathogen of substantial concern to global human and animal health. It is a gram-negative, facultative intracellular, facultative anaerobic organism that can colonize and cause disease in all species of mammals. It is the leading cause of bacterial foodborne disease and is a leading cause of morbidity and mortality in people worldwide.[1] Infection with *Salmonella* arises from consumption of contaminated food or water or direct contact with an infected individual or animal. Infected individuals and animals may shed billions of organisms per gram of feces and represent a major source of environmental contamination. *Salmonella* may live in the soil for up to 300 days in permissive conditions and may colonize plants.[2] As such, this is a tricky pathogen to eradicate and will require a thorough understanding of the organism's biology to develop rational interventions that can interrupt the infection cycle.

Cattle as a Translational Model

Cattle develop disease with clinical features and histologic lesions that are identical to human disease. Cattle develop clinical salmonellosis characterized by fever, inappetence, abdominal pain, and diarrhea that may last as long as 10 days. Histologic lesions are most prominent in the ileum and are characterized by neutrophilic enteritis, submucosal edema, villous blunting, and disruption of the epithelium.[3] Infected animals shed *Salmonella* into the environment for more than 20 days. Most immunocompetent individuals will clear the pathogen with supportive care alone but calves are highly susceptible to systemic spread of the pathogen and may succumb to bacteremia. In addition to clinical disease, cattle may be subclinically infected and these animals are a significant source of *Salmonella* contamination of the food supply. It is estimated that 12–30% of human *Salmonella* disease outbreaks originate from food products of bovine origin.[4,5] The degree of production losses associated with subclinical *Salmonella* infections is unclear. Understanding how *Salmonella* colonizes the bovine intestine therefore serves a dual purpose to create novel interventions to improve both human and animal health.

We use a calf model of infection in our laboratory to understand how *Salmonella* colonizes the intestine. The bovine model is a superior animal model for multiple reasons. First, as already mentioned, cattle develop clinical signs of fever, weight loss, abdominal pain, and diarrhea with histologic features that are identical to that of humans. These common disease features make *Salmonella* pathogenesis research in cattle extremely translatable. Second, in this surgical model, *Salmonella* are injected directly into the lumen of isolated segments of ileum. The large size of calves allows up to 38 different experiments in a single animal. This design is optimal for reducing intra-animal variation, making this model ideal to investigate novel therapeutic targets. Third, in contrast to murine models, neutrophilic inflammation occurs in the presence of intact microbiota in cattle, making this model more representative of the intestinal lumen of people with *Salmonella* infection. Together, these reasons make cattle the ideal host organism to understand the true biology of *Salmonella* in the gut.

Salmonella Pathogenesis

Huge advances have been made in recent decades to elucidate the virulence mechanisms of *Salmonella* using various animal models, including the calf. Salmonellae gain entry into the host through ingestion and must rapidly respond to a wide variety of environmental conditions in the digestive tract: acidic pH in the stomach (abomasum), bile salts, host-derived antimicrobial peptides, high osmolarity of luminal fluid and declining oxygen tensions.[6] The ability to grow in the anaerobic intestinal lumen is critical for *Salmonella* early intestinal colonization.[7] In addition, *Salmonella* competes with the host and resident microflora for continuously changing nutrient sources.[8] The innate immune defenses of the gastrointestinal tract as well as the microbiota-mediated colonization resistance make the intestine an inhospitable environment for incoming pathogens. However, pathogens like *Salmonella* have evolved strategies to overcome these host and microbiota-mediated barriers. One way that *Salmonella* overcomes these barriers is by invading the intestinal epithelium and inducing a neutrophilic enteritis, which it exploits to its benefit.

The *Salmonella* virulence paradigm requires coordinated expression of two type-III secretion systems (TTSS). Both TTSS-1 and TTSS-2 are horizontally acquired genetic loci that are *Salmonella*-specific genes that distinguish *Salmonella* from its close phylogenetic relatives. Acquisition of the TTSS-1 defines the genus *Salmonella* and acquisition of the TTSS-2 defines the species *Salmonella enterica*. This is important because the commercially available PCR for establishment of *Salmonella* infection relies on identification of the gene *invA*, a part of the TTSS-1 gene cluster.

The TTSS is a macromolecular needle that allows for the injection of protein effector molecules directly into the host cell. The cues for expression of the TTSS-1 are present in large abundance in the ileum. *Salmonella* senses environmental cues such as osmolarity of luminal fluid, composition and concentration of microbial-derived short chain fatty acids, oxygen tension and likely many other signals to induce the expression of the TTSS-1.[9] The protein effectors of the TTSS-1 serve multiple purposes.[10] Early secreted effector proteins rearrange the actin cytoskeleton to induce membrane ruffling of non-phagocytic epithelial cells. Next, secreted effector proteins interact with host cell signaling proteins (Rho GTPases) to facilitate alterations to the epithelial cell plasma membrane and to activate transcription factors (NF-kB) that influence expression of inflammatory cytokine genes. Finally, late secreted effector proteins rearrange the plasma membrane to engulf *Salmonella* within a phagosome. The end result of these processes is internalization of *Salmonella* into its *Salmonella*-containing vacuole (SCV) and activation of the inflammatory cascade. Expression of the TTSS-2 occurs within the SCV and prevents phagolysosomal fusion, facilitating intracellular growth and survival.[10] The key to successful infection is passage across the epithelial layer, a process that also requires the TTSS-2 in order to access the lamina propria and induce a strong host inflammatory response.

Pro-inflammatory cytokines and chemokines are induced in response to the TTSS-1 during infection that result in neutrophil trafficking to the intestine. Bovine neutrophils release a wide array of compounds including but not limited to reactive oxygen species (ROS), beta-defensins, serine proteases, metalloproteinases, cathelicidins, and calprotectin when activated.[11] *Salmonella* encounters neutrophil-derived antimicrobial products both within the lumen of the intestine and during transient residence within neutrophils. However, it is relatively resistant to the toxic byproducts of neutrophils through a variety of mechanisms including resistance to ROS and antimicrobial peptides.[12] Interestingly, the study of *Salmonella enteritis* in calves with naturally occurring leukocyte adhesion deficiency (LAD) showed that without neutrophilic influx, *Salmonella* intestinal infection causes less tissue damage and less luminal fluid accumulation.[13] These data suggest that the early stages of neutrophilic infiltration, designed to kill invading pathogens, is to the benefit of the pathogen and not the host.

The host inflammatory response gives *Salmonella* a competitive advantage over resident microflora. Neutrophilic influx creates an oxidative environment, which creates novel nutrients that *Salmonella* can metabolize but normal microflora cannot. The best-characterized example of this is thiosulfate, a by-product of microbial sulfur metabolism, is oxidized to tetrathionate, an alternate anaerobic electron acceptor that *Salmonella* can use but most members of the microflora cannot.[14,15] In addition, effectors of the TTSS-1 directly activate epithelial production of inducible nitric oxide synthase (iNOS) thereby creating nitrate, a preferential terminal electron acceptor that supports metabolism in the intestine. In addition, *Salmonella* uses host-derived nutrients (e.g., ethanolamine, 1,2-propanediol, fructose-asparagine) and scavenges transition metals (e.g., iron, zinc, manganese) to compete with the microbiota during intestinal inflammation.[16,17] It is clear that *Salmonella* benefits from the inflammatory response at the expense of the host, which develops inflammatory diarrhea, expelling *Salmonella* into the environment to repeat the cycle.

Novel Virulence Determinants

Despite the fact that *Salmonella* is a highly studied pathogen and we understand numerous mechanisms of virulence, the incidence of disease has remained unchanged for more than 20 years. One reason why the amazing body of work has not resulted in a novel intervention is that the majority of studies on *Salmonella* pathogenesis utilize poorly translational rodent models of infection. Recent work in our laboratory used a calf model of infection to identify novel mechanisms by which *Salmonella* colonizes the host intestine.[18] We identified 31 novel genes never before described to be important for intestinal colonization. One of these genes is a retron reverse transcriptase that produces a unique molecule called multicopy single-stranded DNA (msDNA). msDNA is a single stranded DNA molecule that is covalently linked to a single-stranded RNA. We demonstrated that msDNA is critical for *Salmonella* to colonize the intestine because it is required for anaerobic growth.[7] Anaerobic growth depends on the ability to alter metabolism to utilize alternate anaerobic electron acceptors. The use of anaerobic electron acceptors is governed by msDNA.[7] Thus msDNA is a novel virulence determinant of *Salmonella* and represents a new area for study that may lead to development of novel therapeutic interventions.

Conclusion

Salmonella is a successful pathogen because it has evolved the ability to thrive during the host inflammatory response, both in the gut and in its intracellular niche. Recent studies have elucidated the mechanisms by which *Salmonella* has evolved to anaerobically respire numerous nutrient sources in the intestine. The understanding of these basic processes is critical to the development of novel therapeutic interventions to disrupt the infection cycle. Novel interventions are critically necessary to reduce the disease burden associated with *Salmonella* to improve both human and animal health.

References

1. Kirk MD, Pires SM, Black RE, *et al.* World Health Organization estimates of the global and regional disease burden of 22 foodborne bacterial, protozoal, and viral diseases, 2010: a data synthesis. *PLoS Med.* 2015;12:e1001921.
2. Ongeng D, Geeraerd AH, Springael D, Ryckeboer J, Muyanja C, Mauriello G. Fate of *Escherichia coli* O157:H7 and *Salmonella enterica* in the manure-amended soil-plant ecosystem of fresh vegetable crops: A review. *Critical Reviews in Microbiology.* 2015;41(3):273–294.
3. Wray C, Sojka WJ. Experimental *Salmonella typhimurium* infection in calves. *Research in Veterinary Science.* 1978;25:139–143.
4. Jackson BR, Griffin PM, Cole D, Walsh KA, Chai SJ. Outbreak-associated *Salmonella enterica* serotypes and food commodities, United States, 1998–2008. *Emerging Infectious Diseases.* 2013;19:1239–1244.
5. Guo C, Hoekstra RM, Schroeder CM, *et al.* Application of Bayesian techniques to model the burden of human salmonellosis attributable to U.S. food commodities at the point of processing: adaptation of a Danish model. *Foodborne Pathogens and Disease.* 2011;8:509–516.
6. Alvarez-Ordonez A, Begley M, Prieto M, *et al. Salmonella* spp. survival strategies within the host gastrointestinal tract. *Microbiology.* 2011;157:3268–3281.
7. Elfenbein JR, Knodler LA, Nakayasu ES, *et al.* Multicopy single-stranded DNA directs intestinal colonization of enteric pathogens. *PLoS Genetics.* 2015;11:e1005472.
8. Faber F, Baumler AJ. The impact of intestinal inflammation on the nutritional environment of the gut microbiota. *Immunology Letters.* 2014;162:48–53.
9. Altier C. Genetic and environmental control of *Salmonella* invasion. *The Journal of Microbiology.* 2005;43 Spec No:85–92.
10. Srikanth CV, Mercado-Lubo R, Hallstrom K, McCormick BA. *Salmonella* effector proteins and host-cell responses. *Cellular and Molecular Life Sciences.* 2011;68:3687–3697.
11. Zanetti M. Cathelicidins, multifunctional peptides of the innate immunity. *Journal of Leukocyte Biology.* 2004;75:39–48.
12. Fang FC. Antimicrobial actions of reactive oxygen species. *MBio.* 2011;2(5).
13. Nunes JS, Lawhon SD, Rossetti CA, *et al.* Morphologic and cytokine profile characterization of *Salmonella enterica* serovar typhimurium infection in calves with bovine leukocyte adhesion deficiency. *Veterinary Pathology.* 2010;47:322–333.
14. Winter SE, Thiennimitr P, Winter MG, *et al.* Gut inflammation provides a respiratory electron acceptor for *Salmonella. Nature.* 2010;467:426–429.

15. Faber F, Tran L, Byndloss MX, *et al.* Host-mediated sugar oxidation promotes post-antibiotic pathogen expansion. *Nature.* 2016;534:697–699.
16. Rivera-Chavez F, Baumler AJ. The pyromaniac inside you: *Salmonella* metabolism in the host gut. *Annual Review of Microbiology.* 2015;69:31–48.
17. Diaz-Ochoa VE, Jellbauer S, Klaus S, Raffatellu M. Transition metal ions at the crossroads of mucosal immunity and microbial pathogenesis. *Frontiers in Cellular Infection Microbiology.* 2014;4:2.
18. Elfenbein JR, Endicott-Yazdani T, Porwollik S, *et al.* Novel determinants of intestinal colonization of *Salmonella enterica* serotype typhimurium identified in bovine enteric infection. *Infection and Immunity.* 2013;81:4311–4320.

acvim228

National Strategy on Antimicrobial Resistance in Switzerland: A Global Approach

Mireille Meylan, DMV, MS, PhD, Dr. Habil., DACVIM, DECBHM
Berne, Switzerland

INTRODUCTION

The discovery and development of antimicrobial drugs belong to the most significant progresses of medicine in the last century. However, excessive and inappropriate antimicrobial drug use has led to the emergence and propagation of resistant bacterial strains, with dramatic consequences for human and animal health. Numerous national and international health organizations (www.ec.europa.eu, www.fda.gov, www.oie.int, www.who.int) have expressed growing concerns regarding antimicrobial resistance (AMR) and published recommendations to limit the spread of AMR via prudent use of antimicrobial drugs. Cephalosporins of 3rd and 4th generation, fluoroquinolones and macrolides are considered as critically important antimicrobials of highest priority for their importance in human medicine. In Switzerland, warnings restricting the use of these three categories of antimicrobial drugs in animals have been included in the product information since several years. They may only be prescribed when either susceptibility testing has shown that the bacteria involved are resistant to other antimicrobials or when previous knowledge of a particular farm/group of animals suggests that a therapeutic success is not to be expected with other drugs.

Antimicrobial use practices in Switzerland, especially in veal calves (the animal group receiving the highest amounts of antimicrobials) have been analyzed in a few studies in the past years.[1-4] While treatment intensity in veal calves was lower than in other European countries (with a mean of 21 days of treatment per calf and year, vs. e.g., 60 days/animal year in Belgium),[3] these studies revealed that a reasonable indication for antimicrobial treatment could not always be found and that the choice of treatment did often not follow the principles of prudent use of antimicrobials. In veal calves, critical antimicrobials were included in most group treatments (macrolides in 55.3% oral treatments and 91.9% parenteral treatments) as well as in individual parenteral treatments (38.3% fluoroquinolones, 13.1% macrolides and 4.7% cephalosporins of 3rd/4th generation).[3] Another study in more intensive veal calf operations showed a rate of 75% individual treatments with critically important antimicrobials (65% fluoroquinolones, 9% cephalosporins of 3rd/4th generation, and 1% macrolides).[1] In dairy calves, 36% of pneumonia cases were treated parenterally with critical antimicrobials (23% fluoroquinolones, 10% macrolides and 3% cephalosporins of 3rd/4th generation), and calf diarrhea was treated with fluoroquinolones in 47% of the cases.[4]

Relatively few data are available on resistance rates in clinical isolates from cattle, but the available data on pathogens isolated from calves with pneumonia point at stable rates of AMR for some antimicrobials groups (penicillins, sulfonamides, tetracyclines), but at increasing rates for others, especially critically important antimicrobials: between 2001 and 2012, resistance rates against tylosin have increased from 0 to 80–100% in Pasteurellaceae, and resistances also appeared against the newer substances tilmicosin and tulathromycin during that time (up to 10% resistant strains). While no resistance to fluoroquinolones was observed in 2001, 8% of the isolated strains of *Mannheimia varigena* were resistant to enrofloxacin and danofloxacin.[5,6] It is further worrisome that the extended-spectrum beta-lactamase (ESBL) gene $bla_{CTX\text{-}M\text{-}14}$ was found in *Escherichia coli* isolated from nasal swabs of calves in two dairy farms where milk containing residues of antimicrobials (used to treat the cows) was regularly fed to calves and where acute mastitis was routinely treated with the 4th generation cephalosporin cefquinome.[7]

Critically important antimicrobials are also used (too) widely in human medicine, in acute care hospitals as well as in outpatient care. Likewise, increasing resistance rates have been observed in isolates from human patients, in particular against fluoroquinolones and 3rd generation cephalosporins in *Escherichia coli* or *Klebsiella pneumoniae*, as well as against fluoroquinolones and macrolides in zoonotic pathogens such as *Campylobacter*. Further reasons for concern are the increasing prevalence of multi-resistant bacteria and the emergence of plasmid-mediated colistin resistances (www.bag.admin.ch). Antimicrobial consumption and resistance in human medicine have been monitored by the Swiss Center for Antimicrobial Resistance since 2006 (www.anresis.ch).

Because the measures taken in the past to limit AMR did not appear to be sufficient, the Swiss government launched a national program for the control and reduction of bacterial resistances to antimicrobial drugs in the fall of 2015, with involvement of all interested stakeholders (i.e., human medicine, veterinary medicine, food safety, agriculture and environment).[8] Groups of experts have developed lists of priorities for measures to limit and optimize the use of antimicrobial drugs, covering all relevant aspects from basic scientific evidence (e.g., investigation of resistance mechanisms or molecular epidemiology) to public information about the hazards associated with excessive and/or inappropriate use of antimicrobials. Different projects and research programs have been initiated to address the priorities identified in the first steps of the project.

National Strategy on Antimicrobial Resistance

The National Strategy Program was launched by two (out of seven) state departments, the Federal Department of Home Affairs and the Federal Department of Economic Affairs, Education and Research. Four Federal Offices participate in the program, the Federal Office of Public Health (FOPH), the Federal Food Safety and Veterinary Office (FSVO), the Federal Office for Agriculture (FOA), and the Federal Office for Environment (FOEN), as well as the state administrations of the 26 Swiss cantons. The participation of all involved stakeholders from the very beginning of the program should ensure the implication of all concerned interest groups in its development, wide acceptance of its results and thus broad support for the implementation of the measures to be proposed in order to minimize antimicrobial resistance in the future.

Objective

The main objective of the National Strategy on Antimicrobial Resistance (StAR) is to preserve antimicrobial drug efficacy to sustainably ensure human and animal health. The strategy was designed to define priority areas for intervention and to develop appropriate measures to reach the goals defined in the project.

Approach

Three workshops with participants from all related interest groups (i.e., human and veterinary medicine, public health, agriculture, and environment) were held to define the contents of the StAR strategy. The first workshop aimed at defining areas where action was needed in priority, the second one was dedicated to the definition of strategic objectives, and the third one to the measures to be developed to achieve the goals of the program. Before and between the workshops, discussions in smaller groups of experts were held around specific topics related to AMR, e.g., for topics around animal health separate groups discussed the situation in dairy cattle, beef cattle, veal calves, swine, and poultry. The aim of the discussions was to identify the most critical areas and to review possible strategies to limit further selection and dissemination of (multi-)resistant bacteria. Eight main fields of action were identified that need to be addressed in priority: surveillance, prevention, prudent use of antimicrobial drugs, reduction of resistances, research and development, cooperation, information and education, and regulatory framework.[8] The objectives related to these eight priority fields are shortly described below.

Surveillance

Systematic monitoring of bacterial resistance and recording of antimicrobial drug use should allow for detailed investigation of the relationships between antimicrobial use and bacterial resistance, as well as for monitoring the effects of control measures designed to reduce antimicrobial use and minimize the selection and propagation of resistant bacteria.

Prevention

Better disease prevention in humans and animals will lead to a reduction of the need for antimicrobial treatment. Since antimicrobial treatment intensity is related to resistance rates, prophylactic measures including e.g., hygiene, vaccination, rapid and reliable diagnostic methods, or improved housing conditions for animals, should allow for a reduction of AMR. The use of antimicrobial drugs to "compensate" for deficient management practices (hygiene, biosecurity, housing, etc.) should be forbidden.

Prudent Use of Antimicrobial Drugs

Excessive and/or inappropriate use of antimicrobial drugs is recognized as one of the main reasons for the selection of bacteria resistant to antimicrobials. Detailed recommendations regarding antimicrobial treatments in humans and animals, in particular the use of critically important or new drugs, are needed to optimize antimicrobial treatments and reduce the selection pressure on bacteria.

Reduction of Resistances

Rapid identification of resistant bacteria allows for measures to prevent their propagation (e.g., in human or animal hospital settings). While prevention is mostly focused on measures to avoid the dissemination of infectious agents, including resistant bacteria, at the herd level for animals, targeted investigations (e.g., upon hospital admission) for human patients at risk of harboring resistant bacteria should allow for minimizing the propagation of AMR.

Research and Development

Understanding causes and interrelations in respect to AMR is urgently needed as the basis for the development and effective implementation of control measures. Targeted research projects must be designed to close the most critical knowledge gaps (e.g., in the field of diagnostic methods).

Cooperation

Bacterial resistance being a complex problem that affects numerous circles, including those related to human and animal health, food hygiene, agriculture and environment, solutions to such problems can only be obtained through collaboration of all involved. This, in turn, requires an efficient structure to coordinate and supervise the efforts undertaken by different research groups. Attention must also be paid to new developments at the international level and to new knowledge becoming available that could influence the further progress of the program.

Information and Prevention

The general population also plays an important role in the implementation of the principles of prudent use of antimicrobial drugs, e.g., through compliance with treatment recommendations (i.e., by not interrupting a prescribed treatment too early or by respecting correct dosages). It is extremely important that the awareness of the problems related to AMR and of the responsibility of each and every individual should be raised through appropriate information campaigns. This is especially true for farmers who may treat large groups of animals with antimicrobials. In addition, continuing education should be provided for professionals to ensure a deep understanding of resistance mechanisms and related factors, prophylactic measures, diagnostic methods and prudent use of antimicrobial drugs. Only this will guarantee appropriate information and education of (human) patients as well as animal owners and caretakers.

Regulatory Framework

An appropriate regulatory framework is necessary to ensure that antimicrobial drugs will retain their efficacy in the future. Regulatory measures should include steps to implement the prudent use of available antimicrobial drugs and to support the development of new therapeutic substances. Incentives should be considered to promote management systems associated with improved animal health and thus reduced need for antimicrobial treatment.

Further critical points identified in the course of the StAR process include the limited availability of first line antibiotics in veterinary medicine (recurrent shortage of products containing penicillins and tetracyclines in the last years), the lack of an economically and ecologically sustainable solution for the disposal of milk from dairy cows during and after (withdrawal times) treatment with antimicrobial drugs, and the lack of reliable statistics on antimicrobial treatment intensity (only data in tons of marketed active substance, but not on the number of treated animals, are available to date) (http://www.bag.admin.ch).

Implementation

The involvement of all stakeholders in the development of the StAR program has allowed for fruitful discussions, which have confirmed that the representatives of the different disciplines (human medicine, veterinary medicine, public health, agriculture, and environment) were not necessarily aware of the difficulties encountered and the efforts already engaged to address them in the other sectors. Thus, the goal of bringing the different actors confronted with the problem of AMR together and of improving the understanding of the overall challenges associated with the topic has already been achieved, or at least first contacts have been made that should improve collaborations in the future.

Since the start of the StAR program, the Swiss legislation on veterinary drugs (Tierarztneimittelverordnung, TAMV, www.admin.ch) has been revised. By law, the use of antimicrobials as growth promoters has been forbidden in Switzerland since 1999 and specific regulations on antimicrobial drug use in food animals have been in place since 2004. The changes implemented in 2016 will ensure a stricter implementation of the principles of prudent use of antimicrobials, e.g., critical antimicrobial drugs may not be given in advance to farmers "for the case an animal should get sick" but will have to be administered exclusively by a veterinarian after clinical assessment of the patient, only if all conditions for choosing a critically important antibiotic are fulfilled. Likewise blanket dry-off strategies are no longer allowed, antibiotic treatment at dry-off may only be used selectively based on milk bacteriology results and/or elevated somatic cell counts. Furthermore, antimicrobial treatment recommendations for food animals (cattle and swine) developed by a panel of experts are now available online for veterinary practitioners, which limit the use of critical antimicrobials to the strict minimum.[9] First results of several research projects launched by the FSVO under the heading "reduction of antimicrobial use and disease prevention in food animals" are expected by the end of this year. Similar initiatives have been launched in human medicine. Furthermore, a National Research Program "Antimicrobial Resistance" (NRP72) approved by the Federal Council in 2015 and conducted by the Swiss National Science Foundation (www.snf.ch) has started with 21 projects (of which two are focused on veterinary medicine) in January 2017. With its holistic and interdisciplinary one-health approach, NRP72 aims to contribute to reducing AMR and complications in treating infectious diseases. It is "expected to increase our knowledge about the origin of resistance genes and their distribution mechanisms, develop swift diagnostic tools, analyses new antimicrobial molecules and propose intervention measures."

Conclusion

The realization of the StAR program represents a challenge because of the large number of research institutions, offices and stakeholders involved. However, the critical situation of ever increasing rates of AMR provides an opportunity for the different interest groups to work together on a public health problem of great actuality and relevance and to share experiences from different perspectives. This should improve reciprocal understanding and lead to more intensive exchanges in the future among the various fields of activity and expertise in order to achieve the goal of limiting AMR in the future.

References

1. Beer G, *et al. Schweiz Arch Tierheilkd.* 2015;157(1):55.
2. Lava M, *et al. Prev Vet Med.* 2016;126:81.
3. Lava M, *et al. Prev Vet Med.* 2016;126:121.
4. Pipoz F, *et al. Schweiz Arch. Tierheilkd.* 2016;158(6):389.
5. Vogel G, *et al. Schweiz Arch. Tierheilkd.* 2001;153(7):341.
6. Rérat M, *et al. Prev Vet Med.* 2012;103:265.
7. Pipoz F, *et al. Schweiz Arch .Tierheilkd.* 2016;158(6):397.
8. https://www.blv.admin.ch/blv/de/home/tiere/tierarzneimittel/antibiotika/nationale-strategie-antibiotikaresistenzen--star--.html
9. https://www.blv.admin.ch/blv/de/home/tiere/tierarzneimittel/antibiotika/nationale-strategie-antibiotikaresistenzen--star--/sachgemaesser-antibiotikaeinsatz.html.

acvim229

Medical Management of Parturition and Dystocia for the Livestock Internist

Robert J. Callan, DVM, MS, PhD, DACVIM
Fort Collins, CO, USA

INTRODUCTION

Medical management of high-risk pregnancies, difficult dystocias, and postpartum complications in the dam is common in livestock practice. Ultrasound examination of the fetus and fetal membranes is an excellent method to monitor the high-risk fetus near parturition. Induction of parturition may be necessary, but appropriate methods vary by species and circumstances. Medical management and support during dystocia can make resolution of a dystocia easier and also improve the likelihood of a favorable outcome for the dam and the fetus. Finally, medical management following parturition can improve recovery following a complicated parturition. This presentation will discuss specific diagnostic and pharmacological management of these complex cases in order to maximize the health of the dam and fetus.

FETAL MONITORING WITH ULTRASOUND

Ultrasound can be used to monitor fetal viability, wellness, and the health of the uterus and placenta. This is most commonly done by transabdominal ultrasound. While a 5 MHz probe will provide helpful information and effectively image the uterine wall and fetal membranes, a 2.5–3 MHz probe is often necessary to locate the heart and determine a fetal heart rate in livestock. Normal fetal heart rates are reported (Table 1).[1-5] In general, mean fetal heart rate decreases during gestation and is at its lowest near parturition. Variation in heart rate is associated with fetal activity and tachycardia during or following fetal activity is a sign of fetal wellbeing if it returns to a normal heart rate at rest. In camelids, fetal heat rates near term that are <50 bpm or >130 bpm suggest fetal stress.[3] Persistent bradycardia or tachycardia is a sign of fetal stress and should be considered an indication for intervention.

Table 1. Normal fetal heart rate near the end of gestation in livestock

Species	Normal fetal HR (bpm)
Bovine	90–125
Caprine	Mean ~140
Ovine	108–126
New World Camelid	80–115 (1.6 to 1.8 times the dam HR)

The uterine wall, placenta, and fetal fluid should also be examined; however, there is limited objective data reported. The allantoic fluid is generally anechoic with some echoic particles.[2,5] The amniotic fluid has slight echogenicity compared to allantoic fluid and more echogenic particles. Increased echogenicity of the allantoic or amniotic fluid is an indication of fetal stress in cattle.[6] The uteroplacental thickness can also be evaluated. The thickness should be <8 mm in camelids.[3] Objective criteria has not been established in ruminants. Increased uteroplacental thickness indicates edema or inflammation and should be considered abnormal. Fetal activity (motion) should also be evaluated.[5,6] Decreased activity and hyperactivity are indicators of fetal stress. However, this can be difficult to fully evaluate as fetal activity varies considerably over time and evaluation should be repeated to get an accurate assessment.

PARTURITION INDUCTION

Parturition may be induced both for elective management choices as well as for non-elective medical treatment for a compromised dam or fetus. Progesterone production during pregnancy occurs at either the corpus luteum or from the placenta, or both. The contributions from each site vary by species. Abortion or parturition in ruminants, swine, llamas, and alpacas can be induced by stopping the production of progesterone using either prostaglandin F2 alpha to lyse the corpus luteum or corticosteroids to inhibit progesterone produced by the placenta. Methods for induction of parturition at term are listed in Table 2.

Table 1. Methods to induce parturition at full term in livestock[7,8]

Species gestation length (days)	CL dependent	Treatment	Comments
Bovine 279–287 Ave 283	No	Dexamethasone 25 mg Dinoprost (Lutalyse), 25 mg Cloprostenol (Estrumate), 500 µg	- Parturition in 3–5 days - CL dependence only lasts for about the first 150 days of gestation. - Glucocorticoids reduce placental progesterone production after 150 days gestation. However, luteal progesterone is still sufficient to maintain pregnancy until the last month of gestation. - Glucocorticoids alone will not induce parturition until the last month of gestation. - Combination prostaglandin and glucocorticoid will induce abortion from 5 days of gestation to term.
Caprine 148–152	Yes	Dinoprost (Lutalyse), 5–10 mg Cloprostenol (Estrumate), 75–100 µg Dexamethasone 15–20 mg	- Good fetal viability when within 2 weeks of expected due date. - PG → Parturition in 28–56 hrs, ave ~31–35 h - Dex → Parturition in 119±30 h
Ovine 142–152 Ave. 147	No	Dexamethasone 15–20 mg	- CL dependence only lasts for about the first 30 days of gestation. - Not reliable prior to 137 days gestation (2 weeks before term). - Good fetal viability when within 1 week of expected due date. - Average time to parturition is ~48 h, range 36–60 h.
Porcine 111–117 Ave. 114–115	Yes	Dinoprost (Lutalyse), 20 mg	- Gestation >116 days should be considered a dystocia. - Induce no earlier than 111 days gestation. - Ave parturition in 30 hours, range 20–36 hours. - Oxytocin (5–10 IU) administered 20 hours after PG will provide closer synchronization of parturition in the next 6 hours. - >3 days prior to due date will result in decreased piglet viability.
New World Camelid (llama & Alpaca) 331–347 Ave. 344	Yes	Cloprostenol (Estrumate), 100–150 µg IM Dinoprost (Lutalyse), 5 mg SQ	- Prostaglandins will induce abortion from day 7 to full term. - Prostaglandins will successfully induce parturition with a viable fetus within 2 weeks of term. - Dosing may need to be repeated at 24 h. - Parturition in 3 to 7 days. - Dinoprost has been associated with acute death in llamas and should be used cautiously. The cause of death is thought to be due to extreme increase in peripheral vascular resistance. This is not reported in alpacas. - Dexamethasone (as little as 0.5 mg) routinely results in

			fetal death when used to induce parturition.

Induction of parturition is sometimes used to help stimulate pulmonary maturation and surfactant production in premature or high risk pregnancies. Early induction of parturition in cattle (258 to 270 days) with either dinoprost or flumethasone resulted in accelerated fetal lung maturation and improved respiratory function after birth.[9] In another study, induction of parturition in cattle with calf survival as early as 230 days gestation was accomplished with a dexamethasone step up protocol. In this study, parturition was induced following the administration of 5 mg dexamethasone IM every 12 hours for 4 days. This was followed by 10 mg dexamethasone every 12 hours on day 5. Cows that had not yet calved received 40 mg dexamethasone and PGF_2 alpha on day 6.[10,11] While most exogenous corticosteroids will consistently induce parturition after 150 days gestation in cattle, isoflupredone (Predef 2X) at doses from 30 to 40 mg IM did not induce abortion in cattle throughout gestation.[12]

Dystocia Decision Making

Dystocia is a complex problem and the decisions made in how to approach this in a patient can have significant impact on the outcome. Consideration of the species, breed, intended use, likely cause, duration, length of gestation, physical status of the dam, value of dam and neonate, costs of intervention, and prognosis are all important factors. Once parturition and dystocia is confirmed, a routine stepwise decision and action plan is helpful to best assist the patient and client (Figure 1).

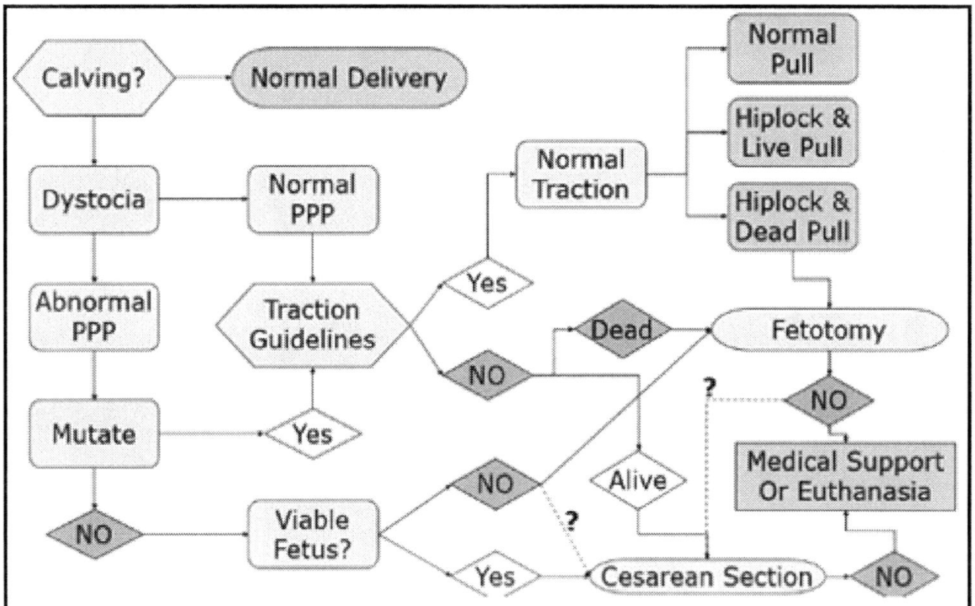

Figure 1. A decision flow chart for dystocia in cattle.

In cattle, one method of traction guidelines is to assess the ability to extend the limbs of the calf beyond the vulva when the cow is recumbent in right lateral recumbency.[13]

- If the calf is in anterior presentation and both forelimb fetlocks can be extended 10 cm (one hand's breadth) beyond the vulva with the force applied from two people, then the shoulders of the calf have fully entered or are past the pelvis of the dam and the calf can be delivered vaginally.
- If the calf is in posterior presentation and the hocks of both hind limbs can be extended past the vulva with the force applied from two people, then the pelvis of the calf is fully engaged or past the pelvis of the dam and the calf can be delivered vaginally.

Positioning the dam in right lateral recumbency during delivery of the neonate is recommended in ruminants and camelids for several reasons:

- Reduces the forces needed to overcome gravity when pulling the neonate up over the pelvis.
- The rumen (or first gastric compartment in camelids) applies pressure to the uterus and aids in expulsion of the neonate.
- Allows for more direct caudal oriented force to be applied by the operator.

Support of the dam should not be overlooked when treating a dystocia. Supplemental oxygen via nasal insufflation may be provided to support both the dam and the fetus, particularly at altitude above 5,000 feet. Dystocia can be a particularly traumatic and painful event. If the dystocia is particularly difficult, non-steroidal anti-inflammatory drugs such as flunixin meglumine should be administered early in the process. If there is significant gross contamination and trauma, systemic antibiotics may be indicated.

UTERINE RELAXATION - TOCOLYTICS

Uterine relaxation can be extremely helpful in cases of dystocia where uterine contraction limits the ability to manipulate the fetus and uterus. Indications to induce uterine relaxation include prolonged dystocia, fetal malposition (particularly breach and retained head), fetotomy, and in some cases when exteriorizing the uterus for caesarean section. There are no tocolytic drugs currently labeled for livestock in the United States. β_2-adrenergic agonists are smooth muscle relaxants and can decrease uterine contractions. While clenbuterol is an effective β_2 agonist and tocolytic, its use in food animal species is illegal in the United States. Epinephrine has β_2-adrenergic activity and can provide effective short-term relaxation of the uterus in ruminants and camelids in active labor. It is administered at a dosage of 1 ml (epinephrine, 1:1000) per 100 lb BW. This can be diluted in saline and administered as an IV bolus to effect. Uterine relaxation is usually appreciated within 2 to 5 minutes and will last about 15 minutes. Epinephrine can also be used to relax the uterus to induce a uterine prolapse to allow for external suturing of a uterine tear.

Non-steroidal anti-inflammatory drugs can also decrease uterine tone during labor by blocking the production of PGF_2 alpha and thus blocking the local PG stimulation of uterine contractions. However, this effect is relatively nominal during parturition when oxytocin levels are high.

REFERENCES

1. Erdogan G. Ultrasonic assessment during pregnancy in goats - a review. *Reprod Domest Anim.* 2012;47:157–163.
2. Ferrer MS, Jones M, Anderson DE, *et al.* Ultrasonographic parameters of fetal well-being and development in alpacas. *Theriogenology.* 2013;79:1236–1246.
3. Tibary A, Rodriguez J, Sandoval S. Reproductive emergencies in camelids. *Theriogenology.* 2008;70:515–534.
4. House JK. The peripartum ruminant. In: Smith BP, ed. *Large Animal Internal Medicine.* 5th ed. St. Louis, MO: Elsevier; 2015.
5. Buczinski S. Ultrasonographic assessment of late term pregnancy in cattle. *Vet Clin North Am Food Anim Pract.* 2009;25:753–765. Table of Contents.
6. Buczinski S, Fecteau G, Comeau G, *et al.* Ultrasonographic fetal well-being assessment, neonatal and postpartum findings of cloned pregnancies in cattle: a preliminary study on 10 fetuses and calves. *Can Vet J.* 2009;50:261–269.
7. Youngquist RS, Threlfall WF. *Current Therapy in Large Animal Theriogenology 2*, 2nd ed. St. Louis, MO: Saunders; 2007.
8. Edmondson MA, Roberts JF, Baird AN, *et al.* Theriogenology of sheep and goats. In: Pugh DG, ed. *Sheep and Goat Medicine.* 2nd ed. Maryland Heights, MO: Elsevier; 2012.
9. Zaremba W, Grunert E, Aurich JE. Prophylaxis of respiratory distress syndrome in premature calves by administration of dexamethasone or a prostaglandin F2 alpha analogue to their dams before parturition. *Am J Vet Res.* 1997;58:404-407.
10. Bleul U. Respiratory distress syndrome in calves. *Vet Clin North Am Food Anim Pract.* 2009;25:179–193, vii.
11. Zerbe H, Zimmermann DK, Bendix A. Neonatal asphyxia in calves: diagnosis, therapy and prophylaxis. *Tierärztliche Praxis Ausgabe G, Grosstiere Nutztiere.* 2008;36:163–169.
12. Mohammedsadegh M. Effect of isoflupredone acetate on pregnancy in cattle. *Vet Rec.* 1994;134:453.
13. Funnell BJ, Hilton WM. Management and prevention of dystocia. *Vet Clin North Am Food Anim Pract.* 2016;32:511–522.

acvim230

BVD: Eradication Programs in Europe

Ingrid Lorenz, DMV, DMV. vet. habil., DECBHM
Poing, Germany

Introduction

Infections with bovine viral diarrhea virus (BVDV) are a significant cause of economic losses to the cattle industry worldwide. Until late in the last century control efforts were limited to vaccination programs in individual herds. With accumulating knowledge of the epidemiology of BVD infection, the focus turned to eradication in several European countries in the 1990s. It is generally accepted that the persistently infected (PI) animal plays the key role in the transmission of BVDV within and between herds. PI animals can develop if a naïve pregnant cow is infected with BVDV in early pregnancy. These animals are immunotolerant to the virus, but will often appear perfectly normal at birth. PI animals will shed large amounts of virus with all body fluids throughout their life. The detection and removal of PI animals stands therefore at the center of all eradication efforts.[1]

Eradication Programs in Europe

The first large scale eradication schemes were launched in the early 1990s in Norway, Sweden, Finland, and Denmark. All those countries are now considered free of BVD. For each country the duration of the program until eradication was completed was about ten years. Austria started a regional and voluntary program in 1997 in Lower Austria following the Scandinavian approach. The Austrian program was extended to the whole country and made compulsory in 2004. Switzerland launched a BVD eradication scheme in 2008 following a different approach, which led to a dramatic drop in BVD prevalence within three years. Germany started an again technically slightly different national campaign in 2011, with its principles then adopted by Ireland in 2012 and Belgium in 2015. The reason for the adoption of different approaches was mainly the difference in BVD prevalence at the start of the campaign and the availability of test systems.[2]

Scandinavian Approach

The common Scandinavian approach, which was also adopted by Austria later on, will be described using the example of Norway.[3] In preparation of the program the serological examination of a limited number of blood samples indicated that 15–20% of dairy cows and 25–30% of dairy herds had been exposed to BVDV, with considerable regional variations throughout the country. Like in all Scandinavian countries the program was triggered by pressure of the farming industry. In Norway it was run by the veterinary authorities with the financial support from the dairy and breeding associations and the meat cooperation.

The program was based on an initial screening of bulk tank milk samples of all dairy herds and examination of pooled blood samples from young animals in beef herds. According to the level of antibodies detected in the bulk tank milk the herds were classified according to the risk of recent exposure to BVDV infection. Herds classified with high antibody levels (absorbance in an indirect ELISA above 0.25; 23% of dairy herds) were subsequently tested for antibody levels in pooled milk from primiparous cows. If this again was positive for BVDV antibodies, a pooled blood sample from about 5 animals 8–12 months of age was examined. In cases where BVDV antibodies were found in pooled blood of young animals the farms were put under trade restrictions since the presence of one or more PI animal was considered probable. Lifting of restrictions was either based on a change in results of these annual routine tests or individual blood testing of all animals of a herd for BVDV and culling of PI animals. Initially, restrictions were placed on 9% of dairy herds and 22% of beef herds.

Swiss Approach

Switzerland used a completely different approach in their national eradication program which was launched in 2008.[4] The reasons were a very high estimated seroprevalence of BVD, a high level of cattle movement and shared summer grazing on mountain pastures. Therefore, testing of individual animals for BVD virus was used in the Swiss eradication scheme. In the first half of 2008 all animals that went on shared summer pastures were tested with the aim to reduce the risk of the generation of new PI calves on pasture. Thereafter, all remaining cattle in the country were tested in the second half of 2008. For this purpose ear notch samples were taken from all animals and were examined either by enzyme-linked immunosorbent-assay (ELISA) or real time reverse transcriptase PCR (rtRT-PCR). Farms were put on movement restrictions until test results were available. If virus-positive animals were found on a farm, pregnant animals were restricted from movement until their calves were born and tested with negative result. In the second phase of the program all newborn calves were tested for BVDV by means of an ear notch sample taken by the farmer. In September 2010 the prevalence of newborn BVD infected calves had dropped to 0.2%.[5] Ear notch testing of newborn calves was discontinued in 2012. Since then BVD free status is monitored serologically by means of bulk milk samples on dairy farms and pooled blood samples on beef farms.

GERMAN APPROACH

Germany introduced a nation-wide BVD eradication campaign from January 2011. It is similar to the Swiss program insofar as it uses antigen testing (mainly in ear notch samples) without serological screening. Also, as opposed to the Scandinavian and Swiss approach, vaccination is permitted as an additional biosecurity measure. The eradication program is based on the mandatory testing of all calves for BVD antigen, initially within the first 6 months of life or prior to movement. The timeframe for testing has recently been shortened to one month. Only certified "unsuspicious animals" (non-PI) can be traded. In this system, it is possible for a cow to receive a referred "unsuspicious" status after the birth of a calf that tested negative for BVDV. PI animals (defined as animals that were tested positive twice within 60 days, or once without retesting within 60 days) have to be culled immediately.[7] With this scheme the prevalence of PI animals has dropped from 0.48% in 2011 to 0.02% in 2016, and the percentage of farms with PI animals from 3.44% to 0.16% in the same time.[8]

DATA HANDLING

It is very obvious that eradication schemes that rely on testing of individual animals can only succeed if animal registration and registration of test results is very tightly handled. Within the European Union a system is implemented in all member states that ensures that every bovine is identified by an individual registration number, which is applied by means of ear tags to both ears. In Germany the national database for the registration of cattle (HI-Tier) is also used for the registration of animal health data including BVD test results. This database serves as virtual link between producers, laboratories, and veterinary authorities. The veterinary authorities will automatically be informed as soon as a laboratory uploaded a positive test result. On the other hand negative test results will be printed on the individual animal's passport that is issued to the owner. Animals can only be traded if they are accompanied by a passport with a negative test result.[9]

The samples are in nearly all cases taken by the farmer, while inserting the identification ear tags. These ear tags are provided by the manufacturer with a tissue sampling system. The sampling tubes are labeled with the animal's unique identification number which can be read into the laboratories' testing systems automatically, so that mistakes during data transfer are minimized.[10]

CONCLUSION

BVDV eradication programs on a regional or country level are successful if they are mandatory, if the farming industry supports them, and if producers are well informed. The sensitivity of the available tests for virus detection is so good, that usually false negatives are only of consideration towards the end of the eradication process. Therefore, outbreak investigations with re-testing of suspicious previously negatively tested animals should be included in any eradication efforts. The availability of a good animal registration system is obligatory, if eradication is attempted by means of individual animal testing.

REFERENCES

1. Stahl K, *et al. Japanese J Vet. Res.* 2012;60(Supplement):S31.
2. Booth R, *et al. Livestock.* 2016;21:338.
3. Waage S, *et al.* In: Proceedings from the World Buiatrics Congress; 1994: 773; Bologna, Italy.
4. Presi P, *et al. Vet Microbiology.* 2010;142:137.
5. Presi P, *et al. Prev Vet Med.* 2011;99:112.
6. Bundesamt für Lebensmittelsicherheit und Veterinärwesen. https://www.blv.admin.ch/blv/de/home/tiere/tiergesundheit/ueberwachung/nationales-ueberwachungsprogramm.html 2017.
7. Schirrmeier H. In: Proceedings from the World Buiatrics Congress; 2014: 245; Cairns, Australia.
8. Wernike K. https://www.fli.de/fileadmin/FLI/IVD/BVD-Statistik2011-2016.pdf. 2017.
9. Carmanns R, *et al.* http://www.tgd-bayern.de/webyep-system/programm/download.php?FILENAME=77-3-3-at-Dokument.pdf&ORG_FILENAME=04_bvd_befundverwaltung_und_ueberwachung.pdf. 2012.
10. Alex M, *et al.* http://www.tgd-bayern.de/webyep-system/programm/download.php?FILENAME=77-3-3-at-Dokument.pdf&ORG_FILENAME=04_bvd_befundverwaltung_und_ueberwachung.pdf. 2014.

acvim231

Bovine Respiratory Coronavirus

Sanjay Kapil, DVM, PhD, DACVM
Stillwater, OK, USA

Bovine respiratory disease (BRD) is the number one cause of mortality in feedlot cattle. Several viruses (including BRCV) and bacteria have been associated with BRD. BCV is also an important cause of calf diarrhea.

Bovine coronavirus (BCV) is a member of antigenic group 2 which also includes other members of ruminant coronaviruses that are genetically and antigenically related to BCV (deer coronavirus). All members of Coronaviridae family BCV have spike-like projections on the surface (Spike and hemagglutinin-esterase).

BCV has world-wide distribution and BECV is responsible for 35% cases of calf scours. The virus is shed at higher titers in young calves up to one month age. In older cattle, the virus is shed at low titers under stress such as snow storm and parturition.

Stress of transportation can cause BCV shedding in nasal cavity. Routes of BCV transmission include feces, aerosols and contact. BCV is stable in organic material such as feces. BCV binds to soil and clay. BCV is susceptible to heat, lipid solvents, soap and pH.

BCV is responsible for both enteric and respiratory diseases. BECV and BRCV are antigenically similar viruses detected in two different compartments of cattle. BRCV is a proven cause of upper respiratory tract disease. Role of BRCV in lower respiratory tract disease needs further investigation. In intestines, BCV multiples in intestinal crypts. A virus related to BCV has been reported in dogs with respiratory disease but role of BCV in canine disease is still unclear? Winter dysentery can be caused by BCV but other agents such as BVD, *Salmonella* can also cause symptoms of hemorrhagic bowel disease.

For confirming the diagnosis of BCV in any clinical syndrome, whole BCV antigen tests will be preferred over nucleic acid amplification techniques for several reasons: BCV antigen tests such as ELISA are semi quantitative; FAT provides evidence of extent of tissue involvement with BCV antigen and BCV-IHC can correlate antigen detection with lesions in formalin-fixed tissues. PCR tests can lead to false-positive diagnosis as small amount of BCV RNA can be detected in clinically normal animals in both feces and nasal swabs. Detection of BCV RNA without clinical disease or lesions is of no clinical significance. However, detection of high copy number (based on Ct values) of BCV RNA in a clinical specimen can be a good correlate with clinical disease. In most real-time RT PCRs, a lower Ct value of BCV RNA such as 15 to 20 can be indicative of role of BCV in clinical disease. Any values between 25 to 30 are intermediate and the clinical value difficult to determine. Values above 30 to 35 are most likely not associated with clinical disease. Recent vaccination of cattle can lead to positive BCV RT PCR (Ct above >30).

Best samples for BCV detection include spiral colon and upper 1/3 of trachea. Samples should be shipped frozen over ice packs over night to a diagnostic laboratory. Based on HA-HI, antigenic variation has been documented in BECV. Guardian vaccine contains both type I and type III BECV. BCV vaccines are available for parenteral and intra nasal administration. Mucosal routes of BCV will provide protection at both enteric and respiratory mucosa due to common mucosal immune systems with rehoming of immune lymphocytes between mucosa. Colostrum immunity can provide 100% protection against calf enteric disease. BCV is highly susceptible virus that is easy to disinfect provided organic material is removed.

References

1. Kapil S *et al*. 1999. *Veterinary Microbiology*. 67:221–230.

1.
acvim232

Mucosal Immune System: Concepts and Applications

Sanjay Kapil, DVM, PhD, DACVM
Stillwater, OK, USA

Mucosal immune system plays an important role in protecting animals and humans from insult on exposed mucosa of respiratory tract, digestive tract, eyes, and reproductive tracts. It should be considered as a common mucosal immune system because of recirculation and homing of lymphocytes between different mucosal sites of animals. Significant developments have happened in the basic concepts of mucosal immunity only in the last 40 years.

In North America, food-producing animals are reared in very high densities, making them more vulnerable to easier spread of infectious diseases by exposure to high concentrations of infectious agents due to lack of hygiene.

Mucosal immune system, like the systemic immunity, has both humoral antibody and cell-mediated components. Mucus lining on mucosa provides a nonspecific physical barrier. However, depending on animal species, the components (antibody isotypes) vary in concentration.

Antibodies at mucosal surfaces (such as IgA) have a secretory component. Similarly, the mucosal lymphocytes have unique surface molecules in different animal species.

I will address the latest developments in mucosal immunity with applications on improvements in health of animals and humans:

1. With the introduction of new laws relating to raising animals with less or no antibiotics in feeds, there is need for increasing applications of prebiotics and probiotics in animal feeds to promote growth.
2. Fecal or gut microbiota transplant for recurrent *Clostridium difficile* diarrhea is highly effective therapy (95% effective) compared to antibiotic treatment. The donors need to be prescreened and gut microbiota properly stored in saline glycerol at -80°C.
3. Mucosal routes of immunization have offered better protection compared to parenteral vaccines in farm animals. The advantages include safety, low cost, advantage of inducing both mucosal and systemic responses.
4. Use of *Lactobacillus* species as vaccine delivery system. These probiotic bacteria are safe and can be used to deliver heterologous antigens for protection against animal and human pathogens. Further, *Lactobacillus* has been found to be stable after oral delivery of the vaccines.

References

1. Wang *et al. Human Vaccines and Immunotherapeutics*. 2014;10:3544–3551.

acvim233

Diagnosis and Health Consequences of Subacute Ruminal Acidosis in Cattle

Edgar F. Garrett, DVM, MS
Urbana, IL, USA

Introduction

Rumen acidosis occurs when a relative excess of non-structural carbohydrates such as starches and sugars are rapidly fermented in the rumen generating organic acids at rates that exceed rumen capacity for absorption and buffering. Ruminal concentrations of organic acids increase resulting in a reduction in rumen pH. The outcome of ingestion of excessive amounts of rapidly fermentable carbohydrates should be viewed as a spectrum of ruminal acidosis ranging from mild to severe with the endpoint determined primarily by the mass of carbohydrate ingested. Mild episodes of ruminal acidosis are commonly referred to as subacute ruminal acidosis (SARA) and severe cases referred to as acute ruminal acidosis. During periods of ruminal acidosis, endotoxins and bacterially derived biogenic amines can accumulate which in turn can lead to upregulation of inflammatory cytokines. Cattle experiencing bouts of SARA will have altered feed intake patterns, have an increased incidence of liver abscesses, be at increased risk for laminitis and have altered milk components. Maximizing cattle health and productivity requires monitoring for signs of ruminal acidosis and adjusting diets and management strategies to minimize the effects of SARA while not compromising the production potential of the animal.

Background on Ruminal Acidosis

Forage adapted animals and grain adapted animals have very different ruminal microbiomes, with forage fed animals having more diverse microbial populations (Makie 1979; Petri 2103; Anderson 2016). When feeding grain, dietary adaptation is necessary to allow an orderly increase in populations of bacteria that utilize starch and sugars and populations of lactate-utilizing bacteria. Equilibrium must be maintained between acid production and acid absorption, utilization, and buffering. Adaptation of the rumen microbiota to a new diet will occur over a period of a few days. When adapting cattle to a high grain diet as in a feedlot, increasing the proportion of grain on a weekly basis allows an orderly progression of microbial adaptation. However, a low proportion of forage in the diet will lead to a loss in diversity in the microbial populations with a decrease in bacteria in the gram-negative phylum Bacteroides and an increase in bacteria in the predominantly gram-positive phylum Firmicutes (Mao 2013). A diet with less than 30% neutral detergent fiber (NDF) on a dry matter basis results in a shift towards the phylum Firmicutes and a loss of bacterial diversity (Anderson 2016). If grain adaptation is done poorly, within 2 to 6 hours after ingestion of rapidly fermentable feed, there is a dramatic increase in the numbers of amylolytic bacteria. *Streptococcus bovis* is of some special interest in this scenario because, given an excess of carbohydrate substrate, *Strep. bovis* can change its fermentation pattern to the production of lactate (Russell 1985). The increased production of lactate has the potential to dramatically change the environment in the rumen because it is a stronger acid than the VFAs that make up the majority of fermentation end products. *Strep. bovis* is thought to play an important role in ruminal acidosis because of its lactate production; however, Wang 2013 did not see an increase in *Strep. bovis* during a SARA challenge suggesting that the precise bacterial changes occurring during SARA are diverse. Protozoa are an important component of the rumen ecology that are overlooked in studies focused on the changes in bacterial populations. Protozoa affect ruminal fermentation by ingesting bacteria and large starch granules. The presence of protozoa in the rumen reduces the number of bacteria in the rumen and moderates the rate of fermentation (Nagaraja TG 1992). Rumen protozoa are pH sensitive and decline significantly as the ruminal falls to pH <5.5.

With a substantial excess of readily fermentable carbohydrate, the changes in bacterial populations and end products of fermentation can lead to a rapid drop in pH resulting in essentially all the normal gram-negative bacteria of the rumen dying off as rumen pH <5.0. In this situation, the homofermentative, lactate producing, acid-tolerant *Lactobacillus* spp. proliferate. The ability of the rumen to buffer acid is overwhelmed and D- and L-lactate accumulate in the rumen. Absorption of these acids through the rumen epithelium results in systemic lactic acidosis. In SARA, the changes in fermentation rate, molar proportions of VFA produced, and rumen pH are temporary and not as severe. In this situation, the supply of carbohydrate is exhausted before the ruminal environment changes drastically. Lactate production increases while the carbohydrate oversupply exists but the additional lactate either does not exceed or marginally exceeds the capacity of the lactate utilizing bacteria to convert lactate to other organic acids. If the concentration of lactate increases, the increase is modest and the increase lasts less than 12 hours (Krause 2005). The reduction in ruminal pH during SARA is the result of the increased rate of production of a wide array of organic acids which exceeds the capacity of the rumen for absorption and buffering. SARA tends to occur cyclically because animals will typically reduce feed intake at rumen pH <5.5; therefore, animals may enter cycles: SARA → decrease intake → normalize rumen → return to feed → SARA. Ruminal pH in SARA ranges from 5.0–5.5. Precise definitions for SARA based on ruminal pH vary but using pH <5.5 for >3 hours or pH <5.8 for >8 hours (Zang 2009) has become well accepted.

Changes in the rumen that occur when high concentrate diets are fed are the most likely cause of changes seen in the health and productivity of the cow; however, in some instances, changes also occur in the hindgut and these changes could contribute to adverse effects on the cow. Starch escaping the rumen can be fermented in the cecum and colon. Changes in microbiome, increased concentration of VFAs, decreased pH, damage to the epithelium of the cecum and colon and local inflammation of the cecum and colon have been demonstrated (Liu 2014; Tao 2014; Li 2012; Gressley 2014).

In addition to microbial adaptation, the rumen undergoes significant morphologic changes as the proportion of grain in the diet increases. Rumen papillae increase in size (surface area) in response to increased intake of fermentable organic matter (Dieho 2016) with the rumen mucosal surface area reaching its maximum by about 3–4 weeks post-calving (Bannnk 2008) and the total mass of the rumen increasing over time by about 20% from early to mid-lactation (Baldwin 2004). The structural changes are most likely a response to increased concentration of VFAs with the effect of the VFAs mediated by hormones and growth factors such as insulin, glucagon, growth hormone, and epidermal growth factor (Penner 2011). The increased surface area facilitates absorption of VFAs and therefore plays a role in homeostasis of the rumen environment. About half of the VFAs are absorbed across the rumen epithelium via a receptor that exchanges a bicarbonate ion for a VFA. A large proportion of rumen buffering is provided by the rumen epithelium (Aschenbach 2011). Thus, the rumen epithelium is a vital component of maintaining a normal rumen environment.

SARA has the potential to cause significant health problems over time because it occurs commonly in both dairies and feedlots. Estimates of prevalence of SARA are 10–26% in dairies (Garrett 1997; Bramley 2008; Kleen 2009; Atkinson 2014) and 15% in the early finishing period and 35% in the late finishing period for feedlot cattle (Castillo-Lopez 2014).

Sequelae to Ruminal Acidosis

Endotoxins and Systemic Inflammatory Response

The major shifts in bacterial populations results in increased ruminal concentration of bacterial endotoxins, lipopolysaccharide (LPS), and lipoteichoic acid (LTA). The LPS and LTA and can induce apoptosis of ruminal epithelial cells which can compromise the barrier function of the rumen epithelium and allow paracellular movement of endotoxins from the rumen to circulation (Eckel 2016). SARA has been shown to cause extensive sloughing of the stratum corneum and decreased cellular tight junctions in the rumen epithelium (Steele 2011). LPS may also be absorbed across the rumen via a transcellular route via TLR-4 during periods of inflammation of the ruminal epithelium when there would be increased expression of TLR-4 on the rumen epithelium (Eckel 2016). Absorption of bacterial endotoxins following SARA leads to systemic inflammatory responses (Gozho 2005; Chang 2015). An extreme systemic inflammatory response to absorbed endotoxins could lead to septic shock but this is not a feature of SARA. However, the inflammatory response associated with absorbed endotoxin may alter other body systems due to changes in lipoprotein trafficking, serum calcium (Eckel 2016), and blood flow in the hoof (Christmann 2002) and these changes could affect animal health. The alteration of the rumen epithelium by SARA can also allow rumen microbes to colonize the damaged epithelial surface resulting in bacterial or fungal rumenitis. Bacteria may cross the damaged rumen epithelium and enter the portal blood leading to bacteremia and liver abscesses (Nagaraja 1998).

Laminitis

Non-infectious lameness in cattle is most commonly associated with sole ulcer, toe ulcer, or white line disease. These entities are often referred to collectively as claw horn disruption lesions and are symptoms of chronic laminitis. A definitive etiology of the lesions has not been determined; however, there are two theories. One theory centers on the importance of the digital cushion lying under the third phalanx. The digital cushion is an important anatomical feature in the hoof of cattle thought to serve as a shock absorbing material to counteract the ground reaction forces transmitted to the hoof during locomotion. Decreased thickness of the digital cushion was associated with increased risk of lameness. The thickness of the digital cushion was associated with body condition score where thin cows had a thinner digital cushion suggesting that diet and energy balance affected the thickness of the digital cushion (Bicahlo 2013). Thus, concussive forces transmitted to the hoof lead to microtrauma that can adversely affect the development of claw horn. This concept fits well with sole ulcer lesions in confined dairy cattle but does not explain laminitic changes such as flaky horn, diffuse sole hemorrhages, and abnormal hoof conformation (i.e., slipper toe) or the occurrence of sole ulcers in well-conditioned beef cattle that are not confined on concrete flooring. The other theory centers on the role of inflammatory cytokines and biogenic amines generated during rumen or hindgut acidosis that adversely affect the growth of the hoof. Collagenases and other matrix metalloproteinases (MMPs) that are involved in the normal growth of the hoof may be abnormally upregulated by some inflammatory cytokines (Clutterbuck 2010) and thus lead to altered hoof growth or impaired epidermal connections of the hoof capsule. Inflammatory cytokines are present in systemic circulation following episodes of acute acidosis and SARA. Increased level of IL-1 was demonstrated in hoof tissue following induction of acute acidosis (Belknap 2002). Additionally, induction of acute acidosis was shown to increase capillary pressure and post-capillary resistance which could promote transvascular movement of fluid and increase tissue pressure (Christmann 2002). Thrombosis, congestion, edema, arteriosclerosis, and hemorrhage are common histologic features of laminitic hooves (Anderson 1980; Boosman 1991; Singh 1992) and are consistent with changes in vascular dynamics as observed by Christman 2002 and Boosman 1989. Similar signs were seen with laminitis following induction of acidosis with oligofructose in dairy heifers (Thoefner 2005). While the etiopathogenesis is still uncertain, studies investigating the effect of varying levels of grain in the diet on lameness and claw lesions supports the role of SARA in laminitis (Manson 1989; Kelly 1990; Bramley 2013). One can conclude that the etiopathogenesis of laminitis in cattle is complex and multifactorial and that carbohydrate feeding and associated SARA contributes to chronic laminitis and lameness as a result of the sequela of chronic laminitis.

The importance of understanding the causes of laminitis are underscored by the remarkably high prevalence lameness in confinement dairy herds. Annual incidence of lameness in dairy herds ranges from 2–50% and averages about 25%. Dairy cows with claw horn lesions are less likely to get pregnant (Hernandez 2001; Sprecher 1997), produce less milk (Hernandez 2002; Warnick 2001) and more likely to be culled from the herd (Hernandez 2002; Booth 2004). The losses associated with claw horn lesions is estimated by the author to be $300 per case. Lameness in dairy herds is an economic and animal welfare priority and can only be controlled by understanding all the facets of the problem.

Liver Abscess

Liver abscess is a recognized consequence of ruminal acidosis. The generally accepted pathogenesis of liver abscesses is that ruminal acidosis causes rumenitis and compromises the barrier function of the ruminal epithelium thus allowing bacteria to colonize the rumen wall and then enter portal circulation. After reaching the liver, bacteria colonize the liver and abscessation follows. The most common bacteria found in liver abscesses are *Fusobacterium necrophorum* and *Trueperella pyogenes. F. necrophorum* is a normal inhabitant of the rumen and is typically found in relatively close association with the wall of the rumen. The numbers of *F. necrophorum* increase significantly as the proportion of grain in the diet increases. *F. necrophorum* is found in higher numbers in abnormal sections of the rumen (i.e., scars from rumenitis) than in normal areas of the rumen. Thus there is a logical link between grain feeding, ruminal acidosis, and the translocation of *F. necrophorum* from the rumen to the liver (Nagaraja 2007).

Liver abscesses are found in 12–32% of grain-fed cattle at slaughter with the range being 1–90%. Animals with category A+ liver abscessation (≥1 large or multiple small active abscesses) have been shown to have reduced average daily gain and feed efficiency and reduced live weight, carcass weight, and dressing percent at slaughter. Tylosin is commonly fed at 90 mg/head/day to help control liver abscesses in feedlot cattle and this has been shown to reduce the incidence of liver abscess by 40–70% and improve average daily gain and feed efficiency. Liver abscesses are considered a financially significant disease process in feedlot cattle because of the loss in carcass value (condemned liver) and loss in productivity (Nagaraja 2007).

Liver abscess is not a commonly considered differential diagnosis in dairy cattle; however, evaluation of the carcasses of adult cattle at slaughter in the Great Lakes region where 87% of the cattle processed were dairy cattle found that 32% of the cattle had one or more liver abscess lesions and 18.5% had A+ lesions (Resac 2014). The effect of liver abscesses on feed intake, feed efficiency, milk production, and risk of culling in dairy cattle is unknown at this time.

Given the high prevalence of liver abscesses in both feedlot and dairy cattle at slaughter, this condition should garner significant attention for control and prevention. The practice of feeding antibiotics to control liver abscesses should be considered tenuous and new approaches to controlling liver abscesses should be developed. Strategies to minimize the risk of ruminal acidosis while maintaining production must be evaluated and other strategies must be developed to minimize the risk or effect of liver abscess when SARA does occur.

Milk Production and Feed Intake

Danscher and coworkers (Danscher 2015) investigated the effect of induced SARA on a wide variety of parameters in lactating dairy cows which provided interesting, missing information about the short-term effects of SARA on lactating dairy cows. During the 3 days of feeding barley and wheat to induce SARA, dry mater intake decreased from 19.7 kg/d to an average of 16.3 kg/d but the cows averaged 14 kg/d for the first day of the SARA challenge and then DMI increased some but did not return to normal until the SARA challenge ended. Surprisingly, milk production was not significantly different during the SARA challenge (22.8 kg/d vs. 20.8 kg/d) but the lack of significance may have been due to the relatively small number of animals in the trial (n=6). However, milk fat% did decrease from 4.5% to 3.5% during the SARA challenge. The change in milk fat associated with ruminal acidosis is due to increased production of conjugated linoleic acid in the rumen which decreases milk fat synthesis in the mammary gland (NRC 2001). Serum calcium decreased significantly during the SARA challenge. This was likely due to endotoxin effect on calcium regulation (Eckel 2016). Rumen contraction rate decreased significantly during the SARA challenge. This could have been due to reduced dry matter intake or a direct effect of the acidosis on the rumen. The detailed study by Danscher 2015 illustrates the potential for significant adverse effects of SARA on animal productivity.

Bloat, Displaced Abomasum, and Ketosis

SARA and the associated systemic inflammation likely increases the risk of other health issues such as displaced abomasum (DA), ketosis, and bloat (Zebeli 2015; Eckel 2016). Chronic bloat is a recognized sequela to acidosis in feedlots and it is assumed to be secondary to inflammation of the rumen that alters the normal rumination cycle. Bloat is not a common feature of SARA in dairy cattle. Displaced abomasum and ketosis are complex metabolic problems driven by many interrelated factors including diet, cow comfort, season, and body condition. SARA may not be a dominant risk factor for DA and ketosis but could play a role in the risk of disease occurrence. The systemic inflammatory response associated with SARA has been shown to alter lipoprotein trafficking and serum calcium levels (Eckel 2016) which could affect the risk of DA and ketosis. In addition, reduction in feed intake associated with SARA would increase the risk for DA and ketosis.

Diagnosing and Monitoring Herds for SARA

SARA is an inconstant feature affecting a subset of animals in a herd which can make diagnosis challenging. The issue is typically approached from the herd perspective rather than trying to identify individuals affected.

For dairies, herd parameters that raise suspicions about SARA include

1. Low bulk tank milk fat percentage.
2. Individual cows with milk protein % > milk fat % on test day.
3. An annual incidence of non-infectious lameness greater than or equal to 15% or a prevalence of greater than or equal to 5% of cows with a locomotion score of 4 out 5.
4. Greater than 5% of cows producing manure with an abnormal consistency, high proportion of large particles or with mucin casts present.

In a feedlot environment, SARA is suspected when feed intake is decreased or erratic, greater than 20% or the group has loose manure or the prevalence of non-infectious lameness in the group is greater than 10%. In both dairy and feedlot, the high risk periods are during the transition from high forage to high grain and at maximum intake. In dairies that would be during the first 3–4 weeks in lactation and then from 90–150 days in milk. In feedlots, it would be during the stepwise adaptation to a high grain diet over the first 3–4 weeks and then during the last 60 days on feed. Animals at these stages of production should be monitored for indications of SARA.

Once the suspicion of SARA has been raised, efforts should be made to confirm or refute the suspicion. Measurement of ruminal pH is the most direct way to diagnose ruminal acidosis. Collection of rumen fluid can be done via a stomach tube or by percutaneous aspiration of rumen fluid (rumenocentesis). Rumenocentesis is the preferred technique because the contamination with saliva complicates interpretation of samples collected with a stomach tube (Duffield 2004). Using rumenocentesis to collect a small volume of rumen fluid from 15 animals in a herd will allow a dichotomous classification of the herd or group into either likely to be affected by SARA or normal. The cut-points for the interpretation of the herd status is greater than or equal to 4 of the 15 animals with a ruminal pH ≤5.5 (Garrett 1999). This approach relies on using the binomial formula to calculate the probability of at least x animals having a pH <5.5 based on data from reference populations. The data cannot be used to make an estimate of prevalence. The rumen is a continuous flow fermenter and the pH changes throughout the day albeit over hours rather than minutes. Given the dynamic nature of the rumen, the point in time measurement with rumenocentesis has limitations that affect the ability to accurately diagnose SARA. Defining SARA as an animal with rumen pH ≤5.5 for greater than 3 hours is a more robust diagnostic cut point. It is only possible to ascertain this data with very frequent sampling via a rumen cannula or an indwelling rumen pH probe. In recent years, several companies (Smaxtec, Austria; Wellcow, Scotland; E-cow, England) began marketing rumen boluses that will collect and store pH and temperature data every 1–10 minutes and then transmit the data wirelessly when activated by a receiving unit. The units are able to measure rumen pH continuously for 4–6 months. Such devices could be used in a subset of animals to accurately determine the effect of dietary changes on the rumen environment. It is important to use multiple animals to evaluate the effect of diet on rumen health because animals vary in their propensity for SARA when fed the same diet (Gao 2014). The pH in the rumen is influence by many variables but two cow-level characteristics that have been recognized are differences in meal size and frequency (Gonzalez 2012) and sorting behavior while eating (Gao 2014).

Fecal pH may be useful as a proxy for SARA. Abnormal digestion kinetics in the rumen can increase the flow of fermentable substrate to the cecum and colon with subsequent fermentation in the cecum and colon. The pH of the feces will decrease if there is sufficient fermentation of substrates in the cecum and colon (Gressley 2014). Normal fecal pH for lactating dairy cows is 6.5–7.0 and cows with SARA have fecal pH ≤6 (Dansche 2015). Data on fecal pH in feedlot cattle is not available. There is no data available to allow interpretation of fecal pH in a group of cows and translate the information into a risk assessment for SARA as was done for rumenocentesis. Measuring fecal pH would be simple, non-invasive, and inexpensive and should be explored for its suitability for a diagnostic test for SARA.

In dairy farming, activity monitors and daily milk weights have been available for the last 2 decades as aides in identifying cows in estrus or experiencing a disease episode. Significant advances have been made in remote monitoring of cow activity that use algorithms to interpret changes in behavioral patterns as indicators of disease. Simple motion sensors have been replaced with accelerometers that can identify cow position in addition to activity which allows evaluation of the important rest periods of cattle. Remote monitoring of rumination activity can be accomplished with either ear mounted accelerometers or acoustic monitors installed in a device attached with a collar around the neck. Some devices can also provide as estimate of time spent eating. Accuracy at monitoring rumination activity was good for the collar mounted device in confined dairy cattle (Schirmann 2009; Byskov 2014) but it was less effective when used in grazing dairy cattle (Ambriz-Vilchis 2015) or feedlot cattle (Goldhawk 2013). The ear tag devices have had mixed results in accurately identifying chewing activities with one study showing a good correlation with ruminating activity (Bikker 2014) and another showing poor correlation with rumination activity (Borchers 2016). Both studies (Bikker 2014; Borchers 2016) indicated only a fair correlation with eating activity. In feedlots, the GrowSafe system (Airdrie, Alberta Canada) allows monitoring of feed intake, drinking bouts, and weight gain to provide continuous data on animal activity and performance which is used as an aide in identifying ill animals.

Data about cow activity acquired by telemetry has been used in dairies to identify cows with metritis, displaced abomasum, ketosis, subclinical ketosis, and indigestion (Schirmann 2016; Stangaferro 2016). Fiber content of the diet is inversely related to risk of ruminal acidosis and directly related to rumination time and SARA increases the risk for lameness associated with laminitis. Therefore, while none of the rumination and activity telemetry systems directly addresses SARA, it is expected that the information from these systems can be adapted to identify cows with SARA with reasonable accuracy. Monitoring cows for SARA in real time with some combination of rumen pH boluses, rumination monitors, and activity monitors will allow early intervention and minimize the adverse effects of SARA on animal health and productivity.

Diet composition is critical in the maintenance of a normal rumen environment. The National Research Council provides guidelines for beef and dairy cattle nutrient requirements. Forages are known to vary substantially depending on soil type, environmental conditions, and maturity at harvest. Forage testing to determine fiber, starch, and protein content is common because of the known variability of the feeds. Grains are often considered uniform products and assumptions are made about their nutrient content and rates of digestion; however, there is substantial variation in rate of starch digestion within grain type. An acidosis index (Black 2008; Lean 2013; Anele 2015) was developed to better characterize the rate of digestion of grains which will provide more accurate diet formulation to minimize the risk of SARA due to unexpectedly rapid fermentation of the grain component of the diet.

Conclusion

The economic demands for production necessitate feeding cattle in a manner that puts them at risk for SARA. There are significant health and production consequences for animals experiencing SARA. Advances in remote monitoring technologies will allow early detection of SARA which will allow fine tuning of feeding practices to limit or avoid SARA while still maintaining high levels of production.

References

1. Ambriz-Vilchis V, Jessop NS, Fawcett RJ, Shaw DJ, Macrae AI. Comparison of rumination activity measured using rumination collars against direct visual observations and analysis of video recordings of dairy cows in commercial farm environments. *J Dairy Sci*. 2015;98:1750–1758.
2. Anderson CL, Schneider CJ, Erickson GE, MacDonald JC, Fernando SC. Rumen bacterial communities can be acclimated faster to high concentrate diets than currently implemented feedlot programs. *J Applied Microbiol*. 2016;120(3):588–599.
3. Anele UY, Swift ML, McAllister TA, Galyean ML, Yang WZ. Evaluation of *in vitro* models for predicting acidosis risk of barley grain in finishing beef cattle. *J Anim Sci*. 2015;93:4852–4859.
4. Aschenbach JR, Penner GB, Stumpff F, Gäbel G. Ruminant nutrition symposium: Role of fermentation acid absorption in the regulation of ruminal pH. *J Anim Sci*. 2011;89:1092–1107.
5. Atkinson O. Prevalence of subacute ruminal acidosis on UK dairy farms. *Cattle Practice*. 2014;22:1.

6. Baldwin RL, McLeod KR, Capuco AV. Visceral tissue growth and proliferation during the bovine lactation cycle. *J Dairy Sci*. 2004;87:2977–2986.
7. Bannink A. Modeling the implications of feeding strategy on rumen fermentation and functioning of the rumen wall. *Anim Feed Sci Technol*. 2008;143:3.
8. Belknap EB, *et al*. Expression of proinflammatory mediators and vasoactive substances in laminitic cattle. In: Proceedings from the 12th International Symposium on Lameness in Ruminants; 2002: 387; Orlando, FL.
9. Bicahlo RC, Oikonomou G. Control and prevention of lameness associated with claw lesions in dairy cows. *Livestock Sci*. 2013;156:96.
10. Bikker JP, van Laar H, Rump P, *et al*. Technical note: Evaluation of an ear attached movement sensor to record cow feeding behavior and activity. *J Dairy Sci*. 2014;97:2974–2979.
11. Black J. Premium grains for livestock program (Australia). Final Report. 2008.
12. Booth CJ, Warnick LS, Gröhn YT, Maizon DO, Guard CL, Janssen D. Effect of lameness on culling in dairy cows. *J Dairy Sci*. 2004;87:4115–4122.
13. Borchers MR, Chang YM, Tsai IC, Wadsworth BA, Bewley JM. A validation of technologies for monitoring dairy cow feeding, ruminating and lying behaviors. *J Dairy Sci*. 2016;99:7458–7466.
14. Bramley E, Lean IJ, Fulkerson WJ, Stevenson MA, Rabiee AR, Costa ND. The definition of acidosis in dairy herds predominantly fed on pasture and concentrates. *J Dairy Sci*. 2008;91:308–321.
15. Bramley E, Costa ND, Fulkerson WJ, Lean IJ. Associations between body condition, rumen fill, diarrhea and lameness and ruminal acidosis in Australian dairy herds. *N Z Vet J*. 2013;61:323–329.
16. Byskov MV, Schulze AK, Weisbjerg MR, Markussen B, Nørgaard P. Recording rumination time by a rumination monitoring system in Jersey heifers fed grass/clover silage and hay at three feeding levels. *J Anim Sci*. 2014;92:1110–1118.
17. Castillo-Lopez E, Wiese BI, Hendrick S, *et al*. Incidence, prevalence, severity, and risk factors for ruminal acidosis in feedlot steers during backgrounding, diet transition, and finishing. *J Anim Sci*. 2014;92:3053–3063.
18. Chang G, Zhuang S, Seyfert HM, *et al*. Hepatic TLR4 signaling is activated by LPS from digestive tract during SARA, and epigenetic mechanisms contribute to enforced TLR4 expression. *Oncotarget*. 2015;6:38578–38590.
19. Christmann U, *et al*. Evaluation of hemodynamics in the normal and laminitic bovine digit. In: Proceedings of the 12th International Symposium on Lameness in Ruminants; 2002: 165; Orlando, FL.
20. Clutterbuck AL, Harris P, Allaway D, Mobasheri A. Matrix metalloproteinases in inflammatory pathologies of the horse. *Vet J*. 2010;183:27–38.
21. Danscher AM, Li S, Andersen PH, Khafinpour E, Kristensen NB, Plaizier JC. Indicators of induced subacute ruminal acidosis (SARA) in Danish Holstein cows. *Acta Vet Scand*. 2015;57:39.
22. Duffield T, Plaizier JC, Fairfield A, *et al*. Comparison of techniques for measurement of rumen pH in lactating dairy cows. *J Dairy Sci*. 2004;87:59–66.
23. Eckel EF, Ametaj BN. Role of bacterial endotoxins in the etiopathogenesis of periparturient diseases of transition dairy cows. *J Dairy Sci*. 2016;99:5967–5990.
24. Garrett EF, *et al*. Cross sectional field study investigating the effect of periparturient dietary management on ruminal pH in early lactation dairy cows. *J Dairy Sci*. 1997;80 (Suppl 1):112.
25. Gao X, Oba M. Relationship of severity of subacute ruminal acidosis to rumen fermentation, chewing activities, sorting behavior and milk production in lactating dairy cows fed a high grain diet. *J Dairy Sci*. 2014;97:3006–3016.
26. Gonzalez LA, *et al*. Ruminal acidosis in feedlot cattle: interplay between feed ingredients, rumen function and feeding behavior (a review). *Anim Sci Feed Tech*. 2012;126:66.
27. Gozho GN, Plaizier JC, Krause DO, Kennedy AD, Wittenberg KM. Subacute ruminal acidosis induces ruminal lipopolysaccharide endotoxin release and triggers and inflammatory response. *J Dairy Sci*. 2005;88:1399–1403.
28. Gressley TF, Hall MB, Armentano LE. Productivity, digestion, and health responses to hindgut acidosis in ruminants. *J Anim Sci*. 2014;89:1120–1130.
29. Hernandez J, Shearer JK, Webb DW. Effect of lameness on the calving to conception interval in dairy cows. *J Am Vet Med Assoc*. 2001;218:1611–1614.
30. Hernandez J, Shearer JK, Webb DW. Effect of lameness on milk yield in dairy cows. *J Am Vet Med Assoc*. 2002;220:640–644.
31. Kelly EF, Leaver JD. Lameness in dairy cattle and the type of concentrate given. *Anim Prod*. 1990;51:221.
32. Khafipour E, Drause DO, Plaizier JC. A grain-based subacute ruminal acidosis challenge causes translocation of lipopolysaccharide and triggers inflammation. *J Dairy Sci*. 2009;92:1060–1070.
33. Kleen JL, Hooijer GA, Rehage J, Noordhuizen JP. Subacute ruminal acidosis in Dutch dairy herds. *Vet Rec*. 2009;164:681–683.
34. Krause KM, Oetzel GR. Inducing subacute ruminal acidosis in lactating dairy cows. *J Dairy Sci*. 2005;88:3633–3639.

35. Lean IJ, *et al.* Impact of nutrition on lameness on claw health in cattle. *Livestock Sci.* 2013;156:71.
36. Lean IJ, Golder HM, Black JL, King R, Rabiee AR. *In vivo* indices for predicting acidosis risk of grains in cattle: comparison with *in vitro* methods. *J Anim Sci.* 2013;91:2823–2835.
37. Li S, Khafipour E, Krause DO, *et al.* Effects of subacute ruminal acidosis challenges on fermentation and endotoxins in the rumen and hindgut of dairy cows. *J Dairy Sci.* 2012;95:294–303.
38. Liu J, Xu T, Zhu W, Mao S. High grain feeding alters caecal bacterial microbiota composition and fermentation and results in caecal mucosal injury in goats. *Br J Nutr.* 2014;112:416–427.
39. Mackie RI, Gilchrist FM. Changes in lactate producing and lactate utilizing bacteria in relation to pH in the rumen of sheep during stepwise adaptation to a high concentrate diet. *Appl Environ Microbiol.* 1979;38:422–430.
40. Manson FJ, Leaver JD. The influence of concentrate amount on locomotion and clinical lameness in dairy cattle. *Anim Prod.* 1988a;47:185.
41. Mao SY, Zhang RY, Wang DS, Zhu WY. Impact of subacute ruminal acidosis (SARA) adaptation on rumen microbiota in dairy cattle using pyrosequencing. *Anaerobe.* 2013;24:12–19.
42. Nagaraja TG, Towne G, Beharka AA. Moderation of ruminal fermentation by ciliated protozoa in cattle fed a high grain diet. *Appl Environ Microbiol.* 1992;58:2410–2414.
43. Nagaraja TG, Chengappa M. Liver abscesses in feedlot cattle: a review. *J Anim Sci.* 1998;76:287–298.
44. Nagaraja TG, Lechtenberg KF. Liver abscesses in feedlot cattle. *Vet Clin North Am Food Anim Pract.* 2007;23:351–369.
45. Penner GB, Steele MA, Aschenbach JR, McBride BW. Molecular adaptation of ruminal epithelia to highly fermentable diets. *J Anim Sci.* 2011;89:1108–1119.
46. Petri RM, Schwaiger T, Penner GB, *et al.* Characterization of the core rumen microbiome in cattle during transition from forage to concentrate as well as during and after an acidotic challenge. *PLoS One.* 2013;8(12):e83424. doi:10.1371/journal.pone.0083424.
47. Rezac DJ, Thomson DU, Siemens MG, Prouty FL, Reinhardt CD, Bartle SJ. A survey of gross pathologic conditions in cull cows at slaughter in the Great Lakes region of the United States. *J Dairy Sci.* 2014;97:4227–4235.
48. Russell JB, Hino T. Regulation of lactate production in *Streptococcus bovis*: A spiraling effect that contributes to rumen acidosis. *J Dairy Sci.* 1985;68:1712–1721.
49. Schirmann K, von Deyserlingk MA, Weary DM, Veira DM, Heuwieser W. Technical note: Validation of a system for monitoring rumination in dairy cows. *J Dairy Sci.* 2009;92:6052–6055.
50. Schirmann K, Weary DM, Heuwieser W, Chapinal N, Cerri RL, von Keyserlingk MA. Rumination and feeding behavior differ between healthy and sick dairy cows during the transition period. *J Dairy Sci.* 2016;99:9917–9924.
51. Sprecher DJ, Hostetler DE, Kaneene JB. A lameness scoring system that uses posture and gait to predict dairy cattle reproductive performance. *Theriogenology.* 1997;47:1179–1187.
52. Stangaferro ML, Wijma R, Caixeta LS, Al-Abri MA, Giordano JO. Use of rumination and activity monitoring for the identification of dairy cows with health disorders: Part 1. Metabolic and digestive disorders. *J Dairy Sci.* 2016;99:7395–7410.
53. Steele MA, Croom J, Kahler M, *et al.* Bovine rumen epithelium undergoes rapid structural adaptations during grain-induced subacute ruminal acidosis. *Am J Physiol Regul Inter Comp Physiol.* 2011;300:R515–R1523.
54. Tao S, Duanmu Y, Dong H, *et al.* High concentrate diet induced mucosal injuries by enhancing epithelial apoptosis and inflammatory response in the hindgut of goats. *PLoS One.* 2014;9(10):e111596.
55. Thoefner MB, Wattle O, Pollitt CC, French KR, Nielsen SS. Histopathology of oligofructose-induced acute laminitis in heifers. *J Dairy Sci.* 2005;88:2774–2782.
56. Wang DS, *et al.* Effects of SARA challenge on fermentation and biogenic amines in the rumen of dairy cows. *Livestock Sci.* 2013;155:262.
57. Warnick LD, Janssen D, Guard CL, Gröhn YT. The effect of lameness on milk production in dairy cows. *J Dairy Sci.* 2001;84:1988–1997.
58. Zang WZ, Beauchemin KA. Increasing physically effective fiber content of dairy cow diets through forage proportion versus forage chop length: Chewing and ruminal pH. *J Dairy Sci.* 2009;92:1603–1615.
59. Zebeli Q, Ghareeb K, Humer E, Metzler-Zebeli BU, Besenfelder U. Nutrition, rumen health and inflammation in the transition period and their role on overall health and fertility in dairy cows. *Res Vet Sci.* 2015;103:126–136.

acvim234

Prevalence of Antimicrobial Resistance in *Mannheimia haemolytica*

Brent Credille, DVM, PhD, DACVIM
Athens, GA, USA

Introduction

Bovine respiratory disease (BRD) is the most common and costly disease of beef cattle in North America. In feedlots, BRD is responsible for 75% of all morbidity and 50–75% of overall mortality. Approximately 16.2% of all cattle entering feedlots will be diagnosed with BRD and 2% will die. In stocker cattle, BRD morbidity occurs far more frequently than what is commonly seen in feedlot cattle and is estimated to be responsible for 90% of all morbidity and mortality in these operations. As a result, it is not unusual to see morbidity risk exceed 75% in certain cohorts of animals. Economically, BRD costs the beef industry $2–4 billion annually, with 20% of total losses due to medication costs and 80% of total losses due to reduced carcass weight and quality. While multiple factors play a role in the development of BRD, bacteria, particularly *Mannheimia haemolytica*, are ultimately responsible for the clinical signs seen in affected cattle. As a result, antimicrobials are a mainstay of therapy and surveys of feedlots across the United States have shown that 100% of cattle diagnosed with BRD are given antimicrobials (oral and/or parenteral) as part of the initial therapeutic regimen. Unfortunately, the stocker and feedlot segments have been encountering issues with antimicrobial resistance in *M. haemolytica* and, based on recently published data, the isolation of strains of *M. haemolytica* that are resistant to multiple antimicrobials is becoming a more frequent occurrence. The purpose of this manuscript is to review the current literature as it pertains to the prevalence of antimicrobial resistance in *M. haemolytica* and how antimicrobial resistance, especially multi-drug antimicrobial resistance, impacts the efficacy of antimicrobial therapy in cattle diagnosed with BRD.

Antimicrobial Susceptibility Testing: A Primer

Bacterial antimicrobial susceptibility is determined *in vitro* using one of several available tests. Disk diffusion, concentration-gradient agar dilution, and broth dilution (macro or micro) have all been used to evaluate susceptibility to antimicrobials. Disk diffusion provides mostly qualitative information (susceptible, intermediate, resistant) while both the broth dilution and concentration-gradient agar diffusion tests provide quantitative data (minimum inhibitory concentration, MIC). With these tests, the MIC is defined as the lowest concentration of antimicrobial that inhibits the growth of target bacteria. It is important to note that inhibition of bacterial growth rather than bacterial killing is the primary endpoint. The designation of a microorganism as susceptible, intermediate, or resistant is determined by comparing the drug's MIC or zone of inhibition to breakpoints established by the Clinical Laboratory Standards Institute (CLSI).

Breakpoints, defined as the concentration of drug above and below which specific bacterial isolates are characterized as either susceptible, intermediate, or resistant, are determined by 3 criteria:

1. Range of *in vitro* MICs of an antimicrobial for a representative population of specific bacterial pathogens
2. PK/PD parameters established on the basis of the relationship between drug concentrations and microbial susceptibility
3. Results of clinical trials in the target species (when available)

When these *in vitro* susceptibility tests are presented to the clinician or researcher, a pathogen will be designated as susceptible, intermediate, or resistant, designations defined as:

- **Susceptible**: An infection caused by the specific isolate may be successfully treated with the recommended dosing regimen of an antimicrobial agent approved for that disease process and infecting microorganism.

- **Intermediate**: An infection by the isolate can be treated at body sites where drugs are physiologically concentrated or when a high dosage can be used. This designation also represents a "buffer zone" that should prevent minor technical factors from causing major discrepancies in interpretations.
- **Resistant**: An infecting isolate is not inhibited by typically achievable concentrations of a specific drug with a standard dosing regimen.

When species-specific breakpoints are used, pathogens with a MIC below an antimicrobial's specific breakpoint have a higher probability of success than organisms with an MIC above the breakpoint. In cattle, 11 antimicrobials have breakpoints established for pathogens associated with BRD (Table 1). For all other antimicrobials breakpoints have been adapted from data collected from plasma or interstitial fluid of humans or other domestic species. Unfortunately, in these situations, there are no data supporting the validity of this extrapolation. For example, the CLSI breakpoint for doxycycline in humans is 4 µg/mL. In horses, administration of doxycycline orally at a dose of 10 mg/kg results in peak concentrations in plasma, synovial fluid, and peritoneal fluid of approximately 0.5 µg/mL.[1] If a pathogen was isolated from one of these sites and if the pathogen had an MIC of 4 µg/mL, it would still be reported as susceptible even though these concentrations are difficult, if not impossible, to achieve in horses. It is also important to note that clinical breakpoints are only relevant for specific bacteria, a specific drug, and a specific organ system. Thus, breakpoints established for ceftiofur against *M. haemolytica* in the respiratory tract are irrelevant when that organism is the cause of disease within another body system (mammary gland, uterus). Therefore, the data obtained from these susceptibility tests must be interpreted with caution. In these situations, knowledge of an infecting organism's MIC combined with pharmacokinetic data ideally describing the concentration of drug within the tissue of interest can assist in predicting efficacy. Today, the literature is littered with publications, old and new, that contain information based on non-standardized testing methods or using interpretive criteria not validated for BRD. As a result, being aware of how susceptibility testing is performed is essential to appropriate interpretation of test results. Therefore, one must be cautious when interpreting antimicrobial susceptibility tests when appropriate methodology is not followed.

Table 1. Antimicrobials with CLSI approved breakpoints for treatment of bovine respiratory disease associated with *M. haemolytica*

Generic name	Antimicrobial class	Approval year	Breakpoint (MIC, µg/ml)	Breakpoint (disc zone, mm)
Oxytetracycline*	Tetracycline	1980	≤4	-
Penicillin*#	β-Lactam	1984	≤25	-
Tilmicosin	Macrolide	1992	≤8	≥14
Florfenicol	Phenicol	1996	≤2	≥19
Ceftiofur	β-Lactam	1998	≤2	≥21
Spectinomycin	Aminocyclitol	1998	≤32	≥14
Enrofloxacin	Fluoroquinolone	1998	≤.25	≥21
Tulathromycin	Macrolide	2005	≤16	≥18
Danofloxacin	Fluoroquinolone	2002	≤25	≥22
Gamithromycin	Macrolide	2011	≤4	≥15
Tildipirosin	Macrolide	2012	≤4	≥20

*Breakpoint valid only when using MICs. No CLSI approved breakpoints exist for these antimicrobials when using Kirby-Bauer methodology.
#Only valid for 22,000 IU/kg dose when given IM every 24 hours.

ANTIMICROBIAL RESISTANCE IN *MANNHEIMIA HAEMOLYTICA*: A BRIEF HISTORY

The earliest published MIC distributions for *M. haemolytica* derived using modern diagnostic laboratory methodology and CLSI approved breakpoints came from a survey of animals that died of BRD over a year period (1988–1992).[2] In this study, 461 *M. haemolytica* isolates were sent from veterinary diagnostic laboratories in Pennsylvania, Wyoming, Iowa, Washington, California, Missouri, Nebraska, Oregon, Kansas, Arizona, Texas, South Dakota, Montana, Minnesota, Oklahoma, Colorado, Utah, Saskatchewan, Alberta, and Quebec to an Upjohn laboratory for MIC determination. The results of this study are reported below in Table 2. It is important to note that the interpretive criteria for tilmicosin have changed since this study was published and the prevalence of resistance to this drug would have been dramatically decreased (>90% susceptible) using currently accepted criteria.

Table 2. MIC data for *M. haemolytica* obtained from the lungs of cattle that died of BRD

Organism	# of isolates	Antimicrobial	MIC50	MIC90	% susceptible
M. haemolytica	461	Tilmicosin	4	8	69.1
		Ceftiofur	<0.03	0.06	100
		Tetracycline	1	32	57
		Spectinomycin	32	64	83.5

In another study, the susceptibility of 390 *M. haemolytica* isolates obtained from the lungs of beef cattle that died from BRD and submitted to the Oklahoma Animal Disease Diagnostic Laboratory was investigated.[3] This study found that susceptibility to tetracycline varied over the course of the study period while susceptibility to ceftiofur and enrofloxacin remained stable (Table 3).

Table 3. Susceptibility of *M. haemolytica* (reported as % susceptibility) obtained from the lungs of feedlot cattle with BRD from 1994–2002

	Year								
Antimicrobial	**1994**	**1995**	**1996**	**1997**	**1998**	**1999**	**2000**	**2001**	**2002**
Ceftiofur	97	98	100	100	98	100	98	96	97
Enrofloxacin	–	–	–	–	–	96	98	89	98
Florfenicol	–	–	100	96	98	97	96	87	90
Spectinomycin	65	49	71	53	55	63	45	29	51
Tilmicosin	90	78	93	83	80	74	85	71	79
Tetracycline	23	46	74	58	42	63	44	34	54

A landmark study from Kansas State University evaluated the prevalence of multi-drug antimicrobial resistance and antimicrobial co-resistance patterns in *M. haemolytica* isolated from the lungs of cattle with BRD over a 3-year period.[4] This work showed that, between 2009 and 2011, the proportion of isolates resistant to 5 or more antimicrobials increased from 5% to 35%. In addition, isolates resistant to either oxytetracycline or tilmicosin were significantly more likely to be resistant to at least one other antimicrobial.

Several studies have begun to investigate the prevalence of antimicrobial resistance in healthy beef cattle. In one study, 10% of animals from 30% of feedlot pens in two feedlots in southern Alberta were submitted for isolation of *M. haemolytica*.[5] Swabs were collected from cattle at the time of feedlot arrival and again within 30 days of feedlot exit. All animals were subjected to standard feedlot practices (treatment of BRD with either ceftiofur or tulathromycin, feeding of ionophores, feeding of chlortetracycline for BRD control, and feeding of chlortetracycline or tylosin for liver abscess control). Over the course of the study, 409 *M. haemolytica* isolates were obtained and resistance to all antimicrobials tested was low.[5]

Another study sampled nearly 5,500 cattle from 4 feedlots in Canada.[6] Deep nasopharyngeal swabs (DNP) were collected from each enrolled animal at the time of arrival and then again at some point prior to feedlot exit. Susceptibility to 21 antimicrobials was evaluated for 2,989 individual *M. haemolytica* isolates. Overall, resistance was rare, with 87% of isolates susceptible to all antimicrobials tested.[6] It should be noted that many of the antimicrobials evaluated in this study do not have CLSI approved breakpoints established for *M. haemolytica* and BRD, making some of the conclusions from this study difficult to interpret.

Work from our lab evaluating the prevalence of resistance in *M. haemolytica* after metaphylaxis with the long-acting macrolide tulathromycin has yielded surprising results. In this study, lightweight, unweaned calves entering a stocker facility in Northeast Georgia were given tulathromycin at the time of arrival to the facility to prevent BRD. DNPs were collected from each animal at the time of arrival and then again 10–14 days later. Susceptibility to ceftiofur, enrofloxacin, tulathromycin, tilmicosin, florfenicol, and gamithromycin was evaluated using disk diffusion methodology. For all of the antimicrobials except ceftiofur, there was a significant increase in the proportion of isolates classified as intermediate or resistant at the time of second sampling compared to samples collected at arrival. Of the 27 calves with *M. haemolytica* cultured at the time of arrival, 14 (51.9%) had only pan-susceptible isolates, 12 (44.4%) had at least one isolate classified as intermediate or resistant to a single antimicrobial class (macrolides), and 1 (3.7%) had at least one isolate classified as intermediate or resistant to two antimicrobial classes (fluoroquinolones and macrolides) and would be considered MDR. Of the 123 calves with *M. haemolytica* cultured at the time of second sampling, 1 (0.8%) had only pan-susceptible isolates, 30 (24.4%) had at least one isolate classified as intermediate or resistant to two antimicrobial classes (fluoroquinolones and macrolides), and 92 (74.8%) had at least one isolate classified as intermediate or resistant to three antimicrobial classes (fluoroquinolones and macrolides in addition to either phenicols or cephalosporins).

Antimicrobial Resistance in *Mannheimia haemolytica*: Impact on Therapy

First treatment success, defined as the proportion of animals successfully responding to a single antimicrobial treatment, is generally high in most populations of cattle on feed. In most situations, first treatment success should range from 80–90%, a value that varies by animal age and weight. Unfortunately, little work has been done to evaluate the impact of antimicrobial resistance on clinical outcome in animals with BRD. The one published study the author is aware of found that 62% of cattle infected with susceptible *M. haemolytica* isolates (n=688) successfully responded to treatment with tilmicosin compared to only 38% of animals with resistant isolates (n=6).[7]

Conclusions

Despite the importance of BRD to the North American cattle industry, there are few well-designed studies that evaluate antimicrobial resistance in bacterial pathogens important to this disease syndrome. Many of these studies are from diagnostic lab submissions and represent dead cattle that have been treated multiple times with many different drugs. Nevertheless, general trends from reports that are available would suggest a decrease in antimicrobial susceptibility in *M. haemolytica* that has occurred over time and this decrease in susceptibility is seen for most of the antimicrobials commonly used to treat cattle diagnosed with BRD. Newer work would suggest that antimicrobial use practices that are commonplace in certain cattle populations might be driving resistance to some extent. As a result, it is important for clinicians working with cattle recognize the importance of antimicrobial resistance in the pathophysiology of BRD and how this might affect treatment of animals with clinical disease.

References

1. Davis JL, *et al. Am J Vet Res.* 2006;67:310.
2. Watts JL, *et al. J Clin Microbiol.* 1994;32:725.
3. Welsh RD, *et al. J Vet Diagn Invest.* 2004;16:426.
4. Lubbers BV, *et al. J Vet Diagn Invest.* 2013;25:413.
5. Klima CL, *et al. Can J Vet Res.* 2014;78:38.
6. Noyes NR, *et al. J Vet Intern Med.* 2015;29:705.
7. McClary DG, *et al. J Am Vet Med Assoc.* 2011;239:129.

acvim235

Molecular Mechanisms of Resistance in *Mannheimia haemolytica*

Brent Credille, DVM, PhD, DACVIM
Athens, GA, USA

INTRODUCTION

Bovine respiratory disease (BRD) is the most common and costly disease of beef cattle in North America. In feedlots, BRD is responsible for 75% of all morbidity and 50-75% overall mortality. Approximately 16.2% of all cattle entering feedlots will be diagnosed with BRD and 2% will die. In stocker cattle, BRD morbidity occurs far more frequently than what is commonly seen in feedlot cattle and is estimated to be responsible for 90% of all morbidity and mortality in these operations. As a result, it is not unusual to see morbidity risk exceed 75% in certain cohorts of animals.[1] Economically, BRD costs the beef industry $2–4 billion annually with 20% of total losses due to medication costs and 80% of total losses due to reduced carcass weight and poor carcass quality. While multiple factors play a role in the development of BRD, bacteria, particularly *Mannheimia haemolytica,* are ultimately responsible for the clinical signs seen in affected cattle. As a result, antimicrobials are a mainstay of therapy. Unfortunately, the stocker and feedlot segments have been encountering issues with antimicrobial resistance in *Mannheimia haemolytica* and, based on recently published data, the isolation of strains of *Mannheimia haemolytica* that are resistant to multiple antimicrobials is becoming a more frequent occurrence.[2] The purpose of this manuscript is to review the current literature as it pertains to the mechanisms of antimicrobial resistance in *Mannheimia haemolytica* and how understanding these mechanisms might help in the development of judicious antimicrobial use strategies in the future.

MECHANISMS OF ANTIMICROBIAL RESISTANCE: A REVIEW

As it relates to antimicrobials, bacteria exhibit two fundamental phenotypes: susceptible and resistant. Bacteria that are phenotypically antimicrobial resistant might be resistant due to intrinsic or acquired resistance mechanisms. Intrinsic antimicrobial resistance mechanisms are natural to all members of a specific taxonomic group and are usually due to structural or biochemical characteristics inherent to the native microorganism. Examples of intrinsic antimicrobial resistance include aminoglycoside resistance in anaerobes, β-lactam resistance in *Mycoplasma* spp., and macrolide resistance in gram-negative enteric organisms (Table 1).

Table 1. Examples of intrinsic antimicrobial resistance

Organism	Intrinsic resistance
Anaerobes	Aminoglycosides
Mycoplasma spp.	β-lactams
Listeria monocytogenes	Oxacillin, cephalosporins, lincosamides
Gram (-) enterics	Macrolides, streptogramins, glycopeptides
Gram (+) organisms	Polymyxins
Pseudomonas spp.	Triclosan

Acquired antimicrobial resistance is associated with the acquisition of phenotypic resistance in a susceptible organism through genetic modification. Otherwise susceptible bacteria might become resistant to antimicrobials through the following mechanisms:

1. Mutation of genes involved in normal cellular processes
2. Acquisition of foreign resistance genes
3. Combinations of mechanisms

The most common means of acquiring antimicrobial resistance is via horizontal gene transfer. Nevertheless, regardless of how resistance is obtained, antimicrobial resistance in manifested in one of several ways and includes the following:

1. Minimizing intracellular concentrations of an antimicrobial
2. Modification of an antimicrobial target
3. Inactivation of the antimicrobial itself

Bacteria may reduce intracellular antimicrobial concentrations by reducing the penetration of antimicrobials into the cell or through the expulsion of an antimicrobial via general or specific efflux pumps. Antimicrobial targets are modified by genetic mutations or post-translational modifications. Inactivation of antimicrobials occurs through the processes of enzymatic hydrolysis or chemical modification. Specific examples of acquired resistance mechanisms are listed in Table 2.

Table 2. Examples of acquired antimicrobial resistance mechanisms *

Antimicrobial class	Resistance mechanism	Genetic determinant
Tetracyclines	Inducible efflux Ribosomal protection	tet(A), tet(B), tet(C) tet(O), tet(M)
Phenicols	Efflux Acetylation	cmlA, floR catA
β-lactams	β-lactamases Alternate penicillin binding proteins	blaTEM, blaCMY mecA
Macrolides	Methylation of rRNA Efflux	erm(42)
Fluoroquinolones	Point mutation in DNA gyrase	gyrA parC

*Not an all-inclusive list

Acquired antimicrobial resistance can be manifested in many different ways. Microbes might be resistant to only a single antimicrobial agent, some antimicrobial agents within a class, an entire class of antimicrobial agents, or antimicrobial agents within multiple classes. It is also important to remember that resistance is best viewed as a continuum and it is not an all or none phenomenon in every circumstance. Microbes might display relative resistance or high-level resistance. With relative resistance, antimicrobial efficacy can be recovered by increasing either drug dose or changing the timing of drug administration. In contrast, high-level resistance implies that there will be no recovery in efficacy with increased drug dosage or exposure.

Mechanisms of Antimicrobial Resistance in *Mannheimia haemolytica* - A Review of Integrative Conjugative Elements

All of the major acquired resistance mechanisms have been reported in *Mannheimia haemolytica* isolates obtained from cattle with BRD. Historically, plasmids were recognized as the most common contributor to resistance associated with horizontal gene transfer.[3] Most recent studies have focused on the role of integrative conjugative elements (ICEs) in the dissemination and propagation of multi-drug antimicrobial resistance (MDR) in *Mannheimia haemolytica*. ICEs are chromosomal mobile genetic elements that can harbor clusters of antimicrobial resistance genes and are found in both gram-negative and gram-positive bacteria.[4] In addition to antimicrobial resistance genes, ICEs also contain genes conferring the ability to excise themselves from the host chromosome, form biofilms, virulence, and genes conferring resistance to heavy metal resistance.[4] These genes are typically acquired through multiple recombination events and, once generated, can establish themselves in a bacterial population and persist through the use of a single antimicrobial agent that selects for the entire gene segment. ICEs were first discovered in BRD pathogens when an ICE was identified in an isolate of *Pasteurella multocida* (ICE *Pmu1*) obtained from a calf that died in a Nebraska feedlot.[5] This ICE contained two regions containing 11 different active antimicrobial resistance genes and an inactive β-lactamase (bla_{OXA-2}, Table 3).[5]

Table 3. Antimicrobial resistance genes (active and inactive) found on ICE *Pmu1*

Gene	Antimicrobial(s) to which resistance is conferred
aadA25, strA, strB, aadB, aphA1	Streptomycin, spectinomycin, gentamicin, neomycin
tetR-tetH	Tetracycline
floR	Florfenicol
sul2	Sulfonamides
erm(42)	Tilmicosin, tildipirosin
msr(E)-mph(E)	Tulathromycin, gamithromycin
bla_{OXA-2}	Inactive (penicillins, 1st and 2nd generation cephalosporins)

Investigations from feedlots in Alberta, Nebraska, and Texas found that 72% of all *Mannheimia haemolytica* were resistance to at least 1 antimicrobial, with 30% (n=18) of these being resistant to 7 different antimicrobial classes.[6] The resistance phenotype of these resistant isolates was found to correlate with PCR amplification of ICE-associated genes. This study also found that these ICE-positive strains of *Mannheimia haemolytica* could successfully transfer the ICEs to other BRD pathogens and *E. coli* via conjugation.[6] More recent work evaluating the genomic signatures of 857 *Mannheimia haemolytica* isolates obtained from cattle found that ICEs harbor up to 12 different antimicrobial resistance genes in 7 different combinations.[3] Interestingly, all isolates harboring ICEs belonged to *Mannheimia haemolytica* isolates in clade 2b. This particular strain of *Mannheimia haemolytica* has been in circulation in North America since at least 2003 and this finding suggests that there has been pressure for selection of this strain over time.[3]

Conclusions

Multiple different mechanisms are responsible for the acquisition of antimicrobial resistance in bacteria. In recent years, ICEs have been recognized as a significant contributor to MDR in pathogens such as *Mannheimia haemolytica.* ICEs can confer resistance to up to 12 different antimicrobials in 7 different classes. Data would suggest that *Mannheimia haemolytica* harboring ICEs belong to a single clade (clade 2b) that has been in circulation in the United States since at least 2003. It is likely that our current patterns of antimicrobial use have selected for this particular strain. This information should allow us to begin developing disease prevention and treatment strategies that place less selection pressure for the overgrowth of MDR strains of *Mannheimia haemolytica.*

References

1. Richeson JT, *et al. J Am Vet Med Assoc.* 2013;243:1035.
2. Lubbers BV, *et al. J Vet Diagn Invest.* 2013;25:413.
3. Clawson ML, *et al.* BMC *Genomics.* 2016;17:982.
4. Wozniak RA, *et al. Nat Rev Microbiol.* 2010;8:552.
5. Michael GB, *et al. J Antimicrob Chemother.* 2012;67:84.
6. Klima CL, *et al. J Clin Microbiol.* 2014;52:438.

1.
acvim236

Impact of Drug Selection and Dosing Regimens on GI Tract Concentrations and Antimicrobial Resistance

Mark G. Papich, DVM; Derek M. Foster, DVM, PhD; Megan E. Jacob
Raleigh, NC, USA

Food animal veterinarians are critical gate-keepers of prudent antimicrobial use, and key to the mitigation of antimicrobial resistance (AMR) development in food animals. The FDA database for cattle shows that most antimicrobials approved are for treatment of bovine respiratory disease (BRD). Thus, this is the most common reason for administration of systemic therapeutic antibiotics, usually by injection. Two of the classes of approved drugs (fluoroquinolones and cephalosporins) have restricted uses because they are crucial to treatment of human infections. Previous studies evaluating the associations between the systemic treatment for BRD and development of AMR in enteric organisms have been variable and inconclusive, likely because the experimental methods and study designs inadequately assess this risk.

Previous studies have attempted to predict drug concentrations in the intestine of cattle, but failed to measure the active, unbound drug concentrations achieved in the gastrointestinal tract (GIT) with systemic drugs. Without this information, one cannot clearly evaluate the effect of antibiotic exposure on enteric bacteria. We have shown in our studies that measurement of active, unbound drug concentrations at the site of action (GIT lumen in these cattle studies) is preferred over plasma concentrations to assess the exposure of these antibiotics on intestinal bacteria.

We have demonstrated in our studies that it is possible using *in vivo* ultrafiltration techniques to directly sample the active drug concentrations (protein unbound) of systemically administered antibiotics. The paper by Warren *et al.* 2014 describes our technique. Our work has focused on two important antibiotics administered to cattle - ceftiofur and enrofloxacin. We have focused on these antibiotics because the FDA and other authorities have expressed concern over the extra-label use of these antibiotics in food-producing animals, and there are current restrictions on the use of these antibiotics in cattle. The extra-label use, or dosage is prohibited. The concern regarding these antimicrobial agents is that antibiotic concentrations in the intestine could potentially select for antimicrobial resistant bacteria that could pose a public health threat. However, these restrictions were placed before there was knowledge of the extent of active antibiotic concentrations excreted into the intestine after injections.

In studies conducted in calves, ceftiofur and enrofloxacin have been administered at different doses, dose intervals, and formulations (Foster *et al.* 2016, and other pending publications). These studies have shown that these antibiotics reach intestinal concentrations after injection that are sufficient to alter the microbial population, but the effects may vary depending on the drug, formulation, and dose.

For enrofloxacin, after a single dose of 7.5 mg/kg SC, there is evidence for passive diffusion into the small intestine. The concentrations in the colon are much higher than ileum, and higher than plasma, which may represent either a concentrating effect in the colon, or active secretion from efflux transporters. In another study in calves with enrofloxacin (publication pending), a high dose of enrofloxacin of 12.5 mg/kg SC once, was compared to three doses at 5 mg/kg SC. In these studies, the concentrations in both ileum and colon were much higher than plasma - reaching concentrations over 400% of plasma concentrations in ileum and colon from the high dose and over 260% and 400% in colon and ileum, respectively. Because of the high concentrations, transport to the intestine by the biliary flow, or efflux transporters is likely. These concentrations are sufficiently high to significantly lower bacteria counts and alter the population of bacteria in the intestine.

When ceftiofur was administered to cattle concentrations in the intestine were lower than plasma (Foster *et al.* 2016). Concentrations in the interstitial fluid (ISF) were low, reflecting the high plasma protein binding (approximately 95%) and it is possible that high protein binding also restricted diffusion into the intestine. Concentrations in the ileum and colon reached concentrations that were approximately 39% and 25% of the plasma concentrations in ileum and colon, respectively. Although these concentrations are lower than plasma, they may be in the range that selects for resistant bacteria, rather than eliminate these organisms.

Subsequent studies (results pending) have focused on the formulation used for ceftiofur administration. In a group of calves, ceftiofur sodium (Naxcel, 2.2 mg/kg) given by repeated dose, was compared to a single dose of ceftiofur crystalline free acid (Excede, 6.6 mg/kg). Because of the different plasma concentration profile produced by these formulations, concentrations produced in the ileum and colon are also affected.

An ideal antibiotic would be one that, after injection, is confined to the plasma and infected tissues (lungs), but restricted from the intestine. Alternatively, an ideal drug would be one that is inactivated in the intestine to mitigate the intestinal antibacterial effects. Such agents - to our knowledge - do not exist for cattle. However, it may be possible to identify the optimum dose, frequency, and dose formulation that will produce the least effect on intestinal bacterial resistance. Our lab will continue these studies to explore these factors.

References

1. Foster DM, Jacob ME, Warren CD, Papich MG. Pharmacokinetics of enrofloxacin and ceftiofur in plasma, interstitial fluid, and gastrointestinal tract of calves after subcutaneous injection, and bactericidal impacts on representative enteric bacteria. *J Vet Pharmacol Ther*. 2016;39(1):62–71.
2. Warren CD, Prange T, Campbell NB, Gerard MP, Martin LG, Jacob ME, Smith GW, Papich MG, Foster DM. Implantation of an ultrafiltration device in the ileum and spiral colon of steers to continuously collect intestinal fluid. *Res Vet Sci*. 2014;97(3):611–615.

acvim238

Challenges of Trace Mineral Diagnosis

Hugues Guyot, DVM, PhD, DECBHM
Liège, Belgium

Introduction

Trace elements are essential in the nutrition of man and animals. They play a major role in health, immunity and reproduction. The production of milk or meat as well as the housing and management conditions of high-performing animals may generate oxidative stress leading to health disorders in cattle. Some trace elements and vitamins have an anti-oxidative action and increase immunity. The most important trace elements in cattle are selenium (Se), copper (Cu), zinc (Zn), iodine (I), cobalt (Co), molybdenum (Mo), manganese (Mn), and iron (Fe). Cobalt is associated with vitamin B12 synthesis in the rumen, so the assessment of Co deficiency can also be done by measuring vitamin B12 blood concentration (Kincaid, Socha 2007). Molybdenum is particularly known for his antagonist effect on Cu absorption. In contrast, Mn and Fe are generally not deficient in cattle but often found in excessive concentrations in feed (in Belgium, unpublished data).

Are Trace Elements Deficiencies Important in Cattle Practice?

During the last 20 years, a discrepancy between the animals' need and the amount of trace elements given with the ration has been observed. The progress in genetics and cattle performance, as well as the impoverishment of soils, has accentuated this disequilibrium. This has consequences on animal's health, reproduction and productivity. Numerous papers have described the negative effects of trace elements deficiencies on cattle's health (Graham 1991). In Europe, trace elements deficiencies are nowadays widely spread in cattle. A study (Guyot *et al.* 2009) displayed the Belgian epidemiology of Se, Zn, Cu and I deficiencies in herds with health problems, compared to healthy herds. As expected, lower blood trace elements concentrations were found in unhealthy herds. The concentrations of trace elements in Belgian herds were compared to those in Ireland and France, which showed similar values (Mee, Rogers 1994; Enjalbert *et al.* 2006). These results can be extrapolated to other European countries, as the depletion in the soils and the nutrition of cattle are similar.

Why Diagnosing?

There is a wide range of symptoms associated with trace element deficiencies and thus the clinical signs are not pathognomonic. Laboratory tests are needed to confirm the diagnosis. Moreover, most of the time, the deficiencies are multiple (Guyot *et al.* 2009) and it is important to know the elements that are deficient in order to adequately correct and avoid competition between minerals. Finally, some trace elements are synergetic, thus the situation might be exacerbated if only one of the 2 minerals is given to face the deficiency (Guyot *et al.* 2011).

How Diagnosing?

Samples

As previously mentioned, clinical signs are not sufficient to make a diagnosis. Laboratory analyses are needed to confirm the diagnosis of deficiency. Moreover, samples must be taken directly from the animals rather than from feedstuffs because the availability and absorption of minerals as well as amount ingested by the animal can widely vary.

Different type of samples can be taken from animals: blood, milk, urine, hair or tissues (principally liver). Liver biopsies are considered as the gold standard for Cu, Zn, Fe, Se, and Co status determination (Ouweltjes *et al.* 2007). This is particularly interesting for animals in a silent phase of hepatic accumulation (Cu) before clinical toxicosis (Lopez-Alonso *et al.* 2006). Nevertheless, even though the procedure is relatively simple, it is not commonly performed in the majority of European countries because most practitioners see this technique as complicated and risky for the animal. Farmers are also generally reluctant to this procedure. Post-mortem sampling (slaughter) can also be an alternative to assess supra-optimal trace mineral exposure. Blood is the most frequently taken and analysed. Blood samples are easily taken and the modern techniques of assay are accurate, precise, sensitive and fast enough to make a good diagnosis. The inductively coupled plasma-mass spectroscopy (ICP-MS) has recently proven its efficacy at this point of view (Miksa *et al.* 2005). Most of the time, blood is centrifuged and plasma or serum are used for ICP-MS assays. Except for Cu where plasma is preferred (plasma>serum: Laven, Smith 2008; Hussein, Staufenbiel 2012), plasma and serum values are similar. The blood level is generally an early indicator of the deficiency phase as serum/plasma represents the transport pool of the mineral. Milk can give an idea of the trace mineral status of a lactating herd. Milk is very easy to sample but analyses are limited to certain elements (Se, I). Urine samples are of low interest, mainly due to the disparity of urine density and thus concentration. Furthermore, reference ranges are missing. Hair is maybe the easiest sample to collect; however, the accuracy of trace elements determination in this sample is poor and subject to external contaminations.

Sampling

It is not the goal to remind the blood sampling or liver biopsy procedure but to focus on the misinterpretation that can arise from wrong samples. The major risk is haemolysis which can falsely increase Se, Cu and Zn plasma/serum concentrations. The delay between sampling and analysis may also lead to false results. A long delay will increase the risk of haemolysis but also the diffusion of Se from erythrocytes (contained in glutathione-peroxidase) to serum/plasma. The activity of enzymes (e.g., glutathione-peroxidase or superoxide dismutase) may also decrease with the time. Finally, some tubes with rubber plugs may contain Zn, leading to a false increase of plasma/serum Zn concentration. Specific plastic tubes, without rubber and free of trace minerals should be used.

Diagnosis of trace elements deficiencies in cattle should be done at a herd level because this is clearly a herd disease. The diagnosis consists in evaluating the nutritional status of the herd. For determining a disease, sick animals should be tested. Thus, for assessing nutritional status, healthy animals should be taken. At a herd level, several healthy animals must be chosen for sampling. The number depends on the deficiency prevalence and the intra-herd variation of trace elements concentration. According to different authors, this number fluctuates from 6 to 15 (Herdt 2000; Herdt *et al.* 2000; Guyot, Rollin 2007; Laven, Nortje 2013). Animals to be sampled must be selected to form a homogeneous group in the target population. Analysis can be made individually or by pool. The pool has the advantage of being less expensive. It is accurate if the number of selected animals is sufficient (about 15 animals) and if the group of sampled animals is homogenous. One pool per target population is necessary. Nevertheless, pooled samples are effective for dosage of plasmatic Se, Zn, Cu, I, but are useless for determining activity of glutathione peroxidase (GPX) in erythrocytes, superoxide dismutase (SOD) or thyroid hormones. While using pool, attention must be paid in the interpretation of results if outliers (heterogeneous group) are present in the sampled population. In general, if the group is homogeneous, an "all-or-nothing" rule is applied as the majority of sampled animals (>70%) are either deficient or healthy (Guyot *et al.* 2009).

Source of Variation

Haemolysis is a huge source of variation as it artificially increases plasmatic Cu, Zn and Se concentrations. GPX and SOD in the red blood cells contain various amounts of these trace elements. The Cu-containing acute-phase protein ceruloplasmin will secondarily increase plasmatic Cu concentration in inflammation and diseases. Concomitantly, a decreased plasmatic Zn concentration will occur while inflammation is present (Milanino *et al.* 1986; Oliva *et al.* 1987). Inflammation has also an impact on Se metabolism and thyroid hormones (e.g., euthyroid sick syndrome; Janosi *et al.* 1998). Other factors, such as gestation, range of lactation, age (calves versus dams) may influence blood trace elements levels. Teat-dipping with I-containing products will interfere with the diagnosis of I deficiency (in milk or plasma) by transcutaneous absorption of I. Some dewormers (closantel, nitroxinil) will also increase the plasmatic I concentrations for a long period after their administration to cattle. The concentration of Se in milk can vary depending on the form of Se given in the ration with more Se transferred to colostrum or milk when cows are fed selenomethionine compared to Na-Selenite. Moreover, at equal dosage of Se in the feed, plasmatic Se and, in a lesser extent, the activity of GPX, reaches higher concentrations if selenomethionine is given in the ration compared to Na-Selenite. The correlation between GPX and blood Se is well described in the literature (Counotte, Hartmans 1989); however, this correlation may change according to the form of Se as well. In any form of Se, following an oral supplementation, plasmatic Se will increase gradually to reach a steady state between 30–60 days, while GPX will continue to progressively increase at least up to 120 days (Guyot *et al.* 2007c).

Analyses and Interpretation

Evaluating trace mineral status on blood or animal's tissues is subject to some limitations and variations. Two different ways are often present: determining nutritional status (e.g., plasmatic or hepatic concentration of the trace element) or functional status (e.g., thyroid hormones, GPX, SOD). Further, the thresholds of these parameters are often discussed. Table 1 proposes different thresholds for assessing nutritional and functional status of the major trace elements Se, Zn, Cu, I and Co. According to Figure 1, they have been chosen considering the concentration of trace elements needed in order to optimize health, welfare, food quality, and productivity and not just avoiding clinical diseases or barely reaching N.R.C. requirements (NRC 2000 and 2001). Besides this, the blood or tissue (e.g., liver) represents different pools of trace elements in the body. When dietary intakes are not sufficient, the pools of storage and transport will decrease, respectively, but they must be depleted before seeing dysfunction and disease (Figure 2). Ideally for the diagnosis, the storage pool must be sampled but in practice, it is not always accessible.

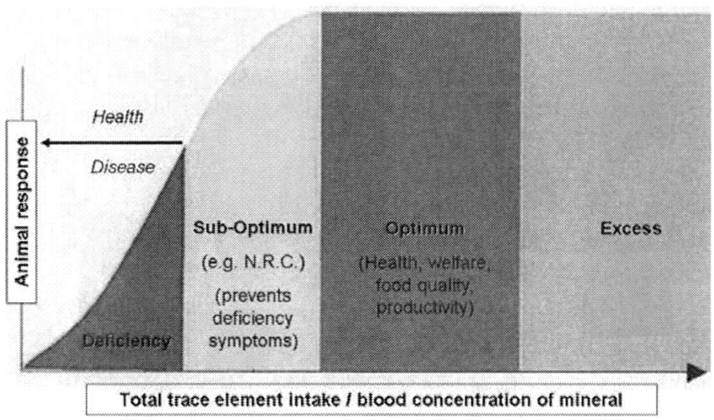

Figure 1. Threshold for determination of correct trace element status in cattle (adapted from Chung 2003)

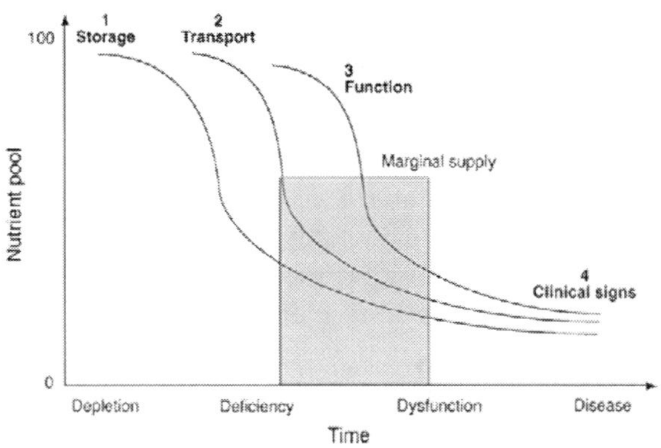

Figure 2. Kinetic of the pools of trace elements, according to the mineral level (from Underwood & Suttle 1999)

Table 1. Threshold for plasma zinc (Zn), plasma copper (Cu), activity of glutathione peroxidase in erythrocytes (GPX), plasma selenium (Se), serum cobalt (Co), plasma thyroxine (T4), plasmatic inorganic iodine (PII) in blood and iodine (I) in milk for adult cattle (Herdt *et al.* 2000; Guyot *et al.* 2007c; Guyot and Rollin 2007; Guyot *et al.* 2009; Guyot *et al.* 2011).

Herd status	Zn (μmol/L)	Cu (μmol/L)	GPX (IU/g Hb)	Se (μg/L)	Co (μg/L)	PII (μg/L)	T4 (nmol/L)	Milk I (μg/L)
Adequate	14–21	13–18	220–600	80–140	0.17–2	45–650	30–120	80
Marginal	8–14	8–13	75–220	65–80	-	15–45	20–30	30–80
Deficient	<8	<8	<75	<65	-	<15	<20	<30

Specific Diagnosis for Se, Cu, Zn, I, and Co

Selenium

Serum/plasma Se is a good indicator of dietary intake. Plasma Se is a nutritional marker that quickly increases (within 2–6 days) after oral supplementation (Ellis *et al.* 1997).

As functional marker and long-term marker, the assay of GPX in red blood cells is often used. Because Se is incorporated in GPX at the time of the erythrocyte's development, GPX activity allows determining Se status of the 100–120 previous days (mean of half-life of red blood cells). As Se is well transferred from dam to calf, the cow in late gestation will show lower Se plasma concentration. After calving, Se levels will progressively increase in the dam. In calves, a higher proportion of Se is in the GPX rather than the plasma, so the GPX activities are higher and plasma Se lower, compared to their dams (Guyot *et al.* 2011). Source of Se fed by cattle (dams or calves) has also an influence on plasma Se and GPX activity. Calves from dams supplemented with Se-yeast (selenomethionine) present higher plasma Se concentrations and GPX activity at birth compared to those supplemented with inorganic sources (Na-Selenite). Iodine supplementation has an impact on Se metabolism with low Se status while animals fed supplementary I, compared with I-deficient diets (Pavlata *et al.* 2005; Guyot *et al.* 2011). Liver biopsy is also a good means to assess Se status, if realisable. Finally, milk Se can also be assessed but is widely influenced by the form of Se given in the ration with higher concentrations when cows are fed with selenomethionine (Guyot *et al.* 2007c).

Copper

The capacity of metallothionein to bind Cu is limited. This reduces the excretion of Cu with the bile and leads to Cu accumulation in the liver (Lopez-Alonso *et al.* 2005). The storage pool for Cu is thus the liver. It reflects the long-term availability of dietary Cu. A decreased liver Cu concentration is an early marker of insufficient Cu dietary intake. The most sensitive way to assess Cu deficiency is therefore the liver biopsy. To accurately interpret the Cu status based on liver samples, the age and production class must be taken into account (Puschner *et al.* 2004a). In some European countries, practitioners and farmers are unenthusiastic to this method of sampling. Instead, plasma Cu is more often used. The transport pool is constituted by the blood and, in a certain extent, also the functional pool. In fact, ceruloplasmin, a protein of inflammation produced by the liver, contains about 80% of the Cu in the blood. Determination of ceruloplasmin activity also seems to be effective in diagnosing Cu status (Laven *et al.* 2007). Plasma Cu assessment seems to be better rather than serum because a variable amount of Cu may be captive in the clot (Laven, Smith 2008; Hussein, Staufenbiel 2012). There is a curvilinear relationship between liver and blood Cu concentration. However, hepatic Cu concentration must fall drastically before a decreased blood Cu concentration can be seen (Underwood, Suttle 1999). Plasma Cu is thus a late (long-term) marker of Cu deficiency but remains useful to be performed at a herd level, with at least 10–15 animals sampled per herd (Herdt 2000). Superoxide dismutase is a Cu-containing enzyme sometimes used in determining Cu status but it is very delicate to interpret. Milk Cu does not present interest as Cu milk concentration is usually poor.

Zinc

The storage pool for Zn is not well defined, making the evaluation of Zn status tricky. Even if hepatic Zn concentration decreases after a long inadequate Zn dietary intake, the liver Zn is not a good means for assessing Zn status in cattle. Moreover, to accurately interpret the Zn status based on liver samples, the age must be taken into account (Puschner *et al.* 2004b). Thus, the plasma/serum Zn concentration is preferred because it is reduced in Zn deficiency. However, there might be dysfunctions (e.g., reduced feed efficiency) before observing a decreased serum Zn concentration. A low serum Zn concentration is indicative of a deficiency while a normal serum Zn does not necessarily plead for a deficiency (Underwood, Suttle 1999).

Iodine

Iodine deficiency can also be assessed by nutritional or functional markers. For the nutritional ones, the most commonly used is plasmatic inorganic iodine (PII). The PII reflects the dietary I intake and its concentration fluctuates very quickly according to the I fed in the ration. For functional markers, the hypothalamus-pituitary-thyroid axis is investigated with thyroid stimulating hormone (TSH) and total thyroid hormones (thyroxine T4 or tri-iodo-thyronine T3) (Guyot *et al.* 2011). Huge variations can physiologically exist (pregnancy, circadian rhythm, birth, age) and are important to be included in the interpretation of the status. Clinical biochemistry laboratories can easily assay thyroid hormones while there is no laboratory able to measure bovine TSH at the moment. TSH, T4 and T3 are usually taken for diagnosing hypothyroidism, which originates in cattle, mainly from I deficiency (Guyot *et al.* 2007a,b). Total blood I is constituted with PII and I from thyroid hormones, setting the interpretation of a short-term and long-term marker not obvious. Milk I can be used for assessing herd I status. It also vary quickly with dietary I, but in a lesser extent than PII. On the field, weighing thyroid gland in (stillborn or goitrous) calves is useful to appreciate I status. The maximal weight of the gland has been determined as follows (Hernandez *et al.* 1972): Y (weight of thyroid gland, in grams) = $0.348 \times Z^{0.944}$ (Z is the body weight, in kg).

Cobalt

Cobalt status can be determined by dosage of Co or vitamin B12 blood concentrations in ruminants. Vitamin B12 is constituted of 4% of Co. The blood concentration of these two elements is very low, necessitating highly performing techniques of dosage (ICP-MS, radioimmunoassay, chemiluminescence, high-performance liquid chromatography). As liver contains the largest concentration of Co, it is the sample of choice and represents the storage pool of Co. Nevertheless, there is a variable proportion of Co that is not associated with vitamin B12, inducing misinterpretation of Co status either in liver or in blood. Correlation between serum Co and serum vitamin B12 is low; however, increased concentration of vitamin B12 is observed in milk and colostrum after Co supplementation (Kincaid, Socha 2007). According to these elements, the assessment of Co status in ruminants, either in liver or blood, is complicated and poorly reliable with routine laboratory assays (Underwood, Suttle 1999: Herdt *et al.* 2000).

Conclusions

Trace elements deficiencies are frequent in cattle. As there are no pathognomonic signs, clinical pathology diagnosis is necessary. Sampling at a herd level increases the reliability of the diagnosis, if samples are taken from a homogeneous group of healthy animals. Blood is most often used because of its higher feasibility in the field. The storage pool is the most effective one for assessing the mineral status but unfortunately is not always represented by the blood. Specific herd thresholds for determining trace elements status in cattle must include health, welfare, food quality and productivity.

References

1. Chung TK. *Poultry Int.* 2003;1:28.
2. Counotte GH, Hartmans. *J Vet Q.* 1989;11:155.
3. Ellis RG, *et al. Am J Vet Res.* 1997;58:760.
4. Enjalbert F, *et al. J Anim Physiol Anim Nutr.* 2006;90:459.
5. Graham TW. *Vet Clin North Am Food Anim Pract.* 1991;7:153.
6. Guyot H, *et al. J Vet Diagn Invest.* 2007a;19:643.
7. Guyot H, *et al. Cattle Pract.* 2007b;15:271.

8. Guyot H, *et al. Livestock Sci.* 2007c;111:259.
9. Guyot H, Rollin F. *Ann Med Vet.* 2007;151:166.
10. Guyot H, *et al.* Epidemiology of trace elements deficiencies in Belgian beef and dairy cattle herds. *J Trace Elem Med Biol.* 2009;23:116.
11. Guyot H, *et al. J Trace Elem Med Biol.* 2011;25:118.
12. Herdt HT. *Vet Clin North Am Food Anim Pract.* 2000;16:387.
13. Herdt HT, *et al. Vet Clin North Am Food Anim Pract.* 2000;16:423.
14. Hernandez MV, *et al. J Anim Sci.* 1972;34:780.
15. Hussein HA, Staufenbiel R. *Biol Trace Elem Res.* 2012;146:47.
16. Janosi S, *et al. Acta Vet Hung.* 1998;46:71.
17. Kincaid RL, Socha MT. *J Dairy Sci.* 2007;90:1880.
18. Laven RA, *et al. N Z Vet J.* 2007;55:171.
19. Laven R, Smith S. *N Z Vet J.* 2008;56:334.
20. Laven RA, Nortje R. *N Z Vet J.* 2013;61:269.
21. Lopez-Alonso M, *et al. Vet J.* 2005;170:332.
22. Lopez-Alonso M, *et al. J Vet Diagn Invest.* 2006;18:71.
23. Mee JF, Rogers PAM. *Irish Vet J.* 1994;47:115.
24. Miksa IR, *et al. J Vet Diagn Invest.* 2005;17:331.
25. Milanino R, *et al. Agents Action.* 1986;19:215.
26. National Research Council (NRC). *Nutrient Requirements of Beef Cattle.* 7th ed. Washington, DC: National Academy Press; 2000: 248p.
27. National Research Council (NRC). *Nutrient Requirements of Dairy Cattle.* 7th ed. Washington, DC: National Academy Press; 2001:408p.
28. Oliva JC, *et al. Esp Fisiol.* 1987;46:25.
29. Ouweltjes W, *et al. Tijdschr Diergeneeskd.* 2007;132:76.
30. Pavlata L, *et al. Vet Med-Czech.* 2005;50:186.
31. Puschner B, *et al. J Vet Diagn Invest.* 2004a;16:382.
32. Puschner B, *et al. J Vet Diagn Invest.* 2004b;16:278.
33. Underwood EJ, Suttle NF. *The Mineral Nutrition of Livestock.* 3rd ed. Oxon: CABI Publishing; 1999.

acvim239

The Use of Inflammatory Markers in Field Cattle Medicine

Hugues Guyot, DVM, PhD, DECBHM
Liège, Belgium

INTRODUCTION

Inflammation is an essential component of innate immunity. Mediators such as cytokines and oxylipids (prostaglandins, leukotriene, and thromboxane) allow amplification of defense mechanisms. They lead to vasodilation and promote migration of immune cells. They regulate the expression of molecules for adhesion of neutrophils. Some of these mediators also have a role in stopping the inflammatory cascade at the end of the aggression. Depending on the situation, these mediators might have a pro- or anti-inflammatory action. Inflammation plays also a role in active immunity as T-lymphocytes are able to synthetize cytokines that can modulate immunity. Immunoglobulins (G and M) can improve phagocytosis capability of neutrophils and macrophages. An adequate inflammatory response is a sign of functional immune system. The "aggression," which triggers an inflammatory process, can be endogenous or exogenous, linked to a pathogen or any other kind of aggression (metabolic, mechanic). The subsequent reaction might be acute (a few hours to several days) or chronic (several weeks, months or years), depending on the resolution of acute inflammation. Inflammation is necessary to control infectious agents, eliminate foreign bodies, repair tissue damage, and inactivate or dilute toxins. Nevertheless, the situation can be deleterious at the moment when inflammation becomes uncontrollable (septic shock, deposition of immune complexes, allergies, autoimmune disease, and fibrosis) or chronic (self-running) and provokes more severe lesions than those directly linked to the aggression itself. Inflammation is a major factor and often a common feature of numerous metabolic and infectious diseases in cattle, related to important economic consequences (mastitis, retention of fetal membranes, metritis, abomasum displacement, and ketosis) (Burvenich *et al.* 2007; Sordillo 2016).

THE DIFFERENT INFLAMMATION BIOMARKERS IN CATTLE PRACTICE

Numerous inflammatory markers are proposed by veterinary laboratories, with different kinetics, sensitivity and specificity according to the species. Some are focused on acute response of inflammation while others are specific for chronic inflammation. There are positive and negative markers. The first ones show an increased concentration in case of inflammation while the others decrease in the same situation. The common acute phase proteins used in ruminant's medicine and gamma-globulins are positive markers. Albumin, thyroxine (T4) and transferrin are negative markers.

Acute Phase Proteins

Dozens of acute phase proteins (APP) exist with major or moderate roles. In cattle practice, the most essentials APP are haptoglobin (Hp) and serum amyloid A (SAA). Fibrinogen (Fibri) and ceruloplasmin (Cp) are also monitored routinely but they have a moderate impact in the inflammatory process. APP are mainly synthetized in the liver but can also be appearing in other organs (mammary gland, uterus, ovaries, respiratory tract, digestive tract, adipose tissues) with a probable role in counterbalancing pro-inflammatory effects within the tissue concerned. Numeration and differentiation of white blood cells can also be useful in some cases. Hp and SAA are early markers (hours) while Fibri will increase later in the course and reaches a peak 10 days after the beginning of the inflammatory process. The increase of these markers is not proportional to the severity of the lesions or the disease.

Serum amyloid A is an alpha-globulin (apolipoprotein) produced mainly by the liver and binds cholesterol and modulates innate immune reactions. However this is the earliest marker of inflammation, its concentration decreases quickly and there is a high variability among animals, making it difficult to interpret and compare. In current practice, this marker won't be used because basal values of healthy animals are not known in advance. SAA increases physiologically around calving (1st week post-partum). In general, healthy animals have SAA concentration below 8.8 mg/L (Jones, Allison 2007).

Derived from SAA, the mammary M-SAA3 is involved in mammary gland defense against Gram+ and Gram- bacteria (Weber *et al.* 2006). SAA can be assessed in serum using ELISA kits.

Haptoglobin is mainly produced by the liver and binds free hemoglobin from damaged red blood cells to protect tissues from the toxicity of free iron. Hp also prevents oxidative damages and has an anti-inflammatory activity during the innate immune response and a bacteriostatic effect. This is an alpha2-globulin that increases 24–48 h after tissue damage. Hp can remain high for 2 weeks. Increased concentration of Hb can be observed in many conditions (metritis, mastitis, enteritis, pneumoniae) and physiologically at the beginning of lactation (Humblet *et al.* 2006). During the first week after calving, Hp must be below 150 mg/L while during the other periods, Hp should be under 30 mg/L (Humblet *et al.* 2006). Hp can be assessed in serum using hemoglobin-binding capacity method adapted to an automated analyzer (Owen *et al.* 1960; Humblet *et al.* 2006) or by ELISA.

Fibrinogen is a beta2-globulin that increases 2 days after initiation of inflammatory process, whatever its origin, and reaches its peak after 10 days. Different methods exist to assess Fibri concentration. In general, a coagulation method is used in most laboratories. Blood must be harvested in Na-Citrate tubes (test done on plasma). Thrombin time (used to assess coagulation troubles) measures the time to convert fibrinogen to fibrin (with a coagulometer). From this test, fibrinogen concentration can be determined. Other methods approximate Fibri concentration by measuring total proteins in plasma and serum after removing fibrinogen by heat precipitation method. It is also feasible to measure total protein in serum (dry tube) and in plasma (Na-heparin tube) and calculate the difference between the 2 tubes that approximates fibrinogen concentration. Fibri concentration should be below 7 g/L. Dehydration can falsely increase Fibri concentration (with relative hyperproteinemia). To avoid this bias and discriminate hyperfibrinogenemia from inflammation and from dehydration, the PP:F ratio method can be used by applying the following formula: PP:F ratio = [plasma total protein] – [Fibrinogen]/[Fibrinogen]. If the PP:F ratio is >15, dehydration should be suspected while if the ratio is <10, inflammation is rather diagnosed. Nevertheless, this method is only valid in adult cattle, without hypoproteinemia and without concomitant inflammation and dehydration. Liver dysfunction may further modulate the Fibri (but also Hp and SAA) concentration.

Ceruloplasmin is an alpha-globulin that increases with inflammation. There is a close relationship between Cp and copper (Cu) while 80% of the latest is contained in Cp. That explains why plasma Cu increases with inflammation. Most of the time, Cp is used to assess Cu deficiency.

Leukogram. As bovines do not have a reserve pool of white blood cells, neutropenia with a predominance of immature forms will be observed (left shift) during acute inflammation (24–48 h). Thereafter, white blood count (WBC) will progressively increase to remain stable and within normal ranges during chronic inflammation.

Chronic Positive Markers: Total Proteins and Globulins

Total proteins (TP) are constituted of albumin and globulins and thus contain most of the inflammation proteins. TP can be measured in the laboratory (Biuret method) or in the clinic or in the field with a refractometer, after centrifugation of the whole blood (plain tube for serum, anticoagulant tube for plasma). Serum total protein (STP) can be measured in adult cattle to determine an inflammatory process, if STP are >81 g/L (Radostits *et al.* 2007). In this case (STP), Fibri will not be taken into account. Unfortunately, there are numerous biases increasing STP artificially: dehydration, icterus, hyperlipemia, hyperglycemia, uremia, and hemolysis. With hyperproteinemia, albumin and/or globulins can increase. If only albumin is increased, dehydration might be suspected. A hyperproteinemia without dehydration is often a hyperglobulinemia. This may happen in e.g., chronic antigenic stimulation or hepatic diseases. The albumin:globulins ratio in healthy animals varies between 0.84–0.94. This ratio will increase in case of intoxication and will decrease with (chronic) inflammatory or infectious process (Russel, Roussel 2007). Hypoproteinemia can be due to hypoalbuminemia, panhypoproteinemia, failure of passive immunity transfer in calves (hypoglobulinemia), excessive losses of albumin (renal amyloidosis), and lack of protein synthesis in the liver (severe and chronic hepatic disease, insufficient absorption, digestion or intakes). STP will be within normal ranges with hypoalbuminemia in case of debilitating diseases. Hypoalbuminemia can also be found in acute or subacute inflammatory process, due to renal and/or gastrointestinal losses. During hemorrhage or severe exudation, panhypoproteinemia will be detected. Eventually, the best way to assess TP is to perform an electrophoresis of total proteins (ETP). This technique will separate all proteins in serum (or plasma) according to their size and concentration in a gel traversed by an electric current. TP are so distributed: Albumin (41–52%), alpha (12–19%, e.g., Hp, Cp) and beta-globulins (11–34%, e.g., Fibri, transferrin), immunoglobulins (or gamma-globulins, synthetized by lymphoid cells, 23–34%). Among immunoglobulins, 70–80% are IgG (gamma-globulins), 10–20% are IgA, 5–10% are IgM, and the remaining in weak concentrations are IgD and IgE. Figure 1 shows a typical picture of an ETP in a healthy adult cow.

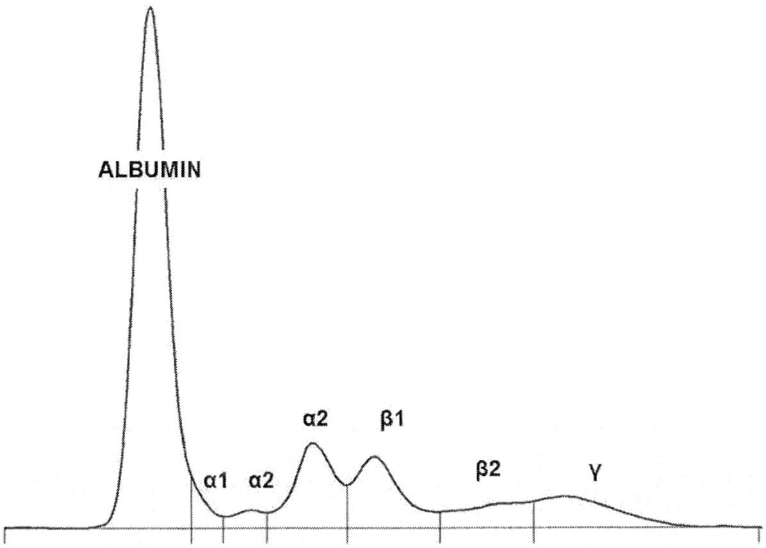

Figure 1. Standard profile of an ETP in a healthy adult cow

Hyper-alpha-globulinemia is observed in acute and subacute phases of inflammation or during viral or bacterial infection. Hyper-beta-globulinemia can be representative of parasitic infestation or, if accompanied with hyper-gamma-globulinemia, the sign of a severe chronic disease (reticulo-peritonitis, hepatic failure). Hyper-gamma-globulinemia is often seen in subacute or chronic inflammatory process. This can be linked to hepatic failure, severe inflammation, infectious diseases (bacterial, viral, parasitic), peritonitis, etc.

Others (Negative Markers)

Albumin represents 40–60% of total serum proteins. It is synthetized by the liver and can bind different ions (Ca, Cu, bilirubin, T4, drugs, etc.). Albumin concentration will decrease in serum during (chronic) inflammation or hypoproteinemia. Its synthesis is in competition with inflammatory markers in the liver. Other proteins such as lipoprotein (cholesterol) or transport proteins (vitamins, trace minerals) also decrease for the same reason during inflammation (Sordillo *et al.* 2009; Trevisi *et al.* 2009).

Transferrin is a beta-globulin produced in the liver, with a half-life of 8 days. This molecule has anti-infectious properties (iron chelation) and decreases while inflammation.

Thyroxine is a hormone produced by the thyroid gland. A drastic decrease might be found during severe infection or inflammation in euthyroid patients. This is also called in ruminants "euthyroid sick syndrome" (Janosi *et al.* 1998). Very low concentration of T4 (<15 nmol/L) may indicate a fatal issue.

Field Tests for Inflammation

Glutaraldehyde Reaction-Test

The method used in this test is the one described by Sandholm (1974). The principle is based on the polymerization (coagulation) of the amine groups present in bovine's whole blood (immunoglobulin [IgG] and fibrinogen) with the aldehyde groups of the glutaraldehyde. Fibrinogen and gamma globulins in contact with glutaraldehyde cause the formation of blood clot. The speed of the formation of the blood clot is proportional to the concentration of these inflammatory proteins (Sandholm 1974; Metzner *et al.* 2007). This test is especially indicated in the diagnosis of chronic inflammatory processes but also during severe acute inflammation. The literature has numerous applications of this test: chronic traumatic reticulo-peritonitis, chronic pericarditis, chronic pneumonia, chronic peritonitis, chronic pleurisy, chronic metritis, chronic hepatitis, major abscesses, chronic arthritis, chronic nephritis and pyelonephritis, and some severe mastitis (Guyot *et al.* 2011; Gaillot *et al.* 2013). The test is quite sensitive but not specific for a particular disease. Clinical examination is essential to determine the origin of inflammation. For very severe acute cases, the fibrinogen concentration may be sufficient to make the test positive. In most of the cases, it takes between 7–10 days after onset of the disease, so that the gamma-globulin concentration is sufficient to trigger a positive test reaction. A field study (Guyot *et al.* 2011) compared the glutaraldehyde-test with laboratory tests (e.g., haptoglobin) and showed a very good correlation. It is important to note that there was no correlation with survival of the animal 2 months after the initial diagnosis. The study by Metzner *et al.* (2007) showed a correlation between the coagulation time of the glutaraldehyde test and the concentration of gamma-globulins and fibrinogen in the blood.

Based on the same principle, it is possible to assess the passive transfer of colostral immunity in calves' blood. The fibrinogen concentration in a young healthy calf is negligible and does not interfere with the test. IgG in blood comes from the absorption of colostrum. According to blood IgG concentration in calves, there will be blood coagulation faster or slower and proportional to the concentration of IgG (Sandholm 1974). This test is indicated in the diagnosis of passive immunity transfer (PIT) in healthy calves between 2 and 6 days of life. In the University of Liège, Faculty of Veterinary Medicine, Bovine Ambulatory Clinic, a field test was developed on that purpose. This test was compared to the gold standard (measurement of IgG by radial immunodiffusion) and presented a sensitivity of 97% and a specificity of 80% (Guyot *et al.* 2015a). The presence of false-negative (tested animals "correct PIT" while they are in failure PIT) is therefore rare. A failure of PIT (FPIT) will have negative effects on health. It will be observed in dairy cattle decreased average daily gain and milk production (1st lactation), increased mortality and culling (1st lactation) (Robinson *et al.* 1988; DeNise *et al.* 1989). In beef herds, a FPIT will result in an increased morbidity (especially respiratory disease) and mortality of calves (pre-weaning), and decreased ADG compared to animals with proper PIT (Wittum, Perino 1995). The Bovine Ambulatory Clinic of Liège continued their investigations with another field test to determine the quality of the colostrum, still based on the same chemical reaction with glutaraldehyde. Distinguishing good from poor quality colostrum allows to adapt the volume administered or to initiate ancillary procedures for a sufficient transfer of IgG. Immunoglobulins (IgG) content of colostrum is highly variable and cannot be predicted. This semi-quantitative field test is indicated to differentiate adequate colostrum from poor quality colostrum (low immunoglobulins concentration) in cattle. The test was also compared to the gold standard (radial immunodiffusion) and showed a sensitivity of 100% and a specificity of 90% (Guyot *et al.* 2015b). To increase test performance and have a global and real idea of colostrum quality in the farm, it is advisable to test the colostrum of at least 3 to 5 cows from the same herd.

Total Proteins in Serum (TPS), Plasma (TPP) and Others

This was already discussed in the paragraph related to Total Proteins. As previously mentioned, it is feasible to approximate the Fibri concentration using a refractometer. Two blood sample tubes must be harvested: one plain tube and one Na-heparin tube. After centrifugation, the plasma and serum are separated and total proteins are measured using a refractometer. The difference between TPP and TPS represents the Fibri. If the concentration observed is ≥7g/L, acute inflammation is suspected. According to a preliminary study (Guyot *et al.* 2011), there is a good correlation (r=0.82) between Fibri measured at the laboratory and estimated Fibri using refractometer. Total protein can also be assessed in cerebrospinal fluid (CSF); however, TP in CSF is too low to be measured with a refractometer. Urinary strips can be useful as they detect weak concentration of TP. Meningitis can be suspected as TP in CSF increases (>1–2 g/L) (Scott 1995). Using a refractometer, TP in joint puncture can help discriminate normal synovial fluid (TP <20 g/L) from septic arthritis (TP >60 g/L). Appearance, consistency, presence of numerous inflammatory cells also add precision to the diagnosis.

To help diagnose peritonitis, TP in peritoneal fluid (paracentesis) can be measured with a refractometer. Exudate is present while TP in peritoneal fluid is above 30 g/L (Wittek *et al.* 2010). Presence of inflammatory cells is also indicative of infection/inflammation.

Eventually, to assess passive immunity transfer for calves, other field tests can be used by measuring TP in serum of calves (must be >56 g/L for adequate immunity transfer) (Vandeputte *et al.* 2011) or TP in colostrum (colostrumeter or refractometer).

Inflammatory Cells

The presence of inflammatory cells can be detected, as an infectious agent is going to attack the host. These cells can be detected in numerous physiological fluids. An EDTA tube will be preferred for sampling to preserve the cells.

While infection in the mammary gland occurs, somatic cells increase in the udder to fight infection. This produces an inflammatory process. Assessment of inflammatory cells in colostrum or milk is an indicator of inflammation/infection in the mammary gland and allows follow-up of the infection or healing. The Californian Mastitis Test (CMT or Schalm's test) (Schalm, Noorlander 1957) is the oldest test used to assess somatic cells in milk. The range of the test is 200,000 to 5,000,000 cells/ml. The principle is based on the formation of gel regarding the number of nucleus (cells), after addition of an anionic surfactant or detergent (reacting with nucleus). The more the aspect of milk looks like jelly, the more the content of somatic cells is important in the milk. Another method is to use automatic cell counters. These automatic counters work with ranges of measurement varying (according to the manufacturer) from 10,000 to 4,000,000 cells/ml.

As in the milk, the presence of somatic cells in urine signs inflammation and/or infection. Urinary strips are not reliable. The CMT and automatic cell counters can be used, depending on the number of cells in the urine.

The normal concentration of cells in the CSF is quite low (<0.012.10^9 cells/L). Somatic cells can increase in the CSF while infection/inflammation (Scott 1995) and can be measured by automatic cell counter or CMT, if the cell concentration is enough.

An increased concentration of somatic cells can be observed while arthritis; however, the usual tests (CMT, automatic cell counter) can be more difficult to interpret, as the articular fluid presents naturally a certain degree of viscosity. The cell concentration in a septic arthritis is above 10,000 cells/ml.

Finally, the normal volume of abdominal liquid is 1 ml/kg body weight. In case of septic peritonitis, the number of somatic cells will also increase to exceed 10,000 cells/µl, as an exudate (Wittek *et al.* 2010). Automatic cell counters and CMT can be used to determine the somatic cell concentration.

Interest of Inflammatory Biomarkers in Cattle Practice

APP can nowadays be used as indicators of the general health of cattle. Researches have raised the interest of these markers in stressful situations, diseases, shipping, and welfare (Ceciliani *et al.* 2012). Inflammatory biomarkers can serve the practitioners in many different situations, as diagnostic or prognostic indicator.

Inflammation and Nutrition

Energy Metabolism

The increased risk of developing pathologies in the early post-partum period is linked to a dysfunction of the immune system and the "physiological" depression of immunity in the post-partum period. At the same time, low reactions of the immune system, as well as a pro/anti-inflammatory imbalance are present, leading to chronic inflammation (Sordillo 2016). During the transition period, the excessive mobilization of fat from tissues is often related to a negative energy balance (NEB). If the situation becomes excessive it may induce a pro-inflammatory process, due to exacerbation of cytokine action by non-esterified fatty acids (NEFA) (Sordillo *et al.* 2009; Trevesi *et al.* 2009). Fat mobilization increases the risks of insulin resistance, fatty liver syndrome and chronic pro-inflammatory state inducing an aggravation of NEB and a higher prevalence of post-partum infections. Raised concentrations of NEFA, beta-hydroxy-butyrate (BOH) or decreased concentration of calcium induce the liberation of pro-inflammatory mediators and reduce the bactericide activity of leucocytes as well as their chemotactical migration and phagocytosis activity. This leads to different common pathologies of the post-partum period: retained fetal membranes, mastitis, metritis, ketosis and left displacement of abomasum (LDA). Recently, a pegylated form of a natural cytokine - bovine granulocyte colony-stimulating factor (bG-CSF) - has been used in peri-parturient dairy cows to fight against post-partum immunity depression by improving neutrophil number and functions (myeloperoxidase release) (Kimura *et al.* 2014). In a natural way, anti-oxidative molecules are specifically expressed around parturition (-7 days to +21 days in milk) (Gessner *et al.* 2013) to fight deleterious effects of inflammation. A deficit in energy and/or hypoglycaemia in dry period will also create a dramatic situation by increasing cytokines and leucocyte synthesis with a less performant immune system, thus, more lesions will be expected in case of infection (Sordillo 2016). Inflammation triggered by periods of metabolic stress is a major factor of complications of post-partum diseases. Liver's health is closely related to inflammatory reactions and pathologies resolution.

Trace Minerals

Inflammatory reactions and oxidations of fatty acids provoke free radical synthesis that induces severe tissue lesions (Sordillo 2016). Specific trace elements and vitamin supplementation help to protect tissues and improve immunity in the post-partum period.

Rumen and Intestines

Other pathologic conditions, such as LDA or subacute ruminal acidosis (SARA), also induce inflammatory process that can be revealed by the assay of APP (Gozho *et al.* 2007; Guzelbektes *et al.* 2010). While increasing the amount of starch in the ration of dairy cows, by-pass starch as well as fermentation in the large intestine also increase, leading to liberation of endotoxins (LPS) which are pro-inflammatory molecules. This can be observed in blood by measuring APP (Plaizier *et al.* 2014). Fibers digestibility of the ration is also a critical point for the inflammatory status of the digestive tract. Intestinal flora drift accompanied with decreased efficacy of fermentations in rumen induce a deterioration of fibers digestibility, predisposing to metabolic troubles (SARA, LDA) that promote inflammation and liberation of APP (Plaizier *et al.* 2008).

Protein Metabolism

The quality of proteins in the ration can influence inflammatory reactions and the resolution of the inflammatory process. For example, methionine and cysteine regulate anti-oxidative and inflammatory protein synthesis. A healthy rumen is nevertheless essential for these amino-acids to be incorporated in specific enzymes and involved in anti-oxidative functions (Osorio *et al.* 2014).

Feedback from the Clinic

Acute and chronic inflammation, besides competing with protein synthesis and energy utilization as well as its interferences with tissues lesions, induce a decrease in milk yield and promote diseases in the herd. Sensitivity to inflammation is individual but the diagnosis needs to be addressed at a herd level. In the University of Liège, 20 dairy herds visited for a drop in milk yields and feed efficiency were assayed for rumen (rumen fluid and faeces sampling) and global health (health score, blood trace elements/vitamins, inflammation biomarkers). In these herds, numerous cases of inflammation were detected and displayed by increased TPS and/or estimated Fibri (without dehydration). Moreover, in the majority of the herds, the albumin/globulin ratio (<0.84) revealed chronic inflammation. Regarding rumen flora, there was no evidence of SARA but instead there was a majority of inactive rumen fluids. There was a positive and significant correlation between inactivity of the flora and chronic inflammation. The origin of this situation might be due to a lack of digestibility of the roughages and/or fermentable energy and/or amino acids with excess of fermentable nitrogen. The consequences could be a reduced synthesis of hepatic proteins (albumin) aggravated by competition with inflammatory proteins synthesis during chronic inflammation. Presence of chronic inflammation is to be taken into account while investigating cattle herds with poor feed efficiency of modifications of milk components (fat and proteins concentration in milk). Monitoring APP in these herds is another efficient way to highlight metabolic and post-partum diseases.

Inflammation and Udder Health

Monitoring the APP in cattle practice is particularly interesting in mastitis, especially in the subclinical form without any evident clinical sign. Hp and SAA are present in milk and are synthetized in the mammary gland during mastitis (Eckersall *et al.* 2001). Moreover, there is a specific isoform of SAA (M-SAA3) that is secreted in milk from cows with mastitis (Weber *et al.* 2006; Molenaar *et al.* 2009). Several studies have shown the rise of APP in milk and serum following udder infection. In case of extra-mammary infection, only serum APP were increased but not in the milk. M-SAA3 could be used as a marker for subclinical mastitis. Gerardi and coworkers (2009) demonstrated a significant correlation between M-SAA3 concentration in milk and somatic cell count. There was no significant correlation between SAA in serum and M-SAA3 in milk, indicating that for subclinical mastitis diagnosis purpose, milk should be used for measuring APP. Using non-steroidal anti-inflammatory drug while treating mastitis has an interest in resolving inflammation and improves reproduction performances (McDougall *et al.* 2016).

Other biomarkers have also been investigated in milk with various results: lactate dehydrogenase, N-acetyl-beta-D-glucosaminidase, alkaline phosphatase, acid phosphatase. They are not specific inflammatory biomarkers.

Inflammation and Reproduction

APP are mainly used for detection and monitoring of metritis. There are evidences of increased concentration of APP during uterine infection. Hp was identified as a potential early marker of metritis in cows. Indeed, cows with high Hp 3 days post-partum are 6.7 more likely to develop metritis, compared to normal range of Hp concentrations (Huzzey *et al.* 2009). Moreover, among pregnant cows, there was a significant higher number of days open in cows with high Hp concentration, compared to cows with lower value of Hp. This shows that subclinical uterine infection can be involved in the prolongation of days open and in a poor conception rate. Efforts must be made in the transition period to avoid acute phase reactions that potentially minimize reproduction performances. Another application of APP in reproduction is to monitor inflammation when progesterone-releasing intravaginal devices are administered to animals, in order to check health and safety of this procedure (Walsh *et al.* 2008).

Inflammation and Other Situations

APP can be used for diagnosing enteritis, peritonitis, pneumonia, endocarditis, fatty liver syndrome, viral, bacterial, and parasitic diseases.

Passive Immunity Transfer & Calves' Health

Newborn calves with low blood IgG concentration in the first 30–60 hours of life, indicating a failure of passive immunity transfer, show higher serum Hp concentration 3–4 weeks after birth. This induces a higher morbidity rate due to respiratory or gastrointestinal tract diseases (Furman-Fratczak *et al.* 2011). The measurement of IgG (laboratory by RID or field-test with glutaraldehyde method) in calves' blood and in colostrum has been previously described in assessing the passive immunity transfer in calves as well as the quality of colostrum.

Studies on respiratory syncytial virus have found that Hp and SAA were sensitive markers of bovine respiratory diseases (Orro *et al.* 2011).

Other researchers have discovered that groups of calves with a higher morbidity rate presented more days with elevated serum Hp and Fibri. Furthermore, the maximum concentration of APP in the sick group of calves was significantly higher compared to healthy group. APP are thus a useful method for monitoring health in calves (Ganheim *et al.* 2007). APP could be assessed in animals entering a new farm, in quarantine conditions, as an early detector of new infections. Nevertheless, measurement of APP in calves arriving in a feedlot is of limited value in identifying potential respiratory diseases and targeting treatments. APP concentration in calves introduced in their new environment had no association with subsequent overall performances (Holland *et al.* 2011). The weaning of calves can cause an increase in serum Hp and SAA 3–5 days post-weaning (Kim *et al.* 2011).

Lameness

Specific claw disorders such as pododermatitis septica, pododermatitis circumscripta, interdigital necrobacillosis and papillomatous digital dermatitis induce a rise in serum Hp. For some of these diseases, a decreased concentration of Hp is seen after treatment (Smith *et al.* 2010).

Stress, Transport, Welfare and Meat Inspection

Studies on animal shipping have found that cattle transported in solitary tie stalls for 4–6 hours showed an increased concentration of serum Hp and SAA; however, in another study, Hp was not significantly increased after transportation (1600 km in 24 hours), while Cp significantly increased on the first day after shipping (Arthington *et al.* 2008; Lomborg *et al.* 2008).

Measurement of APP can be used for assessment of physical welfare (transportation, weaning) and is a prognostic indicator of average daily gain (ADG). ADG decreases as APP concentrations are increasing (Arthington *et al.* 2003).

Finally, the use of APP is suggested for emergency slaughter in cattle. As APP can evaluate early the global health status of cattle, veterinarian could use Hp concentration in the ante-mortem inspection of cattle at a slaughterhouse. Further studies are needed to define useful cut-off values and baseline for being an interesting routine tool in meat inspection (Blagojevic *et al.* 2011).

Conclusions

Nowadays, inflammatory biomarkers are widely analyzed in veterinary laboratories and numerous field tests are also available. This should stimulate practitioners and researchers to use them for clinical diagnosis or for the monitoring of individual and herd health. Close relationship has been established between APP and nutrition, global health, reproduction performances and milk production. Although APP are easily measured, more reference guidelines are needed in order to use them in routine practice and to avoid bias from the use in different conditions.

References

1. Arthington JD, *et al. J Anim Sci.* 2003;81:1120.
2. Arthington JD, *et al. J Anim Sci.* 2008;86:2016.
3. Blagojevic B, *et al. Foodborne Pathog Dis.* 2011;8:1119.
4. Burvenich C, *et al. J Dairy Sci.* 2007;90:E39.
5. Ceciliani F, *et al. J Proteomics.* 2012;75:4207.
6. DeNise SK, *et al. J Dairy Sci.* 1989;72:552.
7. Eckersall PD, *et al. Vet Rec.* 2001;148:35.
8. Furman-Fratczak K, *et al. J Dairy Sci.* 2011;94:5536.
9. Gaillot C, *et al.* In: Proceedings of Buiatrissima, 8th ECBHM symposium; 2013, Bern, Switzerland.
10. Ganheim C, *et al.* Acute phase proteins as indicators of calf herd health. *Vet J.* 2007;173:645.
11. Gerardi G, *et al.* Use of serum amyloid A and milk amyloid A in the diagnosis of subclinical mastitis in dairy cows. *J Dairy Res.* 2009;76:411.
12. Gessner DK, *et al. J Dairy Sci.* 2013;96:1038.
13. Gozho GN, *et al. J Dairy Sci.* 2007;90:856.
14. Guyot H, *et al.* In: Proceeding of the 6th European Congress of Bovine Health Management; 2011, Liège, Belgium.
15. Guyot H, *et al.* In: Proceedings of the XV MEBC & 10th ECBHM Symposium; 2015a, Maribor, Slovenia.
16. Guyot H, *et al.* In: Proceedings of the XV MEBC & 10th ECBHM Symposium; 2015b, Maribor, Slovenia.
17. Guzelbektes H, *et al. J Vet Intern Med.* 2010;24:213.
18. Holland BP, *et al. Am J Vet Res.* 2011;72:1349.
19. Humblet MF, *et al. Vet Clin Pathol.* 2006;35:188.
20. Huzzey J, *et al. J Dairy Sci.* 2009;92:621.
21. Janosi S, *et al. Acta Vet Hung.* 1998;46:71.
22. Jones ML, Allison RW. *Vet Clin North Am Food Anim Pract.* 2007;23:377.
23. Kim MH, *et al. J Vet Sci.* 2011;12:151.
24. Kimura K, *et al. J Dairy Sci.* 2014;97:4842.
25. Lomborg SR, *et al.* Acute phase proteins in cattle after exposure to complex stress. *Vet Res Commun.* 2008;32:575.
26. McDougall S, *et al. J Dairy Sci.* 2016;99:2026.

27. Metzner M, *et al. J Vet Med A Physiol Pathol Clin Med.* 2007;54:449.
28. Molenaar AJ, *et al. Biomarkers.* 2009;14:26.
29. Orro T, *et al. Comp Immunol Microbiol Infect Dis.* 2011;34:23.
30. Owen JA, *et al. J Clin Pathol.* 1960;13:163.
31. Plaizier JC, *et al. Vet J.* 2008;176:21.
32. Plaizier JC, *et al. J Dairy Sci.* 2014;97:7076.
33. Radostits OM, Gay CC, Hinchcliff KW, Constable PD, eds. *Veterinary Medicine.* 10th edition. Philadelphia, PA: Saunders-Elsevier;2007:pp2047.
34. Robinson JD, *et al. J Dairy Sci.* 1988;71-1283.
35. Russel KE, Roussel AJ. *Vet Clin North Am Food Anim Pract.* 2007;23:403.
36. Sandholm M. *Res Vet Sci.* 1974;17:32.
37. Schalm OW, Noorlander DO. *J Am Vet Med Assoc.* 1957;130:199.
38. Scott PR. *Br Vet J.* 1995;151:603.
39. Smith BI, *et al. Vet J.* 2010;186:162.
40. Sordillo LM, *et al. Anim Health Res.* 2009;10:53.
41. Sordillo LM. *J Dairy Sci.* 2016; 99:1.
42. Trevisi E, *et al. J Dairy Res.* 2009;77:310.
43. Vandeputte S, *et al. J Vet Intern Med.* 2011;25:1465.
44. Walsh RB, *et al. Can J Vet Res.* 2008;72:43.
45. Weber A, *et al. Vet Immunol Immunopathol.* 2006;109-79.
46. Wittek T, *et al. Vet Rec.* 2010;166:15.
47. Wittum TE, Perino LJ. *Am J Vet Res.* 1995;56:1149.

1.
acvim244

Current Approaches and Future Directions in Veterinary Education

James N. Moore, DVM, PhD
Athens, GA, USA

Introduction

I suspect that my preparation to become a veterinary educator is similar to that of the majority of faculty members currently teaching in veterinary schools. Namely, after finishing my veterinary education, I completed internship and residency training programs, then did graduate work, and was hired to teach, perform clinical service, and create scholarly publications via research. While I was at home in the research laboratory and clinic, I essentially had no training as an educator. As a result, I patterned my approach after those used by the faculty members that had taught me. In all likelihood, they had similar backgrounds and experiences. Very few, if any, entered veterinary school with a goal of becoming an educator. In my case, I'm convinced it showed.

As I reflect on my initial years as an educator, I realize how poorly prepared I was to lecture. I was ill at ease in front of a group, and I hadn't figured out what they needed to know about a particular topic and what I should leave out. As a result, I believe my lectures were not as well organized as they should have been and likely were far too content-rich for what our students needed. My notes also were excessively in depth in an effort to provide the students as much up-to-date information as possible. In my own naïve way, I was contributing to our students' propensity to memorize whatever information is presented to them. In effect, I was reinforcing the behaviors that I now work constantly to prevent.

During the time I've spent as an educator, I've had ample time to think about the system we currently use to educate veterinary students and how we might improve. In doing so, I've tried to take the same approach I would with any research project I've had the pleasure of working on over the years - 1) identify the problem; 2) hypothesize how to address it; 3) design a study to evaluate the effectiveness of a specific treatment, and that includes appropriate control and treatment groups; 4) collect data from both groups; 5) analyze the data to assess the effectiveness of the treatment; and 6) determine whether or not the results align with the hypothesis. With regard to veterinary education, I believe that we have three intertwined problems that we need to address in a similar manner.

First, the volume of information available for faculty members to share with veterinary students is immense and growing. I used to joke early in my career that it was hard work finding the time to remain minimally behind in my areas of interest. To do so meant trekking up to the library each month to copy and then read as many of the new articles about endotoxemia, colic, laminitis, and circulatory shock in the handful of journals I followed closely, as well as those in a few closely aligned journals. As we all know, today everything is available instantly, with virtually all journals now including manuscripts they've accepted but that have yet to be printed. To see how much new information is out there, I recently performed a PubMed search on 'endotoxemia' for 2016. The response - 341 articles or nearly one new article a day for the year. Clearly, we have to resist the urge to share much of what we know and learn with our students. In essence, just because we can share something, doesn't mean we should.

The second problem we face is that students tend to view every fact as being of equal importance. I believe this is largely because they've been rewarded since they were in elementary school for memorizing and regurgitating facts. As a result, they see every fact as something that might need to be recognized on a multiple-choice test. Simply put, they lack the experience necessary to be able to put individual facts in context in order to identify those that will be important later in their education and those that can be relegated to short-term memory for recall on an impending exam. This problem is intertwined with the problem of information overload, and is exacerbated when faculty members share more facts about a particular topic than necessary. For example, when teaching students about colic in horses, should I spend time describing the different surgical approaches that can be used to treat a horse with an ileocecal intussusception, or should I spend my time telling them how to decide whether or not a horse should be referred for emergency abdominal surgery? This is an easy question to answer; others are more difficult and should be the topic of conversations among faculty members to decide where to draw the line.

The third problem is **siloing**, both on the part of the students and the faculty. For a variety of reasons, many students fail to see how information presented in one course integrates with that in another course, or even within the same course. This certainly is true in the veterinary curriculum, and I've encountered the same problem when I've tutored undergraduate students in physiology. They learn cardiovascular physiology, seemingly master that material for an exam, and then are able to recall very little when the topic shifts to neurophysiology. The end result is what was referred to by Bensley and Ellsworth 25 years ago as **bulimic learning**, a metaphor that speaks for itself. A similar problem can occur when faculty members in the same college are clueless about what the students are being taught in other courses. This can end up with students receiving conflicting information about specific topics, differences that simply may reflect an individual faculty member's personal experience and bias. These discrepancies are amplified when changes occur in the faculty, with new faculty members making independent decisions about the material that needs to be presented in the lectures or courses they've inherited. Things can become murky fairly quickly.

My primary goal in this presentation is not to tell you how to teach, but to encourage you as an educator to think about how you impart new knowledge to veterinary students. It is my hope that this will result in open discussions among faculty members, students, house officers, and staff both within and across veterinary colleges. Rather than view our students' performance on national and state board examinations as evidence of an excellent educational process, we should design studies to assess the effectiveness of interventions to enhance their ability to be problem solvers and critical thinkers.

THE PROS AND CONS OF DIFFERENT APPROACHES

While it may sound from the introductory paragraphs that veterinary curricula today are chaotic and ineffective, this is not the case. In fact, one could easily argue against making any curricular changes based on the well-known fact that more than 95% of veterinary students nationwide pass the North American Veterinary Licensing Examination by the time they graduate. Before we congratulate ourselves on the job we are doing in veterinary education, let's consider the following sample question provided by the International Council for Veterinary Assessment that delivers that examination:

A 5-year-old Quarter horse stallion collapses suddenly and lies covered with sweat in lateral recumbency. When the horse attempts to stand, it exhibits generalized muscle fasciculations, prolapse of the third eyelid, inspiratory stridor, and loose feces. The episode lasts for 25 minutes and then resolves without treatment. The horse appears normal 15 minutes thereafter.

- Which of the following is the most important question to ask the owner?
  a. Does the horse have a history of laminitis?
  b. Has the horse been receiving calcium supplementation?
  c. Has the barn had a recent outbreak of respiratory disease?
  d. What is the bloodline of the horse?
  e. What is the vaccination history of the horse?

Based on our students' collective performance on this examination, most would have answered this question correctly by selecting choice d. This suggests that some of the approaches being used in veterinary schools are working and need to be preserved, that the students we are accepting into our programs are exceptionally well prepared to excel on these types of examinations, or both. While I recognize that examinations such as this likely are necessary at the national level, I question whether we should be using this same approach in our own courses. Does this form of examination allow us to accurately assess the effectiveness of our teaching and our students' ability to solve problems and think critically? Shouldn't we be aiming toward something more along the following lines?

A 5-year-old Quarter horse stallion collapses suddenly and lies covered with sweat in lateral recumbency. When the horse attempts to stand, it exhibits generalized muscle fasciculations, prolapse of the third eyelid, inspiratory stridor, and loose feces. The episode lasts for 25 minutes and then resolves without treatment. The horse appears normal 15 minutes thereafter.

- What is the most important question to ask the owner?

Despite the aforementioned success of our students on the national examination that allows them to practice veterinary medicine, I believe the majority of faculty members in the same veterinary schools would agree that we need to do a better job of preparing our students for life after that examination. Consequently, it is important for us to review what we are doing today, question and assess the validity of our current teaching models, identify our areas of strength and weakness, bolster the areas of strength, devise new teaching approaches to address the areas of weakness, and then assess their efficacy. This may sound easy, but it is tough to do for a variety of reasons - not the least of which is everyone is busy, and tackling these issues takes time and effort.

Lecturing - The Sage on the Stage

One of the most commonly used approaches for imparting new knowledge, particularly in science-based courses, is through the traditional lecture - the instructor speaking to students from the front of the classroom, typically following a set of PowerPoint slides. This 'teaching by telling' approach has its strengths and weaknesses. It is an extremely efficient way to deliver information to a large number of students simultaneously; the instructor is in charge of what material is covered when; and appropriate images, videos, or audio files can be incorporated into the presentation as needed. A well-organized, well-delivered series of lectures is a form of social interaction that often results in a trusting relationship between the instructor and his/her students. This is especially true when the lectures are delivered by the same person; unfortunately, it rarely happens when individual lectures in a course are delivered by a team of instructors.

Lecturing works very well for sharing conceptual and factual information, particularly when there is a large knowledge gap between the instructor and the students. As such, lectures function as a way to ensure that interested, engaged students are exposed to the factual information they need when they begin the more difficult task of posing their own 'what if' scenarios. Furthermore, lectures are perceived as being risk-free by the majority of students; again this is particularly true in a large classroom setting where the chance that an individual student will be called upon to answer a question is extremely small. Furthermore, lectures are perceived as making the learning process easy and effortless for the student; the truth of the matter is that learning requires effort to be effective. Finally, lectures are preferred by students who view themselves as **auditory learners**, although virtually all lectures these days are packed with visual elements.

What about the disadvantages associated with lecturing? A poorly organized, poorly delivered lecture is similar to a poorly told joke - very few, if any, people recall one. It takes work to put together and deliver an engaging lecture, so instructors cannot assume that students are engaged simply because they are sitting in their seats, looking at the screen, and taking notes on their laptops. Lecturing presumes that all students in the class learn at the same pace and have the same background knowledge. Neither of these is true in most classes; one simply has to watch students react when a classmate asks what they perceive to be either an overly basic or complex question.

As mentioned previously, a key feature of lecturing is that it is an excellent way to share factual information with a group. However, it is an extremely poor way to teach higher-order thinking, such as application, analysis, synthesis, or evaluation. A common goal of veterinary education is to promote these skills, and there is convincing evidence that lecturing is not the way to achieve it. Finally, today's presentation software packages often are the bane of a good lecture. Instructors often fill their slides with words or worse yet, put an entire paragraph on a slide and then read it out loud to the students. Many instructors, particularly in the sciences, have little, if any, appreciation for aesthetics. As a result, on-screen colors clash or cause some words to seemingly disappear into a background color, and just because more than one font exists, they all don't need to be used. Finally, these packages are replete with features, such as animations and sound effects, which often detract from the instructor's intended message. A good rule of thumb is to keep things simple, focusing on the key words/concepts you want your students to recall, and avoid having too many distractions.

In many veterinary schools, students are provided electronic copies of the lectures before class so they can follow along with the instructor during the lecture. In some instances, these are copies of the PowerPoint presentations themselves, while in others the students receive pdf versions. Regardless, students typically have versions that allow them to type notes on-screen during the lecture. Students often prefer this approach, believing that taking notes on their laptops far outweighs the potential that the use of laptops could distract them from the material being presented. However, there is considerable evidence that Internet browsing during lectures impairs performance.

While there is considerable evidence that notetaking during lectures enhances retention and learning, much less is known about the potential effects that notetaking on laptops might have on student performance. When notes are taken longhand during lectures, most students are unable to capture entire spoken sentences (i.e., verbatim notetaking). Consequently, they must rely on paraphrasing what has been said. There is evidence in the literature indicating that this method of notetaking has greater effects on retention and learning than verbatim notetaking. In a recent study, Mueller and Oppenheimer (2014) compared the performance of university students asked to take notes during five TED talks covering topics that would be interesting, but not commonly understood by the students. The students were randomly assigned to one of two groups, provided either notebooks or laptops (disconnected from the Internet), and instructed to use their normal notetaking techniques during the lectures. After a 30-minute break, the students were asked a series of factual-recall and conceptual-application questions about the lectures. Students who took notes using the laptop wrote significantly more words and had more verbatim overlap with transcripts of the lectures; however, their performance on conceptual-application questions was significantly less than that of students who took notes longhand. These results suggest that paraphrasing concepts requires greater attention and is superior to notetaking on laptops.

Active Learning Approaches

There are a wide variety of approaches under the umbrella term 'active learning.' They all strive to involve the students more in the learning process than what typically occurs in the more passive style of the lecture. As a result, it also is called learner- or student-centered learning, primarily because it requires more involvement than passively listening and taking notes. The goal is to have students doing something constructive while thinking more deeply about the information at hand. Therefore, active learning has its foundation in the learning theory known as constructivism, which stipulates that the learners must build their own understanding of the material within a relevant context. Ideally, this approach should help students ascend Bloom's taxonomy of learning to analyze, evaluate, and synthesize new ideas.

Examples of active learning approaches range from simple ones that literally take a minute or two of class time and thus can be inserted into a lecture, to flipped classroom activities to complex schemas such as problem-based learning. Some examples of the simple ones include 'clarification pauses,' in which the instructor pauses at various points during a lecture to ask if anyone would like clarification on a key point or topic that will serve as the basis for something later in the lecture or course. This approach works best if the students are advised to review their notes and then to ask questions that will help them clarify what they've written. Another simple approach is called 'think-pair-share' - an activity in which students first try to solve a problem (perhaps deciding which diagnostic test or treatment should be used next in a case study and how to justify the choice) on their own. They are then paired with another student to compare their thought processes and choices, and then share their collective decision with the class. For the past decade or more, many instructors have incorporated 'clicker questions' into their lectures. This approach, whether it involves the use of dedicated student response systems or apps on the students' smartphones, is intended to give all students, particularly the most introverted, a voice. In many instances in veterinary education, this approach is used to give students an opportunity to 'work through' case studies.

I see two drawbacks to the typical use of clicker questions in classrooms. First, it allows introverted students to remain silent, a practice that I don't believe is beneficial to them in a profession that is as people-oriented as veterinary medicine. I would have preferred this approach in school, as just considering asking a question during class caused my adrenals to tremble. In my mind, the typical clicker question approach has another drawback that becomes apparent if/when students are asked to talk through their choices. Consider this scenario: Students are given a short case description and 5 possible treatments (A through E) to consider. Each submits his/her choice electronically, the collective results for the class are displayed on screen at the front of the room, and the answers are split as follows: A - 4%; B - 46%; C - 0%; D - 42%; E - 8%. In the typical classroom approach, students are told to talk with the person next to them and then to re-vote. In all likelihood, there will be a fair number of student pairs comprised of students who voted the same (B-B or D-D). How much do you suspect they get out of this interaction? Essentially nothing. Even if the students are asked to meet in groups of three, there still will be groups comprised of students that made the same choice. Learning Catalytics is a software program that handles this situation in a more logical manner. I've seen it used elsewhere and am hopeful that we will have the opportunity to incorporate it into our classes. This program automatically creates teams of students based on their responses, thereby ensuring that students who voted differently are teamed together to discuss the question at hand.

Another approach that has gained popularity in veterinary schools is the 'flipped classroom,' in which students are assigned material to complete outside of class, and then class time is used by the students to work through case studies in small groups, with a faculty member available nearby to answer questions and clarify misconceptions. Three years ago, Michelle Barton and I decided to take this approach with an equine colic elective class we'd taught the traditional lecture-based way for many years. The number of lectures was reduced to 4 at the beginning of the course, after which the students have a variety of assignments to complete on their own. These include reading through the notes, watching short videos that we provide, and working through cumulative online self-assessments of their knowledge at various points in the course. As we planned the switch, we read a short book about the flipped classroom and agreed with the authors that it makes little sense to videotape lectures for students to watch. As they pointed out, a boring lecture in person is a boring lecture when videotaped. Instead, we took their advice and recorded a series of 15- to 20-minute conversations between the two of us, during which we discussed the clinical application of the material they were assigned to read in the notes. The videos primarily included the audio track of our conversation, interspersed with images or animations to help make our points. We purposefully avoided talking through individual diseases as might occur in a lecture, and often posed questions for them to try to answer. During this course, we meet with students at regular times to discuss the results of the self-assessments, work through case studies in small groups, and have them discuss questions about the material in a clinical-rounds type of approach. Our approach is to encourage them to make mistakes and talk as much as possible so they can get a feeling for what they understand and what they don't.

In 2014, Freeman and colleagues summarized the results of a large study designed to test the hypothesis that lecturing maximizes learning and performance in undergraduate science, technology, and engineering courses, when compared to active learning activities. They performed a meta-analysis of 225 published studies that included examination scores or failure rates. The active learning components varied widely, from the occasional group problem-solving exercise, use of student response systems, and workshop courses. Their results indicated that students in traditional lecture format courses were 1.5 times more likely to fail than students in courses that included active learning activities. Furthermore, they determined that the improvements in performance in courses that included active learning were present regardless of class size, course type, or course level, and that active learning significantly increased student performance on concept inventories. As Freeman and colleagues concluded, this "ask, don't tell" approach may lead to strong increases in student performance. Collectively, these results strongly support the incorporation of active learning opportunities into our courses, particularly early in the veterinary curriculum.

References

1. Barak M, Lipson A, Lerman S. Wireless laptops as means for promoting active learning in large lecture halls. *Journal of Research on Technology in Education*. 2006;38:245–263.

2. Baurelein M. On the benefits of lectures. *The Chronicle of Higher Education*. 2011.
3. Bensley RJ, Ellsworth T. Bulimic learning: a philosophical view of teaching and learning. *Journal of School Health*. 1992;62(8):386–387.
4. Charlton BG. Lectures are an effective teaching method because they exploit human evolved 'human nature' to improve learning - Editorial. *Medical Hypotheses*. 2006;67:1261–1265.
5. Freeman S, Eddy SL, McDonough M, Smith MK, Okoroafor N, Jordt H, Wenderoth MP. Active learning increases student performance in science, engineering and mathematics. *PNAS*. 2014;111(23):8410–8415.
6. Hembrooke H, Gay G. The laptop and the lecture: the effects of multitasking in learning environments. *Journal of Computing in Higher Education*. 2003;15:46–64.
7. Martyn M. Clickers in the classroom: an active learning approach. *Educause Quarterly*. 2007;2:71–74.
8. Mueller PA, Oppenheimer DM. The pen is mightier than the keyboard: advantages of longhand over laptop note taking. *Physiological Science*. 2014;25(6):1159–1168.
9. Skolnick R, Puzo M. Utilization of laptop computers in the school of business classroom. *Academy of Educational Leadership Journal*. 2008;12:1–10.

acvim245

The Chronic Disease of Needing a Gold Standard for Diagnostic Accuracy: Is There a Simple Cure?

Ian A. Gardner, BVSc, MPVM, PhD
Charlottetown, PE, Canada

INTRODUCTION

The belief that a perfect reference test (often termed a gold standard) is required for evaluation of the accuracy of diagnostic tests (measures such as sensitivity, specificity, likelihood ratio of a positive test and area under the receiver-operating characteristic curve) for a designated purpose has been entrenched in the scientific literature for years. Historically, tests such as virus, bacteria, and parasite isolation were considered gold standards when tests such as real-time PCR were validated. When results of a side-by-side comparison in a test validation study showed discordant results (positive-negative or negative-positive), the dilemma became how to interpret PCR-positive results when the isolation results were negative. In a traditional statistical analysis (2 x 2 table with isolation as the reference test), PCR might be deemed to be less specific and less sensitive than isolation for testing clinical samples even though laboratory experiments may have shown PCR to have a lower limit of detection (analytical sensitivity) than isolation.

In the last two decades, the reality that no ante-mortem test in veterinary or human diagnostics is perfect (or is too costly or invasive to use) has led to a search for a "cure." The "cure" is the application of Bayesian latent class models (BLCM) to analyze the joint test results with or without prior information on test sensitivity and specificity. BLCM do not require knowledge of disease status and can be readily implemented using open-source software such as WinBUGS[24]and OpenBugs (www.openbugs.net/w/Downloads). This minimal-harm and well tested "cure" allows users to provide supporting statistical evidence in a diagnostic accuracy study that the PCR-positive, isolation-negative results are likely true positives, as demonstrated in the example that follows.[17]

The main advantages of the BLCM approach are that it can statistically demonstrate superior accuracy of novel technology without the restrictive assumption of a perfect reference standard, that inferences are possible with small sample sizes and that expert scientific opinion about sensitivity (Se) and specificity (Sp) can be incorporated into models. In addition, probability (credible) intervals are generated to capture the uncertainty in parameter estimates. Unlike confidence intervals which are produced in frequentist analyses, the meaning of probability intervals is easier to convey to end-users who may have a role in the selection of tests for a defined purpose (e.g., clinical diagnosis or surveillance). Like all models, certain assumptions need to be met to assure rigor in the estimation process to minimize the risk of bias in Se and Sp estimates. Bayesian inferential procedures require additional care in the data analysis stage because of the incorporation of prior information which can be considered both a strength and a potential weakness.[7, 21]

HISTORICAL PERSPECTIVE

Pioneering work on latent class models (LCM) for estimation of the Se and Sp was done by Hui and Walter[18] in the context of comparison of two skin tests (Tine and Mantoux) for tuberculosis in humans. The basis of the approach was to analyze the distribution of the counts of the joint test results (positive-positive, positive-negative, negative-positive, negative-negative) in 2 populations using maximum likelihood methods. Subject to the underlying assumptions being correct (constant Se and Sp of the tests in the 2 populations, conditional independence of tests[11] and distinct prevalences), the Se and Sp of both tests and prevalences in both populations could be estimated. This work received minimal attention until the mid-1990s in either veterinary or human medicine. In 1996, the first veterinary application of LCM was published comparing the Se and Sp of PCR and impression smears for detection of *Nucleospora salmonis,* a microsporidian parasite of rainbow trout.[14] Results of the analysis showed that PCR was 9 times more sensitive than impression smears, which up until that time had been considered the reference test.

Soon thereafter, a Bayesian approach based on Markov-chain Monte Carlo sampling from the joint posterior distribution was described as an alternative to maximum likelihood in a special issue of *Preventive Veterinary Medicine*.[9] Since 2000 when the paper was published, there has been steady uptake and acceptance of BLCM in the veterinary literature as a valid approach for calculation of Se and Sp, including reanalysis of the Hui and Walter data in a Bayesian framework.[20] In addition, the World Organisation for Animal Health (OIE) lists LCM as an acceptable method for estimation of Se and Sp of OIE-listed tests in the context of "fitness-for-purpose" in its aquatic and terrestrial animal manuals (www.oie.int/international-standard-setting/terrestrial-manual/access-online/) and specifically its chapter on principles of test validation for infectious diseases (www.oie.int/fileadmin/Home/eng/Health_standards/tahm/1.01.06_VALIDATION.pdf).

The simple 2-test in 2-population BCLM which is described in the following example has now been generalized to 3 or more populations and 3 or more tests in a single population, including the comparison of conditionally dependent tests which may require additional information in the form of priors, typically beta distributions for Se and Sp, to ensure that the models are identifiable.[8,15,23] Broadly speaking, lack of identifiability occurs when there are insufficient data to allow unique estimation of model parameters.

The populations from which animals are sampled as part of the experimental design can include those that are known to be free of infection (prevalence=0%), infected populations (prevalence greater than 0 to 100%) or populations of unknown infection status. Models have been extended to use of continuous test results in a receiver-operating characteristic analysis,[6] to consider multiple stages of infection (e.g., clinical and subclinical infections)[19] and to use of semi- and nonparametric models that don't have restrictive parametric assumptions[5]. For brevity, these latter models are not discussed herein. Readers interested in statistical details of BLCM should refer to the textbook by Christensen *et al.*[7]

EXAMPLE

Guthrie *et al.*[17] tested 503 equine blood samples collected from South African horses suspected of having African Horse Sickness (AHS); pyrexic horses with one or more signs typical of AHS including severe respiratory distress, frothy serofibrinous nasal exudate or subcutaneous edema. Two comparison groups of horses, 503 healthy non-vaccinated foals raised in the AHS-free zone in South Africa and 98 adult healthy horses from a race-meet in Dubai, United Arab Emirates. Equine blood samples, collected in EDTA tubes, were tested using a real-time PCR designed to detect all 9 serotypes of AHS virus and virus isolation (VI) was attempted on the blood sample using baby hamster kidney (BHK- 21) cell lines. All testing was done blinded and samples were randomized for laboratory testing Data from the study are presented in Tables 1 and 2 and a detailed description of methods is presented in the published manuscript. All 9 serotypes circulating in South Africa at the time of the study were detected in the 156 VI-positive horses.

Table 1. Horses with clinical signs consistent with AHS

Test results	VI +	VI -	Total
PCR +	156	184	340
PCR -	0	163	163
Total	156	347	503

Table 2. Apparently healthy horses (Dubai, no. in parenthesis)

Test results	VI +	VI -	Total
PCR +	0	0	0
PCR -	0	503 (98)	503 (98)
Total	0	503 (98)	503 (98)

BCLM code for 2 tests in 2 populations was written in WinBUGS using a modification of existing code.[4] Several models were run to compare results based on use of non-informative beta (1,1) priors for the Se and Sp of both tests, an informative beta (9999,1) on the specificity of virus isolation and non-informative priors on the other parameters, and a model based on conditional dependence rather than independence between the tests. Conditional dependence of Se and Sp captures the notion that results of the second test depend on the results of first test in diseased and non-diseased horses, respectively. PCR and VI would potentially be as both are direct indicators of infection. In contrast, a test that directly measures infection would likely be independent of another that indirectly measures infection through use of serological tests. Results from all models run were similar with PCR sensitivity being about 2-fold greater than VI sensitivity and both tests had similar specificities (Table 3)

Table 3. Sensitivity and specificity of PCR and virus isolation (VI) based on use of non-informative priors and conditional independence model based on test results from 503 clinically affected and 503 healthy comparison horses from South Africa

Parameter	Median	95% probability interval
PCR sensitivity	0.996	0.977–0.999
PCR specificity	0.999	0.993–1.0
VI sensitivity	0.458	0.404–0.511
VI specificity	0.999	0.998–1.0

The Se and Sp data were adequate to support a decision to replace virus isolation by PCR for routine clinical diagnosis. However, the comparative Se data would not necessarily be relevant to decisions regarding movements of horses internationally as horses considered for movement would be subclinically infected. A stochastic risk model to estimate the probability of exporting an undetected AHS-infected horse through a vector-protected quarantine facility consistent with OIE trade measures from an infected country used a 5th percentile of 0.708 and a median of 0.978 to capture the possibility that Se of PCR would be lower in a subclinical scenario than in the clinical scenario used to generate the data in Table 3.[27]

Reporting and Design Elements

Reporting and design of test accuracy studies are two inter-related but distinctly different measures of the quality of primary studies, and both need to be addressed in veterinary medicine.[10] In human health, the Standards for the Reporting of Diagnostic Accuracy (STARD) statement (www.stard-statement.org) was first published in 2003 and includes a checklist of 25 key reporting items and a flow diagram to account for losses to follow-up at all stages of testing of subjects.[2] Authors, editors and reviewers of test accuracy studies are encouraged to follow the checklist and submit a flow diagram. In 2015, the checklist was expanded to 30 items and renamed STARD2015[3] and there have been adaptations for paratuberculosis in ruminants[12] and aquatic animal diseases[13] to account for differences in veterinary versus human medicine such as the need to interpret tests at the population level, and the use of experimental challenge studies which are rarely used to study human infectious disease. Because STARD did not consider imperfect reference tests and the use of BLCM to analyze such data, we recently developed a modified version of STARD, named STARD-BLCM, to explicitly address these issues.[23]

QUADAS (Quality of Diagnostic Accuracy Studies) which has been updated to QUADAS-2 (www.bristol.ac.uk/social-community-medicine/projects/quadas/quadas-2/), is a tool for the assessment of methodological quality of primary accuracy studies as part of systematic reviews.[30] QUADAS-2 deals mostly with design issues, potential biases and applicability concerns in human studies. Because systematic reviews are rarely done for infectious diseases of animals, there has been little impetus to adapt these guidelines to animal studies. One source of design guidance from a veterinary epidemiological perspective is available but it is now more than 15 years old and needs updating.[16] In the context of BLCM, selection of appropriate animal populations (e.g., herds) and individual animals within these populations for a designated purpose and standardization of processes to elicit expert opinion about Se and Sp are 2 of many items which would benefit from consensus-based guidelines. A recent review paper of LCMs (frequent and Bayesian) found that nearly 30% of papers failed to describe whether model assumptions were met or assess fit of the data.[29]

Discussion

Many animal health journals are willing to publish test accuracy studies involving the use of Bayesian or frequentist LCM because there are few diseases for which the Se and Sp of ante-mortem tests are close enough to perfect for practical purposes. For brevity, a single example of clinical diagnosis of AHS[17] was used to show the utility of BLCM including the demonstration of the superior sensitivity of PCR over virus isolation. Other examples of the application of BLCM to infectious diseases of animals include feline immunodeficiency virus in cats,[26] *Staphylococcus aureus* intramammary infections of dairy cattle,[25] bovine respiratory disease,[28] and francisellosis in fish,[1] to list a few. The principles outlined in this paper also apply broadly to non-infectious diseases including use of biomarkers for cancer, cardiovascular and musculoskeletal disease.

The availability of easy-to-use computer software has made the use of BLCM far easier than it used to be, but ease of use does not necessarily equate to obtaining the correct parameter estimates (e.g., median and 95% probability intervals). Biased estimates of Se and Sp can occur when the wrong model is used (e.g., conditional independence rather than a dependence model) to compare 2 PCRs where the targeted sequences overlap or 2 variants of bacterial culture (e.g., solid versus liquid culture methods for paratuberculosis). Models may also be non-identifiable without inclusion of prior information about Se and Sp, as described in previous sections of this paper. Therefore, novice users should obtain statistical advice before applying these methods.

Code for many different test accuracy scenarios is available on-line from the School of Veterinary Medicine, U.C. Davis to make it easier for non-specialists to use BLCM (on-line at cadms.ucdavis.edu/diagnostictests/software.html).

References

1. Assis G, *et al. J Fish Dis*. 2016 doi: 10.1111/jfd.12559.
2. Bossuyt PM, *et al. Ann Intern Med*. 2003;138:w1-12.
3. Bossuyt PM, *et al. Radiology*. 2015;277:826-832.
4. Branscum AJ, *et al. Prev Vet Med*. 2005;68:145-163.
5. Branscum AJ, *et al. Stat Med*. 2015;34:2474-2496.
6. Choi Y, *et al. J Agric Biol Environ Stat*. 2006;11:210-229.
7. Christensen R, *et al. Bayesian Ideas and Data Analysis*. Boca Raton, Florida: CRC Press;2010.
8. Dendukuri N, *et al. Stat Med*. 2009;28:441-461.
9. Enoe C, *et al, Prev Vet Med*. 2000;45:61-81.
10. Gardner I. *Prev Vet Med*. 2010;97:136-143.
11. Gardner I, *et al. Prev Vet Med*. 2000;45:107-122.
12. Gardner I, *et al. Prev Vet Med*. 2011;101:18-34.
13. Gardner I, *et al. Dis Aquatic Org*. 2016;118(2):91-111.
14. Georgiadis M, *et al. J Aquat Anim Health*. 1998;10:372-380.
15. Georgiadis M, *et al. J Roy Stat Soc: Applied Stat*. 2003;52:63-76.
16. Greiner M, *et al. Prev Vet Med*. 2000;45:3-22.
17. Guthrie A, *et al. J Virol Methods*. 2013;189:30-35.
18. Hui S, *et al. Biometrics*. 1980;36:167-171.
19. Jafarzadeh R, *et al. Stat Med*. 2015;29:2090-2016.
20. Johnson W, *et al. Am J Epidemiol*. 2001;153:921-924.
21. Johnson W, *et al. Prev Vet Med*. 2009;91:116-121.
22. Jones G, *et al. Biometrics*. 2010;66:855-863.
23. Kostoulas P, *et al. Prev Vet Med*. 2017;138:37-47.
24. Lunn DJ, *et al. Stat Comput*. 2000; 10:325-337.
25. Mahmood Y, *et al. Prev Vet Med*. 2013;1012:309-317.
26. Morton J, *et al. Prev Vet Med*. 2011;104:136-148.
27. Sergeant E, *et al. PLoS One*. 2016;11:e0151757.

28. Timsit E, *et al. Prev Vet Med.* 2016;135:67–73.
29. Van Smeden M, *et al. Am J Epidemiol.* 2014;179:423–431.
30. Whiting P, *et al. BMC Med Res Methodol.* 2003;3:25.

1.
acvim247

Use of Ultrasonography to Assess Pulmonary Lesions in Calves and Foals

Sébastien Buczinski[1], Dr Vét, MSc, DACVIM; Steeve Giguère[2], DVM, PhD, DACVIM
[1]St-Hyacinthe, QC, Canada; [2]Athens, GA, USA

INTRODUCTION

Pneumonia is one of the leading causes of morbidity and mortality in both calves and foals. Successful treatment depends on early recognition of the disease and appropriate antimicrobial therapy. Thoracic ultrasonography has become a widely-used tool for the early detection of pulmonary lesions in foals and calves. This text reviews the applications of thoracic ultrasonography in foals and calves with emphasis on the results of studies critically evaluating the value of this diagnostic modality.

THORACIC ULTRASONOGRAPHY IN FOALS

Although many microorganisms cause respiratory disease in foals, *Rhodococcus equi* is considered the most common cause of severe pneumonia. Pneumonia caused by *R. equi* is often not recognized until it is well advanced and, therefore, difficult to treat. Even severely affected foals may appear to suckle and behave normally to a casual observer. The rationale for screening is the assumption that detecting foals in the early stages of disease and implementing appropriate treatment of affected foals will improve outcome. A variety of screening techniques performed serially have been described, including visual inspection of foals for clinical signs of pneumonia, monitoring rectal temperatures, hematological parameters, serology, and thoracic imaging using either radiography or ultrasonography, with empiric recommendation that screening begin around 3 weeks of age. Studies have shown that serum concentrations of either antibodies against *R. equi*, serum amyloid A, or plasma fibrinogen are not useful screening tests.[1-4] White blood cell (WBC) concentrations performed at monthly intervals had clinically acceptable sensitivity and specificity for early detection of *R. equi* pneumonia at 1 farm.[5] However, serial measurement of white blood cell or neutrophil concentrations resulted in limited performance for prediction of subsequent development of clinically-apparent *R. equi* pneumonia at another farm.[6] In addition to conflicting data regarding the benefit of this approach, additional limitations of using WBC for screening are the time and effort required to collect blood samples from foals, the lag from time of submission to availability of results, and the potential for false-positive results attributable to stress-associated leukocytosis or other infectious or inflammatory conditions that may be common among foals.

In the absence of an effective vaccine, control of infections caused by *R. equi* at endemic farms over the past 15–20 years has relied on early detection of disease using thoracic ultrasonography with antimicrobial therapy for affected foals.[7-9] Ultrasonography of the chest offers several advantages over other screening tests: 1) results are specific for the presence of pulmonary pathology; 2) the procedure can be performed relatively quickly for an individual foal; and 3) results are available immediately. Although controlled studies are lacking, periodic ultrasonography of the chest appears to have decreased mortality due to *R. equi* pneumonia at some farms relative to historical data.[7-9]

Because it is impossible to know which specific foals might recover spontaneously from subclinical disease, and because *R. equi* infections can cause severe disease, many breeding farms elect to treat all foals with ultrasonographic lesions. This approach has resulted in an increased number of foals treated for presumptive *R. equi* pneumonia in the absence of clinical disease. The temporal association between this widespread use of macrolides and rifampin as a result of ultrasonographic screening and the increase in the frequency of detection of resistant isolates,[10] suggest that this practice might not be innocuous. Emergence of macrolide- and rifampin-resistance at a farm after widespread use of these drugs was instituted as part of an ultrasonographic screening program has been documented.[11] On that particular farm, 20%–40% of *R. equi* isolates from pneumonic foals were resistant to macrolides and rifampin.[11] Currently, it is unknown if macrolide- and rifampin- resistance is widespread on many horse farms relying on mass antimicrobial therapy of foals with ultrasonographic lesions or if it is an isolated problem. In addition to the cost and risk for selection for resistant bacteria, unnecessary mass antimicrobial therapy of foals with pulmonary lesions might lead to development of life-threatening adverse reactions (e.g., diarrhea or hyperthermia) in some foals.

Critical Evaluation of Ultrasonographic Screening in Foals

The cumulative incidence of clinical signs of pneumonia attributed to *R. equi* on endemic farms that do not use ultrasonographic screening typically ranges between 5% and 20%[12] whereas the cumulative incidence of foals with ultrasonographic evidence of pulmonary abscessation on *R. equi* endemic farms is generally between 30% and 70%,[7,8,13] suggesting that many sub-clinically affected foals recover without therapy. Indeed, recent double blinded randomized placebo controlled studies have documented that approximately 70–88% of foals with small pulmonary lesions (sum of lesion diameters [or lesion score] of 1–10 cm) recover without antimicrobial therapy.[14,15] In addition, antimicrobial treatment of foals with such small ultrasonographic lesions does not significantly hasten lesion resolution compared to administration of a placebo.[14,15] In contrast, studies performed at the same farm documented that treatment of foals with lesion scores >10 cm provides a significant benefit relative to administration of a placebo.[16,17]

The use of ultrasonography for screening to detect *R. equi* pneumonia was evaluated at another endemic farm with personnel and veterinarians blinded to screening results.[18] Of 270 foals enrolled in the study, 216 (80%) developed sonographically visible pulmonary lesions whereas only 17% of foals developed clinically apparent *R. equi* pneumonia.[18] Resolution without clinically apparent illness or antimicrobial therapy occurred in 79% of foals with ultrasonographic pulmonary lesions.[18] In the aforementioned study, the cumulative sensitivity of ultrasonography was very good (89%) but cumulative specificity was low (62%).[18] Additional studies are needed to establish better criteria to determine the need for therapy in sub-clinically affected foals and to better quantify the risks versus benefits of treating foals with subclinical ultrasonographic lesions.

Thoracic Ultrasonography in Calves

Thoracic ultrasonography has gained in popularity in recent years for assessing lower respiratory tract infections associated with bovine respiratory disease complex in dairy calves.[19,20] When compared with necropsy, thoracic ultrasonography is sensitive and specific for lung lesion detection in chronic[21] and subclinical cases.[22] Ultrasonography has changed the way BRD is diagnosed in dairy calves, revealing the importance of subclinical cases (calves with lesions but without clinical signs). Normal lung parenchyma does not allow penetration of the ultrasound beam which is reflected from the tissue-gas interface with some reverberation artifacts. The main ultrasonographic anomaly associated with BRD in calves is lung consolidation.[19] The alveolar space is filled by exudate and inflammatory deposits associated with lung infection. The affected site can be observed during ultrasonography if adjacent to the pleura. Thoracic ultrasonography can quickly be performed at farms using a systematic screening protocol (1–2 minutes per calf). No specific investment is required other than the ultrasound unit routinely used for pregnancy examinations. The objective is to evaluate the extent of lung consolidation which is associated with 30-day mortality,[20] pre-weaned average daily gain, and increased risk of being culled before the first calving.[23] Different scoring systems exist for scoring/recording lung consolidation in dairy calves.[19]

Thoracic Ultrasonography at the Individual Level

The main interest of thoracic ultrasonography is for diagnostic and prognostic purposes. Thoracic ultrasonography can be used at the individual level to confirm diagnostic in cases where the clinical signs are not obvious. This technique might also help determining the progression or regression of the disease after initiation of therapy. However, the data are scant concerning guidelines for this application. One of the main interests of thoracic ultrasonography at the individual level is for research purposes. Ultrasonographic lesions can be more easily compared from one study to another than clinical signs which are subjectively collected and depend on the observer.[24] Thoracic ultrasonography might therefore be of interest to improve external validity of studies focusing on BRD and could serve as a practical way to standardize case definition.

Thoracic Ultrasonography at the Farm Level

The potential applications of thoracic ultrasonography at the farm level are promising. Calves with lung lesions should have been previously detected as sick by the producer (or at least a majority of them). Prevalence of calves with consolidation could also serve as a way to monitor BRD during preweaning and also serve as a way for monitoring calf health protocol changes.

Limitations of Thoracic Ultrasonography: What We Don't Know

Thoracic ultrasonography is helpful but it is not an infallible tool. Many questions remain unanswered. For example, little is known regarding the specific ultrasonographic pattern characteristics depending on the BRD etiologic agent. The dynamic of lung consolidation is also largely unknown. How much time does it take after an effective treatment to observe resolution of lung consolidation? What proportion of consolidation is reversible vs. non-reversible? Should all calves with consolidation be treated? Would treatment change the outcome of calves with lung consolidation? What is the acceptable farm prevalence of lung lesions in dairy calves? All these questions are unanswered. It is important not to solely rely on ultrasonographic findings when deciding whether therapy should be initiated since a calf without clinical signs might have lung consolidation and inactive lung lesions. Conversely, a calf with severe signs of respiratory disease but without lung lesions should be treated because of its clinical presentation. We are at the beginning of uncovering the applications of ultrasonography for respiratory disease monitoring in dairy calves. We hope that some of the previous questions will be answered soon, thereby helping the clinician to base his/her decision on validated algorithms which include thoracic ultrasonography results.

References

1. Cohen ND, *et al. Equine Vet J.* 2005;37(3):212.
2. Giguère S, *et al. Clin Diagn Lab Immunol.* 2003;10(2):241.
3. Martens RJ, *et al. J Am Vet Med Assoc.* 2002;221(6):825.
4. Giguère S, *et al. J Vet Intern Med.* 2016;30(4):1338.
5. Giguère S, *et al. J Am Vet Med Assoc.* 2003;222(6):775.
6. Chaffin MK, *et al. J Equine Vet Sci.* 2012;32(10S):S20.
7. Slovis NM, *et al. Proc Am Assoc Equine Pract.* 2005;51:274.
8. McCracken JL, *et al. Proc Am Assoc Equine Pract.* 2009;55:38.
9. Venner M, *et al. Vet J.* 2007;174(2):418.
10. Giguère S, *et al. J Am Vet Med Assoc.* 2010;237(1):74.
11. Burton AJ, *et al. Emerg Infect Dis.* 2013;19(2):282.
12. Chaffin MK, *et al. J Am Vet Med Assoc.* 2003;222(4):467.
13. Venner M, *et al. Vet J.* 2009;179(2):301.
14. Venner M, *et al. Vet J.* 2012;192:293.
15. Venner M, *et al. J Vet Intern Med.* 2013;27:171.
16. Hildebrand F, *et al. J Vet Intern Med.* 2015;29(1):333.
17. Venner M, *et al. Vet Rec.* 2013;173(16):397.
18. Chaffin MK, *et al. J Equine Vet Sci.* 2012;32(10S):S20–S21.
19. Ollivett TL, *et al. Vet Clin North Am Food Anim Pract.* 2016;32(1):19.
20. Buczinski S, *et al. J Vet Intern Med.* 2014;28(1):234.
21. Rabeling B, *et al. Vet Rec.* 1998;143(17):468.
22. Ollivett TL, *et al. J Vet Intern Med.* 2015;29(6):1728.
23. Adams EA, *et al. J Dairy Sci.* 2016;99(2):1465.
24. Buczinski S, *et al. N Z Vet J.* 2016;64(4):243.

acvim248

The Science Behind the Emerging Field of Regenerative Medicine in Large Animals

Lindsey Boone, DVM, PhD, DACVS-LA
Auburn, AL, USA

Introduction to Regenerative Medicine

Regenerative medicine is a rapidly developing field of translational biomedicine focused on stimulating or replacing the body's own intrinsic capacity for repair and/or regeneration of damaged, malfunctioning tissues. Scientists and clinicians working within this field of medicine are focused on *in vitro* manipulation of cells, cellular signals (growth factors, cytokines, chemokines), and extracellular matrices for *in vivo* implantation and directed tissue regeneration with the goal of restored tissue function. These agents are typically derived from donor tissues or blood. Blood-derived products, including autologous conditioned serum (Interleukin-1 receptor antagonist protein- IRAP), autologous protein solution, autologous conditioned plasma (also known as platelet rich plasma) and other platelet derived products (platelet lysate, platelet rich fibrin) produce a "soup" of cellular signals that are used to stimulate the healing processes of resident cells in the local wound environment. These products are available to the veterinary market using commercially available kits and centrifuges for point of care processing. However, these products are primarily used in the veterinary market for treatment of musculoskeletal conditions and will not be the focus of these proceedings, but deserve mention. Tissue-derived products include cells (progenitor and stem/stromal cells) and decellularized extracellular matrices for cellular engraftment. Cellular therapy, particularly stem cell therapy, initially gained attention for treatment of musculoskeletal disease, but recently has rapidly gained attention in the field of regenerative medicine for treatment of immune-mediated, inflammatory, and cancerous disease processes.

Modulation of the Wound Environment

The immune system plays a pivotal role in tissue repair and regeneration. After acute injury, microvascular damage results in increased vasopermeability and hemostasis. Stressed or necrotic cells and diseased extracellular matrix along with pathogens within the wound release numerous danger signals (danger associated molecular patters-DAMPS and pathogen associated molecular patterns-PAMPS) that activate chemotaxis of cells of the innate immune system (neutrophils, monocytes, and macrophages) to the wound's microenvironment. Neutrophil infiltration of the wound occurs first resulting in decontamination of the wound and recruitment of other immune and effector cells. Blood and tissue resident macrophages are recruited to the wound and differentiate into a pro-inflammatory phenotype in the early acute inflammatory phase to assist neutrophils with cleanup of the local microenvironment. During this clean-up process, necrotic neutrophils accumulate and are phagocytosed by pro-inflammatory macrophages. This causes a phenotypic switch of the macrophages to a more anti-inflammatory phenotype in the late stage of the acute inflammatory response. The macrophage plays a critical role in the process of wound healing and can be both a blessing and a curse. Depletion of macrophages within the wound results in delayed wound healing and persistent, destructive neutrophilic infiltration.[1] These anti-inflammatory macrophages secrete factors that stimulate cellular recruitment and proliferation of fibroblasts, myofibroblasts, and endothelial cells during the proliferative phase of wound healing. During this stage of wound healing fibroblasts lay down extracellular matrix while and endothelial cells stimulate angiogenesis. This physiologic fibrotic matrix is remodeled overtime through delicate cellular interactions and signals. MSCs play an integral role in regulating these intricate mechanisms of wound healing to ensure adequate tissue repair.

Mesenchymal Stem Cells (MSCS)

In large animal species, tissue-derived cellular therapeutics are most commonly obtained from adipose or bone marrow. These tissues are harvested and processed to concentrate the mononuclear cell population, which can be used in a point of care manner as a cellular concentrate or expanded in culture. These cellular concentrates, including bone marrow aspirate concentrate (BMAC) and adipose derived stromal vascular fraction (ADSVF), contain a heterogenous population of hematopoietic stem, progenitor cells, endothelial cells, erythrocytes, fibroblasts, lymphocytes, monocytes/macrophages, and pericytes and limited mesenchymal stromal cells. Use of these cellular concentrates in large animals has primarily been limited to musculoskeletal related injury, but its use in modulating inflammatory conditions in lab animals has shown promise.[2]

Culture expansion of this mixed mononuclear cell population results in a less heterogenous population of plastic adherent, fibroblast-like cells capable of self-renewal deemed mesenchymal stem/stromal cells (MSCs). MSCs are derived from pericytes or the stroma of the perivascular niche of vascularized mesodermal tissues of fetal or adult origin. MSCs in large animals have been cultured from hematopoietic tissues, adipose, synovial tissues, skin, dental tissues, lung, genital tissues (endometrium), and gestational tissues (placenta and umbilical cord). Standardization of research and scientific laboratory MSC culture methodology is lacking and culture methodology should be tailored to the tissue from which the MSC was derived and the therapeutic need for the MSCs. For example, when clinicians are trying to restore denuded articular cartilage is it better to implant naieve MSC or MSCs that have been pre-differentiated toward the chondrogenic lineage? Is it better to use MSCs obtained from synovial or articular tissues vs. non-joint related tissues? Given the varied phenotypes among MSCs it is important to establish criteria for identifying the MSC phenotype that is being studied, so researchers are comparing apples to apples so to speak rather than apples to oranges. In 2006, the International Society of Cellular Therapy (ISCT) defined the following minimum criteria for definition of an MSC obtained from human bone marrow: adherence to plastic, specific surface antigen expression ($CD105^{hi}$, $CD73^{hi}$, $CD90^{hi}$, $CD45^{lo}$, $CD34^{lo}$, $CD14^{lo}$, $CD11b^{lo}$, $CD79\alpha^{lo}$, or $CD19^{lo}$, and MHC II^{lo}), multipotent differentiation (*in vitro* differentiation into osteoblasts, chondroblasts, and adipocytes).[3] Since this statement, the ICST and the International Federation of Adipose Therapeutics and Sciences has defined minimum criteria for adipose derived MSCs with differences in surface antigen expression ($CD13^{hi}$, $CD73^{hi}$, $CD90^{hi}$, $CD36^{hi}$, $CD11b^{lo}$, $CD45^{lo}$, and $CD106^{lo}$).[4] Interestingly, neither position statement discusses the immunomodulatory functions of MSCs within their minimum criteria, though, ISCT has proposed several key areas of refinement for standardization for defining the immunomodulatory phenotype of MSCs.[5] Full characterization of MSCs in the veterinary literature is difficult due to limited availability of species-specific antibodies and reagents. Surface antigen expression of Equine BMSCs has been studied and proposed phenotype signature would include $MHCI^{hi}$, $CD44^{hi}$, $CD29^{hi}$, $CD90^{hi}$, $CD45RB^{lo}$, $CD11a/CD18^{lo}$ expression; however, consensus in the literature is lacking.[6] For equine ADMSCs the phenotype signature would include $CD29^{hi}$, $CD105^{hi}$, $CD44^{hi}$, $CD90^{hi}$, $CD140b^{hi}$, $CD164^{hi}$, $CD34^{lo}$, $CD45^{lo}$, $CD73^{lo}$, $MHCII^{lo}$.[7] Various surface antigen markers have been reported and/or confirmed for bovine MSCs, but due to the limited number of studies, recommendations for surface antigen expression are not defined.

MSC Immunogenicity

MSCs were previously thought to be immunoprivileged due to their absence of MHC II and costimulatory molecule (CD 40, CD 80, CD 86) expression.[8] This low immunogenic profile combined with the knowledge that patient related factors (age and disease state) could negatively affect the MSC phenotype and concerns related to delayed intervention due to long culture times required for culture expansion of MSCs, prompted clinicians to evaluate the safety and efficacy of allogeneic use of these cells. In large animal species, allogeneic MSCs have been investigated from a safety and/or efficacy perspective following intravenous, intra-arterial, intradermal, intralesional, intra-articular, and intra-ocular.[9-12] The majority of these studies conclude that allogeneic administration of MSCs appears to be safe, however, most of these studies did not characterize the MHC haplotype of the donor and recipient making it difficult to know if these injections were truly allogeneic or syngeneic. This is important because equine MHC II expression has been demonstrated and proven to be dynamic and its expression variable between subject, cell harvests within subject, passage number, and culture conditions.[13] In fact, humoral and cell-mediated immunity has been documented following allogeneic administration in humans and documentation in large animals (horses) is accumulating.[14-17] Owens, *et al*. retrospectively evaluated the serum of horses injected with allogeneic MSCs and found that 37% of horses had developed anti-MSC antibodies.[15] *In vitro*, equine MHC-mismatched MSCs expressing MHCII activated T cells, causing significant increases in T cell proliferation compared to MHC-matched MHC class II negative and MHC matched MSCs proving an immune response.[13] Furthermore, this research group demonstrated cytotoxicity of MSCs when exposed to MHC II mismatched recipient sera. These findings have led to recommendations for haplotype matching prior to allogeneic administration. *In vivo*, MHC-mismatched allogeneic cells do not persist in the tissues as long as allogeneic MHC-matched or autologous MSCs,[18,19] yet allogeneic MHC-mismatched MSCs do persist in tissues longer than other allogeneic cells. This would suggest that these cells are not immunoprivileged, but may be capable of immune evasion long enough to exert their immunomodulatory effects on the environmental niche. The clinical relevance of these alloantibodies in regards to their effect on MSC efficacy is unknown and warrants further investigation.

Regenerative and Reparative Qualities

MSCs were first investigated therapeutically for their "stem" properties, particularly their ability to differentiate into different tissues of mesodermal lineage under certain *in vitro* conditions. Evidence of naieve MSC differentiation following implantation is limited. In fact, demonstration of MSC engraftment and retention at sites of implantation has been disappointing.[20] Despite these findings, however, evidence for improved research and clinical outcomes following MSC therapy is mounting. This has led to the "hit and run" theory, in which scientists believe MSCs are introduced to the environmental niche, they react to this niche via cell-to-cell and paracrine signals, adjust their phenotype resulting in the release of niche appropriate cytokines, chemokines, extracellular vesicles and growth factors. The MSC directly and indirectly interacts temporarily with the environmental niche to establish an environment supportive of tissue repair and regeneration through modulation of site-specific immune and progenitor cells. Stimulation and modulation of progenitor cells is exerted via trophic factors with four main functions: inhibition of ischemia-related apoptosis, inhibition of fibrosis, promotion of angiogenesis and progenitor mitogen activation.[21] These factors include transforming growth factor β (TGFβ), hepatocyte growth factor (HGF), epidermal growth factor (EGF), vascular endothelial growth factor (VEGF), fibroblast growth factor (FGF), platelet derived growth factor (PDGF), insulin-like growth factor (IGF-1), stromal derived factor-1 (SD-1), and angiopoietin-1.

Immunomodulation by MSCs

MSCs have intense immunomodulatory functions, in fact they have been referred to as the "Guardians of Inflammation". Much like a guardian is called to duty when alerted to danger, MSCs must be alerted to danger and activated through inflammatory cytokine priming or ligation of their toll like receptors (TLRs) with danger or pathogen associated molecular patterns (DAMPS and PAMPs). Following activation MSCs modulate the microenvironment through direct cell-cell contact and/or secretion of numerous anti-inflammatory cytokines, chemokines, and adhesion molecules, modulating the surrounding environmental niche toward a more regulatory, immune tolerant environment. Secreted immunomodulatory factors include prostaglandin E2 (PGE2), indoleamine 2, 3 dioxygenase (IDO), nitric oxide (NO), IL-6, TGFβ1, HGF, and tumor necrosis factor inducible gene 6 protein (TSG6). The mediators by which MSCs use to exert their immunomodulatory effects is different between species, tissue source, and targeted immune effector cells. For example, Equine MSCs derived from hematopoietic tissues will secrete NO to induce T cell suppression, whereas MSCs from solid tissue will not secrete NO.[22] Monkey, pig, and human primarily use IDO for T cell suppression whereas rat, mice, rabbit, and hamster NO.[23] In large animals, horses primarily use PGE2 to modulate the inflammatory environment. Additionally, MSCs secrete extracellular vesicles that carry gene regulatory proteins (mRNA and microRNA) and mitochondria that can exert immunosuppressive effects on the target cells.

This activation of MSCs to induce immunoregulation is termed cytokine priming. *In vitro*, activation of MSCs seems to require IFNγ, but other pro-inflammatory cytokines such as TNFα and IL-1β act synergistically with IFNγ. MSCs are plastic and activation is a dynamic process. The inflammatory stimulus, concentration, and duration of exposure to this inflammatory stimulus is important in MSC activation. For example, equine MSCs exposed to 20% allogeneic inflammatory synovial fluid did not alter their immunomodulatory gene expression profile, but exposure to high concentrations (50 ng/ml) of IFNγ and TNFα resulted in significant upregulation of immunomodulatory genes (COX-2, NO, IDO, and IL-6).[24] However, MSC priming with these high concentrations of IFNγ and TNFα reduced MSC viability and impaired trilineage differentiation.[25] When activated with lower concentrations of IFNγ and TNFα for a shorter period of time (12 hours) upregulation of COX-2, NO, IDO, IL-6 was maintained.[26] Additionally, MSC viability and multipotent differentiation was unaffected.[26] Cytokine priming results in a temporary change in the MSC immunomodulatory phenotype, with equine MSCs no longer showing upregulation of COX-2, NO, IDO, and IL-6 expression 7 days after cytokine priming.[26]

It is important to note that under certain conditions, MSCs can actually become pro-inflammatory and act like antigen presenting cells. MSCs can also lose their ability to be activated and induce immunosuppression, this occurs *in vitro* in MSCs at high passage (long culture times), long cryopreservation times, and lack of post thaw equilibration.

Innate Immune System

Neutrophils

Neutrophils kill pathogens through generation of reactive oxygen species, activation of destructive proteases, and generation of extracellular traps. When these mechanisms are prolonged or over activated, marked tissue destruction will occur. Little work has been done investigation the immunomodulatory functions of MSCs on Neutrophils. MSCs inhibit neutrophil migration through secretion of tumor necrosis-factor inducible gene 6 protein (TSG-6). More recently, over activated neutrophils are engulfed by MSCs through ICAM-1/CD18, resulting in MSC secretion of superoxide dismutase (SOD3) dampening the damaging oxidative burst, formation of extracellular traps, and release of destructive proteases.[27]

Monocytes, Macrophages, and Dendritic Cells

MSCs inhibit differentiation and maturation of DCs, modulate their secreted cytokine profile, and downregulate their expression of presentation molecules limiting their ability to stimulate a T cell response.[21] MSCs, therefore, induce a regulatory DC phenotype through PGE2 and IL-10 secretion and STAT3, Notch, and SOCS1 signaling pathways. These regulatory DCs further also suppress the adaptive immune response by induction of T regulatory (Treg) cells discussed below.

Circulating and tissue resident monocytes responds to various environmental stimuli, resulting in differentiation into varying subsets of macrophages. Macrophages activated by TLR ligation and IFNγ are termed classically activated macrophages (M1) whereas macrophages activated by IL-4 and IL-13 are termed alternatively activated macrophages (M2). The M2 macrophages are considered anti-inflammatory and regulatory. They exert their immunosuppressive effect on T cells in an IL-10 dependent manner. Modulation of monocytes by MSCs appears to be dependent on synergistic secretion of PGE2 and IL-6.[28] The Immunomodulation of dendritic cells and monocytes/macrophages by MSCs obtained from large animals has yet to be evaluated.

Natural Killer Cells

MSCs suppress NK cell proliferation, cytokine production, and expression of activation markers via secretion of IDO, HLA-G5, PGE2, TGF-β. Impairment of NK-cell cytotoxicity does require cell to cell contact with MSCs.[29]

Adaptive Immune System

T Cells

MSCs have been shown to suppress T cell proliferation, support survival of quiescent T cell, and support the generation and proliferation of regulatory T cells. MSCs inhibit mitogen activated T cell proliferation through engagement of adhesion molecules including program death receptor (PD-1), program death ligands 1 and 2, intracellular adhesion molecule-1, and vascular adhesion molecule-1 as well as secretion of soluble factors. These factors primarily result in cell cycle arrest at the G0/G1 phase through down regulation of cyclin D2, but induced apoptosis has also been observed.[30] T cell suppression, however, has been shown to be dose dependent, with high MSC: T cell ratios (<1:50) inducing T cell suppression and low MSC:T cell ratios actually stimulating T cell proliferation *in vitro*.[31]

MSCs suppress CD4$^+$ T helper 1 (Th1) differentiation into effector cells and their function indirectly through modulation of DCs and induction of CD4$^+$ regulatory T cells also known as induced regulatory T cells (iTregs). Additionally, MSCs suppress pro-inflammatory CD4$^+$ T helper 17 cells (Th17) by inducing their differentiation into iTreg cells through PGE2.[32] These iTregs are important in prevention of autoimmunity and promotion of an immune tolerant microenvironment. MSCs, on the other hand, encourage the anti-inflammatory CD4$^+$ T helper 2 phenotype. MSCs inhibit cytotoxic CD8$^+$ T (CTL) cell activation reducing CTL mediated allogeneic cell lysis, this serves as one of the mechanisms for immune evasion of MSCs.[21] MSCs can also promote generation of CD8+ T regs through expression of FasLigand and secretion of IL-10.[29] These changes to the CD8$^+$ T cell subset induced by MSCs do not affect CD8$^+$ memory T cells and their immune surveillance capabilities.32 Increased numbers of Tregs have been observed in the spleen following repeat intravenous administration of allogeneic BMSCs, but not ADMSCs in horses.[33]

B Cells

MSCs have been shown to suppress B cell proliferation through induction of cell cycle arrest (G0/G1 phase) and inhibit production of immunoglobulins (IgA, IgM, and IgG).[21] MSCs are thought to mediate B cells through ICAM-1 and IL-6.[21] Regulatory B cells have also been identified and IL-10 is plays a pivotal role in their immunomodulatory effects. B regs primarily exert their effects on T cell subsets through secretion of IL-35 encouraging a Th1 to Th2 shift, suppressing Th17 cells, and encouraging the production of iTregs.[29] MSCs modulation of B cell proliferation and function has not been investigated in large animal species.

In vitro and *in vivo* studies evaluating the immunosuppressive effects of MSCs in large animal species are limited. One of the first *in vitro* studies to evaluate the immunosuppressive effects of MSCs from various tissue sources (adipose, bone marrow, and umbilical cord tissue and blood) on T cell responses identified PGE2 as key immunomodulatory factor in equine MSCs. In this study, unactivated MSCs were shown to have no effect on T cell proliferation, but MSCs significantly suppressed T cell proliferation when activated with allogeneic peripheral blood mononuclear cells (alloantigens) or phytohemagglutinin (PHA) regardless of MSC source. MSCs increased their production of PGE2 and IL-6 following activation and decreased T cell secretion of IFNγ and TNFα. Kynurenine was not secreted following activation of MSCs and therefore IDO was not considered a key regulator in equine MSC induced T cell suppression.[22] Other researchers have debated this finding, however, showing upregulation of IDO gene expression of equine bone marrow derived MSCs following activation with IFNγ and TNFα.[24,26] T cell suppression is induced by both cell-cell contact with MSCs and/or release of soluble factors from activated MSCs. Equine bone marrow MSCs activated by IFNγ suppressed T cell proliferation by 55–56% when cultured in transwell (in-direct contact) or in the presence of conditioned media.[34] Autologous and allogeneic equine BMSCs suppress T cell proliferation equally, supporting allogeneic use for inflammatory conditions provided haplotype matching is peformed.[35] To the author's knowledge, there is only one study in the recent literature that has addressed immunomodulation of bovine MSCs. In this study, MSCs derived from Wharton's jelly induced suppression of activated T cells by 20%.[36]

Our understanding of the mechanisms by which MSCs exert their immunomodulatory function *in vitro* and *in vivo* in large animal species is in its infancy. More work is needed to enhance our understanding of appropriate donor selection, timing of therapy, dose, dosing frequency, proper tissue source for cellular therapy, proper characterization of the MSCs prior to implantation, proper cellular preparation and delivery, etc. However, their use for treatment of inflammatory and/or immune mediated diseases as well as musculoskeletal injury warrant further investigation given their unique biologic properties.

References

1. Sindrilaru A *et al. Adv Wound Care (New Rochelle)*. 2013;2:357–368.

2. Bowles AC *et al. Stem Cells*. 2017;35:532–544.
3. Dominici M *et al. Cytotherapy*. 2006;8.
4. Bourin P *et al. Cytotherapy*. 2013;15:641–648.
5. Krampera M *et al. Cytotherapy*. 2013;15:1054–1061.
6. Radcliffe CH *et al. Stem Cells Dev*. 2010;19:269–282.
7. Marx C *et al. Stem Cells Dev*. 2015;24:803–813.
8. Gao F *et al. Cell Death Dis*. 2016;7:e2062.
9. Williams LB *et al. Equine Vet J*. 2016;48:619–625.
10. Pigott JH *et al. Vet Immunol Immunopathol*. 2013;156:99–106.
11. Pigott JH *et al. Vet Comp Orthop Traumatol*. 2013;26:453–460.
12. Carrade DD *et al. Cytotherapy*. 2011;13:419–430.
13. Schnabel LV *et al. Stem Cell Res Ther*. 2014;5:13.
14. Berglund AK *et al. Equine Vet J*. 2016.
15. Owens SD *et al. Stem Cells Int*. 2016;2016:5830103.
16. Aguiar C *et al. Equine Vet J*. 2015;47:708–714.
17. Bobryshev YV *et al. Cell Tissue Res*. 2015;362:1–7.
18. Zangi L *et al. Stem Cells*. 2009;27:2865–2874.
19. Eliopoulos N *et al. Blood*. 2005;106:4057–4065.
20. Devine SM *et al. Blood*. 2003;101:2999–3001.
21. Regulski MJ *et al. Wounds*. 2017;29:20–27.
22. Carrade DD *et al. Cell Med*. 2012;4:1–11.
23. Su J *et al. Cell Death Differ*. 2014;21:388–396.
24. Barrachina L *et al. Vet Immunol Immunopathol*. 2016;171:57–65.
25. Barrachina L *et al. J Vet Sci*. 2016.
26. Barrachina L *et al. Stem Cells Dev*. 2017;26:15–24.
27. Jiang D *et al. Stem Cells*. 2016;34:2393–2406.
28. Najar M *et al. Cytotherapy*. 2016;18:160–171.
29. Najar M *et al. J Immunother*. 2016;39:45–59.
30. Vellasamy S *et al. Cytotherapy*. 2016;18:1270–1283.
31. Le Blanc K *et al. Lakartidningen*. 2002;99:1318–1321,1324.
32. Duffy MM *et al. Stem Cell Res Ther*. 2011;2:34.
33. Kol A *et al. Stem Cell Res Ther*. 2015;6:73.
34. Paterson YZ *et al. Stem Cell Res Ther*. 2014;5:90.
35. Colbath AC *et al. Stem Cells Dev*. 2016.
36. Cardoso TC *et al. Cell Tissue Res*. 2017;367:243–256.

1\.
acvim249

The Translational Science of the Emerging Field of Regenerative Medicine in Large Animals

John Peroni, DVM, MS, DACVS
Athens, GA, USA

The field of regenerative medicine is defined as the "**process of replacing or regenerating cells, tissues or organs to restore or establish normal function**."[1] This broad field of medicine has gained significant momentum in human and veterinary medicine, thanks to the notion that damaged organs or tissues treated with stem cells can be restored to health, thus improving the quality of life of people and domestic animals. An important clarification should be made in order to specifically define products used with the intent of regenerating tissue. In veterinary medicine, the last decade has seen the application of a plethora of treatments that have been broadly and hastily encompassed within the field of regenerative medicine. It may be more appropriate to consider these treatments as **biological therapies**, given that not all of the products that are used in domestic animals are necessarily meant to regenerate cells or tissue. This is an important clarification to make, especially because the general public has erroneously encompassed any treatment that is biological in nature as tissue regenerating. It is the responsibility of researchers and clinicians to correctly explain and define the use of these treatment options so that the expectations and, in some cases, ethical concerns of the general public are adequately satisfied.

The lack of specific therapies taken from the conventional pharmacopeia aimed at restoring soft tissues to their original makeup has promoted the use of stem cells from a pioneering approach to injury to a common practice recommendation, such that veterinarians regularly include stem cell therapy in the therapeutic options offered to their clients.[6-9] Unfortunately, stem cell therapy has become an accepted part of veterinary practice in the absence of objective data indicating that it actually works.[2,3] This is in large part because stem cell technology has been rapidly commercialized and promoted in the media without appropriate regulation. It should be concerning that this has happened without a thorough understanding of the processes governing the biological effects of this form of therapy. On the other hand, the frequent use of stem cells in veterinary patients can also be viewed as an opportunity. The prospect exists to promote well-designed prospective and controlled studies to be performed rapidly and on a large scale, especially in the field of canine medicine. It should be noted that regulatory bodies that would ensure that appropriate standards of care are followed when using regenerative approaches are still in their infancy in veterinary medicine, with the first Guidance for Industry (GFI) #218 on "Cell-based Products for Animal Use" (www.fda.gov) released in 2015. Furthermore, veterinary species can and have contributed vital information to the development of the regenerative medicine field in people. In fact, naturally occurring domestic animal diseases are routinely used in clinical trials, and veterinary species are frequently used as models to understand the mechanisms and the potential impact of biological therapies in people. As examples, the dog has been used as a model of heart disease[4,5] and spinal cord injury,[6,7] while the horse is a very well-studied model for musculoskeletal disorders such as osteoarthritis and bone repair.[8-14]

Stem cells can promote tissue regeneration, regulate the inflammatory response caused by trauma, and reduce scar formation during the processes of tissue repair. Clinicians, researchers, and members of industry have explored the option of harvesting stem cells from the horse to promote tissue healing and explore their commercial value as a therapeutic agent. As often occurs in the equine industry, this new and exciting treatment has found widespread clinical applications without the development of a corresponding thorough scientific understanding of equine stem cell biology. Although equine researchers and clinicians are expanding our understanding of stem cell use in the horse, most of the biological concepts that currently underpin their use are extrapolated from a wealth of information in the human literature. Researchers have dedicated themselves to understanding the role of stem cells in the maintenance of homeostasis during physiological tissue turnover, to elucidate their phenotypic and genotypic characteristics in a variety of species, and to explain the mechanisms underlying the interactions of these cells with the host environment. In human medicine, the potential therapeutic value of stem cells is supported by the multitude of completed and ongoing clinical trials aimed at resolving many pathologies, including cancer, graft-versus-host disease, multiple sclerosis, and many others (see clinicaltrials.gov, a service of the National Institutes of Health).

Biological therapies encompass a large number of products that can be broadly divided in cell-based or blood-derived preparations. For the purpose of this presentation, we will cover a few aspects of adult stem cell therapy in musculoskeletal disorders, particularly bone repair, and the exciting prospect of using acellular platelet-derived preparations to modulate immune responses.

BONE REPAIR

Methods to induce rapid bone formation in large animals are currently lacking. Developing methods to initiate ossification will be applicable in both fracture healing and in arthrodesis of advanced arthritic joints. Osteoinductive therapeutics can significantly improve treatment outcomes and dramatically reduce recovery intervals. Large animals such as sheep and swine have been used to model fracture healing in order to translate findings to benefit humans, but at the same time help develop strategies that would help heal fractures in large animals as well.

Using an adenovirus-based vector (AdBMP2) for the delivery of mesenchymal stem cells (MSCs) producing BMP2 encapsulated in polyethylene glycol-based hydrogels results in the extremely rapid induction of bone in several animal species. The hydrogel microbeads can be delivered by injection or placed in a surgical wound. Depending on the model and species used, bone is produced within a 7- to 14-day period, and segmental defects are healed within 14 (rodents) or 30 (sheep) days from treatment, resulting in bone containing cortex, trabecular bone, and marrow.

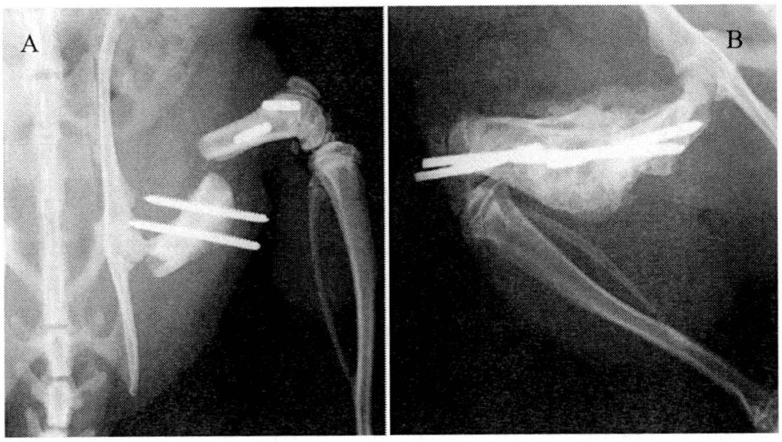

Figure 1. A. Untreated rat femoral defect: no healing at 2 weeks. **B.** AdBMP2-treated rat femoral defect: complete healing at 2 weeks.

Given the success observed in rodents, we tested our construct in large animals. The first objective was to determine if a construct consisting of BMP-2 transduced MSCs contained in a PEG-DA polymer (BMP-2-MSCs) could enhance bone formation in bone defects created in sheep. In these studies, bone formation was observed by 28 days. Bone was seen within the cortical defect and protruding beyond the surface of the bone in an orientation consistent with the downward flow of the injected hydrogel.

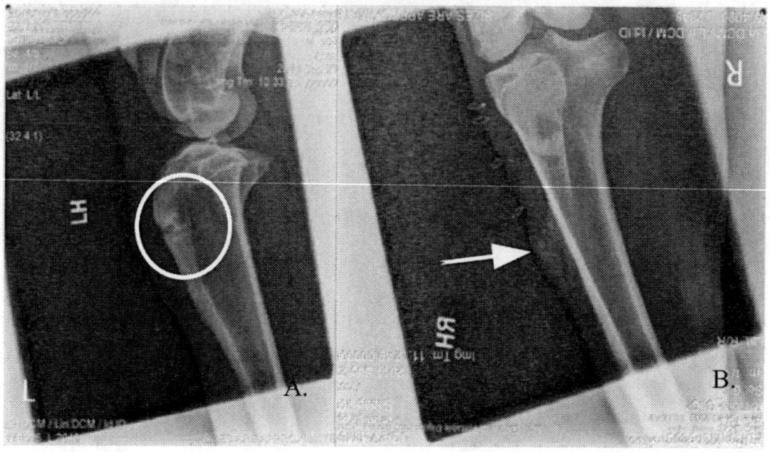

Figure 2. A. L ovine tibia. Unicortical drill hole within white circle with no apparent bone formation. **B.** R ovine tibia. Absence of unicortical defect. Exuberant bone formation indicated by white arrow. Treatment: 28 days after 5 x 10^6 BMP2-MSCs.

Tracking Stem Cells in a Model of Tendonitis

Current knowledge of stem cell migration in tendon lesions is limited. Although GFP tracking[15,16] has been performed, these studies provide a limited perspective that is only able to demonstrate the position and localization of these cells at one time point, and often only in two dimensions. Guest, Smith, and Allen (2008)[15] demonstrated that some of the 1 million cells initially injected into the surgically induced tendon lesion remained at day 10 and 34, and some cells were found engrafted to the surrounding, healthy tissue; however, this study was done over a short time span and did not draw any conclusions with respect to tendon regeneration. In a follow-up study, Guest, Smith, and Allen (2010)[16] showed that of an injected MSC population, less than 5% of the original MSC population could be detected in the lesion after day 10. Again, analysis was not done with respect to overall tissue repair or regeneration. A study by Pacini *et al.* 2007[17] demonstrated a significant increase in tendon tissue density at 30 days following injection of over 1 million stem cells into naturally occurring SDFT lesions. These initial studies indicate that MSCs are capable of altering tendon healing, and further work tracking the migration and presence of MSCs in tendons is vital to furthering this therapeutic treatment.

Work performed in our laboratory has shown that ultrasound may overstate improvements in tendon quality over time.[6] In further studies, we were able to demonstrate that significant differences in healing between MSC-treated and untreated tendons may be evident at 6 weeks from injury but that these differences become less obvious at 28 weeks following the initial trauma. Furthermore, the application of local MSC delivery for acute tendon injuries and the use of MRI for serial monitoring of morphological changes in tendonitis may indicate that MSCs diminish rapidly over the first 2 weeks following injection, and continue to decline over the following 4 weeks. These time points should be considered as opportunities to supplement with a second injection of MSCs.

Platelet Lysate in the Context of Immunomodulation

The techniques used by most laboratories to produce mesenchymal stem cells (MSCs) for research or clinical purposes rely on culture methods that may influence the relationship between MSCs and the host immune system. To date, the majority of cell therapy efforts involving *ex vivo* cell manipulation rely on fetal bovine serum (FBS) to help maintain cell viability and proliferation. By its very nature, however, FBS is xenogeneic. Thus, it has the potential to significantly alter the MSC phenotype rendering cells immunogenic, altering their immune interactivity, and resulting in their destruction by the host immune system, even when autologous tissues are used to produce the MSCs. The aforementioned effects of exposure to FBS during *ex vivo* expansion of equine MSCs may have profound effects on cellular survival and function *in vivo,* resulting in erroneous interpretation of data, unnecessary delays in development of cell-based therapeutics, and ultimately the design of flawed clinical trials. It is, therefore, important to develop expansion media for MSCs that contain recombinant growth supplements or that are sourced in a species-specific manner. Of particular clinical interest are studies showing that platelet lysate is equivalent or superior to FBS in propagating MSCs.[18] There is compelling evidence that platelet lysate can promote rapid MSC growth while maintaining differentiation potential, immune-phenotype and immune-modulatory activities.[19]

In the process of developing platelet lysate, we recently discovered that it has profound effects on immune responses. **Platelet lysate does not activate equine monocytes.** We compared supernatant concentrations of TNF-α and IL-1β from equine monocytes from six horses cultured in the presence of standard monocyte culture media (RPMI-1640) supplemented with either 10% donor horse serum (DHS), 10% fetal bovine serum (FBS), or 10% equine platelet lysate (ePL) manufactured via platelet pheresis and pooled from three donors. After 6 hours of incubation, cell culture supernatants were collected and assayed for the production of TNF-α and IL-1β via ELISA. **Monocytes incubated with FBS exhibited a 40-fold higher baseline production of TNF-α and a 4-fold higher baseline production of IL-1β compared to those incubated with ePL, indicating that platelet lysate does not stimulate the production of inflammatory cytokines, whereas FBS does.**

Platelet lysate dampens inflammatory responses in equine monocytes stimulated with LPS. We proceeded to stimulate equine monocytes with *E. coli* LPS 0111:B4 (50 ng/ml) in regular media (RPMI-1640) supplemented with platelet lysate and/or donor horse serum at 1%, 5%, and 10% concentrations. Donor horse serum (DHS) is important in these experiments, because it provides LPS binding protein that is essential for LPS Toll-like receptor 4 coupling and activation. Experiments were conducted from blood collected from 6 horses and using 3 monocyte cell lines each. After 6 hours of incubation, cell culture supernatants were collected and assayed for the production of TNF-α and IL-1β via ELISA. TNF-α produced by LPS-stimulated monocytes decreased over a thousand-fold in the presence of ePL compared to those incubated in the presence of DHS. Similar results are reported for IL-1β. The production of this cytokine from LPS-stimulated equine monocytes was reduced by over 400-fold in the presence of ePL. More interestingly, for both cytokines, combination of ePL and DHS also resulted in a reduction of TNF-α regardless of the concentration of DHS.

Establishing the efficacy of equine platelet lysate as a media component for the culture of MSCs will have a profound impact on how cell therapy is delivered to our equine patients. Additionally, we are excited to lay the foundation for future work on similar FBS-free media culture techniques for other veterinary species, such as the dog and the cat.

REFERENCES

1. Mason C, *et al. Regen Med.* 2008;3(1):1–5.
2. Becerra P, *et al. J Orthop Res.* 2013;31(7):1096–1102.
3. Watts AE, *et al. Stem Cell Res Ther.* 2011;2(1):4.
4. Fortier LA, *et al. Vet Clin North Am Equine Pract.* 2008;24(1):191–201.
5. Schnabel LV, *et al. J Orthop Res.* 2007;25(2):230–240.
6. Ribitsch I, *et al. Adv Biochem Eng Biotechnol.* 2010.
7. Richardson LE, *et al. Trends Biotechnol.* 2007;25(9):409–416.
8. Welt FG, *et al. Circ Heart Fail.* 2013;6(1):99–106.
9. Rastogi S. *Methods Mol Biol.* 2010;660:111–121.
10. Hiyama A, *et al. J Orthop Res.* 2008;26(5):589–600.
11. Hiyama A, Mochida J, Sakai D. *Cell Mol Biol (Noisy-le-Grand, France).* 2008;54(1):24–32.
12. Frisbie DD, *et al. Equine Vet J.* 2010;42(1):86–89.
13. Zachos, *et al. J Orthop Res.* 2006;24(6):1279–1291.
14. Koch TG, *et al. Can Vet J.* 2008;49(10):1009–1017.
15. Guest DJ, *et al. Equine Vet J.* 40:178–181.
16. Guest DJ, *et al. Equine Vet J.* 42:636–642.
17. Pacini S, *et al. Tissue Eng.* 13:2949–2955.
18. Ben Azouna N, *et al. Stem Cell Res Ther.* 2012;3:6.
19. Capelli C, *et al. Cytotherapy.* 2011;13:786–801.

AUTHOR INDEX

AUTHOR INDEX

AUTHOR INDEX

AUTHOR INDEX

AUTHOR INDEX

American College of Veterinary Internal Medicine (ACVIM)
8301 E. Prentice Avenue, Suite 300
Greenwood Village, CO 80111
www.acvim.org

ISBN 978-1-5108-4279-3

CURRAN ASSOCIATES INC.
proceedings.com